ISBN 978-0-266-95192-6
PIBN 10914630

FB &c Ltd, Dalton House, 60 Windsor Avenue, London, SW19 2RR.
Company number 08720141. Registered in England and Wales.

For support please visit www.forgottenbooks.com

CIRCULATES EVERYWHERE IN CANADA
Also in Great Britain, United States, West Indies, South Africa and Australia.

2018

HARDWARE AND METAL

A Weekly Newspaper Devoted to the Hardware, Metal, Heating and Plumbing Trades in Canada.

Office of Publication, 10 Front Street East, Toronto.

VOL. XIX. MONTREAL, TORONTO, WINNIPEG, JANUARY 5, 1907 *NO. 1.*

See Classified List of Advertisements on Page 67.

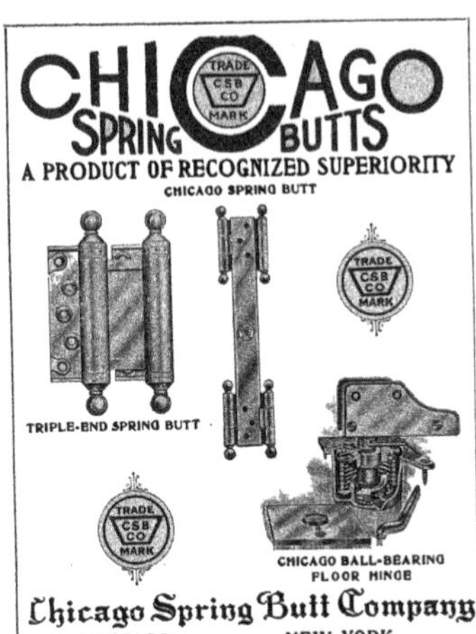
CHICAGO SPRING BUTTS
TRADE CSB CO MARK
A PRODUCT OF RECOGNIZED SUPERIORITY
CHICAGO SPRING BUTT
TRIPLE-END SPRING BUTT
CHICAGO BALL-BEARING FLOOR HINGE
Chicago Spring Butt Company
CHICAGO NEW YORK
CATALOGUE ON REQUEST

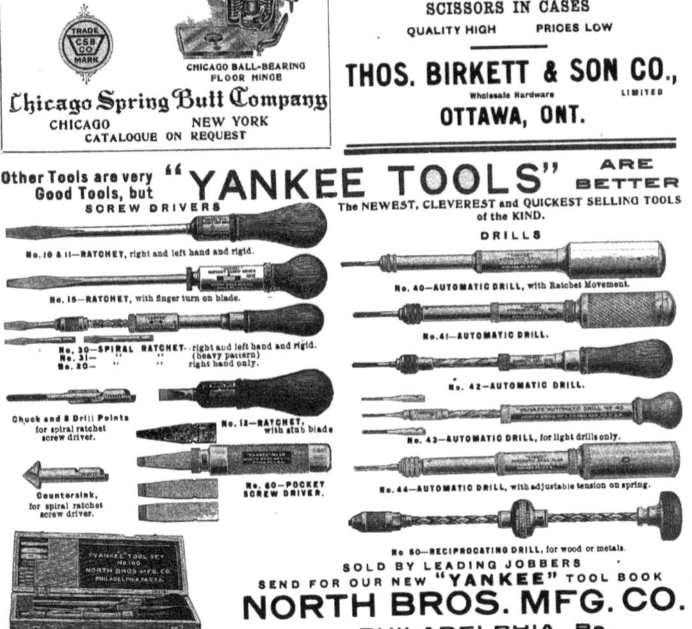
Other Tools are very Good Tools, but "YANKEE TOOLS" ARE BETTER
The NEWEST, CLEVEREST and QUICKEST SELLING TOOLS of the KIND.
SCREW DRIVERS
No. 10 & 11—RATCHET, right and left hand and rigid.
No. 15—RATCHET, with finger turn on blade.
No. 30—SPIRAL RATCHET—right and left hand and rigid.
No. 31— " " (heavy pattern)
No. 20— " " right hand only.
Chuck and 8 Drill Points for spiral ratchet screw driver.
No. 12—RATCHET, with stub blade
Countersink, for spiral ratchet screw driver.
No. 60—POCKET SCREW DRIVER.
DRILLS
No. 40—AUTOMATIC DRILL, with Ratchet Movement.
No. 41—AUTOMATIC DRILL.
No. 42—AUTOMATIC DRILL.
No. 43—AUTOMATIC DRILL, for light drills only.
No. 44—AUTOMATIC DRILL, with adjustable tension on spring.
No 50—RECIPROCATING DRILL, for wood or metals.
SOLD BY LEADING JOBBERS
SEND FOR OUR NEW "YANKEE" TOOL BOOK
NORTH BROS. MFG. CO.
PHILADELPHIA, Pa.

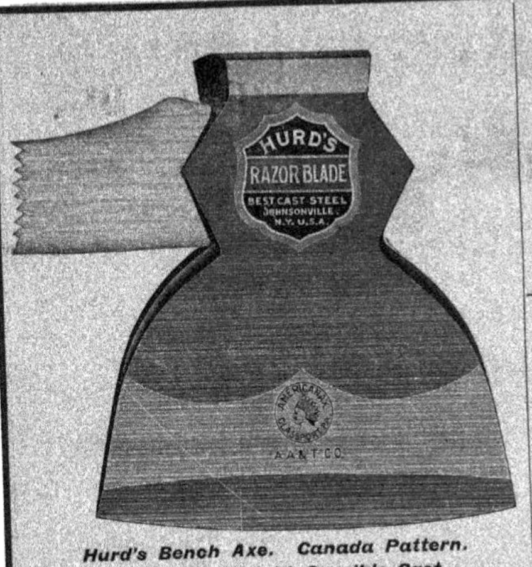

Hurd's Bench Axe. Canada Pattern. Extra Full Weight Crucible Cast. Steel Bit, Polished and Bronzed.

No.	Width cut in.	No.	Width cut in.
2	4½	5	6
3	5	6	6½
4	5½		

HURD'S RAZOR BLADE
JOHNSONVILLE AXE MFG. CO.

Hurd's Broad Axe. Canada Pattern. Bronze Finish. Extra Quality Cast Steel.

Wgt. lbs., 5 to 7½, 8 to 8½, 9 lbs.

QUICK SELLERS

Five Years ago the name "Hurd's" was not known in Canada. To-day there are more ***Hurd's Axes*** sold than any other make.

The quality is superior, and the next surprising thing is, that they are all sold at the same price—or lower—than the Canadian makes.

This is a line that is not cut. There is money in it for you. ***Get wise.***

OTTAWA
TORONTO
VANCOUVER
CALGARY

LEWIS BROS., LTD.

MONTREAL
WINNIPEG

H. S. HOWLAND, SONS & CO. LIMITED

HARDWARE MERCHANTS

Only Wholesale | 138-140 WEST FRONT STREET, TORONTO | Wholesale Only

Horse Clippers and Machines

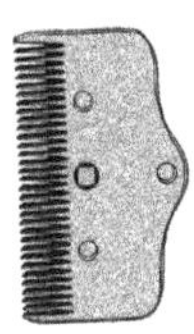

Perfection
Top and Bottom Plates.

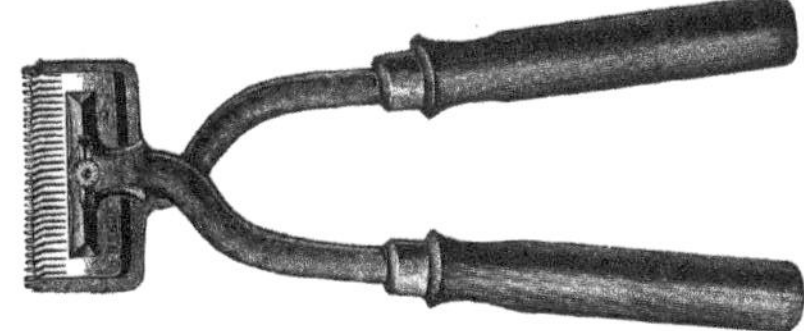

***Boker's* Horse Clippers**
***Burman's Newmarket* " "**

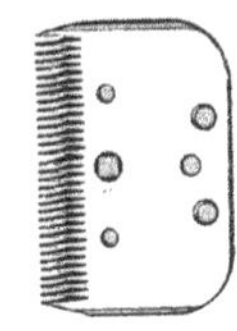

Newmarket
Top and Bottom Plates

Stewart's Patent Knife Plates
Top Plate, One Hole.

Oak Lake Horse Clipper
Three in One. Nickel Plated.

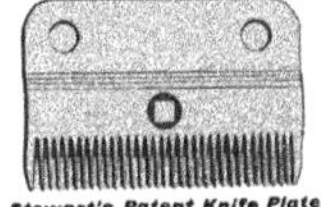

Stewart's Patent Knife Plates
Bottom Plate, One Hole.

20th Century Clipper
Has 12-in. Positive Gear Drive; Steel Flexible Shaft 4½ feet long. Complete with one set Stewart One-nut Tension Knives Weighs15 lbs.

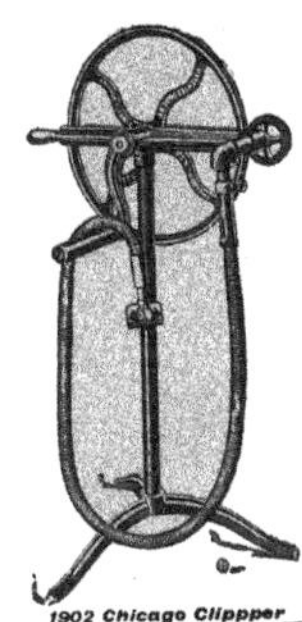

1902 Chicago Clippper
All Gears Cut from Solid Metal.

New Stewart
Inclosed Gear.

For fuller particulars see our Hardware Catalogue

H. S. HOWLAND, SONS & CO., LIMITED

Opposite Union Station.

GRAHAM NAILS ARE THE BEST

Factory: Dufferin Street, Toronto, Ont.

Our Prices are Right | We Ship promptly

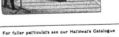

Canada Leads the World in the Manufacture of

HIGH-CLASS SAWS

No. 81

The best and cheapest Hand Saw on the market, quality, temper and finish considered. Every saw warranted. Carved applewood handle. Patented, wave finish blade.

No. 1 Maple Leaf Lance Tooth Cross-Cut Saw, and the No. 0 Racer are tempered under our secret process.

No. 0 Narrow Racer Cross-Cut Saw

If you want saws that will sell and give your customer satisfaction, you will stock the Maple Leaf Saws.

Manufactured by **SHURLY & DIETRICH, - GALT, Ontario**

Brantford—Roofings

This Represents Our

"ASPHALT"
No. 1, 2, 3
"RUBBER"
No. 1, 2, 3

Each Roll is complete with Nails and Cement to lay. Best selling goods in the market.

BRANTFORD ROOFING CO.
LIMITED
BRANTFORD, ONTARIO

SIMPLE

DURABLE

RELIABLE

STYLE F.

THE LEADER

A Name Familiar to Butter Makers in Connection with the

Highest Grade Churns

ITS STRONG POINTS:

Stiff Steel Frame, with malleable iron fittings, Combination Bow Lever and Pedal Drive. Bicycle Ball Bearings. Bolted Trunnions. Cream Breakers. Barrel easily drained and conveniently detatchable for cleaning.

DEALERS: When you handle this line you handle QUALITY.

The Dowswell Manufacturing Co., Limited

W. L. HALDIMAND & SON, Montreal
EASTERN AGENTS

HAMILTON
ONTARIO

GALVANIZED FENCE HOOK FOR FASTENING WOODEN PICKET OR WIRE FENCES

WIRE NAILS, COILED SPRING, BARB and PLAIN FENCE WIRE, OILED and ANNEALED, CLOTHES LINE WIRE, STAPLES, etc.

THE WESTERN WIRE & NAIL CO., Limited. —LONDON, ONT

The Peterborough Lock Manufacturing Company, Limited

Peterborough, Ont.

Manufacturers of all kinds

Rim and Mortise Locks, Inside, Front Door, and Store Door Sets, also full line High-class Builders' Hardware.

Sold by all Leading Jobbers in the Dominion.

Cylinder Night Latch, No. 103.

Four Reasons why INVICTUS FIBRE is the BEST FIBRE PAPER on the Market

1. Invictus Fibre is made from personally selected pulp wood cut on our own limits.
2. This special pulp wood is again rigidly culled on arrival at our mills
3. The pulp from which Invictus Fibre is made is manufactured in our own pulp mill and transferred at once to the paper mill, avoiding all drawbacks of transportation.
4. Invictus Fibre will increase in strength and not grow hard and brittle as does fibre made from sulphite pulp. This is a quality of the peculiar and selected pulp from which Invictus Fibre is made.

A trial order will convince you that we are correct. See that each package bears the words "Invictus Fibre." Every ream contains 480 sheets

TRADE MARK

This design a guarantee of quality.

CANADA PAPER CO., LIMITED.

"Headquarters for high grade Fibre and Manilla Papers"

Toronto. Montreal. Windsor Mills, P.Q.

Don't Stop

to ask your neighbor, lift the load yourself with **The Burr Self-Locking Tackle Block** Saves the labor of two or three men. 600 to 5,000 pounds capacity.

Lewis Bros., Ltd., Montreal
Munro Wire Works, Winnipeg
Vokes Hardware Co., Toronto
F. Hamilton & Co., Hamilton
The Burr Mfg. Co., Cleveland, Ohio

Cap Screws, Set Screws, Machine Screws, Cold Pressed Nuts, Studs, Coupling Bolts, Special Milled Work, etc.

CANADA FOUNDRY COMPANY

LIMITED

HEAD OFFICE and WORKS, TORONTO, ONT.

District Offices: Montreal, Halifax, Ottawa, Winnipeg, Vancouver, Rossland

---

Scrap Iron for Re-working

Better bars for re-working are obtained from selected scrap than from pig iron—have more life and strength.

London Bars are made from selected scrap and come in Flats, Rounds, Squares, Ovals, Half-Ovals, Half-Rounds and Bands.

Large stock always on hand, insuring prompt shipment. 61

London Rolling Mills
London, Canada.

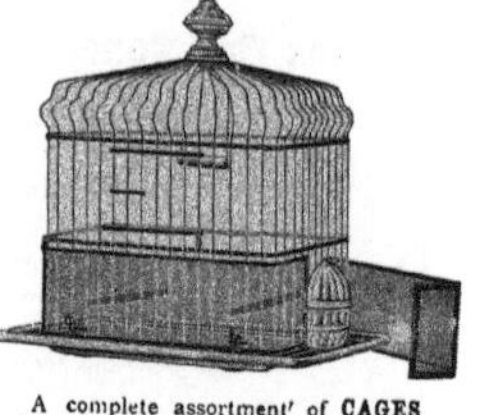

A complete assortment of **CAGES** always in stock; All sizes, styles and prices. Brass and Japanned. With or without guards. Discount on application.

E. T. WRIGHT & CO., - HAMILTON, CAN.

The Bowmanville Foundry Co., Limited
Bowmanville, Ont.

Makers of Good Goods Only.

Expert Bookkeeping

Investigation and auditing of all classes of accounts undertaken. Trading, profit and loss accounts and balance sheets prepared. Sets of books opened and office systems devised in an up-to-date manner. Collection of accounts a specialty.

PERCY P. DAVENPORT

622 McIntyre Block, - WINNIPEG, MAN.

Square your accounts

INTERNATIONAL PORTLAND CEMENT COMPANY, Limited

OTTAWA, - - ONTARIO.

Manufacturers of strictly high-grade Portland Cement made by the most improved method of manufacture in existence. Each barrel contains three hundred and fifty pounds of pure Portland Cement with no adulteration.

The Hanover Portland Cement Co., Limited
HANOVER, ONTARIO

—Manufacturers of the celebrated—

'Saugeen Brand'
OF PORTLAND CEMENT

Prices on application. Prompt shipment

Persons addressing advertisers will kindly mention having seen this advertisement in Hardware and Metal.

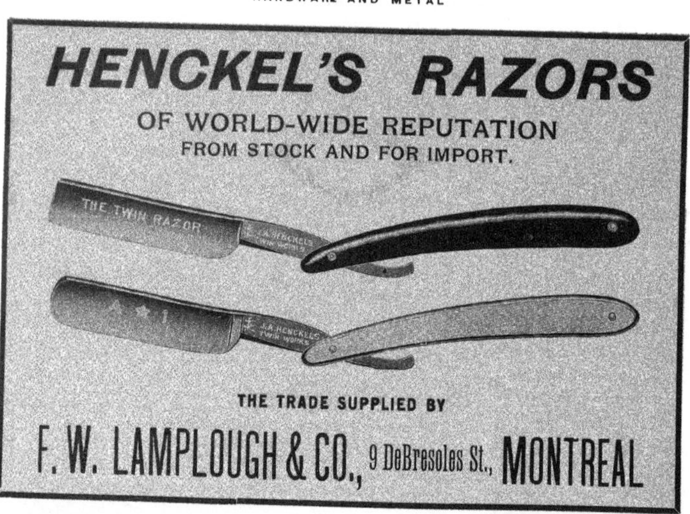

The John Morrow Screw, Limited

MANUFACTURERS OF

Set and Cap Screws, Special Milled Work, Engine Studs Etc. Cold Punched Nuts of every variety of finish.

INGERSOLL, ONTARIO

Covert Mfg. Co.

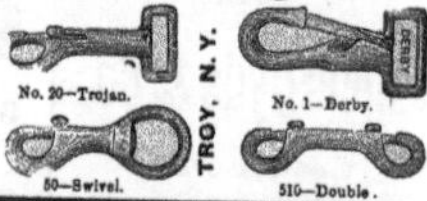

CELEBRATED **HEINISCH** SHEARS

Tailors' Shears,
Trimmers' Shears,
Tinners' Snips, etc.

ACKNOWLEDGED THE BEST.

R. HEINISCH'S SONS CO. NEWARK, N.J., U.S.A. NEW YORK OFFICE, 155 Chambers St

STREET PAVING and SIDEWALKS a SPECIALTY

SILICA BARYTIC STONE CO.

OF ONTARIO, Limited.

Water Proof Floors for Malt Houses, Breweries, Slaughter Houses, Cheese Factories, Cellar, Stable Floors, etc.

Head Office
Ingersoll, Ontario.

Walter Mills, General Manager
Ask for quotations for Septic Tanks.

SEYMOUR SHEARS

HAVE BEEN THE
Standard for over Half a Century.
"QUALITY UNQUESTIONED."
Each pair of our shears bears the above trade mark.

Complete Line *TRIMMERS', BANKERS', BARBERS' and TAILORS' SHEARS, Etc., Etc.*

Latest Catalogue will be sent in exchange for your business card.

Henry T. Seymour Shear Company

WIEBUSCH & HILGER, Limited, NEW YORK, Sole Agents.

"The Tool Holder People"

ARMSTRONG BROS. TOOL CO.

Manufacturers of Armstrong Patent Lathe and Planer Tools and other machine shop specialties.

106 N. Francisco Ave., CHICAGO, ILL.

Persons addressing advertisers will kindly mention having seen this advertisement in Hardware and Metal.

Dennis' Iron Stable Fittings

Sanitary, Indestructible, Economical

A good line for the hardware dealer.
Send for New Stable Fittings Catalogue and Price List "A."

DENNIS IRON & WIRE WORKS, Ltd.
LONDON, - ONTARIO

Buy the Best.

HERCULES

Sash Cord.

Star Brand Cotton Rope
Star Brand Cotton Clothes Lines
Star Brand Cotton Twine

For Sale by all Wholesale Dealers.

B. A. GRAUTOFF & CO.
8 Fenchurch Buildings, London, E.C.
Import and Export.
Old established house with highest connections in Great Britain and on the Continent of Europe.
Cable Address, "Grautoff London," Liebers Code.

A HANDY METHOD for repairing boxes or binding 2 or 3 small ones together, use

Warminton's Clutch Nails

Also strapping, hoop iron and specialties for the shipping dept.

Write for samples to

J. N. WARMINTON
43 Scott St., Toronto 207 St. James St., Montreal

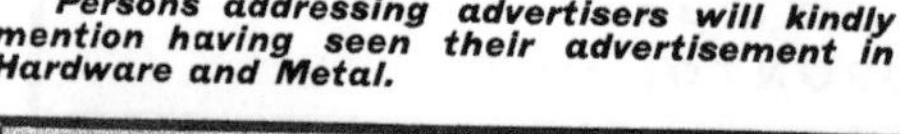

Persons addressing advertisers will kindly mention having seen their advertisement in Hardware and Metal.

SPADES, SHOVELS, Etc.

Entrust your trade to the house that can supply the "just right" **Spades** and **Shovels**.

By getting in touch with us you will become acquainted with the favorite lines:

TRADE MARK

"Peterborough," "Lundy," "Irwin," and 'Hill"

Send in a postcard for our Catalogue and you will see where we stand.

The Peterborough Shovel & Tool Company, Limited, Peterborough, Ont.

Dana

Peerless Freezers

Increase your freezer business.

Takes a woman off her knees

by selling the freezer the women want—because with *it* they make ice-cream with *real* satisfaction. The cook's favorite. Known to all women through advertisements in their favorite magazine.

ASK YOUR JOBBER

THE DANA MFG. CO., Cincinnati. 10 Warren St., New York.

Write for Dana Food Chopper Catalogue

Peerless Woven Wire Fence

Here is a fence that will sell. Once up no need for repairs. It lasts for years—just what you are looking for. It's horse-high, pig-tight and bull-strong. No stock can get over, under or through it. The most unruly animal can't butt it down and there are no barbs to tear and injure stock. It's

WIND AND WEATHER PROOF

The Peerless lock holds it perfectly rigid and the wires can't slip up, down, nor sidewise. Cold weather won't snap it and hot weather won't make it sag. It's the greatest thing out for farmers, stockmen and all other fence users which makes it a red hot proposition for dealers. It comes in rolls all ready to put up. It's easy to handle. Write for prices.

The Banwell Hoxie Wire Fence Company, Ltd.
Dept. J, Hamilton, Ont.

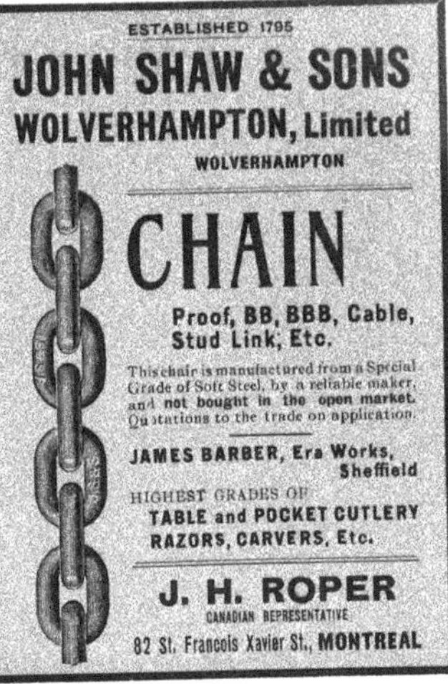

OF MUTUAL INTEREST

A Page of Gossip and Shop Talk

Many complaints have been received from subscribers during the past month about the non-receipt of copies of Hardware and Metal sent them. In every case the fault must rest with the postal department as our mailing has been done promptly each Friday evening and most Ontario subscribers should receive the paper on Saturday. It is annoying to pay for a paper and not get it, but it is also annoying to mail a paper and then be blamed for non delivery.

* * *

The postal department is undoubtedly overtaxed during the holiday season, but when it takes four days to deliver a letter mailed in Toronto to another Toronto address and Christmas parcels received in Toronto before Dec. 25 are not delivered until New Year's Day, there is something radically wrong with the postal service. The desire to conduct the service on too cheap a scale is undoubtedly one reason for the incapacity of the department. No business (public or private) can be conducted successfully by treating its empioyes in such a niggardly way as the country postmasters and the city postman are being treated. They should be paid decent wages and expected to give good service in return. One result of the poor remuneration given country postmasters is their recent demand that they be allowed to compile lists of farmers names for sale to the mail order houses.

* * *

Talking about the postal department brings to mind the agitation Hardware and Metal has helped to foster in opposition to the proposed parcels post c.o.d. collections system, by which country postmasters would become collection agents for the mail order houses—and undertakers for some of the smaller local merchants whose taxes help to keep the smaller municipalities going. Hardware and Metal took up the agitation as soon as it was learned that the mail order houses were sharpening their knives in preparation to take their pound of flesh from the retail trade. As a result of our exposure of the plan dozens of Members of Parliament have promised to guard the interests of their constituents and thousands of letters of protest have been sent to the Postmaster General. Meetings of local merchants have been called and dry goods men, hardware men and other retailers have co-operated by sending a joint letter of protest to Ottawa.

* * *

The Ontario Retail Hardware Association has also done good work by notifying every Ontario hardwareman of the danger of the proposed legislation and the following letter is only one of dozens received showing that the trade is at work and "if everybody hustles success seems easy."

Leamington, Dec. 27, 1906.

Dear Sirs,—On the receipt of Hardware and Metal over a week ago the writer personally called on several business men in town requesting them to write to Mr. Clarke, who is the member in Ottawa representing South Essex. We also personally wrote, covering the whole ground as fully as possible. His home is in Windsor, and he is now home for Christmas holidays. He wrote from the latter place stating he would be down here before he went back to Ottawa, some are looking for him every day. If everybody hustles success seems easy.

Yours tuly,

Greenshield & Moffat.

Leamington, Ont.

That it is the trade newspapers that the retailers must look to to protect their interests has been amply proven again and again. Associations are a splendid thing. They are necessary and no sensible body of men engaged in similar work will remain unorganized. Men go into business for a profit and they should organize to ensure that they get a profit rather than a deficit But no association of retailers can prosper without the support of the trade newspaper.

The subserviency of the daily press to departmental store interests is also beyond doubt but if additional proof was required it is supplied by their suppression and misstatement of the facts in connection with the killing of a woman in an elevator accident in Eaton's Toronto store, and the serious injury of several girls when a crowded elevator fell in Simpson's store a fortnight later.

* * *

Another incident showing the value of the trade newspaper was furnished last week by president A. W. Humphries, of the Ontario Retail Hardware Association. Mr. Humphries wanted to sell a set of tinsmith's tools. He advertised twice in the Globe, got several answers but did not get an offer for the tools at the price asked. He advertised once in Hardware and Metal. The ad. was received too late for classification and Mr. Humphries failed to find it when looking through the paper. He got several replies, however, and looking again found the ad. One who replied was Clayton Brown, Enterprise, Ont. Being told that the price was $250 and assured that they were as represented he immediately sent a cheque for the amount without seeing the tools. When he did receive them he was so delighted that he wrote his thanks to Mr. Humphries, and said he would not re-sell the set for less than $350.

It doesn't need a sledge-hammer to drive home the moral: Hardware and Metal want ads. bring results.

* * *

The editor has some goods things in store for readers of Hardware and Metal. There are some bright and brainy Canadian retailers who know how to use the pen and several of them have promised to give readers of the paper the benefit of their experience. Every retailer will be interested in the article on "Retail Business Methods," by A. E. Provost, Ottawa, in this issue and even more interest will be taken in other articles which are to follow. At present we can announce the early publication of an able article on "The Percentage of Profit," by Fred C. Lariviere, Montreal, two on "Why I Started the Cash System," and "The Cash System in Operation," by A. E. Cote, Morris, Man., and another interesting contribution by M. F. Irwin, the new manager of McClary's branch at St. John, N.B..

Our Ontario man got in conversation with a traveler for a Western Ontario woodenware house the other day in the town of Bolton. After the usual salutation, "What's your line" they started in to discuss trade papers. Our woodenware friend at once remarked that he wished their firm would advertise in Hardware and Metal, as it would make it so much easier for him to sell goods, and would be a great help in opening new accounts. He stated that he would try and induce his manager to take space in Hardware and Metal this year, as he had great faith in the results, and said his house would thank him for the suggestion at the end of next year.

"I am well pleased with the improvements constantly being made in Hardware and Metal," is the way L. Wale, of Stevensville, Ont., expresses himself in a letter in which he renewed his subscription.

I consider Hardware and Metal the hnest trade paper in the Dominion, recently wrote E. P. Paulin, Goderich, one of the active executive members of the Ontario Retail Hardware Association.

"I appreciate the usefulness of Hardware and Metal and as surely as it is an absolute necessity amongst hardwaremen so is good window dressing an absolute necessity for prosperity, and in my opinion the cheapest method of advertising," wrote W. G. Pow, recently. Mr. Pow was on the road for the Hobbs Manufacturing Co. last summer, but is now managing W. R. Hobbs' retail store at Tillsonburg.

THE TURN OF THE YEAR

A Glance at the Past and a View Into the Future — Comments on Trade Conditions by Jobbers and Manufacturers.

"The ——— company is doing the largest business in its history."

The above can be said of almost every business concern in Canada. Fill in the name of any manufacturing company or any jobbing or retail house and a story is told that in all but a very few instances is true, so great is the general prosperity in all parts of the country.

The remarkable volume of trade done in 1905 made it difficult to promise better results in 1906, but with the exception of a few places building and general business has totaled to a greater volume in 1906 than in the twelve months previous.

Looking Forward.

The same conditions would make it hard to forecast the possibilities of 1907 were it not for the heavy booking of orders which manufacturers report, the capacity of many plants having already been contracted for as far ahead as nine months. Prices of nearly every line are very stiff and the tendency is still upward, advance following advance so fast that few are able or desirous of stocking up with goods heavily.

Merchants can well afford to mark up all stocks in hand to present ruling prices as seems unlikely that there will be any decline during 1907, while it is improbable that stocks can be replenished at former prices. Far from affecting the demand for goods adversely, the high prices seem to stimulate trade and encourage even greater buying, all classes appearing to be prosperous and willing to spend money freely. Unless something unforseen happens, the coming year's business will eclipse the splendid records made during the past two years.

Big Year for Building.

Alex. McArthur & Co., the well known paper manufacturers, stated that the past year had been an active one in all the various lines of building paper, owing to the new buildings and extensions which have gone up all over Canada, in keeping with the prosperity which seems to pervade every section. A better appreciation by the public is also noticeable in regard to the insulating advantages of paper and felt on all kinds of buildings, the extremes of heat and cold within them being reduced to a minimum at a trifling cost. The cost of material has advanced and prices of the product are likely to remain firm.

New Factories Ensure Competition.

Adam Taylor, Secretary of the Taylor-Forbes Co., Guelph, states that his firm looked for big business in 1906, and their expectations have been fully realized. The demand for hardware lines and heating goods has been extraordinarily large, and prices have been very low in comparison with the cost of materials and labor. During 1907 higher prices are certain to be the rule, as it is evident that iron will continue high during 1907 and lower prices cannot be expected until well on into 1908. Hot water boilers are likely to advance at once, while quotations on hardware lines are withdrawn, and all orders subject to the prices ruling on the day the order was taken. Mr. Taylor was very enthusiastic about the outlook for trade in Canada in future, but pointed out there is certain to be plenty of competition. The numerous new plants and extensions to old ones, will provide for a very large increase in the output of Canadian factories and foundries.

Paints "Brightening Up."

In discussing the paint and oil situation, William H. Evans of the Canada Paint Company, remarked to "Hardware and Metal" as follows:

"The year 1906 has been characterized by a tremendous enquiry for painting material of every description. This industry has made rapid strides in Canada, and capital has not been "backward in coming forward" to aid and build up the rapidly increasing industries of varnish and paint making. Linseed oil and turpentine have kept somewhat along steady lines, but white lead has met with a sharp advance which has been well maintained and zinc has also felt the world's call for an increased supply of metals in the arts and industries.

Profits to some extent have been curtailed, owing to the prices received from the trade not keeping pace with the enhanced values which prevail for a large number of the chemicals and crude articles which enter into the composition of paint colors and varnishes. Labor is much higher, and all through the lists costs show a marked appreciation which will have to be reckoned with in marking prices for 1907.

Even at the advances which are inevitable, a very large business for 1907 is assured, and the large factories seem to be running on full time right through the holiday season to keep pace with the healthfulness and enthusiasm which is everywhere apparent in Canada."

Heating Goods Active.

W. J. Cluff, of Cluff Bros., Toronto, selling agents for Warden King & Sons, Montreal, is enthusiastic over the amount of business his firm has done during the past year. The demand for all classes of heating supplies has been phenomenal, and their chief difficulty has been in securing prompt deliveries of supplies which are not manufactured by themselves. Mr. Cluff says that conditions will be better in the future, as arrangements are under way to manufacture radiators and other supplies, so that their customers can have the advantage of securing goods of one manufacturer. During the past year, Cluff Bros. have erected a magnificent warehouse in Toronto, and they are now building up their stock in order to keep pace with the increased demand they look for during 1907.

A Year of Prosperity.

"We believe that no such year of prosperity as the one which is past has ever been known in the hardware and metal trades," said W. J. Fuller, president of the Canadian Fairbanks Company, Montreal. "Prices have been constantly advancing, and if the merchants had taken proper advantage of increased values of their stocks they should show very handsome profits at the end of the year.

"During the year the Canadian Fairbanks Co. has put into operation its large plant at Toronto for the manufacture of gas and gasoline engines, steam pumps and other of its specialties, and is endeavoring to become as rapidly as possible, handlers of Canadian-made goods only. They have in view the coming year the establishment of their scale and valve factory, for which a site has already been secured in the Eastern Townships and which will be in operation before June 1st. Every one of the factories which the company represents is filled with orders for months to come, and there is every prospect that the year 1907 will see a much larger volume of business than the year 1906.

"We are carefully on the outlook for any indications which will point to a climb in prices in raw materials so that we may take proper notice and not be caught with stocks of high-priced goods. This is something which every one must realize the importance of."

Prospects for 1907.

"The year just closed has been the best since the branch opened here. As far as I can see the prospects are equally good for 1907 so it will be up to me to keep up the record," writes M. F. Irwin, the new manager of the St. John, N.B., branch of the McClary Manufacturing Company. The McClary business at St. John is steadily growing and the enlargements and changes now being made should mean a considerable increase in their Maritime trade. Being a practical tinsmith, Mr. Irwin should make a success of the new factory being established to supply the eastern trade.

Poor Deliveries Chief Fault.

James G.. Lewis, vice-president of Lewis Bros., Montreal, said the year that has just closed has been the most prosperous in the history of his company, they having found the trade anxious to buy goods and the fault has been the poor deliveries made by the many factories. There is no doubt that the present century will belong to Canada to a larger extent than to any other country and the amount of the purchases made by the trade and the hardware merchants in particular will astonish those who have been in business for the last twenty years. For the present year they recommend all retail merchants to place their orders at once, for, while deliveries may be bad now, they will be worse a few months hence. Those who have goods to sell should ask the market price as they will have no trouble in disposing of their stocks.

Big Year in West.

W. S. Brock, western representative of the Ontario Lead & Wire Company and the Standard Ideal Sanitary Company, who has been in Toronto during the holiday season, states that last year was the best in the history of the plumbing trade in Western Canada. There has been a tremendous demand for all classes of supplies and it has been difficult to keep up with the demand, but Canadian manufacturers have succeeded in doing so with a very small share of the trade going to United States houses. Mr. Brock looks for an equally good trade during 1907; the building industry in the West showing no signs of falling off.

Guessing as to the Future.

The Thos. Davidson Manufacturing Company had a fine year and stated that for the year 1906 the stamped and enamelled ware trade along with other trades using steel as the basis of their supplies have paid advancing prices for their raw materials and had difficulty in obtaining deliveries. The volume of trade was large, but the prices of the products were not in keeping with the advanced cost. There, however, were practically no bad debts, which will compensate to some extent for the low prices. The trend of the business seems to be manufacturing in larger quantities and establishing distributing depots to accomplish quick deliveries in all parts of the country. Besides the advance in the price of steel two of our other raw materials have advanced to prices making a record, viz., pig tin to 43c. per pound and sheet copper to 29c. per pound, so that we are kept guessing as to the future.

Higher Prices Predicted.

"Regarding the outlook for the mixed paint and white lead trade during the coming season," writes John A Straith, president of the Standard Paint and Varnish Works Company, Windsor, Ont. "in our opinion it is exceedingly bright. If we were to judge the future by past experiences it is bound to be good. Our past year's business showed the usual increase on previous years business, and we have no reason for complaint. In our opinion all paint manufacturers along with ourselves will be obliged to further advance prices on mixed paints as well as white leads in the early spring. The present high prices of every commodity entering into the manufacture of a good product, will we think, necessitate this move."

Enlarging Rolling Mills.

The Montreal Rolling Mills Company said in their opinion the year just closed had been one of the busiest and most prosperous in the experience of all classes of industry in this country. As far as they were concerned they had not been able to fill all the orders that had been sent in to them. The great development all over the country, particularly in the Northwest, has created an unprecedented demand for material manufactured by them such as rail fastenings, consisting of spikes, bolts, fish and tie plates, also wire nails, wrought pipe, etc.

Their aim was always to keep abreast of the times in the way of facilities for turning out goods that were most in demand, and in such quantities as were called for; and with this end in view had now in the course of construction additions and alterations to their plant, which they hoped would place them in a position to take care of the wants of their many customers.

Big Demand for Paints.

Mr. Benson, manager of the Sherwin-Williams Toronto branch, reports a very brisk year for 1906. Some of their travelers while at the Cleveland conference estimated that 1907 sales would show an even greater increase. Everywhere building is active and that means a big demand for paints.

Cordage Industry Depressed.

A well known authority on binder twine gave the following more or less gloomy opinions:—

The binder twine, cordage and twine requirements for the past twelve months have naturally increased with the growth of the Dominion; and it is regrettable this expansion, for following reasons, is not benefiting those factories established in Canada:—75 per cent. of the binder twine for Canada's harvest is imported, being on the free list to the world; the greater portion from the United States, into which country Manila binder twine, exported from Canada, must pay 45 per cent. The British preferential tariff permits the importation of much low grade cordage, bearing the name of, but containing little, Manila hemp. The equal of this is not acceptable to buyers from home manufacturers. At two Canadian cities convict labor is employed making binder twine and cordage. Canada is gradually becoming a veritable dumping ground for low grade cordage.

These conditions during the past three years have caused the failure or forced liquidation of seven factories. The dozen existing manufacturing plants, contending as best they can with such serious obstacles, have now by the last Government Budget to face another, the free listing of cordage used by fishermen. Canadian grain, fish and farmers are second to none in the world, and surely do not require protection at the expense of one of the country's leading industries, especially when there is so much competition within the country to regulate prices. With some re-adjustment by the powers that be, the 1907 outlook, so bright for the various industries in our flourishing Dominion, could be considerably improved for the large interests in the cordage manufacture.

A Banner Year.

Rice Lewis & Son, wholesale hardware merchants of Toronto, report a banner year for 1906. As they were at all times taxed to their utmost to fill orders, the manufacturers being largely to blame for delays in shipments as goods could not be turned out quick enough. Stocks are being rapidly repleted, however, and in many cases extensions added to factories to meet with the demand. The prospects for 1907 are even brighter than a year ago, orders coming in steadily.

Excellent Year's Trade.

H. S. Howland, Sons & Co., wholesale hardware merchants of Toronto, have had an excellent year in all branches of their business, orders being received steadily and business being very even throughout the year. The outlook for 1907 is for an even greater volume than in 1906, prices being firm and stocks low owing to the big demand and the scarcity in many lines. The nail situation has for quite a time been very trying to both jobber and retailer. In a great number of cases orders could not be filled, and had to be transferred for spring shipment. Manufacturers are rapidly repleting their stock in view of the large demand looked for during 1907.

TRADE INQUIRIES.

Correspondents desiring to get in touch with any of the firms referred to should quote the reference number when requesting addresses. For information write to Superintendent of Commercial Agencies, Department of Trade and Commerce, Ottawa.

1518. Nuts and bolts—A Lancashire manufacturing firm desires to obtain prices of nuts and bolts from Canadian exporters.

1519. Handles—A Lancashire firm asks for prices of spade and pick handles from Canadian manufacturers of same.

1520. Hardwood stems—A Lancashire firm asks for prices of hardwood stems, 2 feet 8 inches long by 1 1-2 inches diameter at one end, reducing to 1 7-16 inches at other, from Canadian exporters.

1544. A London firm, in the paint, color and varnish trade, contemplates appointing agents in different parts of Canada, and is desirous of hearing from persons who are in a position to entertain such business.

1552. Fireworks—A London firm, manufacturing fireworks of all descriptions, wishes to get into touch with a first-class Canadian firm prepared to introduce and carry stocks of their goods.

1553. Gas stoves—A Birmingham firm, manufacturers of gas stoves, would like to get into touch with Canadian importers of this line of goods.

1554. Motor cars, motor cycles, forges, etc.—A Birmingham firm manufacturers of motor cars, motor cycles, cycles, forges, bellows and anvils, would like to be put in touch with Canadian dealers in these lines of goods.

1555. Electrical lighting apparatus —A Birmingham firm desire to get in touch with Canadian imports of rough electrical lighting apparatus for mines, factories, etc.

1556. Corrugated iron—A Birmingham firm would like to get in touch with several firms in Canada with a view to doing business in corrugated iron.

RETAIL BUSINESS METHODS

An Address by A. E. Provost, Wholesale Grocer, Before the Ottawa Retail Grocers' Association.

"We are not here this evening," said Mr. Provost, "as a mutual admiration society, but as a company of business men met for the very practical purpose of discussing the best methods of carrying on our work. In the few words that I will address to you I wish to speak somewhat as I would to a young man who is about to take charge of a business for me, giving him such points and suggestions as I think might be of advantage to him.

The first consideration for a man embarking in a business is the selection of a stand; but as all of those present are probably located in premises where they expect to remain, I need not do more than mention that point.

Relations With Customers.

Once started in business, the next point you have to consider is what should be your relations with your customers. I would strongly recommend you to make the personal acquaintance of each one of your customers. Take occasion once in a while, if it be only once a week, but whenever you have the chance, of paying some personal attention to each one, waiting upon them, having a little chat with them, or passing the time of day. In this way you will make them feel not only that you are glad to see them in your store, but that they will be glad to come. Your personal acquaintance with your customers and establishing good relations with them in this way, will have the effect of attaching them to you so that you will not be, as some merchants are, largely at the mercy of their employes. For, remember that even your most popular salesman of to-day may be to-morrow in the employ of your business rival, and, but for your personal hold upon your customers, might induce some of them to transfer their trade to his new employer. It is of course well to have salesmen of pleasant and obliging manners, but these are not all the qualities that they ought to possess. It is far more important that they should have that supreme qualification, the ability to sell goods, because that is what they are there for, and not merely to be entertainers of your customers.

Not Too Familiar.

But here I must warn you against being over familiar with your customers, which is almost as great a fault as to be wanting in courtesy. My experience has been that very often a customer with whom you become on too friendly or intimate terms is apt some day to presume on this to ask you for favors which you cannot, in justice to yourself or your business, grant. The true relation between a merchant and his customers, as it appears to me, is something like that which exists in departmental stores between the heads of departments and floor walkers and the public. Thus far, but only thus far, do the proprietors of these great establishments appreciate the importance of personal relations with their customers. That is why we find them engaging for these positions the very merchant whom they have crowded out of business for themselves in the stress of competition.

Good Goods Always.

The next question for consideration is the purchasing of stock. It is very easy to say that the true rule is always to buy the best and buy it right, but as you all know it is pretty hard to carry

A. E. Provost, Ottawa.

that rule out in practice. At the same time, I cannot lay too much stress on the importance of handling the very best goods that are suitable to your particular trade. I must confess that I have a special prejudice against low-grade goods. If a customer tells you that some goods you sold him were not up to the mark, what satisfactory argument or defense have you in reply? No one can defend inferiority. Quality is remembered long after price is forgotten—that is an old and true adage in business of every kind. There is not a man in this audience who complains of the price of his wife's hat if he thinks she looks pretty in it, and in business price is always secondary to quality.

Not Too Much at Once.

Beware of the quantity men who try to induce you to purchase large quantities of one line of goods, using the deceptive argument that five cases are so much cheaper than one. In my comparatively brief experience in the wholesale trade I have been one of the victims of several failures and I have invariably found that these large-quantity goods comprise most of the stock of the men who failed while they had not the means to stock properly with goods more commonly in demand. One of the favorite arguments of these quantity men is the long terms offered—sixty days from the first of next month, and so on; but many a man has found that even after the sixty days had expired and he had to pay for the goods, a large proportion of them were still on his hands.

Speak to All Who Enter.

In the selling of your goods, a word of salutation or greeting to your customers when they come to your store is such an elementary requirement that it is perhaps superfluous to suggest it. But in the press and hurry of business, men are very apt to overlook such little acts of courtesy, and let customers wait for some minutes without noticing them. Even when serving one customer, it is very easy en passent to say a word of greeting to another who has just come in, with the assurance that you will wait on them in a few minutes.

Above all things, tell the truth about the goods you are selling as far as you know it, and avoid exaggeration. By over-praising the qualities of the goods you are selling to-day, you may be unconsciously depreciating the qualities of others which you may have to sell to-morrow. Do not be too anxious to convince your customer of the high qualities of one kind of goods or the inferior qualities of another kind; you can safely leave your customer to find out some of the qualities of the goods he is buying for himself, and some of your customers may use some of the undigested ideas you may have given them to your own disadvantage at some future time.

Credit to New Customers.

Most of you sell largely on the credit system. I would suggest that when a new family comes into your neighborhood and begins to do business with you, it would be well to post yourself upon their financial standing with the dealers in the neighborhood from which they have come. Sometimes they will at first send a child for a small purchase, and you can often learn from a child at such time what you want to know about the family, the father's occupation, etc., so that when the parents come, as they certainly will shortly after, and ask you for credit, you will be in a position to say definitely whether you can give them credit or not. And when you open an account with a new customer, make a definite arrangement as to terms of payment, and see that he lives up to it as far as possible. If on pay day he cannot settle the account, give him to understand that you expect him at least to report and explain the reason why.

With regard to the delivery of goods, it is now a common practice on the part of the larger stores to have from two to four deliveries daily at fixed hours. Experience has proven this to be the very best method. Your customers know the hours of delivery and act accordingly, and you are saved a great deal of expense and trouble.

Watch the Big Advertisers.

As most of you are not large advertisers, I would suggest that you take advantage of the advertisements of the larger stores. Read them regularly, and whenever you find goods advertised that are likely to appeal to your patrons, bring these goods to the front and ticket them at the prices advertised by the large stores. You will be surprised at the effect. For instance, a lady, just after reading in one of these advertisements of 3 pounds of raisins for 25 cents in a big departmental store, walks into your shop; and if she sees a ticket on a box of raisins offering that at the same price, she is at once inspired with confidence in your ability to sell not only raisins, but every other kind of goods as cheaply as the large stores.

Read the Trade Papers.

In this connection, I would recommend you to read the trade papers, which will give you a knowledge of the trade which will enable you to command the market and buy goods at the right time and of course at the right prices.

Just a word about your dealings with your creditors. Do not return a draft because there is some trifling mistake in it, perhaps a consignment has been short by one package. It will be good policy on your part to trust for a few dollars the men who trust you for hundreds, and you will find that you will never have any difficulty about having little errors of this kind rectified. Be careful never to write a cheque which there are no funds to meet. This is simply ruinous to a merchant's credit. These cheques go through the clearing house, and within twenty-four hours after they are written every bank in Ottawa knows that a cheque of yours has not been honored. Some day you may be surprised to find that you are refused a little accommodation by some bank by reason of this neglect on your part.

CANADIAN HARDWARE STORES
No. 10.

BUSY CALGARY STORE.

We are glad to be able to show in this number another of the many fine hardware stores in the Canadian Middle West, that of Linton & Hall, Calgary.

A. T. Linton and E. C. Hall, two Campbellford boys joined forces early in the Fall of 1905 and opened up their hardware store in the "Armstrong Block." Both of these young men being practical in every way and having the advantage of careful training, backed up with the other good qualities necessary to spell the word "Success" viz Perseverance, Energy and Push, their venture from the very first day has been a most successful one.

CANADIAN HARDWARE STORES NO. 10—Linton & Hall's Store at Calgary, Alberta.

Mr. Linton has served with several of the foremost hardware houses in Ontario, beginning as he did with the old pioneer firm of C. S. Gillespie, of Campbellford, Ont. After seven years faithful work there he went with J. W. Walker, of Belleville. Afterwards with John Lewis of the same city and R. R. Casement of Madoc, and finally took a position with the Marks-Clavet-Dobie Co., of Port Arthur, remaining there for several years. Hearing so much of Alberta and trusting much to the great future of the Middle West as a young man's country he came forward at the solicitation of Mr. Hall to join him and begin their career as hardware merchants in the growing city of Calgary in sunny Alberta.

Mr. Hall is not a newcomer by any means, in fact he may be considered something of a pioneer of Calgary. He became identified with the interests of the city eight years ago, and joined the staff of A. McBride & Co., one of the oldest houses in the hardware line in Alberta, remaining with them until the present firm of Linton & Hall came into existence. Mr. Hall is one of the most energetic officers of the Calgary Fire Department, rising from the ranks to the position of Captain of the Aerial Truck Company, besides being a member of several other societies in the town.

Linton & Hall carry a stock of shelf and heavy hardware, paint, oils, glass, etc., and are City Agents for the Gurney Standard Metal Co., of Calgary, being a branch of the Gurney Foundry Co., of Toronto, who make the celebrated "Oxford Chancellor" stoves and ranges. Finding their quarters in the Armstrong Block limited and the necessity of more room crowding them out they sought and found quarters in the "Victoria Block" on the Main Street, a most pretentious building of Alberta sandstone in the very centre of the city. Business expanded with the new move and even now there is no room to spare. Success in any business does not come easily—it means hard persistent work. And both these young men are workers, and they have earned what they possess. —A good business and a good name in the commercial world.

It's the man who never does anything who is always Jonny-on-the-spot when it comes to telling how things should be done.

The Board of Trade of Minnedosa, Man., is at present concentrating its efforts on the organization of the Minnedosa Power Company, for the purpose of developing available water power in the vicinity. The Minnedosa Power Company was formed some few years ago, but owing to insufficiency of capital the charter obtained was never utilized until now, when it is the intention to float the company as speedily as possible. With this object in view the Board of Trade has issued a prospectus inviting applications for stock.

Stoves and Tinware

OVERCOMING STOVE PROBLEMS.

No stove or range ever made has of itself what is called a draft; that must be furnished by the chimney or flue, and even when the draft in chimney or flue is perfect, that is not all that is wanted to insure good work or good bread or biscuits; the other things necessary being proper setting up, good fuel, good material—and last, but not least, a good cook. I had a customer years ago, writes F. P. Hans, a Missouri hardware salesman, who would never guarantee a stove to bake well, and when the lady would ask "Why won't you guarantee the stove to bake well, after saying that you know the store is a good baker?" his answer would be: "I know the stove is a good baker, but I do not know if you know how to bake or not." And there is a lot that hinges on knowing how, in baking as in everything else.

An old-timer told me that when a lady bought a cheap cook stove the chances were that she was going to use it herself, and if the top part of the oven got too hot, would place a paper over the bread or biscuit until enough ashes had gathered on the top oven plate or if the bottom of the oven got too hot, she would put a stove cover under the pan. On the other hand, when a lady bought a high-priced stove the chances were she would have a servant to run it, and if breakfast was late because she had been late in getting up to start the fire, her excuse would be that the stove would not draw. If, on the other hand, she got the oven too hot, or forgot to take out the bread or biscuit in time and they were burnt, the excuse would be the stove could not be regulated and at times I have found these excuses used under conditions mentioned.

Unreasonable Guarantees.

So, in making a sale, do not make any unreasonable guarantees or promises; do not say this stove or range, as the case may be, will do good work set up to any kind of an old chimney and with any kind of fuel; because you know it will not, no matter what may be its name or who may be its maker. To obtain the best results there is needed a good flue or chimney, proper setting up with pipe full size of collar on the stove or range and good fuel, and as stated before, a good cook to run it.

When possible, the dealer should set up every stove or range he sells and the man doing this work should not be the cub or the poorest workman about the place or shop. He should know enough about the business to know if the flue he is going to use has a good draft or not, should so tell the party buying the stove or range, and if possible correct the trouble before leaving the job. If this were always done there would be fewer kicks afterwards.

When the stove or range goes into the country where a man cannot well be sent to set it up, instruct them as far as possible how to set it up and what faults to avoid. Many of the manufacturers have booklets or circulars on this subject that they would be glad to furnish if they knew they would be used.

A good chimney should be 8 x 8 inches inside, and the top of it should be a little higher than the highest part of the comb of the house, and should not have a tree overhanging the chimney, as is sometimes the case, and flue should end from 4 to 6 inches below the opening into which the stove pipe runs, and should not run down to the floor or into the cellar where, perhaps there is an opening into said flue without a cover or stopper of any kind; and remember putting up two or three stoves to one flue is like hitching up two or three wagons to a single team just large enough to handle one wagon easily.

Always Will Be Complaints.

I suppose as long as stoves and ranges are sold there will at times be complaints made and when they come in they should be looked after at once, and it is well to go yourself or send a man who knows the business. The first thing I do in such cases is to fire up the stove or range to see for myself how it burns.

If you find the fire does not burn well, examine the flue and see that it has a good draft. If not, locate the trouble and remedy it if possible, because a good draft is absolutely necessary to good work. Next, look to the fuel and remember you cannot get a hot fire from wet or rotten wood, as the heat used in evaporating the water in green or wet wood is lost to the oven; nor can you get a hot fire from poor coal, especially when mixed with dirt. Then see that the ashes are kept away from the bottom of the fire and from under the grate allowing the air to pass freely through the fuel, as air is as necessary to a fire as is the wood or coal, and a fire can no more burn without any air than you can live without air. All air entering into the fire box should pass under and through the fuel, as any air that passes over the fire checks it and at the same time cools the oven, so I would say always keep the damper slide in the front door closed, and it would be better (in my opinion) if no damper slide was put in the front fire box.

Fire Box Needs Air.

Get all the air necessary into the fire box by drawing out the hearth slide, as in doing so the air is put to the bottom of the fire where it belongs. But while air is necessary to combustion, like in everything else, there may be too much of a good thing. Only one-fourth of the air entering the fire box is oxygen, the only thing that counts, while the other three-fourths is made up of neutral gases that contribute nothing towards combustion, but have to be heated at the expense of the oven; so that while it is necessary to admit all the air into the fire box that is needed to make the fire burn well, no more than is necessary should be admitted to keep the fire burning well. And as no two flues draw exactly alike, nor the same flue draws alike every day, but change as does the weather, the amount of air to be admitted to do the best work can only be learned by experience, as any good cook will tell you, and that is why an oven will get hotter one day than another, using the same kind and same amount of fuel.

HELPS FOR FURNACE MEN.

For years I have set furnaces and in consequence have many customers who need a little instruction every fall to get their heaters operating satisfactorily, writes an American furnace man. They change servants and those who are familiar with the proper management are no longer available. The general and specific charge is that the heater doesn't heat. Frequently we find that the cold air box has the damper shoved in to the fullest extent, making it impossible for the heater to supply the different pipes connected with it. Not infrequently some change in the immediate neighborhood of the cold air inlet affects the air supply, sometimes producing an eddying current, creating a vacuum where there should be a pressure. Often when dampers are put in the hot air pipe at the furnace they may be closed off and it needs a man from my shop to go to the house and open the damper. Frequently in extremely cold weather we are told that even with a good fire the heater fails to do its duty. An investigation finds that the ashpit is banked with ashes and the firepot is more than half full of ashes with a fair fire on the surface, leaving the impression with the house owner, who may go down into the cellar and look in the feed door, that there is a good fire. Here, however, investigation stops. When a $3.50 a day tinsmith or furnaceman spends about an hour removing ashes from the ashpit, clearing the ashes out of the fire and getting a good fire started it is both expensive and vexing, yet I have to make out such bills frequently. Frequently people have automatic draft regulators attached to the smoke pipe and through a misunderstanding of their use and operation they hinder rather than help conditions. After they are once set going properly and some member of the household thoroughly understands, there is no further complaint, and these automatic devices give quite general satisfaction. In fact, in many instances their sale has been both a source of profit to me and of comfort to the owners in keeping up a more regular fire. Too much draft or too much air supply are never the cause of any trouble to me. I can regulate them. But a lack of either will make trouble for any furnaceman. Then every little cause of friction must be removed to help the flow of smoke or air.

The genuine joys of life are to be gotten from useful effort, and to hunt for pleasure is to lose it. Do your work and pleasure will come to you.

RETAIL HARDWARE ASSOCIATION NEWS

Official news of the Ontario and Western Canada Associations will be published in this department. All correspondence regarding association matters should be sent to the secretaries. If for publication send to the Editor of Hardware and Metal, Toronto.

ONTARIO EXECUTIVE MEETS.

The Executive Committee of the Ontario Retail Hardware Association held an important meeting in Toronto on Monday, Dec. 31st, when arrangements were made for the coming March Convention, and action taken on various matters before the Association.

Correspondence was read from Jenkins & Hardy, representing the Canadian Wholesale Hardware Association, and the Sheet Metalware Association in reference to the cartage charges question, the resolution passed at the September Convention recommending that where jobbers sell to persons outside the trade that they protect the retailers, and in reference to the resolution that enamelware manufacturers mark all seconds as such in plain letters. Letters were read from Secretary McRobie of the Western Canada Retail Hardware Association in reference to enamelware price-cutting by departmental stores, and other matters. An interesting communication was also read from Secretary Trowern of the Retail Merchants' Association in reply to a letter advising him of the action being taken by the Retail Hardware Association in the matter of opposing the Parcels Post C.O.D. Collection System. Secretary Trowern agreed that all merchants should cooperate in combating the proposed legislation, and stated that his Association was also opposing other legislation, which they would be glad to have the co-operation of the Hardware Association in combating.

Good Friday Chosen.

It was decided that the date of the annual convention in March would be changed from Tuesday, March 12th, to Thursday and Friday, March 28th and 29th, the latter date being Good Friday when single fares would be available on all railroads. The opening meeting will be on Thursday afternoon, and on Thursday a banquet will be tendered the members of the Association by Col. MacLean, publisher of Hardware and Metal. A very interesting programme will be arranged for Friday morning, afternoon and evening; a "question box" being established in which questions of interest to the trade may be asked by any dealer and replied to and discussed by any others present.

The Executive decided to send a letter to the Railway Commission drawing attention to the unjust conditions existing in connection with the cartage charges imposed on retail hardware merchants, by an arrangement between the jobbing houses, the railway and the cartage companies. The Executive will ask the Railway Commission to eliminate the existing evil, having failed to secure satisfactory action after protesting against the charges to the Wholesale Hardware Association.

The Parcels Post question was discussed at considerable length, and it was decided that a draft letter would be prepared to be published in Hardware and Metal, and that every hardware merchant in Canada be requested to copy the letter on his letterhead, and forward copies to the Post-Master General and his local member of Parliament, mailing the letters so they will reach Ottawa on Feb. 1st next. It was also resolved that the Association views with disfavor the proposed legislation, and that the Executive co-operate with other organizations in securing its defeat. Frank Taylor, of Carleton Place, was appointed to represent the Association if it is found necessary to send a delegation to Ottawa to oppose the measure.

The Secretary reported that new members were already being enrolled for 1907, and many of the old members had forwarded their membership fees for the coming year. It was decided that an active campaign would be conducted in order to increase the membership roll to 300 members if possible, before the March Convention.

Provincial Legislation.

The resolution adopted last May that the "Association petition the Ontario Legislature to so amend the statute law and abolish the exemptions in Section 180 of the Division Court Act, so far as to make small accounts up to $5.00 collectable by garnishment by a procedure in the magistrate's court similar to that for collection of claims for wages, the prosecutor to serve the summons in the case without charge," was discussed and it was decided that the matter would be brought to the attention of the Attorney-General, and each dealer in Ontario requested to urge his local member of the Legislature, to support the proposed amendment.

Complaints were received of jobbers selling direct to persons outside the trade at wholesale prices, and the Secretary was instructed to take the matter up with the representatives of the Wholesale Hardware Association. Two cases discussed were those of jobbers selling cutlery to a School Board at the same price quoted a retail dealer and of jobbers selling skates to young mail order advertisers who have no regular place of business, but who advertise in the daily papers in Stratford, Chatham and other centres, offering to supply skates, etc., at prices which the local trade would be unable to meet. It was felt that jobbers should protect the retail dealers who are their customers in these places, from the competition of those who are under no expense in keeping up places of business.

Better Freight Classification.

The question of freight classification and it was decided that a representative was explained at considerable length of the Executive would meet the railroad freight officials and endeavor to secure a better classification of hardware articles. This matter will be reported on at the March Convention.

As the Western Canada Retail Hardware Association will hold its annual meeting in February, the Executive decided to recommend to that body that a temporary Dominion Retail Hardware Association be formed to be composed of the Presidents and Secretarys of the two existing Associations, until a permanent organization is effected.

Mutual Fire Insurance.

J. Walton Peart, of St. Mary's, Chairman of the Mutual Insurance Committee reported that satisfactory replies had been received from a fair percentage of the members, and he estimated that at least, one million dollars worth of insurance would be placed with the Mutual Fire Insurance Company if one was organized by the Association. Steps are being taken to procure legal advice as to expense of organizing a company, and several old line companies are also being asked to quote a rate at which they will accept fire insurance to the amount of at least one million dollars, if placed with them by the Ontario Hardware Merchants, through the Association.

The Question Box.

At luncheon, when the members of the Executive were entertained by Secretary Wrigley, some very valuable points were brought out in the "round table" discussion, several members saying the pointers received were worth the expense of a trip to Toronto. All agreed that there was no more valuable feature of trade associations than the opportunity convention gatherings gave to members to exchange ideas with men in their own line of trade who are not nearby competitors. Consequently the "Question Box" feature will be made important at the March Convention.

At luncheon on Monday the following questions were suggested:

"Does it pay a dealer doing a $50,000 business to go behind the counter?"

"Is Christmas trade in specialties worth going after?"

"Is it wise to transfer a customer from one clerk to another?"

"Is a parcel boy at 50 cents per day cheaper than one at $1 per day?"

"Does it pay to advertise—and how?"

"Should a portion of a clerk's salary be withheld in order to ensure steady service?"

"Can the shortage of tinsmiths be relieved better by employing helpers?"

The "Question Box" makes a good target for keen thinkers. Who can hit the bull's eye first?

Art follows in the wake of commerce, for without commerce there is neither suplus wealth nor leisure.

HARDWARE AND METAL

Established 1888

The MacLean Publishing Co.
Limited

JOHN BAYNE MACLEAN . . President

Publishers of Trade Newspapers which circulate in the Provinces of British Columbia, Alberta, Saskatchewan, Manitoba, Ontario, Quebec, Nova Scotia, New Brunswick, P.E. Island and Newfoundland.

OFFICES:

MONTREAL, ... 232 McGill Street
Telephone Main 1255
TORONTO, ... 10 Front Street East
Telephones Main 2701 and 2702
WINNIPEG, ... 511 Union Bank Building
Telephone 3726
LONDON, ENG. . . . . . 88 Fleet Street, E.C.
J. Meredith McKim
Telephone, Central 12960

BRANCHES:

CHICAGO, ILL., ... 1001 Teutonic Bldg.
J. Roland Kay
ST. JOHN, N.B. . . . . No. 7 Market Wharf
VANCOUVER, B.C. . . . . Geo. S. B. Perry
PARIS, FRANCE . Agence Havas, 8 Place de la Bourse
MANCHESTER, ENG. . . . . 92 Market Street
ZURICH, SWITZERLAND . . . . Louis Wolf
Orell Fussli & Co.

Subscription, Canada and United States, $2.00
Great Britain, 8s. 6d., elsewhere . 12s

Published every Saturday.

Cable Address { Adscript, London / Adscript, Canada

FIGHTING MAIL ORDER HOUSES.

In Canada the first skirmish in the fight between the mail order houses and the retail trade is now being fought over the proposed parcels post C.O.D. collections system and success seems assured to the retailers if they continue the campaign in the energetic way it has been taken up by the Retail Hardware Association, the dry goods dealers, and others interested.

In the United States, however, the fight has been on for some years and the mail order houses have succeeded in getting rural mail delivery established in many States. The rural carriers are allowed to carry private mail matter from mail order houses but proposals to still further extend these privileges have so far been defeated by the joint fight put up by the hardware and other retail associations. More particulars of these fights will be given in future issues.

This week we desire to draw attention to a proposal recently made that the retail hardware trade fight the mail order houses with their own weapons—a trade catalogue for distribution amongst the farmers and other consumers. The National Hardware Bulletin, organ of the National Retail Hardware Association, has been securing the opinions of manufacturers, jobbers and retailers on the proposition and a result of the replies received has laid it down as an accepted fact that a catalogue would be a good thing if gotten up on right lines as wherever catalogues have been used by dealers they have proved a success. The opinions expressed are strongly against a priced catalogue as prices vary in different sections and even dealers in the same district have not uniform prices. Another strong argument is that uniform prices maintained by a majority of the trade would justify an outcry against a "hardware trust." The Bulletin, in discussing the matter, deduces that such a catalogue cannot be priced, or if priced, the quotations cannot be uniform. It further says:

"Another difficulty presents itself. All dealers do not carry the same goods. A dealer in Nebraska has quarreled with one of the saw manufacturers. Must he perforce advertise that manufacturer's saw? Must he advertise it, when he does not carry the line and would not send away for a sample even to supply an obstinate customer's demand?

"It may be asked why each dealer should not proceed to issue a special catalogue of his own, with local color, showing the lines he is anxious to push. There are two reasons: First, individual catalogues will prove more expensive than when issued by the million. Second, all dealers are not trained in the matter of compiling catalogues nor can they afford to hire a man specially for that purpose. Summarized, the situation is as follows:

"Each dealer should issue some form of catalogue, annually or semi-annually, to people of his community with profit, but the catalogue cannot be priced uniformly, a uniform line of goods cannot be shown, and the work cannot be done by each individual dealer.

"It has been suggested that the National association take up the matter, getting out an individual catalogue for each favorable member. The expense would be minimized by using regulation plates, showing lines each dealer carried and no others. A few pages would be devoted to local matter. The local matter would individualize each catalogue. There would be a country road map, a cut of the principal local buildings, a sketch of the 'oldest inhabitant' and the most popular or successful citizen.

"This catalogue would be as the local dealer wanted it. It would be priced or unpriced as he pleased. It would show no goods he did not carry. It would push no goods he wished to hide. It would cost no more than a general, uniform catalogue, because it would not show so many lines.

"There is one good reason why the National association could not do this. There might be a conflict between two members in the same town. Both might want the catalogue service exclusively. The association could not show favoritism. This objection might be eliminated in some way. The matter ought to be taken up by some private interest."

KEEPING SALESMEN WAITING.

A St. Paul traveling man recently stated that a good buyer never keeps a salesman waiting if it is possible not to, and yet it is true that every salesman losses on an average of two and a half hours each day waiting on buyers. The cost of selling goods is figured in making the selling price and buyers have to pay for it, so that if a salesman's time is lost some one has to stand the loss, and although most buyers would be very much surprised to hear that they pay for it, it is nevertheless true. There are over 100,000 salesmen on the road, whose salary and expense average $10 per day, and their lost time is therefore valued at least at $250,000 a day, which enormous sum must be paid by the buyers, although they get absolutely nothing for it.

KNOCK OFF THE PACKAGE DUTY.

The Government should remove the duty on packages re-imposed under the new tariff. What is the use of it? The excuse that it was imposed because a few importers connived with the shippers, a very remote and impregnant possibility. If they would do this to secure a proportion of their goods free entry, they would do it to secure the same proportion entry at a lower rate of duty. The excuse is a poor one. The duty protects no one and produces no revenue to speak of, but it is a positive nuisance to the trade and a hinderance to business.

KILL THE THING.

Have you sent a postcard to your representative in Parliament asking him to oppose the proposed extension of the parcel post system to assist the big mail order houses? If you haven't do so to-day. It is in the interest of every ordinary retailer that this scheme be killed. What is even more important it is in the interest of the whole country. Tendency to centre trade in the big cities is not in keeping with the successful development of Canada as a nation. A thousand prosperous retailers scattered throughout the country constitute a much greater national asset than a single huge institution in some city and they will better serve the needs of the people.

PROSPERITY AT FLOOD TIDE

A year ago it was thought that the zenith of prosperity had been reached, but it has not taken twelve months to prove that this latter view was incorrect, says Bradstreets of Dec. 29, in a review of Canadian trade conditions. This year, with a steady growth in population and an increasing trade and commerce in all directions at home and abroad, a still greater share of prosperity has fallen to the lot of all classes of this country's inhabitants.

The farmers of Canada have had a splendid year, the country's mines have produced metals and coal as never before, and the output of products of the forest and of the sea compares favorably with that of the previous years. The production of wheat and other grains has been the heaviest in the history of the country. The wheat crop will show an increase of about 10 per cent., for the larger part of which western Canada is responsible. Last year's western wheat crop was in the neighborhood of 85,000,000 bushels, while this year it will run something over 90,000,000 bushels.

One of the most satisfactory features of Canadian trade is that of foreign business. Owing to the fact that government reports are not issued until four or five months after the date reported on, it is impossible to obtain figures covering the latter part of the year, but the report for the fiscal year, which extended half way into 1906, gives a very good indication as to how trade moved, particularly as it is well known that trade in the latter part of this year showed no signs of falling off. The total for the fiscal year was by far the heaviest in the country's history, amounting to $546,929,038, being an increase of $80,616,612 over that of the previous year. It was in export trade that the increase was greatest. The value of the year's exports showed an increase of $44,184,741, the greatest in the history of the country, while the imports increased by $31,646,686. Agricultural exports increased by almost 50 per cent. and those of manufactures by about 20 per cent. To Great Britain, Canada exported 53.96 per cent. of her total exports, against 50.61 last year, and to the United States, 35.68 per cent., against 37.51.

The Golden West.

Canada's present prosperity is largely due to the rapid growth of the western country, the enrichment of western farmers by the high prices for a greater volume of product, and the rapid filling of the country by immigration. Its population has been increased during the past year by well over 100,000 immigrants from Great Britain, the United States and the continent of Europe. The major portion of these have taken up land in the west, and others in the new portions of eastern Canada, but it is in the west that this immigration has most affected the country's advance. The British and American immigration has set a very high average of quality for that of the whole year. The western cattle trade had an exceedingly good year; the cattle were well finished, and shipments were nearly 50 per cent. greater than those of last year. Prices, too, were fairly good. The grain movement was slow owing to the inability of the railroads to properly handle it. At country points, however, it was marketed early, and this had a very good effect upon collections throughout the country, which averaged better this year than they have done for some time. The growth of the manufacturing interests of the west is year by year becoming more pronounced. Winnipeg, among the newer cities, is fast becoming a manufacturing centre, having made considerable progress in this regard during the past year; it is also showing steady growth as a jobbing centre.

Pacific Coast Revival.

The revival in the lumbering and mining industries along the Pacific coast made the year the most remarkable in the history of that part of the country. The lumber mills, particularly, have been unable to keep up with demands even at the high prices ruling, and similar conditions prevail at the mines. The shortage in labor on the coast has been a serious problem. Its cost has advanced greatly, and inferior Asiatic labor has had to be resorted to. The growing of fruit is rapidly becoming an important industry, and a good European trade is hoped for in this connection. The shipping trade has shown a good growth as a result of rapidly increasing trade with the Orient. The crops of Alberta naturally find their outlet on the Pacific Ocean, and flour and grain markets are rapidly being opened in Japan and parts of Africa. Wholesale and retail trade has been good throughout the year and money has been plentiful.

Conditions in Ontario.

Generally speaking, conditions in Ontario and the other older parts of the country have been but little different from those prevailing farther west. Crops generally were good. Fall wheat and most other grains, with the exception of spring wheat, showed an increased total yield. Prices have been good ever since the harvest, and the grain has come forward fairly well. But it was in other lines that farmers were most successful. The dairy season has been an extraordinary one in more than one respect. While the make of cheese was not so heavy as in previous seasons, the high prices prevailing eclipsed anything in the record of the industry. The result was a heavy make of cheese and a light make of butter, in which also prices were high. While cattle prices were not generally good, those for hogs were exceedingly high, and not nearly enough were obtainable. The packing industry suffered on this account. General wholesale trade has had one of the most successful years. The turnover in most lines has been the heaviest on record. The great activity in railroad and general building in all parts of the country has resulted in a scarcity of manufactured steel and iron. Manufacturers have suffered severely in the shortage of labor, and serious delays in making deliveries have occurred.

Canadian banks have had a good year, as shown by annual statements now being published. The one failure, that of the Ontario Bank, was the result of bad management and not of general conditions.

Railway Development.

The scarcity of labor has had some effect upon the work of railway extension. The new Grand Trunk Pacific line is being pushed with all possible speed, and all available men have been working on different parts of it throughout the season. The Canadian Northern, which will be Canada's third transcontinental line, is rapidly being pushed through the west. The same company has completed its line from Toronto north to Sudbury. The Ontario government railway running up into New Ontario has also advanced further during the year. It is rapidly making, tributary to Toronto and other Ontario business centres, a vast area of lands fabulously rich in products of mines and forests. It is along this line the silver mines of Cobalt lie. Where a year ago there was along this line only an occasional lumber or prospecting camp, there are towns and settlements, the existence of which signifies much to older Canada. What the opening of this new country will mean may be gathered from the fact that shipments of ore already run to about $1,000,000 a month. This railroad is also opening up a vast agricultural country well fitted for settlement.

During the past year a good number of settlers have gone into the country, but here is a land extending up to James Bay which will need filling long after western Canada has ceased to be the Mecca of immigrants. It is reasonable to expect its development will do for the industries of eastern and central Canada just what the western expansion is now doing for the country as a whole.

FREIGHT RATES TOO HIGH.

The stove manufacturers of Canada are not satisfied with the new classification made respecting stoves in the recently revised freight rates submitted by the Canadian Freight Agents' Association. The matter will be again discussed by the Freight Association, which is committed to much lower tariffs than formerly.

Don't worry over trifles. If you must worry, pick out something worth while, then get busy.

HARDWARE TRADE GOSSIP

Ontario.

A. Kirpatrick, hardware merchant, Lefroy, called on Toronto jobbers last week.

H. R. Carter, hardware merchant, Chesley, has sold out his business to Davis & Co.

Geo. Praker, of Praker & Son, hardware merchants, Brampton, called on the jobbers in Toronto this week.

Cyrus Birge of the Canadian Screw Company, Hamilton, was a visitor in Toronto, on Thursday of last week.

W. H. White, of Brandon, Man., was a visitor in Toronto this week, calling on the wholesale hardware merchants.

Wm. Harland, of Harland Bros., hardware merchants, Clinton, was a visitor in Toronto last week calling on the wholesale hardware merchants.

Mr. Ballantyne, of the Sherwin-Williams paint staff in Toronto, was on a visit to the head office, Montreal, for several days last week.

The Kelley Hardware Company suffered loss by fire on Dec. 27. Their stock which was valued at $40,000 was destroyed, but partly covered by insurance.

Max Morrell has started at his new position as sales manager for Rice Lewis & Son, W. R. Tait succeeding him as Toronto representative for Lewis Bros., Montreal.

J. R. Marlow, who has been managing the transportation branch of the C.M.A., has resigned, to take charge of the sales department of the Canada Cycle and Motor Company.

John Harold, manager of the Paris Plow Company was presented with a gold-headed cane by the employees last Saturday evening, as a mark of their esteem towards the manager.

The Canadian Iron & Foundry Co. presented each of its 158 Hamilton employees with a fine Christmas turkey. The company did the same for all its employees at six other plants.

Charlie Boyd, manager of the hardware store of the Younge Estate, Gravenhurst, was a visitor in Toronto several days this week calling on the wholesale hardware merchants.

O. S. Hunter, the newly established hardware merchant, Durham, with his manager, W. Coleman, was in Toronto all last week at Rice Lewis & Son, purchasing their stock, which amounted to about $5000.

W. H. Rumball, hardware merchant, of South Woodslee, has sold his business and property. Messrs. Ebbinghaus and A. Rumball, having purchased the stock of hardware and furniture, and will continue the business in the same stand.

Thomas Chamberlain of the Buck Stove Co., Brantford, had a pleasant task Monday afternoon. He presented to each of the men under his supervision a handsome silk handkerchief to show his appreciation of their work and good feeling towards them.

Adam Taylor, of the Taylor-Forbes Company, Guelph, has been elected one of the Guelph directors of the Commercial Travelers' Association of Canada and Arthur F. Hatch, of the Canada Steel Goods Company, Hamilton, a member of the Hamilton Board.

Sir Mackenzie Bowell was showing some visitors around the premises of the Belleville Hardware Company, at Belleville, of which he is president, when he narrowly escaped possibly fatal injuries by being struck by an elevator of the descent of which he was only cognizant just in time. He was standing directly under it.

Alex. Payne, who for some years has been employed on the sample floor for H. S. Howland Sons & Co., of Toronto, has been appointed to the traveling staff, by taking Jack Anderson's place in the Niagara Peninsula. He will no doubt meet with every success, as he is perfectly conversant with every branch of the business.

Sir W. P. Howland, who died New Year's morning in Toronto, was a brother of the late H. S. Howland and uncle of Peleg Howland, president of H. S. Howland & Sons, and George W. Howland of the Graham Nail Works, Toronto. Sir William Howland's life was interwoven in the history of Canada, he having been a "Father of Confederation," and intimately connected with the early development of Western Canada.

Charles Kerr, a 16-year-old Woodstock boy, met with a serious accident a fortnight ago. He was working for Whitney Bros., tinsmiths, and was assisting in the loading of a boiler on a sleigh at the station. The boiler was successfully placed in position, and Kerr stood behind to hold it on. When the sleigh started the boiler slipped, knocking Kerr to the ground, and the boiler fell across the boy. Three fingers were severed, and his back was injured, but not broken.

Quebec.

William Gravel, of Terrebonne, Que., was in Montreal last week.

D. Huette, of St. Hyacinthe, was in Montreal purchasing supplies.

C. Williscraft of Smiths Falls, was a visitor in Montreal last week.

P. Filian, of St. Therese, Que., was in Montreal buying general goods.

W. H. Evans of the Canada Paint Co., Montreal, spent New Year's Day in Ottawa.

Edgar G. Powell has left for Winnipeg, to represent the J. W. Harris Co., Montreal.

L. J. Kirouack, of Upton, Que., called at some of the Montreal wholesale houses last week.

H. W. Aird, of the Canada Paint Co., Montreal, who has been sick for some time, has recovered.

Ernest Bureau, of the traveling staff of Lewis Bros., spent a few days in New York as the guest of his brother-in-law.

Edgar W. Wilkinson, of Harrison Bros. and Howson, cutlery manufacturers, was a visitor at the Montreal office of Hardware and Metal yesterday.

C. C. Ballantyne, manager of Sherwin Williams Co., Montreal, has taken up his duties at the Harbor Board, as one of the three new harbor officials.

F. W. Bronton, Winnipeg, formerly representing the Stewart Manufacturing Co. of Woodstock, Ont., has joined the traveling staff of F. W. Lamplough & Co. He will travel from coast to coast.

At an early hour Thursday morning, Jan. 3, 1907, a disastrous fire cleaned up the best part of a block of Montreal business houses. Among the sufferers were the Montreal factory of the Gillette Safety Razor Co., the Beaver Rubber Clothing Co., the Otis Fensom Elevator Co., and the Dodge Manufacturing Co. of Toronto.

Western.

Burns Bros., hardware merchants, Kawende, Man., will move their stock of hardware, etc., to Portage La Prairie, where they will carry on business.

R. P. McLennan, of McLennan, McFeely & Co., Vancouver, has been nominated as one of the Liberal candidates for Vancouver city in the coming B. C. elections.

McKinleys' hardware store of Ninga, Man., was destroyed by fire last week, damage amounting to $7,000 on stock and $3,000 on building, partly covered by insurance.

Joseph Cahill, one of the oldest commercial travellers in the West, a representative of the Ashdown Hardware Company, died of pneumonia on January 2, in Winnipeg. The body was sent to his former home in Peterboro', for burial.

Smith & Ferguson, hardware merchants of Regina, Sask., have sold their business to Peart Bros., of that place, who control the Western Hardware Company, which will mean an amalgamation of these two powerful business houses. The price paid for the business was in the neighborhood of $125,000 for the business and property. J. W. Smith, proprietor of the Smith & Ferguson business, will continue in the coal business.

Maritime Provinces.

G. L. Munroe, hardware dealer, was a heavy loser in the recent fire at Londonderry, N.S., having only $1,000 insurance.

S. S. Wetmore, travelling representative for A. M. Bell & Company, Halifax, is recovering from a serious attack of influenza.

M. F. Irwin, St. John, N.B., representative of the McClary Manufacturing Company, London, spent the Xmas holidays at London. His family will locate at St. John in January.

L. L. Libby, Halifax, manager of the Canadian Rubber Company, spent the Christmas holidays at his old home at Portland, Maine. Before leaving, the staff gave him a Christmas present in the form of a fine travelling bag.

Markets and Market Notes

(For detailed prices see Current Market Quotations, page 70.)

THE WEEK'S MARKETS IN BRIEF.

TORONTO.

Lock Sets—Now 37½ and 10 per cent.

Hinges—Light and Heavy tee and strap, advancing; also, wrought butts now 70 per cent.; crate hinges and back flaps, 65 and 5 per cent.; hinge hasps, 65 per cent.

Building Paper—E2, D2 and Surprise.

Stanley Tools—20 per cent. advance.

Bar Iron—Advanced to $2.30.

Galvanized Sheets—Appollo advanced 10c.

Boiler Tubes—Advances on some sizes.

Compression and Fuller Work — Lower discounts.

Montreal Hardware Markets

Office of HARDWARE AND METAL,
232 McGill Street,
Montreal, Jan. 4, 1907.

The holidays are past and business has again assumed its usual state.

Jobbers claim that prospects are exceedingly bright for the first three or six months of the year at least.

Prices are still going steadily up.

Sockets firmer, chisels and prune knives have advanced about 10 per cent.

Peterboro and Gurney locks have also advanced, the new discount being 37½ and 10 per cent, instead of 45 and 10 per cent.

Other markets remain about the same.

AXES.—We quote: $7.60 to $9.50 per dozen; double bitt axes, $9.50 to $12 a dozen; handled axes, $7.50 to $9.50; Canadian pattern axes, $7.50 a dozen. follows: No. 3, $1.25; No. 2, $1.50; No. 1, $1.90 a dozen; adze handles, 34-inch, $2.20 a dozen; pick handles, No. 2, $1.70; No. 3, $1.50 a dozen.

LANTERNS—The normal business is being done, at the following prices: "Prism" globes, $1.20; Cold blast, $6.50; No. 0, Safety, $4.00.

COW TIES AND STALL FIXTURES — We still quote: Cow ties discount 40 per cent. off list; stall fixtures, discount 35 per cent. off list.

SLEIGH BELLS.—Some orders are being sent forward this week. We quote: Back Straps, 30c. to $2.50; body straps, 70c. to $3.50; York Eye bells, common, 70c. to $1.50; pear shape, $1.15 to $2; shaft gongs, 20c. to $2.50; Grelots, 35c. to $2; team bells, $1.80 to $5.50; saddle gongs, $1.10 to $2.60.

RIVETS AND BURRS.—Copper is still high and advance in ore expected. We quote: Best iron rivets, section, carriage, and wagon box black rivets, tinned do., copper rivets and tin swede rivets, 60, 10 and 10 per cent.; swede iron burrs are quoted at 60 and 10 and 10 per cent. off new lists; copper rivets with the usual proportion of burrs, 27½ per cent. off, and coppered iron rivets and burrs in 5-lb. carton boxes at 60 and 10 and 10 per cent.; copper burrs alone, 15 per cent., subject to usual charge for half-pound boxes.

HAY WIRE.—Last week's advanced prices are being firmly held. Our quotations are: No. 13, $2.55; No. 14, $2.65; No. 15, $2.80; net cash, f.o.b. Montreal.

MACHINE SCREWS.—Trade is about as usual. We quote: Flat head, iron, 35 per cent.; flat head, brass, 35 per cent.; Felisterhead, iron, 30 per cent.. Felisterhead, brass, 25 per cent.

BOLTS AND NUTS.—Bolts are still scarce, no relief from the shortage having been experienced. Our discounts remain: Carriage bolts, 3 and under 60 and 10; 7-16 and larger, 55 p.c.; fancy carriage bolts, 50 p.c.; sleigh shoe bolts, ⅜ and under, 60 per cent.; 7-16 and over, 50 per cent.; machine bolts, ⅜ and under, 60 per cent.; 7-16 and larger, 55 per cent.

HORSE NAILS.—Sales this week have been large and prices are well maintained. Discounts are as follows: C brand, 40, 10 and 7 per cent.; M.R.M. Co., 55 per cent.; P.B. brand, 55 per cent.

WIRE NAILS.—The market is now firm at $2.30 per keg base f.o.b. Montreal.

CUT NAILS—We continue to quote: $2.50 per 100 lbs.; M.R.M. Co., latest $2.30 per keg base. f.o.b. Montreal.

HORSESHOES.—A steady business is reported. We quote as follows: P.B. new pattern, base price improved pattern iron shoes, light and medium pattern, No. 2 and larger, $3.65; No. 1 and smaller, $3.90; snow pattern, No. 2 and larger, $3.90; No. 1 and smaller, $4.15. Light steel shoes, No. 2 and larger, $4; No. 1 and smaller, $4.25; featherweight, all sizes, No. 0 to 4, $5.60. Toeweight, all sizes, No. 1 to 4, $6.85. Packing, up to three sizes in a keg, 10c. per 100 lbs. More than three sizes, 25c. per 100 lbs extra.

BUILDING PAPER—Conditions remain unchanged.

CEMENT AND FIREBRICK—We are still quoting the following prices: "Lehigh" Portland in wood, $2.54, in cotton sacks, $2.39; in paper sacks, $2.31. Lefarge (non-staining) in wood. $3.40; Belgium, $1.60 to $1.90 per barrel; ex-store, American, $2 to $2.10 ex-cars; Canadian Portland, $2 to $2.05. Firebrick, English and Scotch, $17 to $21. American $30 to $35; White Bros.' English cement. $1.80 in bags. $2.05 in barrels in round lots.

COIL CHAIN.—Last week's advance still holds We are quoting as follows: 5-16 inch, $4.40; 3-8 inch, $3.90; 7-16 inch, $3.70; 1-2 inch, $3.50; 9-16 inch, $3.45; 5-8 inch, $3.35; 3-4 inch, $3.20; 7-8 inch, $3.10; 1 inch, $3.10.

SHOT—Our prices remain as follows: Shot packed in 25-lb. bags, ordinary drop AAA to dust, $7 per 100 lbs.; chilled, No. 1—10, $7.50 per 100 lbs.; brick and seal, $8 per 100 lbs.; ball, $8.50 per 100 lbs. Net list. Bags less than 25 lbs., ½c. per lb. extra, net; f.o.b. Montreal, Toronto, Hamilton, London, St. John, Halifax.

AMMUNITION. — Trade is normal. We are quoting as follows: Loaded with black powder, 12 and 16 gauge, per 1,000, $15; 10 gauge, per 1,000, $18; loaded with smokeless powder, 12 and 16 gauge, per 1,000, $20.50; 10 gauge, $23.50.

GREEN WIRE CLOTH. — Bookings are very satisfactory. Our prices are: In 100 foot rolls, $1.62½ per hundred square feet, in 50 feet rolls, $1.67½ per hundred square feet.

Toronto Hardware Markets

Office of HARDWARE AND METAL,
10 Front Street East,
Toronto Jan 4, 1907.

Hardware markets are very brisk, and several advances are in force this week. The manufacturers of hinges in Canada have been compelled through the rapidly increasing prices of raw material and labor, to again advance prices in hinges, butts and builders' hardware of all kinds. Raw material for these lines has been exceptionally scarce and merchants would do well to order early for their season's requirements as it is possible that when the building season starts hardware stocks will be entirely run out of staple sizes. The demand from the West is so great that Ontario and Quebec jobbers will find it difficult to secure supplies. We are now quoting for, heavy tee and strap, 4 in., $6.25; 5 in. $6.00; 6 in., $5.75; 8 in., $5.50; 10, 12 and 14, $5.25, per 100 lbs. net Light tee and strap hinges now quoting at 65 per cent., formerly 65 and 10. An advance is also made in Peterborough and Gurney locks, advancing from 45 and 10 per cent. to 37½ and 10, present quotation.

Building paper is also quoted at an advance. E.2 and Surprise now 50c. and D.2. at 40c. The cause of the advance of building paper is due to the raise in raw materials, with the ever increasing demand for these goods.

Stanley tools have all advanced 20 per cent. Travelers have again resumed their duties on the road and orders continue coming in very brisk.

AXES AND HANDLES—The business continues very brisk in these lines. Still having large demands from Northern Ontario.

BUILDERS HARDWARE — Orders have started for this line as retailers see the advantage of ordering early. As a shortage is likely to be reported in some lines.

SPORTING GOODS—Hockey sticks and skates continue to sell briskly. A banner year is reported in all lines of these goods.

CHAIN—We still continue to quote: 3/8 in. $3.50; 9-16 in., $3.55; ½ in., $3.60; 7-16 in., $3.85; 3-8 in., $4.10; 5-16 in. $4.70; 1-4 in. $5.10.

RIVETS AND BURRS—No change is anticipated in these articles, and remain as quoted last week, 27½ per cent. on rivets and 15 per cent. on burrs.

SCREWS—Stocks are being rapidly repleted, as manufacturers are giving their best efforts to meet the demand.

BOLTS AND NUTS—Large orders for these are prophesied as a great deal of structural work is being done throughout the country.

TOOLS—Carpenters' tools, hand saws etc., are in great demand. Retail merchants repleting their stock for spring.

SHOVELS—Not much doing in shovels at the present time, but spades, tilling and garden, have started to move.

LANTERNS—The demand still exists and selling $6.50 per dozen.

EXTENSION AND STEP LADDERS —Prices continue as follows: Step ladders at 10c. per foot for 3 to 6 feet, and 11c. per foot for 7 to 10 feet ladders. Waggoner extension ladders, 40 per cent.

POULTRY NETTING—We quote: 2-inch mesh, 19 w.g., discount 50 and 10 per cent. All others 50 per cent.

WIRE FENCING.—Barb and plain twist fencing, galvanized and coiled spring wire are all firm at the advanced prices.

OILED AND ANNEALED WIRE — (Canadian)—Gauge 10, $2.31; gauge 11, $2.37; gauge 12, $2.45 per 100 lbs.

WIRE NAILS—Nails still remain firm at the advanced price, but owing to the high price of raw material, another advance is warranted.

CUT NAILS—We quote $2.30 base f.o.b. Montreal, Toronto 20c higher.

HORSENAILS—"C" brand, 40, 10 and 7½ off list. "M.R.M." brand, 55 per cent. off. "Monarch" brand, 50 and 7½ off. "Peerless," 50 per cent. off.

HORSESHOES—Our quotations continue as follows: P.B. base, $3.65; "M.R.M. Co., latest improved pattern" iron shoes, light and medium pattern, No. 2 and larger, $3.80; No. 1 and smaller, $4.05; snow, No. 2 and larger, $4.05; No. 1 and smaller, $4.30; light steel shoes, No. 2 and larger, $4.15; No. 1 and smaller, $4.40; featherweight, all sizes, 0 to 4, $5.75; toeweight, all sizes 1 to 4, $7. Packing, up to three sizes in a keg, 10c. per 100 lbs. extra; more than three sizes in a keg, 25c. per 100 lbs. extra.

HORSE BLANKETS AND SLEIGH BELLS—Owing to the mild weather lately the demand has fallen off some, but should the cold weather return will liven them up.

BUILDING PAPER—The advance on building paper is owing to the high price of raw material, also the large demand for paper.

CEMENT—We quote: For carload orders, f.o.b. Toronto, Canadian Portland, $1.95. For smaller orders ex warehouse, Canadian Portland, $2.05 upwards.

FIREBRICK—English and Scotch fire brick $27 to $30; American low-grade, $23 to $25; high-grade, $27.50 to $35.

HIDES, WOOL AND FURS.—The supply of hides is in excess of the demand but the course of the market is uncertain. Prices are slightly lower than a week ago.

Hides, inspected, cows and steers, No. 1	..	0 11
" " " " No. 2	....	0 10
Country hides, flat, per lb., cured	....	0 10
" " " green	....	0 09
Calf skins, No. 1, city	...	0 12
" " No. 1, country	....	0 11
Lamb skins	....	1 20
Horse hides, No. 1	3 50	3 75
Rendered tallow, per lb.	0 06½	0 06¾
Pulled wools, super, per lb.	....	0 25
" " extra	....	0 27
Wool, unwashed fleece	....	0 15
" washed fleece	0 24	0 25

FURS.

	No. 1.	Prime
Raccoon	....	1 40
Mink, dark	4 50	5 50
" pale	2 50	3 50
Fox, red	3 00	3 50
" cross	3 00	5 00
Lynx	5 00	6 50
Bear, black	....	1 00
" cubs and yearlings	....	5 00
Wolf, timber	....	2 7
" prairie	....	1 25
Weasel, white	....	0 60
Badger	....	1 10
Fisher, dark	5 00	6 00
Skunk, black	....	1 25
" " short stripe	....	0 90
" " long striped	....	0 50
Marten	3 50	20 00
Muskrat, fall	....	0 16
" winter	....	0 21
" spring	..	0 23
" western	0 12	0 16

Montreal Metal Markets

Office of HARDWARE AND METAL,
232 McGill Street,
Montreal, January 4, 1907

Copper is still advancing. Stocks are light, and deliveries are somewhat hard to get.

Many large metal brokers are unable to quote.

The expected revision in the boiler tube prices has at last taken place.

There have been many little fluctuations during the week, but the tone of the market still remains very strong.

There is a rumor about that a very important advance is about to take place in the British market, but this has not been confirmed.

There have also been several considerable contracts placed for structural steels, etc.

COPPER.—Still further advances have taken place this week, and it is thought that the market will go higher. We are now quoting: Ingot copper, 26c to 26½c; sheet copper, base sizes, 34c.

INGOT TIN.—This metal continues firm, but we still quote: 46½c. to 47c. per lb.

ZINC SPELTER.—Zinc is in a very strong position in the market. We continue to quote: $7.25 to $7.50 per hundred pounds.

PIG LEAD.—Higher prices in England have boosted quotations here and we now quote: $5.50 to $5.60 per hundred pounds.

ANTIMONY—Our price remains 27½ to 28c per pound.

PIG IRON.—We still quote as follows: Londonderry, $24.50; Carron, No. 1, $24.50; Carron, special, $23.50; Summerlee, No. 2, selected, $25.00; Summerlee, No. 3, soft, $23.50.

BOILER TUBES.—The expected new prices have at last come to light, 1½ to 2 inch, $9.50; 2¼ inch, $10.35; 2½ inch, $11.50; 3 inch, $13.00; 3½ inch, $17.00; 4 inch, $21.50; 5 inch, $45.00.

TOOL STEEL—We quote: Colonial Black Diamond, 8c. to 9c.; Sanderson's, 8c. to 45c., according to grade; Jessop's, 13c.; Jonas & Colver's, 10c. to 20c.; "Air Hardening," 65c. per pound; Conqueror, 7½c.; Conqueror high speed steel, 60c.; Jowitt's Diamond J., 6½c. to 7c.; Jowitt's best, 11c. to 11½c.

MERCHANT STEEL.—Our prices are as follows: Sleigh shoe, $2.25; tire, $2.40, spring, $2.75; toecalk, $3.05; machinery iron finish, $2.40 ruled machinery steel, $2.75; mild, $2.25 base and upwards; square harrow teeth $2.40; band steel, $2.45 base. Net cash 30 days. Rivet steel quoted on application.

COLD ROLLED SHAFTING — There is an advance of 25 cents per hundred lbs. over last week's quotations. Present prices are: 3-16 inch to ¼ inch, $7.25; 5-16 inch to 11-32 inch, $6.20; ⅜ inch to 17-32 inch, $5.15; 9-16 inch to 47-64 inch, $4.45; ¾ inch to 17-16 inch, $4.10; 1½ inch to 3 inch, $3.75; 3¼ to 3 7-16 inch, $3.92; 3½ inch to 3 15-16 inch, $4.10; 4 inch to 4 7-16 inch, $4.45; 4½ inch to 4 15-16 inch, $4.80. This is equivalent to 30 per cent. off list.

GALVANIZED IRON.—Firmness prevails in the market. We are still quoting Queen's Head, 28 gauge, $4.60 to $4.85; 26 gauge, $4.45; 22 to 24 gauge, $3.90; 16 to 20 gauge, $3.75; Apollo, 28 gauge, $4.45 to $4.70; 26 gauge, $4.30; 22 and 24 gauge, $3.75; 16 and 20 gauge, $3.60; Comet, 28 gauge, $4.45 to $4.70; 26 gauge, $4.30 to $4.45; 22 and 24, gauge, $3.75 to $4.00; $16 to 20 gauge, $3.60 to $3.85; Fleur-de-Lis, 28 gauge, $4.45 to $4.70; 26 gauge, $4.80; 22 and 24 gauge, $3.75; 16 to 20 gauge, $3.60; Gorbals "Best Best," 28 gauge, $4.45; Colborne Crown, 28 gauge, $4.45; 26 gauge, $4.30; 24 gauge, $3.75. In less than case lots, 25c extra.

Toronto Metal Markets

Office of HARDWARE AND METAL,
10 Front Street East,
Toronto, Jan. 4, 1907

The markets this week are comparitively quiet, few changes being made, and business being interrupted slightly by the holiday week as well as by some houses taking stock. All prices remain very firm, however, the changes made all being increases.

Bar Iron has advanced another peg, and is now at $2.30, with 2 per cent. off. As indicated by us some weeks ago, the turn of the year has strengthened rather than weakened the bar market. Appollo galvanized sheets are also up about ten cents, and there is talk of other sheets advancing. The only other change is in boiler tubes, certain sizes of which show advances.

Copper and lead keep very firm with stocks still short, and deliveries hard to get. Tin is also firm.

Complaints of slow deliveries from Montreal are still coming in, one case being a car load of bar iron which took from November 5, to December 25, to be transported from the wharf to the customer, the rail trip occupying five days.

PIG IRON.—Hamilton, Midland and Londonderry are off the market, and Radnor is quoted at $33 at furnace. Middlesborough is quoted at $24.50, and Summerlee, at $26 f.o.b., Toronto.

BAR IRON.—Bar continues hard to get and we still quote: $2.30, f.o.b., Toronto with 2 per cent. discount.

INGOT TIN—Buying is still lively and prices unchanged at 46 to 46 1-2c per pound.

TIN PLATES—Market continues active and prices firm at recent advances.

SHEETS AND PLATES.—Apollo galvanized sheets has been advanced ten cents, and is now quoted at $4.70 for 10 3-4 ounce. Active buying is reported with prices steady. Stocks are light.

BRASS.—We continue to quote: 26 to 27½c per pound for sheets. Brass has advanced 3c per pound.

COPPER—Shortages are reported with prices still on the upward climb. We now quote: Ingot copper $26 per 100 lbs., and sheet copper $31 to $32 per 100 lbs.

LEAD—Market is active and prices very firm. We quote: $5.40 for imported pig and $5.75 to $6.00 for bar lead.

ZINC SPELTER—Stocks continue light with market firm and active. We quote 7½c. per lb. for foreign and 7c. per lb. for domestic. Sheet zinc is quoted at 8 1-4c. in casks, and 8 1-2c. in part casks.

BOILER PLATES AND TUBES—We quote: Plates, per 100 lbs., ¼ in. to ½ in., $2.50; ⅜ in., $2.35 heads, per 100 lbs., $2.75; tank plates, 3-16 in., $2.65; tubes, per 100 feet, 1½ in., $8.50; 2, $9; 2 1-2, $11.30; 3, $12.50; 3 1-2, $16; 4, $20.00. Terms, 2 per cent. off.

ANTIMONY—Market is active, and stocks scarce. Prices firm at 27c. to 28c.

OLD MATERIAL.—Dealers' buying prices still continue as follows: Heavy copper and wire, 18½c., copper and wire, 18 1-2c. per lb.; light copper 16½ per lb.; heavy red brass, 16½c per lb.; heavy yellow brass, 14½c per lb.; light brass 10½c per lb.; tea lead, $3.70, per 100 lbs.; heavy lead, $4.10 per 100 lbs., scrap zinc, 4½ per lb.; iron No. 1 wrought, $14.00; No. 2 wrought, $6 to $8; machinery cast scrap, $16.50 to $18; stove plate, $13 to $14, malleable and steel, $8; old rubbers, 9c per pound; country mixed rags, $1 to $1.25 per 100 lbs., according to quality.

COAL—The following prices are quoted on coal, slack being short.

Standard Hocking soft coal f.o.b. at mines, lump $1.75, ¾ inch, $1.65, run of mine, $1.40, nut, $1.25; N. & S. $1.10; P. & S., 85c.

Youghiogheny soft coal in cars, bonded at bridges; lump, $2.90; ¾ inch, $2.70; mine run, $2.60; slack, $2.25.

United States Metal Markets

From the Iron Age, Jan 3 1907.

The conviction is general throughout the iron industry that the present tremendous rate of consumption will continue during the first half of the current year, coupled with prevailing generally profitable prices.

There are very many who are acting on the belief that work for full capacity at present or even better prices is assured for the whole of the year 1907. They point to the order books, which on the surface look very encouraging indeed, but which might be turned to a ragged exhibit, after a rush of cancellations such as the trade has witnessed before under similar conditions. There is a disposition to exaggerate the stability of "orders" which it is well to guard against.

The iron industry is passing again through one of its frequent experiences of being caught unawares by a rapidly expanding consumption, followed by a feverish activity to provide adequate producing facilities, which at first fall into line disappointingly slowly and then make themselves felt surprisingly suddenly. That may make its appearance during the second half of 1907.

In its last analysis the course of events in the second half of 1907 will depend first upon the crops, and second upon the ability of our country to finance the betterments and enlargements of its producing and transportation facilities. Time only can tell as to the first, while as to the second we may have ample warning through long continued tightness of the money market.

The holiday week has not sensibly affected the activity which has characterized the markets for some months past. In the Eastern pig iron markets there have been further sales of basic pig iron, and some round lots of foundry and forge iron have been placed. The Buffalo furnaces during the last two weeks of the year booked about 50,000 tons of malleable and foundry iron, and there has been some lively buying for the last half in the Chicago district, with more pending. As yet nothing has been done in the way of purchases of large lots of foreign foundry iron for agricultural interests in the West.

The report is current that two Western roads have placed orders aggregating 90,000 tons for 1908 delivery, which may be interpreted as meaning that so much rolling capacity has been engaged.

Among the sales of structural material is a lot of 4,500 tons for five buildings in San Francisco. It is believed that during the current year a very considerable tonnage will be required for the stricken city. This is in line with the earlier expectations that the true demand would not develop until a considerable time after the disaster.

From the Iron Trade Review, Cleveland, Jan. 3, 1907.

The iron and steel industry enters the new year with the demand for both raw and finished materials of all kinds much stronger than at the beginning of 1906. The increase in consuming capacity, especially in the east, has been tremendous during the past year and there are no indications that production will be greater than the demand. A very active market for pig iron prevailed during the closing days of the year in the east and in Chicago, though in other centres buyers were more conservative. Heavy sales of basic are reported in the east, while foundry consumers of the most conservative class are buying extensively for last half delivery. Foreign iron is reaching further westward, the past few days witnessing the first sale of British iron in Cleveland for nearly five years. The southern pig iron situation is firm, though hampered by the car shortage which shows little improvement.

Two large eastern Pennsylvania rail manufacturers have announced that they will not take any more orders for standard sections for delivery after August, except at $30, an advance of $2 per ton. This is the highest price asked for Bessemer rails since 1901, and the action is significant as indicating an increasing spirit of independence among manufacturers. It is not expected that the leading producer will advance quotations. The action of the independents emphasizes the strength of the finished material market, where the advances have not been as rapid as in crude and semi-finished materials.

In other finished lines, with the exception of structural materials, there is very brisk demand. Plate and sheet specifications are extremely heavy, scarcely a ton of old contracts being unspecified for at the close of the year. Merchant bars are active and the demand for wire products is undiminished.

The holiday season has been productive of sharp decline in sheet quotations at Chicago and other centres, though the market as yet cannot be called weak. There is a serious coke shortage in the Chicago district with some relief being afforded by the output of local by-product ovens.

London, Eng., Metal Markets

From Metal Market Report, Jan 3, 1907.

Cleveland warrants are quoted at 61s. 10½d. and Glasgow Standard warrants at 61s., making price as compared with last week 9d. lower on Cleveland warrants, and 6d. lower on Glasgow Standard warrants.

TIN—Spot Tin opened steady at £193 10s., futures at £194 10s., and after sales of 170 tons of spot and 150 tons of futures closed steady at £193 10s. for spot, £194 10s. for futures making price compared with last week £3 lower on spot and £1 12s. 6d. lower on futures.

COPPER—Spot Copper opened steady at £105 5s., futures £106 12s. 6d., and after sales of 250 tons of spot and 1000 tons of futures, closed steady at £105 5s. for spot, and £106 12s. 6d. for futures making price compared with last week 10s. lower on spot and £1 13s. 6d. higher on futures.

LEAD—The market closed at £19 18s. 9d., making price as compared with last week 1s. 9d. higher.

SPELTER—The market closed at £28, making price as compared with last week unchanged.

N.B. Hardware Trade News

St. John, N. B., Dec. 31.

The close of the year 1906 is the closing of a year of good business for the hardware people of New Brunswick. That seems to be the consensus of opinion, though they are, of course, some of those persons always to be found with words of complaint. The demand for hardware goods has been steady throughout the year, apparently there has been considerable money in circulation and bad debts seem to have become a trifle less numerous, some dealers reporting a marked growth in the tendency of customers to do cash trade. In St. John itself there has been a rather poorer business in builder's hardware than in some other years. The number of new buildings put up has not been what was hoped for and consequently hardware of the type mentioned has not been so strongly in demand as in some other years. The greater part of what may be called the building operations of the year have not come in the erection of new but the remodelling of old structures. Naturally, too, under these circumstances the men of the plumbing department of many of the hardware concerns were not as busily engaged as usual. The mildness of last winter and the indications of mild weather during the coming season have doubtless affected the demands for plumbing. In other lines of the hardware trade, however, 1906 has been a successful year in St. John. Perhaps the most striking feature of the markets was the practically unexampled stiffness of prices. Apparently, too, there is no prospect of any change in price conditions in the immediate future. Indeed the tendency seems to be towards even higher figures than those now prevailing.

The Christmas trade, the dealers say, was excellent. Certainly it seemed to your correspondent that the number of people thronging city stores was greater than usual and that it was only natural to believe that business would be larger as a consequence. Dealers in sporting goods cannot be pleased, however, at the mildness of the weather. A continuance of it must surely mean a poorer trade in winter sporting goods than has been anticipated.

* * *

Once before in this column attention was drawn to the window dressing of St. John hardware stores. At that time your correspondent expressed doubt as to whether or not the hardware men here gave adequate attention to the dressing of their store windows. The doubt still exists, indeed, it has grown stronger. Some of the men apparently give practically no thought to this department of their business; others give but very little. At Christmas time the special efforts put forth—in some cases one is lead to ask: were there any efforts worth the name ?—did not result in the production of any displays worthy of especial note. One or two of the windows attracted a good deal of attention but it is not at all clear that this fact is attachable to skilled window dressing. Surely, St. John hardwaremen should give this matter more attention. The question may be raised, too: Do our merchants give sufficient careful thought to the method of their advertising ? Some of the advertisements seen are certainly not particularly forceful.

* * *

The business done by the stove dealers has on the whole been very satisfactory. City trade has been good and with some firms at least, the sale of stoves to persons in the country has been exceptionally gratifying. The outlook is believed to be for somewhat higher quotations in the Spring than prevailed last season.

Metal prices continue unusually strong. It is thought by local men that further advances are imminent.

* * *

It is reported that a car making plant will shortly be started here. There are many rumors but little definite information has so far been obtainable. Another report had it that the Maritime Rail Works Company has planned the erection here of a much larger plant than it now has. Rolling mills on a generous scale are said to be included in the plan. Mr. S. B. Elkin, secretary-treasurer of the company, has denied the truth of the report. At the same time there is good reason for believing that the company anticipates development here or elsewhere. It may be that Halifax is under consideration as offering a desirable location for an enlarged place for the concern.

* * *

The regular meeting of the Maritime Hardware Association was recently held here. It has not transpired, however, that anything more than routine business was under discussion.

* * *

Mr. Charles Clark, for some years a traveling representative of the Thomas Davidson Company, is to join the staff of the National Cash Register Company.

N.S. Hardware Trade News

Halifax, N.S., Jan. 1, 1907.

The year just closed has been a remarkably prosperous one for the hardware merchants of Halifax. All branches of the trade have been brisk, and the outlook for the year 1907 is most promising. The nail situation now absorbs considerable attention, the record advances having caused the trade to wonder what the next move will be. Nails are now quoted at $2.60 base, for cut, and $2.50 for wire, the highest prices recorded here for a long time.

The yield of gold in Nova Soctia during the past year is estimated at 13,000 ounces, the value of which is $260,000. The gold mining industry has been rather quiet in Nova Scotia for some time past, but the coming year will probably witness a revival of the industry. Several properties have changed hands, and the new owners intend to develop them on an extensive scale. The property at Oldham which has been a big producer in the past is being pumped out, and good returns are anticipated. At Isaac's Harbor and other places considerable development is now going on, and in several gold bearing sections of the Province, new machinery is being installed.

B.C. Hardware Trade News

December 28, 1906.

No class of labor observes the Christmas holiday season more religiously (though not in a religious sense) than the loggers. Every steamer arriving from up-coast points for the past week or more has brought down a full passenger list, nine-tenths of whom were loggers intent on having a really good time during the holiday week in the coast cities. A slight excuse is given these men at this season for as a rule in northern logging districts there is a good deal of snow, even at the coast, and the damp character of snow in the woods close to the ocean makes logging operations difficult. Be that as it may not less than a thousand to fifteen hundred sturdy loggers have left the lumber camps to make merry in the city. The effect is felt directly in the log market for no mills have had a surplus stock on hand and the winter has not found any store ahead as very often occurs.

A result of the temporary reduction in the output is an increase in price and no logs can be bought at less than $10 per thousand at the camps, while it takes $1.50 to tow them to the mill. No wonder then that mill men are already discussing another raise in the price of the sawn lumber. This too, when demand is usually less than at other times. How the mills are going to cope with next season's demands is difficult to see for even from the prairie the call for lumber has not let up as it usually does in the winter. The annual shut-down of most of the mills is likely to be very much reduced and the needed overhaul made in short order this year, for time is valuable when all lumber cut is sold months ahead.

THE MARKETS

Facing a new year, the province of British Columbia finds itself in condition of unprecedented prosperity. In the three chief industries, mining, lumbering and the fisheries, most satisfactory results have marked the year 1906. In the lumber industry more especially, conditions are extraordinary. That the present prices of lumber, high as they are in comparison with a few years ago, are to remain and perhaps increase, is the opinion of many well posted in the trade. Very greatly increased cost of production is the chief reason given for the belief that price of lumber is up to stay. But the great and growing demand in the Canadian Northwest is perhaps the greater factor. Mills simply cannot keep up with their orders. Usually the winter season marks a lessening of demand from the prairie provinces, but the shipments have been so far short of requirements during the past season, that orders have kept up right along, and many orders for next year's business have already been placed.

A good indication of the expansion of the lumber trade in B.C. is taken from the reports of the provincial department of Lands and Works, under which all timber leases are made and dues collected. The returns from this source in 1905 were nearly three-quarters of a million dollars. For 1906, the Chief Commissioner is authority for the statement that the receipts of revenue from timber sources are expected to total $1,000,000. There is but one explanation, and that is that the timber areas of B.C. are becoming valuable in the eyes of the timber men and capitalists, and the big demand for timber is a guarantee that the forests of B.C. will be drawn on indefinitely, and for increasing amounts every year.

Some interest is evinced among timber operators as to the outcome of representations made recently to the government by loggers and mill men for cancellation of reserves for pulp leases. In each case it is the idea to cancel leases which have been held speculatively, and not to hamper a concern actually endeavoring to develop pulp timber reserved. There are a number of large areas so reserved, and so far, with but one exception, that of the Canadian Pacific Sulphite Pulp Co., which is developing its holdings at Swanson Bay, and intends going ahead with a pulp mill, nothing has been done since these leases were granted under special regulations enacted some three or four years ago.

* * *

New Westminster is rapidly coming to the front as a manufacturing centre. Its new glass works, the Crystal Glass Co., being the name of the concern, is rapidly being got into shape, the New Westminster Soap Co., another new industry, is also being put on active basis, while the Schaake Machine Works is enlarging its plant. The Brunette Saw Mill is also erecting a mill for cutting cedar lumber and shingles.

Nanaimo industries are progressing favorably. A new coal company has been formed to operate on veins some 16 miles from the city, the Western Fuel Co. is also contemplating enlarging its works and improving its loading facilities at the wharves. A large cold storage plant is also to be established by a company now being organized, the intention being to furnish storage facilities for storing fruit, fish, dairy products and other perishables, while waiting transportation. The Dominion Government is to be asked for a bonus of $100,000 under existing regulations by which a bonus of 30 per cent. of the cost is given such plants, the rates charged for service, being in return under regulation of the Dominion Department of Agriculture.

The Alley line steamer, Pondo, from New Zealand, reached port on Wednesday with a 2,000 ton cargo, mainly consisting of wool, flax and hardwood, with some hundreds of tons of raw sugar from Fiji for the B.C. Sugar Refinery. The shipments of New Zealand wool go across the country to Montreal. On her outward voyage the Pondo, which is scheduled to sail Jan. 4, will take large consignments of Canadian machinery. Capt. Large of the Pondo saw the New Zealand Exhibition at Christchurch and he states that the Canadian exhibit is the most popular in the whole exhibition.

* * *

The Victoria Trades & Labor Council objects to the contractors for the new Canadian government hydrographic steamer, the B.C. Marine Railway Co., importing the machinery for the new vessel from England instead of making it locally as was the understanding when the contract was awarded. The vessel is to be built at Esquimalt.

* * *

Provincial Mineralogist Robertson, is authority for the estimate that the output of B.C. mines for 1906 will exceed in value the output of 1905 by over 2½ millions of dollars. He bases his figures on the increased prices paid for the principal metals, silver, copper and lead, while admitting that though strikes at Fernie in the coke-producing district, the mines were hampered for fuel and the smelters were also reduced.

According to Mr. Robertsons figures, there were 3,400,000 ounces of silver mined last year, 37,692,000 lbs. of copper, and 56,580,000 of lead. These figures may not be exceeded, but silver has netted producers 7c more per ounce this year, copper 4c per lb. more, and lead 1c per lb. more.

Mr. Robertson looks for a marvellous development in mining in the new Telkwa country, tributary to the Skeena River, and which will be opened up by the G.T.P. Vast deposits of coal as well as very rich veins of copper ore have been located. Several companies have already been formed to open up some of these properties.

* * *

Local hardware market notes are featureless, almost, since the Christmas trade is over. In builders' hardware, nails have gone up to $3.10 for wire nails. Turpentine is now quoted at $10.85 instead of $10.25.

* * *

Extreme activity marks all lines connected with the lumber industry. Within the week, two of the largest mills on Burrard Inlet have changed ownership. In both cases local or at least Canadian capital has been mainly interested, which is a change. Most of the big deals of late years have been made on capital brought in from the United States.

The Pacific Coast Lumber Company, which has had two managers in the year has been turned over to a syndicate headed by Mayor Buscombe of Vancouver and the new owners will take charge at the beginning of the year. Geo. McCormick one of the heaviest stockholders, and his Eastern associates have combined to sell a controlling interest in this splendid mill to the new owners. Geo. McCormick, jr., has been managing since the retirement of W. L. Tait, retires and the new owners will name their manager at the time they take over the mill.

At the same moment that they sold the Pacific Coast mill the McCormick interests were negotiating for the purchase of the Canadian Pacific Lumber Company's mill at Port Moody. This deal was consummated on the same day that the sale of the Pacific Coast mill was completed, though one did not depend on the other. The Port Moody mill was owned by a company comprised of T. W. Paterson, M.P.P., of Victoria, his nephew T. Frank Paterson, of Vancouver and Messrs. Perry D. Roe and Robert Abernethey of Port Moody. These gentlemen will all retire, for the present from active mill-operation on the coast. T. W. Patterson never took an active part in the management of the mill, while T. F. Patterson is now heavily interested in logging operations and handling timber limits. Messrs. Roe and Abernethey are not yet decided as to their movements, but still retain interests on the coast, which will take up their attention.

Another change in mill proprietorship is expected to take place at the end of the year. The McLaren sash and door factory, known as the Vancouver Sash & Door Co., has been bought out by Messrs. Robertson & Hackett, sawmillers, whose mill is right beside the factory on False Creek. Some time ago they had part of their plant burned and the purchase of the factory gives them equipment to replace that which was destroyed.

* * *

Coal demand is very keen on the coast at the present moment. The big Treadwell mine in Alaska has had to shut down its stamp mill for want of fuel, and they are bringing it from Australia. The difficulty is that not sufficient increase in the working staff at the various coast mines has been made to provide for the greatly increased consumption.

With the completion of the Nicola valley branch of the C.P.R. a very short line from the rich Nicola coal fields to the coast has been provided One coal company is getting ready to ship though it will be three months before it will do so. Others are preparing to operate also, now that the rails have been laid.

MANITOBA HARDWARE AND METAL MARKETS

(Market quotations corrected by telegraph up to 12 a.m. Friday, Jan. 4 1907.)

Room 511, Union Bank Building, Office of HARDWARE AND METAL, Winnipeg, Man.

The Christmas trade in all lines of hardware and metal was an exceptionally good one and wholesalers state that they are highly gratified with the business done. With the arrival of the holidays, business became quiet and the past week has been comparatively dull. Most of the local firms are preparing for stock-taking operations and getting ready for the New Year's trade. Prices hold firm at practically no change from the previous week. Judging from the reports of builders the coming year will witness a renewal of last season's remarkable activity in building and all supplies needed in such operations will be in active demand.

LANTERNS. — Quotations are as follows: Cold blast, per dozen, $6.50; coppered cold blast, per dozen, $8.50; cold blast dash, per dozen, $8.50.

WIRE—We quote as follows: Barbed wire, 100 lbs., $2.90; plain galvanized, 6 to 8, $3.39; 10, $3.50; 12, $3.10; 13 $3.20; 14, $3.90; 15, $4.45; 16, $4.60; plain twist, $3; staples, $3.50; oiled unnealed wire, 10, $2.96; 11, $3.02; 12, $3.10; 13, $3.20; 14, $3.30; 15, $3.45. Annealed wires (unoiled) 10c. less.

HORSESHOES — Quotations are as follows: Iron, No. 0 to No. 1, $4.65; No. 2 and larger, $4.40; snowshoes, No. 0 to No. 1, $4.90; No. 2 and larger, $4.65; steel, No. 0 to No. 1, $5; No. 2 and larger, $4.75.

HORSENAILS — Lists and discounts are quoted as follows: No. 10, 20c.; No. 9, 22c.; No. 8, 24c.; No. 7, 26c.; No. 6, 28c.; No. 5, 32c.; No. 4, 40c., per pound. Discounts are quoted as follows: "C" brand, 40, 10 and 7½ per cent., "M" brand and other brands, 55 and 60 per cent. Add 15c. per box.

WIRE NAILS.—Quoted now at $2.70 per keg.

CUT NAILS.—As noted last week cut nails have been advanced to $2.90 per keg.

PRESSED SPIKES — Prices are quoted as follows since the recent advance: ¼ x 5 and 6, $4.75; 5-6 x 5, 6 and 7, $4.40; ⅜ x 6, 7 and 8, $4.25; 7-16 x 7 and 9, $4.15; ½ x 8, 9, 10 and 12, $4.05; ⅝ x 10 and 12, $3.90. All other lengths 25c. extra net.

SCREWS—Discounts are as follows: Flat head, iron, bright, 85 and 10 p.c.; round head, iron, 80 p.c.; flat head, brass, 75 and 10 p.c.; round head, brass 70 and 10 p.c.; coach, 70 p.c.

NUTS AND BOLTS — Discounts are unchanged and continue as follows: Bolts, carriage, ⅜ or smaller, 60 and 5; bolts, carriage, 7-16 and up, 55; bolts, machine, ⅜ and under, 55 and 5; bolts, machine, 7-16 and over, 55; bolts, tire, 65; bolt ends, 55; sleigh shoe bolts, 65 and 10; machine screws, 70; plough bolts, 55; square nuts, case lots, 3; square nuts, small lots, 2½; hex nuts, case lots, 3; hex nuts, smaller lots, 2½ p.c.

RIVETS—Discounts are quoted as follows since the recent advance in the price of copper rivets: Iron, discounts, 60 and 10 p.c.; copper, No. 8, 37c.; copper, No. 10, 40c.; copper, No. 12, 43c.

COIL CHAIN.—Prices have been revised the general effect being an advance. Quotations now are: ¼ inch $7.00; 5-16, $5.35; 3/8, $4.75; 7-16, $4.50; ½, $4.25; 9-16, $4.20; ⅝, $4.25; ¾, $4.10.

SHOVELS—Discounts on spades and shovels continue 40 and 5 p.c.

HARVEST TOOLS — Discounts continue as before, 60 and 5 per cent.

AXE HANDLES—Quoted as follows: Turned, s.g. hickory, doz., $3.15; No. 1, $1.90; No. 2, $1.60; octagon, extra, $2.30; No. 1, $1.60.

AXES.—Quotations are: Bench axes, 40; broad axes, 25 p.c. dis. off list; Royal Oak, per dozen, $6.25; Maple Leaf, $8.25; Model, $8.50; Black Prince, $7.25; Black Diamond, $9.25; Standard flint edge, $8.75; Copper King, $8.25; Columbian, $9.50; handled axes, North Star, $7.75; Black Prince, $9.25; Standard flint edge, $10.75; Copper King, $11 per dozen.

BUTTS—The discount on wrought iron butts is 70 p.c.

CHURNS — The discounts from list continue as before: 45 and 5 per cent.; but the list has been advanced and is now as follows: No. 0, $9; No. 1, $9; No. 2, $10; No. 3, $11; No. 4, $13; No. 5, $16.

CHISELS—Quoted at 70 p.c. off list prices.

AUGER BITS—Discount on "Irwin" bits is 47½ per cent., and on other lines 70 per cent.

BLOCKS—Discount on steel blocks is 35 p.c. off list prices; on wood, 55 p.c.

FITTINGS—Discounts continue as follows: Wrought couplings, 60; nipples, 65 and 10; T's and elbows, 10; malleable bushings, 50; malleable unions, 55 p.c.

GRINDSTONES—As noted last week, the price is now 1½c. per lb., a decline of ¼c.

FORK HANDLES—The discount is 40 p.c. from list prices.

HINGES—The discount on light "T" and strap hinges is 65 p.c. off list prices.

HOOKS—Prices are quoted as follows: Brush hooks, heavy, per doz., $8.75; grass hooks, $1.70.

CLEVISES—Price is now 6½c. per lb.

STOVE PIPES—Quotations are as follows: 6-inch, per 100 feet length, $9; 7-inch, $9.75.

DRAW KNIVES—The discount is 70 per cent. from list prices.

RULES—Discount is 50 per cent.

WASHERS—On small quantities the discount is 35 p.c.; on full boxes it is 40 p.c.

WRINGERS — Prices have been advanced $2 per dozen, and quotations are now as follows: Royal Canadian, $35.00; B.B., $39.75, per dozen.

FILES—Discounts are quoted as follows: Arcade, 75; Black Diamond, 60; Nicholson's, 62½ p.c.

BUILDING PAPER — Prices are as

follows: Plain, Joliette, 40c.; Cyclone, 55c.; Anchor, 55c.; pure fibre, 60c.; tarred, Joliette, 65c.; Cyclone, 80c.; Anchor, 65c.; pure fibre, 80c.

TINWARE, ETC.—Quoted as follows: Pressed, retinned, 70 and 10; pressed, plain, 75 and 2½; pieced, 30; japanned ware, 37½; enamelled ware, Famous, 50; Imperial, 50 and 10; Imperial, one coat, 60; Premier, 50; Colonial, 50 and 10; Royal, 60; Victoria, 45; white 45; Diamond, 50; Granite, 60 p.c.

GALVANIZED WARE. — The discount on pail is now 37½ per cent.; and on other galvanized lines the discount is 30 per cent.

CORDAGE—We quote: Rope, sisal, 7-16 and larger, basis, $11.25; Manila, 7-16 and larger, basis, $16.25; Lathyarn, $11.25; cotton rope, per lb., 21c.

SOLDER—Quoted at 27c. per pound. Block tin is quoted at 45c. per pound.

VISES — Prices are quoted as follows: "Peter Wright," 30 to 34, 14½c.; 35 to 39, 14c.; 48 and larger, 13½c. per lb.

ANVILS—"Peter Wright" anvils are selling at 11c. per lb.

CROWBARS—Quoted now at 4c. per lb.

POWER HORSE CLIPPERS — The "1902" power horse clipper is selling at $12, and the "Twentieth Century" at $6. The "1904" sheep shearing machines are sold at $13.60.

AMMUNITION, ETC.—Quotations are as follows: Cartridges, Dominion R.F. 50 and 5; Dominion, C.F., 33½ C.F., pistol, p.c.; C.F., military, 10 p.c. advance. Loaded shells: Dominion Eley's and Kynoch's soft, 12 gauge, black, $16.50; chilled, 12 gauge, $17.50; soft, 10 gauge, $19.50; chilled, 10 gauge, $20.50. Shot, ordinary, per 100 lbs., $7.25; chilled, $7.75; powder, F.F., keg, Hamilton, $4.75; F.F.G., Dupont's, $5.

IRON AND STEEL.—Quotations are: Bar iron basis, $2.70. Swedish iron basis, $4.95; sleigh shoe steel, $2.75; spring steel, $3.25; machinery steel, $3.50; tool steel, Black Diamond, 100 lbs., $9.50; Jessop, $13.

SHEET ZINC — The price is now $8.50 for cask lots, and $9 for broken lots.

PIG LEAD—Quoted a $5.85 per cwt.

AXLE GREASE—"Mica" axle grease is quoted at $2.75 per case, and "Diamond" at $1.60.

IRON PIPE AND FITTINGS— Revised prices are as follows:—Black pipe, ¼ inch, $2.65; ⅜, $2.80; ½, $3.50; ¾, $4.40; 1, $6.35; 1¼, $8.65; 1½, $10.40; 2, 13.85; 2½, $19.00; 3, $25.00. Galvanized iron pipe, ⅜ inch, $3.75; ½, $4.35; ¾, $5.65; 1, $8.10; 1¼, $11.00; 1½, $13.25; 2, inch, $17.65. Nipples, discounts 70 and 10 per cent.; Unions, couplings; bushings and plugs, 60 per cent.

LEAD PIPE—The price, $7.80, is firmly maintained in view of the advancing lead market.

GALVANIZED IRON—Quoted as follows:—Apollo, 16 gauge, $3.90; 18 and 20, $4.10; 22 and 24, $4.45; 26, $4.40; 28, $4.65, 30 gauge or 10¾ oz., $4.95; Queen's Head, 24, $4.50; 26, $4.65; 28, $5.00.

TIN PLATES—We now quote as follows: IC charcoal, 20 x 28, box, $9.50; IX charcoal, 20 x 28, $11.50; XXI charcoal, 20 x 28, $13.50.

TERNE PLATES—Quoted at $9.

CANADA PLATES—Quoted as follows: Canada plate, 18 x 21, 18 x 24, $3.40; 20 x 28, $3.65; full polished, $4.15.

BLACK SHEETS—Prices are: 10 to 16 gauge, 100 lbs., $3.50; 18 to 22, $3.75; 24, $3.90; 26, $4; 28, $4.10.

PETROLEUM AND GASOLINE.—Silver Star in brls. per gal., 20c.; Sunlight in brls, per gal., 21c.; per case, $2.30; Eocene in brls, per gal., 23c.; per case, $2.50; Pennoline in brls., per gal., 24c.; Crystal Spray, 23c.; Silver Light, 21c.; Engine gasoline in barrels, per gal., 28c., f.o.b. Winnipeg in cases, $2.75

PAINTS, OILS & TURPENTINE — Turpentine is firm at the recent advance White lead, pure, $7; bladder putty in barrels, 2½c.; in kegs, 3¾c.; turpentine, barrel lots, Winnipeg, $1.01; Calgary, $1.08; Lethbridge, $1.08; Edmonton, $1.09. Less than barrel lots 5c. per gallon advance Linseed oil, raw, Winnipeg, 64c.; Calgary, 71c.; Lethbridge, 71c.; Edmonton, 72c.; boiled oil, 3c. per gal. advance on these prices.

WINDOW GLASS—We quote: 16-oz. O.G., single, in 50-ft. boxes—16 to 25 united inches, $2.25; 26 to 40, $2.40; 16-oz. O.G., single, in 100-ft. cases — 16 to 25 united inches, $4; 26 to 40, $4.52; 41 to 50, $4.75; 50 to 60, $5.25; 61 to 70, $5.75 21-oz. C.S., double, in 100-ft. cases—26 to 40 united inches, $7.35; 41 to 50, $8.40; 51 to 60, $9.45; 61 to 70, $10.50; 71 to 80, $11.55; 81 to 85, $12.60; 86 to 90, $14.75; 91 to 95, $17.30.

YEAR'S TRADE REVIEWED.

When Hardware and Metal's young man called upon George E. Davis, secretary-treasurer of Frothingham & Workman, wholesale hardware merchants of Montreal, he found him immersed in a mass of papers which he explained were the day's orders.

In reply to the questions as to the prospects of 1907, judging from the experience of the past year, Mr. Davis replied. "The end of this present prosperity does not yet seem to be in sight and with the swing trade now has there is no reason to anticipate a let-up. Reports from all parts of the country are uniform as to the bountiful crops. Manufacturers also are busy and are using large quantities of goods and supplies in the hardware line. Collections have been easy during the past year, showing that money is circulating freely."

"How about prices?"

"There has been a steady advance in prices, and this not alone in lines of hardware into which iron chiefly enters, but also handles, leather, and rope and such goods. During an advancing market there is generally less cutting of the prices and trade has in consequence, been very satisfactory. The most remarkable thing, however, to note in connection with advancing prices is that consumption has not apparently been checked. There seems no limit to the ability of the country to consume goods of all kinds. It has been very difficult to secure some lines of standard goods, more particularly in the iron line. During all the year it has been almost impossible for a jobber to complete his stock again once it was broken."

"Do you find help hard to obtain?"

"Well, yes. There is more or less difficulty, we have several openings now. There's not much help of an intelligent class offering. This must be another result of the present prosperity, for boys are generally kept longer at school during good times. Hardware and metal does not seem to have much difficulty getting smart fellows," said Mr. Davis, as he shook hands and said good-bye.

BIG AMALGAMATION.

Although details are not yet available we can announce that arrangements are under way for the amalgamation or absorption of the Ontario Tack Company, Hamilton, by the Canada Screw Company, of the same city. The Ontario Tack Company recently purchased the wire nail machinery of the Ontario Lead and Wire Company, Toronto, and has a large interest in the Western Wire Nail Works, at London. The amalgamation of these interests with the Canada Screw Co., which has branches at Toronto, Vancouver and other places, is an important event in the hardware trade, and may help considerably in the movement now under way to bring wire nails back to the profitable prices ruling prior to the price cutting war of the past two years.

TRAVELERS TO MEET.

The Canadian Fairbanks Company's staff of salesmen covering the territory between Montreal and Vancouver will hold a convention in Toronto next week, the headquarters being at the firm's new building on Front street west. Trips of inspection will be made of the Canadian Fairbanks-Morse factory in Toronto, the Union Drawn Steel plant at Hamilton, the Bertram Works at Dundas, and other plants for which the company are sole selling agents.

McCLARY TRAVELERS CHANGES.

The McClary Manufacturing Company have made a number of changes in their staff, to take effect at the end of this year. George Clarke, who has been representing the firm in the western district, has been transferred to Hamilton, where he will be branch manager. Alexander Clarke, manager at Hamilton, will come to London, to be connected with the company's sales department. G. Smith, who has been traveling for the Montreal branch, has been transferred to London to cover the territory heretofore in charge of George Clarke.

WEST INDIAN TRADE.

Pickford & Back, Halifax, N.S., have offered to conduct four representatives from Canada to the British West Indies for the object of increasing Canadian trade there. J. D. Allen will represent the Toronto Board of Trade; A. E. Jones, Halifax Board of Trade; H. B. Schofield, St. John Board of Trade. Montreal has not yet elected its delegate. The party will go on January 15, and be away thirty-eight days. All Canadian manufacturers' catalogues will be taken care of by A. E. Jones, Halifax.

HARDWARE ASSOCIATION ITEMS.

Owing to the illness of his wife, E. P. Paulin, Goderich, was not able to attend the meeting of the Executive Committee in Toronto on Monday.

W. G. Scott, Mt. Forest, was also unable to attend, owing to his being a candidate for re-election as Water Commissioner of his town.

The members present were: President Humphries, Vice-President Hambly, Secretary Wrigley, Treasurer Caslor and Messrs. Brocklebank, Taylor and Peart.

J. Walton Peart, St. Mary's, leaves this month on an extended trip to the Southern States on account of his wife's health. Mrs. Peart is recovering from her recent attack of typhoid fever.

SEASON'S SKATE TRADE.

J. S. Bowbanks, Toronto, representative of the Starr Manufacturing Company of Halifax, N.S., states that the business done in 1906 showed a large increase on 1905. While a great many dealers had skates left over from 1905, it is evident that very little stock will be carried over this season. Mr. Bowbanks travels from Halifax to Vancouver, besides every two years taking a trip to the Old Country, he being the first to sell hockey skates and sticks in England and Scotland where Starr skates have obtained a firm foothold.

An increased business is looked for 1907.

A GOOD INCREASE.

Mr. Patterson, of the Patterson Manufacturing Company, Toronto, reports a very brisk year for their products, building paper, tar felt, wire edge, ready roofing, etc. The increase over 1905 was about 15 per cent., but he looks for a much larger increase for ...

SEASONABLE SUGGESTIONS.

Push ice-scrapers and snow-shovels this month but set a good example by keeping your sidewalks clear. Show that you believe in using the goods you sell.

THOS. DAVIDSON CO.'S BANQUET.

The 10th annual banquet tendered by the President and Directors of the Thos. Davidson Manufacturing Company, to their travellers, office staff, and heads of departments, on December 28, was thoroughly enjoyed by all the participants. The function was held in the club rooms provided for the employees, in the immediate vicinity of the factory.

Owing to the unavoidable absence of Mr. Jas. Davidson, the chair was taken by T. C. Davidson, the Vice-President. The first toast of the evening was "The King," which was right loyally honored. The toast to "Our Travellers," was then proposed by E. Goodwill, who referred to the enormous increase in Canadian trade during the last 25 years. W. H. Morgan, (Central Ontario), and Geo. Mather (B.C.), replied on behalf of the "Knights of the Grip." J. N. Young proposed the "Office Staff and Ware Room," which drew forth a very interesting reply from A. Bindley. W. J. White, K.C., proposed the "Winnipeg Branch," and was followed by J. Taylor Webb, the western manager, who spoke in his usual happy vein. The next toast was "Heads of Departments," proposed by T. C. Davidson, and replied to by Messrs. M. Lachapelle and Jos. Bodfish. The next toast was "The Ladies," proposed by C. D. Koppell, and replied to by T. R. Davidson, on behalf of the fair sex. The toast "Canada," was proposed by R. B. Gray, and replied to by H. B. Chadburn, both of these gentlemen representing parts of the Province of Ontario. After the health of the President and Directors had been enthusiastically drunk, the evening was brought to a close with the singing of the National Anthem.

Those who contributed with songs or recitations included Dr. W. A. Haldimand; Messrs. Jas. Chalmers, C. P. Clarke, A. O. Gee, H. Brisson, W. Johnston, J. O'Dowd, and W. E. Barrat. The item contributed by Jas. Chalmers, (who also designed the invitation cards and menus, which, as usual, were lithographed on tin), was particularly humorous, being an original poem with local hits, which was thoroughly enjoyed.

CATALOGUES AND BOOKLETS

When sending catalogues for review, manufacturers would confer a favor by pointing out the new articles that they contain. It would assist the editor in writing the review.

By mentioning HARDWARE and METAL to show that the writer is in the trade, a copy of these catalogues or other printed matter will be sent by the firms whose addresses are given.

Forty Year's Growth.

The Sherwin-Williams Company, paint manufacturers, Montreal, have issued a pamphlet entitled "Forty Year's Growth," with miniature cuts of their plant in 1866, and of their various large concerns of the present day, including five large plants in Cleveland, Chicago, Montreal, Newark and London, Eng., also their offices and warehouses in fifteen principal cities, scattered from the Atlantic to Pacific Ocean, and from Winnipeg to the Gulf, which demonstrates the great growth of this firm in that space of time.

Novel Souvenir.

Dorken Bros. & Company, of Montreal, manufacturer's agents for H. Bokers & Company, Tree Brand cutlery, have issued a very acceptable little folder wishing the trade A Happy New Year. In the centre is a miniature sketch of the Alps, with a sample sprig of a small flower called "Edeliwiss," which is much sought after by Alpine travellers, and grows only in the Alps and Pyrenees, at a great allitude in situations difficult of access.

Standard Paint Catalogue.

The Standard Paint and Varnish Company of Windsor, are sending out a new catalogue of about 5 1-2 x 8 inches, containing 50 pages fully illustrated, with the products manufactured by them. On the front page are cuts of the present officers of this company, including Jas. A. Straith, President, and general manager; Albert Stolf, factory manager; D. B. Fisher, secretary and assistant manager, Wm. Richards, superintendent dry color department; Wm. Dietzel, superintendent vanish department. Also a cut of Ludger Gravel, the hustling manufacturers' agent, who represents the Standard Paint & Varnish Company in Montreal.

Homes, Healthful and Beautiful.

Under the above caption, the Alabastine Co., of Paris, Ontario, have put out, probably one of the handsomest booklets on home decoration ever issued in Canada. While designed primarily as a catalogue, the book is handsomely illustrated with numerous cuts of the interiors of rooms in colors, and, as well as containing complete information on wall decoration, is filled with many valuable suggestions both pictorial and otherwise on tasteful home furnishing. It is a book which needs only to be seen to be appreciated.

"Homes, Healthful and Beautiful," is gotten out in the form of an edition de luxe, and it is too expensive a book to distribute promiscuously to everyone who might write for it merely out of idle curiosity. The Alabastine Company are therefore making a charge of ten cents for it, which amount, while not nearly covering the cost of the book, is intended to discourage those who would write for it merely to gratity an idle whim. It is well worth reading. A copy will be mailed to any address on receipt of ten cents.

Foundry Supplies.

The Catalogue of the Hamilton, Ont., Facing Mills Company, has been sent out to the trade. This catalogue is of 6 x 9 inches, containing 80 pages, fully illustrated, of their products of foundry supplies, including brushes, sieves, and moulders tools, bellows, shovels, crayons, also foundry facings, foundry equipments, graphite and plumbago. It will be found useful to hardware merchants, in ordering supplies.

Catchy Advertising.

The enterprising firm of James Reid, hardware, harness, wall paper and paint merchants of Perth, Ont., have supplied their customers with a handsome calendar, showing a young lady and a young man. The firm uses the following catchy bit of prose:

"A fellow loves a girl,
That's his business.
A girl loves a fellow,
That's her business.
They get married and need hardware,
That's our business.

Blacksmith's Supplies.

Ludger Gravel, agent for blacksmith's supplies, Montreal, has issued a handsome pastel of a blacksmith surrounded by his tools, and dressed in the regalia of his profession. This pastel is very attractive, and will be given a prominent place by any customer fortunate enough to receive a copy.

FOUNDRY AND METAL INDUSTRIES

The Shipway Iron, Bell and Wire Company, of Toronto, are considering the moving of their plant to Niagara Falls.

A British patent recently issued to M. Gruber of Berlin, Germany, for an aluminum solder, contains 60 parts of tin, 25 parts of zinc, 10 parts of copper, 3 parts of cadmium and 2 parts of aluminum.

The E. A. Pflueger Company, recently incorporated at Akron, Ohio, for the manufacture of fishing tackle and reels, will install a complete plating and polishing plant, and later will put in a brass foundry.

The Percival Plow and Stove Company have organized at Merrickville with the following officers: J. B. Waddell, President; R. C. Percival, Vice-President; E. W. Stickney, managing director and G. S. Seeber, Secretary-Treasurer.

The International Acheson Graphite Company of Niagara Falls, N.Y., are now manufacturing a soft artificial graphite which may be used for electrotyping, lubrication and stove polish. A new discovery by E. G. Acheson has made this possible.

The Nova Scotia Steel Company during November shipped from North Sydney 69,304 tons of coal, or an increase over the same period last year of 10,195 tons. For the 11 months up to November 30, 612,020 tons have been shipped, as against 487,500 tons, or an increase over last year's shipments of 124,453.

A despatch from Glace Bay says: Everything now indicates that the fire in the Hub colliery is out. The temperature of the mine at the bottom of the main shaft is under 100 degrees, and at two other places only 40 and 60 degrees. The Dominion Coal Company are determined to be on the safe side, however, and the Atlantic will continue to pour into the mine for several days. Several powerful pumps have been ordered, and are now on the way to Glace Bay. As soon as they arrive the company will start to pump the mine out.

MAKING SOLDERING FLUX.

Chloride of zinc is very extensively used as a flux in soft soldering. It is an excellent one, too, and nothing has yet been found to take its place. It also has the advantage of cheapness. The action of chloride of zinc in soldering is based upon the fact that it dissolves the oxides of tin and lead upon the solder, and produces a clean surface for uniting with the metal to be soldered.

The mistake that is frequently made in making soldering flux is to add water. The stronger the flux can be made, and still remain in the liquid condition, the better it will be. As it is the chloride of zinc that does the work, the presence of water is a detriment. The water also has the advantage of producing spattering. The action of a weak flux in cleaning the surface of the molten solder is far inferior to that of a strong solution. In order to produce the best results, the flux should be made by dissolving zinc in strong muriatic acid until it will take up no more.

VALUE OF BABBITT METAL.

Babbitt metal is a composition that is used extensively as a substitute for brass in the lining of journal bearings, says the Valve World. To make such a bearing, the shaft for which it is required is placed in position in its recess or cavtiy, and then the melted babbitt metal is poured in around it and allowed to cool, thus forming the bearing for the shaft. The value of babbitt metal for this purpose lies in the fact that it has all the merits of brass as a wearing material, and at the same time melts at a much lower temperature; and also being of a very fluid nature when melted, it may be poured into the cavities around the shaft and the expensive operation of fitting brass to these places thus be dispensed with. What is called the "genuine," or superior babbitt metal is described under the heading of Tin Alloys. But there are a variety of cheaper grades of this material, in which lead is very extensively used alloyed with antimony, or with antimony and tin. While these compositions are not so good as the "genuine," they are of sufficient merit for a large variety of cheap machinery, and are extensively used in the repairing of agricultural implements, and other grades of cheap machinery. An immense quantity of the combination of lead and antimony is produced by the silver smelters, a market for which is found in the production of babbitt metal.

BUILDING AND INDUSTRIAL NEWS

HARDWARE AND METAL would be pleased to receive from any authoritative source building and industrial news of any sort, the formation or incorporation of companies, establishment or enlargement of mills, factories, foundries or other works, railway or mining news. All correspondence will be treated as confidential when desired.

A tannery is being established at Sydney, N.S.

The Red River Metal Company will elect a large warehouse in Winnipeg, to cost $12,000.

The building returns of Calgary and suburbs, for the year 1906, will be in the vicinity of $3,000,000.

The building permits issued in Vancouver for 1906, total $4,084,840, which is more than double those of two years ago.

Gordon Mackay & Company, of Toronto, have been issued a permit for the erection of their new factory at a cost of $60,000.

F. C. Filer, of the Northern Electric & Manufacturing Company, of Montreal, is establishing a branch of his company in Winnipeg.

The Burlington Masonic Hall Company, has been incorporated with a share cpaital of $10,000 for the erection of a hall for the Burlington Masonic Lodge.

The sum of $600,000 has been subscribed in Quebec for the starting of a cement industry. The land has been purchased, and works will be started immediately.

The contract for erecting the Meisel Manufacturing Company's new plant at Port Arthur has been awarded and work will be commenced on the buildings at once.

Another industry has been located at Welland, by John A Reeb, of Port Colborne, and W. J. Sommervile of Welland, who will manufacture wood fibre wall plaster, and cement tile.

A new $100,000 hotel is to be built at St. Johns, Newfoundland. All the stock is held by Newfoundland parties, excepting two Canadian commercial men, Messrs. R. Taylor, of Halifax, and L. Garneau, of Quebec.

The Great Northern Railway has announced a new freight tariff, effective December 27, making voluntary reduction in rates from the south to Winnipeg, Portage la Prairie, Brandon, Prince Albert and Edmonton.

The Winnipeg Foundry Company, has been incorporated with a $20,000 capital; provisional directors: J. T. Hill, machinist; H. Bell, machinist; W. A. Armstrong, agent; A. D. Irish, agent; A. J. Brown, agent, and F. W. Irish, manager, all of Winnipeg.

The Well Machine and Wind Mill Company, has been incorporated with $20,000 capital: provisional directors: C. S. Tyrrell, merchant; J. H. Inkster, agent; H. W. Hutchinson, manager; W. S. Evans, financial agent; J. A. Machray, barrister-at-law, all of Winnipeg.

Montreal shows a large increase in the number of buildings erected, also total value. Permits for new buildings in 1906 are 1,484, as compared with 1,145 for 1905. The value of buildings for 1906 is $7,745,023, compared with $4,779,380, for 1905, an increase of $2,965,643.

Bechtels, Limited, has been incorporated with a share capital of $75,000, for the purpose of manufacturing brick, tile blocks, and cement products. The head office will be at Waterloo, and provisional directors will be: B. E. Bechtel, W. B. Bechtel, W. J. Watson, C. E. Whyard of Waterloo; C. H. Bechtel, of Berlin, and P. A. Watson, of Galt.

The Sydney Cement Company have decided to enlarge their plant, which is situated on ground leased by them from the Dominion Iron & Steel Company, between two railroads, a position of considerable advantage for deliveries of raw material and for shipment of its products. The cement is made from slag which is procured from the blast furnaces of the Dominion Iron & Steel Company.

The first gypsum plaster ever made in Winnipeg, was turned out last week, by the Manitoba Gypsum Company, at their new works. The company first built works on Lake Manitoba, they were burnt down in July last, and it was decided to erect a new factory at Winnipeg, and to take the rock there for treatment. About 70 hands are now employed, and work will continue all the year round.

The Ontario Wind Engine & Pump Company, shipped, on December 26, a consignment of 15 Airmotor Outfits to Egypt. This follows other large shipments made last summer and fall, and these shipments speak loduer than words of the popularity of these Canadian manufacturers in the Levant. It also emphasizes the wisdom of maintaining a reasonable protected tariff to enable Canadian manufacturers to build up not only a home business, but also an export business, for if the manufacturers of this country are not protected they neither can manufacture for the home trade nor for export. The Canadian Airmotors now to be seen in Cyprus, Egypt, Africa, and other distant parts of the world are the best testimony to Canadian Statesmanship.

In Toronto, during 1906, the building permits issued represented the sum of

$13,152,000. Last year the amount was $10,370,000. Application for the construction of several large buildings have recently been received, but the permits have not been issued. Among these are the Public Library and the Dominion Radiator Company's works. The former is $242,000 and the latter $500,000. These, if issued, would bring the amount up to $13,894,000. The number of permits issued during the year was 3,438 and the buildings erected 4,709. Following is a list of the permits issued during the past ten years: 1896, $657,168; 1897, $947,611; 1898, $1,701,630; 1899, $2,010,446; 1900, $1,903,136; 1901, $3,568,883; 1902, $3,854,923; 1903, $4,356,457; 1904, $5,896,120; 1905, $10,370,000.

Galt has given substantial evidence in the way of growth by the erection of 112 dwelling houses. The manufacturing part of the town, has also shown a decided increase, some new firms starting, and a number of old ones adding considerable floor space, 314,710 square feet being added to 15 enterprises, divided viz.: Galt Knitting Company, 87,500 square feet; MacGregor, Gourlay Company, 84,845 square feet; Malleable Iron Company, 36,991 square feet; P. W. Gardiner & Son, 22,400 square feet; Sheldons, Limited, 21,000 square feet; R. McDougall Company, 13,200 square feet; Galt Robe Company, 10,360 square feet; Cowan & Company, 10,230 square feet; McVicker Engine Company, 9,548 square feet; Box Factory, 4,950 square feet; J. J. Stevens Company, 4,100 square feet; Victoria Wheel Works, 3,600 square feet; Peter Hay Knife, 1,600 square feet. The total amount of money expended for building purposes in Galt in 1906, was $560,000.

A well organized and energetic movement is on foot in the Maritime Provinces to build up the steel shipbuilding industry. A statement summing up the argument in favor of a tonnage bonus on steel ships built in Canada was lately submitted to the Tariff Commission, and the Dominion Government has been memorialized to grant such a bonus on a basis of $6 per gross ton. The Boards of Trades and municipal councils of Halifax, Dartmouth and other places have taken the matter up, and much interest is felt regarding the manner in which the request for a bonus will be treated at Ottawa. Years ago in Eastern Canada, shipbuilding assumed very large proportions, but with the advent of steel construction it rapidly declined. Steel vessels can be built cheaper in British yards on account of the cheaper iron, coal, and labor there Those who agitate for a bonus on steel ships point out that all maritime nations give protection to their shipping. There is an increasing demand in Canada for steel vessels for both coastwise and lake trade, but it is claimed that Canadian yards cannot compete with British and foreign builders in this line of construction. It is pointed out that Canadian built tonnage has decreased in less than 30 years from 183,000 to 33,000, while the entries in and out from sea have increased from six to sixteen millions, the coastwise shipping from ten to forty-five millions, and the Great Lakes tonnage has more than doubled. There are now in Canada, besides several smaller firms, four large steel shipbuilding companies—the Algoma Steel Company, Sault Ste. Marie; the Canadian Iron Furnace Company, Montreal; the Collingwood Shipbuilding Company, Collingwood; and the Canadian Shipbuilding Company, Toronto. These yards are not all fully and steadily employed, and it is suggested that while many millions of dollars are being expended in Canada in nearly every industry, including iron and steel, it is time we should build our own ships and do our own carrying trade.

THE TALLEST BUILDING.

Contracts have been awarded for the erection in New York City of what will be the tallest building in the world. This is the proposed new Singer building on Broadway, north of Liberty Street. It will require for the work 6,000 tons of structural steel. The building will have thirty-six regular floors, on top of which will be a dome including four floors, the fortieth floor being 550 feet above the curb. This is a part of the transformation of a city into an abode of cliff-dwellers. The building will be 625 feet high, the tallest skyscraper in the city, and will have wind anchors so that it may be firmly braced against every gale. The wind pressure, on account of the structure's great altitude, will be tremendous, and for that reason the building is to be literally tied to its foundations by an ingenious arrangement of steel rods. They will be three and a half inches in diameter, and descend for nearly fifty feet into the concrete which forms the caissons resting on solid rock eighty-five feet below the curb. The lowest rod has on the end of it a great anchor plate to which it is secured.

LATEST CANADIAN PATENTS.

The following up-to-date list of Canadian patents is reported to us by Egerton R. Case, Solicitor of Patents, Temple Building, Toronto.

Rob. A. Armstrong, Avenmore, Ont., ploughs. Richard H. Castle, Toronto, Ont., oven doors and lamp receptacles therefor. Wm. Hull, Souris, Man., attachments to cultivators. Jas. Wm. McDonald, Winnipeg, nut locks. Frank L. Armstrong, Moose Jaw, Sask., scrapers on disc plows. Wm. Maloney, Smith's Falls, combined cutting knives and distributors for adjusting means for harvester frames. Wm. Maloney, Sherbrooke, threshing harvesters. Harrison C. Frost, Montreal, rubber hose. Wm. F. Kendall, Grimsby, Ont., step ladders. Arthur P. Couture, Winnipeg, window hinges. Alex. Dobson, Beaverton, shockers. Jos. S. Nesbitt, Victoria, B.C., flash signs.

SYSTEM IN FINDING FOUNDRY COSTS.

By D. C. Eggleston

Notwithstanding the advance made in finding costs of production during the last few years it is generally true that years it is generally true that the iron foundry has been neglected Where the attention of cost system experts has been directed toward the iron foundry the tendency has been to try and adapt systems used in other departments. However, the work of the iron foundry is so different in nature from that of other departments that a radical departure must be made from the methods elsewhere employed.

In any iron foundry an analysis of the expense shows that it is incurred in connection with the cupola and several classes of work. The output of the former is in pounds of metal and of the latter in hours of labor. This suggests that the entire cupola expense should be prorated according to the number of pounds of castings produced and the expense against other classes of work on the basis of man-hours worked.

This method requires a division of expense between the cupola and other classes of work. In most cases no difficulty need be experienced in doing this; coke, fire-brick, ladles, lime stone, wood and some water are chargeable to cupola expense The time of the foremen and foremen's clerks should be divided between cupola and other divisions of expense by estimate of the time spent on coke, pig iron, charging sheet and output reports The administration expense not chargeable to the cupola should be distributed to other classes of work on the basis of the number of man hours in each. The rent expense includes all charges on account of grounds, buildings, heating, lighting and fire service. This division of expense is distributed on the basis of the number of square feet of floor space occupied by each class of work. Power expense should be distributed by estimate to the various classes of work.

The foundry clerk should keep a card file on which he records tools such as shovels and sieves received, given out (noting to which class of work), and the number on hand Reference to the file facilitates the distribution of supplies There are often some items concerning which there is doubt as to just which division of work they should be assessed In such cases it is necessary to consult the foundry foreman that as accurate a division may be made as possible

It is the best practice to assign letters indicative of the various rates of expense on the different classes of labor. Thus I F. A. may be used to denote the rate on core work; I. F. B. on crane moulding, and so on. The foreman's clerk notes the proper letters on the workman's time ticket so that the expense can be added simultaneously with the productive labor to the cost of the hob. To the cost of metal the rate per pound for cupola expense can be added, thus giving the indirect expense incurred in connection with the cupola. The sum of productive labor, labor expense, raw material and cupola expense gives the total cost of producing the castings on an order.

In the statement of shop deliveries the productive labor, raw material and expense accounts should be credited and piece parts account is credited and the the casting is used on an order the the piece parts account debited. When proper account debited. Thus the cost system herewith described not only gives accurate costs of production, but also facilitates in keeping a check on the work of the iron foundry. If it is desired to cut down expense a study of the exhibit of iron foundry figures will suggest valuable economies. Comparative figures aid in showing wherein the increase or decrease lies. That the system herein described has been evolved to meet the requirements of the iron foundry in a large manufacturing works where it is in successful operation, recommends it to the attention of iron foundrymen.—Iron Age.

A NOVEL SHAVING MIRROR.

The latest good thing from the ingenious shops of the General Specialty Manufacturing Company, of Philadelphia, is the adjustable shaving and toilet mirror called the "Tipso." This Tipso mirror strikes the average person as just the handiest thing of its kind ever put out.

It is an oval mirror of French plate, bevelled and handsomely mounted in metal. It can be adjusted, by a touch, in any direction, at any angle, so as always to get the best light on the face.

"Tipso" Shaving Mirror.

Each mirror has a heavy ball fixture for screwing into the window frame or wood-work frame near any convenient light, day or night. Stands are provided also for use in dressing tables, supporting the mirror at various heights, a toilet accessory greatly appreciated by women For household use or traveling, the Tipso mirror meets a widely felt want and already it is finding a very large sale.

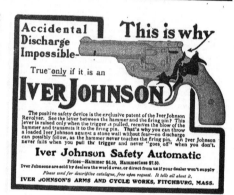

CONDENSED OR "WANT" ADVERTISEMENTS.

Advertisements under this heading 2c. a word first nsertion; 1c. a word each subsequent insertion.

Contractions count as one word, but five figures (as $1,000) are allowed as one word.

Cash remittances to cover cost **must** accompany all advertisements. **In no case** can this rule be overlooked. Advertisements received without remittance cannot be acknowledged.

Where replies come to our care to be forwarded, five cents must be added to cost to cover postage, etc.

SITUATIONS VACANT.

WANTED—An experienced hardwareman at once; write with references and state salary expected to Box 273, Tilsonburg, Ont. [tf]

WANTED an assistant to editor of Hardware and Metal in Montreal—Young man preferred who has had some experience with the hardware or metal trades; Excellent opportunity for advancement; state experience and salary expected; all applications treated confidentially. Address, Manager, HARDWARE AND METAL, Montreal.

TRAVELLER WANTED

TO sell a line of established successful coal heaters, also wood heaters, on commission; must cover territory twice or oftener, say once between January and April and again between May and August; state lines now carried and territory covered; liberal commission given; Box 573, HARDWARE AND METAL, Toronto. [1]

BUSINESS CHANCES.

STOVES—An established, successful working line of patent coal, also wood heaters, for sale; including patent, patterns, materials and some completed stoves; also machinery and dies, if wanted; will go with a line of heaters or ranges, or to sell with some hardware specialty; proprietor has other pressing interests. Address Box 574, HARDWARE AND METAL, Toronto. [1]

WANTED

by January 1st, 1907,

Glass Salesman

Must be experienced and reliable, state references and qualifications, also amount of salary required. Apply only by letter to

Consolidated Plate Glass Co., Limited

Toronto, Canada. (52)

Are you interested in any of the lines that are advertised?

A Post Card will bring you price list and full information.

Don't forget to mention Hardware and Metal.

To Manufacturers' Agents

HARDWARE AND METAL has inquiries from time to time from manufacturers and others wanting representatives in the leading business centres here and abroad.

Firms or individuals open for agencies in Canada or abroad may have their names and addresses placed on a special list kept for the information of inquirers in our various offices throughout Canada and in Great Britain without charge.

Address

Business Manager

HARDWARE AND METAL

Montreal and Toronto

Paint, Oil and Brush Trades

THE SHORTAGE OF GASOLINE.

"Far-reaching are the economic effects of the rising price of gasoline," says the Boston Transcript. "For thousands of scattered factories the power is to-day furnished by gasoline engines, which operate on the explosion principle, made popularly familiar by the operations of the automobile. But the rising price of gasoline is so seriously interfering with the growth, if not the maintenance, of this form of power supply that engine manufacturers are now experimenting on an extensive scale with the less costly distillates in the hope of getting a suitable substitute. It is against this commercial use of gasoline that the present prices will have their first effect. But other of its uses will be affected if its price mounts much higher, notably the development of the autotruck. But for the prospect of diverting some of the demand for gasoline to the other distillates of crude oil, and the likelihood that denatured alcohol will relieve the situation somewhat, gasoline would already have gone much higher than now. The quality of the gasoline is also affected, since it is believed that to-day a proportion of Texas crude oil is often mixed with that from the Pennsylvania wells to make the latter 'go further,' before the process of refining begins. The result is that greater residuum of mineral matter, of which automobilists and others are now complaining.

"America's production of gasoline is apparently not on the increase, although this country's output of crude oil has been steadily, if not sensationally, increasing in volume, and is now larger than ever before. This contrast is due to the character of the Texas and California oil fields, both of which yield a small percentage of gasoline, and also of all other grades of lighter oils. The best fields of these are in Pennsylvania. But that state is not maintaining its proportions of the total supply, and as a result gasoline is falling relatively behind.

"From the oil as it goes through a process of refinement, the first vapors which condense into liquid form are the naphthas, gasoline and kerosene in that order; then come the neutral, or solar, oils which are sold to the gas companies to make the illuminant of our cities. Then follow the lubricating distilates, and finally those heavy oils used in cylinders. In the Pennsylvania field from 60 to 70 per cent. of the oil is turned into kerosene or lighter distillates, while in Texas less than 20 per cent. is thus available. This tells the story. Within the memory of the present generation gasoline sold for 6 or 7 cents a gallon, because the demand at that end of the line was so small. It now retails at about 20 cents. Certain industries, notably the automobile and the power engine run by gasoline, have so focused demand on it that it is sure to increase greatly in price unless relieved by alternative products. It is hoped that by a display of ingenuity the solar oils can yet be used in the factory engines.

"Kerosene, which is still going up in price, although less in proportion than gasoline, is the great illuminant of the world, not only in the rural regions of America, but in all continents, down to the very grade of life which regards artificial light as not worth having. It is a curious fact that the increase in electric-lighting plants apparently increases the demand for kerosene oil, too, often in the very towns where electric works are operated, people accustom themselves to certain degrees of light, forming habits in that regard which, like popular demands in heating and other things, show themselves all along the line. Gas and electric lighting have created a popular taste for more light, and this in turn affects those who depend on kerosene for their illumination.

"The troubles in Russia have greatly affected the situation in Europe today, so that it has been obliged to import large quantities of Texas product, although a few years ago it was with great difficulty that the Texas lubricating oils found any market there. There is a tremendous demand for them now, and one that it is almost impossible adequately to meet. The world might well wonder how it will get along for lubrication of its countless machines, as well as for light, when the stored supply of mineral oils have been exhausted, but for experience with similar worries before. The old lady who expected to have to sit in darkness when all the whales had been driven from the seas expressed a fear which has found many forms. President Dwight, of Yale College, in his celebrated 'Travels,' written a century ago, regarded the outlook of New York state for growth in population and business, limited only by its fuel supplies. He did not think enough wood could be grown there to provide for more than about so many people. Although he visited Niagara he never looked upon it as a substitute for the growing forests.

"Some of the oldest of the Pennsylvania wells are still producing oil, although not so much as formerly. But new regions are constantly discovered, so that the present generation need not worry. The economic price will keep an effective check valve on consumption, lest it run away from supply, just as it is now sending inventors into experimental engines, which may utilize the lower-grade oils."

PAINT AND OIL MARKETS

MONTREAL.

Office of HARDWARE AND METAL,
232 McGill Street.
Montreal, January 4, 1907.

Stock-taking is the order. Some travelers have started out and others are waiting until normal conditions are resumed after the holidays.

Paris Green is being freely advertised this week as usual and prices are normal for the season.

Turpentine and linseed oil remain firm and white lead is firm at the advance of three weeks ago.

LINSEED OIL—We continue to quote: Raw, 1 to 4 barrels, 55c; 5 to 9 barrels, 54c.; boiled, 1 to 4 barrels, 58c.; 5 to 9 barrels, 57c.

TURPENTINE.— We are now quoting: Single barrel, 96c. per gal.; two barrels or over, 95c per gal.; for smaller quantities than barrels, 5c. extra per gal. is charged Standard gallon is 8.40 lbs., f.o.b. point of shipment, net 30 days.

GROUND WHITE LEAD.—The recent advance is being rigidly maintained and grinders have no trouble in getting the following prices: Best brands Government Standard, $7.25 to $7.50; No. 1, $6.90 to $7.15; No. 2. $6.55 to $6.90; No. 3, $6.30 to $6.55 all f.o.b., Montreal.

DRY WHITE ZINC.—We still quote as follows: V.M. Red Seal, 7½c to 8c; Red Seal, 7c to 8c; French V.M., 6c to 7c; Lehigh, 5c to 6c.

WHITE ZINC (ground in oil)—Manufacturers have raised their prices, and we now quote: Pure, 8½c to 9½c.; No. 1, 7c. to 8c.; No. 2, 5¾c. to 6¾c.

PUTTY—A heavy demand is shown. We quote prices: Pure linseed oil, $1.75 to $1.85; bulk in barrels, $1.50; in 25-lb. irons, $1.80; in tins, $1.90; bladder putty in barrels, $1.75.

ORANGE MINERAL—We quote as follows: Casks, 8c.; 100 lb. kegs, 8½c.

RED LEAD—The following quotations are firm: Genuine red lead in casks, $6.00; in 100-lb. kegs, $6.25; in less quantities at the rate of $7 per 100 lbs.; No. 1 red lead, casks, $5.75; kegs $6, and smaller quantities, $6.75.

PARIS GREEN—Prices are as follows: In barrels, about 600 pounds, 23¼c. per lb.; in arsenic kegs, 250 lbs., 23½c.; in 50 lb. drums, 24c.; in 25 lb. drums, 24½c.; in 1 lb. packets, 100 lbs. in case, 25c.; in 1 lb packets, 50 lbs in case, 25½c.; in ½ lb. packets, 100 lbs in case, 27c.; in 1 lb. tins, 26c. f.o.b. Montreal. Term three months net or 2 per cent. 30 days.

SHELLAC—Quotations are as follows: Bleached, in bars or ground, 46c per lb., f.o.b. Eastern Canadian points, Bone Dry, 57c per lb., f.o.b. Eastern Canadian points; T.N. Orange, etc., 48c per lb. f.o.b. New York.

SHELLAC VARNISH—The prices are: Pure white, $2.90 to $2.95; pure orange, $2.70 to $2.75; No. 1 orange, $2.50 to $2.55.

MIXED PAINTS—Prices range from $1.20 to $1.40 per gallon.

CASTOR OIL—We still quote as follows: Firsts in cases, 8½c.; in barrels, 8c.; seconds in cases, 8c.; in barrels, 7½c

BENZINE—We still quote 25 cents in barrels, and 30 cents in smaller quantities.

WINDOW GLASS—Our prices are: First break, 50 feet, $1.85; second break, 50 feet, $1.95; first break, 100 feet, $3.20; second break, 100 feet, $3.40; third break, 100 feet, $3.95; fourth break, 100 feet, $4.15; fifth break, 100 feet, $4.40; sixth break, 100 feet, $4.95. Diamond Star: First break, 50 feet, $2.30; second break, 50 feet, $2.50; first break, 100 feet, $4.40; second break, $4.80; third break, 100 feet, $5.75; fourth break, 100 feet, $6.50; fifth break, 100 feet, $7.50; sixth break, 100 feet, $7.50; seventh break, 100 feet, $8; eighth break, 100 feet, $9. Double Diamond: First break, 50 feet, $3.45; second break, 50 feet, $3.75; first break, 100 feet, $6.75; second break, 100 feet, $7.25; third break, 100 feet, $8.75; fourth break, 100 feet, $10; fifth break, 100 feet, $11.50; sixth break, 100 feet, $12.50; seventh break, 100 feet, $14; eighth break, 100 feet, $16.50; ninth break, 100 feet, $18; tenth break, 100 feet, $20; eleventh break, 100 feet, $24; twelfth break, 100 feet, $28.50. Discount on Diamond Star, 20 per cent.; on Double Diamond, 40 per cent.

TORONTO.

Office of HARDWARE AND METAL,
10 Front Street East.
Toronto, January 4, 1907.

The paint and oil markets remain unchanged, but another increase is looked for in lead and turpentine.

Travelers have again resumed their work on the road, and report an increased demand for paints, etc. Retail merchants will do well to order now, as the large demand will no doubt cause a shortage.

WHITE LEAD—Ex Toronto pure white, $7.40; No. 1, $6.65; No. 2, $6.25; No. 3, $5.90; No. 4, $5.65 in packages of 25 pounds and upwards; 1-2c per pound extra will be charged for 12 1-2 pound packages; genuine dry white lead in casks, $7.00.

RED LEAD—Genuine in casks of 500 lbs., $6.00; ditto, in kegs of 100 lbs., $6.50; No. 1 in casks of 500 lbs., $5.75; ditto, in kegs of 100 lbs., $6.25.

DRY WHITE ZINC— In casks, 7 1-2c.; in 100 lbs., 8c., No. 1, in casks 6 1-2c., in 100 lbs., 7c.

Just try some ordinary green paint on a panel of wood, see how hard it works,—don't cover, lacks body. Then try on another panel

LUCAS IMPERIAL FRENCH GREEN

notice how easy it spreads, how well it covers,—that's body. Makes the cost less for this reason alone, and stays fresh and brilliant. It preserves the wood, keeps the weather out and the color in. It is the **ONE** Green that is free from arsenic or verdigris.

BEWARE OF IMITATIONS

JOHN LUCAS & CO.

NEW YORK, PHILADELPHIA, CHICAGO

"Extry!" "Extry!"

the newsboy's cry.

We all want the "**EXTRA**"—the something out of the ordinary. Someone has said, "It is the 'Extras' that make life worth the living."

The "**EXTRA**" news brings the "**EXTRA**" pennies to the newsboy.

The "**EXTRA**" quality of goods or service brings the "**EXTRA**" dollars to the business man.

EXTRA brings EXTRA

Now, What we have to offer is the Exclusive Agency of

The Martin-Senour Paint

THE PAINT THAT IS

100 per cent pure

The absolute purity of The Martin-Senour House Paint has been the means of obtaining "**EXTRA**" profitable returns to our agents.

The Martin-Senour Paint is not only as pure as the average mixed paint but it is a little "**EXTRA**"—it's 100 per cent. pure.

We would be glad to tell you more about it.

The Martin-Senour Co., Ltd.

"Pioneers of Pure Paints"

142-144 Inspector St., - Montreal

2514-2520 Quarry Street and Archer Avenue
Chicago

(Established 1878)

The Winnipeg Paint & Glass Co.
Winnipeg

The Maritime Paint & Varnish Co.
Halifax, N.S.

WHITE ZINC (ground in oil)—In 25-lb. irons, 8c.; in 12 1-2 lbs., 8 1-2c.

SHINGLE STAIN—In 5-gallon lots, 75c. to 80c. per gallon.

PARIS WHITE—90c. in barrels to $1.25 per 100 lbs.

WHITING—60c. per 100 lbs. in barrels; Gilders' bolted whiting, 90c. in barrels, $1.15 in smaller quantities.

SHELLAC VARNISH—Pure orange in barrels, $2.70; white, $2.82 1-2 per barrel; No. 1 (orange) $2.50; gum shellac, bone dry, 63c., Toronto. T. N. (orange) 51c. net Toronto.

LINSEED OIL—We still quote: Raw, 1 to 4 barrels, 57c.; 5 to 9 barrels, 57c.; boiled, 1 to 4 barrels, 60c.; 5 to 9 barrels, 59c. Toronto, Hamilton, London and Guelph; net 30 days. Advance of 2c. for delivery to outside points.

TURPENTINE—Single barrels, 97c.; 2 to 4 barrels, 96c.; f.o.b. point of shipment, net 30 days. Less than barrels, $1.02 per gallon.

GLUES—French Medal, 12 1-2c. per pound; domestic shee, 10 1-2c. per lb.

PUTTY—Ordinary, 800 casks, $1.50; 100 drums, $1.75, barrels or bladders, $1.75; 100-lb. cases, $1.90;; 25-lb. irons, $1.85; 25-lb. tins (4 to case) $1.90; 12 ½ lb. tins (8 to case) $2.10.

LIQUID PAINTS—Pure, $1.20 to $1.35 per gallon; No. 1, $1.10 per gallon.

BARN PAINTS—Gals., 70c. to 80c.

BRIDGE PAINTS—Gals., 75c. to $1.

CASTOR OIL—English, in cases, 9 1-2c. to 10c. per lb., and 11c. for single tins.

PARIS GREEN—Canadian manufacturers are quoting their base price at 25c and on the English, 25¾ is quoted f.o.b., Toronto.

REFINED PETROLEUM — Dealers are stocking up heavily for the Winter. We still quote: Canadian prime white, 14c.; water white, 16c.; American water white, 16c. to 18c. ex warehouse.

CRUDE PETROLEUM — We quote: Canadian, $1.32; Pennsylvania, $1.58; Ohio, 96c.

SAVE OIL FIELDS FROM TRUST.

If the finds of oil on the Manitoulin Islands are as great as reported, the discovery is of immense importance, or, rather, can be made so if the product can be protected by Government from falling into the maw of the Standard combine, says the Montreal Witness.

Contemporaneous with this discovery is the action of the United States Government, which is taking measures in the South-West for the purpose of preventing discrimination and making the oil trade open to all producers. For this purpose a concession has been granted to the Mellon Company of Pittsburg to erect over reservation lands a pipe from Bartlesville, Indian Territory to points in Texas, with a view to affording an outlet through Gulf ports to the product of the Indian Territory and Oklahoma field.

The officials of the Department of the Interior believe that the independence of operators in the South-Western field is thus secured. Under the government regulations, it will be impossible for the proposed pipe line, which is estimated to cost eight million dollars, ever to fall into the hands of the Standard or any other oil monopoly. The regulations prepared by the Department are hard and fast, and give the Secretary of the Interior authority to cancel on ten days' notice the contract of any individual or company in the event that the terms of the contract are violated. Here is precedent and practice for our Government to follow in the case of the Manitoulin oil wells.

A barrel of crude petroleum containing about forty-three gallons, can be delivered on board ship for the trifle of ten cents, but when it has passed through the hands of the Standard Oil combine, and has gone through a comparatively inexpensive refining process, twenty-five cents a gallon is charged for it. That would be profit with a vengeance. The ten cents would be changed into nearly eleven dollars net. As the oil on the Manitoulin Islands is said to be of such good quality that it would yield a splendid profit even if sold at ten cents a gallon, the Government should do its utmost to secure to the public such a boon.

PLEASE THE WOMEN.

Please a woman with a household novelty and ten to one she will succeed in using it for a variety of things the manufacturer never dreamed of, excepting when he had the nightmare, says Hardware. Disappoint her and the chances are she never afterwards can find a hammer in the whole concern good enough to drive a carpet tack. That isn't the worst of it. She will tell her family so convincingly how she has been swindled that not a member of it will ever trade at such a place again; they would not dare to, lest she find it out.

A man will get outrageously cheated and then will say nothing, but grin sardonically while all of his fellows go in and do likewise. A woman will drag them back by force and call the police if they insist on entering. Perhaps the man may even be cajolled and made to believe that he was not hurt very much; at least that he has been a victim of circumstance rather than of fraud; anyway he wants to go back just to see if it can be done to him again. Swindle a woman and all the apologies in the world won't quiet her. She is after the goods and not excuses. Either she is suited, and all sweet words, or else she isn't and is all words that are not so sweet, and there it ends—only, something else begins.

Certainly it pays to please the women. No other class of customers are more liberal or more loyal; none others half so influential as they in favor of their commercial friends. But an offended woman will get even seven times over if she has to wait ten years to do it, and in the meantime is compelled to organize a special social club or sewing society to accomplish her revenge.

OUR NAME

on a can of Paint or Varnish means that we guarantee its good quality. The name "Sterling" corresponds with the Sterling stamp on silverware—GENUINE—THE BEST.

A storekeeper is to be judged by the quality of the goods he handles.

The best storekeepers sell "Sterling" goods.

Do YOU sell "Sterling" Paint?

There is no better or more satisfactory paint made, and one dealer in your town is going to have the reputation of selling The Best Paint. Our representative will call upon you in a few days to offer a good proposition, which means so much to you that you cannot afford to turn it down. Kindly defer placing your Spring order until he has seen you.

Manufactured by

CANADIAN OIL COMPANY

TORONTO LIMITED

MONTREAL ST. JOHN OTTAWA WINNIPEG VANCOUVER

THE HOUSEKEEPER'S DREAM.

I dreamt that I dwelt in Alabastined halls,
Alabastined ceilings, Alabastined walls,
Where the fairies love to congregate and meet
The air is so etherial, pure and sweet.

Cupids and fairies will visit you at night
In rooms, Alabastined beautiful and bright,
From kalsomines or wall papers tainted with decay
High on upward wing they rise and flee away..

ALABASTINE is the only Wall Coating, which is extensively and persistently advertised to the general public, and is what the people call for when they want anything in the line.

THIS IS NO DREAM

Our publicity campaign for 1907, as planned, is sure to more than double the already large trade for **ALABASTINE.**

Our new illustrated catalogue, "HOMES, HEATHFUL AND BEAUTIFUL" is the sensation of the season in the advertising line.

Get ready to supply the large and growing demand for **ALABASTINE** by ordering a full assortment, either direct or through your nearest jobber, and DO IT NOW.

The ALABASTINE CO., Limited, PARIS, Ont.

Plumbing and Steamfitting

CLOSET CONNECTIONS TO SOIL-PIPE STACK

[Two more letters sent in replying to the question, the prize-winning answer to which was published in HARDWARE AND METAL of December 29.]

We publish herewith two letters received in connection with the last contest, they speak for themselves:

In connection with the sketch by J. R. Wainwright, of Sudbury, Ont., we beg to express the opinion that, as in the case of last week's illustration no vent would be required in the case of any of the fixture connections shewn, for the reasons mentioned in our last issue, referring to the letter by "X" of Ottawa, we think most of the arguments are good, but do not think that ventilation of short fixture connection is necessary, because even if there was no current of air, the natural diffusion of the gases would change the air with sufficient frequency.

Sudbury Plumber's Reply.

In your judgment is it essential that evey closet connection to stack should be provided with a break syphon? Yes. If the closet waste can be called a stack, when it does not continue through the roof. But if the soil pipe has to continue through the roof to be called a stack, then a special break syphon is not essential in the following circumstances:

1. Where only one closet is connected to stack. The stack acting directly as a break syphon. As in sketch No. 1.

2. Where two or more closets are connected to stack, whether it be on the same floor or not. The highest connection does not require a special break syphon. As the condition is the same as in No. 1, sketches 2 and 3.

3. Where two closets are connected one on each side of the stack, neither need a special break syphon, as the condition is the same as in No. 1, sketch No. 4, both connections being the same level.

J. R. Wainwright, Sudbury.

An Ottawa Reply.

With wash down closets adjacent to soil pipes, break syphons are, in my opinion, unwarranted.

To-day, these fixtures have a diameter of about three inches, and the lead bend, to which attachment is made, has a diameter of four inches. Here, then, is a difference of 7 inches in the square of each circle. While the closet is operating, a column of water solid in form, equal to the circle of the closet is in action. Directly it leaves the bowl its syphonic force is broken, by reason of the enlarged channel, the moving column with extended play, is transformed into a splash, in its descent, drawing on the air reserve, that the four inch bend allows, thereby saving the seal of the trap. This argument, would suit a closet any distance from the stack.

This, I think, fully answers your question, but your readers, having in mind the position of the closet, will, perhaps, follow a further reference to the utility of the break syphon. Owing to the two-fold purpose of this tube, it is rather difficult to be clear, and "stick to the point."

To impress my argument forcibly, suppose, the closet was attached to a bend with a diameter no greater than its own, the syphonic force would not be broken until it left the bend. A break syphon then would be of primary importance, but under the conditions which seem general, the break syphon is of only secondary importance.

As a means of thoroughly ventilating a system, its importance is recognized. This, alone, is why it should be connected to, or below every closet tray.

X., Ottawa.

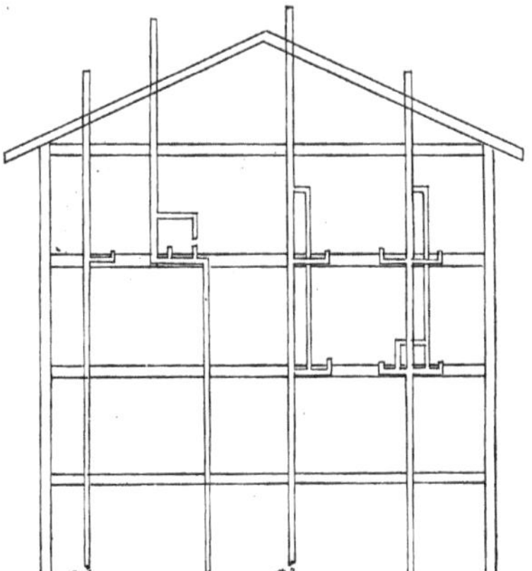

#1. #2 #3 #4

J.R.Wainwright Sudbury, Ont

PLUMBERS IN THE GAS TRADE.

The gas field, as far as the sale of gas fixtures, stoves, piping of houses and other appliances is concerned, was formerly almost entirely controlled by the plumbing trade, says the Metal worker, but since the gas companies have through the introduction of approved appliances and the reduction of their manufacturing expenses found it profitable to sell gas at prevailing rates they have largely monopolized the business through a desire to increase the consumption of gas. And in order that the gas consuming appliances should be brought before the people and be readily purchased they have introduced them at phenomenally low cost. For this reason many plumbers have neglected the trade or else abandoned it altogether.

The gas companies continue to market appliances of this kind at low figures, but their entire aim is to sell material which will consume gas, apparently no attention being paid to the economic combustion of the fuel. This is to their interest and it is to be expected that they would have no special desire to sell gas consuming goods at a low price where the consumption of gas would be low, so long, of course, as it is not particularly wasteful. During the past few years a number of gas heaters, gas burners and appliannes for heating water have been placed on the market which were specially designed for economical combustion of fuel and it would undoubtedly be greatly to the advantage of plumbers to make some study of the subject with a view to establishing a trade in these lines. Many plumbers in cities, particularly those whose places of business are located on the main streets, keep their shops open for an hour or so after the workmen have quit, simply to be in their places of business to receive orders that may come in or to make collections, and some of these have found it profitable to sell gas mantles, shades and the like simply to bring people into the store and the showroom, and thus make them better acquainted with their plumbing wares. It is a form of advertising which pays for itself and is doubtless very advantageous.

DISPOSAL OF HOUSEHOLD WASTES.

In a recent book entitled "Outlines of Practical Sanitation," by Dr. Harvey B. Bashore, inspector of the Department of Health for Pennsylvania, reference is made to the disposal of wastes from the country house. If the dry method of disposing of the human wastes is used there will be certain waste waters from the bath and kitchen sink to be taken care of, the author explains, and this, he says, is best done by some form of surface drain suspended over the garden bed. One such drain shown in the book is made of a 6-inch galvanized roof gutter, pierced every 12 inches by 1-4-inch holes. His discussion of the question is as follows: The gutter allows the filthy water to be distributed evenly over the ground without forming puddles and mud holes.

The solid refuse about country and village houses generally adorns the ash pile or alley. The disposal of these products becomes easy, if the various kinds are collected and kept separate. A good way is to have a series of receptacles for the materials and a certain place for each one, or perhaps have all these receptcles arranged together in a large box near the kitchen door. In one receptacle, which might be a flour sack supported by an iron rack, we would collect the rags, paper, etc. In another tin cans, bottles and such rubbish; then in a suitable can the ashes and in another the garbage—that is, the solid waste from the kitchen.

Now as to the ultimate disposal of this solid waste: The garbage is best disposed of by earth burial—simply put into a shallow furrow in a field and covered with a little earth. If the garden bed is near the kitchen a good way is to have a hole in the bed and practice daily disposal of the garbage. Every evening it should be covered with earth, and in addition a tight board lid should cover the hole during the summer months, else the place may become a breeding place for flies and degenerate into a nuisance. The noncombustible part of the rubbish, such as bottles, tin cans, scraps of metal, etc., can usually be sold to the junk dealer, and the combustible part—rags and paper—if not salable, should be destroyed by fire. Ashes can be used in almost any place for filling, making paths and for foundations under pavements.

WARNER'S GAS CANNON.

The Warner Motor Co., Flatiron Building, New York, are introducing a novelty in the form of a gas cannon, which operates by the explosion of a tiny mixture of carbide gas and air. This gas is made from a small lump of calcium carbide, dropped into water, (carbide can be procured everywhere), and flows into the cannon through a small rubber tube, controlled by a stop cock. This charge is exploded by pulling a lanyard which flashes an electric spark inside the breach, making the operation absolutely like the firing of the big 12 inch guns on a battle ship. Carbide gas when it explodes makes a big noise, but its actual mechanical effect (which is power for doing harm) is extremely small as compared with gun powder. A cork or tight fitting cylinder of wood makes the best projectiles for the cannon. One of these comes with the outfit.

Warner's Gas Cannon.

The cannon is made of cast iron 6½ inches long, neatly mounted on a pressed steel frame which fits over the top of the containing box. The gas generator requires only water and one small lump of calcium carbide to provide ammunition for 20 shots. This carbide can be procured very cheaply anywhere. There is a small pinch-cock on the frame of the cannon by pressing which the gas is admitted to the cannon.

The cannon can be used in many ways and one in operation in a window in holiday seasons should attract all the boys in the neighborhood as well as encouraging many sales.

ABUSES IN THE INSTALLATIONS.

Carelessness in bolting up flanged joints frequently leads to disastrous results, says the Valve World. It often happens that when joints do not come squarely together, the workmen, instead of seeing that the line is made square so that the joints will come together properly, will undertake to spring a very rigid piece of work in the bolting, the result being that if something does not give way under this strain on the bolts, it very likely will do so when the pipe is subjected to expansion and contraction.

Another blunder frequently made by workmen is in not taking pains to tighten the bolts evenly so that they will all be brought up together to the same tension. They will start in and tighten up one side, hard, and then when they come to tighten the opposite side, it throws a very severe strain on the side first tightened. If the flange is not broken by this strain, there is the same danger we have mentioned above, that is, that it will break under the strain of expansion and contraction.

It also frequently happens that when leaks occur after steam is turned on they are not promptly and properly attended to. This, of course, is very bad practice, for every engineer knows that if a leak is allowed to run for any length of time it will injure the material, either the metal or the packing of both. Furthermore, it will then be impossible to tighten it by screwing up the bolts.

Rather than go to the trouble of taking the joint apart, putting in new packing and correcting any injury there may be in the face of the flange, some engineers will keep on in a bullheaded way straining the bolts until the flange either breaks under this strain or later under the expansion and contraction, and then they will seek to place the responsibility on the manufacturer by raising a question as to the quality of the goods or packing.

CLOSET TANK LININGS.

Plumbers quite generally throughout the country are now called on to reline water closet tanks, says the Metal Worker. When the tank is taken out and the lining put in the cost is not endured without complaisance by the customer. Plumbers are willing to confess that they desire to procure goods as cheaply as possible, but are equally desirous that they shall be of such a quality as will not bring complaint from customers. In many of the tanks sent out by the supply houses the lining is of hard rolled copper and very light. The tank lining should have a greater thickness and probably would be better if it were of soft copper. The hard copper draws away from the sides of the tank when its contents are emptied, making a movement at the sharp bends in the corners which not infrequently leads to buckles and breaks. The hard copper is particularly weak, at the seams and at the point of the connection of the outlet piping. Here the natural movement of copper, when the tank is filling and emptying, results in a break around the outlet which experienced plumbers think could be avoided if soft copper was used. If these same plumbers were called upon to reline tanks they would use heavier soft material, and when they do they have little complaint in respect to the service rendered. Some plumbers go so far as to take the thin lining out of a new tank before it goes into the house and reline it with heavier soft material, throwing the old lining into the old metals for sale. This custom is confined to no one section, and the plumbers feel that it is time for the manufacturers to pay less attention to the demand for lower cost and put a lining in their tanks which will insure more satisfactory service.

PLUMBING MARKETS

MONTREAL.

Office of HARDWARE AND METAL,
232 McGill Street,
Montreal, January 4, 1907

"Never busier," is the cry everywhere. This condition of affairs has not been equalled in this city for years. Stock taking has not marred in any way the rush of orders. The old year died with a smiling and prosperous face.

The advance in iron piping is still in the future, but it is believed to be near.

RANGE BOILERS—There is a good demand and the following prices are, well maintained: Iron clad, 30 gallon, $5; 40 gallon, $6.50 net list. Copper, 30 gallon, $25; 35 gallon, $29.50; 40 gallon, $32 net.

LEAD PIPE—Primary markets are still rising, and pipe is firmer. We still quote 5 per cent. discount f.o.b. Montreal, Toronto, St. John, N.B., Halifax: f.o.b., London, 15c per hundred lbs extra; f.o.b. Hamilton, 10c per hundred lbs. extra.

IRON PIPE FITTINGS—The usual trade is being done at unchanged prices. we quote: Discounts on nipples, 1-4 inch to 3 inch, 75 per cent., 3½ inch to 2 inch, 55 per cent.

IRON PIPE. — Supplies continue short and an advance is expected. We still continue to quote as follows: Standard pipe in lots of 100 feet, regular lengths, 1-4 inch, $5.50; 3-8 inch, $5.50; 1-2 inch, $8.50; 3-4 inch, 11.50; 1 inch, $16.50; 1 1-4 inches, $22.50; 1 1-2 inches, $27.00; 2 inches, $36.50, discounts on black pipe, 1-4 inch, 61 per cent.; 3-8 inch, 61 per cent.; 1-2 inch, 70 per cent.; 3-4 inch and upwards, 72 per cent.; Discounts on galvanized pipe: 1-4 inch, 46 per cent.; 3-8 inch, 46 per cent.; 1-2 inch, 60 per cent. Extra heavy pipe of 100 feet lots are quoted as follows: 1-2 inch, $12; 3-4 inch, $15; 1 inch, $22; 1 1-4 inches, $30; 1 1-2 inches, $36; 2 inches, $50. The discount for black pipe is 70 per cent., and for galvanized 60 per cent.

SOIL PIPE AND FITTINGS—Our prices remain as follows: Standard soil pipe, 50 per cent. off list. Standard fittings, 50 and 10 per cent. off list; medium and extra heavy soil pipe. 60 per cent. off. Fittings, 60 per cent. off.

SOLDER—Quotations are unchanged. Our prices are: Bar solder, half-and-half, guaranteed, 25c; No. 2 wiping solder, 22c.

ENAMELWARE.—We still quote: Canadian bath tubs, plate E1, 5 ft., 1st quality, $20.65 special, $18.65; plate E11 and E21 5 ft., 1st quality, $19.50, special $17.15; plate E35, 5 ft., 1st quality, $24.65, special $22.40. American baths, rolled rim, 5 feet 2 1-2 inch rim, $22.25; 3 inch rim, $29.25. Lavatories, discounts, 1st quality, 30 to 30 and 5 per cent.; special, 30 and 10 to 40 per cent. Sinks, 18 x 30 inch, flat rim, 1st quality, $2.60, special, $2.45.

TORONTO.

Office of HARDWARE AND METAL,
10 Front Street East,
Toronto, Jan 4, 18.7

The advance on hot water boilers, spoken of in our report a fortnight ago, has not yet been announced, but the new figures are expected to be made known next week. With prices of iron at their present high pitch, advanced prices on boilers were looked for and would have been made some time ago, except for the keen competition existing between the various producers.

Soil pipe is holding firm at the advanced prices, and there is talk of still higher quotations being made this month.

We report this week additional advances on compression and Fuller work, the advance averaging about 10 per cent.

Iron pipe is still short and there does not seem to be any prospect of stocks being gathered by jobbers during the off season. On the other hand, there is much reason to expect that a shortage will continue throughout the coming year. One Toronto house has secured a large supply of one and one and one-quarter black pipe from United States producers, and are selling it at bare cost to oblige customers who have not been able to secure material for jobs under way.

Trade in both plumbing and heating supplies has been slow, the holiday season and the natural falling off of trade at the beginning of the year, making this condition an expected one.

LEAD PIPE — The discount on lead pipe continues at 5 per cent. off the list price of 7c. per pound. Lead waste, 8c. per pound with 5 off. Caulking lead 5 3-4c. to 6c. per pound. Traps and bends, 50 per cent., discount.

SOIL PIPE AND FITTINGS.—New lists will show advances as follows: Medium and extra heavy pipe and fittings, 60 per cent.; light pipe 50 per cent.; light fittings, 50 and 10 per cent.; 7 and 8 inch pipe, 40 and 5 per cent.

IRON PIPE.—American black pipe has been received in Toronto to relieve the shortage, but it commands higher prices than we are quoting on Canadian goods. 1-inch black pipe is quoted at $4.80 and one-inch galvanized at $6.43. Full list in current market quotations.

IRON PIPE FITTINGS — New lists are as follows: Cast iron elbows, tees, crosses, etc., 62½ per cent.; cast iron plugs and bushings, 62½ per cent.; flange unions, 62½ per cent.; nipples, 70 and 10 per cent.; iron cocks, 55 and 5 per cent.; Canadian malleable, 30 per cent.; malleable unions, 55 and 5 per cent.; malleable bushings, 55 per cent.; cast iron ceiling plates, plain 65 per cent.; cast iron floor, 70 per cent.; hook plates, 60 per cent.; expansion plates, 65 per cent.; headers, 60 per cent, hangers, 65 per cent.; standard list.

GALVANIZED IRON RANGE BOILERS—We quote: 12 gallon capacity, standard, $4.50; extra heavy, $6.50; 18 gallon standard, $4.75; extra heavy, $6.75; 24 gallon, standard, $4.75; extra heavy, $6.75; 30 gallon, standard, $4.75; extra heavy, $7.50; 35 gallon, standard, $5.75; extra heavy, $8.50; 40 gallon, standard, $6.75; 40 gallon, extra heavy, $9.50; 52 gallon, $11; extra heavy, $14; 66 gallon, standard, $18; extra heavy, $20; 82 gallon, standard, $21; extra heavy, $24; 100 gallon, standard, $29; extra heavy, $34; 120 gallon, standard, $34; extra heavy, $40; 144 gallon, standard, $47; extra heavy, $55. Copper range boilers are now net list.

RADIATORS—Prices are very stiff at: Hot water, 47½ per cent.; steam, 50 per cent.; wall radiators, 45 per cent.; specials, 45 per cent. Hot water boilers have not yet advanced.

SOLDER — Quotations remain firm as follows: Bar solder, half-and-half, guaranteed, 27c.; wiping, 23c.

ENAMELWARE—We quote as follows: Standard Ideal, Plate E1, 5 ft., first quality, $21.65; special $20.25; plate E11, 5 ft., first quality, $20.15, special, $19.75. Lavatories, first quality, 20 and 10 to 30 and 10 off; special, 30 and 5 to 30 and 10 per cent. discount Kitchen sinks, plate 300, firsts, 65 and 5 off, specials, 70 per cent. Urinals and range closets, 20 off. Fittings extra. Lower prices in quantities.

HOT-WATER HEATING DEVICE.

Two Danes have taken out American patents for a circulation device for hot-water heating plants, which they describe as follows: Hot-water heating plants in which air is blown into a main rising-pipe in order to increase the circulation of water in the pipe system are well known. The object of the existing patents in this line is to produce a circulation as powerful as possible in proportion to energy expended and in such manner that the sys. does not lose heat and so that the use of air does not cause special difficulty.

WALL RADIATORS POPULAR.

Wall radiation for semi-public places is finding favor with heating contractors, not only on account of the economy of installation, but because of its thermal efficiency as well. It possesses many advantages over the older forms for use in stores, railway stations, and bowling alleys.

THE NEW
"SOLID-ADJUSTABLE"
HAND STOCK AND DIES

At prices that will warrant your inquiries.

I. E. SHANTZ & CO.
BERLIN, - ONTARIO
Selling Agents for Canada

Time and Trouble Saved

by using Armstrong's Ratchet Attachment for threading pipes in cramped positions. Fits all Genuine Armstrong Stocks—is moderate in cost, compact, strong and thoroughly reliable—an invaluable tool for pipe-fitters. Write for catalog and prices.

THE ARMSTRONG MFG. CO.
292 Knowlton St., Bridgeport, Conn.

DISAPPOINTMENTS

Are never attached to orders entrusted to me. I make it my business to be prompt in filling customers wants.

I can be prompt in filling your orders for Enamelware, Heating Goods, etc.

MAKE IT A POINT TO WRITE ME

ORLANDO VICKERY
178-180 Victoria Street TORONTO, CANADA

"Diamond Brand" Fittings

are pronounced the best for many reasons. Here are three:

The Finish is Smooth—The Threads are Perfect—The Material is the Best Obtainable

OSHAWA STEAM & GAS FITTINGS CO., LIMITED
OSHAWA, CANADA

Sharratt & Newth's Glaziers' Diamonds

are unequalled for cutting and wearing qualities.

To be obtained from the principal Dealers in Glass, Hardware, and Painters' Supplies.
Contractors to H. M. Government and the principal English Sheet and Plate Glass Works

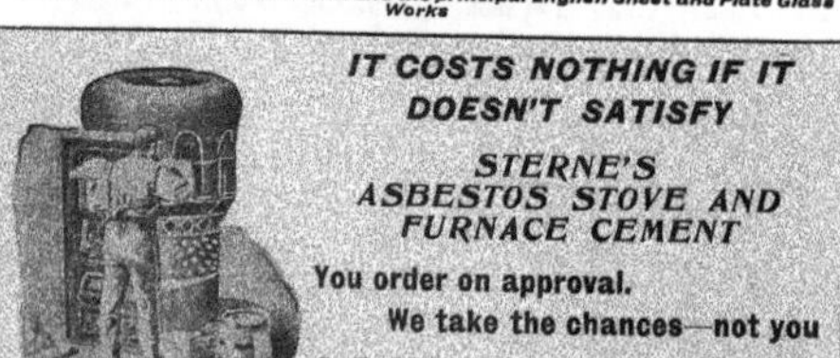

IT COSTS NOTHING IF IT DOESN'T SATISFY

STERNE'S ASBESTOS STOVE AND FURNACE CEMENT

You order on approval.
We take the chances—not you

G. F. STERNE & SON, Manufacturers
BRANTFORD, ONT.

A Certain Sense of Satisfaction

goes with every MERRELL PORTABLE HAND MACHINE, PIPE THREADING and CUTTING MACHINE. We build them that way.

For doing ALL kinds of threading—under ALL conditions—labor trouble and faulty material included—where speed, durability and ease of control count for something, our word to you is to investigate the MERRELL.

Whether it be stationary or portable—hand or power driven—30 DAYS FREE TRIAL must prove conclusively that your choice should be the MERRELL.

This Portable Hand Machine has encased gears, MERRELL standard quick opening and closing die-head, and the latest improved Cutting-off Knife.

Let us tell you more about this machine—The Chasers, Vise and the large range of work covered.

Catalogue for the asking.
Quick shipments.

THE CANADIAN FAIRBANKS CO. Limited
Sole Agents for Canada
MONTREAL, TORONTO, WINNIPEG, VANCOUVER

When You Buy Demand Proof

Don't take any substitute for Glauber Brass goods. Make the salesman show you. Brass has gone up no less than fifteen per cent., and if you don't get proofs before you buy you take chances of getting something less than you pay for.

WE GIVE YOU THE PROOF and ***2 Cocks Free*** for any one sent you defective.

Glauber Brass Mfg. Co.,
Expert Makers of Plumbers' Fine Brass Goods
CLEVELAND, O.

Forwell Foundry Co.

BERLIN, ONT.

Manufacturers of

SOIL PIPE, FITTINGS, and CAST IRON SINKS

Ask Jobbers for "F. F. CO." Brand

An Invitation

is extended to the Trade to inspect our comprehensive stock of Daisy Water Heaters. When your customer experiences the need of a greater Hot Water supply remember that we carry every size and style of heater in stock for immediate shipment.

CLUFF BROTHERS

Selling Agents for WARDEN KING & SON, Limited

Lombard Street, TORONTO

1906 Improved "INTERNATIONAL" BURNER 1907

Complete Exterior View

Complete Sectional View of Burner

Bunsen

TRY A SAMPLE DOZEN

Satisfaction Guaranteed

THREE FINISHES

CLOSED OR OPEN BASKET

	PER DOZ.
Polished Brass	$4 25
Brushed Brass	4 90
Oxidized Copper	4 50
Deck Plates	0 15

The most Scientific Burner on the market to-day. Nothing to equal it in efficiency or appearance.

AN ORNAMENT IN THE MOST ARTISTIC HOME

Leather Packing or Gasket makes it absolutely impossible for any leakage of gas from the needle valve adjustment. Cannot light back into the bunsen or puff out when lit. The accompanying detailed sketch cannot fail to convince the most skeptical of its many advantages over every other burner on the market.

Manufactured expressly for and controlled in Canada, Great Britain and United States by

The International Gas Appliance Co., Limited

535, 537 and 539 Queen St. Esat, TORONTO, CAN.

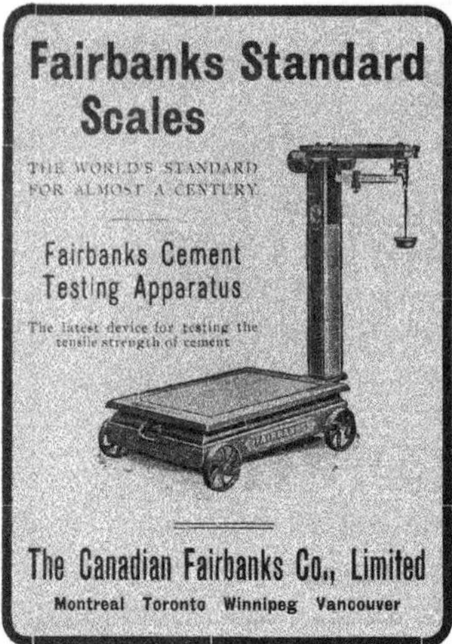

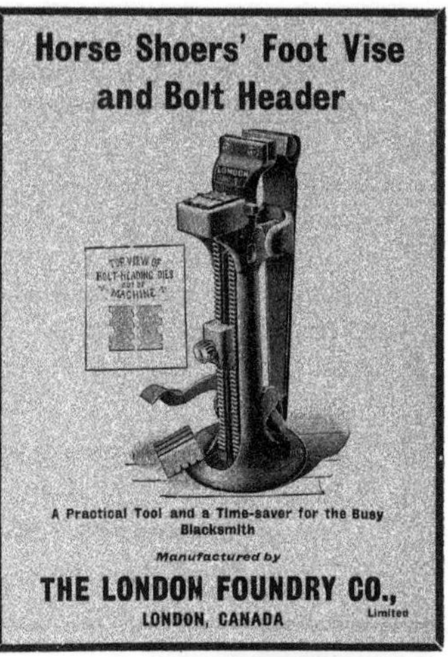

The Ever Ready Dry Battery

FOR AUTOMOBILE and

GASOLINE ENGINE USE

Write for Prices

JOHN FORMAN

248 and 250 Craig St. W., MONTREAL, - Que.

HOTTEST ON EARTH

Our Bonanza Furnaces are known to produce more heat with less fuel than any other make. Jobbers sell at factory prices. Satisfaction guaranteed or money refunded.

No. 38—Price, each, ***$7.80*** net.
No. 39—One size smaller, ***$5.85*** net.

THE TURNER BRASS WORKS

53 MICHIGAN STREET, CHICAGO, U.S.A.

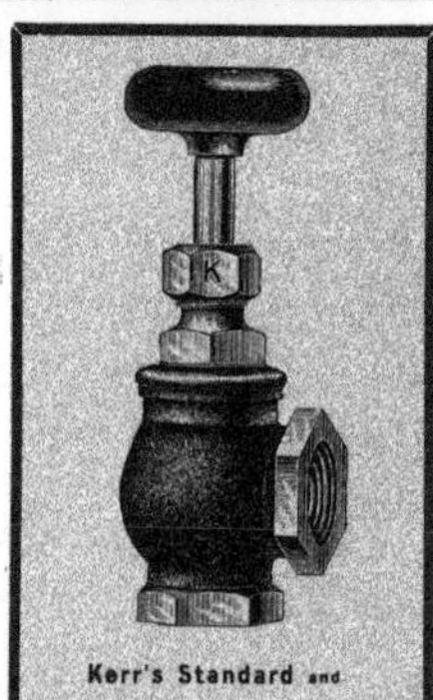

Kerr's Standard and Jenkin Disc Radiator Valves

are perfectly constructed, and of beautiful design. Like all "Kerr" specialties, strictly high-grade

The KERR ENGINE CO.,
Manufacturers Limited

WALKERVILLE - ONTARIO

CURRENT MARKET QUOTATIONS.

Jan. 4, 1907.

These prices are for such qualities and quantities as are usually ordered by retail dealers on the usual terms of credit, the lowest figures being for larger quantities and prompt pay. Large cash buyers can frequently make purchases at better prices. The Editor is anxious to be informed at once of any apparent errors in this list, as the desire is to make it perfectly accurate.

METALS.

ANTIMONY.

Hallett's .... per lb... 0 27½ 0 28

BOILER AND T.K. PITTS.

Plain tinned .... } 35 per cent. off list.
Spun .... }

BABBIT METAL.

Canada Metal Company—Imperial genuine, 60c; Imperial Tough, 56c; White Brass 50c.; Metallic, 35c.; Harris Heavy Pressure, 25c.; Hercules, 25c; White Bronze, 15c.; Star Frictionless, 14c; Alluminoid, 10c.; No. 4, 9c. per lb.

James Robertson Company — Extra Monarch, 45c.; No 1 Monarch 35c; "King" Anti-friction, 20c; Fleur-de-lis Anti-friction 15c; No. 1 Thurber, 12c; Philadelphia 10c; Canadian, 7c; Hardware babbit No. 1, 10c; Hardware babbit No. 2, 8c.; Hardware babbit No. 3, 6½c. Discount on Hardware No. 1, 2, 3, 15 per cent. All others net list.

BRASS.

Rod and Sheet, 14 to 30 gauge, net list.

Sheets, 12 to 14 in.	0 27
Tubing, base, per lb ¼ to 2 in.	0 33

COPPER.

Ingot.		Per 100 lb.
Casting, car lots	25 50	26 00
Bars.		
Cut lengths, round, ¼ to 2 in.	....	32 00
Sheet.		
Plain, 16 oz., 14x48 and 14x60	....	30 00
Plain, 14 oz.	....	31 00
Tinned copper sheet, base	....	33 00
Planished base	....	37 00
Braziers' (in sheets), 4x6 ft., 25 to 30 lb. each, per lb., base	....	0 30

BLACK SHEETS.

	Montreal.	Toronto
8 to 10 gauge	2 60	2 60
12 gauge	2 60	2 65
14 "	2 60	2 60
17 "	2 40	2 50
18 "	2 40	2 55
20 "	2 40	2 55
22 "	2 40	2 70
24 "	2 45	2 75
26 "	2 55	2 85
28 "	2 60	3 00

CANADA PLATES.

Ordinary, 52 sheets	....	3 00
All bright "	....	4 00
Galvanized Canada Plates, 52 sheets		4 35
" " " 60 "		4 60

	Ordinary.	Dom. Crown.
18x24x52	4 25	4
" 60	4 50	4
20x28x80	8 50	8
" 94	9 00	9 85

GALVANIZED SHEETS.

	Fleur-de-Lis.	Gordon Crown.
16 to 20 gauge	3 60	3 85
22 to 24 gauge	3 85	4 00
26 "	4 30	4 20
28 "	4 45	4 50

Apollo.

10¾ oz. (American gauge)	....	4 60
28 gauge "	....	4 40
26 " "	....	4 00
24 " "	....	3 80

	Comet.	Queen's Head.	Bell.
16 to 20 gauge	3 60	3 75	....
22 to 24 gauge	3 75	3 90	....
26 "	4 30	4 45	....
28 "	4 45	4 60	4 45

Less than case lots 10 to 25c. extra.

IRON AND STEEL.

	Montreal.	Toronto.
Common bar, per 100 lb.	2 15	2 25
Forged iron "	2 40	
Refined " "	2 55	70
Horseshoe iron "	2 55	2 70
Hoop steel, 1½ to 3 in. base	....	2 80
Sleigh shoe steel "	2 25	2 30
Tire steel	2 40	2 50
Best sheet cast steel	....	0 12
B. K. Morton "Alpha" high speed		0 65
" annealed	....	0 70
"M" Self-hardening	....	0 50
"J" quality, best warranted	....	0 18
"I" " warranted	....	0 14
"B.C" quality	....	0 09
Jonas & Colver's tool steel	0 10	0 20
" "Novo"		0 65
" " annealed		0 65
Jowett & Sons B P.L. tool steel	....	0 10½

INGOT TIN.

Lamb and Flag and Straits—56 and 28-lb. ingots, 100 lb. $46 50 $47 00

TINPLATES.

Charcoal Plates—Bright

M.L.S., equal to Bradley—		Per box.
I C, 14 x 20 base	....	$6 50
I X, 14 x 20 "	....	8 00
I XX, 14 x 20 base	....	9 50
Famous, equal to Bradley—		
I C, 14 x 20 base	....	6 50
I X, 14 x 20 "	....	8 00
I XX, 14 x 20 base	....	9 50
Raven and Vulture Grades—		
I C, 14 x 20 base	....	5 00
I X "	....	6 00
I XX "	....	7 00
I XXX "	....	8 00
"Dominion Crown Best"—Double Coated, Tissued.		Per box.
I C, 14 x 20 base	....	5 75
I X, 14 x 20 "	....	6 75
I XX, 14 x 20 "	....	7 75
"Allaway's Best"—Standard Quality.		
I C, 14 x 20 base	....	4 50
I X, 14 x 20 "	....	5 25
I XX, 14 x 20 "	....	6 00

Bright Cokes.

Bessemer Steel—		
I.C., 14 x 20 base	...	4 25
20x28, double box	....	8 50

Charcoal Plates—Terne

Dean or J. G. Grade—		
I C., 20x28, 112 sheets	....	8 00
IX., Terne Tin	...	9 50

Charcoal Tin Boiler Plates.

Cookley Grade—
X X, 14x56, 50 sheet bxs. }
" 14x60, " } 7 50
" 14x65, " }

Tinned Sheets.

72x30 up to 24 gauge	....	8 50
" " 26 "	....	9 00

LEAD.

Imported Pig, per 100 lb.	5 40	5 50
Bar, "	5 75	6 00
Sheets, 2½ lb. sq. ft., by roll	....	0 06½
Sheets, 3 to 6 lb. "	....	0 06

NOTE.—Cut sheets ½c per lb., extra. Pipe, by the roll, usual weights per yard, lists at 7c. per lb. and 5 p c. dis. f.o.b. Toronto.

NOTE.—Cut lengths, net price, waste pipe 8-ft lengths, lists at 8c.

SHEET ZINC.

5-cwt. casks	8 00	8 25
Part casks	8 25	8 50

ZINC SPELTER.

Foreign, per 100 lb	7 25	7 50
Domestic	7 00	7 25

PLUMBING AND HEATING

BRASS GOODS, VALVES, ETC.

Standard Compression work, dis. half inch 50 per cent., others 40 per cent.
Cushion work, discount 40 per cent.
Fuller work, ½-inch 60 per cent., others 50 p.c.
Flatway stop and stop and waste cocks, 50 per cent; roundway, 45 per cent.
J.M.T. Globe, Angle and Check Valves, discount 50 per cent.
Standard Globe, Angle and Check Valves discount 52½ per cent.
Kerr standard globes, angles and checks, special, 42½ per cent.; standard, 47½ p.c.
Kerr Jenkins' disc, copper-alloy disc and heavy standard valves, 45 per cent.
Kerr steam radiator valves 60 p.c., and quick-opening hot-water radiator valves, 60 p.c.
Kerr brass, Weber's straightway valves, 42½ per cent.; straightway valves, I.B.B.M., 60 per cent.
J. M. T. Radiator Valves, discount 50 per cent
Standard Radiator Valves, 60 per cent.
Patent Quick-Opening Valves, 65 per cent.
Jenkins' Bros. Globe Angle and Check Valves discount 32½ per cent.

No. 1 compression bath cock	net	2 00
No. 4 " "	"	1 90
No 7 Fuller's	"	2 25
No. 4½ "	"	2 35

Patent Compression Cushion, basin cock, hot and cold, per doz., $16 20
Patent Compression Cushion, bath cock, No. 2208 .... 2 25
Square head brass cocks, discount 50 per cent.
" " iron " " 60 "
Thompson Smoke-test Machine $25.00

BOILERS—COPPER RANGE.

Copper, 30 gallon	25 00
" 35 "	29 00
" 40 "	33 00

Net list.

BOILERS—GALVANIZED IRON RANGE.

Capacity.	Standard.	Extra heavy
30-gallons	4.75	7 50
35 "	5 75	8 50
40 "	6.75	9.50

2 per cent., 30 days

BATH TUBS.

Steel clad copper lined, 15 per cent.

CAST IRON SINKS.

16x24, $1; 18x30, $1; 18x36, $1.30.

ENAMELED CLOSETS AND URINALS

Discount 20 p.c.

ENAMELED BATHS.

Standard Ideal Enameled.

Plate E I, Fittings extra		1st quality	Special
4 and 4½ ft. 3 in. rolled rim		$21 65	20 25
5 feet	"	22 65	21 25
5½ "	"	24 15	22 75
6 "		26 90	25 50
Plate E II			
5 feet	2½ in. "	21 15	19 75
5½ "	2½ " "	22 65	21 25

Closer prices and discounts in quantities.

ENAMELED LAVATORIES.

	1st quality.	Special.
Plate E 100 to E 103	20 & 10 p c.	30 & 5 p c.
" E 104 to E 132	30 & 10 p.c.	40 p.c.

ENAMELED SINKS.

	1st quality.	Special.
Plate E 201, one piece	20 p.c.	20 & 10 p.c.
Plate E, flat iron 300, 65 & 5 p.c.		70 p.c.

LEAD PIPE

Lead Pipe, 7c. per pound, 5 per cent. off.
Lead waste, 8c. per pound, 5 per cent. off.
Caulking lead, 6c. per pound.
Traps and bends, 50 and 10 per cent.

IRON PIPE.

Size (per 100 ft.)	Black.	Size	Galvanized
¼ inch	2 20	¼ inch	3 02
⅜ "	2 20	⅜ "	3 02
½ "	2 63	½ "	3 48
¾ "	3 33	¾ "	4 47
1 "	4 80	1 "	6 43
1¼ "	6 52	1¼ "	8 77
1½ "	7 83	1½ "	10 53
2 "	10 41	2 "	14 04
2½ "	20 59	2½ "	30 36
3 "	27 06	3 "	39 94
3½ "	34 32	3½ "	50 50
4 "	38 63	4 "	57 42

2 per cent. 30 days.

Malleable Fittings—Canadian discount 30 per cent.; American discount 25 per cent.

Cast Iron Fittings 62½; Standard bushings 62½ per cent.; headers, 60; flanged unions 62½, malleable bushings 55; nipples, up to 2 in., 70 and 10 per cent.; up to 6 in., 65 per cent.; malleable lipped unions, 55 and 5 per cent.

SOIL PIPE AND FITTINGS

Medium and Extra heavy pipe and fittings, up to 6 inch, discount 60 per cent.
7 and 8-in. pipe, discount 40 and 5 per cent.
Light pipe, 50 p.c.; fittings, 50 and 10 p.c.

OAKUM.

Plumbers ... per 100 lb. 4 00 4 75

RADIATORS, ETC.

Hot water	47½ p.c.
Steam	50 p.c.
Wall Radiators and Specials	45 p.c.

STOVES, BOILERS, FURNACES, REGISTERS.

Discounts vary from 40 to 70 per cent. according to list.

STOCKS AND DIES.

American discount 25 per cent.

SOLDERING IRONS.

½-lb.	per lb.	0 37
1-lb. or over	"	0 34

SOLDER.

	Montreal	Per lb. Toronto
Bar, half-and-half, guaranteed	0 25	0 27
Wiping	0 22	0 23

PAINTS, OILS AND GLASS.

COLORS IN OIL.

Venetian red, 1-lb. tins, pure.	....	0 08
Chrome yellow "	....	0 15
Golden ochre "	....	0 08
French " "	....	0 08
Marine black "	....	0 04½
Chrome green "	....	0 10
French permanent green "	....	0 13
Signwriters' black "	....	0 15

GLAZIER POINTS.

Discount, 5 per cent.

GLUE.

Domestic sheet	0 10	0 10½
French medal	0 12	0 14

PARIS GREEN.

	Bergers	Canadian
600-lb. casks	0 23½	0 23½
250 lb. drums	0 23½	0 23½
100-lb. "	0 24	0 24
50-lb. "	0 24	0 24
25-lb. "	0 24½	0 24½
1-lb. pkgs, 100 in box	0 25	0 25
1-lb. " 50 "	....	0 25½
1-lb. tins, 100 "	0 27	0 26
½-lb. pkgs	0 27	0 27

2 p. c. 30 days from date of shipment.

PREPARED PAINTS.

Pure, per gallon, in tins......	....	1 20
Second qualities per gallon ..	....	1 00
Barn (in bbls.)............	0 60	0 90
Sherwin-Williams paints, ½ gal	....	1 5?
" " ¼ gal.	....	1 45
" " gal..	....	1 40
Canada Paint Co.'s pure......	....	1 30
Standard Co.'s "New Era,"	....	1 30
Benj. Moore Co.'s "Ark" B'd	....	1 25
" British Navy deck	....	1 50
Brandram-Henderson "Anchor"	....	1 35
Ramsay's paints, Pure, per gal.	....	1 20
" Thistle, "	....	1 00
" Outside, bbls	0 55	0 65
Martin-Senour's 100 p.c. pure, ½ gal.		1 45
" " " " ¼ gal.		1 40
" " " " 1 gal.		1 35
Jamieson's "Crown and Anchor"		1 20
Jamieson's floor enamel	....	1 50
" barn paints, bbls. per gal		0 60
Senour's Floor Paints .....gal.	....	1 20
Sanderson Pearcy's, pure ..	....	1 20
Robertson's pure paints......	....	1 20

PUTTY.

Bulk in bbls..........	1 60
Bulk in less quantity	1 80
Bladders in bbls........	1 80
Bladders in kegs, boxes or loose......	1 95
25-lb. tins........	1 80
12½ lb. tins	2 05
Bladders in bulk or tins less than 100 lb.	1 85
Bulk in 100-lb. irons........	1 80

DRY LEAD.	white	red
Genuine, 560 lb. casks, per cwt	6 75	6 00
Genuine, 100 lb. kegs, "	7 50	6 25
No. 1, 560 lb. casks, per cwt ...	....	5 75
No. 1, 100 lb. kegs, per cwt ..	....	6 00

SHELLAC.

Shellac, bleached, in bars or ground, 46c. per lb., f.o.b. Eastern Canadian Points.
Shellac, bone dry, 57c. per lb., f.o.b., Eastern Canadian Points; 65c., Toronto.
Shellac, T. N. orange, etc., 48c. net, Toronto.

WHITE LEAD GROUND IN OIL.	Per 100 lbs.	
Pure........	7 25	7 50
No. 1	6 90	7 15
No. 2	6 55	6 90
No. 3	6 30	6 65

Above prices are f o b. Montreal.

Munro's Select Flake White..	....	7 50
Elephant and Decorators Pure	....	7 25
Monarch	....	7 40
Standard Decorator's	....	7 15
Essex Genuine	....	6 80
Brandram's B. B. Genuine...	....	8 75
"Anchor," pure	....	6 50
Ramsay's Pure Lead	....	6 40
Ramsay's Exterior	....	6 15
"Crown and Anchor," pure...	....	6 50
Sanderson Pearcy's	....	7 40
Robertson's C.P., lead	....	7 20

WINDOW GLASS.

Size United Inches.	Star	Double Diamond
Under 26	$4 25	$6 25
26 to 40	4 65	6 75
41 to 50	5 10	7 50
51 to 60	5 35	8 50
61 to 70	5 75	9 75
71 to 80	6 25	11 00
81 to 85	7 00	12 50
86 to 90	....	15 00
91 to 95	....	17 50
96 to 100	....	20 50
101 to 105........	....	24 00
100 to 110........	....	27 50

Discount off list, 20 per cent For broken boxes 40 off. All prices per 100 feet.

ZINC—DRY WHITE.

Extra Red Seal	0 07	0 08
French V. M........	0 06	0 25
Lehigh........	0 05	0 06

ZINC—GROUND WHITE.

Pure........	0 08¼	0 09¼
No. 1........	0 07	0 08
No. 2........	0 05¾	0 06¾

VARNISHES.

In 5-gal. lots.	Per gal.	Net.
Carriage, No. 1	1 50	1 60
Pale durable body........	4 10	4 25
" rubbing	2 85	3 20
Gold size, japan........	1 50	1 60
No. 1 brown japan........	0 85	0 90
Elastic oak	....	1 50
Furniture, extra........	1 10	1 25
" No. 1........	0 90	1 00
Hard oil finish........	1 35	1 50
Light oil finish........	1 60	1 70
Damar........	1 75	2 00
Brown japan	1 10	1 20
Black japan	1 10	1 20
" No. 1........	0 85	0 90
Elastilite varnish, 1 gal. can, each..		2 00
Granitine floor varnish, per gal		2 75
Maple Leaf coach enamels; size 1,		1.20
size 2, 70c.; size 3, 40c. each		
Sherwin-Williams' kopal varnish, gal.,		2.50
Canada Paint Co's sun varnish		2 00
"Kyanize" Interior Finish....		2 40
"Flint-Lac," coach.		1 80
"Gold Medal," per gal., in cases..		2 00
Jamieson's Copaline, per gal.		2 00
Sanderson Pearcy's amberine		2 50

BUILDERS' HARDWARE.

BELLS.

Brass hand bells, 60 per cent.
Nickel, 55 per cent.

Gongs, Sargeant's door bells..	5 50	8 00

Peterboro' door bells, discount 37½ and 10 per cent. off new list.

American, house bells, per lb.	0 35	0 40

BUILDING PAPER, ETC.

Tarred Felt, per 100 lb.		2 25
Ready roofing, 2-ply, not under 45 lb. per roll		0 95
Ready roofing, 3-ply, not under 65 lb., per roll		1 20
Carpet Felt ....per ton		55 00
Heavy Straw Sheathing......per ton		35 00
Dry Surprise		0 40
Dry Sheathing.....per roll, 400 sq. ft.		0 50
Tar " ..... " 400 "		0 45
Dry Fibre..... " 400 "		0 55
Tarred Fibre ..... " 400 "		0 65
O. K. & I. X. L. ● " 400 "		0 70
Resin-sized ..... " 400 "		0 45
Oiled Sheathing..... " 600 "		1 00
Oiled " ..... " 400 "		0 70
Root Coating, in barrels ....per gal.		0 17
Roof " small packages		0 25
Refined Tar........per barrel		5 00
Coal Tar		4 00
Coal Tar, less than barrels....per gal.		0 15
Roofing Pitch.......per 100 lb.	0 80	0 90
Slater's felt ....per roll		0 70
Heavy Straw Sheathing f. o. b. St. John and Halifax........		37 50

BUTTS.

Wrought Brass, net revised list.
Wrought Iron, 70 per cent
Cast Iron Loose Pin, discount 60 per cent.
Wrought Steel Fast Joint and Loose Pin, 70 per cent

CEMENT AND FIREBRICK.

"Lafarge" cement in wood		3 40
"Lehigh" cement, in wood		2 54
"Lehigh" cement, cotton sacks.		2 39
"Lehigh" cement, paper sacks		2 31
Fire brick (Scotch) per 1,000....	15 00	24 00
Fire clay (Scotch), net ton		4 95

Paving Blocks per 1,000.

Blue metallic, 9"x4½"x3", ex wharf ..	35 00
Stable pavers, 12"x6"x2", ex wharf ..	50 00
Stable pavers, 9"x4½"x3", ex wharf ..	36 00

DOOR SETS.

Peterboro, 37½ and 10 per cent.

DOOR SPRINGS.

Torrey's Rod........per doz.	....	1 75
Coil, 9 to 11 in........ "	0 95	1 65
English .... "	2 00	4 00

Chicago and Reliance Coil Door Springs, 25 per cent.

STORE DOOR HANDLES.

Per Dozen	1 00	1 50

ESCUTCHEONS.

Discount 50 and 10 per cent., new list
Peterboro, 37½ and 10 per cent.

ESCUTCHEON PINS.

Iron, discount 40 per cent.
Brass, 45 per cent.

HINGES.

Blind, discount 60 per cent.

Heavy T and strap, 4-in., per lb. net .	0 06½
" " 5-in., "	0 06
" " 6-in., "	0 05¾
" " 8-in., "	0 05½
" " 10-in. and larger ..	0 05¼

Light T and strap, discount 65 p.c.
Screw hook and hinge—

under 12 in.... per 100 lb.	....	4 75
over 12 in ...... "	....	3 75
Spring, No. 20, per gro. pairs	....	10 80

Spring, Woodyatt pattern, per gro , No 5, $17.50 No. 10, $18; No. 20, $10 80; No. 12 $20; No. 51, $10; No. 50, $27.50.
Crate hinges and back flaps, 65 and 5 p. c.
Hinge hasps, 65 per cent.

SPRING HINGES.

Chicago Spring Butts and Blanks 12½ per cent.
Triple End Spring Butts, 40 and 5 per cent.
Chicago Floor Hinges, 40 and 5 off
Garden City Fire Engine House Hinges, 12½ per cent.

HOOKS.

Cast Iron.

Bird cage........per doz.	0 50	1 10
Clothes line, No. 61.. "	0 00	0 70
Harness........ "	0 60	12 00
Hat and coat .... per gro.	1 10	10 00
Chandelier ....per doz.	0 50	1 00
Wrought hooks and staples—		
¼ x 5 .... per gross	....	2 65
5-16 x 5 .... "	....	3 30

Bright steel gate hooks and staples, 40 per cent. discount.
Hat and coat wire, discount 62½ per cent.
Screw, bright wire, discount 62½ per cent.

KNOBS.

Door, japanned and N.P., doz	1 50	2 50
Bronze, Berlin....... per doz.	2 75	3 25
Bronze, Genuine .... "	6 00	9 00
Shutter, porcelain, F. & L. screw .... per gross	1 30	2 00
White door knobs ...per doz.	....	2 00

Peterboro knobs, 37½ and 10 per cent.
Porcelain, mineral and jet knobs, net list.

KEYS.

Lock, Canadian dis. 40 to 40 and 10 per cent

Cabinet trunk and padlock American .... per gross	....	0 60

LOCKS.

Peterboro 37½ and 10 per cent.
Russell & Erwin, steel rim $2 50 per doz
Eagle cabinet locks, discount 30 per cent
American padlocks, all steel, 10 to 15 per cent; all brass or bronze, 10 to 25 per cent.

SAND AND EMERY PAPER.

B. & A. sand, discount, 35 per cent
Emery, discount 35 per cent.
Garnet (Rurton's) 5 to 10 per cent. advance on list.

SASH WEIGHTS.

Sectional........per 100 lb.	2 00	2 25
Solid .... "	1 50	1 75

SASH CORD.

Per lb........	....	0 31

BLIND AND BED STAPLES.

All sizes, per lb........	0 07½	0 12

WROUGHT STAPLES.

Galvanized	....	2 75
Plain	....	2 50

Coopers', discount 45 per cent.
Poultry netting staples, discount 40 per cent.
Bright spear point, 75 per cent. discount.

TOOLS AND HANDLES.

ADZES.

Discount 22½ per cent.

AUGERS.

Gilmour's, discount 60 per cent. off list.

AXES.

Single bit, per doz.	5 50	8 50
Double bit, "	10 00	11 00

Bench Axes, 40 per cent.
Broad Axes. 25 per cent.

Hunters' Axes........	5 50	6 00
Boys' Axes	6 25	7 00
Splitting Axes	7 00	12 00
Handled Axes	7 00	9 00
Red Ridge, boys', handled..	....	5 75
" hunters	....	5 25

BITS.

Irwin's auger, discount 47½ per cent.
Gilmour's auger, discount 60 per cent.
Rockford auger, discount 50 and 10 per cent.
Jennings' Gen. auger, net list.
Gilmour's car, 47½ per cent.
Clark's expansive, 40 per cent.

Clark's gimlet, per doz	0 65	0 90
Diamond, Shell, per doz......	1 00	1 50
Nail and Spike, per gross.....	2 25	5 20

BUTCHERS CLEAVERS

German..........per doz.	6 00	9 00
American........ "	12 00	18 00

CHALK.

Carpenters' Colored, per gross	0 45	0 75
White lump..........per cwt.	0 60	0 65

CHISELS.

Warnock's, discount 72½ per cent.
P. S. & W. Extra, discount 72½ per cent

CROSSCUT SAW HANDLES.

S. & D., No. 3	per pair	0 15
S. & D., " 5	"	0 22½
S. & D., " 6	"	0 15
Boynton pattern........	"	0 20

CROWBARS.

3½c. to 4c. per lb.

DRAW KNIVES.

Coach and Wagon, discount 75 per cent.
Carpenters' discount 75 per cent

DRILLS.

Millar's Falls, hand and breast. net list.
North Bros., each set, 50c

DRILL BITS.

Morse, discount 37½ to 40 per cent.
Standard, discount 50 and 5 to 55 per cent.

FILES AND RASPS.

Great Western	75	per cent.
Arcade	75	"
Kearney & Foot	75	"
Disston's........	75	"
American	75	"
J. Barton Smith........	75	"
McClellan	75	"
Eagle	75	"
Nicholson........	66⅔	"
Globe	75	"

Black Diamond, 60, 10 and 5 p.c.
Jowitt's, English list, 27½ per cent.

GAUGES.

Stanley's discount 50 to 60 per cent.

Wire Gauges

Winn's, Nos. 26 to 33 ....each	1 65	2 40

HANDLES.

C & B, fork, 40 per cent., revised list.
C. & B., hoe, 40 per cent, revised list.

American, saw ...... per doz	....	0 25
American, plane ... per gross	3 15	3 75

Canadian, hammer and hatchet 40 per cent
Axe and cant hook handles, 45 per cent.

HAMMERS.

Maydole's, discount 5 to 10 per cent.
Canadian, discount 25 to 27½ per cent.

Magnetic tack ....per doz.	1 10	1 20
Canadian sledge .... per lb.	0 07	0 08½
Canadian ball pean, per lb. ..	0 22	0 25

HATCHETS.

Canadian, discount 40 to 42½ per cent.

Shingle, Red Ridge 1, per doz........	4 40
" " 2, "	4 85
Barrel Underhill........	5 00

MALLETS.

Tinsmiths'..........per doz.	1 25	1 50
Carpenters', hickory, "	1 25	3 75
Lignum Vitae........ "	3 85	5 00
Caulking, each	0 60	2 00

MATTOCKS.

Canadian........... per doz.	5 50	6 00

MEAT CUTTERS.

German, 15 per cent.
American discount, 33⅓ per cent.
Gem ....each 1 15

NAIL PULLERS.

German and American 0 85 2 50
No. 1.......... 0 85
No 1573.......... 0 75

NAIL SETS.

Square, round and octagon, per gross 3 38
Diamond 1 00

PICKS.

Per dozen .... 6 00 9 [illegible]

PLANES.

Wood bench, Canadian discount 40 per cent. American discount 50 per cent.
Wood, fancy Canadian or American 37½ to 40 per cent.
Stanley planes, $1 55 to $3 60, net list prices.

PLANE IRONS.

English .... per doz. 2 00 5 00
Stanley, 2⅝ inch, single 24c., double 39c.

PLIERS AND NIPPERS.

Button's genuine, 37½ to 40 per cent.
Button's imitation....per doz. 5 00 9 00

PUNCHES.

Saddler's ....per doz. 1 00 1 85
Conductor's.......... " 3 00 15 00
Tinners', solid..........per set 0 72
" hollow......per inch 1 00

RIVET SETS.

Canadian, discount 35 to 37½ per cent.

RULES.

Boxwood, discount 70 per cent.
Ivory, discount 20 to 25 per cent.

SAWS.

Atkins, hand and crosscut, 25 per cent.
Disston's Hand, discount 12½ per cent
Disston's Crosscut ....per foot 0 35 0 55
Hack, complete..........each 0 75 2 75
" frame only........each 0 50 1 25

S. & D. solid tooth circular shingle, concave and band, 50 per cent; mill and ice, drag, 30 per cent; cross-cut, 35 per cent; band saws, butcher, 35 per cent.; buck, New Century, $6 25; buck No. 1 Maple Leaf, $5.25; buck, Happy Medium, $4 25, buck, Watch Spring, $1.25; buck, common frame, $4.00.

Spear & Jackson's saws—Hand or rip 26 in., $12 75; 28 in., $11.25; panel, 18 in., $8 25; 20 in., $9, tenon, 10 in., $9.90; 12 in., $10 90; 14 in., $11 50.

SAW SETS.

Lincoln and Whiting 4 75
Hand Sets, Perfect.......... 4 00
X-Cut Sets, " 7 50
Maple Leaf and Premiums saw sets, 40 off.
S. & D. saw swages, 40 off.

SCREW DRIVERS.

Sargent's ....per doz. 0 65 1 00
North Bros., No. 30 . per doz. 16 80

SHOVELS AND SPADES.

Bull Dog, solid neck shovel (No. 2 pol.) $18 50
(Hollow Back) (Reinforced S Scoop)
Moose........ $17 50 .... $16 30
Bear .... 15 00 .... 15 30
Fox.......... 12 50 .... 14 30
Black Cat.... 10 00 .... 13 30
Canadian, discount 45 per cent.

SQUARES.

Iron, discount 20 per cent
Steel, discount 65 and 10 per cent.
Try and Bevel, discount 50 to 52½ per cent.

TAPE LINES.

English, ass skin .....per doz. 2 75 5 00
English, Patent Leather..... 5 50 9 75
Chesterman's ....each 0 90 2 85
" steel........each 0 80 8 00

TROWELS.

Disston's, discount 10 per cent.
S. & D., discount 35 per cent.

FARM AND GARDEN GOODS

BELLS.

American cow bells, 63⅓ per cent.
Canadian, discount 45 and 50 per cent.
American, farm bells, each .. 1 35 3 00

BULL RINGS.

Copper, $1.30 for 2½-inch, and $1.70

CATTLE LEADERS.

Nos. 32 and 33 .. ...per gross 7 50 8 50

BARN DOOR HANGERS.

	doz.	pairs.
Steel barn door..........	8 00	10 00
Stearns wood track	4 50	6 00
Zenith..........	....	9 00
Acme, wood track	5 00	6 50
Atlas	5 00	6 00
Perfect	8 00	11 00
New Milo..........	....	6 50
Steel, covered..........	4 00	11 00
" track, 1 x 3-16 in (100 ft)	....	3 75
" " 1¼ x 3-16 in (100 ft)	....	4 75
Double strap hangers, doz. sets	....	6 40
Standard jointed hangers, "	....	6 40
Steel King hangers "	....	6 25
Storm King and safety hangers	....	7 00
" " rail..........	....	4 25

Chicago Friction, Oscillating and Big Twin Hangers, 5 per cent.

HARVEST TOOLS.

Discount 60 per cent.
S. & D. lawn rakes, Dunn's, 40 off.
" sidewalk and stable scrapers, 40 off.

HAY KNIVES.

Net prices.

HEAD HALTERS.

Jute Rope, ⅜-inch....per gross 9 00
" " ½ " .... " 10 00
" " ⅝ " .... " 12 00
Leather, 1-inch ......per doz. 4 00
Leather, 1¼ " ...... " 5 20
Web.......... " 2 45

HOES.

Garden, Mortar, etc., discount 60 per cent.
Planter.......... per doz. 4 00 4 50

HORSE NAILS.

'C' brand, 40, 10 and 7½ per cent. off list { Oval
M.R.M. Co. brand, 55 per cent. { head

HORSESHOES.

F.O.B. Montreal
M.R.M. Co. brand, base.......... 3 65
Add 15c. Toronto, Hamilton, Guelph.

HORSE WEIGHTS.

Taylor-Forbes, 3½c. per lb.

SCYTHES.

Per doz. net.......... 6 25 9 25

SCYTHE SNATHS.

Canadian, discount 40 per cent.

SNAPS.

Harness, German, discount 25 per cent.
Lock, Andrews' 4 50 11 00

STALL FIXTURES.

Warden King, 35 per cent.

WOOD HAY RAKES.

Ten tooth, 40 and 10 per cent.
Twelve tooth, 45 per cent.

HEAVY GOODS NAILS, ETC.

ANVILS.

Wright's, 80-lb. and over.......... 0 10½
Hay Budden, 80-lb. and over 0 09½
Brook's, 80-lb. and over 0 11¼
Taylor-Forbes, handy 0 05

VISES.

Wright's.......... .... 0 13½
Brook's.......... .... 0 12½
Pipe Vise, Hinge, No. 1...... .... 3 50
" " " No. 2...... .... 5 50
Saw Vise 4 50 5 00
Blacksmiths' (discount) 60 per cent.
" parallel (discount) 45 per cent.

BOLTS AND NUTS

Carriage Bolts, common ($1 list) Per cent.
" " 5/16 and smaller.. 60, 10 and 10
" " 7-16 and up.... 55 and 5
" " Norway Iron ($3 list) 50
Machine Bolts, ⅜ and less .... 60 and 10
Machine Bolts, 7-16 and up..... 55 and 5
Plough Bolts .... 55 and 10
Blank Bolts.......... 55
Bolt Ends.......... 55
Sleigh Shoe Bolts, ⅜ and less .. 60 and 10
" " 7-16 and larger 50 and 5
Coach Screws, conepoint.... 70 and 5
Nuts, square, all sizes, 4c. per lb off.
Nuts, hexagon, all sizes, 4½c. per lb off.
Stove Rods per lb., 5½ to 6c.
Stove Bolts, 75 per cent.

WROUGHT IRON WASHER.

Canadian make discount 40 per cent.

CHAIN.

Proof coil, per 100 lb., 5-16 in., $4.40; ⅜ in., $3 90; 7-16 in., $3.70; ½ in., $3.50; 9-16 in., $3 45; ⅝ in., $3 35; ¾ in., $3.20; ⅞ in., $3.10; 1 in., $3.10.
Halter, kennel and post chains, 40 to 40 and 5 per cent.
Cow ties .... 40 p.c.
Tie out chains .... 65 p.c.
Stall fixtures .... 35 p.c.
Trace chain .... 45 p.c.
Jack chain iron, discount 35 p.c.
Jack chain, brass, discount 40 per cent.

PRESSED SPIKES.

Pressed spikes, ⅝ diameter, per 100 lbs., $3.15

NAILS.	Cut.	Wire.
2d..........	3 80	3 30
3d..........	2 95	2 95
4 and 5d..........	2 70	2 70
6 and 7d..........	2 60	2 60
8 and 9d..........	2 45	2 45
10 and 12d..........	2 40	2 40
16 and 20d..........	2 35	2 35
30, 40, 50 and 60d (base)	2 30	2 30

F.o.b. Montreal. Cut nails, Toronto 20c. higher.
Miscellaneous wire nails, discount 75 per cent
Coopers' nails, discount 40 per cent.

RIVETS AND BURRS.

Iron Rivets, black and tinned, 60, 10 and 10.
Iron Burrs, discount 60 and 10 and 10 p.c.
Copper Rivets, usual proportion burrs, 27½ p.c.
Copper Burrs only, discount 15 per cent.
Extras on Coppered Rivets, ½-lb. packages 1c. per lb.; ¼-lb. packages 2c. lb.
Tinned Rivets, net extra, 4c. per lb.

SCREWS.

Wood, F. H., bright and steel, 87½ per cent.
" R. H., bright, dis. 82½ per cent.
" F. H., brass, dis. 80 per cent.
" R. H., " dis. 75 per cent.
" F. H., bronze, dis. 75 per cent.
" R. H., " dis. 70 per cent.
Drive Screws, dis. 87½ per cent.
Bench, wood ....per doz. 3 25 4 00
" iron 4 25 5 00
Set, case hardened, dis. 60 per cent.
Square Cap, dis. 50 and 5 per cent.
Hexagon Cap, dis. 45 per cent.

MACHINE SCREWS.

Flat head, iron and brass, 35 per cent.
Fellister head, iron, discount 30 per cent.
" " brass, discount 25 per cent.

TACKS, BRADS, ETC.

Carpet tacks, blued .... 80 and 5
" " tinned.......... 80 and 10
" " (in kegs).......... 40
Cut tacks, blued, in dozens only 75 and 10
" ¼ weights 60
Swedes cut tacks, blued and tinned—
In bulk .... 80 and 10
In dozens 75
Swedes, upholsterers', bulk .... 85 and 12½
" brush, blued and tinned bulk 70
Swedes, gimp, blued, tinned and japanned.......... 75 and 12½
Zinc tacks 35
Leather carpet tacks.......... 40
Copper tacks.......... 37½
Copper nails 42½
Trunk nails, black 65
Trunk nails, tinned and blued .. 65
Clout nails, blued and tinned .. 65
Chair nails.......... 35
Patent brads.......... 40
Fine finishing.......... 40
Lining tacks, in papers.......... 10
" " in bulk.......... 15
" " solid heads, in bulk 75
Saddle nails, in papers.......... 10
" " in bulk.......... 15
Tufting buttons, 22 line in dozens only 60
Zinc glaziers' points 5
Double pointed tacks, papers.. 90 and 10
" " " bulk.... 40
Clinch and duck rivets.......... 45
Cheese box tacks, 85 and 5; trunk tacks, 80 and 10.

CUTLERY AND SPORTING GOODS.

AMMUNITION.

B. B. Caps Dominion, 50 and 5 and 25 per cent. American $2.00 per 1000.
C. B. Caps American, $2.60 per 1000

CARTRIDGES.

Rim Fire Cartridges, Dominion, 50 and 5 p.c.
Rim Fire Pistol, discount 30 and 5 per cent., American.
Central Fire, Military and Sporting, American, add 10 per cent. to list. B.B. Caps, discount 40 per cent., American.
Central Fire Pistol and Rifle, list net Amer.
Central Fire Cartridges, pistol and rifle Dominion, 30 and 5 per cent.
Central Fire Cartridges, Sporting and Military, Dominion, 15 per cent., American 10 per cent. advance on list.
Loaded and empty Shells, Crown, 25 and 5; Sovereign, 25, 10 and 10; Regal, 25, 10 and 5; Imperial, 25, 10 and 5. American 20 per cent. discount. Rival and Nitro, 10 per cent. advance on list.
Empty paper shells Dominion, 25 off and empty brass shells, 55 per cent. off. American, 10 per cent. advance on list.
Primers, Dom., 35 per cent.; American $2 05

Wads. per lb.

Best thick brown or grey felt wads, in ¼-lb. bags $0 70
Best thick white card wads, in boxes of 500 each, 12 and smaller gauges 0 29
Best thick white card wads, in boxes of 500 each, 10 gauge.......... 0 35
Thin card wads, in boxes of 1,000 each, 12 and smaller gauges 0 20
Thin card wads, in boxes of 1,000 each, 10gauge.......... 0 25
Chemically prepared black edge grey cloth wads, in boxes of 250 each— Per M.
11 and smaller gauge 0 60
9 and 10 gauges.......... 0 70
7 and 8 " 0 90
5 and 6 " 1 10
Superior chemically prepared pink edge, best white cloth wads in boxes of 250 each—
11 and smaller gauge 1 15
9 and 10 gauges 1 40
7 and 8 " 1 65
5 and 9 " 1 90

SHOT.

Common, $6.50 per 100 lb.; chilled, $7.50 per 100 lb.; buck, seal and ball, $8.50 net list. Prices are f.o.b. Toronto, Hamilton, Montreal, St. John and Halifax. Terms, 2 p.c. for cash in thirty days.

RAZORS.	per doz.	
Elliot's	4 00	18 00
Boker's	7 50	11 00
" King Cutter	13 50	18 50
Wade & Butcher's..........	3 60	10 00
Wilkinson's..........	....	12 50
Lewis Bros.' "Klean Kutter"	8 50	10 50

Clauss Razors and Strops, 50 and 10 per cent.

PLATED GOODS

Holloware, 40 per cent. discount.
Flatware, staples, 40 and 10, fancy 40 and 5 per cent.

SHEARS.

Clauss, nickel, discount 60 per cent.
Clauss, Japan, discount 67½ per cent.
Clauss, tailors, discount 40 per cent.
Seymour's, discount 50 and 10 per cent.

TRAPS (steel.)

Game, Newhouse, discount 30 and 10 per cent.
Game, Hawley & Norton, 50, 10 & 5 per cent.
Game, Victor, 70 per cent.
Game, Oneida Jump (B. & L.) 40 & 2½ [illegible]
Game, steel, 60 and 5 per cent.

SKATES.

Skates, discount 37½ per cent.
Mic Mac hockey sticks, per doz 4 00 5 00
New Rex hockey sticks, per doz 6 25

HOUSE FURNISHINGS.

APPLE PARERS.

Woodyatt Hudson, per doz., net 4 50

BIRD CAGES.

Brass and Japanned, 40 and 10 p. c.

COPPER AND NICKEL WARE.

Copper boilers, kettles, teapots, etc. 30 p.c.
Copper pitts, 33½ per cent.

ENAMELLED WARE.

London, White, Princess, Turquoise, Onyx, Blue and White, discount 50 per cent.
Canada, Diamond, Premier, 50 and 10 p.c.
Pearl, Imperial Crescent, 50 and 10 per cent.
Premier steel ware, 40 per cent.
Star decorated steel and white, 25 per cent.
Japanned ware, discount 50 per cent.
Hollow ware, tinned cast, 35 per cent. off.

KITCHEN NOVELTIES.

Can openers, per doz........... 0 40 0 75
Mincing knives per doz 0 50 0 80
Duplex mouse traps, per doz.... .... 0 65
Potato mashers, wire, per doz... 0 60 0 70
" wood " .. 0 50 0 60
Vegetable slicers, per doz 2 25
Universal meat chopper No. 0, $1; No 1, 1.15.
Enterprise chopper, each 1 30

LAMP WICKS.

Discount, 60 per cent.

LEMON SQUEEZERS.

Porcelain lined...... per doz. 2 20 5 60
Galvanized.......... " 1 87 3 85
King, wood.......... " 2 75 2 90
King, glass.......... " 4 00 4 50
All glass .... " 0 50 0 90

PICTURE NAILS.

Porcelain head ......per gross 1 35 1 50
Brass head.......... " 0 40 1 00
Tin and gilt, picture wire, 75 per cent.

SAD IRONS.

Mrs. Potts, No. 55, polished....per set 0 80
" No. 50, nickle-plated, " 0 92
Common, plain.......................... 4 50
" plated 5 50
Asbestos, per set.............. .... 1 25

TINWARE.

CONDUCTOR PIPE.

2 in. plain or corrugated., per 100 feet. $3.30; 3 in., $4 40; 4 in., $5 80; 5 in., $7.45; 6 in., $9.90.

FAUCETS.

Common, cork-lined, discount 35 per cent.

EAVETROUGHS.

10-inch ....per 100 ft. 3 30

FACTORY MILK CANS.

Discount off revised list, 40 per cent.
Milk can trimmings, discount 25 per cent.

LANTERNS.

No. 2 or 4 Plain Cold Blast....per doz. 6 50
Lift Tubular and Hinge Plain, " 4 75
Better quality at higher prices.
Japanning, 50c. per doz. extra.

OILERS.

Kemp's Tornado and McClary's Model galvanized oil can, with pump, 5 gallon, per dozen 10 00
Davidson oilers, discount 40 per cent
Zinc and tin, discount 50 per cent
Coppered oilers, 20 per cent. off.
Brass oilers, 50 per cent, off.
Malleable, discount 25 per cent

PAILS (GALVANIZED).

Dufferin pattern pails, 40, 10 and 5 per cent.
Flaring pattern, discount 40, 10 and 5 per cent.
Galvanized washtubs 40, 10 and 5 per cent

PIECED WARE.

Discount 40 per cent off list, June, 1899.
10-qt. flaring sap buckets, discount 40 per cent.
6, 10 and 14-qt. flaring pails dis. 40 per cent.
Copper bottom tea kettles and boilers, 35 p.c.
Creamery cans, discount 40 per cent.

STAMPED WARE.

Plain, 75 and 12½ per cent. off revised list.
Retinned, 72½ 10 and 5 per cent. revised list.

SAP SPOUTS.

Bronzed iron with hooks ....per 1,000 7 50
Eureka tinned steel, hooks 8 00

STOVEPIPES.

5 and 6 inch, per 100 lengths 7 00
7 inch........ " " 7 50

STOVEPIPE ELBOWS

5 and 6-inch, common........per doz. 1 32
7-inch........................ " 1 48
Polished, 15c. per dozen extra.

THERMOMETERS.

Tin case and dairy, 75 to 75 and 10 per cent.

TINNERS' SNIPS.

Per doz........................ 3 00 15 00
Clauss, discount 35 per cent.

WIRE.

BRIGHT WIRE.

Discount 62½ per cent.

CLOTHES LINE WIRE.

7 wire solid line, No. 17, $4.90; No. 18, $3.00; No. 19, $2.70; 8 wire solid line, No. 17, $4.45; No. 18, $2.80. No. 19, $2.50. All prices per 1000 ft. measure. F.o.b. Hamilton Toronto, Montreal.

COILED SPRING WIRE.

High Carbon, No. 9, $2 55; No. 11, $3.20; No. 17, $2.80.

COPPER AND BRASS WIRE

Discount 45 per cent.

FINE STEEL WIRE.

Discount 30 per cent. List of extras: In 100-lb. lots: No. 17, $5 — No. 18, $5.50 — No. 19, $6 — No. 20, $6.65 — No. 21, $7 — No. 22, $7 30 — No. 23, $7.65 — No. 24, $8 — No. 25, $9—No. 26, $9 50—No. 27, $10—No. 28, $11—No. 29, $12—No. 30, $13—No. 31, $14—No. 32, $15—No. 33, $16—No. 34, $17. Extras net—tinned wire, Nos. 17-25, $2—Nos. 26-31, $4—Nos. 32-34, $6. Coppered, 75c.—oiling, 10c.—in 25-lb. bundles, 15c.—in 5 and 10-lb. bundles, 25c.—in 1-lb. hanks, 25c. —in ½-lb. hanks, 38c.—in ¼-lb. hanks, 50c.—packed in casks or cases, 15c.—bagging or papering, 10c.

GALVANIZED WIRE.

Per 100 lb.—Nos. 4 and 5, $3.70 — Nos. 6, 7, 8, $3.15 — No. 9, $2.50 — No. 10, $3.20—No. 11, $3.25—No. 12, $2.65 —No. 13, $2.75—No. 14, $3.75—No 15, $4.30 —No. 16, $4.30 from stock. Base sizes, Nos. 6 to 9, $2.35 f.o.b. Cleveland. In carlots 12½c. less.

LIGHT STRAIGHTENED WIRE.

Gauge No.	Over 20 in. per 100 lbs.	10 to 20 in.	5 to 10 in.
0 to 5	$0 50	$0.75	$1 25
6 to 9	0.75	1.25	2 00
10 to 11	1.00	1.75	2.50
12 to 14	1 50	2 25	3 50
15 to 16	2.00	3.00	4.50

SMOOTH STEEL WIRE.

No. 0-9 gauge, $2 25; No. 10 gauge, 6c. extra; No. 11 gauge, 12c extra; No. 12 gauge, 20c. extra; No. 13 gauge, 30c. extra; No 14 gauge, 40c. extra; No. 15 gauge, 55c. extra; No. 16 gauge, 7 c. extra. Add 60c. for coppering and $2 for tinning.

Extra net per 100 lb.—Oiled wire 10c, spring wire $1 25, special hay baling wire 30c., best steel wire 75c., bright soft drawn 15c., charcoal (extra quality) $1.25, packed in casks or cases 15c., bagging and papering 10c., 50 and 100-lb. bundles 10c., in 25-lb. bundles 15c., in 5 and 10-lb. bundles 25c., in 1-lb hanks, 50c., in ½-lb. hanks 75c., in ¼-lb. hanks $1.

POULTRY NETTING.

2 in mesh 19 w.g., discount 50 and 10 per cent. All others 50 per cent.

WIRE CLOTH.

Painted Screen, in 100-ft. rolls, $1 62½, per 100 sq. ft; in 50-ft. rolls, $1.67½, per 100 sq. ft. Terms, 2 per cent. off 30 days.

WIRE FENCING.

Galvanized barb........ .... 2 95
Galvanized, plain twist 3 30
Galvanized barb, f.o.b. Cleveland, $2.70 for small lots and $2.60 for carlots.

WOODENWARE.

CHURNS.

No. 0, $9; No. 1, $9; No. 2, $10; No. 3, $11; No. 4, $13; No. 5, $16; f.o.b. Toronto Hamilton, London and St. Marys, 30 and 30 per cent.; f.o.b. Ottawa, Kingston and Montreal, 40 and 15 per cent. discount, Taylor-Forbes, 30 and 30 per cent.

CLOTHES REELS.

Davis Clothes Reels, dis. 40 per cent.

LADDERS, EXTENSION.

Waggoner Extension Ladders, dis. 40 per cent.

MOPS AND IRONING BOARDS.

"Best" mops..................... 1 25
"900" mops...................... 1 25
Folding ironing boards.......... 12 00 16 50

REFRIGERATORS

Discount, 40 per cent.

SCREEN DOORS.

Common doors, 2, or 3 panel, walnut stained, 4-in. style.........per doz. 7 25
Common doors, 2 or 3 panel, grained only, 4-in. style ....per doz. 7 55
Common doors, 2 or 3 panel, light stair per doz. 9 55

WASHING MACHINES.

Round, re-acting per doz. 60 00
Square " 63 00
Eclipse, per doz 54 00
Dowswell " 39 00
New Century, per doz 75 00
Daisy 54 00

WRINGERS.

Royal Canadian, 11 in., per doz. 34 00
Royal American, 11 in. 36 00
Eze, 10 in., per doz 34 75
Terms, 2 per cent, 30 days.

MISCELLANEOUS.

AXLE GREASE.

Ordinary, per gross......... 6 00 7 00
Best quality..... .... 10 00 12 00

BELTING.

Extra, 60 per cent.
Standard, 60 and 10 per cent.
No. 1, not wider than 6 in., 60, 10 and 10 p.c.
Agricultural, not wider than 4 in., 75 per cent.
Lace leather, per [illegible], 75c.; cut laces, 80c.

BOOT CALKS.

Small and medium, ball ....per M 4 25
Small heel .... " 4 50

CARPET STRETCHERS.

American ....per doz. 1 00 1 50
Bullard's............. " 6 50

CASTORS.

Bed, new list, discount 55 to 57½ per cent.
Plate, discount 52½ to 57½ per cent.

PINE TAR.

½ pint in tins ....per gross ... 7 80
" " 9 60

PULLEYS.

Hothouse ....per doz. 0 55 1 00
Axle.................. " 0 22 0 33
Screw .... " 0 22 1 00
Awning .... " 0 35 2 50

PUMPS.

Canadian cistern 1 40 2 00
Canadian pitcher spout 1 80 3 16

ROPE AND TWINE.

Sisal................................ 0 10½
Pure Manilla 0 15½
"British" Manilla.................... 0 12
Cotton, 3-16 inch and larger... ..0 21 0 23
" 5-32 inch 0 25 0 27
" ¼ inch 0 25 0 28
Russia Deep Sea 0 16
Jute................................. 0 09
Lath Yarn, single 0 10
" double 0 10½
Sisal bed cord, 48 feet........per doz. 0 65
" " 60 feet........ " 0 80
" " 72 feet........ " 0 95

Twine.

Bag, Russian twine, per lb. 0 27
Wrapping, cotton, 3-ply 0 25
" " 4-ply 0 29
Mattress twine per lb. 0 33 0 45
Staging " " 0 27 0 35

SCALES.

Gurney Standard, 40 per cent.
Gurney Champion, 50 per cent.
Burrow, Stewart & Milne—
Imperial Standard, discount 40 per cent.
Weigh Beams, discount 40 per cent.
Champion Scales, discount 50 per cent.
Fairbanks standard, discount 35 per cent.
" Dominion, discount 55 per cent.
" Richelieu, discount 55 per cent.
Warren new Standard, discount 40 per cent.
" " Champion, discount 50 per cent.
" Weighbeams, discount 35 per cent.

STONES—OIL AND SCYTHE.

Washita ....per lb. 0 25 0 37
Hindostan .... " 0 06 0 10
" slip .... " 0 18 0 20
" Axe.......... " 0 10
Deer Creek............ " 0 10
Deerlick .... " 0 25
" Axe .... " 0 15
Lily white .... " 0 42
Arkansas.............. " 1 50
Water-of-Ayr .... " 0 10
Scythe ....per gross 3 50 5 00
Grind, 40 to 200 lb., per ton.... 20 00 22 00
" under 40 lb., " 24 00
" 200 lb. and over 28 00

INDEX TO ADVERTISERS.

CLASSIFIED LIST OF ADVERTISEMENTS.

Auditors.
Davenport, Percy P., Winnipeg.

Babbitt Metal.
Canada Metal Co., Toronto.
Canadian Fairbanks Co., Montreal.
Robertson, Jas. Co., Montreal.

Bath Room Fittings.
Carriage Mounting Co., Toronto.

Belting, Hose, etc.
Gutta Percha and Rubber Mfg. Co. Toronto.

Bicycles and Accessories.
Johnson's, Iver, Arms and Cycle Works Fitchburg, Mass

Binder Twine.
Consumers Cordage Co., Montreal.

Box Strap.
J. N. Warminton, Montreal.

Brass Goods.
Glauber Brass Mfg. Co., Cleveland, Ohio.
Lewis, Rice, & Son., Toronto.
Morrison, Jas., Brass Mfg. Co., Toronto.
Mueller Mfg. Co., Decatur, Ill.
Penberthy Injector Co., Windsor, Ont.
Taylor-Forbes Co., Guelph, Ont.

Bronze Powders.
Canadian Bronze Powder Works, Montreal.

Brushes.
Ramsay, A., & Son Co., Montreal.

Can Openers.
Cumming Mfg. Co. Renfrew.

Cans.
Acme Can Works, Montreal.

Builders' Tools and Supplies.
Covert Mfg Co., West Troy, N.Y.
Frothingham & Workman Co., Montreal.
Howland, H. S., Sons & Co., Toronto.
Hyde, F., & Co., Montreal.
Lewis Bros. & Co , Montreal.
Lewis, Rice, & Son, Toronto.
Lockerby & McComb, Montreal.
Lufkin Rule Co., Saginaw, Mich.
Newman & Sons, Birmingham.
North Bros. Mfg. Co., Philadelphia, Pa.
Stanley Rule & Level Co., New Britain.
Stanley Works, New Britain, Conn.
Stephens, G. F., Winnipeg.
Taylor-Forbes Co., Guelph, Ont.

Carriage Accessories.
Carriage Mountings Co , Toronto.
Covert Mfg. Co., West Troy, N.Y.

Carriage Springs and Axles.
Guelph Spring and Axle Co., Guelph.

Cattle and Trace Chains.
Greening, B., Wire Co., Hamilton.

Churns.
Dowswell Mfg. Co., Hamilton.

Clippers—All Kinds.
American Shearer Mfg. Co., Nashua, N.H.

Clothes Reels and Lines.
Hamilton Cotton Co., Hamilton, Ont.

Cordage.
Consumers' Cordage Co., Montreal.
Hamilton Cotton Co., Hamilton.

Cork Screws.
Erie Specialty Co., Erie, Pa.

Clutch Nails.
J. N. Warminton, Montreal.

Cut Glass.
Phillips, Geo., & Co., Montreal.

Cutlery—Razors, Scissors, etc.
Birkett, Thos., & Son Co., Ottawa.
Clauss Shear Co., Toronto
Dorken Bros. & Co., Montreal.
Heinisch's, R., Sons Co., Newark, N.J.
Howland, H. S. Sons & Co., Toronto.
Phillips, Geo . & Co , Montreal.
Round, John, & Son, Montreal.

Door Hangers.
Door Hanger Co., Hamilton, Ont.

Electric Fixtures.
Canadian General Electric Co., Toronto.
Forman, John, Montreal.
Morrison James, Mfg. Co., Toronto.
Munderloh & Co., Montreal.

Electro Cabinets.
Cameron & Campbell Toronto.

Engines, Supplies, etc.
Kerr Engine Co., Walkerville, Ont.

Files and Rasps.
Barnett Co., G. & H., Philadelphia, Pa.
Nicholson File Co , Port Hope.

Financial Institutions
Bradstreet Co.

Firearms and Ammunition.
Dominion Cartridge Co., Montreal.
Hamilton Rifle Co., Plymouth, Mich.
Harrington & Richardson Arms Co., Worcester, Mass.
Johnson's, Iver, Arms and Cycle Works, Fitchburg, Mass.

Food Choppers.
Enterprise Mfg. Co., Philadelphia, Pa

Galvanizing.
Canada Metal Co., Toronto
Montreal Rolling Mills Co , Montreal
Ontario Wind Engine & Pump Co., Toronto.

Glaziers' Diamonds.
Gibsone, J. B., Montreal.
Pelton, Godfrey S.
Sharratt & Newth, London, Eng.
Shaw, A., & Son, London, Eng.

Hack Saws.
Diamond Saw & Stamping Works, Buffalo

Harvest Tools.
Maple Leaf Harvest Tool Co , Tillsonburg Ont.

Hoop Iron.
Montreal Rolling Mills Co , Montreal.
J. N. Warminton, Montreal.

Horse Blankets.
Heney, E. N., & Co., Montreal.

Horseshoes and Nails.
Canada Horse Nail Co., Montreal.
Montreal Rolling Mills, Montreal.

Hot Water Boilers and Radiators.
Cluff, R. J., & Co., Toronto.
Pease Foundry Co., Toronto.
Taylor-Forbes Co., Guelph.

Ice Cream Freezers.
Dana Mfg. Co., Cincinnati, Ohio.
North Bros. Mfg. Co., Philadelphia, Pa.

Ice Cutting Tools.
Erie Specialty Co., Erie, Pa.
North Bros. Mfg. Co., Philadelphia, Pa.

Injectors—Automatic.
Morrison, Jas., Brass Mfg. Co., Toronto.
Penberthy Injector Co., Windsor, Ont.

Iron Pipe.
Montreal Rolling Mills, Montreal.

Iron Pumps.
McDougall, R., Co., Galt, Ont.

Lanterns.
Kemp Mfg. Co., Toronto.
Ontario Lantern Co., Hamilton, Ont.
Wright, E. T., & Co., Hamilton.

Lawn Mowers.
Birkett, Thos., & Son Co., Ottawa.
Maxwell, D., & Sons, St. Mary's, Ont.
Taylor, Forbes Co., Guelph.

Lawn Swings, Settees, Chairs.
Cumming Mfg. Co., Renfrew.

Ledgers—Loose Leaf.
Business Systems, Toronto.
Copeland-Chatterson Co , Toronto.
Crain, Rolla L., Co., Ottawa.
Universal Systems, Toronto.

Locks, Knobs, Escutcheons, etc.
Peterborough Lock Mfg. Co., Peterborough, Ont.

Lumbermen's Supplies.
Pink, Thos., & Co., Pembroke Ont.

Mantels, Grates and Tiles.
Batty Stove and Hardware Co., Toronto.

Manufacturers' Agents.
Fox, C. H., Vancouver.
Gibb, Alexander, Montreal
Mitchell, David C., & Co., Glasgow, Scot.
Mitchell, H. W., Winnipeg.
Pearce, Frank, & Co. Liverpool, Eng.
Scott, Bathgate & Co., Winnipeg.
Thorne, R. E., Montreal and Toronto.

Metals.
Canada Iron Furnace Co., Midland, Ont.
Canada Metal Co., Toronto.
Eadie, H. G , Montreal.
Gibb, Alexander, Montreal.
Kemp Mfg Co., Toronto
Leslie, A. C., & Co , Montreal.
Lysaght, John, Bristol, Eng.
Nova Scotia Steel and Coal Co., New Glasgow, N.S.
Robertson, Jas., Co., Montreal.
Roper, J. H., Montreal.
Samuel, Benjamin & Co , Toronto.
Stairs, Son & Morrow, Halifax, N S.
Thompson, B. & S. H. & Co. Montreal.

Metal Lath.
Galt Art Metal Co , Galt.
Metallic Roofing Co., Toronto.
Metal Shingle & Siding Co., Preston, Ont.

Metal Polish, Emery Cloth, etc
Oakey, John, & Sons, London, Eng.

Mops.
Cumming Mfg. Co., Refrew.

Mouse Traps.
Cumming Mfg. Co., Renfrew.

Oil Tanks.
Bowser, S. F., & Co., Toronto.

Paints, Oils, Varnishes, Glass.
Bell, Thos , Sons & Co , Montreal.
Canada Paint Co., Montreal.
Canadian Oil Co., Toronto
Consolidated Plate Glass Co., Toronto.
Fenner, Fred., & Co., London, Eng.
Henderson & Potts Co , Montreal.
Imperial Varnish and Color Co., Toronto.
Jamieson, R. C., & Co , Montreal.
McArthur, Corneille & Co., Montreal.
McCaskill, Dougall & Co., Montreal
Montreal Rolling Mills Co , Montreal.
Moore, Benjamin, & Co. Toronto.
Queen City Oil Co , Toronto.
Ramsay & Son, Montreal.
Sherwin-Williams Co., Montreal.
Standard Paint and Varnish Works Windsor, Ont.
Stephens & Co., Winnipeg.
Martin-Senour Co., Chicago.

Perforated Sheet Metals.
Greening, B., Wire Co., Hamilton.

Plumbers' Tools and Supplies.
Borden Co., Warren, Ohio.
Canadian Fairbanks Co., Montreal.
Cluff, R J., & Co., Toronto.
Glauber Brass Co., Cleveland, Ohio.
Jardine, A. B., & Co , Hespeler, Ont.
Jenkins Bros., Boston, Mass.
Lewis, Rice, & Son, Toronto.
Merrell Mfg. Co., Toledo, Ohio.
Montreal Rolling Mills Montreal.
Morrison, Jas., Brass Mfg. Co., Toronto.
Mueller, H., Mfg. Co., Decatur, Ill.
Oshawa Steam & Gas Fitting Co., Oshawa
Robertson Jas., Co. Montreal.
Stairs, Son & Morrow, Halifax, N.S.
Standard Ideal Sanitary Co., Port Hope,
Standard Sanitary Co., Pittsburg.
Stephens, G F., & Co., Winnipeg, Man.
Turner Brass Works, Chicago.
Vickery, Orlando, Toronto.

Portland Cement.
Grey & Bruce Portland Cement Co., Owen Sound.
Hanover Portland Cement Co., Hanover, Ont.
Hyde, F., & Co., Montreal.
Thompson, B. & S. H. & Co., Montreal.

Potato Mashers.
Cumming Mfg Co., Renfrew.

Poultry Netting.
Greening, B., Wire Co., Hamilton, Ont.

Razors.
Clauss Shear Co., Toronto.

Roofing Supplies.
Brantford Roofing Co., Brantford.
McArthur, Alex., & Co., Montreal.
Metal Shingle & Siding Co., Preston, Ont.
Metallic Roofing Co., Toronto.
Paterson Mfg. Co., Toronto & Montreal.

Saws.
Atkins, E. C., & Co., Indianapolis, Ind
Lewis Bros., Montreal.
Shurly & Dietrich, Galt, Ont.
Spear & Jackson, Sheffield, Eng.

Saws—Hack.
Diamond Saw & Stamping Works, Buffalo

Scales.
Canadian Fairbanks Co., Montreal.

Screw Cabinets.
Cameron & Campbell, Toronto.

Screws, Nuts, Bolts.
Montreal Rolling Mills Co., Montreal.
Morrow, John, Machine Screw Co., Ingersoll, Ont.

Sewer Pipes.
Canadian Sewer Pipe Co., Hamilton
Hyde, F., & Co., Montreal.

Shelf Boxes.
Cameron & Campbell, Toronto.

Shears, Scissors.
Clauss Shear Co., Toronto.

Shelf Brackets.
Atlas Mfg. Co., New Haven, Conn

Shellac
Bell, Thos., Sons & Co , Montreal.

Shovels and Spades.
Canadian Shovel Co. Hamilton.
Peterboro Shovel & Tool Co , Peterboro.

Silverware.
Phillips, Geo., & Co., Montreal.
Round, John, & Son, Sheffield, Eng.

Spring Hinges, etc.
Chicago Spring Butt Co., Chicago, Ill.

Steel Rails.
Nova Scotia Steel & Coal Co , New Glasgow, N.S.

Stoves, Tinware, Furnaces
Canadian Heating & Ventilating Co., Owen Sound.
Canada Stove Works, Harriston, Ont.
Clare Bros & Co., Preston.
Davidson, Thos , Mfg. Co., Montreal.
Guelph Stove Co., Guelph.
Gurney Foundry Co , Toronto.
Harris, J. W., Co., Montreal.
Joy Mfg. Co., Toronto.
Kemp Mnfg. Co. Toronto.
McClary Mfg. Co. London.
Pease Foundry Co., Toronto.
Stewart, Jas., Mfg. Co , Woodstock, Ont.
Taylor-Forbes Co., Guelph, Ont.
Wright, E. T., & Co., Hamilton.

Tacks.
Montreal Rolling Mills Co., Montreal.
Ontario Tack Co., Hamilton.

Ventilators.
Harris, J. W., Co., Montreal.
Pearson, Geo. D., Montreal.

Washing Machines, etc
Dowswell Mfg. Co., Hamilton, Ont.
Taylor-Forbes Co., Guelph, Ont.

Wheelbarrows
London Foundry Co., London, Ont.

Wholesale Hardware.
Birkett, Thos., & Sons Co., Ottawa.
Caverhill, Learmont & Co., Montreal.
Frothingham & Workman, Montreal.
Hobbs Hardware Co., London.
Howland, H. S., Sons & Co., Toronto.
Lewis Bros. & Co., Montreal.
Lewis, Rice, & Son, Toronto.

Window and Sidewalk Prisms.
Hobbs Mfg. Co., London, Ont.

Wire Springs.
Guelph Spring Axle Co., Guelph, Ont.
Wallace-Barnes Co., Bristol, Conn.

Wire, Wire Rope, Cow Ties, Fencing Tools, etc.
Canada Fence Co., London.
Dennis Wire and Iron Co., London, Ont.
Dominion Wire Mnfg. Co., Montreal
Greening, B., Wire Co., Hamilton.
Montreal Rolling Mills Co., Montreal.
Western Wire & Nail Co., London, Ont

Woodenware.
Taylor-Forbes Co., Guelph, Ont.

Wrapping Papers.
Canada Paper Co., Toronto.
McArthur, Alex., & Co , Montreal.
Stairs, Son & Morrow, Halifax, N.S.

CIRCULATES EVERYWHERE IN CANADA
Also in Great Britain, United States, West Indies, South Africa and Australia.

HARDWARE AND METAL

A Weekly Newspaper Devoted to the Hardware, Metal, Heating and Plumbing Trades in Canada.
Office of Publication, 10 Front Street East, Toronto.

VOL. XIX. MONTREAL, TORONTO, WINNIPEG, JANUARY 12, 1907 NO. 2.

See Classified List of Advertisements on Page 67.

FIRE SETS and FENDERS

BLACK IRON and BRASS

ANDIRONS

Black Iron Coal Vases

Coal Grates

Brass Coal Scuttles

Coal Tongs

Fenders and Fire Sets

Coal Tongs

WRITE FOR PRICES

RICE LEWIS & SON

LIMITED

TORONTO.

GET OUT AND
HUSTLE

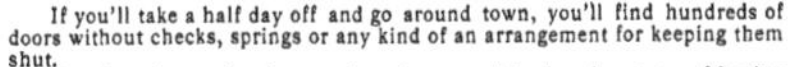

If you'll take a half day off and go around town, you'll find hundreds of doors without checks, springs or any kind of an arrangement for keeping them shut.

Get after the people who own these doors, explain the advantages of having a Liquid Door Check on them, and you'll get a surprise at the number of orders you'll get; then go back to your store and write us for prices and full information about **SARGENT'S LIQUID DOOR CHECK**—the best door check on the market, embodying all the good points of others, with many more which the others lack.

Sargent Liquid Door Check and Spring

SUCCESSFULLY PREVENTS DOORS FROM SLAMMING

Sargent's Liquid Door Check and Spring is suitable for either right or left hand doors.

The Spring is adjustable tension, very compact and powerful, and is made from extra heavy clock-spring steel of the finest quality.

The Check is put up regularly for attaching to the door, but special brackets for attaching in different positions to the soffit and door frame and for any kind of door, can be furnished.

The Check can be instantly varied for light or heavy draft by screwing the regulating thumb screw.

For applying, use the directions packed with each check, and follow them carefully.

Have the cylinder indoors where possible.

Made of cast iron, gold bronze finish.

No. 121. Gold bronze, for light inside doors.
No. 123. " " for medium size inside doors.
No. 125. " " for heavy inside or light outside doors.
No. 127. " " for medium size outside doors.
No. 129. " " for heavy outside doors.

LEWIS BROS., LTD.

MONTREAL

OTTAWA **TORONTO** **WINNIPEG** **VANCOUVER** **CALGARY**

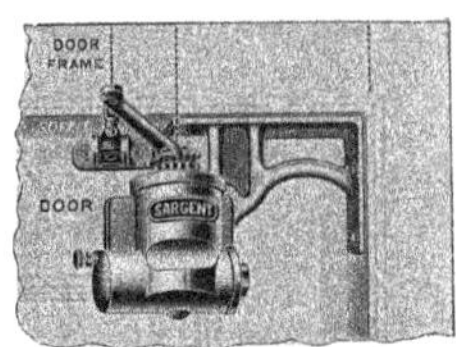

Showing the Check and Spring Applied with Corner Bracket

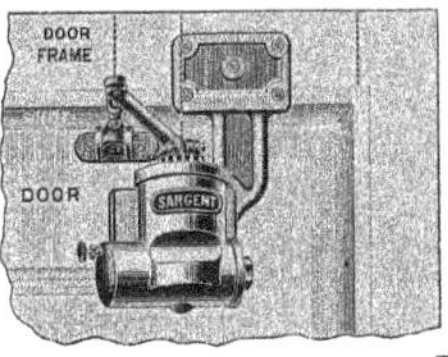

Showing the Check and Spring Applied with Flush Bracket

Showing the Check and Spring Applied with Soffit Bracket

Framing Tools

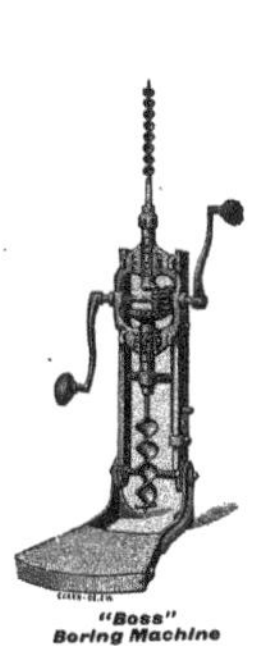

"Boss" Boring Machine

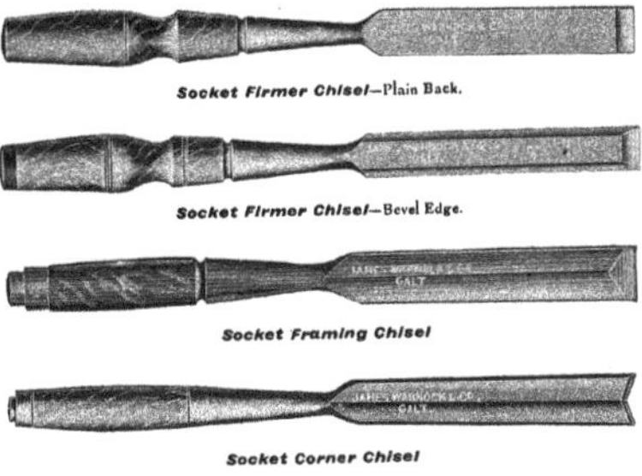

Socket Firmer Chisel—Plain Back.

Socket Firmer Chisel—Bevel Edge.

Socket Framing Chisel

Socket Corner Chisel

Socket Carpenter's Slick

"Miller Falls" Boring Machine

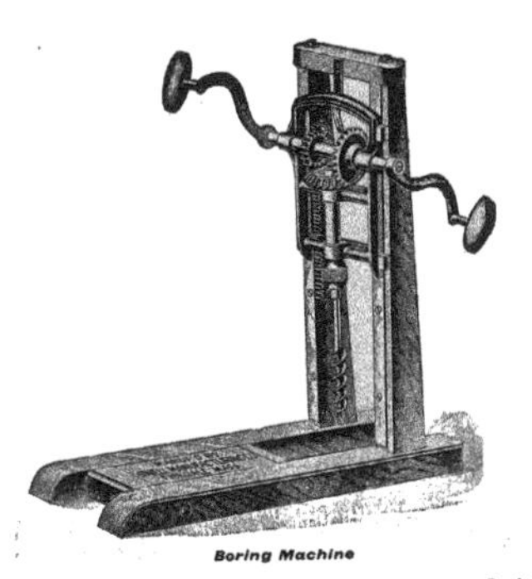

Boring Machine

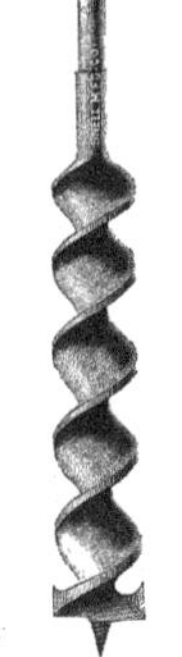

Boring Machine Augur

Angular Boring Machine

A Promise

of good service is made by many an axe with its brightly polished edge, gilt head and brilliant label. But the disappointments are altogether too frequent. When selecting axes for your stock why not choose those which have been known for years for their reliability.

Dunn's "Clipper" Axe is an axe we have sold for 20 years and have always found it reliable and it is still reliable. No better recommendation for an axe is wanted than such a record.

The "Bedford Ebony" is another axe found for many years on our sales records. It is hand-made by expert axe makers and has the natural forge finish.

Stock weights of these axes are 3 to 3½, 3¼ to 3¾, 3½ to 4, 3½ to 4½-lb.

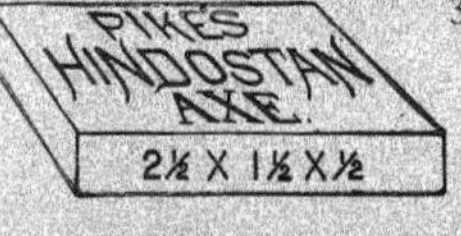

We have Axe Handles
Axe Stones
Axe Wedges
in large variety we ship them promptly. Write us when in a hurry.

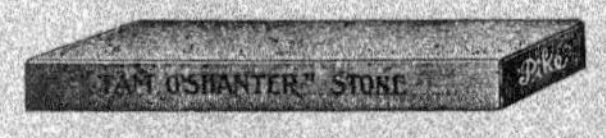

No. 502 TURNER, DAY & WOOLWORTH HANDLE CO. LOUISVILLE Ky

F. & W. Hardware Montreal

FROTHINGHAM & WORKMAN, Limited, MONTREAL, CANADA.

Brantford—Roofings

This Represents Our

"ASPHALT"
No. 1, 2, 3
"RUBBER"
No. 1, 2, 3

Each Roll is complete with Nails and Cement to lay. Best selling goods in the market.

BRANTFORD ROOFING CO.
LIMITED
BRANTFORD, ONTARIO

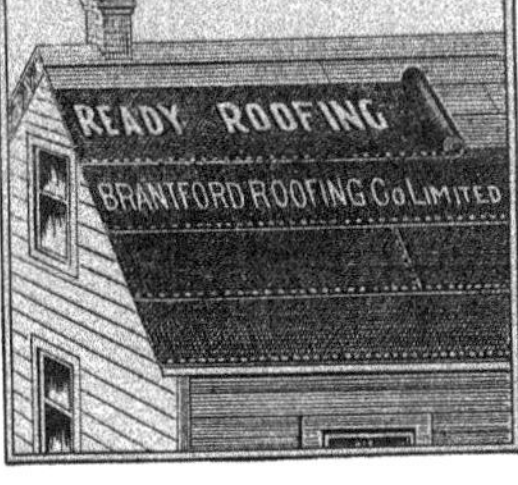

SIMPLE

DURABLE

RELIABLE

STYLE F.

THE LEADER

A Name Familiar to Butter Makers in Connection with the

Highest Grade Churns

ITS STRONG POINTS:

Stiff Steel Frame, with malleable iron fittings, Combination Bow Lever and Pedal Drive. Bicycle Ball Bearings. Bolted Trunnions. Cream Breakers. Barrel easily drained and conveniently detatchable for cleaning.

DEALERS: When you handle this line you handle QUALITY.

The Dowswell Manufacturing Co., Limited

W. L. HALDIMAND & SON, Montreal
EASTERN AGENTS

HAMILTON
ONTARIO

GALVANIZED FENCE HOOK FOR FASTENING WOODEN PICKET ON WIRE FENCES

FENCE HOOK

WIRE NAILS, COILED SPRING, BARB and PLAIN FENCE WIRE, OILED and ANNEALED, CLOTHES LINE WIRE, STAPLES, etc.

THE WESTERN WIRE & NAIL CO., Limited, LONDON, ONT

The Peterborough Lock Manufacturing Company, Limited

Peterborough, Ont.

Manufacturers of all kinds

Rim and Mortise Locks, Inside, Front Door, and Store Door Sets, also full line High-class Builders' Hardware.

Sold by all Leading Jobbers in the Dominion.

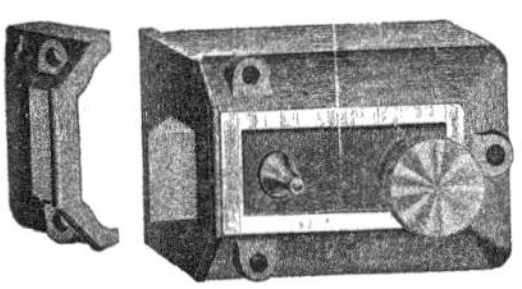

Cylinder Night Latch, No. 103.

Persons addressing advertisers will kindly mention having seen this advertisement in Hardware and Metal.

We are breaking all records in sales of Meat and Food Choppers. If you are not selling more

Enterprise Meat AND Food Choppers

than ever before there is something wrong.

Write us for "ginger" (live advertising matter), to start business bubbling your way.

People are afraid now of packing house products.

They want to chop their own meats.

They are buying Meat Choppers from the dealers who are pushing them. Get in line !

THE ENTERPRISE MFG. CO. OF PA.
PHILADELPHIA, U. S. A.

NEW YORK BRANCH : 10 Warren Street. SAN FRANCISCO BRANCH : 438 Market Street

The John Morrow Screw, Limited

MANUFACTURERS OF
Set and Cap Screws, Special Milled Work, Engine Studs Etc. Cold Punched Nuts of every variety of finish.

INGERSOLL, ONTARIO

CELEBRATED **HEINISCH** SHEARS

Tailors' Shears,
Trimmers' Shears,
Tinners' Snips, etc.

ACKNOWLEDGED THE BEST.

R. HEINISCH'S SONS CO. NEW YORK OFFICE, 155 Chambers St
NEWARK, N.J., U.S.A.

SEYMOUR SHEARS

HAVE BEEN THE
Standard for over Half a Century.
"QUALITY UNQUESTIONED."
Each pair of our shears bears the above trade mark.

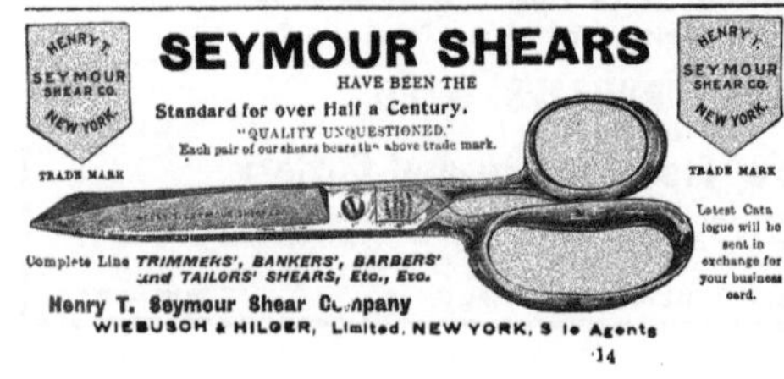

Complete Line TRIMMERS', BANKERS', BARBERS' and TAILORS' SHEARS, Etc., Etc.

Latest Catalogue will be sent in exchange for your business card.

Henry T. Seymour Shear Company
WIEBUSCH & HILGER, Limited, NEW YORK, Sole Agents

Covert Mfg. Co.
Troy, N.Y.

Harness Snaps, Chain, Rope and Web Goods, etc.

For sale by Jobbers at Manufacturers' Prices

20 510 280

STREET PAVING and SIDEWALKS a SPECIALTY

SILICA BARYTIC STONE CO.
OF ONTARIO, Limited.

Water Proof Floors for Malt Houses, Breweries, Slaughter Houses, Cheese Factories, Cellar, Stable Floors, etc.

Head Office
Ingersoll, Ontario.
Walter Mills, General Manager
Ask for quotations for Septic Tanks.

"The Tool Holder People"
ARMSTRONG BROS. TOOL CO.
Manufacturers of Armstrong Patent Lathe and Planer Tools and other machine shop specialties.
106 N. Francisco Ave. CHICAGO, ILL.

GALVANIZING AND TINNING
THE CANADA METAL CO.
TORONTO, ONTARIO.

Persons addressing advertisers will kindly mention having seen this advertisement in Hardware and Metal.

The advertiser talks to you through Hardware and Metal. You should talk to him through correspondence.

OUR LATEST QUICK SELLER

CLOSED

NO. 445
IRONING BOARD

The newest and most convenient.

Insist on your jobber supplying you with No. 445.

Simply raise the top and it can be used as a skirt board.

WE KNOW WHAT THE HOUSEKEEPER WANTS AND WE MAKE THE GOODS

CUMMING'S GOODS ARE GOOD GOODS

—Made by—

The Cumming Manufacturing Co., Limited

Renfrew, Ont.

WINNIPEG: 608 Ashdown Bldg.

Peerless Woven Wire Fence

Here is a fence that will sell. Once up no need for repairs. It lasts for years—just what you are looking for. It's horse-high, pig-tight and bull-strong. No stock can get over, under or through it. The most unruly animal can't butt it down and there are no barbs to tear and injure stock. It's

WIND AND WEATHER PROOF

The Peerless lock holds it perfectly rigid and the wires can't slip up, down, nor sidewise. Cold weather won't snap it and hot weather won't make it sag. It's the greatest thing out for farmers, stockmen and all other fence users which makes it a red hot proposition for dealers. It comes in rolls all ready to put up. It's easy to handle. Write for prices.

The Banwell Hoxie Wire Fence Company, Ltd.
Dept. J, Hamilton, Ont.

ESTABLISHED 1795

JOHN SHAW & SONS
WOLVERHAMPTON, Limited

WOLVERHAMPTON

FUSE

White Countered
Gutta Percha Safety
Single and Double Tape, etc.

Sole Canadian sales agents for

WM. BRUNTON & CO., Wrexham, England

INSIDE PRICES TO THE TRADE

JAMES BARBER, Era Works, Sheffield

Best quality only
TABLE AND POCKET CUTLERY, RAZORS, CARVERS, Etc.

J. H. ROPER

Canadian Representative

82 St. Francois Xavier Street - MONTREAL

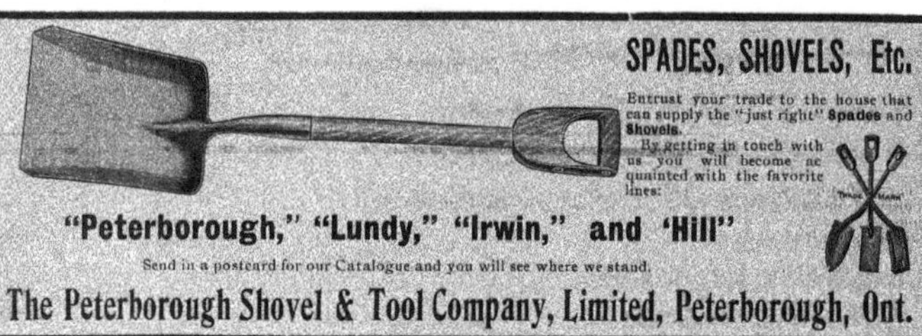

WANTS

MADE KNOWN IN "HARDWARE AND METAL" COLUMNS ARE SOON SATISFIED.

TRY IT YOURSELF

...ESTABLISHED 1849...

BRADSTREET'S

Capital and Surplus, $1,500,000. Offices Throughout the Civilized World

Executive Offices: Nos. 346 and 348 Broadway, New York City, U.S.A.

THE BRADSTREET COMPANY gathers information that reflects the financial condition and the controlling circumstances of every seeker of mercantile credit. Its business may be defined as of the merchants, by the merchants, for the merchants. In procuring, verifying and promulgating information no effort is spared, and no reasonable expense considered too great, that the results may justify its claims as an authority on all matters affecting commercial affairs and mercantile credit. Its offices and connections have been steadily extended, and it furnishes information concerning mercantile persons throughout the civilized world.

Subscriptions are based on the service furnished, and are available only by reputable wholesale, jobbing and manufacturing concerns, and by responsible and worthy financial fiduciary and business corporations. Specific terms may be obtained by addressing the Company at any of its offices. **Correspondence invited.**

—OFFICES IN CANADA—

CALGARY, ALTA.	HAMILTON, ONT.	LONDON, ONT.	MONTREAL, QUE.
HALIFAX, N.S.	QUEBEC, QUE.	ST. JOHN, N.B.	TORONTO, ONT.
OTTAWA, ONT.	VANCOUVER, B.C.	WINNIPEG, MAN.	

THOS. C. IRVING, Gen. Man. Western Canada, Toronto.

KEMP'S FACTORY MILK CAN TRIMMINGS

Are Made to Last

With roll rim bottoms are the kind that stand the wear and tear

Sectional View of Bottom

Prompt attention given letter orders

All Sizes of Tinned Iron in Stock

KEMP MANUFACTURING CO., TORONTO, ONT.

THE KEMP MFG. CO. OF MONTREAL
58 McGill St., Montreal, Que.

THE KEMP MFG. AND METAL CO., Limited
McDermot Ave. E., Winnipeg, Man.

TRAVELLERS AND THEIR SPARE TIME

Some Suggestions for the Improvement of it — The Note Book — Learning Languages — Savings Bank for the ... Small Change of Time.

The profitable use of spare time is a subject of interest to every one ambitious of success. Calvin D. Wilson, writing in The Sample Case, says of it:

The commercial traveler's leisure differs from that of men located at home. He who lives in one place has the evenings in his house or in his office; he has hours that he can count on. The travelling man's unoccupied seasons are in small change, waiting for trains, or on railroad rides or buggy trips across country or in his hotel room.

Yet he has many unemployed minutes, which in a week make hours, as dimes and quarters make dollars. What can he do with these that they may count for something in his success? He can do many things, according to his temperament, health, stage of culture and circumstances. No detailed rules will suit everyone, but general suggestions may prove helpful and each can pick and choose according to his situation. If one does not have a surplus of energy if he has to save up his vital forces in order to concentrate them when in action, it behooves him to seize the idle moments for rest, to sit still in the depot, to doze on the cars, to retire early to his hotel room. Yet even many of the less well-equipped men physically might be benefited by change of mental occupation, according to the present belief that what we often need is not so much repose as the exercise of another set of faculties.

The men with reserves of energy may again be divided into two classes, those who learn chiefly through their senses and such as learn partly by means of reading. The man with a markedly objective mind, who reads principally in the book of men and nature might be diverted from his proper method by advice to become a reader of books and a student.

Keep a Note Book Handy.

We have the impression that traveling men, as well as many others, lose much by often failing to have in their pockets notebooks for ideas, suggestions, plans. Oliver Wendell Holmes said: "No man ever yet caught the reins of a thought save as it galloped by him." It is an invaluable practice to have at hand blank book and pencil, and to get into the habit of seizing the reins of thought as they gallop by. Many of these, perhaps most of them, may prove worthless or impracticable, but in any case you have the ideas written down for examination and scrutiny at leisure. The man with the notebook will soon find he is having more bright ideas than came to him formerly. Then he preserves the facts that come to him by the eye or ear for inspection: he is not trusting to memory which may fail to bring back accurately what he wants when he wants it. The notebook habit causes him to put his ideas into somewhat better shape than if he leaves them unformulated and floating in his mind. The notebook helps him to learn the value of his own thoughts. It gives him a record also of his own ponderings and experiences that may have in the future a larger value than he now perceives. If he saves his thoughts thus, records his observations, while waiting for trains or shut in his hotel room and studies these jottings, he will probably find some diamonds among the pebbles. He may have jotted down in two minutes the idea that proves the germ of a larger future. In times when no new ideas come, the moments may be well spent in looking over the notes already made. Such a course soon comes to have an interest of its own and is a pleasant occupation. It takes away the feeling of wasting time and the spirit of impatience at delays.

E. FIELDING, TORONTO

Who takes care of the cash for the Commercial Travelers' Association of Canada, besides conducting a large business in Turpentine and Linseed Oil.

Study Languages.

There is another line of mental effort, for which odd moments may be utilized, the result of which may be of especial value to traveling men who are able to follow it. This is the learning of a new language. No one needs to be more than reminded that in such a country as ours with its mixed population the knowledge of other languages than English is of every-day value. This is true particularly of traveling men, many of whom are carried by their business into communities where German is commonly spoken. One may be sent by his firm into the French-speaking portion of Canada or Louisiana. His employers may require a man who can talk with the Mexicans in their own tongue, or go into the settlements of the Swedes or Norwegians. His house may have extensive plans that reach out over Europe. In any of these cases the man who is already equipped with a working knowledge of German, French or Spanish may get the plum of the mission. Being able to do what the majority cannot do, he gets a better salary, and if sent abroad gets foreign travel thrown in

Spare minutes tell wonderfully in the acquiring of a new language. The mind comes to it as a change and so with fresh interest. The little learned in a fragmentary way gets in time to be fixed in the memory. Of course some men have no faculty at all for new languages and such, after a fair trial, will have to limit their sphere of business to the people who talk their native tongue. But after all, most men can do far better in learning another language than they think at first thought. Nearly any one can get a speaking acquaintance with German, French or Spanish, if he sets himself about it in the right way. A beginning can be made by the use of such little books as are to be gotten in almost any book store. These appear under some such titles as "How to Speak German Without a Teacher." Any librarian or book dealer can give the information. Something at least may be acquired in this manner and such a book may be carried about and conned anywhere at odd times. A few weeks of this study will bring a man along to a point where he will see what further help he needs. If he can be in one place long enough to get the assistance of a teacher, so much the better. He may in some cases be able to get in touch with night schools or the classes in languages in the Y.M.C.A. Presently he will be reading German or French books on his journeys and talking in these languages on his travels to any who understand them. In due time through use of spare time, he will be equipped for a wide field of employment.

The ambitious traveling man who has gotten a working knowledge of one foreign language will not be likely to stop there. It is a general experience that the learning of other languages becomes much easier after the first one. As is well known, all classes of people in Europe think little of knowing several languages beside their own; of course they often have the advantage of hearing these spoken around about them. Yet what is done by unschooled European peasants ought to prove simple for as bright men as American commercial travelers commonly are.

Spend Time in Systematic Reading.

Reading with a purpose, for some definite end, is of course, a universally desirable use of spare time. This may or may not be apparently of immediately available import, yet may further us greatly. President Roosevelt has stated that he is fond of reading in his leisure hours the history of the dismemberment of the empire of Alexander the Great. This seems a rather remote subject from

American politics; and yet the completed record of the causes and changes that disintegrated one of the greatest empires of ancient times affords a picture in which to study the perils of modern society. The statesman who is familiar with these stories of the past is better able to understand and forestall the dangers of America. Cecil Rhodes, the great empire builder of South Africa, was a constant reader of Gibbon's "Decline and Fall of the Roman Empire," and doubtless thereby had his mind enlarged to understand the scope of the problems of the British Empire.

The intellectual horizon of the traveling man, whose actual business of selling goods may seem quite disconnected from general knowledge, will unquestionably be widened by the reading of history, for example. The wider and stronger his mind, provided he does not sink into a book worm, the better he can do business. He will have a clearer and saner judgment. When reading is connected with daily activity there is little peril of its making men impractical. Roosevelt is one of the best read men of the day and his scholarship has not impaired his practical judgment. The traveling man might well read all American history to advantage. He might take up an elaborate history of business from its beginning to its present developments. Still better, if he aims at culture, would it be to plan a course of reading such as would cover all the most important subjects, history, ethics, political economy, finance and so on. At the end of a few years, by use of spare time alone, he would have furnished himself with all the general facts that any man in any occupation may have gained. He surely should guard against feeding his mind exclusively with the very interesting but scrappy information to be gotten from newspapers and magazines; such reading will at most afford but a superficial knowledge of anything.

How One Man's Study Helped.

If one has a particular bent, he should follow it if it is along lines that promise development. The man of inventive turn may, in his odd moments, note and work out some needed invention. The man of mechanical mind certainly might find pleasure and profit by searching the world as he goes about for the things that are needed or that can be improved. Another man has a turn for mathematics; if he has, he possessses a gift that, worked constantly and to its utmost, may at any time bring forth fruit to his great advantage. We are told that Mr. Hughes, the governor-elect of New York, takes with him on his vacations books on mathematics and such works as Kant's Philosophy. When he had the investigation of insurance matters on his hands, his great skill in mathematics served him well. It seems a long way about for a lawyer to spend time on mathematics and philosophy; yet mathematics in this case, as in many another, worked directly into his practice and Kant's Philosophy doubtless kept him in touch with noble views of human life, and with processes of pure reason. Likewise, the commercial traveler may find any real study while in his leisure. The improvement of mind will certainly serve him and the particular knowledge may at any time fit into a place in his work.

There are still other men who are travelers who have in them, perhaps in an undeveloped state, the gift of writing for the press. The leisure of such may be well employed by writing of their interesting or unusual experiences for the newspapers or magazines, or preparing material for a book, a story, a novel, or a volume on business or on their particular kind of work. Many writers would think they had a gold mine if they were in possession of the impressions and experiences of such commercial travelers as have long made their rounds over a great scope of country.

Reading for Relaxation.

Reading for relaxation has its fit place and seasons and the commercial traveler requires, at least at times, to get away from the actual world into the realm of romance and of poetry and he does well to permit himself the privilege within reason. He surely ought besides to equip himself with some of the vital books that impress the principles of sound faith and wise living. He cannot afford to be without a strong grasp on the truth and his spiritual relations and duties. Otherwise, the spare moments will be the seasons of depression and of darkness. The man who is brave in the actual contact with men and with business is apt to feel unnerved and cheerless in the idle hours.

"'Tis not in the battle nor in the strife
We feel benumbed and wish to be no more
But in the after silence on the shore
When all is lost us but a little life."

In the lonely times, in the lull of the battle, we need anchorage and hope and faith. He is wise who, at such times, has in his grip a book that will rekindle his soul and show him the great truths that remain and are the same both when we are glad and busy and when we are listless and solitary.

The opportunities for exercise and improvement of health in odd moments should not be overlooked. On journeys Talmage used to get off the train during its stops and walk briskly up and down the platform for a few minutes. When he was lecturing, he would take a street car line from his hotel to the country and treat himself to a five or ten minutes' run. The commercial traveler who has ten minutes wait at the depot can step out into the open and fill his lungs with fresh air or snatch a brief walk for a few squares. The spare minutes of each day will do much to invigorate him if he seizes them.

KNOW THE GOODS YOU SELL.

At the recent banquet of the Commercial Travelers' Association of Canada held in Toronto, Thomas Kinnear, a successful wholesale grocer, emphasized the importance of a traveler telling the truth about the goods he sells. "It pays in the long run," he said, "and you may lose an order now and again, but the loss is only a temporary one. Your customer will come back when he finds out that you have dealt honestly with him. Learn to know your goods, study them thoroughly so that you can speak emphatically and authoritatively when selling. If you are thoroughly posted on your goods no competitor can draw a herring across the track and kill your sale." Mr. Kinnear also laid stress upon the improvement in the system of selling goods. It was different now to thirty years ago, when it was the almost unbreakable custom to take your customer out for a drink, before making a sale. The reverse is now the case—it seemed to him that this was nothing more or less than buying trade. He didn't believe in buying trade. It was not necessary. Superior knowledge on the part of the traveler, properly conveyed to his customer, will win out in the end.

SUCCESSFUL JOBBERS AND SALESMEN.

No. 34.

F. W. Lamplough & Co., one of Montreal's manufacturers' agents and jobbers, have recently secured the services of F. G. Brenton as their representative on the road.

T. H. BRENTON,
Representing F. W. Lamplough & Co., Montreal.

Mr. Brenton, although a young man of 30, has already made a name for himself in hardware circles.

He successfully represented the Jas. Stewart Mfg. Co. of Woodstock, Ont., in Western Canada for 3½ years, and in that time, by dint of hard work and a pleasant disposition, worked up a very extensive connection among western hardware dealers.

F. W. Lamplough & Co. have the agency for some of the finest lines on the market to-day, and in view of the fact that Mr. Brenton will cover the field from coast to coast, he should soon be well known to the trade, who we feel sure will receive him in the way he deserves.

DEALERS AND THEIR STORES

IMPROVEMENTS AT PORT ARTHUR

Within the last three months the Marks-Clavet-Dobie Co., Port Arthur, have made extensive improvements in the way of reducing the risk of fire on their premises, having installed nine Bowser Oil Pumps, the tanks of which are located in an absolutely fire proof vault in the basement, which is just completed. They have also installed one of Devs' Time Registers, which is working satisfactorily, their staff numbering 35. It is also their intention in the near future to instal an up-to-date electric cash carrier system and electric elevator.

"Our Xmas business far exceeded any previous years," they write. "We had some good windows but unfortunately were unable to have them photographed, through frost caused by a crack in the plate glass. However, in the future, we hope to send you from time to time, photographs of our displays."

"By the way," continues their letter, "we have just completed our contract for the largest freight sheds in the Dominion, for the C.P.R. at Fort William, built of galvanized corrugated iron, 600 feet long by 60 feet wide."

NEW SPORTING GOODS FIRM.

G. A. Brittain, of the Ashdown Hardware Co., and R. J. McKay of the Hingston-Smith Arms Co., Winnipeg, have joined in a partnership under the name of the Western Sporting Goods Co., and will do business in the Stobart block, on Portage Ave. They expect to open for business by the beginning or the middle of February. Messrs. Britton and McKay have been for the past few years among the foremost of Winnipeg sportsmen.

UP-TO-DATE LISTOWEL STORE.

Messrs. Adolph & Bonnett, Listowel, are removing their stock to their new stand, which they have had specially fitted up for them. When they get everything into position they will have the most up-to-date hardware store in the district, says the Listowel Banner. The new stand will give them more than double the capacity for the display of their large stock. The front has been entirely rebuilt, with new show windows, enclosed, over which is placed prismatic glass, furnishing excellent interior light throughout. On one side of the store have been placed 480 drawers, all carefully sampled, so that goods will be displayed and instantly accessible. Large wall show cases have been built in for the display of silverware on one side and tools on the other, and complete apartments for sorting and handling glass built at the end of the shelves. In fact the classification of the whole stock has been systematized so that any article can be got at at once. In the rear and in part of the upstairs are two store houses where surplus stock will be carried. The workshop for the tinsmithing department is located at the rear upstairs. The basement is floored with cement and well lighted, and will be used for the heavier bulk goods.

ONE OF ORILLIA'S STORES.

The Orillia Hardware Company, which is composed of W. E. Anthony, for 20 years employed with Peaker & Son, Brampton, and S. L. Mullett, who was 11 years in Seaforth, and had four years Western experience, a year ago bought the stock and leased the premises of G. H. White, Orillia.

A new brick workshop 25 x 40 feet, has been built, in which their plumbing steam-fitting and metal work branches are conducted. Noteworthy among their contracts during the past season has been the steam heating plant installed in the Methodist Church at Orillia.

The accompanying cut shows a view of the main floor of their store, which is 35 x 120 feet, and is connected by an arch of about 14 feet, to a room which is used exclusively for the display of stoves, tinware and farming tools. Connected by a fire proof door is the oil house, equipped with the latest improved oil tanks. In the rear are three buildings used for storing old stoves, poultry netting, wire, and cement. Beneath the main shop and stove room is a full sized celler, which contains their stock of rope, woodenware, white lead, glass and nails. The second floor is used to display stoves, and for the storage of unsaleable goods.

Messrs. Mullett & Anthony are enterprising advertisers, and strong believers in up-to-date window dressing.

CANADIAN HARDWARE STORES, No. 11—The Orillia Hardware Company's Store.

BIG CHANGE AT VICTORIA.

As recently announced in these columns, Messrs. Griffin & Greenshaw, of the firms of Eakins & Griffin and Randall & Greenshaw, both of Shoal Lake, Man., have purchased the old established business of Nicholles & Renouf, Victoria, B.C. The new firm are remodeling, enlarging and renovating their store, which is located on a splendid corner only a short block from the business centre of the city.

When the B.C. Hardware Company, the firm's business name, is ready for business, their store will be one of the most modern in British Columbia. Two new windows will be installed, one on Broad St., the other on Yates. The entrance will be large and set off by immense pillars on each side. The store will extend the full length of the structure, the office being placed on the second floor at the rear for the purpose of increasing the accommodation. Thus the alterations planned will improve the lighting, making the general apartment more cheerful than at present. They will also increase the floor space, this being necessary on account of the large amount of new stock ordered. As members of the new firm are both experienced men having plenty of enterprise and every confidence in the future of the Canadian West, their success seems assured.

The St. Lawrence Saw and Steel Works Company has been incorporated with a total capital stock of $40,000, to manufacture saws, bits, shanks, axes, shovels, also machinery of all kinds. The chief place of business will be at Sorel, Que., and the directors are, C. D. Pontbriand, J. Pontbriand, H. M. Pontbriand, of Sorel, and T. D. Pontbriand, of Chicoutimi, Que.

RETAIL HARDWARE ASSOCIATION NEWS

Official news of the Ontario and Western Canada Associations will be published in this department. All correspondence regarding association matters should be sent to the secretaries. If for publication send to the Editor of Hardware and Metal, Toronto.

PROTEST AGAINST PARCELS POST

The executive committee of the Ontario Retail Hardware Association urge the trade throughout Canada to protest against the proposed legislation extending the parcels post to include a c.o.d. collection feature. They suggest that every retail dealer mail a letter to the Postmaster-General so that it will arrive in Ottawa about Feb. 1.

The accompanying draft form of letter should not be copied entire, but is suggested as a basis upon which to outline an argument against the proposed legislation. All letters should be written on the firm's business paper. If the local member of Parliament has not been written yet he should be asked to oppose the measure by a letter or petition signed by every retail merchant in each town or village.

THE PRESS IS MUZZLED.

How many hardware dealers in Canada read accounts recently of serious elevator accidents in Toronto department stores? A woman was killed in Eaton's store, but the daily papers merely announced the fact. A crowded elevator crashed down several stories in Simpson's and several had to be taken to the hospital with broken legs and other injuries. As before, the matter was very briefly referred to by the daily press.

As pointed out in these columns before, the leading daily papers in Canada are controlled by capitalists largely interested in the big mail order houses. This fact constitutes a serious danger as the press is becoming an increasing power for good or ill in the life of the people.

THE QUESTION BOX.

Every hardware merchant in Ontario is invited to send short questions on trade matters to the editor of Hardware and Metal. These will be published on the association page before the convention and then answered at the convention by some one selected by the president, a general discussion following, a report of which will be given in Hardware and Metal for the benefit of those unable to attend.

The questions should be sent in at once so that they can be announced before the convention. If time permits questions may also be asked at the convention but those received beforehand will, of course, receive first attention.

PROFIT IN MUTUAL INSURANCE.

The last quarterly statement of the financial condition of the Retail Hardware Mutual Fire Insurance Company of Minnesota is undoubtedly the most satisfactory from the standpoint of the policy-holders of any statement ever issued by the company.

A healthy increase in business over the amount written during the corresponding period of the previous year, is reported and the company was enabled to add 10 per cent. of the premiums received to the reserve fund and then declare a return dividend for the year 1907, of 45 per cent.

The ratio of losses to premiums was 32 per cent., while the ratio of expenses to premiums was 13 per cent., making the total cost of doing business, including losses and expenses, 45 per cent.

The company now has a cash reserve of upwards of $86,000, which is steadily being increased by the addition of a reasonable percentage of the profits each year. The fact that the expense of doing business was only 13% of the amount of premiums received, speaks volumes for the conservative management which the company has enjoyed and is one of the strongest arguments in favor of the mutual form of insurance, when properly conducted.

Draft Form of Letter.

.... 1907

To the Honorable Mr. Lemieux,
Postmaster General, Ottawa.

Dear Sir,—We understand that you are preparing an amendment to the law providing for the existing parcels post system, the said amendment being intended to provide for the collection on delivery of parcels. Such a system would enable mail order houses to send goods c.o.d. to any part of the country in competition with local retailers.

The large departmental stores are already injuring local merchants to a tremendous extent and any change in the postal laws which would favor them would be a great injustice to the thousands of merchants who have their capital invested in the small towns and cities throughout Canada, aiding by their presence, their energy and their taxes in the up-building of the small communities upon which Canada's future prosperity must depend.

We do not ask for any special privilege and protest against the mail-order houses being given any favors. The proposed legislation would tend to reduce the number of retailers, thus destroying competition, depreciating the value of real estate in the towns and having a similar effect upon the farming districts, as well as tending to decrease the opportunities for pleasure enjoyed by the farmer's family. The deserted farm districts of the New England States are an object lesson to avoid.

We trust that you will see the fairness of our argument that the legislation would be detrimental to the continued progress and development of this fair Dominion and we would ask that you reconsider your announced intention of introducing the proposed amendment.

Yours very truly.

They that stand waiting for fortunes, will never wear out their shoes.

STOVES AND TINWARE

IMPORTANCE OF FIREPLACES IN HEATING.

In discussing the question of heating a village dwelling by means of stoves Dr. Harvey B. Bashore states that the halls are usually cold and, in addition, even in the rooms where stoves are placed the floors are from 6 to 8 degrees colder than the temperature 4 or 5 feet above, a fact easily proved by experiment. As a consequence one's feet are just so much colder than one's head and shoulders. These two defects, cold halls and floors, are certainly factors in producing catarrhal inflammation of the throat and nose, if nothing worse. To remove these defects to a minimum it is necessary to alter somewhat the construction of the rooms. Every one knows the value of the open grate, not so much as a heater, but as an equallizer of room temperature, and herein lies our remedy. Every room should have such a grate, or its equivalent, simply an airshaft connected with the chimney and opening into the room at the floor level. An airshaft so arranged and of suitable dimensions, answers almost as well as an open grate and furnishes the means whereby rooms may be heated very well with ordinary stoves.

When a room which has no fireplace is heated, the heated air rises and spreads along the ceiling in a thick cloud, and if a window is opened the warm air rushes out before it has done much good ; if, on the other hand, there is an open grate, some of the hot air escaping up the chimney creates a partial vacuum ; this, consequently, creates in the room a movement towards the opening, and the upper heated air is more diffused about the room, making the temperature more uniform.

The halls, whether they contain a stove or not, should have an airshaft, for it will assist somewhat in "sucking out" the heated air of the adjoining rooms. A small oil heater placed in the lower hall will be of assistance in keeping the hall temperature at the right point.

TIN CANS UNIVERSAL.

"The American tin can has revolutionized the world," said a millionaire tin can manufacturer, recently. "It has belted the globe and conquered the heathen nations in all lands. Last year in Japan I saw American tomato cans doing service as flower pots. In China I found salmon cans from this country used as soup ladles, and up in the Himalayas I saw painted tin cans used as head dresses for some of the idols."

He may have added that American tin cans are used as hand mirrors by South Sea Island belles, and as ornaments dangling from their necks by some of the Patagonian chiefs. In the Arctic regions Peary found Eskimos using tin cans, that were originally packed with a favorite brand of California peaches, for bird traps. The bright flashing of the tin on the top of a pole attracted the curiosity of the birds, so they could be induced to fly close to the hunters. A queer sort of tambourine or drum was found by one explorer in the Terra del Fuego Islands made out of tin cans. A tribe of Indians on the shores of Behring Strait picked up an American tin can which had been washed ashore, and preserved it as a strange relic of the sea. When a white explorer visited them, the Indians were equally divided as to its origin. Some thought it was the skin of a new kind of sea monster, and others a memento from the gods, with powers for good or evil. It was sacredly preserved and admired until the white visitor presented the Indians with half a dozen similar cans after he had eaten the contents of soup and vegetables from them.

NEW McCLARY TRAVELER.

Gordon Ritchie, for a number of years head of the selling staff of the Sherwin-Williams Company, Montreal branch, has severed his connection with this firm and accepted a position on the traveling staff of the McClary Manufacturing Company.

GORDON RITCHIE.
Who will represent the Montreal Branch of The McClary Mfg. Co. in future.

Salesmen, like poets, are said to be "born, not made" and Mr. Ritchie is peculiarly destined to make a success "on the road." He is alert, well educated, speaks English and French fluently, has a quick eye, full of go and snap, and, withal, is unassuming and extremely popular.

Gordon is a young man who will serve the McClary Manufacturing Company with faithfulness and energy.

A. A. Brown, Montreal manager for the McClary Company is to be congratulated upon this, the latest acquisition to his staff of salesmen.

KITCHEN UTENSILS.

The accompanying cut shows one of the many lines manufactured by John Shaw and Sons, of Wolverhampton, Eng., who have but lately undertaken the manufacture of kitchen utensils in aluminum, tin, steel, brass, and copper, to which they have given their well

"Governor Brand" Kitchen Utensils.

known mark "Governor Brand." They are represented in Canada by J. H. Roper, 82 St. Francois Xavier St., Montreal, who will supply any dealer with information, and a fully illustrated catalogue of these goods.

STOVE SUGGESTIONS.

Hardware merchants are seeing the great need of handling a good line of stoves. In fact heavy hardware is becoming more generally handled every year. Most merchants select some good line and secure a monopoly on it, and take great interest in making it a leader. To do so, a guarantee from the manufacturer should be procured, guaranteeing the baking and cooking quality of the stove. A great many things have to be contended with in the sale of stoves

First, the stove should be carefully studied. A trip to the factory would enable the dealer to talk intelligently on the construction of the stove, also to compare it with any weak points on other prominent stoves.

Second—The place where the stove is to be placed should be studied, seeing that the flue is in perfect condition, which is easily done by lighting a piece of paper and placing it in the pipe hole, to see if the draft is all right.

Third—A great many people want stoves on trial. While this method of selling may be all right with parties with whom you are intimately acquainted, it would not be well to give every person that privilege, as some are never satisfied, and may see a very elaborate looking cheap stove, which they think would suit them much better than your more serviceable one. Unless some agreement is made, whereby if the stove is found O.K., the parties would be made to retain it, the dealer will have to take the stove back, explaining that it will be necessary to sell it as second-hand.

A good plan for this time of year is to have a special stove sale, it being much better to sell surplus stock than having money tied up all year, as it could be placed to much better advantage buying new Spring goods. If two different lines are carried, select one and advertise a special sale, not neglecting to quote prices in your ad , also drawing attention to any special feature of the stove.

HARDWARE AND METAL

Established 1888

The MacLean Publishing Co.
Limited

JOHN BAYNE MACLEAN - President

Publishers of Trade Newspapers which circulate in the Provinces of British Columbia, Alberta, Saskatchewan, Manitoba, Ontario, Quebec, Nova Scotia, New Brunswick, P.E. Island and Newfoundland.

OFFICES:

MONTREAL, ... 232 McGill Street
Telephone Main 1255
TORONTO, ... 10 Front Street East
Telephones Main 2701 and 2702
WINNIPEG, ... 511 Union Bank Building
Telephone 3726
LONDON, ENG. . . . . . 88 Fleet Street, E.C.
J. Meredith McKim
Telephone, Central 12960

BRANCHES:

CHICAGO, ILL., ... 1001 Teutonic Bldg.
J. Roland Kay
ST. JOHN, N.B. . . . . No. 7 Market Whart
VANCOUVER, B.C. . . . . Geo. S. B. Perry
PARIS, FRANCE . Agence Havas, 8 Place de la Bourse
MANCHESTER, ENG. . . . . 92 Market Street
ZURICH, SWITZERLAND . . . . Louis Wolt
Orell Fussli & Co.

Subscription, Canada and United States, $2.00
Great Britain, 8s. 6d., elsewhere - 12s

Published every Saturday.

Cable Address { Adscript, London
Adscript, Canada

PROFIT SHARING IN MONTREAL.

The Jas. Walker Hardware Company of Montreal, have adopted the profit-sharing system in the running of their business. There will not be any material difference in the business itself, with the exception that Messrs. Jas. and D. S. Walker will give over the management to F. Max Hill, who for several years was one of the managers of Weed & Co., of Buffalo, N.Y., and, previous to that, connected with the Yale & Towne Manufacturing Company, of New York City. Mr. Hill has the reputation of being one of the best posted hardware men in the country, and the addition of his experience is counted on being quite a factor to the concern.

Messrs. Jas. and D. S. Walker will not wholly disassociate themselves with the business, but will remain as directors, occupying the positions of president and vice-president respectively. It has always been the policy of this firm to recognize faithful and capable service, and the opportunity to enter the company has been readily accepted by its employes. Several of the oldest employes have been given a block of stock in the new company, and the remainder will participate in the profits.

This is the 50th anniversary of the firm, it having been established in 1857 by Jas. Walker, the father of the above mentioned gentlemen, and strange to say, the business has never moved more than one block from where it started. It at present occupies the building next the Bank of Toronto, on St. James St., and extends right through to Notre Dame St., where the shipping is done. This property has recently been acquired by the firm.

Although the store is well and finely fixed up—being considered one of the best laid out hardware stores on the continent—Mr. Hill, who is of a progressive nature, contemplates many changes, which will not only add greatly to the general appearance of the store, but will facilitate the filling of orders. Paints and several other lines will also be added and it is confidently expected that the united and loyal interest of all connected with the company will insure for the customers of the firm the best service the trade can offer.

EXHIBITION IN MONTREAL.

"Do the Citizens Want an Annual Exhibition Held in Montreal?" is the title of a booklet issued by the Montreal Business Men's League, which was incorporated and established six years ago—one of its objects being to advertise and bring people to Montreal.

An exhibition would no doubt attract a large number of visitors to the city and prove of great material advantage to the citizens. The position of Montreal is more advantageous than that of almost any city in Canada, for the bringing in of visitors from a distance. The Province of Quebec has a distinct link with the densely populated Eastern States. Thousands of the residents of that part of the United States would avail themselves of the incentive offered of visiting Canada upon such an occasion. Interest in the country, relatives and friends in Canada, cheap railway fares and former residence here, would all combine to bring to Montreal a vast number of well-to-do people, whose "ready-spending" characteristics would greatly swell the profits of merchants, hotels, boarding houses, etc. People go to the Toronto exhibition from all over America. Why not Montreal?

An exhibition in Montreal would not alone bring in a direct and important new revenue to the citizens in every conceivable way, but would serve to advertise the city abroad. Toronto exhibition has made that city widely known in all parts of the world, and has, in a great extent, advertised Canada as a progressive country.

Advertising pays, but more advertising pays better. If the Toronto exhibition helps to advertise the country in any way, it's only reasonable to suppose that an exhibition in Montreal would help along the good work the Toronto business men are already doing.

There is no part of the Dominion in which exhibitions in the agricultural districts are more successful, or better attended by the community in general, throughout the province, than in Quebec and it's up to Montreal to organize such an exhibition, to finance and manage it as a public enterprise for the benefit of the city, and of the province and Dominion at large.

SIC 'EM.

"The trouble with some business men," remarked a manufacturer who is well known for his enterprise and business enthusiasm, "is that they are lacking in the quality of energy. A man cannot do anything in business unless he has got some energy about him, and the more he has the better. There are some dogs that will bark at a tramp but the moment the tramp makes a movement towards them they will turn tail and get around the corner of a house. Take a fox terrier, on the other hand, and sic him on, and he gets right down to business and sails right in. When I hear business men talking in gloomy tones and displaying apathy, I just say, "Sic 'em. Sail in and tackle your difficulties."

SIMPLE ACCOUNT SYSTEM.

Although there is no business too small for the loose leaf system to be adapted to some readers may be interested in the simple account system adopted by Bruce and Sanderson, a Toronto retail firm who find it impossible to do a straight cash trade.

By using counter checkbook an itemized statement accompanies each order and a copy is kept. These copies are kept on file in a quarter cut oak filing cabinet. An ordinary Shannon file does the rest. Alphabetically arranged on this file is an ordinary bill head of the firm for each customer and under it a sheet of plain paper. Each morning the totals of the orders sent out the day before are charged on these bill heads and a piece of carbon paper gives a copy on the plain sheet below. Suppose a customer comes in to pay. His account is made and can be produced in an instant. If he pays in full the bill on the file is receipted and handed to him and the copy is destroyed as being of no further use. The copies of orders up to that date are also destroyed as being of no further use. If the customer pays something on account, a credit is made on the bill head and that is given him as a receipt, showing a balance due. A new bill head and clear sheet are put on the file and the balance brought forward. As soon as enough is paid to clear this balance the copy of the old bill is thrown away.

The invoices are kept in a filing cabinet reserved for them. All invoices of each firm from whom they buy are kept together arranged in order of date. In this way the wholesale dealings of the firm for years can be kept in one small cabinet without a scratch of a pen.

THE BUSINESS SITUATION.

By B. E. Walker, President Canadian Bank of Commerce.

While we are enjoying an extraordinary prosperity, there are signs about us of a strain which must bring trouble if they are disregarded. We are a borrowing country, and we cannot be reminded of this too often. As we fix capital in new structures, public or private, railways, buildings, etc., some one must find the capital in excess of what we can ourselves provide out of the saleable products of our labor. The number of countries willing to buy our securities has been steadily increasing, but we must not be blind, as we sometimes seem to be, to the fact that our power to build depends largely on whether these countries have surplus capital to invest. By means of the cable the trading nations of the world have been brought very near together, and while nations of the world have been averted, and the adjustment of capital to the world's needs has been greatly improved, still for the same reason world-wide trouble in the money markets sometimes arises with a suddenness which is alarming to those at least who are not watching for the signs. We are passing through such a period just now, happily without a general breakdown, but unless we mend our ways we are not likely to escape a similar or still worse condition next autumn which may wreck our fair prosperity. Europe is bearing the enormous cost of two great wars, both in the loss of capital actually destroyed and also in the loss to individuals from the decline in the values of the national securities of the countries interested in the wars. And since these wars, losses on an unexampled scale have occurred by earthquakes and fire. The volume of trade and the unusual amount of building in many countries have at the same time vastly increased the amount of capital required. This has been accompanied by a steady rise in prices throughout the world, and by a most pronounced and widespread advance in the scale of personal expenditure. It is true that it has also been accompanied by the greatest production of gold and of other commodities, but the effect of the various influences has naturally been to put upon the money markets a strain which has only just failed to cause a general breakdown of credit. To make the outlook still more serious, the United States, and other less important countries, including Canada, contemplate expenditures on a very large scale for railway and other building. This, then, is a time for every prudent man to survey carefully his financial position. If he has debts he should consider how he will pay them if he should have to face world-wide stringency in money. Has he assets which the world needs for daily use, or assets which will sell only when the sun is shining? If he is happily in easy conditions as to debt, he will, if he is wise, consider every circumstance arising in his business which tends towards debt instead of towards liquidation. As for those who are plunging in real estate at inflated prices and in mining stocks, nothing, we presume, but the inevitable collapse which follows these seasons of mania will do any good.

NO SPECIAL PRIVILEGES.

Both jobbers and retailers are using their influence to induce members of Parliament and the Postmaster General to see the unfairness of the proposed parcels post legislation. To allow country postmasters to become collecting agents for the big mail order houses is to grant a special privilege to merchants who already have great advantages over their smaller competitors.

The retail merchants in the United States have suffered severely from the unfair competition of their big rivals and are up in arms against further extensions of the special privileges given the departmental store by the U.S. Postal Department. They were successful in heading off bills brought into the last Congress by Congressmen Hearst and Henry and are already preparing to make a stiff fight against similar legislation likely to be introduced into Congress this year. The big mail order houses have the cash, but the retail merchants have the votes and influence, which can put out of public life many Congressmen and Senators unless they stand for the interests of their constituents against the big city houses which pay no taxes in country districts and shirk their fair share in their city headquarters.

Two of the most recent developments in the States are the proposals to reduce the rate on parcels of merchandise from one cent per ounce to three cents per pound, and the offer of W. D. Boyce, newspaper publisher, Chicago, to buy out the Postal Department and operate it as a private concern. Boyce represents the big newspaper men who are making millions through departmental store advertising. His proposition is as follows:

"To turn over the postoffice business to a $50,000,000 private corporation under full Government regulation. To reduce by one-half all postal rates, establish rural postal express and apply business methods throughout. To pay the Government rental for postoffice quarters, and charge its regular rates for its postal business. To place in charge a well-known railroad traffic expert to whom the place has been offered at $30,000 annually. To eliminate all sinecures, politics, and the deficit. To pay the Government all profits above 7 per cent. on capital."

If Canadian retailers wish to preserve their rights and prevent the undue growth of the mail order octopus they must be alive and up and doing. In this issue the executive of the Ontario Retail Hardware Association suggest a letter that every dealer should forward to the Postmaster General about Feb. 1, inducing as many retailers as possible to sign the letter or forward a similar one. The suggestion is a good one and should be acted upon.

NEW WHOLESALE AT BRANDON.

The incorporators of the recently organized Hanbury Hardware Company, Brandon, are: John Hanbury, J. Brown, E. Johnston, J. Caldwell, and A. E. Carmichael. The last named was with Marshall Wells Company since their incorporation at Winnipeg and is now manager of the Hanbury Hardware Company. The new company will do wholesale business and have a modern warehouse almost ready for occupancy. The building has five stories, 80x100 feet, being on a spur track. They expect to be open for business March 1st.

MONTREAL BOARD OF TRADE.

George Caverhill has been nominated for the presidency of the Montreal Board of Trade; T. J. Drummond, for the first vice-presidency; C. B. Esdaile, as treasurer; Messrs. J. N. Dougal, and Alex. McLarin, for membership on the Board of Trade Council and Farquhar Robertson for the 2nd vice-presidency. Nominations close Jan. 18, and elections will be held Jan. 29.

BIG YEAR IN SKATES.

Orders are practically all in and filled for skates for this season; in fact travelers are already on the warpath again with new lines, and with improvements here and there, which will be appreciated by hockey players and in fact by all who use skates. A number of orders have already been booked, and skate manufacturers seem to be off to a good start.

Mr. J. S. Bowbanks of the Starr Manufacturing Co., Dartmouth, N.S., stated that many hardware dealers in Ontario had the impression that sales would be very light this season on account of the mild winter last year. He pointed out, however, that their firm did business in all parts of the globe, where they had any skating, and the mild winter in Canada might mean a very cold winter in other countries. He said his firm had made more skates this season than ever before, and had sold practically every one of them.

Hardware Trade Gossip

Ontario.

D. E. Rudd, dealer in creamery supplies, was elected alderman in Guelph, last Monday.

L. A. Whitmore, hardware merchant of Edgley, called on the Toronto jobbers this week.

James T. Stewart, hardware merchant of Manchester, has sold his business to James Young.

John Kelly, hardware merchant, was elected mayor of Oakville on Monday, after an exciting contest.

J. J. Coffee, hardware merchant, Barrie, was elected to the Barrie Council in the elections on Monday.

F. W. Moffatt, stove manufacturer, Weston, was one of those elected to the Weston Town Council, last Monday.

G. A. Binns, hardware merchant, Newmarket, was elected to the Newmarket Council at the elections this week.

Alderman S. Clark, master plumber, Hamilton, was re-elected to the Hamilton City Council at the elections this week.

Joseph J. Bouchard, manager of the Star Iron Company, Montreal, was a visitor in Toronto on Saturday of last week.

J. H. Crow, hardware merchant, Welland, was elected mayor of that booming burg by a substantial majority last Monday.

P. Chappelle, hardware merchant of Brown Hill, was a visitor in Toronto this week, calling on the wholesale hardware trade.

Samuel Ryding, plumber of Toronto Junction, was one of the unsuccessful candidates for the Mayoralty of that town in the elections last Monday.

J. B. Campbell, president of the Acme Can Works of Montreal, passed through Toronto last Saturday on his return east from a trip through Western Canada.

Jos. Wright, of Bennett and Wright, Toronto, who sustained a severe wrench to his leg last Saturday, though still under the doctor's care, is progressing favorably.

H. N. Joy, of the Joy Manufacturing Company, stove manufacturers, Toronto. received serious injury to his head while skating on New Year's night. He is recovering slowly.

T. E. Ransley, representing Greene. Tweed Company, New York, manufacturers of Palmetto packing, attended the conference of Canadian Fairbanks salesmen in Toronto this week.

Mount Forest has decided not to abolish the Water Commission, of which W. G. Scott, hardware merchant, is a leading member. The vote took place at the municipal elections last Monday.

Alderman Samuel Stevely, hardware merchant, and Alderman Wm. Scarlett, of Wortman & Ward, woodenware manufacturers, London, were elected in the municipal elections in that city last Monday.

O. M. Hodson, Toronto representative of H. R. Ives & Company, Montreal, spent several days in Montreal last week, conferring with other salesmen and managers of the company. George Davidson, Woodstock, and Frank Griffin, Stratford, were the other Ontario men at the meeting.

Frank Adams, who has filled the position of secretary for the Peterborough Lock Manufacturing Co. for the past thirteen years, and who severs his connection with that company this week, was last Saturday made the recepient of a handsome gold watch from the directors of the company.

A. G. Buckham, manager of the retail department of the Hudson Bay Company's store at Winnipeg, was a caller on Hardware and Metal in Toronto during the past week. Mr. Buckham reports a very good holiday business in Winnipeg. He has been visiting his old home at Brampton, Ont.

Will Cluff, of Cluff Bros., Ontario selling agents for Warden King & Son, Montreal, met with a rather serious accident a few days ago. While operating his automobile the crank backed up on him and while not breaking any bones, tore the ligament of the arm and put it out of service for two or three weeks time.

Andrew Riddell, stove dealer and plumber, St. Catharines, was re-elected mayor of that city last Monday by about 400 majority. The contest was a very keen one, owing to political disagreements, but Mr. Riddell was supported by the leading Conservatives, as well as Liberals. Mr. Riddell was a visitor in Toronto on Wednesday.

J. Taylor Webb, manager of the Winnipeg branch of the Thos. Davidson Company, Montreal, was a caller on Hardware and Metal, on Tuesday. Mr. Webb is an enthusiastic westerner, and has boundless faith in the trade possibilities of Western Canada. He has been attending a staff conference at Montreal, and incidentally renewing old acquaintanceships in Toronto.

Quebec.

Adolphe Huot, Sorel, was in Montreal during the week.

Geo. Larose, Beloieuil, Que., was seen in Montreal this week.

N. McGillis, of Lancaster, was a visitor in Montreal during the week.

A. G. Eastman, Sutton, Que., was in Montreal on business a few days ago.

O. Pauze, L'Epiphanie, was a visitor at some of the Montreal wholesale houses.

N. Curry, of Rhodes, Curry & Co., Amherst, N.S., was in Montreal during the week.

J. D. McBride, Cranbrook, B.C., is in Montreal, looking up the jobbers with whom he does business.

H. R. Picketts, of Emerson & Fisher, St. John, N.B., has joined the traveling staff of Lewis Bros., Montreal.

J. W. Harris, of J. W. Harris Co., Montreal, who has been confined to the house through illness, has returned to his desk.

R. B. Cherry, representing Sargent & Co., manufacturers of builders' hardware, New York, is calling on the Montreal jobbers.

Mr. Smith, of the Pike Manufacturing Co., Pike, N.H., manufacturers of sharpening stones, is in Montreal on business for his house.

The Canadian branch of The Gillette Safety Razor Co., who were burned out lately, have opened up an office in the Mortimer Building, 2 St. Antoine St., Montreal, where all correspondence should be addressed in future.

G. M. Edwards, of Henderson & Potts, paint manufacturers, Montreal, who has been suffring lately from nervous prostration, has returned from a trip to Boston, where he has been trying to forget business worries.

M. Purvis, of Calgary, spent a few days in Montreal, and, like all western men, speaks most highly of the latter city, but prefers the beautiful climate and pure "ozone" of his western home. He has his office in Calgary with Lintan and Hall.

H. G. Hollis, manager of the New York branch of the Lufkin Rule Co., was a visitor at the Montreal office of Hardware and Metal. Mr. Hollis, although he claims to be 60 years of age, doesn't look the part, is still hale and hearty, and "one of the boys."

Western.

The wife of Louis Moody, general manager of the Canadian Brass Works, Winnipeg, died on Jan. 2nd at that place.

Mr. McKean, foreman in the heavy hardware department of Geo. D. Wood & Co., Winnipeg, was waited on by the members of his department and presented with a handsome gold locket, suitably engraved as a token of the universal esteem and respect in which he is held by his associates.

T. Arthur Kennedy, Winnipeg City, traveler for the McClary Manufacturing Company, before leaving on a trip for the East last week, was waited on by the members of the office and warehouse staff, and after an appropriate address, he was presented with a handsome oak Morris chair.

Markets and Market Notes

(For detailed prices see Current Market Quotations, page 70.)

THE WEEK'S MARKETS IN BRIEF.

MONTREAL.

Iron Pipe—Advanced.
Turpentine—Two cents higher.
Enamelware—New lists show advance.

TORONTO.

Plumbers' Enamelware—Ten per cent. advance.
Iron Pipe—New lists show advances.
Lead Traps and Bends—Advanced to 40 off.
Hot Water Boilers—No immediate change.
Boiler Pitts—Advanced from 35 to 25 per cent.
Furnaces—Five per cent. higher.
Registers—Advanced 5 per cent.
Kitchen Enamelware—Prices withdrawn.
Shovels—Advanced $1 doz. on net price.
Step Ladders—Advanced 1c. per ft.
Rivets and Burrs—Advanced to 25 per cent.
Copper Bottom Boilers and Kettles—Advanced 5 per cent.
Copper Pitts—Advanced 5 per cent.
Factory Milk Cans—Advanced 5 per cent.
Old Material—Several advances.

Montreal Hardware Markets

Office of HARDWARE AND METAL,
232 McGill Street,
Montreal, Jan. 11, 1907.

With the colder weather, business has picked up considerably, and the American travellers are beginning to find their way to our side of the border.

Dealers who place their orders for Spring goods early, are likely to save themselves quite a lot of cash and worry, as manufacturers are away behind in their deliveries, and prices on all lines are expected to go considerably higher within the near future.

Cow ties and sleigh bells are reported among the slowest movers this week.

Sporting goods have also quieted down slightly to what they have been during the holidays.

New prices are daily looked for on building paper.

Wire nails remain firm, but an advance in the price of horse nails is expected shortly.

It is also rumored that steel squares and planes (Stanley's), are in line for an advance.

AXES.—We quote: $7.60 to $9.50 per dozen; double bitt axes, $9.50 to $12 a dozen; handled axes, $7.50 to $9.50; Canadian pattern axes, $7.50 a dozen. follows: No. 3, $1.25; No. 2, $1.50; No. 1½, $1.90 a dozen; adze handles, 34-inch, $2.20 a dozen; pick handles, No. 2, $1.70; No. 3, $1.50 a dozen.

LANTERNS. — Our prices are: "Prism" globes, $1.20; Cold blast, $6.50; No. 0, Safety, $4.00.

COW TIES AND STALL FIXTURES are moving but slowly. We still continue to quote: Cow ties, discount 40 per cent. off list; stall fixtures, discount 35 per cent. off list.

SLEIGH BELLS.—There has been a slight falling off during the past week, Our prices still remain as follows: Back Straps, 30c. to $2.50; body straps, 70c. to $3.50; York Eye bells, common, 70c. to $1.50; pear shape, $1.15 to $2; shaft gongs, 20c. to $2.50; Grelots, 35c. to $2; team bells, $1.80 to $5.50; saddle gongs, $1.10 to $2.60.

RIVETS AND BURRS. — Copper rivets and burrs advance slightly. Following are the quotations: Best iron rivets, section, carriage, and wagon box black rivets, tinned do., copper rivets and tin swede rivets, 60, 10 and 10 per cent.; swede iron burrs are quoted at 60 and 10 and 10 per cent. off new lists; copper rivets with the usual proportion of burrs, 25 per cent. off, and coppered iron rivets and burrs in 5-lb. carton boxes at 60 and 10 and 10 per cent.; copper burrs alone, 15 per cent., subject to usual charge for half-pound boxes.

HAY WIRE.—Prices remain firm. Our quotations are: No. 13, $2.55; No. 14, $2.65; No. 15, $2.80; f.o.b. Montreal.

MACHINE SCREWS.—Trade is about as usual. We quote: Flat head, iron, 35 per cent.; flat head, brass, 35 per cent.; Felisterhead, iron, 30 per cent.. Felisterhead, brass, 25 per cent.

BOLTS AND NUTS—Scarce. Our discounts are: Carriage bolts, 3 and under, 60 and 10; 7-16 and larger, 55 p.c.; fancy carriage bolts, 50 p.c.; sleigh shoe bolts, ⅜ and under, 60 per cent.; 7-16 and over, 50 per cent.; machine bolts, ⅜ and under, 60 per cent.; 7-16 and larger, 55 per cent.

HORSE NAILS.—Advance expected shortly. Discounts are as follows: C brand, 40, 10 and 7 per cent.; M.R.M. Co., 55 per cent.; P.B. brand, 55 per cent.

WIRE NAILS.—The market is now firm at $2.30 per keg base f.o.b. Montreal.

CUT NAILS—We continue to quote: $2.50 per 100 lbs.; M.R.M. Co., latest, $2.30 per keg base. f.o.b. Montreal.

HORSESHOES. — Our prices are: P.B. new pattern, base price, improved pattern iron shoes, light and medium pattern, No. 2 and larger, $3.65; No. 1 and smaller, $3.90; snow pattern, No. 2 and larger, $3.90; No. 1 and smaller, $4.15. Light steel shoes, No. 2 and larger, $4; No. 1 and smaller, $4.25; featherweight, all sizes, No. 0 to 4, $5.60. Toeweight, all sizes, No. 1 to 4, $6.85. Packing, up to three sizes in a keg, 10c. per 100 lbs. More than three sizes, 25c. per 100 lbs. extra.

BUILDING PAPER.—New prices expected daily.

CEMENT AND FIREBRICK.—No change in prices. We are still quoting: "Lehigh" Portland in wood, $2.54, in cotton sacks, $2.39; in paper sacks, $2.31. Lafarge (non-staining) in wood, $3.40; Belgium, $1.60 to $1.90 per barrel; ex-store, American, $2 to $2.10 ex-cars; Canadian Portland, $2 to $2.05. Firebrick, English and Scotch, $17 to $21, American $30 to $35; White Bros.' English cement, $1.80 in bags, $2.05 in barrels in round lots.

COIL CHAIN.—Last week's advance still holds. We are quoting as follows: 5-16 inch, $4.40; 3-8 inch, $3.90; 7-16 inch, $3.70; 1-2 inch, $3.50; 9-16 inch, $3.45; 5-8 inch, $3.35; 3-4 inch, $3.20; 7-8 inch, $3.10; 1 inch, $3.10.

SHOT—Our prices remain as follows: Shot packed in 25-lb. bags, ordinary drop AAA to dust, $7 per 100 lbs.; chilled, No. 1—10, $7.50 per 100 lbs.; brick and seal, $8 per 100 lbs.; ball, $8.50 per 100 lbs. Net list. Bags less than 25 lbs., ½c. per lb. extra, net; f.o.b. Montreal, Toronto, Hamilton, London, St. John, Halifax.

AMMUNITION—The normal business is being done. Our prices are: Loaded with black powder, 12 and 16 gauge, per 1,000, $15; 10 gauge, per 1,000, $18; loaded with smokeless powder, 12 and 16 gauge, per 1,000, $20.50; 10 gauge, $23.50.

GREEN WIRE CLOTH.—We continue to quote: In 100 foot rolls, $1.62½ per hundred square feet, in 50 feet rolls, $1.67½ per hundred square feet.

Toronto Hardware Markets

Office of HARDWARE AND METAL,
10 Front Street East,
Toronto, Jan. 11, 1907.

Jobbers report a great falling off in business during the last week, as many hardware merchants are taking stock. All orders being booked now are for Spring shipment. A great number of merchants were caught on the advance in building paper, and the number of advances which have so rapidly taken place will help the tendency to order goods farther ahead.

A number of minor changes are quoted this week, with a probable further advance in a great many lines. Copper rivets and burrs advanced from 27½ per cent. to 25 per cent. with the market in a very unsteady condition and further advance looked for. Screws are also advancing, but to what extent is not yet known. Step ladders have also advanced 1c. per foot, and are now quoted at 12c. All shovels have advanced $1.00 per dozen on list price, with further advance likely in the near future.

Quotations on all enamelware have been withdrawn. Japanned ware has advanced 5 per cent., copper pitts from 33 1-3 per cent. to 25 per cent., copper bottom tea kettles and boilers from 35 to 30 per cent., and factory milk cans from 40 to 35 per cent., Furnaces and registers have also been advanced 5 per cent. as reported in our metal markets.

AXES AND HANDLES.—Travellers are booking orders for Fall shipment.

GARDEN TOOLS.—While these various lines have nearly all been booked for Spring shipment, quite a number of orders are being received from merchants who had not ordered. An advance is expected.

WASHING MACHINES, ETC. — It is expected more washing machines will be sold this year than ever before, as so many different lines are on the market, a great number of which will be found highly satisfactory. Wringers of Canadian make have so greatly improved of late, that large orders are coming in, business in this line greatly increasing.

SPORTING GOODS.—While the last week very few orders have been received, cold weather will greatly enliven that line.

CHAIN—We still continue to quote: 5/16 in. $3.50; 9-16 in., $3.55; 1/2 in., $3.60; 7-16 in., $3.85; 3-8 in., $4.10; 5-16 in. $4.70; 1-4 in. $5.10.

RIVETS AND BURRS.—An advance has been made on these articles, which are now quoted at 25 per cent. for copper and 60, 10 and 10 for iron and tinned.

SCREWS.—Are advancing, but figures are not yet obtainable.

BOLTS AND NUTS.—Demand is very good, but the supply is very hard to procure.

TOOLS — Carpenters' tools, hand saws, etc., are in great demand, merchants buying largely in view of a further advance.

SHOVELS.—Not much doing at present, but a banner year is reported in the sale of snow shovels.

LANTERNS.—Very little demand at present, as most merchants are stocked up.

EXTENSION AND STEP LADDERS —An advance is quoted on step ladders, which are now 11c. per foot for 3 to 6 feet, and 12c. per foot for 7 to 10 foot ladders. Waggoner extension ladders, 40 per cent.

POULTRY NETTING—We quote: 2-inch mesh, 19 w.g., discount 50 and 10 per cent. All others 50 per cent.

WIRE FENCING.—All galvanized and plain wire is selling very briskly, even at advanced prices.

OILED AND ANNEALED WIRE — (Canadian)—Gauge 10, $2.31; gauge 11, $2.37; gauge 12, $2.45 per 100 lbs.

WIRE NAILS.—No further advance on nails is yet to hand. Stocks are rapidly being repleted.

CUT NAILS—We quote $2.30 base f.o.b. Montreal, Toronto 20c higher.

HORSENAILS—"C" brand, 40, 10 and 7½ off list. "M.R.M." brand, 55 per cent. off. "Monarch" brand, 50 and 7½ off. "Peerless," 50 per cent. off.

HORSESHOES—Our quotations continue as follows: P.B. base, $3.65; "M.R.M. Co., latest improved pattern" iron shoes, light and medium pattern, No. 2 and larger, $3.80; No. 1 and smaller, $4.05; snow, No. 2 and larger, $4.05; No. 1 and smaller, $4.30; light steel shoes, No. 2 and larger, $4.15; No. 1 and smaller, $4.40; featherweight, all sizes, 0 to 4, $5.75; toeweight, all sizes 1 to 4, $7. Packing, up to three sizes in a keg, 10c. per 100 lbs. extra; more than three sizes in a keg, 25c. per 100 lbs. extra.

BUILDING PAPER.—Orders are now being booked at the advanced price for Spring shipments.

CEMENT—We quote: For carload orders, f.o.b. Toronto, Canadian Portland, $1.95. For smaller orders ex warehouse, Canadian Portland, $2.05 upwards.

FIREBRICK—English and Scotch fire brick $27 to $30; American low-grade, $23 to $25; high-grade, $27.50 to $35.

HIDES, WOOL AND FURS.—The supply of hides is in excess of the demand but the course of the market is uncertain. Prices are slightly lower than a week ago.

HIDES, WOOL AND FURS—There are lots of hides coming in, and the market is pretty steady. Nothing is doing in wool, and very few furs are coming. There seems to be a shortage of furs, or the trappers are very late getting to work. Mink is very scarce, and lynx and fox and skunk are also advanced.

Hides, inspected, cows and steers, No. 1	..	0 11
" " " " No. 2	....	0 10
Country hides, flat, per lb., cured	....	0 10
" " " green	....	0 09
Calf skins, No. 1, city	....	0 12
" " No. 1, country		0 11
Lamb skins	1 15	1 25
Horse hides, No. 1	3 50	3 75
Rendered tallow, per lb.	0 05½	0 05½
Pulled wools, super, per lb.	....	0 25
" " extra	....	0 27
Wool, unwashed fleece	....	0 15
" washed fleece	0 24	0 25

FURS

	No. 1.	Prime
Raccoon	....	1 50
Mink, dark	5 00	6 00
" pale	2 50	3 50
Fox, red	8 10	3 50
" cross	3 00	7 00
Lynx	5 00	7 00
Bear, black	....	17 00
" cubs and yearlings	....	5 00
Wolf, timber	....	2 75
" prairie	....	1 25
Weasel, white	....	0 60
Badger	....	1 60
Fisher, dark	5 00	6 00
Skunk, black	....	1 50
" " short striped	....	1 00
" " long striped	...	0 50
Marten,	3 50	20 00
Muskrat., fall	....	0 16
" winter	....	0 20
" spring	..	0 23
" western	0 12	0 16

Montreal Metal Markets

Office of HARDWARE AND METAL,
232 McGill Street,
Montreal, January 11, 1907

Everywhere, we get the same report, business excellent, with no appearance of a let-up, in the near future, at least.

Copper remains unchanged with the market very firm.

Ingot tin weakened slightly during the week, but this was presumably only a fluctuation, as the metal is again back at its old figures.

Zinc spelter shows an advance of 25 cents per 100 pounds over last week's quotations.

Galvanized and black sheets are very firm. All the makers are so full of orders, that it will be difficult to get anything like prompt deliveries next Spring.

COPPER.—Remains unchanged. Market very firm with an advance in the near future. We quote: Ingot copper, 26c to 26½c; sheet copper, base sizes. 34c.

INGOT TIN.—This metal weakened slightly during the week, but is again back at the old figures. We still quote: 46½ to 47c. per pound.

ZINC SPELTER.—Shows an advance of 25 cents per 100 pounds over last week's quotations. Our figures now are: $7.50 to $7.75 per 100 pounds.

PIG LEAD. — No change over last week's prices. We still quote: $5.50 to $5.60 per 100 pounds.

ANTIMONY—Our price remains 27½ to 28 cents per pound.

PIG IRON.—Remains the same. Our prices are: Londonderry, $24.50; Carron, No. 1, $24.50; Carron, special, $23.50; Summerlee. No. 2. selected, $25.00; Summerlee. No. 3, soft, $23.50.

BOILER TUBES.—Still hold at last week's advance. We quote: 1½ to 2 inch. $9.50; 2¼ inch, $10.35; 2½ inch, $11.50; 3 inch. $13.00; 3½ inch, $17.00; 4 inch. $21.50; 5 inch, $45.00.

TOOL STEEL—Stocks are moving very freely. We quote Colonial Black Diamond, 8c. to 9c.; Sanderson's, 8c. to 45c., according to grade; Jessop's, 13c.; Jonas & Colver's, 10c. to 20c.; "Air Hardening," 65c. per pound; Conqueror, 7½c.; Conqueror high speed steel, 60c.; Jowitt's Diamond J., 6½c. to 7c.; Jowitt's best, 11c. to 11½c.

MERCHANT STEEL.—Our prices are as follows: Sleigh shoe, $2.25; tire, $2.40, spring, $2.75; toecalk, $3.05; machinery iron finish, $2.40; reeled machinery steel, $2.75; mild, $2.25 base and upwards; square harrow teeth $2.40; band steel, $2.45 base. Net cash 30 days. Rivet steel quoted on application.

COLD ROLLED SHAFTING — Recent advance still holds good. Present prices are: 3-16 inch to ¼ inch, $7.25; 5-16 inch to 11-32 inch, $6.20; ⅜ inch to 17-32 inch, $5.15; 9-16 inch to 47-64 inch, $4.45; ¾ inch to

17-16 inch, $4.10; 1½ inch to 3 inch, $3.75; 3⅛ to 3 7-16 inch, $3.92; 3½ inch to 3 15-16 inch, $4.10; 4 inch to 4 7-16 inch, $4.45; 4½ inch to 4 15-16 inch, $4.80. This is equivalent to 30 per cent. off list.

GALVANIZED IRON.—Firmness prevails in the market. We are still quoting Queen's Head, 28 gauge, $4.60 to $4.85; 26 gauge, $4.45; 22 to 24 gauge, $3.90; 16 to 20 gauge, $3.75; Apollo, 28 gauge. $4.45 to $4.70; 26 gauge, $4.30; 22 and 24 gauge, $3.75; 16 and 20 gauge, $3.60; Comet, 28 gauge, $4.45 to $4.70; 26 gauge, $4.30 to $4.45; 22 and 24, gauge, $3.75 to $4.00; $16 to 20 gauge, $3.60 to $3.85; Fleur-de-Lis, 28 gauge, $4.45 to $4.70; 26 gauge, $4.30; 22 and 24 gauge, $3.75; 16 to 20 gauge, $3.60; Gorbals "Best Best," 28 gauge, $4.45; Colborne Crown, 28 gauge, $4.45; 26 gauge, $4.30; 24 gauge, $3.75. In less than case lots, 25c extra.

OLD MATERIAL—We quote: Heavy copper, 19c per lb.; light copper, 16c per lb.; heavy red brass, 16½ to 17c per lb.; heavy yellow brass, 14c per lb.; light brass, 10 to 10½c per lb.; tea lead. 4c per lb.; heavy lead, 4½c per lb.; scrap zinc. 4c per lb.; No. 1 wrought. $16.00 to $17.00 per 100 lbs.: No. 2 wrought, $6.00 per 100 lbs.; No. 1 machinery, $17.00 per 100 lbs.; stove plate, $13.00 per 100 lbs.; old rubber, 9½c per lb.; mixed rags, 1 to 1½c per lb.

Toronto Metal Markets

Office of HARDWARE AND METAL,
10 Front Street East,
Toronto, Jan. 11, 1907

Price changes this week are few, the only one to report in sheet, bar or ingot metals being a decline of a cent in ingot tin. Boiler pitts, however, have advanced from 35 to 25 per cent. and several advances have been made in old material, notably in copper, brass, lead and rubber.

Another important change is an advance of 5 per cent. on furnaces and registers, a uniform discount of 45 per cent now covering furnaces and 70 per cent. on registers. Stoves and hot water boilers remain unchanged.

Conditions are similar to those of a week ago. Stock taking is interfering with business, only a seasonable trade being done. Prices remain firm, however, with the exception of ingot tin.

To add spice to the market this week, that old story of the United States Steel Company establishing a Canadian plant at Sandwich has been revived. It is said that 1,000 acres has been bought, and work will be commenced within a week. Most men in touch with the metal markets will take these reports with a grain of salt, however, as the advantages of the corporation coming into the Canadian field are far from apparent. They already have a fair chance after structural steel business, they can ship in tin plate free and the British preference on black and galvanized sheets only amounts to about 3 per cent. when freight rates are deducted. Until work on the new plant has actually commenced, Canadian producers need not get excited over the possibilities of increased competition.

PIG IRON.—Hamilton, Midland and Londonderry are off the market, and Radnor is quoted at $33 at furnace. Middlesborough is quoted at $24.50, and Summerlee, at $26 f.o.b., Toronto.

BAR IRON.—Bar continues hard to get and we still quote: $2.30, f.o.b., Toronto with 2 per cent. discount.

INGOT TIN.—Weakness has developed, and we now quote 45 to 46 cents per pound.

TIN PLATES—Market continues active and prices firm at recent advances.

SHEETS AND PLATES.—Apollo galvanized sheets have been advanced ten cents, and are now quoted at $4.70 for 10 3-4 ounce. Active buying is reported with prices steady. Stocks are light.

BRASS.—We continue to quote: 27½ cents per pound for sheets.

COPPER.—Conditions are unchanged. Stocks are low, buyers need supplies, and prices are very stiff. We now quote: Ingot copper $26 per 100 lbs., and sheet copper $31 to $32 per 100 lbs.

LEAD—Market is active and prices very firm. We quote: $5.40 for imported pig and $5.75 to $6.00 for bar lead.

ZINC SPELTER—Stocks continue light with market firm and active. We quote 7½c. per lb. for foreign and 7c. per lb. for domestic. Sheet zinc is quoted at 8 1-4c. in casks, and 8 1-2c. in part casks.

BOILER PLATES AND TUBES—We quote: Plates, per 100 lbs., ¼ in. to ½ in., $2.50; ⅜ in., $2.35 heads, per 100 lbs., $2.75; tank plates, 3-16 in., $2.65; tubes, per 100 feet, 1½ in., $8.50; 2, $9; 2 1-2, $11.30; 3, $12.50; 3 1-2, $16; 4, $20.00. Terms, 2 per cent. off.

ANTIMONY—Market is active, and stocks scarce. Prices firm at 27c. to 28c.

OLD MATERIAL.—Dealers' buying prices still continue as follows: Heavy copper and wire, 19 cents. per lb.; light copper 17c. per lb.; heavy red brass, 17c. per lb.; heavy yellow brass, 14½c per lb.; light brass 11c. per lb.; tea lead, $3.75 per 100 lbs.; heavy lead, $4.10 per 100 lbs., scrap zinc. 4½ per lb.; iron No. 1 wrought, $14.00; No. 2 wrought, $6 to $8; machinery cast scrap, $16.50 to $18; stove plate, $13 to $14, malleable and steel, $8; old rubbers, 9½c. per pound; country mixed rags, $1 to $1.25 per 100 lbs., according to quality.

COAL.—We continue to quote:

Standard Hocking soft coal f.o.b. at mines, lump $1.75, ¾ inch, $1.65, run of mine, $1.40, nut, $1.25; N. & S. $1.10; P. & S., 85c.

Youghiogheny soft coal in cars, bonded at bridges; lump, $2.90; ¾ inch, $2.70; mine run, $2.60; slack, $2.25.

United States Metal Markets

From the Iron Age, Jan. 10, 1907.

The monthly pig iron statistics collected by The Iron Age show that the output of the coke and anthracite furnaces was 2,236,153 gross tons in December, as compared with 2,187,665 tons in November. The production of the steel works furnaces has broken all records, having reached 1,163,035 tons in December. However, the outlook for maintaining the December output during the current month is not very promising, since the capacity of the furnaces in operation declined from 513,860 tons per week on December 1 to 507,397 tons per week on January 1, 1907.

Scarcity of spot iron is still a marked feature in all the leading pig iron markets, and promises to continue so for some time unless weather conditions are very favorable and transportation facilities improve very materially. The majority of buyers continue to have a good deal of confidence in the second half of the year, but it is only fair to state that an increasing number of consumers have determined to await developments, in view of the high prices prevailing. Sellers generally are very firm, but there are instances cropping up of inducements being made to book orders for delivery during the second half.

In the central west some pretty large orders for foundry iron will probably be placed at an early date, this including 15,000 tons for a pipe founder and 10,000 tons for a machinery manufacturer. The negotiations for a large tonnage for an agricultural implement interest have not yet come to a head. In the East there have been some additional large sales of basic iron at high prices, but otherwise pig iron consumers are growing conservative as to forward deliveries.

In some quarters specifications for structural shapes are not coming in fast enough to employ the full capacity of the mills, thus affording a welcome opportunity to divert the steel into other channels. The eastern plate makers have advanced prices $2 per ton.

A moderate amount of structural work has been placed, this including 8,000 tons of bridge work for the New York Central, and 1,500 tons for the Louisville & Nashville road.

An interesting transaction is the sale of a 5,000-ton lot of steel bars for a reinforced concrete building for a Chicago catalogue house. This is the largest contract for concrete bars yet placed.

From the Iron Trade Review, Jan. 10, 1907.

While the iron buying movement continues extremely heavy in all sections of the country, there are indications of easier conditions, particularly in the east. The Chicago market is still much excited, though all the merchant furnaces in that district have sold two-thirds of their 1907 make. Southern iron is firm with an advancing tendency and deliveries very poor. In New York, Pittsburg and other eastern centres comparatively little iron is being purchased, and concessions are reported on a large purchase of foundry grades. The supply of foreign iron is increasingly scarce, and occasionally foundrymen are compelled to close for a few days at a time. If the market is any less strong it is because of the fact that buyers have ordered far in advance and presents nothing that is alarming.

In certain finished lines easier conditions also prevail. A number of Indiana mills in their eagerness for business have cut prices on bar iron from $4 to $5 per ton below those heretofore prevailing. Quotations of the larger producer are unchanged. Structural materials which so far have constituted the only weak factor of the market are in such light demand that the Illinois Steel Company now operates its mill every alternate week on billets. To counterbalance these facts, however, improvement is noted in a number of lines. San Francisco has finally become an important buyer of building materials, having placed orders for 7,000 tons since January 1, with other large contracts pending. December sales of the American Bridge Company were the heaviest of the year with one exception, and the sales of the year aggregated 660,000 tons. Important bridge and building contracts are shortly to be closed.

An eastern manufacturer, catering to prompt delivery, has announced another advance on plates of $2 per ton, making his price $2.10 Pittsburg, as against $1.60 Pittsburg, quoted by the leading producer. Sheets are strong and firm by reason of the high price of spelter. The American Sheet & Tin Plate Company during the past year shipped in the neighborhood of 1,250,000 tons of sheet products, or 20 per cent. more than 1905.

Eastern furnace interests, unable to secure Lake Superior ore for the coming season, have resorted to foreign ore for which heavy tonnages have been contracted, and will also use local eastern ores. At the close of navigation the amount of ore on Lake Erie docks was about 200,000 tons less than the corresponding period of 1905.

London, Eng., Metal Markets

From Metal Market Report, Jan 10, 1907.

Cleveland warrants are quoted at 60s. 3d., and Glasgow Standard warrants at 56s. 6d., making price as compared with last week 1s. 7½d. lower on Cleveland warrants, and 1s. 6d. lower on Glasgow Standard warrants.

TIN.—Spot tin opened firm at £190 10s., futures at £191, and after sales of 320 tons of spot and 900 tons of futures closed weak at £189 5s. for spot, £189 15s. for futures, making price compared with last week £3 5s. lower on spot and £4 15s. lower on futures.

COPPER—Spot copper opened steady a[illegible] 5s , futures £107 5s., and after sales of 300 tons of spot and 400 tons of futures, closed easy at £106 2s. 6d. for spot, and £107 2s. 6d. for futures, making price compared with last week 17s. 6d. higher on spot and 10s. higher on futures.

LEAD—The market closed at £19 15s., making price as compared with last week 3s. lower.

SPELTER—The market closed at £27 15s., making price as compared with last week unchanged.

N.B. Hardware Trade News

St. John, Jan. 7.

Is there to be a time of marked industrial development in St. John in the near future? The city's growth has been slow of late years, comparatively speaking, but there have been intimations recently that new industries are shortly to come into being. Reference was made in the last issue of Hardware and Metal to the fact that rumor has had it that the Maritime Nail Works concern is contemplating the erection here of a plant much larger than that it is now using. One story has been that the talked-of plant was to be devoted largely to the manufacture of horse shoe nails. Another was that the new works were to be of the rolling mill variety. Mr. S. E. Elkin, of the Maritime Company, has denied the truth of reports. How comes the story, on good authority, that he has secured control of the whole of a large stock of land on what is known as the Marsh Road? What the purpose of this move may be one cannot say. Your correspondent has been informed, however, that Mr. Elkin in seeking a site, is not acting for the Maritime Nail Works Company, as such, though members of the company may be connected with the syndicate for which he is carrying on negotiations.

Another deal about which there is considerable speculation went through a few days ago. At that time a lawyer secured a Leicester Street property for unknown persons. This move, it is generally believed, is taken with a view to establishing a manufacturing industry.

A New Brunswick company that is apparently prospering is the New Brunswick Wire Fence Company, Limited. At their recent annual meeting the directors of the company received excellent reports showing a remarkable growth in business during 1906. Very shortly a new wire weaving machine is to be installed in the company's works. A large annex for warehouse purposes is now being built. Other additional warehouse space and more new machinery, will probably be necessary ere long. The directors for 1907 are: Dr. C. A. Murray, Hon. C. W. Robinson, A. C. Chapman, F. W. Givan and J. T. Hawke, all of Moncton. They have elected Dr. Murray president, Mr. Givan vice-president, and Mr. Chapman, secretary and manager.

It is the same old story asa regards prices, firmness and a tendency to further increase. Ingot tin is at a record making figure; wire and cut nails have taken another jump upwards; brass is quoted at a stronger figure; lead is still at an exceptionally high figure. There is no need to go on. The general statement is sufficient, prices are continuing at remarkable positions. Not for years has there been such a condition of affairs. In some quarters the opinion prevails that a general revision of prices is soon to come here as elsewhere.

The effect of the new tariff on the cordage business is not looked forward to with any degree of complacency by the dealers. The ruling has made the conditions surrounding this branch of trade decidedly uncertain. The free admission of ropes for fishing purposes makes a condition of unsettledness, and opens the way for a variety of difficulties. Then, too, the question comes up: will the new regulations give the United States cordage people too great an advantage?

George McAvity is one of the incorporators of the Dominion Car and Foundry Company, Montreal.

N.S. Hardware Trade News

Halifax, N.S., Jan. 8, 1907.

The hardware trade is very quiet at present. All the jobbers are now busily engaged in taking stock and it will be some time before they get things straightened out again. The weather is very unseasonable, and has a strong tendency to injure business. There is no snow in the woods, and the lumbering business is practically at a standstill. What promised to be a busy season at the outset, has received a serious set back by the weather conditions, and the outlook is not very hopeful. The prices of all lines are very firm, more particularly so with metals, which have an upward tendency. Ingot tin and tin plates are higher now than they have been for some time, and quotations on Paris green are fully 30 per cent. higher than at the same time last year. White lead and all wire products have been advanced by the manufacturers, and it is expected that nails will go still higher. Zinc has been advanced nine cents per cask.

The poor condition of the ice since the opening of the New Year, has caused a marked falling off in the sales of skates, and there is little or no demand for horse shoers' supplies, the streets and roads being entirely free from ice. All the big industries of the province seem to have large orders on hand, and the relighting of the Nova Scotia Steel Company's blast furnaces at Sydney Mines will have good results.

James Simmonds & Co.'s hardware store at Dartmouth, N.S., was damaged by fire last week and the stock was damaged to the extent of $2,574 from water. The members of the firm are very energetic, however, and business is rapidly expanding. The fire did not interfere to any extent with the business, as they made prompt arrangements to fill all the orders on hand.

Calgary Hardware Trade News

Calgary, January 5, 1907.

Dealers here all report a brisk Christmas trade. Local stores made good displays of seasonable stock and reaped the advantage. Special mention may be made of Messrs. Ashdown's display of cutlery and table plate, which by its magnificence suggested a jeweller's store.

The building trade continues busy here, and the demand for the various supplies is consequently brisk. First in importance as regards size and cost, among the jobs under way, is the new Normal School. The British Columbia General Contract Company, who have the contract for the school have just completed a large extension of the C. P.R.'s local machine shop. Built substantially of stone and internally arranged in accordance with the latest designs of the company's engineers, the enlarged shops should enable the daily increasing work to be handled more expeditiously.

Messrs. MacFarlan & Northcott are rapidly getting their new store into shape for business. Unlike the other hardware stores in the city which are all within a radius of a few hundred yards east, their new premises are situated on Eighth Avenue West. With a double window frontage, the interior shows a commodious space for the convenient storage of retail articles, while back premises give accommodation for the heavier branches of the trade. Messrs. MacFarlan and Northcott are also agents for the MacLaughlin carriages.

The reconstruction of the big cement plant in the east of the city is being rapidly pushed on. Mr. Butchart reports the work as nearly completed, but complains of the shortage of labor.

London Hardware Trade News

The elaborate preparations made by the Commercial Travellers' Club of London, for the ball on the 11th inst., gave some indication that the event would be the most successful in the history of the club. Every season this affair is looked upon as one of the greatest events in the city's social life, and this season was no exception. The subscription list was closed more than a week in advance, when an attendance of about 500 was assured. By a specially constructed passage way from the second story of the City Hall to the second story of the Masonic Temple, the two buildings were as one, practically, for the occasion. The decorations of the City Hall gave the suggestion of a big marquee and the same idea was carried out in the decoration of the banqueting hall in the Masonic Temple. Red and black—the club colors—were generously used in the decorating. The officers of the Club are: Hon. C. S. Hyman, Hon.-Pres.; H. E. Buttrey and Donald Ferguson, Hon. Vice-Pres.; C. W. McGuire, President; J. A. Townsend and George Detlor, Vice-Presidents; J. J. Harkness, A. H. Brener, L. C. Johnston, H. W. Lind, A. H. Moran, J. M. Ferguson, T. W. Edwards and F. S. Fisher, members of the Executive Committee. The gentlemen who had in charge the arrangements for the ball were: F. S. Fisher, Honorary Chairman; H. W. Lind, Honorary Secretary; S. F. Glass, convenor, Accommodation and Halls Committee; J. M. Ferguson, convenor, Reception Committee; J. A. Carling, convenor, Refreshment Committee: H. C. McBride, convener, Printing Committee; J. A. Townsend, convener, Finance Committee; A. Tillmann, convenor Decoration Committee, and J. J. Harkness, convener Music Committee. To these be the credit.

CONFERENCE OF TRAVELERS.

About forty managers of branches and traveling salesmen, representing the Canadian Fairbanks Company throughout Canada, have been in conference in Toronto, during the past week, their programme being very instructive to all who participated, including both addresses by men who had been successful in manufacturing and selling the Fairbanks lines, and visits to several of the factories where the Fairbank specialties are produced.

Monday was devoted to a discussion of the merits and best selling methods of scales, gas engines, steam specialties, and railway supplies. On Tuesday gas engines and transmission materials were taken up; a visit to the Canadian Fairbanks-Morse factory being also included. On Wednesday the delegation visited Hamilton and Dundas, to make an inspection of the methods of manufacture in operation in the John Bertram Machine Works at Dundas; the Pratt & Whitney Small Tool Factory at Dundas, and the Union Drawn Steel Co.'s works at Hamilton. Thursday was devoted to valves and steam specialties, a theatre party being given in the evening; and on Friday another trip was made to the Fairbanks-Morse factory and a general recapitulation made of the results of the week's studies.

H. J. Fuller, president of the Canadian Fairbanks Company, is an ardent advocate of conferences of selling staffs, believing that they do much to make the different individuals realize the power of concerted action as well as teaching the new men that it is their business not only to take orders, but to sell goods and know the articles they sell. Mr. Fuller is also a great believer in the "made in Canada" slogan, and is steadily increasing the number of Fairbanks lines made in this country.

Two important announcements can be made this week—one being that the scales and valves at present made by the E. T. Fairbanks Company at their factory at St. Johnsbury, Vermont, will, in future, be made for the Canadian trade at a new factory being established at Beebe Plains, Quebec, only 40 miles from the parent factory in Vermont. The other announcement is that the Canadian Fairbanks Company have contracted with the R. McDougall Co. to take over the entire product of their smaller machine tools, the contract extending for a period of 5 years. The Dart Union Co., of Providence, R.I., also intend to establish a new factory at Toronto. The Fairbanks-Morse gas engines are at present made in Toronto; pumps up to 6 inches are also made in Toronto, the larger sizes being imported, and wood pulleys are made at the Montreal factory. These lines, as well as those made in Hamilton and Dundas, are only a few of those sold by the Fairbanks representatives.

Those persons at the conference included: —

Montreal—H. J. Fuller, D. Guttridge, Gerald Robinson, G. B. Green, J. Falis, R. Miquelon, B. J. Corry, J. S. Bowar, W. S. Howe, J. McLeod, E. J. Holland, G. Drinkwater, M. P. Shea, T. E. Ryder, H. L. De Wolfe.

Toronto—C. J. Brittain, C. S. Hook, F. M. Allen, F. J. Campbell, J. G. Robinson, Geo. Robson, W. J. Sanderson, S. B. Trainer, Geo. Fisher, E. D. Hamilton, P. C. Brooks, J. S. Sanson.

Winnipeg—J. H. Crane, G. K. Tower, P. D. McLaren, A. H. Johnston, S. Jolliff.

Vancouver—G. H. Howard, R. M. Kalberg, F. W. Fisher.

St. John—G. E. Choinier.

Halifax—K. N. Forbes.

New Glasgow—F. A. Lytle.

Chicago—Mr. Jenson.

An important addition to the Montreal staff is W. S. Howe, who has just resigned his position as advertising manager of the S. A. Woods Co., of Boston, manufacturers of wood-working machinery. Mr. Howe has made a splendid reputation for himself in the Eastern States, and should make his way with the company he is now associated with.

MONTREAL FIRM INCORPORATED.

The firm of A. C. Leslie & Company has recently been incorporated under the name of A. C. Leslie & Co., with an authorized capital of $250,000.

This business, which was established by the late A. C. Leslie in 1866, is now one of the oldest in the iron, steel and metal trade in the Dominion, and the rapid growth of its business, combined with the desire to give some of their employes an interest in the business, has led to its incorporation.

No change whatever has been made in the personnel of the management, the directors and officers being as follows: William S. Leslie, president; Albert H. Campbell, vice-president; Thomas H. Jordan, director; Edward H. [illegible], secretary; Frederick H. Foster, treasurer.

Now Made in Canada

HENCEFORTH LOOK FOR THIS LABEL

Blanchite Paint Products First Greet the Canadian Trade in 1907

e are a Purely Canadian Company

including many of the best known business men in Canada, who have purchased valuable rights.

e Manufacture Under a New Process

which does away with the harmful materials paint-makers have tried to eliminate for years.

Do Not Introduce An Experiment

Blanchite has been on the American and European markets and won large contracts right in the greatest centres of competition—London, Eng., and New York, U.S.A.—against all paints made in both countries. When Blanchite won the immense contract for the ***New York Subway,*** 95 paint samples were tested in the chemical laboratories of the Rapid Transit Commission, and also received a 10 months' practical time test under hardest conditions. This page will not hold list of references, which includes ***some of the largest corporations in the world.***

Use Finest Machinery and Purest Raw Material

and we do not adulterate.

Know Our Goods—You Will Know Them Soon

Do not leave your opportunity to your competitor.

We Can Challenge Comparison

With best wishes for New Year,

Yours very truly,

Offices and Factory: 785 King St. W.

TORONTO,

MANITOBA HARDWARE AND METAL MARKETS

(Market quotations corrected by telegraph up to 12 a.m. Friday, Jan. 4, 1907.)

Room 511, Union Bank Building, Office of HARDWARE AND METAL, Winnipeg, Man.

There was no change in prices in the hardware and metal trades this week with the holiday season just ended; business is comparatively slack and will continue to be so for a week or two. Quotations this week are unchanged from the previous week. The trade here looks for a decided activity in business as soon as building operations are under way. Already in Winnipeg there have been several building permits for the coming year taken out and activity in building circles promises to be as pronounced, if not more so, than during the past year when the record was reached here.

LANTERNS. — Quotations are as follows: Cold blast, per dozen, $6.50; coppered cold blast, per dozen, $8.50; cold blast dash, per dozen, $8.50.

WIRE—We quote as follows: Barbed wire, 100 lbs., $2.90; plain galvanized, 6 to 8, $3.39; 10, $3.50; 12, $3.10; 13 $3.20; 14, $3.90; 15, $4.45; 16, $4.60; plain twist, $3; staples, $3.50; oiled annealed wire, 10, $2.96; 11, $3.02; 12, $3.10; 13, $3.20; 14, $3.30; 15, $3.45. Annealed wires (unoiled) 10c. less.

HORSESHOES — Quotations are as follows: Iron, No. 0 to No. 1, $4.65; No. 2 and larger, $4.40; snowshoes, No. 0 to No. 1, $4.90; No. 2 and larger, $4.65; steel, No. 0 to No. 1, $5; No. 2 and larger, $4.75.

HORSENAILS — Lists and discounts are quoted as follows: No. 10, 20c.; No. 9, 22c.; No. 8, 24c.; No. 7, 26c.; No. 6, 28c.; No. 5, 32c.; No. 4, 40c., per pound. Discounts are quoted as follows: "C" brand, 40, 10 and 7½ per cent., "M" brand and other brands, 55 and 60 per cent. Add 15c. per box.

WIRE NAILS.—Quoted now at $2.70 per keg.

CUT NAILS.—As noted last week cut nails have been advanced to $2.90 per keg.

PRESSED SPIKES — Prices are quoted as follows since the recent advance: ¼ x 5 and 6, $4.75; 5-6 x 5, 6 and 7, $4.40; ⅜ x 6, 7 and 8, $4.25; 7-16 x 7 and 9, $4.15: ½ x 8, 9, 10 and 12, $4.05; ⅝ x 10 and 12, $3.90. All other lengths 25c. extra net.

SCREWS—Discounts are as follows: Flat head, iron, bright, 85 and 10 p.c.; round head, iron, 80 p.c.; flat head, brass, 75 and 10 p.c.; round head, brass 70 and 10 p.c.; coach, 70 p.c.

NUTS AND BOLTS — Discounts are unchanged and continue as follows: Bolts, carriage, ⅜ or smaller, 60 and 5; bolts, carriage, 7-16 and up, 55; bolts, machine, ⅜ and under, 55 and 5; bolts, machine, 7-16 and over, 55; bolts, tire, 65; bolt ends, 55; sleigh shoe bolts, 65 and 10; machine screws, 70; plough bolts, 55; square nuts, case lots, 3; square nuts, small lots, 2½; hex nuts, case lots, 3; hex nuts, smaller lots, 2½ p.c.

RIVETS—Discounts are quoted as follows since the recent advance in the price of copper rivets: Iron, discounts, 60 and 10 p.c.; copper, No. 8, 37c.; copper, No. 10, 40c.; copper, No. 12, 43c.

COIL CHAIN.—Prices have been revised the general effect being an advance. Quotations now are: ¼ inch $7.00; 5-16, $5.35; ⅜, $4.75; 7-16, $4.50; ½, $4.25; 9-16, $4.20; ⅝, $4.25; ¾, $4.10.

SHOVELS—Discounts on spades and shovels continue 40 and 5 p.c.

HARVEST TOOLS — Discounts continue as before, 60 and 5 per cent.

AXE HANDLES—Quoted as follows: Turned, s.g. hickory, doz., $3.15; No. 1, $1.90; No. 2, $1.60; octagon, extra, $2.30; No. 1, $1.60.

AXES.—Quotations are: Bench axes, 40; broad axes, 25 p.c. dis. off list; Royal Oak, per dozen, $6.25; Maple Leaf, $8.25; Model, $8.50; Black Prince, $7.25; Black Diamond, $9.25; Standard flint edge, $8.75; Copper King, $8.25; Columbian, $9.50; handled axes, North Star, $7.75; Black Prince, $9.25; Standard flint edge, $10.75; Copper King, $11 per dozen.

BUTTS—The discount on wrought iron butts is 70 p.c.

CHURNS — The discounts from list continue as before: 45 and 5 per cent.; but the list has been advanced and is now as follows: No. 0, $9; No. 1, $9; No. 2, $10; No. 3, $11; No. 4, $13; No. 5, $16.

CHISELS—Quoted at 70 p.c. off list prices.

AUGER BITS—Discount on "Irwin" bits is 47½ per cent., and on other lines 70 per cent.

BLOCKS—Discount on steel blocks is 35 p.c. off list prices; on wood, 55 p.c.

FITTINGS—Discounts continue as follows: Wrought couplings, 60; nipples, 65 and 10; T's and elbows, 10; malleable bushings, 50; malleable unions, 55 p.c.

GRINDSTONES—As noted last week, the price is now 1½c. per lb., a decline of ¼c.

FORK HANDLES—The discount is 40 p.c. from list prices.

HINGES—The discount on light "T" and strap hinges is 65 p.c. off list prices.

HOOKS—Prices are quoted as follows: Brush hooks, heavy, per doz., $8.75; grass hooks, $1.70.

CLEVISES—Price is now 6½c. per lb.

STOVE PIPES—Quotations are as follows: 6-inch, per 100 feet length, $9; 7-inch, $9.75.

DRAW KNIVES—The discount is 70 per cent. from list prices.

RULES—Discount is 50 per cent.

WASHERS—On small quantities the discount is 35 p.c.; on full boxes it is 40 p.c.

WRINGERS — Prices have been advanced $2 per dozen, and quotations are now as follows: Royal Canadian, $35.00; B.B., $39.75, per dozen.

FILES—Discounts are quoted as follows: Arcade, 75; Black Diamond, 60; Nicholson's, 62½ p.c.

BUILDING PAPER — Prices are as follows: Plain, Joliette, 40c.; Cyclone, 55c.; Anchor, 55c.; pure fibre, 60c.; tarred, Joliette, 65c.; Cyclone, 80c.; Anchor, 65c.; pure fibre, 80c.

TINWARE, ETC.—Quoted as follows: Pressed, retinned, 70 and 10; pressed, plain, 75 and 2½; pieced, 30; japanned ware, 37½; enamelled ware, Famous, 50; Imperial, 50 and 10; Imperial, one coat, 60; Premier, 50; Colonial, 50 and 10; Royal, 60; Victoria, 45; white 45; Diamond, 50; Granite, 60 p.c.

GALVANIZED WARE. — The discount on pail is now 37½ per cent.; and on other galvanized lines the discount is 30 per cent.

CORDAGE—We quote: Rope, sisal, 7-16 and larger, basis, $11.25; Manila, 7-16 and larger, basis, $16.25; Lathyarn, $11.25; cotton rope, per lb., 21c.

SOLDER—Quoted at 27c. per pound. Block tin is quoted at 45c. per pound.

VISES — Prices are quoted as follows: "Peter Wright," 30 to 34, 14½c.; 35 to 39, 14c.; 48 and larger, 13½c. per lb.

ANVILS—"Peter Wright" anvils are selling at 11c. per lb.

CROWBARS—Quoted now at 4c. per lb.

POWER HORSE CLIPPERS — The "1902" power horse clipper is selling at $12, and the "Twentieth Century" at $6. The "1904" sheep shearing machines are sol l at $13.60.

AMMUNITION, ETC.—Quotations are as follows: Cartridges, Dominion R.F. 50 and 5; Dominion, C.F., 33⅓ C.F., pistol, p.c.; C.F., military, 10 p.c. advance. Loaded shells: Dominion Eley's and Kynoch's soft, 12 gauge, black, $16.50; chilled, 12 gauge, $17.50; soft, 10 gauge, $19.50; chilled, 10 gauge, $20.50. Shot, ordinary, per 100 lbs., $7.25; chilled, $7.75; powder, F.F., keg, Hamilton, $4.75; F.F.G., Dupont's, $5.

IRON AND STEEL.—Quotations are: Bar iron basis, $2.70. Swedish iron basis, $4.95; sleigh shoe steel, $2.75; spring steel, $3.25; machinery steel, $3.50; tool steel, Black Diamond, 100 lbs., $9.50; Jessop. $13.

SHEET ZINC — The price is now $8.50 for cask lots, and $9 for broken lots.

PIG LEAD—Quoted a $5.85 per cwt.

AXLE GREASE—"Mica" axle grease is quoted at $2.75 per case, and "Diamond" at $1.60.

IRON PIPE AND FITTINGS— Revised prices are as follows:—Black pipe, ¼ inch, $2.65; ⅜, $2.80; ½, $3.50; ¾, $4.40; 1, $6.35; 1¼, $8.65; 1½, $10.40; 2, 13.85; 2½, $19.00; 3, $25.00. Galvanized iron pipe, ⅜ inch. $3.75; ½, $4.35; ¾, $5.65; 1, $8.10; 1¼, $11.00; 1½, $13.25; 2, inch, $17.65. Nipples, discounts 70 and 10 per cent.; Unions, couplings, bushings and plugs, 60 per cent.

LEAD PIPE—The price, $7.80, is firmly maintained in view of the advancing lead market.

GALVANIZED IRON—Quoted as follows:—Apollo, 16 gauge, $3.90; 18 and 20, $4.10; 22 and 24, $4.45; 26, $4.40; 28, $4.65, 30 gauge or 10¾ oz., $4.95; Queen's Head, 24, $4.50; 26, $4.65; 28, $5.00.

TIN PLATES—We now quote as follows: IC charcoal, 20 x 28, box, $9.50; IX charcoal, 20 x 28, $11.50; XXI charcoal, 20 x 28, $13.50.

TERNE PLATES—Quoted at $9.

CANADA PLATES—Quoted as follows: Canada plate, 18 x 21, 18 x 24, $3.40; 20 x 28, $3.65; full polished, $4.15.

BLACK SHEETS—Prices are: 10 to 16 gauge, 100 lbs., $3.50; 18 to 22, $3.75; 24, $3.90; 26, $4; 28, $4.10.

PETROLEUM AND GASOLINE.—Silver Star in brls. per gal., 20c.; Sunlight in brls, per gal., 21c.; per case, $2.30; Eocene in brls, per gal., 23c.; per case, $2.50; Pennoline in brls., per gal., 24c.; Crystal Spray, 23c.; Silver Light, 21c.; Engine gasoline in barrels, per gal., 28c., f.o.b. Winnipeg in cases, $2.75

PAINTS, OILS & TURPENTINE — Turpentine is firm at the recent advance White lead, pure, $7; bladder putty in barrels, 2½c.; in kegs, 3½c.; turpentine, barrel lots, Winnipeg, $1.01; Calgary, $1.08; Lethbridge, $1.08; Edmonton, $1.09. Less than barrel lots 5c. per gallon advance. Linseed oil, raw, Winnipeg, 64c.; Calgary, 71c.; Lethbridge, 71c.; Edmonton, 72c.; boiled oil, 3c. per gal. advance on these prices.

WINDOW GLASS—We quote: 16-oz. O.G., single, in 50-ft. boxes—16 to 25 united inches, $2.25; 26 to 40, $2.40; 16-oz. O.G., single, in 100-ft. cases — 16 to 25 united inches, $4; 26 to 40, $4.52; 41 to 50, $4.75; 50 to 60, $5.25; 61 to 70, $5.75. 21-oz. C.S., double, in 100-ft. cases—26 to 40 united inches, $7.35; 41 to 50, $8.40; 51 to 60, $9.45; 61 to 70, $10.50; 71 to 80, $11.55; 81 to 85, $12.60; 86 to 90, $14.75; 91 to 95, $17.30.

PROMINENT MANUFACTURER DEAD.

Hardwaremen throughout Canada will regret to learn of the rather sudden death a week ago of Leonard McGlashan, of Niagara Falls. With a party of relatives and friends he had gone to California a month ago to enjoy the milder winter climate of the Pacific coast, but on his arrival there was taken ill with typhoid fever, the disease

THE LATE LEONARD McGLASHAN.

which caused his death. Mr. McGlashan was 61 years of age and besides a widow, leaves a daughter and one son, L. Lee McGlashan, of the McGlashan-Clarke company, Niagara Falls, manufacturers of cutlery and silverware.

Mr. McGlashan was probably the wealthiest citizen of Niagara Falls and was one of the first users of power from Niagara River, besides being interested in many enterprises, including mines and railways in the southwestern states and Mexico, and was a large real estate owner. He was until recently the chief owner of the Ontario Silver Company, which had a large factory at Thorold, then at Humberstone and finally located in Niagara Falls, where it is an important industry. Mr. McGlashan sold out his interest to the International Silver Company some time ago, and since then the McGlashan-Clarke Company was formed and built the large works at Niagara Falls, which commenced operations only a month or so ago.

CATALOGUES AND BOOKLETS

When sending catalogues for review, manufacturers would confer a favor by pointing out the new articles that they contain. It would assist the editor in writing the review.

By mentioning HARDWARE and METAL to show that the writer is in the trade, a copy of these catalogues or other printed matter will be sent by the firms whose addresses are given.

A Novel New Year's Card.

We have received from Ludger Gravel, Montreal, a New Year's greeting in the form of an aluminum post card showing a photograph of the sender, and Christmas and New Year's greetings in French and English.

Chrysanthemums.

The Guelph Stove Company have issued a beautiful calendar, it being a reproduction of "Chrysanthemums," a painting by an eminent French artist. The calendar is well worth sending for.

Desk Calendar.

The D. Moore Company, stove manufacturers, of Hamilton, have issued to the trade a very small compact desk calendar, with a cut of their works, which will be found very acceptable by all who receive a copy.

Cordage Calendar.

The Independent Cordage Co., of Toronto, have issued a beautiful calendar for 1907, which will be well received by the trade, and any person not yet receiving one should write mentioning this paper.

A Quebec Calendar.

J. McOuat & Son, hardware merchants, Lachute, Que., have supplied their customers with a very attractive calendar for 1907, a copy of which has been received by Hardware and Metal.

Capewell Horse Nail.

The Capewell Horse Nail Co. have issued a very nice calendar. Also a little booklet, containing history of this nail. Also cuts of various style heads, and sizes. Quite a number of letters from various horse-shoers throughout the country have been received claiming the superior qualities of this nail.

The Metric System.

Lord Kelvin's views on the advantage of the metric system has just been published by the Decimal Association, 605 Salisbury House, London, E.C., and can be procured for the sum of 3d. from these publishers. The book goes extensively into the explanation of the metric system, also verious uses which it is put to.

A Want Book.

The Dominion Wire Mfg. Co., of Montrea,, have issued a "Memo" book for 1907, which is bound in a neat black leather cover, with the company's trade mark on the cover, and contains useful information, which will come in handy to any merchant dealing in wire. Copies may be obtained from the company's salesmen by mentioning Hardware and Metal.

Fairbanks' Calendar.

The Canadian Fairbanks Company, of Montreal, Toronto, Winnipeg, and Vancouver, manufacturers of valves, pipe, and fittings, etc., are sending to the trade a large 1907 calendar, with monthly sheets, containing cuts of different lines manufactured by them. This calendar should prove very useful for store or office use, as the large print makes it very plain and easily read. It will be sent free on request of anyone mentioning this paper.

Year Book for 1907.

Wilkinson, Heywood and Clark, paint manufacturers, Montreal, have issued a very useful little pocket diary and cash account book.

The book is about 3½x6 inches, is bound in a handsome red leather cover with a pocket at the back containing a pencil; it also contains much information of value to paint dealers.

This book is entirely free from advertising matter, with the exception of the firm name, which appears at the top of several of the pages.

This useful little souvenir will, while the supply lasts, be sent by the above firm to any authorized paint dealer who will mention Hardware and Metal in writing.

Ice Cream Freezers.

North Bros. Manufacturing Co., of Philadelphia, have just issued their 1907 catalogue, containing cuts of various lines of freezers made by them. A new feature is the twin freezer, which they have just placed on the market. This freezer is adapted for the making of two different flavors at one time. The can is divided into two partitions, which project above the top of the can and bears against the can lid, so that during freezing nothing can pass from one side to the other. The dasher has a central shaft which fits in the groove of the partition of the can. This dasher carried two sets of scrapers, one set filling in one side of the can and the other set in opposite side. The motion is not the same as the other models on account of the partition in can. It is therefore swung forward and backward, having a rotary motion, the motion being much handier than the old.

Art Metal Work.

The Galt Art Metal Company have issued three fully illustrated catalogues of products manufactured by them. Their Catalogue A. is about 8x12 inches, containing 56 pages, dealing particularly with different designs of ceilings, borders and corners of many kinds. Catalogue B. deals with roofing and siding, also going extensively into the correct methods of laying this roofing, while cuts of their rock, brick and corrugated siding are also displayed. Catalogue C. contains many illustrations of cornices, skylights, finials, ventilators, window caps, etc., with full instructions as to ordering these goods, measurements, etc. Any dealer who has not yet received these catalogues should write and procure them at once in preparation for the coming building season.

Starr Skate Book.

The Starr Manufacturing Company are sending to the trade their Skate Book, which contains illustrations of various grades manufactured by them, as well as valuable information, as to the best methods of placing skates on the boots. The better known lines they manufacture are the Acme (with and without ankle supports), Regal, Mic-Mac "Starr" and Chebucto hockey, Starr tube, ladies' and gentlemen's, Beaver, Starr tube racer, and Starr racer. A new hockey skate has been put on the market known as the "Velox," combining all the good features of the other brands. This skate is meeting with a very large sale. The Starr Company also manufacture hockey sticks, and have just placed on the market two new models, with double-grooved serrated blade, which adds greatly to the shooting qualities, and strength of the stick. In the back of this catalogue will be found the rules of hockey and a table giving comparative sizes of shoes and skates.

LEWIS BROS.' BANQUET.

The travelers of Lewis Bros. staff had their annual re-union on Wednesday, Thursday and Friday of last week.

At the meeting which was held on Thursday, Dec. 27, 1906, short addresses were delivered by Messrs. Jas. G. Lewis, C. M. Strange and C. F. Smallpiece.

The banquet at the Engineers' Club on Friday evening, Dec. 28, 1906, was a great success.

All present seemed to enjoy themselves thoroughly and left with a strong determination to make 1907 the banner year in the history of the firm.

FOUNDRY AND METAL INDUSTRIES

The Pease Furnace Company of Toronto resumed work this week at their factory, after having been closed down for two weeks, for repairs and stock taking.

The Collingwood Shipbuilding Company are about to lay the keel for another big vessel. The steamer will be 490 feet long and is to be built to the order of the Farrar Transportation Company.

The Taylor, Forbes Company's foundry at Guelph suffered $5,000 damage by fire on January 6th. The fire only temporarily stopped the work in the radiator foundry, about 100 men being out of employment for about two days. The loss was fully covered by insurance.

A new fire-clay lining for stoves has been invented by Mr. Werther, formerly superintendent of the open hearth furnaces of the Dominion Iron and Steel Company. The basis of the preparation is Sydney cement, and the cost of this improved lining will be no greater than that of the ordinary iron material.

The annual banquet given by the Pratt & Letchworth Company, malleable iron founders of Brantford, was held last week, a special train bringing O. P. Letchworth, president, and friends from Buffalo, where another branch of their works is situated. The company has grown immensely since their location in 1899, and the entire output now is purely malleable. They started with about 100 men, now they employ 800, and are working night and day. The output for 1906 exceeded 15,000 tons, mostly for use in the construction of cars. The company paid over $400,000 in wages for 1906.

Samples of rock containing copper, found around the Moon River, in the district of Parry Sound, have been sent to New York for examination, and capitalists there lost no time in providing all necessary capital for thoroughly testing the find. Some of the recent assays from the present workings show well the composite character of the vein matter. They gave silver 20 ounces, copper 22 per cent., gold from $10 to $500 per ton, with values in platinum, cobalt and zinc. One of the largest crystals of mica that perhaps has ever been mined has been found in one of these mines, it being about the size of an ordinary door, and required no less than four drill holes to dislodge it. The mica found, though dark in color, contains no iron, and improved in quality with depth.

The year just closed has proved a record one for the output of coal from the collieries of the Nova Scotia Steel and Coal Co., no less than a total of 688,085 tons being raised from three mines. No. 3 led, with a total output of 390,126 tons, No. 1 being second with 257,467, and No. 5, 40,492 tons. Both Nos. 1 and 5 collieries worked single shift, while No. 3 worked double shift and is machine-mined. The total output from the company's collieries at Sydney for the year 1905 was 558,306 tons, and it is the intention of the management to increase last year's output by reaching the million mark for 1907. The output for 1907 is expected to be the largest in the history of the company.

A daily newspaper quotes a director of the Lake Superior Corporation as saying: Notwithstanding the taking off of the bounty July 1, the net earnings for four months since the close of the fiscal year have been fully equal to those for the corresponding period in 1905. The restored bounty went into effect January 1, 1907, and from then on the company receives the benefit, which should still more favorably affect the business as compared with that of a year ago. The steel rail tonnage has been running about 15,000 tons a month and all the subsidiary companies operated, with the exception of one trolley line, are contributing something to the net revenue, as they are all earning their fixed charges and something besides.

BIG FIRMS AMALGAMATE.

Particulars of the amalgamation of the Canada Screw Company, established 1866, and the Ontario Tack Company, established in 1887, two of Hamilton's largest industries, have been announced. Nearly $2,000,000 is involved in the deal. The plants of both concerns are to be considerably enlarged and new lines added to their products. Application has been made for a new charter in which the capital stock will be greatly increased in order to allow for the extension of the works.

The new company will be known as the Canada Screw Co., and the officers will be: Cyrus A. Birge, president; Chas. Alexander, vice-president; F. H. Whitton, general manager; W. F. Coote, secretary-treasurer; James O. Callaghan, director of works. The other directors will be Hon. Senator Gibson and Charles S. Wilcox.

COPPER KING COMING.

It is stated that Senator W. A. Clark, of Montana, the great copper king, will shortly visit Montreal and Quebec in company with his wife. Mrs. Clark, whose maiden name is Lachapelle, is of French-Canadian descent, and has many relatives in Quebec Province.

Anvils and Vises
Machinery Steel
Wire Rope

IN STOCK

A.C. LESLIE & CO.
MONTREAL

HAND DRILLING MACHINERY

We have the best that money can buy.

Our cheaper grades are all good serviceable machines.

The best value in the market whether you want the cheap or the dear kind.

A. B. JARDINE & CO.
HESPELER - - - ONTARIO

NICKEL
The Canadian Copper Co.

NICKEL FOR NICKEL STEEL
The Orford Copper Co.

WRITE US FOR PARTICULARS AND PRICES

General Offices
43 EXCHANGE PLACE
NEW YORK CITY

WIRE NAILS
TACKS
WIRE

Prompt Shipment

The ONTARIO TACK CO.
Limited.
HAMILTON ONT.

"MIDLAND"
BRAND.
Foundry Pig Iron.

Made from carefully selected Lake Superior Ores, with Connellsville Coke as Fuel, "Midland" will rival in quality and grading the very best of the imported brands.

Write for Price to Sales Agents
Drummond, McCall & Co.
MONTREAL, QUE
or to
Canada Iron Furnace Co.
MIDLAND, ONT. Limited

... FULL STOCK ...

Salt Glazed Vitrified

Double Strength Culvert Pipe a Specialty.

THE CANADIAN SEWER PIPE CO.
HAMILTON, ONT. TORONTO, ONT
ST. JOHNS, QUE.

NOVA SCOTIA STEEL & COAL CO., Limited
NEW GLASGOW, N.S.

Manufacturers of

Ferrona Pig Iron
And SIEMENS-MARTIN
OPEN HEARTH STEEL

GET PRICES FOR
GILBERTSON'S
COMET
Brand
GALVANIZED FLAT SHEETS

for any purpose where the best is needed. Wide awake jobbers handle this brand. They are well galvanized true to gauge and especially soft for working up purposes Every sheet bears the name "Gilbertson." That is a guarantee.

MAKERS:
W. GILBERTSON & CO., Limited,
PONTARDAWE, SOUTH WALES
Agent: Alex. Gibb, 13 St. John St., Montreal.

OAKEY'S
The original and only Genuine Preparation for Cleaning Cutlery, 6d. and 1s. Canisters

'WELLINGTON'
KNIFE POLISH

JOHN OAKEY & SONS, Limited
Manufacturers of
Emery, Black Lead, Emery, Glass and Flint Cloths and Papers, etc.

Wellington Mills, London, England
Agent:
JOHN FORMAN, - 644 Craig Street
MONTREAL.

HERE IS A SELLER!

My Patent Cone Rotary Ventilator Sells on Sight. Without a Rival.

No modern building is perfect, from a sanitary standpoint, without it. Sent anywhere under guarantee. Endorsed by best architects. Booklet mailed free. Write to-day. Hundreds of testimonials.

Can. Feb. 27, 1889

THE MONTREAL BOARD OF TRADE
Geo. Hadrill, Secretary

G. D. Pearson, Esq., Montreal: Montreal, February 5th, 1898

Dear Sir,—In reply to your inquiry re the two 18" Rotary Cone Ventilators in use in this building, I can only say that they have now been in operation four years and ten months, during which they have not required attention, and are still in good order. I am quite satisfied with them, thus far.

Yours truly, JOHN YORK, Supt.

Bell Telephone East 2106

GEO. D. PEARSON & CO., Office and Works: **17 Pontiac Street, MONTREAL**

Persons addressing advertisers will kindly mention having seen their advertisement in Hardware and Metal.

BUILDING AND INDUSTRIAL NEWS

HARDWARE AND METAL would be pleased to receive from any authoritative source building and industrial news of any sort, the formation or incorporation of companies, establishment or enlargement of mills, factories, foundries or other works, railway or mining news. All correspondence will be treated as confidential when desired.

Winter doesn't stop building in Canada as is intanced by the large number of permits for new buildings which have been issued during December and the present month in all parts of Canada. For instance, in Toronto alone permits issued totalled to $961,000 in December and from Jan. 1 to 8 the amount was $367,000. The figures given by us in this issue in several cases cover the totals for the past year and are gratifying from every standpoint. Even more pleasing, however, is the certainty that 1907 will show even larger totals than the big records made in 1905 and 1906.

London defeated the waterworks by-law.

Bolton will raise $2,500 for cement walks.

Harriston voted to raise money for school purposes.

The new Normal School at Hamilton will cost $50,000.

Thorold carried a waterworks by-law by a majority of 12.

Paris defeated a by-law to raise $12,000 for school purposes.

A new Public Library will be built at Toronto at the cost of $250,000.

Windsor defeated a $12,000 by-law to improve the electric light plant.

Lindsay defeated the waterworks filtration by-law by a large majority.

Collingwood defeated by-laws to build a new public school and a fire hall.

Mount Forest defeated a by-law to abolish the Water and Light Commission.

Campbellford voted down a by-law to raise $12,000 for improvement of sidewalks.

Building permits for the year for Edmonton totalled $1,868,098, compared with $702,224 in 1905.

The Canadian Bank of Commerce will build a new two-storey, stone bank building in Sydney, N.S.

Stratford defeated by-laws to provide money for a market shelter, trunk sewer, and implement shelter.

At Morrisburg a by-law to provide $20,000 for extension and improvements to the electric power plant, were defeated.

The Waterous Engine Works, of Brantford, have taken out a building permit for a $4,000 addition to their main building.

The amount of customs collected at the port of Toronto eclipses that of 1905 by about $525,000, and is over a million dollars greater than 1904.

A large saw mill will be erected by the Graham Island Lumber Company, at Massett Harbor, B.C., with a cutting capacity of 250,000 feet daily.

The B. F. Graham Lumber Company has purchased the Taylor, Pattison Mill on Victoria Arm. The price paid was in the neighborhood of $50,000.

The Dominion Sewer Pipe Company have acquired property in Hamilton and will erect a factory, for the manufacture of sewer pipes, flue linings, gutter pipe and wall coping.

Peterboro carried a by-law to provide $21,000 additional to the $40,000 already in the hands of the Board of Education to erect a new collegiate institute, costing $61,000.

The Canadian General Electric Company, Peterboro, will shortly let contracts for a two-storey addition, 250 feet long, to their plant. The new addition will be for the lamp and wire departments.

The building permits issued in Victoria, B.C., equal $633,080, nearly $40,000 in advance of those of last year, and it is estimated that the buildings erected in the country immediately surrounding Victoria would easily total in value another $100,000.

Wortman & Ward Manufacturing Company, of London, have asked for a fixed assessment for 10 years at $25,000. The company began business in London 25 years ago, on a very small scale. Now it employs 90 men, and is rapidly increasing its business.

It is said that the Imperial Steel and Wire Company, of Collingwood, will establish a branch in Port Arthur. The factory will have double the capacity of the one at Collingwood, and will employ about 200 men, making wire nails, etc., for the western trade.

A $50,000 factory is to be erected in Toronto by Business Systems, Ltd. This firm has made wonderful progress in the two years of its existence, this being the second time they have moved into larger premises, to meet with the ever increasing demand for their goods.

The contract has been let for the building of the Berg, Sand, Lime and Brick Company, in Brantford. The buildings will consist of a two-storey main building 50 feet square, and another building 100 feet long. The new buildings, when completed, will cost $5,000.

The Galt Metal Works, Galt, manufacturers of architectural sheet metal building machinery, report a 100 per cent. increase in their business during 1906, and with the many new lines they are introducing they expect a corresponding increase during the present year.

The Canada General Electric Company and the Canada Foundry Company who are just completing additions to their buildings at Peterboro and Toronto, are again required to enlarge to meet with the ever increasing demand for their products, which are largely used in construction work.

The Munro Wire Works, in Winnipeg, have extended their plant to meet with the increased business. This new building cost about $30,000, and is set apart for the production of all kinds of wire mattresses and the handling of iron beds. The Northwestern Brass Co. are erecting a steel building, 80x100 feet, at the cost of $40,000.

The Vancouver Portland Cement Company has been incorporated with a share capital of $1,500,000, for the purpose of manufacturing cement, and all kinds of building materials, with head office in Toronto. The directors are J. S. Lovell, W. Bain, R. Towans, E. W. McNeill, W. F. Ralph, H. Chambers and C. H. Blank, all of Toronto.

The Dominion Iron and Steel Co., in order to put through the purchase of extensive coal lands about the Lingan properties, have decided to form a subsidiary company which will issue bonds to cover the purchase of the properties. These bonds will be guaranteed by the Dominion Iron and Steel Co. It is intended that from the start this subsidiary company should have a sinking fund into which will go all moneys above what is required for the interest on the bonds, and in this way the purchase of the properties will be gradually provided for.

Besides electing a Water and Light Commission for 1907 the town of Goderich passed three industrial by-laws, the loans in connection with which aggregate $75,000. The Kensington Furniture Company will now double their plant, and the Goderich Carriage Company, a new concern backed by local men, will take over Walker's carriage works, and will spend $15,000 in enlarging and adding machinery. The Rogers Mfg. Co., who have acquired the plant of the Goderich Engine Works, will also spend a large sum in adding to the present plant, and will go into the manufacture of journal-boxes, and other railway specialties: it is understood that they will take up the construction of railway snow-ploughs.

The development of the agricultural implement trade at Regina continues at a remarkable rate. The fact that six million dollars' worth of implements were sold there in 1906 has apparently spurred the big concerns not now represented to establish warehouses. It is announced that the Nichols-Shepard Co., of Battle Creek, Mich., is about to move its western headquarters from Winnipeg to Regina. It is now negotiating for a large site there and at least two other large concerns have similar action under contemplation. All these will probably move here in the early Spring. In addition to this several important Canadian and American concerns which are now doing business there in only a small way will open large establishments as soon as Spring begins. From present indications the building of warehouses will be one of the features of Regina's constructive programme during 1907.

Montreal will have a busy 1907. During the year the Simplex Car Works, along the Lachine Canal, are to be greatly extended, so as to give double the ordinary capacity. The Canada Car Company have already more orders than they can attend to, and may also have to extend. In the same neighborhood are the Allis-Bullock-Chalmers machinery works, the Wire Works, and the Radiator factory, while the new cotton company, which is to build a factory at Lachine in the Spring or Summer, is materializing. In the east end, the Locomotive Works at Longue Pointe are to be extended at a cost of $1,000,000. In the neighborhood of the Angus shops a new town has grown up, and this will be added to during the Spring and Summer, for the C.P.R. cannot overtake its own orders at these shops, which are the largest and best equipped on the continent. There will be more building in Montreal proper during the coming year than in any previous year, it is predicted.

VOTES ON PUBLIC IMPROVEMENTS

The councils of a large number of Ontario cities and towns submitted by-laws at the municipal elections on Jan. 7, to raise money by way of debentures for various improvements. In the majority of cases these by-laws were carried.

Southampton decided to raise $5,000 to extend the waterworks system.

Napanee passed by-laws to improve the town's sewerage.

Waterloo voted 390 for and 74 against the sewer commission proposal, and also carried a by-law for a $23,000 loan for sewer farm improvements.

St. Catharines voted favorably for extension of the waterworks.

Listowel decided by 17 votes to raise $14,500 to construct new bridges.

Brantford carried sewer extension by-law.

Almonte decided to appoint an electric light commission.

Kincardine carried the Grand Trunk station, Coleman, and Park by-law.

Woodstock endorsed by-laws to establish a parks commission and to raise $20,000 for sewer extensions.

In a quiet vote Guelph took a jump forward in its municipal government by deciding in favor of commissions to manage its waterworks and gas and electric light plants. The vote for the waterworks commission was 1,132 for, 346 against; for gas and electric light commission, 1,104 for 371 against, and to loan $25,000 to enable Morlock Bros. to extend their factory, 966 for, 159 against.

Renfrew, by a vote of 136 to 103, decided to raise $6,000 for sewer extensions, and the county system of good roads was carried, 171 to 65.

Bowmanville decided to have its waterworks managed by a commission.

TO AID INDUSTRIES.

Windsor passed by a majority of 60 an industrial by-law intended to enable the council to arrange with manufacturers seeking locations.

Whitby carried the Ontario Car Works by-law, 359 for and 18 against.

Wingham will loan the Wingham Carriage Co. $5,000.

Only six voted against the by-law submitted at Deseronto, granting the Deseronto Furniture Co. a loan of $10,000.

Welland carried a by-law granting fixed assessment and other concessions to the Robertson Machinery Co.

Goderich decided to aid the Rogers Manufacturing Co., the Furniture Co., and the Carriage Co.

Amherstburg defeated a by-law to grant aid in establishing a canning factory, the required number of favorable votes not being polled.

Orangeville ratified a loan of $15,000 to the Hurndall Novelty Manufacturing Co. by 415 majority.

Only 3 voted nay and 134 yea on the by-law submitted in Cayuga to give the Window Glass Manufacturing Co. 14 acres of land and tax exemption.

The by-law to aid in the rebuilding of the old Dyment Foundry at Barrie secured 173 votes short of the required number.

CONDENSED OR "WANT" ADVERTISEMENTS.

Advertisements under this heading 2c. a word first nsertion; 1c. a word each subsequent insertion.

Contractions count as one word, but five figures (as $1,000) are allowed as one word.

Cash remittances to cover cost **must** accompany all advertisements. **In no case** can this rule be overlooked. Advertisements received without remittance cannot be acknowledged.

Where replies come to our care to be forwarded, five cents must be added to cost to cover postage, etc.

SITUATIONS VACANT.

WANTED—Good practical tinsmith; $18.00 weekly to the right man; steady job. Apply H. H. Gervan, Chilliwack, B.C. (5)

WANTED—Man with two or three thousand dollars to take half interest in old established hardware and tinsmith business in good town. Box 575, HARDWARE AND METAL. (4)

WANTED—Two first-class hardware salesmen, one to travel west and north of London, and one to travel Toronto and vicinity; reply by letter stating experience, references, age and salary, to Box 576, HARDWARE AND METAL, Toronto. (3)

WANTED — Experienced young man, thoroughly conversant with builders' hardware, as traveller for the city. Apply stating references and salary required to E. G. Prior & Co., Ltd., Victoria, B.C. (3)

TRAVELLER WANTED

TO sell a line of established successful coal heaters, also wood heaters, on commission; must cover territory twice or oftener, say once between January and April and again between May and August; state lines now carried and territory covered; liberal commission given; Box 573, HARDWARE AND METAL, Toronto. [1]

BUSINESS CHANCES.

STOVES—An established, successful working line of patent coal, also wood heaters, for sale; including patent, patterns, materials and some completed stoves; also machinery and dies, if wanted; will go with a line of heaters or ranges, or to sell with some hardware specialty; proprietor has other pressing interests. Address Box 574, HARDWARE AND METAL, Toronto. [1]

HARDWARE AND TINWARE business for sale in a thriving town in Eastern Ontario; stock reduced to $1,500; good dairying country, rare chance for new beginner; going out of business on account of ill health. Apply at once to D. Courville, Maxville, Ont. [3]

FOR SALE—A fully equipped Wire Mill and Nail Factory as a going concern, with about 500 tons of wire rods; factory situated on 150 ft. of water-front with trackage optional to buyer. B.C. Wire and Nail Co., Ltd., Vancouver, B.C. (7)

SITUATION WANTED.

HARDWAREMAN (28) open for engagement; west. Box 91. HARDWARE AND METAL, Winnipeg. (4)

Paint, Oil and Brush Trades

CAUSES OF VARNISH CRACKING.

"Techno" in writing upon the subject of the cracking of varnish in The Decorator, says: "So much has already been said on this subject that it will only be necessary to summarize here the causes and their avoidance. Various causes produce cracks of distinct character. The addition of terebine to a varnish for hardening will often cause cracking, especially when exposed to direct sunlight. These cracks at first give the varnish a silky appearance, due to their hair-like fineness and great numbers. Subsequently many of the cracks open out wider under atmospheric variations. But the crack due to terebine is always sharp and clean, and mostly straight, as though cut with a razor edge, crossing the work in all directions. Terebine is sometimes used in graining color and other undercoats prior to varnishing. In such cases the cracks will show their origin to be in undercoats by the depth of every crack, while if the varnish only be at fault the undercoats will in many parts remain unaffected.

The application of any hard, quick-drying coat of paint or varnish on a soft undercoat is liable to cause cracking, and would affect any super-coat likewise. This may sometimes be traceable to a glaze coating prior to varnishing, to a gold-size and turps flatting coat on an oily ground or to a hard-drying varnish on a soft groundwork. Goldsize cracks are distinguishable by their usually lying in the direct line of the brushwork, and having soft round edges, turning inwards, the cracks being less numerous but more open than terebine cracks.

The application of a coat of size upon a hard, non-porous ground prior to varnishing, such as sometimes occurs when re-varnishing old work in cheap jobs, if the size be fairly strong, will sometimes result in cracks, the cracks being notably of polygon shape and the edges having a tendency to curl outwards.

Cracking sometimes occurs only where knots exist in the woodwork. That is generally due to the preliminary use of too much shellac or patent knotting destroying the porous key of the wood, leaving no hold for the priming coat. As no affinity exists between the shellac and the oil paint, the latter cracks by irregular contraction and expansion of the paint and the wood. The remedy is to scrape down to the bare wood and to paint again without fresh knotting, or after a thin coat of reduced strength, in case of new woodwork.

To avoid tendency to cracking, there is no better course than to take care that every coat prior to varnishing, be thin, and allowed to dry hard before applying the following coat. It is important also that no quick-drying medium, such as goldsize or terebine, be employed in painting over a coat mixed with ordinary linseed or boiled oil, though the reverse order may be employed without danger, and in case of quick-drying paint being necessary, employ no oil at all, excepting for the priming coat on new wood, and the finishing varnish may then be elastic or hard as desired without danger of cracking. A hard varnish may be used as an undercoat, and an elastic finishing over-varnish over that. But the reverse order may give rise to the fault under notice."

TURPENTINE.

Turpentine is a product of the pine tree, and in this country almost the entire consumption is produced from the yellow or long-leaf pine of the gulf states.

Formerly vast sections of all the states south of the Virginias and the Ohio river and east of the Mississippi were covered with immense forests of yellow pine, but at the present time the turpentine belt is confined to the gulf states. Though all the available resources in the adjoining states have not been exhausted, the end is plainly in sight unless the devastation, caused by the criminally wasteful methods of production, can be checked.

The chief use of turpentine is in paints and varnishes, where it is employed as a volatile thinning agent and is valued for its penetrating and binding qualities. It evaporates in the right manner leaving no residue. It has the peculiar property of forming ozone, which is practically a condensed form of oxygen, and as oxygen is the cause of the drying of the paints and varnishes, turpentine to this extent serves a double purpose.

About 13 per cent. of the thinning vehicle used in the manufacture of best prepared paint (outside white) is turpentine and japan driers, says Co-operation and Expansion. We have found that this proportion answers the purpose for which it is intended on finishing coats. On the priming coat, however, more turpentine should be added, and especially when painting hard pine, (the source of turpentine in this country), turpentine should constitute from 25 to 40 per cent. of the total amount of thinners used.

Combinations of producers and factors or merchants have had some effect upon prices in recent years; but no combination, however effective, in an industry so widespread as this, could have raised the price of a product to the present current prices for turpentine.

Are you interested in any of the lines that are advertised? A Post Card will bring you price list and full information. Don't forget to mention Hardware and Metal.

PAINT AND OIL MARKETS

MONTREAL.

Office of HARDWARE AND METAL,
232 McGill Street,
Montreal, January 11, 1907

Now that the holidays are over, travellers are brushing up their samples and prices and business has again assumed a more lively aspect. While shipments just now are not very heavy, a good flow of orders is coming in for February and later shipments.

White lead maintains its advances and it looks as if there would be a scarcity in the dry article before the arrival of early spring shipments.

Linseed oil keeps steady, without any special features.

Turpentine in barrels shows an advance of two cents per gallon over last week's prices.

Benzine in barrels drops five cents per gallon.

The Martin-Senour paint has been advanced 5 cents per gallon, and is now quoted at $1.50 pints, $1.45 quarts, and $1.40 gallons.

LINSEED OIL—Remains the same: Raw, 1 to 4 barrels, 55c.; 5 to 9 barrels, 54c.; boiled, 1 to 4 barrels, 58c.; 5 to 9 barrels. 57c.

TURPENTINE—(In barrels) shows an advance of 2 cents per gallon. Smaller quantities remain the same. We quote: Single barrel, 98c. per gal.; two barrels or over, 97c. per gal.; for smaller quantities than barrels, 5c. extra per gal. is charged Standard gallon is 8.40 lbs., f.o b. point of shipment, net 30 days.

GROUND WHITE LEAD—Maintains its advance. Best brands, Government Standard, $7.25 to $7.50; No. 1, $6.90 to $7.15; No. 2. $6.55 to $6.90; No. 3, $6.30 to $6.55 all f.o.b., Montreal.

DRY WHITE ZINC—We still quote as follows: V.M. Red Seal, 7½c. to 8c.; Red Seal, 7c to 8c; French V.M., 6c to 7c; Lehigh, 5c to 6c.

WHITE ZINC—(Ground in oil) Prices remain firm at their recent advance. We quote: Pure, 8½c. to 9½c.; No. 1, 7c. to 8c.; No. 2, 5¾c. to 6¾c.

PUTTY—A heavy demand is shown. We quote prices: Pure linseed oil, $1.75 to $1.85; bulk in barrels, $1.50; in 25-lb. irons, $1.80; in tins, $1.90; bladder putty in barrels, $1.75.

ORANGE MINERAL—Prices are as follows: Casks, 8c.; 100 lb. kegs, 8½c.

RED LEAD—The following quotations are firm: Genuine red lead in casks, $6.00; in 100-lb. kegs, $6.25; in less quantities at the rate of $7 per 100 lbs..; No. 1 red lead, casks, $5.75; kegs $6, and smaller quantities, $6.75.

PARIS GREEN — The following prices still hold good: In barrels, about 600 pounds, 23½c. per lb.; in arsenic kegs, 250 lbs., 23½c.; in 50 lb. drums, 24c.; in 25 lb. drums, 24½c.; in 1 lb. packets, 100 lbs. in case, 25c.; in 1 lb packets, 50 lbs in case, 25½c.; in ½ lb. packets, 100 lbs in case, 27c.; in 1 lb. tins, 26c. f.o.b. Montreal. Term three months net or 2 per cent. 30 days.

SHELLAC—We quote: Bleached, in bars or ground, 46c. per lb., f.o.b. Eastern Canadian points; bone dry, 57c. per lb., f.o-b. Eastern Canadian points; T. N. orange, etc., 48c per lb. f.o.b. New York.

SHELLAC VARNISH—Prices remain: Pure white, $2.90 to $2.95; pure orange, $2.70 to $2.75; No. 1 orange, $2.50 to $2.55.

MIXED PAINTS—Prices range from $1.20 to $1.50 per gallon.

CASTOR OIL—Our prices are: Firsts in cases, 8½c.; in barrels, 8c.; seconds in cases, 8c.; in barrels, 7½c.

BENZINE—(in barrels) Drop 5 cts. We quote: 20 cts. per gallon.

GASOLINE—(In barrels) We quote: 22½ cts. per gallon.

PETROLEUM—(In barrels) Our prices are: American prime white coal, 15½c. per gallon. American water, 17c. per gallon; Pratt's Astral, 19½c. per gallon.

WINDOW GLASS—Our prices are: First break, 50 feet, $1.85; second break, 50 feet, $1.95; first break, 100 feet, $3.20; second break, 100 feet, $3.40; third break, 100 feet, $3.95; fourth break, 100 feet, $4.15; fifth break, 100 feet, $4.40; sixth break, 100 feet, $4.95. Diamond Star: First break, 50 feet, $2.30; second break, 50 feet, $2.50; first break, 100 feet, $4.40; second break, $4.80; third break, 100 feet, $5.75; fourth break, 100 feet, $6.50; fifth break, 100 feet, $7.50; sixth break, 100 feet, $7.50; seventh break, 100 feet, $8; eighth break, 100 feet, $9. Double Diamond: First break, 50 feet, $3.45; second break, 50 feet, $3.75; first break, 100 feet, $6.75; second break, 100 feet, $7.25; third break, 100 feet, $8.75; fourth break, 100 feet, $10; fifth break, 100 feet, $11.50; sixth break, 100 feet, $12.50; seventh break, 100 feet, $14; eighth break, 100 feet, $16.50; ninth break, 100 feet, $18; tenth break, 100 feet, $20; eleventh break, 100 feet, $24; twelfth break, 100 feet, $28.50. Discount on Diamond Star, 20 per cent.; on Double Diamond, 40 per cent.

TORONTO.

Office of HARDWARE AND METAL,
10 Front Street East,
Toronto, Jan 11, 18 7.

Very few changes have to be reported in paint and oil markets this week. We have changed our last week's quotations of raw and boiled oil and now quote: 1 to 3 barrels, raw, 59c; 4 to 7 barrels, raw, 57c; 8 and over, 56c, adding 3 cents for boiled.

Domestic lines, such as aluminum

varnish, etc., are selling well, and book orders for these articles are received daily. Travelers all report the outlook for the business of 1907 even greater than that of 1906.

WHITE LEAD—Ex Toronto pure white, $7.40; No. 1, $6.65; No. 2, $6.25; No. 3, $5.90; No. 4, $5.65 in packages of 25 pounds and upwards; 1-2c per pound extra will be charged for 12 1-2 pound packages; genuine dry white lead in casks, $7.00.

RED LEAD—Genuine in casks of 500 lbs., $6.00; ditto, in kegs of 100 lbs., $6.50; No. 1 in casks of 500 lbs., $5.75; ditto, in kegs of 100 lbs., $6.25.

DRY WHITE ZINC— In casks, 7 1-2c.; in 100 lbs., 8c., No. 1, in casks 6 1-2c., in 100 lbs., 7c.

WHITE ZINC (ground in oil)—In 25-lb. irons, 8c.; in 12 1-2 lbs., 8 1-2c.

SHINGLE STAIN—In 5-gallon lots, 75c. to 80c. per gallon.

PARIS WHITE—90c. in barrels to $1.25 per 100 lbs.

WHITING—60c. per 100 lbs. in barrels; Gilders' bolted whiting, 90c. in barrels, $1.15 in smaller quantities.

SHELLAC VARNISH—Pure orange in barrels, $2.70; white, $2.82 1-2 per barrel; No. 1 (orange) $2.50; gum shellac, bone dry, 63c., Toronto. T. N. (orange) 51c. net Toronto.

LINSEED OIL—We now quote raw, 1 to 3 barrels, 59c; 4 to 7 barrels, 57c; 8 and over, 56c, add 3c to this price for boiled oil. Toronto, Hamilton, London and Guelph net 30 days.

TURPENTINE—Single barrels, 97c.; 2 to 4 barrels, 96c.; f.o.b. point of shipment, net 30 days. Less than barrels, $1.02 per gallon.

GLUES—French Medal, 12 1-2c. per pound; domestic sheet, 10 1-2c. per lb.

PUTTY—Ordinary, 800 casks, $1.50; 100 drums, $1.75, barrels or bladders, $1.75; 100-lb. cases, $1.90;; 25-lb. irons, $1.85; 25-lb. tins (4 to case) $1.90; 12 1 lb. tins (8 to case) $2.10.

LIQUID PAINTS—Pure, $1.20 to $1.35 per gallon; No. 1, $1.10 per gallon.

BARN PAINTS—Gals., 70c. to 80c.

BRIDGE PAINTS—Gals., 75c. to $1.

CASTOR OIL—English, in cases, 9 1-2c. to 10c. per lb., and 11c. for single tins.

PARIS GREEN—Canadian manufacturers are quoting their base price at 25c and on the English, 25¾ is quoted f.o.b., Toronto.

REFINED PETROLEUM — Dealers are stocking up heavily for the Winter. We still quote: Canadian prime white, 14c.; water white, 16c.; American water white, 16c. to 18c. ex warehouse.

CRUDE PETROLEUM — We quote: Canadian, $1.32; Pennsylvania, $1.58; Ohio, 96c.

Fame is delightful, but as collateral it does not rank high.

Be moderate in the use of every thing except fresh air and sunshine. a man a man.

FACTORY FOR ST. JOHN'S, NFLD.

The Canadian trade agent at St. John Nfld, reports that the Standard Manufacturing Company of that place are about to add to their other lines of business, that of making paint. The manager has gone to the United States to look up the necessary machinery and make other arrangements in connection with the proposal. A small factory making copper paint for the use of vessels has been in operation for some years, and having acquired control of the local demand, and driven the American made article out of the market, has not thought it necessary to advertise it in any way. The total quantity of paint used here is not very large, and the latest available figures give the importations as follows:—From Great Britain, $25,942; from Canada, $24,532; from United States, $16,718; total, $66,742.

The duty on paint is 30 per cent. ad valorem, and it is considered that there is a favorable opening under this protection for a successful industry probably amounting to a value of over $100,000 a year.

A SEALING CEMENT.

The composition of a cement to securely seal receptacles closed by screw-cap (or, in fact, by any other method) must necessarily depend upon the nature of the contents of the vessel. If of an alcoholic, oleaginous or resinous nature, for instance, it would require a cement differing from that which would be required for an aqueous solution.

For the first class of articles a cement made with water-soluble gums, gelatin, etc., would be appropriate, while for the other class, rosin, shellac, etc., would answer. For a sealing wax of the first class, kasein dissolved in a 5 per cent. aqueous solution of borax would answer, and so would any of the following: Borax, 1 part; water, 7 parts; shellac, sufficient to make the solution of the desired thickness. A solution of glue would also answer.

For the second class try a mixture of clear rosin, 12 parts; blond shellac, 20 parts; turpentine, 6 parts; dissolved in oil of turpentine. A little experimentation along these lines will enable you to get a satisfactory article.

TURPENTINE FROM PULP.

A Maine company is successfully making turpentine from spruce pulp waste. Heretofore the product has been obtained practically wholly from southern pine. The Industrial Journal, of Bangor, Me., says that "the turpentine obtained as a by-product in the manufacture of spruce pulp answers all purposes of pine turpentine. The ordinary person could never tell the difference between the two."

The Canadian Pacific Railway has completed another huge irrigation project, in Southern Alberta, by which a valley 150 miles in length by 40 miles in width is being brought under cultivation. This block of irrigated lands alone is estimated to have room for half a million people, and it is a significant fact that 95 per cent. of the present settlers in the district are Americans.

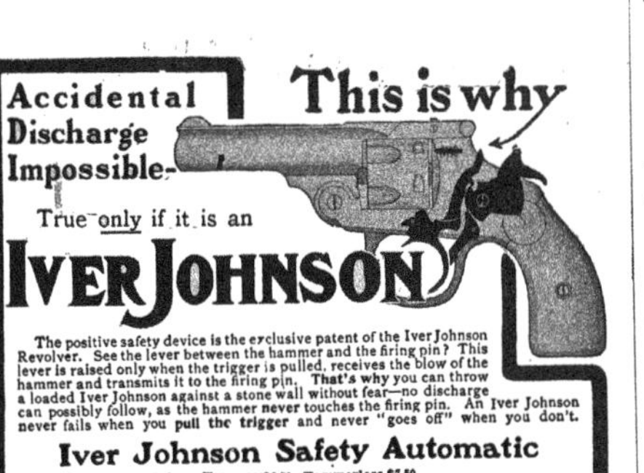

Manufacturers' Agents

CHARLES H. FOX

Vancouver, B.C.

MANUFACTURERS' AGENT

Selling to the Hardware Jobbers Only.

Correspondence Solicited.

ALEXANDER GIBB

Manufacturers' Agent and Metal Broker, 13 St. John Street, Montreal

Representing Canadian, British and American Manufacturers. Correspondence invited from firms wishing to be represented.

H. W. MITCHELL

WINNIPEG, MAN.

Correspondence is solicited from manufacturers wishing a representative in Winnipeg.

Travellers call regularly on the wholesale and retail trade in Western Canada.

Highest references and financial responsibility.

Scott, Bathgate & Co.

BROKERS AND COMMISSION MERCHANTS

Cor. Notre Dame St. East and Victoria St., WINNIPEG, Canada

Storage and Warehousing a Specialty.

First-class facilities in a thoroughly modern brick warehouse centrally located on the railway track.

A Straight Story

We do not manufacture the only dependable paint on earth—but we manufacture **one of the best**—a paint you can rely on and recommend to your best friends—a paint which builds up a paying trade as surely as the ebb and flow of the tide.

Ark Brand Paint

That's it. You practical business men know enough about the science of paint making to know that we have reached the limit of goodness in Ark Brand. There may be a better paint for the Canadian dealer to handle some day—**Ark Brand is the best to-day.**

MURESCO WALL FINISH

Up till now we have failed to discover a better wall finish than **Muresco.** There's nothing to equal it on the market, and it pays the dealer well to handle it.

THE BENJAMIN MOORE CO., LIMITED

FORMERLY THE FRANCIS FROST CO., LIMITED

New York TORONTO, CANADA Chicago

THE STANDARD IDEAL SANITARY CO., LIMITED

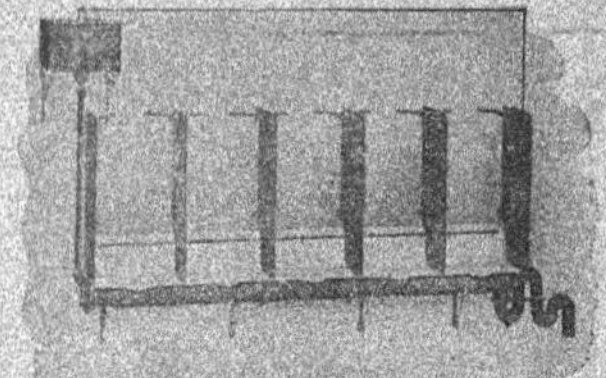

Plate E. 404.—For Public Institutions, Schools, Factories, Asylums, &c.

We manufacture a complete line of

Range-Closets, Range-Urinals, Lavatories and Sanitary Fixtures

Suitable for Public Institutions

The reputation of our goods is unequalled.

Toronto: 50 Colborne Street — ***Montreal: 128 Craig Street West***
Winnipeg: 24 Telfer Building

HOME OFFICE and FACTORIES: PORT HOPE, ONT.

"Valve Quality"
The J.M.T. Valve

The greatest source of satisfaction to the engineer is probably a ***GOOD VALVE***. J.M.T. Valves are the best for all purposes. They're heavy and well made. Have removable discs.

The James Morrison Brass Mfg. Co., Limited

Manufacturers of the largest varieties of Steam Goods in the Dominion.

93-97 West Adelaide St.
TORONTO

Plumbing and Steamfitting

BRANCH VENTILATING PIPES.

The increasing efficiency of sanitary plumbing work during recent years, and the different methods of guarding against the escape of foul air from soil and waste pipes which have been introduced by various inventors and patentees, have been the cause of much discussion among those who advise and direct the work and those who have to find the money for carrying it out.

It was many years before the public in general, and architects in particular, could be persuaded that it was necessary to fix taps in all waste pipes immediately under the fittings, whether they were baths, sinks, or lavatory basins, in order to prevent the foul air from the waste pipes finding its way into the apartments in which these fittings were situated. It was supposed that if the waste pipe discharged over a gulley trap, however foul this may have been, the trap in the waste pipe was not merely unnecessary, but a thing to be avoided. Some, indeed, maintain that traps in waste pipes are a serious obstruction, and describe them as miniature cesspools, while they leave out of account the much larger cesspool, which in many cases is so constructed as to accumulate filth in a dangerous manner. And, strange to say, the local sanitary authorities in many places supported the idea, and actually prohibited the use of traps in the waste pipes. Such a state of things, is, however slowly passing away, and in the modern sanitary by-laws there are clauses which insist upon the fixing of self-cleansing traps in every waste pipe. Moreover, every branch waste pipe near the trap has to be fitted with a branch ventilating pipe, or "puff pipe" as it is sometimes called, leading to external air. This is regarded by many people as unnecessary extravagance. And it must be acknowledged that under some circumstances such a branch ventilating pipe is not really required for retaining the water seal of the trap.

In the case of short waste pipes in connection with flat bottom sinks which are fitted with a waste grating, and not with a plug waste, the necessity does not arise. But where there are several fittings on various floors on one waste pipe every trap must be provided with a branch ventilating pipe in order to preserve the water seal from syphonage, and the local sanitary authority is acting within its rights in insisting on this regulation being carried out. But the point is, have the sanitary authority power to go beyond the letter and spirit of the by-laws and insist on the branch-ventilating pipes being connected to one main ventilating pipe, and either carried above the roof or connected with the upper end of the main waste pipe, and treated in the same way as an anti-syphonage pipe in connection with a stack of soil pipe? This, we understand, is being demanded in several instances, particularly in blocks or residential flats, and, even if it were desirable, has the sanitary authority the power to enforce it, if it is not provided for in the by-laws? Then, if it is done, surely the grating over the gulley into which the main waste pipe discharges should be closed also, because there must be a much larger volume of foul air escape from this grating than is likely to issue from the ends of the small branch ventilating pipes.

A still more important point is that, if it is necessary to provide as complete a fresh air break as possible between the fittings and the drains, then the shorter length of ventilating pipe to the open air the better, beside the importance of having an open grating over the gulley at the foot of the waste pipe. The whole contention of joining-up branch vents is to a large extent based on a misconception, for it can be shown by experiment that instead of the vents discharging air from the waste pipe, at the face of the wall, where they terminate, they are nearly always acting as air inlets, supplying air to the main waste and ventilating pipe which is carried above the roof.—English Plumbers' Review.

CLOSET CONNECTIONS TO SOIL PIPE STACK

[Two more letters sent in replying to the question, the prize-winning answer to which was published in HARDWARE AND METAL of December 29.]

The letters published to-day in answer to our query re break syphons, are self-explanatory, that by Mr. Ross having the decided advantage of being accompanied by an illustration.

The letter by Mr. Julien contains

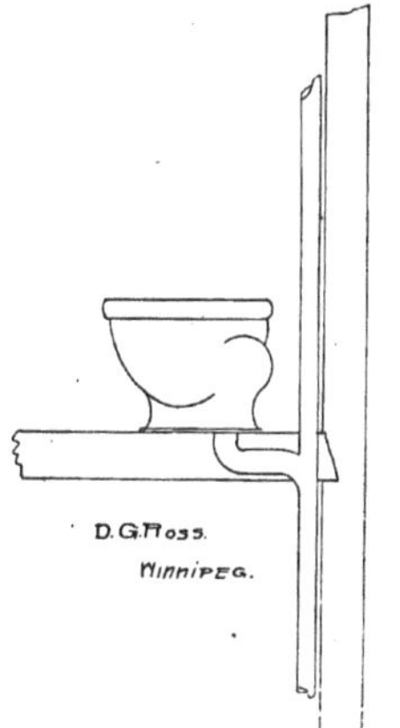

many good arguments, though we cannot agree that under all circumstances a job fitted throughout with pot traps without vents would be perfectly sanitary.

Connection Not Necessary.

In my opinion it is not necessary that every closet connection to soil pipe stack should be fitted with a break syphon. Up-to-date and more sanitary plumbing goods are now on the market which makes it unnecessary for these break syphons to be used The centrifugal lead trap is absolutely non-syphoning and self-cleansing and does away with the vent. There is a disadvantage too where these air vents have been put in, for sooner or later they become choked up, thus causing great inconvenience. If an up-to-date closet is being fixed, the space between bowl and soil pipe stack would be about 18 inches, and a 4-inch lead bend would connect the two. On the accompanying sketch it will be at once seen why a break syphon is unnecessary. The closet cannot syphon, for when it is flushed the discharge of water is not sufficient to fill the short length of pipe or bend used, and the quantity of foul air between bowl and soil pipe stack is reduced to a minimum.

D. G. ROSS, Winnipeg.

I consider that no back vent should be put to closets, when no more than three feet from stack, or even longer, for a horizontal run of 4-inch pipe is never filled with the rush of water from a closet, hence cannot be syphoned. And, as the closet is always put near the stack, I am convinced that no back vent is necessary. I will go still further re back vents, I would call a job perfect if it was done with pot traps.

Do away with back vents, as in large cities. Many people leave their homes for the country, and when they come back they find that their house is filled with sewer gas. There is not much seal in the ordinary trap, with so much circulation of air through the main stack and back vent, that it evaporates the water from said trap.

These are my personal views of modern plumbing, as done in our city at present.

GIL. JULIEN, Ottawa.

Pull down the show window blinds and your clerks can all take a rest.

The merchant who makes money does it by close attention to business.

QUALITIES OF A GOOD SALESMAN

R. T. CRANE, in the Valve World.

Some people seem to believe that a good salesman is one who is able to deceive, humbug, cheat and defraud the buyer, and this, aside from the question of morality, and considered only from the money-making point of view, may be true where the business is such that the seller does not expect to meet or sell to the same person a second time. We claim, however, that any enterprise, which is conducted in that manner is not entitled to the dignity of being classed as a business, for it is nothing more than a swindle.

We maintain that in an ordinary business it is not good policy for salesmen to practice the slightest deception, and one of the first things that we require of our men is that they shall be absolutely honest, frank and fair in their treatment of our customers. We think the value to a salesman of having these qualities appear in his every feature and action and suggested by every word that he utters cannot be overestimated.

Every purchaser is, we believe, desirous of dealing with a house that supplies honest goods at honest prices, that carries stocks in sufficient variety and quantity to take care of his requirements without delay, and where he can be sure of always obtaining the same line of goods.

It is a great relief to a buyer when he finds such a concern, and he should at once see that it is to his interest to confine his dealings to that house, as he is usually so fully occupied in looking after the other portions of his work that he cannot afford to spend his time endeavoring to ascertain whether he has been defrauded either in the quality of goods or in prices.

Trade Only With Reliable Firms.

From our own experience, we know it is a great satisfaction, when purchasing goods, to feel that we are dealing with a house having such a reputation. In fact, we have made it a rule to trade only with concerns of that kind.

For a firm to gain and maintain a reputation for fair dealing, it is not enough for the firm itself to be honest, but the policy of the house must be reflected by the salesmen, who, in many cases, are the only persons connected with the firm the trade ever meet. Hence the importance of having salesmen of good character, who, in soliciting business, make only such statements as the house can live up to. It is apparent, therefore, that an honest concern cannot afford to employ untruthful salesmen, as nothing will cause it greater injury than to have such men out among the trade.

On the other hand, a house selling inferior goods needs salesmen who can and will lie sufficiently to conceal the inferiority beneath a cloak of plausible phrases. Therefore, when we find by experience a salesman whose statements are always correct, and who will not stoop to misrepresentation, we are convinced that his house in honest and reliable, and are disposed to give him our business.

For a salesman to impress the trade in this way involves something more than simply refraining from lying. He must confine himself to what he knows to be true, neither deliberately lying, nor mixing up guesses, opinions and facts in such a way as to deceive or mislead the customer, who, after having discovered a deception, can rarely be sold to again by the salesman, with the result that the house loses a customer.

Salesmen Must Know Goods.

Many salesmen are apt to be slovenly and careless in their replies to questions, sometimes possibly with the intention to deceive, but probably many more times for the purpose of concealing their ignorance on the subject about which they are questioned. This is a practice that we most strongly condemn. We very much prefer that our men shall acknowledge their ignorance to the customer and suggest to him that he apply to the house for information, if the matter is of sufficient importance to take that trouble. While salesmen are not expected to give customers unusual or technical information, still it is exceedingly desirable that they should have an extensive knowledge of the goods they are handling so as to be able to describe their goods and answer questions likely to be asked.

Occasionally a salesman is asked his opinion as to market conditions. Our salesmen are instructed not to give any opinion in the matter of prospective rise or fall in prices, as in our judgment salesmen are not in a position to be reliably informed as to the prospective trend of prices, and we cannot afford to have them deceive or mislead the trade in this way, any more than we can afford to have them misrepresent the quality of our goods. In any event, we do not care to encourage our customers to gamble on the market. However, if customers have contracts on which they would stand to suffer loss if prices were advanced, we feel that it would not be at all improper for our salesmen as a matter of safety to the customers, to recommend that they place orders to cover such contracts.

Cut Out Treating Habit.

One mistake that salesmen frequently make is to urge an audience with a man when he is busy, thereby annoying him so that he will not give them a proper reception. Besides being tactful, it goes without saying that salesmen should be men of good address and affable in manner.

Another feature of the subject which some of the older people in business will remember was a common practice in the past, was the custom of drinking and carousing with customers. In those times some dealers kept a bar in their stores, others would have jugs or bottles of whisky on hand for their customers, and it was looked upon as being rather necessary for salesmen to take people out and entertain them by drinking and carousing with them in order to hold their trade.

We are pleased to be able to state that this practice has almost totally disappeared, business now being transacted more strictly on the basis of merit than was formerly the case. We laid down the rule years ago that if a man's trade could not be obtained without having our salesmen dissipate with him, we did not want it at all.

ATTRACTIVE ANNIVERSARY NUMBER.

The Plumbers' Trade Journal of New York have issued a very attractive anniversary and holiday number, dated January 1st, this being the first issue of their forty-first year. The book comprises 132 pages of well gotten up reading and advertising matter, innumerable illustrations being given to illustrate the remarkable development of the plumbing and heating trades during the lifetime of the paper. To read the various articles and study the illustrations showing the first buildings occupied by industries which are now known to the trade throughout the entire world is very instructive, the accompanying reading matter being also very interesting, showing the lack of sanitary conveniences a quarter of a century ago and the wonderful improvements which have been made since that time. An interesting article to Canadians as well as to the plumbers across the line is one entitled "A Universal Plumbing Code," by Herbert F. Shade, Victoria, B.C., which we will reproduce in a coming issue of Hardware and Metal. The publishers are to be congratulated on the excellence of their anniversary issue.

PLUMBERS' BRASS GOODS.

Catalogue B, for 1907, of the United Brass Manufacturing Company, Cleveland, Ohio, has just been issued to the trade, it being 7x9 inches, containing about 200 pages fully illustrated of their many different lines for water, gas and steam, also all supplies for bath-room and hot water heating. This catalogue will no doubt be found of great value to hardware and plumbing merchants, as the complete lines manufactured by this firm, fully illustrated, will be an invaluable aid in ordering goods, and will be sent to any merchant interested in these goods, by writing, mentioning this paper.

NEW SHOP AT HAILEYBURY.

C. A. McKane & Co. have opened a plumbing, hot water and steam heating shop at Haileybury. Mr. McKane was formerly in business at Newmarket and prior to that time with the Office Specialty Manufacturing Co., and connected with the plumbing and heating trades in Toronto. He should find a good opening for his energies in the rapidly developing north country. Mr. McKane was a caller on Hardware and Metal in Toronto on Tuesday, being now engaged in buying his stocks of material for the coming season.

PLUMBING FAILURE.

The Thorn Plumbing Co., of Toronto has been wound up on the application of the Gurney Foundry Co., who are creditors to the extent of $1,069. The company was incorporated in October, 1905, with a nominal capital of $20,000.

PLUMBING MARKETS

TORONTO.

Office of HARDWARE AND METAL,
10 Front Street East.
Toronto, January 11, 1907

For some weeks we have been looking for an advance on hot water and steam boilers owing to the big jumps which have taken place in the pig iron market since the last changes were made on boiler quotations. Higher prices are certainly warranted by the condition of the iron market, but events during the past week have made it practically certain that present prices will continue unchanged for some time to come.

As predicted by us a month ago, an advance of about 10 per cent. has been announced on enamelware; the new discounts on various lines being quoted below. The condition of the iron market warrants higher prices, and the advanced prices will undoubtedly hold firm during the coming season, as all foundries find it difficult to secure stocks of iron owing to the mills and furnaces being booked so far ahead with orders.

Iron pipe has also been subject to several changes. The new list issued this week of sizes up to 2 1-2 inches, shows an advance, while quotations on larger sizes are also revised. Still another change is in straps and bends, the discount on which has been reduced to 40 per cent. There is also talk of another advance in cast iron fittings.

As is to be expected at this season, trade in both plumbing and heating supplies is none too active. In anticipation, however, of advanced prices, many plumbers are booking their orders ahead for Spring delivery.

LEAD PIPE — The discount on lead pipe continues at 5 per cent. off the list price of 7c. per pound. Lead waste, 8c. per pound with 5 off. Caulking lead 5 3-4c. to 6c. per pound. Traps and bends, 40 per cent., discount.

SOIL PIPE AND FITTINGS.—New lists will show advances as follows: Medium and extra heavy pipe and fittings, 60 per cent.; light pipe 50 per cent.; light fittings, 50 and 10 per cent.; 7 and 8 inch pipe, 40 and 5 per cent.

IRON PIPE.—New lists show advances on all lines up to 2½ inches. 1-inch black pipe is quoted at $4.95, and one-inch galvanized at $6.60. Full list in current market quotations.

IRON PIPE FITTINGS — New lists are as follows: Cast iron elbows, tees, crosses, etc., 62½ per cent.; cast iron plugs and bushings, 62½ per cent.; flange unions, 62½ per cent.; nipples, 70 and 10 per cent.; iron cocks, 55 and 5 per cent.; Canadian malleable, 30 per cent.; malleable unions, 55 and 5 per cent.; malleable bushings, 55 per cent.; cast iron ceiling plates, plain 65 per cent.; cast iron floor, 70 per cent.; hook plates, 60 per cent.; expansion plates, 65 per cent.; headers, 60 per cent, hangers, 65 per cent.; standard list.

GALVANIZED IRON RANGE BOILERS—We quote: 12 gallon capacity, standard, $4.50; extra heavy, $6.50; 18 gallon standard, $4.75; extra heavy, $6.75; 24 gallon, standard, $4.75; extra heavy, $6.75; 30 gallon, standard, $4.75; extra heavy, $7.50; 35 gallon, standard, $5.75; extra heavy, $8.50; 40 gallon, standard, $6.75; 40 gallon, extra heavy, $9.50; 52 gallon, $11; extra heavy, $14; 66 gallon, standard, $18; extra heavy, $20; 82 gallon, standard, $21; extra heavy, $24; 100 gallon, standard, $29; extra heavy, $34; 120 gallon, standard, $34; extra heavy, $40; 144 gallon, standard, $47; extra heavy, $55. Copper range boilers are now net list.

RADIATORS—Prices are very stiff at: Hot water, 47½ per cent.; steam, 50 per cent.; wall radiators, 45 per cent.; specials, 45 per cent. Hot water boilers are still subject to unchanged prices in spite of heavy advances on foundry iron.

SOLDER — Quotations remain firm as follows: Bar solder, half-and-half, guaranteed, 27c.; wiping, 23c.

ENAMELWARE—New lists on enameled baths issued by the Standard Ideal Company on Jan. 3, show a 10 per cent. advance. Lavatories, first quality, 20 and 5 to 20 and 10 off; special, 20 and 10 to 30 and 2½ per cent. discount. Kitchen sinks, plate 300, firsts 60 and 10 off; specials, 65 and 5 per cent. Urinals and range closets, 15 off. Fittings extra.

MONTREAL.

Office of HARDWARE AND METAL,
232 McGill Street.
Montreal, January 11, 1907

General business is excellent. Wholesalers report that the way business has kept up lately is something wonderful and heretofore practically unknown.

The expected advance in iron pipe is no longer a thing of the future—it has arrived.

RANGE BOILERS.—Following prices are well maintained: Iron clad, 30 gallon, $5; 40 gallon, $6.50 net list. Copper, 30 gallon, $25; 35 gallon, $29.50; 40 gallon, $32 net.

LEAD PIPE—Remains firm. We still quote 5 per cent. discount f.o.b. Montreal, Toronto, St. John, N.B., Halifax: f.o.b., London, 15c per hundred lbs extra; f.o.b. Hamilton, 10c per hundred lbs. extra.

IRON PIPE FITTINGS—The usual trade is being done at unchanged prices. we quote: Discounts on nipples, 1-4 inch to 3 inch, 75 per cent., 3½ inch to 2 inch, 55 per cent.

IRON PIPE—Supplies are still short and the looked-for advance has arrived. Standard pipe in lots of 100 feet, regular lengths, 1-4 inch, $5.50; 3-8 inch, $5.50; 1-2 inch, $8.50; 3-4 inch, 11.50; 1 inch, $16.50; 1 1-4 inches, $22.50; 1 1-2 inches, $27.00; 2 inches, $36.50, discounts on black pipe, ¼ inch, 59 per cent.; ⅜ inch, 59 per cent.; ½ inch, 68 per cent.; Discounts on galvanized pipe: 1-4 inch, 44 per cent.; ⅜ inch, 44 per cent.; ½ inch, 58 per cent.; ¾ to 2 inch, 60 per cent. Extra heavy pipe of 100 feet lots are quoted as follows: 1-2 inch, $12; 3-4 inch, $15; 1 inch, $22; 1 1-4 inches, $30; 1 1-2 inches, $36; 2 inches, $50. The discounts on black pipe, ¼ inch, 74 per cent.; ⅜ inch, 69 per cent.; ½ inch to 2 inches, 68 per cent. Galvanized, ¼ inch, 59 per cent.; ⅜ inch, 69 per cent.; ½ to 2 inches 58 per cent.

SOIL PIPE AND FITTINGS—Our prices remain as follows: Standard soil pipe, 50 per cent. off list. Standard fittings, 50 and 10 per cent. off list; medium and extra heavy soil pipe, 60 per cent. off. Fittings, 60 per cent. off.

SOLDER—Prices remain the same: Bar solder, half-and-half, guaranteed 25c.; No. 2 wiping solder, 22c.

ENAMELWARE.—New list issued on Jan. 3 on Canadian ware shows an advance of 10 per cent. Lavatories, discounts, 1st quality, 30 per cent.; special, 30 and 10 per cent. Sinks, 18 x 30 inch, flat rim, 1st quality, $2.60, special, $2.45.

ATTRACTIVE DESK DECORATION.

Cluff Bros., Toronto, selling agents for Warden King & Son, Montreal, have supplied their customers with a very attractive article for desk decoration in the form of an artistically printed and framed quotation of Josh Billings. It is printed in several colors, and contains very valuable tables on the back, giving capacities, dimensions and price list of the Daisy Hot Water Boiler. It will be appreciated by every steam fitter or hardware merchant who receives a copy. Any who have not already received one should forward their name to Cluff Bros. at once before the supply is exhausted. Along with the desk decoration goes a neatly printed souvenir card wishing the trade a prosperous New Year.

TAYLOR FAMILY REUNION.

For the first time in many years the five Taylor brothers, all of whom are well known to the hardware trade throughout Canada, met in Toronto and celebrated their reunion by a dinner at the King Edward Hotel on Tuesday. The five brothers are: Jos. W. Taylor, heating engineer, Johannesburg, South Africa; W. W. Taylor, with Pierce, Butler & Pierce of Syracuse, N.Y., probably America's largest plumbing and heating engineers; John M. Taylor, and Adam Taylor, president and secretary respectively of the Taylor, Forbes Company, of Guelph, manufacturers of Sovereign Hot Water Boilers and Radiators, and James F. Taylor, Toronto, a director of the Taylor, Forbes Company.

THE NEW
"SOLID-ADJUSTABLE"
HAND STOCK AND DIES

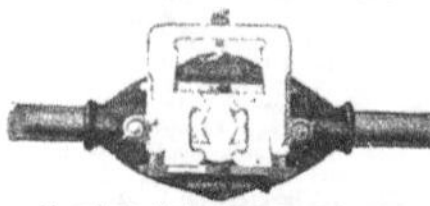

At prices that will warrant your inquiries.

I. E. SHANTZ & CO.

BERLIN, - ONTARIO

Selling Agents for Canada

This is the Mark that Means Something

It is the distinctive brand of genuine Armstrong Stocks and Die and Pipe Fitters' Tools.
No "Armstrong pattern" or "just as good as Armstrong" have this stamp. Don't buy anything in this line that hasn't this brand as it is the mark of genuineness.

THE ARMSTRONG MANUFACTURING CO.

292 Knowlton Street, Bridgeport, Conn.

A Certain Sense of Satisfaction

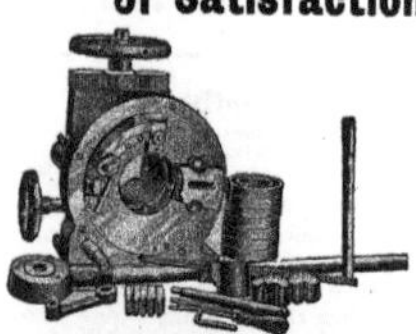

goes with every MERRELL PORTABLE HAND MACHINE, PIPE THREADING and CUTTING MACHINE. We build them that way.

For doing ALL kinds of threading—under ALL conditions—labor trouble and faulty material included—where speed, durability and ease of control count for something, our word to you is to investigate the MERRELL.

Whether it be stationary or portable—hand or power driven—30 DAYS FREE TRIAL must prove conclusively that your choice should be the MERRELL.

This Portable Hand Machine has encased gears, MERRELL standard quick opening and closing die-head, and the latest improved Cutting-off Knife.

Let us tell you more about this machine—The Chasers, Vise and the large range of work covered.

Catalogue for the asking.
Quick shipments.

THE CANADIAN FAIRBANKS CO.

Sole Agents for Canada Limited

MONTREAL, TORONTO, WINNIPEG, VANCOUVER

I Would Like to Know

YOUR WANTS IN ENAMELWARE

that I might prove slow deliveries are **no** part of my business.

It would pay you to test my claims for promptitude.

ORLANDO VICKERY

178-180 Victoria Street TORONTO, CANADA

"Diamond Brand" Fittings

are proclaimed the best by practical men—and they should know.

The Material is first-class—The Threads just right and the Finish as Smooth as can be.

OSHAWA STEAM & GAS FITTINGS CO., LIMITED

OSHAWA, CANADA

When You Buy Demand Proof

Don't take any substitute for Glauber Brass goods. Make the salesman show you. Brass has gone up no less than fifteen per cent., and if you don't get proofs before you buy you take chances of getting something less than you pay for.

WE GIVE YOU THE PROOF and ***2 Cocks Free*** for any one sent you defective.

Glauber Brass Mfg. Co.,

Expert Makers of Plumbers' Fine Brass Goods

CLEVELAND, O.

Sharratt & Newth's Glaziers' Diamonds

are unequalled for cutting and wearing qualities.

To be obtained from the principal Dealers in Glass, Hardware, and Painters' Supplies.

Contractors to H. M. Government and the principal English Sheet and Plate Glass Works

REGISTERED TRADE MARK GRANTED 1682

World's Best Cutlery

You make no mistake in specifying goods made by **JOSEPH RODGERS & SONS, Limited** when ordering cutlery. A reputation of over 200 years at stake, insures the quality of every article.

Write for catalogues from Sole Canadian Agents

JAMES HUTTON & COMPANY, MONTREAL

Forwell Foundry Co.

BERLIN, ONT.

Manufacturers of

SOIL PIPE, FITTINGS, and CAST IRON SINKS

Ask Jobbers for "F. F. CO." Brand

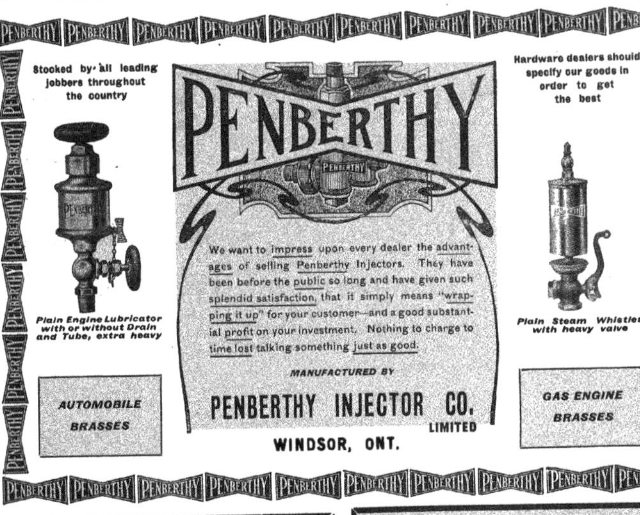

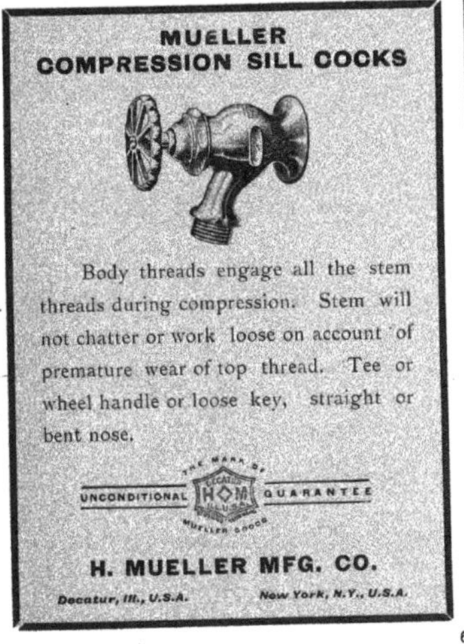

We have pumps duitable to any demand mese upon you by most exatc ing customers.

Our pumps will do anything that a pump can do, and we will be pleased to give the benefit of many years' experience to any one having pump troubles.

McDougall Pumps are Standard Pumps, and with us it is quality first, last and always, while we ask just as little money as we can get along with.

We have a few catalogues for distribution

The R. McDougall Co., Limited

GALT, - CANADA

CURRENT MARKET QUOTATIONS.

Jan. 11, 1907.

These prices are for such qualities and quantities as are usually ordered by retail dealers on the usual terms of credit, the lowest figures being for larger quantities and prompt pay. Large cash buyers can frequently make purchases at better prices. The Editor is anxious to be informed at once of any apparent errors in this list, as the desire is to make it perfectly accurate.

METALS.

ANTIMONY.

Hallett's..............per lb... 0 27½ 0 28

BOILER AND T.K. PITTS.

Plain tinned .... }
Spun.................. } 25 per cent. off list.

BABBIT METAL.

Canada Metal Company—Imperial genuine 60c; Imperial Tough, 65c; White Brass, 50c; Metallic, 35c; Harris Heavy Pressure, 25c; Hercules, 25c; White Bronze, 15c; Star Frictionless, 14c; Alluminoid, 10c; No. 4, 9c. per lb.

BRASS.

Rod and Sheet, 14 to 30 gauge, net list.

Sheets, 12 to 14 in.	0 27
Tubing, base, per lb ⅛-16 to 2 in	0 33
Tubing, ¼ to 3-inch, iron pipe size..	0 33
" 1 to 3-inch, seamless..	0 40

Copper tubing, 5 cents extra.

COPPER.

	Ingot.	Per 100 lb.
Casting, car lots............	25 50	26 00
	Bars.	
Cut lengths, round, ½ to 2 in...	....	32 00
	Sheet.	
Plain, 16 oz., 14x48 and 14x60	....	30 00
Plain, 14 oz..................	....	31 00
Tinned copper sheet, base	....	33 00
Planished base................	....	37 00
Braziers (in sheets) 4x6 ft., 25 to 30 lb. each, per lb., base..	....	0 30

BLACK SHEETS.

	Montreal.	Toronto
8 to 10 gauge	2 60	2 60
12 gauge	2 60	2 65
14 "	2 60	2 60
17 "	2 40	2 50
18 "	2 40	2 55
20 "	2 40	2 53
22 "	2 40	2 70
24 "	2 45	2 75
26 "	2 55	2 85
28 "	2 60	3 00

CANADA PLATES.

Ordinary, 52 sheets	3 00
All bright "	4 01
Galvanized Canada Plates, 52 sheets	4 35
" " " 60 "	4 60

	Ordinary.	Dom. Crown.
18x24x52	4 25	4 35
" 60	4 50	4 60
20x28x80	8 50	8 70
" 94	9 00	9 20

GALVANIZED SHEETS.

	Fleur-de-Lis.	Gordon Crown.
16 to 20 gauge	3 60	3 95
22 to 24 gauge	3 95	4 00
26 "	4 30	4 20
28 "	4 45	4 50

Apollo.	
10¾ oz. (American gauge)........	4 70
28 gauge........ "	4 45
26 " "	4 15
24 " "	3 90

	Comet	Queen's Head.	Bell.
16 to 20 gauge	3 60	3 75	....
22 to 24 gauge	3 75	3 90	....
26 "	4 30	4 45	....
28 "	4 45	4 60	4 45

Less than case lots 10 to 25c. extra.

IRON AND STEEL.

	Montreal.	Toronto.
Common bar, per 100 lb......	2 15	2 30
Forged iron "	2 40	..
Refined " "	2 55	2 70
Horseshoe iron "	2 55	2 70
Hoop steel, 1½ to 3 in. base...	....	2 80
Sleigh shoe steel "	2 25	2 30
Tire steel	2 40	2 50
Best sheet cast steel.........	...	0 12
B. K. Morton "Alpha" high speed.		0 65
" annealed	....	0 70
"M" Self-hardening	....	0 50
"J" quality, best warranted	....	0 18
"I" " warranted	....	0 14
"B.C" quality	....	0 09
Jonas & Colver's tool steel...	0 10	0 20
" "Novo"	....	0 65
" " annealed	....	0 65
Jowett & Sons B P.L. tool steel....		0 10½

INGOT TIN.

Lamb and Flag and Straits—56 and 28-lb. ingots, 100 lb. $46 50 $47 00

TINPLATES.

Charcoal Plates—Bright

		Per box.
M.L.S., equal to Bradley—		
I C, 14 X 20 base	....	$6 50
IX, 14 x 20 "	....	8 00
IXX, 14 x 20 base	....	9 50
Famous, equal to Bradley—		
I C, 14 x 20 base	....	6 50
I X, 14 x 20 "	....	8 00
I X X, 14 x 20 base	....	9 50
Raven and Vulture Grades—		
I C, 14 x 20 base	....	5 00
I X "	....	6 00
I X X "	....	7 00
I X X X "	....	8 00

"Dominion Crown Best"—Double Coated, Tissued.

		Per box.
I C, 14 x 20 base	....	5 75
I X, 14 x 20 "	....	6 75
IXX, 14 x 20 "	....	7 75

"Allaway's Best"—Standard Quality.

I C, 14 x 20 base	....	4 50
I X, 14 x 20	....	5 25
I XX, 14 x 20 "	....	6 00

Bright Cokes.

Bessemer Steel—		
I.C., 14 x 20 base	...	4 25
20x28, double box	....	8 50

Charcoal Plates—Terne

Dean or J. G. Grade—		
I.C., 20x28, 112 sheets	....	8 00
IX., Terne Tin	...	9 50

Charcoal Tin Boiler Plates.

Cookley Grade—
X X, 14x56, 50 sheet bxs. }
" 14x60, " }
" 14x65, " } 7 50

Tinned Sheets.

72x30 up to 24 gauge	....	8 50
" " 26 "	....	9 00

LEAD.

Imported Pig, per 100 lb......	5 40	5 50
Bar, "	5 75	6 00
Sheets, 2½ lb. sq. ft., by roll ..	....	0 06½
Sheets, 3 to 6 lb "	....	0 06

NOTE.—Cut sheets ½c. per lb., extra Pipe, by the roll, usual weights per yard, lists at 7c. per lb and 5 p.c. dis. f o b Toronto.

NOTE.—Cut lengths, net price, waste pipe 8-ft lengths, lists at 8c.

SHEET ZINC.

5-cwt. casks	8 00	8 25
Part casks	8 25	8 50

ZINC SPELTER.

Foreign, per 100 lb	7 25	7 50
Domestic	7 00	7 25

PLUMBING AND HEATING

BRASS GOODS, VALVES, ETC.

Standard Compression work, dis half inch 50 per cent., others 40 per cent.
Cushion work, discount 40 per cent.
Fuller work, ½-inch 70 per cent., others 50 p.c.
Flatway stop and stop and waste cocks, 50 per cent.; roundway, 45 per cent.
J.M.T. Globe, Angle and Check Valves, discount 50 per cent.
Standard Globe, Angle and Check Valves discount 52½ per cent.
Kerr standard globes, angles and checks, special, 42½ per cent; standard, 47½ p c
Kerr Jenkins' disc, copper-alloy disc and heavy standard valves, 45 per cent.
Kerr steam radiator valves 60 p.c., and quick-opening hot-water radiator valves, 60 p.c.
Kerr brass, Weber's straightway valves, 42½ per cent.; straightway valves, I.B.B.M., 60 per cent.
J. M. T. Radiator Valves, discount 50 per cent
Standard Radiator Valves, 60 per cent.
Patent Quick-Opening Valves, 65 per cent.
Jenkins' Bros. Globe Angle and Check Valves discount 32½ per cent.

No. 1 compression bath cock......net		2 00
No. 4 " "	"	1 90
No 7 Fuller's	"	2 25
No. 4½ "	"	2 35

Patent Compression Cushion, basin cock, hot and cold, per doz., $16.20
Patent Compression Cushion, bath cock, No. 2208 .... 2 26
Square head brass cocks, discount 50 per cent.
" " iron " " 60 "
Thompson Smoke-test Machine $25.00

BOILERS—COPPER RANGE

Copper, 30 gallon	25 00
" 35 "	29 00
" 40 "	33 00

Net list.

BOILERS—GALVANIZED IRON RANGE.

Capacity.	Standard.	Extra heavy
30-gallons....	4 75	7 50
35 "	5 75	8 50
40 "	6.75	9.50

2 per cent., 30 days

BATH TUBS.

Steel clad copper lined, 15 per cent.

CAST IRON SINKS.

16x24, $1, 18x30, $1; 18x36, $1.30.

ENAMELED CLOSETS AND URINALS

Discount 15 per cent.

ENAMELED BATHS.

List issued by the Standard Ideal Company Jan. 3, 1907, shows an advance of 10 per cent. over previous quotations.

ENAMELED LAVATORIES.

	1st quality.	Special.
Plate E 100 to E 103 .	20 & 5 p.c.	20 & 10 p.c.
" E 104 to E 132....	20 & 10 p.c.	30 & 2½ p.c.

ENAMELED ROLL RIM SINKS.

	1st quality.	Special.
Plate E 201, one piece	15 & 2½ p.c.	15 & 10 p.c.

ENAMELLED KITCHEN SINKS.

Plate E, flat iron 300, 60 & 10 p.c. 65 & 5 p.c.

LEAD PIPE

Lead Pipe, 7c. per pound, 5 per cent. off.
Lead waste, 8c. per pound, 5 per cent. off.
Caulking lead, 6c. per pound.
Traps and bends, 40 per cent.

IRON PIPE.

Size (per 100 ft.)	Black.		Galvanized
⅛ inch	2 25	⅛ inch	3 08
¼ "	2 25	¼ "	3 08
⅜ "	2 32	⅜ "	3 57
½ "	3 45	½ "	4 60
1 "	4 95	1 "	6 80
1¼ "	6 75	1¼ "	9 00
1½ "	8 10	1½ "	10 80
2 "	10 80	2 "	14 40
2½ "	20 00	2½ "	28 00
3 "	26 00	3 "	36 50
3½ "	33 00	3½ "	45 00
4 "	37 00	4 "	52 50

2 per cent. 30 days.

Malleable Fittings—Canadian discount 30 per cent.; American discount 25 per cent.
Cast Iron Fittings 62½; Standard bushings 62½ per cent.; headers, 60; flanged unions 62½, malleable bushings 55; nipples, up to 2 in., 70 and 10 per cent; up to 6 in., 65 per cent.; malleable lipped unions, 55 and 5 per cent.

SOIL PIPE AND FITTINGS

Medium and Extra heavy pipe and fittings, up to 6 inch, discount 60 per cent.
7 and 8-in. pipe, discount 40 and 5 per cent.
Light pipe, 50 p c.; fittings, 50 and 10 p.c.

OAKUM.

Plumbers ....per 100 lb...... 4 00 4 75

RADIATORS, ETC.

Hot water	47½ p.c.
Steam	50 p.c.
Wall Radiators and Specials	45 p.c.

STOVES, BOILERS, FURNACES, REGISTERS.

Discounts vary from 40 to 70 per cent. according to list.

STOCKS AND DIES.

American discount 25 per cent.

SOLDERING IRONS.

½-lb.	per lb.	0 37
1-lb. or over	"	0 34

SOLDER.

	Montreal	Per lb. Toronto
Bar, half-and-half, guaranteed	0 25	0 27
Wiping	0 22	0 21

PAINTS, OILS AND GLASS.

COLORS IN OIL.

Venetian red, 1-lb. tins, pure.	....	0 09
Chrome yellow "	....	0 15
Golden ochre "	....	0 08
French " "	....	0 08
Marine black "	....	0 04½
Chrome green "	....	0 10
French permanent green "	....	0 13
Signwriters' black "	....	0 15

GLAZIER POINTS.

Discount, 5 per cent.

GLUE.

Domestic sheet	0 10	0 10½
French medal..	0 12	0 14

PARIS GREEN.

	Bergers	Canadian
600-lb. casks	0 23½	0 23½
250 lb. drums	0 23½	0 23½
100-lb. "	0 24	0 24
50-lb. "	0 24	0 24
25-lb. "	0 24½	0 24½
1-lb. pkgs, 100 in box	0 25	0 25
1-lb. " 50 "	....	0 25½
1-lb. tins, 100 "	0 21	0 26
½-lb. pkgs.	0 27	0 27

2 p. c. 30 days from date of shipment.

PREPARED PAINTS.

Pure, per gallon, in tin 1 20
Second qualities, per gallon 1 00
Barn (in bbls.) ... 0 80 0 90
Sherwin-Williams paints, ½ gal .... 1 5.
" " ¼ gal. 1 45
" " gal... .... 1 40
Canada Paint Co.'s pure 1 30
Standard Co.'s "New Era." 1 30
Benj. Moore Co.'s "Ark" B'd 1 25
" British Navy deck 1 50
Brandram-Henderson "Anchor" 1 35
Ramsay's paints, Pure, per gal. 1 20
" Thistle, " 1 00
" Outside, bbls 0 55 0 65
Martin-Senour's 100 p.c pure, ½ gal. 1 45
" " " ¼ gal. 1 40
" " " 1 gal 1 35
Jamieson's "Crown and Anchor" 1 20
Jamieson's floor enamel 1 50
" barn paints, bbls. per gal. 0 60
Senour's Floor Paints ... gal. 1 20
Sanderson Pearcy's, pure 1 20
Robertson's pure paints 1 2)
W. H. & C's English raw, bbls. 0 63
" " boiled, " 0 66

PUTTY.

Bulk in bbls ... 1 50
Bulk in less quantity ... 1 80
Bladders in bbls ... 1 80
Bladders in kegs, boxes or loose ... 1 95
25-lb. tins ... 1 80
12½ lb. tins ... 2 05
Bladders in bulk or tins less than 100 lb. 1 85
Bulk in 100-lb. irons ... 1 80

DRY LEAD.

	white	red.
Genuine, 560 lb. casks, per cwt	6 75	6 00
Genuine, 100 lb. kegs, "	7 50	6 25
No. 1, 560 lb. casks, per cwt ...	....	5 75
No. 1, 100 lb. kegs, per cwt ...	....	6 00

SHELLAC.

Shellac, bleached, in bars or ground, 46c. per lb., f.o.b. Eastern Canadian Points.
Shellac, bone dry, 57c per lb, f o.b., Eastern Canadian Points; 63c., Toronto
Shellac, T N. orange, etc., 48c. net, Toronto

WHITE LEAD GROUND IN OIL. Per 100 lbs

Pure ... 7 25 7 70
No. 1 ... 6 90 7 15
No. 2 ... 6 55 6 90
No. 3 ... 6 30 6 55
Munro's Select Flake White 7 10
Elephant and Decorators' Pure 7 25
Monarch 7 40
Standard Decorator's 7 15
Essex Genuine 6 80
Brandram's B. B. Genuine 8 25
"Anchor," pure 6 50
Ramsay's Pure Lead 6 40
Ramsay's Exterior 6 15
"Crown and Anchor," pure 6 50
Sanderson Pearcy's 7 40
Robertson's C P., lead 7 20
W. H. & C's matured pure English 8 00 8 25

WINDOW GLASS.

Size United Inches	Star	Double Diamond
Under 26	$4 25	$6 25
26 to 40	4 65	6 75
41 to 50	5 10	7 50
51 to 60	5 35	8 50
61 to 70	5 75	9 75
71 to 80	6 25	11 00
81 to 85	7 00	12 50
86 to 90	....	15 00
91 to 95	....	17 50
96 to 100	....	20 50
101 to 105	....	24 00
106 to 110	....	27 50

Discount off list, 3- per cent For broken boxes 40 off. All prices per 100 feet.

ZINC—DRY WHITE.

Extra Red Seal ... 0 07 0 08
French V. M. ... 0 06 0 25
Lehigh ... 0 05 0 06

ZINC—GROUND WHITE.

Pure ... 0 08½ 0 09½
No. 1 ... 0 07 0 08
No. 2 ... 0 05½ 0 06½

VARNISHES.

In 5-gal. lots.	Per gal.	Net.
Carriage, No. 1	1 50	1 60
Pale durable body	4 10	4 25
" rubbing	2 85	3 20
Gold size, japan	1 50	1 60
No. 1 brown japan	0 85	0 90
Elastic oak	....	1 50
Furniture, extra	1 10	1 25
" No. 1	0 90	1 00
Hard oil finish	1 35	1 50
Light oil finish	1 60	1 70
Damar	1 75	2 00
Brown japan	1 10	1 20
Black japan	1 10	1 20
" No. 1	0 85	0 90

Elastilite varnish, 1 gal. can, each .. 2 00
Granitine floor varnish, per gal 2 75
Maple Leaf coach enamels; size 1, 1.20; size 2, 70c.; size 3, 40c. each.
Sherwin-Williams' kopal varnish, gal., 2 50
Canada Paint Co's sun varnish 2 00
"Kyanize" Interior Finish 2 40
"Flint-Lac," coach 1 80
"Gold Medal," per gal., in cases .. 2 00
Jamieson's Copaline, per gal. 2 00
Sanderson Pearcy's amberine 2 50

BUILDERS' HARDWARE.

BELLS.

Brass hand bells, 60 per cent.
Nickel, 55 per cent.
Gongs, Sargeant's door bells .. 5 50 8 00
Peterboro' door bells, discount 37½ and 10 per cent. off new list.
American, house bells, per lb. 0 35 0 40

BUILDING PAPER, ETC.

Tarred Felt, per 100 lb. ... 2 25
Ready roofing, 2-ply, not under 45 lb. per roll ... 0 95
Ready roofing, 3-ply, not under 65 lb., per roll ... 1 20
Carpet Felt ... per ton 55 00
Heavy Straw Sheathing ... per ton 35 00
Dry Surprise ... 0 40
Dry Sheathing ... per roll, 400 sq. ft. 0 50
Tar " ... " 400 " 0 45
Dry Fibre ... " 400 " 0 55
Tarred Fibre ... " 400 " 0 65
O. K. & I. X. L. ... " 400 " 0 70
Resin-sized ... " 400 " 0 45
Oiled Sheathing ... " 600 " 1 00
Oiled " ... " 400 " 0 70
Roof Coating, in barrels ... per gal. 0 17
Roof " small packages ... 0 25
Refined Tar ... per barrel 5 00
Coal Tar ... 4 00
Coal Tar, less than barrels ... per gal. 0 15
Roofing Pitch ... per 100 lb. 0 80 0 90
Slater's felt ... per roll 0 70
Heavy Straw Sheathing f. o. b. St. John and Halifax ... 37 50

BUTTS.

Wrought Brass, net revised list.
Wrought Iron, 70 per cent.
Cast Iron Loose Pin, discount 60 per cent.
Wrought Steel Fast Joint and Loose Pin, 70 per cent

CEMENT AND FIREBRICK

"Lafarge," cement in wood ... 3 40
"Lehigh," cement, in wood ... 2 54
"Lehigh" cement, cotton sacks ... 2 39
"Lehigh" cement, paper sacks ... 2 31
Fire brick (Scotch) per 1,000 ... 15 00 24 00
Fire clay (Scotch), net ton ... 4 95
Paving Blocks per 1,000.
Blue metallic, 9"x4½"x3", ex wharf .. 35 00
Stable pavers, 12"x6"x2", ex wharf .. 50 00
Stable pavers, 9"x4½"x3", ex wharf .. 36 00

DOOR SETS.

Peterboro, 37½ and 10 per cent.

DOOR SPRINGS.

Torrey's Rod ... per doz. 1 75
Coil, 9 to 11 in. ... " 0 95 1 65
English ... " 2 00 4 00
Chicago and Reliance Coil Door Springs, 25 per cent.

STORE DOOR HANDLES.

Per Dozen ... 1 00 1 50

ESCUTCHEONS.

Discount 50 and 10 per cent., new list
Peterboro, 37½ and 10 per cent.

ESCUTCHEON PINS.

Iron, discount 40 per cent.
Brass, 45 per cent.

HINGES.

Blind, discount 60 per cent.
Heavy T and strap, 4-in., per lb. net ... 0 06½
" " 5-in., " ... 0 06
" " 6-in., " ... 0 05¾
" " 8-in., " ... 0 05½
" " 10-in. and larger ... 0 05¼
Light T and strap, discount 65 p.c.
Screw hook and hinge—
under 12 in. ... per 100 lb. 4 75
over 12 in. ... 3 75
Spring, No. 20, per gro. pairs ... 10 80
Spring, Woodyatt pattern, per gro., No. 5, $17.50, No. 10, $18; No. 20, $10.80; No. 12 $20; No. 51, $10; No. 50, $27.50.
Gate hinges and back flaps, 65 and 5 p. c.
Hinge hasps, 65 per cent.

SPRING HINGES.

Chicago Spring Butts and Blanks 12½ per cent.
Triple End Spring Butts, 40 and 5 per cent.
Chicago Floor Hinges, 40 and 5 off.
Garden City Fire Engine House Hinges, 12½ per cent.

HOOKS.

Cast Iron.

Bird cage ... per doz. 0 50 1 10
Clothes line, No. 61 ... " 0 00 0 70
Harness ... " 0 60 12 00
Hat and coat ... per gro. 1 10 10 00
Chandelier ... per doz. 0 50 1 00
Wrought hooks and staples—
1 x 5 ... per gross 2 65
5-16 x 5 ... . 3 30
Bright steel gate hooks and staples, 40 per cent. discount
Hat and coat wire, discount 62½ per cent.
Screw, bright wire, discount 62½ per cent.

KNOBS.

Door, japanned and N.P., doz 1 50 2 50
Bronze, Berlin ... per doz. 2 75 3 25
Bronze, Genuine ... " 6 00 9 00
Shutter, porcelain, F. & L. screw ... per gross 1 30 2 00
White door knobs ... per doz. 2 00
Peterboro knobs, 37½ and 10 per cent.
Porcelain, mineral and jet knobs, net list

KEYS.

Lock, Canadian dis. 40 to 40 and 10 per cent
Cabinet trunk and padlock
American ... per gross 0 60

LOCKS.

Peterboro 37½ and 10 per cent
Russell & Erwin, steel rim $2 50 per doz
Eagle cabinet locks, discount 30 per cent
American padlocks, all steel, 10 to 15 per cent; all brass or bronze, 10 to 25 per cent.

SAND AND EMERY PAPER.

B. & A. sand, discount, 35 per cent
Emery, discount 35 per cent.
Garnet (Rurton's) 5 to 10 per cent. advance on list.

SASH WEIGHTS.

Sectional ... per 100 lb. 2 00 2 25
Solid ... 1 50 1 75

SASH CORD.

Per lb ... 0 31

BLIND AND BED STAPLES.

All sizes, per lb. ... 0 07½ 0 12

WROUGHT STAPLES.

Galvanized ... 2 75
Plain ... 2 50
Coopers', discount 45 per cent.
Poultry netting staples, discount 40 per cent.
Bright spear point, 75 per cent. discount.

TOOLS AND HANDLES.

ADZES.

Discount 22½ per cent.

AUGERS.

Gilmour's, discount 60 per cent. off list.

AXES.

Single bit, per doz. ... 5 50 8 50
Double bit, " ... 10 00 11 00
Bench Axes, 40 per cent.
Broad Axes, 25 per cent.
Hunters' Axes ... 5 50 6 00
Boys' Axes ... 6 25 7 00
Splitting Axes ... 7 00 12 00
Handled Axes ... 7 00 9 00
Red Ridge, boys', handled 5 75
" hunters 5 25

BITS.

Irwin's auger, discount 47½ per cent.
Gilmour's auger, discount 60 per cent.
Rockford auger, discount 50 and 10 per cent.
Jennings' Gen. auger, net list.
Gilmour's car, 47½ per cent.
Clark's expansive, 40 per cent.
Clark's gimlet, per doz ... 0 65 0 90
Diamond, Shell, per doz ... 1 00 1 50
Nail and Spike, per gross ... 2 25 5 20

BUTCHERS CLEAVERS

German ... per doz. 6 00 9 00
American ... " 12 00 18 00

CHALK.

Carpenters' Colored, per gross 0 45 0 75
White lump ... per cwt. 0 60 0 65

CHISELS.

Warnock's, discount 72½ per cent.
P. S. & W. Extra, discount 72½ per cent

CROSSCUT SAW HANDLES.

S. & D., No. 3 ... per pair 0 15
S. & D., " 5 ... " 0 22½
S. & D., " 6 ... " 0 15
Boynton pattern ... " 0 20

CROWBARS.

3½c. to 4c. per lb.

DRAW KNIVES.

Coach and Wagon, discount 75 per cent.
Carpenters discount 75 per cent

DRILLS.

Millar's Falls, hand and breast. net list.
North Bros., each set, 50c.

DRILL BITS.

Morse, discount 37½ to 40 per cent.
Standard, discount 50 and 5 to 55 per cent.

FILES AND RASPS.

Great Western ... 75 per cent.
Arcade ... 75 "
Kearney & Foot ... 75 "
Disston's ... 75 "
American ... 75 "
J. Barton Smith ... 75 "
McClellan ... 75 "
Eagle ... 75 "
Nicholson ... 66⅔ "
Globe ... 75 "
Black Diamond, 60, 10 and 5 p c.
Jowitt's, English list, 27½ per cent.

GAUGES.

Stanley's discount 50 to 60 per cent.

Wire Gauges.

Winn's, Nos. 26 to 33 ... each 1 65 2 40

HANDLES.

C. & B., fork, 40 per cent., revised list.
C. & B., hoe, 40 per cent., revised list
American, saw ... per doz 0 25
American, plane ... per gross 3 15 3 75
Canadian, hammer and hatchet 40 per cent
Axe and cant hook handles, 45 per cent.

HAMMERS.

Maydole's, discount 5 to 10 per cent.
Canadian, discount 25 to 27½ per cent.
Magnetic tack ... per doz. 1 10 1 20
Canadian sledge ... per lb. 0 07 0 08½
Canadian ball pean, per lb. .. 0 22 0 25

HATCHETS.

Canadian, discount 40 to 42½ per cent.
Shingle, Red Ridge 1, per doz ... 4 40
" " 2, " ... 4 85
Barrel Underhill ... 5 00

MALLETS.

Tinsmiths' ... per doz. 1 25 1 50
Carpenters', hickory, " 1 25 3 75
Lignum Vitae ... " 3 85 5 00
Caulking, each ... 0 60 2 00

MATTOCKS.

Canadian ... per doz. 5 50 6 00

MEAT CUTTERS.

German, 15 per cent.
American discount, 33⅓ per cent.
Gem ....each 1 15

NAIL PULLERS.

German and American 0 85 2 50
No. 1 .... 0 85
No 1573 .... 0 75

NAIL SETS.

Square, round and octagon, per gross 3 38
Diamond 1 00

PICKS.

Per dozen 6 00 9 00

PLANES.

Wood bench, Canadian discount 40 per cent, American discount 50 per cent.
Wood, fancy Canadian or American 37½ to 40 per cent.
Stanley planes, $1.55 to $3 60, net list prices.

PLANE IRONS.

English ....per doz. 2 00 5 00
Stanley, 2½ inch, single 24c., double 39c.

PLIERS AND NIPPERS.

Button's genuine, 37½ to 40 per cent.
Button's imitation....per doz. 5 00 9 00

PUNCHES.

Saddlers ....per doz. 1 00 1 85
Conductor's......... " 3 00 15 00
Tinners', solid..........per set 0 72
" hollow......per inch 1 00

RIVET SETS.

Canadian, discount 35 to 37½ per cent.

RULES.

Boxwood, discount 70 per cent.
Ivory, discount 20 to 25 per cent.

SAWS.

Atkins, hand and crosscut, 25 per cent.
Disston's Hand, discount 12½ per cent
Disston's Crosscut ....per foot 0 35 0 55
Hack, complete..........each 0 75 2 75
" frame only.........each 0 50 1 25

S. & D. solid tooth circular shingle, concave and band, 50 per cent; mill and ice, drag, 30 per cent.; cross-cut, 35 per cent; hand saws, butcher, 35 per cent; buck, New Century $6.25, buck No. 1 Maple Leaf, $5.25; buck, Happy Medium $4 25; buck, Watch Spring, $1.25, buck, common frame, $4.00.

Spear & Jackson's saws—Hand or rip 26 in., $12 75; 24 in., $12.25; panel, 18 in., $8.25; 20 in., $9; tenon, 10 in., $9 90; 12 in., $10 9[illegible]; 14 in., $11.50.

SAW SETS.

Lincoln and Whiting 4 75
Hand Sets, Perfect 4 00
X-Cut Sets, " 7 50
Maple Leaf and Premium saw sets, 40 off.
S. & D saw swages, 40 off.

SCREW DRIVERS

Sargent's ....per doz 0 65 1 00
North Bros., No. 31 ..per doz .... 10 8[illegible]

SHOVELS AND SPADES

Bull Dog, solid neck shovel (No. 2 pol).$18 50
(Hollow Back) (Reinforced B Scoop)
Moose.......... $17 50 .... $15 70
Bear .... 15 00 .... 15 30
Fox .... 12 50 .... 14 30
Black Cat..... 10 00 .... 13 30
Canadian, discount 45 per cent.

SQUARES.

Iron, discount 20 per cent.
Steel, discount 65 and 10 per cent.
Try and Bevel, discount 50 to 52½ per cent.

TAPE LINES.

English, ass skin .....per doz. 2 75 5 00
English, Patent Leather 5 50 9 75
Chesterman's ....each 0 90 2 85
" steel..........each 0 80 8 00

TROWELS.

Disston's, discount 10 per cent.
S. & D., discount 35 per cent.

FARM AND GARDEN GOODS

BELLS.

American cow bells, 63⅓ per cent.
Canadian, discount 45 and 50 per cent.
American, farm bells, each .. 1 35 3 00

BULL RINGS.

Copper, $1.30 for 2½-inch, and $1.70

CATTLE LEADERS.

Nos. 32 and 33 .. ...per gross 7 50 8 50

BARN DOOR HANGERS.

		doz. pairs.
Steel barn door	8 00	10 00
Stearns wood track	4 50	6 00
Zenith	....	9 00
Acme, wood track	5 00	6 50
Atlas	5 00	6 00
Perfect	8 00	11 00
New Milo	....	6 50
Steel, covered	4 00	11 00
" track, 1 x 3-16 in (100 ft)	....	3 75
" " 1¼ x 3-16 in (100 ft)	....	4 75
Double strap hangers, doz sets	....	6 40
Standard jointed hangers, "	....	6 40
Steel King hangers "	....	6 25
Storm King and safety hangers	....	7 00
" " rail	....	4 25

Chicago Friction, Oscillating and Big Twin Hangers, 5 per cent.

HARVEST TOOLS.

Discount 60 per cent.
S. & D. lawn rakes, Dunn's, 40 off.
" sidewalk and stable scrapers, 40 off.

HAY KNIVES.

Net prices.

HEAD HALTERS.

Jute Rope, ⅜-inch....per gross 9 00
" " ½ " .... " 10 00
" " ⅝ " .... " 12 00
Leather, 1-inch ......per doz. 4 00
Leather, 1¼ " ...... " 5 20
Web.................. " 2 45

HOES.

Garden, Mortar, etc., discount 60 per cent.
Planter............. per doz. 4 00 4 50

HORSE NAILS.

'C' brand, 40, 10 and 7½ per cent. off list } Oval
M.R.M. Co. brand, 55 per cent. } head

HORSESHOES.

F.O.B. Montreal
M.R.M. Co. brand, base............... 3 65
Add 15c. Toronto, Hamilton, Guelph.

HORSE WEIGHTS

Taylor-Forbes, 3½c. per lb.

SCYTHES.

Per doz. net.................. 6 25 9 25

SCYTHE SNATHS.

Canadian, discount 40 per cent.

SNAPS.

Harness, German, discount 25 per cent.
Lock, Andrews 4 50 11 00

STALL FIXTURES.

Warden King, 35 per cent.

WOOD HAY RAKES.

Ten tooth, 40 and 10 per cent.
Twelve tooth, 45 per cent.

HEAVY GOODS NAILS, ETC.

ANVILS.

Wright's, 80-lb. and over.............. 0 10½
Hay Budden, 80-lb. and over 0 09½
Brook's, 80-lb. and over 0 11½
Taylor-Forbes, handy 0 05

VISES.

Wright's..................... .... 0 13½
Brook's...................... .... 0 12½
Pipe Vise, Hinge, No. 1....... .... 3 50
" " " No. 2....... .... 5 50
Saw Vise 4 50 5 00
Blacksmiths' (discount) 60 per cent.
" parallel (discount) 45 per cent.

BOLTS AND NUTS

Carriage Bolts, common ($1 list) Per cent.
" " ⅜ and smaller.. 60, 10 and 10
" " 7-16 and up .... 55 and 5
" " Norway Iron ($3 list) 50
Machine Bolts, ⅜ and less .... 60 and 10
Machine Bolts, 7-16 and up.... 55 and 5
Plough Bolts .... 55 and 10
Blank Bolts.................... 55
Bolt Ends...................... 55
Sleigh Shoe Bolts, ⅜ and less .. 60 and 10
" " " 7-16 and larger 50 and 5
Coach Screws, conepoint......... 70 and 5
Nuts, square, all sizes, 4c. per off.
Nuts, hexagon, all sizes, 4¼c. per off.
Stove Rods per lb., 5½ to 6c.
Stove Bolts, 75 per cent.

WROUGHT IRON WASHERS.

Canadian make discount 40 per cent.

CHAIN.

Proof coil, per 100 lb., 5-16 in., $4.40; ⅜ in., $3.90; 7-16 in., $3.70; ½ in., $3.50; 9-16 in., $3.45; ⅝ in., $3.35; ¾ in., $3.20; ⅞ in., $3.10; 1 in., $3.10.
Halter, kennel and post chains, 40 to 40 and 5 per cent.
Cow ties .... 40 p.c.
Tie out chains .... 65 p.c.
Stall fixtures .... 35 p.c.
Trace chain .... 45 p.c.
Jack chain iron, discount 35 p.c.
Jack chain, brass, discount 40 per cent.

PRESSED SPIKES.

Pressed spikes, ⅝ diameter, per 100 lbs., $3 15

NAILS.

	Cut.	Wire.
2d	3 80	3 30
3d	2 95	2 95
4 and 5d	2 70	2 70
6 and 7d	2 60	2 60
8 and 9d	2 45	2 45
10 and 12d	2 40	2 40
16 and 20d	2 35	2 35
30, 40, 50 and 60d (base)	2 30	2 30

F.o.b. Montreal. Cut nails, Toronto 20c. higher.
Miscellaneous wire nails, discount 75 per cent.
Coopers' nails, discount 40 per cent.

RIVETS AND BURRS.

Iron Rivets, black and tinned, 60, 10 and 10.
Iron Burrs, discount 60 and 10 and 10 p.c.
Copper Rivets, usual proportion burrs, 27½ p.c.
Copper Burrs only, discount 15 per cent.
Extras on Coppered Rivets, ½-lb. packages 1c. per lb.; ¼-lb. packages 2c. lb.
Tinned Rivets, net extra, 4c. per lb.

SCREWS.

Wood, F. H., bright and steel, 87½ per cent.
" R. H., bright, dis. 82½ per cent
" F. H., brass, dis. 80 per cent.
" R. H., " dis. 75 per cent.
" F. H., bronze, dis. 75 per cent.
" R. H., " dis. 70 per cent.
Drive Screws, dis. 87½ per cent.
Bench, wood ....per doz 3 25 4 00
" iron .... " 4 25 5 00
Set, case hardened, dis. 60 per cent.
Square Cap, dis. 50 and 5 per cent.
Hexagon Cap, dis. 45 per cent.

MACHINE SCREWS.

Flat head, iron and brass, 35 per cent.
Felister head, iron, discount 30 per cent.
" " brass, discount 25 per cent.

TACKS, BRADS, ETC.

Carpet tacks, blued .... 80 and 5
" " tinned .... 80 and 10
" " (in kegs).......... 40
Cut tacks, blued, in dozens only 75 and 10
" ½ weights 60
Swedes cut tacks, blued and tinned—
In bulk .... 80 and 10
In dozens 75
Swedes, upholsterers', bulk .... 85 and 12½
" brush, blued and tinned bulk 70
Swedes, gimp, blued, tinned and japanned......................... 75 and 12½
Zinc tacks 35
Leather carpet tacks.............. 40
Copper tacks...................... 37½
Copper nails 42½
Trunk nails, black 65
Trunk nails, tinned and blued .. 65
Clout nails, blued and tinned .. 65
Chair nails....................... 35
Patent brads...................... 40
Fine finishing.................... 40
Lining tacks, in papers.......... 10
" " in bulk............ 15
" " solid heads, in bulk 75
Saddle nails, in papers.......... 10
" " in bulk............ 15
Tufting buttons, 22 line in dozens only 60
Zinc glaziers' points 5
Double pointed tacks, papers.. 90 and 10
" " " bulk.... 40
Clinch and duck rivets.... 45
Cheese box tacks, 85 and 5; trunk tacks, 80 and 10.

CUTLERY AND SPORTING GOODS.

AMMUNITION.

B. B. Caps Dominion, 50 and 5 and 25 per cent. American $2 00 per 1000.
C. B. Caps American, $2 60 per 1000

CARTRIDGES.

Rim Fire Cartridges, Dominion, 50 and 5 p.c.
Rim Fire Pistol, discount 30 and 5 per cent., American.
Central Fire, Military and Sporting, American, add 10 per cent. to list. B.B. Caps, discount 40 per cent., American.
Central Fire Pistol and Rifle, list net Amer.
Central Fire Cartridges, pistol and rifle Dominion, 30 and 5 per cent.
Central Fire Cartridges, Sporting and Military, Dominion, 15 per cent., American 10 per cent. advance on list.
Loaded and empty Shells, Crown, 25 and 5; Sovereign, 25, 10 and 10; Regal, 25, 10 and 5; Imperial, 25, 10 and 5. American 20 per cent. discount. Rival and Nitro, 10 per cent. advance on list.
Empty paper shells Dominion, 25 off and empty brass shells, 55 per cent. off. American, 10 per cent. advance on list.
Primers, Dom., 35 per cent.; American $2.05

Wads. per lb.

Best thick brown or grey felt wads, in ¼-lb. bags $0 70
Best thick white card wads, in boxes of 500 each, 12 and smaller gauges 0 29
Best thick white card wads, in boxes of 500 each, 10 gauge............ 0 35
Thin card wads, in boxes of 1,000 each, 12 and smaller gauges 0 20
Thin card wads, in boxes of 1,000 each, 10 gauge..................... 0 25
Chemically prepared black edge grey cloth wads, in boxes of 250 each— Per M.
11 and smaller gauge 0 60
9 and 10 gauges.................. 0 70
7 and 8 " 0 90
5 and 6 " 1 10
Superior chemically prepared pink edge, best white cloth wads in boxes of 250 each—
11 and smaller gauge 1 15
9 and 10 gauges 1 40
7 and 8 " 1 65
5 and 9 " 1 90

SHOT

Common, $6.50 per 100 lb.; chilled, $7.50 per 100 lb.; buck, seal and ball, $8.50 net list. Prices are f.o.b. Toronto, Hamilton, Montreal, St. John and Halifax. Terms, 2 p.c. for cash in thirty days.

RAZORS. per doz.

Elliot's	4 00	18 00
Boker's	7 50	11 00
" King Cutter	13 50	18 50
Wade & Butcher's	3 60	10 00
Wilkinson's	....	12 50
Lewis Bros.' "Klean Kutter"	8 50	10 50

Clauss Razors and Strops, 50 and 10 per cent.

PLATED GOODS.

Hollowware, 40 per cent. discount.
Flatware, staples, 40 and 10, fancy 40 and 5 per cent.

SHEARS.

Clauss, nickel, discount 60 per cent.
Clauss, Japan, discount 67½ per cent.
Clauss, tailors, discount 40 per cent.
Seymour's, discount 50 and 10 per cent.

TRAPS (steel.)

Game, Newhouse, discount 30 and 10 per cent.
Game, Hawley & Norton, 50, 10 & 5 per cent.
Game, Victor, 70 per cent.
Game, Oneida Jump (B. & L.) 40 & 2½ p. c.
Game, steel, 60 and 5 per cent.

SKATES.

Skates, discount 37½ per cent.
Mic Mac hockey sticks, per doz 4 00 5 00
New Rex hockey sticks, per doz 6 25

HOUSE FURNISHINGS.

APPLE PARERS.

Woodyatt Hudson, per doz., net 4 50

BIRD CAGES.

Brass and Japanned, 40 and 10 p. c.

COPPER AND NICKEL WARE.

Copper boilers, kettles, teapots, etc. 30 p.c.
Copper pitts, 33⅓ per cent.

ENAMELLED WARE.

London, White, Princess, Turquoise, Onyx, Blue and White, discount 50 per cent.
Canada, Diamond, Premier, 50 and 10 p.c.
Pearl, Imperial Crescent, 50 and 10 per cent.
Premier steel ware, 40 per cent.
Star decorated steel and white, 25 per cent.
Japanned ware, discount 50 per cent.
Hollow ware, tinned cast, 35 per cent. off.

KITCHEN NOVELTIES.

Can openers, per doz............ 0 40 0 75
Mincing knives per doz 0 50 0 80
Duplex mouse traps, per doz.... .... 0 65
Potato mashers, wire, per doz... 0 60 0 70
" wood " .. 0 50 0 60
Vegetable slicers, per doz 2 25
Universal meat chopper No. 0, $1; No. 1, 1.15.
Enterprise chopper, each 1 30

LAMP WICKS.

Discount, 60 per cent.

LEMON SQUEEZERS.

Porcelain lined...... per doz. 2 20 5 60
Galvanized.......... " 1 87 3 85
King, wood.......... " 2 75 2 90
King, glass.......... " 4 00 4 50
All glass .... " 0 50 0 90

PICTURE NAILS.

Porcelain head ......per gross 1 35 1 50
Brass head.......... " 0 45 1 00
Tin and gilt, picture wire, 75 per cent.

SAD IRONS.

Mrs. Potts, No. 55, polished....per set 0 80
" No. 50, nickle-plated, " 0 92
Common, plain........................ 4 50
" plated 5 50
Asbestos, per set............... .... 1 25

TINWARE.

CONDUCTOR PIPE.

2 in. plain or corrugated., per 100 feet, $3.30; 3 in., $4 40; 4 in., $5.80; 5 in., $7.45; 6 in., $9.90.

FAUCETS.

Common, cork-lined, discount 35 per cent.

EAVETROUGHS.

10-inch ....per 100 ft. 3 30

FACTORY MILK CANS.

Discount off revised list, 40 per cent.
Milk can trimmings, discount 25 per cent.

LANTERNS.

No. 2 or 4 Plain Cold Blast....per doz. 6 50
Lift Tubular and Hinge Plain, " 4 75
Better quality at higher prices.
Japanning, 50c. per doz. extra.

OILERS.

Kemp's Tornado and McClary's Model galvanized oil can, with pump, 5 gallon, per dozen 10 00
Davidson oilers, discount 40 per cent.
Zinc and tin, discount 50 per cent.
Coppered oilers, 20 per cent. off.
Brass oilers, 50 per cent. off.
Malleable, discount 25 per cent.

PAILS (GALVANIZED).

Dufferin pattern pails, 40, 10 and 5 per cent.
Flaring pattern, discount 40, 10 and 5 per cent.
Galvanized washtubs 40, 10 and 5 per cent.

PIECED WARE.

Discount 40 per cent off list, June, 1899.
10-qt. flaring sap buckets, discount 40 per cent.
6, 10 and 14-qt. flaring pails dis. 40 per cent.
Copper bottom tea kettles and boilers, 35 p c.
Creamery cans, discount 40 per cent.

STAMPED WARE.

Plain, 75 and 12½ per cent. off revised list.
Retinned, 72½ 10 and 5 per cent. revised list.

SAP SPOUTS.

Bronzed iron with hooks ....per 1,000 7 50
Eureka tinned steel, hooks 8 00

STOVEPIPES.

5 and 6 inch, per 100 lengths 7 00
7 inch........ " " 7 50

STOVEPIPE ELBOWS

5 and 6-inch, common........per doz. 1 32
7-inch........................ " 1 48
Polished, 15c. per dozen extra.

THERMOMETERS.

Tin case and dairy, 75 to 75 and 10 per cent

TINNERS' SNIPS.

Per doz...................... 3 00 15 00
Clauss, discount 35 per cent.

WIRE.

BRIGHT WIRE GOODS

Discount 62½ per cent.

CLOTHES LINE WIRE.

7 wire solid line, No. 17, $4.90; No. 18, $3.00; No. 19, $2.70; 6 wire solid line, No. 17, $4.45; No. 18, $2 80. No. 19, $2.50. All prices per 1000 ft. measure. F.o.b. Hamilton Toronto, Montreal.

COILED SPRING WIRE.

High Carbon, No. 9, $2 55, No. 11, $3.20; No. 12, $2.80.

COPPER AND BRASS WIRE

Discount 45 per cent.

FINE STEEL WIRE.

Discount 30 per cent List of extras: In 100-lb. lots: No. 17, $5 — No. 18, $5.50 — No. 19, $6 — No. 20, $6.65 — No. 21, $7 — No. 22, $7.30 — No 23, $7.65 — No. 24, $8 — No. 25, $9—No. 26, $9.50—No. 27, $10—No. 28, $11—No. 29, $12—No. 30, $13—No. 31, $14—No. 32, $15—No. 33, $16—No. 34, $17. Extras net—tinned wire, Nos. 17-25, $2—Nos. 26-31, $4—Nos. 32-34, $6. Coppered, 75c.—oiling, 10c.—in 25-lb. bundles, 15c.—in 5 and 10-lb. bundles, 25c.—in 1-lb. hanks, 25c. —in ½-lb. hanks, 38c.—in ¼-lb. hanks, 50c.—packed in casks or cases, 15c.—bagging or papering, 10c.

GALVANIZED WIRE.

Per 100 lb.—Nos. 4 and 5, $3 70 — Nos. 6, 7, 8, $3.15 — No 9, $2.50 — No. 10, $3.20 — No. 11, $3.25—No. 12, $2.65 —No. 13, $2.75—No. 14, $3.75—No. 15, $4.30 —No. 16, $4.30 from stock. Base sizes, Nos. 6 to 9, $2.35 f.o.b. Cleveland. In carlots 12½c. less.

LIGHT STRAIGHTENED WIRE.

Gauge No.	Over 20 in. per 100 lbs.	10 to 20 in.	5 to 10 in.
0 to 5	$0 50	$0 75	$1.25
6 to 9	0.75	1 25	2 00
10 to 11	1 00	1 75	2.50
12 to 14	1.50	2.25	3 50
15 to 16	2 00	3 00	4 50

SMOOTH STEEL WIRE.

No. 0-9 gauge, $2 25; No. 10 gauge, 6c extra; No. 11 gauge, 12c extra; No 12 gauge, 20c. extra, No. 13 gauge, 30c. extra; No 14 gauge 40c extra, No 15 gauge, 55c. extra; No 16 gauge, 7 c. extra. Add 60c. for coppering and $2 for tinning

Extra net per 100 lb.—Oiled wire 10c., spring wire $1.25, special hay baling wire 30c., best steel wire 75c., bright soft drawn 15c., charcoal (extra quality) $1.25, packed in casks or cases 15c., bagging and papering 10c., 50 and 100-lb. bundles 10c., in 25-lb. bundles 15c., in 5 and 10-lb. bundles 25c., in 1-lb hanks, 50c., in ½-lb. hanks 75c., in ¼-lb. hanks $1.

POULTRY NETTING.

2 in mesh 19 w. g., discount 50 and 10 per cent. All others 50 per cent.

WIRE CLOTH

Painted Screen, in 100-ft rolls, $1.62½, per 100 sq. ft.; in 50-ft. rolls $1.67½, per 100 sq ft. Terms, 2 per cent. off 30 days.

WIRE FENCING.

Galvanized barb...... .... 2 95
Galvanized, plain twist 3 30
Galvanized barb, f.o.b. Cleveland, $2.70 for small lots and $2 60 for carlot.

WOODENWARE.

CHURNS.

No. 0, $9; No 1, $9; No. 2, $10; No. 3, $11; No. 4, $13; No. 5, $16; f.o.b. Toronto Hamilton, London and St. Marys, 30 and 30 per cent; f o b Ottawa, Kingston and Montreal, 40 and 15 per cent. discount, Taylor-Forbes, 30 and 30 per cent.

CLOTHES REELS.

Davis Clothes Reels, dis. 40 per cent.

LADDERS, EXTENSION.

Waggoner EXtension Ladders, dis. 40 per cent.

MOPS AND IRONING BOARDS.

"Best" mops........................ 1 25
"900" mops........................ 1 25
Folding ironing boards......... 12 00 16 50

REFRIGERATORS

Discount, 40 per cent.

SCREEN DOORS.

Common doors, 2 or 3 panel, walnut stained, 4-in. style...........per doz. 7 25
Common doors, 2 or 3 panel, grained only, 4-in., style ....per doz. 7 55
Common doors, 2 or 3 panel, light stair per doz.............................. 9 55

WASHING MACHINES.

Round, re-acting per doz. 60 00
Square " 63 00
Eclipse, per doz 54 00
Dowswell " 39 00
New Century, per doz 75 00
Daisy................................ 54 00

WRINGERS

Royal Canadian, 11 in., per doz. ... 34 00
Royal American, 11 in. .. 24 00
Eze, 10 in., per doz 34 75
Terms, 2 per cent., 30 days.

MISCELLANEOUS.

AXLE GREASE.

Ordinary, per gross 6 00 7 00
Best quality 10 00 12 00

BELTING.

Extra, 60 per cent.
Standard, 60 and 10 per cent.
No. 1, not wider than 6 in., 60, 10 and 10 p c.
Agricultural, not wider than 4 in., 75 per cent
Lace leather, per side, 75c.; cut laces, 80c

BOOT CALKS.

Small and medium, ball ....per M 4 25
Small heel .... " 4 50

CARPET STRETCHERS.

American ....per doz. 1 00 1 50
Bullard's............. " 6 50

CASTORS.

Bed, new list, discount 55 to 57½ per cent.
Plate, discount 52½ to 57½ per cent.

PINE TAR.

½ pint in tins ....per gross ... 7 80
" " .... " 9 60

PULLEYS.

Hothouse ....per doz. 0 55 1 00
Axle .... " 0 22 0 33
Screw .... " 0 22 1 00
Awning .... " 0 35 2 50

PUMPS.

Canadian cistern 1 40 2 00
Canadian pitcher spout 1 80 3 16

ROPE AND TWINE.

Sisal.................................. 0 10½
Pure Manilla 0 15½
"British" Manilla..................... 0 12
Cotton, 3-16 inch and larger... ..0 21 0 23
" 5-32 inch 0 25 0 27
" ¼ inch 0 25 0 28
Russia Deep Sea 0 16
Jute.................................. 0 09
Lath Yarn, single 0 10
" double 0 10½
Sisal bed cord, 48 feet........per doz. 0 65
" " 60 feet........ " 0 80
" " 72 feet........ " 0 95

Twine.

Bag, Russian twine, per lb. 0 27
Wrapping, cotton, 3-ply 0 25
" " 4-ply 0 29
Mattress twine per lb. 0 33 0 45
Staging " " 0 27 0 35

SCALES.

Gurney Standard, 40 per cent.
Gurney Champion, 50 per cent.
Burrow, Stewart & Milne—
Imperial Standard, discount 40 per cent.
Weigh Beams, discount 40 per cent.
Champion Scales, discount 50 per cent.
Fairbanks standard, discount 35 per cent.
" Dominion, discount 55 per cent.
" Richelieu, discount 55 per cent.
Warren new Standard, discount 40 per cent.
" " Champion, discount 50 per cent.
" Weighbeams, discount 35 per cent.

STONES—OIL AND SCYTHE.

Washita ....per lb. 0 25 0 37
Hindostan .... " 0 06 0 10
" slip .... " 0 18 0 20
" Axe.......... " 0 10
Deer Creek .... " 0 10
Deerlick .... " 0 25
" Axe .... " 0 15
Lily white .... " 0 42
Arkansas.............. " 1 50
Water-of-Ayr .... " 0 10
Scythe ....per gross 3 50 5 00
Grind, 40 to 200 lb., per ton.... 20 00 22 00
" under 40 lb., " 24 00
" 200 lb. and over 28 00

INDEX TO ADVERTISERS.

CLASSIFIED LIST OF ADVERTISEMENTS.

Auditors.
Davenport, Percy P., Winnipeg.

Babbitt Metal.
Canada Metal Co., Toronto.
Canadian Fairbanks Co., Montreal.
Robertson, Jas. Co., Montreal.

Bath Room Fittings.
Carriage Mounting Co., Toronto.

Belting, Hose, etc.
Gutta Percha and Rubber Mfg. Co. Toronto.

Bicycles and Accessories.
Johnson's, Iver, Arms and Cycle Works Fitchburg, Mass

Binder Twine.
Consumers Cordage Co., Montreal.

Box Strap.
J. N. Warminton, Montreal.

Brass Goods.
Glauber Brass Mfg. Co., Cleveland, Ohio.
Lewis, Rice, & Son., Toronto.
Morrison, Jas., Brass Mfg. Co., Toronto.
Mueller Mfg. Co., Decatur, Ill.
Penberthy Injector Co., Windsor, Ont.
Taylor-Forbes Co., Guelph, Ont.

Bronze Powders.
Canadian Bronze Powder Works, Montreal.

Brushes.
Ramsay, A., & Son Co., Montreal.

Can Openers.
Cumming Mfg. Co. Renfrew.

Cans.
Acme Can Works, Montreal.

Builders' Tools and Supplies.
Covert Mfg. Co., West Troy, N.Y.
Frothingham & Workman Co., Montreal.
Howland, H. S., Sons & Co., Toronto.
Hyde, F., & Co., Montreal.
Lewis Bros. & Co., Montreal.
Lewis, Rice, & Son, Toronto.
Lockethy & McComb, Montreal.
Lufkin Rule Co., Saginaw, Mich.
Newman & Sons, Birmingham.
North Bros. Mfg. Co., Philadelphia, Pa.
Stanley Rule & Level Co., New Britain.
Stanley Works, New Britain, Conn.
Stephens, G. F., Winnipeg.
Taylor-Forbes Co., Guelph, Ont.

Carriage Accessories.
Carriage Mountings Co., Toronto.
Covert Mfg. Co., West Troy, N.Y.

Carriage Springs and Axles.
Guelph Spring and Axle Co., Guelph.

Cattle and Trace Chains.
Greening, B., Wire Co., Hamilton.

Churns.
Dowswell Mfg. Co., Hamilton.

Clippers—All Kinds.
American Shearer Mfg. Co., Nashua, N.H.

Clothes Reels and Lines.
Hamilton Cotton Co., Hamilton, Ont.

Cordage.
Consumers' Cordage Co., Montreal.
Hamilton Cotton Co., Hamilton.

Cork Screws.
Erie Specialty Co., Erie, Pa.

Clutch Nails.
J. N. Warminton, Montreal.

Cut Glass.
Phillips, Geo., & Co., Montreal.

Cutlery—Razors, Scissors, etc.
Birkett, Thos., & Son Co., Ottawa.
Clauss Shear Co., Toronto
Dorken Bros. & Co., Montreal.
Heinisch's, R., Sons Co., Newark, N.J.
Howland, H. S. Sons & Co., Toronto.
Phillips, Geo., & Co., Montreal.
Round, John, & Son, Montreal.

Door Hangers.
Door Hanger Co., Hamilton, Ont.

Electric Fixtures.
Canadian General Electric Co., Toronto.
Forman, John, Montreal.
Morrison James, Mfg. Co., Toronto.
Munderloh & Co., Montreal.

Electro Cabinets.
Cameron & Campbell Toronto.

Engines, Supplies, etc.
Kerr Engine Co., Walkerville, Ont.

Files and Rasps.
Barnett Co., G. & H., Philadelphia, Pa.
Nicholson File Co., Port Hope.

Financial Institutions
Bradstreet Co.

Firearms and Ammunition.
Dominion Cartridge Co., Montreal.
Hamilton Rifle Co., Plymouth, Mich.
Harrington & Richardson Arms Co., Worcester, Mass.
Johnson's, Iver, Arms and Cycle Works, Fitchburg, Mass.

Food Choppers.
Enterprise Mfg. Co., Philadelphia, Pa.

Galvanizing.
Canada Metal Co., Toronto.
Montreal Rolling Mills Co., Montreal.
Ontario Wind Engine & Pump Co., Toronto.

Glaziers' Diamonds.
Gibsone, J. B., Montreal.
Pelton, Godfrey S.
Sharratt & Newth, London, Eng.
Shaw, A., & Son, London, Eng.

Hack Saws.
Diamond Saw & Stamping Works, Buffalo

Harvest Tools.
Maple Leaf Harvest Tool Co., Tillsonburg, Ont.

Hoop Iron.
Montreal Rolling Mills Co., Montreal.
J. N. Warminton, Montreal.

Horse Blankets.
Heney, E. N., & Co., Montreal.

Horseshoes and Nails.
Canada Horse Nail Co., Montreal.
Montreal Rolling Mills, Montreal.

Hot Water Boilers and Radiators.
Cluff, R. J., & Co., Toronto.
Pease Foundry Co., Toronto.
Taylor-Forbes Co., Guelph.

Ice Cream Freezers.
Dana Mfg. Co., Cincinnati, Ohio.
North Bros. Mfg. Co., Philadelphia, Pa.

Ice Cutting Tools.
Erie Specialty Co., Erie, Pa.
North Bros. Mfg. Co., Philadelphia, Pa.

Injectors—Automatic.
Morrison, Jas., Brass Mfg. Co., Toronto.
Penberthy Injector Co., Windsor, Ont.

Iron Pipe.
Montreal Rolling Mills, Montreal.

Iron Pumps.
McDougall, R., Co., Galt, Ont.

Lanterns.
Kemp Mfg. Co., Toronto.
Ontario Lantern Co., Hamilton, Ont.
Wright, E. T., & Co., Hamilton.

Lawn Mowers.
Birkett, Thos., & Son Co., Ottawa.
Maxwell, D., & Sons, St. Mary's, Ont.
Taylor, Forbes Co., Guelph.

Lawn Swings, Settees, Chairs.
Cumming Mfg. Co., Renfrew.

Ledgers—Loose Leaf.
Business Systems, Toronto
Copeland-Chatterson Co., Toronto.
Crain, Rolla L., Co., Ottawa.
Universal Systems, Toronto.

Locks, Knobs, Escutcheons, etc.
Peterborough Lock Mfg. Co., Peterborough, Ont.

Lumbermen's Supplies.
Pink, Thos., & Co., Pembroke Ont.

Mantels, Grates and Tiles.
Batty Stove and Hardware Co., Toronto.

Manufacturers' Agents.
Fox, C. H., Vancouver.
Gibb, Alexander, Montreal.
Mitchell, David C., & Co., Glasgow, Scot.
Mitchell, H. W., Winnipeg.
Pearce, Frank, & Co. Liverpool, Eng.
Scott, Bathgate & Co., Winnipeg.
Thorne, R. E., Montreal and Toronto.

Metals.
Canada Iron Furnace Co., Midland, Ont.
Canada Metal Co., Toronto.
Eadie, H. G., Montreal
Gibb, Alexander, Montreal.
Kemp Mfg. Co., Toronto
Leslie, A. C., & Co., Montreal.
Lysaght, John, Bristol, Eng.
Nova Scotia Steel and Coal Co., New Glasgow, N.S.
Robertson, Jas., Co., Montreal.
Roper, J. H., Montreal.
Samuel, Benjamin & Co., Toronto.
Stairs, Son & Morrow, Halifax, N.S.
Thompson, B. & S. H. & Co. Montreal.

Metal Lath.
Galt Art Metal Co., Galt.
Metallic Roofing Co., Toronto.
Metal Shingle & Siding Co., Preston, Ont.

Metal Polish, Emery Cloth, etc
Oakey, John, & Sons, London, Eng.

Mops.
Cumming Mfg. Co., Refrew.

Mouse Traps.
Cumming Mfg. Co., Renfrew.

Oil Tanks.
Bowser, S. F., & Co., Toronto.

Paints, Oils, Varnishes, Glass.
Bell, Thos., Sons & Co., Montreal.
Canada Paint Co., Montreal.
Canadian Oil Co., Toronto
Consolidated Plate Glass Co., Toronto.
Fenner, Fred., & Co., London, Eng.
Henderson & Potts Co., Montreal.
Imperial Varnish and Color Co., Toronto.
Jamieson, R. C., & Co., Montreal.
McArthur, Corneille & Co., Montreal.
McCaskill, Dougall & Co., Montreal.
Montreal Rolling Mills Co., Montreal.
Moore, Benjamin, & Co Toronto.
Queen City Oil Co., Toronto.
Ramsay & Son, Montreal.
Sherwin-Williams Co., Montreal.
Standard Paint and Varnish Works Windsor, Ont.
Stephens & Co., Winnipeg.
Martin-Senour Co., Chicago.

Perforated Sheet Metals.
Greening, B., Wire Co., Hamilton.

Plumbers' Tools and Supplies.
Borden Co., Warren, Ohio.
Canadian Fairbanks Co., Montreal.
Cluff, R. J., & Co., Toronto.
Glauber Brass Co., Cleveland, Ohio.
Jardine, A. B., & Co., Hespeler, Ont.
Jenkins Bros., Boston, Mass.
Lewis, Rice, & Son, Toronto.
Merrell Mfg. Co., Toledo, Ohio.
Montreal Rolling Mills, Montreal.
Morrison, Jas., Brass Mfg. Co., Toronto.
Mueller, H., Mfg. Co., Decatur, Ill.
Oshawa Steam & Gas Fitting Co., Oshawa
Robertson Jas., Co. Montreal.
Stairs, Son & Morrow, Halifax, N.S.
Standard Ideal Sanitary Co., Port Hope.
Standard Sanitary Co., Pittsburg.
Stephens, G. F., & Co., Winnipeg, Man.
Turner Brass Works, Chicago.
Vickery, Orlando, Toronto.

Portland Cement.
Grey & Bruce Portland Cement Co., Owen Sound.
Hanover Portland Cement Co., Hanover, Ont.
Hyde, F., & Co., Montreal.
Thompson, B. & S. H. & Co., Montreal.

Potato Mashers.
Cumming Mfg. Co., Renfrew.

Poultry Netting.
Greening, B., Wire Co., Hamilton, Ont.

Razors.
Clauss Shear Co., Toronto.

Roofing Supplies.
Brantford Roofing Co., Brantford.
McArthur, Alex., & Co., Montreal.
Metal Shingle & Siding Co., Preston, Ont.
Metallic Roofing Co., Toronto.
Paterson Mfg. Co., Toronto & Montreal.

Saws.
Atkins, E. C., & Co., Indianapolis, Ind
Lewis Bros., Montreal.
Shurly & Dietrich, Galt, Ont.
Spear & Jackson, Sheffield, Eng.

Saws—Hack.
Diamond Saw & Stamping Works, Buffalo

Scales.
Canadian Fairbanks Co., Montreal.

Screw Cabinets.
Cameron & Campbell, Toronto.

Screws, Nuts, Bolts.
Montreal Rolling Mills Co., Montreal.
Morrow, John, Machine Screw Co., Ingersoll, Ont.

Sewer Pipes.
Canadian Sewer Pipe Co., Hamilton
Hyde, F., & Co., Montreal.

Shelf Boxes.
Cameron & Campbell, Toronto.

Shears, Scissors.
Clauss Shear Co., Toronto

Shelf Brackets.
Atlas Mfg. Co., New Haven, Conn

Shellac
Bell, Thos., Sons & Co., Montreal.

Shovels and Spades.
Canadian Shovel Co., Hamilton.
Peterboro Shovel & Tool Co., Peterboro.

Silverware.
Phillips, Geo., & Co., Montreal.
Round, John, & Son, Sheffield, Eng.

Spring Hinges, etc.
Chicago Spring Butt Co., Chicago, Ill.

Steel Rails.
Nova Scotia Steel & Coal Co., New Glasgow, N.S.

Stoves, Tinware, Furnaces
Canadian Heating & Ventilating Co., Owen Sound.
Canada Stove Works, Harriston, Ont.
Clare Bros. & Co., Preston.
Davidson, Thos., Mfg. Co., Montreal.
Guelph Stove Co., Guelph.
Gurney Foundry Co., Toronto.
Harris, J. W., Co., Montreal.
Joy Mfg. Co., Toronto.
Kemp Mnfg. Co. Toronto.
McClary Mfg. Co. London.
Pease Foundry Co., Toronto.
Stewart, Jas., Mfg. Co., Woodstock, Ont.
Taylor-Forbes Co., Guelph, Ont.
Wright, E. T., & Co., Hamilton.

Tacks.
Montreal Rolling Mills Co., Montreal.
Ontario Tack Co., Hamilton.

Ventilators.
Harris, J. W., Co., Montreal.
Pearson, Geo. D., Montreal.

Washing Machines, etc
Dowswell Mfg. Co., Hamilton, Ont.
Taylor Forbes Co., Guelph, Ont.

Wheelbarrows
London Foundry Co., London, Ont.

Wholesale Hardware.
Birkett, Thos., & Sons Co., Ottawa.
Caverhill, Learmont & Co., Montreal.
Frothingham & Workman, Montreal.
Hobbs Hardware Co., London.
Howland, H. S., Sons & Co., Toronto.
Lewis Bros. & Co., Montreal.
Lewis, Rice, & Son, Toronto.

Window and Sidewalk Prisms.
Hobbs Mfg. Co., London, Ont.

Wire Springs.
Guelph Spring Axle Co., Guelph, Ont.
Wallace-Barnes Co., Bristol, Conn.

Wire, Wire Rope, Cow Ties, Fencing Tools, etc.
Canada Fence Co., London.
Dennis Wire and Iron Co., London, Ont.
Dominion Wire Mnfg. Co., Montreal
Greening, B., Wire Co., Hamilton.
Montreal Rolling Mills Co., Montreal.
Western Wire & Nail Co., London, Ont

Woodenware.
Taylor-Forbes Co., Guelph, Ont.

Wrapping Papers.
Canada Paper Co., Toronto.
McArthur, Alex., & Co., Montreal.
Stairs, Son & Morrow, Halifax, N.S.

IS CONCEDED BY EXPERTS TO BE THE

BEST BABBITT

For Heavy Slow Machinery
Heavy Line Shafting, etc.

12c. lb. TRY IT.

The James Robertson Co., Limited

144 William St., MONTREAL, QUE.

PFLUEGER'S
Fishing Tackle

Hooks, Flies, Trolls, Spinners, Phantoms, Reels, Furnished Lines.

Everything in Fishing Tackle.

NOTICE—Free to Any Dealer in Sporting Goods, Sent Express Prepaid, 170-Page Illustrated Catalogue No. F25, and Metal Fish Sign in 8-Color Lithograph.

The Enterprise Mfg. Co.
AKRON, Ohio, U.S.A.

WALKER CORK SCREWS

are good holiday souvenirs and live sellers the whole year. Are made well and are made out of good material. That is why we guarantee every one that we put out. When ordering Cork Screws try Walker's

Manufactured only by **ERIE SPECIALTY COMPANY**, ERIE, Pa. U.S.A.

LUFKIN MEASURING TAPES

Steel, Metallic, Linen, Pocket, Ass Skin, Pat. Leather, Bend Leather, Etc.

ARE THE BEST AND MOST POPULAR TAPES IN THE WORLD.
YOUR STOCK IS NOT COMPLETE WITHOUT THEM.

LUFKIN RULE CO., Saginaw, Mich, U.S.A.

London Office and Warehouse—24 and 26 Holborn. New York City Branch—280 Broadway.

For sale by ALL PROMINENT CANADIAN HARDWARE JOBBERS.

CIRCULATES EVERYWHERE IN CANADA

Also in Great Britain, United States, West Indies, South Africa and Australia.

HARDWARE AND METAL

A Weekly Newspaper Devoted to the Hardware, Metal, Heating and Plumbing Trades in Canada.

Office of Publication, 10 Front Street East, Toronto.

VOL. XIX. MONTREAL, TORONTO, WINNIPEG, JANUARY 19, 1907 NO. 3.

See Classified List of Advertisements on Page 67.

WINTER
SPORTING GOODS

Tube Speed Skates

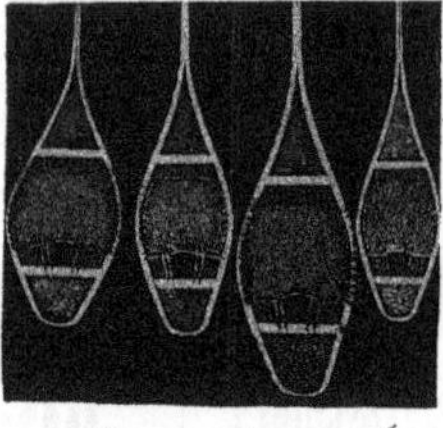
Snow Shoes

Tube Hockey Skates

TOBOGGANS

HAND SLEIGHS

SNOW SHOES

MOCCASINS

Hockey Skates

HOCKEY SUPPLIES

PUCKS
STICKS
SHIN PADS
GAUNTLETS
ANKLE SUPPORTS
ETC.

Clamp Skates

WRITE FOR PRICES

RICE LEWIS & SON
LIMITED

TORONTO.

UNION ADJUSTABLE PLANES

IRON, SMOOTH BOTTOM

All Sizes
Jack
Fore
Jointer

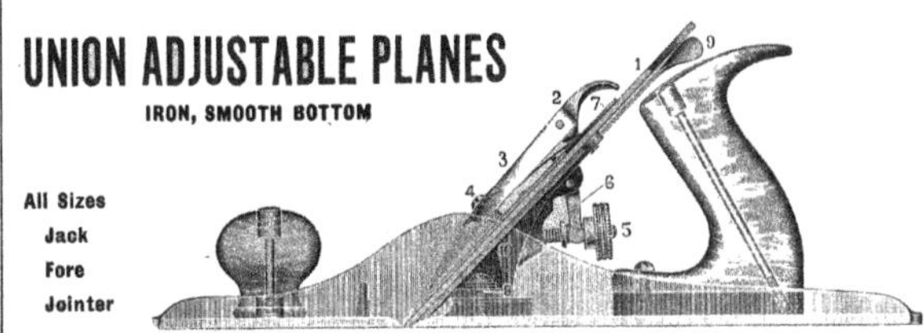

No. 4 U

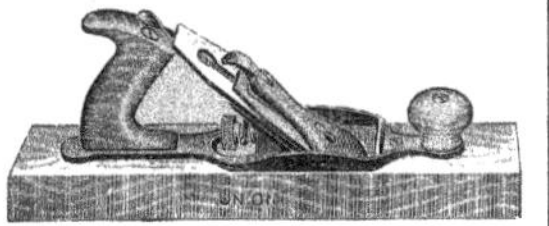

Wood, with Iron Top, No. 26 U. All sizes

Wood, with Iron Top, No. 502. Smooth.
Length, 8 in; Cutter, 1 3-4 in.

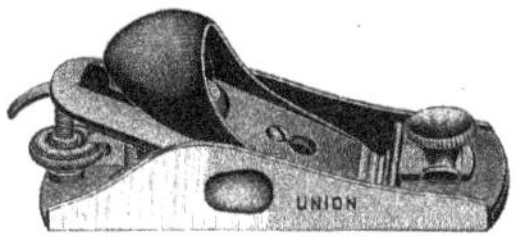

Union Adjustable Block Plane, No. 9 1-2 U.
Length, 6 in.; Cutter, 1 3-4 in.

"ON THE SQUARE"

is about the most appropriate way to express ourselves in regard to

UNION PLANES

They certainly are "on the square."

Not the cheapest goods on the market, nor yet the highest priced, but as regards the quality they take no second place.

Write us for prices and information if you want to get your money's worth in the "plane line" of your business.

LEWIS BROS., LTD.

MONTREAL

OTTAWA
TORONTO
WINNIPEG
VANCOUVER
CALGARY

Consumers Cordage Co., LIMITED

Manufacture every variety of

CORDAGE | PACKING

BINDER TWINE

LATHYARN | SMALL TWINES

From MANILLA, SISAL, ITALIAN, RUSSIAN, JUTE

Tallow Laid Transmission Rope, 3 or 4 Strand

Oil Well Drilling Cables, Lariat Rope, 3 or 4 Strand,
Halyards, Deep Sea Line, Snap Halters, Snap Plow Lines,
Clothes Lines, Hide Cord, Pulp and Paper Cord,
Tarred Rope, Marline, Houseline

Wrapping, Rolling, Seaming, Sewing and Grape Twines

CAREFUL EXECUTION **SPECIAL ORDERS** PROMPT SHIPMENT

Mills: MONTREAL, QUE., and HALIFAX, N.S.

BRANCHES:

F. H. ANDREWS & SON, Quebec. CONSUMERS CORDAGE CO., Limited, St. John, N.B. MacGOWAN & CO., Vancouver.
TEES & PERSSE, Limited, Winnipeg, Calgary and Edmonton. GEO. WOOD, London, England.
MERRICK, ANDERSON & CO., Winnipeg, Binder Twine Distributors for Northwest.

—Supplies for Sugar Makers—

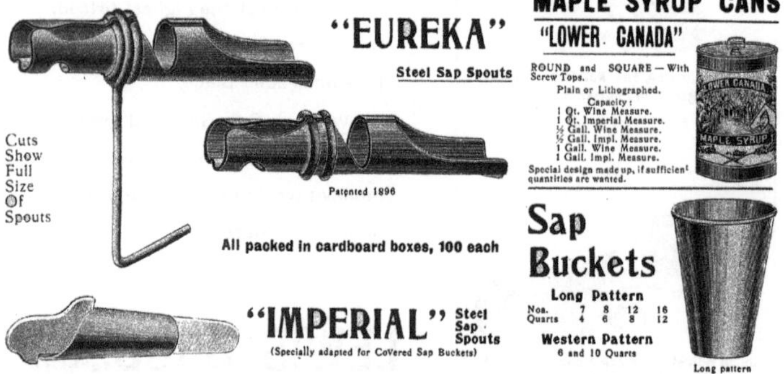

The Thos. Davidson Mfg. Co. Ltd., MONTREAL and WINNIPEG

H. S. HOWLAND, SONS & CO. LIMITED

HARDWARE MERCHANTS.

Only Wholesale — 138-140 WEST FRONT STREET, TORONTO — Wholesale Only

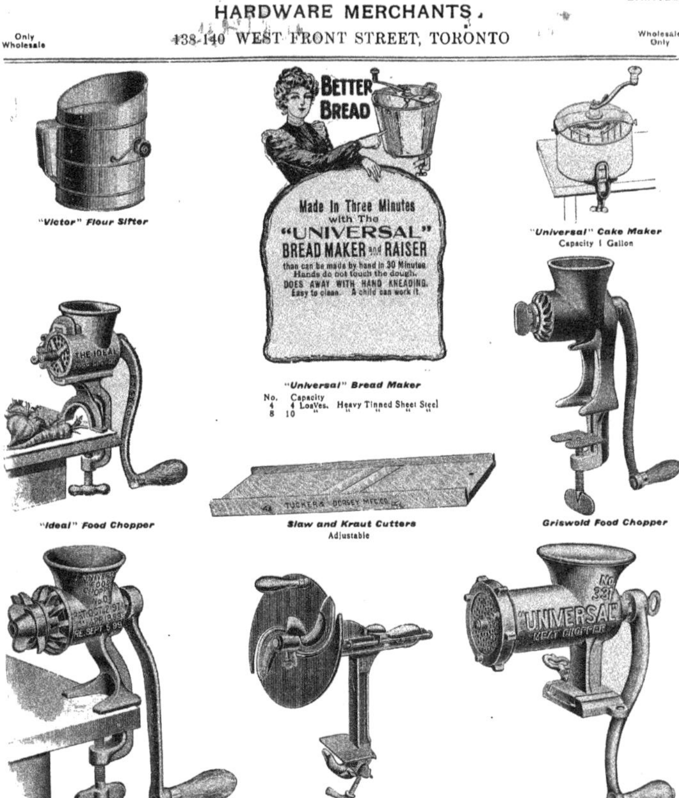
BETTER BREAD
Made In Three Minutes with The
"UNIVERSAL" BREAD MAKER and RAISER
than can be made by hand in 30 Minutes
Hands do not touch the dough.
DOES AWAY WITH HAND KNEADING.
Easy to clean. A child can work it.
THE IDEAL
TUCKER & DORSEY MFG CO.
UNIVERSAL MEAT CHOPPER
No. 331

"Victor" Flour Sifter

"Universal" Cake Maker
Capacity 1 Gallon

"Universal" Bread Maker

No.	Capacity	
4	4 Loaves.	Heavy Tinned Sheet Steel
8	10 "	" " "

"Ideal" Food Chopper

Slaw and Kraut Cutters
Adjustable

Griswold Food Chopper

"Universal" Food Choppers

"Sterling" Vegetable Cutters

"Universal" Meat Choppers

For other Makes of Cutters see our Hardware Catalogue.

H. S. HOWLAND, SONS & CO., LIMITED

Opposite Union Station.

GRAHAM NAILS ARE THE BEST

We Ship promptly — Factory: Dufferin Street, Toronto, Ont. — Our Prices are Right

Brantford—Roofings

This Represents Our

"ASPHALT"
No. 1, 2, 3
"RUBBER"
No. 1, 2, 3

Each Roll is complete with Nails and Cement to lay. Best selling goods in the market.

BRANTFORD ROOFING CO.
LIMITED
BRANTFORD, ONTARIO

SIMPLE

DURABLE

RELIABLE

STYLE F.

THE LEADER

A Name Familiar to Butter Makers in Connection with the

Highest Grade Churns

ITS STRONG POINTS:

Stiff Steel Frame, with malleable iron fittings, Combination Bow Lever and Pedal Drive. Bicycle Ball Bearings. Bolted Trunnions. Cream Breakers. Barrel easily drained and conveniently detatchable for cleaning.

DEALERS: When you handle this line you handle QUALITY.

The Dowswell Manufacturing Co., Limited

W. L. HALDIMAND & SON, Montreal
EASTERN AGENTS

HAMILTON
ONTARIO

GALVANIZED FENCE HOOK FOR FASTENING WOODEN PICKET ON WIRE FENCES

WIRE NAILS, COILED SPRING, BARB and PLAIN FENCE WIRE, OILED and ANNEALED, CLOTHES LINE WIRE, STAPLES, etc.

THE WESTERN WIRE & NAIL CO., Limited, LONDON, ONT

The Peterborough Lock Manufacturing Company, Limited

Peterborough, Ont.

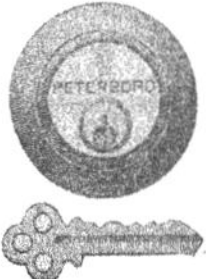

Cylinder Night Latch, No. 103.

Manufacturers of all kinds

Rim and Mortise Locks, Inside, Front Door, and Store Door Sets, also full line High-class Builders' Hardware.

Sold by all Leading Jobbers in the Dominion.

Serviceable Combination Cabinet

SUITABLE FOR HARDWARE TRADE

A COMBINATION THAT EVERYBODY MUST HAVE

CONTENTS

6 Table Knives
6 Dessert Knives
6 Table Forks
3 Table Spoons
6 Dessert Forks
6 Dessert Spoons
6 Tea Spoons
2 Salt Spoons
1 Mustard Spoon

42 Pieces

Price Complete

SILVER PLATED	LIST $22.30
BEST NICKEL SILVER	LIST $14.00

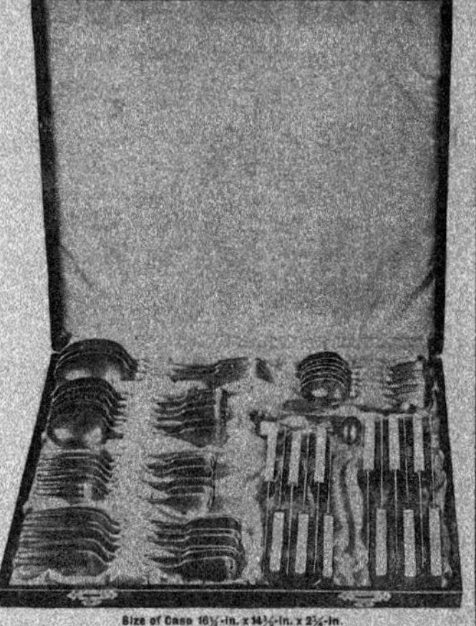

Size of Case 16½-in. x 14½-in. x 2¼-in.

JOHN ROUND & SON, LIMITED

MANUFACTURING

Cutlers and Silversmiths

MONTREAL

Factories SHEFFIELD, Eng.

Established 1847

DOMINION WIRE MANUFACTURING CO. LIMITED

MONTREAL TORONTO

Manufacturers of

Iron and Steel Wire,
Brass Wire,
Copper Wire,
Wire Nails,
Wood Screws,
Staples,
Jack Chain,
Bright Wire Goods,
Spring Cotters,
Steel Wire Barrel Hoops

START THE NEW YEAR RIGHT

No doubt many of our customers throughout the Dominion have completed their stock-taking.

Stocks in many lines must be low.

We would suggest that our customers anticipate their requirements and send us their specifications at an early date, so as to ensure prompt deliveries.

When Ordering PLEASE Specify DOMINION WIRE MFG. CO.'S Goods

The Shrewd Dealer

wastes no floor space. He merely carries the goods his customers demand. Our line of

PIKE POLES

is considered a good deal better than the best marketed by our competitors.

Lumbermen have proven their superiority and always demand them.

We also make a full line of **CANT HOOK, PEAVY** and **AXE HANDLES. NOW'S** a good time to order.

J. H. Still Manufacturing Company, Limited
ST. THOMAS, ONTARIO

: : It Saves Lives : :

A Building Necessity

These are the names that have been made synonymous with the

AEOLIAN VENTILATOR

We are the manufacturers of it and control the patents.

We have placed it on some of the finest buildings in the Dominion.

It is a scientific fact that no modern buildings can afford to neglect sanitation.

We are contractors and sanitary experts. Our ventilator is guaranteed and is placed in position to suit the building as to quality and size.

One Testimony!

MONTREAL, June 12th, 1903.
Canadian Pacific Railway Company.

Messrs. J. W. HARRIS CO., LIMITED,
7 St. Elizabeth St., Montreal, Que.

Gentlemen,—Answering yours of June 1st, I am not personally familiar with the merits of the "AEOLIAN" Ventilators but our Assistant Engineer, Mr. F. Crossley, reports that the Ventilators of this type manufactured by you and used by this Company at the Place Viger Hotel, Chateau Frontenac, Windsor Street Station and elsewhere, have proved satisfactory. Yours truly,

E. H. M[illegible]

The J. W. HARRIS CO., Limited
Successors to LESSARD & HARRIS
General Contractors and Manufacturers
MONTREAL

A★1 Brand Cutlery is Always Worth the Money You Pay for it

Every Knife and Scissor is guaranteed to be of the best quality of steel, right temper, superior finish and finest workmanship. That is why this brand is a favorite in the Canadian Market, and your stock is not complete without the goods of

THE R. BERG CUTLERY CO.

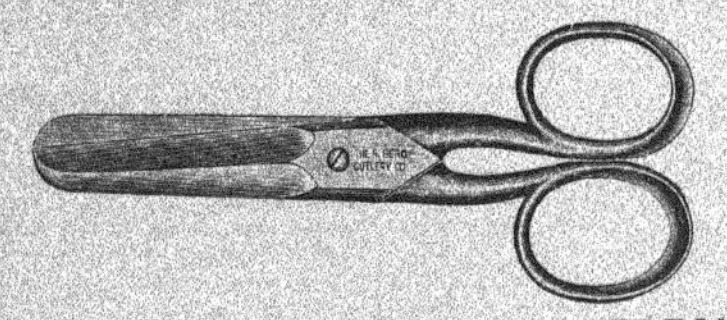

F. W. LAMPLOUGH & CO., MONTREAL

Buy the Best.

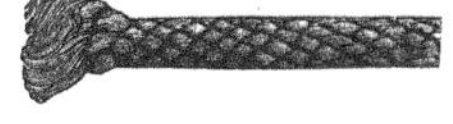

HERCULES

Sash Cord.

Star Brand Cotton Rope
Star Brand Cotton Clothes Lines
Star Brand Cotton Twine

For Sale by all Wholesale Dealers.

B. A. GRAUTOFF & CO.
8 Fenchurch Buildings, London, E.C.
Import and Export.
Old established house with highest connections in Great Britain and on the Continent of Europe.
Cable Address, "Grautoff London." Liebers Code.

GENUINE Pratts Astral Lamp Oil

Sold in all countries and recognized as the highest grade oil manufactured.

WHOLESALE ONLY

THE QUEEN CITY OIL COMPANY Limited, - TORONTO, ONT.

Persons addressing advertisers will kindly mention having seen this advertisement in Hardware and Metal.

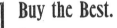

Persons addressing advertisers will kindly mention having seen this advertisement in Hardware and Metal.

Let us Help You Increase Your Business

Why not **go after and get** the fire brick trade in your locality.

Our expert information is at your command and will enable you to make this a paying line.

Make this one of the New Year resolutions and KEEP IT.

Francis Hyde & Co.

Contractors and Foundry Supplies

31 WELLINGTON ST.

MONTREAL

A MATTER OF DOLLARS

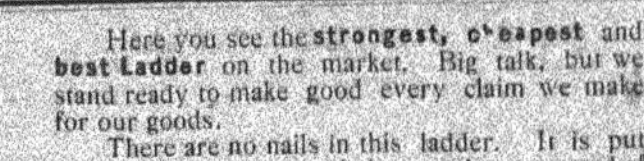

Here you see the **strongest, cheapest** and **best Ladder** on the market. Big talk, but we stand ready to make good every claim we make for our goods.

There are no nails in this ladder. It is put together with screws and rivets and constructed of thoroughly seasoned ash and spruce. Durable, safe and a **money** saver.

That Waggon Jack is a new article on the Canadian market and it already **holds the United States market**. It holds what it raises—because it has double acting dogs.

It is built of steel and malleable iron, there is nothing to get out of order about this Jack. Built to lift all kinds of buggies, heavy waggons and general work.

SEND FOR PARTICULARS NOW.

SHIRREFF MANUFACTURING CO., BROCKVILLE, ONT.

Dennis' Iron Stable Fittings

Sanitary, Indestructible, Economical

A good line for the hardware dealer.

Send for New Stable Fittings Catalogue and Price List "A."

DENNIS IRON & WIRE WORKS, Ltd.
LONDON, - ONTARIO

ELECTRIC SIGNS

advertise you night and day. Without question the cheapest and most effective method of advertising to-day.

We have excellent proposition to make in most Canadian cities for merchants to use electric signs.

Let us hear from you, and we will send you circulars.

THE MARTEL-STEWART CO., Limited
MONTREAL, CANADA.

OUR LINES ARE SELLERS !

Why Not Write for Particulars ?

CABLE ADDRESS "SUPERIOR" — *FARLEY WORKS, SHEFFIELD, ENG*

GEORGE PHILLIPS & CO.

MANUFACTURERS OF

Silverware, Electroplate and Cutlery

We also make the best Razor on the market. Four exclusive brands with advertised names that will help every dealer. It costs a two cent stamp to find out about these solid old country lines which you cannot duplicate elsewhere. ***ALL MONEYMAKERS.***

OFFICE FOR CANADA, PACKARD BUILDING, MONTREAL — TELEPHONE M 557

Persons addressing advertisers will kindly mention having seen this advertisement in Hardware and Metal.

To be continued.

Commercial Success

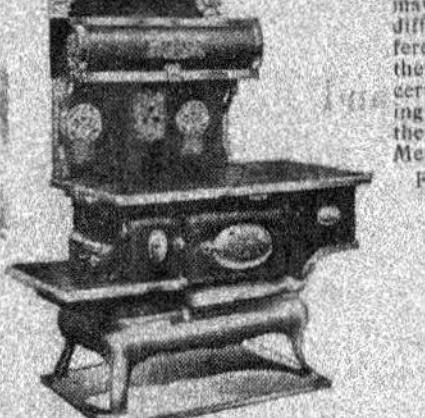

may be attained in different ways by different men; *But*—there is one sure and certain way of cornering a paying trade in the Hardware and Metal Business.

Feature

The

Empire Queen Range

and your success is absolutely assured. The superiority of ***The Empire Queen*** is admitted by nearly all practical men, while its magnificent appearance rivets a customer's attention at once.

The Empire Queen certainly does look well—cooks well—practically sells itself.

The Canadian Heating & Ventilating Co. Limited.

OWEN SOUND, Ontario.

THE CHRISTIE BROS. CO., Limited, 238 King St., Winnipeg, Man., Western Agents.

THE CANADA STOVE AND FURNITURE CO., 126 West Craig St., Montreal, Que., Agents for the Province of Quebec.

A Good Range Saves Worry and Bad Cooking.

A Poor Range is a constant source of Annoyance and Disappointment.

The Joy Malleable Range is among **the best**, to say **the least**, and its sterling malleable iron, double riveted, reinforced, air tight construction makes it the **Longest Liver** in the bunch.

If you want to build up a business institution that your grandchildren can point to with **pride**, adopt The Great Joy Range for your leader—pin your faith right to it and no other result can follow. What think ye?

JOY MFG. CO., 32 Humberside Ave., TORONTO, ONT.

TEES & PERSSE CO., - Western Distributors, Winnipeg

CHAS. H. FOX, - Agent for British Coumbia, Vancouver.

"Samson" Milk Cans, Trimmings, Etc.

"SAMSON"

Now is the time when you should get your stock of these goods into shape for spring business.

If you handle the "Samson" line you know you have something better than your competitor.

You know when you recommend "Samson" Milk Cans and Trimmings to your customers that the goods will bear out your recommendations.

We can ship the finished articles or the trimmings promptly.

"SAMSON"

THE McCLARY MFG. CO.

LONDON, TORONTO, MONTREAL, WINNIPEG, VANCOUVER, ST. JOHN, N.B., HAMILTON, CALGARY.

"EVERYTHING FOR THE TINSHOP"

THE PERCENTAGE OF PROFIT

A discussion invited on an interesting and convincing argument presented in favor of correct bookkeeping by Fred. C. Lariviere, President of Amiot, Lecours & Lariviere, Wholesale and Retail Hardware Merchants, Montreal.

Technically, the profits must be based on cost and the percentage of profits fixed according to the class, quality and appearance of the goods, and this is the only proper course. In practice, however, the percentage of profit may be computed indifferently on the turnover, either on cost or selling values. Provided the expenses are figured on the same basis, the results in dollars and cents would be the same. This is only a transposition of figures with same results as to the amount involved, for nobody would for a moment doubt that dollars and cents are what is looked for in the final results, consequently it should have the precedence over percentage, which is only an accessory. It is a well known fact that bookkeeping must be made to suit the requirements of business, and not business to bookkeeping.

As far as I know it is customary to figure the results of a year's business on the total amount of goods sold at selling prices ; but in using this method it is the business of the merchant, when pricing goods, to apply the following rule : If he wishes to realize a profit of 50 per cent. on selling value he must add 100 per cent. on cash value ; if he wishes to realize 40 per cent. he must add 66 2-3 ; if 33 1-3, he must add 50 ; add 25 ; if 16 2-3 he must add 20.

Provision must be made for a percentage of profit to sales and to capital. There is a wide difference in the two. I will explain by an example taken from actual results. A certain concern I know, started in business with $418.78, the turnover of the year at selling prices being $29,210.14 and the gross profits $6,822.28, or 23.35 per cent. ($22,387.86, cost) or equalling 30 47 per cent. on cost price. The expenses including interest and bad debts were $2,286.60 or 7.96 per cent. on sales at selling prices, or 10.21 per cent. on cost. Now let us see if the results are the same.

The profits on :

$29,210.14, selling value, at 23.35 per cent., equal $6,822.28.

$22,387.86, cost value, at 30.47 per cent., equal $6,822.28. $6,822.28.

The expenses on :

$29,210.14 at 7.96 per cent. equal $2,286.60.

$22,387.86 at 10.21 per cent., equal $2,286.60. $2,286.60. Net profits, $4,535.68.

Therefore the results are the same, the net profits are $4,535,68. Now let us compare this to capital, which was $418 78 at the beginning of business. The result in percentage is 1,120.62, compared to 15.39 per cent. on selling price of sales or 20.26 per cent. on cost price of goods sold. Which is right? Both are correct in any of the ways this problem is figured.

Capital the Only Investment.

The above statement is surely an evidence and an uncontestable proof that the purchase of a certain amount of goods by a merchant in general business is not an investment. The capital necessary for the general administration of a business is the investment, for capital not only provides for the purchase of goods, but also for carrying the book debts, the increase of stock, and meeting the general expenses of administration. If I invest $100 in a business and wait till this amount doubles itself, it is true I will have made 100 per cent. on my capital. This would be the investment, but it would not represent the result of several transactions of buying and selling and the accumulation of profits on each sale. Consequently a purchase cannot, properly speaking, be an investment. Th's 100 per cent. is

FRED. C. LARIVIERE.

not the percentage of profits on my sales either at cost or selling prices unless it would have been made in only one transaction, which is not the case generally in business.

To substantiate this I will submit the following examples :

A. buys $100 worth of goods. A. sells B. $50 worth for $75, making a profit of $25. A. purchases $50 worth of goods. A. sells C. $75 worth for $125, making a profit of $50. A. sells D. balance of stock, $25, for $50, making a profit of $25. Totals : Goods bought $150 ; cost price $150 ; selling price $250 ; profit $100.

A. therefore sold what cost him $150 for $250, realizing a profit of $100, or 100 per cent. on capital ; 66-23 per cent. on cost ; and 40 per cent. on selling prices ; and all these percentages are correct ; hence the different ways, with but one result of a clear profit of $100. It should be clearly evident that the act of buying goods is not what could be termed an investment. The capital placed in the business is the investment.

I hope some other interested parties will send you their views, so that this very important question can be thoroughly studied.

GOOD BUYERS AND GOOD SELLERS.

At the recent hardware association convention across the line, S. Nowell, a St. Louis jobber, delivered a short address in which he made his hearers understand that jobbers have their troubles and emphasized the fact that selling was a much more important operation than buying. He said in part :

"God have mercy on the man who is a good buyer, but who cannot sell the goods he buys. If you are spending 75 per cent. of your time on buying and 25 per cent. on selling, reverse the ratio and devote at least 75 per cent. of your effort to selling. Every head of a mercantile house must be a school teacher. He must be teaching his clerks and his whole force up-to-date ideas on salesmanship. There is not 5 per cent. difference in the cost of the hardware stock, item for item, of the poorest buyer and the best buyer in the State of Wisconsin, but stocks do vary 25 to 50 per cent. in value as to their assortment and character. There is nothing more untrue than the old statement that 'goods well bought are half sold,' if well bought means cheaply."

RULES LEADING TO SUCCESS.

Take as much interest in your employer's business as if it were your own.

Do your work well to-day—you won't have to do any of it over again tomorrow.

Do not expect to get all you can and give nothing. Do a little more work than is demanded.

Do not let your thoughts be always wool-gathering if you expect to earn an increase of salary.

Be prompt. Show that you have an interest in your work above the desire for an extra half hour in bed in the morning. You can't come down half an hour late every morning and impress your employer with the idea that you are a wide-awake, active man or woman with an interest in your work.

Be cheerful and willing. A sullen countenance is not pleasant to look upon by either an employer or a customer. Remember, your pulling power with a customer is one of your assets. The reserve will be your loss. Be courteous. Do not thrust your troubles and inharmony upon those around you It is a poor investment.

WINDOW AND INTERIOR DISPLAYS

PREVENTING FROST ON SHOW WINDOWS.

M. J. Quinn, ventilating expert with Cluff Bros., Toronto, Canadian agents for the Thomas Acme Air Washer for ventilating buildings, in answer to a request for information regarding the prevention of frost on show windows, writes as follows for Hardware and Metal:

"The physical conditions which create frost on window panes are exactly the same as those which create moisture on cold water pipes in summer time, viz., warm air carrying its extreme complement of moisture comes in contact with a cold body and is reduced in temperature, rendering it incapable of carrying the same percentage of moisture, and what is more natural than that the excess of the latter should be deposited where the chilly process takes place, viz., in this case, on the window pane. We anticipate that some considerable difficulty will be experienced in making a window space sufficiently tight from the inner store to render impossible the circulation of warm air from the latter. In this connection we would emphasize that, if absolutely still air can be obtained between the inner and outer glass, very little difficulty will be experienced from frost even on the coldest days, and in this connection it has been found that double glass set in the same sash, leaving a space of say 1½ or 2 inches between the panes, and having the joints all around the four sides made as tight as possible, has given very good satisfaction, and apart from this construction we are not by any means sure that good results can be obtained under other circumstances.

"We do not know of any solution or wash for putting on the window panes to prevent a formation of frost."

DOES THE WINDOW EARN WAGES?

In conducting his business from one year's end to another, the retail dealer must take into account the elemental principles upon which all trade is based—he must keep a strict tabulation of all expenditures, must know exactly at the year's end what his receipts have been, and thus be enabled to arrive at his percentage of profit. Perfect business methods, says the Sporting Goods Dealer, consists in keeping these different transactions or items on a money-making plane. That is, when a certain bill of goods is bought, it should be sold at a figure to allow a certain profit, when a salesman is employed, the proprietor naturally expects him to bring back into the cash box the amount of his salary, and some more, if advertising is carried in the daily paper elsewhere, a full return for money expended is looked for. These various returns often do not come up to expectations, for one cause or another, and this marks the degrees of success attained by different merchants. Now there is one phase of the retail dealer's business that he is apt to be careless about and therefore not realize a full return on the expenditure—that is the display window. The cost of maintaining this exhibit is a just charge to the wage account, exactly as is the salary of a clerk or salesman. Without windows the rent would be much less, so there is every reason for giving special attention to making it a business-bringing feature. The show windows may be considered an employe, and an expensive one. The time of an assistant to dress the window, the cost of working tools and paraphernalia are some of the expenses that may be charged against it. The question remains, does the window sell enough goods to warrant this expense? With the aid of counter books the dealer can keep track of the work of his salesmen and determine what returns they are bringing In the case of the show window this is not so easy. More time and continued experimenting is necessary to investigate the business brought into the house by the outside exhibit.

The proprietor expects his salesmen to be constantly bright, pleasing and wide-awake; he should make his display window the same. There are unlimited possibilities open to the dealer who considers the window dressing proposition in a serious light. By a careful record of the sales of each class of sporting goods during a specified period, and comparing one with the other, bearing in mind when and how the different goods were featured in display, he can form a very satisfactory estimate of the benefits being obtained from this method of publicity. As has been stated time and time again, the display window is probably the dealer's best business getter. Unlike the salesmen and other employes, the window may be asked to work day and night, and there are no holidays or vacations to be considered. But again, unlike good employes, the window needs constant looking after and shaking up. It is apt to get slovenly and out-of-date unless proper attention is bestowed upon it. Its success, therefore, rests with the dealer himself. It is up to him.

THE IMPORTANCE OF DISPLAY.

It is just as necessary that the hardware merchant should dress up his window and put on a "front" as it is for the haberdasher or the dry goods dealer. There is money in the show window, and it rests with each dealer to get it out. "Put your best foot forward" in commercial life as you do in the social world.

Don't, above all things, make an attempt to make a display unless you intend to do it right. Don't bundle a confused mass of articles into your window and "let it go at that." Make it a show window and not a "holy show" window. If you think you haven't the time or you lack the knack of arranging the articles in an artistic manner, encourage one of your clerks to try and develop ability in this line. It will be money well invested. What so quickly engages the public eye as a tastefully arranged store window? It acts as a magnet which draws the elusive coin to your till.

CARD WRITING.

It is not such a difficult matter to write show cards, but many merchants have never made the attempt which might show them that they are capable of producing a really respectable card by means of the pen or brush. Some merchants hardly use cards at all, others rely upon the rubber-stamp outfit. Clerks who might eventually develop great ability in the show-card line are hampered by the instruction of employers who cling to old ideas and fail to see the possibilities of new ones. Then there are the men who attempt the work without any apparent desire for excellence. Their crudity makes them an exhibit which attracts attention—and fails to attract custom. To the first class of card-writers we would say: Learn show-card writing, and to the second class we would say: Learn show-card writing.

By this is meant not necessarily the highest form of card-writing, which at the present day has come to be even a profession in itself, but the proper way to make a card, no matter how plain or simple it is. It is impossible in one lesson to teach show-card writing, but a few primary suggestions may be helpful.

One of the first causes of poor work is a failure to appreciate the value of simplicity. In almost every case the plain card is the best, and it is the first which should be studied. But so-called "fancy" lettering, and scrolls and furbelows have a fascination for some people, and as produced by them they are an abomination to others. They cannot be read at a glance, sometimes, indeed they are next to indecipherable. Legibility is a prime requisite in a show-card, anything hindering it should be most carefully avoided. In the same class with "fancy" schemes may be placed old-fashioned styles of lettering—not the genuine "old styles," which are among the best letters in use to-day, but the freak fashions which appeared in type, roughly speaking, between the years 1850 and 1890, and which even yet, in some localities, the unfortunate country editor is compelled to employ.

Taking it all round, country newspapers although a good thing to support, are a bad place to look for good type faces, but the large city dailies and the magazines which at some time or other reach the most out-of-the-way part of the country, contain models which every dealer having to do with window-cards would do well to study, choose from and copy. Of course in most cases a type printed letter must be modified to a certain extent when used for show-card work. It always pays to do the work well, but it doesn't pay and isn't necessary to spend the time on a temporary price-card that it would take to letter a permanent sign. Show-card writing and sign-writing are two different trades—or professions, for you can be an artist in either of them—and the very best cards are written by some of the fastest workmen.

SPORTING GOODS

BIG YEAR IN SPORTING GOODS.

Dealers from various parts of the country report 1906 to have been the best season in the history of the trade in Canada, the increases in the volume of trade reported, averaging from 10 to 50 per cent, the general prosperity, and the fact that most people have had money to spend on amusements and recreation, being largely responsible for the growth of this branch of trade.

Winter sporting goods sold particularly well during December, one Toronto firm reporting the sale of over 500 toboggans ranging in price from $1.50 to $7.50, the average being about $5.00. Another sporting goods dealer disposed of over 200, selling 25 on one Saturday before Christmas. A feature of the toboggan trade has been the increased demand from the smaller towns and cities where this old time sport is gaining remarkable favor.

Snowshoes have also sold in fair quantities, although the demand has not been so great as a year ago, owing to the large number sold last winter and not used because of the lack of snow. The demand this year has been better in the country districts than in the larger cities. On the other hand, skies have sold better in the larger cities, the customers being chiefly the children of the wealthier citizens.

Ice skates have sold remarkably well, shortages being reported in many districts. Much stock was carried over from 1905, but the demand this season has been so great that stocks will be very light when the present season closes. Sales, will of course, continue until the end of February. Hockey skates have met with the greatest demand. A very profitable side line with many dealers has been boots for hockey skates, one dealer stating that he had sold 1,000 pairs of shoes already this year. Sweaters, moccassins, etc., have also been in active demand. The continued popularity of roller skating has made it possible for many dealers to work up a fair trade in this line, the Harold A. Wilson Company, Toronto, having sold 600 pairs within the last few months. The Henley and Union Hardware Co.'s lines were the chief sellers. The holiday trade in air rifles was also very good, while revolvers continue to have steady sale.

The trap shooting season is now on, and the demand for ammunition is very active, rifles also selling steadily. A line which sells largely in the trap shooting season, is the blue rock imitation birds, manufactured by J. Bowron, Hamilton, Ont. The Dominion Cartridge Co.'s ammunition is also a popular seller. Many dealers are now pushing gun supplies for the different gun clubs which make a specialty of trap shooting, and for the next couple of months, the best possibilities for making trade is in pushing shooting goods.

Gymnasium supplies and fencing goods, have of course, sold well ever since cold weather arrived, there being a steady increase noticeable in both of these branches of the trade.

Dealers are now preparing to get after the business placed by golf, baseball, cricket, lacrosse, football and tennis clubs, the present being an opportune time to circularize and make personal calls on the secretaries of these organizations, quoting prices on the supplies necessary for the coming season's sports. Preparations are also being made for an active trade in boats, canoes, fishing tackle, the season for which commences about the middle of May. Camp supplies will be another line which will demand early attention.

EARLY ENGLISH GUNPOWDER.

A bucket containing bullets and gunpowder has been discovered in the roof of Durham Castle, where it is believed to have been walled up about the year 1641, when the castle was being prepared to withstand a Scottish raid. The bullets are molded spheres of two sizes, and, according to the analyses of Messrs Silberrad and Simpson, consist of a little over 99 per cent. of lead, with iron and silver, and traces of bismuth, arsenic, and antimony. The gunpowder is not granulated like that of the present day, and was evidently prepared by simply mixing the ingredients. It contains about 1 per cent of moisture, and the proportion of the constituents calculated on the dry powder is practically identical with that of the black gunpowder of to-day, viz., nitre, 75 per cent; carbon, 15 per cent; and sulphur, 10 per cent. It is pointed out by Messrs. Silberrad and Simpson that this is a remarkable fact, since the gunpowders made in England at that time contained a considerably larger amount of sulphur. The only gunpowder with the modern proportions in use in the seventeenth century was Prussian musket powder, and hence it is suggested that the Durham powder was probably of Prussian origin.—Knowledge

GUARANTEED FOR THREE YEARS

In the future the Horton Manufacturing Co., Bristol, Conn., have decided to guarantee their "Bristol" steel fishing rods for three years, instead of one year, as heretofore. The three-year guarantee applies to all "Bristol" rods sold since October 1st, 1906, that is, rods which were shipped from their factory since that date. This is an excellent proof of their faith in the rods and should make a good selling argument for the dealers.

AUTO SUPPLIES.

There seems to be a unanimity of opinion as to the advantage of putting in a line of auto accessories or supplies in the hardware store in cities of the right size, says the Hardware Dealers' Magazine, which has been making an investigation of the results secured by United States retail dealers who have stocked the line.

In small towns where there are only one or two automobiles it is obvious that the dealer would not find it advantageous to put in a line, and on the other extreme are large cities where specialty stores control most of this line. Between, however, are many cities wherein the enterprising dealer can gradually feel his way and put in a stock of the smaller supplies. As the automobilists become informed of the dealer carrying such a stock his trade ought to pick up.

As a rule the man who runs a car has money to spend for many of the new things that are being constantly brought out. Something about his car needs replacing, now and then spark plugs give out, batteries become exhausted, tires burst, some of the wrenches or other tools break or get lost, pumps give out, carbide oils and greases are being constantly used up, lamps get in smash-ups, speed indicators frequently get hold of a man's fancy, and possess one he must, and so on with almost no end. The automobile, unlike the bicycle, is not likely to be a fad; every indication points to a large accession to the ranks of autoists. It therefore behooves the wise dealer to get in on the "ground floor" and "grow up" with the business.

CRIME INCREASES BUSINESS.

Retailers of firearms in Pittsburg have for some weeks carried on a tremendous business in firearms and other weapons of defence as a result of a long series of burglaries and other crimes that has visited the community. Revolvers for both home and pocket have been in great demand. It is noticed that the ideas regarding pocket weapons have experienced a change, for the larger calibres have had the largest call. It seems that the 32-calibre revolver is not considered sufficiently effective by the people of Pittsburg who wish to be prepared to receive midnight prowlers.

TO BROWN GUN BARRELS.

Take 1 ounce muriate tincture of steel, 1 ounce spirits of wine, ¼ ounce muriate of mercury, ¼ ounce strong nitric acid, ¼ ounce blue stone, 1 quart water. Mix well and allow to stand 30 days to amalgamate. After the oil or grease has been removed from the barrels by lime the mixture is laid on lightly with a sponge every ten hours. It should be scratched off with a steel wire brush night and morning until the barrels are dark enough; and then the acid is destroyed by pouring on the barrels boiling water, and continuing to rub them until nearly cool. If the barrels are of laminated steel do not dilute the acid so much

RETAIL HARDWARE ASSOCIATION NEWS

Official news of the Ontario and Western Canada Associations will be published in this department. All correspondence regarding association matters should be sent to the secretaries. If for publication send to the Editor of Hardware and Metal, Toronto.

WESTERN DEALERS TO ARMS.

The following circular has been sent to the four hundred members of the Western Retail Hardware Dealers' Association by Secretary McRobie, and if each member does his part as well as the energetic "Mac," the vigorous protests of the "four hundred," following the action of thousands of hardware, dry goods and other merchants in other parts of Canada, should make certain the withdrawal of the Postmaster General's proposed legislation.

A Vigorous Letter.

You have probably read in the trade journals or the daily press of the proposed regulation regarding a new parcel post c.o.d. collection system.

Do you realize what this means to the trade? If you do, then we ask you as merchants whose business is jeopardized to wake up and defeat this iniquitous measure. Such a system, if put in force, would encourage farmers and townspeople to send to the departmental stores and have their purchases sent by mail c o.d., the local postmaster collecting the bill and returning the money.

We want you to write your local member of Parliament urging him to see the Postmaster-General, and oppose the proposed system; also, personally write the Postmaster-General, protesting against the measure; interview the editor of your local paper, and arouse his interest, pointing out to him the loss he will sustain eventually in advertising, etc., if the measure goes through.

The parcel post question is an important matter. Mail order houses in the United States secured similar legislation, but it was repealed owing to the activity of the members of the Retail Hardware and other associations in flooding the Postmaster-General with protests.

Show that you are alive to your interests by protesting now, and by so doing you will not only ward off the present danger, but you will make the law-makers realize that measures affecting the interests of thousands of merchants cannot be sacrificed in order to satisfy the grasping desires of departmental stores.

All merchants should stand firmly together and combat this legislation. We therefore trust that you will take this matter up with your fellow dealers locally, and write both your local member and the Postmaster-General before the 1st of February. (No postage stamp is required on letters sent the Postmaster-General). Yours truly,

J. E. McRobie, Secy.-Treas.
A. J. Falconer, President.

P.S. Don't forget the association's annual meeting which will be held on Tuesday morning, February 12th, at 9 o'clock, in the Scott Block, 271 Main St., Winnipeg.

A GOOD EXAMPLE.

That the hardware trade is alive to the danger of the legislation outlined by the Postmaster-General in Parliament on Dec. 7 last, is proven by the energetic manner in which they have taken up our suggestion that they write letters of protest to their local member of Parliament as well as to the Postmaster-General.

The trade realizes that the proposal to establish a parcels post c.o.d. collection system is inspired by the departmental store millionaires and the retail dealers are not willing to lie down and be trampled upon without showing that they are men capable of upholding their rights.

Numerous hardware dealers in Ontario have written commending the active campaign being conducted by the Ontario Retail Hardware Association, and many are forwarding their membership fees to the organization, which is actively working in their interests.

One of the new firms which, although not in the association in 1906, has forwarded the $3 fee to cover the membership fee for 1907, is J. L. Fenn, Sons, & Co., Bracebridge, a wholesale and retail firm which is not afraid to put its shoulder to the wheel and do their share of pushing in the interests of the trade as a whole. They say:

"We might say that we are strongly opposed to a measure such as is proposed by the Postmaster-General, and we will do all we can to assist the association to protest against such legislation. We wish the association every success in anything they undertake to assist the retail trade."

Draft Form of Letter.

.... 1907

To the Honorable Mr. Lemieux,
Postmaster General, Ottawa.

Dear Sir,—We understand that you are preparing an amendment to the law providing for the existing parcels post system, the said amendment being intended to provide for the collection on delivery of parcels. Such a system would enable mail order houses to send goods c.o.d. to any part of the country in competition with local retailers.

The large departmental stores are already injuring local merchants to a tremendous extent and any change in the postal laws which would favor them would be a great injustice to the thousands of merchants who have their capital invested in the small towns and cities throughout Canada, aiding by their presence, their energy and their taxes in the up-building of the small communities upon which Canada's future prosperity must depend.

We do not ask for any special privilege and protest against the mail-order houses being given any favors. The proposed legislation would tend to reduce the number of retailers, thus destroying competition, depreciating the value of real estate in the towns and having a similar effect upon the farming districts, as well as tending to decrease the opportunities for pleasure enjoyed by the farmer's family. The deserted farm districts of the New England States are an object lesson to avoid.

We trust that you will see the fairness of our argument that the legislation would be detrimental to the continued progress and development of this fair Dominion and we would ask that you reconsider your announced intention of introducing the proposed amendment.

Yours very truly.

Charles Richardson, the manager of the Dominion Cement, Paint and Roofing Company, died last week at his son's residence in Toronto.

EFFECTIVE ADVERTISING

Merely Introductory.

There has been more unadulterated nonsense written about "advertising" during the last decade than on any other subject of interest to mankind—always excepting, of course, "Race Suicide" and "The Annexation of Canada by the United States."

All Advertising Good.

Mr. Retailer: jump in right now and advertise—if you've never done it before! Don't be barred out of the publicity columns of your local papers because you can't write as pretty an ad. as the fellow round the corner. An old Scot had just swallowed a glass of whiskey when a croney inquired: "Is that good whiskey, Sandy ?"

"Good whiskey!" wrathfully ejaculated the old man. "All whiskey is good; but some brands are better than others."

Despite the assertions of many advertising experts I am inclined to think that all advertising is good, but—some ads. are far more effective than others. Even when John Smith merely declares that he has a certain line of goods for sale John Smith is not producing bad advertising. I have studied the methods of advertisers all over the British Empire and the United States, and I have yet to see an ad. which could be considered as damaging to the business of the man who produced it. The moral is double-barrelled: advertise and—keep on advertising.

The Fundamental Essential.

The primary essential for the retailer, who produces his own copy, is —a thorough knowledge of his business —its wants. Some of the best advertising copy I've seen was produced several years ago by a Chicago retailer. That copy was so strong and effective that the gentleman referred to cleared a rich fortune in a few years. He became somewhat careless regarding the management of his business. He lost a knowledge of its needs, but still continued to produce the advertising-good, strong stuff it was, too; but lacking knowledge of the wants of his business, the copy has lost much of its old-time power.

Never Tell a Lie.

Truth is more necessary in your advertising copy than in a preacher's discourse. Don't exaggerate. The merchant who makes exaggerated statements regarding his goods strikes common-sense people in the same way as the fellow who swaggers along the beach in a loud suit. "There's nothing in him," they say—referring to the swaggerer and, well—enough said.

A good way to judge your copy is to leave it aside for a day or two and then revise it. You'll find, as a general rule, that you have spread yourself too much. You'll be able to strengthen your copy in almost every case by cutting out a few sentences.

Study Your Trade.

This is an important point that most retailers seem to overlook. They produce ads. in a kind of happy-go-lucky way without studying their customers at all. Fancy yourself where the other fellow is occasionally.

Women Are Landed Differently.

Be "peppery and to the point" when writing ads. you wish to appeal to men. But don't talk to women that way. A woman wants to see the name of the articles and the price right at a glance. Then a detailed account of the article will interest her after the price has.

Good Intentions.

I intended to criticise two ads. in this article; but these general remarks seemed to be necessary. In another issue we'll have something to say about Jas. Simmonds & Co.'s ad. which we consider just about as good as any we have seen for a long time. Some practical ads. used some time ago by O. M. Hodson will also come in for a word of commendation. We will also reproduce a pretty strong ad. forwarded by D. Brocklebank & Son.

Have any other dealers samples of advertising matter to send for criticism ?

T.J.S.

ORIGINAL ADVERTISING.

An original form of advertising comes from Russia, where a retail dealer recently circulated the following announcement among his customers: "The reason why I have hitherto been able to sell goods so much cheaper than anybody else is that I am a bachelor and do not need to make a profit for the maintenance of a wife and children. It is now my duty to inform the public that this advantage will shortly be withdrawn from them, as I am about to be married. They will, therefore, do well to make their purchases at once at the old rate." The result of this unique proclamation was that such a run on the shop ensued during the next few days that the dealer made money enough to pay the expenses of his wedding on a very lavish scale and leave a snug bank account besides.

Merchants who advertise in their local papers should stir up the editors to do something for the town's business on their own account. The accompanying display is a reduced reproduction of a clation call five inches deep and eight across the top of the front page of a Kentville, N.S., paper. In the same issue the editor said :

"Several of the other towns are holding or have held a merchant's day this season and as will be seen elsewhere in this issue the Kentville merchants have fallen into line and are prepared to meet the outside competition. Every day from now on is Merchants day in Kentville. During the next two weeks we respectfully request our readers to peruse the advertisements in the Chronicle very carefully. Every one will find something interesting."

KENTVILLE MERCHANTS

Are prepared to meet ALL the Outside Competition

All Stores Open Every Evening Until Christmas

Every Day a Merchants Day from now until Xmas

All the Merchants are PUTTING FORTH EVERY EFFORT to supply your wants with the best goods for the least money. Visit the business centre of the Valley and get your money's worth.

This is a good line. The Nova Scotians are to be congratulated and merchants in other parts of Canada should give their local publishers a pointer. This kind of thing shows a town is awake and gives the outsiders an impression that their trade is appreciated. It might also stir up a little latent civic pride and that's what is badly needed in a good many Canadian towns.

GET THERE OR GET LEFT.

"I don't see," said John Jones, the hardware dealer, "how James Smith gets so much more of the furnace business of this locality than I do. I surely am better fitted to take care of it than he is."

"You make the mistake," said his friend, "in this strenuous age, of supposing that business always comes to those who are most capable of taking care of it. On the contrary, they are generally passed in the race, as in your case, by those who make the most strenuous endeavors to secure the same. In other words, present-day commercial success is a case of fight rather than fit."

STOVES AND TINWARE

CLEANING A FURNACE.

The practical work of furnace installation and the designing work—that is, making the plans—are now widely separated in many establishments and often times an engineer, while thoroughly acquainted with the details of furnace work as they are laid out on paper, still knows little of the practical work of installation, and the mistakes that are made not only by the mechanics, but by those who lay out the work. A man who probably spends more of his time over the drafting board than with the hammer or snips, recently told the following story in the Metal Worker, which may be of interest to both branches of the trade.

The Base of the Chimney.

"Several events led up to my taking upon myself the task of cleaning the furnace in my cellar. I had spent a somewhat lengthy vacation at a better watering place than I could afford and was unable to stand the financial strain of having a real authorized furnaceman come in, but I wish I had. After this one the rest of the excuses are unnecessary. I went down to the task with my old clothes on and succeeded in disconnecting the smoke pipe. I found in so doing that the builder had left a square hole in the chimney which when the furnaceman came around he blocked with fire clay. I scraped all of this out and after I had emptied the soot out of the smoke pipe and found that it was sound, turned my attention to removing the accumulation of soot from the chimney. This was not built as I should like to see it, for I certainly believe a chimney would give better service if there were a space at the bottom below the smoke connection so that soot can fall down and be removed through an opening in the chimney near the basement floor. This opening should, of course, be thoroughly blocked up with fire clay when the fire is first started, otherwise the draft is sure to be poor.

Readily Cleaned Register Boxes.

"The next thing to do was to take off all the hot air pipes so that I could remove the bonnet. This was not much of a task, as the different lengths had been soldered together, but the joint between the base of the register box and the elbow of the run had just been shoved together, as was also true of the connection to the bonnet. I find that the furnaceman in hanging the pipes had not been careful to secure as good an alignment as he could and also that the pitch of the pipes varied considerably, in some instances the pitch being greater at the farther end than near the furnace. The register boxes on the first floor were all taken out and cleaned and the accumulation of dirt and dust in them certainly was astonishing. I believe if furnacemen generally knew how much trouble it takes to clean a register box which has the connection to a straight pipe made by cutting and bending over the end of the pipe, instead of double-seaming it, they would be willing to pay the slight additional expense, for the latter is certainly a great improvement and one which is appreciated by the housewife.

Accessible Cold Air Duct.

"When I came to lift the bonnet off I found that instead of using clean good sand in the sand ring, coarse gravel and dirt had been used instead, which when dry blew over the cellar to a considerable extent and called forth remarks on the part of the housewife as to what a dirty thing a furnace is. The castings I found had an accumulation of rust, which, while not thick, still was considerable, and when I thought of brushing this down into the cold air pit at the bottom I wondered how I could ever get it out, as the cold air pit was built in the cellar floor with a concrete top. Not wishing to take the whole furnace down I concluded to let the dirt remain, but I do think it a very unwise practice, particularly in a first-class job, not to leave an opening so that the pit under the furance can be cleaned as well as the cold air box. The greatest objection to furnaces is the dust, and if means were taken for removing this dust periodically there would be much less to complain of. In this furnace I am going to have such an opening made another year, and feel confident that I will be amply repaid for the expense."

STOVE SALESMANSHIP.

What are you in business for? The old answer, "to make money," is too loose and indefinite, and needs to be qualified to be of any benefit. The methods by which you made money last year may seem utter folly to you comparing them with the methods adopted for this year.

A merchant should always be after profits and any proposition not likely to ensure the getting of such profits is not in these days good business. Do not let any condition of competition, local or otherwise, switch you from this idea. At times it seems almost impossible to maintain business on this basis because of some competitive factor, but bear in mind that the successful merchant is he who meets the condition without sacrificing his present or future interest. An easy solution to the problem is to do what others do, but this is oftentimes unprofitable

The next question which would largely aid you in the hunt for perpetual profit is salesmanship. The ordinary conception of salesmanship is the manner and argument which a salesman puts up when confronted by a possible customer. We might think of it as some clever trick which a salesman only displays when a victim arrives. But it is a great mistake, as the class of goods should be studied very closely, comparing them intelligently with several other well known lines: Procure the exclusive agency of some well known stoves and heaters and give your undivided attention to this particular line.

Salesmanship includes everything that tends to advance the business. Supposing there is a family that wants or can be made to want, a stove. You have the stove. Salesmanship is the connecting link between that family and your stove. Whatever exercises an influence, good, bad or indifferent, upon the transfer of your stove to that buyer is the part of the salesmanship of your institution and must be examined, judged and adjusted from that standpoint. A dingy store front discounts the salesman's argument. Many spoil it. A dirty delivery wagon is also a very poor sort of an invitation to call again. A coat of fresh paint often benefits trade more than a cut price. What happens after the buyer reaches the article required—the merit of the thing, the argument, the price and the terms—are the points that clinch the sale.

RENOWN STEEL RANGE.

The Guelph Stove Company will enlarge their plant to accommodate the new line of steel ranges which they are preparing to manufacture. N. L. Stewart manager, and M. Kelly, the superintendent, returned from Detroit last week, after completing a deal for the purchase of the Canadian patents on the modern "Renown" steel range, manufactured in that city by the Independent Stove Company. The range will be made by the Guelph company under the name of the Great Idea steel range. As the Independent Stove Company is composed of men of Canadian birth they have arranged with the Guelph works to handle the Canadian output on all further patents that may be taken out. Walter R. Faulkner, who for some time represented the Independent Stove Co., has been secured by the Guelph Stove Company to represent them in Eastern Ontatio. As Mr. Faulkner is perfectly conversant with this range he will be of great assistance in helping the Guelph company to make this stove a popular seller on the Canadian market.

DEMAND FOR ENAMELED WARE.

An Englishman in China says that the people of even the remote districts are becoming purchasers of foreign cutlery, electroplated ware, and enameled ware. He adds: "Where one used to see a chinaware basin or wooden tub among the odds and ends of a Chinese traveler's baggage on board ship or on land, the enameled is now prominent, and, being unbreakable, it has come to stay and to oust the more fragile article. There should be one in every household in China, and even far away Tibet is importing them across the Indian frontier. The same remarks apply to glass or glassware, lamps, and hardware. There is an ever-increasing desire to possess such goods."

The rebuilding of Manchuria is being carried on so vigorously in repair of the damages of the great war, that the demand for metals and metal goods is enormous, due both to the war and to the evolution among the natives in favor of European instruments of labor. The use of galvanized corrugated iron is also very extensive, mainly for roofing and warehouse purposes.

AMONGST THE SALESMEN

TRAVELERS' BENEFIT.

S. R. Wickett, the new president of the Commercial Travelers' Mutual Benefit Society, has probably had a longer experience on the Board of that institution than any other member, having given his services to this society for upwards of twenty years. He was vice-president in 1898, and again in 1903, and president in 1904. His great business he is a Methodist. He is a member of the S.O.E., and Masonic Orders. Affability has always been a great help to the society. A native of Devonshire, England. He came to Canada in 1855. After a common and High School education, he entered upon the business of leather manufacturing. Coming to Toronto in 1861 he erected a building upon the present location of the extensive factory of Wickett & Craig, of which he was the founder. Mr. Wickett is also a director on several fire and life insurance boards. In politics Mr. Wickett is a National Policy advocate, in religion fable and courteous, Mr. Wickett is one of the best-liked men in the Travelers' Mutual Benefit and in a much wider circle of friends.

Retiring President Joseph Taylor.

Joseph Taylor, the retiring president, has served on the Board as Trustee, Vice-President and President in all about nineteen years, and much of the success of the society is due to his untiring effort in its behalf. He was for fifteen years one of the most active and prominent directors of the Commercial Travelers' Association. He was born in the village of Husthwaite, Yorkshire. On coming to Canada he traveled for Kilgour Bros. for twenty-six years, leaving that firm to return to England, where he spent two years. Returning to this country he engaged with the Canada Paper Co., with whom he has been for the past four years. The world has dealt kindly with him in every way, and it is to be hoped he may yet have many years of activity and usefulness.

PRACTICAL TALK BY SALESMAN.

Geo. West, Toronto, who is the newly elected first vice-president of the Canadian Commercial Travelers' Association, is senior member of West, Taylor, Bickle & Co., manufacturers of brooms, brushes, woodenware, baskets, etc., Norwich, Ont. In conversation with Hardware and Metal last week he said:

"I have had the pleasure of calling on the grocers and hardwaremen of Ontario for some twenty-five years and close observation of the methods of the different men in their lines of business, leads to the conclusion that the man who looks closely after the details (and there are many) is the successful man. This applies equally to office and stock.

President S. R. Wickett.

"As an instance of this: Mr. M—— went into business in 1870 and sold out in 1890, taking enough out of the business to live easily for the balance of his life, but being a comparatively young man, was not satisfied to lead an idle life and went into manufacturing, which proved a failure and he lost what he had made in his first venture, but nothing daunted he procured a stock of hardware and started again in an entirely new district, applying his old methods of looking strictly after the details; and, keeping everything in stock that would likely be called for, from a needle to an anchor, he soon recovered lost ground. The saying goes that if you cannot get an article at any other place in town, go to Mr. M——, he is sure to have it. This opens a new account for him. The customer reasons that if he has the article that is not called for every day he is sure to have the one that is.

"There is nothing that hurts a business so much as being continually out of some line of goods. There are times when it is unavoidable, but this should be the exception and not the rule."

BREAKING IN NEW SALESMEN.

For breaking in new salesmen a Cincinnati wholesale house recently originated a method that has since been adopted with great success by other houses. When a young man is sent out on the road he is allowed to pursue his way for about a week, when he is recalled by a laconic telegram: "Come back at once." When he reports he is taken into the general manager's private office, and there he is immediately taken to task, the general manager opening on him with a volley of questions that would make the ordinary man boil over with anger. The salesman is given a chance to recover his breath and start to explain why he has not done better, when he is interrupted with another volley of questions of a nature to arouse his ire.

If the young man loses his temper he is at once discharged, but if he takes it calmly without trying to get back at the "boss" and without stinging sarcasm or show of temper, he is informed that they have been trying him out to see how he would deal with an irritable customer, and he finishes out his trip and becomes a "regular" on the salesman force.

GEO. WEST, TORONTO
Commercial Travelers' Vice-President.

HARDWARE AND METAL

Established 1888

The MacLean Publishing Co.
Limited

JOHN BAYNE MACLEAN · *President*

Publishers of Trade Newspapers which circulate in the Provinces of British Columbia, Alberta, Saskatchewan, Manitoba, Ontario, Quebec, Nova Scotia, New Brunswick, P.E. Island and Newfoundland.

OFFICES:

MONTREAL, ... 232 McGill Street
Telephone Main 1255
TORONTO, ... 10 Front Street East
Telephones Main 2701 and 2702
WINNIPEG, ... 511 Union Bank Building
Telephone 3726
LONDON, ENG. . . . . . 88 Fleet Street, E.C.
J. Meredith McKim
Telephone, Central 12960

BRANCHES:

CHICAGO, ILL. . . . . . 1001 Teutonic Bldg
J. Roland Kay
ST. JOHN, N.B. . . . . No. 7 Market Wharf
VANCOUVER, B.C. . . . . Geo. S. B. Perry
PARIS, FRANCE - Agence Havas, 8 Place de la Bourse
MANCHESTER, ENG. . . . . 92 Market Street
ZURICH, SWITZERLAND . . . . Louis Wolf
Orell Fussli & Co.

Subscription, Canada and United States, $2.00
Great Britain, 8s. 6d., elsewhere · 12s

Published every Saturday.

Cable Address { Adscript, London / Adscript, Canada

RETAIL METHODS BEST.

"Give us the retail dealers," says Warden Platt, of Kingston Penitentiary, after five years' experience of selling binder twine direct to the farmers who are the consumers. The statement should be recorded, as it proves a strong argument in favor of manufacturers doing business through regular trade channels rather than selling to consumers direct. Here is Warden Platt's statement in full:

"We started in last year with 350 tons and only sold about 200 tons—less than half our possible output. Our twine was as good as any on the market. Our price was two cents below other twine of the same grade. But our terms shut us out, and our terms will always shut us out. Farmers will not send cash for any article they cannot see, and many of them cannot send cash because they have not got it. To get up clubs costs money and increases the price of the twine. Competitors take advantage of our terms and misrepresent the quality of our twine. Altogether the outlook is discouraging. We should be able to run the mill all the year. Give us the retail dealers—give us the market on equal terms with our competitors—and we will sell all the twine we can make if we run our factory night and day every month of the year. If we cannot have the market, why run the factory? All this I say after five years' experience with the present system."

This paper is not an advocate of prison labor—particularly if the government uses the advantage they possess in the way of cheap labor to encourage price cutting by selling twine at two cents per pound less than other manufacturers can market the article at. We reproduce the statement merely for the purpose of drawing attention to the endorsation given the accepted methods of reaching the consumer through the retail trade.

On the surface it would appear that it would be cheaper for manufacturers to sell direct to consumers—that the middleman's profit would be saved if he was done away with. The above, however, is a practical example showthat such is not the case—that, in reality, it costs more to go to the consumer and, in fact, practically presents an insurmountable barrier to trade to attempt to do so when selling goods in quantities.

When business is done through the retailer the farmer deals with a man he knows and he has a guarantee he values. The retailer knows the farmer and can wait till harvest is over for his money if necessity requires. The retailer buys in large quantities and gets lower freight rates by so doing. He, in fact, gets up a farmers' club, takes all the risk, buys in large quantities, knows what he is buying and the men from whom he buys, and then turns over the goods in small quantities, delivering at such times and places as are convenient to the different customers.

In return for this the retailer gets a small percentage of profit, a percentage which would otherwise go towards paying higher freight rates, a bookkeeper's wages for keeping track of the numerous accounts of a business done by making small shipments to consumers. The retailer earns his profit both by his labor and by the use of his credit in acting as the purchasing agent for the consumers. And, as shown in Warden Platt's statement, the local dealer can sell even cheaper than the manufacturer who tries to do a direct trade.

The many seizures of fraudulent binder twine during the past year present another argument why the farmers should do business with the local dealers who stand behind the goods they sell. If twine is not up to the standard it would be far easier to get satisfaction from a business man in the farmer's own locality than from some manufacturer or agent whose place of business is a long distance away.

FOR BETTER FREIGHT HANDLING

Shipping interests in Montreal are taking steps to see what can be done to form a terminal company to handle freight on the wharves. The congestion at the railway terminals in Montreal has come to such a pass that shipping men are preparing now to save themselves the possibility of any loss next season. It is believed that were a company formed to handle all freight there would be no congestion or trouble of any description. The idea at present is to form a company which to all practical intents and purposes would be the Harbor Board. This terminal company would receive the freight arriving in Montreal by the different railways, and charging the railways a sufficient sum to cover operating expenses, would deliver the freight to its destination. Any surplus which might be left at the end of the year would be refunded to the railway companies in amounts varying according to the business they had given the company during the year. This arrangement, it is thought, would be eminently fair to all concerned, and would ensure the wharves being kept reasonably clear of freight at all times.

FREE SCHOOL BOOKS.

The Ontario Government is thinking of adding to the cost of public school education by supplying text books free throughout the province. The socialistic idea is spreading pretty fast just at the moment. It may at first blush look reasonable that if education is to be free the necessary books should also be free; but it is largely a question of cost and if the cost indirectly is going to be greater than at present, where's the gain?

The retail booksellers waited on the Minister of Education at Toronto the other day and protested against the taking away of one of their sources of revenue. They put forward the cogent reason that the cost of Government distribution would not be less than the present retail profits on school books. There are many considerations in this question of cost. For instance, what is the life of a school book? In Toronto where the Board of Education provides free text books an extensive repair department is maintained, which greatly prolongs the usefulness of the average book. In rural districts this would be impossible without doubling the cost of transportation by having books sent to a central depot for repair.

Most general merchants are directly interested in this question and should express their views to their representative in the Legislature.

KEEP UP THE FIGHT.

A leading Hamilton retailer writes that his local M.P. has advised him that "it is not likely that a bill will be introduced in Parliament to amend the parcels post system, as the Postmaster-General has the power to adopt the proposed c.o.d. amendment without introducing a bill in the house."

This emphasizes the necessity of adopting the suggestion that the retail merchants of Canada fairly bombard the Postmaster-General with letters of protest against the proposal. He should be made to understand that by putting into effect the proposed legislation he is benefiting a handful of wealthy mail order merchants at the expense of thousands of retail dealers throughout Canada. Get after the Postmaster-General and camp on his trail until the proposal is withdrawn.

Similar legislation was recently proposed in England, but the retail merchants adopted the suggestion of the trade press and on a certain day fairly flooded the Postmaster-General's office with protests. They kept up the fight and had the satisfaction of seeing the legislation killed. The same tactics have been used successfully in the United States in combatting the ever-greedy mail order merchants and they can win in Canada if the fight is kept up.

To come nearer home for an illustration. A year ago Postmaster-General Aylesworth denied the use of the mails to the "Appeal to Reason," a socialist weekly published in Kansas. Canadian socialists sent thousands of letters to Ottawa, and the article by Eugene V. Debs which had caused the suspension of the paper was reprinted in Canadian papers as well as in circular form. They kept at it till they gained their point and won a victory for the freedom of the press in Canada.

While it must be admitted that the Postmaster General has almost autocratic powers and is to-day acting as a press censor dictating what Canadians can or cannot read, it is doubtful if he will go so far as to attack the thousands of retail merchants and deprive them of the means whereby they earn their livelihood without first securing the consent of Parliament. His announcement in Parliament on Dec. 7 last, in which he promised to fully inform the House when his plans were completed, indicates that he recognizes the rights of the people's representatives in the matter. If he disregards Parliament on this point it is a matter between the members of Parliament and the retail merchants whose interests the former are supposed to conserve.

The moral is plain. The Postmaster General is the one to whom protests should be sent. And it won't do to wait for the legislation to be introduced in Parliament as has been suggested by the Retail Merchants' Association. Protest to the Postmaster General at once in some such letter as is suggested in the Hardware Association page in this issue. Also send a petition signed by every retailer in each town to the local M.P., asking him to see the Postmaster General and urge the withdrawal of the suggested amendment. And, still further, ask the editor of the local paper to point out the dangers of centralizing all wealth, educational advantages, etc., in the big cities and have him send marked copies of his paper to Ottawa. Farmers who own property in town or country, the value of which is likely to be depreciated by the closing of stores, should also be asked to send in their protests.

There is plenty of scope for action Much depends upon the life put into the protest. So far as possible this paper will help to keep the trade posted as to the progress of the agitation. But it rests with each individual retailer to do his part along the lines suggested.

Cutlery, silverware, tools, enamelware, sporting goods, paints, wall paper, stoves, wire netting, and numerous other hardware lines are already being sold in large quantities by the mail order houses. The Postmaster General suggests legislation to further extend this trade. Shall it be allowed to go through without protest?

CALENDAR ADVERTISING.

There are few branches of business in which more money is wasted than in the publication and distribution of calendars and it must be a source of gratification to many to note signs of a decline in the volume of this form of questionable advertising. Much fewer calendars were received at this office for review this year than for several seasons back, those received, however, showing a marked improvement on previous years. If a business concern favors the use of calendars they should get out something worth while and exercise great care in seeing that the distribution is thorough and covers only the field of possible customers. If a calendar is not strong enough to command a position of prominence sufficient to guarantee that it will bring the firm's goods before buyers for 365 days in the year, the expense should be saved.

Both manufacturers and retailers are using calendars more gingerly, the former taking up paper knives, desk decorations, and other forms of business reminders, while many retailers have adopted the pocket-knife souvenir as a more lasting and better business bringing advertisement. One Ontario retailer purchased about 100 two-bladed knives last year and by carefully distributing them has a permanent advertisement in the pockets of a hundred of his best customers. The knives cost about $6 per dozen, and his advertising matter was inserted under the handle on each side. The dealer referred to has found the knife souvenir a paying one and could not be induced to take up calendars again under any consideration.

Another dealer in Ontario got satisfaction out of calendars this year by advertising in the local paper that he would have a nice calendar for distribution, but only one would be given to each household. To secure a calendar a person had to come to the store before Dec. 1 and make a purchase of goods. All names were then written in a calendar book and about Dec. 15 another announcement was made in the local paper that the calendar was ready for distribution and all who had put their names down before Dec. 1 could get their calendar by calling for it at the store. By this plan the retailer succeeded in getting each customer into his store twice before the Christmas holidays. He also succeeded in compiling a first-class mailing list in addition to getting his calendar advertisement into the homes of those he desired as permanent customers. By adopting some such plan as this, expensive calendar advertising can be made to bring good results, but unless similar care is exercised it is far better for the retail dealer to confine his advertising to his local newspapers, changing it often and making a study of how to use the space paid for to the best advantage.

EDITORIAL NOTES.

During the past year excellent progress has been made in the erection of the great railroad and highway cantilever bridge across the St. Lawrence River, near Quebec. Although this structure is by no means the largest bridge in respect of its over-all length, it will contain the largest single span ever erected, the main span over the river measuring 1,800 feet between the towers. The total length from centre to centre of anchorage piers is 2,800 feet, made up of two 500-foot anchor spans, and two 562½-foot cantilever arms, extending over the river and carrying between them a central suspended span of 675 feet. The depth of the trusses over the main piers is 350 feet. The floor system will accommodate two steam railroad tracks, two electric car tracks, two highways for vehicles, and two sidewalks. The bridge is now more than one-half erected, and will be completed during the present year.

HARDWARE TRADE GOSSIP

Ontario.

D. C. Taylor, Lucknow, formerly a hardware merchant, has gone into the lumber business.

L. P. Foucar, hardware merchant, Tottenham, has installed a long distance telephone in his store.

J. D. Murdock and Co., hardware merchants, of Simcoe, Ont., have advertised their business for sale.

G. A. Binns, hardware merchant, of Newmarket, called on the hardware jobbers in Toronto this week.

C. Beirl, hardware merchant, of Markham, was a visitor in Toronto this week, among the hardware jobbers.

Wm. Knight, hardware merchant, of Maple, was in Toronto this week, calling on the wholesale hardware trade.

The hardware store of J. E. Mosley, Huntsville, was partly destroyed by fire last Saturday. Loss covered by insurance.

Geo. A. Beamish, foreman in the plumbing department of W. Doig, hardware merchant, of Russell, Man., has been a visitor in Toronto for the past week.

A. H. Gingerich, hardware merchant, of Woodstock, will move into the new store which he has leased, in about two weeks.

James Young has purchased the Stewart hardware business at Auburn, and will be given possession in a day or two.

John S. Moir, hardware merchant, Arnprior, is having his hardware store enlarged, intending to make it double its present size.

T. A. Kennedy, of the firm of Kennedy Bros., hardware merchants, Winnipeg, is spending a few days with his parents at Westport, Ont.

The F. Hamilton Company, hardware merchants, of Hamilton, have changed their name and will hereafter be known as the Alexander Hardware Co.

G. R. Duncan, of the Canadian Iron and Foundry Company, Fort William, returned last week from Montreal, where he spent the Christmas vacation.

Fred Baker, customs house representative for Rice Lewis & Son, Toronto, has been ill for the past week, and Jack Mills, of the sample floor, is taking his place.

Mr. Johnstone, of Johnstone Bros., hardware merchants, in the new town of Pefferlaw, on the C.N.R., was in Toronto this week placing sorting orders with hardware jobbers.

T. M. Hodgson, Toronto representative of the Dominion Wire Company, was elected to the council of East Toronto last week, heading the poll for ward 1, in which he ran.

Andrew Corbin, who for a long time was first vice-president of C. & F. Corbin, manufacturers, at New Britain, Conn., died at that place on Friday, January 4, in his seventy-fourth year.

Chesney & Smiley, Seaforth, have dissolved partnership as hardware merchants and their business will be continued by Chesney & Wilson. Mr. Smiley has purchased the stock and trade of W. H. Rumball, Aylmer, Ont.

J. T. Carlind, who has been pushing Lunkenheimer valves, etc., in Ontario for some weeks for Cluff Bros., Toronto, will have an exhibition at the convention of the Canadian Stationary Engineers' Association in Ottawa next week.

Geo. J. B. Ramsden, travelling sales manager of the J. H. Still Manufacturing Company, St. Thomas, has been appointed advertising manager, in addition to his other duties with that company.

Thomas Chamberlain, foreman of one of the departments of Buck Stove Works, Brantford, was presented with an address and a beautiful silver butter dish by the men of his department during the holiday week.

Thomas Thomson, aged 78 years, a highly esteemed resident of London, died Jan. 5 at the family residence. Deceased, who was born in Hudson, Que., went to London about fifty-three years ago and engaged in the hardware business.

Two new hardware stores are being started in Toronto. T. A. Whetstone, on College street, and Mr. Franklin, on Bathurst. In one case the order is being held over awaiting the arrival of shelf boxes, which are becoming a recognized part of a hardware's store equipment.

The management of the National Portland Cement Company, of Durham, entertained the employes and their friends, some 300 in number, to a banquet during the holiday week. The company produced 45,000 more barrels of cement than last year. President McWilliams was presented by the employes with a diamond ring, as a small token of respect.

Quebec.

P. Filion, St. Therese, was in Montreal last week.

Felix Cyr, St. Hermas Station, was in Montreal last week on business.

H. G. Eadie, iron and steel merchant, Montreal, is in New York on business.

A. Limoges, Beauharnois, was in Montreal last week buying general goods.

Andrew Forman, of John Forman & Co., electrical supplies, Montreal, is in New York on business.

J. A. Fraser, of Frothingham & Workman, Montreal, has joined the traveling staff of the same firm.

The assets of T. L. Clarke & Company, manufacturers of sleigh bells, Montreal, are advertised for sale.

J. S. McLernon, manager of John Round & Son, Montreal, has gone to the Old Country on a business trip.

Mr. Hudson, of the Russell & Erwin Manufacturing Co., New Britain, Conn., is in Montreal calling on his customers among the wholesalers.

Wm. Paul, Sorel, who is rebuilding the R. & O. steamer, Sovereign, which was destroyed last year by fire at Lachine, is in Montreal purchasing supplies.

A. J. Owen, Canadian representative of Franklin Saunders & Co., Montreal, who has been combining business with pleasure in the Old Country, sailed for home on Jan. 16.

J. B Campbell, of the Acme Can Co., Montreal, has returned to his desk, after a business trip through Western Canada. Mr. Campbell reports "everything lovely" in his line.

R. F. Warren, a young man who has been for the past 7 or 8 years in Frothingham & Workman's Montreal warehouse, has been promoted to the traveling staff and is now calling on the trade.

D. B. Knight, manager of the Frost & Wood Co., Montreal, is severing his connection with that firm, and last week was presented with a gold watch by the office and sales department of that firm.

W. G. How, Montreal representative of the Wilkinson Sword Co., London, England, has been called to the Old Country suddenly owing to the death of his mother. He is expected back about Feb. 15.

A. A. Bittues, manager of the Montreal branch of the Safety Razor Co., has returned from Boston, where he has been on business in connection with the recent fire which destroyed their Canadian factory.

Western Canada.

R. J. Mc Connell, formerly of St. Marys, Ont., but now a prominent hardware merchant of Carman, Man., has erected a fine new two-storey brick store, 30 feet by 80 feet, into which he recently moved. The store is fitted with all the latest improved conveniences.

The employes of the J. H. Ashdown Hardware Company, of Winnipeg, held a large banquet on New Year's eve. The affair was in honor of Messrs. Britton, Sinclair and Thompson, retiring members of the Ashdown staff. G. A. Britton, so long connected with the sporting department of the Ashdown Co., is going into business for himself. C. Sinclair, after a period of nine years' service, is taking a responsible position with the Winnipeg Paint and Glass Co., and F. Thompson has accepted a position with Crane & Ordway. After the menu was served, a presentation was made by W. J. Thorneloe, to each of the retiring members. To Mr. Britton a handsome meerschaum pipe, silver mounted, to Mr. Sinclair, a pair of solid gold cuff links, beautifully engraved, and to Mr. Thompson a gold mounted fountain pen.

Markets and Market Notes

(For detailed prices see Current Market Quotations, page 62.)

THE WEEK'S MARKETS IN BRIEF.

MONTREAL.

Copper Range Boilers—New list shows advances.
Wire Nails—Advanced ten cents per keg.
Building Paper—Advance expected daily.
Old Material—Advanced.

TORONTO.

Copper Range Boilers—New list.
Galvanized Range Boilers—Up 25 cents.
Cast Iron Fittings—Advanced.
Iron Pipe—Changes in large sizes.
J.M.T. Valves—Advanced 5 per cent.
Brass—Higher prices quoted.
Old Materials—Many changes upward.
Stanley Goods—Advanced.
Nails—Advanced 5 cents base price.
Starrett's Tools—Advanced.

Montreal Hardware Markets

Office of HARDWARE AND METAL,
232 McGill Street,
Montreal, Jan. 18, 1907.

Jobbers report that all travelers are now out and orders are coming in freely.

The market remains very firm, with only one advance—wire nails, which have jumped up 10 cents per keg.

Some sizes of screws are beginning to get scarce.

The new building paper prices mentioned last week have not yet come to light, but are expected any moment now.

AXES.—We quote: $7.60 to $9.50 per dozen; double bitt axes, $9.50 to $12 a dozen; handled axes, $7.50 to $9.50; Canadian pattern axes, $7.50 a dozen. follows: No. 3, $1.25; No. 2, $1.50; No. 1, $1.90 a dozen; adze handles, 34-inch, $2.20 a dozen; pick handles, No. 2, $1.70; No. 3, $1.50 a dozen.

LANTERNS. — Our prices are: "Prism" globes, $1.20; Cold blast, $6.50; No. 0, Safety, $4.00.

COW TIES AND STALL FIXTURES —We still quote; Cow ties, discount 40 per cent. off list; stall fixtures, discount 35 per cent. off list.

SLEIGH BELLS—Our prices still remain as follows: Back straps, 30c. to $2.50; body straps, 70c to $3.50; York Eye bells, common, 70c. to $1.50; pear shape, $1.15 to $2; shaft gongs, 20c. to $2.50; Grelots, 35c. to $2; team bells, $1.80 to $5.50; saddle gongs, $1.10 to $2.60.

RIVETS & BURRS—We still quote: Best iron rivets, section carriage, and wagon box black rivets, tinned do., copper rivets and tin swede rivets, 60, 10 and 10 per cent.; swede iron burrs are quoted at 60 and 10 and 10 per cent. off new lists; copper rivets with the usual proportion of burrs, 25 per cent. off, and coppered iron rivets and burrs in 5-lb. carton boxes at 60 and 10 and 10 per cent.; copper burrs alone, 15 per cent., subject to usual charge for half-pound boxes.

HAY WIRE.—Prices remain firm. Our quotations are: No. 13, $2.55; No. 14, $2.65: No. 15, $2.80; f.o.b. Montreal.

MACHINE SCREWS—Some sizes are getting scarce. We quote: Flat head, iron, 35 per cent., flat head brass, 35 per cent., Felisterhead, iron, 30 per cent.. Felisterhead, brass, 25 per cent.

BOLTS AND NUTS — Discounts remain,: Carriage bolts, 3 and under, 60 and 10; 7-16 and larger, 55 p.c.; fancy carriage bolts, 50 p.c.; sleigh shoe bolts, ⅜ and under, 60 per cent.; 7-16 and over, 50 per cent.; machine bolts, ⅜ and under, 60 per cent.; 7-16 and larger, 55 per cent.

HORSE NAILS—New lists have been issued showing several changes, but discounts continue as follows: C brand, 40, 10 and 7 per cent.; M.R.M. Co., 55 per cent.

WIRE NAILS—Advance 10 cents per keg. We quote: $2.40 per keg base f.o.b. Montreal.

CUT NAILS—We continue to quote: $2.50 per 100 lbs.; M.R.M. Co., latest, $2.30 per keg base, f.o.b. Montreal.

HORSESHOES. — Our prices are: P.B. new pattern, base price, improved pattern iron shoes, light and medium pattern, No. 2 and larger, $3.65; No. 1 and smaller, $3.90; snow pattern, No. 2 and larger, $3.90; No. 1 and smaller, $4.15. Light steel shoes, No. 2 and larger, $4; No. 1 and smaller, $4.25; featherweight, all sizes, No. 0 to 4, $5.60. Toeweight, all sizes, No. 1 to 4, $6.85. Packing, up to three sizes in a keg, 10c. per 100 lbs. More than three sizes, 25c. per 100 lbs extra.

BUILDING PAPER—New lists have not yet arrived, but are expected daily.

CEMENT AND FIREBRICK.—No change in prices. We are still quoting: "Lehigh" Portland in wood, $2.54, in cotton sacks, $2.39; in paper sacks, $2.31. Lefarge (non-staining) in wood, $3.40; Belgium, $1.60 to $1.90 per barrel; ex-store, American, $2 to $2.10 ex-cars; Canadian Portland, $2 to $2.05. Firebrick, English and Scotch, $17 to $21, American $30 to $35; White Bros.' English cement, $1.80 in bags, $2 05 in barrels in round lots.

COIL CHAIN—Our quotations remain: 5-16 inch, $4.40; 3-8 inch, $3.90; 7-16 inch, $3.70; 1-2 inch, $3.50; 9-16 inch, $3.45; 5-8 inch, $3.35; 3-4 inch, $3.20; 7-8 inch, $3.10; 1 inch, $3.10.

SHOT—Our prices remain as follows: Shot packed in 25-lb. bags, ordinary drop AAA to dust, $7 per 100 lbs.; chilled, No. 1—10, $7.50 per 100 lbs.; brick and seal, $8 per 100 lbs.; ball, $8.50 per 100 lbs. Net list. Bags less than 25 lbs., ½c. per lb. extra, net; f.o.b. Montreal, Toronto, Hamilton, London, St. John, Halifax.

AMMUNITION — Our prices are: Loaded with black powder, 12 and 16 gauge, per 1,000, $15; 10 gauge, per 1,000, $18; loaded with smokeless powder, 12 and 16 gauge, per 1,000, $20.50; 10 gauge, $23.50.

GREEN WIRE CLOTH.—We continue to quote: In 100 foot rolls, $1.62½ per hundred square feet, in 50 feet rolls, $1.67½ per hundred square feet.

Toronto Hardware Markets

Office of HARDWARE AND METAL,
10 Front Street East,
Toronto, Jan. 18, 1907.

Wholesale hardware merchants report business again very brisk, with prospects of even a brighter year than 1906. In some houses orders are coming in so rapidly that the shipping staff have been working overtime. As predicted by us some time ago, Stanley goods have advanced. A new schedule of prices covering a greater part of their product has been issued by the Stanley Company. Advances are shown in numerous styles of rules, squares, levels and planes, but the feature of the schedule is that it quotes net prices, and discounts have been virtually done away with by the company. Starrett's goods are also quoted at an advance.

NAILS—Have again advanced 5c. per cwt., and are now quoted at $2 35 base price. The great demand still exists for nails, not allowing the manufacturers to catch up with their orders.

AXES AND HANDLES—A great number of new makes of axe handles are on the market, and large orders are being booked for Fall shipment.

GARDEN TOOLS—These various lines are meeting with a good demand. Merchants still continue placing orders.

WASHING MACHINES, ETC. — Are meeting with great demand. As soon as stock-taking is finished merchants will order these goods sent forward. Such a number has been ordered this year that it will be far in excess of anything previous.

SPORTING GOODS — Have again started moving rapidly. The increased cold weather with snow has largely increased the sale of snowshoes and skates. Ammunition is selling well for trap shooting.

CHAIN—Remains at our former quotations: ⅝ in., $3.50, 9-16 in., $3.55; ½ in., $3.60; 7-16 in., $3.85; 3-8 in., $4.10; 5-16 in. $4.70; 1-4 in. $5.10.

RIVETS AND BURRS—Are firm at the advanced price at 25 per cent. for copper, and 60, 10 and 10 for iron and tinned.

SCREWS—The Canada Screw Co. contradict the statement that screws are on an advance. Another well-informed dealer says that the scare is all over, and screws will remain the same as formerly.

BOLTS AND NUTS—Scarcity is very apparent with a large demand coming from all parts.

TOOLS—Stanley goods have all advanced and are now away ahead of all former quotations. Starrett's tools have also experienced a slight advance.

SHOVELS—Snow-shovels are again in great demand and very few will be carried over the season.

LANTERNS—The lantern demand is very quiet at present, all merchants apparently having large stocks on hand.

EXTENSION AND STEP LADDERS—Price remains the same as quoted last week, 11 cents per foot for 3 to 6 feet, and 12 cents per foot for 7 to 10-foot ladders. Waggoner extension ladders 40 per cent.

POULTRY NETTING—We quote: 2-inch mesh, 19 w.g., discount 50 and 10 per cent. All others 50 per cent.

WIRE FENCING—No change in the advanced prices of galvanized and plain wire, and continue selling well.

OILED AND ANNEALED WIRE—(Canadian)—Gauge 10, $2.41; gauge 11, $2.47; gauge 12, $2.55, per 100 lbs.

WIRE NAILS—Another advance of 5 cents on base price of nails has been made and we now quote at $2.35 base.

CUT NAILS—We quote $2.30 base f.o b. Montreal; Toronto, 20c. higher.

HORSENAILS—"C" brand, 40, 10 and 7½ off list. "H.R.M." brand, 60 per cent. off.

HORSESHOES—Our quotations continue as follows: P.B. base, $3.65; "M.R.M. Co., latest improved pattern" iron shoes, light and medium pattern, No. 2 and larger, $3.80; No. 1 and smaller, $4.05; snow, No. 2 and larger, $4.05; No. 1 and smaller, $4.30; light steel shoes, No. 2 and larger, $4.15; No. 1 and smaller, $4.40; featherweight, all sizes, 0 to 4, $5.75; toeweight, all sizes 1 to 4, $7. Packing, up to three sizes in a keg, 10c. per 100 lbs. extra; more than three sizes in a keg, 25c. per 100 lbs. extra.

BUILDING PAPER—Orders are still coming in briskly, the advanced price not affecting the ordering to any extent.

CEMENT—We quote: For carload orders, f.o.b. Toronto, Canadian Portland, $1.95. For smaller orders ex warehouse, Canadian Portland, $2.05 upwards.

FIREBRICK—English and Scotch fire brick $27 to $30; American low-grade, $23 to $25; high-grade, $27.50 to $35.

HIDES, WOOL, FURS—The market is steady with a slight decline in country hides to allow for lack of quality.

Hides, inspected, cows and steers, No. 1	....	0 11½
" " " " No. 2	....	0 10½
Country hides, flat, per lb., cured	0 09½	0 10
" " " green	0 08½	0 09
Calf skins, No. 1, city	....	0 12
" " No. 1, country	....	0 11
Lamb skins	1 15	1 25
Horse hides, No. 1	3 50	3 75
Rendered tallow, per lb.	0 05½	0 06½
Pulled wools, super, per lb.	....	0 25
" " extra	....	0 27
Wool, unwashed fleece	....	0 15
" washed fleece	0 24	0 25

FURS.

	No. 1,	Prime
Raccoon	....	1 50
Mink, dark	5 00	6 00
" pale	2 50	3 50
Fox, red	3 00	3 50
" cross	3 00	7 00
Lynx	5 00	7 00
Bear, black	....	10 00
" cubs and yearlings	....	5 00
Wolf, timber	....	2 75
" prairie	....	1 25
Weasel, white	....	0 60
Badger	....	1 00
Fisher, dark	5 00	6 00
Skunk, black	....	1 25
" " short striped	....	0 90
" " long striped	....	0 50
Marten,	3 50	20 00
Muskrats, fall	....	0 16
" winter	....	0 20
" spring	....	0 23
" western	0 12	0 16

Montreal Metal Markets

Office of HARDWARE AND METAL,
232 McGill Street,
Montreal, January 18, 1907

No startling changes have taken place during the week. Ingot tin is a little easier, having dropped a cent a pound.

Copper remains unchanged and very firm, with no large stocks reported.

Spelter remains unchanged, but is expected to go higher.

Lead and antimony are still firm.

New prices showing an advance in sheet metals are expected next week.

Old material shows several advances.

COPPER—Remains very firm, with the expected advance still a thing of the future. Our prices remain: Ingot copper, 26c to 26½c; sheet copper, base sizes, 34c.

INGOT TIN—Shows a decline of one cent per pound. New prices now are 45½ to 46c per pound.

ZINC SPELTER—Expected to go higher. We still quote: $7.50 to $7.75 per 100 pounds.

PIG LEAD. — No change over last week's prices. We still quote: $5.50 to $5.60 per 100 pounds.

ANTIMONY—Our price remains 27½ to 28 cents per pound.

PIG IRON.—Remains the same. Our prices are: Londonderry, $24.50; Carron, No. 1, $24.50; Carron, special, $23.50; Summerlee, No. 2, selected, $25.00; Summerlee, No. 3, soft, $23.50.

BOILER TUBES—Very firm. We quote: 1½ to 2 in., $9.50; 2½ in., $10.35; 2½ in., $11.50; 3 in., $13.00; 3½ in., $17.00; 4 in. $21.50; 5 in., $45.00.

TOOL STEEL — Our prices remain: Colonial Black Diamond, 8c to 9c; Sanderson's 8c. to 45c., according to grade; Jessop's, 13c.; Jonas & Colver's, 10c. to 20c.; "Air Hardening," 65c. per pound; Conqueror, 7½c.; Conqueror high speed steel, 60c.; Jowitt's Diamond J., 6½c. to 7c.; Jowitt's best, 11c. to 11½c.

MERCHANT STEEL.—Our prices are as follows: Sleigh shoe, $2.25; tire, $2.40, spring, $2.75; toecalk, $3.05; machinery iron finish, $2.40; reeled machinery steel, $2.75; mild, $2.25 base and upwards; square harrow teeth $2.40; band steel, $2.45 base. Net cash 30 days. Rivet steel quoted on application.

COLD ROLLED SHAFTING—Present prices are: 3-16 inch to ¼ inch, $7.25; 5-16 inch to 11-32 inch, $6.20; ⅜ inch to 17-32 inch, $5.15; 9-16 inch to 47-64 inch, $4.45; ¾ inch to 17-16 inch, $4.10; 1½ inch to 3 inch, $3.75; 3⅛ to 3 7-16 inch, $3.92; 3½ inch to 3 15-16 inch, $4.10; 4 inch to 4 7-16 inch, $4.45; 4½ inch to 4 15-16 inch, $4.80. This is equivalent to 30 per cent. off list.

GALVANIZED IRON — We are still quoting Queen's Head, 28 gauge, $4.60 to $4.85; 26 gauge, $4.45; 22 to 24 gauge, $3.90; 16 to 20 gauge, $3.75; Apollo, 28 gauge, $4.45 to $4.70; 26 gauge, $4.30; 22 and 24 gauge, $3.75; 16 and 20 gauge, $3.60; Comet, 28 gauge, $4.45 to $4.70; 26 gauge, $4.30 to $4.45; 22 and 24, gauge, $3.75 to $4.00; $16 to 20 gauge, $3.60 to $3.85; Fleur-de-Lis, 28 gauge, $4.45 to $4.70; 26 gauge, $4.30; 22 and 24 gauge, $3.75; 16 to 20 gauge, $3.60; Gorbals "Best Best," 28 gauge, $4.45; Colborne Crown, 28 gauge, $4.45; 26 gauge, $4.30; 24 gauge, $3.75. In less than case lots, 25c extra.

OLD MATERIAL—We quote: Heavy copper, 19c per lb.; light copper, 16c per lb.; heavy red brass, 17 cents per lb.; heavy yellow brass, 14c per lb.; light brass, 10 to 10½c per lb.; tea lead, 4c per lb.; heavy lead, 4½c per lb.; scrap zinc, 4c per lb.; No. 1 wrought, $16.00 to $17.00 per ton; No. 2 wrought $6.00 per ton; No. 1 machinery, $18.00 per 100 lbs.; stove plate, $13.00 per ton; old rubber, 10 cents per lb.; mixed rags, 1 to 1¼c per lb.

Toronto Metal Markets

Office of HARDWARE AND METAL,
10 Front Street East,
Toronto, Jan 18, 1907

Things are comparatively quiet in metals with only a seasonable trade being done and not many changes to report. Brass has advanced from 27½ cents to 30 cents, and this is the only change, with the exception of old material, in which we quote advances on old copper, brass, lead, iron and rubbers.

There is little to say regarding pig iron. Prices hold very firm, but no changes have been made for the past month. Iron is still hard to get and prospects of relief are anything but bright. Bars hold the recent advance very well. Tin is stronger than it was a week ago and our price remains the same.

Stocks of copper are very low and material is exceedingly hard to get. The electrical manufacturers seem to be absorbing a large percentage of the production and there is talk amongst American buyers of prices soaring higher than ever during 1907. Lead and spelter remain as before—firm, with a fair business being done.

It is announced that the tin plate works at Morrisburg are to be opened

Jan. 24, with a large public demonstration. We will report this matter more fully in another issue.

PIG IRON — Business is quiet, but prices hold very stiff. Hamilton, Midland and Londonderry are off the market, and Radnor is quoted at $33 at furnace. Middlesborough is quoted at $24.50, and Summerlee, at $26 f o.b., Toronto.

BAR IRON—Stocks are easier with prices firm at $2.30 f.o.b. Toronto, with 2 per cent. discount.

INGOT TIN—The market has strengthened and our quotations of 45 to 46 cents a pound are close.

TIN PLATES—Business is active and prices firm with booked orders for Spring and Summer delivery large.

SHEETS AND PLATES—Active buying is reported with prices steady. Stocks are light.

BRASS—A sharp advance of 2½ cents per pound has been made and for sheets we now quote 30 cents per pound.

COPPER—Stocks are very low owing to heavy buying by electrical manufacturers. Prices are very stiff. We now quote: Ingot copper $26 per 100 lbs., and sheet copper $31 to $32 per 100 lbs.

LEAD—Business is fairly active and prices firm. We quote: $5.40 for imported pig and $5.75 to $6.00 for bar lead.

ZINC SPELTER—Stocks continue light with market firm and active. We quote 7½c. per lb. for foreign and 7c. per lb. for domestic. Sheet zinc is quoted at 8 1-4c. in casks, and 8 1-2c. in part casks.

BOILER PLATES AND TUBES—We quote: Plates, per 100 lbs., ¼ in. to ½ in., $2.50; ⅜ in., $2.35 heads, per 100 lbs., $2.75; tank plates, 3-16 in., $2.65; tubes, per 100 feet, 1½ in., $8.50; 2, $9; 2 1-2, $11.30; 3, $12.50; 3 1-2, $16; 4, $20.00. Terms, 2 per cent. off.

ANTIMONY—Market is active, and stocks scarce. Prices firm at 27c. to 28c.

OLD MATERIAL.—Dealers' buying prices still continue as follows: Heavy copper and wire, 20 cents per lb.; light copper, 17½c. per lb.; heavy red brass, 15c. per lb.; heavy yellow brass, 17½c. per lb.; light brass, 11c. per lb; tea lead, $3.75 per 100 lbs.; heavy lead, $4.25 per 100 lbs.; scrap zinc, 4½c. per lb.; No. 1 wrought iron, $15; No. 2 wrought, $8; machinery cast scrap, $17 to $18; stove plate, $14; malleable and steel, $9; old rubbers, 10c. per pound; country mixed rags, $1 to $1.25 per 100 lbs., according to quality.

COAL—The usual difficulties in transportation are being experienced. Slack is still scarce. We quote:

Standard Hocking soft coal f.o.b. at mines, lump $1.75, ¾ inch, $1.65, run of mine, $1.40, nut, $1.25; N. & S. $1.10; P. & S., 85c.

Youghiogheny soft coal in cars, bonded at bridges; lump, $2.90; ¾ inch, $2.70; mine run, $2.60; slack, $2.25.

United States Metal Markets

From the Iron Age, Jan. 17 1907.

The buying movement in pig iron for forward delivery which set in in the Central West since the opening of the year, has made further progress. It is estimated that during the past week there have been sales aggregating 100,000 tons in the Chicago District, including one block of 25,000 tons of malleable iron to one foundry, and 10,000 tons of foundry iron to one machinery manufacturer. Other markets, like Cincinnati and Cleveland, are feeling the movement, the furnaces in the latter district having advanced the price to $22.

Favorable indications are coming from the Birmingham district with regard to the car situation there. It is estimated that there are piled up at the furnaces in the district between 60,000 to 70,000 tons of pig iron, sold and long due to be delivered, which it has been impossible to move.

Pipe makers report heavy inquiries for cast iron pipe, among them being one lot of 12,000 tons for Chicago. The coming letting of 38,000 tons of pipe for New Orleans next week, will have more than local significance. The contract is to be let in six sections, but the bulk of the pipe is for next year's delivery. In view of the relatively high prices it is considered rather doubtful whether the whole of the work will be given out. It will be significant if it is not.

Pittsburg reports that the supply of steel is becoming better, so far as the open market is concerned. On the other hand, tidewater steel works have been able lately to place in the territory west of the Alleghenies some good lots at high prices.

While some large additional steel rail tonnage is pending the largest single order placed was 7,000 tons for the San Antonio & Aransas Pass. It is estimated that all the mills in the country have on their books orders aggregating 2,500,000 tons in addition to about 250,000 tons carried over from last year.

In structural material one order calling for 15,000 tons has been practically closed. Otherwise only fair sized work has come out. One of the eastern mills has advanced its price for shapes to $1.85, at mill, against the usual quotations of $1.85, at tidewater.

From the Iron Trade Review, Jan. 17, 1907

One of the most interesting features in the whole iron and steel business during the week came out when the steel car interests began to make inquiries for plates extending into 1908, and it is announced that inquiries in the neighborhood of 40,000 cars are now in the market, showing that the railroad companies have confidence in the future for practically a year and a half at least, as deliveries on this number of cars could not be made before July, 1908, when the work on hand in this line is taken into consideration.

The demand for sheets continues strong. The principal producer is now behind on deliveries from 14 to 16 weeks. Inability to secure an adequate supply of sheet bars has prevented mills from running to their full capacity. The tremendous demand for plates continues, and mills are overwhelmed with specifications, while mills are having difficulty in obtaining supplies of semi-finished materials.

Owing to the mild weather and offerings of the railroads, old material prices are sagging in nearly all centres. Coke conditions are also somewhat easier, as much better delivery is being given by the railroads.

A decided improvement in the demand for structural material is noted, especially in the Chicago district, where the Illinois Steel Co. has received specifications for about 60,000 tons, of which some 50,000 tons came from a western car builder, and included both plates and shapes. Numerous orders for bridge material aggregating a considerable tonnage, have been placed, and many inquiries for steel for new buildings in New York, Chicago and other cities are pending.

London, Eng., Metal Markets

From Metal Market Report, Jan. 17, 1907.

Cleveland warrants are quoted at 60s. 7½d., and Glasgow standard warrants at 59s. 10d., making price as compared with last week, 4½d. higher on Cleveland warrants, and 3s. 4d. higher on Glasgow standard warrants.

TIN—Spot tin opened steady at £189, futures at £190 5s., and after sales of 360 tons of spot and 180 tons of futures closed easy at £188 15s. for spot, £189 15s. for futures, making price compared with last week 10s. higher on spot and unchanged on futures.

COPPER—Spot copper opened easy at £108 5s., futures, £109, and after sales of 500 tons of spot and 450 tons of futures, closed steady at £108 10s. for spot, and £108 17s. 6d. for futures, making price compared with last week £2 7s. 6d. higher on spot and £1 15s. higher on futures.

LEAD—The market closed at £19 17s. 6d., making price as compared with last week, 2s. 6d. higher.

SPELTER—The market closed at £27 12s. 6d., making price as compared with last week, 2s. 6d. lower.

B.C. Hardware Trade News

Vancouver, Jan. 12, 1907.

It is positively announced that the Chicago & Northwestern Railway, which has now a line under construction from St. Paul to Seattle, will seek entry into Canada at Vancouver, to partake of the coast trade here. Representatives of the road have visited the city from time to time in the past few months with the intent to seek suitable location for terminal facilities, though so far as known no purchases have been made yet.

With the advent of this road, when it ultimately reaches the coast, and the rumored intention of the Canadian Northern also to seek entry to the coast at Vancouver, this city will have earned its title to the name "Terminal City" which it is now usually dubbed. The C. P.R. and the Great Northern are both here, the G.T.P. is coming, and the C. N.R. and the Chicago & Northwestern will make five trunk lines which will some day run into Vancouver.

Before R. G. Macpherson, M.P. left for Ottawa, returning from spending

Christmas at home, he was consulted by the Vancouver Board of Trade on various matters of importance in which this city is interested. Perhaos the most important was the attitude of the Government as to assisting the project of a large terminal elevator built here to encourage the shipment of Alberta grown wheat via this coast. At present a freight rate which is against the western shipping of wheat is in force. But it has been intimated that the C.P.R. would meet the rate question amicably if the Government would provide terminal facilities for storage and shipment of grain similar to that already provided at Montreal. Mr. Macpherson promised to look into the matter fully, and if advisable to have such facilities instead of elevators built and operated by private enterprise he would support the project, in order to see that the shipping of wheat this way was fostered.

It was pointed out at the conference, that the C.P.R. now complained of having to bring empty cars west to accommodate the lumber trade, as there was not traffic enough to the coast to bring out loaded cars in sufficient numbers to supply the coast mills with all the cars they needed for the lumber trade. If the wheat came west the car shortage would be in large measure remedied.

* * *

Another important matter brought to Mr. Macpherson's notice was the Board request sent to the Government for the rescinding of the order in council suspending the coasting laws as far as this coast was concerned. R. P. McLennan, chairman of the meeting, explained that when the suspension of the coasting laws had been asked for it was to give the merchants a chance to ship goods, especially north. Now that the local steamer service had been very greatly improved, it was unnecessary to continue the suspension, and in fact it would be a benefit to the lines now well-established locally, as it would assist them in building up still further. It would also compel other railway companies to improve their terminal facilities if the order were rescinded and the coasting laws enforced.

* * *

Large wharves are to be built at once at Prince Rupert, the Kaien Island terminus of the Grand Trunk Pacific. A large pile-driver was sent north recently and as soon as it reached the new terminus that is to be, it would be put in commission driving piles for the wharves. Recently a small pipe line was installed, the beginning of the water system which will supply the new city. This supply is but temporary to ensure good water for the few cottages and places of business now erected or being erected on the town site. The United Supply & Contracting Co., which has offices in Vancouver, is in charge of much of the improvement going forward at Prince Rupert.

Mining matters in the boundary district have progressed very satisfactorily during 1906. The total shipments of ore for the year as far as reported, were 1,162,034 tons. The three boundary district smelters, the Granby at Grand Forks, the B.C. Copper Co at Boundary Falls, the Dominion Copper Co., at Greenwood, treated during the year 1,180,546 tons.

In Rossland camp, much good work has been accomplished and the output especially of the old mines, Le Roi and Centre Star, greatly increased. The old Northport smelter is now in commission again and as soon as the balance of 30,000 tons still to be sent to Trail has been shipped from the Le Roi mine the rest of the output will go to Northport. Two furnaces have already been blown in at Northport and two more will be blown in as soon as the managoment is sure of the coke supply.

Regarding the coke situation, Manager G. G. S. Lindsey of the Crow's Nest Pass Coal Co., who is in Victoria on business, says that the strike at Fernie is all settled and the mine has resumed its normal activity. The output of coke is again up to the usual amount and no farther trouble is anticipated.

* * *

Many of the coast lumber mills are partially shut down for the annual clean-up and refitting of machinery. Lumber orders are, however, coming in from the Northwest without abatement and the greatest activity is noticeable in the market. Prices have not advanced yet since the beginning of the year, but another increase is quite within the bounds of possibility. Since the holiday season has closed the loggers, minus the proceeds of their December pay-checks, have gone back up coast to the camps, and though snow is heavy in some places, still it is considered that a good deal of timber will now be got out. Recently there were nine tugs, all with large tows of logs, congregated at one time in Secret Cove, a shelter harbor some distance up the coast.

* * *

The possibilities of development of Vancouver's great harbor have been thrown on the screen with startling vividness by the published announcement of the plans of the V.W. & Y. R'y., the coast branch of the Great Northern Railway in this province. Ordinarily the casual observer looks upon Vancouver's harbor as that little fringe of water-front on the south side of Burrard Inlet, facing the city proper, and along which the C.P.R. wharves cover most of the space. But the oft-mooted question of a bridge across what is known as the "Second Narrows," about a mile and a half from the C.P.R. wharves, and a spur line run along the north shore of the inlet for nearly five miles, is now being definitely brought down to a project for immediate execution. This would mean the "arrival," as the French would say, of the greatest harbor in Canada, the whole of Burrard Inlet, with its land-locked security and its great depth of anchorage wide enough also for the merchant fleets of the whole Pacific to swing in, being available for such enterprises as would require dockage and water-front facilities combined with rail connection on land.

The bridge across the Narrows has been looked on as the first step northward, of the V.W. & Y., which holds a charter extending some hundreds of miles north from Vancouver. But the announced intention of the railway to place industrial sites on the north shore of the inlet in the advantageous position of having rail connection, has made the project assume a distinctly new and immediate interest. It is expected that work will begin during the present year on the proposed bridge, which will be a high level structure to permit of shipping passing under it.

* * *

While coal mining on Vancouver Island has long been one of the chief industries, famed far and wide as a grand steaming coal, has been sold in immense quantities in San Francisco and to the many steamers trading to this coast, the operation of mines has been always confined to a limited area. The three chief mines are the Nanaimo, Extension, close to Nanaimo, and Comox, some distance further north on the east coast of the island. Quite recently parties with large capital have become interested in a coal measure of promise at Nanoose Bay, close to Nanaimo also. Now it is reported that some six thousand acres of coal lands at Sooke Bay, close to Victoria, have been leased and work with diamond drill is to commence at once to determine the value of the coal seams under the property, and if results are as expected, there will be another big coal producing mine on the island.

* * *

Wood pulp made in B.C. will be on the market next year, says managing director J. M. Mackinnon, of the Canadian Pacific Sulphite Pulb Co., whose operations at Swanson Bay on the northern coast have been assuming definite shape and increasing activity for some months. Over 83,000 acres of land are owned by the company, which is capitalized at £100,000, most of which is English money. Some leading British pulp men are directly interested. Among them is Mr. John Mackie, managing director of Davidson & Son, of Aberdeen, one of the largest pulp-importing concerns in the old country. Geo. F. Hardy of New York, the best known expert in designing pulp plants, is consulting engineer. It is intended to develop water-power from Swanson river, which is capable of giving from 11,000 to 30,000 horse-power. Many car-loads of machinery for the plant are now in Vancouver, and en route from the manufacturers.

U. S. STEEL PLANT.

The Financial Post, Toronto, refers to the proposal to establish a plant of the United States Steel Corporation at Sandwich as follows: "Daily press despatches from Sandwich say that work will be begun during the early summer by the United States Steel Corporation upon a mammoth steel plant employing 5,000 men. The property is under option and is 1,000 acres in extent, chiefly belonging to the Scotten estate of which Dr. R. Adlinton Newman is the administrator. This statement is at all events premature. The property in question is under option to parties who may or may not represent the U.S. Steel Corporation or any other company. The option has not been exercised as yet. If this somewhat prophetic and prospective announcement were correct, it would be a matter of considerable interest and moment. As a matter of fact the United States Steel Corporation has benn intending to locate in Canada. Another point, Port Colborne, has been mentioned."

AGENTS WANTED.

A United States firm manufacturing a safety clevis are anxious to secure Canadian agents. This clevis has had a large sale in the United States among the hardware trade and it should be a good line for some manufacturer's agent in Canada to take up. Address enquiries to, "Manager," Hardware and Metal, Toronto.

COPPER vs. COAL.

The constantly growing demand for electric power is absorbing every pound of available copper, as substitutes for the metal are being used in practically every other line, says a keen observer of U.S. copper markets.

Sheet copper workers and cornice makers declare that they are using only half the copper they would were the price down to 15 cents. Galvanized iron is taking the place of copper in the building trades ; and the consumption of brass has suffered decidedly in some directions. This is how the advance in of copper to supply the tremendous price has made available a larger amount growth of demand from electrical manufacturers, wire mills, etc.

Copper is coming to be a direct competitor of coal. The world is using mechanical power in a constantly increasing ratio. Every advance in the wages of the coal miners, every increase in the cost of mining supplies and in other mining expenses increases the price of coal and, therefore, advances the cost of creating power.

On the other hand, there have been most decided advances in the science of electrical generation and transmission, and in many places electricity, generated by waterpower, is entering into competition with the isolated steampower plant. While opportunities exist for the creation of electricity by waterpower copper at $1 per pound is cheaper than coal at $7 per ton. Now that electrical power can be transmitted 500 miles or more without any appreciable loss, waterpower privileges are advancing in price ; and the question as to whether it is cheaper to generate electricity at the coal mines and transmit it across country by wire, or to ship the coal to local steam plants, as is done now, is being considered by power economists.

The work of equipping steam railroads with electricity is going ahead much more rapidly than most people realize, and the prospect now is that it will continue as rapidly as is consistent with a reasonable price for copper.

IMPORTING PIG IRON.

The Glasgow Scotsman of recent date says :

"In the course of the last few weeks the American demand for Scottish pig iron has reached dimensions which have not been experienced for years, and which probably not a single member of the trade ever expected to see realized. Years ago the United States were liberal buyers of Scottish brands, and they have always taken more or less, latterly very much less ; but owing to their own enormous production the market has long ceased to be one on which producers in Scotland could reckon. Suddenly, however, a change has come over the situation ; apparent indifference has given place to activity, manifested in a rush of orders from all the consuming centres in the Eastern States, and covering present and prospective wants which sellers on this side find great difficulty in meeting. In fact, they are unable to do so completely.

"It is calculated that since the beginning of the month (November 1 to 19) firms in Boston, Philadelphia, Baltimore, and New York have contracted with Glasgow houses for the shipment of between 30,000 and 40,000 tons, mostly for foundry qualities, but also of hematite. Last week quite 10,000 tons were put through, and orders for many thousands of tons have been declined by makers because of their inability to give delivery. Prompt dispatch is desired in all cases, and as room in the regular steamship lines has been almost wholly taken up, it has been found necessary to charter special steamers, a thing that has not happened in the memory of the present generation of shippers. Besides what has been arranged for on Northern States' account, a cargo of over 4,000 tons has been sold for New Orleans, the first of the kind since 1879.

"If the American demand for Scottish iton is maintained into 190r (and there are strong inquiries to point that way) a condition of things may be created which will prove most embarrassing for home consumers. There are no stocks, public or private, of consequence to fall back upon, and which always were a stand-by in the past. A year ago there were about 92,000 tons in reserve, and at the opening of 1900 there were 277,000 tons."

CIGARS AS ADVERTISEMENT.

J. W. Harris & Co., contractors, Montreal, have adopted an unique form of advertising. They have been sending around to the trade, cigars containing their own hand, marked "J. W. H. & Co."

As these cigars are good quality, it is needless to say that J. W. Harris & Co. are extremely popular just at present.

COPPER FAMINE.

A copper famine is impending owing to the enormous demand for the metal at home and abroad. The demand is due to the great growth of electrical work all over the world. Manufacturers of electrical machinery say that the demand for their products is unprecedented. Authorities in the copper trade say it is doubtful if the output of copper this year will show a material increase over that of 1905. The scarcity of skilled miners has handicapped the larger producers and consequently has reduced production. It is also likely that the Chilian production will be materially reduced by the recent earthquake.

ROGERS BELT PUNCH.

E. C. Atkins & Co., saw manufacturers, Indianapolis, have just closed a deal whereby they secure the entire right to make and sell the Rogers Belt Punch. This little device—for punching holes in belts, straps, etc., is too well known to require an introduction. It is light, convenient, and does its work to perfection. The blade, of sharpened steel, slips into the handle when not in use, so that it is entirely protected. To operate, it is only necessary to pull out a thumb spring which fits into notches in the handle.

The blade is graduated and slightly concave, so that the size of the hole to be cut may be regulated by the amount of blade exposed. By slight pressure, the blade is forced through the belt or strap and by turning the handle around once, an even hole of uniform size is cut. It sells for fifty cents and should and will be in the pocket of every one having to cut uniform holes in belting, straps, and leather. Many of the best jobbers carry them, or it can be bought through E. C. Atkins & Co., 77 Adelaide Street east, Toronto.

BRADSTREET'S FAILURE RECORD

Bradstreet's record of failures in Canada for 27 years past is as follows :

	Number.	Assets.	Liabilities.
1906	1,232	$4,258,310	$9,540,915
1905	1,424	6,550,331	13,837,176
1904	1,177	4,136,618	10,018,299
1903	956	3,832,197	8,328,362
1902	1,092	3,597,220	8,328,658
1901	1,370	5,196,951	11,656,937
1900	1,333	4,244,932	10,786,276
1899	1,285	4,507,608	11,077,891
1898	1,427	4,085,722	9,644,100
1897	1,907	5,191,647	13,147,929
1896	2,179	6,724,535	16,208,460
1895	1,923	6,299,177	15,793,359
1894	1,873	11,947,253	23,985,283
1893	1,781	7,388,692	15,690,404
1892	1,682	4,848,005	11,603,210
1891	1,846	6,014,000	14,884,000
1889	1,616	6,119,585	13,147,910
1888	1,730	7,178,744	15,498,242
1887	1,315	8,407,000	17,054,000
1886	1,186	5,566,474	11,240,025
1885	1,286	4,201,831	9,210,334
1884	1,363	9,074,000	17,126,000
1883	1,464	12,367,000	22,155,000
1882	755	3,948,000	8,139,000
1881	607	3,278,475	6,122,208
1880	839	4,700,372	9,340,929

The above statistics go far to confirm the advices of large trade and good profits in Canada in the past year. There were 1,232 failures in the Dominion in 1906, a decrease of 13.4 per cent. from 1905, while the liabilities were $9,540,915, a decrease of 31 per cent. from the preceding year.

NEW GILLETTE FACTORY.

A. A. Bittues, manager of the Montreal factory of the Gillette Safety Razor Co., has returned from a trip to the head office in Boston, where he has been reporting conditions in connection with the fire which recently put their Canadian branch out of business. Mr. Bittues explains that the board have seen fit to allow him to find another factory in which to continue this end of the business within the shortest possible time.

The firm is at present trying to make arrangements with the Ottawa Government to allow them to bring in razors in such condition that they can be completed by some Canadian firm, as the bulk of the machinery used in their factory was built by themselves—it can't be bought in the open market and takes 4 or 5 months to make. However, Mr. Bittues reports that there will be no unnecessary delay in getting things going again.

The firm, although in Montreal, has been treated extremely well by the hardware trade in general, having had many kind offers of assistance.

MANITOBA HARDWARE AND METAL MARKETS

(Market quotations corrected by telegraph up to 12 a m. Friday, Jan 11, 1907.)

Room 511, Union Bank Building, Office of HARDWARE AND METAL, Winnipeg, Man.

The past week has been a comparatively quiet one in the hardware and metal trades, but all wholesalers predict that the next few weeks will see a decided improvement and they expect the coming Spring to show a better volume of trade than any previous period. The congested condition of the roads owing to the recent unprecedented heavy snow storms has doubtless had the effect of limiting business, but orders are coming in fairly well notwithstanding, though shipping will not commence to any extent for another month.

Prices in some lines have advanced, notably copper rivets, wrought butts, wire nails, barb wire and shovels. Lead also shows an advance, and prices in the paint and oil departments hold firm, with linseed oil 3 cents higher.

LANTERNS. — Quotations are as follows: Cold blast, per dozen, $6.50; coppered cold blast, per dozen, $8.50; cold blast dash, per dozen, $8.50.

WIRE—We quote as follows: Barbed wire, 100 lbs., $2.90; plain galvanized, 6 to 8, $3.39; 10, $3.50; 12, $3.10; 13 $3.20; 14, $3.90; 15, $4.45; 16, $4.60; plain twist, $3.45; staples, $3.25; oiled annealed wire, 10, $2.96; 11, $3.02; 12, $3.10; 13, $3.20; 14, $3.30; 15, $3.45. Annealed wires (unoiled) 10c. less.

HORSESHOES — Quotations are as follows: Iron, No. 0 to No. 1, $4.65; No. 2 and larger, $4.40; snowshoes, No. 0 to No. 1, $4.90; No. 2 and larger, $4.65; steel, No. 0 to No. 1, $5; No. 2 and larger, $4.75.

HORSENAILS — Lists and discounts are quoted as follows: No. 10, 20c.; No. 9, 22c.; No. 8, 24c.; No. 7, 26c.; No. 6, 28c.; No. 5, 32c.; No. 4, 40c., per pound. Discounts are quoted as follows: "C" brand, 40, 10 and 7½ per cent., "M" brand and other brands, 55 and 60 per cent. Add 15c. per box.

WIRE NAILS—Quoted now at $2.80 per keg.

CUT NAILS.—As noted last week cut nails have been advanced to $2.90 per keg.

PRESSED SPIKES — Prices are quoted as follows since the recent advance: ¼ x 5 and 6, $4.75; 5-6 x 5, 6 and 7, $4.40; ⅜ x 6, 7 and 8, $4.25; 7-16 x 7 and 9, $4.15; ½ x 8, 9, 10 and 12, $4.05; ⅝ x 10 and 12, $3.90. All other lengths 25c. extra net.

SCREWS—Discounts are as follows: Flat head, iron, bright, 85 and 10 p.c.; round head, iron, 80 p.c.; flat head, brass, 75 and 10 p.c.; round head, brass 70 and 10 p.c.; coach, 70 p.c.

NUTS AND BOLTS — Discounts are unchanged and continue as follows: Bolts, carriage, ⅜ or smaller, 60 and 5; bolts, carriage, 7-16 and up, 55; bolts, machine, ⅜ and under, 55 and 5; bolts, machine, 7-16 and over, 55; bolts, tire, 65; bolt ends, 55; sleigh shoe bolts, 65 and 10; machine screws, 70; plough bolts, 55; square nuts, case lots, 3; square nuts, small lots, 2½; hex nuts, case lots, 3; hex nuts, smaller lots, 2½ p.c.

RIVETS—Discounts are quoted as follows since the recent advance in the price of copper rivets: Iron, discounts, 60 and 10 p.c.; copper, No. 7, 38c.; No. 8, 38½c.; copper, No. 10, 41½c.; copper, No. 12, 44½c.; assorted, No. 8 and 10, 39½c. and 42½c.

COIL CHAIN.—Prices have been revised, the general effect being an advance. Quotations now are: ¼ inch $7.00; 5-16, $5.35; ⅜, $4.75; 7-16, $4.50; ½, $4.25; 9-16, $4.20; ⅝, $4.25; ¾, $4.10.

SHOVELS—List has advanced $1 per dozen on all spades, shovels and scoops.

HARVEST TOOLS — Discounts continue as before, 60 and 5 per cent.

AXE HANDLES—Quoted as follows: Turned, s.g. hickory, doz., $3.15; No. 1, $1.90; No. 2, $1.60; octagon, extra, $2.30; No. 1, $1.60.

AXES.—Quotations are: Bench axes, 40; broad axes, 25 p.c. dis. off list; Royal Oak, per dozen, $6.25; Maple Leaf, $8.25; Model, $8.50; Black Prince, $7.25; Black Diamond, $9.25; Standard flint edge, $8.75; Copper King, $8.25; Columbian, $9.50; handled axes, North Star, $7.75; Black Prince, $9.25; Standard flint edge, $10.75; Copper King, $11 per dozen.

BUTTS—The discount on wrought iron butts is 65 and 5 p.c.

CHURNS — The discounts from list continue as before: 45 and 5 per cent.; but the list has been advanced and is now as follows: No. 0, $9; No. 1, $9; No. 2, $10; No. 3, $11; No. 4, $13; No. 5, $16.

CHISELS—Quoted at 70 p.c. off list prices.

AUGER BITS—Discount on "Irwin" bits is 47½ per cent., and on other lines 70 per cent.

BLOCKS—Discount on steel blocks is 35 p.c. off list prices; on wood, 55 p.c.

FITTINGS—Discounts continue as follows: Wrought couplings, 60; nipples, 65 and 10; T's and elbows, 10; malleable bushings, 50; malleable unions, 55 p.c.

GRINDSTONES—As noted last week, the price is now 1½c. per lb., a decline of ½c.

FORK HANDLES—The discount is 40 p.c. from list prices.

HINGES—The discount on light "T" and strap hinges is 65 p.c. off list prices.

HOOKS—Prices are quoted as follows: Brush hooks, heavy, per doz., $8.75; grass hooks, $1.70.

CLEVISES—Price is now 6½c. per lb.

STOVE PIPES—Quotations are as follows: 6-inch, per 100 feet length, $9; 7-inch, $9.75.

DRAW KNIVES—The discount is 70 per cent. from list prices.

RULES—Discount is 50 per cent.

WASHERS—On small quantities the discount is 35 p.c.; on full boxes it is 40 p.c.

WRINGERS — Prices have been advanced $2 per dozen, and quotations are now as follows: Royal Canadian, $35.00; B.B., $39.75, per dozen.

FILES—Discounts are quoted as follows: Arcade, 75; Black Diamond, 60; Nicholson's, 62½ p.c.

BUILDING PAPER — Prices are as follows: Plain, Joliette, 40c.; Cyclone, 55c.; Anchor, 55c.; pure fibre, 60c.; tarred, Joliette, 65c.; Cyclone, 80c.; Anchor, 65c.; pure fibre, 80c.

TINWARE, ETC.—Quoted as follows: Pressed, retinned, 70 and 10; pressed, plain, 75 and 2½; pieced, 30; japanned ware, 37½; enamelled ware, Famous, 50; Imperial, 50 and 10; Imperial, one coat, 60; Premier, 50; Colonial, 50 and 10; Royal, 60; Victoria, 45; white 45; Diamond, 50; Granite, 60 p.c.

GALVANIZED WARE. — The discount on pail is now 37½ per cent.; and on other galvanized lines the discount is 30 per cent.

CORDAGE—We quote: Rope, sisal, 7-16 and larger, basis, $11.25; Manila, 7-16 and larger, basis, $16.25; Lathyarn, $11.25; cotton rope, per lb., 21c.

SOLDER—Quoted at 27c. per pound. Block tin is quoted at 45c. per pound.

VISES — Prices are quoted as follows: "Peter Wright," 30 to 34, 14½c.; 35 to 39, 14c.; 48 and larger, 13½c. per lb.

ANVILS—"Peter Wright" anvils are selling at 11c. per lb.

CROWBARS—Quoted now at 4c. per lb.

POWER HORSE CLIPPERS — The "1902" power horse clipper is selling at $12, and the "Twentieth Century" at $6. The "1904" sheep shearing machines are sold at $13.60.

AMMUNITION, ETC.—Quotations are as follows: Cartridges, Dominion R.F. 50 and 5; Dominion, C.F., 33⅓ C.F., pistol, p.c.; C.F., military, 10 p.c. advance. Loaded shells: Dominion Eley's and Kynoch's soft, 12 gauge, black, $16.50; chilled, 12 gauge, $17.50; soft, 10 gauge, $19.50; chilled, 10 gauge, $20.50. Shot, ordinary, per 100 lbs., $7.25; chilled, $7.75; powder, F.F., keg, Hamilton, $4.75; F.F.G., Dupont's, $5.

IRON AND STEEL.—Quotations are: Bar iron basis, $2.70. Swedish iron basis, $4.95; sleigh shoe steel, $2.75; spring steel, $3.25; machinery steel, $3.50; tool steel, Black Diamond, 100 lbs., $9.50; Jessop, $13.

SHEET ZINC — The price is now $8.50 for cask lots, and $9 for broken lots.

PIG LEAD—Quoted a $5.85 per cwt.

AXLE GREASE—"Mica" axle grease is quoted at $2.75 per case, and "Diamond" at $1.60.

IRON PIPE AND FITTINGS— Revised prices are as follows:—Black pipe, ¼ inch, $2.65; ⅜, $2.80; ½, $3.50; ¾, $4.40; 1. $6.35; 1¼, $8.65; 1½, $10.40; 2, 13.85; 2½, $19.00; 3, $25.00. Galvanized iron pipe, ⅜ inch. $3.75; ½, $4.35; ¾, $5.65; 1, $8.10; 1¼, $11.00; 1½, $13.25; 2, inch, $17.65. Nipples, discounts 70 and 10 per cent.; Unions, couplings, bushings and plugs, 60 per cent.

LEAD PIPE—The price, $7.80, is firmly maintained in view of the advancing lead market.

GALVANIZED IRON—Quoted as follows:—Apollo, 16 gauge, $3.90; 18 and 20, $4.10; 22 and 24, $4.45; 26, $4.40; 28, $4.65, 30 gauge or 10¾ oz., $4.95; Queen's Head, 24, $4.50; 26, $4.65; 28, $5.00.

TIN PLATES—We now quote as follows: IC charcoal, 20 x 28, box, $9.50; IX charcoal, 20 x 28, $11.50; XXI charcoal, 20 x 28, $13.50.

TERNE PLATES—Quoted at $9.

CANADA PLATES—Quoted as follows: Canada plate, 18 x 21, 18 x 24, $3.40; 20 x 28, $3.65; full polished, $4.15.

BLACK SHEETS—Prices are: 10 to 16 gauge, 100 lbs., $3.50; 18 to 22, $3.75; 24, $3.90; 26, $4; 28, $4.10.

PETROLEUM AND GASOLINE.—Silver Star in brls. per gal., 20c.; Sunlight in brls, per gal., 21c.; per case, $2.30; Eocene in brls, per gal., 23c.; per case, $2.50; Pennoline in brls., per gal., 24c.; Crystal Spray, 23c.; Silver Light, 21c.; Engine gasoline in barrels, per gal., 28c., f.o.b. Winnipeg in cases, $2.75.

PAINTS, OILS AND TURPENTINE—Turpentine is firm at the recent advance. White lead, pure, $6.50 to $7.50, according to brand; bladder putty in barrels, 2½c.; in kegs, 3¼c.; turpentine, barrel lots, Winnipeg, $1.01; Calgary, $1.08; Lethbridge, $1.08; Edmonton, $1.09. Less than barrel lots 5c. per gallon advance. Linseed oil, raw, Winnipeg, 67c.; Calgary, 74c.; Lethbridge, 74c.; Edmonton, 75c.; boiled oil, 3c. per gal. advance on these prices.

WINDOW GLASS—We quote: 16-oz. O.G., single, in 50-ft. boxes—16 to 25 united inches, $2.25; 26 to 40, $3.40; 16-oz. O.G., single, in 100-ft. cases — 16 to 25 united inches, $4; 26 to 40, $4.52; 41 to 50, $4.75; 50 to 60, $5.25; 61 to 70, $5.75. 21-oz. C.S., double, in 100-ft. cases—26 to 40 united inches, $7.35; 41 to 50, $8.40; 51 to 60, $9.45; 61 to 70, $10.50; 71 to 80, $11.55; 81 to 85, $12.60; 86 to 90, $14.75; 91 to 95 $17.30.

CATALOGUES AND BOOKLETS

When sending catalogues for review, manufacturers would confer a favor by pointing out the new articles that they contain. It would assist the editor in writing the review.

By mentioning HARDWARE and METAL to show that the writer is in the trade, a copy of these catalogues or other printed matter will be sent by the firms whose addresses are given.

Letter Opener.

The London Rolling Mill Company, London, Ont., have supplied their customers with a very neat paper knife for the office desk. The novelty is a very useful and certain to be a lasting reminder. Customers who have not received samples should mention this paper when asking for one.

Contractors' Supply Calendar.

Amiot, Lecours & Lariviere, Montreal, dealers in contractors' supplies, are sending to the trade a 1907 calendar, containing illustrations of several of the lines carried by them. This calendar is of a good size for office use and will be sent to anyone who mentions this paper when writing.

Memoranda Calendar.

A very neat calendar of a most sensible sort has been issued by Sybry, Searls & Co., Cannon Steel Works, Sheffield, England. The upper part of the card shows three views of the interior and exterior of the works, and the calendar portion has the business man's week from Monday to Saturday on separate sheets with small blanks for memoranda each day. The reverse side contains a great deal of information of interest to metal workers. No doubt the publishers would be glad to send copies to any of our readers who make application.

Carborundum Booklet.

The Carborundum Co., Niagara Falls, N.Y., in a recent booklet, illustrate and list a few of the standard forms of carborundum stones, such as razor hones, oil stones, knife sharpeners, pocket stones, etc. Mention this paper if a copy is desired.

New List on Horse Nails.

The Canada Horse Nail Company, Montreal, have issued under date Jan. 1st, a revised hardware trade price list, for their hot-forged "C" brand horse nails. As several important alterations have been made, which are of interest to that portion of the trade dealing in horseshoe nails, we herewith reproduce the list in full: No. 4, $4.50; No. 5, $3.75; No. 6, $3.50; No. 7, $3.25; No. 8, $3.00; No. 9, $3.00; No. 10 and larger, $2.75 per box. All patterns, oval and countersunk, are the same price. They direct the attention of the trade, in making comparisons, to the fact that their standard of lengths are unchanged, and remain the same as adopted by them for many years, these differing materially from the lengths for the same numbers used by other manufacturers. The "C" brand No. 9 is the same length as other number eights and the "C" brand No. 12 is the same length as other number tens, etc. Particulars as to discounts, terms and conditions of sale may be had from the makers direct on application.

DIVIDING THE PROFITS.

News comes from Minnesota of a strange agreement drawn up, signed and lived up to by G. H. O. Roberts, a hardwareman conducting a store at No. 108 Western avenue, Minneapolis, Minn. Distrusting his fellow men as partners, he drew up a document on May 12, 1893, and inscribed the same upon the flyleaf of his ledger, repeating it each year since. It ran as follows:

"I promise that, as the Lord shall prosper me, I shall act as His steward and give to Him as follows:

"'If I make $1,000 annually, $100; $1,500, $250; $2,000, $400; $2,500, $625; $3,000, $1,000; $5,000, $2,500.'"

Mr. Roberts claims that he has prospered by carrying out his agreement until he has reached the last-mentioned figures. He says he has no desire to go back to his old method of doing business before he made the agreement, and, in fact, would be afraid to do so, for he is certain that his present business would never have grown to its proportions if he had continued along the old lines.

FOUNDRY AND METAL INDUSTRIES

The Sandwich location offers many advantages over others suggested, but it is too early to make any definite announcement regarding the corporation's intentions.

The plant of the Goderich Engine and Bicycle Co. has been purchased by the Rogers Manufacturing Co., of Toronto. General foundry and railway specialty work will be done in the shops.

Promoters from Harvey, Illinois, have been at Windsor looking over a site on which they may establish a big manufacturing plant if the United States Steel Company carries out its proposal to build near Sandwich.

New York capitalists are in communication with the municipal council of Goderich, with a view to building a $5,000,000 steel plant. The concern would employ fifteen hundred hands, and would occupy one hundred acres of land.

Luke Thompson, engineer, has brought a suit against the Ontario Sewer Pipe Company for $10,000. Thompson was employed as engineer at the company's plant at Mimico, and was badly scalded by the breaking of an alleged defective connection.

A big deposit of high gradé iron ore has been discovered near Desbarats, Ont., a few miles below the Soo. It is said to be as good as any in the Lake Superior country. It lies close to the water, and can be shipped to southern lake ports easily, making a much shorter haul than from Lake Superior.

The Hamilton and Fort William Navigation Company of Hamilton, have placed an order with the Canadian Shipbuilding Company for an 8,500-ton freight steamér, to be built for the iron ore, coal, and wheat trade. This steamer, which will be one of the largest on the lakes, will be built at the shipbulding company's Port Colborne shipyard. The machinery will be built in Toronto.

The cost of copper refining by electricity has been reduced greatly in recent years in the American refineries by the introduction of mechanical devices for casting the anode slabs of crude copper and for charging and discharging the vats. According to an article by John. B. C. Kershaw in Cassier's Magazine, the expenditure on hand labor has thus been greatly reduced, and the time during which vats are laid off for recharging and cleaning has been curtailed. The current density used has also been greatly increased by the use of improved methods of circulating the electrolyte, and by the addition of a very small percentage of hydrochloric acid to the copper sulphate solution.

BUILDING AND INDUSTRIAL NEWS

HARDWARE AND METAL would be pleased to receive from any authoritative source building and industrial news of any sort, the formation or incorporation of companies, establishment or enlargement of mills, factories, foundries or other works, railway or mining news. All correspondence will be treated as confidential when desired.

A new Y.M.C.A. building at St. John, N.B., will be erected at the cost of $60,000.

John McPherson Company, Hamilton, is increasing the capital from $100,000 to $500,000.

Building permits have been issued in Toronto from January 8 to 12, 1907, totalling $55,000.

The Merchants Bank, of Toronto, have procured premises at Streetsville, and will erect a brick bank in the Spring.

The Cyclone Woven Wire Fence Company, of Toronto, have increased their capital stock from $75,000 to $100,000.

The National Spring & Wire Company have leased a factory in St. Catharines and will get into operation in the near future.

The Imperial Coal Company, of Beersville, N.B., are planning to establish a brick making plant with a capacity of 20,000 per day.

The Winton Automobile Company will establish a branch factory in Toronto. It is also intended to establish factories in England.

The plants of the Tilson Manufacturing Company and the Standard Fitting and Valve Company, of Guelph, will be rapidly pushed to completion in the early Spring.

The Temiskaming & Northern Ontario Railway Commission is asking for tenders for the construction of general offices at North Bay, at the cost of $25,000 or $30,000.

Cudahy Company, manufacturers of Dutch Cleanser, and H. B. Johnstone Company, leather manufacturers, of Toronto, sustained a total loss of $12,000 by fire, which occurred on Wednesday.

The pipe works of New Westminster, B.C., are rapidly nearing completion. The main building is 138 feet by 68 feet. It is expected that the company will be turning out pipe by the middle of next month.

There were erected in Montreal during 1906, 1,484 new buildings, valued at $7,745,023, as compared with 1,145 valued at $4,779,380 for the previous year. Ten structures exceeded in estimated cost $100,000 each.

Large consignments of sisal have been landed at Halifax this season by the Mexican Liner Sokoto. It is estimated that one hundred carloads per month will be imported via Halifax for Ontario manufacturers of rope.

Mr. Menzies and R. S. McLean, doing business under the name of the Sydney Electric and Construction Company, have dissolved partnership, and business will be carried on by Mr. Menzies and J. Morley, under the old name.

The Hugh Johnson Syndicate, of Ottawa, have procured premises on Queen street, Toronto, opposite the city hall. The price paid for this property was in the neighborhood of $175,000, and contracts have been let for a $6,000 brick building.

The contract for the locomotive shops at Moncton, N.B., for the Intercolonial Railway, to replace those recently burned, has been awarded to a Montreal firm, at a price of about half a million dollars, and will be built entirely of concrete and steel.

It is feared that the West Indian Electric Company, which operates tram cars, telephones and an electric light system in Kingston, Jamaica, must have lost heavily by the earthquake. The company is owned in Montreal. The capital out is $800,000 stock and $600,000 bonds.

The Ontario Wind Engine and Pump Co., of Winnipeg, has been awarded the contract for a 75,000-gallon water tank on a 50-foot tower for the waterworks system of the town of Arcola. The water is being brought to the town from four or five miles distant and forced into the tower for domestic use and fire protection.

The Toronto Plastering and Supply Company assigned to Osler Wade last week. This company was incorporated about a year ago, but it made a poor start, locking up all the capital in buildings and plant. No money was left for running expenses, and an effort to procure capital for this purpose was not successful.

Nine new industries have been located in Toronto during 1906. These are the Kindall Bed and Mattress Co., St. Louis; Blanchite Process Paint Co., New York; Canada Bolt and Nut Co., Boston; De Sauga Silk Co., St. Etienne, France; Chemical Laboratories; Dominion Carriage Works, Berlin Electrical Manufacturing Company.

The Sherbrooke Novelty Company has been incorporated with a share capital of $20,000 for the purpose of manufacturing electrical and mechanical appliances, household utensils and smallwares. The head office will be at Sherbrooke, Que., and the directors will be E. R. Ebbitt and A. C. Snowden, Montreal; R. A. Wright, A. T. Boydell and J. A. Swan, Sherbrooke.

The ratepayers of Port Arthur have voted to guarantee $75,000 worth of bonds for the Meisel Manufacturing Co., and to loan the Seaman-Kent Co. $15,000. The Meisel Co. will establish works for the manufacture of mill machinery and heavy harvesting implements. The Seaman-Kent Co.'s factory will be devoted to the manufacture of hardwood flooring, mantels, etc.

A large brick building at Point Edward, formerly used as a warehouse and pattern shop by the Canada Machinery Company, was destroyed by fire on Jan. 15, with a loss of fifteen thousand dollars. The building contained patterns formerly used by the company valued at five thousand dollars. The origin of the fire is unknown. The main part of the plant was destroyed by fire two years ago, and only the pattern shop and office stood. The latter was threatened, and was saved with difficulty. The loss is partly covered by insurance.

The Schultz Manufacturing Company has been incorporated with a share capital of $50,000 to purchase the business of Ernest Schultz, known as The Schultz Manufacturing Company, makers of stamped, pressed and spun metal goods, castings, lamps, lanterns, etc., with head office at Hamilton and directors to be E. P. Schultz, A. H. Brittain and E. A. Schultz, of Hamilton.

The Western Canada Development Company has been incorporated with a $1,000,000 capital, to carry on business as miners, coal miners, oil producers, refiners and gas makers, in any part of the Dominion of Canada with head office to be at Winnipeg, incorporating J. S. Hough, A. C. Ferguson, C. Williams, E. B. Lindsay and W. M. Graham, all of Winnipeg.

A wire cloth factory, which will give employment to about 700 people, will be established in Montreal in the spring. The plant will also manufacture wire screens and many kinds of wire novelties. The new company, which will be largely composed of American capitalists will be capitalized at $200,000. A factory for the manufacture of spokes, wheels and other articles of wood used in the building of vehicles will also be established. Montreal will be the distributing point for the various provinces.

A NOVEL DOOR LOCK.

A door lock of decidedly unique form has recently been invented by Peter Ebbeson, of St. Paul, Neb. While the construction of this lock is not complicated, yet it has been ingeniously designed to prevent operation with a false key. Furthermore, it comprises a latch of such form as to prevent shaking or rattling of the door. The lock consists of three disk-like sections, A, B, and C, the disk B being stationary and the others revolvable. The disks are mounted in a socket in the door and project from opposite sides thereof. The latch is operated by a pair of knobs at opposite sides of the door, which are respectively secured to the disks A and C. In the face of the disk A is an eccentric slot, adapted to receive a stud projecting from the door frame. By operating the knob of disk A the latter may be turned to engage the stud in the eccentric slot, thus locking the door. The disk C is connected with the disk A by a series of bevel gears, so that by operating the knob of disk C, it is possible to rotate the disk A to latch or unlatch the door. In order to lock the door a novel mechanism has been provided in the central disk B, a barrel being mounted in this disk. This barrel is provided with a bolt, which is adapted to engage a slot in one of the bevel gears, and thus prevent rotation of the other two disks. In the barrel is a tumbler, which is carried on a short shaft mounted to slide in slots in the barrel. This tumbler is provided with a projection at its upper end adapted normally to register with the central one of three flanges, projecting from a block above. Now, in order to unlock the latch, a key is inserted in the barrel, and this presses the tumbler, when the projection thereon clears the central projection, and the barrel may be rotated to move the bolt clear of the bevel gears. The tumbler is held in normal position by a pair of springs which bear against its shaft. It will be observed that the ends of the tumbler are of odd form, which the key must fit to prevent the tumbler from tilting on its axis when being pushed clear of the projection. If a false key is introduced, the tumbler will be tilted up into engagement with the outer projections.

48-STOREY BUILDING.

The Metropolitan Life Co. will erect a tower in New York, as an annex to their main building. The tower when complete will be the tallest structure on earth. Height above sidewalk 690½ ft.; number of stories below sidewalk, 2; grand total floor area of new Metropolitan building, 1,035,663 square feet (about 25 acres.)

A REMARKABLE WRECK.

An extraordinary and expensive cement advertisement was furnished by a recent shipwreck. The ship Socoa, bound for San Francisco with a cargo of cement for use in the rebuilding of the city, was wrecked off the Lizard on the Cornwall coast. The ship struck a rock, which tore a large hole in her side, and remained fastened as upon a pivot. When the salvage crew arrived to see about taking the Socoa from her perilous position, says The Cement Age, the men found a remarkable condition of affairs. The water had entered the hold and its action upon the cargo had caused the cement to set. It had accommodatingly set hard around the rock that pierced the side of the ship, which now remains fastened there permanently in its unique position. The entire cargo has become as hard as stone, and nothing can be done with the ship except to dismantle as much of the wood as can be removed. The hull will remain there for many years as a conspicuous advertisement for the cement manufacturers.

CAPEWELL SALESMEN'S DINNER.

The first annual dinner of the Ontario salesmen of the Capewell Horse Nail Company, of Toronto, was held at McConkey's restaurant, Toronto, on Jan 7th, and proved to be a very successful gathering, of a nature half social and half business. The chair was taken by the manager, Chas. H. Fleming, and for over three hours an interesting talk ensued regarding trade conditions and the progress the company had made during the past year. It is pleasant to record that the prevailing note of the discussion was one of decided optimism, and all the salesmen present reported favorably on conditions in their respective territories, and that there was a generally expressed determination that 1907 should eclipse all previous records. An attractive menu card was prepared, each dish bearing some reference to the company's product. Those present were Chas. H. Fleming, manager; F. E. Mews, chief accountant, and Messrs. J. Clothers, W. T. Crosby, F. May, C. Andres, W. A. Dean, G. W. West and J. R. Myers, traveling salesmen.

THE FINANCIAL POST.

To fill one of the greatest journalistic openings yet presenting itself in Canada, The Financial Post has appeared, under the direction of Lt.-Col. J. B. MacLean, of the MacLean Publishing Co., Montreal, Toronto and Winnipeg, and Mr Stewart Houston, a well-known financial writer, of Toronto. By its first number, issued on Saturday last, the paper proves its ability to fill the position. The aim of the publication as set forth by its publishers is to present to the public in a popular manner accurate information relating to the financial interests and the legitimate investments of Canada—not only news on these subjects, but the results of special enquiry from the best authoritative sources by expert writers In the first number appear articles on the Hudson's Bay Company, the Grand Trunk Pacific, Banks in the Northwest, Cobalt Comparisons, and an especially interesting collection of financial authorities' opinions regarding C.P.R., 25 years ago and at the present day. The Post begins under auspicious circumstances, in the fact that it has behind it a well-tried publishing firm, and the early struggles of the ordinary periodical will be obviated by an organization already perfected. The MacLean Publishing Co., which was established 20 years ago, is now the largest concern of its kind in Canada, and owns six trade papers, The Busy Man's Magazine and a number of provincial dailies and weeklies—twenty-one in all.

ST. JOHN JOBBERS MEET.

The annual meeting of the St. John Iron and Hardware Association of N.B., was held on January 8, and the following officers were elected: President, W. S. Fisher; vice-president, J. M Robertson; secretary-treasurer, John J. Barry; directors, W H. Thorne, Thomas McAvity, M. E. Agar.

Executive committee, J. M Robertson, W. S Fisher, W. H Thorne, Thomas McAvity, M. E. Agar, John J. Barry.

Wholesale committee, Emerson & Fisher, W. H. Thorne & Co., Ltd., S. Hayward Co., Ltd., M. E. Agar, James Robertson Co., Limited; T. McAvity & Sons, Robertson, Foster & Smith; T. & E. R. Burpee, McClary Mfg. Co., Ltd.

Manufacturers' committee, James Pender, George McAvity, George W. Ketchum, R. B. Emerson, John Keeffe, S. Elkin, M. F. Irwin.

CONDENSED OR "WANT" ADVERTISEMENTS.

Advertisements under this heading 2c. a word first insertion; 1c. a word each subsequent insertion.

Contractions count as one word, but five figures (as $1,000) are allowed as one word.

Cash remittances to cover cost must accompany all advertisements. In no case can this rule be overlooked. Advertisements received without remittance cannot be acknowledged.

Where replies come to our care to be forwarded, five cents must be added to cost to cover postage, etc.

SITUATIONS VACANT.

WANTED—Good practical tinsmith; $18.00 weekly to the right man; steady job. Apply H. H. Gervan, Chilliwack, B.C. (5)

WANTED—Man with two or three thousand dollars to take half interest in old established hardware and tinsmith business in good town. Box 575, HARDWARE AND METAL. (4)

WANTED—Two first-class hardware salesmen, one to travel west and north of London, and one to travel Toronto and vicinity; reply by letter stating experience, references, age and salary, to Box 576, HARDWARE AND METAL, Toronto. (3)

WANTED — Experienced young man, thoroughly conversant with builders' hardware, as traveller for the city. Apply stating references and salary equired to E. G. Prior & Co., Ltd., Victoria, B.C. (3)

WANTED—An experienced hardware man at once apply stating wages required to The Geo. Taylor Hardware Co., Ltd., New Liskeard, Ont. (4)

WANTED—Experienced pricer for wholesale hardware; none but experienced man need apply. Box 578, HARDWARE AND METAL, Toronto.

BUSINESS CHANCES.

HARDWARE stock about $7,000, including set of tinners' tools; stock is all new and situated on main line C.P.R. in Saskatchewan town of 1,800 population; building can be leased, is brick, two storeys and basement; net profit $4,000 yearly. Box 577, HARDWARE AND METAL, Toronto. (7)

HARDWARE BUSINESS established 17 years for sale—including stoves, ranges and tinware; a good plumbing and tinshop in connection; stock about $6,000. J. D. Murdoch & Co., Simcoe, Ont. (6)

FOR SALE — Hardware and tinware business in Western Ontario town of about 3,000 population; one opposition; turnover seventeen to twenty thousand dollars. Apply Box 679, HARDWARE AND METAL Toronto.

HARDWARE AND TINWARE business for sale in a thriving town in Eastern Ontario; stock reduced to $1,500; good dairying country, rare chance for new beginner; going out of business on account of ill health. Apply at once to D. Courville, Maxville, Ont. (3)

FOR SALE—A fully equipped Wire Mill and Nail Factory as a going concern, with about 500 tons of wire rods; factory situated on 150 ft. of water-front with trackage optional to buyer B.C. Wire and Nail Co., Ltd., Vancouver, B.C. (7)

SITUATION WANTED.

HARDWAREMAN (28) open for engagement; west. Box 91. HARDWARE AND METAL, Winnipeg. (4)

Persons addressing advertisers will kindly mention having seen their advertisement in Hardware and Metal.

Paint, Oil and Brush Trades

RAW AND BOILED OIL.

For some purposes raw oil is preferable to boiled. The raw oil does not harden as quickly as the boiled, consequently is more in evidence for outside work. In speaking on the subject of oils an old veteran of the brush, quoted by the Canadian Painter, said :—"Raw linseed oil generally finds most favor. It is clear, works freely, and in white lead paint on outside work requires the addition of little driers. It allows the paint to cover well, and is durable. Mixed in raw oil a paint dries with a glossy surface, which is a factor of durability. Boiled oil, on the other hand, is for white lead paints, less satisfactory than raw. Though a paint mixed with it may go further, it does not cover so well, is apt to discolor light tints, is harder to spread, less durable, and a constant source of blistering. Where the use of boiled oil comes in is in mixing dark colors for outside work, such as Brunswick green, Indian red, bronze green and such colors, obtained by a painter in a dry state. The comparatively thick condition of boiled oil assures a better gloss under these circumstances, and assures their drying under ordinary conditions without the addition of driers."

TURPENTINE MARKETS.

The following is an extract from the Savannah Naval Stores Review, on the cause of the advance of turpentine : "So far this Winter there have been but three or four days that could be called cold, and there has been an absence of rain equally noticeable. As a result of this state of the weather the turpentine farms have practically cleaned up everything in sight. The boxes, as one man remarked, have been "wiped dry," and everyone is looking for an almost total cessation of receipts at any time. The impression prevails in trade circles that the late receipts of the season of 1906-07 and the early receipts of 1907-08 will be noticeably small. As a result the feeling prevails that conditions are in favor of a continuance of high values, and that even better values than at present prevailing are not unlikely in both spirits, turpentine and lower grades of rosins. Buyers seem desirous to gather in stocks of turps while there are any to be had, although they naturally do not like to run the market up at their own expense. Stocks are not heavy and as there are probably three months ahead during which the consumers must be fed from the stuff now in hand and the small lots to arrive there appears to be good basis for the bullish feeling that is beginning to show itself, and which may bring important developments at any time."

PROFIT IN PAINTS

In the modern, well-stocked hardware store ready-mixed and ready-to-use paint is one of the important staples. It is a convenient commodity, a good seller and a profit-maker. The enterprising retailer will, therefore, see to it that his paint stock is kept complete and fresh and that it is given its full share of display in the store.

The great bulk of the ready-mixed paints carried in stock by hardware dealers is the product of well-known and responsible firms whose guarantee of weights and measures may be relied upon. The quantities required for any given amount of space to be covered may therefore be learned from the dealer with accuracy, whereby waste and unnecessary expense are avoided and considerable annoyance to both dealer and consumer is obviated.

In buying ready-mixed paints prepared by one of the well-established paint houses the purchaser has the satisfaction of knowing that he is getting much better paint that can be prepared by hand, finely ground and perfect for all purposes in its completed state.

Another advantage of the ready-mixed paints handled by the hardware dealer is that the colors are accurate and uniform at all times, so that when any color has to be duplicated it can be done with exactitude and confidence. In the manufacture of such paints high-salaried chemists and scientific experts superintend the various processes for the makers, and the finest machinery being used, under the factory system, the results obtained are at all times uniform and exact. Manufacturing in large quantities also makes the product cheaper than hand-mixed paints, as well as more reliable and convenient.

In wearing and lasting qualities ready-mixed paints of good brands are superior to any others, while varieties of colors and shades can be obtained that are unknown or impossible with hand-mixing.

Not the least important advantage gained by the use of ready-mixed paints is the saving of time. These paints are always ready for use and there is true economy in their use.

Millions of dollars are invested in the paint business of Canada, and each maker strives to obtain the highest perfection of his product in all meritorious qualities. The result is that the retailer can buy stocks with confidence, knowing that he is securing paints that will give every possible satisfaction to the consumer.

Ready-mixed paints are a profitable line for the hardware dealer to handle and a large business can often be worked up in special brands. In communities where such paints are not in general use a little systematic advertising of their merits will often bring about their use in large quantities to the advantage of all concerned.

PAINT AND OIL MARKETS

MONTREAL.

Office of HARDWARE AND METAL,
232 McGill Street,
Montreal, January 18, 1907

The placing of orders for Spring shipment, alluded to last week, is assuming such proportions as to indicate an extremely brisk trade when the season actually opens. Manufacturers expect that from now on their mills will be kept busy.

Prices remain unchanged, and the only special feature is the very firm position of turpentine.

LINSEED OIL—We still quote: Raw, 1 to 4 barrels, 55c.; 5 to 9 barrels, 54c.; boiled, 1 to 4 barrels, 58c.; 5 to 9 barrels, 57c.

TURPENTINE—(In barrels)—The recent advance still holds, with a very firm market. Small quantities remain the same. We quote; Single barrel, 98c. per gal; two barrels or over, 97c. per gal.; for smaller quantities than barrels, 5c. extra per gal. is charged Standard gallon is 8.40 lbs., f.o.b. point of shipment, net 30 days.

GROUND WHITE LEAD—Still maintains its advance: Best brands, Government Standard, $7.25 to $7.50; No. 1, $6.90 to $7.15; No. 2. $6.55 to $6.90; No. 3, $6.30 to $6.55, all f.o.b., Montreal.

DRY WHITE ZINC—We still quote as follows: V.M. Red Seal, 7½c. to 8c.; Red Seal, 7c to 8c; French V.M., 6c to 7c; Lehigh, 5c to 6c.

WHITE ZINC—(Ground in oil) Prices remain firm at their recent advance. We quote: Pure, 8½c. to 9½c.; No. 1, 7c. to 8c.; No. 2, 5¾c. to 6¾c.

PUTTY.—Our prices are: Pure linseed oil, $1.75 to $1.85; bulk in barrels, $1.50; in 25-lb. irons, $1.80; in tins, $1.90; bladder putty in barrels, $1.75.

ORANGE MINERAL—Prices are as follows: Casks, 8c.; 100 lb. kegs, 8½c.

RED LEAD—The following quotations are firm: Genuine red lead in casks, $6.00; in 100-lb. kegs, $6.25; in less quantities at the rate of $7 per 100 lbs.; No. 1 red lead, casks, $5.75; kegs $6, and smaller quantities, $6.75.

PARIS GREEN — The following prices still hold good: In barrels, about 600 pounds, 23½c. per lb.; in arsenic kegs, 250 lbs., 23½c.; in 50 lb. drums, 24c.; in 25 lb. drums, 24½c.; in 1 lb. packets, 100 lbs. in case, 25c.; in 1 lb packets, 50 lbs in case, 25½c.; in ½ lb. packets, 100 lbs in case, 27c.; in 1 lb. tins, 26c. f.o.b. Montreal. Term three months net or 2 per cent. 30 days.

SHELLAC—We quote: Bleached, in bars or ground, 46c. per lb., f.o.b. Eastern Canadian points; bone dry, 57c. per lb., f.o-b. Eastern Canadian points; T. N. orange, etc., 48c. per lb. f.o.b. New York.

SHELLAC VARNISH—Prices remain: Pure white, $2.90 to $2.95; pure orange, $2.70 to $2.75; No. 1 orange, $2.50 to $2.55.

MIXED PAINTS—Prices range from $1.20 to $1.50 per gallon.

CASTOR OIL—Our prices are: Firsts in cases, 8½c.; in barrels, 8c.; seconds in cases, 8c.; in barrels, 7½c.

BENZINE— In bbls.)—We still quote: 20 cents per gallon.

GASOLINE—(In bbls.)—Our price is 22½ cents per gallon.

PETROLEUM—(In barrels) Our prices are: American prime white coal, 15½c. per gallon. American water, 17c. per gallon; Pratt's Astral, 19½c. per gallon.

WINDOW GLASS—Our prices are: First break, 50 feet, $1.85; second break, 50 feet, $1.95; first break, 100 feet, $3.20; second break, 100 feet, $3.40; third break, 100 feet, $3.95; fourth break, 100 feet, $4.15; fifth break, 100 feet, $4.40; sixth break, 100 feet, $4.95. Diamond Star: First break, 50 feet, $2.30; second break, 50 feet, $2.50; first break, 100 feet, $4.40; second break, $4.80; third break, 100 feet, $5.75; fourth break, 100 feet, $6.50; fifth break, 100 feet, $7.50; sixth break, 100 feet, $7.50; seventh break, 100 feet, $8; eighth break, 100 feet, $9. Double Diamond: First break, 50 feet, $3.45; second break, 50 feet, $3.75; first break, 100 feet, $6.75; second break, 100 feet, $7.25; third break, 100 feet, $8.75; fourth break, 100 feet, $10; fifth break, 100 feet, $11.50; sixth break, 100 feet, $12.50; seventh break, 100 feet, $14; eighth break, 100 feet, $16.50; ninth break, 100 feet, $18; tenth break, 100 feet, $20; eleventh break, 100 feet, $24; twelfth break, 100 feet, $28.50. Discount on Diamond Star, 20 per cent.; on Double Diamond, 40 per cent.

TORONTO.

Office of HARDWARE AND METAL,
10 Front Street East,
Toronto, Jan. 18, 1907.

Paint manufacturers report a very brisk season at present, as all hardware merchants are ordering before the spring advances.

Paris green is experiencing a great trade as the majority of orders are taken for English green which will arrive about the first of May. While the new shipping terms will not encourage the early buying of Canadian green, merchants will generally see the advantage of ordering early in view of the shortage expected.

The Carter White Lead Company, of Montreal, have closed down their corroding mill, partly owing to being unable to procure enough raw material, and partly to internal troubles.

Turpentine has again advanced owing to the great scarcity in Savannah.

We are now quoting $1.02 for single barrels and $1.07 in small quantities. Raw and boiled oil remain at last week's quotations.

WHITE LEAD—Ex Toronto pure white, $7.40; No. 1, $6.65; No. 2, $6.25; No. 3, $5.90; No. 4, $5.65 in packages of 25 pounds and upwards; 1-2c per pound extra will be charged for 12 1-2 pound packages; genuine dry white lead in casks, $7.00.

RED LEAD—Genuine in casks of 500 lbs., $6.00; ditto, in kegs of 100 lbs., $6.50, No. 1 in casks of 500 lbs., $5.75; ditto, in kegs of 100 lbs., $6.25.

DRY WHITE ZINC— In casks, 7 1-2c.; in 100 lbs., 8c., No. 1, in casks 6 1-2c., in 100 lbs., 7c.

WHITE ZINC (ground in oil)—In 25-lb. irons, 8c.; in 12 1-2 lbs., 8 1-2c.

SHINGLE STAIN—In 5-gallon lots, 75c. to 80c. per gallon.

PARIS WHITE—90c. in barrels to $1.25 per 100 lbs.

WHITING—60c. per 100 lbs. in barrels; Gilders' bolted whiting, 90c. in barrels, $1.15 in smaller quantities.

SHELLAC VARNISH—Pure orange in barrels, $2.70; white, $2.82 1-2 per barrel; No. 1 (orange) $2.50; gum shellac, bone dry, 63c., Toronto. T. N. (orange) 51c. net Toronto.

LINSEED OIL—We continue to quote: Raw, 1 to 3 barrels, 59c; 4 to 7 barrels, 57c; 8 and over, 56c, add 3c to this price for boiled oil. Toronto, Hamilton, London and Guelph net 30 days.

TURPENTINE—As predicted by us, turps have again advanced and we now quote: Single barrels, $1.05; 2 to 4 barrels, $1.04, f.o.b. point of shipment, net 30 days. Less than barrels, $1.07 per gallon.

GLUES—French Medal, 12 1-2c. per pound; domestic shee., 10 1-2c. per lb.

PUTTY—Ordinary, 800 casks, $1.50; 100 drums, $1.75; barrels or bladders, $1.75; 100-lb. cases. $1.90;; 25-lb. irons, $1.85; 25-lb. tins (4 to case) $1.90; 12 ½ lb. tins (8 to case) $2 10.

LIQUID PAINTS—Pure, $1.20 to $1.35 per gallon; No. 1, $1.10 per gallon.

BARN PAINTS—Gals., 70c. to 80c.

BRIDGE PAINTS—Gals., 75c. to $1.

CASTOR OIL—English, in cases, 9 1-2c. to 10c. per lb., and 11c. for single tins.

PARIS GREEN—On Canadian green, the base price should be 25¼c instead of 25c. English remains at former quotations, 25¾ base f.o.b. Toronto.

REFINED PETROLEUM — Dealers are stocking up heavily for the Winter. We still quote: Canadian prime white, 14c.; water white, 16c.; American water white, 16c. to 18c. ex warehouse.

CRUDE PETROLEUM — We quote: Canadian, $1.32; Pennsylvania, $1.58; Ohio, 96c.

MIXED PAINT ADVANCES.

Recent advances on mixed paints were largely owing to the increased cost of raw materials, tin cans, boxes and labor. Manufacturers hesitated individually to make a move that might meet with strong disfavor on the part of their customers, but a few of the stronger manufacturers made the advance and their example was followed by the others. In the majority of cases, customers received the announcement more favorably than had been expected, and the demand for brands of paint enjoying a good reputation was not lessened by a slight increase in price. In fact the advances had been expected by the retailer, as the high prices of raw material warranted the advance.

That the logic of events is strongly in favor of the advance is sufficiently proved by the mere fact that it has been made by practically every paint manufacturer. It would be impossible to bring all the manufacturers together into a concerted movement to this effect by mere urging. Not until each manufacturer saw that the life of his business depended on getting a better price for his product, commensurate with its increased cost, was he willing to declare an advance.

LEAVES PAINT FOR STOVES.

Gordon Ritchie, who recently accepted a position with the Montreal branch of the McClary Manufacturing Company, was not head of the selling staff of the Sherwin-Williams Company at Montreal, as reported. For the past six years he held various positions on the S.-W. office staff, but it was an injustice to both Mr. Ritchie and his former employers to connect his name with the position he was said to have resigned. Mr. Ritchie has many friends in the paint trade who wish him every success in the stove end of the business.

LATEST CANADIAN PATENTS.

The following up-to-date list of Canadian patents is reported to us by Egerton R. Case, solicitor of patents, and expert in patent causes, Temple Building, Toronto:—Isaac Peabody, St. Mary's, N.B., conveyors; J. J. Holmes, Chatham, Ont., refrigerating machines; Jas. F. Latimer, Toronto, apparatus for separating and refining graphite; Geo. A. Bennett, Winnipeg, cushions for bicycles; Thos. M. Morgan, Longue Point, Que., apparatus for burning cement; Thos. Wilson, Ottawa, adjustable roofs; Wm. J. Watson, Ladysmith, B.C., smelting furnaces; Hiram W. Hixon, Victoria Mines, Ont., blast furnaces; Goldsmith English, Windsor, Ont., the construction of electrical annunciators; Peter Stoddart, Copper Cliff, tapping jackets for blast furnace settling wells or for hearths; Fred. J. Gilmore, Montreal, apparatus for building submerged concrete works from above the surface of the water; Albert Belair, Ahuntsic, Que., burial caskets; Wm. C. Gurney, Toronto, gas ranges and stoves; Robt. N. Gundy, Guelph, cooking stoves and ranges.

THE STANDARD IDEAL SANITARY CO., LIMITED

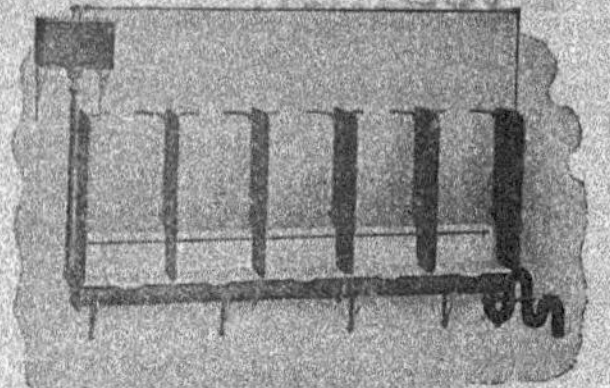

Plate E. 404.—For Public Institutions, Schools, Factories, Asylums, &c.

We manufacture a complete line of

Range-Closets, Range-Urinals, Lavatories and Sanitary Fixtures

Suitable for Public Institutions

The reputation of our goods is unequalled.

Toronto: 50 Colborne Street *Montreal: 128 Craig Street West*
Winnipeg: 24 Telfer Building

HOME OFFICE and FACTORIES: PORT HOPE, ONT.

The James Morrison Brass Mfg. Co., Limited

PLUMBERS' SUPPLIES AND SPECIALTIES

93-95 Adelaide St. West, - - - TORONTO

Plumbing and Steamfitting

THE PLUMBING OF A COTTAGE

By J. A. F. Cardiff in the National Builder, Chicago

Article XIV.

Refrigerator Wastes.

In the modern cottages the refrigerator waste is supplanting the old style drip pan as a means of carrying away the drippings from the ice. This improvement is not only a great convenience to the housekeeper, but is also a more sanitary arrangement as the water which drips from the ice chamber is more or less slimy and unhealthy.

Under the refrigerator a safe, about fifteen inches square, is constructed by nailing hardwood beveled strips to the floor so as to form a pan and lining the pan with six-pound sheet lead turned over and tacked to the wood strips in the manner shown in Fig. 59. An outlet is left in the centre of the safe for the connection of the waste pipe. Over the outlet a removable strainer is placed so as to prevent any foreign substance from entering the pipe, and the space immediately about the outlet is countersunk. Sheet copper is sometimes used instead of sheet lead for the lining.

The piping should discharge over a sink in the cellar and should never be directly connected to the drainage system. There are times when the refrigerator is temporarily out of use and consequently there is not enough water to maintain the seal in the trap, so that if the waste was connected to the drainage system the foul air and gases would enter the house through the refrigerator waste. Discharging the pipe over a sink does away with this evil.

Fig. 58

The sink should be supplied with water and should be properly connected up to the drainage system, and the trap on the sink waste should be back vented. The sink should always be placed in a location where it will be used at frequent intervals so that the seal of the sink trap will be properly maintained.

The piping is usually of galvanized wrought iron one and one-quarter inches in diameter, with galvanized wrought iron recessed and threaded drainage fittings.

All horizontal offsets should be made accessible by means of a cleanout as at "A," Fig. 58.

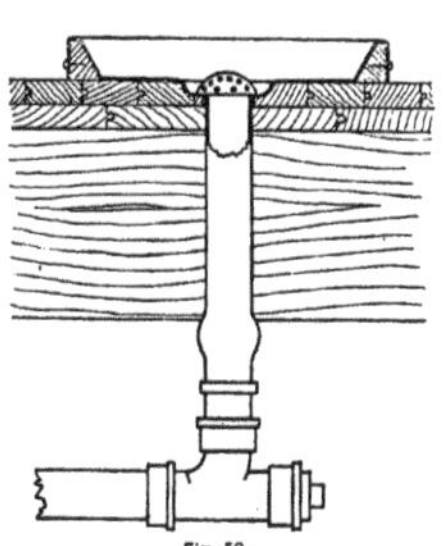

Fig. 59.

Near the outlet of the pipe a light swing check valve should be provided to prevent the cellar air from entering the rooms above through the waste pipe. No trap is required on the refrigerator waste when installed in this manner, but frequently the check valve is omitted and a trap formed by bending the pipe as at "B." This trap serves the same purpose as the check valve, but is not so good for the reason that the seal is liable to be broken by evaporation and the trap forms an inaccessible pocket for obstructions which may enter the pipe.

The connection between the safe and the iron piping is usually made with a short piece of lead pipe, soldered to the safe lining and wiped or soldered to a brass ferrule or nipple screwed into the hub of the iron fitting.

LUNKENHEIMER'S BIG CALENDAR.

The Lunkenheimer Company, Cincinnati, Ohio, who are represented in Canada by Cluff Bros., Toronto, have supplied their customers with a monster daily calendar which, besides telling at a glance the day of the year, constantly reminds buyers of the varied line of valves and engineering specialties manufactured by them. Steamfitters and engineers desiring one of the calendars should mention this paper when writing the firm at Cincinnati.

INTERESTING APPLICATION OF AN AIR WASHER

The following description of an air washing apparatus recently installed in the Illinois Steel Company's drafting room in Chicago has been supplied by the Buffalo Forge Company, whose system was used.

"In the past the only objection to the fan system of indirect heating and ventilation has been found in the introduction of dust and soot and occasionally foul gases due to the proximity of the ventilation intake being necessarily near to the street level or a neighboring chimney. This feature would be noticeably unpleasant in a drafting room because of the deposit of dust which would occur on the drawings to the annoyance of the draftsmen.

"The general layout and scheme of distribution of the ventilation is carefully shown on the accompanying drawing. The entering air passes into the tempering coils through a grating placed near the corner of the building. The tempering coils contain 1,000 feet of 1-inch pipe and serve in cold weather to raise the temperature of the air to 70 or 80 degrees F., whence it proceeds directly to an air washer.

"The washing chamber, in which we will include all the apparatus provided for washing and afterward removing the free moisture, has a rated capacity of 10,600 cubic feet of air per minute and is 4 feet 10 inches wide, 4 feet 8 inches high and 7 feet long.

"The washing itself is accomplished by spraying the incoming air through 76 specially designed nozzles, the direction of water projection being parallel to that of the air travel. Each of these nozzles will atomize, it is claimed, between 1½ and 2 pounds of water per minute as supplied by the pump. The water passing through the nozzle is atomized by centrifugal force acquired by its own circular path in leaving the nozzle exit.

"The washing is completed and the free moisture removed by drawing the air through what the writer will choose to call eliminator plates, a system of vertical baffle plates. The first part of each plate presents to the air merely a water covered surface to which the solid particles originally present in the air will cling and be left by the air as it passes to the second or rear portion of the plates which are provided with gutters running lengthwise to break the continuity of the water film and prevent it from being blown from the far side of the plates by the on-rushing air. The water removed from the air runs to a settling tank built beneath the eliminator in the concrete foundation and above which the eliminator plates are supported on 4-inch galvanized I beams. Provision is made in the settling tank for drawing off the "settlings" and then the clearer water is led to a well with which the suction of the small brass centrifugal pump belt driven from a 2 h.p. 440 volt A.C. motor, is connected and then this water, after straining, is supplied again to the nozzles. The nozzles are readily taken apart by removing the knurled cap with one's fingers to remove any scum that may have been deposited there.

"The humidity control is through the agency of a themostat, a heat indicating instrument. The amount of moisture which the air will absorb in the form of water vapor has been found after careful test to be controlled more by the temperature of the water exposed to the air, than through the air temperature. Therefore if the thermostat be set and calibrated by trial and it controls a steam jet playing into the water supplied to the spray nozzles, it can control the spray water temperature and hence the humidity which is kept at a constant percentage of the saturation point at all temperatures.

"The washing chamber and eliminator plate, which are of galvanized iron, are housed in the same materials, stiffly raced, riveted and soldered so as to be water-tight. In the washer casing is placed a water-tight door and two plate glass windows to allow inspection of its action while in operation. Connecting with this housing on the washer end is the inlet conduit and the other end connects with the heater coils.

"After leaving the heater coils which supply radiation from 1,500 feet of 1-inch pipe, and passing through a 70-inch lower provided with a double set of curved blades, the air is forced through the system of ducts shown by the building plan to the various parts of the drafting room. The blower is belt driven from a Westinghouse 5 h.p. 440 volt 3 phase A.C. motor and makes 475 revolutions per minute at a capacity of 10,600 cubic feet of air per minute."

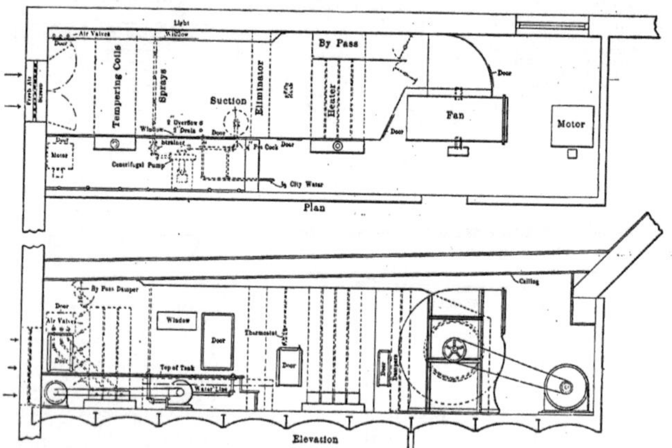

Air Washer System in the Illinois Steel Company's Drafting Room.

PLUMBING MARKETS

MONTREAL.

Office of HARDWARE AND METAL,
232 McGill Street,
Montreal, January 17, 1907

Business here remains excellent. There has never been a winter when general conditions have been so good. This is due, no doubt, to the great amount of building going on at present. Heretofore building operations are generally suspended with the arrival of Jack Frost, but this year from some cause or other, they have continued, and contractors are pushing things just as though the thermometer registered 80 degrees in the shade.

With this unexpected activity, jobbers are having a great deal of trouble and find it almost impossible to get deliveries.

Copper range boilers have taken a jump and the discount is now only 5 per cent.

Other prices remain the same.

RANGE BOILERS—We now quote: Iron clad, 30 gallon, $5; 40 gallon, $6.50 net list. Copper, 30 gallon, $33; 35 gallon, $38; 40 gallon, $43; 45 gallon, $50; 50 gallon, $58; 60 gallon, $68; 70 gallon, $80; 80 gallon, $95; 90 gallon, $115; 100 gallon, $150 net.

LEAD PIPE—Remains firm. We still quote 5 per cent. discount f.o.b. Montreal, Toronto, St. John, N.B., Halifax: f.o.b., London, 15c per hundred lbs extra; f.o.b. Hamilton, 10c per hundred lbs. extra.

IRON PIPE FITTINGS—We still quote: Discounts on nipples, ¼ inch to 3 inch, 75 per cent., 3½ inch to 2 inch, 55 per cent.

IRON PIPE—Supplies still short with last week's advance holding firm. We quote: Standard pipe in lots of 100 feet, regular lengths, ¼ inch, $5.50; ⅜ inch, $5.50; ½ inch, $8.50; ¾ inch, 11.50; 1 inch, $16.50; 1 1-4 inches, $22.50; 1 1-2 inches, $27.00; 2 inches, $36.50, discounts on black pipe, ¼ inch, 59 per cent.; ⅜ in. 59 per cent.; ½ in. 68; ¾ to 2, 70. Discounts on galvanized pipe: ¼ inch, 44 per cent.; ⅜ inch, 44 per cent.; ½ inch, 58 per cent.; ¾ to 2 inch, 60 per cent. Extra heavy pipe of 100 feet lots are quoted as follows: 1-2 inch, $12; 3-4 inch, $15; 1 inch, $22; 1 1-4 inches, $30; 1 1-2 inches, $36; 2 inches, $50. The discounts on black pipe, ¼ inch, 74 per cent.; ⅜ inch, 69 per cent.; ½ inch to 2 inches, 68 per cent. Galvanized, ¼ inch, 59 per cent.; ⅜ inch, 69 per cent.; ½ to 2 inches 58 per cent.

SOIL PIPE AND FITTINGS—Our prices remain as follows: Standard soil pipe, 50 per cent. off list. Standard fittings, 50 and 10 per cent. off list; medium and extra heavy soil pipe, 60 per cent. off. Fittings, 60 per cent. off.

SOLDER—We quote: Bar solder, half and half, guaranteed, 25c; No. 2, wiping solder, 22c.

SOLDER—Prices remain the same: Bar solder, half-and-half, guaranteed 25c.; No. 2 wiping solder, 22c.

ENAMELWARE — Recent advance on Canadian ware still holds. We quote: Canadian baths, see Jan. 3, 1907 list. Lavatories, discounts, 1st quality, 30 per cent.; special, 30 and 10 per cent. Sinks, 18 x 30 inch, flat rim, 1st quality, $2.60, special, $2.45.

TORONTO.

Office of HARDWARE AND METAL,
10 Front Street East.
Toronto, January 18, 190

Advances have taken place this week on both galvanized iron and copper range boilers, the advance on the first mentioned amounting to 25 cents on each boiler, 30-gallon Standard being now quoted at $5, and extra heavy at $7.75. On copper range boilers a new list has been issued as a result of the many sharp advances which have taken place in the last few months in the copper market. The old lists quoted $25, $29, and $33 on 30, 35 and 40-gallon boilers, while the new figures are $33, $38 and $43 on the same sizes. On the old quotations, the net list was quoted, while now special discounts are given to cover the difference in the advance on the list over the real advance in the cost of manufacturing raw material.

Other changes this week include the advance on J.M.T. globe, angle and check valves from 50 to 45 per cent.; the jump in cast iron fittings from 62½ to 60 per cent., and an advance on American enamelware sufficient to cover the recent advance in duty. It is also understood that an advance of $1 per tub is to be made on American baths. In addition to these changes, we have revised our quotation on iron pipe in sizes from 2½ inches upward.

While business is necessarily slow at this season, the volume of business that is being done is very satisfactory, both plumbing and heating contractors keeping most of their men at work on jobs which are being rushed to completion during the cold weather.

LEAD PIPE — The discount on lead pipe continues at 5 per cent. off the list price of 7c. per pound. Lead waste, 8c. per pound with 5 off. Caulking lead 5 3-4c. to 6c. per pound. Traps and bends, 40 per cent. discount.

SOIL PIPE AND FITTINGS — Recent advances are held to firmly with demand seasonable. Medium and extra heavy pipe and fittings, 60 per cent.; light pipe 50 per cent.; light fittings, 50 and 10 per cent.; 7 and 8 inch pipe, 40 and 5 per cent.

IRON PIPE.—We are revising last week's figures on large sizes. One-inch black pipe is quoted at $4.95, and one-inch galvanized at $6.60. Full list in current market quotations.

IRON PIPE FITTINGS—Cast iron fittings show several advances. Cast iron elbows, tees, crosses, etc., 60 per cent., cast iron plugs and bushings, 60 per cent.; flange unions, 60 per cent.; nipples, 60 and 10 per cent.; iron cocks, 55 and 5 per cent.; Canadian malleable, 30 per cent.; malleable unions, 55 and 5 per cent.; malleable bushings, 55 per cent.; cast iron ceiling plates, pl..n 65 per cent.; cast iron floor, 70 per cent.; hook plates, 60 per cent.; expansion plates, 65 per cent.; headers, 60 per cent, hangers, 65 per cent.; standard list.

RANGE BOILERS—An advance of 25 cents has been made on all sizes of galvanized iron and we now quote: Galvanized iron, 30 gallon, standard, $5; extra heavy, $7.75; 35 gallon, standard, $6; extra heavy, $8.75; 40 gallon, standard, $7; 40 gallon, extra heavy, $9.75, net list. Copper range boilers—New lists quote: 30 gallon, $33; 35 gallon, $38; 40 gallon, $43. Discounts up to 15 per cent.

RADIATORS—Prices are very stiff at: Hot water, 47½ per cent.; steam, 50 per cent.; wall radiators, 45 per cent.; specials, 45 per cent. Hot water boilers are unchanged.

SOLDER — Quotations remain firm as follows: Bar solder, half-and-half, guaranteed, 27c.; wiping, 23c.

ENAMELWARE—New lists on enameled baths issued by the Standard Ideal Company on Jan. 3, show a 10 per cent. advance. Lavatories, first quality, 20 and 5 to 20 and 10 off; special, 20 and 10 to 30 and 2½ per cent. discount. Kitchen sinks, plate 300, firsts 60 and 10 off; specials, 65 and 5 per cent. Urinals and range closets, 15 off. Fittings extra.

A NEW EXHAUST HEAD.

In modern engineering the removal of water and grease from the waste steam of the power plant is considered quite essential to good operation. For this purpose exhaust heads are used, and the new head just put on the market by the Canadian Buffalo Forge Co. is the result of the applied experience of almost one-third of a century in building power plant equipment. This head depends for its operation on the excessive inertia of the water or grease over that of steam vapor, water weighing approximately seventeen hundred times as much as the vapor, volume for volume.

Exhaust steam, rising into the head, strikes the inverted cone deflector, and the water and grease, by the change of direction, are thrown to the outer shell, where they find their way to the drip pipe. On account of its small weight per cubic foot the steam readily changes its direction of flow and finds its way through the outlet to the outside air. The shell of this apparatus is carefully constructed of heavy galvanized sheet iron, which is strongly riveted at the joints and soldered, the drip pipes from the deflector and to the outside air being held in place by machine-threaded clamp nuts, so that there is no possibility of their working loose or breaking off.

THE NEW

"SOLID-ADJUSTABLE" HAND STOCK AND DIES

At prices that will warrant your inquiries.

I. E. SHANTZ & CO.

BERLIN, - ONTARIO

Selling Agents for Canada

This is the Mark that Means Something

It is the distinctive brand of genuine Armstrong Stocks and Die and Pipe Fitters' Tools.

No "Armstrong pattern" or "just as good as Armstrong" have this stamp. Don't buy anything in this line that hasn't this brand as it is the mark of genuineness.

THE ARMSTRONG MANUFACTURING CO.

292 Knowlton Street, Bridgeport, Conn.

A Certain Sense of Satisfaction

goes with every MERRELL PORTABLE HAND MACHINE, PIPE THREADING and CUTTING MACHINE. We build them that way.

For doing ALL kinds of threading—under ALL conditions—labor trouble and faulty material included—where speed, durability and ease of control count for something, our word to you is to investigate the MERRELL.

Whether it be stationary or portable—hand or power driven—30 DAYS FREE TRIAL must prove conclusively that your choice should be the MERRELL.

This Portable Hand Machine has encased gears, MERRELL standard quick opening and closing die-head, and the latest improved Cutting-off Knife.

Let us tell you more about this machine—The Chasers, Vise and the large range of work covered.

Catalogue for the asking.
Quick shipments.

THE CANADIAN FAIRBANKS CO.

Sole Agents for Canada Limited

MONTREAL, TORONTO, WINNIPEG, VANCOUVER

Send Off That Order Now

TEST MY CLAIMS TO-DAY

I want to know your **ENAMELWARE WANTS** so that I may prove to you how promptly shipments can be made. You'll find the **Quality** and **Price** just right.

ORLANDO VICKERY

178-180 Victoria Street TORONTO, CANADA

"THE BEST ON EARTH"

That's what practical men say about

"Diamond Brand" Fittings

and they should know. We guarantee that the ***finish is as smooth*** as ***can be***, the ***threads just right*** and the ***material first-class***.

OSHAWA STEAM & GAS FITTINGS CO.,

OSHAWA, CANADA LIMITED

When You Buy Demand Proof

Don't take any substitute for Glauber Brass goods. Make the salesman show you. Brass has gone up no less than fifteen per cent., and if you don't get proofs before you buy you take chances of getting something less than you pay for.

WE GIVE YOU THE PROOF and ***2 Cocks Free*** for any one sent you defective.

Glauber Brass Mfg. Co.,

Expert Makers of Plumbers' Fine Brass Goods

CLEVELAND, O.

Sharratt & Newth's Glaziers' Diamonds

are unequalled for cutting and wearing qualities.

To be obtained from the principal Dealers in Glass, Hardware, and Painters' Supplies.

Contractors to H. M. Government and the principal English Sheet and Plate Glass Works

World's Best Cutlery

You make no mistake in specifying goods made by

JOSEPH RODGERS & SONS, Limited

when ordering cutlery. A reputation of over 200 years at stake, insures the quality of every article.

Write for catalogues from Sole Canadian Agents

JAMES HUTTON & COMPANY, MONTREAL

Forwell Foundry Co.

BERLIN, ONT.

Manufacturers of

SOIL PIPE, FITTINGS, and CAST IRON SINKS

Ask Jobbers for "F. F. CO." Brand

Horse Shoers' Foot Vise and Bolt Header

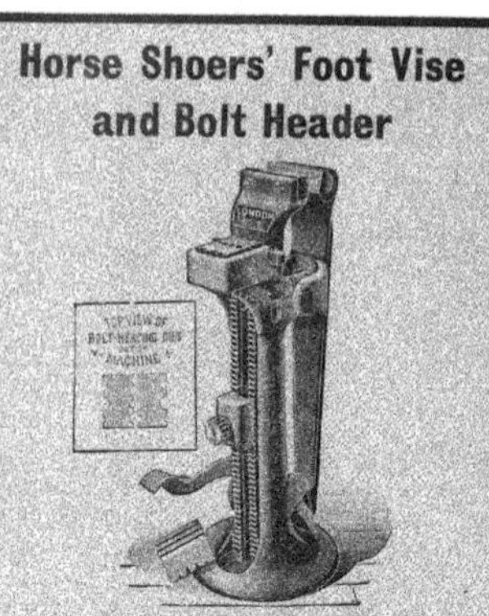

A Practical Tool and a Time-saver for the Busy Blacksmith

Manufactured by

THE LONDON FOUNDRY CO., Limited
LONDON, CANADA

Hardware and Plumber—

We have pumps suitable to any demand made upon you by most exacting customers.

Our pumps will do anything that a pump can do, and we will be pleased to give the benefit of many years' experience to any one having pump troubles.

McDougall Pumps are Standard Pumps, and with us it is quality first, last and always, while we ask just as little money as we can get along with.

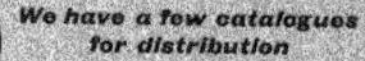

We have a few catalogues for distribution

The R. McDougall Co., Limited
GALT, - CANADA

The Long and the Short of It

"IT" MEANS

THE "LONG" is the time it wears. It stands the test of heat and frost, wind and rain, and give your customers many years of solid satisfaction.

THE "SHORT" part is the time needed to put it on. A great deal of time—which means a great deal of money—is saved to the man who uses

SHIELD BRAND
READY ROOFING

LOCKERBY & McCOMB
65 SHANNON STREET
MONTREAL

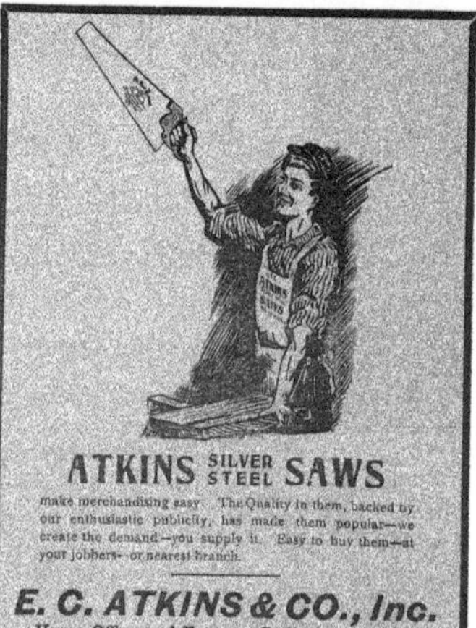

ATKINS SILVER STEEL SAWS

make merchandising easy. The Quality in them, backed by our enthusiastic publicity, has made them popular—we create the demand—you supply it. Easy to buy them—at your jobbers—or nearest branch.

E. C. ATKINS & CO., Inc.
Home Office and Factory, Indianapolis, Ind.
Canadian Branch, 77 Adelaide St. E., TORONTO

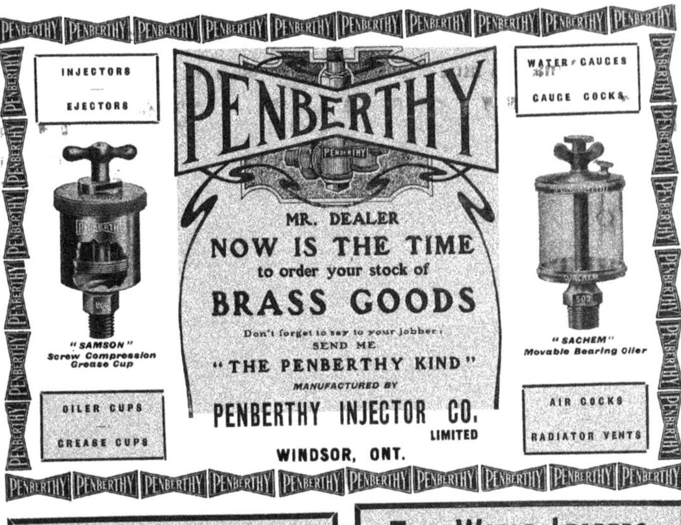
INJECTORS
EJECTORS
PENBERTHY
WATER GAUGES
GAUGE COCKS
"SAMSON"
Screw Compression Grease Cup
MR. DEALER
NOW IS THE TIME
to order your stock of
BRASS GOODS
Don't forget to say to your jobber:
SEND ME
"THE PENBERTHY KIND"
MANUFACTURED BY
PENBERTHY INJECTOR CO.
LIMITED
WINDSOR, ONT.
"SACHEM"
Movable Bearing Oiler
OILER CUPS
GREASE CUPS
AIR COCKS
RADIATOR VENTS

MUELLER FULLER BIBBS
STANDARD PATTERN
No rough or sharp seat edges to cut the ball. Deep seat with rounded edges and surface curved to conform to the shape of the ball. Plain or for hose and all styles of pipe connection.
UNCONDITIONAL GUARANTEE
H. MUELLER MFG. CO.
Decatur, Ill., U.S.A.
New York, N.Y., U.S.A.

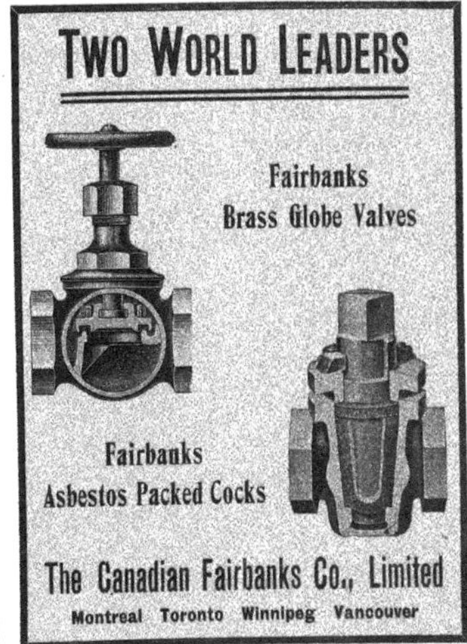
TWO WORLD LEADERS
Fairbanks
Brass Globe Valves
Fairbanks
Asbestos Packed Cocks
The Canadian Fairbanks Co., Limited
Montreal Toronto Winnipeg Vancouver

CURRENT MARKET QUOTATIONS.

Jan. 18, 1907.

These prices are for such qualities and quantities as are usually ordered by retail dealers on the usual terms of credit, the lowest figures being for larger quantities and prompt pay. Large cash buyers can frequently make purchases at better prices. The Editor is anxious to be informed at once of any apparent errors in this list, as the desire is to make it perfectly accurate.

METALS.

ANTIMONY.

Hallett's..............per lb... 0 27½ 0 28

BOILER AND T K. PITTS.

Plain tinned / Spun ... 25 per cent. off list.

BABBIT METAL.

Canada Metal Company—Imperial genuine, 60c.; Imperial Tough, 66c.; White Brass 50c.; Metallic, 35c.; Harris Heavy Pressure, 25c.; Hercules, 25c.; White Bronze, 15c.; Star Frictionless, 14c.; Alluminoid, 10c.; No. 4, 9c per lb.

BRASS.

Rod and Sheet, 14 to 30 gauge, net list

Sheets, 12 to 14 in.	0
Tubing, base, per lb. 16 to 2 in	0
Tubing, 2 to 3-inch, iron pipe size..	0 27
" 1 to 3-inch, seamless	0 30

Copper tubing, 5 cents extra

COPPER.

	Ingot.	Per 100 lb.
Casting, car lots	25 50	26 00
	Bars.	
Cut lengths, round, ½ to 2 in.	....	32 00
	Sheet.	
Plain, 16 oz., 14x48 and 14x60	....	30 00
Plain, 14 oz.	....	31 00
Tinned copper sheet, base	....	33 00
Planished base	....	37 00
Braziers' (in sheets) 4x6 ft., 25 to 30 lb. each, per lb., base..	....	0 30

BLACK SHEETS.

	Montreal.	Toronto
8 to 10 gauge	2 60	2 60
12 gauge	2 60	2 65
14 "	2 60	2 60
17 "	2 40	2 50
18 "	2 40	2 55
20 "	2 40	2 53
22 "	2 40	2 70
24 "	2 45	2 75
26 "	2 55	2 85
28 "	2 60	3 00

CANADA PLATES.

Ordinary, 52 sheets	3 00
All bright "	4 00
Galvanized Canada Plates, 52 sheets	4 35
" " " 60 "	4 60

	Ordinary.	Dom. Crown.
18x24x52	4 25	4 35
" 60	4 50	4 60
20x28x80	8 50	8 70
" 94	9 00	9 20

GALVANIZED SHEETS.

	Fleur-de-Lis	Gordon Crown
16 to 20 gauge	3 60	3 95
22 to 24 gauge	3 95	4 00
26 "	4 30	4 20
28 "	4 45	4 50

Apollo.

10¾ oz. (American gauge)	4 70
28 gauge "	4 45
26 " "	4 15
24 " "	3 9

	Comet	Queen's Head.	Bell.
16 to 20 gauge	3 60	3 75	....
22 to 24 gauge	3 75	3 90	....
26 "	4 30	4 45	....
28 "	4 45	4 60	4 45

Less than case lots 10 to 25c. extra.

IRON AND STEEL.

	Montreal.	Toronto
Common bar, per 100 lb.	2 15	2 30
Forged iron "	2 40	
Refined " "	2 55	2 70
Horseshoe iron "	2 55	2 70
Hoop steel, 1½ to 3 in. base		2 80
Sleigh shoe steel "	2 25	2 30
Tire steel	2 40	2 50
Best sheet cast steel	...	0 12
B. K Morton "Alpha" high speed.	....	0 65
" annealed	....	0 70
"M" Self-hardening	....	0 50
"J" quality, best warranted	....	0 18
"I" " warranted	....	0 14
"B.C" quality	....	0 09
Jonas & Colver's tool steel..	0 10	0 20
" "Novo"	....	0 65
" " annealed	....	0 65
Jowett & Sons B P.L. tool steel	....	0 10½

INGOT TIN.

Lamb and Flag and Straits—56 and 28-lb. ingots, 100 lb. $46 50 $47 00

TINPLATES.

Charcoal Plates—Bright

	Per box.
M.L.S., equal to Bradley—	
I C, 14 x 20 base	$6 50
I X, 14 x 20 "	8 00
I XX, 14 x 20 base	9 50
Famous, equal to Bradley—	
I C, 14 x 20 base	6 50
I X, 14 x 20 "	8 00
I X X, 14 x 20 base	9 50
Raven and Vulture Grades—	
I C, 14 x 20 base	5 00
I X "	6 00
I X X "	7 00
I X X X "	8 00

"Dominion Crown Best"—Double Coated, Tissued.

	Per box.
I C, 14 x 20 base	5 75
I X, 14 x 20 "	6 75
I XX, 14 x 20 "	7 75

"Allaway's Best"—Standard Quality.

I C, 14 x 20 base	4 50
I X, 14 x 20 "	5 25
I XX, 14 x 20 "	6 00

Bright Cokes.

Bessemer Steel—	
I.C., 14 x 20 base	4 25
20x28, double box	8 50

Charcoal Plates—Terne

Dean or J. G. Grade—	
I.C., 20x28, 112 sheets	8 00
IX., Terne Tin	9 50

Charcoal Tin Boiler Plates.

Cookley Grade—
X X, 14x56, 50 sheet bxs. / " 14x60, " / " 14x65, " } 7 50

Tinned Sheets.

72x30 up to 24 gauge	8 50
" " 26 "	9 00

LEAD.

Imported Pig, per 100 lb.	5 40	5 50
Bar, "	5 75	6 00
Sheets, 2½ lb. sq. ft., by roll..	....	0 06½
Sheets, 3 to 6 lb "	....	0 06

NOTE.—Cut sheets ½c per lb., extra. Pipe, by the roll, usual weights per yard, lists at 7c. per lb. and 5 p.c. dis. f.o.b. Toronto.

NOTE.—Cut lengths, net price, waste pipe 8-ft lengths, lists at 8c.

SHEET ZINC.

5-cwt casks	8 00	8 25
Part casks	8 25	8 50

ZINC SPELTER.

Foreign, per 100 lb	7 25	7 50
Domestic	7 00	7 25

PLUMBING AND HEATING

BRASS GOODS, VALVES, ETC.

Standard Compression work, dis half inch 50 per cent., others 40 per cent.
Cushion work, discount 40 per cent.
Fuller work, ½-inch 60 per cent., others 50 p.c.
Flatway stop and stop and waste cocks, 50 per cent.; roundway, 45 per cent.
J.M.T. Globe, Angle and Check Valves, discount 45 per cent.
Standard Globe, Angle and Check Valves discount 52½ per cent.
Kerr standard globes, angles and checks, special, 42½ per cent., standard, 47½ p.c.
Kerr Jenkins' disc, copper-alloy disc and heavy standard valves, 45 per cent.
Kerr steam radiator valves 60 p.c., and quick-opening hot-water radiator valves, 60 p.c.
Kerr brass, Weber's straightway valves, 42½ per cent., straightway valves, I.B.B.M., 60 per cent
J. M. T. Radiator Valves, discount 50 per cent
Standard Radiator Valves, 60 per cent.
Patent Quick-Opening Valves, 65 per cent.
Jenkins' Bros. Globe Angle and Check valves discount 32½ per cent.

No. 1 compression bath cock......net	2 00
No. 4 " "	1 90
No 7 Fuller's "	2 25
No. 4½ " "	2 35

Patent Compression Cushion, basin cock, hot and cold, per doz., $16.20
Patent Compression Cushion, bath cock, No 2208 .... 2 25
Square head brass cocks, discount 50 per cent.
" " iron " " 60 "
Thompson Smoke-test Machine $25.00

BOILERS—COPPER RANGE.

Copper, 30 gallon	33 00
" 35 "	38 00
" 40 "	43 00

Discounts on new list.

BOILERS—GALVANIZED IRON RANGE.

Capacity	Standard.	Extra heavy
30-gallons	5.00	7.75
35 "	6 00	8 75
40 "	7 00	9.75

2 per cent., 30 days

BATH TUBS.

Steel clad copper lined, 15 per cent.

CAST IRON SINKS.

16x24, $1; 18x30, $1; 18x36, $1.30.

ENAMELED CLOSETS AND URINALS

Discount 15 per cent.

ENAMELED BATHS.

List issued by the Standard Ideal Company Jan. 3, 1907, shows an advance of 10 per cent over previous quotations.

ENAMELED LAVATORIES.

	1st quality	Special.
Plate E 100 to E 103	20 & 5 p c.	20 & 10 p c
" E 104 to E 132	20 & 10 p c.	30 & 2½ p c.

ENAMELED ROLL RIM SINKS.

	1st quality.	Special.
Plate E 201, one piece.	15 & 2½ p.c.	15 & 10 p.c.

ENAMELLED KITCHEN SINKS.

Plate E, flat iron 300, 60 & 10 p.c. 65 & 5 p.c.

LEAD PIPE

Lead Pipe, 7c. per pound, 5 per cent. off
Lead waste, 8c. per pound, 5 per cent. off.
Caulking lead, 6c. per pound.
Traps and bends, 40 per cent.

IRON PIPE.

Size (per 100 ft.)	Black.		Galvanized
¼ inch	2 25	¼ inch	3 08
⅜ "	2 25	⅜ "	3 08
½ "	2 72	½ "	3 57
¾ "	3 45	¾ "	4 60
1 "	4 95	1 "	6 60
1¼ "	6 75	1¼ "	9 00
1½ "	8 10	1½ "	10 80
2 "	10 80	2 "	14 40
2½ "	18 00	2½ "	24 00
3 "	23 50	3 "	31 50
3½ "	30 00	3½ "	40 00
4 "	34 00	4 "	45 00

2 per cent. 30 days.

Malleable Fittings—Canadian discount 30 per cent.; American discount 25 per cent.
Cast Iron Fittings 60; Standard bushings 60 per cent.; headers, 60; flanged unions 60, malleable bushings 55; nipples, up to 3 in., 65 and 10 per cent; up to 6 in., 65 per cent; malleable lipped unions, 55 and 5 per cent.

SOIL PIPE AND FITTINGS

Medium and Extra heavy pipe and fittings, up to 6 inch, discount 60 per cent.
7 and 8-in. pipe, discount 40 and 5 per cent.
Light pipe, 50 p.c.; fittings, 50 and 10 p.c.

OAKUM.

Plumbers ....per 100 lb..... 4 00 4 75

RADIATORS, ETC.

Hot water	47½ p.c.
Steam	50 p.c.
Wall Radiators and Specials	45 p.c.

STOVES, BOILERS, FURNACES, REGISTERS.

Discounts vary from 40 to 70 per cent. according to list.

STOCKS AND DIES.

American discount 25 per cent.

SOLDERING IRONS.

½-lb.	per lb.	0 37
1-lb. or over	"	0 34

SOLDER.

	Montreal	Per lb. Toronto
Bar, half-and-half, guaranteed	0 25	0 27
Wiping	0 22	0 23

PAINTS, OILS AND GLASS.

COLORS IN OIL.

Venetian red, 1-lb. tins, pure.	....	0 08
Chrome yellow	"	0 15
Golden ochre	"	0 08
French "	"	0 08
Marine black	"	0 04½
Chrome green	"	0 10
French permanent green	"	0 13
Signwriters' black	"	0 15

GLAZIER POINTS.

Discount, 5 per cent.

GLUE.

Domestic sheet	0 10	0 10½
French medal	0 12	0 14

OILS AND TURPENTINE.

W. H. & C's English raw oil, bbls.	0 63
" " boiled "	0 66

PARIS GREEN.

	Bergers	Canadian
600-lb casks	0 23½	0 23½
250 lb. drums	0 23½	0 23½
100-lb "	0 24	0 24
50-lb. "	0 24	0 21
25-lb. "	0 24½	0 24½
1-lb. pkgs, 100 in box	0 25	0 25
1-lb. " 50 "	....	0 25½
1-lb. tins, 100 "	0 2[illegible]	0 26
½-lb. pkgs	0 27	0 27

2 p. c. 30 days from date of shipment.

PREPARED PAINTS.

Pure, per gallon, in tin	....	1 20
Second qualities, per gallon	....	1 00
Barn (in bbls.)	0 60	0 90
Sherwin-Williams paints, ½ gal	....	1 50
" " ¼ gal.	....	1 45
" " gal.	....	1 40
Canada Paint Co.'s pure	....	1 30
Standard Co.'s "New Era"	....	1 35
Benj. Moore Co.'s "Ark" B'd	....	1 25
" British Navy deck	....	1 50
Brandram-Henderson "Anchor"	....	1 35
Ramsay's paints, Pure, per gal.	....	1 30
" Thistle, "	....	1 00
" Outside, bbls	0 55	0 65
Martin-Senour's 100 p.c. pure, ½ gal.	....	1 45
" " " " ¼ gal.	....	1 40
" " " " 1 gal.	....	1 35
Jamieson's "Crown and Anchor"	....	1 20
Jamieson's floor enamel	....	1 50
" barn paints, bbls per gal.	....	0 60
Senour's Floor Paints ... gal.	....	1 20
Sanderson Pearcy's pure	....	1 20
Robert on's pure paints	....	1 20

PUTTY.

Bulk in bbls.	1 50
Bulk in less quantity	1 80
Bladders in bbls	1 80
Bladders in kegs, boxes or loose	1 95
25-lb. tins	1 80
12½ lb. tins	2 05
Bladders in bulk or tins less than 100 lb.	1 85
Bulk in 100-lb. irons	1 80

DRY LEAD.

	white	red.
Genuine, 560 lb. casks, per cwt	6 75	6 00
Genuine, 100 lb. kegs, "	7 50	6 25
No. 1, 560 lb. casks, per cwt	....	5 75
No. 1, 100 lb. kegs, per cwt	....	6 00

SHELLAC.

Shellac, bleached, in bars or ground, 44c per lb, f.o.b., Eastern Canadian points

Shellac, bone dry, 57c per lb. f.o.b., Eastern Canadian Points; 63c, Toronto.

Shellac, T. N. orange, etc., 48c. net, Toronto

WHITE LEAD GROUND IN OIL. Per 100 lbs.

Pure	7 25	7 50
No. 1	6 90	7 15
No. 2	6 55	6 90
No. 3	6 30	6 55
Munro's Select Flake White	....	7 50
Elephant and Decorators Pure	....	7 25
Monarch	....	7 40
Standard Decorator's	....	7 15
Essex Genuine	....	6 80
Brandram's B. B. Genuine	....	8 25
" Anchor," pure	....	6 50
Ramsay's Pure Lead	....	6 40
Ramsay's Exterior	....	6 15
"Crown and Anchor," pure	....	6 50
Sanderson Pearcy's	....	7 40
Robertson's C P., lead	....	7 20
W H. & C's matured pure English	8 00	8 25

WINDOW GLASS.

Size United Inches.	Star	Double Diamond
Under 26	$4 25	$6 25
26 to 40	4 65	6 75
41 to 50	5 10	7 50
51 to 60	5 35	8 50
61 to 70	5 75	9 75
71 to 80	6 25	11 00
81 to 85	7 00	12 50
86 to 90	....	15 00
91 to 95	....	17 50
96 to 100	....	20 50
101 to 105	....	24 00
106 to 110	....	27 50

Discount off list, 25 per cent For broken boxes 40 off. All prices per 100 feet

ZINC—DRY WHITE.

Extra Red Seal	0 07	0 08
French V. M.	0 06	0 25
Lehigh	0 05	0 06

ZINC—GROUND WHITE.

Pure	0 08½	0 09½
No. 1	0 07	0 08
No. 2	0 05½	0 06½

VARNISHES

In 5-gal. lots.	Per gal.	Net.
Carriage, No. 1	1 50	1 60
Pale durable body	4 10	4 25
" rubbing	2 85	3 20
Gold size, japan	1 50	1 60
No. 1 brown japan	0 85	0 90
Elastic oak	....	1 50
Furniture, extra	1 10	1 25
" No. 1	0 90	1 00
Hard oil finish	1 35	1 50
Light oil finish	1 60	1 70
Damar	1 75	2 00
Brown japan	1 10	1 20
Black japan	1 10	1 20
" No. 1	0 85	0 90
Elastilite varnish, 1 gal. can, each		2 00
Granitine floor varnish, per gal		2 75
Maple Leaf coach enamels, size 1, size 2, 70c.; size 3, 40c. each		1.20
Sherwin-Williams kopal varnish, gal.		2.50
Canada Paint Co's sun varnish		2 00
"Kyanize" Interior Finish		2 40
"Flint-Lac," coach		1 80
"Gold Medal," per gal. in cases		2 00
Jamieson's "opaline, per gal.		2 00
Sanderson Pearcy's amberine	....	2 50

BUILDERS' HARDWARE.

BELLS.

Brass hand bells, 60 per cent.

Nickel, 55 per cent.

Gongs, Sargeant's door bells	5 50	8 00
Peterboro door bells, discount 37½ and 10 per cent. off new list.		
American, house bells, per lb.	0 35	0 40

BUILDING PAPER, ETC.

Tarred Felt, per 100 lb.		2 25
Ready roofing, 2-ply, not under 45 lb. per roll		0 95
Ready roofing, 3-ply, not under 65 lb., per roll		1 20
Carpet Felt	per ton	55 00
Heavy Straw Sheathing	per ton	35 00
Dry Surprise		0 40
Dry Sheathing	per roll, 400 sq. ft.	0 50
Tar "	" 400 "	0 45
Dry Fibre	" 400 "	0 55
Tarred Fibre	" 400 "	0 65
O. K. & I. X. L.	" 400 "	0 70
Resin-sized	" 400 "	0 45
Oiled Sheathing	" 600 "	1 00
Oiled "	" 400 "	0 70
Roof Coating, in barrels	per gal.	0 17
Roof " small packages	"	0 25
Refined Tar	per barrel	5 00
Coal Tar	"	4 00
Coal Tar, less than barrels	per gal	0 15
Roofing Pitch	per 100 lb. 0 80	0 90
Slater's felt	per roll	0 70
Heavy Straw Sheathing f. o. b. St. John and Halifax		37 50

BUTTS.

Wrought Brass, net revised list.

Wrought Iron, 70 per cent.

Cast Iron Loose Pin, discount 60 per cent

Wrought Steel Fast Joint and Loose Pin, 70 per cent.

CEMENT AND FIREBRICK.

"Lafarge" cement in wood		3 40
"Lehigh" cement, in wood		2 54
"Lehigh" cement, cotton sacks		2 39
"Lehigh" cement, paper sacks		2 31
Fire brick (Scotch) per 1,000	15 00	24 00
Fire clay (Scotch), net ton		4 95

Paving Blocks per 1 000.

Blue metallic, 9"x4½"x3", ex wharf	35 00
Stable pavers, 12"x6"x2", ex wharf	50 00
Stable pavers, 9"x4½"x ", ex wharf	36 00

DOOR SETS.

Peterboro, 37½ and 10 per cent

DOOR SPRINGS.

Torrey's Rod	per doz.	....	1 75
Coil, 9 to 11 in.	"	0 95	1 65
English	"	2 00	4 00

Chicago and Reliance Coil Door Springs, 25 per cent

STORE DOOR HANDLES

Per Dozen	1 00	1 50

ESCUTCHEONS.

Discount 50 and 10 per cent., new list

Peterboro, 37½ and 10 per cent.

ESCUTCHEON PINS.

Iron, discount 40 per cent.

Brass, 45 per cent.

HINGES.

Blind, discount 60 per cent.

Heavy T and strap, 4-in., per lb. net	0 06¼
" " 5-in., "	0 06
" " 6-in., "	0 05¾
" " 8-in., "	0 05½
" " 10-in. and larger	0 05¼

Light T and strap, discount 65 p.c.

Screw hook and hinge—

under 12 in. ... per 100 lb.	....	4 75
over 12 in. "	....	3 75
Spring, No. 20, per gro. pairs	....	10 80

Spring, Woodyatt pattern, per gro., No. 5, $17.50 No. 10, $18; No. 20, $10 80; No. 12 $20; No. 51, $10; No. 50, $27.50.

Crate hinges and back flaps, 65 and 5 p. c

Hinge hasps, 65 per cent.

SPRING HINGES

Chicago Spring Butts and Blanks 12½ per cent

Triple End Spring Butts, 40 and 5 per cent.

Chicago Floor Hinges, 40 and 5 off

Garden City Fire Engine House Hinges, 12½ per cent

HOOKS.

Cast Iron.

Bird cage	per doz.	0 50	1 10
Clothes line, No. 61	"	0 00	0 70
Harness	"	0 60	12 00
Hat and coat	per gro.	1 10	10 00
Chandelier	per doz.	0 50	1 00
Wrought hooks and staples—			
¼ x 5	per gross	....	2 65
5-16 x 5	"	....	3 30

Bright steel gate hooks and staples, 40 per cent. discount

Hat and coat wire discount 62½ per cent

Screw, bright wire, discount 62½ per cent.

KNOBS.

Door, japanned and N. P., doz	1 50	2 50
Bronze, Berlin ... per doz	2 75	3 25
Bronze, Genuine "	6 00	9 00
Shutter, porcelain, F. & L screw ... per gross	1 30	2 00
White door knobs ... per doz		2 00

Peterboro knobs, 37½ and 10 per cent

Porcelain mineral and jet knobs, net list.

KEYS.

Lock, Canadian dis. 40 to 40 and 10 per cent

Cabinet trunk and padlock American ... per gross	....	0 60

LOCKS.

Peterboro 37½ and 10 per cent

Russell & Erwin, steel rim $2 50 per doz

Eagle cabinet locks, discount 30 per cent

American padlocks, all steel, 10 to 15 per cent; all brass or bronze, 10 to 25 per cent.

SAND AND EMERY PAPER.

B. & A. sand, discount, 35 per cent

Emery, discount 35 per cent.

Garnet (Rurton's) 5 to 10 per cent. advance on list.

SASH WEIGHTS.

Sectional ... per 100 lb.	2 00	2 25
Solid "	1 50	1 75

SASH CORD.

Per lb.	....	0 31

BLIND AND BED STAPLES

All sizes, per lb.	0 07½	0 12

WROUGHT STAPLES.

Galvanized	....	2 75
Plain	....	2 50

Coopers', discount 45 per cent.

Poultry netting staples, discount 40 per cent.

Bright spear point, 75 per cent. discount

TOOLS AND HANDLES.

ADZES.

Discount 22½ per cent.

AUGERS.

Gilmour's, discount 60 per cent. off list.

AXES.

Single bit, per doz.	5 50	8 50
Double bit, "	10 00	11 00
Bench Axes, 40 per cent.		
Broad Axes, 25 per cent.		
Hunters' Axes	5 50	6 00
Boys' Axes	6 25	7 00
Splitting Axes	7 00	12 00
Handled Axes	7 00	9 00
Red Ridge, boys', handled	....	5 75
" hunters	....	5 25

BITS.

Irwin's auger, discount 47½ per cent.

Gilmour's auger, discount 60 per cent.

Rockford auger, discount 50 and 10 per cent.

Jennings' Gen. auger, net list.

Gilmour's car, 47½ per cent

Clark's expansive, 40 per cent.

Clark's gimlet, per doz	0 65	0 90
Diamond, Shell, per doz.	1 00	1 50
Nail and Spike, per gross	2 25	5 20

BUTCHERS CLEAVERS

German ... per doz.	6 00	9 00
American "	12 00	18 00

CHALK.

Carpenters' Colored, per gross	0 45	0 75
White lump ... per cwt.	0 60	0 65

CHISELS.

Warnock's, discount 72½ per cent.

P. S. & W. Extra, discount 72½ per cent

CROSSCUT SAW HANDLES.

S. & D., No. 3	per pair	0 15
S. & D., " 5	"	0 22½
S. & D., " 6	"	0 15
Boynton pattern	"	0 20

CROWBARS.

3½c to 4c. per lb

DRAW KNIVES

Coach and Wagon, discount 75 per cent.

Carpenters discount 75 per cent

DRILLS.

Millar's Falls, hand and breast net list.

North Bros, each set, 50c

DRILL BITS.

Morse, discount 37½ to 40 per cent.

Standard, discount 50 and 5 to 55 per cent.

FILES AND RASPS.

Great Western	75 per cent.
Arcade	75 "
Kearney & Foot	75 "
Disston's	75 "
American	75 "
J. Barton Smith	75 "
McClellan	7 "
Eagle	75 "
Nicholson	66⅔ "
Globe	75 "

Black Diamond, 60, 10 and 5 p.c.

Jowitt's, English list, 27½ per cent.

GAUGES.

Stanley's discount 50 to 60 per cent.

Wire Gauges

Winn's, Nos. 26 to 33 ... each	1 65	2 40

HANDLES.

C. & B., fork, 40 per cent., revised list.

C. & B., hoe, 40 per cent., revised list.

American, saw ... per doz	....	0 25
American, plane ... per gross	3 15	3 75

Canadian, hammer and hatchet 40 per cent

Axe and cant hook handles, 45 per cent.

HAMMERS.

Maydole's, discount 5 to 10 per cent.

Canadian, discount 25 to 27½ per cent.

Magnetic tack ... per doz.	1 10	1 20
Canadian sledge ... per lb.	0 07	0 08½
Canadian ball pean, per lb.	0 22	0 25

HATCHETS.

Canadian, discount 40 to 42½ per cent.

Shingle, Red Ridge 1, per doz	4 40
" " 2, "	4 85
Barrel Underhill	5 00

MALLETS.

Tinsmiths' ... per doz.	1 25	1 50
Carpenters', hickory, "	1 25	3 75
Lignum Vitae "	3 85	5 00
Caulking, each	0 60	2 00

MATTOCKS.

Canadian ... per doz.	5 50	6 00

MEAT CUTTERS.

German, 15 per cent.
American discount, 33⅓ per cent.
Gem ....each 1 15

NAIL PULLERS.

German and American 0 85 2 50
No. 1.......................... 0 85
No 1573.......................... 0 75

NAIL SETS.

Square, round and octagon, per gross 3 38
Diamond 1 00

PICKS.

Per dozen 6 00 9 00

PLANES.

Wood bench, Canadian discount 40 per cent. American discount 50 per cent.
Wood, fancy Canadian or American 37½ to 40 per cent.
Stanley planes, $1 55 to $3 60, net list prices.

PLANE IRONS.

English ....per doz. 2 00 5 00
Stanley, 2⅝ inch, single 24c., double 39c.

PLIERS AND NIPPERS.

Button's genuine, 37½ to 40 per cent.
Button's imitation....per doz. 5 00 9 00

PUNCHES.

Saddler's ....per doz. 1 00 1 85
Conductor's.......... " 3 00 15 00
Tinners', solid..........per set 0 72
" hollow......per inch 1 00

RIVET SETS.

Canadian, discount 35 to 37½ per cent.

RULES.

Boxwood, discount 70 per cent.
Ivory, discount 20 to 25 per cent.

SAWS.

Atkins, hand and crosscut, 25 per cent.
Disston's Hand, discount 12½ per cent
Disston's Crosscut ....per foot 0 35 0 55
Hack, complete..........each 0 75 2 75
" frame only..........each 0 50 1 25

S & D. solid tooth circular shingle, concave and band, 50 per cent; mill and ice, drag, 30 per cent; cross-cut, 35 per cent; hand saws, butcher, 35 per cent; buck, New Century $6.25; buck No. 1 Maple Leaf, $5.25; buck, Happy Medium, $4.25; buck, Watch Spring, $4.25, buck, common frame, $4.00.

Spear & Jackson's saws—Hand or rip, 26 in., $12.75; 24 in., $11.25, panel, 18 in., $8.25; 20 in., $9, tenon, 10 in., $9.90, 12 in., $10.90; 14 in., $11.50.

SAW SETS.

Lincoln and Whiting 4 75
Hand Sets, Perfect 4 00
X-Cut Sets, " 7 50
Maple Leaf and Premium saw sets, 40 off.
S & D saw swages, 40 off.

SCREW DRIVERS.

Sargent's ....per doz. 0 65 1 00
North Bros., No. 30 ..per doz. 16 80

SHOVELS AND SPADES

Bull Dog, solid neck shovel (No. 2 pol.) $18 50

(Hollow Back)		(Reinforced S Scoop)
Moose	$17 50	$16 30
Bear	15 00	15 30
Fox	12 50	14 30
Black Cat	10 00	13 30

Canadian, discount 45 per cent.

SQUARES.

Iron, discount 20 per cent.
Steel, discount 65 and 10 per cent.
Try and Bevel, discount 50 to 52½ per cent.

TAPE LINES.

English, ass skin .....per doz. 2 75 5 00
English, Patent Leather 5 50 9 75
Chesterman's ....each 0 90 2 85
" steel ....each 0 80 8 00

TROWELS.

Disston's, discount 10 per cent.
S. & D., discount 35 per cent.

FARM AND GARDEN GOODS

BELLS.

American cow bells, 63⅓ per cent.
Canadian, discount 45 and 50 per cent.
American, farm bells, each .. 1 35 3 00

BULL RINGS.

Copper; $1.30 for 2½-inch, and $1.70

CATTLE LEADERS.

Nos. 32 and 33 .. ...per gross 7 50 8 50

BARN DOOR HANGERS.

	doz.	pairs.
Steel barn door	8 00	10 00
Stearns wood track	4 50	6 00
Zenith	....	9 00
Acme, wood track	5 00	6 50
Atlas	5 00	6 00
Perfect	8 00	11 00
New Milo	....	6 50
Steel, covered	4 00	11 00
" track, 1 x 3-16 in (100 ft)	....	3 75
" " 1¼ x 3-16 in (100 ft)	....	4 75
Double strap hangers, doz. sets	....	6 40
Standard jointed hangers, "	....	6 40
Steel King hangers "	....	6 25
Storm King and safety hangers	....	7 00
" " rail	....	4 25

Chicago Friction, Oscillating and Big Twin Hangers, 5 per cent.

HARVEST TOOLS.

Discount 60 per cent.
S. & D. lawn rakes, Dunn's, 40 off.
" sidewalk and stable scrapers, 40 off.

HAY KNIVES.

Net prices.

HEAD HALTERS.

Jute Rope, ⅜-inch....per gross 9 00
" " ½ " .... " 10 00
" " ⅝ " .... " 12 00
Leather, 1-inch ......per doz. 4 00
Leather, 1¼ " ...... " 5 20
Web.................. " 2 45

HOES.

Garden, Mortar, etc., discount 60 per cent.
Planter.............. per doz. 4 00 4 50

HORSE NAILS.

'C' brand, 40, 10 and 7½ per cent. off list } Oval
M.R.M Co. brand, 55 per cent. } head

HORSESHOES.

F.O.B. Montreal
M R.M 'Co brand, base.................. 3 65
Add 15c Toronto, Hamilton, Guelph

HORSE WEIGHTS

Taylor-Forbes, 3½c. per lb

SCYTHES.

Per doz. net.................... 8 25 9 25

SCYTHE SNATHS.

Canadian, discount 40 per cent.

SNAPS.

Harness, German, discount 25 per cent.
Lock, Andrews' 4 50 11 00

STALL FIXTURES

Warden King, 35 per cent.

WOOD HAY RAKES.

Ten tooth, 40 and 10 per cent
Twelve tooth, 45 per cent.

—

HEAVY GOODS NAILS, ETC.

ANVILS.

Wright's, 80-lb. and over.............. 0 10½
Hay Budden, 80-lb. and over 0 09½
Brook's, 80-lb. and over 0 11½
Taylor-Forbes, handy................ 0 05

VISES.

Wright's...................... .... 0 13½
Brook's...................... .... 0 12½
Pipe Vise, Hinge, No. 1...... .... 3 50
" " " No. 2...... .... 5 50
Saw Vise 4 50 5 00
Blacksmiths' (discount) 60 per cent.
" parallel (discount) 45 per cent.

BOLTS AND NUTS

	Per cent.
Carriage Bolts, common ($1 list)	
" " ⅜ and smaller..	60, 10 and 10
" " 7-16 and up	55 and 5
" " Norway Iron ($3 list)	50
Machine Bolts, ⅜ and less	60 and 10
Machine Bolts, 7-16 and up....	55 and 5
Plough Bolts	55 and 10
Blank Bolts	55
Bolt Ends	55
Sleigh Shoe Bolts, ⅜ and less ..	60 and 10
" " " 7-16 and larger	50 and 5
Coach Screws, conepoint	70 and 5

Nuts, square, all sizes, 4c. per off.
Nuts, hexagon, all sizes, 4¼c. per off.
Stove Rods per lb., 5½ to 6c.
Stove Bolts, 75 per cent.

WROUGHT IRON WASHER.

Canadian make discount 40 per cent.

CHAIN.

Proof coil, per 100 lb., 5-16 in., $4.40; ⅜ in., $3.90; 7-16 in., $3.70; ½ in., $3.50; 9-16 in., $3.45; ⅝ in., $3.35; ¾ in., $3.20, ⅞ in., $3.10; 1 in., $3.10.
Halter, kennel and post chains, 40 to 40 and 5 per cent.
Cow ties .... 40 p.c.
Tie out chains .... 65 p.c.
Stall fixtures .... 35 p.c.
Trace chain .... 45 p.c.
Jack chain iron, discount 35 p.c.
Jack chain, brass, discount 40 per cent.

PRESSED SPIKES

Pressed spikes, ⅝ diameter, per 100 lbs., $3 15

NAILS.	Cut.	Wire.
2d	3 80	3 30
3d	2 95	2 95
4 and 5d	2 70	2 70
6 and 7d	2 60	2 60
8 and 9d	2 45	2 45
10 and 12d	2 40	2 40
16 and 20d	2 35	2 35
30, 40, 50 and 60d (base)	2 30	2 30

F.o.b. Montreal. Cut nails, Toronto 20c. higher.
Miscellaneous wire nails, discount 75 per cent
Coopers' nails, discount 40 per cent.

RIVETS AND BURRS

Iron Rivets, black and tinned, 60, 10 and 10
Iron Burrs, discount 60 and 10 and 10 p.c.
Copper Rivets, usual proportion burrs, 27½ p.c.
Copper Burrs only, discount 15 per cent.
Extras on Coppered Rivets, ½-lb packages 1c. per lb.; ¼-lb packages 2c. lb
Tinned Rivets, net extra, 4c. per lb.

SCREWS.

Wood, F. H., bright and steel, 87½ per cent.
" R. H., bright, dis. 82½ per cent
" F. H., brass, dis. 80 per cent.
" R. H., " dis. 75 per cent.
" F. H., bronze, dis. 75 per cent.
" R. H., " dis. 70 per cent.
Drive Screws, dis. 87½ per cent.
Bench, wood ....per doz. 3 25 4 00
" iron .... " 4 25 5 00
Set, case hardened, dis. 60 per cent
Square Cap, dis. 50 and 5 per cent.
Hexagon Cap, dis. 45 per cent.

MACHINE SCREWS.

Flat head, iron and brass, 35 per cent.
Felister head, iron, discount 30 per cent.
" " brass, discount 25 per cent

TACKS, BRADS, ETC.

Carpet tacks, blued .... 80 and 5
" " tinned............ 80 and 10
" " (in kegs).......... 40
Cut tacks, blued, in dozens only 75 and 10
" ½ weights 60
Swedes cut tacks, blued and tinned—
In bulk.......................... 80 and 10
In dozens 75
Swedes, upholsterers', bulk .... 85 and 12½
" brush, blued and tinned bulk 70
Swedes, gimp, blued, tinned and japanned........................ 75 and 12½
Zinc tacks 35
Leather carpet tacks.............. 40
Copper tacks...................... 37½
Copper nails 42½
Trunk nails, black 65
Trunk nails, tinned and blued .. 65
Clout nails, blued and tinned .. 65
Chair nails........................ 35
Patent brads...................... 40
Fine finishing.................... 40
Lining tacks, in papers.......... 10
" " in bulk............ 15
" " solid heads, in bulk 75
Saddle nails, in papers.......... 10
" " in bulk............ 15
Tufting buttons, 22 line in dozens only 60
Zinc glaziers' points 5
Double pointed tacks, papers.. 90 and 10
" " " bulk.... 40
Clinch and duck rivets.......... 45
Cheese box tacks, 85 and 5; trunk tacks, 80 and 10.

—

CUTLERY AND SPORTING GOODS.

AMMUNITION.

B. B. Caps Dominion, 50 and 5 and 25 per cent. American $2.00 per 1000.
C. B. Caps American, $2 60 per 1000

CARTRIDGES

Rim Fire Cartridges, Dominion, 50 and 5 p.c.
Rim Fire Pistol, discount 30 and 5 per cent., American.
Central Fire, Military and Sporting, American, add 10 per cent. to list. B.B. Caps, discount 40 per cent., American.
Central Fire Pistol and Rifle, list net Amer.
Central Fire Cartridges, pistol and rifle Dominion, 30 and 5 per cent.
Central Fire Cartridges, Sporting and Military, Dominion, 15 per cent., American 10 per cent. advance on list.
Loaded and empty Shells, Crown, 25 and 5; Sovereign, 25, 10 and 10; Regal, 25, 10 and 5; Imperial, 25, 10 and 5. American 20 per cent. discount. Rival and Nitro, 10 per cent. advance on list.
Empty paper shells Dominion, 25 off and empty brass shells, 55 per cent. off. American, 10 per cent. advance on list.
Primers, Dom., 35 per cent.; American $2 05

Wads. per lb.

Best thick brown or grey felt wads, in ½-lb. bags $0 70
Best thick white card wads, in boxes of 500 each, 12 and smaller gauges 0 29
Best thick white card wads in boxes of 500 each, 10 gauge............ 0 35
Thin card wads, in boxes of 1,000 each, 12 and smaller gauges 0 20
Thin card wads, in boxes of 1,000 each, 10 gauge...................... 0 25
Chemically prepared black edge grey cloth wads, in boxes of 250 each— Per M.
11 and smaller gauge 0 60
9 and 10 gauges.................. 0 70
7 and 8 " 0 90
5 and 6 " 1 10
Superior chemically prepared pink edge, best white cloth wads in boxes of 250 each—
11 and smaller gauge 1 15
9 and 10 gauges 1 40
7 and 8 " 1 65
5 and 6 " 1 90

SHOT.

Common, $6.50 per 100 lb.; chilled, $7.50 per 100 lb.; buck, seal and ball, $8.50 net list. Prices are f.o.b. Toronto, Hamilton, Montreal, St. John and Halifax. Terms, 5 p.c. for cash in thirty days.

RAZORS.	per doz.	
Elliot's	4 00	18 00
Boker's	7 50	11 00
" King Cutter	13 50	18 50
Wade & Butcher's	3 60	10 00
Wilkinson's	....	12 50
Lewis Bros.' "Klean Kutter"	8 60	10 50

Clauss Razors and Strops, 50 and 10 per cent

PLATED GOODS

Holloware, 40 per cent. discount.
Flatware, staples, 40 and 10, fancy 40 and 5 per cent.

SHEARS.

Clauss, nickel, discount 60 per cent.
Clauss, Japan, discount 67½ per cent.
Clauss, tailors, discount 40 per cent.
Seymour's, discount 50 and 10 per cent.

TRAPS (steel.)

Game, Newhouse, discount 30 and 10 per cent
Game, Hawley & Norton, 50, 10 & 5 per cent.
Game, Victor, 70 per cent
Game, Oneida Jump (B. & L.) 40 & 2½ p. c.
Game, steel, 60 and 5 per cent.

SKATES.

Skates, discount 37½ per cent.
Mic Mac hockey sticks, per doz 4 00 5 00
New Rex hockey sticks, per doz 6 25

HOUSE FURNISHINGS.

APPLE PARERS.

Woodyatt Hudson, per doz., net 4 50

BIRD CAGES.

Brass and Japanned, 40 and 10 p. c.

COPPER AND NICKEL WARE.

Copper boilers, kettles, teapots, etc. 30 p.c.
Copper pitts, 33⅓ per cent.

ENAMELLED WARE.

London, White, Princess, Turquoise, Onyx, Blue and White, discount 50 per cent.
Canada, Diamond, Premier, 50 and 10 p c.
Pearl, Imperial Crescent, 60 and 10 per cent.
Premier steel ware, 40 per cent.
Star decorated steel and white, 25 per cent.
Japanned ware, discount 50 per cent.
Hollow ware, tinned cast, 35 per cent. off.

KITCHEN NOVELTIES.

Can openers, per doz............ 0 40 0 75
Mincing knives per doz 0 50 0 80
Duplex mouse traps, per doz 0 65
Potato mashers, wire, per doz... 0 60 0 70
" wood " .. 0 50 0 60
Vegetable slicers, per doz 2 25
Universal meat chopper No. 0, $1; No 1, 1.15.
Enterprise chopper, each 1 30

LAMP WICKS.

Discount, 60 per cent.

LEMON SQUEEZERS.

Porcelain lined......	per doz.	2 20	5 60
Galvanized..........	"	1 87	3 85
King, wood..........	"	2 75	2 90
King, glass..........	"	4 00	4 50
All glass	"	0 50	0 90

PICTURE NAILS.

Porcelain head ......per gross 1 35 1 50
Brass head.......... " 0 40 1 00
Tin and gilt, picture wire, 75 per cent.

SAD IRONS.

Mrs. Potts, No. 55, polished....per set 0 80
" No. 50, nickle-plated, " 0 92
Common, plain......................... 4 50
" plated 5 50
Asbestos, per set.............. .... 1 25

TINWARE.

CONDUCTOR PIPE.

2 in. plain or corrugated., per 100 feet, $3 30; 3 in., $4 40; 4 in, $5 80; 5 in., $7.45; 6 in., $9.90.

FAUCETS.

Common, cork-lined, discount 35 per cent.

EAVETROUGHS.

10-inch ....per 100 ft. 3 30

FACTORY MILK CANS.

Discount off revised list, 40 per cent.
Milk can trimmings, discount 25 per cent.

LANTERNS.

No. 2 or 4 Plain Cold Blast....per doz. 6 50
Lift Tubular and Hinge Plain, " 4 75
Better quality at higher prices.
Japanning, 50c. per doz. extra.

OILERS.

Kemp's Tornado and McClary's Model galvanized oil can, with pump, 5 gallon, per dozen 10 00
Davidson oilers, discount 40 per cent
Zinc and tin, discount 50 per cent.
Coppered oilers, 20 per cent. off.
Brass oilers, 50 per cent, off.
Malleable, discount 25 per cent

PAILS (GALVANIZED).

Dufferin pattern pails, 40, 10 and 5 per cent.
Flaring pattern, discount 40, 10 and 5 per cent.
Galvanized washtubs 40, 10 and 5 per cent.

PIECED WARE.

Discount 40 per cent off list, June, 1899.
10-qt. flaring sap buckets, discount 40 per cent.
6, 10 and 14-qt. flaring pails dis. 40 per cent.
Copper bottom tea kettles and boilers, 35 p.c.
Creamery cans, discount 40 per cent.

STAMPED WARE.

Plain, 75 and 12½ per cent. off revised list.
Retinned, 72½ 10 and 5 per cent revised list.

SAP SPOUTS.

Bronzed iron with hooks ....per 1,000 7 50
Eureka tinned steel, hooks 8 00

STOVEPIPES.

5 and 6 inch, per 100 lengths 7 00
7 inch........ " " 7 50

STOVEPIPE ELBOWS

5 and 6-inch, common........per doz. 1 32
7-inch.......................... " 1 48
Polished, 15c. per dozen extra.

THERMOMETERS.

Tin case and dairy, 75 to 75 and 10 per cent

TINNERS' SNIPS.

Per doz...................... 3 00 15 00
Clauss, discount 35 per cent.

WIRE.

BRIGHT WIRE GOODS

Discount 62½ per cent.

CLOTHES LINE WIRE.

7 wire solid line, No. 17, $4.90; No. 18, $3.00; No. 19, $2.70; 6 wire solid line, No. 17, $4.45; No. 18, $2.80 No. 19, $2.50. All prices per 1000 ft. measure. F.o.b. Hamilton Toronto, Montreal.

COILED SPRING WIRE.

High Carbon, No. 9, $2 55, No. 11, $3.20; No. 12, $2 80.

COPPER AND BRASS WIRE

Discount 45 per cent.

FINE STEEL WIRE.

Discount 30 per cent List of extras: In 100-lb. lots: No. 17, $5 — No. 18, $5.50 — No. 19, $6 — No. 20, $6.65 — No. 21, $7 — No. 22, $7.30 — No. 23, $7.65 — No. 24, $8 — No. 25, $9—No. 26, $9 50—No. 27, $10—No. 28, $11—No. 29, $12—No. 30, $13—No. 31, $14—No. 32, $15—No. 33, $16—No. 34, $17. Extras net—tinned wire, Nos. 17-25, $2—Nos. 26-31, $4—Nos 32-34, $6. Coppered, 75c.—oiling, 10c.—in 25-lb. bundles, 15c.—in 5 and 10-lb. bundles, 25c.—in 1-lb. hanks, 25c. —in ½-lb. hanks, 38c.—in ¼-lb. hanks, 50c.—packed in casks or cases, 15c.—bagging or papering, 10c.

GALVANIZED WIRE.

Per 100 lb.—Nos. 4 and 5, $3 70 — Nos. 6, 7, 8, $3.15 — No. 9, $2.50 — No. 10, $3.20 — No. 11, $3.25—No. 12, $2 65 —No. 13, $2 75—No. 14, $3.75—No 15, $4.30 —No. 16, $4.30 from stock. Base sizes, Nos. 6 to 9, $2.35 f.o.b. Cleveland. In carlots 12½c. less.

LIGHT STRAIGHTENED WIRE.

Gauge No.	Over 20 in. per 100 lbs.	10 to 20 in.	5 to 10 in.
0 to 5	$0.50	$0 75	$1 25
6 to 9	0.75	1 25	2 00
10 to 11	1 00	1.75	2 50
12 to 14	1 50	2 25	3 50
15 to 16	2.00	3.00	4 50

SMOOTH STEEL WIRE.

No. 0-9 gauge, $2 25; No. 10 gauge, 6c. extra; No. 11 gauge, 12c extra; No. 12 gauge, 20c. extra; No. 13 gauge, 30c. extra; No 14 gauge, 40c. extra; No. 15 gauge, 55c. extra; No. 16 gauge, 70c extra Add 60c. for coppering and $2 for tinning.

Extra net per 100 lb.—Oiled wire 10c, spring wire $1.25, special hay baling wire 30c, best steel wire 75c., bright soft drawn 15c., charcoal (extra quality) $1.25, packed in casks or cases 15c., bagging and papering 10c., 50 and 100-lb. bundles 10c., in 25-lb. bundles 15c., in 5 and 10-lb. bundles 25c., in 1-lb hanks, 50c., in ½-lb. hanks 75c., in ¼-lb. hanks $1

POULTRY NETTING.

2 in mesh 19 w. g., discount 50 and 10 per cent. All others 50 per cent.

WIRE CLOTH

Painted Screen, in 100-ft rolls, $1 62½, per 100 sq. ft; in 50-ft. rolls $1 67½, per 100 sq. ft. Terms, 2 per cent. off 30 days.

WIRE FENCING.

Galvanized barb...... .... 2 95
Galvanized, plain twist 3 30
Galvanized barb, f.o.b. Cleveland, $2 70 for small lots and $2 60 for carlot.

WOODENWARE.

CHURNS

No. 0, $9; No. 1, $9; No. 2, ; No. 3, $11; No. 4, $13; No. 5, $16; f.o.b. Toronto Hamilton, London and St. Marys, 30 and 30 per cent; f.o.b Ottawa, Kingston and Montreal, 40 and 15 per cent discount, Taylor-Forbes, 30 and 30 per cent.

CLOTHES REELS.

Davis Clothes Reels, dis. 40 per cent.

LADDERS, EXTENSION.

Waggoner Extension Ladders, dis. 40 per cent.

MOPS AND IRONING BOARDS.

"Best" mops........................ 1 25
"900" mops.......................... 1 25
Folding ironing boards......... 12 00 16 50

REFRIGERATORS

Discount, 40 per cent.

SCREEN DOORS.

Common doors, 2 or 3 panel, walnut stained, 4-in. style..........per doz. 7 25
Common doors, 2 or 3 panel, grained only, 4-in., style ....per doz. 7 55
Common doors, 2 or 3 panel, light stair per doz.......................... 9 55

WASHING MACHINES.

Round, re-acting per doz. 60 00
Square " 63 00
Eclipse, per doz 54 00
Dowswell " 39 00
New Century, per doz 75 00
Daisy................................ 54 00

WRINGERS.

Royal Canadian, 11 in., per doz. ... 34 00
Royal American, 11 in. ... 34 00
Eze, 10 in., per doz 37 75
Terms, 2 per cent., 30 days.

MISCELLANEOUS.

AXLE GREASE.

Ordinary, per gross 6 00 7 00
Best quality 10 00 12 00

BELTING.

Extra, 60 per cent.
Standard, 60 and 10 per cent.
No. 1, not wider than 6 in., 60, 10 and 10 p c.
Agricultural, not wider than 4 in., 75 per cent.
Lace leather, per side, 75c; cut laces, 80c.

BOOT CALKS.

Small and medium, ball ....per M 4 25
Small heel .... " 4 50

CARPET STRETCHERS.

American ....per doz. 1 00 1 50
Bullard's............ " 6 50

CASTORS.

Bed, new list, discount 55 to 57½ per cent.
Plate, discount 52½ to 57½ per cent.

PINE TAR.

½ pint in tins ....per gross ... 7 80
" " .... " 9 60

PULLEYS.

Hothouse	per doz.	0 55	1 00
Axle	"	0 22	0 33
Screw	"	0 22	1 00
Awning	"	0 35	2 50

PUMPS.

Canadian cistern 1 40 2 00
Canadian pitcher spout ... 1 80 3 16

ROPE AND TWINE.

Sisal..................................		0 10½
Pure Manilla		0 15½
"British" Manilla..................		0 12
Cotton, 3-16 inch and larger...	0 21	0 23
" 5-32 inch	0 25	0 27
" ⅛ inch	0 25	0 28
Russia Deep Sea		0 16
Jute...................................		0 09
Lath Yarn, single		0 10
" double		0 10½
Sisal bed cord, 48 feet........per doz.		0 65
" " 60 feet........ "		0 80
" " 72 feet........ "		0 95

Twine.

Bag, Russian twine, per lb.		0 27
Wrapping, cotton, 3-ply		0 25
" " 4-ply		0 29
Mattress twine per lb.	0 33	0 45
Staging " "	0 27	0 35

SCALES.

Gurney Standard, 40 per cent.
Gurney Champion, 50 per cent.
Burrow, Stewart & Milne—
Imperial Standard, discount 40 per cent.
Weigh Beams, discount 40 per cent.
Champion Scales, discount 50 per cent.
Fairbanks standard, discount 35 per cent.
" Dominion, discount 55 per cent.
" Richelieu, discount 55 per cent.
Warren new Standard, discount 40 per cent.
" " Champion, discount 50 per cent.
" Weighbeams, discount 35 per cent.

STONES—OIL AND SCYTHE.

Washita	per lb.	0 25	0 37
Hindostan	"	0 06	0 10
" slip	"	0 18	0 20
" Axe..........	"	....	0 10
Deer Creek	"	....	0 10
Deerlick	"	....	0 25
" Axe	"	....	0 15
Lily white	"	....	0 42
Arkansas..............	"	....	1 50
Water-of-Ayr	"	....	0 10
Scythe	per gross	3 50	5 00
Grind, 40 to 200 lb	per ton....	20 00	22 00
" under 40 lb.,	"	....	24 00
" 200 lb and over	....	....	28 00

KEARNEY & FOOT

The Best Rasp Made

The Rasp that sells

MADE BY

Nicholson File Co., Dominion Wks. **Port Hope, Ont.**

Wrought Iron Pipe and Cast Iron Fittings

Our Stock in these lines is complete. Can fill all orders promptly. Be supplied before the Fall rush commences.

WM. STAIRS, SON & MORROW, Limited, HALIFAX, N.S.

INDEX TO ADVERTISERS

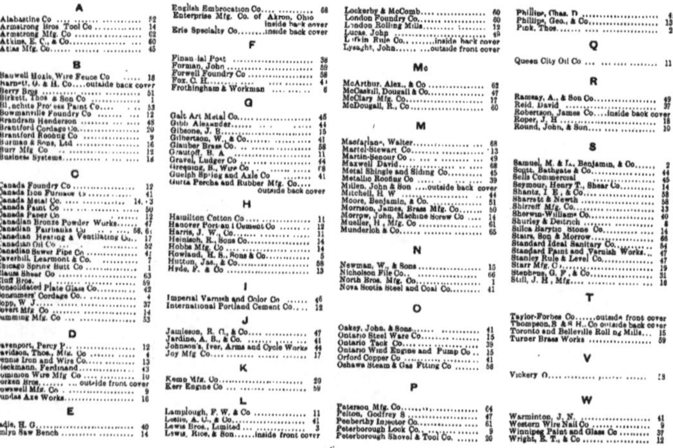

CLASSIFIED LIST OF ADVERTISEMENTS.

Auditors.
Davenport, Percy P., Winnipeg.

Babbitt Metal.
Canada Metal Co., Toronto.
Canadian Fairbanks Co., Montreal.
Robertson, Jas. Co., Montreal.

Bath Room Fittings.
Carriage Mounting Co., Toronto.

Belting, Hose, etc.
Gutta Percha and Rubber Mfg. Co. Toronto.

Bicycles and Accessories.
Johnson's, Iver, Arms and Cycle Works Fitchburg, Mass

Binder Twine.
Consumers Cordage Co, Montreal.

Box Strap.
J. N. Warminton, Montreal.

Brass Goods.
Glauber Brass Mfg. Co., Cleveland, Ohio.
Lewis, Rice, & Son., Toronto.
Morrison, Jas., Brass Mfg. Co., Toronto.
Mueller Mfg Co., Decatur, Ill.
Penberthy Injector Co., Windsor, Ont.
Taylor-Forbes Co., Guelph, Ont.

Bronze Powders.
Canadian Bronze Powder Works, Montreal.

Brushes.
Ramsay, A., & Son Co., Montreal.

Can Openers.
Cumming Mfg. Co. Renfrew

Cans.
Acme Can Works, Montreal

Builders' Tools and Supplies.
Covert Mfg. Co., West Troy, N.Y.
Frothingham & Workman Co., Montreal.
Howland, H. S., Sons & Co., Toronto.
Hyde, F., & Co., Montreal.
Lewis Bros. & Co., Montreal.
Lewis, Rice, & Son, Toronto.
Lockerby & McComb, Montreal.
Lufkin Rule Co., Saginaw, Mich.
Newman & Sons, Birmingham.
North Bros. Mfg. Co., Philadelphia, Pa.
Stanley Rule & Level Co., New Britain.
Stanley Works, New Britain, Conn.
Stephens, G. F., Winnipeg.
Taylor-Forbes Co., Guelph, Ont.

Carriage Accessories.
Carriage Mountings Co., Toronto.
Covert Mfg. Co., West Troy, N.Y.

Carriage Springs and Axles.
Guelph Spring and Axle Co., Guelph.

Cattle and Trace Chains.
Greening, B., Wire Co., Hamilton.

Churns.
Dowswell Mfg. Co., Hamilton.

Clippers—All Kinds.
American Shearer Mfg. Co., Nashua, N.H.

Clothes Reels and Lines.
Hamilton Cotton Co., Hamilton, Ont.

Cordage.
Consumers' Cordage Co., Montreal.
Hamilton Cotton Co., Hamilton.

Cork Screws.
Erie Specialty Co., Erie, Pa.

Clutch Nails.
J. N. Warminton Montreal.

Cut Glass.
Phillips, Geo., & Co., Montreal.

Cutlery—Razors, Scissors, etc.
Birkett, Thos., & Son Co., Ottawa.
Clauss Shear Co., Toronto
Dorken Bros. & Co., Montreal.
Heinisch's, R., Sons Co., Newark, N.J.
Howland, H. S. Sons & Co., Toronto.
Phillips, Geo., & Co., Montreal.
Round, John, & Son, Montreal.

Electric Fixtures.
Canadian General Electric Co., Toronto.
Forman, John, Montreal.
Morrison James, Mfg. Co., Toronto.
Munderloh & Co., Montreal.

Electro Cabinets.
Cameron & Campbell Toronto.

Engines, Supplies, etc.
Kerr Engine Co., Walkerville, Ont.

Files and Rasps.
Barnett Co., G. & H., Philadelphia, Pa.
Nicholson File Co., Port Hope

Financial Institutions
Bradstreet Co.

Firearms and Ammunition.
Dominion Cartridge Co., Montreal.
Hamilton Rifle Co., Plymouth, Mich.
Harrington & Richardson Arms Co., Worcester, Mass.
Johnson's, Iver, Arms and Cycle Works, Fitchburg, Mass.

Food Choppers.
Enterprise Mfg. Co., Philadelphia, Pa.
Shirreff Mfg Co., Brockville, Ont.

Galvanizing.
Canada Metal Co., Toronto.
Montreal Rolling Mills Co., Montreal.
Ontario Wind Engine & Pump Co., Toronto.

Glaziers' Diamonds.
Gibsone, J. B., Montreal.
Pelton, Godfrey S.
Sharratt & Newth, London, Eng.
Shaw, A., & Son, London, Eng.

Handles.
Still, J. H., Mfg. Co.

Harvest Tools.
Maple Leaf Harvest Tool Co., Tillsonburg Ont.

Hoop Iron.
Montreal Rolling Mills Co., Montreal.
J. N. Warminton, Montreal

Horse Blankets.
Heney, E. N., & Co., Montreal.

Horseshoes and Nails.
Canada Horse Nail Co., Montreal
Montreal Rolling Mills, Montreal.

Hot Water Boilers and Radiators.
Cluff, R. J., & Co., Toronto
Pease Foundry Co., Toronto.
Taylor-Forbes Co., Guelph.

Ice Cream Freezers.
Dana Mfg. Co., Cincinnati, Ohio.
North Bros. Mfg. Co., Philadelphia, Pa.

Ice Cutting Tools.
Erie Specialty Co., Erie, Pa.
North Bros. Mfg. Co., Philadelphia, Pa.

Injectors—Automatic.
Morrison, Jas., Brass Mfg. Co., Toronto.
Penberthy Injector Co., Windsor, Ont.

Iron Pipe.
Montreal Rolling Mills, Montreal.

Iron Pumps.
McDougall, R., Co., Galt, Ont.

Lanterns.
Kemp Mfg. Co., Toronto.
Ontario Lantern Co., Hamilton, Ont.
Wright, E. T., & Co., Hamilton.

Lawn Mowers.
Birkett, Thos., & Son Co., Ottawa.
Maxwell, D., & Sons, St. Mary's, Ont.
Taylor, Forbes Co., Guelph.

Lawn Swings, Settees, Chairs.
Cumming Mfg. Co., Renfrew.

Ledgers—Loose Leaf.
Business Systems, Toronto.
Copeland-Chatterson Co., Toronto.
Crain, Rolla L., Co., Ottawa.
Universal Systems, Toronto.

Locks, Knobs, Escutcheons, etc.
Peterborough Lock Mfg. Co., Peterborough, Ont.

Lumbermen's Supplies.
Pink, Thos., & Co., Pembroke Ont.

Manufacturers' Agents.
Fox, C. H., Vancouver.
Gibb, Alexander, Montreal.
Mitchell, David C., & Co., Glasgow, Scot.
Mitchell, H. W., Winnipeg.
Pearce, Frank, & Co. Liverpool, Eng.
Scott, Bathgate & Co., Winnipeg.
Thorne, R. E., Montreal and Toronto.

Metals.
Canada Iron Furnace Co., Midland, Ont.
Canada Metal Co., Toronto.
Eadie, H. G., Montreal.
Gibb, Alexander, Montreal.
Kemp Mfg. Co., Toronto
Leslie, A. C., & Co., Montreal.
Lysaght, John, Bristol, Eng.
Nova Scotia Steel and Coal Co., New Glasgow, N.S.
Robertson, Jas., Co., Montreal
Roper, J. H., Montreal
Samuel, Benjamin & Co., Toronto.
Stairs, Son & Morrow, Halifax, N.S.
Thompson, B. & S. H. & Co. Montreal.

Metal Lath.
Galt Art Metal Co., Galt
Metallic Roofing Co., Toronto.
Metal Shingle & Siding Co., Preston, Ont.

Metal Polish, Emery Cloth, etc
Oakey, John, & Sons, London, Eng.

Mops.
Cumming Mfg. Co., Refrew.

Mouse Traps.
Cumming Mfg. Co., Renfrew

Oil Tanks
Bowser, S. F., & Co., Toronto.

Paints, Oils, Varnishes, Glass.
Bell, Thos., Sons & Co., Montreal.
Canada Paint Co., Montreal.
Canadian Oil Co., Toronto
Consolidated Plate Glass Co., Toronto.
Fenner, Fred., & Co., London, Eng.
Henderson & Potts Co., Montreal.
Imperial Varnish and Color Co., Toronto.
Jamieson, R. C., & Co., Montreal.
McArthur, Corneille & Co., Montreal.
McCaskill, Dougall & Co., Montreal.
Montreal Rolling Mills Co., Montreal.
Moore, Benjamin, & Co. Toronto.
Queen City Oil Co., Toronto.
Ramsay & Son, Montreal.
Sherwin-Williams Co., Montreal.
Standard Paint and Varnish Works Windsor, Ont.
Stephens & Co., Winnipeg.
Martin-Senour Co., Chicago.

Perforated Sheet Metals.
Greening, B., Wire Co., Hamilton.

Plumbers' Tools and Supplies.
Bordan Co., Warren, Ohio.
Canadian Fairbanks Co., Montreal.
Cluff, R. J., & Co., Toronto
Glauber Brass Co., Cleveland, Ohio.
Jardine, A. B., & Co., Hespeler, Ont.
Jenkins Bros., Boston, Mass.
Lewis, Rice, & Son, Toronto.
Merrell Mfg. Co., Toledo, Ohio.
Montreal Rolling Mills Montreal.
Morrison, Jas., Brass Mfg. Co., Toronto.
Mueller, H., Mfg. Co., Decatur, Ill.
Oshawa Steam & Gas Fitting Co., Oshawa
Robertson Jas., Co. Montreal.
Stairs, Son & Morrow, Halifax, N.S.
Standard Ideal Sanitary Co., Port Hope,
Standard Sanitary Co., Pittsburg
Stephens, G. F., & Co., Winnipeg, Man.
Turner Brass Works, Chicago.
Vickery, Orlando, Toronto.

Portland Cement.
International Portland Cement Co., Ottawa, Ont
Hanover Portland Cement Co., Hanover, Ont.
Hyde, F., & Co., Montreal.
Thompson, B. & S. H. & Co., Montreal.

Potato Mashers.
Cumming Mfg Co., Renfrew.

Poultry Netting.
Greening, B., Wire Co., Hamilton, Ont.

Razors.
Clauss Shear Co., Toronto.

Roofing Supplies.
Brantford Roofing Co., Brantford.
McArthur, Alex., & Co., Montreal.
Metal Shingle & Siding Co., Preston, Ont.
Metallic Roofing Co., Toronto.
Paterson Mfg. Co., Toronto & Montreal.

Saws.
Atkins, E. C., & Co., Indianapolis, Ind
Lewis Bros., Montreal.
Shurly & Dietrich, Galt, Ont
Spear & Jackson, Sheffield, Eng.

Scales.
Canadian Fairbanks Co., Montreal.

Screw Cabinets.
Cameron & Campbell. Toronto

Screws, Nuts, Bolts.
Montreal Rolling Mills Co., Montreal.
Morrow, John, Machine Screw Co., Ingersoll, Ont

Sewer Pipes.
Canadian Sewer Pipe Co., Hamilton
Hyde, F., & Co., Montreal.

Shelf Boxes.
Cameron & Campbell, Toronto

Shears, Scissors.
Clauss Shear Co., Toronto.

Shelf Brackets.
Atlas Mfg. Co., New Haven, Conn

Shellac
Bell, Thos., Sons & Co., Montreal.

Shovels and Spades.
Canadian Shovel Co., Hamilton.
Peterboro Shovel & Tool Co., Peterboro.

Silverware.
Phillips, Geo., & Co., Montreal.
Round, John, & Son, Sheffield, Eng

Spring Hinges, etc.
Chicago Spring Butt Co., Chicago, Ill.

Steel Rails.
Nova Scotia Steel & Coal Co., New Glasgow, N.S.

Stoves, Tinware, Furnaces
Canadian Heating & Ventilating Co., Owen Sound.
Canada Stove Works, Harriston, Ont.
Clare Bros. & Co., Preston.
Davidson, Thos., Mfg. Co., Montreal,
Guelph Stove Co., Guelph.
Gurney Foundry Co., Toronto.
Harris, J. W., Co., Montreal.
Joy Mfg. Co., Toronto.
Kemp Mnfg. Co. Toronto.
McClary Mfg Co. London.
Pease Foundry Co., Toronto.
Stewart, Jas., Mfg. Co., Woodstock, Ont.
Taylor-Forbes Co., Guelph, Ont.
Wright, E. T., & Co., Hamilton.

Tacks.
Montreal Rolling Mills Co., Montreal.
Ontario Tack Co., Hamilton.

Ventilators.
Harris, J. W., Co., Montreal.
Pearson, Geo. D., Montreal.

Washing Machines, etc
Dowswell Mfg Co., Hamilton, Ont.
Taylor-Forbes Co., Guelph, Ont.

Wheelbarrows
London Foundry Co., London, Ont.

Wholesale Hardware.
Birkett, Thos., & Sons Co., Ottawa.
Caverhill, Learmont & Co., Montreal.
Frothingham & Workman, Montreal.
Hobbs Hardware Co., London.
Howland, H. S., Sons & Co., Toronto.
Lewis Bros. & Co., Montreal.
Lewis, Rice, & Son, Toronto.

Window and Sidewalk Prisms.
Hobbs Mfg. Co., London, Ont.

Wire Springs.
Guelph Spring Axle Co., Guelph, Ont.
Wallace-Barnes Co., Bristol, Conn.

Wire, Wire Rope, Cow Tics, Fencing Tools, etc.
Canada Fence Co., London
Dennis Wire and Iron Co., London, Ont.
Dominion Wire Mnfg. Co., Montreal
Greening, B., Wire Co., Hamilton.
Montreal Rolling Mills Co., Montreal.
Western Wire & Nail Co., London, Ont

Woodenware.
Taylor-Forbes Co., Guelph, Ont.

Wrapping Papers.
Canada Paper Co., Toronto.
McArthur, Alex., & Co., Montreal.
Stairs, Son & Morrow, Halifax, N.S.

FAVORITE CHURN

More sold in the Dominion than all others combined

In 8 sizes, churning from ½ to 30 gallons.
IMPROVED STEEL FRAME. PATENT FOOT and LEVER DRIVE.
PATENT STEEL ROLLER BEARINGS

Easy to operate. Bolts throughout in place of wood screws. Superior in workmanship and finish. **No other just as good. Accept no substitute.** If not offered by your jobber write direct to us.

MAXWELL LAWN MOWER

Made in 4 different heights of wheel and widths, ranging from 8 to 25-in. in plain or ball bearings as required. Improved for coming season. Not equalled anywhere for clean cutting, durability and finish.

DAVID MAXWELL & SONS
ST. MARY'S, ONT.

AGENTS
W. PORTEOUS JACK, 1256 Robson, Vancouver
GEO. PEATT & CO., P.O. Box 466, Winnipeg, Man.
BAXTER, PATERSON & CO., 102 St. Antoine St., Montreal.

Trade Mark

"LION" BRAND

PUTZ TRIPLE EXTRACT

METAL POLISH

LIQUID and PASTE

For polishing all kinds of metals "Lion" Brand Putz Triple Extract Metal Polish has no equal. Gives a lasting lustre. Put up in liquid and paste form. Has a ready sale. If not in stock send a postal for exceptional terms. Something that will give you handsome profits.

English Embrocation Co.,
Montreal, Quebec

Perforated Sheet Metals

IN

BRASS, COPPER, STEEL, Etc.

ALL SIZES OF PERFORATIONS AND THICKNESS OF METALS

FOR

MINERS' USE
GRAIN-CLEANING MACHINERY
BEE KEEPERS
MALT KILN FLOORS, Etc.

THE B. GREENING WIRE CO.
LIMITED
HAMILTON, Ont. MONTREAL, Que.

THURBER

Babbitt Metal

IS CONCEDED BY EXPERTS TO BE THE

BEST BABBITT

For Heavy Slow Machinery
Heavy Line Shafting, etc.

12c. lb. TRY IT

The James Robertson Co., Limited
144 William St., MONTREAL, QUE.

PFLUEGER'S
Fishing Tackle

Hooks, Flies, Trolls, Spinners, Phantoms, Reels, Furnished Lines.

Everything in Fishing Tackle.

NOTICE—Free to Any Dealer in Sporting Goods, Sent Express Prepaid, 170-Page Illustrated Catalogue No. F25, and Metal Fish Sign in 8-Color Lithograph.

The Enterprise Mfg. Co.
AKRON, Ohio, U.S.A.

WALKER CORK SCREWS

are good holiday souvenirs and live sellers the whole year. Are made well and are made out of good material. That is why we guarantee every one that we put out. When ordering Cork Screws try **Walker's**

Manufactured only by ERIE SPECIALTY COMPANY, ERIE, Pa. U.S.A.

LUFKIN MEASURING TAPES

Steel, Metallic, Linen, Pocket, Ass Skin, Pat. Leather, Bend Leather, Etc.

ARE THE BEST AND MOST POPULAR TAPES IN THE WORLD.
YOUR STOCK IS NOT COMPLETE WITHOUT THEM.

LUFKIN RULE CO., Saginaw, Mich, U.S.A.

London Office and Warehouse—24 and 26 Holborn. New York City Branch—280 Broadway.

For sale by ALL PROMINENT CANADIAN HARDWARE JOBBERS.

"Redstone"

High Pressure

Sheet Packing

A packing that will hold. For use in highest pressures for steam, hot or cold water and air. Packs equally well for all.

From actual tests, we believe that this packing is the most durable and satisfactory of any on the market. Try a sample lot and see for yourself.

Manufactured Solely by

THE GUTTA PERCHA & RUBBER MFG. CO.

of TORONTO, LIMITED

HEAD OFFICES,

47 Yonge Street, Toronto.

Branches: Montreal, Winnipeg, Vancouver.

SHEET ZINC,

SHEET COPPER,

SPELTER,

INGOT TIN,

BRASS and COPPER TUBES.

For lowest prices send your enquiries to

B. & S. H. THOMPSON & Co.

LIMITED

53 ST. SULPICE STREET, MONTREAL

SHELBY SEAMLESS STEEL TUBES

INVALUABLE IN

- General Machine Shop Work
- Bicycle Repairs
- Hydraulic Work
- Boiler Work

We carry the largest stock of steel tubes in the Dominion. We can ship promptly almost any size or quantity called for.

CANADIAN DISTRIBUTORS

JOHN MILLEN & SON, LTD.

321 St. James St.
MONTREAL

132 Bay Street
TORONTO

CIRCULATES EVERYWHERE IN CANADA
Also in Great Britain, United States, West Indies, South Africa and Australia.

HARDWARE AND METAL

A Weekly Newspaper Devoted to the Hardware, Metal, Heating and Plumbing Trades in Canada.

Office of Publication, 10 Front Street East, Toronto.

VOL. XIX. MONTREAL, TORONTO, WINNIPEG, JANUARY 26, 1907 NO. 4.

See Classified List of Advertisements on Page 71.

SPORTING GOODS

TARGET RIFLES

Gun Cases

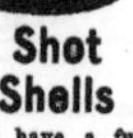

Shot Shells

We have a full stock of shells loaded with Smokeless or Black Powder.

SHELLS

STEVENS RIFLES

SAVAGE RIFLES

Write for Prices

RICE LEWIS & SON

LIMITED

TORONTO.

Wouldn't you, Mr. Store-keeper, be worth more if you could do 100 per cent. more work?

That means that you can do ten hours work in five, because you can enter up your accounts and make up your customer's bill with one writing. You can do your york better, more quickly, and more accurately by the use of

THE CRAIN CONTINUOUS SYSTEMS

than by the old out-of-date methods. Then, again, the Crain binders are more durable, and better made than any other, and do not have metal parts protruding to scratch your desks. They will hold one or a thousand sheets, as you wish.

Drop us a card about any system you are interested in.

The Rolla L. Crain Co., Limited

OTTAWA, CANADA

TORONTO: 18 Toronto Street
MONTREAL: Alliance Bldg., 107 St. James St.
WINNIPEG: 11 Nanton Block

WE CARRY IN STOCK

Horse-shoers' Clip Horn Anvils
Prospectors' Anvils and Bellows
Prospectors' Picks, Shovels and Stoves
The Famous Blacksmith Blower No. 400
The Grand Bay Blower No. 100
Challenge Forges with Hood, Extra Strong Frame
Hero Forges With and Without Hood
Navy Wheelbarrows with Steel or Wood Trays
Steel Pulley Blocks, Single, Double or Triple

Write for prices. Prompt shipment.

THOS. BIRKETT & SON CO., LIMITED

Wholesale Hardware

OTTAWA, ONT.

Other Tools are very Good Tools, but **"YANKEE TOOLS"** ARE BETTER

The NEWEST, CLEVEREST and QUICKEST SELLING TOOLS of the KIND.

SCREW DRIVERS

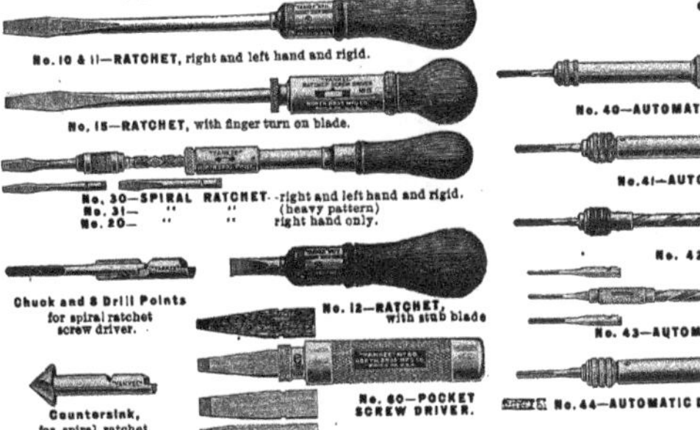

No. 10 & 11—RATCHET, right and left hand and rigid.

No. 15—RATCHET, with finger turn on blade.

No. 30—SPIRAL RATCHET—right and left hand and rigid.
No. 31— " " (heavy pattern)
No. 20— " " right hand only.

Chuck and 8 Drill Points for spiral ratchet screw driver.

No. 12—RATCHET, with stub blade

No. 60—POCKET SCREW DRIVER.

Countersink, for spiral ratchet screw driver.

DRILLS

No. 40—AUTOMATIC DRILL, with Ratchet Movement.

No. 41—AUTOMATIC DRILL.

No. 42—AUTOMATIC DRILL.

No. 43—AUTOMATIC DRILL, for light drills only.

No. 44—AUTOMATIC DRILL, with adjustable tension on spring.

No. 50—RECIPROCATING DRILL, for wood or metals.

SOLD BY LEADING JOBBERS

SEND FOR OUR NEW "YANKEE" TOOL BOOK

NORTH BROS. MFG. CO.

PHILADELPHIA, Pa.

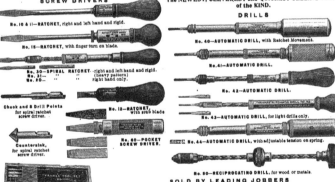

Pruning Saws and Shears

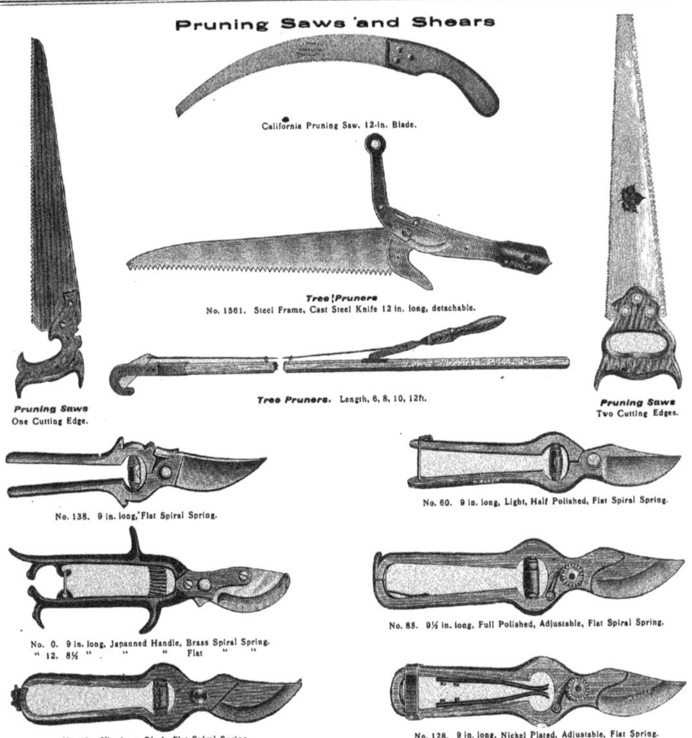

California Pruning Saw. 12-in. Blade.

Tree Pruners
No. 1561. Steel Frame, Cast Steel Knife 12 in. long, detachable.

Tree Pruners. Length, 6, 8, 10, 12ft.

Pruning Saws
One Cutting Edge.

Pruning Saws
Two Cutting Edges.

No. 138. 9 in. long, Flat Spiral Spring.

No. 60. 9 in. long, Light, Half Polished, Flat Spiral Spring.

No. 0. 9 in. long, Japanned Handle, Brass Spiral Spring.
" 12. 8½ " " " Flat "

No. 85. 9½ in. long, Full Polished, Adjustable, Flat Spiral Spring.

No. 40. 8 in. long. Black, Flat Spiral Spring.

No. 128. 9 in. long, Nickel Plated, Adjustable, Flat Spring.

For other Pruners see our Hardware Catalogue.

SAFE REVOLVERS

SAFETY HAMMER REVOLVERS
.22, .32, .38 Calibres

SAFETY HAMMERLESS REVOLVERS
.22, .32, .38 Calibres

Iver Johnson's grip can be furnished on these Revolvers at an extra charge.

No article should be selected with more care than a Revolver.

For defence it must be sure and reliable.

For safety it must be so constructed that it will not "Fire" at the wrong time.

Dealers handling Iver Johnson Revolvers know that each of these weapons is absolutely safe. Accidental discharge is impossible.

THIS IS WORTH CONSIDERING

With these safety features the cost is no greater than for Revolvers without them.

Caverhill Learmont & Co

Wholesale Distributors
MONTREAL and WINNIPEG

Canada Leads the World in the Manufacture of

HIGH-CLASS SAWS

No. 81

The best and cheapest Hand Saw on the market, quality, temper and finish considered. Every saw warranted. Carved applewood handle. Patented, wave finish blade.

No. 1 Maple Leaf Lance Tooth Cross-Cut Saw, and the No. 0 Racer are tempered under our secret process.

No. 0 Narrow Racer Cross-Cut Saw

If you want saws that will sell and give your customer satisfaction, you will stock the Maple Leaf Saws.

Manufactured by **SHURLY & DIETRICH, - GALT, Ontario**

Brantford—Roofings

This Represents Our

"ASPHALT"
No. 1, 2, 3
"RUBBER"
No. 1, 2, 3

Each Roll is complete with Nails and Cement to lay. Best selling goods in the market.

BRANTFORD ROOFING CO.
LIMITED
BRANTFORD, ONTARIO

SIMPLE

DURABLE

RELIABLE

STYLE F.

THE LEADER

A Name Familiar to Butter Makers in Connection with the

Highest Grade Churns

ITS STRONG POINTS:

Stiff Steel Frame, with malleable iron fittings, Combination Bow Lev r and Pedal Drive. Bicycle Ball Bearings. Bolted Trunnions. Cream Breakers. Barrel easily drained and conveniently detatchable for cleaning.

DEALERS: When you handle this line you handle QUALITY.

The Dowswell Manufacturing Co., Limited

W. L. HALDIMAND & SON, Montreal
EASTERN AGENTS

HAMILTON
ONTARIO

GALVANIZED FENCE HOOK FOR FASTENING WOODEN PICKET ON WIRE FENCES

WIRE NAILS, COILED SPRING, BARB and PLAIN FENCE WIRE, OILED and ANNEALED, CLOTHES LINE WIRE, STAPLES, etc.

THE WESTERN WIRE & NAIL CO., Limited. ——LONDON, ONT.

The Peterborough Lock Manufacturing Company, Limited

Peterborough, Ont.

Manufacturers of all kinds

Rim and Mortise Locks, Inside, Front Door, and Store Door Sets, also full line High-class Builders' Hardware.

Sold by all Leading Jobbers in the Dominion.

Cylinder Night Latch, No. 103.

Serviceable Combination Cabinet

SUITABLE FOR HARDWARE TRADE

A COMBINATION THAT EVERYBODY MUST HAVE

CONTENTS

6 Table Knives
6 Dessert Knives
6 Table Forks
3 Table Spoons
6 Dessert Forks
6 Dessert Spoons
6 Tea Spoons
2 Salt Spoons
1 Mustard Spoon
42 Pieces

Price Complete

SILVER PLATED LIST $22.30

BEST NICKEL SILVER LIST $14.00

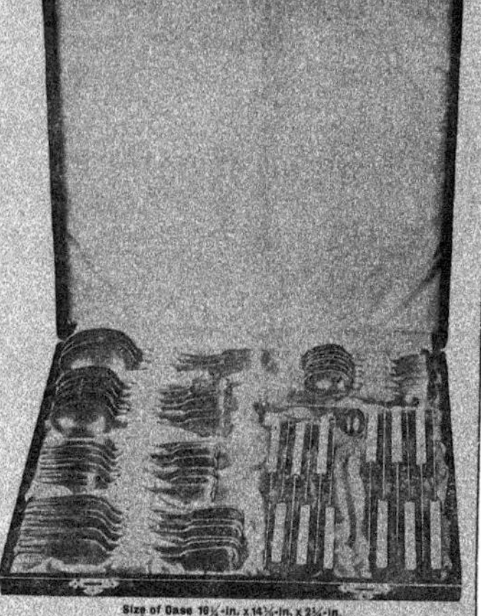

Size of Case 16¼-in. x 14½-in. x 2¼-in.

JOHN ROUND & SON, LIMITED

MANUFACTURING

Cutlers and Silversmiths

MONTREAL

Factories SHEFFIELD, Eng. *Established* 1847

Dame Quality at Our Bidding Waits

Our ***Handsaws*** are made of "top-notch" Crucible Cast Steel, they are tempered in oil and carefully tested. Every one is guaranteed.

The illustration shows one of our Patent Ground, Hollow-Back, Highly-Tempered, Highly-Finished ***Handsaws.*** It is praised the world over.

We also make Back Saws, Circular Saws, Hack Saws, Mill Saws, Canada Webs, Files, Edge Tools, etc.

The sun never sets on Spear & Jackson's Wares

A POSTCARD SENT TO US MAY MEAN A PILE OF BUSINESS SENT TO YOU.

SPEAR & JACKSON ÆTNA WORKS **Sheffield, England**

Telegraphic Address: "Spear," Sheffield

THE
SPECIALLY ROLLED STEEL
used in making
Hamilton Rifles

gives the maximum strength with the minimum weight—the condition par excellence in rifle-making.

HAMILTON RIFLES are made in three models—Nos. 15, 19, 23.

No. 15 sells at $2.00
No. 19 sells at $2.50
No. 23 sells at $3.50

ORDER FROM THE JOBBER

THE HAMILTON RIFLE CO.
Dept. 71 *Plymouth, Mich.*

Are you interested in any of the lines that are advertised? A Post Card will bring you price list and full information. Don't forget to mention Hardware and Metal.

Working Iron in Common or Refined Bars

London Bars are made in the following sizes:

Flats, 7/16 inch to 4 inch x 3/16 inch to 1¼ inch
Round Edge Tire, 1 inch to 3 inch
Rounds, 3/16 inch to 2¼ inch
Squares, ¼ " to 2¼ "
Ovals, ½ " to 1½ "
Half Ovals, ½ " to 2½ "
Half Rounds, ½ " to 2¼ "

54

London Rolling Mills
London, Canada

WRIGHT'S
NO. 10 TRACTION ENGINE HEAD LIGHT

Best of its kind on the market, Brass Nickel Plated conical Burner, Heavy Copper Nickel Plated Reflector, throws a strong brilliant steady light over a large area.

Supplied with without. Brackets, as required. Full stock of wicks chimneys, etc. always on hand

Write us for illustrations and prices.

MANUFACTURED BY
E. T. WRIGHT & CO., - HAMILTON, CAN.

Will Hold up a Shelf

That's what a shelf brackets for.

For this purpose there can be NOTHING BETTER, NOTHING CHEAPER than the BRADLEY STEEL BRACKET. It is well Japanned, Strong and Light. The saving on freight is a good profit aside from the lower price at which the goods are sold. Order direct or through your jobbers.

ATLAS MFG. CO.,
New Haven, Conn.,

Expert Bookkeeping

Investigation and auditing of all classes of accounts undertaken. Trading, profit and loss accounts and balance sheets prepared. Sets of books opened and office systems devised in an up-to-date manner. Collection of accounts a specialty.

PERCY P. DAVENPORT
622 McIntyre Block, - WINNIPEG, MAN.

Square your accounts

INTERNATIONAL PORTLAND CEMENT COMPANY, Limited
OTTAWA, - - ONTARIO.

Manufacturers of strictly high-grade Portland Cement made by the most improved method of manufacture in existence. Each barrel contains three hundred and fifty pounds of pure Portland Cement with no adulteration.

The Hanover Portland Cement Co., Limited
HANOVER, ONTARIO
—Manufacturers of the celebrated—
'Saugeen Brand"
OF PORTLAND CEMENT
Prices on application. Prompt shipment

Persons addressing advertisers will kindly mention having seen this advertisement in Hardware and Metal.

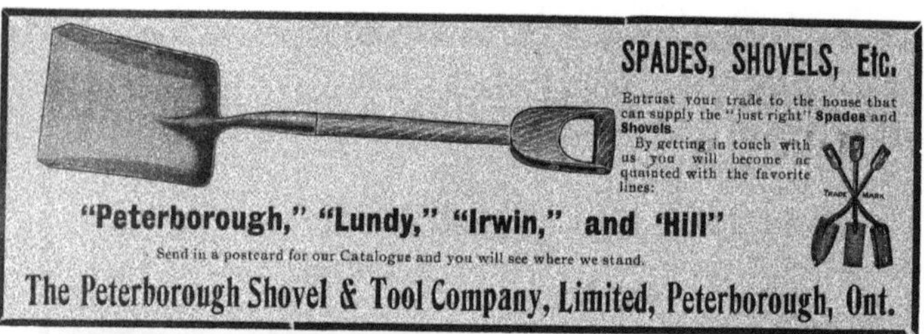

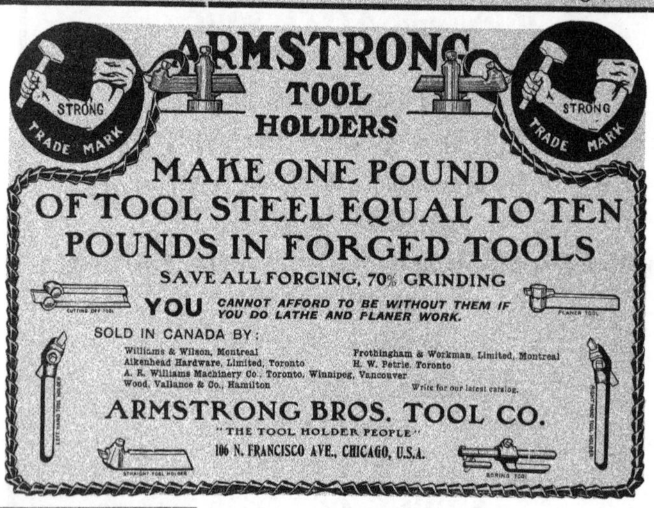

A BOY'S POCKET MONEY

Any boy can earn money in his spare time after school or on Saturday.

We want a bright, intelligent boy, between 9 and 16 years old, in every town and city and village in Canada.

We will start him in business selling our useful publications.

You will make a profit on each sale and earn a watch as well for good work.

Simply send your name to us.

Address

The MacLean Pub. Co.

10 Front St. East, Toronto

The Canadian Bronze Powder Works

R. E. THORNE & CO.

The only bronze powder works under the British flag.

High Grade bronze powders and bronze liquids.

Can fill all orders at short notice.

MONTREAL — TORONTO

WORKS AT VALLEYFIELD. ***NO ORDER TOO LARGE***

Let us Help You Increase Your Business

Why not **go after and get** the fire brick trade in your locality.

Our expert information is at your command and will enable you to make this a paying line.

Make this one of the New Year resolutions and KEEP IT.

Francis Hyde & Co.

Contractors and Foundry Supplies

31 WELLINGTON ST.

MONTREAL

A MATTER OF DOLLARS

Here you see the **strongest, cheapest** and **best Ladder** on the market. Big talk, but we stand ready to make good every claim we make for our goods.

There are no nails in this ladder. It is put together with screws and rivets and constructed of thoroughly seasoned ash. Durable, safe and a **money** saver.

That Waggon Jack is a new article on the Canadian market and it already **holds the United States market**. It holds what it raises—because it has double acting dogs.

It is built of steel and malleable iron, there is nothing to get out of order about this Jack. Built to lift all kinds of buggies, heavy waggons and general work.

SEND FOR PARTICULARS NOW.

SHIRREFF MANUFACTURING CO., BROCKVILLE, ONT.

Don't Stop

to ask your neighbor, lift the load yourself with **The Burr Self-Locking Tackle Block** Saves the labor of two or three men. 600 to 5,000 pounds capacity.

Lewis Bros., Ltd., Montreal
Munro Wire Works, Winnipeg
Vokes Hardware Co., Toronto
F. Hamilton & Co., Hamilton
The Burr Mfg. Co., Cleveland, Ohio

Cap Screws, Set Screws, Machine Screws, Cold Pressed Nuts, Studs, Coupling Bolts, Special Milled Work, etc.

CANADA FOUNDRY COMPANY

LIMITED

HEAD OFFICE and WORKS, TORONTO, ONT.

District Offices: Montreal, Halifax, Ottawa, Winnipeg, Vancouver, Rossland

OUR LINES ARE SELLERS !

Why Not Write for Particulars ?

CABLE ADDRESS "SUPERIOR" *FARLEY WORKS, SHEFFIELD, ENG*

GEORGE PHILLIPS & CO.

MANUFACTURERS OF

Silverware, Electroplate and Cutlery

We also make the best Razor on the market. Four exclusive brands with advertised names that will help every dealer. It costs a two cent stamp to find out about these solid old country lines which you cannot duplicate elsewhere. ***ALL MONEYMAKERS.***

OFFICE FOR CANADA, PACKARD BUILDING, MONTREAL **TELEPHONE M 557**

Talking Points That Tell
H. & R. REVOLVERS

Excellence of Construction
Accuracy
Rapidity of Action
Absolute Safety

No matter which model or at what price it may be sold, every **H. & R. Revolver** undergoes the most rigid inspection both as to material and workmanship. Every part is accurate to the thousandth part of an inch and absolutely without a flaw.

H. & R. POLICE PREMIER REVOLVERS

combine more good features than any revolvers of other make selling for twice the money. The **safety hammer** enables them to be discharged quicker than any other hammer revolver. **Automatic Ejector, Independent Cylinder Stop** (found only on other make revolvers of much higher price). **Light Weight. Beautiful in Design and Finish** are the points that make them sell rapidly.

Our Catalog tells about our full line and gives valuable selling points.

HARRINGTON & RICHARDSON ARMS CO. WORCESTER, MASS.

314 Park Ave.

ALL STEEL AXE

There are those who not only want a good Axe, but a good looker, one that they can take a pride in, and take care of. There is nothing in the variety of painted axes that looks quite as well as the All Steel Full Polished Axe. Our All Steel Axe is as good as it looks and is an easy seller.

The Dundas Axe Works

DUNDAS, ONT. To be continued.

The Tank that Saves

The loss of from 15% to 50%, which experts say is customary when oil is stored in faulty barrels and cans, is entirely stopped by the Bowser.

A Bowser Oil Storage Outfit prevents evaporation, leakage, waste, spilling and dripping and provides a safe, clean, economical, convenient and altogether satisfactory method of storing oil.

If you want a clean store, a safe store, and bigger profits on your oil business, write for catalog V. It tells how.

Cut 15

S. F. BOWSER & CO., Inc.

66-68 Fraser Ave. Toronto, Ont.

Jowitt's Cast Steel

All Standard Sizes in Steel.

SPECIAL PRICES FOR IMPORT.

H. G. EADIE,

22 St. John St. Montreal

Peerless Woven Wire Fence

Here is a fence that will sell. Once up no need for repairs. It lasts for years—just what you are looking for. It's horse-high, pig-tight and bull-strong. No stock can get over, under or through it. The most unruly animal can't butt it down and there are no barbs to tear and injure stock. It's

WIND AND WEATHER PROOF

The Peerless lock holds it perfectly rigid and the wires can't slip up, down, nor sidewise. Cold weather won't snap it and hot weather won't make it sag. It's the greatest thing out for farmers, stockmen and all other fence users which makes it a red hot proposition for dealers. It comes in rolls all ready to put up. It's easy to handle. Write for prices.

The Banwell Hoxie Wire Fence Company, Ltd.
Dept. J, Hamilton, Ont.

ESTABLISHED 1795

JOHN SHAW & SONS
WOLVERHAMPTON, Limited
WOLVERHAMPTON

Soil Pipe, Linseed Oil,
Canada Plates, Tin Plates,
Galvanized Sheets, Black Sheets,
Sheet Zinc.

JAMES BARBER, Era Works, Sheffield

Best quality only
TABLE AND POCKET CUTLERY,
RAZORS, CARVERS, Etc.

J. H. ROPER
Canadian Representative
82 St. Francois Xavier Street - MONTREAL

SPECIAL SKATES FOR SPECIAL PURPOSES

Have you ever thought of the value of being able to meet every taste, or whim, or fancy, or need of your customers? Of course, it's a pretty hard thing to do, but you can at least do it in SKATES if you have our complete line in stock. COMPLETENESS AND QUALITY are the invincible combination that have made our line so successful.

We draw your attention to a few illustrations of special skates which we make.

Figure skating is coming in again, which is evidenced by the demand we have for figure skates this year, and you will find the STARR FIGURE SKATE the most satisfactory you can put on your counters.

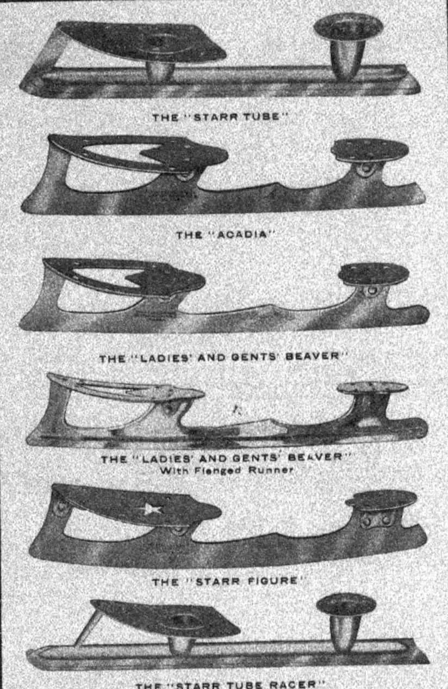

THE "STARR TUBE"

THE "ACADIA"

THE "LADIES' AND GENTS' BEAVER"

THE "LADIES' AND GENTS' BEAVER" With Flanged Runner

THE "STARR FIGURE"

THE "STARR TUBE RACER"

Not only are our skates of various sorts,—self-fastening, hockey, fancy-skating, rink-skating, racing, etc. — popular with skaters, but our hockey sticks are popular with hockey players, and enjoy the very highest reputation. What more could the dealer want behind him than such a reputation?

Our representative is now on the road soliciting orders for 1907-08, and will call on you shortly; but in case you would rather not wait for him, drop us a line; we are always pleased to answer letters.

The "Mic-Mac" Hockey Stick.

The "Rex" Hockey Stick.

HIGHEST AWARDS WHEREVER EXHIBITED AT INTERNATIONAL AND OTHER EXHIBITIONS

GOLD MEDAL AT THE **DOMINION EXHIBITION 1906**

THE **STARR MFG. Co.,** LIMITED

Head Office and Works, DARTMOUTH, N.S.
Branch Office, 126 Wellington St. W., TORONTO

What to Do Next Month

February is a short month, but for the live hardwareman there is a lot to do. If stocktaking has not been done in January, it should be tackled at once as it is not only a big job in itself but it suggests many other important things to be done. An inventory sale should be advertised, short lines should be stocked up, departments which have not paid their share of the profits and expenses should be reorganized or done away with and the results of the inventory should be carefully studied for lessons showing how better results can be procured during the coming year.

* * *

Stocktaking is probably the most disliked job of the year. It is a mean job, but it is necessary and if it is gone after in the proper spirit the results can be made of far greater value than if it is tackled gingerly with kid gloves. Go at stocktaking as though it was determined to get something out of it. Keep an accurate memorandum of every article in stock and clean up everything that is dirty or shopworn.

* * *

An Eastern Ontario dealer recently bought a $4,000 stock of hardware and put ten tons of the stock in the scrap-iron pile. He didn't pay for the scrap—the man who had to stand the loss was the fellow who had evidently scamped his stocktaking. He had probably gone on year after year making an inventory of goods and fooling himself into the belief that he had several thousand dollars worth of saleable stock. Instead of having an annual inventory sale he kept hoarding up the old unsaleable goods, allowing them to become more unsaleable all the time. He evidently didn't figure that the space in his store, on his shelves and in his warehouse was given to goods which ought to be gotten rid of for anything above the scrap-iron price over the counter or for the old material figure out of the back door. He didn't count up the lost time in repeated handling of the stuff during the year and at stocktaking. But finally he learned his lesson—or if he didn't he ought to. He sold to a dealer with modern methods, a man who will have nothing but saleable goods taking up shelf room or floor space.

* * *

Another concern we have heard of went on paying splendid dividends for a number of years. It made a big name for itself, but one day there came a bump and when creditors put in an appearance it was found that the dividends had been paid on the strength of a wind-filled inventory. The concern thought they were sound and borrowed on their credit. The day came when the balloon burst, creditors flocked in and the house of cards crumbled to pieces.

* * *

Many a merchant who thought himself worth ten or twenty thousand dollars has found that instead of having it in cash or its equivalent it was in old and worthless stock. Make sure that in the inventory taken this year that the figures totaled up represent real dollars and not wind.

When a man takes up stocktaking his aim should be to find out what he is worth and how much has been made during the past year. In too many cases he is anxious to make himself believe he is worth more than he really is. He wants a favorable result and the tendency is for him to make the stock sheet tell too favorable a story. But some day this man will have to close out his stock and try and make his imaginary personal worth measure up with the real selling figures. If a merchant wants to know what stock he has he should know it exactly, figuring a fair percentage for depreciation. A wind-filled inventory won't fool the credit man in a wholesale house. The only person it is apt to fool is the dealer himself.

If a system of stocktaking cannot be adopted showing at the end of each day the real condition of the business, try, by all means, to have an accurate inventory showing the actual situation at least once in each year.

* * *

After stocktaking has been completed and everything has been cleaned up an "inventory sale" will be in order to clean out any goods which are deteriorating in value. Many dealers have such a sale before stocktaking in order to bring down stocks to as low a point as possible. Both seasons are opportune times to push off such goods which the merchant has not found profitable and which can be replaced by better selling lines. It will be far better to stand a small loss now than a larger one later on, particularly as it brings cash into the till during the dull season.

* * *

There is no better season of the year to plan a new arrangement of the store than at stocktaking time. Stocks are at their lowest point. Holiday and winter goods are largely out of the way and spring shipments have not yet arrived. More time can be given now to a "spring housecleaning" or to installing a new silent salesman, cabinet of shelf boxes, etc., than later on, so the dealer who aims to make his store move with the times will get busy at once on securing quotations from the manufacturers of sehlf boxes, silent salesmen and similar fixtures.

Much depends upon the arrangement of the stock, and any dealer who has found during stocktaking that the space under the counters, or the top shelving or in out of the way corners in the warehouse has become a catch-all for odd articles, should devise some means whereby a better system can be followed in future. The more modern fixtures there are installed, the less opportunity there is for slovenly clerks to dispose of odd things. There should be a place for everything and everything should be kept in its place. But to keep things in their places a modern stock card system should be used in connection with modern store fixtures.

* * *

While stocktaking time is not usually a season when extensive buying is done it behooves every dealer to keep a keen eye upon the market reports, particularly this year, when nearly every line has either advanced sharply or is likely to move up higher. By buying early the dealer not only has the advantage of making certain the delivery of goods which are likely to be scarce, but he gets the advantage of present prices, often with a protection against a decline. From no quarter does there appear to be any indications, however, of lower prices during 1907 on the most used metals and materials which govern the great bulk of articles sold in the hardware store.

* * *

Take stoves, for instance. The cold winter has run stocks in the hands of manufacturers and retailers down to the lowest point in years. Pig iron holds firm at sky-high prices with stocks hard to procure in foreign markets, all Canadian furnaces being booked up for about a year ahead. Stove founders have already advanced stove prices once and furnaces and registers twice. Some have had to close down on account of the shortage of iron. A further advance seems certain. Under the circumstances it would appear that the dealer who books orders early stands a better chance of making money than the one who holds off in the possible hope that prices of iron may drop or that stove manufacturers will continue to disagree on the question of an advance.

* * *

Paints and white lead are other lines which should be booked early. Paint materials have gone up in price and the stiffness in the lead market and growing shortage of white lead on both home and foreign markets would point to further advances in the spring. Ever since the Government put the duty on white lead the tendency has been upward and the turning point appears to be a long distance off yet.

* * *

And in staple hardware lines the conditions are about the same. Jobbers find their chief difficulty to be the matter of securing goods to fill orders. Manufacturers are loaded up with business and with the increase in population the demand is steadily increasing. Prices, too, are on the advance, although most lines have already been marked up. It can be safely said of all shelf and heavy goods, as well as of metals, that the dealer who books his orders earliest this year is the one who stands the best chance of showing a margin on the right side of the ledger when another stocktaking season comes around.

Canadian Retailers Stirred to Action

Thousands of letters of protest being sent to the Postmaster General and Members of Parliament—Strong Arguments against proposed Parcels Post legislation Extracts from copies of a few of the letters written by Retailers.

It is doubtful if anything has ever stirred the retailers of Canada so deeply as the announcement made in Parliament on Dec. 7th last, by Postmaster-General Lemieux, that he was preparing legislation to amend the Parcels Post Law, the proposal including provision for a c. o. d. system, whereby goods could be sent to any point and collection made by the post office. It can easily be realized how this would work in favor of the retail catalogue houses, by obviating the necessity of sending the purchase price with every small order.

The announcement came just when Christmas activity was at its greatest, and consequently few retailers noticed the incident, particularly as the large daily newspapers which carry pages of mail order advertising took care not to emphasize or draw attention to the announcement, which was of particular interest to their thousands of retail readers.

Trade Papers Take Action.

The MacLean trade newspapers, however, were alive to the interests of the retailers and promptly investigated the report and suggested a method by which the trade could effectively oppose the measure. We urged that each one communicate at once with the local member of Parliament denouncing the proposal, and stating that he would be expected to cast his vote against it.

We also urged dealers to write the Postmaster-General direct, giving reasons why the legislation should not be passed and further suggested that the editors of the country newspapers be asked to support the position taken by their local advertisers whose business is endangered. The Ontario and Western Canada Retail Hardware Associations backed this up by circularizing every dealer in the four provinces which their organization covers, hardware dealers being urged to take prompt and united action in bringing their influence to bear on the Postmaster-General in opposition to his proposed amendments to the Parcels Post Law.

From every part of the Dominion the retail trade has responded promptly, and thousands of letters of protest have been forwarded to Ottawa. In many cases petitions containing the name of every retailer in a town have been sent to the members of Parliament to be forwarded to the Postmaster-General. This is as it should be. The retail trade should not lie down and be trampled upon without protest.

In one town close to Toronto the retail trade has suffered severely from mail order competition and yet have failed to take action, taking the pessimistic view that their efforts would be futile. That this is an exception, however, is clear from the large number of dealers who have sent to us copies of the letters they have forwarded to Ottawa. As will be seen from the following extracts from a few of them nearly every section of the Dominion is represesnted in the correspondence:

Extracts from Merchants' Letters.

"Now I am decidedly opposed to any such legislation. We have express companies who forward parcels with promptitude and at reasonable rates. Such a bill would build up catalogue houses and tear away many local businesses. If this Dominion is to prosper it must be by making trade conditions of such a nature as to give the greatest good to the greatest number. This bill would handicap such conditions."—Ontario.

* * *

"We believe such legislation would tend to favor a few monopolists to the injury of a multitude of local merchants who, as a rule, rank amongst our best citizens and pay heavy taxes for the benefit of the localities in which they do business. Competition is sufficiently keen among local merchants to protect the consumer from over-charge. The parties soliciting this legislation aim at securing a good portion of the mercantile business in sections of the country where they pay no taxes."—Ontario. This was signed by a number of merchants and a gentleman signing himself an ex-M.P.

Has Hurt Trade in Britain.

"We learn that a bill will be introduced into Parliament at the present session to establish a cheap parcel post system in Canada and whereby the post office will collect on all parcels, that is, provide a c.o.d. service. This means the introduction of the British parcel post system, a system that has ruined the business of the local dealer outside of the large centres in that country. The merchants of Canada will oppose this measure in the most emphatic manner and we ask you to use your influence to have this bill defeated, as it will seriously injure, if not destroy, in many places the enterprises that do most towards promoting a vigorous and prosperous growth in local towns and villages."—Ontario. This communication was signed by all the leading merchants of the town.

* * *

"The passage of such a bill would put a great many local retailers out of business, and put a large amount of extra work and expense on the postmasters. I hope you will oppose the measure."—Ontario. This merchant is also a postmaster.

* * *

"I would ask you to earnestly oppose this measure, as, in my opinion, it would be a great detriment to at least three-fourths of the people engaged in mercantile business. As it tends towards the centralization of trade in one or two cities, it would injuriously affect hundreds of other cities, towns and villages. It would also tend to the further burdening of the mails, on account of which the public is already suffering. I cannot see that there would be any general benefit resulting from such a measure, and in the interests of a large proportion of the people of Canada, and especially of this city and district, I would again ask you to oppose it."—Ontario.

* * *

"The growth and prosperity of this Western Canada surely depends upon the energy, prosperity and success of her local citizens."—Manitoba.

Will Affect Farmers Adversely.

"If this becomes law it will greatly facilitate the progress of Toronto and Montreal catalogue houses, which have already become a menace to the prosperity of our town and county. A consideration of this growing evil shows that it is unfair to local merchants, and in the end will be disastrous to those who encourage it. Local support is necessary to ensure good, healthy business, and progressive merchants are necessary to the growth and success of their towns. Without them there is no market, the largest taxpayers disappear, there is little or no demand for labor, and as a direct result, farmers' lands decrease in value, as an active market is necessary for the profitable sale of their produce. Merchants in small towns, particularly in a farming community, are obliged to give their customers long credits, often tide them over a period of misfortune, and handle at a loss much inferior produce, consequently they are placed at a disadvantage when brought into competition with catalogue concerns, whose business is conducted on a strictly pay-in-advance basis, and who are not called upon to pay anything locally."—Nova Scotia.

* * *

"I feel that we already have enough competition from this class of traders, who pay nothing in the way of local taxation, and I also feel that the general public would not benefit by the proposed change. People can buy as cheaply at home as from the catalogue houses, but they are influenced by the advertising features of the latter."—Quebec.

"You are, of course, aware of the immense amount of money sent out of the small towns every day to the catalogue houses for goods that command a good profit, leaving the local dealer to supply the staples at a very small profit, and pay the taxes."—Quebec.

Would Kill Local Trade.

"Not only would such legislation be unfair to the merchants, but every business would be adversely affected. The tendency would be to kill local trade, lessen local competition and enterprise, and depreciate real estate values. In order to make our ideas as plain as possible we will view as an illustration this village, with a population of about 800, six general, five grocery and two hardware stores, two harness shops, three merchant tailors, and a number of smaller businesses besides. Undoubtedly it may be admitted that competition is keen enough here. We conduct one of these general stores, carry a stock of from $12,000 to $15,000, employ ten hands the year round, pay rent, taxes and insurance. We take an interest in the progress of the village and foster the growth of our market. The result is good prices and accommodation for the farmers from the surrounding country. They cannot do without the local market and stores; it is impossible for them to drive to Toronto with their butter, eggs, fowl, etc. It is clear that prosperous towns and villages tend to create prosperous farmers. One cannot well do without the other and this condition is a natural one that exists all over Canada. We purchase the product of the farm for which we pay prices established by local competition. During the past few weeks we paid out for fowl alone over $5,000. By the combination of retailing goods and handling farm produce we are able to conduct and sustain a fairly profitable business, and at least four other stores here are conducted on similar lines. We have nothing to fear from fair competition, but should be protected from legislation giving the mammoth mail order houses any unfair advantages."—Ontario.

Detrimental to Progress.

"Such a bill, should it become law, would be a menace to our interests in proportion as it would benefit a class of merchants who reap where they do not sow, and who do business in communities to whose progress they do not contribute by their presence, their energy or their taxes. We would be sorry to oppose a measure that would contribute to the comforts and conveniences of the people as a whole, but in opposing the measure in question—which we will do most determinedly, and which we ask you to do as representing us—we would point out that a policy which tends to the centralization of business, and the nurture of large concerns, cannot but be detrimental to that progress in country communities which we must all desire to see."—Alberta.

* * *

"We could fill pages giving our individual reasons for asking you to oppose such a bill, but have confidence in your ability to grasp the full effect that it would have without our going further into the matter."—Saskatchewan. This was signed by every business man in the town.

"We would ask you to keep tab on this matter, and oppose it to the extent of your ability, as it seriously menaces trade in every place east of Montreal, especially in Prince Edward Island. . . . Every dollar that goes out of the province impoverishes it just that much."—Prince Edward Island.

* * *

"I respectfully call your attention to the stand which I take in the matter of the bill about to be introduced re extension of the parcel post system. I believe this is merely a scheme of the departmental stores. If adopted it would be a source of injury to every rising town in the west. Our country needs population, not monopoly—legislation that will fill the west with business centres instead of crushing out the small dealer."—Alberta.

In the Interests of the Few.

"This legislation would be in the interests of the few. It would be unjust to the local retailer, who would, in effect, have the Government in league against him. It would be another step toward centralization, and another favor given to large corporations. It is of great importance to the farmers that large and up-to-date stores be found at their very doors. The legislation referred to would not help this, but operate against it."—Ontario.

* * *

"This measure, I think, would be very unfair to town, village and country merchants. We are here on the ground where we have to pay our taxes, and where we have an interest in the welfare of our particular locality. We have a right to be protected so far as possible from what would certainly be unfair means employed by mail order houses, who have no interest whatever in our community, except to get our trade from us. If you can see your way clear to oppose this measure I think you would be conferring a favor on not only the business men, but the whole community in general."—Ontario.

* * *

"Progressive communities the country over are almost exclusively those who have the spirit of, and practice, home protection. Progressive towns always increase the land value of the adjacent territory. The large retail catalogue houses, by absorbing the trade of the smaller towns and villages, act as a detriment to their welfare, and also that of the country adjacent."—Saskatchewan.

Merchants Pay the Taxes.

"For the Government to step in and assist mail order houses to distribute their goods, at a nominal charge, would be most unfair. . . . We pay taxes to keep up our roads, streets, bridges, etc., but if the mail order houses, without assisting us in our municipal expenditures, are to be enabled to take our business away—the means provided by the Government—what benefit do we get from these taxes?"—Ontario.

* * *

"Mail order houses do great injury to our town and country stores, and are no benefit to those who purchase goods from them. I have compared the goods with what we sell, and see no difference."—Nova Scotia.

Manifestly Unfair.

"We do not fear any competitor if he is placed on an equal footing with ourselves, but it is so manifestly unfair that we should have to pay a business tax, support our schools and our churches and be expected to respond to every local demand, and the catalogue houses come in without paying any of these things, that we shall expect you to oppose the measure when it comes up."—Ontario.

* * *

"By passing such a bill the Government would certainly, whether consciously or not, be playing into the hands of the catalogue houses."—Quebec.

* * *

"Quite a number of our people are thoughtlessly sending a great deal of their money to the city stores, thereby cutting their own throats, ruining their home town, decreasing value of property, etc. They are building up Toronto and other cities at the expense, and to the stagnation, of their own community. If people would spend their money at home it would come back to them in a better town, better schools, better churches, better stores, better social advantages, and increased value of property. The city stores, on an average, don't sell any cheaper than we do. They don't help pay taxes, etc. They are great and clever advertisers, and, you know, fields are green far away. Mail order competition is harmful enough in the smaller towns without giving them extra legislation."—Ontario.

No Special Favors.

The retail trade do not seek any special favors from the Postal Department, and are quite right in protesting against special privileges being given the large mail order houses. The Toronto Globe, in commenting editorially on the Postmaster-General's statement in Parliament, said: "In this promised reform there is nothing revolutionary and the moral sense of the people, as well as of the Government, gives an assurance that no private interest will be treated unfairly." The real object of the bill, however, is the unfair treatment of private interests, and behind it are the T. Eaton Co. and other mail order concerns. The editorial comment referred to, whether intentionally or not, tends to disguise as a public benefaction a measure which will benefit one section of merchants at the expense of another section.

In order to guarantee that "no private interest will be treated unfairly" the retail dealers must be up and doing. If there are any who have not yet written their M.P. and the Postmaster-General they should do so at once. And in every town from Halifax to Vancouver a meeting of the merchants should be called to make certain that the campaign will be continued until the objectionable features of the proposed legislation have been withdrawn.

Doing Business on Cash Basis

Sensible Article by E. Thurston in the National Hardware Bulletin—How a Credit Business was transferred to a cash system.

Perhaps I can interest you and some of your readers in a few words about myself and my business, as my method of doing business is somewhat distinctive from the general run of hardware store policies in the fact that I conduct a hardware store on a cash basis and keep so close to cash in every transaction, both in store and shop, that I feel that I can rightly claim the distinction that I do a cash business. History is always dull unless it comes home to the reader of it. I hope to get near home to some of my fellow hardwaremen, and not only interest but be of some benefit to them in point out a surer and pleasanter path to pursue.

A few years ago I bought out a small tinshop that was being run in connection with a general hardware store and lumber yard, and as I was handy at all kinds of mechanical work and having a pretty thorough knowledge of the tinner's trade I got along fairly well, managing to save up a few dollars by close application to business and still closer to collections for work done.

Things went well along this line for about two years when for lack of room in my then limited quarters, and what appeared to me then a splendid opportunity to engage in the merchandise business, I bought out a small hardware stock and added it to my shop work, also leased the building for a term of years.

Tinsmith Became Merchant.

At that moment I became a merchant instead of a tradesman in the eyes of the public, and then and there I was expected to conform to the rules of the town and the customs of the other merchants of the town. While I worked at my trade my customers seemed to believe I had need for the money earned and payments came readily, but as a merchant all I needed was to sell them goods on credit, so in my timidity in business decision and lest I should offend some of my customers, I decided to do a little credit business in a very conservative way.

Well, things went about as you would suppose. Before the first year had closed I discovered that in spite of all my conservatism and as I thought shrewd business tact, my books showed more goods scattered out over the country on unlimited time than I had in the store. I realized the situation fully, I saw that a change must be made speedily otherwise there would be a case of "failure of health" and a "splendid business opportunity for sale," so I took counsel with more experienced and older business heads and through their friendly advice and due consideration of the situation myself, on January 1st following, I issued in print in our town paper this proclamation to the public:

A New Year's Resolution.

After due consideration we have decided from this date forward to do a cash business in selling goods. We do not wish to humiliate anyone by refusing them credit, neither do we wish to become embarrassed when our bills become due and we have nothing to pay with. Every favor and accommodation consistent with good business principles will be accorded our customers in buying goods. We do not invite open accounts. By selling for cash we can and will sell cheaper, will have more friends, will be able to promptly meet our obligations, to present no duns and to do more business. Shopwork same terms as heretofore.

(Signed.) E. Thurston.

Such was the determined step that we took several years ago and such is our policy to-day, and proud we are that we have departed from that annoying and vexatious credit plan, and that we have placed our business on as near a strictly cash basis as it is possible to do so in a business with its varied make up such as we have done.

In our shopwork we have never attempted to do a spot cash business as the very nature of it would not permit us to do so. It would be disastrous to any shop that depends on public patronage, and quite a hardship on the patrons to demand of them a cash settlement the moment the work was done. This would be utterly impractical for often in case of a breakdown repairs must be made at once, regardless of cost or terms.

We have in connection with the store both a harness and tinshop. Now some may conclude this method of conducting shopwork in connection with store would tend to depreciate the cash status of the store, but not so. It works the other way, the rigid policy of the store tends to draw the shop to its high level.

A Man with a Principle.

Perhaps you would like to know what we did about those customers who said they "would not spend their money where they could not get credit." Well we let them go their way, refusing to discuss our method with them in open conversation, but fighting our battle unceasingly through the town paper, and in most cases they came back and traded with us and really admired our pluck. The world admires a man for having a principle and sticking to it.

With a great many our plan of doing business served as an advertiser as it singled us out as distinct from the other merchants of the town. We have long since found that no one man can sell all the goods of the town nor can he sell to everybody. It is our opinion, based largely on the last few years' experience and close observation, that the time is ripe for radical changes in methods of doing business. Credit is an old fogy and will prove disastrous sooner or later to the advocate of it. The catalogue house gets the cash anywhere from a week to months before the goods appear. Surely we can demand it when the goods are delivered to our customers. The poor merchant has permitted himself to be made a pack horse until everyone feels that he is invited to ride because he goes easy.

Telephone Calls.

Now as to the price question on goods sold under the cash plan of doing business: Well, we have leaders, as we have always had, with very close prices, usually on staple articles, but the great bulk of hardware goods can be pushed on quality to a better advantage than on price. When we changed from credit to cash we made a slight change in price but rather defended our position on the theory that the goods were the equivalent of the money and that we were justly entitled to the money for our goods.

We keep no customers' books—only a little memorandum book in which we note goods sent out in response to telephone orders or otherwise. These are usually settled in a day or two. If the goods are not settled for promptly and the party does not treat the transaction as a favor shown him, his call is passed by next time.

This we believe to be as near a practical cash business as it is possible to do under the existing surroundings. We live up to our motto in dealing with our customers. We pay them cash on the spot, rarely giving cheques, only for large sums, thus forging our way by example as well as precept.

A Pertinent Question

It has frequently been stated that in order to make it possible to do business on a cash basis it is necessary for all the dealers in a town to agree to do away with the credit system at once. This week we publish the experience of a retail hardwareman who had the nerve to stand on his own shoe leather and who made a success of the cash system in spite of the opposition of dealers who continued to give credit.

In the Western Provinces where the transportation problem makes it hard for farmers to get cash for their grain without selling at low prices it is said that the "Cash System" cannot be adopted. We know, however, of several dealers in the West who have successfully changed their store policy. Why should farmers hold their grain for months and make big money while at the same time borrowing money in the form of goods from the merchant, who in many cases does not even charge interest? Too many farmers leave money in the bank drawing interest while at the same time running long accounts at the hardware stores. If the merchant loses it is largely his own fault.

The editor of Hardware and Metal wants to hear from the hardware dealers in Canada who have been progressive enough to put their business on a cash basis. It is hoped that some helpful suggestions can be given if a list of "We Sell for Cash Only" hardwaremen is compiled. Travellers and merchants are, therefore, requested to forward the names of dealers who do not give credit.

Catalogue Houses.

The effect of the catalogue house is still felt on our trade to some degree, but they are doing less business here now than a few years ago. In most cases persons buying from them have realized that they did not get any bargains over what they could get at home. Many delayed and badly damaged shipments from the catalogue house in the past has proved a damper to their patronage here. Occasionally we are called upon to meet the price on some particular article, which we can usually do to the satisfaction of our customers, but withal the catalogue house business seems to be on the wane in this locality.

Our local town paper made a vigorous fight in a high handed style against the "sending off" way of buying and for the trading at home idea, and it certainly has had its effect. I find my customers quite liberal with the question; ninety-nine per cent. prefer to buy at home if they can do as well, and ninety-five per cent. of my customers will buy at home with slight odds in price against them.

Quality Versus Price.

While salesmanship ranks first as a necessary adjunct in marketing goods, a large assortment well displayed, prompt service honestly rendered adds greatly to any business. A few very cheap and inferior staple articles of merchandise on hand we have found to be an advantage. We use these cheap articles as stool pigeons to induce the purchase of a better grade of goods which in the majority of cases works out all right, but when the price is the main item of consideration we can then meet any price named, holding the trade of the customer and proving to him by the goods that the cheap articles are dear at any price.

We sell cheap goods only when we have to and not because we want to, for it is a settled fact that a better and a far more satisfactory business can be built up on a first-class high grade of goods than on a cheap spurious line.

Thorough mastership of a business comes through a concentration of forces which fully qualifies one for the work at hand. No one can be too thorough in the knowledge of the goods that he offers for sale. President Bush of the National Retail Hardware Association struck the nail on the head when he stated that manufacturers should be more explicit in the description of their wares which they place on the market through the regular course of merchandising. Our stores throughout the country, and especially the country stores, are too much operated by clerks and order takers instead of qualified salesmen.

Credit Drives off Customers.

In concluding this article I will say if any of my fellow hardwaremen are contemplating changing from credit to cash they should decide first that it is the right thing to do, and once decided gather up all your nerve and will power and go into it for dear life, for it is no harder to say no to a friend than to say, pay me and not get it. Later see him going to your competitors to avoid you after you feel that you befriended him by letting him walk off with your goods. It is our honest opinion, based on years of experience and observation, that credit will drive away more good customers from a place of business than it will bring in or hold. Of course the dead beat is not expected to come back to a store to trade after he is fully "loaded," neither will your best friends stick to you sometimes if they owe you an obligation that they cannot meet. Credit is dangerous as it frequently places a merchant where he can neither go ahead nor get out of business. Have you ever seen such cases?

The great catalogue houses have cut the cash plan pattern for us large and clear, and they are doing everything possible to sustain this pattern before the buying public. Can we not come in for some of the good that is mixed with the evil that flows from the great catalogue house fountain? Shall we have more cash houses and less "failing health?"

THE SIEGE OF BATTLEBURG.

For many years the peaceful little town of Battleburg had lain in its pretty valley among the mountains overlooked apparently by the Scouts of The Catalogue House Army. Its inhabitants were happy and contented and the goods they bought from the old veteran general merchant filled their every want.

The General had worked up a good business and won the esteem of the people, but he scented trouble from afar and knew that sooner or later the enemy would appear and bombard the town with tons upon tons of large paper shells called catalogues, the ammunition used by the Catalogue House Army in its new "Hot Air Battery," a recent invention of one of its clever and efficient officers.

He, therefore, like a good general, worked out his plan of campaign, posted his sentries and sat down to await the coming of the enemy.

The enemy came, and came with a rush that nearly put the general out of business, old campaigner that he was.

He soon recovered himself, however, and got down to business. He secured one of their shells, or catalogues, looked it over long and carefully, then sent for one of his aids, Captain Dimes, a young but efficient officer.

The interview lasted some time before the captain emerged and proceeded at once to his quarters at the Battleburg Bulletin, the flourishing daily paper of the town.

That evening the columns of the Bulletin contained an appeal from General Merchant to the citizens to rally around him and offering them as good terms as the enemy.

The population laughed; they were being treated very well, they thought, by the enemy. He was making them fine promises and showing them pictures of the beautiful articles he was going to sell them for half the price the General would charge, and they thought they couldn't do better than stick to him.

The Weapons That Won.

The old General didn't despair, but kept right on with his campaign. Every night he attacked the enemy through the columns of the Bulletin. He showed up all his schemes, invited the public to call at his (the Gerenal's) store and see the kind of goods the catalogue houses would send them, and ended up by offering to duplicate any article shown in the catalogue at the same price, plus expressage or freight.

Thus the fight went on with the advantage on the side of the Catalogue Houses, until the unfaithful ones who had been the first deserters began to receive their shipments.

Then came the General's opportunity. He explained how the great Catalogue Houses, by illustrations, clever descriptions and low prices induced the country people to send away their hard earned money for goods they hadn't even seen. He explained how these concerns had such an enormous line of customers all over the country that a few dissatisfied ones here and there made no material difference. If the goods were not satisfactory, who was going to lose sleep over it? Surely not the manager of the Catalogue House. It was the money they were after, and as they got that before the customer saw the goods, the deal, so far as they were concerned, was closed.

He showed them by actual comparison that the goods they received from the Catalogue House were inferior to what he had been selling them all along, but told them if that was the quality of goods they wanted he would cheerfully supply them at Catalogue House terms.

It was a long, hard fight, and the General went to bed tired every night, but he stuck to it, and soon his heart was gladdened by the sight of his old customers coming back faster and faster and many new ones besides.—Westmount, in Iron Age.

METAL AND HARDWARE MEN DINE.

The annual dinner of the metal and hardware section of the Montreal Board of Trade was held at the Canada Club, Montreal, on the 17th inst., when about 40 members of this branch of manufacture sat down to a most sumptuous repast.

Among the guests were: James Davidson, Geo. Caverhill, A. E. Hanna, C. F. Hibbert, H. Walter Dorken, George Boyd, W. H. Farrell, C. M. Strange, A. H. Campbell, C. H. Godfrey, J. R. Kinghorn, R. Starke, Geo. A. Childs, A. Jeannotte, F. C. Lariviere, Geo. R. Muirhead, G. C. Seybold, J. B. Peck, J. A. Fuller, W. Hudson, T. H. Jordan, T. P. Howard, A. Clift, T. Hutchison, C. B. Rittenhouse, E. G. Jackson, A. Gordon McPherson, Wilfrid Lauriault and others.

A POPULAR CALENDAR.

Many requests have been made for the finely lithographed calendar of the Capewell Horse Nail Co. which represents the advantage of the horse and buggy over the automobile from the girls' point of view. The company have still some copies of this calendar left, and they have offered to mail a copy to any reader of Hardware and Metal.

HARDWARE AND METAL

Established 1888

The MacLean Publishing Co.
Limited

JOHN BAYNE MACLEAN - President

Publishers of Trade Newspapers which circulate in the Provinces of British Columbia, Alberta, Saskatchewan, Manitoba, Ontario, Quebec, Nova Scotia, New Brunswick, P.E. Island and Newfoundland.

OFFICES:

Montreal, - - - - - 232 McGill Street, Telephone Main 1255
Toronto, - - - - - 10 Front Street East, Telephones Main 2701 and 2702
Winnipeg, - - - 511 Union Bank Building, Telephone 3726
London, Eng. - - - - 88 Fleet Street, E.C., J. Meredith McKim, Telephone, Central 12960

BRANCHES:

Chicago, Ill. - - - - 1001 Teutonic Bldg, J. Roland Kay
St. John, N.B. - - - No. 7 Market Wharf
Vancouver, B.C. - - - - Geo. S. B. Perry
Paris, France - Agence Havas, 8 Place de la Bourse
Manchester, Eng. - - - 92 Market Street
Zurich, Switzerland - - - Louis Wolf, Orell Fussli & Co.

Subscription, Canada and United States, $2 00
Great Britain, 8s. 6d., elsewhere - 12s

Published every Saturday.

Cable Address { Adscript, London / Adscript, Canada

ALLEGED HOSTILITY TO U.S. MAGAZINES.

For some time past enquiries have been received from United States manufacturers and others regarding our position on the question of an alleged movement in Canada to keep out United States magazines. Regarding this we wish to state very emphatically that there is no such organized movement and any action that has been taken has not originated with publishers in Canada. Regarding our own position in the matter we are very much opposed to any movement reacting against the best publications in the United States, believing that Canadians should have the most liberal access to the best reading possible. A portion of a letter written by Col. J. B. MacLean, president of the MacLean Publishing Co., publishers of Hardware and Metal, who is himself a member of the executive of the Canadian Manufacturers' Association, to Mr. J. P. Murray, chairman of the Commercial Intelligence Committee of the C.M.A., will show the stand he takes in the matter.

"There is a matter which I think might be brought before the Commercial Intelligence Committee. The Canadian Government propose shutting out U.S. publications, trade newspapers, magazines and that sort of thing. I would probably benefit by such a policy more than anyone else in the country, owning as I do so many newspapers, but I think it is a very short-sighted policy on the part of the Government. By all means let them shut out the cheap and useless magazines, but I think it is to the advantage of our people, particularly of our business men, manufacturers, and their employes, to read the best that is published in the United States and Great Britain that they may be kept right up to the times in every particular in every department of manufacturing and selling of goods. We are doing a lot in this way in our own papers, but we cannot cover all the fields and we never hesitate to recommend or advise any of our readers anywhere in Canada to subscribe for and read the best publications in the United States.

"I think this is a matter the association should deal with and point out to the Government the inadvisability of such a course."

Considerable activity has been noticeable of late in the Post Office Department of both the United States and Canada, which has led up to the present condition of affairs. The most important move has been the abrogation by Canada of the post convention between the United States and Canada in so far as that convention relates to second-class matter. This will take effect next Spring. The original instrument was ratified in 1888, but it was amended two years ago so as to provide "for the right of each administration to decline to transmit through its mails, except when duly prepaid by stamps affixed in the country of origin at the rate applicable to miscellaneous printed matter, such newspapers and periodocals as it would decline to transmit through its mails under the statutory newspaper and periodical privileges accorded to publishers and newsdealers, if such newspapers and periodicals were published in its own country."

In abrogating the convention, Canada notifies the United States that, if by legislation new regulations are framed controlling the second-class matter in the United States, it will be prepared to enter upon negotiations for another convention relating to this class of matter.

Thus it will be seen that Canada is still ready to make another treaty as soon as the United States Government, by legislation, frames new regulations controlling second-class matter. It is over this point that the whole difficulty has apparently arisen, giving effect that Canada is antagonistic to United States publications. She is merely waiting until the United States Government frames new laws in this matter, at which time there is not the shadow of a doubt but that a suitable treaty will be made permitting of the entrance into Canada of reputable United States magazines on the same terms as they are received through the mails of the United States.

More than this, we have reason to believe on the authority of an official of the Canadian Post Office Department that it was on the suggestion of a high official in the U.S. Post Office Department at Washington that the Canadian Government has taken the action referred to above. This was evidently done to assist the U.S. authorities in some scheme which has not yet been made public.

RETAINING EXPERIENCED HELP.

In small towns the general idea among male clerks is that when they have received a certain amount of experience they will go away in search of larger opportunities. This applies to almost every line of business. They leave just at the time when they are most valuable to the merchant with whom their apprenticeship has been served and very often their places have to be filled by the promotion of less competent men. So numbers of small stores go on year after year acting as training schools for salesmen, and continually changing the personnel of their staffs.

Such was the case, up till a couple of years ago, in a store located in a town of about two thousand population. It had two competitors, who had always claimed the larger share of the business, although the store in question was going ahead splendidly. Besides the two members of the firm three salesmen were employed, and at the time we speak of they had resolved to seek a larger field for their efforts. Instead, however, of allowing them to go, the firm organized a joint stock company and took them into the business. All three remained and worked with an energy that has brought big results.

Ask a resident of the town which store appears to be doing the most prosperous trade and he is more than likely to mention this one. He may also comment on the excellence of the service which it gives. We were able to detect no peculiar circumstances that would contribute to the company's success ; it seemed due entirely to the fact that all members of the staff were enthusiastic in endeavoring to bring it to the fore. Other merchants may abstract from this a solution of the problem of retaining experienced help.

SHORTAGE IN WHITE LEAD.

There is a shortage of white lead, owing to most firms depending on heavy supplies from the local white lead corroding firm, which has been forced to close down for a couple of months owing to its inability to procure enough raw material.

This company, being in fact the only producers of dry white lead in Canada at present, the lead grinders were somewhat chary in importing too much dry white lead last fall, especially in view of the high figures which prevailed at the corroding plants in the Old Country.

Now, we understand, the Canadian firm which converts the lead from pig into the dry, or powdered form, have had to close down for a couple of months in mid-winter. Consequently, they have not been able to turn out dry white lead, and the result, as is naturally to be expected, creates a shortage which is difficult to overcome, during the close of navigation to Montreal and interior points.

SENSIBLE TALK.

It is seldom that a manufacturer comes out in opposition to the protectionist policy, but R. T. Crane, of Chicago, takes this unusual position in a recent article, discussing U.S. Senator J. J. Hill's recent plea for reciprocity in Canada. Mr. Crane is probably the largest manufacturer of valves and heating supplies in America, one of his company's many warehouses being located in Winnipeg.

Mr. Crane contends that "a large amount of trade with the Canadians has already been lost, and that much more is likely to be lost, by the stupid and short-sighted policy of our Congress in making the Dingley tariff law a bar to reciprocal trade relations, not only with Canada, but with every other foreign country. We deliberately took a course opposed to reciprocity at a time when Canada appeared to be entirely satisfied with its trade relations with this country, and thus lost the enormous advantage we then possessed, and it is now too late for us to try to re-establish our former position, much less to take advantage of the greatly improved conditions in the Dominion. In this case we have antagonized a neighbor that, by its geographical situation, ought to have been one country with the United States. There is no more sense in putting up trade barriers between Canada and this country than there would be in putting them up between Illinois and Indiana. Once we had committed ourselves to the Dingley tariff, we antagonized the Canadians, and they then swung sharply toward a fiscal policy of protection. This policy is now thoroughly and generally approved, and there is no prospect that the Canadians will change it."

Continuing, Mr. Crane points out that to change Canada's fiscal policy would be unfair to both Canadian manufacturers and to Americans, who have established factories in Canada, with the expectation that the policy would be permanent. He further argues that the high protective policy in the States is forcing manufacturers there to pay high prices for raw materials, this increase in manufacturing costs placing American goods at a disadvantage in foreign markets.

MERCHANTS AND TRAVELERS.

The travelling representative of the manufacturer and jobber is ubiquitous, and sometimes he claims that his reception by merchants is not as cordial as it should be. The merchant in defence says that he would be able to do very little else if he gave any time to every commercial man who called upon him. He recognizes thoroughly the value of examining as many samples as possible, for the purpose of comparing prices and qualities, but is forced to draw the line somewhere.

There is one type of drummer who not only does a great deal of harm to himself, but excites a prejudice against his fellows. He's the man who is persistent in the wrong place. A good illustration of this point was given by a merchant who called upon the editor a few days ago. He was writing out an order in his private office, and had only about five minutes to catch the mail, when the door opened and in walked a commercial man, without even knocking. He was told that it would be impossible to talk to him just then, but he persisted. Very much annoyed, the merchant took him by the shoulders and pushed him outside. Fortunately there are few men of this class on the road.

On the other hand, there are times when travelers have a right to expect a little more consideration than they get. We met the representative of a Toronto house in an Ontario city one day, where he had gone in response to a special request from one of the merchants. This was the only call that he had to make there, and he naturally expected to get through in the morning, so that he could leave on the first afternoon train. He arrived about 9 a.m., and it was about 4 p.m. before he was given an opportunity to show his samples. Now we have reason to know that the merchant was not busy in the meantime. His reason for acting as he did we do not know, but, thoughtless or otherwise, he wasted a half day of the traveler's valuable time.

SHORTAGE OF CARS.

The Great West has been hit hard during the past few months, first by the shortage of cars to move the big crop garnered by the farmers, and second by the shortage of coal, exceptionally cold Winter, bad railway service and, in some cases, actual shortage of provisions. But despite these drawbacks it would be hard to find anyone who has any kick to register about the country, much though they criticize the railways and other institutions.

One protest has, however, been registered by the Gazette, of Francis, Sask., which through the courtesy of one of our subscribers, reproduced, in its issue of Jan. 4 last, our recent editorial on "Returned Drafts." The Gazette says western retailers fully appreciate the advantage of taking every discount possible and making every saving on interest. It says further:

"Retailers are practically at the mercy of the wholesalers at the present moment owing to shortage of cars. No wheat is being moved from the west, the elevators are full, barns of every description are full, and grain is even lying on the ground. When the farmer has no money to pay his bills, how can the wholesaler expect the retailer to pay his, much as he would like to get that discount!"

The point of our recent criticism was that retailers should hesitate before returning drafts for trivial causes. "Shortage of cars," however, is not a trivial cause and we entirely agree with the Gazette that the influence of every branch of trade ought to be used to induce the railroads to supply the necessary number of cars to move the big crops of the west upon the turning into cash of which so much depends.

THE BLACK LIST.

The merchants in some towns throughout Ontario have taken steps to round up the "dead beats" and force them to cast off their practice of "doing" and accept an honest way of living or go wanting. (This is a good time of the year to put this rule into force).

Every business man has his troubles and losses with this class of people who are cunning in their way. For the first few weeks the bills are paid. Next a dollar or two is carried over and then, like a cloud-burst, the orders come in until a bill of $30 or $40 is on the book, when they suddenly disappear and start a new cash account at another store.

To try and collect the debt is a waste of time, although the cases are nothing less than theft.

In various towns the mercantile men have joined hands, each handing in his list of "dead beats," and under the heading of "Black List" are hung in the windows announcing that those whose names appear on the card will be refused credit.

In the past the Government law somewhat protected the dead beats, but it is understood that at the next sitting of the Legislature a change will be made, giving magistrates a limited power to deal with cases of this kind.

Markets and Market Notes

(For detailed prices see Current Market Quotations, page 66.)

THE WEEK'S MARKETS IN BRIEF.

MONTREAL.

Ingot Copper—Advanced ½ cent.
Galvanized Iron—One line shows advance.
Old Material—Several lines show slight advances.
Paris Green—Advanced one cent per pound.
Turpentine—Advance of one cent per gallon.
Copper Rivets and Burrs—Advanced.
Shovels and Scoops—Advanced $1 per dozen (list.)

TORONTO.

Paris Green—One cent per lb. advance.
Copper Rivets and Burrs—Ten per cent. advance.
Old Material—Lead and rubber higher.

Montreal Hardware Markets

Office of HARDWARE AND METAL,
232 McGill Street,
Montreal, Jan. 25, 1907.

Business in hardware circles this week is reported excellent. In fact, most jobbers say that they don't know which way to turn. The goods are going out faster than they are coming in, and one firm has such a floor full of shelf goods laid out (all big orders) that it's almost impossible to get around them.

Many lines are beginning to run short, and unless some relief comes, there are likely to be quite a number of advances in the near future.

Screen doors and many other spring lines are to be seen among stocks.

Copper rivets have again advanced, the new discount being 15 per cent.

All shovels and scoops advanced $1.00 per doz. (list).

Machine screws are also among the week's advances, the new list being 70 per cent.

AXES — A slight increase in the sales is reported over last week. We quote: $7.60 to $9.50 per dozen; double bitt axes, $9.50 to $12 a dozen; handled axes, $7.50 to $9.50; Canadian pattern axes, $7.50 a dozen.

LANTERNS. — Our prices are: "Prism" globes, $1.20; Cold blast, $6.50; No. 0, Safety, $4.00.

COW TIES AND STALL FIXTURES —We still quote; Cow ties, discount 40 per cent. off list; stall fixtures, discount 35 per cent. off list.

SLEIGH BELLS—Moving, but very slowly. Our prices remain: Back straps, 30c. to $2.50; body straps, 70c. to $3.50; York Eye bells, common, 70c. to $1.50; pear shape, $1.15 to $2; shaft gongs, 20c. to $2.50; Grelots, 35c. to $2; team bells, $1.80 to $5.50; saddle gongs, $1.10 to $2.60.

RIVETS AND BURRS — Copper rivets and burrs advanced. New discount 15 per cent. Our prices now are: Best iron rivets, section carriage, and wagon box black rivets, tinned do., copper rivets and tin swede rivets, 60, 10 and 10 per cent.; swede iron burrs are quoted at 60 and 10 and 10 per cent. off new lists; copper rivets with the usual proportion of burrs, 25 per cent. off, and coppered iron rivets and burrs in 5-lb. carton boxes at 60 and 10 and 10 per cent.; copper burrs alone, 15 per cent., subject to usual charge for half-pound boxes.

HAY WIRE.—Prices remain firm. Our quotations are: No. 13, $2.55; No. 14, $2.65; No. 15, $2.80; f.o.b. Montreal.

MACHINE SCREWS — Still scarce, with no prospects of a near relief. We quote: Flat head, iron, 35 per cent., flat head brass, 35 per cent., Felisterhead, iron, 30 per cent.. Felisterhead, brass, 25 per cent.

BOLTS AND NUTS—Scarce, discounts remain: Carriage bolts, 3 and under, 60 and 10; 7-16 and larger, 55 per cent.; fancy carriage bolts, 50 per cent.; sleigh shoe bolts, ⅜ and under, 60 per cent.; 7-16 and over, 50 per cent.; machine bolts, ⅜ and under, 60 per cent.; 7-16 and larger, 55 per cent.

HORSE NAILS—Fair demand. Discounts remain as follows: C brand, 40, 10 and 7 per cent.; M.R.M. Co., 55 per cent.

WIRE NAILS—No change from last week's advance. We still quote: $2.40 per keg base f.o.b. Montreal.

CUT NAILS—We continue to quote: $2.50 per 100 lbs.; M.R.M. Co., latest, $2.30 per keg base, f.o.b. Montreal.

HORSESHOES. — Our prices are: P.B. new pattern, base price, improved pattern iron shoes, light and medium pattern, No. 2 and larger, $3.65; No. 1 and smaller, $3.90; snow pattern, No. 2 and larger, $3.90; No. 1 and smaller, $4.15. Light steel shoes, No. 2 and larger, $4; No. 1 and smaller, $4.25; featherweight, all sizes, No. 0 to 4, $5.60. Toeweight, all sizes, No. 1 to 4, $6.85. Packing, up to three sizes in a keg, 10c. per 100 lbs. More than three sizes, 25c. per 100 lbs. extra.

BUILDING PAPER—New lists show advances.

CEMENT AND FIREBRICK.—No change in prices. We are still quoting: "Lehigh" Portland in wood, $2.54, in cotton sacks, $2.39; in paper sacks, $2.31. Lefarge (non-staining) in wood, $3.40; Belgium, $1.60 to $1.90 per barrel; ex-store, American, $2 to $2.10 ex-cars; Canadian Portland, $2 to $2.05. Firebrick, English and Scotch, $17 to $21, American $30 to $35; White Bros.' English cement, $1.80 in bags, $2.05 in barrels in round lots.

COIL CHAIN—Our quotations remain: 5-16 inch, $4.40; 3-8 inch, $3.90; 7-16 inch, $3.70; 1-2 inch, $3.50; 9-16 inch, $3.45; 5-8 inch, $3.35; 3-4 inch, $3.20; 7-8 inch, $3.10; 1 inch, $3.10.

SHOT—Our prices remain as follows: Shot packed in 25-lb. bags, ordinary drop AAA to dust, $7 per 100 lbs.; chilled, No. 1—10, $7.50 per 100 lbs.; brick and seal, $8 per 100 lbs.; ball, $8.50 per 100 lbs. Net list. Bags less than 25 lbs., ¼c. per lb. extra, net; f.o.b. Montreal, Toronto, Hamilton, London, St. John, Halifax.

AMMUNITION — Slow. Very little sold. Our prices remain: Loaded with black powder, 12 and 16 gauge, per 1,000, $15; 10 gauge, per 1,000, $18; loaded with smokeless powder, 12 and 16 gauge, per 1,000, $20.50; 10 gauge, $23.50.

GREEN WIRE CLOTH.—We continue to quote: In 100 foot rolls, $1.62½ per hundred square feet, in 50 feet rolls, $1.67½ per hundred square feet.

Toronto Hardware Markets

Office of HARDWARE AND METAL,
10 Front Street East,
Toronto, Jan. 24, 1907.

Hardware jobbers report business very brisk with very few advancers quoted this week. Travellers all report business very brisk with exceptionally large orders for spring. At present some are selling lines of lawn mowers, hose, etc., which are generally meeting with good demand although the bulk of the trade placed their orders in the fall.

Copper rivets with usual proportion copper burrs, that is to say, weight of copper burrs not exceeding ⅓ of combines weight of copper rivets and copper burrs, have advanced 10 per cent. and are now quoted at 15 per cent. Copper burrs in excess of ⅓ of combined weight of copper rivets net list, also copper burrs net list. A large amount of copper has been used of late for street railways, and other electrical enterprises, and the price has rapidly advanced.

No further advance is quoted on nails and they still remain at $2.35 base, but another advance is looked for in a short time. The advance on wood screws has failed to materialize, the manufacturers claiming that stocks are rapidly rounding into shape. Were the expected advance a large one shipments from the States would be encouraged.

AXES AND HANDLES—The numerous new lines of axe handles which have been placed on the market are rapidly being sold with large orders for fall shipment.

GARDEN TOOLS—Small book orders for these articles continue to be received.

WASHING MACHINES—At present these goods are very prominent with large orders. The majority of merchants are placing these goods in conspicuous positions for an early sale.

SPORTING GOODS — Skates are again in great demand and the increased fall of snow throughout the country will help the sale of snow shoes. Trap shooting is becoming very popular throughout the country, and will greatly increase the sale of ammunition.

CHAIN—Remains at our former quotations: ⅝ in., $3.50; 9-16 in., $3.55; ½ in., $3.60; 7-16 in., $3.85; 3-8 in., $4.10; 5-16 in. $4.70; 1-4 in. $5.10.

RIVETS AND BURRS—An advance of 10 per cent. has been made on these articles, and are now quoted at 15 per cent. on copper rivets and burrs. All burrs in excess of ⅓ weight of rivets are net list.

SCREWS—The expected advance has not materialized and it does not look as though prices will be changed for the present although raw material has gone up.

BOLTS AND NUTS—The scarcity on these articles continues with large orders continually being received.

TOOLS—The various advances on tools do not tend to diminish the sale, as large orders are taken for spring shipments.

SHOVELS—Snow-shovels are again in great demand and very few will be carried over the season.

LANTERNS—The lantern demand is very quiet at present, all merchants apparently having large stocks on hand.

EXTENSION AND STEP LADDERS—Price remains the same as quoted last week, 11 cents per foot for 3 to 6 feet, and 12 cents per foot for 7 to 10-foot ladders. Waggoner extension ladders 40 per cent.

POULTRY NETTING—We quote, 2 inch mesh, 19 w.g., discount 50 and 10 per cent, all others 50 per cent.

WIRE FENCING—Galvanized and plain wire remain at advanced prices, but continue selling well.

OILED AND ANNEALED WIRE—(Canadian)—Gauge 10, $2.41; gauge 11, $2.47; gauge 12, $2.55, per 100 lbs.

WIRE NAILS—Remain firm at last week's quotation of $2 35 base, with a probable further advance in the near future.

CUT NAILS—We quote $2.30 base f.o.b. Montreal; Toronto, 20c. higher.

HORSENAILS—Capewell brand, list per pound, No. 5, 15cts; No. 6, 14cts; Nos. 7 and 8, 13cts; No. 9 to 12, 12cts; f.o.b. Toronto and Winnipeg, discounts on application. "C" brand, 40, 10 and 7½ off list. "M.R.M." brand, 60 per cent. off.

HORSESHOES—Our quotations continue as follows: P.B. base, $3.65; "M.R.M. Co., latest improved pattern" iron shoes, light and medium pattern, No. 2 and larger, $3.80; No. 1 and smaller, $4.05; snow, No. 2 and larger, $4.05; No. 1 and smaller, $4.30; light steel shoes, No. 2 and larger, $4.15; No. 1 and smaller, $4.40; featherweight, all sizes, 0 to 4, $5.75; toeweight, all sizes 1 to 4, $7. Packing, up to three sizes in a keg, 10c. per 100 lbs. extra; more than three sizes in a keg, 25c. per 100 lbs. extra.

BUILDING PAPER—Orders are still coming in briskly, the advanced price not affecting the ordering to any extent.

CEMENT—We quote: For carload orders, f.o.b. Toronto, Canadian Portland, $1.95. For smaller orders ex warehouse, Canadian Portland, $2.05 upwards.

FIREBRICK—English and Scotch fire brick $27 to $30; American low-grade, $23 to $25; high-grade, $27.50 to $35.

HIDES, WOOL, FURS—The market is steady with a slight decline in country hides to allow for lack of quality.

Hides, inspected, cows and steers, No. 1	....	0 11½
" " " " No. 2	....	0 10½
Country hides, flat, per lb., cured	0 09½	0 10
" " green	0 08½	0 09
Calf skins, No. 1, city	....	0 12
" " No. 1, country	....	0 11
Lamb skins	1 20	1 25
Horse hides, No. 1	3 50	3 75
Rendered tallow, per lb.	0 05½	0 06½
Pulled wools, super, per lb.	....	0 25
" " extra	....	0 27
Wool, unwashed fleece	....	0 15
" washed fleece	0 24	0 25

FURS.

	No. 1,	Prime
Raccoon	....	1 50
Mink, dark	5 00	6 00
" pale	2 50	3 50
Fox, red	3 00	3 50
" cross	3 00	7 00
Lynx	5 00	7 00
Bear, black	....	12 00
" cubs and yearlings	....	5 00
Wolf, timber	....	2 75
" prairie	....	1 25
Weasel, white	....	0 60
Badger	....	1 60
Fisher, dark	5 00	6 00
Skunk, black	....	1 25
" " short striped	....	0 90
" " long striped	....	0 56
Marten	3 50	20 00
Muskrat., fall	....	0 10
" winter	....	0 26
" spring	....	0 23
" western	0 12	0 10

Montreal Metal Markets

Office of HARDWARE AND METAL,
232 McGill Street,
Montreal, January 25, 1907

The markets this week show several advances, among which ingot copper again has a place, being extremely short just at present, with not much offering.

Tin is firming up again and will advance if the market goes any higher, which it will very likely do next week.

Zinc spelter remains firm, with no changes at present. The same refers to lead and antimony.

Galvanized iron shows an advance.

A fine business is being done in old material, and our prices this week show quite a few advances.

Structural iron and steel is somewhat scarce; the big mills in the States are running full capacity, and promising deliveries in from three to six months; therefore, nearly all large buildings in course of construction are likely to be held up more or less, and contractors should look well ahead for their requirements.

COPPER — Extremely scarce, with very little appearing. Ingot copper shows an advance of half a cent over last week's figures. Our prices now are: Ingot copper, 26½ to 27 cts.; sheet copper, base sizes, 34 cts.

INGOT TIN—Firming up again and expected to advance shortly. We still quote: 45½ to 46c. per pound.

ZINC SPELTER—Prices have not yet advanced. We are still quoting: $7.50 to $7.75 per 100 lbs.

PIG LEAD—Very firm. Our prices remain: $5.50 to $5.60 per 100 lbs.

ANTIMONY—No change over last week's prices, we still quote: 27½ to 28 cts. per pound.

PIG IRON—Very hard to get at present, with prospects of remaining so for some time. We are quoting: Londonderry, $24.50; Carron, No. 1, $24.50; Carron, special, $23.50; Summerlee, No. 2 selected, $25.00; Summerlee, No. 3 soft, $23.50.

BOILER TUBES—Very firm. We quote: 1½ to 2 in., $9.50; 2¼ in., $10.35; 2½ in., $11.50; 3 in., $13.00; 3½ in., $17.00; 4 in. $21.50; 5 in., $45.00.

TOOL STEEL — Our prices remain: Colonial Black Diamond, 8c to 9c; Sanderson's 8c. to 45c., according to grade; Jessop's, 13c.; Jonas & Colver's, 10c. to 20c.; "Air Hardening," 65c. per pound; Conqueror, 7½c.; Conqueror high speed steel, 60c.; Jowitt's Diamond J., 6¼c. to 7c.; Jowitt's best, 11c. to 11½c.

MERCHANT STEEL.—Our prices are as follows: Sleigh shoe, $2.25; tire, $2.40, spring, $2.75; toecalk, $3.05; machinery iron finish, $2.40; reeled machinery steel, $2.75; mild, $2.25 base and upwards; square harrow teeth $2.40; band steel, $2.45 base. Net cash 30 days. Rivet steel quoted on application.

COLD ROLLED SHAFTING—Present prices are: 3-16 inch to ¼ inch, $7.25; 5-16 inch to 11-32 inch, $6.20; ⅜ inch to 17-32 inch, $5.15; 9-16 inch to 47-64 inch, $4.45; ¾ inch to 17-16 inch, $4.10; 1½ inch to 3 inch, $3.75; 3⅛ to 3 7-16 inch, $3.92; 3½ inch to 3 15-16 inch, $4.10; 4 inch to 4 7-16 inch, $4.45; 4½ inch to 4 15-16 inch, $4.80. This is equivalent to 30 per cent. off list.

GALVANIZED IRON — One line shows an advance of 10 cents (Gorbal's "Best Best" 28 gauge, now selling at $4.55). We now quote: Queen's Head, 28 gauge, $4.60 to $4.85;

26 gauge, $4.45; 22 to 24 gauge, $3.90; 16 to 20 gauge, $3.75; Apollo, 28 gauge, $4.45 to $4.70; 26 gauge, $4.30; 22 and 24 gauge, $3.75; 16 and 20 gauge, $3.60; Comet, 28 gauge, $4.45 to $4.70; 26 gauge, $4.30 to $4.45; 22 and 24 gauge, $3.75 to $4.00; $16 to 20 gauge, $3.60 to $3.85; Fleur-de-Lis, 28 gauge, $4.45 to $4.70; 26 gauge, $4.30; 22 and 24 gauge, $3.75; 16 to 20 gauge, $3.60; Gorbals "Best Best," 28 gauge, $4.55; Colborne Crown, 28 gauge, $4.45; 26 gauge, $4.30; 24 gauge, $3.75. In less than case lots, 25c extra.

OLD MATERIAL — Several lines show slight advances. Our prices are: Heavy copper, 20c per lb.; light copper, 17c per lb.; heavy red brass, 17c per lb.; heavy yellow brass, 14c per lb.; light brass, 10 to 10½c per lb.; tea lead, 4¼c per lb.: heavy lead, 4½c. per lb.; scrap zinc, 4¼c. per lb.; No. 1 wrought $17.00 per ton; No. 2 wrought $6.00 per ton; No. 1 machinery, $18.00 per ton; stove plate, $13.00 per ton; old rubber, 10½ cents per lb.; mixed rags, 1 to 1¾c per lb.

Toronto Metal Markets

Office of HARDWARE AND METAL,
10 Front Street East,
Toronto, Jan 25, 1907

While prices on all basic metals are the same as a week ago, conditions have strengthened somewhat, and ingot tin is gaining in strength. Copper continues the great feature, its great firmness and scarcity being considered entirely free from speculative dealings by close followers of the market. Lead is increasing in activity and scarcity, and higher prices will probably result. In old materials tea lead and old rubbers have shown advances and in coal easier conditions are reported with revised quotations.

Business is being transacted in large volume, the movement of stocks being much greater than usual at this season. The high prices increase the totals very materially, as with the same tonnage of metals the cash receipts will show totals fully double those of a year or so ago. The difficulty in securing stocks in large quantities also makes necessary the placing of more orders, this aiding in making business more lively.

Manufacturers who use tubes and bars still experience difficulty in securing shipments of supplies, and screws, wire nails and similar goods continue scarce as a result.

PIG IRON — Prices hold very stiff, but there is not much activity at this season. Hamilton, Midland and Londonderry are off the market, and Radnor is quoted at $33 at furnace. Middlesborough is quoted at $24.50, and Summerlee, at $26 f.o.b., Toronto.

BAR IRON—Quotations remain as before with a fair trade being done. We quote: $2.30 f.o.b. Toronto, with 2 per cent. discount.

INGOT TIN — Speculation has strengthened the market, but we are still quoting 45 to 46 cents a pound.

TIN PLATES—Orders for future delivery are still being placed but most dealers have bought. Prices are firm.

SHEETS AND PLATES—Active buying is reported with prices steady. Stocks are light and prices advancing.

BRASS—Along with copper this metal holds a very firm position. We quote 30 cents per pound.

COPPER — Stocks are very low and still prices are very stiff. We still quote: Ingot copper $26 per 100 lbs., and sheet copper $31 to $32 per 100 lbs.

LEAD — Material is scarce with a very active demand from buyers who are short. Prices keep very firm. We quote: $5.40 for imported pig and $5.75 to $6.00 for bar lead.

ZINC SPELTER—Stocks continue light with the market active. We quote 7½c. per lb. for foreign and 7c. per lb. for domestic. Sheet zinc is quoted at 8 1-4c. in casks, and 8 1-2c. in part casks.

BOILER PLATES AND TUBES—We quote: Plates, per 100 lbs., ¼ in. to ½ in., $2.50; ⅜ in., $2.35 heads, per 100 lbs., $2.75; tank plates, 3-16 in., $2.65; tubes, per 100 feet, 1½ in., $8.50; 2, $9; 2 1-2, $11.30; 3, $12.50; 3 1-2, $16; 4, $20.00. Terms, 2 per cent. off.

ANTIMONY—Stocks are in fair condition with prices unchanged at 27c. per pound.

OLD MATERIAL—Dealers report a good demand with supplies coming in well. Their buying prices are: Heavy copper and wire, 20 cents per lb.; light copper, 17½c. per lb.; heavy red brass, 17½c. per lb.; heavy yellow brass, 15c. per lb.; light brass, 11c. per lb.; tea lead, $4.00 per 100 lbs.; heavy lead, $4.25 per 100 lbs.; scrap zinc, 4½c. per lb.; No. 1 wrought iron, $15; No. 2 wrought, $8; machinery cast scrap, $17 to $18; stove plate, $14; malleable and steel, $9; old rubbers, 10½c. per pound; country mixed rags, $1 to $1.25 per 100 lbs., according to quality.

COAL—Stocks are in good shape, with the scarcity of slack being relieved. We quote:

Standard Hocking soft coal f.o.b. at mines, lump $1.50, ¾ inch, $1.40; run of mine, $1.30; nut, $1.25; N. & S., $1; P. & S. 85c.

Youghiogheny soft coal in cars, bonder at bridges: lump, $2.60; ¾ inch, $2.50; mine run, $2.40; slack, $2.15.

United States Metal Markets

From the Iron Age, Jan. 24, 1907.

There is distinct evidence, notably in the New York market, of an eagerness on the part of some sellers of pig iron to secure business for spot and for early delivery, and the market is weaker, with lower prices accepted for what little business there is. Unless weather conditions prove adverse there is the prospect that the famine is over. It appears, too, as though shipments from the Birmingham district are improving, at least to tidewater markets. To points north of the Ohio the floods have checked the movement.

Considerable foreign iron is still coming forward, but so far as we can learn there has been no buying of round lots abroad for some time, and it looks as though the end of the import movement is pretty well in sight.

Buying for delivery during the second half has continued on quite a good scale, particularly in the territory west of the Allegheny Mountains, but a feeling of conservatism is spreading among founders and steel makers who are now partly covered and are inclined to await developments before committing themselves further.

In order to compensate for the high price of spelter the price of galvanized sheets will be advanced 10 cents per 100 lb. by the leading interest. On blue annealed sheets there will be an advance of 5 cents per 100 lb., prices hitherto having been out of harmony with other departments of the sheet trade.

A significant report is that a fair sized business has been done in tin plate for delivery during the third quarter, a premium of 10 cents per box being paid.

From the Iron Trade Review, Jan. 21 1907.

Pig iron is the absorbing topic, and its steady movement and upward trend are dominating features of the present situation. Transportation facilities have been greatly marred by the storms of the past week, and an unprecedented demand for spot iron is one of the contingencies expected.

Specifications for finished material show no evidence of declining. They are abnormally large for plates, steel bars and sheets, in each of which mills are months behind in deliveries. Structural steel is improving, though the increased tonnage is almost wholly from railroads and carbuilders. Building work is seasonably dull, and specifications from contractors naturally are light. There is a good inquiry for light and heavy rails, and a few small sales of the latter to traction enterprises have been made at $30 mill. Scrap continues weak and prices are lower.

Inquiry for cast iron pipe product is heavier than in years. Municipal wants are slow in coming forward, but specifications from gas companies and other private interests are unusually brisk.

The action of the Pennsylvania railroad in organizing a $100,000,000 car trust, following the announcement of a new issue of capital stock and bonds amounting to $200,000,000 has aroused much interest, and is generally accepted as indicating unprecedented expenditures for freight cars. Although $25,000,000 of the trust certificates will be for payment on cars ordered for 1906 and $19,000,000 for cars already ordered for 1907, further heavy ordering at an early date is expected. The principal car interest is now in the market for 30,000 tons of pig iron for the last half of the year and has placed an order for 5,000 tons of shapes, angles and plates.

London, 'Eng., Metal Markets

From Metal Market Report, Jan. 21, 1907.

Cleveland warrants are quoted at 59s. 10½d., and Glasgow Standard warrants at 59s., making price as compared with last week 9d. lower on Cleveland warrants, and 10d. lower on Glasgow Standard warrants.

TIN—Spot tin opened strong at £192 10s., futures at £192 15s., and after sales of 250 tons of spot and 550 tons of futures, closed easy at £192 5s. for spot, £192 10s. for futures, making price compared with last week £3 10s. higher on spot and £2 15s. higher on futures.

COPPER—Spot copper opened strong at £107, futures £108 3s., and after sales of 500 tons of spot and 800 tons of futures, closed easy at £106 15s. for spot, and £108 for futures, making price compared with last week £1 15s. lower on spot and 17s. 6d. lower on futures.

LEAD.—The market closed at £19 15s. making price as compared with last week 2s. 6d. lower.

SPELTER — The market closed at £23 17s. 6d., making price as compared with last week 15s. lower.

London Hardware Trade News

The prominence attained by London last week, when the published clearing house returns showed an increase of 73.4 per cent. over the figures of a year ago, is significant. This increase was by far the largest experienced by any clearing house in Canada, the increase in Vancouver, the second city on the list, being 51.9 per cent. London's merchants and manufacturers are enjoying a vast increase in business.

The past year was marked by rapid advancement in industrial London. A total of $1,250,000 was spent on new buildings, including 300 houses and many factories. The list of new business premises and additions erected during 1906 includes: McClary Manufacturing Co., Trafalgar Street, 100 x 230 feet, five stories high; Wortman & Ward Manufacturing Co., York Street, extension 50 x 100 feet, 2 stories; London Foundry Co., Thames Street, extension 54 x 36 feet, 2 stories; George White & Sons Co., east end warehouse, 240 x 140 feet, 1 storey; Dennis Wire and Iron Co., Dundas Street extension 24 x 45 feet, 1 storey.

The number of hands now employed in London factories is 9,000. The McClary Mfg. Co. employs over 800 people, London Rolling Mills 250, E. Leonard & Sons 200, Labatt Mfg. Co. 175, George White & Sons 160, London Bolt and Hinge Works 100, Wortman & Ward Mfg. Co. 90, and many others in the hardware and metal trade are among the important factories in the city.

There were 1,259 sales of real estate for the year, the consideration being $2,033,486. The assessment of these properties totaled $1,447,822, being but 71 per cent. of the actual value. So the city taxes on a basis of sale value are less than 15 mills on the dollar. Building lots in many sections have more than doubled in value in the past five years in London. The city cannot have a "boom." At least such a thing never seemed within the range of possibilities. But beyond question the city is having an era of prosperity hitherto unprecedented.

Edgar M. Bogart, of Calgary, who has been visiting relatives in London during the past month, left on Jan. 17 for the west. Mr. Bogart has resigned his position as traveler for a western hardware firm, to engage in business with N. McCutcheon, of Toronto. Many friends in this city will wish him every success in his new venture in the west.

N.B. Hardware Trade News

St. John, Jan. 21.

There have been plenty of rumors of late about plans for industrial development in St. John. Some of these reports, in all of which there seems some foundation of truth, have been touched upon in this column in previous issues. One that has excited quite a bit of interest is that to the effect that large car shops are to be established here on a stretch of property along the Marsh Road. Men whose names were associated with the affair were of the stamp of J. A. Likely and T. H. Estabrooks, leading citizens. Now the car shops story is quiescent. In its place, however, has come another report. In this the names of some of the gentlemen said to be connected with the car shop plans are also used. If this latter story be correct there are plans on foot to have a large carriage factory put up on the site above mentioned. There is room for such a plant here, but of course there are many things to be considered in the connection. Definite information is not obtainable, but the rumors may be picked up on the street.

* * *

For the past couple of weeks business has been a trifle on the quiet side with wholesale houses. Preparations for the spring trade have been going on steadily but the actual sales recorded have naturally been comparatively small though for the season of the year they have been satisfactory. It has perhaps been noticeable that while houses dealing in other lines have had or are having "clearing up" sales, the hardware people have not been holding such special sales. One would think that, to a large degree at least, what holds good in other lines would be true of hardware. But steady going, even sales seem to be satisfactory to the hardware houses at this as at other times.

* * *

High prices still reign. The forecasts are: Further advances. Copper is so scarce that quite a number of dealers will not offer quotations at all. Moreover, there is no indication of an easing of market conditions as regards this metal in the near future. Poultry netting generally is higher too than at this time last year. Wire fencing may be referred to in the same way. This is due, of course, largely to the advances in steel rods. Piece tinware, too, has jumped up of late. White lead has become even firmer than formerly and additional advances are expected. As regards cordage, however, there is a different tale to tell. The market is practically paralyzed so far as cordage is concerned. For this the conditions of the new tariff are accountable.

At the annual meeting of the St. John Iron and Hardware Association held recently, it was decided that the association's annual dinner should be held on January 30th.

B.C. Hardware Trade News

Vancouver, Jan. 19, 1907.

From official figures the totals of ore shipments in the Kootenay in 1906 have been compiled. There were in all 1,570,148 tons shipped. This was divided approximately as follows:

Boundary shipments, total 1,135,138 tons, including Grandy mines, 801,598; Mother Lode, 101,539; Brooklyn-Stemwinder, 142,625; Sunset, 41,128; Rawhide, 26,646; Emma, 12,662. There were 27 shippers in this district.

Rossland district, 281,711 tons; including, Centre Star, 114,972; Le Roi, 126,252; Le Roi No. 2, 22,166, Smelted, 13,200. Twelve shippers.

Slocan district, 133,299 tons, including, St. Eugene, 26,068; Sullivan, 27,320; Ymir, 12,451; La Plata, 14,630; Eva, 10,314; Queen, 7,401; Hunter V, 4,188; North Star, 3,334. Number of shippers, 125.

* * *

The first shipments of the machinery for the new dredge to be built on False Creek, to be used in Vancouver harbor, have been shipped west. The machinery is all the work of the Polson Iron Works, Toronto, and some unavoidable delays have occurred in making delivery. It was originally intended that the machinery would be here last Fall, so that the new dredge would be at work in the Spring. The cost of the plant complete will be about $100,000. It is of the suction type, long pipes laid on pontoons delivering the mud excavated from the bed of the harbor to any point desired. The whole dredge is being shipped in the knock-down and will be put together here.

* * *

Chandelar, the newest mining camp in the golden north, is the latest discovery of Wada, the hardy little Japanese, who first reported the Tanana country, though his finds in the latter never brought him much direct gain. The little brown man, who is considered one of the hardiest men in the far north, dropped into Dawson four years ago this winter, when the thermometer was 50 below. His tales of the wonderful riches of the Tanana country given with such circumstantial detail and precision, induced a great stampede to the new district, though the weather was cruelly severe at the time. The stampede is still recalled as about the greatest in the history of Dawson. At the time the stampeders were disappointed in not finding Wada's fairy tales verified at once. They were indeed so angry that they first decided to hang the unlucky little Jap, but relented. Later, the finds in the Tanana district more than verified the biggest tales Wada had told, of the wealth of the district. This time the little Jap is reaping the reward by getting big results for himself from his claim which is one of the richest in the Chandelar field.

PUBLICATION DELAYED.

Owing to the storms of the past week many trains have been delayed and considerable matter for publication has necessarily been crowded out. We trust, therefore, that our subscribers will understand the difficulties of publication under the circumstances and will know why the paper is delayed if it is not received at the usual time.

MANITOBA HARDWARE AND METAL MARKETS

(Market quotations corrected by telegraph up to 12 a.m. Friday, Jan. 25, 1907.)

Room 511, Union Bank Building, Office of HARDWARE AND METAL, Winnipeg, Man.

Storms have demoralized train services, and many orders are being delayed in shipment.

LANTERNS. — Quotations are as follows: Cold blast, per dozen, $6.50; coppered cold blast, per dozen, $8.50, cold blast dash, per dozen, $8.50.

WIRE—We quote as follows: Barbed wire, 100 lbs., $2.90; plain galvanized, 6 to 8, $3.39; 10, $3.50; 12, $3.10; 13 $3.20; 14, $3.90; 15, $4.45; 16, $4.60, plain twist, $3.45; staples, $3.25; oiled annealed wire, 10, $2.96; 11, $3.02; 12, $3.10; 13, $3.20; 14, $3.30; 15, $3.45. Annealed wires (unoiled) 10c. less.

HORSESHOES — Quotations are as follows: Iron, No. 0 to No. 1, $4.65; No. 2 and larger, $4.40; snowshoes, No. 0 to No. 1, $4.90; No. 2 and larger, $4.65; steel, No. 0 to No. 1, $5; No. 2 and larger, $4.75.

HORSENAILS — Lists and discounts are quoted as follows: Capewell brand, list, per lb., Nos. 5, 15c.; 6, 14c.; 7 and 8, 13c.; 9 to 12, 12c.; f.o.b. Winnipeg. Discounts on application. No. 10, 20c.; No. 9, 22c.; No. 8, 24c.; No. 7, 26c.; No. 6, 28c.; No. 5, 32c.; No. 4, 40c., per pound. Discounts are quoted as follows: "C" brand, 40, 10 and 7½ per cent., "M" brand and other brands, 55 and 60 per cent. Add 15c. per box.

WIRE NAILS—Quoted now at $2.80 per keg.

CUT NAILS.—As noted last week cut nails have been advanced to $2.90 per keg.

PRESSED SPIKES — Prices are quoted as follows since the recent advance: ¼ x 5 and 6, $4.75; 5-6 x 5, 6 and 7, $4.40; ⅜ x 6, 7 and 8, $4.25; 7-16 x 7 and 9, $4.15; ½ x 8, 9, 10 and 12, $4.05; ⅝ x 10 and 12, $3.90. All other lengths 25c. extra net.

SCREWS—Discounts are as follows: Flat head, iron, bright, 85 and 10 p.c.; round head, iron, 80 p.c.; flat head, brass, 75 and 10 p.c.; round head, brass 70 and 10 p.c.; coach, 70 p.c.

NUTS AND BOLTS — Discounts are unchanged and continue as follows: Bolts, carriage, ⅜ or smaller, 60 and 5; bolts, carriage, 7-16 and up, 55; bolts, machine, ⅜ and under, 55 and 5; bolts, machine, 7-16 and over, 55; bolts, tire, 65; bolt ends, 55; sleigh shoe bolts, 65 and 10; machine screws, 70; plough bolts, 55; square nuts, case lots, 3; square nuts, small lots, 2½; hex nuts, case lots, 3; hex nuts, smaller lots, 2½ p.c.

RIVETS—Discounts are quoted as follows since the recent advance in the price of copper rivets: Iron, discounts, 60 and 10 p.c.; copper, No. 7, 38c.; No. 8, 38½c.; copper, No. 10, 41½c.; copper, No. 12, 44½c.; assorted, No. 8 and 10, 39½c. and 42½c.

COIL CHAIN.—Prices have been revised, the general effect being an advance. Quotations now are: ¼ inch $7.00; 5-16, $5.35; ⅜, $4.75; 7-16, $4.50; ½, $4.25; 9-16, $4.20; ⅝, $4.25; ¾, $4.10.

SHOVELS—List has advanced $1 per dozen on all spades, shovels and scoops.

HARVEST TOOLS — Discounts continue as before, 60 and 5 per cent.

AXE HANDLES—Quoted as follows: Turned, s.g. hickory, doz., $3.15; No. 1, $1.90; No. 2, $1.60; octagon, extra, $2.30; No. 1, $1.60.

AXES.—Quotations are: Bench axes, 40; broad axes, 25 p.c. dis. off list; Royal Oak, per dozen, $6.25; Maple Leaf, $8.25; Model, $8.50; Black Prince, $7.25; Black Diamond, $9.25; Standard flint edge, $8.75; Copper King, $8.25; Columbian, $9.50; handled axes, North Star, $7.75; Black Prince, $9.25; Standard flint edge, $10.75; Copper King, $11 per dozen.

BUTTS—The discount on wrought iron butts is 65 and 5 p.c.

CHURNS — The discounts from list continue as before: 45 and 5 per cent.; but the list has been advanced and is now as follows: No. 0, $9; No. 1, $9; No 2, $10; No. 3, $11; No. 4, $13; No. 5, $16.

CHISELS—Quoted at 70 p.c. off list prices.

AUGER BITS—Discount on "Irwin" bits is 47½ per cent., and on other lines 70 per cent.

BLOCKS—Discount on steel blocks is 35 p.c. off list prices; on wood, 55 p.c.

FITTINGS—Discounts continue as follows: Wrought couplings, 60; nipples, 65 and 10; T's and elbows, 10; malleable bushings, 50; malleable unions, 55 p.c.

GRINDSTONES—As noted last week, the price is now 1½c. per lb., a decline of ½c.

FORK HANDLES—The discount is 40 p.c. from list prices.

HINGES—The discount on light "T" and strap hinges is 65 p.c. off list prices.

HOOKS—Prices are quoted as follows: Brush hooks, heavy, per doz., $8.75; grass hooks, $1.70.

CLEVISES—Price is now 6½c. per lb.

STOVE PIPES—Quotations are as follows: 6-inch, per 100 feet length, $9; 7-inch, $9.75.

DRAW KNIVES—The discount is 70 per cent. from list prices.

RULES—Discount is 50 per cent.

WASHERS—On small quantities the discount is 35 p.c.; on full boxes it is 40 p.c.

WRINGERS — Prices have been advanced $2 per dozen, and quotations are now as follows: Royal Canadian, $35.00; B.B., $39.75, per dozen.

FILES—Discounts are quoted as follows: Arcade, 75; Black Diamond, 60; Nicholson's, 62½ p.c.

BUILDING PAPER — Prices are as follows: Plain, Joliette, 40c.; Cyclone, 55c.; Anchor, 55c.; pure fibre, 60c.; tarred, Joliette, 65c.; Cyclone, 80c.; Anchor, 65c.; pure fibre, 80c.

TINWARE, ETC.—Quoted as follows: Pressed, retinned, 70 and 10; pressed, plain, 75 and 2½; pieced, 30; japanned ware, 37½; enamelled ware, Famous, 50; Imperial, 50 and 10; Imperial, one coat, 60; Premier, 50; Colonial, 50 and 10; Royal, 60; Victoria, 45; white 45; Diamond, 50; Granite, 60 p.c.

GALVANIZED WARE. — The dis-

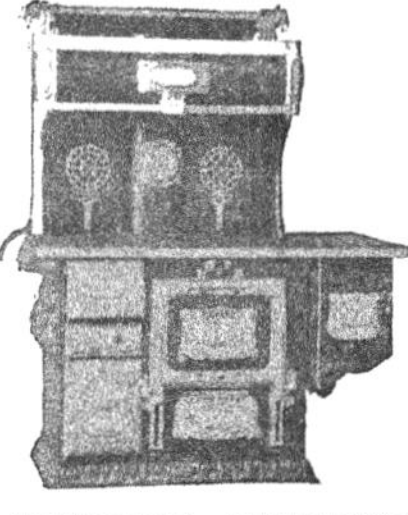

count on pail is now 37½ per cent.; and on other galvanized lines the discount is 30 per cent.

CORDAGE—We quote: Rope, sisal, 7-16 and larger, basis, $11.25; Manila, 7-16 and larger, basis, $16.25; Lathyarn, $11.25; cotton rope, per lb., 21c.

SOLDER—Quoted at 27c. per pound. Block tin is quoted at 45c. per pound.

VISES — Prices are quoted as follows: "Peter Wright," 30 to 34, 14½c.; 35 to 39, 14c.; 48 and larger, 13½c. per lb..

ANVILS—"Peter Wright" anvils are selling at 11c. per lb.

CROWBARS—Quoted now at 4c. per lb.

POWER HORSE CLIPPERS — The "1902" power horse clipper is selling at $12, and the "Twentieth Century" at $6. The "1904" sheep shearing machines are sold at $13.60.

AMMUNITION, ETC.—Quotations are as follows: Cartridges, Dominion R.F. 50 and 5; Dominion, C.F., 33⅓ C.F., pistol, p.c.; C.F., military, 10 p.c. advance. Loaded shells: Dominion Eley's and Kynoch's soft, 12 gauge, black, $16.50; chilled, 12 gauge, $17.50; soft, 10 gauge, $19.50;

chilled, 10 gauge, $20.50. Shot, ordinary, per 100 lbs., $7.25; chilled, $7.75; powder, F.F., keg, Hamilton, $4.75; F.F.G., Dupont's, $5.

IRON AND STEEL.—Quotations are: Bar iron basis, $2.70. Swedish iron basis, $4.95; sleigh shoe steel, $2.75; spring steel, $3.25; machinery steel, $3.50; tool steel, Black Diamond, 100 lbs., $9.50; Jessop, $13.

SHEET ZINC — The price is now $8.50 for cask lots, and $9 for broken lots.

PIG LEAD—Quoted a $5.85 per cwt.

AXLE GREASE—"Mica" axle grease is quoted at $2.75 per case, and "Diamond" at $1.60.

IRON PIPE AND FITTINGS— Revised prices are as follows:—Black pipe, ¼ inch, $2.65; ⅜, $2.80; ½, $3.50; ¾, $4.40; 1, $6.35; 1¼, $8.65; 1½, $10.40; 2, 13.85; 2½, $19.00; 3, $25.00. Galvanized iron pipe, ⅜ inch, $3.75; ½, $4.35; ¾, $5.65; 1, $8.10; 1¼, $11.00; 1½, $13.25; 2, inch, $17.65. Nipples, discounts 70 and 10 per cent.; Unions, couplings, bushings and plugs, 60 per cent.

LEAD PIPE—The price, $7.80, is firmly maintained in view of the advancing lead market.

GALVANIZED IRON—Quoted as follows:—Apollo, 16 gauge, $3.90; 18 and 20, $4.10; 22 and 24, $4.45; 26, $4.40; 28, $4.65, 30 gauge or 10¾ oz., $4.95; Queen's Head, 24, $4.50; 26, $4.65; 28, $5.00.

TIN PLATES—We now quote as follows: IC charcoal, 20 x 28, box, $9.50; IX charcoal, 20 x 28, $11.50; XXI charcoal, 20 x 28, $13.50.

TERNE PLATES—Quoted at $9.

CANADA PLATES—Quoted as follows: Canada plate, 18 x 21, 18 x 24, $3.40; 20 x 28, $3.65; full polished, $4.15.

BLACK SHEETS—Prices are: 10 to 16 gauge, 100 lbs., $3.50; 18 to 22, $3.75; 24, $3.90; 26, $4; 28, $4.10.

PETROLEUM AND GASOLINE.—Silver Star in brls. per gal., 20c.; Sunlight in brls, per gal., 21c.; per case, $2.30; Eocene in brls, per gal., 23c.; per case, $2.50; Pennoline in brls., per gal., 24c.; Crystal Spray, 23c.; Silver Light, 21c.; Engine gasoline in barrels, per gal., 28c., f.o.b. Winnipeg in cases, $2.75.

PAINTS, OILS AND TURPENTINE—Turpentine is firm at the recent advance. White lead, pure, $6.50 to $7.50, according to brand; bladder putty in barrels, 2½c.; in kegs, 3¾c.; turpentine, barrel lots, Winnipeg, $1.01; Calgary, $1.08; Lethbridge, $1.08; Edmonton, $1.09. Less than barrel lots 5c. per gallon advance. Linseed oil, raw, Winnipeg, 67c.; Calgary, 74c.; Lethbridge, 74c.; Edmonton, 75c.; boiled oil, 3c. per gal. advance on these prices.

WINDOW GLASS—We quote: 16-oz. O.G., single, in 50-ft. boxes—16 to 25 united inches, $2.25; 26 to 40, $2.40; 16-oz. O.G., single, in 100-ft. cases — 16 to 25 united inches, $4; 26 to 40, $4.52; 41 to 50, $4.75; 50 to 60, $5.25; 61 to 70, $5.75. 21-oz. C.S., double, in 100-ft. cases—26 to 40 united inches, $7.35; 41 to 50, $8.40; 51 to 60, $9.45; 61 to 70, $10.50; 71 to 80, $11.55; 81 to 85, $12.60; 86 to 90, $14.75; 91 to 95, $17.30.

BOARD OF TRADE ELECTIONS.

The Montreal Board of Trade nominations are now closed. The expected nomination to the candidates for positions on the executive did not materialize, and George Caverhill is elected president of the Montreal Board of Trade for 1907 by acclamation. Thomas J. Drummond and Peter Lyall are the candidates for the position of second vice-president; Farquhar Robertson is elected by acclamation to the position of second vice-president, and C. B. Esdaile treasurer by acclamation.

A UNIQUE PORTABLE PUNCH AND SHEAR.

The illustration herewith shows a new type of punch and shear being placed on the market by the Canadian Buffalo Forge Co., Montreal, Que. The frame is made of one piece of extremely heavy rolled steel armor plate. All fittings are machined from drop forgings, and at every point in the design and

Portable Punch and Shear.

construction precaution has been taken to supply a machine which is practically indestructible. This tool is remarkable as a high power portable punch and shear. As an example, six-inch angle iron can be sheared in two cuts. The armour-plate frame of the tool makes it so light that it is easily rolled about the shop, or from place to place for odd jobs. It is a style of machine that blacksmiths and repair-shop craftsmen can use to advantage. The punches and dies furnished with this machine are ⅜, ½, ⅝, ¾-inch. The machine will shear a 6x⅝-inch bar, depth of throat 7 inches, height 58 inches, weight with truck 850 lbs.

BURMAN HORSE CLIPPERS.

One of the first manufacturers to make horse clippers was the firm of Burman & Sons, Birmingham, England, which was established in 1871, they being the original patentees of the reversible double action clipper. They have since extended the scope of business by manufacturing toilet clippers for barbers' use, power horse clipping machines and sheep shearing machinery, both for hand and power and for complete equipment for shearing stations. They have also absorbed by purchase the clipper departments of the following firms, during the last few years: Hipkiss & Co., Birmingham; Phillips & Bullows, Walsall; W. Bown, clipper department, Birmingham; The Kimberley Tool Co., clipper department, Birmingham. The works have been rebuilt and added to from time to time to cope with the increasing business.

The firm's best known horse clipper is the Newmarket, which has had a large trade in Canada for the last 20 years. Other well known machines are the Grand, Goodwood and Handicap, while amongst toilet clippers the Despatch is firmly established as a popular selling line at moderate price. A new feature in high-class barbers' clippers is the push spring toilet, which will shortly be placed on the market. The power horse clipper is slowly but surely ousting the hand clipper from the market, and now that the power clipper is offered at about the same price as half dozen hand clippers, it is not to be wondered at that they are becoming increasingly popular.

Burman & Sons have been represented in Canada for many years by B. & S. H. Thompson & Co., Montreal, who hold a stock of their goods. A new booklet for the year 1907 has just been issued and copies will be sent to any dealer mentioning this paper.

CAUSES OF FAILURE.

Bradstreets of January 22, analyzes the failures of 1906 in the Dominion of Canada and Newfoundland, as follows:

Failures due to	No.	Assets.
Incompetence	203	$ 878,185
Inexperience	41	250,238
Lack of capital	626	2,266,775
Unwise credits	13	1 90,100
Failures of others	14	101,200
Extravagance	9	52,175
Neglect	41	52,064
Competition	9	12,218
Specific conditions	168	392,766
Speculation	7	26,600
Fraud	108	182,760
Totals	1,239	$4,305,076

PREMIUMS HELP TRADE.

R. F. Lund, general sales agent for the Dover Manufacturing Company, Canal Dover, Ohio, was a caller at the Toronto office of Hardware and Metal on Monday last. Mr. Lund looks after the sale of asbestos sad irons as premiums for newspapers and says that experience has shown the sale of special irons in this way to be very helpful to the trade in the regular goods sold exclusively through the retail hardware trade. In Philadelphia, for instance, the "North American" used twenty-one carloads of the small French laundry irons as premiums, making a house-to-house canvass of the city and district, "Asbestos" irons being thus given a tremendous amount of free advertising. As a result of the premium campaign the demand for the other sizes of irons was increased fully 1000 per cent.

The Toronto Daily Star has contracted for 10,000 French sets of "Asbestos" irons to be used as premiums and the results of their canvass in Toronto will be watched with interest by the trade.

CATALOGUES AND BOOKLETS

When sending catalogues for review, manufacturers would confer a favor by pointing out the new articles that they contain. It would assist the editor in writing the review.

By mentioning HARDWARE and METAL to show that the writer is in the trade a copy of these catalogues or other printed matter will be sent by the firms whose addresses are given.

Artistic and Dramatic.

The James Robertson Company, Montreal and Toronto, have supplied their plumbing and paint customers with a beautiful dramatic calendar, a companion piece to the one issued a year ago. The calendar is a very expensive one, suitable for drawingroom decoration, containing, as it does splendid artistic reproductions of the likenesses of four of the best known actors and actresses, each picture being mounted separately.

Power and Hand Clippers.

Burman and Sons, Birmingham, Eng., have just issued a catalogue of 7x10 inches, containing 30 pages fully illustrated with various lines of horse and sheep clippers, for both hand and power use, as well as other devices which are manufactured by them for the sharpening of knives for these instruments.

Handsomely Framed Picture.

Shurly & Dietrich, Galt manufacturers of the famous Maple Leaf saws, are supplying the trade with a large, beautiful frame, enclosing a handsome advertisement, showing a maple leaf with a hand-saw in the centre. The prominent place given these pictures by the trade in former years, will no doubt be continued. The picture will prove an ornament to any hardware store, and any customer not having received a sample should immediately send and procure one for store decoration.

TAXES ON TRAVELERS.

The British Board of Trade has issued a useful bluebook, which gives the following list of British colonies where there are regulations or taxes upon travelers, with the annual amount payable:

New Zealand (£5).
Cape Colony (£25).
Natal (£10).
Transvaal (£10).
Orangia (£20).
Prince Edward Island ($20).
Quebec ($50 up).
British Columbia ($100).
British Honduras ($10).
Bechuanaland ($10).

In New Zealand the fee is held as a guarantee that the income tax due on the business done in the colony will be repaid. In Prince Edward Island and British Columbia the licenses are much higher for travelers selling intoxicating liquors or cigars.

In all other parts of the British Empire commercial travelers can carry on their work without hindrance.

Appended is a list of foreign countries which impose taxes on the ambassadors of trade, together with the annual amount charged:

Argentine Republic (varies in different provinces).
Bolivia ($300).
Brazil (varies according to province).
Bulgaria (£6).
Congo Free State (£8).
Denmark (£9).
Germany (1s).
Mexico (varies according to province).
Norway (£5 11s. per month).
Paraguay (£10 to £15).
Russia (£26 7s. 5d.).
Sweden (£5 11s. per month).
Uruguay (£21).

The commercial traveler in Russia has no fewer than four separate payments to make every year.

GOVERNMENT TO MAKE WIRE NETTING.

The Government of Victoria, Australia, has accepted a local bid for the supply of eight machines at the price of $5,000, for the manufacture of wire netting. These machines are for the purpose of establishing the industry in the penitentiary at Melbourne and supplying prison-made wire netting to landowners at cost price on long terms of repayment to enable them to cope with the rabbit pest.

Put yourself in your employer's place and figure out what kind of an employe you would hire to get the most out of your business. Then set yourself to try to be that employe.

A BETTER QUALITY OF LINSEED OIL FOR CANADA.

The manufacture of Strictly Pure Screw Press Linseed Oil in Canada by The Sherwin-Williams Co. marks a new standard of quality for linseed oil in the Dominion. By the "screw press process", which they control exclusively in Canada, they are able to extract from the flaxseed a quality of linseed oil not heretofore possible. Their "Screw Press Oil", as the new brand is called, is clearer, purer and lighter than other brands of strictly pure linseed oils.

A Battery of Screw Presses.

The "screw press process" extracts the oil from the flaxseed at a low temperature. In all other processes the seed is heated to a high temperature to secure a satisfactory yield of oil per bushel of seed. Where the seed is highly heated, the albumen and mucilaginous matter are freed and allowed to get into the oil. This foreign matter in the oil assists in the disintegration of the paint or varnish film. The "screw press process" leaves this matter in the meal.

The Company's new linseed oil mill is a part of their manufacturing plant at Patrick and Centre Streets, Montreal. They have erected special buildings there for their screw presses, filter presses, storage tanks, etc., and have built a large elevator for the handling of the seed, and steel tanks of large storage capacity for seed storage.

The New Linseed Oil Mill of The Sherwin-Williams Co., at Montreal.

The line of linseed oils The Sherwin-Williams Co. is now manufacturing in Canada includes the following:

The S-W. Strictly Pure Screw Raw Linseed Oil.

The S-W. Strictly Pure Screw Press Kettle Boiled Linseed Oil.

The S-W. Strictly Pure Screw Press Bleached or Refined Linseed Oil.

The Company will be glad to send further information and prices to those interested in *quality* linseed oil. Address, The Sherwin-Williams Co., 639 Centre St., Montreal, Que.

FOUNDRY AND METAL INDUSTRIES

The Dyment Foundry Company, of Barrie, have a few men working to complete some orders which were on hand at time of fire. When these are finished the men will be let go and the buildings closed up.

The citizens of Sydney have voted to grant a bonus of $50,000 to the new Rolling Mills Company. The company will begin building operations as soon as possible.

The Ontario Iron and Steel Company's plant, which has been under construction for some months at Welland, will be a much larger industry than was expected. The plans now are for an expenditure of $300,000, and the works will employ about five hundred men. It is expected they will be completed by June 1st.

The Niagara Falls Machine and Foundry Company have secured a large contract to supply cast and wrought iron work for the Michigan Central's Detroit river tunnel from the contractors, the Butler Bros. & Hoff Company. They have another contract to supply fire hydrants for the city of Toronto. The company's foundry is to be enlarged at once

A charter has been granted by the Manitoba Government to the Manitoba Rolling Mills Company, with a capital of $100,000, to take over and operate the Kirkwood Iron & Steel Rolling Mills, of Winnipeg. These mills were built by T. M. Kirkwood, Toronto, who has lately associated with him the president and officers of the United States Horseshoe Company, of Erie, Pa., Mr. Kirkwood will be the vice-president of the company.

R. L. Worthington, who has severed his connections with the Vulcan Iron Works, of Winnipeg, entertained the foreman and office staff at a tea on Jan. 7, when he was presented with a beautiful combination desk and book case. Mr. Worthington is resigning his position as draughtsman after six years with the Vulcan Works, to accept a similar position with Kelly Bros. & Mitchell, Winnipeg.

Another car construction plant is to be established at Montreal. It will be the largest steel and wood car factory in Canada. This week a number of capitalists and railway men, including Fred Eaton, president of the American Car and Foundry Company, Berwick, Pa., were shown over the city by Messrs. Frank H. Dunn and O. C. Kahler, of Berwick, Pa. Both of these gentlemen recently resigned from the service of the Dominion Car Company. A site has been selected between the Canada Car Company's premises and that of the Dominion Car Company, near Montreal West. The new works are to have a capacity of fifty to sixty completed cars a day. The estimated cost is five million dollars. The formality of organizing the company has yet to be attended to, and even the name is not yet decided upon, but operations will be started at once, it is stated, and it is promised that the works will be turning out cars early in the coming autumn.

POINTS ABOUT NEEDLES

One needle is a pretty small item, but the daily consumption of something like 3,000,000 needles all over the world makes a pretty big total. Every year the women of the United States break, lose, and use about 300,000,000 of these little instruments, writes H R Christy in Scientific American.

Our needles are the finished products of American ingenuity, skill and workmanship, and yet how many people, threading a needle or taking a stitch, have ever given a thought to the various processes through which the wire must pass ere it comes out a needle? The manufacture of a single needle includes some twenty-one or twenty-two different processes, as follows: Cutting the wire into lengths; straightening, by rubbing while heated, pointing the ends on grind-stones; stamping impression for the eyes; grooving; eying, the eye being pierced by screw presses; splitting, threading the double needle by the eyes on short lengths of fine wire; filing, removing the "cheek" left on each side of the eye by stamping; breaking, separating the two needles on the one length of wire; heading, heads filed and smoothed to remove the burr left by stamping and breaking; hardening in oil, the needle is thus made brittle; tempering; picking, separating those crooked in hardening; straightening the crooked ones; scouring and polishing; bluing, softening the eyes by heat; drilling or cleaning out the sides of the eye; head-grinding; point-setting, or the final sharpening; final polishing; then papering, and finally, labeling. For wrapping, purple paper is used, because it prevents rusting.

The needle, as we see it to-day, is the evolved product of centuries of invention. In its primitive form it was made of bone, ivory, or wood. The making of Spanish needles was introduced into England during the reign of Queen Elizabeth. Point by point the manufacture has improved, until the little instrument is one of the highly-finished products of nineteenth century machinery and skill.

SEWER PIPE

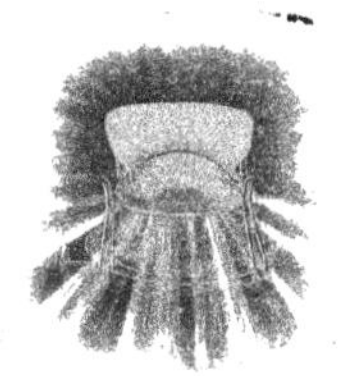

BUILDING AND INDUSTRIAL NEWS

HARDWARE AND METAL would be pleased to receive from any authoritative source building and industrial news of any sort, the formation or incorporation of companies, establishment or enlargement of mills, factories, foundries or other works, railway or mining news. All correspondence will be treated as confidential when desired.

The new fire station in Toronto will cost nearly $90,000.

The new Union Station to be built in Toronto will cost $2,000,000.

The Bank of Montreal will erect a fine office building in Vancouver.

A new theatre will be built in London, Ont., by A. J. Small, Toronto.

A new collegiate institute will be erected at Brandon, Man., at the cost of $65,000.

The William Rogers Company have commenced the plating of silverware at Niagara Falls.

The Mickle Dyment Lumber Company will erect a new factory in Brantford at the cost of $10,000.

J. V. Jenkins, proprietor of the Quinte Hotel, Belleville, which was burnt some time ago, will rebuild the hotel.

A new church will be erected in Regina at the cost of $40,000, having a seating capacity of 1,000 people.

Building permits to the amount of $84,900 have been issued in Toronto for the week ending January 19.

The Canada Screw Company will begin extensive building operations in the spring, doubling the capacity of their plant in Hamilton.

The Phillips Manufacturing Company of Toronto have plans out for a large factory, which will be erected for them in the spring.

The Ontario Lamp and Lantern Company, of Hamilton, have taken out a permit for an addition to their factory to cost $2,500.

The Dominion Express Company will erect a large and commodious shipping and receiving building near the C.P.R. depot in Winnipeg.

Owing to the scarcity of brick the new telephone building at Edmonton is to be of concrete up to the street level, and of cement blocks above that.

A factory for the manufacture of automatic fire shutters will be established at Niagara Falls shortly. The company has many orders on hand now.

The Montreal Street Railway Company are asking the Quebec Government for authority to increase their capitalization from $10,000,000 to $18,000,000.

A large metal manufacturing company is negotiating to locate in Calgary. They offer to build a plant costing $100,000 and employ from sixty to seventy hands.

The M. C. Galarneau Company, of Montreal, who were burned out in the late disastrous fire at that place, have decided to erect a large warehouse for their leather business.

Three hundred houses and a number of factories were erected in London, Ont., last year. The total cost of the dwellings is estimated at $750,000, and of the factories and public buildings $450,000, or $1,200,000 in all.

The Standard Fuel Manufacturing Company of Fairview, Halifax, have decided to move their works to Amherst, N.S. The company manufactures a fire lighter composed of sawdust and a chemical mixture formed into briquettes.

The Metal Shingle and Siding Company, Preston, are erecting a fine new two-storey office building. The building is being metal-sheeted in the Ionic style, and the interior will have metal walls and ceilings.

The Canadian Northern Railway has purchased about 110 acres of land on the Don flats, Toronto, to be used for roundhouses, car sheds, repair shops, cold storage plant and other requirements of the railway.

Marble is being used in the construction of new buildings and the remodelling of old to a greater extent than ever before. The use of this material is a valuable indication of the strength of the building market, and of the increasing prosperity of the country.

If the by-law granting the Plumbers' Supplies Co., of Galt, $15,000 is passed, they will erect a factory containing 44,000 square feet of floor space. The company must agree to expend $10,000 on buildings and plant and to employ at the end of 1908 at least 50 hands.

The largest and heaviest single block of granite ever sent from the United States into Canada has just been shipped to Montreal and will be fashioned into a memorial to be erected in honor of the late Hon. Raymond Prefontaine, Minister of Marine and Fisheries.

The Northern Turpentine Company has been incorporated with a share capital of $250,000, for the purpose of manufacturing turpentine, tar, charcoal, etc. The head office will be at Ottawa, the directors being A. H. Edwards, E. A. Reid, and R. G. Code, of Ottawa.

Owing to the remarkable electrical development in Canada, the Canadian General Electric Company, of Peterboro, has found it necessary to erect still another large machine shop, which will mean that between three and four acres of floor space will be occupied by this department.

A meeting of the creditors of the Toronto Plaster and Supply Company, who have a factory in Toronto, and which assigned about a week ago, was held last Saturday. The firm was given a week in which to make an offer for the property. Its liabilities are about $4,200, and there is a surplus of about $2,000.

The Safety Explosives Company has been incorporated with a share capital stock of $300,000 for the purpose of manufacturing explosives and chemicals of every nature. The head office will be at Montreal, Que., the directors being W. H. Evans, R. W. Withycomb, W. A. Wier, W. J. Wright and A. W. C. Macalister of Montreal.

The Producers Torpedo Company has been incorporated with a share capital of $15,000, for the purpose of manufacturing nitro-glycerine, dynamite and other high explosives. The head office will be at Leamington, the provisional directors being G. W. Benson, E. Wigle and E. Winter, of Leamington, and Wm. Fleming, of Kingsville.

Tudhope, Anderson & Company have been incorporated with a share capital of $300,000 for the purpose of manufacturing all kinds of machinery, farm implements, carriages, waggons, harness, etc. The head office will be at Winnipeg, the directors being H. F. Anderson, J. J. Bryan, S. H. Ross and E. A. Stulter, of Winnipeg, and Jas. B. Tudhope, of Orillia.

A Brantford man has secured the contract for the erection of Normal Schools for the Ontario Government at the following places and prices: North Bay, $54,000; Peterborough, $53,800; Stratford, $52,050; Hamilton, $52,050. The schools will be uniform in size, with 78 feet frontage and a depth of 110 feet, built of stone and brick. Including the basement there are four floors.

The Canadian Nut & Bolt Company have decided to build a factory at Niagara Falls, Ont., commencing work early in February. They have a contract to supply a large quantity of nuts and bolts for railway switches to the Canadian Ramapo Iron Works Company, which is building a big plant there. About fifty men will be employed in the nut and bolt works. The buildings will be of steel and concrete.

Geo. W. Reid and Company have been incorporated with a capital stock of $150,000, for the purpose of manufacturing concrete, cement, asphalt, sheet metal roofing preparations, and other materials which can be used by contractors, builders or roofers. Head office will be at Montreal, and directors will be C. T. Williams, F. H. Barwick, E. C. Barwick and K. N. Church, of Montreal, and J. K. McNutt, of Westmount.

Do not shirk your work and be always thinking of the money side of the proposition. Give good value for the money you receive, and you will be sure to succeed.

C.M.A. EXECUTIVE MEETING.

For the first time in the history of the association, the executive committee of the Canadian Manufacturers' Association met in Montreal, January 17, 1907. Up to the present time these meetings have been held in Toronto, and have been well attended by representatives from Montreal, Quebec and other eastern points. The principal idea in holding the meeting in Montreal is to allow representatives to attend who are at a long distance from Toronto, on which account they have heretofore been prevented from taking an active part in the work of the association.

The members, sixty-two in number, were entertained to a luncheon at the Canada Club, and afterwards held their meeting, when the recent tariff changes, technical education, and other matters were discussed at length.

The following members of the executive council were present: J. S. N. Dougall, William Bramley, E. Tougas, Alderman Sadler, George E. Drummond, D. J. Fraser, S. W. Ewing, Fred. Birks, Louis Simpson, J. C. Casavant, Robert Munro, Col. Burland, James Davidson, R. C. Wilkins, J. H. Sherrard, Hon. J. D. Rolland, J. J. McGill, W. T. Whitehead, Wm. Smaill, R. J. Younge, D. Morrice, T. Esmond Peck, C. C. Ballantyne, P. Hamill, William Cauldwell, J. R. Kinghorn, S. S. Boxer, Joseph Horsfall, J. M. H. Robertson, F. W. Fairman, Montreal; J. P. Murray, T. A. Staunton, J. H. Housser, George Brigden, Col. J. B. MacLean, G. Frank Beer, F. A. Ralph, A. S. Rogers, W. B. Tindall, W. K. George, George Booth, J. M. Sinclair, Toronto; George E. Amyot, G. A. Vaudry, Quebec; Lloyd Harris, Harry Cockshutt, C. H. Waterous, Brantford; William Levis, Halifax; F. L. Haszard, Charlottetown; J. Hewton, Kingston; Jas. Playfair, Midland; A. Saunders, Goderich; T. J. Storey, Brockville; C. S. G. Wilson, Ingersoll; J. M. Jenckes, Sherbrooke; R. J. White, Smith's Falls, D. Wilson, Collingwood.

ARTISTIC TILLSONBURG WINDOW

W. R. Hobbs, Tillsonburg, had a most effective Christmas display of seasonable goods. "Passing down the street at night," writes E. Norman, London, an observing traveling salesman, "there appeared to be myriads of multi-colored lights moving and flashing in the store window. Curiosity aroused, the sightseer drew closer, and saw tastefully dressed oblong trays carrying silver and plated goods, cutlery and other Christmas presents, coming into view and receding therefrom, the trays being held horizontally while making a circular revolution. The suggestion conveyed by the advancing goods to the onlooker's acquisitive desires, though mute, was forcible, and we learn that coin of the realm was freely exchanged as a result. The window dresser is to be congratulated on the artistic and business effects he obtained from two cycle wheels, a few laths, much bunting, a home-made water motor, and a piece of string used as an invisible driving belt. Electric light lamps of red, green, orange, and blue, alternately flashing, furnished the myriad light ensemble, skilfully used to attract the distant passer-by to become a close inspector and final purchaser."

BARGAINS IN SPORTING GOODS.

In almost every line of business it has been found to be both profitable and good policy to hold low-priced sales, or bargain days, at certain times when conditions and circumstances seem to justify them. This method is an excellent one, says the Sporting Goods Dealer, for clearing out old stock and making room for new models. Sporting goods dealers have for some time been followers of the practice and these sales are being held with increasing frequency. A bargain department as a permanent fixture of the store where sporting goods are retailed is hardly a wise policy, because it inevitably results in a general reputation for cheapness which naturally produces a slight distrust in the quality of goods. However, there are certain classes of goods, such as fishing tackle, that can be offered at reduced prices without affecting the reputation of the store, while at the same time serving as a bait for other business. The bargain department in the sporting goods trade is seldom a source of profit and its purpose is mainly to rid the shelves of superfluous goods; the profit is always incidental and indirect rather than primary. In disposing of "stale" goods at low-priced sales, it is a good plan to mix in some articles of higher grade and established prices in order to present a showing that is not entirely marked by cheapness. In pricing bargains it should be with the idea of getting rid of them at any cost and not with the idea that a profit of any amount is to be forthcoming. Out-of-date goods should have no place on the shelves of a first-class store and the money they will bring is much better invested in new goods. The retailer of sporting goods really has many opportunities to present goods at bargain prices. New models in guns, changed styles in sportsmen's clothing, fresh tackle and other goods are constantly being received. Some of this material finds a ready sale at its appointed prices during the season of its usefulness, and much of it does not "take" but remains behind to accumulate dust and occupy space that can be ill afforded. It is not only as a means of removing stale goods that the bargain sale is convenient and beneficial to the retail merchant. A few choice articles that will insure a fair profit can be placed prominently in the store and advertised at stated periods. They will help to draw customers to the place and in every way call attention to the fact that there is a store where sporting goods are sold and whose manager is wide awake and after business.

OTTAWA STOVE FOUNDRY.

The Canada Stove Company, Ottawa, which succeeded the Ottawa Furnace & Foundry Company, Ottawa, several months ago, are now manufacturing "Standard" stoves and ranges and "National" hot water boilers and hot air furnaces. They have made several improvements on the old lines made by the Ottawa company as well as extensive improvements to the plant. They have also installed the latest sanitary improvements for the comfort and convenience of the office staff and workmen. The Canada Stove Company is working under a provincial incorporation, the capitalization being $90,000.

CONDENSED OR "WANT" ADVERTISEMENTS.

Advertisements under this heading 2c. a word first nsertion; 1c. a word each subsequent insertion.

Contractions count as one word, but five figures (as $1,000) are allowed as one word.

Cash remittances to cover cost **must** accompany all advertisements. **In no case** can this rule be overlooked. Advertisements received without remittance cannot be acknowledged.

Where replies come to our care to be forwarded, five cents must be added to cost to cover postage, etc.

SITUATIONS VACANT.

WANTED—Good practical tinsmith; $18.00 weekly to the right man; steady job. Apply H. H. Gervan, Chilliwack, B.C. (5)

WANTED—Man with two or three thousand dollars to take half interest in old established hardware and tinsmith business in good town. Box 575, HARDWARE AND METAL. (4)

WANTED—An experienced hardware man at once apply stating wages required to The Geo. Taylor Hardware Co., Ltd., New Liskeard, Ont. [4]

WANTED—Experienced pricer for wholesale hardware; none but experienced man need apply. Box 578, HARDWARE AND METAL, Toronto.

BRIGHT, intelligent boy wanted in every town and village in Canada; good pay besides a gift of a watch for good work. Write The MacLean Publishing Company, 10 Front Street E., Toronto. [tf]

BUSINESS CHANCES.

HARDWARE stock about $7,000, including set of tinners' tools; stock is all new and situated on main line C.P.R. in Saskatchewan town of 1,800 population; building can be leased, is brick, two storeys and basement; net profit $4,000 yearly. Box 577, HARDWARE AND METAL, Toronto. [7]

HARDWARE BUSINESS established 17 years for sale—including stoves, ranges and tinware; a good plumbing and tinshop in connection; stock about $6,000. J. D. Murdoch & Co., Simcoe, Ont. [6]

FOR SALE — Hardware and tinware business in Western Ontario town of about 3,000 population; one opposition; turnover seventeen to twenty thousand dollars. Apply Box 679, HARDWARE AND METAL Toronto.

FOR SALE—A fully equipped Wire Mill and Nail Factory as a going concern, with about 500 tons of wire rods; factory situated on 150 ft. of water-front with trackage optional to buyer. B.C. Wire and Nail Co., Ltd., Vancouver, B.C. (7)

SITUATION WANTED.

HARDWAREMAN (28) open for engagement; west. Box 91. HARDWARE AND METAL, Winnipeg. (4)

FOR SALE.

FOR SALE—One three-foot geared three-inch roller, in good condition. Apply St. Mary's Hardware Limited. [4]

Persons addressing advertisers will kindly mention having seen their advertisement in Hardware and Metal.

Paint, Oil and Brush Trades

FLAX GROWERS' BURDEN.

Deputations recently waited on Finance Minister Fielding, with petitions requesting an increased duty on flax products The request was refused, however, as it was shown that the change would have a far-reaching effect and heavily burden the makers of paint who use linseed oil, and through these the industries that use paint extensively. The furniture men would also suffer, and other extensive users of tow, which is the rough, green fibre of flax.

During the past fiscal year we imported 1,170,012 gallons of linseed oil valued at $438,937, and on this the duty was $77,945 The rate was 25 per cent., but 1,027,730 gallons valued at $381,170 were imported under the preferential tariff. There were 951 cwt. of flax and raw tow imported free, the value being $5,978, also large quantities of oil cake and flax seed, both being on the free list.

The deputation which waited on the Minister wanted the old duty of $1.25 per cwt. restored, a duty of ten cents per bushel imposed on flax seed, and $10 per ton on green tow. While flax-growing is yet in its infancy in Canada, it is carried on quite profitably in sections of Ontario and the west, and those engaged in it are called upon to bear a share of the burden that certainly must fall somewhere, as the suggested duties would increase the cost of oil cake to the stock-raisers, in addition to increasing the cost of all paints by making linseed oil dearer.

AN INAUGURAL DINNER

The Standard Paint Co. of Canada, Montreal, gave a banquet at the Canada Club on Friday evening, the 18th inst., on the occasion of the opening of their works at Highlands, near Montreal.

After the good things had been disposed of, Major Lockerby, vice chairman proposed a toast to "The King." and the "President of the United States" "Dominion Parliament" was replied to by the Hon J. B Rolland, who, after many suitable remarks, ended up by saying he was glad to see the Standard Paint Co. of Canada had not only located in the Province of Quebec, but in Canada's largest city.

The "City of Montreal" was proposed by Peter Lyall, and suitably responded to by Alderman Sadler.

"Canadian Industries" was proposed by D. M. Stewart of the Sovereign Bank, and responded to by Messrs. Dougall, Hood, and Lariviere.

"Our Guests" by Felix Jellinck, responded to by Messrs. Decarie, Place, and Black

"Our Company" by J. L. Decarie, responded to by R. L. Shainwald.

The banquet itself was a sumptuous affair, and served to bring the Standard Paint Co. of Canada and its officers well before the trade in Montreal.

Among those present were: R. L. Shainwald, New York, president of the company; vice-chairman Major D. M. Lockerby, of Montreal, one of the directors; Louis C. Rugen, secretary; J. N. Richards, sales manager; Max Drey, vice-president; Felix Jellinck, treasurer; H. F. Gillespie, assistant superintendent American works; E. H. Morris, Philadelphia; Senator J. B. Rolland; D. A. Campbell; D. J. Munn, of Alex. McArthur & Co.; C. A. Lockerby; Chas. Pope Tucker; John Hugh McComb; J. W. McGregor, of J. C. Wilson & Co.; Howard McComb; John Fisher; L. C. Rugen, Jr., superintendent new works, Montreal; Major R. Starke; E. Lamontague; J. H. Bell; M. Birkett; C. Smallpeice, of Lewis Bros., Montreal; G. R. Burton, of the R. E. T. Pringle Co., Montreal; John Black, Cobalt; A. F. Ross; R. W. Garth; D. Farrell; John Smith; Mr. Belanger; E. F. Casey, of the Montreal street railway; N. Greyburn; A. H. Edwards, of Montreal Light & Power Co; Andrew Dawes; Alderman Sadler, acting mayor; Lt.-Col D. M. Stewart, mgr. Sovereign Bank; A. R. Oughtred, K.C.; Peter Lyall; McLea Wallbank; J. S. R. Dougall, president Manufacturers' Association; Geo. Head, president builders' exchange; J. Gagnon, vice-president builders' exchange; J. Lorne, secretary builders' exchange; F. C. Lariviere; Geo. Esplin; A. Scott Robinson; J. L. Decarie, mayor Notre Dame de Grace; A. Ramsay, and many others.

MANITOULIN OIL FIELDS.

If the finds of oil on the Manitoulin Islands are as great as reported the discovery is of immense importance or rather can be made so if the product can be protected from falling into combine hands. A barrel of crude petroleum containing about forty-three gallons can be delivered on board ship for the trifle of ten cents.

Be conscientious. Don't have a relation die too often. Funerals sometimes grow too monotonous to an employer during the baseball season or on matinee afternoons.

Business Grows with the Dealer

who makes it his business to feature a good reliable paint with some lasting life about it—a paint well tried and tested and proven a sure trade builder. That is

NEW ERA PAINT **stands in a class by itself for Beauty, Durability and Economy. It holds its color and is impervious to all weather conditions**

Note—You are not giving yourself a square deal unless you see our catalogue before making up your order list for Spring deliveries.

STANDARD PAINT AND VARNISH WORKS CO., LIMITED
WINDSOR . ONTARIO

YOU MUST BE TIRED

of selling STICKY FLOOR PAINTS. Your customers are certainly tired of USING THEM.

THE KITCHEN IS SO IMPORTANT a room that no one can afford to put it out of commission while the ordinary floor paint is DRYING.

THAT IS WHY YOU SEE SO MANY KITCHENS WITH THE PAINT PEELED OFF AND RUBBED OFF—almost ready for another coat—but still STICKY.

THE STRONG POINT about

JAMIESON'S Floor Enamels

is that they dry over night, leaving a BRILLIANT AND LASTING FINISH.

Write for particulars

R. C. JAMIESON & CO.
LIMITED

Sole Manufacturers, MONTREAL

McCaskill, Dougall & Co.

Manufacturers

RAILWAY, CARRIAGE AND BOAT **VARNISHES.**
HIGH-GRADE FURNITURE and HOUSE VARNISHES

MONTREAL.

Louden's Double Strap Barn Door Hangers

Are the Standard

They Have Case Hardened Bearings, Track Scrapers, Revolving Washers.

They hold the door closer to the track, hang perfectly plumb, and allow the door to hang closer to the wall than other hangers do, while they are just as flexible, and are the easiest running hangers made.

Manufactured by

LOUDEN MACHINERY CO.
GUELPH, ONT.

We manufacture 15 different styles of Hay Carriers, 10 different styles of Barn Door Hangers, also Feed and Litter Carriers, Cow Stanchions, Barn Door Latches, etc.

GLAZERS' DIAMONDS OF PROVED WORTH

Having a Century's Reputation for Reliability.

MADE BY

A. SHAW & SON, London, Eng.

CANADIAN AGENT

GODFREY S. PELTON, 201 Coristine Building, Montreal

MONEYMAKERS FOR TINSMITHS

Can. Feb. 27, 1889

That's what my PATENT CONE ROTARY VENTILATORS are.

It's strong argument to say architects recommend them; it's stronger to have the guarantee not only of the makers but of thousands of users. Let us send you booklet. It will tell you new thoughts on ventilation. Write to-day. Here is one testimony:

St. John, N.B., Oct. 31st, 1889

G. D. Pearson, Esq.,

The Cone Rotary Ventilator you placed on chimney of house, to prevent its smoking. I am pleased to say I have not been annoyed by smoke since using it Before using it I could only burn wood, now I burn hard coal, and I would recommend it to all parties in need of such.

JAMES McALLISTER,
James Grant's House, City Road.

Bell Telephone East 2106

GEO. D. PEARSON & CO., Office and Works: **17 Pontiac Street, MONTREAL**

PAINT AND OIL MARKETS

MONTREAL.

Office of HARDWARE AND METAL,
232 McGill Street,
Montreal, January 2?, 1907

Business has fallen off somewhat during the past week, owing to travelers being impeded by storms and extreme weather conditions, therefore receipts of orders has been somewhat of an erratic description.

Sales of white lead have been slightly checked by a new disposition on the part of purchasers not to pay the advances, but from all appearances, still higher figures are inevitable, owing to the shortage caused by the local white lead company (the only one in Canada at present) having closed down.

The only change this week is in turpentine, which has advanced one cent per gallon.

LINSEED OIL—Very firm. We still quote: Raw, 1 to 4 barrels, 55c.; 5 to 9 barrels, 54c.; boiled, 1 to 4 barrels, 58c.; 5 to 9 barrels, 57c.

TURPENTINE—(In barrels.) Another advance of one cent per gallon has taken place. We quote: Single barrel, 99 cents per gal.; two barrels or over, 98c. per gal.; for smaller quantities than barrels, 5c. extra per gal. is charged. Standard gallon is 8.40 lbs., f.o.b. point of shipment, net 30 days.

GROUND WHITE LEAD—Prices are still firm, but an advance is expected shortly, owing to the shortage caused by the closing down of the local white lead works. Our prices are: Best brands Government Standard, $7.25 to $7.50; No. 1, $6.90 to $7.15; No. 2, $6.55 to $6.90; No. 3, $6.30 to $6.55, all f.o.b., Montreal.

DRY WHITE ZINC—Prices remain very firm with no tendency to advance in the near future. We still quote: V.M. Red Seal, 7½c. to 8c.; Red Seal, 7c to 8c; French V.M., 6c to 7c; Lehigh, 5c to 6c.

WHITE ZINC—Our quotations are: Pure, 8½c. to 9½c.; No. 1, 7c. to 8c.; No. 2, 5¾c. to 6¾c.

PUTTY.—Our prices are: Pure linseed oil, $1.75 to $1.85; bulk in barrels, $1.50; in 25-lb. irons, $1.80; in tins, $1.90; bladder putty in barrels, $1.75.

ORANGE MINERAL—Prices are as follows: Casks, 8c.; 100 lb. kegs, 8½c.

RED LEAD—The following quotations are firm: Genuine red lead in casks, $6.00; in 100-lb. kegs, $6.25; in less quantities at the rate of $7 per 100 lbs.; No 1 red lead, casks, $5.75; kegs $6, and smaller quantities, $6.75

PARIS GREEN—Prices remain the same and some jobbers report a little more activity among buyers. In barrels, about 600 pounds, 23½c. per lb.; in arsenic kegs, 250 lbs., 23½c.; in 50 lb. drums, 24c.; in 25 lb. drums, 24½c; in 1 lb packets, 100 lbs. in case, 25c.; in 1 lb packets, 50 lbs in case, 25½c.; in ½ lb. packets, 100 lbs in case, 27c.; in 1 lb. tins, 26c. f.o.b. Montreal. Term three months net or 2 per cent. 30 days.

SHELLAC—We quote: Bleached, in bars or ground, 46c. per lb., f.o.b. Eastern Canadian points; bone dry, 57c. per lb., f.o.b. Eastern Canadian points; T. N. orange, etc., 48c per lb. f.o.b. New York.

SHELLAC VARNISH—Prices remain: Pure white, $2.90 to $2.95; pure orange, $2.70 to $2.75; No. 1 orange, $2.50 to $2.55.

MIXED PAINTS—A fair business is being done at the same prices. We still quote: $1.20 to $1.50 per gal.

CASTOR OIL—Our prices are: Firsts in cases, 8½c.; in barrels, 8c.; seconds in cases, 8c.; in barrels, 7½c.

BENZINE—(In bbls.)—We still quote: 20 cents per gallon.

GASOLINE—(In bbls.)—Our price is 22½ cents per gallon.

PETROLEUM—(In barrels) Our prices are: American prime white coal, 15½c. per gallon. American water, 17c. per gallon, Pratt's Astral, 19½c. per gallon.

WINDOW GLASS — Prices remain firm. We still quote: First break, 50 feet, $1.85; second break, 50 feet, $1.95; first break, 100 feet, $3.20; second break, 100 feet, $3.40; third break, 100 feet, $3.95; fourth break, 100 feet, $4.15; fifth break, 100 feet, $4.40; sixth break, 100 feet, $4.95. Diamond Star: First break, 50 feet, $2.30; second break, 50 feet, $2.50; first break, 100 feet, $4.40; second break, $4.80; third break, 100 feet, $5.75; fourth break, 100 feet, $6.50; fifth break, 100 feet, $7.50; sixth break, 100 feet, $7.50; seventh break, 100 feet, $8; eighth break, 100 feet, $9. Double Diamond: First break, 50 feet, $3.45; second break, 50 feet, $3.75; first break, 100 feet, $6.75; second break, 100 feet, $7.25; third break, 100 feet, $8.75; fourth break, 100 feet, $10; fifth break, 100 feet, $11.50; sixth break, 100 feet, $12.50; seventh break, 100 feet, $14; eighth break, 100 feet, $16.50; ninth break, 100 feet, $18; tenth break, 100 feet, $20; eleventh break, 100 feet, $24; twelfth break, 100 feet, $28.50. Discount on Diamond Star, 20 per cent.; on Double Diamond, 40 per cent.

TORONTO.

Office of HARDWARE AND METAL,
10 Front Street East,
Toronto, Jan. 25, 1907

At present the paint trade is very brisk, with large orders being booked daily, for Spring shipments. Hardware merchants and industrial concerns such as, carriage and furniture factories, are placing good-sized orders.

While white lead is still unchanged, another general advance is looked for, it having been quoted at a 15-cent advance in eastern markets.

Boiled oil and turpentine remain the same as quoted last week.

Paris green has experienced another advance, it now being 1 cent higher per lb. on both English and Canadian green, with further advances probable. The demand has been very large for this line, and owing to the very high price and great demand of copper, which is extensively used as an ingredient, will make possible a further advance.

Litharge has also advanced from 5½ to 6½ cents per lb. We have received new figures on English castor oil, and are now quoting at 10½ to 11 cents per lb., in cases, and 12 cents for single tins.

WHITE LEAD—Ex Toronto pure white, $7.40; No. 1, $6.65; No. 2, $6.25; No. 3, $5.90; No. 4, $5.65 in packages of 25 pounds and upwards; 1-2c per pound extra will be charged for 12 1-2 pound packages; genuine dry white lead in casks, $7.00.

RED LEAD—Genuine in casks of 500 lbs., $6.00; ditto, in kegs of 100 lbs., $6.50; No. 1 in casks of 500 lbs., $5.75; ditto, in kegs of 100 lbs., $6.25. Litharge, 6½ cents lb.

DRY WHITE ZINC— In casks, 7 1-2c.; in 100 lbs., 8c., No. 1, in casks 6 1-2c., in 100 lbs., 7c.

WHITE ZINC (ground in oil)—In 25-lb. irons, 8c.; in 12 1-2 lbs., 8 1-2c.

SHINGLE STAIN—In 5-gallon lots, 75c. to 80c. per gallon.

PARIS WHITE—90c. in barrels to $1.25 per 100 lbs.

WHITING—60c. per 100 lbs. in barrels; Gilders' bolted whiting, 90c. in barrels, $1.15 in smaller quantities.

SHELLAC VARNISH—Pure orange in barrels, $2.70; white, $2.82 1-2 per barrel; No. 1 (orange) $2.50; gum shellac, bone dry, 63c., Toronto. T. N. (orange) 51c. net Toronto.

LINSEED OIL — Remaining steady at last week's quotations: Raw, 1 to 3 barrels, 58c.; 4 to 7 barrels, 57c.; 8 and over, 56c., add 3c. to this price for boiled oil. Toronto, Hamilton, London and Guelph net 30 days.

TURPENTINE—Remains steady at the advanced price and we quote single barrels $1 to $1.02 per gallon f.o.b point of shipment, net 30 days Less than barrels, $1 07 per gallon

GLUES—French Medal, 12 1-2c. per pound; domestic sheet, 10 1-2c. per lb.

PUTTY—Ordinary, 800 casks, $1.50; 100 drums, $1.75, barrels or bladders, $1.75; 100-lb. cases, $1.90;; 25-lb. irons, $1.85; 25-lb. tins (4 to case) $1.90; 12 ½ lb. tins (8 to case) $2.10.

LIQUID PAINTS—Pure, $1.20 to $1.35 per gallon; No. 1, $1.10 per gallon.

BARN PAINTS—Gals., 70c. to 80c.

BRIDGE PAINTS—Gals., 75c. to $1.

CASTOR OIL—English, in cases, 10½ to 11 cents per lb., and 12 cents for single tins.

PARIS GREEN—Another advance of 1 cent per lb. is now quoted on green. And Canadian is now listed at 26½c. base price, English at 26¾c. base f.o.b. Toronto.

REFINED PETROLEUM — Dealers are stocking up heavily for the Winter. We still quote: Canadian prime white, 14c.; water white, 16c.; American water white, 16c. to 17c. ex warehouse.

CRUDE PETROLEUM — We quote: Canadian, $1.32; Pennsylvania, $1.58; Ohio, 96c.

New Paint Factory near Montreal.

PROSPECTS BRIGHT IN PAINT TRADE.

Although January is ordinarily a quiet month in the paint trade, business this month among the manufacturers and jobbers is very satisfactory. Many hardware merchants have taken stock of their paints first, so as to enable them to get their orders placed before the general advance in the spring. Others are holding their orders for white lead on account of the present prevailing high prices, but it is doubtful if they will be able to do better than at present.

The general outlook for the paint trade for 1907 is brighter than any previous year. Large numbers of buildings erected last year, many of which are not finished, will require painting. The bright prospects and large number of permits issued to date should encourage manufacturers to turn out an even larger supply than in previous years, especially as spring orders from manufacturers of carriages and farm implements are very large in volume.

Hardware merchants will do well to place their orders for paint as soon as possible this spring, as many manufacturers have definitely announced another advance to take place in the spring.

THE STANDARD IDEAL SANITARY CO., LIMITED

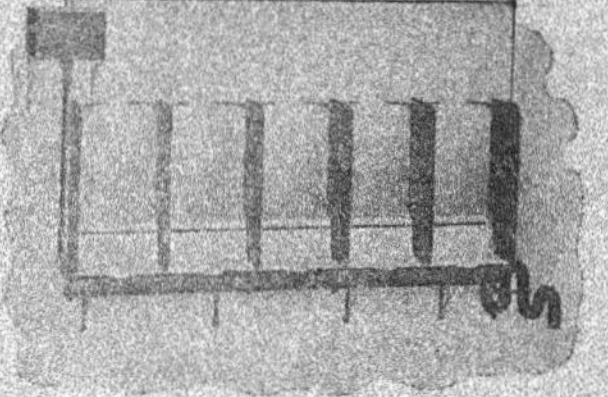
Plate E. 404.—For Public Institutions, Schools, Factories, Asylums, &c.

We manufacture a complete line of

Range-Closets, Range-Urinals, Lavatories and Sanitary Fixtures

Suitable for Public Institutions

The reputation of our goods is unequalled.

Toronto: 50 Colborne Street ***Montreal: 128 Craig Street West***
Winnipeg: 24 Telfer Building

HOME OFFICE and FACTORIES: PORT HOPE, ONT.

The James Morrison Brass Mfg. Co., Limited

PLUMBERS' SUPPLIES AND SPECIALTIES

93-95 Adelaide St. West, - - - TORONTO

Plumbing and Steamfitting

A UNIVERSAL PLUMBING CODE

A Plea for Uniformity—By Herbert F. Shade, Victoria, B.C.

There is no question of such prime importance in the minds of those interested in the proper and efficient disposal of household wastes by water carriage,

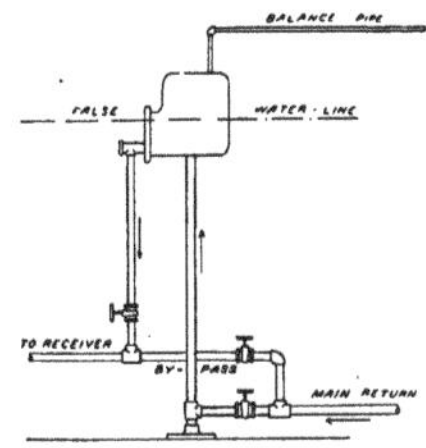

Fig. 1—Steam Trap Connected for Use as a Water Line Trap.

as is the establishment and the strict maintenance of a universal plumbing code, one that will give all the structural precautions necessary to be observed in order to prevent the ingress of sewer air into our homes and buildings, and such that will provide the immediate removal of all household wastes. As outlined in the Plumbers' Trade Journal by Mr. Shade, the main essentials are as follows :

Class of Fixtures.—That the receptacle where the water is soiled shall be of an impervious nature, and so formed that it will be readily cleansed by a minimum amount of water.

Waste Pipes.—That the pipes which convey the wastes away from the fixtures should be of such diameter that they would be self-cleansing in their action, and of such material that their efficiency would not be impaired by the action of the use for which they are intended to serve.

Traps.—That there should be placed as close to the fixture as possible a barrier in the form of a water seal, to prevent the air contained in the pipes coming in contact with the air in the room.

Vents.—That the fixture trap should be supplied with an air pipe, attached to the uppermost portion of the sewer side of the trap, and carried to the outer air, so as to prevent the loss of the trap seal by syphonic action, and to establish a current of fresh air through the drainage system at all times.

Main Drain.—That the whole system should be perfectly sound and tight, and of one continuous grade to the outlet without any impediment whatever, save the fixture trap.

These are the chief features of all plumbing regulations, and are so acknowleged by almost every one connected with the profession, but the construction of plumbing is of such a complex nature that it could not be governed by these few and simple rules, and it is in the complication of detail where a wide difference of opinion prevails, and to these questions, only, that are most at issue I will express my views.

Methods of Back Venting.

The necessity of the trap at the fixture is an undisputed fact, but in the application of the vent pipe in its proper form there has lately been considerable discussion, and while my opinion may be simply a repetition of what has been said along these lines, it is the result of practical observations following the use of the vent pipe. To be better understood I will refer to Fig. 1, which illustrates the ordinary method of back venting, the demerits of which are as follows, viz.:

1. That the opening of the vent pipe at (a) will be eventually closed by the adhesion of grease contained in waste matter to the sides of the vent pipe. (This I have found to be true by actual observations).
2. That the opening at (a) in the majority of cases is smaller in area than the vent pipe, as this is solely dependant upon the mechanic who actually performs the work.
3. That the vent connection being of such close proximity to the water seal that it adds additional danger of the loss of the seal by evaporation.
4. That any scale, silt or other matter that falls down the vertical vent pipe, would lodge at (b) and thereby destroy its purpose.

The economic and effective remedies of the defects of Fig. 1 are eliminated by the method shown in Fig. 2, which is almost in general use in the west.

1. That it is practically impossible to close up the opening of vent at (c) by grease, hair, lint. etc.
2. That the area of the opening of the vent pipe at (c) is as large, and in most cases larger than the vent pipe.
3. That the circulation of air through the vent system is farther removed from the water seal, which reduces the liability of evaporation.
4. That any scale, silt or other matter that falls down the vertical vent pipe is carried direct to the waste pipe, and thus removed from the system.

Should Vent Pipe Be Abolished ?

Much has been said regarding the abolishment of the vent pipe and the general use of non-syphon traps, but the better knowledge due to considerable familiarity with plumbing has proved to my satisfaction that the vent pipe cannot be done away with, unless we look upon the actual cost of trap venting in preference to the proper safeguards for our health.

Most regulations allow vent pipes to converge into one main vent within a certain radius, provided that the branch vents connect to the main vent above the fixture it is intended to serve. In Victoria this is allowable provided that the branch vent does not deviate from an angle less than 45 degrees from vertical (as shown in the sketch of our system). In few cases this cannot practically be done, but in this event the horizontal portion must connect to a main vent leading direct to a waste pipe; this removes the liability of the horizontal portion becoming obstructed, save that which it would collect in its own length. This is what is termed "continuous venting," which increases the life of vent pipes indefinitely, as they cannot be blocked under anything like ordinary conditions. For this class of work there are now special fittings on the market which have given excellent results by test.

The much debated question of the trap on the main drain is still an open one, some authorities being strenuous in their advocacy of its adoption, while

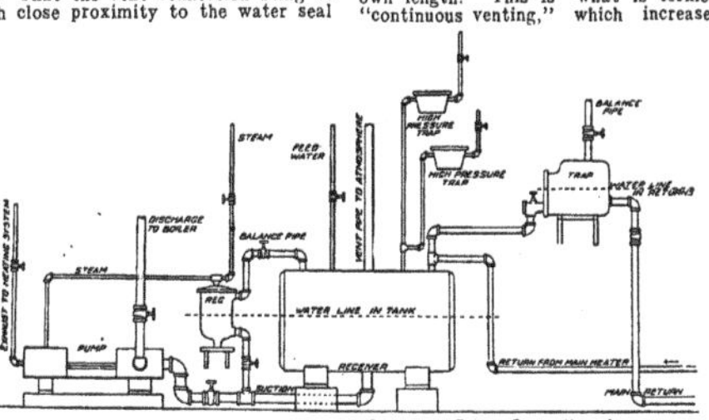

Fig. 2—General Method of Making the Main Supply and Return Connections for an Exhaust Steam Heating System.

others are just as strong in their opinion to the contrary. The writer previous to entering upon his present duties was a supporter of the use of the trap, but by actual observations and experiments, and results obtained, I am now convinced that the omission of the trap is a step in the right direction, and by its use the proper ventilation of our sewerage system cannot be attained. We have in Victoria 16 miles of sewers where the use of the trap is general, and about 14 miles where the connections are direct without the trap, the difference in the efficiency between the trapped and the non-trapped drain has proven to me that the latter is far the most sanitary of the two methods.

Ventilating Rooms.

The theory of ventilating a room by a continuous inflow of fresh air, and a constant outflow of foul and exhausted air ; precisely the same requirements are necessary for your sewerage and drainage systems. On this point almost all authorities agree and that it should be carried out with simplicity and without mechanical aid.

Now as there is no doubt as to the desirability of establishing a current of air through the whole system, and under ordinary conditions there would be a better movement of air if the inlet was at the opening in the street, and the outlet at the soil pipe above the roof, this cannot possibly be secured with trapped house connections, therefore it is necessary to remove the trap and make every connection to the sewer a vent for the same, causing the openings on the street to be inlets for the admission of fresh air, instead of foul air outlets as they are when the trap is used.

Whenever the omission of the trap is anything like general the ventilation secured through the connections reduces the percentage of gas in our sewers to a minimum, and there is no other way in the writer's opinion that sewers can be so effectively ventilated as through house drains, and I cannot see where there can be the slightest objection to this use being made of them.

If the connection between the house and the sewer is without interruption, all wastes that are discharged into it are carried away to a safe distance from the house before putrefaction sets in. If a trap is placed in its course, its seal will collect more or less accumulation of sewage matter, and it is now a recognized fact that sewer gas to a large per cent. is home made, its production taking place within the house drains themselves, and as a rule with few exceptions the air contained in the house drain is far more offensive than that contained in the sewer where the trap on the house drain is used, therefore it seems that by the omission of the interceptor trap the best possible solution of how to ventilate sewers is reached.

The objection generally advanced against this system of sewer ventilation is the cry that sewer gas is laid on to the house, and that contagious diseases existing in other houses will communicate their infection with the houses that are not directly cut off by the use of the trap. This theory has long since been an exploded one, and it has been demonstrated by researches of eminent scientists that pathogenic or disease breeding bacteria have never been found in sewer air, and that in sewage itself they do not and are not likely to thrive.

FEATURES OF EXHAUST STEAM HEATING

Charles L. Hubbard, in The Heating and Ventilating Magazine

Steam, after being usesd in an engine, contains the greater part of its heat, and if not condensed or used for other purposes it can usually be employed for heating without affecting to any great extent the power of the engine.

In general, we may say that it is a matter of economy to use the exhaust for heating, although various factors must be considered in each case to determine to what extent this is true. The more important considerations bearing upon the subject are: the relative quantities of steam required for power and for heating, the length of the heating season, the type of engine used, the pressure carried, and, finally, whether the plant under consideration is entirely new or whether, on the other hand, it involves the adapting of an old heating system to a new plant.

The first use to be made of the

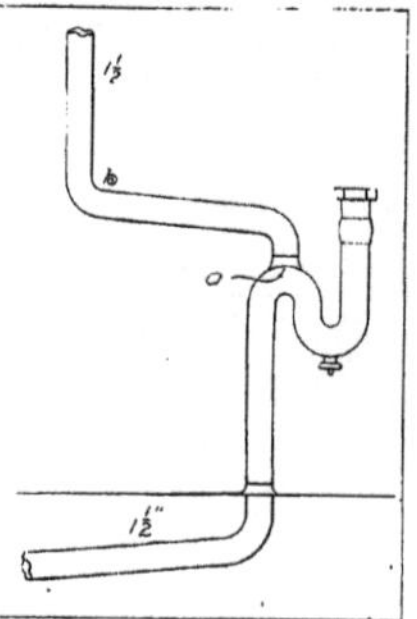

Fig. 1—Ordinary Method of Back Venting.

exhaust steam is, of course, the heating of the feed-water, as this effects a constant saving both summer and winter, and can be done without materially increasing the back pressure on the engine.

Under ordinary conditions about 1-6 of the steam supplied to the engine can be used in this way, or more nearly 1-5 of the exhaust discharged by the engine.

We may assume in average practice that about 80 per cent. of the steam supplied to an engine is discharged in the form of steam at a lower pressure, the remaining 20 per cent. being partly converted into work, and partly lost through condensation.

The latent heat in a pound of steam at atmospheric pressure is 966; therefore, out of each pound of live steam furnished to the engine there will be 966 x 8 = 773 heat units available for heating purposes. It requires practically 210 - 50 = 160 heat units to raise one pound of feed-water from 50 degrees to 210 degrees; and this is

$$\frac{160}{773} = \frac{1}{4.8}$$

or about 1-5 of the available exhaust. Taking this into account, there remains for other heating purposes 8 x 4-5 = .64 of the entire quanity of steam supplied to the engine.

When the quantity of steam required for heating is small compared with the total amount supplied to the engines, or where the heating season is short, it is often more economical to run the engines condensing and use live steam for heating. This can be determined in any particular case by computing the saving in fuel by the use of a condenser, taking into account the interest and depreciation on the first cost of the condensing apparatus, and the cost of water, if it must be purchased, and comparing it with the cost of heating with live steam.

Usually, however, in the case of office buildings and institutions, and commonly in the case of shops and factories, especially in northern latitudes, it is advantageous to use the exhaust for heating, even if a condenser is installed for summer use.

The principal objection to the use of exhaust steam for heating has been the higher back pressure required on the engine. There are two ways of offsetting this loss; one by raising the initial or boiler pressure, and the other by increasing the cut-off of the engine.

Engines are usually designed to work most economically at a given cut-off, so that in most cases it is undesirable to change the cut-off to any extent. Raising the boiler pressure, on the other hand, is not so objectionable if the increase amounts to only a few pounds.

A modern heating plant with ample mains and branches should circulate freely on a pressure not exceeding 2 pounds, and good results are often obtained on one pound or even less.

The systems of steam heating in common use are those in which the water of condensation flows back to the boiler by gravity. When exhaust steam is used the pressure is much below that of the boiler and the condensation must be returned either by a return trap or pump.

As the exhaust steam is often insufficient to supply the entire heating system, it must be supplemented with live steam taken directly from the boiler.

This must first pass through a regulating valve in order to reduce the pressure to that carried on the heating system.

An engine does not deliver steam continuously but at regular intervals at the end of each stroke, and further the amount will vary with the work done.

When the work is light very little

steam will be admitted to the engine, and for this reason the supply available for heating will vary somewhat, depending upon the use made of the power. In mills the amount of exhaust steam is practically constant; in office buildings where the exhaust is obtained from a lighting plant, the variation is greater, especially if power is also supplied for running elevators.

In order to make the steam supply automatic, the system must be equipped with both a reducing valve and a relief or back-pressure valve—the former for admitting live steam when the exhaust is not sufficient to maintain the required pressure, and the latter to allow the surplus to pass outboard when the quantity is greater than can be condensed in the heating system. Before entering the heating main the exhaust should be passed through a good form of oil separator, and it is well to still further purify the return by the use of a settling chamber.

Any of the usual systems of piping may be used, but great care should be taken that as little resistance as possible is introduced at bends and fittings; and the mains and branches should be of ample size. Usually the best results are obtained from the systems in which the main steam pipe is carried directly to the top of the building, the distributing pipes run from that point, and the radiating surfaces supplied by a down-flowing current of steam.

Pipe sizes should generally be based on a drop in pressure of ¼ pound in 200 feet length of run.

In order to secure the benefit of a sealed return, a "false water-line," so called, is established by the use of a water-line trap. This consists of a constant discharge float trap of large size, mounted at such an elevation as may be necessary to produce the desired depth of water-seal.

The return pipe should be brought into the bottom of the trap instead of the top as is usually the case, and a balance pipe for equalizing the pressure connected with the top.

This pipe should be, if possible, not less than 15 or 20 feet in length and provided with a globe valve near its connection with the heating main.

The reason for this is to allow the space above the water in the trap to become filled with air, after which the valve may be opened just enough to allow the steam pressure to act upon the air cushion in the trap without driving it out. If the steam is allowed to enter the trap, its contact with the cooler return water will cause a rapid condensation, thus producing a partial vacuum within the trap which will often result in draining the entire system of sealed returns, and causing at the same time violent surging and water hammer.

Fig. 1 shows a Curtis trap connected up for use as a water-line trap.

The general method of making the main supply and return connections for an exhaust steam heating system are shown in Fig. 2.

When it becomes necessary to connect an old heating system having small or poorly graded mains and returns with a power plant, different methods must be resorted to, as pressures of 10 pounds or more are often required to produce proper circulation in the coldest weather.

Under these circumstances it is generally necessary to employ one of the so-called vacuum systems by means of which a good circulation of steam can be produced without raising the back pressure on the engine at all, and in many cases the pressure may even be reduced below that of the atmosphere.

One point to be borne in mind in the design of any steam plant, is its simplicity; and, when mechanical appliances such as pumps, traps, automatic valves, etc., can be dispensed with without interfering with the proper

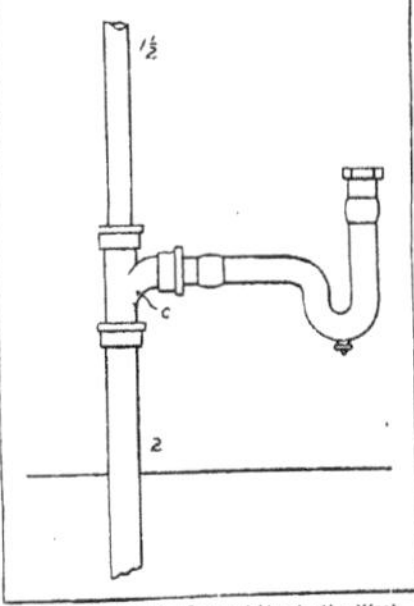

Fig. 2—Method in General Use in the West.

working of a plant, this should be done.

With properly proportioned and well-graded pipes, the pressure required for circulating the steam through a heating system, even of large size, should not exceed 2 pounds. Now, the back pressure of an engine exhausting into the atmosphere is usually about two pounds, so that practically no additional load is put upon the egnine when the exhaust is used in a well-designed heating system; the system simply acts as a large condenser in which condensation takes place under a slight pressure instead of in a vacuum.

We find the increase in boiler pressure required to raise the back pressure from 2 pounds to 5 pounds is only 5 pounds, an amount so small that it can involve no harmful or wasteful results. So far as the efficiency of the radiators is concerned, it makes no difference whether the air is drawn out or forced out, provided that the radiators are steam-filled; and, as the temperature increases with the steam pressure, the heating capacity of a radiator filled with steam under pressure is slightly greater than when steam below atmospheric pressure is used.

Under the pressure system, air valves that are poorly constructed or not well-adjusted, may prevent the radiators from freeing themselves of air. Vacuum systems, on the other hand, employ these or similar devices. It is as necessary in one case as in the other, that the valves be properly adjusted and cared for.

The vacuum system has a large field of its own, and is of much value in cases calling for the special features that it possesses. In modern plants with pipes of ample size, however, it would seem preferable to rely as far as possible upon steam pressure for removing air from radiators, and upon gravity for returning the condensation to the receiving tank rather than to install special mechanical means for accompanying the same results.

CHANGE OF NAME.

The Ontario Lead & Wire Company, Toronto, has announced its change of name to Somerville, Limited. The business will be conducted as before, under the presidency of A. F. Somerville, with no change in the staff or business address. It will be remembered that the company recently disposed of their wire nail and white lead business, with the intention of going into the plumbing supply trade exclusively.

NEW ENAMELWARE CATALOGUE.

The Seamless Steel Bathtub Company, of Detroit, Mich., have just issued their 1907 catalogue, which contains a short description of the plant where the seamless steel bathtub is made, and also contains a number of views of the inside of the works, illustrative of the peculiarly interesting process used by them. The book is a very handsome one indeed and any plumber would do well to write for a copy mentioning this paper.

A LEAD PIPE BUG.

Electric engineers and fire underwriters interested in the Chicago stock yards have become alarmed over the advent of unidentified larvae swarming certain sections of the packing plant's, and insisting on feeding upon the lead insulation of electric wires. These brown, hairy little wigglers, each five-eighths of an inch long, are moving through the hoof storage houses at the yards, gnawing irregular patches of lead, often cutting through the cloth and rubber insulation and short-circuiting the electric current. Holes an inch long and half an inch wide have been cut through one-tenth inch thickness of lead pipe. "The lead pipe cinch" bug is the facetious designation given the creature.

PLUMBING PARAGRAPHS.

W. G. Minns is establishing a plumbing shop at 562 College street, Toronto.

A new plumbing shop is being established at 626 Bloor street west, Toronto, by Thomas Mulvihill

Electrical and Lighting

ELECTRICAL PROGRESS.

More and more is that subtle and powerful natural agency that no man can describe, and that we speak of as electricity, becoming a force that aids in doing the work of the world. Its appliances are so numerous that it would take pages to name them. Its presence in the office and the store for purposes of light or as the motive power for the telephone is so common that it no longer attracts attention. It is rapidly gaining headway in the shops and factories as a source of heat and power. In the modern equipped machine shop here are some of the purposes of power to which it is applied: The automatic oiling devise, magnetic chucks, lifting magnet, relining engine cylinders, whip and hoist, crane, plate planer, bending and straightening machine, shears, flanging machine, bead shaper, chain maker, welding machine, portable hammer, portable riveting machine, centering tool, double slotting machine, boring mill, press, multiple drill press, milling machine, tube drilling machine, portable drill, steel saw, grindstone, blollows, blower, fan, boiler tube cleaner.

For heat there are the soldering iron, solder pot, forging soldering iron, tempering of tools, brazers. In the plumbing and tin shops, in steel mills and blast furnaces, in all kinds of factories, electricity is now doing the work once done by steam and still earlier by hand.

NEW ENGLISH GAS MANTLE.

An English writer describes an improved British gas mantle said to be as remarkable as the newly-invented German mantle, which uses a form of copper cellulose impregnated with certain salts. The new English mantle is dipped in a solution of thorium and cerium, in which is added an ingredient called "laddite," which so adds to its strength and life that it has been uninjured after burning two thousand five hundred hours.

ELECTRIC ILLUMINATION.

Never has greater interest been displayed in new methods of electric illumination than during the past year, says the Scientific American. If the promises which are held out by the inventors of metallic filament lamps are fulfilled we may soon see the passing of the carbon filament bulb.

Although the Nernst lamp, on which great hopes were based because it requires only half as much current as the carbon filament, has proved too costly and the Osmium lamp has been found wanting for the same reason and for the additional reason that its voltage of 47 is too low for ordinary circuits, the tantalum and tungsten lamps seem likely successors of the standard incandescent lamp. The tantalum consumes about as much energy as the Osmium lamp, but its long filament renders its use possible on a 110-volt circuit and on circuits of even higher voltage. Its useful life of 400 to 600 hours and its maximum life of 1,000 hours and more compare favorably with those of the best electric incandescent lamps in use. The filament is very delicate but able to stand greater variations in voltage than the carbon filament. When broken the ends readily fuse, so that the tantalum's usefulness, although impaired, is not utterly destroyed. The present low cost of construction (about 50 cents), coupled with its high voltage, give it a decided advantage over the Osmium filament.

Guelcher's iridium lamp is made only for low tensions (24 volts); it consumes it is claimed, only 1 to 1.5 watts per candle power, and costs about 87 cents. What its life may be it is impossible to state, inasmuch as no figures have been published. It is open to many of the objections leveled at the osmium lamp.

More promising is the tungsten lamp, which is now made by four European firms using as many different processes. The normal tungsten lamp of Just and Hanamann seems to give about 30 to 40 candles at 110 volts and consumes 1.1 watts per candle. Kuzel's tungsten lamp is said to show an efficiency of 1 to 1.25 watts per candle for 19 to 32-candle lamps, with a useful life of 1,000 hours, at the end of which the loss in candle power is said to be but 10 or 15 per cent. When broken the filament automatically welds together as in the tantalum lamp. The Osmium tungsten lamps have shown from 54.7 to 55.6 candles and from 1.026 to 1.047 watts per candle at 110 volts.

Whether these new lamps will fulfill the hopes placed in them can of course be determined only by thorough tests under conditions approximating those of actual service. At present the metallic filament lamp is in its experimental stage. The necessity of using the tungsten lamp in the inverted vertical position may perhaps be regarded as a defect; yet quite recently the vertical incandescent gas mantle has invaded an extensive field hitherto monopolized by the electric light.

UNSCIENTIFIC HEATING APPLIANCES.

The elementary principles of such branches of physical science as sound, light and heat have been so universally taught during the last half century, especially since the revival in technical education of the last few years, that one might reasonably imagine it would be impossible now to find up-to-date specialties, of domestic or public utility, which are constructed in flagrant violation of the elementary laws on which the efficiency of such articles depends, says a writer in an exchange, who recently saw a window full of beautiful specimens of the metal workers' art in copper, brass and iron. All were highly finished in various styles and generally, no doubt, in harmony with the purpose in view; but among them were beautifully polished and nickel-plated radiators.

So flagrant a breach of the laws of radiation might pass with a smile were it an isolated case, but judging from the number exhibited in certain quarters the glitter seems to have caught on and the growing anomaly calls for correction.

The use of radiators in various forms is extending so rapidly that it is important for both maker and dealer to understand the elementary laws which tend toward efficiency. Makers and fitters alike will be forwarding their own best interests by leading clients in this matter. Clients look to the trade for efficient service from the apparatus applied. Let them know that radiation and polish are antagonistic; that a rough, unpolished surface is the best radiator, and that a smooth, bright surface is the very worst. Every stroke of labor put upon a radiator in the direction of polish is money spent in destroying its heating power. A polished, plated surface will not give one-third the radiant heat which is given from the same radiator in the rough, thinly coated with a flatted color.

A simple way of testing this is to place a thermometer at a given distance from an urn full of boiling water. Take the temperature first from the polished side, and afterward from the other, which should be coated with whiting or lampblack; the difference will convince the skeptic. How often has one touched a hot polished surface and got burnt because it gave no warning by radiation.

An instance of misuse of the science of heat on the opposite extreme of application is seen in the cheap, single-cased gas cookers now so extensively used. Here the object should be to conserve the heat within the cooker, allowing none to escape by radiation, instead of which the outer sides of the cooker are specially adapted for efficient radiation, thus throwing away heat which should be helping the cooking, and causing discomfort by heating a room which is required to be as cool as possible.

NEW PLUMBERS' TEXT-BOOK.

The announcement has been made that the series of articles written by J. J. Crosgrove and published in "Modern Sanitation" during the past two years, are to be published in book form as a text book for plumbers, architects, sanitary engineers, technical institutes, and plumbing inspectors. The articles have excited a great deal of interest amongst the trade and many requests have been made for their permanent publication. This has now been decided upon by the Standard Sanitary Manufacturing Company, publishers of "Modern Sanitation," they being careful to state that the publication of the book is not intended to be in any way an advertisement for the publishers, their only object being to advocate the advancement of plumbing and sanitation along the most modern and approved lines. In order to have the first edition large enough to supply the demand the publishers are asking for advance orders. Publication will be commenced at once and Canadian plumbers who have not already ordered are urged to send their names at once to The MacLean Publishing Company, Toronto. The price of the book will be $3, and will comprise about 400 pages, on enameled book paper, profusely illustrated, thoroughly indexed and handsomely bound in cloth, with gilt edges.

PLUMBING MARKETS

MONTREAL.

Office of HARDWARE AND METAL,
232 McGill Street,
Montreal, January 25, 1907

There are no changes in prices this week, all lines remaining very firm.

Jobbers report that orders are now coming in from their travelers, and that general business conditions seem to be very good throughout the province of Quebec. No trouble is being experienced with collections; merchants all over the province seeming to share in the general prosperity.

Owing to the difficulty in getting delivery of some lines, building operations in Montreal are not progressing as rapidly as some contractors would like it, but on the whole there seems nothing to complain of and everybody in expecting a banner year in plumbing goods.

RANGE BOILERS—Prices remain the same, but there will likely be another advance shortly on the copper article, if the present shortage of copper continues. Our prices at present are: Iron clad, 30 gallon, $5; 40 gallon, $6.50 net list. Copper, 30 gallon, $33; 35 gallon, $38; 40 gallon, $43; 45 gallon, $50; 50 gallon, $58; 60 gallon, $68; 70 gallon, $80; 80 gallon, $95; 90 gallon, $115; 100 gallon, $150 net.

LEAD PIPE—Remains firm. We still quote 5 per cent. discount f.o.b. Montreal, Toronto, St. John, N.B., Halifax: f.o.b., London, 15c per hundred lbs extra; f.o.b. Hamilton, 10c per hundred lbs. extra.

IRON PIPE FITTINGS—A very brisk business is being done at the same figures. We still quote: Discounts on nipples, ¼ inch to 3 inch, 75 per cent., 3½ inch to 2 inch, 55 per cent.

IRON PIPE — The recent advance is still maintained and some sizes are reported short. We quote: Standard pipe in lots of 100 feet, regular lengths, ¼ inch, $5.50; ⅜ inch, $5.50; ½ inch, $8.50; ¾ inch, 11.50; 1 inch, $16.50; 1 1-4 inches, $22.50; 1 1-2 inches, $27.00; 2 inches, $36.50, discounts on black pipe, ¼ inch, 59 per cent.; ⅜ in. 59 per cent.; ½ in. 68; ¾ to 2, 70. Discounts on galvanized pipe: ¼ inch, 44 per cent.; ⅜ inch, 44 per cent.; ½ inch, 58 per cent.; ¾ to 2 inch, 60 per cent. Extra heavy pipe of 100 feet lots are quoted as follows: 1-2 inch, $12; 3-4 inch, $15; 1 inch, $22; 1 1-4 inches, $30; 1 1-2 inches, $36; 2 inches, $50. The discounts on black pipe, ¼ inch, 74 per cent.; ⅜ inch, 69 per cent.; ½ inch to 2 inches, 68 per cent. Galvanized, ¼ inch, 59 per cent.; ⅜ inch, 69 per cent.; ½ to 2 inches 58 per cent.

SOIL PIPE AND FITTINGS—Our prices remain as follows: Standard soil pipe, 50 per cent. off list. Standard fittings, 50 and 10 per cent. off list; medium and extra heavy soil pipe, 60 per cent. off. Fittings, 60 per cent. off.

SOLDER—Prices remain the same: Bar solder, half-and-half, guaranteed 25c.; No. 2 wiping solder, 22c.

ENAMELWARE — Recent advance on Canadian ware still holds. We quote: Canadian baths, see Jan. 3 1907 list. Lavatories, discounts, 1st quality, 30 per cent.; special, 30 and 10 per cent. Sinks, 18 x 30 inch, flat rim, 1st quality, $2.60, special, $2.45.

AN EASY QUESTION.

Send in a reply to the question by March 1, and take a chance after the valuable prize offered.

If a customer asked the following question: "What are the advantages of hot water heating over steam heating in a private residence?" what arguments would you use in reply? Also outline what advantages steam heating possesses over water for this same class of work.

* * *

The above is the third of a series of questions Hardware and Metal is asking the trade in Canada to answer. Some interesting replies were received to the preceding questions and the simplicity of the above should encourage a large number of replies, the publication of which should be helpful to readers of this paper who engage in heating work.

For the best reply we offer a clothbound copy of J. J. Cosgrove's splendid text-book "Principles and Practice of Plumbing."

TORONTO.

Office of HARDWARE AND METAL,
10 Front Street East
Toronto, January 25, 1907

With no price changes during the week there is little to report other than that travelers are on the road and getting fair business. Plumbers, like other business men, are getting more in the habit of placing their orders well ahead in order to get deliveries promptly. Building is going on actively and some good orders are being placed—one Galt job, for which supplies have just been ordered, is for nineteen tubs, closets, laundry tubs, etc., and 38 lavatories.

Jobbers are well satisfied with collections and are looking for a good Spring trade. Pipe supplies are still low, but skelp is said to be easier to procure. A shortage is looked for, however, in the Spring when big orders are placed.

LEAD PIPE — The discount on lead pipe continues at 5 per cent. off the list price of 7c. per pound. Lead waste, 8c. per pound with 5 off. Caulking lead 5 3-4c. to 6c. per pound. Traps and bends, 40 per cent., discount.

SOIL PIPE AND FITTINGS — Recent advances are held to firmly with demand seasonable. Medium and extra heavy pipe and fittings, 60 per cent.; light pipe 50 per cent.; light fittings, 50 and 10 per cent.; 7 and 8 inch pipe, 40 and 5 per cent.

IRON PIPE.—We are revising last week's figures on large sizes. One-inch black pipe is quoted at $4.95, and one-inch galvanized at $6.60. Full list. in current market quotations.

IRON PIPE FITTINGS—Cast iron fittings show several advances. Cast iron elbows, tees, crosses, etc., 60 per cent.; cast iron plugs and bushings, 60 per cent.; flange unions, 60 per cent.; nipples, 60 and 10 per cent.; iron cocks, 55 and 5 per cent.; Canadian malleable, 30 per cent.; malleable unions, 55 and 5 per cent.; malleable bushings, 55 per cent.; cast iron ceiling plates, plain 65 per cent.; cast iron floor, 70 per cent.; hook plates, 60 per cent.; expansion plates, 65 per cent.; headers, 60 per cent, hangers, 65 per cent.; standard list.

RANGE BOILERS—An advance of 25 cents has been made on all sizes of galvanized iron and we now quote: Galvanized iron, 30 gallon, standard, $5; extra heavy, $7.75; 35 gallon, standard, $6; extra heavy, $8.75; 40 gallon, standard, $7; 40 gallon, extra heavy, $9.75, net list. Copper range boilers—New lists quote: 30 gallon, $33; 35 gallon, $38; 40 gallon, $43. Discounts 5 to 15 per cent.

RADIATORS—Prices are very stiff at: Hot water, 47½ per cent.; steam, 50 per cent.; wall radiators, 45 per cent.; specials, 45 per cent. Hot water boilers are unchanged.

SOLDER — Quotations remain firm as follows: Bar solder, half-and-half, guaranteed, 27c.; wiping, 23c.

ENAMELWARE—New lists on enameled baths issued by the Standard Ideal Company on Jan. 3, show a 10 per cent. advance. Lavatories, first quality, 20 and 5 to 20 and 10 off; special, 20 and 10 to 30 and 2½ per cent. discount. Kitchen sinks, plate 300, firsts 60 and 10 off; specials, 65 and 5 per cent. Urinals and range closets, 15 off. Fittings extra.

THE NEW
"SOLID-ADJUSTABLE"
HAND STOCK AND DIES

At prices that will warrant your inquiries.

I. E. SHANTZ & CO.
BERLIN, - ONTARIO
Selling Agents for Canada

This is the Mark that Means Something

It is the distinctive brand of genuine Armstrong Stocks and Die and Pipe Fitters' Tools.

No "Armstrong pattern" or "just as good as Armstrong" have this stamp. Don't buy anything in this line that hasn't this brand as it is the mark of genuineness.

THE ARMSTRONG MANUFACTURING CO.
292 Knowlton Street, Bridgeport, Conn.

CUT OUT ALL WORRY

As far as your Enamelware wants are concerned you should certainly do so. Promptitude is the key-note of my business and my name is a sufficient guarantee for the quality of the goods I deliver.

ORLANDO VICKERY
178-180 Victoria Street TORONTO, CANADA

"DIAMOND BRAND" FITTINGS

No better fittings were ever placed on the Canadian market and all practical men know it.

Materials—The Best Procurable.
Workmanship—First-class.
Threads—Just as smooth as smooth can be.

OSHAWA STEAM & GAS FITTINGS CO., LIMITED
OSHAWA, CANADA

Sharratt & Newth's Glaziers' Diamonds

are unequalled for cutting and wearing qualities.

To be obtained from the principal Dealers in Glass, Hardware, and Painters' Supplies.
Contractors to H. M. Government and the principal English Sheet and Plate Glass Works

World's Best Cutlery

You make no mistake in specifying goods made by **JOSEPH RODGERS & SONS, Limited** when ordering cutlery. A reputation of over 200 years at stake, insures the quality of every article.

Write for catalogues from Sole Canadian Agents
JAMES HUTTON & COMPANY, MONTREAL

A Certain Sense of Satisfaction

goes with every MERRELL PORTABLE HAND MACHINE, PIPE THREADING and CUTTING MACHINE. We build them that way.

For doing ALL kinds of threading—under ALL conditions—labor trouble and faulty material included—where speed, durability and ease of control count for something, our word to you is to investigate the MERRELL.

Whether it be stationary or portable—hand or power driven—30 DAYS FREE TRIAL must prove conclusively that your choice should be the MERRELL.

This Portable Hand Machine has encased gears, MERRELL standard quick opening and closing die-head, and the latest improved Cutting-off Knife.

Let us tell you more about this machine—The Chasers, Vise and the large range of work covered.

Catalogue for the asking.
Quick shipments.

THE CANADIAN FAIRBANKS CO. Limited
Sole Agents for Canada
MONTREAL, TORONTO, WINNIPEG, VANCOUVER

When You Buy Demand Proof

Don't take any substitute for Glauber Brass goods. Make the salesman show you. Brass has gone up no less than fifteen per cent., and if you don't get proofs before you buy you take chances of getting something less than you pay for.

WE GIVE YOU THE PROOF and ***2 Cocks Free*** for any one sent you defective.

Glauber Brass Mfg. Co.,
Expert Makers of Plumbers' Fine Brass Goods
CLEVELAND, O.

Forwell Foundry Co.

BERLIN, ONT.
Manufacturers of
SOIL PIPE, FITTINGS, and CAST IRON SINKS
Ask Jobbers for "F. F. CO." Brand

THE LINE TO PUSH

There are undoubtedly many people in your town whose business or home-demands require an unusual quantity of hot water. These Heaters are adapted for warming offices, conservatories and small dwellings, or supplying hot water for factories, stables, barber shops, etc.

It is a good time of year to push this line and keep the men busy during the dull season.

ALL SIZES KEPT IN STOCK

CLUFF BROTHERS

Selling Agents for

WARDEN KING & SON, Limited, - - Lombard Street, TORONTO

The Ever Ready Dry Battery

FOR AUTOMOBILE and

GASOLINE ENGINE USE

Write for Prices

JOHN FORMAN

248 and 250 Craig St. W., MONTREAL, - Que.

No. 30 King. Price each $4.55 net

HOTTEST ON EARTH

We manufacture the largest line of mechanical tools using gasoline, kerosene and alcohol as fuel. Your jobber will supply you at factory prices. Guaranteed to please you or money refunded.

THE TURNER BRASS WORKS

53 MICHIGAN STREET, - - - CHICAGO, U.S.A.

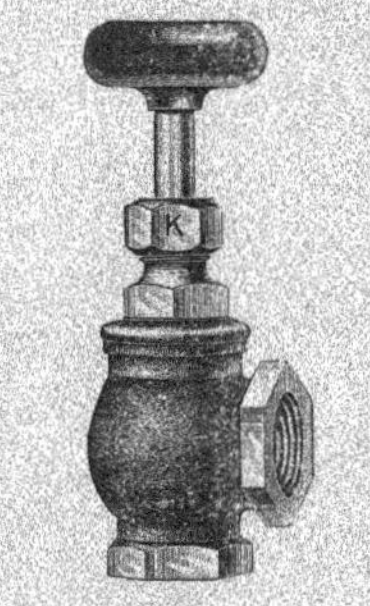

Kerr's Standard and Jenkin Disc Radiator Valves

are perfectly constructed, and of beautiful design. Like all "Kerr" specialties, strictly high-grade

The KERR ENGINE CO.,

Manufacturers Limited

WALKERVILLE - ONTARIO

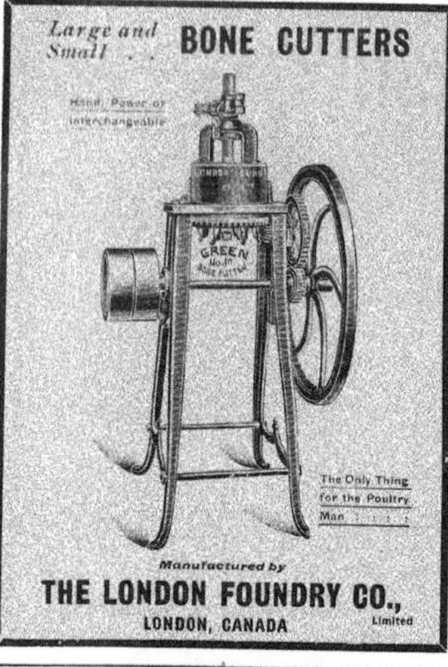

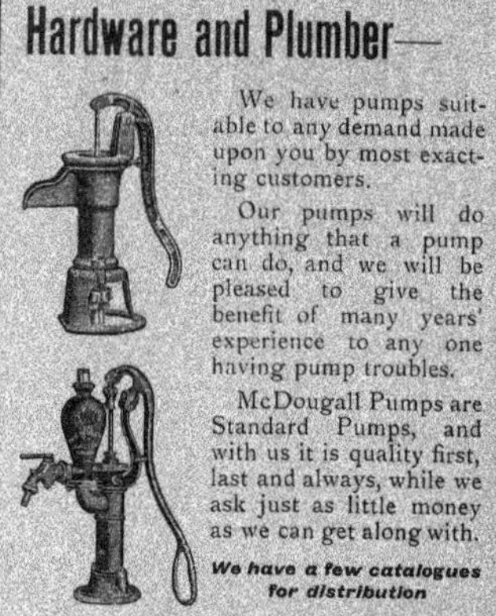

Hardware and Plumber—

We have pumps suitable to any demand made upon you by most exacting customers.

Our pumps will do anything that a pump can do, and we will be pleased to give the benefit of many years' experience to any one having pump troubles.

McDougall Pumps are Standard Pumps, and with us it is quality first, last and always, while we ask just as little money as we can get along with.

We have a few catalogues for distribution

The R. McDougall Co., Limited

CALT, - CANADA

The Long and the Short of It

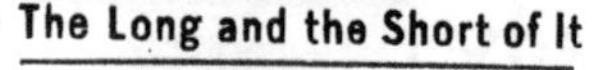

" IT " MEANS

THE "LONG"

is the time it wears. It stands the test of heat and frost, wind and rain, and give your customers many years of solid satisfaction.

THE "SHORT"

part is the time needed to put it on. A great deal of time—which means a great deal of money—is saved to the man who uses

SHIELD BRAND READY ROOFING

LOCKERBY & McCOMB
65 SHANNON STREET
MONTREAL

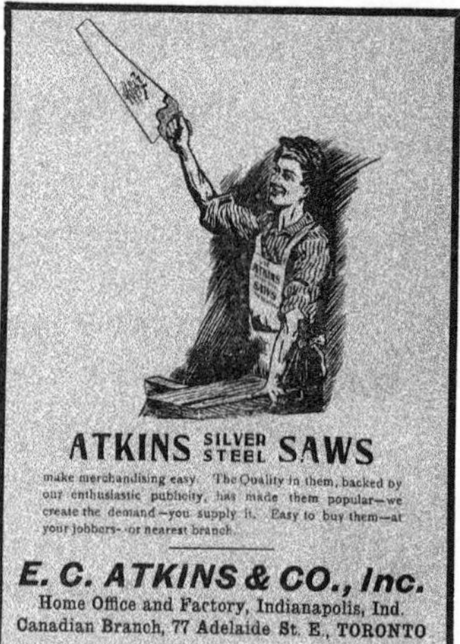

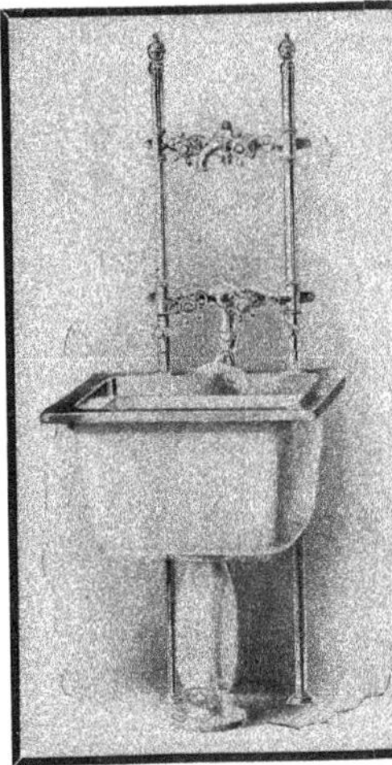

OUR ADVERTISING IS YOUR SALESMAN

Our extensive national advertising has made "Standard" Porcelain Enameled Ware the favorite with the House-owners in every part of the country. "Standard" Fixtures are easiest to specify for Kitchen, Laundry, Bathroom or Pantry because the House-owner is already prejudiced in their favor.

"Standard" Plate P1369—Porcelain Enameled Slop Sink with Nickel-plated Improved Flushing Rim, on Trap Standard Clean-out and 3-inch Nickel-plated Vent Connection to Wall, Nickel-plated Strainer, Compression Supply and Flush Fittings complete with Supply Pipes to Floor, Nickel-plated Unions, Tees, Pipe Straps, Air Chambers and Spun Cup—enameled inside or all over.

Standard Sanitary Mfg. Co.
PITTSBURGH, Pa., U.S.A.

Mechanical Draft

An application of fans for securing the highest efficiency of fuel and the greatest steaming capacity of boilers—Fiske.

THERE is nothing new or irrational about Mechanical Draft—It dates back from the first use of bellows, and is best appreciated in the latest and largest power plants.

Buffalo Electric "B" Volume Exhauster

MECHANICAL DRAFT means a positive control over your boilers. No dependence on weather conditions. No fixed limit to boiler capacity. Quick response to demands for steam.

A typical Buffalo Induced Draft Plant.

Left Hand Bottom Horizontal Discharge Fan with Cylinder below shaft engine.

MECHANICAL DRAFT means no tall, expensive chimneys. Perfect combustion of low-grade fuels. Realizes, in connection with fuel economizers, the highest possible boiler efficiency.

MECHANICAL DRAFT has long ceased to be an experiment, but is well defined and accepted engineering practice. The best results are only possible when installations are designed and executed by specialists in this service. Bring your draft problems to us.

CANADIAN BUFFALO FORGE CO., Limited
MONTREAL, QUE.

Strictly Brand

IS A FAVORITE

Solder

A Specialty of and **PACKED** by

THE CANADA METAL CO., Toronto

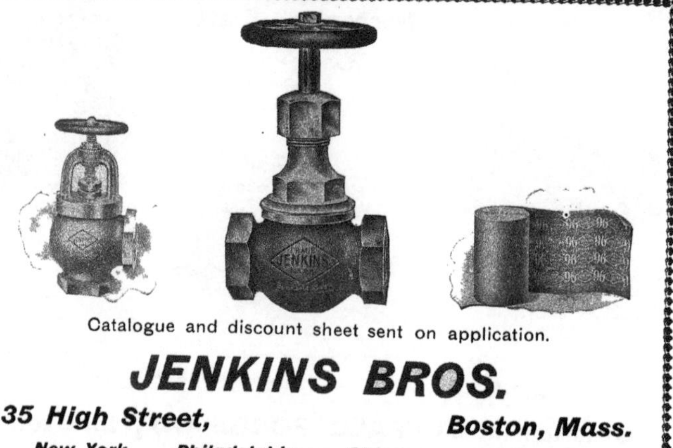

Catalogue and discount sheet sent on application.

JENKINS BROS.

35 High Street, Boston, Mass.

New York Philadelphia Chicago London, E. C.

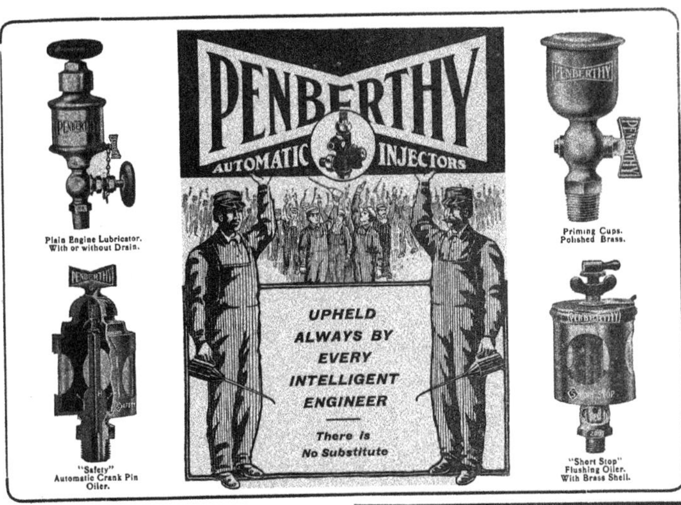
PENBERTHY
AUTOMATIC INJECTORS
UPHELD ALWAYS BY EVERY INTELLIGENT ENGINEER
There is No Substitute
Plain Engine Lubricator. With or without Drain.
"Safety" Automatic Crank Pin Oiler.
Priming Cups. Polished Brass.
"Short Stop" Flushing Oiler. With Brass Shell.

Mueller Water Strainer
PATENTED
C—109
Foreign matter can't cling to the screen in this strainer and choke it. A baffle plate at the inlet receives the first impact of foreign matter and deflects it past the screen into the pocket. Made in regular pipe sizes up to 3-inch and in larger upon order.
UNCONDITIONAL GUARANTEE
THE MARK OF MUELLER GOODS
H. MUELLER MFG. CO.
Decatur, Ill., U.S.A. New York, N.Y., U.S.A.

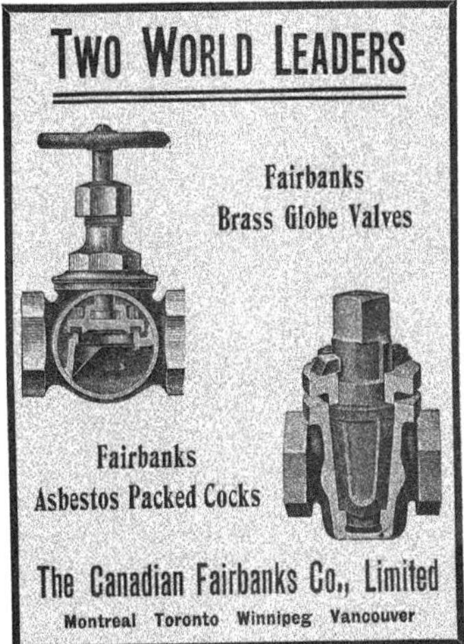
TWO WORLD LEADERS
Fairbanks Brass Globe Valves
Fairbanks Asbestos Packed Cocks
The Canadian Fairbanks Co., Limited
Montreal Toronto Winnipeg Vancouver

CURRENT MARKET QUOTATIONS.

Jan 25, 1907.

These prices are for such qualities and quantities as are usually ordered by retail dealers on the usual terms of credit, the lowest figures being for larger quantities and prompt pay. Large cash buyers can frequently make purchases at better prices. The Editor is anxious to be informed at once of any apparent errors in this list, as the desire is to make it perfectly accurate.

METALS.

ANTIMONY.

Hallett's..............per lb... 0 27½ 0 28

BOILER AND T.K. PITTS.

Plain tinned. .... } 25 per cent. off list
Spun.................. }

BABBIT METAL.

Canada Metal Company—Imperial genuine 60c.; Imperial Tough, 60c; White Brass, 50c; Metallic, 35c; Harris Heavy Pressure, 25c; Hercules, 25c; White Bronze, 15c; Star Frictionless, 14c; Alluminoid, 10c; No. 4, 9c per lb

BRASS.

Rod and Sheet, 14 to 30 gauge, net list.

Sheets, 12 to 14 in.	0 27
Tubing, base, per lb ⅛ to 2 in	0 33
Tubing, ⅛ to 3-inch, iron pipe size	0 33
" ⅛ to 3-inch, seamless	0 40

Copper tubing, 5 cents extra

COPPER.

	Ingot.	Per 100 lb.
Casting, car lots	26 50	27 00
Bars.		
Cut lengths, round, ½ to 2 in	....	32 00
Sheet.		
Plain, 16 oz., 14x48 and 14x60	....	30 00
Plain, 14 oz.	....	31 00
Tinned copper sheet, base	....	33 00
Planished base	....	37 00
Braziers' (in sheets), 4x6 ft., 25 to 30 lb. each, per lb., base	....	0 30

BLACK SHEETS.

	Montreal	Toronto
8 to 10 gauge	2 70	2 60
12 gauge	2 70	2 65
14 "	2 70	2 60
17 "	2 50	2 50
18 "	2 50	2 55
20 "	2 50	2 55
22 "	2 50	2 70
24 "	2 55	2 75
26 "	2 65	2 85
28 "	2 70	3 00

CANADA PLATES.

Ordinary, 52 sheets	2 75	2 90
All bright "	3 80	3 90
Galvanized, 52 sheets	4 35	4 45
" 60 "	4 60	4 70

	Ordinary.	Dom. Crown.
18x24x52	4 25	4 35
" 60	4 50	4 60
20x28x80	8 50	8 70
" 94	9 00	9 20

GALVANIZED SHEETS.

	Fleur-de-Lis.	Gordon Crown
16 to 20 gauge	3 80	3 95
22 to 24 gauge	4 05	4 00
26 "	4 30	4 20
28 "	4 55	4 50

Apollo	
10¾ oz. (American gauge)	4 70
28 gauge "	4 45
26 " "	4 15
24 " "	3 90

	Comet	Queen's Head.	Bell.
16 to 20 gauge	3 60	3 95	....
22 to 24 gauge	3 75	4 20	....
26 "	4 30	4 45	....
28 "	4 45	4 70	4 45

Less than case lots 10 to 25c. extra.

IRON AND STEEL.

	Montreal.	Toronto
Common bar, per 100 lb.	2 15	2 30
Forged iron "	2 40	..
Refined " "	2 55	2 70
Horseshoe iron "	2 55	2 70
Hoop steel, 1½ to 3 in. base	..	2 80
Sleigh shoe steel "	2 25	2 30
Tire steel	2 40	2 50
Best sheet cast steel	...	0 12
B. K. Morton "Alpha" high speed.		0 65
" annealed	....	0 70
"M" Self-hardening	....	0 50
"J" quality, best warranted	....	0 18
"I" " warranted	....	0 14
"B.C" quality	....	0 09
Jonas & Colver's tool steel	0 10	0 20
" "Novo"	....	0 65
" " annealed	....	0 65
Jowett & Sons B P.L. tool steel	....	0 10½

INGOT TIN.

Lamb and Flag and Straits—56 and 28-lb. ingots, 100 lb. $45 50 $46 00

TINPLATES.

Charcoal Plates—Bright

M.L.S., equal to Bradley—		Per box.
I C, 14 x 20 base	....	$6 50
I X, 14 x 20 "	....	8 00
I XX, 14 x 20 base	....	9 50
Famous, equal to Bradley—		
I C, 14 X 20 base	....	6 50
I X, 14 x 20 "	....	8 00
I X X, 14 x 20 base	....	9 50
Raven and Vulture Grades—		
I C, 14 x 20 base	....	5 00
I X "	....	6 00
I X X "	....	7 00
I X X X "	....	8 00

"Dominion Crown Best"—Double Coated, Tissued.		Per box.
I C, 14 x 20 base	5 50	5 75
I X, 14 x 20 "	6 50	6 75
I XX, 14 x 20 "	7 50	7 75
"Allaway's Best"—Standard Quality.		
I C, 14 X 20 base	....	4 50
I X, 14 x 20 "	....	5 25
I XX, 14 x 20 "	....	6 00

Bright Cokes.

Bessemer Steel—		
I C., 14 x 20 base	...	4 25
20x28, double box	....	8 50

Charcoal Plates—Terne

Dean or J. G. Grade—		
I C., 20x28, 112 sheets	7 25	8 00
IX., Terne Tin	...	9 50

Charcoal Tin Boiler Plates.

Cookley Grade—
X X, 14x56, 50 sheet bxs. }
" 14x60, " } 7 50
" 14x65, " }

Tinned Sheets.

7⅞x30 up to 24 gauge	....	8 50
" " 26 "	....	9 00

LEAD.

Imported Pig, per 100 lb.	5 50	5 60
Bar, "	5 75	6 00
Sheets, 2½ lb. sq. ft., by roll	....	0 06½
Sheets, 3 to 6 lb. "	....	0 06

NOTE.—Cut sheets ½c. per lb., extra. Pipe, by the roll, usual weights per yard, lists at 7c. per lb. and 5 p.c. dis. f.o.b. Toronto.

NOTE.—Cut lengths, net price, waste pipe 8-ft lengths, lists at 8c.

SHEET ZINC.

5-cwt casks	8 00	8 25
Part casks	8 25	8 50

ZINC SPELTER.

Foreign, per 100 lb	7 50	7 75
Domestic	7 00	7 25

PLUMBING AND HEATING

BRASS GOODS, VALVES, ETC.

Standard Compression work, dis half inch 50 per cent., others 40 per cent.
Cushion work, discount 40 per cent.
Fuller work, ½-inch 60 per cent., others 50 p.c.
Flatway stop and stop and waste cocks, 50 per cent; roundway, 45 per cent.
J.M.T. Globe, Angle and Check Valves, discount 45 per cent.
Standard Globe, Angle and Check Valves discount 52½ per cent.
Kerr standard globes, angles and checks, special, 42½ per cent.; standard, 47½ p.c.
Kerr Jenkins' disc, copper-alloy disc and heavy standard valves, 45 per cent.
Kerr steam radiator valves, 60 p.c., and quick-opening hot-water radiator valves, 60 p.c.
Kerr brass, Weber's straightway valves, 42½ per cent.; straightway valves, I.B.B.M., 60 per cent.
J. M. T. Radiator Valves, discount 50 per cent
Standard Radiator Valves, 60 per cent.
Patent Quick-Opening Valves, 65 per cent.
Jenkins' Bros Globe Angle and Check Valves discount 32½ per cent.

No. 1 compression bath cock......net	2 00
No. 4 " " "	1 90
No 7 Fuller's "	2 25
No. 4½ " "	2 35

Patent Compression Cushion, basin cock, hot and cold, per doz., $16.20
Patent Compression Cushion, bath cock, No. 2208 .... 2 25
Square head brass cocks, discount 50 per cent.
" " iron " " 60 "
Thompson Smoke-test Machine $25.00

BOILERS—COPPER RANGE.

Copper, 30 gallon	33 00
" 35 "	38 00
" 40 "	43 00

Discounts on new list.

BOILERS—GALVANIZED IRON RANGE.

Capacity.	Standard.	Extra heavy
30 gallons	5 00	7.75
35 "	6 00	8 75
40 "	7 00	3.75

2 per cent, 30 days

BATH TUBS.

Steel clad copper lined, 15 per cent.

CAST IRON SINKS.

16x24, $1; 18x30, $1; 18x36, $1.30.

ENAMELED CLOSETS AND URINALS

Discount 15 per cent.

ENAMELED BATHS.

List issued by the Standard Ideal Company Jan 3, 1907, shows an advance of 10 per cent. over previous quotations.

ENAMELED LAVATORIES.

	1st quality	Special.
Plate E 100 to E 103	20 & 5 p c	20 & 10 p c.
" E 104 to E 132	20 & 10 p c.	30 & 2½ p.c

ENAMELED ROLL RIM SINKS.

	1st quality.	Special.
Plate E 201, one piece.	15 & 2½ p.c.	15 & 10 p c.

ENAMELLED KITCHEN SINKS.

Plate E, flat rim 300, 60 & 10 p c. 65 & 5 p.c.

HEATING APPARATUS

Stoves and Ranges—Discounts vary from 40 to 70 per cent according to list.
Furnaces—45 per cent.
Registers—70 per cent.
Hot Water Boilers—Discounts vary.
Hot Water Radiators—47½ per cent.
Steam Radiators—50 per cent.
Wall Radiators and Specials—45 per cent.

LEAD PIPE

Lead Pipe, 7c. per pound, 5 per cent. off.
Lead waste, 8c. per pound, 5 per cent. off.
Caulking lead, 6c. per pound.
Traps and bends, 40 per cent.

IRON PIPE.

Size (per 100 ft.)	Black.		Galvanized
¼ inch	2 25	¼ inch	3 08
⅜ "	2 25	⅜ "	3 18
½ "	2 72	½ "	3 57
¾ "	3 45	¾ "	4 60
1 "	4 93	1 "	6 60
1¼ "	6 75	1¼ "	9 00
1½ "	8 10	1½ "	10 80
2 "	10 80	2 "	14 40
2½ "	18 00	2½ "	24 00
3 "	23 50	3 "	31 50
3½ "	30 00	3½ "	40 00
4 "	34 00	4 "	45 00

2 per cent. 30 days.

Malleable Fittings—Canadian discount 30 per cent.; American discount 25 per cent.
Cast Iron Fittings 60; Standard bushings 60 per cent; headers, 60; flanged unions 60, malleable bushings 55; nipples, up to 3 in., 65 and 10 per cent; up to 6 in., 65 per cent; malleable lipped unions, 55 and 5 per cent.

SOIL PIPE AND FITTINGS

Medium and Extra heavy pipe and fittings, up to 6 inch, discount 60 per cent.
7 and 8-in. pipe, discount 40 and 5 per cent.
Light pipe, 50 p c.; fittings, 50 and 10 p c.

OAKUM.

Plumbers ....per 100 lb..... 4 00 4 75

STOCKS AND DIES.

American discount 25 per cent.

SOLDERING IRONS.

½-lb. per lb.	....	0 37
1-lb. or over "	....	0 34

SOLDER.

	Montreal	Per lb. Toronto
Bar, half-and-half, guaranteed	0 25	0 27
Wiping	0 22	0 23

PAINTS, OILS AND GLASS.

BRUSHES

Paint and household, 70 per cent..

COLORS IN OIL.

Venetian red, 1-lb. tins, pure.	....	0 08
Chrome yellow "	....	0 15
Golden ochre "	....	0 08
French " "	....	0 08
Marine black "	....	0 04½
Chrome green "	....	0 10
French permanent green "	....	0 13
Signwriters' black "	....	0 15

GLUE.

Domestic sheet	0 10	0 10½
French medal	0 12	0 14

OILS AND TURPENTINE.

W. H. & C's English raw oil, bbls.		0 63
" " boiled. "	....	0 66
Southern turpentine, single bbls	0 98	1 05
Ben ine, bbls, per gal	....	0 20

PARIS GREEN.

	Berger's	Canadian
600-lb. casks	0 24½	0 24½
250 lb. drums	0 24½	0 24½
100-lb. "	0 25	....
50-lb. "	0 25	0 25½
1-lb. pkgs, 100 in box	0 26	0 26½
½-lb. "	0 28	0 28½
1-lb tins, 100 in box	0 27	0 27½
½-lb. pkgs.	....	0 23

PREPARED PAINTS.

Pure, per gallon, in tins 1
Second qualities per gallon 1
Barn (in bbls.) 0 60 0
Sherwin-Williams paints, ½ gal. 1
" " ¼ gal. 1
" " gal... 1
Canada Paint Co.'s pure 1
Standard P. & V. Co.'s "New Era" 1
Benj. Moore Co.'s "Ark" B'd 1
" British Navy deck 1
Brandram-Henderson "Anchor" 1
Ramsay's paints, Pure, per gal. 1
" Thistle, " 1
" Outside, bbls 0 55 0
Martin-Senour's 100 p.c. pure, ½ gal. 1
" " " " ¼ gal. 1
" " " " 1 gal. 1
Senour's Floor Paints.....gal. 1
Jamieson's "Crown and Anchor" 1
Jamieson's floor enamel 1
" barn paints, bbls, per gal. 0
Sanderson Pearcy's, pure 1 20
Robertson's pure paints 1 20

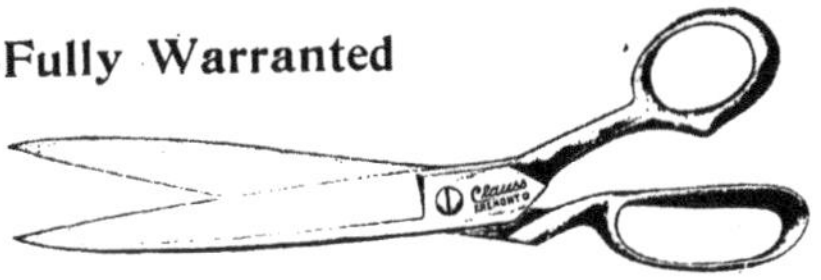

PUTTY.

Bulk in bbls. 1 50
Bulk in less quantity 1 80
Bladders in bbls 1 80
Bladders in kegs, boxes or loose 1 95
25-lb. tins 1 80
12½ lb. tins 2 05
Bladders in bulk or tins less than 100 lb. 1 85
Bulk in 100-lb. irons 1 80

DRY LEAD.

	white	red.
Genuine, 560 lb. casks, per cwt	6 75	6 00
Genuine, 100 lb. kegs, "	7 50	6 25
No. 1, 560 lb. casks, per cwt ..	....	5 75
No. 1, 100 lb. kegs, per cwt ..	....	6 00

SHELLAC.

Shellac, bleached, in bars or ground, 46c. per lb., f o.b. Eastern Canadian Points.
Shellac, bone dry, 57c per lb., f o.b. Eastern Canadian Points; 63c, Toronto.
Shellac, T. N. orange, etc, 48c. net, Toronto

TURPENTINE AND OIL

Turpentine, single barrels	0 98	1 03
Linseed Oil, " raw	0 55	0 59
" " boiled	0 58	0 62
Benzine, " per gal	0 20	0 20

WHITE LEAD GROUND IN OIL. Per 100 lbs.

Pure	7 25	7 50
No. 1	6 90	7 15
No. 2	6 55	6 90
No. 3	6 30	6 55
Munro's Select Flake White..	....	7 50
Elephant and Decorators' Pure	....	7 25
Monarch	....	7 40
Standard Decorator's	....	7 15
Essex Genuine	....	6 80
Brandram's B. B. Genuine	....	8 75
"Anchor," pure	....	6 50
Ramsay's Pure Lead	....	6 40
Ramsay's Exterior	....	6 15
"Crown and Anchor," pure	....	6 50
Sanderson Pearcy's	....	7 40
Robertson's C.P., lead	....	7 20
W.H.&C' matured pure English	8 00	8 21

WINDOW GLASS.

Size United Inches.	Star	Double Diamond
Under 26	$4 25	$6 25
26 to 40	4 65	6 75
41 to 50	5 10	7 50
51 to 60	5 35	8 50
61 to 70	5 75	9 75
71 to 80	6 25	11 00
81 to 85	7 00	12 50
86 to 90	....	15 00
91 to 95	....	17 50
96 to 100	....	20 50
101 to 105	....	24 00
106 to 110	....	27 50

Discount off list, 20 per cent For broken boxes 40 off. All prices per 100 feet.

ZINC—DRY WHITE.

Extra Red Seal	0 07	0 08
French V. M.	0 06	0 25
Lehigh	0 05	0 06

ZINC—GROUND WHITE.

Pure	0 08½	0 09½
No. 1	0 07	0 08
No. 2	0 05½	0 06½

VARNISHES.

In 5-gal. lots.	Per gal.	Net.
Carriage, No. 1	1 50	1 60
Pale durable body	4 10	4 25
" rubbing	2 85	3 20
Gold size, japan	1 50	1 60
No. 1 brown japan	0 85	0 90
Elastic oak	....	1 50
Furniture, extra	1 10	1 25
" No. 1	0 90	1 00
Hard oil finish	1 35	1 0
Light oil finish	1 60	1 70
Damar	1 75	2 00
Brown japan	1 10	1 20
Black japan	1 10	1 20
" No. 1	0 85	0 90
Elastilite varnish, 1 gal. can, each..		2 00
Granitine floor varnish, per gal		2 10
Maple Leaf coach enamels, size 1, size 2, 70c.; size 3, 40c. each.		1 20
Sherwin-Williams' kopal varnish, gal.,		2 50
Canada Paint Co's sun varnish		2 00
"Kyanize" Interior Finish....		2 40
"Flint-Lac," coach		1 80
"Gold Medal," per gal., in cases ..		2 00
Jamieson's Copaline, per gal.		2 00
Sanderson Pearcy's amberine		2 50

BUILDERS' HARDWARE.

BELLS.

Brass hand bells, 60 per cent.
Nickel, 55 per cent.
Gongs, Sargeant's door bells.. 5 50 8 00
American, house bells, per lb. 0 35 0 40
Peterboro' door bells, discount 37½ and 10 per cent. off new list.

BUILDING PAPER, ETC.

Tarred Felt, per 100 lb. 2 25
Ready roofing, 2-ply, not under 45 lb. per roll 0 95
Ready roofing, 3-ply, not under 65 lb., per roll 1 20
Carpet Felt ....per ton 55 00
Heavy Straw Sheathing......per ton 35 00
Dry Surprise 0 40
Dry Sheathing.....per roll, 400 sq. ft. 0 50
Tar " ..... " 400 " 0 45
Dry Fibre......... " 400 " 0 55
Tarred Fibre ..... " 400 " 0 65
O. K. & I. X. L.... " 400 " 0 70
Resin-sized .... " 400 " 0 45
Oiled Sheathing.... " 600 " 1 00
Oiled " ..... " 400 " 0 70
Root Coating, in barrels......per gal. 0 17
Roof " small packages " 0 23
Refined Tar..........per barrel 5 00
Coal Tar " 4 00
Coal Tar, less than barrels....per gal. 0 15
Roofing Pitch ...... per 100 lb. 0 80 0 90
Slater's felt ....per roll 0 70
Heavy Straw Sheathing f. o. b. St. John and Halifax 37 50

BUTTS.

Wrought Brass, net revised list.
Wrought Iron, 70 per cent.
Cast Iron Loose Pin, discount 60 per cent.
Wrought Steel Fast Joint and Loose Pin, 70 per cent.

CEMENT AND FIREBRICK.

"Lafarge" cement in wood 3 40
"Lehigh" cement, in wood 2 54
"Lehigh" cement, cotton sacks 2 39
"Lehigh" cement, paper sacks 2 31
Fire brick (Scotch) per 1,000 .... 15 00 24 00
Fire clay (Scotch), net ton 4 95
Paving Blocks per 1,000
Blue metallic, 9"x4½"x3", ex wharf .. 36 00
Stable pavers, 12"x6"x2", ex wharf .. 50 00
Stable pavers, 9"x4½"x3", ex wharf .. 36 00

DOOR SETS

Peterboro, 37½ and 10 per cent

DOOR SPRINGS.

Torrey's Rod .... per doz. 1 75
Coil, 9 to 11 in. " 0 95 1 65
English .... " 2 00 4 00
Chicago and Reliance Coil Door Springs, 25 per cent.

STORE DOOR HANDLES.

Per Dozen 1 00 1 50

ESCUTCHEONS.

Discount 50 and 10 per cent., new list
Peterboro, 37½ and 10 per cent.

ESCUTCHEON PINS.

Iron, discount 40 per cent.
Brass, 45 per cent.

HINGES.

Blind, discount 60 per cent.
Heavy T and strap, 4-in., per lb. net.. 0 06¼
" " 5-in., " 0 06
" " 6-in., " 0 05¾
" " 8-in., " 0 05½
" " 10-in. and larger .. 0 05¼
Light T and strap, discount 65 p.c.
Screw hook and hinge—
under 12 in.... per 100 lb. 4 75
over 12 in....... " 3 75
Spring, No. 20, per gro. pairs 10 80
Spring, Woodyatt pattern, per gro., No. 5, $17.50 No. 10, $18; No. 20, $10 80; No. 12 $20; No. 51, $10; No. 50, $27.50.
Crate hinges and back flaps, 65 and 5 p. c.
Hinge hasps, 65 per cent.

SPRING HINGES

Chicago Spring Butts and Blanks, 12½ per cent.
Triple End Spring Butts, 40 and 5 per cent.
Chicago Floor Hinges, 40 and 5 off
Garden City Fire Engine House Hinges, 12½ per cent.

HOOKS.

Cast Iron.

Bird cage	per doz.	0 50	1 10
Clothes line, No. 61..	"	0 00	0 70
Harness	"	0 60	12 00
Hat and coat	per gro.	1 10	10 00
Chandelier	per doz.	0 50	1 00

Wrought hooks and staples—
¼ x 5 .... per gross 2 65
5-16 x 5 .... " 3 30
Bright steel gate hooks and staples, 40 per cent discount.
Hat and coat wire, discount 62½ per cent.
Screw, bright wire, discount 62½ per cent.

KNOBS.

Door, japanned and N.P., doz 1 50 2 50
Bronze, Berlin........ per doz. 2 75 3 25
Bronze, Genuine .... " 6 00 9 00
Shutter, porcelain, F. & L. screw .... per gross 1 30 2 00
White door knobs....per doz. 2 00
Peterboro knobs, 37½ and 10 per cent.
Porcelain, mineral and jet knobs, net list.

KEYS.

Lock, Canadian dis. 40 to 40 and 10 per cent
Cabinet trunk and padlock American .... per gross 0 60

LOCKS.

Peterboro, 37½ and 10 per cent.
Russell & Erwin, steel rim $2 50 per doz
Eagle cabinet locks, discount 30 per cent
American padlocks, all steel, 10 to 15 per cent; all brass or bronze, 10 to 25 per cent.

SAND AND EMERY PAPER.

B. & A. sand, discount, 35 per cent.
Emery, discount 35 per cent.
Garnet (Rurton's). 5 to 10 per cent. advance on list.

SASH WEIGHTS.

Sectional..........per 100 lb. 2 00 2 25
Solid 1 50 1 75

SASH CORD.

Per lb 0 31

BLIND AND BED STAPLES.

All sizes, per lb 0 07½ 0 12

WROUGHT STAPLES.

Galvanized 2 75
Plain 2 50
Coopers', discount 45 per cent.
Poultry netting staples, discount 40 per cent.
Bright spear point, 75 per cent. discount.

TOOLS AND HANDLES.

ADZES.

Discount 22½ per cent.

AUGERS.

Gilmour's, discount 60 per cent. off list.

AXES.

Single bit, per doz 5 50 8 50
Double bit, ... 10 00 11 00
Bench Axes, 40 per cent.
Broad Axes. 25 per cent.

Hunters' Axes	5 50	6 00
Boys' Axes	6 25	7 00
Splitting Axes	7 00	12 00
Handled Axes	7 00	9 00
Red Ridge, boys', handled..	....	5 75
" hunters	....	5 25

BITS.

Irwin's auger, discount 47½ per cent.
Gilmour's auger, discount 60 per cent.
Rockford auger, discount 50 and 10 per cent.
Jennings' Gen. auger, net list.
Gilmour's car, 47½ per cent.
Clark's expansive, 40 per cent.
Clark's gimlet, per doz 0 65 0 90
Diamond, Shell, per doz...... 1 00 1 50
Nail and Spike, per gross..... 2 25 5 20

BUTCHERS CLEAVERS

German............per doz. 6 00 9 00
American.......... " 12 00 18 00

CHALK.

Carpenters' Colored, per gross 0 45 0 75
White lump..........per cwt. 0 60 0 65

CHISELS.

Warnock's, discount 72½ per cent.
P. S. & W. Extra, discount 72½ per cent

CROSSCUT SAW HANDLES.

S. & D., No. 3 ....per pair 0 15
S. & D., " 5 .... " 0 22½
S. & D., " 6 .... " 0 15
Boynton pattern................ " 0 20

CROWBARS.

3½c. to 4c. per lb.

DRAW KNIVES.

Coach and Wagon, discount 75 per cent.
Carpenters' discount 75 per cent

DRILLS.

Millar's Falls, hand and breast. net list.
North Bros., each set, 50c.

DRILL BITS.

Morse, discount 37½ to 40 per cent.
Standard, discount 50 and 5 to 55 per cent.

FILES AND RASPS.

Great Western .... 75 per cent.
Arcade .... 75 "
Kearney & Foot .... 75 "
Disston's................ 75 "
American .... 75 "
J Barton Smith................ 75 "
McClellan .... 75 "
Eagle .... 75 "
Nicholson................ 66⅔ "
Globe .... 75 "
Black Diamond, 60, 10 and 5 p c.
Jowitt's, English list, 27½ per cent.

GAUGES.

Stanley's discount 50 to 60 per cent.

Wire Gauges

Winn's, Nos. 26 to 33 ....each 1 65 2 40

HANDLES.

C. & B., fork, 40 per cent., revised list.
C. & B., hoe, 40 per cent., revised list.
American, saw ...... per doz. 0 25
American, plane ... per gross 3 15 3 75
Canadian, hammer and hatchet 40 per cent
Axe and cant hook handles, 45 per cent.

HAMMERS.

Maydole's, discount 5 to 10 per cent.
Canadian, discount 25 to 27½ per cent.
Magnetic tack......per doz. 1 10 1 20
Canadian sledge ... per lb. 0 07 0 08½
Canadian ball pean, per lb. .. 0 22 0 25

HATCHETS.

Canadian, discount 40 to 42½ per cent.
Shingle, Red Ridge 1, per doz.......... 4 40
" " 4 85
Barrel Underhill.......... 5 00

MALLETS.

Tinsmiths'..........per doz. 1 1
Carpenters', hickory, " 1 3
Lignum Vitae........ " 3 85 5 00
Caulking, each 0 60 2 00

MATTOCKS.

Canadian.......... per doz. 5 50 6 00

MEAT CUTTERS.

German, 15 per cent.
American discount, 33⅓ per cent.
Gem ....each 1 15

NAIL PULLERS.

German and American 0 85 2 50
No. 1...................... 0 85
No 1573.................... 0 75

NAIL SETS.

Square, round and octagon, per gross 3 38
Diamond 1 00

PICKS.

Per dozen 6 00 9 00

PLANES.

Wood bench, Canadian discount 40 per cent. American discount 50 per cent.
Wood, fancy Canadian or American 37½ to 40 per cent.
Stanley planes, $1 55 to $3 60, net list prices.

PLANE IRONS.

English ....per doz. 2 00 5 00
Stanley, 2½ inch, single 24c., double 39c.

PLIERS AND NIPPERS.

Button's genuine, 37½ to 40 per cent.
Button's imitation....per doz. 5 00 9 00

PUNCHES.

Saddler's ....per doz. 1 00 1 85
Conductor's.......... " 3 00 15 00
Tinners', solid..........per set 0 72
" hollow......per inch 1 00

RIVET SETS.

Canadian, discount 35 to 37½ per cent.

RULES.

Boxwood, discount 70 per cent.
Ivory, discount 20 to 25 per cent.

SAWS.

Atkins, hand and crosscut, 25 per cent.
Disston's Hand, discount 12½ per cent
Disston's Crosscut ....per foot 0 35 0 55
Hack, complete..........each 0 75 2 75
" frame only........each 0 50 1 25

S. & D. solid tooth circular shingle, concave and band, 50 per cent; mill and ice, drag, 30 per cent.; cross-cut, 35 per cent.; hand saws, butcher, 35 per cent; buck, New Century, $6.25; buck, No. 1 Maple Leaf, $5.25; buck, Happy Medium, $4.25; buck, Watch Spring, $4.25; buck, common frame, $4.00.

Spear & Jackson's saws—Hand or rip, 26 in., $12 75; 28 in., $14.25; panel, 18 in., $8 25; 20 in., $9; tenon, 10 in., $9.90; 12 in., $10 90; 14 in., $11.50.

SAW SETS.

Lincoln and Whiting 4 75
Hand Sets, Perfect.................. 4 00
X-Cut Sets. " 7 50
Maple Leaf and Premiums saw sets, 40 off.
S. & D. saw swages, 40 off.

SCREW DRIVERS.

Sargent's ....per doz. 0 65 1 00
North Bros., No. 30 ..per doz. 16 80

SHOVELS AND SPADES.

Bull Dog, solid neck shovel (No. 2 pol.). $18 50

	(Hollow Back)	(Reinforced S Scoop)
Moose.........	$17 50	$16 30
Bear..........	15 00	15 30
Fox...........	12 50	14 30
Black Cat....	10 00	13 30

Canadian, discount 45 per cent.

SQUARES.

Iron, discount 20 per cent.
Steel, discount 65 and 10 per cent.
Try and Bevel, discount 50 to 52½ per cent.

TAPE LINES.

English, ass skin.....per doz. 2 75 5 00
English, Patent Leather..... 5 50 9 75
Chesterman's............each 0 90 2 85
" steel........each 0 80 8 00

TROWELS.

Disston's, discount 10 per cent.
S. & D., discount 35 per cent.

FARM AND GARDEN GOODS

BELLS.

American cow bells, 63⅓ per cent.
Canadian, discount 45 and 50 per cent.
American, farm bells, each .. 1 35 3 00

BULL RINGS.

Copper, $1.30 for 2½-inch, and $1.70

CATTLE LEADERS.

Nos. 32 and 33 .. ...per gross 7 50 8 50

BARN DOOR HANGERS.

	doz.	pairs.
Steel barn door..............	8 00	10 00
Stearns wood track	4 50	6 00
Zenith.......................	....	9 00
Acme, wood track	5 00	6 50
Atlas........................	5 00	6 00
Perfect	8 00	11 00
New Milo.....................	....	6 50
Steel, covered...............	4 00	11 00
" track, 1 x 3-16 in(100 ft)	....	3 75
" " 1½ x 3-16 in(100 ft)	....	4 75
Double strap hangers, doz. sets	....	6 40
Standard jointed hangers, "	....	6 40
Steel King hangers "	....	6 25
Storm King and safety hangers	....	7 00
" " rail............	....	4 25

Chicago Friction, Oscillating and Big Twin Hangers, 5 per cent.

HARVEST TOOLS.

Discount 60 per cent.
S. & D. lawn rakes, Dunn's, 40 off.
" sidewalk and stable scrapers, 40 off.

HAY KNIVES.

Net list

HEAD HALTERS.

Jute Rope, ⅜-inch....per gross 9 00
" " ½ " .... " 10 00
" " ⅝ " .... " 12 00
Leather, 1-inch ......per doz. 4 00
Leather, 1¼ " ...... " 5 20
Web................ " 2 45

HOES.

Garden, Mortar, etc., discount 60 per cent.
Planter.............. per doz. 4 00 4 50

SCYTHES.

Per doz. net.................. 6 25 9 25

SCYTHE SNATHS.

Canadian, discount 40 per cent.

SNAPS.

Harness, German, discount 25 per cent.
Lock, Andrews' 4 50 11 00

STALL FIXTURES.

Warden King, 35 per cent.

WOOD HAY RAKES.

Ten tooth, 40 and 10 per cent
Twelve tooth, 45 per cent.

—

HEAVY GOODS NAILS, ETC.

ANVILS.

Wright's, 80-lb. and over.............. 0 10½
Hay Budden, 80-lb. and over 0 09½
Brook's, 80-lb. and over 0 11¼
Taylor-Forbes, handy.................. 0 05

VISES.

Wright's....................... .... 0 13½
Brook's........................ .... 0 12½
Pipe Vise, Hinge, No. 1 .... 3 50
" " " No. 2...... .... 5 50
Saw Vise 4 50 5 00
Blacksmiths' (discount) 60 per cent.
" parallel (discount) 45 per cent.

BOLTS AND NUTS

	Per cent.
Carriage Bolts, common ($1 list	
" " ⅜ and smaller..	60, 10 and 10
" " 7-16 and up	55 and 5
" " Norway Iron ($3 list)	50
Machine Bolts, ⅜ and less	60 and 10
Machine Bolts, 7-16 and up....	55 and 5
Plough Bolts	55 and 10
Blank Bolts.......	55
Bolt Ends......................	55

Sleigh Shoe Bolts, ⅜ and less .. 60 and 10
" " " 7-16 and larger 50 and 5
Coach Screws, conepoint...... 70 and 5
Nuts, square, all sizes, 4c. per off.
Nuts, hexagon, all sizes, 4¼c. per off.
Stove Rods per lb., 5½ to 6c.
Stove Bolts, 75 per cent.

CHAIN.

Proof coil, per 100 lb., 5-16 in., $4.40; ⅜ in., $3.90; 7-16 in., $3.70; ½ in., $3.50; 9-16 in., $3.45; ⅝ in., $3.35; ¾ in., $3.20; ⅞ in., $3.10; 1 in., $3.10.
Halter, kennel and post chains, 40 to 40 and 5 per cent.
Cow ties .... 40 p.c.
Tie out chains .... 65 p.c.
Stall fixtures .... 35 p.c.
Trace chain .... 45 p.c.
Jack chain iron, discount 35 p.c.
Jack chain, brass, discount 40 per cent.

HORSE NAILS.

'C' brand, 40, 10 and 7½ per cent. off list { Oval
M.R.M. Co. brand, 55 per cent. { head
Capewell brand, Nos. 9 to 12, base per pound, 12c. Discounts on application.

HORSESHOES.

F.O.B. Montreal
M R.M. Co. brand, base.............. 3 65
Add 15c. Toronto, Hamilton, Guelph.

HORSE WEIGHTS.

Taylor-Forbes, 3½c. per lb.

NAILS.	Cut.	Wire.
2d..............................	3 80	3 30
3d..............................	2 95	2 95
4 and 5d........................	2 70	2 70
6 and 7d	2 60	2 60
8 and 9d	2 45	2 45
10 and 12d	2 40	2 40
16 and 20d......................	2 35	2 35
30, 40, 50 and 60d (base)	2 30	2 30

F.o.b. Montreal. Cut nails, Toronto 20c. higher.
Miscellaneous wire nails, discount 75 per cent
Coopers' nails, discount 40 per cent.

PRESSED SPIKES.

Pressed spikes, ⅝ diameter, per 100 lbs, $3.15

RIVETS AND BURRS.

Iron Rivets, black and tinned, 60, 10 and 10.
Iron Burrs, discount 60 and 10 and 10 p.c.
Copper Rivets, usual proportion burrs, 27½ p.c.
Copper Burrs only, discount 15 per cent.
Extras on Coppered Rivets, ½-lb. packages 1c. per lb.; ¼-lb. packages 2c. lb.
Tinned Rivets, net extra, 4c. per lb.

SCREWS.

Wood, F. H., bright and steel, 87½ per cent.
" R. H., bright, dis. 82½ per cent.
" F. H., brass, dis. 80 per cent.
" R. H., " dis. 75 per cent.
" F. H., bronze, dis. 75 per cent.
" R. H., " dis. 70 per cent.
Drive Screws, dis. 87½ per cent.
Bench, wood ....per doz. 3 25 4 00
" iron 4 25 5 00
Set, case hardened, dis. 60 per cent.
Square Cap, dis. 50 and 5 per cent.
Hexagon Cap, dis. 45 per cent.

MACHINE SCREWS.

Flat head, iron and brass, 35 per cent.
Fellister head, iron, discount 30 per cent.
" " brass, discount 25 per cent.

TACKS, BRADS, ETC.

Carpet tacks, blued .... 80 and 5
" tinned .... 80 and 10
" (in kegs)........... 40
Cut tacks, blued, in dozens only 75 and 10
" ½ weights 60
Swedes cut tacks, blued and tinned—
In bulk.......................... 80 and 10
In dozens 75
Swedes, upholsterers', bulk .... 85 and 12½
" brush, blued and tinned bulk 70
Swedes, gimp, blued, tinned and japanned.......................... 75 and 12½
Zinc tacks 35
Leather carpet tacks............... 40
Copper tacks....................... 37½
Copper nails 42½
Trunk nails, black 65
Trunk nails, tinned and blued .. 65
Clout nails, blued and tinned .. 65
Chair nails........................ 35
Patent brads....................... 40
Fine finishing..................... 40
Lining tacks, in papers............ 10
" " in bulk............ 15
" " solid heads, in bulk 75
Saddle nails, in papers............ 10
" " in bulk............ 15
Tufting buttons, 22 line in dozens only 60
Zinc glaziers' points 5
Double pointed tacks, papers.. 90 and 10
" " " bulk.... 40
Clinch and duck rivets............ 45
Cheese box tacks, 85 and 5; trunk tacks, 80 and 10.

WROUGHT IRON WASHERS.

Canadian make, discount 40 per cent.

—

CUTLERY AND SPORTING GOODS.

AMMUNITION.

B. B. Caps Dominion, 50 and 5 and 25 per cent. American $2 00 per 1000.
C. B. Caps American, $2.60 per 1000

CARTRIDGES.

Rim Fire Cartridges, Dominion, 50 and 5 p.c.
Rim Fire Pistol, discount 30 and 5 per cent., American.
Central Fire, Military and Sporting, American, add 10 per cent. to list. B.B. Caps, discount 40 per cent., American.
Central Fire Pistol and Rifle, list net Amer.
Central Fire Cartridges, pistol and rifle Dominion, 30 and 5 per cent.
Central Fire Cartridges, Sporting and Military, Dominion, 15 per cent., American 10 per cent. advance on list.
Loaded and empty Shells, Crown, 25 and 5; Sovereign, 25, 10 and 10; Regal, 25, 10 and 5; Imperial, 25, 10 and 5. American 20 per cent. discount. Rival and Nitro, 10 per cent. advance on list.
Empty paper shells Dominion, 25 off and empty brass shells, 55 per cent. off. American, 10 per cent. advance on list.
Primers, Dom., 35 per cent.; American $2.05

Wads. per lb.

Best thick brown or grey felt wads, in ¼-lb. bags $0 70
Best thick white card wads in boxes of 500 each, 12 and smaller gauges 0 29
Best thick white card wads, in boxes of 500 each, 10 gauge............ 0 35
Thin card wads, in boxes of 1,000 each, 12 and smaller gauges 0 20
Thin card wads, in boxes of 1,000 each, 10 gauge...................... 0 25
Chemically prepared black edge grey cloth wads, in boxes of 250 each— Per M.
11 and smaller gauge 0 60
9 and 10 gauges.................... 0 70
7 and 8 " 0 90
5 and 6 " 1 10
Superior chemically prepared pink edge, best white cloth wads in boxes of 250 each—
11 and smaller gauge 1 15
9 and 10 gauges 1 40
7 and 8 " 1 65
5 and 9 " 1 90

SHOT

Common, $6.50 per 100 lb.; chilled, $7.50 per 100 lb.; buck, seal and ball, $8.50 net list. Prices are f.o.b. Toronto, Hamilton, Montreal, St. John and Halifax. Terms, 2 p.c. for cash in thirty days.

RAZORS.	per doz.	
Elliot's	4 00	18 00
Boker's	7 50	11 00
" King Cutter	13 50	18 50
Wade & Butcher's............	3 60	10 00
Lewis Bros.' "Klean Kutter"	8 50	10 50

Clauss Razors and Strops, 50 and 10 per cent

PLATED GOODS

Hollowware, 40 per cent discount.
Flatware, staples, 40 and 10, fancy 40 and 5 per cent.

SHEARS.

Clauss, nickel, discount 60 per cent.
Clauss, Japan, discount 67½ per cent.
Clauss, tailors, discount 40 per cent.
Seymour's, discount 50 and 10 per cent.

TRAPS (steel.)

Game, Newhouse, discount 30 and 10 per cent
Game, Hawley & Norton, 50, 10 & 5 per cent.
Game, Victor, 70 per cent.
Game, Oneida Jump (B. & L.) 40 & 2½ p. c.
Game, steel, 60 and 5 per cent.

SKATES.

Skates, discount 37½ per cent.
Mic Mac hockey sticks, per doz 4 00 5 00
New Rex hockey sticks, per doz 6 25

HOUSE FURNISHINGS.

APPLE PARERS.

Woodyatt Hudson, per doz., net 4 50

BIRD CAGES.

Brass and Japanned, 40 and 10 p. c.

COPPER AND NICKEL WARE.

Copper boilers, kettles, teapots, etc. 35 p.c.
Copper pitts, 25 per cent.

ENAMELLED WARE.

London, White, Princess, Turquoise, Onyx, Blue and White, discount 50 per cent.
Canada, Diamond, Premier, 50 and 10 p c.
Pearl, Imperial Crescent, 60 and 10 per cent.
Premier steel ware, 40 per cent.
Star decorated steel and white, 25 per cent.
Japanned ware, discount 45 per cent.
Hollow ware, tinned cast, 35 per cent. off.

KITCHEN SUNDRIES.

Can openers, per doz.	0 40	0 75
Mincing knives per doz	0 50	0 80
Duplex mouse traps, per doz....		0 65
Potato mashers, wire, per doz...	0 60	0 70
" wood " ..	0 50	0 60
Vegetable slicers, per doz		2 25

Universal meat chopper No. 0, $1; No.1, 1 15.
Enterprise chopper, each 1 30
Spiders and fry pans, 50 per cent.

LAMP WICKS.

Discount, 60 per cent.

LEMON SQUEEZERS.

Porcelain lined......	per doz.	2 20	5 60
Galvanized..........	"	1 87	3 85
King, wood..........	"	2 75	2 90
King, glass..........	"	4 00	4 50
All glass	"	0 50	0 90

PICTURE NAILS.

Porcelain head ......per gross 1 35 1 50
Brass head.......... 1 00
Tin and gilt, picture wire, 75 per cent.

SAD IRONS.

Mrs. Potts, No. 55, polished....per set 0 80
" No. 50, nickle-plated, " 0 92
Common, plain........................ 4 50
" plated 5 50
Asbestos, per set............... .... 1 25

TINWARE.

CONDUCTOR PIPE.

2 in. plain or corrugated., per 100 feet, $3.30; 3 in., $4 40; 4 in., $5.80; 5 in., $7.45; 6 in., $9.90.

FAUCETS.

Common, cork-lined, discount 35 per cent.

EAVETROUGHS.

10-inch ....per 100 ft. 3 30

FACTORY MILK CANS.

Discount off revised list, 35 per cent.
Milk can trimmings, discount 25 per cent.
Creamery Cans, 45 per cent.

LANTERNS.

No. 2 or 4 Plain Cold Blast....per doz. 6 50
Lift Tubular and Hinge Plain, " 4 75
Better quality at higher prices.
Japanning, 50c. per doz. extra.

OILERS.

Kemp's Tornado and McClary's Model galvanized oil can, with pump, 5 gallon, per dozen 10 00
Davidson oilers, discount 40 per cent
Zinc and tin, discount 50 per cent
Coppered oilers, 20 per cent. off.
Brass oilers, 50 per cent. off.
Malleable, discount 25 per cent

PAILS (GALVANIZED).

Dufferin pattern pails, 40, 10 and 5 per cent.
Flaring pattern, discount 40, 10 and 5 per cent.
Galvanized washtubs 40, 10 and 5 per cent

PIECED WARE.

Discount 35 per cent off list, June, 1899.
10-qt. flaring sap buckets, discount 40 per cent.
6, 10 and 14-qt. flaring pails dis. 40 per cent.
Copper bottom tea kettles and boilers, 30 p.c.
Coal hods, 40 per cent.

STAMPED WARE.

Plain, 75 and 12½ per cent. off revised list.
Retinned, 72½ 10 and 5 per cent. revised list.

SAP SPOUTS.

Bronzed iron with hooks ....per 1,000 7 50
Eureka tinned steel, hooks 8 00

STOVEPIPES.

5 and 6 inch, per 100 lengths 7 00
7 inch........ " " 7 50
Nestable, discount 40 per cent.

STOVEPIPE ELBOWS

5 and 6-inch, common........per doz. 1 32
7-inch........................ " 1 48
Polished, 15c. per dozen extra.

THERMOMETERS.

Tin case and dairy, 75 to 75 and 10 per cent

TINNERS' SNIPS.

Per doz...................... 3 00 15
Clauss, discount 35 per cent.

TINNERS' TRIMMINGS.

Discount, 45 per cent.

WIRE.

BRIGHT WIRE GOODS

Discount 62½ per cent.

CLOTHES LINE WIRE.

7 wire solid line, No. 17, $4.90; No. 18, $3.00; No. 19, $2.70; 6 wire solid line, No. 17, $4.45; No. 18, $3.80. No. 19, $3.50. All prices per 1000 ft. measure. F.o.b. Hamilton Toronto, Montreal.

COILED SPRING WIRE.

High Carbon, No. 9, $2 55, No. 11, $3.20; No. 12, $2.80.

COPPER AND BRASS WIRE.

Discount 45 per cent.

FINE STEEL WIRE.

Discount 30 per cent. List of extras: In 100-lb. lots: No. 17, $5 — No. 18, $5.50 — No. 19, $6 — No. 20, $6.65 — No. 21, $7 — No. 22, $7.30 — No. 23, $7.65 — No. 24, $8 — No. 25, $9 — No. 26, $9.50 — No. 27, $10 — No. 28, $11 — No. 29, $12 — No. 30, $13 — No. 31, $14 — No. 32, $15 — No. 33, $16 — No. 34, $17. Extras net — tinned wire, Nos. 17-25, $2 — Nos. 26-31, $4 — Nos. 32-34, $6. Coppered, 75c. — oiling, 10c. — in 25-lb. bundles, 15c. — in 5 and 10-lb. bundles, 25c. — in 1-lb. hanks, 25c. — in ½-lb. hanks, 38c. — in ¼-lb. hanks, 50c. — packed in casks or cases, 15c. — bagging or papering, 10c.

GALVANIZED WIRE.

Per 100 lb. — Nos. 4 and 5, $3.70 — Nos. 6, 7, 8, $3.15 — No. 9, $2.50 — No. 10, $3 20 — No. 11, $3.25 — No. 12, $2 65 — No. 13, $2.75 — No. 14, $3.75 — No. 15, $4.30 — No. 16 $4.30 from stock. Base sizes, Nos. 6 to 9, $2.35 f.o.b. Cleveland. In carlots 12½c. less.

LIGHT STRAIGHTENED WIRE.

Gauge No.	Over 20 in. per 100 lbs.	10 to 20 in.	5 to 10 in.
0 to 5	$0.50	$0 75	$1 25
6 to 9	0 75	1 25	2 00
10 to 11	1.00	1.75	2 50
12 to 14	1.50	2.25	3 50
15 to 16	2.00	3.00	4.50

SMOOTH STEEL WIRE.

No. 0-9 gauge, $2 25; No. 10 gauge, 6c. extra; No. 11 gauge, 12c extra; No. 12 gauge, 20c. extra; No. 13 gauge, 30c. extra; No 14 gauge, 40c. extra; No. 15 gauge, 55c. extra; No 16 gauge, 70c extra. Add 60c. for coppering and $2 for tinning.

Extra net per 100 lb. — Oiled wire 10c., spring wire $1.25, special hay baling wire 30c., best steel wire 75c., bright soft drawn 15c., charcoal (extra quality) $1.25, packed in casks or cases 15c., bagging and papering 10c., 50 and 100-lb. bundles 10c., in 25-lb. bundles 15c., in 5 and 10-lb. bundles 25c., in 1-lb hanks, 50c., in ½-lb. hanks 75c., in ¼-lb. hanks $1.

POULTRY NETTING.

2 in mesh 19 w. g., discount 50 and 10 per cent. All others 50 per cent.

WIRE CLOTH.

Painted Screen, in 100-ft. rolls, $1 62½, per 100 sq. ft; in 50-ft. rolls, $1.67½, per 100 sq. ft. Terms, 2 per cent. off 30 days.

WIRE FENCING.

Galvanized barb...... .... 2 95
Galvanized, plain twist 3 30
Galvanized barb, f.o.b. Cleveland, $2 70 for small lots and $2.60 for carlot.

WOODENWARE.

CHURNS.

No. 0, $9; No. 1, $9; No. 2, ; No. 3, $11; No. 4, $13; No. 5, $16.; f.o.b. Toronto Hamilton, London and St. Marys, 30 and 30 per cent; f.o.b Ottawa, Kingston and Montreal, 40 and 15 per cent. discount, Taylor-Forbes, 30 and 30 per cent.

CLOTHES REELS.

Davis Clothes Reels, dis. 40 per cent.

LADDERS, EXTENSION.

Waggoner Extension Ladders, dis. 40 per cent.

MOPS AND IRONING BOARDS.

"Best" mops................................ 1 25
"900" mops................................. 1 25
Folding ironing boards............ 12 00 16 50

REFRIGERATORS

Discount, 40 per cent.

SCREEN DOORS.

Common doors, 2 or 3 panel, walnut stained, 4-in. style..........per doz. 7 25
Common doors, 2 or 3 panel, grained only, 4-in., style ....per doz. 7 55
Common doors, 2 or 3 panel, light stair per doz. 9 55

WASHING MACHINES.

Round, re-acting per doz.	60 00
Square " "	63 00
Eclipse, per doz	54 00
Dowswell "	39 00
New Century, per doz	75 00
Daisy..	54 00

WRINGERS.

Royal Canadian, 11 in., per doz. ... 34 00
Royal American, 11 in. ... 34 00
Eze, 10 in., per doz 3 75
Terms, 2 per cent., 30 days.

MISCELLANEOUS.

AXLE GREASE.

Ordinary, per gross	6 00	7 00
Best quality	10 00	12 00

BELTING.

Extra, 60 per cent.
Standard, 60 and 10 per cent.
No. 1, not wider than 6 in., 60, 10 and 10 p.c.
Agricultural, not wider than 4 in., 75 per cent
Lace leather, per side, 75c; cut laces, 80c.

BOOT CALKS.

Small and medium, ball ....per M 4 25
Small heel .... " 4 50

CARPET STRETCHERS.

American ....per doz. 1 00 1 50
Bullard's............ " 6 50

CASTORS.

Bed, new list, discount 55 to 57½ per cent.
Plate, discount 52½ to 57½ per cent.

PINE TAR.

½ pint in tins ....per gross ... 7 80
" " 9 60

PULLEYS.

Hothouse	per doz.	0 55	1 00
Axle	"	0 22	0 33
Screw	"	0 22	1 00
Awning	"	0 35	2 50

PUMPS.

Canadian cistern	1 40	2 00
Canadian pitcher spout	1 80	3 16

ROPE AND TWINE.

Sisal..................................		0 10½
Pure Manilla		0 15½
"British" Manilla....................		0 12
Cotton, 3-16 inch and larger...	0 21	0 23
" 5-32 inch	0 25	0 27
" ¼ inch	0 25	0 28
Russia Deep Sea		0 16
Jute.......................................		0 09
Lath Yarn, single		0 10
" double		0 10½
Sisal bed cord, 48 feet.......per doz.		0 65
" " 60 feet....... "		0 80
" " 72 feet....... "		0 95

Twine.

Bag, Russian twine, per lb.		0 27
Wrapping, cotton, 3-ply		0 25
" " 4-ply		0 29
Mattress twine per lb.	0 33	0 45
Staging " "	0 27	0 35

SCALES.

Gurney Standard, 40 per cent.
Gurney Champion, 50 per cent.
Burrow, Stewart & Milne—
Imperial Standard, discount 40 per cent.
Weigh Beams, discount 40 per cent.
Champion Scales, discount 50 per cent.
Fairbanks standard, discount 35 per cent.
" Dominion, discount 55 per cent.
" Richelieu, discount 55 per cent.
Warren new Standard, discount 40 per cent.
" " Champion, discount 50 per cent.
" Weighbeams, discount 35 per cent.

STONES—OIL AND SCYTHE.

Washita	per lb.	0 25	0 37
Hindostan	"	0 06	0 10
" slip	"	0 18	0 20
" Axe.........	"	....	0 10
Deer Creek............	"	....	0 10
Deerlick	"	....	0 25
" Axe	"	....	0 15
Lily white	"	....	0 42
Arkansas...............	"	....	1 50
Water-of-Ayr	"	....	0 10
Scythe	per gross	3 50	5 00
Grind, 40 to 200 lb., per ton....		20 00	22 00
" under 40 lb., "		....	24 00
" 200 lb. and over		....	28 00

INDEX TO ADVERTISERS.

CLASSIFIED LIST OF ADVERTISEMENTS.

Auditors.
Davenport, Percy P., Winnipeg.

Babbitt Metal.
Canada Metal Co., Toronto.
Canadian Fairbanks Co., Montreal.
Robertson, Jas. Co., Montreal.

Bath Room Fittings.
Carriage Mounting Co., Toronto.

Belting, Hose, etc.
Gutta Percha and Rubber Mfg. Co. Toronto.

Bicycles and Accessories.
Johnson's, Iver, Arms and Cycle Works Fitchburg, Mass.

Binder Twine.
Consumers Cordage Co., Montreal.

Box Strap.
J. N. Warminton, Montreal.

Brass Goods.
Glauber Brass Mfg. Co., Cleveland, Ohio.
Lewis, Rice, & Son., Toronto.
Morrison, Jas., Brass Mfg. Co., Toronto.
Mueller Mfg. Co., Decatur, Ill.
Penberthy Injector Co., Windsor, Ont.
Taylor-Forbes Co., Guelph, Ont.

Bronze Powders.
Canadian Bronze Powder Works, Montreal.

Brushes.
Ramsay, A., & Son Co., Montreal.

Can Openers.
Cumming Mfg. Co. Renfrew.

Cans.
Acme Can Works, Montreal.

Builders' Tools and Supplies.
Covert Mfg. Co., West Troy, N.Y.
Frothingham & Workman Co., Montreal.
Howland, H. S., Sons & Co., Toronto.
Hyde, F., & Co., Montreal.
Lewis Bros. & Co., Montreal.
Lewis, Rice, & Son, Toronto.
Lockerby & McComb, Montreal.
Lufkin Rule Co., Saginaw, Mich.
Newman & Sons, Birmingham.
North Bros. Mfg. Co., Philadelphia, Pa.
Stanley Rule & Level Co., New Britain.
Stanley Works, New Britain, Conn.
Stephens, G. F., Winnipeg.
Taylor-Forbes Co., Guelph, Ont.

Carriage Accessories.
Carriage Mountings Co., Toronto.
Covert Mfg. Co., West Troy, N.Y.

Carriage Springs and Axles.
Guelph Spring and Axle Co., Guelph.

Cattle and Trace Chains.
Greening, B., Wire Co., Hamilton.

Churns.
Dowswell Mfg. Co., Hamilton.

Clippers—All Kinds.
American Shearer Mfg. Co., Nashua, N.H.

Clothes Reels and Lines.
Hamilton Cotton Co., Hamilton, Ont.

Cordage.
Consumers' Cordage Co., Montreal.
Hamilton Cotton Co., Hamilton.

Cork Screws.
Erie Specialty Co., Erie, Pa.

Clutch Nails.
J. N. Warminton, Montreal.

Cut Glass.
Phillips, Geo., & Co., Montreal.

Cutlery—Razors, Scissors, etc.
Birkett, Thos., & Son Co., Ottawa.
Clauss Shear Co., Toronto
Dorken Bros. & Co., Montreal.
Heinisch's, R., Sons Co., Newark, N.J.
Howland, H. S. Sons & Co., Toronto.
Phillips, Geo. & Co., Montreal.
Round, John, & Son, Montreal.

Electric Fixtures.
Canadian General Electric Co., Toronto.
Forman, John, Montreal.
Morrison James, Mfg. Co., Toronto.
Munderloh & Co., Montreal.

Electro Cabinets.
Cameron & Campbell Toronto.

Engines, Supplies, etc.
Kerr Engine Co., Walkerville, Ont.

Files and Rasps.
Barnett Co., G. & H., Philadelphia, Pa.
Nicholson File Co., Port Hope.

Financial Institutions
Bradstreet Co.

Firearms and Ammunition.
Dominion Cartridge Co., Montreal.
Hamilton Rifle Co., Plymouth, Mich.
Harrington & Richardson Arms Co., Worcester, Mass.
Johnson's, Iver, Arms and Cycle Works, Fitchburg, Mass.

Food Choppers.
Enterprise Mfg. Co., Philadelphia, Pa.
Shirreff Mfg. Co., Brockville, Ont

Galvanizing.
Canada Metal Co., Toronto.
Montreal Rolling Mills Co., Montreal.
Ontario Wind Engine & Pump Co., Toronto.

Glaziers' Diamonds.
Gibsone, J. B., Montreal.
Pelton, Godfrey S.
Sharratt & Newth, London, Eng.
Shaw, A., & Son, London, Eng.

Handles.
Still, J. H., Mfg. Co.

Harvest Tools.
Maple Leaf Harvest Tool Co., Tillsonburg, Ont.

Hoop Iron.
Montreal Rolling Mills Co., Montreal.
J. N. Warminton, Montreal.

Horse Blankets.
Heney, E. N., & Co., Montreal.

Horseshoes and Nails.
Canada Horse Nail Co., Montreal.
Montreal Rolling Mills, Montreal.
Capewell Horse Nail Co., Toronto

Hot Water Boilers and Radiators.
Cluff, R. J., & Co., Toronto.
Pease Foundry Co., Toronto.
Taylor-Forbes Co., Guelph.

Ice Cream Freezers.
Dana Mfg. Co., Cincinnati, Ohio.
North Bros. Mfg. Co., Philadelphia, Pa.

Ice Cutting Tools.
Erie Specialty Co., Erie, Pa.
North Bros. Mfg. Co., Philadelphia, Pa.

Injectors—Automatic.
Morrison, Jas., Brass Mfg. Co., Toronto.
Penberthy Injector Co., Windsor, Ont.

Iron Pipe.
Montreal Rolling Mills, Montreal.

Iron Pumps.
McDougall, R., Co., Galt, Ont.

Lanterns.
Kemp Mfg. Co., Toronto.
Ontario Lantern Co., Hamilton, Ont.
Wright, E. T., & Co., Hamilton.

Lawn Mowers.
Birkett, Thos., & Son Co., Ottawa.
Maxwell, D., & Sons, St. Mary's, Ont.
Taylor, Forbes Co., Guelph.

Lawn Swings, Settees, Chairs.
Cumming Mfg. Co., Renfrew.

Ledgers—Loose Leaf.
Business Systems, Toronto.
Copeland-Chatterson Co., Toronto.
Crain, Rolla L., Co., Ottawa.
Universal Systems, Toronto.

Locks, Knobs, Escutcheons, etc.
Peterborough Lock Mfg. Co., Peterborough, Ont.

Lumbermen's Supplies.
Pink, Thos., & Co., Pembroke Ont.

Manufacturers' Agents.
Fox, C. H., Vancouver.
Gibb, Alexander, Montreal.
Mitchell, David C., & Co., Glasgow, Scot.
Mitchell, H. W., Winnipeg.
Pearce, Frank, & Co., Liverpool, Eng.
Scott, Bathgate & Co., Winnipeg.
Thorne, R. E., Montreal and Toronto.

Metals.
Canada Iron Furnace Co., Midland, Ont.
Canada Metal Co., Toronto
Eadie, H. G., Montreal.
Gibb, Alexander, Montreal.
Kemp Mfg. Co., Toronto
Leslie, A. C., & Co., Montreal.
Lysaght, John, Bristol, Eng.
Nova Scotia Steel and Coal Co., New Glasgow, N.S.
Robertson, Jas., Co., Montreal
Roper, J. H., Montreal
Samuel, Benjamin & Co., Toronto.
Stairs, Son & Morrow, Halifax, N.S.
Thompson, B. & S. H. & Co. Montreal.

Metal Lath.
Galt Art Metal Co., Galt
Metallic Roofing Co., Toronto.
Metal Shingle & Siding Co., Preston, Ont.

Metal Polish, Emery Cloth, etc
Oakey, John, & Sons, London, Eng.

Mops.
Cumming Mfg. Co., Refrew.

Mouse Traps.
Cumming Mfg. Co., Renfrew

Oil Tanks.
Bowser, S. F., & Co., Toronto.

Paints, Oils, Varnishes, Glass.
Bell, Thos., Sons & Co., Montreal.
Canada Paint Co., Montreal.
Canadian Oil Co., Toronto.
Consolidated Plate Glass Co., Toronto.
Fenner, Fred., & Co., London, Eng.
Henderson & Potts Co., Montreal.
Imperial Varnish and Color Co., Toronto.
Jamieson, R. C., & Co., Montreal.
McArthur, Corneille & Co., Montreal.
McCaskill, Dougall & Co., Montreal.
Montreal Rolling Mills Co., Montreal.
Moore, Benjamin, & Co. Toronto.
Queen City Oil Co., Toronto.
Ramsay & Son, Montreal.
Sherwin-Williams Co., Montreal.
Standard Paint and Varnish Works Windsor, Ont
Stephens & Co., Winnipeg.
Martin-Senour Co., Chicago.

Perforated Sheet Metals.
Greening, B., Wire Co., Hamilton.

Plumbers' Tools and Supplies.
Borden Co., Warren, Ohio.
Canadian Fairbanks Co., Montreal.
Cluff, R. J., & Co., Toronto.
Glauber Brass Co., Cleveland, Ohio.
Jardine, A. B., & Co., Hespeler, Ont.
Jenkins Bros., Boston, Mass.
Lewis, Rice, & Son, Toronto.
Merrell Mfg. Co., Toledo, Ohio.
Montreal Rolling Mills, Montreal.
Morrison, Jas., Brass Mfg. Co., Toronto.
Mueller, H., Mfg. Co., Decatur, Ill.
Oshawa Steam & Gas Fitting Co., Oshawa
Robertson Jas., Co. Montreal.
Stairs, Son & Morrow, Halifax, N.S.
Standard Ideal Sanitary Co., Port Hope,
Standard Sanitary Co., Pittsburg.
Stephens, G. F., & Co., Winnipeg, Man.
Turner Brass Works, Chicago
Vickery, Orlando, Toronto.

Portland Cement.
International Portland Cement Co., Ottawa, Ont
Hanover Portland Cement Co., Hanover, Ont.
Hyde, F., & Co., Montreal.
Thompson, B. & S. H. & Co., Montreal.

Potato Mashers.
Cumming Mfg. Co., Renfrew.

Poultry Netting.
Greening, B., Wire Co., Hamilton, Ont.

Razors.
Clauss Shear Co., Toronto.

Roofing Supplies.
Brantford Roofing Co., Brantford.
McArthur, Alex., & Co., Montreal.
Metal Shingle & Siding Co., Preston, Ont.
Metallic Roofing Co., Toronto.
Paterson Mfg. Co., Toronto & Montreal.

Saws.
Atkins, E. C., & Co., Indianapolis, Ind
Lewis Bros., Montreal.
Shurly & Dietrich, Galt, Ont.
Spear & Jackson, Sheffield, Eng.

Scales.
Canadian Fairbanks Co., Montreal.

Screw Cabinets.
Cameron & Campbell, Toronto.

Screws, Nuts, Bolts.
Montreal Rolling Mills Co., Montreal
Morrow, John, Machine Screw Co., Ingersoll, Ont.

Sewer Pipes.
Canadian Sewer Pipe Co., Hamilton
Hyde, F., & Co., Montreal.

Shelf Boxes.
Cameron & Campbell, Toronto.

Shears, Scissors.
Clauss Shear Co., Toronto.

Shelf Brackets.
Atlas Mfg. Co., New Haven, Conn

Shellac
Bell, Thos., Sons & Co., Montreal.

Shovels and Spades.
Canadian Shovel Co., Hamilton.
Peterboro Shovel & Tool Co., Peterboro.

Silverware.
Phillips, Geo., & Co., Montreal.
Round, John, & Son, Sheffield, Eng.

Skates.
Canada Cycle & Motor Co., Toronto.

Spring Hinges, etc.
Chicago Spring Butt Co., Chicago, Ill.

Steel Rails.
Nova Scotia Steel & Coal Co., New Glasgow, N.S.

Stoves, Tinware, Furnaces
Canadian Heating & Ventilating Co., Owen Sound
Canada Stove Works, Harriston, Ont.
Clare Bros & Co., Preston.
Davidson, Thos., Mfg. Co., Montreal.
Guelph Stove Co., Guelph.
Gurney Foundry Co., Toronto.
Harris, J. W., Co., Montreal.
Joy Mfg. Co., Toronto.
Kemp Mnfg. Co. Toronto.
McClary Mfg. Co. London.
Pease Foundry Co., Toronto
Stewart, Jas., Mfg. Co., Woodstock, Ont.
Taylor-Forbes Co., Guelph, Ont.
Wright, E. T., & Co., Hamilton.

Tacks.
Montreal Rolling Mills Co., Montreal.
Ontario Tack Co., Hamilton.

Ventilators.
Harris, J. W., Co., Montreal.
Pearson, Geo. D., Montreal.

Washing Machines, etc.
Dowswell Mfg. Co., Hamilton, Ont.
Taylor-Forbes Co., Guelph, Ont.

Wheelbarrows.
London Foundry Co., London, Ont.

Wholesale Hardware.
Birkett, Thos., & Sons Co., Ottawa.
Caverhill, Learmont & Co., Montreal.
Frothingham & Workman, Montreal.
Hobbs Hardware Co., London.
Howland, H. S., Sons & Co., Toronto.
Lewis Bros. & Co., Montreal.
Lewis, Rice, & Son, Toronto.

Window and Sidewalk Prisms.
Hobbs Mfg. Co., London, Ont.

Wire Springs.
Guelph Spring Axle Co., Guelph, Ont.
Wallace-Barnes Co., Bristol, Conn.

Wire, Wire Rope, Cow Ties, Fencing Tools, etc.
Canada Fence Co., London.
Dennis Wire and Iron Co., London, Ont.
Dominion Wire Mnfg. Co., Montreal
Greening, B., Wire Co., Hamilton.
Montreal Rolling Mills Co., Montreal.
Western Wire & Nail Co., London, Ont.

Woodenware.
Taylor-Forbes Co., Guelph, Ont.

Wrapping Papers.
Canada Paper Co., Toronto.
McArthur, Alex., & Co., Montreal.
Stairs, Son & Morrow, Halifax, N.S.

CIRCULATES EVERYWHERE IN CANADA

Also in Great Britain, United States, West Indies, South Africa and Australia.

HARDWARE AND METAL

A Weekly Newspaper Devoted to the Hardware, Metal, Heating and Plumbing Trades in Canada.

Office of Publication, 10 Front Street East, Toronto.

VOL. XIX. MONTREAL, TORONTO, WINNIPEG, FEBRUARY 2, 1907 NO. 5.

See Classified List of Advertisements on Page 67.

NICKEL-PLATED BATH-ROOM FITTINGS

- Paper Holders
- Tumbler Holders
- Brush Holders
- Soap Holders
- Towel Bars

- Robe Hooks
- Match Holders
- Bath Seats
- Mirrors
- Bath Sprays

TUMBLER AND SOAP HOLDERS

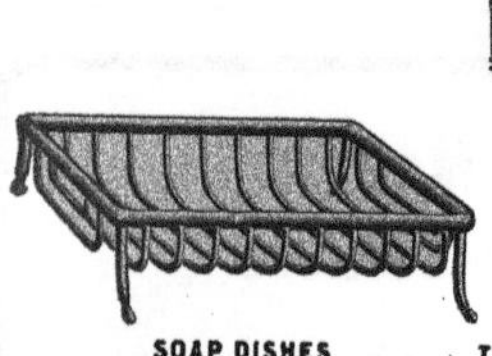

SOAP DISHES

TUMBLER AND SOAP HOLDERS

SOAP DISHES

Everything in Bath-room Fittings

TOWEL BARS

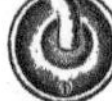

Everything in Bath-room Fittings

Write for Prices

RICE LEWIS & SON

LIMITED

TORONTO.

CHICAGO SPRING BUTTS
TRADE CSB CO MARK
A PRODUCT OF RECOGNIZED SUPERIORITY
CHICAGO SPRING BUTT
TRIPLE-END SPRING BUTT
CHICAGO BALL-BEARING FLOOR HINGE
Chicago Spring Butt Company
CHICAGO NEW YORK
CATALOGUE ON REQUEST

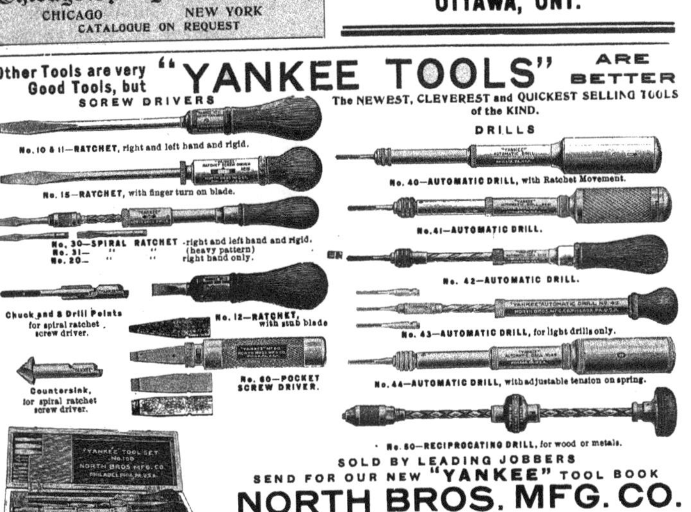
Other Tools are very Good Tools, but "YANKEE TOOLS" ARE BETTER
SCREW DRIVERS
The NEWEST, CLEVEREST and QUICKEST SELLING TOOLS of the KIND.
DRILLS
No. 10 & 11—RATCHET, right and left hand and rigid.
No. 15—RATCHET, with finger turn on blade.
No. 30—SPIRAL RATCHET -right and left hand and rigid.
No. 31— " " (heavy pattern)
No. 20— " " right hand only.
Chuck and 8 Drill Points for spiral ratchet screw driver.
No. 12—RATCHET, with stub blade
Countersink, for spiral ratchet screw driver.
No. 60—POCKET SCREW DRIVER.
No. 40—AUTOMATIC DRILL, with Ratchet Movement.
No. 41—AUTOMATIC DRILL.
No. 42—AUTOMATIC DRILL.
No. 43—AUTOMATIC DRILL, for light drills only.
No. 44—AUTOMATIC DRILL, with adjustable tension on spring.
No. 50—RECIPROCATING DRILL, for wood or metals.
"YANKEE" TOOL SET No. 100 NORTH BROS. MFG. CO. PHILADELPHIA, PA. U.S.A.
SOLD BY LEADING JOBBERS
SEND FOR OUR NEW "YANKEE" TOOL BOOK
NORTH BROS. MFG. CO.
PHILADELPHIA, Pa.

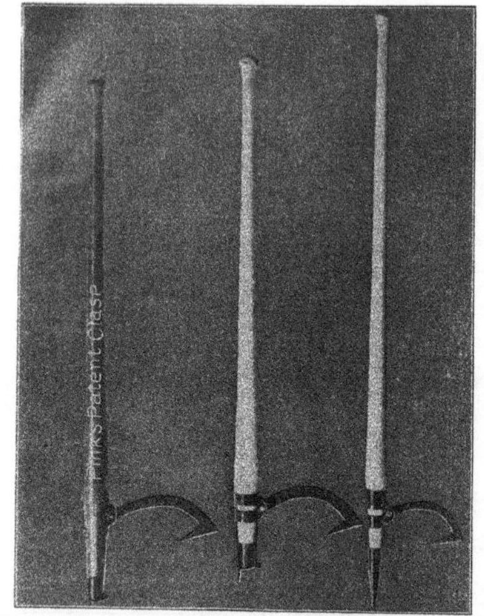

Pink's Lumbering Tools

MADE IN CANADA

Send for Catalogue and Price List

THE STANDARD TOOLS

in every Province of the Dominion, New Zealand, Australia, Etc.

We manufacture all kinds of Lumber Tools

Pink's Patent Open Socket Peaveys.
Pink's Patent Open Socket Cant Dogs.
Pink's Patent Clasp Cant Dogs, all Handled with Split Rock Maple.

These are light and durable tools.

Sold throughout the Dominion by all Wholesale and Retail Hardware Merchants

MANUFACTURED BY

Long Distance Phone No. 87 **THOMAS PINK**
Pembroke, Ont., Canada.

ESTABLISHED 1855

HEADQUARTERS FOR

PIG IRON

ENGLISH and SCOTCH

Prompt and forward delivery. Interesting prices.

Warehouse: 54-56-58 Front St. West and 1 Piper St.

M. & L. Samuel, Benjamin & Co.

TORONTO

SAMSON
EIGHT LEVER
CORBIN
CORBIN
TITAN
SAMSON-CORBIN
No. 2916¼—2½ in., bright finish steel case and shackle, eight secure levers, two double-bitted steel keys.
No. 412, YALE
2 in., Bower-Barff steel case, bright steel shackle, brass locking mechanism, 2 flat steel keys.
$ PROFITABLE STOCK $
NO LINE PAYS A BIGGER PROFIT THAN THIS
PADLOCKS AT LEWIS BROS.' PRICES
No company carries a bigger range than we do.
ALL THE POPULAR MAKES
Corbin, Yale & Towne, Miller Lock Co., Slaymaker, Mallory, Wheeler, etc.
WRITE FOR PRICES
LEWIS BROS., Ltd.
OTTAWA
TORONTO
MONTREAL
WINNIPEG
VANCOUVER
CALGARY
PROTECTOR, JR.
MILLER
No. 16B—1 5/16 in., heavy steel case, brass plated; steel levers, wards and keys; brass forged shackle.
No. 916, SLAYMAKER
2 in., all parts solid bronze metal, attractive design, richly finished, quick shackle action, 2 corrugated keys.
No. 4167¼
EAGLE LOCK CO.
2 in., brassed steel case and shackle, spring shackle, 2 flat steel keys.
EAGLE LOCK CO
U.S.A.
TERRYVILLE CT
SLAYMAKER-LANCASTER, PA.
SLAYMAKER
NE PLUS
ULTRA

H. S. HOWLAND, SONS & CO. LIMITED

HARDWARE MERCHANTS

Only Wholesale — 138-140 WEST FRONT STREET, TORONTO — Wholesale Only

Coil Chain, Proof and BBB Qualities

"Cant Hooks"—4-4½-ft. Handles

"Peavies" 5-5½-ft. Handles

Samson's Calks

Timber Carriers or Log Hooks

5-Ring Calks

Repair Links

Lipscomb's Ice Calks

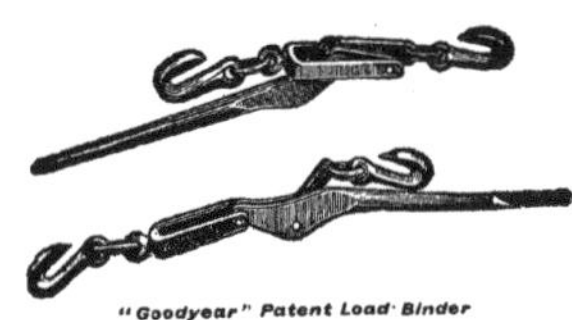

"Goodyear" Patent Load Binder

Lipscomb's Lumbermen's Calks

Grab and Slide Hook

Pike Poles—12 and 14-feet Handles

For fuller particulars see our Hardware Catalogue.

H. S. HOWLAND, SONS & CO., LIMITED

Opposite Union Station.

GRAHAM NAILS ARE THE BEST

Factory: Dufferin Street, Toronto, Ont.

We Ship promptly — Our Prices are Right

Morse Twist Drills are perfect drills. They are made scientifically by the people who have made the name **"Morse"** synonymous with the words **"twist drill."** Many discriminating users will have no other. The largest machine shops in all parts of Canada use them.

The **Morse Twist Drill** can be had in a greater variety of sizes and styles than any other make of drills. If you have a twist drill trade sell **Morse Drills.**

No. 102 No. 109 No. 104

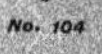

Morse Twist Drills are put up in attractive and convenient sets. They will sell to the machinists in your town.

Ask our travellers for prices, or write direct.

Five sets of drills can be had in these indexed cases. Cases are also sold empty

No. 14. Set of Bit Stock Drills, 9 sizes

F. & W. Hardware Montreal

FROTHINGHAM & WORKMAN, Limited, MONTREAL, CANADA.

SAFE
REVOLVERS
Hammer the Hammer
SAFETY HAMMER REVOLVERS
.22, .32, .38 Calibres
SAFETY HAMMERLESS REVOLVERS
.22, .32, .38 Calibres
Iver Johnson's grip can be furnished on these Revolvers at an extra charge.
No article should be selected with more care than a Revolver.
For defence it must be sure and reliable.
For safety it must be so constructed that it will not "Fire" at the wrong time.
Dealers handling Iver Johnson Revolvers know that each of these weapons is absolutely safe. Accidental discharge is impossible.
THIS IS WORTH CONSIDERING
With these safety features the cost is no greater than for Revolvers without them
Caverhill Learmont & Co
Wholesale Distributors
MONTREAL and WINNIPEG

Serviceable Combination Cabinet

SUITABLE FOR HARDWARE TRADE

A COMBINATION THAT EVERYBODY MUST HAVE

CONTENTS

6 Table Knives
6 Dessert Knives
6 Table Forks
3 Table Spoons
6 Dessert Forks
6 Dessert Spoons
6 Tea Spoons
2 Salt Spoons
1 Mustard Spoon
42 Pieces

Price Complete

SILVER PLATED LIST $22.30

BEST NICKEL SILVER LIST $14.00

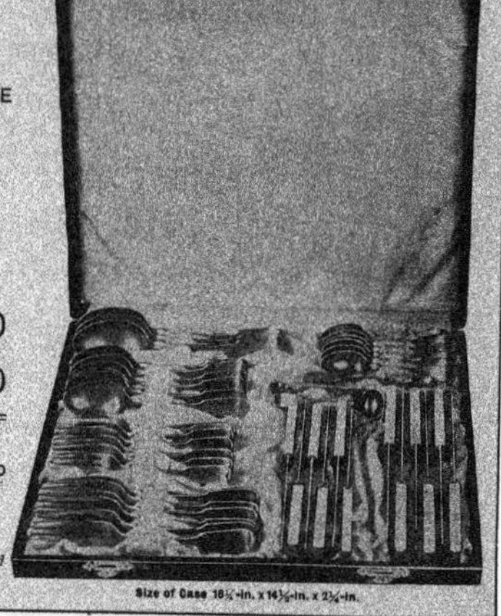

Size of Case 18½-in. x 14½-in. x 2¾-in.

JOHN ROUND & SON, LIMITED

MANUFACTURING

Cutlers and Silversmiths

MONTREAL

Factories SHEFFIELD, Eng.

Established 1847

DREADNOUGHT

A HIGH-GRADE ENAMELLED WARE

Write for Catalogue

ONTARIO STEEL WARE, LIMITED

115-121 Brock Ave. and 79-91 Florence St.

TORONTO, ONT.

PERFECT AXE

Another Axe that is a good-looker is the Perfect. It is like the All Steel Full Polished, with, in addition, the Phantom Bevil painted red. This Bevil by some makers is produced by a drop hammer, which spoils the shape of the axe to a certain extent. We grind it out, which, though it costs more, produces a much better effect and leaves the axe full where fullness is an advantage for bursting a chip, yet our price is not higher. It pays to get our line every time.

The Dundas Axe Works

To be continued. **DUNDAS, ONT.**

CHECKED HEAD
TRADE MARK

OUR FACTORY: The best equipped and most up-to-date in the country for the manufacture of HORSE NAILS.

OUR PROCESS: Unequalled, and exclusively our own. Insuring absolute uniformity of temper and shape.

OUR REFERENCE: Any horse-shoer.

OUR FACILITIES: Include prompt shipment; from Toronto or from Winnipeg.

OUR PRODUCT: "The best in the world." Imitated but never equalled.

OUR REPUTATION: World-wide.

OUR OBJECT: To place the horses of the country on a better footing.

OUR DISCOUNTS: YOURS for the asking.

OUR POPULAR 1907 CALENDAR: A few copies left. Have you got YOURS?

The Capewell Horse Nail Co.

TORONTO, ONT.

Western Branch: 212 Nena St., WINNIPEG, MAN.

DOMINION WIRE MANUFACTURING CO.
LIMITED

MONTREAL TORONTO

Manufacturers of

Iron and Steel Wire
Barb Wire
Galvanized Wire
Coiled Spring Fencing
Brass Wire
Copper Wire
Wire Nails
Wood Screws
Staples
Jack Chain
Bright Wire Goods
Spring Cotters
Steel Wire Barrel Hoops

SEASONABLE REMINDERS:

Barb Wire

Galvanized Wire

Galvanized Coiled Spring Fencing

Fence Staples

Wire Nails——Wood Screws

To ensure prompt deliveries send us your orders at an early date.

When **Horse Clippers** are mentioned the name **BURMAN** soon crops up.

It spells Quality and Reliability, and carries with it the weight of 36 years experience.

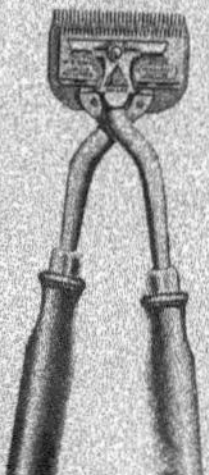

Newmarket

A Favourite for over Thirty Years.

A large stock of Hand Clippers Newmarket, Grand, Handicap, etc., on hand, also a Consignment of our new No. 13 Power Horse Clipper.

Just the machine for the owner of one horse.

Packs into case 17 x 11 x 7. Weight 34 lbs.
All cogs machine-cut from solid.

No. 13 Power Clipper

Agents in Canada (holding stock):

B. & S. H. Thompson & Co., Ltd., Montreal

BURMAN & SONS, Ltd., Lee Bank Works, BIRMINGHAM

Four Reasons why INVICTUS FIBRE is the BEST FIBRE PAPER on the Market

1. Invictus Fibre is made from personally selected pulp wood cut on our own limits.
2. This special pulp wood is again rigidly culled on arrival at our mills
3. The pulp from which Invictus Fibre is made is manufactured in our own pulp mill and transferred at once to the paper mill, avoiding all drawbacks of transportation.
4. Invictus Fibre will increase in strength and not grow hard and brittle as does fibre made from sulphite pulp. This is a quality of the peculiar and selected pulp from which Invictus Fibre is made.

A trial order will convince you that we are correct. See that each package bears the words "Invictus Fibre." Every ream contains 480 sheets.

TRADE MARK

This design a guarantee of quality.

CANADA PAPER CO., LIMITED,
"Headquarters for high grade Fibre and Manilla Papers"

Toronto. Montreal. Windsor Mills, P.Q.

A TOOL THAT'S STAMPED
STANLEY
IS THE BEST OF IT'S KIND

IMPROVED CARPENTER'S TOOLS

Sold by all Hardware Dealers

STANLEY RULE & LEVEL CO.
NEW BRITAIN., Conn., U.S.A.

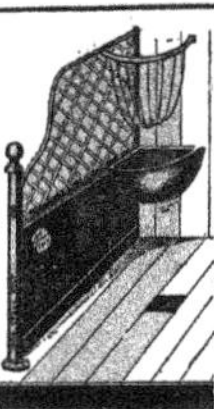

Dennis' Iron Stable Fittings

Sanitary, Indestructible, Economical

A good line for the hardware dealer.

Send for New Stable Fittings Catalogue and Price List "A."

DENNIS IRON & WIRE WORKS, Ltd.
LONDON, - ONTARIO

E. T. WRIGHT & CO.
HAMILTON, ONT.

DAIRY PAILS
STRAINER PAILS
MILK PANS
SAP BUCKETS
CALF WEANERS
MILK CANS AND TRIMMINGS
SAP SPOUTS
SYRUP CANS

Send for our Illustrated Catalogue

Will Hold up a Shelf

That's what a shelf brackets for.

For this purpose there can be NOTHING BETTER NOTHING CHEAPER than the BRADLEY STEEL BRACKET. It is well Japanned, Strong and Light. The saving on freight is a good profit aside from the lower price at which the goods are sold. Order direct or through your jobbers.

ATLAS MFG. CO.,
New Haven, Conn..

Expert Bookkeeping

Investigation and auditing of all classes of accounts undertaken. Trading, profit and loss accounts and balance sheets prepared. Sets of books opened and office systems devised in an up-to-date manner. Collection of accounts a specialty.

PERCY P. DAVENPORT
622 McIntyre Block, - WINNIPEG, MAN.

Square your Accounts

INTERNATIONAL PORTLAND CEMENT COMPANY, Limited
OTTAWA, - - ONTARIO.

Manufacturers of strictly high-grade Portland Cement made by the most improved method of manufacture in existence. Each barrel contains three hundred and fifty pounds of pure Portland Cement with no adulteration.

The Best Door Closer is . . .

NEWMAN'S INVINCIBLE FLOOR SPRING

Will close a door silently against any pressure of wind. Has many working advantages over the ordinary spring, and has twice the wear. In use throughout Great Britain and the Colonies. Gives perfect satisfaction. Made only by

W. NEWMAN & SONS,
Hospital St., - BIRMINGHAM

GET YOUR SHARE OF THE PROFITS

THERE'S MONEY IN HOCKEY STICKS, AND

NOW'S YOUR TIME TO CORNER YOUR SHARE

"EMPIRE" BRAND

is in great demand because that brand is just as perfect as Hockey Sticks can be made. Only the best Rock Elm and White Ash is used in the production of the "EMPIRE" brand.

You can depend on prompt shipment.

And it does pay to place your order with the largest manufacturers in Canada.

J. H. Still Manufacturing Company, Limited

ST. THOMAS, ONTARIO

Business Getters for Tinsmiths

Are there any defective chimneys in your locality?

Then it is your chance to get business.

THE

ZEPHYR Ventilator

saves the expense of building a high chimney.

It is so constructed as to guarantee a perfect draft and, therefore, easy combustion.

By using the ZEPHYR it is possible to burn soft coal with the best results.

The smoke is easily drawn off, enabling an even heat to be generated.

THE ZEPHYR IS STRONGLY PUT UP AND WILL RESIST ANY STORM

This is what others think of it:

Montreal, February 20th, 1901.

Messrs. HARRIS Co., Montreal. Providence Mother House.

Dear Sirs,

It is with pleasure that I can say that your "AEOLIAN" Ventilator is the best Ventilator we have used on our establishment so far. We had eight (8) of the large sizes put up on our Mother House and on the Hospice Gamelin, and they gave us entire satisfaction.

Yours respectfully, SISTER MADELEINE, Dep.

We manufacture and guarantee

The Har is Filter — The Zephyr Ventilator

The Aeolian Venti ator — Expansion Conductor Pipe

The J. W. HARRIS COMPANY, Limited

Successors to LESSARD & HARRIS

Contraactors *Montreal*

Peerless Woven Wire Fence

Here is a fence that will sell. Once up no need for repairs. It lasts for years—just what you are looking for. It's horse-high, pig-tight and bull-strong. No stock can get over, under or through it. The most unruly animal can't butt it down and there are no barbs to tear and injure stock. It's

WIND AND WEATHER PROOF

The Peerless lock holds it perfectly rigid and the wires can't slip up, down, nor sidewise. Cold weather won't snap it and hot weather won't make it sag. It's the greatest thing out for farmers, stockmen and all other fence users which makes it a red hot proposition for dealers. It comes in rolls all ready to put up. It's easy to handle. Write for prices.

The Banwell Hoxie Wire Fence Company, Ltd.
Dept. J, Hamilton, Ont.

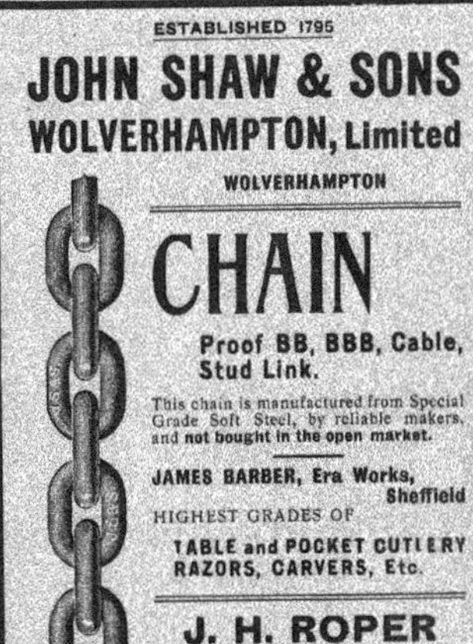

ANOTHER DEALER

has proven the solidity of our claims regarding

The
Empire Queen Range

We knew that THE EMPIRE QUEEN was just as pretty and as good as a good range could be, but the constant endorsement of our assertions by practical men is pleasing in the extreme.

Quality always tells and always will.

The margin of profit on this range pleases the dealer, and it goes without saying that the Empire Queen gives the utmost satisfaction to the lady in the case.

We have an interesting booklet for you.

The Canadian Heating & Ventilating Co. Limited,
OWEN SOUND, Ontario.

THE CHRISTIE BROS. CO., Limited, 238 King St., Winnipeg, Man., Western Agents.
THE CANADA STOVE AND FURNITURE CO., 126 West Craig St., Montreal, Que., Agents for the Province of Quebec.

Some **FEATURES** of

The Joy Malleable Steel Range

A six-hole polished steel range with malleable tops, doors, dampers, etc., for anthracite, bituminous and lignite coal, or wood.

Copper-lined, steel-encased, flush-top reservoir, heating by contact.

Removable duplex grates, in no place weakened by cotter-pin holes.

Heavy, durable firebox linings.

They pay good dividends to the dealer who handles them and return Satisfaction to the user.

JOY MFG. CO., 32 Humberside Ave., TORONTO, ONT.

TEES & PERSSE CO., - Western Distributors, Winnipeg
C. H. FOX, Agent - Vancouver, B.C.

SAP TIME IS COMING

Just a few short weeks until you need these goods; why not order now and be sure to have them when needed?

"EUREKA" CAST IRON AND STEEL SAP SPILES

You know that these spiles will not load you up with complaints from your customers.

SAP PAILS

Made in six sizes, in both straight and flaring patterns. We carry sap-pan materials in both tinned and galvanized Iron as follows—

FLARING PATTERN

Tinned Iron	Galvanized-Iron
48x96x20	48x96x22
48x96x22	36x96x22
48x96x24	36x96x24
36x84x22	
36x84x24	

Prompt
Shipment
Guaranteed

STRAIGHT PATTERN

THE McCLARY MFG. CO.

LONDON, TORONTO, MONTREAL, WINNIPEG, VANCOUVER, ST. JOHN, N.B., HAMILTON, CALGARY.

"EVERYTHING FOR THE TINSHOP"

BETTER THAN EVER

THE FOUR MAPLE LEAF BRANDS

BINDER TWINE FOR 1907

GILT EDGE—650 ft. Pure Manila.
GOLD LEAF—600 ft. Manila.
SILVER LEAF—550 ft. Standard Manila.
MAPLE LEAF—500 ft. Standard.

Good Material; Bright, Clean, Smooth, Even CORD, of unusual Length and Strength. Specially Treated to make it "Insect Proof."

Agents wanted in localities where we are not now represented.

Write us for SAMPLES and our new ROPE and CORDAGE price list.

BINDER TWINE

THE BRANTFORD CORDAGE CO., Ltd.

BRANTFORD, CANADA.

Manufacturers' Agents

CHARLES H. FOX

Vancouver, B.C.

MANUFACTURERS' AGENT

Selling to the Hardware Jobbers Only.

Correspondence Solicited.

ALEXANDER GIBB

Manufacturers' Agent and Metal Broker,
13 St. John Street, Montreal

Representing Canadian, British and American Manufacturers. Correspondence invited from firms wishing to be represented.

H. W. MITCHELL

WINNIPEG, MAN.

Correspondence is solicited from manufacturers wishing a representative in Winnipeg.

Travellers call regularly on the wholesale and retail trade in Western Canada.

Highest references and financial responsibility.

Scott, Bathgate & Co.

BROKERS AND COMMISSION MERCHANTS

Cor. Notre Dame St. East and Victoria St., WINNIPEG, Canada

Storage and Warehousing a Specialty.

First-class facilities in a thoroughly modern brick warehouse centrally located on the railway track.

"THE EMLYN" SAW BENCH

Made in 6 sizes. Best value obtainable. Specially designed for export. With or without "Emlyn" Patent Guard. Sole maker—

CHARLES D. PHILLIPS,

Cables—
"Machinery," Newport.

Emlyn Engineering Work
NEWPORT, MON., ENGLAND

B. A. GRAUTOFF & CO.

8 Fenchurch Buildings, London, E.C.

Import and Export.

Old established house with highest connections in Great Britain and on the Continent of Europe.

Cable Address, "Grautoff London." Liebers Code.

Our Metal Ceilings never leak, crack or fall. They are a real protection against fire, dirt and vermin.

This would make a fine ceiling in your store. It is one of our Louis XIV Designs.

Write for our Literature

The Metal Shingle and Siding Co., Ltd.

Montreal Branch:
Cor. St. Catherine and DeLorimier

Preston, Ont.

Sales Agents

ST. JOHN, N.B., Emerson & Fisher, QUEBEC, J. A. Bernard, WINNIPEG, Clare & Brockest, CALGARY, Ellis & Grogan.

Quit Taking Chances

Every time you recommend any old kind of cement to a customer you're taking chances. It does'nt pay to do so. Best clinch your trade by recommending

"SAUGEEN BRAND" OF PORTLAND CEMENT

all the time. Prices are right. Prompt shipment you can depend on. This is a good time to write.

The Hanover Portland Cement Co., Ltd., Hanover, Ont.

The Canadian Bronze Powder Works

R. E. THORNE & CO.

The only bronze powder works under the British flag.

High Grade bronze powders and bronze liquids.

Can fill all orders at short notice.

MONTREAL — TORONTO

WORKS AT VALLEYFIELD. *NO ORDER TOO LARGE*

No Trade Like the Hardware

TO RECOGNIZE A GOOD THING QUICKLY
AND HERE'S THE GOOD THING

Canada's Famous
"AUTOMOBILE" AND "CYCLE" SKATES
New 1907 Designs Just Out

YOU'LL WANT THEM — WAIT FOR TRAVELLER

Canada Cycle & Motor Co., Limited
TORONTO JUNCTION

KEMP'S FACTORY MILK CANS

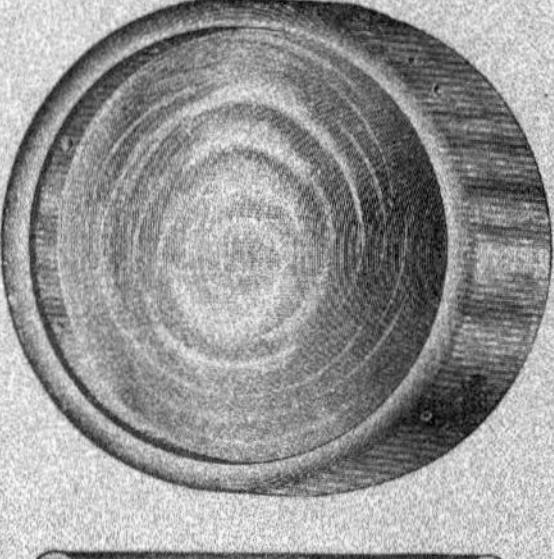

Roll rim

Factory Milk Can Trimmings

1 set consists of—
- 1 roll rim bottom
- 1 body hoop
- 1 top hoop
- 1 cover hoop
- 1 pair side handles
- 1 pair top handles

Made in six standard sizes, 10, 15, 20, 25, 30, 40 gallons

Factory milk cans always supplied with

KEMP'S ROLL RIM BOTTOM

are the kind that stand the test

Matters of Mutual Interest

Some Shop Talk Interesting Alike to Publishers and Subscribers.

The value of the trade newspaper has been emphasized in the agitation going on against the proposed parcels post legislation. Without a live trade press the retailers would have been kept in ignorance of the blow which was being aimed at retail merchandizing. The daily newspapers evidently consider the interests of their mail order advertisers paramount and those of their thousands of retailer subscribers of secondary importance. Not only would the parcels post express service hit the retailer but it would also do injury to the jobbing and manufacturing interests, as the big mail order corporations largely ignore the jobber and, where possible, do their own manufacturing or control the output of subsidiary factories. Manufacturers, wholesalers and retailers all have their trade associations, but the connecting link is the trade newspaper which, while not controlled by any branch of the trade, realizes fully that the retail trade is the one section upon whose success or failure all other branches depend.

A Big Advertising Contract.

Trade paper publicity is more and more regarded as a factor of permanent value in marketing, and while it is not yet common practice to do so, it is by no means unusual for manufacturers to anticipate their publicity needs as they do their requirements in raw materials, and contract in advance for a supply sufficient for two or three and sometimes five years' consumption.

A recent and notable case of this kind is that of the Oneida Community which has contracted with the Jewelers' Circular-Weekly, for one page in a special position in every issue for five years, at a yearly rate of $5,300. This will be in addition to space in another part of the paper which will cost about $1,000 more per year, making a total of $31,500.

It is particularly interesting to note that this contract, which is said to be the largest ever given to a commercial trade paper, was not sought by the publisher and, therefore, was not secured through the strong personal solicitation which sometimes over-persuades the advertiser.

Spring Special Number.

Active preparations are being made for the annual spring number of Hardware and Metal, to be issued on March 23 next, and it is safe to say that the 1907 issue will be even more interesting than any previous number. This is promising a great deal, as L. J. A. Surveyer, one of Montreal's most successful merchants, on April 11 last wrote:

"I take the liberty to write and congratulate you for the splendid spring number of Hardware and Metal we received last week. For some months past we have noticed a marked improvement in your paper, but this spring edition is the finest and has surpassed any other hardware publication we know of. Hardware and Metal is always interesting to read and also instructive, and we all read it from cover to cover."

In the reading matter a resume will be given of the progress of the past year and a comparison made of the most important market changes. The departments devoted to "The Merchant and His Store," "How Publicity Helps Selling," "With the Men Who Sell," "About Goods Hardwaremen Handle," will be as practical, authoritative and educative as it is possible to make them, and the whole issue will be one which the trade will look for and keep for reference after it has been published.

As in 1906 all advertising will be kept before and after the reading matter, the general appearance of the number having been greatly improved by the change. Practically every representative manufacturing and jobbing firm doing business with the hardware and kindred trades in Canada will have interesting announcements to make regarding their 1907 goods, and the advertising pages will be read with interest by every retail buyer.

How Travelers Help.

Freeman W. Davis, one of the Ontario traveling salesmen for Sanderson Pearcy & Co., when forwarding an interesting item of news recently, said: "I think it is a traveler's duty to forward any suggestions of this kind to our trade paper, as well as selling goods." Mr. Davis drew attention to the plan of blacklisting deadbeats that many Ontario merchants have adopted. The matter was treated editorially, thanks to Mr. Davis' suggestion. Anything which helps to make the retailer's financial position stronger will help the manufacturer and jobber. Mr. Davis recognizes this and realizes that the trade paper is the connecting link binding together the different branches of the trade.

J. H. Lyons, Ontario sales representative for M. & L. Samuel, Benjamin & Co., Toronto, drew our attention to a rising stove foundry only recently established and, like many other travelers, supplied us with some good items of news.

W. L. Lambert, who has been out selling "Automobile" and "Cycle" skates for the Canada Cycle & Motor Co., Toronto Junction, traveled with our circulation man for a few hours and in conversation said he could be quoted as saying that he placed Hardware and Metal first as an advertising medium in Canada for sporting goods. Mr. Lambert said that his office had been in receipt of numerous requests for their "Three-in-one" pictorial hanger as a result of a brief notice in Hardware and Metal.

Clerk Makes Good Suggestion.

D. G. Ross, a Winnipeg clerk, who has sent in able replies to the questions asked in our plumbing department, says he has profited greatly by these questions, not only in trying to solve them but by reading the opinions of others. He is however, more interested in hardware than plumbing, and suggests the establishment of a column devoted to such questions, say once a month, which would benefit and interest the hardware clerk.

"There are hundreds of clerks," writes Mr. Ross, "who do not read any trade paper, in fact I have worked in houses where not one of the employes thought of reading or subscribing to any trade journal. Now I would suggest that this matter be taken up and every clerk worthy of the name equipped with Hardware and Metal. Let it be his lesson book from which he learns something each week. If knowledge is power we want that knowledge, for is it not the man who carries the most trade knowledge and uses it, who is most fitted for the best job, and draws the biggest salary. Then let our trade journal be the source through which we obtain that knowledge. I venture to say too that if hardware dealers recognized the trade journal as the best advertising medium for clerks, etc., it would be to their own advantage as well as that of the clerk."

Mr. Ross' suggestion of a column for questions by clerks is a good one, and we invite the men behind the counter to forward their queries for information. Let Mr. Ross lead off and the column will be started. Initials only will be printed, but it is advisable that the name of the writer should be signed to each communication

Britishers Watching Canada.

Word comes from our London office that there is an ever-increasing demand for copies of Hardware and Metal, occasioned, no doubt, by the growing interest of Britishers in the growth and development of Canada. Many of these enquiries are from manufacturers and merchants who apply for copies in their own names, and numerous copies are supplied to unknown enquirers through book-stalls and news agents. Our London manager states that the secretaries of some of the most important exchanges and commercial rooms in Great Britain are subscribing for the paper for the benefit of members, and in this way our connection with the old land is extending more widely than we can estimate. This is a point to be remembered by those who wish to establish a connection amongst Old Country houses. Many copies are enquired for by manufacturers, who hope to obtain from our pages the names of possible representatives.

Two Recent Commendations.

Bent & Cahoon, North Sydney, N.S., who call themselves by the English term "ironmongers," wrote last month: "We have been reading your paper for 19 years and look upon it as an old chum."

Magladery Bros., New Liskeard, Ont., also sent the following self-explanatory letter: "Kindly send our Hardware and Metal to our address at New Liskeard instead of to Parkhill as formerly. We have purchased the hardware business of S. Eplett here and we sure want Hardware and Metal. It helped to make our Parkhill business the success that it was and we want it to do us another good turn."

AMONGST THE SALESMEN

LONDON TRAVELERS' SOCIAL SIDE.

The Travelers' Club of London is a social organization. London is also the headquarters of the Western Ontario Commercial Travelers' Association, which is business from the drop of the hat, but those members of the association who are members of the club, cast aside all business cares when they enter the club-rooms even from the drop of the hat on the club hat-rack. "Hang hats. Hang care," might be their thought. They have a large clubhouse on Richmond St., and have a good time there all the time. Sometimes they let their friends know what good fellows they are. They are not so very clannish. They can mingle some and when they undertake the part of hosts there is no lack of guests.

H. N. Lind — G. H. Detlor, 2nd Vice-President — C. M. McGuire, President — J. S. Townsend, 1st Vice-President

J. J. Harkness — A. H. Brener — H. E. Buttrey — F. Fisher, Treasurer

Three weeks ago Al. Davis, who travels for the Carling Brewing & Malting Co., had in charge a social evening at the club, when refreshments were served and a delightful programme of music presented. Over two hundred attended. More recently the club entertained twice that number of guests and the event was pronounced to be the most successful social func on of the season there. On this occasion a ball, such as the club gives annually, was held in the City Hall with the Masonic Temple used as an annex. All arrangements had been perfected to the slightest detail. The decorating of the hall was done by Messrs. A. Screaton & Co., and they are not likely to suffer any loss of business in consequence of credit being given where credit is due. Satisfactory music was supplied by an orchestra of twelve pieces under the direction of Tony Cortese and I do not know the name of the caterer who had charge of the midnight supper in the banqueting hall of the Masonic Temple, but he deserves credit, too. As for the work of the various committees who had charge of the affair it may have been hard or it may have been easy, but the result entitles them to an increase in popularity.

Mr. Fred S. Fisher was the honorary chairman, while Mr. H. W. Lind was the secretary. The conveners of the various committees were:

Finance—J. S. Townsend.
Reception—J. M. Ferguson.
Decoration—A. Tillmann.
Accommodation and halls—S. F. Glass.
Refreshment—J. A. Carling.
Printing—H. C. McBride.
Music and programme—J. J. Harkness.

The lady patronesses were: Mrs. Adam Beck. Mrs. Jeffrey Hale, Mrs. Charles S. Hyman, Mrs. H. W. Lind, Mrs. Arthur Little, Mrs. R. C. Macfie, Mrs H. C. McBride. Mrs. C. W. McGuire, Mrs. Edwin Paull, Mrs. E. B. Smith. Mrs. E. R. C. Struthers, and Mrs. A. Tillmann.

The stewards were: James A. Angus. R. Arkell, Dr. A. V. Becher, J. M. Ferguson, George S. Gibbons, A. O. Graydon, A. F. Kerrigan, F. J. Lind, A. M. Masuret, C. Mathewson, A. H. Moran, Dr. A. Scott, W. Wright, A. Zimmerman.

Three-fourths of the officers of the Travelers' Club of London are Mystic Shriners, members of Rameses Temple. They are: H. E. Buttrey, Donald Ferguson, C. W. McGuire, Geo. Detlor, J. J. Harkness, A. H., Brener, L. C. Johnston, H. W. Lind and T. W. Edwards.

At the adjourned annual meeting of the Travelers' Club of London, Jan. 12, Vice-president J. S. Townsend was presented with a set of silverware and cutlery of 51 pieces as a token of the club's regard for him.

• • •

Donald Ferguson, manager of the Ferguson Lumber Company and president of the Maple Leaf Automobile & Electric Company, was for three years a director of the Western Ontario Commercial Travelers' Association. He is an ardent horseman and has for several years been prominent in promoting the interests of the London Turf Association.

ISLANDERS GETTING AHEAD.

Writing from St. John's, Newfoundland, E. D. Arnaud, Canadian Commercial Agent, says: "A western correspondent of a local newspaper reports that travellers representing St. John's firms, are gradually supplanting the travellers for Canadian houses in that part of the colony where they formerly came in considerable numbers, and have captured the bulk of the trade. This it is now asserted is being lost to Canada in consequence of the enterprise displayed by several firms in this city, and the energetic work done by their representatives in the field."

MARITIME DEALERS ORGANIZE.

For some time there has been a quiet movement going on in Nova Scotia and last week it culminated in the formation of the Nova Scotia Retail Hardware Association, with about two dozen of the best retail houses in the province as members.

Further particulars will be given next week. In the meantime we extend our congratulations.

GAS AT 36 CENTS.

The Paul Automatic Gas Co., Montreal, has been incorporated, with a stated capital of $145,000, to manufacture a machine to generate gasoline gas in a most economical way.

The company claims that 2 gallons of gasoline will produce 1,000 feet of a most brilliant gas, which is superior for light, heating or cooking, to ordinary gas, or that produced by any other machine, at a cost of only 36 cents per thousand cubic feet. They also claim that this machine is perfectly safe, and insurance companies recognizing this fact, have permitted its installation without extra charge.

The company have quarters at 965 St. Catherine street east, Montreal, and will be glad to send interested parties who will mention this paper when writing, a booklet containing full information regarding their machine.

PUBLIC SPIRITED HARDWARE MERCHANTS

NO. 7

Last December the city of Regina elected by acclamation to the office of mayor one of her best known and most highly respected business men, J. W. Smith. Mr. Smith is an old hardware-man, head of the firm of Smith & Ferguson, which has recently been bought out by the Western Hardware Co., of Regina, and as such he has a wide acquaintance among the hardware trade.

Although a busy business man who has in recent years built up one of the biggest hardware trades in the west, Mr. Smith has always found time to take an active interest in municipal politics, and the citizens of Regina were making no experiment when they elected him mayor by acclamation a few weeks ago.

J. W. SMITH
Mayor of Regina.

In 1884 Mr. Smith was elected to the first municipal council that Regina ever had, and from that date to 1904 he was almost continuously a member of the council or the school board. He was mayor in 1889 at the time of a hot dispute between the city and the trustees of the Town Site Company as to the amount of taxes due by the latter. So well satisfied were the citizens of Regina with the way in which Mr. Smith handled this question that at the close of his term he was offered re-election by acclamation. The honor was declined as Mr. Smith could not spare the time from his business that year.

In 1902 the Board of Trade—of which Mr. Smith was then president—decided that a new and progressive policy was urgently required by the city and requested Mr. Smith to stand for mayor in order that the new policy might be carried into effect. He was elected and occupied the mayor's chair during 1902 and 1903, and during his term of office Regina was incorporated as a city and the water-works contract was made.

Last December Mr. Smith was again urged to accept a nomination and he was honored with an election by acclamation. A business man is urgently required at the helm of the civic ship of state as the city is now to work under a new charter, a feature of which is a provision for the introduction of the ward system.

Regina is growing fast and provision must needs be made for a city many times its present size. A far-sighted business man is the sort of mayor required in this growing time and Regina is fortunate in having secured the services of J. W. Smith.

BIG BUSINESS CHANGE.

On Jan. 15 the Smith & Ferguson Company, of Regina, Sask., turned over their business to Peart Bros., who have been doing business in Regina under the name of the Western Hardware Company, the price paid for the premises, stock and goodwill being about $125,000. J. W. Smith, the owner of the Smith & Ferguson business, will continue his coal business. The land and building occupied a site 54 x 175 feet, and was figured as worth $60,000, the stock being expected to figure at $65,000. The new owners have also leased the old company's warehouse with an option on purchasing. They still have the big warehouse erected by the Western Hardware Company during the past year. Since the fire which burned the Western Hardware Block several months ago business has been transacted in the Donohue Block, but this store will be closed out and all the trade carried on in the newly purchased store.

The Regina Standard says: "Peart Brothers are among the most successful young business men in the city. T. W. Peart came here just ten years ago from his home in St. Mary's, Ont., to take a position with the Smith & Ferguson Co. After remaining in the employ of that firm for six years he started in business on his own account four years ago and has been trading since under the name of the Western Hardware Co., building up an important and valuable business in a comparatively short space of time." The other member of the firm is Milton Peart, who joined his brother in the Regina enterprise several years ago. Another brother is J. Walton Peart, manager of the St. Mary's Hardware Company, and a leading member of the Ontario Retail Hardware Association, he also having an interest in the Regina enterprise.

CONTRACT AWARDED.

The contract for plumbing and heating appliances, in the Government immigration home, in Toronto, has been awarded to J. Armstrong, Toronto, the price being $1,998

OUR LETTER BOX

Correspondence on matters of interest to the hardware trade is solicited. Manufacturers, jobbers, retailers and clerks are urged to express their opinions on matters under discussion.

Any questions asked will be promptly answered. Do you want to buy anything, want some shelving, a silent salesman, any special line of goods, anything in connection with the hardware trade? Ask us. We'll supply the necessary information.

Furnace for Fuel Oil.

The Ontario Nickel Co., Worthington, Ont., desire to secure a Canadian made furnace in which fuel oil can be burned.

Answer—Vanzant & Faird, Petrolea, have a furnace of this kind, and another is being manufactured in the Welland district about which we have been unable to learn very much. Can any reader supply the desired information? —Editor.

Book on Window Dressing.

A clerk who reads Hardware and Metal wants to secure a book on window dressing, particularly adapted to the hardware trade. There is no book or magazine making a specialty of hardware windows "The American Hardware Store," price $3, and sold in Canada by the MacLean Pub. Co., Toronto, is the nearest we can suggest.—Editor.

U. S. Duties on Binder Twine.

"A Subscriber" asks: "What is the duty, if any, payable on binder twine, 650 feet per lb., going into the United States from Canada? Has any change been made recently?"

Answer.—The only definite source of information in reference to the U. S. tariff is from the Treasury Department. The U. S. consul at Toronto states unofficially that on manilla twine there is a duty of 45 per cent., and roughly speaking all other twine is free. There is, however, a regulation that all twine over 600 feet in length is dutiable. While the general U. S. tariff has been unchanged for many years there have been several changes in the regulations. —Editor.

Cobbler's Outfits.

A. McBride & Co., Calgary, Alta., asks for the name of the manufacturers of the Economy Cobbler's Outfit.

Answer.—Root Bros. & Co., Plymouth, Ohio, make the line referred to.—Editor.

ELECTRICAL CONTRACTORS MEET

The regular meeting of the Electrical Contractors' Association of the Province of Quebec was held at Montreal on Thursday, the 24th ult., when the following officers were elected for the coming year: President, J. E. Scott, representative to Builders' Exchange, E. W. Sayer; secretary, J. H. Lauer; Executive Committee, W. R. Shaw, F. J. Parsons, D. McQuaid, N. Simoneau, R. Moncel, W. J. O'Leary and James Bennett.

It was decided at this meeting to hold the annual banquet on Feb. 6.

Retail Hardware Association News

Official news of the Ontario and Western Canada Associations will be published in this department. All correspondence regarding Association matters should be sent to the secretaries. If for publication send to the Editor of Hardware and Metal, Toronto.

Officers Retail Hardware and Stove Dealers' Association of Western Canada:

President—A. J. Falconer, Deloraine
First Vice-President—J. B. Curran, Brandon.
Second Vice-President—W. M. Gordon, Winnipeg
Secretary-Treasurer—J. E. McRobie, Winnipeg
Executive—Alberta, A. E. Clements, Olds; C. F. Comer, Calgary; A. R. Auger, Okotoks.
Manitoba—H. S. Price, Boissevain; A. P. Macdonald, Winnipeg; O. Gilmer, Winnipeg
Saskatchewan—G. K. Smith, Moose Jaw; S. A. Clark, Saskatoon, J. R. Fox, Weyburn.
Officers Western Board (elected by general merchants and hardwaremen in joint session) President, W. G. McLaren, Souris, Man., vice-president, G. K. Smith, Moose Jaw, Sask.; H. C. Hamelin, Winnipeg, Man; secretary, W. A. Coulson, Winnipeg, Man; treasurer, J. E. McRobie, Winnipeg, Man., auditor J. A. Lindsay, Winnipeg Man.
Association offices, 53 Scott building, Main street, Winnipeg

Officers Ontario Retail Hardware and Stove Dealers' Association:

President—A. W. Humphries, Parkhill
1st Vice-President—W. G. Scott, Mount Forest.
2nd Vice-President—J. R. Hambly, Barrie.
Treasurer—John Caslor, Toronto.
Secretary—Weston Wrigley, 10 Front St. East, Toronto.
Executive Committee—The officers and J. W. Peart, St. Mary's; M. R. Marshall, Dunnville; D. Brocklebank, Arthur; F. W. Jeffery, Midland; E. P. Paulin, Goderich, and Frank Taylor, Carleton Place
Auditors—J. W. Peacock and C. F. Moorhouse, Toronto.

WESTERN CONVENTION.

The Western Retail Hardware Association will hold their annual meeeting in the Association's board room in the Scott Building, Winnipeg, on Tuesday, February 12th. Notices have been sent to the trade notifying them of the meeting and a big attendance is expected. The meeting is being held during the Bonspiel week, and consequently the visiting dealers will have the advantage of special railway rates. Important business will come up for discussion and a good attendance is expected and desired.

A WEAK OBJECTION.

In a letter written to a St. Catharines hardwareman, E. A. Lancaster, M.P., states that he appreciates the position of retail merchants on the parcels post matter but argues that the measure would be for the benefit of the great majority of the people as an attack on the monopoly of the express companies and the exorbitant rates charged by them.

The objection he makes is a weak one, as if he or any section of the people desire to attack the monopoly of the express companies and reduce the exorbitant rates they are said to be charging, the proper method to adopt is to bring the matter before the Railway Commission or introduce legislation to rearrange and make more equitable the said express rates. Instead of doing this, Mr. Lancaster's argument is that the express companies should be gotten after in a roundabout way through the postal department, by means of legislation which would benefit a few wealthier departmental store merchants at the expense of thousands of retailers scattered throughout the country.

The Parcels Post Collections system would not benefit the farmers in any way, as they would have to pay the same charges for their mail order purchases as at the present time, plus the c. o. d. collection charges. On the other hand, the proposed legislation would injure the farmer very materially by putting out of business numerous stores in the smaller towns and villages, reducing the number of purchasers of farm products and lessening the value of real estate in business centres, besides having a similar effect upon the real estate in the farming districts. The most intelligent farmers look forward to their home life being made less isolated by the extension of the electric railways, farm telephone systems, etc., and by the small country towns being made educational and amusement centres for the farmer and his family. The proposed legislation will work against this ideal and make it farther from realization, as it will tend to centralize wealth more and more in the hands of the multi-millionaires in the larger cities.

Mr. Lancaster is an able advocate of the farmer's position, and as he promises to try and prevent any injustice being done the smaller local merchants for the benefit of the few large monopolistic department stores in the large centres, it is hoped that he will see the justice of the agitation against the parcels post c.o.d. collections system.

QUESTIONS FOR DISCUSSION.

Very few dealers have forwarded to the secretary of the Ontario Association questions on trade matters for discussion at the annual convention on Good Friday next.

What is wanted is short, terse questions from dealers who have been puzzled at times as to how best to advertise, to sell or to buy, what new lines to take up, how to keep and get the best results from clerks, tinsmiths or other employes, what new bookkeeping appliances or store fixtures have given the best results, and similar subjects, a discussion of which would prove helpful to every hardware merchant.

Think over some question you would like discussed and forward it to the secretary at 10 Front street east, Toronto, at once.

The question box feature of the convention offers possibilities of great value to every one attending. Everyone present at the May convention will remember that the discussion was brought out by Mr. Brocklebank's query about oil storage tanks. The only difficulty then was that the discussion was brought on just before adjournment, but this will be remedied at the Good Friday convention.

Send in the questions.

PEDDLERS' COMPETITION.

A Washington judge has decided that the law licensing peddlers is unconstitutional, and the general trend of legal opinion seems to be along the same line. A careful study of the lay of the land will afford many a merchant insight into peddlers' methods and light for meeting their business in a way adequate and satisfactory. Nine times out of ten, local merchants can secure the business where their goods and prices are compared with those of the peddler. The advantage the latter individual has is in being on the ground with the goods. There is nothing which will take the place of personal contact, and goods and price comparisons. This method will not only defeat the aims of the peddlers but will sell your own goods as well.

MEMBERSHIP BRINGS RETURNS.

The Ontario Retail Hardware Association has been proven a good thing by more than one member. One prominent member shortly after the September convention found a jobbing house selling goods in his town to persons outside the trade. He immediately "cut out" doing business with them and notified them why. This didn't stop the jobber selling to consumers, but it did an equally good thing—it forced him to protect the retailer.

Unsolicited credit notes have been sent to the retailer on several occasions since then, the total being many times the amount of the membership fee in the Association.

If the Association can be made a help to the retailer in this way it can be of even greater assistance if the membership is doubled or trebled. When a retailer feels that he has two or three hundred dealers behind him in adjusting a grievance over some trade evil he is going to take a firmer stand, and it is only by taking a strong stand—and being in the right—that he can succeed. Join the Association if you haven't already done so.

HARDWARE TRADE GOSSIP

Ontario.

Bishop & Ball, hardware merchants, Wingham, have dissolved partnership.

John O'Neill, a well known tinsmith of Hamilton, was found dead in bed on Friday, Jan. 11.

D. Courville, hardware merchant, of Maxville, has advertised his hardware and tinshop for sale.

C. L. Macnab, of Macnab Bros., hardware merchants, Orillia, was a visitor in Toronto last week.

L. I. Hunt & Son, hardware merchants of Alvinston, have sold their business to Telford Bros.

Wm. McArthur, tinsmith, Almonte, died recently after a little over a week's illness of pneumonia. He was 62 years of age.

Alex. McAllister, of Carleton Place, is retiring from the paint business, having accepted a position as traveler for a wholesale firm.

B. Tedford, of the firm of Tedford Bros., hardware merchants, Alvinston, is in Blenheim this week, moving his family to Alvinston.

R. J. Cluff, of Cluff Bros., Toronto, was in Montreal last week on business for Warden King & Son, whose Ontario selling agents his firm are.

McKinnon & Walker, hardware merchants, of Cobalt, have dissolved partnership, and business in future will be carried on by W. T. Walker.

A. E. Lewis of Jno. Lewis & Co., Belleville, stove dealers, has been a visitor in Toronto during the week, attending the curling tournament.

The Simonds Canada Saw Company, Montreal, have removed their office and factory in Toronto from 265 King St. West to 105 Adelaide St. East.

The D. Pike Company, extensive dealers in sportsmen's supplies, Toronto, suffered loss by fire last week to the extent of $2,500, covered by insurance.

W. J. Cluff, of Cluff Bros., Toronto, has returned from a trip to Chicago and New York, where he was on business in connection with their jobbing trade in heating supplies.

B. S. Jackes, proprietor of the Model hardware store, Dundas Street, Toronto, will remove to his Yonge Street store, where he will concentrate all his efforts in future.

A. P. Horsman, manager of the Gurney Standard Metal Company, of Calgary, Alta., has been in Toronto during the past week, conferring with the head office officials.

Geo. Merrick of Toronto has been added to the traveling staff of the Pedlar People, Oshawa. Mr. Merrick has had several years' experience on the road and should make good in his new position.

Thomas Black, Winnipeg representative of the Pedlar People, metal roofing manufacturers, Oshawa, visited the office of the Pedlar People at Toronto, Oshawa and Montreal during the past fortnight.

Charlie Overholt, manager of the hardware store of Prince & Company, Bloor Street West, Toronto, has accepted a position with Lewis Bros., Montreal, and will commence work in his new position next week.

H. N. Joy, of the Joy Manufacturing Company, makers of the Joy malleable ranges, Toronto, who was severely injured by falling while skating on New Year's Day, is recovering and expects to be back at work shortly.

On the occasion of his recent marriage, the office staff of the Kemp Manufacturing Company, Toronto, presented W. J. Virtue, of the purchasing department of the company, with a handsome bronze clock, and a massive leather-covered chair.

A. A. Brown, manager of the McClary Mfg. Co.'s branch at Montreal, and A. D. Kenelly, manager of the Toronto branch, spent several days in London recently, conferring with the heads of the company regarding the coming year's business.

The employes of the McClary Manufacturing Company of London presented Thomas A. Greer, on the eve of his departure from that city, with a handsome suitcase and gentleman's companion. Mr. Greer has been with McClary's for twelve years.

Milton A. Adams, lately employed by A. W. Moore, hardware merchant, St. Catharines, and previously with R. Hadden, Picton, has accepted a position as traveling salesman with Lewis Bros., Montreal, and will represent them in Western Ontario.

Charles Array, who for the past four years has been traveling for the United Factories, Toronto, has left for New York to accept a similar position with J. J. Adams & Company, one of the largest brush manufacturing concerns in the United States.

E. Holt Gurney, of the Gurney Foundry Company, Toronto, will return next week from a visit to the hot springs at Banff, Alta., where he has been staying with James Drew, manager of the Winnipeg branch of the Gurney Foundry Company.

The hardware store of H. Jones, Uxbridge, is displaying two nice, attractive windows this week, one being of washing machines, wringers, etc., which are arranged very artistically, intermingled with other household lines, giving life to an otherwise dry exhibit.

Lewis Bros., wholesale hardware merchants, of Montreal, are having extensive improvements made to their Toronto office: the old stock will be cleaned out, and a complete new line of samples installed. When completed they will be very handsome and inviting sample rooms.

Peleg Howland, head of the firm of H. S. Howland, Sons & Company, wholesale hardware merchants, Toronto, has been unable to attend to business during the past week, but is around again. H. A. Gunn and several other members of the staff have also been suffering from the grippe.

A Denny, formerly with the James Smart and Gurney Tilden companies, has accepted a position with D. Moore & Company, Hamilton, and will cover the territory between Toronto and Windsor. Mr. Denny replaces Clark Yuker, who has resigned and is now wintering in California.

G. P. Graham, the new leader of the Liberal party in the Province of Ontario, received his early business education in the hardware stores of J. A. Carman & Co., Iroquois, and M. F. Beach & Co., Winchester, Ont. He was born in 1859 and in 1880 forsook the hardware for the newspaper business. For the past nine years he has represented Brockville in the Legislature as well as publishing the Brockville Recorder.

Mrs. Wm. Hodson, mother of Oscar M. Hodson, Toronto representative of the H. R. Ives Co., Montreal, and mother-in-law of J. R. Hambly, vice-president of the Ontario Retail Hardware Association, died at the home of Mr. Hambly last week, and was buried at Drayton. She was 80 years of age.

Quebec.

H. Sibley, L'Annonciation, was in Montreal this week on business.

J. A. Paquin, St. Eustache, was in Montreal on business this week.

L. H. Goudry, Montreal, has left for the Old Country on a business trip

A. Sweet, Winchester, Ont., visited some of the Montreal wholesalers last week.

H. H. Bourke, Windsor Mills, was a caller on some of the Montreal wholesalers this week.

W. R. Dalton, of Dalton & Sons, Kingston, Ont., was in Montreal last week on business.

Chas. A. Alexander, of A. E. Alexander & Son, Campbellton, N.B., was in Montreal this week.

Mr. Wilson, of Madole & Wilson, Napanee, called on some of the Montreal wholesalers this week.

Philemon Morency, hardware and paint merchant, of Montreal, has been succeeded by Morency and Cote.

Ludger Gravel, Montreal, has left for Detroit and other points west to see some of the big firms he represents here.

Among the firms registered in Montreal last week was the Dominion Asbestos Co., Harold H. Robertson, president.

A. O. Campbell, Vancouver, representative of Lewis Bros, Montreal, is in Montreal on his annual visit. Mr. Campbell goes as far as St. John, N.B.

I. C. Stewart, president of the Imperial Publishing Co., Halifax, publishers of the Maritime Merchant, called at the Montreal office of Hardware and Metal last week.

Alex. McArthur & Co., manufacturers of building paper, Montreal, held their annual meeting on January 17th, 1907, when a very satisfactory statement of the year's business was submitted.

The firm of J. H. Bell & Co., of Haileybury, is now known as the "Bell, Rochester Hardware Co," of Haileybury. The members of the firm remain as before, that is, J. H. Bell and G. H. Rochester.

The Ross Rifle Company, of Quebec, have entered an action for $10,000 against Le Nationaliste, of Montreal, for the publication of alleged libellous correspondence, which, besides criticizing the Ross Rifle Company in regard to their location in Quebec, went on to adversely criticize the rifle and steel material.

Alexander A. Wilson, one of the most active of the old-time business men of Montreal, passed away on Saturday afternoon, the 18th inst., after an illness of about ten days, brought about by an attack of paralysis. Mr. Wilson was a native of Coteau du Lac, Que. He was born in 1835, and was therefore in his 72nd year. A little over fifty years ago he opened a large hardware and paint business on St. Paul street, and was known from one end of Canada to the other.

HARDWARE AND METAL

Established 1888

The MacLean Publishing Co.

Limited

JOHN BAYNE MACLEAN - President

Publishers of Trade Newspapers which circulate in the Provinces of British Columbia, Alberta, Saskatchewan, Manitoba, Ontario, Quebec, Nova Scotia, New Brunswick, P.E. Island and Newfoundland.

OFFICES:

MONTREAL, ... 232 McGill Street, Telephone Main 1255
TORONTO, ... 10 Front Street East, Telephones Main 2701 and 2702
WINNIPEG, . . . 511 Union Bank Building, Telephone 3726
LONDON, ENG. . . . . 88 Fleet Street, E.C., J. Meredith McKim, Telephone, Central 12960

BRANCHES:

CHICAGO, ILL. . . . . 1001 Teutonic Bldg., J. Roland Kay
ST. JOHN, N.B. . . . No. 7 Market Wharf
VANCOUVER, B.C. . . . Geo. S. B. Perry
PARIS, FRANCE - Agence Havas, 8 Place de la Bourse
MANCHESTER, ENG. . . . 92 Market Street
ZURICH, SWITZERLAND . . . Louis Wolt, Orell Fussli & Co.

Subscription, Canada and United States, $2.00
Great Britain, 8s. 6d., elsewhere - 12s

Published every Saturday.

Cable Address { Adscript, London / Adscript, Canada

WESTERN SNOW BLOCKADES.

That a serious condition of affairs exists in western Canada at the present time owing to the severity of the Winter, the shortage of fuel, the blockades of passenger and freight trains and the general inability of the railways to afford adequate transportation facilities in the face of unusual difficulties it would be folly to deny. At the same time it is very easy to exaggerate the gravity of the situation, and it is to be feared that alarming reports sent east and south by Winnipeg newspaper correspondents anxious to grind out "good copy," have created a wrong impression in eastern Canada and in the United States.

Business in the west has suffered of course from the general tie-up of the railways and the severity of the Winter. Inclement weather has hurt the retail trade and the train blockades have hindered the distribution of goods to the retailers. Shortage of cars last Fall retarded the movement of grain, the number of cars shipped from Sept. 1st, 1906, to Dec. 12th 1906, when navigation closed on the great lakes, being 2,443 less than for the corresponding period last year. In a country where the consumer is almost entirely dependent for his ready money on the sale of his wheat this car shortage is bound to have an effect which reacts upon general business conditions. It is not strange that there should be some complaint from the wholesale houses of the slowness of collections.

In the far west the ranching industry has suffered from the unusual severity of the Winter. Owing to the unusual depth of the snow the cattle have been without food and the loss has been heavy. Newspaper despatches estimate the loss at 25 per cent., but it is too early yet for definite statements.

Nevertheless in spite of conditions which are admittedly unfavorable there is no occasion for undue alarm. Ontario, Quebec and the Maritime Provinces survived snow blockades in 1903 and 1904 which were much worse than those now being endured in the west. There was no undue alarm at the time because it was known and recognized that these conditions were unusual and the prosperity of those provinces was on a sound basis. Similarly it must be recognized that the prosperity of the west rests upon too sound a basis to allow of serious alarm because of temporarily unfavorable conditions. The troubles of this Winter are but the growing pains of the west and should be so regarded. They are due mainly to the inability of the railways to provide adequate transportation facilities. The country has in short grown too fast for the railways; they have not been able to keep up with the procession. They are, however, doing their best and with the expenditure of many additional millions during the present year on rolling stock and equipment the railways should next season be in better shape to handle the output of the west. In the meantime those who know western Canada best are losing no sleep through anxiety as to the outcome.

PUBLISHERS HELP RETAILERS.

The following editorial from Printer and Publisher, the trade newspaper published by the MacLean Publishing Company for Canadian printers, shows that some country publishers are already active in support of the retailers in their fight against the proposed parcels post C.O.D. system.

The action taken by the Forest Free Press should be copied by the publisher of every Canadian newspaper which does not depend for its support upon departmental store advertising. Retail merchants should not hesitate to ask their local editors to use their pens freely in writing articles endorsing the position taken by the Forest Free Press.

The editorial from Printer and Publisher reads as follows:

"Strenuous efforts should be made by the readers of this paper to defeat a bill which is to be introduced during the present session of Parliament, and which has for its object, the amending of the parcel post law so as to enable postmasters, in their official capacity, to deliver C.O.D. parcels, to collect the money due on them and to remit the same to the senders. Provision is also to be made for cheaper parcel postage rates.

"The commendable attitude upon this bill taken by the Forest Free Press, should be studied by all persons who are interested not only in their own advancement, but in the prosperity of the country at large. It points out some facts which should be taken into consideration by every member of Parliament before voting on the proposed bill. It rightly contends that a C.O.D. parcel post system would have a strong tendency to impair the proper and legitimate functions of the postoffice. A C.O.D. parcel post delivery would give those engaged in the mail order business an unfair advantage over country town merchants, which would undoubtedly result in a decrease in the business done by the latter, without any particular advantage accruing to the consumer.

"It will be generally conceded that the presence in the country of a large number of thriving towns and villages is more to be desired than one or two mammoth cities. By the passage of such a bill as the one in question, the smaller towns and villages of the land would be greatly injured. No effort should be spared therefore by the publishers of this country to defeat the bill, which is detrimental to themselves and the country at large, and to say the least of it, looks exceedingly monopolistic."

ADVANCE IN PARIS GREEN.

In referring to our quotations, it will be noticed that pure paris green has been marked up $1.00 per 100 lbs. This is inevitable owing to the thrice repeated story that all chemicals entering into the composition of paris green are far higher in price than has yet been experienced in recent years.

It is not thought that these high prices are due to any cornering or manipulation, but are simply owing to the tremendous demand in the metal and chemical market for the base articles used in making this insecticide.

Only a limited quantity of paris green can be made in Canada during the winter months, and as it has been found that in spreading the manufacture into the warmer seasons, it is very injurious to the workmen, there is liable to be a considerable shortage experienced this year, therefore it behooves the trade generally to make their purchases early, as from the reasons given above, a superabundance of supply cannot possibly be counted upon.

A WINNIPEG TRIUMPH.

Winnipeg retailers have an organization that does things; they are not content with holding meetings and making speeches which result in no definite action. They are organized because they believe that things can be accomplished by united effort which would be impossible without it; and they have proved the reasonableness of their belief

Reference has been made before in these columns to the fact that Winnipeg retailers—and particularly grocers and butchers—have been put to great expense to have their garbage removed. Garbage is removed by the city from all private residences, but the dealers were forced to remove it themselves. They were held up by the scavengers and several stores were paying $4 to $5 per week for service that was very unsatisfactory. This grievance was taken up by the retail associations in the city and some time ago Secretary Coulson and a large deputation of dealers waited on the Winnipeg city council and asked that the city remove all garbage from stores and apartment blocks free of charge. The request was taken into serious consideration, and now the announcement is made that the city will remove garbage free of charge for three months. If the cost should be found to be too great to charge against the regular business tax a small special tax may be charged. In any event the city will assume the full responsibility for the removal of garbage and this was the main contention. Dealers were unable to get satisfactory scavenger service and were liable to fines in consequence. All this has been changed by united effort. The moral is obvious.

GEO. CAVERHILL
Caverhill, Learmont & Co., President Montreal of Trade.

HEATING PROBLEMS.

During the extreme cold weather of last week the natural gas supply at Hamilton, Ont., failed. Dozens of houses depended upon the gas supply for heat, while many restaurants and supplies had no other means of cooking food Much suffering and inconvenience resulted, therefore, and there was a rush for coal stoves and heaters.

In western Canada there has also been much suffering from the extreme cold because of the shortage of coal resulting from the strikes of coal miners and the failure of the railroads to meet the demands of the country. In the Pacific coast cities the usual mild climate has not forced all householders to equip their premises with furnaces or boilers and the present cold snap has given stove dealers a big trade as well as loading up plumbers with work from frozen water pipes.

THOMAS J. DRUMMOND
Drummond, McCall & Co., 1st Vice-President, Montreal Board of Trade.

RAILROADS BACK DOWN.

That Canadian railways have much to learn yet in the proper conduct of their business has been illustrated by their actions during the past couple of months.

The new customs tariff went into effect at the end of November and it gave an increased preference on imports from Great Britain, particularly on metals. Here was an opportunity for graft, thought the railways, and they proceeded to put into force an increased tariff on freight carried from Atlantic ports, the increases being most noticeable on the articles which were given an increased preference in the new customs tariff The new regulations were put into force at once without regard for the interests of merchants.

Protests were immediately made against the disturbance of freight rates in the middle of a season, after all contracts had been made for the supply of

J. S. N. DOUGALL
McCaskill, Dougall & Co., Member Executive Montreal Board of Trade.

goods for several months to come. It was rightly contended that due notice should be given by the railroad companies when rates are to be advanced, the notice to be several months ahead in order to allow the merchants to figure freight charges correctly on import orders.

The railroads were forced to take back water. Another notice has been issued countermanding the previous order and the old rates are in force. The incident is interesting, showing the greed of the railway corporations and how it was frustrated for once.

MONTREAL BOARD OF TRADE ELECTIONS.

The adjourned meeting of the Montreal Board of Trade took place at noon on January 29, 1907, F. H. Mathewson, the retiring president, occupying the chair for the last time.

The results of the elections for first vice-president, members of the council, and members of the board of arbitration, were announced to the large concourse of members present by the secretary, and were as follows:—

1st vice-pres., Thos. J. Drummond.

Members of Council: J. R. Binning. A. A. Ayer, Geo. L. Cains, W. W. Craig, J. S. N. Dougall, Geo. A. Kohl. J. P. Mullarkey, J. L. McCulloch, Edgar McDougall, Alex. McLaurin. Alex Orsali, and J. A. Richardson.

Members Arbitration Board: James Carruthers, Sir Geo. A. Drummond, Geo. F. Drummond, Wm. I. Greer, E. B. Greenshields, Arthur J. Hodgson, F. H. Mathewson, R. W. McDougall, Alex. McPhee, John McKergow, Alex. Ramsay, James Thom.

Immediately after the reading of the returns Mr. Mathewson introduced his successor, Geo. Caverhill, and the latter was installed as president for 1907.

Markets and Market Notes

(For detailed prices see Current Market Quotations, page 62.)

THE WEEK'S MARKETS IN BRIEF.

MONTREAL.

Galvanized Iron—Some lines advanced.
Pig Iron—Shows advances.
Copper Wire Coat and Hat Hooks — Advance 10 cents per gross for 3-inch

TORONTO.

Jenkins Air Vents—50 cents higher.
Soldering Irons—Quoted higher.
Plumbers' Oakum—Now $4.75.
Kerr Weber Straightway Valves—Advanced.
Window Glass—New list, lower prices.
Wire Coat Hooks—Advanced.
Nails—Advanced 5c. base.
Black and Galvanized Sheets—Higher.
Boiler Plates—Advance expected.

Montreal Hardware Markets

Office of HARDWARE AND METAL,
232 McGill Street,
Montreal, Feb 1, 1907.

Although last week's pace has quieted down slightly, there is still plenty doing in hardware circles, and almost every mail brings in orders for spring shipment.

Owing to heavy storms in the West, orders from that quarter have fallen off slightly during the week.

Copper wire coat and hat hooks have advanced 10 cents per gloss on the 3-inch size, being now quoted at 60c instead of 50 cents.

The new list on building paper, mentioned last week, is dated January 14, 1907, and shows advances of 5 cents on several lines.

AXES. — Selling well. We still quote: $7.60 to $9.50 per dozen; double bitt axes, $9.50 to $12 a dozen; handled axes, $7.50 to $9.50; Canadian pattern axes, $7.50 a dozen.

LANTERNS. — Our prices are: "Prism" globes, $1.20; Cold blast, $6.50; No. 0, Safety, $4.00.

COW TIES AND STALL FIXTURES—We still quote; Cow ties, discount 40 per cent. off list; stall fixtures, discount 35 per cent. off list.

SLEIGH BELLS—Moving, but very slowly. Our prices remain: Back straps, 30c. to $2.50; body straps, 70c. to $3.50; York Eye bells, common, 70c. to $1.50; pear shape, $1.15 to $2; shaft gongs, 20c. to $2.50; Grelots, 35c. to $2; team bells, $1.80 to $5.50; saddle gongs, $1.10 to $2.60.

RIVETS AND BURRS.—Last week's advance on the copper article still holds good. We quote: Best iron rivets, section carriage, and wagon box black rivets, tinned do., copper rivets and tin swede rivets, 60, 10 and 10 per cent.; swede iron burrs are quoted at 60 and 10 and 10 per cent. off new lists; copper rivets with the usual proportion of burrs, 25 per cent. off, and coppered iron rivets and burrs in 5-lb. carton boxes at 60 and 10 and 10 per cent.; copper burrs alone, 15 per cent., subject to usual charge for half-pound boxes.

HAY WIRE.—Prices remain firm. Our quotations are: No. 13, $2.55; No. 14, $2.65; No. 15, $2.80; f.o.b. Montreal.

MACHINE SCREWS.—Stocks seem to be getting into shape and the scarcity is not so noticeable as last week. Our prices remain: Flat head, iron, 35 per cent., flat head brass, 35 per cent., Felisterhead, iron, 30 per cent., Felisterhead, brass, 25 per cent.

BOLTS AND NUTS—Scarce, discounts remain: Carriage bolts, 3 and under, 60. and 10; 7-16 and larger, 55 per cent.; fancy carriage bolts, 50 per cent.; sleigh shoe bolts, 3/8 and under, 60 per cent.; 7-16 and over, 50 per cent.; machine bolts, 3/8 and under, 60 per cent.; 7-16 and larger, 55 per cent.

HORSE NAILS.—Discounts remain: C brand, 40, 10 and 7 per cent.; M.R.M. Co., 35 per cent.

WIRE NAILS.—Our prices remain: $2.40 per keg base f.o.b. Montreal.

CUT NAILS—We continue to quote: $2.50 per 100 lbs.; M.R.M. Co., latest, $2.30 per keg base, f.o.b. Montreal.

HORSESHOES. — Our prices are: P.B. new pattern, base price, improved pattern iron shoes, light and medium pattern, No. 2 and larger, $3.65; No. 1 and smaller, $3.90; snow pattern, No. 2 and larger, $3.90; No. 1 and smaller, $4.15. Light steel shoes, No. 2 and larger, $4; No. 1 and smaller, $4.25; featherweight, all sizes, No. 0 to 4, $5.60. Toeweight, all sizes, No. 1 to 4, $6.85. Packing, up to three sizes in a keg, 10c. per 100 lbs. More than three sizes, 25c. per 100 lbs. extra.

BUILDING PAPER. — See list of January 14, 1907 (advance of 5 cents on several lines).

CEMENT AND FIREBRICK.—No change in prices. We are still quoting: "Lehigh" Portland in wood, $2.54, in cotton sacks, $2.39; in paper sacks, $2.31. Lefarge (non-staining) in wood, $3.40; Belgium, $1.60 to $1.90 per barrel; ex-store, American, $2 to $2.10 ex-cars; Canadian Portland, $2 to $2.05. Firebrick, English and Scotch, $17 to $21; American $30 to $35; White Bros.' English cement, $1.80 in bags, $2.05 in barrels in round lots.

COIL CHAIN—Our quotations remain: 5-16 inch, $4.40; 3-8 inch, $3.90; 7-16 inch, $3.70; 1-2 inch, $3.50; 9-16 inch, $3.45; 5-8 inch, $3.35; 3-4 inch, $3.20; 7-8 inch, $3.10; 1 inch, $3.10.

SHOT—Our prices remain as follows: Shot packed in 25-lb. bags, ordinary drop AAA to dust, $7 per 100 lbs.; chilled, No. 1—10, $7.50 per 100 lbs.; brick and seal, $8 per 100 lbs.; ball, $8.50 per 100 lbs. Net list. Bags less than 25 lbs., ½c. per lb. extra, net; f.o.b. Montreal, Toronto, Hamilton, London, St. John, Halifax.

AMMUNITION — Slow. Very little sold. Our prices remain: Loaded with black powder, 12 and 16 gauge, per 1,000, $15; 10 gauge, per 1,000, $18; loaded with smokeless powder, 12 and 16 gauge, per 1,000, $20.50; 10 gauge, $23.50.

GREEN WIRE CLOTH.—We continue to quote: In 100 foot rolls, $1.62½ per hundred square feet, in 50 feet rolls, $1.67½ per hundred square feet.

Toronto Hardware Markets

Office of HARDWARE AND METAL,
10 Front Street East,
Toronto, February. 1, 1907.

Hardware business continues very brisk and in every way gives prospects of as bright a year as last.

Looking at it from the point of view of one prominent hardwareman, connected with a leading wholesale house in Toronto, it is questionable, even with present large orders if the business will be as large for 1907, as last year. Owing to the remarkably open Winter, building operations then went on all Winter, and of course demanded material to keep going. This year the weather having been more severe, building has not been so active, but there has been no apparent effect on the ordering of goods. The reason given for this is the many advances which have lately taken place, and the prospects of further advances which have the tendency to encourage the hardware merchant to order early.

Innumerable price changes are being made by jobbers, and while a great number of these are of minor interest, they are all, nevertheless, of importance to the hardware merchant. France, Germany, and Great Britain have all advanced their prices in hardware, owing to the great demand and high price of all raw materials.

Nails have again advanced and are now quoted at $2.40 base.

All copper goods may be expected to advance further owing to the great demand and high price of copper.

AXES AND HANDLES—Orders for these articles have been nearly all taken, the numerous lines meeting with general satisfaction.

WASHING MACHINES —Merchants who have received these goods are placing them in a prominent position, a great many making window displays to a very effective advantage for Spring trade.

SPORTING GOODS — Hockey sticks are still selling briskly. Skates are also very brisk, but in many cases sizes cannot be procured.

CHAIN—Remains at our former quotations: ⅝ in., $3.50; 9-16 in., $3.55; ½ in., $3.60; 7-16 in., $3.85; 3-8 in., $4.10; 5-16 in., $4.70, 1-4 in., $5.10.

RIVETS AND BURRS—Remain at the 10 per cent. advance as quoted last week, but a further advance may take place again shortly, as the raw material is so high.

SCREWS—Stocks seem to be in good shape, and nothing further has been heard of an advance. All American manufacturers of screws have lately advanced their products 5 points, but it will have no material difference on Canadian products.

BOLTS AND NUTS—Large orders are being received for these articles daily, and the demand cannot be coped with. Since the factory burning down here a year ago the demand has never been caught up with.

TOOLS—Large orders are being sent out daily, the advance on Stanley and Starrett's goods not affecting the ordering for Spring shipment.

SHOVELS—A number of new lines have been placed on the market and are meeting with quite a demand.

LANTERNS—The demand is very quiet, merchants apparently having large stocks on hand.

EXTENSION AND STEP LADDERS —Price remains the same as quoted last week, 11 cents per foot for 3 to 6 feet, and 12 cents per foot for 7 to 10-foot ladders. Waggoner extension ladders 40 per cent.

POULTRY NETTING—We quote, 2-inch mesh, 19 w.g., discount 50 and 10 per cent, all others 50 per cent.

WIRE FENCING — Galvanized and plain wire remain at advanced prices, but continue selling well.

OILED AND ANNEALED WIRE—(Canadian)—Gauge 10, $2.41; gauge 11, $2.47; gauge 12, $2.55, per 100 lbs.

WIRE NAILS—A further advance of 5 cents per cwt. has been made, and we now quote $2.40 base.

CUT NAILS—We quote $2.30 base f.o.b. Montreal; Toronto, 20c. higher.

HORSENAILS—Capewell brand, list per pound, No. 5, 15c.; No. 6, 14c.; Nos. 7 and 8, 13c; Nos. 9 to 12, 12c.; f.o.b. Toronto and Winnipeg, discounts on application. "C" brand, 40, 10 and 7½ off list. "M.R.M." brand, 60 per cent. off.

HORSESHOES—Our quotations continue as follows: P.B. base, $3.65; "M.R.M. Co., latest improved pattern" iron shoes, light and medium pattern, No. 2 and larger, $3.80; No. 1, and smaller, $4.05; snow, No. 2 and larger, $4.05; No. 1 and smaller, $4.30; light steel shoes, No. 2 and larger, $4.15; No. 1 and smaller, $4.40; featherweight, all sizes, 0 to 4, $5.75; toeweight, all sizes, 1 to 4, $7. Packing, up to three sizes in a keg, 10c. per 100 lbs. extra; more than three sizes in a keg, 25c. per 100 lbs. extra.

BUILDING PAPER—At present very little business is done in this line, but all goods remain firm at advanced prices.

CEMENT—We quote: For carload orders, f.o.b. Toronto, Canadian Portland, $1.95. For smaller orders ex warehouse, Canadian Portland, $2.05 upwards.

FIREBRICK—English and Scotch firebrick, $27 to $30; American low-grade, $23 to $25; high-grade, $27.50 to $35.

HIDES, WOOL AND FURS — Hides are lower, lamb skins higher, and tallow is slightly advanced Furs also continue to advance.

Hides, inspected, cows and steers, No. 1	....	0 11
" " " " No. 2	....	0 10
Country hides, flat, per lb., cured	....	0 09½
" " " green	....	0 08½
Calf skins, No. 1, city	....	0 12
" " No. 1, country		0 11
Lamb skins	1 25	1 35
Horse hides, No. 1	3 50	3 75
Rendered tallow, per lb.	0 05½	0 05¾
Pulled wools, super, per lb.	....	0 25
" " extra	....	0 27
Wool, unwashed fleece	....	0 15
" washed fleece	0 24	0 25

FURS.

	No. 1,	Prime
Raccoon	....	1 50
Mink, dark	5 00	7 00
" pale	2 50	4 50
Fox, red	3 00	4 00
" cross	3 00	10 00
Lynx	5 00	8 00
Bear, black	....	12 00
" cubs and yearlings	....	5 00
Wolf, timber	....	2 75
" prairie	....	1 25
Weasel, white	0 10	0 65
Badger	0 75	1 75
Fisher, dark	6 00	8 00
Skunk, No. 1	....	1 50
Marten,	3 50	20 00
Muskrat, fall	....	0 17
" winter	..	0 23
" western	0 12	0 17

Montreal Metal Markets

Office of HARDWARE AND METAL,
232 McGill Street,
Montreal, February 1, 1907

Metals, like sheep, seem to have a tendency to follow a leader. Last week ingot copper started the procession skywards, and now, as though through sympathy—or jealousy—other metals must do likewise. Pig iron is one of the culprits this week to add to the trials of manufacturers who depend upon the prices of raw material to keep the family larder well stocked and at the same time not pull too hard on the string attached to the long-suffering consumers' pocket book. Some lines of galvanized iron also have a liking for the atmosphere up in the region of the clouds, and if we may take anything from present symptoms, they seem bent on reaching their destination without any unnecessary delay.

Ingot copper remains scarce and very firm in price.

Ingot tin dropped slightly during the week, but has again stiffened up.

COPPER.—Still very short, with no changes over last week's prices. Our quotations remain: Ingot copper, 26½ to 27 cents; sheet copper, base sizes, 34 cents.

INGOT TIN.—Dropped slightly during the week, but has firmed up again. Our prices remain: 45½ to 46 cents per pound.

ZINC SPELTER. — No changes yet. We are quoting: $7.50 to $7.75 per 100 lbs.

PIG LEAD—Very firm. Our prices remain: $5.50 to $5.60 per 100 lbs.

ANTIMONY—No change over last week's prices, we still quote: 27½ to 28 cts. per pound.

PIG IRON. — Some lines show advances. We now quote: Londonderry, $24.50; Carron, No. 1, $27.00; Carron, special, $25.50; Summerlee, No. 2, selected, $25.00; Summerlee, No. 3, soft, $23.50.

BOILER TUBES. — Remain very firm. Our prices are: 1½ to 2 in., $9.50; 2¼ in., $10.35; 2½ in., $11.50; 3 in., $13.00; 3½ in., $17.00; 4 in., $21.50; 5 in., $45.00.

TOOL STEEL — Our prices remain: Colonial Black Diamond, 8c to 9c; Sanderson's 8c. to 45c., according to grade; Jessop's, 13c.; Jonas & Colver's, 10c. to 20c.; "Air Hardening," 65c. per pound; Conqueror, 7½c.; Conqueror high speed steel, 60c.; Jowitt's Diamond J., 6½c. to 7c.; Jowitt's best, 11c. to 11½c.

MERCHANT STEEL.—Our prices are as follows: Sleigh shoe, $2.25; tire, $2.40, spring, $2.75; toecalk, $3.05; machinery iron finish, $2.40; reeled machinery steel, $2.75; mild, $2.25 base and upwards; square harrow teeth $2.40; band steel, $2.45 base. Net cash 30 days. Rivet steel quoted on application.

COLD ROLLED SHAFTING—Present prices are: 3-16 inch to ¼ inch, $7.25; 5-16 inch to 11-32 inch, $6.20; ⅜ inch to 17-32 inch, $5.15; 9-16 inch to 47-64 inch, $4.45; ¾ inch to 17-16 inch, $4.10; 1¼ inch to 3 inch, $3.75; 3⅛ to 3 7-16 inch, $3.92; 3½ inch to 3 15-16 inch, $4.10; 4 inch to 4 7-16 inch, $4.45; 4½ inch to 4 15-16 inch, $4.80. This is equivalent to 30 per cent. off list.

GALVANIZED IRON. — Shows advances. We now quote: Queen's Head, 28 gauge, $4.70; 26 gauge, $4.45; 22 to 24 gauge, $4.20; 16 to 20 gauge, $3.95; Apollo, 28 gauge, $4.45; 26 gauge, $4.30; 22 and 24 gauge, $3.75; 16 and 20 gauge, $3.60; Comet, 28 gauge, $4.45; 26 gauge, $4.35; 22 and 24 gauge, $3.75; 16 to 20 gauge, $3.60; Fleur-de-Lis, 28 gauge, $4.55; 26 gauge, $4.30; 22 and 24 gauge, $4.05; 16 to 20 gauge, $3.80; Gorbals "Best Best," 28 gauge, $4.55; Colborne Crown, 28 gauge, $4.45; 26 gauge, $4.30; 24 gauge, $3.75. In less than case lots, 25c extra.

OLD MATERIAL. — Remains the same. Our prices are: Heavy copper, 20c. per lb.; light copper, 17c. per lb.; heavy red brass, 17c. per lb.; heavy yellow brass, 14c per lb.; light brass, 10 to 10½c per lb.; tea lead, 4½c per lb.; heavy lead, 4½c. per lb.; scrap zinc, 4½c. per lb; No. 1 wrought $17.00 per ton; No. 2 wrought $6.00 per ton; No. 1 machinery, $18.00 per ton; stove plate, $13.00 per ton; old rubber, 10½ cents per lb.; mixed rags, 1 to 1¼c per lb.

Toronto Metal Markets

Office of HARDWARE AND METAL,
10 Front Street East,
Toronto, Feb 1, 1907

The same old story is to be told about ingot metals. With the exception of tin they are all stiff in price and exceedingly hard to get.

Copper, particularly, is scarce and high in price, there being practically a famine in this metal. The Amalgamated is extending its power over the entire sources of supply and little change can be looked for, therefore, as it will not be to their interest to weaken the market. Brass is also difficult to obtain, with stocks very light.

Lead is more active, but supplies continue hard to get. Tin has declined a cent but black and galvanized sheets have advanced ten cents and boiler plates will do so during the coming week it is expected.

The many reports of metals being stolen indicate that the light-fingered gentry have been reading the market reports.

The volume of business transacted continues to total at large figures in spite of the difficulty experienced in filling orders in some metals. Bookings are large and with nearly all mills filled up for months ahead there appears to be every possibility of iron, copper and other high priced metals advancing to even higher figures.

PIG IRON. — The market continues strong with a good book business doing. Present delivery orders are seasonably light. Hamilton, Midland and Londonderry are off the market, and Radnor is quoted at $33 at furnace. Middlesborough is quoted at $24.50, and Summerlee, at $26 f o b., Toronto.

BAR IRON.—No change has taken place since last week. Prices are firm and we still quote: $2.30 f.o.b. Toronto, with 2 per cent. discount.

INGOT TIN — Ingots have declined one cent owing to a decreased demand, and we are now quoting 44 to 45 cents a pound.

TIN PLATES. -- Conditions are unchanged with prices firm. Only a seasonable business is being done.

SHEETS AND PLATES.—Both black and galvanized sheets have advanced 10 cents. Stocks are low and the market active.

BRASS. — Light stocks are reported everywhere with deliveries hard to obtain. We continue to quote 30 cents per pound.

COPPER. — Almost a famine exists and the high prices are expected to go even higher. We continue to quote: Ingot copper $26 per 100 lbs., and sheet copper $31 to $32 per 100 lbs.

LEAD.—Stocks are light with a very lively demand from buyers who are short. Prices keep very firm. We quote: $5.40 for imported pig and $5.75 to $6.00 for bar lead.

ZINC SPELTER—Stocks continue light with the market active. We quote 7½c. per lb. for foreign and 7c. per lb. for domestic. Sheet zinc is quoted at 8 1-4c. in casks, and 8 1-2c. in part casks.

BOILER PLATES AND TUBES.—An advance is looked for but we still quote: Plates, per 100 lbs., ¼ in. to ½ in., $2.50; ⅜ in., $2.35 heads, per 100 lbs., $2.75; tank plates, 3-16 in., $2.65; tubes, per 100 feet, 1½ in., $8.50; 2, $9; 2 1-2, $11.30; 3, $12.50; 3 1-2, $16; 4, $20.00. Terms, 2 per cent. off.

ANTIMONY. — More activity is reported, but stocks are plentiful. Prices are unchanged at 27c. per pound.

OLD MATERIAL—Dealers report a good demand with supplies coming in well. Their buying prices are: Heavy copper and wire, 20 cents per lb.; light copper, 17½c. per lb.; heavy red brass, 17½c. per lb.; heavy yellow brass, 15c. per lb.; light brass, 11c. per lb.: tea lead, $4.00 per 100 lbs.; heavy lead, $4 25 per 100 lbs.; scrap zinc, 4½c. per lb.; No. 1 wrought iron, $15; No. 2 wrought, $8; machinery cast scrap, $17 to $18; stove plate, $14; malleable and steel, $9; old rubbers, 10½c. per pound; country mixed rags, $1 to $1.25 per 100 lbs., according to quality.

COAL—Stocks are in good shape, with the scarcity of slack being relieved. We quote:

Standard Hocking soft coal f.o.b. at mines, lump $1.50, ¾ inch, $1.40; run of mine, $1.30; nut, $1.25; N. & S., $1; P. & S. 85c.

Youghiogheny soft coal in cars, bonder at bridges: lump, $2.60; ¾ inch, $2.50; mine run, $2.40; slack, $2.15.

United States Metal Markets

From the Iron Age, Jan. 31, 1907.

Deliveries of pig iron to foundries and steel works have been better, particularly north of the Potomac and east of the Allegheny Mountains, and the demand for spot and prompt iron has much lessened. The territory more directly dependent upon Southern iron seems still to be suffering from inadequate shipments Generally speaking, founders have been much exasperated during the past few months over the inability to get iron on old lower price contracts, while a stiff premium when paid for spot brought the same iron to the foundry. Founders, too, are exercised all over the country over the action of the railroads in demanding an advance of 25s. per ton freight on pig iron from Southern furnaces, effective Feb. 1, when the very iron was due them months ago, and did not reach them because the railroads could not handle it.

Prices abroad have weakened and Middlesbrough to-day was reported at 56s. 10½d. This means that No. 3 Middlesbrough can be laid down, duty paid, at $19.50 to $20. Last week some foreign iron was sold on arrival at a concession in preference to storing. There is still considerable foreign iron to come, practically all of it sold. There is at least one large inquiry for foreign iron in the market, but that is for a consumer who would use it for export goods.

So far as the pig iron market for the second half of the year is concerned, there has been quite a movement in the Central West. Cincinnati reports sales of 15,000 tons for the second half to a large implement maker, a total of 10,000 tons to two malleable foundries, and 8,000 tons to a car builder. In New England a number of melters have purchased an aggregate of about 5,000 tons at a shade under $24. Pittsburgh reports the market for Bessemer pig a shade easier for the second and third quarters, quoting Valley iron at $22 at furnace Ferromanganese is weaker.

There has been a sharp advance in iron skelp, due to the fact that leading consumers have made some large purchases during the past two weeks. The manufacturers of merchant pipe and of tubes have again advanced prices $2 per ton. The demand is active.

From the Iron Trade Review, Jan. 31, 1907

Advances in finished materal were plentiful this week. The National Tube Co. raised merchant pipe and boiler tubes one point, or $2 a ton each, and the American Sheet and Tin Plate Co. advanced. blue annealed sheets $1 a ton, galvanized sheets $2 a ton, and galvanized corrugated roofing 10 cents a square. Store prices were also advanced like amounts. Independent tin plate manufacturers also advanced their prices 10 cents per base box for third quarter deliveries. Jobbers' prices on tubes are unchanged, but an advance of $2 a ton was made on warehouse shipments of plates.

In spite of the fact that buyers are showing more deliberation in placing orders for the last half of the year, and that slightly lower quotations are made in some cases, the pig iron market continues very strong. The importance of some small concessions recently made in the east under unusual conditions has been exaggerated. Six months ago a great many furnaces found themselves hopelessly oversold and have since been very conservative in taking orders. Owing to this policy, and also to weather conditions being favorable to a large production, these furnace interests now find themselves more comfortably fixed, and in a few cases they are able to offer spot iron Notwithstanding concessions on small lots, the prices in the east have remained firm, sales having been in good volume and in some instances quotations have been advanced. Taken as a whole, the pig iron conditions are easier but furnish absolutely no cause for belief that there is to be any decided decline.

The demand for structural material has improved in most centres, but competition is brisk, and the principal producer has recently lost a number of orders to independent interests which were eager for business. There is, however, plenty of work in sight with plates and specifications for operation requiring big tonnage, but financial considerations prevent immediate action.

London, Eng., Metal Markets

From Metal Market Report, Jan. 30, 19.7.

Cleveland warrants are quoted at 56s. 9d., and Glasgow Standard warrants at 56s., making price as compared with last week, 2s. 3d. lower on Cleveland warrants and 3s. lower on Glasgow Standard warrants.

TIN—Spot tin opened weak at £189 5s., futures at £189 5s., and after sales of 220 tons of spot and 550 tons of futures closed weak at £189 for spot, £189 for futures, making price compared with last week, £3 5s. lower on spot and £3 10s. lower on futures.

COPPER—Spot copper opened weak at £106 12s. 6d., futures, £107 17s. 6d., and after sales of 400 tons of spot and 300 tons of futures, closed easy at £106 5s. for spot, and £107 12s. 6d. for futures, making price compared with last week, 10s. lower on spot and £1 12s. 6d. lower on futures.

LEAD—The market closed at £19 15s., making price as compared with last week unchnged.

SPELTER—The market closed at £26 15s., making price as compared with last week, £2 7s. 6d. higher.

Calgary Hardware Trade News

Jan. 19, 1907.

Messrs. Cushing Brothers, the largest sash and door manufacturers of the west, are reported to be contemplating the vacation of their premises and the erection of a large factory on the brewery flat in the east of the city. Their present location covers a very valuable portion in the centre of the city. This firm have been kept working at their fullest capacity in all departments, the demand for the stock sizes of doors and sashes having been enormous for the city and large shipments have also been made to Banff, Laggan, and the other busy towns on the line out to British Columbia. Special outfits for banks, hotels and stores have also been greatly in demand

At the first meeting of the new city council there were brought forward several matters of moment to the commercial interests of Calgary. A recommendation from the local board of trade regarding civic telephones was read and it was agreed to again refer the matter to the Fire Committee, which had the matter in charge last year. They are to investigate the matter, secure all information for the installing and maintenance of a municipal telephone system and report to the council A proposal was also submitted from the Calgary Power and Street Railway Co., a syndicate that has been formed to supply the city with electric power and to operate a street car system throughout the city. It is proposed to generate electricity at Radnor and transmit to sub-stations at Calgary, to be sold to the city 50 per cent. lower than the city can produce. A 35-year franchise is asked for a street railway, the city to have the option, after 10 years, of purchasing at cost, less depreciation or plus 10 per cent. If, after 10 years, the city does not care to buy the system, the company will allow them one-third of the net revenue. A first-class system is promised, with a 5-cent tariff and special fares for workmen. It has been resolved to obtain more details of the proposal and if approved by the council the question is to be submitted to the ratepayers for decision.

The Canadian agent of the Perfection power block machines, Minneapolis, has been in the city making arrangements for the installation of a plant to be operated by a local syndicate. They will manufacture power pressed stone, which can be used for general building purposes, being able, it is said, to stand a pressure of over one hundred tons per square foot.

It is also reported that a Winnipeg wholesale house are to open a branch here, to be followed afterwards by a manufacturing plant.

B.C. Hardware Trade News

Vancouver, Jan. 18, '07.

Building trades have never in six or seven years had such a stoppage. The recent cold weather has been such that very few men cared to work, and many unfinished buildings have been left till the cold wave has passed over. The thermometer has risen to some 15 degrees above zero and a snow fall of 6 inches last night has made a change. Indications are now that a January thaw of the old-fashioned sort is coming.

In the interior the smelters and some of the big mines have had to shut down for lack of coke and coal. Heavy snows have so interfered with transportation that the collieries of the Crow's Nest Pass, on which the whole interior depends for fuel, have not been able to get coke and coal forward.

A most interesting deal in power was consummated last week in the interior, when Lorne A. Campbell, manager of the West Kootenay Power & Light Co., closed a deal giving his company control of the South Kootenay Water Power Co. This enables the West Kootenay Power & Light Co. to enter the district of Yale, including the boundary towns and mines, and supply power. Last year this company sought amendments to its charter to empower it to do this, but the Cascade Power Co., operating at Cascade, in the boundary district, successfully fought the application. Under the charter of the South Kootenay Co. just acquired, the West Kootenay Co. has ample powers to supply electric energy in the district. Hitherto this company, which has its power station on the Kootenay river between Nelson and Robson, has confined its operations to the Rossland and Nelson camps and surrounding distrcts. The deal just closed means a great alteration in the boundary district.

The lumber situation is likely to be as much of a problem as ever in the approaching season, for orders placed for shipment right along have been badly tied up for lack of cars, while weather conditions have tied up the railway and thus delayed shipments still further. The stormy weather has stopped almost all logging in the coast district till the deep snow melts off, and towing has been almost impossible. In many of the small bays and harbors up-coast, where log booms are gathered, there has been so much ice that tugs could not get out with their tows. The mills have practically all closed down for the present owing to the difficulty of running in the frost, and the reduction in cut is from the same cause.

In hardware circles a few changes are reported. Bar iron base has been advanced by 20c., to $3.10. White lead is up again, an advance of $1.00 on previous prevailing price. Wire nails, imported, have been put up 10c. base. Wire of all grades is away up, and quotations from the English manufacturers quite keep pace with the increases quoted on the American side of the line, where conditions at the mills are such that no further orders are desired.

Vancouver, Jan. 24, 1907.

Members of the B.C. Lumbermen's Association express themselves as having nothing to fear from the investigations to be made by the select committee of the Dominion House, which by resolution was decided upon this week, to inquire into the working of an alleged lumber combine of unlawful nature. The assertions that fines are imposed on members for selling to non-members of the Northwest Lumber Dealers Association, and that expulsion is also resorted to, are denied by men well posted in the workings of the association. The general charge made by the Alberta member, Mr. Herron, who moved for the special committee, that an unlawful combine exists is specifically denied.

Actual conditions in the lumber industry in this province, so the lumbermen claim, will bear out the prices charged for lumber at the present time. The extreme high price paid for logs, of some $10 per thousand on the average for the past few months, the scarcity of labor and the high wages prevailing, are all factors in the regulation of the price. The millmen assert that the railways get the lion's share of the price paid by the Northwest consumers for lumber.

While the Dominion House of Commons was dealing with the motion to appoint a select committee, the Mountain Lumbermen's Association, 125 strong, was holding its quarterly meeting at Nelson, B.C. This association, while it is affiliated with the coast association, is practically confined to the millmen operating in the interior, or, mountain section of B. C. The session closed last Friday decided to increase prices of lumber in the cheaper grades by 50c. per thousand. This was the only change in the price list. The labor question was discussed and cheap rates from the east and from Europe were advocated as a means of getting in increased help for mills and lumber camps. The advertising of British Columbia in Europe was also discussed and approved, in a general way.

On the coast the log famine is getting worse, and would be severely felt if all the mills were in full running condition. But most of them are still shut down, all having been compelled to close by the culmination of the cold snap last week. The bays and inlets on the northern coast, where logs are usually collected and boomed ready to be brought to the mills, have all been frozen over, so that the tugs could not work. The lumber camps are, many of them, still shut down owing to snow in the woods. Many logs were fast in the ice, so that it was impossible to move booms. The warm weather this week has made a rapid change, but the trade has not recovered and logs are still slow coming in. If conditions had been made to order, they could not be more ideal for creating a stringency in the lumber supply and consequent upward tendency of prices, for the demand from the Northwest is still insistent, for the lumber. The inability of the railways to supply cars and the difficulty to keep the line open, let alone move through freight, have tended to protect the millmen but as soon as there are cars to be had, and the line is open for shipment of freight, the question of price will again be considered.

Mr. Robert Lowe, of White Horse, a prominent pioneer merchant of the Yukon, a member of the Yukon Council (elective), from its inception, and the man popularly supposed to have had the refusal of the governorship of the Yukon as successor to W. B. McInnes, spent a few days in Vancouver this week on his way home from a brief eastern visit. Mr. Lowe is deeply interested in the Great North. His faith in its resources is being tangibly rewarded, for he is one of the heaviest operators in coal and copper lands in the White Horse distrist. He states that much activity has been shown in copper claims for the past year, and that the coming summer will see large development, for many prominent practical mining men from Spokane, Montana, and other copper mining sections have become interested in the copper properties of the north. Mr. Lowe sold one copper property when out on this trip, and he has others. He has also a promising coal property, which he has under development. He recently paid a fancy figure to his other two partners for their interests in it, and now owns it solely. He expects to have it developed on a large basis this season.

There is some certainty of a smelter being built at or near White Horse which will at once assure a number of the near-by copper properties becoming shippers. The general condition of the Yukon is steady and satisfactory. The Guggenheims taking hold of the Klondike mines has given a stability to the operations there, though, of course, a change of business basis has followed. The identification of Wm. Mackenzie, of the Canadian Northern, with the Conrad Consolidated Copper Mines at Conrad City, near White Horse, has done a great deal to impress the permanency and value of the district upon the business public. Mr. Lowe left for the north on the steamer Dolphin on Tuesday.

Giving as a reason for the increase, the unprecedented demand for coal from the mines of this coast, the Western Fuel Co., operating the Nanaimo collieries, has granted voluntarily an increase of 5 per cent. in their wages to the miners. This is for January, and for February the increase is to be 10 per cent. The Dunsmuir mines, at Ladysmith, have also granted a 10 per cent. increase to their men. At the same time the price of coal has gone up 50c. per ton, so that the consumer pays the whole shot—as usual.

FIRE IN PILLOW & HERSEY PLANT

The building on Mill St., Montreal, which was recently occupied by The Pillow & Hersey Manufacturing Co., was damaged on the evening of the 25th ult. to the extent of $2,000 by a fire which broke out in the machine shop. There was little machinery in the shop at the time, as most of it had already been removed to make room for the Ogilvie Flour Mills Co., who recently bought the building.

NEW HORSESHOE FACTORY.

The Toronto and Belleville Rolling Mills, which were recently overhauled and thoroughly equipped with the latest machinery, are now being used in making a fine brand of horseshoes. The company announces that they have installed improved machinery and have secured the services of expert workmen to operate this part of their plant under the supervision of P. J. Smith, known to horseshoers in all parts of Canada. The new shoes are fully guaranteed and should meet with a large sale.

WESTERN TRADE NOTES.

J. A. N. Collinge joined the traveling staff this week of Woods Western.

J. A. Fraser, hardware merchant, of Russell, Man., will shortly move into his new store in the Wright block.

C. D. Sparrow, hardware merchant, of Fairfax, Man., has sold out his business to Tufts Bros.

Landon, Hellier and Heartnell, tinsmiths, of Saskatoon, have been succeeded in business by W. P. Landon.

W. W. La Chance, architect, Saskatoon, was in Winnipeg this week registered at the Royal Alexandra. He was on his return journey from a holiday trip to the east.

Philip Walker, of Grose & Walker, Winnipeg, manufacturers' agents, is in Chicago this week in consultation with several of the firms whom he and Mr. Grose represent in the west.

J. A. Gilhaly, hardware merchant, of Arden, Man., was in Winnipeg this week calling on his host of friends in the trade. He was on his way home from a holiday trip to Ontario.

On leaving the employ of the Vulcan Iron Works, Winnipeg, to accept a position with Kelly Bros. & Mitchell, George Duross was presented by the Vulcan employes with an address and handsome Morris chair.

Arthur Britton, of Winnipeg, late of the James Ashdown Hardware Company, stayed at Port Hope on his way to Montreal to purchase goods for his store, known as the Western Sporting Goods Company, which he will open in Winnipeg.

Gordon McKeag, of Winnipeg, formerly of Fergus, has been appointed manager of the Anchor Wire Fence Company, of that city. Mr. McKeag has represented this firm on the road for the past seven years, and is deserving of his promotion.

C. D. Waldon, sales manager of the Pease-Waldon Co., Winnipeg, returned home last week by way of Chicago from an extended business and pleasure trip in eastern Canada. He was accompanied by W. G. Jones, treasurer of the Pease Foundry Co., Toronto.

Geo. D. Wood, head of Geo. D. Wood & Co., left Winnipeg on Saturday afternoon of last week for Hot Springs, Arkansas. Mr. Wood has been confined to his home for the last four months by a severe attack of sciatica, and it is hoped that he may receive permanent relief from the curative waters and balmy climate of Hot Springs.

One of the most disastrous fires in the history of Saskatoon occurred on Jan. 17th, in the hardware store of Oliver & Kempthorne, on Second avenue. The blaze began in the furnace room upstairs. The early man had just started the morning fire and had gone away. Mr. Kempthorne, who, with his sister, occupied rooms in the front of the building, was aroused by the smoke and made a hurried exit. His sister was soon afterwards taken out of the front window of the burning building by firemen. The total loss is estimated at $54,000, and the insurance is placed at $28,000.

MARITIME TRADE NOTES.

The stock of Crump and Perriers, plumbers and gasfitters, of Halifax, was destroyed by fire on January 18.

Ernest T. Dryden, of Moncton, N.B., has accepted a position with the John Watson Hardware Company, of Houlton, Maine.

UNITED STATES TRADE NOTES

The Atlas Manufacturing Co., New Haven, Conn., manufacturers of Bradley steel shelf brackets, wire closet hooks, etc., have increased their capitalization from $5,000 to $70,000, and are planning to increase their output as well as enlarge their line by adding several new sizes of brackets, coat and hat hooks and tinned spoons. The Atlas Company are well known to the trade in Canada and their growth will be noted with interest.

MANITOBA HARDWARE AND METAL MARKETS

(Market quotations corrected by telegraph up to 12 a.m. Friday, Feb. 1, 1907.)

Room 511, Union Bank Building, Office of HARDWARE AND METAL, Winnipeg, Man.

Although the railways continue to accept all classes of freight for delivery there are complaints from retailers in all parts of the west that they cannot get their goods. Snow blockades and unusually severe weather have combined to tie up the railways, and it will be some little time before a normal condition of affairs is restored.

Discounts have been reduced on Peterborough and Gurney locks, and prices have been advanced on building paper and rivets.

LANTERNS. — Quotations are as follows: Cold blast, per dozen, $6.50; coppered cold blast, per dozen, $8.50, cold blast dash, per dozen, $8.50.

WIRE.—Prices on barb wire for 1907 shew a sharp advance as compared with last year. Revised prices are as follows: Barbed wire, 100 lbs., $3.22½; plain galvanized, 6, 7 and 8, $3.70; No. 9, $3.25; No. 10, $3.70; No. 11, $3.80; No. 12, $3.45; No. 13, $3.55; No. 14, $4.00; No. 15, $4.25; No. 16, $4.40; plain twist, $3.45; staples, $3.50; oiled annealed wire, 10, $2.96; 11, $3.02; 12, $3.10; 13, $3.20; 14, $3.30; 15, $3.45 Annealed wires (unoiled) 10c. less

HORSESHOES — Quotations are as follows: Iron, No. 0 to No. 1, $4.65; No. 2 and larger, $4.40; snowshoes, No. 0 to No. 1, $4.90; No. 2 and larger, $4.65; steel, No. 0 to No. 1, $5; No. 2 and larger, $4.75.

HORSENAILS — Lists and discounts are quoted as follows: Capewell brand, list, per lb., Nos. 5, 15c.; 6, 14c.; 7 and 8, 13c.; 9 to 12, 12c.; f.o.b. Winnipeg. Discounts on application. No. 10, 20c.; No. 9, 22c.; No. 8, 24c.; No. 7, 26c.; No. 6, 28c.; No. 5, 32c.; No. 4, 40c., per pound. Discounts are quoted as follows: "C" brand, 40, 10 and 7½ per cent., "M" brand and other brands, 55 and 60 per cent. Add 15c. per box.

WIRE NAILS—Quoted now at $2.80 per keg.

CUT NAILS.—Recently advanced to $2.90 per keg.

PRESSED SPIKES — Prices are quoted as follows since the recent advance: ¼ x 5 and 6, $4.75; 5-6 x 5, 6 and 7, $4.40; ⅜ x 6, 7 and 8, $4.25; 7-16 x 7 and 9, $4.15; ½ x 8, 9, 10 and 12, $4.05; ⅝ x 10 and 12, $3.90. All other lengths 25c. extra net.

SCREWS—Discounts are as follows: Flat head, iron, bright, 85 and 10 p.c.; round head, iron, 80 p.c.; flat head, brass, 75 and 10 p.c.; round head, brass 70 and 10 p.c.; coach, 70 p.c.

NUTS AND BOLTS — Discounts are unchanged and continue as follows: Bolts, carriage, ⅜ or smaller, 60 and 5; bolts, carriage, 7-16 and up, 55; bolts, machine, ⅜ and under, 55 and 5; bolts, machine, 7-16 and over, 55; bolts, tire, 65; bolt ends, 55; sleigh shoe bolts, 65 and 10; machine screws, 70; plough bolts, 55; square nuts, case lots, 3; square nuts, small lots, 2½; hex nuts, case lots, 3; hex nuts, smaller lots, 2½ p.c.

RIVETS.—Prices on copper rivets have been increased and revised quotations are: Iron, discounts, 60 and 10 p.c.; copper, No. 7, 43c; No. 8, 43½c.; No. 9, 45½c.; copper, No. 10, 47c.; copper, No. 12, 50½c.; assorted, No. 8, 44½c., and No. 10, 48c.

COIL CHAIN.—Prices have been revised, the general effect being an advance. Quotations now are: ¼ inch $7.00; 5-16, $5.35; ⅜, $4.75; 7-16, $4.50; ½, $4.25; 9-16, $4.20; ⅝, $4.25; ¾, $4.10.

SHOVELS—List has advanced $1 per dozen on all spades, shovels and scoops.

HARVEST TOOLS — Discounts continue as before, 60 and 5 per cent.

AXE HANDLES—Quoted as follows: Turned, s.g. hickory, doz., $3.15; No. 1, $1.90; No. 2, $1.60; octagon, extra, $2.30; No. 1, $1.60.

AXES.—Quotations are: Bench axes, 40; broad axes, 25 p.c. dis. off list; Royal Oak, per dozen, $6.25; Maple Leaf, $8.25; Model, $8.50; Black Prince, $7.25; Black Diamond, $9.25; Standard flint edge, $8.75; Copper King, $8.25; Columbian, $9.50; handled axes, North Star, $7.75; Black Prince, $9.25; Standard flint edge, $10.75; Copper King, $11 per dozen.

BUTTS—The discount on wrought iron butts is 65 and 5 p.c.

CHURNS — The discounts from list continue as before: 45 and 5 per cent.; but the list has been advanced and is now as follows: No. 0, $9; No. 1, $9; No. 2, $10; No. 3, $11; No. 4, $13; No. 5, $16.

CHISELS—Quoted at 70 p.c. off list prices.

AUGER BITS—Discount on "Irwin" bits is 47½ per cent., and on other lines 70 per cent.

BLOCKS—Discount on steel blocks is 35 p.c. off list prices; on wood, 55 p.c.

FITTINGS—Discounts continue as follows: Wrought couplings, 60; nipples, 65 and 10; T's and elbows, 10; malleable bushings, 50; malleable unions, 55 p.c.

GRINDSTONES.—Quoted now at 1½c. per lb.

FORK HANDLES—The discount is 40 p.c. from list prices.

HINGES—The discount on light "T" and strap hinges is 65 p.c. off list prices.

HOOKS—Prices are quoted as follows: Brush hooks, heavy, per doz., $8.75; grass hooks, $1.70.

CLEVISES—Price is now 6½c. per lb.

STOVE PIPES—Quotations are as follows: 6-inch, per 100 feet length, $9; 7-inch, $9.75.

DRAW KNIVES—The discount is 70 per cent. from list prices.

RULES—Discount is 50 per cent.

WASHERS—On small quantities the discount is 35 p.c.; on full boxes it is 40 p.c.

WRINGERS — Prices have been advanced $2 per dozen, and quotations are now as follows: Royal Canadian, $35.00; B.B., $39.75, per dozen.

FILES—Discounts are quoted as follows: Arcade, 75; Black Diamond, 60; Nicholson's, 62½ p.c.

LOCKS.—The discount on Peterborough and Gurney locks is now 40 per cent.

BUILDING PAPER. — Prices have been advanced and quotations are now as follows: Anchor, plain, 66c.; tarred, 69c; Victoria, plain, 71c; tarred, 84c.

TINWARE, ETC.—Quoted as follows: Pressed, retinned, 70 and 10; pressed, plain, 75 and 2½; pieced, 30; japanned ware, 37½; enamelled ware, Famous, 50; Imperial, 50 and 10; Imperial, one coat, 60; Premier, 50; Colonial, 50 and 10; Royal, 60; Victoria, 45; white 45; Diamond, 50; Granite, 60 p.c.

GALVANIZED WARE. — The discount on pail is now 37½ per cent.; and on other galvanized lines the discount is 30 per cent.

CORDAGE—We quote: Rope, sisal, 7-16 and larger, basis, $11.25; Manila, 7-16 and larger, basis, $16.25; Lathyarn, $11.25; cotton rope, per lb., 21c.

SOLDER—Quoted at 27c. per pound. Block tin is quoted at 45c. per pound.

VISES — Prices are quoted as follows: "Peter Wright," 30 to 34, 14½c.; 35 to 39, 14c.; 48 and larger, 13½c. per lb.

ANVILS—"Peter Wright" anvils are selling at 11c. per lb.

CROWBARS—Quoted now at 4c. per lb.

POWER HORSE CLIPPERS — The "1902" power horse clipper is selling at $12, and the "Twentieth Century" at $6. The "1904" sheep shearing machines are sold at $13.60.

AMMUNITION, ETC.—Quotations are as follows: Cartridges, Dominion R.F. 50 and 5; Dominion, C.F., 33⅓ C.F., pistol, p.c.; C.F., military, 10 p.c. advance. Loaded shells: Dominion Eley's and Kynoch's soft, 12 gauge, black, $16.50; chilled, 12 gauge, $17.50; soft, 10 gauge, $19.50; chilled, 10 gauge, $20.50. Shot, ordinary, per 100 lbs., $7.25; chilled, $7.75; powder, F.F., keg, Hamilton, $4.75; F.F.G., Dupont's, $5.

IRON AND STEEL.—Quotations are: Bar iron basis, $2.70. Swedish iron basis, $4.95; sleigh shoe steel, $2.75; spring steel, $3.25; machinery steel, $3.50; tool steel, Black Diamond, 100 lbs., $9.50; Jessop. $13.

SHEET ZINC — The price is now $8.50 for cask lots, and $9 for broken lots.

PIG LEAD.—The price is advancing and the range of quotations among Winnipeg houses makes it difficult to quote with exactitude. The average price is about $6.00.

AXLE GREASE—"Mica" axle grease is quoted at $2.75 per case, and "Diamond" at $1.60.

IRON PIPE AND FITTINGS— Revised prices are as follows:—Black pipe, ¼ inch, $2.65; ⅜, $2.80; ½, $3.50; ¾, $4.40; 1. $6.35; 1¼, $8.65; 1½, $10.40; 2, 13.85; 2½, $19.00; 3, $25.00. Galvanized iron pipe, ⅜ inch. $3.75; ½, $4.35; ¾, $5.65; 1, $8.10; 1¼, $11.00; 1½, $13.25; 2, inch, $17.65. Nipples, discounts 70 and 10 per cent.; Unions, couplings, bushings and plugs, 60 per cent.

LEAD PIPE—The price, $7.80, is firmly maintained in view of the advancing lead market.

GALVANIZED IRON—Quoted as follows:—Apollo, 16 gauge, $3.90; 18 and 20, $4.10; 22 and 24, $4.45; 26, $4.40; 28, $4.65, 30 gauge or 10¾ oz., $4.95; Queen's Head, 24, $4.50; 26, $4.65; 28, $5.00.

TIN PLATES—We now quote as follows: IC charcoal, 20 x 28, box, $9.50; IX charcoal, 20 x 28, $11.50; XXI charcoal, 20 x 28, $13.50.

TERNE PLATES—Quoted at $9.

CANADA PLATES—Quoted as follows: Canada plate, 18 x 21, 18 x 24, $3.40; 20 x 28, $3.65; full polished, $4.15.

BLACK SHEETS—Prices are: 10 to 16 gauge, 100 lbs., $3.50; 18 to 22, $3.75; 24, $3.90; 26, $4; 28, $4.10.

PETROLEUM AND GASOLINE.—Silver Star in brls. per gal., 20c.; Sunlight in brls, per gal., 21c.; per case, $2.30; Eocene in brls, per gal., 23c.; per case, $2.50; Pennoline in brls., per gal., 24c.; Crystal Spray, 23c.; Silver Light, 21c.; Engine gasoline in barrels, per gal., 28c., f.o.b. Winnipeg in cases, $2.75.

PAINTS, OILS AND TURPENTINE.—The trade are expecting another advance in turpentine. We quote: White lead, pure, $6.50 to $7.50, according to brand; bladder putty in barrels, 2½c.; in kegs, 3¾c.; turpentine, barrel lots, Winnipeg, $1.01; Calgary, $1.08; Lethbridge, $1.08; Edmonton, $1.09. Less than barrel lots 5c. per gallon advance. Linseed oil, raw, Winnipeg, 67c.; Calgary, 74c.; Lethbridge, 74c.; Edmonton, 75c.; boiled oil, 3c. per gal. advance on these prices.

WINDOW GLASS—We quote: 16-oz. O.G., single, in 50-ft. boxes—16 to 25 united inches, $2.25; 26 to 40, $2.40; 16-oz. O.G., single, in 100-ft. cases — 16 to 25 united inches, $4; 26 to 40, $4.52; 41 to 50, $4.75; 50 to 60, $5.25; 61 to 70, $5.75. 21-oz. C.S., double, in 100-ft. cases—26 to 40 united inches, $7.35; 41 to 50, $8.40; 51 to 60, $9.45; 61 to 70, $10.50; 71 to 80, $11.55; 81 to 85, $12.60; 86 to 90, $14.75; 91 to 95, $17.30.

See a well-dressed window, and you will see people looking in it.

The world is full of wealth and possibilities, the hustlers are after both.

Don't stop to sing "Bringing in the Sheaves," but go out and get a bunch.

FOUNDRY AND METAL INDUSTRIES

An English syndicate will absorb many large steel plants in B.C., and also take over great coal and iron areas.

An employee of the Canada Foundry, Toronto, was working on a crane doing some riveting on Wednesday, when he fell twenty feet, fracturing his skull. He is expected to recover.

The production of lead in Canada for the year 1906 amounts to 26,000 tons. The production for 1905 was 27,000 tons. The record year was 1900, when a total of 30,000 tons was reached.

The rail mill of the Lake Superior Corporation has been turning out 700 to 750 tons a day. The rail output this month is expected to run 18,000 tons against 16,700 tons the previous high record.

Another rich discovery of iron ore has been made near Port Arthur, which has proven to be the most extensive and highest grade ore found in that section. Samples of the red-ore will assay 60 per cent. iron.

The total output of coal in Nova Scotia for 1906 amounted in round numbers to 5,170,000 tons, an increase of 488,000 over 1905. The Cape Breton mines gave an increase of 352,000 tons, Picton 88,500. Cumberland decrease, 34,506, Inverness 80,000.

Geo. S. Wynne has been appointed acting general manager of the Lake Superior Corporation in place of Mr. Sawyer, who has returned to Pittsburg. The general superintendent will be in full charge of the subsidiary companies and report to Mr. Wynne. The executive are still on the lookout for a general manager for the whole concern.

An 18-foot seam of iron ore, located at Barachois, C.B., has lately been discovered. Samples of the ore were supplied the Dominion Iron and Steel Company for analysis, but their report as to the quality of the rock was not satisfactory to the promoters, and test loads were then sent to Sydney Mines for a like purpose. The report from there was that the ore was of a good quality and unique for the purpose of mixing with the Wabana ore in the manufacture of bessemer pig.

HARD GENUINE BABBITT.

As made by the inventor, Isaac Babbitt, genuine babbitt metal was a soft mixture and was really nothing but pewter. The mixture which he used was composed of 50 parts of tin, 5 parts of antimony and 1 part of copper. This mixture, while suitable for many uses, utterly fails to meet the requirements of others. Even the so-called standard mixture of the present time, which consists of 96 parts of tin, 8 parts of antimony, and 4 parts of copper does not always serve the purpose.

A hard genuine babitt which will give the best service that it is possible to obtain, except in cases of severe pounding, is made of the following: Tin, 100 lbs.; antimony, 8 lbs.; copper, 8 lbs.

This mixture will stand hammering into the recesses of the box and will wear without cracking.—Brass World.

PONDEROUS STEAM SHOVELS.

As an illustration of the rapid rate at which railroad building is going forward in Canada might be mentioned the contracts for steam shovels recently placed with the Canada Foundry Company by the Canadian Pacific Railway, the Canadian Northern, and several firms of railway contractors. The building of large steam shovels is a new industry, as far as this country is concerned, the shovels heretofore used on heavy construction work having been imported. The Canada Foundry Company, however, has established a special department for the construction of steam shovels. These implements are known as the Bucyrus steam shovels, which are ponderous machines weighing 90 tons apiece, and three loads of the dipper more than fill a flat car.

A photograph in the possession of the Canada Foundry Company tells more graphically than words the speed at which the Bucyrus shovels excavate the earth. The picture was taken as a string of flat cars was backing down a track beside one of the monster shovels, which kept on working as the cars passed it, and on each car was placed a load from the dipper.

TO PRESERVE IRON AGAINST RUST.

To preserve iron against rust, immerse it for a few minutes in a solution of blue vitriol, then in a solution of hyposulphite of soda, acidulated with chlorhydric acid, says the Inland Printer. This gives a blue-black coating, which neither air nor water will affect.

Persons addressing advertisers will kindly mention having seen this advertisement in Hardware and Metal.

BUILDING AND INDUSTRIAL NEWS

HARDWARE AND METAL would be pleased to receive from any authoritative source building and industrial news of any sort, the formation or incorporation of companies, establishment or enlargement of mills, factories, foundries or other works, railway or mining news All correspondence will be treated as confidential when desired.

A new Y.M.C.A. building will be erected in Montreal at the cost of $300,000

The Vancouver Portland Cement Company, operating at Tod Inlet, B.C., are enlarging their plant.

The Mickle-Dyment planing factory at Brantford will have a $10,000 extension made in the Spring.

The Dominion Radiator Company has taken out a permit for the erection of a plant to cost $160,000, in Toronto.

Tenders are asked for the erection of a large factory building on Carlaw Ave., Toronto, for the Phillips Manufacturing Company.

The work which the Canada Foundry Company, of Toronto, have been doing on the blast furnace at Port Arthur is completed.

The Dominion Carriage Company, Truro, N.S., recently organized, with a capital of $250,000, will erect new building in the Spring.

Weyburn leads all towns in southern Saskatchewan in building operations during 1906, and is fourth in the whole province, the total amount being $303,000.

The statement for the foreign trade of Canada for the six months ending with December 31, shows an increase in exports of $12,890,000, and in imports of $30,124,000.

The county of Wentworth has decided to build a house of refuge. The building will cost between $35,000 and $40,000, and it is proposed to have a farm of 125 acres in connection.

The Toronto Electric Light Company will apply to the Ontario Government for an increase in its capital stock of $1,000,000, which will bring the total capitalization to $4,000,000.

Arcola, Sask., has, in keeping with the general progress of the west, made rapid strides during the past year. Something over $190,000 has been put in buildings and other improvements.

The Canada Cycle and Motor Company, of Toronto Junction, has purchased property in Ottawa, and will erect showrooms, and repair shop, expecting to be open for business in the Spring.

A large bank building will be erected at Victoria, B.C., by the Royal Bank of Canada. The Bank of Nova Scotia have also purchased property on King street, Toronto, and will erect a bank building.

The Canadian Pacific Railway Co. will build a new depot at Calgary, Alta. It is estimated that the structure will cost $200,000, and outside of Winnipeg will be the finest on their western lines, and one of the finest on the whole system.

The new building just completed for the Toronto Type Foundry Company, Toronto, is one of the largest in the city, being 168x85 feet. The whole structure is built largely of concrete, more than 500 barrels of cement being used.

The Galt Art Metal Company have succeeded in carrying off the contracts for all sheet metal work, such as cornices, ceiling, etc., on all four Normal schools to be erected at Hamilton, Stratford, Peterboro and North Bay, amounting to between $16,000 and $18,000.

When the Galt Malleable Iron Works opened in October last 25 men were employed. Last pay-day at the works 132 regularly employed workmen received their pay envelopes, an increase of over 400 per cent. in the number of employes.

A grant of $50,000 has been made for the erection of a hygienic institute at London by the Provincial Government. Besides the $50,000 for building, an annual grant for five years of $5,000 a year will be made for maintenance by the Government.

The plant of the Canadian Glass Works, located on the outskirts of Montreal, and employing about 150 hands, was burned to the ground on Wednesday. The loss is in the neighborhood of $20,000 on the plant and equipment and is covered by insurance.

Messrs. Ney, Camp and Company have been incorporated with a capital of $40,000 for the purpose of manufacturing furniture, etc. The head office will be at Stratford, and provisional directors being, W. J. Ney, N. W. Camp and J. H. Bamber, Stratford.

The Imperial Steel and Barb Wire Co., of Collingwood, is looking for a suitable point in the west to establish a plant for the manufacture of barb wire and nails. The company is looking for inducements, guaranteeing to employ 100 to 200 men all the year round.

A large, four-storey building in Hamilton will be erected in the Spring, and will be sub-let to small manufacturers. This proposition should meet with great success, as small manufacturers are frequently handicapped by being unable to get suitable buildings in starting out.

The Collingwood Shipping Co. has been incorporated with a capital of $90,000, for the purpose of constructing elevators, wharves, docks, warehouses, etc. The head office will be at Collingwood, and provisional directors being, W. T. Allan, M. Brophy and W. A. Hogg, of Collingwood.

The Dominion Bearing Company, of St. Catharines, are manufacturing automobile equipment, engines, running gear and bearings for supply to a selling concern. The company will shortly place on the market an improved roller bearing. About eighteen men are now employed in the works.

The Standard Concrete Construction Company, Toronto, has been incorporated with a capital of $100,000, for the purpose of general building and construction. The head office will be at Toronto, and provisional directors being, F. Rielly, J. B. Bartram and E. A. Scott, Toronto.

The Brandon and Robertson Manufacturing Company has been incorporated with a capital of $75,000, for the purpose of manufacturing implements, machinery, etc., the head office to be at Brandon, Man., and provisional directors being, W. Brandon, J. A. Robertson and R. J. Brandon, of Brandon, Man.

The Ham & Nott Company, of Brantford, are about to build a large factory in Ottawa. The new factory will manufacture bee-keeping supplies, spring beds, and screen doors, and will employ over one hundred hands while in full operation. Arrangements are now being made with regard to a suitable site and other features.

The Montreal Wood Mosaic Flooring Company has been incorporated, with a capital stock of $5,000, for the purpose of manufacturing wood flooring, window and door screens, steel mats, etc. The head office will be at Montreal, the directors being, A. McLean, Buffalo, and D. H. McLennan, C. Stewart and R. W. Barclay, of Montreal.

The North Shore Transportation and Wreckage Company, Quebec, has been incorporated, with a capital stock of $250,000, for the purpose of manufacturing vessels, steamboats, machinery, etc., and to construct wharves, piers, warehouses, etc. The head office will be in Quebec, the directors being, J. A. Fafard, O. C. Bernier and A. Tegnon, of Quebec.

An agreement has been signed with the town council of Welland by E. Billings, of the Spencer Company, of Hartford, Conn., the largest manufacturers of drop forgings in the United States, to locate there. The branch will be known as the Canadian Billings and Spencer Company, and plans for a plant to cost about $100,000 will be prepared at once.

The Dominion Graphite Company, of New York, will erect a large mill, representing an outlay of nearly a million dollars for the treatment of graphite, to be mined in the Laurentian ranges near Buckingham, Que. Not long ago the Dry Concentrating Company, of New York, announced their intention of erecting a smelter with the same object in view, and both plants are now in course of construction.

The recently instituted Carleton Place Board of Trade, has completed organization, with Frank Taylor, of Taylor Bros., hardware merchants, as secretary, and W. J. Muirhead, another hardware dealer, as one of the executive. Forty members have already allied themselves to the new board. A resolution was passed endorsing the proposed by-law to loan Bates & Innes $10,000 for ten years to assist them in starting the Gillies woolen mill.

A new apartment block will be erected in Winnipeg which will accommodate one hundred families and will be furnished throughout with every known modern device for the comfort and convenience of its tenants, including electric elevators, window cleaning, janitor service, telephone in every suite, giving free service over the entire building, and connecting with outside service by means of a private switchboard; mail chutes in every corridor, auto garage in basement, bell boys, dance hall, and roof garden, all free to the use of the tenants.

The manufacturing and retail trade interests of Quebec held a mass meeting of citizens recently after church service to protest against the proposed conversion of the Plains of Abraham and Coverfields into a national park, as it would mean the strangulation of the Ross rifle factory, the laboratories of the Quebec cartridge factory, and a blow to Quebec manufacturing industries in general. The committee who wish to propagate the national park in question are expecting one million dollars aid from the Dominion Government. George E. Amyot, president of the Quebec Board of Trade, presided. Resolutions were adopted requesting the Dominion Government to grant the request of the Ross Rifle Company for an additional area of land, about three acres, in the Coverfields. About fourteen hundred were present.

The acquisition of the Qu'Appelle, Long Lake and Saskatchewan Railway by the Canadian Northern Company has already resulted in substantial reductions in passenger, freight and express rates. The Canadian Pacific Railway, which formerly operated the line, charged 4 cents per mile for single fares, and this has been reduced to 3½ cents by the C. N.R. The reduction in freight rates is about 20 per cent., while the express charges have been reduced 10 to 33 per cent. The service they have maintained so far has been unsatisfactory, both to the public and to the C.N.R. Passenger trains have on occasions since Jan. 1, taken two to three days to travel from Prince Albert to Regina, a distance of 254 miles. Snow blockades and the lack of coal have been the principal hindrances, while the continued cold weather has played havoc with the motive power supplied for the division. It is understood that the company is making some progress in its efforts to clear the line and restore the normal service. But in the meantime there is great inconvenience and in some cases actual hardship among the residents in towns and villages along the route.

NEW ZEALAND'S EXHIBITION.

The New Zealand international exhibition, now being held at Christchurch, New Zealand, is of much larger proportions than was anticipated, and the Government is well pleased with the response to the invitation to participate. The largest section of the exhibition is that occupied by Canada. It contains a floor area of 17,000 square feet, while the surrounding walls afford a space of 5,000 square feet, which have been handsomely decorated with bunting, pictures, charts, etc., all of which will prove an interesting and educational feature. The following are some of the Canadian manufacturers who are represented in in the Canadian court: The Canadian Rubber Company, Victoria Wheel Works, Pacific Coast Pipe Company, The Dodge Mfg. Co., The Montreal Rolling Mills, Dowswell Mfg. Co., Berlin Rubber Company, R. MacDougall Co., Stauntons, Chestnut Canoe Co., Peterboro Canoe Co., E. T. Wright & Co., D. Maxwell & Co., Canada Cycle and Motor Co., McClary Mfg. Co., Metallic Roofing Co., The Waggoner Ladder Company, Imperial Oil Company, Hart Corundum Company. The mineral exhibit is attracting great attention, particularly the exhibits of asbestos, mica, corundum and nickel, these being to a certain extent new minerals to New Zealanders. This exhibit occupies a space of 3,491 feet, and consists of thirty-five pyramids and stands. Among the more prominent exhibits are gold quartz from Nova Scotia, Ontario and British Columbia; nuggets from Quebec, British Columbia and the Yukon; gold-copper ores and silver lead ores, British Columbia; copper ores, Nova Scotia, Ontario and British Columbia; zinc ores, Ontario and British Columbia ; mercury, British Columbia ; antimony, Nova Scotia and New Brunswick ; nickel and copper ores, Sudbury ; iron ores from Nova Scotia, Quebec, Ontario and British Columbia ; metallurgical products from the Trail smelter, Trail, British Columbia ; exhibit of ferro-chromo, ferro-silicon and ferro-phosphorus, by the Electric Reduction Company, of Buckingham, Quebec ; manganese, Nova Scotia ; chromic iron, Quebec ; molyodenite, Quebec, Ontario and British Columbia ; tungsten, British Columbia ; coal from Nova Scotia, Alberta and British Columbia ; petroleum and products, Imperial Oil Co., Sarnia, Ont.; apatite, Quebec ; iron ochre and paints, Canada Paint Company, Montreal, salt, Ontario ; corundum, Craig mine, Raglan, Ontario, corundum wheels, Ontario ; asbestos, from the Thetford and Black Lake mines, Quebec ; mica, Quebec, Ontario and British Columbia ; graphite, from Ontario and Quebec ; felspar, Quebec and Ontario ; building stones, granite,syenite, sandstone, limestone and dolomite; ornamental stones, serpentine, marble, jaspas, conglomerate, sodalite, etc., marls, clays, infusorial earth, barite, magnesite, cement, talc.

CANADIAN PATENTS.

F. W. Moore, Montreal, has secured a Canadian patent for a combined tea and coffee pot.

A patent for blanks for forming five-tined pitch forks has been issued to O. K. Janson, Tillsonburg, Ont.

CONDENSED OR "WANT" ADVERTISEMENTS.

Advertisements under this heading 2c. a word first insertion; 1c. a word each subsequent insertion.

Contractions count as one word, but five figures (as $1,000) are allowed as one word.

Cash remittances to cover cost **must** accompany all advertisements. **In no case** can this rule be overlooked. Advertisements received without remittance cannot be acknowledged.

Where replies come to our care to be forwarded, five cents must be added to cost to cover postage, etc.

SITUATIONS VACANT.

WANTED—Good practical tinsmith; $18.00 weekly to the right man; steady job. Apply H. H. Gervan, Chilliwack, B.C. (5)

WANTED—Experienced pricer for wholesale hardware; none but experienced man need apply. Box 578, HARDWARE AND METAL, Toronto.

BRIGHT, intelligent boy wanted in every town and village in Canada; good pay besides a gift of a watch for good work. Write The MacLean Publishing Company, 10 Front Street E., Toronto. [tf]

WANTED—An experienced hardwareman at once. Apply, stating wages required, to Thomas Oliver, Hardware Merchant, Copper Cliff, Ont. [6]

BUSINESS CHANCES.

HARDWARE stock about $7,000, including set of tinners' tools; stock is all new and situated on main line C.P.R. in Saskatchewan town of 1,800 population; building can be leased, is brick, two storeys and basement; net profit $4,000 yearly. Box 577, HARDWARE AND METAL, Toronto. [7]

HARDWARE BUSINESS established 17 years for sale—including stoves, ranges and tinware; a good plumbing and tinshop in connection; stock about $6,000. J. D. Murdoch & Co., Simcoe, Ont. [6]

FOR SALE—A fully equipped Wire Mill and Nail Factory as a going concern, with about 500 tons of wire rods; factory situated on 150 ft. of waterfront with trackage optional to buyer. B.C. Wire and Nail Co., Ltd., Vancouver, B.C. (7)

HARDWARE business, established 32 years in town of 2700 population, no opposition; stock of about $4,000, consisting of hardware, paints and oils, tinware and graniteware, annual turn over $13,000 (for 1906), owners going into manufacturing; building owned by present owners, who will rent reasonably. Box U, Alexander, Ont. [5]

FOR SALE.

CASH CARRIER SYSTEM in first-class order for sale at a bargain; nine stations; will sell separately or complete. Box 580, HARDWARE AND METAL, Toronto. [5]

HARDWARE Business for sale in one of the best towns in Canada, best of reasons for selling. Box 582 HARDWARE AND METAL, Toronto. (5)

ARTICLES WANTED.

WANTED—A set of tinsmith tools, second-hand; state full particulars, H. A. Hoar, Barrie. [6]

WANTED—Eight foot cornice; state price and particulars. J. E. Hussey, Melbourne, Ont. [8]

The advertiser talks to you through Hardware and Metal. You should talk to him through correspondence.

A★1 Brand Cutlery is Always Worth the Money You Pay for it

Every Knife and Scissor is guaranteed to be of the best quality of steel, right temper, superior finish and finest workmanship. That is why this brand is a favorite in the Canadian Market, and your stock is not complete without the goods of

THE R. BERG CUTLERY CO.

F. W. LAMPLOUGH & CO., MONTREAL

Hardware Dealers Should Wake Up—

to the fact that **SHEET METAL BUILDING GOODS** could be made a large and ever increasing factor in their business. Farmers, manufacturers and all owners and builders are daily demanding these goods because they fill satisfactorily the gap which is being left by the diminishing supply of good and cheap timber. Some dealers have already established a considerable trade in this line, proving that it can be done, but it's "only as a drop in the bucket" compared to what's to come. If the **HARDWARE DEALERS** and **TINSMITHS** don't look after this valuable trade, they will live to regret it. **Some one else will reap the harvest.**

Don't say—"**There's nothing in it for me**"—because there certainly is, providing you have the **RIGHT GOODS** and the **RIGHT FIRM** behind them. It doesn't matter about the manufacturers who sell direct to the consumer. One good, square firm, who will **PROTECT** and **HELP** a good orginization of agents, can, with those agents, do more business than three or four of the other kind.

We can show you—all we need to prove to you how this combination will work in 1907 is some more good agents—we've got the goods and the other qualifications. One of our representatives will be along your way soon after New Years to talk to you on the subject. If you are interested drop us a line at once and make certain that you will not be missed.

The GALT ART METAL CO., Limited, GALT, ONT.

Western Distributing Agents, **THE McCLARY MFG. CO.**, Winnipeg and Calgary.

PLAIN TROJAN HOOP IRON

made in four widths, $\frac{1}{2}$, $\frac{5}{8}$, $\frac{3}{4}$ and 1 inch, put up as cut in rolls of 2,000 feet, continuous, packed in barrels of 500 pounds.

STRONGER THAN ORDINARY HOOP AND CHEAPER

J. N. WARMINTON

43 Scott St., Toronto — 207 St. James St., Montreal

Paint, Oil and Brush Trades

METHODS OF MAKING WHITE LEAD.

White lead, or carbonate of lead, is made from the metal by two methods, the second of which is merely a development of the first. The first, or old Dutch method, is still widely used, and has been the way of manufacture perhaps since the Roman times, the Dutch centuries ago learning it from the Venetians.

The metallic lead is cast into the form of star gratings, or thin perforated sheets, in such a way as to render quite easy its conversion into the carbonate. These pieces of lead are placed in small earthenware jars resembling flower pots about six inches deep, and wider at the top than at the bottom. A piece of wood is placed in the pot so as to confine the lead to the upper portions thereof. Some acetic acid (vinegar) is then poured into the bottom of the vessel, the lead placed in the upper part occupying about two-thirds of the jar, and kept from the vinegar by the wood, and a slab of lead is placed on top. These jars are then arranged in parallel rows, forming a square tier of about 20 feet each way, or 400 altogether, and the whole covered with tan bark or stable litter; another tier is placed on the top of the first and treated likewise. This process is repeated until there is a pile eight or ten tiers high formed, all embedded in the litter.

Left to themselves for six or eight weeks, the fermentation of the litter keeps up a suitable temperature and a constant supply of carbonic acid. The vinegar in the jars is converted into gas by the heat, and this aided by the imprisoned air, attacks the lead, which becomes thickly coated with a white substance (basic acetate.) The carbonic acid given off from the hotbed in turn converts this basic acetate into the carbonate of white lead.

During the whole of the six or eight weeks this constant action and reaction continues until the whole of the metallic lead is changed into the white lead. After this is completed the pots are emptied and the newly made carbonate of lead broken up and ground in water to a fine paste. It is then placed in little conical moulds and allowed to dry. It is then ready for use, and is what is called dry white lead.

Some years since this was an extensive industry in Holland and numerous factories had their existence around Rotterdam. But of late years white lead manufacturing is being conducted extensively in Prussia, Great Britain, and the United States, causing the Dutch industry to dwindle down to smaller proportions. In Saxony and Prussia, and also in Great Britain and the United States, the second method, which is merely a development of the first, is followed. The factory is a two-storey building the ground floor of which is occupied by a furnace. The floor of the upper chamber is composed of boards loosely arranged so that the heat may readily pass through from the furnace beneath. Square or rectangular watertight boxes with numerous openings on top are placed on the first floor. Above these are erected frameworks reaching to the roof, to which short sticks are attached in such a way as to enable them to sustain the bars of thin sheets of lead above the boxes. Some unfermented grape or raisin juice or cider is placed in the boxes, and this is made to ferment by the action of the heat from the furnace. The fumes from the grape juice together with the oxygen in the air decompose the lead, changing it to the carbonate. This falls in flakes to the floor, where it can be gathered by the workmen, broken into fine particles, ground in water and dried in the moulds; it is then ready for shipment.

NEW PAINT FACTORY IN B.C.

The purchase of the Dominion Paint Company's premises, at Victoria, B.C., by W. E. Staneland a short time ago, promises to add another flourishing industry to the commerce of that city.

Mr. Staneland, who has been with P. McQuade & Co. for seven years, and with paint businesses all his life, will at once extend the business carried on by W. T. Andrews. While retaining the latter's agency for Burrell & Co., of London, Eng., Mr. Staneland will go into the manufacture of high-grade pure prepared paints, which he will put on the market as the Staneland paints. He is adding seventy-five feet to his warehouse, and is installing an iron and tin plant.

A NEW USE FOR GLUE.

An application for burns can be made, it is said, by taking fifteen ounces of the best glue, breaking it into small pieces and adding two pints of water. This having become soft, should be dissolved by means of a water bath; two ounces of glycerine and six drachms of carbolic acid should be added, and the heat continued until the whole is thoroughly dissolved. On cooling this mixture hardens to an elastic mass, covered with a shining parchment-like skin, and may be kept for any length of time. When required for use it is placed for a few minutes in a water bath until sufficiently liquid, and applied by means of a brush. In about two minutes it forms a shining, smooth, flexible and nearly transparent skin.

ANOTHER RUBBER MERGER.

The Canadian Consolidated Rubber Company, Montreal, will in the near future absorb the Merchants and Berlin Rubber Companies, of Berlin, which have a capital of $500,000, and employ 600 hands. The two companies have been purchased by interests friendly to the Consolidated Company, and officials of the latter do not deny that the ultimate object is the consolidation of the two companies named with the larger concern.

Paint and Oil Markets

MONTREAL.

Office of HARDWARE AND METAL,
232 McGill Street,
Montreal, February 1, 1907

The only change in prices to be recorded this week is the advance in Paris green of one cent per pound, which was received too late for publication in our issue of Jan. 26th. This advance, instead of acting as a detriment to sales, has given feeling to the market, and orders are pouring in, both for immediate and spring shipment.

General lines are not over active as far as shipping is concerned, but manufacturers report a good volume of trade for later shipments, and the market generally may be noted as firm and strong.

LINSEED OIL—No change in prices. We still quote: Raw, 1 to 4 barrels, 55c; 5 to 9 barrels, 54c.; boiled, 1 to 4 barrels, 58c.; 5 to 9 barrels, 57c.

TURPENTINE—(In barrels) Last week's advance of one cent per gallon still holds. Our prices are: Single barrel, 98c. per gal.; for smaller quantities than barrels, 5c. extra per gal. is charged. Standard gallon is 8.40 lbs., f.o.b. point of shipment, net 30 days.

GROUND WHITE LEAD—Expected advance has not yet arrived. Our prices remain: Best brands Government Standard, $7.25 to $7.50; No. 1, $6.90 to $7.15; No. 2, $6.55 to $6.90; No. 3, $6.30 to $6.55, all f.o.b., Montreal.

DRY WHITE ZINC—Still firm. We quote: V.M. Red Seal, 7½c to 8c; Red Seal, 7c to 8c; French V.M., 6c to 7c; Lehigh, 5c to 6c.

WHITE ZINC—Our quotations are: Pure, 8¼c. to 9½c.; No. 1, 7c. to 8c.; No. 2, 5¾c. to 6¾c.

PUTTY.—Our prices are: Pure linseed oil, $1.75 to $1.85; bulk in barrels, $1.50; in 25-lb. irons, $1.80; in tins, $1.90; bladder putty in barrels, $1.75.

ORANGE MINERAL—The following prices still hold: Casks, 8c; 100 lb. kegs, 8½c.

RED LEAD—The following quotations are firm: Genuine red lead in casks, $6.00; in 100-lb. kegs, $6.25; in less quantities at the rate of $7 per 100 lbs.; No. 1 red lead, casks, $5.75; kegs $6, and smaller quantities, $6.75.

PARIS GREEN—Advanced one cent per pound, and orders pouring in. Our prices now are: In barrels, about 600 pounds, 23¼c. per lb.; in arsenic kegs, 250 lbs., 23½c.; in 50 lb. drums, 24c.; in 25 lb. drums, 24½c.; in 1 lb. packets, 100 lbs. in case, 25c.; in 1 lb packets, 50 lbs in case, 25½c.; in ½ lb. packets, 100 lbs in case, 27c.; in 1 lb. tins, 26c. f.o.b. Montreal. Term three months net or 2 per cent. 30 days.

SHELLAC—Prices remain: Bleached in bars or ground, 46c. per lb., f.o.b. Eastern Canadian points; bone dry, 57c per lb., f.o.b. Eastern Canadian points; T. N. orange, etc., 48c per lb. f.o.b. New York.

SHELLAC VARNISH—Very firm. We still quote: Pure white, $2.90 to $2.95; pure orange, $2.70 to $2.75; No. 1 orange, $2.50 to $2.55.

MIXED PAINTS—Business good; no change in prices. We still quote: $1.20 to $1.50 per gallon.

CASTOR OIL—Our prices are: Firsts in cases, 8½c.; in barrels, 8c.; seconds in cases, 8c.; in barrels, 7½c.

BENZINE— In bbls.)—We still quote: 20 cents per gallon.

GASOLINE—(In bbls.)—Our price is 22½ cents per gallon.

PETROLEUM—(In barrels) Our prices are: American prime white coal, 15½c. per gallon. American water, 17c. per gallon; Pratt's Astral, 19½c. per gallon.

WINDOW GLASS — Prices remain firm. We still quote: First break, 50 feet, $1.85; second break, 50 feet, $1.95; first break, 100 feet, $3.20; second break, 100 feet, $3.40; third break, 100 feet, $3.95; fourth break, 100 feet, $4.15; fifth break, 100 feet, $4.40; sixth break, 100 feet, $4.95. Diamond Star: First break, 50 feet, $2.30; second break, 50 feet, $2.50; first break, 100 feet, $4.40; second break, $4.80; third break, 100 feet, $5.75; fourth break, 100 feet, $6.50; fifth break, 100 feet, $7.50; sixth break, 100 feet, $7.50; seventh break, 100 feet, $8; eighth break, 100 feet, $9. Double Diamond: First break, 50 feet, $3.45; second break, 50 feet, $3.75; first break, 100 feet, $6.75; second break, 100 feet, $7.25; third break, 100 feet, $8.75; fourth break, 100 feet, $10; fifth break, 100 feet, $11.50; sixth break, 100 feet, $12.50; seventh break, 100 feet, $14; eighth break, 100 feet, $16.50; ninth break, 100 feet, $18; tenth break, 100 feet, $20; eleventh break, 100 feet, $24; twelfth break, 100 feet, $28.50. Discount on Diamond Star, 20 per cent.; on Double Diamond, 40 per cent.

TORONTO.

Office of HARDWARE AND METAL,
10 Front Street East.
Toronto, February 1, 1907

The paint trade continues very brisk with continued large orders for Spring shipment. The orders for paint this season are much more numerous than previous years, and much larger quantities are being ordered.

White lead is also selling very rapidly. While the recent 15c. advance of lead in Montreal has not been followed here as yet, merchants throughout the country are ordering very briskly in view of the advance in prices which must take place.

Turpentine remains very firm. Very little ordering is being done for turps at present, which are quoted at an advance of ½c. per gallon in Savannah over

last week's price. Merchants will no doubt have to watch very carefully for the cheaper and adulterated grades of turpentine which will be placed on the market as a result of the advanced prices.

The sale of Paris green continues large in spite of the high prices.

The Standard Oil Company have announced an advance of a quarter-cent a gallon on all grades of refined oil, naphtha and gasolene in barrels, owing to the higher price of barrels.

WHITE LEAD—Ex Toronto pure white, $7.40; No. 1, $6.65; No. 2, $6.25; No. 3, $5.90; No. 4, $5.65 in packages of 25 pounds and upwards; 1-2c per pound extra will be charged for 12 1-2 pound packages; genuine dry white lead in casks, $7.00.

RED LEAD—Genuine in casks of 500 lbs., $6.00; ditto, in kegs of 100 lbs., $6.50, No. 1 in casks of 500 lbs., $5.75; ditto, in kegs of 100 lbs., $6.25. Litharge, 6½ cents lb.

DRY WHITE ZINC— In casks, 7 1-2c.; in 100 lbs., 8c., No. 1, in casks 6 1-2c., in 100 lbs., 7c.

WHITE ZINC (ground in oil)—In 25-lb. irons, 8c.; in 12 1-2 lbs., 8 1-2c.

SHINGLE STAIN—In 5-gallon lots, 75c. to 80c. per gallon.

PARIS WHITE—90c. in barrels to $1.25 per 100 lbs.

WHITING—60c. per 100 lbs. in barrels; Gilders' bolted whiting, 90c. in barrels, $1.15 in smaller quantities.

SHELLAC VARNISH—Pure orange in barrels, $2.70; white, $2.82 1-2 per barrel; No. 1 (orange) $2.50; gum shellac, bone dry, 63c., Toronto. T. N. (orange) 51c. net Toronto.

LINSEED OIL—No advance has been made lately. Very little selling at present, prices remaining at former quotations: Raw, 1 to 3 barrels, 58c.; 4 to 7 barrels, 57c.; 8 and over, 56c, add 3c. to this price for boiled oil. Toronto, Hamilton, London and Guelph net 30 days.

TURPENTINE — Has experienced a small advance of ½c. per gallon in Savannah, but this will have no present effect here. Very little is being sold at present, but prices remain steady at last week's quotations. Single barrels, $1.02 per gallon f.o.b. point of shipment, net 30 days. Less than barrels, $1.07 per gallon.

GLUES—French Medal, 12 1-2c. per pound; domestic sheet, 10 1-2c. per lb.

PUTTY—Ordinary, 800 casks, $1.50; 100 drums, $1.75; barrels or bladders, $1.75; 100-lb. cases, $1.90;; 25-lb. irons, $1.85; 25-lb. tins (4 to case) $1.90; 12 ½ lb. tins (8 to case) $2.10.

LIQUID PAINTS—Pure, $1.20 to $1.35 per gallon; No. 1, $1.10 per gallon.

BARN PAINTS—Gals., 70c. to 80c.

BRIDGE PAINTS—Gals., 75c. to $1.

CASTOR OIL—English, in cases, 10½ to 11 cents per lb., and 12 cents for single tins.

PARIS GREEN—Last week's advance of 1 cent per pound on green has encouraged many dealers who were holding off to buy now. The great majority of merchants, however, had stock ordered.

REFINED PETROLEUM — Dealers are stocking up heavily for the Winter. We still quote: Canadian prime white, 14c.; water white, 16c.; American water white, 16c. to 18c. ex warehouse.

CRUDE PETROLEUM — We quote: Canadian, $1.32; Pennsylvania, $1.58; Ohio, 96c.

CENTRALIZATION HELPS TO SUCCESS.

Centralization is the keynote of business to-day. It is shown as much in the retail trade of the rural village as in the big commercial organizations of the country. In most instances it is the solution of the problem of diminishing trade and profits. To compete successfully in the market the merchant of the village, of the town, or of the city, must carry as few lines as possible. He must cut down his stock to the products of as few manufacturers in the same line as possible. He must centralize his selling force. In this way only, can he put the necessary energy and life into the sale of goods. This centralization is not effected by cutting down the number of classes of goods carried by the merchant, but in cutting down the number of lines of each class of goods. Instead of handling half a dozen or more half lines, he carries one full line. The merchant with one full line of each class of goods with the products of but one manufacturer, always providing, of course, that the products will give his customers satisfaction, is in a much stronger position that the merchant loaded down with the goods of various makers.

He can much easier and much better advertise one full line than several short lines. Every effort made to sell one article, every sale made, advertises all the other products of the line in his stock. And, the very fact that he carries a full line shows that he has confidence in the goods, and is an argument to the prospective buyer of every article in the line. One full line requires less investment and minimizes the likelihood of carrying over stock. The merchant who confines his stock to one full line avoids delay in shipping, shortage, errors in filling orders, etc. He finds it easier to take care of his trade.

The manufacturer can deal much more liberally with the merchant who carries his products only in stock. Much closer co-operation is possible in going after trade. The manufacturer is assured of the merchant's loyalty—their interests are common—and he can advertise thoroughly in the dealer's territory without the risk of promoting the sale of competitor's goods. The paint trade affords a very good instance of the tendency towards centralization of interests and of the benefits derived. The large paint concerns are appointing agents in each town, giving them a monopoly on their complete line, working hand-in-hand with them for the sale of their goods. The plan has proved generally successful, the dealer recognizing the material advantage it affords.

Never stop to think if you expect to win the race.

The Long and the Short of It

"IT" MEANS

THE "LONG"

is the time it wears. It stands the test of heat and frost, wind and rain, and give your customers many years of solid satisfaction.

THE "SHORT"

part is the time needed to put it on. A great deal of time—which means a great deal of money—is saved to the man who uses

SHIELD BRAND READY ROOFING

LOCKERBY & McCOMB
65 SHANNON STREET
MONTREAL

ALABASTINE

A familiar household word. A necessity in every home.	"Homes Healthful and Beautiful." Health and beauty go hand in hand.
The best advertised, the most saleable Wall Coating.	No taint, no smell, no vitiated air in rooms with Alabastined walls and ceilings.
Has no competitor. There is no other "just-as-good."	Ever more and more in popular favor and demand.
Wins golden opinions from Painters and Decorators.	Ready for use with Cold Water. Simple and easy of application.
Order NOW, so as to be ready for the Spring rush.	Direct or through your jobber.

THE ALABASTINE CO., Limited

PARIS, ONT.

The Handy Fruit and Vegetable Slicer

A Quick Seller

from the word "go"

THE SENSATION OF THE DAY IN

SIMPLE IN OPERATION

MARVELLOUS IN RESULTS

Kitchen Furnishings

Full directions with each Slicer.

Ask your jobber for them.

NOTHING in recent years has been devised that adds so much to dainty cookery. It slices all kinds of Fruit and Vegetables in an infinite variety of fancy and attractive designs, absolutely

WITHOUT WASTE

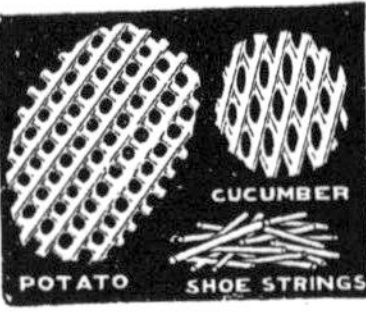

THESE

are a few of the designs, and there is absolutely no limit to the varieties possible to produce. No trick about it. Anyone can use it.

Made by

The Cumming Mfg. Co., Ltd.

Renfrew, Ont.

WINNIPEG: 608 Ashdown Bldg.

THE STANDARD IDEAL SANITARY CO., LIMITED

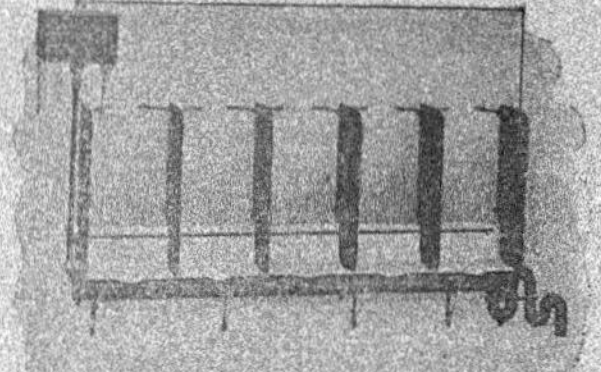

Plate E. 404.—For Public Institutions, Schools, Factories, Asylums, &c.

We manufacture a complete line of

Range-Closets, Range-Urinals, Lavatories and Sanitary Fixtures

Suitable for Public Institutions

The reputation of our goods is unequalled.

Toronto: 50 Colborne Street ***Montreal: 128 Craig Street West***
Winnipeg: 24 Telfer Building

HOME OFFICE and FACTORIES: PORT HOPE, ONT.

Plumbing and Steamfitting

THE PLUMBING OF A COTTAGE

By J. A. F. Cardiff in the National Builder, Chicago

Article XV.

With this instalment we will commence the consideration of the water supply system of a cottage.

Figure 60 represents a section of a modern cottage having the usual amount of plumbing fixtures, and shows a system of piping for the supply and distribution of cold water, the supply being taken from the city water main. In this case it is assumed that the pressure in the mains is ample to raise the water to the highest fixture in the building.

The illustration provides for a simple system of piping such as is commonly required. The street main is tapped and a one and one-half-inch service pipe is run in through the front foundation wall and through the cellar to the bases of the various risers. Just inside the front wall a stop and waste cock is provided to shut off the entire supply. All piping is graded to this cock so that the entire system may be drained dry. Two risers are carried up to the two stacks of fixtures and are provided with stop cocks at their bases so that either riser may be shut off independently of the other. These risers decrease in size as the various branches are taken off, and at the top of each, where marked "A," the pipe is extended so as to form an air chamber to prevent water hammer. A similar air chamber is provided near where the service pipe passes through the front wall.

Branches to all fixtures, except those in the attic, are provided with individual stop cocks so that any one fixture may be cut off for repairs, etc. The branch supply to the attic fixtures has one stop cock for the two fixtures. Stop cocks on all branches are very desirable, as they greatly facilitate repairs. They are particularly desirable on the supply to a water closet tank, as the ball cocks or tank valves generally require repairs more frequently than any other fixtures.

A more detailed description of the work will be given in future articles.

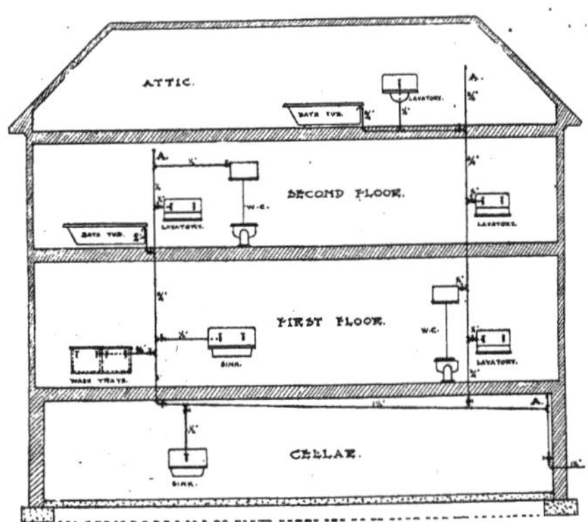

Fig 60

A NOTABLE SEWER CONNECTION.

For a number of years a hotel occupied largely during the summer season and equipped with some 250 fixtures, carried its sewage for some distance in a 10-in. pipe, where it was allowed to discharge without further attention. Recently the owners of the property on which the sewage was discharged notified the hotel owners that some other disposition must be made of the sewage. The old outlet was within about 600ft. of a sewer, and when the hotel owners called upon the plumber for a solution of their problem he found a right of way for a sewer connection and fortunately the ground sloped toward the sewer with rather a sharp fall in the last 150 or 200 feet. The item of expense was an important one, and to carry a 10-in. pipe would have entailed a considerable cost and such a connection would have been rather large in comparison with the size of the branch sewer with which it was to be connected.

Reducing the Pipe Size Successively.

The plumber was one of a class who studied the various problems in his field of enterprise and was a wide reader of the trade press and the books published on all phases of the plumbing question. From his reading he was acquainted with the Waring system of sewerage, particularly as applied to Memphis, Tenn., where the feature is the small size of the sewers used, says the Metal Worker. He was of the opinion that the 10-in. pipe could be reduced in size if it was done judiciously and a considerable saving in expense effected. He submitted his proposition to the hotel owners, and was given the order to try the experiment. He began by lowering the far end of the 10-in. sewer, and connected it by means of a tapering joint with an 8-in. pipe, which was carried some distance. Then this was reduced by means of a tapering joint to a 6-in. pipe, great care being taken to have the inside pipe at the joint where the reduction was made perfectly smooth, so that nothing could lodge to effect an accumulation which might eventually stop the pipe. After running some distance with a 6-in. pipe the piping was reduced to 4-in., this connection being made at the top of a sharp declivity, and the 4-in. pipe continued the remainder of the 600 feet, where it was connected with the sewer.

One Trouble With Stoppage.

This unusual sewer has now been in operation several years with only one complaint. A stoppage was noticed on account of the sewage backing up in the lowest fixtures of the building. Immediately the plumber's assitance was sought to discover the point where the stoppage was located and to remove it. He was of the opinion that this must be somewhere near the point of connection of the 6-in. pipe with the 4-in. pipe, probably at the bottom of the declivity where the 4-in. pipe began running with a slight fall. The sewer pipe was opened at this point and immediately spectators were entertained by the discharge of a fountain of filth much to the disgust of the workmen who made the opening. This experience gave evidence that the stoppage was further along the line. Without any means of discovering the exact point the trench was opened some distance beyond and fortunately immediately where the stoppage occurred. Here it was found that a small tin mustard box had passed through some plumbing fixture into the large sewer and passed through the various reduced mains and well along in the 4-in. line until by an unfortunate tilt it caught in the top of the sewer and became wedged fast and caused the accumulation which eventually stopped up the sewer. The obstruction was removed and the sewer line put in perfect repair, investigation showing that the velocity of the discharge had left the bore of the 4-in. sewer free from any accumulation and quite clean, as sewers go.

Since this one experience the sewer has been doing service for a few years, and up to the present time there has been no second complaint of an obstruction, notwithstanding that some fixture in the hotel is frequently found stopped and the usual variety of tooth brushes, hand scrubs, wash cloths, towels and other matters never intended to pass into the sewer compressed in one mass.

KING RADIATOR CO. SECURE SITE.

Arrangements have practically been completed for the organization of the King Radiator Co., with a capitalization of a quarter of a million dollars, with Cluff Bros., Toronto, as the chief promoters; associated with them being the Warden King interests, of Montreal. Fred Somerville, of the Ontario Lead & Wire Company, Toronto, acted for these interests in securing their new site. The promoters have ordered their machinery, and have their patterns already prepared; the most experienced man they could obtain in one of the large plants in the United States, having been secured as superintendent of the proposed works.

A site for the factory has been secured on the Ashbridge's Bay marsh, in the eastern part of Toronto, a lease having been recommended by the Board of Control on the land, comprising 3 acres, with 11 additional acres of water lots. The site is at the foot of Cherry street, and westward to the line of the proposed new concrete breakwater. About $150,000 will be spent during the next six months in the erection of foundry buildings, etc., and from 100 to 150 men will be employed in the new works. The matter has to be ratified by the city council before it is finally settled.

In addition to a full line of radiators, the new company may also manufacture boilers at the Toronto works, and it has been definitely decided that soil pipe and range boilers will be produced.

The location of this industry at Toronto will have an important bearing upon the steamfitting and heating trades, and Cluff Bros., the promoters, must be congratulated upon the rapid development of their business in Ontario, which has made possible the present enlargement. They have already one of the finest warehouses and office buildings on the continent, and with the foundry, which will be in operation by August 1st next, they will be splendidly equipped to look after the interests of their customers.

BRITISH COLUMBIA PLUMBERS BUSY.

Plumbers and stove dealers have had their innings in B.C. for the past fortnight. Business in these lines is usually very slack in the first month of the year, and the time is as a rule taken up with stock-taking. But the cold wave which has blown down from the north-east and slopped over in the whole Pacific Coast region, even to sunny California, has made the balmy climate of B.C. a dream for the present. Fuel is at a premium, not because of its absolute scarcity but because it is difficult to deliver it quickly enough. The unwonted icy weather has made the streets and hillside one glare, and the effectiveness of delivery teams is greatly reduced.

At New Westminster, where the coal supply is all brought in by barges up the Fraser River, the coal shortage has become alarming as the river is so badly frozen over it is no longer possible for steamers to break their way through, let alone tow barges of coal. At Nanaimo and Ladysmith, the big coal towns on the island, orders have been to stop the export of coal until local demands have been supplied. But even this order has not sufficed to deliver the coal as promptly as desired. In Vancouver, coal dealers do not undertake to deliver at all. And as to taking orders, the rule is to limit the amount to one person at one time, to a quarter-ton lot. Even at that the only way to get it is to secure a dray and send it after the coal. Bad roads in from the country have made it difficult to team in wood, and as nearly every sawmill on Burrard Inlet is now closed down, the supply of mill wood is out, so that the shortage while not serious, is a very great inconvenience temporarily. But with the thermometer never lower than 2.3 below zero, which was the lowest reached and that dip for only one night, the weather is not sufficient to alarm people used to or prepared for it.

* * *

The most serious phase of the situation is the condition of the water pipes connecting to a large percentage of the houses in Vancouver. In many places not sufficient protection has been given where the service pipe enters the building and freezing has been common. In several localities, notably where street grades have been cut down, the service connections have been left near the surface or exposed and the inevitable result has followed. In many places the lucky householder who has kept his water service unfrozen is the object of envy and the source of water supply for the whole neighborhood.

The electric thawing device so commonly used in the east has never before been required or brought into requisition in Vancouver. The superintendent of waterworks has, however, arranged with the electric company for a current and for the necessary machine for generating the heat, so that frozen mains are being thawed out. The electric company also advertises that private parties may have their service pipes thawed out for the charge of $5.00 in cases where the connection is all iron, and $8 where the connection is partly lead pipe.

There have been several instances where water heaters connected with kitchen ranges or furnaces have blown up through negligence on the part of people in control, fires being set in stoves before connecting pipes were thawed. No fatalities have occurred, but in one case, that of the home of the Children's Aid Society, three little chaps were badly hurt, one of them having a leg broken through the explosion of the water jacket.

It is needless to say that the plumber is the busiest man in the city at the present moment, and the most sought after. Day and night the staff of every plumbing shop is busy, and new work is simply at a standstill till the insistent demand for repairs is attended to. There is one feature of the disagreeable experience which will lead to benefit, and that is, it has brought into prominence defects in the system of plumbing allowed in the city, where due care is not taken to protect plumbing from cold weather, and in future provision will be made in the regulations for meeting the unexpected.

BORDEN COMPANY EXPANDING.

B. T. Borden, of the Borden Company, Warren, Ohio, expects to establish a factory shortly in Canada for the manufacture of pipe threading machinery and dies. The Borden Company has broken ground at Warren for the election of a new plant in that city, to be devoted exclusively to the manufacture of hand and power pipe threading machinery. The buildings will be constructed of brick and iron, and hence absolutely fireproof. They will be all one-storey high with the exception of the office building, which will be a two-storey structure. Work will be pushed along rapidly, as the company expects to have the plant ready for occupancy early in the spring. The new plant will be equipped with up-to-date machinery.

NEW INVENTIONS.

F. W. Kingsbury, Evansville, Ind., has secured two new U.S. patents, one being for an improvement in water-closet tanks, which has for its object to provide novel means for supporting the tank and for holding it by the plumbing connections in interlocked engagement with the supporting means. The construction dispenses with the expensive and objectionable back plate or board and brackets ordinarily employed. The other is for an improvement in water-closet seats, and has for its object the provision of a seat which will present no unusual appearance, will be strong and durable, and will have no sockets or other openings in its exposed faces to be filled by putty, litharge, cement, or the like.

J. E. H. Paddon, Montreal, has secured a Canadian patent on a new waste pipe trap.

METAL THIEVES AT MORRISON'S.

Detectives captured two thieves in the premises of the James Morrison Brass Manufacturing Co., Toronto, on Wednesday night. Several hundred pounds of copper and brass had been stolen and the police got wind that another haul was to be made.

PLUMBING MARKETS

MONTREAL.

Office of HARDWARE AND METAL,
232 McGill Street,
Montreal, February 1, 1907

"Nothing to report this week" is the cry that greeted our representative on his weekly visit to Montreal's largest plumbing supply houses, and the gentlemen from whom we gather our information, in all cases, seemed to have an unusual desire to be back among their papers, which goes to prove that the banner year, which we predicted last week, is beginning to command their attention already. In fact it is only necessary for an "outsider" to step inside the door of any large house to see for himself, without asking any questions, that there's something doing.

The markets remain very firm, with no changes received—or expected as yet—over last week's quotations. Orders continue to pour in for spring shipment which goes to show that the plumber is a wise man and knows how to anticipate his wants, as well as anybody else.

RANGE BOILERS—The expected advance in the copper article is yet a thing of the future, but as copper is reported short from all quarters, it will assuredly come sooner or later. For the present, out prices remain: Iron clad, 30 gallon, $5; 40 gallon, $6.50 net list. Copper, 30 gallon, $33; 35 gallon, $38; 40 gallon, $43; 45 gallon, $50; 50 gallon, $58; 60 gallon, $68; 70 gallon, $80; 80 gallon, $95; 90 gallon, $115; 100 gallon, $150 net.

LEAD PIPE—Very firm. No change in prices. We still quote 5 per cent. discount f.o.b. Montreal, Toronto, St. John, NfB., Halifax; f.o.b. London, 15c. per hundred lbs extra; f.o.b. Hamilton, 10c per hundred lbs. extra.

IRON PIPE FITTINGS—A very brisk business is being done at the same figures. We still quote: Discounts on nipples, ¼ inch to 3 inch, 75 per cent., 3½ inch to 2 inch, 55 per cent.

IRON PIPE — We still quote: Standard pipe in lots of 100 feet, regular lengths, ¼-inch, $5.50; ⅜ inch, $5.50; ½ inch, $8.50; ¾ inch, 11.50; 1 inch, $16.50; 1 1-4 inches, $22.50; 1 1-2 inches, $27.00; 2 inches, $36.50, discounts on black pipe, ¼ inch, 59 per cent.; ⅜ in. 59 per cent.; ½ in. 68; ¾ to 2, 70. Discounts on galvanized pipe: ¼ inch, 44 per cent.; ⅜ inch, 44 per cent.; ½ inch, 58 per cent.; ¾ to 2 inch, 60 per cent. Extra heavy pipe of 100 feet lots are quoted as follows: 1-2 inch, $12; 3-4 inch, $15; 1 inch, $22; 1 1-4 inches, $30; 1 1-2 inches, $36; 2 inches, $50. The discounts on black pipe, ½ inch, 74 per cent.; ¾ inch, 69 per cent.; 1 inch to 2 inches, 68 per cent. Galvanized, ½ inch, 59 per cent.; ¾ inch, 69 per cent.; 1 to 2 inches 58 per cent.

SOIL PIPE AND FITTINGS—Our prices remain as follows: Standard soil pipe, 50 per cent. off list. Standard fittings, 50 and 10 per cent. off list; medium and extra heavy soil pipe, 60 per cent. off. Fittings, 60 per cent. off.

SOLDER—Prices remain the same: Bar solder, half-and-half, guaranteed 25c.; No. 2 wiping solder, 22c.

ENAMELWARE—A good business is being done at the same prices. We continue to quote: Canadian baths, see Jan. 3 1907 list. Lavatories, discounts, 1st quality, 30 per cent.; special, 30 and 10 per cent. Sinks 18 x 30 inch, flat rim, 1st quality, $2.60, special, $2.45.

AN EASY QUESTION.

Send in a reply to the question by March 1, and take a chance after the valuable prize offered.

If a customer asked the following question: "What are the advantages of hot water heating over steam heating in a private residence?" what arguments would you use in reply? Also outline what advantages steam heating possesses over water for this same class of work.

* * *

The above is the third of a series of questions Hardware and Metal is asking the trade in Canada to answer. Some interesting replies were received to the preceding questions and the simplicity of the above should encourage a large number of replies, the publication of which should be helpful to readers of this paper who engage in heating work.

For the best reply we offer a clothbound copy of J. J. Cosgrove's splendid text-book "Principles and Practice of Plumbing."

TORONTO.

Office of HARDWARE AND METAL,
10 Front Street East,
Toronto, Feb 1, 1907

The approach of Feb. 4th is looked upon with considerable interest by jobbers of heating goods, this being practically the big settling day of the year. There has been a tremendous volume of work done in all parts of Ontario during the past year, and with the exception of Toronto, good prices have been secured. In the Queen City, however, there has been considerable cutting, and while it is not expected that any firms will go under, some of the price-cutters will have some difficulty in meeting their payments.

The trade in both heating goods and plumbing supplies is rather slow, as the jobbers are not very anxious to book orders for future shipment owing to the expectation of higher prices. Plumbers are learning to look farther ahead than formerly, however, and from now on, will place orders considerably ahead of the time when they will require their shipments.

A few slight readjustments have been made in prices, but no big lines have changed. Some houses are advancing Jenkins air vents from $5.00 to $5.50, and Kerr Webber straightway valves are being advanced from 42½ to 40 per cent. Soldering irons are also quoted higher, and plumbers' oakum is firm at $4.75. American enamelware has been advanced, but the new lists are not yet available. It is certain, however, that the new figures will be $1.00 to $2.00 higher on each bath tub.

LEAD PIPE — The discount on lead pipe continues at 5 per cent. off the list price of 7c. per pound. Lead waste, 8c. per pound with 5 off. Caulking lead 5 3-4c. to 6c. per pound. Traps and bends, 40 per cent., discount.

SOIL PIPE AND FITTINGS — Recent advances are held to firmly with demand seasonable. Medium and extra heavy pipe and fittings, 60 per cent.; light pipe 50 per cent.; light fittings, 50 and 10 per cent.; 7 and 8 inch pipe, 40 and 5 per cent.

IRON PIPE.—We are revising last week's figures on large sizes. One-inch black pipe is quoted at $4.95, and one-inch galvanized at $6.60. Full list in current market quotations.

IRON PIPE FITTINGS—Cast iron fittings show several advances. Cast iron elbows, tees, crosses, etc., 60 per cent.; cast iron plugs and bushings, 60 per cent.; flange unions, 60 per cent.; nipples, 60 and 10 per cent.; iron cocks, 55 and 5 per cent.; Canadian malleable, 30 per cent.; malleable unions, 55 and 5 per cent.; malleable bushings, 55 per cent.; cast iron ceiling plates, plain 65 per cent.; cast iron floor, 70 per cent.; hook plates, 60 per cent.; expansion plates, 65 per cent.; headers, 60 per cent, hangers, 65 per cent.; standard list.

RANGE BOILERS—An advance of 25 cents has been made on all sizes of galvanized iron and we now quote: Galvanized iron, 30 gallon, standard, $5; extra heavy, $7.75; 35 gallon, standard, $6; extra heavy, $8.75; 40 gallon, standard, $7; 40 gallon, extra heavy, $9.75, net list. Copper range boilers—New lists quote: 30 gallon, $33; 35 gallon, $38; 40 gallon, $43. Discounts 5 to 15 per cent.

RADIATORS—Prices are very stiff at. Hot water, 47½ per cent.; steam, 50 per cent.; wall radiators, 45 per cent.: specials, 45 per cent. Hot water boilers are unchanged.

SOLDER — Quotations remain firm as follows: Bar solder, half-and-half, guaranteed, 27c.; wiping, 23c.

ENAMELWARE—New lists on enameled baths issued by the Standard Ideal Company on Jan. 3, show a 10 per cent. advance. Lavatories, first quality, 20 and 5 to 20 and 10 off; special, 20 and 10 to 30 and 2½ per cent. discount. Kitchen sinks, plate 300, firsts 60 and 10 off; specials, 65 and 5 per cent. Urinals and range closets, 15 off. Fittings extra.

THE NEW

"SOLID-ADJUSTABLE" HAND STOCK AND DIES

At prices that will warrant your inquiries.

I. E. SHANTZ & CO.

BERLIN, - ONTARIO

Selling Agents for Canada

This is the Mark that Means Something

It is the distinctive brand of genuine Armstrong Stocks and Die and Pipe Fitters' Tools.

No "Armstrong pattern" or "just as good as Armstrong" have this stamp. Don't buy anything in this line that hasn't this brand as it is the mark of genuineness.

THE ARMSTRONG MANUFACTURING CO.

292 Knowlton Street, Bridgeport, Conn.

RELY ON ME

Hundreds of dealers do, as far as ENAMELWARE wants are concerned, because they have proven it good business policy to do so. Orders are attended to with care and dispatch. The Quality of the goods I deliver has to be disputed for the first time.

ORLANDO VICKERY

178-180 Victoria Street TORONTO, CANADA

Threads—Just as smooth as smooth can be.
Workmanship—First-class.
Materials—The Best Procurable.

These brief statements tell the whole story of

"Diamond" brand Fittings

We would like to mail you an interesting catalogue.

OSHAWA STEAM & GAS FITTINGS CO., LIMITED

OSHAWA, CANADA

Sharratt & Newth's Glaziers' Diamonds

are unequalled for cutting and wearing qualities.

To be obtained from the principal Dealers in Glass, Hardware, and Painters' Supplies.

Contractors to H. M. Government and the principal English Sheet and Plate Glass Works

World's Best Cutlery

You make no mistake in specifying goods made by **JOSEPH RODGERS & SONS, Limited** when ordering cutlery. A reputation of over 200 years at stake, insures the quality of every article.

Write for catalogues from Sole Canadian Agents

JAMES HUTTON & COMPANY, MONTREAL

A Certain Sense of Satisfaction

goes with every MERRELL PORTABLE HAND MACHINE, PIPE THREADING and CUTTING MACHINE. We build them that way.

For doing ALL kinds of threading—under ALL conditions—labor trouble and faulty material included—where speed, durability and ease of control count for something, our word to you is to investigate the MERRELL.

Whether it be stationary or portable—hand or power driven—30 DAYS FREE TRIAL must prove conclusively that your choice should be the MERRELL.

This Portable Hand Machine has encased gears, MERRELL standard quick opening and closing die-head, and the latest improved Cutting-off Knife.

Let us tell you more about this machine—The Chasers, Vise and the large range of work covered.

Catalogue for the asking.
Quick shipments.

THE CANADIAN FAIRBANKS CO.

Sole Agents for Canada — Limited

MONTREAL, TORONTO, WINNIPEG, VANCOUVER

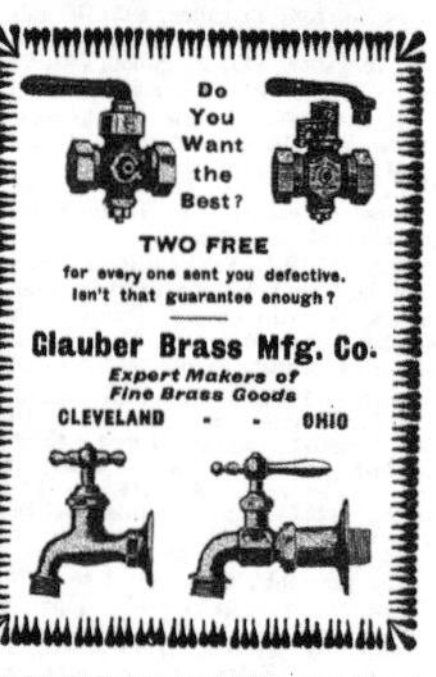

Forwell Foundry Co.

BERLIN, ONT.

Manufacturers of

SOIL PIPE, FITTINGS, and CAST IRON SINKS

Ask Jobbers for "F. F. CO." Brand

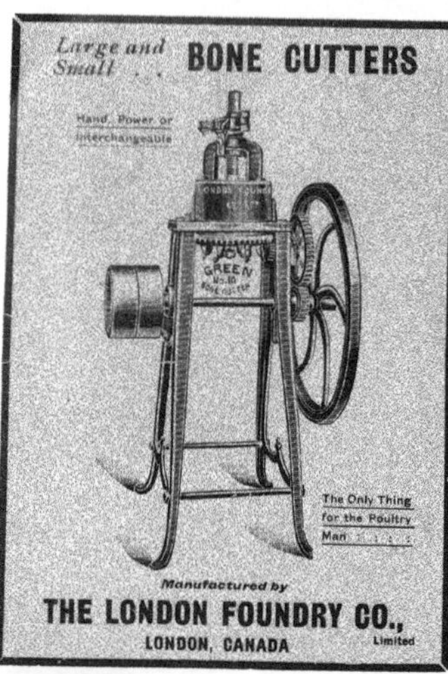

The Ever Ready Dry Battery

FOR AUTOMOBILE and

GASOLINE ENGINE USE

Write for Prices

JOHN FORMAN

248 and 250 Craig St. W., MONTREAL, - Que.

HOTTEST ON EARTH

Our Bonanza Furnaces are known to produce more heat with less fuel than any other make. Jobbers sell at factory prices. Satisfaction guaranteed or money refunded.

No. 38—Price, each, ***$7.80*** net.
No. 39—One size smaller, ***$5.85*** net.

THE TURNER BRASS WORKS

53 MICHIGAN STREET, - - - CHICAGO, U.S.A.

LEGAL CARDS

ATWATER, DUCLOS & CHAUVIN
Advocates, Montreal
Albert W. Atwater, K.C., Consulting Counsel for City of Montreal. Chas. A. Duclos. Henry N. Chauvin.

ROBINSON & GREEN
Barristers, Solicitors, Etc.
John A. Robinson, John R. Green, Solicitors for the Imperial Bank of Canada, the Southern Loan & Savings Co., St. Thomas, Ont.

BOY WANTED

ANY BOY in every town and village in Canada can earn many dollars by selling our publications among his neighbors and relatives and business men.

YOU can begin at once. No money needed to start. We will provide you with capital. A watch to the boy who does good work, as special prize.

Write at once to

THE MACLEAN PUBLISHING CO.
10 Front St. E., TORONTO

CURRENT MARKET QUOTATIONS.

Feb. 1, 1907.

These prices are for such qualities and quantities as are usually ordered by retail dealers on the usual terms of credit, the lowest figures being for larger quantities and prompt pay. Large cash buyers can frequently make purchases at better prices. The Editor is anxious to be informed at once of any apparent errors in this list, as the desire is to make it perfectly accurate.

—

METALS.

ANTIMONY.

Cookson's ....per lb... 0 27½ 0 28

BOILER AND T.K. PITTS.

Plain tinned .... } 25 per cent. off list.
Spun.................. }

BABBIT METAL.

Canada Metal Company—Imperial genuine, 60c.; Imperial Tough, 60c; White Brass, 50c.; Metallic, 35c.; Harris Heavy Pressure, 25c.; Heroules, 25c.; White Bronze, 15c.; Star Frictionless, 14c; Alluminoid, 10c; No. 4, 9c. per lb.

BRASS.

Rod and Sheet, 14 to 30 gauge, net list.

Sheets, 12 to 14 in.	0 27
Tubing, base, per lb. ⅛ to 2 in	0 33
Tubing, 1 to 3-inch, iron pipe size..	0 31
" 1 to 3-inch, seamless..	0 36

Copper tubing, 6 cents extra

COPPER.

	Ingot.	Per 100 lb.
Casting, car lots............	26 50	27 00
Bars.		
Cut lengths, round, ½ to 2 in...	....	32 00
Sheet.		
Plain, 16 oz., 14x48 and 14x60	....	31 00
Plain, 14 oz....................	....	32 00
Tinned copper sheet, base	....	34 00
Planished base................	....	38 00
Braziers' (in sheets), 4x6 ft., 25 to 30 lb. each, per lb., base..	....	0 30

BLACK SHEETS.

	Montreal.	Toronto
8 to 10 gauge	2 70	2 70
12 gauge......................	2 70	2 60
14 "	2 70	2 55
17 "	2 50	2 65
18 "	2 50	2 65
20 "	2 50	2 65
22 "	2 50	2 75
24 "	2 55	2 80
26 "	2 65	2 85
28 "	2 70	3 00

CANADA PLATES.

Ordinary, 52 sheets	2 75	2 90
All bright "	3 80	3 90
Galvanized, 52 sheets	4 35	4 45
" 60 "	4 60	4 70

	Ordinary.	Dom. Crown
18x24x52	4 25	4 35
" 60....................	4 50	4 60
20x28x80	8 50	8 70
" 94....................	9 00	9 20

GALVANIZED SHEETS.

	Fleur-de-Lis.	Gordon Crown.
16 to 20 gauge	3 80	3 95
22 to 24 gauge	4 05	4 00
26 "	4 30	4 20
28 "	4 55	4 50

Apollo.

10¾ oz (American gauge)........	....	4 85
28 gauge.......... "	....	4 55
26 " .... "	4 15	4 30
24 " .... "	3 90	4 30

	Comet	Queen's Head.	Bell.
16 to 20 gauge	3 60	3 95	....
22 to 24 gauge	3 75	4 20	....
26 "	4 30	4 45	....
28 "	4 45	4 70	4 45

Less than case lots 10 to 25c. extra.

IRON AND STEEL.

	Montreal.	Toronto.
Common bar, per 100 lb......	2 15	2 30
Forged iron "	2 40	.
Refined " "	2 55	2 70
Horseshoe iron "	2 55	2 70
Hoop steel, 1½ to 3 in. base...	....	2 80
Sleigh shoe steel "	2 25	2 30
Tire steel	2 40	2 50
Best sheet cast steel..........	...	0 12
B. K. Morton "Alpha" high speed.		0 65
" annealed	....	0 70
"M" Self-hardening.........	....	0 50
"J" quality, best warranted	....	0 18
"I" " warranted......	....	0 14
"B.C" quality	....	0 09
Jonas & Colver's tool steel...	0 10	0 20
" "Novo".......	....	0 65
" " annealed	....	0 65
Jowett & Sons B P.L. tool steel....		0 10½

INGOT TIN.

Lamb and Flag and Straits—
56 and 28-lb. ingots, 100 lb. $45 50 $46 00

TINPLATES.

Charcoal Plates—Bright

M.L.S., equal to Bradley—		Per box.
I C, 14 x 20 base	....	$6 50
I X, 14 x 20 "	....	8 00
IXX, 14 x 20 base	....	9 50
Famous, equal to Bradley—		
I C, 14 x 20 base..........	....	6 50
I X, 14 x 20 "	....	8 00
I X X, 14 x 20 base..........	....	9 50
Raven and Vulture Grades—		
I C, 14 x 20 base	....	5 00
I X "	....	6 00
I X X "	....	7 00
I X X X "	....	8 00

"Dominion Crown Best"—Double Coated, Tissued.

		Per box.
I C, 14 x 20 base..........	5 50	5 75
I X, 14 x 20 "	6 50	6 75
IXX x 20 "	7 50	7 75

"Allaway's Best"—Standard Quality.

I C, 14 x 20 base..........	....	4 50
I X, 14 x 20 "	....	5 25
I XX, 14 x 20 "	....	6 00

Bright Cokes.

Bessemer Steel—		
I.C., 14 x 20 base	...	4 25
20x28, double box	....	8 50

Charcoal Plates—Terne

Dean or J. G. Grade—		
I C., 20x28, 112 sheets	7 25	8 00
IX., Terne Tin	...	9 50

Charcoal Tin Boiler Plates.

Cookley Grade—
X X, 14x56, 50 sheet bxs }
" 14x60, " } 7 50
" 14x65, " }

Tinned Sheets.

72x30 up to 24 gauge...... ...	....	8 50
" " 26 "	....	9 00

LEAD.

Imported Pig, per 100 lb......	5 50	5 60
Bar, "	5 75	6 00
Sheets, 2½ lb. sq. ft., by roll ..	....	0 06½
Sheets, 3 to 6 lb. "	....	0 06

NOTE.—Cut sheets ½c. per lb., extra. Pipe, by the roll, usual weights per yard, lists at 7c. per lb. and 5 p.c. dis. f.o.b. Toronto.

NOTE.—Cut lengths, net price, waste pipe 8-ft lengths, lists at 8c.

SHEET ZINC.

5-cwt. casks	8 00	8 25
Part casks....................	8 25	8 50

ZINC SPELTER.

Foreign, per 100 lb	7 50	7 75
Domestic	7 00	7 25

—

PLUMBING AND HEATING

BRASS GOODS, VALVES, ETC.

Standard Compression work, dis half inch 50 per cent., others 40 per cent.

Cushion work, discount 40 per cent.

Fuller work, ½-inch 60 per cent., others 50 p.c.

Flatway stop and stop and waste cocks, 50 per cent.; roundway, 45 per cent.

J.M.T. Globe, Angle and Check Valves, discount 45 per cent.

Standard Globe, Angle and Check Valves discount 52½ per cent.

Kerr standard globes, angles and checks, special, 42½ per cent.; standard, 47½ p c.

Kerr Jenkins' disc, copper-alloy disc and heavy standard valves, 45 per cent.

Kerr steam radiator valves, 60 p.c., and quick-opening hot-water radiator valves, 60 p.c.

Kerr brass, Weber's straightway valves, 40 per cent.; straightway valves, I B.B.M., 60 per cent.

J. M. T. Radiator Valves, discount 50 per cent

Standard Radiator Valves, 60 per cent.

Patent Quick-Opening Valves, 65 per cent.

Jenkins' Bros. Globe Angle and Check Valves discount 32½ per cent.

No. 1 compression bath cock......net		2 00
No. 4 " " ...... "		1 90
No 7 Fuller's.............. .... "		2 25
No. 4½, " .... "		2 35

Patent Compression Cushion, basin cock, hot and cold, per doz., $16.20

Patent Compression Cushion, bath cock, No. 2208 .... 2 25

Square head brass cocks, discount 50 per cent.
" " iron " " 60 "

Thompson Smoke-test Machine $25.00

BOILERS—COPPER RANGE.

Copper, 30 gallon............	33 00
" 35 "	38 00
" 40 "	43 00

5 to 15 per cent discount

BOILERS—GALVANIZED IRON RANGE.

Capacity	Standard.	Extra heavy
30-gallons ...	5.00	7.75
35 "	6 00	8 75
40 "	7 00	9.75

2 per cent., 30 days

BATH TUBS.

Steel clad copper lined, 15 per cent.

CAST IRON SINKS.

16x24, $1; 18x30, $1; 18x36, $1.30.

ENAMELED CLOSETS AND URINALS

Discount 15 per cent.

ENAMELED BATHS.

List issued by the Standard Ideal Company Jan 3, 1907, shows an advance of 10 per cent. over previous quotations.

ENAMELED LAVATORIES.

	1st Quality.	Special.
Plate E 100 to E 103 ..	20 & 5 p.c.	20 & 10 p.c.
" E 104 to E 132....	20 & 10 p.c.	30 & 2½ p.c.

ENAMELED ROLL RIM SINKS.

	1st quality.	Special.
Plate E 901, one piece.	15 & 2½ p.c.	15 & 10 p c.

ENAMELLED KITCHEN SINKS.

Plate E, flat rim 300, 60 & 10 p.c. 65 & 5 p.c

HEATING APPARATUS.

Stoves and Ranges—Discounts vary from 40 to 70 per cent. according to list.

Furnaces—45 per cent.

Registers—70 per cent.

Hot Water Boilers—Discounts vary.

Hot Water Radiators—47½ per cent.

Steam Radiators—50 per cent.

Wall Radiators and Specials—45 per cent.

LEAD PIPE

Lead Pipe, 7c. per pound, 5 per cent. off.

Lead waste, 8c. per pound, 5 per cent. off.

Caulking lead, 6c. per pound.

Traps and bends, 40 per cent.

IRON PIPE.

Size (per 100 ft.)	Black.		Galvanized
[illegible] inch........	2 25	[illegible] inch........	3 08
[illegible] "	2 25	[illegible] "	3 08
[illegible] "	2 72	[illegible] "	3 57
[illegible] "	3 45	[illegible] "	4 60
1 "	4 95	1 "	6 60
1¼ "	6 75	1¼ "	9 00
1½ "	8 10	1½ "	10 80
2 "	10 80	2 "	14 40
2½ "	18 00	2½ "	24 00
3 "	23 50	3 "	31 50
3½ "	30 00	3½ "	40 00
4 "	34 00	4 "	45 00

2 per cent. 30 days.

Malleable Fittings—Canadian discount 30 per cent.; American discount 25 per cent.

Cast Iron Fittings 60%; Standard bushings 60 per cent.; headers, 60; flanged unions 60, malleable bushings 55; nipples, up to 3 in., 65 and 10 per cent.; up to 6 in., 65 per cent.; malleable lipped unions, 55 and 5 per cent.

SOIL PIPE AND FITTINGS

Medium and Extra heavy pipe and fittings, up to 6 inch, discount 60 per cent.

7 and 8-in. pipe, discount 40 and 5 per cent.

Light pipe, 50 p.c.; fittings, 50 and 10 p.c.

OAKUM.

Plumbers ... per 100 lb... .. .... 4 75

STOCKS AND DIES.

American discount 25 per cent.

SOLDERING IRONS.

½-lb.	per lb.	....	0 38
1-lb. or over	"	....	0 35

SOLDER.

	Montreal	Per lb. Toronto
Bar, half-and-half, guaranteed	0 25	0 27
Wiping........................	0 22	0 23

—

PAINTS, OILS AND GLASS.

BRUSHES

Paint and household, 70 per cent..

COLORS IN OIL.

Venetian red, 1-lb. tins	pure.	....	0 08
Chrome yellow	"	....	0 15
Golden ochre	"	....	0 08
French "	"	....	0 08
Marine black	"	....	0 04½
Chrome green	"	....	0 10
French permanent green	"	....	0 13
Signwriters' black	"	....	0 15

GLUE.

Domestic sheet	0 10	0 10½
French medal..	0 12	0 14

PARIS GREEN.

	Berger's	Canadian
600-lb. casks	0 24½	0 24½
250 lb. drums	0 24½	0 24½
100-lb. "	0 25	....
50-lb. "	0 25	0 25½
1-lb. pkgs, 100 in box	0 26	0 26½
½-lb. "	0 28	0 28½
1-lb. tins, 100 in box	0 27	0 27½
½ lb pkg	....	0 29

PREPARED PAINTS.

Pure, per gallon, in tins	....	1 20
Second qualities, per gallon	....	1 00
Barn (in bbls.)	0 80	0 90
Sherwin-Williams paints, ½ gal.	....	1 50
" " ¼ gal.	....	1 45
" " gal.	....	1 40
Canada Paint Co.'s pure	....	1 30
Standard P. & V. Co.'s "New Era."	....	1 30
Benj. Moore Co.'s "Ark" B'd	....	1 25
" British Navy deck	....	1 50
Brandram-Henderson "Anchor"	....	1 35
Ramsay's paints, Pure, per gal.	....	1 20
" Thistle, "	....	1 00
" Outside, bbls	0 55	0 65
Martin-Senour's 100 p.c. pure, ½ gal.		1 50
" " " ¼ gal.		1 45
" " " 1 gal.		1 40
Senour's Floor Paints.....gal.	....	1 20
Jamieson's "Crown and Anchor"		1 20
Jamieson's floor enamel	....	1 50
" barn paints, bbls per gal.		0 60
Sanderson, Pearcy's, pure	....	1 20
Robertson's pure paints	....	1 20

PUTTY

Bulk in bbls.	1 50
Bulk in less quantity	1 80
Bladders in bbls	1 80
Bladders in kegs, boxes or loose	1 95
25-lb. tins	1 80
12½ lb. tins	2 05
Bladders in bulk or tins less than 100 lb.	1 85
Bulk in 100-lb. irons	1 80

SHELLAC.

Shellac, bleached, in bars or ground, 46c. per lb., f.o.b. Eastern Canadian points

Shellac, bone dry, 57c. per lb., f.o.b., Eastern Canadian Points; 63c, Toronto.

Shellac, T. N. orange, etc., 48c. net, Toronto

TURPENTINE AND OIL.

Benzine, per gal	....	0 20
Turpentine single barrels	0 98	1 05
Linseed Oil, " raw	0 55	0 59
" " boiled	0 58	0 62
W. H. & C's English raw oil, bbls		0 63
" " boiled	....	0 66

WHITE LEAD GROUND IN OIL. Per 100 lbs.

Pure	7 25	7 50
No. 1	6 90	7 15
No. 2	6 55	6 90
No. 3	6 30	6 55
Munro's Select Flake White	....	7 50
Elephant and Decorators' Pure	....	7 25
Monarch	....	7 40
Standard Decorator's	....	7 15
Essex Genuine	....	6 80
Brandram's B. B. Genuine	....	8 25
" Anchor," pure	....	6 50
Ramsay's Pure Lead	....	6 40
Ramsay's Exterior	....	6 15
"Crown and Anchor," pure	....	6 50
Henderson Pearcy's	....	7 40
Robertson's C P., lead	....	7 21
W H. & C' matured pure English	8 00	8 25

WHITE AND RED DRY LEAD.

	white	red.
Genuine, 560 lb. casks, per cwt	6 75	6 00
Genuine, 100 lb. kegs, "	7 50	6 25
No. 1, 560 lb. casks, per cwt	....	5 75
No. 1, 100 lb. kegs, per cwt	....	6 00

WINDOW GLASS.

Size United Inches.	Star	Double Diamond
Under 26	$4 25	$6 25
26 to 40	4 65	6 75
41 to 50	5 10	7 50
51 to 60	5 35	8 50
61 to 70	5 75	9 75
71 to 80	6 25	11 00
81 to 85	7 00	12 50
86 to 90	....	15 00
91 to 95	....	17 50
96 to 100	....	20 50
101 to 105	....	24 00
106 to 110	....	27 50

Discount—16-oz., 75 per cent.; 21 oz 20 per cent. per 100 feet. Broken boxes 50 per cent. off list.

ZINC—DRY WHITE.

Extra Red Seal	0 07	0 08
French V. M.	0 06	0 25
Lehigh	0 05	0 06

ZINC—GROUND WHITE.

Pure	0 08½	0 09½
No. 1	0 07	0 08
No. 2	0 05½	0 06½

VARNISHES.

In 5-gal. lots.	Per gal.	Net.
Carriage, No. 1	1 50	1 60
Pale durable body	4 10	4 25
" rubbing	2 85	3 20
Gold size, japan	1 50	1 60
No. 1 brown japan	0 85	0 90
Elastic oak	....	1 50
Furniture, extra	1 10	1 25
" No. 1	0 90	1 00
Hard oil finish	1 35	1 50
Light oil finish	1 60	1 70
Damar	1 75	2 00
Brown japan	1 10	1 20
Black japan	1 10	1 20
" No. 1	0 85	0 90
Elastilite varnish, 1 gal. can, each		2 00
Granitine floor varnish, per gal		2 10
Maple Leaf coach enamels; size 1, size 2, 70c.; size 3, 40c. each.		1 20
Sherwin-Williams' kopal varnish, gal.		2 50
Canada Paint Co's sun varnish		2 00
"Kyanize" Interior Finish		2 40
"Flint-Lac," coach		1 80
"Gold Medal," per gal., in cases		2 00
Jamieson's Copaline, per gal.		2 00
Sanderson Pearcy's amberine		2 50

BUILDERS' HARDWARE.

BELLS.

Brass hand bells, 60 per cent.

Nickel, 55 per cent.

Gongs, Sargeant's door bells	5 50	8 00
American, house bells, per lb.	0 35	0 40

Peterboro' door bells, discount 37½ and 10 per cent. off new list.

BUILDING PAPER, ETC.

Tarred Felt, per 100 lb.	2 25
Ready roofing, 2-ply, not under 45 lb. per roll	1 00
Ready roofing, 3-ply, not under 65 lb., per roll	1 25
Carpet Felt.....per ton	60 00
Heavy Straw Sheathing.....per ton	35 00
Dry Surprise	0 45
Dry Sheathing.....per roll, 400 sq. ft.	0 50
Tar " " 400 "	0 45
Dry Fibre " 400 "	0 55
Tarred Fibre " 400 "	0 65
O. K. & I. X. L. " 400 "	0 70
Resin-sized " 400 "	0 45
Oiled Sheathing " 600 "	1 00
Oiled " " 400 "	0 70
Roof Coating, in barrels.....per gal.	0 17
Roof " small packages "	0 25
Refined Tar.....per barrel	5 00
Coal Tar "	4 00
Coal Tar, less than barrels.....per gal.	0 15
Roofing Pitch.....per 100 lb. 0 80	0 90
Slater's felt.....per roll	0 70
Heavy Straw Sheathing f. o. b. St. John and Halifax	37 50

BUTTS.

Wrought Brass, net revised list.

Wrought Iron, 70 per cent.

Cast Iron Loose Pin, discount 60 per cent.

Wrought Steel Fast Joint and Loose Pin, 10 per cent.

CEMENT AND FIREBRICK.

"Lafarge" cement in wood		3 40
"Lehigh" cement, in wood		2 54
"Lehigh" cement, cotton sacks		2 39
"Lehigh" cement, paper sacks		2 31
Fire brick (Scotch) per 1,000	15 00	24 00
Fire clay (Scotch), net ton		4 93

Paving Blocks per 1,000.

Blue metallic, 9"x4½"x3", ex wharf	35 00
Stable pavers, 12"x6"x2", ex wharf	50 00
Stable pavers, 9"x4½"x", ex wharf	36 00

DOOR SETS.

Peterboro, 37½ and 10 per cent.

DOOR SPRINGS.

Torrey's Rod.....per doz.	....	1 75
Coil, 9 to 11 in. "	0 95	1 65
English "	2 00	4 00

Chicago and Reliance Coil Door Springs, 25 per cent.

STORE DOOR HANDLES.

Per Dozen	1 00	1 50

ESCUTCHEONS.

Discount 50 and 10 per cent., new list

Peterboro, 37½ and 10 per cent.

ESCUTCHEON PINS.

Iron, discount 40 per cent.

Brass, 45 per cent.

HINGES.

Blind, discount 60 per cent.

Heavy T and strap, 4-in., per lb. net	0 06¼
" " 5-in., "	0 06
" " 6-in., "	0 05¾
" " 8-in., "	0 05½
" " 10-in. and larger	0 06¼

Light T and strap, discount 65 p.c.

Screw hook and hinge—

under 12 in.... per 100 lb.	4 75
over 12 in. "	3 75
Spring, No. 20, per gro. pairs	10 80

Spring, Woodyatt pattern, per gro., No. 5, $17.50 No. 10, $18; No. 20, $10 80; No. 12 $20; No. 51, $10; No. 50, $27 50.

Crate hinges and back flaps, 65 and 5 p. c

Hinge hasps, 65 per cent.

SPRING HINGES.

Chicago Spring Butts and Blanks 12½ per cent.

Triple End Spring Butts, 40 and 5 per cent.

Chicago Floor Hinges, 40 and 5 off.

Garden City Fire Engine House Hinges, 12½ per cent.

HOOKS.

Cast Iron.

Bird cage.....per doz.	0 50	1 10
Clothes line, No. 61 "	0 00	0 70
Harness "	0 60	12 00
Hat and coat.....per gro.	1 10	10 00
Chandelier.....per doz.	0 50	1 00
Wrought hooks and staples—		
¼ x 5 .....per gross	....	2 65
5-16 x 5 "	....	3 30

Bright steel gate hooks and staples, 40 per cent discount

Hat and coat wire, discount 62½ per cent.

Screw, bright wire, discount 65 per cent.

KNOBS.

Door, japanned and N.P., doz	1 50	2 50
Bronze, Berlin.....per doz.	2 75	3 25
Bronze, Genuine "	6 00	9 00
Shutter, porcelain, F. & L. screw.....per gross	1 30	2 00
White door knobs.....per doz.	....	2 00

Peterboro knobs, 37½ and 10 per cent.

Porcelain, mineral and jet knobs, net list.

KEYS.

Lock, Canadian dis. 40 to 40 and 10 per cent

Cabinet trunk and padlock American.....per gross	0 60

LOCKS.

Peterboro 37½ and 10 per cent

Russell & Erwin, steel rim $2 50 per doz

Eagle cabinet locks, discount 30 per cent

American padlocks, all steel, 10 to 15 per cent; all brass or bronze, 10 to 25 per cent.

SAND AND EMERY PAPER.

B. & A. sand, discount, 35 per cent

Emery, discount 35 per cent.

Garnet (Burton's). 5 to 10 per cent. advance on list.

SASH WEIGHTS.

Sectional.....per 100 lb.	2 00	2 25
Solid "	1 50	1 75

SASH CORD.

Per lb.	....	0 31

BLIND AND BED STAPLES

All sizes, per lb.	0 07½	0 12

WROUGHT STAPLES.

Galvanized	....	2 75
Plain	....	2 50

Coopers', discount 45 per cent.

Poultry netting staples, discount 40 per cent.

Bright spear point, 75 per cent. discount.

TOOLS AND HANDLES.

ADZES.

Discount 22½ per cent.

AUGERS.

Gilmour's, discount 60 per cent. off list.

AXES.

Single bit, per doz.	5 50	8 50
Double bit, "	10 00	11 00

Bench Axes, 40 per cent.

Broad Axes, 25 per cent.

Hunters' Axes	5 50	6 00
Boys' Axes	6 25	7 00
Splitting Axes	7 00	12 00
Handled Axes	7 00	9 00
Red Ridge, boys', handled	....	5 75
" hunters	....	5 25

BITS.

Irwin's auger, discount 47½ per cent.

Gilmour's auger, discount 60 per cent.

Rockford auger, discount 50 and 10 per cent.

Jennings' Gen. auger, net list.

Gilmour's car, 47½ per cent.

Clark's expansive, 40 per cent.

Clark's gimlet, per doz	0	0
Diamond, Shell, per doz.	1 65	1 90
Nail and Spike, per gross	3 00	5 [illegible]

BUTCHERS CLEAVERS

German.....per doz.	6 00	9 00
American "	12 00	18 00

CHALK.

Carpenters' Colored, per gross	0 45	0 75
White lump.....per cwt.	0 60	0 65

CHISELS.

Warnock's, discount 72½ per cent.

P. S. & W. Extra, discount 72½ per cent

CROSSCUT SAW HANDLES.

S. & D., No. 3.....per pair	0 15
S. & D., " 5 "	0 22½
S. & D., " 6 "	0 15
Boynton pattern "	0 20

CROWBARS.

3½c. to 4c. per lb.

DRAW KNIVES.

Coach and Wagon, discount 75 per cent.

Carpenters' discount 75 per cent

DRILLS.

Millar's Falls, hand and breast net list

North Bros., each set, 50c.

DRILL BITS.

Morse, discount 37½ to 40 per cent.

Standard, discount 50 and 5 to 55 per cent.

FILES AND RASPS.

Great Western	75 per cent.
Arcade	75 "
Kearney & Foot	75 "
Disston's	75 "
American	75 "
J. Barton Smith	75 "
McClellan	75 "
Eagle	75 "
Nicholson	66⅔ "
Globe	75 "

Black Diamond, 60, 10 and 5 p.c.

Jowitt's, English list, 27½ per cent.

GAUGES.

Stanley's discount 50 to 60 per cent.

Wire Gauges.

Winn's, Nos. 26 to 33.....each	1 65	2 40

HANDLES.

C. & B., fork, 40 per cent., revised list.

C. & B., hoe, 40 per cent., revised list.

American, saw.....per doz.	....	0 25
American, plane.....per gross	3 15	3 75

Canadian, hammer and hatchet 40 per cent

Axe and cant hook handles, 45 per cent.

HAMMERS.

Maydole's, discount 5 to 10 per cent.

Canadian, discount 25 to 27½ per cent.

Magnetic tack.....per doz.	1 10	1 20
Canadian sledge.....per lb.	0 07	0 08½
Canadian ball pean, per lb.	0 22	0 25

HATCHETS.

Canadian, discount 40 to 42½ per cent.

Shingle, Red Ridge 1, per doz.	4 40
" " 2, "	4 85
Barrel Underhill	5 00

MALLETS.

Tinsmiths'.....per doz.	1 25	1 50
Carpenters', hickory, "	1 25	3 75
Lignum Vitae "	3 85	5 00
Caulking, each	0 60	2 00

MATTOCKS.

Canadian.....per doz.	5 50	6 00

MEAT CUTTERS.

German, 15 per cent.
American discount, 33½ per cent.
Gem ....each 1 15

NAIL PULLERS.

German and American 0 85 2 50
No. 1........................ 0 85
No 1573...................... 0 75

NAIL SETS.

Square, round and octagon, per gross 3 38
Diamoud 1 00

PICKS.

Per dozen 6 00 9 00

PLANES.

Wood bench, Canadian discount 40 per cent. American discount 50 per cent.
Wood, fancy Canadian or American 37½ to 40 per cent.
Stanley planes, $1 55 to $3 60, net list prices

PLANE IRONS.

English ....per doz. 2 00 5 00
Stanley, 2⅝ inch, single 24c., double 39c.

PLIERS AND NIPPERS.

Button's genuine, 37½ to 40 per cent.
Button's imitation....per doz. 5 00 9 00

PUNCHES.

Saddler's ....per doz. 1 00 1 85
Conductor's.......... " 3 00 15 00
Tinners', solid..........per set 0 72
" hollow......per inch 1 00

RIVET SETS.

Canadian, discount 35 to 37½ per cent.

RULES.

Boxwood, discount 70 per cent.
Ivory, discount 20 to 25 per cent.

SAWS.

Atkins, hand and crosscut, 25 per cent.
Disston's Hand, discount 12½ per cent
Disston's Crosscut ....per foot 0 35 0 55
Hack, complete..........each 0 75 2 75
" frame only.........each 0 50 1 25

S. & D. solid tooth circular shingle, concave and band, 50 per cent; mill and ice, drag, 30 per cent.; cross-cut, 35 per cent; band saws, butcher, 35 per cent, buck, New Century $6 25, buck No. 1 Maple Leaf, $5 25; buck, Happy Medium, $4.25; buck, Watch Spring, $1.25, buck, common frame, $4.00.

Spear & Jackson's saws—Hand or rip, 26 in., $12 75; 24 in., $11.25; panel, 18 in., $8 25; 20 in., $9; tenon, 10 in., $9.90; 12 in., $10 90; 14 in., $11.50.

SAW SETS.

Lincoln and Whiting 4 75
Hand Sets, Perfect 4 00
X-Cut Sets " 7 50
Maple Leaf and Premiums saw sets, 40 off.
S. & D. saw swages, 40 off.

SCREW DRIVERS.

Sargent's ....per doz. 0 65 1 00
North Bros., No. 30 ..per doz. 16 80

SHOVELS AND SPADES.

Bull Dog, solid neck shovel (No. 2 pol.) $18 50
(Hollow Back) (Reinforced S Scoop.)
Moose........ $17 50 .... $16 30
Bear .... 15 00 .... 15 30
Fox .... 12 60 .... 14 30
Black Cat.... 10 00 .... 13 30
Canadian, discount 45 per cent.

SQUARES.

Iron, discount 20 per cent.
Steel, discount 65 and 10 per cent.
Try and Bevel, discount 50 to 52½ per cent.

TAPE LINES.

English, ass skin .....per doz. 2 75 5 00
English, Patent Leather 5 50 9 75
Chesterman's ....each 0 90 2 85
" steel........each 0 80 8 00

TROWELS.

Disston's, discount 10 per cent.
S. & D., discount 35 per cent.

FARM AND GARDEN GOODS

BELLS.

American cow bells, 63½ per cent.
Canadian, discount 45 and 50 per cent.
American, farm bells, each .. 1 35 3 00

BULL RINGS.

Copper, $1.30 for 2½-inch, and $1.70

CATTLE LEADERS.

Nos. 32 and 33 ... per gross 7 50 8 50

BARN DOOR HANGERS. doz. pairs.

Steel barn door 8 00 10 00
Stearns wood track 4 50 6 00
Zenith 9 00
Acme, wood track 5 00 6 50
Atlas 5 00 6 00
Perfect 8 00 11 00
New Milo 6 50
Steel, covered 4 00 11 00
" track, 1 x 3-16 in (100 ft) 3 75
" 1¼ x 3-16 in (100 ft) 4 75
Double strap hangers, doz. sets 6 40
Standard jointed hangers, " 6 40
Steel King hangers 6 25
Storm King and safety hangers 7 00
" rail 4 25
Chicago Friction, Oscillating and Big Twin Hangers, 5 per cent.

HARVEST TOOLS.

Discount 60 per cent.
S. & D. lawn rakes, Dunn's, 40 off.
" sidewalk and stable scrapers, 40 off.

HAY KNIVES.

Net list

HEAD HALTERS.

Jute Rope, ⅜-inch ...per gross 9 00
" " ½ " 10 00
" " " 12 00
Leather, 1-inch ......per doz. 4 00
Leather, 1¼ " 5 20
Web .... " 2 45

HOES.

Garden, Mortar, etc., discount 60 per cent.
Planter .... per doz. 4 00 4 50

SCYTHES.

Per doz. net 6 25 9 25

SCYTHE SNATHS.

Canadian, discount 40 per cent.

SNAPS.

Harness, German, discount 25 per cent.
Lock, Andrews' 4 50 11 00

STABLE FITTINGS.

Warden King, 35 per cent.

WOOD HAY RAKES.

Ten tooth, 40 and 10 per cent.
Twelve tooth, 45 per cent.

—

HEAVY GOODS, NAILS, ETC.

ANVILS.

Wright's, 80-lb. and over............. 0 10½
Hay Budden, 80-lb and over........... 0 09½
Brook's, 80-lb. and over 0 11½
Taylor-Forbes, handy 0 05

VISES.

Wright's.......................... .... 0 13½
Brook's........................... .... 0 12½
Pipe Vise, Hinge, No. 1 .... 3 50
" No. 2 .... 5 50
Saw Vise 4 50 5 00
Blacksmiths' (discount) 60 per cent.
" parallel (discount) 45 per cent.

BOLTS AND NUTS

Carriage Bolts, common ($1 list Per cent.
" " ⅜ and smaller.. 60, 10 and 10
" " 7-16 and up .... 55 and 5
" " Norway Iron ($3 list) 50
Machine Bolts, ⅜ and less .... 60 and 10
Machine Bolts, 7-16 and up .... 55 and 5
Plough Bolts .... 55 and 10
Blank Bolts 55
Bolt Ends 55
Sleigh Shoe Bolts, ⅜ and less .. 60 and 10
" " " 7-16 and larger 50 and 5
Coach Screws, conepoint...... 70 and 5
Nuts, square, all sizes, 4c. per off.
Nuts, hexagon, all sizes, 4¼c. per off.
Stove Rods per lb., 5½ to 6c.
Stove Bolts, 75 per cent.

CHAIN.

Proof coil, per 100 lb., 5-16 in., $4.40; ⅜ in., $3 90; 7-16 in., $3.70; ½ in., $3 50; 9-16 in., $3 45; ⅝ in., $3 35; ¾ in., $3.20; ⅞ in., $3 10; 1 in. $3.10.
Halter, kennel and post chains, 40 to 40 and 5 per cent.
Cow ties .... 40 p.c.
Tie out chains .... 65 p.c.
Stall fixtures .... 35 p.c.
Trace chain .... 45 p.c.
Jack chain iron, discount 35 p.c.
Jack chain, brass, discount 40 per cent.

HORSE NAILS.

'C' brand, 40, 10 and 7½ per cent. off list } Oval
M.R.M. Co. brand, 55 per cent. } head
Capewell brand, Nos. 9 to 12, base per pound, 12c. Discounts on application.

HORSESHOES.

F.O.B. Montreal
M.R.M. Co. brand, base.................. 3 65
Add 15c. Toronto, Hamilton, Guelph.

HORSE WEIGHTS

Taylor-Forbes, 3½c. per lb.

NAILS.	Cut.	Wire
2d	3 80	3 30
3d	2 95	2 95
4 and 5d	2 70	2 70
6 and 7d	2 60	2 60
8 and 9d	2 45	2 45
10 and 12d	2 40	2 40
16 and 20d	2 35	2 35
30, 40, 50 and 60d (base)	2 30	2 30

F.o.b. Montreal. Cut nails, Toronto 20c. higher.
Miscellaneous wire nails, discount 75 per cent
Coopers' nails, discount 40 per cent.

PRESSED SPIKES.

Pressed spikes, ⅝ diameter, per 100 lbs, $3.15

RIVETS AND BURRS.

Iron Rivets, black and tinned, 60, 10 and 10
Iron Burrs, discount 60 and 10 and 10 p.c.
Copper Rivets, usual proportion burrs, 15 p.c.
Copper Burrs only, net list.
Extras on Coppered Rivets, ½-lb. packages 1c. per lb.; ¼-lb. packages 2c. lb.
Tinned Rivets, net extra, 4c. per lb.

SCREWS.

Wood, F. H., bright and steel, 87½ per cent.
" R. H., bright, dis. 82½ per cent.
" F. H., brass, dis. 80 per cent.
" R. H., " dis. 75 per cent.
" F. H., bronze, dis. 75 per cent.
" R. H., " dis. 70 per cent.
Drive Screws, dis. 87½ per cent.
Bench, wood ....per doz. 3 25 4 00
" iron .... " 4 25 5 00
Set, case hardened, dis. 60 per cent.
Square Cap, dis. 50 and 5 per cent.
Hexagon Cap, dis. 45 per cent.

MACHINE SCREWS.

Flat head, iron and brass, 35 per cent.
Felister head, iron, discount 30 per cent.
" " brass, discount 25 per cent.

TACKS, BRADS, ETC.

Carpet tacks, blued .... 80 and 5
" " tinned............ 80 and 10
" " (in kegs)......... 40
Cut tacks, blued, in dozens only 75 and 10
" ½ weights 60
Swedes cut tacks, blued and tinned—
In bulk........................ 80 and 10
In dozens 75
Swedes, upholsterers', bulk .... 85 and 12½
" brush, blued and tinned bulk 70
Swedes, gimp, blued, tinned and japanned........................ 75 and 12½
Zinc tacks 35
Leather carpet tacks............... 40
Copper tacks....................... 37½
Copper nails 42½
Trunk nails, black 65
Trunk nails, tinned and blued .. 65
Clout nails, blued and tinned .. 65
Chair nails........................ 35
Patent brads....................... 40
Fine finishing..................... 40
Lining tacks, in papers.......... 10
" " in bulk........... 15
" " solid heads, in bulk 75
Saddle nails, in papers.......... 10
" " in bulk........... 15
Tufting buttons, 22 line in dozens only 60
Zinc glaziers' points 5
Double pointed tacks, papers.. 90 and 10
" " " bulk.... 40
Clinch and duck rivets........... 45
Cheese box tacks, 85 and 5; trunk tacks, 80 and 10.

WROUGHT IRON WASHERS.

Canadian make, discount 40 per cent.

—

CUTLERY AND SPORTING GOODS.

AMMUNITION.

B. B. Caps Dominion, 50 and 5 and 25 per cent. American $2.00 per 1000.
C. B. Caps American, $2.60 per 1000

CARTRIDGES.

Rim Fire Cartridges, Dominion, 50 and 5 p.c.
Rim Fire Pistol, discount 30 and 5 per cent., American.
Central Fire, Military and Sporting, American, add 10 per cent. to list. B.B. Caps, discount 40 per cent., American.
Central Fire Pistol and Rifle, list net Amer.
Central Fire Cartridges, pistol and rifle Dominion, 30 and 5 per cent.
Central Fire Cartridges, Sporting and Military, Dominion, 15 per cent, American 10 per cent. advance on list.
Loaded and empty Shells, Crown, 25 and 5; Sovereign, 25, 10 and 10; Regal, 25, 10 and 5; Imperial, 25, 10 and 5. American 20 per cent. discount. Rival and Nitro, 10 per cent. advance on list.
Empty paper shells Dominion, 25 off and empty brass shells, 55 per cent. off. American, 10 per cent. advance on list.
Primers, Dom., 35 per cent.; American $2 05

Wads. per lb.

Best thick brown or grey felt wads, in ¼-lb. bags $0 70
Best thick white card wads, in boxes of 500 each, 12 and smaller gauges 0 29
Best thick white card wads, in boxes of 500 each, 10 gauge............. 0 35
Thin card wads, in boxes of 1,000 each, 12 and smaller gauges 0 20
Thin card wads, in boxes of 1,000 each, 10 gauge...................... 0 25
Chemically prepared black edge grey cloth wads, in boxes of 250 each— Per M.
11 and smaller gauge 0 60
9 and 10 gauges.................. 0 70
7 and 8 " 0 90
5 and 6 " 1 10
Superior chemically prepared pink edge, best white cloth wads in boxes of 250 each—
11 and smaller gauge 1 15
9 and 10 gauges 1 40
7 and 8 " 1 65
5 and 9 " 1 90

SHOT.

Common, $6.50 per 100 lb.; chilled, $7.50 per 100 lb.; buck, seal and ball, $8.50 net list. Prices are f.o.b. Toronto, Hamilton, Montreal, St. John and Halifax. Terms, 2 p.c. for cash in thirty days.

RAZORS. per doz.

Elliot's 4 00 18 00
Boker's 7 50 11 00
" King Cutter 13 50 18 50
Wade & Butcher's............ 3 60 10 00
Lewis Bros.' "Klean Kutter" 8 50 10 50
Clauss Razors and Strops, 50 and 10 per cent

PLATED GOODS

Hollowware, 40 per cent. discount.
Flatware; staples, 40 and 10, fancy 40 and 5 per cent

SHEARS.

Clauss, nickel, discount 60 per cent.
Clauss, Japan, discount 67½ per cent.
Clauss, tailors, discount 40 per cent.
Seymour's discount 50 and 10 per cent.

TRAPS (steel.)

Game, Newhouse, discount 30 and 10 per cent
Game, Hawley & Norton, 50, 10 & 5 per cent.
Game, Victor, 70 per cent.
Game, Oneida Jump (B. & L.) 40 & 2½ p. c.
Game, steel, 60 and 5 per cent.

SKATES.

Skates, discount 37½ per cent.
Mic Mac hockey sticks, per doz 4 00 5 00
New Rex hockey sticks, per doz 6 25

HOUSE FURNISHINGS.

APPLE PARERS.

Woodyatt Hudson, per doz., net 4 50

BIRD CAGES.

Brass and Japanned, 40 and 10 p. c.

COPPER AND NICKEL WARE.

Copper boilers, kettles, teapots, etc. 35 p.c
Copper pitts, 25 per cent.

ENAMELLED WARE.

London, White, Princess, Turquoise, Onyx, Blue and White, discount 50 per cent.
Canada, Diamond, Premier, 50 and 10 p c.
Pearl, Imperial Crescent, 50 and 10 per cent.
Premier steel ware, 40 per cent.
Star decorated steel and white, 25 per cent.
Japanned ware, discount 45 per cent.
Hollow ware, tinned cast, 35 per cent. off.

KITCHEN SUNDRIES.

Can openers, per doz.	0 40	0 75
Mincing knives per doz	0 50	0 80
Duplex mouse traps, per doz	....	0 65
Potato mashers, wire, per doz.	0 60	0 70
" wood "	0 50	0 60
Vegetable slicers, per doz	....	2 25
Universal meat chopper No. 0, $1; No.1,		1 15.
Enterprise chopper, each	...	1 30

Spiders and fry pans, 50 per cent.

LAMP WICKS.

Discount, 60 per cent.

LEMON SQUEEZERS.

Porcelain lined	per doz.	2 20	5 60
Galvanized	"	1 87	3 85
King, wood	"	2 75	2 90
King, glass	"	4 00	4 50
All glass	"	0 50	0 90

PICTURE NAILS.

Porcelain head ......per gross	1 5	1 50
Brass head		1 00

Tin and gilt, picture wire, 75 per cent.

SAD IRONS.

Mrs. Potts, No. 55, polished....per set	0 80
" No. 50, nickle-plated, "	0 9)
Common, plain	4 50
" plated	5 50
Asbestos, per set	1 25

TINWARE.

CONDUCTOR PIPE.

2 in. plain or corrugated., per 100 feet, $3 30; 3 in., $4 40; 4 in., $5.80; 5 in., $7.45; 6 in., $9.9).

FAUCETS.

Common, cork-lined, discount 35 per cent.

EAVETROUGHS.

10-inch ....per 100 ft. 3 30

FACTORY MILK CANS.

Discount off revised list, 35 per cent.
Milk can trimmings, discount 25 per cent.
Creamery Cans, 45 per cent.

LANTERNS.

No. 2 or 4 Plain Cold Blast....per doz. 6 50
Lift Tubular and Hinge Plain, " 4 75
Better quality at higher prices.
Japanning, 50c. per doz. extra.

OILERS.

Kemp's Tornado and McClary's Model galvanized oil can, with pump, 5 gallon, per dozen 10 00
Davidson oilers, discount 40 per cent
Zinc and tin, discount 50 per cent.
Coppered oilers, 20 per cent. off.
Brass oilers, 50 per cent. off.
Malleable, discount 25 per cent

PAILS (GALVANIZED).

Dufferin pattern pails, 40, 10 and 5 per cent.
Flaring pattern, discount 40, 10 and 5 per cent.
Galvanized washtubs 40 10 and 5 per cent.

PIECED WARE.

Discount 35 per cent off list, June, 1899.
10-qt. flaring sap buckets, discount 40 per cent.
6, 10 and 14-qt. flaring pails dis. 40 per cent.
Copper bottom tea kettles and boilers, 30 p.c.
Coal hods, 40 per cent.

STAMPED WARE.

Plain, 75 and 12½ per cent. off revised list.
Retinned, 72½ per cent. revised list.

SAP SPOUTS.

Bronzed iron with hooks ....per 1,000 7 50
Eureka tinned steel, hooks 8 00

STOVEPIPES.

5 and 6 inch, per 100 lengths 7 00
7 inch " " 7 50
Nestable, discount 40 per cent.

STOVEPIPE ELBOWS

5 and 6-inch, common........per doz. 1 32
7-inch " 1 48
Polished, 15c. per dozen extra.

THERMOMETERS.

Tin case and dairy, 75 to 75 and 10 per cent.

TINNERS' SNIPS.

Per doz........................ 3 00 15
Clauss, discount 35 per cent.

TINNERS' TRIMMINGS

Discount, 45 per cent.

WIRE.

BRIGHT WIRE GOODS

Discount 62½ per cent.

CLOTHES LINE WIRE.

7 wire solid line, No. 17 $4.90; No. 18, $3.00; No. 19, $2.70; 4 wire solid line, No. 17, $4.45; No. 18, $2.80. No. 19, $2.50. All prices per 1000 ft. measure. F.o.b. Hamilton Toronto, Montreal.

COILED SPRING WIRE.

High Carbon, No. 9, $2 55, No. 11, $3.20; No. 12, $2.80.

COPPER AND BRASS WIRE.

Discount 45 per cent.

FINE STEEL WIRE.

Discount 30 per cent. List of extras: In 100-lb. lots: No. 17, $5 — No. 18, $5.50 — No. 19, $6 — No. 20, $6.65 — No. 21, $7 — No. 22, $7.30 — No. 23, $7.65 — No. 24, $8 — No. 25, $9 — No. 26, $9 50 — No. 27, $10 — No. 28, $11 — No. 29, $12 — No. 30, $13 — No. 31, $14 — No. 32, $15 — No. 33, $16 — No. 34, $17. Extras net — tinned wire, Nos. 17-25, $2 — Nos. 26-31, $4 — Nos. 32-34, $6. Coppered, 75c. — oiling, 10c. — in 25-lb. bundles, 15c. — in 5 and 10-lb. bundles, 25c. — in 1-lb. hanks, 25c. — in ½-lb. hanks, 38c. — in ¼-lb. hanks, 50c. — packed in casks or cases, 15c. — bagging or papering, 10c.

GALVANIZED WIRE.

Per 100 lb. — Nos. 4 and 5, $3.70 — Nos. 6, 7, 8, $3.15 — No. 9, $2.50 — No. 10, $3 20 — No. 11, $3.25 — No. 12, $2 65 — No. 13, $2.75 — No. 14, $3.75 — No 15, $4.30 — No. 16, $4.30 from stock. Base sizes, Nos. 6 to 9, $2 35 f.o.b. Cleveland. In carlots 12½c. less.

LIGHT STRAIGHTENED WIRE.

Gauge No.	Over 20 in. per 100 lbs.	10 to 20 in.	5 to 10 in.
0 to 5	$0.50	$0 75	$1.25
6 to 9	0.75	1 25	2 00
10 to 11	1.00	1 75	2 50
12 to 14	1.50	2.25	3 50
15 to 16	2.00	3.00	4.50

SMOOTH STEEL WIRE.

No. 0-9 gauge, $2.25; No. 10 gauge, 6c extra; No. 11 gauge, 12c extra; No. 12 gauge, 20c. extra; No. 13 gauge, 30c extra, No 14 gauge, 40c. extra. No. 15 gauge, 55c. extra; No. 16 gauge, 7 c. extra. Add 60c. for coppering and $2 for tinning.

Extra net per 100 lb. — Oiled wire 10c, spring wire $1.25, special hay baling wire 30c, best steel wire 75c., bright soft drawn 15c., charcoal (extra quality) $1.25, packed in casks or cases 15c., bagging and papering 10c., 50 and 100-lb. bundles 10c., in 25-lb. bundles 15c., in 5 and 10-lb. bundles 25c., in 1-lb hanks, 50c., in ½-lb. hanks 75c., in ¼-lb. hanks $1.

POULTRY NETTING.

2 in mesh 19 w. g., discount 50 and 10 per cent. All others 50 per cent.

WIRE CLOTH

Painted Screen, in 100-ft rolls, $1.62½, per 100 sq. ft.; in 50-ft. rolls $1.67½, per 100 sq. ft Terms, 2 per cent. off 30 days.

WIRE FENCING.

Galvanized barb...... .... 2 95
Galvanized, plain twist 3 30
Galvanized barb, f.o b. Cleveland, $2.70 for small lots and $2 60 for carlot.

WOODENWARE.

CHURNS.

No. 0, $9; No 1, $9; No. 2, ; No. 3, $11; No. 4, $13; No. 5, $16.; f.o.b. Toronto Hamilton, London and St. Marys, 30 and 30 per cent; f.o.b. Ottawa, Kingston and Montreal, 40 and 15 per cent discount, Taylor-Forbes, 30 and 30 per cent.

CLOTHES REELS.

Davis Clothes Reels, dis. 40 per cent.

LADDERS, EXTENSION.

Waggoner Extension Ladders, dis. 40 per cent.

MOPS AND IRONING BOARDS.

"Best" mops		1 25
"900" mops		1 25
Folding ironing boards	12 00	16 50

REFRIGERATORS

Discount, 40 per cent.

SCREEN DOORS.

Common doors, 2 or 3 panel, walnut stained, 4-in. style...........per doz. 7 25
Common doors, 2 or 3 panel, grained only, 4-in., style ....per doz. 7 55
Common doors, 2 or 3 panel, light stair per doz. 9 55

WASHING MACHINES.

Round, re-acting per doz.	60 00
Square "	63 00
Eclipse, per doz	54 00
Dowswell "	39 00
New Century, per doz	75 00
Daisy	54 00

WRINGERS.

Royal Canadian, 11 in., per doz.	34 00
Royal American, 11 in.	34 00
Eze, 10 in, per doz	3 75

Terms, 2 per cent, 30 days.

MISCELLANEOUS.

AXLE GREASE.

Ordinary, per gross	6 00	7 00
Best quality	10 00	12 00

BELTING.

Extra, 60 per cent.
Standard, 60 and 10 per cent.
No. 1, not wider than 6 in., 60, 10 and 10 p.c.
Agricultural, not wider than 4 in., 75 per cent
Lace leather, per side, 75c.; cut laces, 80c.

BOOT CALKS.

Small and medium, ball ....per M 4 25
Small heel .... " 4 50

CARPET STRETCHERS.

American ....per doz.	1 00	1 50
Bullard's "	....	6 50

CASTORS.

Bed, new list, discount 55 to 57½ per cent.
Plate, discount 52½ to 57½ per cent.

PINE TAR.

½ pint in tins ....per gross ... 7 80
" " " ... 9 60

PULLEYS.

Hothouse	per doz.	0 55	1 00
Axle	"	0 22	0 33
Screw	"	0 22	1 00
Awning	"	0 35	2 50

PUMPS.

Canadian cistern	1 40	2 00
Canadian pitcher spout	1 80	3 16

ROPE AND TWINE.

Sisal		0 10½
Pure Manilla		0 15½
"British" Manilla		0 12
Cotton, 3-16 inch and larger	0 21	0 23
" 5-32 inch	0 25	0 27
" ¼ inch	0 25	0 28
Russia Deep Sea		0 16
Jute		0 09
Lath Yarn, single		0 10
" double		0 10½
Sisal bed cord, 48 feet........per doz.		0 65
" " 60 feet........ "		0 80
" " 72 feet........ "		0 95

Twine.

Bag, Russian twine, per lb.	....	0 27
Wrapping, cotton, 3-ply	....	0 25
" " 4-ply	....	0 29
Mattress twine per lb.	0 33	0 45
Staging " "	0 27	0 35

SCALES.

Gurney Standard, 40 per cent.
Gurney Champion, 50 per cent.
Burrow, Stewart & Milne—
Imperial Standard, discount 40 per cent.
Weigh Beams, discount 40 per cent.
Champion Scales, discount 50 per cent.
Fairbanks standard, discount 35 per cent.
" Dominion, discount 55 per cent.
" Richelieu, discount 55 per cent.
Warren new Standard, discount 40 per cent.
" " Champion, discount 50 per cent.
" Weighbeams, discount 35 per cent.

STONES—OIL AND SCYTHE.

Washita	per lb.	0 25	0 37
Hindostan	"	0 06	0 10
" slip	"	0 18	0 20
" Axe	"	....	0 10
Deer Creek	"	....	0 10
Deerlick	"	....	0 25
" Axe	"	....	0 15
Lily white	"	....	0 42
Arkansas	"	....	1 50
Water-of-Ayr	"	....	0 10
Scythe	per gross	3 50	5 00
Grind, 40 to 200 lb., per ton	....	20 00	22 00
" under 40 lb., "	....	....	24 00
" 200 lb. and over	....	....	23 00

INDEX TO ADVERTISERS.

CLASSIFIED LIST OF ADVERTISEMENTS.

Auditors
Davenport, Percy P., Winnipeg.

Babbitt Metal.
Canada Metal Co , Toronto.
Canadian Fairbanks Co., Montreal.
Robertson, Jas. Co., Montreal.

Bath Room Fittings.
Carriage Mounting Co., Toronto.

Belting, Hose, etc.
Gutta Percha and Rubber Mfg. Co Toronto.

Bicycles and Accessories.
Johnson's, Iver, Arms and Cycle Works Fitchburg, Mass

Binder Twine.
Consumers Cordage Co., Montreal.

Box Strap.
J N. Warminton, Montreal.

Brass Goods.
Glauber Brass Mfg. Co., Cleveland, Ohio.
Lewis, Rice, & Son., Toronto.
Morrison, Jas , Brass Mfg. Co., Toronto.
Mueller Mfg. Co., Decatur, Ill.
Penberthy Injector Co., Windsor, Ont.
Taylor-Forbes Co., Guelph, Ont.

Bronze Powders.
Canadian Bronze Powder Works, Montreal

Brushes.
Ramsay, A., & Son Co., Montreal.

Can Openers.
Cumming Mfg. Co. Renfrew.

Cans.
Acme Can Works, Montreal.

Builders' Tools and Supplies.
Covert Mfg. Co., West Troy, N.Y.
Frothingham & Workman Co., Montreal.
Howland, H. S., Sons & Co., Toronto.
Hyde, F., & Co., Montreal.
Lewis Bros. & Co., Montreal.
Lewis, Rice, & Son, Toronto.
Lockerby & McComb, Montreal.
Lufkin Rule Co., Saginaw, Mich.
Newman & Sons, Birmingham.
North Bros. Mfg. Co., Philadelphia, Pa.
Stanley Rule & Level Co., New Britain.
Stanley Works, New Britain, Conn.
Stephens, G. F., Winnipeg.
Taylor-Forbes Co., Guelph Ont.

Carriage Accessories.
Carriage Mountings Co., Toronto.
Covert Mfg. Co., West Troy, N.Y.

Carriage Springs and Axles.
Guelph Spring and Axle Co., Guelph.

Cattle and Trace Chains.
Greening, B., Wire Co., Hamilton.

Churns.
Dowswell Mfg. Co., Hamilton.

Clippers—All Kinds.
American Shearer Mfg. Co., Nashua, N.H

Clothes Reels and Lines.
Hamilton Cotton Co., Hamilton, Ont.

Cordage.
Consumers' Cordage Co., Montreal.
Hamilton Cotton Co., Hamilton.

Cork Screws.
Erie Specialty Co., Erie, Pa.

Clutch Nails.
J. N. Warminton, Montreal.

Cut Glass.
Phillips, Geo., & Co., Montreal.

Cutlery—Razors, Scissors, etc.
Birkett, Thos., & Son Co., Ottawa.
Clauss Shear Co., Toronto
Dorken Bros. & Co., Montreal.
Heinisch's, R., Sons Co., Newark, N.J.
Howland, H. S. Sons & Co., Toronto.
Phillips, Geo., & Co., Montreal.
Round, John, & Son, Montreal.

Electric Fixtures.
Canadian General Electric Co., Toronto.
Forman, John, Montreal.
Morrison James, Mfg. Co., Toronto.
Munderloh & Co., Montreal.

Electro Cabinets.
Cameron & Campbell Toronto.

Engines, Supplies, etc.
Kerr Engine Co., Walkerville, Ont.

Files and Rasps.
Barnett Co., G. & H., Philadelphia, Pa
Nicholson File Co., Port Hop

Financial Institutions
Bradstreet Co.

Firearms and Ammunition.
Dominion Cartridge Co., Montreal.
Hamilton Rifle Co., Plymouth, Mich
Harrington & Richardson Arms Co., Worcester, Mass.
Johnson's, Iver, Arms and Cycle Works, Fitchburg, Mass.

Food Choppers.
Enterprise Mfg Co., Philadelphia, Pa.
Shirreff Mfg. Co , Brockville, Ont.

Galvanizing.
Canada Metal Co., Toronto.
Montreal Rolling Mills Co., Montreal.
Ontario Wind Engine & Pump Co , Toronto.

Glaziers' Diamonds.
Gibsone, J. B., Montreal.
Pelton, Godfrey S.
Sharratt & Newth, London, Eng.
Shaw, A., & Son, London, Eng.

Handles
Still, J H , Mfg Co.

Harvest Tools.
Maple Leaf Harvest Tool Co , Tillsonburg Ont.

Hoop Iron.
Montreal Rolling Mills Co , Montreal.
J. N. Warminton, Montreal.

Horse Blankets.
Heney, E. N., & Co., Montreal.

Horseshoes and Nails.
Canada Horse Nail Co., Montreal.
Montreal Rolling Mills, Montreal.
Capewell Horse Nail Co , Toronto.

Hot Water Boilers and Radiators.
Cluff, R. J., & Co., Toronto.
Pease Foundry Co., Toronto.
Taylor-Forbes Co., Guelph.

Ice Cream Freezers.
Dana Mfg. Co., Cincinnati, Ohio.
North Bros. Mfg. Co., Philadelphia, Pa.

Ice Cutting Tools.
Erie Specialty Co., Erie, Pa.
North Bros. Mfg. Co., Philadelphia. Pa

Injectors—Automatic.
Morrison, Jas., Brass Mfg. Co., Toronto.
Penberthy Injector Co., Windsor, Ont.

Iron Pipe.
Montreal Rolling Mills, Montreal.

Iron Pumps.
McDougall, R., Co., Galt, Ont.

Lanterns.
Kemp Mfg. Co., Toronto.
Ontario Lantern Co., Hamilton, Ont.
Wright, E. T., & Co., Hamilton.

Lawn Mowers.
Birkett, Thos., & Son Co., Ottawa.
Maxwell, D., & Sons, St. Mary's, Ont.
Taylor, Forbes Co., Guelph.

Lawn Swings, Settees, Chairs.
Cumming Mfg. Co., Renfrew.

Ledgers—Loose Leaf.
Business Systems, Toronto.
Copeland-Chatterson Co , Toronto.
Crain, Rolla L., Co., Ottawa.
Universal Systems, Toronto.

Locks, Knobs, Escutcheons, etc.
Peterborough Lock Mfg. Co., Peterborough, Ont.

Lumbermen's Supplies.
Pink, Thos., & Co., Pembroke Ont.

Manufacturers' Agents.
Fox, C. H., Vancouver.
Gibb, Alexander, Montreal.
Mitchell, David C , & Co., Glasgow, Scot.
Mitchell, H. W., Winnipeg.
Pearce, Frank, & Co. Liverpool, Eng.
Scott, Bathgate & Co., Winnipeg.
Thorne, R. E., Montreal and Toronto.

Metals.
Canada Iron Furnace Co., Midland, Ont.
Canada Metal Co., Toronto.
Eadie, H. G., Montreal.
Gibb, Alexander, Montreal.
Kemp Mfg. Co., Toronto
Leslie, A. C., & Co., Montreal.
Lysaght, John, Bristol, Eng.
Nova Scotia Steel and Coal Co., New Glasgow, N.S.
Robertson, Jas., Co , Montreal
Roper, J. H., Montreal.
Samuel, Benjamin & Co., Toronto.
Stairs, Son & Morrow, Halifax, N.S.
Thompson, B. & S. H. & Co. Montreal.

Metal Lath.
Galt Art Metal Co., Galt.
Metallic Roofing Co , Toronto.
Metal Shingle & Siding Co., Preston, Ont.

Metal Polish, Emery Cloth, etc
Oakey, John, & Sons, London, Eng.

Mops.
Cumming Mfg. Co., Refrew.

Mouse Traps.
Cumming Mfg. Co., Renfrew.

Oil Tanks.
Bowser, S. F., & Co , Toronto.

Paints, Oils, Varnishes, Glass.
Bell, Thos , Sons & Co , Montreal.
Canada Paint Co., Montreal.
Canadian Oil Co., Toronto
Consolidated Plate Glass Co., Toronto.
Fenner, Fred., & Co., London, Eng.
Henderson & Potts Co , Montreal.
Imperial Varnish and Color Co., Toronto.
Jamieson, R. C., & Co , Montreal.
McArthur, Corneille & Co., Montreal.
McCaskill, Dougall & Co., Montreal.
Montreal Rolling Mills Co., Montreal.
Moore, Benjamin, & Co. Toronto.
Queen City Oil Co., Toronto.
Ramsay & Son, Montreal.
Sherwin-Williams Co., Montreal.
Standard Paint and Varnish Works Windsor, Ont
Stephens & Co , Winnipeg.
Martin-Senour Co., Chicago.

Perforated Sheet Metals.
Greening, B., Wire Co., Hamilton.

Plumbers' Tools and Supplies.
Borden Co., Warren, Ohio.
Canadian Fairbanks Co., Montreal.
Cluff, R. J., & Co., Toronto.
Glauber Brass Co., Cleveland, Ohio.
Jardine, A. B., & Co , Hespeler, Ont.
Jenkins Bros., Boston, Mass.
Lewis, Rice, & Son, Toronto.
Merrell Mfg. Co., Toledo, Ohio.
Montreal Rolling Mills Montreal.
Morrison, Jas., Brass Mfg. Co., Toronto.
Mueller, H., Mfg. Co., Decatur, Ill.
Oshawa Steam & Gas Fitting Co., Oshawa
Robertson Jas., Co. Montreal.
Stairs, Son & Morrow, Halifax, N.S.
Standard Ideal Sanitary Co., Port Hope,
Standard Sanitary Co., Pittsburg.
Stephens, G F., & Co., Winnipeg, Man.
Turner Brass Works, Chicago.
Vickery, Orlando, Toronto.

Portland Cement.
International Portland Cement Co., Ottawa, Ont.
Hanover Portland Cement Co., Hanover, Ont.
Hyde, F., & Co., Montreal.
Thompson, B. & S. H. & Co., Montreal.

Potato Mashers.
Cumming Mfg Co., Renfrew.

Poultry Netting.
Greening, B., Wire Co., Hamilton, Ont.

Razors.
Clauss Shear Co., Toronto.

Roofing Supplies.
Brantford Roofing Co., Brantford.
McArthur, Alex., & Co., Montreal.
Metal Shingle & Siding Co., Preston, Ont.
Metallic Roofing Co., Toronto.
Paterson Mfg. Co., Toronto & Montreal.

Saws.
Atkins, E. C., & Co., Indianapolis, Ind
Lewis Bros., Montreal.
Shurly & Dietrich, Galt, Ont.
Spear & Jackson, Sheffield, Eng.

Scales.
Canadian Fairbanks Co., Montreal.

Screw Cabinets.
Cameron & Campbell, Toronto.

Screws, Nuts, Bolts.
Montreal Rolling Mills Co., Montreal.
Morrow, John, Machine Screw Co., Ingersoll, Ont.

Sewer Pipes.
Canadian Sewer Pipe Co., Hamilton
Hyde, F., & Co., Montreal.

Shelf Boxes.
Cameron & Campbell, Toronto

Shears, Scissors.
Clauss Shear Co., Toront .

Shelf Brackets.
Atlas Mfg. Co., New Haven, Conn

Shellac
Bell, Thos , Sons & Co., Montreal.

Shovels and Spades.
Canadian Shovel Co., Hamilton.
Peterboro Shovel & Tool Co , Peterboro.

Silverware.
Phillips, Geo., & Co., Montreal.
Round, John, & Son, Sheffield, Eng.

Skates.
Canada Cycle & Motor Co , Toronto.

Spring Hinges, etc.
Chicago Spring Butt Co., Chicago, Ill

Steel Rails.
Nova Scotia Steel & Coal Co., New Glasgow, N.S.

Stoves, Tinware, Furnaces
Canadian Heating & Ventilating Co., Owen Sound.
Canada Stove Works, Harriston, Ont.
Clare Bros. & Co., Preston.
Davidson, Thos., Mfg. Co., Montreal,
Guelph Stove Co., Guelph.
Gurney Foundry Co., Toronto.
Harris, J. W., Co., Montreal.
Joy Mfg. Co., Toronto.
Kemp Mnfg. Co. Toronto.
McClary Mfg. Co. London.
Pease Foundry Co., Toronto.
Stewart, Jas., Mfg. Co., Woodstock, Ont.
Taylor-Forbes Co., Guelph, Ont.
Wright, E. T., & Co., Hamilton.

Tacks.
Montreal Rolling Mills Co., Montreal.
Ontario Tack Co., Hamilton.

Ventilators.
Harris, J. W., Co., Montreal.
Pearson, Geo. D., Montreal.

Washing Machines, etc
Dowswell Mfg Co., Hamilton, Ont.
Taylor-Forbes Co., Guelph, Ont.

Wheelbarrows.
London Foundry Co., London, Ont.

Wholesale Hardware.
Birkett, Thos., & Sons Co., Ottawa.
Caverhill, Learmont & Co., Montreal.
Frothingham & Workman, Montreal.
Hobbs Hardware Co., London.
Howland, H. S., Sons & Co., Toronto.
Lewis Bros. & Co., Montreal.
Lewis, Rice, & Son, Toronto.

Window and Sidewalk Prisms.
Hobbs Mfg. Co., London, Ont.

Wire Springs.
Guelph Spring Axle Co., Guelph, Ont.
Wallace-Barnes Co., Bristol, Conn.

Wire, Wire Rope, Cow Ties, Fencing Tools, etc.
Canada Fence Co., London.
Dennis Wire and Iron Co., London, Ont.
Dominion Wire Mnfg. Co., Montreal
Greening, B., Wire Co., Hamilton.
Montreal Rolling Mills Co., Montreal.
Western Wire & Nail Co., London, Ont

Woodenware.
Taylor-Forbes Co., Guelph, Ont.

Wrapping Papers.
Canada Paper Co., Toronto.
McArthur, Alex., & Co , Montreal.
Stairs, Son & Morrow, Halifax, N.S.

Babbitt Metal

IS CONCEDED BY EXPERTS TO BE THE

BEST BABBITT

For Heavy Slow Machinery
Heavy Line Shafting, etc.

12c. lb. TRY IT.

The James Robertson Co., Limited
144 William St., MONTREAL, QUE.

PFLUEGER'S

Fishing Tackle

Hooks, Flies, Trolls, Spinners, Phantoms, Reels, Furnished Lines.

Everything in Fishing Tackle.

NOTICE—Free to Any Dealer in Sporting Goods, Sent Express Prepaid, 170-Page Illustrated Catalogue No. F25, and Metal Fish Sign in 8-Color Lithograph.

The Enterprise Mfg. Co.
AKRON, Ohio, U.S.A.

WALKER CORK SCREWS

are good holiday souvenirs and live sellers the whole year. Are made well and are made out of good material. That is why we guarantee every one that we put out. When ordering Cork Screws try Walker's

Manufactured only by ERIE SPECIALTY COMPANY, ERIE, Pa. U.S.A.

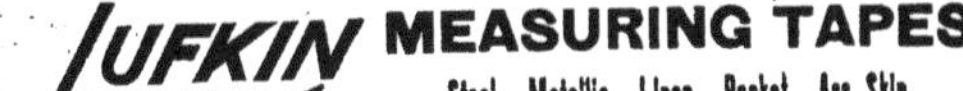

Steel, Metallic, Linen, Pocket, Ass Skin, Pat. Leather, Band Leather, Etc.

ARE THE BEST AND MOST POPULAR TAPES IN THE WORLD.
YOUR STOCK IS NOT COMPLETE WITHOUT THEM.

LUFKIN RULE CO., Saginaw, Mich, U.S.A.

London Office and Warehouse—24 and 26 Holborn. New York City Branch—280 Broadway.

For sale by ALL PROMINENT CANADIAN HARDWARE JOBBERS.

CIRCULATES EVERYWHERE IN CANADA

Also in Great Britain, United States, West Indies, South Africa and Australia.

HARDWARE AND METAL

A Weekly Newspaper Devoted to the Hardware, Metal, Heating and Plumbing Trades in Canada.

Office of Publication, 10 Front Street East, Toronto.

VOL. XIX. MONTREAL, TORONTO, WINNIPEG, FEBRUARY 9, 1907 NO. 6.

See Classified List of Advertisements on Page 71.

WORK BENCHES AND TOOL BOXES

NEW TOOL BOXES

No. 1 Chest, 19 x 9 x 6¾
contains 16 tools

No. 3 Chest, 26 x 12 x 12
contains 37 tools

No. 5 Chest, 33 x 16 x 16
contains 52 tools

TOOL BOXES ALL SIZES

TOOL CABINETS

No. 1 Cabinet,
2 ft. 8 in. x 1 ft. 8 in. x 5½ in.
contains 26 tools

No. 3 Cabinet,
2 ft. 8 in. x 1 ft. 8 in. x 5½ in.
contains 30 tools

*No. 5 Cabinet,
2 ft. 8 in. x 1 ft. 8 in. x 5½ in.
contains 44 tools

SCROLL SAWS and LATHES

WORK BENCHES and CARVING BENCHES

WORK BENCHES 4—6½ and 7 feet

Write for Prices

RICE LEWIS & SON

LIMITED

TORONTO.

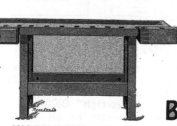

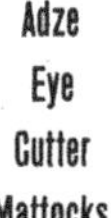

CAST STEEL
First quality, polished cutters, weight 5½ lbs.

It's a little early yet to talk this line, but the present is a good time to buy, as the Spring will soon be here and the prices we are making now, are a good deal lower than they will be then.

We are headquarters for everything in the pick line including Clay, Rock, Stone, Coal Drifting, Prospectors, Tamping, etc.

If you want to save something on your pick trade, write us now.

CROW BARS

We stock all weights and sizes, assorted, 10 to 35 lbs.

As a usual thing, this line is sold by the retailers with about the same profit as nails. If you know where to buy, you will find it would pay you a better profit.

TRY US

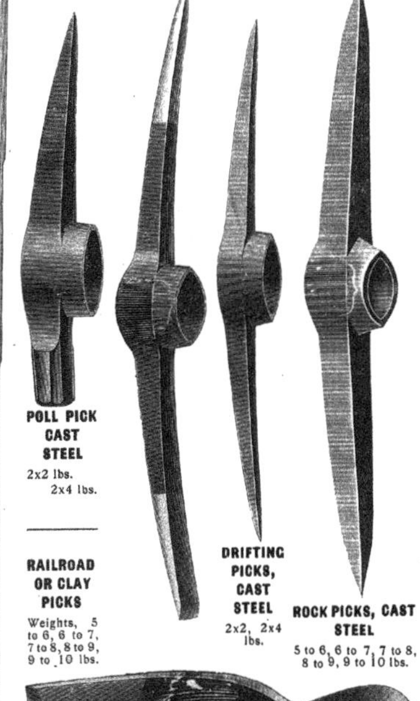

GRUB HOES, CAST STEEL

No.	1	2
Average width, inches	3¾	4
" weight, lbs.	3	3½

LEWIS BROS. Ltd.

MONTREAL
WINNIPEG
OTTAWA TORONTO
VANCOUVER CALGARY

Consumers Cordage Co., LIMITED

Manufacture every variety of

CORDAGE — PACKING

BINDER TWINE

LATHYARN — SMALL TWINES

From Manilla, Sisal, Italian, Russian, Jute

Tallow Laid Transmission Rope, 3 or 4 Strand

Oil Well Drilling Cables, Lariat Rope, 3 or 4 Strand,
Halyards, Deep Sea Line, Snap Halters, Snap Plow Lines,
Clothes Lines, Hide Cord, Pulp and Paper Cord,
Tarred Rope, Marline, Houseline

Wrapping, Rolling, Seaming, Sewing and Grape Twines

Careful Execution **SPECIAL ORDERS** Prompt Shipment

Mills: MONTREAL, QUE., and HALIFAX, N.S.

BRANCHES:

F. H. ANDREWS & SON, Quebec. CONSUMERS CORDAGE CO., Limited, St. John, N.B. MacGOWAN & CO., Vancouver.
TEES & PERSSE, Limited, Winnipeg, Calgary and Edmonton. GEO. WOOD, London, England.
MERRICK, ANDERSON & CO., Winnipeg, Binder Twine Distributors for Northwest.

Have You Tried SOLDERALL ?

Tinsmiths should make a test of this New Anticorrosive Fluid—It is the best thing in the market to-day for tinning and soldering all metals.

Unsurpassed for Cleanliness of Work and Strength of Joint made.

No Noxious Fumes to Injure the Health or Affect the Skin.

Economises Time and Labor, Saves Solder

No rust or discoloration. The Solder will follow the fluid like a magnet

PUT UP FOR SALE IN ½-GAL., 1-GAL. AND 5-GAL. CANS.

Solderall is always sent out **full strength** and may be diluted according to the class of work for which it may be required.

SOLE AGENTS AND MANUFACTURERS

The Thos. Davidson Mfg. Co. Ltd., MONTREAL and WINNIPEG

H. S. HOWLAND, SONS & CO. LIMITED

HARDWARE MERCHANTS

Only Wholesale — 138-140 WEST FRONT STREET, TORONTO — Wholesale Only

GAME TRAPS

"Star," with and without chains

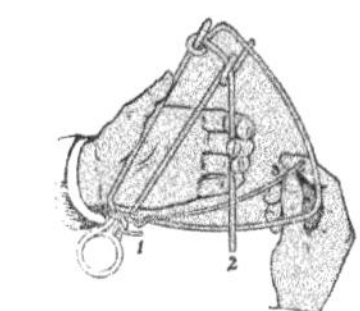

Stop Thief Trap
For catching small fur-bearing animals

"Victor"—Single Spring
Nos. 0, 1, 1½—with and without chains

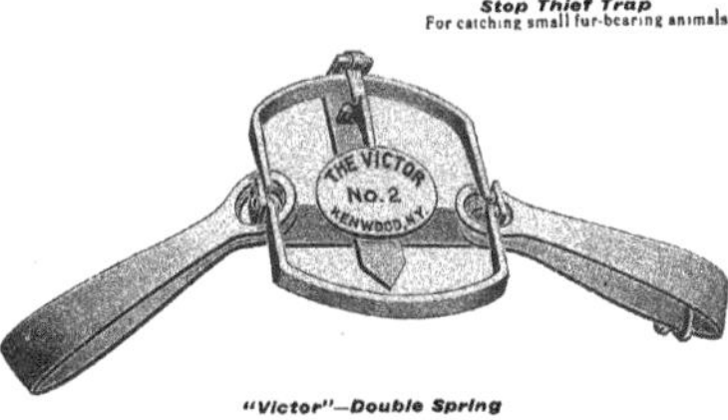

"Victor"—Double Spring
Nos. 2, 3, 4—with and without chains

Hawley and Norton—Single Spring
Nos. 0, 1, 1½—with and without chains

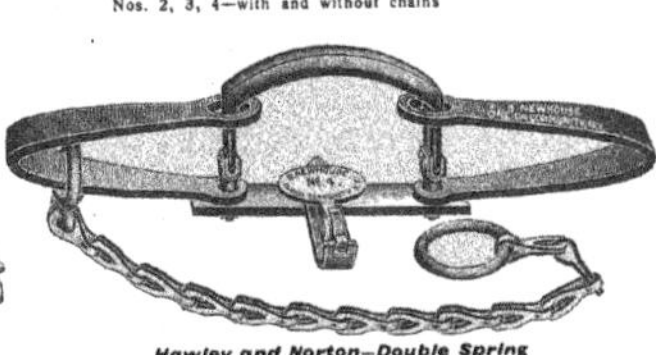

Hawley and Norton—Double Spring
Nos. 2, 3, 4,—with and without chains

BEAR TRAP

"Newhouse" Bear Traps

Fuller description see our Hardware Catalogue.

H. S. HOWLAND, SONS & CO., LIMITED

Opposite Union Station.

GRAHAM NAILS ARE THE BEST

We Ship promptly — Factory: Dufferin Street, Toronto, Ont. — Our Prices are Right

Morse Twist Drills are perfect drills. They are made scientifically by the people who have made the name **"Morse"** synonymous with the words **"twist drill."** Many discriminating users will have no other. The largest machine shops in all parts of Canada use them.

The **Morse Twist Drill** can be had in a greater variety of sizes and styles than any other make of drills. If you have a twist drill trade sell **Morse Drills.**

No. 102 No. 109 No. 104

Five sets of drills can be had in these indexed cases. Cases are also sold empty

Morse Twist Drills are put up in attractive and convenient sets. They will sell to the machinists in your town.

Ask our travellers for prices, or write direct.

No. 14. Set of Bit Stock Drills, 9 sizes

F. & W. Hardware Montreal

FROTHINGHAM & WORKMAN, Limited, MONTREAL, CANADA.

Brantford—Roofings

This Represents Our

"ASPHALT"
No. 1, 2, 3
"RUBBER"
No. 1, 2, 3

Each Roll is complete with Nails and Cement to lay. Best selling goods in the market.

BRANTFORD ROOFING CO.
LIMITED
BRANTFORD, ONTARIO

SIMPLE

DURABLE

RELIABLE

STYLE F.

THE LEADER

A Name Familiar to Butter Makers in Connection with the

Highest Grade Churns

ITS STRONG POINTS:

Stiff Steel Frame, with malleable iron fittings, Combination Bow Lever and Pedal Drive. Bicycle Ball Bearings. Bolted Trunnions. Cream Breakers. Barrel easily drained and conveniently detatchable for cleaning.

DEALERS: When you handle this line you handle QUALITY.

The Dowswell Manufacturing Co., Limited

W. L. HALDIMAND & SON, Montreal
EASTERN AGENTS

HAMILTON
ONTARIO

GALVANIZED FENCE HOOK FOR FASTENING WOODEN PICKET OR WIRE FENCES

WIRE NAILS, COILED SPRING, BARB and PLAIN FENCE WIRE, OILED and ANNEALED, CLOTHES LINE WIRE, STAPLES, etc.

THE WESTERN WIRE & NAIL CO., Limited, LONDON, ONT.

The Peterborough Lock Manufacturing Company, Limited

Peterborough, Ont.

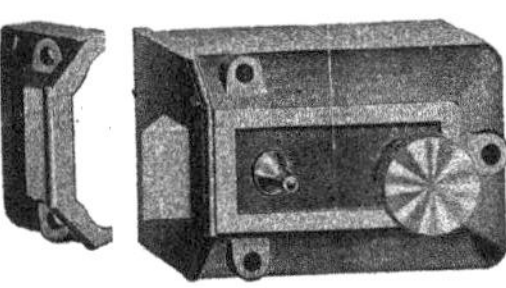

Cylinder Night Latch, No. 103.

Manufacturers of all kinds

Rim and Mortise Locks, Inside, Front Door, and Store Door Sets, also full line High-class Builders' Hardware.

Sold by all Leading Jobbers in the Dominion.

CHALLENGE AXE

This is a good every day Axe and will challenge any other Axe in the market to meet it in quality and finish at the price. Our orders for this season are far in excess of any former year, and as our travellers are on the road at present it will pay you to wait and see them, if they have not already called on you. You might as well have the best for the money in the market.

The Dundas Axe Works

To be continued. **DUNDAS, ONT.**

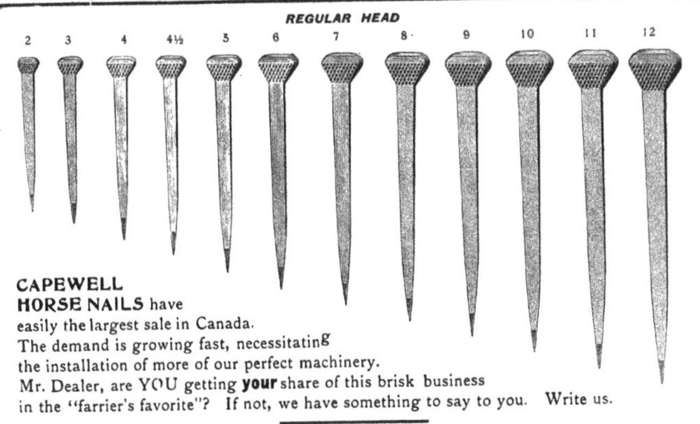

CAPEWELL HORSE NAILS have easily the largest sale in Canada. The demand is growing fast, necessitating the installation of more of our perfect machinery. Mr. Dealer, are YOU getting **your** share of this brisk business in the "farrier's favorite"? If not, we have something to say to you. Write us.

THE CAPEWELL HORSE NAIL CO., Toronto, Ont.

Western Branch: 212 Nena St., WINNIPEG, MAN.

THE

SPECIALLY ROLLED STEEL

used in making

Hamilton Rifles

gives the maximum strength with the minimum weight—the condition par excellence in rifle-making.

HAMILTON RIFLES are made in three models—
Nos. 15, 19, 23.

No. 15 sells at $2.00
No. 19 sells at $2.50
No. 23 sells at $3.50

ORDER FROM THE JOBBER

THE HAMILTON RIFLE CO.

Dept. 71 ***Plymouth, Mich.***

Scrap Iron for Re-working

Better bars for re-working are obtained from selected scrap than from pig iron—have more life and strength.

London Bars are made from selected scrap and come in Flats, Rounds, Squares, Ovals, Half-Ovals, Half-Rounds and Bands.

Large stock always on hand, insuring prompt shipment. 51

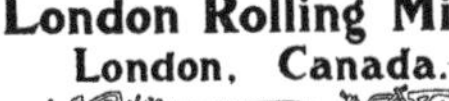

London Rolling Mills
London, Canada.

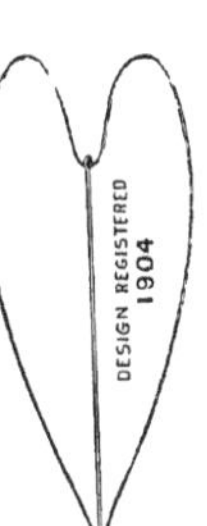

THE

ANTON CARPET BEATER

Has double the beating surface of any carpet beater on the market

LIGHT, yet STRONG and DURABLE; STIFF, yet PLIABLE and EASY TO HANDLE.

Centre rod made of heavy spring steel wire. Outside wires go through the handle and are clinched at bottom end making it impossible for handle to come off; handle filled with lead, thereby making it a perfectly balanced article.

FOR SALE BY

E. T. WRIGHT & CO.
HAMILTON, ONT.

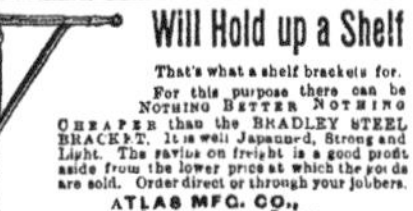

Will Hold up a Shelf

That's what a shelf brackets for.

For this purpose there can be NOTHING BETTER NOTHING CHEAPER than the BRADLEY STEEL BRACKET. It is well Japanned, Strong and Light. The saving on freight is a good profit aside from the lower price at which the goods are sold. Order direct or through your jobbers.

ATLAS MFG. CO.,
New Haven, Conn.,

Expert Bookkeeping

Investigation and auditing of all classes of accounts undertaken. Trading, profit and loss accounts and balance sheets prepared. Sets of books opened and office systems devised in an up-to-date manner. Collection of accounts a specialty.

PERCY P. DAVENPORT
622 McIntyre Block, - WINNIPEG, MAN.

Square your Accounts

INTERNATIONAL PORTLAND CEMENT COMPANY, Limited
OTTAWA, - - ONTARIO.

Manufacturers of strictly high-grade Portland Cement made by the most improved method of manufacture in existence. Each barrel contains three hundred and fifty pounds of pure Portland Cement with no adulteration.

Don't Forget the Name...

NEWMAN'S INVINCIBLE FLOOR SPRINGS

Strong, Quick, Reliable, Effective.

Will close a door against any pressure of wind. Far ahead of ordinary door springs, pneumatic or otherwise. Ask your wholesaler.

W. NEWMAN & SONS, Birmingham.

Money for You, Lots of it

Yes, there are more dollars in retailing Window Glass than some dealers think—and **most dealers know it.**

You certainly are not giving yourself a square deal—if you have not seen our prices for spring deliveries. Besides, **you do save money** by importing direct.

These strong cases assure delivery in good order, while our facilities assure promptitude.

We are always glad to quote prices.

Have you our catalogue?

The Hobbs Manufacturing Co., Limited

LONDON TORONTO WINNIPEG

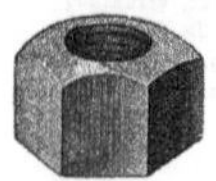

The John Morrow Screw, Limited

MANUFACTURERS OF

Set and Cap Screws, Special Milled Work, Engine Studs Etc. Cold Punched Nuts of every variety of finish.

INGERSOLL, ONTARIO

CELEBRATED **HEINISCH** SHEARS

Tailors' Shears,
Trimmers' Shears,
Tinners' Snips, etc.

ACKNOWLEDGED THE BEST.

R. HEINISCH'S SONS CO. NEW YORK OFFICE, 155 Chambers St
NEWARK, N.J., U.S.A.

HENRY T. SEYMOUR SHEAR CO. NEW YORK.
TRADE MARK

SEYMOUR SHEARS

HAVE BEEN THE

Standard for over Half a Century.

"QUALITY UNQUESTIONED."

Each pair of our shears bears the above trade mark.

HENRY T. SEYMOUR SHEAR CO. NEW YORK.
TRADE MARK

Latest Catalogue will be sent in exchange for your business card.

Complete Line *TRIMMERS', BANKERS', BARBERS' and TAILORS' SHEARS, Etc., Etc.*

Henry T. Seymour Shear Company

WIEBUSCH & HILGER, Limited, NEW YORK, Sole Agents

Covert Mfg. Co.

TROY, N. Y.

Trojan Strap No. 20

Rein Chains

Harness Snaps, Chain, Rope and Web Goods, etc. For sale by Jobbers, at Manufacturers' prices.

STREET PAVING and SIDEWALKS a SPECIALTY

SILICA BARYTIC STONE CO.

OF ONTARIO, Limited.

Water Proof Floors for Malt Houses, Breweries, Slaughter Houses, Cheese Factories, Cellar, Stable Floors, etc.

Head Office
Ingersoll, Ontario.
Walter Mills, General Manager
Ask for quotations for Septic Tanks.

"The Tool Holder People"

ARMSTRONG BROS. TOOL CO.

Manufacturers of Armstrong Patent Lathe and Planer Tools and other machine shop specialties.

106 N. Francisco Ave., CHICAGO, ILL.

Persons addressing advertisers will kindly mention having seen this advertisement in Hardware and Metal.

It Doesn't Sell Itself

BUT—if ever any article manufactured by man came near doing so, it's

The

Empire Queen Range

The lady in the case always decides on the sort of range for the home. **The Empire Queen's** beautiful appearance rivets a woman's attention at once, and while you briefly point out the inner perfections of this range, the man in the case is getting down into his pocket or reaching for his cheque book.

The Canadian Heating & Ventilating Co. Limited.

OWEN SOUND, Ontario.

THE CHRISTIE BROS. CO., Limited, 238 King St., Winnipeg, Man., Western Agents.
THE CANADA STOVE AND FURNITURE CO., 126 West Craig St., Montreal, Que., Agents for the Province of Quebec.

The Wellesville Blue Polished Steel used in making the

Joy Malleable and Steel Range

gives and keeps the finest appearance with the smallest expense of labor.

No blacking or Japan needed on it. A little kerosene with a cloth does the trick.

And every part rivetted

GAS TIGHT
AIR TIGHT
DUST TIGHT and
STAY TIGHT

JOY MFG. CO., 32 Humberside Ave., TORONTO, ONT.

TEES & PERSSE CO., - Western Distributors, Winnipeg
C. H. FOX, Agent - Vancouver, B.C.

SAP TIME IS COMING

Just a few short weeks until you need these goods; why not order now and be sure to have them when needed?

"EUREKA" CAST IRON AND STEEL SAP SPILES

You know that these spiles will not load you up with complaints from your customers.

SAP PAILS

FLARING PATTERN

Made in six sizes, in both straight and flaring patterns. We carry sap-pan materials in both tinned and galvanized Iron as follows—

Tinned Iron	Galvanized Iron
48x96x20	48x96x22
48x96x22	36x96x22
48x96x24	36x96x24
36x84x22	
36x84x24	

Prompt Shipment Guaranteed

STRAIGHT PATTERN

THE McCLARY MFG. CO.

LONDON, TORONTO, MONTREAL, WINNIPEG, VANCOUVER, ST. JOHN, N.B., HAMILTON, CALGARY.

"EVERYTHING FOR THE TINSHOP"

Reach the Right People!

Hardware and Metal want ads reach the right people. It is the only paper in Canada that will reach the hardware-men, plumbers, manufacturers and travellers from the Atlantic to the Pacific.

For a mere trifle you can talk to all the men in Canada who are interested in the hardware trade.

Are you taking advantage of this?

Somebody in Canada wants just what you have to sell: we can find that somebody for you.

Condensed advertisements in Hardware and Metal cost 2c. per word for first insertion, 1c. per word for subsequent insertions. Box number 5c. extra. Send money with advertisement. Write or phone our nearest office.

Hardware and Metal

Toronto **Montreal** **Winnipeg**

Business Getters for Tinsmiths

Are there any defective chimneys in your locality?

Then it is your chance to get business.

THE

ZEPHYR Ventilator

saves the expense of building a high chimney.

It is so constructed as to guarantee a perfect draft and, therefore, easy combustion.

By using the ZEPHYR it is possible to burn soft coal with the best results.

The smoke is easily drawn off, enabling an even heat to be generated.

THE ZEPHYR IS STRONGLY PUT UP AND WILL RESIST ANY STORM

This is what others think of it:

Montreal, February 20th, 1901.
Messrs. HARRIS Co., Montreal. Providence Mother House.
Dear Sirs,
It is with pleasure that I can say that your "AEOLIAN" Ventilator is the best ventilator we have used on our establishment so far. We had eight (8) of the large sizes put up on our Mother House and on the Hospice Gamelin, and they gave us entire satisfaction.
Yours respectfully, SISTER MADELEINE, Dep.

We manufacture and guarantee

The Harris Filter — The Zephyr Ventilator
The Aeolian Ventilator — Expansion Conductor Pipe

The J. W. HARRIS COMPANY, Limited

Successors to LESSARD & HARRIS

Contractors ***Montreal***

Peerless Woven Wire Fence

Here is a fence that will sell. Once up no need for repairs. It lasts for years—just what you are looking for. It's horse-high, pig-tight and bull-strong. No stock can get over, under or through it. The most unruly animal can't butt it down and there are no barbs to tear and injure stock. It's

WIND AND WEATHER PROOF

The Peerless lock holds it perfectly rigid and the wires can't slip up, down, nor sidewise. Cold weather won't snap it and hot weather won't make it sag. It's the greatest thing out for farmers, stockmen and all other fence users which makes it a red hot proposition for dealers. It comes in rolls all ready to put up. It's easy to handle. Write for prices.

The Banwell Hoxie Wire Fence Company, Ltd.

Dept. J, Hamilton, Ont.

A Booklet about the Best Wire Fencing in Canada ought to be sought by Hardwaremen

It costs nothing—i. e., costs **you** nothing.

It tells much—i.e., tells **you** much.

It leads to enquiry—i.e., leads the **enterprising dealer** to enquiry.

It cannot go unless it is addressed—i.e., cannot go unless it has **your** address.

Speak Up!

THE OWEN SOUND WIRE FENCE CO., Limited

OWEN SOUND, ONT.

Sold by { Messrs. Caverhill, Learmont & Co., Montreal
Messrs. Christie Bros. Co., Limited, Winnipeg

Persons addressing advertisers will kindly mention having seen their advertisement in this paper.

TRADE MARK A★1 THE R. BERG HARDWARE CO.

LOCKS

of all kinds and descriptions and of

FIRST CLASS QUALITY

Ask your jobber for prices on our well-known goods, comprising all lines of tools, general hardware, cutlery, etc., etc. You will find they are all right and the goods are too

F. W. LAMPLOUGH & CO., - Montreal

Can. Feb. 27, 1889

Clean, Pure Air in Houses

at all times can be obtained only by ventilators. Mine is guaranteed to work all the time and give satisfaction. Fix up your chimneys. Write to-day for particulars about the best seller on the market.

CAUTION—Beware of imitations; any with Fan or Wheels in the inside to block the interior not genuine

St. John, N.B., December 28th, 1889.

G. D. Pearson, Esq.
The Cone Rotary Ventilator you put on the chimney of my house, Elliott Row, to prevent it smoking, answers the purpose perfectly.

HENRY GILBERT,
Clinton House, Germain St.

Bell Telephone East 2106

GEO. D. PEARSON & CO., Office and Works: **17 Pontiac Street, MONTREAL**

CARRIAGE SPRINGS & AXLES

ANCHOR BRAND

THE GUELPH SPRING & AXLE CO. LIMITED
GUELPH, ONT.

Hockey Sticks Hockey Sticks

Empire, Imperial and **Champion** are the names of the three best brands on the market. Why! Because they are made of the best Rock Elm and White Ash that can be secured.

We have been making Hockey Sticks for over Twenty Years and are the largest manufacturers of them in Canada.

Our sticks are used by all the best players. Our 1907 samples are now ready.

Before placing your next order write us for prices.

J. H. STILL MFG. COMPANY, Limited

ST. THOMAS, ONTARIO

Persons addressing advertisers will kindly mention having seen this advertisement in Hardware and Metal.

BOY WANTED

ANY BOY in every town and village in Canada can earn many dollars by selling our publications among his neighbors and relatives and business men.

YOU can begin at once. No money needed to start. We will provide you with capital. A watch to the boy who does good work, as special prize.

Write at once to

THE MACLEAN PUBLISHING CO.
10 Front St. E., TORONTO

Four Reasons why INVICTUS FIBRE is the BEST FIBRE PAPER on the Market

1. Invictus Fibre is made from personally selected pulp wood cut on our own limits.
2. This special pulp wood is again rigidly culled on arrival at our mills
3. The pulp from which Invictus Fibre is made is manufactured in our own pulp mill and transferred at once to the paper mill, avoiding all drawbacks of transportation.
4. Invictus Fibre will increase in strength and not grow hard and brittle as does fibre made from sulphite pulp. This is a quality of the peculiar and selected pulp from which Invictus Fibre is made.

A trial order will convince you that we are correct. See that each package bears the words "Invictus Fibre." Every ream contains 480 sheets

TRADE MARK

This design a guarantee of quality.

CANADA PAPER CO., LIMITED,
"Headquarters for high grade Fibre and Manilla Papers"

Toronto. Montreal. Windsor Mills, P.Q.

Are you interested in any of the lines that are advertised?

A Post Card will bring you price list and full information.

Don't forget to mention Hardware and Metal.

The "Automobile" Skate has won more praise than any other skate ever made in Canada. And you are not likely to forget it, for

"It's the Skate Everybody is Boosting"

Don't buy until you see the new designs and then **COMPARE PRICES, DESIGN and FINISH.**

Canada Cycle & Motor Co., Limited

TORONTO JUNCTION

Branches: Winnipeg, Man., Vancouver, B.C., Melbourne, Aust.

WRITE FOR CATALOGUE.

ARE YOU GETTING READY

FOR SPRING'S ENQUIRIES FOR SAP SUPPLIES?

SAP PAILS

FLARING

6 and 10 qts.

No. 8, 12, 16
6, 8, 12 qt.
Straight and extra deep

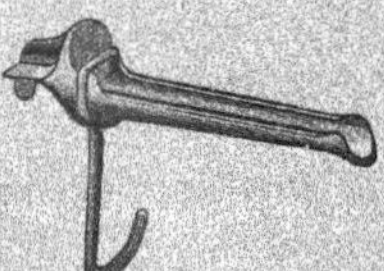

Sap Spouts, Eureka, Maple leaf and E.T. Three favorites with the trade and will give satisfaction to your customers.

SYRUP CANS, QUEBEC, ROUND AND SQUARE

1/4, 1/2, 1 gallons, plain and decorated stencilled.

Write for Prices

KEMP MANUFACTURING CO., TORONTO, ONT

THE KEMP MFG. CO. OF MONTREAL,
Montreal, Que.

THE KEMP MFG. AND METAL CO.
McDermot Ave. E., Winnipeg, Man

P.M.G. May Withdraw C.O.D. Clause

Postmaster General Understood to be Prepared to Withdraw -Proposal Unpopular in Parliament - Success Seems Assured—Keep Up the Fight.

Letters written by members of Parliament to retail dealers indicate that the campaign of petitions and letters of protest conducted by the Retail Hardware Associations and the trade newspapers during the past six or eight weeks has borne fruit and Postmaster-General Lemeiux, in view of the unpopularity of the proposed changes in the postal regulations amongst the members of Parliament and the thousands of letters which have been sent to the P.M.G.'s office, has practically decided to withdraw the c.o.d. feature which is so objectionable to the retailers throughout Canada. This news will be gratifying to every reader of Hardware and Metal and to each member of the Ontario and Western Canada Retail Hardware Associations, both of which organizations took prompt action towards protecting the interests of the trade as a whole. But the fight is not yet won. Things appear to be coming our way, but to let up in the campaign now would be suicidal.

P. M. G. Has Powers of Czar.

In Hardware and Metal of Jan. 19 we stated that a Hamilton retailer had been advised by his local M.P. that "it is not likely that a bill will be introduced in Parliament to amend the parcels post system, as the Postmaster-General has the power to adopt the proposed c.o.d. amendment without introducing a bill in the House."

At that time we referred to the exclusion of certain newspapers from the Canadian mails, in one case the Postmaster-General having been [illegible] withdraw a ruling after enforcing it for a brief period owing to the thousands of letters of protest written him. We urged that every retailer write the Postmaster-General, as well as the members of Parliament. If the publishers of a Socialist weekly could force the Postal Department to take back water after action had been taken, surely Canadian retail merchants should be able to head off the striking of a blow at their heads.

To make sure, however, that the Postmaster-General could so materially damage the business interests of thousands of taxpaying retailers by introducing new postal regulations without consulting Parliament Postmaster-General Lemieux was written to on January 17 as follows:

"We have been informed that you have the power to adopt the parcels post system, which you referred to in the House a short time ago, without introducing a bill Will you please inform us if this is so." Common courtesy would have warranted the sending of a reply to this letter, but to date it is unanswered

On January 31, Dr. Sproule, M.P., representing East Grey, inquired from the floor of the House if the Government intended introducing a bill this session for carrying post parcels c.o.d "No legislation is necessary," was Postmaster-General Lemieux's answer, "it is a departmental regulation and is now under consideration."

Inactivity Means Defeat.

The Postmaster-General, therefore, can not only exercise the powers of a press censor, can say what Canadians should or should not read, but he can cause the ruination of innumerable small business men by a simple revision of the postal regulations. These are wide powers, such as are only exercised by Czars and autocrats. They are not in keeping with our theories of constitutional government, but it must be remembered that the people still have the ballot and were a Liberal or Conservative Government to go too far in the way of initiating dangerous and unpopular postal regulations the people have the power to mete out punishment to the guilty ones, and there are enough retail merchants in any constituency to change its political complexion.

With a clear knowledge of the powers of the Postmaster-General, and a knowing that he is still "considering" the revision of the parcels post regulations, it behooves every retailer to "get busy" and "keep busy."

It isn't enough to have already written one letter. That letter did good. Follow it up by another letter to the Postmaster-General, sending a copy to the local member of Parliament and another copy to us. Pass the letter around amongst the retailers in your town and have them sign it before forwarding to Ottawa. And don't fail to show the letter or petition to the editor of your local paper, so that he can write an editorial and send copies of the paper to the P.M.G. and M.P.

Keep up the fight until we advise you that success has been achieved It will be far batter [illegible] law-makers at Ottawa with letters than to send too few.

If the politicians see that Canadian retailers are alive to their interests and in fighting mood against the capitulation of the Postal Department to the mail order millionaires it can be taken for granted that they will think twice before letting the hammer fall.

A Conspiracy of Silence.

While the retailers have been active in sending mail bags full of letters of protest to Ottawa, there has been no indication, outside of the articles in the MacLean trade newspapers and the circular letters of the Retail Hardware Associations that the retail trade is in danger of being delivered a solar plexis blow by the departmental store interests

Were it not for the Postmaster-General's reply to Dr. Sproule, M.P., in Parliament on January 31, many dealers might rightly think that there was "much ado about nothing."

Surprise has been expressed that the daily papers, even those of the highest standing, have made no attempt to inform the public as to what the effect of the proposed legislation would be. Their silence on a question of such national importance is significant, and is to be condemned with all possible severity. Throughout the country we find their attitude interpreted by retail merchants as good evidence that they are under the thumb of the department store interests.

What other reason can they advance to explain their inaction? Some might defend themselves by saying that the proposal is still in very indefinite form, and that it would be unwise to turn public opinion one way or the other until an idea can be given of just what it will embrace. That would be a poor excuse. They know the intention is that the post office shall carry much larger parcels than at present, at a cheaper rate than the express companies now give—virtually the establishment of a postal express service; that provision shall be made for the sending of parcels from any point to any other point c.o.d., collection to be made by the postmaster at destination; that with this system in operation the Government would, as a matter of course, absorb the express companies; that the post-office would be expected to carry parcels at about cost; that the big retail catalogue houses would be enabled to enter more directly into competition with every local merchant in Canada, with the natural result that great numbers of the latter would be forced into liquidation or a position of insignificance, as mere vendors of staple lines, that the wholesale trade would suffer in accordance with the effect upon retailers; and that, viewed from every standpoint—if they take the trouble to examine the situation—the enactment of such legislation would be one of the worst things that could happen to the mercantile life of Canada at the present time.

Effect of Proposed Changes.

The adoption of the proposed change in the postal system would mean the concentration of the great volume of trade in the large cities, to the impoverishment of the smaller cities, towns and villages. It would act as a powerful deterrent to the general progress of the country, for in the building up of a nation the small merchant plays an important part. He has constituted one of the most potent factors in the development of great areas in the West, through the extension of long credit to settlers, thereby permitting them to devote their cash capital to the improvement of their farms. He sometimes got his money after the first crop; he sometimes had to wait until the second. In this way he tided thousands of families over the crucial period of their pioneer existence.

The catalogue houses cared nothing about this trade, for the reason that payments were not spot cash. The local merchant catered to it without opposition from outside, and now that circumstances for the settler have improved, and he has more money at his disposal, the big city stores value his business. With the assistance of parcel post and special express rates they are able to get a good deal of it at the present time. Now it is proposed to equip them with advantages that would mean the attraction to their hands of all, or nearly all, the cash trade, while the local merchant would be left to take care of the credit business and shoulder bad debts besides. This would preclude the possibility of the latter being able to buy his goods for cash and sell as cheaply as he otherwise would. In calculating his percentage of profits he would be obliged to make provision for terms exacted by manufacturers and wholesalers on long dating and also for bad debts. As the country became more prosperous and people were able to buy more and more for cash, and recognized the advantage of so doing—the tendency in this direction is becoming stronger every day—his trade would gradually be depleted and that of the catalogue houses strengthened.

Cash to City, Credit at Home.

Throughout the older-settled portions of the Dominion it would be difficult to find a community from which mail orders are not sent. Often it is found in small towns, villages or cross-road settlements, where a good proportion of the buying public is composed of farmers, that many people send all their ready cash, to the catalogue houses and look to the local merchants to extend them good long credit at any time, and go even farther at seasons when crops are backward. Then, in the general stores, farmers find a market for their produce, with keen competition to assure them of good prices. Kill off the general stores and you kill off this market, as well as deprive the farmer of the credit accommodation that he often finds necessary.

It does not require any extraordinary perception to see what effect the decrease of activity in local markets would have on the towns, villages and surrounding communities with respect to real estate values, etc. The effect would extend from one of Canada to the other.

We see no season for the silence of the metropolitan press other than that it is influenced by the advertising patronage of the catalogue houses. The conclusion that this is so is forced upon us, though we would like to believe differently. Not one of them gave any prominence to the matter in its news columns, and only one that we have seen took it up editorially, then making no mention whatever of the injury that would be done to local merchants is such a change were adopted. Another, in an obscure paragraph of seven or eight lines, did go so far as to say that great benefit would accrue to mail order houses. Even the question of Dr. Sproule and the answer of the Postmaster-General, quoted in the opening paragraphs of this article, were ignored by all Toronto papers but one in their parliamentary reports. Their representatives in the press gallery must have been acting under instructions or else a strict censorship in this regard is exercised over their copy at the offices of publication.

Jobbers Should Assist Retailers.

Every wholesaler should use his influence with both members of Parliament and the Postmaster-General to aid in defeating any further extensions of the parcels post regulations which are intended to switch business from the country dealer to the city mail order house.

As already pointed out, such legislation will be certain to hurt the jobbers in the long run, even if jobbers sell to departmentals, as the tendency is for the mail order houses to secure control of the output of factories owned or controled by them. Dry goods jobbers have taken up the agitation actively and wholesalers in other lines might well follow their example.

MORE LETTERS FROM DEALERS.

Following are a few additional extracts from the thousands of letters being forwarded to the Postmaster-General and members of Parliament by retail dealers in all parts of Canada in opposition to the proposed extension of the parcels post system:

* * *

"It would be a great injustice to the many local merchants doing business to-day and paying business taxes to allow the mail order houses to step in and get the cream of country trade through the medium of this c.o.d. privilege. . . . I think mail order houses should be taxed on the amount of business they get in the country."—Ontario.

* * *

"In 99 cases out of 100 the consumer can purchase just as cheaply at home as from the department stores, if he takes into consideration the postage or express. I trust you will do all you can to oppose this measure. If put through its effects would be far reaching and exceedingly injurious."—Ontario.

"If passed, this bill would place a great many local merchants in such a position that the mail, order houses would practically ruin their businesses."—Nova Scotia.

* * *

"This bill has, no doubt, been initiated by the retail catalogue houses. If passed they would inundate the towns and villages of Canada with parcels, to the injury of local merchants, who pay the local taxes and keep the wheels of commerce moving."—Nova Scotia.

* * *

"For some years past the mail order houses have been industriously circulating their catalogues throughout the Island, and thousands of dollars have been sent to them each year for merchandise that might have been purchased just as cheaply at home. This is a serious loss to the merchants, and the province generally, and if the proposed bill is allowed to pass, mail order business will be greatly increased."—Prince Edward Island.

"You are aware that in all towns the business portion is the one which pays the bulk of the taxes, and that for the privilege of doing business. Any legislation that would allow outsiders easy access to towns, without contributing to the maintenance of said towns, would be unjust. We ask no favors, and do not want the other party to get any either."—Ontario.

* * *

"Being a merchant yourself you will understand the matter fully. There is no doubt that the large mail order houses are at the bottom of the proposal."—Prince Edward Island.

* * *

"I would ask you to oppose this bill and try to secure the co-operation of other members as well."—New Brunswick.

* * *

"We have been informed that a bill will be introduced into Parliament at the present session to establish a cheap parcel post system in Canada, whereby the post office shall collect on parcels. A moment's reflection will, we think, show you the injustice of the proposal. It would mean placing at the disposal of the few mail order houses a great public utility, to the support of which these concerns contribute not a dollar in comparison to the hundreds of dollars contributed by those whom such a move would injure. No doubt this ridiculously unjust and unbusinesslike proposal is instigated by the mail order houses. They are not satisfied that the postal service carries their trainloads of catalogues into every corner of the Dominion, thus placing them in a position to unfairly compete with the local merchant, who has to bear his share of local expenses, while the catalogue house goes free; they now want the post office to carry their wares cheaper, and do their collecting to boot—in short, to transact business for them at the expense of the country. We are confident that, should this bill appear, you will lend a hand in passing it to the place where it belongs—the waste basket."—Quebec.

RETAILER CHARGED WITH CONSPIRACY.

Montgomery, Ward & Co., Chicago, have begun action in the circuit court at Sioux Falls, S.D., to secure an injunction against the South Dakota Retail Merchants' Association on the ground that the association has conspired to prevent jobbers and manufacturers from selling goods to so-called mail order houses. This shows that the mail order houses are hard hit. There is little question but what the merchants will win, as there has been no conspiracy. The jobbers and manufacturers are simply protecting their own interests in their efforts to supply retail merchants in preference to mail order houses. There is no conspiracy, and never has been. Montgomery, Ward & Co. will be unable to prove their case, and simply hope to stem the tide that is setting in against them.

Do not make the same mistake twice. The show window is the merchant's great ad.

Window and Interior Display

Photos of Window Displays and Sketches or Photos of Store Interiors or Display Fixtures Solicited.

PRIZES AWARDED FOR DISPLAYS.

The results of our recent competition of photos of window displays of Christmas goods was not as satisfactory as it should have been. About a dozen clerks and dealers sent photos, but there seemed to be a desire to do it on a cheap basis. One of the finest displays came in on a very small amateur photo, so small and unsatisfactory that it would be a waste of money to have an engraving made from it. Had it been taken properly it would at least have won second prize.

A Lesson for Clerks.

This should be a lesson to window dressers. When a display is arranged that one feels proud of, a photo should be taken of it—not a poor photo, but a good one which will do the display justice. Even if it costs a dollar or so the expenditure will be repaid in many ways. A clerk who earns a reputation as a window dresser can command a higher salary than one who is only good behind the counter. His services will be in demand, and it is his own fault if he does not get the publicity necessary to widen his reputation. Excessive modesty, like parsimonious photo taking, does not pay.

It is equally necessary for the dealer to make a reputation for his store window displays. He should encourage his clerks to develop their talents allowing them to go to a little expense for papers for backgrounds, window cards, etc. And it will pay him to give the local photographer from 50 cents to $1 for a photo of a good display. If the display has merit and a photo is supplied us, we will have photo engravings made for reproduction in Hardware and Metal, and where a merchant desires to have the cut used in his local paper we will loan the engraving for this purpose. The local publisher will be only too glad to give space free to any advertiser for such a cut and accompanying article.

Winnipeg Clerk Wins First Prize.

W. J. Illsey, manager of the retail cutlery department for the J. H. Ashdown Hardware Company, Winnipeg, who was awarded the first prize of $6 cash, forwarded a very clear and well taken photo of a double window containing cutlery, carving sets, case goods of flat plated ware, hollow silver ware, cut glass and lighting fixtures, all timely lines for holiday sale. The background was of white silk, the upper portion being taken up with rows of electric bulbs artistically covered with holly. The photo will be reproduced in November next when the thoughts of dealers will be centred upon preparations for the next holiday season. These window displays were backed up with well prepared advertisements in the daily papers, copies of which were forwarded along with an interesting letter containing a description of the arrangement of the display and the results secured.

Second Honors to Halifax Traveler.

S. S. Wetmore, Newfoundland traveler for A. M. Bell & Co., Halifax, whose novel window displays have made him known to the hardware trade in all parts of Canada, was awarded the second cash prize of $4. His display was specially designed for publicity, the idea being to attract attention to the store, reliance being made upon other window displays of purely hardware goods to induce possible customers to enter the store and buy. From this standpoint the display did not really come under the conditions of our contest, but the display and the photo were so far in advance of the others that Mr. Wetmore's contribution had to be given second place.

The Other Prize Winners.

For window displays A. H. Gingrich, Woodstock, and R. W. Craigie, Goderich, the later a clerk with C. C. Lee, prizes valued at $2 were awarded, while a fountain pen was also given to L. H. Fleischauer, clerk in Peter Hymmen's big store at Berlin, for the best advertisements advertising lines shown in the Christmas windows.

Experience seems to show that photos of window displays can best be taken at night with the window lights supplying the light. In some cases flashlights can be used, but the advice of the photographer would have to be taken on this point. In the case of the photo supplied by Mr. Gingrich a flashlight was used and, there being no glass dividing the window from the rest of the store, the photo is peculiar in showing both the window display and the store interior.

FOR THE FROSTED SHOW WINDOW.

In northern portions of Russia, where zero weather is sufficiently common, experience has taught the owners of show windows that the only effective protection is a 3-inch air-space between two panes of glass The outer sash is rendered as nearly tight as possible by calking the chinks and pasting strips of paper over the crevices The glass is then carefully cleaned and dried on a clear, mild day, and a second sash, fitted with the same care to prevent all circulation of air, is inserted about three inches within the first. The double panes are said to obstruct the view very little. The physical cause of the deposit of moisture and ice upon windows is the difference in temperature between the surface of the glass and the air bearing a relatively high proportion of moisture, which comes in contact with it.

TELL THE STORY PLAINLY.

Too many trimmers, especially those who are somewhat new at the work, begin on a window with the feeling that they must do something great and wonderful. They try to, and the results are not always artistic or effective.

It is the same way with persons who are not used to writing for publication. Asked to write "something for the paper," they feel that it is up to them to produce some sort of ornate literary masterpiece which will be a gem of its kind and will attract widespread attention. They do not realize that all that is necessary is for them to tell in simple language and straightforward fashion what they know or think about the subject at hand, and that an article of this sort, absolutely devoid of any frills or furbelows, will be twice as readable and twice as strong as anything else.

Instead of starting out with the idea that he must produce a wonderful and remarkable trim and getting somewhat "rattled" in his valiant endeavor to do so, the trimmer should bear in mind that it is necessary for him only to display the goods at his disposal in strong, simple fashion, so that people will be attracted into the store to buy.

Of course, this does not mean that the trimmer should make no effort to produce an artistic and beautiful window, but simplicity is the thing to strive for and the fewer frills and furbelows there are in the window the more effective it will be.

It has been said that after looking at a man who is really well dressed one will not be able to tell what he had on; in other words, that the man who is really well dressed wears nothing that obtrudes itself upon the attention and that the things he has on are so perfectly in harmony that the whole effect, while restful to the eye, is not particularly noticeable.

It is so with the window. It should have a strong, direct effect upon the beholder, but he should not be made aware that there has been any effort made to entrap his attention or to force him to look. The highest art is the concealment of art.

TIMOTHY EATON, THE MAN

Sketch of the Great Merchant Who Passed Away Last Week—His Own Fortune-maker—An Irishman of Scotch Descent.

By the death of Timothy Eaton on Thursday of last week Canada loses its greatest merchant. He began as a retailer and remained a retailer and his monument is the great business he built up. Mr. Eaton came to Canada in 1857 with £100, a common school education and the memory of a good mother. He has left a business that is among the two or three largest retail stores in the world. In proportion to the country and population, it is the largest in the world by long odds. Marshall Field's, Chicago, is larger, but not nearly as much larger as the population of Chicago is larger than Toronto's or that of the United States larger than Canada's. A member of a government survey called at a settler's shanty in the back woods of northern Ontario. He wanted to see the settler, but the wife only was at home. When he told her he was from Toronto she became interested at once; asked him in and began to question him about Eaton's. She told him they had been getting their supplies from Eaton's for several years and were greatly pleased with the service but she had never been in the store. She wanted to know everything he knew about the store and its founder.

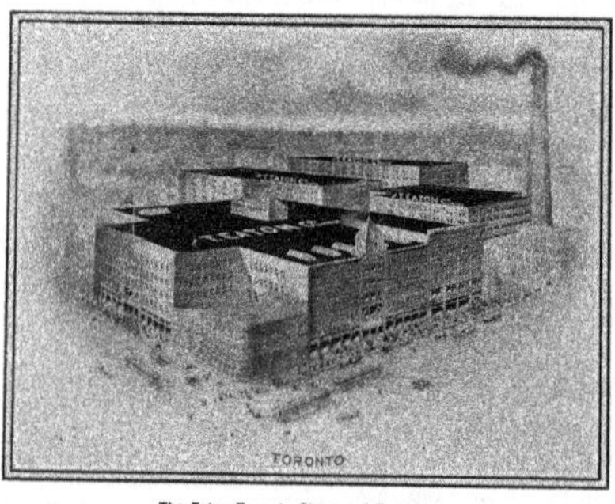
The Eaton Toronto Store and Factories.

Eaton the Man.

The Eaton store is known to most Canadians, but the man behind it has been little in the limelight of publicity. Few men knew his politics. In commerce or finance outside his business he was practically unheard of. He had a passion for merchandising, not for money making. He said to his son not very long ago: "If anyone had told me years ago that I would control as much as a million dollars, I would have laughed at him." His mind was not on money making, but his ambition was wrapped up in his business. Every faculty of his alert and powerful intellect, every ounce of his abundant physical strength was directed along the lines of building up the Eaton store. His manner was abrupt to bluntness. Personal considerations weighed little or nothing with him when the welfare of the store was at the other end of the scale. But his vision was far-seeing; his politics were broad, liberal, and founded in justice and honesty. "The greatest good to the greatest number" was a motto of his. He was a man of strong individuality and great self-reliance. His mind worked with such rapidity that his decisions, even in important matters, were often instantaneous.

At one time there was a vacancy at the head of an important department that had to do with the whole store. Two good men in the department wanted the place and a good deal of feeling developed among the other managers. It came to the ears of Mr. Eaton. He asked for an explanation. "Let them both go," was his immediate decision. One of them is now one of the highest-salaried men of his class in America, and the other is manager of a big store in another city.

A Just Man.

An incident of many years ago will show how, while his sense of justice could be appealed to, his first consideration was the business.

When Mr. Eaton was located on Yonge St., south of Queen, he employed for the print counter a young Scotchman just arrived from the Old Country. A couple of weeks after a methodist clergyman and his wife were in the store. She bought some print and moved on with Mr. Eaton to another counter. Her husband had gone to the methodist bookroom just around the corner. When she went to pay Mr. Eaton for something she had bought

THE LATE TIMOTHY EATON.

from him she missed her purse, and saying she left it at the print counter went to get it. The Scotch clerk told her there had been no one at the counter and if the purse was left there it would be here yet. He began to search but could not find it and she accused him of having it. Mr. Eaton came and the clerk said to him what he had said to the customer. Mr. Eaton said he would have him searched and called the foreman. The Scotch clerk put his back against the shelves and said to his employer, "send for a policeman, and I'll stand here, but no man in this house

JOHN C. EATON,
The Present Head of the Business.

will put a hand on me." No one wanted to tackle the angry Scot, and while they waited for the officer the minster came back. His wife explained that she had lost her purse and he replied, "oh, no, my dear, you haven't. Don't you remember you gave it to me ?"

Demanded Apology.

The customers were moving away when the clerk interrupted with "this woman has accused me of a crime, and if she doesn't apologize it'll be my turn to send for an officer." The minister came back to apologize, but that wouldn't do. He insisted upon his accuser doing so and she did. When they had gone Mr. Eaton called him to the office, paid him off and told him to go. When he had his money the clerk said:

Mr. Eaton, if you had a son 3,000 miles from home and he was accused falsely of theft and did not stand up for his honor, what would you think of him?"

"But you've lost me that customer," growled Mr. Eaton.

"Better lose a thousand customers than that a son should sully his honor," burst forth the Scotchman.

"Em. Go back to your department," was all Mr. Eaton said, and that ended the matter.

A Blunt Man.

Mr. Eaton's bluntness was positive, assertive. His intensely practical mind, keyed by nature to rapid action, could not brook parleying. He made up his mind quickly and acted promptly. Like other people he made mistakes, but he was never ashamed to admit one as soon as he saw a better way. With him nothing was finished till he thought it right. He had no patience with makeshifts. He might order up a stairway in the store to-day and to-morrow morning if he saw a better arrangement he would order it down again with just as little hesitation. His ambition never halted at the biggest or best store. His aim was ever a better store. Restless progressiveness was a dominant characteristic.

Judging Men.

One not inconsiderable factor in his success was his faculty for estimating men: He was going through the store once and came upon a big Irishman, lately employed as floorwalker. "How long have you been here ?" he asked abruptly. "It's none of your business" responded the Irishman cheerfully, not knowing his employer in person. "Well, what are you doing ?" persisted Mr. Eaton. "I'm minding my own business, and that's more than you're doing," returned the Irishman, testily. Mr. Eaton's comment afterwards was that he wished he had several hundred more like that in the store.

Another time a manager of a department affecting the whole business, made arrangements for a week off. In the course of the week something went wrong and the operations of the department were practically suspended for a couple of days by Mr. Eaton's order. Shortly after the manager returned Mr. Eaton was in the private lunchroom when this man entered. He called him over and in his characteristically brusque style accosted him :

"So you're to blame for all this trouble we've had."

Without hesitation the manager shouldered the blame.

"Yes sir, I guess that's right."

Just as gruffly Mr. Eaton said it shouldn't have happened, and the conversation was ended, but next pay day the manager found an advance of $10 a week in his salary. Mr. Eaton had wanted to see if the man was big enough for his place and would shoulder his responsibility as manager, present or absent.

Treatment of Employes.

Talking about Mr. Eaton with men in the employ of the store to-day or with men who left it many years ago, they all render his conversation in a tone as brusque and abrupt as one can imagine a man to speak. Yet in the treatment of his employes he was large hearted. There has never been a strike on account of wages or conditions. A manager told the writer he knew he was getting more money than he could get anywhere else. If an employe is away a day, two days at most, someone is sent to see if he or she is ill. Should that be the case, it is seen that he or she has proper medical attendance, or if necessary is sent to a hospital, and the hospital bill is paid. Before the business assumed such tremendous proportions Mr. Eaton would go himself in cases of sickness. He took pride in his staff. Speaking to his pastor once after a store picnic he exclaimed, "It would do your heart good to see them."

Early Closing.

More than any other man, perhaps more than any score, Mr. Eaton has helped on the movement for a shorter workday in Canada. One Christmas Eve, years ago, there was such a crush in the store that the crowds were almost unmanageable. Mr. Eaton declared he would never keep his store open at night again. He inaugurated the 5 o'clock closing, first during the summer and extended it to the whole year. He next closed the store at one o'clock on Saturdays during the hot months and his intention was to extend that month by month. He often said the time would come, not in his lifetime but his sons', that stores would close all of Saturday all the year. His theory was that if people had one clear day in which to enjoy themselves, they would attend more regularly to their religious duties on Sunday and would be the better for it, both as employes and as citizens.

In religion Mr. Eaton was a Methodist, and a stout adherent of the old-fashioned orthodoxy, but he was not an aggressively devout man. He took his religion as a matter of course and his business as a matter of fact.

His Ancestors.

Mr. Eaton's forbears were farmers near Belfast, Ireland, and he was descended from a Scotch settlement of 300 years ago when Cromwell was making Ireland over. His father died before he was born, but his mother was a brave, ambitious woman, and despite their poverty, secured to her children the national school education available at the little village of Clogher. After the potato famine of 1847 the oldest boy came to Canada and Timothy was apprenticed to a draper at Portglenone, a small market town. After five years he took his wages, £100, and followed his elder brother. He started in business in Kirkton, Huron County, in 1857, where supplies had to be brought in by wagon. Later he removed to St. Mary's where he was in partnership with his brother. The field was not large enough for him and he had become possessed of a desire to do business on a cash basis. He said to his brother, "You take the money and I'll take the business, or I'll take the money and you take the business."

The latter solution was adopted and he came to Toronto in 1869 and opened a small store on Yonge St., south of King. Then he removed to a store between Queen and Richmond and some time later to 190 Yonge St.

Principles of Business.

Early in his independent business career Mr. Eaton laid down three principles, no credit, no misrepresentations, one price to all. On these by the force of his ability, his courage and his dogged perseverance, he built his great success. All three qualities were often needed in the early days to pull him through. At one time a big dry goods house in Toronto guaranteed all his bills and he paid five per cent. for it. A Scotch house once had an account of $29,000 against him, and becoming nervous sent a man to collect it. Mr. Eaton told him as brusquely as usual that he would pay it whenever he got ready. But the agent entered suit and an agreement was made for the payment at the rate of $1,500 a month. The payments were all met promptly. But for many years the Eaton store has bought for cash and its buyers circumnavigate the globe.

No rule of business could be adhered to more strictly than Mr. Eaton's to sell for cash. It is stated that Mrs. Eaton has asked a clerk to send an article home for her without the cash and been refused. The wealthiest customers pay the cash or get their goods c.o.d.

One is inclined to seek the reason for Mr. Eaton's great success. It is to be found only in the man himself. He is one of the few very rich men whose wealth is quite independent of any government aid or public privilege. He made his money and built his success as a trader and the business he built up amply entitles him to be named a merchant prince.

LAWN MOWER GRINDER.

In placing the 1907 pattern of the Ideal lawn mower grinder upon the market, Root Bros., Plymouth, O., assert that they have now made the machine as nearly perfect as it can be built. They state that owing to the great interest taken in automobiles a large number of repair shops have sprung up, and that not one of these ought to be deemed complete if it does not possess an Ideal grinder. and the hardware dealer who does such work as sharpening lawn mowers, is another man who will be interested, as will many blacksmiths. Think of doing in 15 minutes—and with very little exertion even if the grinder is run by hand instead of power—what formerly required an hour of downright hard work with a file, and even then one was not always sure that the bevel on each blade was accurate, and so much depends on this bevel ! Skates, tools, etc., can be sharpened as well as lawnmowers on the grinder. Any desired information as to price and terms can be secured from the manufacturers by mentioning this paper and writing on the firm's business paper.

HARDWARE AND METAL

Established 1888

The MacLean Publishing Co.

Limited

JOHN BAYNE MACLEAN . *President*

Publishers of Trade Newspapers which circulate in the Provinces of British Columbia, Alberta, Saskatchewan, Manitoba, Ontario, Quebec, Nova Scotia, New Brunswick, P.E. Island and Newfoundland.

OFFICES:

MONTREAL, ... 232 McGill Street
Telephone Main 1255
TORONTO, ... 10 Front Street East
Telephones Main 2701 and 2702
WINNIPEG, ... 511 Union Bank Building
Telephone 3726
LONDON, ENG . . . . . 88 Fleet Street, E.C.
J. Meredith McKim
Telephone, Central 12960

BRANCHES:

CHICAGO, ILL. . . . . . 1001 Teutonic Bldg.
J. Roland Kay
ST. JOHN, N.B . . . . No. 7 Market Wharf
VANCOUVER, B.C. . . . . Geo. S. B. Perry
PARIS, FRANCE - Agence Havas, 8 Place de la Bourse
MANCHESTER, ENG. . . . . 92 Market Street
ZURICH, SWITZERLAND . . . . Louis Wolf
Orell Fussli & Co.

Subscription, Canada and United States, $2.00
Great Britain, 8s. 6d., elsewhere . 12s

Published every Saturday.

Cable Address { Adscript, London
Adscript, Canada

STEEL COMPANIES MAKING MONEY.

It has been customary to hear hard luck stories from the Canadian steel companies, owing probably to the repeated reports regarding the proposed establishment in Canada of a branch plant of the United States Steel Corporation. Through these stories the public has been educated to the belief that there is no money in the steel business in Canada, and that the promoters are philanthropists in building up Canadian industry and in providing work for our working people.

It is refreshing, therefore, to read that at least one concern, the Hamilton Steel and Iron Company, is enjoying a large measure of prosperity. According to the Toronto Daily News the company's "stock has jumped from 110 to 135 within the last few days as a result of an effort being made by a few of the largest shareholders to buy in all of the stock. The reason for this action is said to be that $500,000 worth of bonus stock is to be distributed pro rata among the present shareholders. The plant is to be doubled, but the money will be taken out of the reserve fund.'

A Montreal paper also says that the common stock of the Nova Scotia Steel Company will be put on a 6 per cent. basis for 1907, and that a dividend of 6 per cent. for the six months ended March 31, will be paid in April. The dividend will take only $75,000 out of the net earnings for 1906, which are estimated at over $900,000, and which, after the payment of the bond interest, would leave about $450,000, so that after the payment of the dividend there would be left over $100,000 for extensions and improvements. It is expected from the unfilled orders at hand that the company's net earning for 1907 will exceed those of 1906.

With conditions such as are indicated in these extracts existing it looks as though the United States Steel Corporation would be making no mistake in entering the Canadian field, with furnaces at some lake port in Ontario. It is interesting to know that the corporation is well thought of in the States, where it is best known, the American Metal Market Report of Feb. 4, having the following to say about them:

"The general feeling in the trade here and also in leading financial circles is that the commission's report will give the steel corporation a clean bill of health, although their reports on some other corporations have been of a different character. No matter the revolution and changes the Steel Corporation caused in the iron and steel trade, in many cases so distressing to hundreds of individuals, in many cases leading to the complete elimination of some of the most active, worthy, and at one time, leading members of the iron and allied industries, still it must be acknowledged that during the past two years the United States Steel Corporation has not excited in the trade the feeling of distrust and animosity created by other trusts, but instead has steadily gained friends. We believe the reason for this has been the fairness and frankness with which the public has been treated by the Steel Corporation, in the free publicity of their reports, and the management of its business; their absolutely uniform fairness towards buyers, both on prices, and opportunity to make their purchases, for definite deliveries at definite prices, and the courtesy and consideration with which the powers they possess have been exercised.

"Contrast the feeling in the trade that exists to-day towards the American Smelting & Refining Company, (the Pig Lead Trust) caused by their dictatorial use, and we think abuse, of the powers that they enjoy through the 2½c. per pound duty that protects them in their monopoly, and one sees then at a glance why the U.S. Steel Corporation has won the confidence of the trade, become a settled institution, respected, and one might say popular."

WINNIPEG DELINQUENT LIST.

Winnipeg retailers are setting a worthy example to retailers in all parts of the country in the harmonious way in which they are working together and the intelligent manner in which they are reaping the benefits of organization. Every dealer who gives credit at all has his troubles with dead beats and slow pay customers. Often he is imposed upon by customers who ask for credit at a time when they owe heavy store bills to other merchants, bills which they are unable to pay or which they have no intention of paying. If he knew of this heavy indebtedness he would hesitate to supply goods on credit, but under ordinary circumstances he has no means of knowing about it.

Winnipeg retailers have grappled with this problem in a practical way. Grocers, butchers, dry goods men, druggists, etc., have been organized in Winnipeg in separate associations by secretary W. A. Coulson. The hardware men have long had an efficient organization and their secretary is J. E. McRobie, secretary of the Western Canada Retail Hardware Association. The members of these various branch associations furnish Mr. Coulson and Mr. McRobie with the names and addresses of "delinquents" who refuse to pay their accounts and with whom no satisfactory settlement can be made; with the names and addresses of slow pay customers and with the names of delinquent and slow pay customers who have changed their addresses and cannot now be located by those dealers to whom they are in debt. The secretaries keep secret the sources of their information, but, from the data supplied, they compile a list which is supplied to association members in good standing. This list is revised every fortnight and is thus kept up to date in every respect.

The writer has before him the list dated January 15th, 1907. It contains the names and addresses of 144 "delinquent" customers and of 75 slow pay customers. It also gives the names and former addresses of 50 delinquent customers who cannot now be located.

All this is valuable information and must, in the course of a year, save the members many times their small membership fee. The list is in no sense a "black list." It is not posted up and the members are under no obligation not to sell or give credit to those who are listed. The dealer has the list for his own information and if he gives credit to a customer whom other dealers report as a "delinquent" he runs the risk and walks into danger with his eyes open.

MONTREAL HARDWARE MERGER.

To have spent thirty-eight years in active, strenuous life as a hardware merchant and to have lived with success and honor, and now to retire as a director in one of the largest commercial mergers in the hardware trade of Montreal—tells briefly the life of J. P. Seybold, head of the Seybold, Sons & Co., which was the other day merged with the Starke Hardware Co., into the Starke-Seybold, Ltd. Away back in 1869 a little hardware store was opened by J. P. Seybold at the corner of Murray and St. Joseph streets—the latter is now Notre Dame street. Young Seybold had saved a small sum of money. That was all he had in the world, except the capital of grit and determination and character. He stocked to the extent of about $1,600, a nervy thing to do in those days when credits were all six months. Mr. Seybold adopted a principle, which in telling, he said would be worth while impressing upon the retail merchants of the land to-day. It was, that he never lived on the wholesaler's money. That is, he never permitted his credits to go past what he could pay at any moment with his own money; he with his small capital, coupled with the cash he took in and the profits made, was able to pay his bills so that inside of 18 months he was getting his five per cent. discount. From that day it was plain sailing, although the difficulties were many.

At the unheard-of hour of a quarter to six every morning this young man was at his store. He was the boss and the staff and he looked after everyone. Often he said that by eight o'clock he had his expenses made for that day. The store never closed before 9 to 9.30 each evening and on Saturdays 11.30. Imagine what the young men of to-day would say if their employers "dared" to ask them to follow this routine for years "There is only one secret about success," said Mr. Seybold, "and that is hard work and the concentration of effort upon one spot."

There were, however, bigger things looming up, and in 1878 a company was formed for wholesale trade under the caption, Seybold & Son Co., located on St. Paul street, just west of the new Coristine building. These premises were fitted for the wholesale trade and the little shop on Murray street passed into other hands. In 1901, at the time of the big Board of Trade fire, the Seybold building was completely destroyed. The head of the firm had resources and ability, and shortly on the opposite side of St. Paul street was erected the handsome structure of seven stories. This has been occupied for five years and the wholesale business has grown steadily and two sons, Gordon and Herbert, were taken in, giving an extra "S" to the name, as Seybold, Sons & Co. These handsome premises are now for sale or to let, as the merger occasions removal to the Starke Co.'s headquarters on St. Peter street.

The new directorate will include the Seybold sons Gordon and Herbert, and two Starke brothers, Robert and William, also J. P. Seybold, who will not be in active business. The new company will be capitalized at $400,000

Strong Personality.

A glance at J. P. Seybold would see a man inclined to be medium of stature, but well built, erect, full-chested, and well dressed, with the air of an Old Country business man. Fifty-nine years, thirty-eight of which have been filled with hard, devoted work, have stamped their mark upon a broad, well-shaped head—have marked it with reflectiveness, but kept its serenity. The silvery hair is no indication of age, for he was that

J. P. SEYBOLD, MONTREAL
Who has Retired from Business After 38 Years of Active Business Life as a Hardware Merchant.

way at thirty. The rest of the face might be taken for a young man of thirty—clean shaven, rosy, as if the owner had never lost a night's sleep or a day's enjoyment.

With a steady, clear eye, Mr. Seybold states proudly that he never speculated in his life outside of his own business His well modulated voice and general avoidance of extravagance at once impels one to the idea that investments of one's earnings would be safe in any venture where such a man had control.

Asked how he came to get such a grasp of the French trade, he said, laconically: "I speak French; I learned it myself down on Murray street." That, however, is not the secret. He knows how to solve the French character as he knows other characters and happened to concentrate his mind on Quebec.

His fad is flowers. A devoted lover of botany and of floral culture, J. P. Seybold has a handsome conservatory attached to his home. Now that he will have time to spend, he claims his flowers will be one of his greatest pleasures. Beneficent work has not escaped him, for he is a life governor of the Montreal general hospital.

He declares that three-quarters of the failures in business are due to extravagant living and too much ambition on the part of young men.

RETAIL HARDWARE CONVENTIONS.

The convention of the Western Canada Retail Hardware Association in Winnipeg next week is an important one, and should be largely attended. It is feared, however, that owing to the railway blockades many dealers will be unable to visit Winnipeg for the gathering. Reports of disagreements over price arrangements have been received and this matter should be discussed at length by the delegates. Complaints are heard, also, of members being slow in paying their fees. This is short-sighted, if not careless action, as the secretary and executive cannot be expected to conduct the business of the association without funds. Even if a member disagrees with some actions taken he should consider the general advantages of organization and realize that it would be suicidal to allow the work of organization so far accomplished to go to naught owing to inactivity and shortness of funds. Stay inside and oppose the matters you do not favor, but pay your dues and have an organization to fight the mail order houses or other enemies when occasion requires.

The convention held at Halifax recently was merely a preliminary gathering. It was decided to organize a Nova Scotia Retail Hardware Association and over two dozen representative firms have already enrolled. At an early date another convention will be held at which officers will be elected.

Preparations for the annual convention of the Ontario Retail Hardware Association, to be held on Good Friday, March 29, are being made, and indications point to it being a large and enthusiastic gathering.

When a note comes due at a wholesale house and you have failed to satisfy it, you have delivered a sledge hammer blow against your own reputation.

HARDWARE TRADE GOSSIP

Quebec

Leblanc Joliette was in Montreal last week on business.

E. Monbleau, St. Johns, Que., was in Montreal on business last week.

C. O. Gervais, St. Johns, Que., was in Montreal last week on business.

A. J. Silberstein, of New York, was in Montreal on business this week.

August Tanguay, of St. Johns, Que., was in Montreal on business last week.

J. Thompson, Sherbrooke, Que., was in Montreal last week purchasing supplies.

A. T. Hunter, of Howick, Que., was among the busy men in Montreal last week.

J. A. Plourde, Watton, Que., called on some of the Montreal jobbers last week.

Mr. Cuzner, of McDougall & Cuzner, Ottawa, called on the Montreal jobbers last week.

J. Beverley Robinson, metal manufacturers' agent, Montreal, has returned from a busines strip to Europe.

Mr. Stackhouse, of Stackhouse and McElroy, Kinburn, Ont., called on some of the Montreal wholesalers last week.

Geo. Brown, Eastern Townships, traveler for Caverhill, Learmont & Co., Montreal, was a caller at the Montreal office of Hardware and Metal last week.

Mr. Abrams, of Rostern, Sask., is in Montreal at present, after interviewing the Ottawa Government about several matters in connection with his business.

Jack Botrell, representing the Canadian Fairbanks Co., in Winnipeg, has successfully avoided snow drifts and railway tie-ups, and arrived in Montreal.

J. W .Harris, head of the J. W. Harris Co., Montreal, spent a portion of this week in New York. He leaves shortly for Winnipeg and other western points.

Robert Munro, managing director of the Canada Paint Co., Montreal, has left for Pine Hurst, North Carolina, where he will remain for the next month.

Wm. Hall, of the Hall Engineering Works, Montreal, returned this week from a trip through the large manufacturing centres of Great Britain, and reported several new agencies, as well as great prosperity everywhere.

F. Hammar, of the firm of Schuchardt & Schutte, New York, visited Montreal this week and was introduced to the trade by J. J. Sophus, Montreal, representative of the firm.

Harry A. Norton, vice-president of the Norton Jack Co., of Boston, was a caller at the Montreal office of Hardware and Metal last week. Mr. Norton has just returned from a trip through India, Egypt, and the Holy Land.

J. J. Sophus, Montreal, representative of Schuchardt & Schutte, New York, spent a portion of the week in Quebec, visiting the trade. Mr. Sophus is going to be married next week, and will take up his residence permanently in Montreal.

Chas. Dietrich, of Shurly and Dietrich, saw manufacturers, of Galt, is in Montreal. Business in their three leading lines is exceedingly good all through Quebec, and he says the story of the Key and the San Francisco disaster, is being fired at him from all sides.

H. G. Eadie, and H. P. Douglas, of Montreal, have formed a partnership under the name of Eadie and Douglas, to carry an extensive building and contracting business. Mr. Douglas was formerly with the Canadian White Co., and with his partner will make a strong business team.

Ontario.

Dreaney Bros., hardware merchants of Cobalt, have opened a branch store at Englehart.

H. Munderloh, of Munderloh & Co., Montreal, was a visitor in Toronto during the past week.

Fred Somerville, of the Toronto plumbing supply house which bears his name, is on a business trip to Winnipeg.

The tin shop of Robert Maines, tinsmith and plumber, Orillia, was completely destroyed by fire on Monday last.

C. L. Macnab, of Macnab Bros., hardware merchants, Orillia, called on Toronto jobbers last week on his way to New York.

James Stewart, hardware merchant of Pembroke, was quietly married on Wednesday of last week to Miss Charlotte Cameron, also of Pembroke.

Chas. P. Brown, formerly of McMurtry's hardware store, Galt, has taken a position with the Taylor Hardware Company of New Liskeard.

C. S. E. Saunders, of London, died at his home last week. Mr. Saunders had been in the employ of James Wright and Company, hardware merchants of London, since boyhood, having done the buying for his department for years.

Clark (Baldy) Graham, who for a number of years was employed with Rice Lewis and Sons, Toronto, has accepted a position with Lewis Bros, Montreal, as their travelling representative in Toronto.

Andrew Archibald, for many years with Chesney & Smiley, Seaforth, has joined the firm, which will be known in future as Chesney & Archibald instead of Chesney & Wilson, as previously announced. The new proprietors are both young men and have a prosperous business. Mr. Smiley, who came from St. Thomas five years ago, is now at Aylmer taking over the Rumball business at that town.

Western Canada.

Hintz & Lynch, hardware merchants of Yonda, Sask., have dissolved partnership.

Schwanz Bros., hardware merchants, of Rocanville, Man., have dissolved partnership.

James & Hughes, hardware merchants of Napinka, Man., have dissolved partnership.

The Stovel Hardware Company, of Edmonton, Alberta, will change hands in a short time.

Messrs. Hockin & Siddoner, Moose Jaw, Sask., will erect a large hardware store shortly.

W. B. Shannon & Company, hardware merchants, of Carberry, Man., are retiring from business.

Rattray, Cameron & Company, dealers in hardware specialties, Winnipeg, suffered loss by fire last week.

Owing to shortage of coal the Sherwin-Williams paint factory of Winnipeg had to close this week.

Siemens & Wiele, hardware merchants of Herbert, Sask., have dissolved partnership. Business will be carried on by G. P. Siemens.

The Brandon Hardware Company has made application to the Manitoba Government for power to increase its capital stock to $150,000.

The Peart Bros. Company, hardware merchants of Regina, will open for business this week in the old stand of Smith & Ferguson Company.

R. P. Maclennan, hardware merchant, Vancouver, headed the Liberal ticket in the recent B.C. elections, but five Conservatives romped in ahead of him.

J. G Rattray, of Virden, Man., has been nominated by the Liberals of Virden as their candidate for the Legislature in the approaching Manitoba elections. Mr. Rattray is proprietor of a hardware store in Pipestone.

Maritime Provinces.

The members of the St. John Iron and Hardware Association held the thirteenth annual dinner at the Union Club on January 30. Elaborate preparations had been made for the banquet, which was a very enjoyable function. The menu was in the form of a horseshoe, and formed an appropriate souvenir of the occasion.

TORONTO BOARD OF TRADE.

R. C. Steele, of the Steele, Briggs Company, seed merchants, has been chosen president of the Toronto Board of Trade to succeed Peley Howland, of H. S. Howland Sons & Co. For the executive council, Peleg Howland, Sigmund Samuel, of Samuel, Benjamin & Co.; A D .Fisher, skate manufacturer, and Mark H. Irish, of the Monarch Brass Manufacturing Company, are amongst those nominated, while A. Burdette Lee, of Rice Lewis & Son, is one of the ten candidates for representatives on the Exhibition Board.

Markets and Market Notes

(For detailed prices see Current Market Quotations, page 66.)

THE WEEK'S MARKETS IN BRIEF.

MONTREAL.

Damar Varnish—Advanced.
Brass Safety Chain—Advanced 10 per cent.
Flat Irons—Several lines show a five per cent. advance.
Hay Wire—Advanced five cents.
Wire Staples—Five cents higher.
Wire Nails—Advanced five cents.

TORONTO.

Cast Iron Fittings—Now 57½ per cent.
Ingot Tin—Reaction and higher prices.
Copper—Both ingots and sheets advanced.
Old Metals—Several lines higher.
Damar Varnish—Advanced.
Bluestone—Quoted higher.

Montreal Hardware Markets

Office of HARDWARE AND METAL,
232 McGill Street,
Montreal, Feb. 8, 1907.

Business still holds good in hardware circles, but heavy storms are delaying shipments in all quarters and the spring is likely to find the railways with more freight on hand than they will be able to handle comfortably. This means that the merchant who leaves his purchasing till the last minute will find himself short on several lines. Order now and get your goods started, and you will be more likely to have goods on hand when called for.

Brass safety chain has advanced 10 per cent., also some lines of flat irons show a 5 per cent. advance. Hay wire, wire staples and wire nails have all advanced 5 cents over last week's price.

AXES — We still quote: $7.60 to $9.50 per dozen; double bitt axes, $9.50 to $12 a dozen; handled axes, $7.50 to $9.50; Canadian pattern axes, $7.50 a dozen.

LANTERNS. — Our prices are: "Prism" globes, $1.20; Cold blast, $6.50; No. 0, Safety, $4.00.

COW TIES AND STALL FIXTURES —We still quote; Cow ties, discount 40 per cent. off list; stall fixtures, discount 35 per cent. off list.

SLEIGH BELLS—Our prices remain: Back straps, 30c to $2.50; body straps, 70c to $3.50; York Eye bells, common, 70c. to $1.50; pear shape, $1.15 to $2; shaft gongs, 20c. to $2.50; Grelots, 35c. to $2; team bells, $1.80 to $5.50; saddle gongs, $1.10 to $2.60.

RIVETS AND BURRS — —We are still quoting: Best iron rivets, section carriage, and wagon box black rivets, tinned do.,copper rivets and tin swede rivets, 60, 10 and 10 per cent.; swede iron burrs are quoted at 60 and 10 and 10 per cent. off new lists; copper rivets with the usual proportion of burrs, 25 per cent. off, and coppered iron rivets and burrs in 5-lb. carton boxes at 60 and 10 and 10 per cent.; copper burrs alone, 15 per cent., subject to usual charge for half-pound boxes.

HAY WIRE—Prices have advanced 5 cents. We now quote: No. 13, $2.60; No. 14, $2.70; No. 15, $2.85; f.o.b., Montreal.

MACHINE SCREWS—No advance yet. We still quote: Flat head, iron, 35 per cent., flat head brass, 35 per cent., Felisterhead, iron, 30 per cent.. Felisterhead, brass, 25 per cent.

BOLTS AND NUTS—Scarce, discounts remain: Carriage bolts, 3 and under, 60 and 10; 7-16 and larger, 55 per cent.; fancy carriage bolts, 50 per cent.; sleigh shoe bolts, ⅜ and under, 60 per cent.; 7-16 and over, 50 per cent.; machine bolts, ⅜ and under, 60 per cent.; 7-16 and larger, 55 per cent.

HORSE NAILS.—Discounts remain: C brand, 40, 10 and 7 per cent.; M.R.M. Co., 55 per cent.

WIRE NAILS—Advanced 5 cents per keg. Our price is now $2.45 per keg f.o.b., Montreal.

CUT NAILS—We still quote: $2.30 per keg base f.o.b., Montreal.

HORSESHOES—Our prices remain: P.B. new pattern, base price, $3.50 per 100 lbs.; M.R.M. Co., latest improved pattern iron shoes, light and medium pattern, No. 2 and larger, $3.65; No. 1 and smaller, $3.90; snow pattern, No. 2 and larger, $3.90; No. 1 and smaller, $4.15. Light steel shoes, No. 2 and larger, $4; No. 1 and smaller, $4.25; featherweight, all sizes, No. 0 to 4, $5.60. Toeweight, all sizes, No. 1 to 4, $6.85. Packing, up to three sizes in a keg, 10c. per 100 lbs. More than three sizes, 25c. per 100 lbs extra.

BUILDING PAPER—No change in last week's prices.

CEMENT AND FIREBRICK.—No change in prices. We are still quoting: "Lehigh" Portland in wood, $2.54, in cotton sacks, $2.39; in paper sacks, $2.31. Lefarge (non-staining) in wood, $3.40; Belgium, $1.60 to $1.90 per barrel; ex-store, American, $2 to $2.10 excars; Canadian Portland, $2 to $2.05. Firebrick, English and Scotch, $17 to $21, American $30 to $35; White Bros.' English cement, $1.80 in bags, $2.05 in barrels in round lots.

COIL CHAIN—Our quotations remain: 5-16 inch, $4.40; 3-8 inch, $3.90; 7-16 inch, $3.70; 1-2 inch, $3.50; 9-16 inch, $3.45; 5-8 inch, $3.35; 3-4 inch, $3.20; 7-8 inch, $3.10; 1 inch, $3.10.

SHOT—Our prices remain as follows: Shot packed in 25-lb. bags, ordinary drop AAA to dust, $7 per 100 lbs.; chilled, No. 1—10, $7.50 per 100 lbs.; brick and seal, $8 per 100 lbs.; ball, $8.50 per 100 lbs. Net list. Bags less than 25 lbs., ½c. per lb. extra, net; f.o.b. Montreal, Toronto, Hamilton, London, St. John, Halifax.

AMMUNITION — Slow. Very little sold. Our prices remain: Loaded with black powder, 12 and 16 gauge, per 1,000, $15; 10 gauge, per 1,000, $18; loaded with smokeless powder, 12 and 16 gauge, per 1,000, $20.50; 10 gauge, $23.50.

GREEN WIRE CLOTH.—We continue to quote: In 100 foot rolls, $1.62½ per hundred square feet, in 50 feet rolls, $1.67½ per hundred square feet.

Toronto Hardware Markets

Office of HARDWARE AND METAL,
10 Front Street East,
Toronto, February 8, 1907

Business for the past week has been quiet, some firms having had their travelers off the road helping finish up stock taking and also filling orders which were taken the previous week.

Jobbers whose year ended in December managed to have their stock taking finished between Christmas and New Years.

Orders continue to steadily arrive, even with travelers off the road, as all retail merchants are anxious to stock up before any further advance.

No price changes are quoted this week and everything remains steady at former quotations.

AXES AND HANDLES—Many large orders for these articles were taken some time ago, but the trade appears to be greatly interested in the new lines which are exhibited for various jobbers by their traveling representatives.

WASHING MACHINES—These articles, along with various household devices, are meeting with an ever increasing demand for spring trade.

SPORTING GOODS—The remarkably fine winter which we have so far been favored with has greatly added to the demand for many lines. The shortage in skates was greatly in evidence some time ago, and at present some lines are not procurable at any price. While the snowshoe demand is very good in northern Ontario it is not above the ordinary in the eastern or western sections. Trap shooting is at present greatly indulged in and, in consequence, ammunition is selling very well.

CHAIN—Remains at our former quotations: ½ in., $3.50; 9-16 in., $3.55; ½ in., $3.60; 7-16 in., $3.85; 3-8 in., $4.10, 5-16 in., $4.70; 1-4 in., $5.10.

RIVETS AND BURRS—Remain firm at the advanced price. Copper being at such a high price, and so scarce, another advance is probable in a short time.

SCREWS—Exceptionally large orders have been received by jobbers during the past couple of weeks, and in every case orders have been satisfactorily filled, which goes to show the large demand was anticipated.

BOLTS AND NUTS—These articles continue to be very largely asked for, but very hard to procure.

TOOLS—A good demand still continues for these various articles and many merchants are displaying a full line in their show windows.

SHOVELS—Demand is very good. Orders which were taken are now being shipped so as to be ready fou the opening of spring, when large quantities will be used in connection with the many buildings to be erected.

EXTENSION AND STEP LADDERS—These remain at former quotations: 11 cents per foot for 3 to 6 feet, and 12 cents per foot for 7 to 10-foot ladders. Waggoner extension ladders 40 per cent.

POULTRY NETTING—Orders are beginning to be filled, and shipments made daily. We still quote: 2-inch mesh, 19 w.g., discount 50 and 10 per cent., all others 50 per cent.

WIRE FENCING—The same applies to this line, as all booked orders are being rapidly filled and prices remain as before.

OILED AND ANNEALED WIRE—(Canadian)—Gauge 10, $2.41; gauge 11, $2.47; gauge 12, $2.55, per 100 lbs.

WIRE NAILS—Remain firm at the advanced price of $2.40 base, with continued large orders from every district.

CUT NAILS—We quote $2.30 base f.o.b. Montreal; Toronto, 20c. higher.

HORSENAILS—"C" brand 40, 10 and 7½ o fflist; "M.R.M. brand 60 per cent. off.

HORSESHOES — The demand for horseshoes has been very good during the last couple of months, large quantities having been disposed of in northern Ontario. Quotations remain at former prices. P. B. base, $3.65; "M.R.M. Co., latest improved pattern" iron shoes, light and medium pattern, No. 2 and larger, $3.80; No. 1, and smaller, $4.05; snow, No. 2 and larger, $4.05; No. 1 and smaller, $4.30; light steel shoes, No. 2 and larger, $4.15; No. 1 and smaller, $4.40; featherweight, all sizes, 0 to 4, $5.75; toeweight, all sizes, 1 to 4, $7. Packing, up to three sizes in a keg, 10c. per 100 lbs. extra; more than three sizes in a keg, 25c. per 100 lbs. extra.

BUILDING PAPER — Business in picking up considerably in this line. As soon as spring opens things will largely increase the sale.

CEMENT—For carload orders, f.o.b. Toronto, we quote: Canadian Portland, $1.95. For smaller orders ex warehouse, Canadian Portland, $2.05 upwards.

FIREBRICK—English and Scotch firebrick, $27 to $30, American low-grade, $23 to $25; high-grade, $27.50 to $35.

HIDES, WOOL AND FURS — Hides are slightly lower. Some furs, cross fox, timber wolf and winter and western rats are higher. Tallow is higher.

Hides, inspected, cows and steers, No. 1	....	0 10½
" " " " No. 2	....	0 09½
Country hides, flat, per lb., cured	....	0 09
" " " green	....	0 08
Calf skins, No. 1, city	....	0 12
" " No. 1, country	....	0 11
Lamb skins	1 25	1 35
Horse hides, No. 1	3 50	3 75
Rendered tallow, per lb.	0 05½	0 06
Pulled wools, super, per lb.	....	0 25
" " extra	....	0 27
Wool, unwashed fleece	....	0 15
" washed fleece	0 24	0 25

FURS.

	No. 1.	Prime
Raccoon	....	1 50
Mink, dark	5 00	7 00
" pale	2 50	4 50
Fox, red	3 00	4 00
" cross	2 00	20 00
Lynx	5 00	8 00
Bear, black	....	12 00
" cubs and yearlings	....	5 00
Wolf, timber	....	3 25
" prairie	....	1 25
Weasel, white	0 10	0 65
Badger	0 75	1 75
Fisher, dark	6 00	8 00
Skunk, No 1	....	1 10
Marten,	3 50	20 00
Muskrat., fall	....	0 17
" winter	..	0 25
" western	0 12	0 18

Montreal Metal Markets

Office of HARDWARE AND METAL,
232 McGill Street,
Montreal, February 8, 1907

No changes are reported in the metal markets this week, although copper and tin are firming up and jobbers generally predict that another advance, in copper at least, may be expected very shortly. All other metals remain practically the same, with no outward appearance of advancing just at present.

Sellers of old material report a steady business, all lines moving well. The American market has weakened slightly in scrap iron, and although this does not affect us here, it is pleasant to hear of a lower price on anything, even though it is across the border.

COPPER — Wherever there's copper you'll find the long arm of "Amalgamated" working overtime, with the result that this metal is still very short, and although no new advances are reported the market is very firm. Our quotations remain: Ingot copper, 26½ to 27 cents; sheet copper, base sizes, 34 cents.

INGOT TIN—Market again firming up. We still quote: 45½ to 46 cents per pound.

ZINC SPELTER — We still quote: $7.50 to $7.75 per 100 pounds.

PIG LEAD—Very firm. Our prices remain: $5.50 to $5.60 per 100 lbs.

ANTIMONY—No change over last week's prices, we still quote: 27½ to 28 cts. per pound.

PIG IRON. — Some lines show advances. We now quote: Londonderry, $24.50; Carron, No. 1, $27.00; Carron, special, $25.50; Summerlee, No. 2, selected, $25.00; Summerlee, No. 3, soft, $23.50.

BOILER TUBES. — Remain very firm. Our prices are: 1½ to 2 in., $9.50; 2¼ in., $10.35; 2½ in., $11.50; 3 in., $13.00; 3½ in., $17.00; 4 in., $21.50; 5 in., $45.00.

TOOL STEEL — Our prices remain: Colonial Black Diamond, 8c to 9c; Sanderson's 8c. to 45c., according to grade; Jessop's, 13c.; Jonas & Colver's, 10c. to 20c.; "Air Hardening," 65c. per pound; Conqueror, 7½c.; Conqueror high speed steel, 60c.; Jowitt's Diamond J., 6½c. to 7c.; Jowitt's best, 11c. to 11½c.

MERCHANT STEEL.—Our prices are as follows: Sleigh shoe, $2.25; tire, $2.40, spring, $2.75; toecalk, $3.05; machinery iron finish, $2.40; reeled machinery steel, $2.75; mild, $2.25 base and upwards; square harrow teeth $2.40; band steel, $2.45 base. Net cash 30 days. Rivet steel quoted on application.

COLD ROLLED SHAFTING—Present prices are: 3-16 inch to ¼ inch, $7.25; 5-16 inch to 11-32 inch, $6.20; ⅜ inch to 17-32 inch, $5.15; 9-16 inch to 47-64 inch, $4.45; ¾ inch to 17-16 inch, $4.10; 1½ inch to 3 inch, $3.75; 3⅛ to 3 7-16 inch, $3.92; 3½ inch to 3 15-16 inch, $4.10; 4 inch to 4 7-16 inch, $4.45; 4½ inch to 4 15-16 inch, $4.80. This is equivalent to 30 per cent. off list.

GALVANIZED IRON. — Shows advances. We now quote: Queen's Head, 28 gauge, $4.70; 26 gauge, $4.45; 22 to 24 gauge, $4.20; 16 to 20 gauge, $3.95; Apollo, 28 gauge, $4.45; 26 gauge, $4.30; 22 and 24 gauge, $3.75; 16 and 20 gauge, $3.60; Comet, 28 gauge, $4.45; 26 gauge, $4.35; 22 and 24 gauge, $3.75; 16 to 20 gauge, $3.60; Fleur-de-Lis, 28 gauge, $4.55; 26 gauge, $4.30; 22 and 24 gauge, $4.05; 16 to 20 gauge, $3.80; Gorbals "Best Best," 28 gauge, $4.55; Colborne Crown, 28 gauge, $4.45; 26 gauge, $4.30; 24 gauge, $3.75. In less than case lots, 25c extra.

OLD MATERIAL — American market weakened slightly during the week on scrap iron, but has not affected prives here. We still quote: Heavy copper, 20c. per lb.; light copper, 17c. per lb.; heavy red brass, 17c. per lb.; heavy yellow brass, 14c per lb.; light brass, 10 to 10½c per lb.; tea lead. 4¾c per lb.: heavy lead, 4½c. per lb.; scrap zinc, 4½c. per lb.; No. 1 wrought $17.00 per ton; No. 2 wrought $6.00 per ton; No. 1 machinery, $18.00 per ton; stove plate, $13.00 per ton; old rubber, 10½ cents per lb.; mixed rags, 1 to 1¼c per lb.

The Niagara Falls Machine and Foundry Company, Niagara Falls, Ont., have secured a large contract to supply cast and wrought iron work for the Michigan Central's Detroit river tunnel from the contractors, the Butler Bros., and Hoof Company. They also have a contract to supply fire hydrants for the city of Toronto. The company will enlarge their foundry at once.

Toronto Metal Markets

Office of HARDWARE AND METAL,
10 Front Street East.
Toronto, Feb 8, 1907

February, usually a dull month in the metal business, is a busy one this year and indications are that, like January, it will show a record-breaking volume of business transacted. Buyers are conservative and, while covering their requirements for the coming season, are not being carried off their feet by the speculative possibilities of the market.

Pig iron is, according to cable reports, much easier in Great Britain Beginning last August Cleveland Warrants rose steadily from 51s. 6d. to 64s. early in December, since when they have dropped to 57s.

It is hard to believe that this movement had any other foundation than the prospective—and afterwards realized—American demand. The warrant market opened in 1906 at 55s., and dropped to a low point of 48s. 6d. late in March. There was a slight advance thereafter, gliding into the heavier advance noted above as commencing in August. Conditions in the British iron trade have not so changed. It has been uniformly prosperous throughout the year and the varying quotations are the result of speculation. On the best English and Scotch grades the quotations we give are close. As before, Canadian furnaces are booked ahead for two or three quarters, while a fair amount of British iron is being imported.

Copper and tin are the leading figures on the bill this week, copper being advanced half a cent, and tin having firmed up to the figures maintained for some time past. The market in Toronto is practically bare of copper and as it is predicted that the production of the Lake Superior region in 1907 will be smaller than in 1906, conditions seem favorable for even higher prices. The Amalgamated Copper Company now has the market in its own hands and as orders for future delivery are not being accepted it is plain that lower prices cannot be looked for.

There is talk of further advances on plates and sheets, but we are not changing our figures this week. Sheet lead is quoted higher by some jobbers.

PIG IRON—Import orders are coming through satisfactory. Hamilton, Midland and Londonderry are off the market, and Radnor is quoted at $33 at furnace. Middlesborough No. 1 is quoted at $24.50, and Summerlee No. 1 at $26 f.o.b., Toronto.

On bar iron prices are firm and we still quote: $2.30 f.o.b., Toronto, with 2 per cent. discount.

TIN—Ingots have experienced a sharp reaction and prices are again firm at 45c with more asked in some cases. Plates are firm and unchanged.

SHEETS AND PLATES—No further advances have been made on black sheets or Canada plates, although there is talk of higher prices. Galvanized sheets are in good demand.

BRASS—Light stocks are reported everywhere with deliveries hard to obtain. We continue to quote 30 cents per pound.

COPPER—The market is bare and higher prises seem probable. We continue to quote: Ingot copper, $26.50 per 100 lbs., and sheet copper, $32 per 100 lbs.

LEAD—Stocks are light with a very lively demand, and prices very firm. We quote: $5.40 for imported pig and $5.75 to $6.00 for bar lead. For sheet lead $7 is asked.

ZINC SPELTER—Stocks continue tight with the market active. We quote 7½c. per lb. for foreign and 7c. per lb. for domestic. Sheet zinc is quoted at 8 1-4c. in casks, and 8 1-2c. in part casks.

BOILER PLATES AND TUBES.—An advance is looked for but we still quote: Plates, per 100 lbs., ¼ in. to ½ in., $2.50; ⅜ in., $2.35 heads, per 100 lbs., $2.75; tank plates, 3-16 in., $2.65; tubes, per 100 feet, 1½ in., $8.50; 2, $9; 2 1-2, $11.30; 3, $12.50; 3 1-2, $16; 4, $20.00. Terms, 2 per cent. off.

ANTIMONY. — More activity is reported, but stocks are plentiful. Prices are unchanged at 27c. per pound.

OLD MATERIAL.—Dealers report a good demand with supplies coming in well. Their buying prices are: Heavy copper and wire, 20 cents per lb.; light copper, 18c per lb.; heavy red brass, 18c per lb.; heavy yellow brass, 15c. per lb.; light brass, 11c. per lb.; tea lead, $4.25 per 100 lbs.; heavy lead, $4.50 per 100 lbs.; scrap zinc, 4½c. per lb.; No. 1 wrought iron, $15; No. 2 wrought, $8; machinery cast scrap, $18; stove plate, $14; malleable and steel, $9; old rubbers, 10½c. per pound; country mixed rags, $1 to $1.25 per 100 lbs., according to quality.

COAL — No shortages are reported. We still quote: Standard Hocking soft coal f.o.b. at mines, lump $1.50, ¾ inch, $1.40; run of mine, $1.30; nut, $1.25; N. & S., $1; P. & S. 85c. Youghiogheny soft coal in cars, bonder at bridges: lump, $2.60; ¾ inch, $2.50; mine run, $2.40; slack, $2.15.

United States Metal Markets

From the Iron Age, Feb 7, 1907.

In the markets east of the Allegheny mountains there has been little movement during the past week, after the heavy business done in January. The demand, both for early and late delivert, is rather light, and a canvass of the trade in different sections indicates considerable indifference at the present level of prices Considerable foreign iron is coming in and is still due, but prices are lower, one cargo to arrive in the Delaware having sold at a shade over $21. In the central west and in the Chicago district there is a good deal of inquiry for the second half, notably for malleable Bessemer Pittsburg reports that the Jones & Laughlin Steel Company has purchased a considerable quantity of Bessemer pig to make up for inadequate output of its own furnaces, some of which are out of blast

The tonnage placed for plates is stated to have been unprecedented, and the leading mills are now crowded with work for the next four or five months.

Specifications for structural material have been coming in at a satisfactory rate during the past few weeks.

There has been a tremendous pressure for tubular goods. Among the orders recently placed is one lot of 1,000 tons for the Pennsylvania Company. Quite a number of moderate-sized contracts for cast iron pipe are cropping up.

The demand for sheets and for steel bars continues lively, and premiums of 10c. per box are being paid for tin plate for delivery during the third quarter.

From the Iron Trade Review, Feb. 7, 1907

Pig iron is the centre of attraction because of the many adverse currents affecting it. The agitation for the repeal of the advance in southern freight rates, the loud protest against the exasperating delays to shipments, and the efforts of several large consuming interests to secure lower prices, have produced a situation containing an element of uncertainty. This is manifested by a disposition on the part of smaller melters to postpone purchases until it is more definitely known what the railroads will accomplish in the next thirty days towards moving the large tonnage of iron that has accumulated on southern yrds.

Following the publication about two weeks ago of reports of weakness in the American pig iron market, a determined bear assault was made on warrant iron in England, and at least one large American consumer was able to buy a round tonnage at about $16.50 New Orleans. The report of this transaction has been noised about within a few days, and has caused hesitancy on the part of some buyers, but all indications are that it would be extremely difficult to buy any considerable tonnage of foreign at so low a price, and it is not probable that the importation of iron will result in demoralizing prices in this country to any great extent. Practically all of the foreign iron which can be delivered before April 7 has been sold and the amount that can be landed in this country after that date does not promise to be very large. Taken as a whole, the pig iron market is easier, but with little, if any evidence of weakness.

London, Eng., Metal Markets

From Metal Market Report, Feb 6, 19.7.

Cleveland warrants are quoted at 56s. 9d, and Glasgow Standard warrants at 56s., making price, as compared with last week, unchanged on Cleveland warrants and Glasgow Standard warants.

TIN—Spot tin opened firm at £193 5s., futures at £192 15s., and after sales of 380 tons of spot and 560 tons of futures closed easy at £193 for spot., £192 7s. 6d. for futures, making price compared with last week £4 higher on spot and £3 7s. 6d. higher on futures.

COPPER—Spot copper opened steady at £107 10s., futures £108 10s., and after sales of 300 tons of spot and 600 tons of futures closed quiet at £107 7s. 6d. for spot, and £108 7s. 6d. for futures, making price compared with last week £1 2s. 6d. higher on spot and 15s. higher on futures.

LEAD—The market closed at £19 12s. 6d., making price as compared with last week 3s. 6d. lower.

SPELTER—The market closed at £26, making price as compared with last week 15s. lower.

Calgary Hardware Trade News

Feb. 2, 1907.

The questions of cheap power and street cars are still being thrashed out between the council and the syndicate. The power proposition is one of considerable importance to manufacturers and great interest is being taken in the matter localty. It is expected, if the offer of the syndicate is not accepted, that the municipality will tackle the matter on a larger scale than they have so far attempted.

G. F. Stephens & Co., paint manufacturers are having a large warehouse erected on Eighth Avenue, the main busithoroughfare of the city. The block, which is almost completed, is built of brick terra cotta finished front. Facing to Eighth Avenue there is a large covered entrance suitable for shipping, while the block extends to the spur track at the rear where there is a large receiving platform at which goods can be unloaded direct from the cars. The block consists of a deep basement with three floors above, electric elevators will be fitted to run from basement to top. The front is fitted with two large segment head windows and the designs for the interior work here, show a tasteful partition in glass, giving suitable accommodation for show room and offices. Messrs. Stephens will occupy the east half, and the west part has been leased for five years to Ames, Holden & Co., wholesale boot and shoe manufacturers, of Montreal.

Manson, Campbell & Co., of Chatham, Ontario, manufacturers of fanning mills, incubators, etc., also propose building a large warehouse here. J. I. Campbell, the Alberta and B.C. manager, has made Calgary his headquarters for the past three years, but with the rapid expansion of the West, their present premises have become too small. The new block will be in the centre of the city where a suitable site has been acquired.

Very little progress has been made in the building line during the past two months owing to the almost incessant severe frost. Builders also complain of the prohibitive price of lumber, but present tendencies point to stiff prices for some time to come.

London Hardware Trade News

A hardware store will be opened next week in the new Sovereign Bank building, at the corner of Dundas and Adelaide Streets. Mr. William Kilpatrick, a well known hardware salesman of this city, has leased the premises and will sever his connection with the Purdom-Gillespie Hardware Company, by whom he is now employed, on the 9th instant. Mr. Kilpatrick has been with the Purdom-Gillespie Hardware Co. for the past three years. He began his experience in the trade as an employe of James Wright & Co., and later worked for Mr. James Reid. His knowledge of the business is extensive and his ability is such that his friends are confident of the success of his venture.

* * *

The interest shown by the public generally in the progress of the Hardware Hockey League is an indication that the various teams are composed of good players. Two evenings a week games are played at the Simcoe Street rink by league teams and there is a large attendance at every game. The dealers represented in the league are: The Purdom-Gillespie Co., James Cowan, A. Westman, Geo. Taylor & Son and McLean's. The staffs of the last two houses on the list were drawn upon to make up one team and the others put in a team each. One series of games has been played and the second series is in progress, with the Purdom-Gillespie's in the lead. Mr. Phelps of the Hobbs Hardware Co., is the efficient official referee.

* * *

A. J. Morgan, now in business on Dundas Street near Wellington, is soon to locate in the store now occupied by Clark & Smith on Dundas Street near Talbot. Mr. Morgan deals in garden supplies, poultry and pet animal supplies, etc. The premises newly leased by him are larger and much nearer the market square than his present stand. Mr. Morgan was with the Hobbs Hardware Co. before he went into business for himself.

* * *

Ex-Mayor George Taylor who established the Geo. Taylor Hardware Co. at New Liskeard a few years ago, has made several fortunate investments in mining property. He is one of the pioneers of New Ontario. His company has a branch store at Cobalt and last week the mining right in the lot on which the company's stable stands was sold for over $5,000.

B.C. Hardware Trade News

Vancouver, January 31, 1907.

With politics in the air to suffcation, and bad weather, there has not been the building activity during the month of January that would otherwise have been recorded. Therefore trade in builders' hardware has been lighter. The sawmills have nearly all been shut down more or less, and many of the upcoast logging camps have hardly operated at all in the month, so that trade in the lumber supply line has been light for the hardware men. Nevertheless none complain, as business for the first of the year is fully as good as expected. Stock-taking occupied most of the dealers at the first of the month, and that being over, the decks are cleared ready for a busy season. And the prospects are that the incoming building season will eclipse all previous records.

* * *

Though building operations have not been very active, for the causes mentioned, nevertheless another record for Vancouver has been made by the number of permits issued, the total value being for the month, $187,265. This is the largest on record for any January. In 1905 the total permits issued was $185,000. Last year it was but $176,425.

* * *

Steamer Tottenham, which has been delayed in loading because of ice in the Fraser river, has finished her big lumber cargo at this port. She loaded the first portion of the cargo at Millside, Fraser river, but her agents moved her to Vancouver for fear of the river freezing, which it afterwards did. The remainder of the cargo had to be lightered to the big tramp steamer. She is taking out railroad ties and timbers and her total is to be over 4,000,000 feet, which is the record for any vessel loading at a Pacific coast port. The Tottenham already holds the record, having taken 3,800,000 feet from Portland on a previous occasion. Her present cargo goes to Port San Marques, Mexico.

* * *

Five carloads of machinery for the big pulp mill the Canadian Pacific Sulphite Pulp & Paper Co. are building at Swanson Bay, have reached this port and will be sent forward as soon as space can be secured on one of the coast steamers. Other cars of machinery are en route and more is to be shipped yet for the plant, the first of its kind in the province. J. M. Mackinnon, managing director of the company, has gone north to inspect work now under way at the site of the mill. A sawmill is in operation getting out timbers for the various buildings, a wharf has already been built and other buildings are in process of construction. The company is amply provided with capital, the entire stock having been subscribed for in England. A very large area of choice spruce timber has been secured and a water power is to be developed for operating the mill.

* * *

J. H. Pilisbury, C.E., engineer-in-charge of survey work on the townsite of Prince Rupert, the G.T.P. terminal city up the coast, has gone north to conduct active operations in the field. He left by the steamer Amur a few days ago. By the same steamer the representatives of the United Supply & Contracting Co. went north to look after contracts they have secured from the G.T.P. for clearing portions of the townsite. The logging engines and other machinery for the clearing work were shipped north on the freighting steamer Venture, and when they arrive a large gang of men is to be put to work at once. The townsite is not to be placed in the market till next Fall, by which time a great deal of clearing will have been done.

* * *

Iron ores of Vancouver Island in the vicinity of Quatsino Sound are to be carried to Irondale, Wash., where James A. Moore, of Seattle, has bought the old blast furnaces and steel plant, which he intends to refit, enlarge and operate on a large scale. It is announced that a number of steamers have been purchased to be put in commission in the ore-carrying trade, exclusively, from the mines of the northern part of Vancouver Island, where Mr. Moore holds a number of valuable deposits. The old German steamer Mariechen, badly wrecked on the west coast of Vancouver Island, has also been purchased by Mr. Moore, who intends fitting it up as a barge to be put in the ore-carrying trade as well.

Car shortage is still affecting the smelter industry of the interior because there are not enough cars available to carry coal and coke from the mines to the smelters. There is reported to be sufficient coke in readiness, but it is difficult, if not impossible, to secure cars to load. Some of the smelters in the boundary district have been only running part of their furnaces because of this coke shortage. The Trail smelter has also been running only part of its plant. The Northport smelter, in Washington, which is once more running on Le Roi ores, has been running full time, but the big surplus of ore accumulated during the coal miners' strike has not been reduced, and with fresh shipments daily, wil not be materially lessened.

* * *

Turbines or horizontal type are to be used in developing the water power of Stave Lake Power Co., was in charge is under construction. This is to be the initial installation, and there is room for a vastly larger power when demand warrants the capital expenditure on installing the plant to generate it. The horizontal type turbines have been installed at Niagara and at the Kakabeka Falls plant of the Kamin' iquia Power Co., near Ft. William. t both these plants, Wm. Kennedy, .E., of Montreal, who is consulting gineer for the Stave Lake Power Co., was in charge of installation of the plant.

Stave Lake is but 38 miles from Vancouver, so that the prospects of power supply for industries for the terminal city are very bright. It is expected that the power company will have their dam installed before high water in the spring and tenders for the turbines and generating plant will be called for in a few weeks, so that construction will proceed quickly. Three electric generators, of 5,000 kilowatt capacity each, will be installed in the power house. These will likely be made by one of the large Canadian manufacturers of electrical machinery in the east.

Work on construction of the new electric line to be built by the B.C. Electric Railway from New Westminster to Eburne, along the north shore of the Fraser River, is proceeding as fast as weather permits. Nearly all the grading has been done and rails will be laid early in the spring. This line will traverse a section but little settled up so far, and the entire distance is being divided in acreage of 3, 4, 5 and 10 acre blocks, most of which is being bought by people who intend making their homes on the land and raising fruit, etc. The line will also serve a large portion of the delta lands of the Fraser not now in touch with any railway. The whole district is subsidiary to the cities of New Westminster and Vancouver.

NEW STEEL CHEESE VAT.

The Steel Trough and Machine Company, Tweed, Ont., are introducing a steel vat support which should sell well to dairymen. They will supply the supports to tinsmiths or hardwaremen desiring to make their own tinwork, or will supply both tanks and supports when desired for twin or single cream vats, curd sinks, skim milk vats or water tanks The steel vat support is built on such a principle that the stock frame work can be knocked down and bundled for shipment, which makes freight on it very low. The legs are made of steel channel like what is used is bridge or structural work, giving the greatest strength to the least weight; they are held together by cross sections firmly bolted to them. Bars of steel run lengthwise in cross sections to which they are bolted to support the bottom of pan. The tank itself is made of heavy galvanized steel riveted and soldered and reinforced around the top edge with heavy angle steel which extends outwardly over the top of the legs where it is bolted. Thus almost the entire weight of vat rests directly on angle steel and legs. The tank is not riveted or bolted to legs or framework otherwise than at angle steel as described above so that the twisting of the frame work would not in any way injure the tank or cause it to leak. By the removal of the bolts at top of legs the tank can be lifted out entirely independent of the frame work.

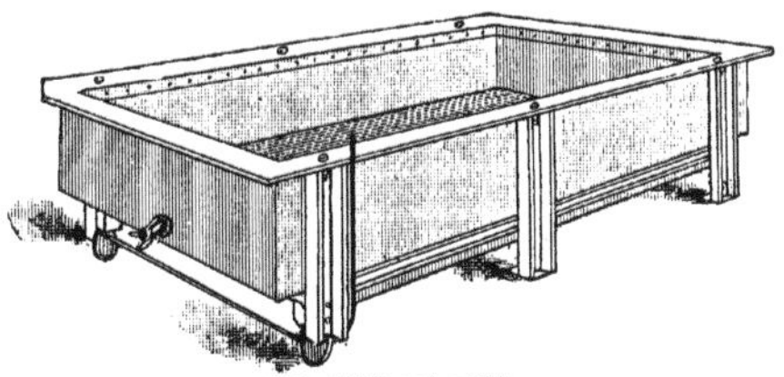

All Steel Vat Support and Vat.

CLEAN OUT HORSE BLANKETS.

Are the robes and blankets selling as they should? The season for their sale is passing and they do not make the best line in the world to carry over. Never again will they look so attractive as when new, their colors uninjured by dust or a degree of fading; besides there are mice and moths to guard against. There is still enough cold weather ahead to render them of interest, although the time for it is shortenning up enough to justify season sales and sale prices.

Find out the people who will be in need of blankets by the beginning of another season and give them enough of an inducement to buy now to pay them for the trouble of carrying their purchase over. They can afford to carry it cheaper than you can because with them it is only a purchase in advance of something they are sure to need in the future; with you it is so much old stock to be carried into next year's tag ends on which, more than likely, a reduction will be expected because it is old. Better make the reduction now and let the other fellow do the carrying. Probably you have already sold enough blankets from the lot to afford a cut on the remainder to about cost and still make money. more than would be made in carrying the stuff over.

If the stock in hand is large it might pay to go into quite an elaborate campaign in order to close it out promptly; but do not fail to make it bring in customers and sell other goods. There need be no real loss in an occasional well-conducted at-cost or below cost sale. Take a lesson from the mail-order people, who always accompany each sale with an enticing descriptive circular of something else. Frequently their greatest profit is in the subsequent sales.

Get your blanket advertisement busy in the newspaper of your town, remembering that a moderate-sized advertisement containing one good honest reason that will interest the average horse owner is worth several times its space in meaningless type gymnastics. Make your advertisements talk and do not make them shout; but make them talk blanket, not general hardware. Set the store window displaying blanket attractions. Make the whole affair as much a blanket gala day as though offering holiday goods before Christmas.

Hardware suggests that after the blanket sale is over, and there will be no more demand for them until another season, do not fail to gather up those that remain and store, carefully folded, in a tight, dry, mouse-proof and moth-proof chest that contains nothing else and will not be disturbed. Do not allow them to kick around in the way upon the counters half the Summer, or to lie in a tumbled heap in some dusty corner, as frequently is done to a greater or a less extent in many hardware stores. Even though the freshness of their appearance must be lost to some extent, there are few other things in the hardware store that require more considerate treatment or that, considering their nature, are more frequently made unsalable through careless handling.

MANITOBA HARDWARE AND METAL MARKETS

(Market quotations corrected by telegraph up to 12 a.m. Friday, Feb. 8, 1907.)

Room 511, Union Bank Building, Office of HARDWARE AND METAL, Winnipeg, Man.

Owing to snow blockades on the railways both mail and shipments of goods have been delayed. Dealers are complaining, especially those on the Canadian Northern, over which through trains between Edmonton and Winnipeg have not been running for three weeks. Conditions are equally bad in the Northwestern States. A combination of circumstances is making this winter a hard one for both settlers and merchants.

LANTERNS. — Quotations are as follows. Cold blast, per dozen, $6.50; coppered cold blast, per dozen, $8.50, cold blast dash, per dozen, $8.50.

WIRE.—Prices on barb wire for 1907 shew a sharp advance as compared with last year. Revised prices are as follows: Barbed wire, 100 lbs., $3.22½; plain galvanized, 6, 7 and 8, $3.70; No. 9, $3.25; No. 10, $3.70; No. 11, $3.80; No. 12, $3.45; No. 13, $3.55; No. 14, $4.00; No. 15, $4.25; No. 16, $4.40; plain twist, $3.45; staples, $3.50; oiled annealed wire, 10, $2.96; 11, $3.02; 12, $3.10; 13, $3.20; 14, $3.30; 15, $3.45. Annealed wires (unoiled) 10c. less.

HORSESHOES — Quotations are as follows: Iron, No. 0 to No. 1, $4.65; No. 2 and larger, $4.40; snowshoes, No. 0 to No. 1, $4.90; No. 2 and larger, $4.65; steel, No. 0 to No. 1, $5; No. 2 and larger, $4.75.

HORSENAILS — Lists and discounts are quoted as follows: No. 10, 20c.; No. 9, 22c.; No. 8, 24c.; No. 7, 26c.; No. 6, 28c.; No. 5, 32c.; No. 4, 40c., per pound. Discounts are quoted as follows: "C" brand, 40, 10 and 7½ per cent., "M" brand and other brands, 55 and 60 per cent. Add 15c. per box.

WIRE NAILS—Quoted now at $2.80 per keg.

CUT NAILS.—Recently advanced to $2.90 per keg.

PRESSED SPIKES — Prices are quoted as follows since the recent advance: ¼ x 5 and 6, $4.75; 5-16 x 5, 6 and 7, $4.40; ⅜ x 6, 7 and 8, $4.25; 7-16 x 7 and 9, $4.15; ½ x 8, 9, 10 and 12, $4.05; ⅝ x 10 and 12, $3.90. All other lengths 25c. extra net.

SCREWS—Discounts are as follows: Flat head, iron, bright, 85 and 10 p.c.; round head, iron, 80 p.c.; flat head, brass, 75 and 10 p.c.; round head, brass 70 and 10 p.c.; coach, 70 p.c.

NUTS AND BOLTS — Discounts are unchanged and continue as follows: Bolts, carriage, ⅜ or smaller, 60 and 5; bolts, carriage, 7-16 and up, 55; bolts, machine, ⅜ and under, 55 and 5; bolts, machine, 7-16 and over, 55; bolts, tire, 65; bolt ends, 55; sleigh shoe bolts, 65 and 10; machine screws, 70; plough bolts, 55; square nuts, case lots, 3; square nuts, small lots, 2½; hex nuts, case lots, 3; hex nuts, smaller lots, 2½ p.c.

RIVETS.—Prices on copper rivets have been increased and revised quotations are: Iron, discounts, 60 and 10 p.c.; copper, No. 7, 43c; No. 8, 43½c.; No. 9, 45½c.; copper, No. 10, 47c.; copper, No. 12, 50½c.; assorted, No. 8, 44½c., and No. 10, 48c.

COIL CHAIN.—Prices have been revised, the general effect being an advance. Quotations now are: ¼ inch $7.00; 5-16, $5.35; ⅜, $4.75; 7-16, $4.50; ½, $4.25; 9-16, $4.20; ⅝, $4.25; ¾, $4.10.

SHOVELS—List has advanced $1 per dozen on all spades, shovels and scoops.

HARVEST TOOLS — Discounts continue as before, 60 and 5 per cent.

AXE HANDLES—Quoted as follows: Turned, s.g. hickory, doz., $3.15; No. 1, $1.90; No. 2, $1.60; octagon, extra, $2.30; No. 1, $1.60.

AXES.—Quotations are: Bench axes, 40; broad axes, 25 p.c. dis. off list; Royal Oak, per dozen, $6.25; Maple Leaf, $8.25; Model, $8.50; Black Prince, $7.25; Black Diamond, $9.25; Standard flint edge, $8.75; Copper King, $8.25; Columbian, $9.50; handled axes, North Star, $7.75; Black Prince, $9.25; Standard flint edge, $10.75; Copper King, $11 per dozen.

BUTTS—The discount on wrought iron butts is 65 and 5 p.c.

CHURNS — The discounts from list continue as before: 45 and 5 per cent.; but the list has been advanced and is now as follows: No. 0, $9; No. 1, $9; No. 2, $10; No. 3, $11; No. 4, $13; No. 5, $16.

CHISELS—Quoted at 70 p.c. off list prices.

AUGER BITS—Discount on "Irwin" bits is 47½ per cent., and on other lines 70 per cent.

BLOCKS—Discount on steel blocks is 35 p.c. off list prices; on wood, 55 p.c.

FITTINGS—Discounts continue as follows: Wrought couplings, 60; nipples, 65 and 10; T's and elbows, 10; malleable bushings, 50; malleable unions, 55 p.c.

GRINDSTONES.—Quoted now at 1½c. per lb.

FORK HANDLES—The discount is 40 p.c. from list prices.

HINGES—The discount on light "T" and strap hinges is 65 p.c. off list prices.

HOOKS—Prices are quoted as follows: Brush hooks, heavy, per doz., $8.75; grass hooks, $1.70.

CLEVISES—Price is now 6½c. per lb.

STOVE PIPES—Quotations are as follows: 6-inch, per 100 feet length, $9; 7-inch, $9.75.

DRAW KNIVES—The discount is 70 per cent. from list prices.

RULES—Discount is 50 per cent.

WASHERS—On small quantities the discount is 35 p.c.; on full boxes it is 10 p.c.

WRINGERS — Prices have been advanced $2 per dozen, and quotations are now as follows: Royal Canadian, $35.00; B.B., $39.75, per dozen.

FILES—Discounts are quoted as follows: Arcade, 75; Black Diamond, 60; Nicholson's, 62½ p.c.

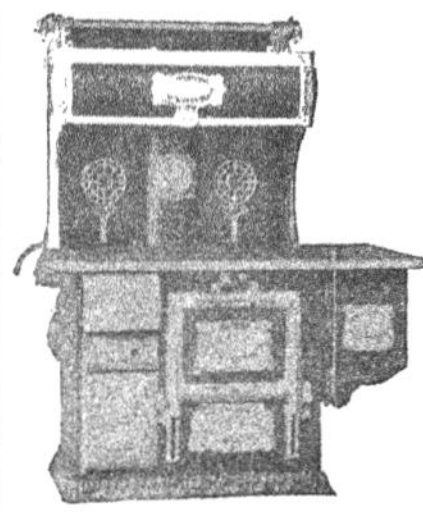

LOCKS.—The discount on Peterborough and Gurney locks is now 40 per cent.

BUILDING PAPER. — Prices have been advanced and quotations are now as follows: Anchor, plain, 66c.; tarred, 69c; Victoria, plain, 71c; tarred, 84c

TINWARE, ETC.—Quoted as follows. Pressed, retinned, 70 and 10; pressed, plain, 75 and 2½; pieced, 30; japanned ware, 37½; enamelled ware, Famous, 50; Imperial, 50 and 10; Imperial, one coat, 60; Premier, 50; Colonial, 50 and 10; Royal, 60; Victoria, 45; white 45; Diamond, 50; Granite, 60 p.c.

GALVANIZED WARE. — The discount on pail is now 37½ per cent.; and on other galvanized lines the discount is 30 per cent.

CORDAGE—We quote: Rope, sisal, 7-16 and larger, basis, $11.25; Manila, 7-16 and larger, basis, $16.25; Lathyarn, $11.25; cotton rope, per lb, 21c.

SOLDER—Quoted at 27c. per pound Block tin is quoted at 45c. per pound.

VISES — Prices are quoted as follows: "Peter Wright," 30 to 34, 14½c; 35 to 39, 14c.; 48 and larger, 13½c. per lb.

ANVILS—"Peter Wright" anvils are selling at 11c. per lb.

CROWBARS—Quoted now at 4c. per lb.

POWER HORSE CLIPPERS — The "1902" power horse clipper is selling at $12, and the "Twentieth Century" at $6. The "1904" sheep shearing machines are sold at $13 00

AMMUNITION, ETC.—Quotations are as follows: Cartridges, Dominion R.F. 50 and 5; Dominion, C.F., 33½ C.F., pistol, p.c.; C.F., military, 10 p.c. advance. Loaded shells: Dominion Eley's and Kynoch's soft, 12 gauge, black, $16.50; chilled, 12 gauge, $17.50; soft, 10 gauge, $19.50, chilled, 10 gauge, $20.50. Shot, ordinary, per 100 lbs., $7.25; chilled, $7.75; powder, F.F., keg. Hamilton, $4.75; F.F.G., Dupont's, $5.

IRON AND STEEL.—Quotations are. Bar iron basis, $2 70. Swedish iron basis, $4.95; sleigh shoe steel, $2.75; spring steel, $3.25; machinery steel, $3.50; tool steel, Black Diamond, 100 lbs, $9 50; Jessop, $13.

SHEET ZINC — The price is now $8.50 for cask lots, and $9 for broken lots.

PIG LEAD.—The price is advancing and the range of quotations among Winnipeg houses makes it difficult to quote with exactitude. The average price is about $6.00.

AXLE GREASE—"Mica" axle grease is quoted at $2.75 per case, and "Diamond" at $1.60.

IRON PIPE AND FITTINGS— Revised prices are as follows:—Black pipe, ¼ inch, $2.65; ⅜, $2.80; ½, $3.50; ¾, $4.40; 1. $6.35; 1¼, $8.65; 1½, $10.40; 2, 13.85; 2½, $19.00; 3, $25.00. Galvanized iron pipe, ⅜ inch. $3.75; ½, $4.35; ¾, $5.65; 1, $8.10; 1¼, $11.00; 1½, $13.25; 2, inch, $17.65. Nipples, discounts 70 and 10 per cent.; Unions, couplings, bushings and plugs, 60 per cent.

LEAD PIPE—The price, $7.80, is firmly maintained in view of the advancing lead market.

GALVANIZED IRON—Quoted as follows:—Apollo, 16 gauge, $3.90; 18 and 20, $4.10; 22 and 24, $4.45; 26, $4.40; 28, $4.65, 30 gauge or 10¾ oz., $4.95; Queen's Head, 24, $4.50; 26, $4.65; 28, $5.00.

TIN PLATES—We now quote as follows: IC charcoal, 20 x 28, box, $9.50; IX charcoal, 20 x 28, $11.50; XXI charcoal, 20 x 28, $13.50.

TERNE PLATES—Quoted at $9.

CANADA PLATES—Quoted as follows: Canada plate, 18 x 21, 18 x 24, $3.40; 20 x 28, $3.65; full polished, $4.15.

BLACK SHEETS—Prices are: 10 to 16 gauge, 100 lbs., $3.50; 18 to 22, $3.75; 24, $3.90; 26, $4; 28, $4.10.

PETROLEUM AND GASOLINE.—Silver Star in brls. per gal., 20c.; Sunlight in brls, per gal., 21c.; per case, $2.30; Eocene in brls, per gal., 23c.; per case, $2.50; Pennoline in brls., per gal., 24c.; Crystal Spray, 23c.; Silver Light, 21c.; Engine gasoline in barrels, per gal., 28c., f.o.b. Winnipeg in cases, $2.75.

PAINTS, OILS AND TURPENTINE.—The trade are expecting another advance in turpentine. We quote: White lead, pure, $6.50 to $7.50, according to brand; bladder putty in barrels, 2½c.; in kegs, 3¼c.; turpentine, barrel lots, Winnipeg, $1.01; Calgary, $1.08; Lethbridge, $1.08; Edmonton, $1.09. Less than barrel lots 5c. per gallon advance. Linseed oil, raw, Winnipeg, 67c.; Calgary, 74c.; Lethbridge, 74c.; Edmonton, 75c.; boiled oil, 3c. per gal. advance on these prices.

WINDOW GLASS—We quote: 16-oz. O.G., single, in 50-ft. boxes—16 to 25 united inches, $2.25; 26 to 40, $2.40; 16-oz. O.G., single, in 100-ft. cases — 16 to 25 united inches, $4; 26 to 40, $4.52; 41 to 50, $4.75; 50 to 60, $5.25; 61 to 70, $5.75. 21-oz. C.S., double, in 100-ft. cases—26 to 40 united inches, $7.35; 41 to 50, $8.40; 51 to 60, $9.45; 61 to 70, $10.50; 71 to 80, $11.55; 81 to 85, $12.60; 86 to 90, $14.75; 91 to 95, $17.30.

STOVE MAKERS BANQUETED.

The employes of the Copp foundry at Fort William on Friday of last week were entertained to their annual banquet by Mr. and Mrs. H. Copp, served in the capacious showrooms of the firm.

SUIT FOR "RAKE-OFF" WITHDRAWN.

The suit brought by the Bennett and Wright Co. against W. J. Gage and Co., Toronto, to recover $9,613 for plumbing done in the latter firm's warehouse, has been finally settled, after hanging fire in the courts for over a year. Gage and Co. pay the plumbing firm $9,981, the increase of $300 including some other small accounts and interest. Mr. Gage preferred to settle the suit rather than fight it out in the courts, considering that the amount of the supposed "rake-off" was not worth the costs of the litigation.

NEW PAINT FACTORY.

The Standard Manufacturing Company, of St. John, Newfoundland, have just completed the construction and fitting up of their new building which is to be used as a paint factory. Part of the machinery for the production of both lead and liquid paint is now being installed. Experts have been engaged to carry on the manufacture of these commodities.

BUSINESS CHANCES.

HARDWARE business, established 32 years in town of 2700 population, no opposition; stock of about $4,000, consisting of hardware, paints and oils, tinware and graniteware, annual turn over $13,000 (for 1906), owners going into manufacturing; building owned by present owners, who will rent reasonably. Box U, Alexandria, Ont. [5]

FOUNDRY AND METAL INDUSTRIES

The Peterboro Steel Rolling Mills Company, Peterboro, are considering the erection of a new plant.

The C.P.R. will erect car-shops at London, Ont., and will probably employ about five hundred hands.

Nathaniel Dyment, proprietor of the Dyment Foundry Company, Barrie, died at his home in Barrie this week.

The Northwest Brass Company, Winnipeg, Man., are erecting a new steel building, 100x80 feet, at a cost of about $10,000.

John E. Wilson, of St. John, N.B., will erect a large addition of 40x100 feet to his foundry, and will go extensively into the manufacture of stoves.

Archibald W. Duff, of Montreal, on leaving for Winnipeg to take charge of a branch establishment of the Canadian Westinghouse Company, was the recipient of a gold watch from his Montreal friends.

The Dominion Steel Company's Sydney rail mill in January broke the record for the greatest monthly production of steel rods, Their mill turned out 7,960 tons, the former record being 7,000 tons.

The moulding shop in connection with the business of the Wilson Scale Works, Toronto, was gutted by fire last Saturday afternoon. Valuable patterns of some machinery was destroyed. The loss, about $3,000, is covered by insurance.

Iron sheets coated with aluminum are now being manufactured in considerable quantities, and have been found to be very durable under long exposure These aluminum-coated sheets will probably supplant galvanized iron for many purposes.

The main building of the new plant which Jenkins Bros., New York, valve manufacturers, are erecting in Montreal, will be 200x50 feet. There will also be a foundry 150x64 feet, and an engine and boiler room 60x45 feet. Later, the company expects to build a foundry 100x60 feet.

PROTECTING CUTLERY MARKS.

A decided effort is being made in Sheffield to prevent the marking of foreign made cutlery in such a way as will convey the impression that it is of the Sheffield manufacture. The plan under consideration is to so mark the Sheffield-made goods as will distinguish the genuine from the imitation, but it is pointed out by authorities in England that it would be very difficult to discover any mark which would show conclusively that the goods had been made in England, and would still be out of reach of infringers or imitators. The manufacturers of cutlery in Sheffield are facing a marked advance of certain raw materials which may affect seriously the price at which the finished article can be placed upon the market. Celluloid has advanced about 10 per cent., and even more serious than this is the recent advances in the price of ivory. Within a short time past this material has advanced £12 per hundredweight in price. For a number of years past the Sheffield buyers who have attended the ivory auction sales have declined to bid because they could not afford the prices paid for ivory by American manufacturers of piano keys, who were apparently in need of all the stock that was offered. It is said, on good authority, that in the past three months Sheffield has cut up for use the smallest amount of ivory ever before used.

LAST YEAR'S METAL PRODUCTION

In its issue of January 5th the Engineering and Mining Journal publishes the statistics of the production of the principal metals and mineral substances in 1906, the figures for 1905 being given in comparison. The year which has closed was one of great prosperity in the mining districts, and in most branches of mineral and metal production. The production of gold in the world is estimated at $404,649,685, against $379,867,373 in 1905. The production of copper in North America (including the United States, Canada and Mexico) was 1,097,300,000 lbs., in 1906, against 1,063,582,700 lbs. in 1905 Most of these statistics are based on direct reports from the producers.

PROMPT PAY.

Settle on the day. Your reputation for meeting your obligations promptly is your best asset. Every day does something toward making or unmaking your reputation as a business man, says the Hardware Trade.

When your bills come due and you assume it is not necessary for you to make the proper arrangements with the wholesaler, letting that slide by the day of payment, you have registered with the wholesaler, and with any others whom he may give the information to, the fact that you are "unsatisfactory" pay. Never take anything for granted when a business reputation is at stake.

Do business on business principles. When obligations are due, they are due. Many merchants do not need to be told this. Many retailers who have not yet become very good merchants fail to realize these truths.

LETTER BOX.

P. S. Stewart & Son, Renfrew, asks for the name of the manufacturer of the "Winnipeg" Hockey Skates.

Ans.—We believe that J. K. McCullough & Co., Winnipeg, are the manufacturers. If not, can any reader supply the correct address?—Editor.

"CARRON"
"AYRESOME"
"CLARENCE"

PIG IRON

Finest Quality.
Prompt Shipments.

A.C. LESLIE & CO. Limited
MONTREAL

HAND DRILLING MACHINERY

We have the best that money can buy.

Our cheaper grades are all good serviceable machines.

The best value in the market whether you want the cheap or the dear kind.

A. B. JARDINE & CO.
HESPELER - - - ONTARIO

NICKEL
The Canadian Copper Co.

NICKEL FOR NICKEL STEEL
The Orford Copper Co.

WRITE US FOR PARTICULARS AND PRICES

General Offices
43 EXCHANGE PLACE
NEW YORK CITY

OAKEY'S The original and only Genuine Preparation for Cleaning Cutlery, 6d. and 1s. Canisters

'WELLINGTON' KNIFE POLISH

JOHN OAKEY & SONS, Limited
Manufacturers of
Emery, Black Lead, Emery, Glass and Flint Cloths and Papers, etc.

Wellington Mills, London, England

Agent:
JOHN FORMAN, - 644 Craig Street
MONTREAL.

"MIDLAND"
BRAND.
Foundry Pig Iron.

Made from carefully selected Lake Superior Ores, with Connellsville Coke as Fuel, "Midland" will rival in quality and grading the very best of the imported brands.

Write for Price to Sales Agents

Drummond, McCall & Co.
MONTREAL, QUE
or to
Canada Iron Furnace Co. Limited
MIDLAND, ONT.

...FULL STOCK...

Salt Glazed Vitrified

Double Strength Culvert Pipe a Specialty.

THE CANADIAN SEWER PIPE CO.
HAMILTON, ONT. TORONTO, ONT
ST. JOHNS, QUE.

GET PRICES FOR
GILBERTSON'S
COMET
Brand
GALVANIZED FLAT SHEETS

for any purpose where the best is needed. Wide awake jobbers handle this brand. They are well galvanized, true to gauge and especially soft for working up purposes. Every sheet bears the name "Gilbertson." That is a guarantee.

MAKERS:
W. GILBERTSON & CO., Limited.
PONTARDAWE, SOUTH WALES
Agent: Alex. Gibb, 13 St. John St., Montreal.

NOVA SCOTIA STEEL & COAL CO., Limited
NEW GLASGOW, N.S.

Manufacturers of

Ferrona Pig Iron

And SIEMENS-MARTIN
OPEN HEARTH STEEL

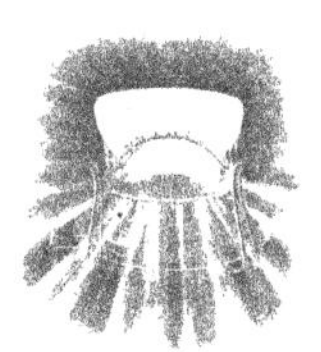

50% MORE LIGHT

BING GLASS CONE BURNERS

MANUFACTURED IN CANADA
BY
Ontario Lantern & Lamp Co.
HAMILTON, ONT.

BUILDING AND INDUSTRIAL NEWS

HARDWARE AND METAL would be pleased to receive from any authoritative source building and industrial news of any sort, the formation or incorporation of companies, establishment or enlargement of mills, factories, foundries or other works, railway or mining news. All correspondence will be treated as confidential when desired.

An opera house will be erected at New Liskeard, Ont.

A market building will be erected at Berlin, Ont., at a cost of about $20,000.

Plans are being prepared for the erection of a new hotel at Kingston to cost $120,000.

An up-to-date hotel building is going to be erected at Woodstock at the cost of $60,000.

The Metallic Roofing Company, of Toronto, will enlarge their factory on King and Dufferin streets.

The Ottawa Steel Casting Company have applied for an increase of $100,000 to their capital stock, making it $350,-000.

A new skating rink will be erected at Ottawa to replace the one destroyed by fire a week ago. The new rink will cost about $75,000.

Plans have been completed for a handsome new three-storey building to be erected by the Oddfellows of Strathcona, Alberta.

A new theatre will be erected on Richmond street, Toronto, at a cost of about $75,000. The promoters include Rust and Weber of New York city.

The Meisel shops at Port Arthur will be erected almost entirely of glass. Preparations are being made for rushing the construction work in the spring.

Negotiations are being carried on for the establishment of a match factory in Lindsay. J. D. Mantion, formerly of Hull, Que., is promoter of the industry.

The National Car Company, Halifax, N.S., recently incorporated with a capital of $1,000,000 intend erecting a plant for the construction of from 15 to 25 cars per day.

A new company has been incorporated with a capital of $100,000, to be known as The Plimley Automobile Company, and will erect a new brick garage at Victoria, B.C.

Plans are now ready for the new dam and bridge at Buckhorn, on the Trent Canal. Tenders will be called in a short time, and it is estimated that the work will cost about $35,000.

Building permits to the value of $750,000 have been issued in Toronto for January as against $450,000 during January last year.

The Canadian-Mexican Pacific Steamship service will carry samples of general merchandise free of charge to the various ports touched by them in order to show what Canada and Mexico have to offer each other.

A new grain elevator with a capacity of about 10,000,000 bushels will be erected at Port Arthur for the Grand Trunk Pacific Railway. The contract has been awarded to Barnett and Mcueen of Port Arthur.

The Standard Glass Company has been incorporated with a share capital of $40,000, for the purpose of manufacturing glass and glassware. The head office is to be in Toronto, the provisional directors being: J. Hurst, W. Hand, W. Jeffery and W. F. Oliver, of Toronto.

The Gamer Manufacturing Company has been incorporated with a share capital of $50,000 for the purpose of manufacturing bedsteads, etc. The head office will be at Chesley, the directors being: Arthur Gamer, Ada Gamer, A. Harrod, T. E. Devitt and F. Gillon of Weston, Ont.

The Wingold Stove Company has been incorporated with a capital stock of $40,000, for the purpose of manufacturing stoves, ranges, furnaces, etc. The head office will be at Winnipeg, the directors being: F. B. Blanchard, O. Gensmer, R. Kellow, L. Blanchard and I. A. Mackay, of Winnipeg.

The Kelly Island Lime Company has been incoeporated with a share capital of $40,000 for the purpose of manufacturing lime, mortar, cement, sewer pipe, etc. The head office will be in Windsor, the provisional directors being: A. R. Bartlet, N. A. Bartlet and A. W. MacGregor of Windsor.

Love Brothers, has been incoruorated with a share capital of $150,000 for the purpose of manufacturing building materials and supplies. The head office will be at Toronto, the provisional directors being: P. Love, H. W. Love, E. G. Long, F. Orford, W. J. Coulter, and E. J. Barton of Toronto.

The Eastern Construction Company has been incorporated with a share capital of $1,000,000, for the purpose of manufacturing machinery for construction purposes. The head office will be at Ottawa, the directors being: J. Gillespie, H. H. Short, F. H. Honeywell, S. B. Johnston and J. B. Prendergast of Ottawa.

The M. McKenzie Company has been incorporated with a capital stock of $75,000 for the purpose of manufacturing mill, marine, railway and contractors' supplies. The head office will be in Montreal, the directors being: U. D. Hamilton, A. A. Lunan, A. Dunn and L. Lahaye of Montreal, and Alex. Lunan of Huntingdon, Que.

The Crescent Machine Company has been incorporated with a share capital of $20,000 for the purpose of manufacturing machinery, implements, rolling stock and hardware. The head office will be at Montreal, the directors being: C. M. Gardiner, C. D. Drabble, W. A. Patterson. H. S. Williams and W. Bovey, of Montreal.

A new building will be erected in Regina for stock show purposes. The new building will be 144 feet by 101 feet, to be used as a show and sale ring and for stabling accommodation.

In the centre will be a show ring 80 x 39 feet, and surrounding the arena is seating accommodation for over 2,000. The building will cost in the neighborhood of $15,000.

A school building will be erected at Cobalt at the cost of about $8,000.

A new hotel will be erected at Kenora, Ont., at the cost of $100,000.

The Canadian Pacific Railway Company have purchased a site for their new ticket office in Montreal, the city freight office, a local telegraph office, and other offices of the railway company. The Dominion Express Company have secured room in part of the building. The price of the property was a little over half a million dollars.

Cameron & Company has been incorporated with a share capital of $100,000 for the purpose of manufacturing furniture, doors, sash, pulp, turpentine, wood alcohol, etc. The head office will be at Ottawa, the provisional directors being: U. A. Cameron, R. G. Cameron, G. C. Edwards, G. Cameron of Ottawa, J. H. Gates of Burlington, and Hugh McLean of Buffalo, N.Y.

The Consolidated Bicycle and Motor Company, Winnipeg, Man., has been incorporated, with a capital of $60,000, for the purpose of manufacturing bicycles, motor cars, boats, pumps, etc. The head office will be at Winnipeg, provisional directors being C. T. Cruikshank, J. A. Hudson and J. S. St. Mars, all of Winnipeg.

To spend nearly half a million dollars during the present year in building some sixty new school rooms to relieve the overcrowding, and to erect a big building at the corner of York and Richmond streets, to serve as Board of Education offices, board rooms and store rooms, and to make extensive improvements and enlargements is the policy of the Property Committee of the Board of Education, Toronto.

GOODS OUT OF STOCK.

In a recent interview on the treatment of customers by the retailer of sporting goods, a writer in the Sportings Goods Dealer has the following to say: "No doubt in common with a great many others I am often asked for articles which I do not have in stock, and I claim that my acquaintance with this industry is as large as that of most retailers. Under these circumstances the best course to pursue with customers, is a question which naturally suggests itself. Peronally, I frankly state at the outset that I do not hold the particular line required, but suggest that I have a very similar article which should answer just as well. Mind, I do not force this substitute on my customer, or say that it is better than the article required. This, in my opinion, is by no means advisable, and, in fact, according to the letter of law, comes very nearly within the definition of substitution. However, there can be no possible objection taken to pointing out that one has an equally good line in stock, and more often than not a sale will result. If, however, my customer is particularly set on obtaining a certain article, it is always polite to offer to procure it for him, even if the resulting sale does not provide a very large profit, or if it involves some considerable trouble to obtain. The dealer has at least prevented the customer from applying at some other establishment, and has created a good effect by his evident anxiety to please the client. If I am not acquainted with the maker of the particular line desired, I usually drop a note to my trade paper, and in nine cases out of ten the paper furnishes me with the information and puts me in the proper channel for obtaining a supply. There is some information, however, with regard to certain lines, which even the editorial knowledge can not supply, and this fact leads me to consider seriously whether such a line is worth the retailers attention at all. In my opinion, unless a trader is unable to learn or is not familiarized by advertisements with all particulars of any really sound line, he is quite justified in refusing to handle it. This may seem a somewhat sweeping assertion, but it must be borne in mind that if a firm have sufficient confidence in the merits of their goods, they are always willing to go to some expense to notify the trade of its particular advantages."

NO THINKING HERE.

A prominent business man told me recently that his great weakness was his inability to stop thinking after retiring, writes O. S. Marden in Success. In this way he is robbed of so much sleep that he feels all used up the next day.

I advised him to cultivate the habit of closing the door of his business brain at the same time that he closed the door of his business office. "You should," I said, "insist on changing the current of your thoughts when you leave your business for the day just as you change your environment or as you change your dress for dinner when you go home in the evening. Turn your thoughts to your wife and children, to their joys and cares; talk to them; play games with them; read some humorous or entertaining story, or some strong, interesting book that will lift you, in spite of yourself, out of your business rut. Go out for a long walk or a ride; fill your lungs with strong, sweet, fresh air; look about you and observe the beauties of nature, or have a hobby of some kind to which you can turn for recreation and refreshment when you quit your regular business. Be master of your mind; learn to control it instead of allowing it to control you and tyrannize over you.

Hang up in your bed chamber, in a conspicuous place, where you can always see it, a card bearing in bold illuminated characters this motto: NO THINKING HERE.

There is no short, easy road to success.

CATALOGUES AND BOOKLETS

When sending catalogues for review, manufacturers would confer a favor by pointing out the new articles that they contain. It would assist the editor in writing the review.

By mentioning HARDWARE and METAL to show that the writer is in the trade, a copy of these catalogues or other printed matter will be sent by the firms whose addresses are given

Artistic Pastel.

James Smart, manufacturers of hardware and tools, stoves, ranges and furnaces, Brockville and Winnipeg, are sending to their customers a handsome calendar in the shape of a pastel, tinted in gold, it making a very beautiful ornament for the office or shop.

Money Saving Oil Tanks.

A small booklet issued by S. F. Bowser & Company, Toronto, entitled, "How to Retail Gasoline," should prove very interesting to hardware merchants who handle gasoline. By installing the Bowser long distance outfit, loss can be turned into profit besides doing away with any possible danger of explosion or fire. Any merchant who has not yet received the booklet should write for one, mentioning this paper.

Atkins' Anniversary Souvenir.

E. C. Atkins & Co., Indianapolis and Toronto, manufacturers of Atkins Saws, have issued a handsome book in commemoration of their fiftieth anniversary, entitled "The Saw, its Ancient and Modern Development." The binding is a clever imitation of a piece of sawed timber, while the grain of the wood correctly brought out, while high grade paper, cuts, type and artistic arrangement combine to make the production harmonious and effective in a high degree

A very interesting sketch of the world's industrial development leading up to the invention of the circular saw, is given, followed by a sketch of the ancestry, youth and business experience of E. C. Atkins, founder of the company. This introduces a history of the company, with brief sketches of the men who have been active in its management and are now directing its affairs. Some description of the present plant is accompanied by numerous handsome illustrations, while the concluding pages of portraits show the company's appreciation of its efficient clerical, mechanical and selling staff.

The company is to be congratulated on their half century of progress, as well as on the beauty and good taste of its souvenir We do not know how generally the Atkins company can distribute their souvenir book, but any dealer who secures a copy is bound to value it as a decided acquisition to his library.

Rolling Mills Price List.

The price list of bolts, nuts, hinges, etc., manufactured by John White, London Bolt and Hinge Works, London, Ont., also containing prices of bar iron and steel washers, etc., manufactured by the London Rolling Mills Company, has been sent to the trade, and should prove very handy to hardware merchants, as the number of various bolts, in 100 lbs. is given, as well as cuts of various styles of nuts, hinges, screw hooks for various purposes, harrow teeth, etc. The last ten pages contain quotations of the London Rolling Mill Company's product of iron and steel bars and bands, sleigh, shoe steel, and washers, and is particularly valuable.

Imperial Varnish Catalogue.

The trade sales catalogue, No. 10 of the Imperial Varnish and Color Company of Toronto, has been sent out to the trade, containing illustrations of the many lines manufactured by them. Suggestions of numerous methods for increasing the sale of this paint are also supplied.

Carriage Castings.

Catalogue No. 7 of malleable iron goods manufactured by P. Kyle, Merrickville, Ont., has been issued, it containing views and sizes of carriage and wagon castings, such as whiffletree hooks, centre irons, clevises, plates, couplings, bolster plates, neck yoke tips, wear irons, step pads, etc., hame trimmings, and various tools and fittings for agricultural workf Any merchant not receiving this catalogue should write for a copy, mentioning this paper. The many illustrations will be found of great value in ordering goods.

Marlin's New Catalogue Ready.

A handsome hanger in colors is being distributed by the Marlin Fire Arms Company, New Haven, Conn. It is really an announcement of the new catalogue of this firm, for across the top of the hanger are the words, "New Catalogue Now Ready." The hanger is very pretty and will ornament the walls of the store or the show window. The new catalogue is also a handsome trade publication and will be interesting to all dealers in firearms. Mention this paper.

HOW GERMANS ENCOURAGE CASH.

In the German town of Glauchau a novel plan has been adopted by merchants for the encouragement of cash payments. The plan is to give coupons for cash sales and where purchases aggregate 110 marks the customer is entitled to 5 marks or 4½% on his purchases. The method of carrying it out is German. An association of relatives has been formed and each pays an entrance fee of 50 marks (23.8 cents.) The local bank handles the funds and redeems the coupons. To enable it to do that the merchants buy the coupons from the bank; and to cover expenses pay a little more than the redemption of the stamps call for. The bank's renumeration is the handling of the funds. The association's surplus amounting this year to 500 marks, is distributed as prizes is amounts of 5 marks each to the persons whose names appear on the redeemed books first drawn at random from the files of the current year.

CONDENSED OR "WANT" ADVERTISEMENTS.

Advertisements under this heading 2c. a word first insertion; 1c. a word each subsequent insertion.

Contractions count as one word, but five figures (as $1,000) are allowed as one word.

Cash remittances to cover cost **must** accompany all advertisements. **In no case** can this rule be overlooked. Advertisements received without remittance cannot be acknowledged.

Where replies come to our care to be forwarded, five cents must be added to cost to cover postage, etc.

SITUATIONS VACANT.

WANTED—Experienced pricer for wholesale hardware; none but experienced man need apply. Box 578, HARDWARE AND METAL, Toronto.

BRIGHT, intelligent boy wanted in every town and village in Canada; good pay besides a gift of a watch for good work. Write The MacLean Publishing Company, 10 Front Street E., Toronto. [tf]

WANTED—An experienced hardwareman at once. Apply, stating wages required, to Thomas Oliver, Hardware Merchant, Copper Cliff, Ont. [6]

BUSINESS CHANCES.

HARDWARE stock about $7,000, including set of tinners' tools; stock is all new and situated on main line C.P.R. in Saskatchewan town of 1,800 population; building can be leased, is brick, two storeys and basement; net profit $4,000 yearly. Box 577, HARDWARE AND METAL, Toronto. [7]

HARDWARE BUSINESS established 17 years for sale—including stoves, ranges and tinware; a good plumbing and tinshop in connection; stock about $6,000. J. D. Murdoch & Co., Simcoe, Ont. [6]

FOR SALE—A fully equipped Wire Mill and Nail Factory as a going concern, with about 500 tons of wire rods; factory situated on 150 ft. of water-front with trackage optional to buyer. B.C. Wire and Nail Co., Ltd., Vancouver, B.C. (7)

FOR SALE.

FOR SALE—Hardware, tinsmithing and stove business; stock about $3,000, doing a good steady business; expenses low. Apply Box 582 HARDWARE AND METAL, Toronto. (8)

HARDWARE Business; including stoves, tinware and tinsmith tools, in thriving town in West Ontario peninsular, stock about $5,000; building can be leased if desired, dwelling also. Box 583 HARDWARE AND METAL, Toronto. (17)

ARTICLES WANTED.

WANTED—A set of tinsmith tools, second-hand; state full particulars, H. A. Hoar, Barrie. [6]

WANTED—8 foot cornice brake; state price and particulars. J. E. Hussey, Melbourne, Ont. (9)

ANY hardwaremen having a stock of cut nails they wish to dispose of at reduced prices should write at once to Box 501 HARDWARE & METAL, Toronto.

To Manufacturers' Agents

HARDWARE AND METAL has inquiries from time to time from manufacturers and others wanting representatives in the leading business centres here and abroad.

Firms or individuals open for agencies in Canada or abroad may have their names and addresses placed on a special list kept for the information of inquirers in our various offices throughout Canada and in Great Britain without charge.

Address
Business Manager
HARDWARE AND METAL
Montreal and Toronto

Persons addressing advertisers will kindly mention having seen their advertisement in Hardware and Metal.

Paint, Oil and Brush Trades

THE LURE OF THE WEST.

Still another of the Canada Paint Company's staff has been called to their branch works in Manitoba. A few days ago Robert Bremner, jr., received orders to proceed to Winnipeg, and, as he is a universal favorite, the opportunity was seized upon to make him a little presentation. This took the form of a neat chamois purse generously lined with some golden beauties. On behalf of the staff, Mr. Robert Munro, the managing director of the company, made the presentation and, in wishing Mr. Bremner farewell, alluded to the corp d'esprit and bond of harmony which existed between the staff generally and the various branches in Montreal, Toronto and Winnipeg.

On Thursday last at the Canadian Pacific Railway station in Montreal a large crowd also assembled to say "good-bye" to Mr. Bremner. In addition to a strong contingent from the Canada Paint Company's works there was a good assemblage of old chums, aunties, sisters and cousins (Bob, like a tru Irishman, counts his cousins by the dozens.) As the train pulled out there was hand-shaking, cheers, and not a few tears, one old Salt bound for Esquimalt, remarking as the lachrymal flow came in a perceptible stream: "Vast heaving there—Belay making Eye (high) water on the poop deck or you'll soon have a bloomin' skiting rink!" The night was cold but the send-off was warm. Success to Mr. Bremner, jr., in the Prairie City.

TURPENTINE VS. BENZINE.

Now that turpentine has reached such high prices and become so scarce, benzine is being largely used as a substitute. For thinning paint deodorized benzine replaces spirits of turpentine to a great extent in the manufacture of low priced varnishes and ready mixed paints, and almost to the exclusion of turpentine by manufacturers of agricultural implements, farm wagons and shade cloth. Wherever a low priced solient or volatile article is desired, it finds a ready market. And wherever it is simply necessary to cover a surface with paint, that is to dry rapidly, afterwards to be protected by varnish, it serves the purpose, because of its rapid and complete evaporation. While it is a more reliable thinner for paint than fatty or adulturated turpentine, it can by no means replace turpentine, as it leaves oil paints on drying, more porous and less binding, and cannot be employed with satisfaction by the coach painter, or decorator for interior work.

Try it for yourself by mixing color fairly strong with oil and drier, then thin one half of the paint with turpentine, and the other half with the same measure of benzine, apply the two mixtures side by side, on an old painted board, and expose the board for a week or so and see what happens.

RAW AND BOILED OIL.

For some purposes raw oil is preferable to boiled. The raw oil does not harden as quickly as the boiled, consequently is more in evidence for outside work. In speaking on the subject of oils, an old veteran of the brush, quoted by an exchange, said:

"Raw linseed oil generally finds most favor. It is clear, works freely, and in white lead paint on outside work requires the addition of little driers. It allows the paint to cover well, and is durable. Mixed in raw oil a paint dries with a glossy surface, which is a factor of durability. Boiled oil, on the other hand, is for white lead paints, less satisfactory than raw. Though a paint mixed with it may go further, it does not cover so well, is apt to discolor light tints, is harder to spread, less durable, and a constant source of blistering. Where the use of boiled oil comes in is in mixing dark colors for outside work, such as Brunswick green, Indian red, bronze green and such colors obtained by a painter in a dry state. The comparatively thick condition of boiled oil assures a better gloss under these circumstances, and assures their drying under ordinary conditions without the addition of dryers."

WOOD STAIN INSTRUCTOR.

S. C. Johnstone and Son, of Racine, Wis., have recently published an extremely interesting and instructive handbook upon their specialty. It furnishes valuable information regarding the different kinds of wood—hard, medium and soft—and the best methods to follow in order to preserve and bring out the full beauty of grain and texture.

The different varieties of each wood are dealt with separately, and full instructions are given of the methods used for obtaining all the various effects which are now so much in demand. While the book advertises more especially the materials manufactured by this firm, it contains a vast amount of information which will greatly aid hardware merchants in intelligently selling stains, varnishes, etc. It is well illustrated and well printed, and will be well received by hardware merchants.

YOUR PAINT AND VARNISH WANTS

No manufacturer in Canada can so readily and efficaciously supply your paint and varnish wants and we wish you to test our Business Promptitude as well as the Quality of our goods, if you have never done so before.

NEW ERA PAINT

makes friends
makes dollars
and life-long customers

N.B.—You are not giving yourself a square deal, unless you have our catalogue.

STANDARD PAINT AND VARNISH WORKS CO., LIMITED
WINDSOR · ONTARIO

Your CUSTOMERS ARE HUNGRY

for something really satisfactory for their Floors.

Give them

JAMIESON'S FLOOR ENAMEL

and they get what, up to the present, they have hoped and longed for in vain.

DRIES HARD OVER NIGHT
DOES NOT PEEL OFF
IS NEVER STICKY
MIRROR GLOSS
HARD AS IRON
STANDS REPEATED WASHING

Manufactured by

B. C. JAMIESON & CO.
LIMITED
16-30 Nazareth St., MONTREAL

Write for prices and show card

McCaskill, Dougall & Co.

Manufacturers

RAILWAY, CARRIAGE AND BOAT **VARNISHES.**
HIGH-GRADE FURNITURE and HOUSE VARNISHES

MONTREAL.

A. RUDD & CO., St. Helens, Lancashire, England

Manufacturers of highest grade

GLAZIERS' DIAMONDS In all patterns. Also Circle Boards, Beam Compasses, Gauge Glass Cutters, Boxwood Squares and Laths, Plate Glass Nippers, Atmospheric Soldering Irons, etc., etc., etc.,

Canadian Agent: J. B. GIBSONE, P.O. Box No. 478, MONTREAL

GLAZERS' DIAMONDS OF PROVED WORTH

Having a Century's Reputation for Reliability.

MADE BY

A. SHAW & SON, London, Eng.

CANADIAN AGENT

GODFREY S. PELTON, - 201 Coristine Building, Montreal

Louden's Double Strap Barn Door Hangers

Are the Standard

They Have Case Hardened Bearings, Track Scrapers, Revolving Washers.

They hold the door closer to the track, hang perfectly plumb, and allow the door to hang closer to the wall than other hangers do, while they are just as flexible, and are the easiest running hangers made.

Manufactured by

LOUDEN MACHINERY CO.
GUELPH, ONT.

We manufacture 15 different styles of Hay Carriers, 10 different styles of Barn Door Hangers, also Feed and Litter Carriers, Cow Stanchions, Barn Door Latches, etc.

PAINT AND OIL MARKETS

MONTREAL.

Office of HARDWARE AND METAL,
232 McGill Street,
Montreal, February 8, 1907

There is nothing very startling to report in the markets this week.

Turpentine is reported to be very firm with higher prices at the stills, and it is thought that much inclement weather is put up with at the pine woods, figures may soar even five or ten cents per gallon. Local figures, however, have not been changed, but an advance may be recorded at any moment.

With the exception of Paris green, which is now being in fair quantities, with orders marked for February delivery, there is no special rush in the paint and oil department.

In view of a sharp advance in damar gum, all damar varnishes have been advanced about 50 cents per gallon. Varnish and Japans generally, owing to the higher price of turpentine, are being firmly held.

LINSEED OIL—Still remains at the same prices. We quote: Raw, 1 to 4 barrels, 55c; 5 to 9 barrels, 54c; boiled, 1 to 4 barrels, 58c; 5 to 9 barrels, 57c.

TURPENTINE—(In barrels). Higher prices prevail at the stills, and if much inclement weather is met with at the pine woods, prices are likely to advance 5 or 10c per gallon. Our prices are: Single barrel, 98c per gal.; for smaller quantities than barrels, 5c extra per gal. is charged. Standard gallon is 8.40 lbs., f.o.b. point of shipment, net 30 days.

GROUND WHITE LEAD—Our prices remain: Best brands Government Standard, $7.25 to $7.50; No. 1, $6.90 to $7.15; No. 2, $6.55 to $6.90; No. 3, $6,30 to $6.55, all f.o.b., Montreal.

DRY WHITE ZINC—Still firm. We quote: V.M. Red Seal, 7½c to 8c; Red Seal, 7c to 8c; French V.M., 6c to 7c; Lehigh, 5c to 6c.

WHITE ZINC—Our quotations are: Pure, 8½c. to 9½c.; No. 1, 7c. to 8c.; No. 2, 5¾c. to 6¾c.

PUTTY.—Our prices are: Pure linseed oil, $1.75 to $1.85; bulk in barrels, $1.50; in 25-lb. irons, $1.80; in tins, $1.90; bladder putty in barrels, $1.75.

ORANGE MINERAL—The following prices still hold: Casks, 8c; 100 lb. kegs, 8½c.

RED LEAD—The following quotations are firm: Genuine red lead in casks, $6.00; in 100-lb. kegs, $6.25; in less quantities at the rate of $7 per 100 lbs.; No. 1 red lead, casks, $5.75; kegs $6, and smaller quantities, $6.75.

PARIS GREEN—We continue to quote: In barrels, about 600 lbs., 23¼c per lb.; in arsenic kegs, 250 lbs., 23½c.; in 50 lb. drums, 24c.; in 25 lb. drums, 24½c.; in 1 lb. packets, 100 lbs. in case, 25c.; in 1 lb packets, 50 lbs in case, 25½c.; in ½ lb. packets, 100 lbs in case, 27c.; in 1 lb. tins, 26c. f.o.b. Montreal. Term three months net or 2 per cent. 30 days.

SHELLAC—Prices remain: Bleached in bars or ground, 46c. per lb., f.o.b. Eastern Canadian points; bone dry, 57c per lb., f.o.b. Eastern Canadian points; T. N. orange, etc., 48c per lb. f.o.b. New York.

SHELLAC VARNISH—Very firm. We still quote: Pure white, $2.90 to $2.95; pure orange, $2.70 to $2.75; No. 1 orange, $2.50 to $2.55.

MIXED PAINTS—No change in prices. We still quote: $1.20 to $1.50 per gallon.

PETROLEUM—(In barrels) Our prices are: American prime white coal, 15½c. per gallon. American water, 17c. per gallon; Pratt's Astral, 19½c. per gallon.

WINDOW GLASS—Prices remain firm. We still quote: First break, 50 feet, $1.85; second break, 50 feet, $1.95; first break, 100 feet, $3.20; second break, 100 feet, $3.40; third break, 100 feet, $3.95; fourth break, 100 feet, $4.15; fifth break, 100 feet, $4.40; sixth break, 100 feet, $4.95. Diamond Star: First break, 50 feet, $2.30; second break, 50 feet, $2.50; first break, 100 feet, $4.40; second break, $4.80; third break, 100 feet, $5.75; fourth break, 100 feet, $6.50; fifth break, 100 feet, $7.50; sixth break, 100 feet, $7.50; seventh break, 100 feet, $8; eight break, 100 feet, $9. Double Diamond: First break, 50 feet, $3.45; second break, 50 feet, $3.75; first break, 100 feet, $6.75; second break, 100 feet, $7.25; third break, 100 feet, $8.75; fourth break, 100 feet, $10; fifth break, 100 feet, $11.50; sixth break, 100 feet, $12.50; seventh break, 100 feet, $14; eighth break, 100 feet, $16.50; ninth break, 100 feet, $18; tenth break, 100 feet, $20; eleventh break, 100 feet, $24; twelfth break, 100 feet, $28.50. Discount on Diamond Star, 20 per cent.; on Double Diamond, 40 per cent.

TORONTO.

Office of HARDWARE AND METAL,
10 Front Street East.
Toronto, February 8, 1907

The paint trade still continues very brisk, with large orders being received for future shipment.

Very few advances are quoted this week, the principal one being on damar, which has advanced 50c. per gallon. Clear damar is now quoted at $2.80, and No. 1 at $2.50 per gallon. This will have a tendency to cause further advances on varnishes. Bluestone has also advanced, owing partly to the large quantities required for the manufacture of paris green.

White lead is in good demand and large orders are being booked for present shipment. No advance is quoted

this week but merchants are ordering in large quantities, in view of a further advance, which seems certain to take place in the near future.

Turpentine remains at last week's prices, but is much firmer in Europe and Savannah. No change is made in raw or boiled oil, very little ordering being done for either oils or turps at present. No let-up is reported in the sale of paris green, book orders being received daily for this article, which has practically no substitute.

WHITE LEAD—Ex Toronto pure white, $7.40; No. 1, $6.65; No. 2, $6.25; No. 3, $5.90; No. 4, $5.65 in packages of 25 pounds and upwards; 1-2c per pound extra will be charged for 12 1-2 pound packages; genuine dry white lead in casks, $7.00.

RED LEAD—Genuine in casks of 500 lbs., $6.00; ditto, in kegs of 100 lbs., $6.50; No. 1 in casks of 500 lbs., $5.75; ditto, in kegs of 100 lbs., $6.25.

DRY WHITE ZINC.—In casks, 7½c.; in 100 lbs., 8c., No. 1, in casks 6½c. in 100 lbs., 7c. (Ground in oil)—In 25 lb. irons, 8c.; in 12½ lbs., 8½c.

SHELLAC VARNISH—Pure orange in barrels, $2.70; white, $2.82 1-2 per barrel; No. 1 (orange) $2.50; gum shellac, bone dry, 63c., Toronto. T. N. (orange) 51c. net Toronto.

LINSEED OIL.—Very little is being sold at present, with no advances quoted for this week. Prices remain at: Raw, 1 to 3 barrels, 58c.; 4 to 7 barrels, 57c.; 8 and over, 56c, add 3c. to this price for boiled oil. Toronto, Hamilton, London and Guelph net 30 days.

TURPENTINE.—The markets at Savannah, and Europe are very firm. The demand for the season's supply of turps has not started as yet. We still quote: Single barrels, $1.02 per gallon f.o.b. point of shipment, net 30 days. Less than barrels, $1.07 per gallon.

PARIS GREEN.—The demand remains extra good, in view of the further advances expected. Merchants are remembering their difficulties in securing second supplies last year, even at the advanced price. We now quote base price per lb. on Canadian 26½c., and English 26¾c., f.o.b. Toronto.

PETROLEUM—For refined we still quote: Canadian prime white, 14c.; water white, 16c.; American water white, 16c. to 18c. ex. warehouse. On crude the prices are: Canadian, $1.32; Pennsylvania, $1.58; Ohio, 96c.

For additional quotations see current prices at back of paper.

PAINT COMPANY MOVES.

Benjamin Moore & Company, formerly the Francis Frost Company, have removed from their old factory on Queen street east, Toronto, to their fine new $30,000 factory building at the corner of Cawthra and Grove avenues, Toronto Junction, and in future all mail should be sent to the new address. The new factory is 50 x 150 feet, and has three stories and basement, with room for further expansion as the demand for Muresco wall finish and Ark brand paint makes necessary further enlargements.

CARE OF BRUSHES.

One conspicuously famous brush maker has declared the art of brush making to be "an art preservative." The practical carriage and wagon painter is deeply concerned in the achievements of that art, because every distinct advancement made therein makes possible an equally distinct advancement in the art of painting. To a greater extent, perhaps, than any other class of painters, the carriage and wagon painter should be interested in making up his brush equipment of tools of the best quality.

The brush made of reliable stock, having the proper "hand" and point, and which balances live a "thoroughbred," is an excellent tool to buy, regardless of the price. The vehicle painter requires a brush made scientifically, by the outlay of honest workmanship, and of material that is wholly above suspicion. A brush that has simply the price to recommend it is usually an unreliable article and worketh evil, like a thief in the night, unexpectedly. In making choice of a brush for putting on priming, lead, and roughstuff, and for such other features of general use as requires a round or oval bristle brush, the painter may properly look at the filling of the tool. Deception, if practised at all, is usually placed where it shows least.

The first-class brush is distinctively the brush that shows quality—uniform quality—from centre to outside, says the Painter and Decorator. Other things being equal, the brush that is made up uniformly as to its bristle equipment will develop a good point, and all carriage painters are alive to the importance of this virtue in both paint and varnish brushes. Much of the usefulness of a brush depends upon the manner of caring for it when it comes into the paint shop. The bristle brushes used for priming, lead, and roughstuff require bridling until worn down somewhat.

There are many patent brush bridles now procurable at a nominal cost which tend to give a brush better shape than the shop-made bridle. If these are not at hand, the painter can take "tufting cord" (our friends the carriage trimmers keep it) and wind the brush securely, but not too tightly; or he can take a piece of lightweight rubber cloth and, extending the piece well down to the handle, tie it at the proper distance around the bristles, the rubber side of the piece should be fastened next the bristles. Then from where it is tied around the bristles, fold the piece back into the handle and tie securely. Trim off, and a bridle is furnished that is perfectly water and paint-proof, the cloth sides of the rubber being folded inside. For a shop-made bridle this will be found very serviceable. After bridling drop a little oil paint into the heel of the brush and set it away in a dust-proof compartment for a few days.

Construction work on the blast furnace plant at Port Arthur is now completed, and the date of blowing-in depends on the arrival of ore from the Atikokan mines.

The Long and the Short of It

"IT" MEANS

THE "LONG"
is the time it wears. It stands the test of heat and frost, wind and rain, and give your customers many years of solid satisfaction.

THE "SHORT"
part is the time needed to put it on. A great deal of time—which means a great deal of money—is saved to the man who uses

SHIELD BRAND READY ROOFING

LOCKERBY & McCOMB
65 SHANNON STREET
MONTREAL

ANTICIPATE
THE MOST ACTIVE AND PROFITABLE

Spring Season
IN

WALL PAPER

AND BE FULLY PREPARED FOR IT

We have our best-selling styles in all grades for season of 1907 in stock. We know them to be reliable—attractive—decorative, and the BEST VALUE IN THE MARKET;

Grasp This Fact Good and Hard.

A post-card application will bring you our EXPRESS SAMPLE BOOKS, charges prepaid, to enable you to make selections for stock.

Borders at Flat Prices.

WRITE TO-DAY. DON'T WAIT.

STAUNTONS LIMITED
Manufacturers of Superior Wall Papers
TORONTO

Not a Pipe Dream

"BEST"

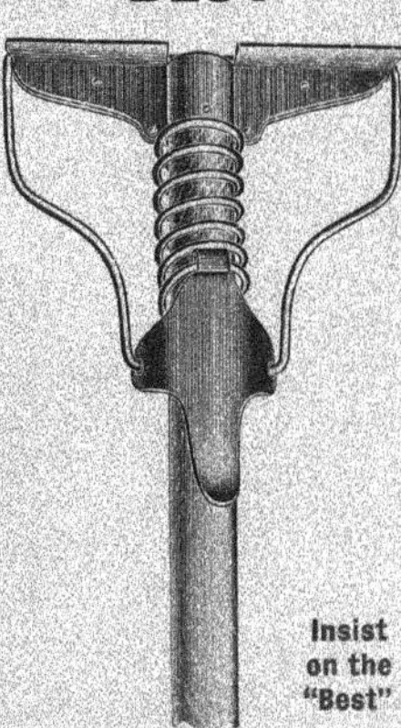

AN ACTUAL FACT

The "Best" is the

Strongest
Simplest
Most Durable

Better get a stock of the "Best," and be prepared for the Spring trade.

THEY ARE SOLD BY ALL THE LEADING JOBBERS

"Cumming's Goods are Good Goods"

MADE BY
THE CUMMING MFG. CO., Limited
RENFREW, ONT.

Western Representatives:
F. A. REINHARDT, J. J. MAGUIRE } *608 Ashdown Bldg., WINNIPEG, MAN.*

Insist on the "Best"

Plumbing and Steamfitting

RETURN TUBULAR BOILERS FOR LOW PRESSURE STEAM HEATING

By W. B. McKay, Toronto.

It is not the object of this article to discuss the relative merits of the return tubular and the many makes of both steel and cast sectional boilers, but rather to set forth such information in as clear and as succinct manner as possible, as may be useful to heating engineers, architects and contractors when the use of such boilers is required.

One point necessary to the successful installation of a steam heating plant is to have ample boiler power, and it is for this reason that a careful examination of the table in fig. 1 will enable you to specify the proper size for any requirements. The data in table fig. 1 was collected from the actual practice of to-day, and is standard for Canada. Throughout the United States a slight modification of this same type of boiler is used, the difference being that the head is set back in to make a smoke chamber on the front end of the boiler. There is no advantage in doing this, and, in fact, some boiler men consider it a disadvantage, as it leaves a dry end of the plates exposed to the hot gases. We need only devote our time, however, to the first-mentioned of these types as this is the one most used in low pressure heating to-day, throughout Canada—the type we are most interested in. The information in the table is laid down in columns under specific headings, thus enabling us to dwell on each, briefly, and more to the point.

Boiler Horse Power.

The term "Horse power" in column 1 of the table, is really only the builder's rating of the boiler, calculated on a basis of the heating surface tabulated in column 2. It will be seen that, approximately, fifteen square feet of heating surface in this table is equivalent to a boiler horse power. The reason of this will be observed from the following remarks in detail. First of all, it must be understood what constitutes a boiler horse power and what is figured upon as being heating surface in a boiler of this type.

A boiler horse power, corectly speaking, is not a unit of so many square feet of heating surface (heating surface being a variable quantity, horse power,

RETURN TUBULAR BOILERS FOR LOW PRESS. STEAM HEATING.

HP	SQ FT	SHELL		TUBES			THICKNESS		GRATES		S.V.	PIPES	BRICK REQ		WEIGHT
Horse Power	Heating Surface	Diam. Inch.	Length Feet	Number	Diam.	Length.	Shell	Heads	Width	Length	W&L Safety Valve	Blow Off	Fire Brick	Common Brick	Weight—Full Front Fixtures &c. Complete
12	177	30	7	30	2½	7	¼	⅜	30	24	1½	1¼	550	4600	3310
14	202	30	8	30	2½	8	¼	⅜	30	30	2	1¼	650	5000	3700
16	254	36	8	38	2½	8	¼	⅜	36	30	2	1¼	700	5300	4550
20	287	36	10	28	3	10	¼	⅜	36	36	2	1¼	750	5500	5013
25	376	42	10	38	3	10	¼	⅜	42	36	2½	1¼	750	6700	6750
35	451	42	12	38	3	12	¼	⅜	42	42	2½	1¼	850	7600	7100
40	596	48	12	52	3	12	¼	7/16	48	48	3	1½	1100	10900	8300
45	695	48	14	52	3	14	¼	7/16	48	54	3	1½	1150	11800	9200
50	722	54	12	64	3	12	5/16	7/16	54	48	3½	1½	1250	12900	10800
60	844	54	14	64	3	14	5/16	7/16	54	54	3½	1½	1400	13800	11800
70	1035	60	14	80	3	14	11/32	7/16	60	60	4	1½	1550	16700	14200
80	1085	60	16	62	3½	16	11/32	7/16	60	66	4	1½	1600	18300	15400

Fig. 1

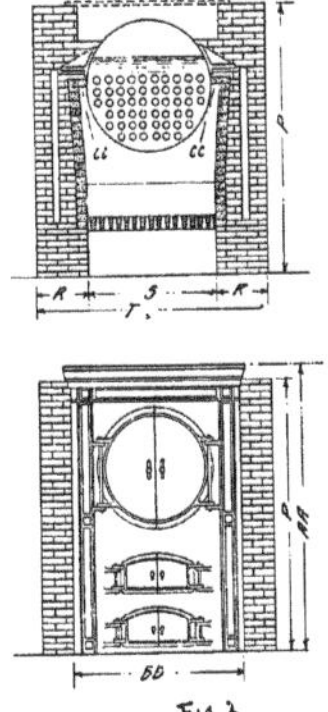

Fig. 2

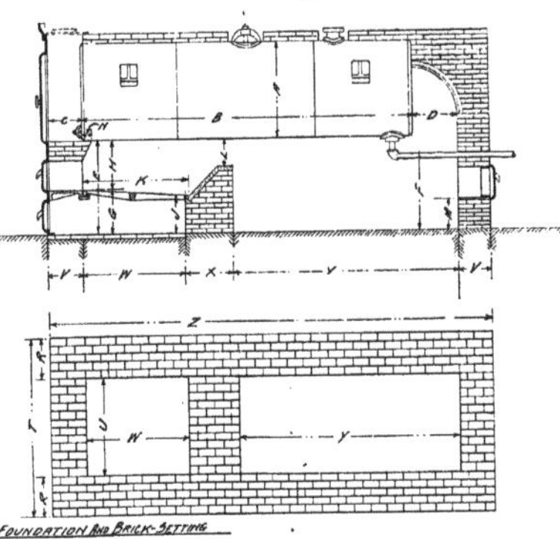

FOUNDATION AND BRICK-SETTING
STANDARD RETURN TUBULAR BOILERS
FOR LOW PRESSURE STEAM HEATING.

then, would also vary), but is the evaporation of thirty pounds of water at a temperature of 100 degrees (Fahr.) feeding into the boiler, into steam at 70 pounds per square inch, gauge pressure, its equivalent being the evaporation of 34.5 pounds from water at 212 degrees into steam at 212 degrees, or 0 pounds pressure, and the heating surface referred to here and in column 2 of the table is the result of the following computation: Take two-thirds of the circumference of the shell, multiplied by its length (both dimensions in inches); the circumference of one of the tubes, multiplied by the common length, which would be the length of the boiler in inches, by the number of tubes; two-thirds the area of both the heads, less twice the area of the ends of all the tubes. Adding these three together and dividing by 144 brings the result into square feet of heating surface in a boiler of this type. From this, it will be noticed that some of the surface figured as heating surface is very inefficient, and, while this surface cannot be ignored, certain allowances must be made to offset this inefficiency.

Liberal Allowance.

As before stated, fifteen square feet of heating surface to the horse power has been adopted in the compilation of this table and in doing so ample allowance had been made to offset many adverse conditions met with in the operation and installation of this particular type of boiler. When we recognize the fact that a boiler horse power may be developed from 11.5 square feet of heating surface it will readily be seen that the allowance is a liberal one. The conditions under which a horse power was developed, on a test, from 11.5 square feet of heating surface were, of course, most favorable, inasmuch as hard coal was used and very careful attention was given the fires. It might also be noted here that, during the test mentioned, the temperature of the fire in gases passing through the combustion chamber and tubes was abnormally high, making the temperature of the escaping gases in the chimney rather higher than is to be desired in low pressure heating. The conditions surrounding this test vary widely from those to be met with in the average low pressure steam heating plant. It is the desire, in the design of a steam heating plant, to provide boiler power sufficient to operate the entire system at a moderately slow rate of combustion, this providing for inferior grades of fuel and lack of care and attention necessary to economical results. From the foregoing it will be seen that it is advisable to choose a boiler having a low capacity rating, or in other words, with an under-estimated capacity.

Estimating Radiation.

With the information given in the first two columns of our table it becomes an easy matter to intelligently select a boiler for any requirements. Further note may be made of these two columns, resolving the information therein contained into terms more easily understood by the men engaged in low pressure heating. A boiler power is generally conceded as being equivalent to 100 square feet of gross radiation (mains, etc., included as radiating surface). Some heating men use as a basis in figuring, 85 square feet of actual radiator surface to one boiler horse power, in this way making an allowance for mains, etc. This gives good satisfaction, and may be used in conjunction with this table (fig. 1).

Passing on to the other columns in this table, which are, for the most part, explanatory in themselves, very little comment is required. Under the heading of "Shell" are the figures designating the size of the boiler (diameter in inches—by length in feet). For this reason it is customary in designating a boiler to mention the size according to the dimensions rather than by the rated horse power. Thus we may say a 42-inch x 10-ft. boiler, instead of a 25-horse power boiler. The thickness of plates in column 5 are designed for 100 pounds pressure and the boiler must stand a test of 150 pounds pressure (cold water) before being released from the manufacturer's hands, no boilers being constructed with plates thinner than those mentioned. The grate area is proportioned to the heating surface to give the best possible results and agrees with authorities on heating boilers, assuming the ration of one square foot of grate surface to, approximately, 36 square feet of heating surface.

Estimating Brick Work.

Under the heading of "brick required" will be found information of value to contractors for heating, which can be used for estimating purposes, simplifying the computation of the cost of bricking in, prices on brickwork usually being laid down in various localities as so much per thousand bricks, laid. The last column contains the approximate weights of the various sizes of boilers, complete with all fixtures, front, grate-bars, trimmings, etc.

Figures 2 and 3 are almost self-explanatory. Figure 2 shows general dimensioned drawings of various views of the brick setting of this type of boiler, and, together with the dimensions tabulated in fig. 3 should furnish information sufficient for the erection of these boilers.

The table and dimensions are for the setting of a single boiler. If two or more boilers are to be set in a battery, one wall between two boilers is all that is necessary but this dividing wall should, of course, be not less than six inches wider than the wall on either side of the boilers. It may be often necessary to deviate slightly from the information laid down in this table to meet existing conditions, but, under no circumstances, should the thickness of the walls be any less than given. The drawing shown shows the brickwork carried up to the full height of the boiler. This is not absolutely necessary, although it makes a very much neater construction, but, in any case, the upper part of the boiler shell exposed should be arched over with brick, two courses laid on edge, to prevent loss of heat by radiation.

STANDARD DIMENSIONS OF SETTING ETC OF RETURN TUBULAR BOILERS FOR LOW PRESSURE STEAM HEATING

HP	SIZE A	B	C	D	E	F	G	H	J	K	L	M	N	P	R	S	T	U	V	W	X	Y	Z	AA	BB	CC
12	30	7	12	20						24	8		5		16	30	62	30	16	22½	20	61½	11-5			3
14	30	8	12	20						30	8		5		16	30	62	30	16	28½	20	67½	12-0		52	3
16	30	8	12	20	41	40	20	21	18	30	10	18	5	83	16	36	68	36	16	28½	20	67½	12-0	7-4½	52	3
20	30	10	12	20	41	40	20	21	18	36	10	18	5	83	16	36	68	36	16	34½	24	81½	14-0	7-4½	52	3
25	42	10	14	20	46	45	22	24	20	36	10	22	5	96	16	42	74	42	16	34½	24	81½	14-7	8-6	58	3
35	42	12	14	20	46	45	22	24	20	42	10	22	5	96	16	42	74	42	16	40½	28	95½	16-2	8-6	58	3
40	48	12	16	24	46	45	22	24	20	48	12	22	5	99	20	48	88	48	16	46½	28	95½	16-8	8-9	64	3
45	48	14	16	24	46	45	22	24	20	54	12	22	5	99	20	48	88	48	16	52½	32	107½	18-8	8-9	64	3
50	54	12	16	24	50	47	24	26	22	48	12	24	5	114	20	54	94	54	20	46½	28	95½	17-0	9-11½	72	4
60	54	14	16	24	50	47	24	26	22	54	12	24	5	114	20	54	94	54	20	52½	32	107½	19-0	9-11½	72	4
70	60	14	18	24	51	50	24	27	22	60	12	24	5	112	24	60	108	60	20	58	32	102	19-2	10-0	84	4
80	60	16	18	24	51	50	24	27	22	66	12	24	5	112	24	60	118	66	20	64	36	116	21-2	10-0	84	4

Fig. 3.

INVENTOR MAKES BIG CLAIMS.

To merely turn on a gas jet in the cellar and forthwith to have a continuous supply of hot water heating for a large house as well as hot air for the registers is the claim made by Wm. Howard, 248 Macdonell avenue, Toronto, for an invention which he demonstrated to some friends on Wednesday. Not content with supplying hot water, hot air and steam heat from the same little furnace, he claims that he can utilize the waste heat to cook delicacies in the kitchen range.

This combination heater can be operated by either coal, gas or electricity. The heater is made in six different sizes. A four-horse power size boiler, the inventor claims, is big enough for a six-storey building. Its gas consumption is only about thirty-five cubic feet an hour, or about equal to an ordinary kitchen burner.

PLUMBING MARKETS

MONTREAL.

Office of HARDWARE AND METAL,
239 McGill Street,
Montreal, February 8, 1907

Prices on almost everything necessary to human existence have been advancing so rapidly lately, that it is somewhat of a novelty to have a whole week go by without having a single advance to record. However, that is the condition here at present, and dealers who have purchased supplies during this state of tranquility may feel that they have done themselves a good turn, if prices on any lines should advance in the near future. Unluckily, these favorable conditions are not likely to last for any lengthy period, as past experiences go to show that an article cannot remain at the same price very long. Some sort of a change is due, if only to keep people guessing, and if the prophesies of the "wise ones" count for anything, this change will not mean lower prices.

General conditions remain practically unchanged, business keeping up about the same pace as reported last week.

RANGE BOILERS—Copper boilers have not shown any advance over last week's quotations. We still quote: Iron clad, 30 gallon, $5; 40 gallon, $6.50 net list. Copper, 30 gallon, $33; 35 gallon, $38; 40 gallon, $43.

LEAD PIPE—Very firm. No change in prices. We still quote 5 per cent. discount f.o.b. Montreal, Toronto, St. John, N.B., Halifax; f.o.b. London, 15c per hundred lbs extra; f.o.b. Hamilton, 10c per hundred lbs. extra.

IRON PIPE FITTINGS—Good business reported, with no change in prices. We still quote. Discounts on nipples, ¼ inch to 3 inch, 75 per cent., 3½ inch to 2 inch, 55 per cent.

IRON PIPE — We still quote: Standard pipe in lots of 100 feet, regular lengths, ¼-inch, $5.50; ⅜ inch, $5.50; ½ inch, $8.50; ¾ inch, 11.50; 1 inch, $16.50; 1 1-4 inches, $22.50; 1 1-2 inches, $27.00; 2" inches, $36.50, discounts on black pipe, ¼ inch, 59 per cent.; ⅜ in. 59 per cent.; ½ in. 68; ¾ to 2, 70. Discounts on galvanized pipe: ¼ inch, 44 per cent.; ⅜ inch, 44 per cent.; ½ inch, 58 per cent.; ¾ to 2 inch, 60 per cent. Extra heavy pipe of 100 feet lots are quoted as follows: 1-2 inch, $12; 3-4 inch, $15; 1 inch, $22; 1 1-4 inches, $30; 1 1-2 inches. $36; 2 inches, $50. The discounts on black pipe, ¼ inch, 74 per cent.; ⅜ inch, 69 per cent.; ½ inch to 2 inches, 68 per cent. Galvanized, ¼ inch, 59 per cent.; ⅜ inch, 69 per cent.; ½ to 2 inches 58 per cent.

SOIL PIPE AND FITTINGS—Our prices remain: Standard soil pipe, 50 per cent. off list. Standard fittings, 50 and 10 per cent. off list; medium and extra heavy soil pipe, 60 per cent. off. Fittings, 60 per cent. off.

SOLDER—Prices remain the same: Bar solder, half-and-half, guaranteed 25c.; No. 2 wiping solder, 22c.

ENAMELWARE—A good business is being done at the same prices. We continue to quote: Canadian baths, see Jan. 3 1907 list. Lavatories, discounts, 1st quality, 30 per cent.; special, 30 and 10 per cent. Sinks 18 x 30 inch, flat rim, 1st quality, $2.60, special, $2.45.

TORONTO.

Office of HARDWARE AND METAL,
10 Front Street East,
Toronto, Feb. 8, 1907.

Collections are reported to have been very satisfactory with paper due on Feb. 1 well met by the trade. The season was a particularly good one, with the exception of the shortage in iron pipe, which held back many jobs from being completed. Some of the small firms which indulged in price cutting are said to have run close to the wind, but there have been fewer notes extended than expected.

Some in the trade are looking for a reaction in prices and orders for supplies are being held over as long as possible. There does not appear to be any ground for this hope, however, as quotations on iron, copper and lead are all either very firm at present prices or are still on the advance with stocks hard to get and large quantities sold for future delivery.

This week the only advance is another jump upward on cast iron fittings from 60 to 67½ per cent. It is understood, however, that iron body values are also likely to go higher.

Business is seasonable with one inch black pipe still short. Pipe makers are still finding it hard to get skelp, as producers in the States have found it more profitable to keep their furnaces busy on other lines. It is freely predicted that pipe will be even more scarce during the coming summer than in 1906.

LEAD PIPE—Buying is seasonable with the discount unchanged at 5 per cent. off the list price of 7c. per pound. Lead waste, 8c. per pound with 5 off. Caulking lead 5 3-4c. to 6c. per pound. Traps and bends, 40 per cent., discount.

SOIL PIPE AND FITTINGS—No changes in prices have been made since our last report and we still quote: Medium and xtra heavy pipe and fittings, 60 per cent.; light pipe 50 per cent.; light fittings, 50 and 10 per cent.; 7 and 8 inch pipe, 40 and 5 per cent.

IRON PIPE—One inch black pipe continues scarce and many are buying ahead in order to ensure deliveries later on. One-inch black pipe is quoted at $4.95, and one-inch galvanized at $6.60. Full list in current market quotations.

IRON PIPE FITTINGS—An advance has been made on cast iron fittings, the change being from 60 to 57½ per cent. We now quote: Cast iron elbows, tees, crosses, etc., 57½ per cent.; cast iron plugs and bushings, 57½ per cent.; flange unions, 57½ per cent.; nipples, 60 and 10 per cent.; iron cocks, 55 and 5 per cent.; Canadian malleable, 30 per cent.; malleable unions, 55 and 5 per cent.; malleable bushings, 55 per cent.; cast iron ceiling plates, plain 65 per cent.; cast iron floor, 70 per cent.; hook plates, 60 per cent.; expansion plates, 65 per cent.; headers, 60 per cent, hangers, 65 per cent.; standard list.

RANGE BOILERS — Sales are fairly large, as many inside jobs are being completed. We still quote: Galvanized iron, 30 gallon, standard, $5, extra heavy, $7.75; 35 gallon, standard, $6; extra heavy, $8.75; 40 gallon, standard, $7; 40 gallon, extra heavy, $9.75, net list. Copper range boilers—New lists quote: 30 gallon, $33; 35 gallon, $38; 40 gallon, $43. Discounts 5 to 15 per cent.

RADIATORS — Buying continues good for this season, with prices unchanged at: Hot water, 47½ per cent.; steam, 50 per cent.; wall radiators, 45 per cent.; specials, 45 per cent. Hot water boilers are unchanged.

SOLDER—Tin is weaker, but lead is advancing. We quote: Bar solder, half-and half, guaranteed, 27c; wiping, 23c.

ENAMELWARE—No changes have been made with the January list still prevailing on Canadian baths. We quote: Lavatories, first quality, 20 and 5 to 20 and 10 off; special, 20 and 10 to 30 and 2½ per cent. discount. Kitchen sinks, plate 300, firsts 60 and 10 off; specials, 65 and 5 per cent. Urinals and range closets, 15 off. Fittings extra.

THE NEW
"SOLID-ADJUSTABLE"
HAND STOCK AND DIES

At prices that will warrant your inquiries.

I. E. SHANTZ & CO.
BERLIN, - ONTARIO
Selling Agents for Canada

This is the Mark that Means Something

It is the distinctive brand of genuine Armstrong Stocks and Die and Pipe Fitters' Tools.

No "Armstrong pattern" or "Just as good as Armstrong" have this stamp. Don't buy anything in this line that hasn't this brand as it is the mark of genuineness.

THE ARMSTRONG MANUFACTURING CO.
292 Knowlton Street, Bridgeport, Conn.

My Business is Yours

At least my business is to attend to your ***enamel-ware wants,*** and most dealers have found it good business policy to depend on me The quality of the goods I deliver please most and it will please you.

ORLANDO VICKERY
178-180 Victoria Street TORONTO, CANADA

PRACTICAL MEN DECLARE that
"Diamond" brand Fittings
are incomparably the best on the Canadian market.

THREADS WORKMANSHIP and MATERIAL ALL PERFECT

OSHAWA STEAM & GAS FITTINGS CO., LIMITED
OSHAWA, CANADA

Sharratt & Newth's Glaziers' Diamonds

are unequalled for cutting and wearing qualities.

To be obtained from the principal Dealers in Glass, Hardware, and Painters' Supplies.
Contractors to H. M. Government and the principal English Sheet and Plate Glass Works

---

World's Best Cutlery

You make no mistake in specifying goods made by **JOSEPH RODGERS & SONS, Limited** when ordering cutlery. A reputation of over 200 years at stake, insures the quality of every article.

Write for catalogues from Sole Canadian Agents

JAMES HUTTON & COMPANY, MONTREAL

A Certain Sense of Satisfaction

goes with every MERRELL PORTABLE HAND MACHINE, PIPE THREADING and CUTTING MACHINE. We build them that way.

For doing ALL kinds of threading—under ALL conditions—labor trouble and faulty material included—where speed, durability and ease of control count for something, our word to you is to investigate the MERRELL.

Whether it be stationary or portable—hand or power driven—30 DAYS FREE TRIAL must prove conclusively that your choice should be the MERRELL.

This Portable Hand Machine has encased gears, MERRELL standard quick opening and closing die-head, and the latest improved Cutting-off Knife.

Let us tell you more about this machine—The Chasers, Vise and the large range of work covered.

Catalogue for the asking.
Quick shipments.

THE CANADIAN FAIRBANKS CO. Limited
Sole Agents for Canada
MONTREAL, TORONTO, WINNIPEG, VANCOUVER

Forwell Foundry Co.
BERLIN, ONT.

Manufacturers of
SOIL PIPE, FITTINGS, and CAST IRON SINKS
Ask Jobbers for "F. F. CO." Brand

KERR'S GLOBE AND GATE VALVES
STRICTLY HIGH GRADE. TESTED & PACKED
THE KERR ENGINE CO. LIMITED
VALVE AND HYDRANT MANUFACTURERS
WALKERVILLE, ONT.

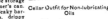

Can You answer the question on Hot Water vs. Steam Heating asked on page 57?

CURRENT MARKET QUOTATIONS.

Feb 8, 1907.

These prices are for such qualities and quantities as are usually ordered by retail dealers on the usual terms of credit, the lowest figures being for larger quantities and prompt pay. Large cash buyers can frequently make purchases at better prices. The Editor is anxious to be informed at once of any apparent errors in this list, as the desire is to make it perfectly accurate.

METALS.

ANTIMONY.

Cookson's ....per lb... 0 27½ 0 28

BOILER AND T.K. PITTS.

Plain tinned .... } 25 per cent. off list.
Spun .... }

BABBIT METAL.

Canada Metal Company—Imperial, genuine, 60c; Imperial Tough, 56c., White Brass, 50c; Metallic, 35c; Harris Heavy Pressure, 25c.; Hercules, 25c; White Bronze, 15c.; Star Frictionless, 14c; Alluminoid, 10c; No. 4, 9c per lb

BRASS.

Rod and Sheet, 14 to 30 gauge, net list.

Sheets, 12 to 14 in.	0 27
Tubing, base, per lb ⅛ to 2 in	0 33
Tubing, ¼ to 3-inch, iron pipe size..	0 31
" 1 to 3-inch, seamless..	0 36

Copper tubing, 6 cents extra

James Robertson Co.—Extra and genuine Monarch, 60c; Crown Monarch, 50c; No 1 Monarch, 40c, King, 30c, Fleur-de-lis, 20c; Thurber, 15c; Philadelphia, 12c; Canadian, 10c; hardware, No 1, 15c; No. 2, 12c, No 3, 10c per lb.

COPPER

Ingot.		Per 100 lb.
Casting, car lots	26 50	27 00
Bars.		
Cut lengths, round, ½ to 2 in...	....	32 00
Sheet.		
Plain, 16 oz., 14x48 and 14x60	....	31 00
Plain, 14 oz.	....	32 00
Tinned copper sheet, base	....	34 00
Planished base	....	38 00
Braziers' (in sheets), 4x6 ft., 25 to 30 lb. each, per lb., base..	....	0 30

BLACK SHEETS.

	Montreal.	Toronto
8 to 10 gauge	2 70	2 70
12 gauge	2 70	2 50
14 "	2 70	2 55
17 "	2 50	2 65
18 "	2 50	2 65
20 "	2 50	2 65
22 "	2 50	2 75
24 "	2 55	2 80
26 "	2 65	2 85
28 "	2 70	3 00

CANADA PLATES.

Ordinary, 52 sheets	2 75	2 90
All bright "	3 80	3 90
Galvanized, 52 sheets	4 35	4 45
" 60 "	4 60	4 70

	Ordinary.	Dom. Crown
18x24x52	4 25	4 35
" 60	4 50	4 60
20x28x80	8 50	8 70
" 94	9 00	9 20

GALVANIZED SHEETS.

	Fleur-de-Lis.	Gordon Crown
16 to 20 gauge	3 80	3 95
22 to 24 gauge	4 05	4 00
26 "	4 30	4 20
28 "	4 55	4 50

Apollo.

10¾ oz. (American gauge)	....	4 85
28 gauge .... "	....	4 55
26 " .... "	4 15	4 20
24 " .... "	3 85	4 90

	Comet	Queen's Head.	Bell.
16 to 20 gauge	3 60	3 95	....
22 to 24 gauge	3 75	4 20	....
26 "	4 30	4 45	....
28 "	4 45	4 70	4 45

Less than case lots 10 to 25c. extra.

IRON AND STEEL.

	Montreal.	Toronto.
Common bar, per 100 lb......	2 15	2 30
Forged iron "	2 40	..
Refined " "	2 55	2 70
Horseshoe iron "	2 55	2 70
Hoop steel, 1½ to 3 in. base...	....	2 80
Sleigh shoe steel "	2 25	2 30
Tire steel	2 40	2 50
Best sheet cast steel..........	...	0 12
B. K. Morton "Alpha" high speed.		0 65
" annealed	....	0 70
"M" Self-hardening	....	0 50
"J" quality, best warranted	....	0 18
"I" " warranted......	....	0 14
"B.C" quality	....	0 09
Jonas & Colver's tool steel....	0 10	0 20
" "Novo"......	....	0 65
" " annealed	....	0 65
Jowett & Sons B.P.L. tool steel....		0 10½

INGOT TIN.

Lamb and Flag and Straits—
56 and 28-lb. ingots, 100 lb. $44 50 $45 00

TINPLATES.

Charcoal Plates—Bright

M.L.S., equal to Bradley—		Per box.
I C, 14 x 20 base	....	$6 50
I X, 14 x 20 "	....	8 00
I XX, 14 x 20 base	....	9 50
Famous, equal to Bradley—		
I C, 14 x 20 base..........	....	6 50
I X, 14 x 20 "	...	8 00
I X X, 14 x 20 base........	....	9 50
Raven and Vulture Grades—		
I C, 14 x 20 base	....	5 00
I X "	....	6 00
I X X "	....	7 00
I X X X "	....	8 00

"Dominion Crown Best"—Double Coated, Tissued.

		Per box.
I C, 14 x 20 base..........	5 50	5 75
I X, 14 x 20 "	6 50	6 75
I XX x 20 "	7 50	7 75

"Allaway's Best"—Standard Quality.

I C, 14 x 20 base.....	....	4 50
I X, 14 x 20 "	....	5 25
I XX, 14 x 20 "	....	6 00

Bright Cokes.

Bessemer Steel—		
I C., 14 x 20 base	...	4 25
20x28, double box	....	8 50

Charcoal Plates—Terne

Dean or J. G. Grade—		
I C., 20x28, 112 sheets	7 25	8 00
IX., Terne Tin	...	9 50

Charcoal Tin Boiler Plates.

Cookley Grade—
X X, 14x56, 50 sheet bxs. }
" 14x60, " } 7 50
" 14x65, " }

Tinned Sheets.

72x30 up to 24 gauge...... ...	....	8 50
" " 26 "	....	9 00

LEAD.

Imported Pig, per 100 lb......	5 50	5 60
Bar, "	5 75	6 00
Sheets, 2½ lb. sq. ft., by roll ..	....	0 07½
Sheets, 3 to 6 lb. "	....	0 07

NOTE.—Cut sheets ½c. per lb., extra Pipe, by the roll, usual weights per yard, lists at 7c. per lb. and 5 p.c. dis. f.o.b. Toronto.

NOTE.—Cut lengths, net price, waste pipe 8-ft lengths, lists at 8c.

SHEET ZINC.

5-cwt. casks	8 00	8 25
Part casks....................	8 25	8 50

ZINC SPELTER.

Foreign, per 100 lb	7 50	7 75
Domestic	7 00	7 25

PLUMBING AND HEATING

BRASS GOODS, VALVES, ETC.

Standard Compression work, dis half inch 50 per cent., others 40 per cent.
Cushion work, discount 40 per cent.
Fuller work, ½-inch 60 per cent., others 50 p.c.
Flatway stop and stop and waste cocks, 50 per cent.; roundway, 45 per cent.
J.M.T. Globe, Angle and Check Valves, discount 45 per cent.
Standard Globe, Angle and Check Valves discount 52½ per cent.
Kerr standard globes, angles and checks, special, 42½ per cent.; standard, 47½ p.c.
Kerr Jenkins' disc, copper-alloy disc and heavy standard valves, 45 per cent
Kerr steam radiator valves, 60 p.c., and quick-opening hot-water radiator valves, 60 p.c.
Kerr brass, Weber's straightway valves, 40 per cent.; straightway valves, I.B.B.M., 60 per cent.
J. M. T. Radiator Valves, discount 50 per cent
Standard Radiator Valves, 60 per cent.
Patent Quick-Opening Valves, 65 per cent.
Jenkins' Bros. Globe Angle and Check Valves discount 32½ per cent.

No. 1 compression bath cock......net	2 00
No. 4 " " ...... "	1 90
No 7 Fuller's.............. .... "	2 25
No. 4½, " .... "	2 35

Patent Compression Cushion, basin cock, hot and cold, per doz., $16.20
Patent Compression Cushion, bath cock, No. 2208 .... 2 25
Square head brass cocks, discount 50 per cent.
" " iron " " 60 "
Thompson Smoke-test Machine $25.00

BOILERS—COPPER RANGE.

Copper, 30 gallon............	33 00
" 35 "	38 00
" 40 "	43 00

15 per cent. discount

BOILERS—GALVANIZED IRON RANGE.

Capacity	Standard.	Extra heavy
30-gallons ...	5.00	7 75
35 "	6 00	8 75
40 "	7.00	9 75

2 per cent., 30 days.

BATH TUBS.

Steel clad copper lined, 15 per cent.

CAST IRON SINKS

16x24, $1; 18x30, $1; 18x36, $1.30.

ENAMELED CLOSETS AND URINALS

Discount 15 per cent.

ENAMELED BATHS.

List issued by the Standard Ideal Company Jan. 3, 1907, shows an advance of 10 per cent. over previous quotations.

ENAMELED LAVATORIES.

	1st quality.	Special.
Plate E 100 to E 103 .	20 & 5 p.c.	20 & 10 p.c.
" E 104 to E 132....	20 & 10 p.c.	30 & 2½ p.c.

ENAMELED ROLL RIM SINKS.

	1st quality.	Special
Plate E 201, one piece.	15 & 2½ p.c.	15 & 10 p.c.

ENAMELED KITCHEN SINKS.

Plate E, flat rim 300, 60 & 10 p.c. 65 & 5 p.c

HEATING APPARATUS

Stoves and Ranges—Discounts vary from 40 to 70 per cent. according to list
Furnaces—45 per cent.
Registers—70 per cent.
Hot Water Boilers—Discounts vary
Hot Water Radiators—47½ per cent
Steam Radiators—50 per cent
Wall Radiators and specials—45 per cent.

LEAD PIPE

Lead Pipe, 7c. per pound, 5 per cent. off
Lead waste, 8c. per pound, 5 per cent. off.
Caulking lead, 6c. per pound.
Traps and bends, 40 per cent.

IRON PIPE.

Size (per 100 ft.)	Black.		Galvanized
¼ inch........	2 25	¼ inch........	3 08
⅜ "	2 25	⅜ "	3 08
½ "	2 72	½ "	3 57
¾ "	3 45	¾ "	4 60
1 "	4	1 "	6 60
1¼ "		1¼ "	9 00
1½ "		1½ "	10 80
2 "		2 "	14 40
2½ "		2½ "	24 00
3 "	6 95	3 "	31 50
3½ "	[illegible]	3½ "	40 00
4 "	34 00	4 "	45 00

2 per cent. 30 days.

Malleable Fittings—Canadian discount 30 per cent.; American discount 25 per cent.
Cast Iron Fittings 60; Standard bushings 60 per cent.; headers, 60; flanged unions 60, malleable bushings 55; nipples, up to 3 in., 65 and 10 per cent; up to 6 in., 65 per cent; malleable lipped unions, 55 and 5 per cent.

SOIL PIPE AND FITTINGS

Medium and Extra heavy pipe and fittings, up to 6 inch, discount 60 per cent.
7 and 8-in. pipe, discount 40 and 5 per cent.
Light pipe, 50 p.c.; fittings, 50 and 10 p.c.

OAKUM.

Plumbers ....per 100 lb... .. .... 4 75

STOCKS AND DIES.

American discount 25 per cent.

SOLDERING IRONS.

½-lb. .... per lb.	....	0 38
1-lb. or over .... "	....	0 35

SOLDER.

	Montreal	Per lb. Toronto
Bar, half-and-half, guaranteed	0 25	0 27
Wiping....	0 22	0 23

PAINTS, OILS AND GLASS

BRUSHES

Paint and household, 70 per cent.

CHEMICALS

	In casks per lb.
Sulphate of copper (bluestone or blue vitrol)	0 09
Litharge, ground	0 06
" flaked	0 06½
Green copperas (green vitrol)	0 01
Sugar of lead	0 08
Lump olive..........	0 01½

COLORS IN OIL.

Venetian red, 1-lb. tins pure.	....	0 08
Chrome yellow "	....	0 15
Golden ochre "	....	0 08
French " "	....	0 08
Marine black "	....	0 04½
Chrome green "	....	0 10
French permanent green "	....	0 13
Signwriters' black "	....	0 15

GLUE.

Domestic sheet	0 10	0 10½
French medal..	0 12	0 12½

PARIS GREEN.

	Berger's	Canadian
600-lb. casks	0 24½	0 24½
250 lb. drums	0 24½	0 24½
100-lb. "	0 25	....
50-lb. "	0 25	0 25½
1-lb. pkgs, 100 in box	0 26	0 26½
½-lb. "	0 28	0 28½
1-lb. tins, 100 in box	0 27	0 27½
½-lb. pkg	....	0 29

PARIS WHITE.

In bbls 0 90

PREPARED PAINTS.
Pure, per gallon, in tins 1 30
Second qualities per gallon 1 10
Barn (in bbls.) 0 85 0 90
Sherwin-Williams paints, ½ gal. 1 50
" " ¼ gal. 1 45
" " gal... .... 1 40
Canada Paint Co.'s pure 1 30
Standard P. & V. Co.'s "New Era." 1 30
Benj. Moore Co.'s "Ark" B'd 1 25
" British Navy deck 1 50
Brandram-Henderson "Anchor" 1 35
Ramsay's paints, Pure, per gal. 1 20
" Thistle, " 1 00
Outside, bbls 0.55 0 65
Martin-Senour's 100 p.c. pure, ½ gal. 1 50
" " " " ¼ gal. 1 45
" " " " 1 gal. 1 40
Senour's Floor Paints..... gal. 1 20
Jamieson's "Crown and Anchor" 1 20
Jamieson's floor enamel 1 50
" barn paints, bbls. per gal. 0 60
Sanderson Pearcy's, pure 1 20
Robertson's pure paints 1 20

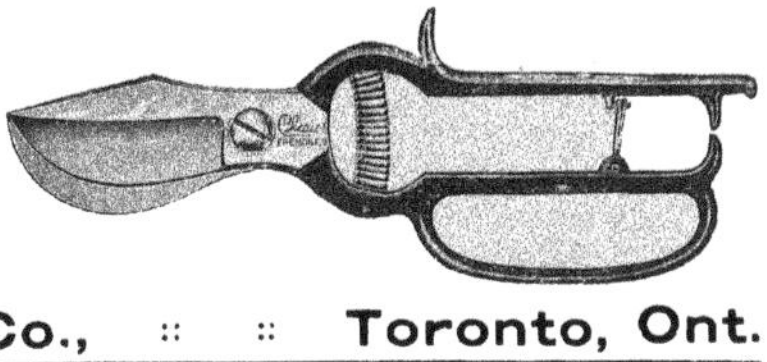

PUTTY.
Bulk in bbls. 1 50
Bulk in less quantity 1 80
Bladders in bbls. 1 80
Bladders in kegs, boxes or loose...... 1 95
25-lb. tins 1 80
12½ lb. tins 2 05
Bladders in bulk or tins less than 100 lb. 1 85
Bulk in 100-lb. irons 1 80

SHINGLE STAINS.
In 5-gallon lots 0 75 0 87

SHELLAC.
White 0 65
Fine orange 0 60
Medium orange 0 55
F.o.b. Montreal or Toronto.

TURPENTINE AND OIL
Castor oil 0 08 0 10
Gasoline 0 2½
Benzine, per gal 0 20
Turpentine single barrels 0 98 1 05
Linseed Oil, " raw.. 0 55 0 59
" " boiled 0 58 0 62
W. H. & C's English raw oil, bbls .. 0 63
" " boiled " 0 66

WHITE LEAD GROUND IN OIL Per 100 lbs
Canadian pure 7 15
No. 1 Canadian 6 80
Munro's Select Flake White.. .... 7 40
Elephant and Decorators' Pure ... 7 41
Monarch 7 40
Standard Decorator's 7 15
Essex Genuine 6 80
Brandram's B. B. Genuine 8 25
"Anchor," pure 6 50
Ramsay's Pure Lead 6 40
Ramsay's Exterior 6 15
"Crown and Anchor," pure 6 50
Sanderson Pearcy's 7 40
Robertson's O.P., lead 7 20
W. H. & C's matured pure English 8 00 8 25

WHITE AND RED DRY LEAD.

	white	red.
Genuine, 560 lb. casks, per cwt	6 75	6 00
Genuine, 100 lb. kegs, "	7 50	6 50
No. 1, 560 lb. casks, per cwt ...	6 25	5 75
No. 1, 100 lb. kegs, per cwt ..	7 00	6 25

WINDOW GLASS.

Size United Inches.	Star	Double Diamond
Under 26	$4 25	$6 25
26 to 40	4 65	6 75
41 to 50	5 10	7 50
51 to 60	5 35	8 50
61 to 70	5 75	9 75
71 to 80	6 25	11 00
81 to 85	7 00	12 50
86 to 90	....	15 00
91 to 95	....	17 50
96 to 100	....	20 50
101 to 105	....	24 00
100 to 110	....	27 50

Discount—16-oz., 25 per cent.; 21 oz. 30 per cent. per 100 feet. Broken boxes 50 per cent. off list

WHITING
Plain, in bbls 0 60
Gilders bolted in bands........ .... 0 90

WHITE DRY ZINC.
Extra Red Seal, V.M... ... 0 07½ 0 08¼

WHITE GROUND ZINC.
Pure, in 25-lb. irons 0 08¼
No. 1, " " 0 07
No. 2, " " 0 05½

VARNISHES.
In 5-gal. lots. Per gal. cans.
Carriage, No. 1 .... 1 50
Pale durable body........ .... 3 50
" hard rubbing 3 00
Finest elastic gearing 3 00
Elastic oak 1
Furniture, polishing 2 0
Furniture, extra 1 1
" " No. 1........ .... 0 9
" " union 0 80
Light oil finish 1 40
Gold size japan 1 80
Brown japan 0 95
No 1 brown japan 0 90
Baking black japan 1 35
No. 1 black japan 0 90
Benzine black japan 0 70
Crystal Damar 2 80
No. 1 " 2 50
Pure asphaltum 1 40
Oilcloth 1 50
Lightning dryer 0 70
Elastilite varnish, 1 gal. can, each.. 2 00
Granitine floor varnish, per gal 2 10
Maple Leaf coach enamels; size 1, 1 20
size 2, 70c.; size 3, 40c. each.
Sherwin-Williams' kopal varnish, gal., 2 50
Canada Paint Co's sun varnish 2 00
"Kyanize" Interior Finish.... .... 2 40
"Flint-Lac," coach........ .... 1 80
"Gold Medal," per gal., in cases.. 2 00
Jamieson's Copaline, per gal. 2 00

BUILDERS' HARDWARE

BELLS.
Brass hand bells, 60 per cent.
Nickel, 55 per cent.
Gongs, Sargeant's door bells.. 5 50 8 00
American, house bells, per lb. " 35 0 40
Peterboro' door bells, discount 37½ and 10 per cent. off new list.

BUILDING PAPER, ETC.
Tarred Felt, per 100 lb. 2 25
Ready roofing, 2-ply, not under 45 lb. per roll 1 00
Ready roofing, 3-ply, not under 65 lb., per roll 1 25
Carpet Felt........ per ton 60 00
Heavy Straw Sheathing....... per ton 35 00
Dry Surprise........ 0 40
Dry Sheathing.....per roll, 400 sq. ft. 0 50
Tar " 400 " 0 45
Dry Fibre........ " 400 " 0 55
Tarred Fibre .... " 400 " 0 65
O. K. & I. X. L.... " 400 " 0 70
Resin-sized .... " 400 " 0 45
Oiled Sheathing.... " 600 " 1 00
Oiled " .... " 400 " 0 70
Roof Coating, in barrels.. ...per gal. 0 17
Roof " small packages 0 25
Refined Tar .... per barrel 5 00
Coal Tar " 4 00
Coal Tar, less than barrels.... per gal. 0 15
Roofing Pitch....... per 100 lb. 0 80 0 90
Slater's felt .... per roll 0 70
Heavy Straw Sheathing f. o. b. St. John and Halifax........ 37 50

BUTTS.
Wrought Brass, net revised list.
Wrought Iron, 70 per cent.
Cast Iron Loose Pin, discount 60 per cent
Wrought Steel Fast Joint and Loose Pin, 70 per cent.

CEMENT AND FIREBRICK.
"Lafarge" cement in wood 3 40
"Lehigh" cement, in wood 2 54
"Lehigh" cement, cotton sacks. 2 39
"Lehigh" cement, paper sacks 2 31
Fire brick (Scotch) per 1,000.... 15 00 24 00
Fire clay (Scotch), net ton 4 95
Paving Blocks per 1,000
Blue metallic, 9"x4½"x3", ex wharf .. 35 00
Stable pavers, 12"x6"x2", ex wharf .. 50 00
Stable pavers, 9"x4½"x3", ex wharf .. 36 00

DOOR SETS.
Peterboro, 37½ and 10 per cent.

DOOR SPRINGS.
Torrey's Rod .... per doz. 1 75
Coil, 9 to 11 in. .... " 0 95 1 65
English .... " 2 00 4 00
Chicago and Reliance Coil Door Springs, 25 per cent.

STORE DOOR HANDLES.
Per Dozen 1 00 1 50

ESCUTCHEONS.
Discount 50 and 10 per cent., new list
Peterboro, 37½ and 10 per cent.

ESCUTCHEON PINS.
Iron, discount 40 per cent.
Brass, 45 per cent.

HINGES.
Blind, discount 60 per cent.
Heavy T and strap, 4-in., per lb. net.. 0 06½
" " 5-in., " 0 06
" " 6-in., " 0 05¾
" " 8-in., " 0 05¼
" " 10-in. and larger . 0 05¼
Light T and strap, discount 65 p.c.
Screw hook and hinge—
under 12 in.... per 100 lb. 4 75
over 12 in. " 3 75
Spring, No. 20, per gro. pairs 10 80
Spring, Woodyatt pattern, per gro., No. 5, $17.50. No. 10, $18; No. 20, $10 80; No 12 $20; No 51, $10; No. 50, $27.50.
Crate hinges and back flaps, 65 and 5 p. c.
Hinge hasps, 65 per cent.

SPRING HINGES.
Chicago Spring Butts and Blanks 12½ per cent.
Triple End Spring Butts, 40 and 5 per cent.
Chicago Floor Hinges, 40 and 5 off.
Garden City Fire Engine House Hinges, 12½ per cent

HOOKS.
Cast Iron.
Bird cage .... per doz. 0 50 1 10
Clothes line, No. 61.. " 0 00 0 70
Harness .... " 0 60 12 00
Hat and coat .... per gro. 1 10 10 00
Chandelier .... per doz. 0 50 1 00
Wrought hooks and staples—
⅜ x 5 .... per gross 2 65
5-16 x 5 " 3 30
Bright steel gate hooks and staples, 40 per cent. discount
Hat and coat wire, discount 62½ per cent.
Screw, bright wire, discount 65 per cent.

KNOBS.
Door, japanned and N. P., doz 1 50 2 50
Bronze, Berlin .... per doz. 2 75 3 25
Bronze, Genuine .... " 6 00 9 00
Shutter, porcelain, F. & L. screw .... per gross 1 30 2 00
White door knobs ...per doz 2 00
Peterboro knobs, 37½ and 10 per cent.
Porcelain, mineral and jet knobs, net list.

KEYS
Lock, Canadian dis. 40 to 40 and 10 per cent
Cabinet trunk and padlock
American .... per gross 0 60

LOCKS.
Peterboro 37½ and 10 per cent
Russell & Erwin, steel rim $2 50 per doz
Eagle cabinet locks, discount 30 per cent
American padlocks, all steel, 10 to 15 per cent; all brass or bronze, 10 to 25 per cent.

SAND AND EMERY PAPER.
B. & A. sand, discount, 35 per cent
Emery, discount 35 per cent.
Garnet (Burton's) 5 to 10 per cent. advance on list.

SASH WEIGHTS.
Sectional........ per 100 lb. 2 00 2 25
Solid .... " 1 50 1 75

SASH CORD.
Per lb. 0 31

BLIND AND BED STAPLES
All sizes, per lb. 0 07½ 0 12

WROUGHT STAPLES.
Galvanized 2 75
Plain 2 50
Coopers', discount 45 per cent.
Poultry netting staples, discount 40 per cent.
Bright spear point, 75 per cent discount.

TOOLS AND HANDLES.

ADZES.
Discount 22½ per cent.

AUGERS.
Gilmour's, discount 60 per cent. off list

AXES.
Single bit, per doz. 5 50 8 50
Double bit, " 10 00 11 00
Bench Axes, 40 per cent.
Broad Axes, 25 per cent.
Hunters' Axes.............. 5 50 6 00
Boys' Axes 6 25 7 00
Splitting Axes 7 00 12 00
Handled Axes 7 00 9 00
Red Ridge, boys', handled.. .... 5 75
" hunters 5 25

BITS.
Irwin's auger, discount 47½ per cent.
Gilmour's auger, discount 60 per cent.
Rockford auger, discount 50 and 10 per cent
Jennings' Gen. auger, net list.
Gilmour's car, 47½ per cent.
Clark's expansive, 40 per cent.
Clark's gimlet, per doz 0 65 0 90
Diamond, Shell, per doz...... 1 00 1 50
Nail and Spike, per gross..... 2 25 5 20

BUTCHERS CLEAVERS
German.......... per doz. 6 00 9 00
American.......... " 12 00 18 00

CHALK.
Carpenters' Colored, per gross 0 45 0 75
White lump.......... per cwt. 0 60 0 65

CHISELS.
Warnock's, discount 72½ per cent.
P. S. & W. Extra, discount 72½ per cent

CROSSCUT SAW HANDLES.
S. & D., No. 3 .... per pair 0 15
S. & D., " 5 .... " 0 22½
S. & D., " 6 .... " 0 15
Boynton pattern.............. " 0 20

CROWBARS.
2½c. to 4c. per lb.

DRAW KNIVES.
Coach and Wagon, discount 75 per cent.
Carpenters' discount 75 per cent

DRILLS.
Millar's Falls, hand and breast net list
North Bros., each set, 50c.

DRILL BITS.
Morse, discount 37½ to 40 per cent.
Standard, discount 50 and 5 to 55 per cent.

FILES AND RASPS.
Great Western .... 75 per cent.
Arcade .... 75 "
Kearney & Foot .... 75 "
Disston's.............. 75 "
American .... 75 "
J. Barton Smith .... 75 "
McClellan .... 75 "
Eagle .... 75 "
Nicholson.............. 66⅔ "
Globe .... 75 "
Black Diamond, 60, 10 and 5 p.c.
Jowitt's, English list, 27½ per cent.

GAUGES.
Stanley's discount 50 to 60 per cent.
Wire Gauges.
Winn's, Nos. 26 to 33 ....each 1 65 2 40

HANDLES.
C. & B., fork, 40 per cent., revised list.
C. & B., hoe, 40 per cent., revised list.
American, saw ...... per doz. 0 25
American, plane ... per gross 3 15 3 75
Canadian, hammer and hatchet 40 per cent
Axe and cant hook handles, 45 per cent

HAMMERS.
Maydole's, discount 5 to 10 per cent.
Canadian, discount 25 to 27½ per cent.
Magnetic tack .... per doz. 1 10 1 20
Canadian sledge .... per lb. 0 07 0 08½
Canadian ball pean, per lb. .. 0 22 0 25

HATCHETS.
Canadian, discount 40 to 42½ per cent
Shingle, Red Ridge 1, per doz.......... 4 40
" " 2, " 4 85
Barrel Underhill 5 00

MALLETS.

Tinsmiths'............per doz. 1 25 1 50
Carpenters', hickory, " 1 25 3 75
Lignum Vitae........ " 3 85 5 00
Caulking, each 0 60 2 00

MATTOCKS.

Canadian............ per doz. 5 50 6 00

MEAT CUTTERS.

German, 15 per cent.
American discount, 33⅓ per cent.
Gem ....each 1 15

NAIL PULLERS.

German and American 0 85 2 50
No. 1 .... 0 85
No 1573...................... 0 75

NAIL SETS.

Square, round and octagon, per gross 3 38
Diamond 1 00

PICKS.

Per dozen 6 00 9 00

PLANES.

Wood bench, Canadian discount 40 per cent. American discount 50 per cent.
Wood, fancy Canadian or American 37½ to 40 per cent.
Stanley planes, $1 55 to $3 60, net list prices.

PLANE IRONS.

English ....per doz. 2 00 5 00
Stanley, 2⅝ inch, single 24c, double 39c.

PLIERS AND NIPPERS.

Button's genuine, 37½ to 40 per cent.
Button's imitation....per doz. 5 00 9 00

PUNCHES.

Saddler's ....per doz. 1 00 1 85
Conductor's.......... " 3 00 15 00
Tinners', solid..........per set 0 85
" hollow......per inch 1 00

RIVET SETS.

Canadian, discount 35 to 37½ per cent.

RULES.

Boxwood, discount 70 per cent.
Ivory, discount 20 to 25 per cent.

SAWS.

Atkins, hand and crosscut, 25 per cent.
Disston's Hand, discount 12½ per cent.
Disston's Crosscut.....per foot 0 35 0 55
Hack, complete..........each 0 75 2 75
" frame only.........each 0 50 1 25
S. & D solid tooth circular shingle, concave and band, 50 per cent; mill and ice, drag, 30 per cent.; cross-cut, 35 per cent; hand saws, butcher, 35 per cent; buck, New Century, $6.25; buck, No. 1 Maple Leaf, $5.25; buck, Happy Medium, $4.25; buck, Watch Spring, $4.25, buck, common frame, $4.00.
Spear & Jackson's saws—Hand or rip, 26 in., $12.75, 28 in., $14.25; panel, 18 in., $8.25; 20 in., $9, tenon, 10 in., $9.90; 12 in., $10.90; 14 in., $11.50.

SAW SETS.

Lincoln and Whiting 4 75
Hand Sets, Perfect 4 00
X-Cut Sets, " 7 50
Maple Leaf and Premium saw sets, 40 off.
S. & D. saw swages, 40 off.

SCREW DRIVERS.

Sargent's ....per doz. 0 65 1 00
North Bros., No. 30 ..per doz. 16 80

SHOVELS AND SPADES.

Bull Dog, solid neck shovel (No. 2 pol.) $18 50
(Hollow Back) (Reinforced S Scoop)
Moose......... $17 50 .... $16 30
Bear.......... 15 50 .... 15 30
Fox........... 12 50 .... 14 30
Black Cat..... 10 00 .... 13 30
Canadian, discount 45 per cent.

SQUARES.

Iron, discount 20 per cent.
Steel, discount 65 and 10 per cent.
Try and Bevel, discount 50 to 52½ per cent.

TAPE LINES.

English, ass skin......per doz. 2 75 5 00
English, Patent Leather..... 5 50 9 75
Chesterman's..........each 0 90 2 85
" steel..........each 0 80 8 00

TROWELS.

Disston's, discount 10 per cent.
S. & D., discount 35 per cent.

FARM AND GARDEN GOODS

BELLS.

American cow bells, 63⅓ per cent.
Canadian, discount 45 and 50 per cent.
American, farm bells, each .. 1 35 3 00

BULL RINGS.

Copper, $1.30 for 2½-inch, and $1.70

CATTLE LEADERS.

Nos. 32 and 33 .. ...per gross 7 50 8 50

BARN DOOR HANGERS.

doz. pairs.
Steel barn door.............. 8 00 10 00
Stearns wood track 4 50 6 00
Zenith....................... .... 9 00
Acme, wood track 5 00 6 50
Atlas........................ 5 00 6 00
Perfect 8 00 11 00
New Milo..................... .... 6 50
Steel, covered............... 4 00 11 00
" track, 1 x 3-16 in(100 ft) 3 75
" " 1¼ x 3-16 in(100 ft) 4 75
Double strap hangers, doz. sets.... 6 40
Standard jointed hangers, " 6 40
Steel King hangers " 6 25
Storm King and safety hangers 7 00
" " rail.................. .... 4 25
Chicago Friction, Oscillating and Big Twin Hangers, 5 per cent.

HARVEST TOOLS.

Discount 60 per cent.
S. & D. lawn rakes, Dunn's, 40 off.
" sidewalk and stable scrapers, 40 off.

HAY KNIVES.

Net list.

HEAD HALTERS.

Jute Rope, ⅜-inch.....per gross 9 00
" " ½ " 10 00
" " ⅝ " 12 00
Leather, 1-inch .....per doz. 4 00
Leather, 1¼ " ..... " 5 20
Web.................. " 2 45

HOES.

Garden, Mortar, etc., discount 50 per cent.
Planter.............. per doz. 4 00 4 50

SCYTHES.

Per doz. net.................. 6 25 9 25

SCYTHE SNATHS.

Canadian, discount 40 per cent.

SHAPS.

Harness, German, discount 25 per cent.
Lock, Andrews'.............. 4 50 11 00

STABLE FITTINGS.

Warden King, 35 per cent.

WOOD HAY RAKES.

Ten tooth, 40 and 10 per cent.
Twelve tooth, 45 per cent.

HEAVY GOODS, NAILS, ETC.

ANVILS.

Wright's, 80-lb. and over.......... 0 10½
Hay Budden, 80-lb. and over 0 09½
Brook's, 80-lb. and over 0 11½
Taylor-Forbes, handy............... 0 05

VISES.

Wright's......................... .... 0 13½
Brook's.......................... .... 0 12½
Pipe Vise, Hinge, No. 1 .... 3 50
" " No. 2 .... 5 50
Saw Vise 4 50 5 00
Blacksmiths' (discount) 60 per cent.
" parallel (discount) 45 per cent.

BOLTS AND NUTS

Carriage Bolts, common ($1 list Per cent.
" " ⅜ and smaller.. 60, 10 and 10
" " 7-16 and up.... 55 and 5
" " Norway Iron ($3 list) 50
Machine Bolts, ⅜ and less .... 60 and 10
Machine Bolts, 7-16 and up..... 55 and 5
Plough Bolts .... 55 and 10
Blank Bolts..................... 55
Bolt Ends....................... 55
Sleigh Shoe Bolts, ⅜ and less .. 60 and 10
" " " 7-16 and larger 50 and 5
Coach Screws, conepoint...... 70 and 5
Nuts, square, all sizes, 4c. per cent. off
Nuts, hexagon, all sizes, 4¼c. per cent. off.
Stove Rods per lb., 5½ to 6c.
Stove Bolts, 75 per cent.

CHAIN.

Proof coil, per 100 lb., 5-16 in., $4.40; ⅜ in., $3.90; 7-16 in., $3.70; ½ in., $3.50; 9-16 in., $3.45; ⅝ in., $3.35; ¾ in., $3.20; ⅞ in., $3.10; 1 in., $3.10.
Halter, kennel and post chains, 40 to 40 and 5 per cent.
Cow ties .... 40 p.c.
Tie out chains .... 65 p.c.
Stall fixtures .... 35 p.c.
Trace chain .... 45 p.c.
Jack chain, iron, discount 35 p.c.
Jack chain, brass, discount 40 per cent.

HORSE NAILS.

'C' brand, 40, 10 and 7½ per cent. off list } Oval
M.R.M. Co. brand, 55 per cent. } head

HORSESHOES.

F.O.B. Montreal
M.R.M. Co. brand, base...................... 65
Add 15c. Toronto, Hamilton, Guelph.

HORSE WEIGHTS.

Taylor-Forbes, 3½c. per lb.

NAILS. Cut. Wire.
2d........................... 3 80 3 35
3d........................... 2 95 3 00
4 and 5d..................... 2 70 2 75
6 and 7d..................... 2 60 2 65
8 and 9d..................... 2 45 2 50
10 and 12d................... 2 40 2 45
16 and 20d................... 2 35 2 40
30, 40, 50 and 60d (base) ... 2 30 2 35
F.o.b. Montreal.
Cut nails, Toronto 20c. higher.
Miscellaneous wire nails, discount 75 per cent.
Coopers' nails, discount 40 per cent.

PRESSED SPIKES.

Pressed spikes, ⅝ diameter, per 100 lbs. $3.15

RIVETS AND BURRS.

Iron Rivets, black and tinned, 60, 10 and 10.
Iron Burrs, discount 60 and 10 and 10 p.c.
Copper Rivets, usual proportion burrs, 15 p.c.
Copper Burrs only, net list.
Extras on Coppered Rivets, ½-lb. packages 1c. per lb.; ¼-lb. packages 2c. lb.
Tinned Rivets, net extra, 4c. per lb.

SCREWS.

Wood, F. H., bright and steel, 87½ per cent.
" R. H., bright, dis. 82½ per cent.
" F. H., brass, dis. 80 per cent.
" R. H., " dis. 75 per cent.
" F. H., bronze, dis. 75 per cent.
" R. H., " dis. 70 per cent.
Drive Screws, dis. 87½ per cent.
Bench, wood ....per doz. 3 25 4 00
" iron 4 25 5 00
Set, case hardened, dis. 60 per cent.
Square Cap, dis. 50 and 5 per cent.
Hexagon Cap, dis. 45 per cent.

MACHINE SCREWS.

Flat head, iron and brass, 35 per cent.
Felister head, iron, discount 30 per cent.
" " brass, discount 25 per cent.

TACKS, BRADS, ETC.

Carpet tacks, blued .... 80 and 5
" " tinned.............. 80 and 10
" " (in kegs)............ 40
Cut tacks, blued, in dozens only 75 and 10
" ¼ weights 60
Swedes cut tacks, blued and tinned—
In bulk......................... 80 and 10
In dozens 75
Swedes, upholsterers', bulk .... 85 and 12½
" brush, blued and tinned bulk 70
Swedes, gimp, blued, tinned and japanned......................... 75 and 12½
Zinc tacks 35
Leather carpet tacks............ 40
Copper tacks.................... 37½
Copper nails 42½
Trunk nails, black 65
Trunk nails, tinned and blued .. 65
Clout nails, blued and tinned .. 65
Chair nails..................... 35
Patent brads.................... 40
Fine finishing.................. 40
Lining tacks, in papers......... 10
" " in bulk........... 15
" " solid heads, in bulk 75
Saddle nails, in papers......... 10
" " in bulk........... 15
Tufting buttons, 22 line in dozens only 60
Zinc glaziers' points 5
Double pointed tacks, papers.. 90 and 10
" " " bulk.... 40
Clinch and duck rivets.... .. 45
Cheese box tacks, 85 and 5; trunk tacks, 80 and 10.

WROUGHT IRON WASHERS.

Canadian make, discount 40 per cent.

CUTLERY AND SPORTING GOODS.

AMMUNITION.

B. B. Caps American $2.00 per 1000.
C. B. Caps American, $2 60 per 1000

CARTRIDGES.

Rim Fire Cartridges, 50 and 5 p.c.
Rim Fire Pistol, 30 and 5 per cent., American.
Central Fire, Military and Sporting American, 10 per cent. advance, B.B. Caps, discount 40 per cent., American.
Central Fire Pistol and Rifle, net Amer. list
Loaded and empty Shells, American 20 per cent. discount. Rival and Nitro, 10 per cent. advance.
Empty paper shells American, 10 per cent. advance.
Primers, American $2 03

Wads. per lb.
Best thick brown or grey felt wads, in ¼-lb. bags $0 70
Best thick white card wads, in boxes of 500 each, 12 and smaller gauges 0 29
Best thick white card wads, in boxes of 500 each, 10 gauge............ 0 35
Thin card wads, in boxes of 1,000 each, 12 and smaller gauges 0 20
Thin card wads, in boxes of 1,000 each, 10 gauge...................... 0 25
Chemically prepared black edge grey cloth wads, in boxes of 250 each— Per M.
11 and smaller gauge 0 60
9 and 10 gauges.................. 0 70
7 and 8 " 0 90
5 and 6 " 1 10
Superior chemically prepared pink edge, best white cloth wads in boxes of 250 each—
11 and smaller gauge 1 15
9 and 10 gauges 1 40
7 and 8 " 1 65
5 and 9 " 1 90

SHOT.

Common, $6 50 per 100 lb.; chilled, $7.50 per 100 lb.; buck, seal and ball, $8.50 net list. Prices are f.o.b. Toronto, Hamilton, Montreal, St. John and Halifax. Terms, 2 p.c. for cash in thirty days.

RAZORS. per doz.
Elliot's 4 00 18 00
Boker's 7 50 11 00
" King Cutter 13 50 18 50
Wade & Butcher's.............. 3 60 10 00
Lewis Bros' "Klean Kutter" 8 50 10 50
Clauss Razors and Strops, 50 and 10 per cent.

PLATED GOODS

Holloware, 40 per cent discount.
Flatware, staples, 40 and 10, fancy, 40 and 5.

SHEARS.

Clauss, nickel, discount 60 per cent.
Clauss, Japan, discount 67½ per cent.
Clauss, tailors, discount 40 per cent.
Seymour's, discount 50 and 10 per cent.

TRAPS (steel.)

Game, Newhouse, discount 30 and 10 per cent
Game, Hawley & Norton, 50, 10 & 5 per cent.
Game, Victor, 70 per cent.
Game, Oneida Jump (B. & L.) 40 & 2½ p. c.
Game, steel, 60 and 5 per cent.

SKATES.

Skates, discount 37½ per cent.
Mic Mac hockey sticks, per doz 4 00 5 00

HOUSE FURNISHINGS.

APPLE PARERS.

Woodyatt Hudson, per doz., net 4 50

BIRD CAGES.

Brass and Japanned, 40 and 10 p. c.

COPPER AND NICKEL WARE.

Copper boilers, kettles, teapots, etc. 35 p.c.
Copper pitts, 25 per cent.

ENAMELED WARE.

London, White, Princess, Turquoise, Onyx, Blue and White, discount 50 per cent.
Canada, Diamond, Premier, 50 and 10 p c.
Pearl, Imperial Crescent, 60 and 10 per cent.
Premier steel ware, 40 per cent.
Star decorated steel and white, 25 per cent.
Japanned ware, discount 45 per cent.
Hollow ware, tinned cast, 35 per cent. off.

KITCHEN SUNDRIES.

Can openers, per doz.	0 40	0 75
Mincing knives per doz	0 50	0 80
Duplex mouse traps, per doz.	....	0 65
Potato mashers, wire, per doz.	0 60	0 70
" wood "	0 50	0 60
Vegetable slicers, per doz	....	2 25
Universal meat chopper No. 0, $1; No. 1,		1 15.
Enterprise chopper, each	...	1 30

Spiders and fry pans, 50 per cent.

LAMP WICKS.

Discount, 60 per cent.

LEMON SQUEEZERS.

Porcelain lined	per doz.	2 20	5 60
Galvanized	"	1 87	3 85
King, wood	"	2 75	2 90
King, glass	"	4 00	4 50
All glass	"	0 50	0 90

PICTURE NAILS.

Porcelain head ......per gross	1 5	1 50
Brass head		1 00

Tin and gilt, picture wire, 75 per cent.

SAD IRONS.

Mrs. Potts, No. 55, polished....per set	0 80
" No. 50, nickle-plated, "	0 92
Common, plain	4 50
" plated	5 50
Asbestos, per set	1 25

TINWARE.

CONDUCTOR PIPE.

2 in. plain or corrugated, per 100 feet. $3 30; 3 in, $4 40; 4 in., $5.80; 5 in., $7.45; 6 in., $9.90.

FAUCETS.

Common, cork-lined, discount 35 per cent.

EAVETROUGHS.

10-inch ....per 100 ft. 3 30

FACTORY MILK CANS.

Discount off revised list, 35 per cent.
Milk can trimmings, discount 25 per cent.
Creamery Cans, 45 per cent.

LANTERNS.

No. 2 or 4 Plain Cold Blast ..per doz. 6 50
Lift Tubular and Hinge Plain " 4 75
Better quality at higher prices.
Japanning, 50c. per doz. extra.

OILERS.

Kemp's Tornado and McClary's Model galvanized oil can, with pump, 5 gallon, per dozen 10 00
Davidson oilers, discount 40 per cent
Zinc and tin, discount 50 per cent.
Coppered oilers, 20 per cent. off.
Brass oilers, 50 per cent, off.
Malleable, discount 25 per cent

PAILS (GALVANIZED).

Dufferin pattern pails, 40, 10 and 5 per cent.
Flaring pattern, discount 40, 10 and 5 per cent.
Galvanized washtubs 40, 10 and 5 per cent

PIECED WARE.

Discount 35 per cent off list, June, 1899.
10-qt. flaring sap buckets, discount 40 per cent.
6, 10 and 14-qt. flaring pails dis. 40 per cent.
Copper bottom tea kettles and boilers, 30 p c.
Coal hods, 40 per cent.

STAMPED WARE.

Plain, 75 and 12½ per cent. off revised list.
Retinned, 72½ per cent revised list.

SAP SPOUTS.

Bronzed iron with hooks ....per 1,000 7 50
Eureka tinned steel, hooks 8 00

STOVEPIPES.

5 and 6 inch, per 100 lengths 7 00
7 inch........ " " 7 50
Nestable, discount 40 per cent.

STOVEPIPE ELBOWS

5 and 6-inch, common........per doz. 1 32
7-inch........................ " 1 48
Polished, 15c. per dozen extra.

THERMOMETERS.

Tin case and dairy, 75 to 75 and 10 per cent.

TINNERS' SNIPS.

Per doz....................... 3 00 15
Clauss, discount 35 per cent.

TINNERS' TRIMMINGS

Discount, 45 per cent.

WIRE.

BRIGHT WIRE GOODS

Discount 62½ per cent.

CLOTHES LINE WIRE

7 wire solid line, No. 17, $4.90; No. 18, $3.00; No. 19, $2.70; 4 wire solid line, No. 17, $4.45; No. 18, $2.80. No. 19, $2.50. All prices per 1000 ft. measure. F.o.b. Hamilton Toronto, Montreal.

COILED SPRING WIRE

High Carbon, No. 9, $2 90, No. 11, $3 45; No. 12, $3 15.

COPPER AND BRASS WIRE

Discount 37½ per cent.

FINE STEEL WIRE.

Discount 25 per cent. List of extras: In 100-lb. lots: No. 17, $5 — No. 18, $5 50 — No. 19, $6 — No. 20, $6 65 — No. 21, $7 — No. 22, $7.30 — No 23, $7.65 — No 24, $8 — No. 25, $9—No. 26, $9 50—No. 27, $10—No. 28, $11—No. 29, $12—No. 30, $13—No. 31, $14—No. 32, $15—No 33, $16—No. 34, $17. Extras net—tinned wire, Nos. 17-25, $2—Nos. 26-31, $4—Nos. 32-34, $6. Coppered, 75c —oiling, 10c.—in 25-lb. bundles, 15c.—in 5 and 10-lb. bundles, 25c.—in 1-lb. hanks, 25c. —in ½-lb. hanks, 38c.—in ¼-lb. hanks, 50c.—packed in casks or cases, 15c.—bagging or papering, 10c.

GALVANIZED WIRE.

Per 100 lb.—Nos. 4 and 5, $3.70 — Nos. 6, 7, 8, $3.15 — No. 9, $2 50 — No. 10, $3 20 — No. 11, $3 25 — No. 12, $2.65 —No. 13, $2.75—No. 14. $3.75—No 15, $4.30 —No. 16, $4.30 from stock. Base sizes, Nos. 6 to 9, $2 35 f.o.b. Cleveland. In carlots 12½c. less.

LIGHT STRAIGHTENED WIRE

Gauge No.	Over 20 in. per 100 lbs.	10 to 20 in.	5 to 10 in.
0 to 5	$0.50	$0.75	$1 25
6 to 9	0 75	1 25	2 00
10 to 11	1 00	1 75	2 50
12 to 14	1.50	2.25	3 50
15 to 16	2 00	3.00	4 50

SMOOTH STEEL WIRE.

No. 0-9 gauge, $2 30; No. 10 gauge, 6c. extra; No. 11 gauge, 12c extra, No 12 gauge, 20c. extra; No. 13 gauge, 30c. extra; No 14 gauge, 40c. extra; No 15 gauge, 55c. extra; No 16 gauge, 70c extra Add 60c for coppering and $2 for tinning

Extra net per 100 lb.—Oiled wire 10c, spring wire $1 25, special hay baling wire 30c., best steel wire 75c., bright soft drawn 15c., charcoal (extra quality) $1.25, packed in casks or cases 15c., bagging and papering 10c., 50 and 100-lb. bundles 10c., in 25-lb. bundles 15c., in 5 and 10-lb. bundles 25c., in 1-lb hanks, 50c., in ½-lb. hanks 75c., in ¼-lb. hanks $1.

POULTRY NETTING.

2 in mesh 19 w. g., discount 50 and 10 per cent. All others 50 per cent.

WIRE CLOTH.

Painted Screen, in 100-ft rolls, $1 62½, per 100 sq. ft; in 50-ft. rolls, $1.67½, per 100 sq ft. Terms, 2 per cent. off 30 days.

WIRE FENCING.

Galvanized barb............... .... 2 95
Galvanized, plain twist 3 30
Galvanized barb, f.o b. Cleveland, $2.70 for small lots and $2 60 for carlots.

WOODENWARE.

CHURNS.

No. 0, $9; No. 1, $9; No 2, ; No 3, $11; No 4, $13, No. 5, $16.; f.o b. Toronto Hamilton, London and St. Marys, 30 and 30 per cent; f o b. Ottawa, Kingston and Montreal, 40 and 15 per cent. discount, Taylor-Forbes, 30 and 30 per cent.

CLOTHES REELS.

Davis Clothes Reels, dis. 40 per cent.

LADDERS, EXTENSION.

Waggoner Extension Ladders, dis. 40 per cent.

MOPS AND IRONING BOARDS.

"Best" mops..................... .. 1 25
"900" mops 1 25
Folding ironing boards......... 12 00 16 50

REFRIGERATORS

Discount, 40 per cent

SCREEN DOORS.

Common doors, 2 or 3 panel, walnut stained, 4-in. style..........per doz. 7 25
Common doors, 2 or 3 panel, grained only, 4-in., style ....per doz. 7 55
Common doors, 2 or 3 panel, light stain per doz. 9 55

WASHING MACHINES.

Round, re-acting per doz.	60 00
Square "	63 00
Eclipse, per doz	54 00
Dowswell "	39 00
New Century, per doz	75 00
Daisy	54 00

WRINGERS

Royal Canadian, 11 in., per doz. ... 34 00
Royal American, 11 in. ... 34 00
Eze, 10 in., per doz 3 75
Terms, 2 per cent., 30 days.

MISCELLANEOUS.

AXLE GREASE.

Ordinary, per gross	6 00	7 00
Best quality	10 00	12 00

BELTING.

Extra, 60 per cent.
Standard, 60 and 10 per cent.
No. 1, not wider than 6 in., 60, 10 and 10 p c.
Agricultural, not wider than 4 in., 75 per cent
Lace leather, per side, 75c., cut laces, 80c.

BOOT CALKS.

Small and medium, ball ....per M 4 25
Small heel .... " 4 50

CARPET STRETCHERS.

American ....per doz.	1 00	1 50
Bullard's .... "	....	6 50

CASTORS.

Bed, new list, discount 55 to 57½ per cent.
Plate, discount 52½ to 57½ per cent.

PINE TAR.

½ pint in tins ....per gross ... 7 80
" " .... " 9 60

PULLEYS.

Hothouse ....per doz.	0 55	1 00
Axle "	0 22	0 33
Screw "	0 22	1 00
Awning "	0 35	2 50

PUMPS.

Canadian cistern	1 40	2 00
Canadian pitcher spout	1 80	3 16

ROPE AND TWINE.

Sisal		0 10½
Pure Manilla		0 15½
"British" Manilla		0 12
Cotton, 3-16 inch and larger	0 21	0 23
" 5-32 inch	0 25	0 27
" ¼ inch	0 25	0 28
Russia Deep Sea		0 16
Jute		0 09
Lath Yarn, single		0 10
" double		0 10½
Sisal bed cord, 48 feet........per doz.		0 65
" " 60 feet........ "		0 80
" " 72 feet........ "		0 95

Twine.

Bag, Russian twine, per lb.	....	0 27
Wrapping, cotton, 3-ply	....	0 25
" " 4-ply	....	0 29
Mattress twine per lb.	0 33	0 45
Staging " "	0 27	0 35

SCALES.

Gurney Standard, 40 per cent.
Gurney Champion, 50 per cent.
Burrow, Stewart & Milne—
Imperial Standard, discount 40 per cent.
Weigh Beams, discount 40 per cent.
Champion Scales, discount 50 per cent.
Fairbanks standard, discount 35 per cent
" Dominion, discount 55 per cent.
" Richelieu, discount 55 per cent.
Warren new Standard, discount 40 per cent.
" " Champion, discount 50 per cent.
" Weighbeams, discount 35 per cent.

STONES—OIL AND SCYTHE.

Washita	per lb.	0 25	0 37
Hindostan	"	0 06	0 10
" slip	"	0 18	0 20
" Axe	"	....	0 10
Deer Creek	"	....	0 10
Deerlick	"	....	0 25
" Axe	"	....	0 15
Lily white	"	....	0 42
Arkansas	"	....	1 50
Water-of-Ayr	"	....	0 10
Scythe	per gross	3 50	5 00
Grind, 40 to 200 lb.,	per ton	20 00	22 00
" under 40 lb.,	"	....	24 00
" 200 lb. and over		....	28 00

INDEX TO ADVERTISERS.

CLASSIFIED LIST OF ADVERTISEMENTS.

Auditors
Davenport, Percy P., Winnipeg.

Babbitt Metal.
Canada Metal Co., Toronto.
Canadian Fairbanks Co., Montreal.
Robertson, Jas. Co,, Montreal.

Bath Room Fittings.
Carriage Mounting Co., Toronto.

Belting, Hose, etc.
Gutta Percha and Rubber Mfg. Co Toronto.

Bicycles and Accessories.
Johnson's, Iver, Arms and Cycle Works Fitchburg, Mass

Binder Twine
Consumers Cordage Co., Montreal.

Box Strap.
J. N. Warminton, Montreal.

Brass Goods.
Glauber Brass Mfg. Co, Cleveland, Ohio.
Lewis, Rice, & Son., Toronto.
Morrison, Jas., Brass Mfg. Co., Toronto.
Mueller Mfg. Co., Decatur, Ill.
Penberthy Injector Co., Windsor, Ont.
Taylor-Forbes Co., Guelph, Ont.

Bronze Powders.
Canadian Bronze Powder Works, Montreal.

Brushes.
Ramsay, A., & Son Co, Montreal.

Can Openers.
Cumming Mfg. Co. Renfrew.

Cans.
Acme Can Works, Montreal.

Builders' Tools and Supplies.
Covert Mfg. Co., West Troy, N.Y.
Frothingham & Workman Co., Montreal.
Howland, H. S., Sons & Co., Toronto.
Hyde, F., & Co., Montreal.
Lewis Bros. & Co., Montreal.
Lewis, Rice, & Son, Toronto.
Lockerby & McComb, Montreal.
Lufkin Rule Co., Saginaw, Mich.
Newman & Sons, Birmingham.
North Bros. Mfg. Co., Philadelphia, Pa.
Stanley Rule & Level Co., New Britain.
Stanley Works, New Britain, Conn.
Stephens, G. F., Winnipeg.
Taylor-Forbes Co., Guelph, Ont.

Carriage Accessories.
Carriage Mountings Co., Toronto.
Covert Mfg. Co., West Troy, N.Y.

Carriage Springs and Axles.
Guelph Spring and Axle Co., Guelph.

Cattle and Trace Chains.
Greening. B., Wire Co., Hamilton.

Churns.
Dowswell Mfg. Co., Hamilton.

Clippers—All Kinds.
American Shearer Mfg. Co., Nashua, N.H.

Clothes Reels and Lines.
Hamilton Cotton Co., Hamilton, Ont.

Cordage.
Consumers' Cordage Co., Montreal.
Hamilton Cotton Co., Hamilton.

Cork Screws.
Erie Specialty Co., Erie, Pa.

Clutch Nails.
J. N. Warminton, Montreal.

Cut Glass.
Phillips, Geo., & Co., Montreal.

Cutlery—Razors, Scissors, etc.
Birkett, Thos., & Son Co., Ottawa.
Clauss Shear Co., Toronto
Dorken Bros. & Co., Montreal.
Heinisch's, R., Sons Co., Newark, N.J.
Howland, H. S. Sons & Co., Toronto.
Phillips, Geo.. & Co., Montreal.
Round, John, & Son, Montreal.

Electric Fixtures.
Canadian General Electric Co., Toronto.
Forman, John, Montreal.
Morrison James, Mfg. Co., Toronto.
Munderloh & Co., Montreal.

Electro Cabinets.
Cameron & Campbell Toronto.

Engines, Supplies, etc.
Kerr Engine Co., Walkerville, Ont.

Fencing—Woven Wire
McGregor-Banwell Fence Co., Walkerville, Ont
Owen Sound Wire Fence, Co., Owen Sound
Banwell Hoxie Wire Fence Co., Hamilton.

Files and Rasps.
Barnett Co., G. & H., Philadelphia, Pa.
Nicholson File Co, Port Hope

Financial Institutions
Bradstreet Co.

Firearms and Ammunition.
Dominion Cartridge Co., Montreal.
Hamilton Rifle Co., Plymouth, Mich.
Harrington & Richardson Arms Co., Worcester, Mass.
Johnson's, Iver, Arms and Cycle Works, Fitchburg, Mass.

Food Choppers.
Enterprise Mfg. Co., Philadelphia, Pa.
Shirreff Mfg Co, Brockville, Ont.

Galvanizing.
Canada Metal Co., Toronto.
Montreal Rolling Mills Co, Montreal
Ontario Wind Engine & Pump Co., Toronto.

Glaziers' Diamonds.
Gibsone, J. B., Montreal.
Pelton, Godfrey S.
Sharratt & Newth, London, Eng.
Shaw, A., & Son, London, Eng

Handles
Still, J. H, Mfg. Co.

Harvest Tools.
Maple Leaf Harvest Tool Co, Tillsonburg Ont.

Hoop Iron.
Montreal Rolling Mills Co. Montreal.
J. N. Warminton, Montreal.

Horse Blankets.
Heney, E. N., & Co., Montreal.

Horseshoes and Nails.
Canada Horse Nail Co., Montreal.
Montreal Rolling Mills, Montreal
Capewell Horse Nail Co., Toronto.

Hot Water Boilers and Radiators.
Cluff, R. J., & Co., Toronto.
Pease Foundry Co., Toronto.
Taylor-Forbes Co., Guelph.

Ice Cream Freezers.
Dana Mfg. Co., Cincinnati, Ohio.
North Bros. Mfg. Co., Philadelphia, Pa

Ice Cutting Tools.
Erie Specialty Co., Erie, Pa.
North Bros. Mfg. Co., Philadelphia, Pa.

Injectors—Automatic.
Morrison, Jas., Brass Mfg. Co., Toronto.
Penberthy Injector Co., Windsor, Ont.

Iron Pipe.
Montreal Rolling Mills, Montreal.

Iron Pumps.
McDougall, R., Co., Galt, Ont.

Lanterns.
Kemp Mfg. Co., Toronto.
Ontario Lantern Co., Hamilton, Ont.
Wright, E. T., & Co., Hamilton.

Lawn Mowers.
Birkett, Thos., & Son Co., Ottawa.
Maxwell, D., & Sons, St. Mary's, Ont.
Taylor, Forbes Co., Guelph.

Lawn Swings, Settees, Chairs.
Cumming Mfg. Co., Renfrew.

Ledgers—Loose Leaf.
Business Systems Toronto.
Copeland-Chatterson Co, Toronto.
Grain, Rolla L., Co., Ottawa.
Universal Systems, Toronto.

Locks, Knobs, Escutcheons, etc.
Peterborough Lock Mfg. Co., Peterborough, Ont.

Lumbermen's Supplies.
Pink, Thos., & Co., Pembroke Ont.

Manufacturers' Agents.
Fox, C. H., Vancouver.
Gibb, Alexander, Montreal.
Mitchell, David C, & Co, Glasgow, Scot.
Mitchell, H. W., Winnipeg.
Pearce, Frank, & Co. Liverpool, Eng.
Scott, Bathgate & Co., Winnipeg.
Thorne, R. E., Montreal and Toronto.

Metals.
Canada Iron Furnace Co., Midland, Ont.
Canada Metal Co., Toronto.
Eadie, H. G., Montreal.
Gibb, Alexander, Montreal.
Kemp Mfg Co., Toronto
Leslie, A. C., & Co, Montreal.
Lysaght, John, Bristol, Eng.
Nova Scotia Steel and Coal Co., New Glasgow, N S.
Robertson, Jas, Co, Montreal
Roper, J H., Montreal.
Samuel, Benjamin & Co, Toronto.
Stairs, Son & Morrow, Halifax, N.S.
Thompson, B. & S. H. & Co. Montreal.

Metal Lath.
Galt Art Metal Co., Galt.
Metallic Roofing Co., Toronto.
Metal Shingle & Siding Co., Preston, Ont.

Metal Polish, Emery Cloth, etc
Oakey, John, & Sons, London, Eng.

Mops.
Cumming Mfg. Co., Refrew.

Mouse Traps.
Cumming Mfg. Co., Renfrew.

Oil Tanks.
Bowser, S. F., & Co., Toronto.

Paints, Oils, Varnishes, Glass.
Bell, Thos, Sons & Co, Montreal
Canada Paint Co., Montreal.
Canadian Oil Co., Toronto.
Consolidated Plate Glass Co., Toronto.
Fenner, Fred., & Co., London, Eng.
Henderson & Potts Co, Montreal.
Imperial Varnish and Color Co., Toronto.
Jamieson, R C., & Co, Montreal.
McArthur, Corneille & Co, Montreal.
McCaskill Dougall & Co, Montreal.
Montreal Rolling Mills Co., Montreal.
Moore Benjamin, & Co. Toronto.
Queen City Oil Co., Toronto.
Ramsay & Son, Montreal.
Sherwin-Williams Co., Montreal.
Standard Paint and Varnish Works Windsor, Ont.
Stephens & Co, Winnipeg.
Martin-Senour Co., Chicago.

Perforated Sheet Metals.
Greening, B., Wire Co., Hamilton.

Plumbers' Tools and Supplies.
Borden Co., Warren, Ohio.
Canadian Fairbanks Co., Montreal.
Cluff R. J., & Co., Toronto.
Glauber Brass Co., Cleveland, Ohio
Jardine, A. B., & Co, Hespeler, Ont.
Jenkins Bros., Boston, Mass.
Lewis, Rice, & Son, Toronto.
Merrell Mfg. Co., Toledo, Ohio.
Montreal Rolling Mills Montreal.
Morrison, Jas., Brass Mfg. Co., Toronto.
Mueller, H., Mfg. Co, Decatur, Ill.
Oshawa Steam & Gas Fitting Co., Oshawa
Robertson Jas., Co. Montreal.
Stairs, Son & Morrow, Halifax, N.S.
Standard Ideal Sanitary Co, Port Hope,
Standard Sanitary Co., Pittsburg.
Stephens, G F., & Co, Winnipeg, Man.
Turner Brass Works, Chicago.
Vickery, Orlando, Toronto.

Portland Cement.
International Portland Cement Co. Ottawa, Ont
Hanover Portland Cement Co., Hanover, Ont.
Hyde, F., & Co., Montreal.
Thompson, B. & S. H. & Co., Montreal.

Potato Mashers.
Cumming Mfg Co., Renfrew.

Poultry Netting.
Greening, B., Wire Co., Hamilton, Ont.

Razors.
Clauss Shear Co., Toronto.

Roofing Supplies.
Brantford Roofing Co., Brantford.
McArthur, Alex., & Co., Montreal.
Metal Shingle & Siding Co., Preston, Ont.
Metallic Roofing Co, Toronto.
Paterson Mfg. Co., Toronto & Montreal.

Saws.
Atkins, E. C., & Co., Indianapolis, Ind
Lewis Bros., Montreal.
Shurly & Dietrich, Galt, Ont
Spear & Jackson, Sheffield, Eng.

Scales.
Canadian Fairbanks Co., Montreal.

Screw Cabinets.
Cameron & Campbell, Toronto.

Screws, Nuts, Bolts.
Montreal Rolling Mills Co., Montreal.
Morrow, John, Machine Screw Co., Ingersoll, Ont.

Sewer Pipes.
Canadian Sewer Pipe Co., Hamilton
Hyde, F., & Co., Montreal.

Shelf Boxes.
Cameron & Campbell, Toronto

Shears, Scissors.
Clauss Shear Co., Toront.

Shelf Brackets.
Atlas Mfg. Co., New Haven, Conn

Shellac
Bell, Thos, Sons & Co., Montreal

Shovels and Spades.
Canadian Shovel Co., Hamilton.
Peterboro Shovel & Tool Co, Peterboro.

Silverware.
Phillips, Geo., & Co., Montreal.
Round, John, & Son, Sheffield, Eng.

Skates.
Canada Cycle & Motor Co, Toronto.

Spring Hinges, etc.
Chicago Spring Butt Co., Chicago, Ill.

Steel Rails.
Nova Scotia Steel & Coal Co., New Glasgow, N.S.

Stoves, Tinware, Furnaces
Canadian Heating & Ventilating Co., Owen Sound.
Canada Stove Works, Harriston, Ont.
Clare Bros. & Co, Preston.
Davidson, Thos., Mfg. Co., Montreal.
Guelph Stove Co., Guelph.
Gurney Foundry Co., Toronto.
Harris, J. W., Co., Montreal.
Joy Mfg. Co., Toronto.
Kemp Mnfg. Co. Toronto.
McClary Mfg. Co. London.
Pease Foundry Co., Toronto.
Stewart, Jas., Mfg. Co., Woodstock, Ont.
Taylor-Forbes Co., Guelph, Ont.
Wright, E. T., & Co., Hamilton.

Tacks.
Montreal Rolling Mills Co., Montreal.
Ontario Tack Co., Hamilton.

Ventilators.
Harris, J. W., Co., Montreal.
Pearson, Geo. D, Montreal.

Washing Machines, etc.
Dowswell Mfg. Co., Hamilton, Ont.
Taylor Forbes Co., Guelph, Ont.

Wheelbarrows.
London Foundry Co, London, Ont.

Wholesale Hardware.
Birkett, Thos., & Sons Co., Ottawa.
Caverhill, Learmont & Co., Montreal.
Frothingham & Workman, Montreal.
Hobbs Hardware Co., London.
Howland, H. S., Sons & Co., Toronto.
Lewis Bros. & Co., Montreal.
Lewis, Rice, & Son, Toronto.

Window and Sidewalk Prisms.
Hobbs Mfg. Co., London, Ont.

Wire Springs.
Guelph Spring Axle Co., Guelph, Ont.
Wallace-Barnes Co., Bristol, Conn.

Wire, Wire Rope, Cow Ties, Fencing Tools, etc.
Canada Fence Co., London.
Dennis Wire and Iron Co., London, Ont.
Dominion Wire Mnfg. Co., Montreal
Greening, B., Wire Co., Hamilton.
Montreal Rolling Mills Co., Montreal.
Western Wire & Nail Co., London, Ont

Woodenware.
Taylor-Forbes Co., Guelph, Ont.

Wrapping Papers.
Canada Paper Co., Toronto.
McArthur, Alex., & Co, Montreal.
Stairs, Son & Morrow, Halifax, N.S.

WIRE ROPE

"Acme" brand

Highest grade of hoisting rope made.
Extra tensile strength for heavy work.

Use Greening's Rope Grease for Lubrication.

THE B. GREENING WIRE CO.
LIMITED

HAMILTON, Ont. **MONTREAL, Que.**

TRADE MARK

Est. 1868. Inc. 1895.

Black Diamond File Works

G. & H. Barnett Company

PHILADELPHIA

Twelve Medals

Awarded

By **JURORS** at

International Expositions

Special Prize

Gold Medal at Atlanta, 1895

Copy of catalogue sent free to any interested file user upon application.

"Redstone"

High Pressure

Sheet Packing

A packing that will hold. For use in highest pressures for steam, hot or cold water and air. Packs equally well for all.

From actual tests, we believe that this packing is the most durable and satisfactory of any on the market. Try a sample lot and see for yourself.

Manufactured Solely by

THE GUTTA PERCHA & RUBBER MFG. CO.

of TORONTO, LIMITED

HEAD OFFICES,

47 Yonge Street, Toronto.

Branches: Montreal, Winnipeg, Vancouver.

SHEET ZINC,

SHEET COPPER,

SPELTER,

INGOT TIN,

BRASS and COPPER TUBES.

For lowest prices send your enquiries to

B. & S. H. THOMPSON & Co.

LIMITED

53 ST. SULPICE STREET, MONTREAL

CIRCULATES EVERYWHERE IN CANADA
Also in Great Britain, United States, West Indies, South Africa and Australia.

HARDWARE AND METAL

A Weekly Newspaper Devoted to the Hardware, Metal, Heating and Plumbing Trades in Canada.
Office of Publication, 10 Front Street East, Toronto.

VOL. XIX. MONTREAL, TORONTO, WINNIPEG, FEBRUARY 16, 1907 NO. 7.

See Classified List of Advertisements on Page 71.

WE CARRY IN STOCK

Horse-shoers' Clip Horn Anvil
Prospectors' Anvils and Bellows
Prospectors' Picks, Shovels and Stoves
The Famous Blacksmith Blower No. 400
The Grand Bay Blower No. 100
Challenge Forges with Hood, Extra Strong Frame
Hero Forges With and Without Hood
Navy Wheelbarrows with Steel or Wood Trays
Steel Pulley Blocks, Single, Double or Triple

Write for prices. Prompt shipment.

THOS. BIRKETT & SON CO.,
Wholesale Hardware LIMITED
OTTAWA, ONT.

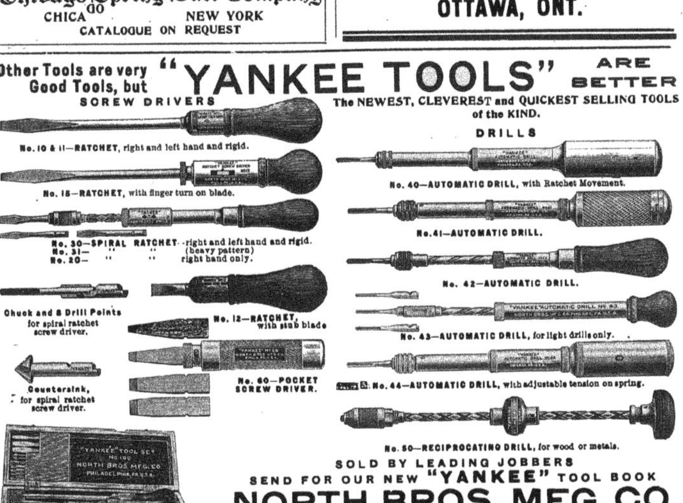

GOOD PROFIT ON THESE

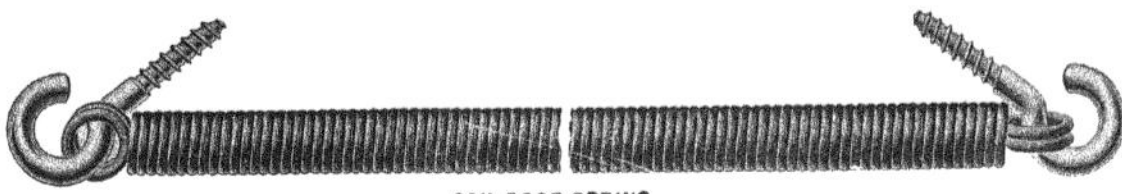

COIL DOOR SPRING
Length 17 inches: Including Screw Hooks, Japanned.
All sizes, Nos. 1 to 6.

There are many little things in the Hardware Trade that some Merchants don't pay enough attention to.

Door Springs, although a very small item, will pay you a handsome profit.

Our annual sales for the lines illustrated, run into hundreds of gross, which is the best possible evidence that we've struck the goods the trade desire, and know how to price them satisfactorily.

FOR FURTHER INFORMATION, ASK OUR TRAVELERS, OR WRITE US.

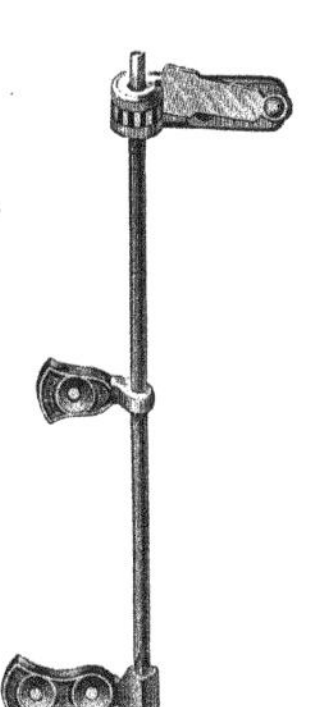

TORREY DOOR SPRING
39 in., Japanned Steel, Rod Japanned, Malleable Fixtures.

CARTLAND'S HELICAL DOOR SPRING
JAPANNED.
5, 6, 7 or 8 inch.

RELIANCE COIL DOOR SPRING RATCHET TENSION ADJUSTMENT.

LEWIS BROS., LTD.

MONTREAL
WINNIPEG
OTTAWA VANCOUVER
TORONTO CALGARY

Consumers Cordage Co., LIMITED

Manufacture every variety of

CORDAGE | PACKING

BINDER TWINE

LATHYARN | SMALL TWINES

From MANILLA, SISAL, ITALIAN, RUSSIAN, JUTE

Tallow Laid Transmission Rope, 3 or 4 Strand

Oil Well Drilling Cables, Lariat Rope, 3 or 4 Strand,
Halyards, Deep Sea Line, Snap Halters, Snap Plow Lines,
Clothes Lines, Hide Cord, Pulp and Paper Cord,
Tarred Rope, Marline, Houseline

Wrapping, Rolling, Seaming, Sewing and Grape Twines

CAREFUL EXECUTION **SPECIAL ORDERS** PROMPT SHIPMENT

Mills: MONTREAL, QUE., and HALIFAX, N.S.

BRANCHES:

F. H. ANDREWS & SON, Quebec. CONSUMERS CORDAGE CO., Limited, St. John, N.B. MacGOWAN & CO., Vancouver.
TEES & PERSSE, Limited, Winnipeg, Calgary and Edmonton. GEO. WOOD, London, England.
MERRICK, ANDERSON & CO., Winnipeg, Binder Twine Distributors for Northwest.

Have You Tried SOLDERALL?

Tinsmiths should make a test of this New Anticorrosive Fluid—It is the best thing in the market to-day for tinning and soldering all metals.

Unsurpassed for Cleanliness of Work and Strength of Joint made.

No Noxious Fumes to Injure the Health or Affect the Skin.

Economises Time and Labor, Saves Solder

No rust or discoloration. The Solder will follow the fluid like a magnet

PUT UP FOR SALE IN ½-GAL., 1-GAL. AND 5-GAL. CANS.

Solderall is always sent out **full strength** and may be diluted according to the class of work for which it may be required.

SOLE AGENTS AND MANUFACTURERS

The Thos. Davidson Mfg. Co. Ltd., MONTREAL and WINNIPEG

H. S. HOWLAND, SONS & CO. LIMITED

HARDWARE MERCHANTS

Only Wholesale — 138-140 WEST FRONT STREET, TORONTO — Wholesale Only

NICHOLSON AND KARNEY & FOOTE

For a thoroughly reliable File use Nicholson or Kearney & Foote.

For a severe test in Filing try Nicholson or Kearney & Foote.

Taper Files—Ordinary and slim, 3 to 8 inches long.

Round Files—Bastard and smooth, 3 to 16 inches long.

Square Files—3 to 16 inches long.

Cross Cut Saw Files.

Flat Files—Second cut and smooth, 3 to 14 inches long.

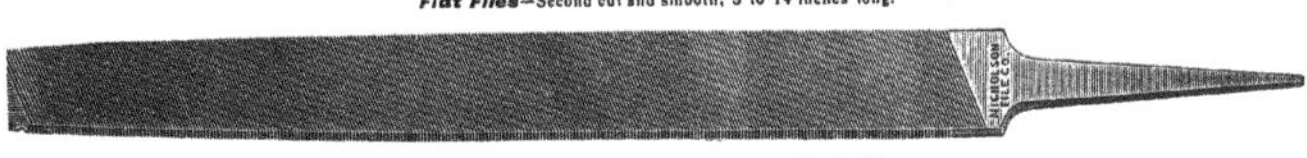

Flat Bastard—Hand and Bastard cut, 4 to 16 inches long.
Half Round—Bastard and smooth, 4 to 16 inches long.

For
Cabinet Files
" Rasps
Shoe "
Mower Files
See our catalogue

For
Planer Knife Files
Pit Saw Files
Auger Bit "
Double ender
See our catalogue.

Horse Rasps—Plain and Tanged, 12 to 16 inches long

FOR A FULLER LINE SEE OUR HARDWARE CATALOGUE

H. S. HOWLAND, SONS & CO., LIMITED

Opposite Union Station.

GRAHAM NAILS ARE THE BEST

We Ship promptly — Factory: Dufferin Street, Toronto, Ont. — Our Prices are Right

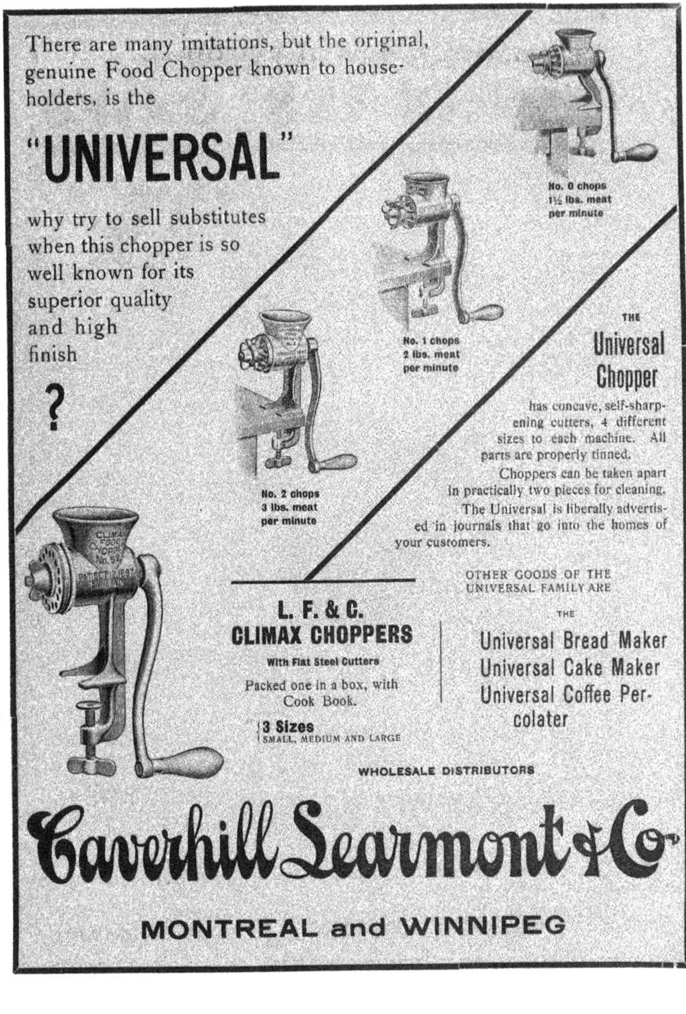
There are many imitations, but the original, genuine Food Chopper known to householders, is the
"UNIVERSAL"
why try to sell substitutes when this chopper is so well known for its superior quality and high finish
?
No. 0 chops 1½ lbs. meat per minute
No. 1 chops 2 lbs. meat per minute
No. 2 chops 3 lbs. meat per minute
THE
Universal Chopper
has concave, self-sharpening cutters, 4 different sizes to each machine. All parts are properly tinned.
Choppers can be taken apart in practically two pieces for cleaning.
The Universal is liberally advertised in journals that go into the homes of your customers.
L. F. & C.
CLIMAX CHOPPERS
With Flat Steel Cutters
Packed one in a box, with Cook Book.
3 Sizes
SMALL, MEDIUM AND LARGE
OTHER GOODS OF THE UNIVERSAL FAMILY ARE
THE
Universal Bread Maker
Universal Cake Maker
Universal Coffee Percolater
WHOLESALE DISTRIBUTORS
Caverhill Learmont & Co.
MONTREAL and WINNIPEG

Canada Leads the World in the Manufacture of

HIGH-CLASS SAWS

No. 81

The best and cheapest Hand Saw on the market, quality, temper and finish considered. Every saw warranted, Carved applewood handle. Patented, wave finish blade.

No. 1 Maple Leaf Lance Tooth Cross-Cut Saw, and the No. 0 Racer are tempered under our secret process.

No. 0 Narrow Racer Cross-Cut Saw

If you want saws that will sell and give your customer satisfaction, you will stock the Maple Leaf Saws.

Manufactured by **SHURLY & DIETRICH, - GALT, Ontario**

Persons addressing advertisers will kindly mention having seen their advertisement in Hardware and Metal.

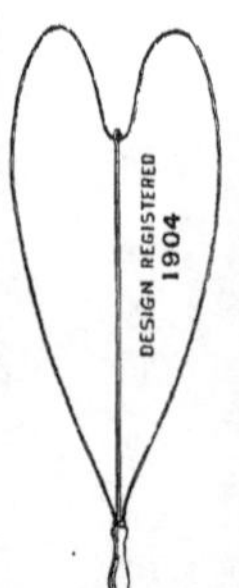

SPADES, SHOVELS, Etc.
Entrust your trade to the house that can supply the "just right" Spades and Shovels.
By getting in touch with us you will become acquainted with the favorite lines:
"Peterborough," "Lundy," "Irwin," and 'Hill"
Send in a postcard for our Catalogue and you will see where we stand.
The Peterborough Shovel & Tool Company, Limited, Peterborough, Ont.

PRATT & WHITNEY
DRILLS REAMERS, TAPS, DIES, ETC.
are made within the finest limits of precision and of the best material.
They have great strength—wear longer and the finish is better than the ordinary.
PRATT & WHITNEY SMALL TOOLS ARE THE STANDARD THE WORLD OVER.
NOW MADE IN CANADA
CANADIAN FAIRBANKS CO., Limited
SOLE AGENTS
Montreal Toronto Winnipeg Vancouver

The Cost of Installing Business Systems.
SYSTEM THE SECRET OF SUCCESS
The cost of installing Business Systems is in proportion to the size of the business.
A small business requires a more or less simple system of accounting the expense is almost nothing compared to a big business whose arrangements are more complex.
Our Monthly Account System is especially adapted to the retail business.
Our Monthly Account System does it all in one operation makes the entries in your books and prepares your bills at the same time.
Our Monthly Account System stops the work from piling up at the end of the month and enables you to have each customer's bill ready at a moment's notice.
Let us tell you how Business Systems can be adapted to your business.
It costs you one cent to know.
A post card will bring you the information.
BUSINESS SYSTEMS
LIMITED
84 SPADINA AVE.
TORONTO, CANADA

Money for You,
Lots of it

Yes, there are more dollars in retailing Window Glass than some dealers think—and **most dealers know it.**

You certainly are not giving yourself a square deal—if you have not seen our prices for spring deliveries. Besides, **you do save money** by importing direct.

These strong cases assure delivery in good order, while our facilities assure promptitude.

We are always glad to quote prices.

Have you our catalogue?

The Hobbs Manufacturing Co., Limited

LONDON TORONTO WINNIPEG

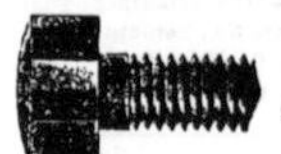

The John Morrow Screw, Limited

MANUFACTURERS OF

Set and Cap Screws, Special Milled Work, Engine Studs Etc. Cold Punched Nuts of every variety of finish.

INGERSOLL, ONTARIO

CELEBRATED **HEINISCH** SHEARS

Tailors' Shears,
Trimmers' Shears,
Tinners' Snips, etc.

ACKNOWLEDGED THE BEST.

R. HEINISCH'S SONS CO. NEW YORK OFFICE, 155 Chambers St NEWARK, N.J., U.S.A.

HENRY T. SEYMOUR SHEAR CO. NEW YORK. TRADE MARK

SEYMOUR SHEARS

HAVE BEEN THE

Standard for over Half a Century.

"QUALITY UNQUESTIONED."

Each pair of our shears bears the above trade mark.

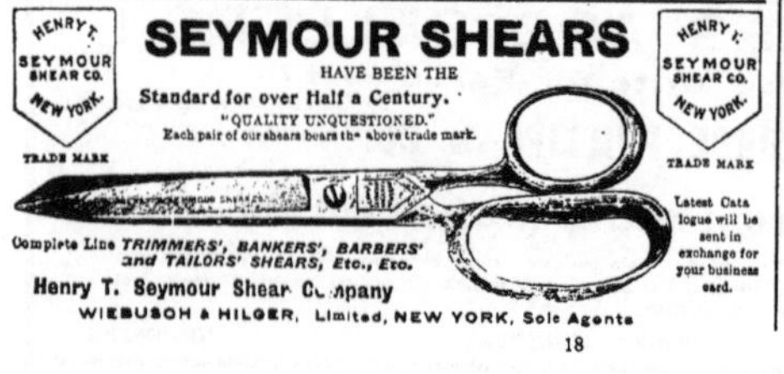

Latest Catalogue will be sent in exchange for your business card.

Complete Line *TRIMMERS', BANKERS', BARBERS' and TAILORS' SHEARS, Etc., Etc.*

Henry T. Seymour Shear Company

WIEBUSCH & HILGER, Limited, NEW YORK, Sole Agents

Covert Mfg. Co.

TROY, N. Y.

Harness Snaps, Chain, Rope and Web Goods, etc. For sale by Jobbers, at Manufacturers' prices.

STREET PAVING and SIDEWALKS a SPECIALTY

SILICA BARYTIC STONE CO.

OF ONTARIO, Limited.

Water Proof Floors for Malt Houses, Breweries, Slaughter Houses, Cheese Factories, Cellar, Stable Floors, etc.

Head Office *Ingersoll, Ontario.*

Walter Mills, General Manager

Ask for quotations for Septic Tanks.

"The Tool Holder People"

ARMSTRONG BROS. TOOL CO.

Manufacturers of Armstrong Patent Lathe and Planer Tools and other machine shop specialties.

106 N Francisco Ave. CHICAGO, ILL.

Persons addressing advertisers will kindly mention having seen this advertisement in Hardware and Metal.

The Profit That Tells

Yes, the profit always tells in the sale of any article. All that the manufacturer can do is to assure lasting quality as we have done in

The **Empire Queen Range**

We say without the least fear of contradiction, that **The Empire Queen** is as perfect as any range manufactured at any price. Most Canadian dealers know that and we are willing to forward proofs of our statement to those who don't.

Let's show you just how profitable it is to push The Empire Queen. A post-card will bring interesting particulars.

The Canadian Heating & Ventilating Co. Limited.
OWEN SOUND, Ontario.

THE CHRISTIE BROS. CO., Limited, 238 King St., Winnipeg, Man., Western Agents.
THE CANADA STOVE AND FURNITURE CO., 126 West Craig St., Montreal, Que., Agents for the Province of Quebec.

THE IDEAL LAWN-MOWER GRINDER

In offering the Ideal Grinder to the hardware fraternity, the repair men and blacksmiths of Canada this season, we do so with the utmost confidence in the reliability of the machine. We have made some improvements over previous patterns, and now claim the Grinder to be as nearly perfect as can be made. A study of the illustration will show the principle upon which it works. Any style of lawn-mower may be inserted, and within 15 minutes each blade will possess not only a keen edge, but an EXACT BEVEL, and it is this exact bevel which is so necessary to the perfect operation of the mower. One can obtain the same price for sharpening a mower that he did when he used the "file and elbow grease," and spent an hour or more in doing the work now requiring 15 minutes. Fitted for operation by power when needed, and can be used for tools, skates, etc., making it an all-around utensil. Send for interesting circular.

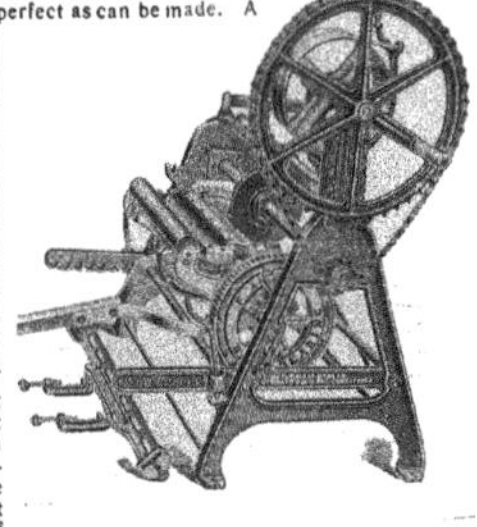

Ideal Lawn Mower Grinder.

ROOT BROS. CO., PLYMOUTH, OHIO.

Are You Ready for the Sap Season Trade?

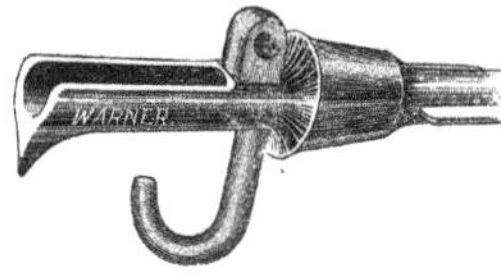

"WARNER" SPILE

If not, you should place your order now; if you do you are sure to have the goods in good time. We can supply you with both

"Eureka" Cast Iron and Steel Sap Spiles

and the famous

Warner Malleable Tin Spile

SAP PAILS

FLARING PATTERN

Made in six sizes, in both straight and flaring patterns. We carry sap-pan materials in both tinned and galvanized Iron as follows—

Tinned Iron	Galvanized Iron
48x96x20	48x96x22
48x96x22	36x96x22
48x96x24	36x96x24
36x84x22	
36x84x24	

Prompt Shipment Guaranteed

STRAIGHT PATTERN

THE McCLARY MFG. CO.

LONDON, TORONTO, MONTREAL, WINNIPEG, VANCOUVER, ST. JOHN, N.B., HAMILTON, CALGARY

"EVERYTHING FOR THE TINSHOP"

The Long and the Short of It

"IT" MEANS

THE "LONG"

is the time it wears. It stands the test of heat and frost, wind and rain, and give your customers many years of solid satisfaction.

THE "SHORT"

part is the time needed to put it on. A great deal of time—which means a great deal of money—is saved to the man who uses

SHIELD BRAND READY ROOFING

LOCKERBY & McCOMB
65 SHANNON STREET
MONTREAL

Ventilation is Important

as a feature of modern buildings.

There are many buildings and residences whose occupants are afflicted with drowsiness and headaches, because the ventilation is imperfect or has been altogether neglected.

THE AEOLIAN VENTILATOR

is unexcelled as a means of producing pure air in large buildings, closet rooms, vaults of churches, schools, factories, stables, etc.

As will be seen from the accompanying illustration, **The Aeolian Ventilator** is very ornamental in appearance, while also possessing exceptional lasting qualities.

We like to talk about the "Aeolian" Will you write us for particulars?

Read this testimonial:

Ottawa, June 4th, 1903.

Messrs. The J. W. Harris Co., Ltd., Montreal.

Gentlemen,—In reply to your letter of 1st of June, inst., beg to say "Aeolian" Ventilators we got from you have always proved satisfactory to us and we can recommend them when such an article is required.

Yours respectfully,
McKINLEY & NORTHWOOD.

The Price is Inconsiderable in Comparison with the Results

MANUFACTURED BY
THE J. W. HARRIS COMPANY, LIMITED
SUCCESSORS TO LESSARD & HARRIS, CONTRACTORS
MONTREAL

THE AEOLIAN VENTILATOR
(Can be supplied in copper, if so desired.)

Peerless Woven Wire Fence

Here is a fence that will sell. Once up no need for repairs. It lasts for years—just what you are looking for. It's horse-high, pig-tight and bull-strong. No stock can get over, under or through it. The most unruly animal can't butt it down and there are no barbs to tear and injure stock. It's

WIND AND WEATHER PROOF

The Peerless lock holds it perfectly rigid and the wires can't slip up, down, nor sidewise. Cold weather won't snap it and hot weather won't make it sag. It's the greatest thing out for farmers, stockmen and all other fence users which makes it a red hot proposition for dealers. It comes in rolls all ready to put up. It's easy to handle. Write for prices.

The Banwell Hoxie Wire Fence Company, Ltd.
Dept. J, Hamilton, Ont.

Per Rod, You Make More Profit on the "Dillon"

than any other fence. For just this reason: The "Dillon" **is all fence,** no locks, ratchets, pickets, etc., to increase the consumer's price and reduce the dealer's profit.

Rolled out in your store those No. 9 High Carbon wires, which are 500 lbs. stronger than ordinary fence wire, show their value. And mind you, all the profit isn't on your side. Your customer shares it with you. But then **that's your profit** after all, isn't it?

THE OWEN SOUND WIRE FENCE CO., Limited
OWEN SOUND, ONT.

Sold by { Messrs. Caverhill, Learmont & Co., Montreal
Messrs. Christie Bros. Co., Limited, Winnipeg

Persons addressing advertisers will kindly mention having seen their advertisement in this paper.

Persons addressing advertisers will kindly mention having seen their advertisement in Hardware and Metal.

Young Man! Do You Want to Make a Profitable Change?

"A rolling stone gathers no moss," but it acquires a shine that enhances its value—where it has value to begin with.

Knowledge wins your business. Knowledge is the result of intelligent experience. To gain enough experience to make one's services valuable it is necessary to make a change occasionally. But take no chances. State what sort of position you desire in the want columns of Hardware and Metal, and you are practically assured of a profitable change.

Condensed advertisements in Hardware and Metal cost 2c. per word for first insertion and 1c. per word for subsequent insertion. Box numbers 5 cents extra. Write or phone our nearest office.

HARDWARE and METAL, Toronto, Montreal, Winnipeg

GOTHIC CEILING PLATE NO. 5002

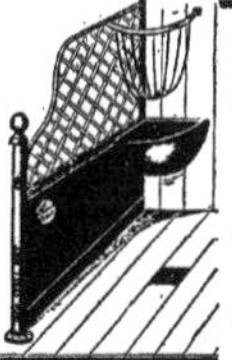

Some Present Trade Requirements

Tight Money for a Time—Law of Averages Calls for Lean Years—Corruption Follows Paternalism—Iron Bonuses Unfair to Ontario—Relations With the Mother Country Should be Undisturbed—Railway Commission has Raised Freight Rates—Should Have More Export Help.

By Peleg Howland, Retiring President of The Toronto Board of Trade.

The large increase in deposits in the chartered banks would indicate that the people of Canada generally are thrifty and saving, notwithstanding the extraordinary inducements offered for speculation in mining and other enterprises. All these deposits, however, find ready employment in the expansion of business as shown by the increase in current loans. The difficulty among bankers now would seem to be to keep their reserves in safe proportion to their liabilities and find means for the constantly-increasing demands of their customers; in consequence a period of dear money is probable for at least some time to come, with the possibility of curtailment in a degree of the prevailing activity.

While the railway construction going on not only in the new provinces but in the older portions of the Dominion and the constantly-increasing immigration must help towards continued prosperity, it must be remembered that we have had a series of unusually good crops, upon which we are largely dependant, and that the law of average would lead to the expectation of lean years; that the prices of nearly all commodities, including labor, are dangerously near the point of inflation; that our national expenditure is becoming extraordinarily large with immediate prospects of increase; that we are a debtor nation, and that the drain upon us for interest on our foreign borrowings, national, provincial, municipal and private, must be very heavy. It would not, therefore, seem out of place to echo the note of warning already given out by financial men and urge reasonable caution.

The United States have during the past few years been experiencing a prosperity perhaps proportionately greater than our own, they have had, however in the process of developing their almost limitless resources, periods of extreme depression. These in the past used to precede ours almost a year; whether that will continue to be the case cannot be foretold. It would be well, though, to keep a careful watch upon the conditions there for any sign of change.

Too Much Reliance on Law.

Comment is timely on the deplorable revelations made during the past year by the investigations into some of our commercial, financial, insurance and political affairs. While doubtless periods of prosperity and speculation must have accompanying evils of this kind, the conclusion cannot be avoided that they are also fostered by paternalism, whether taking the form of protection by customs duties, by bonuses, national or other, or of sumptuary laws that tend to make the individual lean on government help for his rate of wages, his profits or for the safety of his investments, to the weakening of the sturdy independence, self-reliance and toughness of fibre which is of such essential importance. The lesson taught in my opinion, is against any attempt to effect a cure by further adding to the criminal code, increasing the list of restrictions or by making government still more responsible.

Opposed to Government Bonuses.

I am strongly opposed to government bonuses generally as being one of the worst forms of class legislation, and I desire to call attention to the fact that the manufacture of the base products of iron and steel in Canada is now being fostered by a series of bonuses accompanied by customs restrictions, and very large and increasing sums are being exacted annually from the whole people for the benefit chiefly of an industry located in the far east, which, if capable of producing as represented when established, should be able to stand alone. Ontario is naturally situated to draw

PELEG HOWLAND, TORONTO.

its raw materials for the more finished products in iron and steel from Pennsylvania, Ohio and West Virginia, and if allowed to do so would have chances of very large growth in all the smaller industries, which are really of so much more benefit to a community than the immense establishments, to a large degree monopolies, which produce the coarser materials. It would seem unwise on our part not to avail ourselves of our neighbor's coal and iron as long as they remain the cheapest products of the kind, rather than tax ourselves to upbuild an industry a thousand miles away. We are, as yet, the largest consumers of this products, and bear the greater portion of the taxation, but are handicapped by the double freights in getting back into the east, which has as well an advantage proportionately in freights westward. Ontario must sooner or later awaken to the injustice of the present policy, which takes from it the chance of supremacy that naturally belongs to it, with the accompanying danger of sectional difference."

Opposed to Preferential Treatment.

It may perhaps be proper for me to say here that personally I do not favor the resolution urging preferential treatment of our products in the old country, nor that indicating that Canada should contribute to the defence of the empire, feeling as I do that we are able to progress without favors, and believing that entanglements must arise from any attempt to alter our relations to the mother country or to make contributions without representation, with the grave danger of our either breaking away entirely or losing that measure of independence which is of so much importance to us.

In regard to representations made before the Railway Commission that an investigation should be made into the freight rates of Ontario and that there is a discrimination in favor of Montreal in the rates to points both east and west of Toronto, no judgment has yet been given.

The conclusion is forced that the commission has not the necessary expert assistance to get through its work, and I would advocate representations to the government for such changes as will make the commission more efficient, including the extension of its jurisdiction to Dominion Government railways, which are now in my opinion, improperly exempt. The question of rates and of discrimination is of more importance to the people at large than the rights of one railroad to cross another, or matters of a similar nature which seem to chiefly occupy the attention of the commissioners.

The railroads have recently submitted to the commission schedules of rates for merchandise bound inward from many places in the United States; they are substantially higher than those now existing, and their adoption is suggested in order to overcome the decrepancy in the rates existing from American and Canadian places on the border, those from the latter being now much the higher. The co-operation of the manufacturers is sought by placing certain of their raw materials on a special basis, which will mean practically no advance Generally speaking, however, their adoption will mean higher freights for merchants and consumers on inward stuff from the United States. A protest has been entered against these being considered at all until a general and thorough inquiry be made into the fairness of the rates now existing throughout Ontario. Any adjustment of rates upward should be resisted by every possible means.

Commission Raises Rates.

I would also call attention to the arrangements existing between railroads which prevent them from competing in rates with one another. Such combinations are surely as harmful as arrangements affecting the prices of commodities, and are as much entitled to be forbidden by law. The Railway Commission has undoubtedly up to the present been an aid to these freight associations, and has served to increase rather than reduce the rates, thus helping to add to the revenue of the railroads, and it is questionable if the uniformity obtained is a compensation to the people generally for the additional cost of the service.

Unfair Discrimination.

The custom of charging the ordinary shipper from various centres in the northwest in some instances more than double the rates given to wholesale-dealers located in those centres shipping to traders, and which is a factor in the many difficulties experienced by the eastern merchant in getting into that country with his goods, still exists, and should have the attention of the commission at once. An eastern merchant shipping to—I will say Calgary—a carload of goods to a distributing agent to be broken up and reshipped to various consignees at points in the neighborhood, must be on a list as carrying a stock in Calgary, and the consignees must be traders before these consignees can get the benefit of the so-called traders' rates, otherwise being charged the ordinary class rates. The following examples of these rates, given as existing on January 5 last, will give an idea of the differences between the traders' rates and the so-called class rates which are paid by the ordinary shipper or consignee, the figures being per hundred pounds:—

Calgary to Revelstoke—Class rates—First, $1.35; second, $1.13; third 90c.; fourth, 68c. Trades rates—First class—64c.; second, 55c.; third, 45c; fourth, 37c. Calgary to Fernie, B. C.—Class rates—First, $1.27; second, $1.06; third, 85c.; fourth, 64c. Traders' rates—First class, 61c.; second, 53c.; third, 44c.; fourth, 37c.

Favors Local Merchant.

Similar conditions exist at Winnipeg, Brandon, Regina, etc., though the differences are not so extreme. This system meets with the favor of the local wholesale merchant, as it protects him from his eastern competitor, and also of the retailer in the smaller towns, as he is in his turn protected, the tendency being to force the consumer to patronize the nearby merchant. It is improper, however, and contrary to the principle upon which charters are granted It is not practised in the east, the tendency here being rather to give the adventage if any, to the more distant place. It should be stopped.

Reciprocal Demurrage.

The country suffers every year from a shortage of cars and from the inability of the railroads to move promptly the rolling stock they are able to supply. This year the trouble has been particularly serious, and has been felt by all classes of the business community, being one of the causes of the scarcity of money. The railroads now very properly enforce demurrage charges on customers detaining cars over a certain specified time, but are under no penalty themselves if shipments remain in their possession longer than is reasonable. Any system of reciprocal demurrage will be vigorously opposed by the railroads and many strong arguments will be used against it. It seems reasonable, however, to think that what works well in one way should in the other, and that its adoption might lead to such exertion on the part of the companies as will overcome, at least to some extent the evil complained of. The system has already been advocated by board, and it is well worthy of further consideration.

Anti-Dumping Law Unfair.

The attention of the Toronto Board of Trade Council has recently been called to the anti-dumping clause in the custom tariff, the application of which is being extended at the present session of Parliament. 'I cannot urge too strongly that the matter be thoroughly investigated. A regulation that restricts a merchant's right to buy as cheap as his ability and capital will permit, and penalize him if he does exercise such ability by requiring, besides the regular duty, the payment in the shape of special duty of any concession he can obtain in price below what the authorities fix as the home value of the goods in the country of export up to fifteen per cent. of such home value, and compels the taking of such an oath as the following—"That no arrangement or understanding affecting the purchase price of the said goods has been or will be made or entered into between the said importer and the exporter, or by anyone on behalf of either of them, other than as shown in the said invoices, either by way of discount, rebate, salary, compensation or in any other manner whatsoever " which forms a part of the oath taken on the entry of goods for consumption—approaches too nearly to despotism, surely, to be borne by this community, and is besides calculated, as the oath suggests, to encourage fraud and dishonesty, and is not in the interest of the consumer. In my opinion, as a restriction on trade, no combination can compare with it. If the manufacturer requires more protection it should be afforded him in some less objectionable way.

NEW INVENTIONS.

A hydrocarbon-burner, invented by J. N. Blain and O. H. Smith, Ottawa, Kan., is said to present efficient means for carbureting the air let into the burner. A further object is to construct the fire-pan so as to enable the same to be readily inserted in an ordinary stove or furnace and to provide the same with special means for facilitating the gasifying of the fuel when fed thereto.

A newly patented spring fish-hook, invented by A. S. Martin, Geneseo, Ill., relates to hooks of the type in which a spring-actuated auxiliary hook is released when the fish strikes at the bait, thus allowing a plurality of hooks to obtain a firm hold upon the fish and thereby prevent its escape. The device is especially useful in the catching of quick-striking fish, such as trout or bass. The hook is only operated by actual contact.

A. Allen, Lead, S.D., has a new invention which refers especially to that class of saw frames for receiving a detachable blade. The object is not only to improve the construction of saw-blades of this character, making them more convenient to handle, cheapening and simplifying them, but also to provide a novel and easily-operated means for stretching the saw-blade in the frame after it is applied thereto.

B. Stoll, Gardena, N.D., has a new carpenters' tool, the object of which is the provision of a new and improved carpenters' tool more especially designed for pressing floor-boards, sheathing-boards, and the like, into proper position for nailing. It is very simple and durable in construction and can be cheaply manufactured.

A patent has been granted J. A. R. Damonte, New Orleans, La., for a magnetic tack hammer in which the tacks are placed in the magazine indiscriminately, and when the hammer is brought up to a striking position it causes the tacks to scatter, and on the outward swing or striking movement they find their way one at a time through the tube and slot in the handle and down into the slideway.

W. McCormick, Hillyard, Wash., has secured a patent for a bolt-extractor, designed, primarily, for the removal of crown-bolts from the crown-sheets of locomotives, although its use is not limited to this particular class of work, since it will be found to be an effective means for removing bolts in other relations, especially those with round or other forms of heads on which an ordinary type of wrench cannot obtain a purchase.

D. W. Patton, Moberly, Mo., has invented a nut lock which when in use the nut is screwed on the bolt the distance desired and turned so that the flattened end of the bolt lies parallel to the grooves in the nut. The staple is inserted in the grooves, thus holding the flattened end portion of the bolt between the two arms thereof and preventing its rotation. Means are adopted to engage the edges of the nut and prevent accidental displacement of the locking member. The latter being preferably of wire, its outer ends may be bent out of alignment after its insertion, thus serving as an additional means for holding the same in place

---

WIRE ROPE FOR HAY FORKS.

The B. Greening Wire Company, Hamilton, report that during the last few years they have sold wire rope to be used as hauling rope on hay forks, and last year while the sale was not large, they sold more than they had ever sold before, this indicating that wire rope is a success for this purpose. In a circular letter the company points out that as Manilla rope is generally sold on a very fine margin, the hardware trade should be interested in substituting a wire rope that will bear the merchant a better profit for a line that is cut so fine. An opportune time to look into this subject is before the farmers get busy with their spring work. Wire rope is also used for the track, to take the place of rod, but this is a different make of rope altogether to the hauling rope:

PROPOSED CHANGES IN THE NEW TARIFF

Hon. Mr. Fielding has given notice of about 120 changes in the now tariff schedule brought down in Parliament on November 29th last. No official explanation of the changes will be given until the resolutions are moved in the House. For the most part they are of a technical or relatively unimportant character, designed to meet the objections of manufacturers and others on minor points.

Changes in Iron Duties.

Some important changes are made in the duties in iron and steel. A new item is introduced comprising flat-eye bar blanks, not punched or drilled or rolled ; edge plates of steel over twelve inches wide for use in steel structural work or car construction, the preferential tariff being fixed at $2 per ton, the intermediate at $2.75, and the general at $3.

On scrap iron and similar material the British preference has been reduced to 50 cents per ton. The duty on rolled iron or steel beams and other rolled shapes other than railway bars or rails has been fixed at $2 per ton under preferential, $2.75 intermediate and $3 general. Boiler plate not more than thirty inches wide, wrought seamless tubes for boilers, wire rigging, and steel wire to be used in the manufacture of rope have been placed on the free list.

The changes, as they affect the hardware trade, are as follows :

Item 261 is amended to place spirits of turpentine on the free list instead of dutiable at 5 per cent.

Item 270 provides for a reduction in the general tariff on crude petroleum, gas oils other than naphtha, benzine and gasoline, the duty to be one and a half cents a gallon instead of two and a half cents.

Item 281 is changed to read :—"Fire brick of a class and kind not made in Canada, preferential free, others 5 per cent. each." The present tariff rates are 5, 7½, and 10 per cent.

Item 316A is a new one. It reads :—"Incandescent lamp globes for use in the manufacture of incandescent lamps, and mantle stocking for gas lights, 5, 7½ and 10 per cent."

Duties on Glass.

Item 326 adds to the dutiable articles of glass of all kinds, bottles, lamp chimneys, globes, etc., the words, "blown glass, table and other cut glass ware."

Item 344 changed to read :—"Tinware, japanned or not, and all manufactures of tin, n. o. p., 15, 22½ and 25 per cent."

Item 355, which provides for the free importation of Britannia metal and German silver, strikes out the words "or bars" when it applies to the free importation of German silver in bars.

Item 374, on iron or steel scrap, wrought, being waste or refuse, including punchings, cuttings, or clippings of iron or steel plates or sheets having been in actual use ; crop ends of tin plate, bars, or of blooms, or of rails, the same not having been in actual use, the British preferential has been reduced from seventy to fifty cents.

Item 379 has been changed to read "rolled iron or steel beams, channels, angles and other rolled shapes of iron, steel not punched, drilled or further manufactured than rolled, weighing not less than thirty-five pounds per lineal yard, not being square, flat, oval or round shapes, and not being railway bars or rails, per ton preferential, $2 ; intermediate, $2.75 ; general, $3."

Item 379 is a new item, as follows : "Flat eye bar blanks, not punched or drilled, and universal mill or rolled plates of steel over twelve inches wide, for use exclusively in the manufacture of bridges or of steel structural work or of car construction, per ton, preferential, $2 ; intermediate, $2 75 ; general, $3.

Boiler Plate Free to Manufacturers.

Item 380 is changed to read as follows :—"Boiler plate of iron or steel, not less than thirty inches in width and not less than a quarter of an inch in thickness, for use exclusively in the manufacture of boilers, under regulations by the Minister of Customs, is made free in all the tariffs."

Item 381 has been changed to read :—"Rolled iron or steel plates, not less than thirty inches in width and not less than ¼ of an inch in thickness, n. o. p., per ton, preferential, $2 ; intermediate, $2.75 ; general $3."

The item formerly was "Plates of 48 inches preferential, 5 per cent.' intermediate, 10 per cent., and general, 10 per cent."

On item 384 "strips, polished or not," have been added after "steel sheets." Preferential, free ; intermediate, 5 per cent.; general, 7½ per cent. "Flat galvanized iron or steel sheets" have been struck out.

Item 386 has been made to read :—"Rolled iron or steel and cast steel, in bars, bands, hoops, scroll, strip, sheet or plate, of any size, thickness or width, galvanized or coated with any material or not, and steel blanks for the manufacture of milling cutters, when of greater value than 3½ cents per pound : preferential, free ; intermediate and general, 5 per cent. each."

Item 387 is a new one, and reads :—"Steel in bars or sheets, to be used exclusively in the manufacture of shovels, when imported by manufacturers of shovels : preferential, 10 per cent.; intermediate, 12½ per cent ; general, 15 pet cent."

Item 395—Wrought or seamless iron or steel tubes for boilers, n o.p., under regulations prescribed by Minister of Customs, flues and corrugated tubes for marine boilers, are made free in all three tariffs. It was : Preferential, free; intermediate and general, 5 per cent. each.

Item 397—Tubes of rolled iron or steel not joined or welded, not more than 1½ inches in diameter, n.o.p., have been made free in all tariffs.

Increase on Coil Chain.

Item 410—Coil chain, coil chain links and chain shackles of iron or steel five-sixteenths of an inch in diameter and over, preferential, 5 per cent. as at present; intermediate increased from 5 per cent. to 7½ per cent , and general from 5 per cent. to 10 per cent.

Item 411—Malleable sprocket or link belting chain is made free in all the tariffs. It was 15 per cent., intermediate 17½, and general 20 per cent.

Item 151A is new. "Stoves, urns of metal and dove tails, chaplets and hinge tubes of tin for use in the manufacture of stoves, preferential 5 per cent., intermediate 7½ per cent , general 10 per cent."

Item 460—After the words "processes in" the words "iron or copper" are struck out and the word "metals" inserted instead.

Item 461A is new. "Iron or steel pipes, not butt or lap welded, and wire-bound wooden pipe not less than 30 inches internal diameter, when for use exclusively in alluvial gold-mining, preferential 5 per cent., intermediate 7½ per cent., general 10 per cent.

Item 462, which formerly included only "blast furnace slag trucks," is amended to include blowers of steel of a class or kind not made in Canada, for use in the smelting of ores or in the reduction, separation or refining of metals, rotary kilns, revolving roasters and furnaces, of metals of a class or kind not made in Canada, designed for roasting ore, mineral, rock or clay, blast furnace, slag trucks and slag pots of a kind not made in Canada, made free in all tariffs.

Item 524A is new. Seamless cotton or linen duck, in circular form of a class or kind not made in Canada, for use in the manufacture of hose pipe, made free in all tariffs.

Item 682—After the words Manila rope insert "not exceeding 1½ inches in circumference"; before the words "fishing nets" strike out the words "deep sea."

Drawback Increased to 99 Per Cent.

Schedule "B" covering goods subject to drawback for home consumption is amended so as to make the portion of duty (not including special or dumping duty) payable as drawback, 99 per cent. instead of 95 per cent.

Item 1,002 in this schedule is changed to read : "Malleable iron castings and pig iron," instead of "rolled iron, rolled steel and pig iron."

In item 1,009 all the words after augur bits are struck out, and bit braces is added to the other items in which drawback is payable.

The following items are also added to schedule "B" as subject to payment of drawbacks :

1,014—Nickel, nickel silver and German silver in bars, rods, strips, sheets and plates, when used in the manufacture of spoons and cutlery, a drawback of 65 per cent.

Item 1,015—Rolled angles of iron or steel, nine and ten gauge, not over one and a half inches wide, and used in the manufacture of bedsteads, are subject to a drawback of 99 per cent.

Item 1,017—Lap-welded tubing of iron or steel, not less than two and a half inches diameter, threaded and coupled or not, testing 1,000 pounds pressure to the square inch, when used in oil or natural gas wells, and for transmission of natural gas under high pressure from gas wells to points of distribution, drawback 99 per cent.

Item 1,018—Machinery imported prior to July 1st, 1908 and other articles not machinery, when entering into the cost of tin plate manufactured in Canada, drawback 99 per cent.

Item 1,019—Bituminous coal, when imported by proprietors of smelting works and converted at the works into coke for the smelting of metals from ores, drawback 99 per cent.

Retail Hardware Association News

Official news of the Ontario and Western Canada Associations will be published in this department. All correspondence regarding Association matters should be sent to the secretaries. If for publication send to the Editor of Hardware and Metal, Toronto.

Officers Retail Hardware and Stove Dealers' Association of Western Canada:

President—A. J Falconer, Deloraine
First Vice-President—J B Curran, Brandon
Second Vice-President—W. M Gordon, Winnipeg
Secretary-Treasurer—J. E McRobie, Winnipeg.
Executive—Alberta, A. E Clements, Olds; C F Comer, Calgary, A. R. Auger, Okotoks.
Manitoba—H S Price, Boissevain; A. P. Macdonald, Winnipeg; O Gilmer, Winnipeg
Saskatchewan—G K. Smith, Moose Jaw, S A. Clark, Saskatoon; J. R. Fox, Weyburn.
Officers Western Board (elected by general merchants and hardwaremen in joint session). President, W. G. McLaren, Souris, Man., vice-president, G. K. Smith, Moose Jaw, Sask.; H C. Hamelin, Winnipeg, Man.; secretary, W. A Coulson, Winnipeg, Man.; treasurer, J E McRobie, Winnipeg, Man., auditor J. A Lindsay, Winnipeg Man.
Association offices, 53 Scott building, Main street, Winnipeg.

Officers Ontario Retail Hardware and Stove Dealers' Association:

President—A. W Humphries, Parkhill
1st Vice-President—W G Scott, Mount Forest.
2nd Vice-President—J R. Hambly, Barrie
Treasurer—John Caslor, Toronto
Secretary—Weston Wrigley, 10 Front St. East, Toronto.
Executive Committee—The officers and J. W. Peart, St Mary's; M R Marshall, Dunnville; D Brocklebank, Arthur, F. W Jeffery, Midland, E P. Paulin, Goderich, and Frank Taylor, Carleton Place
Auditors—J W Peacock and C F. Moorhouse, Toronto

NOVA SCOTIA ORGANIZING.

The date for the convention of the new Nova Scotia Retail Hardware Association has not yet been fixed, but announcement will be made in these columns as soon as it is decided upon. At a preliminary meeting held at Halifax, on Jan. 29, the following representative firms were enrolled personally or by letter:

C. P. Moore, Sydney.
A. J. Walker & Son, Truro.
H. A. West, Annapolis.
Douglas & Co., Amherst.
Dunlap Bros. Co., Amherst.
Dakin Bros., Digby.
George McLaren, Liverpool.
C. V. Mackintosh, Liverpool.
Thompson & Sutherland, New Glasgow.
Bent & Cohoon, New Glasgow.
Middleton Hardware Co., Middleton.
J. H. Cameron, Parrsboro.
D. G. Kirk, Antigonish.
K. Sweet & Co., Antigonish.
Frank Powers, Lunenburg.
J. T. Powers & Co., Bridgewater.
A. S. Austen, Halifax.
A. L. Melvin, Halifax.
E. K. Spinney, Yarmouth.
S. A. Crowell & Son, Yarmouth.
Wilcox Bros., Windsor.
Karl Freeman, Bridgetown.
T. P. Calkin & Co., Kentville.
H. A. Patton, Oxford.

Frank Powers, Lunenburg, one of the most progressive stove dealers and heating contractors in the Maritime provinces, has been an active spirit in the work of organizing, according to W. H. T. Spinney, Yarmouth, who writes to the editor of Hardware and Metal: "I thank you for the interest you take in the Maritime provinces. While our business is mostly wholesale, yet we are interested in the retail merchants and wish to assist them all we can that both the jobber and retailer may participate in profits which are due their respective branches."

WESTERN HARDWARE DEALERS PROTEST.

A strong deputation of hardwaremen, representing the Retail Hardware and Stove Dealers' Association of Western Canada appeared before the Law Amendments committee of the Manitoba Legislature last week to protest against some proposed legislation in reference to druggists and early closing. For some time the Winnipeg druggists have been in trouble about the early

FRANK POWERS, LUNENBURG N. S.
Who is taking an active part in the work of organizing the retail hardware trade in Nova Scotia.

closing by-law. Under the Winnipeg city by-law all retail stores are compelled to close at 6 p.m. Under the strict interpretation of the law the drug stores are also supposed to close, although they may open for emergency calls; in practice it is a well understood thing that the drug stores may remain open for the sale of drugs and medicines. Now, the drug stores carry many articles which the also stocked by the hardware and stationery stores and there have been complaints that they have been selling these lines after hours. One or two prosecutions have resulted and the druggists have claimed that they were being prosecuted in a petty manner.

Appeals to the city council accomplished nothing and they accordingly went to the Manitoba Legislature for an Act enabling drug stores to remain open after hours.

The hardware men were alive to a danger in the situation, and accordingly a strong deputation waited on the Law Amendments committee last week to point out the unfairness of allowing drug stores to remain open after 6 p.m. to sell razors, cutlery and other lines of goods also carried in hardware stores that are compelled by law to close at 6 p.m. Among those in the delegation were A. J. Faulkner, president of the Western Retail Hardware Association; J. E. McRobie, secretary; O. Gilmer, J. E. Thomas, A. P. Macdonald, W. A. Templeton, J. E. Riley, A. Dykes, F. W. Weir, E. B. Lennox, W. M. Gordon, C. C. Faulkner, Neil Lightley, Robert Wyatt and Chas. W. Graham.

J. W. E. Armstrong, solicitor for the association, explained the grievances of the hardwaremen. Druggists stock such goods as razors, gasoline, sporting goods, all sorts of oil, cutlery, benzine, etc., all of which are legitimate hardware lines. It would be unfair to allow the drug stores to sell these lines after hours.

It was finally decided to leave the matter with the local councils, who may specify classes of goods which may not be sold in drug stores after hours.

LETTER FROM BARRIE.

The following letter has been sent for publication on this page by one of the members of the Ontario Association:

Sir,—Allow me to congratulate you and the executive offcers of our association for the excellent work done since its formation less than a year ago. I have read with a great deal of interest and benefit of the good work accomplished, as reported in the columns of that excellent journal: Hardware and Metal. In order to keep abreast of the times and to be thoroughly posted in our business it is very essential that we should be strongly organized and kept posted continually by such a valued journal as the Hardware and Metal.

Why, would you believe, when the petition to be sent the Postmaster-General re the Parcel Post Delivery was taken to the business men of our town by the hardware dealers, the other merchants do not know the reason for the action. That in itself is sufficient to convince any sane man alive to his interests how necessary it is to have an organization, with live officers, continually looking after our interests and protecting the same. Then it behooves every hardware merchant to become a member of our organization and strengthen the power of our vigilant officers by sending their application for membership to our secretary and also by reading Hardware and Metal every week.

CONVENTION AT WINNIPEG.

In spite of the very adverse conditions and snow blockades in many districts which prevented a large percentage of the members of the Western Canada Retail Hardware and Stove Dealers' Association from attending the annual meeting, which was held in Winnipeg on Tuesday and Wednesday of this week, there was a fair number on hand to take part in the business transacted at the convention. All of the old officers were elected, as follows:

President, A. J. Falconer, Deloraine.

First vice-president, J. B. Curran, Brandon.

Second vice-president, W. M. Gordon, Winnipeg.

Secretary-treasurer, J. E. McRobie, Winnipeg.

Executive, Alberta—A. E. Clements, Olds; C. F. Comer, Calgary; A. R. Auger, Okotoks.

Manitoba—H. S. Price, Boissevain; A. P. Macdonald, Winnipeg; O. Gilmer, Winnipeg.

Saskatchewan—G. K. Smith, Moose Jaw; S. A. Clark, Saskatoon; J. R. Fox, Weyburn.

Dominion Association Endorsed.

After some discussion, in which many members expressed themselves as gratified with the extension of the retail hardware association work into Ontario and the Maritime Provinces, the suggestion made by the executive of the Ontario Association that a temporary organization be effected to cover the Dominion, by appointing the presidents and secretaries of the Ontario and Western Canada Associations as a provisional committee to act until a permanent Dominion association is formed. This action is, of course, to be endorsed by the convention in Toronto on Good Friday, but it can be taken for granted with a provisional committee composed of President A. J. Falconer and Secretary J. E. McRobie, appointed by the Western Association, the movement in Ontario will fall into line and perfect the proposed organization. Definite steps can then be taken to secure affiliation of the new Nova Scotia Association, and arrange for the organization of the trade in other provinces.

Parcels Post Legislation Condemned.

A strong resolution was passed by the convention calling upon the Postmaster-General to withdraw the legislation which he has outlined in providing for an extension of the parcels post system and the incorporation of the c.o.d. feature in the postal regulations. The sentiment of the delegates was entirely opposed to the measure under consideration by the Postmaster-General, as it would tend to centralize trade in the larger cities and deal a heavy blow at hundreds of small towns and cities in the western provinces, making it much harder for the farmers and other settlers to secure the goods they require in any quantity and with any promptitude. It was felt that the measure would particularly hurt the retail trade and the farmers in the west, and steps will be taken to see that each western member of Parliament is thoroughly informed as to the meaning of the proposed legislation.

Trade Press Excluded.

For the first time in the history of the western association the representatives of the trade press were excluded from the meetings of the convention. This action came as a surprise, as it was almost entirely due to the assistance given by the trade newspapers that the association had been able to organize itself so strongly in such a short period of time. At the time of writing, actions taken by the convention at its later sessions, are not known. It is understood, however, that a lengthy discussion took place on the subject of price agreements covering the trade in the west. This feature has been emphasized considerably, and much difference has resulted between the members of th association in recent months over disagreements and price-cutting. In some states on the other side, price agreements are in force covering the large districts, but it would seem that it is impossible to arrange such matters satisfactorily in Canada, especially on a large scale. In Ontario it has been felt that all price agreements should be entirely arranged by the local merchants in any one town or district, the provincial association having nothing to do whatever with the arrangement of minimum prices. If the association succeeds in preventing the jobbers from selling to the consumers without protecting the retail trade, it has more than justified its existence.

Association Funds.

A financial report will be published next week along with a full report of the convention's doings, as supplied by the proposed committee. It is understood, however, that a large percentage of the members have been very slow in forwarding their fees to the secretary, and it has been felt by all who have the interests of the trade at heart that this is a serious mistake, as the trade stands to give much through the existence of the asociation, certainly far more than the membership fee. It would be a serious mistake for the members to allow the work of the association to be handicapped in any way by lack of funds.

MAIL ORDER HOUSE LOSES.

A decision has just been rendered by Judge Carland, of the United States Court, Sioux Falls, S.D., denying the application of Montgomery Word & Co., Chicago, for a temporary injunction restraining the officers and directors of the South Dakota Retail Merchants' Association and the editor of the Commercial News from continuing their alleged boycott against the catalogue house. Because some jobbers and manufacturers refused to sell goods to the catalogue house the company alleged that it had been damaged through being unable to purchase the goods required by it to fill the orders of its customers, and accordingly brought suit against the alleged offending parties, as has been noted in our columns. The importance of the decision is the fact that had the plaintiff won in this case there was a general belief that the intention of the company was to institute similar actions in other states where merchants' associations are waging a warfare against the encroachments of the mail order houses. The association made up a list of the jobbers and manufacturers who would not agree to sell to the retail trade exclusively, and circulated this list amongst about 1,500 retail dealers. The mail order house claimed the list to be a "black list," but the court decision indicates that the retail trade can continue to decide who they shall or shall not buy from—as well as allowing them to take joint action in this matter.

WRITE EARLY AND OFTEN.

Everyone knows that the mail order business done by the few large department stores is enormous, and that it has effected great injury to the small merchants throughout the length and breadth of the country. However, it is small compared with what would be done by them, and exclusive mail order houses that would spring up, if the Postmaster-General undertook to facilitate it by putting in force the parcels post c.o.d. legislation still being considered by him.

That the catalogue houses are behind the extension of the parcels post system there is little doubt, and it must not be supposed that they would be satisfied to have the Government make a profit out of the delivery business. Of course, it will be argued that lower rates would do the most good to the greatest number of people, a thin disguise to the real object in view. The mail order people care nothing so long as their own ends are served. What they want is the extension of their operations at the expense of the national treasury and the retail merchants.

Consumers are not asking for this change in the postal system, and they do not need it. The country is better without it. Better far, in the interest of the national life, a thousand independent business men than one business man with a thousand or five thousand employes. The movement, if successful, means the pulling down of small institutions all over the Dominion to build up a few mammoth houses in the big cities. Surely the cities are draining the country too fast already.

All that any local merchant asks is a fair chance, and he would certainly not be getting it under the provisions of this bill. He has it in his power to turn aside the blow which it aims at him. Let him do so by seeing that a stream of letters of protest continues to flow into the office of the Postmaster-General until the word is given that the scheme is as dead as it deserves to be.

Write three or four times if necessary. Postage costs nothing, as the letters O.H.M.S. are as good as a two-cent stamp when the letter is written on matters of Government legislation.

Dealers and Their Stores

Retail Dealers are Requested to Supply Information Regarding Business Changes, Sketches of the Development of Large Retail Businesses, Etc., Accompanied by Photos or Floor Plans and an Outline of the Methods or Systems used in Conducting Their Business.

ARTHUR'S DEPARTMENTAL STORE.

It is doubtful if there are many stores in the smaller towns in Canada more generally and favorably known that that of D. Brocklebank & Son, at Arthur, Ont. We hope soon to be able to publish photos and floor plans of the building, the dimensions of which are 46 x 100 feet with a floor space totaling 125,000 feet.

The store is light and airy with splendid ventilation. The ceilings and partitions are metallic, while the establishment throughout is heated by steam and lighted with electricity (comprising nearly 100 incandescents). Modern lavatories, both up-tsairs and down for the use and convenience of customers have been installed. A splendid cash carrier system, converging in different directions to the cash office, is most serviceable to waiting customers. In the rear is an elevator for taking goods up and down stairs, while a separate cement building, 24 x 36 feet, is used for storing oils and heavy iron, and for a tin shop. By a self-measuring tank system oil is carried from the storehouse into the main building. This materially reduces insurance rates. Another storehouse, 24 x 30 feet, is situated further in the rear.

In the basement a salesroom, 30 x 50 feet, is devoted to crockery, lamps, churns, washing machines and woodenware, stocks of glass, nails, rope, etc., occupying the remainder of the cellar. The first floor is handsomely fitted up with plate glass silent salesmen and modern shelving. On the left side a varied stock of general hardware, cutlery, silverware, clocks, and paint, while crockery, tinware and enamelled ware occupies the centre of the floor, these household lines being in charge of a bright saleslady, who has made her department attractive and popular with the women of the town who buy kitchen goods. The right side of the main floor is given over to large stocks of clothing and men's furnishings, while in the rear a complete stock of harness is carried. Upstairs there is a display of stoves and heating goods comprising several of the leading makes, and, in addition, a line of sewing machines and bicycles. Furnaces and hot water boilers are carried in the tin and work shop in the rear. All the buying is done by E. W. Brocklebank, the junior member of the firm, and in keeping with the progressive methods of the company credit is eschewed and the cash system in force.

D. Brocklebank, whose likeness is reproduced, is the head of the firm, he being well known to the trade as one of the executive officers of the Ontario Retail Hardware Association. He was born, 51 years ago, at Malton, Peel county, his experience comprising four years as a school teacher, two years with Beatty Bros., Fergus, and from 1884 to 1892 in the hardware business at Arthur. For a year he published a paper in Arthur, then spending three years in the dry goods business at Simcoe, in 1896 returning to Arthur and buying out H. W. Gillrie's hardware store. In 1905 he built the Brocklebank block and took his son into partnership.

Mr. Brocklebank is also part owner of the Arthur and Wingham flour mills and has given his fellow citizens three

D. BROCKLEBANK, ARTHUR
Member of the Executive Committee of the Ontario Retail Hardware Association.

years service in the town council and six on the public school board. Business interests have prevented him devoting much time to public life, but those who know him, either as a citizen of Arthur, as a business man, or as an officer of the Retail Hardware Association, realize that he possesses the qualities which would make it possible for him to serve with credit to himself in any capacity his friends placed him.

MITCHELL STORE ENLARGED.

According to travelers, one of the most attractive stores in Western Ontario is that of F. A. Campbell. The property on the east side, lately purchased from the Burret estate, has been connected with the old store by large openings in the dividing wall and both sides have been thoroughly refitted in the most up-to-date style, the facing of the shelving and drawers being hardwood. The large business done by Mr. Campbell left him cramped for room, but hereafter he will have no reason to complain in this respect, for with a frontage of 40 feet and a depth of 100 feet, with two stores in the rear and a large cellar, he will have all the accommodation required for his growing business. In addition to hardware he will hereafter carry a full line of tin and graniteware, and in fact almost everything tht may be found in city stores.

Mr. Campbell is one of the most prosperous and wide awake hardwaremen to be found anywhere; is deserving of success, for his aim has always been to handle the best, buy for cash and please his patrons.

ENTERPRISING YOUNG MEN.

The Peterborough Hardware Company will in future be under new management, four prominent young business men of about 30 years of age who have been connected with this firm for a number of years, having taken control. The new company will be composed of V. Eastwood, president; Manson Comstock, managing director; Harry Morgan, vice-president; C. W. Blewitt, secretary-treasurer. These young men are all thoroughly acquainted with every detail of the business, having started in as clerks in the store when young boys and by hard and close attention to business combined with natural ability, climbed to their present position.

The Peterborough Hardware Company is one of the best known retail and jobbing businesses in Eastern Ontario, having been established in 1893 by the amalgamation of the McKee & Davidson and Fortye & Phalen firms.

During the recent cold snap at Vancouver a peculiar acident occurred to J. A. Flett & Co.'s new hardware store on Westminster avenue. The store is located in a brick and stone block and the store windows have as a foundation long stones laid in the ground. The frost raised these stones, and with the weight of the upper story pressing from above and the frost driving the stones upward the large plate glass windows buckled and suddenly crashed into a thousand fragments. Manager Owens feared that the front was in danger, and promptly got out carpenters and propped up the entrance. The remaining plate glass has been protected by being loosened so as to be capable of instant removal, and the owners of the building notified. The loss in the glass is fully $150.

Effective Hardware Advertising

We purposely reproduce D. Brocklebank & Son's ad. this week because it contains the elements of strength and weakness in a peculiar degree. The man who wrote this ad. is capable of turning out first-class copy—copy strong and effective enough to fill his store pretty regularly with eager buyers, and there is no reason under the sun why he should not do so. He has grasped the primary fact of all effective advertising, viz.: that reason why copy is the essential thing. The border and the set-up stand in the same relation to a good ad. as the frame and point-of-view to a good picture. A Turner landscape remains a Turner landscape it matters not how dilapidated the canvas and frame may be.

Reasons Why You Should Buy at This Store.

That's good stuff—just the sort of stuff the economic man or woman looks for in an ad. Both say: "Mr. So-and-So advertises all the time; but he doesn't give us any very substantial reasons for dealing at his store."

This sort of buyer—male and female—is a pretty numerous party in every community. "Reasons why" appearing under the head of Brocklebank's store news attracted him. And Mr. Brocklebank can rest assured that his ad., in this particular instance, was read. That's the half of the fight. You would all be content to worry along, particularly if you were convinced that your ads. were read.

* * *

"Because everything is new and up-to-date." That is a pretty attractive reason; but your next reason, Mr. Brocklebank, is in no way illuminating, or fetching. Your prices may be "always right," but when it is possible for a man to state exactly what these "right prices" are in round figures he should always do so. Suppose one of your citizens had wanted a stove about the time your ad. appeared. You have advertised standard stoves. But if a competitor had advertised the same line of stoves—specifying the make and prices in every case—you can rest assured that the fellow on the stove hunt would not have called on you first, and probably—very probably, not at all. Stoves are thoroughly and systematically advertised by the makers. The people know stove values. And they would have appreciated your range and stove talk much more if you had stated the prices.

This fault holds all through the ad. When I want an article I look through the publicity columns of the newspapers and I go to that dealer who frankly states the price. And, mind you, I seldom go to the man who has an article marked down too much. Money does talk. But human nature makes me avoid the low-prices man just as much as I avoid the no-price man.

* * *

Let me state right here that I have touched the point where the big mail order houses get ahead of the country and small city dealers every time. The mail order house sends out an attractive catalogue once or twice a year and backs it up with circulars and other printed matter of a high order. That catalogue describes an article attractively and states the price in almost every case. The buyer knows exactly what he has to pay for the article he wants. You can't afford to produce either a handsome catalogue or an attractive circular, and I am quite sure that the publicity columns of your daily papers will pay you best at any rate—if properly used.

* * *

For instance, Mr. Brocklebank advertises a good line of sewing machines.

Brocklebank's Store News

Reasons Why You Should Buy at This Store!

Why? Because Everything is New and Up to Date
Reason The Prices are Always Right
Hardware Largest Assortment in Town
Stoves and Ranges A Most Complete Selection to Choose from
Ready-Made Clothing Always the Best Quality and the Newest Styles
Crockery Largest Assortment in Town

Ranges—Do you need a Range this winter? Better buy one at once, whether you buy from us or not. All Ranges have been advanced 5%, this means from 2 to 3 dollars on a range. We have quite a few in stock [illegible] about 20 but they are moving out quickly one or two a day. These sold up goes the price. You can find an excellent assortment here including Happy Thought, Empire Queen, [illegible], [illegible], Pandora. Our prices will stand comparison.

Harness—[illegible] compare prices to see how much you can save by buying your harness here? Ever enquire from someone who knows whether or not it isn't just a little better than you've been buying? Ever look through our harness department and see the kind of stock and workmanship in them? [illegible] We guarantee every strap and buckle we sell, we sell everything in harness that you might want.

Coal—You ought to have your bin full. We have Scranton Coal. The clean kind.

General Hardware—You can find as complete and as well assorted a stock of hardware as can be seen in Ontario. Everything that is usually found in a hardware store is here. We invite comparison of prices. Quality is our first consideration always.

Coal Heaters—Same thing happened them as the Ranges. Up [illegible]. You might as well save a few dollars by buying now, before the other fellow gets the snap. No doubt about finding what you want here. [illegible] of choice. Look em over.

Coal or Wood Heater—We have lots of these. You'll have no trouble finding the one you want. We also have several wood stoves in new and second hand cooks and heaters.

Sewing Machines—You can see the most up-to-date sewing machines on the market here. The White Sewing Machine has some points about it you will find in no other. You may have one on trial. [illegible] for 5 years.

Crockery—Have you looked through our crockery yet? You are welcome whether you buy or not. Most ladies are interested in nice crockery. You can see it here. Everything in crockery at the lowest prices you ever heard of.

Ready-to-Wear Clothing—There is nothing in the shape of mens ready-to-wear clothing you cannot find here. You can save from 3 to 8 dollars on a suit, proportionately on an overcoat. Why not see it. Every person is pleased who bought. Everything in Men's furnishings.

D. Brocklebank & Son

Every young matron wants a sewing machine. She wants to know "the points" about this particular machine—wherein it differs from other makes and if these points will make it a more useful instrument than Mrs. Smith's machine is to her. And she wants to know the price before she'll ready your descriptive matter—because she has a handsome catalogue at home which states the prices quite fearlessly and describes several good machines so thoroughly and interestingly that she feels like borrowing the money to secure one of these catalogue machines right away.

* * *

The dealer must fight the catalogue house through the publicity columns of his local papers. And he must use the same kind of ammunition as the mail order men, if he wants to exist at all. That ammunition is—money-talking, descriptive advertising. "Brocklebank's Store News" is effective advertising; but I am convinced that a man with Mr. Brocklebank's apparent grip of the subject he can make it about ten times more effective.

Advertising Schemes.

Another dealer has sent in several ads. for criticism. The ads. are deliberately printed with typographical errors and a prize of $3 worth of goods is offered to the one sending in the largest list of errors occurring "in this space between now and August 21st."

Who do you suppose is going to pick out mistakes in your ads. for several months, all for the sum of $3, or $2, which is the amount of the second prize?

However, it is not a bad scheme, even if it isn't advertising. Why not offer a handsome study lamp, a set of knives and forks, or some useful present or presents for Christmas?

This dealer can accomplish better advertising. The brilliancy of his scheme does not eclipse the fact that ninety-nine schemes out of every hundred are the poorest kind of advertising and always bad business policy.

* * *

Now, Mr. Hardwareman, send along your ads. for criticism. You'll get my honest opinion, and that's just exactly what you want—isn't it?

T. J. STEWART.

A FRIENDLY CHAT.

Every storekeeper ought to occasionally call his clerks together and talk to them somewhat as follows:

"You and I want to make more money in this store, but in order to do this, we must do more busines. Now I want suggestions from all of you about anything you think we can do to increase trade and gain new customers. Or if you see any waste going on that we might save I want you to tell me. I want each one of you to keep your eyes open, use your brains and think of something that will improve this business. If there is anyone who does not care to try to improve this business we don't want him here. I would rather get rid of him and divide his salary among those who do care. Now, as soon as a suggestion comes to your mind, let us have it, and we will see if all of us cannot make more money this year than we did last."

HARDWARE AND METAL

Established 1888

The MacLean Publishing Co.
Limited

JOHN BAYNE MACLEAN . President

Publishers of Trade Newspapers which circulate in the Provinces of British Columbia, Alberta, Saskatchewan, Manitoba, Ontario, Quebec, Nova Scotia, New Brunswick, P.E. Island and Newfoundland.

OFFICES:

MONTREAL, ... 232 McGill Street, Telephone Main 1255
TORONTO, ... 10 Front Street East, Telephones Main 2701 and 2702
WINNIPEG, ... 511 Union Bank Building, Telephone 3726
LONDON, ENG . . . . . 88 Fleet Street, E.C., J. Meredith McKim, Telephone, Central 12960

BRANCHES:

CHICAGO, ILL. . . . . . 1001 Teutonic Bldg, J. Roland Kay
ST. JOHN, N.B. . . . . No. 7 Market Wharf
VANCOUVER, B.C. . . . . Geo. S. B. Perry
PARIS, FRANCE - Agence Havas, 8 Place de la Bourse
MANCHESTER, ENG. . . . . 92 Market Street
ZURICH, SWITZERLAND . . . . Louis Wolf, Orell Fussli & Co.

Subscription, Canada and United States, $2.00
Great Britain, 8s. 6d., elsewhere - 12s

Published every Saturday.

Cable Address { Adscript, London / Adscript, Canada

CURB PARTY BIAS.

It is invigorating to have a man like Peleg Howland come out boldly and say what he thinks about measures passed by the Government, of which he is a recognized supporter. He says, plump and plain, that the Railway Commission has helped to increase freight rates by recognizing the traffic associations. He thinks and says the anti-dumping clause of the tariff is a mistake, because it interferes with the merchant's inherent right to buy in the cheapest market. Too many men in this country have put political clamps on their independence and if they can't praise their leaders and their works, sit silent. If the commercial leaders would curb their party bias and give vent occasionally to their honest views, it would be a fine thing for Canada What this Dominion needs most is men to speak out, regardless of what party idol is blasphemed.

A SUBSERVIENT DAILY PRESS.

Over in Chicago there is a serious epidemic of scarlet fever and hundreds of children are dying. A milk company brought in supplies from an infected district and the daily papers suppressed the name of the company. Finally fever was discovered in one of the sweatshops where clothing is made for one of the leading departmental stores. So bad did the epidemic become that in spite of powerful influence, the city officials had to act.

It was decreed that crowds should not be allowed to congregate anywhere, in order to stop the spread of the disease. That hit the big stores, whose proprietors control the daily papers, and the latter at once announced that the fever was decreasing in spite of overwhelming figures to the contrary. The crowds are allowed to gather in the stores, but small handfuls of people skating in the parks have been moved on by the police.

As in Chicago, so in Canada. There is legislation under consideration at Ottawa which, if adopted, will strike a hard blow at every branch of retail trade, the chief beneficiaries being the mail order concerns, which own and control Canada's leading daily papers. Have the papers come out in opposition to the proposed legislation, or have they advised their retail subscribers of their danger? On the contrary, they have remained silent, the few words they have printed being in support of the measures.

Retailers who are subscribers to the big daily papers should let them know that their actions are being watched.

MERCHANTS' EXCURSIONS.

The Inter-State Commerce Commission has issued a ruling that merchants' excursions are legal under the new rate law recently put in operation across the border, and these excursions, which have been so popular with both jobbers and retailers in the States will, therefore, be continued.

This week a circular was received from the Merchants' Association of New York, announcing that merchants' rates from the central west, which is designated by the railroads as Central Passenger Association Territory, will be in effect to New York City on February 9th-13th, inclusive, and March 2nd-5th, inclusive, with a thirty-day return limit. The special rate is a fare and one-third for the round trip, the reduction being granted under the certificate plan. The association will follow this circular up with maps and street directories of New York City and invitations to make the Merchants' Association building a headquarters while in the metropolis. These arrangements are of great advantage to thousands of merchants who desire to visit jobbing and manufacturing houses in order to get in touch with the most up-to-date lines on the market.

A year or so ago an effort was made to establish similar merchants' excursions to Toronto and other Canadian centres, but the plan was blocked by the railway companies. The semi-annual conventions of the Retail Hardware Associations are, therefore, the only organized movements encouraging the retail trade to visit the jobbing centres but they do not provide sufficient time for the two branches of trade to get together and become thoroughly acquainted.

The matter of merchants' excursions along the lines in operation across the border might to advantage be taken up again by the Boards of Trade as well as the wholesale and retail associations.

COBALT—A DISEASE.

The United States Consul in the Transvaal in a report to the Department of Trade and Commerce, Washington, says: "Stock speculation on the part of everybody and wild cat gold and diamond mines have taken away money from everything save the barest necessities and in many instances reduced men to bankruptcy and poverty who two years ago counted their sterling in many thousands." Canada has a similar trouble. The gambling disease seems latent in the blood everywhere and a year or two hence many men will be cursing Cobalt as the cause of their troubles. It isn't the man who can count his sterling by thousands one need worry about; their case is sad, but they are comparatively few. The men to be pitied are the hundreds and thousands whose small savings or business go the way of speculation. Many a good business has been consumed by the disease and families in comparative independence because of a paying business are swept into the street. Of course here and there someone starts with a few hundred dollars and makes a fortune; but look at the losses that go to make his gains.

FARMERS' BOARDS OF TRADE.

Farmers' Boards of Trade are institutions the merchants should encourage. Organization will benefit agriculture and the improvement of agriculture means more merchandizing and better returns for the merchants. The first farmers' Board of Trade in Canada was organized in King's county, N.S., twelve years ago. It has done excellent work. At its meetings the farmers from all over the county get together and talk things over. It keeps them informed. The informed man is the progressive, the successful man in any line of endeavor. The merchant who looks beyond the bounds of his own business and takes a patriotic interest in the betterment of his community is, other things being equal, the best merchant.

PUBLIC SPIRITED HARDWARE MERCHANTS

NO. 8

At the present time no town in Ontario is attracting more attention from an industrial standpoint than Welland. The past couple of years has seen a remarkable awakening, and with the wonderful possibilities of electrical power from Niagara Falls, combined with the shipping facilities of the town at the head of deep water navigation and the foot of transportation on the Great Lakes, the future of Welland is bound to be one of marked industrial expansion and prominence.

It is fitting that at this stage in the development of the town, when questions of building restrictions, sanitary improvement, concessions to new enterprises, etc., are to be decided, that a level-headed and successful hardware dealer and plumber should be at the helm. For 21 years J. H. Crow has been in the hardware business in Welland, and in the recent municipal elections was chosen by a decisive majority

J. H. CROW
Recently elected Mayor of Welland Ont.

over two rivals, as the mayor of the town for 1907.

Mr. Crow was born in Pelham Township, Welland County in 1856, and after receiving his education in the public, high and normal schools, spent eight years as a teacher in Welland and Lincoln Counties. In 1886 he entered into the hardware business at Welland with A. E. Taylor and ten years later bought out his partner, the business, which included plumbing and gasfitting, steadily increasing until 1906, which, on account of the extensive building, industrial and natural gas operations, was a banner year. Mr. Crow qualified for the mayoralty by sitting for five years in the town council. He is also prominent in Masonic and Methodist church work.

In a letter to Hardware and Metal, the new mayor says Welland will boom again in 1907 as the Plymouth Cordage Company intend to duplicate their plant at a probable cost of half a million dollars, besides erecting forty additional houses for their employes. There is not a vacant house in town, and those in course of erection have been secured by the employes of the Ontario Iron and Steel Company.

CHANGE FOR THE GOOD.

The Ontario Government's proposal to employ prison labor on farm work in future instead of having the convicts make binder twine, brooms, wooden-ware, etc., is a step in the right direction. It will temporarily affect the various manufacturers who have been utilizing the facilities of the central prison to secure stocks of these goods, but this will be easily offset by the removal of this class of competition which other manufacturers have had to meet.

To remove the prison out into the country and have the convicts get next to nature by learning how to cultivate the soil while enjoying the wholesome air and surroundings of the country, should have excellent results in the way of reforming many first offenders. City life is more productive of crime than farm life and anything which would tend to make it possible for the unfortunates to begin life again as farm laborers or homesteaders, should be encouraged.

One problem which will have to be considered, of course, is the guarding of the prisoners. The old system of confinement within brick walls paced by armed guards could hardly be made to work satisfactorily if the prison was removed to the country. But could not the prisoners be made to guard themselves? It is said that in the State of Oregon the convicts are put to work on the farms practically unguarded. If a prisoner escapes the term of the remaining prisoners is lengthened while, on the other hand, if all work faithfully an allowance for good service is made.

Criminals as a rule are the creatures of their surroundings. If born under degraded circumstances or forced to live surrounded by temptation the weak can hardly be blamed for falling. Society is largely at fault and it is the duty of our Governments to endeavor to overcome this fault both in the way of removing the cause and in remedying the effects The suggested removal of the Central Prison from Toronto appears to be in line with a recognition of this duty and all thinking citizens will hope to hear of good results following the adoption of the proposal.

MONTREAL AGENT MARRIES.

A wedding of considerable interest to the Montreal trade took place Thursday afternoon when J. Sophus, Canadian representative of Schuchardt & Schutte of New York and Berlin, was married to Miss Jennie Colquhoun of Manchester, England. The ceremony took place at St. Thomas' Church, and was performed by the Rev. Canon Renaud. While distinctly private, a large number of immediate friends attended, and many handsome gifts were received. The young couple left on a short tour to the Southern States and on their return will take up their residence in Montreal.

J. Sophus has been in Montreal a little more than a year and a half during which time he has made steady progress. Formerly he was superintendent of large works in Manchester, England, and before that again had been engaged as a traveler for machinery and metals in Europe, visiting Germany and Russia as well as other countries. Gifted as a linguist he has exceptional advantages, and as Canadian representative of Schuchardt & Schutte he deserves congratulations. This firm have branches in China, Japan, Russia, France, India, and other places, having also immense headquarters in New York City. They have confidence in the Canadian market and are spending large sums this year to promote their interests. Hardware and Metal supplements the many good wishes extended to Mr. Sophus on his new step.

OPPOSE PARCEL POST.

At a meeting of the council of the Portage la Prairie board of trade, held last week it was decided to protest

J. SOPHUS
An Enterprising Young Montreal Manufacturer's' Agent who was married a few days ago.

against the Dominion Government's new arrangement re the parcel post, which it proposed to bring in, and the secretary wired the local member, J. Crawford, to this effect.

It is maintained that the regulation will discriminate against smaller and local merchants in favor of the departmental stores, as it would facilitate the latter's shipments of goods through the mails.

NERVOUS PROSTRATION AMONG CLERKS.

Nervous prostration comes from letting the work chase you. When you chase the work, you eat, sleep and laugh, and the man who can do these three things is immune from everything from jiggers to paresis.

HARDWARE TRADE GOSSIP

Ontario.

R. N. Youmans, plumber, Parliament street, Toronto, has put in a stock of hardware.

M. R. Summerfeldt, hardware merchant of Mt. Albert, called on Toronto jobbers this week.

J. Henry & Son, Orono, have bought out their opposition, D. H. Walsh, and have taken over the stock.

W. S. Piper, hardware, stove and furniture dealer of Fort William, has advertised his business for sale.

James Johnston, manager of the retail department of Rice Lewis & Son, Toronto, has been away ill for the past week.

Slimman Bros. have taken over the business of Messrs. McLaughlin, Sturtbridge & Company, hardware merchants, of Drayton.

O. B. Manville, hardware merchant of Prince Albert, Sask., was a visitor in Toronto this week calling on the hardware jobbers.

John Armstrong & Company, hardware merchants of Brigden, will enlarge their store to meet with the requirements of the business.

Louis Payette, of Warden King & Son, and R. J. Cluff, of Cluff Bros., Toronto, returned on Tuesday from a ten days' trip to Winnipeg.

McMaster's hardware store at Ridgetown, was destroyed by fire on Thursday. The loss will be over $13,000, with only $7,000 insurance.

James D. Smith, Baysville, is remodelling his store. Mr. Smith was elected to the municipal council last month, being second in the poll.

H. N. Joy, malleable range manufacturer, Toronto, has returned to his office after six weeks' illness following a blow on his head received while skating.

James Thomson, saleman in E. J. Torrens' hardware store, Tillsonburg, has taken a position as traveler. Mr. Willings, of Niagara Falls, has taken the vacancy with Mr. Torrens.

The police are after William Roach, a cashier in Bond and Co.'s hardware store, Guelph, who is alleged to have left for Uncle Sam's domains with some $700 of the firm's money. It is said that there is a woman or two in the case.

George James, of James & Reid, Perth, joined the army of benedicts last week. When asked as to where he was going for his honeymon he replied that he is too busy to take on now but expected to take a good one in the summer time.

F. A. Howard, for many years a member of the firm of Turnbull, Howard & Company, hardware merchants, Brantford, has severed his connection with the firm, which will in future be conducted by W. R. Turnbull and E. B. Cutcliffe.

F. Y. W. Braithwaite, Blind River, was a caller at the Toronto office of Hardware and Metal on Thursday. Mr. Braithwaite is making a semi-annual buying trip and reports conditions on the north shore of Georgian Bay to be excellent.

The annual meeting of the Peterborough Lock Manufacturing Company was held on Monday last. The following directors were reelected: Hon. J. R. Stratton, W. T. Morrow, T. H. Fortye, A. L. Davis and Wm. Irwin. The Hon. J. R. Stratton was reelected president and W. T. Morrow vice-president. The company commenced business in Peterborough 21 years ago and have made steady and satisfactory progress and at present pay out in wages to employes $45,000 per annum.

Quebec.

J. Paquin, St. Eustache, was in Montreal on business this week.

C. O. Gervais, St. John, Que., was in Montreal this week on business.

Mr. Aubin, Glen Robertson, called on some of the Montreal wholesalers this week.

R. H. Cowan of the Traders Company, Buckingham, was in Montreal this week buying goods.

A. Ault, of Ault Bros., Montreal, is in the Royal Victoria Hospital at present, with typhoid fever.

The Standard Paint Co., of Montreal, have opened up nice quarters in the Board of Trade Building.

J. B. Campbell, president of the Acme Can Co., Montreal, is in Buffalo attending the canners' convention.

Cunin Bros., stovepipe elbow manufacturers, of Montreal, suffered loss by fire last week, damaged to the extent of $2,000.

J. S. McLernan, Montreal representative of John Round & Son, cutlery manufacturers, Sheffield, England, has returned from a business trip to the Old Country.

Lewis Bros., Montreal, are making some changes in their warehouse, which, when completed, will bring the "Buyers' Office" from the fourth floor, where it is now located, up to the seventh, with the offices and counting rooms.

The annual general meeting of the shareholders of Frothingham & Workman was held on Tuesday, and a satisfactory report and statement of the business for 1906 laid before the meeting. The following were re-elected directors for the ensuing year: Messrs. E. Archibald, C. D. Monk, W. G. Lemesurict, E. C. Eaton and W. C. Davis.

J. S. Laurier, hardware merchant, of Montreal, died at the General Hospital this week. An attack of appendicitis necessitated an operation, which was performed on February 2. A complication of peritonitis carried him off. Mr. Laurier was born at Joliette, Que., in 1861. He opened his commercial career with a course at the commercial college of his native town. Coming to Montreal a short time after, he became a clerk in the hardware store of Mr. Lallemand and later entered the J. L. Lafleur hardware store. After a few years' service there, he opened the hardware store at the corner of St. James and Inspector Streets, where by his energy and activity he built up a prosperous business.

J. Sophus, Canadian representative of Schuchardt & Schutte, formerly with headquarters at 13 St. John St., has decided to move to 102 St. Antoine St., where commodious warerooms will be employed. The company have made rapid strides in this country under the direction of Mr. Sophus, and to meet the expansion they have given over the sales end of the heavy machinery department for the province of Quebec to Baxter, Paterson & Co., an enterprising firm, who have many other large agencies and whose headquarters are at the same address. This will give Mr. Sophus greater scope, and at the same time will be a valuable supplementary force for the Schuchardt & Schutte Co., who have branches in all parts of the world. Large stocks will be kept on hand, both of tools and hardware lines.

Western.

W. H. Sparling, Minnedosa, was in Winnipeg last week on business.

J. W. Bacon, Enderby, B.C., has secured a Canadian patent on a new baking pan.

Stockwell and Perry, hardware merchants, Estevan, Sask., have dissolved partnership.

W. Stewart Plum, Coulee, Man., called at several of the Winnipeg wholesale houses last week.

The Marshall-Wells Hardware Co. expect to move into their new warehouse about March 1st.

It was Thomas Dallas and not Thos. Wallace, as misprinted, who recently established a hardware store at Lamont, Alta.

A. J. Falconer, Deloraine, president of the Western Retail Hardware Association, was in Winnipeg last week on business.

The firm of Anderson and Thomas, hardware merchants, Winnipeg, Man., has been dissolved by mutual consent. The business will be continued by J. E. Thomas.

Alf. Kelley, of the firm of Hislop, Kelley and Younge, hardware merchants, High River Alberta, is at present in Toronto visiting paint, oil, and hardware jobbers.

Edgar G. Powell, of J. W. Harris Co., Montreal, is in Winnipeg at present on business for his firm. He called at the Winnipeg office of Hardware and Metal on Monday of last week.

W. A. Kenning has resigned his position as sales manager of the Miller-Morse Hardware Co., Winnipeg, and he has been succeeded by Louis J. Blackwood, who was formerly with Merrick, Anderson & Co.

J. A. Lindsay, former president of the Western Retail Hardware and Store Dealers' Association has returned to Winnipeg from an extended holiday tour in Scotland. Mr. Lindsay's health is much improved.

The following merchants of Cartright, Man., are amalgamated under style of the Merchants, Limited: Moore & Hills, general store; J. A. McKenzie, F. & F. M. Macklin, hardware; Cannon & Pickle, general store; L. H. Phillips, hardware, and Wm. Gemmell, grocer.

Markets and Market Notes

(For detailed prices see Current Market Quotations, page 66.)

THE WEEK'S MARKETS IN BRIEF.

MONTREAL.

Plumbers' Brass Goods—Advanced 10 to 15 per cent., and in some cases more.

TORONTO.

Plumbers' Brass Goods—Valves advanced.

Cast Iron Fittings—Last week's advance now general.

Linseed Oil—Advanced one cent.

Turpentine—Declined, owing to change in tariff.

Coil Chain—Advancing.

Coil Spring Wire—Advanced 5c. per cwt.

Paris Green—One cent. higher.

Montreal Hardware Markets

Office of HARDWARE AND METAL,
232 McGill Street,
Montreal, Feb 15, 1907.

Trade still remains good in hardware circles, and although collections at present are not all they might be, jobbers in general are very well satisfied with their year's business, and express wonder at the way things have kept going.

Owing to the low temperature and biting wind experienced here during the week, many city salesmen have been remaining indoors as much as possible and have taken this opportunity to get their inside work up to date. The severe storms in Western Ontario have also held many travelers back a day or two.

Although there are no changes to record in prices at present, it is expected that owing to the recent advances in raw material entering into the manufacture of ammunition, prices in this line will be higher than last season.

Axes are reported to be moving very well, but the sales of lanterns have fallen off slightly. Cow ties and sleigh bells are very quiet. Rivets and burrs, hay wire, horseshoes and horse nails are moving very well. There is a heavy demand for bolts and nuts, and some sizes of machine screws are still scarce.

AXES.—Moving very well. We still quote: $7.60 to $9.50 per dozen; double bitt axes, $9.50 to $12 a dozen; handled axes, $7.50 to $9.50; Canadian pattern axes, $7.50 a dozen.

LANTERNS.—Sales have fallen off slightly. Our prices remain: "Prism" globes, $1.20; Cold blast, $6.50; No. 0, Safety, $4.

COW TIES AND STALL FIXTURES—Very quiet. We are still quoting: Cow ties, discount 40 per cent. off list; stall fixtures, discount 35 per cent. off list.

RIVETS AND BURRS.—Our prices are: Best iron rivets, section carriage, and wagon-box black rivets, tinned do., copper rivets and tin swede rivets, 60, 10 and 10 per cent.; swede iron burrs are quoted at 60 and 10 and 10 per cent. off new lists; copper rivets with the usual proportion of burrs, 25 per cent. off, and coppered iron rivets and burrs in 5 lb. carton boxes at 60 and 10 and 10 per cent.; copper burrs alone, 15 per cent., subject to usual charge for half-pound boxes.

HAY WIRE.—Beginning to move faster. Prices remain: No. 13, $2.60; No. 14, $2.70; No. 15, $2.85; f.o.b. Montreal.

MACHINE SCREWS.—Some sizes still short. Prices remain same as last week: Flat head, iron, 35 per cent., flat head brass, 35 per cent., Felisterhead, iron, 30 per cent., Felisterhead, brass, 25 per cent.

BOLTS AND NUTS.—Demand is very heavy. Discounts remain: Carriage bolts, 3 and under, 60 and 10; 7-16 and larger, 55 per cent.; fancy carriage bolts 50 per cent.; sleigh shoe bolts, ⅜ and under, 60 per cent.; 7-16 and over, 50 per cent.; machine bolts, ⅜ and under, 60 per cent.; 7-16 and larger, 55 per cent.

HORSE NAILS.—Good demand at present. Discounts remain: C brand, 40, 10 and 7 per cent.; M.R.M. Co., 55 per cent.

WIRE NAILS.—We still quote: $2.45 per keg, f.o.b. Montreal.

CUT NAILS.—Prices remain: $2.30 per keg, base, f.o.b. Montreal.

HORSESHOES.—Sales increasing. We still quote: P.B. new pattern, base price, $3.50 per 100 lbs.; M.R.M. Co., latest improved pattern iron shoes, light and medium pattern, No. 2 and larger, $3.65; No. 1 and smaller, $3.90; snow pattern, No. 2 and larger, $3.90; No. 1 and smaller, $4.15. Light steel shoes, No. 2 and larger, $4; No. 1 and smaller, $4.25; featherweigh, all sizes, No. 0 to 4, $5.60. Toeweight, all sizes, No. 1 to 4, $6.85. Packing up to three sizes in a keg, 10c. per 100 lbs. More than three sizes, 25c. per 100 lbs. extra.

BUILDING PAPER.—No change in prices.

CEMENT AND FIREBRICK.—Prices remain: "Lehigh" Portland in wood, $2.54, in cotton sacks, $2.39; in paper sacks, $2.31. Lefarge (non-staining) in wood, $3.40; Belgium, $1.60 to $1.90 per barrel; ex-store, American, $2 to $2.10, ex-cars; Canadian Portland, $2 to $2.05. Firebrick, English and Scotch $17 to $21. American $30 to $35; White Bros.' English cement, $1.80 in bags, $2.05 in barrels in round lots.

COIL CHAIN.—Our quotations remain: 5-16 inch, $4.40; ⅜ inch, $3.90; 7-16 inch, $3.70; ½ inch, $3.50; 9-16 inch $3.45; 5-8 inch, $3.35; ¾ inch, $3.20; ⅞ inch, $3.10; 1 inch, $3.10.

SHOT.—We are still quoting: Shot packed in 25-lb. bags, ordinary drop AAA to dust, $7 per 100 lbs.; chilled, No. 1—10, $7.50 per 100 lbs.; brick and seal, $8 per 100 lbs.; ball, $8.50 per 100 lbs. Net list. Bags less than 25 lbs., ½c. per lb. extra, net; f.o.b. Montreal, Toronto, Hamilton, London, St. John, Halifax.

AMMUNITION.—Owing to the recent advances in raw material, prices are likely to be higher this season than last, but our prices still remain: Slow. last, but our prices still remain: Loaded with black powder, 12 and 16 gauge, per 1,000, $15; 10 gauge, per 1,000, $18; loaded with smokeless powder, 12 and 16 gauge, per 1,000, $20,50; 10 gauge, $23.50.

GREEN WIRE CLOTH.—We continue to quote: In 100-foot rolls, $1.62½ per hundred square feet, in 50-feet rolls, $1.67½ per hundred square feet.

Toronto Hardware Markets

Office of HARDWARE AND METAL,
10 Front Street East,
Toronto February 15, 1907

While business is not as brisk as formerly it remains fairly steady, with numerous orders for small staple lines being received daily. All other lines for spring and summer, having been ordered some time ago, are being shipped with these small orders, so as to make up a shipment.

Very few changes have to be made this week, the most important being the 5 per cent. advance duty on American steel chain, which will naturally increase the price on the Canadian market.

An advance of 5 cents per cwt. has been made on coil spring wire. No doubt this advance has been expected, as the high price of rods, and of raw material has warranted it, but this will not materially effect the hardware merchant, as all these lines will be ordered in the fall, and will soon be shipped to the purchaser.

Paint brushes and utensils for spring housecleaning are meeting with good demand. Graniteware is selling very rapidly at present, with a view of an advance in the near future, as the ever-increasing cost of raw material makes this likely.

AXES AND HANDLES—All the business for the season is practically over in this line for wholesale merchants. But retailers continue doing a brisk business. Many display full lines of these articles, accompanied by saws, in their windows.

WASHING MACHINES—These articles, along with various appliances for housecleaning, are meeting with great demand. All orders which have been taken, are being rapidly filled.

SPORTING GOODS—Many marchants have been buying large quantities of fishing poles, tackle, etc.

CHAIN—The 5 per cent advance on the duty will naturally cause an advance on chain. We, however, continue to quote: ⅝ in., $3.50; 9-16 in., $3.55, ½ in., $3.60; 7-16 in., $3.85; 3-8 in., $4.10; 5-16 in., $4.70; 1-4 in., $5.10.

RIVETS AND BURRS—Remain firm at the advanced price, with great numbers being sold

SCREWS—Quite a large number are being sold at present, with stocks in a very good condition.

BOLTS AND NUTS—The demand for these articles has eased off somewhat, allowing stocks to be repleted.

TOOLS—Still continue selling very freely, merchants stocking up and arranging their windows with displays of various kinds of tools

SHOVELS—The demand still continues good for all lines. All book orders are being rapidly sent forward, as the demand will increase with the opening of spring.

EXTENSION AND STEP LADDERS—The sale continues very brisk, as larger numbers of these will be used for spring work. We continue to quote: 11 cents per foot for 3 to 6 feet, and 12 cents per foot for 7 to 10-foot ladders. Waggoner extension ladders 40 per cent.

POULTRY NETTING—Orders which have been booked are being sent forward. We continue to quote: 2-inch mesh, 19 w g., discount 50 and 10 per cent., all others 50 per cent.

WIRE FENCING—Coiled spring wire has advanced 5 cents per cwt., which will likely have an influence towards advancing woven wire fencing.

OILED AND ANNEALED WIRE—(Canadian)—Gauge 10, $2.41; gauge 11, $2 47; gauge 12, $2 55, per 100 lbs.

WIRE NAILS—No further advance is quoted on wire nails, but the feeling is that the market has not yet reached its highest point. We continue to quote $2.10 base.

CUT NAILS—We quote $2.30 base f.o.b. Montreal; Toronto, 20c. higher.

HORSENAILS—"C" brand 40, 10 and 7½ off list; "M.R.M." brand 60 per cent. off.

HORSESHOES—The demand still continues good, the scarcity of snow, and slippery roads, aiding greatly in ordering of shoes. Quotations remain at former prices: P. B. base, $3.65; "M. R M. Co., latest improved pattern" iron shoes, light and medium pattern, No. 2 and larger, $3.80; No. 1, and smaller, $4.05; snow, No. 2 and larger, $4.05; No. 1 and smaller, $4.30; light steel shoes, No. 2 and larger, $4.15; No. 1 and smaller, $4.40; featherweight, all sizes, 0 to 4, $5 75; toeweight, all sizes, 1 to 4, $7. Packing, up to three sizes in a keg, 10c. per 100 lbs. extra; more than three sizes in a keg, 25c. per 100 lbs. extra.

BUILDING PAPER—The opening of spring will see this line revived, but very little is doing at present.

CEMENT—For carload orders, f.o.b. Toronto, we quote: Canadian Portland, $1.95. For smaller orders ex warehouse, Canadian Portland, $2.05 upwards.

FIREBRICK—English and Scotch firebrick, $27 to $30; American low-grade, $23 to $25; high-grade, $27.50 to $35.

HIDES, WOOL AND FURS — Prices are unchanged. Trade is quiet.

Hides, inspected, cows and steers, No. 1		0 10
" " " " No. 2		0 09½
Country hides, flat, per lb., cured		0 09
" " " green		0 08
Calf skins, No. 1, city		0 12
" " No. 1, country		0 11
Lamb skins	1 25	1 35
Horse hides, No. 1	3 50	3 75
Rendered tallow, per lb.	0 05½	0 06
Pulled wools, super, per lb.		0 25
" " extra		0 27
Wool, unwashed fleece		0 15
" washed fleece	0 24	0 25

FURS.

	No. 1.	Prime
Raccoon	....	1 50
Mink, dark	5 00	7 00
" pale	2 50	4 50
Fox, red	3 00	4 00
" cross	2 00	20 00
Lynx	5 00	8 00
Bear, black	....	12 00
" cubs and yearlings	....	5 00
Wolf, timber	....	3 25
" prairie	....	1 25
Weasel, white	0 10	0 65
Badger	0 75	1 75
Fisher, dark	6 00	8 00
Skunk, No 1	....	1 10
Marten	3 50	20 00
Muskrat, fall	....	0 17
" winter	....	0 25
" western	0 12	0 18

Montreal Metal Markets

Office of HARDWARE AND METAL,
232 McGill Street,
Montreal, February 15, 1907

One of Montreal's foremost metal brokers, who is in the Old Country at present, cables that the markets there are exceedingly strong and although there have been several temporary fluctuations lately, the outlook is for higher prices in the near future.

About the same condition prevails here this week as in the past fortnight, that is, there are no changes of any consequence to report. Zinc eased off slightly during the week, but this is only temporary as orders are already in for spring delivery, and prices are sure to be very firm when that season arives.

Pig lead is a little weaker in tone with a tendency to lower prices. Copper remains the same with the market stronger than ever. Spelter remains firm at last weeks' price. Pig iron is firming up.

COPPER—Prices remain the same with market stronger than ever. We still quote: Ingot copper, 26½ to 27c.; sheet copper, base sizes, 34c.

INGOT TIN — Remains firm. Our prices remain: 45½ to 46c. per pound.

ZINC SPELTER — Eased off temporarily during the week. We still quote: $7.50 to $7.75 per 100 pounds.

PIG LEAD — A little weaker in tone, with a tendency to lower prices. We still quote: $5.50 to $5.60 per 100 lbs.

ANTIMONY — We still quote: 27½ to 28c. per pound.

PIG IRON — Firming up. We still quote: Londonderry, $24.50; Carron, No. 1, $27; Carron, special, $25.50; Summerlee, No. 2, selected, $25; Summerlee, No. 3, soft, $23.50.

BOILER TUBES—Remain very firm. Our prices are: 1½ to 2 in., $9.50; 2¼ in., $10.35; 2½ in., $11.50; 3 in., $13; 3½ in., $17; 4 in., $21.50; 5 in., $45.

TOOL STEEL — Our prices remain: Colonial Black Diamond, 8c. to 9c.; Sanderson's, 8c. to 45c., according to grade; Jessop's, 13c.; "Air Hardening," 65c. per pound; Conqueror, 7½c.; Conqueror, high speed steel, 60c.; Jewitt's Diamond J., 6½c. to 7c.; Jowitt's best, 11c. to 11½c.

COLD ROLLED SHAFTING—Present prices are: 3-16 inch to ¼ inch, $7.25; 5-16 inch to 11-32 inch, $6.20; ⅜ inch to 17-32 inch, $5.15; 9-16 inch to 47-64 inch, $4.45; ¾ inch to 17-16 inch, $4.10; 1½ inch to 3 inch, $3.75; 3⅛ inch to 3 7-16 inch $3.92; 3½ inch to 3 15-16 inch, $4.10; 4 inch to 4 7-16 inch, $4.45; 4½ inch to 4 15-16 inch, $4.80. This is equivalent to 30 per cent. off list.

GALVANIZED IRON — Market strong, with no changes to report.

OLD MATERIAL — We still quote: Heavy copper, 20c. per pound; light copper, 17c. per pound; heavy red brass, 17c. per pound; heavy yellow brass, 14c. per pound; light brass, 10 to 10½c. per pound; tea lead, 4¾c. per pound; heavy lead, 4½c. per pound; scrap zinc, 4¼c. per pound; No. 1 wrought, $17 per ton; No. 2 wrought, $6 per ton; No. 1 machinery, $18 per ton; stove plate, $13 per ton; old rubber, 10½c. per pound; mixed rags, 1 to 1¼c. per pound.

Toronto Metal Markets

Office of HARDWARE AND METAL,
10 Front Street East,
Toronto, Feb. 15, 1907

Sheet copper shows the only advance for the week, it being marked up to 35 cents per pound in response to an advance on the other side, where the market is governed by an association. Stocks are very low everywhere and the demand is increasing from electrical and other quarters. Production is also being hampered in the lake region by the unprecedented cold weather. With producers sold for months ahead, the continued unsatisfied daily inquiry for copper shows demand is in excess of supply. It is only by refusing to quote at present on far futures that the market is kept quiet, steady and at the present basis. We are confident, says the Metal Market Report, that if the large producers would agree to sell May, June and July at ½c. per pound over the supposed nominal market price, it would quickly be taken, and an immediate advance made on spot and all down the line. This advance might be a great mistake, hence we congratulate producers on their present attitude of keeping out of the market and keeping control of some of their supplies.

Locally the important topic is the revised tariff, the tendency of which is to make matters easier for the trade. Boiler tubes will be even better than before, but boiler plates are left far from satisfactory, as merchants will be forced to carry two separate stocks in order to do business with manufacturers and general buyers. For the former there is no duty, while for the latter $3 per ton is charged. The cutting out of the duty of 5 per cent. on tubes will not affect the price, as the price advanced the same amount last week in the producing market. All sizes of plates will now be rated at the same quotation.

Business has been very lively and one firm which does business in all parts of Canada reports a 20 per cent. gain in January over the same month last year.

In reviewing the statistics of pig iron production in the U. S. for 1906, which show a ten per cent. increase and a total of over 25 million tons, the Metal Market Report says: "When will this rapid increase in pig iron production halt? Certainly not this year; we shall certainly break the record again, as we are making at the rate of more than 27,-000,000 tons a year at the present time, with new furnaces to come in, and the iron is sold, either in the form of pig iron or of finished steel products, for more than half the remainder of the year."

PIG IRON—In foreign markets conditions have been easier but locally things are unchanged with prices the same. Buying continues brisk. Hamilton, Midland and Londonderry are off the market, and Radnor is quoted at $.53 at furnace. Middlesborough No. 1 is quoted at $24.50, and Summerlee No. 1 at $26 f.o.b., Toronto.

BAR IRON — Stocks are picking up but buying is active. Steady prices rule and we still quote: $2.30 f.o.b., Toronto, with 2 per cent. discount.

TIN — Both ingots and plates have been fluctuating in foreign markets but locally ingots are firm at 45c. and charcoal plates, base, at $6.50. Prices are again firm at 45c. with more asked in some cases. Plates are firm and unchanged.

SHEETS AND PLATES—No further advances have been made on black sheets or Canada plates, although there is talk of higher prices. Galvanized sheets are in good demand.

BRASS—Conditions are not improving and stocks continue light. We still quote 30 cents per pound on sheets.

COPPER—Sheets have advanced in response to higher prices on the other side. We now quote: Ingot copper, $26.50 per 100 lbs., and sheet copper, $35 per 100 lbs.

LEAD—Stocks continue low. Buying is fair with prices unchanged. We quote: $5.40 for imported pig and $5.75 to $6.00 for bar lead. For sheet lead $7 is asked.

ZINC SPELTER—A fair business is doing but stocks are short. We quote 7½c. per lb. for foreign and 7c. per lb. for domestic. Sheet zinc is quoted at 8 1-4c. in casks, and 8 1-2c. in part casks.

BOILER PLATES AND TUBES.—An advance is looked for, but we still quote: Plates, per 100 lbs., ¼ in. to ½ in., $2.50; heads, per 100 lbs., $2.75; tank plates, 3-16 in., $2.65; tubes, per 100 feet, 1½ in., $8.50; 2, $9; 2 1-2, $11.30; 3, $12.60; 3 1-2, $16; 4, $20.00. Terms, 2 per cent. off.

ANTIMONY—The market continues lively with prices unchanged at 27c. per pound.

OLD MATERIAL.—Dealers report a good demand with prices continuing on the upward trend. Their buying prices are: Heavy copper and wire, 20c. per lb.; light copper, 18c per lb.; heavy red brass, 18c per lb.; heavy yellow brass, 15c. per lb.; light brass, 11c. per lb.; tea lead, $4.25 per 100 lbs.; heavy lead, $4.50 per 100 lbs.; scrap zinc, 4½c. per lb.; No. 1 wrought iron, $15; No. 2 wrought, $8; machinery cast scrap, $18; stove plate, $14; malleable and steel, $9; old rubbers, 10½c. per pound; country mixed rags, $1 to $1.25 per 100 lbs., according to quality.

COAL—While there is plenty of soft coal, small sizes of anthracite are very short in all parts of Ontario. We still quote: Standard Hocking soft coal f.o.b. at mines, lump, $1.50, ¾ inch, $1.40; run of mine, $1.30; nut, $1.25; N. & S., $1; P. & S. 85c. Youghiogheny soft coal in cars, bonder at bridges: lump, $2.60; ¾ inch, $2.50; mine run, $2.40; slack, $2.15.

United States Metal Markets

From the Iron Age, Feb 14, 1907.

The blast furnaces have not been working well lately and the output has not been as large as it should be. During January the production of coke and anthracite pig iron amounted to 2,205,407 gross tons, as compared with 2,235,306 tons in December, the production of the steel works stacks having been 1,406,-397 and 1,445,528 tons, respectively. The decrease in output was therefore due entirely to the latter. The merchant furnaces held their own. It is interesting to note that three new stacks were added to their number during January, the Josephine, Federal and second Toledo having made their first iron in that month. We entered February with an active furnace capacity of 492,359 tons per week, as compared with 507,-322 tons per week on January 1.

The announcement has just been made that the contract for 38,000 tons of water pipe for New Orleans has been let on the basis of $33 per net ton for the 1907 and $32 for the 1908 delivery. This shows that the fears were not justified, that present high prices for material might cause the closing of contracts for next year's part of the work to be postponed.

In the steel rail trade an interesting event has been the sale, for export, of a lot of 50,000 tons of 80-lb. rails, for delivery during the second half of this year. It is worthy of special mention that the price realized for these rails nets the mill more than is being obtained for domestic business.

From the plate, bar, sheet, tin plate, tube and wire mills comes the story of a rush of specifications on old orders and of new orders. Among the larger contracts recently closed with tube mills is one for 400 miles of pipe for the Southern Pacific, for a new oil line in California.

From the Iron Trade Review, Feb. 14, 1907

The pig iron market shows indication of lower prices this week in nearly all parts of the country. The United Stases Steel Corporation and other large interests are of the opinion that if the prices of finished products are further advanced, consumption will be decreased and the continuance of prosperity will be imperilled. Hence, strong influences are being exerted in favor of conservatism. Steel makers are declining to buy pig iron at the present prices, saying that they cannot dispose of their own products if they must pay such high prices for the crude material. Buyers of foundry iron are also making a determined effort to obtain lower prices. Although some furnace interests are very firm, and are declining to make any concessions, the general disposition is to quote lower prices for future delivery. Furnace interests in the Mahoning and Shenango valleys have sold practically no steel, making iron for the last half of the year, but are not disposed to sell at $20, which is being offered by some steelmakers. At present it is a waiting game, and it will be extremely interesting to discover whether the buyers or the sellers will win. The present probability seems to be that the price on Bessemer iron for last half delivery will be pounded down to $20.

The freight situation in the South does not improve. When the railroads decided to postpone the advance in freight rates 30 days, it was hoped that there would be a heavy movement of pig iron from the south during February, but with nearly half of the present month past, there is no indication that the situation will be relieved to any great extent. It will not be surprising if the shippers renew the agitation in opposition to the advance in freight rates.

N.B. Hardware Trade News

St. John, Feb 11.—The annual dinners of the St. John Iron and Hardware Association are always most enjoyable but none of the thirteen so far held has equalled for pleasure that of a few days ago. It was held at the Union Club, forty sitting down to prettily decorated tables. During the evening banjo numbers and vocal solos added to the enjoyment. After dinner came the customary round of toasts. The King was proposed by the president, W. S. Fisher. Our Association, proposed by J. H. McAvity, was responded to by John Keeffe and M. E. Agar. The City and Board of Trade, proposed by the president, was responded to by Mayor Sears, J. H. McRobbie and Ald. Rowan. The Iron and Hardware Manufacturers, proposed by M. E. Agar, was responded to by C. McDonald, S. E. Elkin, J. A. McAvity and T. Stewart. Our Guests, proposed by John Keeffe, was responded to by F. W. Sumner, of Moncton, S. H. White, of Sussex, Wm. Kerr and H. N. Stevens. The Ladies, proposed by G. McDonald, was replied to by J. G. Harrison and Fred Foster.

The menu card was novel and attractive. It was well printed and closed about with a nickeled horse shoe.

* * *

J. E. Wilson is preparing to make noteworthy advances in his business. At the present time he and his foreman, Walter Thompson, are visiting Upper Canadian and United States foundries, inspecting the latest improved machinery and seeking new ideas in the manufacture of stoves, stovefitting, etc. On their return men will begin the erection of an addition to the Brussels St. foundry. The new structure will be one hundred by forty feet. Here much new machinery will be installed as soon as posible. Thus the capacity of the Wilson foundry will be considerably increased. In the rear of the Sydney street store Mr. Wilson will also put up a large brick building of three stories. Here will be found needed room for the carrying on of his sheet metal and other business.

* * *

A factory for the manufacture of the Security Placket Fastener is shortly to be opened here. The man behind this scheme is H. R. McLellan, who has already ordered the machinery needed for the factory. Mr. McLellan plans on putting nearly a thousand of the fasteners on the market every year. The plan is to sell them wherever an opening is to be seen or can be made. The factory will be located on Queen street.

* * *

The markets are continuing strong. The metal advances, however, have not been quite so marked of late as a few weeks ago. All previous advances have been maintained. The stove men have given notice of another advance. Roofing papers have gone up slightly. The paper has gone up five cents per roll, sheathing five cents per hundred pounds.

B.C. Hardware Trade News

Vancouver, B.C., Feb. 8, 07.

Action officially has not yet been made public by the B.C. Lumbermen's Association, or by the Mountain Lumbermen's Association, regarding the investigation to be held before a select committee of the House of Commons into the allegation that the lumbermen are operating an illegal combine. Individuals, however, have made some emphatic protests against the assertions which have been made by members of the Dominion House. G. P. Wells, secretary of the Mountain Lumbermen's Association, denies most emphatically that such a combine exists. He asserts that the millmen will welcome the investigation and the sooner the committee is appointed the better it will suit the lumbermen of the Province.

Mr. Wells describes the association as a body which meets at intervals to discuss matters relative to the industry. In so far as the accepted meaning of trust or combine goes, Mr. Wells says, the association is far removed from any such character.

Discussing actual conditions and how the lumbermen deal with them, Mr. Wells instances the recent annual meeting of the association. They found, he said, that stocks in the Northwest retail yards were very light at present, and that at most of the big mills the stocks on hand were also small. At the present moment orders which would absorb every foot of lumber now in stock have been placed, so that practically a lumber famine may be looked for when the season opens. Notwithstanding this acute condition, the Mountain Lumbermen in the recent convention made no general advances, and no changes of any great importance. On the contrary, they evinced a desire to hold the present prices without further increases. This, too, in face of large increase of cost, from higher priced logs and higher cost of milling. The restricting of output is not even thought of. In fact the output increases from year to year, 1906 being 20 per cent. greater than 1905.

* * *

Vancouver Island is to be developed company has been formulating plans ever since purchasing the E. & N. Railway from the Dunsmuir interests. The huge land grant, comprising some of the best parts of the island, which went with the railway, is to be cleared in districts and put on the market by the C.P.R. and some of this work goes on at once. The climate of Vancouver Island is very mild and salubrious and many districts are admirably adapted to fruit-growing. The company expects to secure a large influx of settlers by means of advertising its cleared lands for sale.

Another important undertaking which the railway has announced its intention of going on with at once, is a branch of the island railway to go from the east coast at or near Nanaimo, to Alberni on the west coast. Whether this foreshadows the establishment of a port on the west coast for the Japan-China line of steamers, is not yet known. But it will in any event open up a large district, and also bring into imediate touch, the isolated areas of the west coast of the island, which have heretofore depended on a steamer service at weekly intervals.

* * *

Discussing coal mining and that industry generally, Thos. R. Stockett, recently expressed the opinion that the hardships in the Northwest would result in good to the people themselves and to the coal mining companies. Mr. Stockett, who is now manager of the Western Fuel Co.'s mines at Nanaimo, was formerly in charge of the Crow's Nest Pass Coal Co.'s mines at Fernie, so that he is in a position to know conditions as they are. He points out that even now, the mines of which he is manager, are slackening off in orders. And that in the summer season it is almost a case of shut-down because the demand is so light. This is even more emphatically the case with the mines of the Crow's Nest. Then when winter comes on, with the extremely doubtful conditions of transportation, everyone is rushing in orders. If consumers, especially in the prairie provinces, where the winter needs are accurately known always, would fill up their coal cellars and coal sheds in the summer and fall months, when railway transportation was good, and before the grain rush begins and bad weather stops transportation, there would be no trouble.

Speaking of conditions at Nanaimo, Mr. Stockett said that his company was now producing as much coal as ever the New Vancouver Coal Co., its predecessor in the Nanaimo mines, ever did in its palmiest days. Three years ago Nanaimo was considered at its lowest ebb, but now there was solid prosperity and quiet progress, exciting not the least comment, though, as he said, the amount of coal being raised equalled that of the best boom days of the island coal city.

* * *

The unprecedented winter conditions still prevail across the Northwest and in the mountains, interrupting even mail service. For days at a time no trains arrive at the coast, and mails are not received at regular intervals. Shipments of lumber have practically ceased until the weather across the continent moderates. Cars are hard to get, impossible almost, and once loaded there is no surety that they will be moved.

QUEBEC MAN HONORED.

W. H. Wiggs, of the Mechanics Supply Co., Quebec, has been in Toronto this week attending the 35th annual convention of the Ontario and Quebec Young Men's Christian Associations. Mr. Wiggs is one of the most ardent workers in Y.M.C.A. work in Canada, and was honored by being elected president of the Ontario and Quebec Associations at the convention.

PRESENTATIONS TO W. C. GURNEY

On Saturday morning last the staffs of the Gurney Foundry Co., at Toronto and the Junction, assembled at the head office to show their appreciation of their vice-president, William Cromwell Gurney, who for many years has been active in the work of the company, and who was married on Tuesday to Miss Meyer, of Toronto. First vice-president Carrick, on behalf of the joint staffs of the head office and the branches at Montreal, Winnipeg, Calgary and Vancouver, in a few fitting and well chosen words tendered their good wishes for Mr. Gurney's life as a benedict along with a substantial reminder in the form of a cabinet of silverware, comprising over 300 pieces. The superintendents and foremen, representing the mechanical staffs, also joined in the good wishes, accompanied by a Morris reclining chair upholstered in leather as well as several others to match for hall and library use. The presentations are an indication of the excellent good feeling existing between employers and employes in this old established and steadily growing business.

NOVELTIES FOR HARDWAREMEN

Readers desiring the addresses of the manufacturers or jobbers of the articles described, or any additional information, can secure the same from the Editor of HARDWARE & METAL.

PERFECT HANDLE SCREW DRIVER.

H. D. Smith & Co., Plantsville, Conn., have recently introduced to the Canadian trade their perfect handle screwdrivers, illustrated herewith. The

Pocket for Wooden Handle in Screw Driver.

wooden handle is pocketed in each side of the drop forging in such a manner, and so securely, that by test, a handle split with a cold chisel had to be further split to splinters and pried out before it could be removed.

The handles are treated with a proofing process, which renders them impervious to water. They are finished in their natural color—a rich red. The handle of the driver may be used as a hammer to force the point of the screw into the wood, owing to its partial metal construction. It may also be used with a monkey wrench for tightening or loosening cap screws, set screws, etc. The user may put all the power he wishes back of the wrench, and the manufacturer assures him of an unqualified guarantee back of the screwdriver.

The blade, bolster, handle flat and hammer head are drop forged of one piece of steel. The oval shaped handle gives greater leverage and fits the hand better. The screwdriver is well balanced and this means much to one who uses screwdrivers extensively.

The company will forward catalogues on application.

HOUSEHOLD CONVENIENCES.

The H. W. Johns-Manville Co., N.Y., represented in Canada by A. E. Patton, 54 Columbia Ave., Westmount, Montreal, are placing on the market a number of asbestos household conveniences, two of which are illustrated.

The "Milwaukee" iron holder is a soft, heat-resisting pad, doubly faced with asbestos cloth which is so fastened to the back that a pocket is made

Asbestos Toaster. "Milwaukee" Iron Holder.

for the hand, which makes possible a more secure hold on the iron and prevents the possibility of scorching the hands. This holder is thickly padded, positively resists heat, and is inodorous.

The Asbestos Toaster No. 1 consists of a heavy asbestos mat, backed with backed with sheet steel, covered with a fine wire mesh and fitted with a cold handle. It is especially designed for toasting, since it diffuses the heat over the entire surface, thereby rendering the article equally brown all over.

FLUID LEVEL INDICATOR.

The Turner Brass Works, Franklin and Michigan Sts., Chicago, Ill., are manufacturing the "Ross" Fluid Level Indicator, which is especially designed for gasoline tanks on automobiles and launches. It is a combined indicator and filler cap, and it is rigid and simple in construction. A brass socket is fitted

Machine Screw Driver in operation.

with six screw holes for fastening on to the tank. The filler cap contains the brass dial casting and indicator, which screws on to the brass socket. A leather washer intervenes between the brass socket and filler cap to make the same air tight. The filler cap contains a magnifying glass which makes it easy at all times to read the dial. Into the dial casting are screwed two rigid steel guide rods on which the float slides up and down, the centre screw revolving the pointer over the dial plate. The dial used is of special composition and is not affected by gasoline. It is secured in a groove at the bottom of the dial casting, and is easily removed to change the style of dial. The makers furnish dials marked E (empty), ¼, ½, ¾, and F (full); or 1, 2, 3, etc., in inches or in gallons when a large number of indicators is wanted for a certain size and style of tank, in which case the depth of the fluid at each gallon is to be specified by the purchaser. The device lists at $6 for a 14-inch indicator and filler socket for flat top tank, and $5 for indicator alone. For each additional 2-inch depth of a tank over 14 inches deep, an extra charge of 50 cents is added. For a cylindrical tank an extra charge of $1 is made for the larger and heavier brass tank socket.

For boat type indicators of small size an extra charge of $2 is made for a 6-inch tube, which runs from the deck to the tank, and 30 cents for each additional inch thereof. The device is made for all styles of tanks, whether cylindrical, square, oblong or otherwise.

FRUIT AND VEGETABLE PARER.

The fruit and vegetable parer shown in the accompanying cut is stamped from a single piece of steel firmly held in a substantial wooden handle. The curved blade fits the circular fruit and takes off a thin, wide paring twice as wide and twice as thin as a paring made with a knife, it is said. The parer is very economical of the fruit or vegetables, as it is made on the principle of a safety razor. The knife on the front edge can be used when necessary. The parer is pulled straight down with the right hand, turning the fruit with the left. To remove eyes, the thumb is placed on the point of the parer and by a twist of the wrist the eye is taken out. The parer is put on the market by Phoenix Hardware Mfg. Company, Homer, N.Y.

Guest's Patent Fruit and Vegetable Parer.

WHAT ASBESTOS REALLY IS.

In the important work of protecting life and property from fire and heat there is a growing appreciation of the value of asbestos, and a constant increase in its use. It has a combination of properties unlike that of any other substance found in nature. No other product as yet discovered could take its place. It has been called mineral wool, and also the connecting link between the mineral and the vegetable kingdoms. After the fibres of asbestos have been separated from their mother rock they have a fluffy softness and whiteness much like that of wool or cotton, and by a process very similar to that of ordinary weaving, they are converted into cloth. It is a cloth, however, which, owing to its mineral origin, is impervious to fire, and herein lies its value.

The clerk who believes that he is not being paid as much as he deserves, generally deserves less than he is getting. The other kind either get an advance or look elsewhere for a position.

MANITOBA HARDWARE AND METAL MARKETS

(Market quotations collected by telegraph up to 12 a.m. Friday, Feb. 15, 1907.)

Room 511, Union Bank Building, Office of HARDWARE AND METAL, Winnipeg, Man.

Milder weather and the attractions of the Bonspiel have combined to bring a good number of hardware merchants to town. The wholesale houses are busy entertaining their customers and selling them spring hardware. Travelers are able now to reach points on the branch lines which had been cut off for some time and business is showing a marked improvement in consequence.

Wire nails have been advanced to $2.40 f.o.b. Fort William and $2.85 f.o.b. Winnipeg. Turpentine has been advanced to $1.10 in Winnipeg. A few other minor changes will be noted.

LANTERNS.—Quotations are as follows: Cold blast, per dozen, $6.50; coppered, cold blast, per dozen, $8.50; cold blast dash, per dozen, $8.50.

WIRE — New prices on soft copper wire and brass spring wire have just been announced. Quotations are as follows. Barbed wire, 100 lbs., $3.22½; plain galvanized, 6, 7 and 8, $3.70; No. 9, $3.25; No. 10, $3.70; No. 11, $3.80; No. 12, $3.45; No. 13, $3.55; No. 14, $4.00; No. 15, $4.25; No. 16, $4.40; plain twist, $3.45; staples, $3.50; oiled annealed wire, 10, $2.96; 11, $3.02; 12, $3.10; 13, $3.20; 14, $3.30; 15, $3.45. Annealed wires (unoiled) 10c. less; soft copper wire, base, 36c.; brass spring wire, base, 30c.

HORSESHOES — Quotations are as follows: Iron, No. 0 to No. 1, $4.65; No. 2 and larger, $4.40; snowshoes, No 0 to No. 1, $4.90; No. 2 and larger, $4.65; steel, No. 0 to No. 1, $5; No 2 and larger, $4.75.

HORSENAILS — Lists and discounts are quoted as follows: Capewell brand, list, per lb., Nos. 5, 15c., 6, 14c.; 7 and 8, 13c.; 9 to 12 12c.; f.o.b. Winnipeg. Discounts on application. No. 10, 20c.; No. 9, 22c.; No. 8, 24c.; No. 7, 26c., No. 6, 28c.; No. 5, 32c.; No. 4, 40c., per pound. Discounts are quoted as follows: "C" brand, 40, 10 and 7½ per cent., "M" brand and other brands, 55 and 60 per cent. Add 15c. per box.

WIRE NAILS—Quoted now at $2.85 f.o.b. Winnipeg, and $2.40 f.o.b. Fort William.

CUT NAILS.—Recently advanced to $2.90 per keg.

PRESSED SPIKES — Prices are quoted as follows since the recent advance: ¼ x 5 and 6, $4.75; 5-6 x 5, 6 and 7, $4.40; ⅜ x 6, 7 and 8, $4.25; 7-16 x 7 and 9, $4.15; ½ x 8, 9, 10 and 12, $4.05; ⅝ x 10 and 12, $3.90. All other lengths 25c. extra net.

SCREWS—Discounts are as follows: Flat head, iron, bright, 85 and 10 p.c.; round head, iron, 80 p.c.; flat head, brass, 75 and 10 p.c.; round head, brass 70 and 10 p.c.; coach, 70 p.c.

NUTS AND BOLTS — Discounts are unchanged and continue as follows: Bolts, carriage, ⅜ or smaller, 60 and 5; bolts, carriage, 7-16 and up, 55; bolts, machine, ⅜ and under, 55 and 5; bolts, machine, 7-16 and over, 55; bolts, tire, 65; bolt ends, 55; sleigh shoe bolts, 65 and 10; machine screws, 70; plough bolts, 55; square nuts, case lots, 3; square nuts, small lots, 2½; hex nuts, case lots, 3; hex nuts, smaller lots, 2½ p.c.

RIVETS—Since the recent advance in copper rivets prices are quoted as follows: Iron, discounts, 60 and 10 p.c.; copper, No. 7, 43c; No. 8, 43½c.; No. 9, 45½c.; copper, No. 10, 47c.; copper, No. 12, 50½c.; assorted, No. 8, 44½c., and No. 10, 48c.

COIL CHAIN.—Prices have been revised, the general effect being an advance. Quotations now are: ¼ inch $7.00; 5-16, $5.35; ⅜, $4.75; 7-16, $4.50; ½, $4.25; 9-16, $4.20; ⅝, $4.25; ¾, $4.10.

SHOVELS—List has advanced $1 per dozen on all spades, shovels and scoops.

HARVEST TOOLS — Discounts continue as before, 60 and 5 per cent.

AXE HANDLES—Quoted as follows: Turned, s.g. hickory, doz., $3.15; No. 1, $1.90; No. 2, $1.60; octagon, extra, $2.30; No. 1, $1.60.

AXES.—Quotations are: Bench axes, 40; broad axes, 25 p.c. dis. off list; Royal Oak, per dozen, $6.25; Maple Leaf, $8.25; Model, $8.50; Black Prince, $7.25; Black Diamond, $9.25; Standard flint edge, $8.75; Copper King, $8.25; Columbian, $9.50; handled axes, North Star, $7.75; Black Prince, $9.25; Standard flint edge, $10.75; Copper King, $11 per dozen.

BUTTS—The discount on wrought iron butts is 65 and 5 p.c.

CHURNS — The discounts from list continue as before: 45 and 5 per cent.; but the list has been advanced and is now as follows: No. 0, $9; No. 1, $9; No. 2, $10; No. 3, $11; No. 4, $13; No. 5, $16.

CHISELS— uoted at 70 p.c. off list prices.

AUGER BITS—Discount on "Irwin" bits is 47½ per cent., and on other lines 70 per cent.

BLOCKS—Discount on steel blocks is 35 p.c. off list prices; on wood, 55 p.c.

FITTINGS—Discounts continue as follows: Wrought couplings, 60; nipples, 65 and 10; T's and elbows, 10; malleable bushings, 50; malleable unions, 55 p.c.

GRINDSTONES.—Quoted now at 1½c. per lb.

FORK HANDLES—The discount is 40 p.c. from list prices.

HINGES—The discount on light "T" and strap hinges is 65 p.c. off list prices.

HOOKS—Prices are quoted as follows: Brush hooks, heavy, per doz., $8.75; grass hooks, $1.70.

CLEVISES—Price is now 6½c. per lb.

STOVE PIPES—Quotations are as follows: 6-inch, per 100 feet length, $9; 7-inch, $9.75.

DRAW KNIVES—The discount is 70 per cent. from list prices.

RULES—Discount is 50 per cent.

WASHERS—On small quantities the discount is 35 p.c.; on full boxes it is 40 p.c.

WRINGERS — Prices have been ad-

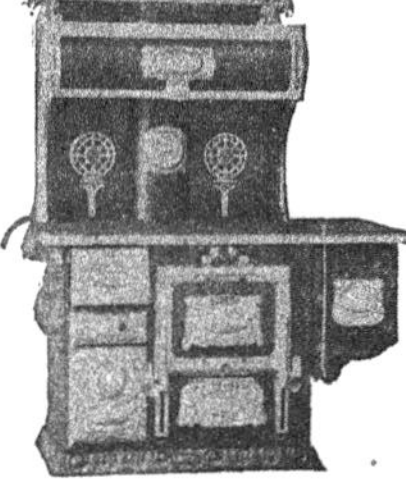

TINWARE, ETC.—Quoted as follows: Pressed, retinned, 70 and 10; pressed, plain, 75 and 2½; pieced, 30; japanned ware, 37½; enamelled ware, Famous, 50; Imperial, 50 and 10; Imperial, one coat, 60; Premier, 50; Colonial, 50 and 10; Royal, 60; Victoria, 45; white 45; Diamond, 50; Granite, 60 p.c.

GALVANIZED WARE. — The discount on pail is now 37½ per cent.; and on other galvanized lines the discount is 30 per cent.

CORDAGE—We quote: Rope, sisal, 7-16 and larger, basis, $11.25; Manila, 7-16 and larger, basis, $16.25; Lathyarn, $11.25; cotton rope, per lb., 21c.

SOLDER—Quoted at 27c. per pound. Block tin is quoted at 45c. per pound.

VISES — Prices are quoted as follows: "Peter Wright," 30 to 34, 14½c.; 35 to 39, 14c.; 48 and larger, 13½c. per lb.

ANVILS—"Peter Wright" anvils are selling at 11c. per lb.

CROWBARS—Quoted now at 4c. per lb.

POWER HORSE CLIPPERS — The "1902" power horse clipper is selling at $12, and the "Twentieth Century" at $6. The "1904" sheep shearing machines are sold at $13.60.

WRINGERS — Prices have been advanced $2 per dozen, and quotations are now as follows: Royal Canadian, $35.00; B.B., $39.75, per dozen.

FILES—Discounts are quoted as follows: Arcade, 75; Black Diamond, 60; Nicholson's, 62½ p.c.

LOCKS.—The discount on Peterborough and Gurney locks is now 40 per cent.

BUILDING PAPER. — Prices have been advanced and quotations are now as follows: Anchor, plain, 66c.; tarred, 69c; Victoria, plain, 71c; tarred, 84c

AMMUNITION, ETC.—Quotations are as follows: Cartridges, Dominion R.F. 50 and 5; Dominion, C.F., 33⅓ C.F., pistol, p.c.; C.F., military, 10 p.c. advance. Loaded shells: Dominion Eley's and Kynoch's soft, 12 gauge, black, $16.50; chilled, 12 gauge, $17.50; soft, 10 gauge, $19.50; chilled, 10 gauge, $20.50. Shot, ordinary, per 100 lbs., $7.25; chilled, $7.75; powder, F.F., keg, Hamilton, $4.75; F.F.G., Dupont's, $5.

IRON AND STEEL.—Quotations are: Bar iron basis, $2.70. Swedish iron basis, $4.95; sleigh shoe steel, $2.75; spring steel, $3.25; machinery steel, $3.50; tool steel, Black Diamond, 100 lbs., $9.50; Jessop. $13.

SHEET ZINC — The price is now $8.50 for cask lots, and $9 for broken lots.

PIG LEAD.—The price is advancing and the range of quotations among Winnipeg houses makes it difficult to quote with exactitude. The average price is about $6.00.

AXLE GREASE—"Mica" axle grease is quoted at $2.75 per case, and "Diamond" at $1.60.

IRON PIPE AND FITTINGS— Revised prices are as follows:—Black pipe, ¼ inch, $2.65; ⅜, $2.80; ½, $3.50; ¾, $4.40; 1. $6.35; 1¼, $8.65; 1½, $10.40; 2, 13.85; 2½, $19.00; 3, $25.00. Galvanized iron pipe, ⅜ inch. $3.75; ½, $4.35; ¾, $5.65; 1, $8.10; 1¼, $11.00; 1½, $13.25; 2, inch, $17.65. Nipples, discounts 70 and 10 per cent.; Unions, couplings, bushings and plugs, 60 per cent.

LEAD PIPE—The price, $7.80, is firmly maintained in view of the advancing lead market.

GALVANIZED IRON—Quoted as follows:—Apollo, 16 gauge, $3.90; 18 and 20, $4.10; 22 and 24, $4.45; 26, $4.40; 28, $4.65, 30 gauge or 10¾ oz., $4.95; Queen's Head, 24, $4.50; 26, $4.65; 28, $5.00.

TIN PLATES—We now quote as follows: IC charcoal, 20 x 28, box, $9.50; IX charcoal, 20 x 28, $11.50; XXI charcoal, 20 x 28, $13.50.

TERNE PLATES—Quoted at $9.

CANADA PLATES—Quoted as follows: Canada plate, 18 x 21, 18 x 24, $3.40; 20 x 28, $3.65; full polished, $4.15.

BLACK SHEETS—Prices are: 10 to 16 gauge, 100 lbs., $3.50; 18 to 22, $3.75; 24, $3.90; 26, $4; 28, $4.10.

PETROLEUM AND GASOLINE.—Silver Star in brls. per gal., 20c.; Sunlight in brls, per gal., 21c.; per case, $2.30; Eocene in brls, per gal., 23c.; per case, $2.50; Pennoline in brls., per gal., 24c.; Crystal Spray, 23c.; Silver Light, 21c.; Engine gasoline in barrels, per gal., 28c., f.o.b. Winnipeg in cases, $2.75.

PAINTS, OILS AND TURPENTINE—Turpentine has been advanced to $1.10. We quote: White lead, pure, $6.50 to $7.50, according to brand; bladder putty in barrels, 2½c.; in kegs, 3¾c.; turpentine, barrel lots, Winnipeg, $1.01; Calgary, $1.08; Lethbridge, $1.08; Edmonton, $1.09. Less than barrel lots 5c. per gallon advance. Linseed oil, raw, Winnipeg, 67c.; Calgary, 74c.; Lethbridge, 74c.; Edmonton, 75c.; boiled oil, 3c. per gal. advance on these prices.

WINDOW GLASS—We quote: 16-oz. O.G., single, in 50-ft. boxes—16 to 25 united inches, $2.25; 26 to 40, $2.40; 16-oz. O.G., single, in 100-ft. cases — 16 to 25 united inches, $4; 26 to 40, $4.52; 41 to 50, $4.75; 50 to 60, $5.25; 61 to 70, $5.75. 21-oz. C.S., double, in 100-ft. cases—26 to 40 united inches, $7.35; 41 to 50, $8.40; 51 to 60, $9.45; 61 to 70, $10.50; 71 to 80, $11.55; 81 to 85, $12.60; 86 to 90, $14.75; 91 to 95, $17.30.

STOVES AND TINWARE

CHANCES OF THE STOVE TRADE.

For some time both the stove manufacturer and the stove dealer have shared the disadvantages resulting from tradesmen in other lines taking up stoves as a side-line, and selling them to the great annoyance of the manufacturer; and to make inroads on the profits of the stove dealer. Some stove manufacturers, on whom responsibility rest lightly, have shown little insight as to the effect new methods of marketing their product would have upon the regular stove dealer.

Stoves sold by these parties are never properly set up or intelligently connected, and in numerous instances the final purchaser suffers no small discomfort and irritation from the ignorance on the part of those from whom the stove was purchased. Manufacturers should protect the stove dealer, who has his establishment open every day in the week to give assistance to a cuctomer for some manufacturer's stove.

In the course of a year stove dealers receive many visits from prospective buyers, who ask a great many questions about different lines of stoves, the answer in all cases is gladly given, and when the stove is purchased from the dealer, any little wrong has to be adjusted. Stoves which have been purchased from department stores, furniture dealers, installment houses, and mail order concerns, are too often the cheapest goods that the stove manufacturers of the country have to offer. What they lack in meritorious construction is, to deceive the buyer, covered up with gorgeous nickel trimmings, and it is a matter of common knowledge that this class of goods seldom includes in its line of samples the goods made by those stove manufacturers who are best known to the legitimate stove dealers.

There are some who are close observers of the conditions who express the opinion that this demoralizing method of marketing the stove product has reached the zenith, and that a change for the better is at hand.

Stove dealers should handle a good reliable line, also carrying all the accessories. and complete line of repairs. Nothing discourages a customer so much as having to wait until repairs are ordered from the factory. It will also give customers the opinion that either you are selling very few stoves, or are very slack in your methods, by not having all repairs parts on hand.

—

PROTECT THE CUSTOMER.

"If I was a hardware dealer," grumbled a disappointed customer, recently, "I would try and keep the commoner repairs for the goods I handled. Last month I purchased an expensive hard coal stove and yesterday by accident got a hole punched through the door. To-day I came in here on purpose for the necessary repairs and was met by the calm assurance that 'we do not keep mica.' For my part I would either keep it or quit selling stoves that require it. It's too much like telling a man to go to thunder as soon as you have made a sale and taken his money," says Hardware.

There is a good deal of justice in the camplaint and a good suggestion, for every retailer; keep in stock a reasonable supply of repairs for your own line of goods and do not fail to include in that list the parts most likely to become broken that would most effectively disable the article sold. In the case of stoves the necessary repairs would be a few sheets of mica, an assortment of stove bolts, pipe, including elbows; dampers. knobs and hinges, with a supply of furnace cement; these will give your customers a reasonable protection against unreasonable inconvenience from break-downs, whether caused by defects in the stove or by some accident. Such an outfit will be found to cost little and will render almost any stove usable until the heavier and more expensive repairs can be ordered for it from the factory. Because a fire-pot happens to crack is no reason why the purchaser must suffer the inconvenience of changing stoves or the chills of inclement weather while a new one is in transit. More than likely a few cents worth of furnace cement will run him until the new piece comes.

Of course the principle holds in other lines; so far as possible protect the customer against break-downs and when a break-down does occur protect him as effectually from serious loss and inconvenience by having at hand something to tide him over until the new piece can be obtained. The effort will be appreciated. though you may be told it through continued patronage rather than by words.

—

THE STOVE DEPARTMENT.

"The stove department in the hardware business does not," insists a retailer in an exchange, "pay a cent of profit to the average dealer, though it certainly should. A stock of stoves runs into money, and large expense is incurred in handling them, and how often sales are made at 10 per cent. margin when the actual expense is more. In my opinion a stove should never be sold at less than 25 per cent. margin, and the custom of throwing in this and that article should be stopped.

"Bazaar and department stores do not usually handle stoves because of the investment and labor involved. They want us to handle all the undesirable lines and they pirate upon our shelf and case goods, where the investment and labor are less and the profits more. Do you think we should stand with out neighbor hardwareman in maintaining satisfactory prices on all these goods which we only keep, thus enabling us to meet their price cutting on shelf and other goods? How often we pay a man $10 per week to black and set up stoves and our charge for doing so is so small that the dry goods clerk, getting $5 per week, feels he cannot afford to soil his hands in blacking a stove when he can get the hardwareman to do it for perhaps the cost of the blacking material. I speak of this to illustrate the many things we do at a less margin than we can afford to, and no one to blame but ourselves, competition not forcing us to do so—simply one hardwareman cutting the other's throat, and the chickens always come bome to roost."

CHANGE FOR THE BETTER.

Stove manufacturers are going back 25 years to the plain and more sensible ornamentation of stoves, which is more in keeping with the use and intent of a cooking stove. Nickel is not used so much, and it is hoped will not be revived. Plain goods are the innovation.

With the prosperity that this country has been blessed with during the past year, there is no doubt that the tendency has been for everyone to buy a better grade, and higher priced articles of every kind, and this certainly applied to the stove business.

The tendency of the trade is in the direction of buying good goods, not necessarily the highest grades, but honestly made stoves, that have the reputation of an experienced manufacturer behind them. Stove dealers are more inclined to connect their reputation with goods that give the best of satisfaction, and are most highly commended by those who use them. They consider that it is more important to make a sale that lays the foundation for future business, than to make a sale that will prevent future business.

Consequently the many stove manufacturers who are devoting themselves to improve the quality of their goods are experiencing the most satisfactory to improve the quality of their goods is ever becoming more pronounced, although the low-grade meets with large demand in some sections of the country.

GET AFTER STOVE TRADE.

Now when business is slack is the time to get out and hustle. There are many lines which offer good returns for such enterprise. Among these, stoves are worthy of considerable attention. Many householders need to be convinced that they should get a new stove, that an old, out of date one is a cause of loss, using much more fuel to do the same work than does a new stove of good pattern.

How can these people be reached? Many hardwaremen content themselves with drawing the attention of buyers by advertising. Some try window displays to interest the public. Others again issue circulars for distribution to their customers. An increasing number of stove dealers are however more insistent. They make a point of personally interviewing everyone whom they feel might possibly be persuaded to get a new stove. The latter plan is well worth trying.

NEW ONTARIO PROSPEROUS.

James Purvis, of Purvis Bros., Sudbury, who was one of the big Algoma delegation which waited on the Ontario Government this week, called at the Toronto office of Hardware and Metal on Thursday. Mr. Purvis says last year was a splendid one for all New Ontario hardwaremen, but 1907 bids fair to be even better. If the lumbermen had called a convention and prevailed upon the weather man to allow them to run his business this winter, they could not have made a better job of it and conditions could not have been more favorable. All winter and lumber supply lines have sold in large quantities, and with the selection of Sudbury as a new district judicial centre much new building would result, in addition to the natural development of the town.

LETTER BOX.

The Ontario Nickel Company, Worthington, Ont., writes: "Please inform us who makes the Spooner Babbit Metal?"

Answer.—Alonzo W. Spooner, Port Hope, Ont.

Taylor Bros., Carleton Place, ask: "Can you get us a quotation in Toronto on slate blackboards 3 x 15 feet?"

Answer.—Steinberger & Hendry Company, 37 Richmond Street West, Toronto, have supplied the information by letter.

Gordon Ritchie has just returned from his initial trip for the McClary Manufacturing Co. One well known French Canadian merchant in the eastern townships greeted him with: "Bon jour Monsieur Reechay, I seed your picture in Hardware and Mettall. I hexpee you since long taaime."

FOUNDRY AND METAL INDUSTRIES

The production of copper in the United States has increased from 27,000 tons in 1880 to 430,000 in 1906, and the United States now furnishes over 57 per cent. of the world's supply.

The railway construction works and rolling mills, which have been proposed for Sydney, C. B., will be erected in the Spring, as sufficient capital for the financing of the scheme is available.

A $10,000,000 corporation is in process of formation with a view to entering into competition with the Pittsburg Reduction Co., the heaviest producer of aluminum in the United States. It is reported that the enterprise will erect a plant along the Cumberland River, in Kentucky, at a cost approximating $3,500,000.

A delegation composed chiefly of members of the legal profession waited on the Hon. Frank Cochrane, Minister of Lands, Forests and Mines, on behalf of their clients, to disapprove of a plan to impose a tax per acre on mining lands. He intimated that at least part of the revenue derived from taxation of the mines of the province would be applied to encourage the smelting and refining of ores in Ontario, as it would increase in the value of mining properties, besides allowing their owners to work deposits of ore of lower grade than was profitable at present.

The Canadian branch of the American Locomotive Works at Montreal, is largely extending its structural steel plant, and has so many orders booked for this kind of steel that even the enlarged plant will be kept running to its fullest capacity throughout the year.

The Ottawa Car Company has in view an extensive expansion of its business. It will conduct steam and freight cars and railway rolling stock generally, in addition to its present business in electric, as well as carriage work of all kinds, including ordnance and other wagons required for the army service. New works, covering several acres, will be built.

A branch establishment of the Campagnie des Metaux Unital will likely be located in Toronto. Two of the Jacques, of Paris, France, sons of the proprietor of the company, are at present in Toronto. They intend to exploit their product, a metal for the manufacture of mining drills and for tools of all kinds The claim is made that by a patent process in connection with a special ore secured in Algeria, it is far superior to steel, wearing a great deal longer, and being much more easily handled by blacksmiths. It is so hard that it will cut glass like a diamond, yet will batter out without breaking. The durability of this product will be tested at the Cobalt mines. Should the factory locate in Toronto, employment will be provided for 2,000 or more men.

CRANE COMPANY'S ANNUAL GIFT.

The close of 1906 rounded another successful year for the Crane Company, Chicago, and, following a custom adopted in 1900, the company generously shared the fruits of their success with its employes. On a basis of 10 per cent of the earnings of employes for the year, there was distributed as a Christmas present, in round numbers, $330,000. This was $25,000 more than the distribution of 1905, and makes a total of $1,500,000 thus distributed to date. Every employe, irrespective of his position or length of service, received 10 per cent. of his year's earnings in a lump sum, accompanied by a card bearing the good wishes of the company and a bit of wholesome advice as to the wisdom of preparing for the inevitable "rainy day." Six thousand employes were beneficiaries of the latest gift of the Crane Company.

STEAM ECONOMY IN ROLLING MILLS.

A Rateau exhaust steam accumulator and turbine has been installed at the Hallside Works of the Steel Company of Scotland to use the exhaust steam from one high pressure cogging machine with two cylinders, each 40 x 60 in., one finishing train engine with two cylinders, each 42 x 60 in., two small mill engines and four steam hammers, delivering a total of about 41,000 ℔. of steam per hour. The power is used for lighting and to drive rolls, saws, sand blast apparatus and machine tools. The accumulator, which has two compartments, separates from the water any oil carried over, and maintains the water in energetic circulation. It has a diameter of 11 ft. 6 in. and a length of 34 in. The turbine is rated at 700 h.p. and has 11 sets of blades, with wheels 40 in. in diameter. It develops its full power at a speed of 1,500 revolutions per minute, and when working at an overload of 10 per cent. the inlet pressure is never over 12 ℔. absolute. The vacuum is 28 in. The generator was designed to deliver 2,000 amperes, and has carbon brushes and a commutator provided with special ventilating ducts through the centre. The ventilation is set up by means of a fan blade attached to each bar. Tests of the unit have shown a consumption of 66.4 ℔. of steam per kilowatt hour at one-tenth load, and with an absolute inlet pressure under 3 ℔. per square inch. At 200 k.w. load and an absolute admission pressure of 5.35 ℔., the consumption was 42 ℔. per unit. A half load and an admission pressure of 8.25 ℔. absolute, the steam used was 37 ℔. per unit. With a load of 450 k.w. and an admission pressure of 11.4 ℔. absolute, the figure was 36.6 ℔. The vacuum fell during these tests from 28.7 in. to 27.9 in., as the load increased.

SEWER PIPE

BUILDING AND INDUSTRIAL NEWS

HARDWARE AND METAL would be pleased to receive from any authoritative source building and industrial news of any sort, the formation or incorporation of companies, establishment or enlargement of mills, factories, foundries or other works, railway or mining news. All correspondence will be treated as confidential when desired.

Universal Systems, Toronto, have decided to erect a plant in Montreal.

A new Baptist church will be built on Pape Avenue, Toronto, at a cost of $4,500.

A new hotel and depot will be built by the Grand Trunk Railway in Ottawa, to cost from a million to a million and a quarter dollars.

A company has been formed in Kingston for the manufacture of brick. It will be called the Perfect Brick and Tile Company, having a capital of $50,000.

The contract for the erection of the Peterborough armouries has been let. The successful tenderer is Geo. H. Proctor, Sarnia, and the figure of the contract signed is $125.190.

The Ottawa Car Company will extend their plant and branch out into the manufacture of steam railway rolling stock construction. The new shops will cover several acres of land.

Permits have been issued in Toronto for the erection of a grand stand and horticultural buildings at the Exhibition grounds. The former will cost $216,465 and the horticultural building $90,000.

The A. Workman and Company has been incorporated with a share capital of $60.000 for the purpose of manufacturing and dealing in hardware. The head office will be at Ottawa, the directors being T. Workman, A. Workman, and A. A. Whillians, of Ottawa.

M. McKenzie & Company have been incorporated with a capital of $75,000 for the purpose of manufacturing railway, marine and contractors' supplies. The head office will be at Montreal, the directors being, W. D. Hamilton, A. A. Lunan, A. Dunn, L. Lahaye and A. Lunan of Montreal.

The King Edward Hotel, Toronto, is to be enlarged this spring when a two-storey addition will be added. This will provide 200 more rooms, 100 on each flat. A lot of new machinery will be installed for lighting and other purposes, and for power to run a number of up-to-date labor-saving machines.

The Dominion Smelters, Limited, has been incorporated with a share capital of $1.000.000 for the purpose of carrying on business as miners and refiners. The head office will be in Sault Ste. Marie, Ont., the provisional directors being H. H. Muggley, W. S. Peters and T. J. MacCune, of Oshkosh, Wisconsin.

The Manitoba and Saskatchewan Coal Company has been incorporated with a capital stock of $1.000.000 for the purpose of carrying on a mining business, also to manufacture brick, tile, pipe, pottery, earthenware, etc. The head office will be at Winnipeg, the directors being W. W. McMillan, E. Thomson, A. S. Swinford and J. C. Thomson, of Winnipeg, Man.

The Manitoba Linseed Oil Mills has been incorporated with a capital of $200,000 and the company has a site in Winnipeg. They will commence building operations in the Spring and intend erecting a thoroughly well equipped and up-to-date plant for the manufacture of raw and boiled linseed oil and refined varnish oils. It is intended to have the plant in operation to handle the crop of 1907.

The Theodore Lefebvre and Company has been incorporated with a capital stock of $80,000, for the purpose of manufacturing all kinds of paints, oils, chemicals and drugs. The head office will be at Montreal, the directors being C. A. M. Lefebvre, L. M. T. Lefebvre, R. B. Lefebvre, and L. S. Lefebvre, of Montreal.

The Foreign Rail Joint Company has been incorporated with a share capital of $50,000 for the purpose of manufacturing rails, rail joints, angle bars, and all kinds of railway supplies and railway contractors' supplies. The head office will be at Toronto, the provisional directors being J. S. Lovell, W. Bain, R. Gowans, and E. W. McNeill.

The Georgian Bay Oil Company has been incorporated with a share capital of $1,000,000 for the purpose of carrying on business as gas producers and petroleum oil refiners. The head office will be in Fort Erie, the provisional directors being S. Johnston, F. B. Johnston, F. R. MacKelcan, A. J. Thomson, and R. H. Parmenter, of Toronto.

The Canada Arms and Rifle Sights Company has been incorporated with a share capital of $200,000 for the purpose of manufacturing guns, rifles, and other small arms; also to acquire the business of the Mitchell Rifle Sight Company. The head office will be at Toronto, the provisional directors being H. Dixon, T. Cocking, and R. Staite, of Toronto.

Building this coming spring will be much more expensive than was the case last season. The reason for this is mainly owing to an advance in the prices of woodwork materials, which have rapidly increased in every branch, lumber having advanced from $2 to $3 on each 1,000 feet. Laths and shingles are also away in advance of former prices.

The Sanderson-Harold Company, of Paris, manufacturers of refrigerators, screen doors, etc., contemplate building a large addition to their factory, making it double the capacity of the present one. The move is compulsory in order to keep pace with the growth of business. 70 hands are now employed and the factory has been running day and night for months.

The Brockville Cement Pressed Brick and Concrete Company has been incorporated with a share capital of $10,000, to manufacture cement bricks, cement blocks, cement tile, and all building material made from cement. The head office will be at Brockville, the provisional directors being B. Dillon, A. T. Wilgress, R. N. Horton, J. C. Yanwood, and H. A. Stewart, of Brockville.

"Warden King, Limited," has been incorporated with a capital stock of $1,000,000 for the purpose of manufacturing all kinds of heating apparatus, soil pipes, steam fittings, builders', plumbers', and steamfitters' supplies. The head office will be in Montreal, the directors being J. C. King, L. A. Payette, R. C. McMichael, of Montreal, and W. Grieg and F. G. Bush, of Westmount, Que.

The Canadian Rand Company, has been incorporated with a capital stock of $500,000, for the purpose of taking over the Canadian Rand Drill Company and continue in the manufacture of air compressors, rock drills, pumps, pneumatic tools, etc. The head office will be at Sherbrooke, Que., the directors being, G. Doubleday, New York; E. Webber, of Montreal; S. W. Jenckes, W. Fanwell and H. D. Lawrence, of Sherbrooke, Que.

The Starke-Seybold Company Montreal, has been incorporated with a capital stock of $400,000, having amalgamated the Starke Hardware Company, composed of W. Starke, G. R. Starke, and R. Starke, and the Seybold and Sons Company, composed of J. P. Seybold, G. C. Seybold and H. B. Seybold, to carry on business as importers, manufacturers and dealers in hardware, metals, paints, oils, etc.

Ham and Nott, manufacturers of refrigerators and screen doors, Brantford, will likely build another plant, double the size of present one. Owing to being unable to make satisfactory arrangements with Ottawa it is probable they will not go there. The company at present employ 125 hands and are working day and night. In the event of the management maintaining their head plant at Brantford, the number of employees will be doubled.

It is said that the cement works at Longue Point, near Montreal, operated by James Morgan, have been sold to the Fordwick Co., of Virginia, which has decided to establish a second plant in the vicinity of Montreal, costing in the neighborhood of $2,000,000, with a capacity of 6,000 barrels per day. The Fordwick Co. will also build large docks to store coal from Sydney. The engineering firm of W. S. Barstow & Co., of New York, are preparing the plans and specifications, and have given the building contract to the Canadian White Co., of Montreal. The new plant will employ from 1,000 to 2,000 men.

A meeting of the Starr Manufacturing Company, of Dartmouth, was held last Thursday in Halifax, N.S., when several important changes were contemplated and decided on. The capital stock was increased to $1,000,000 and it was also decided to amalgamate the Dartmouth Rolling Mills Company with the Starr Manufacturing Company. The works of the combined companies will be enlarged. The skate business of the Starr factory has been gratifying during the past year and the orders already booked for the year 1907 from the western provinces are largely in excess of any preceding year in the history of the company.

The commerce through the Canadian and American canals at Sault Ste. Marie in the navigation season of 1906 aggregated 51,751,080 net tons, an increase of approximately 7,500,000 tons over the preceding season, and of nearly 20,000,000 over that of 1904. The United States canal cared for 45,180,-292 tons, and the canal in Canadian territory for 6,570,788, while of the entire movement, 41,584,905 tons moved in an easterly and 10,166,175 in a westerly direction. The principal items in the former movement were 138,607,764 bushels of grain, 6,484,754 barrels of flour, and 35,357,042 net tons of iron ore, while of the latter, 8,739,630 net net tons of coal and 984,265 net tons of general merchandise composed the bulk of the tonnage. As compared with like like shipments in 1905, grain gained over 30,000,000 bushels, flour more than 700,000 barrels, iron ore approximately 4,000,000 tons, and coal over 2,000,-000 tons.

The recent declaration of an extra dividend of 1 per cent. by the directors of the International Silver Co., of Niagara Falls, N.Y., on the preferred stock, is considered by many financial interests as equivalent to placing the stock on a 5 per cent. dividend basis, as it is believed that the corporation will continue to pay such an extra dividend of 1 per cent. every year in addition to its regular 4 per cent. per annum, payable quarterly. The last quarterly dividend of 1 per cent. was paid on Jan. 1, and the next is due on April 1. The extra dividend of 1 per cent. will be paid on March 1. There is now outstanding $6,607,500 of preferred stock, upon which the directors of the company eventually intend to pay 7 per cent. yearly. On Jan. 1, 1903, the company issued scrip for 21½ per cent. unpaid dividends on the preferred stock, and it is now estimated that the corporation is in arrears to the extent of 36½ per cent. on unpaid preferred dividends.

READY FOR BUSINESS.

The new silverware factory at Niagara Falls of the McGlashan, Clark Co., is now in full running order. It consists of a two-storey solid concrete building, floors and all, inside dimensions, 300x 70. It is situated on Palmer avenue and extends to the main line of the Michigan Central Railway.

This factory has been fitted out with all the latest machinery and new devices for producing the latest designs in cutlery, electro-plate, and solid nickel-silver flatware, and no expense has been spared in completing this modern factory.

Lee McGlashan, the manager, who has had many years' experience in the silverware business, is a son of the late Leonard McGlashan, well-known to the trade throughout Canada in connection with this line of business. A number of skilled workmen, who were trained by him, have assumed charge of the different departments in connection with the new factory. An invitation is extended to the trade to visit this factory when at the Falls.

CATALOGUES AND BOOKLETS

When sending catalogues for review, manufacturers would confer a favor by pointing out the new articles that they contain. It would assist the editor in writing the review.

By mentioning HARDWARE and METAL to show that the writer is in the trade, a copy of these catalogues or other printed matter will be sent by the firms whose addresses are given.

Paterson Roofing Calendar.

The Paterson Company, manufacturers of felt paper, tarred felt, wire edge ready roofing, etc., of Toronto and Montreal, have issued a beautiful calendar for 1907 practically devoid of advertising, but drawing attention to the different lines of goods manufactured by them.

Electric Conveniences.

Electric heating is the title of the book which the Canadian General Electric Company, Toronto, are sending to the trade. This book contains cuts of many useful articles which are used for traveling and can be attached to electric lights. In the many articles illustrated some will be found very useful for the hardware stock. Mention this paper when writing.

Carriage Trimmings.

H. D. Smith & Company, of Plantsville, Conn., will send to the trade their etxtensive catalogue, fully illustrated with carriage trimmings manufactured by them. They also make the perfect handle screwdriver, double screwdriver bits and lineman's climbers, all good lines for hardware merchants.

Sprayers.

The Barnes Manufacturing Company, of Mansfield, Ohio, have issued their spray pump catalogue illustrating their many different lines of pumps for various uses. A few handy formulas are quoted in the back of the book for the making of different solutions, for spraying trees, potato plants, etc., also a very handy calendar telling when to use different solutions, and for what purpose intended. This catalogue will be found very instructive by the hardware merchants.

Electric Fixtures.

R. E. T. Pringle & Company, Montreal and Toronto, are sending several pamphlets, containing cuts and descriptions of new electrical lines manufactured by them. Greenwood's Safety Guard, for electric lamps, is constructed of bars of flat steel, making it very rigid and strong. A very strong feature of this guard is the arrangement of inner guard wires, with coil springs at the top, which, when the bulb is enclosed in the guard, keep the bulb in a central position, greatly reducing the possibility of breaking. Copies can be secured on application by mentioning this paper.

1866-1906.

Forty Years' Record is the title of the book issued by A. C. Leslie & Co., Montreal, giving a very interesting account of the life of the late Alex. C. Leslie, founder of the firm. Likenesses of the present members, and sales staff of the firm, and views of their enlarged premises, which they have been continually adding to, are also given. Copies will be sent to all dealers in metals who write, mentioning Hardware and Metal.

Refrigerator Catalogue.

We have received from C. P. Fabien, 3167-3169 Notre Dame St., Montreal, a very handsome little catalogue, about 6 x 9 inches in size. This book is printed on very fine glazed paper, and the handsome half-tone cuts show a complete line of refrigerators, from the small article used in every house, to the largest hotel refrigerator. The catalogue also gives the sizes, ice capacity, shipping weights and prices of the different lines. Mr. Fabien will be pleased to forward one of his booklets to any person who will mention Hardware and Metal when writing.

Fine Gas Range Catalogue.

The McClary Manufacturing Company of London have just issued their first complete catalogue of gas ranges, having decided to go into the gas range business extensively. The general get-up of the book is striking, the cover being particularly attractive, "Caloric Gas Ranges, Catalogue 68," standing out clearly in white letters on a green background. The gas stove business is steadily increasing in importance, and if gas is available in the district, no dealer should neglect this branch of the trade. In the cities, gas is steadily pushing coal into the background as a fuel for cooking both in summer and winter, although as yet it is not an important factor in heating.

The catalogue comprises 32 pages and cover, innumerable illustrations being given showing the construction of the "Caloric," the many styles in which it is made, as well as a fairly complete line of utensils specially serviceable for use on gas stoves. List prices are also quoted.

Regular customers have been supplied with the book, but its utility is such as to warrant everyone in the trade who handles heating or lighting goods to send for a copy. Use the firm's letter-head and mention this paper when writing.

WORLD-WIDE TRADE.

A very interesting advertisement appeared in the special merchant shippers' edition of "The Ironmonger" of Jan. 19th. This was a double page display illustrating this company saying, "The sun never sets on Lysaght's galvanized iron." It showed twenty steamers on which shipments has recently been made from English ports to merchants in Asia, Africa, Australia, North and South America, and many islands of the seas. The idea of having an illustration of a steamer in each case making two lines of ten each, made a most striking way of showing what a firm hold on the world's trade Britain has in this line.

CONDENSED OR "WANT" ADVERTISEMENTS.

SITUATIONS VACANT.

WANTED—Experienced pricer for wholesale hardware; none but experienced man need apply. Box 578, HARDWARE AND METAL, Toronto.

BRIGHT, intelligent boy wanted in every town and village in Canada; good pay besides a gift of a watch for good work. Write The MacLean Publishing Company, 10 Front Street E., Toronto. (tf)

BUSINESS CHANCES.

HARDWARE stock about $7,000, including set of tinners' tools; stock is all new and situated on main line C.P.R. in Saskatchewan town of 1,800 population; building can be leased, is brick, two storeys and basement; net profit $4,000 yearly. Box 577, HARDWARE AND METAL, Toronto. [7]

FOR SALE—A fully equipped Wire Mill and Nail Factory as a going concern, with about 500 tons of wire rods; factory situated on 150 ft. of water-front with trackage optional to buyer. B.C. Wire and Nail Co., Ltd., Vancouver, B.C. (7)

FOR SALE—Established profitable furnace, stove and tinware business in live growing manufacturing city in old Ontario; excellent stock, good stand with long lease; business growing rapidly (last year nearly $20,000); stock and tools about $5,000. Box 585, HARDWARE AND METAL, Toronto. [7]

FOR SALE—Established, profitable furnace stove and tinware business, in live growing manufacturing city in old Ontario. Excellent stock. Good stand with long lease; business growing rapidly (last year nearly $20,000). Stock and tools about $5,000. Box 588, HARDWARE AND METAL, Toronto. [7]

HARDWARE business established fifty years, for sale, including stoves, ranges and tinware; good clean stock, reduced to about $2,000. Box 35, Newcastle, Ont. [7]

PLUMBING, Steam, Hot Water, Gasfitting and Tinsmithing business for sale; $15,000 business done last year, and only $3,000 or less invested; in one of the best towns in Canada. Waterworks and sewage. Owner retiring. Box 587, HARDWARE AND METAL, Toronto. [10]

HARDWARE and Harness Business for sale in good town. Owner retiring for good reason. Stock about $2,500, in excellent shape. This is worth investigating, for particulars. apply Richard Tew, 23 Scott Street, Toronto.

FOR SALE—In Eastern Ontario, hardware, tinsmithing and stove business, stock reduced to $1,500; doing a good steady business, expenses low; must sell ot once on account of ill-health. Building, including dwelling, for sale or rent. Apply to Box 590, HARDWARE AND METAL, Toronto. [8]

FOR SALE.

FOR SALE—Hardware, tinsmithing and stove business; stock about $3,000, doing a good steady business; expenses low. Apply Box 582 HARDWARE AND METAL, Toronto. (8)

HARDWARE Business; including stoves, tinware and tinsmith tools, in thriving town in West Ontario peninsular, stock about $5,000; building can be leased if desired, dwelling also. Box 583 HARDWARE AND METAL, Toronto. (17)

ARTICLES WANTED.

WANTED—8 foot cornice brake; state price and particulars. J. E. Hussey, Melbourne, Ont. (9)

ANY hardwaremen having a stock of cut nails they wish to dispose of at reduced prices should write at once to Box 501 HARDWARE & METAL, Toronto.

WANTED.

WANTED—By a competent and reliable man, a small set of books to keep in spare time, accounts made out, etc. References. Address Bookkeeper, HARDWARE AND METAL, Montreal. [7]

WANTED—Six-foot brake and thirty-inch tinsmith's grover. Box 586, HARDWARE AND METAL. [8]

It will pay you

We give herewith the experience of one of our subscribers. There is no reason why you cannot obtain just as good results.

THE PEOPLE'S HARDWARE STORE.

HARDWARE AND METAL
Toronto, Ont.
Essex, Jan. 27, 1906

Dear Sirs—Within one month I have "bought and sold" with the best of satisfaction—the result of $1.68 worth of condensed ads. in HARDWARE AND METAL.

In the future when I want to know anything in the hardware business I will seek your paper.

Yours truly, IVESON L. COOKE.

Remember your ad. will be read by Hardware Dealers in all parts of Canada.

2 cents per word for first insertion
1 cent " " subsequent insertions

Send cash with advertisement, 5c. additional for Box number

HARDWARE AND METAL

MONTREAL TORONTO WINNIPEG

Paint, Oil and Brush Trades

BLANCHITE SELLING AGENTS.

The Blanchite Process Paint Company, which, for a young concern, has started business in Canada with a vigor which presages success, has arranged as follows for the placing of their paints and enamels with the retail trade:

Central Ontario will be handled by H. Lyn Hudson; Western Ontario will be handled by G. W. Pinkerton; Eastern Ontario, including Ottawa, will be handled by the General Supply Company of Canada, who will have representatives in the city of Ottawa and two travelers on the road; the city of Montreal will be handled by Kenelm M. Trigge, of the General Supply Company of Canada; the city of Winnipeg and the Province of Manitoba will be handled by H. W. Glassco, who has been for years connected with the firms of J. H. Ashdown & Company, and Geo. D. Wood & Company.

Rice Lewis & Son, Toronto, are now handling Blanchite paints exclusively, and their traveling representatives will also sell the paints throughout Ontario.

NEW WINNIPEG COMPANY.

The Manitoba Linseed Oil Mills has been incorporated with a capital of $200,000, and have purchased a site in Winnipeg. They will commence building operations in the early spring, and intend erecting a thoroughly well equipped and up to date plant for the manufacture of raw and boiled linseed oil and refined varnish oils. It is intended to have the plant in operation to handle the crop of 1907.

MONTREAL FIRM ENLARGES.

The Canadian Economic Lubricant Company, of Montreal, manufacturers of oils and grease, have recently enlarged their premises by taking up the entire block on Wellington street, their numbers now being from 23 to 29. W. B. Converse, formerly of the Consumers' Cordage Company, and now managing director of the Colonial Cordage Company, Toronto, is largely interested in the lubricant company.

USES FOR TURPENTINE.

There are few housekeepers that are not familiar with some of the numerous uses of turpentine, says the Drug Review, and as its odor is clean and wholesome it has the advantage of many remedies whose odors are offensive.

Turpentine and soap will remove ink stains from linen.

A few drops added to water in which clothes are boiled will whiten them.

It will exterminate roaches if sprinkled about.

Moths will leave if it is sprinkled about.

from furniture cause by water.

Turpentine will remove wheel grease, pitch and tar stains.

Clean gilt frames with a sponge moistened in turpentine.

Ivory knife handles that have become yellow can be restored to their former whiteness by rubbing with turpentine.

Carpets can be cleaned and colors restored by going over occasionally with a broom dipped in warm water to which a little turpentine is added.

An equal mixture of turpentine and linseed oil will remove white marks

AN OILY CUSTOMER.

"Please, sir," piped the tiny customer, whose head scarcely reached the counter, "father wants some oak varnish."

"How much does your father want, my little man?" asked the shopman.

"Father says, will you fill this?" said the little fellow, handing him over a half-gallon can.

It was duly filled and handed over.

"Father will pay you on Saturday," said the recipient, casually. And then the face of the shopman grew dark.

"We don't give credit here," he said. "Gimme back the can."

Meekly the little lad handed back the can, which was emptied, and handed back to him with a scowl.

"Thank you, sir," he said. "Father said you'd be sure to leave enough in the can, round the sides, for him to finish the job he wants to do, and I think you 'ave, sir."

ARGENTINA'S SEED SUPPLY.

Argentina, as a source for the supply of linseed, has not lately been doing well. The coming yield is put at 50,000 tons less than the quantity forecasted by the Government Agricultural Department. The exports up to now—about 350,000 tons—are 150,000 tons less than the quantity for the same period last year. Touching the next harvest, prospects happily were better, as the drought, which had lasted in the linseed areas almost six months, had come to an end with a copious rainfall.

Wm. O. Greenway, representing a number of British color and other manufactures, has opened an office at 13 St. John St., Montreal. Mr. Greenway is up-to-date, and a hustler, and although he has only been established here for a couple of weeks, he has already called on most of the important men in the trade.

Paint and Oil Markets

MONTREAL.

Office of HARDWARE AND METAL,
232 McGill Street,
Montreal, February 15, 1907

Not a change to report this week in paints or oils. All prices are reported very firm. Owing to the extremely severe weather experienced here at present, orders have fallen off slightly, but this is only thought to be a temporary lull, and from present indications, the spring will bring at least a good business in all lines.

Linseed oil and turpentine remain very firm, and a good flow of orders are coming in for Paris green.

LINSEED OIL—Very firm. We still quote: Raw, 1 to 4 barrels, 55c; 5 to 9 barrels, 54c.; boiled, 1 to 4 barrels, 58c.; 5 to 9 barrels, 57c.

TURPENTINE — Prices remain at: Single barrel, 98c. per gal.; for smaller quantities than barrels, 5c. extra per gal. is charged. Standard gallon is 8.40 lbs., f.o.b. point of shipment, net 30 days.

GROUND WHITE LEAD — We still quote: Best brands, Government standard, $7.25 to $7.50; No. 1, $6.90 to $7.15; No. 2, $6.55 to $6.90; No. 3, $6.30 to $6.55, all f.o.b., Montreal.

DRY WHITE ZINC—Still firm. We quote; V.M., Red Seal, 7½c. to 8c.; Red Seal, 7c. to 8c.; French V.M., 6c. to 7c.; Lehigh, 5c. to 6c.

White ZINC—Our quotations are: Pure, 8½c. to 9½c.; No. 1, 7c. to 8c.; No. 2, 5¾c. to 6¾c.

PUTTY—Our prices are: Pure linseed oil, $1.75 to $1.85, bulk in barrels, $1.50; in 25-lb. irons, $1.80; in tins, $1.90; bladder putty in barrels, $1.75.

ORANGE MINERAL—The following prices still hold: Casks, 8c.; 100-lb. kegs, 8½c.

RED LEAD—The following quotations are firm: Genuine red lead, in casks, $6; in 100-lb. kegs, $6.25; in less quantities at the rate of $7 per 100 lbs.; No 1 red lead, casks, $5.75; kegs, $6, and smaller quantities, $6.75.

PARIS GREEN — We continue to quote: In barrels, about 600 lbs., 23¼c. per lb.; in arsenic kegs, 250 lbs, 23½c., in 50-lb. drums, 24c.; in 25-lb. drums, 24½c.; in 1-lb packets, 100 lbs. in case, 27c.; in 1-lb. packets, 50 lbs. in case, 27c.; in ½-lb. packets, 100 lbs. in case, 27c.; in 1-lb. tins, 26c. f.o.b Montreal Terms, three months net or 2 per cent. 30 days.

SHELLAC—Prices remain, Bleached, in bars or ground, 46c. per lb., f.o.b Eastern Canadian points; bone dry, 57c. per lb., f.o.b. Eastern Canadian points, T.N. orange, etc., 48c. per lb. f.o.b New York

SHELLAC VARNISH—Very firm. We still quote: Pure white, $2.90 to $2.95; pure orange, $2.70 to $2.75; No. 1 orange, $2.50 to $2.55.

MIXED PAINTS — No change in prices. We still quote; $1.20 to $1.50 per gallon.

PETROLEUM — (In barrels) — Our prices are: American prime white coal, 15½c. per gallon, American water, 17c. per gallon; Pratt's Astral, 19½c. per gallon.

WINDOW GLASS—Prices remain firm. We still quote: First break, 50 feet, $1.85; second break, 50 feet, $1.95; first break, 100 feet, $3.20; second break, 100 feet, $3.40; third break, 100 feet, $3.95; fourth break, 100 feet, $4.15; fifth break, 100 feet, $4.40; sixth break, 100 feet, $4.95. Diamond Star: First break, 50 feet, $2.30; second break, 50 feet, $2.50; first break, 100 feet, $4.40; second break, $4.80; third break, 100 feet, $5.75; fourth break, 100 feet, $6.50; fifth break, 100 feet, $7.50; sixth break, 100 feet, $7.50; seventh break, 100 feet, $8; eight break, 100 feet, $9. Double Diamond: First break, 50 feet, $3.45; second break, 50 feet, $3.75; first break, 100 feet, $6.75; second break, 100 feet, $7.25; third break, 100 feet, $8.75; fourth break, 100 feet, $10; fifth break, 100 feet, $11.50; sixth break, 100 feet, $12.50; seventh break, 100 feet, $14; eighth break, 100 feet, $16.50; ninth break, 100 feet, $18; tenth break, 100 feet, $20; eleventh break, 100 feet, $24; twelfth break, 100 feet, $28.50. Discount on Diamond Star, 20 per cent.; on Double Diamond, 40 per cent.

TORONTO.

Office of HARDWARE AND METAL
10 Front Street East,
Toronto Feb. 15, 1907.

Orders continue very brisk with spring business starting to move, turpentine and linseed oil selling very briskly.

Owing to the five per cent. duty being removed from turpentine, it is now being quoted at a lower price. The Savannah prices are higher, however, and the forest fires which have been raging in the turpentine district for the last few weeks and have destroyed many thousand trees boxed for turpentine, will prevent any decrease in price. Barrel quantities are now quoted at 99 cents per gallon, and small lots at $1 04 per gallon.

Linseed Oil has been advanced 2c. per gal. in England, and an advance of 1c. has been made here. The high price of flax seed in all countries is the general cause of the advance.

No change is made on white lead, which remains at former quotations.

Paris Green has experienced a further advance of 1c. per lb., this being the second advance which has been made this year. It is expected it will continue to advance with every further order of any quantity which the manufacturer receives, as practically all the output is ordered in advance. The base price is now 27¼.

WHITE LEAD—Ex Toronto, pure white, $7.40; No. 1, $6.65; No. 2, $6.25; No. 3, $5.90; No. 4, $5.65 in pack-

ages of 25 pounds and upwards; ½c. per pound extra will be charged for 12½ pound packages; genuine dry white lead in casks, $7.

RED LEAD.—Genuine in casks of 500 lbs., $6; ditto, in kegs of 100 lbs., $6.50; No. 1 in casks of 500 lbs., $5.75; ditto, in kegs of 100 lbs., $6.25.

DRY WHITE ZINC.—In casks, 7½c.; in 100 lbs., 8c., No. 1, in casks 6½c. in 100 lbs., 7c. Ground in oil—In 25 lb. irons, 8c.; in 12½ lbs., 8½c.

SHELLAC VARNISH.—Pure orange in barrels, $2.70; white, $2.82½ per barrel; No. 1 (orange) $2.50; gum shellac, bone dry, 63c. Toronto. T. N. (orange) 51c. net Toronto.

LINSEED OIL.—Starting to move very freely, and quoted at a 1c. advance over last week's price, and now quote. Raw, 1 to 3 barrels, 59c.; 4 to 7 barrels, 58; 8 and over, 57c.; add 3c. to this price for boiled oil, f.o.b. Toronto, Hamilton, London and Guelph, net 30 days.

TURPENTINE.—Owing to the 5 per cent. duty being removed from turps., we now quote: Single barrels at 99c. per gal., f.o.b., point of shipment, net 30 days; less than barrels, $1.04 per gallon.

PARIS GREEN.—Another advance of 1c. per lb. is now quoted, and the demand still continues very brisk. The last advance is caused largely owing to the jobbers having to replete their stock and the general high price of raw materials from which it is manufactured. We now quote base price on both English and Canadian, 27½c. per lb.

PETROLEUM.—For refined we still quote: Canadian prime white, 14c.; water white, 16c.; American water white, 16c. to 18c. ex. warehouse. On crude the prices are: Canadian, $1.32; Pennsylvania, $1.58; Ohio, 96c.

For additional quotations, and revised list on Paris Green, see current prices at back of paper.

BOILED LINSEED OIL.

There is a general distrust of boiled oil. It is seen in the directions for the use of mixed paints, which say: "Thin only with raw linseed oil," or "Use no boiled oil." The reason, of course, is that much of the modern so-called "boiled oil" is not boiled at all, but has been treated with chemicals or litharge. Many decorators scarcely use it at all, pinning their faith to a pure raw oil, and the addition, when necessary, of driers to the paint mixed therewith.

It may be asked whether the decline of the use of boiled oil amongst decorators is not connected with the question of quality, says the Oil and Color Trades Journal. Many articles in the paint trade vary a good deal in quality, partly by the very nature of the article, though more often by sophistication; but the point is whether or not the quality of boiled oil varies more than is absolutely necessary. Of course, we are putting aside the adulterated boiled oils and the innumerable substitutes for the moment, and merely considering whether honest, straightforward manufacturers of the article need vary so much in the quality of the oil they turn out. It is a matter of very considerable importance, for in all probability a uniform fine quality in boiled oil would lead to an increased use, which would far more than repay the trouble taken to ensure a standard of high excellence.

The enormous practical difference to a painter between a badly boiled oil and a well boiled oil is, perhaps, only recognized by those engaged in the decorator's business; but manufacturers may rest assured that it is very great indeed. As all the paint world knows, there is little satisfaction in supplying raw linseed oil, for the business has been so cut that no living profit can be obtained from it; but in boiled oil there is usually a margin of profit for the manufacturer, and a substantial increase in the trade is worth some trouble to obtain. The number of drying ingredients used in the different makes of boiled oil is very considerable, and no authority has yet arisen to tell us which is the absolute best. It is not in the nature of things that they should all be alike the best, and there is no doubt that some oils are inferior by reason of the character of the driers used. Boiled oils were certainly better and more reliable in quality in the older days, when litharge was the recognized drier incorporated with the oil; but this fact does not, of course, prove that litharge is now the best drier, because at that time fire-boiled oils were the usual thing, whereas now the steam boiled is more frequently made.

It may be that the transition from fire-boiling has damaged the quality of the oil, or at any rate made it subject to variations; but probably the more specific cause for variation in the quality of boiled oil is difference in the practice of manufacturers with regard to the age or maturity of the linseed oil used for boiling purposes. Any carelessness in this respect is sure to result in an inferior article. A fine boiled oil is almost equal to a varnish in its effect upon the paint, whereas some oils that we have seen are merely brown liquids having a diluent effect—and that only—upon the stiff paints with which they are mixed.

The question arises whether it would be possible to establish some standard of quality in boiled oils upon which the user might rely; say that the drier should be fixed and openly stated, the raw oil used to be of a certain definite maturity, and the temperature and time of boiling to be guaranteed, says our contemporary. There would remain only the question of packing, and this is not unimportant, for we have known the finest boiled oil practically ruined by careless packing in damp casks. The packing of any boiled oil in casks or barrels is indeed of questionable wisdom, although it is so prevalent in the trade. Probably the adoption of the system of packing in drums would in the end soon pay for itself, for the convenience of handling would be greatly increased, and in many cases the retailer, who sells to the painter, could then sell his oil without breaking bulk. Thousands of working painters could buy a five or ten-gallon drum of boiled oil who cannot in any case take a barrel of forty-two gallons. In India oil is always sold in these small drums and if the Indian painter can afford the expense it seems reasonable to believe that the British painter could do the same. The boiled oil business certainly needs some reform if it is not to decline still further, and eventually to become quite worthless to the manufacturer.

Plumbing and Steamfitting

RE-SEALING AND ANTI-SYPHON TRAPS VERSUS BACK VENTING *

As the American Society of Inspectors of Plumbing and Sanitary Engineers is organized primarily to bring about uniform laws governing the installation and maintenance of plumbing and house drainage, and methods of their enforcement, the question of re-sealing and anti-syphon traps should be studied by them in their relation to their legitimate permissible or compulsory use in a code of building sanitation.

First of all, it should be borne in mind that a code of building sanitation to stand the test of its legality in the courts, must be drawn so as to require only the minimum of safety; otherwise it is likely to be declared illegal, should any one be disposed to attack it. It should be possible to give a valid and consistent reason for every requirement contained therein. The requirement that every fixture directly connected with the drainage system should be trapped, and a safe method of maintaining the trap seal against syphonage, and evaporation, for a reasonable length of time is unquestioned. But there is a decided difference of opinion as to which method of maintaining the trap seal is the most reliable—back venting or re-sealing and anti-syphonic trap construction—and also a difference of opinion as to the volume of water necessary to guard against complete evaporation of the trap seal, during the average length of absence in the summer months.

I propose to demonstrate that venting of fixture traps is not absolutely reliable, as to its first installation, and that after it has been used for a period of a few years it is absolutely unreliable; also to demonstrate that the vertical soil and waste pipe stack with the lateral branches extending not to exceed ten feet with waste pipe undiminished in size, so as to admit of the greatest self-cleaning action, with the use of re-sealing and anti-syphon traps, is a safer method of construction than plain traps back vented.

Back Venting Insufficient.

First, back venting, though the greatest care may be taken to make the connection to the waste pipe in the most approved manner, viz.: taken off at an angle of 45 degrees beyond the crown of the trap, in its installation, it is liable to be rendered non-efficient from the following causes: Where galvanized iron pipe is used the galvanizing may form a film which would completely stop the pipe. In the use of lead or cast iron, the molten metal may enter the pipe through the joint when it is being made, and partially or completely fill it. Oakum placed in the pipe to prevent building material from falling into it may be forgotten and allowed to remain, thus completely stopping the pipe, and if the ends of the vents are not protected during construction, building material may fall into them, and lodging in the elbows completely stop them. All this presupposes carelessness on the part of the plumber, but constitute a factor which cannot be ignored, as it often exists.

After the vent pipes have been installed, there are conditions over which the plumber has no control that operate to render the vent pipe non-efficient, viz.: the freezing of the roof outlets in protracted cold weather, the building of nests by the birds, and the falling of leaves into the roof outlets The greasy vapor from the discharge of kitchen sink waste rising in the vent pipe, and the discharge of greasy sink waste where the waste pipe is partially or entirely clogged rising in the vent pipe to the height of the water in the sink leaves a deposit of grease on the cold vent pipe, gradually stops it up entirely.

50 Gallon Tank Placed 10 feet Above Trap Connection.

There is also the proposition of the average plumber clearing the sink waste pipe by means of an opening into the waste pipe and leaving the vent untouched, or of trying to force the obstruction from the waste pipe and forcing what might be in the vent pipe up against an elbow beyond the point of accessibility, or of using a blind washer on the back vent union and forgetting to take it out. Then there is the added danger of piping through which no water passes to indicate a leak should one exist, and those of us who have made smoke tests of buildings that have been in use for a few years know to what an alarming extent this danger exists; and it is the part of wisdom to reduce it to a minimum.

As the fixture outlet is an open end plus the trap seal, and the back vent pipe cannot be relied upon, it naturally follows that to be on the safe side every trap ought to be a re-sealing trap, if placed where the discharge of fixtures above it does not exceed more than two-thirds of the area of stack into which it discharges, and the vertical discharge is only of sufficient length to permit of a re-seal to a standard depth and volume; or, an anti-syphon tray where the discharge of fixtures above it exceeds more than two-thirds of the area of the stack.

Standard Trap Construction.

It has been very gratifying to me in discussing this proposition with intelligent, up-to-date plumbers throughout the country, to find so many who believe that properly constructed and tested re-sealing and anti-syphon traps at a limit of ten feet from the stack, without back vent, is a much better construction than other traps back vented. This leads us up to the proposition of the standard of trap construction which is applicable to all forms of traps, viz.: They must be free from secret partitions through which, in case of defect, the trap seal could leak into the waste pipe unobserved, or sewer air leak into the room above the trap seal; they must he self-cleaning—that means free from deflectors or pockets that would retard the natural flow of the waste water. It is conceded by those who have given the matter careful study, that a trap that would be satisfactory under a wash bowl, bath tub, wash trays, urinals, etc., would not be satisfactory for self-cleaning action under a kitchen sink discharging greasy waste water, so that when the trap placed under a kitchen sink is other than an approved grease trap, or a plain trap, it should be equal in self-cleaning action to the plain trap. This can be accomplished by using only a drum trap with a tangential inlet. It is apparent that there should be a fixed standard of diameters and inlets for drum traps, and the following is in my judgment a satisfactory one:

Every inlet to a drum trap shall be either through the bottom, or on the side on a line with the bottom, so as to prevent any sediment pocket, and shall be either tangetial or bell-shaped. The diameter of a drum trap provided with a bottom or side inlet shall not be greater than twice the diameter of the inlet. The clearance between the outer shell of a drum trap and the stand pipe forming a centre outlet through the bottom of the drum, shall not exceed the diameter of the inlet.

The next question is the standard for a depth of seal, or equivalent in vol-

* A paper read by H. J. Luff, Sanitary Engineer, Cleveland, at the second annual meeting of the American Society of Inspectors of Plumbing and Sanitary Engineers.

ume, and in connection with this proposition it is necessary to take into consideration the fact that many people leave their homes in the summer time for from one to three months, with no thought of the effect of evaporation on the seals of the traps under their plumbing fixtures. In the winter time, if they leave home, they usually remember the effect of freezing and take the necessary precaution. I have found that a safe estimate of evaporation is one inch per month, so that it is advisable that a positive re-seal equal to the three-inch seal in a plain trap be the standard, and the depth of such re-seal should never be less than one inch.

Another feature in trap construction on which a standard should be adopted is the minimum amount of superficial surface. It is a well known fact that a re-sealing and anti-syphon trap must have considerably more superficial surface than a plain trap back vented to prevent syphonage. I have found by experiment that this amount can be kept below three times the superficial surface of a plain trap with a three-inch seal, and a space between the trap equal to the diameter of the trap, so that it would be reasonable to limit the superficial surface in a trap to three times the same size trap with a three-inch seal and space between trap equal to the diameter of trap. While it is desirable if practicable to place the cleanout of the trap in the trap seal or on the inlet side of trap, it is by no means as essential as some maintain, but when located on the outlet of the trap it should be restricted to a maximum distance of the diameter of the inlet from the top of the outlet to the under side of cleanout screw. This would permit of a pressure on the cleanout screw sufficient to indicate a leak should one exist. While it is possible to prevent trap syphonage by increasing the size of the waste pipe, and many plumbers resort to this method, the largely increased fouling surface thereby added, with the possibility of a gradually decreasing diameter from the deposits on its sides, would suggest the advisability of using re-sealing and anti-syphon trap construction and the same size waste pipe.

This brings us up to the proposition of a satisfactory test to require of a re-sealing and anti-syphon trap. If we establish the distance that a trap may be placed from the ventilating stack as ten feet and allow that ten feet to be horizontal or vertical, and re-sealing trap to be used where the discharge of the fixtures above does not exceed two-thirds of the area of the stack, it naturally follows that the following test is a safe one for the re-sealing of traps:

Safe Test for Re-sealing Traps.

Place on the trap inlet a bowl-shaped tank sufficient to hold water enough to fill thirteen feet of pipe the size of trap to be tested. The outlet from tank to be full bore, trap to have a vertical discharge of the same size as the trap, and ten feet in length. After the discharge of tank contents, the seal retained should not be less than one inch in depth and the volume of water not less than the full three-inch seal in a plain trap. No movable parts will be allowed in trap when test is being made. If an anti-syphon trap is to be used where the discharge of fixtures above may completely fill the stack, it naturally follows that the test required should be as severe as the severest condition that is likely to exist in drainage construction. We will suppose that sets of wash trays without overflow are placed above each other and discharging into the same stack, the roof outlet is frozen over or clogged from other causes, the discharge of the two topmost fixtures uniting are sufficient to fill the stack, and passing by the fixtures trap below, produce a strong syphonic action on this trap; and that at this time the plugs are placed in the wash trays above simultaneously, the water in the trap below will be thrown into violent action differing from that caused by the steady downward pull with an open end above. While this condition may seem improbable, it nevertheless is not impossible, so that the test required for an anti-syphon trap should be severe enough to cover such a possibility and the following test would, I believe, prove a safe one.

Place a 50-gallon tank 10 feet above the trap connection, connect the trap to a Y connection on a vertical pipe no smaller than the trap outlet and drop from this point thirty feet. The entire pipe shall be vertical and full bore throughout. Place a quick opening valve on the pipe above trap inlet. The entire contents of the tank, 50 gallons of water, shall be discharged through this pipe by opening and closing the valve every five seconds, after which test the seal retained shall not be less than 1 inch in depth and the volume of water not less than the 3-inch seal in a plain trap. No movable parts will be allowed in trap during the time that test is being made.

BRASS WORKS FOR GALT.

The Galt Brass Manufacturing Company has been incorporated with a share capital of $40,000 for the purpose of manufacturing all kinds of brass goods and plumbers', gas-fitters', steam-fitters' and engineers' supplies. The head office will be at Galt, the provisional directors being J. Scott, B. F. Bennett and A. Groff, of Galt. The officers will be James H. Smith, president; Peter H. Cowan, vice-president; W. D. Sheldon, secretary-treasurer, and B. F. Bennett, managing director. The company is backed by some of Galt's wealthiest manufacturers.

ROBERTSON TRAVELERS CHANGE.

E. A. Rogers, city traveler for the Toronto branch of the James Robertson Company, has resigned his position in order to go into business for himself, and J. R. Foster, who has been representing the firm in Western Ontario, will take Mr. Rogers' place. L. J. Martin is to take Mr. Foster's territory, and W. H. Cunningham will cover Mr. Martin's ground in future.

NEW BRASS GOODS FACTORY.

Following the recent changes in the organization of the Ontario Lead & Wire Co., in dropping their white lead and wire nail business, and changing their name to Somerville, Limited, another important development is to take place. A new company is to be formed and about $100,000 invested in machinery and plant for a plumbers' brass goods foundry. Some of the machinery has already been purchased, and Mr. Somerville is now away arranging for the remainder. The plant will be established in Toronto, probably in the east end.

PROTECTING PIPES FROM FREEZING.

The means generally employed to prevent pipes from freezing consist in the use of coatings which protect against cold, says an exchange, and non-conductors of heat, such as straw, cork, and oakum. There are, however, more effective agents, also practicable for use in thawing frozen pipes. The pipes are first covered with a thin layer of straw, sawdust or tanbark. Pieces of unslacked lime as large as the fist are then packed around them, and enveloped in another layer of some non-conducting material, straw, oakum, or cork, and the whole is held firmly together by means of a wrapping of coarse linen. The first layer is for the purpose of protecting the pipes from the action of the fresh lime, which would cause the metal to rust. The lime draws moisture from the air and the materials surrounding it, and is made warm by means of the chemical reaction. The outer covering allows only a small amount of atmospheric air to pass through, so that the lime remains unslacked to keep up the temperature during an entire winter.

This method, with slight variations, can be applied to the thawing out of frozen pipes. For this purpose somewhat more lime is to be packed around the pipes, and water poured over it. The heat generated will melt the ice in the pipes. The ground in winter can also be thawed out in this way, when, it is desired to lift paving stones without breaking.

PLUMBING PARAGRAPHS.

The Calgary Plumbing & Heating Co., Calgary, Alta., have dissolved.

The Northern Plumbing & Heating Company, Winnipeg, Man., has been dissolved and the business in the future will be carried on by C. H. Rivercomb and E. T. Mitchell.

W. H. Meadows, J. N. O'Neil, and Benjamin Kirk, plumbing inspectors, Toronto, have petitioned for an increase in their salaries. Although employed by the city for 19 years, they say, they are now getting only $75 more than at that time.

Two hundred and fifty people were present at the euchre and dance given on Feb. 7 by the Montreal Master Plumbers' Association, and the function proved more than ordinarily enjoyable. The earlier part of the evening was given over to the euchre, for which there were sixty tables set. Thereafter the members of the association with their guests and lady friends sat down to a charming supper. From midnight until a late hour dancing was carried on with spirit and gusto. Among the officers of the association present were: Messrs. N. Turcot, president; J. A. Godon, first vice-president; J. A. Marier, second vice-president; J. E. Walsh, secretary; John Watson, treasurer, Jos. Thibeault, Jos. Lamarche, P. C. Ogilvie and Jos. Laurier, along with the officers, forming the committee.

Plumbing Markets

MONTREAL.

Office of HARDWARE AND METAL,
232 McGill Street.
Montreal, February 15, 1907

The feature of the market this week is the advance in plumbers' brass goods; all lines of taps, stop cocks, etc., having taken their place in the limelight by jumping up 10 to 15 per cent., and in some cases even higher. The reason this advance has been delayed so long is due to the fact that manufacturers have been standing aside to see what height the metal market would reach. Now that there seems to be more steadiness in that quarter, they consider that the time has arrived when they may safely draw up their new schedule of prices.

Cast iron fittings are reported very hard to get here, but no advance has yet been made over our last quotations.

Considerable uncertainty seems to prevail in regard to what trend the markets may take in the near future, some jobbers predicting that prices have about reached the limit, while others say that many lines are yet due to advance.

Trade in all lines of plumbing supplies remains excellent, and from present indications seems due to stay so.

RANGE BOILERS—An exceedingly good business is being done in this line, and prices remain very firm. We still quote: Iron clad, 30 gal., $5; 40 gal., $6.50 net list. Copper, 30 gal., $33; 35 gal., $38; 40 gal., $43.

LEAD PIPE—Very firm. No change over last week's quotations. Our prices are: 5 per cent. discount f.o.b. Montreal, Toronto, St. John, N.B., Halifax; f.o.b. London, 15c. per hundred lbs. extra; f.o.b. Hamilton, 10c. per hundred lbs. extra.

IRON PIPE FITTINGS — Hard to get, but no advance reported as yet. We still quote: Discounts on nipples, ¼ inch to 3 inch, 75 per cent., 3½ inch to 2 inch, 55 per cent.

IRON PIPE — Expected that supplies will be even harder to get this year than last. Our prices are: Standard pipe in lots of 100 feet, regular lengths, ¼ inch, $5.50; ⅜ inch, $5.50; ½ inch, $8.50; ¾ inch, $11.50; 1 inch, $16.50; 1¼ inch, $22.50; 1½ inch, $27.; 2 inches, $36.50. Discounts on black pipe: ¼ inch, 59 per cent.; ⅜ inch, 59 per cent.; ½ inch, 68; ¾ to 2, 70. Discounts on galvanized pipe: ¼ inch, 44 per cent.; ⅜ inch, 44 per cent.; ½ inch, 58 per cent.; ¾ to 2 inch, 60 per cent. Extra heavy pipe of 100 feet lots are quoted as follows: ½ inch, $12; ¾ inch, $15; 1 inch, $22; 1¼ inch, $30, 1½ inch, $36; 2 inches, $50. The discounts on black pipe: ¼ inch, 74 per cent.; ⅜ inch, 69 per cent.; ½ inch to 2 inches, 68 per cent. Galvanized: ¼ inch, 59 per cent.; ⅜ inch, 69 per cent.; ½ to 2 inches, 58 per cent.

SOIL PIPE FITTINGS — Prices remain: Standard soil pipe, 50 per cent. off list. Standard fittings, 50 and 10 per cent. off list; medium and extra heavy soil pipe, 60 per cent. off. Fittings, 60 per cent. off.

SOLDER — No change in prices. We still quote: Bar solder, half-and-half, guaranteed, 25c.; No. 2 wiping solder, 22c.

ENAMELWARE — We quote: Canadian baths, see Jan. 3, 1907, list. Lavatories, discounts, 1st quality, 30 per cent.; special, 30 and 10 per cent. Sinks, 18x30 inch, flat rim, 1st quality, $2.60; special, $2.45.

TORONTO.

Office of HARDWARE AND METAL,
10 Front Street East
Toronto, February 15, 1907

Last week's advance on cast iron fittings has been followed generally by the trade this week, and, as predicted, an upward move has been made on iron body valves, the discount having been changed from 60 to 55 and 5 per cent. on Standard, and from 55 and 5 to 50 on Jenkins' disc. These have been accompanied by similar changes on saving checks, Webber gate valves and safety valves. Likewise advances of from 2½ to 5 per cent. have been made on all globe, angle and check valves. Some minor changes have also been made in both cast iron and malleable fittings.

Some booking for future delivery is being quietly done by some shrewd buyers who realize that both iron and copper are bound to retain their present strength and are more likely to go even higher than take a slump during 1907. With the exception of such orders, however, business is slow and is not likely to brighten up for two or three weeks.

Rumors of a new supply house to be established in Toronto are received from well informed parties.

LEAD PIPE — Trade is dull and the discount continues at 5 per cent. off the list price of 7c. per pound. Lead waste, 8c. per pound, with 5 off. Caulking lead, 5¾c. to 6c. per pound. Traps and bends, 40 per cent. discount.

SOIL PIPE AND FITTINGS—Buying is seasonable. We still quote: Medium and extra heavy pipe and fittings, 60 per cent.; light pipe, 50 per cent.; light fittings, 50 and 10 per cent.; 7 and 8 inch pipe, 40 and 5 per cent.

IRON PIPE — The scarcity in the most used sizes continues, with booked orders fairly large. One inch black pipe is quoted at $4.95, and 1 inch galvanized at $6.60. Full list in current market quotations.

IRON PIPE FITTINGS—Last week's advance has now been generally followed. We now quote: Cast iron fittings, 57½ per cent.; cast iron plugs and bushings, 60 per cent.; flange unions, 60 per cent.; nipples, 70 and 10 per cent.; iron cocks, 55 and 5 per cent.; Canadian malleable, 30 per cent.; malleable unions, 55 and 5 per cent.; malleable bushings, 55 per cent.; cast iron ceiling plates, plain, 65 per cent.; cast iron floor, 70 per cent.; hook plates, 60 per cent.; expansion plates, 65 per cent.; headers, 60 per cent.; hangers, 65 per cent.; standard list.

RANGE BOILERS — Trade is seasonable with little buying being done. We quote: Galvanized iron, 30 gallon, standard, $5; extra heavy, $7.75; 35 gallon, standard, $6; extra heavy, $8.75; 40 gallon, standard, $7; 40 gallon, extra heavy, $9.75, net list. Copper range boilers—New lists quote: 30 gallon, $33; 35 gallon, $38; 40 gallon $43. Discounts 5 to 15 per cent.

RADIATORS—At this season buying is not active. Prices continue unchanged at: Hot water, 47½ per cent.; steam, 50 per cent.; wall radiators, 45 per cent.; specials, 45 per cent. Hot water boilers are unchanged.

SOLDER — We quote: Bar solder, half-and-half, guaranteed, 27c.; wiping, 23c.

ENAMELWARE — No changes have been made, with the January list still prevailing on Canadian baths. We quote: Lavatories, first quality, 20 and 5 to 20 and 10 off; special, 20 and 10 to 30 and 2½ per cent. discount. Kitchen sinks, plate 300, firsts, 60 and 10 off; specials, 65 and 5 per cent. Urinals and range closets, 15 off. Fittings extra.

The Steamfitter

that pushes the Daisy Hot Water Boiler knows that he is handling a Heater of unquestioned merit. It has stood the test of time and its installation means perfect satisfaction to the customer and a substantial profit to the fitter.

CLUFF BROTHERS

Lombard Street, TORONTO

Selling Agents for - - - WARDEN KING & SON, Limited

The Oil Tank That Saves

Unless you have a Bowser Equipment, you are constantly losing oil through evaporation, dripping, leakage and waste. These losses are often enough to eat away your entire profit. Don't put up with these losses any longer. Instal a Bowser and make a good profit on your oil business.

The Bowser is absolutely leak and evaporation proof.

The Bowser pump will not allow dripping.

The accurate self-measuring feature prevents waste and loss through careless handling.

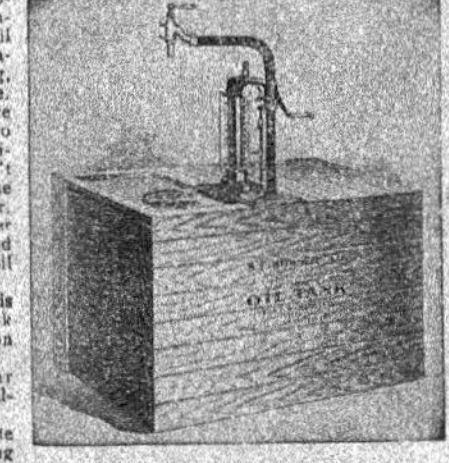

Cut 19. First Floor Outfit for Non-Lubricating Oils

The Bowser handles the heaviest oils without gumming.

The Bowser will save and earn money for you because it will stop all losses.

Send for Catalog V which explains the Bowser Paint and Non-Lubricating Oil Outfits.

S. F. BOWSER & CO., Inc.

66-68 Fraser Ave. Toronto, Can.

Can You answer the question on Hot Water vs. Steam Heating asked on page 61?

PENBERTHY PENBERTHY PENBERTHY PENBERTHY PENBERTHY PENBERTHY PENBERTHY PENBERTHY PENBERTHY PENBERTHY

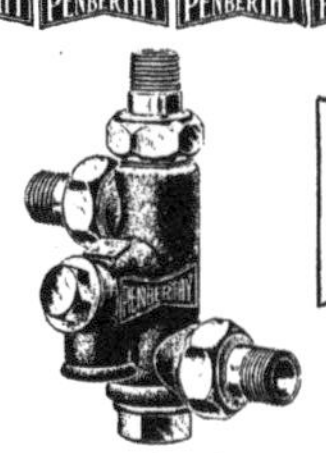

Automatic Injectors

Some dealers know a good article and always stock a full line of Penberthy Injectors and Brass Specialties. They have found from past experience that it is a* POPULAR LINE *and* SELLS WELL. *No time lost in talking goods as

Penberthy Brasses Speak for Themselves

Why not stock the best? Drop us a postal and we will mail you our new 1907 catalog

We manufacture

INJECTORS — OIL CUPS — AIR COCKS — WHISTLES
WATER GAUGES — GREASE CUPS — RADIATOR VENTS — GAS ENGINE
GAUGE COCKS — LUBRICATORS — CYLINDER COCKS — SUPPLIES, Etc.

PENBERTHY INJECTOR CO., LTD., WINDSOR, ONT.

DRILLING MACHINES

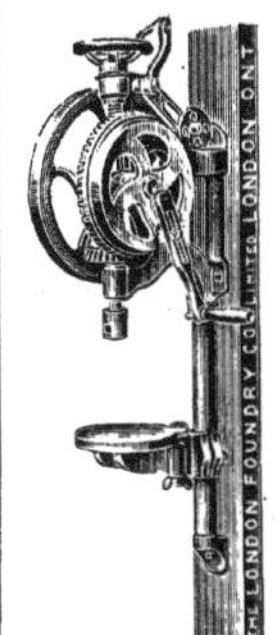

FOR ALL PURPOSES

ALSO

Blacksmith Machine Tools,

Tuyere Irons,

Toe Calk Blocks,

Wagon Skeins,

Foot Vises,

Etc.

Manufactured by

THE LONDON FOUNDRY CO., Limited
LONDON, CANADA

ATKINS SILVER STEEL SAWS

are the most profitable Saws to sell. Because they satisfy your customer and bring him back to YOU. Good mechanics appreciate them and buy them. The best Dealers everywhere sell them. Write for details and catalogue.

E. C. ATKINS & CO., Inc.

The Silver Steel Saw People,

Home Office and Factory, Indianapolis, Ind.

Branches—Toronto, Atlanta, Chicago, Memphis, Minneapolis, New Orleans, New York City, Portland, San-Francisco, Seattle.

CURRENT MARKET QUOTATIONS.

Feb. 15, 1907.

These prices are for such qualities and quantities as are usually ordered by retail dealers on the usual terms of credit, the lowest figures being for larger quantities and prompt pay. Large cash buyers can frequently make purchases at better prices. The Editor is anxious to be informed at once of any apparent errors in this list, as the desire is to make it perfectly accurate.

METALS.

ANTIMONY.

Cookson's ....per lb... 0 27½ 0 28

BOILER AND T.K. PITTS.

Plain tinned.........} 25 per cent. off list.
Spun..................}

BABBIT METAL.

Canada Metal Company—Imperial, genuine, 60c; Imperial Tough, 60c; White Brass, 50c; Metallic, 35c; Harris Heavy Pressure, 25c; Hercules, 25c; White Bronze, 15c; Star Frictionless, 14c; Alluminoid, 10c; No 4, 9c per lb.

James Robertson Co—Extra and genuine Monarch, 60c, Crown Monarch, 50c; No. 1 Monarch, 40c; King, 30c; Fleur-de-lis, 20c; Thurber, 15c; Philidelphia, 12c; Canadian, 10c; hardware, No. 1, 15c; No. 2, 12c; No. 3, 10c per lb.

BRASS.

Rod and Sheet, 14 to 30 gauge, net list.

Sheets, 12 to 14 in.	0 27
Tubing, base, per lb 5-16 to 2 in	0 33
Tubing, ½ to 3-inch, iron pipe size	0 31
" 1 to 3-inch, seamless	0 36

Copper tubing, 6 cents extra.

COPPER.

Ingot.		Per 100 lb.
Casting, car lots	26 50	27 00
Bars.		
Cut lengths, round, ½ to 2 in	....	32 00
Sheet.		
Plain, 16 oz., 14x48 and 14x60	....	31 00
Plain, 14 oz.	....	32 00
Tinned copper sheet, base	....	34 00
Planished base	....	38 00
Braziers' (in sheets), 4x6 ft., 25 to 30 lb. each, per lb., base	....	0 30

BLACK SHEETS.

	Montreal.	Toronto
8 to 10 gauge	2 70	2 70
12 gauge	2 70	2 80
14 "	2 70	2 55
17 "	2 50	2 65
18 "	2 50	2 65
20 "	2 50	2 65
22 "	2 50	2 75
24 "	2 55	2 80
26 "	2 65	2 85
28 "	2 70	3 00

CANADA PLATES.

Ordinary, 52 sheets	2 75	2 90
All bright "	3 80	3 90
Galvanized, 52 sheets	4 35	4 45
" 60 "	4 60	4 70
	Ordinary.	**Dom. Crown**
18x24x52	4 25	4 35
" 60	4 50	4 60
20x28x80	8 50	8 70
" 94	9 00	9 20

GALVANIZED SHEETS.

	Fleur-de-Lis.	Gordon Crown.
16 to 20 gauge	3 80	3 95
22 to 24 gauge	4 05	4 00
26 "	4 30	4 20
28 "	4 55	4 50

Apollo.

10¾ oz. (American gauge)	....	4 85
28 gauge "	....	4 65
26 " "	4 15	4 20
24 " "	3 85	3 90

	Comet	Queen's Head.	Bell.
16 to 20 gauge	3 60	3 95	....
22 to 24 gauge	3 75	4 20	....
26 "	4 30	4 45	....
28 "	4 45	4 70	4 45

Less than case lots 10 to 25c. extra.

IRON AND STEEL.

	Montreal.	Toronto
Common bar, per 100 lb.	2 15	2 30
Forged iron "	2 40	..
Refined " "	2 55	2 70
Horseshoe iron "	2 55	2 70
Hoop steel, 1½ to 3 in. base	....	2 80
Sleigh shoe steel "	2 25	2 30
Tire steel	2 40	2 50
Best sheet cast steel	...	0 12
B. K. Morton "Alpha" high speed.		0 65
" annealed	....	0 70
"M" Self-hardening	....	0 50
"J" quality, best warranted	....	0 18
"I" " warranted	....	0 14
"B.C" quality	....	0 09
Jonas & Colver's tool steel	0 10	0 20
" "Novo"	....	0 65
" " annealed	....	0 65
Jowett & Sons B P.L tool steel	....	0 10½

INGOT TIN.

Lamb and Flag and Straits—
56 and 28-lb. ingots, 100 lb. $45 00 $45 50

TINPLATES.

Charcoal Plates—Bright

M.L.S., equal to Bradley—		Per box.
I C, 14 x 20 base	....	$6 50
I X, 14 x 20 "	....	8 00
I XX, 14 x 20 base	....	9 50
Famous, equal to Bradley—		
I C, 14 x 20 base	....	6 50
I X, 14 x 20 "	....	8 00
I X X, 14 x 20 base	....	9 50
Raven and Vulture Grades—		
I C, 14 x 20 base	....	5 00
I X "	....	6 00
I X X "	....	7 00
I X X X "	....	8 00

"Dominion Crown Best"—Double Coated, Tissued.

		Per box.
I C, 14 x 20 base	5 50	5 75
I X, 14 x 20 "	6 50	6 75
I XX x 20 "	7 50	7 75

"Allaway's Best"—Standard Quality.

I C, 14 x 20 base	....	4 50
I X, 14 x 20 "	....	5 25
I XX, 14 x 20 "	....	6 00

Bright Cokes.

Bessemer Steel—		
I C., 14 x 20 base	...	4 25
20x28, double box	....	8 50

Charcoal Plates—Terne

Dean or J. G. Grade—		
I C., 20x28, 112 sheets	7 25	8 00
IX., Terne Tin	...	9 50

Charcoal Tin Boiler Plates.

Cookley Grade—
X X, 14x56, 50 sheet bxs. }
" 14x60, " } 7 5)
" 14x65, " }

Tinned Sheets.

72x30 up to 24 gauge	....	8 50
" " 26 "	....	9 00

LEAD.

Imported Pig, per 100 lb.	5 50	5 60
Bar, "	5 75	6 00
Sheets, 2½ lb. sq. ft., by roll	....	0 07½
Sheets, 3 to 6 lb. "	....	0 07

NOTE.—Cut sheets ½c. per lb., extra. Pipe, by the roll, usual weights per yard, lists at 7c. per lb. and 5 p.c. dis f.o.b Toronto.

NOTE.—Cut lengths, net price, waste pipe 8-ft lengths, lists at 8c.

SHEET ZINC.

5-cwt. casks	8 00	8 25
Part casks	8 25	8 50

ZINC SPELTER.

Foreign, per 100 lb	7 50	7 75
Domestic	7 00	7 25

PLUMBING AND HEATING

BRASS GOODS, VALVES, ETC.

Standard Compression work, dis half inch 50 per cent., others 40 per cent.

Cushion work, discount 40 per cent.

Fuller work, ½-inch 50 per cent., others 50 p.c.

Flatway stop and stop and waste cocks, 50 per cent.; roundway, 45 per cent.

J.M.T. Globe, Angle and Check Valves, discount 45 per cent.

Standard Globe, Angle and Check Valves discount 52½ per cent.

Kerr standard globes, angles and checks, special, 42½ per cent.; standard, 47½ p.c.

Kerr Jenkins' disc, copper-alloy disc and heavy standard valves, 45 per cent.

Kerr steam radiator valves, 60 p.c., and quick-opening hot-water radiator valves, 60 p.c.

Kerr brass, Weber's straightway valves, 40 per cent.; straightway valves, I.B.B.M., 60 per cent.

J. M. T. Radiator Valves, discount 50 per cent

Standard Radiator Valves, 60 per cent.

Patent Quick-Opening Valves, 65 per cent.

Jenkins' Bros. Globe Angle and Check Valves discount 27½ per cent.

No. 1 compression bath cock ......net	2 00
No. 4 " " "	1 90
No 7 Fuller's "	2 25
No. 4½ " "	2 35

Patent Compression Cushion, basin cock, hot and cold, per doz., $16.20

Patent Compression Cushion, bath cock, No. 2208 ...... 2 25

Square head brass cocks, 50 per cent.
" " iron " 60 "

Thompson Smoke-test Machine 25.00

BOILERS—COPPER RANGE.

Copper, 30 gallon	33 00
" 35 "	38 00
" 40 "	43 00

15 per cent.

BOILERS—GALVANIZED IRON RANGE.

Capacity	Standard.	Extra heavy
30-gallons	5 00	7.75
35 "	6 00	8 75
40 "	7.00	9.75

2 per cent., 30 days.

BATH TUBS.

Steel clad copper lined, 15 per cent.

CAST IRON SINKS.

16x24, $1; 18x30, $1; 18x36, $1.30.

ENAMELED CLOSETS AND URINALS

Discount 15 per cent.

ENAMELED BATHS.

List issued by the Standard Ideal Company Jan. 3, 1907, shows an advance of 10 per cent. over previous quotations.

ENAMELED LAVATORIES.

	1st quality.	Special.
Plate E 100 to E 103	20 & 5 p.c.	20 & 10 p.c.
" E 104 to E 132	20 & 10 p.c.	30 & 2½ p.c.

ENAMELED ROLL RIM SINKS.

	1st quality.	Special.
Plate E 201, one piece.	15 & 2½ p.c.	15 & 10 p.c.

ENAMELED KITCHEN SINKS.

Plate E, flat rim 300, 60 & 10 p.c. 65 & 5 p.c

HEATING APPARATUS

Stoves and Ranges—Discounts vary from 40 to 70 per cent. according to list

Furnaces—45 per cent

Registers—70 per cent

Hot Water Boilers—Discounts vary.

Hot Water Radiators—47½ per cent.

Steam Radiators—50 per cent

Wall Radiators and Specials—45 per cent.

LEAD PIPE

Lead Pipe, 7c. per pound, 5 per cent. off.

Lead waste, 8c. per pound, 5 per cent. off.

Caulking lead, 6c. per pound.

Traps and bends, 40 per cent.

IRON PIPE.

Size (per 100 ft.)	Black.		Galvanized
¼ inch	2 25	¼ inch	3 08
⅜ "	2 25	⅜ "	3 08
½ "	2 72	½ "	3 57
¾ "	3 45	¾ "	4 60
1 "	4 95	1 "	6 84
1¼ "	6 75	1¼ "	9 00
1½ "	8 10	1½ "	10 80
2 "	10 80	2 "	14 40
2½ "	18 00	2½ "	24 00
3 "	23 50	3 "	31 50
3½ "	30 00	3½ "	40 00
4 "	34 00	4 "	45 00

2 per cent. 30 days.

Malleable Fittings—Canadian discount 30 per cent.; American discount 25 per cent.

Cast Iron Fittings 57½; Standard bushings 57½ per cent.; headers, 57½; flanged unions 57½, malleable bushings 55; nipples, 70 and 10 per cent.; malleable lipped unions, 55 and 5 per cent.

SOIL PIPE AND FITTINGS

Medium and Extra heavy pipe and fittings, up to 6 inch, discount 60 per cent.

7 and 8-in. pipe, discount 40 and 5 per cent.

Light pipe, 50 p.c.; fittings, 50 and 10 p.c.

OAKUM.

Plumbers ....per 100 lb... .. .... 4 75

STOCKS AND DIES.

American discount 25 per cent.

SOLDERING IRONS.

½-lb.	per lb.	0 38
1-lb. or over	"	0 35

SOLDER.

Per lb.	Montreal	Toronto
Bar, half-and-half, guaranteed	0 25	0 27
Wiping	0 22	0 23

PAINTS, OILS AND GLASS

BRUSHES

Paint and household, 70 per cent.

CHEMICALS.

	In casks per lb.
Sulphate of copper (bluestone or blue vitrol)	0 09
Litharge, ground	0 06
" flaked	0 06½
Green copperas (green vitrol)	0 01
Sugar of lead	0 08
Lump olive	0 01½

COLORS IN OIL.

Venetian red, 1-lb. tins pure.	....	0 08
Chrome yellow "	....	0 15
Golden ochre "	....	0 09
French " "	....	0 08
Marine black "	....	0 04½
Chrome green "	....	0 10
French permanent green "	....	0 13
Signwriters' black "	....	0 15

GLUE.

Domestic sheet	0 10	0 10½
French medal	0 12	0 12½

PARIS GREEN.

	Berger's	Canadian
600-lb. casks	0 25½	0 25½
250 lb. drums	0 25½	0 25½
100-lb. "	0 26½	....
50-lb. "	0 26½	0 26½
1-lb. pkgs, 100 in box	0 27½	0 27½
½-lb. "	0 29½	0 29½
1-lb. tins, 100 in box	0 28½	0 28½
½-lb. pkg	0 30½	0 30

F.o b. Toronto.

PARIS WHITE.

In bbls 0 9)

PREPARED PAINTS.

Pure, per gallon, in tin 1 30
Second qualities per gallon 1 10
Barn (in bbls.) 0 65 0 90
Sherwin-Williams paints, 1 gal. 1 40
Canada Paint Co.'s pure 1 30
Standard P. & V. Co.'s "New Era." 1 30
Benj. Moore Co.'s "Ark" B'd 1 25
" British Navy deck 1 50
Brandram-Henderson's "English" 1 45
Ramsay's paints, Pure, per gal. 1 20
" Thistle, " 1 00
" Outside, bbls 0 55 0 65
Martin-Senour's 100 p.c. pure, ½ gal. 1 50
" " " ¼ gal. 1 45
" " " 1 gal. 1 40
Senour's Floor Paints ....gal. 1 20
Jamieson's "Crown and Anchor" 1 20
Jamieson's floor enamel 1 50
" barn paints, bbls, per gal. 0 60
Sanderson Pearcy's, pure 1 20
Robertson's pure paints 1 20

PUTTY.

Bulk in bbls. 1 50
Bulk in less quantity 1 80
Bladders in bbls 1 80
Bladders in kegs, boxes or loose 1 95
25-lb. tins 1 80
12½ lb. tins 2 05
Bladders in bulk or tins less than 100 lb. 1 85
Bulk in 100-lb. irons 1 80

SHINGLE STAINS

In 5-gallon lots 0 75 0 80

SHELLAC.

White 0 65
Fine orange 0 60
Medium orange 0 55
F.o.b Montreal or Toronto.

TURPENTINE AND OIL.

Castor oil 0 08 0 10
Gasoline 0 2¼
Benzine, per gal 0 20
Turpentine, single barrels 0 98 1 05
Linseed Oil, " raw 0 55 0 59
" " boiled 0 58 0 62
W. H. & C's English raw oil, bbls .. 0 63
" " boiled, " 0 66

WHITE LEAD GROUND IN OIL. Per 100 lbs

Canadian pure 7 15
No. 1 Canadian 6 80
Munro's Select Flake White 7 40
Elephant and Decorators' Pure 7 40
Monarch 7 40
Standard Decorator's 7 15
Essex Genuine 6 80
Brandram's B. B. Genuine 8 40
"Anchor," pure 7 00
Ramsay's Pure Lead 6 40
Ramsay's Exterior 6 15
"Crown and Anchor," pure 6 50
Sanderson Pearcy's 7 40
Robertson's C P. lead 7 20
W H. & C's matured pure English 8 00 8 25

WHITE AND RED DRY LEAD.

	white	red.
Genuine, 560 lb. casks, per cwt	6 75	6 00
Genuine, 100 lb. kegs, "	7 50	6 50
No. 1, 560 lb. casks, per cwt	6 25	5 75
No. 1, 100 lb. kegs, per cwt	7 00	6 25

WINDOW GLASS.

Size United Inches.	Star	Double Diamond
Under 26	$4 25	$6 25
26 to 40	4 65	6 75
41 to 50	5 10	7 50
51 to 60	5 35	8 50
61 to 70	5 75	9 75
71 to 80	6 25	11 00
81 to 85	7 00	12 50
86 to 90	....	15 00
91 to 95	....	17 50
96 to 100	....	20 50
101 to 105	....	24 00
106 to 110	....	27 50

Discount—16-oz., 25 per cent; 21-oz. 30 per cent. per 100 feet. Broken boxes 50 per cent. off list.

WHITING.

Plain, in bbls 0 60
Gilders bolted in bands 0 90

WHITE DRY ZINC.

Extra Red Seal, V M 0 07½ 0 08½

WHITE GROUND ZINC.

Pure, in 25-lb. irons 0 08½
No. 1, " " 0 07
No. 2, " " 0 05½

VARNISHES.

In 5-gal. lots. Per gal cans

Carriage, No. 1 .... 1 50
Pale durable body 3 50
" hard rubbing 3 00
Finest elastic gearing 3 00
Elastic oak 1 50
Furniture, polishing 2 00
Furniture, extra 1 15
" " No. 1 .... 0 90
" union 0 80
Light oil finish 1 40
Gold size japan 1 80
Brown japan 1 95
No 1 brown japan 0 90
Baking black japan 1 35
No 1 black japan 0 90
Benzine black japan 0 70
Crystal Damar 2 80
No 1 " 2 40
Pure asphaltum 1 40
Oilcloth 1 50
Lightning dryer 0 70
Elastilite varnish, 1 gal. can, each .. 2 00
Granitine floor varnish, per gal 2 10
Maple Leaf coach enamels; size 1, 1 20
Sherwin-Williams' kopal varnish, gal., 2 50
Canada Paint Co's sun varnish 2 00
"Kyanize" Interior Finish 2 40
"Flint Lac," coach 1 80
B H. Co's "Gold Medal," in cases 2 00
Jamieson's Copaline, per gal. 2 00

BUILDERS' HARDWARE.

BELLS.

Brass hand bells, 60 per cent.
Nickel, 55 per cent.
Gongs, Sargeant's door bells .. 5 50 8 00
American, house bells, per lb. 0 35 0 40
Peterboro' door bells, discount 37½ and 10 per cent. off new list.

BUILDING PAPER, ETC.

Tarred Felt, per 100 lb. 2 25
Ready roofing, 2-ply, not under 45 lb. per roll 1 00
Ready roofing, 3-ply, not under 65 lb., per roll 1 25
Carpet Felt ....per ton 60 00
Heavy Straw Sheathing ....per ton 35 00
Dry Surprise 0 41
Dry Sheathing ....per roll, 400 sq. ft. 0 50
Tar " " 400 " 0 45
Dry Fibre " 400 " 0 55
Tarred Fibre " 400 " 0 65
O. K. & I. X. L. " 400 " 0 70
Resin-sized " 400 " 0 45
Oiled Sheathing " 600 " 1 00
Oiled " " 400 " 0 70
Roof Coating, in barrels ....per gal. 0 17
Roof " small packages " 0 25
Refined Tar ....per barrel 5 00
Coal Tar " 4 00
Coal Tar, less than barrels ....per gal. 0 15
Roofing Pitch ....per 100 lb. 0 80 0 90
Slater's felt ....per roll 0 70
Heavy Straw Sheathing f. o. b. St. John and Halifax 37 50

BUTTS.

Wrought Brass, net revised list
Wrought Iron, 70 per cent.
Cast Iron Loose Pin, discount 60 per cent
Wrought Steel Fast Joint and Loose Pin, 70 per cent.

CEMENT AND FIREBRICK.

"Lafarge" cement in wood 3 40
"Lehigh" cement, in wood 2 54
"Lehigh" cement, cotton sacks 2 39
"Lehigh" cement, paper sacks 2 31
Fire brick (Scotch) per 1,000 .... 15 00 24 00
Fire clay (Scotch), net ton 4 95

Paving Blocks per 1 000.

Blue metallic, 9"x4½"x3", ex wharf .. 35 00
Stable pavers, 12"x6"x2", ex wharf .. 50 00
Stable pavers, 9"x4½"x3", ex wharf .. 36 00

DOOR SETS.

Peterboro, 37½ and 10 per cent.

DOOR SPRINGS.

Torrey's Rod ....per doz. 1 75
Coil, 9 to 11 in. " 0 95 1 65
English " 2 00 4 00
Chicago and Reliance Coil Door Springs, 25 per cent.

STORE DOOR HANDLES.

Per Dozen 1 00 1 50

ESCUTCHEONS.

Discount 50 and 10 per cent., new list
Peterboro, 37½ and 10 per cent.

ESCUTCHEON PINS.

Iron, discount 40 per cent.
Brass, 45 per cent.

HINGES.

Blind, discount 60 per cent.
Heavy T and strap, 4-in., per lb. net .. 0 06½
" " 5-in., " 0 06
" " 6-in., " 0 05¾
" " 8-in., " 0 05¼
" " 10-in. and larger .. 0 05¼
Light T and strap, discount 65 p.c.
Screw hook and hinge— under 12 in. .... per 100 lb. 4 75
over 12 in. 3 75
Spring, No. 20, per gro. pairs 10 80
Spring, Woodyatt pattern, per gro., No. 5, $17 50 No. 10, $18; No. 20, $10 80; No. 12 $20; No. 51, $10; No. 50, $27 50.
Crate hinges and back flaps, 65 and 5 p. c.
Hinge hasps, 65 per cent.

SPRING HINGES.

Chicago Spring Butts and Blanks 12½ per cent.
Triple End Spring Butts 40 and 5 per cent.
Chicago Floor Hinges, 40 and off.
Garden City Fire Engine House Hinges, 12½ per cent.

CAST IRON HOOKS.

Bird cage ....per doz. 0 50 1 10
Clothes line, No. 61 .. " 0 00 0 70
Harness " 0 60 12 00
Hat and coat ....per gro. 1 10 10 00
Chandelier ....per doz. 0 50 1 00
Wrought hooks and staples—
¼ x 5 ....per gross 2 65
5-16 x 5 .... " 3 30
Bright steel gate hooks and staples, 40 per cent. discount
Hat and coat wire discount 67½ per cent.
Screw, bright wire, discount 65 per cent.

KNOBS.

Door, japanned and N.P., doz 1 50 2 50
Bronze, Berlin ....per doz. 2 75 3 25
Bronze, Genuine 6 00 9 00
Shutter, porcelain, F. & L. screw ....per gross 1 30 2 00
White door knobs ....per doz. 2 00
Peterboro knobs, 37½ and 10 per cent.
Porcelain, mineral and jet knobs, net list.

KEYS.

Lock, Canadian dis 40 to 40 and 10 per cent
Cabinet trunk and padlock American ....per gross 0 60

LOCKS.

Peterboro 37½ and 10 per cent
Russell & Erwin, steel rim $2 50 per doz.
Eagle cabinet locks, discount 30 per cent
American padlocks, all steel, 10 to 15 per cent; all brass or bronze, 10 to 25 per cent.

SAND AND EMERY PAPER.

B & A. sand, discount, 35 per cent
Emery, discount 35 per cent.
Garnet (Rurton's) 5 to 10 per cent. advance on list.

SASH WEIGHTS.

Sectional ....per 100 lb. 2 00 2 25
Solid 1 50 1 75

SASH CORD.

Per lb 0 31

BLIND AND BED STAPLES.

All sizes, per lb 0 07½ 0 12

WROUGHT STAPLES.

Galvanized 2 75
Plain 2 50
Coopers', discount 45 per cent.
Poultry netting staples, discount 40 per cent.
Bright spear point, 75 per cent. discount.

TOOLS AND HANDLES.

ADZES.

Discount 22½ per cent.

AUGERS.

Gilmour's, discount 60 per cent. off list.

AXES.

Single bit, per doz. 5 50 8 50
Double bit, " 10 00 11 00
Bench Axes, 40 per cent.
Broad Axes 25 per cent.
Hunters' Axes 5 50 6 00
Boys' Axes 6 25 7 00
Splitting Axes 7 00 12 00
Handled Axes 7 00 9 00
Red Ridge, boys', handled 5 75
" hunters 5 25

BITS.

Irwin's auger, discount 47½ per cent.
Gilmour's auger, discount 60 per cent.
Rockford auger, discount 50 and 10 per cent.
Jennings' Gen. auger, net list.
Gilmour's car, 47½ per cent.
Clark's expansive, 40 per cent.
Clark's gimlet, per doz 0 65 0 90
Diamond, Shell, per doz. 1 00 1 50
Nail and Spike, per gross 2 25 5 20

BUTCHERS CLEAVERS.

German ....per doz. 6 00 9 00
American .... " 12 00 18 00

CHALK.

Carpenters' Colored, per gross 0 45 0 75
White lump ....per cwt. 0 60 0 65

CHISELS.

Warnock's, discount 72½ per cent.
P. S. & W. Extra, discount 72½ per cent.

CROSSCUT SAW HANDLES.

S. & D., No. 3 ....per pair 0 15
S. & D., " 5 .... " 0 22½
S. & D., " 6 .... " 0 15
Boynton pattern .... " 0 20

CROWBARS.

3½c. to 4c. per lb.

DRAW KNIVES.

Coach and Wagon, discount 75 per cent.
Carpenters' discount 75 per cent

DRILLS.

Millar's Falls, hand and breast, net list.
North Bros., each set, 50c.

DRILL BITS.

Morse, discount 37½ to 40 per cent.
Standard, discount 50 and 5 to 55 per cent.

FILES AND RASPS.

Great Western .... 75 per cent.
Arcade .... 75 "
Kearney & Foot .... 75 "
Disston's .... 75 "
American .... 75 "
J. Barton Smith .... 75 "
McClellan .... 75 "
Eagle .... 75 "
Nicholson .... 66⅔ "
Globe .... 75 "
Black Diamond, 60, 10 and 5 p c.
Jowitt's, English list, 27½ per cent.

GAUGES.

Stanley's discount 50 to 60 per cent.

Wire Gauges.

Winn's, Nos. 26 to 33 ....each 1 65 2 40

HANDLES.

C. & B., fork, 40 per cent., revised list.
C. & B., hoe, 40 per cent., revised list.
American, saw ....per doz 0 25
American, plane ....per gross 3 15 3 75
Canadian, hammer and hatchet 40 per cent
Axe and cant hook handles, 45 per cent.

HAMMERS.

Maydole's, discount 5 to 10 per cent.
Canadian, discount 25 to 27½ per cent.
Magnetic tack ....per doz. 1 10 1 20
Canadian sledge ....per lb. 0 07 0 08½
Canadian ball pean, per lb. .. 0 22 0 25

HATCHETS.

Canadian, discount 40 to 42½ per cent.
Shingle, Red Ridge 1, per doz 4 40
" " 2, " 4 60
Barrel Underhill 5 15

MALLETS.

Tinsmiths'............per doz. 1 25 1 50
Carpenters', hickory, " 1 25 3 75
Lignum Vitae........ " 3 85 5 00
Caulking, each 0 60 2 00

MATTOCKS.

Canadian............ per doz. 5 50 6 00

MEAT CUTTERS.

German, 15 per cent
American discount, 33⅓ per cent.
Gem ....each 1 15

NAIL PULLERS.

German and American 0 85 2 50
No. 1....................... 0 85
No 1573..................... 0 75

NAIL SETS.

Square, round and octagon, per gross 3 38
Diamond 1 00

PICKS.

Per dozen 6 00 9 00

PLANES.

Wood bench, Canadian discount 40 per cent. American discount 50 per cent.
Wood, fancy Canadian or American 37½ to 40 per cent.
Stanley planes, $1.55 to $3 60, net list prices.

PLANE IRONS.

English ....per doz. 2 00 5 00
Stanley, 2½ inch, single 24c., double 39c.

PLIERS AND NIPPERS.

Button's genuine, 37½ to 40 per cent.
Button's imitation....per doz. 5 00 9 00
Berg's wire fencing......... .... 1 72 5 50

PUNCHES

Saddler's ....per doz. 1 00 1 85
Conductor's.......... " 3 00 15 00
Tinners', solid..........per set 0 72
" hollow......per inch 1 00

RIVET SETS.

Canadian, discount 35 to 37½ per cent.

RULES.

Boxwood, discount 70 per cent.
Ivory, discount 20 to 25 per cent.

SAWS.

Atkins, hand and crosscut, 25 per cent.
Disston's Hand, discount 12½ per cent
Disston's Crosscut ....per foot 0 35 0 55
Hack, complete..........each 0 75 2 75
" frame only.........each 0 50 1 25

S. & D. solid tooth circular shingle, concave and band, 50 per cent; mill and ice, drag, 30 per cent; cross-cut, 35 per cent.; band saws, butcher, 35 per cent; buck, New Century, $6.25; buck No. 1 Maple Leaf, $5.25; buck, Happy Medium, $4.25; buck, Watch Spring, $4.25; buck, common frame, $4 00.

Spear & Jackson's saws—Hand or rip, 26 in., $12.75; 24 in., $11 25, panel, 18 in., $8 25; 20 in., $9; tenon, 10 in., $9 90; 12 in., $10 90; 14 in., $11.50

SAW SETS.

Lincoln and Whiting 4 75
Hand Sets, Perfect 4 00
X-Cut Sets, " 7 50
Maple Leaf and Premiums saw sets, 40 off.
S. & D. saw swages, 40 off.

SCREW DRIVERS

Sargent's ....per doz. 0 65 1 00
North Bros., No. 30 . per doz 16 80

SHOVELS AND SPADES.

Bull Dog, solid neck shovel (No. 2 pol.). $18 50
(Hollow Back) (Reinforced S Scoop.)
Moose......... $17 50 .... $16 30
Bear .... 15 00 .... 15 30
Fox .... 12 50 .... 14 30
Black Cat.... 10 00 .... 13 30
Canadian, discount 45 per cent.

SQUARES.

Iron, discount 20 per cent.
Steel, discount 65 and 10 per cent.
Try and Bevel, discount 50 to 52½ per cent.

TAPE LINES.

English, ass skin .....per doz. 2 75 5 00
English, Patent Leather 5 50 9 75
Chesterman's... ....each 0 90 2 85
" steel ....each 0 80 8 00
Berg's, each...... .... 0 75 12 50

TROWELS.

Disston's, discount 10 per cent.
S. & D., discount 35 per cent.
Berg's, brick, 9¼x11 .... 4 00
" pointing, 9¼x5.......... .... 2 10

FARM AND GARDEN GOODS

BELLS.

American cow bells, 63⅓ per cent.
Canadian, discount 45 and 50 per cent.
American, farm bells, each .. 1 35 3 00

BULL RINGS.

Copper, $1.30 for 2½-inch, and $1.70

CATTLE LEADERS.

Nos. 32 and 33 .. ...per gross 7 50 8 50

BARN DOOR HANGERS.

doz. pairs.
Steel barn door................ 8 00 10 00
Stearns wood track 4 50 6 00
Zenith............................ 9
Acme, wood track 5 00 6
Atlas.............................. 5 00 6
Perfect 8 00 11
New Milo........................ 6
Steel, covered.................. 4 00 11
" track, 1 x 3-16 in(100 ft) 3
" " 1¼ x 3-16 in(100 ft) 4
Double strap hangers, doz. sets.... 6
Standard jointed hangers, " 6
Steel King hangers " 6
Storm King and safety hangers 7 00
" " rail.............. .... 4 50
Chicago Friction, Oscillating and Big Twin Hangers, 5 per cent.

HARVEST TOOLS.

Discount 60 per cent.
S. & D. lawn rakes, Dunn's, 40 off.
" sidewalk and stable scrapers, 40 off.

HAY KNIVES.

Net list

HEAD HALTERS.

Jute Rope, ⅜-inch....per gross 9 00
" " ½ " .... " 10 00
" " ⅝ " .... " 12 00
Leather, 1-inch ......per doz. 4 00
Leather, 1¼ " ...... " 5 20
Web.................. " 2 45

HOES.

Garden, Mortar, etc., discount 60 per cent.
Planter.............. per doz. 4 00 4 50

SCYTHES.

Per doz. net.................... 8 25 9 25

SCYTHE SNATHS.

Canadian, discount 40 per cent.

SNAPS.

Harness, German, discount 25 per cent.
Lock, Andrews' 4 50 11 00

STABLE FITTINGS

Warden King, 35 per cent

WOOD HAY RAKES.

Ten tooth, 40 and 10 per cent.
Twelve tooth, 45 per cent.

HEAVY GOODS, NAILS, ETC.

ANVILS

Wright's, 80-lb. and over...... .... 0 10½
Hay Budden, 80-lb. and over 0 09½
Brook's, 80-lb. and over 0 11¼
Taylor-Forbes, handy 0 05
Columbia Hardware Co., per lb. 0 09½

VISES.

Wright's... .... 0 13½
Berg's, per lb 0 12½
Brook's........................ .... 0 12½
Pipe Vise, Hinge, No. 1....... .... 3 50
" " " No. 2....... .... 5 50
Saw Vise 4 50 5 00
Blacksmiths' (discount) 60 per cent.
" parallel (discount) 45 per cent.

BOLTS AND NUTS

Carriage Bolts, common ($1 list Per cent.
" " ⅜ and smaller.. 60, 10 and 10
" " 7-16 and up .... 55 and 5
" " Norway Iron ($3 list) 50

Machine Bolts, ⅜ and less .... 60 and 10
Machine Bolts, 7-16 and up.... 55 and 5
Plough Bolts .... 55 and 10
Blank Bolts....... 55
Bolt Ends....................... 55
Sleigh Shoe Bolts, ⅜ and less .. 60 and 10
" " " 7-16 and larger 50 and 5
Coach Screws, conepoint...... 70 and 5
Nuts, square, all sizes, 4c. per cent. off
Nuts, hexagon, all sizes, 4¼c. per cent off
Stove Rods per lb., 5½ to 6c.
Stove Bolts, 75 per cent.

CHAIN.

Proof coil, per 100 lb., 5-16 in., $4.40; ⅜ in., $3 90; 7-16 in., $3.70; ½ in., $3 50; 9-16 in., $3 45; ⅝ in., $3 35; ¾ in., $3.20; ⅞ in., $3.10; 1 in., $3.10.
Halter, kennel and post chains, 40 to 40 and 5 per cent.
Cow ties .... 40 p.c.
Tie out chains .... 65 p.c.
Stall fixtures .... 35 p.c.
Trace chain .... 45 p.c.
Jack chain iron, discount 35 p.c.
Jack chain, brass, discount 40 per cent.

HORSE NAILS.

'C' brand, 40, 10 and 7½ per cent. off list } Oval
M.R.M. Co. brand, 55 per cent. } head
Capewell brand, discounts on application.

HORSESHOES.

F.O.B. Montreal
M.R.M. Co. brand, base 65
Add 15c. Toronto, Hamilton, Guelph.

HORSE WEIGHTS

Taylor-Forbes, 3½c. per lb.

NAILS.	Cut.	Wire.
2d	3 80	3 45
3d	2 95	3 10
4 and 5d	2 70	2 85
6 and 7d	2 60	2 75
8 and 9d	2 45	2 60
10 and 12d	2 40	2 55
16 and 20d	2 35	2 50
30, 40, 50 and 60d (base)	2 30	2 45

F.o.b. Montreal.
Cut nails, Toronto 20c. higher.
Miscellaneous wire nails, discount 75 per cent
Coopers' nails, discount 40 per cent.

PRESSED SPIKES.

Pressed spikes, ⅝ diameter, per 100 lbs, $3 15

RIVETS AND BURRS.

Iron Rivets, black and tinned, 60, 10 and 10.
Iron Burrs, discount 60 and 10 and 10 p.c.
Copper Rivets, usual proportion burrs, 15 p.c.
Copper Burrs only, net list.
Extras on Coppered Rivets, ½-lb. packages 1c. per lb.; ¼-lb. packages 2c. lb.
Tinned Rivets, net extra, 4c. per lb.

SCREWS.

Wood, F. H., bright and steel, 87½ per cent.
" R. H., bright, dis. 82½ per cent
" F. H., brass, dis. 80 per cent.
" R. H., " dis. 75 per cent.
" F. H., bronze, dis. 75 per cent.
" R. H., " dis. 70 per cent.
Drive Screws, dis. 87½ per cent.
Bench, wood ....per doz. 3 25 4 00
" iron .... " 4 25 5 00
Set, case hardened, dis. 60 per cent.
Square Cap, dis. 50 and 5 per cent.
Hexagon Cap, dis. 45 per cent.

MACHINE SCREWS.

Flat head, iron and brass, 35 per cent.
Felister head, iron, discount 30 per cent.
" " brass, discount 25 per cent.

TACKS, BRADS, ETC.

Carpet tacks, blued .... 80 and 5
" " tinned............ 80 and 10
" " (in kegs).......... 40
Cut tacks, blued, in dozens only 75 and 10
" ½ weights 60

Swedes cut tacks, blued and tinned—
In bulk.......................... 80 and 10
In dozens 75
Swedes, upholsterers', bulk .... 85 and 12½
" brush, blued and tinned bulk 70
Swedes, gimp, blued, tinned and japanned.......................... 75 and 12½
Zinc tacks 35
Leather carpet tacks............ 40
Copper tacks..................... 37½
Copper nails 42½
Trunk nails, black 65
Trunk nails, tinned and blued .. 65
Clout nails, blued and tinned .. 65
Chair nails........................ 35
Patent brads....................... 40
Fine finishing..................... 40
Lining tacks, in papers.......... 10
" " in bulk............ 15
" " solid heads, in bulk 75
Saddle nails, in papers.......... 10
" " in bulk........... 15
Tufting buttons, 22 line in dozens only 60
Zinc glaziers' points 5
Double pointed tacks, papers.. 90 and 10
" " " bulk.... 40
Clinch and duck rivets.......... 45
Cheese box tacks, 85 and 5; trunk tacks, 80 and 10.

WROUGHT IRON WASHERS.

Canadian make, discount 40 per cent.

SPORTING GOODS.

AMMUNITION.

B. B. Caps American $2.00 per 1000.
C. B. Caps American, $2.60 per 1000

CARTRIDGES.

Rim Fire Cartridges, 50 and 5 p.c.
Rim Fire Pistol, 30 and 5 per cent., American.
Central Fire, Military and Sporting American, 10 per cent. advance. B.B. Caps, discount 40 per cent., American.
Central Fire Pistol and Rifle, net Amer. list
Loaded and empty Shells, American 20 per cent. discount. Rival and Nitro, 10 per cent. advance.
Empty paper shells American, 10 per cent. advance.
Primers, American $2 05

Wads. per lb.
Best thick brown or grey felt wads, in ¼-lb. bags $0 70
Best thick white card wads, in boxes of 500 each, 12 and smaller gauges 0 29
Best thick white card wads, in boxes of 500 each, 10 gauge............ 0 35
Thin card wads, in boxes of 1,000 each, 12 and smaller gauges 0 20
Thin card wads, in boxes of 1,000 each, 10 gauge...................... 0 25
Chemically prepared black edge grey cloth wads, in boxes of 250 each— Per M.
11 and smaller gauge 0 60
9 and 10 gauges...................... 0 70
7 and 8 " 0 90
5 and 6 " 1 10
Superior chemically prepared pink edge, best white cloth wads in boxes of 250 each—
11 and smaller gauge 1 15
9 and 10 gauges 1 40
7 and 8 " 1 65
5 and 9 " 1 90

SHOT.

Common, $6.50 per 100 lb.; chilled, $7.50 per 100 lb.; buck, seal and ball, $8.50 net list. Prices are f.o.b. Toronto, Hamilton, Montreal, St. John and Halifax. Terms, 2 p.c. for cash in thirty days.

TRAPS (steel.)

Game, Newhouse, discount 30 and 10 per cent.
Game, Hawley & Norton, 50, 10 & 5 per cent.
Game, Victor, 70 per cent.
Game, Oneida Jump (B. & L.) 40 & 2½ p. c.
Game, steel, 60 and 5 per cent.

SKATES.

Skates, discount 37½ per cent.
Mic Mac hockey sticks, per doz 4 00 5 00

CUTLERY AND SILVERWARE.

RAZORS. per doz.

Elliot's 4 00 18 00
Boker's 7 50 11 00
" King Cutter 13 50 18 50
Wade & Butcher's.......... 3 60 10 00
Lewis Bros.' "Klean Kutter" 8 50 10 50
Henckel's 7 50 20 00
Berg's.......... 7 50 20 00
Clauss Razors and Strops, 50 and 10 per cent

KNIVES.

Farriers-Stacey Bros., doz 3 50

PLATED GOODS

Holloware, 40 per cent. discount.
Flatware, staples, 40 and 10, fancy, 40 and 5.

SHEARS.

Clauss, nickel, discount 60 per cent.
Clauss, Japan, discount 67½ per cent.
Clauss, tailors, discount 40 per cent.
Seymour's, discount 50 and 10 cent
Berg's.......... per 6 00 12 00

HOUSE FURNISHINGS.

APPLE PARERS.

Woodyatt Hudson, per doz., net 4 50

BIRD CAGES.

Brass and Japanned, 40 and 10 p c.

COPPER AND NICKEL WARE.

Copper boilers, kettles, teapots, etc. 35 p.c.
Copper pitts, 25 per cent.

ENAMELED WARE.

London, White, Princess, Turquoise, Onyx, Blue and White, discount 50 per cent.
Canada, Diamond, Premier, 50 and 10 p c.
Pearl, Imperial Crescent, 50 and 10 per cent.
Premier steel ware, 40 per cent.
Star decorated steel and white, 25 per cent.
Japanned ware, discount 45 per cent.
Hollow ware, tinned cast, 35 per cent. off.

KITCHEN SUNDRIES.

Can openers, per doz.......... 0 40 0 75
Mincing knives per doz 0 50 0 80
Duplex mouse traps, per doz.... .... 0 65
Potato mashers, wire, per doz... 0 60 0 70
" wood " .. 0 50 0 60
Vegetable slicers, per doz 2 25
Universal meat chopper No. 0, $1; No. 1, 1.15.
Enterprise chopper, each 1 30
Spiders and fry pans, 50 per cent.
Star A.1 chopper 5 to 32 .... 1 35 4 10
" " 100 to 103.......... 1 35 2 00
Kitchen hooks, bright.......... .... 0 62½

LAMP WICKS.

Discount, 60 per cent.

LEMON SQUEEZERS.

Porcelain lined...... per doz. 2 20 5 60
Galvanized.......... " 1 87 3 85
King, wood.......... " 2 75 2 90
King, glass.......... " 4 00 4 50
All glass .... " 0 50 0 90

METAL POLISH.

Tandem metal polish paste...... .... 6 00

PICTURE NAILS.

Porcelain head ......per gross 1 35 1 50
Brass head.......... 0 40 1 00
Tin and gilt, picture wire, 75 per cent.

SAD IRONS.

Mrs. Potts, No. 55, polished....per set 0 80
" No. 50, nickle-plated, " 0 9½
Common, plain.......... 4 50
" plated 5 50
Asbestos, per set.......... .... 1 25

TINWARE.

CONDUCTOR PIPE.

2-in. plain or corrugated., per 100 feet. $3.30; 3 in., $4 40; 4 in., $5.80; 5 in., $7.45; 6 in., $9.90.

FAUCETS.

Common, cork-lined, discount 35 per cent.

EAVETROUGHS.

10-inch .... per 100 ft. 3 30

FACTORY MILK CANS.

Discount off revised list, 35 per cent
Milk can trimmings, discount 25 per cent.
Creamery Cans, 45 per cent.

LANTERNS.

No. 2 or 4 Plain Cold Blast....per doz. 6 50
Lift Tubular and Hinge Plain, " 4 75
Better quality at higher prices.
Japanning, 50c. per doz. extra.

OILERS.

Kemp's Tornado and McClary's Model galvanized oil can, with pump, 5 gallon, per dozen 11 00
Davidson oilers, discount 40 per cent
Zinc and tin, discount 50 per cent.
Coppered oilers, 20 per cent. off.
Brass oilers, 50 per cent. off.
Malleable, discount 25 per cent

PAILS (GALVANIZED).

Dufferin pattern pails, 45 per cent.
Flaring pattern, discount 40, 10 and 5 per cent.
Galvanized washtubs 40 10 and 5 per cent.

PIECED WARE.

Discount 35 per cent off list, June, 1899.
10-qt. flaring sap buckets, discount 40 per cent.
6, 10 and 14-qt. flaring pails dis. 40 per cent.
Copper bottom tea kettles and boilers, 30 p.c.
Coal hods, 40 per cent.

STAMPED WARE.

Plain, 75 and 12½ per cent. off revised list.
Retinned, 72½ per cent revised list.

SAP SPOUTS.

Bronzed iron with hooks ....per 1,000 7 50
Eureka tinned steel, hooks 8 00

STOVEPIPES.

5 and 6 inch, per 100 lengths 7 00
7 inch .... " " 7 50
Nestable, discount 40 per cent.

STOVEPIPE ELBOWS

5 and 6-inch, common..........per doz. 1 32
7-inch.......... " 1 48
Polished, 15c. per dozen extra.

THERMOMETERS.

Tin case and dairy, 75 to 75 and 10 per cent.

TINNERS' SNIPS.

Per doz.......... 3 00 15
Clauss, discount 35 per cent.

TINNERS' TRIMMINGS

Discount, 45 per cent.

WIRE.

ANNEALED CUT HAY BAILING WIRE.

No. 12 and 13, $4; No. 13½, $4 10; No. 14, $4.20; No 15, $4 50; in lengths 6' to 11', 25 per cent.; other lengths 20c. per 100 lbs extra; if eye or loop on end add 25c. per 100 lbs. to the above.

BRIGHT WIRE GOODS

Discount 62½ per cent.

CLOTHES LINE WIRE.

7 wire solid line, No. 17, $4.90; No. 18, $3.00; No. 19, $2.70; 8 wire solid line, No. 17, $4.45; No. 18, $2.80. No. 19, $2.50. All prices per 1000 ft. measure. F.o.b. Hamilton Toronto, Montreal.

COILED SPRING WIRE.

High Carbon, No. 9, $2.90, No. 11, $3.45; No. 17, $3.15.

COPPER AND BRASS WIRE.

Discount 37½ per cent.

FINE STEEL WIRE.

Discount 25 per cent. List of extras: In 100-lb. lots: No. 17, $5 — No. 18, $5.50 — No. 19, $6 — No. 20, $6.65 — No. 21, $7 — No. 22, $7.30 — No. 23, $7.65 — No. 24, $8 — No. 25, $9 — No. 26, $9.50 — No. 27, $10 — No. 28, $11 — No. 29, $12 — No. 30, $13 — No. 31, $14 — No. 32, $15 — No. 33, $16 — No. 34, $17. Extras net—tinned wire, Nos. 17-25, $2 — Nos. 26-31, $4 — Nos. 32-34, $6. Coppered, 75c.—oiling, 10c.—in 25-lb. bundles, 15c.—in 5 and 10-lb. bundles, 25c.—in 1-lb. hanks, 25c. —in ½-lb. hanks, 38c.—in ¼-lb. hanks, 50c.—packed in casks or cases, 15c.—bagging or papering, 10c.

FENCE STAPLES.

Bright .... 2 70 Galvanized.... 3 15

HAY WIRE IN COILS.

No. 13, $2 60; No. 14, $2.70; No. 15, $2.85; f.o b., Montreal.

GALVANIZED WIRE.

Per 100 lb.—Nos. 4 and 5, $3.70 — Nos. 6, 7, 8, $3.15 — No. 9, $2.50 — No. 10, $3.20 — No. 11, $3.25 — No. 12, $2.65 — No. 13, $2.75 — No. 14, $3.75 — No. 15, $4 30 — No. 16, $4.30 from stock. Base sizes, Nos. 6 to 9, $2.35 f.o.b. Cleveland. In carlots 12½c. less.

LIGHT STRAIGHTENED WIRE.

Gauge No.	Over 20 in. per 100 lbs.	10 to 20 in.	5 to 10 in.
0 to 5	$0.50	$0 75	$1 25
6 to 9	0.75	1 25	2 00
10 to 11	1.00	1 75	2 50
12 to 14	1.50	2 25	3.50
15 to 16	2.00	3.00	4 50

SMOOTH STEEL WIRE.

No. 0.9 gauge, $2.30; No. 10 gauge, 6c. extra; No. 11 gauge, 12c extra; No 12 gauge, 20c. extra; No. 13 gauge, 30c. extra; No 14 gauge, 40c. extra; No 15 gauge, 55c. extra; No 16 gauge, 70c extra. Add 60c. for coppering and $2 for tinning.

Extra net per 100 lb.—Oiled wire 10c, spring wire $1.25, bright soft drawn 15c., charcoal (extra quality) $1.25, packed in casks or cases 15c., bagging and papering 10c., 50 and 100-lb. bundles 10c., in 25-lb. bundles 15c., in 5 and 10-lb. bundles 25c., in 1-lb hanks, 50c., in ½-lb. hanks 75c., in ¼-lb. hanks $1.

POULTRY NETTING.

2 in mesh 19 w. g., discount 50 and 10 per cent. All others 50 per cent.

WIRE CLOTH.

Painted Screen, in 100-ft rolls, $1 62½, per 100 sq. ft.; in 50-ft. rolls, $1.67½, per 100 sq. ft. Terms, 2 per cent. off 30 days.

WIRE FENCING.

Galvanized barb.......... .... 2 95
Galvanized, plain twist 3 30
Galvanized barb, f.o b. Cleveland, $2.70 for small lots and $2.60 for carlots

WOODENWARE.

CHURNS.

No. 0, $9; No. 1, $9; No. 2, ; No. 3, $11; No. 4, $13; No. 5, $16; f.o b. Toronto Hamilton, London and St Marys, 30 and 30 per cent; f o b. Ottawa, Kingston and Montreal, 40 and 15 per cent. discount, Taylor-Forbes, 30 and 30 per cent.

CLOTHES REELS.

Davis Clothes Reels, dis. 40 per cent.

LADDERS, EXTENSION.

Waggoner Extension Ladders, dis 40 per cent.

MOPS AND IRONING BOARDS.

"Best" mops.......... .... 1 25
"900" mops.......... .... 1 25
Folding ironing boards.......... 12 00 16 50

REFRIGERATORS

Discount, 40 per cent.

SCREEN DOORS.

Common doors, 2 or 3 panel, walnut stained, 4-in. style..........per doz. 7 25
Common doors, 2 or 3 panel, grained only, 4-in., style..........per doz. 7 55
Common doors, 2 or 3 panel, light stair per doz.......... 9 55

WASHING MACHINES.

Round, re-acting per doz. 60 00
Square " 63 00
Eclipse, per doz 54 00
Dowswell " 39 00
New Century, per doz 75 00
Daisy.......... 54 00

WRINGERS.

Royal Canadian, 11 in., per doz. ... 34 00
Royal American, 11 in. ... 34 00
Eze, 10 in., per doz 3. 75
Terms, 2 per cent., 30 days.

MISCELLANEOUS.

AXLE GREASE.

Ordinary, per gross 6 00 7 00
Best quality 10 00 12 00

BELTING.

Extra, 60 per cent.
Standard, 60 and 10 per cent.
No. 1, not wider than 6 in., 60, 10 and 10 p c.
Agricultural, not wider than 4 in., 75 per cent.
Lace leather, per side, 75c; cut laces, 80c.

BOOT CALKS.

Small and medium, ball ....per M 4 25
Small heel .... " 4 50

CARPET STRETCHERS.

American ....per doz. 1 00 1 50
Bullard's.......... " 6 50

CASTORS.

Bed, new list, discount 55 to 57½ per cent.
Plate, discount 52½ to 57½ per cent.

PINE TAR.

½ pint in tins ....per gross ... 7 80
" " .... " 9 60

PULLEYS.

Hothouse ....per doz. 0 55 1 00
Axle .... " 0 22 0 33
Screw .... " 0 22 1 00
Awning .... " 0 35 2 50

PUMPS.

Canadian cistern 1 40 2 00
Canadian pitcher spout ... 1 80 3 16
Berg's wing pump, 75 per cent.

ROPE AND TWINE.

Sisal.......... 0 10½
Pure Manilla 0 15½
"British" Manilla.......... 0 12
Cotton, 3-16 inch and larger... ..0 21 0 23
" 5-32 inch 0 25 0 27
" ¼ inch 0 25 0 28
Russia Deep Sea 0 16
Jute.......... 0 09
Lath Yarn, single 0 10
" double 0 10½
Sisal bed cord, 48 feet........per doz. 0 65
" " 60 feet........ " 0 80
" " 72 feet........ " 0 95

Twine.

Bag, Russian twine, per lb. 0 27
Wrapping, cotton, 3-ply 0 25
" " 4-ply 0 29
Mattress twine per lb. 0 33 0 45
Staging " " 0 27 0 35

SCALES.

Gurney Standard, 40 per cent.
Gurney Champion, 50 per cent.
Burrow, Stewart & Milne—
Imperial Standard, discount 40 per cent.
Weigh Beams, discount 40 per cent.
Champion Scales, discount 50 per cent.
Fairbanks standard, discount 35 per cent.
" Dominion, discount 55 per cent.
" Richelieu, discount 55 per cent.
Warren new Standard, discount 40 per cent.
" " Champion, discount 50 per cent.
" Weighbeams, discount 35 per cent.

STONES—OIL AND SCYTHE.

Washita ....per lb. 0 25 0 37
Hindostan .... " 0 06 0 10
" slip .... " 0 18 0 20
" Axe.......... " 0 10
Deer Creek.......... " 0 10
Deerlick .... " 0 25
" Axe .... " 0 15
Lily white .... " 0 42
Arkansas.......... " 1 50
Water-of-Ayr .... " 0 10
Scythe ....per gross 3 50 5 00
Grind, 40 to 200 lb., per ton.... 20 00 22 00
" under 40 lb., " 24 00
" 200 lb. and over 28 00

KEARNEY & FOOT

The Best Rasp Made

The Rasp that sells

MADE BY

Nicholson File Co., Dominion Wks. **Port Hope, Ont.**

Wrought Iron Pipe and Cast Iron Fittings

Our Stock in these lines is complete. Can fill all orders promptly. Be supplied before the Fall rush commences.

WM. STAIRS, SON & MORROW, Limited, HALIFAX, N.S.

INDEX TO ADVERTISERS.

CLASSIFIED LIST OF ADVERTISEMENTS.

Auditors
Davenport, Percy P., Winnipeg.

Babbitt Metal.
Canada Metal Co., Toronto.
Canadian Fairbanks Co., Montreal.
Robertson, Jas. Co., Montreal.

Bath Room Fittings.
Carriage Mounting Co., Toronto.

Belting, Hose, etc.
Gutta Percha and Rubber Mfg. Co. Toronto.

Bicycles and Accessories.
Johnson's, Iver, Arms and Cycle Works Fitchburg, Mass

Binder Twine
Consumers Cordage Co., Montreal.

Box Strap.
J. N. Warminton, Montreal.

Brass Goods.
Glauber Brass Mfg. Co., Cleveland, Ohio.
Lewis, Rice, & Son., Toronto.
Morrison, Jas., Brass Mfg. Co., Toronto.
Mueller Mfg. Co., Decatur, Ill.
Penberthy Injector Co., Windsor, Ont.
Taylor-Forbes Co., Guelph, Ont.

Bronze Powders.
Canadian Bronze Powder Works, Montreal.

Brushes.
Ramsay, A., & Son Co., Montreal.

Can Openers.
Cumming Mfg. Co. Renfrew.

Cans.
Acme Can Works, Montreal.

Builders' Tools and Supplies.
Covert Mfg. Co., West Troy, N.Y.
Frothingham & Workman Co., Montreal.
Howland, H. S., Sons & Co., Toronto.
Hyde, F., & Co., Montreal.
Lewis Bros. & Co., Montreal.
Lewis, Rice, & Son, Toronto.
Lockerby & McComb, Montreal.
Lufkin Rule Co., Saginaw, Mich.
Newman & Sons, Birmingham.
North Bros. Mfg. Co., Philadelphia, Pa.
Stanley Rule & Level Co., New Britain.
Stanley Works, New Britain, Conn.
Stephens, G. F., Winnipeg.
Taylor-Forbes Co., Guelph, Ont.

Carriage Accessories.
Carriage Mountings Co., Toronto.
Covert Mfg. Co., West Troy, N.Y.

Carriage Springs and Axles.
Guelph Spring and Axle Co., Guelph.

Cattle and Trace Chains.
Greening, B., Wire Co., Hamilton.

Churns.
Dowswell Mfg. Co., Hamilton.

Clippers—All Kinds.
American Shearer Mfg. Co., Nashua, N.H.

Clothes Reels and Lines.
Hamilton Cotton Co., Hamilton, Ont.

Cordage.
Consumers' Cordage Co., Montreal.
Hamilton Cotton Co., Hamilton.

Cork Screws.
Erie Specialty Co., Erie, Pa.

Clutch Nails.
J. N. Warminton, Montreal.

Cut Glass.
Phillips, Geo., & Co., Montreal.

Cutlery—Razors, Scissors, etc.
Birkett, Thos., & Son Co., Ottawa.
Clauss Shear Co., Toronto
Dorken Bros. & Co., Montreal.
Heinisch's, R., Sons Co., Newark, N.J.
Howland, H. S. Sons & Co., Toronto.
Lamplough, F. W., & Co., Montreal.
Phillips, Geo., & Co., Montreal.
Round, John, & Son, Montreal.

Electric Fixtures.
Canadian General Electric Co., Toronto.
Forman, John, Montreal.
Morrison James, Mfg. Co., Toronto.
Munderloh & Co., Montreal.

Electro Cabinets.
Cameron & Campbell Toronto.

Engines, Supplies, etc.
Kerr Engine Co., Walkerville, Ont.

Fencing—Woven Wire.
McGregor-Banwell Fence Co., Walkerville, Ont
Owen Sound Wire Fence, Co., Owen Sound
Banwell Hoxie Wire Fence Co., Hamilton.

Files and Rasps.
Barnett Co., G. & H., Philadelphia, Pa.
Nicholson File Co., Port Hope

Financial Institutions
Bradstreet Co.

Firearms and Ammunition.
Dominion Cartridge Co., Montreal.
Hamilton Rifle Co., Plymouth, Mich
Harrington & Richardson Arms Co., Worcester, Mass.
Johnson's, Iver, Arms and Cycle Works, Fitchburg, Mass.

Food Choppers
Enterprise Mfg Co., Philadelphia, Pa
Lamplough, F. W., & Co., Montreal.
Shirreff Mfg. Co., Brockville, Ont.

Galvanizing.
Canada Metal Co., Toronto.
Montreal Rolling Mills Co., Montreal
Ontario Wind Engine & Pump Co., Toronto.

Glaziers' Diamonds.
Gibsone, J. B., Montreal.
Pelton, Godfrey S.
Sharratt & Newth, London, Eng.
Shaw, A., & Son, London, Eng.

Handles
Still, J. H., Mfg. Co.

Harvest Tools.
Maple Leaf Harvest Tool Co., Tillsonburg Ont.

Hoop Iron.
Montreal Rolling Mills Co Montreal.
J. N. Warminton, Montreal.

Horse Blankets.
Heney, E. N., & Co., Montreal.

Horseshoes and Nails.
Canada Horse Nail Co., Montreal.
Montreal Rolling Mills, Montreal.
Capewell Horse Nail Co., Toronto.

Hot Water Boilers and Radiators.
Cluff, R. J., & Co., Toronto
Pease Foundry Co., Toronto.
Taylor-Forbes Co., Guelph.

Ice Cream Freezers.
Dana Mfg. Co., Cincinnati, Ohio.
North Bros. Mfg. Co., Philadelphia, Pa.

Ice Cutting Tools.
Erie Specialty Co., Erie, Pa.
North Bros. Mfg. Co., Philadelphia, Pa.

Injectors—Automatic.
Morrison, Jas., Brass Mfg. Co., Toronto.
Penberthy Injector Co., Windsor, Ont.

Iron Pipe.
Montreal Rolling Mills, Montreal.

Iron Pumps.
Lamplough, F. W., & Co., Montreal.
McDougall, R., Co., Galt, Ont.

Lanterns.
Kemp Mfg. Co., Toronto.
Ontario Lantern Co., Hamilton, Ont.
Wright, E. T., & Co., Hamilton.

Lawn Mowers.
Birkett, Thos., & Son Co., Ottawa.
Maxwell, D., & Sons, St. Mary's, Ont.
Taylor, Forbes Co., Guelph.

Lawn Swings, Settees, Chairs.
Cumming Mfg. Co., Renfrew.

Ledgers—Loose Leaf.
Business Systems Toronto.
Copeland-Chatterson Co., Toronto.
Crain, Rolla L., Co. Ottawa.
Universal Systems, Toronto.

Locks, Knobs, Escutcheons, etc.
Peterborough Lock Mfg. Co., Peterborough, Ont.

Lumbermen's Supplies.
Pink, Thos., & Co., Pembroke Ont.

Manufacturers' Agents.
Fox, C. H., Vancouver.
Gibb, Alexander, Montreal.
Mitchell, David C., & Co., Glasgow, Scot.
Mitchell, H. W., Winnipeg.
Pearce, Frank, & Co. Liverpool, Eng.
Scott, Bathgate & Co., Winnipeg.
Thorne, R. E., Montreal and Toronto.

Metals.
Canada Iron Furnace Co., Midland, Ont.
Canada Metal Co., Toronto.
Eadie, H. G., Montreal
Gibb, Alexander, Montreal.
Kemp Mfg Co., Toronto
Leslie, A. C., & Co., Montreal.
Lysaght, John, Bristol, Eng.
Nova Scotia Steel and Coal Co., New Glasgow, N S.
Robertson, Jas., Co., Montreal
Roper, J. H., Montreal.
Samuel, Benjamin & Co., Toronto.
Stairs, Son & Morrow, Halifax, N S.
Thompson, B. & S. H. & Co. Montreal.

Metal Lath.
Galt Art Metal Co Galt
Metallic Roofing Co., Toronto.
Metal Shingle & Siding Co., Preston, Ont.

Metal Polish, Emery Cloth, etc
Oakey, John, & Sons, London, Eng

Mops.
Cumming Mfg Co., Refrew.

Mouse Traps.
Cumming Mfg Co., Renfrew

Oil Tanks
Bowser, S. F., & Co., Toronto.

Paints, Oils, Varnishes, Glass
Bell, Thos., Sons & Co., Montreal.
Canada Paint Co., Montreal.
Canadian Oil Co. Toronto
Consolidated Plate Glass Co., Toronto
Fenner, Fred., & Co., London, Eng.
Henderson & Potts Co., Montreal.
Imperial Varnish and Color Co., Toronto.
Jamieson, R. C., & Co., Montreal
McArthur, Corneille & Co., Montreal.
McCaskill, Dougall & Co., Montreal.
Montreal Rolling Mills Co., Montreal.
Moore Benjamin, & Co. Toronto.
Queen City Oil Co., Toronto.
Ramsay & Son, Montreal.
Sherwin-Williams Co., Montreal.
Standard Paint and Varnish Works Windsor, Ont.
Stephens & Co., Winnipeg.
Martin-Senour Co., Chicago.

Perforated Sheet Metals.
Greening, B., Wire Co., Hamilton.

Plumbers' Tools and Supplies.
Borden Co., Warren, Ohio.
Canadian Fairbanks Co., Montreal.
Cluff, R. J., & Co., Toronto.
Glauber Brass Co., Cleveland, Ohio.
Jardine, A. B., & Co., Hespeler, Ont.
Jenkins Bros., Boston, Mass.
Lewis, Rice, & Son, Toronto.
Merrell Mfg. Co., Toledo, Ohio.
Montreal Rolling Mills Montreal.
Morrison, Jas., Brass Mfg. Co., Toronto.
Mueller, H., Mfg. Co., Decatur, Ill.
Oshawa Steam & Gas Fitting Co., Oshawa
Robertson Jas., Co. Montreal.
Stairs, Son & Morrow, Halifax, N.S.
Standard Ideal Sanitary Co., Port Hope.
Standard Sanitary Co., Pittsburg.
Stephens, G. F., & Co., Winnipeg, Man.
Turner Brass Works, Chicago.
Vickery, Orlando, Toronto.

Portland Cement.
International Portland Cement Co., Ottawa, Ont.
Hanover Portland Cement Co., Hanover, Ont.
Hyde, F., & Co., Montreal.
Thompson, B. & S. H. & Co., Montreal.

Potato Mashers.
Cumming Mfg Co., Renfrew.

Poultry Netting.
Greening, B., Wire Co., Hamilton, Ont.

Razors.
Clauss Shear Co., Toronto.

Roofing Supplies.
Brantford Roofing Co., Brantford.
McArthur, Alex., & Co., Montreal.
Metal Shingle & Siding Co., Preston, Ont.
Metallic Roofing Co., Toronto.
Paterson Mfg. Co., Toronto & Montreal.

Saws.
Atkins, E. C., & Co., Indianapolis, Ind
Lewis Bros., Montreal.
Shurly & Dietrich, Galt, Ont.
Spear & Jackson, Sheffield, Eng.

Scales.
Canadian Fairbanks Co., Montreal.

Screw Cabinets.
Cameron & Campbell, Toronto.

Screws, Nuts, Bolts.
Montreal Rolling Mills Co., Montreal.
Morrow, John, Machine Screw Co., Ingersoll, Ont.

Sewer Pipes.
Canadian Sewer Pipe Co., Hamilton
Hyde, F., & Co., Montreal.

Shelf Boxes.
Cameron & Campbell, Toronto.

Shears, Scissors.
Clauss Shear Co., Toronto.

Shelf Brackets.
Atlas Mfg. Co., New Haven, Conn

Shellac
Bell, Thos., Sons & Co., Montreal.

Shovels and Spades.
Canadian Shovel Co Hamilton.
Peterboro Shovel & Tool Co., Peterboro.

Silverware.
Phillips, Geo., & Co., Montreal
Round, John, & Son, Sheffield, Eng

Skates.
Canada Cycle & Motor Co., Toronto.

Spring Hinges, etc.
Chicago Spring Butt Co., Chicago, Ill.

Steel Rails.
Nova Scotia Steel & Coal Co., New Glasgow, N.S.

Stoves, Tinware, Furnaces
Canadian Heating & Ventilating Co., Owen Sound.
Canada Stove Works, Harriston, Ont.
Clare Bros. & Co., Preston.
Davidson, Thos., Mfg. Co., Montreal.
Guelph Stove Co., Guelph.
Gurney Foundry Co., Toronto.
Harris, J. W., Co., Montreal.
Joy Mfg Co., Toronto.
Kemp Mnfg. Co. Toronto.
McClary Mfg. Co. London.
Pease Foundry Co., Toronto.
Stewart, Jas., Mfg. Co., Woodstock, Ont.
Taylor-Forbes Co., Guelph, Ont.
Wright, E. T., & Co., Hamilton.

Tacks.
Montreal Rolling Mills Co., Montreal.
Ontario Tack Co., Hamilton.

Ventilators.
Harris, J. W., Co., Montreal.
Pearson, Geo. D., Montreal.

Washing Machines, etc.
Dowswell Mfg. Co., Hamilton, Ont.
Taylor-Forbes Co., Guelph, Ont.

Wheelbarrows.
London Foundry Co., London, Ont.

Wholesale Hardware.
Birkett, Thos., & Sons Co., Ottawa.
Caverhill, Learmont & Co., Montreal.
Frothingham & Workman, Montreal.
Hobbs Hardware Co., London.
Howland, H. S., Sons & Co., Toronto.
Lamplough, F. W., & Co., Montreal.
Lewis Bros. & Co., Montreal.
Lewis, Rice, & Son, Toronto.

Window and Sidewalk Prisms.
Hobbs Mfg. Co., London, Ont.

Wire Springs.
Guelph Spring Axle Co., Guelph, Ont.
Wallace-Barnes Co., Bristol, Conn.

Wire, Wire Rope, Cow Ties, Fencing Tools, etc.
Canada Fence Co., London.
Dennis Wire and Iron Co., London, Ont.
Dominion Wire Mnfg. Co., Montreal
Greening, B., Wire Co., Hamilton.
Montreal Rolling Mills Co., Montreal.
Western Wire & Nail Co., London, Ont

Woodenware.
Taylor-Forbes Co., Guelph, Ont.

Wrapping Papers.
Canada Paper Co., Toronto.
McArthur, Alex., & Co., Montreal.
Stairs, Son & Morrow, Halifax, N.S.

CIRCULATES EVERYWHERE IN CANADA
Also in Great Britain, United States, West Indies, South Africa and Australia.

HARDWARE AND METAL

A Weekly Newspaper Devoted to the Hardware, Metal, Heating and Plumbing Trades in Canada.

Office of Publication, 10 Front Street East, Toronto.

VOL. XIX. MONTREAL, TORONTO, WINNIPEG, FEBRUARY 23, 1907 **NO. 8.**

See Classified List of Advertisements on Page 71.

TORCHES — WE CAN SHIP PROMPTLY — TORCHES

These Torches are indispensible for all kinds of mechanics, Electricians, Dentists and Plumbers will find them most useful for their business.

PLUMBERS' HEATER

Torches of all kinds in stock, suitable for Painters, Jewellers, Opticians, Gas and Steamfitters, Pattern-makers, etc.

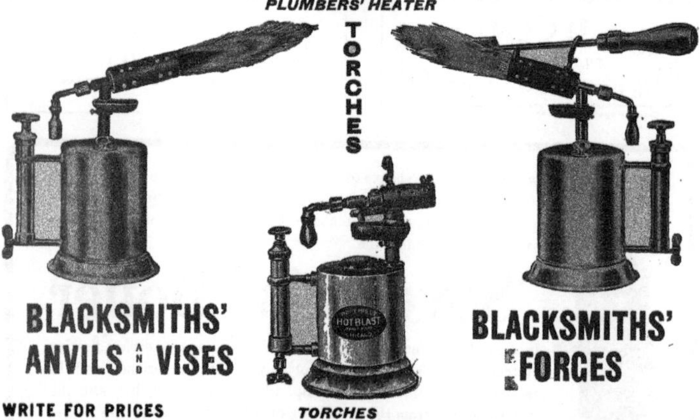

BLACKSMITHS' ANVILS AND VISES

BLACKSMITHS' FORGES

TORCHES

WRITE FOR PRICES

RICE LEWIS & SON

LIMITED

TORONTO.

Your Business at Your Finger Tips

The merchant, if he is wise, no longer keeps his accounts upon tickets, in memorandum books, or in a loose way, but has a system by which at a glance, he can tell what he is doing

THE CRAIN MONTHLY ACCOUNT SYSTEM

has been in use by successful business men for years. It does the work in one-half the time, and in a way that is much more satisfactory to his customers and more pleasant to himself than plodding along in the old way. In this system, too, you require a binder, such as the Crain, with an unlimited capacity, which binds on one sheet as easily as it does 1,000, not one that you must have a certain number of sheets in before it will bind, and then only allows you to extend to a limited number of accounts.

Write us for particulars

The Rolla L. Crain Co., Limited

OTTAWA, CANADA

TORONTO: 18 Toronto Street
MONTREAL: Alliance Bldg., 107 St. James St.
WINNIPEG: 11 Nanton Block

WE CARRY IN STOCK

Horse-shoers' Clip Horn Anvil
Prospectors' Anvils and Bellows
Prospectors' Picks, Shovels and Stoves
The Famous Blacksmith Blower No. 400
The Grand Bay Blower No. 100
Challenge Forges with Hood, Extra Strong Frame
Hero Forges With and Without Hood
Navy Wheelbarrows with Steel or Wood Trays
Steel Pulley Blocks, Single, Double or Triple

Write for prices. Prompt shipment.

THOS. BIRKETT & SON CO., LIMITED

Wholesale Hardware

OTTAWA, ONT.

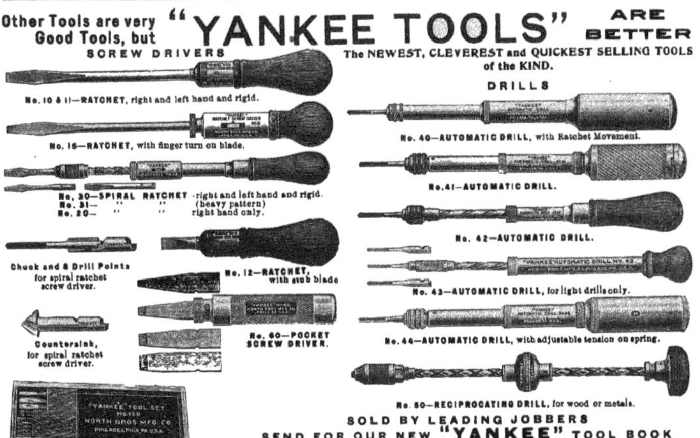

SOLD BY LEADING JOBBERS

SEND FOR OUR NEW "YANKEE" TOOL BOOK

NORTH BROS. MFG. CO.

PHILADELPHIA, Pa.

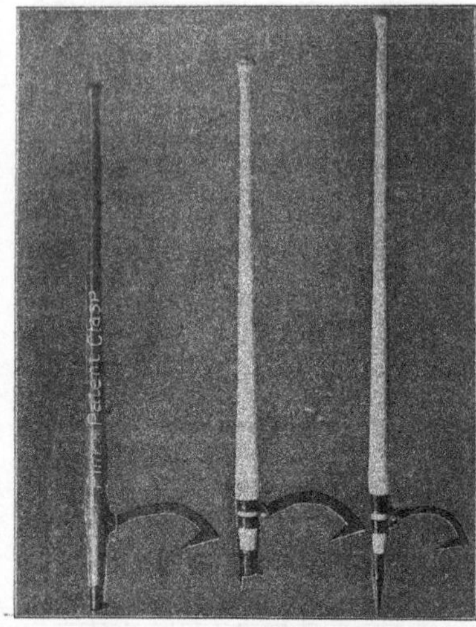

Pink's Lumbering Tools

MADE IN CANADA

Send for Catalogue and Price List

THE STANDARD TOOLS

in every Province of the Dominion, New Zealand, Australia, Etc.

We manufacture all kinds of Lumber Tools

Pink's Patent Open Socket Peaveys.
Pink's Patent Open Socket Cant Dogs.
Pink's Patent Clasp Cant Dogs, all Handled with Split Rock Maple.

These are light and durable tools.

Sold throughout the Dominion by all Wholesale and Retail Hardware Merchants

MANUFACTURED BY

Long Distance Phone No. 87

THOMAS PINK
Pembroke, Ont., Canada.

METALS

ANTIMONY
COPPER
LEAD
TIN
ZINC

Large Stocks — Lowest Prices

M. & L. Samuel, Benjamin & Co.

TORONTO

Consumers Cordage Co., LIMITED

Manufacture every variety of

CORDAGE

PACKING

BINDER TWINE

LATHYARN

SMALL TWINES

From MANILLA, SISAL, ITALIAN, RUSSIAN, JUTE

Tallow Laid Transmission Rope, 3 or 4 Strand

Oil Well Drilling Cables, Lariat Rope, 3 or 4 Strand, Halyards, Deep Sea Line, Snap Halters, Snap Plow Lines, Clothes Lines, Hide Cord, Pulp and Paper Cord, Tarred Rope, Marline, Houseline

Wrapping, Rolling, Seaming, Sewing and Grape Twines

CAREFUL EXECUTION **SPECIAL ORDERS** PROMPT SHIPMENT

Mills: MONTREAL, QUE., and HALIFAX, N.S.

BRANCHES:

F. H. ANDREWS & SON, Quebec. CONSUMERS CORDAGE CO., Limited, St. John, N.B. MacGOWAN & CO., Vancouver
TEES & PERSSE, Limited, Winnipeg, Calgary and Edmonton. GEO. WOOD, London, England.
MERRICK, ANDERSON & CO., Winnipeg, Binder Twine Distributors for Northwest.

Have You Tried SOLDERALL?

Tinsmiths should make a test of this New Anticorrosive Fluid—It is the best thing in the market to-day for tinning and soldering all metals.

Unsurpassed for Cleanliness of Work and Strength of Joint made.

No Noxious Fumes to Injure the Health or Affect the Skin.

Economises Time and Labor, Saves Solder

No rust or discoloration. The Solder will follow the fluid like a magnet

PUT UP FOR SALE IN ½-GAL., 1-GAL. AND 5-GAL. CANS.

Solderall is always sent out **full strength** and may be diluted according to the class of work for which it may be required.

SOLE AGENTS AND MANUFACTURERS

The Thos. Davidson Mfg. Co. Ltd., MONTREAL and WINNIPEG

CAVERHILL LEARMONT & Company

The New No. 44 Genuine

UNIVERSAL BREAD MAKER

Can be clamped to the table, so that the operator can use it either sitting or standing. Capacity, 2 to 6 loaves. "Recipe" Book with each machine.

Half Dozen in a Case

The Universal mixes and kneads the dough with scientific accuracy, thoroughly and evenly mixing all the ingredients. Thus, every particle of starch is burst by the heat, because every bit of flour is covered with a film of moisture, and by expanding is more easily permeated by the heat in baking.

Genuine

UNIVERSAL CAKE MAKERS

Mix Cake Batter, Whip Cream or Beat Eggs

A household necessity. Neatly tinned, strong, and will make better cake than old methods.

Remember, Universal goods are well known to your customers now through extensive advertising in magazines which go into their homes.

Capacity : 4 Quarts Batter.

Wholesale Distributors:

Caverhill Learmont & Co

Montreal and Winnipeg

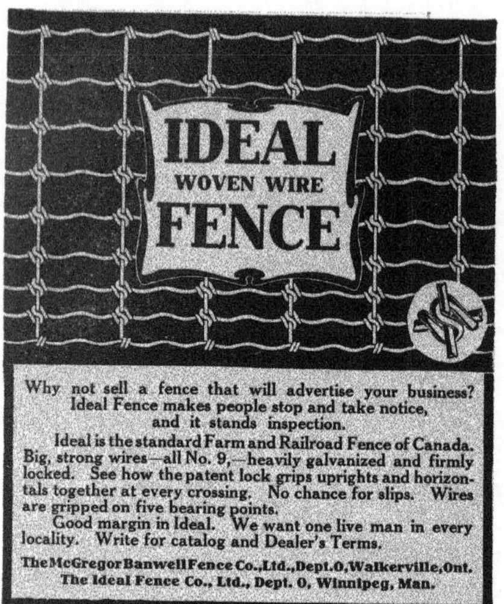

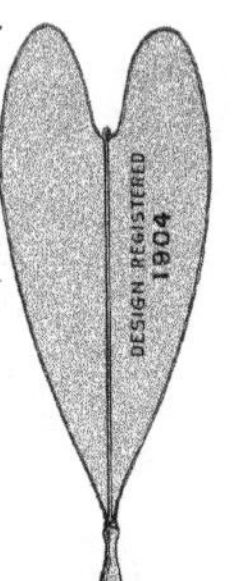

THE

ANTON CARPET BEATER

Has double the beating surface of any carpet beater on the market.

LIGHT, yet STRONG and DURABLE; STIFF, yet PLIABLE and EASY TO HANDLE.

Centre rod, made of heavy spring steel wire. Outside wires go through the handle and are clinched at bottom end making it impossible for handle to come off; handle filled with lead, thereby making it a perfectly balanced article.

FOR SALE BY

E. T. WRIGHT & CO.
HAMILTON, ONT.

Will Hold up a Shelf

That's what a shelf brackets for.

For this purpose there can be NOTHING BETTER, NOTHING CHEAPER than the BRADLEY STEEL BRACKET. It is well Japanned, Strong and Light. The saving on freight is a good profit aside from the lower price at which the goods are sold. Order direct or through your jobbers.

ATLAS MFG. CO.
New Haven, Conn.

Expert Bookkeeping

Investigation and auditing of all classes of accounts undertaken. Trading, profit and loss accounts and balance sheets prepared. Sets of books opened and office systems devised in an up-to-date manner. Collection of accounts a specialty.

Davenport, Pickup & Co.
622 McIntyre Block and 422 Ashdown Block
WINNIPEG MAN.

Square your Accounts

INTERNATIONAL PORTLAND CEMENT COMPANY, Limited
OTTAWA, - - ONTARIO.

Manufacturers of strictly high-grade Portland Cement made by the most improved method of manufacture in existence. Each barrel contains three hundred and fifty pounds of pure Portland Cement with no adulteration.

The Best Door Closer is . . .
NEWMAN'S INVINCIBLE FLOOR SPRING
Will close a door silently against any pressure of wind. Has many working advantages over the ordinary spring, and has twice the wear. In use throughout Great Britain and the Colonies. Gives perfect satisfaction. Made only by
W. NEWMAN & SONS,
Hospital St., - - BIRMINGHAM

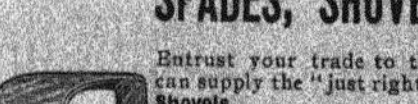

SPADES, SHOVELS, Etc.

Entrust your trade to the house that can supply the "just right" **Spades** and **Shovels**.

By getting in touch with us you will become acquainted with the favorite lines:

TRADE MARK

"Peterborough," "Lundy," "Irwin," and 'Hill"

Send in a postcard for our Catalogue and you will see where we stand.

The Peterborough Shovel & Tool Company, Limited, Peterborough, Ont.

PRATT AND WHITNEY SMALL TOOLS

Insure

A finer grade of steel—superior workmanship and better cutting and wearing qualities at the price of ordinary tools. They produce better work and cut the manufacturing cost to the lowest figure.

SPECIFY PRATT AND WHITNEY IN YOUR NEXT ORDER AND INSIST UPON GETTING TOOLS STAMPED "P. and W."

TAPS, DIES, REAMERS, MILLING CUTTERS, DRILLS AND SPECIAL TOOLS

NOW MADE IN CANADA

THE CANADIAN FAIRBANKS CO., LTD.

CANADIAN SALES AGENTS

MONTREAL TORONTO WINNIPEG VANCOUVER

Dame Quality at Our Bidding Waits

Our ***Handsaws*** are made of "top-notch" Crucible Cast Steel, they are tempered in oil and carefully tested. Every one is guaranteed.

The illustration shows one of our Patent Ground, Hollow-Back, Highly-Tempered, Highly-Finished ***Handsaws.*** It is praised the world over.

We also make Back Saws, Circular Saws, Hack Saws, Mill Saws, Canada Webs, Files, Edge Tools, etc.

The sun never sets on Spear & Jackson's Wares

A POSTCARD SENT TO US MAY MEAN A PILE OF BUSINESS SENT TO YOU.

SPEAR & JACKSON ÆTNA WORKS **Sheffield, England**

Telegraphic Address: "Spear," Sheffield

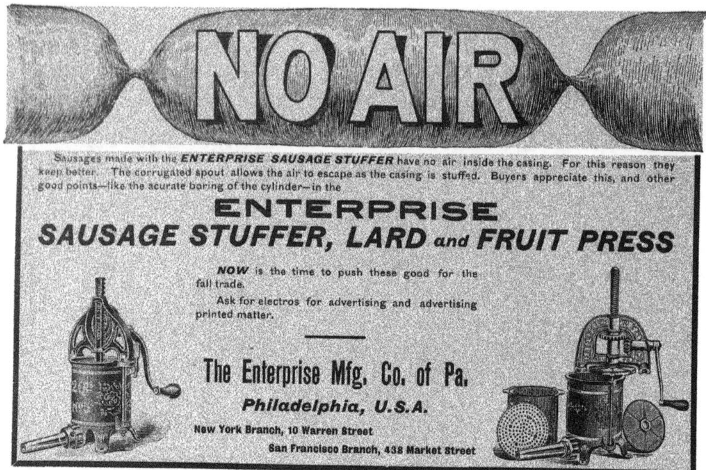

Talking Points That Sell
H. & R. REVOLVERS

Excellence of Construction
Accuracy
Rapidity of Action
Absolute Safety

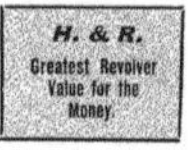

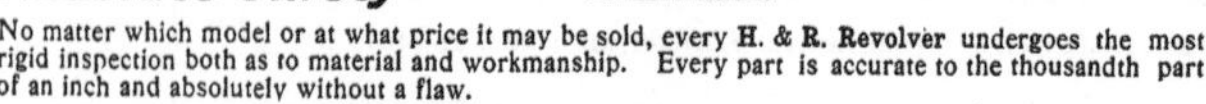

No matter which model or at what price it may be sold, every **H. & R. Revolver** undergoes the most rigid inspection both as to material and workmanship. Every part is accurate to the thousandth part of an inch and absolutely without a flaw.

H. & R. POLICE PREMIER REVOLVERS

combine more good features than any revolvers of other make selling for twice the money. The safety hammer enables them to be discharged quicker than any other hammer revolver. **Automatic Ejector, Independent Cylinder Stop** (found only on other make revolvers of much higher price), **Light Weight, Beautiful in Design and Finish** are the points that make them sell rapidly.

Our Catalog tells about our full line and gives valuable selling points.

HARRINGTON & RICHARDSON ARMS CO., WORCESTER, MASS.

314 Park Ave.

Persons addressing advertisers will kindly mention having seen this advertisement in Hardware and Metal.

WE ARE ADVERTISED

by our loving friends who are using

"Automobile" and "Cycle" Skates

The skate everybody is boosting

NEW CATALOGUE JUST OUT. ***WRITE FOR IT.***

Canada Cycle & Motor Co., Limited

TORONTO JUNCTION, CANADA

Branches: Winnipeg, Vancouver, Melbourne, Aust.

Kemp's Diamond Ware

Enameled Covers

Every Piece Guaranteed Perfect

KEMP MANUFACTURING CO.,

TORONTO, CAN.

THE KEMP MFG. CO. OF MONTREAL,
39 St. Antoine St. Montreal, Que.

THE KEMP MFG. AND METAL CO., Ltd.
McDermot Ave. E., Winnipeg, Man.

What To Do Next Month

March is an important month for the hardwareman, as it is the beginning of the spring season; the month that makes the average man's blood tingle and fills the mind of his good wife with thoughts of the housecleaning that is to be done in the near future.

* * *

Stock-taking is over and you know just how you came through your last year's business—or should, if you are a wide-awake merchant. You can therefore go into this month with a clear brain and steady nerve, and pay strict attention to the noble art of getting the best you can out of your worthy neighbors.

* * *

As spring and housecleaning go hand in hand, let us first turn our attention to the articles necessary for the successful carrying on of this hunt for dirt and germs.

Paints, Varnishes, Wallpapers. Etc.

Of course head the list, and together with carpet sweepers, brooms, pails, mops, feather dusters, whisks carpet beaters, carpet stretchers, mop wringing pails, etc., would make a very appropriate display, and if a little ingenuity is exercised, a very attractive exhibit can be made. Floor stains and Kalsomine, also come among the needs of housecleaners.

Of course the vacuum cleaners, have in the large cities, greatly facilitated housecleaning, and therefore taken a great deal of trade from the hardwareman, but in towns where these machines have not yet made their appearance, the retailer can, with a little longheadedness, reap a veritable harvest from the above lines.

Not only should a good display be made of these lines, but advertising should be run continually in the daily or weekly papers. There are many catch phrases which can be successfully used, such as "Helps to Housecleaning," "Get After the Dirt," "Cleanliness is Godliness," "Housecleaning Simplified," "Make Housecleaning a Pleasure," "Lighten Woman's Work," etc. To make a success out of your housecleaning advertising you must always bear in mind that 90 per cent. of this work is done by women, and your ads. must always be headed with phrases that will attract a woman's attention. Its always necessary when getting after the fair sex, to describe your goods fully, and, don't forget a woman must see prices in an ad. or she's not all interested. This is a most important fact for you to remember.

* * *

Sugaring time must also be given some thought and all during March, you should be after the farmers. Sap spouts have a habit of getting lost during the winter months, and sap pails will not last for ever, so get into the game. A little advertising wouldn't hurt this line either. Tell the farmers to look over their stocks of spouts and pails early in order not to be left when the rush comes. A good heading for such an ad. would be "Time and Sap Wait for No Man," or something after that style.

* * *

Again, "In the spring a young man's fancy, etc." Watch for the June brides that are to be. A nice fat order for kitchen utensils would repay you for your trouble, etc. Don't think over it too long, your competitor is also a smart man. With the breaking up of winter, houses begin to go up in large numbers and several contracts for builders' hardware, building paper, etc., will help to swell the bank account.

A GREAT VICTORY

During the past eight weeks Hardware and Metal, in conjunction with the Ontario and Western Canada Retail Hardware Associations, has conducted an active campaign against the proposed parcels post c.o.d. system.

With the exception of the MacLean Trade Newspapers, particularly the Dry Goods Review and Printer and Publisher, we are not aware of any other hands being raised in opposition to the P.M.G.'s proposal. Certain it is that the Retail Merchants' Association were invited to join in the fight and declined.

The hardware trade has reason to congratulate itself upon its ability to act unitedly in time of trouble, upon having live associations in several provinces ready to take the lead in opposing injurious legislation—and on having a medium which keeps the retail trade in every part of Canada posted upon all matters of general interest when their interests are betrayed by the great metropolitan daily papers.

Success has crowned the labors of those who took up the agitation against the extension of the parcels post system as desired by the mail order houses. In Parliment last Friday night (Feb. 14) Hon. Mr. Lemieux, Postmaster-general announced that because of the unanimous opposition expressed to the scheme on both sides of the House he would drop the proposal. He said he still considered the scheme a good one, however, and the plan migh be carried into effect later on. "At present,' he said, "we are fighting against wind mills."

The victory is a notable one. The Retail Hardware Associations and the trade press has shown the power they can wield. Strengthen the Associations—extend their membership—and their powers for good work will be so much greater.

As this paper covers the whole of Canada, its rather hard for us to advise you of all the goods that you might dispose of in the early spring, owing to the different climates prevailing from coast to coast, but we give a line that is used generally, and leave it to your own good judgment to add to it. For instance, in Montreal, March is the great sidewalk cleaning month and naturally the dealer turns over quite a lot of money on sidewalk cleaners, shovels, etc., whereas in Vancouver these articles would have no sale at all.

Lawn mowers, lawn and garden tools, are also lines what may be displayed as soon as the snow begins to leave.

Sporting goods leave the dealer a good profit, and if you have not already done so you should lose no time in getting after the managers of football, baseball, lacrosse, fishing and shooting clubs. Circulars may be sent out, but we would advise you to send these gentlemen personal letters asking them to give you a chance before placing their orders elsewhere, and offering to procure for them, any lines you do not stock regularly. It might also be to your advantage to offer special club rates to any clubs or proposed clubs in your district.

* * *

Among the lines you should order from your jobber during March are ice cream freezers, screen doors and windows, paints, oils, glass, etc., wringers and washing machines, churns, enamelware, fishing tackle, Paris green and all kinds of tools.

Owing to the difficulty the jobbers are experiencing in getting deliveries of goods, it would be well perhaps for you to place your orders for all lines a little earlier than heretofore, in order to receive your shipments in time.

As already stated, this is just a page of suggestions and reminders, and does not begin to cover the thousand and one ways by which you can add to your sales during this season, but if you'll sit down some evening and think it over, you'll be surprised at the number of practical ideas that will come to you. Don't be content to sit and let trade come to you. Use your brain and get after trade. Remember the world's most successful men are the men who use their brains more than their hands.

* * *

The dealer who possesses foresight will also begin to lay plans in March for his summer campaign. Where his store is located in or near a tourist centre, this preparation will have an important bearing on the success of the summer business. Once more we would reiterate the advice for each individual dealer to profit by his previous experiences. He is now in a position to know where he fell short in previous years and he is able to remedy the mistakes then made. A rearrangement of store fixtures and of stock can be recommended. It is better for the summer visitor to find an altered store than to come back and see the same old arrangement. Alterations are not difficult to make and they indicate life and progressiveness. These, in turn, have an influence on customers. March is a good time to attend to this item.

Western Retail Hardware Association

Attendance None Too Large, Owing to Snow Blockades—Keen Interest Shown—Trade Press Excluded—Parcels Post Condemned—Western Board Dissolved—Hardwaremen Stand Alone—Dominion Association Organized.

On Tuesday, February 12th, the Western Retail Hardware and Stove Dealers' Association held their annual meeting in the association board room in Winnipeg. Owing to the blockades on the railways the attendance from country points was not so large as it would otherwise have been, but from all parts of the west came cheering messages from dealers who were unable to attend and the enthusiasm of the meeting more than compensated for the small attendance.

In little more than two years and a half the Western Retail Hardware Association has developed from its small but enthusiastic nucleus in the city of Winnipeg into a strong organization embracing the great bulk of the retail hardware trade of the three Provinces of Manitoba, Saskatchewan and Alberta. A few impatient members have expressed dissatisfaction with the association because they have not seen the immediate, direct and tangible results from membership which their enthusiasm led them to expect, but the majority have the sound common sense to recognize that there are many indirect benefits from membership and that the work of perfecting the organization is necessarily slow.

An important step was the appointment of the president and secretary as a committee to act with the president and secretary of the Ontario Association as a temporary executive of a new Dominion Association, to embrace both organizations and any others which may be formed in the other provinces and which may be willing to unite.

A. J. FALCONER, DELORAINE
Re-elected President

On the C.O.D. Parcel Post Bill the association expressed itself in no uncertain way and the president and secretary were instructed to prepare a resolution on the subject to be forwarded to the Postmaster-General.

The Register.

An examination of the register shows the following names of members in attendance who did not neglect the formality of presenting the association with their autographs: Wm. M. Gordon, E. B. Lennox, N. H. Lightly, J. E. Riley, C. C. Falconer, S. Drewe, F. E. McFeely, A. P. Macdonald, Oliver Gilmer, W. A. Templeton, I. Woods, J. A. Lindsay, C. A. Baskerville, C. A. B. Whiting, J. E. Thomas, A. E. Dykes, R. Wyatt, Winnipeg; W. Montgomery, Stonewall, Man.; F. Anderson, Teulon, Man.; J. G. Rattray, Pipestone, Man.; R. W. Birch, Roland, Man.; W. Chalmers, Deloraine, Man.; J. A. Decosse, Somerset, Man.; W. P. Sparling, Melita, Man.; W. K. Cherry, Waskada, Man.; J. Estlin, Melita, Man.; J. A. Mountjoy, Heward, Sask.; J. B. Curran, Brandon, Man.; G. K. Smith, Moose Jaw, Sask.; R. J. McConnell, Carman, Man.; D. J. Fummerton, Reston, Man.; A. E. Jones, Milestone, Sask.; O. Kuss, rep. Neudorf Trading Co., Neudorf, Sask.; M. Isbister, Saskatoon, Sask.; E. A. Coleman, Darlingford, Man.; G. M. Brown, Stonewall, Man.; E. A. Walker, Grenfell, Sask.; H. S. Price, Boissevain, Man.; A. J. Falconer, Deloraine, Man.; George Houston, Cypress River, Man.; J. C. Stewart, Grayson, Sask.; D. I. Stewart, Rosenfeld, Man.

J. B. CURRAN, BRANDON
Re-elected 1st Vice-President

Letters of Regret.

Letters of regret at inability to attend the convention were read from C. F. Comer, Calgary; G. L. Foerster, Neepawa; R. O. Bird, Tofield; D. Bradley, Holmfield; G. R. McLean, Waskada; F. Babb, Portage La Prairie; J. A. Gilhooly, Arden; Lambert & Earle, Elkhorn; Linton & Hall, Calgary; and J. Drewe, manager Calgary branch J. H. Ashdown Hardware Co.

Mr. Comer, in the course of a very cordial letter, said: "Our association has had a very successful season and everything has been harmonious. We certainly appreciate the efforts of yourself and the association generally in the interests of the hardware and store trade in the west, and we think it is the duty of every hardware merchant doing business in the territory covered by the association to become one of its members and thus help the good work along. We are in receipt of your circular letter regarding the proposed Parcel Post regulations and we have added our protest to the number that we hope may have been sent to the Postmaster-General and the different members of Parliament."

President's Address.

President A. J. Falconer of Deloraine, Man., occupied the chair and in opening the convention delivered the following address:

"Gentlemen: I extend to you a very hearty welcome to this our third annual meeting. While it is not quite as large as we would like to have it, the uncertainty of railway traveling at the present time has prevented a great many members from meeting with us to-day; however, this does not apply to the Winnipeg members. Although the meeting is small I feel that a lot of good will be accomplished.

Wm. GORDON, WINNIPEG
Re-elected 2nd Vice-President

"The by-laws have been somewhat changed from former times, and I see the first order of business is the President's address.

"The secretary has received a number of letters from members who were unable to attend, in which they pledged their support and goodwill to the association in the good work we are carrying on for the benefit of the retail hardware and stove dealers throughout this great west of ours; a work which your officers and executive feel is not receiving the support financially from quite a number of members that it is entitled to, and which will be more fully and clearly shown to you when the report of the executive is read.

"The primary object in forming this association was to foster and encourage local associations. A great many of the local boards were formed, and in the majority of cases so gratifying were the results that, I am sure, when the secretary visits where friction exists and explains the advantages already gained by the dealers at other points, they will immediately organize their local boards. But right here let me say, gentlemen, that it rests with each individual hardwareman to make that local association a success; the secretary cannot do it all. You will have to have that confidence one in another that cannot be destroyed by falsehoods circulated for selfish gain.

"At the inception of this association the relationship existing between the retailers and jobbers was not as satisfactory as it is at present. Through the efforts of your secretary the jobbers are confining their trade to the legitimate hardwaremen, and the jobbers recognize that the Retail Hardware Association is a body capable of guarding their own interests.

"In regard to legislation, this is clearly proven by the private action taken by the secretary and executive when the Postmaster-General announced his proposed C.O.D. Parcel Post Bill, a

J. E. McROBIE, WINNIPEG
Re-elected Secretary-Treasurer

bill which is clearly in the interests of departmental and other large stores, and detrimental to the small retailer. In this connection I may say that the secretary and myself sent a letter to the Postmaster-General and the secretary also wrote you all asking you to take action in the matter, and a great many of you have done so. We have the assurance that the Postmaster-General is prepared to withdraw his proposed legislation. This is gratifying, if we had not taken this action, this would have gone through.

"Now there was another matter that concerned you very much and which was sprung on us at the last moment and it was the proposed amendment to the Shops Regulation Act which the druggists asked Parliament to pass, clearly in their own interests. A few of the Winnipeg dealers and other merchants at once saw the proposed bill and immediately got to work to look after their own interests, and other merchants, to see that this should not be passed, with the result that a limitation clause has been inserted, through our effects.

"In addition to the election of officers for 1907 you will be asked to consider carefully a number of questions of importance to the association

"1st—The general use of collection

C. F. COMER, CALGARY
One of Alberta's Representatives on the Executive.

forms for bad debts. This we think is a good thing but not used enough.

"2nd—The advisability of publishing a delinquent list for the exclusive use of every paid-up member of the association. It would be a good thing for every individual member to have a list of his bad debts sent to the secretary and the secretary prepare same and send one to every hardwareman.

"Now, before the election of officers takes place, let me thank the executive and members for the loyal support accorded me at all times, and I know that the same support will be given my successor.

"Now, in conclusion, let me appeal to you to continue the good work which is so vital to the success of the hardwaremen in this great west of ours."

Secretary's Report.

The report of the secretary was as follows:

"Gentlemen: I beg to place before you the report of the association for the past six months. By it you will see that the association has not been as successful financially as it should. This is accounted for by the failure and indifference shown by a number of members in not remitting their dues, although repeatedly requested to do so. Many replies to these requests were to

G. K. SMITH, MOOSE JAW
One of Saskatchewan's Representatives on the Executive

the effect that the association was a good thing, but had not done them any good so far, and until it did they would not pay their dues.

"It is unfortunate that these dealers have not exercised a little patience in order to allow of the association arriving at that stage where it could show them visible results, there is no doubt but that they, with others, have shared equally in any benefits the trade have so far obtained

"A striking contrast to these is the evidence of one of our far north members, who, in conversation with me the other day, spoke as follows: 'When the association was formed I did not see at first how it could possibly benefit me, situated as I was, so far away. On thinking it over I came to the conclusion that as the best hardwaremen in the country were at the head of the movement, there must be something in it, so decided to become a member. On my return home after the first meeting I decided as one of the objects of the association was the eliminating of price cutting, and as I had heretofore been compelled, as I thought, to cut prices on representation of customers that my opposition was doing likewise, that I would live up to this feature and test its sincerity. I did so, and the result has been that during the past year I have sold goods at a better profit, have done a larger and more profitable business, and have no doubt but that in a few years every retailer will have benefited to the extent of thousands of dollars through the influence of our association.'

"I have letters from many others testifying to the good the association has accomplished Don t you think it would be suicide on your part to allow the success secured during the past two years to be lost now? I venture to say that it you do that within two months you will regret it. Every day matters of interest to your business are cropping up. You have the mail order and catalogue houses reaching out more and more after your trade. They are

H. S. PRICE, BOISSEVAIN
One of Manitoba's Representatives on the Executive

now trying to secure legislation to enable them to still further encroach on your territory; are you going to allow them to do so without a protest? If not, how do you expect to combat them if not united?

"A better understanding between the manufacturers, jobbers and dealers has been secured by the association. You are receiving better protection to-day than a year ago; much more, however, remains to be done along these lines, and it is only by your active support of the association that this can be accomplished.

"The question before you to-day, and which you will be called upon to decide, is the continuance of this association on the same broad lines as at present. If you decide to do so, measures will have to be adopted to place it on a sound financial basis.

"This could be done by appointing a member in each locality whose duty it would be to see that every member in his district paid his dues. If this could be accomplished, it would mean the saving of hundreds of dollars to the association and allow of the executive pushing the work more vigorously.

"I would advise that Alberta and part of Saskatchewan be requested to form local associations, which could work with you in all matters requiring legislation or affecting the general welfare of the trade. They are too far away, and the cost to give them the attention they require is too great; the feeling among some of them is that Winnipeg is too far away to properly adjust matters locally.

"Gentlemen, I trust that you will give these matters to-day the careful consideration they deserve; a climax in your association's history is with you, weigh carefully the welfare of your business before deciding. Remember, should you cast aside what you have already accomplished, it will take double the work to overcome, should you ever attempt it again. Not only that, but you will have acknowledged that as hardwaremen who are credited in the commercial arena as the best business heads it contains—have not the energy to accomplish what others have found necessary to do for their own protection."

J. E. McROBIE,
Sec.-Treas.

Dominion Association Executive.

A communication was read from Weston Wrigley, secretary of the Ontario Retail Hardware Association, regarding the proposed Dominion Retail Hardware Association, the principle of which was endorsed at the meeting last July. Mr. Wrigley suggested that the president and secretary of each association be appointed to act as the member of a temporary executive for the Dominion Association until a permanent organization could be effected. The suggestion was received with enthusiasm, and was acted upon, President A. J. Falconer and Secretary J. E. McRobie being appointed to act with the

A. P. MacDONALD, WINNIPEG
One of Manitoba's Representatives on the Executive

president and secretary of the Ontario Retail Hardware Association.

Parcel Post Bill.

The proposed parcel post legislation came up for discussion and was roundly condemned by the association members, a number taking part in the general discussion. It was contended that this legislation was wanted only by the mail order houses and that it would be detrimental to the interests of the retail trade. The president and secretary were instructed to prepare a strong resolution on the subject and forward it to Ottawa.

Reciprocal Demurrage.

The efforts of the Winnipeg Union to secure from the railways a system of reciprocal demurrage charges were warmly endorsed. It was pointed out that the present system is unfair. The merchant is forced to pay demurrage charges if he delays to unload his goods from the car, but the railway company is not compelled to compensate the dealers for delay in delivering goods. Oliver Gilmer moved the following resolution, seconded by J. E. Riley:

"That this association express its unqualified approval of the action that has been taken by the Jobbers' Union to have proper reciprocal demurrage charges which would be beneficial to all parties concerned. And that a copy of this resolution be sent to the Jobbers' Union." Carried unanimously.

Officers Elected.

The officers for last year were re-elected by acclamation. Four members of the executive elected last year for only one year were re-elected last week for a term of two years. There were C. F. Comer, Calgary; A. E. Clements, Olds; A. R. Auger, Okotoks; and J. R. Fox, Weyburn. The other members of the executive were elected last year for a term of two years. The officers and executive committee for the ensuing year are therefore as follows:

President—A. J. Falconer, Deloraine.

First Vice-President—J. B. Curran, Brandon.

Second Vice-President—W. M. Gordon, Winnipeg.

Secretary-Treasurer—J. E. McRobie, Winnipeg.

Executive—Alberta, A. E. Clements, Olds; C. F. Comer, Calgary; A. R. Auger, Okotoks.

Manitoba—H. S. Price, Boissevain; A. P. Macdonald, Winnipeg; O. Gilmer, Winnipeg.

Saskatchewan—G. K. Smith, Moose Jaw; S. A. Clark, Saskatoon; J. R. Fox, Weyburn.

Western Board Disbands.

A year ago the "Western Board" was formed at the request of E. M. Trowern, Toronto, as a connecting link between the Retail Hardware Association of Western Canada on the one hand, and the Retail Merchants' Association of Canada (of which Mr. Trowern of Toronto is secretary) on the other hand. On Thursday morning, February 14th,

O. GILMER, WINNIPEG
One of Manitoba's Representatives on the Executive

the Western Board met in secret session and at the conclusion of their deliberations the following statement was handed out:

"A meeting of the Western Board of the association was held in the association rooms Thursday morning, February 14th, at 9 o'clock.

"After considerable discussion on our connection with the Retail Merchants' Association of Canada, (headquarters Toronto, E. M. Trowern, secretary) it was moved by Mr. Coulson and seconded by Mr. McRobie that we sever our connection with the above body and that the secretary be instructed to advise the secretary of the Retail Merchants' Association of Canada (E. M. Trowern, Toronto) to this effect. Carried.

"G. K. Smith, Moose Jaw, occupied the chair in the absence of W. G. McLaren, who had written that he was unable to attend owing to sickness in his family.

"It was felt by the members present that our connection with the association in the east had hampered us considerably in handling western trade evils, and the feeling was general that we should be willing to work in conjunction with the eastern association on any matters of legislation pertaining to the trade as a whole.

"Moved by Mr. McRobie and seconded by Mr. Coulson, that this Western Board be dissolved. Carried."

This action of the Western Board disposes for good of the vexed question of the relations of the western associations with Mr. Trowern's association in Toronto. The agreement entered into last February has been cancelled entirely and the Western Retail Hardware and Stove Dealers' Association is now on exactly the same footing as when it was first organized. The dissolution of the Western Board also disposes of the affiliation with the Retail Merchants' Association of Western Canada. The two western associations are now as separate, independent and distinct as when they were first organized.

A CORRECTION.

In the summarized report of the convention published last week reference was made to a lengthy discussion which it was understood took place on the subject of price agreements covering the trade in the west. Secretary McRobie states positively that this subject was not dealt with at the convention. The exclusion of the trade press from the convention is responsible for the reference to the discussion at the convention, a brief telegraphic report being all that was procurable by our reporter in time for last week's issue.

Some differences have undoubtedly existed in certain districts, but the fact that the matter did not come up at the convention shows that in all but a few districts the work of the association is meeting with the hearty co-operation of all in the trade. Where differences exist we would urge that the proper method to pursue is to bring the matter to the attention of the executive so that an adjustment can be arrived at promptly. It is much preferable to try and arrange matters on the inside to withdrawing from membership and indulging in open price-cutting warfare.

There are times when an ounce of ingenuity discounts a ton of energy.

PUBLIC SPIRITED HARDWARE MERCHANTS

NO. 9

To be elected mayor of a thriving Canadian city, without having had previous aldermanic experience, and then to be re-elected mayor over another strong candidate by the largest majority in the history of the city, is an honor to be appreciated by any man. And it is appreciated by the subject of this sketch, Mayor Andrew Riddell, of St. Catharines, known to the stove and plumbing trades throughout Ontario as one of the live men in that branch of mercantile activity.

Mr. Riddell, as will be seen by the engraving, is a kindly but determinedly visaged gentleman of mature years, and the success that has crowned his efforts as a merchant has made possible the splendid endorsation he has received as mayor at the hands of his fellow-citizens.

Although not having sat as an alderman, Mr. Riddell has not been without

ANDREW RIDDELL, ST. CATHARINES.
A Prominent Plumber and Stove Dealer, recently re-elected Mayor by several hundred majority.

public experience, he having been for seventeen years one of the water commissioners of the city, as well as a member, deputy and chief of the fire brigade for many years. He retired from these public positions in 1905 in order to devote all his time to business, but in 1906 he was prevailed upon to accept the nomination for mayor, with the result already outlined. As no mayor had ever been elected previously without having aldermanic experience the result was very complimentary to Mr. Riddell.

Born in Scotland in 1839, coming to Canada in 1850, and locating in St. Catharines in 1857, Mr. Riddell took up the trade of tinsmithing, going into business for himself in 1873, building up a large trade as a plumber and metal worker, stove dealer and heating contractor. His eldest son has been associated with the business for many years, having charge of the mechanical work.

St. Catharines has been enjoying an era of prosperity for some time, all lines of manufacturing being rushed with orders. Situated on the Lake Ontario end of the Welland canal, only a few miles from the big power plants at Niagara Falls, and in the natural gas district, the future of the city is assured and it is safe to say that during Mr. Riddell's regime as mayor, the building for the future will be sound and substantial.

PIONEER STOVE FOUNDER DEAD.

Thomas Lang Moffat, one of the best known stove founders in Canada, died at his residence, Weston, Ont., last Tuesday, after an illness of some months. Deceased was well known throughout the trade and has put the stove foundry business on a broad and sound basis. Mr. Moffat was a native of Scotland, but has lived in Canada for quite a number of years. Previous to coming to Canada he was connected with the Marine Works on the Clyde at Glasgow.

The Moffat Stove Company at Weston, founded fourteen years ago by Mr. Moffat, has grown into a large industry doing a successful business in all parts of Canada. Mr. Moffat's ancestry dates back to the first iron master of Scotland, the family having been in the iron business for many generations. It is not remarkable, therefore, that all his sons are directly connected with the stove industry, having been brought up by their father "in the way they should go."

Deceased was seventy years of age, he being survived by his wife, five sons and one daughter. The sons are: J. K. Moffat, T. L. Moffat, F. W. Moffat and A. B. Moffat, all of Weston, and C. L. Moffat of Winnipeg; Mrs. Matheson, of Toronto, is the only daughter.

NEW FREIGHT CLASSIFICATION.

At a conference recently held between representatives of the Canadian Freight Association and the Canadian stove manufacturers, a number of reductions were decided upon, which practically gives a third-class rating for stoves in less than carloads, as asked for by the stove manufacturers. The matter is being followed up with the Railway Commission with a view to having it embodied in supplement No. 8 to classification No. 12 which is now before them.

FISHING LINE FOR COD.

(From the Minneapolis Journal.)

"A fishing line worth $2,000?"

"Yes, sir."

"I don't believe it."

"It's the truth. It's a codfish line. It's one of those lines to which you owe your Sunday morning fishballs and your less appetizing, equally helpful, cod liver oil. These codfish lines, you see, are frequently eight miles long. They have 4,680 hooks. They'll often land 2,500 cod. No wonder they cost $2,000, eh?"

HARDWARE AND METAL

Established 1888

The MacLean Publishing Co.
Limited

JOHN BAYNE MACLEAN - President

Publishers of Trade Newspapers which circulate in the Provinces of British Columbia, Alberta, Saskatchewan, Manitoba, Ontario, Quebec, Nova Scotia, New Brunswick, P.E. Island and Newfoundland.

OFFICES:

MONTREAL, ... 232 McGill Street, Telephone Main 1255
TORONTO, ... 10 Front Street East, Telephones Main 2701 and 2702
WINNIPEG, ... 511 Union Bank Building, Telephone 3726
LONDON, ENG. . . . . . 88 Fleet Street, E.C. J. Meredith McKim, Telephone, Central 12960

BRANCHES:

CHICAGO, ILL. - . . . . 1001 Teutonic Bldg. J. Roland Kay
ST. JOHN, N.B. . . . . No. 7 Market Wharf
VANCOUVER, B.C. . . . . Geo. S. B. Perry
PARIS, FRANCE - Agence Havas, 8 Place de la Bourse
MANCHESTER, ENG. . . . . 92 Market Street
ZURICH, SWITZERLAND . . . - Louis Wolf, Orell Fussli & Co.

Subscription, Canada and United States, $2.00
Great Britain, 8s. 6d., elsewhere - 12s

Published every Saturday.

Cable Address { Adscript, London / Adscript, Canada

NEW ADVERTISERS.

Howard, W. M., Toronto.
Standard Paint Works, Montreal.

WESTERN ASSOCIATION.

Radical and far-reaching changes were made last week by the retail conventions which met in Winnipeg. It will be remembered that a year ago the Retail Hardware Association of the west and the Retail Merchants' Association of Western Canada decided to affiliate for the accomplishment of certain common objects and for that reason elected in joint session a common executive to be known as the Western Board. This Western Board in turn affiliated with the Retail Merchants' Association of Canada, an organization whose headquarters are in Toronto.

The conventions held in Winnipeg last week swept all these affiliations out of existence. In the first place the Western Board met and repudiated the affiliation with the Retail Merchants' Association of Canada. It also decided to disband. In the second place the conventions approved of the action of the Western Board and it was decided that the affiliation between the two western associations should be at an end. The two western associations are therefore absolutely independent from this time on. Their relations are friendly and no doubt they will often act together to attain common objects, the fact that their head offices are in the same building making this easy and convenient. But in some important matters their interests conflict and for this reason absolute independence has been decided upon.

But while the retail hardwaremen of the west have thought it advisable to be independent of the general storemen with whom they disagree on some points they have welcomd the prospect of closer relations with the Ontario Retail Hardware Association. They have appointed their president and secretary to act with the president and secrtary of the Ontario organization as a joint temporary executive of a Dominion Retail Hardware Association.

KEEN COMPETITION EXPECTED.

Both manufacturers and users of heating apparatus are looking for an interesting season during 1907 owing to the introduction into the trade of new factors.

During the past couple of years the Taylor-Forbes Company has come into the market with a very large boiler and radiator plant at Guelph and their advertising has been conducted vigorously. The American Radiator Company have established a plant at Brantford but they have not yet gone after the steamfitters trade. The Pease Foundry Company have erected larger foundries at Toronto and the Gurney Foundry have also been forced to extend their plant. The Canada Radiator Company have also moved their works from Port Hope to Montreal. Further extensions under way at present include: a doubling of the boiler plant of the Star Iron Works, at Montreal, additions to the foundry of the H. R. Ives Company, Montreal, and the erection of the big radiator plant of the King Radiator Co., at Toronto, to work in connection with the Warden King Works, at Montreal, where the capacity for manufacturing Daisy boilers has been doubled. Another expansion is the erection of the new plant of the Dominion Radiator Company at Toronto.

All of these changes have been outlined in our news columns from time to time, but when connected together they make an extended list. A similar expansion has been taking place in the plumbers' brass goods field, new plants having been established during the past year or so at London and Port Colborne and new industries being now projected at Galt and Toronto.

As a result of this development in the manufacturing industry, coupled with the aggressive advertising done by the Taylor-Forbes Company and more recently by Cluff Brothers, the other firms are preparing to pay more attention to their publicity department and some lively advertising will undoubtedly be done during 1907. The competition will not stop there, however, as the rush for business will be participated in by each house, both by personal canvess in the cities and by the traveling salesmen on the road.

Prices, too, will feel the effect of the situation, as they have already done, boilers and radiators not having advanced in price in keeping with the higher quotations ruling for pig iron, etc., and the advanced prices of stoves. The one factor which saves the situation is the steady expansion of the market caused by the continued building activity. Every new plant under construction will probably be kept busy by this increased call for goods, but it would appear that the requirements of the trade will not require much further expansion for a year or so at least.

PARCELS POST PROPOSALS KILLED.

As announced on another page a great victory has been won in the agitation against the proposed parcels post C.O.D. system. Every credit should be given the Ontario and Western Retail Hardware Associations, they being the only organized bodies to take up the agitation and back up the work done by the MacLean trade newspapers.

As has already been pointed out, the daily press representatives at Ottawa practically suppressed all information regarding the proposals and their probable effect. To have, in the face of this obstacle, succeeded in heading off the mail order house legislation is, therefore, a source of satisfaction to the trade and to us.

Full credit should be given Hon. Mr. Lemieux, Postmaster-General, for his prompt action in dropping the proposals as soon as he learned the strength of public opinion against the legislation. Mr. Lemieux is a level-headed and progressive legislator, as is shown by his work in bringing down legislation aiming to rid the country of the evil effects of strikes by providing for a public investigation into the controversial points at issue between employers and employes. It is probable that his parcels post proposals were made on the advice of department heads to whom he, as a new head of the department, must look to for guidance on matters of postal regulations." Certain it is that the postal department is being made more efficient and the staff is not so starved as under former administrations.

The Postmaster-General, in considering future legislation affecting the parcels post system, will be sure to consider the interests of the retail merchant throughout the country. The point has been driven home that the retail trade does not want any special favors at the hands of the government and rightfully complain when legislation is proposed tending to divert trade from their hands into the departmental store channels.

The lesson of solidarity learned by the retailers should not be lost. Unorganized the merchants are not well equipped to conduct an offensive or defensive campaign in matters of legislation. Organized into associations for mutual benefit united and prompt action can be taken.

Retail hardwaremen in Ontario and the Western Provinces who have not yet joined the association have much to thank this organization for as if the present attack on retail interests had not been frustrated, more and heavier blows would have been struck. No firm should begrudge the amount of a year's membership fee in the Retail Hardware Association, and provinces where associations have not yet been formed should take up the work of organization without delay.

WESTERN RETAIL HARDWARE CONVENTION.

Last week for the first time in its history the Western Retail Hardware Association excluded from its annual convention the representatives of the trade press. At all previous conventions a representative of Hardware and Metal has been present, and from the reports in this paper the great bulk of the membership who are unable to attend a meeting held hundreds of miles from their homes have been able to keep in touch with the work of the association. In this issue we are able to present only a bare summary of the proceedings for the reason that the executive of the association decided to do their own reporting. Now the members of the executive are business men, not reporters, and they delegated their duties to the secretary, who, although animated by the best intentions, is likewise a business man and not a reporter. Reporting is a business just as selling hardware is a business. A reporter would not make a good hardware merchant and a hardware merchant does not make a good reporter. Each might learn in time, but neither can do work unfamiliar to him and do it well at a moment's notice.

For this sudden and unexpected action of the association executive no reasons or explanations worthy of serious consideration or intended to be taken seriously have been advanced. The association has no reason to complain of the character of the reports of previous conventions; they have been fair, accurate, impartial and complete, and the association executive have made no complaints. Their action is, therefore, all the harder to understand.

At the meeting last week there were present a full delegation of members from the city of Winnipeg, and a few members from the country. That there were not more present from the country was due to the snow blockades on the railways. A merchant leaving his business to attend the convention under the conditions that prevail this winter could not be sure when he could return home, and it is not strange that only a few venturesome spirits braved the dangers and inconveniences of travel. The members who were unable to attend have a right to expect a report in their trade paper. Moreover, distances are so great in the west that it costs money to attend the annual convention. One member, speaking at the convention, last week, is reported to have said that his trips to the two conventions cost him $60 each, which, with his membership fee of $10, amounts to $130 per year. Not many association members are prepared to spend that amount of money to attend the conventions, and if the executive hope to keep up the interest and enthusiasm of the country members they should not deprive those members of their only means of knowing what is being done. The bare minutes of the meeting dished up in skeleton form are not likely to awaken much interest or arouse much enthusiasm.

If the association had anything to hide, the action of its executive could be explained, but unless some radical changes of policy were decided upon last week the association has everything to gain and nothing to lose from the fullest publicity—in the strictly trade press. With the aims and objects of the association as we understand them this paper is in hearty accord, and we would be false to the best interests of the retail trade were we, because of the recent action of the association executive, to cease to advocate the organization of the retail hardware trade of the west and the energetic prosecution of the work of the association. We merely regret that owing to the action of the association executive we are unable in this issue to give our retail friends in the country who were unable to attend the convention, the full and impartial report to which we think they are entitled and which they have always received from this paper in the past.

A BACKWARD STEP.

Postal matters seem to be absorbing a good deal of attention at present, the latest to come to the front being the announcement that after May 1st next the Canadian Postal Department will not accept for delivery any United States publications mailed as second-class matter.

This will exclude from the Canadian mails all magazines, daily papers, etc., and will force weekly periodicals and smaller monthlies to pay postage at regular rates in order to secure delivery to subscribers on this side of the line. Monthly magazines will be able to secure a circulation through booksellers, but such trade papers as the "Iron Age" will be practically excluded from the country as they do not look for a general circulation and their weight is such as to prevent them paying the eight cent per pound rate which will be substituted for the one cent charge after May 1st next.

It is inconceivable that public opinion in Canada will be so narrow as to allow the new regulation to become permanent. Canadian publications should be protected against the sensational yellow journalism of the States, and British periodicals should be encouraged in securing a larger circulation in Canada, but the workings of the new regulation will be so drastic as to cause wide discontent.

If we are to have a press censorship at Ottawa, and Canadians are to be told what they shall or shall not read, it is time to establish a definite tribunal which will consider each case upon its merits. The right of trial should not be taken away from publishers. A censorship may be necessary in excluding immoral publications, but scientific, political and industrial publications should have the free use of the mails subject to a democratic, not an autocratic, censorship.

Hardware dealers should protest against such a backward step as has been decided upon. Had the postal rates been raised slightly the reason for complaint would not have been so great. Regulations, however, which are so drastic as to exclude representative trade newspapers of high standing should not be allowed to be put into force unchallenged.

Markets and Market Notes

(For detailed prices see Current Market Quotations, page 66.)

THE WEEK'S MARKETS IN BRIEF.

MONTREAL.

Jenkin's Valves—Advanced about 5 per cent. Discount is now 27½ per cent.

TORONTO.

Plumbers' Brass Goods—Some lines lower.

Old Material—Lead and iron slightly weaker. Rubbers a little higher.

Oiled and Annealed Wire—5 cents higher.

Montreal Hardware Markets

Office of HARDWARE AND METAL,
232 McGill Street,
Montreal, Feb 22, 1907.

Conditions remain practically unchanged here this week, most lines continuing to move well, but money is rather tight, and considering the volume of business that has been done during the past two or three months, jobbers are at a loss to understand why collections are not better than they are.

Most jobbers are suffering through delays on the railroads. Car lots on different lines are anywhere from two to four weeks overdue. This makes a shortage in some lines.

Prices on building paper and wire nails are extremely stiff.

Wood screws are very scarce, some of the jobber's shelves being practically empty. One by six are particularly difficult to get.

The only change in prices to record this week is on Jenkins' valves, which have advanced about five per cent, the discount now being 27½ per cent.

AXES — Although the jobbers' season for this line is practically over, sales are still reported to be very good. This is due, no doubt, to the severe weather experienced here this winter. Prices remain: $7.60 to $9.50 per dozen; double bitt axes, $9.50 to $12 a dozen; handled axes, $7.50 to $9.50; Canadian pattern axes, $7.50 a dozen.

LANTERNS — Moving, but very slowly. We still quote: Prism globes, $1.20; cold blast, $6.50; No. 0, safety, $4.

COW TIES AND STALL FIXTURES.—Very quiet. We are still quoting: Cow ties, discount 40 per cent. off list; stall fixtures, discount 35 per cent. off list.

RIVETS AND BURRS.—Our prices are: Best iron rivets, section carriage, and wagon-box black rivets, tinned do., copper rivets and tin swede rivets, 60, 10 and 10 per cent.; swede iron burrs are quoted at 60 and 10 and 10 per cent. off new lists; copper rivets with the usual proportion of burrs, 25 per cent. off, and coppered iron rivets and burrs in 5 lb. carton boxes at 60 and 10 and 10 per cent.; copper burrs alone, 15 per cent., subject to usual charge for half-pound boxes.

HAY WIRE — Waking up. Prices are. No. 13, $2.60; No. 14, $2.70; No. 15, $2.85; f.o.b. Montreal.

MACHINE SCREWS — The shortage not yet relieved. Prices remain: Flat head, iron, 35 per cent., flat head brass, 35 per cent., Felisterhead, iron, 30 per cent., Felisterhead, brass, 25 per cent.

BOLTS AND NUTS.—Demand is very heavy. Discounts remain: Carriage bolts, 3 and under, 60 and 10; 7-16 and larger, 55 per cent.; fancy carriage bolts 50 per cent.; sleigh shoe bolts, ⅜ and under, 60 per cent.; 7-16 and over, 50 per cent.; machine bolts, ⅜ and under, 60 per cent.; 7-16 and larger, 55 per cent.

HORSE NAILS — Demand increasing. Discounts remain: C brand, 40, 10 and 7 per cent.; M.R.M. Co., 55 per cent.

WIRE NAILS.—We still quote: $2.45 per keg, f.o.b. Montreal.

CUT NAILS.—Prices remain: $2.30 per keg, base, f.o.b. Montreal.

HORSESHOES — Moving well. We still quote: P.B. new pattern, base price, $3.50 per 100 lbs.; M.R.M. Co., latest improved pattern iron shoes, light and medium pattern, No. 2 and larger, $3.65; No. 1 and smaller, $3.90; snow pattern, No. 2 and larger, $3.90; No. 1 and smaller, $4.15. Light steel shoes, No. 2 and larger, $4; No. 1 and smaller, $4.25; featherweigh, all sizes, No. 0 to 4, $5.60. Toeweight, all sizes, No. 1 to 4, $6.85. Packing up to three sizes in a keg, 10c. per 100 lbs. More than three sizes, 25c. per 100 lbs. extra.

BUILDING PAPER.—No change in prices.

BUILDING PAPER—Warmer weather creating a good demand, but no change in prices yet.

CEMENT AND FIREBRICK.—Prices remain: "Lehigh" Portland in wood, $2.54, in cotton sacks, $2.39; in paper sacks, $2.31. Lefarge (non-staining) in wood, $3.40; Belgium, $1.60 to $1.90 per barrel; ex-store, American, $2 to $2.10, ex-cars; Canadian Portland, $2 to $2.05. Firebrick, English and Scotch $17 to $21, American $30 to $35; White Bros.' English cement, $1.80 in bags, $2.05 in barrels in round lots.

COIL CHAIN.—Our quotations remain: 5-16 inch, $4.40; ⅜ inch, $3.90; 7-16 inch, $3.70; ½ inch, $3.50; 9-16 inch $3.45; 5-8 inch, $3.35; ¾ inch, $3.20; ⅞ inch, $3.10; 1 inch, $3.10.

SHOT.—We are still quoting: Shot packed in 25-lb. bags, ordinary drop AAA to dust, $7 per 100 lbs.; chilled, No. 1—10, $7.50 per 100 lbs.; brick and seal, $8 per 100 lbs.; ball, $8.50 per 100 lbs. Net list. Bags less than 25 lbs., ½c. per lb. extra, net; f.o.b. Montreal, Toronto, Hamilton, London, St. John, Halifax.

AMMUNITION — Our prices are: Loaded with black powder, 12 and 16 gauge, per 1,000, $15; 10 gauge, per 1,000, $18; loaded with smokeless powder, 12 and 16 gauge, per 1,000, $20.50; 10 gauge, $23.50.

GREEN WIRE CLOTH.—We continue to quote: In 100-foot rolls, $1.62½ per hundred square feet, in 50-feet rolls, $1.67½ per hundred square feet.

Toronto Hardware Markets

Office of HARDWARE AND METAL,
10 Front Street East,
Toronto, February 22, 1907.

Wholesale dealers are not doing a very brisk business, although, of course, they are holding their own. Retailers are doing a good business in the staple goods with the approach of spring and house-cleaning activity. Orders with the wholesalers are being booked well in advance this year, a plan which should be carried out by all retailers.

Practically all the jobbers report a scarcity in screws; one company stating that they could not be gotten "for love or money." One screw manufactory reports a very low stock, not having even their usual six months' stock on hand.

Paint brushes and all cleansing utensils are meeting with a good demand, in anticipation of the spring housecleaning activities.

Special attention is called to our quotation of prices on binder twine which have just been announced by some of the manufacturers. The prices given below practically cover the Ontario market.

TWINE — International and Plymouth, 500 ft. Sisal, 9½-9⅝; 500 ft. Standard, 9½-9⅝; 550 ft. Standard Manilla, 10¾; 600 ft. Manilla, 12¼; 650 ft. Pure Manilla, 13¾-14. Car lots ¼-cent less, five ton lots ⅛-cent less central delivery.

AXES AND HANDLES — The demand for these is practically over as far as the wholesale merchants are concerned. Retailers report quite a brisk business.

SCREWS—Wholesalers report a great scarcity of these, some firms not having a sufficient supply to fill an order.

BOLTS AND NUTS — Business in these is heavy. Demand greatly exceeds the supply, the stock being low with no immediate prospects of repletion.

CHAIN — No change in prices, which we quote as before: ⅜-in., $3.50; 9-16-in.,

$3.55; ½-in., $3.60; 7-16-in., $3.85; ⅝-in., $4.10; 5-16-in., $4.70; ¾-in., $5.10.

RIVETS AND BURRS — Prices remain firm and unchanged, with large orders.

TOOLS—Are in good demand. Merchants are completing their stock and a great business in these is looked for during the coming summer, which will be unprecedented for building.

SHOVELS — The demand for these continues strong with prospects of enlivenment with the opening of spring.

EXTENSION AND STEP LADDERS—The demand is strong, prices remaining unchanged: 11 cents per foot for 3 to 6 feet, and 12 cents per foot for 7 to 10-foot ladders. Waggoner extension ladders 40 per cent.

POULTRY NETTING — This line in good demand, all outstanding orders being rapidly and promptly filled. We continue to quote: 2-inch mesh, 19 w.g., discount 50 and 10 per cent., all others 50 per cent.

OILED AND ANNEALED WIRE—We have revised our quotations on all Canadian wire: gauge 10, $2.46; gauge 11, $2.52; gauge 12, $2.61 per 100 lbs.

WIRE NAILS — No changes in price. Demand is good. We continue to quote $2.40 base.

BUILDING PAPER — The business so far in this is slow. However, with the opening of spring and renewed building operations the demand will be strong.

CUT NAILS—We quote $2.30 base f.o.b. Montreal; Toronto, 20c. higher.

HORSENAILS—"C" brand 40, 10 and 7½ off list; "M.R.M." brand 60 per cent. off. Capewell brand, quotations on application.

HORSESHOES — The demand is strong, large orders coming from New Ontario, especially from railroad companies and lumber camps. Quotations remain the same: P.B. base, $3.65; "M. cr prices: P. B. base, $3.65; "M. R.M. Co., latest improved pattern" iron shoes, light and medium pattern, No. 2 and larger, $3.80; No. 1, and smaller, $4.05; snow, No. 2 and larger, $4.05; No. 1 and smaller, $4 30; light steel shoes, No. 2 and larger, $4.15; No. 1 and smaller, $4.40; featherweight, all sizes, 0 to 4, $5.75; toeweight, all sizes, 1 to 4, $7. Packing, up to three sizes in a keg, 10c. per 100 lbs. extra; more than three sizes in a keg, 25c. per 100 lbs. extra.

CEMENT—For carload orders, f.o.b. Toronto, we quote: Canadian Portland, $1.95. For smaller orders ex warehouse, Canadian Portland, $2:05 upwards.

FIREBRICK—English and Scotch firebrick, $27 to $30; American low-grade, $23 to $25; high-grade, $27.50 to $35.

HIDES, WOOL, FURS — Trade is very quiet at unchanged prices.

Hides, inspected, cows and steers, No. 1	....	0 10½
" " " " No. 2	....	0 09½
Country hides, flat, per lb., cured	....	0 09
" " " green	....	0 08
Calf skins, No. 1, city	....	0 12
" " No. 1, country		0 11
Lamb skins	1 25	1 35
Horse hides, No. 1	3 50	3 75
Rendered tallow, per lb.	0 05½	0 06
Pulled wools, super, per lb.	....	0 25
" " extra	....	0 27
Wool, unwashed fleece	....	0 15
" washed fleece	0 24	0 26

FURS.

	No 1.	Prime
Raccoon	....	1 50
Mink, dark	5 00	7 00
" pale	2 50	4 50
Fox, red	3	4
" cross	3 00	20
Lynx	5 00	
Bear, black	....	8
" cubs and yearlings	....	12
Wolf, timber	....	3 00
" prairie	....	1 00
Weasel, white		0
Badger	0 10	1
Fisher, dark	3 00	8
Skunk, No 1	....	1 55
Marten	3 50	20 00
Muskrat, fall	....	0 17
" winter	....	0 2
" western	0 12	0 15

Montreal Metal Markets

Office of HARDWARE AND METAL,
232 McGill Street,
Montreal, February 22, 1907

"Copper again," not another advance yet, but just something to talk about. It's higher in New York, with a market shortage, and practically none at all to be had here. This looks as if we might expect another "boost" in prices here. Consumption has completely wiped out all reserve stocks, and there is practically nothing at present to show that there will be any relief, or any decline in prices for some time to come.

Spelter here is firmer. Pig iron is a little easier in England, for future deliveries. This, however, has not affected the Montreal market.

Galvanized iron remains the same as reported last week.

Ingot tin and antimony remain firm, with no change in prices.

COPPER—Prices remain the same here, but are higher in New York. All reserve stocks are wiped out and there doesn't seem to be any chance of immediate relief. We still quote: Ingot copper, 26½ to 27c; sheet copper, base sizes, 34c.

INGOT TIN—Remains the same (very firm, with no change in prices.) We still quote: 45½ to 46c per lb.

ZINC SPELTER—Firmer if anything. We still quote: $7.50 to $7.75 per 100 lbs.

PIG LEAD—We reported last week that this article was a little weaker, with a tendency to lower prices. This was no doubt only a fluctuation, as no decline is reported yet. Our prices remain: 5.50 to $5.60 per 100 lbs.

ANTIMONY—We are still quoting: 27½ to 28c per lb.

PIG IRON—The English market is a little easier for future deliveries, but things here remain unchanged. We still quote: Londonderry, $24.50; Carron, No. 1, $27; Carron, special, $25.50; Summerlee, No. 2, selected, $25; Summerlee, No. 3, soft, $23.50.

BOILER TUBES—Remain very firm. Our prices are: 1½ to 2 in., $9.50; 2½ in., $10.35; 2¾ in., $11.50; 3 in., $13; 3½ in., $17; 4 in., $21.50; 5 in., $45.

TOOL STEEL—Our prices remain: Colonial Black Diamond, 8c. to 9c.; Sanderson's, 8c. to 45c., according to grade; Jessop's, 13c.; "Air Hardening," 65c. per pound; Conqueror, 7½c.; Conqueror, high speed steel, 60c.; Jewitt's Diamond J., 6½c. to 7c.; Jowitt's best, 11c. to 11½c.

COLD ROLLED SHAFTING—Present prices are: 3-16 inch to ¼ inch, $7.25; 5-16 inch to 11-32 inch, $6.20; ⅜ inch to 17-32 inch, $5.15; 9-16 inch to 47-64 inch, $4.45; ¾ inch to 17-16 inch, $4.10; 1½ inch to 3 inch, $3.75; 3⅛ inch to 3 7-16 inch $3.92; 3½ inch to 3 15-16 inch, $4.10; 4 inch to 4 7-16 inch, $4.45; 4½ inch to 4 15-16 inch, $4.80. This is equivalent to 30 per cent. off list.

GALVANIZED IRON — Market strong, with no changes to report.

OLD MATERIAL — We still quote: Heavy copper, 20c. per pound; light copper, 17c. per pound; heavy red brass, 17c. per pound; heavy yellow brass, 14c. per pound; light brass, 10 to 10½c. per pound; tea lead, 4½c. per pound; heavy lead, 4½c. per pound; scrap zinc, 4½c. per pound; No. 1 wrought, $17 per ton; No. 2 wrought, $6 per ton; No. 1 machinery, $18 per ton; stove plate, $13 per ton; old rubber, 10½c. per pound; mixed rags, 1 to 1½c. per pound.

Toronto Metal Markets

Office of HARDWARE AND METAL,
10 Front Street East,
Toronto, Feb 22, 1907

Transportation problems are uppermost at present, complaints of long delays in receiving shipments being heard on every hand. The railways seem to be utterly unable to handle the large business under way and in consequence of cars being lost between tidewater and Toronto some factories are forced to bank their fires. In one case seventy men will be idle if shipments of pig iron are not hurried along.

During the past week a temporary weakness has developed in the iron markets across the line and some have been quick to predict that the "great divide" had been passed and prices of iron, like water, would run down hill from now on. It is said that some buyers have asked producers to delay shipments in view of the weakness. Locally, however, little stock is taken in these bearish reports, they being looked upon as the work of speculators. Stocks of pig iron in Ontario are far from satisfactory and even though prices declined somewhat furnaces would keep busy for months ahead on present orders. It is not to be expected, of course, that prices will keep up to their present notch for all time, and a decline must come some time, but with quotations $10 higher than those ruling a year or so ago a drop of a dollar or two would not affect the market much.

Business done during February has been fully double that of a year ago and there has been no let-up whatever from

January. Jobbers, therefore, are looking for an even greater volume of trade to be done when navigation opens up.

Copper continues to be a feature with stocks in no better shape than for some time past. Early deliveries are difficult to obtain. Tin continues strong and lead unchanged.

PIG IRON—Speculation has demoralized outside markets this week, but no weakness has been manifested locally. Buying continues active and deliveries slow. Hamilton, Midland, and Londonderry are of the market, and Radnor is quoted at $33 at furnace. Middlesborough No. 1 is quoted at $24.50, and Summerlee No. 1 at $26 f.o.b., Toronto.

BAR IRON—The market is firm and continues active. Stocks are improving slightly. Steady prices rule and we still quote: $2.30 f.o.b., Toronto, with 2 per cent. discount.

TIN—Demand continues very active with prices steady at 45c. and charcoal plates, base, at $6 50.

SHEETS AND PLATES—In the face of light stocks a brisk demand exists for Canada plates, and black sheets. Galvanized sheets are in good demand.

BRASS—Conditions are not improving and stocks continue light. We still quote 30 cents per pound on sheets.

COPPER—Prices keep stiff with stocks at a minimum. Deliveries for present use are exceedingly hard to get. We now quote: Ingot copper, $26.50 per 100 lbs., and sheet copper, $35 per 100 lbs.

LEAD—Buying is fair with prices unchanged. We quote: $4.50 for imported pig and $5.75 to $6 for bar lead. For sheet lead $7 is asked.

ZINC SPELTER—business is fairly active for this season with stocks light. We quote 7½c per lb. for foreign and 7c. per lb. for domestic. Sheet zinc is quoted at 8 1-4c. in casks, and 8 1-2c. in part casks.

BOILER PLATES AND TUBES—We still quote: Plates, per 100 lbs., ¼ in to ½ in., $2.50; heads, per 100 lbs., $2.75; tank plates, 3-16 in., $2.65; tubes, per 100 feet, 1½ in., $8.50; 2, $9; 2 1-2, $11.30; 3, $12.50; 3 1-2, $16; 4, $20.00. Terms, 2 per cent. off.

ANTIMONY—The market continues unchanged at 27c. per pound.

OLD MATERIAL—Iron is lower, but rubbers have advanced. Buying prices are: Heavy copper and wire, 20c. per lb.; light copper, 18c per lb.; heavy red brass, 18c per lb.; heavy yellow brass, 15c. per lb.; light brass, 11c. per lb.; tea lead, $4.25 per 100 lbs.; heavy lead, $4.40 per 100 lbs.; scrap zinc, 4½c. per lb.; No. 1 wrought iron, $14; No. 2 wrought, $8, machinery cast scrap, $19.50; stove plate, $12; malleable and steel, $9; old rubbers, 11c. per pound; country mixed rags, $1 to $1.25 per 100 lbs., according to quality.

COAL — The shortage of hard coal continues. We still quote:—Standard Hocking soft coal, f.o.b. at mines, lump, $1.50, ¾ inch, $1.40; run of mine, $1.30; nut, $1.25; N. & S., $1; P. & S. 85c. Youghiogheny soft coal in cars, bonder at bridges: lump, $2.60; ¾ inch, $2.50; mine run, $2.40; slack, $2.15.

London, Eng., Metal Markets

From Metal Market Report, Feb 22, 1907.

Cleveland warrants are quoted at 55s. 10½d., and Glasgow standard warrants at 55s., making price as compared with last week unchanged on Cleveland warrants and Glasgow standard warrants.

TIN—Spot tin opened quiet at £191 15s., futures at £191, and after sales of 150 tons of spot, and 210 tons of futures, closed easy at £191 5s. for spot, £190 10s. for futures, making price compared with last week, £1 5s. lower on spot and 7s. 6d. lower on futures.

COPPER—Spot copper opened steady at £107 17s. 6d., futures £108 17s. 6d., and after sales of 150 tons of spot and 450 tons of futures, closed quiet at £107 17s. 6d. for spot, and £109 for futures, making price compared with last week, 17s. 6d. higher on spot and £1 17s. 6d. higher on futures.

LEAD—The market closed at £19 10s., making price as compared with last unchanged.

The market closed at £26 2s. 6d., making price as compared with last week, 2s. 6d. higher.

United States Metal Markets

From the Iron Age, Feb 21, 1907.

The pig iron markets throughout the country have been quiet and in some quarters show some easing off.

The English market has receded to 54½ shillings, but there is no new buying for shipment by consumers. It is considered probable, however, that one interest may arrange for March and April shipments to this country under special circumstances.

We note sales to a leading interest of 9,300 tons of low phosphorus pig iron by eastern Pennsylvania and New York furnaces, partly for shipment to Pittsburg, at prices which are lower than those recently prevailing. There is more business pending from other quarters in low phosphorus iron.

Reports from the finished iron and steel trade are uniformly cheerful and in some branches, notably in bars and in plates, the pressure for deliveries is enormous.

The manufacturers of cast iron pipe report a continuance of heavy business at a comparatively early period of the year. A Pittsburg concern which controls a number of waterworks in different parts of the country is in the market for 7,900 tons, partly for new construction and partly for season requirements of existing plants.

From the Iron Trade Review, Feb. 21, 1907.

Ten days remain of the one month's continuance of the present freight rate between Birmingham and the Ohio river granted by southern railroads. At the time the postponement was announced, it was intimated that a united effort would be made during the following thirty days to move the iron that had accumulated on furnace yards. In the three weeks that have elapsed there has been no distribution commensurate with the gravity of the situation. Stocks on furnace yards in the Birmingham district are greater to-day than they were at the opening of the month. The effort of the railroads, if any was made, has been a signal failure. Despite frantic appeals of officials of all the furnaces in the affected district, the supply of cars has been wholly inadequate. Furnaces are not moving their daily output, much less the accumulated tonnage. For ten days past one producer with a stock of 30,000 tons and a daily output of 1,150 tons reports that it has had an average of only one car a day from all the railroads in the district. Another large maker states that for three weeks it has not received a single car from the largest trunk line in the south. Others report that the only cars received were those consigned to them loaded with freight. These are not isolated instances, but in one form or another, the car shortage is the substance of every report from the south. Under the circumstances it is futile to expect early relief from the present congestion.

The demand for finished material is unabated. Daily sales continually exceed the daily output. Specifications on mills are the heaviest in history; but instead of overtaking them, producers are steadily falling further behind. Jobbers are having trouble in replenishing their stocks, and because of the uncertainty of mill shipments, they are forced to the recourse of picking up odd lots wherever obtainable. The only price change is in boiler tubes which have been placed to move the market, either upward or downward. Scrap is without improvement and prices are lower.

After an absence of many months from the market, the United States Steel Corporation has placed an order aggregating 9,300 tons of low phosphorus pig iron for delivery in June and August. The purchase is not of great importance, but is significant as showing that the corporation stocks are needed for ordinary grades and cannot be spared to make the low phosphorous iron which is used in the manufacturing of armour plate.

N.S. Hardware Trade News

Halifax, N.S., Feb. 19.

Just at the present time the hardware trade in the province is a little on the quiet side. The jobbers in Halifax are now busy placing their shelf goods and marking them. The prospects for a big spring trade are excellent, and some small orders are already coming in. The lumber promises to reap a harvest this season and this will be of immense benefit to the trade. Prices on all lines particularly staples, have been advanced. Among the most notable is a big increase in hand wood planes. Lines that formerly sold for 95c. are now marked $1.15, and those that sold for 75c. last year are now $1. The manufacturers have increased the price of padlocks 20 per cent. The manufacturers of nails have also added 5c. to the list, bringing the price up to $2.50 per keg. So far the jobbers have made no change. The nail situation is the cause of considerable comment among the trade. The Elder-Dempster line steamer Dahomey, which arrived on Saturday last from Vera Cruz and Mexican ports, brought a big cargo, which included five thousand bales of sisal, some of which is for the Consumers' Cordage Co., and the rest for London and Liverpool and points in Quebec. The steamer also brought one hundred tons of bitumen for London.

* * *

Hugh Sutherland, of New Glasgow, is interested in a large tin area in Lunenburg. There is a large deposit of ore there, and prosperity has been going on for over a score of years. The assays of ore go 11.42 per cent., which is counted very good. Mr. Sutherland is showing a wonderful rock crystal of some twenty-five pounds, perfect hexagon in shape. A monster one weighing 150 pounds was sent to the provincial museum at Halifax. These were found embedded in a very fine deposit of china-clay as white as chalk and containing no grit, in the tin area in Lunenburg.

* * *

It is reported that Messrs. Rhodes, Curry & Co. are about to start a nail mill in Amherst. So far there are only rumors in circulation, but it would not be surprising if they took definite shape.

* * *

The Department of Railways and Canals has just awarded the contract for the car repair shops and planing mill in connection with the new round-house at Halifax to Rhodes, Curry & Co., of Amherst. The firm's tender was the lowest, and the price is about $80,000. M. E. Keefe, of Halifax, has the contract for the round-house.

* * *

The employes of the hardware firm of William Stairs, Son, & Morrow had a most enjoyable sleigh drive last week. Seated in a sleigh drawn by four horses the party drove about the city, through Dutch Village, and then proceeded to Popple's hotel, on the Herring Cone road. During the afternoon all kinds of games, dancing and singing were indulged in. At seven o'clock the party sat down and did full justice to a bountiful repast prepared by Host Popple in grand style. There were no toasts, the men being all temperate. After speeches by William J. Stairs and Captain Douglas and a few others, the games and singing were continued. The party reached the city shortly before midnight, after having one of the best sleigh drives for years.

B.C. Hardware Trade News

Vancouver, B.C., Feb. 15, 1907.

The construction work now being carried on in the southern interim of B.C. by the Great Northern Railway is being pushed rapidly, steel being laid to Oro and by the end of the month it is expected that it will reach Keremeos, the centre of the rich agricultural district of the Similkameen. With rail transpart that far, the question of cost of supplies will be much simplified. The contractors had to team every pound of supplies so far over mountain trails that it was a losing proposition. When the rails reach Keremeos the whole valley as far as Princeton will be reached at moderate cost for transport, and construction will be just that much accelerated.

The final stages of the line from Princeton, across Hope Mountains to the Fraser valley, and so to the coast are to be definitely decided on this year and the last link in the short line from the coast to the interior will then be proceeded with. From Cloverdale the Fraser valley section is being pushed eastward up the Fraser valley to the Hope Mountains. At Vancouver, the V. W. & Y. Railway's plans are said to include a high level bridge across Burrard Inlet, at the Second Narrows, just east of the city boundary, thus giving rail access to North Vancouver. That line put in, the construction to the northern interior of the province is to go on, a route by Howe Sound and thence across the Pemberton Meadows and Lillooet to northern Cariboo, being the general line as foreshadowed by the railway people.

* * *

Other coast railway construction is being looked for, pricipally on Vancouver Island, where the C.P.R. has inaugurated vast plans for development of the princely domain which has heretofore lain untouched. The extension of the E. & N. Railway north from Nanaimo to the northern part of the island with a branch to Alberni, on the west coast, or rather at the head of Alberni Canal, which is an inlet nearly dividing the island and running in from the west coast, the clearing and opening for settlement of large areas of the E. & N. Railway land grant, and the development of the mining and timber resources of the island are among the items on the plans of the C.P.R. and to better further these plans the company has recently promoted A. Marpole, general superintendent, to the position of chief executive in the west, with headquarters at Vancouver. The island improvements will be largely under his care.

* * *

Activity continues in the movement toward Prince Rupert, the G.T.P.'s northern terminal. Several parties are now organized to leave this month for the new city, to be in on the ground floor. The reported commencement of construction in the B.C. portions of the new railway has had a marked influence on the movement north. The coast cities are looking to a great trade to spring up from which they shall benefit during the construction days of the new town and of the railway from the coast end.

* * *

The British steamer Trafalgar which is loading at Hastings saw mill, a part cargo for Australia brought half a cargo of California red wood, which also goes to Australia.

The Holt line steamer Titan, due about the end of February from British ports, has about 2,500 tons of general cargo for B.C. ports. The next steamer of this service to arrive will be the Cyclops, which left Liverpool in January and is due in March. She will have over 3,00 tons of cargo, mainly heavy metals and cement.

* * *

In mining the advent of the Guggenheims, the great New York operators in the field on Vancouver Island, is expected to start quite a boom in copper-gold properties, of which there are numbers close to the city of Victoria, in the south-western end of the island. These properties have never been actively developed, though so close to the capital. A bond has been given on a group of claims at Sooke, to the Guggenheims.

* * *

English miners ariving at Nanaimo tell of an expected arrival of large numbers of their co-workers from England. They say an influx of Chinese miners into British coal mines has begun, and will be a threatening menace to the white miners. Over 300 of these Orientals arrived the day the miners left Liverpool for B.C. It has just been announced that the C.P.R. has made a concession in rates third-class from England, making it $40 to British Columbia, as it now is to Calgary. Many skilled operatives and other British laborers are expected to take advantage of the reduction and come to this Province.

FIRE IN LEAD WORKS.

About four o'clock Friday afternoon, February 15th, fire broke out in the works of the Jas. Robertson Co., Montreal, which would have proven most disastrous had it not been for the effective work of the firemen. As it was, the damage amounted to about $60,000, covered by insurance.

The fire broke out in what is known as the central portion of the works, and did greatest damage in what is known as the shot tower and the first flat of the building through which the tower runs. At one time this tower, which reaches a height of about 140 feet, presented the appearance of an immense torch.

In the portion destroyed were the designs and patterns of the company, and the brass moulding works, together with warehouses in which were stored tin plate and enamelled ware.

The Robertson Company employ about 125 hands, forty of whom were in the building affected by the flames.

The shot tower was used for moulding shot by gravity, and constituted a grave danger, as at the top there was a huge caldron used for melting lead, which would hold about two tons of the molten metal. It was feared that this pot would break loose and fall on the firemen, but it was supported by steel beams and held its place.

It is not known what started the fire, but it originated in a part of the factory that was used as a warehouse, and from there worked its way to the shot tower.

MANITOBA HARDWARE AND METAL MARKETS

(Market quotations corrected by telegraph up to 12 a.m. Friday, Feb. 22, 1907.)

Room 511, Union Bank Building, Office of HARDWARE AND METAL, Winnipeg, Man.

Normal weather conditions prevail once more in the west and there is now some approach at normal transportation conditions. The branch lines are all open again and dealers are getting delivery of long delayed shipments of freight. Except for an advance in petroleum market prices are unchanged.

LANTERNS.—Quotations are as follows: Cold blast, per dozen, $6.50; coppered, $8.50; dash, $8.50.

WIRE — Prices are as follows: Barbed wire, 100 lbs., $3.22½; plain galvanized, 6, 7 and 8, $3.70; No. 9, $3.25; No. 10, $3.70; No. 11, $3.80; No. 12, $3.45; No. 13, $3.55; No. 14, $4.00; No. 15, $4.25; No. 16, $4.40; plain twist, $3.45; staples, $3.50; oiled annealed wire, 10, $2.96; 11, $3.02; 12, $3.10; 13, $3.20; 14, $3.30; 15, $3.45. Annealed wires (unoiled) 10c. less; soft copper wire, base, 36c.; brass spring wire, base, 30c.

HORSESHOES — Quotations are as follows: Iron, No. 0 to No. 1, $4.65; No. 2 and larger, $4.40; snowshoes, No. 0 to No. 1, $4.90; No. 2 and larger, $4.65; steel, No. 0 to No. 1, $5; No. 2 and larger, $4.75.

HORSENAILS — Lists and discounts are quoted as follows: Capewell brand, quotations on application. No. 10, 20c.; No. 9, 22c.; No. 8, 24c.; No. 7, 26c.; No. 6, 28c.; No. 5, 32c.; No. 4, 40c., per lb. Discounts: "C" brand, 40, 10 and 7½ per cent., "M" brand and other brands, 55 and 60 per cent. Add 15c. per box.

WIRE NAILS—Quoted now at $2.85 f.o.b. Winnipeg, and $2.40 f.o.b. Fort William.

CUT NAILS—Now $2.90 per keg.

PRESSED SPIKES — Prices are: ¼ x 5 and 6, $4.75; 5-16 x 5, 6 and 7, $4.40; ⅜ x 6, 7 and 8, $4.25; 7-16 x 7 and 9, $4.15; ½ x 8, 9, 10 and 12, $4.05; ⅝ x 10 and 12, $3.90. All other lengths 25c. extra net.

SCREWS—Discounts are as follows: Flat head, iron, bright, 85 and 10 p.c.; round head, iron, 80 p.c.; flat head, brass, 75 and 10 p.c.; round head, brass 70 and 10 p.c.; coach, 70 p.c.

NUTS AND BOLTS— Discounts are: Bolts, carriage, ⅜ or smaller, 60 and 5; bolts, carriage, 7-16 and up, 55; bolts, machine, ⅜ and under, 55 and 5; bolts, machine, 7-16 and over, 55; bolts, tire, 65; bolt ends, 55; sleigh shoe bolts, 65 and 10; machine screws, 70; plough bolts, 55; square nuts, case lots, 3; square nuts, small lots, 2½; hex nuts, case lots, 3; hex nuts, smaller lots, 2½ p.c.

RIVETS—Iron, discounts, 60 and 10 p.c.; copper, No. 7, 43c.; No. 8, 43½c.; No. 9, 45¼c.; copper, No. 10, 47c.; copper, No. 12, 50½c.; assorted, No. 8, 44½c., and No. 10, 48c.

COIL CHAIN—Quotations now are: ¼ in., $7.00; 5-16, $5.35; ⅜, $4.75; 7-16, $4.50; ½, $4.25; 9-16, $4.20; ⅝, $4.25; ¾, $4.10.

SHOVELS—List has advanced $1 per dozen on all spades, shovels and scoops.

HARVEST TOOLS — Discounts continue at 60 and 5 per cent.

AXE HANDLES—Quoted as follows: Turned, s.g. hickory, doz., $3.15; No. 1, $1.90; No. 2, $1.60; octagon, extra, $2.30; No. 1, $1.60.

AXES.—Quotations are: Bench axes, 40; broad axes, 25 p.c. dis. off list; Royal Oak, per dozen, $6.25; Maple Leaf, $8.25; Model, $8.50; Black Prince, $7.25; Black Diamond, $9.25; Standard flint edge, $8.75; Copper King, $8.25; Columbian, $9.50; handled axes, North Star, $7.75; Black Prince, $9.25; Standard flint edge, $10.75; Copper King, $11 per dozen.

BUTTS—Wrought iron, 65 and 5 p.c.

CHURNS—Discounts continue at 45 and 5 per cent.; but the list has been advanced and is now as follows: No. 0, $9; No. 1, $9; No. 2, $10; No. 3, $11; No. 4, $13; No. 5, $16.

CHISELS—70 p.c. off list prices.

AUGER BITS—"Irwin" bits, 47½ per cent., and other lines 70 per cent.

BLOCKS—Discount on steel blocks is 35 p.c. off list prices; on wood, 55 p.c.

FITTINGS—Discounts continue as follows: Wrought couplings, 60; nipples, 65 and 10; T's and elbows, 10; malleable bushings, 50; malleable unions, 55 p.c.

GRINDSTONES—1½c. per lb.

FORK HANDLES—Discount is 40 p.c.

HINGES—The discount on light "T" and strap hinges is 65 p.c.

HOOKS—Prices are quoted as follows: Brush hooks, heavy, per doz., $8.75; grass hooks, $1.70.

STOVE PIPES—Quotations are as follows: 6-inch, per 100 feet length, $9; 7-inch, $9.75.

DRAW KNIVES—70 per cent.

RULES—Discount is 50 per cent.

WASHERS—On small quantities discount is 35 p.c.; on full boxes 40 p.c.

TINWARE, ETC.—Quoted as follows: Pressed, retinned, 70 and 10; pressed, plain, 75 and 2½; pieced, 30; japanned ware, 37½; enamelled ware, Famous, 50; Imperial, 50 and 10; Imperial, one coat, 60; Premier, 50; Colonial, 50 and 10; Royal, 60; Victoria, 45; white 45; Diamond, 50; Granite, 60 p.c.

GALVANIZED WARE—Pails, 37½ per cent.; other galvanized lines 30 per cent.

CORDAGE—We quote: Rope, sisal, 7-16 and larger, basis, $11.25; Manila, 7-16 and larger, basis, $16.25; Lathyarn, $11.25; cotton rope, per lb., 21c.

SOLDER—Quoted at 27c. per pound. Block tin is quoted at 45c. per pound.

VISES — "Peter Wright," 30 to 34, 14½c.; 35 to 39, 14c.; 48 and larger, 13½c. per lb.

Anvils—"Peter Wright," 11c. per lb crow bars, 4c. per lb.

POWER HORSE CLIPPERS—"1902" power clipper, $12; "Twentieth Cen-

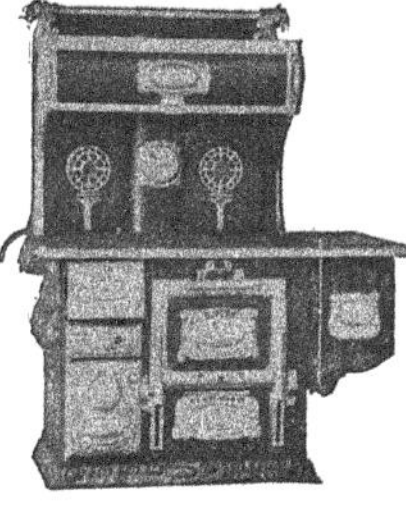

tury" $6. "1904" sheep shearing machines, $13.60.

WRINGERS—Royal Canadian, $35; B B., $39.75, per dozen.

FILES—Arcade, 75; Black Diamond, 60; Nicholson's, 62½ p.c.

LOCKS—Peterboro and Gurney, 40 per cent.

BUILDING PAPER—Anchor, plain, 66c.; tarred, 69c.; Victoria, plain, 71c.; tarred, 84c.

AMMUNITION, ETC. — Cartridges, rim fire, 50 and 5; central fire, 33½ p.c.; military, 10 p.c. advance. Loaded shells: 12 guage, black, $16.50; chilled, 12 gauge, $17.50; soft, 10 gauge, $19.50; chilled, 10 gauge, $20.50. Shot, ordinary, per 100 lbs., $7.25; chilled, $7.75; powder, F.F., keg, Hamilton, $4.75; F.F.G., Dupont's, $5.

IRON AND STEEL—Bar iron basis, $2.70. Swedish iron basis, $4.95; sleigh shoe steel, $2.75; spring steel, $3.25; machinery steel, $3.50; tool steel, Black Diamond, 100 lbs., $9.50, Jessop, $13.

SHEET ZINC—We quote $8.50 for cask lots, and $9 for broken lots.

PIG LEAD—The average price is about $6.

AXLE GREASE—"Mica" axle grease $2.75 per case; "Diamond," $1.60.

IRON PIPE AND FITTINGS—Black pipe, ¼ inch, $2.65, ⅜, $2.80, ½, $3.50; ¾ $4.40; 1, $6.35; 1¼, $8.65; 1½, $10.40; 2, 13.85; 2½, $19.00; 3, $25.00. Galvanized iron pipe, ⅜ inch, $3.75; ½, $4.35; ¾, $5.05; 1, $8.10; 1¼, $11.00; 1½, $13.25; 2, inch, $17.65. Nipples, discounts 70 and 10 per cent.; Unions, couplings, bushings and plugs, 60 per cent.

LEAD PIPE—Market is firm at $7.80.

GALVANIZED IRON — Apollo, 16 gauge, $3.90; 18 and 20, $4.10; 22 and 24, $4.45; 26, $4.40; 28, $4.65, 30 gauge or 10¾ oz., $4.95; Queen's Head, 24, $4.50; 26, $4.65; 28, $5.00.

TIN PLATES—IC charcoal, 20x28, box $9.50; IX charcoal, 20x28, $11.50; XXI charcoal, 20x28, $13.50.

TERNE PLATES—Quoted at $9.

CANADA PLATES—18x21, 18x24, $3.40; 20 x 28, $3.65; full polished, $4.15.

BLACK SHEETS—10 to 16 gauge, 100 lbs., $3.50; 18 to 22, $3.75; 24, $3.90; 26, $4; 28, $4.10.

PETROLEUM AND GASOLINE.—There has been an advance in petroleum and prices are now quoted as follows: Silver Star in brls, per gal., 21c.; Sunlight in brls. per gal., 23c.; per case, $2.30; Eocene in brls, per gal., 24c.; per case, $2.50; Pennoline in brls., per gal., 24c.; Crystal Spray, 23c.; Silver Light, 21c.; Engine gasoline in barrels, per gal., 27c., f.o.b. Winnipeg in cases, $2.75.

PAINTS AND OILS — White lead, Pure, $6.50 to $7.50, according to brand; bladder putty, in bbls., 2½c.; in kegs, 3¾c.; turpentine, barrel lots, Winnipeg, $1.01; Calgary, $1.08; Lethbridge, $1.08; Edmonton, $1.09. Less than barrel lots 5c. per gallon advance. Linseed oil, raw, Winnipeg, 67c.; Calgary, 74c.; Lethbridge, 74c.; Edmonton, 75c.; boiled oil, 3c. per gal. advance on these prices.

WINDOW GLASS — 16-oz. O.G., single, in 50-ft. boxes — 16 to 25 united inches, $2.25; 26 to 40, $2.40; 16-oz. O.G., single, in 100-ft. cases — 16 to 25 united inches, $4; 26 to 40, $4.52; 41 to 50, $4.75; 50 to 60, $5.25; 61 to 70, $5.75. 21-oz. C.S., double, in 100-ft. cases—26 to 40 united inches, $7.35; 41 to 50, $8.40; 51 to 60, $9.45; 61 to 70, $10.50; 71 to 80, $11.55; 81 to 85, $12.60; 86 to 90, $14.75; 91 to 95, $17.30.

HARDWARE TRADE GOSSIP

Ontario.

N. B. Howden, hardware merchant, Watford, died last week.

C. W. Hancock and Co., hardware merchants, Wallaceburg, have advertised their dissolution of partnership.

D. Brocklebank, of D. Brocklebank & Son, Arthur, visited Toronto during the week and took in the horse show.

Mr. Ketley, hardware merchant, of Welland was a visitor in Toronto last week calling on the hardware jobbers.

J. H. Vandusen, of the firm of McDonald and Vandusen, hardware merchants, Tara, died at his home last week.

Harvey Lewis Hymmen, son of Peter Hymmen, hardware merchant, Berlin, was married to Miss Dorothy Boehmer at Berlin, on Wednesday, the 20th inst.

J. J. Foote, London, sales manager of the McClary Manufacturing Co., spent Wednesday in Toronto attending the Stove Manufacturers' monthly meeting.

W. Boyle, of Boyle & Son, Napanee, has not been very well of late and has gone on a visit to friends in Buffalo. Mrs. Boyle accompanies him on the trip.

Benjamin Moore, of New York, the discoverer of "Muresco," has been in Toronto this week looking over the new plant at Toronto Junction, of Benjamin Moore & Co.

Mr. Jamieson, jr., of Bowes, Jamieson Stove Works, Hamilton, was in Toronto on Wednesday, attending the monthly meeting of the Stove Manufacturers' Association.

Madole & Wilson, hardware merchants, Napanee, have desolved partnership. M. S. Madole continues the business. Mr. Wilson intends going to the Northwest very shortly.

J. T. Sheridan, president of the Pease Foundry Co., Toronto, has gone to Naples, Italy, for a six weeks' holiday. Mr. Sheridan's friends expect that his health will be much restored by this pleasant remedy.

Arthur Durnan, city traveler for Rice, Lewis and Son, Toronto, was presented last week with a silver tea service, on the eve of his marriage. Chas. Elliston made the presentation on behalf of the employes of the company.

W. W. Near, general manager; Mr. Rooke, treasurer; and Messrs. Ogg, Mitchell and Taylor, of the bookkeeping staff of the Page-Hersey Iron, Tube and Lead Company, are removing from Guelph to Toronto, owing to the transfer of the firm's headquarters to the Temple Building, Toronto.

Charles Stewart, of the James Stewart Manufacturing Company, Woodstock; John T. Tilden, of the Gurney-Tilden Company, Hamilton; Wm. Burrows, jr., of Burrows, Stewart & Milne, Hamilton; and Stanley Robinson, of the D. More Company, Hamilton, were visitors in Toronto on Wednesday attending the monthly meeting of the Stove Manufacturers' Association.

Ross Mackinnon, who travels for the Pease Foundry Co., Toronto, received a pleasant surprise on Friday the 15th inst., when the other men of the travelling and office staff presented him with a handsome cut glass bowl. Mr. Mackinnon is about to assist in annexing the United States by taking charge of a fair daughter of Raleigh, N.C., whom he has found willing to halve his sorrows and double his joys, for an indefinite period.

Gerry Bros., Fort William who have affiliated stores at Brussels and Blyth, Ont., and at Indian Head, Sask., are earnest believers in the future of their town, which, they say, has splendid prospects for extensive building developments during 1907. They draw attention to an error in a recent issue in which Port Arthur was given credit for a new grain elevator which is to be erected at the Grand Trunk Pacific terminal point—Fort William. J. T. Armstrong, manager of the Northern Engineering & Supply Company has also written pointing ou that Barnett-McQueen, the contractors for the new elevator, live at Fort William and the work will be done in that town.

M. W. Connor & Son., Madoc, have been successful in winning three prizes offered by the J. Stevens, Arms & Tool Co., Chicopee Falls, Mass., for the best written and most striking advertisement, winning the following prizes: 1 double barrel hammer shotgun, 1 44 ideal rifle, 1 115 single barrel shotgun. It is exceedingly creditable to our Canadian hardware firms that they should be so successful against keen competition from all over the United States and Canada. Messrs. Connor & Son appear to be all alive from the keen northern air of Madoc, which they say is impregnated with molecules of iron from the Marmoras, which causes it to act as a tonic and sharpen the wits of the inhabitants there abouts.

Quebec.

The Montreal Hardware Co., Montreail, was slightly damaged by fire on the 19th inst.

Dwight Brainerd, of Hamilton Powder Co., Montreal, is going to England shortly on business.

Mr. Tongas, head of P. D. Dods & Co., paint manufacturers, Montreal, has been laid up for the past week.

F. O. Lewis, president of Lewis Bros., Montreal, accompanied by Mrs. Lewis, sailed for England on the 22nd inst.

The board of directors of the Northern Electric and Manufacturing Co., of Montreal, has been increased from seven to nine.

Geo. Edwards, manager of Henderson & Potts, Montreal, has been very sick for several months, and is going south for his health.

A. Macfarlane, manufacturers' agent, Montreal, spent a portion of the week in Quebec. He said that business was flourishing in the ancient capital.

J. B. Barduas, St. Charles, Richelieu, was in Montreal last week on business.

J. E. Majean, L'Epiphanie, was in Montreal this week buying goods.

E. G. M. Cape, who for some time past has been connected with C. E. Deakin, general contractor, Montreal, has left to go in business for himself.

J. R. Baxter, of Baxter, Paterson & Co., Montreal, called on the trade in St. Hyacinthe and other Quebec towns this week and reports business booming.

S. H. Lewis and W. G. Hamilton, of the eastern traveling staff of the Canada Paint Co., were in Montreal for a few days this week renewing their samples.

M. Johnson, formerly in charge of the sample room of Lewis Bros., Montreal, has severed his connection with the firm, in order to seek his fortune in the Northwest.

F. Wilkinson, manager of B. and S. H. Thompson & Co., hardware and metal brokers, Montreal, is going to Europe shortly on business. He will remain there about five weeks.

Col. Cole, representing M. Hartley Co., of New York, is in Montreal at present. Col. Cole, who is a great man for finding bull's-eyes with a rifle, is taking a few crack shots at the trade.

J. M. H. Robertson and J. A. Robertson, of the James Robertson Company, Montreal, returned last week from Baltimore, where they had been paying their annual pilgrimage to the headquarters of the firm in the United States.

J. W. Harris, general contractor; F. C. Gariviere, hardware merchant, and Alex. Prudhomme, hardware merchant, of Montreal, were added to the council membership of the Chambre de Commerce, at the general meeting held on Feb. 14, 1907.

Maritime.

A. F. Armstrong, of Dimock & Armstrong, hardware merchants, Windsor, N.S., was re-elected mayor of that city in the elections in Nova Scotia a fortnight ago.

H. F. Cohoon, of Bent & Cohoon, North Sydney, C.B., an old subscriber to Hardware and Metal, is at present enjoying a holiday tour in England. He favored our London office with a call the morning he arrived, and seemed pleased to see the familiar buff cover so many miles away from home. He is to return to Canada by the Tunisian on February 22nd.

Western.

John Hanbury, president of the new Hanbury Hardware Co., of Brandon, was in Winnipeg last week on business.

O. Landen and C. A. Whiting, hardware merchants, Winnipeg, will hereafter be known as the Whiting Hardware Company.

The hardware store belonging to the estate of Edward Crouter, Gladstone, Man., has been sold to W. H. Squair, of that place.

James Stepler, of Carveth & Stepler, hardware merchants, Fielding, Sask., is in Winnipeg consulting a physician, being threatened with appendicitis.

T. L. Waldon, of the Marchall-Wells Hardware Co., has returned to business after an extended trip in the east. The Marchall-Wells Co. expect to move into their new warehouse about the first of the month.

John Isbister, of Saskatoon, was one of the enthusiastic devotees of the "roarin' game" during the Winnipeg Bonspiel, but he found time during his stay of about ten days in the city to look up his friends in the hardware and heating trade.

W. G. Jones, treasurer of the Pease Foundry Co., Toronto, is registered at the Royal Alexandra hotel, Winnipeg, this week. Mr. Jones has returned from a business trip in the western provinces and evidently found traveling this winter hard on his health as he is confined to his room with a severe cold.

FOUNDRY AND METAL INDUSTRIES

The Lake Superior Corporation has obtained the contract to supply 65,000 tons of rails to the Temiskaming and Northern Ontario Railway.

The North American Refining and Smelting Company will likely erect their smelter in Thorold, instead of St. Catharines, as at first intended

The Lake Superior Corporation steel rail mill made a new high record in the month of January, turning out 19,285 tons, the previous high record being 16,800 tons.

The last year has been a very prosperous one for the Lake Superior iron region, as the output approximates 38,400,000 tons of ore, which is far greater than any previous one in the history of that district.

The total amount paid to date in iron and steel bounties in Canada is $8,814,833, of which $998,000 went to the Soo Company; $1,416,469 to the Hamilton companies; $151,095 to the Deseronto company; $1,369,158 to the Nova Scotia Steel Co., at Sydney, and $3,466,519 to the Dominion Steel Co.

The Nova Scotia Steel and Coal Company will carry out their original intention of erecting another furnace. Already orders have been received which far exceed the capacity of the plant, and in order to keep up with the increasing demand the company will have to enlarge their works by another furnace.

The Madison Williams foundry, which is the latest acquisition to Lindsay's industries, is now ready for business. A complete equipment of machinery and patterns and moulding apparatus is in place. Besides a general foundry business, the company will specialize on water-wheels. Pulleys, beltings and tie cutting machines will also be manufac-

JONES BROS. NEW PLANT.

Jones Bros., Bracondale, Ont., manufacturers of stoves linings, fittings, etc., who were burnt out a short time ago, have almost completed the building of a new plant. The kiln shed is 37x33 feet, covered with sheet iron. The machine shop will be 115x28 feet, one story in height. They expect to be ready for business again by April 1.

MANITOBA ROLLING MILLS.

The United States Horse Shoe Company, Erie, Pa., which was recently incorporated as the Manitoba Rolling Mill Company, with a capital of $100,000, and will operate the Kirkwood Rolling Mills, at Winnipeg, Man., have the works now equipped with one 9-in. guide mill, with five stands of housings, one 16-in-bar mill, with three stands of housings, and the necessary heating furnaces, engines, shears, etc. It is the intention to materially enlarge the plant by adding a number of busheling furnaces, squeezer, shears, railroad spike machines and horse shoe machines, and engage in the manufacture of bar iron, railroad spikes and horse shoes, adding other commodities as trade conditions warrant. Contracts have already been placed for the additional equipment, and the company expects to have the plant in active operation within the next four months. T. M. Kirkwood, Toronto, is vice-president.

SEWER PIPE

BUILDING AND INDUSTRIAL NEWS

HARDWARE AND METAL would be pleased to receive from any authoritative source building and industrial news of any sort, the formation or incorporation of companies, establishment or enlargement of mills, factories, foundries or other works, railway or mining news. All correspondence will be treated as confidential when desired.

A permit is to be taken out by George Manuel for a warehouse at Edmonton, to cost $2,690.

A new hotel will be built on the corner of York and Front streets, Toronto, by G. Percival, Toronto.

An addition is to be made to the Sylvester Feed Co.'s block on Yates St., Victoria. It will cost about $6,000.

The Corundum Wheel Company, of Hamilton, have taken out a permit for an addition to their factory to cost $1,000.

The Hamilton Incubator Company have secured a permit for the erection of a brick factory in Hamilton, at the cost of $6,000.

James Woods, of the Woods Company, manufacturers of lumberman's supplies, Ottawa, will erect a $200,000 factory at Winnipeg.

Kurze and McLean Company have procured the lease of a factory in Stratford, and will go into the manufacture of acetylene gas plants.

Rhodes, Curry & Co., of Amherst, N.S., have been awarded the contract for the new round-house at Halifax The price is about $80,000.

Building permits have been taken out by Fred Perkins of Edmonton for the erection of three houses on the corner of Seventh and Victoria Sts.

The Bennett Theatrical Enterprise Company intend erecting a factory in London, Ont., for the manufacture of films for moving picture machines.

A. M. Orpen has bought a block of land in the Avenue Road district, Toronto, and may spend this spring $100,000 in the erection of new houses.

The Grand Trunk Pacific Company has called for tenders at Victoria, B.C., for the erection of a new hotel at Prince Rupert, costing in the neighborhood of $50,000.

The David Spencer Company will erect an eight-storey building in Vancouver, B.C., to be used for manufacturing purposes. The cost estimated at about $200,000.

The National Car Company, Toronto, have accepted the site of 5 acres offered by the town of Whitby, for their car works. Work will be begun on the new shops at once.

A new Y.M.C.A. building will be erected in Ottawa, containing 97 dormitories. It will be of steel construction and fire-proof throughout, costing in the neighborhood of $200,000.

The Nichols-Shepard Threshing Machines Company, of Battle Creek, Mich., will erect a warehouse in Regina. Regina headquarters will be a source of supply for 150 agents.

The Massey-Harris Company, of Brantford, are making extensive additions to their plant by building a 100-foot addition to their foundry, and also enlarging their blacksmith shop.

The Iron and Brass Manufacturing Company, of Weston, have now located in Chesley. A new building 300 feet long has been erected, and the company will employ 75 hands.

John Mann Son and Company, of Brantford, will establish a plant for the manufacture of sand lime brick in that city. The plant will have a capacity of 30,000 bricks per day.

A sawmill with a capacity of 40,000 feet a day will be erected by the B.C. Mills, Timber & Trading Co. in Burnaby Lake, B.C. The contract is let to Mitchell & Ferris, of Vancouver.

The Security Light Company, Toronto, has been incorporated, with a capital of $40,000, to manufacture gas, gasoline, etc. The provisional directors are, H. E. Pearce, A. Gate and T. Minton, Toronto.

A $110,000 hotel will be erected by E. J. Fader, of the Torpedo Touring and Tug Company, at New Westminster. The new building will be known as the Hotel Russel and is to be completed by October, 1907.

The Gamer Manufacturing Company, Chesley, Ont., has been incorporated with a capital of $50,000, to manufacture bedsteads, etc. The provisional directors include A. Gamer, A. Harrod and T. E. Devitt, Weston, Ont.

"Electro Metals, Limited," is the name of a New York concern which purposes to build and operate an electric ore-reducing plant in the Niagara Peninsula. English capital from New Castle, Eng., is largely interested.

At the annual meeting of the International Portland Cement Co., held in Ottawa this week, the shareholders approved the recommendation of the directors to double the capac[illegible]lant at Hull. At present it is turning out 2,000 barrels of cement per day.

The Wingold Stove Company, of Winnipeg, Man., has been incorporated with a capital of $40,000, to manufacture stoves, ranges, furnaces, machinery, wagons, carriages, bicycles, etc. The provisional directors are: E. B. Blanchard, O. Gensmer, R. Kello, Winnipeg, Man.

The Excelsior Constructing and Paving Company has been incorporated with a share capital of $30,000, to manufacture, deal and take contracts in cement and concrete works. The head office will be at Toronto, the provisional directors being J. G. Murphy, W. Mitchell, and O. Murphy, Toronto.

A shipbuilding company is to be formed at Kingston, with a site on the west side of the dry dock. The company will likely have a capital of $100,000. Already the companies existing here are full of business, and there is easily a field for increased and larger operations. The new company would be handy for repairing vessels that enter the dock.

Wm. Stones, Sons and Company has been incorporated with a share capital of $100,000, for the purpose of dealing in wool, hides, skins, greases, raw furs, salt, lime and cement. The head office will be at Brantford, the directors being, W. Stone, Sr., A. W. Stone, and J. F. Stone, of Woodstock; E. E. Thornton and P. S. Thornton, of Brantford.

Trussed Concrete Steel Company of Canada, has been incorporated with a share capital of $200,000, to manufacture materials used in connection with concrete reinforcement. The head office will be at Walkerville, the directors being G. Kahn, D. C. Raymond and L. Wiman, of Toronto.

One of the largest sawmills on the Pacific coast is to erected shortly at Vancouver. The contract for the machinery of the new plant has been let to the Prescott Co., of Menominee, and the contract is let to E. H. Heaps & Co. The new plant will have a capacity to handle logs ten feet in diameter and 110 feet long.

The Montreal Steel Works made a record showing in the fiscal year ending Dec. 31. The financial statement submitted at the annual meeting shows that the management has placed the company in a very strong position. Net earnings for the year amounted to $196,097, and after the payment of dividends on preferred stock there was earned over 20 per cent. on the common stock.

Montreal is to have a big new cotton mill. W. T. Whitehead, formerly of the Dominion Textile Company, and for many years identified with the cotton interests, has been successful in forming a company with a capital of $1,500,000, under the title of the Mount Royal Spinning Company. All the capital has been subscribed. The new mill, which will be situated either at Maissoneuve or on the Lachine Canal, will, it is expected, be in operation in eighteen months.

The Regina Board of Trade held its annual meeting on Feb. 7, electing the following officers: President, P. Cooper; vice-president, A. E. Whitmore; secretary-treasurer, A. E. Boyle; council, P. McAra, Jr., A. T. Hunter, J. M. Young, H. F. Mytton, J. F. Bole, A. E. Jarvis, D. J. Taylor, Robert Sinton and E. A. McCallum. Secretary Boyle is also publicity commissioner and his work during 1906 came in for much praise. For advertising during the year $1,669 was spent to make this thriving town more favorably known as a manufacturing, trade and residental centre.

WATER PURIFICATION.

When a marked industrial growth is taking place in a new country and many factors are exerting their influence either for or against the development of different localities or districts, it is very interesting to note the rivalry that often exists between various towns and cities, any of which may be selected as the site for a coming industry, says Canadian Machinery.

When a manufacturer happens to be looking for a location for his plant, he naturally desires to select a locality which will offer him the cheapest production, and he is more often influenced by such a consideration than by any artificial stimulus that may be offered. Therefore, after such questions as raw materials, market and transportation have been considered, the effect of local conditions on the cost of manufacture is of vital consequence.

As compared with the other natural advantages of a manufacturing region, the availability of a pure and cheap water supply is of the greatest economic importance. The amount of water that is used in some industries is hard to conceive when thinking of it in gallons and it is only when water has to be purchased that this factor is thought of at all, as there are many industries that would hardly exist at all if they were obliged to consider the water they use as an item of expense. However, it is these very manufacturers that are most seriously concerned with the quality of the water that they use, for a very small content of an injurious ingredient may seriously affect the quality of the material treated when such large quantities of water are used.

We have many localities in Canada where the water is relatively soft owing to the insolubility of the rocks and soil over which the streams and rivers pass. Yet these same waters often carry in solution ferruginous and oganic matters which are equally deleterious to many industries. On the other hand are many streams whose course lies mainly through a limestone region and these always contain what are commonly called hard waters.

In Eastern Canada the quality of our average river water is not considered bad from an industrial standpoint and this is one of the reasons that the manufacturer so often ignores the fact that even moderately bad water may do a lot of injury when used in sufficient quantity.

Boiler Feed Waters.

There is one use for water which nearly all manufacturers have in common and that is in the production of steam. In this operation a more or less impure water is continually coming into the boiler while nothing but pure water is passing out in the form of steam and it is common experience that what we consider to be pretty good water will soon leave behind considerable residue. The question to determine from the standpoint of boiler economy is, how bad will the water have to be before it will pay to treat it for the removal of the scale-forming solids prior to its use in steam production. This of course depends upon the relation between the loss sustained from using the natural water and the cost of purification, and whenever the matter is treated as being worth consideration the probabilities are always in favor of the methods of purifying.

OUR LETTER BOX

Correspondence on matters of interest to the hardware trade is solicited. Manufacturers, jobbers, retailers and clerks are urged to express their opinions on matters under discussion.

Any questions asked will be promptly answered. Do you want to buy anything, want some shelving, a silent salesman, any special line of goods, anything in connection with the hardware trade? Ask us. We'll supply the necessary information

Electric Cooking Utensils.

K. Stairs, of William Stairs, Son & Morrow, Halifax, asks for the name of some firm manufacturing cooking utensils heated by electricity.

Answer—The Simplex Electric Company, Boston, Mass., and the Promethus Electric Supply Company, New York, make complete lines of these goods, the former being represented in Canada by the Canadian General Electric Company, Toronto, and the latter by the R. E. T. Pringle Company, Montreal. There are many other concerns making electric sad irons, etc., but for complete lines the above are the best known manufacturers.—Editor.

Car Seats.

Gerry Bros., Fort William, ask for the names of manufacturers or jobbers of car seats in Canada.

Answer—The Crosson Car Company, Cobourg, the Ottawa Car Company, Ottawa, and the Toronto Railway Company, Toronto, all make some seats for their own use but the largest part of the seats used are imported, Wakefield Bros., Philadelphia, sending a large number of seats to Canada.—Editor.

Diamond Drills.

Arthur Cote, Verner, Ont., wants the address of a manufacturer of diamond drills.

Answer—The Canadian Rand Co., Sherbrooke, Que., are the only Canadian manufacturers.—Editor.

Series of Questions.

The Ontario Nickel Company, Worthington, Ont., inquire for the names of jobbers of carborundum in sheets ½-inch thick and about 18 inches long by 6 inches wide; also of Canadian makers of emery wheels and dealers in electrical supplies.

Answer—The Carborundum Company, Niagara Falls, N.Y., represented in Canada by Rice Lewis & Son, Toronto, and other jobbers, can make the sizes desired. It is not stocked by the jobbers referred to. The Canadian Corundum Company, Hamilton; Hart Emery Wheel Co., Hamilton, and the Prescott Emery Wheel Company, Prescott, make emery wheels. The Canadian General Electric Company, Toronto; Munderloh & Co., Montreal; John Millen & Co., Montreal; John Forman, Montreal; Sayer Electric Company, Montreal; R. E. T. Pringle & Co., Montreal, all handle electrical supplies.

CONDENSED OR "WANT" ADVERTISEMENTS.

Advertisements under this heading 2c. a word first insertion; 1c. a word each subsequent insertion.

Contractions count as one word, but five figures (as $1,000) are allowed as one word.

Cash remittances to cover cost **must** accompany all advertisements. **In no case** can this rule be overlooked. Advertisements received without remittance cannot be acknowledged.

Where replies come to our care to be forwarded, five cents must be added to cost to cover postage, etc.

SITUATIONS VACANT.

BRIGHT, intelligent boy wanted in every town and village in Canada; good pay besides a gift of a watch for good work. Write The MacLean Publishing Company, 10 Front Street E., Toronto. [tf]

EXPERIENCED Hardware Clerk Wanted—One to take charge, must be good salesman and stock-keeper; state experience and salary expected. Address G. McLean (Hardware), London. 9

BUSINESS CHANCES.

PLUMBING, Steam, Hot Water, Gasfitting and Tinsmithing business for sale; $15,000 business done last year, and only $3,000 or less invested; in one of the best towns in Canada. Waterworks and sewage. Owner retiring. Box 587, HARDWARE AND METAL, Toronto. [10]

HARDWARE and Harness Business for sale in good town. Owner retiring for good reason. Stock about $2,500, in excellent shape. This is worth investigating, for particulars. apply Richard Tew, 23 Scott Street, Toronto.

FOR SALE—In Eastern Ontario, hardware, tinsmithing and stove business, stock reduced to $1,500; doing a good steady business, expenses low; must sell at once on account of ill-health. Building, including dwelling, for sale or rent. Apply to Box 590, HARDWARE AND METAL, Toronto. [8]

FOR SALE—Foundry and stove works on the Pacific Coast, fully equipped and up-to-date. Apply P.O. Box 760, Victoria, B.C. [10]

HARDWARE Business—Including stoves, tinware, etc; in best location of the city. Apply Box 261, Vancouver, B.C. 8

FOR SALE.

HARDWARE Business; including stoves, tinware and tinsmith tools, in thriving town in West Ontario peninsular, stock about $5,000; building can be leased if desired, dwelling also. Box 583 HARDWARE AND METAL, Toronto. (17)

ARTICLES WANTED.

WANTED—8 foot cornice brake; state price and particulars. J. E. Hussey, Melbourne, Ont. (9)

ANY hardwaremen having a stock of cut nails they wish to dispose of at reduced prices should write at once to Box 501 HARDWARE & METAL, Toronto.

WANTED.

WANTED—Six-foot brake and thirty-inch tinsmith's grover. Box 586, HARDWARE AND METAL. [8]

CORNICE FOREMAN WANTED. Must be capable of taking full charge of Metal Cornice Department. Permanent position to right man. State experience and wages expected. Galt Art Metal Co., Limited, Galt, Ont. (9)

BOOKKEEPER WANTED

WANTED—A good bookkeeper; stating experience, reference and salary. Cunningham Hardware Co., Westminster, B.C. [13]

Get Our Catalogs

and Price Lists.—That's the surest way to be "in the game." Your Opposition will have to "step fast" if he beats you and Our Line. We are using our every endeavor to make Our Line interesting to "the trade." We want good representation in every locality. Write us about it to-day.

Western Distributing Agents:

THE McCLARY MFG. CO.
Winnipeg and Calgary

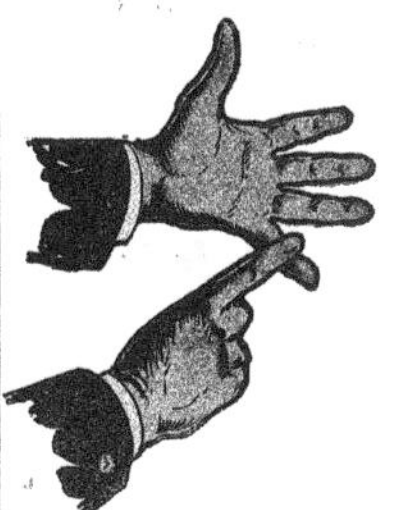

Five good reasons why you should handle "M.R.M." Horse Shoes

MATERIAL is the best that can be procured for the purpose.

SHAPE is such that the Farrier has less fitting than is required with most makes of shoes.

CREASE is so placed as to give the proper setting to the nail.

FINISH is clean and smooth.

THEY are asked for by all the best Farriers.

MADE BY

THE MONTREAL ROLLING MILLS CO.

Are you interested in any of the lines that are advertised? A Post Card will bring you price list and full information. Don't forget to mention Hardware and Metal.

Paint, Oil and Brush Trades

PAINT ECONOMY.

"It sounds paradoxical to say that money spent for paint is money saved, yet with certain limitations it is true. Paint is used on structures for two purposes—to preserve and to beautify them Beauty itself has a cash value, and as the materials protected by good paint are far more costly than the paint itself, it is evident on both counts that money invested in good paint is money saved.

The selection of good paint is not so difficult as it has been made to appear, but the painter's judgment on this point is not always unbiased. His natural impulse is to recommend the paint that affords him the largest profit. Probably the most economical type of paint is prepared or ready-mixed paint. It is manufactured on a large scale by rigid standards and formulas in thoroughly equipped factories under the supervision of experienced specialists. To obtain the best results such paint should be applied by a practical painter, who understands the adaptation of consistency to the condition of surface. The leading brands of paints of this class are so widely used that it is seldom difficult for the intending purchaser to obtain from his own neighborhood evidence of their quality and behavior in service.

There is room for choice here as elsewhere, of course; but the choice—excluding manifestly fraudulent products—depends far less upon the composition of the paint than upon the method of manufacture. The most highly approved materials insufficiently ground or improperly combined will not yield results comparable to those obtained from second-grade materials scientifically prepared. Paint on a building should be examined every spring and fall, and as soon as deterioration becomes apparent a fresh coat should be applied. In this way its efficiency and beauty may be maintained at a very low average annual expense almost indefinitely; and if a good ready-mixed paint is used this repainting should not ordinarily be necessary more than once in about five years. The important requisite is to apply the fresh coat before the old one begins to "let go." This is the practice with the successful trust estates, and will be found by any property-owner who adopts it to be most economical."

PREPARED PAINT BEST.

Many progressive painstaking painters have tried to make in their own shops better paint than straight lead and oil mixtures. They have failed because they have not the mill equipment necessary to paint making, and these failures are largely responsible for the painters' prejudice against combination paints. They are not disposed to acknowledge that the unsatisfactory results are due to the methods employed rather than to the pigments used. With the proper machinery, and necessary technical information, anyone can make good paint, but no painter, not even the largest contractor, consumes enough paint to warrent the expense of efficient laboratory and mill equipment.

If a white lead and oil job goes wrong, the painter blames the oil. If a prepared paint or combination lead job does not wear well, the painter blames the pigments. As a matter of fact, the painter is sometimes to blame. He applies the paint in a way, or at a time that makes dissatisfaction inevitable.

Paint is often applied immediately after a rain, or in the early morning after a heavy dew or frost. This explains why three sides of a building will sometimes be in a good condition, while on the one side, or a portion of it, the paint has peeled. Paint will not hold on a damp surafce. Natural moisture (as found in green lumber) is even worse than the dampness formed by rain, dew and frost: green lumber therefore should never be painted, nor should painting be resumed after a rain, nor in the morning before the surface has had ample time to dry.

Nothing will make paint hold on pitchy boards. The heat of the sun draws the pitch to the surface; the paint is loosened, and peeling results. Knots pitchy surfaces that are given a coat of good orange shellac will hold paint if there is not too much pitch.

MAKE YOUR WINDOWS SELL PAINT.

Do you use your windows to the best advantage? There is no better way of your keeping paint constantly before the public than by neat and attractive window displays, and no time more opportune than now.

A properly dressed window is a valuable advertising asset; not only is it a forceful advertisement, but it is invariably taken as an indication of the condition of the store and the stock.

At this season a well trimmed window, is particularly effective. It is a great aid to your spring advertising campaign. Draw attention to paint as an aid to housecleaning at once and keep up the campaign for several weeks.

The Standard Paint Co., of Canada, are now thoroughly established in their new Montreal factory at Highlands suburb, and have a suite of well-equipped offices at room 11n Board of Trade building. They have a unique and novel method of advertising their Ruberoid and P & B brands of goods, one a roofing and flooring, and the other a waterproof preservative allied to paint. Tablets, with descriptive slips attached, showing a slanting shape of a roof with the exact article, Ruberoid, on it, are distributed, and serve as convincing arguments for salesmen,

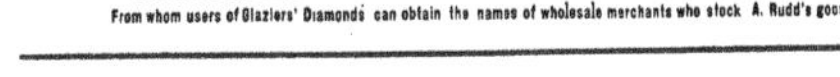

PAINT AND OIL MARKETS

MONTREAL.

Office of HARDWARE AND METAL,
232 McGill Street,
Montreal, February 22, 1907

Firmness characterizes the market this week, in all lines, but particularly in white lead, an article which is in pretty good demand, while stocks of dry in maintaining their quotations.

There is less cutting reported on all the staples than was formerly the case, indeed, with a brisk business, and a shortage of crude supplies, the manufacturers seem to have little difficulty in maintaining theit quotations.

Business remains good, and no advances are reported in any lines.

LINSEED OIL—Remains very firm. We still quote: Raw, 1 to 4 barrels, 55c.; 5 to 9 barrels, 54c.; boiled, 1 to 4 barrels, 58c.; 5 to 9 barrels, 57c.

TURPENTINE — No change in prices. Our quotations remain: Single barrel, 98c. per gal.; for smaller quantities than barrels, 5c. extra per gal. is charged. Standard gallon is 8.40 lbs., f.o.b. point of shipment, net 30 days.

GROUND WHITE LEAD — We still quote: Best brands, Government standard, $7.25 to $7.50; No. 1, $6.90 to $7.15; No. 2, $6.55 to $6.90; No. 3, $6.30 to $6.55, all f.o.b., Montreal.

DRY WHITE ZINC — Very firm. We still quote: V.M., Red Seal, 7½c. to 8c.; Red Seal, 7c. to 8c.; French V.M., 6c. to 7c.; Lehigh, 5c. to 6c.

White ZINC—Our quotations are: Pure, 8½c. to 9½c.; No. 1, 7c. to 8c.; No. 2, 5¾c. to 6¾c.

PUTTY—Our prices remain; Pure linseed oil, $1.75 to $1.85; bulk in bbls., $1.50; in 25-lb. irons, $1 80; in tins, $1.90; bladder putty in barrels, $1.75.

ORANGE MINERAL—The following prices still hold: Casks, 8c.; 100-lb. kegs, 8½c.

RED LEAD—The following quotations are firm: Genuine red lead, in casks, $6; in 100-lb. kegs, $6.25; in less quantities at the rate of $7 per 100 lbs.; No. 1 red lead, casks, $5.75; kegs, $6, and smaller quantities, $6.75.

PARIS GREEN—We revise our quotations: In barrels, about 600 lbs., 25½c. per lb.; in arsenic kegs, 250 lbs., 25¾c.; in 50-lb. drums, 26¼c.; in 25-lb. drums, 26¾c.; in 1-lb. packets, 100 lbs. in case, 27¼c.; in 1-lb. packets, 50 lbs. in case, 27¾c.; in ½-lb. packets, 100 lbs. in case, 29¼c.; in 1-lb. tins, 28¼c. f.o.b. Montreal. Terms, three months net or 2 per cent. 30 days.

SHELLAC—Prices remain; Bleached, in bars or ground, 46c. per lb., f.o.b Eastern Canadian points; bone dry, 57c. per lb., f.o.b. Eastern Canadian points; T.N. orange, etc., 48c. per lb. f.o.b. New York.

SHELLAC VARNISH—Very firm. We still quote: Pure white, $2.90 to $2.95; pure orange, $2.70 to $2.75; No. 1 orange, $2.50 to $2.55.

MIXED PAINTS — No change in prices. We still quote; $1.20 to $1.50 per gallon.

PETROLEUM — (In barrels) — Our prices are: American prime white coal, 15½c. per gallon.; American water, 17c. per gallon; Pratt's Astral, 19½c. per gallon.

WINDOW GLASS—Prices remain firm. We still quote: First break, 50 feet, $1.85; second break, 50 feet, $1.95; first break, 100 feet, $3.20; second break, 100 feet, $3.40; third break, 100 feet, $3.95; fourth break, 100 feet, $4.15; fifth break, 100 feet, $4.40; sixth break, 100 feet, $4.95. Diamond Star: First break, 50 feet, $2.30; second break, 50 feet, $2.50; first break, 100 feet, $4.40; second break, $4.80; third break, 100 feet, $5.75; fourth break, 100 feet, $6.50; fifth break, 100 feet, $7.50; sixth break, 100 feet, $7.50; seventh break, 100 feet, $8; eight break, 100 feet, $9. Double Diamond: First break, 50 feet, $3.45; second break, 50 feet, $3.75; first break, 100 feet, $6.75; second break, 100 feet, $7.25; third break, 100 feet, $8.75; fourth break, 100 feet, $10; fifth break, 100 feet, $11.50; sixth break, 100 feet, $12.50; seventh break, 100 feet, $14; eighth break, 100 feet, $16.50; ninth break, 100 feet, $18; tenth break, 100 feet, $20; eleventh break, 100 feet, $24; twelfth break, 100 feet, $28.50. Discount on Diamond Star, 20 per cent.; on Double Diamond, 40 per cent.

TORONTO.

Office of HARDWARE AND METAL
10 Front Street East,
Toronto Feb 22, 1907.

While very few orders are being taken at present for paint, orders which have been booked are being sent forward in fair quantities. The high prices of raw materials continue and manufacturers may have to advance their products again.

The following is an extract from the Savannah Naval Stores Review, on the causes of the high prices of turpentine: "So far this winter there have been but three or four days that could be called cold, and there has been an absence of rain equally noticeable. As a result of this state of the weather the turpentine farms have practically cleaned up everything in sight. The boxes, as one man remarked, have been "wiped dry, and everyone is looking for an almost total cessation of receipts at any time. The impression prevails in trade circles that the late receipts of the season of 1906-07 and the early receipts of 1907-08 will be noticeably small. As a result the feeling prevails that conditions are in favor of a continuance of high values, and that even better values than at present prevailing are not unlikely in both spirits, turpentine and lower grades of rosins. Buyers seem desirous to gather in stocks of

Ramsay's Paints

Can you hold up your pot of paint, when you offer it for sale, and honestly say you have the best? Everything is in the quality for future trade—for the man you want to come back. With Ramsay's Paints you get the best. We guarantee it, and you can give that confidently to your customer. With our aids to trade, you can secure business if you take advantage of what we offer. With our price, we believe we offer the highest grade at the lowest figure of any paint made to-day. Listen !

A. RAMSAY & SON COMPANY

Est. 1842 MONTREAL PAINT MAKERS

Just try some ordinary green paint on a panel of wood, see how hard it works,—don't cover, lacks body. Then try on another panel

LUCAS IMPERIAL FRENCH GREEN

notice how easy it spreads, how well it covers,—that's body. Makes the cost less for this reason alone, and stays fresh and brilliant. It preserves the wood, keeps the weather out and the color in. It is the **ONE** Green that is free from arsenic or verdigris.

BEWARE OF IMITATIONS

JOHN LUCAS & CO.

NEW YORK, PHILADELPHIA, CHICAGO

Painting Profit for the Dealer

Every stroke of the brush works for the dealer's advantage when Martin-Senour Paint is being put on. The way it covers and flows tickles the painter and he orders more for the next job. Its fine appearance and wearing qualities make it in demand when more painting is needed and herein lies the secret of its

Success as a Trade Builder

It is uniformly excellent in quality—100 per cent pure is the standard of its manufacture. The public knows where it gets the best value for its money. Dealers who handle Martin-Senour Paint are the ones who reap the benefit of future orders from present sales. Martn-Senour Paint is the kind that brings buyers back over and over.

Write to us for our special proposition to dealers.

The MARTIN-SENOUR CO., Ltd.

"Pioneers of Pure Paints"

142-144 Inspector Street, Montreal.
2514-2520 Quarry Street and Archer Avenue, Chicago. (Est. 1878.)

The Winnipeg Paint and Glass Co., Ltd. Winnipeg. The Maritime Paint and Varnish Co., Ltd. Halifax, N. S.

turps while there are any to be had, although they naturally do not like to run the market up at their own expense. Stocks are not heavy and as there are probably three months ahead during which the consumers must be fed from the stuff now in hand and the small lots to arrive there appears to be good basis for the bullish feeling that is beginning to show itself, and which may bring important developments at any time."

Linseed oil is firm at the 1c per gallon advance last week. White lead remains firm at the old price, and is meeting with good demand. No further advance is quoted on Paris green, which remains at 27¼c base price.

WHITE LEAD—Ex Toronto pure white, $7.40 ; No. 1, $6.65 ; No. 2, $6.25 ; No. 3, $5.90 ; No. 4, $5.65 in packages of 25 pounds and upwards ; ½c. per pound extra will be charged for 12½ pound packages ; genuine dry white lead in casks, $7.

RED LEAD—Genuine in casks of 500 lbs., $6 ; ditto, in kegs of 100 lbs., $6.50 ; No. 1 in casks of 500 lbs., $5.75 ; ditto, in kegs of 100 lbs., $6.25.

DRY WHITE ZINC.—In casks, 7½c.; in 100 lbs., 8c-, No. 1, in casks 6½c. in 100 lbs., 7c. Ground in oil—In 25 lb. irons, 8c.; in 12½ lbs., 8½c.

SHELLAC VARNISH.—Pure orange in barrels, $2.70 ; white, $2.82½ per barrel ; No. 1 (orange) $2.50 ; gum shellac, bone dry, 63c. Toronto. T. N. (orange) 51c. net Toronto.

LINSEED OIL—Remains firm at the 1c advance and is selling very freely. We continue to quote: Raw, 1 to 3 barrels, 59c; 4 to 7 barrels, 58 ; 8 and over, 57c.; add 3c. to this price for boiled oil, f.o.b. Toronto, Hamilton, London and Guelph, net 30 days.

TURPENTINE—Stocks on Southern markets are low and the market generally remains firm. We quote: Single barrels at 99c per gal., f.o.b. point of shipment, net 30 days: less than barrels, $1.04 per gallon.

PARIS GREEN—Quotations remain at last week's figures, and most merchants are understood to have placed their orders. We now quote base price on both English and Canadian at 27¼c per lb.

For additional quotations see current quotations at back of paper.

OIL CRUSHING AT WINNIPEG.

We reproduce a cut of the Canada Paint Company's elevator and linseed oil mills at Winnipeg, where the most modern appliances are installed to crush the Manitoba flax seed and to produce linseed oil from the celebrated Manitoba seed. The product is a beautifully bright, plump oil, free from all fibrous or mucilaginous matter—the color of a pale sherry—"clear as a bell and free from sediment."

Canada Paint Company's Oil Mills at Winnipeg.

A ready market is found in Canada for all the linseed oil which this mill will produce, but the residuary product of linseed meal and cake is largely exported. Linseed meal besides being a pharmacopeiae remedy of great value is a valuable food for cattle eagerly bought by stock-raisers, especially in the Old Country, where food stuffs are naturally more expensive than in Canada.

In addition to the oil-works to the trade as the Winnipeg linseed oil mills, the Canada Paint Company, under the management of Edward Barry, has a complete paint and color factory in Winnipeg, fitted up with great completeness and detail.

"LET EVERY TUB STAND ON ITS OWN BOTTOM"

The great trouble, however, is that a lot of Tubs are so made that they can't stand long enough to warrant their expense. The wood is not properly kiln-dried, the hoops loosely rivetted and the finishing so poorly done that they warp, hoops drop off and they go to pieces.

A "Cane-made" Tub or a "Cane-made" Pail

has every detail of manufacture so closely looked after that ***a perfect article*** is the result. **Pin your faith to "CANE'S BRANDS"**

UNITED FACTORIES, LIMITED

TORONTO, CAN.

The Necessity of Featuring the Best Paint

100 per cent. durability

You would not try to cross the Atlantic in an open boat. Might as well as try to corner a good paying trade in your line without considering paint and its results. Paint makes or mars a business. Every customer your paint defects cost you means the loss of ten probable customers.

ARK BRAND PAINT

differs from all other paints on the market. The paint difference—a difference which hundreds of dealers appreciate—is that it is made from pure linseed oil and the most durable pigments known.

MURESCO WALL FINISH

has restored and improved more retail businesses than any other article that could be mentioned. Quality tells and always will. Muresco is easily applied, durable and strongly endorsed by the best decorators in Canada.

BENJAMIN MOORE & CO., LIMITED

(Formerly The Francis-Frost Co., Limited)

NEW OFFICES AND WORKS:

TORONTO JUNCTION, CANADA

New York, Chicago, Cleveland, Carteret, N.J.

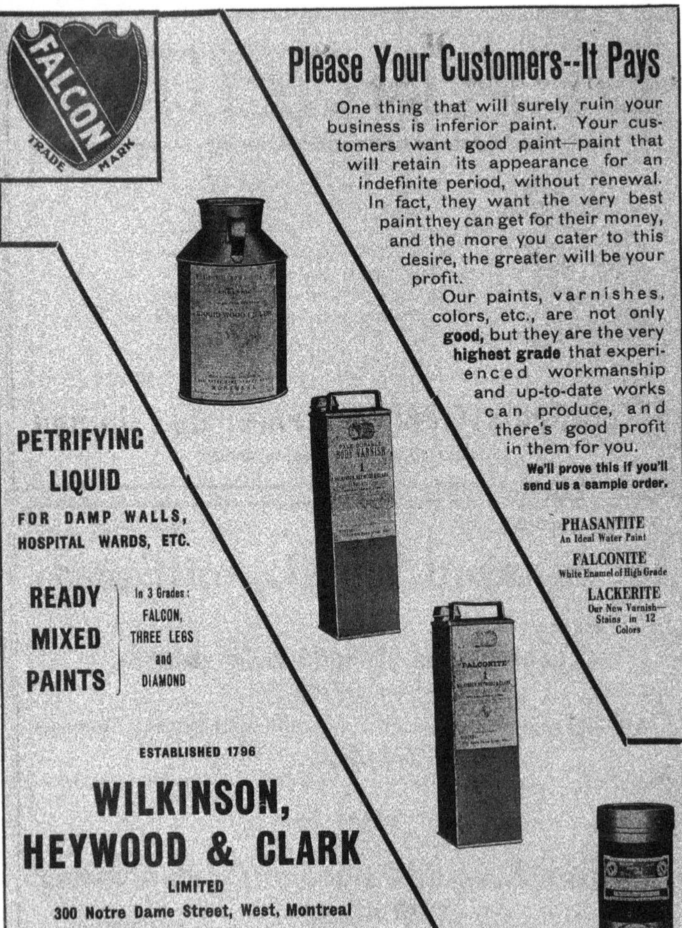
FALCON
TRADE MARK
Please Your Customers--It Pays
One thing that will surely ruin your business is inferior paint. Your customers want good paint—paint that will retain its appearance for an indefinite period, without renewal. In fact, they want the very best paint they can get for their money, and the more you cater to this desire, the greater will be your profit.
Our paints, varnishes, colors, etc., are not only good, but they are the very highest grade that experienced workmanship and up-to-date works can produce, and there's good profit in them for you.
We'll prove this if you'll send us a sample order.
PHASANTITE
An Ideal Water Paint
FALCONITE
White Enamel of High Grade
LACKERITE
Our New Varnish—Stains in 12 Colors
PETRIFYING LIQUID
FOR DAMP WALLS, HOSPITAL WARDS, ETC.
READY MIXED PAINTS
In 3 Grades: FALCON, THREE LEGS and DIAMOND
ESTABLISHED 1796
WILKINSON, HEYWOOD & CLARK
LIMITED
300 Notre Dame Street, West, Montreal
Branches and Agencies all over the world
FRANCIS C. REYNOLDS
Canadian Manager

Please mention HARDWARE AND METAL when writing.

Plumbing and Steamfitting

PUBLICITY FOR THE PLUMBER AND FITTER

By P. A. Smith, manager Publicity Department of Cluff Bros., Toronto.

"Advertising is to business what steam is to machinery—the great propelling power."—Lord Macauley.

There are few men in the commercial world to-day who will dispute that statement. Successful business men realize that in this era of strenuous competition that judicious advertising is the great propelling force that makes large enterprises possible. It is no longer looked upon as a hazardous speculation or an expense—but as a safe, sure business investment. Money properly invested in advertising will just as certainly produce profits as money invested in any other line.

The subject under discussion, however, is how advertising effects the plumber and steam fitter. The average member of the trade undoubtedly believes in it. Its efficacy is a concrete fact which has been driven home through the success of many large advertisers, and perhaps there are some in his own town. It is likely that his next door neighbor, the grocer, gives generous publicity to his business; perhaps the clothier or hardwareman has made money through unstinted use of the local newspaper. He looked upon these examples as a matter of course and contents himself by saying, "My business is different." Different in some characteristics, yes; but in essential principles it is just the same. It is quite true the plumber cannot offer the special inducement of a sale of sugar, or of hats, or of saws, but he has goods to market just the same, and the quality of his workmanship is a feature to be advertised, the value of which few plumbers and steamfitters thoroughly appreciate.

In the naturally dull winter months our friends are satisfied to sit in their workshops and wait until the spring opens with its attendent building. Little do they realize that power of the art of increasing trade, of gaining prestige, good will and added profits by means of that merchant-idea of educating the public mind to a proper appreciation of the value of modern goods and excellent workmanship. The average sanitary or heating engineer is content to energetically pursue the contract for work on every new building in his neighborhood, and is quite satisfied if he secures 15 or 20 per cent. of all he bids upon. He is ever keen and alert to master every new feature which tends to enhance his mechanical knowledge, but how little study does he give the art of showing and selling his goods; how little does he realize the great possibilities of creating a demand for a modern sanitary or heating apparatus.

Let us cite an example of how this can be done. It is curiously true that the vast majority of owners of old homes are under the impression that the work of installing a modern hot water or steam heating apparatus can be accomplished successfully only in a new building where special provision is made for the purpose. If they understood the truth of the matter there is no doubt what they would do, and this work would not be the target of all the bids in that locality. It is true also that few people thoroughly appreciate the wonderful strides that have been made in the construction of warming apparatus during the past few years.

What are you doing to convince the layman of the great benefits of modern warming methods in which result added convenience, cleanliness, comfort, health and economy. What means are you using to attract the new-comer to your place of business? What efforts are you making to place your name and your business before the great consuming public.

We would make a passing word to those of the trade who attempt in some measure to advertise. It has been the experience of the writer that the great majority of plumbers merely fill their space with a business card and leave it there unchanged until they receive notice of the expiration of their contract. They are throwing their money away. A space 3 inches deep and double column wide shows up well and allows room to display.

A standing advertisement, be it ever so small, soon loses its power to attract, and it is quite necessary, therefore, that a frequent change be made each setting forth some single indisputable fact why you are "the plumber" or why a particular system of heating should be adopted. You must not forget that in inserting a business card in the newspaper, there are other plumbers and steamfitters whose names are just as euphonious as yours, and that merely telling people who you are will not sell goods for you. Change your advertisements frequently.

If you are not already an advertiser, invest $25 or $50 in your local paper, mailing cards, or in various other means which are so valuable an aid in stirring the public to a proper realization of the advantages of your business

MONTREAL M.P.A.

The annual election of the Master Plumbers' Association of Montreal held last week resulted as follows:

President, Ald. Nap. Turcot; 1st vice-president, John A. Gordon; 2nd vice-president, J. A. Marier; secretary, James E. Walsh; trustees, Jos. Thibeault, John Watson, David Ouimet.

The choice of the new president was a very popular one and caused general satisfaction, whilst a hearty vote of thanks was accorded the retiring president, P. C. Ogilvie.

PRESENTATION TO TRAVELER.

An interesting event took place las Saturday when E. A. Rogers, who was retiring from the position of city traveler for the James Robertson Company, Toronto, was entertained at dinner at the St. Charles Hotel, by H. S. Harwood, manager, Geo. F. Clare, cashier, and the traveling staff of the Toronto branch of the Robertson Company, he being presented with a beautiful gold locket and watch fob, engraved with the monogram of the recipient and the initials of the givers. The dinner was enlivened with many complimentary references to Mr. Rogers' eight years good work as traveler for the company and it evidenced a fine spirit of good fellowship, particularly as it is Mr. Rogers' intention to go into business for himself in Toronto in the plumbing supply line. Previous to his connection with the Robertson Company, Mr. Rogers was for eleven years on the road in Toronto and Ontario for the James Morrison Brass Company, Toronto.

A NEW FLEXIBLE STEAM PACKING.

The modern practice of using steam at high pressures and temperatures has brought with it certain problems which did not confront the engineer of ten years ago with his low-pressure plant. The chief of these problems is undoubtedly the question of suitable packings, a question which grows rapidly more serious as steam pressures are increased. Practically the only high-pressure packings hitherto available have been either metallic, hence rigid and unyielding, requiring constant regrindings, or of rubber, which is not absolutely impervious to steam, has not the necessary wearing qualities, and cannot withstand high temperatures. Recently, a new type of packing has been invented by Frederick M. Ekert, which seems to overcome the defects of previous packings. The material is very tough and tenacious, and it is sufficiently flexible or plastic to conform itself to all irregularities, thus absolutely preventing leakage. It is composed principally of rubber and asbestos fibres with which certain pore-filling substances are mixed. The packing is absolutely impervious to water or steam, and is a non-conductor of heat. Furthermore, it is self-lubricating, owing to the presence of graphite in its composition. It is made into valve disks, which will withstand any pressure up to 450 pounds continuous service, and also in sheets for use on pumps, cylinders, steam chest covers, manhole covers, and the like. In addition to these, a nickel-protected disk is made for superheated steam, which is adapted to withstand temperatures up to 900 deg. Fahrenheit.

E. A. ROGERS, TORONTO

Who has been calling on the Plumbing and Steamfitting trade in Ontario for the past twenty years, representing the James Morrison and James Robertson Companies

A similar material, in which cotton fibres are used in place of asbestos, Mr. Ekert provides for the manufacture of puncture-proof automobile tires, mattings and the like.

SASKATOON WATER WORKS.

Saskatoon is asking for tenders before March 20 for the following work:

Contract "A"—Labor on pipelaying for waterworks and sewerage comprising about 27,000 feet of trenching. Contract "C"—Concrete foundation for water tower housing for water supply. Contract "L"—Sedimentation basin.

STORING WATER IN HOUSES.

J. J. Cosgrove, author of "Principles and Practice of Plumbing," and an eminent sanitary authority, has the following to say in "Modern Sanitation" in relation to the construction of house tanks:

House tanks are used to store water for the supply of buildings and should be located at least 10 feet above the level of the highest fixture to be supplied. There are two kinds of tanks commonly used—wooden tanks and iron tanks. When located outside of buildings on roofs or in other exposed positions, wooden tanks are generally used; when located inside of buildings iron tanks are generally used. During warm weather moisture condenses on the outside of iron tanks and if not cared for will drip to the floor and wet both floor and ceiling below. To prevent this a drip pan should be placed under all iron tanks and a drip pipe from the pan extended to some convenient sink or connected to the overflow pipe from the tank.

Objections to Wooden Lined Tanks.

Lead-lined wooden tanks were formerly extensively used and in some localities are still, to a limited extent; but owing to the liability of carbonates or sulphates of lead being dissolved from the lining and poisoning the water, lead should not be used for tank linings, particularly in localities where the water is soft.

Copper-lined wooden tanks are sometimes used. From a chemical standpoint copper linings are not so objectionable as lead, particularly when the copper is tinned; however, copper linings present so many joints and seams that some of them are liable to leak, and, in some waters, soldered copper joints disintegrate owing to a chemical or galvanic action of the metals.

In extremely tall buildings fixtures on the lower floors are supplied with water direct from the street mains; the upper floors are supplied with water from the house tank on the roof, and intermediate tanks are installed, so that not more than eight floors of the building are supplied with water from any one tank. In such installations the house supply from the roof tank should be cross connected to the house supply from all the intermediate tanks and to the house supply for the lower floors, so that in case of necessity the entire building can be supplied with water from the house tank, which can be filled by pumping from the suction tank.

Size of Overflow Pipes.

Storage tanks should be provided with overflow pipes of sufficient capacity to safely carry off the greatest quantity of water likely to be discharged by the supply pipe. It is a safe rule to allow for the overflow pipe twice the diameter or four times the sectional area of the supply pipe. Overflow pipes from tanks located on roofs of buildings may discharge onto the roof. Overflow pipes from tanks located inside of buildings should discharge into a properly tapped and water-supplied sink. Under no circumstances should they connect directly to the drainage system.

The size of storage tanks depends upon the number of people to be supplied. They should have sufficient storage capacity for one day's supply to tide over possible periods of breakdown of pump or boiler. When figuring the capacity of storage tanks 100 gallons of water per day per capita should be allowed in

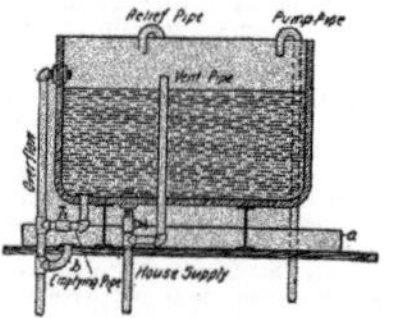

hotels, hospitals, apartment houses and public institutions.

Tank Connections.

The general arrangement of pipe connections to a house tank is shown in the accompanying engraving. The clean-out or emptying pipe is valved and connected to the overflow pipe. The house supply extends a few inches above the bottom of the tank to prevent sediment entering the pipe. Below the valve that controls the house supply is connected a vent pipe to admit air to the house supply and permit it to empty when the valve is shut off. A vapor or relief pipe from the highest point in the hot water supply system bends over the tank and thus permits the escape of steam. The pump may discharge into the house tank in the manner indicated when the pump is not controlled automatically. When it is, the pump pipe should enter the tank through the bottom and be controlled by a balanced float valve. A drip pan, (a), under the tank and extending a few inches on all sides of it catches the water condensation and discharges it through the waste pipe, (b), into the overflow pipe. When a tank is supplied with water by a pump that is not automatic in operation a telltale pipe should run from a point in the tank about 2 inches below the level of the overflow pipe to the engineer's sink. Water flowing through the pipe then notifies the engineer when the tank is full.

PLUMBING MARKETS

Office of HARDWARE AND METAL,
232 McGill Street,
Montreal, February 22, 1907

Nothing startling has happened this week in plumbing circles. The volume of business done this year, has far surpassed anything for the same period of previous years.

For the past while everything that is used by the plumber has been steadily advancing. Even the wood used for closet seats, cisterns, etc., is no exception, and plumbers are buying goods very freely in order to be on the safe side, should raw material go higher still. When asked by our representative when things were going to stop going up, one prominent jobber said, "It's hard to say. Different people take different views of the situation. Personally, I don't think it's due to manipulation, so much as increasing demand, but there's no saying. The trusts could, no doubt, by making the output a trifle short of the demand, boost up prices, but the shortage in many lines is no doubt due to the tremendous demand of a growing continent.

RANGE BOILERS—An excellent business is being done in this line. We still quote: Iron clad, 30 gal., $5; 40 gal., $6.50 net list. Copper, 30 gal., $33; 35 gal., $38; 40 gal., $43.

LEAD PIPE — Still firm. No change in prices. We still quote: 5 per cent. discount f.o.b. Montreal, Toronto, St. John, N.B., Halifax; f.o.b. London, 15c. per hundred lbs. extra; f.o.b. Hamilton, 10c. per hundred lbs. extra.

IRON PIPE FITTINGS—A shortage is reported in some lines, but we still quote: Discounts on nipples, ¼ inch to 3 inch, 65 per cent., 3½ to 2 inch, 67½ per cent.

IRON PIPE—Some lines are reported short, and it is expected that they will remain so. Our prices remain: Standard pipe in lots of 100 feet, regular lengths, ¼ inch, $5.50; ⅜ inch, $5.50; ½ inch, $8.50; ¾ inch, $11.50; 1 inch, $16.50; 1¼ inch, $22.50; 1½ inch, $27.; 2 inches, $36.50. Discounts on black pipe: ¼ inch, 59 per cent.; ⅜ inch, 59 per cent.; ½ inch, 68; ¾ to 2, 70. Discounts on galvanized pipe: ¼ inch, 44 per cent.; ⅜ inch, 44 per cent.; ½ inch, 58 per cent.; ¾ to 2 inch, 60 per cent. Extra heavy pipe of 100 feet lots are quoted as follows: ½ inch, $12; ¾ inch, $15; 1 inch, $22; 1¼ inch, $30, 1½ inch, $36; 2 inches, $50. The discounts on black pipe: ½ inch, 74 per cent.; ¾ inch, 69 per cent.; ½ inch to 2 inches, 68 per cent. Galvanized: ½ inch, 59 per cent.; ¾ inch, 69 per cent.; ½ to 2 inches. 58 per cent.

SOIL PIPE FITTINGS — Prices remain: Standard soil pipe, 50 per cent. off list. Standard fittings, 50 and 10 per cent. off list; medium and extra heavy soil pipe, 60 per cent. off. Fittings, 60 per cent. off.

SOLDER — No change in prices. We still quote: Bar solder, half-and-half, guaranteed, 25c.; No. 2 wiping solder, 22c.

ENAMELWARE — We quote: Canadian baths, see Jan. 3, 1907, list. Lavatories, discounts, 1st quality, 30 per cent.; special, 30 and 10 per cent. Sinks, 18x30 inch, flat rim, 1st quality, $2.60; special, $2.45.

TORONTO.

Office of HARDWARE AND METAL,
10 Front Street East.
Toronto, February 22, 1907

A peculiar condition exists this week in that we are forced to revise our quotations on plumbers' brass goods, reducing the price on account of the keen competition for business. In spite of the continued high quotations on copper and the almost certainty that prices will go even higher it appears that recent advances on brass goods have been too large and the discounts have not been firmly maintained, some houses frankly admitting that in order to get business they have had to increase their discounts. The changes we note this week include an extra 5 per cent. on compression and fuller work and both flatway and roundway goods.

The shortage on iron pipe continues, and with the opening of the season close at hand stocks of pipe are almost bare. This situation presages even worse conditions than existed last summer and the firms which have placed their business early will be the best off in the long run.

In the States the opinion is expressed that pig iron has passed its high point but as all the furnaces are booked ahead until about November future business is all that will be affected. This may make ie easier to procure skelp for iron pipe in the fall.

Trade in other lines remains normal for this season, plus a fair percentage of booked orders placed by far-seeing firms which do not want to be delayed in completing contracts secured.

LEAD PIPE—Business continues fair for this season. The discount continues at 5 per cent. off the list price of 7c. per pound. Lead waste, 8c. per pound, with 5 off. Caulking lead, 5¾c. to 6c. per pound. Traps and bends, 40 per cent. discount.

SOIL PIPE AND FITTINGS—Many orders are being placed for future delivery. We still quote: Medium and extra heavy pipe and fittings, 60 per cent.; light pipe, 50 per cent.; light fittings, 50 and 10 per cent.; 7 and 8 inch pipe, 40 and 5 per cent.

IRON PIPE—Stocks are remarkably low for so close to the opening of the building season. One inch black pipe is quoted at $4.95, and 1 inch galvanized at $6.60. Full list in current market quotations.

IRON PIPE FITTINGS—Plenty of fittings are available and business is being placed for future delivery. We now quote: Cast iron fittings, 57½ per cent.; cast iron plugs and bushings, 60 per cent.; flange unions, 60 per cent.; nipples, 70 and 10 per cent.; iron cocks, 55 and 5 per cent.; Canadian malleable, 30 per cent.; malleable unions, 55 and 5 per cent.; malleable bushings, 55 per cent.; cast iron ceiling plates, plain, 65 per cent.; cast iron floor, 70 per cent.; hook plates, 60 per cent.; expansion plates, 65 per cent.; headers, 60 per cent.; hangers, 65 per cent.; standard list.

RANGE BOILERS — A fair amount of business is being placed. We quote: Galvanized iron, 30 gallon, standard, $5; extra heavy, $7.75; 35 gallon, standard, $6; extra heavy, $8.75; 40 gallon, standard, $7; 40 gallon, extra heavy, $9.75, net list. Copper range boilers—New lists quote: 30 gallon, $33; 35 gallon, $38; 40 gallon $43. Discounts 5 to 15 per cent.

RADIATORS—At this season buying is not active. Prices continue unchanged at: Hot water, 47½ per cent.; steam, 50 per cent.; wall radiators, 45 per cent.; specials, 45 per cent. Hot water boilers are unchanged.

SOLDER — We quote: Bar solder, half-and-half, guaranteed, 27c.; wiping, 23c.

ENAMELWARE — No changes have been made, with the January list still prevailing on Canadian baths. We quote: Lavatories, first quality, 20 and 5 to 20 and 10 off; special, 20 and 10 to 30 and 2½ per cent. discount. Kitchen sinks, plate 300, firsts, 60 and 10 off; specials, 65 and 5 per cent. Urinals and range closets, 15 off. Fittings extra.

PLUMBING PARAGRAPHS.

Somerville, Limited, Toronto, are to spend $100,000 in establishing a new plumbers' brass goods plant in Toronto. The plans show a factory 100x150 and work will be commenced shortly. A separate company will not be formed as originally announced.

The employes of Purdy, Mansell, and Co., Toronto, held their 18th annual sleighing party and banquet to East Toronto last Friday night.

The Ottawa Plumbers' and Steamfitters' Union, have sent a written request to the Master Plumbers' Association of Ottawa, asking for an increase in wages and for Saturday afternoons off the year round. They ask that this change go into force on May 1st.

To Manufacturers' Agents

HARDWARE AND METAL has inquiries from time to time from manufacturers and others wanting representatives in the leading business centres here and abroad.

Firms or individuals open for agencies in Canada or abroad may have their names and addresses placed on a special list kept for the information of inquirers in our various offices throughout Canada and in Great Britain without charge.

Address
Business Manager
HARDWARE AND METAL
Montreal and Toronto

This is the Mark that Means Something

It is the distinctive brand of genuine Armstrong Stocks and Die and Pipe Fitters' Tools.

No "Armstrong pattern" or "just as good as Armstrong" have this stamp. Don't buy anything in this line that hasn't this brand as it is the mark of genuineness.

THE ARMSTRONG MANUFACTURING CO.,
292 Knowton Street, Bridgeport, Conn.

A Certain Sense of Satisfaction

goes with every MERRELL PORTABLE HAND MACHINE, PIPE THREADING and CUTTING MACHINE. We build them that way.

For doing ALL kinds of threading—under ALL conditions—labor trouble and faulty material included—where speed, durability and ease of control count for something, our word to you is to investigate the MERRELL.

Whether it be stationary or portable—hand or power driven—30 DAYS FREE TRIAL must prove conclusively that your choice should be the MERRELL.

This Portable Hand Machine has encased gears, MERRELL standard quick opening and closing die-head, and the latest improved Cutting-off Knife.

Let us tell you more about this machine—The Chasers, Vise and the large range of work covered.

Catalogue for the asking.
Quick shipments.

THE CANADIAN FAIRBANKS CO.
Sole Agents for Canada Limited
MONTREAL, TORONTO, WINNIPEG, VANCOUVER

When I stop talking about the quality of my goods I start talking about prompt deliveries. Then I demonstrate both quality and promptitude, by filling the **Enamelware Wants** of any man, anywhere. Send on your order.

ORLANDO VICKERY

178-180 Victoria Street *TORONTO, CANADA*

LASTING FITTINGS

Every practical man, capable of delivering an unbiased opinion, declares that

"Diamond" brand Fittings

are the most durable ever placed on the market. Prove the statement by sending in that test order to-day.

OSHAWA STEAM & GAS FITTINGS CO.,
OSHAWA, CANADA LIMITED

Do You Want the Best?

TWO FREE
for every one sent you defective.
Isn't that guarantee enough?

Glauber Brass Mfg. Co.
Expert Makers of Fine Brass Goods
CLEVELAND - - OHIO

Sharratt & Newth's Glaziers' Diamonds

are unequalled for cutting and wearing qualities.

To be obtained from the principal Dealers in Glass, Hardware, and Painters' Supplies.

Contractors to H. M. Government and the principal English Sheet and Plate Glass Works

World's Best Cutlery

You make no mistake in specifying goods made by

JOSEPH RODGERS & SONS, Limited

when ordering cutlery. A reputation of over 200 years at stake, insures the quality of every article.

Write for catalogues from Sole Canadian Agents

JAMES HUTTON & COMPANY, MONTREAL

Forwell Foundry Co.

BERLIN, ONT.

Manufacturers of

SOIL PIPE, FITTINGS, and CAST IRON SINKS

Ask Jobbers for "F. F. CO." Brand

The Daisy Way

Of hot-water heating has been the subject of our conversation with you for a good many years. We are still talking of the Daisy Boiler and have convinced the vast majority of our listeners that it is the boiler to handle.

CLUFF BROTHERS

Lombard Street, TORONTO

Selling Agents for - - - WARDEN KING & SON, Limited

Satisfactory Oil Storage—The Proof

THE RUSSILL HARDWARE CO.
126 E. King St.

Toronto, Can., January 3, 1907

Messrs. S. F. Bowser & Co.
66-68 Fraser Ave.
Toronto, Canada

Dear Sirs:—

Just five years ago, you installed for us ten of your Bowser Self Measuring Oil Tanks, the aggregate gallonage of same amounting to twelve hundred gallons.

We wish to express the unqualified satisfaction which the outfit has given us, and we feel that it would be absolutely impossible for us to handle the volume of oil business which we enjoy without the assistance of such an outfit.

We further consider, the money expended for the purchase of same was the very best investment ever made by us in the way of store fittings.

Trusting that your goods are meeting with the success which their merit entitles them to, we are.

Yours truly,

THE RUSSILL HARDWARE CO.
Per Frank Russill, Mgr.

Catalogue V describes the Bowser Outfits which do the work. Sent free on request.

S. F. BOWSER & CO., Inc.

66-68 Fraser Ave. Toronto, Can.

We have pumps suitable to any demand made upon you by most exacting customers.

Our pumps will do anything that a pump can do, and we will be pleased to give the benefit of many years' experience to any one having pump troubles.

McDougall Pumps are Standard Pumps, and with us it is quality first, last and always, while we ask just as little money as we can get along with.

We have a few catalogues for distribution

The R. McDougall Co., Limited

CALT, - CANADA

Oakey's Emery,
Oakey's Emery Cloth,
Holley Carburetors,
Gas Engine Ignition Supplies,
Fixtures and Shades, etc.

WRITE FOR PRICES

JOHN FORMAN

248 and 250 Craig St. W., - MONTREAL, QUE.

HOTTEST ON EARTH

Our Bonanza Furnaces are known to produce more heat with less fuel than any other make. Jobbers sell at factory prices. Satisfaction guaranteed or money refunded.

No. 38—Price, each, ***$7.80*** net
No. 39—One size smaller, ***$5.85*** net

THE TURNER BRASS WORKS

53 MICHIGAN STREET, - CHICAGO, U.S.A.

Manufacturers' Agents

CHARLES H. FOX

Vancouver, B.C.

MANUFACTURERS' AGENT

Selling to the Hardware Jobbers Only.

Correspondence Solicited.

ALEXANDER GIBB

Manufacturers' Agent and Metal Broker, 13 St. John Street, Montreal

Representing Canadian, British and American Manufacturers. Correspondence invited from firms wishing to be represented.

H. W. MITCHELL

WINNIPEG, MAN.

Correspondence is solicited from manufacturers wishing a representative in Winnipeg.

Travellers call regularly on the wholesale and retail trade in Western Canada.

Highest references and financial responsibility.

Scott, Bathgate & Co.

BROKERS AND COMMISSION MERCHANTS

Cor. Notre Dame St. East and Victoria St., WINNIPEG, Canada

Storage and Warehousing a Specialty.

First-class facilities in a thoroughly modern brick warehouse centrally located on the railway track.

Babbitt, Solder, Lead Pipe, Etc.

The
Canada Metal Co.
Limited
TORONTO

"Standard" Plate P-851

PARAGON "STANDARD" Enameled Washout Closet—adapted for work requiring least possible outlay, but demanding a thoroughly sanitary and effective fixture. Enameled inside and outside with 2-inch Brass Flush Connection and Improved Roll-Top flushing rim. Woods and fittings Quartered Oak, Cherry, Walnut, High Dutch Oak, English Oak, Ox Blood, Bird's Eye Maple, White Enamel and Mahogany.

"Standard" FIXTURES GIVE PERMANENT SATISFACTION

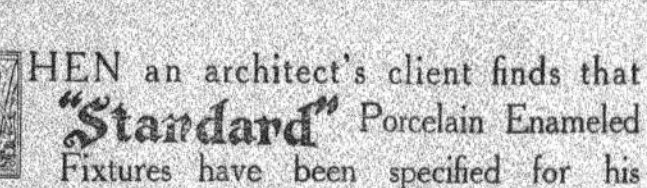

WHEN an architect's client finds that "Standard" Porcelain Enameled Fixtures have been specified for his property he is pleased, because he has been taught by actual experiences to regard "Standard" Fixtures as the most beautiful, the most sanitary and the most satisfactory equipment that he can secure. The Architect on the other hand knows that his customer has the right impression of "Standard" fixtures and that the test of use will increase his satisfaction.

Standard Sanitary Mfg. Co.
PITTSBURGH, PA.

Buffalo Cupola Blowers

For All High Pressure Blast Service

B VOLUME BLOWERS FOR LARGE QUANTITIES OF AIR AT MODERATE PRESSURE

"The **running record** of a blower embraces its **effectiveness**, its **consumption of power**, its **maintenance cost**, and its **period of usefulness**, and should be the determining factor in its choice."

A lively appreciation of these requirements has made **BUFFALO BLOWERS** unequalled for the

ECONOMIC HANDLING OF AIR UNDER ANY CONDITIONS

BUFFALO CUPOLA BLOWER

MADE IN CANADA BY

THE CANADIAN BUFFALO FORGE CO., Limited, Montreal, Canada

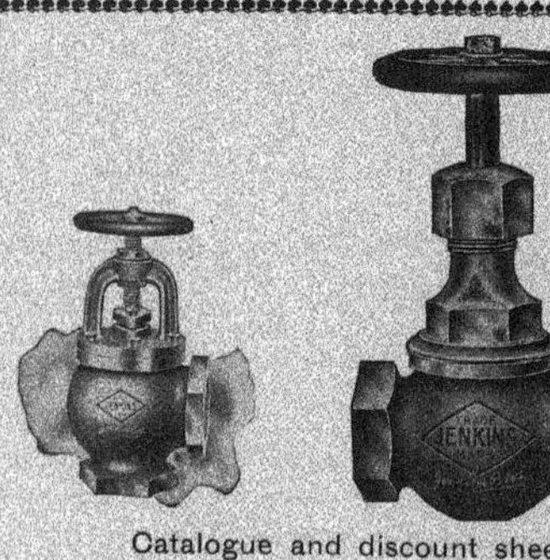

Catalogue and discount sheet sent on application.

JENKINS BROS.

35 High Street, Boston, Mass.

New York Philadelphia Chicago London, E. C.

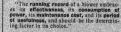

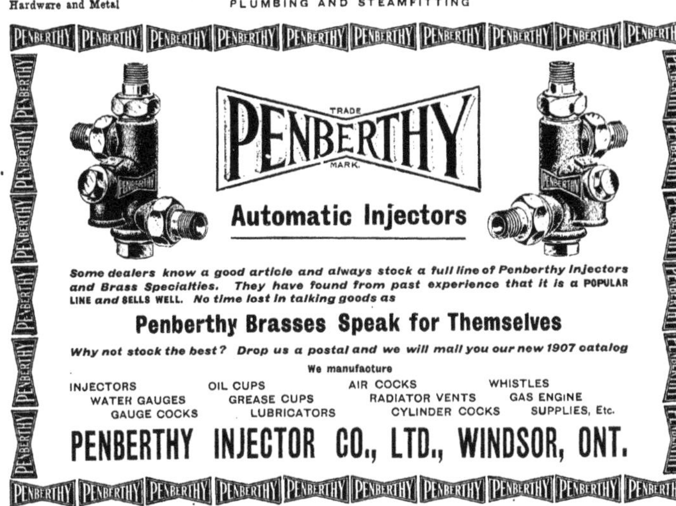
PENBERTHY
TRADE PENBERTHY MARK.
Automatic Injectors
Some dealers know a good article and always stock a full line of Penberthy Injectors and Brass Specialties. They have found from past experience that it is a POPULAR LINE and SELLS WELL. No time lost in talking goods as
Penberthy Brasses Speak for Themselves
Why not stock the best? Drop us a postal and we will mail you our new 1907 catalog
We manufacture
INJECTORS OIL CUPS AIR COCKS WHISTLES
WATER GAUGES GREASE CUPS RADIATOR VENTS GAS ENGINE
GAUGE COCKS LUBRICATORS CYLINDER COCKS SUPPLIES, Etc.
PENBERTHY INJECTOR CO., LTD., WINDSOR, ONT.

CURRENT MARKET QUOTATIONS.

Feb 22, 1907.

These prices are for such qualities and quantities as are usually ordered by retail dealers on the usual terms of credit, the lowest figures being for larger quantities and prompt pay. Large cash buyers can frequently make purchases at better prices. The Editor is anxious to be informed at once of any apparent errors in this list, as the desire is to make it perfectly accurate.

METALS.

ANTIMONY.

Cookson's ....per lb... 0 27½ 0 28

BOILER AND T K. PITTS.

Plain tinned......... } 25 per cent off list.
Spun.................. }

BABBIT METAL.

Canada Metal Company—Imperial, genuine 60c; Imperial Tough, 60c.; White Brass, 50c; Metallic, 35c.; Harris Heavy Pressure, 25c.; Hercules, 25c. White Bronze, 15c., Star Frictionless, 14c.; Alluminoid, 10c.; No. 4, 9c. per lb.

James Robertson Co.—Extra and genuine Monarch, 60c, Crown Monarch, 50c., No 1 Monarch, 40c; King, 30c; Fleur-de-lis, 20c; Thurber, 15c. Philadelphia, 12c.; Canadian, 10c; hardware, No. 1, 15c.; No. 2, 12c; No. 3, 10c per lb

BRASS.

Rod and Sheet, 14 to 30 gauge, 25 p c advance

Sheets, 12 to 14 in.	0 30
Tubing, base, per lb 5-16 to 2 in	0 33
Tubing, 1 to 3-inch, iron pipe size..	0 31
" 1 to 3-inch, seamless......	0 36

Copper tubing, 6 cents extra

COPPER.

Ingot.

		Per 100 lb.
Casting, car lots..........	26 50	27 00

Bars.

Cut lengths, round, ½ to 2 in...	....	35 00

Sheet.

Plain, 16 oz., 14x48 and 14x60	....	35 00
Plain, 14 oz...........	....	36 00
Tinned copper sheet, base	....	38 00
Planished base..............	....	39 00
Braziers' (in sheets), 4x6 ft., 25 to 30 lb. each, per lb., base..	....	0 35

BLACK SHEETS.

	Montreal.	Toronto
8 to 10 gauge	2 70	2 70
12 gauge.........	2 70	2 80
14 "	2 70	2 55
17 "	2 50	2 65
18 "	2 50	2 65
20 "	2 50	2 65
22 "	2 50	2 75
24 "	2 55	2 80
26 "	2 65	2 85
28 "	2 70	3 00

CANADA PLATES

Ordinary, 52 sheets	2 75	2 90
All bright "	3 80	3 90
Galvanized, 52 sheets	4 35	4 45
" 60 "	4 60	4 70

	Ordinary.	Dom. Crown
18x24x52	4 25	4 35
" 60........	4 50	4 60
20x28x80	8 50	8 70
" 94	9 00	9 20

GALVANIZED SHEETS

	Fleur-de-Lis.	Gordon Crown.	Colborne Crown.
16 to 20 gauge..	3 80	3 95	
22 to 24 gauge..	4 05	4 00	3 75
26 " ..	4 30	4 20	4 30
28 " ..	4 55	4 50	4 45

Apollo

10¾ oz. (American gauge)........	....	4 85
24 gauge........ "	4 45	4 55
26 " .... "	4 15	4 30
4 " .... "	3 85	3 90

	Gorbal's Best.	Comet.	Queen's Head.	Bell.
16 to 20 gauge	....	3 60	3 95	....
22 to 24 gauge	....	3 75	4 20	....
26 "	..	4 30	4 45	....
28 "	4 55	4 45	4 70	4 45

Less than case lots 10 to 25c. extra.

IRON AND STEEL.

	Montreal.	Toronto
Common bar, per 100 lb......	2 15	2 30
Forged iron "	2 40	..
Refined " "	2 55	2 70
Horseshoe iron "	2 55	2 70
Hoop steel, 1½ to 3 in. base...	....	2 80
Sleigh shoe steel "	2 25	2 30
Tire steel	2 40	2 50
Best sheet cast steel.........	...	0 12
B. K. Morton "Alpha" high speed.		0 65
" annealed	....	0 70
"M" Self-hardening	....	0 60
"J" quality, best warranted	....	0 18
"I" " warranted......	....	0 14
"B C" quality	....	0 09
Jonas & Colver's tool steel....	0 10	0 20
" "Novo"......	....	0 65
" " annealed	....	0 65
Jowett & Sons B P.L. tool steel....		0 10½

INGOT TIN.

Lamb and Flag and Straits—

56 and 28-lb. ingots, 100 lb.	$45 00	$45 50

TINPLATES.

Charcoal Plates—Bright

			Per box.
M.L.S., equal to Bradley—			
I C, 14 x 20 base		....	$6 50
I X, 14 x 20 "		....	8 00
I X X, 14 x 20 base		....	9 50
Famous, equal to Bradley—			
I C, 14 x 20 base........		....	6 50
I X, 14 x 20 "		....	8 00
I X X, 14 x 20 base........		....	9 50
Raven and Vulture Grades—			
I C, 14 x 20 base		....	5 00
I X "		....	6 00
I X X "		....	7 00
I X X X "		....	8 00
"Dominion Crown Best"—Double Coated, Tissued.			Per box.
I C, 14 x 20 base........		5 50	5 75
I X, 14 x 20 "		6 50	6 75
I X X x 20 "		7 50	7 75
"Allaway's Best"—Standard Quality.			
I C, 14 x 20 base.....		....	4 50
I X, 14 x 20 "		....	5 25
I X X, 14 x 20 "		....	6 00

Bright Cokes.

Bessemer Steel—		
I C, 14 x 20 base	...	4 25
20x28, double box	....	8 50

Charcoal Plates—Terne

Dean or J G. Grade—		
I C., 20x28, 112 sheets	7 25	8 00
IX., Terne Tin	...	9 50

Charcoal Tin Boiler Plates.

Cookley Grade— X X, 14x56, 50 sheet bxs. " 14x60, " " 14x65,	....	7 50

Tinned Sheets.

72x30 up to 24 gauge...... ...	....	9 50
" " 26 "	....	9 00

LEAD.

Imported Pig, per 100 lb......	5 50	5 60
Bar, "	5 75	6 00
Sheets, 2½ lb. sq. ft., by roll ..	....	0 07½
Sheets, 3 to 6 lb "	....	0 07

NOTE.—Cut sheets ½c. per lb., extra. Pipe, by the roll, usual weights per yard, lists at 7c. per lb. and 5 p.c. dis f.o.b. Toronto.

NOTE.—Cut lengths, net price, waste pipe 8-ft lengths, lists at 8c.

SHEET ZINC.

5-cwt. casks	8 00	8 25
Part casks..............	8 25	8 50

ZINC SPELTER.

Foreign, per 100 lb	7 50	7 75
Domestic	7 00	7 25

PLUMBING AND HEATING

BRASS GOODS, VALVES, ETC.

Standard Compression work, dis. half inch 50 per cent., others 50 per cent.
Cushion work, discount 40 per cent.
Fuller work, ½-inch 65 per cent., others 55 p.c.
Flatway stop and stop and waste cocks, 55 per cent; roundway, 45 per cent.
J.M.T. Globe, Angle and Check Valves, discount 45 per cent.
Standard Globe, Angle and Check Valves discount 52½ per cent.
Kerr standard globes, angles and checks, special, 42½ per cent.; standard, 47½ p.c.
Kerr Jenkins' disc, copper-alloy disc and heavy standard valves, 45 per cent.
Kerr steam radiator valves, 60 p.c., and quick-opening hot-water radiator valves, 60 p.c.
Kerr brass, Weber's straightway valves, 40 per cent.; straightway valves, I B B.M, 60 per cent.
J. M. T. Radiator Valves, discount 50 per cent.
Standard Radiator Valves, 60 per cent.
Patent Quick-Opening Valves, 65 per cent.
Jenkins' Bros. Globe Angle and Check Valves discount 27½ per cent.

No. 1 compression bath cock......	net	2 00
No. 4 "	"	1 90
No. 7 Fuller's..............	"	2 25
No. 4½ "	"	2 35

Patent Compression Cushion, basin cock, hot and cold, per doz., $16.20
Patent Compression Cushion, bath cock, No. 2208 .... 2 25
Square head brass cocks, 50 per cent.; " iron " 60 "
Thompson Smoke-test Machine 25.00

BOILERS—COPPER RANGE.

Copper, 30 gallon........	......	33 00
" 35 "	......	38 00
" 40 "	......	43 00

15 per cent

BOILERS—GALVANIZED IRON RANGE.

Capacity	Standard.	Extra heavy
30-gallons ...	5 00	7 75
35 "	6.00	8 75
40 "	7.00	9 75

2 per cent, 30 days

BATH TUBS.

Steel clad copper lined, 15 per cent.

CAST IRON SINKS.

16x24, $1, 18x30, $1; 18x36, $1.30.

ENAMELED CLOSETS AND URINALS

Discount 15 per cent.

ENAMELED BATHS.

List issued by the Standard Ideal Company Jan 3, 1907, shows an advance of 10 per cent over previous quotations.

ENAMELED LAVATORIES.

	1st quality.	Special.
Plate E 100 to E 103 ..	20 & 5 p c.	20 & 10 p.c.
" E 104 to E 132....	20 & 10 p c.	30 & 2½ p.c

ENAMELED ROLL RIM SINKS.

	1st quality.	Special.
Plate E 201, one piece.	15 & 2½ p c.	15 & 10 p c.

ENAMELED KITCHEN SINKS.

Plate E, flat rim 300, 60 & 10 p c. 65 & 5 p.c

HEATING APPARATUS

Stoves and Ranges—Discounts vary from 40 to 70 per cent according to list.
Furnaces—45 per cent
Registers—70 per cent
Hot Water Boilers—Discounts vary.
Hot Water Radiators—47½ per cent.
Steam Radiators—50 per cent
Wall Radiators and Specials—45 per cent.

LEAD PIPE

Lead Pipe, 7c. per pound, 5 per cent. off.
Lead waste, 8c. per pound, 5 per cent. off.
Caulking lead, 6c. per pound.
Traps and bends, 40 per cent.

IRON PIPE.

Size (per 100 ft.)	Black.		Galvanized
⅛ inch........	2 25	⅛ inch........	3 08
¼ "	2 25	¼ "	3 08
⅜ "	2 72	⅜ "	3 57
½ "	3 45	½ "	4 80
¾ "	4 95	¾ "	6 60
1 "	6 75	1 "	9 00
1¼ "	8 10	1¼ "	10 80
1½ "	10 80	1½ "	14 40
2 "	18 00	2 "	24 00
2½ "	23 50	2½ "	31 50
3 "	30 00	3 "	40 00
3½ "	34 00	3½ "	45 00

2 per cent. 30 days.

Malleable Fittings—Canadian discount 30 per cent.; American discount 25 per cent.
Cast Iron Fittings 57½; Standard bushings 57½ per cent; headers, 57½; flanged unions 57½, malleable bushings 55; nipples, 70 and 10 per cent.; malleable lipped unions, 55 and 5 per cent.

SOIL PIPE AND FITTINGS

Medium and Extra heavy pipe and fittings, up to 6 inch, discount 60 per cent.
7 and 8-in. pipe, discount 40 and 5 per cent.
Light pipe, 50 p c.; fittings, 50 and 10 p.c.

OAKUM.

Plumbers ....per 100 lb... .. .... 4 75

STOCKS AND DIES.

American discount 25 per cent.

SOLDERING IRONS.

½-lb. .... per lb.	....	0 38
1-lb. or over	....	0 35

SOLDER.

	Per lb. Montreal.	Toronto
Bar, half-and-half, guaranteed	0 25	0 27
Wiping................	0 22	0 23

PAINTS, OILS AND GLASS.

BRUSHES

Paint and household, 70 per cent.

CHEMICALS.

	In casks per lb
Sulphate of copper (bluestone or blue vitriol)	0 09
Litharge, ground	0 06
" flaked..............	0 06½
Green copperas (green vitriol)......	0 01
Sugar of lead................	0 09
Lump alum..................	0 01½

COLORS IN OIL.

Venetian red, 1-lb. tins pure.	....	0 08
Chrome yellow "	....	0 15
Golden ochre "	....	0 08
French " "	....	0 06
Marine black "	....	0 04½
Chrome green "	....	0 10
French permanent green "	....	0 13
Signwriters' black "	....	0 15

GLUE.

Domestic sheet	0 10	0 10½
French medal..............	0 12	0 12½

PARIS GREEN.

	Berger's	Canadian
600-lb. casks	0 23½	0 25½
250 lb. drums	0 25½	0 25½
100-lb. "	0 26½	....
50-lb. "	0 26½	0 26½
1-lb. pkgs, 100 in box	0 27½	0 27½
½-lb. "	0 29½	0 29½
1-lb. tins, 100 in box	0 28½	0 28½
½-lb. pkg	0 30½	0 30

F.o.b. Toronto.

PARIS WHITE.

In bbls 0 90

PREPARED PAINTS.

'ure, per gallon, in tins .... 1 3[illegible]
Second qualities, per gallon 1 10
Barn (in bbls.) 0 85 0 90
Sherwin-Williams paints, 1 gal. 1 40
Canada Paint Co.'s pure 1 30
Standard P. & V. Co.'s "New Era." 1 30
Benj. Moore Co.'s "Ark" B'd 1 25
" British Navy deck 1 50
Brandram-Henderson's "English" 1 45
Ramsay's paints, Pure, per gal. 1 20
" Thistle, " 1 00
" Outside, bbls 0 55 0 65
Martin-Senour's 100 p.c. pure, ½ gal. 1 50
" " " " ¼ gal. 1 45
" " " " 1 gal. 1 40
Senour's Floor Paints .....gal. 1 20
Jamieson's "Crown and Anchor" 1 20
Jamieson's floor enamel 1 50
" barn paints, bbls, per gal. 0 60
Sanderson Pearcy's, pure 1 20
Robertson's pure paints 1 20

PUTTY.

Bulk in bbls 1 50
Bulk in less quantity 1 80
Bladders in bbls 1 80
Bladders in kegs, boxes or loose 1 95
25-lb. tins 1 80
12½ lb. tins 2 05
Bladders in bulk or tins less than 100 lb. 1 85
Bulk in 100-lb. irons 1 80

SHINGLE STAINS.

In 5-gallon lots 0 75 0 80

SHELLAC.

White 0 65
Fine orange 0 60
Medium orange 0 55
F.o.b Montreal or Toronto.

TURPENTINE AND OIL.

Castor oil 0 08 0 10
Gasoline 0 27½
Benzine, per gal 0 20
Turpentine, single barrels 0 98 1 05
Linseed Oil, " raw.. 0 55 0 59
" " boiled 0 58 0 62
W. H. & C's English raw oil, bbls.. .. 0 63
" " boiled, " 0 66

WHITE LEAD GROUND IN OIL. Per 100 lbs

Canadian pure 7 15
No. 1 Canadian 6 80
Munro's Select Flake White.. .... 7 40
Elephant and Decorators' Pure .... 7 4[illegible]
Monarch 7 40
Standard Decorator's 7 15
Essex Genuine 6 80
Brandram's B. B. Genuine... .... 8 40
"Anchor," pure 7 00
Ramsay's Pure Lead 6 40
Ramsay's Exterior 6 15
"Crown and Anchor," pure... .... 6 50
Sanderson Pearcy's 7 40
Robertson's O P., lead 7 20
W.H.& C's matured pure English 8 00 8 25

WHITE AND RED DRY LEAD.

	white	red.
Genuine, 560 lb. casks, per cwt	6 75	6 00
Genuine, 100 lb. kegs, "	7 50	6 50
No. 1, 560 lb. casks, per cwt ...	6 25	5 75
No. 1, 100 lb. kegs, per cwt ..	7 00	6 25

WINDOW GLASS.

Size United Inches.	Star	Double Diamond
Under 26	$4 25	$6 25
26 to 40	4 65	6 75
41 to 50	5 10	7 50
51 to 60	5 35	8 50
61 to 70	5 75	9 75
71 to 80	6 25	11 00
81 to 85	7 00	12 50
86 to 90	....	15 00
91 to 95	....	17 50
96 to 100	....	20 50
101 to 105	....	24 00
106 to 110	....	27 50

Discount—16-oz, 25 per cent.; 21-oz 30 per cent. per 100 feet. Broken boxes 50 per cent. off list.

WHITING.

Plain, in bbls 0 60
Gilders bolted in bands .... 0 90.

WHITE DRY ZINC

Extra Red Seal, V M 0 07½ 0 08½

WHITE GROUND ZINC.

Pure, in 25-lb. irons 0 08½
No. 1, " " 0 07
No. 2, " " 0 05½

VARNISHES.

In 5-gal. lots. Per gal cans.

Carriage, No. 1 .... 1 50
Pale durable body 3 50
" hard rubbing 3 00
Finest elastic gearing 3 00
Elastic oak 1 50
Furniture, polishing 2 00
Furniture, extra 1 15
" " No 1 .... 0 90
" " union 0 80
Light oil finish 1 40
Gold size japan 1 80
Brown japan 0 95
No. 1 brown japan 0 90
Baking black japan 1 35
No. 1 black japan 0 90
Benzine black japan 0 70
Crystal Damar 2 80
No. 1 " 2 50
Pure asphaltum 1 40
Oilcloth 1 50
Lightning dryer 0
Elastilite varnish, 1 gal. can, each.. 2
Granitine floor varnish, per gal 2
Maple Leaf coach enamels; size 1, 1
Sherwin-Williams' kopal varnish, gal., 2
Canada Paint Co's sun varnish 2
"Kyanize" Interior Finish 2
"Flint-Lac," coach 1
B H Co's "Gold Medal," in cases 2 70
Jamieson's Copaline, per gal. .... 2 [illegible]

BUILDERS' HARDWARE.

BELLS.

Brass hand bells, 60 per cent.
Nickel, 55 per cent.
Gongs, Sargeant's door bells.. 5 50 8 00
American, house bells, per lb. 0 35 0 40
Peterboro' door bells, discount 37½ and 10 per cent. off new list.

BUILDING PAPER, ETC.

Tarred Felt, per 100 lb. 2 25
Ready roofing, 2-ply, not under 45 lb. per roll 1 00
Ready roofing, 3-ply, not under 65 lb., per roll 1 25
Carpet Felt ....per ton 60 00
Heavy Straw Sheathing ....per ton 35 00
Dry Surprise 0 40
Dry Sheathing .....per roll, 400 sq. ft. 0 50
Tar " ..... " 400 " 0 45
Dry Fibre .... " 400 " 0 55
Tarred Fibre ...... " 400 " 0 65
O. K. & I. X. L .... " 400 " 0 70
Resin-sized .... " 400 " 0 45
Oiled Sheathing .... " 600 " 1 00
Oiled " .... " 400 " 0 70
Root Coating, in barrels.. ...per gal. 0 17
Roof " small packages " 0 25
Refined Tar ....per barrel 5 00
Coal Tar .... " 4 00
Coal Tar, less than barrels ...per gal. 0 15
Roofing Pitch ....per 100 lb. 0 80 0 90
Slater's felt ....per roll 0 70
Heavy Straw Sheathing f o. b. St. John and Halifax 37 50

BUTTS.

Wrought Brass, net revised list.
Wrought Iron, 70 per cent.
Cast Iron Loose Pin, discount 60 per cent.
Wrought Steel Fast Joint and Loose Pin, 70 per cent.

CEMENT AND FIREBRICK.

"Lafarge" cement in wood 3 40
"Lehigh" cement, in wood 2 54
"Lehigh" cement, cotton sacks. 2 39
"Lehigh" cement, paper sacks 2 31
Fire brick (Scotch) per 1,000 .... 15 00 24 00
Fire clay (Scotch), net ton 4 95

Paving Blocks per 1,000.

Blue metallic, 9"x4½"x3", ex wharf .. 35 00
Stable pavers, 12"x6"x2", ex wharf .. 50 00
Stable pavers, 9"x4½"x3", ex wharf .. 36 00

DOOR SETS.

Peterboro, 37½ and 10 per cent.

DOOR SPRINGS.

Torrey's Rod ....per doz. 1 75
Coil, 9 to 11 in .... " 0 95 1 65
English .... " 2 00 4 00
Chicago and Reliance Coil Door Springs, 25 per cent.

STORE DOOR HANDLES

Per Dozen 1 00 1 50

ESCUTCHEONS.

Discount 50 and 10 per cent., new list
Peterboro, 37½ and 10 per cent.

ESCUTCHEON PINS.

Iron, discount 40 per cent.
Brass, 45 per cent.

HINGES.

Blind, discount 60 per cent.
Heavy T and strap, 4-in., per lb. net . 0 06½
" " 5-in., " 0 06
" " 6-in., " 0 05¾
" " 8-in., " 0 05½
" " 10-in. and larger .. 0 05¼
Light T and strap, discount 65 p.c.
Screw hook and hinge—
under 12 in .... per 100 lb. 4 75
over 12 in .... " 3 75
Spring, No. 20, per gro. pairs 10 80
Spring, Woodyatt pattern, per gro., No 5, $17.50 No. 10, $18; No. 20, $10 80; No. 12 $20; No. 51, $10; No. 50, $27 50.
Crate hinges and back flaps, 65 and 5 p. c.
Hinge hasps, 65 per cent.

SPRING HINGES.

Chicago Spring Butts and Blanks 12½ per cent.
Triple End Spring Butts, 40 and 5 per cent.
Chicago Floor Hinges, 40 and 5 off
Garden City Fire Engine House Hinges, 12½ per cent.

CAST IRON HOOKS.

Bird cage ....per doz. 0 50 1 10
Clothes line, No. 61.. " 0 00 0 70
Harness .... " 0 60 12 00
Hat and coat .... per gro. 1 10 10 00
Chandelier ....per doz. 0 50 1 00
Wrought hooks and staples—
¼ x 5 .... per gross 2 65
5-16 x 5 .... " 3 30
Bright steel gate hooks and staples, 40 per cent. discount
Hat and coat wire, discount 62½ per cent.
Screw, bright wire, discount 65 per cent.

KNOBS.

Door, japanned and N.P., doz 1 50 2 50
Bronze, Berlin .... per doz. 2 75 3 25
Bronze, Genuine .... " 6 00 9 00
Shutter, porcelain, F. & L. screw .... per gross 1 30 2 00
White door knobs ....per doz. 2 00
Peterboro knobs, 37½ and 10 per cent.
Porcelain, mineral and jet knobs, net list.

KEYS.

Lock, Canadian dis. 40 to 40 and 10 per cent
Cabinet trunk and padlock American .... per gross 0 60

LOCKS.

Peterboro 37½ and 10 per cent
Russell & Erwin, steel rim $2 50 per doz
Eagle cabinet locks, discount 30 per cent
American padlocks, all steel, 10 to 15 per cent; all brass or bronze, 10 to 25 per cent.

SAND AND EMERY PAPER.

B & A. sand, discount, 35 per cent
Emery, discount 35 per cent.
Garnet (Burton's). 5 to 10 per cent. advance on list.

SASH WEIGHTS.

Sectional ....per 100 lb. 2 00 2 25
Solid .... " 1 50 1 75

SASH CORD.

Per lb 0 31

BLIND AND BED STAPLES

All sizes, per lb 0 07½ 0 12

WROUGHT STAPLES.

Galvanized 2 75
Plain 2 50
Coopers', discount 45 per cent.
Poultry netting staples, discount 40 per cent.
Bright spear point, 75 per cent. discount.

TOOLS AND HANDLES.

ADZES.

Discount 22½ per cent.

AUGERS.

Gilmour's, discount 60 per cent. off list.

AXES.

Single bit, per doz. 5 50 8 50
Double bit, " 10 00 11 00
Bench Axes, 40 per cent.
Broad Axes 25 per cent.
Hunters' Axes 5 50 6 00
Boys' Axes 6 25 7 00
Splitting Axes 7 00 12 00
Handled Axes 7 00 9 00
Red Ridge, boys', handled.. .... 5 75
" hunters 5 25

BITS.

Irwin's auger, discount 47½ per cent
Gilmour's auger, discount 60 per cent.
Rockford auger, discount 50 and 10 per cent.
Jennings' Gen. auger, net list.
Gilmour's car, 47½ per cent.
Clark's expansive, 40 per cent.
Clark's gimlet, per doz 0 65 0 90
Diamond, Shell, per doz 1 00 1 50
Nail and Spike, per gross 2 25 5 20

BUTCHERS CLEAVERS

German ....per doz. 6 00 9 00
American .... " 12 00 18 00

CHALK.

Carpenters' Colored, per gross 0 45 0 75
White lump ....per cwt. 0 60 0 65

CHISELS.

Warnock's, discount 72½ per cent.
P. S. & W. Extra, discount 72½ per cent

CROSSCUT SAW HANDLES.

S. & D., No. 3 ....per pair 0 15
S. & D., " 5 .... " 0 22½
S. & D., " 6 .... " 0 15
Boynton pattern .... " 0 20

CROWBARS.

3½c. to 4c. per lb.

DRAW KNIVES.

Coach and Wagon, discount 75 per cent.
Carpenters' discount 75 per cent

DRILLS.

Millar's Falls, hand and breast. net list.
North Bros., each set, 50c.

DRILL BITS.

Morse, discount 37½ to 40 per cent.
Standard, discount 50 and 5 to 55 per cent.

FILES AND RASPS.

Great Western .... 75 per cent.
Arcade .... 75 "
Kearney & Foot .... 75 "
Disston's .... 75 "
American .... 75 "
J. Barton Smith .... 75 "
McClellan .... 75 "
Eagle .... 75 "
Nicholson .... 66⅔ "
Globe .... 75 "
Black Diamond, 60, 10 and 5 p c.
Jowitt's, English list, 27½ per cent.

GAUGES.

Stanley's discount 50 to 60 per cent.

Wire Gauges

Winn's, Nos. 26 to 33 ....each 1 65 2 40

HANDLES.

C. & B., fork, 40 per cent., revised list.
C. & B., hoe, 40 per cent., revised list.
American, saw ...... per doz. 0 25
American, plane ... per gross 3 15 3 75
Canadian, hammer and hatchet 40 per cent
Axe and cant hook handles, 45 per cent.

HAMMERS.

Maydole's, discount 5 to 10 per cent.
Canadian, discount 25 to 27½ per cent.
Magnetic tack ....per doz. 1 10 1 20
Canadian sledge .... per lb. 0 07 0 08½
Canadian ball pean, per lb. .. 0 22 0 25

HATCHETS.

Canadian, discount 40 to 42½ per cent.
Shingle, Red Ridge 1, per doz 4 40
" " 2, " 4 80
Barrel Underhill 5 05

MALLETS.

Tinsmiths' ....per doz. 1 25 1 50
Carpenters', hickory, " 1 25 3 75
Lignum Vitae........ " 3 85 5 00
Caulking, each 0 60 2 00

MATTOCKS.

Canadian............ per doz. 5 50 6 00

MEAT CUTTERS.

German, 15 per cent.
American discount, 33⅓ per cent.
Gem ....each 1 15

NAIL PULLERS.

German and American 0 85 2 50
No. 1........................ 0 85
No 1573...................... 0 75

NAIL SETS.

Square, round and octagon, per gross 3 38
Diamond 1 00

PICKS.

Per dozen 6 00 9 00

PLANES.

Wood bench, Canadian discount 40 per cent. American discount 50 per cent.
Wood, fancy Canadian or American 37½ to 40 per cent
Stanley planes, $1.55 to $3 60, net list prices.

PLANE IRONS.

English ....per doz. 2 00 5 00
Stanley, 2⅜ inch, single 24c, double 39c.

PLIERS AND NIPPERS.

Button's genuine, 37½ to 40 per cent.
Button's imitation....per doz. 5 00 9 00
Berg's wire fencing.......... .... 1 72 5 50

PUNCHES.

Saddlers ....per doz. 1 00 1 85
Conductor's.......... " 3 00 15 00
Tinners', solid..........per set 0 72
" hollow......per inch 1 00

RIVET SETS.

Canadian, discount 35 to 37½ per cent.

RULES.

Boxwood, discount 70 per cent.
Ivory, discount 20 to 25 per cent

SAWS.

Atkins, hand and crosscut, 25 per cent.
Disston's Hand, discount 12½ per cent
Disston's Crosscut ....per foot 0 35 0 55
Hack, complete..........each 0 75 2 75
" frame only..........each 0 50 1 25
S. & D. solid tooth circular shingle, concave and band, 50 per cent; mill and ice, drag, 30 per cent; cross-cut, 35 per cent; hand saws, butcher, 35 per cent; buck, New Century, $6.25, buck No. 1 Maple Leaf, $5.25; buck, Happy Medium, $4.25; buck Watch Spring, $4.25, buck, common frame, $4.00.
Spear & Jackson's saws—Hand or rip, 26 in., $12 75; 24 in., $12 25; panel, 18 in., $8 25; 20 in., $9; tenon, 10 in., $9.90; 12 in., $10 9; 14 in., $11.50

SAW SETS.

Lincoln and Whiting 4 75
Hand Sets, Perfect 4 00
X-Cut Sets, " 7 50
Maple Leaf and Premiums saw sets, 40 off.
S. & D saw swages, 40 off.

SCREW DRIVERS.

Sargent's ....per doz. 0 65 1 00
North Bros., No. 30 ..per doz. 16 80

SHOVELS AND SPADES.

Bull Dog, solid neck shovel (No. 2 pol) $18 50
(Hollow Back) (Reinforced S Scoop)
Moose........ $17 50 .... $16 30
Bear .... 15 00 .... 15 30
Fox........... 12 50 .... 14 30
Black Cat.... 10 00 .... 13 30
Canadian, discount 45 per cent.

SQUARES.

Iron, discount 20 per cent.
Steel, discount 65 and 10 per cent.
Try and Bevel, discount 50 to 52½ per cent.

TAPE LINES

English, ass skin .....per doz. 2 75 5 00
English, Patent Leather..... 5 50 9 75
Chesterman's............each 0 90 2 85
" steel........each 0 80 8 00
Berg's, each...................... 0 75 2 50

TROWELS.

Disston's, discount 10 per cent.
S & D., discount 35 per cent.
Berg's, brick, 924x11 .... 4 00
" pointing, 924x5........... .... 2 10

FARM AND GARDEN GOODS

BELLS.

American cow bells, 63⅓ per cent.
Canadian, discount 45 and 50 per cent.
American, farm bells, each .. 1 35 3 00

BULL RINGS.

Copper, $1.30 for 2½-inch, and $1.70

CATTLE LEADERS.

Nos. 32 and 33 .. ...per gross 7 50 8 50

BARN DOOR HANGERS

		doz. pairs.
Steel barn door................	8 00	10 00
Stearns wood track	4 50	6 00
Zenith........................	....	9 00
Acme, wood track	5 00	6 50
Atlas........................	5 00	6 00
Perfect	8 00	11 00
New Milo....................	....	6 50
Steel, covered................	4 00	11 00
" track, 1 x 3-16 in(100 ft)	....	3 75
" 1¼ x 3-16 in(100 ft)	....	4 75
Double strap hangers, doz. sets....	....	6 40
Standard jointed hangers, "	....	6 40
Steel King hangers	....	6 25
Storm King and safety hangers	....	7 00
" " rail................	....	4 25

Chicago Friction, Oscillating and Big Twin Hangers, 5 per cent.

HARVEST TOOLS.

Discount 60 per cent.
S. & D. lawn rakes, Dunn's, 40 off.
" sidewalk and stable scrapers, 40 off.

HAY KNIVES.

Net list

HEAD HALTERS.

Jute Rope, ⅜-inch....per gross 9 00
" " ½ " .... " 10 00
" " ⅝ " .... " 12 00
Leather, 1-inch ....per doz. 4 00
Leather, 1¼ " ...... " 5 20
Web................... " 2 45

HOES.

Garden, Mortar, etc., discount 60 per cent.
Planter.......... ... per doz. 4 00 4 50

SCYTHES.

Per doz. net.................. 6 25 9 25

SCYTHE SNATHS.

Canadian, discount 40 per cent.

SNAPS.

Harness, German, discount 25 per cent.
Lock, Andrews'.............. 4 50 11 00

STABLE FITTINGS

Warden King, 35 per cent

WOOD HAY RAKES.

Ten tooth, 40 and 10 per cent
Twelve tooth, 45 per cent.

HEAVY GOODS, NAILS, ETC.

ANVILS.

Wright's, 80-lb. and over............ 0 10½
Hay Budden, 80-lb. and over 0 09½
Brook's, 80-lb. and over 0 11½
Taylor-Forbes, handy 0 05
Columbia Hardware Co , per lb. 0 09½

VISES.

Wright's..................... .... 0 13½
Berg's, per lb.................. .... 0 12½
Brook's........................ .... 0 12½
Pipe Vise, Hinge, No. 1...... .... 3 50
" " " No. 2...... .... 5 50
Saw Vise 4 50 5 00
Blacksmiths' (discount) 60 per cent.
" parallel (discount) 45 per cent.

BOLTS AND NUTS

Carriage Bolts, common ($1 list Per cent.
" " ⅜ and smaller.. 60, 10 and 10
" " 7-16 and up .... 55 and 5
" " Norway Iron ($3 list) 50

Machine Bolts, ⅜ and less .... 60 and 10
Machine Bolts, 7-16 and up.... 55 and 5
Plough Bolts .... 55 and 10
Blank Bolts................... 55
Bolt Ends...................... 55
Sleigh Shoe Bolts, ⅜ and less .. 60 and 10
" " 7-16 and larger 50 and 5
Coach Screws, conepoint .... 70 and 5
Nuts, square, all sizes, 4c. per cent. off
Nuts, hexagon, all sizes, 4⅛c. per cent. off.
Stove Rods per lb., 5½ to 6c.
Stove Bolts, 75 per cent.

CHAIN.

Proof coil, per 100 lb., 5-16 in., $4.40; ⅜ in., $3 93; 7-16 in., $3.70; ½ in., $3 50; 9-16 in., $3 45; ⅝ in., $3.35; ¾ in., $3 20; ⅞ in., $3.10; 1 in., $3.10.
Halter, kennel and post chains, 40 to 40 and 5 per cent.
Cow ties .... 40 p.c.
Tie out chains .... 65 p.c.
Stall fixtures .... 35 p.c.
Trace chain .... 45 p.c.
Jack chain iron, discount 35 p.c.
Jack chain, brass, discount 40 per cent.

HORSE NAILS.

'C' brand, 40, 10 and 7½ per cent. off list { Oval
M.R.M. Co. brand, 55 per cent. } head
Capewell brand, quotations on application.

HORSESHOES.

F.O.B. Montreal
M.R.M. Co. brand, base.................. 65
Add 15c. Toronto, Hamilton, Guelph.

HORSE WEIGHTS

Taylor-Forbes, 3c. per lb.

NAILS.	Cut.	Wire.
2d	3 90	3 45
3d	2 95	3 10
4 and 5d	2 70	2 85
6 and 7d	2 60	2 75
8 and 9d	2 45	2 60
10 and 12d	2 40	2 55
16 and 20d	2 35	2 50
30, 40, 50 and 60d (base)	2 30	2 45

F.o.b. Montreal.
Cut nails, Toronto 20c. higher.
Miscellaneous wire nails, discount 75 per cent
Coopers' nails, discount 40 per cent.

PRESSED SPIKES

Pressed spikes, ⅝ diameter, per 100 lbs, $3.15

RIVETS AND BURRS.

Iron Rivets, black and tinned, 60, 10 and 10.
Iron Burrs, discount 60 and 10 and 10 p.c.
Copper Rivets, usual proportion burrs, 15 p.c.
Copper Burrs only, net list
Extras on Coppered Rivets, ½-lb. packages 1c. per lb.; ¼-lb. packages 2c. lb
Tinned Rivets, net extra, 4c. per lb.

SCREWS.

Wood, F. H., bright and steel, 87½ per cent.
" R. H., bright, dis. 82½ per cent
" F. H., brass, dis. 80 per cent.
" R. H., " dis. 75 per cent
" F. H., bronze, dis. 75 per cent.
" R. H., " dis. 70 per cent.
Drive Screws, dis. 87½ per cent.
Bench, wood ....per doz. 3 25 4 00
" iron 4 25 5 00
Set, case hardened, dis. 60 per cent.
Square Cap, dis. 50 and 5 per cent.
Hexagon Cap, dis. 45 per cent.

MACHINE SCREWS.

Flat head, iron and brass, 35 per cent.
Felister head, iron, discount 30 per cent.
" " brass, discount 25 per cent.

TACKS, BRADS, ETC.

Carpet tacks, blued .... 80 and 5
" " tinned .... 80 and 10
" " (in kegs) 40
Cut tacks, blued, in dozens only 75 and 10
" ½ weights 60
Swedes cut tacks, blued and tinned—
In bulk......................... 80 and 10
In dozens 75
Swedes, upholsterers', bulk .... 85 and 12½
" brush, blued and tinned bulk 70
Swedes, gimp, blued, tinned and japanned....................... 75 and 12½
Zinc tacks 35
Leather carpet tacks............ 40
Copper tacks.................... 37½
Copper nails 42½
Trunk nails, black 65
Trunk nails, tinned and blued .. 65
Clout nails, blued and tinned .. 65
Chair nails....................... 35
Patent brads..................... 40
Fine finishing................... 40
Lining tacks, in papers.......... 10
" " in bulk............ 15
" " solid heads, in bulk 75
Saddle nails, in papers.......... 10
" " in bulk............ 15
Tufting buttons, 22 line in dozens only 60
Zinc glaziers' points 5
Double pointed tacks, papers.. 90 and 10
" " " bulk.... 40
Clinch and duck rivets.......... 45
Cheese box tacks, 85 and 5; trunk tacks, 80 and 10.

WROUGHT IRON WASHERS.

Canadian make, discount 40 per cent.

SPORTING GOODS.

AMMUNITION.

B. B. Caps American $2.00 per 1000.
C. B. Caps American, $2.60 per 1000

CARTRIDGES.

Rim Fire Cartridges, 50 and 5 p.c.
Rim Fire Pistol, 30 and 5 per cent, American.
Central Fire, Military and Sporting American, 10 per cent. advance, B.B. Caps, discount 40 per cent., American.
Central Fire Pistol and Rifle, net Amer list
Loaded and empty Shells, American 20 per cent. discount. Rival and Nitro, 10 per cent. advance
Empty paper shells American, 10 per cent. advance.
Primers, American $2 05

Wads. per lb.

Best thick brown or grey felt wads, in ¼-lb. bags $0 70
Best thick white card wads, in boxes of 500 each, 12 and smaller gauges 0 29
Best thick white card wads in boxes of 500 each, 10 gauge............ 0 35
Thin card wads, in boxes of 1,000 each, 12 and smaller gauges 0 20
Thin card wads, in boxes of 1,000 each, 10gauge....................... 0 25
Chemically prepared black edge grey cloth wads, in boxes of 250 each— Per M.
11 and smaller gauge 0 60
9 and 10 gauges.................. 0 70
7 and 8 " 0 90
5 and 6 " 1 10
Superior chemically prepared pink edge, best white cloth wads in boxes of 250 each—
11 and smaller gauge 1 15
9 and 10 gauges 1 40
7 and 8 " 1 65
5 and 9 " 1 90

SHOT.

Common, $6.50 per 100 lb.; chilled, $7.50 per 100 lb.; buck, seal and ball, $8.50 net list. Prices are f.o.b. Toronto, Hamilton, Montreal, St. John and Halifax. Terms, 2 p.c. for cash in thirty days.

TRAPS (steel.)

Game, Newhouse, discount 30 and 10 per cent.
Game, Hawley & Norton, 50, 10 & 5 per cent.
Game, Victor, 70 per cent.
Game, Oneida Jump (B. & L.) 40 & 2½ p. c.
Game, steel, 60 and 5 per cent.

SKATES.

Skates, discount 37½ per cent.
Mic Mac hockey sticks, per doz 4 00 5 00

CUTLERY AND SILVERWARE.

RAZORS.

	per doz.	
Elliot's	4 00	18 00
Boker's	7 50	11 00
" King Cutter	13 50	18 50
Wade & Butcher's	3 60	10 00
Lewis Bros.' "Klean Kutter"	8 50	10 50
Henckel's	7 50	20 00
Berg's	7 50	20 00

Clauss Razors and Strops, 50 and 10 per cent

KNIVES.

Farriers-Stacey Bros., doz 3 50

PLATED GOODS

Hollowware, 40 per cent. discount.
Flatware, staples, 40 and 10, fancy, 40 and 5.

SHEARS.

Clauss, nickel, discount 60 per cent.
Clauss, Japan, discount 67½ per cent.
Clauss, tailors, discount 40 per cent.
Seymour's, discount 50 and 10 per cent.
Berg's 6 00 12 00

HOUSE FURNISHINGS.

APPLE PARERS.

Woodyatt Hudson, per doz., net 4 50

BIRD CAGES.

Brass and Japanned, 40 and 10 p c.

COPPER AND NICKEL WARE.

Copper boilers, kettles, teapots, etc. 35 p.c.
Copper pitts, 25 per cent.

ENAMELED WARE.

London, White, Princess, Turquoise, Onyx, Blue and White, discount 50 per cent.
Canada, Diamond, Premier, 50 and 10 p.c.
Pearl, Imperial Crescent, 50 and 10 per cent.
Premier steel ware, 40 per cent.
Star decorated steel and white, 25 per cent.
Japanned ware, discount 45 per cent.
Hollow ware, tinned cast, 35 per cent. off.

KITCHEN SUNDRIES.

Can openers, per doz	0 40	0 75
Mincing knives per doz	0 50	0 80
Duplex mouse traps, per doz	....	0 65
Potato mashers, wire, per doz	0 60	0 70
" wood "	0 50	0 60
Vegetable slicers, per doz	....	2 25
Universal meat chopper No. 0, $1; No. 1, 1.15.		
Enterprise chopper, each	....	1 30
Spiders and fry pans, 50 per cent.		
Star Al chopper 5 to 32	1 35	4 10
" " 100 to 103	1 33	2 00
Kitchen hooks, bright	....	0 62½

LAMP WICKS.

Discount, 60 per cent.

LEMON SQUEEZERS.

Porcelain lined	per doz.	2 20	5 60
Galvanized	"	1 87	3 85
King, wood	"	2 75	2 90
King, glass	"	4 00	4 50
All glass	"	0 50	0 90

METAL POLISH.

Tandem metal polish paste 6 00

PICTURE NAILS.

Porcelain head ......per gross 1 35 1 50
Brass head 0 40 1 00
Tin and gilt, picture wire, 75 per cent.

SAD IRONS.

Mrs. Potts, No. 55, polished....per set 0 80
" No. 50, nickle-plated, " 0 92
Common, plain 4 50
" plated 5 50
Asbestos, per set 1 25

TINWARE.

CONDUCTOR PIPE.

2-in. plain or corrugated., per 100 feet, $3.30; 3 in., $4 40; 4 in., $5.80; 5 in., $7.45; 6 in., $9.9).

FAUCETS.

Common, cork-lined, discount 35 per cent.

EAVETROUGHS.

10-inch ....per 100 ft. 3 30

FACTORY MILK CANS

Discount off revised list, 35 per cent
Milk can trimmings, discount 25 per cent.
Creamery Cans, 45 per cent.

LANTERNS.

No. 2 or 4 Plain Cold Blast....per doz. 6 50
Lift Tubular and Hinge Plain, " 4 75
Better quality at higher prices.
Japanning, 50c. per doz. extra.

OILERS.

Kemp's Tornado and McClary s Model galvanized oil can, with pump, 5 gallon, per dozen 10 92
Davidson oilers, discount 40 per cent
Zinc and tin, discount 50 per cent
Coppered oilers, 20 per cent. off.
Brass oilers, 50 per cent. off.
Malleable, discount 25 per cent

PAILS (GALVANIZED).

Dufferin pattern pails, 45 per cent.
Flaring pattern, discount 45 per cent.
Galvanized washtubs 40 per cent

PIECED WARE.

Discount 35 per cent off list, June, 1899.
10-qt. flaring sap buckets, discount 35 per cent.
6, 10 and 14-qt. flaring pails dis. 35 per cent.
Copper bottom tea kettles and boilers, 30 p.c.
Coal hods, 40 per cent.

STAMPED WARE.

Plain, 75 and 12½ per cent. off revised list.
Retinned, 72½ per cent revised list.

SAP SPOUTS.

Bronzed iron with hooks ....per 1,000 7 50
Eureka tinned steel, hooks 8 00

STOVEPIPES.

5 and 6 inch, per 100 lengths 7 64 7 91
7 inch .... " " 8 18
Nestable, discount 40 per cent.

STOVEPIPE ELBOWS

5 and 6-inch, common........per doz. 1 32
7-inch .... " 1 48
Polished, 15c. per dozen extra.

THERMOMETERS.

Tin case and dairy, 75 to 75 and 10 per cent.

TINNERS' SNIPS.

Per doz........ 3 00 15
Clauss, discount 35 per cent.

TINNERS' TRIMMINGS

Discount, 45 per cent.

WIRE.

ANNEALED CUT HAY BAILING WIRE.

No. 12 and 13, $4; No 13½, $4 10; No. 14, $4 20; No 15, $4 50; in lengths 6' to 11', 25 per cent.; other lengths 20c. per 100 lbs extra; if eye or loop on end add 25c per 100 lbs. to the above

BRIGHT WIRE GOODS

Discount 62½ per cent.

CLOTHES LINE WIRE.

7 wire solid line, No. 17. $4.90; No. 18, $3.00; No. 19, $2.70; 6 wire solid line, No. 17, $4.45; No. 18, $2.80. No. 19, $2.50. All prices per 1000 ft. measure. F.o.b. Hamilton Toronto, Montreal.

COILED SPRING WIRE.

High Carbon, No. 9, $2 90, No. 11, $3.45; No. 19, $3.15.

COPPER AND BRASS WIRE

Discount 37½ per cent.

FINE STEEL WIRE.

Discount 25 per cent. List of extras: In 100-lb. lots: No. 17, $5 — No. 18, $5.50 — No. 19, $6 — No. 20, $6.65 — No. 21, $7 — No. 22, $7.30 — No. 23, $7.65 — No. 24, $8 — No. 25, $9 — No. 26, $9.50 — No. 27, $10 — No. 28, $11 — No. 29, $12 — No. 30, $13 — No. 31, $14 — No. 32, $15 — No. 33, $16 — No. 34, $17. Extras net—tinned wire, Nos. 17-25, $2 — Nos. 26-31, $4 — Nos. 32-34, $6. Coppered, 75c.—oiling, 10c.—in 25-lb. bundles, 15c.—in 5 and 10-lb. bundles, 25c.—in 1-lb. hanks, 25c. —in ½-lb. hanks, 38c.—in ¼-lb. hanks, 50c.—packed in casks or cases, 15c.—bagging or papering, 10c.

FENCE STAPLES.

Bright........ 2 70 Galvanized.... 3 15

HAY WIRE IN COILS.

No. 13, $2.60; No. 14, $2.70; No. 15, $2.85, f.o.b., Montreal.

GALVANIZED WIRE.

Per 100 lb.—Nos. 4 and 5, $3.70—Nos. 6, 7, 8, $3.15 — No. 9, $2.50 — No. 10, $3.20 — No. 11, $3.25 — No. 12, $2.65 — No. 13, $2.75 — No. 14, $3.75 — No 15, $4.30 — No. 16, $4 30 from stock. Base sizes, Nos. 6 to 9, $2.35 f.o.b. Cleveland. In carlots 12½c. less.

LIGHT STRAIGHTENED WIRE.

Gauge No.	Over 20 in. per 100 lbs.	10 to 20 in.	5 to 10 in.
0 to 5	$0 50	$0 75	$1 25
6 to 9	0.75	1 25	2 00
10 to 11	1 00	1.75	2 50
12 to 14	1.50	2.25	3.50
15 to 16	2.00	3.00	4.50

SMOOTH STEEL WIRE.

No. 0-9 gauge, $2 30; No. 10 gauge, 6c. extra; No. 11 gauge, 12c extra; No 12 gauge, 20c. extra; No. 13 gauge, 30c. extra; No 14 gauge, 40c. extra; No 15 gauge, 55c. extra; No 16 gauge, 70c. extra. Add 60c. for coppering and $2 for tinning.

Extra net per 100 lb.—Oiled wire 10c., spring wire $1.25, bright soft drawn 15c., charcoal (extra quality) $1.25, packed in casks or cases 15c., bagging and papering 10c., 50 and 100-lb. bundles 10c., in 25-lb. bundles 15c., in 5 and 10-lb. bundles 25c., in 1-lb hanks, 50c., in ½-lb. hanks 75c., in ¼-lb. hanks $1.

POULTRY NETTING.

2 in mesh 19 w. g., discount 50 and 10 per cent. All others 50 per cent.

WIRE CLOTH.

Painted Screen, in 100-ft rolls, $1 62½, per 100 sq. ft.; in 50-ft. rolls $1.67½, per 100 sq ft. Terms, 2 per cent. off 30 days.

WIRE FENCING.

Galvanized barb 2 95
Galvanized, plain twist 3 30
Galvanized barb, f.o.b. Cleveland, $2.70 for small lots and $2.60 for carlo's

WOODENWARE.

CHURNS.

No. 0, $9; No. 1, $9; No. 2, $10; No. 3, $11; No. 4, $13; No. 5, $16; f.o.b. Toronto Hamilton, London and St. Marys, 30 and 30 per cent; f o b Ottawa, Kingston and Montreal, 40 and 15 per cent discount. Taylor-Forbes, 30 and 30 per cent.

CLOTHES REELS.

Davis Clothes Reels, dis. 40 per cent.

LADDERS, EXTENSION.

Waggoner Extension Ladders, dis. 40 per cent.

MOPS AND IRONING BOARDS.

"Best" mops 1 25
"900" mops 1 25
Folding ironing boards 12 00 16 50

REFRIGERATORS

Discount, 40 per cent.

SCREEN DOORS.

Common doors, 2 or 3 panel, walnut stained, 4-in. style ....per doz. 7 25
Common doors, 2 or 3 panel, grained only, 4-in., style ....per doz. 7 55
Common doors, 2 or 3 panel, light stair per doz. 9 85

WASHING MACHINES.

Round, re-acting per doz 60 00
Square " 63 00
Eclipse, per doz 54 00
Dowswell " 39 00
New Century, per doz 75 00
Daisy 54 00

WRINGERS.

Royal Canadian, 11 in., per doz. ... 34 00
Royal American, 11 in. ... 34 00
Eze, 10 in., per doz ... 3 75
Terms, 2 per cent., 30 days.

MISCELLANEOUS.

AXLE GREASE.

Ordinary, per gross 6 00 7 00
Best quality 10 00 12 00

BELTING.

Extra, 60 per cent.
Standard, 60 and 10 per cent.
No. 1, not wider than 6 in., 60, 10 and 10 p.c.
Agricultural, not wider than 4 in., 75 per cent
Lace leather, per side, 75c; cut laces, 80c

BOOT CALKS.

Small and medium, ball ....per M 4 25
Small heel .... " 4 50

CARPET STRETCHERS.

American ....per doz. 1 00 1 50
Bullard's 6 50

CASTORS.

Bed, new list, discount 55 to 57½ per cent.
Plate, discount 52½ to 57½ per cent.

PINE TAR.

½ pint in tins ....per gross ... 7 80
1 " " 9 60

PULLEYS.

Hothouse	per doz.	0 55	1 00
Axle	"	0 22	0 33
Screw	"	0 22	1 00
Awning	"	0 35	2 50

PUMPS.

Canadian cistern 1 40 2 00
Canadian pitcher spout 1 80 3 16
Berg's wing pump, 75 per cent.

ROPE AND TWINE.

Sisal		0 10½
Pure Manilla		0 15½
"British" Manilla		0 12
Cotton, 3-16 inch and larger	0 21	0 23
" 5-32 inch	0 25	0 27
" ¼ inch	0 25	0 28
Russia Deep Sea		0 16
Jute		0 09
Lath Yarn, single		0 10
" double		0 10½
Sisal bed cord, 48 feet ....per doz.		0 65
" " 60 feet		0 80
" " 72 feet		0 95

Twine.

Bag, Russian twine, per lb.	....	0 27
Wrapping, cotton, 3-ply	....	0 25
" " 4-ply	....	0 29
Mattress twine per lb.	0 33	0 45
Staging "	0 27	0 35

SCALES.

Gurney Standard, 40 per cent.
Gurney Champion, 50 per cent.
Burrow, Stewart & Milne—
Imperial Standard, discount 40 per cent.
Weigh Beams, discount 40 per cent.
Champion Scales, discount 50 per cent.
Fairbanks standard, discount 35 per cent.
" Dominion, discount 55 per cent.
" Richelieu, discount 55 per cent.
Warren new Standard, discount 40 per cent.
" " Champion, discount 50 per cent.
" Weighbeams, discount 35 per cent.

STONES—OIL AND SCYTHE.

Washita	per lb.	0 25	0 37
Hindostan	"	0 06	0 10
" slip	"	0 10	0 20
" Axe	"	....	0 10
Deer Creek	"	....	0 10
Deerlick	"	....	0 25
" Axe	"	....	0 15
Lily white	"	....	0 42
Arkansas	"	....	1 50
Water-of-Ayr	"	....	0 10
Scythe	per gross	3 50	5 00
Grind, 40 to 200 lb., per ton		20 00	22 00
" under 40 lb.	"	....	24 00
" 200 lb. and over	"	....	28 00

INDEX TO ADVERTISERS.

CLASSIFIED LIST OF ADVERTISEMENTS.

Auditors.
Davenport, Pickup & Co., Winnipeg.

Babbitt Metal.
Canada Metal Co , Toronto.
Canadian Fairbanks Co., Montreal.
Robertson, Jas. Co,, Montreal.

Bath Room Fittings.
Carriage Mounting Co., Toronto.

Belting, Hose, etc.
Gutta Percha and Rubber Mfg. Co. Toronto.

Bicycles and Accessories.
Johnson's, Iver, Arms and Cycle Works Fitchburg, Mass

Binder Twine.
Consumers Cordage Co., Montreal.

Box Strap.
J. N. Warminton, Montreal.

Brass Goods.
Glauber Brass Mfg. Co., Cleveland, Ohio.
Lewis, Rice, & Son., Toronto.
Morrison, Jas., Brass Mfg. Co., Toronto.
Mueller Mfg. Co., Decatur, Ill.
Penberthy Injector Co., Windsor, Ont.
Taylor-Forbes Co., Guelph, Ont.

Bronze Powders.
Canadian Bronze Powder Works, Montreal.

Brushes.
Ramsay, A., & Son Co., Montreal.

Can Openers.
Cumming Mfg. Co. Renfrew.

Cans.
Acme Can Works, Montreal.

Builders' Tools and Supplies.
Covert Mfg. Co., West Troy, N.Y.
Frothingham & Workman Co., Montreal.
Howland, H. S., Sons & Co., Toronto.
Hyde, F., & Co., Montreal.
Lewis Bros. & Co., Montreal.
Lewis, Rice, & Son, Toronto.
Lockerby & McComb, Montreal.
Lufkin Rule Co., Saginaw, Mich.
Newman & Sons, Birmingham.
North Bros. Mfg. Co., Philadelphia, Pa.
Stanley Rule & Level Co., New Britain.
Stanley Works, New Britain, Conn.
Stephens, G. F., Winnipeg.
Taylor-Forbes Co., Guelph, Ont.

Carriage Accessories.
Carriage Mountings Co., Toronto.
Covert Mfg. Co., West Troy, N.Y.

Carriage Springs and Axles.
Guelph Spring and Axle Co., Guelph.

Cattle and Trace Chains.
Greening, B., Wire Co., Hamilton.

Churns.
Dowswell Mfg. Co., Hamilton.

Clippers—All Kinds.
American Shearer Mfg. Co., Nashua, N.H.

Clothes Reels and Lines.
Hamilton Cotton Co., Hamilton, Ont.

Cordage.
Consumers' Cordage Co., Montreal.
Hamilton Cotton Co., Hamilton.

Cork Screws.
Erie Specialty Co., Erie, Pa.

Clutch Nails.
J. N. Warminton, Montreal.

Cut Glass.
Phillips, Geo., & Co., Montreal.

Cutlery—Razors, Scissors, etc.
Birkett, Thos., & Son Co., Ottawa.
Clauss Shear Co., Toronto
Dorken Bros. & Co., Montreal.
Heinisch's, R., Sons Co., Newark, N.J.
Howland, H. S. Sons & Co., Toronto.
Lamplough, F. W., & Co , Montreal.
Phillips, Geo., & Co., Montreal.
Round, John, & Son, Montreal.

Electric Fixtures.
Canadian General Electric Co., Toronto.
Forman, John, Montreal.
Morrison James, Mfg. Co., Toronto.
Munderloh & Co., Montreal.

Electro Cabinets.
Cameron & Campbell Toronto.

Engines, Supplies, etc.
Kerr Engine Co., Walkerville, Ont.

Fencing—Woven Wire.
McGregor-Banwell Fence Co , Walkerville, Ont
Owen Sound Wire Fence, Co., Owen Sound
Banwell Hoxie Wire Fence Co , Hamilton.

Files and Rasps.
Barnett Co., G. & H., Philadelphia, Pa.
Nicholson File Co., Port Hope

Financial Institutions
Bradstreet Co.

Firearms and Ammunition.
Dominion Cartridge Co., Montreal.
Hamilton Rifle Co., Plymouth, Mich
Harrington & Richardson Arms Co., Worcester, Mass.
Johnson's, Iver, Arms and Cycle Works, Fitchburg, Mass.

Food Choppers
Enterprise Mfg Co., Philadelphia, Pa.
Lamplough, F. W., & Co , Montreal.
Shirreff Mfg. Co , Brockville, Ont.

Galvanizing.
Canada Metal Co., Toronto.
Montreal Rolling Mills Co., Montreal
Ontario Wind Engine & Pump Co., Toronto.

Glaziers' Diamonds.
Gibsone, J. B , Montreal.
Pelton, Godfrey S.
Sharratt & Newth, London, Eng.
Shaw, A., & Son, London, Eng.

Handles
Still, J. H., Mfg. Co

Harvest Tools.
Maple Leaf Harvest Tool Co , Tillsonburg Ont.

Hoop Iron.
Montreal Rolling Mills Co , Montreal.
J N. Warminton, Montreal.

Horse Blankets.
Heney, E. N., & Co., Montreal

Horseshoes and Nails.
Canada Horse Nail Co., Montreal.
Montreal Rolling Mills, Montreal
Capewell Horse Nail Co , Toronto

Hot Water Boilers and Radiators.
Cluff, R. J., & Co , Toronto
Pease Foundry Co., Toronto.
Taylor-Forbes Co., Guelph.

Ice Cream Freezers.
Dana Mfg. Co., Cincinnati, Ohio.
North Bros. Mfg. Co., Philadelphia, Pa

Ice Cutting Tools.
Erie Specialty Co., Erie, Pa.
North Bros. Mfg. Co., Philadelphia, Pa.

Injectors—Automatic.
Morrison, Jas., Brass Mfg. Co., Toronto.
Penberthy Injector Co , Windsor, Ont.

Iron Pipe.
Montreal Rolling Mills, Montreal.

Iron Pumps.
Lamplough, F. W., & Co , Montreal.
McDougall, R., Co., Galt, Ont.

Lanterns.
Kemp Mfg. Co., Toronto.
Ontario Lantern Co., Hamilton, Ont.
Wright, E. T., & Co., Hamilton.

Lawn Mowers.
Birkett, Thos., & Son Co., Ottawa.
Maxwell, D., & Sons, St. Mary's, Ont.
Taylor, Forbes Co., Guelph.

Lawn Swings, Settees, Chairs.
Cumming Mfg. Co., Renfrew.

Ledgers—Loose Leaf.
Business Systems Toronto.
Copeland-Chatterson Co , Toronto
Crain, Rolla L., Co , Ottawa.
Universal Systems, Toronto.

Locks, Knobs, Escutcheons, etc.
Peterborough Lock Mfg. Co., Peterborough, Ont.

Lumbermen's Supplies.
Pink, Thos , & Co., Pembroke Ont.

Manufacturers' Agents.
Fox, C. H., Vancouver.
Gibb, Alexander, Montreal.
Mitchell, David C , & Co , Glasgow, Scot.
Mitchell, H. W., Winnipeg.
Pearce, Frank, & Co. Liverpool, Eng.
Scott, Bathgate & Co., Winnipeg
Thorne, R. E., Montreal and Toronto.

Metals.
Canada Iron Furnace Co., Midland, Ont.
Canada Metal Co., Toronto.
Eadie, H. G., Montreal.
Gibb, Alexander, Montreal.
Kemp Mfg Co., Toronto
Leslie, A C , & Co , Montreal.
Lysaght, John, Bristol, Eng.
Nova Scotia Steel and Coal Co., New Glasgow, N S.
Robertson, Jas , Co , Montreal.
Roper, J H., Montreal.
Samuel, Benjamin & Co , Toronto.
Stairs, Son & Morrow, Halifax, N S
Thompson, B & S. H. & Co. Montreal

Metal Lath.
Galt Art Metal Co., Galt
Metallic Roofing Co., Toronto.
Metal Shingle & Siding Co , Preston, Ont.

Metal Polish, Emery Cloth, etc
Oakey, John, & Sons, London, Eng

Mops.
Cumming Mfg. Co., Refrew

Mouse Traps.
Cumming Mfg. Co., Renfrew

Oil Tanks
Bowser, S F., & Co., Toronto.

Paints, Oils, Varnishes, Glass.
Bell, Thos , Sons & Co , Montreal.
Canada Paint Co., Montreal.
Canadian Oil Co., Toronto
Consolidated Plate Glass Co., Toronto.
Fenner, Fred , & Co., London, Eng.
Henderson & Potts Co., Montreal.
Imperial Varnish and Color Co., Toronto.
Jamieson, R. C., & Co , Montreal.
McArthur, Corneille & Co., Montreal.
McCaskill, Dougall & Co., Montreal.
Montreal Rolling Mills Co., Montreal.
Moore, Benjamin, & Co Toronto.
Queen City Oil Co., Toronto.
Ramsay & Son, Montreal.
Sherwin-Williams Co., Montreal.
Standard Paint and Varnish Works Windsor, Ont.
Stephens & Co , Winnipeg.
Martin-Senour Co., Chicago.

Perforated Sheet Metals.
Greening, B., Wire Co , Hamilton.

Plumbers' Tools and Supplies.
Borden Co., Warren, Ohio.
Canadian Fairbanks Co., Montreal.
Cluff, R. J., & Co., Toronto.
Glauber Brass Co., Cleveland, Ohio.
Jardine, A. B., & Co , Hespeler, Ont.
Jenkins Bros., Boston, Mass.
Lewis, Rice, & Son, Toronto.
Merrell Mfg. Co., Toledo, Ohio.
Montreal Rolling Mills Montreal.
Morrison, Jas., Brass Mfg. Co., Toronto.
Mueller, H., Mfg Co , Decatur, Ill.
Oshawa Steam & Gas Fitting Co., Oshawa
Robertson Jas , Co. Montreal.
Stairs, Son & Morrow, Halifax, N S.
Standard Ideal Sanitary Co., Port Hope,
Standard Sanitary Co , Pittsburg.
Stephens, G F , & Co , Winnipeg, Man.
Turner Brass Works, Chicago.
Vickery, Orlando, Toronto.

Portland Cement.
International Portland Cement Co., Ottawa, Ont
Hanover Portland Cement Co., Hanover, Ont.
Hyde, F., & Co., Montreal.
Thompson, B. & S. H. & Co., Montreal.

Potato Mashers.
Cumming Mfg Co., Renfrew.

Poultry Netting.
Greening, B., Wire Co., Hamilton, Ont.

Razors.
Clauss Shear Co., Toronto.

Roofing Supplies.
Brantford Roofing Co., Brantford.
McArthur, Alex., & Co., Montreal.
Metal Shingle & Siding Co., Preston, Ont.
Metallic Roofing Co , Toronto.
Paterson Mfg. Co., Toronto & Montreal.

Saws.
Atkins, E. C., & Co., Indianapolis, Ind
Lewis Bros., Montreal.
Shurly & Dietrich, Galt, Ont.
Spear & Jackson, Sheffield, Eng,

Scales.
Canadian Fairbanks Co., Montreal.

Screw Cabinets.
Cameron & Campbell, Toronto.

Screws, Nuts, Bolts.
Montreal Rolling Mills Co., Montreal,
Morrow, John, Machine Screw Co., Ingersoll, Ont.

Sewer Pipes.
Canadian Sewer Pipe Co , Hamilton
Hyde, F., & Co., Montreal.

Shelf Boxes.
Cameron & Campbell, Toronto.

Shears, Scissors.
Clauss Shear Co., Toront .

Shelf Brackets.
Atlas Mfg. Co , New Haven, Conn

Shellac
Bell, Thos , Sons & Co , Montreal

Shovels and Spades.
Canadian Shovel Co , Hamilton.
Peterboro Shovel & Tool Co , Peterboro.

Silverware.
Phillips, Geo , & Co., Montreal.
Round, John, & Son, Sheffield, Eng.

Skates.
Canada Cycle & Motor Co., Toronto.

Spring Hinges, etc.
Chicago Spring Butt Co., Chicago, Ill.

Steel Rails.
Nova Scotia Steel & Coal Co., New Glasgow, N.S.

Stoves, Tinware, Furnaces
Canadian Heating & Ventilating Co., Owen Sound
Canada Stove Works, Harriston, Ont.
Clare Bros. & Co , Preston.
Davidson, Thos., Mfg. Co , Montreal,
Guelph Stove Co., Guelph.
Gurney Foundry Co., Toronto.
Harris, J. W., Co., Montreal.
Joy Mfg. Co., Toronto.
Kemp Mnfg. Co. Toronto.
McClary Mfg. Co. London.
Pease Foundry Co., Toronto.
Stewart, Jas., Mfg. Co., Woodstock, Ont.
Taylor-Forbes Co., Guelph, Ont.
Wright, E. T., & Co., Hamilton.

Tacks.
Montreal Rolling Mills Co., Montreal.
Ontario Tack Co., Hamilton.

Ventilators.
Harris, J. W., Co , Montreal.
Pearson, Geo. D., Montreal.

Washing Machines, etc
Dowswell Mfg. Co., Hamilton, Ont.
Taylor-Forbes Co., Guelph, Ont.

Wheelbarrows.
London Foundry Co , London, Ont.

Wholesale Hardware.
Birkett, Thos., & Sons Co., Ottawa.
Caverhill, Learmont & Co., Montreal.
Frothingham & Workman, Montreal.
Hobbs Hardware Co., London.
Howland, H. S., Sons & Co., Toronto.
Lamplough, F. W., & Co , Montreal.
Lewis Bros. & Co., Montreal.
Lewis, Rice, & Son, Toronto.

Window and Sidewalk Prisms.
Hobbs Mfg. Co., London, Ont.

Wire Springs.
Guelph Spring Axle Co., Guelph, Ont.
Wallace-Barnes Co., Bristol, Conn.

Wire, Wire Rope, Cow Ties, Fencing Tools, etc.
Canada Fence Co., London.
Dennis Wire and Iron Co., London, Ont.
Dominion Wire Mnfg. Co., Montreal
Greening, B., Wire Co., Hamilton.
Montreal Rolling Mills Co., Montreal.
Western Wire & Nail Co., London, Ont

Woodenware.
Taylor-Forbes Co., Guelph, Ont.

Wrapping Papers.
Canada Paper Co., Toronto.
McArthur, Alex., & Co , Montreal.
Stairs, Son & Morrow, Halifax, N.S.

FAVORITE CHURN

More sold in the Dominion than all others combined

In 8 sizes, churning from ½ to 30 gallons.

IMPROVED STEEL FRAME. PATENT FOOT and LEVER DRIVE.
PATENT STEEL ROLLER BEARINGS

Easy to operate. Bolts throughout in place of wood screws. Superior in workmanship and finish. **No other Just as good. Accept no substitute.** If not offered by your jobber write direct to us.

MAXWELL LAWN MOWER

Made in 4 different heights of wheel and widths, ranging from 8 to 25-in. in plain or ball bearings as required. Improved for coming season. Not equalled anywhere for clean cutting, durability and finish.

DAVID MAXWELL & SONS
ST. MARY'S, ONT.

D. C. Mitchell & Co.

GLASGOW

STEEL PLATES,
SHEETS, ANGLES,
BARS,
PIG IRON,

REPRESENTATIVE

Eadie-Douglas Company
22 St. John St., MONTREAL

IMPROVED STEEL WIRE TRACE CHAINS

Every Chain Guaranteed.
Gives Universal Satisfaction.

THE
B. GREENING WIRE CO., LIMITED

Hamilton, Ont. Montreal, Que.

HARDWARE BABBITTS

No. 1, No. 2, No. 3

The wearing quality of our

HARDWARE BABBITTS

is the secret of the large orders the

HARDWARE MERCHANTS

are sending us. If not on our list

GET IN LINE

LIBERAL DISCOUNTS. TERMS RIGHT

The James Robertson Co., Limited

144 William St., MONTREAL, QUE.

PFLUEGER'S Fishing Tackle

Hooks, Flies, Trolls, Spinners, Phantoms, Reels, Furnished Lines.

Everything in Fishing Tackle.

NOTICE—Free to Any Dealer in Sporting Goods, Sent Express Prepaid, 170-Page Illustrated Catalogue No. F25, and Metal Fish Sign in 8-Color Lithograph.

The Enterprise Mfg. Co.

AKRON, Ohio, U.S.A.

WALKER CORK SCREWS

are good holiday souvenirs and live sellers the whole year. Are made well and are made out of good material. That is why we guarantee every one that we put out. When ordering Cork Screws try **Walker's**

Manufactured only by **ERIE SPECIALTY COMPANY**, ERIE, Pa. U.S.A.

LUFKIN MEASURING TAPES

Steel, Metallic, Linen, Pocket, Ass Skin, Pat. Leather, Bend Leather, Etc,

ARE THE BEST AND MOST POPULAR TAPES IN THE WORLD.
YOUR STOOK IS NOT COMPLETE WITHOUT THEM.

LUFKIN RULE CO., Saginaw, Mich, U.S.A.

London Office and Warehouse—24 and 26 Holborn. New York City Branch—280 Broadway.

For sale by ALL PROMINENT CANADIAN HARDWARE JOBBERS.

CIRCULATES EVERYWHERE IN CANADA

Also in Great Britain, United States, West Indies, South Africa and Australia.

HARDWARE AND METAL

A Weekly Newspaper Devoted to the Hardware, Metal, Heating and Plumbing Trades in Canada.

Office of Publication, 10 Front Street East, Toronto.

VOL. XIX. MONTREAL, TORONTO, WINNIPEG, MARCH 2, 1907 NO. 9.

See Classified List of Advertisements on Page 71.

No. 1, 2, 3, 4, 4½
No. 17, 19
No. 5, 5½, 6, 7, 8
Think—
Skilled mechanics always ask for "STANLEY'S PLANES."
Is it not far more profitable to sell them the "genuine STANLEY," which gives every satisfaction and holds them, than to sell them an inferior substitute and have them go elsewhere for the next tool they require?
The irons in these good planes are made from the best English steel, of equal thickness and temper throughout.
Each tool passes a rigid examination before leaving the works, and is fully guaranteed against defect in either material or workmanship.
For household and other purposes we have another make of Iron Planes made in all the leading styles and sizes, known as "Hunts," which sells at a low price.
Caverhill, Learmont & Co.
PLANE AND TOOL HEADQUARTERS
Montreal and Winnipeg
He Sells Most
Who Offers
Stanley's Tools
124
Varieties
in Stock to
Select From.
A Plane
for Every Purpose.
No. 113
Corrugated Bottom
Same price as plain
Nos. 26 to 34

Brantford—Roofings

This Represents Our

"ASPHALT"
No. 1, 2, 3
"RUBBER"
No. 1, 2, 3

Each Roll is complete with Nails and Cement to lay. Best selling goods in the market.

BRANTFORD ROOFING CO.
LIMITED
BRANTFORD, ONTARIO

SIMPLE

DURABLE

RELIABLE

STYLE F.

THE LEADER

A Name Familiar to Butter Makers in Connection with the

Highest Grade Churns

ITS STRONG POINTS:

Stiff Steel Frame, with malleable iron fittings, Combination Bow Lever and Pedal Drive. Bicycle Ball Bearings. Bolted Trunnions. Cream Breakers. Barrel easily drained and conveniently detatchable for cleaning.

DEALERS: When you handle this line you handle QUALITY.

The Dowswell Manufacturing Co., Limited

W. L. HALDIMAND & SON, Montreal
EASTERN AGENTS

HAMILTON
ONTARIO

GALVANIZED FENCE HOOK FOR FASTENING WOODEN PICKET ON WIRE FENCES

WIRE NAILS, COILED SPRING, BARB and PLAIN FENCE WIRE, OILED and ANNEALED, CLOTHES LINE WIRE, STAPLES, etc.

THE WESTERN WIRE & NAIL CO., Limited, ——LONDON, ONT.

The Peterborough Lock Manufacturing Company, Limited

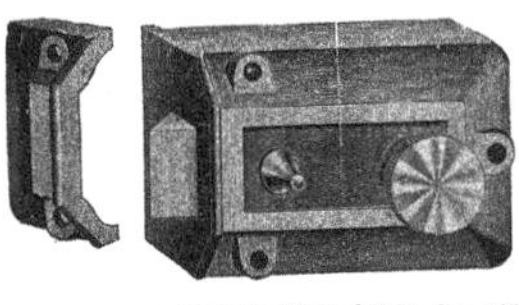

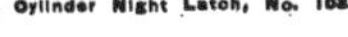

Cylinder Night Latch, No. 108.

Peterborough, Ont.

Manufacturers of all kinds

Rim and Mortise Locks, Inside, Front Door, and Store Door Sets, also full line High-class Builders' Hardware.

Sold by all Leading Jobbers in the Dominion.

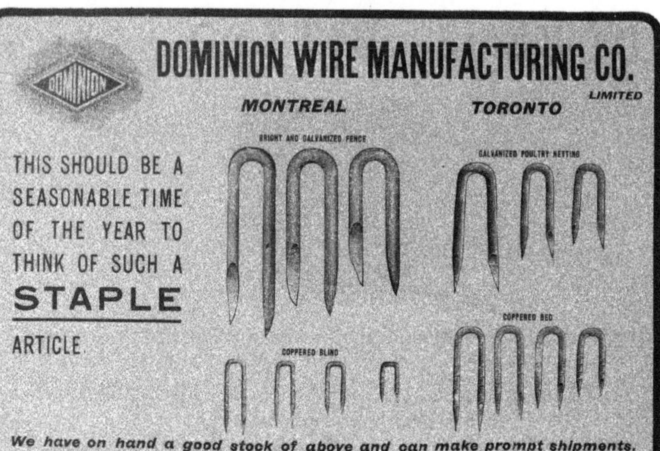

Expert Bookkeeping

Investigation and auditing of all classes of accounts undertaken. Trading, profit and loss accounts and balance sheets prepared. Sets of books opened and office systems devised in an up-to-date manner. Collection of accounts a specialty.

Davenport, Pickup & Co.

622 McIntyre Block and 422 Ashdown Block
WINNIPEG MAN

Square your Accounts

INTERNATIONAL PORTLAND CEMENT COMPANY, Limited

OTTAWA, - - ONTARIO.

Manufacturers of strictly high-grade Portland Cement made by the most improved method of manufacture in existence. Each barrel contains three hundred and fifty pounds of pure Portland Cement with no adulteration.

Don't Forget the Name. . .

NEWMAN'S INVINCIBLE FLOOR SPRINGS

Strong, Quick, Reliable, Effective.

Will close a door against any pressure of wind. Far ahead of ordinary door springs, pneumatic or otherwise. Ask your wholesaler.

W. NEWMAN & SONS, Birmingham.

"Perfection" Stoves and Ranges

The "Home Perfection" entirely new for 1907.

The "Home Perfection" is bound to be a most popular line, where wood is the fuel.

The "Home Perfection" will be made in both stove and range construction, in four sizes.

HOME PERFECTION

DISTINCTIVE FEATURES: Large Square Oven; Long and roomy Fire-box; Removable Oven Bottom; Deep Ash Pit with Drop Hearth; Rounded Bottom Copper Tank; Handsome Ornamentation; ***Strictly First-class.***

DIMENSIONS

No.	Fire-box	Oven
No. 9—20	Fire-box 24 inches	Oven 20 x 20 x 12 inches
" 9—22	" 26 "	" 22 x 22 x 13 "
" 9—24	" 28 "	" 24 x 24 x 14 "
" 9—26	" 30 "	" 26 x 26 x 15 "

PROGRESSIVE DEALERS cannot afford to be without the "Home Perfection" line on their floor.

☞ OUR PRICES ARE INTERESTING

The James Smart Mfg. Co.

LIMITED

Western Branch: **WINNIPEG, MAN.** **BROCKVILLE, ONT.**

Let us Help You Increase Your Business

Why not **go after and get** the fire brick trade in your locality.

Our expert information is at your command and will enable you to make this a paying line.

Make this one of the New Year resolutions and KEEP IT.

Francis Hyde & Co.

Contractors and Foundry Supplies

31 WELLINGTON ST.

MONTREAL

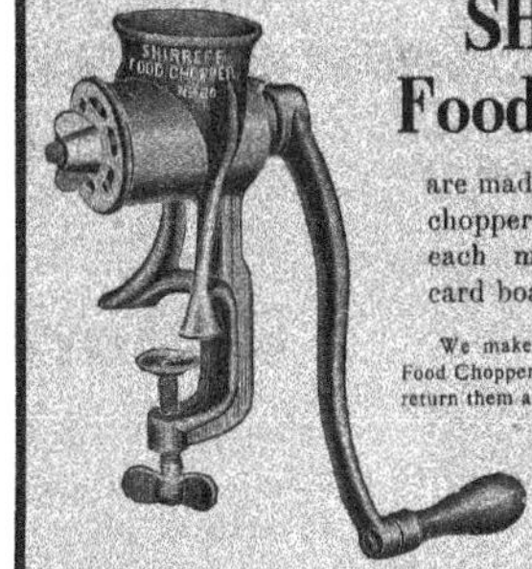

SHIRREFF Food Choppers

are made in four sizes, every chopper has steel k nives and each machine is packed in card board box.

We make you a fair proposition. If our Food Choppers are not satisfactory, you may return them at our expense.

Write for price list before placing orders for Fall delivery.

SHIRREFF MANUFACTURING CO., BROOKVILLE ONT.

Don't Stop

to ask your neighbor, lift the load yourself with **The Burr Self-Locking Tackle Block** Saves the labor of two or three men. 600 to 5,000 pounds capacity.

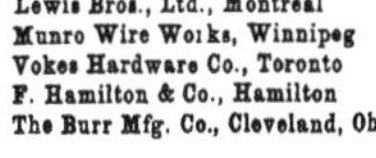

Lewis Bros., Ltd., Montreal
Munro Wire Works, Winnipeg
Vokes Hardware Co., Toronto
F. Hamilton & Co., Hamilton
The Burr Mfg. Co., Cleveland, Ohio

Cap Screws, Set Screws, Machine Screws, Cold Pressed Nuts, Studs, Coupling Bolts, Special Milled Work, etc.

CANADA FOUNDRY COMPANY

LIMITED

HEAD OFFICE and WORKS, TORONTO, ONT.

District Offices: Montreal, Halifax, Ottawa, Winnipeg Vancouver, Rossland

OUR LINES ARE SELLERS!

Why Not Write for Particulars?

CABLE ADDRESS "SUPERIOR" *FARLEY WORKS, SHEFFIELD, ENG.*

GEORGE PHILLIPS & CO.

MANUFACTURERS OF

Silverware, Electroplate and Cutlery

We also make the best Razor on the market. Four exclusive brands with advertised names that will help every dealer. It costs a two cent stamp to find out about these solid old country lines which you cannot duplicate elsewhere. *ALL MONEYMAKERS.*

OFFICE FOR CANADA, PACKARD BUILDING, MONTREAL **TELEPHONE M 557**

A Perpetual Demand

There is a perpetual demand for window glass of all shades and designs and most shrewd dealers are alive to that fact. Have you considered just how much in actual dollars and cents a good stock of glass may mean to you? It does not matter how extraordinary and varied the demands of your customers may be, we are always ready to fill your orders promptly. Our Catalogue is a panoramic view of

Window Glass

plain or artistic and coloured as you please. It also contains suggestions for

Fan Lights, Door Lights, Hall Windows, Side-Lights, Etc., made in Copper or German Silver Bars.

You cannot possibly secure an order which we cannot supply. This is just the time of year to talk glass with your friends, because they talk of building. Experienced packers and ultra strong cases assure delivery in perfect condition.

The Hobbs Manufacturing Co., Limited

LONDON TORONTO WINNIPEG

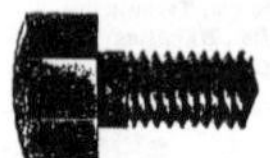

The John Morrow Screw, Limited

MANUFACTURERS OF
Set and Cap Screws, Special Milled Work, Engine Studs Etc. Cold Punched Nuts of every variety of finish.

INGERSOLL, ONTARIO

Covert Mfg. Co.
Troy, N.Y.

Harness Snaps, Chain, Rope and Web Goods, etc.

For sale by Jobbers at Manufacturers' Prices

20 510 280

CELEBRA.ED **HEINISCH** SHEARS

Tailors' Shears,
Trimmers' Shears,
Tinners' Snips, etc.

ACKNOWLEDGED THE BEST.

R. HEINISCH'S SONS CO. NEW YORK OFFICE, 155 Chambers St
NEWARK, N.J., U.S.A.

STREET PAVING and SIDEWALKS a SPECIALTY

SILICA BARYTIC STONE CO.

OF ONTARIO, Limited.

Water Proof Floors for Malt Houses, Breweries, Slaughter Houses, Cheese Factories, Cellar, Stable Floors, etc.

Head Office
Ingersoll, Ontario.
Walter Mills, General Manage
Ask for quotations for Septic Tanks.

HENRY T. SEYMOUR SHEAR CO. NEW YORK
TRADE MARK

SEYMOUR SHEARS

HAVE BEEN THE
Standard for over Half a Century.
"QUALITY UNQUESTIONED."
Each pair of our shears bears the above trade mark.

HENRY T. SEYMOUR SHEAR CO. NEW YORK
TRADE MARK

Latest Catalogue will be sent in exchange for your business card.

Complete Line *TRIMMERS', BANKERS', BARBERS' and TAILORS' SHEARS, Etc., Etc.*

Henry T. Seymour Shear Company

WIEBUSCH & HILGER, Limited, NEW YORK Sole Agents

"The Tool Holder People"

ARMSTRONG BROS. TOOL CO.

Manufacturers o Armstrong Patent Lathe and Planer Tools and other machine shop specialties.

106 N. Francisco Ave. CHICAGO, ILL.

Persons addressing advertisers will kindly mention having seen this advertisement in Hardware and Metal.

"WE WORK WITH THE TRADE"

Your most fastidious customer will be pleased with our

METAL SPANISH TILES

For private residences and buildings where ornamental effects are desired, there is no prettier or more durable roof-covering obtainable.

We are glad to help you estimate prices on all prospective jobs, and we can make you solid with your customers by our **prompt shipments** and **liberal treatment.** Our price is right, too.

TELL US WHAT YOU NEED

The Metal Shingle & Siding Co. Limited

PRESTON, ONT.

MONTREAL FACTORY, Cor. St. Catherine and Delorimier Sts.

SALES AGENCIES

ST. JOHN, N.B. Emerson & Fisher — QUEBEC J. A. Bernard — WINNIPEG Clare & Brockest — CALGARY Ellis & Grogan

The Man You Know

talks pretty regularly these days about building in the spring. His wife looks forward to decorating the kitchen of the new home with

The Empire Queen Range

not just because it's beautiful, but because all her friends owning one, declare that **THE EMPIRE QUEEN** makes cooking a pleasure. All practical men know the reason why—oven efficiency explains all. There's just as much for you in selling **THE EMPIRE QUEEN** as in selling the ordinary range, and then your reputation is assured.

The Canadian Heating & Ventilating Co. Limited.

OWEN SOUND, Ontario.

THE CHRISTIE BROS. CO., Limited, 238 King St., Winnipeg, Man., Western Agents.

THE CANADA STOVE AND FURNITURE CO., 126 West Craig St., Montreal, Que., Agents for the Province of Quebec.

ALL No. 9 WIRE

Peerless Woven Wire Fence

Here is a fence that will sell. Once up no need for repairs. It lasts for years—just what you are looking for. It's horse-high, pig-tight and bull-strong. No stock can get over, under or through it. The most unruly animal can't butt it down and there are no barbs to tear and injure stock. It's

WIND AND WEATHER PROOF

The Peerless lock holds it perfectly rigid and the wires can't slip up, down, nor sidewise. Cold weather won't snap it and hot weather won't make it sag. It's the greatest thing out for farmers, stockmen and all other fence users which makes it a red hot proposition for dealers. It comes in rolls all ready to put up. It's easy to handle. Write for prices.

The Banwell Hoxie Wire Fence Company, Ltd.
Dept. J, Hamilton, Ont.

You Can Sell Real Good Fencing at a Profit.

Just another way of saying you can sell DILLON HINGE-STAY FIELD FENCE and be handsomely paid for doing so.

The "Dillon", you understand, is made not only to look well, but to have an awful lot of strength and common sense under its looks.

This quality is what makes it a friend maker and sale maker.

More information for he who wants it.

THE OWEN SOUND WIRE FENCE CO., Limited

OWEN SOUND, ONT.

Sold by { **Messrs. Caverhill, Learmont & Co., Montreal**
Messrs. Christie Bros. Co., Limited, Winnipeg

Transient Traders and Pedlars

They do a Good Deal of Harm to Legitimate Merchants— A By-law Which is Used in Ontario.

Numbers of Ontario municipalities have by-laws and see that they are enforced, with results that are satisfying to the merchants. Others have passed by-laws, but they exist in name only. In one town our representative visited recently we were told that transient traders and pedlars were much more injurious to local trade interests than the catalogue houses. "We have a by-law," said one dealer, "but the authorities pay no attention to it. As a matter of fact, one of the members of our council rents premises to a good many transients who come in with stocks."

If this merchant and the others in his town get together and go after the municipal council they can secure the enforcement of the by-law. An eye should be kept on the constable to see that he is on the alert for transients.

* * *

Following are some extracts from the by-law which has been passed by a number of Ontario municipalities:

1. No person shall carry on the business mentioned in the next succeeding section of this by-law unless and until he shall procure a license so to do, and every person so licensed shall be subject to the provisions of this by-law.

2. There shall be taken out by:

(a) All hawkers, pedlars, petty chapmen or other persons carrying on petty trades or who go from place to place or to other men's houses on foot or with any animal bearing or drawing any goods, wares or merchandise, for sale or in or with any boat, vessel or other craft, otherwise carry goods, wares or merchandise for sale (except that no such license shall be required for hawking, peddling or selling from any vehicle or other conveyance goods, wares or merchandise to any retail dealer or for hawking or peddling goods, wares or merchandise, the growth, produce or manufacture of this province, not being liquors within the meaning of the law relating to taverns or tavern licenses, if the same are being hawked or peddled by the manufacturer or producer of such goods, wares or merchandise, or by his bona fide servants or employes having written authority in that behalf and such servant or employe shall produce and exhibit his written authority when required so to do by any municipal or peace officer, and provided that in any prosecution for a violation of any such by-law against any hawker, pedlar or petty chapman or other person mentioned in this sub-section on the ground that any such person has not obtained a license in pursuance of any by-law passed thereunder and the defence is set up that such person does not require any such license by reason of the fact that he is peddling goods, wares or other merchandise to a retail dealer, or is hawking or peddling goods, wares or merchandise the growth, produce of manufacture of this province, not being liquors within the meaning of the law relating to taverns or tavern licenses, and is the manufacturer or producer thereof or the bona fide servant or employe of such manufacturer or producer thereof having written authority in that behalf, then and in such case it shall not be necessary for the complainant to show affirmatively that the person so prosecuted does not come within the defence so set up, but the onus of proving that he does come within such defence shall rest upon the person so prosecuted and in the event of his failing to establish at his trial that he does so come within such defence he may be convicted of a violation of this sub-section; but no such license shall be required from any tinker, cooper, glazier, harness-mender or any person usually trading or mending kettles, tubs, household goods or umbrellas, or going about carrying with him proper materials for such mending.

(b) The word "Hawkers" in this clause shall include all persons who, being agents for persons not resident within the county, sell or offer for sale tea, dry goods, watches, plated ware, silverware, furniture, carpets, upholstery, millinery or jewelry, or carrying and expose samples or patterns of any such goods to be afterwards delivered within the county to any person not being a wholesale or retail dealer in such goods, wares or merchandise.

(c) Every transient trader or other person whose name has not been duly entered on the assessment roll in respect of income or business assessment for the then current year, and who may offer goods or merchandise of any description for sale by auction or in any other manner, conducted by himself or by a licensed auctioneer, or by his agent or otherwise; but no license to any transient trader shall affect, apply to or restrict the sale of the stock of an insolvent estate which is being sold or disposed of within the city or town in which the insolvent carried on business therewith at the time of the issue of a writ of attachment or of the execution of an assignment.

(d) The words "Transient Trader" wherever they occur shall extend to and include any person commencing in the municipality the business in the said clauses mentioned who has not resided continuously in such municipality for a period of at least three months next preceding the time of the commencement by him of such business therein.

8. DISPLAYING BADGES — Every hawker, pedlar, transient trader or other person licensed under this by-law who carries on his business or calling with any wagon, cart or other vehicle, shall, at the time of the issue of his license, receive from the chief constable two plates bearing a number, which shall be affixed on a prominent place on the right and left side of the outside of such wagon, cart or other vehicle, and shall remain thereon during the period for which the license is granted; and no other devise displaying a number shall be exhibited upon the outside of such wagon, cart or other vehicle, and such plates shall be returned to the chief constable at the expiration of the term of the license; and every such licensed person shall have his name and address legibly painted on each side of his vehicle in letters at least three inches long.

11. FEES—There shall be levied and collected from the applicant for every license granted for the business or object are specified in this by-law, a license fee as follows:

(1) For a license for a hawker, pedlar or petty chapman selling on foot, any fee you deem advisable.

(2) For a license for a hawker, pedlar or petty chapman selling with a vehicle or with a horse or horses or other animal with a vehicle, any fee you deem advisable.

(3) For a license to carry on the business of a transient trader selling from stationary premises or on foot or with a vehicle or with a horse or horses or other animal with a vehicle, not to exceed in cities and towns $250, and in other municipalities $100.

Providing that the sum so paid for a license shall be credited to the transient trader paying the same upon and on account of taxes for the unexpired portion of the then current year, as well as any subsequent taxes, should such trader remain in the municipality a sufficient time for taxes to become due and payable by him and in any other event to be taken and used by the municipality as a portion of the license fund of such municipality.

Provided, nevertheless, that the license fee imposed by any by-law of any village, situated within any territorial district may be a sum not exceeding $200.

12. PENALTY—Except as hereinafter provided, any person convicted of a breach of any of the provisions of this by-law shall forfeit and pay, at the discretion of the convicting magistrate, a penalty not exceeding the sum of fifty dollars for each offense exclusive of costs; and in default of payment of the said penalty and costs forthwith, the said penalty and costs, or costs only, may be levied by distress and sale of the goods and chattels of the offender; and case of there being no distress found out of which such penalty can be levied, the convicting magis'rate may commit the offender to the common gaol of the city or town of . . . . with or without hard labor, for any period not exceeding six calendar months, unless the said penalty and costs be sooner paid.

* *

The by-law as it stands can be enforced against a merchant going into a place and purchasing a business. He may be regarded as a transient until he has resided there three months. His license fee, in case he remains, will apply on taxes. This clause will be amended at the present session of the Ontario Legislature so that legitimate retailers cannot be subjected to such annoyance.

Retail Hardware Association News

Official news of the Ontario and Western Canada Associations will be published in this department. All correspondence regarding Association matters should be sent to the secretaries. If for publication send to the Editor of Hardware and Metal, Toronto.

Officers Retail Hardware and Stove Dealers' Association of Western Canada:

President—A. J. Falconer, Deloraine.
First Vice-President—J. B Curran, Brandon.
Second Vice-President—W. M. Gordon, Winnipeg.
Secretary-Treasurer—J. E. McRobie, Winnipeg.
Executive—Alberta, A E Clements, Olds; C F. Comer, Calgary; A. R. Auger, Okotoks.
Manitoba—H S. Price, Boissevain, A. P Macdonald, Winnipeg; O Gilmer, Winnipeg.
Saskatchewan—G K. Smith, Moose Jaw; S A. Clark, Saskatoon, J. R. Fox, Weyburn.
Association offices, 53 Scott building, Main street, Winnipeg.

Officers Ontario Retail Hardware and Stove Dealers' Association:

President—A. W Humphries, Parkhill
1st Vice-President—W G. Scott, Mount Forest.
2nd Vice-President—J R Hambly, Barrie.
Treasurer—John Caslor, Toronto.
Secretary—Weston Wrigley, 10 Front St East, Toronto.
Executive Committee—The officers and J W Peart, St. Mary's, M. R Marshall, Dunnville, D Brocklebank, Arthur; F W. Jeffery, Midland, E. P. Paulin, Goderich, and Frank Taylor, Carleton Place
Auditors—J. W. Peacock and C F. Moorhouse, Toronto.

ONTARIO CONVENTION.

The Ontario Retail Hardware Association annual convention will be held in Labor Temple, Toronto, on March 28 and 29. Single fare on all railroads.

GET READY FOR CONVENTION.

Good Friday will soon be here and every hardwareman in Ontario should be making preparations to attend the convention of the Ontario Retail Hardware Association. The holiday will allow delegates to secure the single fare rate, and as the gathering provides an excellent opportunity for retailers to get together and exchange ideas no up-to-date merchant should miss the treat.

The association's success in forcing the Postmaster-General to withdraw the proposed parcels post c.o.d. legislation has given the organization a big boost, and both new members and renewals are being forwarded in encouraging numbers. There is every indication of the association being an even greater success in 1907 than in its first year of existence.

As the association has no organizer in the field and, in fact, no paid official of any kind, each merchant should co-operate by forwarding his membership fee without waiting for it to be called for. If each one does his part in this matter as well as was done in the parcels post agitation, the association will be in splendid shape to take up a strong campaign against other evils.

OPPOSITION IN NOVA SCOTIA.

Word comes from Nova Scotia to the effect that an effort is being made by certain parties to prevent the formation of the Nova Scotia Retail Hardware Association. The attempt should not succeed. Organization should help both jobbers and retailers and any who oppose this modern method of doing business should be made to realize that they are opponents of progress.

ORGANIZE AGAINST CREDIT EVIL.

Oak Point, Man., retailers are determined to fight the credit system and they have signed a hard and fast agreement to sell only for cash for one year from March 1st. Each dealer has paid in $300 as a guarantee of good faith. For the first breach of the agreement the fine is $50, for the second $100, and for the third, $150. Fines are to be devoted to charitable purposes.

BIG DIVIDENDS.

James Purvis, of Purvis Bros., Sudbury, who was a visitor in Toronto a couple of weeks ago, told the secretary of the Ontario Association that he had just received a credit note for $6.36 for cartage paid on a carload shipment from one of the jobbing houses. Mr. Purvis says this one item will pay his membership fee for two years and he cannot understand the diffidence some of the retailers show about joining the association.

Mr. Purvis was the first to urge the trade in Ontario to organize an association, but has so far been unable to attend any of the conventions. He will be on hand on Good Friday, however, and, being one of the largest retailers in northern Ontario, should be a valuable addition to the delegates' roll.

If large firms, like Purvis Bros., find it an advantage to belong to the association and can by working hand in hand with their fellow retailers make their membership profitable, there is no reason why retailers, whose interests are not as extensive, should not enjoy the benefits of membership when they can be secured for only $3 per annum.

DISTRICT ASSOCIATIONS.

President Humphries expects to have something interesting to report to the Good Friday convention regarding the formation of a district association. The matter is now in hand in his district and the trade seem to be falling in line with the suggestion with avidity.

In London there is a strong local association which the retailers have used to good advantage. Up to the present, however, all the members have not seen fit to affiliate with the Ontario Association.

In Niagara Falls a movement is also on foot to form a local association covering one or two towns. A meeting is to be arranged at once and a definite plan worked out. As it is, the trade are all members of the provincial association and are working together harmoniously, so it should not be hard to get a local or district association going.

It is hoped that some good reports can be made to the convention on district organization so that other sections can take up the matter during the year.

PARCELS POST IN PARLIAMENT.

An interesting discussion on the subject of the C.O.D. parcels post took place in Parliament at Ottawa on Feb. 8, but as the daily press combined to maintain silence on this matter, the public has been kept in ignorance on the subject. Through Richard Blain, M.P., a hardware merchant at Brampton, Ont., however, copies of Hansard containing the report have been received and any reader desiring copies can secure them on request of Mr. Blain or their local M.P.

One of the interesting features of the discussion was the presentation of a petition (the form letter suggested in Hardware and Metal of Jan. 19) signed by Ald. J. J. Coffey, hardware merchant, and about 50 other retail merchants at Barrie. Messrs. A. Martin, Queens, P.E.I.; J. J. Hughes, Kings, P.E.I.; A. McLennan, Inverness, N.S.; G. O. Alcorn, Prince Edward; H. Lennox, South Simcoe; H. S. Clements, West Kent; H. H. Miller, South Grey; and A. A. McLean, Queens, P.E.I.; were the members who took part in the discussion.

B.C. ORGANIZING.

The annual meeting of the B. C. Wholesale Hardware Dealers' Association was held in Victoria recently, delegates being present from New Westminster, Vancouver, and Victoria. Officers were elected as follows: President, C. Swanger, of E. G. Prior & Co., Victoria; vice-president, John Boyd, of Boyd, Burns & Co., Vancouver; secretary-treasurer, John Burns, Vancouver. It was decided to hold the annual meeting in New Westminster. A meeting of the retailers' association was also held, the wholesalers entertaining the retailers at a banquet.

WILL HELP EVERY RETAILER.

"We are pleased to note that the parcels post legislation was defeated. It was through your suggestion that we acted in this matter," write Hose & Canniff, one of the new members of the association.

The defeat of this legislation should be worth a dozen times $3 per year to every hardwareman in Ontario. How many will contribute one $3 to strengthen the association which took up and successfully led the fight against this departmental store game?

RETAILERS UNITE TO MEET MAIL ORDER COMPETITION

Six retail stores in Cartwright, Man., have amalgamated to form a joint stock company, the objects in view being to restrict the credit system and to enable the Cartwright stores to meet the competition of the big catalogue houses. This is a new way of dealing with a couple of problems with which almost every dealer is confronted, and the trade in all parts of the Dominion will be interested in watching the measure of success or failure which may attend this novel experiment.

Cartwright is a small Manitoba village of about 500 population, serving a rich and prosperous farming community. Its merchants would have found business conditions quite satisfactory were it not for two things, viz., they could not escape allowing long credits to their customers, and many of these long credit customers were sending their cash to Winnipeg and Toronto to the catalogue houses. It is an old story, a story which may be duplicated in almost any part of Canada. The local store gives credit to its customers and they, being ungrateful and unappreciative, send their cash, when they have it, to the catalogue houses in the city. In one week it is said that one catalogue house took $1,200 in cash from the little village of Cartwright.

Confronted with these conditions, the Cartwright dealers have taken a bold step, the results of which it is too early yet to predict. Six firms—Moore & Hills, McKenzie & Byce, L. H. Phillips, Wm. Gemmill, W. Macklem and Cannon & Pickell—have formed a joint stock company, to be known as "Merchants Limited." In a circular letter to their customers these firms announce that the new company is capitalized at $50,000, divided into shares of $100 each, and that any farmer or private individual may purchase these shares to the extent of $500. R. F. Moore is president and general manager; Wm. Gemmill 1st vice-president; L. H. Phillips, 2nd vice-president; and J. J. Hill secretary treasurer.

To Fight Department Stores.

In the circular to their customers announcing their new departure the promoters do not disguise the fact that one of their main objects is to fight the city department store. To quote from their circular:

"We would impress strongly upon our patrons that this company is not formed for the purpose of inflating prices or restraining trade; on the other hand, it is formed expressly to enable us to compete successfully with catalogue houses.

"On account of the merger we shall be enabled to buy on the very cheapest available markets, and to save local freights, as practically all our stocks can be bought in car lots. This fact will allow us to sell goods very much cheaper than under present conditions. We would recommend any person who entertains any doubt on this point to from time to time compare our prices with those ruling in adjacent towns."

The Department Idea.

The new company—Merchants Limited—is in fact a new department store, nothing more, nothing less. Its departments are not all under one roof, but it is not impossible that that may come in the future. For the present, stocks have been consolidated and rearranged, groceries being in one store, hardware in another, and so on. To fight the mail order department of the department store in Winnipeg, another department store has been formed in Cartwright. That is the gist of the whole matter.

Against Credits.

But the department store in Cartwright is not to be conducted on exactly the same lines as a city department store. For example, while it is out to fight the credit system it is not insisting upon a strictly cash system of doing business. Credit is being given but only according to a given system. A customer who wants credit must secure a coupon book good for a stated amount, say $10 or $25. In order to get this book he must give Merchants Limited his note for the amount; he is allowed credit in no other way. When he buys

R. F. MOORE, PRESIDENT

goods on credit he pays for them with his coupons. Bookkeeping expenses are thus eliminated and in their advertising the new firm are making a strong feature of this point. Reduced expenses, they say, will allow a reduction in retail prices.

The Produce Question.

Evidently the new company purposes grappling with the produce question as it presents itself to every country store. This is a problem which never confronts the city department store and the country department store has to blaze a trail of its own.

Every country dealer, whether in Western Canada or Eastern Canada knows the difficulty of handling butter and eggs and other farm produce satisfactorily. Competition among rival stores has often forced dealers to pay for butter and eggs a higher price than they can get from the wholesale produce houses. Of course payment is made in goods upon which there is a profit, but the practice is suicidal nevertheless. The customer who pays cash should get the discounts, not the customer who pays in produce which costs time, trouble, and money to handle. Moreover there is seldom in the country store any efficient method of grading butter, and because he does not want to offend an otherwise good customer many a merchant pays for poor butter as high a price as he pays for the first class article.

Merchants Limited announce that all butter is to be graded and paid for according to quality; and it may be presumed quite safely that prices for produce will not be more than the market value. A big, strong concern can grapple with a problem of this kind and solve it satisfactorily.

Probable Outcome.

What will be the outcome of this venture? What will be the attitude towards it of the farmer and the general public? These are moot questions and time will give the only satisfactory answer.

In the meantime it seems that the general public in Cartwright are disposed to regard the new move with approval. Merchants Limited are making a strong bid for local sympathy and they are appealing to local self interest in a convincing way. They state that they are prepared to compete with the catalogue houses and that they invite comparison of prices. In the course of their circular announcement they say:

"We feel that our patrons will agree with us that these departmental stores are drawing on the resources of the small towns and building up the cities. This it is evident cannot be for the weal of people residing in the country and on farms, for it is generally admitted that a good town helps materially to enhance the value of adjacent farm lands; therefore we would ask your co-operation and we will prove to you that you will be benefited in dollars and cents."

The obvious danger is that the public will view the elimination of competition in Cartwright with profound distrust. Everything will depend upon the manner in which the consolidated business is managed. This paper will follow the new venture and from time to time will publish news of its success or failure.

MAKE YOUR STORE ATTRACTIVE.

Realization of changed trade requirements and determination to fit yourself to them can hardly show in any one direction more resultfully than in the way you display your goods.

Make your whole store attractive. People generally like that way of storekeeping well enough to go blocks out of their way to buy even a trifling thing where goods are kept in nice, clean, attractive style and where customers are made to feel at home and to understand that their business is appreciated.

They cannot buy on sight in your store unless they do see your goods. If you keep your wares way up near the ceiling in dust proof boxes, don't be disappointed because they stay there.

Of two evils choose neither; both will come home to roost.

It's a wise boy who knows enough to laugh at his father's jokes.

AMONGST THE SALESMEN

GOOD SALESMANSHIP.

They have in Boston a salesmanship class in connection with one of the night schools. Leading merchants have appeared before these young people who desire to succeed in business, and each has emphasized some special line of effort that shall mean advancement in the chosen occupation.

Mr. Ferris, of the Gilchrist Co., in his address last week gave as his motto: "Make up your mind what you want; go after it and keep after it until you get it." "I did not come here to-night to teach you how to sell goods or how to trim windows, but how to get along and reach the top of the ladder," he said. "There is lots of room at the top, the bottom is much crowded. Great opportunities are always open for energetic men and women. Ambition accomplishes many things and will always make people better. The surest test of a man's mettle is his ambition. Ambition makes friends and makes enemies, and makes happiness and unhappiness, and the sooner you get ambition the better it will be for you.

"If you have ambition and do not get discouraged you will surely get along. If you desire to be a Rockefeller or a Morgan, you cannot smother it, you will be such a man.

"Take the best job that is offered to you and have ambition, but be sure always to back it up with honest methods. Keep at all times from getting a big head. If you have ambition and determination you will succeed. Every man who starts with a determination can be his own master.

"If I should lose my position to-morrow I would start in as a salesman and I would at the outset get the names of all people who come into the store and try to please them. Try to excel. Be optimistic. If you are worth getting a raise in salary you will get it."

NEW COLOR AGENCY.

The stories of the "slowness" of the Englishman in business are numerous, but as business men most of them "arrive." In fact, it is said that few have ever been able to beat the British trader at his own business. W. O. Greenway, 13 St. John Street, Montreal, is a good example of the bright, up-to-date Britisher. He arrived in Montreal recently and opened well equipped offices and sample rooms at the address noted, selling a varied list of British and Continental goods along such lines as fine colors and lakes of all kinds for all purposes; ochres, siennas, umbers, oxides, purple browns, turkey and venetian reds, ivory blacks, crocus, rottenstone, etc.; also varnish gums, gambia lac for knottings, polishes etc., hat manufacturing, insulating, etc., English filling-up powder (pink and grey) lithopone, zinc white, barytes, sulphate of copper, flake and powdered litharge, also red lead and many other kindred specialties.

Mr. Greenway is a "hustler" and has been 27 years in the trade. He will cover the whole Dominion and parts of the United States. His varied experience in the color and allied trades both in Britain and on the European continent, also by previous traveling in United States and Canada, give him accomplishments that mark him as one with rare possibilities before him in Canada. Mr. Greenway is not only seeking business in the lines named, all of which he can offer to advantage, but is also looking for certain Canadian produce for regular shipment to England.

His well printed noteheads bear a unique picture at the top. It embraces a map, showing in red the British Islands and Canada. London and Montreal are in bold black type and sweeping, as if in the form of an ocean liner's track or as a connecting link is the name Wm. O. Greenway. Altogether a striking and original example of letter heading.

TRAVELERS' TAXES IN NEW ZEALAND.

J. S. Larke, Canada's commercial agent in Australia, writes: "In New

WM. O GREENWAY, Montreal.

Zealand the tax of fifty pounds levied a few years ago was changed to an income tax on the goods sold. It was considered that a profit of five per cent. had been made on all sales, and the Government levied a tax of two and a half per cent. upon this profit. This amounted to one-eighth of one per cent., or twelve and a half cents on the hundred dollars of orders taken. It is impossible for the majority of travelers to know how much their sales have been as the larger firms commonly send in their orders through their London or New York agents, and this furnished a means of evading the act and of often escaping it. The Government now requires every traveler to make a deposit of five pounds on entering the colony and pay duty on his samples also. On leaving it he is required to make a declaration of his sales, when, if the deposit exceeds the tax due, the excess is refunded to him, with the duty on the samples he may take away with him. He may enter at any port or take his departure from any port of the colony, a certificate of the deposit and duty paid being given him at the port of entry, which is honored at the port of departure."

A TRAVELER'S MISTAKE.

"A salesman came into my office the other day," said a retailer of hardware, "whom I had never seen before, but of whose goods I had formed a pleasant impression.

"As he came up to my desk I held out my hand, and was prepared to give him a cordial reception, but before I had a chance to say more than 'Glad to see you,' he broke out something like this: 'Never thought I would get here with that poky old local. Is that the best you can do in the way of trains? Then the rain and mud nearly finished me. Never struck worse roads in my life. Nasty weather for any man to be out, anyway.'

"It took all the cordiality out of me, and threw a chill over the interview. He was a good salesman so far as some things went, but a mighty poor one in others. It took him a half-hour to thaw out the ice that his ill-timed salutation and his growl at our town had formed over my buying impulses."

A LITTLE DIPLOMACY.

I was in search of my first job and had been scanning the "Help Wanted" columns every day for a month before school closed, when at last I read the words: "Boy Wanted—Call at No.— Market Street, at 8.00 o'clock a.m.; References," writes a man who is now manager of a large business.

I showed it to the principal at school, who readily gave me a "To whom it may concern" reference, setting forth all my good qualities, and, armed with this, I presented myself at No.— Market Street at 7.30 a.m., when lo! to my dismay there stood twenty others ahead of me.

I took my position in the rapidly increasing line, however, and had waited fifteen minutes when a young man appeared to sweep the sidewalk. While watching him a plan occurred to me. Why not send my letter in ahead?

Calling the boy, I asked him if he would take my "references" in if I would finish sweeping the sidewalk, to which he readily agreed. Upon his explaining matters to the "Chief" the latter immediately sent for me, "took me on trial" and dismissed the line of applicants.

I am still employed by the same firm, but as a salesman now, having given up my broom six years ago to another.

When you die the world will keep right on going around.

A man never fully realizes how homely he really is until he had his picture taken in a group.

Sporting Goods and Supplies

CANOES AND CAMP OUTFITS

E. T. Keyser Gives Some Practical Advice Regarding Stocking and Selling Canoes and Camping Supplies.

The hardware dealer who makes any pretence at handling sporting goods and who has watched conditions closely, during the last few seasons, could scarcely help noticing that the canoeist is gradually assuming the position, as a commercial proposition, which the bicyclist did in the palmy days of the wheel.

The canoeist is not, and probably never will be, as numerous as the bicyclist once was, but as an individual purchaser, he ranks far higher. When the bicyclist had bought a wheel, ranging in price from $150 downward to $70 or $80, a cyclometer lamp, a repair kit, bell and a few other sundries, he was tolerably dead as a customer, and the balance of his money was spent in board and lodging at wayside inns. These latter expenditures amounted to no small item, but did the dealer in bicycle supplies no earthly good.

The canoeist of to-day, on the other hand, expends from $25 to $170 for his canoe and rigging and is then in the market for cushions, back rests, paddles, etc. Before he has used his canoe many weeks, his afternoon trips have lengtehened to Saturday afternoon and Sunday cruises, which necessitate, in addition to his outfit, a tent, a cot, cooking utensils, camp stove, a lantern, a folding chair, waterproof clothing bags, etc., until he has gradually accumulated an equipment, the retail price of which has amounted to fully as much, if not more, than the amount which he paid for his craft.

Then, he discovers that his short cruises allow opportunities for fishing, shooting, rifle and pistol practice and amateur photography that no other class of outdoor sportsman can enjoy within such a short distance from home. Incidentally, such articles as camp blankets, ponchos, bathing suits, sweaters and other articles of clothing are always in demand by him.

Don't jump at the conclusion that, because a complete canoe equipment embraces a large and various assortment of sporting goods, with its value well up in the hundreds, that the canoeist must necessarily be a rich man, and therefore few and far between.

The canoeist has this advantage over any other class of outdoor lover. The money which he spends for his equipment takes the form of an investment. He cooks his own food in camp and pays no rent for his tent site, and the difference between the amount which he spends in a year for the food which he, himself, prepares and that which the bicyclist bought at a restaurant, and the saving afforded by pitching his tent with no one to demand rent over the cost of a room, at even a moderate-priced hotel, will about even things up in expenditures as far as canoeist and bicyclist are concerned. Remember also, and this is a main point of my argument, that the money which the bicyclist spent, generally went into the till of some wayside boniface, while the dollars, which the canoeist puts in circulation, go into the bank account of the merchant who has sufficient enterprise to handle goods which the canoeist needs.

A membership in a local canoe club will pay an enterprising dealer many hundred per cent. dividends on his initiation fees and dues, and he is thereby in a position to get into close touch with the canoeist in his vicinity, and also be on the qui vive for the trade of the individual who is just about to purchase his first craft.

There is a fair profit in canoes, and a good one in camp outfits. The sale of one naturally leads to the sale of the other, and the man, who purchases a satisfactory craft, is very apt to return to the same establishment, when he desires to accumulate his camp outfit. It therefore behooves the dealer to use care in the selection of his stock.

The canoeist's trade is worth having, as I have stated before, his original purchase of craft bears a good profit to the dealer, and there never was a canoeist yet, no matter what the size of his equipment, who was not in the market every successive season for additions to it. There are to-day thousands of towns, throughout the country, where men are now sending their orders for craft and outfits to outside parties, and where an establishment, at which they might procure their supplies, would secure a profitable trade. There are thousands of other towns where facilities for canoeing exist, and where a single canoe, put in commission, would induce the sale of dozens more. These are points well worth consideration by the up-to-date retailer who wishes to increase his business, and increase it in a paying direction.

Cut prices have not yet struck the canoeing trade and, from all present indications, it will be some time before they do so. Of course, ordinary common sense must be used in the selection of one's stock. In a well wooded country, where dead wood is plentiful, one of the light wood-burning camp stoves will be found the article most in demand for cooking purposes. Where wood is scarce, a vapor stove burning kerosene is a better proposition.

Then, too, it must be borne in mind that a large, heavy tent, although it may be admirably adapted for a camping party, staying in the same vicinity for many days or weeks, would not appeal to the canoeist in the slightest degree. The canoeist's tent must be light, portable and easy to pitch.

The canoeist, who makes a Saturday and Sunday camp, and whose tent remains in the same spot for but one night, does not care to go to the trouble involved, in properly settling a wall tent, with its numerous guy ropes and pegs, and the addition of a fly, with its extra bulk and weight, is not to be thought of. Therefore a marquee tent, with a single pointed pole, or a pyramid tent, both of waterproof material, and furnished with a waterproof floor cloth, and of comparatively light weight and giving, when set up an area of from 7 by 7 to 7½ by 7½ feet are the styles of tents which can be sold to best advantage to the canoeing trade. It is not necessary for the dealer to lay in an elaborate and costly stock of cooking and eating outfits. He can make these up from his own stock of tin or enamelware, cooking utensils, in various combinations, packing the various assortments in tin buckets of various sizes, accordingly as they are designed for one, two or more parties. He can have made up, to his special order, bags of waterproof canvas with double tops, the inside one tieing up like a half empty salt bag and the outside one closing by a draw string, running through brass grommets. These will last longer and keep clothing just as dry and pay a better profit to the retailer than the rubber clothing bags, which chip, scale and crack on exposure to the heat of the sun and the rough usage of camp life.

SPRINGS ON SKATES.

The absence of springs on ice skates has been more or less of a mystery. That they are needed is an undisputed fact, as the skater has no means of lessening the jarring naturally incident to the uneven surface of the ice. A New York inventor, noticing the total absence of springs on ice skates, experimented with them and found them desirable in every way. The springs are placed between the steel runner and the plates for the reception of the toe and the heel of the shoe. The addition of the springs also tends to ease the feet while skating, besides greatly adding to the enjoyment of the sport.

NOVEL FISH HOOK.

W. J. Evans, Minneapolis, Minn., has invented a fish hook on which the bait is fully exposed at all times, but cannot escape from the hook or be detached by the fish. At the same time no part of the hook is passed through the body of the bait and the frog, the bait preferably used, may swim about, and even rise to the surface and breathe with nearly as great freedom as though the hook were not attached. The swimming bait attracts the fish and cruelty to live bait is obviated.

STOVES AND TINWARE

STOVES OR BOILERS.

There are many homes in which the right kind of stoves will be more satisfactory to the occupants than some of the heating apparatus which is more persistently urged by salesmen representing other lines, says a writer in the Metal Worker. A short time ago I sold two parlor double heaters because they were cheaper and would be more satisfactory than a hot water heating boiler and radiators in the same building. I would not have made the sale had I not learned from the stove dealer that he had lost the sale of some stoves because the people were thinking of putting in a central heating plant. I asked the address of the people so that I might go and see them, and I reminded the man and his wife, who were familiar with the use of stoves, how pleasant it is to see the fire and feel the direct heat of a hot fire in cold weather. They were familiar from long experience with the comfort derived from a stove in the room, and when I further explained that a hot air pipe could be run up to a register in the floor to heat one or more bedrooms from each stove, they decided the arrangement was just what they wanted. I did not attempt to sell them a cheap stove or an unnecessarily fancy one. I selected a good common sense type of parlor double heater that I knew would do strong heating on the upper floor. I feel sure that as the result of this sale my customer will have demand for similar stoves next winter. I have explained to him how he should bring to the attention of other possible buyers the pleasure and comfort which this customer is deriving so that the good stoves would be doing good advertising for him as long as they are in use.

I believe that very little is gained from selling cheap stoves, and regret that some stove manufacturers do not design their heating stoves with a view to more practical efficiency and pay less attention to the unnecessary gaudy decorators. Stove dealers will do well to be careful in selecting their stock to get only the most efficient heating stoves that are attractive without being so covered with nickel plate as to make their presence too conspicuous in an otherwise tastefully furnished home. The trend in range ornamentation and the ease with which they sell is evidence that merit and chaste design are more popular and profitable than stoves which depend on show to secure the attention of the buyer and be forever after an unsatisfactory purchase, though the owner may not know the real cause.

STORING STOVES.

"What in the world to do with the stoves," is a question that is now bothering many a woman, says the Stoves and Hardware Reporter. If an appeal to the man of the house is made he will generally say, "Oh, leave 'em up; you'll find them there all right in the fall." But that doesn't suit at housecleaning time and the stove dealer who caters to this business can, with a little push, obtain considerable of it to do. An advantage to the dealer is that he generally gets the job of putting the stoves up again, furnishing repairs, new pipes, etc., and often a new stove. He gains an entrance to the houses of customers, has an opportunity to make friends with the women folks, and if he is wide-awake and alert to his business this will lead to many sales. By all means, store all the stoves you can and make the charges reasonable, depending on future business.

HELP CLERK TO KNOW GOODS.

The average retailer is a busy man, and often says, "If I spent the time the average traveler thinks I should spend with him I would not accomplish anything," writes W. L. Leckie, a stove traveler. You are in business to make money, and in order to sell a thing successfully, the only way to learn about them is to go to the factory where they are made and see them made, which many of you have not the time to do, or listen to the traveler.

Another thing, is that it is as necessary for the clerk to know about your goods as it is for you. How many of you take your clerks into your confidence, or give them all the information you know about what you are dealing in? How is a clerk, if you do not do that, going to explain your goods to the customers? I believe the best thing the retailer can do is to educate the clerk on any article that you wish him to sell. And I believe it is just as essential for the clerk to try and learn, and pick up all the knowledge he can, as it is for his employer to inform him.

Another very necessary thing in the selling of stoves, or any other article, but more especially in stoves, is that of display. I am sorry to say that I go into many stores where the proprietor, though he has been successful in many ways and has made money, is a little bit negligent in the display of his goods. I have gone into stores where I have seen stoves covered up with blankets or harnesses or dishpans and nothing to see about the stove but perhaps the legs, or the extreme top, and this man will tell me that he has not been as successful in seling stoves at Jones over there, that he cannot sell them; and one of the great reasons why he does not sell them is because he does not display them. I believe when a lady walks into a store to buy a range or any other article, it is half sold when it appeals to or strikes her fancy, and in order to do that it ought to be clean and polished up and made attractive, but it should not be covered up with a lot of other goods entirely foreign to stoves.

Don't get your gasoline and kerosene cans mixed. That error cost two lives in Ohio last year.

NEW WINNIPEG MANAGER.

W. L. Helliwell, formerly Toronto city traveler for the Gurney Foundry Co., has been apointed manager of the Winnipeg branch of this company, Mr. Helliwell, accompanied by E. Holt Gurney, left on Monday to assume his new duties.

Mr. Drewe, formerly manager at Winnipeg, is ill at Banff.

SOME GASOLINE DON'TS.

Some statistics on the use of gasoline have recently been prepared by H. B. Davis, the state fire marshal of Ohio, who has given some very interesting suggestions regarding lighting protection as noted in these columns some time ago. The information given is as follows:

Don't allow too much fluid to flow into the burner or fail to close it tight when putting the fire out.

Don't slop the stuff—it is more dangerous than powder. Three-fourths of the accidents occur while filling the reservoir.

Don't keep gasoline in any jug or in a can larger than 2 gal., because it is difficult to pour the thin stuff from either without spilling it.

Finally, don't hunt the source of an odor of gasoline with a light. The result of finding it is always instantaneous and disastrous.

Don't fill the can quite full, for gasoline expands much more than water when it becomes warm and is likely to force open a seam in the reservoir.

Don't fail to turn the burner shut before filling the reservoir, for the fluid leaking through it will make a vapor which will set on fire one who strikes a match to light the stove.

Don't pour gasoline from one vessel to another in a room in which there is a fire or light, because the invisible vapor of gasoline will be drawn to any nearby fire, lamp, candle or gas jet.

Don't fill the stove reservoir while the burner is alight. Vapor of gasoline being heavier than air will reach the flame, and the flash will so frighten the filler that more gasoline will be spilled and the room instantly filled with flame.

Don't leave any gasoline can open, because currents of air draw out the vapor. All gasoline stoves should have bottom and three sides closed to prevent combustible material from reaching the flame, and the main burner grates should be two feet from the floor.

Don't fail to watch closely for leaks in reservoir or burner, because gasoline, being but two-thirds as dense as water, will exude through a smaller hole. Remember, too, that when the leak is small there is no drop or damp spot anywhere to show its existence, because the gasoline vaporizes as fast as it exudes.

AN INGENIOUS LIGHT.

An ingenious beacon is located at Arnish Rock, Stornway Bay, in the Hebrides, Scotland. It is a cone of cast-iron plates, surmounted by an arrangement of prisms and a mirror which reflect the light from the lighthouse on Lewis Island, 500 feet distant across the channel.

HARDWARE TRADE GOSSIP

Ontario.

H. Dakin, Galt, was in Toronto this week.

J. H. Vandusen, hardware merchant, od Tara, is dead.

J. H. Hedley, Markdale, is opening up a tinsmith and repair shop.

Moore and Nunn are opening a new hardware store at Lansdowne.

W. A. Hillhouse, Shelburne, has sold out to Button, Spilker & Co.

Chas. R. Banks, Peterboro, has opened a new sporting goods store.

W. Mitchell, Port Stanley, has sold his hardware stock to H. M. Ellison.

F. Grice, of A. J. Ross Co., Cargill, was a buyer in Toronto last Saturday.

B. Werner, hardware merchant, of Innerkip, has sold his business to go west.

Young & McBurney, Wingham, have dissolved partnership, Mr. McBurney retiring.

Jelly & White, tinware and stoves, Shelburne, have dissolved partnership, R. A. Jelly continuing.

Fred Schaeffer, Beaslau, was in Galt last week buying machinery for his brick plant from the Goldie-McCulloch Co.

Col. W. M. Gartshore, manager of the McClary Mfg. Co., London, has left with his wife and daughter for Florida.

Brooks, Sanford Hardware Co., Toronto, have sent A. E. West to Hamilton to represent them as traveler for their branch in that city.

George H. Beaudin, representing J. Wiss & Sons, shear manufacturers, Newark, N.J., called at the Toronto office of Hardware and Metal on Saturday.

B. A. Perry, of the James Morrison Brass Mfg. Co., superintendent of fixture department, was in the train wreck at Guelph this week and had his back injured.

J. B. Reade, for many years a valued employe of H. S. Howland, Sons & Co., has accepted the position of buyer for the Kennedy Hardware Co. George H. Smith, formerly with the Kennedy Company, has joined the staff of Rice Lewis & Son.

J. W. Searle, London, Eng., called at the Toronto office of Hardware and Metal on Tuesday. Mr. Searle is paying a business visit to Canada and United States in the interests of Moffatt's Co., of London, Eng., manufacturers of the "Lucas" incandescent gas lamp.

Amongst the Ontario plumbers seen in Toronto this week were; Ald. Chas. Bull, St. Thomas; Mr. Peters, of Peters & Sylvester, Stratford; Geo. Scott, of Scott & Bennett, Galt; Jack Brady and Jas. Struthers, of Butchart & Co., Owen Sound; Hugh Wallace, of Hamilton; Jas. Boxall, of Boxall & Matthie, Lindsay; and Will Clark, of Adam Clark & Son, of Hamilton.

D. W. Rose, formerly buyer for Van Tuyl & Fairbank, Petrolea, has accepted a position with H. S. Howland Co., Toronto, and will leave for the west next week. He will represent this firm in the district between Winnipeg and Regina. Thos. Wright, formerly covering the district from Winnipeg to the mountains, will travel between Regina and Edmonton in the future.

Quebec.

A. Forget & Co., plumbers, Montreal, are registered.

E. Manbleau, St. John's, Que., is in Montreal buying goods.

The Philbin Specialty Co., hardware, Montreal, has been registered.

A meeting of the creditors of A. D. Armand, hardware dealer, Montreal, was held Feb. 25th.

Wm. Smaill, of the Canada Horse Nail Co., Montreal, has returned from a business trip to Toronto.

C. C. Race, representing Reveillon Freres, of Edmonton, is calling on the Montreal jobbers this week.

John P. Scott, metal merchant, Duchess St., Toronto, has taken out a permit for a new warehouse.

A. W. Stanley, of the Stanley Rule and Level Co., New Britain Conn., is in Montreal calling on the trade.

The Fife, Searle Co., Montreal, have moved from St. Paul St., corner of St. Peter St., to three doors east on St. Paul St.

R. G. Howe, representing the Wilkinson Sword Co., in Montreal, has returned from a trip to the head office of the firm in London, Eng.

A Curry, who lost his eye while working in the Angus shops of the C.P.R., Montreal, has entered suit for $5,000 against that company.

Ex-Mayor O'Donohue, of Stratford, who was killed this week in the railroad wreck near Guelph was connected with the Owen Sound Cement Company for many years.

J. H. L. Pettier, of the Maison, Jean, Paquette, hardware dealers, Montreal, is calling on the trade in Granby and other Quebec towns, and reports business excellent.

S. D. Pearson Co., Montreal, manufacturers of patent cone rotary ventilators, have secured the contract for fitting up the new "La Patrie" building, on St. Catherine St. east.

J. Sophus, representing Schuehardt & Schutte, in Canada, has returned from his wedding trip, and will take up his residence in Montreal. Mr. Sophus has taken up business quarters with Baxter and Patterson, 102 St. Antoine St.

The Canadian branch of the Gillette Safety Razor Co., Montreal, which was destroyed by fire some time ago, has taken up quarters at 622 St. Paul St., Montreal, where correspondence should be addressed.

R. F. Harrison, of London, Eng., is in Montreal to take over the sales managership of the F. W. Reddaway Belting Co. Mr. Harrison comes here with a well-established record in the Old Country, and he will have complete control of the Canadian business.

Geo. T. Moss has severed his connection with the Canada Screw Factory, Hamilton, going to the Canadian Westinhouse Co. Before leaving he was presented with a gold locket by the employes of the screw factory.

V. Eastwood, who has severed his connection, as manager of the Peterborough Hardware Company, to take the management of the Royal Bank in that city, was recently waited on by the officers and staff of the hardware company, and presented with a handsome gold-headed cane.

The Seybold Co., have now taken up their quarters with the Starke Hardware Co., on St. Peter St., Montreal. The long established Dominion prestige of the Starke Company, coupled with the excellent Quebec clientele of the Seybold Co., should help to make the Starke-Seybold Hardware Co. one of the strongest concerns of its kind in Canada.

Frothingham & Workman, Montreal, recently stated in an advertisement of "Lowmoor" bar iron, that they had sold this brand of iron for the past 30 years. At the time of writing the advertisement there was no positive information as to how long time they had sold this brand of iron. There was no doubt, however, that they had sold it for as long a period as the advertisement stated. An early edition of this old firm's catalogue has just been found which shows that in 1867 Frothingham & Workman were importing "Lowmoor" bar iron. This 1867 catalogue, as well as that of 1865, shows that at those dates Frothingham & Workman were selling lines of English goods which they still sell in large quantities, among them being Thomas Jowitt & Sons' files and rasps, and Armitage mousehole forge anvils.

Maritime.

In a fire at Pictou, N.S., Feb. 21, S. H. Hinnison and E. Cameron, hardware merchants, lost heavily.

Western.

T. Weir & Son, Minitonas, Man., hardware and implement dealers, have dissolved.

James Sutherland, Moose Jaw, has sold his hardware and tinsmithing business to A. E. MacKenzie.

J. J. Heinrichs and Jno. Driedger, hardware dealers, Osler, Sask., lost heavily in a fire which visited that town Feb. 17.

L. G. McKam, of New Westminster, B.C., has invented a hand saw called the Twentieth Century hand saw and rips, cuts-off and mitres, practically doing the work of three saws in one.

W. J. Simpson, manager and buyer of hardware and crockery for Robinson & Co., Winnipeg, left on Feb. 16 for a month's purchasing tour of Chicago, Pittsburg and eastern points.

Among the hardware men who visited the Winnipeg wholesale houses last week were: S. Schwanz, Rocanville; J. H. Dulmage, Battleford; A. Johnson, Fort Frances; Alex. Anderson, Saltcoats; Jas. Anderson, Binscarth; L. C. Porteous, Carlyle; J. Sutherland, Kenora, and D. Donahy, Belmont.

HARDWARE AND METAL

Established 1888

The MacLean Publishing Co.
Limited

JOHN BAYNE MACLEAN - President

Publishers of Trade Newspapers which circulate in the Provinces of British Columbia, Alberta, Saskatchewan, Manitoba, Ontario, Quebec, Nova Scotia, New Brunswick, P. E. Island and Newfoundland.

OFFICES:

MONTREAL, ... 232 McGill Street, Telephone Main 1255
TORONTO, ... 10 Front Street East, Telephones Main 2701 and 2702
WINNIPEG, ... 511 Union Bank Building, Telephone 3726
LONDON, ENG . . . . . 88 Fleet Street, E.C., J. Meredith McKim, Telephone, Central 12960

BRANCHES:

CHICAGO, ILL. . . . . . 1001 Teutonic Bldg, J. Roland Kay
ST. JOHN, N.B . . . . No. 7 Market Wharf
VANCOUVER, B.C. . . . . Geo. S. B. Perry
PARIS, FRANCE . Agence Havas, 8 Place de la Bourse
MANCHESTER, ENG. . . . . 92 Market Street
ZURICH, SWITZERLAND . . . . Louis Wol, Orell Fussli & Co.

Subscription, Canada and United States, $2.00
Great Britain, 8s. 6d., elsewhere - 12s

Published every Saturday.

Cable Address { Adscript, London / Adscript, Canada

NORMAL CONDITIONS IN WEST.

Once again normal conditions prevail in the west. It would be idle to deny that up to the first week in February the winter was exceptionally severe—it was in fact the worst in twenty years—and the resulting losses to the ranches in the Far West and to all classes through the tie-up of the railways have been heavy. At the same time people in the east and south should take with much more than the proverbial pinch of salt the stories which have been telegraphed from Winipeg by a few unscrupulous correspondents anxious to offer saleable "copy"; and they should know that during the month of February mild weather has prevailed and the railways are now grappling successfully with the tie-up in transportation. They were caught unprepared by the worst winter in a generation and the disasters which followed will surely result in provision for adequate equipment in future.

With a return to mild weather the branch lines have been opened up and trains on the main lines are running almost on schedule time. The fuel scarcity is rapidly being relieved in all parts of the west and merchants are getting delivery of their goods from the wholesale houses. A great amount of money is necessarily tied up in the country through the inability of the railways to move more than a small proportion of the 1906 wheat crop. It is evident that the railways will be busy with the 1906 crop almost up to the date when the 1907 crop commences to move; but with the opening of the lake and rail routes heavy shipments of grain will commence and money will again be plentiful from the Great Lakes to the Rockies. Eastern wholesalers and manufacturers should not permit themselves to be alarmed by foolish newspaper stories for they who know the conditions in the west best are the men who have most confidence in its immediate future. There is no reason for alarm. Normal conditions again prevail.

COBALT SPECULATION.

Hardware jobbers informed us that quite a number of retailers throughout Ontario have invested in Cobalt mining stocks, and in consequence are experiencing difficulty in meeting their accounts. It is a fact that about ninety per cent. of the mining companies, so-called, are nothing but stock-selling propositions, frauds, if ever there was a fraud. Mining is a highly specialized form of industry. The merchant of training and experience who knows the pitfalls and difficulties of the business, thinks the man a fool who dashes into it as a cat might under pressure of a dog-hunt take to the water But it is much easier for the uninformed to make money in retailing than in mining; for the man who knows it is infinitely easier.

If the dealers who have put money into Cobalt had kept it in their business and had taken as much interest in it as they have in the mining market their aggregate profits would have been immensely larger. Not only would their profits have been larger, but their businesses would have been worth more.

There are retailers with money to invest that they do not need in their business. If they are in doubt what to do with it let them ask their banker or a reputable broker. But they should avoid the mining market as they would contagion. Even in the few cases where the mining proposition is a legitimate undertaking, as soon as the public have put in enough and there appears to be a possibility of dividends, a few sharp speculators who know the game are likely to form a combination, "buy" the market and dispossess the rest of the shareholders of all interest in the property, save an unpleasant recollection. A man's business is the place for his money. If he has more than his business requires, there are fields of legitimate investment where his money can browse in safety and not keep him awake at nights or take his attention from his business. Next week we will publish from The Financial Post a statement showing the comparative money earning capacity of various lines of investment.

MALLEABLE INDUSTRY THREATENED.

The hardware trade is united in its strong denunciation of the amendment of the revised customs tariff whereby the duty of 17½ per cent., formerly 20 per cent. on malleable iron castings is to be removed, thus ruining the malleable iron industry in Canada located now in six leading towns, Oshawa, Galt, St. Catharines, Smith's Falls, Brantford, and Walkerville, by placing castings on the free list and leaving Canadian manufacturers helpless against the competition of American firms. If the proposed amendments are carried out, it will practically ruin the malleable industry in Canada, and, moreover, deal a hard blow to the towns in which these industries have been located, for if they have to close down, which is altogether probable, the taxation of the towns will be doubled and the trade decreased. Several thousand men are engaged in this industry in Ontario. We can hardly conceive of a government, by its legislation, deliberately and ruthlessly putting these men out of the way of earning a living in a trade which they spent many years in becoming proficient in.

We ask all hardware merchants and all those directly or indirectly interested in the malleable industry to co-operate in strong endeavors to urge the Government to a serious reconsideration of the proposed measure. If any industry is to suffer a change in the duties let it be the implement industry which has for so long been nursed by the Government or the iron and steel industries, the bounties to which are sufficient to allow of enough change to save the malleable industry from destruction.

NEW POSTAL REGULATIONS.

Attention was drawn last week to the abrogation of the postal convention between the Canadian and United States Governments, and the possibility of the exclusion of all magazines and trade publications, along with a host of other publications, owing to the imposition of the rate of 8c. per pound, in place of the one cent per pound rate to be discontinued.

Regarding the new regulations the Woodstock Sentinel-Review says:

"Speaking of the proposed intellectual preference for the encouragement of British periodical literature in this country, a contemporary points out that the rate from Canada to Great Britain on newspapers and periodicals is only one-fifteenth the rate from Britain to Canada. The British postal authorities have persistently refused to modify the rate for the reason that it would mean too great a loss of revenue.

"It would appear from this that the first step towards encouraging the circulation of British periodical literature in Canada should be taken by Great Britain, not by Canada.

"Associated with this talk of an intellectual preference we hear references to the advisability of discouraging the yellow literature of the United States. But why speak exclusively of the yellow literature of the United States? Not surely because there is no yellow literature published in Great Britain.

"The fact is there is yellow literature and bad literature coming from both countries, and the badness that bears a British imprint is not a bit better than the badness that bears an American imprint.

"This is a matter in regard to which the tastes of the Canadian people will probably have the last say. There are a great many cheap magazines printed in America that find their way into this country because the Canadian people cannot get anything else that suits their tastes and their purses any better. It is nonsense to condemn these magazines because they are cheap. They ought to be judged by the character of their contents as well as by their price.

"The fact is that many of the cheap American magazines are very creditable productions and perfectly harmless in their effect on either the morality or the loyalty of the readers.

"Some of the most pernicious publications sold in this town come across the ocean, not across the boundary line.

"This is a matter in regard to which a little more common sense and a little less prejudice would be desirable. By all means let anything be done that may properly be done to facilitate the circulation of wholesome periodical literature of Great Britain in this country; the more wholesome literature the better, whatever country it comes from; but if the British publishers expect to gain ground in this country they must undertake first to learn something about Canada and the needs and tastes of Canadian readers. Canadians are fond of reading the cheap American magazines, partly because they are cheap, partly because they deal with problems and phases of life with which Canadians are more or less familiar, and partly because there is always the possibility of coming across the work of a Canadian writer."

A. BURDETTE LEE'S DEATH.

The hardware trade in Toronto was shocked on Tuesday last on hearing of the sudden death of Major A. Burdette Lee, president of Rice Lewis & Son, and one of the best known men in the trade in the city. Mr. Lee had been ill with la grippe for about ten days, but the trouble was not considered serious until a couple of days before his death, when symptoms of pneumonia developed, and the patient lost strength rapidly. The deceased was the eldest son of A. B. Lee, who died in July, 1904, and who for a generation had been president of the old-established firm of Rice Lewis & Son. On the death of his father, A. Burdette Lee became president of the company, and retained that office until he passed away on Monday last.

Major Lee was in his 46th year, having been born in Toronto in 1861. With the exception of about a year spent in the fur business in Winnipeg, he had been associated with Rice Lewis & Son all his life. In 1885 he, as a member of the Queen's Own Rifles, served as a captain in the Northwest rebellion, rising from the ranks and being pay-master on

THE LATE MAJOR A. BURDETTE LEE
President of Rice Lewis & Son, Toronto.

retiring from active service in 1905. He was educated at Upper Canada College and Trinity College School, Port Hope, and was a member of the Granite and Albany clubs, and Military Institute, he being an enthusiastic bowler and curler.

The deceased leaves a widow and two sons, Stanley Burdette Lee, and Norman Barker Lee, the former having been serving in the sporting goods department of Rice Lewis & Son for the past few months. One brother and four sisters survive, the brother being Gordon Lee, a successful rancher at Marshall, Sask., and the sisters, Mrs. H. C. Wilson, of Edmonton; Mrs. W. C. Crowther, Toronto; Mrs. D. MacMurchie, Toronto, and Mrs Harold Mara, Toronto. Victor Lee, a brother, who was also associated with Rice Lewis & Son, died in August, 1905, there being, therefore, three deaths, of father and two sons, within the past three years.

The funeral took place on Thursday afternoon, the wholesale and retail stores being closed down in respect for the deceased president, who was particularly well liked by all the inside employes and traveling men.

LICENSE TRANSIENT TRADERS.

In a great many communities the merchants are troubled more or less by transient traders and pedlars operating without licenses, in some cases carrying away thousands of dollars in a year, without expending a cent by way of fee to the municipalities where they get the business. They enter directly into competition with local retailers who are obliged to pay heavy taxes for the privilege of selling goods to the public.

Where such conditions exist in Ontario the fault lies with the merchants themselves for an Act passed at the last session of the Provincial Legislature provides a means whereby they can have transients put on such footing that their competition need not be feared. All municipalities are empowered to pass by-laws imposing fees up to a certain substantial amount. Other provinces may secure legislation of like nature if the merchants present the matter to their local House.

The imposition of good, stiff licenses on transients is in no sense a restriction of trade, but rather the protection of legitimate traders. Heavy taxes or rent enter into a merchant's cost of doing business, and must be covered by the profit which he makes on his goods. Transients, in the majority of cases, have no expense of this kind to reckon with, and can afford to undersell the resident merchant. They take many thousands of dollars annully from the customers of retail stores. A fee which places them on equal footing with the latter is prohibitive to most of them, as it should be.

MARKETS IN DAILY PAPERS.

President Humphries, of the Ontario Retail Hardware Assn. has taken wise action in writing the editor of the Mail and Empire, Toronto, requesting that the publication of paint, oil and glass markets be discontinued on its financial page. One of the leading daily papers in Canada recently discontinued the publication of such market reports on Mr. Humphries' request and he should be successful in this instance. If the reports are not discontinued hardware merchants who are subscribers to the Mail and Empire should write the editor pointing out the injustice of giving wholesale quotations to retail buyers, who do not figure on freight charges, and similar incidentals.

Markets and Market Notes

(For detailed prices see Current Market Quotations, page 66.)

THE WEEK'S MARKETS IN BRIEF.

MONTREAL.

Door Checks and Springs—Discount on some lines, now 10 per cent.

Pipe Stocks and Dies—No's 1 to 2, 75 and 2½ discount.

Shot—New selling basis—advanced.

Old Material—Some lines advanced.

TORONTO.

Shot—New list with 5 per cent. off.

Galvanized Sheets—Gordon Crown advanced.

Montreal Hardware Markets

Office of HARDWARE AND METAL,
232 McGill Street,
Montreal, March 1, 1907.

In spite of the continuous cold weather wholesalers are doing a very brisk business, but money is still reported tight.

The shortage in wood screws is still very marked, especially ½x6, and 2x14, many jobbers having scarcely any on hand at all.

Rivets and burrs, machine screws, bolts and nuts, horse shoes and horse nails, and coil chain, are all selling well. Some shipments of green wire cloth are already going forward

Door checks and springs, (some lines) have advanced, the discount now being 10 per cent. An advance is also reported in pipe stocks and dies, Nos. 1 to 2, 75 and 2½ per cent. discount. Shot has also advanced 5 per cent., and higher prices on ammunition are looked for daily.

We have received the latest prices on binder twine, lawn mowers and poultry netting, and show same below.

TWINE—"Standard," 500 ft., sisal, 9½c.; "Tiger," 550 ft., manilla, 9½c.; "Red Cap," 600 ft., manilla, 12½c., "Blue Ribbon," 650 ft., pure manilla, 13¾ to 14c. Car lots ¼ less; thousand pound lots, ⅛c. less.

AXES—Axes and handles still continue to sell well here, in spite of the fact that the jobbers' season is practically over.

LANTERNS—Sales are few and far between, and this line can now be counted among the dead ones as far as wholesalers are concerned.

COW TIES AND STALL FIXTURES—This line is another of the "has-beens," moving very slowly now.

RIVETS AND BURRS—Selling well. Prices remain firm, with no changes.

HAY WIRE—Orders are now coming in freely. No change in prices: No. 13, $2.60; No. 14, $2.70; No 15, $2.85; f.o.b Montreal

WIRE NAILS.—We still quote: $2.45 per keg, f.o.b. Montreal.

CUT NAILS.—Prices remain: $2.30 per keg, base, f.o.b. Montreal.

MACHINE SCREWS—Good demand. No change in prices.

BOLTS AND NUTS—All factories are running full blast. Demand exceedingly heavy. Shortage in some lines. No change in prices.

HORSE NAILS—Change in the seasons is starting to create a heavy demand and dealers are trying to get their orders in in good time.

HORSE SHOES—Good demand. No change in prices.

BUILDING PAPER—Prices are very stiff, but no higher.

CEMENT AND FIREBRICK — Good business is being done at the same prices.

COIL CHAIN—Good demand Prices stiff with no changes.

SHOT—Advanced .5 pet cent Prices now are: Shot, packed in 25-lb. bags, ordinary drop AAA to dust, $7.35 per 100lbs.; chilled, No. 1—10. $7.88 per 100 lbs.; brick and seal, $8.40 per 100 lbs.; hall, $8.93 per 100 lbs 5 per cent. off. Bags less than 25 lbs., ½c. per lb. extra, net; f o.b. Montreal, Toronto, Hamilton London St. John, Halifax.

AMMUNITION—Jobbers are now paying a little more, but have not yet raised their prices. We still quote: Loaded with black powder, 12 and 16 gauge, per 1,000, $15; 10 gauge, per 1,000, $18; loaded with smokeless powder, 12 and 16 gauge, per 1,000, $20.50; 10 gauge, $23.50.

GREEN WIRE CLOTH—Some shipments are being made. No change in prices: In 100-foot rolls, $1.62½ per hundred square feet, in 50-feet rolls, $1.67½ per hundred square feet.

LAWN MOWERS—Quite a few good orders are now being received. Our prices are: Low wheel, 12, 14 and 16 in., $2.30 each; 9-in. wheel by 12 inch, $2.85 each; 14-inch, $3 each; 16-inch, $3.12½ each; high wheel, 12-inch, $4.05 each; 14-inch, $4.25 each; 16-inch, $4-50 each; 18-inch, $4.75 each.

POULTRY NETTING—2x2 mesh 50 and 5 per cent. discount, other sizes 50 per cent. off.

Toronto Hardware Markets

Office of HARDWARE AND METAL,
10 Front Street East,
Toronto, March 1, 1907

Hardware trade conditions have improved considerably, prices strengthening, with some lines an upward tendency. The approach of spring and moderation of the climate are the main causes for the improvement in trade. Orders with the wholesalers are coming in rapidly, especially from the west and Northern Ontario. In spite of J. J. Hill's prediction of a depression business is very active.

Binder twine prices are firm, with reports of a probable advance in sisal.

Axes and handles still have a good demand in the retail trade. The wholesale trade is practically ended.

Screws are firm in price with a heavy demand. Business in bolts and nuts is heavy; the demand, however, exceeds the supply, building operations becoming more extensive, the stock being low with little immediate prospects of repletion.

Mechanics tools are in good demand. Carpenters' tool prices are firm and business active with the retailers and with some of the wholesalers, as some retailers are still getting in stock. The demand for shovels continues with the approach of spring and railroad activities.

Axes and handles, still have a good demand in the retail trade. Retail business is heavy.

BINDER TWINE—500 ft. Sisal, 9½-9⅝; 500 ft. Standard, 9½-9⅝; 550 ft. Standard Manilla, 10¾; 600 ft. Manilla, 12¼; 650 ft. pure Manilla, 13¾-14. Car lots ¼c less, five ton lots ⅛c less, central delivery. There are prospects of advance in prices of Sisal.

SCREWS—Wholesalers report a good business, with active prices.

BOLTS AND NUTS—These continue to experience a good demand, which exceeds the supply considerably.

CHAIN—No change in prices.

RIVETS AND BURRS—Prices are firm and unchanged, with heavy demand.

MECHANICS' TOOLS—The demand for these is heavy, especially carpenters. Shovels continue to experience a good demand.

EXTENSION AND STEP LADDERS—The demand is strong, prices remaining unchanged: 11 cents per foot for 3 to 6 feet, and 12 cents per foot for 7 to 10-foot ladders. Waggoner extension ladders 40 per cent.

POULTRY NETTING—A good demand prevails for this line, retailers getting their spring stock rapidly filled. Prices are unchanged: 2-inch mesh, 19 w.g. discount 50 and 10 per cent., all others 50 per cent.

OILED AND ANNEALED WIRE—Prices in these are unchanged. We continue to quote: Canadian wire: gauge, 10, $2.46; gauge 11, $2.52; gauge 12, $2.61 per 100 lbs.

WIRE NAILS—The demand continues heavy, with prices firm and unchanged. We continue to quote $2.40 base.

HORSESHOES AND HORSENAILS—The demand is heavy, especially in Northern Ontario and in all centres of railroad building operations. Prices remain firm and unchanged.

BUILDING PAPER—The business in this as yet is not active, but should in a few weeks enliven.

HIDES, WOOL, FURS—Business is quiet. For all prices in these see The Canadian Grocer.

Montreal Metal Markets

Office of HARDWARE AND METAL,
232 McGill Street,
Montreal, March 1, 1907

There is no advance as yet over last week's prices on copper, but the shortage is just as great as ever. Some well informed gentlemen say that copper will come down, while others just as well informed, say it will go up. One Montreal jobber says, "None of us know where it is going, to tell the truth." This man is also well informed. Judging from the above, we are pretty safe in saying that when the present shortage is relieved, copper will go down, but until then, "nothing doing." Spelter is very firm, with the tone of the market higher. Ingot tin is firmer than last week, minimum price firmly held. Lead has been steady for the past week, with no change in prices. Some lines of old material have advanced slightly.

COPPER.—Conditions remain as stated last week. Ingot copper, 26¼ to 27c.; sheet copper, base sizes, 34c.

INGOT TIN.—Stronger than last week. Prices remain: 45½ to 46c. per lb.

ZINC SPELTER.—Firm at 7.50 to 7.75 per 100 lbs.

PIG LEAD.—Steady for the past week. Prices are, 5.50 to 5.60 per 100 lbs.

ANTIMONY.—Firm at 27½ to 28c. per pound.

PIG IRON.—Unchanged: Londonderry, $24.50; Carron, No. 1, 27.00; Carron special, 25.50; Summerlee No. 2, selected, 25; Summerlee No. 3 soft, $23.50.

BOILER TUBES.—Very firm at the same prices.

TOOL STEEL.—A fair business is reported, with no change in prices.

COLD ROLLED SHAFTING — No changes to report.

GALVANIZED IRON. — Market strong. Same prices prevail.

OLD MATERIAL.—Several lines advanced. Our prices now are: Heavy copper 20c. per pound; light copper 17c. per pound; heavy red brass, 18½c. per pound; heavy yellow brass, 15c. per pound; light brass, 11 to 11½c. per pound; tea lead, 4¼ to 4½c per pound; heavy lead, 4½c. per pound; scrap zinc, 4½c. per pound; No. 1 wrought, $17 per ton; No. 2 wrought, $6 per ton; No. 1 machinery, $18 per ton; stove plate, $14 per ton; old rubber, 11½c. per pound; mixed rags, 1 to 1¼c. per pound.

Carry a full line and let your customers look to your store for everything in the paint line.

Toronto Metal Markets

Office of HARDWARE AND METAL,
10 Front Street East,
Toronto, March 1, 1907

Tariff matters have attracted considerable attention in the metal markets this week, the danger which threatens Canada's malleable iron industry being realized by every metal merchant. Everywhere the sentiment is expressed that the Government should reconsider their action in throwing this market open to United States producers. The malleable industry has developed into an important branch of the metal trade and should not be squeezed out of existence to benefit implement manufacturers or steel producers.

The statement that the 1,000 employes of the Lancashire Steel Company was kept going during the shipping strike last summer solely because of the Canadian tariff has also been noted with interest.

American markets, which have been full of bearish reports during the past fortnight, are strengthening again and the break which has been so freely predicted is apparently not due yet. It is generally realized that an end has been reached to the upward movement in iron but a gradual easing off is looked for rather than a sharp break. A desire to create business by slightly lowering prices is now said to be developing. Advices, however, report consumption continuing at a record pace and this should keep prices firm. An intelligent view taken is that when underlying conditions are sound, as they are to-day, adverse reports do not precipitate trouble, but rather that the dangerous time is the time when there are no notes of warning, and everyone feels that there are no troubles ahead. Time and again in this period of extreme prosperity there have been little waves of pessimistic talk and they have vanished overnight.

Copper keeps both scarce and firm. Receipts of copper from the mines are slow and refineries are, therefore, unable to forward orders with any promptitude. Better conditions are looked for with the opening of spring.

Regarding tin plate a foreign view is that a falling off in demand need hardly be feared during the next six months, as many consumers have been out of the market for some time and are only just coming to the front again. This particularly applies to China and Japan, which have been working off the enormous supplies which accumulated during the war, and have only resumed buying during the last few weeks. In Russia also trade has been affected by the political crisis, and is only beginning to recover.

Complaints of shipping delays continue, two carloads of pig iron being forwarded from the Old Country to a Western Ontario user, being lost somewhere between Boston and here, it having been shipped from that point on Feb. 4. In such cases, and they are numerous to-day, jobbers are put to endless worry, trouble and expense in correspondence and telegraphing in order to endeavor to satisfy customers. When an order is taken it is by no means filled.

Lead and smelter continue active with a good value of trade doing. Galvanized steets are strong with an advance noted on gordon crown, 26 gauge being moved from $4.20 to $4.40 and 28 from $4.40 to $4.60. There is also talk of higher prices on brass tubing and sheet brass.

PIG IRON—The market is stronger outside. Railroad blockades are causing great distress among the foundries. Buying continues active and deliveries slow. Hamilton, Midland, and Londonderry are off the market, and Radnor is quoted at $33 at furnace. Middlesborough No. 1 is quoted at $24.50, and Summerlee No. 1 at $26 f.o.b., Toronto.

BAR IRON—Stocks are improving slightly. Steady prices rule and we still quote: $2.30 f.o.b., Toronto, with 2 per cent. discount.

TIN—Prices are steady with a fair amount of buying at 45 cents for ingots. The base on charcoal plates continues at $6.50.

SHEETS AND PLATES—Gordon Crown galvanized sheets have advanced. A good trade is being done in sheets and Canada Plates.

BRASS—We still quote 30 cents per pound on sheets, with some talk of an advance on sheets and on brass tubing.

COPPER — Deliveries for present use are exceedingly hard to get. We now quote: Ingot copper, $26.50 per 100 lbs., and sheet copper, $35 per 100 lbs.

LEAD—The market is active with prices unchanged. We quote, $4.50 for imported pig and $5.75 to $6 for bar lead. For sheet lead $7 is asked.

ZINC SPELTER—Stocks are light but trade is good at former prices which are 7½c per lb. for foreign and 7c per lb. for domestic. Sheet zinc is quoted at 8¼c in casks, and 8½c in part casks.

ANTIMONY—A fair trade is reported with price unchanged at 27c per pound.

OLD MATERIAL — Buying prices are: Heavy copper and wire, 20c. per lb.; light copper, 18c per lb.; heavy red brass, 18c per lb.; heavy yellow brass, 15c. per lb.; light brass, 11c. per lb.; tea lead, $4.25 per 100 lbs.; heavy lead, $4.40 per 100 lbs; scrap

ONTARIO STEEL

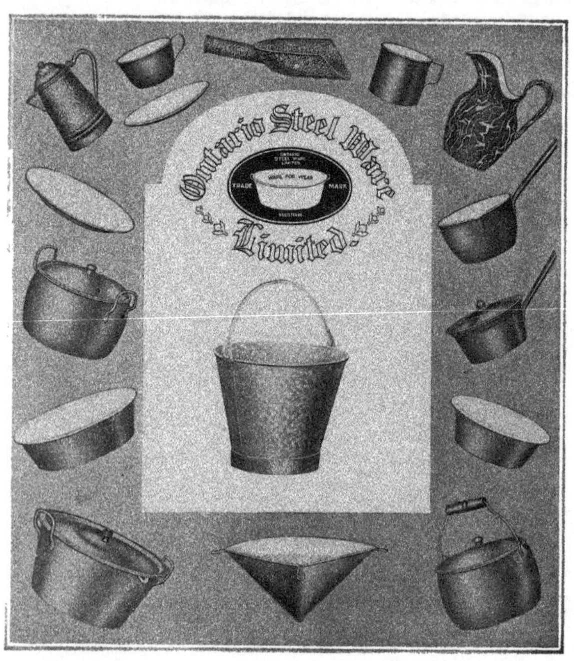

115-121 Brock Avenue, - 79-91

zinc, 4½c. per lb.; No. 1 wrought iron, $13.50; No. 2 wrought, 8; machinery cast scrap, $17.50; stove plate, $12; malleable and steel, $8.50, old rubbers, 11c. per pound; country mixed rags, $1 to $1.25 per 100 lbs., according to quality.

COAL — The shortage of hard coal continues. We still quote :—Standard Hocking soft coal, f.o.b. at mines, lump, $1.50, ¾ inch, $1.40; run of mine, $1.30; nut, $1.25; N. & S., $1; P. & S. 85c. Youghiogheny soft coal in cars, bonder at bridges: lump, $2.60; ¾ inch, $2.50; mine run, $2.40; slack, $2.15.

London, Eng., Metal Markets

From Metal Market Report, F b 26, 1907.

PIG IRON—Cleveland warrants are quoted at 54s. 6d., and Glasgow standard warrants at 53s. 9d., 1s. 4½d. lower, making price as compared with last week on Cleveland warrantes 1s. 3d. lower.

TIN—Spot tin opened quiet at £192 15s., futures at £191, and after sales of 310 tons of spot, and 80 tons of futures closed weak at £192 for spot, £190 10s. for futures, making price compared with last week 15s. higher on spot and unchanged on futures.

COPPER—Spot copper opened firm at £107 15s, futures £109, and after sales of 300 tons of spot and 500 tons of futures, closed easy at £107 7s. 6d. for spot, and £108 10s. for futures, making price compared with last week 10s. lower on spot and 10s. lower on futures.

LEAD—The market closed at £19 10s., making price as compared with last week unchanged.

SPELTER—Best selected closed at £114 10s., making price as compared with last week 10s. higher.

United States Metal Markets

From the Iron Age, Feb 28, 1907.

So far as the finished branches of the iron and steel industry are concerned business is developing in a very satisfactory manner. The mills are full of work, and specifications are coming in steadily. During the first three weeks of the current month the new orders for the United States Steel Corporation have come in at a daily rate which is practically equivalent to the full capacity of the plants. In other words, the corporation, even in a month which is usually quiet, as February is, has not been gaining on its extraordinary accumulation of orders. Shipments during the first half of February were for obvious reasons not quite up to the output of the mills.

The sale of 50,000 tons of steel rails to the Manchurian road, to which we alluded some weeks since, has been followed by contracts for bridge work amounting to 8,600 tons. It is estimated that the February tonnage of the American Bridge Company will aggregate 40,000 tons. This includes 3,200 tons for the new shops of the Grand Trunk Western and 7,500 tons for the Waverly warehouse of the Steel Corporation. Among the contracts pending are 6,500 tons for the Chicago Corn Exchange Bank and 2,000 tons additional for the New Haven road.

In the steel market the most interesting development is the purchase by the leading interest of what tonnage was offered in the market by a new open hearth steel plant in the Buffalo district about to begin production.

The pressure in the wire and tube trades for an advance in prices has been persistent recently. It is thoroughly understood, however, that the most powerful interests are adverse to any raising of prices.

The eastern bar iron markets are strong and makers are now quoting on the basis of their own mills, rather than on the Pittsburgh base.

The pig iron markets are featureless. The spot markets are controlled entirely by local and temporary conditions, with a tendency toward a lowering of premiums. Middlesbrough iron has been offered in cargo lots at $20.50, at tidewater, but for the future business is confined to purchases made for export work. For forward delivery of foundry irons negotiations are pending east and south, but buyers and sellers seem too far apart. Prompt basic is scarce, although the mills have a good deal of tonnage due to them on old contracts.

From the Iron Trade Review, Feb. 28 1907.

The centre of attraction is pig iron. Keen interest is displayed in every move and rumor regarding the situation. The air is full of bearish talk and authenticated reports are current that southern prices are being shaded for other deliveries besides the last half.

Considering the number of rumors afloat, which would tend to increase any inherent weakness, remarkable strength is shown. One event which has been widely referred to in newspaper reports as an indication of weakness, was the action of a large independent steel maker in refusing, on account of what he considered the excessive price, to take about 50,000 tons of Bessemer and basic pig iron on a sliding scale contract which permitted such action in case the price reached a certain point. The ease with which this tonnage, suddenly thrown upon the market, was absorbed is, however, proof of the strength of present conditions. At the time the iron was refused one interest which makes basic iron was badly in need of iron for its customers and bought a part of the tonnage at about $22, and a large steel making interest, whose production of iron had been curtailed by the blowing out of several of its furnaces, was another buyer.

Indications now are that shipments of iron ore from the Lake Superior ranges next season may amount to 42,000,000 tons, about 2,000,000 tons more than has been estimated.

The situation regarding steel bars has now reached an acute stage and delivery promises are about on a par with those on steel plates If all contracts now on the books are fully specified, mills will operate uninterruptedly for the next seven months. No definite promises are being made on new business and current specifications are so heavy that the leading producers have inaugurated a system of apportionment by which their steel bar output will be divided as equitably as possible, this plan now being in operation on plate distribution.

THE PRICE OF COPPER.

"What causes such high prices to prevail in the metal market" is a question we have on every side, and the answers we get when we ask the question, ourselves, are various. Many people, for instance, think the high prices are due to manipulation, but the great majority claim that the trouble arises from the tremendous demand of a growing continent. This after all seems to be the more rational explanation.

Of course some people will ask why there should be a shortage in copper, for instance, unless it is brought about by "amalgamated" holding up the production. "Amalgamated" certainly could do so, did they desire, but do they? We do not think so.

In the first place, these great enterprises depend for their existence, on doing a tremendous business at a small margin of profit. The very existence of the trusts proves this. The greater the volume of business done, the less the expense proportionately.

Suppose there were three manufacturers in one city, making the same class of goods, and doing about the same amount of business. The heads of two of these concerns get together and decide to amalgamate. What's the result? By manufacturing on a larger scale, the cost of production is reduced, therefore the product can be sold by the larger firm at a much lower figure than the smaller man can hope to meet, without lowering in any way the quality of the goods, therefore, the smaller man must join forces with his rivals, or get out of the game. Thus, in order to keep down the manufacturing cost, the sales must be kept up and the plant running full blast, else there would soon be no profit, and if the price were raised enough that a small independent manufacturer could make a profit on his goods, wouldn't he rather stay out of any combine and stick to himself? A man would rather be the whole thing in a small business than only half the show in a larger one, but rather than go to the wall, he joins the combine.

Just so soon as a combine or trust raises its prices to a figure that allows competition, its power begins to fade away.

If the world's copper mines were nearly exhausted, there would be some object in the great producers holding back for the highest possible prices in order to make the best profit they could while the metal lasted, but with the great mineral wealth of the North America continent, this is unnecessary. There is copper enough for many years to come, and therefore no reason that we can see for any shortage, except the want of labor.

Metal is taking the place of wood in many articles of everyday use, and the strides that all branches of electrical manufacture have taken lately have used up practically all reserve stocks. In fact metals are used now almost as fast as they can be mined, and the only relief seems to be in increasing the output of the mines, and this can only be done by increasing the number of laborers employed.

B.C. Hardware Trade News

Vancouver, B.C., Feb. 22, '07.

The lumbermen will be represented before the select committee of the House of Commons which is to hear evidence regarding the charge made on the floor of the House that a combine exists among the lumbermen, by which prices are unduly enhanced. All along the millmen have asserted that they welcomed such an investigation, as it will give them opportunity to go on record in a manner impossible otherwise, as to their position and as to the fairness of present lumber prices. R. H. H. Alexander, secretary of the B. C. Lumbermen's Association is to go to Ottawa and give evidence, and other lumbermen are expected to be in attendance there also.

* * *

From all accounts prices of lumber from the coast mills are not at all likely to be less. In fact a well-defined rumor has it that on April 1st the list is to be increased materially. There has been no raise in lumber prices from the coast since Dec. 13, last, when a raise of $1.00 went into effect. The Mountain Lumbermen's Association at their meeting in January increased their schedule on rough grades but allowed the higher classes of lumber to remain. It is said that they too will be increased when the coast prices go up. Owing to their situation the mills of the interior are able to sell somewhat cheaper than coast mills for the good reason that the freight rate is lower. The quality too of the timber or rather the class, is very different from that of the coast, so there are two distinct schedules of prices on B.C. lumber. The fact is the coast and interior lumber does not compete in very many classes, and certainly not in the clear grades of fir and cedar.

* * *

To meet the growth of the city the Vancouver Gas Co. is planning large increase of its plant. The present location of the producing plant is so restricted that a new site has to be secured for the proposed larger plant. The company now has an expert on the ground, L. L. Merrifield, M. Inst.M.E., C.G.S., of Toronto, and this gentleman is to go over the whole situation and make recommendations for the enlargement of the present capacity of the gas plant.

* * *

The cry is going up from many small towns and outlying districts of the province for "more light." The interruption in freight traffic on the railways has not only practically stopped shipments out from the distributing centres of the province, but the steady drain on stocks of oil has reduced them till there is no coal oil in Vancouver and Victoria to meet demands. Other lines of hardware are also badly broken up by non-receipt of goods. It is expected that much delayed freight will soon be coming forward. The C.P.R. has been making a better record for the last week, their line not having been altogether closed at any time. The Great Northern, which was completely blocked for a fortnight or more, has announced that its lines are once more cleared and open for traffic, and that freight may be expected forward at once.

* * *

The greatest interest is being taken by the hardware trade in the altered schedule of freight rates, caused by the recent decision of the Railway Commission, giving Vancouver shippers a rate to Nelson and common points, which means the interior of B.C., of 62 per cent. of the rate from Winnipeg to the same points. As pointed out in connection with the grocery trade last week, the rate is generally satisfactory, but there are special cases in which a special commodity rate, not based on class is of course withdrawn by the new ruling. In such cases there will be an actual increase, by sticking to the letter of the ruling.

* * *

Coast building operations have opened up very strongly since the cold snap has been over. Many large blocks for Vancouver have already been begun. Others are projected, and the usual large percentage of increase in the number of residences is also to be made. Building is very active in Victoria, several blocks being projected, and a large number of residences. Many interior towns have experienced a satisfactory growth too, owing to the influx of settlers, all the available land suitable for farming or fruit-growing now being sought eagerly. The building of the Grand Trunk's new city of Prince Rupert is bound to go on this summer. Many buildings are already erected and with the rush of people who plan to go there almost at once, there is bound to be activity. This is having its effect already on stocks of builders' hardware, plumbing supplies, etc. Much more so is it having its effect on the building trades. The carpenters are said to have served notice on the contractors that at the expiry of the present wage schedule agreement, on April, there must be a very material increase in the rates of pay. Other building trades are also restive, and there may be some return to the disturbances of three or four years ago.

* * *

The Robertson Lumber Co., formerly Kamloops Lumber Co., has rebuilt and enlarged the capacity of its plant at Enderby, on the Okanagan branch railway, at a cost of $75,000. The new plant is to have a capacity of nearly 100,000 feet per day. It will be ready for operation in about two months.

N.S. Hardware Trade News

Halifax, N.S., Feb. 26, 1907.

The bad condition of the roads during the past week has had a very detrimental effect on business, and just at present there is not much doing in the hardware line. Some small orders for spring goods have come in, but it is a little early yet for any rush. Prices now appear to be pretty steady throughout the list, though the jobbers look upon the markets with the utmost uncertainty.

* * *

The new hardware store of L. E. Wamboldt, of Lunenburg, N.S., is up-to-date in every respect. The building is of the flat-iron style, being 63 feet in length, 43 feet in width in the front and tapering to 18 feet in the rear. The store has large plate glass windows and is well lighted. All the ceilings are of metal of a handsome design, and the building is heated throughout with hot air. Mr. Wamboldt is one of the hustling merchants of the Gloucester of Nova Scotia, and his business is most prosperous.

* * *

The employes of Crump & Perrier, plumbers and steam fitters and dealers in shelf hardware, held their annual sleigh-ride to Popple's, on the Herring Cove road, last week. Both members of the firm accompanied the party, which went out on two four-horse teams. After games, songs, etc., the party sat down to a bountiful dinner, and all did it full justice. The toast list was, "The King," proposed by G. A. Perrier; "The Firm," proposed by Charles Rent, the bookkeeper, was responded to by Messrs. Crump and Perrier. The last toast was proposed by Mr. Perrier, "The Employes," and responded to by Mr. Rant for the men, and A. Gale for the boys. The good feeling that exists between employer and employe in this firm is very manifest, both Messrs. Crump and Perrier taking a deep interest in all matters of pleasure which the employes desire to carry out.

* * *

In the destruction by fire of the Henderson block, at Pictou, last week, H. Higginson's hardware store, with all contents, was destroyed. The store of E. Cameron, dealer in stoves and tinware, also situated in the same block, was completely destroyed. Both stores were partially insured, but the loss to the owners will be considerable.

* * *

James Cummings, of the staff of H. H. Fuller & Co., has gone to Winnipeg, where he has accepted a position with the wholesale hardware firm of Miller, Morse & Co., Mr. Cummings was very popular in Halifax, and his many friends regret his departure.

Arthur McAdams, formerly traveler for H. H. Fuller & Co., has joined the staff of Austin Bros.

Thomas Graham has accepted a position with H. H. Fuller & Co., and is now on the road. Mr. Graham has been identified with the hardware trade of the Maritime Provinces for many years. He was employed by Black Bros., up to the time that firm went out of business, and he then joined the staff of Emmerson & Fisher, St. John. Later he went to Glace Bay, being connected with the firm of F. E. Moir & Co., of that town.

MANITOBA HARDWARE AND METAL MARKETS

Market quotations corrected by telegraph up to 12 a m. Friday, March 1. , 1907.)

Room 511, Union Bank Building, Office of HARDWARE AND METAL, Winnipeg, Man.

With a return to normal weather conditions the branch lines on the railways are all open again and the C.P.R. claim to have delivered all delayed freight. Business shows renewed activity, as retail stocks were allowed to run low during the snow blockades and the travellers are now reaching customers whom they were forced to neglect a few weeks ago.

LANTERNS—Cold blast, per dozen, $6.50 ; coppered, $8.50 ; dash, $8.50.

WIRE—Barbed wire, 100 lbs., $3.22½ ; plain galvanized, 6, 7 and 8, $3.70; No. 9 $3.25; No. 10, $3.70; No. 11, $3.80; No. 12, $3.45; No. 13, $3.55; No. 14, $4.00; No. 15, $4.25; No. 16, $4.40; plain twist, $3.45; staples, $3.50; oiled annealed wire, 10, $2.96 ; 11, $3.02 ; 12, $3.10 ; 13, $3.20 ; 14, $3.30 ; 15, $3.45. Annealed wires (unoiled) 10c. less; soft copper wire, base, 36c.; brass spring wire, base, 30c.

HORSESHOES—Iron No. 0 to No. 1, $4.65; No. 2 and larger, $4.40; snowshoes, No. 0 to No. 1, $4.90; No. 2 and larger, $4.65 ; steel, No. 0 to No. 1, $5 ; No. 2 and larger, $4.75.

HORSENAILS — Capewell brand, quotations on application. No. 10, 20c.; No 9, 22c. ; No. 8, 24c. ; No. 7, 26c. ; No. 6, 28c. ; No. 5, 32c. ; No. 4, 40c., per lb. Discounts: "C" brand, 40, 10 and 7½ per cent., "M" brand and other brands, 55 and 60 per cent. Add 15c. per box.

WIRE NAILS—$2.85 f.o.b. Winnipeg, and $2.40 f o.b. Fort William.

CUT NAILS—Now $2.90 per keg.

PRESSED SPIKES — ¼ x 5 and 6, $4.75 ; 5-6 x 5, 6 and 7, $4.40; ⅜ x 6, 7 and 8, $4.25; 7-16 x 7 and 9, $4.15; ½ x 8, 9, 10 and 12, $4.05; ⅝ x 10 and 12, $3.90. All other lengths 25c. extra net.

SCREWS—Flat head, iron, bright, 85 and 10 p.c. round head, iron, 80 p.c.; flat head, brass, 75 and 10 p.c.; round head, brass, 70 and 10 p.c.; coach, 70 p.c.

NUTS AND BOLTS — Bolts, carriage, ⅜ or smaller, 60 and 5 ; bolts, carriage, 7-16 and up, 55 ; bolts, machine, ⅜ and under, 55 and 5 ; bolts, machine, 7-16 and over, 55 ; bolts, tire, 65 ; bolt ends, 55 ; sleigh shoe bolts, 65 and 10 ; machine screws, 70 ; plough bolts, 55 ; square nuts, case lots, 3 ; square nuts, small lots, 2½ ; hex nuts, case lots, 3 ; hex nuts, smaller lots, 2½ p.c.

RIVETS — Iron, 60 and 10 p.c.; copper, No. 7, 43c., No. 8, 42½c.; No. 9, 45¼c.; copper, No. 10, 47c.; copper, No. 12, 50½c.; assorted, No. 8, 44½c., and No. 10, 48c.

COIL CHAIN—¼ in., $7 ; 5-16, $5.35 ; ⅜, $4.75 ; 7-16, $4.50 ; ½, $4.25 ; 9-16, $4.20 ; ⅝, $4.25 ; ¾, $4.10.

SHOVELS—List has advanced $1 per dozen on all spades, shovels and scoops.

HARVEST TOOLS—60 and 5 p.c.

AXE HANDLES—Turned, s.g. hickory doz., $3.15 ; No. 1, $1.90. No. 2, $1.60 ; octagon, extra, $2.30 ; No. 1, $1.60.

AXES — Bench axes, 40 ; broad axes, 25 p.c. dis. off list ; Royal Oak, per dozen, $6.25; Maple Leaf, $8.25 ; Model, $8.50 ; Black Prince, $7.25 ; Black Diamond, $9.25 ; Standard flint edge, $8.75 ; Copper King, $8.25 ; Columbian, $9.50 ; handled axes, North Star, $7.75 ; Black Prince, $9.25 ; Standard flint edge, $10.75 ; Copper King, $11 per dozen.

CHURNS—45 and 5 per cent.; list as follows ; No. 0, $9 ; No. 1, $9 ; No. 2, $10 ; No. 3, $11 ; No. 4, $13 ; No. 5, $16. ?

AUGER BITS—"Irwin" bits, 47½ per cent., and other lines 70 per cent.

BLOCKS—Steel blocks, 35 p.c.; wood 55 p.c.

FITTINGS—Wrought couplings, 60 ; nipples, 65 and 10 ; T's and elbows, 10; malleable bushings, 50 ; malleable unions, 55 p.c.

HINGES—Light "T" and strap, 65 p.c.

HOOKS—Brush hooks, heavy, per doz., $8.75 ; grass hooks, $1.70.

STOVE PIPES—6-inch, per 100 feet length, $9 ; 7-inch, $9.75.

TINWARE, ETC. — Pressed, retinned, 70 and 10 ; pressed, plain, 75 and 2½ ; pieced, 30 ; japanned ware, 37½ ; enamelled ware, Famous, 50 ; Imperial, 50 and 10 ; Imperial, one coat, 60 ; Premier, 50 ; Colonial, 50 and 10 ; Royal, 60 ; Victoria, 45 ; white 45 ; Diamond, 50 ; Granite, 60 p.c.

GALVANIZED WARE—Pails, 37½ per cent.; other galvanized lines 30 per cent.

CORDAGE — Rope, sisal, 7-16 and larger, basis, $11.25 ; Manilla, 7-16 and larger, basis, $16.25 ; Lathyarn, $11.25 ; cotton rope, per lb., 21c.

SOLDER—Quoted at 27c. per pound. Block tin is quoted at 45c. per pound.

POWER HORSE CLIPPERS—"1902" power clipper, $12 ; "Twentieth Century" $6. "1904" sheep shearing machines, $13.60.

WRINGERS—Royal Canadian, $35 ; B.B., $39.75, per dozen.

FILES—Arcade, 75 ; Black Diamond, 60 ; Nicholson's, 62½ p.c.

LOCKS—Peterboro and Gurney, 40 per cent.

BUILDING PAPER—Anchor, plain, 66c.; tarred, 69c.; Victoria, plain, 71c.; tarred, 84c.

AMMUNITION, ETC. — Cartridges, rim fire, 50 and 5 ; central fire, 33⅓ p.c.; military, 10 p.c. advance. Loaded shells: 12 guage, black, $16.50 ; chilled, 12 gauge, $17.50; soft, 10 gauge, $19.50; chilled, 10 gauge, $20.50. Shot, ordinary, per 100 lbs., $7.25; chilled, $7.75; powder, F.F., keg, Hamilton, $4.75 ; F.F.G., Dupont's, $5.

IRON AND STEEL—Bar iron basis, $2.70. Swedish iron basis, $4.95 ; sleigh shoe steel, $2.75 ; spring steel, $3.25 ; machinery steel, $3.50 ; tool steel, Black Diamond, 100 lbs., $9.50 ; Jessop, $13.

SHEET ZINC—$8.50 for cask lots, and $9 for broken lots.

PIG LEAD—Average price is $6.

AXLE GREASE—"Mica," 1-lb. tins, $11 per gross; 3-lb. tins, $2.40 per doz.; $1.60 per doz. case; 10-lb. iron pails, 60c. each; 15-lb. pails, 80c.; 25-lb. pails, $1.25; "Diamond," 1-lb. wooden boxes, $6.40 per gross.

IRON PIPE AND FITTINGS—Black pipe, ¼ inch, $2.65; ⅜, $2.80; ½, $3.50; ¾ $4.40; 1, $6.35; 1¼, $8.65; 1½, $10.40; 2, 13.85; 2½, $19.00; 3, $25.00. Galvanized iron pipe, ⅜ inch, $3.75; ½, $4.35; ¾, $5.65; 1, $8.10; 1¼, $11.00; 1½, $13.25; 2, inch, $17.65. Nipples, 70 and 10 per cent.; unions, couplings, bushings and plugs, 60 per cent.

LEAD PIPE—Market is firm at $7.80.

GALVANIZED IRON—Apollo, 16 gauge, $3.90; 18 and 20, $4.10; 22 and 24, $4.45; 26, $4.40; 28, $4.65, 30 gauge or 10¾ oz., $4.95; Queen's Head, 24, $4.50; 26, $4.65; 28, $5.00.

TIN PLATES—IC charcoal, 20x28, box $9.50; IX charcoal, 20x28, $11.50; XXI charcoal, 20x28, $13.50.

TERNE PLATES—Quoted at $9.

CANADA PLATES—18x21, 18x24, $3.40, 20 x 28, $3.65, full polished, $4.15.

BLACK SHEETS—10 to 16 gauge, 100 lbs., $3.50; 18 to 22, $3.75; 24, $3.95; 26, $4; 28, $4.10.

PETROLEUM AND GASOLINE.—Advanced prices are: Astral, 20c. per gallon in barrel, per case, $2.65; Silver Star in bbls, per gal., 21c., Sunlight in bbls. per gal., 23c., per case, $2.50; Eocene in bbls, per gal., 24c., per case, $2.80; Pennoline in bbls., per gal., 24c.; Crystal Spray, 23c., Silver Light, 21c., Engine gasoline in barrels, per gal., 27c., f.o.b. Winnipeg in cases, $2.75; lamp gasoline, per gal., 33c., f.o.b. Winnipeg, in cases, $3.25.

PAINTS AND OILS — White lead, Pure, $6.50 to $7.50, according to brand; bladder putty, in bbls., 2¾c.; in kegs, 3¾c.; turpentine, barrel lots, Winnipeg, $1.01; Calgary, $1.08; Lethbridge, $1.08, Edmonton, $1.09. Less than barrel lots 5c. per gallon advance. Linseed oil, raw, Winnipeg, 67c., Calgary, 74c., Lethbridge, 74c., Edmonton, 75c.; boiled oil, 3c. per gal. advance on these prices.

WINDOW GLASS — 16-oz. O.G., single, in 50-ft. boxes — 16 to 25 united inches, $2.25; 26 to 40, $2.40, 16-oz. O.G., single, in 100-ft. cases — 16 to 25 united inches, $4; 26 to 40, $4.52; 41 to 50, $4.75; 50 to 60, $5.25; 61 to 70, $5.75. 21-oz. C.S., double, in 100-ft. cases—26 to 40 united inches, $7.50; 41 to 50, $8.40; 51 to 60, $9.45, 61 to 70, $10.50; 71 to 80, $11.55, 81 to 85, $12.60; 86 to 90, $14.75; 91 to 95, $17.50.

FURS PLENTIFUL IN NORTH.

Experienced fur dealers expect to get a large consignment of fur from the far north next summer, as the winter, so far, has been good for the traffic. According to the signs timber wolves are plentiful in the Peace River district, for the north, the moose travel south, and when the wolves are very numerous in old-timers say they never remember the moose to have been so plentiful as this season. Marten and lynx also reported to be very plentiful. They live on rabbits, and this year the rabbit supply is abundant.

The scarcity of muskrats is due, it is said, to the deep snow. They live in burrows, and when the snow is very deep it covers up their homes. The trappers set their traps on top of the muskrat holes and when the snow is very deep they do not know where to find them.

The coyotes are very plentiful this winter, and on account of the snow being so deep they find it hard to get footing. Farmers in the surrounding district report that the coyotes come close up to their dwellings in search of food.

Altogether the dealers are expecting a prosperous season for the fur trappers in the north, and next summer they expect the number and size of the packs of fur to be larger and better than ever. The cold winter will make a thick growth of fur on the animals, so that the pelts should be of the very best quality. As Edmonton is practically the last town to the north on the railway, and is a well-known traders' post, this fur industry is of considerable importance to it, about a million dollars' worth of furs passing through each year.

PIG IRON PRODUCTION IN 1906.

The American Iron and Steel Association's statistics just published show that the total production of pig iron in Canada in 1906 amounted to 541,957 gross tons, against 468,003 tons in 1905, an increase of 73,954 tons, or over 15 per cent. Of the total production in 1906, 525,716 tons was made with coke, 16,-021 tons with charcoal, and 220 tons with electricity. The production of basic pig iron in Canada in 1906 amounted to 246,223 tons, against 172,102 tons in 1905, and the production of Bessemer pig iron was 165,609 tons, against 149,203 tons in 1905. Basic pig iron was made in 1906 by three companies owning six furnaces, and Bessemer pig iron by two companies owning three coke furnaces. The basic and Bessemer pig iron was all made with coke, Canada has not made spiegeleisen or ferro manganese since 1899.

On December 31, 1906, Canada had 15 completed blast furnaces, of which eight were in blast and seven were idle. Of the total 12 usually use coke for fuel and three use charcoal. In addition one furnace to use coke was building, and three coke furnaces were partly erected on December 31. Work on the partly erected furnaces was suspended, however, some time ago.

SHUTTING OFF CREDIT IN CARTWRIGHT.

R. F. Moore, Cartwright, Man., president of Merchants', Limited, the new retail consolidation to which extended reference was made in a recent issue of this paper, was in Winnipeg last week and in the course of an interview gave an interesting account of the first week's experiences. It will be remembered that the dealers in Cartwright have formed a joint stock company in order that by united action they may combat the credit evil and the competition of the mail order houses.

"How did you come to think of this scheme?" asked The Canadian Grocer.

"I have had it in mind for two or three years," was the reply. "One Friday night a month or so ago, we merchants in Cartwright met to pass a resolution on the parcels post matter. From that we naturally drifted into a discussion of the credit troubles and the competition of the mail order houses. I put my propositions before the others in rough form and the next Monday night we met and decided definitely to go on with the deal."

"How did the public receive the news?"

"Not very favorably at first. Most of the farmers shouted "combine" and said they would send to other towns and to Eatons; some talked of starting another store. But we have met and are meeting all these arguments and I think most people are disposed now to give us a fair trial. I don't think there will be any trouble when people see that we are not enhancing prices."

"I put it to them this way: You will have no kick if you get your goods as cheap as you are getting them now, will you? To that they must answer "no." Well then, I add, we expect to be able to sell cheaper to you than we are selling to you now for the reason that we can buy in large quantities."

"You are introducing the coupon system of giving credit, are you not?"

"Yes, we are introducing coupons, both for cash and credit sales. In the first place a customer has to settle all amounts owing before we give him credit at all and in this way we are getting in our accounts. Then credit is given only by the sale of coupon books for which the customer gives his note. This note doesn't bear interest until after a date in the fall when the farmers should have plenty of ready cash."

"How is the scheme working?"

"Very well, but of course there is trouble at the start. One of the wealthiest men in our town sent up the first day for a gallon of coal oil. The messenger had no money with him. We refused to let him take the oil without paying for it by cash or coupon. Of course the account was perfectly safe but we had to maintain our system.

"The man came to the store, very much annoyed, but soon saw the justice of our position."

"You mentioned cash coupon books, Mr. Moore."

"Oh yes, we are selling cash coupon books as well as credit coupons. On a $10 book we allow 2 per cent. discount, on a $25 book 3 per cent. discount, and on a $50 book 5 per cent. discount.

"It is early yet to speak of our experiment," concluded Mr. Moore, "as we have been running now only one week, but so far I am more than pleased with the results. Our cash sales the first week were very satisfactory and our credit sales were a smaller proportion than usual of the total amount."

ORGANIZE FOR PROTECTION.

The retailers in Redvers; Sask., have recently organized a Trades' Protection Association in order to deal with the evils attending the long credit system. The car shortage has caused a scarcity of ready cash and outstanding accounts have become so large that the Redvers merchants have decided the time has come for drastic measures. Under the circumstances the strictly cash method of doing business is not considered practicable, but some curtailment of credit is considered absolutely necessary. At a formal meeting of the Redvers dealers the following resolution was passed:

"That we form the Redvers Mutual Protective Association.

"That the following be elected officers pro tem: President, John Carter; secretary-treasurer, R. Douglas; executive committee, T. King, John Dodds and Richard Curle.

"That the object of this association shall be to overcome to some degree the evils of the credit system and protect ourselves against the abuse of it.

"That each member doing credit business shall furnish monthly to the secretary-treasurer a list of customers' names which shall also show the indebtedness of each party named.

"That the secretary is to record in his books, to which all members shall have access at any time, the total indebtedness of members debtors, which has been incurred with any or all of the members.

"That a special committee be appointed to draft a circular which shall define the amount due by each individual who has been placed on the book of the association, and who shall receive by mail a copy of said circular.

"That we communicate with merchants doing business in other surrounding districts with the view to their co-operation and also to forward the extension of the association to other towns and places in Saskatchewan."

NEWSPAPER ADVERTISING.

(From "The Autobiography of a Business Man," in Everybody's Magazine.)

For a number of years I advertised only in my windows and in some of the street cars, because I did not feel that I could afford to advertise in the daily papers. Two years ago last September I was having a cravanette coat sale, and I succeeded in selling for a couple of weeks about fifty coats a day. I thought I would try a column ad. in one of the evening papers. The next day this column ad. appeared in one of the evening papers, and by the bye, it was not the one that has the largest circulation in Chicago; I selected the paper that this ad. appeared in because they gave me a low rate, but they agreed to give my ad. a good position in the paper. The result was that the next day the sales, which formerly had been about 50 coats a day, jumped to 142, and in 50 days I had sold over 3,500 raincoats. For the year following that sale I continued to advertise in this one paper. Last fall I felt that I could afford to invest say, about $5,000 in advertising in some one of the other papers. I used three morning papers, and three evening papers, the best in Chicago. The results have been something phenomenal. I did not have to invest the $5,000. The profits came back from the newspaper advertising before their bills came in, and I do not figure to-day that I have a dollar invested in advertising.

MR. HUMPHRIES NEW STORE.

President A. W. Humphries, of the Ontario Retail Hardware Association, has moved his stock into his new store at Parkhill and has been joined in the business by his son, Ernest, formerly an official at the Institute for the Blind, at Brantford. The new store is more central than the old, being only four stores from the market and next door to the largest dry goods store in town. It was formerly occupied by Magladery Bros.

MONEY IN COPPER.

In British Columbia it is predicted that in 1907 the copper smelters at Granby will produce 25,000,000 pounds. This should be produced at a cost of 9c. per lb., and if it can be sold for 25c. per lb., the Granby Company should earn this year $4,000,000, which is equal to $29.62 per share. Every one-cent per pound fluctuation in the price of copper makes a difference of $250,000 a year in the earnings of the company, equal to $1.85 per share. On a declining copper market, earnings per share would be as follows, figuring on an output of 25,000,000 lbs. per annum:

Price copper.	Earnings per share.
25c.	$29.62
20c.	20.37
15c.	11.11

The Granby Company is now paying dividends at the rate of $12 per share per annum, so that copper would have to be maintained at above 15c. per lb. to permit of the continuation of these dividends.

NEW INVENTIONS.

R. Parker, Lakewood, N.J., has invented an improved pipe clamp embodying in its construction a plurality of jaws, which are universally adjustable, adapting them to support pipes of irregular forms, branch joints, and any kind of pipe-fitting. The nature of the construction is such that it may be folded to occupy a small compass, enabling the clamp to be conveniently carried from place to place, and manufactured at a small cost.

C. F. Smith, New York, N.Y., has secured a patent for a shovel for use in shifting ashes before the same are removed from the stove. The invention is particularly directed to a form of detachable bottom for the shovel and a novel device for securing the same in place, the device being of such construction and placed in such position as not to interfere with the use of the shovel in the ordinary manner.

FOUNDRY AND METAL INDUSTRIES

J. E. Leonard, M.P., president of the Montreal Reduction and Smelting Co., has been for the past year directing the erecting and equipping of a smelting and refining plant at Trout Creek, Ont. It is near North Bay, is the first and only custom smelter for the Ontario mining districts, and will soon be ready to receive ore.

LEAD SMELTING IN ONTARIO.

D. G. Kerr, Toronto, states that the Stanley Smelting Works have been quietly at work for the past four years opening out and developing the lead resources of their property in Eastern Ontario. Present development on two large veins ten to twelve feet wide, fully warrants them in making a large increase in their plants, and they have placed orders with three prominent Canadian machinery manufacturing companies for all the necessary equipment, which is estimated to cost, with buildings, $75,000. Their mine promises Mr. Kerr says, with suitable plant, a large production of lead ore of 10 per cent. to keep a mill running of a capacity of 200 to 500 tons of ore per day. The demand for lead, he states, cannot be supplied, orders are three months behind, the price is going up, and at present it is $112 per ton for refined pig lead.

STANDARD METALS.

An interesting book, which is being quoted from considerably of late is "Standard Metals," written by R. T. Crane, of the Crane Company, Chicago, the large heating supply house. The work was published some years ago and is a recognized authority in technical schools. It describes where the various metals are found, the manner of their production, their character and their uses in mechanics.

SIFTING SKIMMINGS.

Brass skimmings should never be washed without first sifting them, says the Brass World. Many brass founders believe that the white material in yellow skimmings is nothing but oxide of zinc and carries no copper. As a matter of fact it will frequently run from 20 to 30 per cent. in copper, and if washed, this will be wasted.

All brass skimmings should first be sifted through a rather coarse riddle, say a 20 mesh, and in this manner the fine material can be separated from the metal. The fine, white oxide of zinc which passes through the riddle may be sold for copper contents.

in the riddle should be hand-picked and the large pieces of metal removed. The remainder may be washed in a washing barrel, or otherwise treated.

It is a noteworthy fact that although yellow brass contains less copper than red metal, the skimmings are much richer in copper. The oxide of zinc which escapes seems to carry copper with it. A lot of fine, yellow brass skimmings which have a grey appearance and look as if they contained nothing but oxide of zinc and a little charcoal, may actually contain over 30 per cent. of copper. Repeated tests have demonstrated that this is the case.

Persons addressing advertisers will kindly mention having seen this advertisement in Hardware and Metal.

BUILDING AND INDUSTRIAL NEWS

HARDWARE AND METAL would be pleased to receive from any authoritative source building and industrial news of any sort, the formation or incorporation of companies, establishment or enlargement of mills, factories, foundries or other works, railway or mining news. All correspondence will be treated as confidential when desired.

The Bank of Nova Scotia will build a branch in Winnipeg.

A branch of the Toronto Bank has been opened at Colborne.

The steam saw-mill at Rothesay, N. B., is to be sold by public auction.

A shipbuilding company is to be formed at Kingston, with a capital of $100,000.

The congregation of St. Giles church, Winnipeg, will erect a new church to cost $50,000.

A $50,000 Baptist church is to be erected in Vancouver, with seating capacity of 950.

McCamus, McKelvie & Dowser, New Liskeard, will open a sash and door factory at Englehart.

G. R. Charleson, Minnedosa, has closed his machine shop, to build a new and more capacious one.

Two additional storeys will be built to Stobart, Sons & Co.'s warehouse in Winnipeg this spring.

An apartment block to contain one hundrex suites is to be erected at a cost of $300,000 in Winnipeg.

Scott and Jamieson, meat and produce dealers, Renfrew, are having a cold storage building erected.

The Ellison Milling & Elevator Co. is calling for tenders for a $75,000 mill and elevator for Calgary.

A $15,000 or $20,000 apartment house is to be erected on Queen street east by McCarthy & Co., Toronto.

Two sawmills, with a total capacity of 160,000 feet of lumber, are being built on False Creek, Vancouver.

The Canadian Westinghouse Co., Hamilton, will build a large extension to its present plant this year.

The Canadian Flax Cordage Co. have asked St. Marys for a loan of $5,000 to extend their plant in that town.

A. Haggett & Co., craftsmen, decorators and importers, Victoria, will erect a three-storey building, 80x25 feet.

Nerlich & Co., Toronto, have been granted a permit for a two-storey addition to their warehouse, to cost $35,000.

A permit has been granted to the A. J. Burton Saw Co., Vancouver, for the erection of a $3,300 addition to its plant.

The International Portland Cement Co. have decided to duplicate their plant at Hull. The new capacity will be 4,500 barrels.

J. W. Woods, of Ottawa, has purchased the Redford block in Winnipeg for the Woods Western Co., at a cost of over $100,000.

In the spring one of the biggest private owner car shops in Canada will be erected a few miles east of Medicine Hat, Assa.

Work has been commenced on a magnificent apartment block, 130x197 feet, in Winnipeg. The architects are A. and W. Melville.

A big permit was issued in Winnipeg last week for the erection of a $95,000 warehouse on Princess street for the Fairchilds Co.

The building of the four new Normal schools to be erected at Hamilton, North Bay, Stratford and Peterboro, was commenced on Monday.

An American syndicate has an option on a central block in Ottawa, facing the Government buildings, for the erection of a 300-room hotel.

The Grand Trunk Ry. are making big extensions to the Point St. Charles works, with a view to manufacturing their own rolling stock.

The Imperial Bank has taken out a permit to erect a branch bank on the northwest corner of Bloor street and Lansdowne avenue, Toronto.

The Macgregor-Gourlay Co., Galt, have let the contract for an additional building to their plant, for a pattern shop. Sibley, Miller & Nicol are the contractors.

The Pacific Whaling Co. intend building a barrel factory at Nanaimo. Barrels, not only for their own trade, but also for the general trade, will be manufactured.

Plans for the new Saskatoon office of the National Trust Co. have been received at the Toronto office. It will be a two-storey building, costing about $25,000.

The Rat Portage Lumber Co. will install a $100,000 plant in the $250,000 mill at Harrison, B.C., purchased three years ago and cutting will commence in the spring.

The T., H. & B. Ry. will replace the workshops, recently destroyed by fire, at Hamilton, with a brick building to cost $10,000. George E. Mills has the contract.

A syndicate of American capitalists has secured options on properties in Toronto, with a view to building a large hotel of about 300 rooms, and seven storeys high.

The Alberta Brick Co. has been formed at Medicine Hat, with a capital of $20,000. The president of the company is W. C. Harris, with W. A. Wyatt as secretary-treasurer.

The new Bessemer converter being installed at the Dominion Iron and steel works at Sydney, is nearing completion. This converter will increase the output by 50 000 or 60,000 tons a month.

The Canadian White Co., Montreal, have received a contract from the Fordwick Co., New York, for the construction at Kingston of a million-dollar cement plant, with an initial capacity of 2,500 barrels per day.

The Standard Sand & Machine Co., Cleveland, O., is building a complete plant for handling sand for continuous moulding. This is to be installed in the plant of the Standard Ideal Co., Port Hope, Canada.

The Canada Screw Co., Hamilton, has been incorporated, with a Dominion charter, and a capital of $2 500,000. The incorporators are C. A. Birge, C. Alexander, F. H. Witton, J. O. Callaghan, W. F. Coote, C. S. Wilcox and Hon. Wm. Gibson.

Adam Zimmerman, M.P., is at the head of a big company organized to manufacture balbriggan underwear at

Hamilton. The company is capitalized at $300,000, the Lawrence Co., of Lowell, Mass., being interested. Building operations will begin at once.

The Hamilton and Fort William Nav. Co., have placed an order with the Canadian Shipbuilding Co. for an 8,000-tons steamer, for the upper lake coal and ore trade. The vessel will be 460 feet long, 53 feet beam, and a 29 feet draught. Quadruple expansion engines and Scotch boilers will be installed.

So far this session there have been no less than 65 petitions received by the Canadian Parliament against a continuance of the bounties on steel and iron after June 30 of this year. The petitions, which are all identical in form, are signed almost wholly by farmers of Ontario, the average number of names on each petition being about fifty. They point out that about ten million dollars has already been paid in bounties to the iron and steel industries on Canada, and that if the period of paying bounties is extended four years from June 30 next, as proposed, the total amount paid in bounties will probably reach twenty-five millions.

USES OF ALUMINUM.

It is only a few years since we first began to hear of aluminum, and of its wonderful combination of strength and lightness. The stir occasioned by its discovery passed after a time, as is the case with everything when the novelty wears off. Practical men, however, did not overlook the advantages it had to offer, and soon it was being tested on a great many articles where reduction of weight meant increased efficiency.

A good example of its uses can be seen any day at the Canada Cycle & Motor Co.'s works at Toronto Junction. Here is a separate foundry for moulding the special alloys of aluminum they require for automobile engine crank and transmission cases and plates for skates. By introducing this metal into motor construction the weight of the car has been reduced several hundred pounds and the strength of these parts has not been impaired thereby. Rather has the strength of the car been increased, for the other parts have consequently less weight and strain to carry.

Experience with this metal in this way opened up new avenues for its usefulness Probably the latest adaptation is for skate tops, for which this company has secured patents. This application reduces the weight of the skates by at least one-third, and at the same time admits of a very unique and elegant design. These "Automobile" and "Cycle" skates, as they are called, have been on the market now for over two seasons and have demonstrated one more practical use of aluminum.

THE IRON BOUNTIES.

In a letter to the Globe, J. H. Sinclair, M.P., quotes from the annual address delivered before the Toronto Board of Trade recently by the retiring president, Peleg Howland, and takes up the cudgels on behalf of the iron and steel industry in Nova Scotia, which he felt had been treated unfairly in the address. Mr. Sinclair says:

I have no desire to make any personal attack on Mr. Peleg Howland, but, judging by the above paragraph, I cannot help wondering how he ever became President of a Board of Trade in a great Canadian city. He refers to Nova Scotia as "the far east." To him it is evidently a foreign country. He can see no reason why the people of Ontario should tax themselves to build up an industry a thousand miles away. I suppose it is our misfortune in Nova Scotia that we are a thousand miles away from the place where Mr. Howland lives but he must remember that Canada is a large country, and that a considerable portion of it lies outside the boundaries of Ontario, and that the people of those remote regions are his fellow-countrymen.

When Mr. Howland complains that the people of Ontario are being taxed to build up an outside industry he evidently forgets that Sault Ste. Marie, Hamilton and Midland are within the bounds of his own province. I suppose Mr. Howland will be surprised when I tell him that out of a total sum of $6,037,350 paid by the Canadian Government for bounties to the manufacturers of iron and steel, no less than $2,570,831, or nearly one-half of the whole amount, was paid not to the far east at all, but to manufacturers in the Province of Ontario. Again, when Mr. Howland says that Ontario is naturally situated to draw its raw material such as iron and steel from the United States. I wish to remind him that the Maritime Provinces likewise are so situated that they could purchase with even greater advantage their manufactured goods from the New England States rather than from the manufacturers of Ontario. The one proposal is just as reasonable as the other. The Maritime Provinces are to-day the best outside market open to the Ontario manufacturer. It has been his hunting ground ever since confederation, and has done not a little to make him the prosperous man that he is to-day. In Nova Scotia we export our lumber, apples and agricultural products to Great Britain. We send our fish to the West Indies, and with the proceeds we purchase flour and manufactured goods largely from Ontario. The only article that the people of Nova Scotia sell to Ontario is steel, and Mr. Howland would stop that, too, if he could. We in the Maritime Provinces are quite well aware that we are subjected to heavy taxation to build up industries in the other provinces, but we are not complaining. We hope we are broad enough to recognize that our countrymen both east and west have equal rights with ourselves, and that the prosperity of any one section or province means the prosperity of the whole. If we are to build up a great country north of the 49th parallel our business men must be broad enough to feel interested in the advancement of the other localities even if they are a thousand miles away.

It's as difficult to find a friend as it is to lose an enemy.

Distance lends enchantment to the view of a friend who is always in need.

Always listen to an honest opinion—if for no other reason than because it's different from your own.

CATALOGUES AND BOOKLETS

When sending catalogues for review, manufacturers would confer a favor by pointing out the new articles that they contain. It would assist the editor in writing the review.

By mentioning HARDWARE and METAL to show that the writer is in the trade, a copy of these catalogues or other printed matter will be sent by the firms whose addresses are given.

From Nova Scotia.

We have received from Rhodes, Curry & Co., Amherst, N.S., a very useful little souvenir, in the form of a card case and memo book. This is made of handsome maroon leather, and lined with tan calfskin. The cover is free from advertising, with the exception of the firm's name and address, which appears in gold lettering. The above firm will be pleased, as long as the supply lasts, to send one to any in the trade who will mention Hardware and Metal when writing.

Samples of Roofing.

The tremendous increase in the price of lumber in the last few years has made shingles entirely out of the question on most farm buildings. In order that every reader of this paper may become familiar with a roofing which is very much better than shingles and is rapidly taking their place, the Barrett Mfg. Company offer to send free a sample of their Amatite Roofing to every reader of this paper. The feature of this roofing is a real mineral surface on top, which not only makes coating and painting unnecessary, but is a splendid fire-retardent. We suggest that every reader write at once for a free sample and illustrated booklet. The request should be adressed to the Barrett Manufacturing Company, New York.

Maple Sugar Industry.

The Department of Agriculture at Ottawa have just issued a very effective and instructive book dealing on the making maple syrup and sugar which should prove very interesting to hardware merchants making a specialty of sap pails and spouts. It will be sent to anyone writing the Department of Agriculture, Ottawa, Canada.

LETTER BOX.

Enamelware Manufacturers.

J. M. Bothwell, Allandale, writes: "Could you give me the address of any Canadian firms who are in the market for graniteware? I am wanting to purchase, and thought you could give me the name of some such firm."

Answer.—The older firms in the graniteware business are the Kemp Manufacturing Co., Toronto; The McClary Manufacturing Co., London; The Thomas Davidson Manufacturing Co., Montreal. A new firm which has come into the market recently is the Ontario Steelware, Limited, Brock Ave., Toronto. All of the above companies are advertising with Hardware and Metal. --Editor.

Wire Nail Machines.

Edward Norman, London, Ont., writes: "Can you give me the names and addresses of Canadian makers of wire nail machines?"

Answer.—There is no Canadian manufacturer of wire nail machines, but the following American firms manufacture them: H. J. Miller's Sons, Bridgewater, Mass.; National Machinery Co., Tiffin, Ohio; Perkins, Henry Co., Bridgewater, Mass.; Turner, Vaughn & Taylor Co., Cuyahoga Falls, Ohio; and Walton & Macke, Kokomo, Ind.—Editor.

NOTES.

G. Leblanc, Joliette, was in town last week buying supplies.

H. J. Elliott, South Durham, was in Montreal buying goods this week.

J. A. Paquin, St. Eustache, called on the Montreal jobbers this week.

W. H. Clapperton, Maria Capes, Que., was in Montreal this week buying goods.

F. Gasselin, St. Alexandre d'Iberville, visited some of the Montreal jobbers this week.

W. W. Chown, of Belleville, now in the hardware business in Edmonton, reports good trade, notwithstanding adverse weather conditions.

T. L. Clark & Co., bell factory, 208 Papineau Ave., Montreal, was burnt out recently, the damage amounting to about $5,000. The fire started on the top floor where the brass foundry is located, and although the firemen succeeded in confining it to that flat, the two lower floors were flooded with water.

A. B. Manship, of the Canadian Wire and Steel Company, died at his home in Hamilton, last Wednesday. The deceased was 56 years of age and had been a resident of Hamilton for five or six years, coming from Cleveland to establish the steel company.

The annual meeting of the Williams Manufacturing Co., was held in Montreal, on Feb. 13th. The old board of directors was re-elected, and Bartlett McLennan was chosen as president and Chas. W. Davis vice-president, and managing director. Alex. Dube was appointed secretary.

The Vancouver Power Co., which at present is the only organization supplying electric energy in the mainland coast cities, is figuring on developing its present installation of 12,000 h.p. by adding three monster units of 10,000 h.p. each. One of these will go in at once and all will be installed in little over a year from date.

We have been asked to state by Pilchers, Limited, London, England, that there is no connection between their house and that of the Mr. Pilcher, whose case of Pilcher versus Price & Co., has recently been reported in various trade papers. The plaintiff in the case refer-

CONDENSED OR "WANT" ADVERTISEMENTS.

Advertisements under this heading 2c. a word first nsertion; 1c. a word each subsequent insertion.

Contractions count as one word, but five figures (as $1,000) are allowed as one word.

Cash remittances to coVer cost must accompany all adVertisements. In no case can this rule be overlooked. AdVertisements receiVed without remittance cannot be acknowledged.

Where replies come to our care to be forwarded, five cents must be added to cost to coVer postage, etc.

SITUATIONS VACANT.

BRIGHT, intelligent boy wanted in eVery town and Village in Canada; good pay besides a gift of a watch for good work. Write The MacLean Publishing Company, 10 Front Street E., Toronto. (tf)

EXPERIENCED Hardware Clerk Wanted—One to take charge, must be good salesman and stockkeeper; state experience and salary expected. Address G. McLean (Hardware), London. 9

WANTED—Hardware Clerk, must be good stock keeper, state age, experience and salary expected. Peart Bros., Hardware Co., Limited, Regina, Saskatchewan. (12)

BUSINESS CHANCES.

PLUMBING, Steam, Hot Water, Gasfitting and Tinsmithing business for sale; $15,000 business done last year, and only $3,000 or less inVested; in one of the best towns in Canada. Waterworks and sewage. Owner retiring. Box 587, HARDWARE AND METAL, Toronto. (10)

HARDWARE and Harness Business for sale in good town. Owner retiring for good reason. Stock about $2,500, in excellent shape. This is worth inVestigating, for particulars apply Richard Tew, 23 Scott Street, Toronto.

FOR SALE—Foundry and stoVe works on the Pacific Coast, fully equipped and up-to-date. Apply P.O. Box 760, Victoria, B.C. (10)

HARDWARE Business; including stoVes, tinware and tinsmith tools, in thriVing town in West Ontario peninsular, stock about $5,000; building can be leased if desired, dwelling also. Box 583 HARDWARE AND METAL, Toronto. (17)

HARDWARE Business Stock about $12,000.00 is in good condition, could be reduced to suit purchaser if necessary. Apply George Taylor & Son, Hardware, London, Ont. (11)

FOR SALE—Stock of Hardware and Groceries in live village in central Ontario. Stock in first-class condition. Situation best in village. Reason for selling ill health. Box 591 HARDWARE AND METAL. (9)

FOR SALE.

FOR SALE—Five h. p. Campbell Gas Engine, first-class condition, capable of deVeloping eight h. p. Can be seen in operation. 1-8 ft Cornice brake. Apply R. & W. Kerr, Limited, Montreal. (10)

ARTICLES WANTED.

WANTED—8 foot cornice brake; state price and particulars. J. E. Hussey, Melbourne, Ont. (9)

ANY hardwaremen having a stock of cut nails they wish to dispose of at reduced prices should write at once to Box 501 HARDWARE & METAL, Toronto.

WANTED—1 second hand power punch suitable for range making. Must be in first-class condition with 16 or 18 inch throat. R. & W. Kerr, Limited, Montreal. (10)

WANTED.

CORNICE FOREMAN WANTED. Must be capable of taking full charge of Metal Cornice Department. Permanent position to right man. State experience and wages expected. Galt Art Metal Co., Limited, Galt, Ont. (9)

BOOKKEEPER WANTED

WANTED—A good bookkeeper; stating experience, reference and salary. Cunningham Hardware Co., Westminster, B.C. (13)

Galt Sure Grip Shingles

make the HANDSOMEST and most DURABLE roof. Are the EASIEST and QUICKEST to either SELL or put on. See them and get acquainted.

GALT ART METAL CO. Ltd.
GALT, ONT.

Get in Shape

To giVe your customers intelligent information and trade-winning prices on the most up-to-the-minute and best-selling line of Sheet Metal Building Goods. This is an important matter to the Hardware and Tinsmith Trades. There will be an enormous demand for these goods this year. Builders and Owners from your Very locality are now enquiring about this line. Speak up—And let them know "you're in the game."

Get Our Catalogs

and Price Lists. That's the surest way to be "in the game." Your Opposition will haVe to "step fast" if he beats you and Our Line. We are using our eVery endeaVor to make Our Line interesting to "the trade." We want good representation in eVery locality. Write us about it to-day.

Western Distributing Agents:

THE McCLARY MFG. CO.
Winnipeg and Calgary

Galt Steel Siding

is made in original designs. Sharp, bold patterns. Close-fitting and invisable joints make them great sellers. People won't have the ordinary kinds once they see ours.

GALT ART METAL CO. Ltd.
GALT, ONT.

"QUALITY UNSURPASSED"

BELLEVILLE BRAND

TRADE B MARK

HORSE SHOES

MANUFACTURED BY

TORONTO AND BELLEVILLE ROLLING MILLS

LIMITED

BELLEVILLE, ONT.

Are you interested in any of the lines that are advertised? A Post Card will bring you price list and full information. Don't forget to mention Hardware and Metal.

Paint, Oil and Brush Trades

PAINT BUSINESS ENLARGING.

While many hardware merchants have for some time handled paints there is undoubtedly a growing tendency to increase the attention given to this line, and to carry in stock on a larger or a smaller scale some suitable brand of mixed paints, says the Iron Age. This tendency and the great possibilities for the sale of this class of goods in the hardware store are nowhere better recognized than by the paint manufacturers themselves, many of whom are putting forth every effort to secure prestige in the hardware trade by being among the first to take advantage of the tendency toward enlargement of the variety of goods handled by merchants in this branch of business. It is a noteworthy fact that at the conventions of retail hardwaremen held throughout the United States last year paint manufacturers were prominently represented. At one of these conventions, indeed, five different manufacturers reserved expensive suites in the hotel which was the convention headquarters, where they installed elaborate exhibits and distributed unlimited quantities of advertising matter among the 300 merchants present. It may be presumed that at the conventions which are being held this spring the same tendency will be observed, and merchants thus be given an opportunity to get into close touch with paint manufacturers and consider intelligently and practically the advisabilityof taking up the sale of this line of goods.

GET READY FOR SPRING.

Spring paint selling will soon be in full blast. Upon the business you do—the orders you fill—during the present months depend to a very large degree your paint profits for the year. To make this spring's business exceed all past records every paint dealer must maintain its fighting attitude, must "work its plan," must call into service every aid at its command, says an exchange.

"Keeping everlastingly at it" will win. Don't let up in your efforts because things are coming your way, and you feel you are getting a fair proportion of local business. You want more than a fair share. You want ever so much more than you got last year because it's only by progress and continual advances that you can maintain your lead in the paint field.

To "keep ahead, always," you must go after orders, large and small, little and big. Never overlook the specialites. Give a good share of your time and attention to the profitable field of balls, schools, churches, offices and store buildings. If you see a building going up or have a customer whom you cannot quite close with, call on your manufacturer for aid with forceful letters and strong follow ups. Continue to use all the advertising helps and be a "full line" advertiser.

Keep your newspaper space filled with good "copy." Color cards, hangers, posters, fence and field signs are invariably effective when properly distributed.

LUMINOUS PAINT.

Luminous paint is made in the following manner, according to a recipe printed in the Engineering and Mining Journal. Take oyster shells, wash and clean them in warm water; put the shells into a fire for half an hour; after taking them out and allowing them to cool, they are finely powdered by pounding, and the gray parts which are of no use, are thrown away. The powder is then put into a crucible in alternate layers with flowers of sulphur, the lid is put on and the crucible sealed up with a paste made of sand and beer. After this is dry, the crucible is put on the fire and baked for one hour, then cooled, and the cover taken off; all gray parts are now separated, and the remaining substance is mixed into a thin paint with a mastic varnish. Before applying the luminous paint, the article should be given two coats of white lead and turpentine, to form a body ground for the luminant.

This luminous paint, which is really a polysulphide of calcium, has the property of emiting, in darkness, light which was previously absorbed, and when kept in darkness for 12 or 15 hours without being subject to any light, loses its luminosity. Luminous paint of this sort has been used for clock dials, lanterns, etc.; however, its practical application can hardly be said to have met with success.

In connection with the use of luminous paint in mines, it may be stated that when subjected to the light from electric bulbs or an electric arc light, such paint will retain its luminosity and emit light when placed in darkness. For this reason, it is possible that where arrows are painted on the sides of an entry that has been lighted by electricity, should the light suddenly go out, the arrows painted with luminous paint would still emit the light which was previously absorbed.

The early bird catches the spring paint business. Be the early bird.

Time spent in talking cheap paint is time wasted—you'll never have another chance with the same customer.

Let your Paint be a Business Promoter!

How many Hardwaremen are as particular about their "silent salesman" as they might be? There are many Dealers shrewd enough to be careful, mighty careful about the paint they supply their customers. We know that by the orders which come our way for

NEW ERA PAINT the most effective "silent salesman" ever manufactured.

Have you our catalogue? You owe it to yourself to secure it before the spring rush comes. A post card brings it.

STANDARD PAINT AND VARNISH WORKS CO., LIMITED

WINDSOR - ONTARIO

Floors to be Proud of

are produced by using

JAMIESON'S FLOOR ENAMEL

We show a fine range of colors, all of which give a finish which cannot be equalled by any **Floor Paint** sold.

They wear well, never stick after drying, and do not rub off by washing: In short, they have none of the defects of the common floor paints.

Try a Sample order of these goods and you'll want some more.

R. C. JAMIESON & CO.
LIMITED

16-30 Nazareth St., MONTREAL

McCaskill, Dougall & Co.

Manufacturers — RAILWAY, CARRIAGE AND BOAT **VARNISHES.** HIGH-GRADE FURNITURE and HOUSE VARNISHES

MONTREAL.

A. RUDD & CO., St. Helens, Lancashire, England

Manufacturers of highest grade

GLAZIERS' DIAMONDS in all patterns. Also Circle Boards, Beam Compasses, Gauge Glass Cutters, Boxwood Squares and Laths, Plate Glass Nippers Atmospheric Soldering Irons, etc., etc., etc.,

Canadian Agent: J. B. GIBSONE, P.O. Box No. 476 MONTREAL

GLAZIERS' DIAMONDS OF PROVED WORTH

Having a Century's Reputation for Reliability.

MADE BY

A. SHAW & SON, London, Eng.

CANADIAN AGENT

GODFREY S. PELTON, - 201 Coristine Building, Montreal.

BEFORE ordering colors think of **THE VIA GELLIA COLOR CO.**

Raisers and Levigators of Mineral Colors

Mines and works at

Matlock-Bath, Cromford and Bonsall (Eng) and Ovoca (Ireland)

Our specialities are

ENGLISH and TURKEY UMBERS (raw and burnt) TURKEY RED, RED AND PURPLE OXIDES, PURPLE BROWNS, YELLOW OCHRES, ROTTEN STONE (lump and powder) and Superior Ivory Blacks.

These are exclusive lines and are strong money-makers in the trade. Write for samples and prices to

WM. O. GREENWAY, 13 St. John St., Montreal

It will pay you.

Persons addressing advertisers kindly mention having seen their advertisement in Hardware and Metal.

PAINT AND OIL MARKETS

MONTREAL.

Office of HARDWARE AND METAL,
232 McGill Street,
Montreal, March 1, 1907

There is nothing of special interest to report in paint and oil circles at present, all prices remaining the same as quoted last week.

Although sales have fallen off slightly of late, the market remains fairly firm, with a fair demand for stocks.

The quitness in the retail and jobbing trade, is looked upon as being the lull before the storm, and not much is expected to be doing now, until the regular spring season opens. The only people who are rushing now, are the manufacturers, who are all running their plants full blast, in anticipation of a good spring trade.

LINSEED OIL—Still firm at the old prices: Raw, 1 to 4 barrels, 55c.; 5 to 9 barrels, 54c.; boiled, 1 to 4 barrels, 58c.; 5 to 9 barrels, 57c.

TURPENTINE—Fair business is being done. Prices remain the same as last week: Single barrel, 98c per gal.; for smaller quantities than barrels, 5c extra per gal. is charged. Standard gallon is 8.40 lbs., f.o.b. point of shipment, net 30 days.

GROUND WHITE LEAD—The following prices have remained steady now for some time: Best brands, Government standard, $7.25 to $7.50; No. 1, $6.90 to $7.15; No. 2, $6.55 to $6.90; No. 3, $6.30 to $6.55, all f.o.b., Montreal.

DRY WHITE ZINC—Still firm at the following prices: V.M., Red Seal, 7½ to 8c; Red Seal, 7c to 8c; French V.M., 6c to 7c; Lehigh, 5c to 6c.

WHITE ZINC—Pure, 8½ to 9½c; No. 1, 7c to 8c; No. 2, 5¾c to 6¾c.

PUTTY—Prices are as follows: Pure linseed oil, $1.75 to $1.85; bulk in bbls., $1.50; in 25-lb. irons, $1.80; in tins, $1.90; bladder putty in barrels, $1.75.

ORANGE MINERAL—The following prices still hold: Casks, 8c.; 100-lb. kegs, 8½c.

RED LEAD—Is firm at the following figures: Genuine red lead, in casks, $6; in 100-lb. kegs, $6.25; in less quantities at the rate of $7 per 100 lbs.; No. 1 red lead, casks, $5.75; kegs, $6, and smaller quantities, $6.75.

PARIS GREEN—Last week's revision holds good: In barrels, about 600 lbs., 25½c per lb.; in arsenic kegs, 250 lbs., 25¾c; in 50-lb. drums, 26¼c; in 25-lb. drums, 26¾c; in 1-lb. packets, 100 lbs. in case, 27¼c; in 1-lb. packets, 50 lbs. in case, 27¾c; in ½-lb. packets, 100 lbs. in case, 29¼c; in 1-lb. tins, 28¼c f.o.b. Montreal. Terms, three months net or 2 per cent. 30 days.

SHELLAC—Bleached, in bars or ground, 46c per lb., f.o.b. Eastern Canadian points; bone dry, 57c per lb., f.o.b. Eastern Canadian points; T.N. orange, etc., 48c per lb. f.o.b. New York.

SHELLAC VARNISH—Firm at last week's figures: Pure white, $2.90 to $2.95; pure orange, $2.70 to $2.75; No. 1 orange, $2.50 to $2.55.

PETROLEUM — American prime white coal, 15½c per gallon; American water, 17c per gallon; Pratt's Astral, 19½c per gallon.

WINDOW GLASS—First break, 50 feet, $1.85; second break, 50 feet, $1.95; first break, 100 feet, $3.20; second break, 100 feet, $3.40; third break, 100 feet, $3.95; fourth break, 100 feet, $4.15; fifth break, 100 feet, $4.40; sixth break, 100 feet, $4.95. Diamond Star: First break, 50 feet, $2.30; second break, 50 feet, $2.50; first break, 100 feet, $4.40; second break, $4.80; third break, 100 feet, $5.75; fourth break, 100 feet, $6.50; fifth break, 100 feet, $7.50; sixth break, 100 feet, $7.50; seventh break, 100 feet, $8; eight break, 100 feet, $9. Double Diamond: First break, 50 feet, $3.45; second break, 50 feet, $3.75; first break, 100 feet, $6.75; second break, 100 feet, $7.25; third break, 100 feet. $8.75; fourth break, 100 feet, $10; fifth break, 100 feet, $11.50; sixth break, 100 feet, $12.50; seventh break, 100 feet, $14; eighth break, 100 feet, $16.50; ninth break, 100 feet, $18; tenth break, 100 feet, $20; eleventh break, 100 feet, $24; twelfth break, 100 feet, $28.50. Discount on Diamond Star, 20 per cent.; on Double Diamond, 40 per cent.

TORONTO.

Office of HARDWARE AND METAL,
10 Front Street East.
Toronto, March 1, 1907

The trade in paints and oils is improving Prices generally are strengthening, with good prospects for advance, and all dealers are looking for milder weather, as they are in all other branches of trade, to enliven the business.

Paris green is experiencing a good demand, a large majority of the orders being booked. Turpentine prices are firm, but unchanged. Linseed oil prices are strengthening, with prospects of an advance. Manufacturers are understood to have already advanced prices to jobbers. White lead prices are stronger, with a tendency to rise, the quotations, however, are still unchanged.

WHITE LEAD — Pure white, $7.40, No. 1, $6.65; No. 2, $6.25; No. 3, $5.90; No. 4, $5.65 in packages of 25 lbs and upwards; ½c. per lb. will be charged extra for 12½ lbs. packages; genuine dry white lead in casks, $7.

RED LEAD.—Genuine in casks of 500 lbs., $6; ditto, in kegs of 100 lbs., $6.50; No. 1 in casks of 500 lbs., $5.75; ditto, in kegs of 100 lbs., $6.25.

DRY WHITE ZINC.—In casks, 7½c.; in 100 lbs., 8c., No. 1, in casks, 6½c., in 100 lbs., 7c. Ground in oil—In 25 lb. irons, 8c.; in 12½ lbs., 8½c.

SHELLAC VARNISH.—Pure orange in barrels, $2.70 ; white, $2.82½ per barrel ; No. 1 (orange) $2.50 ; gum shellac, bone dry, 63c., Toronto. T. N. (orange) 51c. net Toronto.

LINSEED OIL—The prices in this are strengthening, with a quite probable advance. We continue to quote : Raw, 1 to 3 barrels, 59c; 4 to 7 barrels, 58 ; 8 and over, 57c.; add 3c. to this price for boiled oil, f.o.b. Toronto, Hamilton, London and Guelph, net 30 days.

TURPENTINE—Prices are firm, but unchanged : Single barrels at 99c. per gal., f.o.b. point of shipment, net 30 days ; less than barrels, $1.04 per gallon.

PARIS GREEN—Trade in Paris green has been brisk, with a large number of orders booked. We continue to quote English and Canadian at 27½c. per lb.

PETROLEUM—The trade in all by-products is steady and improving. We quote : Prime white, 13c.; water white, 14c.; Pratt's astral, 18c.

For additional quotations see current quotations at back of paper.

HOUSECLEANING TIME APPROACHING.

"The best way to clean house is to paint," is a timely motto containing more of fact than fiction, says an exchange. Paint is a great renovator, not simply for brightening up, but literally making clean. Further, it is a little recognized insecticide, destroying both egg and adult ; and those unfortunate in getting into apartments infested with vermin will find it a more satisfactory remedy than carbon bisulphide.

Convince a woman on one or two of these points, and you have an interested listener. Few women who have never substituted the paint brush for the scrub brush are aware that they can do the most of their own painting and varnishing. With prepared paint and a few simple directions regarding care and selection of brushes, which the dealer should be prepared to throw in, creditable work is not beyond the reach of the most inexperienced. But the successful salesman should familiarize himself with the common problems of the housewife, anticipating her needs and showing a cordial interest. He should know what colors harmonize—for the color cards are not fully satisfying to the novice ; how to prepare sizing for the kitchen wall if it is not hard finished plaster ; to warn her against using anything but iron paint on tin. These and other "kinks" will be appreciated and prove profitable on both sides ; for the real home-maker delights in using her own hands in every possible way in beautifying her home, and the newly acquired manual skill will find two ways to use paint where one was found before.

CANADIAN OIL CO.'S EXTENDING.

The Canadian Oil Co., Toronto, manufacturers of "Sterling Brand" oils, paints, and varnishes, are adding a second story over a part of their buildings. This increase of floor space will allow for enlargement of office room, and improvement in storing and shipping facilities. Storage batteries will be installed so that their own steam power can be used for lighting purposes. The recent enlargements of this company's refineries at Petrolea will relieve the Toronto plant of most of the work of compounding oils, except for special orders. The Toronto plant will now be devoted chiefly to the manufacture of paints and varnishes.

FOUR WAYS OF SELLING.

In their weekly market report of a recent date, Strasbaugh, Silver & Co., of Aberdeen, Md., go a little afield from the market and under "Points Worth Remembering" quote a successful manufacturer as follows :

"The desire to buy cheap, anything, is a mania with some people.

"To favor an article simply because its first cost is low is not only unfair, but a poor business policy.

"To buy at a fair price is shrewd, but it is false economy to forget that below a certain limit, cheapness is suicidal should be shunned.

"There are four ways to sell goods :

"1 The Dishonest Method—To sell so low that you can't earn a living, and the sheriff will finally wind up your affairs, and your creditors carry your losses.

"2 The Misrepresentation Method — Palm off upon your customers unfair goods and persuade them that they are the best.

"3 The Suicidal Method — Employ workmen at less than living wages ; buy the cheapest materials ; pare everything down to the lowest notch, and patch results.

"4 The Straight Business Method—Buy the best ; employ skilled labor ; thoroughly know your own business and business values. Provide special facilities for the execution of the greatest amount of high-grade production at a minimum cost. Avoid extravagant management, expensive methods, and have your dealings with customers who appreciate treatment that is not one way.

PRICE CARD SUGGESTION.

The store that does not use a generous number of price cards is not doing the business it ought to do. Everybody knows that mouse traps sell for five cents and there is a good profit in them at that price. Take a ten cent bushel basket and put in a false bottom about four inches from the top. It can be quickly done by using strings and weaving a network into the basket. Throw a gross of these traps—print a large card something like this :

> "That mouse kept you awake last night. Here's the remedy, 5c."

Tack it to the basket, set the basket just outside the door and that gross of traps will fairly melt away.

What will make them sell so fast? Surely not the price, because you have always sold them for five cents—what is it? If you aren't a believer in price cards try this plan. Put the goods in sight where they will sell themselves, and keep changing both the card and the goods.

BERRY BROTHERS' VARNISHES

are one grand story of success

BECAUSE THEY ARE THE BEST

They are the safest goods to handle,
and the most reliable goods to use.

Write for Catalogue, Information and Prices.

BERRY BROTHERS, Limited

NEW YORK
BOSTON
PHILADELPHIA
BALTIMORE

Varnish Manufacturers
DETROIT
Canadian Factory—WALKERVILLE ONT.

CHICAGO
CINCINNATI
ST. LOUIS
SAN FRANCISCO

Let's Advertise Your Business This Season, Mr. Dealer

100 per cent. durability

We mean that, every word of it. We are prepared to advertise your business and practically prepared to make a success of it. The business not advertised cannot live in the twentieth century. Hardwaremen cannot afford to advertise in the ordinary way. Here is our plan. We claim a good deal for

ARK BRAND PAINT

That paint will endure. That paint cannot fade as other paints do, because it's made different. There is lasting life and beauty in every drop of **ARK BRAND** and so confident are we of it's excellence and superiority that we are willing to insert a **money back clause** in ou agreement with any practical man. Use **ARK BRAND PAINT**, feature it, push it this season and if it dces not advertise your business as never before we want to know about it.

MURESCO WALL FINISH

is nearly as famous and as much of a favorite as Ark Brand Paint. Our aim is to excel in all things. Because our ideals are high we have produced the best paint and the best wall paper finish ever placed on the market. Hundreds of dealers in Canada say so. We would like you to know their opinion. A post card brings full particulars.

BENJAMIN MOORE & CO., LIMITED

(Formerly The Francis-Frost Co., Limited)

NEW OFFICES AND WORKS:

TORONTO JUNCTION, CANADA

New York, Chicago, Cleveland, Carteret, N. J.

Standard Ideal Ware is manufactured after a process which makes the porcelain enameling virtually a part of the iron. It will not chip, crack or discolor. Durability, Purity, and Beauty of Design have served to give **Standard Ideal** fixtures a reputation which is a full warrant to any architect or builder to not only suggest but recommend this ware for residences or public institutions.

We make Baths, Lavatories, Sinks, and other Plumbing equipment.

Send for Catalogue

The Standard Ideal Co., Limited
Port Hope, Ontario

Aubin's Patent Refrigerators

GROCERS should not be without one

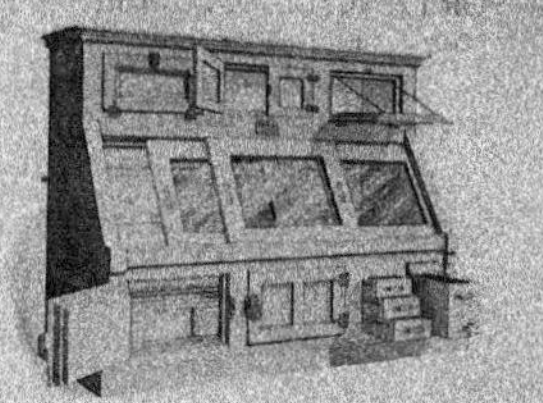

Perfect Dry Air Circulation System
Highest Testimonials from Leading Grocers in Canada

Made in 10 styles

Silver Medal Quebec Exhibition
Diplomas—Ottawa, Montreal and Toronto

WRITE FOR CATALOGUE

C. P. FABIEN, Proprietor and Manufacturer

OFFICE AND FACTORY
3167 to 3171 Notre Dame St.
MONTREAL, CANADA

WAREROOMS
4 to 8 Fabien Ave.
31 to 45 William St.

Branches at Ottawa, Winnipeg and Windsor, N.S.

MINCING KNIVES

DON'T TAKE CHANCES

WOODENWARE OF ALL KINDS

Stock the No. 445 FOLDING IRONING BOARD

"Cumming's Goods are Good Goods"

A Spring Seller in great demand.
The one that will **sell** and sell **easy.**
The one that will please your customers.
Ask Your Jobber For Them

WESTERN REPRESENTATIVES:
F. A. REINHARDT
J. J. MAGUIRE
608 Ashdown Bldg., Winnipeg

CAN OPENERS

MADE BY

THE CUMMING MANUFACTURING CO., LIMITED
RENFREW, ONTARIO

Plumbing and Steamfitting

IMPORTANCE OF PLUMBING AND HEATING

Position of the Trade Unique—Influence and Responsibility Not Generally Realized.

Written for Hardware and Metal by C. E. Oldacre, Toronto.

The plumber, steam and hot water fitter occupies an unique and more important economic and industrial position in his relations to his fellow-beings than for which he probably takes credit to himself.

Upon him to-day all are largely dependent for the state of their health and of their bodily comfort. A properly and orderly-arranged plumbing system is essential to every modern house or building to-day, as also is a properly heated one. Without either one or both the building, if intended for habitation, is incomplete. It is quite necessary, that, no matter how unpretentious or how extensive these systems may be, they shall be of good quality and able to serve the purpose for which they are intended.

Country Growing Rapidly.

Any one acquainted with conditions in Canada to-day knows it is experiencing a most unprecedented growth and one of great expansion in every direction—new country is constantly being opened up, new mills, new warehouses, new mines, new railroads are everywhere. There is a constant streaming influx of people from other lands—additional facilities being added for both railroad and water transportations.

All this means new towns—new homes, new churches, new schools, new post offices, new fire halls, new banks, new opera houses, new business blocks.

And that means that the plumber, steam and hot water fitter has a constantly expanding field before him and new problems to consider every day. And how to meet all these conditions and their attendant emergencies is the question which continually faces him day by day.

As in all new and expanding countries there are situations to be met and many vexatious annoyances from one source or another that present themselves, that must be overcome, that do not occur to those in old and well-settled communities. Take the one located a considerable distance from the more important present trade centres; he is subjected quite frequently to annoying delays, without seeming reason, in getting the requisite materials, due to the inadequate transportation facilities—both rail and water.

Order Goods Early.

One thing that might well be touched on here, however, is that knowing these conditions one should close his contracts as early as possible, urging upon the customer the great necessity of so doing, and place his requisitions with the manufacturer definitely as long ahead of the actual requirement of the goods as possible, foreseeing one's wants in any particular line as well as can be, so as to avoid the usual delays occurring during periods when there are naturally the greatest demands, when there is the greatest call upon manufacturers, transportation companies and tradesmen.

The present is the time of year when we have the greatest amount of time for preparing for the coming season's business—to look back over the good working and profitable work and at probably some not so well working plants. But there is now time to consider the causes and reasons that led to the success or failure of one or other. It is well known that any poor and unsuccessful work is heralded far and near and soon becomes the thorn in one's side and is often used to good advantage by one's competitors.

Heating Knowledge Necessary.

How to accomplish good heating work is the question that should be ever before the steamfitter. As complete a knowledge of the work in all details so far as possible, a little more learned every day, is the surest safeguard, and a correct installation in accordance with this knowledge and experience is sure to produce the best results. Do not put too much confidence in 'rule of thumb" and cut and dry methods, or the policy of "that's good enough."

It is first essential to have a well laid out plan of the work in hand, and then to thoroughly execute the same in accordance with the best practice of the trade.

Is it not too true that too many fitters depend on others entirely too much for planning their work and getting out material lists and costs?

Would it not be much better and more profitable if the fitter did this to a very large extent for himself and would thoroughly check all details of radiation, piping, boiler sizes and cost before submitting tenders on any work.

Would it not be a great advantage to him to so acquaint himself with the real details of his work so that he would determine for himself the radiations required under all conditions and all temperatures?

Would it not prevent going back over work that has been considered complete and a good many complaints from poor working jobs, with their consequent loss of money, time and prestige.

Every Job Different.

Even the best laid out plans of a heating system can not and do not show all the minor details of the work, but it is necessary that the fitter shall be able to adjust the actual work to any and all local conditions so the plant may be correctly installed. To do this, there are many seemingly small details that have to be thoroughly looked to in the course of the erection of the work. If these things were zealously looked to we know that many fitters would have a good many more dollars in their pockets at the end of a good season of business.

Speaking of failures, we all know that heating apparatus as a rule goes into the hands of many people utterly devoid of any mechanical knowledge and if there is any way that a plant can be operated wrongly they will surely find that wrong way sooner or later.

But from some pieces of work inspected we cannot deny that faulty work is too often to blame. The faults are often too glaring to be pardonable at this day.

Importance of Plumbing and Heating.

We too often find boilers too small,—too little radiators, badly placed radiators, too small piping, too little pitch to pipes—improperly connected branches and radiators. To illustrate a few cases are cited.

Some Defective Work.

A steam radiator is connected to a main by a six foot arm or branch and in this arm there is a drop of nearly two inches and the radiator does not give good results,—it throws water from the air valve,—it hammers at times,—it does not heat easily without other radiators.

In a building there is 1,200 feet of radiation and to the main there is connected a boiler that at the best rating that could be given it is not capable of supplying over 1,200 feet gross radiation in zero weather. When the thermometer is down to 20 degrees below zero, even though the boiler is stoked with wood as fast as it can be got through the fire, not a pound of steam will show on the gauge and some of the radiators will remain cold, or partly cold.

Then there's a good sized printing office requiring to properly heat it 2,000 feet of radiation and some one has put in 1,500 feet of radiation, and a small boiler, and he put it in because he could tender lower than someone that would have put in the proper amount.

What are the consequences? The typesetters can hardly get their fingers loosened up in the morning before 9 or 10 o'clock, on account of the cold, and when it comes to run off the paper, the ink will not work right, the paper sticks and the pressman says: "Perhaps there's too much electricity." Would not a properly laid out and carefully executed piece of work at a fair price have been the cheapest?

Then think of the days lost by school children from poor working heating plants due to one cause and another. Can you recall any such cases this winter?

Here's a case where numbers of the radiators are full, so to speak, of water. The branches carrying these radiators are taken out of the side of the steam main near its end when they could just as easy have been taken from the top; only a nipple and an elbow was saved.

Poorly Heated Hotel.

Here's an hotel heated with about 3,000 feet of hot water radiation. There is as good a fire in the heater as could be asked for. The main drops two inches in about thirty feet and in this distances the branches taken off the main are more than equal in size to the main. The radiators of course do not heat equally, and some not at all.

Then here is a building in which a

steam radiator is located close to the entrance, where the door is constantly opening and closing. The radiator has 48 square feet of surface; it is connected to the main by a 1-inch branch 15 feet long which is dropped 1¼ inches in its run and is not relieved at the end next the radiator. The radiator naturally doesn't work very well.

Then there is a living room that has 2,400 cubic feet of contents and 70 feet of glass surface. When the temperature gets down to 20 degrees below zero this room is about 55 to 58 degrees with 80 feet of hot water radiation. It is a dining room and needs to be warm. If it had about 120 feet of radiation then a temperature around 70 degrees might have been expected.

Here's a hall that has 40 feet of hot water radiation with a temperature around 50 degrees when the thermometer outside shows 20 degrees below. 70 square feet of radiation would have been nearer the mark.

Scores of cases similar to these could be cited, only accentuating the necessity of good and careful work.

Mistakes to Avoid.

Above all, don't put in too little radiation or too small a boiler on any of your work. With hot water heating particularly there is very little or no danger of getting in too much radiation. With plenty you can carry a lower temperature in the water and a lower fire and if a real good cold snap comes you have the reserve to meet it. You have a pleased customer and one that will recommend your work.

Get a good rule for calculating your quantities of radiation and with your experience use it and don't put in too little radiation when correct judgment tells you that you should put in more. Size up the situation, the cubic contents, the glass surface, the construction of the building and the purposes of the building or rooms that are to be heated.

Safe Rule to Follow.

It is safe to say that if you want 70 degrees in ground floor rooms and 60 degrees in upper rooms of houses when the thermometer is at zero outside under ordinary exposures and ordinary amount of glass you do not want to count on less than 1 square foot of radiation for each 25 cubic feet of space downstairs and 35 cubic feet upstairs, and if you figure for 20 degrees below zero outside allow this amount for 20 and 30 cubic feet respectively. If colder weather is to be met and temperatures inside as noted are required, it will demand a still larger amount of radiating surface.

With steam, it requires about five-eighths as much radiation as with hot water heating, but of course steam is not so well adapted to heating requirements in private houses.

Competition may be said to be the life of trade—but sometimes it is the death of it as well. In Canada there is plenty of trade for all. Do not take a contract on a piece of work for the glory there may be in it and then cut the work down or slight it and find there is neither glory nor profit in the work.

Let some one else take that kind of work in preference for there is more profit in good work and a great deal more credit in the work in the end. Good work always brings trade and ensures a fair profit and far less trouble. Work that is well performed speaks for itself and although you may not hear so much from it directly as you will of unsatisfactory work, the net results are far greater and wider reaching.

THE CLEVELAND SANITARY CODE

A notable improvement over the ordinary sanitary code of cities may be cited in the ordinance governing plumbing, lately compiled for the city of Cleveland, says Modern Sanitation. This code is remarkable not so much for the direction as for the length it goes towards providing an adequate and suitable supply of water, and for the provision made to properly dispose of household wastes, particularly the liquid portion known as sewage, from buildings isolated from public sewers.

No doubt the action of the Cleveland commission will be followed by commissions of other cities when drafting laws to govern sanitation of their communities, and in the near future as many safeguards will be placed around the citizens to protect them from polluted water as there are now to protect them from the less harmful "sewer gas." In addition to providing a community with a sterile water supply, sanitary codes will see that the distributing systems are so proportioned that an adequate supply of water can be had simultaneously at all fixtures, and that the annoyance of one fixture robbing another is made impossible.

In the matter of sewage purification and disposal, the Cleveland code has gone further than have most other sanitary laws and has specified under what conditions septic tanks and filter beds are to be used. This part of the code, however, lacks much of the mandatory directness of the rest of the laws, which specify size and proportions so any architect or engineer can design a drainage system; whereas, in the section devoted to sewage purification no definite data is given that would help an individual in designing a purification works. This lack of data on sewage purification, however, in no way reflects on the completeness of the code compiled by Commissioner Eisemann. The practice of sewage purification for the past ten years has been in a transitory stage, and out of the chaos that reigned definite principles are only now being formulated. Perhaps when another code is drafted, the principles underlying the practice of sewage purification will be sufficiently understood so that portion of the code can be enlarged upon.

There are many excellent provisions in this ordinance that it would be well to incorporate in other city codes, and in order to bring them to the notice of interested persons and in the interest of sanitary progress, various sections will be published from time to time and commented on in this magazine.

A very excellent provision of the Cleveland sanitary code is section 5 of title IX, which provides:

"When inside or outside (rain) conductors are made gas and water tight, and do not open near doors, windows or other openings, within the distance prescribed for soil and waste vents, traps may be omitted."

The distance prescribed for waste vents is 3 feet above and 12 feet away from any door, window, scuttle or air shaft, and 6 feet away from a chimney or smoke flue.

There is no logical reason why the old practice should be adhered to of trapping rain leaders that open above the roof and well away from all openings leading to the interior of buildings. One of the objects aimed at in designing a drainage system is to provide for circulation of air through the pipes to maintain the air within the system comparatively free from deleterious gases. To do so, special pipes are extended to points above the roof, where they are left open to the atmosphere, and if rain conductors can be made to serve the dual purpose of rain conductor and vent pipe, a much purer air can be maintained within the drainage system. Such being the case, interposing a trap, instead of bettering the work from a sanitary standpoint, is actually detrimental to the system.

A JAPANESE BATHROOM.

A tiny space four by six feet. In it were four objects—a stool to sit upon when washing one's self before getting into the bath, a shining brass wash basin, a wooden pail and dipper, in which to fetch the bath water, and the tub. The tub, like most private baths, was round, casket shaped, and made of white wood. It was, perhaps, thirty inches in diameter and twenty-seven inches high. A copper funnel or tube passing through the bottom went up inside close to the edge. This, filled with lighted charcoal, supplied heat for the water. The pipe was higher than the tub, so the water could not leak inside. A few tarnsverse bars of wood fitted into grooves and formed a protection so the bather could kneel in the tub without coming in contact with the hot pipe. The walls of the room were of white wood, with a pretty grain, the floor of pine, laid with a slight slope and grooved so the water might flow into a gutter and through a bamboo pipe to the yard. A moon-shaped lattice window high up let in air and light. As a provision for more ventilation the two outside walls for a foot below the ceiling were lattice of bamboo slats.

As my eye traveled from object to object I quickly sized up the cost. For the tub 8 yen, and it would last indefinitely; 2 yen for the brass basin; 50 sen for the pail and dipper and 25 sen for the stool. Eleven yen would fit up my bathroom, and I asked for nothing nicer.—The Craftsman.

We are weaving character every day, and the way to weave the best character is to be kind and to be useful. And character is the result of but two things—our mental attitude and the way we spend our time.

Plumbing Markets

MONTREAL.

Office of Hardware and Metal,
232 McGill Street,
Montreal, March 1, 1907

High prices do not seem to have affected the plumbing business to any great extent, people continue to buy in spite of the high figures that prevail in all lines. Most jobbers were a little scary when they started their travelers out this year, and wouldn't have been greatly surprised if some of them had been "handed a lemon," therefore the good, generous orders most plumbing salesmen have turned over to their respective houses were happy surprises. There was no doubt as to the ultimate outcome of affairs, only slight anxiety as to what the public would do until they became educated to the prevailing prices. However, the test is over and business remains excellent. This is due no doubt to the building boom which has been going on here lately. Things are higher than previous years, in fact they are very high, but we must have them just the same.

There are no changes in prices, everything remaining as quoted last week.

RANGE BOILERS—Prices on the copper articles are now up in keeping with the prevailing price of copper, and unless an advance takes place in the sheet article, no higher figures are likely to be quoted for some time to come. Prices are: Iron clad, 30 gal., $5; 40 gal., $6.50 net list. Copper, 30 gal., $33; 35 gal., $38; 40 gal., $43.

LEAD PIPE—Firm at 5 per cent. discount f.o.b. Montreal, Toronto, St. John, N.B., Halifax; f.o.b. London, 15c per hundred lbs. extra; f.o.b. Hamilton, 10c per hundred lbs. extra.

IRON PIPE FITTINGS—Some lines still short: Discounts on nipples, ¼ inch to 3 inch, 65 per cent., 3½ to 2 inch, 67½ per cent.

IRON. PIPE — Standard pipe in lots of 100 feet, regular lengths, ¼ inch, $5.50; ⅜ inch, $5.50; ½ inch, $8.50; ¾ inch, $11.50; 1 inch, $16.50; 1¼ inch, $22.50; 1½ inch, $27.; 2 inches, $36.50. Discounts on black pipe: ¼ inch, 59 per cent.; ⅜ inch, 59 per cent.; ½ inch, 68; ¾ to 2, 70. Discounts on galvanized pipe: ¼ inch, 44 per cent.; ⅜ inch, 44 per cent.; ½ inch, 58 per cent.; ¾ to 2 inch, 60 per cent. Extra heavy pipe of 100 feet lots are quoted as follows: ½ inch, $12; ¾ inch, $15; 1 inch, $22; 1¼ inch, $30, 1½ inch, $36; 2 inches, $50. The discounts on black pipe: ½ inch, 74 per cent.; ¾ inch, 69 per cent.; 1 inch to 2 inches, 68 per cent. Galvanized: ½ inch, 59 per cent.; ¾ inch, 69 per cent.; 1 to 2 inches, 58 per cent.

SOIL PIPE AND FITTINGS—Standard soil pipe, 50 per cent. off list. Standard fittings, 50 and 10 per cent. off list; medium and extra heavy soil pipe, 60 per cent. off. Fittings, 60 per cent. off.

SOLDER — No change in prices. We still quote: Bar solder, half-and-half, guaranteed, 25c.; No. 2 wiping solder, 22c.

ENAMELWARE — We quote: Canadian baths, see Jan. 3, 1907, list. Lavatories, discounts, 1st quality, 30 per cent.; special, 30 and 10 per cent. Sinks, 18x30 inch, flat rim, 1st quality, $2.60; special, $2.45.

TORONTO.

Office of Hardware and Metal
10 Front Street East,
Toronto March 1, 1907.

Last week we referred to the price-cutting going on in plumbers' brass goods, a peculiar situation in view of the steadily rising copper market. We are again revising our quotations on these goods, but no figures can be quoted as definite in view of the cutting going on. As a general rule the trade would prefer to have reasonable and steady prices than an unsettled market.

Business in plumbing supplies is picking up and orders for roughing in goods are being placed more generally. Heating goods, are, of course, out of season and little is being done with them. Prices on other lines are stationary.

LEAD PIPE — The discount continues at 5 per cent. off the list price of 7c. per pound. Lead waste, 8c. per pound, with 5 off. Caulking lead, 5¾c. to 6c. per pound. Traps and bends, 40 per cent. discount.

SOIL PIPE AND FITTINGS—Business is increasing We quote: Medium and extra heavy pipe and fittings, 60 per cent.; light pipe, 50 per cent.; light fittings, 50 and 10 per cent.; 7 and 8 inch pipe, 40 and 5 per cent.

IRON PIPE—Stocks are low, with one-inch black pipe quoted at $4.95, and one-inch galvanized at $6.60. Full list in current market quotations.

IRON PIPE FITTINGS—The following quotations rule. Cast iron fittings, 57½ per cent.; cast iron plugs and bushings, 60 per cent.; flange unions, 60 per cent.; nipples, 70 and 10 per cent.; iron cocks, 55 and 5 per cent.; Canadian malleable, 30 per cent.; malleable unions, 55 and 5 per cent.; malleable bushings, 55 per cent.; cast iron ceiling plates, plain, 65 per cent.; cast iron floor, 70 per cent.; hook plates, 60 per cent.; expansion plates, 65 per cent.; headers, 60 per cent.; hangers, 65 per cent.; standard list.

RANGE BOILERS—Galvanized iron, 30-gal., standard, $5; extra heavy, $7.75, 35-gal. standard, $6, extra heavy, $8.75, 40-gal., standard, $7; 40 gallon, extra heavy, $9.75, net list. Copper range boilers—New lists quote: 30 gallon, $33; 35 gallon, $38; 40 gallon $43. Discounts 5 to 15 per cent.

RADIATORS—Hot water, 47½ per cent., steam, 50 per cent.; wall radiators 45 per cent.; specials, 45 per cent. Hot water boilers are subject to an open market.

SOLDER—Bar solder, half-and-half, guaranteed, 27c., wiping, 23c.

ENAMELWARE—The January list still prevails on Canadian baths. We quote: Lavatories, first quality, 20 and 5 to 20 and 10 off; special, 20 and 10 to 30 and 2½ per cent. discount. Kitchen sinks, plate 300, firsts, 60 and 10 off; specials, 65 and 5 per cent. Urinals and range closets, 15 off. Fittings extra.

ELECTRIC LAMP THAT TALKS.

Two new and unique inventions have been put on the market. One of these is an incandescent lamp that lights when it is spoken to; the other is a talking arc lamp. They are the inventions of Frank Moulan, star in "The Grand Mogul." While there is no practical use to which either of these inventions may be put they demonstrate the wonders of electricity.

The incandescent lamp that talks, however, is a reality. The device is a simple one—just an ordinary telephone transmitter with a highly sensitive diaphram, which vibrates to such an extent when in contact with the human voice that it touches the wire that starts the motor for the lights.

The speaking arc lamp is also a reality. It is wired from a highly sensitive telephone transmitter into which the voice is injected, the sounds being audibly reproduced from the sputtering carbons of the lamp.

EXCELLENT POINTS

The Daisy Hot Water Boiler embodies features in its construction that place it in a position by itself. It is easily handled, durable and efficient and the fitter handling them never fails to acquire a reputation for proficiency and good practice.

CLUFF BROTHERS

Lombard Street, TORONTO

Selling Agents for - - - WARDEN KING & SON, Limited

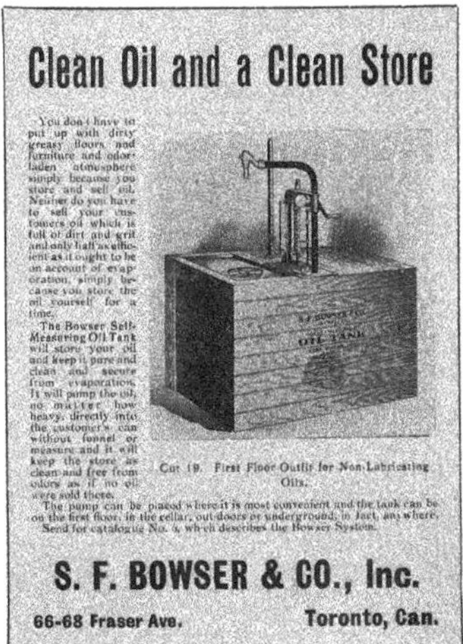

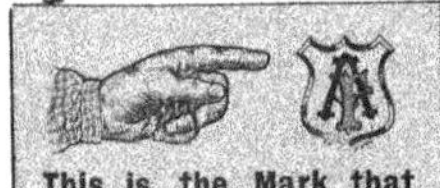

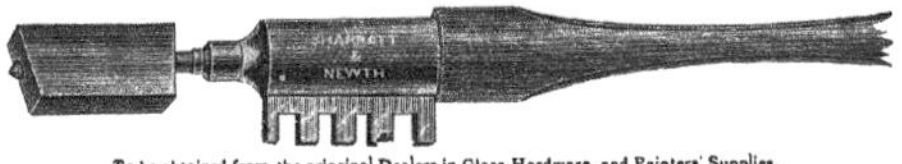
SHARRATT
&
NEWTH

REGISTERED TRADE MARK
GRANTED 1682

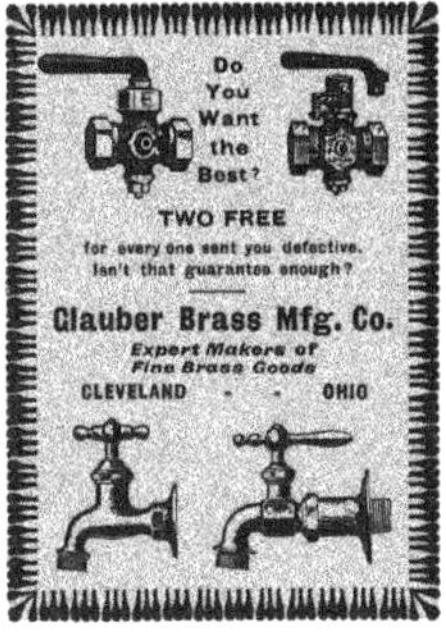
Do
You
Want
the
Best?
TWO FREE
for every one sent you defective.
Isn't that guarantee enough?
Glauber Brass Mfg. Co.
Expert Makers of
Fine Brass Goods
CLEVELAND - - OHIO

CURRENT MARKET QUOTATIONS.

Mar. 1. 1907.

These prices are for such qualities and quantities as are usually ordered by retail dealers on the usual terms of credit, the lowest figures being for larger quantities and prompt pay. Large cash buyers can frequently make purchases at better prices. The Editor is anxious to be informed at once of any apparent errors in this list, as the desire is to make it perfectly accurate.

METALS.

ANTIMONY.

Cookson's ....per lb...	0 27	0 27½

BOILER PLATES AND TUBES.

Plates, ¼ to ½ inch, per 100 lb..	....	2 50
Heads, per 100 lb	....	2 75
Tank plates 3-16 inch	....	2 65
Tubes per 100 feet, 1½ inch ..	8 50	....
" " 2 " ..	9 00	9 50
" " 2½ " ..	11 30	11 50
" " 3 " ..	12 50	13 00
" " 3½ " ..	16 00	17 00
" " 4 " ..	20 00	21 50

2 per cent off

BOILER AND T.K. PITTS.

Plain tinned.........}
Spun...................} 25 per cent. off list.

BABBIT METAL.

Canada Metal Company—Imperial, genuine 60c; Imperial Tough, 60c; White Brass, 50c.; Metallic. 35c.; Harris Heavy Pressure, 25c.; Hercules, 25c; White Bronze, 15c.; Star Frictionless, 14c.; Alluminoid, 10c.; No. 4, 9c. per lb.

James Robertson Co.—Extra and genuine Monarch, 60c., Crown Monarch, 50c; No. 1 Monarch, 40c; King, 30c; Fleur-de-lis, 20c; Thurber, 15c; Philadelphia, 12c; Canadian, 10c; hardware, No 1, 15c; No. 2, 12c; No. 3, 10c. per lb.

BRASS.

Rod and Sheet, 14 to 30 gauge, 25 p c. advance.

Sheets, 12 to 14 in.	0 30
Tubing, base, per lb 5-16 to 2 in	0 33
Tubing, ¼ to 3-inch, iron pipe size..	0 31
" ¼ to 3-inch, seamless..	0 35

Copper tubing, 6 cents extra.

COPPER.

Ingot.		Per 100 lb.
Casting, car lots..............	26 50	27 00
Bars.		
Cut lengths, round. ½ to 2 in...	....	35 00
Sheet		
Plain, 16 oz., 14x48 and 14x60	....	35 00
Plain, 14 oz.	....	36 00
Tinned copper sheet, base	....	38 00
Planished, base	....	39 00
Braziers (in sheets), 4x6 ft., 25 to 30 lb. each, per lb., base .	....	0 35

BLACK SHEETS.

	Montreal.	Toronto
8 to 10 gauge	2 70	2 70
12 gauge	2 70	2 80
14 "	2 70	2 55
17 "	2 50	2 65
18 "	2 50	2 65
20 "	2 50	2 65
22 "	2 50	2 75
24 "	2 55	2 80
26 "	2 65	2 85
28 "	2 70	3 00

CANADA PLATES.

Ordinary, 52 sheets	2 75	2 90
All bright "	3 80	3 93
Galvanized, 52 sheets	4 35	4 45
" 60 "	4 60	4 70
	Ordinary.	**Dom. Crown**
18x24x52	4 25	4 35
" 60	4 50	4 60
20x28x80	8 50	8 70
" 94	9 00	9 20

GALVANIZED SHEETS.

	Fleur-de-Lis.	Gordon Crown.	Colborne Crown.
16 to 20 gauge..	3 80	3 95	....
22 to 24 gauge ..	4 05	4 00	3 75
26 " ..	4 30	4 20	4 30
28 " ..	4 55	4 50	4 45

Apollo.

10¾ oz. (American gauge)........	....	4 85
28 gauge........	4 45	4 55
26 "	4 15	4 30
24 "	3 85	3 90

	Gorbal's Best.	Comet.	Queen's Head.	Bell.
16 to 20 gauge		3 60	3 95	....
22 to 24 gauge	....	3 75	4 20	....
26 "	....	4 30	4 45	....
28 "	4 55	4 45	4 70	4 45

Less than case lots 10 to 25c. extra.

IRON AND STEEL.

	Montreal.	Toronto.
Common bar, per 100 lb......	2 15	2 30
Forged iron "	2 40	
Refined "	2 55	2 70
Horseshoe iron "	2 55	2 70
Hoop steel, 1½ to 3 in. base....	....	2 80
Sleigh shoe steel "	2 25	2 30
Tire steel	2 40	2 50
Best cast steel	....	0 12
B. K. Morton "Alpha" high speed	....	0 65
" annealed	....	0 70
"M" Self-hardening	....	0 50
"J" quality, best warranted	....	0 18
"I" " warranted	....	0 14
"B C" quality	....	0 09
Colonial black diamond	0 08	0 09
Sanderson's	0 08	0 45
Jessop's	0 12	0 13
Air hardening	0 60	0 65
Conqueror	0 07½	0 50
Jowett's diamond J	0 06½	0 07
Jonas & Colver's tool steel....	0 10	0 20
" "Novo"	....	0 65
" annealed	....	0 65
Jowett & Sons B.P.L. tool steel	0 10½	0 11

COLD ROLLED SHAFTING.

9-16 to 11-16 inch	....	0 06
¾ to 1 7-16 "	....	0 05½
1 7-16 to 3 "	....	0 05

Montreal 30, Toronto 35 to 40 per cent.

INGOT TIN.

Lamb and Flag and Straits—
56 and 28-lb. ingots, 100 lb. $45 00 $45 50

TINPLATES.

Charcoal Plates—Bright

M.L.S., equal to Bradley—		Per box.
I C, 14 x 20 base	....	$6 50
I X, 14x20 "	....	8 00
I XX, 14 x 20 base	....	9 50
Famous, equal to Bradley—		
I C, 14 x 20 base	....	6 50
I X, 14 x 20 "	....	8 00
I XX, 14 x 20 base	....	9 50
Raven and Vulture Grades—		
I C, 14 x 20 base	....	5 00
I X "	....	6 00
I X X "	....	7 00
I X X X "	....	8 00
"Dominion Crown Best"—Double Coated, Tissued.		Per box.
I C, 14 x 20 base	5 50	5 75
I X, 14 x 20 "	6 50	6 75
I XX " 20 "	7 50	7 75
"Allaway's Best"—Standard Quality.		
I C, 14 x 20 base	....	4 50
I X, 14 x 20 "	....	5 25
I XX, 14 x 20 "	....	6 00

Bright Cokes.

Bessemer Steel—		
I C, 14 x 20 base	....	4 25
20x28, double box	....	8 50

Charcoal Plates—Terne.

Dean or J. G. Grade—		
I C, 20x28, 112 sheets	7 25	8 00
I X, Terne Tin	....	9 50

Charcoal Tin Boiler Plates.

Cookley Grade—
X X, 14x56, 50 sheet bxs. }
" 14x60, " }
" 14x65, " } 7 50

Tinned Sheets.

72x30 up to 24 gauge	....	8 50
" " 26 "	....	9 00

LEAD.

Imported Pig, per 100 lb......	5 50	5 60
Bar, ...	5 75	6 00
Sheets, 2½ lb. sq. ft., by roll ..	....	0 07½
Sheets, 3 to 6 lb. "	....	0 07

NOTE.—Cut sheets ½c. per lb., extra. Pipe, by the roll, usual weights per yard, lists at 7c. per lb. and 5 p.c. dis. f.o.b. Toronto.

NOTE.—Cut lengths, net price, waste pipe 8-ft lengths, lists at 8c.

SHEET ZINC.

5-cwt. casks	8 00	8 25
Part casks	8 25	8 50

ZINC SPELTER.

Foreign, per 100 lb	7 50	7 75
Domestic	7 00	7 25

PLUMBING AND HEATING

BRASS GOODS, VALVES, ETC.

Standard Compression work, 55 per cent.

Cushion work, discount 40 per cent.

Fuller work, 65 per cent.

Flatway stop and stop and waste cocks, 60 per cent.; roundway, 55 per cent.

J.M.T. Globe, Angle and Check Valves, 45 per cent.

Standard Globe, Angle and Check Valves 55 per cent.

Kerr standard globes, angles and checks, special, 42½ per cent.; standard, 47½ p.c

Kerr Jenkins' disc, copper-alloy disc and heavy standard valves, 45 per cent.

Kerr steam radiator valves, 60 p.c., and quick-opening hot-water radiator valves, 60 p.c.

Kerr brass, Weber's straightway valves, 40 per cent.; straightway valves, I.B.B.M., 60 per cent.

J. M. T. Radiator Valves 50 per cent.

Standard Radiator Valves, 60 per cent.

Patent Quick-Opening Valves, 65 per cent.

Jenkins' Bros. Globe Angle and Check Valves 27½ per cent.

No. 1 compression bath cock......net	2 00
No. 4 " " ...... "	1 90
No 7 Fuller's.............. .... "	2 25
No. 4½, " .... "	2 35
Patent Compression Cushion, basin cock, hot and cold, per dz., ...	$16.20
Patent Compression Cushion, bath cock, No. 2208	2 25

Square head brass cocks, 50 per cent.
" " iron " 60 "

Thompson Smoke-test Machine 25.00

BOILERS—COPPER RANGE.

Copper, 30 gallon............	33 00
" 35 "	38 00
" 40 "	43 00

15 per cent

BOILERS—GALVANIZED IRON RANGE.

Capacity.	Standard.	Extra heavy
30-gallons....	5.00	7.75
35 "	6 00	8 75
40 "	7.00	9.75

2 per cent., 30 days.

BATH TUBS.

Steel clad copper lined, 15 per cent.

CAST IRON SINKS.

16x24, $1; 18x30, $1; 18x36, $1.30.

ENAMELED CLOSETS AND URINALS

Discount 15 per cent.

ENAMELED BATHS.

List issued by the Standard Ideal Company Jan. 3, 1907, shows an advance of 10 per cent. over previous quotations.

ENAMELED LAVATORIES.

	1st quality	Special.
Plate E 100 to E 103 ..	20 & 5 p c.	20 & 10 p c.
" E 104 to E 132....	20 & 10 p.c.	30 & 2½ p.c.

ENAMELED ROLL RIM SINKS.

	1st quality.	Special
Plate E 201, one piece.	15 & 2½ p.c.	15 & 10 p.c.

ENAMELED KITCHEN SINKS.

Plate E, flat rim 300, 60 & 10 p.c. 65 & 5 p.c

HEATING APPARATUS.

Stoves and Ranges—Discounts vary from 40 to 70 per cent. according to list.

Furnaces—45 per cent.

Registers—70 per cent.

Hot Water Boilers—Discounts vary.

Hot Water Radiators—47½ per cent.

Steam Radiators—50 per cent

Wall Radiators and Specials—45 per cent.

LEAD PIPE

Lead Pipe, 7c. per pound, 5 per cent. off.

Lead waste, 8c. per pound, 5 per cent. off.

Caulking lead, 6c. per pound.

Traps and bends, 40 per cent.

IRON PIPE.

Size (per 100 ft.)	Black.		Galvanized
¼ inch........	2 25	¼ inch........	3 08
⅜ "	2 25	⅜ "	3 08
½ "	2 72	½ "	3 57
¾ "	3 45	¾ "	4 60
1 "	4 95	1 "	6 80
1¼ "	6 75	1¼ "	9 00
1½ "	8 10	1½ "	10 80
2 "	10 80	2 "	14 40
2½ "	18 00	2½ "	24 00
3 "	23 50	3 "	31 50
3½ "	30 00	3½ "	40 00
4 "	34 00	4 "	45 00

2 per cent. 30 days.

Malleable Fittings—Canadian discount 30 per cent.; American discount 25 per cent.

Cast Iron Fittings 57½; Standard bushings 57½ per cent.; headers, 57½; flanged unions 57½, malleable bushings 55; nipples, 70 and 10 per cent.; malleable lipped unions, 55 and 5 per cent.

SOIL PIPE AND FITTINGS

Medium and Extra heavy pipe and fittings, up to 6 inch, discount 60 per cent.

7 and 8-in. pipe, discount 40 and 5 per cent.

Light pipe, 50 p c.; fittings, 50 and 10 p.c.

OAKUM.

Plumbers ....per 100 lb... .. .... 4 75

STOCKS AND DIES.

American discount 25 per cent.

SOLDERING IRONS.

½-lb.................. per lb.	....	0 38
1-lb. or over .... "	....	0 35

SOLDER.

Per lb.	Montreal	Toronto
Bar, half-and-half, guaranteed	0 25	0 27
Wiping	0 22	0 23

PAINTS, OILS AND GLASS.

BRUSHES

Paint and household, 70 per cent..

CHEMICALS.

	In casks per lb.
Sulphate of copper (bluestone or blue vitrol)	0 09
Litharge, ground	0 06
" flaked	0 06½
Green conperas (green vitrol)	0 01
Sugar of lead	0 08
Lump olive	0 01½

COLORS IN OIL.

Venetian red, 1-lb. tins	pure.	....	0 08
Chrome yellow	"	....	0 15
Golden ochre	"	....	0 08
French "	"	....	0 08
Marine black	"	....	0 04½
Chrome green	"	....	0 10
French permanent green	"	....	0 13
Signwriters' black	"	....	0 15

GLUE.

Domestic sheet	0 10	0 10½
French medal..	0 12	0 12½

PARIS GREEN.

	Berger's	Canadian
600-lb. cask	0 25½	0 25½
250 lb. drums	0 25¾	0 25¾
100-lb. "	0 26¼	...
50-lb. "	0 26¼	0 26¼

1-lb. pkgs, 100 in box 0 27½ 0 27½
½-lb. " 0 29½ 0 29½
1-lb. tins, 100 in box 0 28½ 0 28½
½-lb. pkg 0 30½ 0 30
F.o.b. Toronto.

PARIS WHITE.
In bbls 0 90

PREPARED PAINTS.
Pure, per gallon, in tins 1 30
Second qualities, per gallon 1 10
Barn (in bbls.) 0 65 0 90
Sherwin-Williams paints, 1 gal. 1 40
Canada Paint Co.'s pure 1 30
Standard P. & V. Co.'s "New Era." 1 30
Benj. Moore Co.'s "Ark" B'd 1 25
" British Navy deck 1 50
Brandram-Henderson's "English" 1 45
Ramsay's paints, Pure, per gal. 1 20
" Thistle, " 1 00
" Outside, bbls 0 55 0 65
Martin-Senour's 100 p.c. pure, ½ gal. 1 50
" " " " ¼ gal. 1 45
" " " " 1 gal. 1 40
Senour's Floor Paints.....gal. 1 20
Jamieson's "Crown and Anchor" 1 20
Jamieson's floor enamel...... .... 1 50
" barn paints, bbls, per gal. 0 60
Sanderson Pearcy's, pure 1 20
Robertson's pure paints...... .... 1 20

PUTTY.
Bulk in bbls........ 1 50
Bulk in less quantity 1 80
Bladders in bbls........ 1 80
Bladders in kegs, boxes or loose...... 1 95
25-lb. tins........ 1 80
12½ lb. tins 2 05
Bladders in bulk or tins less than 100 lb. 1 85
Bulk in 100-lb. irons........ 1 80

SHINGLE STAINS.
In 5-gallon lots 0 75 0 80

SHELLAC.
White........ .... 0 65
Fine orange........ .... 0 60
Medium orange........ .... 0 55
F.o.b. Montreal or Toronto.

TURPENTINE AND OIL.
Castor oil........ 0 08 0 10
Gasoline 0 22½
Benzine, per gal 0 17 0 20
Turpentine, single barrels 0 98 1 05
Linseed Oil, " raw.. 0 57 0 59
" " boiled 0 60 0 62
W. H. & C's English raw oil, bbls... 0 63
" " boiled, " 0 66

WHITE LEAD GROUND IN OIL. Per 100 lbs
Canadian pure 7 15
No. 1 Canadian 6 80
Munro's Select Flake White.. .... 7 40
Elephant and Decorators' Pure 7 40
Monarch 7 40
Standard Decorator's 7 15
Essex Genuine 6 80
Brandram's B. B. Genuine... .... 8 40
"Anchor," pure 7 00
Ramsay's Pure Lead 6 40
Ramsay's Exterior 6 15
"Crown and Anchor," pure... .... 6 50
Sanderson Pearcy's 7 40
Robertson's C P., lead........ .... 7 20
W. H. & C's matured pure English 8 00 8 25

WHITE AND RED DRY LEAD. white red.
Genuine, 560 lb. casks, per cwt 6 75 6 00
Genuine, 100 lb. kegs, " 7 50 6 50
No. 1, 560 lb. casks, per cwt ... 6 25 5 75
No. 1, 100 lb. kegs, per cwt .. 7 00 6 25

WINDOW GLASS.

Size United Inches.	Star	Double Diamond
Under 26	$4 25	$6 25
26 to 40	4 65	6 75
41 to 50	5 10	7 50
51 to 60	5 35	8 50
61 to 70	5 75	9 75
71 to 80	6 25	11 00
81 to 85	7 00	12 50
86 to 90	....	15 00
91 to 95	....	17 50
96 to 100	....	20 50
101 to 105........	....	24 00
106 to 110........	....	27 50

Discount—16-oz., 25 per cent.; 21-oz. 30 per cent per 100 feet. Broken boxes 50 per cent. off list.

WHITING.
Plain, in bbls 0 60
Gilders bolted in barrels........ .... 0 90

WHITE DRY ZINC.
Extra Red Seal, V.M. 0 07½ 0 08½

WHITE GROUND ZINC.
Pure, in 25-lb. irons 0 08½
No. 1, " 0 07
No. 2, " 0 05½

VARNISHES.
In 5-gal. lots. Per gal. cans.
Carriage, No. 1 .... 1 50
Pale durable body 3 50
" hard rubbing 3 00
Finest elastic gearing 3 00
Elastic oak 1 50
Furniture, polishing 2 00
Furniture, extra 1 15
" " No. 1 .... 0 90
" " union 0 80
Light oil finish 1 40
Gold size japan 1 80
Brown japan 0 95
No. 1 brown japan 0 90
Baking black japan 1 35
No. 1 black japan 0 90
Benzine black japan 0 70
Crystal Damar 2 80
No. 1 " 2 50
Pure asphaltum........ .... 1 40
Oilcloth........ .... 1 50
Lightning dryer 0 70
Elastilite varnish, 1 gal. can, each.. 2 00
Granitine floor varnish, per gal 2 50
Maple Leaf coach enamels; size 1, 1 20
Sherwin-Williams' kopal varnish, gal., 2 50
Canada Paint Co's sun varnish 2 00
"Kyanize" Interior Finish.... .... 2 40
"Flint-Lac," coach........ .... 1 80
B H. Co's "Gold Medal," in cases 2 00
Jamieson's Copaline, per gal. 2 00

BUILDERS' HARDWARE.

BELLS.
Brass hand bells, 60 per cent.
Nickel, 55 per cent.
Gongs, Sargeant's door bells.. 5 50 8 00
American, house bells, per lb. 0 35 0 40
Peterboro' door bells, discount 37½ and 10 per cent. off new list.

BUILDING PAPER, ETC.
Tarred Felt, per 100 lb. 2 25
Ready roofing, 2-ply, not under 45 lb. per roll 1 00
Ready roofing, 3-ply, not under 65 lb., per roll........ 1 25
Carpet Felt ....per ton 60 00
Heavy Straw Sheathing.......per ton 35 00
Dry Surprise........ 0 40
Dry Sheathing.....per roll, 400 sq. ft. 0 50
Tar " ..... " 400 " 0 45
Dry Fibre........ " 400 " 0 55
Tarred Fibre ..... " 400 " 0 65
O. K. & I. X. L.... " 400 " 0 70
Resin-sized .... " 400 " 0 45
Oiled Sheathing.... " 600 " 1 00
Oiled " .... " 400 " 0 70
Roof Coating, in barrels.. ...per gal. 0 17
Roof " small packages " 0 25
Refined Tar........per barrel 5 00
Coal Tar .... " 4 00
Coal Tar, less than barrels....per gal. 0 15
Roofing Pitch....... per 100 lb. 0 80 0 90
Slater's felt ....per roll 0 70
Heavy Straw Sheathing f. o. b. St. John and Halifax........ 37 50

BUTTS.
Wrought Brass, net revised list.
Wrought Iron, 70 per cent
Cast Iron Loose Pin, discount 60 per cent.
Wrought Steel Fast Joint and Loose Pin, 70 per cent.

CEMENT AND FIREBRICK.
Canadian Portland........ 2 00 2 10
Belgium........ 1 60 1 90
White Bros. English 1 80 2 05
"Lafarge" cement in wood 3 40
"Lehigh" cement, in wood 2 54
"Lehigh" cement, cotton sacks. 2 39
"Lehigh" cement, paper sacks 2 31
Fire brick, Scotch, per 1,000 .... 27 00 30 00
" English........ 17 00 21 00
" American, low 23 00 25 00
" " high 27 50 35 00
Fire clay (Scotch), net ton 4 95
Paving Blocks per 1,000
Blue metallic, 9"x4½"x3", ex wharf .. 35 00
Stable pavers, 12"x6"x2", ex wharf .. 50 00
Stable pavers, 9"x4½"x3", ex wharf .. 36 00

DOOR SETS.
Peterboro, 37½ and 10 per cent.

DOOR SPRINGS.
Torrey's Rod........per doz. 1 75
Coil, 9 to 11 in........ " 0 95 1 65
English .... " 2 00 4 00
Chicago and Reliance Coil Door Springs, 25 per cent.

STORE DOOR HANDLES
Per Dozen 1 00 1 50

ESCUTCHEONS.
Discount 50 and 10 per cent., new list
Peterboro, 37½ and 10 per cent.

ESCUTCHEON PINS.
Iron, discount 40 per cent.
Brass, 45 per cent.

HINGES.
Blind, discount 60 per cent.
Heavy T and strap, 4-in., per lb. net.. 0 06½
" " 5-in., " 0 06
" " 6-in., " 0 05¾
" " 8-in., " 0 05½
" " 10-in. and larger.. 0 05¼
Light T and strap, discount 65 p.c.
Screw hook and hinge—
under 12 in.... per 100 lb. 4 75
over 12 in........ " 3 75
Spring, No. 20, per gro. pairs 10 80
Spring, Woodyatt pattern, per gro., No. 5, $17.50 No. 10, $18; No. 20, $10 80; No. 12 $20; No. 51, $10; No. 50, $27 50.
Crate hinges and back flaps, 65 and 5 p. c.
Hinge hasps, 65 per cent.

SPRING HINGES.
Chicago Spring Butts and Blanks 12½ per cent.
Triple End Spring Butts, 40 and 5 per cent.
Chicago Floor Hinges, 40 and 5 off
Garden City Fire Engine House Hinges, 12½ per cent.

CAST IRON HOOKS.
Bird cage........per doz. 0 50 1 10
Clothes line, No. 61.. " 0 00 0 70
Harness........ " 0 60 12 00
Hat and coat ....per gro. 1 10 10 00
Chandelier ....per doz. 0 50 1 00
Wrought hooks and staples—
¼ x 5 .... per gross 2 65
5-16 x 5 .... " 3 30
Bright steel gate hooks and staples, 40 per cent. discount.
Hat and coat wire, discount 62½ per cent.
Screw, bright wire, discount 65 per cent.

KNOBS.
Door, japanned and N.P., doz 1 50 2 50
Bronze, Berlin........per doz. 2 75 3 25
Bronze, Genuine .... " 6 00 9 00
Shutter, porcelain, F. & L. screw .... per gross 1 30 2 00
White door knobs. ..per doz. 2 00
Peterboro knobs, 37½ and 10 per cent.
Porcelain, mineral and jet knobs, net list.

KEYS.
Lock, Canadian dis. 40 to 40 and 10 per cent
Cabinet trunk and padlock
American .... per gross 0 60

LOCKS.
Peterboro, 37½ and 10 per cent.
Russell & Erwin, steel rim $2.50 per doz.
Eagle cabinet locks, discount 30 per cent
American padlocks, all steel, 10 to 15 per cent.; all brass or bronze, 10 to 25 per cent.

SAND AND EMERY PAPER.
B. & A. sand, discount, 35 per cent
Emery, discount 35 per cent.
Garnet (Rurton's). 5 to 10 per cent. advance on list.

SASH WEIGHTS.
Sectional........per 100 lb. 2 00 2 25
Solid .... " 1 50 1 75

SASH CORD.
Per lb........ .... 0 31

BLIND AND BED STAPLES.
All sizes, per lb........ 0 07½ 0 12

WROUGHT STAPLES.
Galvanized 2 75
Plain 2 50
Coopers', discount 45 per cent.
Poultry netting staples, discount 40 per cent.
Bright spear point, 75 per cent. discount.

TOOLS AND HANDLES.

ADZES.
Discount 22½ per cent.

AUGERS.
Gilmour's, discount 60 per cent. off list.

AXES.
Single bit, per doz. 5 50 8 50
Double bit, " 10 00 11 00
Bench Axes, 40 per cent.
Broad Axes, 25 per cent.
Hunters' Axes........ 5 50 6 00
Boys' Axes 6 25 7 00
Splitting Axes 7 00 12 00
Handled Axes 7 00 9 00
Red Ridge, boys', handled.. .... 5 75
" hunters 5 25

BITS.
Irwin's auger, discount 47½ per cent.
Gilmour's auger, discount 60 per cent.
Rockford auger, discount 50 and 10 per cent.
Jennings' Gen. auger, net list.
Gilmour's car, 47½ per cent.
Clark's expansive, 40 per cent.
Clark's gimlet, per doz 0 65 0 90
Diamond, Shell, per doz...... 1 00 1 50
Nail and Spike, per gross..... 2 25 5 20

BUTCHERS CLEAVERS
German........net doz. 6 00 9 00
American........ " 12 00 18 00

CHALK.
Carpenters' Colored, per gross 0 45 0 75
White lump........per cwt. 0 60 0 65

CHISELS.
Warnock's, discount 72½ per cent.
P. S. & W. Extra, discount 72½ per cent

CROSSCUT SAW HANDLES.
S. & D., No. 3........per pair 0 15
S. & D., " 5........ " 0 22½
S. & D., " 6........ " 0 15
Boynton pattern........ " 0 20

CROWBARS.
3½c. to 4c. per lb.

DRAW KNIVES.
Coach and Wagon, discount 75 per cent.
Carpenters' discount 75 per cent

DRILLS.
Millar's Falls, hand and breast, net list.
North Bros., each set, 50c.

DRILL BITS.
Morse, discount 37½ to 40 per cent.
Standard, discount 50 and 5 to 55 per cent.

FILES AND RASPS.
Great Western .... 75 per cent.
Arcade .... 75 "
Kearney & Foot .... 75 "
Disston's........ 75 "
American .... 75 "
J. Barton Smith........ 75 "
McClellan .... 75 "
Eagle .... 75 "
Nicholson........ 66⅔ "
Globe .... 75 "
Black Diamond, 60, 10 and 5 p.c.
Jowitt's, English list, 27½ per cent.

GAUGES.
Stanley's discount 50 to 60 per cent.

Wire Gauges.
Winn's, Nos. 26 to 33 ....each 1 65 2 40

HANDLES.
C. & B., fork, 40 per cent., revised list.
C. & B., hoe, 40 per cent., revised list.
American, saw ...... per doz. 0 25
American, plane .... per gross 3 15 3 75
Canadian, hammer and hatchet 40 per cent.
Axe and cant hook handles, 45 per cent.

HAMMERS.
Maydole's, discount 5 to 10 per cent.
Canadian, discount 25 to 27½ per cent.
Magnetic tack........per doz. 1 10 1 20
Canadian sledge .... per lb. 0 07 0 08½
Canadian ball pean, per lb. .. 0 22 0 25

HATCHETS.
Canadian, discount 40 to 42½ per cent.
Shingle, Red Ridge 1, per doz........ 4 40
" " " 2 .... 4 80
Barrel Underhill........ 5 05

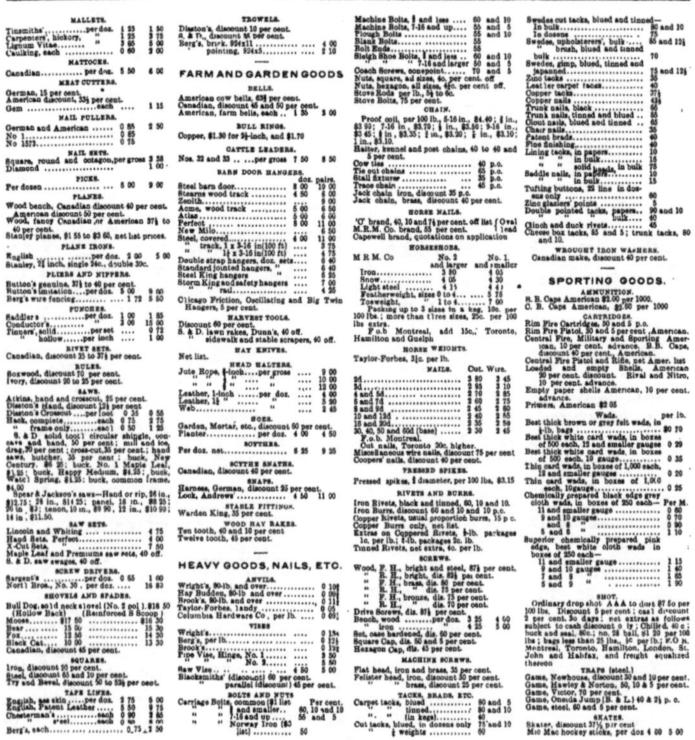

MALLETS.
Tinsmiths' ... per doz. 1 25 1 50
Carpenters', hickory, " 1 25 3 75
Lignum Vitae ... " 3 85 5 00
Caulking, each ... 0 60 2 00

MATTOCKS.
Canadian ... per doz. 5 50 6 00

MEAT CUTTERS.
German, 15 per cent.
American discount, 33⅓ per cent.
Gem ... each 1 15

NAIL PULLERS.
German and American ... 0 85 2 50
No 1 ... 0 85
No 1573 ... 0 75

NAIL SETS.
Square, round and octagon, per gross 3 38
Diamond ... 1 00

PICKS.
Per dozen ... 6 00 9 00

PLANES.
Wood bench, Canadian discount 40 per cent. American discount 50 per cent.
Wood, fancy Canadian or American 37½ to 40 per cent.
Stanley planes, $1 55 to $3 60, net list prices.

PLANE IRONS.
English ... per doz. 2 00 5 00
Stanley, 2⅝ inch, single 24c., double 39c.

PLIERS AND NIPPERS.
Button's genuine, 37½ to 40 per cent.
Button's imitation ... per doz. 5 00 9 00
Berg's wire fencing ... 1 72 5 50

PUNCHES.
Saddler's ... per doz. 1 00 1 85
Conductor's ... " 3 00 15 00
Tinners', solid ... per set 0 72
" hollow ... per inch 1 00

RIVET SETS.
Canadian, discount 35 to 37½ per cent.

RULES.
Boxwood, discount 70 per cent.
Ivory, discount 20 to 25 per cent.

SAWS.
Atkins, hand and crosscut, 25 per cent.
Disston's Hand, discount 12½ per cent
Disston's Crosscut ... per foot 0 35 0 55
Hack, complete ... each 0 75 2 75
" frame only ... each 0 50 1 25
S. & D. solid tooth circular shingle, concave and band, 50 per cent; mill and ice, drag, 30 per cent; cross-cut, 35 per cent.; hand saws, butcher, 35 per cent; buck, New Century, $6 25; buck, No. 1 Maple Leaf, $1.25; buck, Happy Medium, $4.25; buck, Watch Spring, $1.25; buck, common frame, $4.00.
Spear & Jackson's saws—Hand or rip, 26 in., $12.75; 28 in., $11 25; panel, 18 in., $8 25; 20 in., $9; tenon, 10 in., $9 90, 12 in., $10 90; 14 in., $11.50.

SAW SETS.
Lincoln and Whiting ... 4 75
Hand Sets, Perfect ... 4 00
X-Cut Sets, " ... 7 50
Maple Leaf and Premium saw sets, 40 off.
S. & D. saw swages, 40 off.

SCREW DRIVERS.
Sargent's ... per doz. 0 65 1 00
North Bros., No. 30, per doz. 16 80

SHOVELS AND SPADES.
Bull Dog, solid neck shovel (No. 2 pol.). $18 50
(Hollow Back) (Reinforced S Scoop)
Moose ... $17 50 ... $16 30
Bear ... 15 00 ... 15 30
Fox ... 12 50 ... 14 30
Black Cat ... 10 00 ... 13 30
Canadian, discount 45 per cent.

SQUARES.
Iron, discount 20 per cent.
Steel, discount 65 and 10 per cent.
Try and Bevel, discount 50 to 52½ per cent.

TAPE LINES.
English, ass skin ... per doz. 2 75 5 00
English, Patent Leather ... 5 50 9 75
Chesterman's ... each 0 90 2 85
" steel ... each 0 80 8 00
Berg's, each ... 0 75 2 50

TROWELS.
Disston's, discount 10 per cent.
S. & D., discount 35 per cent.
Berg's, brick, 9¼x11 ... 4 00
" pointing, 9¼x5 ... 2 10

FARM AND GARDEN GOODS

BELLS.
American cow bells, 63⅓ per cent.
Canadian, discount 45 and 50 per cent.
American, farm bells, each .. 1 35 3 00

BULL RINGS.
Copper, $1.30 for 2½-inch, and $1.70

CATTLE LEADERS.
Nos. 32 and 33 .. ... per gross 7 50 8 50

BARN DOOR HANGERS. doz. pairs.
Steel barn door ... 6 00 10 00
Stearns wood track ... 4 50 6 00
Zenith ... 9 00
Acme, wood track ... 5 00 6 50
Atlas ... 5 00 6 00
Perfect ... 8 00 11 00
New Milo ... 6 50
Steel, covered ... 4 00 11 00
" track, 1 x 3-16 in (100 ft) ... 3 75
" 1¼ x 3-16 in (100 ft) ... 4 75
Double strap hangers, doz. sets ... 6 40
Standard jointed hangers, " ... 6 40
Steel King hangers " ... 6 25
Storm King and safety hangers ... 7 00
" rail ... 4 25
Chicago Friction, Oscillating and Big Twin Hangers, 5 per cent.

HARVEST TOOLS.
Discount 60 per cent.
S. & D. lawn rakes, Dunn's, 40 off.
" sidewalk and stable scrapers, 40 off.

HAY KNIVES.
Net list.

HEAD HALTERS.
Jute Rope, ¾-inch ... per gross 9 00
" " ⅝ " 10 00
" " ½ " 12 00
Leather, 1-inch ... per doz. 4 00
Leather, 1¼ " 5 20
Web 2 45

HOES.
Garden, Mortar, etc., discount 50 per cent.
Planter ... per doz. 4 00 4 50

SCYTHES.
Per doz. net ... 5 25 9 25

SCYTHE SNATHS.
Canadian, discount 40 per cent.

SNAPS.
Harness, German, discount 25 per cent.
Lock, Andrews' ... 4 50 11 00

STABLE FITTINGS.
Warden King, 35 per cent.

WOOD HAY RAKES.
Ten tooth, 40 and 10 per cent
Twelve tooth, 45 per cent.

HEAVY GOODS, NAILS, ETC.

ANVILS.
Wright's, 80-lb. and over ... 0 10½
Hay Budden, 80-lb and over ... 0 09½
Brook's, 80-lb. and over ... 0 11¼
Taylor-Forbes, 1andy ... 0 05
Columbia Hardware Co, per lb. 0 09½

VISES
Wright's ... 0 13½
Berg's, per lb ... 0 12½
Brook's ... 0 12½
Pipe Vise, Hinge, No. 1 ... 3 50
" " " No. 2 ... 5 50
Saw Vise ... 4 50 5 00
Blacksmiths' (discount) 60 per cent.
" parallel (discount) 45 per cent.

BOLTS AND NUTS
Carriage Bolts, common ($1 list) Per cent.
" " ⅜ and smaller.. 60, 10 and 10
" " 7-16 and up ... 55 and 5
" " Norway Iron ($3 list) ... 50
Machine Bolts, ⅜ and less .... 60 and 10
Machine Bolts, 7-16 and up .... 55 and 5
Plough Bolts ... 55 and 10
Blank Bolts ... 55
Bolt Ends ... 55
Sleigh Shoe Bolts, ⅜ and less .. 60 and 10
" " " 7-16 and larger 50 and 5
Coach Screws, conepoint ... 70 and 5
Nuts, square, all sizes, 4c. per cent. off
Nuts, hexagon, all sizes, 4¼c. per cent off.
Stove Rods per lb., 5½ to 6c.
Stove Bolts, 75 per cent.

CHAIN.
Proof coil, per 100 lb., 5-16 in., $4.40; ⅜ in., $3 90; 7-16 in., $3.70; ½ in., $3.50; 9-16 in., $3 45; ⅝ in, $3.35; ¾ in., $3.30; ⅞ in., $3.10; 1 in., $3.10.
Halter, kennel and post chains, 40 to 40 and 5 per cent.
Cow ties ... 40 p.c.
Tie out chains ... 65 p.c.
Stall fixtures ... 35 p.c.
Trace chain ... 45 p.c.
Jack chain, iron, discount 35 p.c.
Jack chain, brass, discount 40 per cent.

HORSE NAILS.
'C' brand, 40, 10 and 7½ per cent. off list } Oval
M.R.M. Co. brand, 55 per cent. } head
Capewell brand, quotations on application

HORSESHOES.

M R M. Co	No. 2 and larger	No. 1 and smaller
Iron	3 80	4 05
Snow	4 05	4 30
Light steel	4 15	4 40
Featherweight, sizes 0 to 4		5 75
Toeweight, " 1 to 4		7 00

Packing up to 3 sizes in a keg, 10c. per 100 lbs.; more than three sizes, 25c. per 100 lbs extra.
F.o.b Montreal, add 15c., Toronto, Hamilton and Guelph

HORSE WEIGHTS.
Taylor-Forbes, 3½c. per lb.

NAILS.	Cut.	Wire.
2d	3 80	3 45
3d	2 95	3 10
4 and 5d	2 70	2 85
5 and 7d	2 60	2 75
8 and 9d	2 45	2 60
10 and 12d	2 40	2 55
16 and 20d	2 35	2 50
30, 40, 50 and 60d (base)	2 30	2 45

F.o.b. Montreal.
Cut nails, Toronto 20c. higher.
Miscellaneous wire nails, discount 75 per cent
Coopers' nails, discount 40 per cent.

PRESSED SPIKES.
Pressed spikes, ⅝ diameter, per 100 lbs, $3.15

RIVETS AND BURRS.
Iron Rivets, black and tinned, 60, 10 and 10.
Iron Burrs, discount 60 and 10 and 10 p.c.
Copper Rivets, usual proportion burrs, 15 p c.
Copper Burrs only, net list.
Extras on Coppered Rivets, ½-lb. packages 1c. per lb.; ¼-lb. packages 2c. lb.
Tinned Rivets, net extra, 4c. per lb.

SCREWS.
Wood, F. H., bright and steel, 87½ per cent.
" R. H., bright, dis. 82½ per cent.
" F. H., brass, dis. 80 per cent.
" R. H., " dis. 75 per cent.
" F. H., bronze, dis. 75 per cent.
" R. H., " dis. 70 per cent.
Drive Screws, dis. 87½ per cent.
Bench, wood ... per doz. 3 25 4 00
" iron ... 4 25 5 00
Set, case hardened, dis. 60 per cent.
Square Cap, dis. 50 and 5 per cent.
Hexagon Cap, dis. 45 per cent.

MACHINE SCREWS.
Flat head, iron and brass, 35 per cent.
Felister head, iron, discount 30 per cent.
" " brass, discount 25 per cent.

TACKS, BRADS, ETC.
Carpet tacks, blued ... 80 and 5
" tinned ... 80 and 10
" " (in kegs) ... 40
Cut tacks, blued, in dozens only 75 and 10
" ¼ weights ... 60
Swedes cut tacks, blued and tinned—
In bulk ... 80 and 10
In dozens ... 75
Swedes, upholsterers', bulk ... 85 and 12½
" brush, blued and tinned bulk ... 70
Swedes, gimp, blued, tinned and japanned ... 75 and 12½
Zinc tacks ... 35
Leather carpet tacks ... 40
Copper tacks ... 37½
Copper nails ... 42½
Trunk nails, black ... 65
Trunk nails, tinned and blued .. 85
Clout nails, blued and tinned .. 65
Chair nails ... 35
Patent brads ... 40
Fine finishing ... 40
Lining tacks, in papers ... 10
" " in bulk ... 15
" " solid heads, in bulk 75
Saddle nails, in papers ... 10
" " in bulk ... 15
Tufting buttons, 22 line in dozens only ... 60
Zinc glaziers' points ... 5
Double pointed tacks, papers.. 90 and 10
" " " bulk ... 40
Clinch and duck rivets ... 45
Cheese box tacks, 85 and 5; trunk tacks, 80 and 10.

WROUGHT IRON WASHERS.
Canadian make, discount 40 per cent.

SPORTING GOODS.

AMMUNITION.
B. B. Caps American $2.00 per 1000.
C. B. Caps American, $2.60 per 1000

CARTRIDGES.
Rim Fire Cartridges, 50 and 5 p.c.
Rim Fire Pistol, 30 and 5 per cent, American.
Central Fire, Military and Sporting American, 10 per cent. advance. B.B. Caps, discount 40 per cent., American.
Central Fire Pistol and Rifle, net Amer. list
Loaded and empty Shells, American 20 per cent. discount. Rival and Nitro, 10 per cent. advance.
Empty paper shells American, 10 per cent. advance.
Primers, American $2 05

Wads. per lb.
Best thick brown or grey felt wads, in ¼-lb. bags ... $0 70
Best thick white card wads, in boxes of 500 each, 12 and smaller gauges 0 29
Best thick white card wads, in boxes of 500 each, 10 gauge ... 0 35
Thin card wads, in boxes of 1,000 each, 12 and smaller gauges ... 0 20
Thin card wads, in boxes of 1,000 each, 10 gauge ... 0 25
Chemically prepared black edge grey cloth wads, in boxes of 250 each— Per M.
11 and smaller gauge ... 0 60
9 and 10 gauges ... 0 70
and 8 " ... 0 90
5 and 6 " ... 1 10
Superior chemically prepared pink edge, best white cloth wads in boxes of 250 each—
11 and smaller gauge ... 1 15
9 and 10 gauges ... 1 40
7 and 8 " ... 1 65
5 and 9 " ... 1 90

SHOT.
Ordinary drop shot AAA to dust $7 50 per 100 lbs. Discount 5 per cent; cash discount 2 per cent. 30 days; net extras as follows subject to cash discount only; Chilled, 40 c; buck and seal, 80c.; no. 28 ball, $1 20 per 100 lbs; bags less than 25 lbs., ¼c per lb.; F.O B. Montreal, Toronto, Hamilton, London, St. John and Halifax, and freight equalized thereon

TRAPS (steel.)
Game, Newhouse, discount 30 and 10 per cent.
Game, Hawley & Norton, 50, 10 & 5 per cent.
Game, Victor, 70 per cent.
Game, Oneida Jump (B. & L.) 40 & 2½ p. c.
Game, steel, 60 and 5 per cent.

SKATES.
Skates, discount 37½ per cent
Mic Mac hockey sticks, per doz 4 00 5 00

CUTLERY AND SILVERWARE.

RAZORS.

	per doz.	
Elliot's	4 00	18 00
Boker's	7 50	11 00
" King Cutter	13 50	18 50
Wade & Butcher's	3 60	10 00
Lewis Bros. "Klean Kutter"	8 50	10 50
Henckel's	7 50	20 00
Berg's	7 50	20 00

Clauss Razors and Strops, 50 and 10 per cent

KNIVES.

Farriers-Stacey Bros., doz 3 50

PLATED GOODS

Hollowware, 40 per cent. discount.
Flatware, staples, 40 and 10, fancy, 40 and 5.

SHEARS.

Clauss, nickel, discount 60 per cent.
Clauss, Japan, discount 67½ per cent.
Clauss, tailors, discount 40 per cent.
Seymour's, discount 50 and 10 per cent.
Berg's 6 00 12 00

HOUSE FURNISHINGS.

APPLE PARERS.

Woodyatt Hudson, per doz, net 4 50

BIRD CAGES.

Brass and Japanned, 40 and 10 p. c.

COPPER AND NICKEL WARE.

Copper boilers, kettles, teapots, etc. 35 p.c.
Copper pitts, 25 per cent.

ENAMELLED WARE.

London, White, Princess, Turquoise, Onyx, Blue and White, discount 50 per cent.
Canada, Diamond, Premier, 50 and 10 p c.
Pearl, Imperial Crescent, 60 and 10 per cent
Premier steel ware, 40 per cent.
Star decorated steel and white, 25 per cent.
Japanned ware, discount 45 per cent.
Hollow ware, tinned cast, 35 per cent. off.

KITCHEN SUNDRIES.

Can openers, per doz	0 40	0 75
Mincing knives per doz	0 50	0 80
Duplex mouse traps, per doz		0 65
Potato mashers, wire, per doz.	0 60	0 70
" wood "	0 60	0 60
Vegetable slicers, per doz		2 25
Universal meat chopper No. 0, $1; No. 1, 1.15.		
Enterprise chopper, each		1 30
Spiders and fry pans, 50 per cent.		
Star Al chopper 5 to 32	1 35	4 10
" " 100 to 103	1 35	2 00
Kitchen hooks, bright		0 62½

LAMP WICKS.

Discount, 60 per cent.

LEMON SQUEEZERS.

Porcelain lined	per doz.	2 20	5 60
Galvanized	"	1 87	3 85
King, wood	"	2 75	2 90
King, glass	"	4 00	4 50
All glass	"	0 50	0 90

METAL POLISH.

Tandem metal polish paste 6 00

PICTURE NAILS.

Porcelain head	per gross	1 25	1 50
Brass head		0 40	1 00

Tin and gilt, picture wire, 75 per cent.

SAD IRONS.

Mrs. Potts, No. 55, polished....per set	0 80
" No. 50, nickle-plated, "	0 94
Common, plain	4 50
" plated	5 40
Asbestos, per set	1 25

TINWARE.

CONDUCTOR PIPE

2-in. plain or corrugated, per 100 feet, $3 30; 3 in, $4 40; 4 in, $5.80; 5 in, $7.45; 6 in., $9.90.

FAUCETS.

Common, cork-lined, discount 35 per cent.

EAVETROUGHS.

10-inch ......per 100 ft. 3 30

FACTORY MILK CANS.

Discount off revised list, 35 per cent
Milk can trimmings discount 25 per cent.
Creamery Cans, 45 per cent.

LANTERNS.

No. 2 or 4 Plain Cold Blast....per doz.		6 50
Lift Tubular and Hinge Plain,	"	4 75
No 0, safety	"	4 00

Better quality at higher prices.
Japanning, 50c. per doz. extra.
Prism globes, per doz., $1 20.

OILERS.

Kemp's Tornado and McClary's Model galvanized oil can, with pump, 5 gallon, per dozen 10 92
Davidson oilers, discount 40 per cent
Zinc and tin, discount 50 per cent.
Coppered oilers, 20 per cent. off.
Brass oilers, 50 per cent. off.
Malleable, discount 25 per cent

PAILS (GALVANIZED).

Dufferin pattern pails, 45 per cent.
Flaring pattern, discount 45 per cent.
Galvanized washtubs 40 per cent.

PIECED WARE.

Discount 35 per cent off list, June, 1899.
10-qt. flaring sap buckets, discount 35 per cent.
6, 10 and 14-qt. flaring pails dis. 35 per cent.
Copper bottom tea kettles and boilers, 30 p.c.
Coal hods, 40 per cent.

STAMPED WARE.

Plain, 75 and 12½ per cent. off revised list.
Retinned, 72½ per cent revised list.

SAP SPOUTS.

Bronzed iron with hooks ....per 1,000 7 50
Eureka tinned steel, hooks 8 00

STOVEPIPES.

5 and 6 inch, per 100 lengths	7 64	7 91
7 inch	"	8 18

Nestable, discount 40 per cent.

STOVEPIPE ELBOWS

5 and 6-inch, common....per doz.	1 32
7-inch "	1 48

Polished, 15c. per dozen extra.

THERMOMETERS.

Tin case and dairy, 75 to 75 and 10 per cent.

TINNERS' SNIPS.

Per doz 3 00 15
Clauss, discount 35 per cent.

TINNERS' TRIMMINGS.

Discount, 45 per cent.

WIRE.

ANNEALED CUT HAY BAILING WIRE

No. 12 and 13, $4; No 13½, $4 10; No. 14, $4 25; No 15, $4 50; in lengths 6' to 11', 25 per cent.; other lengths 20c. per 100 lbs extra; if eye or loop on end add 25c. per 100 lbs. to the above.

BRIGHT WIRE GOODS

Discount 62½ per cent.

CLOTHES LINE WIRE.

7 wire solid line, No. 17, $4.90; No. 18, $3.00; No. 19, $2.70; 6 wire solid line, No. 17, $4.45; No. 18, $2.80. No. 19, $2.50. All prices per 1000 ft. measure. F.o.b. Hamilton Toronto, Montreal.

COILED SPRING WIRE

High Carbon, No. 9, $2 90, No. 11, $3 45; No. 17, $3 15.

COPPER AND BRASS WIRE.

Discount 37½ per cent.

FINE STEEL WIRE.

Discount 25 per cent. List of extras: In 100-lb. lots: No. 17, $5 — No. 18, $5.50 — No. 19, $6 — No. 20, $6.65 — No. 21, $7 — No. 22, $7.30 — No 23, $7.65 — No. 24, $8 — No. 25, $9—No. 26, $9 50—No. 27, $10—No. 28, $11—No. 29, $12—No. 30, $13—No. 31, $14—No. 32, $15—No. 33, $16—No. 34, $17. Extras net—tinned wire, Nos. 17-25, $2—Nos. 26-31, $4—Nos. 32-34, $6. Coppered, 75c.—oiling, 10c.—in 25-lb. bundles, 15c.—in 5 and 10-lb. bundles, 25c.—in 1-lb. hanks, 25c. —in ½-lb. hanks, 38c.—in ¼-lb. hanks, 50c.—packed in casks or cases, 15c.—bagging or papering, 10c.

FENCE STAPLES.

Bright.......... 2 75 Galvanized.... 3 15

HAY WIRE IN COILS.

No. 13, $2 60; No. 14, $2 70; No. 15, $2.85; f.o b., Montreal.

GALVANIZED WIRE.

Per 100 lb.—Nos. 4 and 5, $3.70 — Nos. 6, 7, 8, $3.15 — No. 9, $2.60 — No. 10, $3 20 — No. 11, $3 25—No. 12, $2 65 —No. 13, $2.75—No. 14, $3.75—No 15, $4 30 —No. 16, $4.30 from stock. Base sizes, Nos. 6 to 9, $2.35 f.o.b. Cleveland. In carlots 12½c. less.

LIGHT STRAIGHTENED WIRE.

Gauge No.	Over 20 in. per 100 lbs.	10 to 20 in.	5 to 10 in.
0 to 5	$0 50	$0 75	$1 25
6 to 9	0.75	1.25	2 00
10 to 11	1.00	1 75	2 50
12 to 14	1.50	2 25	3 50
15 to 16	2.00	3.00	4.50

SMOOTH STEEL WIRE.

No. 0-9 gauge, $2.30; No. 10 gauge, 6c. extra; No. 11 gauge, 12c extra; No 12 gauge, 20c. extra; No. 13 gauge, 30c. extra; No 14 gauge, 40c. extra; No. 15 gauge, 55c. extra, No 16 gauge, 70c extra. Add 60c. for coppering and $2 for tinning

Extra net per 100 lb.—Oiled wire 10c., spring wire $1.25, bright soft drawn 15c., charcoal (extra quality) $1.25, packed in casks or cases 15c., bagging and papering 10c., 50 and 100-lb. bundles 10c., in 25-lb. bundles 15c., in 5 and 10-lb. bundles 25c., in 1-lb hanks, 50c., in ½-lb. hanks 75c., in ¼-lb. hanks $1.

POULTRY NETTING.

2 in mesh 19 w. g., discount 50 and 10 per cent. All others 50 per cent.

WIRE CLOTH.

Painted Screen, in 100-ft rolls, $1 62½, per 100 sq. ft.; in 50-ft. rolls, $1.67½, per 100 sq ft. Terms, 2 per cent. off 30 days.

WIRE FENCING.

Galvanized barb...... 2 95
Galvanized, plain twist 3 30
Galvanized barb, f.o b. Cleveland, $2.70 for small lots and $2.60 for carlots.

WOODENWARE.

CHURNS.

No. 0, $9; No 1, $9; No. 2, $10; No. 3, $11; No. 4, $13; No. 5, $16; f.o b. Toronto Hamilton, London and St Marys, 30 and 30 per cent; f o b Ottawa, Kingston and Montreal, 40 and 15 per cent. discount, Taylor-Forbes, 30 and 30 per cent.

CLOTHES REELS.

Davis Clothes Reels, dis. 40 per cent.

LADDERS, EXTENSION.

Waggoner Extension Ladders, dis. 40 per cent.

MOPS AND IRONING BOARDS.

"Best" mops		1 25
"900" mops		1 25
Folding ironing boards	12 00	16 50

REFRIGERATORS

Discount, 40 per cent.

SCREEN DOORS.

Common doors, 2 or 3 panel, walnut stained, 4-in. style.........per doz. 7 25
Common doors, 2 or 3 panel, grained only, 4-in., style ....per doz. 7 55
Common doors, 2 or 3 panel, light stair per doz. 9 56

WASHING MACHINES.

Round, re-acting per doz.	60 00
Square "	63 00
Eclipse, per doz	54 00
Dowswell "	39 00
New Century, per doz	75 00
Daisy	54 00

WRINGERS.

Royal Canadian, 11 in., per doz.	34 00
Royal American, 11 in.	34 00
Eze, 10 in., per doz	3 75

Terms, 2 per cent., 30 days.

MISCELLANEOUS.

AXLE GREASE.

Ordinary, per gross	6 00	7 00
Best quality	10 00	12 00

BELTING.

Extra, 60 per cent.
Standard, 60 and 10 per cent.
No. 1, not wider than 6 in., 60, 10 and 10 p.c.
Agricultural, not wider than 4 in., 75 per cent
Lace leather, per side, 75c.; cut laces, 80c.

BOOT CALKS.

Small and medium, ball	per M	4 25
Small heel	"	4 50

CARPET STRETCHERS.

American ....per doz.	1 00	1 50
Bullard's "	....	6 50

CASTORS.

Bed, new list, discount 55 to 57½ per cent.
Plate, discount 52½ to 57½ per cent.

PINE TAR.

½ pint in tins ....per gross	...	7 80
" "	....	9 60

PULLEYS.

Hothouse	per doz.	0 55	1 00
Axle	"	0 22	0 33
Screw	"	0 22	1 00
Awning	"	0 35	2 50

PUMPS.

Canadian cistern	1 40	2 00
Canadian pitcher spout	1 80	3 16

Berg's wing pump, 75 per cent.

ROPE AND TWINE.

Sisal		0 10½
Pure Manilla		0 15½
"British" Manilla		0 12
Cotton, 3-16 inch and larger	0 21	0 23
" 5-32 inch	0 25	0 27
" ¼ inch	0 25	0 28
Russia Deep Sea		0 16
Jute		0 09
Lath Yarn, single		0 10
" double		0 10½
Sisal bed cord, 48 feet........per doz.		0 65
" " 60 feet........ "		0 80
" " 72 feet........ "		0 95

Twine.

Bag, Russian twine, per lb.	....	0 27
Wrapping, cotton, 3-ply	....	0 25
" " 4-ply	....	0 29
Mattress twine per lb.	0 33	0 45
Staging " "	0 27	0 35

SCALES.

Gurney Standard, 40 per cent.
Gurney Champion, 50 per cent.
Burrow, Stewart & Milne—
Imperial Standard, discount 40 per cent.
Weigh Beams, discount 40 per cent.
Champion Scales, discount 50 per cent.
Fairbanks standard, discount 35 per cent.
" Dominion, discount 55 per cent.
" Richelieu, discount 55 per cent.
Warren new Standard, discount 40 per cent.
" " Champion, discount 50 per cent.
" Weighbeams, discount 35 per cent.

STONES—OIL AND SCYTHE.

Washita	per lb.	0 25	0 37
Hindostan	"	0 06	0 10
" slip	"	0 18	0 20
" Axe	"	....	0 10
Deer Creek	"	....	0 10
Deerlick	"	....	0 25
" Axe	"	....	0 15
Lily white	"	....	0 42
Arkansas	"	....	1 50
Water-of-Ayr	"	....	0 10
Scythe	per gross	3 50	5 00
Grind, 40 to 200 lb., per ton		20 00	22 00
" under 40 lb.,	"	....	24 00
" 200 lb. and over		....	26 00

INDEX TO ADVERTISERS.

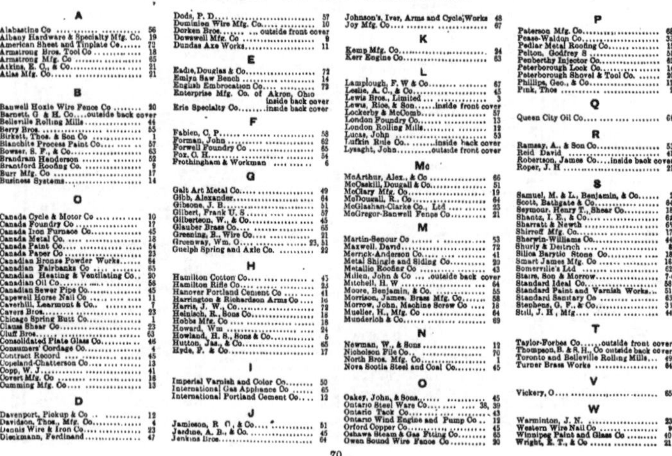

CLASSIFIED LIST OF ADVERTISEMENTS.

Auditors.
Davenport, Pickup & Co., Winnipeg.

Babbitt Metal.
Canada Metal Co., Toronto.
Canadian Fairbanks Co., Montreal.
Robertson, Jas. Co., Montreal.

Bath Room Fittings.
Carriage Mounting Co., Toronto.

Belting, Hose, etc.
Gutta Percha and Rubber Mfg. Co. Toronto.

Bicycles and Accessories.
Johnson's, Iver, Arms and Cycle Works Fitchburg, Mass

Binder Twine.
Consumers Cordage Co., Montreal.

Box Strap.
J. N. Warminton, Montreal.

Brass Goods.
Glauber Brass Mfg. Co., Cleveland, Ohio.
Lewis, Rice, & Son., Toronto.
Morrison, Jas., Brass Mfg. Co., Toronto.
Mueller Mfg. Co., Decatur, Ill.
Penberthy Injector Co., Windsor, Ont.
Taylor-Forbes Co., Guelph, Ont.

Bronze Powders.
Canadian Bronze Powder Works, Montreal.

Brushes.
Ramsay, A., & Son Co., Montreal.

Can Openers.
Cumming Mfg. Co. Renfrew.

Cans.
Acme Can Works, Montreal.

Builders' Tools and Supplies.
Covert Mfg. Co., West Troy, N.Y.
Frothingham & Workman Co., Montreal.
Howland, H. S., Sons & Co., Toronto.
Hyde, F., & Co., Montreal.
Lewis Bros. & Co., Montreal.
Lewis, Rice, & Son, Toronto.
Lockerby & McComb, Montreal.
Lufkin Rule Co., Saginaw, Mich.
Newman & Sons, Birmingham.
North Bros. Mfg. Co., Philadelphia, Pa.
Stanley Rule & Level Co., New Britain.
Stanley Works, New Britain, Conn.
Stephens, G. F., Winnipeg.
Taylor-Forbes Co., Guelph, Ont.

Carriage Accessories.
Carriage Mountings Co., Toronto.
Covert Mfg. Co., West Troy, N.Y.

Carriage Springs and Axles.
Guelph Spring and Axle Co., Guelph.

Cattle and Trace Chains.
Greening, B., Wire Co., Hamilton.

Churns.
Dowswell Mfg. Co., Hamilton.

Clippers—All Kinds.
American Shearer Mfg. Co., Nashua, N.H.

Clothes Reels and Lines.
Hamilton Cotton Co., Hamilton, Ont.

Cordage.
Consumers' Cordage Co., Montreal.
Hamilton Cotton Co., Hamilton.

Cork Screws.
Erie Specialty Co., Erie, Pa.

Clutch Nails.
J. N. Warminton, Montreal.

Cut Glass.
Phillips, Geo., & Co., Montreal.

Cutlery—Razors, Scissors, etc.
Birkett, Thos., & Son Co., Ottawa.
Clauss Shear Co., Toronto
Dorken Bros. & Co., Montreal.
Heinisch's, R., Sons Co., Newark, N.J.
Howland, H. S. Sons & Co., Toronto.
Lamplough, F. W., & Co., Montreal.
Phillips, Geo., & Co., Montreal.
Round, John, & Son, Montreal.

Electric Fixtures.
Canadian General Electric Co., Toronto.
Forman, John, Montreal.
Morrison James, Mfg. Co., Toronto.
Munderloh & Co., Montreal.

Electro Cabinets.
Cameron & Campbell Toronto.

Engines, Supplies, etc.
Kerr Engine Co., Walkerville, Ont.

Fencing—Woven Wire.
McGregor-Banwell Fence Co., Walkerville, Ont
Owen Sound Wire Fence Co, Owen Sound.
Banwell Hoxie Wire Fence Co., Hamilton.

Files and Rasps.
Barnett Co., G. & H., Philadelphia, Pa.
Nicholson File Co., Port Hope

Financial Institutions
Bradstreet Co.

Firearms and Ammunition.
Dominion Cartridge Co., Montreal.
Hamilton Rifle Co., Plymouth, Mich.
Harrington & Richardson Arms Co., Worcester, Mass.
Johnson's, Iver, Arms and Cycle Works, Fitchburg, Mass.

Food Choppers
Enterprise Mfg. Co., Philadelphia, Pa.
Lamplough, F. W., & Co., Montreal.
Shirreff Mfg. Co, Brockville, Ont.

Galvanizing.
Canada Metal Co., Toronto.
Montreal Rolling Mills Co., Montreal.
Ontario Wind Engine & Pump Co., Toronto.

Glaziers' Diamonds.
Gibsone, J. B., Montreal.
Pelton, Godfrey S.
Sharratt & Newth, London, Eng.
Shaw, A., & Son, London, Eng.

Handles.
Still, J. H, Mfg. Co.

Harvest Tools.
Maple Leaf Harvest Tool Co, Tillsonburg, Ont.

Hoop Iron.
Montreal Rolling Mills Co, Montreal.
J. N. Warminton, Montreal.

Horse Blankets.
Heney, E. N., & Co., Montreal.

Horseshoes and Nails.
Canada Horse Nail Co., Montreal.
Montreal Rolling Mills, Montreal.
Capewell Horse Nail Co., Toronto

Hot Water Boilers and Radiators.
Cluff, R. J., & Co., Toronto.
Pease Foundry Co., Toronto.
Taylor-Forbes Co., Guelph.

Ice Cream Freezers.
Dana Mfg. Co., Cincinnati, Ohio.
North Bros. Mfg. Co., Philadelphia, Pa.

Ice Cutting Tools.
Erie Specialty Co., Erie, Pa.
North Bros. Mfg. Co., Philadelphia, Pa.

Injectors—Automatic.
Morrison, Jas., Brass Mfg. Co., Toronto.
Penberthy Injector Co., Windsor, Ont.

Iron Pipe.
Montreal Rolling Mills, Montreal.

Iron Pumps.
Lamplough, F. W., & Co, Montreal.
McDougall, R., Co., Galt, Ont.

Lanterns.
Kemp Mfg. Co., Toronto.
Ontario Lantern Co., Hamilton, Ont.
Wright, E. T., & Co., Hamilton.

Lawn Mowers.
Birkett, Thos., & Son Co., Ottawa.
Maxwell, D., & Sons, St. Mary's, Ont.
Taylor, Forbes Co., Guelph.

Lawn Swings, Settees, Chairs.
Cumming Mfg. Co., Renfrew.

Ledgers—Loose Leaf.
Business Systems Toronto.
Copeland-Chatterson Co, Toronto.
Crain, Rolla L., Co., Ottawa.
Universal Systems, Toronto.

Locks, Knobs, Escutcheons, etc.
Peterborough Lock Mfg. Co., Peterborough, Ont.

Lumbermen's Supplies.
Pink, Thos., & Co., Pembroke Ont.

Manufacturers' Agents.
Fox, C. H., Vancouver.
Gibb, Alexander, Montreal.
Mitchell, David C., & Co., Glasgow, Scot.
Mitchell, H. W., Winnipeg.
Pearce, Frank, & Co. Liverpool, Eng.
Scott, Bathgate & Co., Winnipeg.
Thorne, B. E., Montreal and Toronto.

Metals.
Canada Iron Furnace Co., Midland, Ont.
Canada Metal Co., Toronto.
Eadie, H. G., Montreal.
Gibb, Alexander, Montreal.
Kemp Mfg Co., Toronto
Leslie, A. C., & Co, Montreal.
Lysaght, John, Bristol, Eng.
Nova Scotia Steel and Coal Co., New Glasgow, N.S.
Robertson, Jas, Co., Montreal.
Roper, J. H., Montreal.
Samuel, Benjamin & Co., Toronto.
Stairs, Son & Morrow, Halifax, N S.
Thompson, B. & S. H. & Co. Montreal.

Metal Lath.
Galt Art Metal Co., Galt.
Metallic Roofing Co., Toronto.
Metal Shingle & Siding Co., Preston, Ont.

Metal Polish, Emery Cloth, etc
Oakey, John, & Sons, London, Eng.

Mops.
Cumming Mfg. Co., Refrew.

Mouse Traps.
Cumming Mfg Co., Renfrew.

Oil Tanks.
Bowser, S. F., & Co., Toronto.

Paints, Oils, Varnishes, Glass.
Bell, Thos, Sons & Co, Montreal.
Canada Paint Co., Montreal.
Canadian Oil Co., Toronto.
Consolidated Plate Glass Co., Toronto.
Fenner, Fred., & Co., London, Eng.
Henderson & Potts Co, Montreal.
Imperial Varnish and Color Co., Toronto
Jamieson, R. C., & Co., Montreal.
McArthur, Corneille & Co., Montreal.
McCaskill, Dougall & Co., Montreal.
Montreal Rolling Mills Co, Montreal.
Moore, Benjamin, & Co. Toronto.
Queen City Oil Co., Toronto.
Ramsay & Son, Montreal.
Sherwin-Williams Co., Montreal.
Standard Paint and Varnish Works Windsor, Ont.
Stephens & Co, Winnipeg.
Martin-Senour Co., Chicago.

Perforated Sheet Metals.
Greening, B., Wire Co., Hamilton.

Plumbers' Tools and Supplies.
Borden Co., Warren, Ohio.
Canadian Fairbanks Co., Montreal.
Cluff, R. J., & Co., Toronto.
Glauber Brass Co., Cleveland, Ohio.
Jardine, A. B., & Co, Hespeler, Ont.
Jenkins Bros., Boston, Mass.
Lewis, Rice, & Son, Toronto.
Merrell Mfg. Co., Toledo, Ohio.
Montreal Rolling Mills Montreal.
Morrison, Jas., Brass Mfg. Co., Toronto.
Mueller, H., Mfg. Co, Decatur, Ill.
Oshawa Steam & Gas Fitting Co., Oshawa
Robertson Jas, Co. Montreal.
Stairs, Son & Morrow, Halifax, N.S.
Standard Ideal Sanitary Co, Port Hope.
Standard Sanitary Co., Pittsburg.
Stephens, G. F, & Co., Winnipeg, Man.
Turner Brass Works, Chicago.
Vickery, Orlando, Toronto.

Portland Cement.
International Portland Cement Co., Ottawa, Ont.
Hanover Portland Cement Co., Hanover, Ont.
Hyde, F., & Co., Montreal.
Thompson, B. & S. H. & Co., Montreal.

Potato Mashers.
Cumming Mfg. Co., Renfrew.

Poultry Netting.
Greening, B., Wire Co., Hamilton, Ont.

Razors.
Clauss Shear Co., Toronto.

Roofing Supplies.
Brantford Roofing Co., Brantford.
McArthur, Alex., & Co., Montreal.
Metal Shingle & Siding Co., Preston, Ont.
Metallic Roofing Co, Toronto.
Paterson Mfg. Co., Toronto & Montreal.

Saws.
Atkins, E. C., & Co., Indianapolis, Ind
Lewis Bros., Montreal.
Shurly & Dietrich, Galt, Ont.
Spear & Jackson, Sheffield, Eng.

Scales.
Canadian Fairbanks Co., Montreal.

Screw Cabinets.
Cameron & Campbell, Toronto.

Screws, Nuts, Bolts.
Montreal Rolling Mills Co., Montreal.
Morrow, John, Machine Screw Co., Ingersoll, Ont.

Sewer Pipes.
Canadian Sewer Pipe Co., Hamilton
Hyde, F., & Co., Montreal.

Shelf Boxes.
Cameron & Campbell, Toronto.

Shears, Scissors.
Clauss Shear Co, Toront .

Shelf Brackets.
Atlas Mfg. Co., New Haven, Conn

Shellac
Bell, Thos., Sons & Co., Montreal.

Shovels and Spades.
Canadian Shovel Co., Hamilton.
Peterboro Shovel & Tool Co, Peterboro.

Silverware.
Phillips, Geo., & Co., Montreal.
Round, John, & Son, Sheffield, Eng.

Skates.
Canada Cycle & Motor Co., Toronto.

Spring Hinges, etc.
Chicago Spring Butt Co, Chicago, Ill.

Steel Rails.
Nova Scotia Steel & Coal Co., New Glasgow, N.S.

Stoves, Tinware, Furnaces
Canadian Heating & Ventilating Co., Owen Sound.
Canada Stove Works, Harriston, Ont.
Clare Bros. & Co., Preston.
Davidson, Thos, Mfg. Co., Montreal,
Guelph Stove Co., Guelph.
Gurney Foundry Co., Toronto.
Harris, J. W., Co., Montreal.
Joy Mfg. Co., Toronto.
Kemp Mnfg. Co. Toronto.
McClary Mfg. Co. London.
Pease Foundry Co., Toronto.
Stewart, Jas., Mfg. Co., Woodstock, Ont.
Taylor-Forbes Co., Guelph, Ont.
Wright, E. T., & Co., Hamilton.

Tacks.
Montreal Rolling Mills Co., Montreal.
Ontario Tack Co., Hamilton.

Ventilators.
Harris, J. W., Co., Montreal.
Pearson, Geo. D., Montreal.

Washing Machines, etc.
Dowswell Mfg Co., Hamilton, Ont.
Taylor-Forbes Co., Guelph, Ont.

Wheelbarrows.
London Foundry Co., London, Ont.

Wholesale Hardware.
Birkett, Thos., & Sons Co., Ottawa.
Caverhill, Learmont & Co., Montreal.
Frothingham & Workman, Montreal.
Hobbs Hardware Co., London.
Howland, H. S., Sons & Co., Toronto.
Lamplough, F. W., & Co., Montreal.
Lewis Bros. & Co., Montreal.
Lewis, Rice, & Son, Toronto.

Window and Sidewalk Prisms.
Hobbs Mfg. Co., London, Ont.

Wire Springs.
Guelph Spring Axle Co., Guelph, Ont.
Wallace-Barnes Co., Bristol, Conn.

Wire, Wire Rope, Cow Ties, Fencing Tools, etc.
Canada Fence Co, London.
Dennis Wire and Iron Co., London, Ont.
Dominion Wire Mnfg. Co., Montreal
Greening, B., Wire Co., Hamilton.
Montreal Rolling Mills Co., Montreal.
Western Wire & Nail Co., London, Ont

Woodenware.
Taylor-Forbes Co., Guelph, Ont.

Wrapping Papers.
Canada Paper Co., Toronto.
McArthur, Alex., & Co, Montreal.
Stairs, Son & Morrow, Halifax, N.S.

is more deadly than desease itself. Bad air is the curse of crowded living, in all public places such as asylums, clubs, big offices, schools and theatres. **PURE AIR CAN BE OBTAINED ONLY BY OUR SYSTEM OF VENTILATION.**

It has stood the tests of scientific experts.

It is in use in the finest buildings in Canada.

Its price is reasonable and its benefits lasting.

Testimony:

MONTREAL, June 7th, 1901

Messrs. J. W. HARRIS CO., LIMITED, Montreal

Dear Sirs,

I have been using your "AEOLIAN" Ventilators on several houses in this city with the best results. The tenants of these houses declared that your Ventilators are of great advantage to them. I also installed two (2) 18 inch on my stable and I can say that your Ventilator cannot be surpassed by any to take out the sweating or bad smell in any room.

F. D. MONK

Ask for booklet re Aeolian and Zephyr Ventilators, also booklet on Filter, etc.

The J. W. HARRIS CO., Limited

Contractors and Manufacturers

MONTREAL

D. C. Mitchell & Co.

GLASGOW

STEEL PLATES, SHEETS, ANGLES, BARS, PIG IRON,

REPRESENTATIVE

Eadie-Douglas Company

22 St. John St., MONTREAL

SUPPORT CANADIAN INDUSTRIES

FREE
TO
HARDWARE DEALERS

We are supplying electros of the cut in this ad. for use in your local advertising, besides offering special terms for immediate orders for

"LION" Brand

PUTZ TRIPLE EXTRACT

METAL POLISH

IN LIQUID OR PASTE

The most reliable metal polish in the market. Once tried always used.

Support Canadian Industries

Manufactured by

English Embrocation Co., Montreal

Specify the Following Products

Manufactured by

American Sheet and Tin Plate Company

PITTSBURG, PA.

If you want the best.

"Apollo Best Bloom," Galvanized Sheets,
Blue Annealed and American Bessemer Black Sheets,
Wellsville Polish, Range Steel,
Wood's "A" Planished Iron,
Black Steel specialties for all purposes,
Tinplates, Coke and Charcoal, all grades,
Terne Plates, Old Style and New Method Processes.

Place your order with

B. & S. H. THOMPSON & CO.
LIMITED

Dominion of Canada Sales Agents

MONTREAL, CANADA

HARDWARE BABBITTS

No. 1, No. 2, No. 3

The wearing quality of our

HARDWARE BABBITTS

is the secret of the large orders the

HARDWARE MERCHANTS

are sending us. If not on our list

GET IN LINE

LIBERAL DISCOUNTS. TERMS RIGHT

The James Robertson Co., Limited

144 William St., MONTREAL, QUE.

PFLUEGER'S Fishing Tackle

Hooks, Flies, Trolls, Spinners, Phantoms, Reels, Furnished Lines.

Everything in Fishing Tackle.

NOTICE—Free to Any Dealer in Sporting Goods, Sent Express Prepaid, 170-Page Illustrated Catalogue No. F25, and Metal Fish Sign in 8-Color Lithograph.

The Enterprise Mfg. Co.

AKRON, Ohio, U.S.A.

WALKER CORK SCREWS

are good holiday souvenirs and live sellers the whole year. Are made well and are made out of good material. That is why we guarantee every one that we put out. When ordering Cork Screws try **Walker's**

Manufactured only by **ERIE SPECIALTY COMPANY,** ERIE, Pa. U.S.A.

LUFKIN MEASURING TAPES

Steel, Metallic, Linen, Pocket, Ass Skin, Pat. Leather, Bend Leather, Etc.

ARE THE BEST AND MOST POPULAR TAPES IN THE WORLD. YOUR STOCK IS NOT COMPLETE WITHOUT THEM.

LUFKIN RULE CO., Saginaw, Mich, U.S.A.

London Office and Warehouse—24 and 26 Holborn. New York City Branch—280 Broadway.

For sale by ALL PROMINENT CANADIAN HARDWARE JOBBERS.

Awarded
By **JURORS** at
International Expositions
Special Prize
Gold Medal at Atlanta, 1895

Copy of catalogue sent free to any interested file user upon application.

"Redstone"

High Pressure

Sheet Packing

A packing that will hold. For use in highest pressures for steam, hot or cold water and air. Packs equally well for all.

From actual tests, we believe that this packing is the most durable and satisfactory of any on the market. Try a sample lot and see for yourself.

Manufactured Solely by

THE GUTTA PERCHA & RUBBER MFG. CO.
of TORONTO, LIMITED
HEAD OFFICES,
47 Yonge Street, Toronto.
Branches: Montreal, Winnipeg, Vancouver.

SHEET ZINC,
SHEET COPPER,
SPELTER,
INGOT TIN,
BRASS and COPPER TUBES.

For lowest prices send your enquiries to

B. & S. H. THOMPSON & Co.
LIMITED
53 ST. SULPICE STREET, MONTREAL

"Cyrus" Bicycles
"Adlake" Bicycles

Two wheels that sell readily on their merits

Every cent of their price is represented in the wheel.

Write for our catalogue. Get our trade price on these lines.

JOHN MILLEN & SON,
Limited

321 St. James St.	**132 Bay Street**
MONTREAL	**TORONTO**

CIRCULATES EVERYWHERE IN CANADA
Also in Great Britain, United States, West Indies, South Africa and Australia.

HARDWARE AND METAL

A Weekly Newspaper Devoted to the Hardware, Metal, Heating and Plumbing Trades in Canada.
Office of Publication, 10 Front Street East, Toronto.

VOL. XIX. MONTREAL, TORONTO, WINNIPEG, MARCH 9, 1907. NO. 10.

See Classified List of Advertisements on Page 71.

Would you like to know each evening as you close your store that your accounts are up-to-date and ready to render?

It is the proper way to keep books, and the nearest approach to a cash business—to have a system such as

THE CRAIN MONTHLY ACCOUNT SYSTEM

where accounts are always ready to render. No other labor-saving device on the market is just as good as the Crain. We manufacture this system in different sizes and styles of binding, ranging in price from $14.25 up with 1,000 sheets of invoices.

If you are interested write for catalogue E, which will give you full information.

The Rolla L. Crain Co., Limited

OTTAWA, CANADA

TORONTO: 18 Toronto Street
MONTREAL: Alliance Bldg., 107 St. James St.
WINNIPEG: 11 Nanton Block

WE CARRY IN STOCK

Horse-shoers' Clip Horn Anvil
Prospectors' Anvils and Bellows
Prospectors' Picks, Shovels and Stoves
The Famous Blacksmith Blower No. 400
The Grand Bay Blower No. 100
Challenge Forges with Hood, Extra Strong Frame
Hero Forges With and Without Hood
Navy Wheelbarrows with Steel or Wood Trays
Steel Pulley Blocks, Single, Double or Triple

Write for prices. Prompt shipment.

THOS. BIRKETT & SON CO., LIMITED

Wholesale Hardware

OTTAWA, ONT.

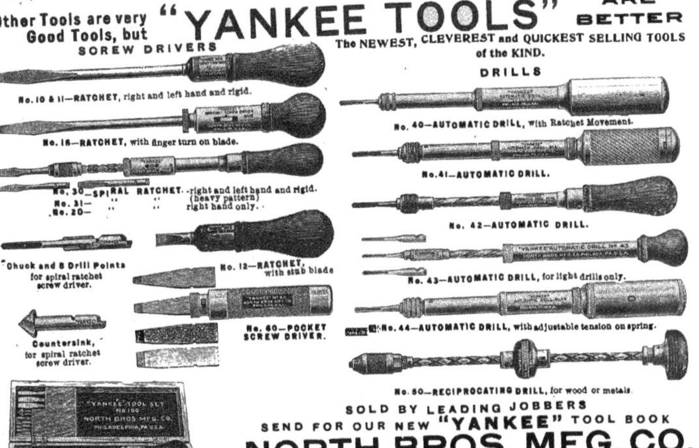

Wrenches and Pipe Tools

We carry the largest range of of any HARDWARE House in CANADA **PIPE TOOLS**

It will pay you to get our prices on EVERYTHING pertaining to this LINE

STILLSON WRENCH

6-18 inch—Wooden Handle
24-36 inch—Steel Handle

VULCAN WRENCH

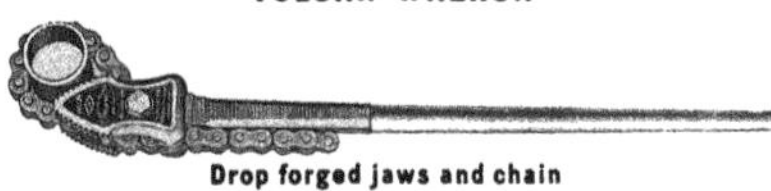

Drop forged jaws and chain
All parts interchangeable

We stock Stillson's Wrenches
We stock Trimo Wrenches **ALL SIZES**

And Bemis & Call Pipe Wrenches, Bemis & Call Combination Wrenches, V. and B. Pipe Wrenches, Wescott Pipe Wrenches, Bonney's Combination Pipe Wrenches and Pipe Cutters

TRIMO PIPE CUTTER

No. 1—⅛—1¼
No. 2—½—2
No. 3—1¼—3

We carry Barnes Pipe Cutters, all numbers, and Saunders Pipe Cutters, all numbers.

BROWN'S ADJUSTABLE PIPE TONGS

No. 1—⅛—¾
No. 1½—⅜—1
No. 2—½—1¼

No. 3—1—2
No. 4—1½—3
No. 5—2½—4

Special attention given MAIL ORDERS.
Our specialty is prompt shipments.

LEWIS BROS., Ltd.

OTTAWA
TORONTO

MONTREAL
WINNIPEG

VANCOUVER
CALGARY

Consumers Cordage Co.,
LIMITED

Manufacture every variety of

CORDAGE — PACKING

BINDER TWINE

LATHYARN — SMALL TWINES

From Manilla, Sisal, Italian, Russian, Jute

Tallow Laid Transmission Rope, 3 or 4 Strand

Oil Well Drilling Cables, Lariat Rope, 3 or 4 Strand,
Halyards, Deep Sea Line, Snap Halters, Snap Plow Lines,
Clothes Lines, Hide Cord, Pulp and Paper Cord,
Tarred Rope, Marline, Houseline

Wrapping, Rolling, Seaming, Sewing and Grape Twines

Careful Execution **SPECIAL ORDERS** Prompt Shipment

Mills: MONTREAL, QUE., and HALIFAX, N.S.

BRANCHES:

F. H. ANDREWS & SON, Quebec. CONSUMERS CORDAGE CO., Limited, St. John, N.B. MacGOWAN & CO., Vancouver
TEES & PERSSE, Limited, Winnipeg, Calgary and Edmonton. GEO. WOOD, London, England.
MERRICK, ANDERSON & CO., Winnipeg, Binder Twine Distributors for Northwest.

Insure Prompt Shipment by Placing Your Orders For OIL STOVES Now.

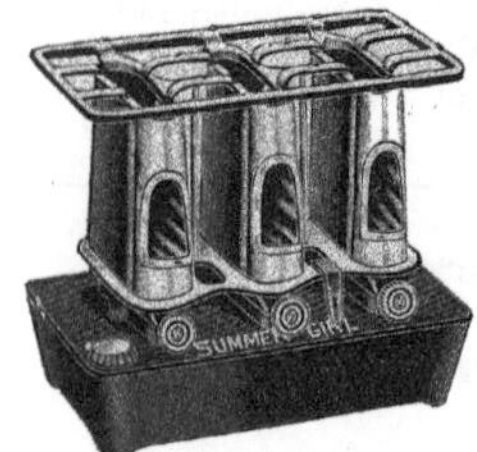

THE

Summer Girl

Width of flame 4 inches

Cast Iron Bottom.
1, 2 and 3 Burners.

German

Glass Bottom.
1, 2 and 3 Burners

"Eagle" **Tin Bottom, 2 and 3 Burners**

We have on hand a good stock of the above lines at right prices, and can make prompt shipment.

We are also Sole Agents in Canada for the

"Standard New Process" Oil and Gasoline Stoves

The Cleanest, Safest and Most Efficient Stoves in the market to-day

The Thos. Davidson Mfg. Co. Ltd., MONTREAL and WINNIPEG

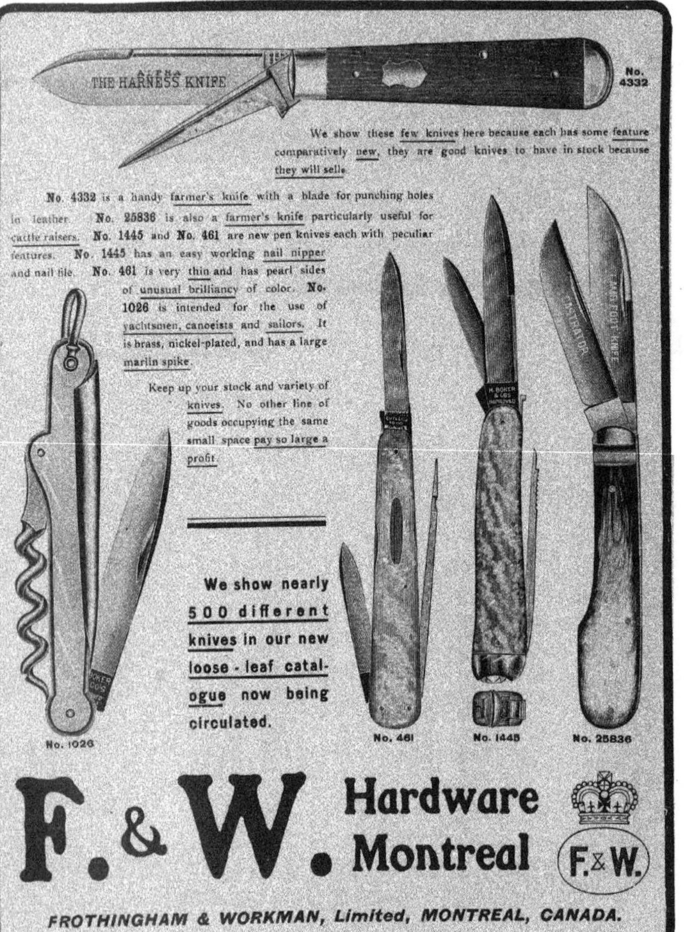
THE HARNESS KNIFE
No. 4332
We show these few knives here because each has some feature comparatively new, they are good knives to have in stock because they will sell.
No. 4332 is a handy farmer's knife with a blade for punching holes in leather. No. 25836 is also a farmer's knife particularly useful for cattle raisers. No. 1445 and No. 461 are new pen knives each with peculiar features. No. 1445 has an easy working nail nipper and nail file. No. 461 is very thin and has pearl sides of unusual brilliancy of color. No. 1026 is intended for the use of yachtsmen, canoeists and sailors. It is brass, nickel-plated, and has a large marlin spike.
Keep up your stock and variety of knives. No other line of goods occupying the same small space pay so large a profit.
We show nearly 500 different knives in our new loose-leaf catalogue now being circulated.
No. 1026
No. 461
No. 1445
No. 25836
F. & W.
Hardware Montreal
F. & W.
FROTHINGHAM & WORKMAN, Limited, MONTREAL, CANADA.

PARKER'S
Low Set Jaw, Semi-Steel
VISES

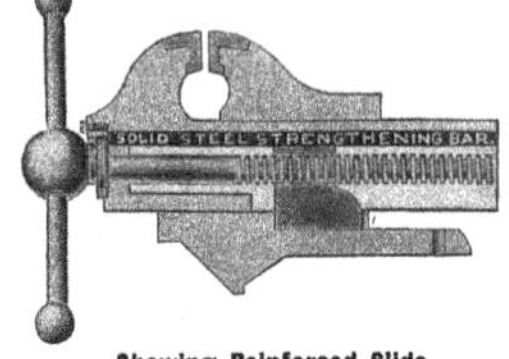

Showing Reinforced Slide

SEMI-STEEL

No.					
No. 229	With the	Reinforced	Slide	3¼-in.	Jaws
" 239	"	"	"	3¾ "	"
" 249	"	"	"	4¼ "	"
" 259	"	"	"	4¾ "	"

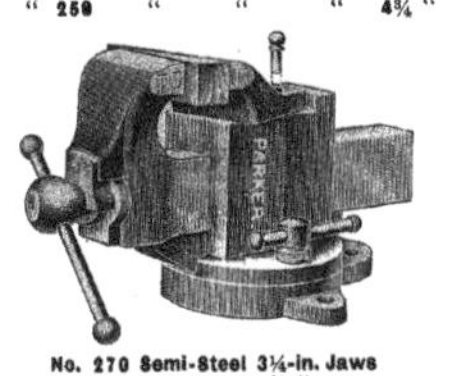

No.			
No. 270	Semi-Steel	3¼-in.	Jaws
" 271	"	3⅝ "	"
" 272	"	4½ "	"
" 273	"	5 "	"

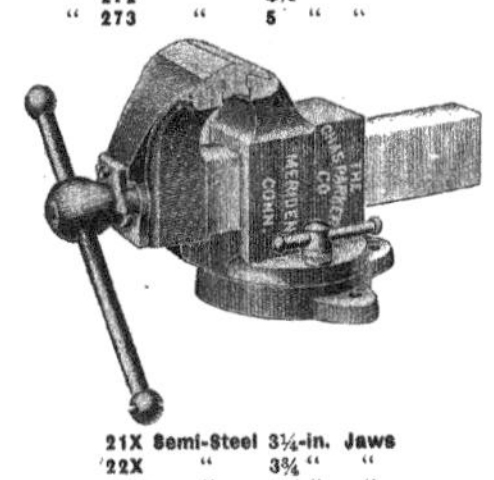

No.			
21X	Semi-Steel	3¼-in.	Jaws
22X	"	3¾ "	"
23X	"	4¼ "	"
24X	"	4¾ "	"

The Solid Steel Strengthening Bar as used in the Slide of Parker's New Vises, combined with the blending of steel and best iron in the making of the castings (known as Semi-Steel) make them practically unbreakable.

This is why the machinist will have no other.

Then the Steel Jaws can be removed and replaced when necessary.

The low set jaws make them the most desirable vise for the mechanic to use.

To Sum Up

Parker's Vises are the best that money can buy.

They are sold at a reasonable price.

They are taken in preference to others in the leading machine and railroad shops.

Caverhill Learmont & Co.

WHOLESALE DISTRIBUTORS

MONTREAL and WINNIPEG

Canada Leads the World in the Manufacture of

HIGH-CLASS SAWS

No. 51

The best and cheapest Hand Saw on the market, quality, temper and finish considered. Every saw warranted. Carved applewood handle. Patented, wave finish blade.

No. 1 Maple Leaf Lance Tooth Cross-Cut Saw, and the No. 0 Racer are tempered under our secret process.

No. 0 Narrow Racer Cross-Cut Saw

If you want saws that will sell and give your customer satisfaction, you will stock the Maple Leaf Saws.

Manufactured by **SHURLY & DIETRICH, - GALT, Ontario**

Brantford—Roofings

This Represents Our

"ASPHALT"
No. 1, 2, 3
"RUBBER"
No. 1, 2, 3

Each Roll is complete with Nails and Cement to lay. Best selling goods in the market.

BRANTFORD ROOFING CO.
LIMITED
BRANTFORD, ONTARIO

SIMPLE

DURABLE

RELIABLE

STYLE F.

THE LEADER

A Name Familiar to Butter Makers in Connection with the

Highest Grade Churns

ITS STRONG POINTS:

Stiff Steel Frame, with malleable iron fittings, Combination Bow Lever and Pedal Drive. Bicycle Ball Bearings. Bolted Trunnions. Cream Breakers. Barrel easily drained and conveniently detatchable for cleaning.

DEALERS: When you handle this line you handle QUALITY.

The Dowswell Manufacturing Co., Limited

W. L. HALDIMAND & SON, Montreal
EASTERN AGENTS

HAMILTON
ONTARIO

GALVANIZED FENCE HOOK FOR FASTENING WOODEN PICKET ON WIRE FENCES

WIRE NAILS, COILED SPRING, BARB and PLAIN FENCE WIRE, OILED and ANNEALED, CLOTHES LINE WIRE, STAPLES, etc.

THE WESTERN WIRE & NAIL CO., Limited, —LONDON, ONT.

The Peterborough Lock Manufacturing Company, Limited

Cylinder Night Latch, No. 108.

Peterborough, Ont.

Manufacturers of all kinds

Rim and Mortise Locks, Inside, Front Door, and Store Door Sets, also full line High-class Builders' Hardware.

Sold by all Leading Jobbers in the Dominion.

RIDGED AXES

There has been a general tendency to make axes flat for some time, and although an axe is seldom used now to chop a tree down, still there are sections of the country where a well-ridged axe—an axe that will burst a chip—sells best. To meet this demand we make the *Forest Queen* in Square Bit and the *Peerless* in Round Bit.

GIVE YOUR CUSTOMERS WHAT THEY WANT. IT SAVES BOTH TIME AND TROUBLE

The Dundas Axe Works

Dundas, Ont.

CAPEWELL HORSE NAILS

"THE BEST IN THE WORLD"

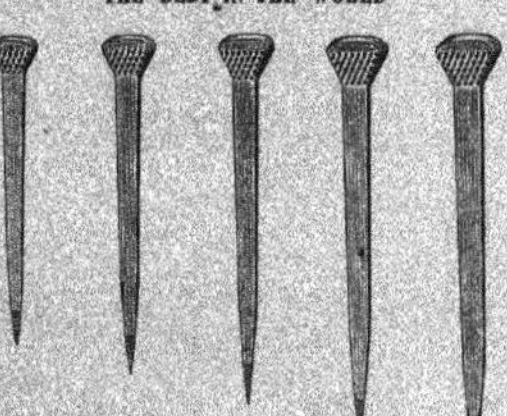

Two Samples of what Our Files Contain

DUNNVILLE, ONT., FEB. 9, '07.

This is to certify that I HAVE USED HORSE NAILS FOR FIFTY YEARS, and I consider the CAPEWELL NAIL the best ever made.

JOHN HAITH.

LADYSMITH, B.C., FEB. 25, '07.

I have used your nails ON ALL KINDS OF HORSES. I find them SUPERIOR TO ALL OTHERS, and I have used some dozens of brands.

J. M. MCMILLAN.

A satisfied customer is the best advertisment in the world. Keep Your customers satisfied by selling them the best—the **CAPEWELL.**

The CAPEWELL HORSE NAIL CO., TORONTO, ONT.

Branch, 212 Nena Street, Winnipeg, Man.

Dana

Peerless Freezers

Increase your freezer business.

Takes a woman off her knees

by selling the freezer the women want—because with *it* they make ice-cream with *real* satisfaction. The cook's favorite. Known to all women through advertisements in their favorite magazine.

ASK YOUR JOBBER

THE DANA MFG. CO., Cincinnati. 10 Warren St., New York.

Write for Dana Food Chopper Catalogue

WHEELBARROWS

ALL KINDS

Railroad, Dock, Concrete Work, Farm and Garden Barrows

SEND US YOUR ENQUIRIES FOR

Cider Mills and Presses
Drilling Machines
Forges
Trucks, for all purposes
Washing Machines
Butter Workers and Shipping Boxes
Blacksmith Machine Tools
Wagon Skeins

THE LONDON FOUNDRY CO., Limited

LONDON, CANADA

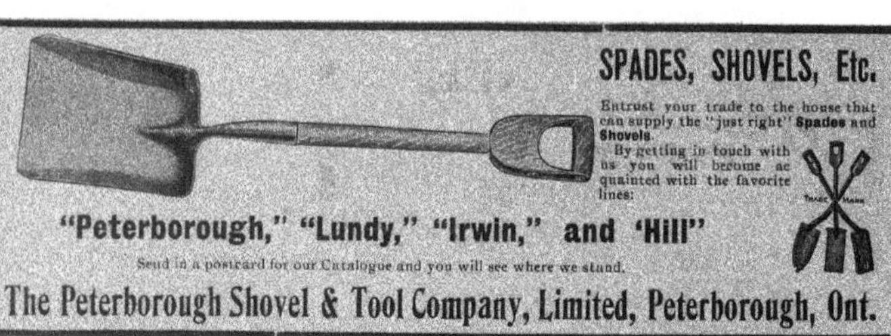

SPADES, SHOVELS, Etc.

Entrust your trade to the house that can supply the "just right" **Spades** and **Shovels**.

By getting in touch with us you will become acquainted with the favorite lines:

"Peterborough," "Lundy," "Irwin," and 'Hill"

Send in a postcard for our Catalogue and you will see where we stand.

The Peterborough Shovel & Tool Company, Limited, Peterborough, Ont.

NORTON GRINDING WHEELS

MADE OF ALUNDUM

are better than others because of their fast cutting qualities and uniformity.

ALUNDUM

s the hardest, sharpest and most durable abrasive known. **ALUNDUM WHEELS** are furnished in any size or shape and in the various grades of hardness.

If you are finishing your work in a lathe we can save you money, and get better results with **ALUNDUM WHEELS**.

CANADIAN FAIRBANKS CO., Limited

MACHINERY AND SUPPLIES

MONTREAL TORONTO WINNIPEG VANCOUVER

PERPETUAL CATALOGUES

are the most essential feature of modern hardware selling---you can get your shelf goods together---your heavy specials together and in any order---it is what facilitates buying and gets money on the flow to you.

UNIVERSAL SYSTEMS LIMITED

8-10 ADELAIDE STREET W.,
TORONTO, ONT.

14 BANK OF OTTAWA BLDG.,
MONTREAL, QUE.

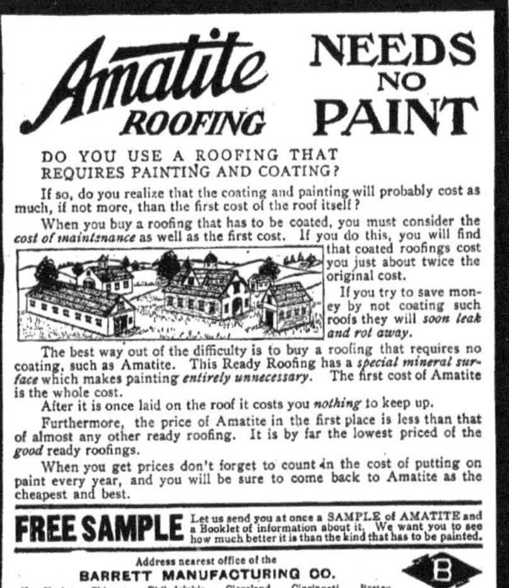

Roofing troubles! End them with Congo.

Leave the old roof there and just lay Congo Roofing over it. It won't cost much; it won't take long.

Everybody who tries Congo likes it the first year, likes it better the second year, is proud of it the third year and gets more enthusiastic after that the longer he waits for it to wear out.

A sample is pretty good evidence. We will send you one free. Ask for it. Address any of the following—

Buchanan-Foster Co., Philadelphia, Pa.; Geo. E. Smith & Co., Halifax; S. Hayward Co., St. John, N.B.; Mechanics Supply Co., Quebec; The Starke Hardware Co., Montreal; H. S. Howland, Sons & Co., Limited, Toronto

QUICK

A want ad in **Hardware and Metal** will reach practically every Hardware Merchant in Canada, from Halifax to Vancouver, within four days.

If you are a Clerk and want to secure another position; if you are a Hardware Dealer and want to secure a clerk—in fact, if you want anything in which the Hardwaremen, Clerks, Travellers or Manufacturers are interested,

Use our Condensed Ad column

RATES

2c. per word................1st insertion
1c. " " ......subsequent insertions

HARDWARE & METAL

Montreal Toronto Winnipeg

MR. DEALER, this is a good time for you to indulge in profitable conversations. Lots of your friends talk of building.

Recommend

"SAUGEEN BRAND" OF PORTLAND CEMENT

and you not only secure lasting friendships but also corner a paying trade. Our dispatch is almost as well known as the quality of "Saugeen Brand."

The Hanover Portland Cement Co., Ltd., Hanover, Ont.

combines **Safety and Efficiency** that cannot be found in any hammer revolver of other makes.

IT can be drawn from the pocket and fired with great rapidity because the hammer is sunk in the frame and CANNOT CATCH ON THE CLOTHING, nor be accidentally discharged, yet can be readily cocked for deliberate aim when desired. These safety features are not found in revolvers of other makes. Shells are automatically ejected—making it easy to reload. Made of the best materials, finely finished, and with proper care will last a lifetime, and gives perfect satisfaction to your customers. Write for catalog.

HARRINGTON & RICHARDSON ARMS CO.

Makers of the celebrated H & R Single Guns

314 Park Ave., Worcester, Mass.

Millions of Tons Annually

There are millions of tons of Window Glass sold annually. Are you getting your share of the profits?

The demand for Window Glass is increasing with startling rapidity. That demand must go on increasing until there is an imperishable brand of glass manufactured. Are you placing yourself in a position to meet the ever-increasing demand? Our catalogue not only shows

Window Glass

plain, and in all the variations of the glass manufacturers' art; but it also contains suggestions for

Fan Lights, Door Lights, Hall Windows, Side-Lights, Etc., made in Copper or German Silver Bars.

This catalogue is a source of money to every dealer who owns it. We have one for you. Can you afford to do without it when your name and address on a postcard will bring it to you?

The Hobbs Manufacturing Co., Limited

LONDON TORONTO WINNIPEG

The John Morrow Screw, Limited

MANUFACTURERS OF

Set and Cap Screws, Special Milled Work, Engine Studs Etc. Cold Punched Nuts of every variety of finish.

INGERSOLL, ONTARIO

CELEBRATED **HEINISCH** SHEARS

Tailors' Shears,
Trimmers' Shears,
Tinners' Snips, etc.

ACKNOWLEDGED THE BEST.

R. HEINISCH'S SONS CO. NEW YORK OFFICE, 155 Chambers St NEWARK, N.J., U.S.A.

HENRY T. SEYMOUR SHEAR CO. NEW YORK — TRADE MARK

SEYMOUR SHEARS

HAVE BEEN THE

Standard for over Half a Century.

"QUALITY UNQUESTIONED."

Each pair of our shears bears the above trade mark.

Latest Catalogue will be sent in exchange for your business card.

Complete Line *TRIMMERS', BANKERS', BARBERS' and TAILORS' SHEARS, Etc., Etc.*

Henry T. Seymour Shear Company

WIEBUSCH & HILGER, Limited, NEW YORK, Sole Agents

COVERT MFG. CO.

TROY, N.Y.

DERBY SNAP

With Plated Rust Proof and Guarded Spring

"The Latest and Best."

For Sale by all Jobbers at Manufacturer's Price

STREET PAVING and SIDEWALKS a SPECIALTY

SILICA BARYTIC STONE CO.

OF ONTARIO, Limited.

Water Proof Floors for Malt Houses, Breweries, Slaughter Houses, Cheese Factories, Cellar, Stable Floors, etc.

Head Office
Ingersoll, Ontario.
Walter Mills, General Manage
Ask for quotations for Septic Tanks.

Persons addressing advertisers will kindly mention having seen this advertisement in Hardware and Metal.

IMPROVED CARPENTER'S TOOLS

Sold by all Hardware Dealer

STANLEY RULE & LEVEL CO.

NEW BRITAIN., Conn., U.S.A.

Four Reasons why INVICTUS FIBRE is the BEST FIBRE PAPER on the Market

1. Invictus Fibre is made from personally selected pulp wood cut on our own limits.
2. This special pulp wood is again rigidly culled on arrival at our mills
3. The pulp from which Invictus Fibre is made is manufactured inlour own pulp mill and transferred at once to the paper mill, avoiding all drawbacks of transportation.
4. Invictus Fibre will increase in strength and not grow hard and brittle as does fibre made from sulphite pulp. This is a quality of the peculiar and selected pulp from which Invictus Fibre is made.

A trial order will convince you that we are correct. See that each package bears the words "Invictus Fibre." Every ream contains 480 sheets

TRADE MARK

This design a guarantee of quality.

CANADA PAPER CO., LIMITED,

"Headquarters for high grade Fibre and Manilla Papers"

Toronto. Montreal. Windsor Mills, P.Q

HOW TO BUILD A GOOD FENCE

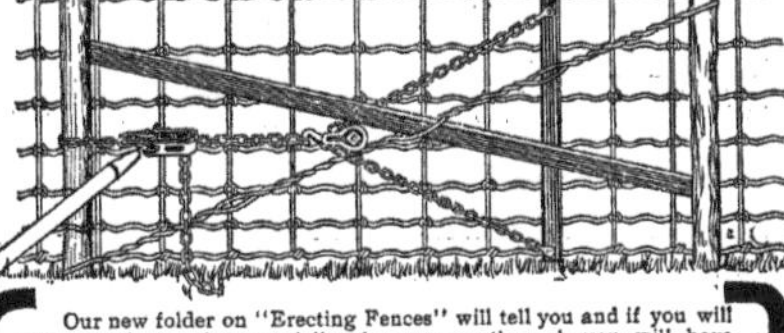

Our new folder on "Erecting Fences" will tell you and if you will follow the instructions carefully when you are through you will have a good job.

It's full of valuable and interesting information on fence building and tells how to erect woven wire fencing in the quickest and most substantial manner.

No farmer, fence man or any one interested in fence construction should fail to write for a copy. It gives all the information required for building fences and we send it

FREE!

In addition there is also a complete and very interesting description of the manufacture of fence wire. Persons who have never had the privilege of visiting a wire mill will find this article of especial interest.

It also has an article quoted from a bulletin issued by the U. S. Department of Agriculture on the manufacture of concrete fence posts, showing how posts made of this most useful and durable material can be manufactured at home. Don't fail to write for a copy today. Ask for our folder called, "**Erecting Fences.**" Remember it's free. Address

THE BANWELL HOXIE WIRE FENCE CO., LTD., Dept. J, Hamilton, Ont.

Want Help?

If you are located in a small city or country town, a want ad. placed in the local papers means that possibly half-a-dozen young men will read it.

If you are located in even a big city, a want ad. in any metropolitan daily costs money. You are paying for a circulation of 100,000, and there is no possibility of more than 200 of the men you want to reach seeing that ad.

Doesn't that look like throwing money away? It does. Because you are paying for 99,800 of a circulation that is absolutely worthless as far as your wants are concerned.

Every Hardware clerk sufficiently interested in his business to pay you as an employee, reads

Hardware and Metal

When you want a man your ad. should appear in our Want columns. It certainly will pay you well to consider this.

Condensed advertisements in Hardware and Metal cost 2c. per word for first insertion, and 1c. a word for subsequent insertions. Box number, 5 cents extra. Write or phone our nearest office.

HARDWARE AND METAL

Toronto Montreal Winnipeg

I WILL TALK

to practically every Hardware merchant in Canada from the Atlantic to the Pacific. I cannot do it all in one day, but during the first twenty-four hours I will deliver your message to every Hardware merchant in Ontario. I travel all day Sunday and on Monday morning there will not be a village within the limits of Halifax in the East, and Brandon in the West, into which I will not have penetrated.

I cannot go any further East, so I now devote all my energies to the West, and so many new towns are springing up here each week that I haven't as much time as I used to have to enjoy the scenery. But I like talking to hardwaremen, clerks, travellers and manufacturers, especially as they are always glad to see me and hear the news I have to tell them. Tuesday noon I am at Calgary, Wednesday noon at Kamloops, and by Thursday morning I reach Vancouver, having been in all the mining towns and all through the fruit districts of British Columbia.

I have been eighteen years on the road and I have a pretty good connection. I never intrude when a man is busy, but just bide my time, because I know men pay far more attention to what you have to say if you catch them when they have a few moments to spare. So I often creep into their pocket when they are going home at night, and when supper is over Mr. Hardwareman usually finds me. He must be glad to see me, because he listens to what I have to say for an hour or more.

I try to always tell the truth, and men put such confidence in what I say that I would feel very sorry to deceive them even inadvertently. Probably some other week I will tell you about the different classes of people I meet. In the meantime if you want a message delivered to HARDWAREMEN, PLUMBERS, CLERKS, MANUFACTURERS or TRAVELLERS—and want it delivered quickly—I'm your man.

THE WANT AD MAN

Condensed Advertisements in Hardware and Metal cost 2c. per word for first insertion, 1c. per word for subsequent insertions. Box number 5c. extra. Send money with advertisement. Write or phone our nearest office

Hardware and Metal

MONTREAL TORONTO WINNIPEG

Persons addressing advertisers will kindly mention having seen this advertisement in Hardware and Metal.

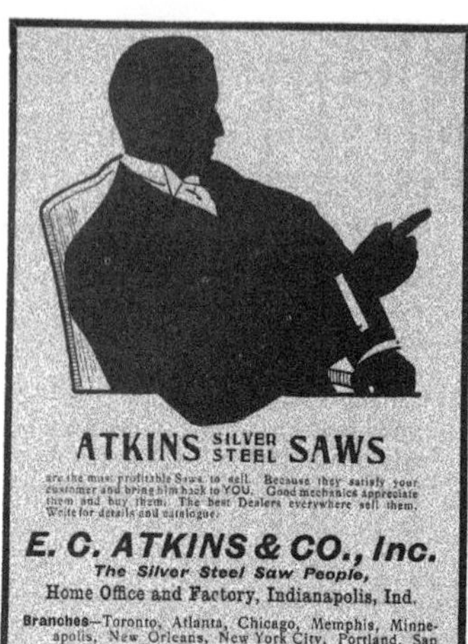
ATKINS SILVER STEEL SAWS
are the most profitable Saws to sell. Because they satisfy your customer and bring him back to YOU. Good mechanics appreciate them and buy them. The best Dealers everywhere sell them. Write for details and catalogue.
E. C. ATKINS & CO., Inc.
The Silver Steel Saw People,
Home Office and Factory, Indianapolis, Ind.
Branches—Toronto, Atlanta, Chicago, Memphis, Minneapolis, New Orleans, New York City, Portland, San Francisco, Seattle.

The Long and the Short of It
"IT" MEANS
SHIELD BRAND
READY ROOFING
THE "LONG"
is the time it wears. It stands the test of heat and frost, wind and rain, and give your customers many years of solid satisfaction.
THE "SHORT"
part is the time needed to put it on. A great deal of time—which means a great deal of money—is saved to the man who uses
SHIELD BRAND
READY ROOFING
LOCKERBY & McCOMB
65 SHANNON STREET
MONTREAL

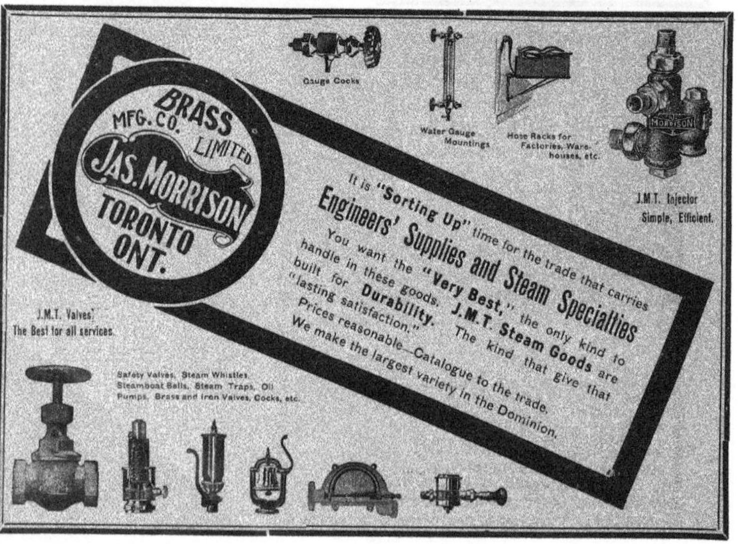
BRASS MFG. CO. LIMITED
JAS. MORRISON
TORONTO ONT.
Gauge Cocks
Water Gauge Mountings
Hose Racks for Factories, Warehouses, etc.
MORRISON
J.M.T. Injector
Simple, Efficient.
It is "Sorting Up" time for the trade that carries
Engineers' Supplies and Steam Specialties
You want the "Very Best," the only kind to handle in these goods. J.M.T. Steam Goods are built for Durability. The kind that give that "lasting satisfaction."
Prices reasonable—Catalogue to the trade.
We make the largest variety in the Dominion.
J.M.T. Valves
The Best for all services.
Safety Valves, Steam Whistles, Steamboat Bells, Steam Traps, Oil Pumps, Brass and Iron Valves, Cocks, etc.

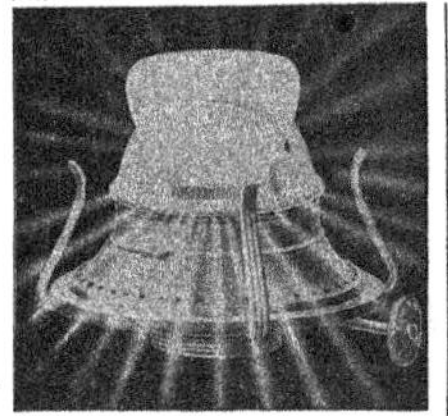

It's the

GLASS CONE

that makes this new burner a great success.

It gives full benefit of flame down to the very wick top.

FOR SALE BY ALL PROMINENT DEALERS.

MANUFACTURED BY

The Ontario Lantern & Lamp Co.
LIMITED
HAMILTON, ONT.

AutoSpray

Compressed-Air Automatic

Rest hand sprayer made. An absolute necessity for every farm and garden. Will repay its cost in one season for potatoes alone. Saves time, labor and material. A boy can do the work. Will run continuously for 6 to 10 minutes.

CAVERS BROS., GALT. ONT.

Sole Agents for the Dominion

THE

SPECIALLY ROLLED STEEL

used in making

Hamilton Rifles

gives the maximum strength with the minimum weight—the condition par excellence in rifle-making.

HAMILTON RIFLES are made in three models—Nos. 15, 19, 23

No. 15 sells at $2.00
No. 19 sells at $2.50
No. 23 sells at $3.50

ORDER FROM THE JOBBER

THE HAMILTON RIFLE CO.

Dept. 71 ***Plymouth, Mich.***

The New Concern is Ready for Business

The McGlashan, Clarke Co., Limited

Manufacturers of

Cutlery, Electro Plate and Solid Nickel-Silver Flatware

Experience, modern ideas, new and latest machinery enable us to assert our motto—

QUALITY, FINISH and DISPATCH

The McGlashan, Clarke Co. Ltd., NIAGARA FALLS, ONT.

Stuffy Rooms and Deadly Air

Can. Feb. 27, 1889

can be avoided by ventilators. Mine is guaranteed to work all the time and give satisfaction. Fix up your chimneys. Write to-day for particulars about the best seller on the market.

CAUTION—Beware of imitations; any with Fan or Wheels in the inside to block the interior not genuine

Montreal, Aug. 1st, 1894.

G. D. Pearson Co., Montreal.

Dear Sir,—The three eighteen-inch Cone Rotary Ventilators you put on the stable and storage have given the best satisfaction and can safely recommend them to anyone wanting a really good ventilator.

JOHN ATKIN, Brewer,
per C. H. Wallace.

Bell Telephone East 2106

GEO. D. PEARSON & CO., Office and Works **17 Pontiac Street, MONTREAL**

Reputation and Profit

The shrewd dealer so orders his business that he clinches both at once. There are several exclusive features about

The Howard Heater

which makes it immensely superior to any other system ever placed on the Canadian market.

The Howard eliminates tin pipes from your cellar, assures pure air and is easily controlled. The Heater burns any kind of fuel, and contracts can be filled on the shortest notice, a guarantee being given with each heater.

All goods are "Union Made," Mr. Howard himself having been a "union man" for some years.

WM. HOWARD - 248 MACDONNELL AVE. PHONE PARK 2633 - **Toronto**

High Quality Galvanized Ware

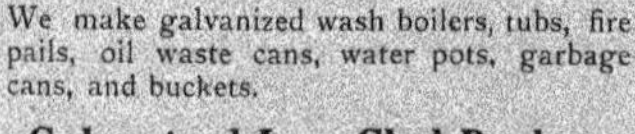

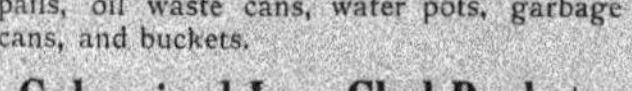

We make galvanized wash boilers, tubs, fire pails, oil waste cans, water pots, garbage cans, and buckets.

Galvanized Iron Clad Buckets
WITH ROLL RIM BOTTOM

Galvanized Garbage Cans
NOS. 1, 2, 3, 4 and 5.

Galvanized Strt. Buckets
12—14—16

Galvanized Flaring Pails
11—12—13—14

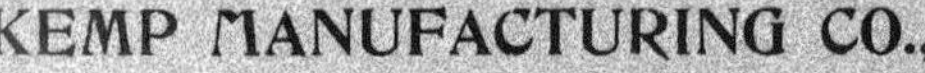

KEMP MANUFACTURING CO.,

TORONTO, CAN.

THE KEMP MFG. AND METAL CO., Ltd.
McDermot Ave. E., Winnipeg, Man.

THE KEMP MFG. CO. OF MONTREAL,
39 St. Antoine St., Montreal, Que.

Parcels Post in Parliament

How the Different Members Expressed Themselves—Postmaster General Threatens to Bring the Matter Up Again.

On the evening of Feb. 15th, Hon. Mr. Lemieux, Postmaster-General, announced in Parliament his intention to withdraw the proposal to introduce a parcels post C.O.D. system. It would have been a most pernicious piece of legislation and would have jeopardized the interests of innumerable merchants throughout the Dominion by giving preference and aid to the large departmental stores in carrying parcels at less than cost price and allowing them to collect cash on delivery. Hon. Mr. Lemieux said he still considered the scheme a good one; and it might be carried into effect later. Another fight may be looked for next session, and it will, therefore, be interesting to review the discussion.

The postal express legislation was first referred to in a speech delivered by the member for South Grey (Mr. Miller) on Nov. 30, 1906, which, as the Postmaster-General said, foreshadowed an arrangement for the arrangement and delivery of post parcels C.O.D. Hon. Mr. Lemieux, referring to it, then said: "It is contemplated that the post office should be the means by which parcels could be despatched and collections of any charges thereon made by the department on some definite scale."

Members Declare Themselves.

The next mention of the proposed legislation was made on Feb. 8, 1907, by Alex. Martin (Queen's, P.E.I.) He said "The country postmasters are the poorest paid officials in the public service, and I congratulate the Postmaster-General that he intends to increase their salaries. But I understand it is the intention of the hon. gentleman to introduce a bill this session which will enable postal parcels to be delivered C.O.D., and to carry them at a less rate than the actual cost. I understand also that the Postmaster-General bases the increase in the salaries of the postmasters on the fact that an extra amount of work will thus be thrown on them. Well, I do not believe that the postal service of this country should be used for the purpose of carrying parcels at unremunerative rates, to the advantage of some large business concerns in the cities but to the detriment of the storekeepers throughout the country. Our country merchants are doing an honest business and bearing their share of the municipal and provincial taxation in their several localities, and I do not think it is any business of the Post Office Department to endeavor to injure their trade by such a proposal."

Following this statement of Mr. Martin's came comments upon the proposal by various members, among them Mr. Logan, J. J. Hughes, Mr. Avery, and Mr. Henderson. Mr. Martin in his statement, deplored the poor salaries paid to postmasters. Mr. Logan qualified this by adding that if the postmasters' salaries were poor, the mail carriers' were even more so, and the legislation proposed would incur more work for these with an incommensurate advance of salaries. J. J. Hughes (King's, P.E.I.) expressed his decided opposition to any such proposal to give "additional facilities to the large order houses in the cities for sending their goods all over the country to the detriment of the retail merchants." Retail merchants do their duty as citizens and are at present subjected to very keen competition from the large order houses without having this condition aggravated by the enforcement of such a measure as the one under consideration. They (the large order houses) send out circulars stating that they can furnish all kinds of goods from their departmental stores and they can do that very cheaply under present conditions. But those who give orders have to pay the money in advance, and I know that in many cases when the goods are received there is disappointment. If, therefore, the facilities for sending out these goods be increased, the evil will be increased. I would impress upon the Postmaster-General as earnestly as I can that it will not be a wise policy, in the public interest generally, to tax a whole people in order that a few large order houses or departmental stores may be enabled to send their goods more cheaply throughout the country and thus make it a delivery wagon of our post office system. Mr. Hughes continued by observing that in the United States the system had been tried and found to be enormously expensive. Would it be wise, then, for a country very much weaker, financially, to assume such burdens?

Quotations From Hansard.

We quote from Hansard of that day (Feb. 8) reports of the discussions by various members:

Mr. McLennan (Inverness): "I desire to say a word or two before the hon. the Postmaster-General (Mr. Lemieux) will have time to carry out the various sugestions made to him for doing away with the surplus. I must add my protest against the C.O.D. system being established in connection with the parcel post service for the reason given by other speakers, that the postal service in Canada would be merely an express agency for the benefit of a few departmental stores in the great cities of Canada, to the great detriment of the local trader in the country towns and villages. I may say that when the people of Canada compare notes as to the success of the administration of the present Postmaster-General the question uppermost with them will not be how large a surplus he has rolled up but how much he has advanced the mail and postal service of the people of Canada. In the United States they have an annual deficit of several million dollars in the mail and postal service and no protest will be found in the newspapers from end to end of that country, so eager are the people of that country to secure an efficient mail service; and I am sure the people of Canada are just as eager to secure an efficient service as they are and would be perfectly content to see the surplus wholly wiped out in order to secure a more efficient mail and postal service than at present prevails in a great many rural districts in Canada. I might cite an instance in the county which I have the honor to represent. Three or four mining towns are springing up in it and trade with the country districts around is springing up in various directions from these mining centres, and when I assure the Postmaster-General, who is comparatively new to the department, that the postal service in Inverness does not cost as much to-day by nearly $5,000 as it did ten years ago, I think he will be inclined not to turn down so many of my applications for post offices as he has done since he took office. I must do like the member for Cumberland (Mr. Logan.)"

Petition From Barrie.

H. Lennox (South Simcoe) said: "May I be permitted to call attention to another matter, and that is the proposal to inaugurate, by departmental regulation, a system of parcel post or C.O.D. delivery? I have had many communications addresed to me on the subject. One of them is as follows: 'Kindly present this petition to the Postmaster-General and oblige the foregoing subscribers.' As a matter of fact, it is rather a letter than a petition which I have received and it expresses, in very concise language, and very clearly, the grievance under which follows:

Barrie, January 19, 1907.

To the Hon. Mr. Lemieux,
Postmaster-General, Ottawa.

Dear Sir,—We understand that you are preparing an amendment to the law providing for the existing parcels post system, the said amendment being intended to provide for the collection on delivery of parcels. Such a system would enable mail order houses to send goods C.O.D. to any part of the country in competition with local retailers. The large departmental stores are already injuring local merchants to a tremendous extent, and any change in the postal laws which would favor them would be a great injustice to the thousands of merchants who have their capital invested in the small towns and cities throughout Canada, and who are aiding by their presence, their energy and their taxes, in the upbuilding of the small communities upon which Canada's future prosperity must depend.

We do not ask for any special privilege and protest against the mail order houses being given any favors. The proposed legislation would tend to reduce the number of retailers thus destroying competition, depreciating the value of real estate in the towns and having a similar effect upon the farming districts, as well as tending to decrease the opportunities for pleasure enjoyed by the farmer's family. The deserted farm districts of the New England States are an object lesson to avoid.

We trust that you will see the fairness of our argument that the legislation would be detrimental to the continued progress and development of this fair Dominion, and we would ask that you reconsider your announced intention of introduceding the proposed amendment.

This letter is signed by John J. Coffey, hardware merchant, and some fifty merchants and business men in Barrie and surrounding district."

Mr. Lennox added: "Were this matter one which the people generally understood, there would be probably a greater number of petitions sent in against the proposed change. Most people are under the impression that what is proposed is the introduction of a bill; but as it is a change that can be effected simply by departmental regulation, I take this opportunity of bringing the subject to the attention of the hon. minister in the hope that he will very seriously consider its effect before he concludes to take the course suggested, one of the effects of which would be to injure the trade of the merchants in the rural parts and detract from the general prosperity of Canada ———"

H. S. Clements (West Kent): "I think that the c.o.d. system would work to the detriment of the rural storekeepers and not to the advantage of the public."

A Novel Suggestion.

H. H. Miller (South Grey): "I do not think that it was the intention of the Postmaster-General, in introducing the c.o.d. system, to assist the department stores at the expense of the stores in the towns and villages of Canada. It has occurred to me that, perhaps, a c.o.d. system might be established, that, instead of bringing grist to the department store mill will help the stores in smaller places to hold their own against the departmental stores. I think there a provision already by which newspapers pay a certain rate within a certain distance, and a certain increased rate at a longer distance from the place of publication. I have wondered whether it would not be possible to bring into effect the proposed parcel post system on somewhat the same plan. Nobody appears to have any objection to it except to the c.o.d. system. Could you not make the c.o.d. part of the system only applicable to mails within a certain distance, say ten miles from the point of departure? That would not bring in the large departmental store, but it would procure a great advantage to people in the rural sections who, by means of the rural telephone system, could procure parcels by mail from the neighboring town or village not more than ten miles away."

A. A. McLean (Queen's County, P. E. I.): "It is not in the interest of the retail merchants of the different portions of Canada that the large departmental stores should receive the benefit of that c.o.d. delivery in any portion of the country. I think that if it were limited, as was suggested by my hon. friend to my left, it might perhaps be of some benefit, but to allow the large departmental stores in Toronto and Montreal to send their goods c.o.d. at very small cost all over Canada is not in the interest of trade."

Postmaster-General Declares Hmiself.

Hon. A. Lemieux (Postmaster-General): "I do not know, I never dreamt of that great rivalry which seems to exist between the deparemental stores and the country stores, but I quite ag with hon. gentlemen on both sides of the House that there is certainly to-day an undercurrent against the establishment of this parcel post system on account of the injury it might cause to the country merchants. For my
I have been biased in favor of cities, having been born and brought up in a city, but it was not my intention to hurt at all the interests or the feelings of country merchants when I thought that I should perhaps give this country the benefit of that system, because my hon. friends on both sides must not forget that the parcel post system is not at all a new system; it exists in Europe, it exists in England; in fact, in England you can send practically anything through His Majesty's mails, and the English mail system is considered to be the most modern and best equipped in the world. There they have the parcel post system and the c.o.d. system. We thought that the Canadian mails could perhaps be used to bring to terms the express companies of this country. We would have afforded additional channels for the public and at the same time we would, perhaps, have put an end to what is practically a monopoly. But, however, I find public opinion, that is to say public opinion framed by the country merchants, against the proposed system, and I may say frankly that I would not run against public opinion in the present case, but I think that if public opinion was educated, public opinion would back up the postoffice department in the proposed establishment of this c.o.d. system."

A. Martin (Queen's County, P.E.I.): "Now, let me tell the Postmaster-General that he never should impose on this country such a scheme as the free parcel post delivery. I am as anxious as anyone could be that people should get cheap goods, but I don't believe that the departmental stores of this country should be able to sell cheap goods at the expense of the postal service."

T. Martin (North Wellington): "I am glad the minister has taken time to give very careful consideration to this question of the c.o.d. delivery. There is certainly a feeling in the country against that proposal. I trust that the matter can be arranged."

Discussion Continued.

The next mention of the c.o.d. delivery proposal was made in the House on Monday (Feb. 11), when. F. L. Schaffner (Souris, Man.) asked:

1. Is the government aware that there is a strong and universal agitation in the province of Manitoba against the extension of the postal system to enable parcels to be carried on the c.o.d. plan?

2. Has the government received any protests from the province of Manitoba against the extension of the postal system to enable parcels to be carried on the c.o.d. plan?

Hon. Rodolphe Lemieux (Postmaster-General):

1. Yes, a strong but not universal agitation has been observed in the province of Manitoba.

2. Yes.

A Dead Issue.

On Feb. 15, Hon. J. G. Haggart asked the Postmaster-General: "Are you entering into a new arrangement for the carrying of parcels, a new system of parcel post?"

Mr. Lemieux: "My own personal view until lately was that we should establish a parcel post system in Canada just as they have in England, but, in order to avoid all useless discussion, I may say at once that I have yielded to the practically unanimous opinion expressed by hon. members on both sides of the House."

Mr. Haggart: "Is it the express companies that have stirred up that opposition?"

Mr. Lemieux. "I am afraid the express companies have stirred up all that trouble and have sent out postal cards all over the country to arouse the country merchants. At all events, although personally I believe that a good parcel post system on a c.o.d. basis as they had in England and all continental countries would be a good system, and that in the lang run a well educated public in this country would sustain the government in such a service, but for the present I am fighting against windmills."

Mr. Herron: "I see an article in the Retail Merchants' Journal of Montreal, with reference to the parcel post system—

Mr. Lemieux: "It is a dead issue for the present."

Retail Hardware Association News

Official news of the Ontario and Western Canada Associations will be published in this department. All correspondence regarding Association matters should be sent to the secretaries. If for publication send to the Editor of Hardware and Metal, Toronto.

THE CONVENTION QUESTION BOX.

A leading feature of the coming convention of the Ontario Retail Hardware Association in Toronto on March 28 and 29 is to be the "Question Box" discussion in which every dealer is invited to express his views. Already over two dozen questions have been suggested for discussion and the success of the feature is assured. There is still time, however, for a few more practical questions on subjects of interest to any branch of the retail hardware trade.

The program which is being mapped out will probably include a meeting of the executive on the morning of March 28. The convention will open in the Labor Temple at 1 p.m. on Thursday and as much business as possible transacted during the afternoon. In the evening the delegates will be tendered a banquet by Lt.-Col. J. B. MacLean, publisher of Hardware and Metal. Two sessions will be held on Good Friday, the election of officers and concluding business being gotten out of the way as soon as possible in order to take up the "Question Box."

Here are the questions so far suggested for discussion:

Should large retailers be able to buy goods as cheaply as jobbers or, in other words, is the quantity system of buying the most satisfactory manufacturers can adopt?

Should there be a difference between cash and credit prices? If so, how much? How can this best be handled?

What percentage of a merchant's turnover should his expenses be? What percentage should be spent in advertising alone?

Should goods be marked in plain figures or ciphers?

Should some novelty be given away each year?

Do small remembrances to special customers at Christmas time do more good than harm?

Assuming that the jobber is willing to pay exchange on local cheques, what is the minimum cheque on which we should ask him to pay exchange?

When remitting money orders should the cost of the order be taken out of the remittance?

What do we know of a satisfactory system of filing catalogues from a two-page price list to the regular hardware catalogue?

To what extent does it pay a hardware dealer to advertise in the local newspaper? What forms of advertising have been found the most profitable? Is it profitable to advertise in programs, etc.

Does it pay a dealer doing a $50,000 business to go behind the counter?

Is it wise to interfere with a clerk or transfer a customer from one clerk to another?

Is a parcel boy at 50c. a day cheaper than one at $1 per day?

Can the shortage of tinsmiths be relieved by employing helpers?

Does it pay to weigh freight and check freight bills?

Do 5 and 10 cent counters pay in small towns?

What is the best method of handling stove repairs and storing parts?

Does it pay to borrow money to discount bills?

What is the cheapest and best method of handling oil?

If a hardware dealer has not been handling stoves, what is the best selling system he can adopt to secure publicity and make profits?

Should the retail hardwareman confine his buying to representatives of jobbing houses?

What is the best method of dealing with goods substituted by jobbers for lines not in stock?

In a retail business of $25,000 to $50,000 per year, what is the best system of bookkeeping?

How can unsaleable lines be disposed of satisfactorily without cutting prices?

Should a portion of a clerk's salary be withheld or does it pay to offer bonuses to clerks to secure faithful service and prevent them leaving on short notice?

Does it pay to employ female clerks to look after the housefurnishing department and lady customers?

Does it pay to handle special lines for holiday trade? If so, what lines pay best?

WESTERN ASSOCIATION AT WORK

J. E. McRobie, secretary of the Western Association, has addressed the following circular letter to the western hardware trade:—

The annual meeting of the above association was held on Tuesday, February 12th. Unfortunately weather conditions prevented the attendance of a great number, who wrote expressing their regret at their inability to attend and wishing the association every success.

The association has in the past year secured a better understanding between jobbers and retailers, you are receiving better protection to-day than ever before. It has been recognized that the association is a body capable of guarding its own interests It remains with you to retain the hold which has been secured; this can be done only by each dealer giving the association his support, don't wait until these benefits has been secured to you, before becoming a member, come in at once and assist your fellow dealers who have been struggling during the past two years in your interest.

Your association together with others, have been instrumental in having the proposed parcel post c.o.d. bill withdrawn, a measure, which had it gone through, would have interfered materially with dealers in all parts of the country.

We are now working in conjunction with the eastern association towards the formation of a Dominion Hardware Association, with a view of impressing on the manufacturers and jobbers the disastrous results to the retail trade if they do not take steps to protect the trade against the catalogue and mail order houses, and the selling to consumers; also to assist the trade generally in overcoming indiscriminate price cutting. This has been secured by the hardware dealers on the other side; benefitting by their success there is no reason to doubt that the same good results can be obtained here.

I again beg to call your attention to the collecting forms which are furnished you free of charge. They have been successful in securing bad accounts by those who have used them; it is your own loss if you fail to take advantage and secure the same results.

We are issuing a delinquent list, a copy of which will be furnished you, such a list in your hands is a valuable protection against dead-beats. I want you to send me a list of all bad accounts on your books before the 1st of April, giving names in full, their occupations and address so they may be included in the list.

The failure of some members to pay their last year's dues has caused a deficit, which is to be regretted, and your executive trusts and expect that every member will pay his dues promptly this year.

J. E. McROBIE, Secy.-Treas.

Statement to Jan. 31, 1907.

Receipts—		
As per July report		$1,823.94
Membership fees		480.00
Notes discounted		150.00
		$2.453.94
Disbursements—		
As per July report		$1,552.93
Salaries	$659.82	
Rent	90.00	
Postage	40.50	
Printing	37.65	
Travelling Exp.	25.40	
Telephone ac.	27.45	
Exchange	3.15	
Light	3.75	
Sundries	5.90	
Furniture	26.80	
		956.42
		$2,509.35
Assets—		
Membership fees due		$1,003.00
Furniture, etc.		130.83
		$1,133.83
Liabilities—		
Rent	$ 67.50	
Sec. salary	459.12	
Rent typewriter	30.00	
Henderson Ptg. Co.	4.55	
A. L. Simpson	19.50	
Imperial Bank (nts.)	150.00	
Overdraft	55.41	
		$786.08
Surplus over liabilities if all dues were paid	347.75	
		$1,133.83

J. E. McROBIE,
Secretary-Treasurer.

HARDWARE AND METAL

Established 1888

The MacLean Publishing Co.
Limited

JOHN BAYNE MACLEAN . President

Publishers of Trade Newspapers which circulate in the Provinces of British Columbia, Alberta, Saskatchewan, Manitoba, Ontario, Quebec, Nova Scotia, New Brunswick, P.E. Island and Newfoundland.

OFFICES:

MONTREAL, ... 232 McGill Street
Telephone Main 1255
TORONTO, ... 10 Front Street East
Telephones Main 2701 and 2702
WINNIPEG, ... 511 Union Bank Building
Telephone 3726
LONDON, ENG . . . . . 88 Fleet Street, E.C.
J. Meredith McKim
Telephone, Central 12960

BRANCHES:

CHICAGO, ILL. . . . . . 1001 Teutonic Bldg.
J. Roland Kay
ST. JOHN, N B. . . . . No. 7 Market Wharf
VANCOUVER, B.C. . . . . Geo. S. B. Perry
PARIS, FRANCE . Agence HaVas, 8 Place de la Bourse
MANCHESTER, ENG. . . . . 92 Market Street
ZURICH, SWITZERLAND . . . . Louis Wolf
Orell Fussli & Co.

Subscription, Canada and United States, $2 00
Great Britain, 8s. 6d., elsewhere . 12s

Published eVery Saturday.

Cable Address { Adscript, London
Adscript, Canada

NEW ADVERTISERS.

Barrett Mfg. Co., New York.
Bird, T. W., & Son, East Walpole, Mass.
Buchanan Foster Co., Philadelphia.
Ontario Metal NoVelty Co., Toronto.
Sanderson, Pearcy Co., Toronto.

THE PARCELS POST QUESTION.

Elsewhere in this issue we reproduce a full report of the discussions in Parliament on the parcels post c.o.d. matter so that readers of Hardware and Metal may know who their friends were in the House of Commons. The discussion is well worth reading and can be referred to in case the matter is brought up again as the Postmaster General suggests.

It is amusing to read the conclusion drawn by Hon. Mr. Haggert and Hon. Mr. Lemieux that the express companies have been behind the agitation conducted against this measure. The honorable gentlemen know better, but if it is necessary the retail merchants of Canada can disillusion them if the "dead issue" is brought to life again.

Another interesting incident was the reading in Parliament of one of the form letters suggested in Hardware and Metal. Hundreds of these were forwarded to Ottawa, but this one was backed up by half a hundred signatures secured by a hardware merchant in Barrie.

The discussion was closed by Mr. Herron, (Alberta) stating that he understood the Postmaster General to say in the House that he could put this system into effect without any enabling legislation, but he had seen an article in the Retail Merchants Journal of Montreal from which he quoted: "We are pleased to say on the authority of the Postmaster General, as we anticipated, that no such legislation can be put into operation unless it is presented before the House in a regular manner, and if any change is proposed in the present system he will give us ample opportunity to be heard." To this Mr. Lemieux replied. "This parcel post system could be introduced under the regulations of the department and without any legislation, but in view of the almost unanimous opposition from members on both sides I would not think of putting it into effect. I can tell my hon. friend that no such scheme will be adopted for the present."

The calm assurance of the editor of the Journal that he could go to sleep until the Postmaster General wakened him is on a par with the action of Dominion Secretary Trowern, of the Retail Merchants' Association of Toronto who when asked if he would co-operate with the Ontario Retail Hardware Association in opposing the parcels post c.o.d. legislation replied on Dec. 29 as follows:

"In this matter of the proposed amendment to the Parcel Post System which is of very great importance to Retail Merchants, we are keeping our eye closely on it, but our Legislative Committee, some of whom are constantly at Ottawa during the entire session, feel that it is always inadvisable for organized bodies to attack any legislation until it is presented in proper form before the House, because it has the effect of weakening our influence by putting us in a position of raising objections before the facts are properly before us. When the proposal comes properly before the House, you can rest assured that we will throw our whole weight against it, if it has any of the features that were incorporated proper form before the House, because in the measure introduced in the United States, or any other features that are injurious to the Retail interest."

If the Retail Hardware Associations had laid down and waited until the "legislation" was brought down in the House as suggested by the Retail Merchants' Association it can be taken for granted that the Parcels Post regulations would be in force before an agitation could have been worked up against it. The time to prevent a snowball from developing into an avalanche is to stop its progress while yet in the snowball stage.

The Retail Hardware Associations have the satisfaction of knowing that they were the only organized national or provincial bodies to oppose the mail order houses in their attempts to secure this extension of the postal regulations. They were aided by some local Boards of Trade, a few local branches of the Retail Merchants Association and by Hardware and Metal, the Dry Goods Review, Printer and Publisher and a few other trade newspapers. No other aid was forthcoming, however, and the credit for the great victory must, therefore, rest largely with the hardware associations, which first took up the fight and carried it to a successful issue.

AN EDITOR WITH BACKBONE.

It isn't always that a daily paper has the courage to stand up against a big advertiser and insist that he must not attempt to interfere with the news and editorial departments of the paper. In fact it is charged, and the charge seems to have substantial foundation, that a considerable part of the daily press in the big Canadian cities dare not publish news items or make editorial comments which are distasteful to the management of the big departmental stores. However that may be there is at least one editor of a big paper who refuses to submit to dictation from an advertiser.

It was some months ago when one day shortly before the evening edition of a certain city daily went to press the telephone rang and a reporter answered it.

"This is Mr. Dash, of Blank's, speaking. We have had an accident in our building and a couple of men are badly hurt. Now we don't want one line of this to appear in the paper."

The reporter was a new man and he didn't want to take any responsibility, especially as he knew that Dash was the manager of the big department store. So he brought the "main guy" to the telephone.

Mr. Dash repeated his demand to the editor-in-chief.

"What's this?" replied the editor. "You know we are running a newspaper don't you?"

"I don't care about that," was the reply. "This is Mr. Dash, of Blank's, speaking, and I insist that I don't want a line in your paper about this accident."

"We are running a newspaper and this accident is news," was the reply. "We are going to publish it of course. If you have any suggestions to make as to the way you want the story told, we'll be glad to hear them, but we're going to publish the story."

"Look here, young man," was the irate reply. "I don't know who you are, but I want you to know that I am Mr. Dash, of Blank's, and if one line of this accident goes in the paper your people will hear about it when they come around for a renewal of their advertising contract."

"I don't care whether it is Mr. Dash, of Blank's, that's speaking, or the Emperor William of Germany," was the reply. "We are publishing a newspaper and we are going to print the story. What's more, when the time comes to renew the advertising you will find that Blank's will need their advertisement in this paper more than this paper will need Blank's advertising."

That was all. The story was published and the advertising was not cancelled.

RIDICULOUS PRICE CUTTING.

A traveler in western Ontario has forwarded the following communication, which speaks for itself:

"A glaring case of ridiculous undercutting was named to me at St. Thomas where it is alleged churns and washing machines are sold for 25c. profit, and in some cases at cost. The same appears at Aylmer. My suggestion to the complainant that retailers and manufacturers would find their best interests mutually served by uniting to put out the evil, was met with the reply that the retailers are not strong or united enough for this work."

This situation is regrettable and should not be. Hardware merchants are not in business for their health. It is folly for dealers to sell goods at such a close margin, as it doesn't allow for carrying on business, and if continued is bound to result in both the retailer and the jobber coming out short.

Manufacturers could undoubtedly assist in maintaining the price, but there is nothing so important in this work as the existence of a good feeling amongst the retailers in each town. In St. Thomas practically all the hardwaremen are in the Retail Hardware Association. Of course, existing laws prevent the arrangement of prices, but if the trade would only get together occasionally and talk over trade matters it would not be long before such ruinous price-cutting would be discontinued.

In one town in northern Ontario where most of the trade are members of the Ontario Association, the best of feeling exists locally and the several dealers act as a unit in different matters. For instance, not long ago a retailer secured an order from a small consumer for some wire rope and forwarded it to a manufacturer for shipment. Later on, however, he received a telephone message to cancel the order, as a traveler representing a jobbing house had called and quoted a lower price. The jobbing house got the business, but its travelers have got the cold shoulder from every hardwareman in the town since that incident happened.

Associations have a value far beyond the mere arrangement of prices. Standing alone and apart there is every encouragement given to shysters who spread reports of the lower prices offered by the other fellow. United, a more friendly feeling exists, profits are more sure and conditions are better all round.

SHAKING HANDS WITH THE ASSIGNEE.

Upon the man who has got his attention fixed on the possibility of sudden and increased profit in mining stocks a single warning has little effect. So again we venture to remark that the likelihood of a man permanently augmenting his bank account by mining stock speculation is much more remote than his chance for making money in the grocery business, providing he gives it as earnest attention as the man gambling in mining stocks gives to the mining situation. A man with limited capital who ventures into the mining market, accompanied only by his shadow, has a fine disregard of danger. It is almost impossible for him to know the actual value of the stock offered him and the man who is trusting to luck to make money for him has the assignee on one side and the bailiff on the other. We promised to reproduce this week from The Financial Post an article showing the comparative earning power of various lines of investments. For lack of space we have had to hold it over a week.

ATTENTION TO DETAILS.

Nowadays, when a dealer's store is so full of a diversity of goods and new lines are being added so frequently, it is undoubtedly hard to bestow the necessary attention on each new article that is offered for sale. Yet a dealer loses by not specializing on each particular class of his new goods. Novelty wears off rapidly and unless new goods are merchandized quickly, they become an encumbrance.

It is advisable, therefore, to plan a retailing campaign for each novelty secured and, as far as possible, have plans cut and dried beforehand. This may seem burdensome and even impossible but, once the habit is cultivated, it will be seen to work wonders.

Take your clerks into your confidence and discuss plans with them. For instance, you expect a consignment of fishing tackle, rods or some other sporting goods novelty. Your old way of retailing them would be to show them in your window as soon as they arrive. By this means you would of course dispose of a good many. But surely there are other methods that will bring you in greater returns. For a day or two you must concentrate your efforts on selling them. Let your whole staff devise means to this end. Then, having cleared out the bulk of the supply, you are ready to bring forward other goods.

GEO. D. WOOD NOT DEAD.

Someone should suffer for the circulation of the report this week of the death of Mr. Geo. D. Wood, the successful Winnipeg wholesale hardware merchant There is no truth in the report and there was no foundation for it

For some months Mr. Wood has not been in good health, his trouble being diagnosed as sciatica. He recently went to Hot Springs, Arkansas, for treatment and was there told that spinal disease was the enemy he had to fight. A fortnight ago he came east to Hamilton, and after about a week's rest went on to Winnipeg with his family.

Mr. Wood's numerous friends will read with pleasure the contradiction of the report of his death and will be glad to know that his ultimate recovery is looked for. We trust that this hope will be realized.

NEW WHOLESALE FOR REGINA.

J. W. Smith, Regina, is organizing a wholesale hardware company in that city to be known as the J. W. Smith Hardware Co. It will be remembered that a few weeks ago the Smith & Fergusson Co. was dissolved and the retail business sold to Peart Bros., proprietors of the Western Hardware Co. At that time Mr. Smith had no intention of starting a wholesale business, but it is hard for an old hardwareman to give up the line of business to which he has devoted most of his life, and Mr. Smith, although young in years, is old in the business. So the announcement has recently been made of the proposed organization of the J. W. Smith Hardware Co. Associated with Mr. Smith as secretary-treasurer of the company will be W. O. Lott, who for so long a time was connected with the Smith & Fergusson Co

At least $250,000 of the total capitalization of $500,000 will be paid up at once, the capital being subscribed in Regina and in Eastern Canada. A large three-storey warehouse 75x110 feet will be erected at once and Mr. Smith hopes that the new firm will be ready for business by the fall. They are already in negotiations with eastern manufacturers with regard to agencies for special lines.

The hardwareman who does not attend the convention of his retail association is neglecting an opportunity to make a better merchant of himself

Make a fair start and keep up the pace until the finish. Waning enthusiasm never helped any merchant over the rough spots.

Markets and Market Notes

(For detailed prices see Current Market Quotations, page 66.)

THE WEEK'S MARKETS IN BRIEF.

MONTREAL.

Linseed Oil—Advanced 5c.

TORONTO.

Linseed Oil—Three cents higher owing to new flax seed duty.

Planished Copper—Now 43 cents.

Copperware — Advanced from 35 to 30 per cent.

Copper Pitts—Changed from 25 to 20 per cent.

Ingot Copper—Half a cent higher.

Montreal Hardware Markets

Office of HARDWARE AND METAL,
232 McGill Street,
Montreal, March 8, 1907.

The hardware trade is now in full swing here, and jobbers are getting ready their orders for spring shipment. There seems to be no immediate prospect of the shortage in screws being relieved, especially in some lines. The heavy demand for bolts and nuts is still maintained, and some sizes of wrought washers are beginning to run short. Jobbers state that deliveries of sash weights are very hard to get, manufacturers being very slow in filling their orders. There is an exceedingly good demand at present for all lines of carpenters', machinists' and lumbering tools.

The growth of the Northwest has greatly helped along the sales of binding twine, manufacturers having booked most of their business already. There is a tendency on the part of some manufacturers to hold out for a flat price of 14 cents on the 650 feet article, but local makers report that they see no reason why business should not be done at 13¾c.

TWINE—There are no changes to report this week, although some manufacturers are trying to hold out for a 14-cent fiprice on the 650-feet article. "Standard," 500 ft., sisal, 9⅝c.; "Tiger," 550 ft., manilla, 9¾c.; "Red Cap," 600 ft., manilla, 12¾c.; "Blue Ribbon," 650 ft., pure manilla, 13¾ to 14c. Car lots ¼ less; thousand pound lots, ⅛c. less.

AXES—The severe winter has evidently lengthened the jobbers' season on axes and handles, as they are still reported to be moving very well.

RIVETS AND BURRS—Sales continue good and prices firm, with no changes.

HAY WIRE—The usual spring business is now being done and prices remain : No. 13, $2.60 ; No. 14, $2.70 ; No. 15, $2.85, f o.b Montreal.

WIRE NAILS—Sales are very good, and prices stiffer than last week, but quotations remain $2.45 per keg f o.b Montreal

CUT NAILS—Moderate demand. No extraordinary features. We still quote $2.30 per keg f.o.b. Montreal.

MACHINE SCREWS—Demand brisk. Prices remain unchanged.

BOLTS AND NUTS — Manufacturers are running their plants up to the limit, but the demand for some sizes still exceeds the supply.

HORSE NAILS—Bad roads are helping along the demand. No change in prices.

HORSE SHOES—The same applies here as to horse nails. Most of the blacksmith shops have all the business they can handle.

BUILDING PAPER—Very active with stiff prices, but no change.

CEMENT AND FIREBRICK— Moving well. No change in prices.

POULTRY NETTING—2x2 mesh, 50 and 5 per cent. discount ; other sizes 50 per cent. off.

GREEN WIRE CLOTH—In 100 ft. rolls, $1.62½ per 100 square feet ; in 50 foot rolls, $1.67½ per 100 square feet.

LAWN MOWERS—Demand good. Prices unchanged ; Low wheel, 12, 14 and 16 in., $2.30 each ; 9-in. wheel by 12 inch, $2.85 each ; 14-inch, $3 each ; 16-inch, $3.12½ each ; high wheel, 12-inch, $4.05 each ; 14-inch, $4.25 each ; 16-inch, $4.50 each ; 18-inch, $4.75 each.

COIL CHAIN—Prices firm and unchanged. Demand good.

SHOT—Last week's advance is well maintained: Shot, packed in 25-lb. bags, ordinary drop AAA to dust, $7.35 per 100lbs.; chilled, No. 1—10, $7.88 per 100 lbs.; brick and seal, $8.40 per 100 lbs.; ball, $8.93 per 100 lbs 5 per cent. off. Bags less than 25 lbs., 1c. per lb. extra, net ; f.o.b. Montreal, Toronto, Hamilton London St. John, Halifax.

AMMUNITION—There seems to be a tendency to higher prices, but no advance has taken place as yet.: Loaded with black powder, 12 and 16 gauge, per 1,000, $15: 10 gauge, per 1,000, $18 : loaded with smokeless powder, 12 and 16 gauge, per 1,000, $20.50; 10 gauge, $23.50.

Toronto Hardware Markets

Office of HARDWARE AND METAL,
10 Front Street East,
Toronto, March 8, 1907.

Trade conditions in hardware are becoming much better. The approach of spring, with its railroad operations and housecleaning activities, is doing much to better these conditions. Wholesalers are getting prepared to ship their spring orders which this year are large owing to increased building operations.

The situation in screws has not yet been relieved. Big demands prevail with little supply to fill them, and with no immediate prospects of repletion.

The demand for bolts and nuts continues strong, prices fair and unchanged.

Although the prices of Sisal have advanced binder twine prices remain firm and unchanged. Good business is being done, a large number of orders being booked and more coming.

All copper manufactures have advanced in price; copper kettles from 35 to 30, copper pitts from 25 to 20.

The demand for wire nails is good, with firm and unchanged prices.

BINDER TWINE — All the prices in raw material have advanced, twine prices remain unchanged. We continue to quote: 500 ft. Sisal, 9½-9⅝; 500 ft. standard, 9½-9⅝; 550 ft. standard manilla, 10⅝; 600 ft. manilla, 12¼; 650 ft. pure manilla, 13¾-14. Car lots ¼c. less, five ton lots ⅛c. less, central delivery.

SCREWS — The wholesale business is good. They experience a strong demand, but stocks are low with little prospects of being repleted.

BOLTS AND NUTS — There is a good demand for these, the supply, however, being inadequate.

CHAIN — Prices remain firm and unchanged, with good demand.

RIVETS AND BURRS — The demand is steady, with unchanged prices.

MECHANICS' TOOLS — These experience a good demand owing to the increased activity in building operations. This summer will witness a big business in tools of every description as an enormous amount of building will be done—stores, houses and railroads.

EXTENSION AND STEP LADDERS —Demand remains strong, with unchanged prices: 11 cents per foot for 3 to 6 feet, and 12 cents per foot for 7 to 10-foot ladders. Waggoner extension ladders, 40 per cent.

POULTRY NETTING — The demand for this continues strong, the retailers' stock being well filled up. Prices remain unchanged: 2-inch mesh, 19 w.g. discount 50 and 10 per cent., all others 50 per cent.

OILED AND ANNEALED WIRE — Demand continues good with firm and unchanged prices: Canadian wire: gauge, 10, $2.46; gauge, 11, $2.52; gauge 12, $2.61 per 100 lbs.

WIRE NAILS — A good demand continues for these. Prices remain firm and unchanged, with a limited stock. We continue to quote: $2.40 base.

HORSESHOES AND HORSENAILS —Prices remain firm and unchanged, with good demand, especially from New Ontario and the newly opened up districts in the west.

BUILDING PAPER — Business is as yet quiet, but should enliven with the renewal of building operations.

WASHERS, WRINGERS AND CARPET SWEEPERS — The business in these is commencing with good demand.

LAWN MOWERS — The demand for these is increasing, retailers getting their stock rapidly filled.

HARVEST AND GARDEN TOOLS—Business in these is becoming active among wholesalers, the merchants getting their spring stock completed.

HIDES, WOOLS AND FURS — The business in these is about all over.

Hides, inspected, cows and steers, No. 1	....	0 10½
" " " " No. 2	....	0 09½
Country hides, flat, per lb., cured	....	0 09
" " " green	....	0 08
Calf skins, No. 1, city	....	0 13
" " No. 1, country	....	0 12
Sheep skins	1 35	1 50
Horse hides, No. 1	3 50	3 75
Rendered tallow, per lb.	....	0 06
Pulled wools, super, per lb.	....	0 25
" " extra	....	0 27
Wool, unwashed fleece	....	0 15
" washed fleece	0 24	0 25

FURS.

	No. 1,	Prime
Raccoon	....	1 50
Mink, dark	5 00	7 00
" pale	2 50	4 50
Fox, red	3 00	4 00
" cross	2 00	20 00
Lynx	5 00	8 00
Bear, black	....	12 00
" cubs and yearlings	....	5 00
Wolf, timber	....	3 25
" prairie	....	1 25
Weasel, white	0 10	0 65
Badger	0 75	1 75
Fisher, dark	6 00	8 00
Skunk, No. 1	....	1 50
Marten	3 50	20 00
Muskrat, fall	....	0 17
" winter	....	0 25
" western	0 12	0 18

Montreal Metal Markets

Office of HARDWARE AND METAL,
232 McGill Street,
Montreal, March 8, 1907

The metal situation this week remains practically unchanged. In fact there seems to be a "waiting attitude" in the air, as if people were standing aside to see how things were going to turn out.

The shortage of copper is still very marked and from present prospects there seems to be no chances of any relief for some time to come. This is not only the case in Canada, but across the border also, a report being received here, that some of the largest American consumers have been compelled to buy small lots of 5,000 or 10,000 lbs. from retail dealers, and pay exceptionally high prices. It is also said that for the first time since the United States became large producers of copper, the imports for February exceeded the exports which were for the first 26 days of the month only 7,379 tons.

Pig iron is weaker in England and deliveries are very much harder to get, but so far, this has made no material difference in the market here.

Tin is very much firmer and although it costs a little more, prices have not yet been raised. Lead also advanced a little during the past 3 or 4 days. This, although not enough to change the prices, shows the standing of the market. Lead is selling in the States for $6.30 spot cash, in minimum car lots of 50,000 lbs. Boiler tubes are unchanged at present, but next week will very likely see new prices.

COPPER.—This metal is till extremely short both in this country and the United States, but no advance has taken place. Prices remain: Ingot copper 26½ to 27c.; sheet copper, base sizes, 34c.

INGOT TIN.—The market is firmer, with a higher tendency, but prices remain: 45½ to 46c. per pound.

ZINC SPELTER.—Still firm at 7.58 to 7.75 per 100 lbs.

PIG LEAD.—Has been advancing for the past three or four days, but not enough to affect prices. We are still quoting $5.50 to $5 60 per 100 lbs.

ANTIMONY.—Remains firm at 27¼ to 28c. per lb.

GALVANIZED IRON.—Market continues strong at the same prices.

PIG IRON.—Weaker in England and deliveries are very much harder to get, but this has not had any effect on the market here, prices remaining as quoted last week: Londonderry, $24.50; Carron, No. 1, $27; Carron special, $25.50; Summerlee No. 3 soft, $23.50.

BOILER TUBES.—The old prices still prevail, but next week will very likely see an advance.

TOOL STEEL.—Prices remain unchanged.

COLD ROLLED SHAFTING.— No changes to report.

OLD MATERIAL.—Prices are: Heavy copper 20c. per pound; light copper 17c. per pound; heavy red brass, 18½c. per pound; heavy yellow brass, 15c. per pound; light brass, 11 to 11½c. per pound; tea lead, 4¼ to 4½c per pound; heavy lead, 4½c. per pound; scrap zinc, 4½c. per pound; No. 1 wrought, $17 per ton; No. 2 wrought, $6 per ton; No. 1 machinery, $18 per ton; stove plate, $14 per ton; old rubber, 11½c. per pound; mixed rags, 1 to 1½c. per pound.

Toronto Metal Markets

Office of HARDWARE AND METAL,
10 Front Street East,
Toronto, March 8, 1907

Ingot Copper continues to be the feature of the market with a half-cent advance noted this week. Stocks in Toronto continue bare with supplies coming very slow. A leading New York copper interest is quoted as follows:—"The average price for copper for the entire twelve months of 1907 will probably be 25 cents a pound, and if it falls short of this figure it will be by an exceedingly small margin; and there is a possibility that an average quotation even higher will result. Indications are that the market will hold steady at the prevailing prices, but should manufacturers sell their product in excess of their copper supplies and be forced to cover their future requirements in competition with each other, a scramble would ensue (such as forced the price of copper from 20 to 25 cents a pound) in which advances to 30 cents, or even higher, would undoubtedly result. It is up to the consumers to use judgment in obtaining supplies if they would not see higher copper prices." It is said that the Calumet and Hecla Company has sold 14,000,000 pounds of their brand for July delivery at 26½ cents and the same quantity for September at 26 cents. With these conditions outside the local market cannot be anything else than strong.

Pig iron is firmer in Great Britain and in spite of much talk the market has not yet broken in the States. A reaction across the line would not have very much effect in the Canadian market, however, for some time. Conditions in the Old Country are more interesting to Canadians and the increasing strength there indicates a steady market for some time to come. The recent weakness was caused, it is understood by the high rentals and dear money, these forcing speculators to unload some of the hundreds of thousands of tons of Middlesborough and other iron stored on warrants. With speculators wanting to sell the market was bound to weaken. In the United States the situation is a complicated one. The producers of steel making pig have in the past depended largely upon the large steel making interests to take a considerable proportion of their output. They have lost these interests as customers; the large steel interests who have not been buying for months will not buy at present high prices; should prices gradually recede, in the next three months, six months, or twelve months, they could hardly count upon large purchases by such interests, unless for the purpose of supporting the market, because these interests are building many additional blast furnaces of their own, to come in at various times. In connection with this it is interesting to note the revival of the story that the Steel Corporation are to establish a furnace at Sandwich, Ont. The Corporation is at the present time building no less than 16 blast furnaces (including the 8 at Gary, Indiana), which will give it more than 2,500,000 tons of pig iron a year.

Gadvanized iron and Canada plates are selling well just now and jobbers' stocks, which have been low have been strengthened by good receipts during the past week. In order to secure early deliveries jobbers asked producers to ship in March goods wanted in June and material wanted in November next is being asked for in June. With iron and tin firm in prices there is no indication of any weakness in sheets and plates.

PIG IRON.—Firmer conditions are reported from Great Britain and locally all prices keep firm. Buying is lively with the railroads still slow on delivery Hamilton, Midland, and Londonderry are off the market, and Radnor is quoted at $33 at furnace. Middlesborough No. 1 is quoted at $24.50, and Summerlee No. 1 at $26 f.o.b. Toronto.

BAR IRON—Bar iron is moving very freely with prices firm. We quote: $2.30 f.o.b., Toronto, with 2 per cent. discount.

TIN—Outside markets have strengthened and higher prices are asked for. Locally prices are unchanged at 45c for ingots The base on charcoal plates continues at $6.50.

SHEETS AND PLATES—There is a large movement of both Canada plates and galvanized iron. Black sheets are also in fair demand. Prices keep firm with no sign of weakness evident.

BRASS—We still quote 30 cents per pound on sheets. At present prices buying is none too active.

COPPER—Another advance has been made and we are quoting 50 cents higher on ingots. Planished sheets have also advanced to 43 cents. We now quote: Ingot copper, $27.00 per 100 lbs. and sheet copper, $35 per 100 lbs.

LEAD—Primary markets are very strong and conditions locally are very active. An advance is looked for in the States, but the duty keeps the two markets entirely apart. We quote: $5.50 for imported pig and $5.75 to $6 for bar lead. For sheet lead $7 is asked.

ZINC SPELTER—Stocks are light but trade is good at former prices which are 7½c per lb. for foreign and 7c per lb. for domestic. Sheet zinc is quoted at 8¼c in casks, and 8½c in part casks. Sheet zinc manufacturers attribute the present demand for zinc to the existing high price of sheet coper.

ANTIMONY—Prices unchanged at 27c per pound.

OLD MATERIAL — Buying prices are: Heavy copper and wire, 20c. per lb.; light copper, 18c per lb.; heavy red brass, 18c per lb.; heavy yellow brass, 15c. per lb.; light brass, 11c. per lb.; tea lead, $4.25 per 100 lbs.; heavy lead, $4.40 per 100 lbs.; scrap zinc, 4½c. per lb.; No. 1 wrought iron, $13.50; No. 2 wrought, 8; machinery cast scrap, $17.50; stove plate, $12; malleable and steel, $8.50, old rubbers, 11c. per pound; country mixed rags, $1 to $1.25 per 100 lbs., according to quality.

COAL — We still quote:— Standard Hocking soft coal, f.o.b. at mines, lump, $1.50, ¾ inch, $1.40; run of mine, $1.30; nut, $1.25; N. & S., $1; P. & S. 85c. Youghiogheny soft coal in cars, bonder at bridges: lump, $2.60; ¾ inch, $2.50; mine run, $2.40; slack, $2.15.

United States Metal Markets

From the Iron Age, March 7, 1907.

There is continued activity in nearly all branches of finished iron and steel, and some of the reports are almost buoyant In the lighter lines, in sheets, hoops and bands, and tin plate, the mills are far behind their orders, and premiums are being paid for early delivery. The orders for iron and steel bars during February were very heavy, and in the wire industry the spring trade is just opening up with much promise.

Not very much has been done lately in steel rails, but a number of good sales for export have been effected. There has been some cancellation of railroad rolling stock, the Gould system having countermanded a total of 4,000 cars and 100 locomotives. The shops, however, are overloaded with work.

For structural material the prospective demand is very large. It is estimated by a leading interest that there is a tonnage of 150,000 to 175,000 tons in the bids to be opened in March, and in the awards expected to be made in that month on bids which have already been opened. It is noted that there is considerable railroad bridge work in this which indicates that in this branch at least the railroad requirements continue heavy.

The steel market is in a somewhat perplexing condition. There is still a marked scarcity, the sale of 5,000 tons by a new plant in the Buffalo district to a large interest having been based upon the necessity on the part of the latter to make long delayed deliveries on old contracts. On the other hand, in the territory east of the Allegheny mountains lower prices are being offered for forward delivery, and it is even reported that a southern plant has appeared as a seller. An interesting fact is that there has been a sale of 6,000 to 7,000 tons of sheet bars for export for the last quarter of this year, with further important negotiations pending.

A good deal of significance appears to be assigned in the trade to the purchase by the Steel Corporation of some lots of Bessemer pig, aggregating between 5,000 and 6,000 tons for delivery during this month. The reported price is $22 at Valley furnace. It is well known that in spite of having all its furnaces except 2 small stacks in operation the Steel Corporation is short of iron, but there have been no purchases for many months, and it is not clear that this latest transaction foreshadows any systematic buying campaign. It must be noted, however, that probably later on the Steel Corporation will be forced to blow out some of its stacks. While winter weather lasts repairs are deferred, but they must be made in the spring.

While there has been a fair amount of buying of foundry iron throughout the country, buyers and sellers are generally somewhat apart, so far as contracting for forward delivery is concerned.

From the Iron Trade Review, March 7, 1907.

Whatever lull has been experienced in the iron trade since the first of the year has been attributed to the alleged retrenchment policies adopted by some of the leading trunk lines. Postponements of rail and car deliveries by a few of the roads have been magnified by those who have been forecasting a slump in iron and steel values, and the reports that have eminated from certain financial centres have been indicative of a general curtailment of purchases and a suspension of shipments of material already ordered. In fact, however, only two roads of prominence have concluded that the retrenchment period has arrived—the Wabash, which has temporarily suspended operations on all new undertakings, and a western road, which in its mad rush to the coast, has been expending an average of $1,000,000 monthly. Despite this curtailment the latter system has already promulgated a list of machine tool requirements which will cost approximately $100,000, and which indicates that its more important extension work along has been affected.

The Grand Trunk is proceeding with the erection of shops at Battle Creek, Mich., which, when completed, will represent an expenditure of $5,000,000, machine tool builders only recently having been asked to figure on the equipment requirements. The Big Four is likewise going ahead with the erection of its new shops at Indianapolis, which will require at least four years to complete.

It is frankly admitted by the rail makers that there has been some readjustment of shipments, due to the postponement of deliveries for a few of the roads, but in no instances have the manufacturers been embarassed, as other roads anxiously awaited an opportunity to step into the breach. On March 1, the rail bookings of the Carnegie Steel Co. for 1907 were within 80,000 tons of the total ordered last year, and the indications are that the phenomenal record of 1906 will be surpassed in the remaining ten months.

Nor is the situation, as reported, regarding the car builders, borne out by the facts. As in rails, there has been a readjustment of shipments, but these deferred deliveries have in no degree resulted in a curtailment of the output of any of the car plants. Most of the roads were anxious to have their orders brought forward, the car shortage still figuring as the all-important factor preventing the satisfactory movement of the freight.

During the first two months of this year, it is estimated that contracts for no less than 40,000 cars were placed, one manufacturer alone having been awarded 22,000 of this total. In further refutation of the curtailment charges it can be reliably stated that the majority of these cars are for delivery this year and only a few of the builders are in position to make reservations before December. Time of delivery rather than price, governed the placing of practically all of these orders, and instead of reducing the accumulation of car orders, the builder reports that his total was increased 5,000 during the months of January and February.

London, Eng., Metal Markets

From Metal Market Report, March 5, 19.7.

PIG IRON.—Cleveland warrants are quoted at 54s. 7½d. and Glasgow standard warrants at 54s., making prices as compared with last week on Cleveland warrants 1½d. higher; on Glasgow warrants 3d. higher.

TIN.—Spot tin opened easy at £191 15s., futures at £191, and after sales of 80 tons of spot, and 70 tons of futures closed quiet at £191, 15s. for spot,

£190, 15s. for futures, making price compared with last week £1 higher on spot and 5s. higher on futures.

COPPER.—Spot copper opened easy at £109 15s. futures £110 17s. 6d, and after sales of 300 tons of spot and 500 tons of futures, closed easy at £109 10s. for spot and £110 12s. 6d. for futures making price compared with last week £2 2s. 6d. higher on spot and £2 2s. 6d. higher on futures.

LEAD.—The market closed at £19 16s. 3d. making price as compared with last week 6s. 3d. higher.

SPELTER.—Best selected closed at £116 making price as compared with last week £1 10s. higher.

Western Ontario News

By F. T. Yealland.

London, March 6, 1907.

London hardware merchants, in common with those of other cities of the province, are enjoying a prosperous trade, and look for a continuance, if not an increase, of that prosperity. The cause of this optimism is found in the fact that London is a fitting-out point for intending Northwest settlers of no mean importance, and as the approaching spring bids fair to see a large migration from this district westward, a lively trade in settlers' supplies is naturally looked for, as it is felt by many that people can buy much cheaper here than they can in the west. Another favorable feature of the situation is that the coming season is expected to show building activity locally even greater than last year's record-breaker, which, of course, is likewise good for the hardware business.

* * *

In metal manufacturing lines all branches are kept busy. The Wortman & Ward Company, London, are adding largely to their premises and plant.

The London Brass Works Co., having found their business to have outgrown their premises, on Carling street, have purchased a site in the East End for a new building.

The Scott Machine Co., one of the latest additions to London's manufacturing concerns, is rapidly building up a large business, and there is every indication that before long they, too, will have to look for more extensive quarters.

The big concerns, such as McClary's, Leonard's, White's and other, continue to more than hold their own, and taken all in all the hardware and metal trades in London are in an eminently satisfactory state.

Nova Scotia News

By James Hickey.

Halifax, N.S., March 5.

It is reported that a steel shipbuilding industry is about to be established at Handsport, N.S., and that work will be commenced this Spring. John Churchill, a well known builder of modern ships is a leading figure in the new industry.

* * *

Rhodes, Curry & Co., of Amherst, have received an order to build forty three-ton store cars for the Dominion Iron and Steel Company. The Wallace Stone Company has also placed an order with the same firm for eleven cars of the same type. Big gangs of men are working in the iron ore mines at Bell Island, Nfld., and large quantities of ore are being taken out and placed in the stock piles for shipment in the Spring to Sydney. The first steamers will arrive at the Island in May, and there will be a larger output of ore than at any time since the mines were opened. Some 900,000 tons were shipped last year, but next season over one million tons will be sent away.

* * *

The following shipments have been made by the Robb Engineering Co.: 1 vertical boiler to the Canadian Portland Cement Co. at Marlbank, Ont.; to James Valentine & Co., Winnipeg; 2 10 h.p. and 2 20 h.p. vertical boilers; to Joseph Hayens, Montreal, 1 90 h.p. tandem compound engine; to the Allis Chalmers Bullock Co., 1 15 h.p. engine, and to Jenkins Bros., Montreal, 1 125 h.p. Robb-Armstrong engine.

* * *

David MacKay, one of Bridgewater's oldest merchants, is going out of business, and the store has been taken over by W. O. Bates and W. B. Freeman, who will carry on an extensive business which will include paints and oils, and ship chandelry.

* * *

James Simmonds & Co., hardware merchants of Dartmouth, N.S., have almost completed the refitting of their store, which was recently damaged by fire. A free use of paint, and the installation of a metallic ceiling makes a great improvement in the appearance of the shop, and in its new dress, it will present a marked change over the place before the fire.

* * *

The Dartmouth Cordage Company contemplate entering on extensive building operations shortly. It is planned to erect about thirty cottages in the north end of Dartmouth for the employes of the company, following out the plan so successfully carried out by the Acadia Sugar Refining Co. at Woodside.

* * *

Dr. Kendall, M.P.P. of Sydney, has introduced a bill in the Nova Scotia Legislature to regulate the prices of electric light and power, and he has also asked for a commission to inquire into the cost of coal, etc., as follows:

1. To enquire into the practice that prevails in promoting the formation of companies to operate coal areas in Nova Scotia.
2. To enqqire into the methods of mining to ascertain whether in important operations large quantities of coal are being lost which should be saved to the province.
3. To enquire into the desirability of Government supervision of the development of coal properties.
4. To enquire into the conditions on which coal is supplied to consumers in Nova Scotia.
5. To enquire into whether a colliery or collieries owned and operated by the Government would be in the interests of the people of Nova Scotia.

The above is considered to have a very important bearing on mining interests and a spirited debate is anticipated when it comes before the House for consideration.

British Columbia News

By C. S. D. Perry.

Vancouver, B.C., March 1, 07.

There has been some feeling of unrest in the mining centres of Kootenay and the Boundary districts, caused by the discussions between coal operators and operatives as to the making of another agreement. The meeting to be held in Calgary is expected to result in an amicable arrangement, and so far as present indications show, it is likely there will not be much change from that now in force. The old agreement is out in less than a month and is to be renewed at the meeting at Calgary. As there have been such fierce fights in the past, in Crow's Nest Pass coal mines, the nervousness felt in the interior mining districts was natural. The Crow's Nest Pass collieries supply the whole smelting industry of the Kootenay and Boundary with coke and coal, and a tie-up of the mines is a serious evil. Just at present, with all metals at record prices, the condition of mining is very prosperous, every producing mine being worked steadily.

One prominent smelterman, in discussing the situation, said that while he had a big bunch of coke and coal on the way, to guard, as far as possible, against accidents, he had not much fear of a strike. He very shrewdly pointed out that the miners had not yet got over the effect of the last prolonged struggle, and they were now so hard up that they were not in shape to stand another strike.

* * *

Nelson Board of Trade has taken a strong stand in urging on the Dominion government the continuance of the tariff on lead and lead products. A unanimous resolution was forwarded from a meeting of the board held a few days ago, protesting against any lowering of the present duties on these lines. Any disturbance of present conditions would destroy the confidence which has hugely been created by the Government's course in paying lead bounties, and serious injury to the lead-mining industry would result. Ministers and local members have been advised of the stand taken.

* * *

Fluctuations in hardware prices are all of upward tendency at the present time. Shortage of stocks, through non-arrival of freight is the universal complaint. Some recent increases in quotations by jobbers are: Wire nails, advanced from $3.10 to $3.50, with hardly any in the market here; putty, advanced from $3.50 to $3.75; tarred building paper, from 80c. to $1; turpentine, from $10.85 to $11 per case; asphalt, California, is now quoted from $26 to $30 per ton in large quantities, with likelihood of this high price continuing. Several local industries are affected by this increase, which is about 60 per cent. over prices prevailing less than a year ago.

MANITOBA HARDWARE AND METAL MARKETS

Market quotations corrected by telegraph up to 12 a.m. Friday, March 8, 1907)

F R. Munro, Western Manager HARDWARE AND METAL, Room 511, Union Bank Building, Winnipeg, Man.

As illustrating the disadvantages under which wholesale houses are laboring this winter it is worthy of note that one of the Winnipeg houses has at the present time some fifty-odd cars of hardware at Fort William waiting for shipment to various western points.

Except for an advance in shot and a change in the discounts on Penberthy injectors there are no important market changes this week. Discounts on Penberthy injectors are now 70 and 10 per cent.

LANTERNS—Cold blast, per dozen, $6.50 ; coppered, $8.50 ; dash, $8.50.

WIRE—Barbed wire, 100 lbs., $3.22½ ; plain galvanized, 6, 7 and 8, $3.70; No. 9 $3.25; No. 10, $3.70; No. 11, $3.80; No. 12, $3.45; No. 13, $3.55; No. 14, $4.00; No. 15, $4.25; No. 16, $4.40; plain twist, $3.45; staples, $3.50; oiled annealed wire, 10, $2.96 ; 11, $3.02 ; 12, $3.10 ; 13, $3.20 ; 14, $3.30 ; 15, $3.45. Annealed wires (unoiled) 10c. less; soft copper wire. base, 36c.; brass spring wire, base, 30c.

HORSESHOES—Iron No. 0 to No. 1, $4.65; No. 2 and larger, $4.40; snowshoes, No. 0 to No. 1, $4.90; No. 2 and larger, $4.65 ; steel, No. 0 to No. 1, $5 ; No. 2 and larger, $4.75.

HORSENAILS — Capewell brand, quotations on application. No. 10, 20c.; No. 9, 22c. ; No. 8, 24c. ; No. 7, 26c. ; No. 6, 28c. ; No. 5, 32c. ; No. 4, 40c., per lb. Discounts: "C" brand, 40, 10 and 7½ per cent., "M" brand and other brands, 55 and 60 per cent. Add 15c. per box.

WIRE NAILS—$2.85 f.o.b. Winnipeg, and $2 40 f o.b. Fort William.

CUT NAILS—Now $2.90 per keg.

PRESSED SPIKES — ¼ x 5 and 6, $4.75 ; 5-6 x 5, 6 and 7, $4.40; ⅜ x 6, 7 and 8, $4.25; 7-16 x 7 and 9, $4.15; ½ x 8, 9, 10 and 12, $4.05; ⅝ x 10 and 12, $3.90. All other lengths 25c. extra net.

SCREWS—Flat head, iron, bright, 85 and 10 p.c.: round head, iron, 80 p.c.; flat head, brass, 75 and 10 p.c.; round head, brass, 70 and 10 p.c.; coach, 70 p.c.

NUTS AND BOLTS — Bolts, carriage, ⅜ or smaller, 60 and 5 ; bolts, carriage, 7-16 and up, 55 ; bolts, machine, ⅜ and under, 55 and 5 ; bolts, machine, 7-16 and over, 55 ; bolts, tire, 65 ; bolt ends, 55 ; sleigh shoe bolts, 65 and 10 ; machine screws, 70 ; plough bolts, 55 ; square nuts, case lots, 3 ; square nuts, small lots, 2½ ; hex nuts, case lots, 3 ; hex nuts, smaller lots, 2½ p.c.

RIVETS — Iron, 60 and 10 p.c.; copper, No. 7, 43c., No. 8, 42½c.; No. 9, 45½c.; copper, No. 10, 47c.; copper, No. 12, 50½c.; assorted, No. 8, 44½c., and No. 10, 48c.

COIL CHAIN—¼ in., $7 ; 5-16, $5.35 ; ⅜, $4.75 ; 7-16, $4.50 ; ½, $4 25 ; 9-16, $4 20 ; ⅝, $4 25 ; ¾, $4.10.

SHOVELS—List has advanced $1 per dozen on all spades, shovels and scoops.

HARVEST TOOLS—60 and 5 p.c.

AXE HANDLES—Turned, s g. hickory doz., $3.15 ; No 1, $1.90, No. 2, $1.60 ; octagon, extra, $2.30 ; No. 1, $1.60.

AXES — Bench axes, 40 ; broad axes, 25 p c. dis. off list ; Royal Oak, per dozen, $6.25; Maple Leaf, $8.25 ; Model, $8.50 ; Black Prince, $7.25 ; Black Diamond, $9.25 ; Standard flint edge, $8.75 ; Copper King, $8.25 ; Columbian, $9.50 ; handled axes, North Star, $7.75 ; Black Prince, $9.25 ; Standard flint edge, $10.75 ; Copper King, $11 per dozen.

CHURNS—45 and 5 per cent.; list as follows ; No. 0, $9 ; No. 1, $9 ; No. 2, $10 ; No. 3, $11 ; No. 4, $13 ; No. 5, $16. ?

AUGER BITS—"Irwin" bits, 47½ per cent., and other lines 70 per cent.

BLOCKS—Steel blocks, 35 p.c.; wood 55 p.c.

FITTINGS—Wrought couplings, 60 ; nipples, 65 and 10 ; T's and elbows, 10; malleable bushings, 50 ; malleable unions, 55 p.c.

HINGES—Light 'T' and strap, 65 p.c.

HOOKS—Brush hooks, heavy, per doz., $8.75 ; grass hooks, $1.70.

STOVE PIPES—6-inch, per 100 feet length, $9 ; 7-inch, $9.75.

TINWARE, ETC. — Pressed, retinned, 70 and 10 ; pressed, plain, 75 and 2½ ; pieced, 30 ; japanned ware, 37½ ; enamelled ware, Famous, 50 ; Imperial, 50 and 10 ; Imperial, one coat, 60 ; Premier, 50 ; Colonial, 50 and 10 ; Royal, 60 ; Victoria, 45 ; white 45 ; Diamond, 50 ; Granite, 60 p.c.

GALVANIZED WARE—Pails, 37½ per cent.; other galvanized lines 30 per cent.

CORDAGE — Rope, sisal, 7-16 and larger, basis, $11.25 ; Manilla, 7-16 and larger, basis, $16.25 ; Lathyarn, $11.25 ; cotton rope, per lb., 21c.

SOLDER—Quoted at 27c. per pound. Block tin is quoted at 45c. per pound.

POWER HORSE CLIPPERS—"1902" power clipper, $12 ; "Twentieth Century" $6. "1904" sheep shearing machines, $13.60.

WRINGERS—Royal Canadian, $35 ; B.B., $39.75, per dozen.

FILES—Arcade, 75 ; Black Diamond, 60 ; Nicholson's, 62½ p.c.

LOCKS—Peterboro and Gurney, 40 p.c.

BUILDING PAPER—Anchor, plain, 66c.; tarred, 69c.; Victoria, plain, 71c.; tarred, 84c.

AMMUNITION, ETC.—Shot has been advanced 50 cents per 100 pounds in sympathy with the advanced lead market. We quote : Cartridges, rim fire, 50 and 5 ; central fire, 33½ p.c.; military, 10 p.c. advance. Loaded shells: 12 guage, black, $16.50 ; chilled, 12 gauge, $17.50; soft, 10 gauge, $19.50; chilled, 10 gauge, $20.50. Shot, ordinary, per 100 lbs., $7.75 ; chilled, $8.10 ; powder, F.F., keg, Hamilton, $4.75 ; F.F.G., Dupont's, $5.

IRON AND STEEL—Bar iron, basis, $2.70. Swedish iron basis, $4.95 ; sleigh shoe steel, $2.75 ; spring steel, $3.25 ; machinery steel, $3.50 ; tool steel, Black Diamond, 100 lbs., $9.50 ; Jessop, $13.

SHEET ZINC—$8.50 for cask lots, and $9 for broken lots.

PIG LEAD—Average price is $6.

AXLE GREASE—"Mica," 1-lb. tins, $11 per gross ; 3-lb. tins, $2.40 per doz.; $4.60 per doz. case ; 10-lb. iron pails, 60c. each ; 15-lb. pails, 80c.; 25-lb. pails, $1.25 ; "Diamond," 1-lb. wooden boxes, $6.40 per gross.

IRON PIPE AND FITTINGS—Black pipe, ¼ inch, $2.65; ⅜, $2.80; ½, $3.50; ¾ $4.40; 1, $6.35; 1¼, $8.65; 1½, $10.40; 2, 13.85; 2½, $19.00; 3, $25.00. Galvanized iron pipe, ⅜ inch, $3.75; ½, $4.35; ¾, $5.65; 1, $8.10; 1¼, $11.00; 1½, $13.25; 2, inch, $17.65. Nipples, 70 and 10 per cent.; unions, couplings, bushings and plugs, 60 per cent.

LEAD PIPE—Market is firm at $7.80.

GALVANIZED IRON — Apollo, 16 gauge, $3.90; 18 and 20, $4.10; 22 and 24, $4.45; 26, $4.40; 28, $4.65, 30 gauge or 10¾ oz., $4.95; Queen's Head, 24, $4.50; 26, $4.65; 28, $5.00.

TIN PLATES—IC charcoal, 20x28, box $9.50 ; IX charcoal, 20x28, $11.50 ; XXI charcoal, 20x28, $13.50.

TERNE PLATES—Quoted at $9.

CANADA PLATES—18x21, 18x24, $3.40; 20 x 28, $3.65; full polished, $4.15.

BLACK SHEETS—10 to 16 gauge, 100 lbs., $3.50 ; 18 to 22, $3.75 ; 24, $3.-90 ; 26, $4 ; 28, $4.10.

LUBRICATING OILS—Prices of lubricating oils have been advanced and quotations are now as follows : 600W. cylinders, 80c., capital cylinders, 50c.; solar red engine, 30c.; Atlantic red engine, 29c.; heavy castor, 28c.; medium castor, 27c.; ready harvester, 28c.; standard hand separator oil, 35c.; standard gas engine oil, 35c. per gallon.

PETROLEUM and GASOLINE—Quoted as follows since the recent advances : Silver Star in brls, per gal., 21c.; Sunlight in brls. per gal., 23c.; per case, $2.30; Eocene in brls, per gal., 24c.; per case, $2.50; Pennoline in brls., per gal., 24c.; Crystal Spray, 23c.; Silver Light, 21c.; Engine gasoline in barrels, gal. 27c., f.o.b. Winnipeg in cases, $2.75.

PAINTS AND OILS — White lead, Pure, $6.50 to $7.50, according to brand ; bladder putty, in bbls., 2½c.; in kegs, 3¾c.; turpentine, barrel lots, Winnipeg, $1.01; Calgary, $1.08; Lethbridge, $1.08; Edmonton, $1.09. Less than barrel lots

5c. per gallon advance. Linseed oil, raw, Winnipeg, 67c.; Calgary, 74c.; Lethbridge, 74c.; Edmonton, 75c.; boiled oil, 3c. per gal. advance on these prices.

WINDOW GLASS — 16-oz. O.G., single, in 50-ft. boxes — 16 to 25 united inches, $2.25; 26 to 40, $3.40; 16-oz. O.G., single, in 100-ft. cases — 16 to 25 united inches, $4; 26 to 40, $4.52; 41 to 50, $4.75; 50 to 60, $5.25; 61 to 70, $5.75. 21-oz. C.S., double, in 100-ft. cases—26 to 40 united inches, $7.35; 41 to 50, $8.40; 51 to 60, $9.45; 61 to 70, $10.50; 71 to 80, $11.55; 81 to 85, $12.60; 86 to 90, $14.75; 91 to 95, $17.30.

HARDWARE TRADE GOSSIP

Ontario.

The Field Hardware Co., of Cobourg, is selling out.

T. N. Healey, of Rice, Lewis Co., is on the sick list.

J. H. Hedley, Markdale, is opening a new tinsmith and repair shop.

J. Wright, has opened up a hardware store on Avenue Road, Toronto.

The Massive Corundum Co., of Niagara Falls, has received a charter.

J. W. Macdonald, of Macdonald & Hay, North Bay, was in Toronto this week.

J. L. Flanders, manufacturer of fences, Ottawa, has assigned to R. W. Grant.

J. W. Cornell, Dundalk, has sold his hardware and harness business to W. J. and R. Russell.

A. Whittaker, manufacturer of stoves and furniture, Windsor, has sold to Whittaker & Teahan.

G. Smith, formerly buyer for the Kennedy Hardware Company, Toronto, has joined the Rice, Lewis Co.'s staff.

The Capewell Horse Nail Co., of Toronto, has juht added $3,000 worth of machinery and shafting to its plant.

C. L. Lightfoot, Vancouver, representative of the Gurney Foundry Co., has been a visitor in Toronto this week.

L. R. Greene, manager of Department of publicity, for the Sherwin-Williams Co., in Cleveland was in Toronto this week.

The Dominion Radiator Co., Toronto, is suing Cartier & Cote, of Sturgeon Falls, to recover $1,098 for goods sold and delivered.

F. L. Kelly & Co., tinware and crockery merchants, Ottawa, have assigned to W. A. Cole, a meeting of the creditors being held this week.

J. A. Bostwick, of the office staff of the Canadian Oil Co., Toronto, left this week for Kansas City, where he will engage in the sewer pipe business.

Mel. Canfield, connected with the hardware store of Watt & Squire, at Brantford, has been appointed manager of J. R. Myers' hardware store at Stratford.

Amongst the hardwaremen in Toronto last week were: R. Shannon, Tara; W. T. Walker, Cobalt; Mr. Phillips, of Phillips Bros., Havelock, and Mr. Carter, Fessenton.

Mr. Rice, of Rice & Barnes, Whitby; Mr. Spright, Acton; E. Brasford, Welland: J. P. Noonan, Mount Forest; R. Campbell, of Simmonds & Campbell, Sault Ste. Marie; and F. Bailes, of John Bailes, Oshawa, were visitors in Toronto this week.

A remarkable cure of lockjaw is reported at Nicholls Hospital, Peterboro. John Haris, foreman polisher in the Peterboro Shovel Factory, received a trival injury some weeks ago. For eighteen days his jaws were tightly locked and his only sustenance was in liquid form, taken by a tube inserted where a tooth had been removed. Circulation of the blood also was stopped for a short time. He resumed his duties at the works last week.

Western Canada.

Porteous Bros., hardware merchants, Carlyle, Sask., have sold to H. Stephens.

Brown & Mitchell, hardware merchants, Brandon, have dissolved, Mr. Brown retiring.

The Marshall-Wells Co. have removed to their big new warehouse on Market St., Winnipeg.

John Dunfield has arrived at Fox-Warren, Man., to take charge of the hardware business carried on by his son.

Geo. D. Wood has returned to Winnipeg from his trip to the Southern States, in search of better health. His friends will be sorry to learn that his recovery has not been so nearly complete as was hoped for.

W. G. Jones, treasurer of the Pease Foundry Co., Toronto, left Winipeg for the Southern States early in the week. Mr. Jones had been confined to his room in the Royal Alexandra for more than a fortnight owing to a severe cold.

Quebec.

H. Dagenais succeeds Sauve & Dagenais, hardware merchants, Montreal.

George Phillip & Co., Montreal, wholesale silverware dealers, are registered.

The Montreal Rag & Metal Co. has dissolved, a new partnership being registered.

OUR LETTER BOX

Correspondence on matters of interest to the hardware trade is solicited Manufacturers, jobbers, retailers and clerks are urged to express their opinions on matters under discussion.

Any questions asked will be promptly answered. Do you want to buy anything want some shelving, a silent salesman, any special line of goods anything in connection with the hardware trade? Ask us We'll supply the necessary information.

Fire Escapes.

P. S. Stewart, hardware merchant, Renfrew, writes: "Can you supply us with names of makers of fire escapes for public buildings?"

Answer: The following are the names of manufacturers of fire escapes: A. N. Cameron, Perth; McGregor & McIntyre, Toronto; Alex. McKay, Montreal; A. E. Whitehouse, Montreal; Shipway Iron Bell and Wire Mfg. Co., Toronto; Toronto Fence and Ornamental Iron Works, Toronto; Wm. Wallace, Three Rivers; National Fence Co., Merrickville.—Editor.

Corks for Bottles.

Taylor Bros. & Co., hardware merchants, Carleton Place, write: "Will you kindly give us the address of a firm in Toronto who make a specialty of corks for bottles?"

Answer: We give the names of two firms in Toronto: Dominion Crown Cork Co., 73 Adelaide Street West, and the Freysing Cork Co., 71 Sumach St. —Editor.

Palmetto Packing.

R. McV., Southampton, Ont., writes: "Please tell me, through your valuable paper, who makes or where a steam packing called "palmetto" can be procured at once."

Answer: The Canadian Fairbanks Co. Front Street West, Toronto, are the sole agents for Canada.—Editor.

Labels for Bolt Rack.

The Alameda Farmers' Elevator and Trading Co., hardware merchants, Alameda, Sask., write: "Can you inform us where we can procure a full set of printed labels for a bolt rack?"

Answer: We do not know where you could procure a set of these and we would suggest that you cut the labels off your boxes and use them.—Editor.

Solution of Discount Problem.

A commercial traveler writes: "I would like to see the solution of the following proposition worked out in your journal, it is very simple, but it is surprising how many different answers are given:

"A buys goods at 70 and 10 off list price.

B buys goods at 60 and 10 off list price.

"How much cheaper does A buy his goods than B or how much dearer does B pay for his goods than A?"

Answer: This is a very simple proposition and solutions forwarded to us will be published in this department.—Editor.

CATALOGUES AND BOOKLETS

When sending catalogues for review, manufacturers would confer a favor by pointing out the new articles that they contain. It would assist the editor in writing the review.

By mentioning HARDWARE and METAL to show that the writer is in the trade, a copy of these catalogues or other printed matter will be sent by the firms whose addresses are given.

New Penberthy Lines.

The Penberthy Injector Co., Windsor, have issued their spring catalogue, which is quite an improvement over previous issues.

The company has added to its already comprehensive lines of manufactures several articles, such as Standard Steam Whistles, Plain Engine Lubricators and Penberthy Steam Lubricators. This last article is a new idea and different in construction from the ordinary steam lubricator on the market. This lubricator has a patent combination filler, gauge glass and vent arm. On ordinary lubricators the gauge glass and sight feed glass is a long tube similar to a water gauge glass. The Co.'s new lubricator is fitted with round tempered glass discs, which are out of the way, thus precluding any possibility of breakage—a decided advantage over the long tube style. The company expects to have this new article on the market this coming summer, having already applied for a patent. This is a device with a bright future, as anything new from the hands of this firm, which is patronized by all the leading dealers in Canada, is sure to be.

All firms desiring copies of the new spring catalogue will do well to write at once.

Valuable Selling Helps.

The American Fork and Hoe Co., Cleveland, O., have issued a 48-page catalogue with an additional booklet entitled "Selling Pointers," containing valuable advice to salesmen and clerks. The catalogue contains cuts of the Co.'s articles manufactured and made attractive by some reproductions of photographs of agricultural scenes. Dealers in gardeners' implements will do well to send for copies, as, while the company does not endeavor to sell to the trade in Canada, the "Selling Pointers" will be appreciated by every dealer and clerk.

Attractive Announcement.

The Standard Sanitary Manufacturing Co., Pittsburg, Pa., have published in the form of a neat booklet the announcement of the opening of their new showrooms in that city.

New Enamelware.

The Ontario Steel Ware Co., Brock avenue, Toronto, have just issued a splendid catalogue of eighty-seven pages. This company is new and if this booklet is at all a criterion of their work, their success is already guaranteed.

As an introduction they state that they have decided that the question of most importance to their patrons is conciseness and brevity of description. Their decision has been admirably borne out in the compilation of the catalogue. The first page contains a reproduction of a letter to the firm from Dr. Allan, of Toronto University, who has analyzed their enamel and testifies to its absolute purity, thus avoiding all danger of poisoning.

The engravings throughout the catalogre are perfect representations both in color and in form of the company's manufactures. The paper throughout is of the best quality, and the printing is perfect. The last two pages contain an index, alphabetically arranged, of the entire contents. Dealers who wish to keep posted regarding the development of the kitchen enamelware in Canada should send for a copy of the book, which can be had if this paper is mentioned and the firms' letter paper used.

Snowflake Axle Grease.

The Snowflake Axle Grease Company, of Fitchburg, Mass., have published a handsome booklet, called "The Ambassador," presenting in a forcible style the superiority of their goods

FOUNDRY AND METAL INDUSTRIES

Young Bros., Almonte, have sold out their foundry business.

The Grand Trunk has placed an order for 3,000 tons of steel for new shops with the American Bridge Co

The German Canadian Smelting and Refining Co. has been incorporated under provincial charter with a capital of $1,000,000.

The T. & No. O. Railway Commission let the contract to the Montreal Locomotive Works for six new locomotives. The aggregate cost will be about $125,000. Delivery must be made by the end of October.

R. Bigley, manufacturer of stoves and furnaces, and Toronto agent for the Buck Stove Co., Brantford, has purchased land on the Weston Road, near Toronto Junction, and purposes building a factory to cost $7,000, employing at first 50 men, afterwards 120 men.

An important transfer of property took place last week, when Frank Oliver manager of the Perrin Plow Co.; B. Knapp, Morrisburg, and R. S. Harder, Aurora, purchased the plant of the Rideau foundry, Smith's Falls, and all stock on hand. The new company will commence operations at once in the manufacture of stoves. The stock on hand amounts to $10,000, including about 500 completed stoves.

OPPORTUNITY FOR INVENTORS.

Fame and fortune waits the lucky individual who can rediscover the combination of metals from which the Egyptians, the Aztecs, and the Incas of South America made their tools and arms. Though each of these nations reached a high state of civilization, none of them ever discovered iron, in spite of the fact that the soil of all three countries was largely impregnated with it. Their substitute for it was a combination of metals which had the temper of steel Despite the greatest efforts, the secret of this composition has baffled scientists, and has become a lost art. The greatest explorer, Humboldt, tried to discover it from an analysis of a chisel found in an ancient Inca silver mine, but all that he could find out was that it appeared to be a combination of a small portion of tin with copper. This combination will not give the hardness of steel, so it is evident that tin and copper could not have been its only component parts. Whatever might have been the nature of the metallic combination, these ancient races were able so to prepare pure copper that it equalled in temper the finest steel produced at the present day by the most scientifically approved process. With their bronze and copper instruments they were able to quarry and shape the hardest stones, such as granite and porphyry, and even cut emeralds and like substances. A rediscovery of the lost art might revolutionize many trades in which steel at present holds the monopoly. If copper could thus be tempered now its advantage over steel would be very great, and it would no doubt be preferred to the latter in numerous industries. It is a curious fact that though this lost secret still baffles modern scientists, it must have been discovered independently by the three races that made use of it so long ago.

ALL INGOT METALS IN STOCK

A. C. LESLIE & CO. Limited
MONTREAL

No. 7 Hand Drill

Cut gearing, change of speed for light or heavy work. Ball bearings, Rack and pinion feed, either hand or automatic. Treadle feed to lower the spindle to the work. A weighted lever raises the spindle instantly as soon as the feed is disengaged. Incomparably the best hand drill in the market, and also a first-class wood boring machine. Fitted for power if desired without extra charge. Weight 375 lbs. Send for circular.

It pays to use the best tools.

A. B. JARDINE & CO.
HESPELER - - - ONTARIO

NICKEL

The Canadian Copper Co.

NICKEL FOR NICKEL STEEL

The Orford Copper Co.

WRITE US FOR PARTICULARS AND PRICES

General Offices
43 EXCHANGE PLACE
NEW YORK CITY

OAKEY'S The original and only Genuine Preparation for Cleaning Cutlery, 6d. and 1s. Canisters

'WELLINGTON' KNIFE POLISH

JOHN OAKEY & SONS, Limited
Manufacturers of
Emery, Black Lead, Emery, Glass and Flint Cloths and Papers, etc.
Wellington Mills, London, England

Agent:
JOHN FORMAN, - 644 Craig Street
MONTREAL.

"MIDLAND"
BRAND.
Foundry Pig Iron.

Made from carefully selected Lake Superior Ores, with Connellsville Coke as Fuel, "Midland" will rival in quality and grading the very best of the imported brands.

Write for Price to Sales Agents

Drummond, McCall & Co.
MONTREAL, QUE.
or to
Canada Iron Furnace Co.
MIDLAND, ONT. Limited

...FULL STOCK...

Salt Glazed Vitrified

Double Strength Culvert Pipe a Specialty.

THE CANADIAN SEWER PIPE CO.
HAMILTON, ONT. TORONTO, ONT
ST. JOHNS, QUE.

GET PRICES FOR
GILBERTSON'S COMET Brand
GALVANIZED FLAT SHEETS

for any purpose where the best is needed. Wide awake jobbers handle this brand. They are well galvanized, true to gauge and especially soft for working up purposes. Every sheet bears the name "Gilbertson." That is a guarantee.

MAKERS:
W. GILBERTSON & CO., Limited.
PONTARDAWE, SOUTH WALES
Agent: Alex. Gibb, 13 St. John St., Montreal.

NOVA SCOTIA STEEL & COAL CO., Limited
NEW GLASGOW, N.S.

Manufacturers of
Ferrona Pig Iron
And SIEMENS-MARTIN
OPEN HEARTH STEEL

Buy the Best.

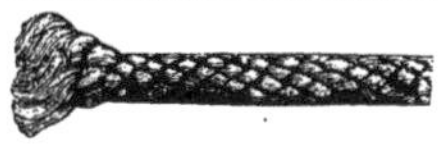

HERCULES
Sash Cord.

Star Brand Cotton Rope
Star Brand Cotton Clothes Lines
Star Brand Cotton Twine

For Sale by all Wholesale Dealers.

TRADE WITH ENGLAND

Every Canadian who wishes to trade successfully with the Old Country should read

"Commercial Intelligence"
(The address is 168 Fleet St., London, England.)

The cost is only 6c. per week. (Annual subscription, including postage, $4.80.) Moreover, regular subscribers are allowed to advertise without charge in the paper. See the rules.

B. A. GRAUTOFF & CO.
8 Fenchurch Buildings, London, E.C.
Import and Export.
Old established house with highest connections in Great Britain and on the Continent of Europe.
Cable Address, "Grautoff London," Liebers Code.

Persons addressing advertisers will kindly mention having seen this advertisement in Hardware and Metal.

BUILDING AND INDUSTRIAL NEWS

HARDWARE AND METAL would be pleased to receive from any authoritative source building and industrial news of any sort, the formation or incorporation of companies, establishment or enlargement of mills, factories, foundries or other works, railway or mining news. All correspondence will be treated as confidential when desired.

A new furniture factory employing 40 or 50 men will be erected at Elmira.

The Ford Fire Shutter Co., Welland, has booked a $2,000 order from a Toronto firm.

A new Y.M.C.A. building will be erected in Woodstock the coming summer to cost $25,000.

Lendon Bros, Leamington, are extending their premises by the erection of a new warehouse.

A permit has been issued for the erection of a new post office at Edmonton to cost $250,000.

Christie, Brown & Co., Toronto, will extend their premises by a six-storey and basement factory.

The Ottawa Car Co. are looking for a suitable site at Hintonburg for the location of a new factory.

D. Gavin, Vancouver, has been granted a permit for the erection of a $16,000 brick warehouse in that city.

A permit has been granted to the Ottawa Wine Vault Co. for the erection of a stone building at a cost of $28,000.

D. Whitney will have a large brick and stone business block erected at Lethbridge, costing from $10,000 to $15,000.

A large extension will be made to the car works planing mill of the Rhodes, Curry & Co., Amherst, N.S., in the spring.

J. W. Gallagher has been awarded the contract for the construction of J. M. Young's large new $10,000 residence in Regina.

The Canadian Northern and Mackenzie & Mann interests are looking for a suitable site for their car works at Hamilton.

A new public library is to be ereceed the coming summer at Woodstock costing $20,000. The architects are Chadwick & Beckett, Toronto.

The A. Darling Co., Toronto, has secured a permit to erect a nine-storey warehouse at the corner of Spadina Ave. and Adelaide St., to cost $100,000.

W. E. Bonner, of the Non-resident Owners' Realty Co., Winnipeg, will erect a new hotel at Strathcona this summer, costing between $10,000 and $15,000.

The congregation of the Central Presbyterian church, Toronto, has about completed arrangements for raising $150,000 required for their new church.

The Scientific Brick Co., Toronto, is applying for a winding-up order in the affairs of the Modern Brick and Stone Co. The latter Co. was capitalized at $200,000.

Chas. Rogers & Sons have been granted a permit to erect a three-storey factory adjacent to their present premises at the corner of Defoe and Tecumseh streets, Toronto.

A factory for the manufacture of automobiles and bicycles will be erected at Port Arthur. McKinnon Bros. have secured a site and building operations will commence at once.

The National Spring and Wire Co., of the State of Michigan, has secured the right to do business in Ontario. A. W. Marquis, of St. Catharines, has been appointed as their attorney for Ontario.

The J. W. Woods Mfg. Co.'s plant, Ottawa, will be moved to Hull, Que. Mr. Woods has granted a $40,000 contract with Holbrook and Sutherland, who will undertake the erection of the new factory.

The McFarlane and Douglas Co. have been incorporated under provincial charter, with a capital of $100,000, to carry on the business of founders and machinists and the manufacture of builders' supplies.

The Toronto Plate Glass Importing Co. has been incorporated with a capital of $250,000, to carry on business as a glass merchant and manufacturer. S. J. Rutherford, E. Hill and J. G. Hutchison are the provisional directors of the company.

The amalgamated business of the Canada Screw Co. and the Ontario Tack Co. may be removed from Hamilton to Welland. Negotiations are on for the change but the company state that no definite decision has yet been arrived at.

A canning and preserving plant will be erected at Pelham, Ont., to cost about $12,000. Following are the officers of the company: George Arnold, Ridgeville, president; A. Armbrust, Pelham, vice-president; G. J. McCormick, Welland, secretary. The capital stock is $40,000.

Seventeen independent millings firms in Manitoba and Saskatchewan have merged under the name of the Canadian Consolidated Flour Mills Co., with a capital of $2,000,000. They have a combined output of 3,000 barrels a day, and an elevator capacity of 1,500,000 bushels.

The Dominion Quarry Co., with headquarters at Montreal, has been incorporated with a capital of $20,000/to carry on a general quarry business throughout Canada. The promoters are: L. C. Rivard, R. Delfausse, J. G. Avard, E. L. Rivard, and A. B. Dufresne, all of Montreal.

The Dominion Government will enlarge its accommodation at Ottawa by the erection of two buildings on the eastern side of the canal. One building will have a floor area of 300,000 square feet, and cost $2,000,000; the other an area of 100,000 square feet, costing $750,000. The land for the site of the new buildings has been secured at a cost of $500,000.

The International Veneer and Lumber Co., of Philadelphia, will establish a factory at Arnprior if given by the town a free site of two acres, a subvention of $2,000, exemption from taxation for a period of ten years, free water during a period of ten years for a 250 horse-power boiler, and assistance from the council in placing $8,000 stock.

The Muirhead Milling Co. will erect a 200-barrel flour mill at Port Arthur to cost $100,000.

TRANSPORTATION IN THE WEST: NEED FOR RECIPROCAL DEMURRAGE

The Winnipeg Jobbers and Shippers Association which has recently been organized under the presidency of Mayor J. H. Ashdown, is actively interesting itself in the transportation requirements of the west. G. E. Carpenter, the manager of the freight department of the association, writes this paper as follows:

Much is being said regarding the severe conditions under which the railway companies are laboring in their endeavors to meet the transportation requirements of the public, and justly too, because one and all agree that the weather conditions west of the Great Lakes have been the worst in years. At the same time the present situation would have been much better if the railway companies had had a clean sheet to work on when the Winter commenced. On the contrary they were very far behind.

Therefore the present congested condition is due to more than the exceptional weather conditions.

If a law had been in effect last Summer providing for reciprocal demurrage it is safe to say that the railway companies would have kept abreast of the requirements of the times by obtaining the necessary locomotives and cars to move freight promptly. Then we would have had no such accumulation when Winter broke upon us.

In several of the States to the south of us reciprocal demurrage is in effect. In the State of Virginia a very fair law is in force, which, if applied here, would do much to overcome the serious delay of freight, and present car shortage.

Under the ruling of the Corporation Commission of the Virginian State Legislature, when a shipper makes verbal or written application to a railroad company for car or cars to be loaded with any kind of freight embraced in the tariff of said company, stating in said application the character of the freight, and its final destination, the railroad company shall furnish same within four days of seven o'clock a.m. the day following such application.

Or, when shipper making such application specifies a future day on which he wishes to make a shipment, giving not less than four days notice thereof, computing from seven o'clock a.m. the day following such application, the company shall furnish such car or cars on the day specified in the application.

For failure to comply with this rule, the company so offending shall forfeit and pay to the shipper applying the sum of $1 per car per day, or fraction of a day's delay after expiration of free time, upon demand in writing, made within thirty days thereafter by the shipper.

When freight in carloads or less is tendered to a railroad company, and correct shipping instructions given, the railroad agent must immediately receive the same for shipment, and issue bills of lading therefor, and whenever such shipments have been so received by any railroad company, they must be carried forward at the rate of not less than fifty miles per day of twenty-four hours, computing from seven o'clock a.m. the day following receipt of shipment, and for failure to receive and transport such shipments within the time prescribed, the railroad company so offending shall forfeit and pay to the shipper the sum of $1 per car per day, or fraction thereof, on all carload freight, and one cent per hundred pounds per day on freight in less than carloads, with minimum charge of five cents for any one package, upon demand in writing by the shipper, or other party whose interest is affected by such delay; provided that in computing the time of freight in transit there shall be allowed twenty-four hours at each point where transferring from one railroad to another, or rehandling of freight is involved.

The period during which the movement of freight is suspended on account of accident, or any cause not within the power of the railroad company to prevent, shall be added to the free time allowed in this rule, and counted as additional free time.

RULE III.

Railroad companies shall within twenty-four hours after arrival of shipments, give notice, by mail or otherwise, to consignee of the arrival of shipments, together with weight and amount of freight charges thereon, and where goods or freight in carload quantities arrive, such notice shall contain also identifying numbers, letters and initials of the car or cars, and, if transferred in transit, the number and initials of the car in which originally shipped. Any railroad company failing to give such notice shall forfeit, and pay to the shipper or other party whose interest is affected, the sum of $1 per car per day, or fraction of a day's delay, on all carload shipments, and one cent per hundred pounds per day or fraction thereof on freight in less than car loads, with minimum charge of five cents for any one package, after the expiration of the said twenty-four hours; provided that not more than one dollar per day be charged for any one consignment not in excess of a carload.

This rule is also applicable to steamboat and steamship lines.

FREIGHT CAR ORDERS.

The reports of freight car construction the past two years illustrates the strenuous efforts the railroads are making to handle the remarkable increase of traffic. Reports from the car construction plants, and estimates of the freight cars built by the railroads in their own shops, show that no less than 460,000 freight cars were constructed in the United States during the past two years. This compares with less than 300,000 cars in the years 1903 and 1904, and 350,000 cars in 1901 and 1902. In Canada there has also been an enormous output, with several new plants established and more under way. It is not, however, so much a scarcity of freight cars, as of locomotives, which has been troubling the railroads and causing havoc amongst shippers generally.

FIRE IN ROOFING FACTORY.

A fire which broke out on the morning March 1, 1907, in the factory of the Alex. McArthur Co., manufacturers of roofing materials, corner of Harbor and Logan Sts., Montreal, caused considerable excitement while it lasted, as well as completely gutting the factory in which it started. The excitement was caused by the fire threatening to spread across Harbor St. and seize a holder of the east end gas plant of the Montreal Light, Heat and Power Co. The heat was so intense that the bricks of the holder were quite hot, half an hour after the fire had been beaten down to a smouldering heap.

The McArthur plant consists of a one story brick building, while behind it are sheds for storage purposes, the boiler house, and another big shed built of galvanized iron.

The fire started in the factory which is used for the manufacture of roofing materials and increased with great rapidity, the large amount of inflammable material in the place aiding its spread. Within a few minutes after the start of the fire, there was an explosion that blew out the front wall of the building. This was presumably caused by a tar tank that stood in the corner of the building catching on fire. Tar on fire, and confined, generates gas.

The fire burned fiercely for an hour, before the firemen got the better of it. The loss on the factory was well covered by insurance.

SOME INDUSTRIAL NONSENSE.

The Canada Tin Plate & Sheet Steel Co., Morrisburg, has been continuing to supply the public through the daily press with a lot of inflated matter regarding their industry. They have made big claims and made demands for preposterous and impossible concessions, but on their side nothing has materialized. They announced a great opening celebration for Jan 26, on which day flags would fly and all of Morrisburg's citizens would be out to celebrate the inception of a new industry. From information we have gathered, no celebration was held, in fact, the existent conditions could never have permitted it. It was impossible to hold a celebration for the following reasons :

1. There was no machinery there at the time the announcement was made of any account.

2. The foundations for the machinery were not completed.

Not one pound of material of any kind was on the ground for the making of tin plate (nor is there at the present time.)

4. Skilled workmen were not there, nor were there any places for them to live in.

5. The power plant for supplying power to the company was not completed (nor is it yet completed).

Such was the condition of things at the time the company had announced for its opening. Conditions have not materially improved up to date. We can hardly conceive of a company making such large promises and falling so far short of ever fulfilling those promises.

"SHEFFIELD" GOODS.

The Canadian agent at Leeds, Eng., says he is informed that a very large amount of German cutlery, razors, etc., is shipped from Germany to Sheffield, and is then sent to Canada as Sheffield goods.

CONDENSED OR "WANT" ADVERTISEMENTS.

Advertisements under this heading 2c. a word first nsertion; 1c. a word each subsequent insertion.

Contractions count as one word, but five figures (as $1,000) are allowed as one word.

Cash remittances to coVer cost **must** accompany all adVertisements. **In no case** can this rule be oVerlooked AdVertisements receiVed without remittance cannot be acknowledged.

Where replies come to our care to be forwarded, five cents must be added to cost to coVer postage, etc.

SITUATIONS VACANT.

BRIGHT, intelligent boy wanted in eVery town and Village in Canada; good pay besides a gift of a watch for good work. Write The MacLean Publishing Company, 10 Front Street E., Toronto. (tf)

WANTED—First-class tinsmith to take charge of shop; must be good furnace man and understand giVing estimates. Address Box 593, HARDWARE AND METAL, Toronto. [13]

WANTED—Hardware Clerk, must be good stock keeper, state age, experience and salary expected. Peart Bros., Hardware Co., Limited, Regina, Saskatchewan. [12]

EXPERIENCED Hardware Clerk Wanted—One to take change, must be good salesman and stock-keeper. state experience and salary expected. Box 594. HARDWARE AND METAL, Toronto. [12]

SITUATION WANTED

WANTED—By a competent and reliable man, a small set of books to keep in spare time, accounts made out, etc. References. Address Bookkeeper, HARDWARE AND METAL, Montreal. [11]

HARDWARE Clerk with four years' experience is open for engagement April 1st in Alberta, Calgary preferred. A1 reference. Box 592, HARDWARE AND METAL, Toronto. [11]

BUSINESS CHANCES.

PLUMBING, Steam, Hot Water, Gasfitting and Tinsmithing business for sale; $15,000 business done last year, and only $3,000 or less inVested; in one of the best towns in Canada. Waterworks and sewage. Owner retiring. Box 587, HARDWARE AND METAL, Toronto. [10]

HARDWARE and Harness Business for sale in good town. Owner retiring for good reason. Stock about $2,500, in excellent shape. This is worth inVestigating, for particulars apply Richard Tew, 23 Scott Street, Toronto.

FOR SALE—Foundry and stoVe works on the Pacific Coast, fully equipped and up-to-date. Apply P.O. Box 760, Victoria, B.C. [10]

HARDWARE Business; including stoVes, tinware and tinsmith tools, in thriving town in West Ontario peninsular, stock about $5,000; building can be leased if desired, dwelling also. Box 583 HARDWARE AND METAL, Toronto. (17)

HARDWARE Business Stock about $12,000.00 is in go d condition, could be reduced to suit purchaser if necessary. Apply George Taylor & Son, Hardware. London, Ont. (11)

FOR SALE—Established, profitable furnace stoVe and tinware business, in liVe growing manufacturing city in old Ontario. Excellent stock. Good stand with long lease; business growing rapidly (last year nearly $20,000). Stock and tools about $5,000. Box 588, HARDWARE AND METAL, Toronto. [10]

FOR SALE.

FOR SALE—FiVe h. p. Campbell Gas Engine, first-class condition, capable of deVeloping eight h. p. Can be seen in operation. 1-8 ft Cornice brake. Apply R & W. Kerr, Limited, Montreal. [10]

CANADIAN Patent for Sale—Metal "Cramp Plate" for locking horizontal wires to upright stays in wire fence construction; a positive lock; easily manufactured; good investment. P. P. Kee, Brandon, Man. [12]

ARTICLES WANTED.

WANTED—1 second hand power punch suitable for range making. Must be in first-class condition with 16 or 18 inch throat. R. & W. Kerr, Limited, Montreal. (10)

BOOKKEEPER WANTED

WANTED—A good bookkeeper; stating experience, reference and salary. Cunningham Hardware Co., Westminster, B.C. [13]

Galt Sure Grip Shingles

make the HANDSOMEST and most DURABLE roof. Are the EASIEST and QUICKEST to either SELL or put on. See them and get acquainted.

GALT ART METAL CO. Ltd.
GALT, ONT

Get in Shape

To giVe your customers intelligent information and trade-winning prices on the most up-to-the-minute and best-selling line of Sheet Metal Build'ng Goods. This is an important matter to the Hardware and Tinsmith Trades. There will be an enormous demand for these goods this year. Bu lders and Owners from your Very locality are now enquiring about this line. Speak up—And let them know "you're in the game."

Get Our Catalogs

and Price Lists. That's the surest way to be "in the game." Your Opposition will haVe to "step fast" if he beats you and Our Line. We are using our eVery endeavor to make Our Line interesting to "the trade." We want good representation in eVery locality. Write us about it to-day.

Western Distributing Agents:

THE McCLARY MFG. CO.
Winnipeg and Calgary

Galt Steel Siding
GALT ART METAL CO. Ltd.
GALT, ONT.

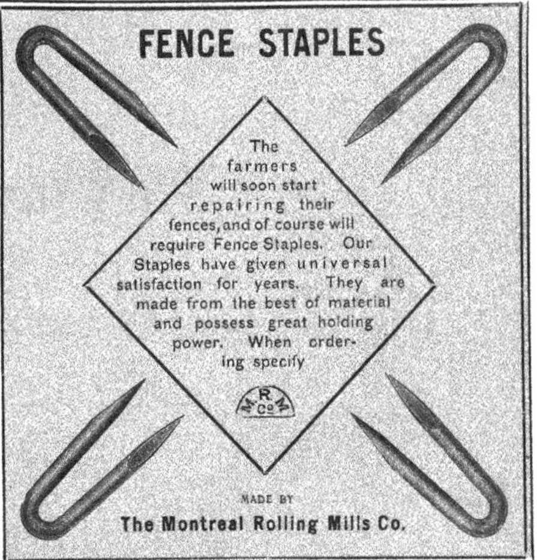
FENCE STAPLES
The farmers will soon start repairing their fences, and of course will require Fence Staples. Our Staples have given universal satisfaction for years. They are made from the best of material and possess great holding power. When ordering specify
M R M Co
MADE BY
The Montreal Rolling Mills Co.

Persons addressing advertisers will kindly mention having seen their advertisement in Hardware and Metal.

Paint, Oil and Brush Trades

SOAP IN PAINTS.

It might not appear at first sight that the introduction of soap into paints would be likely to be attended by satisfactory results, the very nature of soap appearing on the surface to be entirely at variance with those qualities which we look for in a well-finished ground color. While this holds good in a general way, it is undoubtedly the case that soap in small quantities, and with due regard to the special nature of the soap employed, into paints, is occasionally attended with success.

A number of paint grinders have found that, in the grinding of white lead in oil, cases occur where the pigment with difficulty amalgamates with the medium, owing probably to the presence of a small trace of moisture in the white lead. This gives rise to a more or less crumbly paint, a difficulty which is surmounted in the ordinary way by the introduction of a larger proportion of oil. This method, however, is not to be recommended, as it usually happens that when the paint is left for some time it assumes a sloppy condition owing to the excess of oil tending to separate out.

Under the circumstances we have described, says the Oil and Color Trades Journal, the introduction of quite a small proportion of pale neutral soft soap is attended with good results. The proportion ought not to exceed 2 to 4 oz. per cwt. of paint, and the effect appears to be to bind the pigment and the oil more closely together. It must not be understood that we recommend this addition of soft soap as a general practice. On theoretical grounds alone it is not to be defended. The point we are drawing attention to is that under certain adverse conditions the addition of this material enables the paint grinder to get a passable product.

Cases also occur where through a certain stringent conditions in specifications it is necessary to grind pigments with what the practical paint grinder would term an insufficient quantity of oil. Under such conditions the introduction of a small quantity of soap enables the proportion of oil to be cut down considerably. This is not infrequently done in the grinding of zinc paints for certain foreign markets, where the taste or fancy of the purchaser of these materials leads him to demand that the zinc paint shall be of a very firm consistency.

This brings us to the question of the employment of what the chemist terms "soaps" in other branches of paint manufacture. A soap in general chemical language is, we need hardly remind our technical readers, the combination of a metallic base with a fatty acid. Consequently the oleates and linoleates of lead, manganese, and other metals are chemically speaking, just as strictly soaps as the oleates or linoleates of soda and potash. The use of the linoleates of the heavier metals is spreading in the paint trade, and it is probably true that at the present time the uses and application of these materials have not been fully studied.

It is usual in trade literature to describe as soaps the compounds formed by the union of resin acids and metals, these compounds being spoken of as resin soaps. This is not strictly correct, the similarity being more of the nature of analogy than strict chemical resemblance All the same, use of the so-called resin soaps has increased to an enormous extent, and many of the more specialized products of the paint manufacturers' art contain these bodies in varying proportions.

SUBSTITUTES FOR LINSEED OIL.

It is usual to consider that linseed oil is "par excellence" the fluid medium in which to grind all pigments without exception, says the Oil and Color Trades Journal. This, however, is not the case, there being certain pigments which do not give satisfactory results when ground in this medium. Take, for example, Chinese and Prussian blues and their derivatives, Brunswick and Celestial blues. Every practical paint grinder knows that if these articles are ground in linseed oil pure and simple, the paint is liable, in the course of time, to assume an indiarubber-like consistency. With boiled oil the tendency is even more marked, but as the phenomenon has been frequently observed in connection with raw oil it is reasonable to suppose that the action is primarily due to the oil. In such cases the whole or part of the linseed oil should be replaced with an oil whose fatty acids consist largely of oleic acid. Whale oil and olive oil are chiefly used for fine pigments such as Chinese blue and Prussian blue, while for cheaper materials a part of the linseed oil may be replaced by cotton oil. Poppy oil is also used in grinding certain fine colors. It is an interesting fact in this connection to note that on the continent where some of the grades of white lead commonly employed contain a preponderating proportion of hydrate of lead, poppy oil is used instead of linseed oil in the grinding of the paint, it being found that when linseed oil is employed the product is too crisp, and is liable to set up to a hard mass in the kegs, besides tending to go off color.

Hardwaremen *all over this wide Dominion declare that we have about reached the limit in our famous brand. The lustre stays where*

NEW ERA PAINT the most effective "silent salesman" on the market

is used. You can recommend ***NEW ERA*** *all the time and be confident of its lasting qualities. You should have our catalogue. A post card brings it.*

STANDARD PAINT AND VARNISH WORKS CO., LIMITED
WINDSOR - ONTARIO

Floors to be Proud of

are produced by using

JAMIESON'S FLOOR ENAMEL

We show a fine range of colors, all of which give a finish which cannot be equalled by any **Floor Paint** sold.

They wear well, never stick after drying, and do not rub off by washing. In short, they have none of the defects of the common floor paints.

Try a Sample order of these goods and you'll want some more.

R. C. JAMIESON & CO.
LIMITED
16-30 Nazareth St., MONTREAL

A POINTER

Pearcy's Pure Prepared Paints

have been the longest in the market and have won the approval of all dealers. Prompt deliveries a specialty.

Sanderson Pearcy & Co., Ltd.
Toronto, Canada

A. RUDD & CO., St. Helens, Lancashire, England
Manufacturers of highest grade

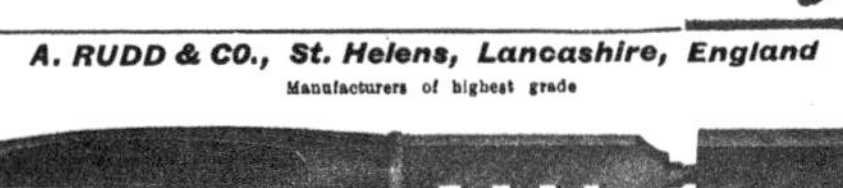

GLAZIERS' DIAMONDS in all patterns. Also Circle Boards, Beam Compasses, Gauge Glass Cutters, Boxwood Squares and Laths, Plate Glass Nippers Atmospheric Soldering Irons, etc., etc., etc.,

Canadian Agent: J. B. GIBSONE, P.O. Box No. 478 MONTREAL

GLAZIERS' DIAMONDS OF PROVED WORTH
Having a Century's Reputation for Reliability.
MADE BY
A. SHAW & SON, London, Eng.
CANADIAN AGENT
GODFREY S. PELTON, - 201 Coristine Building, Montreal

WE SPECIALLY CATER FOR
PAINT GRINDERS.
Our Pure Greens, Permanent Reds, Blues, Lakes, etc. are worth ***attention.*** All qualities, with ***prices that talk!***
Write to-day for samples to
WM. O. GREENWAY, 13 St. John St., MONTREAL

W. J. COWAN & SONS
Expert Color Manufacturers,
BETHNAL GREEN, LONDON, E. (England)

PAINT AND OIL MARKETS

MONTREAL.

Office of HARDWARE AND METAL,
232 McGill Street,
Montreal, March 8, 1907

Trade is beginnig to pick up wonderfully, due no doubt to the warm weather which has at last made its appearance. There are no changes in prices to report, all staples remaining practically unchanged.

The high prices on white lead seem to act as a tonic to the sales of ready mixed paints, and it is reported that while in some circles the demand for ground white lead is decidedly sluggish the heavy demand for ready mixed paints stands out in bold relief.

The congestion on the railways seems to be relieved, and supplies of crude material from all parts are coming forward and shipments of the finished product will shortly go off with a quick step.

Owing to rumors of a duty on flax seed, a sharp advance has taken place on linseed oil, both the raw and boiled article jumping up 5 cents per gallon.

LINSEED OIL — Very firm, with an upward tendency, but no change as yet from the old prices, which are: Raw, 1 to 4 barrels, 60c.; 5 to 9 barrels, 59c.; boiled. 1 to 4 barrels, 63c.; 5 to 9 barrels. 62c.

TURPENTINE — A moderate business is being done at the following prices, which are very firm: Single barrel, 98c. per gallon; for smaller quantities than barrels, 5c extra per gal. is charged. Standard gallon is 8.40 lbs., f.o.b. point of shipment. net 30 days.

GROUND WHITE LEAD—There is a fair inquiry, but business is slightly hampered by the high prices: Best brands, Government standard, $7.25 to $7.50; No. 1, $6.90 to $7.15; No. 2, $6.55 to $6.90; No. 3, $6.30 to $6.55; all f.o.b. Montreal.

DRY WHITE ZINC — Very quiet. Prices remain unchanged and firm. V. M., Red Seal, 7½c. to 8c.; Red Seal, 7c. to 8c.; French V. M., 6c. to 7c.; Lehigh, 5c. to 6c.

WHITE ZINC — Demand spasmodic and the following prices firm: Pure, 8½c. to 9½c.; No. 1, 7c. to 8c.; No. 2, 5¾c. to 6¾c.

PUTTY — Owing to the excessively severe climatic conditions experienced lately, the demand is easy. Prices are: Pure linseed oil, $1.75 to $1.85; bulk in bbls., $1.50; in 25-lb. irons, $1.80; in tins, $1.90; bladder putty in bbls., $1.75.

ORANGE MINERAL — Is becoming scarce, but not enough so as yet to warrant any change in quotations, which remain: Casks, 8c.; 100-lb. kegs, $8½c.

RED LEAD — Stocks are light, and higher prices are likely to prevail before the opening of navigation, but for the present they remain: Genuine red lead, in casks, $6; in 100-lb. kegs, $6.25; in less quantities at the rate of $7 per 100 lbs.; No. 1 red lead, casks, $5.75; kegs, $6, and smaller quantities, $6.75.

PARIS GREEN—No special features, with the exception that figures are well maintained: In barrels, about 600 lbs., 25½c per lb.; in arsenic kegs, 250 lbs., 25¾c; in 50-lb. drums, 26¼c; in 25-lb. drums, 26¾c; in 1-lb. packets, 100 lbs. in case, 27¼c; in 1-lb. packets, 50 lbs. in case, 27¾c; in ½-lb. packets, 100 lbs. in case, 29¼c; in 1-lb. tins, 28¼c f.o.b. Montreal. Terms, three months net or 2 per cent. 30 days.

SHELLAC GUMS — Prices keep steady. Some fair orders have lately been placed. We still quote: Bleached, in bars or ground, 46c. per lb., f.o.b. Eastern Canadian points; bone dry. 57c. per lb., f.o.b. Eastern Canadian points; T. N. orange, etc., 48c. per lb., f.o.b. New York.

SHELLAC VARNISH—Responds to brisk trade conditions, and is actively inquired for: Pure white $2.90 to $2.95; pure orange, $2.70 to $2.75; No. 1 orange, $2.50 to $2.55.

PETROLEUM — American prime white coal, 15½c per gallon; American water, 17c per gallon; Pratt's Astral, 19½c per gallon.

WINDOW GLASS—First break, 50 feet, $1.85; second break, 50 feet, $1.95: first break, 100 feet, $3.20; second break, 100 feet, $3.40; third break, 100 feet, $3.95; fourth break, 100 feet. $4.15; fifth break, 100 feet, $4.40; sixth break, 100 feet, $4.95. Diamond Star: First break, 50 feet, $2.30; second break. 50 feet, $2.50; first break, 100 feet. $4.40; second break, $4.80; third break. 100 feet, $5.75; fourth break, 100 feet, $6.50; fifth break, 100 feet, $7.50; sixth break, 100 feet, $7.50; seventh break, 100 feet, $8; eight break, 100 feet, $9. Double Diamond: First break, 50 feet, $3.45; second break, 50 feet, $3.75; first break, 100 feet, $6.75; second break, 100 feet, $7.25; third break, 100 feet, $8.75; fourth break, 100 feet, $10; fifth break, 100 feet, $11.50; sixth break, 100 feet, $12.50; seventh break, 100 feet, $14; eighth break, 100 feet, $16.50: ninth break, 100 feet, $18; tenth break, 100 feet, $20; eleventh break, 100 feet. $24; twelfth break, 100 feet, $28.50. Discount on Diamond Star, 20 per cent.; on Double Diamond, 40 per cent.

TORONTO.

Office of HARDWARE AND METAL,
10 Front Street East,
Toronto March 8, 1907.

Trade conditions in paints and oils are practically unchanged, except that prices are strengthening. One important change, however, has been made in

the price of linseed oil owing to a raise in the duty of flax seed.

Hon. Mr. Fielding on Wednesday in the House of Commons announced that the tariff brought down in November had made flaxseed free in all tariffs. The duty on linseed oil, lead ground in oil, and other articles which had flaxseed as their basis, had been held, at the request of several delegations, which had urged that the production of linseed oil was an important one, and that a duty on flaxseed was desirable. The Government had decided that the request was reasonable, and that instead of being admitted free, a duty would be placed on it of 7½c. a bushel, British preference; 10 cents intermediate and general tariff.

Following this advance in the duty of flaxseed has come the advance in the price of linseed oil of 3c. to rule at present, with a possibility of being advanced perhaps to 5c.

Turpentine prices are unsteady, the range of prices being lengthened.

Water white petroleum prices have advanced ½c.

The orders for all lines of paints and oils have been booked, and little change in conditions will come until the early part of the Summer when orders are being repeated.

WHITE LEAD — Pure white, $7.40; No. 1, $6.65; No. 2, $6.25; No. 3, $5.90; No. 4, $5.65 in packages of 25 lbs. and upwards; ½c. per lb. will be charged extra for 12½ lbs. packages; genuine dry white lead in casks, $7.

RED LEAD.—Genuine in casks of 500 lbs., $6; ditto, in kegs of 100 lbs., $6.50; No. 1 in casks of 500 lbs., $5.75; ditto, in kegs of 100 lbs., $6.25.

DRY WHITE ZINC.—In casks, 7½c.; in 100 lbs., 8c., No. 1, in casks, 6½c., in 100 lbs., 7c. Ground in oil—In 25 lb. irons, 8c.; in 12½ lbs., 8½c.

SHELLAC VARNISH.—Pure orange in barrels, $2.70; white, $2.82½ per barrel; No. 1 (orange) $2.50; gum shellac, bone dry, 63c. Toronto. T. N. (orange) 51c. net Toronto.

LINSEED OIL.—There is a general advance of 3c. owing to change of tariff. We quote: Raw, 1 to 3 barrels, 62c. to 65c.; 4 to 7 barrels, 61c. to 64c.; 8 and over, 60c.; add 3c. to this price for boiled oil, f.o.b. Toronto, Hamilton, London and Guelph, net 30 days.

TURPENTINE.—Prices are variable this week; trade conditions unchanged, with a possible advance owing to scarcity of labor and the drought in Southern States. We quote: Single barrels at 97c. to $1 per gal., f.o.b. point of shipment, net 30 days; less than barrels, $1.02 to $1.05 per gallon.

PARIS GREEN.—The trade is steady most of the orders, however, being booked up. We continue to quote: English and Canadian at 27½c. per lb.

PETROLEUM.—Trade is steady with moderate demands. We quote: Prime white, 13c.; water white, 14½c.; Pratt's astral, 18c.

For additional quotations see current quotations at back of paper.

DANGEROUS SUBSTITUTES.

Considerable consternation has been excited among those engaged in the paint and turpentine trade recently by the frequency with which substitutes for turpentine are being offered as the price of true turpentine advances. Speaking to Hardware and Metal this week, John R. Walsh dealer in spirits, turpentine and naval stores, Savannah, Ga., who is in Toronto for a short time, said, "It's a dangerous business. Only insignificant firms could safely perpetrate such a thing. Our business is conducted under a rigid government inspection and we can't do it."

The cause for the offering of substitutes by certain firms is the upward tendency of turpentine prices owing to the scarcity of labor in the southern States and the very dry weather being experienced there. So dry is it that Mr. Walsh says forest fires have been raging in Georgia. The price for futures, Mr. Walsh states, is 65c. at Savannah, with a tendency still higher.

Regarding the increased substitution, the National Fire Protection Association of the United States has been and is sending out warning to its members of the dangers of it. A chemical analysis of some of these substitutes which are sold under such various and misleading names as special spirits, paint thinner, turpoline, and mineral turpentine, show that they are far from being pure, and in some cases extremely dangerous to handle. It is found that they are usually mixtures of true turpentine and naphtha.

The trade in Canada will do well to go slow in the matter of buying these substitutes. We understand that the trade is being asked to buy turpentine made in Cleveland, the absurbity of which is self-evident, as the pine belt is in Georgia and not in Ohio.

WALL PAPER.

Last month wall paper manufacturers completed their orders, and are now busy on next year's new set of samples. Warehouses are filled and shipping will be general during the next few months. What will be shown next year is the question holding the attention of the trade. A still further improvement is promised in all lines. In these days of hard competition goods must possess the necessary qualities to sell on their merits; they must possess artistic value and quality to stand against, and compare favorably with, the best. The best goods sell, and the retailer will go far to find the best values. Therefore, it is incumbent upon each manufacturer to produce the best line of samples that he has ever done, for to reduce his range, or to repeat too frequently the stereotyped patterns, would mean ruin. On novelty and freshness of patterns depends the result of the season.

This sampling season will bring out some handsome effects in English designs. This means rich floral pattern of artistic qualities, open and highly colored. Ingrain has to a large extent been displaced by heavy figure and floral effects and also the landscape borders and woodland side-walls. These have taken well with the trade.

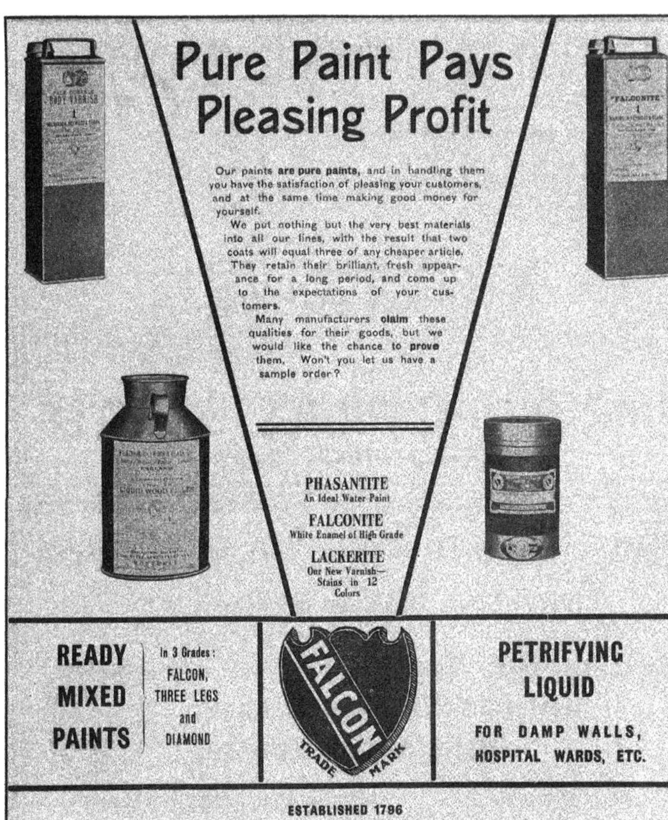

Pure Paint Pays
Pleasing Profit
Our paints are pure paints, and in handling them you have the satisfaction of pleasing your customers, and at the same time making good money for yourself.
We put nothing but the very best materials into all our lines, with the result that two coats will equal three of any cheaper article. They retain their brilliant, fresh appearance for a long period, and come up to the expectations of your customers.
Many manufacturers claim these qualities for their goods, but we would like the chance to prove them. Won't you let us have a sample order?
PHASANTITE
An Ideal Water Paint
FALCONITE
White Enamel of High Grade
LACKERITE
Our New Varnish-Stains in 12 Colors
READY MIXED PAINTS
In 3 Grades: FALCON, THREE LEGS and DIAMOND
FALCON
TRADE MARK
PETRIFYING LIQUID
FOR DAMP WALLS, HOSPITAL WARDS, ETC.
ESTABLISHED 1796
WILKINSON, HEYWOOD & CLARK
LIMITED
300 Notre Dame Street West, Montreal
Branches and Agencies all over the world
FRANCIS C. REYNOLDS
Canadian Manager

Plumbing and Steamfitting

THE PLUMBING OF A COTTAGE

By J. A. F. Cardiff in the National Builder, Chicago

Article XVI.

Water Supply.

Continuing the consideration of the water supply of a cottage, we illustrate a plumbing section showing a system of piping for the supply and distribution of cold water, taking the supply from the city water main. The illustration is intended to serve as an example for the installation of the piping of a modern cottage having the usual amount of plumbing fixtures, and also includes the use of a pressure regulator and a water meter.

The pipe is galvanized wrought iron, except the service connection. Lead pipe for water supply is no longer used.

Taking up the work in detail, we indicate the city water main in the street at "A." The connection to this main is made by an employe of the city or company that owns the water system. This is done by drilling the street main and screwing in a brass corporation cock, "B," to which the service pipe, "C," is connected. The drilling of the water main and the insertion of the corporation cock is accomplished without shutting off the water in the mains, by means of a device especially made for the purpose.

The size of the corporation cock which may be inserted in a street main is usually determined by the municipality water department or the water company. In New York city the largest tap permitted, except when water meters are installed, is ⅝-inch in diameter, and in other cities seldom exceeds ¾-inch in diameter. The diameter of the service pipe, however, should not be governed by the size of the tap, but should be determined solely by the amount of water required for the building.

The service connection, "C," should be of lead pipe and bent in a wavy form, as shown, so as to be flexible and thereby give and take with any settlement of the building or the street main. The plumber should furnish the mason with an iron pipe sleeve to build in the wall at the point where the service connection will pass through it. This is much preferable to leaving a hole in the wall or cutting one when the piping is being installed.

At "D" is shown the valve which shuts off the entire piping system. At "E" is provided an air chamber to absorb all shocks that occur in the pipes, and protect the meter, "F," from them. Water meters are used to measure the quantity of water which passes through the service pipe to the building. They are required in many cities where the water tax is determined by the amount of water consumed.

At "G" is shown a branch taken off the main house supply to feed a sill cock located on the outside surface of the building and used for cleaning sidewalks, sprinkling lawns, etc. The branch is fitted inside the wall with a shut-off valve with a drip, and when the valve is closed the water in the pipe between the valve and the sill cock passes out through the drip, leaving the pipe empty. If this water was allowed to remain in the pipe it would be likely to freeze, because of its exposed location.

At "H" is shown a pressure regulator, which is a device required when the minimum pressure in the street main is greater than is required to raise the water to the highest fixtures. By means of this regulator the excessive pressure may be reduced so that it is just sufficient to supply the highest fixtures. A greater pressure than this would cause considerable annoyance and wear and tear on the piping.

When a pressure regulator is installed on a water supply system, a relief or safety valve should always be installed on the hot water supply boiler to provide relief for the excessive pressure generated by the water heating apparatus. This is necessary for the reason that the pressure regulator is similar to a check valve, in that it would prevent any excessive pressure generated within the building from finding relief through the service connection and street main.

The sizes of pipe are determined in general by the number of fixtures and the pressure of water in the main. For a low pressure, that is, a pressure not exceeding 20 pounds per square inch, the sizes given in the illustration are about right. The pipe is reduced in size in the manner indicated as the various branches are taken off.

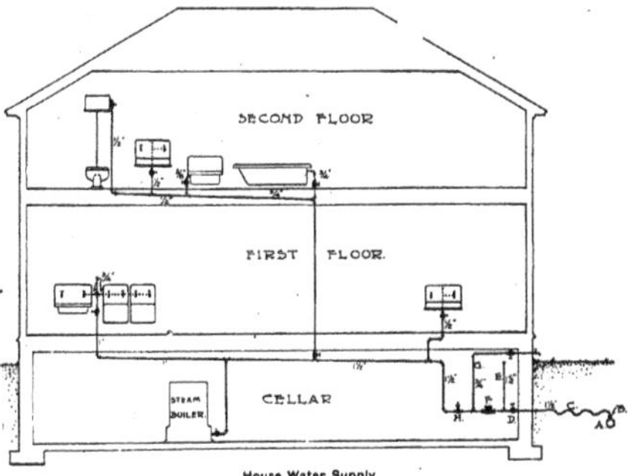

House Water Supply

PLUMBING CHANGE AT BELLEVILLE.

A big change has just been made in the firm of John Lewis & Co., plumbers and steamfitters, Belleville, although the business will be conducted under the old name. Arthur E. Lewis has sold his interest in the business to M. R. Doyle and Jas. H. Reeves, two practical young men who are thoroughly in touch with all the work done by the firm.

Arthur Lewis left on Monday for Calgary, where he has accepted a position in a large wholesale hardware firm. During his long residence at Belleville he made a host of friends by his honesty, integrity and fair dealing principles and has done much to place the firm of John Lewis & Co. in the front rank.

The firm is now composed of Mrs. John Lewis, of Belleville, Jones L. Lewis, editor of the Herald, Hamilton, and Messrs. Doyle and Reeves. Mr. Doyle opened up the plumbing business with the late John Lewis about 15 years ago, and he has been a faithful employee of the firm ever since. He will have charge of the mechanical end of the business, while Mr. Reeves will look after the business end of the firm, he having had a long experience in all branches of the hardware business.

PLUMBING PARAGRAPHS.

Lewis Bros., plumbers and stove merchants, Belleville, are to remove to Calgary.

Geo. Ecclestone, hardware merchant, Bracebridge, has now an up-to-date plumbing department added to his business.

Port Arthur is to have another plumbing establishment—a branch of a Winnipeg concern. Already there are three good firms established.

Moore & Brown, hardware merchants, Sault Ste. Marie, have established a plumbing department, with F. G. Sweetzer, formerly of Portsmouth, N.S., in charge.

APPLICATION OF HEAT UNIT TO BUILDINGS.

Address by Geo. G. Bennett before the Ohio Society Mechanical, Electric and Steam Engineers.

When estimating the heat required for the warming of buildings, the following methods may be used. Determine the loss in heat units through the walls and windows and make some allowance for the loss of heat b. leakage of cold air, due to the faulty construction of the room or building. The total loss in heat units for the entire building is found by the following method, which will be found to give a very close approximation. It has been determined by practical experiment that the loss in heat units through the change of air in the room or building, caused by bad construction when not otherwise ventilated, is about 1.430 heat units per square foot of surface per hour. It will require 0.238 heat units to raise one cubic foot of air one degree from absolute zero, but we are only figuring on zero on the Fehrenheit scale, and the specific heat of air at this temperature is 0.0864; therefore, to find the loss in heat units, we must multiply 0.230 by 0.0864, which equals 0.0205. Now, this is equal to the loss of heat in one cubic foot for one hour for each degree difference of temperature, and the room or building must be heated to 70 degrees, with an outside temperature of zero. To find the total loss we must multiply 0.0205 by 70 degrees, which equal 1.439 heat units for one hour.

Outside walls will require 0.223 heat units per hour per degree difference of temperature between the inside and outside temperatures. The difference is 70 degreess, which, multiplied by 0.223, equals 15.61 heat units, which gives the loss for one square foot for one hour. The next operation is to find the loss in heat units through one square foot of glass. The glass being the coldest part of the building, it requires a greater number of heat units ta counteract the cooling effect of the outside temperature. The glass will lose 0.543 heat units per hour per degree of difference of temperature. Now, as in the former example, the difference between the inside and outside temperatures is 70 degrees, and 0.543 multiplied by 70 degrees equals 38.01 heat units, which is the loss for one square foot of glass. The total loss per hour will be about as follows:

	Heat Units
Loss for 1 cubic foot of air	14.39
Loss for 1 sq. ft. of outside wall	15.61
Loss for 1 sq. ft. of glass	38.01
Total	55.059

I will now give a practical example of the above methods of determining the amount of radiation for a given sized room. The room is 16 ft. square by 10 ft. high, and has four windows containing 32 square feet of glass each. The total number of heat units necessary to counteract the cooling effect of glass, outside walls and cubical contents is as follows:

For the cubical contents we have 16 x 16 x 10—2,560 cubic feet, which is the cubical contents. This, multiplied by 1.439 equals 3,683.84, the heat units. The square feet of outside wall equals 16 x 10 x 2—320, and this multiplied by 15.61 equals 4,995.20 heat units lost through the outside walls. The square feet of glass equals 32 x 4—128, and this multiplied by 38.01—4.864 heat units lost through the glass.

The total amount is equal to:

	Heat Units
Cubical contents	3,683.840
Outside wall	4,995.20
Glass.	4,864
Total....	13,542

The total loss in heat units for the above room is now represented by 13,-542.04.

The number of square feet of pipe or other radiating surface that will be needed to counteract this loss is determined by the aid of the difference between the temperature of the steam pipe and the temperature at which the room is kept. We will assume the temperature of the steam to be 228 degrees —5 pounds gauge pressure. The difference between the temperature of the steam pipe and the temperature of the room will be the number of heat units given off in the room. The room is to be kept at 70 degrees, then 228—70—158 heat units—i.e., each square foot of heating surface will give off to the air in the room 158 heat units, so if we divide 13,542 by 158 the quotient will be the square feet of pipe or radiating surface, which is equal to 85.7.

If it is desired to heat the same room by hot water the process is somewhat similar, with the exception that the temperature of the water will be considerably less than that of the steam, so that the square feet of heating surface will be increased that much. In hot water heating the temperature of the water very rarely gets above 180 degrees, and it will be found safer to estimate the temperature of the water at about 140 degrees. If the outside temperature should get below zero the temperature of the water could be raised to 180 or 200 degrees.

Harrison & Robertson, plumbers and steamfitters, 389 Spadina avenue, Toronto, have dissolved partnership. R. W. Harrison will continue the business at the same location.

The Canadian Buffalo Forge Co., Montreal, have the contract for the heating and ventilation work on the new Molsons Bank building, on Bay street, Toronto. Fiddes & Hogarth, Toronto, are doing the heating work for the contractors. The ventilation is on the fan system and the heating hot water combination.

Ed. Holer, a plumber, employed by W. H. Jones, hardware merchant, Ingersoll, is suffering from scalp wounds sustained while at work at the John Morrow screw works. He fell from a ladder, his head coming in contact with a machine. Forty stitches were necessary to close the wounds.

RADIATOR EFFICIENCY.

Short, vertical cylinders are presumably the best that can be devised for giving off heat. If they be increased in height, say two or three times, they would do less duty because the air in contact with the upper parts would have been warmed by the lower part as it passes upward, and, therefore, is not capable of extracting as much heat. The higher a radiator is, the lower its efficiency per square foot of surface, and 36 or 38 inches has been established as a fair limit of height, says the Engineering Review.

Prof. R. C. Carpenter, in a series of tests at Sibley College, demonstrated the fact that "with radiators of the same height and same dimensions no difference could be traced, which was due either to material or to thickness. With radiators of the same material and the same height, but of different depths, or with varying numbers of tubes, we find invariably that higher results are obtained from the thinner radiator or those with the fewest rows of tubes.

NATURAL WATERWORKS SYSTEM

The artesian well belt of South Dakota is now attracting special attention because of the recent development of a number of wells containing hot water sufficient for the operation of hot water heating plants. This feature in addition to the cold water artesian wells for irrigation purposes in the arid sections of the state, which have also shown a bountiful suply, has enthused the residents of the commonwealth to a high degree. It is declared to be evident that South Dakota is underlaid with a natural waterworks system. The promise is held forth that in addition to abundance of water for irrigation, homes, live stock and crops, hot water for domestic purposes and heat is soon to be added. Water power for the driving of reasonably heavy machinery is, among the possibilities. In fact on many farms to-day artesian water power is utilized to operate corn shellers, separators, shredders and other rural machinery.

Recently contractors who sank an artesian well for the Chicago & Northern Railway Company at Midland found that the water which came to the surface with great force had a temperature of 135 degrees. This water, piped to residences, where it would go by its own force, would furnish all the heat desired. There are now in South Dakota about thirty artesian wells, 400 feet deep, sixty-five over 1,000 feet deep and two 2,500 feet deep. At Aberdeen the sewage is pumped away by power obtained from an artesian well, and elsewhere in the state electricity is generated for power and light purposes.

A. A. Horton, district hydrographer of the United States goelogical survey, says: "Artesian wells are certainly doing great things for the Dakotas and Montana. But I fear the flow of water will never reach sufficient power to operate heavy machinery. There is hot water in the belt now, lots of it. But no one can tell how long it will last."

PLUMBING MARKETS

MONTREAL.

Office of HARDWARE AND METAL,
232 McGill Street,
Montreal, March 8, 1907

In spite of the high prices prevailing at present, business in plumbing circles is booming. In fact it was never better.

Although there are no changes in prices to record this week, there is a tendency in the market to higher prices on several lines. Next week will likely see an advance in the price of iron pipe, as high as 3-inch. No figures have been received by the jobbers as yet, however, and prices are merely withdrawn.

Jobbers seem to be very much in the dark, in regard to the future prices that will prevail, one large house in fact, give all their quotations with the understanding that they are for immediate acceptance, and refuse to hold an offer over for even 24 hours.

RANGE BOILERS—Very good business is being done in both the copper and iron clad article, and prices on both are very firm; Iron clad, 30-gal., $5, 40-gal., $6.50 net list. Copper, 30 gal., $33; 35 gal., $38; 40 gal., $43.

LEAD PIPE—Owing to the heavy building operations there is a good demand for this article, and the market is firm at last week's quotations. Discount is: 5 p.c. f.o b. Montreal, Toronto, St. John, N.B., Halifax; f.o.b. London, 15c per hundred lbs. extra; f.o.b. Hamilton, 10c per hundred lbs. extra.

IRON PIPE FITTINGS—Shortage in some lines, and very firm prices, are the prevailing features in connection with this line. Discounts remain: On nipples, ¼-inch to 3-inch, 65 per cent.; 3½ to 2-inch, 67½ per cent.

IRON PIPE—Prices are withdrawn but the new list will likely be distributed by next week. Prices will be higher on all lines up to three-inch.

SOIL PIPE AND FITTINGS—Business is very good in this line We still quote: Standard soil pipe, 50 p.c. off list. Standard fittings, 50 and 10 per cent. off list; medium and extra heavy soil pipe, 60 per cent. off. Fittings, 60 per cent. off.

SOLDER — No change in prices. We still quote: Bar solder, half-and-half, guaranteed, 25c.; No. 2 wiping solder. 22c.

ENAMELWARE — We quote: Canadian baths, see Jan. 3, 1907, list. Lavatories, discounts, 1st quality, 30 per cent.; special, 30 and 10 per cent. Sinks, 18x30 inch, flat rim, 1st quality, $2.60; special, $2.45.

TORONTO.

Office of HARDWARE AND METAL,
10 Front Street East.
Toronto, March 8, 1907

The plumbing supplies business is picking up, although trade is not yet by any means as brisk as might be wished for. One thing is certain, conditions are better than they were a fortnight ago.

During the past week there have been many reports of price-cutting in plumbing lines, both enamelware and brass goods. In the latter line particularly the market has been unsteady for a fortnight or more. While plumbers are keen buyers and not over anxious to pay higher prices than necessary, they are, nevertheless, believers in a steady market, not favoring price-cutting on contracts and preferring to have jobbers and manufacturers maintain their prices on an equitable basis. Probably more than any other section of business men, they look upon price-cutting in any branch

of the trade as demoralizing to all other branches.

Orlando Vickery, Toronto, representative of the Amherst Foundry Co, who has developed quite a business in plumbing supplies during the past eighteen months, disclaims any price-cutting in his lines, and states that no matter what others may do he will not sell below the regular figure, based on the cost of manufacture and a fair margin of profit. The recent advance of 7½ per cent. on "Beaver" enamelware is being held to firmly.

LEAD PIPE—The discount is still 5 per cent off the list price of 7c. per lb Lead waste, 8c per lb, with 5 off. Caulking lead, 5¼c. to 6¾c. per lb. Traps and bends, 40 per cent. off.

SOIL PIPE AND FITTINGS—Business is increasing We quote: Medium and extra heavy pipe and fittings, 60 per cent.; light pipe, 50 per cent.; light fittings, 50 and 10 per cent.; 7 and 8 inch pipe, 40 and 5 per cent.

IRON PIPE—Prices have been withdrawn and an advance is looked for daily. Stocks are low, with prospects of scarcity for the future. We continue to quote. 1-inch black pipe, $4.95, 1-inch galvanized at $6.60.

PIPE IRON FITTINGS—Prices are unsteady. Some jobbers have lowered the price from 57½ to 60 per cent. We continue to quote, however at 57½ per cent Cast iron fittings, 57½ per cent.; cast iron plugs and bushings, 60 per cent.; flange unions, 60 per cent.; nipples, 70 and 10 per cent.; iron cocks, 55 and 5 per cent.; Canadian malleable, 30 per cent.; malleable unions, 55 and 5 per cent.; malleable bushings, 55 per cent.; cast iron ceiling plates, plain, 65 per cent.; cast iron floor, 70 per cent.; hook plates, 60 per cent.; expansion plates, 65 per cent.; headers, 60 per cent.; hangers, 65 per cent.; standard list.

RANGE BOILERS—Galvanized iron, 30-gal., standard, $5; extra heavy, $7.75; 35-gal. standard, $6; extra heavy, $8.75, 40-gal., standard, $7; 40 gallon, extra heavy, $9.75, net list. Copper range boilers—New lists quote: 30 gallon, $33; 35 gallon, $38; 40 gallon $43. Discounts 5 to 15 per cent.

RADIATORS—The demand for these is brisk. Keen business is being done, orders being booked ahead in view of the erection of a new factory. Prices are unchanged. We continue to quote: Hot water, 47½ per cent; steam, 50 per cent.; wall radiators, 15 per cent.; specials, 45 per cent. Hot water boilers are subject to an open market.

SOLDER—Bar solder, half-and-half, guaranteed, 27c.; wiping, 23c.

ENAMELWARE—Price-cutting in this line is reported. We continue to quote, however, the January prices, as they are about the average prices Lavatories, first quality, 20 and 5 to 20 and 10 off; special, 20 and 10 to 30 and 2½ per cent. discount. Kitchen sinks, plate 300, firsts, 60 and 10 off; specials, 65 and 5 per cent. Urinals and range closets, 15 off. Fittings extra.

PLUMBING PARAGRAPHS.

The Northwestern Brass Co., Winnipeg, representing an investment of $10,000, will be turning out its products in less than two months and will employ from 60 to 75 hands.

The King Radiator Co, Toronto, who secured a site for their radiator works in Ashbridge Bay, Toronto, will commence building operations at once

The Canadian Brass Mfg. Co., recently located at Galt, will be established by July 1st.

The Wolverine Brass Works, of Grand Rapids, have decided to locate a branch in Toronto. The manager, who was in Toronto last week, says the duty levied on United States Goods makes it impossible to export at a profit to Canada and hence the decision to open a branch here.

TRADE
PENBERTHY
MARK
AUTOMATIC
INJECTORS
PENBERTHY
RELIABILITY
SIMPLICITY
DURABILITY

Synonomous with excellence and completeness in hot water heating is

The Daisy Hot Water Boiler

For over twenty years the Daisy has represented what is best in Hot Water Boiler construction, and hundreds of steamfitters attribute the cause of their success in heating to the unquestioned satisfaction the installation of the Daisy has given.

CLUFF BROTHERS

Lombard Street, TORONTO

Selling Agents for - - - WARDEN KING & SON, Limited

No Evaporation Means No Danger

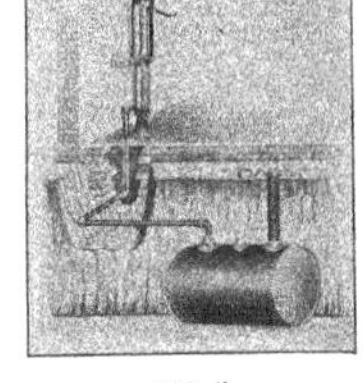

Cut No. 42
Long Distance Outfit for Gasolene

Gasolene will not explode while in liquid form. It will remain in its liquid form until exposed to the air, when it will evaporate and form a dangerous vapor. A system of gasolene storage that prevents evaporation is a safe system.

With the Bowser, gasolene is not exposed to the air from the time it first enters the storage tank until it is used by your customer.

If you want to know how this is done, send for our Catalog V. It explains the system.

S. F. BOWSER & CO., Inc.

66-68 Fraser Ave. Toronto, Can.

Ask any Pump Man

who has ever used our Fig. 92, and if he does not say "it is the best ever," then we will acknowledge "the exception which proves the rule." There is no use looking for anything better than we have in our catalogue and prices must, of course, be right, as well. Let us have a little talk together.

THE

R. McDougall Company, LIMITED

GALT, CANADA

Are you interested in any of the lines that are advertised?
A Post Card will bring you price list and full information.
Don't forget to mention Hardware and Metal.

The Canadian Bronze Powder Works

R. E. THORNE & CO.

The only bronze powder works under the British flag.
High Grade bronze powders and bronze liquids.
Can fill all orders at short notice.

MONTREAL — TORONTO

WORKS AT VALLEYFIELD. ***NO ORDER TOO LARGE***

HOTTEST ON EARTH

Our Bonanza Furnaces are known to produce more heat with less fuel than any other make. Jobbers sell at factory prices. Satisfaction guaranteed or money refunded.

No. 38. Price, each, ***$7.80,*** net.
No. 39. One size smaller, ***$5.85,*** net.

THE TURNER BRASS WORKS

53 MICHIGAN STREET, - CHICAGO, U.S.A.

Manufacturers' Agents

CHARLES H. FOX
Vancouver, Canada
MANUFACTURERS' AGENT

Selling to the Hardware Jobbers Only.

Representing Canadian, British and American Manufacturers, correspondence invited. Cable Address, Chasfox Vancouver, P.O.Box 1105.

ALEXANDER GIBB

Manufacturers' Agent and Metal Broker,
13 St. John Street, Montreal

Representing Canadian, British and American Manufacturers. Correspondence invited from firms wishing to be represented.

H. W. MITCHELL
WINNIPEG, MAN.

Correspondence is solicited from manufacturers wishing a representative in Winnipeg.
Travellers call regularly on the wholesale and retail trade in Western Canada.
Highest references and financial responsibility.

Scott, Bathgate & Co.

BROKERS AND COMMISSION MERCHANTS

Cor. Notre Dame St. East and Victoria St., WINNIPEG, Canada
Storage and Warehousing a Specialty.

First-class facilities in a thoroughly modern brick warehouse centrally located on the railway track.

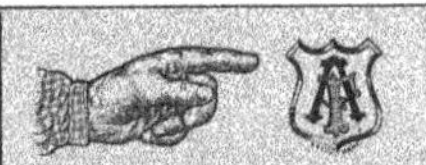

This is the Mark that Means Something

It is the distinctive brand of genuine Armstrong Stocks and Die and Pipe Fitters' Tools.

No "Armstrong pattern" or "just as good as Armstrong" have this stamp. Don't buy anything in this line that hasn't this brand as it is the mark of genuineness.

THE ARMSTRONG MANUFACTURING CO.,
292 Knowlton Street, Bridgeport, Conn.

A Certain Sense of Satisfaction

goes with every MERRELL PORTABLE HAND MACHINE, PIPE THREADING and CUTTING MACHINE. We build them that way.

For doing ALL kinds of threading—under ALL conditions—labor trouble and faulty material included—where speed, durability and ease of control count for something, our word to you is to investigate the MERRELL.

Whether it be stationary or portable—hand or power driven—30 DAYS FREE TRIAL must prove conclusively that your choice should be the MERRELL.

This Portable Hand Machine has encased gears, MERRELL standard quick opening and closing die head, and the latest improved Cutting-off Knife.

Let us tell you more about this machine. The Chasers, Vise and the large range of work covered.

Catalogue for the asking.
Quick shipments.

THE CANADIAN FAIRBANKS CO.
Sole Agents for Canada Limited
MONTREAL, TORONTO, WINNIPEG, VANCOUVER

FACTS ARE STUBBORN THINGS

Most shrewd dealers are wise to the way I conduct my business. My customers' interests are never lost sight of for a single moment during business hours. **"Quality, Dispatch,** and then **More Quality."** That's how one man describes my goods and methods.

ORLANDO VICKERY
178-180 Victoria Street *TORONTO, CANADA*

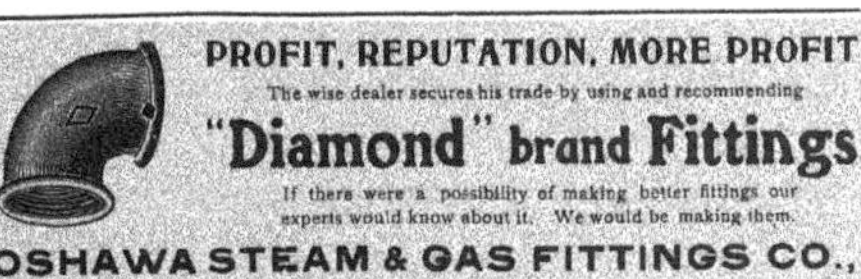

OSHAWA STEAM & GAS FITTINGS CO., LIMITED
OSHAWA, CANADA

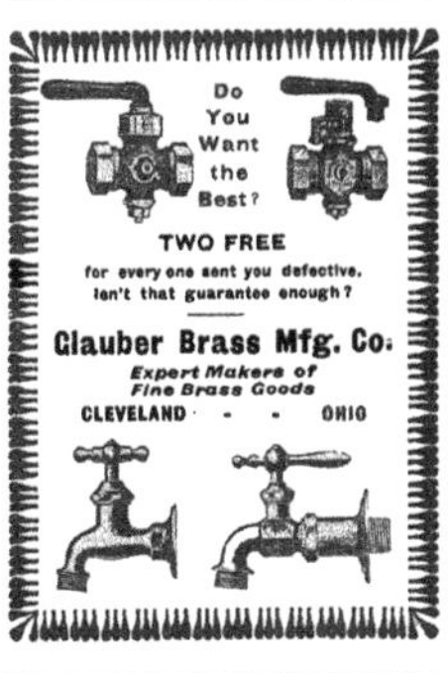

Sharratt & Newth's Glaziers' Diamonds

are unequalled for cutting and wearing qualities.

To be obtained from the principal Dealers in Glass, Hardware, and Painters' Supplies.

Contractors to H. M. Government and the principal English Sheet and Plate Glass Works

World's Best Cutlery

You make no mistake in specifying goods made by **JOSEPH RODGERS & SONS, Limited** when ordering cutlery. A reputation of over 200 years at stake, insures the quality of every article.

Write for catalogues from Sole Canadian Agents

JAMES HUTTON & COMPANY, MONTREAL

Forwell Foundry Co.
BERLIN, ONT.
Manufacturers of
SOIL PIPE, FITTINGS, and CAST IRON SINKS
Ask Jobbers for "F. F. CO." Brand

CURRENT MARKET QUOTATIONS.

Mar. 8, 1907.

These prices are for such qualities and quantities as are usually ordered by retail dealers on the usual terms of credit, the lowest figures being for larger quantities and prompt pay. Large cash buyers can frequently make purchases at better prices. The Editor is anxious to be informed at once of any apparent errors in this list, as the desire is to make it perfectly accurate.

METALS.

ANTIMONY.

Cookson's ....per lb...	0 27	0 27½

BOILER PLATES AND TUBES

Plates, ¾ to ½ inch, per 100 lb.	....	2 50
Heads, per 100 lb.	....	2 75
Tank plates 3-16 inch	....	2 65
Tubes per 100 feet, 1½ inch ..	8 50	....
" " 2 " ..	9 00	9 50
" " 2½ " ..	11 30	11 50
" " 3 " ..	12 50	13 00
" " 3½ " ..	16 00	17 00
" " 4 " ..	20 00	21 50

2 per cent off

BOILER AND T K. PITTS.

Plain tinned.........} 25 per cent off list.
Spun...............}

BABBIT METAL.

Canada Metal Company—Imperial, genuine 60c; Imperial Tough, 60c.; White Brass, 50c; Metallic, 35c.; Harris Heavy Pressure, 25c.; Hercules, 25c.; White Bronze, 15c.; Star Frictionless, 14c; Alluminoid, 10c; No. 4, 9c. per lb.

James Robertson Co.—Extra and genuine Monarch, 60c.; Crown Monarch, 50c; No 1 Monarch, 40c; King, 30c; Fleur-de-lis, 20c; Thurber, 15c; Philidelphia, 12c; Canadian, 10c, hardware, No 1, 15c; No. 2, 12c; No 3, 10c per lb.

BRASS

Rod and Sheet, 14 to 30 gauge, 25 p c advance.

Sheets, 12 to 14 in.	0 30
Tubing, base, per lb 5-16 to 2 in	0 33
Tubing, ¾ to 3-inch, iron pipe size..	0 31
" 1 to 3-inch, seamless..	0 36

Copper tubing, 6 cents extra.

COPPER.

Ingot.		Per 100 lb.
Casting, car lots	26 50	27 00
Bars.		
Cut lengths, round, ½ to 2 in...	....	35 00
Sheet.		
Plain, 16 oz., 14x48 and 14x60	....	35 00
Plain, 14 oz.	....	36 00
Tinned copper sheet, base	....	38 00
Planished base	....	39 00
Braziers' (in sheets) 4x6 ft., 25 to 30 lb. each, per lb., base..	....	0 35

BLACK SHEETS.

	Montreal	Toronto
8 to 10 gauge	2 70	2 70
12 gauge	2 70	2 80
14 "	2 70	2 55
17 "	2 50	2 65
18 "	2 50	2 65
20 "	2 50	2 65
22 "	2 50	2 75
24 "	2 55	2 80
26 "	2 65	2 85
28 "	2 70	3 00

CANADA PLATES.

Ordinary, 52 sheets	2 75	2 90
All bright "	3 80	3 90
Galvanized, 52 sheets	4 35	4 45
" 60 "	4 60	4 70
	Ordinary.	Dom. Crown
18x24x52	4 25	4 35
" 60	4 50	4 60
20x28x80	8 50	8 70
" 94	9 00	9 20

GALVANIZED SHEETS

	Fleur-de-Lis.	Gordon Crown	Colborne Crown
16 to 20 gauge..	3 80	3 95	....
22 to 24 gauge ..	4 05	4 00	3 75
26 " ..	4 30	4 40	4 30
28 " ..	4 55	4 60	4 45

Apollo.

10¾ oz. (American gauge)	....	4 85
28 gauge "	4 45	4 55
26 " "	4 15	4 30
24 " "	3 85	3 90

	Gorbal's Best.	Comet	Queen's Head.	Bell.
16 to 20 gauge	....	3 60	3 95	....
22 to 24 gauge	....	3 75	4 20	....
26 "	....	4 30	4 45	....
28 "	4 55	4 45	4 70	4 45

Less than case lots 10 to 25c. extra.

IRON AND STEEL.

	Montreal.	Toronto.
Common bar, per 100 lb.	2 15	2 30
Forged iron "	2 40	..
Refined " "	2 55	2 70
Horseshoe iron "	2 55	2 70
Hoop steel, 1½ to 3 in. base...	....	2 80
Sleigh shoe steel "	2 25	2 30
Tire steel	2 40	2 50
Best sheet cast steel	...	0 12
B. K. Morton "Alpha" high speed.		0 65
" annealed	....	0 70
"M" Self-hardening	....	0 50
"J" quality, best warranted	....	0 18
"I" " warranted	....	0 14
"B.C" quality		0 09
Colonial black diamond.	0 08	0 09
Sanderson's	0 08	0 45
Jessop's	0 12	0 13
Air hardening	0 60	0 65
Conqueror	0 07½	0 60
Jowett's diamond J	0 06½	0 07
Jonas & Colver's tool steel...	0 10	0 20
" "Novo"	..	0 65
" " annealed	....	0 65
Jowett & Sons B P.L. tool steel	0 10½	0 11

COLD ROLLED SHAFTING.

9-16 to 11-16 inch	....	0 06
¾ to 1 7-16 "	....	0 05½
1 7-16 to 3 "	....	0 05

Montreal 30, Toronto 35 to 40 per cent.

INGOT TIN.

Lamb and Flag and Straits—56 and 28-lb. ingots, 100 lb.	$45 00	$45 50

TINPLATES.

Charcoal Plates—Bright

			Per box.
M.L.S., equal to Bradley—			
I C, 14 x 20 base		....	$6 50
IX, 14 x 20 "		....	8 00
IXX, 14 x 20 base		....	9 50
Famous, equal to Bradley—			
I C, 14 x 20 base		....	6 50
I X, 14 x 20 "		....	8 00
I X X, 14 x 20 base		....	9 50
Raven and Vulture Grades—			
I C, 14 x 20 base		....	5 00
I X "		....	6 00
I X X "		....	7 00
I X X X "		....	8 00
"Dominion Crown Best"—Double Coated, Tissued			Per box.
I C, 14 x 20 base		5 50	5 75
I X, 14 x 20 "		6 50	6 75
IXX x 20 "		7 50	7 75
"Allaway's Best"—Standard Quality.			
I C, 14 x 20 base		....	4 50
I X, 14 x 20 "		....	5 25
I XX, 14 x 20 "		....	6 00
Bright Cokes			
Bessemer Steel—			
I C, 14 x 20 base		...	4 25
20x28, double box		....	8 50
Charcoal Plates—Terne			
Dean or J. G. Grade—			
I C, 20x28, 112 sheets		7 25	8 00
IX., Terne Tin		...	9 50
Charcoal Tin Boiler Plates.			
Cookley Grade—			
X X, 14x56, 50 sheet bxs. } " 14x60, " } " 14x65, " }		...	7 50
Tinned Sheets.			
72x30 up to 24 gauge		....	8 50
" " 26 "		....	9 00

LEAD.

Imported Pig, per 100 lb.	5 50	5 60
Bar, "	5 75	6 00
Sheets, 2½ lb. sq. ft., by roll ..	....	0 07½
Sheets, 3 to 6 lb. "	....	0 07

NOTE.—Cut sheets ½c. per lb., extra. Pipe, by the roll, usual weights per yard, lists at 7c. per lb. and 5 p.c. dis. f.o.b. Toronto.

NOTE.—Cut lengths, net price, waste pipe 8-ft lengths, lists at 8c.

SHEET ZINC.

5-cwt. casks	8 00	8 25
Part casks	8 25	8 50

ZINC SPELTER.

Foreign, per 100 lb	7 50	7 75
Domestic	7 00	7 25

PLUMBING AND HEATING

BRASS GOODS, VALVES, ETC.

Standard Compression work, 55 per cent.
Cushion work, discount 40 per cent.
Fuller work, 65 per cent.
Flatway stop and stop and waste cocks, 60 per cent.; roundway, 55 per cent.
J.M.T. Globe, Angle and Check Valves, 45 per cent.
Standard Globe, Angle and Check Valves 55 per cent.
Kerr standard globes, angles and checks, special, 42½ per cent.; standard, 47½ p.c.
Kerr Jenkins' disc, copper-alloy disc and heavy standard valves, 45 per cent.
Kerr steam radiator valves 60 p.c., and quick-opening hot-water radiator valves, 60 p.c.
Kerr brass, Weber's straightway valves, 40 per cent.; straightway valves, I.B.B.M., 60 per cent.
J. M. T. Radiator Valves 50 per cent.
Standard Radiator valves, 60 per cent.
Patent Quick-Opening Valves, 65 per cent.
Jenkins' Bros Globe Angle and Check Valves 27½ per cent.

No. 1 compression bath cock......net	2 00
No. 4 " " ...... "	1 90
No 7 Fuller's .... "	2 25
No. 4½ " .... "	2 35
Patent Compression Cushion, basin cock, hot and cold, per doz., ...	$16.20
Patent Compression Cushion, bath cock, No. 2208	2 25

Square head brass cocks, 50 per cent.
" " iron, " 60 "
Thompson Smoke-test Machine 25.00

BOILERS—COPPER RANGE.

Copper, 30 gallon	33 00
" 35 "	38 00
" 40 "	43 00

15 per cent

BOILERS—GALVANIZED IRON RANGE.

Capacity	Standard.	Extra heavy
30-gallons ...	5 00	7.75
35 "	6 00	8 75
40 "	7 00	9 75

2 per cent., 30 days

BATH TUBS.

Steel clad copper lined, 15 per cent.

CAST IRON SINKS.

16x24, $1, 18x30, $1; 18x36, $1.30.

ENAMELED BATHS.

List issued by the Standard Ideal Company Jan. 3, 1907, shows an advance of 10 per cent over previous quotations.

ENAMELED CLOSETS AND URINALS

Discount 15 per cent.

ENAMELED LAVATORIES.

	1st quality.	Special.
Plate E 100 to E 103 .	20 & 5 p.c.	20 & 10 p.c.
" E 104 to E 132....	20 & 10 p.c.	30 & 2½ p.c.

ENAMELED ROLL RIM SINKS.

	1st quality.	Special.
Plate E 201, one piece.	15 & 2½ p.c.	15 & 10 p.c.

ENAMELED KITCHEN SINKS.

Plate E, flat rim 300, 60 & 10 p.c. 65 & 5 p.c.

HEATING APPARATUS.

Stoves and Ranges—Discounts vary from 40 to 70 per cent. according to list.
Furnaces—45 per cent.
Registers—70 per cent
Hot Water Boilers—Discounts vary.
Hot Water Radiators—47½ per cent.
Steam Radiators—50 per cent
Wall Radiators and Specials—45 per cent.

LEAD PIPE

Lead Pipe, 7c. per pound, 5 per cent. off.
Lead waste, 8c. per pound, 5 per cent. off.
Caulking lead, 6c per pound.
Traps and bends, 40 per cent.

IRON PIPE.

Size (per 100 ft.)	Black.		Galvanized
¼ inch	2 25	¼ inch	3 08
⅜ "	2 25	⅜ "	3 08
½ "	2 72	½ "	3 57
¾ "	3 45	¾ "	4 60
1 "	4 95	1 "	6 64
1¼ "	6 75	1¼ "	9 00
1½ "	8 10	1½ "	10 80
2 "	10 80	2 "	14 40
2½ "	18 00	2½ "	24 00
3 "	23 50	3 "	31 50
3½ "	30 00	3½ "	40 00
4 "	34 00	4 "	45 00

2 per cent. 30 days.
Malleable Fittings—Canadian discount 30 per cent.; American discount 25 per cent.
Cast Iron Fittings 57½; Standard bushings 57½ per cent.; headers, 57½; flanged unions 57½, malleable bushings 55; nipples, 70 and 10 per cent; malleable lipped unions, 55 and 5 per cent.

SOIL PIPE AND FITTINGS

Medium and Extra heavy pipe and fittings, up to 6 inch, discount 60 per cent.
7 and 8-in. pipe, discount 40 and 5 per cent
Light pipe, 50 p.c.; fittings, 50 and 10 p.c

OAKUM.

Plumbers ....per 100 lb...	4 75

STOCKS AND DIES

American discount 25 per cent.

SOLDERING IRONS.

½-lb. .... per lb.	0 38
1-lb. or over .... "	0 35

SOLDER.

Per lb.	Montreal	Toronto
Bar, half-and-half, guaranteed	0 25	0 27
Wiping	0 22	0 23

PAINTS, OILS AND GLASS.

BRUSHES

Paint and household, 70 per cent.

CHEMICALS.

	In casks per lb.
Sulphate of copper (bluestone or blue vitrol)	0 09
Litharge, ground	0 06
" flaked	0 06½
Green copperas (green vitrol)	0 01
Sugar of lead	0 08
Lump olive	0 01¾

COLORS IN OIL.

Venetian red, 1-lb. tins pure.	0 08
Chrome yellow "	0 15
Golden ochre "	0 08
French " "	0 08
Marine black "	0 04½
Chrome green "	0 10
French permanent green "	0 13
Signwriters' black "	0 15

GLUE.

Domestic sheet	0 10	0 10½
French medal	0 12	0 12½

PARIS GREEN.

	Berger's	Canadian
600-lb. cask	0 25½	0 25½
250-lb. drums	0 25¾	0 25¾
100-lb. "	0 26½	....
50-lb. "	0 26¾	0 26¾

1-lb. pkgs, 100 in box 0 27½ 0 27½
½-lb. " 0 29½ 0 29½
1-lb. tins, 100 in box .... 0 28½ 0 2?½
½-lb pkg 0 30½ 0 30½
F.o.b Toronto.

PARIS WHITE.

In bbls 0 90

PREPARED PAINTS.

Pure, per gallon, in tins 1 30
Second qualities, per gallon 1 10
Barn (in bbls.) 0 85 0 90
Sherwin-Williams paints, 1 gal. 1 40
Canada Paint Co.'s pure 1 30
Standard P. & V. Co.'s "New Era." 1 30
Benj. Moore Co.'s "Ark" B'd 1 25
" British Navy deck 1 50
Brandram-Henderson's "English" 1 45
Ramsay's paints, Pure, per gal. 1 20
" Thistle, " 1 00
" Outside, bbls 0 55 0 65
Martin-Senour's 100 p.c. pure, ½ gal. 1 50
" " " " ¼ gal. 1 45
" " " " 1 gal. 1 40
Senour's Floor Paints ... gal. 1 20
Jamieson's "Crown and Anchor" 1 20
Jamieson's floor enamel 1 50
" barn paints, bbls, per gal. 0 60
Sanderson Pearcy's, pure 1 20
Robertson's pure paints 1 20

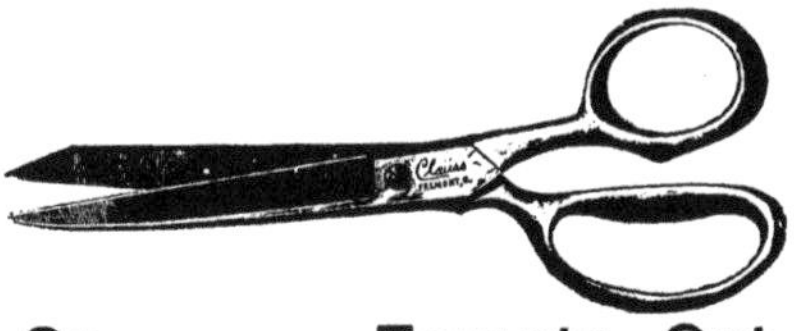

PUTTY.

Bulk in bbls 1 50
Bulk in less quantity 1 80
Bladders in bbls 1 80
Bladders in kegs, boxes or loose 1 95
25-lb. tins 1 80
12½ lb. tins 2 05
Bladders in bulk or tins less than 100 lb. 1 85
Bulk in 100-lb. irons 1 80

SHINGLE STAINS.

In 5 gallon lots 0 75 0 80

SHELLAC.

White 0 65
Fine orange 0 60
Medium orange 0 55
F.o.b. Montreal or Toronto.

TURPENTINE AND OIL.

Castor oil 0 08 0 10
Gasoline 0 22½
Benzine, per gal 0 17 0 20
Turpentine, single barrels 0 98 1 05
Linseed Oil, raw .. 0 57 0 59
" boiled 0 60 0 62
W. H. & C's English raw oil, bbls 0 63
" boiled, " 0 66

WHITE LEAD GROUND IN OIL. Per 100 lbs

Canadian pure 7 15
No. 1 Canadian 6 80
Munro's Select Flake White 7 40
Elephant and Decorators' Pure 7 40
Monarch 7 40
Standard Decorator's 7 15
Essex Genuine 6 80
Brandram's B. B. Genuine 8 40
"Anchor," pure 7 00
Ramsay's Pure Lead 6 40
Ramsay's Exterior 6 15
"Crown and Anchor," pure 6 50
Sanderson Pearcy's 7 40
Robertson's C.P. lead 7 20
W. H. & C's matured pure English 8 00 8 25

WHITE AND RED DRY LEAD. white red.

Genuine, 560 lb. casks, per cwt 6 75 6 00
Genuine, 100 lb. kegs, " 7 50 6 50
No. 1, 560 lb. casks, per cwt .. 6 25 5 75
No. 1, 100 lb. kegs, per cwt .. 7 00 6 25

WINDOW GLASS.

Size United Inches.	Star	Double Diamond
Under 26	$4 25	$6 25
26 to 40	4 65	6 75
41 to 50	5 10	7 50
51 to 60	5 35	8 50
61 to 70	5 75	9 75
71 to 80	6 25	11 00
81 to 85	7 00	12 50
86 to 90	....	15 00
91 to 95	....	17 50
96 to 100	....	20 50
101 to 105	....	24 00
106 to 110	....	27 50

Discount—16-oz., 75 per cent.; 21-oz. 30 per cent. per 100 feet Broken boxes 50 per cent.

WHITING.

Plain, in bbls 0 60
Gilders bolted in bands 0 90

WHITE DRY ZINC.

Extra Red Seal, V.M. 0 07½ 0 08½

WHITE GROUND ZINC.

Pure, in 25-lb. irons 0 08½
No. 1, " 0 07
No. 2, " 0 05½

VARNISHES.

In 5-gal. lots. Per gal. cans.
Carriage, No. 1 .... 1 50
Pale durable body 3 50
" hard rubbing 3 00
Finest elastic gearing 3 00
Elastic oak 1 50
Furniture, polishing 2 00
Furniture, extra 1 15
" No. 1 .... 0 90
" union 0 80
Light oil finish 1 40
Gold size japan 1 80
Brown japan 0 95
No. 1 brown japan 0 90
Baking black japan 1 35
No. 1 black japan 0 90
Benzine black japan 0 70
Crystal Damar 2 80
No. 1 " 2 50
Pure asphaltum 1 40
Oilcloth 1 50
Lightning dryer 0 70
Elastilite varnish, 1 gal. can, each .. 2 00
Granitine floor varnish, per gal 2 10
Maple Leaf coach enamels; size 1, 1 20
Sherwin-Williams' kopal varnish, gal., 2 50
Canada Paint Co's sun varnish 2 00
"Kyanize" Interior Finish 2 40
"Flint-Lac," coach 1 80
B.H. Co's "Gold Medal," in cases 2 00
Jamieson's Copaline, per gal 2 00

BUILDERS' HARDWARE.

BELLS.

Brass hand bells, 60 per cent.
Nickel, 55 per cent.
Gongs, Sargeant's door bells .. 5 50 8 00
American, house bells, per lb. 0 35 0 40
Peterboro' door bells, discount 37½ and 10 per cent. off new list.

BUILDING PAPER, ETC.

Tarred Felt, per 100 lb. 2 25
Ready roofing, 2-ply, not under 65 lb. per roll 1 00
Ready roofing, 3-ply, not under 65 lb., per roll 1 25
Carpet Felt .... per ton 60 00
Heavy Straw Sheathing .... per ton 35 00
Dry Surprise 0 40
Dry Sheathing .... per roll, 400 sq. ft. 0 50
Tar " 400 " 0 45
Dry Fibre " 400 " 0 55
Tarred Fibre " 400 " 0 65
O. K. & I. X. L. " 400 " 0 70
Resin-sized " 400 " 0 45
Oiled Sheathing " 600 " 1 00
Oiled " 400 " 0 70
Roof Coating, in barrels .... per gal. 0 17
Roof " small packages " 0 25
Refined Tar .... per barrel 5 00
Coal Tar " 4 00
Coal Tar, less than barrels .... per gal. 0 15
Roofing Pitch .... per 100 lb. 0 80 0 90
Slater's felt .... per roll 0 70
Heavy Straw Sheathing f. o. b. St. John and Halifax 37 50

BUTTS.

Wrought Brass, net revised list.
Wrought Iron, 70 per cent.
Cast Iron Loose Pin, discount 60 per cent.
Wrought Steel Fast Joint and Loose Pin, 70 per cent.

CEMENT AND FIREBRICK.

Canadian Portland 2 00 2 10
Belgium 1 50 1 90
White Bros English 1 80 2 03
"Lafarge" cement in wood 3 40
"Lehigh" cement, in wood 2 54
"Lehigh" cement, cotton sacks 2 39
"Lehigh" cement, paper sacks 2 31
Fire brick, Scotch, per 1,000 .... 27 00 30 00
" English 17 00 21 00
" American, low 21 00 25 00
" " high 27 00 35 00
Fire clay (Scotch), net ton 4 95
Paving Blocks per 1,000.
Blue metallic, 9"x4½"x3", ex wharf .. 35 00
Stable pavers, 12"x6"x2", ex wharf .. 50 00
Beaubie pavers, 9"x4½"x3", ex wharf .. 36 00

DOOR SETS.

Peterboro, 37½ and 10 per cent.

DOOR SPRINGS.

Torrey's Rod .... per doz. 1 75
Coil, 9 to 11 in. .... " 0 95 1 65
English .... " 2 00 4 00
Chicago and Reliance Coil 25 per cent.

STORE DOOR HANDLES.

Per Dozen 1 00 1 50

ESCUTCHEONS.

Discount 50 and 10 per cent., new list
Peterboro, 37½ and 10 per cent.

ESCUTCHEON PINS.

Iron, discount 40 per cent.
Brass, 45 per cent.

HINGES.

Blind, discount 60 per cent.
Heavy T and strap, 4-in., per lb. net .. 0 06½
" " 5-in., " 0 06
" " 6-in., " 0 05¾
" " 8-in., " 0 05¾
" " 10-in. and larger .. 0 05¾
Light T and strap, discount 65 p.c.
Screw hook and hinge—
under 12 in .... per 100 lb. 4 75
over 12 in. .... " 3 75
Spring, No. 20, per gro. pairs 10 80
Spring, Woodyatt pattern, per gro., No. 5, $17.50 No. 10, $18; No. 20, $10 80; No. 12 $20; No. 51, $10; No. 50, $27 50.
Crate hinges and back flaps, 65 and 5 p. c
Hinge hasps, 65 per cent.

SPRING HINGES.

Chicago Spring Butts and Blanks 12½ per cent.
Triple End Spring Butts, 40 and 5 per cent.
Chicago Floor Hinges, 40 and 5 off.
Garden City Fire House Hinges, 12½ p c

CAST IRON HOOKS.

Bird cage .... per doz. 0 50 1 10
Clothes line, No. 61 .. " 0 00 0 70
Harness .... " 0 60 12 00
Hat and coat .... per gro. 1 10 10 00
Chandelier .... per doz. 0 50 1 00
Wrought hooks and staples—
¼ x 5 .... per gross 2 65
5-16 x 5 .... " 3 30
Bright steel gate hooks and staples, 40 p.c.
Hat and coat wire, discount 67½ per cent.
Screw, bright wire, discount 65 per cent

KNOBS.

Door, japanned and N.P., doz 1 50 2 50
Bronze, Berlin .... per doz. 2 75 3 25
Bronze, Genuine .... " 6 00 9 00
Shutter, porcelain, F. & L. screw .... per gross 1 30 2 00
White door knobs .... per doz. 2 00
Peterboro knobs, 37½ and 10 per cent.
Porcelain, mineral and jet knobs, net list.

KEYS.

Lock, Canadian dis. 40 to 40 and 10 per cent
Cabinet trunk and padlock American .... per gross 0 60

LOCKS.

Peterboro 37½ and 10 per cent
Russell & Erwin steel rim $2 50 per doz.
Eagle cabinet locks, discount 30 per cent
American padlocks, all steel, 10 to 15 per cent; all brass or bronze, 10 to 25 per cent.

SAND AND EMERY PAPER.

B. & A. sand, discount, 35 per cent
Emery, discount 35 per cent.
Garnet (Rurton's) 5 to 10 per cent. advance

SASH WEIGHTS.

Sectional .... per 100 lb. 2 00 2 25
Solid .... " 1 50 1 75

SASH CORD.

Per lb 0 31

BLIND AND BED STAPLES.

All sizes, per lb. 0 07½ 0 12

WROUGHT STAPLES.

Galvanized 2 75
Plain 2 50
Coopers', discount 45 per cent.
Poultry netting staples, discount 40 per cent.
Bright spear point, 75 per cent. discount.

TOOLS AND HANDLES.

ADZES.

Discount 22½ per cent.

AUGERS.

Gilmour's, discount 60 per cent. off list.

AXES.

Single bit, per doz. 5 50 8 50
Double bit, " 10 00 11 00
Bench Axes, 40 per cent.
Broad Axes, 25 per cent.
Hunters' Axes 5 50 6 00
Boys' Axes 6 25 7 00
Splitting Axes 7 00 12 00
Handled Axes 7 00 9 00
Red Ridge, boys', handled 5 75
" hunters 5 25

BITS.

Irwin's auger, discount 47½ per cent.
Gilmour's auger, discount 60 per cent.
Rockford auger, discount 50 and 10 per cent.
Jennings' Gen. auger, net list.
Gilmour's car, 47½ per cent.
Clark's expansive, 40 per cent.
Clark's gimlet, per doz 0 65 0 90
Diamond, Shell, per doz. 1 00 1 50
Nail and Spike, per gross 2 25 5 20

BUTCHERS CLEAVERS.

German .... per doz. 6 00 9 00
American .... " 12 00 18 00

CHALK.

Carpenters' Colored, per gross 0 45 0 75
White lump .... per cwt. 0 60 0 65

CHISELS.

Warnock's, discount 72½ per cent.
P. S. & W. Extra, discount 72½ per cent

CROSSCUT SAW HANDLES.

S. & D., No. 3 .... per pair 0 15
S. & D., " 5 .... " 0 22½
S. & D., " 6 .... " 0 15
Boynton pattern .... " 0 20

CROWBARS.

3½c. to 4c. per lb.

DRAW KNIVES.

Coach and Wagon, discount 75 per cent.
Carpenters' discount 75 per cent

DRILLS.

Millar's Falls, hand and breast, net list
North Bros., each set, 50c.

DRILL BITS.

Morse, discount 37½ to 40 per cent.
Standard, discount 50 and 5 to 55 per cent.

FILES AND RASPS.

Great Western .... 75 per cent.
Arcade .... 75 "
Kearney & Foot .... 75 "
Disston's .... 75 "
American .... 75 "
J Barton Smith .... 75 "
McClellan .... 75 "
Eagle .... 75 "
Nicholson .... 66⅔ "
Globe .... 75 "
Black Diamond, 60, 10 and 5 p.c.
Jowitt's, English list, 27½ per cent.

GAUGES.

Stanley's discount 50 to 60 per cent.

Wire Gauges.

Winn's, Nos. 26 to 33 .... each 1 65 2 40

HANDLES.

C. & B., fork and hoe, 40 p. c., revised list.
American, saw .... per doz. 0 25
American, plane .... per gross 3 15 3 75
Canadian, hammer and hatchet 40 per cent.
Axe and cant hook handles, 45 per cent.

HAMMERS.

Maydole's, discount 5 to 10 per cent.
Canadian, discount 25 to 27½ per cent.
Magnetic tack .... per doz. 1 10 1 20
Canadian sledge .... per lb. 0 07 0 08½
Canadian ball pean, per lb. .. 0 22 0 25

HATCHETS.

Canadian, discount 40 to 42½ per cent.
Shingle, Red Ridge 1, per doz. 4 40
" " 2, " 4 80
Barrel Underhill 5 05

MALLETS.

Tinsmiths'	per doz.	1 25	1 50
Carpenters', hickory,	"	1 25	3 75
Lignum Vitae........	"	3 85	5 00
Caulking, each		0 60	2 00

MATTOCKS.

Canadian .... per doz. 5 50 6 00

MEAT CUTTERS.

German, 15 per cent.
American discount, 33⅓ per cent.
Gem .... each 1 15

NAIL PULLERS.

German and American	0 85	2 50
No. 1	0 85	
No 1573	0 75	

NAIL SETS.

Square, round and octagon, per gross 3 38
Diamond ....

PICKS.

Per dozen 6 00 9 00

PLANES.

Wood bench, Canadian discount 40 per cent. American discount 50 per cent.
Wood, fancy Canadian or American 37½ to 40 per cent.
Stanley planes, $1 55 to $3 60, net list prices

PLANE IRONS.

English .... per doz. 2 00 5 00
Stanley, 2½ inch, single 24c., double 39c.

PLIERS AND NIPPERS.

Button's genuine, 37½ to 40 per cent.
Button's imitation....per doz. 5 00 9 00
Berg's wire fencing 1 72 5 50

PUNCHES.

Saddlers'	per doz.	1 00	1 85
Conductor's........	"	3 00	15 00
Tinners', solid........	per set	....	0 72
" hollow........	per inch	....	1 00

RIVET SETS.

Canadian, discount 35 to 37½ per cent.

RULES.

Boxwood, discount 70 per cent.
Ivory, discount 20 to 25 per cent

SAWS.

Atkins, hand and crosscut, 25 per cent.
Disston's Hand, discount 12½ per cent
Disston's Crosscut ....per foot 0 35 0 55
Hack, complete........each 0 75 2 75
" frame only........each 0 50 1 25
S & D, solid tooth circular shingle, concave and band, 50 per cent; mill and ice, drag, 30 per cent.; crosscut, 35 per cent.; hand saws, butcher, 35 per cent.; buck, New Century, $6.25, buck, No 1 Maple Leaf, $5.25, buck, Happy Medium, $4.25; buck, Watch Spring, $4.25, buck, common frame, $4.00.
Spear & Jackson's saws—Hand or rip 26 in., $12 75; 28 in., $14.25, panel, 18 in., $8 25; 20 in., $9; tenon, 10 in., $9.90; 12 in., $10 90; 14 in., $11.50.

SAW SETS.

Lincoln and Whiting	4 75
Hand Sets, Perfect	4 00
X-Cut Sets	7 50

Maple Leaf and Premium saw sets, 40 off.
S. & D. saw swages, 40 off.

SCREW DRIVERS.

Sargent's ....per doz. 0 65 1 00
North Bros., No. 30 ..per doz. 16 80

SHOVELS AND SPADES.

Bull Dog, solid neck shovel (No 2 pol.) $18 50

	(Hollow Back)	(Reinforced S Scoop)
Moose........	$17 50	$16 30
Bear	15 50	15 30
Fox	12 50	14 30
Black Cat....	10 00	13 30

Canadian, discount 45 per cent.

SQUARES.

Iron, discount 20 per cent.
Steel, discount 65 and 10 per cent.
Try and Bevel, discount 50 to 52½ per cent.

TAPE LINES.

English, ass skin....	per doz.	2 75	5 00
English, Patent Leather		5 50	9 75
Chesterman's	each	0 90	2 85
" steel........	each	0 80	8 00
Berg's, each		0.75	2 50

TROWELS.

Disston's, discount 10 per cent.
S. & D., discount 35 per cent.
Berg's, brick, 9¼x11 .... 4 00
" pointing, 9¼x5 .... 2 10

FARM AND GARDEN GOODS

BELLS.

American cow bells, 63⅓ per cent.
Canadian, discount 45 and 50 per cent.
American, farm bells, each .. 1 35 3 00

BULL RINGS.

Copper, $1.30 for 2½-inch, and $1.70

CATTLE LEADERS.

Nos. 32 and 33 .. ...per gross 7 50 8 50

BARN DOOR HANGERS.

	doz. pairs.	
Steel barn door........	8 00	10 00
Stearns wood track	4 50	6 00
Zenith		9 00
Acme, wood track	5 00	6 50
Atlas	5 00	6 00
Perfect	8 00	11 00
New Milo........		6 50
Steel, covered	4 00	11 00
" track, 1 x 3-16 in (100 ft)	....	3 75
" 1¼ x 3-16 in (100 ft)	....	4 75
Double strap hangers, doz. sets	....	6 40
Standard jointed hangers, "	....	6 40
Steel King hangers "	....	6 25
Storm King and safety hangers	....	7 00
" " rail........	....	4 25

Chicago Friction, Oscillating and Big Twin Hangers, 5 per cent.

HARVEST TOOLS

Discount 60 per cent.
S. & D. lawn rakes, Dunn's, 40 off.
" sidewalk and stable scrapers, 40 off.

HAY KNIVES.

Net list.

HEAD HALTERS.

Jute Rope, ¾-inch....	per gross	....	9 00
" " ⅝ "	"	....	10 00
" " ½ "	"	....	12 00
Leather, 1-inch	per doz.	....	4 00
Leather, 1½ "	"	....	5 20
Web........	"	....	2 45

HOES.

Garden, Mortar, etc., discount 50 per cent.
Planter........per doz. 4 00 4 50

SCYTHES.

Per doz. net........ 6 25 9 25

SCYTHE SNATHS.

Canadian, discount 40 per cent.

SNAPS.

Harness, German, discount 25 per cent.
Lock, Andrews' 4 50 11 00

STABLE FITTINGS

Warden King, 35 per cent

WOOD HAY RAKES.

Ten tooth, 40 and 10 per cent
Twelve tooth, 45 per cent.

HEAVY GOODS, NAILS, ETC.

ANVILS.

Wright's, 80-lb. and over........	0 10½
Hay Budden, 80-lb. and over........	0 09½
Brook's, 80-lb. and over	0 11¼
Taylor-Forbes, handy	0 05
Columbia Hardware Co., per lb	0 09½

VISES.

Wright's........	....	0 14
Berg's, per lb	....	0 12½
Brook's........	....	0 12½
Pipe Vise, Hinge, No. 1........	....	3 50
" No. 2........	....	5 50
Saw Vise	4 50	5 00

Blacksmiths' (discount) 60 per cent. parallel (discount) 45 per cent.

BOLTS AND NUTS

	Per cent.
Carriage Bolts, common ($1 list) ⅜ and smaller..	60, 10 and 10
" " 7-16 and up ..	55 and 5
" " Norway Iron ($3 list)	50
Machine Bolts, ⅜ and less ...	61 and 10
Machine Bolts, 7-16 and up....	55 and 5
Plough Bolts	55 and 10
Blank Bolts........	55
Bolt Ends........	55
Sleigh Shoe Bolts, ⅜ and less ..	60 and 10
" " " 7-16 and larger	50 and 5
Coach Screws, conepoint.......	70 and 5

Nuts, square, all sizes, 4c. per cent. off
Nuts, hexagon, all sizes, 4¼c. per cent. off
Stove Rods per lb., 5½ to 6c.
Stove Bolts, 75 per cent.

CHAIN.

Proof coil, per 100 lb., 5-16 in., $4.40; ⅜ in., $3 90; 7-16 in., $3.70; ½ in., $3 50; 9-16 in., $3 45; ⅝ in., $3 35; ¾ in., $3.20; ⅞ in., $3.10; 1 in., $3.10.
Halter, kennel and post chains, 40 to 40 and 5 per cent.

Cow ties	40 p.c.
Tie out chains	65 p.c.
Stall fixtures	35 p.c.
Trace chain	45 p.c.

Jack chain iron, discount 35 p.c.
Jack chain, brass, discount 40 per cent

HORSE NAILS.

'C' brand, 40, 10 and 7½ per cent. off list } Oval
M.R.M. Co. brand, 55 per cent. } head
Capewell brand, quotations on application

HORSESHOES.

M R M Co	No. 2 and larger	No. 1 and smaller
Iron........	3 80	4 05
Snow........	4 05	4 30
Light steel	4 15	4 40
Featherweight, sizes 0 to 4		5 75
Toeweight, " 1 to 4........		7 00

Packing up to 3 sizes in a keg, 10c per 100 lbs; more than three sizes, 25c. per 100 lbs extra. F o b. Montreal, add 15c, Toronto, Hamilton and Guelph

HORSE WEIGHTS.

Taylor-Forbes, 3½c. per lb.

NAILS.	Cut.	Wire.
2d........	3 80	3 45
3d........	2 95	3 10
4 and 5d........	2 70	2 85
6 and 7d	2 60	2 75
8 and 9d	2 45	2 60
10 and 12d	2 40	2 55
16 and 20d........	2 35	2 50
30, 40, 50 and 60d (base)	2 30	2 45

F.o.b. Montreal. Cut nails, Toronto 20c. higher.
Miscellaneous wire nails, discount 75 per cent
Coopers' nails, discount 40 per cent.

PRESSED SPIKES.

Pressed spikes, ⅝ diameter, per 100 lbs, $3 15

RIVETS AND BURRS.

Iron Rivets, black and tinned, 60, 10 and 10.
Iron Burrs, discount 60 and 10 and 10 p.c.
Copper Rivets, usual proportion burrs, 15 p c.
Copper Burrs only, net list
Extras on Coppered Rivets, ½-lb. packages 1c. per lb.; ¼-lb. packages 2c. lb
Tinned Rivets, net extra, 4c. per lb.

SCREWS.

Wood, F H., bright and steel, 87½ per cent.
" R. H., bright, dis. 82½ per cent
" F. H., brass, dis. 80 per cent.
" R. H., " dis. 75 per cent.
" F. H., bronze, dis. 75 per cent.
" R. H., " dis. 70 per cent.
Drive Screws, dis. 87½ per cent.

Bench, wood	per doz.	3 25	4 00
" iron	"	4 25	5 00

Set, case hardened, dis. 60 per cent.
Square Cap, dis. 50 and 5 per cent.
Hexagon Cap, dis. 45 per cent.

MACHINE SCREWS

Flat head, iron and brass, 35 per cent.
Felister head, iron, discount 30 per cent.
" " brass, discount 25 per cent

TACKS, BRADS, ETC.

Carpet tacks, blued	80 and 5
" " tinned........	80 and 10
" " (in kegs)........	40
Cut tacks, blued, in dozens only	75 and 10
" ½ weights	60
Swedes cut tacks, blued and tinned—	
In bulk........	80 and 10
In dozens	75
Swedes, upholsterers', bulk	85 and 12½
" " brush, blued and tinned bulk	70
Swedes, gimp, blued, tinned and japanned........	75 and 12½
Zinc tacks	35
Leather carpet tacks........	40
Copper tacks........	37½
Copper nails	42½
Trunk nails, black	65
Trunk nails, tinned and blued ..	65
Clout nails, blued and tinned ..	65
Chair nails........	35
Patent brads........	40
Fine finishing........	40
Lining tacks, in papers........	10
" " in bulk........	15
" " solid heads, in bulk	75
Saddle nails, in papers........	10
" " in bulk........	15
Tufting buttons, 22 line in dozens only	60
Zinc glaziers' points	5
Double pointed tacks, papers..	90 and 10
" " " bulk....	40
Clinch and duck rivets........	45

Cheese box tacks, 85 and 5; trunk tacks, 80 and 10.

WROUGHT IRON WASHERS.

Canadian make, discount 40 per cent.

SPORTING GOODS.

AMMUNITION.

B. B. Caps American $2 00 per 1000.
C. B. Caps American, $2 60 per 1000

CARTRIDGES.

Rim Fire Cartridges, 50 and 5 p.c.
Rim Fire Pistol, 30 and 5 per cent., American.
Central Fire, Military and Sporting American, 10 per cent. advance. B.B. Caps, discount 40 per cent., American.
Central Fire Pistol and Rifle, net Amer. list
Loaded and empty Shells, American 20 per cent. discount. Rival and Nitro, 10 per cent. advance
Empty paper shells American, 10 per cent advance.
Primers, American, $2.05

Wads.	per lb.
Best thick brown or grey felt wads, in ¼-lb. bags	$0 70
Best thick white card wads, in boxes of 500 each, 12 and smaller gauges	0 29
Best thick white card wads, in boxes of 500 each, 10 gauge........	0 35
Thin card wads, in boxes of 1,000 each, 12 and smaller gauges	0 20
Thin card wads, in boxes of 1,000 each, 10 gauge........	0 25
Chemically prepared black edge grey cloth wads, in boxes of 250 each—	Per M
11 and smaller gauge	0 60
9 and 10 gauges........	0 70
7 and 8 "	0 90
5 and 6 "	1 10
Superior chemically prepared pink edge, best white cloth wads in boxes of 250 each—	
11 and smaller gauge	1 15
9 and 10 gauges	1 40
7 and 8 "	1 65
5 and 6 "	1 90

SHOT

Ordinary drop shot, AAA to dust $7.50 per 100 lbs. Discount 5 per cent; cash discount 2 per cent. 30 days; net extras as follows subject to cash discount only; Chilled, 40c; buck and seal, 80c; no 28 ball, $1 20 per 100 lbs.; bags less than 25 lbs., ½c. per lb; F.O.B. Montreal, Toronto, Hamilton, London, St. John and Halifax, and freight equalized thereon.

TRAPS (steel.)

Game, Newhouse, discount 30 and 10 per cent.
Game, Hawley & Norton, 50, 10 & 5 per cent.
Game, Victor, 70 per cent.
Game, Oneida Jump (B & L) 40 & 2½ p. c.
Game, steel, 60 and 5 per cent.

SKATES

Skates, discount 37½ per cent.
Mic Mac hockey sticks, per doz 4 00 5 00

CUTLERY AND SILVERWARE.

RAZORS. per doz.

Elliot's 4 00 18 00
Boker's 7 50 11 00
" King Cutter 13 50 18 50
Wade & Butcher's............ 3 60 10 00
Lewis Bros.' "Kleen Kutter' 8 50 10 50
Henckel's 7 50 20 00
Berg's 7 50 20 00
Clauss Razors and Strops, 50 and 10 per cent

KNIVES.

Farriers-Stacey Bros., doz 3 50

PLATED GOODS

Hollowware, 40 per cent discount.
Flatware, staples, 40 and 10, fancy, 40 and 5.

SHEARS.

Clauss, nickel, discount 60 per cent.
Clauss, Japan, discount 67½ per cent.
Clauss, tailors, discount 40 per cent.
Seymour's, discount 50 and 10 per cent.
Berg's............ .... 6 00 12 00

HOUSE FURNISHINGS.

APPLE PARERS.

Woodyatt Hudson, per doz., net 4 50

BIRD CAGES.

Brass and Japanned, 40 and 10 p. c.

COPPER AND NICKEL WARE.

Copper boilers, kettles, teapots, etc. 35 p.c.
Copper pitts, 25 per cent.

ENAMELED WARE.

London, White, Princess, Turquoise, Onyx, Blue and White, discount 50 per cent.
Canada, Diamond, Premier, 50 and 10 p c.
Pearl, Imperial Crescent, 50 and 10 per cent.
Premier steel ware, 40 per cent.
Star decorated steel and white, 25 per cent.
Japanned ware, discount 45 per cent.
Hollow ware, tinned cast, 35 per cent. off.

KITCHEN SUNDRIES

Can openers, per doz.............. 0 40 0 75
Mincing knives per doz 0 50 0 80
Duplex mouse traps, per doz 0 65
Potato mashers, wire, per doz... 0 60 0 70
" wood " .. 0 50 0 60
Vegetable slicers, per doz 2 25
Universal meat chopper No. 0, $1; No.1, 1 15.
Enterprise chopper, each 1 30
Spiders and fry pans, 50 per cent.
Star Al chopper 5 to 32 .... 1 35 4 10
" " 100 to 103......... 1 35 2 00
Kitchen hooks, bright 0 62½

LAMP WICKS.

Discount, 60 per cent.

LEMON SQUEEZERS.

Porcelain lined..... per doz. 2 20 5 60
Galvanized.......... " 1 87 3 85
King, wood.......... " 2 75 2 90
King, glass.......... " 4 00 4 50
All glass .... " 0 50 0 90

METAL POLISH.

Tandem metal polish paste 6 00

PICTURE NAILS.

Porcelain head ......per gross 1 75 1 50
Brass head.......... 0 40 1 00
Tin and gilt, picture wire, 75 per cent.

SAD IRONS.

Mrs. Potts, No. 55, polished....per set 0 80
" No. 50, nickle-plated, " 0 9½
Common, plain......................... 4 50
" plated 4 50
Asbestos, per set 1 25

TINWARE.

CONDUCTOR PIPE.

2-in. plain or corrugated, per 100 feet, $3.30; 3 in, $4 40; 4 in, $5 80; 5 in, $7 45; 6 in., $9.9 .

FAUCETS.

Common, cork-lined, discount 35 per cent.

EAVETROUGHS.

10-inch ....per 100 ft. 3 30

FACTORY MILK CANS

Discount off revised list, 35 per cent
Milk can trimmings, discount 25 per cent.
Creamery Cans, 45 per cent.

LANTERNS.

No. 2 or 4 Plain Cold Blast ...per doz. 6 50
Lift Tubular and Hinge Plain, " 4 75
No 0, safety................. " 4 00
Better quality at higher prices.
Japanning, 50c. per doz extra.
Prism globes, per doz, $1 20.

OILERS.

Kemp's Tornado and McClary's Model galvanized oil can, with pump, 5 gallon, per dozen 10 92
Davidson oilers, discount 40 per cent
Zinc and tin, discount 50 per cent
Coppered oilers, 20 per cent. off.
Brass oilers, 50 per cent. off.
Malleable, discount 25 per cent

PAILS (GALVANIZED).

Dufferin pattern pails, 45 per cent.
Flaring pattern, discount 45 per cent.
Galvanized washtubs 40 per cent

PIECED WARE.

Discount 35 per cent off list, June, 1899.
10-qt. flaring sap buckets, discount 35 per cent.
6, 10 and 14-qt. flaring pails dis. 35 per cent.
Copper bottom tea kettles and boilers, 30 p.c.
Coal hods, 40 per cent.

STAMPED WARE.

Plain, 75 and 12½ per cent. off revised list.
Retinned, 72½ per cent revised list.

SAP SPOUTS.

Bronzed iron with hooks ....per 1,000 7 50
Eureka tinned steel, hooks 8 00

STOVEPIPES.

5 and 6 inch, per 100 lengths 7 64 7 91
7 inch .... " " 8 18
Nestable, discount 40 per cent.

STOVEPIPE ELBOWS

5 and 6-inch, common.........per doz. 1 32
7-inch......................... " 1 48
Polished, 15c. per dozen extra.

THERMOMETERS.

Tin case and dairy, 75 to 75 and 10 per cent

TINNERS' SNIPS.

Per doz...................... 3 00 15
Clauss, discount 35 per cent.

TINNERS' TRIMMINGS

Discount, 45 per cent.

WIRE.

ANNEALED CUT HAY BAILING WIRE.

No. 12 and 13, $4; No 13½, $4 10; No 14, $4 2 ; No 15 $4 50, in lengths 6' to 11', 25 per cent., other lengths 20c per 100 lbs extra; if eye or loop on end add 25c per 100 lbs. to the above

BRIGHT WIRE GOODS

Discount 62½ per cent.

CLOTHES LINE WIRE

7 wire solid line, No. 17 $4 90; No. 18, $3.00; No. 19, $2.70; 2 wire solid line, No. 17, $4.45; No. 18, $2.80, No. 19, $2.50. All prices per 1000 ft. measure. F.o b. Hamilton Toronto, Montreal

COILED SPRING WIRE

High Carbon, No. 9, $2 90, No 11, $3 45; No. 12, $3 15.

COPPER AND BRASS WIRE.

Discount 37½ per cent

FINE STEEL WIRE.

Discount 25 per cent. List of extras: In 100-lb. lots: No. 17, $5 — No. 18, $5.50 — No. 19, $6 — No. 20, $6.65 — No. 21, $7 — No. 22, $7.30 — No 23, $7.65 — No. 24, $8 — No. 25, $9—No. 26, $9.50—No. 27, $10—No. 28, $11—No. 29, $12—No 30, $13—No. 31, $14—No. 32, $15—No. 33, $16—No. 34, $17. Extras net—tinned wire, Nos. 17-25, $2—Nos. 26-31, $4—Nos. 32-34, $6. Coppered, 75c.—oiling, 10c.—in 25-lb. bundles, 15c.—in 5 and 10-lb. bundles, 25c.—in 1-lb. hanks, 25c. —in ½-lb. hanks, 38c.—in ¼-lb. hanks, 50c.—packed in casks or cases, 15c.—bagging or papering, 10c.

FENCE STAPLES.

Bright.......... 2 75 Galvanized..... 3 15.

HAY WIRE IN COILS.

No. 13, $2 60; No. 14, $2 70; No. 15, $2.85; f.o b., Montreal.

GALVANIZED WIRE

Per 100 lb.—Nos. 4 and 5, $3.70 — Nos. 6, 7, 8, $3.15 — No 9, $2.50 — No. 10, $3 20 — No. 11, $3 25—No. 12, $2.65 —No. 13, $2 75—No. 14, $3.75—No 15, $4.30 —No. 16, $4.30 from stock. Base sizes, Nos. 6 to 9, $2.35 f.o.b. Cleveland. In carlots 12½c. less.

LIGHT STRAIGHTENED WIRE.

Gauge No.	Over 20 in. per 100 lbs.	10 to 20 in.	5 to 10 in.
0 to 5	$0.50	$0 75	$1.25
6 to 9	0.75	1.25	2 00
10 to 11	1 00	1.75	2.50
12 to 14	1 50	2.25	3 50
15 to 16	2.00	3.00	4 50

SMOOTH STEEL WIRE.

No. 0-9 gauge, $2 30; No 10 gauge, 6c. extra; No. 11 gauge, 12c extra; No 12 gauge, 20c. extra; No. 13 gauge, 30c. extra; No 14 gauge, 40c. extra; No. 15 gauge, 55c. extra; No. 16 gauge, 70c extra. Add 60c. for coppering and $2 for tinning.

Extra net per 100 lb.—Oiled wire 10c., spring wire $1 25, bright soft drawn 15c., charcoal (extra quality) $1.25, packed in casks or cases 15c., bagging and papering 10c., 50 and 100-lb. bundles 10c., in 25-lb. bundles 15c., in 5 and 10-lb. bundles 25c., in 1-lb hanks, 50c., in ½-lb. hanks 75c., in ¼-lb. hanks $1.

POULTRY NETTING.

2 in mesh 19 w. g., discount 50 and 10 per cent. All others 50 per cent.

WIRE CLOTH.

Painted Screen, in 100-ft. rolls, $1 62½, per 100 sq. ft.; in 50-ft. rolls $1 67½, per 100 sq ft. Terms, 2 per cent. off 30 days.

WIRE FENCING.

Galvanized barb...... .... 2 95
Galvanized, plain twist 3 30
Galvanized barb, f o b, Cleveland, $2.70 for small lots and $2.60 for carlots.

WOODENWARE.

CHURNS.

No. 0, $9; No. 1, $9; No. 2, $10; No. 3, $11; No. 4, $13; No. 5, $16; f.o b. Toronto Hamilton, London and St. Marys, 30 and 30 per cent; f o b Ottawa, Kingston and Montreal, 40 and 15 per cent discount, Taylor-Forbes, 30 and 30 per cent.

CLOTHES REELS.

Davis Clothes Reels, dis. 40 per cent.

LADDERS, EXTENSION.

Waggoner Extension Ladders, dis. 40 per cent.

MOPS AND IRONING BOARDS.

"Best" mops.......................... 1 25
"900" mops 1 25
Folding ironing boards......... 12 00 16 50

REFRIGERATORS

Discount, 40 per cent.

SCREEN DOORS.

Common doors, 2 or 3 panel, walnut stained, 4-in. style..........per doz. 7 25
Common doors, 2 or 3 panel, grained only, 4-in., style ....per doz. 7 55
Common doors, 2 or 3 panel, light stair per doz........................... 9 55

WASHING MACHINES.

Round, re-acting per doz. 60 00
Square " 63 00
Eclipse, per doz 54 00
Dowswell " 39 00
New Century, per doz 75 00
Daisy............................... 54 00

WRINGERS

Royal Canadian, 11 in., per doz. ... 34 00
Royal American, 11 in. .. 34 00
Eze, 10 in., per doz 37 75
Terms, 2 per cent., 30 days.

MISCELLANEOUS.

AXLE GREASE.

Ordinary, per gross......... 6 00 7 00
Best quality..... .... 10 00 12 00

BELTING.

Extra, 60 per cent.
Standard, 60 and 10 per cent.
No. 1, not wider than 6 in., 60, 10 and 10 p.c.
Agricultural, not wider than 4 in, 75 per cent
Lace leather, per side, 75c; cut laces, 80c.

BOOT CALKS.

Small and medium, ball ....per M 4 25
Small heel .... " 4 50

CARPET STRETCHERS.

American ....per doz. 1 00 1 50
Bullard's............. " 6 50

CASTORS.

Bed, new list, discount 55 to 57½ per cent.
Plate, discount 52½ to 57½ per cent.

PINE TAR.

½ pint in tins ....per gross ... 7 80
" " .... " 9 60

PULLEYS.

Hothouse ....per doz. 0 55 1 00
Axle.................... " 0 22 0 33
Screw .... " 0 22 1 00
Awning .... " 0 35 2 50

PUMPS.

Canadian cistern 1 40 2 00
Canadian pitcher spout ... 1 80 3 16
Berg's wing pump, 75 per cent.

ROPE AND TWINE.

Sisal.................................. 0 10½
Pure Manilla 0 15½
"British" Manilla..................... 0 12
Cotton, 3-16 inch and larger......0 21 0 23
" 5-32 inch 0 25 0 27
" ¼ inch 0 25 0 28
Russia Deep Sea 0 16
Jute................................. 0 09
Lath Yarn, single 0 10
" double 0 10½
Sisal bed cord, 48 feet........per doz. 0 65
" " 60 feet........ " 0 80
" " 72 feet........ " 0 95

Twine.

Bag, Russian twine, per lb. 0 27
Wrapping, cotton, 3-ply 0 25
" " 4-ply 0 29
Mattress twine per lb. 0 33 0 45
Staging " " 0 27 0 35

SCALES.

Gurney Standard, 40 per cent.
Gurney Champion, 50 per cent.
Burrow, Stewart & Milne—
Imperial Standard, discount 40 per cent.
Weigh Beams, discount 40 per cent.
Champion Scales, discount 50 per cent.
Fairbanks standard, discount 35 per cent.
" Dominion, discount 55 per cent.
" Richelieu, discount 55 per cent.
Warren new Standard, discount 40 per cent.
" " Champion, discount 50 per cent.
" Weighbeams, discount 35 per cent.

STONES—OIL AND SCYTHE.

Washita ....per lb. 0 25 0 37
Hindostan .... " 0 06 0 10
" slip .... " 0 18 0 20
" Axe.......... " 0 10
Deer Creek............ " 0 10
Deerlick .... " 0 25
" Axe .... " 0 15
Lily white .... " 0 42
Arkansas.............. " 1 50
Water-of-Ayr .... " 0 10
Scythe ....per gross 3 50 5 00
Grind, 40 to 200 lb., per ton.... 20 00 22 00
" under 40 lb., " 24 00
" 200 lb. and over 26 00

INDEX TO ADVERTISERS.

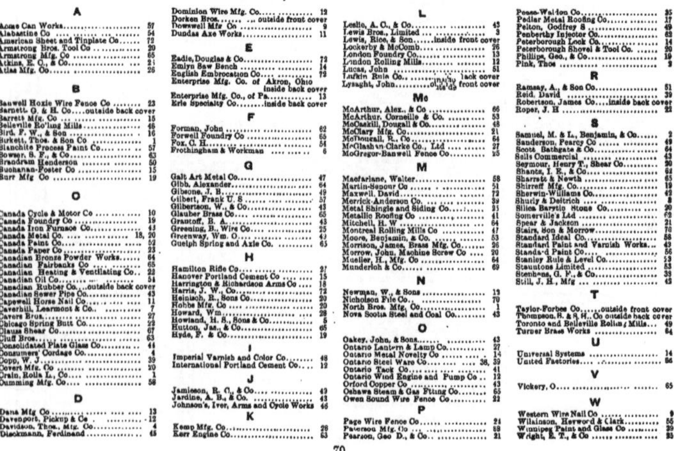

CLASSIFIED LIST OF ADVERTISEMENTS.

Auditors.
Davenport, Pickup & Co., Winnipeg.

Babbitt Metal.
Canada Metal Co., Toronto.
Canadian Fairbanks Co., Montreal.
Robertson, Jas. Co., Montreal.

Bath Room Fittings.
Carriage Mounting Co , Toronto.

Belting, Hose, etc.
Gutta Percha and Rubber Mfg. Co. Toronto.

Bicycles and Accessories.
Johnson's. Iver, Arms and Cycle Works Fitchburg, Mass

Binder Twine.
Consumers Cordage Co., Montreal.

Box Strap.
J. N. Warminton, Montreal.

Brass Goods.
Glauber Brass Mfg. Co., Cleveland, Ohio.
Lewis, Rice, & Son., Toronto.
Morrison, Jas., Brass Mfg. Co., Toronto.
Mueller Mfg. Co., Decatur, Ill.
Penberthy Injector Co., Windsor, Ont.
Taylor-Forbes Co., Guelph, Ont.

Bronze Powders.
Canadian Bronze Powder Works, Montreal.

Brushes.
Ramsay, A., & Son Co., Montreal.

Can Openers.
Cumming Mfg. Co. Renfrew.

Cans.
Acme Can Works, Montreal.

Builders' Tools and Supplies.
Covert Mfg. Co., West Troy, N.Y.
Frothingham & Workman Co , Montreal.
Howland, H. S., Sons & Co., Toronto.
Hyde, F., & Co., Montreal.
Lewis Bros. & Co., Montreal.
Lewis, Rice, & Son, Toronto.
Lockerby & McComb, Montreal.
Lufkin Rule Co., Saginaw, Mich.
Newman & Sons, Birmingham.
North Bros. Mfg. Co., Philadelphia, Pa.
Stanley Rule & Level Co., New Britain.
Stanley Works, New Britain, Conn.
Stephens, G. F., Winnipeg.
Taylor-Forbes Co., Guelph, Ont.

Carriage Accessories.
Carriage Mountings Co., Toronto.
Covert Mfg. Co., West Troy, N.Y.

Carriage Springs and Axles.
Guelph Spring and Axle Co., Guelph.

Cattle and Trace Chains.
Greening. B., Wire Co., Hamilton.

Churns.
Dowswell Mfg. Co., Hamilton.

Clippers—All Kinds.
American Shearer Mfg. Co.,Nashua,N.H.

Clothes Reels and Lines.
Hamilton Cotton Co., Hamilton, Ont.

Cordage.
Consumers' Cordage Co., Montreal.
Hamilton Cotton Co., Hamilton.

Cork Screws.
Erie Specialty Co., Erie, Pa.

Clutch Nails.
J. N. Warminton, Montreal.

Cut Glass.
Phillips, Geo., & Co., Montreal.

Cutlery—Razors, Scissors, etc.
Birkett, Thos., & Son Co., Ottawa.
Clauss Shear Co., Toronto
Dorken Bros. & Co., Montreal.
Heinisch's, R., Sons Co., Newark, N.J.
Howland, H. S. Sons & Co., Toronto.
Lamplough, F. W., & Co , Montreal.
Phillips, Geo. & Co., Montreal.
Round, John, & Son, Montreal.

Electric Fixtures.
Canadian General Electric Co., Toronto.
Forman, John, Montreal.
Morrison James, Mfg. Co., Toronto.
Munderloh & Co., Montreal.

Electro Cabinets.
Cameron & Campbell Toronto.

Engines, Supplies, etc.
Kerr Engine Co., Walkerville, Ont.

Fencing—Woven Wire.
McGregor-Banwell Fence Co., Walkerville, Ont.
Owen Sound Wire Fence Co., Owen Sound
Banwell · Hoxie Wire Fence Co , Hamilton.

Files and Rasps.
Barnett Co., G. & H., Philadelphia, Pa.
Nicholson File Co., Port Hope

Financial Institutions
Bradstreet Co.

Firearms and Ammunition.
Dominion Cartridge Co., Montreal.
Hamilton Rifle Co , Plymouth, Mich.
Harrington & Richardson Arms Co., Worcester, Mass.
Johnson's, Iver, Arms and Cycle Works, Fitchburg, Mass.

Food Choppers
Enterprise Mfg. Co., Philadelphia, Pa.
Lamplough, F. W., & Co., Montreal.
Shirreff Mfg Co., Brockville, Ont.

Galvanizing.
Canada Metal Co., Toronto
Montreal Rolling Mills Co., Montreal
Ontario Wind Engine & Pump Co., Toronto.

Glaziers' Diamonds.
Gibsone, J. B., Montreal.
Pelton, Godfrey S.
Sharratt & Newth, London, Eng.
Shaw, A., & Son, London, Eng.

Handles.
Still, J. H., Mfg. Co.

Harvest Tools.
Maple Leaf Harvest Tool Co , Tillsonburg, Ont.

Hoop Iron.
Montreal Rolling Mills Co , Montreal.
J. N. Warminton, Montreal.

Horse Blankets.
Heney, E. N., & Co., Montreal.

Horseshoes and Nails.
Canada Horse Nail Co., Montreal
Montreal Rolling Mills, Montreal.
Capewell Horse Nail Co., Toronto.

Hot Water Boilers and Radiators.
Cluff, R. J., & Co Toronto.
Pease Foundry Co. Toronto.
Taylor-Forbes Co., Guelph.

Ice Cream Freezers.
Dana Mfg. Co , Cincinnati, Ohio.
North Bros. Mfg. Co., Philadelphia, Pa

Ice Cutting Tools.
Erie Specialty Co., Erie, Pa.
North Bros. Mfg. Co., Philadelphia, Pa.

Injectors—Automatic.
Morrison, Jas., Brass Mfg Co., Toronto.
Penberthy Injector Co., Windsor, Ont.

Iron Pipe.
Montreal Rolling Mills, Montreal.

Iron Pumps.
Lamplough, F. W., & Co., Montreal.
McDougall, R., Co., Galt, Ont.

Lanterns.
Kemp Mfg. Co., Toronto.
Ontario Lantern Co., Hamilton, Ont.
Wright. E. T., & Co., Hamilton.

Lawn Mowers.
Birkett, Thos., & Son Co., Ottawa.
Maxwell. D., & Sons, St. Mary's, Ont.
Taylor, Forbes Co., Guelph.

Lawn Swings, Settees, Chairs.
Cumming Mfg. Co., Renfrew.

Ledgers—Loose Leaf.
Business Systems T ronto
Copeland-Chatterson Co , Toronto.
Crain, Rolla L., Co., Ottawa.
Universal Systems, Toronto.

Locks, Knobs, Escutcheons, etc.
Peterborough Lock Mfg. Co., Peterborough, Ont.

Lumbermen's Supplies.
Pink, Thos., & Co., Pembroke Ont.

Manufacturers' Agents.
Fox, C. H., Vancouver.
Gibb, Alexander, Montreal.
Mitchell, David C , & Co., Glasgow, Scot.
Mitchell, H. W., Winnipeg.
Pearce, Frank, & Co. Liverpool, Eng.
Scott, Bathgate & Co., Winnipeg.
Thorne, R. E., Montreal and Toronto.

Metals.
Canada Iron Furnace Co., Midland, Ont.
Canada Metal Co., Toronto.
Eadie, H. G., Montreal.
Gibb, Alexander, Montreal.
Kemp Mfg Co., Toronto
Leslie, A. C., & Co., Montreal.
Lysaght, John, Bristol, Eng.
Nova Scotia Steel and Coal Co , New Glasgow, N.S.
Robertson, Jas , Co., Montreal.
Roper, J. H , Montreal.
Samuel, Benjamin & Co , Toronto.
Stairs, Son & Morrow, Halifax, N.S.
Thompson, B. & S. H. & Co. Montreal.

Metal Lath.
Galt Art Metal Co., Galt.
Metallic Roofing Co., Toronto.
Metal Shingle & Siding Co., Preston, Ont.

Metal Polish, Emery Cloth, etc
Oakey, John, & Sons, London, Eng.

Mops.
Cumming Mfg Co., Refrew.

Mouse Traps.
Cumming Mfg. Co., Renfrew.

Oil Tanks.
Bowser, S. F., & Co., Toronto.

Paints, Oils, Varnishes, Glass.
Bell, Thos., Sons & Co., Montreal.
Canada Paint Co., Montreal.
Canadian Oil Co., Toronto
Consolidated Plate Glass Co., Toronto.
Fenner, Fred., & Co., London, Eng.
Henderson & Potts Co., Montreal.
Imperial Varnish and Color Co., Toronto
Jamieson, R. C., & Co., Montreal.
McArthur, Corneille & Co., Montreal.
McCaskill, Dougall & Co., Montreal.
Montreal Rolling Mills Co., Montreal.
Moore, Benjamin, & Co. Toronto.
Queen City Oil Co., Toronto.
Ramsay & Son, Montreal.
Sherwin-Williams Co., Montreal.
Standard Paint and Varnish Works Windsor, Ont.
Stephens & Co , Winnipeg.
Martin-Senour Co., Chicago.

Perforated Sheet Metals.
Greening, B., Wire Co., Hamilton.

Plumbers' Tools and Supplies.
Borden Co., Warren, Ohio.
Canadian Fairbanks Co., Montreal.
Cluff, R. J., & Co., Toronto.
Glauber Brass Co., Cleveland, Ohio.
Jardine, A. B., & Co , Hespeler, Ont.
Jenkins Bros., Boston, Mass.
Lewis, Rice, & Son, Toronto.
Merrell Mfg. Co., Toledo, Ohio.
Montreal Rolling Mills Montreal.
Morrison, Jas., Brass Mfg. Co., Toronto.
Mueller, H., Mfg. Co., Decatur, Ill.
Oshawa Steam & Gas Fitting Co., Oshawa
Robertson Jas., Co. Montreal.
Stairs, Son & Morrow, Halifax, N.S.
Standard Ideal Sanitary Co., Port Hope.
Standard Sanitary Co., Pittsburg.
Stephens, G. F., & Co., Winnipeg, Man.
Turner Brass Works, Chicago.
Vickery, Orlando, Toronto.

Portland Cement.
International Portland Cement Co., Ottawa, Ont.
Hanover Portland Cement Co., Hanover, Ont.
Hyde, F., & Co., Montreal.
Thompson, B. & S. H. & Co., Montreal.

Potato Mashers.
Cumming Mfg. Co., Renfrew.

Poultry Netting.
Greening, B., Wire Co., Hamilton, Ont.

Razors.
Clauss Shear Co., Toronto.

Refrigerators.
Fabien, C. P., Montreal.

Roofing Supplies.
Brantford Roofing Co., Brantford.
McArthur, Alex., & Co., Montreal.
Metal Shingle & Siding Co., Preston, Ont.
Metallic Roofing Co , Toronto.
Paterson Mfg. Co., Toronto & Montreal.

Saws.
Atkins, E. C., & Co., Indianapolis, Ind
Lewis Bros., Montreal.
Shurly & Dietrich, Galt, Ont.
Spear & Jackson, Sheffield, Eng.

Scales.
Canadian Fairbanks Co., Montreal.

Screw Cabinets.
Cameron & Campbell. Toronto.

Screws, Nuts, Bolts.
Montreal Rolling Mills Co., Montreal.
Morrow, John, Machine Screw Co., Ingersoll, Ont.

Sewer Pipes.
Canadian Sewer Pipe Co., Hamilton
Hyde, F., & Co., Montreal.

Shelf Boxes.
Cameron & Campbell, Toronto.

Shears, Scissors.
Clauss Shear Co., Toront .

Shelf Brackets.
Atlas Mfg. Co., New Haven, Conn

Shellac
Bell, Thos , Sons & Co , Montreal.

Shovels and Spades.
Canadian Shovel Co , Hamilton.
Peterboro Shovel & Tool Co., Peterboro.

Silverware.
Phillips, Geo., & Co., Montreal.
Round, John, & Son, Sheffield, Eng.

Skates.
Canada Cycle & Motor Co., Toronto.

Spring Hinges, etc.
Chicago Spring Butt Co., Chicago, Ill.

Steel Rails.
Nova Scotia Steel & Coal Co., New Glasgow, N.S.

Stoves, Tinware, Furnaces
Canadian Heating & Ventilating Co., Owen Sound.
Canada Stove Works, Harriston, Ont.
Clare Bros. & Co., Preston.
Davidson, Thos., Mfg. Co., Montreal,
Guelph Stove Co., Guelph.
Gurney Foundry Co., Toronto.
Harris, J. W., Co., Montreal.
Joy Mfg. Co., Toronto.
Kemp Mnfg. Co. Toronto.
McClary Mfg. Co. London.
Pease Foundry Co., Toronto.
Stewart, Jas., Mfg. Co., Woodstock, Ont.
Taylor-Forbes Co., Guelph, Ont.
Wright, E. T., & Co., Hamilton.

Tacks.
Montreal Rolling Mills Co., Montreal.
Ontario Tack Co., Hamilton.

Ventilators.
Harris, J. W., Co., Montreal.
Pearson, Geo. D., Montreal.

Washing Machines, etc.
Dowswell Mfg. Co., Hamilton, Ont.
Taylor-Forbes Co., Guelph, Ont.

Wheelbarrows.
London Foundry Co., London, Ont.

Wholesale Hardware.
Birkett, Thos., & Sons Co., Ottawa.
Caverhill, Learmont & Co., Montreal.
Frothingham & Workman, Montreal.
Hobbs Hardware Co., London.
Howland, H. S., Sons & Co., Toronto.
Lamplough, F. W., & Co., Montreal.
Lewis Bros. & Co., Montreal.
Lewis, Rice, & Son, Toronto.

Window and Sidewalk Prisms.
Hobbs Mfg. Co., London, Ont.

Wire Springs.
Guelph Spring Axle Co., Guelph, Ont.
Wallace-Barnes Co., Bristol, Conn.

Wire, Wire Rope, Cow Ties, Fencing Tools, etc.
Canada Fence Co., London.
Dennis Wire and Iron Co., London, Ont.
Dominion Wire Mnfg. Co., Montreal
Greening, B., Wire Co., Hamilton.
Montreal Rolling Mills Co., Montreal.
Western Wire & Nail Co., London, Ont.

Woodenware.
Taylor-Forbes Co., Guelph, Ont.

Wrapping Papers.
Canada Paper Co., Toronto.
McArthur, Alex., & Co , Montreal.
Stairs, Son & Morrow, Halifax, N.S.

That Stuffy Room

is more deadly than disease itself. Bad air is the curse of crowded living. In all public places such as asylums, clubs, big offices, schools and theatres, **PURE AIR CAN BE OBTAINED ONLY BY OUR SYSTEM OF VENTILATION.**

It has stood the tests of scientific experts.

It is in use in the finest buildings in Canada.

Its price is reasonable and its benefits lasting.

Testimony:

MONTREAL, June 7th, 1901.

MESSRS. J. W. HARRIS CO., LIMITED, Montreal

Dear Sirs:

I have been using your "AEOLIAN" Ventilators on several houses in this city with the best results. The tenants of these houses declared that your Ventilators are of great advantage to them. I also installed two (2) 18 in. on my stable and I can say that your Ventilator cannot be surpassed by any to take out the sweating or bad smell in any room.

F. D. MONK.

Ask for booklet re Aeolian and Zephyr Ventilators, also booklet on Filter, etc.

The J. W. HARRIS CO., Limited

Contractors and Manufacturers - MONTREAL

D. C. Mitchell & Co.

GLASGOW

STEEL PLATES, SHEETS, ANGLES, BARS, PIG IRON,

REPRESENTATIVE

Eadie-Douglas Company

22 St. John St., MONTREAL

SUPPORT CANADIAN INDUSTRIES

FREE

TO

HARDWARE DEALERS

We are supplying electros of the cut in this ad. for use in your local advertising, besides offering special terms for immediate orders for

"LION" Brand

PUTZ TRIPLE EXTRACT

METAL POLISH

IN LIQUID OR PASTE

The most reliable metal polish in the market. Once tried always used.

Support Canadian Industries

Manufactured by

English Embrocation Co., Montreal

Specify the Following Products

Manufactured by

American Sheet and Tin Plate Company

PITTSBURG, PA.

If you want the best.

"Apollo Best Bloom," Galvanized Sheets,
Blue Annealed and American Bessemer Black Sheets,
Wellsville Polish, Range Steel,
Wood's "A" Planished Iron,
Black Steel specialties for all purposes,
Tinplates, Coke and Charcoal, all grades,
Terne Plates, Old Style and New Method Processes.

Place your order with

B. & S. H. THOMPSON & CO.
LIMITED

Dominion of Canada Sales Agents

MONTREAL, CANADA

HARDWARE BABBITTS

No. 1, No. 2, No. 3

The wearing quality of our

HARDWARE BABBITTS

is the secret of the large orders the

HARDWARE MERCHANTS

are sending us. If not on our list

GET IN LINE

LIBERAL DISCOUNTS. TERMS RIGHT

The James Robertson Co., Limited

144 William St., MONTREAL, QUE.

PFLUEGER'S Fishing Tackle

Hooks, Flies, Trolls, Spinners, Phantoms, Reels, Furnished Lines.

Everything in Fishing Tackle.

NOTICE—Free to Any Dealer in Sporting Goods, Sent Express Prepaid, 170-Page Illustrated Catalogue No. F25, and Metal Fish Sign in 8-Color Lithograph.

The Enterprise Mfg. Co.

AKRON, Ohio, U.S.A.

WALKER CORK SCREWS

are good holiday souvenirs and live sellers the whole year. Are made well and are made out of good material. That is why we guarantee every one that we put out. When ordering Cork Screws try **Walker's**

Manufactured only by **ERIE SPECIALTY COMPANY,** ERIE, Pa. U.S.A.

LUFKIN MEASURING TAPES

Steel, Metallic, Linen, Pocket, Ass Skin, Pat. Leather, Bend Leather, Etc.

ARE THE BEST AND MOST POPULAR TAPES IN THE WORLD.
YOUR STOCK IS NOT COMPLETE WITHOUT THEM.

LUFKIN RULE CO., Saginaw, Mich, U.S.A.

London Office and Warehouse—24 and 26 Holborn. New York City Branch—280 Broadway.

For sale by ALL PROMINENT CANADIAN HARDWARE JOBBERS.

Special Prize

Gold Medal at Atlanta, 1895

Copy of catalogue sent free to any interested file user upon application.

"Redstone"

High Pressure

Sheet Packing

A packing that will hold. For use in highest pressures for steam, hot or cold water and air. Packs equally well for all.

From actual tests, we believe that this packing is the most durable and satisfactory of any on the market. Try a sample lot and see for yourself.

Manufactured Solely by

THE GUTTA PERCHA & RUBBER MFG. CO.

of TORONTO, LIMITED

HEAD OFFICES,

47 Yonge Street, Toronto.

Branches: Montreal, Winnipeg, Vancouver.

SHEET ZINC,

SHEET COPPER,

SPELTER,

INGOT TIN,

BRASS and COPPER TUBES.

For lowest prices send your enquiries to

B. & S. H. THOMPSON & Co.

LIMITED

53 ST. SULPICE STREET, MONTREAL

Our Packing is Different—and Better

THE CANADIAN RUBBER CO. OF MONTREAL LIMITED

Mark of Quality.

155 Granville Street, - Halifax, N.S.
Imperial Bank Bldg., St James Street } Montreal, Que.
Front & Yonge Sts., - Toronto, Ont.
89 Princess St., - Winnipeg, Man.
403 Cordova St., - Vancouver, B.C.
Dewdney St., - - Regina, Sask.
First St. W., - - Calgary, Alta.
Wharf St., - - - Victoria, B.C.

CIRCULATES EVERYWHERE IN CANADA

Also in Great Britain, United States, West Indies, South Africa and Australia.

HARDWARE AND METAL

A Weekly Newspaper Devoted to the Hardware, Metal, Heating and Plumbing Trades in Canada.

Office of Publication, 10 Front Street East, Toronto.

VOL. XIX. MONTREAL, TORONTO, WINNIPEG, MARCH 16, 1907 NO. 11.

See Classified List of Advertisements on Page 67.

CEMENT TOOLS

THE NEW CEMENT ERA

is as yet in its infancy, and with its rapid growth we will assist the cement worker in perfecting with ease his most troublesome and symmetrical, as well as plain, work in the building of cement (concrete) walks, steps, gutters, curbing, drives, etc.

Our Tools are of Special Hard Bronze, which cannot rust—saves cleaning the tools after using them.

Each Tool is highly polished and fitted with a Hardwood Handle.

EDGER

RIB ROLLER

CURB EDGER

CORNER TROWEL

CURB EDGER

WRITE FOR PRICES

RICE LEWIS & SON

LIMITED

TORONTO.

GET YOUR SHARE

of the Horse Clipper Business this season. BUSINESS IS SURE when you stock up with our line of HAND CLIPPERS, FETLOCK CLIPPERS and SINGEING LAMPS.

Hurd's No. 274A Clipper.
Boker's No. 1669 "
" " 1668 "
" " 1704 "
Burman's No. 294 Clipper.
Brown's Newmarket "
"Hunter" Fetlock "
No. 1596 Fetlock Shears.
No. 160 Singeing Lamp.
No. 165 " "
No. 156 " "
Eclipse " "

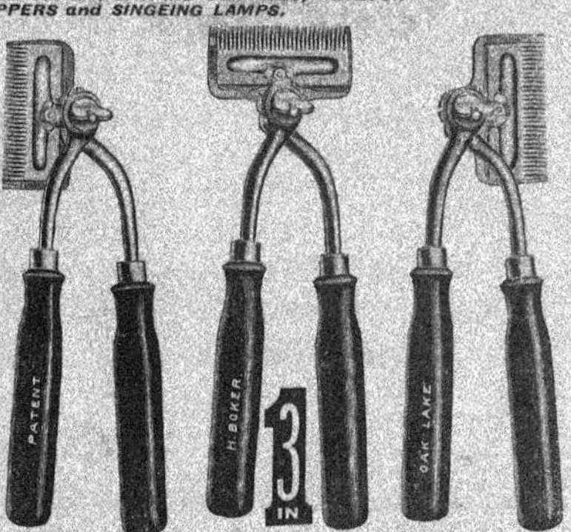

Boker's No. 4093 Clipper, Reversible, Nickel Plated, Ebonized Handles, Steel Ferrules.

Stewart No. 1 (enclosed type).

THE STEWART No. 1 HAND POWER CLIPPER

is a new clipping machine, sturdy, compact and perfect in every detail of material and workmanship, and at a price within the reach of every farmer and horseman in the land.

This machine converts the slow motion of the crank handle to the high speed required for the knives with practically no frictional loss. All moving parts are cut from the solid steel bar and hardened, and, being thoroughly encased and protected from dust and dirt, will wear longer than any other machine made. It has an oil tempered, flexible shaft and the highest grade clipper knives, is mounted on a suitable tripod stand and packed in a case 6x11x12½, weight boxed 36 lbs.

20th Century Clipper in operation.

We also carry in stock the Newmarket and 20th Century Power Clippers.

The Stewart Patent New Model 1904 Sheep Shearing Machine

is a marvel of simplicity and strength. The power required to drive it is practically nothing. All parts are interchangeable. It does not weary the operator, tire the boy, or torture the sheep.

IT SAVES WOOL, TIME, LABOR AND SHEEP.

Anyone who raises wool for the market and owns but a few head of sheep, will find this machine a self-paying investment.

WRITE US FOR PRICES.

LEWIS BROS., LIMITED

Wholesale Hardware, MONTREAL,

and at ***Ottawa, Toronto, Winnipeg, Calgary and Vancouver.***

Consumers Cordage Co., LIMITED

Manufacture every variety of

CORDAGE PACKING

BINDER TWINE

LATHYARN SMALL TWINES

From Manilla, Sisal, Italian, Russian, Jute

Tallow Laid Transmission Rope, 3 or 4 Strand

Oil Well Drilling Cables, Lariat Rope, 3 or 4 Strand,
Halyards, Deep Sea Line, Snap Halters, Snap Plow Lines,
Clothes Lines, Hide Cord, Pulp and Paper Cord,
Tarred Rope, Marline, Houseline

Wrapping, Rolling, Seaming, Sewing and Grape Twines

Careful Execution **SPECIAL ORDERS** Prompt Shipment

Mills: MONTREAL, QUE., and HALIFAX, N.S.

BRANCHES:

F. H. ANDREWS & SON, Quebec. CONSUMERS CORDAGE CO., Limited, St. John, N.B. MacGOWAN & CO., Vancouver
TEES & PERSSE, Limited, Winnipeg, Calgary and Edmonton. GEO. WOOD, London, England.
MERRICK, ANDERSON & CO., Winnipeg, Binder Twine Distributors for Northwest.

It is advisable to place your orders for Refrigerators now, as the demand will be great later.

Size 4.

OLD PROBS. says:—"Expect a sizzling hot summer after such an old-fashioned cold winter."

OURS—THE "NORTHERN"—

REFRIGERATORS on the DRY COLD-AIR PRINCIPLE.

NO INTERIOR SWEATING.

They are made from Well-seasoned Hardwood with Antique Flow Finish.

ZINC LINED THROUGHOUT.
NEW STYLE SWING BASE.
EXTRA LARGE ICE BOX.

AIR SPACE AND INSULATED WITH PAPER AND WOOD.

Also fitted with Improved Galvanized Iron Shelves.
Raised Panels. Fancy Brass Locks and Hinges.

In fact, they are SPLENDID VALUE.

Made in 3 Sizes.

No. 1 and 2 with SINGLE DOORS.
No. 4 " DOUBLE DOORS.

We can ship promptly. Write for dimensions and prices.

The Thos. Davidson Mfg. Co. Ltd., MONTREAL and WINNIPEG

H. S. HOWLAND, SONS & CO. LIMITED

HARDWARE MERCHANTS

Only Wholesale — 138-140 WEST FRONT STREET, TORONTO — Wholesale Only

HAND SEED SOWERS

Simple, Strong and Practical.

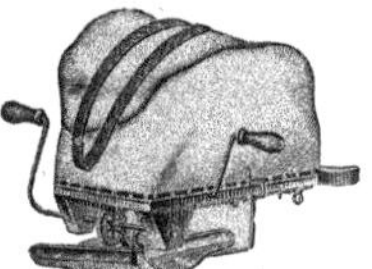

The "Little Wonder" Seed Sower

It is constructed on a single bottom. The novel feed agitating device ensures a uniform flow of seed and in any desired quality.

These machines are well known for their high standard of perfection and accuracy of seeding.

"Champion, Jr." Seed Sower.

Anyone seeking a low priced seeder will find convenience and quality in the "Champion, J." found in no other machine of the kind.

Corn Planters

The "Premier" Seed Sower

The ***"Premier"*** will sow any seed that is sown broadcast, such as oats, wheat, rye, rice, flax, millet clover, timothy, grass seed, etc.

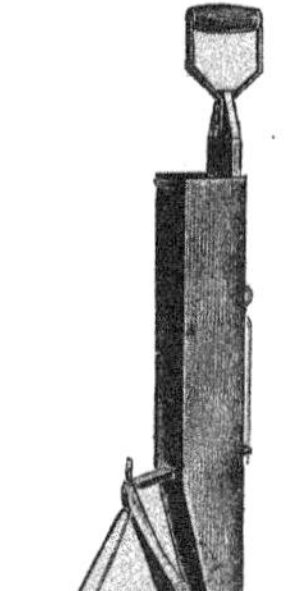

The "Cyclone" Seed Sower

The "Cyclone" will sow timothy, clover, wheat, oats, rye, buckwheat, bonedust, turnip, millet, corn, cotton and all other grain and grass seed. It distributes evenly and works perfectly. The operator can regulate it to sow any desired quantity per acre, and sow forty to sixty acres per day.

Kent's Genuine Triumph.
Corn Planter.

The "Essex," Corn Planter.
One hand Corn Planter.

Kent's Genuine Triumph
Corn Planters with Pumkin seed attachment.

H. S. HOWLAND, SONS & CO., LIMITED

Opposite Union Station.

GRAHAM NAILS ARE THE BEST

We Ship promptly — Factory: Dufferin Street, Toronto, Ont. — Our Prices are Right

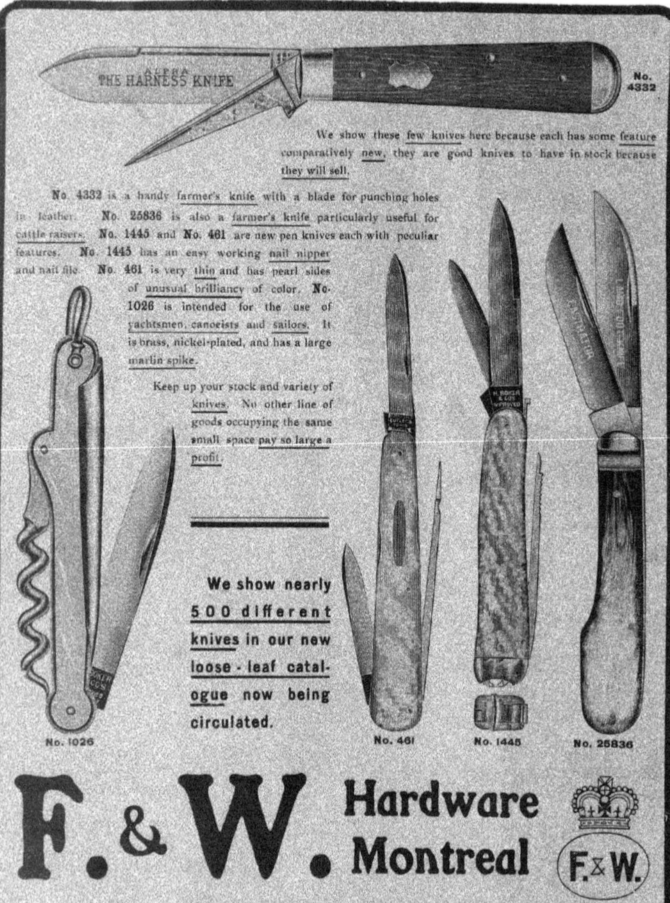

ALPHA
THE HARNESS KNIFE
No. 4332
We show these few knives here because each has some feature comparatively new, they are good knives to have in stock because they will sell.
No. 4332 is a handy farmer's knife with a blade for punching holes in leather. No. 25836 is also a farmer's knife particularly useful for cattle raisers. No. 1445 and No. 461 are new pen knives each with peculiar features. No. 1445 has an easy working nail nipper and nail file. No. 461 is very thin and has pearl sides of unusual brilliancy of color. No. 1026 is intended for the use of yachtsmen, canoeists and sailors. It is brass, nickel-plated, and has a large marlin spike.
Keep up your stock and variety of knives. No other line of goods occupying the same small space pay so large a profit.
We show nearly 500 different knives in our new loose-leaf catalogue now being circulated.
No. 1026
No. 461
No. 1445
No. 25836
F. & W.
Hardware
Montreal
F. & W.
FROTHINGHAM & WORKMAN, Limited, MONTREAL, CANADA.

WARM DAYS will be here before you realize it.

The ice crop all over Canada has been good this year. Cheap ice will mean a great demand for Freezers.

This would suggest getting your order for Freezers in to us now.

We will give you the best Freezers made at a price no greater than you would pay for poor ones.

The tubs of our Freezers are cedar and have electric welded hoops, which will not drop off. The cans are the best to be had, thoroughly tinned, made for wear.

They freeze the cream quickly and use the least possible quantity of ice.

All sizes in stock from 1 quart for small families up to the largest hotel size.

THE AMERICAN TWIN FREEZER will freeze two distinct flavors at one operation without extra effort on the part of the operator.

It appeals to the careful housekeeper, as it gives her the advantage of having in effect two freezers at a cost a trifle over the old style single flavor freezer.

THE GEM FREEZER is double acting. It freezes quickly, making rich, smooth dainties for dessert, such as Ice Cream, Water Ice, Sherbets, Frozen Puddings, etc.

THE BLIZZARD FREEZER is the best low-priced freezer made. It is single acting, easy running and easily cleaned.

Complete Recipe Book With Each Freezer

Caverhill Learmont & Co

WHOLESALE DISTRIBUTORS } MONTREAL and WINNIPEG

WHITCOMB

STEEL BARN DOOR CATCH

It is convenient; it is strong: it is durable; it is adapted to any thickness of door; and harness, etc., will not catch on it. Fully guaranteed.

Get a sample and see what an excellent Latch it is.

ALBANY HARDWARE SPECIALTY MFG. CO.

Manufacturers of PATENTED SPECIALTIES BY CONTRACT

ALBANY, WISCONSIN, U.S.A.

CROWN JEWEL

An axe like the Crown Jewel, that has a reputation, is the axe that most merchants want to sell. It gives the customer satisfaction, which keeps him your friend. There are, however, people who want an axe, but not having much to do with it they want a cheap axe. We have a line that meets this demand, and is the best line in the market at the price. If you are not familiar with it, write us.

The Dundas Axe Works

Dundas, Ont.

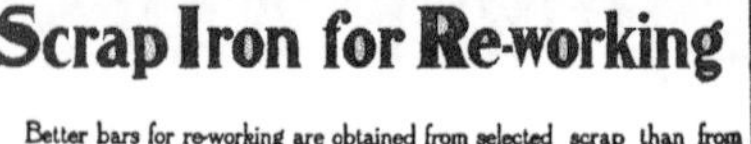

Better bars for re-working are obtained from selected scrap than from pig iron—have more life and strength.

London Bars are made from selected scrap and come in Flats, Rounds, Squares, Ovals, Half-Ovals, Half-Rounds and Bands.

Large stock always on hand, insuring prompt shipment. 51

London Rolling Mills
London, Canada.

Expert Bookkeeping

Investigation and auditing of all classes of accounts undertaken. Trading, profit and loss accounts and balance sheets prepared. Sets of books opened and office systems devised in an up-to-date manner. Collection of accounts a specialty.

Davenport, Pickup & Co.
622 McIntyre Block and 422 Ashdown Block
WINNIPEG MAN.

Square your Accounts

INTERNATIONAL PORTLAND CEMENT COMPANY, Limited
OTTAWA, - - ONTARIO.

Manufacturers of strictly high-grade Portland Cement made by the most improved method of manufacture in existence. Each barrel contains three hundred and fifty pounds of pure Portland Cement with no adulteration.

The Best Door Closer is . . .
NEWMAN'S INVINCIBLE FLOOR SPRING
Will close a door silently against any pressure of wind. Has many working advantages over the ordinary spring, and has twice the wear. In use throughout Great Britain and the Colonies. Gives perfect satisfaction. Made only by
W. NEWMAN & SONS,
Hospital St., - BIRMINGHAM

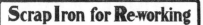

There is only ***One Best, One Lightest, One Strongest, One Keenest.***

"Automobile" and "Cycle" Skates

Their Elegance of Design and Simplicity of Construction at once place them **IN A CLASS BY THEMSELVES.**

Aluminum Foot Plates and Nickel Steel Blades, the lightest and strongest combination possible.

WRITE FOR CATALOGUE

CANADA CYCLE AND MOTOR CO., Limited

Branches—*OTTAWA* *VANCOUVER* *WINNIPEG* *MELBOURNE, AUST.*

TORONTO JUNCTION, CAN.

Young Man! Do You Want to Make a Profitable Change?

"A rolling stone gathers no moss," but it acquires a shine that enhances its value—where it has value to begin with.

Knowledge wins your business. Knowledge is the result of intelligent experience. To gain enough experience to make one's services valuable it is necessary to make a change occasionally. But take no chances. State what sort of position you desire in the want columns of Hardware and Metal, and you are practically assured of a profitable change.

Condensed advertisements in Hardware and Metal cost 2c. per word for first insertion and 1c. per word for subsequent insertion. Box numbers 5 cents extra. Write or phone our nearest office.

HARDWARE and METAL, Toronto, Montreal, Winnipeg

THE LONDON FOUNDRY CO., Limited

LONDON, CANADA

SCRAPERS

Wheelbarrows

for Railroad, Dock, Concrete, Farm and Garden

Also

Trucks, for all purposes
Drilling Machines
Forges
Blacksmith Foot Vises, etc.
Wagon Skeins
Bone Cutters
Butter Workers and Shipping Boxes
Cider Mills and Presses

PAROID ROOFING

"THE TIME TESTED KIND"

Plant Steamship Co's, Wharves, Halifax, N.S., covered with Paroid.

Dealers all over Canada and the United States doubled their sales on Paroid last year. Our new advertising plan promises to make a bigger increase for them this year. Paroid is used by the leading railroad systems, manufacturers and farmers throughout Canada. It has stood the test of time and that's the test that tells.

Let us tell you about our advertising plan which will bring customers to your store for other things as well as Paroid. Sign your name to a postal to-day—the best roofing season of the year will soon be here.

F. W. BIRD & SON, Makers

Established in U.S.A. 1817 ***Factory at HAMILTON, ONT.***

Offices at Winnipeg, Man., and Hamilton, Ont.

$2 FOR THIS SMALL SUM THE $2

MANUFACTURER and SUPPLY MERCHANT

may keep posted on new openings for trade.

The CANADIAN CONTRACT RECORD

reports weekly all projected building and other construction works throughout Canada as well as new business enterprises.

Send your name and address with $2 for a year's subscription to

Canadian Contract Record

Toronto, Montreal, Winnipeg, and Vancouver

$2 $2

Auto Spray

Compressed-Air Automatic

Best hand sprayer made. An absolute necessity for every farm and garden. Will repay its cost in one season for potatoes alone. Saves time, labor and material. A boy can do the work. Will run continuously for 6 to 10 minutes.

CAVERS BROS., GALT, ONT.

Sole Agents for the Dominion

The "WAYNE" CARPET BEATER

Handy, Useful and Durable Household Article. "Not a Seasonable Toy"

Is now made in three styles—to retail at popular prices.

Patented Sept. 8, 1903

The "WAYNE" is the best whip upon the market for beating clothes, carpets, rugs and furniture. Constructed of the best coppered steel wire obtainable and mounted on a seasoned hardwood handle.

The wire is fastened to the handle in such a manner that it is impossible to work it loose. A large ferule pressed over the handle prevents the latter from splitting.

The 'WAYNE" beater being made of spring steel wire always retains its shape, also eliminates the danger of puncturing or tearing article being cleaned.

This beater is so well balanced that you get full stroke and best results with very little effort. The "WAYNE CARPET BEATER" is a trade winner. Its usefulness is recognized on sight and sells wherever shown. Place your order now.

Manufactured by

Ontario Metal Novelty Mfg. Co.

Limited

111 King Street East, - Toronto, Ont.

Manufacturers of Bath Room Fixtures and Metal Novelties. Dies, Tools, Etc. Look for something new in every week's paper. All our own make. Made in Canada. Send for prices.

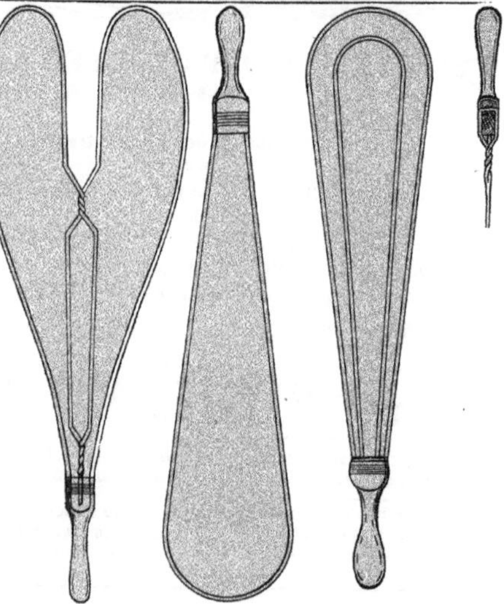

Are You Losing Money?

Rather a personal question; but you do not need to have the silver fever in order to lose money. If you are not securing all the dollars you might make ***you are losing money*** — actually throwing dollars away. There ***are*** some men in the Hardware business, bright fellows, too, who fail to see — or seem to fail to see — the money that's in

Window Glass

for them. The demand for Glass of all shades and variations ***must*** increase. There's just one way of limiting the demand for Glass — all sorts of Glass — and that way is to manufacture an imperishable brand. We'll do that some day, but in the meantime we wish to know if you have our large and beautifully illustrated catalogue? You are losing money every day you do without it. Send off that post card now.

The Hobbs Manufacturing Co., Limited

LONDON TORONTO WINNIPEG

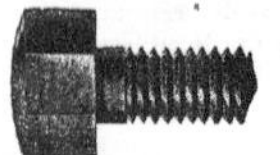

The John Morrow Screw, Limited

MANUFACTURERS OF

Set and Cap Screws, Special Milled Work, Engine Studs Etc. Cold Punched Nuts of every variety of finish.

INGERSOLL, ONTARIO

COVERT MFG. CO.

Troy, N.Y.

Steel Carriage and Wagon Jacks

Harness Snaps, Chain, Rope and Web Goods, etc.

SOLD BY ALL LEADING JOBBERS

CELEBRATED HEINISCH SHEARS

Tailors' Shears,
Trimmers' Shears,
Tinners' Snips, etc.

ACKNOWLEDGED THE BEST.

R. HEINISCH'S SONS CO. NEW YORK OFFICE, 155 Chambers St. NEWARK, N.J., U.S.A.

STREET PAVING and SIDEWALKS a SPECIALTY

SILICA BARYTIC STONE CO. OF ONTARIO, Limited.

Water Proof Floors for Malt Houses, Breweries, Slaughter Houses, Cheese Factories, Cellar, Stable Floors, etc.

Head Office *Ingersoll, Ontario.*

Walter Mills, General Manager

Ask for quotations for Septic Tanks.

Persons addressing advertisers will kindly mention having seen this advertisement in Hardware and Metal.

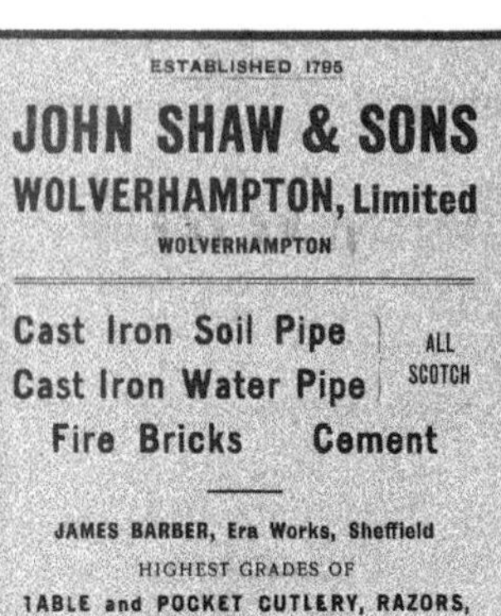

ESTABLISHED 1795

JOHN SHAW & SONS
WOLVERHAMPTON, Limited
WOLVERHAMPTON

Cast Iron Soil Pipe
Cast Iron Water Pipe } ALL SCOTCH
Fire Bricks Cement

JAMES BARBER, Era Works, Sheffield

HIGHEST GRADES OF

TABLE and POCKET CUTLERY, RAZORS, CARVERS, Etc.

J. H. ROPER
CANADIAN REPRESENTATIVE

82 St. Francois Xavier St., - MONTREAL

IT'S DOUBTFUL—IF

a better range than Empire Queen will be devised during this century. But it is certain that a better than

The
Empire Queen Range

is not anywhere in sight. That's big talk; but hundreds of practical men will endorse every claim we make for THE EMPIRE QUEEN and do that right over their signatures.

Yes, The Empire Queen looks well, cooks well and practically sells itself. We like to mail proofs of our assertions to sceptical men.

The Canadian Heating & Ventilating Co. Limited.
OWEN SOUND, Ontario.

THE CHRISTIE BROS. CO., Limited, 238 King St., Winnipeg, Man., Western Agents.

THE CANADA STOVE AND FURNITURE CO., 126 West Craig St., Montreal, Que., Agents for the Province of Quebec.

Australasian Hardware and Machinery.
The Organ of the Hardware, Machinery and Kindred trades of the Antipodes.

SUBSCRIPTION $1.25 PER ANNUM,

post free to any part of the world.

PUBLISHING OFFICES:
Melbourne - Fink's Buildings.
Sydney - Post Office Chambers.

BRITISH OFFICES:
London, - 42 Cannon St., E.C.

CANADIAN AND AMERICAN ENQUIRIES will receive prompt attention if addressed to the LONDON OFFICE, 42 CANNON STREET, E.C.

Specimen Copies Free on Application.

Strand Wires in the "Dillon" You Can Bet They are Sound

and sounder than users usually find some other strands. Do you know the reason? Here it is:

"Dillon"

Strand Wires, whether heavy or light are High Carbon Hard Spring Wire—yes, every one of them. That accounts for the good deal of difference in the orders you can get—in the results "order-getting" "Dillon" can give your patrons.

THE OWEN SOUND WIRE FENCE CO., Limited
OWEN SOUND, ONT.

Sold by { Messrs. Caverhill, Learmont & Co., Montreal; Messrs. Christie Bros. Co., Limited, Winnipeg

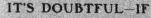

I WILL TALK

to practically every Hardware merchant in Canada from the Atlantic to the Pacific. I cannot do it all in one day, but during the first twenty-four hours I will deliver your message to every Hardware merchant in Ontario. I travel all day Sunday and on Monday morning there will not be a village within the limits of Halifax in the East, and Brandon in the West, into which I will not have penetrated.

I cannot go any further East, so I now devote all my energies to the West, and so many new towns are springing up here each week that I haven't as much time as I used to have to enjoy the scenery. But I like talking to hardwaremen, clerks, travellers and manufacturers, especially as they are always glad to see me and hear the news I have to tell them. Tuesday noon I am at Calgary, Wednesday noon at Kamloops, and by Thursday morning I reach Vancouver, having been in all the mining towns and all through the fruit districts of British Columbia.

I have been eighteen years on the road and I have a pretty good connection. I never intrude when a man is busy, but just bide my time, because I know men pay far more attention to what you have to say if you catch them when they have a few moments to spare. So I often creep into their pocket when they are going home at night, and when supper is over Mr. Hardwareman usually finds me. He must be glad to see me, because he listens to what I have to say for an hour or more.

I try to always tell the truth and men put such confidence in what I say that I would feel very sorry to deceive them even inadvertently. Probably some other week I will tell you about the different classes of people I meet. In the meantime if you want a message delivered to HARDWAREMEN, PLUMBERS, CLERKS, MANUFACTURERS or TRAVELLERS—and want it delivered quickly—I'm your man.

THE WANT AD MAN

Condensed Advertisements in Hardware and Metal cost 2c. per word for first insertion, 1c. per word for subsequent insertions. Box number 5c. extra. Send money with advertisement. Write or phone our nearest office

Hardware and Metal

MONTREAL **TORONTO** WINNIPEG

Persons addressing advertisers will kindly mention having seen this advertisement in Hardware and Metal.

ATKINS SILVER STEEL SAWS
are the most profitable Saws to sell. Because they satisfy your customer and bring him back to YOU. Good mechanics appreciate them and buy them. The best Dealers everywhere sell them. Write for details and catalogue.
E. C. ATKINS & CO., Inc.
The Silver Steel Saw People,
Home Office and Factory, Indianapolis, Ind.
Branches—Toronto, Atlanta, Chicago, Memphis, Minneapolis, New Orleans, New York City, Portland, San Francisco, Seattle.

The Long and the Short of It
"IT" MEANS
SHIELD BRAND
READY ROOFING
LOCKERBY & McCOMB
THE "LONG"
is the time it wears. It stands the test of heat and frost, wind and rain, and give your customers many years of solid satisfaction.
THE "SHORT"
part is the time needed to put it on. A great deal of time—which means a great deal of money—is saved to the man who uses
SHIELD BRAND
READY ROOFING
LOCKERBY & McCOMB
65 SHANNON STREET
MONTREAL

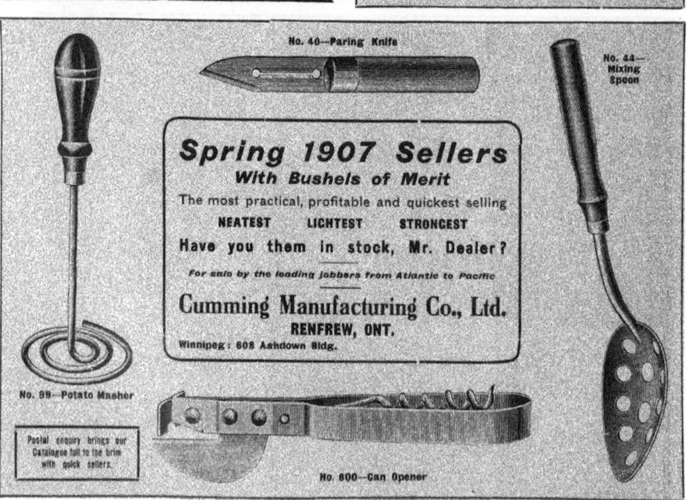
No. 40—Paring Knife
No. 44—Mixing Spoon
Spring 1907 Sellers
With Bushels of Merit
The most practical, profitable and quickest selling
NEATEST LIGHTEST STRONGEST
Have you them in stock, Mr. Dealer?
For sale by the leading jobbers from Atlantic to Pacific
Cumming Manufacturing Co., Ltd.
RENFREW, ONT.
Winnipeg: 608 Ashdown Bldg.
No. 99—Potato Masher
Postal enquiry brings our Catalogue full to the brim with quick sellers.
No. 600—Can Opener

QUICK

A want ad in **Hardware and Metal** will reach practically every Hardware Merchant in Canada, from Halifax to Vancouver, within four days.

If you are a Clerk and want to secure another position; if you are a Hardware Dealer and want to secure a clerk—in fact, if you want anything in which the Hardwaremen, Clerks, Travellers or Manufacturers are interested,

Use our Condensed Ad column

RATES

2c. per word..................1st insertion
1c. " " ....subsequent insertions

HARDWARE & METAL

Montreal Toronto Winnipeg

Hardwaremen, Attention!

Your primary object is—to corner dollars. Can you afford to recommend any old heating system when the best Heater on the market is so easy priced? All practical men, who have noted the exclusive features of

The Howard Heater

unhesitatingly declare it to be second to none manufactured anywhere at any price. Practical men like yourselves say so. I do not ask you to accept unverified statements.

The Howard is superior in that it is easily controlled, eliminates pipes from your cellar and assures pure air.

Let us mail you interesting facts which every dealer in business for money should know.

WM. HOWARD - 248 MACDONNELL AVE. PHONE PARK 2633 - **Toronto**

Kemp's Cold Blast Lanterns

Plain and Japanned

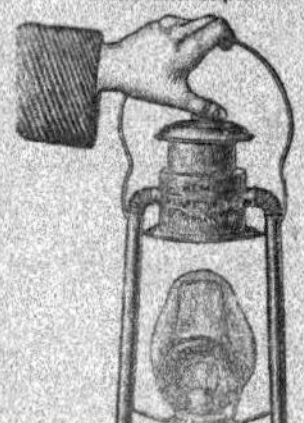

Dash Board

Copper-plated

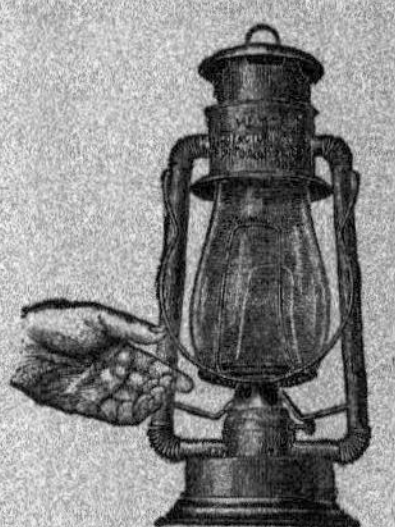

The light that never fails

KEMP MANUFACTURING CO.,

TORONTO, CAN.

THE KEMP MFG. AND METAL CO., Ltd.
McDermot Ave. E., Winnipeg, Man.

THE KEMP MFG. CO. OF MONTREAL
39 St. Antoine St., Montreal, Que.

Building a Retail Business

H. N. Higinbotham, in Saturday Evening Post.

Attentiveness to customers should be instant, in season and out of season, from the moment the customer crosses the threshold of the store until he is out on the street again. There is nothing which customers more invariably resent than indifference on the part of the storekeeper or his clerk. To go into a store and be forced to stand about without being recognized is a kind of treatment which goes against the grain of human nature, and will not be tolerated by the average American. Of course, there are at times in almost every store when there is much business to be done in a small time, and it is absolutely necessary to keep several customers waiting. This does not mean, however, that there is any necessity for the proprietor or his assistant to fail to give a pleasant word of greeting to those who are obliged to await their turn.

It is an easy matter for the storekeeper or the clerk to give a nod of welcome and a word of excuse to those who cannot be immediately served. The customer who has been pleasantly greeted and told that he will receive attention in a few minutes will wait cheerfully, whereas he would leave the store in ill-humor if his presence were entirely ignored. The ability to keep waiting customers in good humor and make them feel that they are not neglected is one of the most valuable qualities a storekeeper or a clerk can possibly possess. Just ordinary courtesy and attentiveness demand very little time, and they are far more valuable than flattery or argument in the sale of goods.

Tell the Truth.

Perhaps it will be thought that a word of caution with regard to the over-representation of goods is entirely superfluous and ungracious. I wish this were the case, but I fear that the observation is not at all out of order. Very often the zeal and ambition of the clerk carries him further in this direction than his employer would wish him to go. Again, there are those in the merchandizing business, as in every other business, who allow their desire for gain to impart an elastic quality to their consciences, and who lost sight of the fact that nothing can be more fatal to their final success than misrepresentation.

As a sheer matter of policy, it is far better for the merchant to understate than to overstate the merits of his goods When a merchant gains a reputation in his community for never misrepresenting his goods he is on the high road to success. He will command the trade of his neighbors and he will hold it against obstacles.

In selecting his clerks every merchant should give careful consideration to securing those who have tact, pleasing manners, and all that is included in the term "a good personal address." At the same time, it will be well for him to remember that a clerk may have all of these qualities and still be a comparative failure. Many a merchant employing quite a force of clerks has been perplexed by the problems presented by this matter of the personal equation. Here is a clerk, for example, who outclasses all his fellow-workers in personal popularity with customers. His graces of manner attract the patrons of the store to him in a degree which marks him as an exceptional favorite. In spite of this, the totals of his sales from week to week fall below several of his co-laborers at the counter. He puts in as many hours as they do and works as diligently as they. Here, on the face of things, is a clear contradiction which is decidedly puzzling to the merchant. What is the difficulty?

Popular Clerk Lacked Decision.

In nearly all cases of this kind, and they are so numerous as to be almost general, I believe the cause of the trouble lies in the fact that the popular clerk, who has decidedly the lead over all his fellows, lacks in decision. Not long ago I saw an interview on this very subject in which the superintendent of one of the big State street retail stores, in Chicago, hit the nail squarely on the head as far as this point is concerned. He related how he had been puzzled to reconcile the fact that the most popular clerk in the fancy dress-goods department took about third rank, so far as the volume of his sales was concerned. The young man had a decided charm of manner which naturally attracted customers to him, and it was apparent to the management that this clerk had practically first call on a large percentage of the customers who came to the department to buy.

At last the superintendent decided to give his personal attention to the solution of this very interesting and practical riddle. To this end he stationed himself near the fancy dressgoods counter and began to watch the young man at his work. A young matron came down the aisle, passed a clerk who was at liberty to wait on her, and went direct to the young man in question. It so happened that the superintendent was acquainted with the woman, and knew, from having waited on her himself before being promoted to an executive position, that she was not especially difficult to please.

The Fatal Fault.

She handed the clerk a sample of goods of which she was having a skirt made, and told him she wished to get the material for a waist which would be appropriate to wear with the skirt. He looked at the sample, turned around to the shelves and started to take out a bolt of goods, hesitated, went to another, and then to still another. Then he drew out two or three bolts, looked at them and finally shoved them back into place. After making several false motions of this kind, he at length returned to the first bolt of goods which his hand had touched, drew it out and displayed it on the counter.

To the keen eyes of the watchful superintendent it was apparent that the indecision of the clerk had communicated itself to the customer, and, after examining the goods in a somewhat indifferent way, she asked if he had not something else more attractive. Then he took down the bolts which he had glanced at in his hesitating pilgrimage along the shelves, showed them to his customer, and discussed with her their relative merits. Still, she was apparently unsatisfied, and he once more ranged the shelves and brought down other patterns. There was more conversation and further comparison of goods, and for the third time he turned about and began to forage in a somewhat hopeless and hesitating way.

More than once the customer seemed to be on the very point of decision, but finally she offered an apology, and, saying that she would come again, she left the counter without buying.

The Secret of Success.

In the meantime, the clerk whom she had passed by had waited upon three customers and sold to every one of them. His manner of work was this: He gave very careful attention to the description of the goods wanted by his customer, and in each case asked one or two questions to bring out more clearly the desires of his patron. But in each instance after he had learned this he turned round and his hand went unhesitatingly to just the right goods. There was decision in his manner, and he placed the bolt of cloth upon the counter with the remark: "I think this is just what you wish." The watchful eye of the superintendent had taken in his movements as well as those of the popular clerk, and he was convinced of two things. first, that the taste or judgment of the clerk who made the three sales was no better than that of the popular clerk who failed to make his sales; second, that the decision with which the less popular clerk acted was the secret of his success.

The conclusions of this superintendent were, to my mind, entirely correct. He found out why the man who attracted the fewer customers was able to make the greater number of sales. Incidentally it may be added that the popular clerk remained at the fancy dressgoods counted while the other was promoted to the head of the department.

Retail Hardware Association News

Official news of the Ontario and Western Canada Associations will be published in this department. All correspondence regarding Association matters should be sent to the secretaries. If for publication send to the Editor of Hardware and Metal, Toronto.

BRITISH COLUMBIA HARDWARE CONVENTIONS

The following report of the recent conventions of the British Columbia Retail Hardware and Wholesale Hardware Associations has been forwarded for publication by John Burns, hardware manufacturers' agent, Vancouver, he being secretary-treasurer of both organizations :

Wholesalers Oppose Parcels Post.

The annual meeting of the B.C. Wholesale Hardware Association was held in

C. P. W. SCHWENGERS, Victoria, B.C.
Of E. G. Prior & Co., Victoria, and President B. C. Wholesale Hardware Association

Victoria, B.C., on Friday, Feb. 8th. There was a full attendance and the greatest interest was manifested in the proceedings by the members, among whom the greatest harmony was shown to exist and a general feeling seemed to be evident that the association was proving a benefit to all parties interested either directly or indirectly. Two sessions were held, the morning one being devoted to the reports of the retiring officers and elections for the ensuing year.

W. G. MacKenzie, of Wood, Vallance & Leggat, Vancouver, the retiring president, in his report spoke of the good work the association was doing and the great assistance the members had been to the officers in carrying out necessary work of the year. The importance of taking action to oppose the new postal arrangements of the Government, etc., etc. The secretary-treasurer's report showed the financial affairs to be in excellent condition.

Election of Officers.

The election of officers then took place and resulted as follows :

President—C. P. W. Schwengers, of E. G. Prior & Co., Ltd , Victoria.

Vice-President—John Boyd, of Boyd, Burns & Co., Vancouver.

Secretary-Treasurer — John Burns, Vancouver.

Executive Committee—W. G. MacKenzie, J. Elliott, A. Cunningham.

Pipe Committee—C. A. Godson, W. A. Akhurst, J. Boyd, C. P. W. Schwengers.

There was no balloting, all nominations being unanimous. The afternoon session was devoted to detail work and laying plans for the year's business, every effort being made to extend the usefulness of the association.

Retailers and Jobbers Dine.

On the following evening the Wholesale Association entertained the members of the British Columbia Retail Hardware Association at a banquet, there being a large attendance in the city on account of the annual meeting of the latter organization taking place on that day. It was the first function of its kind and was indicative of the very cordial relations existing between the two organizations. Mr. Schwengers, the president, occupied the chair, and took occasion to remark on this, and that it would be the aim of the officers of the association and himself to try and further cement this good feeling and lend their best endeavors to keep the retail hardware business on a sound basis. The same hopeful note was also struck by Mr. MacKenzie, who spoke to the same point. In accordance with the general desire, the function was made as informal as possible and was voted by one and all a distinct success.

Officers of Retail Association.

The officers of the Retail Dealers' Association are :

L. B. Lusby, president.
C. Snell, vice-president.
John Burns, secretary-treasurer.
Executive Committee—Messrs. Kirk, Ogilvie and Stearman.

There was a good attendance of members from New Westminster, Vancouver and Victoria, the principal matter of discussion being the question of parcel post C.O.D. system. A committee was appointed to get up a petition against the proposal.

The next annual meeting will be held in New Westminster.

DISTRICT ASSOCIATION FORMED.

A well attended meeting of the retail hardware merchants of North Middlesex, North Lambton and South Middlesex was held at Ailsa Craig on Thursday, March 7 and the North Middlesex Hardware Association formed to include the hardware merchants in Strathroy, Lucan, Forest, Exeter, Parkhill Ailsa Craig, Thedford, Arkona, Crediton and Dashwood. In all fifteen firms have signified their intention of joining the district organization. Those attending the first meeting were : A. W. Humphries, Parkhill ; J. E. Westcott, Ailsa Craig ; W. Scott, R. F. Scott and Wm. Lawrie, Forest; Wm. Kennedy, Ailsa Craig; Young Bros., Crediton; and Mr. Harrison, of Brewer & Harrison, Parkhill. Several others wrote, regretting their inability to be present.

An interesting discussion took place on the subject of cash and credit prices the general opinion being that dealers should avoid price cutting and secure a larger margin of profit on goods sold on credit than on articles sold for cash.

Election of Officers.

It was decided that the membership fee would be $1 per annum and the following officers were chosen to carry on the business of the new association :

President—J. E. Westcott, Ailsa Craig.

Secretary—Wm. Lawrie, Forest.

Executive Committee—Messrs. Haney, Strathroy; Heaman, Exeter, and Harrison, Parkhill.

Hardwaremen in other parts of Ontario who desire to secure additional particulars regarding the work of the district association should write to Secretary Lawrie for the information.

SHOULD TAKE A BROAD VIEW.

The Iron Age points out that one of the dangers which lie in the path of the retail associations is the disposition sometimes found to take a low and narrow view of the retailer's business, and to discourage the growth of retail houses in the direction of a local job-

A. R. AUGER, OKOTOKS, ALBERTA
One of Alberta's Representatives on the Executive of the Western Retail Hardware Association.

bing trade, which in course of time may assume larger dimensions. Such houses who are principally retail in their scope, but who serve an eminently useful purpose as distributors to the cross roads and other small trade in their immediate vicinity, are most desirable customers for either the jobber or the manufacturer, and in view of their enterprise and enlarging business relations are to be given an honorable place among retail merchants. They should be indeed regarded as exceedingly desirable members of retail hardware associations. Their identification with such an organization adds strength to it. If, however, on the other hand they should be regarded as out of place in such associations simply because with a retail business they are also doing some jobbing, their absence from the membership will weaken the association and rob it of a dignity and influence which it would otherwise possess. This matter came up in the Illinois convention and the subject was there disposed of in an eminently wise and sensible manner, as the association decided that persons doing a retail business and branching out into wholesale lines were entitled to membership. If this broad gauged view of the question were not taken there would be danger that the associations would become limited to the smaller and less progressive houses.

CONGRATULATIONS TO WM. GORDON.

As showing the respect in which the Retail Hardware Association of Western Canada is held among the manufacturers and wholesale men, the following letter, received by W. M. Gordon, Winnipeg, is highly significant :

Toronto, Feb. 23, 1907;

Wm. Gordon,
Winnipeg, Man.,

Dear Sir,—The writer was very pleased indeed to see by Hardware and Metal of this week that you have been re-elected second vice-president of the western association and wishes to congratulate you and also the association on being able to bring men of your class to the head of their affairs. From the work you and the other officers are doing in that association I fell that an organization of that kind is a help to every man throughout western Canada.

Assuring you again that I am very pleased to hear of your re-election, I am,

Yours very truly,
NORMAN A. WYLIE,
Kemp Mfg. Co., Ltd.

ACTIVITY IN THE EAST.

As indicative of the interest taken in this question in the Maritime Provinces, the following editorial from a recent issue of the "Telephone," of Bear River, N.S., is worth reproducing :

"United action is being taken against a proposed legislation providing for the collection on delivery of parcels by post. This measure, if carried, would give added impetus to the system of mail order trading which is already jeopardizing the interests of the local dealers in our provincial towns.

"The attention of Bridgetown merchants was directed to the proposed amendment some weeks ago and they immediately sent in a petition through the local member, S. W. W. Pickup, requesting him to use his influence in preventing the measure to be carried into effect. They also enlisted the attention and caused to be added to their petition the protests of Middleton, Lawrencetown and Paradise merchants.

The Annapolis Board of Trade, ever alive to the interests of their town, have made a point of forwarding protests to the Postmaster General and to Mr. Pickup, and the Bear River Board of Trade will also appeal for the prevention of proposed amendment to the postal delivery law. Many other towns in the province have adopted the same tactics and it is probable the purpose of the measure will be thwarted.

'The interests of the local merchants are one with the interests of the general public, and we are sure their efforts will have the sympathy of all but the few wealthy mail order merchants who would benefit by the proposed legislation."

QUESTION BOX AT CONVENTION.

The interest manifested in the "Question Box" feature of the Good Friday convention of the Ontario Retail Hardware Association has been such as to make certain remarkably interesting morning and afternoon sessions of the convention on the holiday when this feature comes up. No lengthy addresses will be delivered at the convention and the discussions will be practical and to the point, if the officers of the association can carry their wishes into effect.

CONVENTION PROGRAMME.

The second annual convention of the Ontario Retail Hardware Association will be held in Room 2, Labor Temple, Church street, Toronto, on Thursday and Friday, March 28 and 29, the program being as follows :

Thursday, March 28.

11 a.m.—Executive committee meeting.

1 p.m.—Roll call, president's address, secretary's report, committee reports, etc.

7 p.m.—Complimentary banquet tendered the members of the association by Lt.-Col. MacLean, publisher of Hardware and Metal.

Friday, March 29.

9.30 a.m.—Election of officers, general business and question box discussion.

2 p.m.—Unfinished business and question box discussion. Adjournment.

The attendance at the convention ought to be representative of every part of the province. The association has been successful in its work during the past year and the program outlined under another heading should encourage all live hardwaremen to attend.

The banquet will be the big social feature of the convention, and a number of prominent hardwaremen have been invited to reply to toasts and speak on practical subjects. Richard Blain, M.P., who is a retail hardware merchant himself, has been invited to give a short account of the parcels post discussion in Parliament, as well as to explain other legislation (such as the new co-operative act) which will be injurious to the retail trade.

A few additional questions have been submitted for the "Question Box," which now includes the following topics for discussion :

Should large retailers be able to buy goods as cheaply as jobbers or, in other words, is the quantity system of buying the most satisfactory manufacturers can adopt ?

Should there be a difference between cash and credit prices ? If so, how much ? How can this best be handled ?

What percentage of a merchant's turnover should his expenses be ? What percentage should be spent in advertising alone ?

Should goods be marked in plain figures or ciphers ?

Should some novelty be given away each year ?

Do small remembrances to special customers at Christmas time do more good than harm ?

Assuming that the jobber is willing to pay exchange on local cheques, what is the minimum cheque on which we should ask him to pay exchange ?

When remitting money orders should the cost of the order be taken out of the remittance ?

What do we know of a satisfactory system of filing catalogues from a two-page price list to the regular hardware catalogue ?

To what extent does it pay a hardware dealer to advertise in the local newspaper ? What forms of advertising have been found the most profitable ? Is it profitable to advertise in programs, etc.

Does it pay a dealer doing a $50,000 business to go behind the counter ?

Is it wise to interfere with a clerk or transfer a customer from one clerk to another ?

Is a parcel boy at 50c. a day cheaper than one at $1 per day ?

Can the shortage of tinsmiths be relieved by employing helpers ?

Does it pay to weigh freight and check freight bills ?

Do 5 and 10 cent counters pay in small towns ?

What is the best method of handling stove repairs and storing parts ?

Does it pay to borrow money to discount bills ?

What is the cheapest and best method of handling oil ?

If a hardware dealer has not been handling stoves, what is the best selling system he can adopt to secure publicity and make profits ?

Should the retail hardwareman confine his buying to representatives of jobbing houses ?

What is the best method of dealing with goods substituted by jobbers for lines not in stock ?

In a retail business of $25,000 to $50,000 per year, what is the best system of bookkeeping ?

How can unsaleable lines be disposed of satisfactorily without cutting prices ?

Should a portion of a clerk's salary be withheld or does it pay to offer bonuses to clerks to secure faithful service and prevent them leaving on short notice ?

Does it pay to employ female clerks to look after the housefurnishing department and lady customers ?

Does it pay to handle special lines for holiday trade ? If so, what lines pay best ?

Does it pay to allow customers 2 per cent. if they pay accounts before the first of the month, for the preceding month ?

HARDWARE TRADE GOSSIP

Ontario.

D. P. Warren, of Waterford, has opened up a new hardware store.

L. Payette, of Warden, King & Son, Montreal, was in Toronto this week.

A. J. Ross, Owen Sound, hardware merchant, has assigned to J. Pringle.

J. H. Plummer, president of the Dominion Steel Co., is in Toronto this week.

Geo. Lasher & Son, hardware merchants, Toronto, have sold to W. Robertson.

D. McGregor, of Johnston & McGregor, stove manufacturers, Smith's Falls, is dead.

W. H. Wiggs, of the Mechanics' Supply Co., Quebec, was a visitor in Toronto last Friday.

George Taylor & Son, hardware merchants, London, are advertising their business for sale.

Ballantyne Bros., hardware merchants, Kincardine, have sold their business to C. J. McAllister.

Richardson & Sneyd, hardware jobbers, Toronto, have dissolved, J. E. Richardson continuing.

W. Conn, hardware merchant, Aylmer, has sold out; D. C. Davis being the provisional assignee.

C. A. Birge, president of the Canada Screw Co., was in Toronto this week looking after the interests of his company.

W. Lundy, an employee of the Madison, Williams Mfg. Co., Lindsay, was choked with gas while working at the cupola.

Frank Taylor, of Taylor Bros., Carleton Place and Almonte, was a caller at the Toronto office of Hardware and Metal last Saturday.

J. T. Lawler, of the Lawler Regulator Co., New York, and author of Lawler's Hand-book on Plumbing, was a visitor in Toronto this week.

Peleg Hamilton, of the H. S. Howland, Son & Co., Toronto, left last week for Latona, Florida, where he will holiday for about a month.

Mr. Armstrong, of the Blanchite Process Paint Co., is able to be around again after being laid up for about a month with one of his legs injured by a street car.

Among those who were in Toronto this week were: J. B. Skelton, Palmerston; H. Christie, of Christie Bros., Owen Sound; A. Dundas, Cobourg; C. A. Binns, Newmarket; Alex. Hamilton, Herb. Parr, Blackstock.

W. B. Riley, of Arthur street, Toronto, was tried in the Sessions Court this week for receiving and disposing of stolen property. During the trial, a witness, Fred Corner, was taken into custody on a charge of perjury. The goods which Riley is charged with receiving were six gas stoves. Corner swore that the day the detectives searched Riley's premises the box of gas stoves was seen by the police and by him. The officers who made the search swore Corner's statements were false, and he was taken into custody. His trial has not yet taken place. The jury declared Riley not guilty.

Quebec.

Mr. King, of F. Hyde & Co., Montreal, is in the Lower Provinces on business.

Mr. Doherty, of the Sussex Mfg. Co., Sussex, N.B., was a recent visitor in Montreal.

J. G. Lewis, vice-president of Lewis Bros., Montreal, has been made a justice of the peace.

H. Moulden, of the Taylor, Forbes Co., Guelph, Ont., was in Montreal last week on business.

J. A. Tetreault, of the Thomas Robertson Co., was on the sick list for a few days last week.

J. Fowler, of the Fowler Axe Works, St. John, N. B., was in Montreal last week calling on the trade.

Thos. Blaikie, city traveler for the Dominion Wire Co., Montreal, was laid up for a few days last week.

B. and S. H. Thompson, Montreal, are moving from their present premises on St. Sulpice street to St. Paul street, Montreal.

B. S. Saunders, one of Lewis Bros. salesmen, has been presented with a little Saunders, to follow in his father's footsteps.

Wm. H. Newton, Ottawa, representative of the Canada Paper Co., was at headquarters in Montreal for a few days recently.

Tom Fortye, representing the Peterboro Lock Co., Peterboro, Ont., was calling on some of the Montreal jobbers last week.

Chas. Dietrich, eastern traveler for Shurly & Dietrich, Galt, was a caller at the Montreal office of Hardware and Metal last week.

J. C. McCartney, of the Can. Shovel Co. of Hamilton and New York, called at the Montreal office of Hardware and Metal last week.

Robert Munro, president of the Canada Paint Co., Montreal, has returned from a short holiday spent at Palm Beach, North Carolina.

The Metal Shingle and Siding Company have fitted up new premises at the corner of Delormier avenue and St. Catherine street, Montreal.

C. A. McNoun, of Vankleek Hill, was in Montreal last week on business. He has purchased the stock of J. S. McIntosh, hardware dealer, of Vankleek Hill.

E. Tougas, proprietor and general manager of P. D. Dodds & Co., paint manufacturers, Montreal, was confined to the house for a few days this week with a severe cold, but has now recovered.

Wayland Williams, well known to the trade in Montreal, in connection with Walter H. Laurie & Co., is now looking after the advertising of seven Montreal firms. He still retains his old position with the Laurie Company.

Frank Bishop, senior partner of Bishop & Son, Brantford, Ont., was in town last week visiting the leading manufacturers and wholesale merchants. He spent Sunday with his old time friend, W. H. Evans, of the Canada Paint Co.

Geo. Caverhill, the new president of the Montreal Board of Trade, entertained his fellow-officers, the members of the council, and a few other guests, to dinner at the St. James Club last week. Among those present were:—J. B. Learmont, T. H. Newman, J. A. Richardson, Geo. A. Kohl, Geo. E. Drummond, T. J. Drummond and others.

Western Canada.

G. C. Truman, hardware merchant, Fairlight, Sask., has sold out.

S. Taylor, Yellow Grass, Sask., has sold out to D. Gunn and A. Jaques, both of Manitou, Man., who will continue the hardware business formerly done by S. Taylor under the name of Gunn & Jaques.

As mentioned in a recent issue, Anderson & Thomas, Winnipeg, has dissolved partnership, Mr. Anderson retiring owing to advancing years, and Mr. Thomas continuing the business under the old firm name. Hardware and Metal is in receipt of a circular letter from Mr. Thomas announcing the change.

Hastings & Willoughby are a new hardware and heating firm starting business in Regina. For a number of years Mr. Hastings was foreman of the Smith & Ferguson Co., and he has a wide acquaintance among the trade and an enviable reputation as a heating engineer. The new firm are making arrangements for a number of agencies and have been appointed agents for the Pease-Waldon Co.'s line of warm air furnaces.

PLATE GLASS COMPANY'S FIRE.

The premises of the Toronto Plate Glass Importing Company, Toronto, was partially destroyed by fire on Saturday, March 9, the cause of the fire being unknown. The building was valued at $40,000, and insured for $30,000; the stock being valued at $125,000 and insured for $100,000. The company are now building a factory on the Don roadway and anticipate no difficulty in filling their orders as usual, as they have two other premises in addition to the one damaged by fire. All the books, etc., were saved and the temporary offices have been opened in the Crown Bank building, corner of Victoria and Queen streets, and next door to the former headquarters.

HARDWARE AND METAL

Established 1888

The MacLean Publishing Co.
Limited

JOHN BAYNE MACLEAN . President

Publishers of Trade Newspapers which circulate in the Provinces of British Columbia, Alberta, Saskatchewan, Manitoba, Ontario, Quebec, Nova Scotia, New Brunswick, P.E. Island and Newfoundland.

OFFICES:

MONTREAL, ... 232 McGill Street
Telephone Main 1255
TORONTO, ... 10 Front Street East
Telephones Main 2701 and 2702
WINNIPEG, ... 511 Union Bank Building
Telephone 3726
LONDON, ENG. . . . . .. 88 Fleet Street, E.C.
J. Meredith McKim
Telephone, Central 12960

BRANCHES:

CHICAGO, ILL. . . . . . 1001 Teutonic Bldg.
J. Roland Kay
ST. JOHN, N.B. . . . . No. 7 Market Wharf
VANCOUVER, B.C. . . . . Geo. S. B. Perry
PARIS, FRANCE . Agence Havas, 8 Place de la Bourse
MANCHESTER, ENG. . . . . 92 Market Street
ZURICH, SWITZERLAND . . . . Louis Wolf
Orell Fussli & Co

Subscription, Canada and United States, $2.00
Great Britain, 8s. 6d., elsewhere . . 12s

Published every Saturday.

Cable Address { Adscript, London
Adscript, Canada

TURPENTINE SUBSTITUTES.

An interesting situation is developing in the turpentine markets. The price has been climbing upward steadily under the supervision of the manipulators who are understood to represent the Standard Oil Company, and, as predicted by Mr. Walsh of Savannah, in his interview with Hardware and Metal a week ago, the price is likely to advance still farther, even $1.25 being a possibility before 1907 is over. The price of labor has increased and the present pine groves cannot last much longer, it being predicted that five years is the length of life of the sources of supply under the control of the Rockefeller interests.

This being the case and it being a well known fact that painters in all parts of the country are substituting benzine for turps in ever increasing quantities, all interested in the paint trade would undoubtedly welcome the introduction of a species of turpentine which will answer the same purpose and do the same work as the present standard spirits of turpentine. If by any process of manufacture such an article can be produced and marketed the trade will have an alternative to turn to in case the foregoing market predictions materialize.

In Maine, in British Columbia and in Ontario efforts are being made to produce turpentine from pine stumps but so far results do not seem to be satisfactory. "Brazilian" and "Russian" turpentines have not won a place for themselves on the world's market as reliable substitutes and the trade, much though it desires to avoid paying exhorbitant prices for the standard article has rightly become suspicious of all substitutes.

A new article which offers hope of proving the alternative desired is now being introduced in Canada and several analysis, have been made by buyers, some paint and carriage manufacturers claiming that it has given satisfactory results. We understand that the importing company, the Defiance Manufacturing and Supply Company, Toronto, a branch of the American Lino-oil Company, of Cleveland, have asked the customs department to have an analysis made in the Dominion Government's Labratory at Ottawa, several samples having ben taken from shipments passing through the customs at Toronto. The result of these tests will be looked for with interest and Hardware and Metal will supply the information as soon as it is obtainable.

MANITOBA ELECTIONS.

It is regrettable that in the recent Manitoba elections so few business men were candidates; the legal and agricultural interests were well represented, but only a few business men sought the suffrages of the electorate. It is, however, gratifying to note that the few business men elected are substantial men of affairs who are certain to give a good account of themselves in the Legislature and to bring to bear upon the discussion of the business of the province the business experience and training which they have gained in the successful conduct of their own affairs.

J. T. Gordon, the member-elect for South Winnipeg, is the head of the immense Gordon, Ironsides & Fares business, and one of the ablest and most substantial business men in western Canada. In Portage La Prairie the Liberal leader, who is a prominent business man, met defeat, but at the hands of another business man, Hugh Armstrong. The Legislature would be much the stronger if both Mr. Armstrong and Mr. Brown were in it, and it is to be hoped that the opposition will find another seat for Mr. Brown. In Virden J. G. Rattray, a prominent hardware merchant, was defeated by the Provincial Treasurer. In Morden, Geo. Ashdown, another particularly able business man, brother of Mayor Ashdown of Winnipeg, was also defeated.

The defeats of good business men are to be regretted since there were so few in the field, for the Legislature contains too many politicians and too few practical men of business affairs. It is, however, gratifying to know that the business men who have been elected are men who will through sheer force of their ability make their influence felt.

VALUE OF A FIRM'S NAME.

Recent business changes in Regina, Sask., raise a question as to the value of a well-established and widely known firm name. Last fall the big hardware store of the Western Hardware Co., Regina, was totally destroyed by fire. In December the business of the Smith & Fergusson Co. was sold to Peart Bros., the former proprietors of the Western Hardware Co. As a result two widely known firms—the Smith & Fergusson Co and the Western Hardware Co.—have disappeared from the hardware trade entirely. Mr. Smith is continuing his coal business under the old name of the Smith & Fergusson Co., but his new wholesale hardware business is to be incorporated as the J. W. Smith Hardware Co.

These changes raise the question of the value of firm names. The Smith & Fergusson Co., and the Western Hardware Co., were widely and favorably known. Did not Mr. Smith and Peart Bros. sacrifice valuable assets when in the course of a few business changes they dropped the old firm names ?

REPORT LIKELY GROUNDLESS.

The Canadian agent at Leeds, Eng., stated about ten days ago that he was informed that a very large amount of German cutlery, razors and knives, is being shipped from Germany to Sheffield to Canada under the name of Sheffield goods. From information we have gathered we are led to believe that the report is not correct as the present trade conditions in Germany are so favorable that there is no inducement to do such a thing, and, moreover, English traders are not apt to connive with such schemes and too rigid an inspection of imported goods is maintained to allow smuggling. If such a report was true and goods were being brought into Canada under an assumed name, protests and complaints would have been heard before this. Buyers with an average knowledge of cutlery are able to distinguish German and English goods, not that the German manufacturers are generally inferior to the English, but that certain grades are

In successful stores the floor is not used as a waste-basket, nor the counter for a catch-all.

Markets and Market Notes

(For detailed prices see Current Market Quotations, page 62.)

THE WEEK'S MARKETS IN BRIEF.

MONTREAL.

Putty—Advanced 20 cents per 100 lbs.
Zinc Spelter—Prices reduced 25 cts. per 100 lbs.

TORONTO.

Turpentine—Higher prices quoted.
Varnishes—Change in duty means advances.
Pig Iron English iron quoted easier.
Old Materials—Brass higher.

Montreal Hardware Markets

Office of HARDWARE AND METAL,
232 McGill Street,
Montreal, March 15, 1907.

The spring rush continues and jobbers are finding no difficulty in disposing of their heavy stocks. The seasons seems to be further advanced at present than for the corresponding period last year, and on the average, trade is also a little better. This may be caused by the retailers ordering earlier than they have been in the habit of doing in the past, so as to be on the safe side should any unexpected advances take place. This is a good plan and should be given more attention by retailers, some of whom are prone to hold back their orders till their shelves are almost empty of the desired article, and then blame the jobber for not being able to deliver the goods immediately.

Retailers are doing quite a business now in all lines of sidewalk cleaning implements.

Some lines of chain and coat screws are reported to be running short.

TWINE—No special features to report. Prices remain unchanged, and orders are now mostly booked up for the coming season. "Standard" 500 ft., sisal, 9¼c.; "Tiger," 550 ft., manilla, 9¾c.; "Red Cap," 600 ft., manilla, 12¼c.; "Blue Ribbon," 650 ft., pure manilla, 13¾ to 14c. Car lots ¼ less; thousand pound lots, ⅛c. less.

AXES—Axes and handles are still reported to be moving fairly well.

RIVETS AND BURRS—Are in good demand. Prices unchanged and very firm.

HAY WIRE—The demand is good, and there are no changes to report in prices.

WIRE NAILS—The demand for these still continues good, and prices remain stiff and unchanged at $2.45 per keg f.o.b. Montreal.

CUT NAILS—No extraordinary features. The demand is moderate at $2.30 per keg f.o.b. Montreal.

MACHINE SCREWS—A very brisk business is being done at the old prices.

BOLTS AND NUTS—Sales this year so far exceed those of the same season last year. Prices remain unchanged, and the shortage is still very marked in some lines.

HORSE NAILS—This line is reported to be moving extremely well just at present.

HORSE SHOES—Large sales are reported, and the output is well taken care of.

BUILDING PAPER—With the warmer weather, the sales are increasing.

POULTRY NETTING—Dealers are buying freely. 2x2 mesh, 50 and 5 per cent.; other sizes, 50 per cent. off.

GREEN WIRE CLOTH—Orders have mostly been booked now for the season. Prices remain: In 100 ft. rolls, $1.62½ per 100 square feet; in 50 foot rolls, $1.67½ per 100 square feet.

LAWN MOWERS—What is said of green wire cloth, also applies to lawn mowers. Low wheel, 12, 14 and 16 in., $2.30 each; 9-in. wheel by 12 inch, $2.85 each; 14-inch, $3 each; 16-inch, $3.12½ each; high wheel, 12-inch, $4.05 each; 14-inch, $4.25 each; 16-inch, $4.50 each; 18-inch, $4.75 each.

COIL CHAIN—Some sizes are getting scarce.

SHOT — Not in great demand at present. We still quote.: Shot, packed in 25-pound bags, ordinary drop AAA to dust, $7.35 per 100lbs.; chilled, No. 1—10, $7.88 per 100 lbs.; brick and seal, $8.40 per 100 lbs.; ball, $8.93 per 100 lbs. 5 per cent. off. Bags less than 25 lbs., ¼c. per lb. extra, net; f.o.b. Montreal, Toronto, Hamilton London St. John, Halifax.

AMMUNITION — Moving moderately. No change in Prices: Loaded with black powder, 12 and 16 gauge, per 1,000, $15; 10 gauge, per 1,000, $18; loaded with smokeless powder, 12 and 16 gauge, per 1,000, $20.50; 10 gauge, $23.50.

Toronto Hardware Markets

Office of HARDWARE AND METAL,
10 Front Street East,
Toronto, March 15, 1907.

Hardware trade conditions continue good, most of the seasonable lines experiencing strong demand. Wholesale dealers report big orders, especially for garden tools, and mechanics' outfits in view of the greatly extended building operations the coming summer.

The scaccity in screws continues with no prospects of relief. Orders are coming in rapidly, and the shippers are unable to fill them. If relief does not soon come the situation will become grave; the scarcity is even now sufficient to excite alarm.

A scarcity also prevails in wire nails, with no prospects of repletion. The demand is heavy, prices remaining unchanged.

Sporting goods, fishing tackle and guns are moving, quite a number of orders for fishing tackle having been received.

The moderation in the climate this week has helped the trade very much, and wholesalers are preparing to get out their spring shipments.

BINDER TWINE—Twine prices remain firm and unchanged. Most of the orders have been booked and shipments will commence early in April. We continue to quote:—500 feet Sisal, 9½-9⅝; 500 feet, standard, 9½-9⅝; 550 ft. standard manilla, 10¾; 600 ft. manilla, 12¼; 650 ft. pure manilla, 13¾-14. Car lots ¼c. less, five ton lots ⅛c. less, central delivery.

SCREWS—Business continues brisk, orders be,ng booked tapidly. The supply, however, is very low with no immediate prospects of repletion.

BOLTS AND NUTS—Business in these is brisk, with good demand, but inadequate supply.

CHAIN—Business is steady with firm and unchanged prices and good demand.

RIVETS AND BURRS—The demand is good, with unchanged prices.

MECHANICS' TOOLS—The demand for these is increasing. The moderation in the climate has increased building operations. The present congested housing conditions should do much towards the strengthening of the trade in all tools.

EXTENSION AND STEP LADDERS—Demand continues strong, with unchang(d prices. We continue to quote: 11 cents per foot for 3 to 6 feet, and 12 cents per foot for 7 to 10-foot ladders. Waggoner extension ladders 40 per cent.

POULTRY NETTING — The demand continues good, with unchanged prices: 2-inch mesh, 19 w.g., discount 50 and 10 per cent., all others 50 per cent.

OILED AND ANNEALED WIRE — Business in these is good, demand strong with firm and unchanged prices. We continue to quote: Canadian wire: gauge, 10, $2.46; gauge, 11, $2.52; gauge 12, $2.61 per 100 lbs.

WIRE NAILS—The demand continues strong. The supply is becoming small, and those who have their supply of wire nails in may congratulate themselves for there are no immediate prospects of repletion. Prices are firm and unchanged We continue to quote: $2.40 base.

HORSESHOES AND HORSENAILS—Demand continues strong, especially from newly-opened-up districts. Prices remain firm and unchanged.

BUILDING PAPER—Business in this is enlivening as building operations commence.

WASHERS, WRINGERS AND CARPET SWEEPERS—The demand for these is increasing with the opening up of spring.

LAWN MOWERS—Business in these is increasing, retailers getting their stocks completed.

HARVEST AND GARDEN TOOLS—Business is increasing in these and should in a few weeks be strong. Demand is good, retailers completing their stock as soon as possible.

HIDES, WOOLS AND FURS—We continue to quote the old prices, as consignments of furs are still coming in.

HIDES AND RAW FURS — Country hides are lower. Sheepskins are higher.

Hides, inspected, cows and steers, No. 1	....	0 10½
" " " " No. 2	....	0 09½
Country hides, flat, per lb., cured	....	0 09
" " " green	....	0 08
Calf skins, No. 1, city	....	0 13
" " No. 1, country	0 11	0 12
Sheep skins	1 45	1 60
Horse hides, No. 1	3 50	3 75
Rendered tallow, per lb.	0 05½	0 06
Pulled wools, super, per lb.	....	0 25
" " extra	....	0 27

FURS.

	No. 1.	Prime
Raccoon	....	1 50
Mink, dark	5 00	7 00
" pale	2 50	4 50
Fox, red	3 00	4 00
" cross	2 00	20 00
Lynx	5 00	8 00
Bear, black	....	12 00
" cubs and yearlings	....	5 00
Wolf, timber	....	3 00
" prairie	....	1 00
Weasel, white	0 10	0 65
Badger	0 75	1 75
Fisher, dark	6 00	8 00
Skunk, No. 1	....	1 40
Marten,	3 50	20 00
Muskrat., No. 1	0 28	0 30
" " 2	....	0 25
" " 3	....	0 18
" " 4 and kits	....	0 05

Montreal Metal Markets

Office of HARDWARE AND METAL,
232 McGill Street,
Montreal, March 15, 1907

Quietness seems to reign this week as far as metals are concerned. One large jobber says: "Things are jogging along in the same old way." Business remains good and prices in most cases firm and unchanged.

Copper is still extremely scarce and the prices are firmest of any metal reported this week, and, although an advance was reported in the Toronto markets last week, prices here remain unchanged. Tin is firmer than last week, with an upward tendency, and there will probably be an advance in prices next week. Zinc spelter is a little easier at present than when last quoted. Pig lead is very firm, and it is reported that prices during the past week reached the high water mark. The expected advance in the price of boiler tubes failed to connect, and quotations remain unchanged.

All kinds of structural iron and steel are in good demand owing to the extensive building operations going on here at present, and the mills and works have orders booked for some time to come.

COPPER — This metal is still extremely short, and it's a practical impossibility to obtain any large quantities on the local market. Although an advance was reported in the Toronto markets last week, no change has taken place here, and prices remain extremely firm. Ingot copper is quoted at 26½ to 27c., and sheet copper, base sizes, 34c.)per pound.

INGOT TIN—Although there are no changes this week the market is very much firmer and next week will probably see an advance. For the present we quote: 45½ to 46c. per pound.

ZINC SPELTER — The market has been a little easier during the week and we have reduced our prices 25c. per 100 lbs. We now quote: $7.25 to $7.50 per 100 lbs.

PIG LEAD—Remains very firm, and prices during the week reached the high water mark. We are still quoting $5.50 to $5.60 per 100 lbs.

ANTIMONY—Prices remain firm and unchanged at the following figures: 27½ to 28c. per lb.

GALVANIZED IRON—Heavy building operations have created a good demand, and prices remain firm and unchanged.

PIG IRON—No special features, the following prices remaining firm: Londonderry, $24.50; Carron, No. 1, $27; Carron, special, $25.50; Summerlee, No. 3, soft $23.50.

BOILER TUBES—The advance which was expected to take place this week failed to materialize and the old list still prevails.

TOOL STEEL — Prices remain unchanged.

COLD ROLLED SHAFTING — No changes to report.

OLD MATERIAL.—Prices are: Heavy copper 20c. per pound; light copper 17c. per pound; heavy red brass, 18½c. per pound; heavy yellow brass, 15c. per pound; light brass, 11 to 11½c. per pound; tea lead, 4¼ to 4½c per pound; heavy lead, 4½c. per pound; scrap zinc, 4½c. per pound; No. 1 wrought, $17 per ton; No. 2 wrought, $6 per ton; No. 1 machinery, $18 per ton; stove plate, $14 per ton; old rubber, 11½c. per pound; mixed rags, 1 to 1¼c. per pound.

Toronto Metal Markets

Office of HARDWARE AND METAL,
10 Front Street East,
Toronto, March 18, 1907

While there is a good business being done in most metals, matters are running along with little change. English pig iron is quoted a little easier, but Canadian furnaces are still booked ahead for months and no iron can be procured for present delivery without paying a premium. The market in the States keeps firm.

Copper continues very scarce and last week's advances are held to firmly. As illuserating the condition in foreign markets the American Metal Market Report says that a new high record was made in London last Monday and for deliveries in three months from now, buyers in London are paying for G M.B. £111 10s. equal to about ¼c. per pound more than the spot price. G.M.B., means, "Good merchantable brands," and is the term under which the London copper speculative trading is done. The market continues to demonstrate its remarkable intrinsic strength, in the face of conditions in the American and foreign stock markets that ordinarily would create lack of confidence and ultra conservatism. But present untoward conditions have no effect on copper or the copper trade here or abroad. It does not enable our producers to get any more copper mined and smelted, and does not seem to affect orders for the manufactured article. The question is how to supply enough metal to meet the demands of the world.

Tin has been fluctuating in outside markets, owing to speculation, but the local continues firm. Business in galvanized iron and black sheets is very good while lead continues to manifest great strength. Zinc spelter is a little weaker but prices remain as before. Old materials show a few changes, brass being higher and iron and rubbers a trifle weaker.

PIG IRON—Recent weakness in English markets has made slightly lower prices possible on English iron, but the market is in good condition. Hamilton and Midland are off the market, Radnor is quoted at $33 at furnace, Londonderry at $26, Middlesborough No. 1 at $23,50, and Summerlee No. 1 at $25.50 f.o.b. Toronto.

BAR IRON—Bar iron is moving freely, with prices steady. We quote: $2.30 f.o.b., Toronto, with 2 per cent. discount.

TIN—Prices are unchanged at 45c. for ingots. The base on charcoal plates continues at $6.50.

SHEETS AND PLATES — Canada plates and galvanized iron are being bought freely and black sheets are also in fair demand. Prices keep firm with no sign of weakness evident.

BRASS—We still quote 30 cents per pound on sheets.

COPPER—The market retains its firmness in spite of last week's advance and we continue to quote: Ingot copper, $27 per 100 lbs., and sheet copper, $35 per 100 lbs.

LEAD—There is talk of even higher prices and the firmness of the market gives warrant for this feeling. The same quotations stand: $5.50 for imported pig and $5.75 to $6 for bar lead. For sheet lead $7 is asked.

ZINC SPELTER—Spelter is reported weaker, but we still quote:— 7½c per lb. for foreign and 7c per lb. for domestic. Sheet zinc is quoted at 8¼c in casks, and 8½c in part casks.

ANTIMONY — Market is fairly strong with the ruling price 27c. per pound.

OLD MATERIAL — Buying prices are: Heavy copper and wire, 20c. per lb.; light copper, 18c per lb.; heavy red brass, 18½c. per lb.; heavy yellow brass, 15c. per lb.; light brass, 12c per lb.; tea lead, $4.25 per 100 lbs.; heavy lead, $4 40 per 100 lbs; scrap zinc, 4½c. per lb.; No. 1 wrought iron, $13; No. 2 wrought, $8; machinery cast scrap, $17.50; stove plate, $12; malleable and steel, $8.50, old rubbers, 11c. per pound; country mixed rags, $1 to $1.25 per 100 lbs., according to quality.

COAL—Standard Hocking soft coal, f. o b. mines, lump, $1 50, ¾-in, $1 40, run of mine, $1.30; nut, $1.25; N. & S., $1; P. & S. 85c, Youghiogheny soft coal in

cars, bonder at bridges: lump, $2.60; 3/4 inch. $2.50; mine run, $2.40; slack, $2.15.

London, Eng., Metal Markets

From Metal Market Report, March 13, 1907.

PIG IRON—Cleveland warrants are quoted at 53s. 10½d., and Glasgow standard warrants at 53s., making prices as compared with last week on Cleveland warrants 9d. lower, on Glasgow warrants 1s. lower.

TIN—Spot tin opened firm at £192 5s., futures at £191, and after sales of 260 tons of spot, and 180 tons for futures closed easy at £192 5s. for spot, and £190 15s. for futures, making price compared with last week 10s. higher on spot and unchanged on futures.

COPPER—Spot copper opened steady at £110 7s. 6d., futures £111 12s. 6d., and after sales of 500 tons of spot and 300 tons of futures, closed quiet at £110 7s. 6d. for spots, and £111 10s. for futures, making price compared with last week 17s. 6d. higher on spot and 17s. 6d. higher on futures.

LEAD—The market closed at £20, making price as compared with last week £3 17s. higher.

SPELTER—The market closed at £26 10s., making price as compared with last week 10s. higher.

British Columbia News

By G. S. B. Perry.

Vancouver, B.C., March 8, 1907.

New industries, in fact all industries, established in Vancouver are to be encouraged by exemption from taxation on all improvements, as buildings, machinery, where the industry is strictly a manufacturing industry. This action is the result of some agitation through the Markets and Industries Committee, a special committee of the city council which has been engaged in promoting and extending business in the city. It was first proposed to limit the industries receiving the benefit to those that employed white labor only. On it being pointed out by one member, a mill man, that of a monthly pay roll of $15,000 he only paid some $4,000 or $5,000 to Orientals, the limitation was removed and the recommendation unanimously adopted by the committee.

Water rates are to be reduced to manufacturing industries also. The present rate is 7c. per hundred cubic feet, and it is proposed to reduce this to 3c. per hundred cubic feet. One gentleman presented a bill which his company had paid, amounting to $227.-89 for one month, and then but three boilers were in operation. In his new plant fourteen boilers were to be installed, and at the same rate the cost would be appalling. The fear was advanced in this, as in the question of taxation, that industries would seek suburban locations, thus depriving the city of the benefit of their presence and the distribution of their wage sheets, and other disbursements.

The big Blue Funnel liner Titan, 500 ft. long, and 9,000 tons gross, reached port last Saturday from Liverpool. She first discharged cargo at Victoria and then came on here. The vessel had 2,100 tons of miscellaneous merchandise for Vancouver, principally metals and cement.

The lumber manufacturers will not by any means let the investigation of allegations regarding the industry go by default. When the House of Commons select committee begins its sittings on March 15th, the lumbermen will be on hand with data to show increased cost all along the line, from the logging camps to the loading in cars. The advances in wages average from 50c. to $1 per day over rates formerly in effect. At the same time, men are hard to get. The shortage in the number of men who can be secured for the work is also a serious factor as it has had its influence in putting up the price of logs where they are bought in the open market. There are not nearly enough logs now being marketed to keep the price down.

Coal measures on Queen Charlotte Islands are to be exploited. A syndicate of Victoria men has negotiated a sale of several thousand acres at $20 per acre. Extensive tests are to be made during the summer.

Coal operators at Edmonton, where a number of mines of good quality lignite coal have been opened for years, are out with a resolution endorsing the recommendation of the Edmonton Board of Trade that freight rates on coal be reduced during the months of August and September. It is pointed out that at that time many cars are coming west loaded with machinery and other merchandise. These are not in demand for return shipments and go back empty. Later on, the wheat movement blocks the single track railways, and coal cannot be moved, so that if the coal were got out earlier than the wheat it could be stored at various points along the line. This would serve a double purpose as farmers marketing their wheat could take home a stock of coal. It is the aim that the experience of the winter just ending shall not be repeated.

The British Columbia-Mexican steamer service is now in full running order direct from the ports of this province to Salinas Cruz, the Pacific terminal of the Tehuantepec Railway, giving direct connection from Pacific to Atlantic across the Isthmus. Three steamers, the Woodford, the Georgia and the Lonsdale have been acquired. The first to reach B. C. ports will be the Georgia, which is due to sail from Salinas Cruz on March 15 for Victoria and Vancouver.

Water power for electric generation continues to excite interest in this province. Many streams of considerable dimensions in the coast district have been recorded for power purposes and others are being sought. It is stated that the B. C. Electric Railway Co., owning and operating all electric lines at Victoria, Vancouver and New Westminster, may, if sufficient power can be secured, electrify the E. & N. Railway, now owned by the C.P.R. Co., and running from Victoria to Nanaimo. The change which electric power would make, in case the electric company and the C.P.R. came to terms, can hardly be calculated. A district now sparsely settled, adaptable for occupation by a large number of people on small holdings, is served by the railway, and rapid and frequent transit as provided by an electric line, would make a vast improvement and provide for the numerous incoming settlers who wish to locate on land.

On the mainland several large water powers have been recorded and some of them are to be developed this year.

N.S. Hardware Trade News

Halifax, N.S., March 12, 1907.

The travelers who are now on the road are sending in orders for spring lines, and the opening of the business appears unusually bright. Industrial development is expanding in many districts and everything gives promise of a big trade. All lines of goods are very firm, and some advances are looked for, but at the present time the jobbers have not increased their prices very materially.

* * *

The weather is very favorable for lumbering operations, and some record cuts are anticipated. The Sissiboo Lumber Company is cutting over a million feet per month.

* * *

D. G. Kirk, of Antigonish, has added another to his list of stores, having recently purchased the store, merchandise, etc., from W. M. Strople, on Bayfield road. A new stock of goods will be ordered and Mr. Strople will continue the management of the store.

* * *

Bruce McKay, of Parrsboro, formerly a member of the firm of the Fox River Lumber Company, will leave shortly for Calgary, Alta., where he will engage in business.

* * *

W. Curry, of Rhodes, Curry & Co., Amherst, has gone on a tour of the West Indies.

* * *

James Dempster, proprietor of Dempster's Mills, Halifax, and Robie Davidson, lumber merchant, both of Halifax, have gone to the West Indies for the benefit of their health.

* * *

R. W. W. Buck, of St. John, representing the Dayton Computing Scale Co., of Ohio, is touring Prince Edward Island, in the interests of his firm.

* * *

On the completion of the Dominion Iron and Steel Company's Bessemer controverter at Sydney, the company expect to increase its output by five or six thousand tons a month. It is expected that it will be in operation next month.

MALLEABLE IRON DUTIES REVISED.

The Government has backed down from the position it originally took regarding the malleable iron industry. All malleables that were 25 per cent. before will now be dutiable at 27½ per cent., thus putting the industry in a better position than before. Malleable castings, formerly 17½, are now 20 per cent. The Globe commenting on the subject editorially says that a prominent Ontario manufacturer negotiated unsuccessfully with no less than five malleable iron founders in the province, who were unable to book orders, as the capacities of the foundries was not equal to any greater output. An application to the Finance Minister for leave to import during the stringent season was refused, as he has no authority to levy or remit except as the Government approves. Our foundries are taxed to their utmost capacity and a demand unsupplied will not last.

An important change in the tariff is in the matter of sprocket chains, formerly dutiable at 15, 17½ and 20 per cent., which now go on the free list.

PRESENTATION AT WINNIPEG.

Fred A. Sermon has resigned his position with The Miller-Moorse Hardware Co., Winnipeg, and is going to St. Louis to take a position with a big hardware house there. A little less than two years ago Mr. Sermon came to Winnipeg from Montreal, where he had been with the firm of A. Macfarlane & Co., he thus has a wide acquaintance among the hardware trade of Canada.

The other night Mr. Sermon was entertained at the Mariaggi Hotel, Winnipeg, by a number of his friends and business associates. Mr. Fred A. Reinhardt occupied the chair and in a few well chosen remarks expressed the sentiments of the company when he said that their regret in losing Mr. Sermon's presence among them was only surpassed by the joy it afforded them to know that in making this move he was climbing one step higher on the ladder of commercial importance. A pleasant evening was spent and Mr. Sermon carries with him the very best wishes of his many friends in Winnipeg.

TO TEST STEEL BOX CARS.

After experimenting with steel flat cars and steel passenger, postal, baggage and Pullman cars, railroad officials are devoting their attention to steel box cars, and it is likely contracts will be awarded shortly by some of the leading railroads of the country for a large number of such cars. Local railroad men are in the west inspecting the new steel box car built by the Union Pacific Railroad Company, which is the first car of its kind ever constructed. One or two other roads which have been experimenting with the steel box car have instructed their motive power officials to watch carefully the experiments with the new car built by the Union Pacific, and upon the result of this investigation will depend largely what action is to be taken by the various roads in the construction of the new equipment.

The new box car is regarded by motive power experts as the extreme development of the steel car, and they think it strange that the experiment of building a box car with a steel body has not been tried before, for it is only a short step from the high steel gondola to the box car. Some railroad officials have not thought it important that the steel construction should be carried above the underframe, but they now realize that this can be done to good advantage.

It is claimed that the new box car weighs 37,800 pounds, or about 12 per cent. less than the standard box car used on the Harriman lines. The two cars have practically the same dimensions, the difference in cubic capacity being only about 50 cubic feet. This saving of weight is accomplished largely by using only one centre sill and designing the side frames as efficient plate girders of considerable length.

The body framing of a wooden box car has always been more or less of a failure as a truss to carry that part of the load between bolsters. The combination of green timber and imperfect fitting usually results in a side frame, which carries little or none of the load after a few months of service, but the steel box car, the sides having all the members rigidly attached to each other, supports a great deal of the load which would otherwise be carried by the longitudinal sills.

Railroad officials realize that it will cost more to repair a steel box car than the present equipment in case the car is damaged, but they figure that repairs will not be required nearly so often.

MANITOBA HARDWARE AND METAL MARKETS

Market quotations corrected by telegraph up to 12 a.m. Friday, March 15, 1907.)

F. R. Munro, Western Manager HARDWARE AND METAL, Room 511, Union Bank Building, Winnipeg, Man.

Wholesale houses still complain of the difficulty in getting through shipments from Fort William to western points, and report that many cars of hardware are held up in that town.

An advance of five cents per gallon in linseed oils and a decline in petroleum are the features of the market this week. The discount on copperware is now 25 per cent.

LANTERNS—Cold blast, per dozen, $6.50 ; coppered, $8.50 ; dash, $8.50.

WIRE—Barbed wire, 100 lbs., $3.22½ ; plain galvanized, 6, 7 and 8, $3.70; No. 9 $3.25; No. 10, $3.70; No. 11, $3.80; No. 12, $3.45; No. 13, $3.55; No. 14, $4.00; No. 15, $4.25; No. 16, $4.40; plain twist, $3.45; staples, $3.50; oiled annealed wire, 10, $2.96 ; 11, $3.02 ; 12, $3.10 ; 13, $3.20 ; 14, $3.30 ; 15, $3.45. Annealed wires (unoiled) 10c. less; soft copper wire, base, 36c.; brass spring wire, base, 30c.

HORSESHOES—Iron No. 0 to No. 1, $4.65; No. 2 and larger, $4.40; snowshoes, No. 0 to No. 1, $4.90; No. 2 and larger, $4.65 ; steel, No. 0 to No. 1, $5 ; No. 2 and larger, $4.75.

HORSENAILS — Capewell brand, quotations on application. No. 10, 20c.; No. 9, 22c. ; No. 8, 24c. ; No. 7, 26c. ; No. 6, 28c. ; No. 5, 32c. ; No. 4, 40c. per lb. Discounts: "C" brand, 40, 10 and 7½ p.c.; "M" brand and other brands, 55 to 60 p.c. Add 15c. per box.

WIRE NAILS—$2.85 f.o.b. Winnipeg, and $2.40 f.o.b. Fort William.

CUT NAILS—Now $2.90 per keg.

PRESSED SPIKES — ¼ x 5 and 6, $4.75 ; 5-6 x 5, 6 and 7, $4.40; ⅜ x 6, 7 and 8, $4.25; 7-16 x 7 and 9, $4.15; ½ x 8, 9, 10 and 12, $4.05; ⅝ x 10 and 12, $3.90. All other lengths 25c. extra net.

SCREWS.—Flat head, iron, bright, 85 and 10; round head, iron, 80; flat head, brass, 75 and 10; round head, brass, 70 and 10; coach, 70.

NUTS AND BOLTS — Bolts, carriage, ⅜ or smaller, 60 and 5 ; bolts, carriage, 7-16 and up, 55 ; bolts, machine, ⅜ and under, 55 and 5 ; bolts, machine, 7-16 and over, 55 ; bolts, tire, 65 ; bolt ends, 55 ; sleigh shoe bolts, 65 and 10 ; machine screws, 70 ; plough bolts, 55; square nuts, cases, 3; square nuts, small lots, 2½; hex nuts, cases, 3; hex nuts, small lots, 2½ p.c. Stone bolts, 70 and 10 p.c.

RIVETS — Iron, 60 and 10 p.c.; copper, No. 7, 43c., No. 8, 42½c.; No. 9, 45¼c.; copper, No. 10, 47c.; copper, No. 12, 50½c.; assorted, No. 8, 44½c., and No. 10, 48c.

COIL CHAIN—¼ in., $7 ; 5-16, $5.35 ; ⅜, $4.75 ; 7-16, $4.50 ; ½, $4.25 ; 9-16, $4.20 ; ⅝, $4.25 ; ¾, $4.10.

SHOVELS—List has advanced $1 per dozen on all spades, shovels and scoops.

HARVEST TOOLS—60 and 5 p.c.

AXE HANDLES—Turned. s.g. hickory doz., $3.15 ; No. 1, $1.90. No. 2, $1.60 ; octagon, extra, $2.30 ; No. 1, $1.60.

AXES — Bench axes, 40 ; broad axes, 25 p.c. dis. off list ; Royal Oak, per dozen, $6.25: Maple Leaf. $8.25 ; Model, $8.50 ; Black Prince, $7.25 ; Black Diamond, $9.25 ; Standard flint edge, $8.75 ; Copper King, $8.25 ; Columbian, $9.50 ; handled axes, North Star, $7.75 ; Black Prince, $9.25 ; Standard flint edge, $10.75 ; Copper King, $11 per dozen.

CHURNS.—45 and 5; list as follows: No. 0, $9; No. 1, $9; No. 2, $10; No. 3, $11; No. 4, $13; No. 5, $16.

AUGER BITS—"Irwin" bits, 47½ per cent., and other lines 70 per cent.

BLOCKS—Steel blocks, 35; wood, 55.

FITTINGS—Wrought couplings, 60 ; nipples, 65 and 10 ; T's and elbows, 10; malleable bushings, 50 ; malleable unions, 55 p.c.

HINGES—Light "T" and strap, 65.

HOOKS—Brush hooks, heavy, per doz., $8.75 ; grass hooks, $1.70.

STOVE PIPES—6-inch, per 100 feet length, $9 ; 7-inch, $9.75.

TINWARE, ETC. — Pressed, retinned, 70 and 10 ; pressed, plain, 75 and 2½ ; pieced, 30 ; japanned ware, 37½ ; enamelled ware, Famous, 50 ; Imperial, 50 and 10 ; Imperial, one coat, 60 ; Premier, 50 ; Colonial, 50 and 10 ; Royal, 60 ; Victoria, 45 ; white 45 ; Diamond, 50 ; Granite, 60 p.c.

GALVANIZED WARE—Pails, 37½ per cent.; other galvanized lines 30 per cent.

CORDAGE — Rope, sisal, 7-16 and larger, basis, $11.25 ; Manilla, 7-16 and larger, basis, $16.25 ; Lathyarn, $11.25 ; cotton rope, per lb., 21c.

SOLDER—Quoted at 27c. per pound. Block tin is quoted at 45c. per pound.

POWER HORSE CLIPPERS—"1902" power clipper, $12 ; "Twentieth Century" $6. "1904" sheep shearing machines, $13.60.

WRINGERS—Royal Canadian, $35 ; B.B., $39.75, per dozen.

FILES—Arcade, 75 ; Black Diamond, 60 ; Nicholson's, 62½ p.c.

LOCKS—Peterboro and Gurney, 40 p.c.

BUILDING PAPER—Anchor, plain, 66c.; tarred, 69c.; Victoria, plain, 71c.; tarred, 84c.

AMMUNITION, ETC.:—Cartridges, rim fire, 50 and 5 ; central fire, 33⅓ p.c.; military, 10 p.c. advance. Loaded shells: 12 guage, black, $16.50 ; chilled, 12 gauge, $17.50; soft, 10 gauge, $19.50; chilled, 10 gauge, $20.50. Shot, ordinary, per 100 lbs., $7.75 ; chilled, $8.10 ; powder, F.F., keg, Hamilton, $4.75 ; F.F.G., Dupont's, $5.

IRON AND STEEL—Bar iron basis, $2.70. Swedish iron basis, $4.95 ; sleigh shoe steel, $2.75 ; spring steel, $3.25 ; machinery steel, $3.50 ; tool steel, Black Diamond, 100 lbs., $9.50 : Jessop, $13.

SHEET ZINC—$8.50 for cask lots, and $9 for broken lots.

PIG LEAD—Average price is $6.

AXLE GREASE—"Mica," 1-lb. tins, $11 per gross ; 3-lb. tins, $2.40 per doz.;

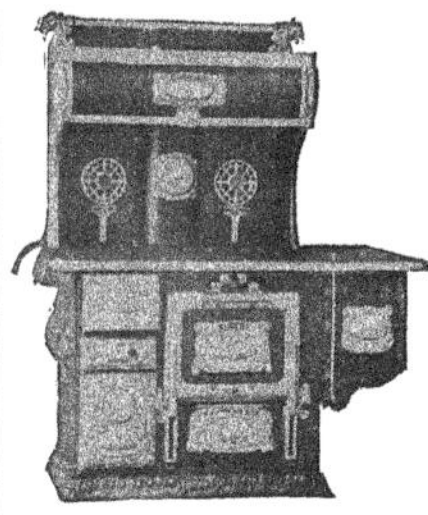

$4.60 per doz. case; 10-lb. iron pails, 60c. each; 15-lb. pails, 80c.; 25-lb. pails, $1.25; "Diamond," 1-lb. wooden boxes, $6.40 per gross.

COPPER — Planished copper, 44c. per lb.; plain, 39c.

IRON PIPE AND FITTINGS—Black pipe, ¼ inch, $2.65; ⅜, $2.80; ½, $3.50; ¾ $4.40; 1, $6.35; 1¼, $8.65; 1½, $10.40; 2, 13.85; 2½, $19.00; 3, $25.00. Galvanized iron pipe, ⅜ inch. $3.75; ½, $4.35; ¾, $5.65; 1, $8.10; 1¼, $11.00; 1½, $13.25; 2, inch, $17.65. Nipples, 70 and 10 per cent.; unions, couplings, bushings and plugs, 60 per cent.

LEAD PIPE—Market is firm at $7.80.

GALVANIZED IRON — Apollo, 16 gauge, $3.90; 18 and 20, $4.10; 22 and 24, $4.45; 26, $4.40; 28, $4.65, 30 gauge or 10¾ oz., $4.95; Queen's Head, 24, $4.50; 26, $4.65; 28, $5.00.

TIN PLATES—IC charcoal, 20x28, box $9.50; IX charcoal, 20x28, $11.50; XXI charcoal, 20x28, $13.50.

TERNE PLATES—Quoted at $9.

CANADA PLATES—18x21, 18x24, $3.40; 20 x 28, $3.65; full polished, $4.15.

BLACK SHEETS—10 to 16 gauge, 100 lbs., $3.50; 18 to 22, $3.75; 24, $3.90; 26, $4; 28, $4.10.

LUBRICATING OILS—600W. cylinders, 80c.; capital cylinders, 50c.; solar red engine, 30c.; Atlantic red engine, 29c.; heavy castor, 28c.; medium castor, 27c.; ready harvester, 28c.; standard hand separator oil, 35c.; standard gas engine oil, 35c. per gallon.

PETROLEUM AND GASOLINE — Silver Star in brls., per gal., 20c.; Sunlight in brls. per gal., 22c.; per case, $2.35; Eocene in brls, per gal., 24c.; per case, $2.50; Pennoline in brls., per gal., 24c.; Crystal Spray, 23c.; Silver Light, 21c.; Engine gasoline in barrels, gal. 27c., f.o.b. Winnipeg in cases, $2.75.

PAINTS AND OILS — White lead, Pure, $6.50 to $7.50, according to brand; bladder putty, in bbls., 2½c.; in kegs, 3¾c.; turpentine, barrel lots, Winnipeg, $1.01; Calgary, $1.08; Lethbridge, $1.08; Edmonton, $1.09. Less than barrel lots 5c. per gallon advance. Linseed oil, raw, Winnipeg, 72c.; Calgary, 79c.; Lethbridge, 79c.; Edmonton, 80c.; boiled oil, 3c. per gal. advance on these prices.

WINDOW GLASS — 16-oz. O.G., single, in 50-ft. boxes — 16 to 25 united inches, $2.25; 26 to 40, $2.40; 16-oz. O.G., single, in 100-ft. cases — 16 to 25 united inches, $4; 26 to 40, $4.52; 41 to 50, $4.75; 50 to 60, $5.25; 61 to 70, $5.75. 21-oz. C.S., double, in 100-ft. cases—26 to 40 united inches, $7.36; 41 to 50, $8.40; 51 to 60, $9.45; 61 to 70, $10.50; 71 to 80, $11.55; 81 to $17.30.

United States Metal Markets

From the Iron Age, March 14, 1907.

The monthly pig iron statistics show that the output for February, a short month, of coke and anthracite pig iron amounted to 2,045,068 tons, as compared with 2,205,607 gross tons in January. This is an increase of 1,880 tons daily.

In pig iron both sides maintain a defensive attitude, and there is comparatively little business doing for delivery during the second quarter.

The pipe and tube trade is in an exceptional position. The National Tube Company has been forced to withdraw prices, and has notified the trade that no more orders will be booked for shipment prior to June 1, prices to be fixed at the time of shipment of the order. Outside mills have withdrawn their prices and are quoting an advance of $2 to $4 per ton.

The sheet and tin plate mills are running under tremendous pressure. There is some talk of advances in prices, but it is well understood that the leading interest is opposed to any such action.

The implement makers and the mills rolling bars and merchant shapes are still negotiating over the fixing of prices on season contracts. Last year the mills made concessions, but they show little disposition to do so this year.

From the Iron Trade Review, March 14, 1907

Despite the absence of noteworthy transactions, the iron and steel market gained considerable strength during the week. Structural fitters are being urged by the steam roads to hurry forward material. The structural mills of the Carnegie Steel Co. this week received specifications from the American Bridge Co. for 26,000 tons of material for rolling within the next sixty days, and plates for 1,000 steel cars were placed by the western plant of the Pressed Steel Car Co.

The demand for wire nails has reached abnormal proportions, most of the mills being from four to six weeks behind in their deliveries. So heavy were the requirements of jobbers during the winter months that few of the manufacturers were able to accumulate stocks for the spring trade, while shipments are again be pro-rated to meet the needs of the greatest number. An early advance, not only on wire nails, but wire products generally, is anticipated, some of the independent mills already asking premiums for prompt deliveries.

Persons addressing advertisers will kindly mention having seen this advertisement in Hardware and Metal.

BUILDING AND INDUSTRIAL NEWS

HARDWARE AND METAL would be pleased to receive from any authoritative source building and industrial news of any sort, the formation or incorporation of companies, establishment or enlargement of mills, factories, foundries or other works, railway or mining news. All correspondence will be treated as confidential when desired.

A waterworks plant will be erected at Welland, to cost $70,000.

A new Methodist church is to be erected at Victoria at a cost of $25,000.

Work has been commenced on the addition to the Galt Malleable Iron works.

The Berlin (Ont.) Hotel Co. has been incorporated, with a capital of $250,000.

Wentworth County Council will erect a House of Refuge at Hamilton, to cost $40,000.

The W. R. Brock Co., Calgary, will open in a few weeks their large new warehouse.

The Washburn Crosby Co., of Minneapolis, will erect an 8,000-barrel mill at Keewatin.

The congregation of Central church, Hamilton, will build a new church, to cost $165,000.

The Dominion Smelters Co. will build a customs smelter at Parry Sound, to cost $1,500,000.

A new Anglican church will be erected at Fort William to cost between $35,000 and $40,000.

The Merchants Bank of Canada will establish another branch at Montreal to cost $100,000.

Jas. Coristine Co., Montreal, will erect a seven-storey office building, to cost about $82,000.

Public Works Department, Ottawa, is asking for tenders for erection of a post office at Fernie, B.C.

A company is being formed for the purpose of building a skating rink at Thornhill, to cost $1,500.

Kelly Bros. & Co., Kenora, have received the contract to erect a $100,000 round-house for the C.P.R. at Kenora.

Bryan & Henderson have been awarded the contract for the erection of the Carnegie Hall at Lucknow, to cost $9,475.

Damage to the extent of $600 was done to the brass works of W. Coulter & Sons, Toronto, on Tuesday, March 12.

The Carnegie library, municipal buildings and fire hall, Sault Ste. Marie, were completely destroyed by fire last week.

The Lethbridge, Alta., Knights of Pythias, in conjunction with a Calgary syndicate, will erect a new brick block in Lethbridge.

Smith Bros. & Wilson have been awarded the contract for the erection of Ald. T. Wilkinson's new warehouse in Regina to cost $12,000.

The Polson Iron Works have been awarded the contract for the construction of a $12,000 hydraulic pump for the city of Toronto.

J. & J. McDiarmid Co. has been awarded the contract for the erection at St. Boniface, Man., of a new post office, to cost $20,255.

S. F. Whitham, Brantford, contractor for the new Normal school there, has taken out a permit for the building, which will cost $52,000.

The L. & N.O. railway commission has contracted to the Montreal Locomotive Works for six new locomotives, the aggregate cost to be $125,000.

The T. Eaton Co., Toronto, have been granted a permit for the erection of an eight-storey building, at the corner of Yonge and Albert streets.

The Northern Turpentine Co. has been incorporated with a capital of $250,000 for the establishment of a large turpentine plant at Killaloe, Ont.

W. R. Brock & Co., Toronto, have been granted a permit for the erection of a two-storey addition to the Piper street wing of their factory.

A cordage company will be established at Valleyfield, Que., arrangements having been made with the Montreal Cotton Co. for supply of electric power.

An immense cement plant will be erected at Kendall, B.C., to cost $1,000,000, employing 700 men, and having a daily capacity of 5,000 barrels.

The Vancouver Board of Trade has passed a resolution calling on the Dominion Government to build a 250,000-bushel elevator in that city to store Alberta wheat.

The Tidman Silver and Aluminum Co., Toronto, has been incorporated, with a capital of $40,000, to manufacture and deal in aluminum ware, silverware and metal novelties. Provisional directors of the company are: F. A. Lewis, E. Gillis and M. W. Mayhr.

Two Guelph men, W. W. Hadley and John Grey, have secured options on zinc ore lands in Frontenac, from Richardson & Sons, and will build a smelting and refining plant, provided the council grant them land and exemption from general taxes for ten years.

A bill to give the Dominion Iron and Steel Company the right to form subsidiary companies to carry on any trade or business within the limits of the company, and to guarantee the stock of these companies, has been introduced in the Nova Scotia Legislature.

The King Electrical Works, Montreal, has been incorporated, with a capital of $30,000, to carry on the business of manufacturers of and dealers in electrical machinery, brass, copper and other metals. The promoters are: E. F. Surveyer, A. Chase-Casgrain, J. W. Weldon, E. M. McDougall, and S. J. Le-Huray, all of Montreal.

The Imperial Steel & Wire Company offer to construct at Hamilton a wire drawing mill with a capacity of one hundred tons of wire daily, at an approximate cost of $200,000, if the city will guarantee the principal and interest of the first $100,000 and grant a site of ten acres of land with a river frontage of 150 feet.

One of the new factories lately contributed to Canada by Uncle Sam is the Anthony Fence Company, of Walkerville, Ont. This new concern is an outgrowth of a similar business established at Tecumseh, Michigan, and although one of the youngest of American fence companies, having been under way only about three years, it now ranks as one of the most important of its kind. The Walkerville business is financed largely by Canadians.

B. C. TRADE NEWS.

By May 1st it is expected that the new terminal city of the coast, Prince Rupert, is to have its first electric light plant in operation. The B. C. Tie & Timber Co., of which D. H. Hays, brother of President Hays, of the G.T.P., is manager, is installing at its mill a plant capable of generating electricity for 1,200 lights. Tris will be run by steam power, using the waste from the saw mill for fuel. The Canadian General Electric Co. is supplying the plant.

* * *

A pole line from Vancouver to the location of the Stave Lake Power Co.'s generating plant, a distance of over 38 miles, is to be surveyed this year. The company is now receiving tenders for turbines and electrical equipment sufficient for the generation of ten thousand H.P. This portion of the development will be completed this season, and the transmission of power, it is expected, will be possible before many months. The entire installation will develop much more than the initial 10,000 h.p.

* * *

There will be two or more coal mines added to the shipping list in B.C. this year, it is likely, as a reslt of the establishing of rail transportation into the Similkameen and Nicola districts. The Diamond Vale Coal & Iron Co. expects to be shipping by May 1st, its property being near Coutlee on the Spences Bridge-Nicola branch of the C.P.R. The company is about to take over this line from the contractors, who have been operating it under construction. The last few miles, added last season, have now been completed. The inspection visit of the representative of the Department of Railways is all that is now necessary to allow of the road being formally taken over.

* * *

The Alien Labor law of Canada has been demonstrated for once, and the unlucky subject of the demonstration is S. A. Mundy, the millionaire head of the Mundy Lumber Co., which has mills at Three Valley Lake, a few miles west of Revelstoke on the main line of the C.P.R. Mr. Mundy is an American, a former resident of Bradford, Pa. In attempting to operate his mill Mr. Mundy found himself up against the shortage of labor, which is chronic in this province. He inserted advertisements in papers back in the Quaker state offering not only to pay transportation of willing laborers, but provide them a lunch basket as well if they would go to work in his mill in far B. C. He secured several employes in this way, and the trouble began when an employe getting higher rate of pay was let out to make room for all those brought in. He instigated the action, and a decision has just been handed down fining Mr. Mundy $200 and costs.

It was held that the advertisement offering to engage men under the terms stated, formed, together with the subsequent acts of sending men on the basis offered, a "contract" within the meaning of the Act. The trial judge, Justice Morrison in imposing the smaller penalty took occasion to comment that he was not sure, from the course defendant adopted, that he was aware of the provisions of the Alien Labor Act, or had deliberately ignored it. This is about the first case in which a conviction under the Canadian Alien Labor law has been secured. The difficulty, as a rule, lies in proving that a contract has been entered into in a foreign country.

* * *

One of the largest timber and milling transfers made in this province for many months has just been announced, M. Carlin, F. W. Jones and Hugh Sutherland have sold their one-third interest in the Columbia River Lumber Co., for close to one millions dollars cash. The purchasers are American capitalists, headed by J. D. McCormick, of Duluth. This gentleman will in future reside in Vancouver. The company's principal mill is at Golden, B.C., and they have several other r- on the main line of the C.P.R. between there and Kamloops. Wm. Mackenzie, of the Canadian Northern Railway, holds the other two-thirds interest in the property.

* * *

Car chortage is causing trouble among the shingle men, who have been trying all the month of February to get down to business after their six weeks' close down. But car shortage is not all. They cannot get shingle bolts, most of the mills cutting from bolts instead of from large timber. There are no heavy stocks on hand, and if more timber is not soon secured the stocks will remain light.

A NEW MINING JOURNAL.

The first number of the Canadian Mining Journal has just been issued. This journal takes the place of what was formerly known as The Canadian Mining Review, which has been incorporated with the new paper. The front cover contains an engraving of a beautiful specimen of Cobalt nickel silver ore. The articles throughout are entertaining, well illustrated, and the paper should secure a large circulation.

NEW PATENTS.

O. Katsenberger, San Antonio, Texas, has recently secured a patent on a door lock arranged so as to be opened by a key from the outside of the door and also to be opened from the inside without employing a key.

C. F. Smith, New York City, has secured a patent on a new tin dipper. The inventor overcomes the objection of not being able to dip milk from an almost empty can, by providing means to swing the cup to a vertical or horizontal position, a means operable by the hand holding the dipper.

OUR LETTER BOX

Correspondence on matters of interest to the hardware trade is solicited. Manufacturers, jobbers, retailers and clerks are urged to express their opinions on matters under discussion.

Any questions asked will be promptly answered. Do you want to buy anything, want some shelving, a silent salesman, any special line of goods, anything in connection with the hardware trade? Ask us. We'll supply the necessary information

Fire Escape Ladders.

The Coulogne Hardware Co., Coulogne, Que., writes : "Would you kindly let us know who manufactures iron fire escape ladders?"

Answer—Last week we gave a list of manufacturers of iron fire escapes and ladders : The Dennis Wire and Iron Works, London, however, are the most extensive manufacturers in Canada, their brands being sold throughout the country from Halifax to Vancouver, and their men being engaged at present at Calgary erecting fire escape and iron stairway. Since the beginning of the year shipments have been made to Vancouver, B.C., Rimouski, Que., Stanstead, Que., Edmonton, Alta., Brandon, Man., Halifax, N.S., and Wolfville, N. S.

Boot Calks.

E. Chown & Son, Kingston, write : "Can you tell us who makes the Sampson Boot Calks? We are anxious to purchase some and would like to know where they are made."

The "Samson" brand of boot calks, shovels, etc., is marketed through H. S. Howland, Sons & Co., Toronto.

Cement Tile Moulds.

John Hillhouse, Clifford, Ont., writes: "Can you direct me to makers of moulds for cement tile?"

Answer—We do not know of any Canadian manufacturers of cement tile moulds, but H C. Baird & Son, Parkhill, Ont., are makers of moulds for common tile.

Problem in Discounts.

Fred C. Lariviere, of Amiot, Lecours & Lariviere, Montreal, writes : Problem : A buys at 70, 10 per cent. off price list ; B buys at 60, 10 per cent. off price list.

1st question—How much cheaper does A buy his goods than B? Answer—A buys his goods 25 per cent. cheaper than B, or 33 1-3 per cent. if his own cost is considered.

2nd question—How much dearer does B pay for his goods than A? Answer—B pays 33 1-3 per cent. dearer than A, or 25 per cent., if his own cost is considered.

Furniture Manufacturers.

W. Holborn, Sutton West, writes : "Would like you to send me list of furniture manufacturers in Ontario."

Answer : The following are manufacturers in Ontario of furniture : Ellis Furniture Co, Ingersoll, M Erb & Co., Berlin; John Ferguson & Sons, London; Charles Rogers & Sons Co., Toronto; Snyder, Ross & Co., Waterloo ; Canada Furniture Manufacturers, Toronto. The entire list is too long for insertion here, but the above are among the most important.—Editor.

B. C. REPEALS TRAVELERS' TAX.

The British Columbia Government has introduced a bill to repeal the tax on commercial travelers from other provinces, which has awakened so much protest. The only exception is the case of travelers for tobacco and liquor firms, who have no branch houses in the province.

CONDENSED OR "WANT" ADVERTISEMENTS.

Advertisements under this heading 2c. a word first nsertion; 1c. a word each subsequent insertion.

Contractions count as one word, but five figures (as $1,000) are allowed as one word.

Cash remittances to cover cost must accompany all advertisements. In no case can this rule be overlooked. Advertisements received without remittance cannot be acknowledged.

Where replies come to our care to be forwarded, five cents must be added to cost to cover postage, etc.

SITUATIONS VACANT.

BRIGHT, intelligent boy wanted in every town and village in Canada; good pay besides a gift of a watch for good work. Write The MacLean Publishing Company, 10 Front Street E., Toronto. (tf)

WANTED—First-class tinsmith to take charge of shop; must be good furnace man and understand giving estimates. Address Box 593, HARDWARE AND METAL, Toronto. (13)

WANTED—Hardware Clerk, must be good stock keeper, state age, experience and salary expected. Peart Bros., Hardware Co., Limited, Regina, Saskatchewan. (12)

EXPERIENCED Hardware Clerk Wanted—One to take change, must be good salesman and stock-keeper, state experience and salary expected, Box 594, HARDWARE AND METAL, Toronto. (12)

WANTED—Experienced Hardware Clerk—Must be good stock-keeper, and strictly sober; references required. Hose & Caniff, Kenora, Ont. (13)

TINSMITH—With 3 or 4 years experience, and with some knowledge of plumbing; state wages; steady job. W. B. Clifton, Alliston, Ont. (14)

WANTED—First-class tinsmith; married, age 35 to 40; steady work; none other need apply. Philip Drench, Walkerville, Ont. (13)

WANTED—Immediately, reliable hardware clerk for B.C.; must be good buyer, stockkeeper and salesman of general hardware, etc.; state habits, capabilities, reasons for changing and salary expected to Box 596, HARDWARE AND METAL, Toronto.

SITUATION WANTED

WANTED—By a competent and reliable man, a small set of books to keep in spare time, accounts made out, etc. References. Address Bookkeeper, HARDWARE AND METAL, Montreal. (11)

HARDWARE Clerk with four years' experience is open for engagement April 1st in Alberta, Calgary preferred. A1 reference. Box 592, HARDWARE AND METAL, Toronto. (11)

WANTED—By experienced hardware clerk and bookkeeper, position as head clerk in retail. Box 598, HARDWARE AND METAL, Toronto.

WANTED—Hardware clerk with 7 years experience open for engagement for April 1st; A1 reference; full charge if necessary. Box 599, HARDWARE AND METAL, Toronto. (12)

FOR SALE.

CANADIAN Patent for Sale—Metal "Cramp Plate" for locking horizontal wires to upright stays in wire fence construction; a positive lock; easily manufactured; good investment. P. P. Kee, Brandon, Man. (12)

FOR SALE—A good set of tinner's tools, including eight-foot brake. Apply Box 597, HARDWARE AND METAL, Toronto. (11)

BUSINESS CHANCES.

HARDWARE and Harness Business for sale in good town. Owner retiring for good reason. Stock about $2,500, in excellent shape. This is worth investigating, for particulars apply Richard Tew, 23 Scott Street, Toronto.

HARDWARE Business; including stoves, tinware and tinsmith tools, in thriving town in West Ontario peninsular, stock about $5,000; building can be leased if desired, dwelling also. Box 583 HARDWARE AND METAL, Toronto. (17)

HARDWARE Business Stock about $12,000.00 is in good condition, could be reduced to suit purchaser if necessary. Apply George Taylor & Son, Hardware, London, Ont. (11)

PLUMBING, Heating, Stoves and Furnishings business for sale—Town of [illegible]; first-class farming district; $3,500. Box [illegible]95, HARDWARE AND METAL, Toronto. (12)

Paint, Oil and Brush Trades

PROFITS IN PREPARED PAINT

By S. W. F. Davis, of Sanderson Pearcy & Co., Toronto.

In the first place, it is necessary to get the agency for a pure, prepared paint, which you can rely on, as mixed paint is one of the most profitable lines handled in a hardware store. Try and instil in the minds of your clerks how to talk intelligently on mixed paint. The chief fault is in the salesman not knowing his business.

A customer comes in and asks for a quart, or a half gallon of paint, as the case may be, and finds he has not enough to finish the job. Then he comes for another pint and finds the color a little different. Then there is a complaint and he says the paint is no good. If the salesman had known his business he would have sold him enough in the first place.

Always sell a little more than the customer requires and be sure and tell him to keep his paint well stirred. Shake the tin of paint thoroughly, and if the customer is not going to use it at once have him turn it upside down and let it stand thus until it is required. On opening it should be thinned down and stirred thoroughly until it is of equal body throughout. Also advise the customer to stir it occasionally while using to keep the pigments thoroughly in suspension.

Use a good brush and spread evenly allowing ample time to dry between first and second coats, the time depending greatly on the weather, and whether for inside or outside use. The more coats applied the better, but be sure the paint is thin or it is sure to peel off. Shellac all knots and sappy wood as heat in time will otherwise draw sap through the paint.

Never paint over wood that is not perfectly dry. Do not wash a floor before painting. Sweep it and if there are any grease spots take benzine or turpentine and clean them off, as paint will not dry over grease, water or pitch. Do not put paint on too thick or in damp or rainy weather.

It is the "mystery mixtures" that have prejudiced many minds against prepared paints. "The burned child dreads the fire."

Scientific Paint Making.

At the outset it should be explained that paint making is not a mere mixing of solids and liquids, but a scientific process involving technical knowledge, skilled labor and expensive and powerful machinery. Good paint is made chiefly of linseed oil, in which solid coloring matter or pigments are ground — not simply mixed. The pigments serve a triple purpose—they thicken the paint (oil alone would run off), they beautify surfaces, and protect the oil from the sun and rain. Turpentine, dryers, varnish, &c., are added to paint to serve a special, rather than a general purpose.

Linseed oil is so important a part of paint that it is often called "the life of paint." It may well be so regarded, for just so long as the oil lasts the paint has life. When the oil, from chemical action or long exposure, is no longer present, the paint is dead. Dead paint, a lustreless, dry, chalky coating, has no protective qualities and disintegrates rapidly. The pigments have no cohesive attraction after the decomposition of the oil.

The property which gives linseed oil its peculiar paint value is that it absorbs oxygen from the air when it is exposed in thin films, as in painting. This film, from oxidation, begets the valuable properties of being practically neutral in its reaction, more or less transparent, and quite elastic. The property of absorbing oxygen is increased by heating linseed oil, and it is still further increased by adding to the oil while it is being heated, certain bodies, known as dryers. While boiled oil dries more quickly than raw oil, and leaves a more lustrous surface, the film is less elastic and is liable to crack.

Anything added to the paint to unduly hasten its drying, will destroy the elasticity and toughness of the oil film. Therefore all dryers should be used sparingly and with caution.

Furthermore, any pigment chemically active, will not make an enduring paint. If used at all, they should be in combination with more stable or inert pigments, to neutralize as far as possible, the chemical action on the oil.

It is this chemical activity that lessens to a marked degree the value of white lead as a paint pigment. Strictly pure white lead carries comparatively little oil, and the little it does carry soon ceases to be linseed oil, by reason of the chemical change that takes place when the lead and oil are united and exposed to the elements. To correct this serious defect in the lead and oil mixture, modern paint makers have added various pigments in varying formula, and years of experience have proven that of these combinations those

(Continued on page 50.)

SOME day a genius may discover a better process of manufacturing paint than we know; but that day seems far away. Our expert analysts are certain that

NEW ERA PAINT the most effective "silent salesman" on the market

will hold first place against all comers anywhere in sight. You can recommend New Era with full assurance that your customer will advertise your business right along.

STANDARD PAINT AND VARNISH WORKS CO., LIMITED

WINDSOR · ONTARIO

AN AID TO CLEANLINESS

Kitchen floors are hard to keep clean when covered with the ordinary **floor paints,** which accumulate dirt, and rub off when washed.

JAMIESON'S Floor Enamels

dry very hard, in a short time, and leave a gloss which prevents the collection of dirt. They wear like iron, and are not affected by washing.

These are sure business-getters. Made in all the popular colors.

BY

R. C. JAMIESON & CO.

LIMITED

16-30 Nazareth St., MONTREAL

The wise dealer takes no chances. He recommends

Pearcy's Pure Prepared Paints

because he has tried and tested them in every way.

PROMPT DELIVERIES A SPECIALTY

SANDERSON PEARCY & CO., Ltd.

Toronto, Canada

A. RUDD & CO., St. Helens, Lancashire, England

Manufacturers of highest grade

GLAZIERS' DIAMONDS

IN ALL PATTERNS.

Made only by thoroughly experienced workmen, 10 years as glass cutter and 16 years as chooser and setter being the experience of our head setter.

Canadian Agent: J. B. GIBSONE, P.O. Box No. 478 MONTREAL

From whom users of Glaziers' Diamonds can obtain the names of wholesale merchants who stock A. Rudd's goods.

GLAZIERS' DIAMONDS OF PROVED WORTH

Having a Century's Reputation for Reliability.

MADE BY

A. SHAW & SON, London, Eng.

CANADIAN AGENT

GODFREY S. PELTON, - 201 Coristine Building, Montreal

BEFORE ordering colors think of **THE VIA GELLIA COLOR CO.**

Raisers and Levigators of Mineral Colors

Mines and works at

Matlock-Bath, Cromford and Bonsall (Eng) and Ovoca (Ireland)

Our specialities are

ENGLISH and TURKEY UMBERS (raw and burnt) TURKEY RED, RED AND PURPLE OXIDES, PURPLE BROWNS, YELLOW OCHRES, ROTTEN STONE (lump and powder) and Superior Ivory Blacks.

These are exclusive lines and are strong money-makers in the trade. Write for samples and prices to

WM. O. GREENWAY, 13 St. John St., Montreal

It will pay you.

Paint and Oil Markets

MONTREAL.

Office of HARDWARE AND METAL,
232 McGill Street,
Montreal, March 15, 1907

A splendid activity is maintained in all lines of colors, paints and varnishes.

The duty of 10 cents per bushel imposed on flaxseed last week, naturally raised the cost of linseed oil 5 cents per gallon, and the increase took effect immediately.

It is only natural to assume that the higher price will effect all productions in paints and varnishes. In fact already the leading manufacturers are conning their lists with special interest, in order to mark up future quotations. It is not necessary to tell even a novice that advances will be absolutely necessary. Putty also feels the high price of oil, and has been advanced 20 cents per 100 pounds.

It is now thought that the tariff is absolutely finished for this session, and that any fluctuations for the balance of the year will be due to trade conditions, and not from tariff readjustment.

The advance in linseed oil created no little interest among dealers. Those who held stocks naturally felt very comfortable, and those who were short were just the opposite.

LINSEED OIL—The demand has been somewhat checked by the sharp increase which took place last week, and which is firmly maintained: Raw 1 to 4 barrels, 60c.; 5 to 9 barrels, 59c.; boiled, 1 to 4 barrels, 63c.; 5 to 9 barrels, 62c.

TURPENTINE—The demand is fairly active, with no change in prices, which remain: Single barrel, 98c. per gallon; for smaller quantities than barrels, 5c extra per gal. is charged. Standard gallon is 8.40 lbs., f.o.b. point of shipment. net 30 days.

GROUND WHITE LEAD—With the advent of spring weather (especially in the west) large quantities are now being shipped, and prices are exceedingly strong. Best brands, Government standard, $7.25 to $7.50; No. 1, $6.90 to $7.15; No. 2, $6.55 to $6.90; No. 3, $6.30 to $6.55; all f.o.b. Montreal.

DRY WHITE ZINC — In fair demand only. Prices remain: V. M., Red Seal, 7½c. to 8c.: Red Seal, 7c. to 8c.; French V. M., 6c. to 7c.; Lehigh, 5c. to 6c.

WHITE ZINC GROUND IN OIL—Inquiry moderate, prices firm. Pure, 8½c. to 9½c.; No. 1, 7c. to 8c.; No. 2, 5¾c. to 6¾c.

PUTTY—Advanced 20 cents per 100 lbs. Shipments are somewhat larger than last week. We now quote: Pure linseed oil, $1.75 1-5 to $1.85 1-5; bulk in bbls., $1.50 1-5; in 25-lb irons, $1.80 1-5; in tins, $1.90 1-5; bladder putty in bbls., $1.75 1-5.

ORANGE MINERAL—This article is still short, but there is not much call for it at present, so prices remain the same: Casks, 8c.; 100-lb. kegs, 8½c.

RED LEAD—Moving slowly. Stocks being light are carefully husbanded. Genuine red lead, in casks, $6: in 100-lb. kegs, $6.25; in less quantities at the rate of $7 per 100 lbs.; No. 1 red lead, casks, $5.75; kegs, $6, and smaller quantities, $6.75.

PARIS GREEN—The demand seems to have been met for the present, and there is no change in quotations: In barrels, about 600 lbs., 25½c per lb.; in arsenic kegs, 250 lbs., 25¾c; in 50-lb. drums, 26¼c; in 25-lb. drums, 26¾c; in 1-lb. packets, 100 lbs. in case, 27¼c; in 1-lb. packets, 50 lbs. in case, 27¾c; in ½-lb. packets, 100 lbs. in case, 29¼c; in 1-lb. tins, 28¼c f.o.b. Montreal. Terms, three months net or 2 per cent. 30 days.

SHELLAC GUMS—In fair demand. No change in prices: Bleached, in bars or ground, 46c. per lb., f.o.b. Eastern Canadian points; bone dry. 57c. per lb., f.o.b. Eastern Canadian points; T. N. orange, etc., 48c. per lb., f.o.b. New York.

SHELLAC VARNISH—Trade is exceedingly brisk and some good shipments are being made. Prices remain: Pure white, $2.90 to $2.95; pure orange, $2.70 to $2.75; No. 1 orange, $2.50 to $2.55.

PETROLEUM — American prime white coal, 15½c per gallon; American water, 17c per gallon; Pratt's Astral. 19½c per gallon.

WINDOW GLASS—First break, 50 feet, $1.85; second break, 50 feet, $1.95: first break, 100 feet, $3.20; second break, 100 feet, $3.40; third break, 100 feet, $3.95; fourth break, 100 feet. $4.15; fifth break, 100 feet, $4.40; sixth break, 100 feet, $4.95. Diamond Star: First break, 50 feet, $2.30; second break. 50 feet, $2.50; first break, 100 feet. $4.40; second break, $4.80; third break. 100 feet, $5.75; fourth break, 100 feet. $6.50; fifth break, 100 feet, $7.50; sixth break, 100 feet, $7.50; seventh break, 100 feet, $8; eight break, 100 feet, $9. Double Diamond: First break, 50 feet, $3.45; second break, 50 feet, $3.75; first break, 100 feet, $6.75; second break, 100 feet, $7.25; third break, 100 feet, $8.75; fourth break, 100 feet, $10; fifth break, 100 feet, $11.50; sixth break, 100 feet, $12.50; seventh break, 100 feet, $14; eighth break, 100 feet, $16.50; ninth break, 100 feet, $18; tenth break, 100 feet, $20; eleventh break, 100 feet, $24; twelfth break, 100 feet, $28.50. Discount on Diamond Star, 20 per cent.; on Double Diamond, 40 per cent.

TORONTO.

Office of HARDWARE AND METAL,
10 Front Street East.
Toronto, March 15, 19 7

The paint and oil trade is brisk, having improved during the last week with the moderation in climate and opening of spring.

The demand for the leads is strengthening while for linseed oil it continues strong despite the advance of 5c. a gal-

ion in duty and a consequent advance in price.

Trade in turpentine is unchanged but prices have advanced owing to the scarcity of labor in Georgia, the prices being the highest in years in Savannah.

The orders for oils, paints and paris green are practically all booked up, the wholesalers awaiting spring freight.

An advance has been made in the duty on varnish amounting to 5c. a gallon, and some price changes are looked for next week.

It is stated that the prices in mixed paints will be affected by the advance in oil.

WHITE LEAD.—We continue to quote: Pure white, $7.40; No. 1, $6.65; No. 2, $6.25; No. 3, $5.90; No. 4, $5.65 in packages of 25 lbs. and upwards; ½c. per lb. will be charged extra for 12½ lbs. packages; genuine dry white lead in casks, $7.

RED LEAD.—Prices remain unchanged, with good demand: Genuine in casks of 500 lbs., $6; ditto, in kegs of 100 lbs., $6.50; No. 1 in casks of 500 lbs., $5.75; ditto, in kegs of 100 lbs., $6.25.

DRY WHITE ZINC.—In casks, 7½c.; in 100 lbs., 8c., No. 1, in casks, 6½c., in 100 lbs., 7c. Ground in oil—In 25 lb. irons, 8c.; in 12½ lbs., 8½c.

SHELLAC VARNISH.—Pure orange in barrels, $2.70; white, $2.82½ per barrel; No. 1 (orange) $2.50; gum shellac, bone dry, 63c. Toronto. T. N. (orange) 51c. net Toronto.

LINSEED OIL.—As announced a week ago, there is a general advance of 3c. owing to the change of tariff, with prospects of a still greater advance to 5c. We quote: Raw, 1 to 3 barrels, 64c.; 4 to 7 barrels, 63c.; 8 barrels and over, 62c.; add 3c. to this price for boiled oil, f.o.b. Toronto, Hamilton, London and Guelph, net 30 days.

TURPENTINE.—Prices continue variable, the tendency, of course, being upward. Prices this week have advanced owing to scarcity of labor and the dought in Georgia. We quote: Single barrels at $1.02, f.o.b. point of shipment, net 30 days; less than barrels, $1.07 per gallon.

PARIS GREEN.—Trade continues steady, the greater part of the orders, however, being booked. We continue to quote: English and Canadian at 27½c. base.

PETROLEUM.—Trade is good, the demand increasing. Prices remain firm and unchanged. We quote: Prime white, 13c.; water white, 14½c.; Pratt's astral, 18c.

For additional quotations see current quotations at back of paper.

WHERE COLORS COME FROM.

The cochineal bug furnished many of the most brilliant colors, including the bright carmine, crimson, purple lake, and scarlet. The cuttlefish gives the sepia, and Indian yellow comes from the camel. Ivory chips produce ivory black, and the exquisite Persian blue was discovered accidentally by fusing horses' hoofs and other refuse animal matter with impure potassium carbonate. Crimson lake comes from the roots and barks of certain trees; blue-black from the charcoal of the vine chalk; and Turkey red comes from the root of the madder plant found in Hindustan. India ink is made from burned camphor by the Chinese.—Ex.

TO MAKE WATERPROOF GLUE.

Waterproof glue is manufactured of gum shellac three parts and india rubber one part by weight, these constituents being dissolved in separate vessels in ether, free from alcohol, subject to a gentle heat. When thoroughly dissolved the two solutions are mixed and kept for some time in a vessel tightly sealed.

PROFITS IN PREPARED PAINTS.

(Continued from page 46.)

carrying the greater amount of oil make the better paint when they have good opacity or covering qualities.

Advantages of White Lead.

White lead has some good points of excellence not possessed by any other pigment. It has remarkable covering qualities, is a good dryer, and works easily under the brush. A good paint without lead would be like good bread without flour. Unfortunately, lead's qualities are such that it continues to dry in linseed oil until finally no oil, as such, remains, and what paint is left on the painted surface is like chalk and is blown or washed from the building.

The best paint authorities agree that white lead should not be used alone. Zinc has great affinity for linseed oil, carrying from 60 to 70 per cent. more oil than white lead. This gives it exceptional spreading capacity. Zinc, unlike lead, is not affected by chemical action when united with oil, nor is it injured by gases and atmospheric elements which so quickly destroy lead. Zinc itself would be an ideal paint pigment were it not that it dries too hard and that it lacks porosity.

Painters have learned from practical experience that a combination of lead and other more enduring pigments makes a better paint than lead alone, but some of them have yet to learn that it is quite impossible for them, even with the knowledge of proportions to mix by hand a paint so composed. The very nature of pigments is such that they require a thorough mixing and grinding in their dry state and again in their paste form, and again and again in their semi-liquid and liquid forms. This mixing, chasing, grinding and regrinding is done in expensive mills especially constructed to handle the pigments used by this or that manufacturer. This explains why makers of prepared paints have acquired success and reputation, while painters have found their home-made mixtures of similar pigments to be unsatisfactory.

In the spring the dealer should advertise his line freely in his local papers and place circulars, printed matter and cards in the hands of every possible customer. If printed matter is included in parcels delivered to customers they will take them home and look them over so that when the time comes for painting they will have had time to study over what they require.

Good window displays of all the goods should also be made as the window has been proven to be one of the best selling mediums for this branch of the hardwareman's business.

BERRY BROTHERS' VARNISHES

are one grand story of success

BECAUSE THEY ARE THE BEST

They are the safest goods to handle,
and the most reliable goods to use.

Write for Catalogue, Information and Prices.

BERRY BROTHERS, Limited

NEW YORK
BOSTON
PHILADELPHIA
BALTIMORE

Varnish Manufacturers
DETROIT
Canadian Factory—WALKERVILLE ONT.

CHICAGO
CINCINNATI
ST. LOUIS
SAN FRANCISCO

SOME DEALERS WOO DISASTER

100°/₀ durability

from a business standpoint simply because they don't discriminate between rational evidence and hot air. Take paint for instance. Every dealer knows that he simply cannot afford to sell or reccommend bad paint—or even ordinary paint. He must sell and reccommend the best. Make your leader

ARK BRAND PAINT

and you can rest assured of commercial success. We do not say stock **ARK BRAND** because we say it's the best paint. We say you should reccommend and sell **ARK BRAND** because hundreds of practical men have proven it's superiority. They say so right over their signatures, and we want to mail the proofs to you.

MURESCO

Well here's the opinion of practical men

Sarnia, March 12th, 1907

Gentlemen:—Replying to yours of yesterday we have no intention of dropping Muresco. In fact we expect larger sales for it this year than ever. We have sold Muresco for about seven years and have never heard a complaint about it. It gives our customers the best of satisfaction.

Yours truly, (Signed), ROUGVIE BROS.

BENJAMIN MOORE & CO., LIMITED

(Formerly The Francis-Frost Co., Limited)

NEW OFFICES AND WORKS:

TORONTO JUNCTION, CANADA

New York, Chicago, Cleveland, Carteret, N.J.

Another great victory for the PRESERVATIVE and SANITARY qualities of

BLANCHITE

PAINTS and ENAMELS—they are used exclusively on the immense

Filtration Plant of Pittsburg

Our Cold Water Paint is used as a filler over the concrete surface. No. 41 Pearl Grey Preservative Paint as damp-proof primer, and No. 26 Cream Preservative as finisher. All tanks and pipes are enamelled.

Write for Information. A Few Agencies Left.

THE BLANCHITE PROCESS PAINT CO., Limited

785 King Street West, TORONTO, ONT.

Two Things to Keep in Mind when Ordering Paint

The Brand of Quality

"Island City"

and then

EIGHT HOURS

Think of it! **EIGHT HOURS** is all that is necessary for **ISLAND CITY** brand of **House** or **Floor** paints to dry.

Guaranteed never to **peel** or **crack** after drying.

These are the strong talking points.

If your Spring stock is lacking, let us know.

Our Travellers get all the credit of orders sent in from their ground.

P. D. DODS & CO., - MONTREAL

VANCOUVER TORONTO WINNIPEG

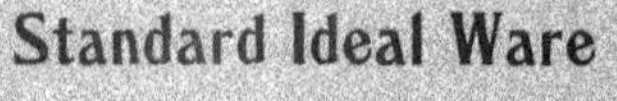

THE WARE OF QUALITY

STANDARD IDEAL WARE is enameled after a process that makes the porcelain as a part of the iron. Its one-piece construction and smooth, snowy, opaque appearance ensures beauty, durability and purity.

We make Bath Tubs, Lavatories, Sinks, Urinals, Slop Hoppers, Laundry Tubs and other necessary plumbing equipment.

Write for Catalogue

The Standard Ideal Co., Limited

Sales Offices and Sample Rooms, 50 Colborne Street, TORONTO, 128 West Craig Street, MONTREAL and 24 Telfer Block, WINNIPEG

Home Office and Factories, - PORT HOPE, Ont.

SOLVED AT LAST

Your problem is solved at last. The **Cleveland Wall Paper Cleaner** has captured the U.S. markets because of merit and merit alone. There is no other wall paper cleaner nearly as good.

Guaranteed To Keep For Two Years

Not Dangerous. Does Not Contain Acids

The Cleveland Wall Paper Cleaner

does clean wall paper and frescoed walls. It does not streak or spot them, and with ordinary care, almost any one can use it. Let us forward you particulars regarding **your** margin of profit.

Put up in cases of 3 doz. one pound cans
Bulk—10, 25 and 50 pound cans. Prompt Shipments

FRANK U. S. GILBERT

1224 Woodland Ave. CLEVELAND, OHIO

PLUMBING GOODS

And Sanitary Specialties

"High Grade"

Closet Combinations

You may have clients who demand something better than the ordinary—Something that has genuine merit both in efficiency and appearance.

The "**Astoria**" outfit combines these essentials—a Square Back Vitrious-China Bowl with extra large Water Seal, large Noiseless Tank and Saddle Seat—all woodwork Piano Polish.

Also furnished with "Nethery" Patent Flush Valve for concealed Tank supply or direct pressure from main.

Sectional View of Syphon-Jet Closet with Nethery Valve.

The "Astoria" Low Tank Combination.

Enameled Baths *Enameled Lavatories* *Enameled Sinks and Laundry Tubs*

The James Morrison Brass Mfg. Co., Limited, 93-97 Adelaide St. W., TORONTO

Manufacturers of the Largest Variety in the Dominion

Plumbing and Steamfitting

EXCELLENT WORK IN BRITISH COLUMBIA

Some work approved by President Davis, of the Plumbing Inspectors' Association of America, has encouraged "A. S.," of Victoria, B.C., to write describing a job done by him in that city, the work differing materially from that approved by Mr. Davis, although being of a similar nature. The job was in a tenement building in Victoria, where the system of venting is what is termed venting in the rough, and from the Victoria point of view is the best yet adopted by any city on the continent.

A Practical Letter.

"According to the plumbing and sewerage by-laws all main and branch sewers under a building must be of a soil pipe termed extra heavy. The sizes of same will be given in the plan of building just completed and approved by our plumbing inspector, H. Shade, who is a practical man for the position he occupies, having passed a strict examination in practical and theoretical work, including plan drawing, before receiving his appointment. Under his supervision we have a set of plumbing by-laws which the practical plumber is proud of. According to them all plumbing work must be subject to the following tests:

"At the completion of the roughing in of the work all ends of wastes must be plugged and a water test of not less than fifteen pounds be given to the work before calling in the inspector for the first examination. If everything is tight you may then start in to instal your fixtures, and as soon as this is done the final test is given by using the smoke test. At the completion of this a certificate is issued from the inspector's office giving a guarantee to the property owner that the work is in first-class condition, and is in compliance with the city plumbing and sewerage by-laws.

"I am giving this general routine of work so as to compare same with work in other cities, for, after studying some of the working plans in plumbing, I have come to the conclusion that something is radically wrong with the system in having so many different varieties of work, as anyone who studies the problem must realize that there is one goal and that is perfection. How much easier it would be for the plumber who travels from north to south, east to west, to be familiar with his trade, so that should he be in Chicago one week and land in Victoria the next he could take his position along with his fellow employees and go to work without having to study the by-laws and learn the trade over again.

"From my own experience as an employer of labor I find that a universal by-law is what is needed, so that we may be able to kill the sewer pipe monopolists, et al., who are holding Chicago by the throat at the present day for the lust of gold. 'Let the people die from bad plumbing and sewer work, the plumber will get the blame anyway; and we can sit in our armchairs and gather the shekels in,' says the sewer pipe combine of Chicago.

"But there are other questions nearer home to the plumber which ought to be considered seriously, and one is, that when a building is being erected, generally speaking, no provision is made for the piping unless you can squeeze everything between the studding or ex-

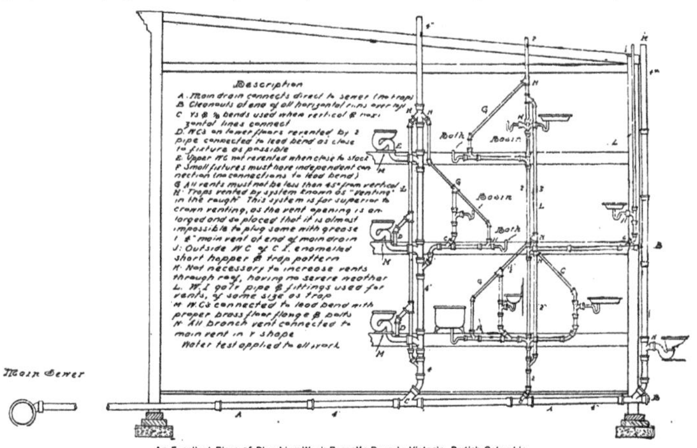
Description
A. Main drain connects direct to sewer (no trap)
B Cleanouts at end of all horizontal runs over top
C Ys & ⅛ bends used when vertical & horizontal lines connect
D. W.C's on lower floors revented by 2 pipe connected to lead bend as close to fixture as possible
E. Upper W.C not revented when close to stack
F Small fixtures must have independent connection (no connections to lead bend)
G All vents must not be less than 45° from vertical
H: Traps vented by system known as "venting in the rough" This system is far superior to crown venting, as the vent opening is enlarged and so placed that it is almost impossible to plug same with grease
I 4" main vent at end of main drain
J: Outside W.C of C.I. enamelled short hopper & trap pattern
K: Not necessary to increase vents through roof, having no severe weather
L. W.I galv pipe & fittings used for vents, of same size as trap
M W.C's connected to lead bend with proper brass floor flange & bolts
N All branch vent connected to main vent in Y shape
Water test applied to all work
Main Sewer
Bath
Basin

An Excellent Piece of Plumbing Work Recently Done in Victoria, British Columbia.

pose the piping on the outside. In milder climates, such as we have in Victoria, the latter course is generally adopted, but in the eastern states I have noticed the soil pipes and wastes are exposed inside the building, which gives a very bad impression to the owner. The architect provides a place for the fixtures, but forgets that there are soil pipes to be taken to them. Were the plumbers to put their heads together this would not be the case, and better satisfaction would be given the owner and the trade in general raised a notch higher, for it is every move that counts.

Vent Pipes Clean Themselves.

"In my accompanying plan you will notice that all venting is done perpendicularly or at an angle of 45 degrees. The reason for this is to give all vent pipes an opportunity of cleaning themselves directly into the main soil pipes, and I think you will agree with me that this is a good law. In most of the cities of the United States I find that they allow horizontal runs of vent pipes, which must be condemned by anyone who studies the question of venting.

"Did ever any of the trade take the trouble to inquire into old work that they have taken out, and see whether it was full of rust or scales, generally at the elbows? If not, try it the next time. The coal fumes and acids from factories may have a good deal to do with this, but as long as the easy plumber gives those factories the opportunity to do so by this system of horizontal venting just so long will the defect remain. It is one thing to do a job and have it give satisfaction for the time being, but it is another to have the same job give satisfaction for all time to come.

"Another point I might touch upon is the system of direct connection to the sewer versus the intercepting trap. Some engineers take exception to the direct system, but I have yet to see the city adopting such that ever went back to the intercepting trap, and I have been informed of many who have adopted the direct after experiencing the indirect. Those who have adopted the intercepting trap must have found out by this time that the stoppage in any sewer is generally from this, and then the smell from the streets where the vents for flushing the main sewer are is enough to knock you down. If it is good to vent and take off the smell from your own sewerage until it passes your property, is it not just as well to vent until it passes out to the sea?

"Think it out, fellow craftsmen, and let us have your opinion. That is what I am writing this article for. My ambition is to see the plumbing fraternity recognized as men of culture in whatever community they may reside, in company with doctors, lawyers and preachers, using the Golden Rule at all times for the benefit of their less informed fellow citizens."

A Discussion Suggested.

This is an interesting letter and we will be glad to give space to the comments of anyone in the trade on the system outlined and views advanced. The Plumbers' Trade Journal of New York, by whom the letter was published a month ago, added the following comment, which covered the several points raised very completely and concisely:

"As we look at it, the elevation of plumbing work as done in Victoria, which correspondent submits, shows an excellent piece of work. The system which he calls 'venting in the rough' is certainly, he suggests, the best that has yet been adopted. This system has been referred to in the columns of the Journal as 'continuous venting,' and the advantages of the system and many illustrations of it have often been brought to the attention of our readers. It appears to us that it is only question of time before our cities will adopt this method very extensively.

"The correspondent's remarks on the architect bring up again the question which has often filled every plumber with dismay. It is an old subject, one upon which much has been written, but we do not see that much headway is being made, and it strikes us that it is a difficult matter to adjust. In the layout of our large public buildings architects are often employed who have in their service experts on the various lines of construction, and under such such conditions it is usually to be observed that the arrangements for properly installing the piping are far better than in those buildings designed by the architect who employs no such assistants, and who has little or no knowledge of practical plumbing installation. Until the architect is educated along this line we do not see that conditions will change, unless, indeed, the architect can ever be persuaded to come down from his dignity sufficiently to allow himself to consult with tbt plumber regarding the layout of this part of the work.

Hard to Harmonize.

"As to a universal plumbing ordinance, we believe that it would be a splendid thing; but probably as unattainable as it is splendid. In the first place, our correspondent would not want to see the intercepting trap appear in that universal law. The other fellow would want it there. Our correspondent does not increase his roof pipe, while in another part of the country it is necessary.

"The question is along the same lines as a universal religion. Custom has a great deal to do with these things, and is often stronger than law. Various local conditions often make necessary certain changes in the plumbing ordinance. In other words, it appears to us that while it is a pleasing thing to consider, there would be about as much chance of bringing the various elements together on one common ground as there is of bringing our two great political parties together in one common party. However, we can see where much advantage might be derived if, for instance, smaller sections of country, each state or province, for instance, could be governed by a single plumbing ordinance."

WATERWORKS BY-LAWS.

Wetaskiwin, Alta., ratepayers have voted favorably on the following bylaws: To provide for the raising of the sum of $140,000 for the construction of a municipal system of waterworks and sewers. For the raising of the sum of $30,000, for improving and extending the municipal electric light and power plant. For $2,500 for buying a hospital site, and $10,000 for the erection of a hospital.

CHAIN FOR PIPE HANGING.

The Oneida Community Company are manufacturing a chain for pipe hanging which consists of two parts. An adjustment screws solidly into the ceiling, on which is hung a strong chain which holds the pipe. This chain may be shortened or lengthened at will, thus allowing the pipe to be hung at any distance from the ceiling. This is a very simple and convenient device and is becoming very popular. Readers may obtain additional information by writing the Oneida Community Co., Oneida, N. Y., and mentioning the Plumber and Steamfitter.

PLUMBING MARKETS

MONTREAL.

Office of HARDWARE AND METAL,
232 McGill Street,
Montreal, March 15, 1907

Things are booming now in all branches of the plumbing trade, both wholesale and retail. Wholesalers are working at top speed, in order to keep their orders up to date, and the usual spring rush is starting with the practical men of the trade.

The prices were withdrawn on iron pipe last week, but so far, the new schedule is not made up. We can only say that there will be an advance of about 4 per cent. on the old list, on all lines up to 3 inch.

RANGE BOILERS.—The demand remains good, with no change in prices; which are very firm at the following figures: Iron clad, 30-gal., $5; 40-gal. $6.50 net list. Copper, 30 gal., $33; 35 gal., $38; 40 gal., $43.

LEAD PIPE.—Business is very brisk, and prices remain firm with no changes. Discount is: 5 p.c. f.o.b. Montreal, Toronto, St. John, N.B., Halifax; f.o.b. London, 15c per hundred lbs. extra; f.o.b. Hamilton, 10c per hundred lbs. extra.

IRON PIPE FITTINGS.—The demand in some lines still exceeds the supply, and the following prices are very firm: Discounts on nipples, ¼-inch to 3-inch, 65 per cent.; 3½ to 2-inch, 67½ per cent.

IRON PIPE.—The new prices are not yet made up, but the advance will be about 4 per cent. on all lines up to 3 inch.

SOIL PIPE AND FITTINGS.—Business good and prices firm. We quote: Standard soil pipe, 50 p.c. off list. Standard fittings, 50 and 10 per cent. off list; medium and extra heavy soil pipe, 60 per cent. off. Fittings, 60 per cent. off.

SOLDER.—No changes to report. Prices remain: Bar solder, half-andhalf, guaranteed, 25c.; No. 2 wiping solder, 22c.

ENAMELWARE — We quote: Canadian baths, see Jan. 3, 1907, list. Lavatories, discounts, 1st quality, 30 per cent.; special, 30 and 10 per cent. Sinks, 18x30 inch, flat rim, 1st quality, $2.60; special, $2.45.

TORONTO.

Office of HARDWARE AND METAL,
10 Front Street East,
Toronto March 15, 1907.

The feature of the plumbing markets in Toronto at present is the slashing in prices on plumbers' brass goods, enamelware and closets. Trade has been quiet owing to the dull winter season, but is beginning to enliven with the moderation in the climate and opening up of spring.

The cut in brass is especially out of place just now, with copper making another advance. The brass goods manufacturers say they cannot get deliveries of copper in less than eight weeks' time and their cost of production is constantly advancing.

The new list on iron pipe is out with about a ten per cent. advance. Pipe continues to be scarce.

LEAD PIPE—The discount is still 5 per cent. off the list price of 7c. per lb. Lead waste, 8c. per lb., with 5 off. Caulking lead, 5¾c. to 6¾c. per lb. Traps and bends, 40 per cent. off.

SOIL PIPE AND FITTINGS — Business continues good. We continue to quote: Medium and extra heavy pipe and fittings, 60 per cent; light pipe, 50 per cent.; light fittings, 50 and 10 per cent.; 7 and 8 in. pipe, 40 and 5 per cent.

IRON PIPE—We give the new prices in iron pipe. Stocks are low, with prospects of scarcity. We quote: 1-inch black pipe, $5.12; 1-inch galvanized, $6.-77.

PIPE IRON FITTINGS—Prices continue unsteady. We continue to quote: Cast iron fittings, 57½ per cent.; cast iron plugs and bushings, 60 per cent.; flange unions, 60 per cent.; nipples, 70 and 10 per cent.; iron cocks, 55 and 5 per cent.; Canadian malleable, 30 per cent.; malleable unions, 55 and 5 per cent.; malleable bushings, 55 per cent.; cast iron ceiling plates, plain, 65 per cent.; cast iron floor, 70 per cent.; hook plates, 60 per cent.; expansion plates, 65 per cent.; headers, 60 per cent.; hangers, 65 per cent.; standard list.

RANGE BOILERS — We continue to quote: Galvanized iron, 30-gal., standard, $5; extra heavy, $7.75; 35-gal. standard, $6; extra heavy, $8.75; 40-gal., standard, $7; 40 gallon, extra heavy, $9.75, net list. Copper range boilers—New lists quote: 30 gallon, $33; 35 gallon, $38; 40 gallon $43. Discounts 5 to 15 per cent.

RADIATORS—The demand continues brisk, orders being booked ahead in view of the erection of a new factory. We continue to quote: Hot water, 47½ per cent.; steam, 50 per cent; wall radiators, 45 per cent.; specials, 45 per cent. Hot water boilers are subject to an open market.

SOLDER—Bar solder, half-and-half, guaranteed, 27c.; wiping, 23c.

ENAMELWARE— Price-cutting continues in these in some quarters. We continue to quote January prices; Lavatories, first quality, 20 and 5 to 20 and 10 off; special, 20 and 10 to 30 and 2½ per cent. discount. Kitchen sinks, plate, 300, firsts, 60 and 10 off; specials, 65 and 5 per cent. Urinals and range closets, 15 off. Fittings extra.

"What are the Advantages of Hot Water over Steam Heating in a Private Residence?"

What arguments would you use in reply if a customer asked you the above question.

Send us an answer and also outline what advantages steam heating possesses over water for this same class of work.

For the best reply we offer a cloth bound copy of J. J. Cosgrove's splendid textbook, "Principles and Practice of Plumbing."

Address the Plumbing Editor

Hardware and Metal

Montreal Toronto Winnipeg

PRICE OF ILLUMINATING GAS IN ENGLAND.

The price of illuminating gas in Widnes, Lancashire, England, is now 32 cents to small consumers, but will be reduced to 30 cents on July 1. Large consumers will pay from 22 to 26 cents. This is claimed to be the cheapest gas in the world. The town has about 30,-000 population. The price of gas is remarkably low everywhere in Great Britain, whether under public or private control, the general range of price being 40 to 70 cents.

ELECTRIC HEATING FOR RESIDENCES.

In a paper read before the Atlantic City convention of the National Electric Light Association, Jas. I. Ayer stated that in cooking, gas at $1 per thousand cubic feet is equivalent to electricity at 2½ cents per kilowatt hour, and that while electric cooking is practicable at a higher rate, it will require a 5 to 3 cent rate to make it an important competitor of gas. A 10-gallon jacketed hot water boiler can be heated to 150 degrees Fahrenheit with approximately 2½ kilowatt hours. Radiators for occasional use in ordinary-sized bedrooms should have at least a capacity of one kilowatt, while for heating a bathroom a heater of at leats two kilowatts should be used, in order to heat the room quickly—fifteen to twenty minutes. The cost of doing the average family ironing with an electric iron is given at $1 to $1.25. Instantaneous hot-water heaters are not recommended, although for small quantities of water the cost of operation is not excessive.

CRAIN CO. EXPANDS.

The Rolla L. Crain Co., Ottawa, have been auhorized by the Kalamezoo Loose Leaf Binder Co. to manufacture their products in Canada.

Fuel Consumption

How much fuel will the

Daisy Hot Water Boiler

consume ?

It is doubtful if the steamfitter is asked a question quite so frequently, and there is no more difficult one to answer accurately. But this fact can be truthfully emphaized : there is no boiler made which has such an area of fire surface, in proportion to grate area; such ample provision for positive and accurate circulation, and perfect fuel consumption. It is not hard to sell the Daisy when you convince your prospective customer of these positive truths.

CLUFF BROTHERS

Lombard Street, TORONTO

Selling Agents for - - - WARDEN KING & SON, Limited

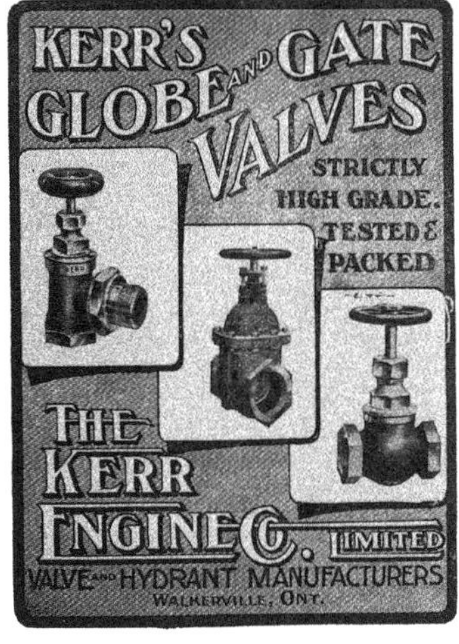

It's a Pleasure to Sell Oil

The Bowser Self-Measuring Oil Storage Outfit makes it a positive pleasure to sell oil. Why? Because it is convenient. Just a few steps to the pump, one simple stroke and the can is filled. It's just as easy as selling a hammer or saw. Because it is clean. All the oil is kept where it belongs, in the tank or in the customer's can. Because it is safe. No danger of fire or explosion. Because it is economical. It pays for itself in one year by its own saving of time, labor and oil.

If you handle oil with the old style tank, you can realize what the Bowser gives you. Catalog V tells more about the Bowser. Send for it to-day.

Cut 15. CELLAR OUTFIT
For non-lubricating oils.

S. F. BOWSER & CO., Inc.

66-68 Fraser Ave. Toronto, Can.

Don't Be Misled

Users of Mueller Brass Coods are Guaranteed Absolute Immunity From All Liability on Account of Patent Suits.

The Glauber Company in recent advertisements in trade journals seeks to make business capital of the fact that they have instituted a suit against the H. Mueller Manufacturing Company, alleging "imitating and infringing supplies."

It is not the desire of the H. Mueller Manufacturing Company to try this Suit in the public prints or to be drawn into a newspaper controversy, but in justification to itself and the trade, feels called upon to direct attention of dealers and consumers to the fact that they will be protected.

The H. Mueller Manufacturing Company is financially responsible and has always stood behind its goods with an *Unconditional Guarantee.* It will continue to do so, but in addition at this time wishes to assure the trade and the public absolute immunity from any liability of whatsoever kind arising from the use of its goods because of this or any similar suit.

Whether the action brought possesses merit is a question for the courts to determine. The mere bringing of the suit carries with it no more weight as to its merit than a bald and unsupported statement of an individual. The H. Mueller Manufacturing Company asks the trade and the public to withhold judgment and not be misled by any statements printed in paid advertisements.

We will continue to manufacture the supplies in question and will protect all sellers and users of same.

H. MUELLER MFG. CO.

Decatur, Ill., U.S.A. *New York City, U.S.A.*

ASK ANY PUMP MAN

who has ever used our Fig. 92, and if he does not say "it is the best ever," then we will acknowledge "the exception which proves the rule." There is no use looking for anything better than we have in our catalogue and prices must, of course, be right, as well. Let us have a little talk together.

THE

R. McDougall Company, LIMITED

GALT, CANADA

Are you interested in any of the lines that are advertised?
A Post Card will bring you price list and full information.
Don't forget to mention Hardware and Metal.

The Canadian Bronze Powder Works

R. E. THORNE & CO.

The only bronze powder works under the British flag.
High Grade bronze powders and bronze liquids.
Can fill all orders at short notice.

MONTREAL — TORONTO

WORKS AT VALLEYFIELD. *NO ORDER TOO LARGE*

HOTTEST ON EARTH

We manufacture the largest line of mechanical tools using gasoline, kerosene, and alcohol as fuel. Your jobber will supply you at factory prices. Guaranteed to please you or money refunded.

THE TURNER BRASS WORKS

53 MICHIGAN STREET, - CHICAGO, U.S.A.

No. 30 King. Price, each $5.20 net.

Manufacturers' Agents

CHARLES H. FOX
Vancouver, Canada
MANUFACTURERS' AGENT

Selling to the Hardware Jobbers Only.

Representing Canadian, British and American Manufacturers, correspondence invited. Cable Address, Chasfox Vancouver, P.O.Box 1105.

ALEXANDER GIBB

Manufacturers' Agent and Metal Broker,
13 St. John Street, Montreal

Representing Canadian, British and American Manufacturers. Correspondence invited from firms wishing to be represented.

H. W. MITCHELL
WINNIPEG, MAN.

Correspondence is solicited from manufacturers wishing a representative in Winnipeg.
Travellers call regularly on the wholesale and retail trade in Western Canada.
Highest references and financial responsibility.

Scott, Bathgate & Co.

BROKERS AND COMMISSION MERCHANTS

Cor. Notre Dame St. East and Victoria St., WINNIPEG, Canada
Storage and Warehousing a Specialty.

First-class facilities in a thoroughly modern brick warehouse centrally located on the railway track.

PAGE ACME White Fences

Any height to 8 ft. Any length you say. From 16 cents a foot. Gates to match, from $2.25. Last longer. Easy to put up. Get booklet.

PAGE WIRE FENCE CO., Limited

Walkerville - Toronto - Montreal
St. John - Winnipeg 211

A Certain Sense of Satisfaction

goes with every MERRELL PORTABLE HAND MACHINE PIPE THREADING and CUTTING MACHINE. We build them that way.

For doing ALL kinds of threading—under ALL conditions—labor trouble and faulty material included—where speed, durability and ease of control count for something, our word to you is to investigate the MERRELL.

Whether it be stationary or portable—hand or power driven—30 DAYS FREE TRIAL must prove conclusively that your choice should be the MERRELL.

This Portable Hand Machine has encased gears, MERRELL standard quick opening and closing die-head, and the latest improved Cutting-off Knife.

Let us tell you more about this machine—The Chasers, Vise and the large range of work covered.

Catalogue for the asking.
Quick shipments.

THE CANADIAN FAIRBANKS CO.
Sole Agents for Canada Limited
MONTREAL, TORONTO, WINNIPEG, VANCOUVER

Your Interests Are Mine

That explains the increasing volume of my business from month to month. Then, Quality always tells, and, as you know, you can only secure First Quality Enamelware by placing your orders with me. Prompt delivery guaranteed.

ORLANDO VICKERY
178-180 Victoria Street TORONTO, CANADA

There may be be better fittings placed on the market some day, but to-day

"Diamond" brand Fittings

hold the market because of their sterling and lasting worth. Remember the brand. Wholesale only.

OSHAWA STEAM & GAS FITTINGS CO., LIMITED
OSHAWA, CANADA

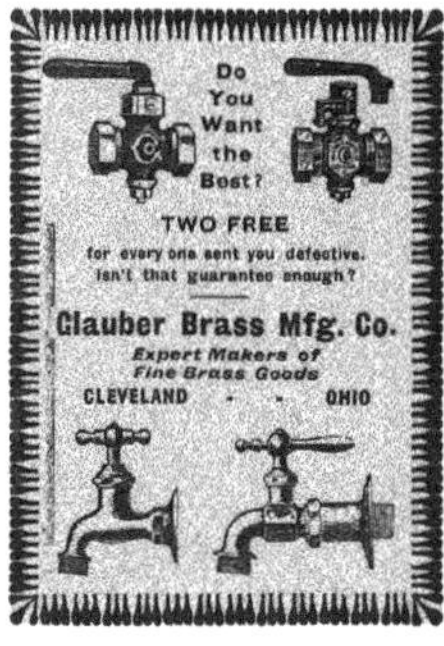

Sharratt & Newth's Glaziers' Diamonds

are unequalled for cutting and wearing qualities.

To be obtained from the principal Dealers in Glass, Hardware, and Painters' Supplies

Contractors to H. M. Government and the principal English Sheet and Plate Glass Works

World's Best Cutlery

You make no mistake in specifying goods made by **JOSEPH RODGERS & SONS, Limited** when ordering cutlery. A reputation of over 200 years at stake, insures the quality of every article.

Write for catalogues from Sole Canadian Agents

JAMES HUTTON & COMPANY, MONTREAL

Forwell Foundry Co.

BERLIN, ONT.

Manufacturers of

SOIL PIPE, FITTINGS, and CAST IRON SINKS

Ask Jobbers for "F. F. CO." Brand

Persons addressing advertisers will kindly mention having seen their advertisement in Hardware and Metal.

CURRENT MARKET QUOTATIONS.

Mar. 15, 1907.

These prices are for such qualities and quantities as are usually ordered by retail dealers on the usual terms of credit, the lowest figures being for larger quantities and prompt pay. Large cash buyers can frequently make purchases at better prices. The Editor is anxious to be informed at once of any apparent errors in this list, as the desire is to make it perfectly accurate.

METALS.

ANTIMONY.

Cookson's ....per lb...	0 27	0 27½

BOILER PLATES AND TUBES

Plates, ¼ to ½ inch, per 100 lb		2 50
Heads, per 100 lb		2 75
Tank plates 3-16 inch		2 65
Tubes per 100 feet, 1½ inch ..	8 50	....
" " 2 " ..	9 00	9 50
" " 2½ " ..	11 30	11 50
" " 3 " ..	12 50	13 00
" " 3½ " ..	16 00	17 00
" " 4 " ..	20 00	21 50

2 per cent off

BOILER AND T.K. PITTS.

Plain tinned .... }
Spun .... } 25 per cent. off list.

BABBIT METAL.

Canada Metal Company—Imperial, genuine 60c ; Imperial Tough, 60c ; White Brass, 50c ; Metallic, 35c ; Harris Heavy Pressure, 25c ; Hercules, 25c ; White Bronze, 15c ; Star Frictionless, 14c ; Alluminoid, 10c ; No. 4, 9c per lb.

James Robertson Co.—Extra and genuine Monarch, 60c ; Crown Monarch, 50c ; No. 1 Monarch, 40c ; King 30c ; Fleur-de-lis, 20c ; Thurber, 15c ; Philadelphia, 12c ; Canadian, 10c.; hardware, No. 1, 15c.; No. 2, 12c ; No. 3, 10c per lb.

BRASS.

Rod and Sheet, 14 to 30 gauge, 25 p.c. advance.

Sheets, 12 to 14 in.	0 30
Tubing, base, per lb 5-16 to 2 in	0 33
Tubing, ½ to 3-inch, iron pipe size..	0 31
" ½ to 3-inch, seamless..	0 36

Copper tubing, 6 cents extra

COPPER.

	Ingot.	Per 100 lb.
Casting, car lots............	....	27 00
	Bars	
Cut lengths, round, ½ to 2 in...	....	35 00
	Sheet.	
Plain, 16 oz., 14x48 and 14x60	....	35 00
Plain, 14 oz	....	36 00
Tinned copper sheet, base	....	38 00
Planished base..............	....	43 00
Braziers' (in sheets), 4x6 ft., 25 to 30 lb each, per lb., base..	....	0 35

BLACK SHEETS.

	Montreal.	Toronto
8 to 10 gauge	2 70	2 70
12 gauge........................	2 70	2 60
14 "	2 70	2 55
17 "	2 50	2 65
18 "	2 50	2 65
20 "	2 50	2 65
22 "	2 50	2 75
24 "	2 55	2 80
26 "	2 65	2 85
28 "	2 70	3 00

CANADA PLATES

Ordinary, 52 sheets	2 75	2 90
All bright "	3 80	3 90
Galvanized, 52 sheets	4 35	4 45
" 60 "	4 60	4 70
	Ordinary.	Dom. Crown
18x24x52	4 25	4 35
" 60....................	4 50	4 60
20x28x80	8 50	8 70
" 94	9 00	9 20

GALVANIZED SHEETS.

	Fleur-de-Lis.	Gordon Crown	Colborne Crown
16 to 20 gauge	3 80	3 95	....
22 to 24 gauge ..	4 05	4 00	3 75
26 " ..	4 30	4 40	4 30
28 " ..	4 55	4 60	4 45

Apollo.

10¾ oz. (American gauge)	....	4 85
28 gauge.......... "	4 45	4 55
26 " .... "	4 15	4 30
24 " .... "	3 85	3 90

	Gorbal's Best.	Comet	Queen's Head.	Bell.
16 to 20 gauge	....	3 80	3 95	....
22 to 24 gauge	....	3 75	4 20	....
26 "	....	4 30	4 45	....
28 "	4 55	4 45	4 70	4 45

Less than case lots 10 to 25c. extra.

IRON AND STEEL.

	Montreal.	Toronto.
Common bar, per 100 lb......	2 15	2 30
Forged iron "	2 40	
Refined " "	2 55	2 70
Horseshoe iron "	2 55	2 70
Hoop steel, 1½ to 3 in. base...	....	2 80
Sleigh shoe steel "	2 25	2 30
Tire steel	2 40	2 50
Best sheet cast steel..........	....	0 12
B. K. Morton "Alpha" high speed.	....	0 65
" annealed	....	0 70
"M" Self-hardening........	....	0 50
"J" quality, best warranted	....	0 18
" " warranted......	....	0 14
"B.C" quality	....	0 09
Colonial black diamond......	0 08	0 09
Sanderson's	0 08	0 45
Jessop's	0 12	0 13
Air hardening	0 60	0 65
Conqueror	0 07½	0 60
Jowett's diamond J	0 06½	0 07
Jonas & Colver's tool steel....	0 10	0 20
" "Novo"	....	0 65
" " annealed ..	....	0 65
Jowett & Sons B P.L. tool steel	0 10½	0 11

COLD ROLLED SHAFTING.

9-16 to 11-16 inch	....	0 06
¾ to 1 7-16 "	....	0 05½
1 7-16 to 3 "	....	0 05

Montreal 30, Toronto 35 to 40 per cent.

INGOT TIN.

Lamb and Flag and Straits—
56 and 28-lb. ingots, 100 lb. $45 00 $45 50

TINPLATES.

Charcoal Plates—Bright

		Per box.
M.L.S., equal to Bradley—		
I C, 14 x 20 base	....	$6 50
IX, 14 x 20 "	....	8 00
IXX, 14 x 20 base	....	9 50
Famous, equal to Bradley—		
I C, 14 x 20 base..........	....	6 50
I X, 14 x 20 "	....	8 00
I X X, 14 x 20 base.........	....	9 50
Raven and Vulture Grades—		
I C, 14 x 20 base	....	5 00
I X "	....	6 00
I X X "	....	7 00
I X X X "	....	8 00

"Dominion Crown Best"—Double Coated, Tissued.

		Per box.
I C, 14 x 20 base...........	5 50	5 75
I X, 14 x 20 "	6 50	6 75
IXX " x 20 "	7 50	7 75

"Allaway's Best"—Standard Quality.

I C, 14 x 20 base..........	....	4 50
I X, 14 x 20 "	....	5 25
I XX, 14 x 20 "	....	6 00

Bright Cokes.

Bessemer Steel—		
I.C., 14 x 20 base	....	4 25
20x28, double box	....	8 50

Charcoal Plates—Terne

Dean or J G Grade—		
I.C., 20x28, 112 sheets	7 25	8 00
IX., Terne Tin	....	9 50

Charcoal Tin Boiler Plates.

Cookley Grade— X X, 14x56, 50 sheet bxs } " 14x60, " } " 14x65, " }	....	7 50

Tinned Sheets.

72x30 up to 24 gauge.......	....	8 50
" " 26 "	....	9 00

LEAD.

Imported Pig, per 100 lb......	5 50	5 60
Bar, "	5 75	6 00
Sheets, 2½ lb. sq. ft., by roll ..	....	0 07½
Sheets, 3 to 6 lb. "	....	0 07

NOTE.—Cut sheets ½c. per lb., extra. Pipe, by the roll, usual weights per yard, lists at 7c. per lb. and 5 p.c. dis. f.o.b. Toronto.

NOTE.—Cut lengths, net price, waste pipe 8-ft lengths, lists at 8c.

SHEET ZINC.

5-cwt. casks	8 00	8 25
Part casks......................	8 25	8 50

ZINC SPELTER.

Foreign, per 100 lb	7 50	7 75
Domestic	7 00	7 25

PLUMBING AND HEATING

BRASS GOODS, VALVES, ETC.

Standard Compression work, 55 per cent.
Cushion work, discount 40 per cent.
Fuller work, 65 per cent.
Flatway stop and stop and waste cocks, 60 per cent ; roundway, 55 per cent.
J.M.T. Globe, Angle and Check Valves, 45 per cent.
Standard Globe, Angle and Check Valves 55 per cent.
Kerr standard globes, angles and checks, special, 42½ per cent.; standard, 47½ p.c.
Kerr Jenkins' disc, copper-alloy disc and heavy standard valves, 40 per cent.
Kerr steam radiator valves, 60 p.c., and quick-opening hot-water radiator valves, 60 p.c.
Kerr brass, Weber's straightway valves, 40 per cent. ; straightway valves, I B B.M., 60 per cent
J. M. T. Radiator Valves 50 per cent.
Standard Radiator Valves, 60 per cent.
Patent Quick-Opening Valves, 65 per cent.
Jenkins' Bros. Globe Angle and Check Valves 27½ per cent.

No. 1 compression bath cock......net	2 00
No. 4 " " ...... "	1 90
No 7 Fuller's .... "	2 25
No. 4½ " .... "	2 35
Patent Compression Cushion, basin cock, hot and cold, per doz., ...	$16.20
Patent Compression Cushion, bath cock, No. 2208	2 25

Square head brass cocks, 50 per cent.
" " iron " 60 "
Thompson Smoke-test Machine 25.00

BOILERS—COPPER RANGE.

Copper, 30 gallon............	33 00
" 35 "	38 00
" 40 "	43 00

15 per cent

BOILERS—GALVANIZED IRON RANGE.

Capacity.	Standard.	Extra heavy
30-gallons ...	5 00	7 75
35 "	6 00	8 75
40 "	7 00	9 75

2 per cent , 30 days.

BATH TUBS.

Steel clad copper lined, 15 per cent.

CAST IRON SINKS.

16x24, $1, 18x30, $1 ; 18x36, $1.30.

ENAMELED BATHS.

List issued by the Standard Ideal Company Jan. 3, 1907, shows an advance of 10 per cent over previous quotations.

ENAMELED CLOSETS AND URINALS

Discount 15 per cent.

ENAMELED LAVATORIES.

	1st quality.	Special.
Plate E 100 to E 103 ..	20 & 5 p.c.	20 & 10 p.c.
" E 104 to E 132....	20 & 10 p.c.	30 & 2½ p.c.

ENAMELED ROLL RIM SINKS.

	1st quality.	Special.
Plate E 201, one piece.	15 & 2½ p.c.	15 & 10 p.c.

ENAMELED KITCHEN SINKS.

Plate E, flat rim 300, 60 & 10 p.c. 65 & 5 p.c

HEATING APPARATUS.

Stoves and Ranges—Discounts vary from 40 to 70 per cent. according to list.
Furnaces—45 per cent.
Registers—70 per cent.
Hot Water Boilers—Discounts vary.
Hot Water Radiators—47½ per cent.
Steam Radiators—50 per cent.
Wall Radiators and Specials—45 per cent.

LEAD PIPE.

Lead Pipe, 7c. per pound, 5 per cent. off.
Lead waste, 8c. per pound, 5 per cent. off.
Caulking lead, 6c. per pound.
Traps and bends, 40 per cent.

IRON PIPE.

Size (per 100 ft.)	Black.	Galvanized
¼ inch........	2 25	3 08
⅜ "	2 25	3 07
½ "	2 72	3 87
¾ "	3 57	4 72
1 "	5 12	6 77
1¼ "	6 98	9 23
1½ "	8 37	11 07
2 "	11 16	14 76
2½ "	18 00	24 00
3 "	23 50	31 50
3½ "	30 00	40 00
4 "	34 00	45 00

2 per cent 30 days.

Malleable Fittings—Canadian discount 30 per cent.; American discount 25 per cent.

Cast Iron Fittings 57½; Standard bushings 57½ per cent ; headers, 57½, flanged unions 57½, malleable bushings 55; nipples, 70 and 10 per cent ; malleable lipped unions, 55 and 5 per cent.

SOIL PIPE AND FITTINGS.

Medium and Extra heavy pipe and fittings, up to 6 inch, discount 60 per cent.
7 and 8-in. pipe, discount 40 and 5 per cent.
Light pipe, 50 p.c.; fittings, 50 and 10 p.c.

OAKUM.

Plumbers .....per 100 lb... .. .... 4 75

STOCKS AND DIES.

American discount 25 per cent.

SOLDERING IRONS.

½-lb. .... per lb.	0 38
1-lb. or over	0 35

SOLDER.

	Montreal	Toronto
Bar, half-and-half, guaranteed	0 25	0 27
Wiping	0 22	0 21

PAINTS, OILS AND GLASS.

BRUSHES.

Paint and household, 70 per cent..

CHEMICALS.

	In casks per lb.
Sulphate of copper (bluestone or blue vitrol)	0 09
Litharge, ground	0 06
" flaked	0 06½
Green copperas (green vitrol)	0 01
Sugar of lead..................	0 08
Lump alum	0 01½

COLORS IN OIL.

Venetian red, 1-lb. tins pure.	0 08
Chrome yellow "	0 15
Golden ochre "	0 08
French " "	0 08
Marine black "	0 04½
Chrome green "	0 10
French permanent green "	0 13
Signwriters' black "	0 15

GLUE

Domestic sheet	0 10	0 10½
French medal..............	0 12	0 12½

PARIS GREEN.

	Berger's	Canadian
600-lb. cask	0 25½	0 25½
250-lb. drums	0 25½	0 25½
100-lb. "	0 26	....
50-lb. "	0 26½	0 26½

1-lb. pkgs, 100 in box	0 27½	0 27½
½-lb. "	0 29½	0 29½
1-lb. tins, 100 in box	0 28½	0 28½
½-lb pkg	0 30½	0 30

F.o.b. Toronto.

PARIS WHITE.

In bbls		0 90

PREPARED PAINTS

Pure, per gallon, in tins	....	1 30
Second qualities per gallon ..	....	1 10
Barn (in bbls.)	0 65	0 90
Sherwin-Williams paints, 1 gal.	....	1 40
Canada Paint Co.'s pure	....	1 30
Standard P. & V. Co.'s "New Era"		1 30
Benj. Moore Co.'s "Ark" B'd	....	1 25
" British Navy deck	....	1 50
Brandram-Henderson's "English"		1 45
Ramsay's paints, Pure, per gal.	....	1 20
" Thistle, "	....	1 00
" Outside, bbls	0 55	0 65
Martin-Senour's 100 p c. pure, ½ gal.		1 50
" " " " ¼ gal.		1 45
" " " " 1 gal.		1 40
Senour's Floor Paints.....gal.	....	1 20
Jamieson's "Crown and Anchor"		1 20
Jamieson's floor enamel	....	1 50
" barn paints, bbls. per gal.		0 60
Sanderson Pearcy's, pure ..	....	1 20
Robertson's pure paints	....	1 20

PUTTY.

Bulk in bbls	1 50
Bulk in less quantity	1 80
Bladders in bbls	1 80
Bladders in kegs, boxes or loose	1 95
25-lb. tins	1 80
12½ lb. tins	2 05
Bladders in bulk or tins less than 100 lb.	1 85
Bulk in 100-lb. irons	1 80

SHINGLE STAINS.

In 5-gallon lots	0 75	0 80

SHELLAC.

White	0 65
Fine orange	0 60
Medium orange	0 55

F.o.b. Montreal or Toronto.

TURPENTINE AND OIL.

Castor oil	0 08	0 10
Gasoline	....	0 22½
Benzine, per gal	0 17	0 20
Turpentine, single barrels	0 98	1 05
Linseed Oil, " raw ..	0 60	0 62
" " boiled	0 63	0 65
W. H. & C's English raw oil, bbls ..		0 63
" " boiled, "		0 66

WHITE LEAD GROUND IN OIL		Per 100 lbs
Canadian pure	....	7 15
No. 1 Canadian	....	6 80
Munro's Select Flake White ..	....	7 40
Elephant and Decorators' Pure	....	7 40
Monarch	....	7 40
Standard Decorator's	....	7 15
Essex Genuine	....	6 80
Brandram's B. B. Genuine ...	....	8 40
"Anchor," pure	....	7 00
Ramsay's Pure Lead	....	6 40
Ramsay's Exterior	....	6 15
"Crown and Anchor," pure ...	....	6 50
Sanderson Pearcy's	....	7 40
Robertson's C.P., lead	....	7 20
W. H. & C's matured pure English	8 00	8 25

WHITE AND RED DRY LEAD.	white	red.
Genuine, 560 lb. casks, per cwt	6 75	6 00
Genuine, 100 lb. kegs, "	7 50	6 50
No. 1, 560 lb. casks, per cwt ...	6 25	5 75
No. 1, 100 lb. kegs, per cwt ..	7 00	6 25

WINDOW GLASS.

Size United Inches.	Star	Double Diamond
Under 26	$4 25	$6 25
26 to 40	4 65	6 75
41 to 50	5 10	7 50
51 to 60	5 35	8 50
61 to 70	5 75	9 75
71 to 80	6 25	11 00
81 to 85	7 00	12 50
86 to 90	....	15 00
91 to 95	....	17 50
96 to 100	....	20 50
101 to 105	....	24 00
106 to 110	....	27 50

Discount—16-oz., 25 per cent.; 21-oz. 30 per cent. per 100 feet. Broken boxes 50 per cent.

WHITING.

Plain, in bbls	0 60
Gilders bolted in bands	0 90

WHITE DRY ZINC.

Extra Red Seal, V.M.	0 07½	0 08½

WHITE GROUND ZINC.

Pure, in 25-lb. irons	0 08½
No. 1, "	0 07
No. 2, "	0 05½

VARNISHES. In 5-gal. lots.	Per gal. cans
Carriage, No. 1	1 50
Pale durable body	3 50
" hard rubbing	3 00
Finest elastic gearing	3 00
Elastic oak	1 50
Furniture, polishing	2 00
Furniture, extra	1 15
" No. 1	0 90
" union	0 80
Light oil finish	1 40
Gold size japan	1 80
Brown japan	0 95
No. 1 brown japan	0 90
Baking black japan	1 35
No. 1 black japan	0 90
Benzine black japan	0 70
Crystal Damar	2 80
No. 1	2 50
Pure asphaltum	1 40
Oilcloth	1 50
Lightning dryer	0 70
Elastilite varnish, 1 gal. can, each.	2 00
Granitine floor varnish, per gal	2 50
Maple Leaf coach enamels; size 1,	1 20
Sherwin-Williams' kopal varnish, gal.,	2 50
Canada Paint Co's sun varnish	2 00
"Kyanize" Interior Finish	2 40
"Flint-Lac," coach	1 80
B H Co's "Gold Medal," in cases	2 00
Jamieson's Copaline, per gal.	2 00

BUILDERS' HARDWARE.

BELLS.

Brass hand bells, 60 per cent.
Nickel, 55 per cent.

Gongs, Sargeant's door bells ..	5 50	8 00
American, house bells, per lb.	0 35	0 40

Peterboro' door bells, discount 37½ and 10 per cent. off new list.

BUILDING PAPER, ETC.

Tarred Felt, per 100 lb.		2 25
Ready roofing, 2-ply, not under 45 lb., per roll		1 00
Ready roofing, 3-ply, not under 65 lb., per roll		1 25
Carpet Felt ...... per ton		60 00
Heavy Straw Sheathing ...... per ton		35 00
Dry Surprise		0 40
Dry Sheathing ..... per roll, 400 sq. ft.		0 50
Tar " ..... " 400 "		0 45
Dry Fibre ...... " 400 "		0 55
Tarred Fibre ...... " 400 "		0 65
O. K. & I. X. L. .... " 400 "		0 70
Resin-sized ...... " 400 "		0 45
Oiled Sheathing .... " 600 "		1 00
Oiled " .... " 400 "		0 70
Roof Coating, in barrels ...... per gal.		0 17
Roof " small packages "		0 25
Refined Tar ...... per barrel		5 00
Coal Tar ...... "		4 00
Coal Tar, less than barrels .... per gal.		0 15
Roofing Pitch ...... per 100 lb.	0 80	0 90
Slater's felt ...... per roll		0 70
Heavy Straw Sheathing f. o. b. St. John and Halifax		37 50

BUTTS.

Wrought Brass, net revised list.
Wrought Iron, 70 per cent.
Cast Iron Loose Pin, discount 60 per cent.
Wrought Steel Fast Joint and Loose Pin, 70 per cent.

CEMENT AND FIREBRICK.

Canadian Portland	2 00	2
Belgium	1 60	1
White Bros. English	1 80	2
"Lafarge" cement in wood		3
"Lehigh" cement, in wood		2
"Lehigh" cement, cotton sacks.		2
"Lehigh" cement, paper sacks ..		2
Fire brick, Scotch, per 1,000	27	0
" English	17	1
" American, low	23	5 10
" " high	27 60	35 [illegible]
Fire clay (Scotch), net ton		4 95

Paving Blocks per 1,000.

Blue metallic, 9"x4½"x3", ex wharf ..	00
Stable pavers, 12"x6"x2", ex wharf ..	5[illegible] 00
Stable pavers, 9"x4½"x3", ex wharf ..	36 00

DOOR SETS.

Peterboro, 37½ and 10 per cent.

DOOR SPRINGS.

Torrey's Rod .... per doz.	....	1 75
Coil, 9 to 11 in. ...... "	0 95	1 65
English .... "	2 00	4 00

Chicago and Reliance Coil 25 per cent.

STORE DOOR HANDLES

Per Dozen	1 00	1 50

ESCUTCHEONS.

Discount 50 and 10 per cent., new list
Peterboro, 37½ and 10 per cent.

ESCUTCHEON PINS.

Iron, discount 40 per cent.
Brass, 45 per cent.

HINGES.

Blind, discount 60 per cent.

Heavy T and strap, 4-in., per lb. net.	0 06½
" " 5-in., "	0 06
" " 6-in., "	0 05½
" " 8-in., "	0 05½
" " 10-in. and larger ..	0 05½

Light T and strap, discount 65 p.c.

Screw hook and hinge— under 12 in. ... per 100 lb.	....	4 75
over 12 in. ...... "	....	3 75
Spring, No. 20, per gro. pairs	....	10 80

Spring, Woodyatt pattern, per gro., No. 5, $17.50 No. 10, $18; No. 20, $10 80; No. 12 $20; No. 51, $10; No. 50, $27 50.
Crate hinges and back flaps, 65 and 5 p. c.
Hinge hasps, 65 per cent.

SPRING HINGES.

Chicago Spring Butts and Blanks, 12½ per cent.
Triple End Spring Butts, 40 and 5 per cent.
Chicago Floor Hinges, 40 and 5 off.
Garden City Fire House Hinges, 12½ p c.

CAST IRON HOOKS.

Bird cage ...... per doz.	0 50	1 10
Clothes line, No. 61 .. "	0 00	0 70
Harness ...... "	0 60	12 00
Hat and coat ...... per gro.	1 10	10 00
Chandelier ...... per doz.	0 50	1 00
Wrought hooks and staples— ¼ x 5 ...... per gross	....	2 65
5-16 x 5	....	3 30

Bright steel gate hooks and staples, 40 p.c.
Hat and coat wire, discount 62½ per cent.
Screw, bright wire, discount 65 per cent.

KNOBS.

Door, japanned and N.P., doz	1 50	2 50
Bronze, Berlin ...... per doz.	2 75	3 25
Bronze, Genuine .... "	6 00	9 00
Shutter, porcelain, F. & L. screw ...... per gross	1 30	2 00
White door knobs .. per doz.		2 00

Peterboro knobs, 37½ and 10 per cent.
Porcelain, mineral and jet knobs, net list.

KEYS.

Lock, Canadian dis. 40 to 40 and 10 per cent

Cabinet trunk and padlock American ...... per gross	....	0 60

LOCKS.

Peterboro, 37½ and 10 per cent
Russell & Erwin, steel rim $2 50 per doz.
Eagle cabinet locks, discount 30 per cent
American padlocks, all steel, 10 to 15 per cent; all brass or bronze, 10 to 25 per cent.

SAND AND EMERY PAPER.

B. & A. sand, discount, 35 per cent
Emery, discount 35 per cent.
Garnet (Rurton's) 5 to 10 per cent. advance

SASH WEIGHTS.

Sectional ...... per 100 lb.	2 00	2 25
Solid ...... "	1 50	1 75

SASH CORD.

Per lb.	....	0 31

BLIND AND BED STAPLES.

All sizes, per lb.	0 07½	0 12

WROUGHT STAPLES.

Galvanized	....	2 75
Plain	....	2 50

Coopers', discount 45 per cent.
Poultry netting staples, discount 40 per cent.
Bright spear point, 75 per cent. discount.

TOOLS AND HANDLES.

ADZES.

Discount 22½ per cent.

AUGERS.

Gilmour's, discount 60 per cent. off list.

AXES.

Single bit, per doz.	5 50	8 50
Double bit, "	10 00	11 00

Bench Axes, 40 per cent.
Broad Axes, 25 per cent.

Hunters' Axes	5 50	6 00
Boys' Axes	6 25	7 00
Splitting Axes	7 00	12 00
Handled Axes	7 00	9 00
Red Ridge, boys', handled ..	....	5 75
" hunters	....	5 25

BITS.

Irwin's auger, discount 47½ per cent.
Gilmour's auger, discount 60 per cent.
Rockford auger, discount 50 and 10 per cent.
Jennings' Gen. auger, net list.
Gilmour's car, 47½ per cent.
Clark's expansive, 40 per cent.

Clark's gimlet, per doz	0 65	0 90
Diamond, Shell, per doz.	1 00	1 50
Nail and Spike, per gross	2 25	5 20

BUTCHERS CLEAVERS

German ...... per doz.	6 00	9 00
American ...... "	12 00	18 00

CHALK.

Carpenters' Colored, per gross	0 45	0 75
White lump ...... per cwt.	0 60	0 65

CHISELS.

Warnock's, discount 72½ per cent.
P. S. & W. Extra, discount 72½ per cent

CROSSCUT SAW HANDLES.

S. & D., No. 3	per pair	0 15
S. & D., " 5	"	0 22½
S. & D., " 6	"	0 15
Boynton pattern	"	0 20

CROWBARS.

3½c. to 4c. per lb.

DRAW KNIVES.

Coach and Wagon, discount 75 per cent.
Carpenters' discount 75 per cent

DRILLS.

Miller's Falls, hand and breast, net list
North Bros, each set, 50c.

DRILL BITS.

Morse, discount 37½ to 40 per cent.
Standard, discount 50 and 5 to 55 per cent.

FILES AND RASPS.

Great Western	75	per cent.
Arcade	75	"
Kearney & Foot	75	"
Disston's	75	"
American	75	"
J. Barton Smith	75	"
McClellan	75	"
Eagle	75	"
Nicholson	66⅔	"
Globe	75	

Black Diamond, 60, 10 and 5 p.c.
Jowitt's, English list, 27½ per cent.

GAUGES.

Stanley's discount 50 to 60 per cent.

Wire Gauges.

Winn's, Nos. 26 to 33 .... each	1 65	2 40

HANDLES.

C. & B., fork and hoe, 40 p. c., revised list.

American, saw ...... per doz.	....	0 25
American, plane ... per gross	3 15	3 75

Canadian, hammer and hatchet 40 per cent
Axe and cant hook handles, 45 per cent.

HAMMERS.

Maydole's, discount 5 to 10 per cent.
Canadian, discount 25 to 27½ per cent

Magnetic tack ...... per doz.	1 10	1 20
Canadian sledge .... per lb.	0 07	0 08½
Canadian ball pean, per lb. ..	0 22	0 25

HATCHETS.

Canadian, discount 40 to 42½ per cent.

Shingle, Red Ridge 1, per doz.	4 40
" " 2, "	4 80
Barrel Underhill	5 06

MALLETS.

Tinsmiths' ....per doz. 1 25 1 50
Carpenters', hickory, " 1 25 3 75
Lignum Vitae .... " 3 85 5 00
Caulking, each 0 60 2 00

MATTOCKS.

Canadian............ per doz. 5 50 6 00

MEAT CUTTERS.

German, 15 per cent.
American discount, 33⅓ per cent.
Gem ....each 1 15

NAIL PULLERS.

German and American 0 85 2 50
No. 1.............................. 0 85
No 1573........................ 0 75

NAIL SETS.

Square, round and octagon, per gross 3 38
Diamond 1 00

PICKS.

Per dozen 6 00 9 00

PLANES.

Wood bench, Canadian discount 40 per cent.
American discount 50 per cent.
Wood, fancy Canadian or American 37½ to 40 per cent.
Stanley planes, $1.55 to $3 60, net list prices.

PLANE IRONS.

English ....per doz. 2 00 5 00
Stanley, 2½ inch, single 24c., double 39c.

PLIERS AND NIPPERS.

Button's genuine, 37½ to 40 per cent.
Button's imitation....per doz. 5 00 9 00
Berg's wire fencing.............. 1 72 5 50

PUNCHES.

Saddler's ....per doz. 1 00 1 85
Conductor's.......... " 3 00 15 00
Tinners', solid...........per set 0 72
" hollow......per inch 1 00

RIVET SETS.

Canadian, discount 35 to 37½ per cent.

RULES.

Boxwood, discount 70 per cent.
Ivory, discount 20 to 25 per cent.

SAWS.

Atkins, hand and crosscut, 25 per cent.
Disston's Hand, discount 12½ per cent
Disston's Crosscut ....per foot 0 35 0 55
Hack, complete..........each 0 75 2 75
" frame only........each 0 50 1 25

S. & D. solid tooth circular shingle, concave and band, 50 per cent; mill and ice, drag, 30 per cent; cross-cut, 35 per cent.; hand saws, butcher, 35 per cent.; buck, New Century, $6.25; buck, No 1 Maple Leaf, $5 25; buck, Happy Medium, $4.25; buck Watch Spring, $4.25; buck, common frame, $4.00.

Spear & Jackson's saws—Hand or rip 26 in., $12.75; 24 in., $12.25; panel, 18 in., $8 25; 20 in., $9; tenon, 10 in., $9 90; 12 in., $10 90; 14 in., $11.50.

SAW SETS.

Lincoln and Whiting 4 75
Hand Sets, Perfect................... 4 00
X-Cut Sets. " 7 50
Maple Leaf and Premium saw sets, 40 off.
S. & D. saw swages, 40 off.

SCREW DRIVERS

Sargent's ....per doz. 0 65 1 00
North Bros., No. 30 ..per doz. 16 80

SHOVELS AND SPADES.

Bull Dog, solid neck shovel (No. 2 pol.). $18 50
(Hollow Back) (Reinforced S Scoop.)
Moose........ $17 50 .... $16 30
Bear .... 15 00 .... 15 30
Fox............ 12 50 .... 14 30
Black Cat.... 10 00 .... 13 30
Canadian, discount 45 per cent.

SQUARES.

Iron, discount 20 per cent.
Steel, discount 65 and 10 per cent.
Try and Bevel, discount 50 to 52½ per cent.

TAPE LINES.

English, ass skin.....per doz. 2 75 5 00
English, Patent Leather 5 50 9 75
Chesterman's..............each 0 90 2 85
" steel.........each 0 80 8 00
Berg's, each......................... 0 75 2 50

TROWELS.

Disston's, discount 10 per cent.
S & D., discount 35 per cent.
Berg's, brick 9¼x11 .... 4 00
" pointing, 9¼x5............ .... 2 10

FARM AND GARDEN GOODS

BELLS.

American cow bells, 63⅓ per cent.
Canadian, discount 45 and 50 per cent.
American, farm bells, each .. 1 35 3 00

BULL RINGS.

Copper, $1.30 for 2½-inch, and $1.70

CATTLE LEADERS.

Nos. 32 and 33 .. ...per gross 7 50 8 50

BARN DOOR HANGERS.

	doz.	pairs.
Steel barn door................	6 00	10 00
Stearns wood track	4 50	6 00
Zenith.......................	....	9 00
Acme, wood track	5 00	6 50
Atlas............................	5 0)	6 00
Perfect	8 00	11 00
New Milo........................	....	6 50
Steel, covered.......................	4 00	11 00
" track, 1 x 3-16 in(100 ft)	....	3 75
" " 1½ x 3-16 in(100 ft)	....	4 75
Double strap hangers, doz. sets	....	6 40
Standard jointed hangers, "	....	6 40
Steel King hangers "	....	6 25
Storm King and safety hangers	....	7 00
" " rail..................	....	4 25

Chicago Friction, Oscillating and Big Twin Hangers, 5 per cent.

HARVEST TOOLS.

Discount 60 per cent.
S. & D. lawn rakes, Dunn's, 40 off.
" sidewalk and stable scrapers, 40 off.

HAY KNIVES.

Net list.

HEAD HALTERS.

Jute Rope, ⅜-inch....per gross 9 00
" " ½ " .... " 10 00
" " ⅝ " .... " 12 00
Leather, 1-inch ....per doz. 4 00
Leather, 1¼ " ...... " 5 20
Web.................. " 2 45

HOES.

Garden, Mortar, etc., discount 60 per cent.
Planter.............. per doz. 4 00 4 50

SCYTHES.

Per doz. net.................. 6 25 9 25

SCYTHE SNATHS.

Canadian, discount 40 per cent.

SNAPS.

Harness, German, discount 25 per cent.
Lock, Andrews' 4 50 11 00

STABLE FITTINGS.

Warden King, 35 per cent.

WOOD HAY RAKES.

Ten tooth, 40 and 10 per cent.
Twelve tooth, 45 per cent.

HEAVY GOODS, NAILS, ETC.

ANVILS.

Wright's, 80-lb. and over............... 0 10½
Hay Budden, 80-lb. and over 0 09½
Brook's, 80-lb. and over 0 11½
Taylor-Forbes, Dandy 0 05
Columbia Hardware Co., per lb. 0 09½

VISES.

Wright's.......................... .... 0 13½
Berg's, per lb...................... 0 12½
Brook's.......................... .... 0 12½
Pipe Vise, Hinge, No. 1...... .... 3 50
" " " No. 2...... .... 5 50
Saw Vise 4 50 5 00
Blacksmiths' (discount) 60 per cent.
" parallel (discount) 45 per cent.

BOLTS AND NUTS

Carriage Bolts, common ($1 list — Per cent
" " ⅜ and smaller.. 60, 10 and 10
" " 7-16 and up.... 55 and 5
" " Norway Iron ($3 list) 50
Machine Bolts, ⅜ and less .... 60 and 10
Machine Bolts, 7-16 and up.... 55 and 5
Plough Bolts .... 55 and 10
Blank Bolts.................... 55
Bolt Ends........................ 55
Sleigh Shoe Bolts, ⅜ and less .. 60 and 10
" " " 7-16 and larger 50 and 5
Coach Screws, conepoint...... 70 and 5
Nuts, square, all sizes, 4c. per cent. off
Nuts, hexagon, all sizes, 4¼c. per cent. off
Stove Rods per lb., 5½ to 6c.
Stove Bolts, 75 per cent.

CHAIN.

Proof coil, per 100 lb., 5-16 in., $4.40; ⅜ in., $3 90; 7-16 in., $3 70; ½ in., $3 50; 9-16 in., $3 45; ⅝ in., $3.35; ¾ in., $3.20; ⅞ in., $3.10; 1 in., $3.10.
Halter, kennel and post chains, 40 to 40 and 5 per cent.
Cow ties .... 40 p.c.
Tie out chains .... 65 p.c.
Stall fixtures .... 35 p.c.
Trace chain .... 45 p.c.
Jack chain iron, discount 35 p.c.
Jack chain, brass, discount 40 per cent

HORSE NAILS.

'C' brand, 40, 10 and 7½ per cent. off list } Oval
M.R.M. Co. brand, 55 per cent. } head
Capewell brand, quotations on application

HORSESHOES.

M.R.M. Co	No. 2 and larger	No. 1 and smaller
Iron..................	3 80	4 05
Snow..................	4 05	4 30
Light steel	4 15	4 40
Featherweight, sizes 0 to 4...	....	5 75
Toeweight, " 1 to 4........		7 00

Packing up to 3 sizes in a keg, 10c. per 100 lbs.; more than three sizes, 25c. per 100 lbs extra. F.o.b. Montreal, add 15c, Toronto, Hamilton and Guelph.

HORSE WEIGHTS.

Taylor-Forbes, 3½c. per lb.

NAILS.	Cut.	Wire.
2d..............................	3 80	3 45
3d..............................	2 95	3 10
4 and 5d........................	2 70	2 85
6 and 7d	2 60	2 75
8 and 9d	2 45	2 60
10 and 12d	2 40	2 55
16 and 20d.......................	2 35	2 50
30, 40, 50 and 60d (base)	2 30	2 45

F.o.b. Montreal. Cut nails, Toronto 20c, higher.
Miscellaneous wire nails, discount 75 per cent
Coopers' nails, discount 40 per cent.

PRESSED SPIKES.

Pressed spikes, ⅝ diameter, per 100 lbs, $3.15

RIVETS AND BURRS.

Iron Rivets, black and tinned, 60, 10 and 10.
Iron Burrs, discount 60 and 10 and 10 p.c.
Copper Rivets, usual proportion burrs, 15 p.c.
Copper Burrs only, net list.
Extras on Coppered Rivets, ½-lb. packages 1c. per lb.; ¼-lb. packages 2c. lb
Tinned Rivets, net extra, 4c. per lb.

SCREWS.

Wood, F. H., bright and steel, 87½ per cent.
" R. H., bright, dis. 82½ per cent.
" F. H., brass, dis. 80 per cent.
" R. H., " dis. 75 per cent.
" F. H. bronze, dis. 75 per cent.
" R. H., " dis. 70 per cent.
Drive Screws, dis. 87½ per cent.
Bench, wood ....per doz. 3 25 4 00
" iron .... " 4 25 5 00
Set, case hardened, dis. 60 per cent.
Square Cap, dis. 50 and 5 per cent.
Hexagon Cap, dis. 45 per cent.

MACHINE SCREWS.

Flat head, iron and brass, 35 per cent.
Felister head, iron, discount 30 per cent.
" " brass, discount 25 per cent.

TACKS, BRADS, ETC.

Carpet tacks, blued .... 80 and 5
" " tinned............ 80 and 10
" " (in kegs)........... 40
Cut tacks, blued, in dozens only 75 and 10
" ½ weights 60
Swedes cut tacks, blued and tinned—
In bulk.......................... 80 and 10
In dozens 75
Swedes, upholsterers', bulk .... 85 and 12½
" brush, blued and tinned bulk 70
Swedes, gimp, blued, tinned and japanned.......................... 75 and 12½
Zinc tacks 35
Leather carpet tacks............. 40
Copper tacks...................... 37½
Copper nails 42½
Trunk nails, black 65
Trunk nails, tinned and blued .. 85
Clout nails, blued and tinned .. 65
Chair nails........................ 35
Patent brads...................... 40
Fine finishing..................... 40
Lining tacks, in papers.......... 10
" " in bulk........... 15
" " solid heads, in bulk 75
Saddle nails, in papers.......... 10
" " in bulk............ 15
Tufting buttons, 22 line in dozens only 60
Zinc glaziers' points 5
Double pointed tacks, papers.. 90 and 10
" " " bulk.... 40
Clinch and duck rivets.......... 45
Cheese box tacks, 85 and 5; trunk tacks, 80 and 10.

WROUGHT IRON WASHERS.

Canadian make, discount 40 per cent.

SPORTING GOODS.

AMMUNITION.

B. B. Caps American $2.00 per 1000.
C. B. Caps American, $2.60 per 1000

CARTRIDGES.

Rim Fire Cartridges, 50 and 5 p.c.
Rim Fire Pistol, 30 and 5 per cent., American.
Central Fire, Military and Sporting American, 10 per cent. advance. B.B. Caps, discount 40 per cent., American.
Central Fire Pistol and Rifle, net Amer. list
Loaded and empty Shells, American 20 per cent. discount. Rival and Nitro, 10 per cent. advance.
Empty paper shells American, 10 per cent. advance.
Primers, American $2.05

Wads. — per lb.

Best thick brown or grey felt wads, in ½-lb. bags $0 70
Best thick white card wads, in boxes of 500 each, 12 and smaller gauges 0 29
Best thick white card wads, in boxes of 500 each, 10 gauge............. 0 35
Thin card wads, in boxes of 1,000 each, 12 and smaller gauges 0 20
Thin card wads, in boxes of 1,000 each, 10 gauge........................ 0 25
Chemically prepared pink edge grey cloth wads, in boxes of 250 each— Per M.
11 and smaller gauge 0 60
9 and 10 gauges................... 0 70
7 and 8 " 0 90
5 and 6 " 1 10
Superior chemically prepared pink edge, best white cloth wads in boxes of 250 each—
11 and smaller gauge 1 15
9 and 10 gauges 1 40
7 and 8 " 1 65
5 and 6 " 1 90

SHOT.

Ordinary drop shot, AAA to dust $7.50 per 100 lbs. Discount 5 per cent; cash discount 2 per cent. 30 days; net extras as follows subject to cash discount only; Chilled, 40 c.; buck and seal, 80c.; no. 28 ball, $1 20 per 100 lbs.; bags less than 25 lbs., ½c per lb; F.O.B. Montreal, Toronto, Hamilton, London, St. John and Halifax, and freight equalized thereon.

TRAPS (steel.)

Game, Newhouse, discount 30 and 10 per cent.
Game, Hawley & Norton, 50, 10 & 5 per cent.
Game, Victor, 70 per cent.
Game, Oneida Jump (B. & L.) 40 & 2½ p. c.
Game, steel, 60 and 5 per cent.

SKATES.

Skates, discount 37½ per cent.
Mic Mac hockey sticks, per doz 4 00 5 00

CUTLERY AND SILVERWARE.

RAZORS.

	per doz.	
Elliot's	4 00	18 00
Boker's	7 50	11 00
" King Cutter	13 50	18 50
Wade & Butcher's	3 60	10 00
Lewis Bros.' "Klean Kutter"	8 50	10 50
Henckel's	7 50	20 00
Berg's	7 50	20 00

Clauss Razors and Strops, 50 and 10 per cent

KNIVES.

Farriers-Stacey Bros., doz 3 50

PLATED GOODS

Hollowware, 40 per cent. discount.
Flatware, staples, 40 and 10, fancy, 40 and 5.

SHEARS.

Clauss, nickel, discount 60 per cent.
Clauss, Japan, discount 67½ per cent.
Clauss, tailors, discount 40 per cent.
Seymour's, discount 50 and 10 per cent.
Berg's 6 00 12 00

HOUSE FURNISHINGS.

APPLE PARERS.

Woodyatt Hudson, per doz., net 4 50

BIRD CAGES.

Brass and Japanned, 40 and 10 p. c.

COPPER AND NICKEL WARE.

Copper boilers, kettles, teapots, etc. 30 p.c.
Copper pitts, 20 per cent.

ENAMELED WARE.

London, White, Princess, Turquoise, Onyx, Blue and White, discount 50 per cent.
Canada, Diamond, Premier, 50 and 10 p c.
Pearl, Imperial Crescent, 50 and 10 per cent.
Premier steel ware, 40 per cent.
Star decorated steel and white, 25 per cent.
Japanned ware, discount 45 per cent.
Hollow ware, tinned cast, 35 per cent. off.

KITCHEN SUNDRIES.

Can openers, per doz	0 40	0 75
Mincing knives per doz	0 50	0 80
Duplex mouse traps, per doz	....	0 65
Potato mashers, wire, per doz	0 60	0 70
" wood "	0 50	0 60
Vegetable slicers, per doz	....	2 25

Universal meat chopper No. 0, $1; No. 1, 1 15.

Enterprise chopper, each	...	1 30

Spiders and fry pans, 50 per cent.

Star Al chopper 5 to 32	1 35	4 10
" " 100 to 103	1 35	2 00
Kitchen hooks, bright	....	0 62½

LAMP WICKS.

Discount, 60 per cent.

LEMON SQUEEZERS.

Porcelain lined	per doz.	2.20	5 60
Galvanized	"	1 87	3 85
King, wood	"	2 75	2 90
King, glass	"	4 00	4 50
All glass	"	0 50	0 90

METAL POLISH.

Tandem metal polish paste 6 00

PICTURE NAILS.

Porcelain head ......per gross	1 35	1 50
Brass head	0 40	1 00

Tin and gilt, picture wire, 75 per cent.

SAD IRONS.

Mrs. Potts, No. 55, polished....per set	0 80
" No. 50, nickle-plated, "	0 91
Common, plain	4 50
" plated	5 50
Asbestos, per set	1 25

TINWARE.

CONDUCTOR PIPE.

2-in. plain or corrugated., per 100 feet, $3.30; 3 in., $4.40; 4 in., $5.80; 5 in., $7.45; 6 in., $9.90.

FAUCETS.

Common, cork-lined, discount 35 per cent.

EAVETROUGHS.

10-inch ....per 100 ft. 3 30

FACTORY MILK CANS.

Discount off revised list, 35 per cent
Milk can trimmings, discount 25 per cent.
Creamery Cans, 45 per cent.

LANTERNS.

No. 2 or 4 Plain Cold Blast	per doz.	6 50
Lift Tubular and Hinge Plain,	"	4 75
No. 0, safety	"	4 00

Better quality at higher prices.
Japanning, 50c. per doz. extra.
Prism globes, per doz, $1 20.

OILERS.

Kemp's Tornado and McClary's Model galvanized oil can, with pump, 5 gallon, per dozen 10 92
Davidson oilers, discount 40 per cent
Zinc and tin, discount 50 per cent.
Coppered oilers, 20 per cent. off.
Brass oilers, 50 per cent. off.
Malleable, discount 25 per cent

PAILS (GALVANIZED).

Dufferin pattern pails, 45 per cent.
Flaring pattern, discount 45 per cent.
Galvanized washtubs 40 per cent

PIECED WARE.

Discount 35 per cent off list, June, 1899.
10-qt. flaring sap buckets, discount 35 per cent.
6, 10 and 14-qt. flaring pails dis. 35 per cent.
Copper bottom tea kettles and boilers, 30 p.c.
Coal hods, 40 per cent.

STAMPED WARE.

Plain, 75 and 12½ per cent. off revised list.
Retinned, 72½ per cent revised list.

SAP SPOUTS.

Bronzed iron with hooks ....per 1,000	7 50
Eureka tinned steel, hooks	8 00

STOVEPIPES.

5 and 6 inch, per 100 lengths	7 64	7 91
7 inch " "	....	8 18

Nestable, discount 40 per cent.

STOVEPIPE ELBOWS

5 and 6-inch, common ....per doz.	1 32
7-inch "	1 48

Polished, 15c. per dozen extra.

THERMOMETERS.

Tin case and dairy, 75 to 75 and 10 per cent.

TINNERS' SNIPS.

Per doz 3 00 15
Clauss, discount 35 per cent.

TINNERS' TRIMMINGS

Discount, 45 per cent.

WIRE.

ANNEALED CUT HAY BAILING WIRE.

No 12 and 13, $4; No. 13½, $4 10; No. 14, $4.2; No. 15 $4 50; in lengths 6' to 11', 25 per cent.; other lengths 20c. per 100 lbs extra; if eye or loop on end add 25c. per 100 lbs. to the above.

BRIGHT WIRE GOODS

Discount 62½ per cent.

CLOTHES LINE WIRE.

7 wire solid line, No. 17, $4.90; No. 18, $3.00; No. 19, $2.70; 6 wire solid line, No. 17, $4.45; No. 18, $2.80, No. 19, $2.50. All prices per 1000 ft. measure. F.o.b. Hamilton Toronto, Montreal.

COILED SPRING WIRE.

High Carbon, No. 9, $2.90, No. 11, $3.45; No. 12, $3 15.

COPPER AND BRASS WIRE.

Discount 37½ per cent.

FINE STEEL WIRE.

Discount 25 per cent. List of extras: In 100-lb. lots: No. 17, $5 — No. 18, $5.50 — No. 19, $6 — No. 20, $6.65 — No. 21, $7 — No. 22, $7.30 — No 23, $7.65 — No. 24, $8 — No. 25, $9 — No. 26, $9.50 — No. 27, $10 — No. 28, $11 — No. 29, $12 — No. 30, $13 — No. 31, $14 — No. 32, $15 — No. 33, $16 — No. 34, $17. Extras net — tinned wire, Nos. 17-25, $2 — Nos. 26-31, $4 — Nos. 32-34, $6. Coppered, 75c. — oiling, 10c. — in 25-lb. bundles, 15c. — in 5 and 10-lb. bundles, 25c. — in 1-lb. hanks, 25c. — in ½-lb. hanks, 38c. — in ¼-lb. hanks, 50c. — packed in casks or cases, 15c. — bagging or papering, 10c.

FENCE STAPLES.

Bright .... 2 75 Galvanized 3 15

HAY WIRE IN COILS.

No. 13, $2.60; No. 14, $2.70; No. 15, $2.85; f.o.b., Montreal.

GALVANIZED WIRE.

Per 100 lb. — Nos. 4 and 5, $3.70 — Nos. 6, 7, 8, $3.15 — No. 9, $2.50 — No. 10, $3 20 — No. 11, $3 25 — No. 12, $2.65 — No. 13, $2.75 — No. 14, $3.75 — No 15, $4.30 — No. 16, $4.30 from stock. Base sizes, Nos. 6 to 9, $2.35 f.o.b. Cleveland. In carlots 12½c. less.

LIGHT STRAIGHTENED WIRE.

Gauge No.	Over 20 in. per 100 lbs.	10 to 20 in.	5 to 10 in.
0 to 5	$0.50	$0 75	$1.25
6 to 9	0.75	1.25	2 00
10 to 11	1.00	1 75	2.50
12 to 14	1 50	2.25	3.50
15 to 16	2.00	3.00	4.50

SMOOTH STEEL WIRE.

No. 0-9 gauge, $2 30; No. 10 gauge, 6c. extra; No. 11 gauge, 12c extra; No. 12 gauge, 20c. extra; No. 13 gauge, 30c. extra; No 14 gauge 40c extra; No. 15 gauge, 55c. extra; No 16 gauge, 70c. extra. Add 60c. for coppering and $2 for tinning.

Extra net per 100 lb. — Oiled wire 10c., spring wire $1 25, bright soft drawn 15c., charcoal (extra quality) $1.25, packed in casks or cases 15c., bagging and papering 10c., 50 and 100-lb. bundles 10c., in 25-lb. bundles 15c., in 5 and 10-lb. bundles 25c., in 1-lb hanks, 50c., in ½-lb. hanks 75c., in ¼-lb. hanks $1.

POULTRY NETTING.

2 in mesh 19 w. g., discount 50 and 10 per cent. All others 50 per cent.

WIRE CLOTH.

Painted Screen, in 100-ft rolls, $1.62½, per 100 sq. ft.; in 50-ft. rolls $1.67½, per 100 sq. ft. Terms, 2 per cent. off 30 days.

WIRE FENCING.

Galvanized barb	2 95
Galvanized, plain twist	3 30

Galvanized barb, f.o.b. Cleveland, $2.70 for small lots and $2.60 for carlots.

WOODENWARE.

CHURNS.

No. 0, $9; No. 1, $9; No. 2, $10; No. 3, $11; No. 4, $13; No. 5, $16; f.o.b. Toronto Hamilton, London and St. Marys, 30 and 30 per cent; f.o.b. Ottawa, Kingston and Montreal, 40 and 15 per cent discount. Taylor-Forbes, 30 and 30 per cent.

CLOTHES REELS.

Davis Clothes Reels, dis. 40 per cent.

LADDERS, EXTENSION.

Waggoner Extension Ladders, dis. 40 per cent.

MOPS AND IRONING BOARDS.

"Best" mops		1 25
"900" mops		1 25
Folding ironing boards	12 00	16 50

REFRIGERATORS

Discount, 40 per cent.

SCREEN DOORS.

Common doors, 2 or 3 panel, walnut stained, 4-in. style ....per doz.	7 25
Common doors, 2 or 3 panel, grained only, 4-in., style ....per doz.	7 55
Common doors, 2 or 3 panel, light stair per doz.	9 55

WASHING MACHINES.

Round, re-acting per doz.	60 00
Square "	63 00
Eclipse, per doz	54 00
Dowswell "	39 00
New Century, per doz	75 00
Daisy	54 00

WRINGERS.

Royal Canadian, 11 in., per doz.	34 00
Royal American, 11 in.	34 00
Eze, 10 in., per doz	37 75

Terms, 2 per cent., 30 days.

MISCELLANEOUS.

AXLE GREASE.

Ordinary, per gross	6 00	7 00
Best quality	10 00	12 00

BELTING.

Extra, 60 per cent.
Standard, 60 and 10 per cent.
No. 1, not wider than 6 in., 60, 10 and 10 p.c.
Agricultural, not wider than 4 in., 75 per cent
Lace leather, per side, 75c.; cut laces, 80c.

BOOT CALKS.

Small and medium, ball	per M	4 25
Small heel	"	4 50

CARPET STRETCHERS.

American	per doz.	1 00	1 50
Bullard's	"	....	6 50

CASTORS.

Bed, new list, discount 55 to 57½ per cent.
Plate, discount 52½ to 57½ per cent.

PINE TAR.

½ pint in tins	per gross	...	7 80
1 " "	"	....	9 60

PULLEYS.

Hothouse	per doz.	0 55	1 00
Axle	"	0 22	0 33
Screw	"	0 22	1 00
Awning	"	0 35	2 50

PUMPS.

Canadian cistern	1 40	2 00
Canadian pitcher spout	1 80	3 16

Berg's wing pump, 75 per cent.

ROPE AND TWINE.

Sisal		0 10½
Pure Manilla		0 15½
"British" Manilla		0 12
Cotton, 3-16 inch and larger	0 21	0 23
" 5-32 inch	0 25	0 27
" ⅛ inch	0 25	0 28
Russia Deep Sea		0 16
Jute		0 09
Lath Yarn, single		0 10
" double		0 10½
Sisal bed cord, 48 feet ....per doz.		0 65
" " 60 feet "		0 80
" " 72 feet "		0 95

Twine.

Bag, Russian twine, per lb.	....	0 27
Wrapping, cotton, 3-ply	....	0 25
" " 4-ply	....	0 29
Mattress twine per lb.	0 33	0 45
Staging "	0 27	0 35

SCALES.

Gurney Standard, 40 per cent.
Gurney Champion, 50 per cent.
Burrow, Stewart & Milne—
Imperial Standard, discount 40 per cent.
Weigh Beams, discount 40 per cent.
Champion Scales, discount 50 per cent.
Fairbanks standard, discount 35 per cent.
" Dominion, discount 55 per cent.
" Richelieu, discount 55 per cent.
Warren new Standard, discount 40 per cent.
" " Champion, discount 50 per cent.
" Weighbeams, discount 35 per cent.

STONES—OIL AND SCYTHE.

Washita	per lb.	0 25	0 37
Hindostan	"	0 06	0 10
" slip	"	0 18	0 20
" Axe	"	....	0 10
Deer Creek	"	....	0 10
Deerlick	"	....	0 25
" Axe	"	....	0 15
Lily white	"	....	0 42
Arkansas	"	....	1 50
Water-of-Ayr	"	....	0 10
Scythe	per gross	3 50	5 00
Grind, 40 to 200 lb., per ton		20 00	22 00
" under 40 lb., "		....	24 00
" 200 lb. and over		....	20 00

INDEX TO ADVERTISERS.

CLASSIFIED LIST OF ADVERTISEMENTS.

Auditors.
Davenport, Pickup & Co., Winnipeg.

Babbitt Metal.
Canada Metal Co, Toronto.
Canadian Fairbanks Co., Montreal.
Robertson, Jas. Co., Montreal.

Bath Room Fittings.
Carriage Mounting Co., Toronto.

Belting, Hose, etc.
Gutta Percha and Rubber Mfg. Co. Toronto.

Bicycles and Accessories.
Johnson's Iver, Arms and Cycle Works Fitchburg, Mass.

Binder Twine.
Consumers Cordage Co, Montreal.

Box Strap.
J. N. Warminton, Montreal.

Brass Goods.
Glauber Brass Mfg Co, Cleveland, Ohio.
Kerr Engine Co., Walkerville, Ont.
Lewis, Rice, & Son., Toronto.
Morrison, Jas., Brass Mfg. Co., Toronto.
Mueller Mfg. Co, Decatur, Ill.
Penberthy Injector Co., Windsor, Ont.
Taylor-Forbes Co., Guelph, Ont.

Bronze Powders.
Canadian Bronze Powder Works, Montreal.

Brushes.
Ramsay, A., & Son Co., Montreal.

Can Openers.
Cumming Mfg. Co. Renfrew.

Cans.
Acme Can Works, Montreal.

Builders' Tools and Supplies.
Covert Mfg. Co., West Troy, N.Y.
Frothingham & Workman Co., Montreal.
Howland, H. S., Sons & Co., Toronto.
Hyde, F., & Co., Montreal.
Lewis Bros. & Co., Montreal.
Lewis, Rice, & Son, Toronto.
Lockerby & McComb, Montreal.
Lufkin Rule Co., Saginaw, Mich.
Newman & Sons, Birmingham.
North Bros. Mfg. Co., Philadelphia, Pa.
Stanley Rule & Level Co., New Britain.
Stanley Works, New Britain, Conn.
Stephens, G. F., Winnipeg.
Taylor-Forbes Co., Guelph, Ont.

Carriage Accessories.
Carriage Mountings Co, Toronto.
Covert Mfg. Co., West Troy, N.Y.

Carriage Springs and Axles.
Guelph Spring and Axle Co., Guelph.

Cattle and Trace Chains.
Greening, B., Wire Co., Hamilton.

Churns.
Dowswell Mfg. Co., Hamilton.

Clippers—All Kinds.
American Shearer Mfg. Co., Nashua, N.H.

Clothes Reels and Lines.
Hamilton Cotton Co., Hamilton, Ont.

Cordage.
Consumers' Cordage Co., Montreal.
Hamilton Cotton Co., Hamilton.

Cork Screws.
Erie Specialty Co., Erie, Pa.

Clutch Nails.
J. N. Warminton, Montreal.

Cut Glass.
Phillips, Geo., & Co., Montreal.

Cutlery—Razors, Scissors, etc.
Birkett, Thos., & Son Co., Ottawa.
Clauss Shear Co., Toronto.
Dorken Bros. & Co., Montreal.
Heinisch's, R., Sons Co., Newark, N.J.
Howland, H. S. Sons & Co., Toronto.
Hutton, Wm., & Sons, Ltd., London, Eng.
Lamplough, F. W., & Co., Montreal.
Phillips, Geo. & Co, Montreal.
Round, John, & Son, Montreal.

Electric Fixtures.
Canadian General Electric Co, Toronto.
Forman, John, Montreal.
Morrison James, Mfg. Co., Toronto.
Munderloh & Co., Montreal.

Electro Cabinets.
Cameron & Campbell Toronto.

Engines, Supplies, etc.
Kerr Engine Co., Walkerville, Ont.

Fencing—Woven Wire.
McGregor-Banwell Fence Co., Walkerville, Ont.
Owen Sound Wire Fence Co, Owen Sound.
Banwell Hoxie Wire Fence Co, Hamilton.

Files and Rasps.
Barnett Co., G. & H., Philadelphia, Pa.
Nicholson File Co., Port Hope.

Financial Institutions.
Bradstreet Co.

Firearms and Ammunition.
Dominion Cartridge Co., Montreal.
Hamilton Rifle Co, Plymouth, Mich.
Harrington & Richardson Arms Co., Worcester, Mass.
Johnson's, Iver, Arms and Cycle Works, Fitchburg, Mass.

Food Choppers.
Enterprise Mfg Co., Philadelphia, Pa.
Lamplough, F. W., & Co., Montreal.
Shirreff Mfg. Co., Brockville, Ont.

Galvanizing.
Canada Metal Co., Toronto.
Montreal Rolling Mills Co., Montreal.
Ontario Wind Engine & Pump Co., Toronto.

Glaziers' Diamonds.
Gibsone, J. B., Montreal.
Pelton, Godfrey S.
Sharratt & Newth, London, Eng.
Shaw, A., & Son, London, Eng.

Handles.
Still, J. H., Mfg. Co.

Harvest Tools.
Maple Leaf Harvest Tool Co, Tillsonburg, Ont.

Hoop Iron.
Montreal Rolling Mills Co., Montreal.
J. N. Warminton, Montreal.

Horse Blankets.
Heney, E. N., & Co., Montreal.

Horseshoes and Nails.
Canada Horse Nail Co., Montreal.
Montreal Rolling Mills, Montreal.
Capewell Horse Nail Co., Toronto.

Hot Water Boilers and Radiators.
Cluff, R. J., & Co Toronto.
Pease Foundry Co, Toronto.
Taylor-Forbes Co., Guelph.

Ice Cream Freezers.
Dana Mfg. Co., Cincinnati, Ohio.
North Bros. Mfg. Co., Philadelphia, Pa.

Ice Cutting Tools.
Erie Specialty Co., Erie, Pa.
North Bros. Mfg. Co., Philadelphia, Pa.

Injectors—Automatic.
Morrison, Jas., Brass Mfg. Co., Toronto.
Penberthy Injector Co., Windsor, Ont.

Iron Pipe.
Montreal Rolling Mills, Montreal.

Iron Pumps.
Lamplough, F. W., & Co, Montreal.
McDougall, R., Co., Galt, Ont.

Lanterns.
Kemp Mfg. Co., Toronto.
Ontario Lantern Co., Hamilton, Ont.
Wright, E. T., & Co., Hamilton.

Lawn Mowers.
Birkett, Thos., & Son Co., Ottawa.
Maxwell, D., & Sons, St. Mary's, Ont.
Taylor, Forbes Co., Guelph.

Lawn Swings, Settees, Chairs.
Cumming Mfg. Co., Renfrew.

Ledgers—Loose Leaf.
Business Systems Toronto.
Copeland-Chatterson Co, Toronto.
Crain, Rolla L., Co., Ottawa.
Universal Systems, Toronto.

Locks, Knobs, Escutcheons, etc.
Peterborough Lock Mfg. Co, Peterborough, Ont.

Lumbermen's Supplies.
Pink, Thos., & Co., Pembroke Ont.

Manufacturers' Agents.
Fox, C. H., Vancouver.
Gibb, Alexander, Montreal.
Mitchell, David C., & Co, Glasgow, Scot.
Mitchell, H. W., Winnipeg.
Pearce, Frank, & Co., Liverpool, Eng.
Scott, Bathgate & Co., Winnipeg.
Thorne, R. E., Montreal and Toronto.

Metals.
Canada Iron Furnace Co., Midland, Ont.
Canada Metal Co., Toronto.
Eadie, H. G., Montreal.
Gibb, Alexander, Montreal.
Kemp Mfg Co, Toronto
Leslie, A. C, & Co., Montreal.
Lysaght, John, Bristol, Eng
Nova Scotia Steel and Coal Co., New Glasgow, N S.
Robertson, Jas., Co., Montreal.
Roper, J. H., Montreal.
Samuel, Benjamin & Co, Toronto.
Stairs, Son & Morrow, Halifax, N.S.
Thompson, B. & S. H. & Co. Montreal.

Metal Lath.
Galt Art Metal Co., Galt.
Metallic Roofing Co., Toronto.
Metal Shingle & Siding Co., Preston, Ont.

Metal Polish, Emery Cloth, etc
Oakey, John, & Sons, London, Eng.

Mops.
Cumming Mfg. Co., Refrew.

Mouse Traps.
Cumming Mfg Co., Renfrew.

Oil Tanks.
Bowser, S F., & Co., Toronto.

Paints, Oils, Varnishes, Glass.
Bell, Thos, Sons & Co, Montreal.
Canada Paint Co., Montreal.
Canadian Oil Co., Toronto.
Consolidated Plate Glass Co., Toronto.
Fenner, Fred, & Co., London, Eng.
Henderson & Potts Co., Montreal.
Imperial Varnish and Color Co., Toronto.
Jamieson, R. C, & Co, Montreal.
McArthur, Corneille & Co., Montreal.
McCaskill, Dougall & Co, Montreal.
Montreal Rolling Mills Co, Montreal.
Moore, Benjamin, & Co. Toronto.
Queen City Oil Co, Toronto.
Ramsay & Son, Montreal.
Sherwin-Williams Co., Montreal.
Standard Paint and Varnish Works Windsor, Ont.
Stephens & Co, Winnipeg.
Martin-Senour Co., Chicago.

Perforated Sheet Metals.
Greening, B, Wire Co., Hamilton.

Plate Glass.
Toronto Plate Glass Co, Toronto.

Plumbers' Tools and Supplies.
Borden Co., Warren, Ohio.
Canadian Fairbanks Co., Montreal.
Cluff, R. J., & Co., Toronto.
Glauber Brass Co., Cleveland, Ohio.
Jardine, A. B., & Co, Hespeler, Ont.
Jenkins Bros., Boston, Mass.
Kerr Engine Co, Walkerville, Ont.
Lewis, Rice, & Son, Toronto.
Merrell Mfg. Co., Toledo, Ohio.
Montreal Rolling Mills Montreal.
Morrison, Jas, Brass Mfg. Co., Toronto.
Mueller, H., Mfg. Co., Decatur, Ill.
Oshawa Steam & Gas Fitting Co., Oshawa
Robertson Jas., Co Montreal.
Stairs, Son & Morrow, Halifax, N.S.
Standard Ideal Sanitary Co., Port Hope.
Standard Sanitary Co., Pittsburg.
Stephens, G F, & Co, Winnipeg, Man.
Turner Brass Works, Chicago.
Vickery, Orlando, Toronto.

Portland Cement.
International Portland Cement Co., Ottawa, Ont
Hanover Portland Cement Co., Hanover, Ont.
Hyde, F., & Co., Montreal.
Thompson, B. & S. H. & Co., Montreal.

Potato Mashers.
Cumming Mfg Co., Renfrew.

Poultry Netting.
Greening, B., Wire Co., Hamilton, Ont.

Razors.
Clauss Shear Co., Toronto.

Refrigerators.
Fabien, C. P., Montreal.

Roofing Supplies.
Brantford Roofing Co, Brantford.
McArthur, Alex., & Co., Montreal.
Metal Shingle & Siding Co., Preston, Ont.
Metallic Roofing Co., Toronto.
Paterson Mfg. Co., Toronto & Montreal.

Saws.
Atkins, E. C., & Co., Indianapolis, Ind
Lewis Bros., Montreal.
Shurly & Dietrich, Galt, Ont.
Spear & Jackson, Sheffield, Eng.

Scales.
Canadian Fairbanks Co., Montreal.

Screw Cabinets.
Cameron & Campbell, Toronto.

Screws, Nuts, Bolts.
Montreal Rolling Mills Co., Montreal.
Morrow, John, Machine Screw Co., Ingersoll, Ont.

Sewer Pipes.
Canadian Sewer Pipe Co., Hamilton
Hyde, F., & Co., Montreal.

Shelf Boxes.
Cameron & Campbell, Toronto.

Shears, Scissors.
Clauss Shear Co., Toronto.

Shelf Brackets.
Atlas Mfg. Co., New Haven, Conn

Shellac
Bell, Thos., Sons & Co, Montreal.

Shovels and Spades.
Canadian Shovel Co, Hamilton.
Peterboro Shovel & Tool Co., Peterboro.

Silverware.
Hutton, Wm., & Sons, Ltd, London, Eng.
Phillips, Geo, & Co., Montreal.
Round, John, & Son, Sheffield, Eng.

Skates.
Canada Cycle & Motor Co., Toronto.

Spring Hinges, etc.
Chicago Spring Butt Co., Chicago, Ill.

Steel Rails.
Nova Scotia Steel & Coal Co., New Glasgow, N.S.

Stoves, Tinware, Furnaces
Canadian Heating & Ventilating Co., Owen Sound.
Canada Stove Works, Harriston, Ont.
Clare Bros. & Co., Preston.
Davidson, Thos., Mfg. Co., Montreal.
Guelph Stove Co., Guelph.
Gurney Foundry Co., Toronto.
Harris, J. W., Co., Montreal.
Joy Mfg. Co., Toronto.
Kemp Mnfg. Co. Toronto.
McClary Mfg. Co. London.
Pease Foundry Co., Toronto.
Stewart, Jas., Mfg. Co., Woodstock, Ont.
Taylor-Forbes Co., Guelph, Ont.
Wright, E. T., & Co., Hamilton.

Tacks.
Montreal Rolling Mills Co., Montreal.
Ontario Tack Co., Hamilton.

Ventilators.
Harris, J. W., Co., Montreal.
Pearson, Geo. D., Montreal.

Washing Machines, etc.
Dowswell Mfg. Co., Hamilton, Ont.
Taylor-Forbes Co., Guelph, Ont.

Wheelbarrows.
London Foundry Co., London, Ont.

Wholesale Hardware.
Birkett, Thos., & Sons Co., Ottawa.
Caverhill, Learmont & Co., Montreal.
Frothingham & Workman, Montreal.
Hobbs Hardware Co., London.
Howland, H. S., Sons & Co., Toronto.
Lamplough, F. W., & Co., Montreal.
Lewis Bros. & Co., Montreal.
Lewis, Rice, & Son, Toronto.

Window and Sidewalk Prisms.
Hobbs Mfg. Co., London, Ont.

Wire Springs.
Guelph Spring Axle Co., Guelph, Ont.
Wallace-Barnes Co., Bristol, Conn.

Wire, Wire Rope, Cow Ties, Fencing Tools, etc.
Canada Fence Co., London.
Dennis Wire and Iron Co., London, Ont.
Dominion Wire Mnfg. Co., Montreal
Greening, B., Wire Co., Hamilton.
Montreal Rolling Mills Co., Montreal.
Western Wire & Nail Co., London, Ont

Woodenware.
Taylor-Forbes Co., Guelph, Ont.

Wrapping Papers.
Canada Paper Co., Toronto.
McArthur, Alex., & Co, Montreal.
Stairs, Son & Morrow, Halifax, N.S.

: : It Saves Lives : :

A Building Necessity

These are the names that have been made synonymous with the

AEOLIAN VENTILATOR

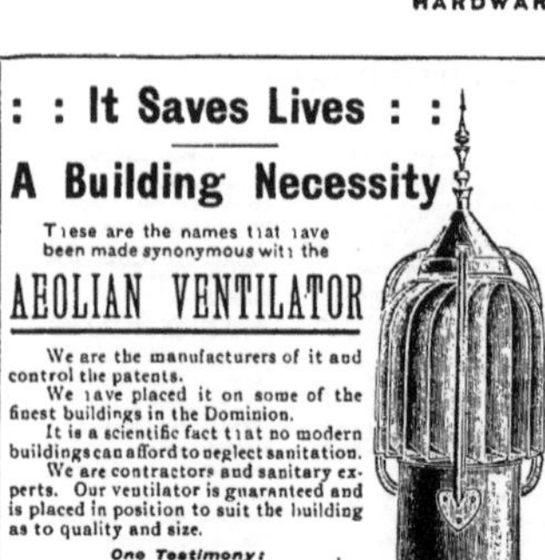

We are the manufacturers of it and control the patents.

We have placed it on some of the finest buildings in the Dominion.

It is a scientific fact that no modern buildings can afford to neglect sanitation.

We are contractors and sanitary experts. Our ventilator is guaranteed and is placed in position to suit the building as to quality and size.

One Testimony:

Montreal, June 12th, 1903

Canadian Pacific Railway Company.

Messrs. J. W. Harris Co., Limited,
7 St. Elizabeth St., Montreal, Que.

Gentlemen,—Answering yours of June 1st, I am not personally familiar with the merits of the "Aeolian" Ventilators but our Assistant Engineer, Mr. F. Crossley, reports that the Ventilators of this type manufactured by you and used by this Company at the Place Viger Hotel, Chateau Frontenac, Windsor Street Station and elsewhere, have proved satisfactory. Yours truly,

E. H. McHENRY, Chief Engineer.

The J. W. HARRIS CO., Limited
Successors to LESSARD & HARRIS
General Contractors and Manufacturers
MONTREAL

D. C. Mitchell & Co.

GLASGOW

STEEL PLATES,
SHEETS, ANGLES,
BARS,
PIG IRON,

REPRESENTATIVE

Eadie-Douglas Company
22 St. John St., MONTREAL

SUPPORT CANADIAN INDUSTRIES

FREE
TO
HARDWARE DEALERS

We are supplying electros of the cut in this ad. for use in your local advertising, besides offering special terms for immediate orders for

"LION" Brand
PUTZ TRIPLE EXTRACT
METAL POLISH
IN LIQUID OR PASTE

The most reliable metal polish in the market. Once tried always used.

Support Canadian Industries

Manufactured by

English Embrocation Co., Montreal

Specify the Following Products

Manufactured by

American Sheet and Tin Plate Company
PITTSBURG, PA.

If you want the best.

"Apollo Best Bloom," Galvanized Sheets,
Blue Annealed and American Bessemer Black Sheets,
Wellsville Polish, Range Steel,
Wood's "A" Planished Iron,
Black Steel specialties for all purposes,
Tinplates, Coke and Charcoal, all grades,
Terne Plates, Old Style and New Method Processes.

Place your order with

B. & S. H. THOMPSON & CO.
LIMITED

Dominion of Canada Sales Agents
MONTREAL, CANADA

Awarded
By **JURORS** at
International Expositions
Special Prize
Gold Medal at Atlanta, 1895

Copy of catalogue sent free to any interested file user upon application.

"Redstone"

High Pressure

Sheet Packing

A packing that will hold. For use in highest pressures for steam, hot or cold water and air. Packs equally well for all.

From actual tests, we believe that this packing is the most durable and satisfactory of any on the market. Try a sample lot and see for yourself.

Manufactured Solely by

THE GUTTA PERCHA & RUBBER MFG. CO.
of TORONTO, LIMITED

HEAD OFFICES,
47 Yonge Street, Toronto.
Branches: Montreal, Winnipeg, Vancouver.

Burrow, Stewart & Milne(a)

SHEET ZINC,

SHEET COPPER,

SPELTER,

INGOT TIN,

BRASS and COPPER TUBES.

For lowest prices send your enquiries to

B. & S. H. THOMPSON & Co.
LIMITED

53 ST. SULPICE STREET, MONTREAL

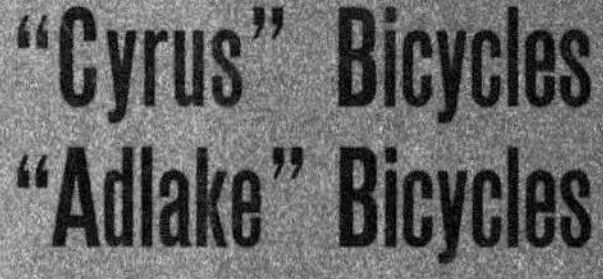

Two wheels that sell readily on their merits

Every cent of their price is represented in the wheel.

Write for our catalogue. Get our trade price on these lines.

JOHN MILLEN & SON,
Limited

321 St. James St. MONTREAL	132 Bay Street TORONTO

RDWARE
& METAL

CANADA

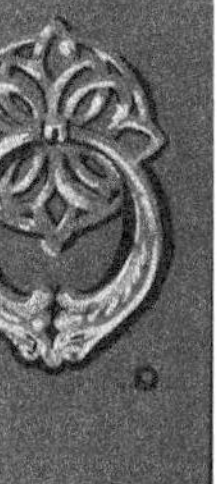

THE ONLY PAPER
PUBLISHED IN CANADA
DEVOTED EXCLUSIVELY
TO THE HARDWARE METAL &
PLUMBING TRADES.

MADE IN CANADA

No. 1 Pocket Flash Light.

Bell Battery

Vest Pocket Flash Light.

Our Formula of Success

Put a Heart in Your Engine

No. 6
2½ x 6
22 Amperes

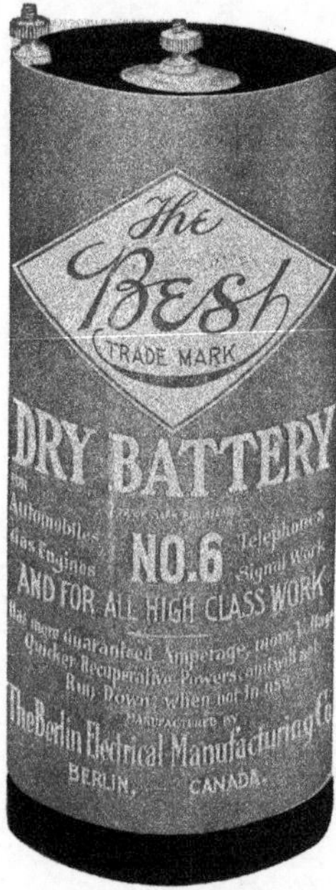

The Motor Trades Best Friend

No. 7
3 x 7
27 Amperes

Dry Battery users have focused prejudice and criticism on "Best" Batteries this past year. Now it is their Standard. Quality is bound to win out, always.

HARDWARE AND METAL

A Weekly Newspaper Devoted to the Hardware, Metal, Heating and Plumbing Trades in Canada.

Office of Publication, 10 Front Street East, Toronto.

VOL. XIX. MONTREAL, TORONTO, WINNIPEG, MARCH 23, 1907 NO. 12.

Hardware Notice

THE year 1907 will be the greatest in the history of Canada, and the live hardware merchant will be the first to feel the benefit, as the contractor and builder get their supplies from him. Our facilities for supplying all kinds of building material and tools are of the very best, and our aim is to supply goods promptly. Our stock will be found most complete in all requirements of the trade. Write for prices.

LIMITED

NT

SHIP CH ND

Wood Pulley BLOCKS

We carry in stock regular inside iron strapped blocks, loose hooks and beckets, with wood or iron sheaves, plain or patent roller bushed, all sizes, from 3 to 10 inch in single or double blocks, also steel tackle blocks for rope from $\frac{5}{8}$ to 2-in. diameter in one, two, or three sheave, or snatch blocks.

Chain Anchors

WHITE LEAD

Rope Oakum

GALVANIZED RIGGING

WRITE FOR PRICES

A Good Hardware Catalogue

should contain

a line of goods that sell, correct information and adequate description.

These objects we have kept in mind in making our new loose-leaf catalogue which we are now sending to our customers.

The goods shown in the catalogue have been chosen for their selling qualities. They are correctly described and sufficient information is given as to material, finish, size, weight and packages to enable you to order without seeing the goods.

They are priced so that you can show this book to your customers without disclosing your cost price.

We show a few leaves from the catalogue on the opposite page.

You ought to get in line for a copy of this catalogue.

F. & W. Hardware Montreal

FROTHINGHAM & WORKMAN, Limited, MONTREAL, CANADA.

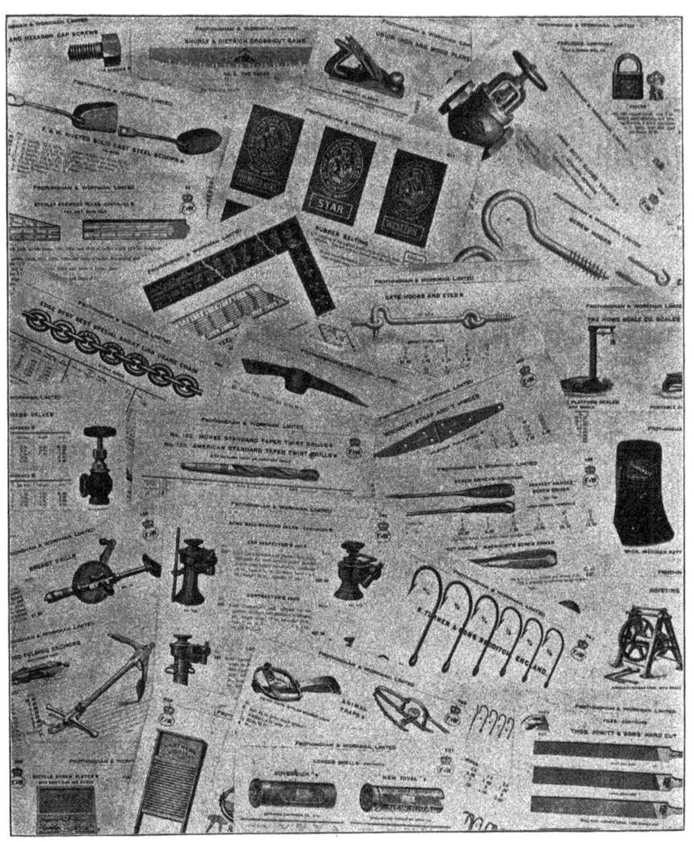

F. & W. Ha ware n rea

F. & W.

FROTHINGHAM & WORKMAN, Limited, MONTREAL, CANADA

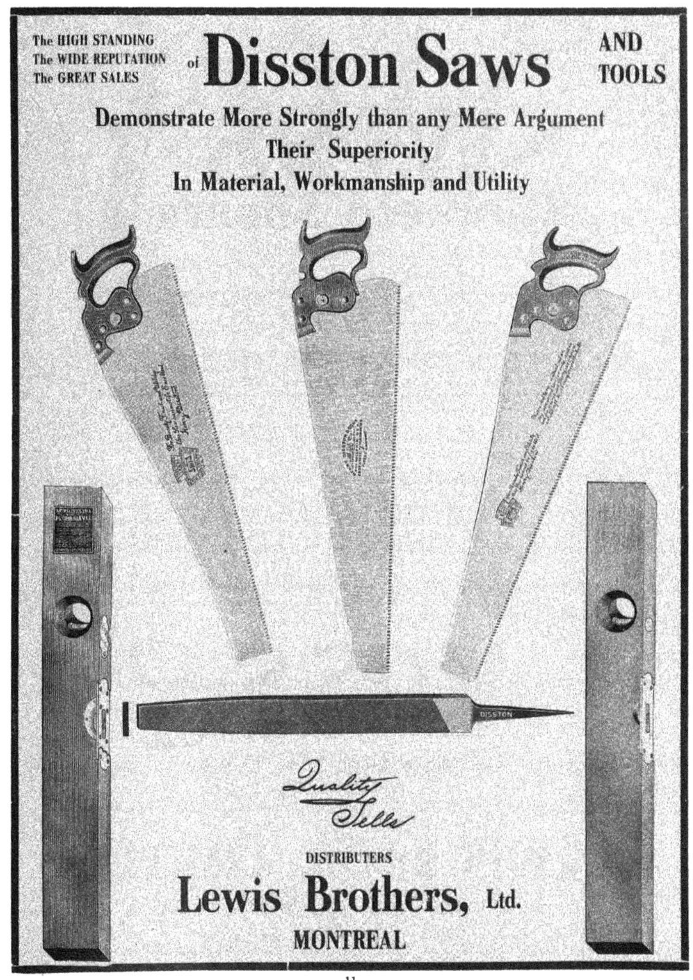
The HIGH STANDING
The WIDE REPUTATION
The GREAT SALES
of Disston Saws
AND TOOLS
Demonstrate More Strongly than any Mere Argument
Their Superiority
In Material, Workmanship and Utility
DISSTON
Quality Tells
DISTRIBUTERS
Lewis Brothers, Ltd.
MONTREAL

"Stielweld" Shears

The exclusive Wiss "Stielweld" process insures a cutting edge that will stay sharp forever, and also a frame proof against breakage.

WHY take all this trouble to make a Wiss Shear blade, when it is possible to make a casting of iron, "convert" it into so-called "steel," give it a coat of enamel or nickel and have it look almost like a genuine Wiss "Stielweld?"

BECAUSE we make Shears not alone for looks, but for perfect and lasting cutting power.

Wiss Shears are adjusted to a hair's breadth. They cut evenly from heel to point of blade, and never pinch or chew the cloth, even after a lifetime of constant use.

If your dealer can't show you the name "Wiss" stamped on the blade, go to another store.

All shears so stamped are *guaranteed* to give satisfaction. If for any reason they fail, your dealer will exchange them, or we will.

OUR BOOK, "POINTED SHARPNESS," MAILED FREE

It illustrates, describes and gives prices of 150 Varieties of shears and scissors, designed for all conceivable uses. Write for it.

J. WISS & SONS CO., 15-33 LITTLETON AVENUE, NEWARK, N. J., U. S. A.

WISS HELPS TO SELL WISS GOODS by enormous advertising through an educational campaign now being read by 25,000,000 people every month. This campaign teaches the public that this brand, whether it is on shears, scissors, razors, tinner snips or pruning shears, means that the article will give absolute satisfaction. No if's or and's about it.

J. WISS & SONS
NEWARK, N.J.
U.S.A.

INCREASE YOUR CUTLERY SALES AND YOUR CUTLERY PROFITS. We wonder sometimes why some dealers bother with the shear business at all, they pay so little attention in selecting the line they are to sell. If you only sell a shear or two every week it is because you have't the right shear or do not go at all in the right way. Without the right quality you will not get the kind of shear and scissors trade that will increase with time—Trade that sticks—Trade that knows your quality is right—Trade that pays a good price for good goods. In other words—Wiss Trade. The Wiss name will get good business for you. The Wiss quality is so far ahead of any other—it is the standard the world over.

OUR SYSTEM OF INSPECTION is such that we are able to guarantee to you that when you sell an article bearing our trade mark you should have no fear of your customer returning it as unsatisfactory. Wiss goods have been sold under this guarantee for 60 years.

OF NEWARK, N.J., U.S.A.

Established 1848.

Send a Postal to

LEWIS BROS., LIMITED

14-28 Bleury Street, MONTREAL.

One of their travelers will show you samples of the entire Wiss line.

The Man Who Writes an Ad.

saying he is a general dealer in all kinds of goods, and then proceeds to ennumerate them, had better retire from business, for he will have no business left if he depends on such advertisements to help him out. It is important that you specialize and attact attention to one line or department.

We attribute no small portion of the success of our business to

Guns and Ammunition

We carry the biggest range, have the most attractive patterns (and prices) of any house in the trade.

These Two Guns

have been winners from the start. There is a good demand for them **NOW**, and a good profit in them for you. **Next year** other jobbers will have them, so will your competitors. We are working on our new Sporting Goods Catalogue for this year. Mail us a card and will enter your name for one just as soon as it is issued.

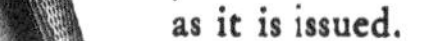

No. 470 x 12 Ga.

No. 469 x 12 Ga.

Goods sold to the trade only---All enquiries received from people who are not Storekeepers or Sporting Goods Dealers are referred to our nearest customer.

LEWIS BROS., Limited

MONTREAL

TORONTO OTTAWA WINNIPEG CALGARY VANCOUVER

UNIVERSITY GROUNDS, TORONTO

H. S. HOWLAND, SONS & CO. LIMITED

HARDWARE MERCHANTS

Only Wholesale — 138-140 WEST FRONT STREET, TORONTO — Wholesale Only

R. WALLACE—SILVER PLATED WARE

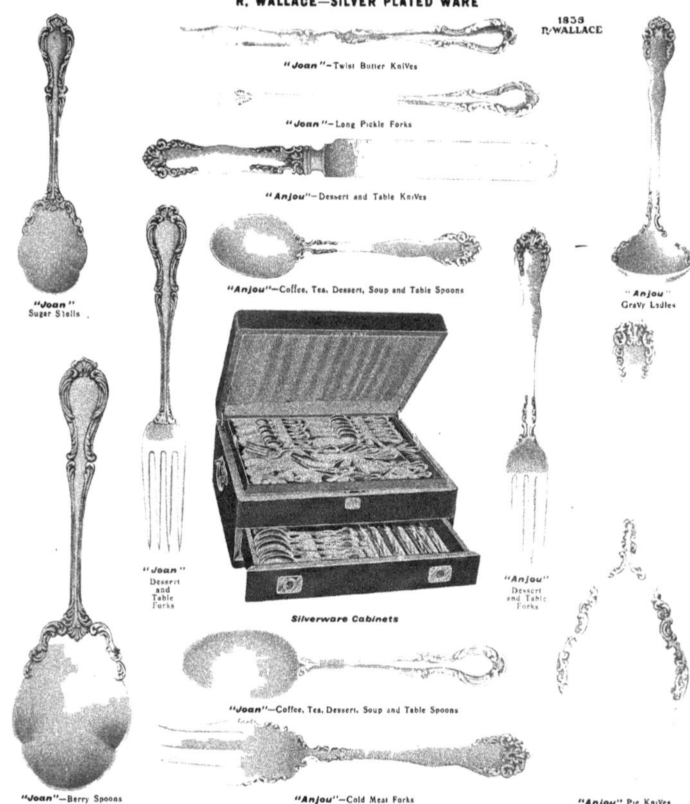

"Joan"—Twist Butter Knives

"Joan"—Long Pickle Forks

"Anjou"—Dessert and Table Knives

"Anjou"—Coffee, Tea, Dessert, Soup and Table Spoons

"Joan" Sugar Shells

"Anjou" Gravy Ladles

"Joan" Dessert and Table Forks

Silverware Cabinets

"Anjou" Dessert and Table Forks

"Joan"—Coffee, Tea, Dessert, Soup and Table Spoons

"Joan"—Berry Spoons

"Anjou"—Cold Meat Forks

"Anjou" Pie Knives

Sterling Silverware quoted on application

H. S. HOWLAND, SONS & CO., LIMITED

We Ship Promptly — Opposite Union Station — Our Prices are Right

The Shirreff Lines Assure the Paying Trade

It will pay every dealer in this Dominion—pay him well—to push the famous dollar-making Shirreff lines, because our goods are reliable, durable and easy-priced. Every article we market has demonstrated it superiority over all competitive lines, and the wise hardwareman will cut out this page for future reference.

Dollars for You

Just figure it out in your own way and if you can come to any other conclusion we would like to know about it.

The "General" Food Chopper is practically a necessity in every kitchen, because particular housewives, knowing its virtues, insist on it. The "General" will last a lifetime. It's a self-cleaner, a self-sharpener, and a money-maker. You are losing money every day you fail to carry a stock of "General" Food Choppers.

The Young Matron's Favorite

It would be practically impossible to sell any other bread-maker to an intelligent woman who knows **The "General" Bread-maker** and its perfections. The "General" is simply an easier, quicker and cleaner way of making bread. It accomplishes as much work in 3 minutes, as far as kneading is concerned, as could be accomplished by hand in half-an-hour. The women of Canada demand the "General" Bread-maker. Can you supply the demand?

The Eureka Wagon Jack

is a new article on the Canadian market, although it has already captured the United States market. It holds whatever it raises—because it has double-acting dogs.

This jack is built of steel and malleable iron. There is nothing to get out of order about the Eureka. Built to lift all kinds of buggies, heavy wagons, and general work.

There you see the ***strongest, cheapest*** and ***best Ladder*** on the market. Big talk, but we stand ready to prove any claim we make for our goods.

There are no nails in this ladder. It is put together with screws and rivets and constructed of thoroughly seasoned ash. Durable, safe, and ***a money-saver.***

We claim that our goods are better than competitive lines, and none the worse because they are "Made in Canada."

Every article we market must give utter and complete satisfaction, because we back its goodness with a money-back guarantee.

Write now for our catalogues and literature.

THE SHIRREFF MANUFACTURING CO., - Brockville, Ont.

No. 1, 2, 3, 4, 4½
No. 17, 19
No. 5, 5½, 6, 7, 8
Think—
Skilled mechanics always ask for "STANLEY'S PLANES."
Is it not far more profitable to sell them the "genuine STANLEY" which gives every satisfaction and holds them, than to sell them an inferior substitute and have them go elsewhere for the next tool they require?
The irons in these good planes are made from the best English steel, of equal thickness and temper throughout.
Each tool passes a rigid examination before leaving the works, and is fully guaranteed. Any tool showing defect in either material or workmanship, if brought to our notice, will be replaced without cost to the owner.
For household and other purposes we have another make of Iron Planes made in all the leading styles and sizes, known as "Hunts" which sells at a low price.
Caverhill, Learmont & Co.
PLANE AND TOOL HEADQUARTERS
Montreal and Winnipeg
He Sells Most
Who Offers
Stanley's Tools
124
Varieties
in Stock to
Select From.
A Plane
for Every Purpose.
No. 113
Corrugated Bottom
Same price as plain
Nos. 26 to 34

anufacturing Co.

Toronto

Highest Grade Enameled Ware

Corner River and Gerrard Streets

Kemp Manufacturing Co.

Toronto

OUR NEW PREMISES IN WINNIPEG
KEMP
KEMP MANUFACTURING & METAL CO., McDermot Ave. E., WINNIPEG
OUR NEW PREMISES IN MONTREAL
THE KEMP M'F'G CO OF MONTREAL
KEMP MANUFACTURING CO. OF MONTREAL 39 St. Antoine St., MONTREAL

The Ladies' Desire

You will find that the average lady knows more about **"Universal Choppers"** than you ever dreamed of.

When buying, she does not merely ask for a "chopper" but emphasizes the **Universal** kind.

WHY?

She has seen the maker's extensive advartising in the journals that enter the home.

Dozens of her friends are now satisfied users of the **Universal,** they recommend it highly and she must have one also.

REMEMBER

The **Universal Chopper** is inexpensive, yet will last a lifetime.

It has four cutters which will chop all kinds of food from very fine to very coarse, as desired. These cutters are so made that any wear tends to keep their edges keen. Whole chopper is easily cleaned, comes apart without trouble of any kind. All parts are neatly tinned.

We carry **Universal Choppers for Butchers' Use** as well as for the home, in all sizes.

No matter what luck you have had with other choppers you will find that handling the **Universal** will bring increased sales, increased sales will mean **more profit for you. Try it.**

CAVERHILL, LEARMONT & CO.

WHOLESALE DISTRIBUTORS

MONTREAL and WINNIPEG

MY BUSINESS

is to talk each week to hardware merchants, plumbers, clerks, manufacturers and travellers. I find out what they want and then I satisfy those wants.

Sometimes I find a man in Halifax who wants to buy a set of tinner's tools and I may find a purchaser for him in Vancouver.

Frequently hardware merchants buy new show cases and then they want to sell the old ones. Well, no person in their own town wants to buy them and they can't afford to start out on the road for they can't travel as cheaply as I can. So I find a purchaser for them, it may be in a nearby town or it may be half way across the continent. My charge is the same in each case.

Hardware merchants are always wanting clerks, and clerks are always wanting new positions, and when you know all these people it isn't so hard to introduce them and let them come to terms if they can.

I had some work last week that kept me pretty busy. A hardware merchant in Eastern Ontario wanted to buy some cut nails, so he told me to go into all the hardware stores in Canada and deliver this message:—

> ANY hardwaremen having stock of cut nails they wish to dispose of at reduced prices should write at once to Box 501 HARDWARE & METAL, Toronto.

Over sixty dealers left their work at once and wrote offering to sell their stock. A number of others seemed to be interested and will probably write later. I charged him half a dollar for my work. Do you think this was too much?

I am always glad to carry any message no matter what it is. You see, I am a sort of Jack-of-all-Trades. I'm an employment bureau and a second-hand store although I sell a good many new articles as well.

I do my work well because I have been going over the ground for over 18 years and know practically every hardware merchant, plumber, stove and tinware dealer in Canada.

Have you a message to be delivered to any of these people?

Do you want it delivered quickly?

Condensed Advertisements in Hardware and Metal cost 2c. per word for first insertion, 1c. per word for subsequent insertions. Box number 5c. extra. Send money with advertisement. Write or phone our nearest office

Hardware and Metal

MONTREAL TORONTO WINNIPEG

It will pay you

We give herewith the experience of one of our subscribers. There is no reason why you cannot obtain just as good results.

THE PEOPLE'S HARDWARE STORE.

HARDWARE AND METAL, Toronto Ont. — Essex, Jan. 27, 1906

Dear Sirs—Within one month I have "bought and sold" with the best of satisfaction—the result of $1.68 worth of condensed ads. in HARDWARE AND METAL.
In the future when I want to know anything in the hardware business I will seek your paper.

Yours truly, IVESON L. COOKE.

Remember your ad. will be read by Hardware Dealers in all parts of Canada.

2 cents per word for first insertion
1 cent " " subsequent insertions

Send cash with advertisement, 5c. additional for Box number

HARDWARE AND METAL
MONTREAL TORONTO WINNIPEG

THE

McGlashan, Clarke Co.,

Limited

Have Expert Workmen,
Latest Machinery,
Practical Experience
and Facilities Unsurpassed

MANUFACTURERS

CUTLERY, ELECTRO PLATE and SOLID NICKEL SILVER FLATWARE

DOMINION WIRE MANUFACTURING CO.

LIMITED

MONTREAL TORONTO

Manufacturers of

Iron and Steel Wire
Barb Wire
Galvanized Wire
Coiled Spring Fencing
Brass Wire
Copper Wire
Wire Nails
Wood Screws
Staples
Jack Chain
Bright Wire Goods
Spring Cotters
Steel Wire Barrel Hoops

WIRE NAILS

FIRST COME FIRST SERVED

You have still time this month to send us your specifications for

WIRE NAILS

If you want prompt deliveries we would suggest

DO IT NOW

N. B. How about— ***FENCE STAPLES***

ESTABLISHED 1847

Fitted Cabinets of all sizes. All Union Made

The most practical cabinet for the Canadian Market

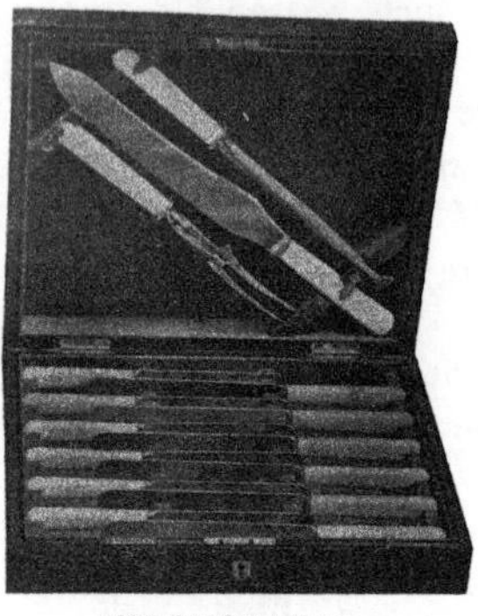

R618—Union Cabinet—15 pieces.

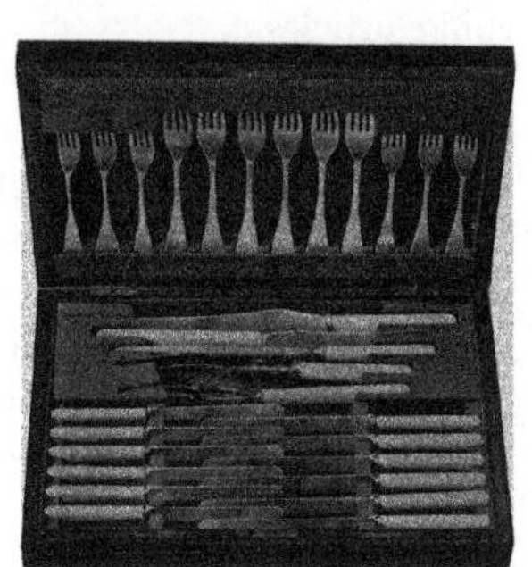

R619—Union Cabinet—29 pieces.

Our Cabinets are all Union-made, and are guaranteed to stand any climate. Fitted with finest quality flatware, and hand forged steel cutlery.

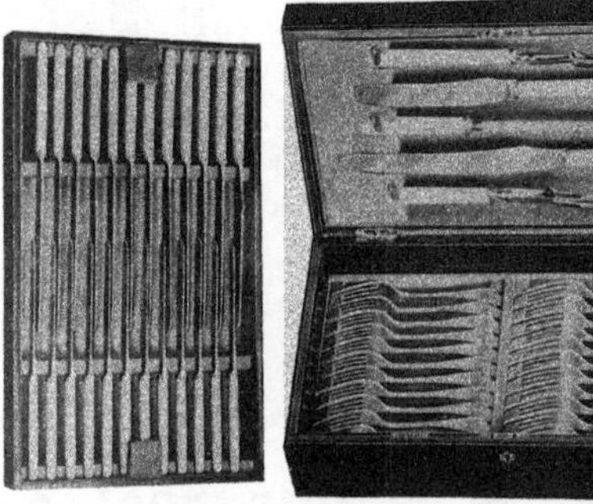

R620—Union Cabinet—53 pieces.

JOHN ROUND AND SON, LIMITED

MONTREAL

WRITE FOR QUOTATIONS.

FACTORIES—SHEFFIELD, ENG.

THE PRECISION OF

STANDARD DRILLS

Were we asked to tell in two words the secret of the great increase in sales of

STANDARD DRILLS it would be

"Precision" and "Quality"

MADE BY MECHANICS ESPECIALLY SKILLED IN THEIR PARTICULAR BRANCHES, AND EXPERT IN THE USE OF PRECISION GAUGES, WITH WHICH ALL MEASUREMENTS IN THE COURSE OF MANUFACTURE ARE MADE.

WITH USING THE HIGHEST GRADE OF MATERIAL AND MODERN MACHINERY AND APPLIANCES THE BEST TO BE HAD IN DRILLS IS THE RESULT.

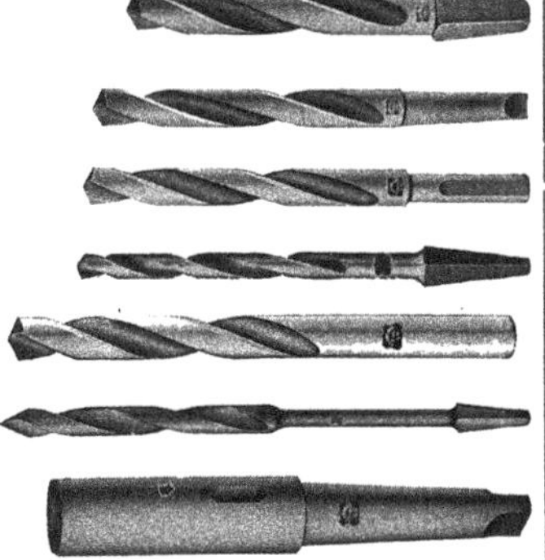

Is it any wonder then that STANDARD DRILLS are now used by the leading

RAILWAYS, CAR MAKERS and LARGE MACHINE SHOPS

When you need Drills let us hear from you; we will gladly co-operate with you with a view to furnishing the best tools adapted to any particular purpose.

Standard Catalogue sent to dealers on request

CAVERHILL, LEARMONT & CO.

Montreal and Winnipeg

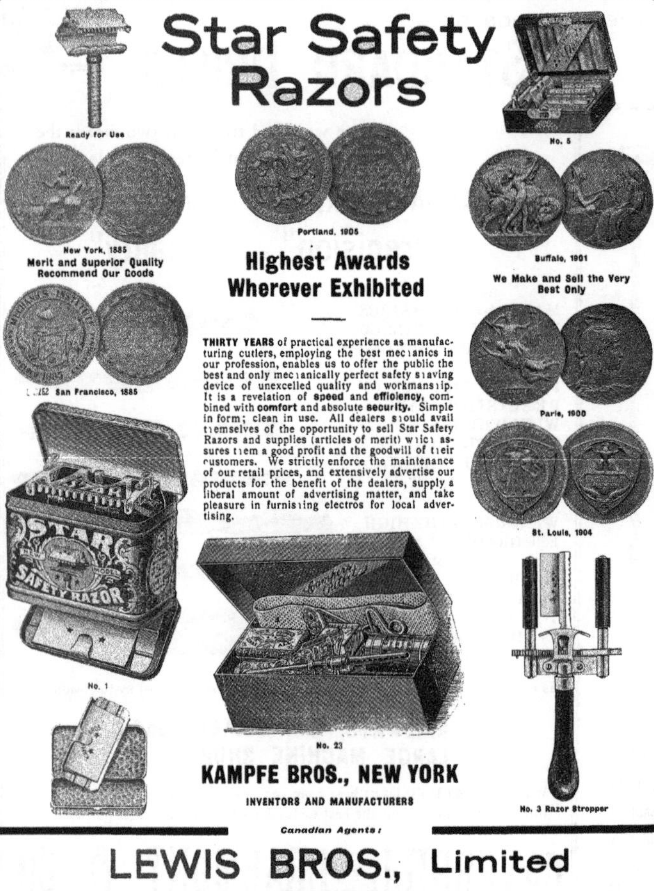

Star Safety Razors
Ready for Use
No. 5
Portland, 1905
New York, 1885
Merit and Superior Quality Recommend Our Goods
Highest Awards Wherever Exhibited
Buffalo, 1901
We Make and Sell the Very Best Only
San Francisco, 1885
THIRTY YEARS of practical experience as manufacturing cutlers, employing the best mechanics in our profession, enables us to offer the public the best and only mechanically perfect safety shaving device of unexcelled quality and workmanship. It is a revelation of speed and efficiency, combined with comfort and absolute security. Simple in form; clean in use. All dealers should avail themselves of the opportunity to sell Star Safety Razors and supplies (articles of merit) which assures them a good profit and the goodwill of their customers. We strictly enforce the maintenance of our retail prices, and extensively advertise our products for the benefit of the dealers, supply a liberal amount of advertising matter, and take pleasure in furnishing electros for local advertising.
Paris, 1900
St. Louis, 1904
STAR SAFETY RAZOR
No. 1
No. 23
KAMPFE BROS., NEW YORK
INVENTORS AND MANUFACTURERS
No. 3 Razor Stropper
Canadian Agents:
LEWIS BROS., Limited
MONTREAL
TORONTO OTTAWA WINNIPEG CALGARY VANCOUVER

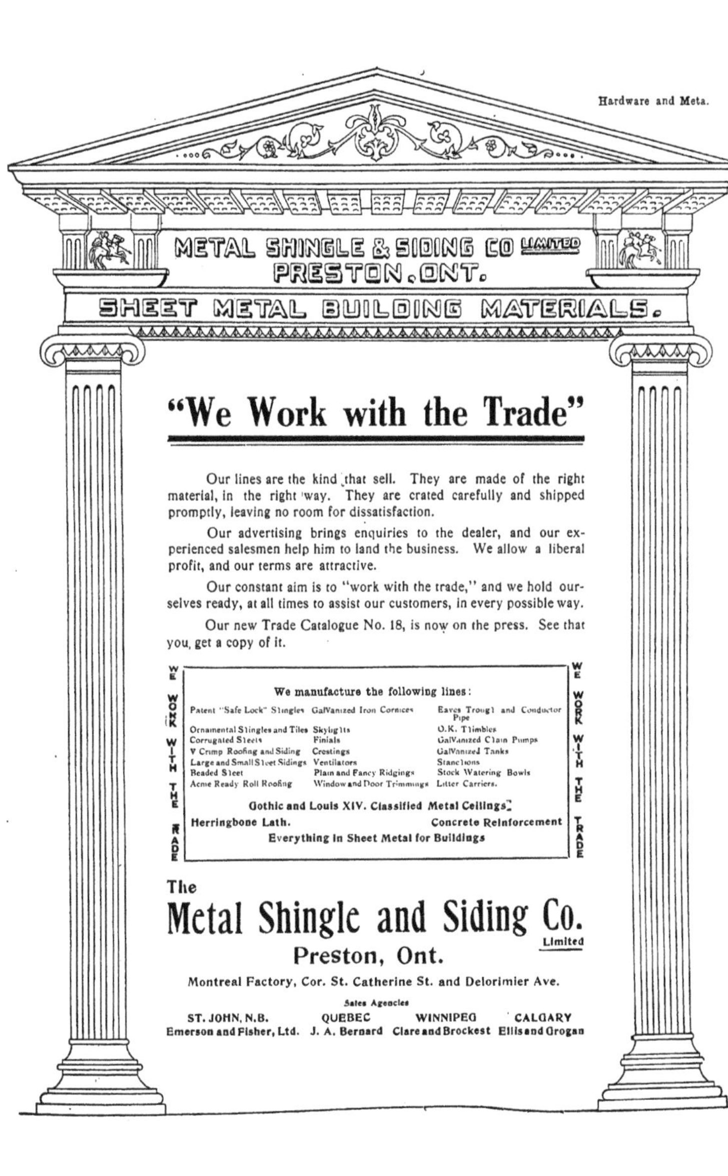
METAL SHINGLE & SIDING CO LIMITED
PRESTON, ONT.
SHEET METAL BUILDING MATERIALS.
"We Work with the Trade"
Our lines are the kind that sell. They are made of the right material, in the right way. They are crated carefully and shipped promptly, leaving no room for dissatisfaction.
Our advertising brings enquiries to the dealer, and our experienced salesmen help him to land the business. We allow a liberal profit, and our terms are attractive.
Our constant aim is to "work with the trade," and we hold ourselves ready, at all times to assist our customers, in every possible way.
Our new Trade Catalogue No. 18, is now on the press. See that you, get a copy of it.
WE WORK WITH THE TRADE
We manufacture the following lines:
Patent "Safe Lock" Shingles
Ornamental Shingles and Tiles
Corrugated Sheets
V Crimp Roofing and Siding
Large and Small Sheet Sidings
Beaded Sheet
Acme Ready Roll Roofing
Galvanized Iron Cornices
Skylights
Finials
Crestings
Ventilators
Plain and Fancy Ridgings
Window and Door Trimmings
Eaves Trough and Conductor Pipe
O.K. Thimbles
Galvanized Chain Pumps
Galvanized Tanks
Stanchions
Stock Watering Bowls
Litter Carriers.
Gothic and Louis XIV. Classified Metal Ceilings.
Herringbone Lath.
Concrete Reinforcement
Everything in Sheet Metal for Buildings
WE WORK WITH THE TRADE
The
Metal Shingle and Siding Co.
Limited
Preston, Ont.
Montreal Factory, Cor. St. Catherine St. and Delorimier Ave.
Sales Agencies
ST. JOHN, N.B.
Emerson and Fisher, Ltd.
QUEBEC
J. A. Bernard
WINNIPEG
Clare and Brockest
CALGARY
Ellis and Grogan

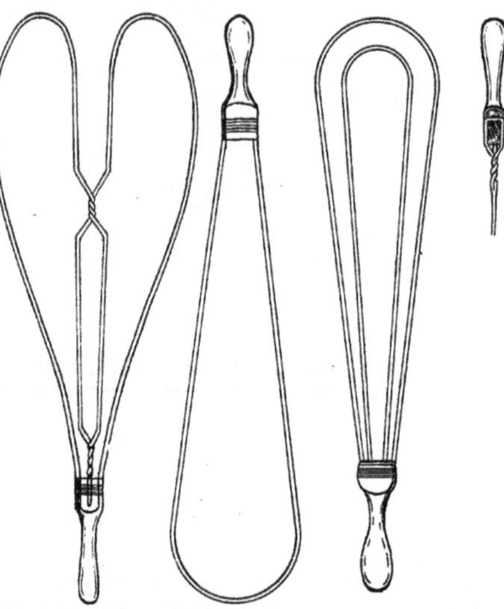

Are You Handling the Proper Handles?

There is a reasonable doubt—unless you are carrying the Handles made by the J. H. Still Manufacturing Co. Because it has been demonstrated beyond the shadow of a doubt, that the **J. H. Still lines** are not only the best Handles but also **the most profitable** for the dealer to handle.

The Pick and Axe Handles shown in the cut are made of Southern Hickory, and of first-class Southern Hickory at that.

The Adze Handles, Whiffletrees and Neck Yokes are made from Hickory and second growth Ash of the very best quality. Second-class material cannot by any chance find its way into any Handle we make and the workmanship is A 1.

Surface or Railway Pick

House Adze

Miner Pick

Heavy Team Neck Yoke

THEY ALL SPELL DOLLARS FOR YOU

Besides these lines we also manufacture Fork, Hoe and Rake Handles out of the best Ash money can buy.

Our Broad Axe, Bench and Hunters' Axe Handles made St. Thomas famous, while our Hammer, Cant Hook and Peavy Handles have set a pace for genuine goodness which our competitors strive to equal in vain.

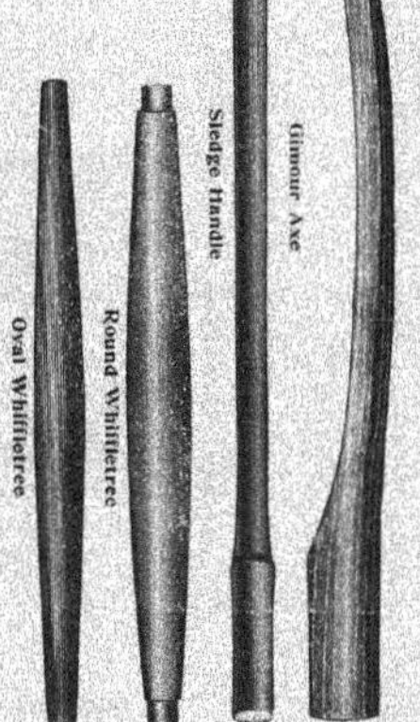

The immense superiority of the J. H. Still lines is readily and practically universally granted. Owning our own huge timber reserves helps some. But **our system of doing business conserves the dealer's profit and every hardwareman in Canada should know our "Profit Story." Write to-day.**

J. H. Still Mfg. Co., - St. Thomas, Ont.

ICE CREAM
FREEZERS
THAT SELL
SEND FOR
1907
FREEZER
BOOK
FREE
AMERICAN
TWIN FREEZER
SHEPARD'S
LIGHTNING
THEY ARE
Well Advertised
In Demand
Easily Sold
Satisfactory in Use
Of Known Reputation
THEY EXCEL IN
Easy Running
Quick Freezing
Economy
Convenience
Practical Results
The Gem Freezer
Blizzard Freezer
Our line of Freezers comprises the largest variety made in any one factory. They embody all that is best and of intrinsic merit in a Freezer. Beside lines illustrated above we make large sizes with fly wheels or pulleys, and Machine Freezers, Ice Breakers, etc.
THEY ARE SOLD BY
LEADING JOBBERS
EVERYWHERE
NORTH BROS. MFG. CO., PHILADELPHIA, Pa., U. S. A.

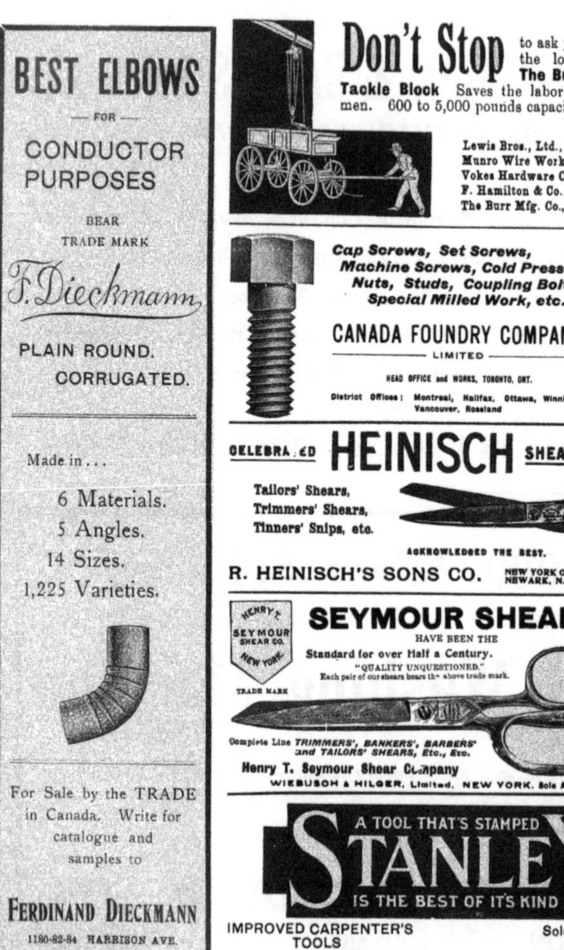
BEST ELBOWS
FOR
CONDUCTOR PURPOSES
BEAR TRADE MARK
F. Dieckmann
PLAIN ROUND.
CORRUGATED.
Made in . . .
6 Materials.
5 Angles.
14 Sizes.
1,225 Varieties.
For Sale by the TRADE in Canada. Write for catalogue and samples to
FERDINAND DIECKMANN
1180-82-84 HARRISON AVE.
CINCINNATI, O., U.S.A.
Don't Stop to ask your neighbor, lift the load yourself with The Burr Self-Locking Tackle Block Saves the labor of two or three men. 600 to 5,000 pounds capacity.
Lewis Bros., Ltd., Montreal
Munro Wire Works, Winnipeg
Vokes Hardware Co., Toronto
F. Hamilton & Co., Hamilton
The Burr Mfg. Co., Cleveland, Ohio
Cap Screws, Set Screws, Machine Screws, Cold Pressed Nuts, Studs, Coupling Bolts, Special Milled Work, etc.
CANADA FOUNDRY COMPANY
LIMITED
HEAD OFFICE and WORKS, TORONTO, ONT.
District Offices: Montreal, Halifax, Ottawa, Winnipeg Vancouver, Rossland
CELEBRATED HEINISCH SHEARS
Tailors' Shears,
Trimmers' Shears,
Tinners' Snips, etc.
ACKNOWLEDGED THE BEST.
R. HEINISCH'S SONS CO.
NEW YORK OFFICE, Chambers St
NEWARK, N.J., U.S.A.
HENRY T. SEYMOUR SHEAR CO. NEW YORK.
TRADE MARK
SEYMOUR SHEARS
HAVE BEEN THE
Standard for over Half a Century.
"QUALITY UNQUESTIONED."
Each pair of our shears bears the above trade mark.
Complete Line TRIMMERS', BANKERS', BARBERS' and TAILORS' SHEARS, Etc., Etc.
Henry T. Seymour Shear Company
WIEBUSCH & HILGER, Limited, NEW YORK, Sole Agents
Latest Catalogue will be sent in exchange for your business card.
A TOOL THAT'S STAMPED
STANLEY
IS THE BEST OF IT'S KIND
IMPROVED CARPENTER'S TOOLS
Sold by all Hardware Dealer
STANLEY RULE & LEVEL CO.
NEW BRITAIN., Conn., U.S.A.

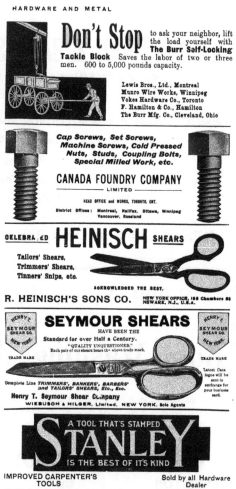

THE ECLIPSE CURRY COMB

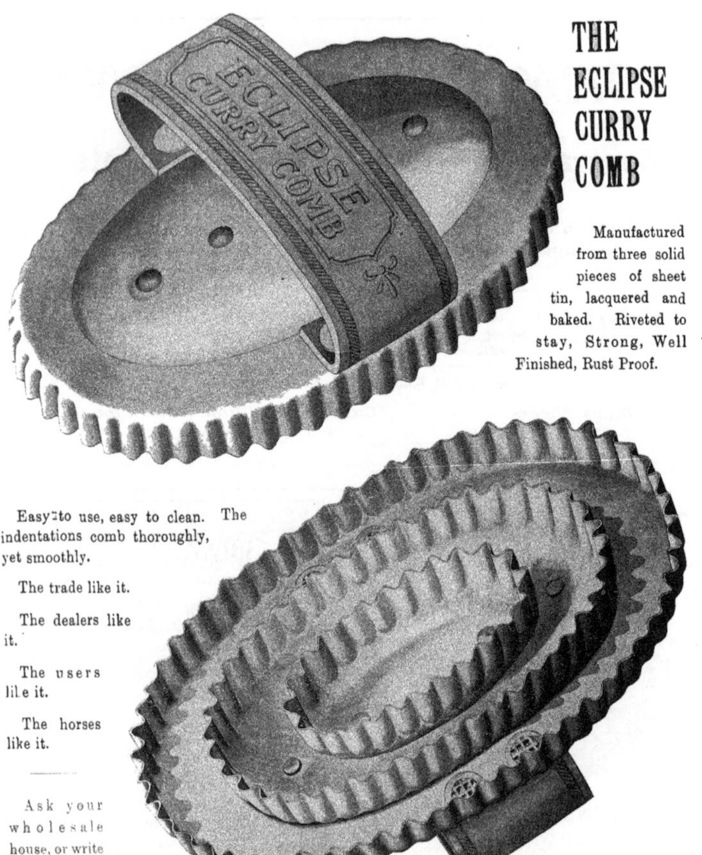

Manufactured from three solid pieces of sheet tin, lacquered and baked. Riveted to stay, Strong, Well Finished, Rust Proof.

Easy to use, easy to clean. The indentations comb thoroughly, yet smoothly.

The trade like it.

The dealers like it.

The users like it.

The horses like it.

Ask your wholesale house, or write direct to

THE ECLIPSE MANUFACTURING CO., Limited,

OTTAWA, ONT.

The "Manitoba" Grain Scoop

Largely used for the rapid handling of grain. Every Farmer, Miller and Storekeeper wants the "Manitoba" Scoop.

For sale by all Wholesale Hardware Houses in Winnipeg

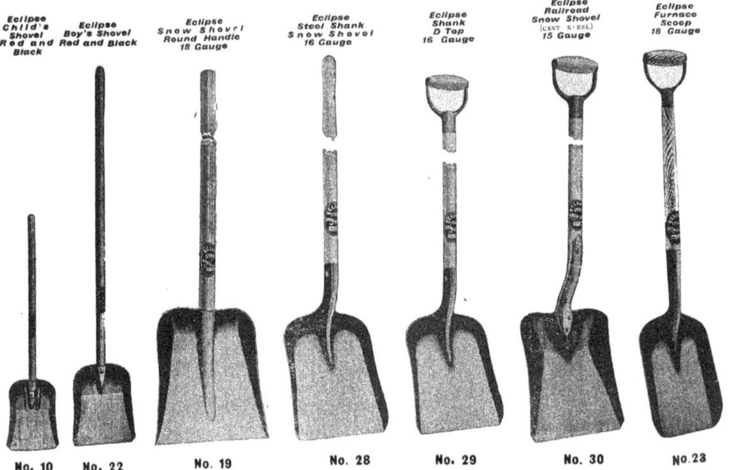

THE ECLIPSE MANUFACTURING CO., LIMITED,

Manufacturers of STEEL SNOW SHOVELS, GRAIN SHOVELS, Etc.

OTTAWA, ONT.

TRADE MARK
Est. 1863.
Inc. 1895.
Black Diamond File Works
G. & H. Barnett Company
PHILADELPHIA
Twelve
TRADE MARK
Medals
Awarded
By JURORS at
International Expositions
Special Prize
Gold Medal at Atlanta, 1895
Copy of catalogue sent free to any interested file user upon application.
TRADE MARK

SUPPORT CANADIAN INDUSTRIES
FREE
TO
HARDWARE DEALERS
PUTZ TRIPLE EXTRACT LIQUID
METAL POLISH
MANUFACTURED BY
ENGLISH EMBROCATION Co
Montreal, Canada.
We are supplying electros of the cut in this ad. for use in your local advertising, besides offering special terms for immediate orders for
"LION" Brand
PUTZ TRIPLE EXTRACT
METAL POLISH
IN LIQUID OR PASTE
The most reliable metal polish in the market. Once tried always used.
Support Canadian Industries
Manufactured by
English Embrocation Co., Montreal

A. G. HEWTON
Traveller Eastern and Western Ontario

The Product

Nothing but the very highest grade of Imported English Oak Tanned Stock enters into the manufacture of our belting.

J. C. FAIR
Traveller Maritime Provinces.

The Firm

The J. C. McLaren Belting Co., of Montreal are the pioneer manufacturers of leather belting in Canada, having been established by the late J. C. McLaren in 1856, and later on the manu- of card clothing and reeds was taken up. After the death of founder of the firm, the business was successfully carried on by his son D. W. McLaren, until his death recently, when the management came under the hands of F. A, Johnson, the present head of the firm.

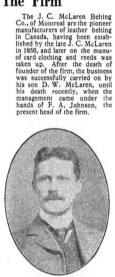

J. R. MILFORD
Superintendent Card Clothing Department

Agencies and Branches

St. John, N.B.—Canadian Oil Co.,

Hamilton—Alexander Hdw. Co.,

Ottawa—W. F. Colston & Co.,

Toronto (Branch) — 50 Colborne Street.

London – A. Westman

St. Thomas – Ingram & Davey

STANLEY J. GONIN
Traveller Montreal and vicinity

The J. C. McLaren Belting Co., Montreal

Toronto · Winnipeg
St. John · Vancouver

WM. LINTON
Traveller Central Ontario

W. S. BROCK
Manager Winnipeg Branch

The Result

The result is, our belts will meet all reasonable demands made upon them.

The lives of your employees depend upon the belt you use, see that you got a **belt that's fully guaranteed. Ours are.**

We have no hesitation in fully guaranteeing our

Extra Oak Brand

A DOUGLAS McARTHUR
Manager Toronto Branch

Agencies and Branches

Winnipeg (Branch)—156 Lombard Street.
Chatham, Ont.—Park Bros.
Calgary—Great West Saddlery Co.,
Edmonton—Great West Saddlery Co.,
Regina—H. W. Laird & Co.,
Prince Albert—A. H. Woodman,
Vancouver — Canadian Fairbanks Co., Limited

F. A. JOHNSON
Director and General Manager,

The General Manager

F. A. Johnson, since he has been head of the firm has developed the business very rapidly. In the early eighties, he entered the lumber business and for seven years followed this line. For the next few years he engaged at mining and milling of ores in California, and then returned Canada, having accepted the position of assistant manager of the English Portland Cement Co. Martbank, Ont. He next came to Montreal in connection with the Standard Asbestos Co., of which concern he became a director and secretary-treasurer. He later became associated with the J. C. Mc Laren Belting Co., of Montreal, succeeding to the position of director and general manager on the death of D. W. McLaren. It is evident Mr. Johnson has had long and varied experience in the using as well as the manufacture of leather belting.

JOHN EDWARDS
Factory Superintendent

The J. C. McLaren Belting Co., Montreal

Leather and Rubber Belting, Card Clothing and Reed Manufacturers; Lacing and General Mill Supplies

Toronto Winnipeg
St. John Vancouver

WM. S. CAMPBELL
Accountant

MR. DEALER:—Do you know that the

H & R
Greatest Revolver
Value for the
Money

H&R POLICE AUTOMATIC

combines **Safety and Efficiency** that cannot be found in any hammer revolver of other makes.

IT can be drawn from the pocket and fired with great rapidity because the hammer is sunk in the frame and CANNOT CATCH ON THE CLOTHING, nor be accidentally discharged, yet can be readily cocked for deliberate aim when desired. These safety features are not found in revolvers of other makes. Shells are automatically ejected—making it easy to reload. Made of the best materials, finely finished, and with proper care will last a lifetime, and gives perfect satisfaction to your customers. Write for catalog.

HARRINGTON & RICHARDSON ARMS CO.

Makers of the celebrated H & R Single Guns

314 Park Ave., Worcester, Mass.

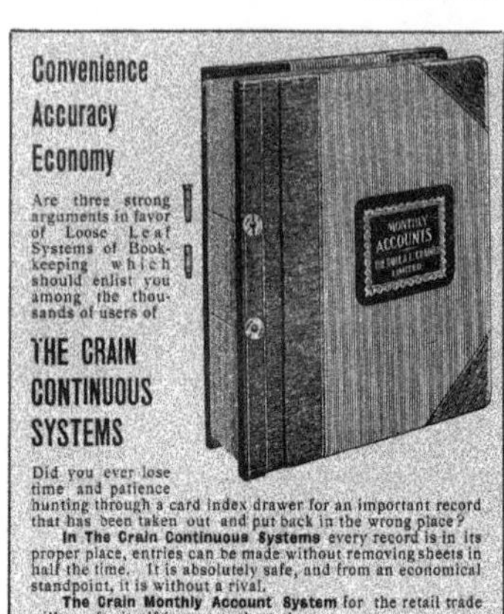
Convenience
Accuracy
Economy
Are three strong arguments in favor of Loose Leaf Systems of Book-keeping which should enlist you among the thousands of users of
THE CRAIN CONTINUOUS SYSTEMS
MONTHLY ACCOUNTS
Did you ever lose time and patience hunting through a card index drawer for an important record that has been taken out and put back in the wrong place?
In The Crain Continuous Systems every record is in its proper place, entries can be made without removing sheets in half the time. It is absolutely safe, and from an economical standpoint, it is without a rival.
The Crain Monthly Account System for the retail trade will pay for itself in a very short time.
Write for catalogue E which will give you full information
THE ROLLA L. CRAIN CO., Limited
OTTAWA, CANADA
TORONTO, 18 Toronto St.
MONTREAL, Alliance Bldg.
WINNIPEG, Nanton Block.

CHICAGO SPRING BUTTS
TRADE CSB CO MARK
A PRODUCT OF RECOGNIZED SUPERIORITY
CHICAGO SPRING BUTT
TRIPLE-END SPRING BUTT
CHICAGO BALL-BEARING FLOOR HINGE
Chicago Spring Butt Company
CHICAGO
NEW YORK
CATALOGUE ON REQUEST

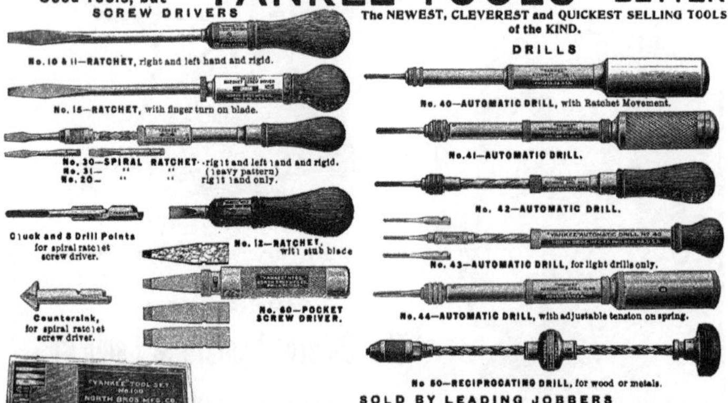
Other Tools are very Good Tools, but
"YANKEE TOOLS" ARE BETTER
The NEWEST, CLEVEREST and QUICKEST SELLING TOOLS of the KIND.
SCREW DRIVERS
No. 10 & 11—RATCHET, right and left hand and rigid.
No. 15—RATCHET, with finger turn on blade.
No. 30—SPIRAL RATCHET—right and left hand and rigid.
No. 31— " " (heavy pattern)
No. 20— " " right hand only.
Chuck and 8 Drill Points for spiral ratchet screw driver.
No. 12—RATCHET, with stub blade
Countersink, for spiral ratchet screw driver.
No. 60—POCKET SCREW DRIVER.
"YANKEE" TOOL SET No. 100 NORTH BROS MFG. CO. PHILADELPHIA, PA. U.S.A.
DRILLS
No. 40—AUTOMATIC DRILL, with Ratchet Movement.
No. 41—AUTOMATIC DRILL.
No. 42—AUTOMATIC DRILL.
No. 43—AUTOMATIC DRILL, for light drills only.
No. 44—AUTOMATIC DRILL, with adjustable tension on spring.
No. 50—RECIPROCATING DRILL, for wood or metals.
SOLD BY LEADING JOBBERS
SEND FOR OUR NEW "YANKEE" TOOL BOOK
NORTH BROS. MFG. CO.
PHILADELPHIA, Pa.

CAPEWELL

HORSE NAILS

The Best in the World and "Made in Canada"

CITY HEAD

2 3 4 4½ 5 6 7 8 9 10 11

MADE WELL
HOLD WELL
DRIVE WELL } **CAPEWELL**
LOOK WELL
SELL WELL

CAPEWELL HORSE NAILS

are known, in the Horse Nail World, as the Standard of Perfection among all competitors. "Imitation is the sincerest flattery," and we acknowledge the compliment paid us by the manufacturer who marks his boxes "Also called City Head." We are the **ONLY** makers in Canada of the genuine City Head Horse Nail, as illustrated above. Both our Regular and City Heads are the approved half-countersunk pattern—the best adapted for **ALL** classes of shoeing. The material of which **CAPEWELL HORSE NAILS** are made is the very highest grade of Swedish iron, made specially for us, and imported by us direct from Sweden. The magnitude of our contracts, which exceed those of all other makers combined, insures the most minute care being taken in the preparation of our material. With the best rods obtainable in the markets of the world and the "**CAPEWELL**" process, which is controlled **EXCLUSIVELY** by us, we are able to produce a horse nail which is absolutely unequalled. We invite your comparison, in any particular and in any test. The best and most conclusive test of all is in service in the hoof. The horse-shoers of Canada are rendering their verdict in no unmistakable voice, and the business of this Company has increased by leaps and bounds. We know our product is what we claim for it, "**THE BEST IN THE WORLD**," and we shall always strive to merit your patronage.

PROMPT SHIPMENTS. **EVERY NAIL GUARANTEED.**

Samples, Prices, Catalogue and 1907 Calendar free on application.

THE CAPEWELL HORSE NAIL CO., TORONTO, ONT.

Branch: WINNIPEG, MAN.

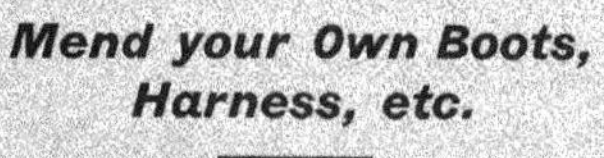

Mend your Own Boots, Harness, etc.

"ALL-U-WANT."

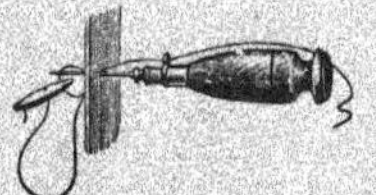

The **ALL-U-WANT** is being advertised extensively to the Farmer, and Hardwaremen will find a ready sale for it if they have a sample to show their customers.

G. Parsons & Son, Limited

79 FRONT STREET E.,
TORONTO.

Counter Check Books

ALL STYLES MADE.
"GET THE BEST."

Ask to see

The "DUPLEX."

Samples on application, or Traveller will call.

The Carter-Crume Company, Limited,
Toronto, Ont.

THE AJAX MANUFACTURING CO., LTD.

MAKERS OF

"AJAX" EMERY WHEELS.
"AJAX" SHARPENING STONES.
"AJAX" WALL TILING.
"AJAX" FLOOR TILING.
"AJAX" PLASTIC FLOORING.

Agents for "Metile" Wall Tiling and "Flexible Veneer."

Office, 333 Elgin Avenue, Winnipeg, Man.

PHONE 6456.

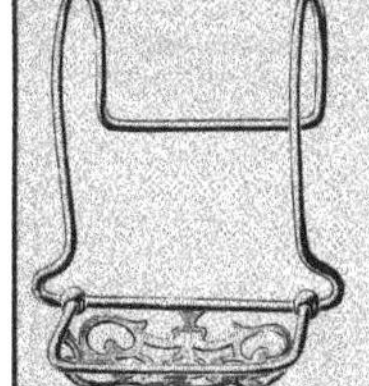

SOAP DISHES

BATH ROOM FIXTURES

OF THE HIGHEST QUALITY
ARE MANUFACTURED IN
CANADA BY

THE CARRIAGE MOUNTINGS CO., Ltd.

TORONTO

Write for Catalog "B2"

PAPER FIXTURES

TOWEL BARS

PRIVATE AND CONFIDENTIAL

TO THE HARDWARE DEALER

We doubt if you are aware of the possibilities of remunerative business in

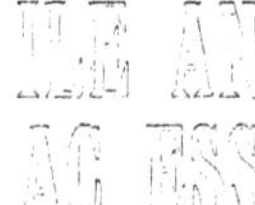

MOT R

IES

The increasing use of auto cars and boats ensures a constant demand. When an item is needed it is wanted quick, and **will be bought from the nearest man who has it to sell.** Then (this is confidential) the profits are large and the stock need only be small, as we carry the reserve stock—the largest in Canada of auto accessories.

Here, for instance, is one of many makes of Spark Plugs we carry . . . . .

The Rajah Spark Plug

Sells for $1.50

We sell to you for (Figure by mail)

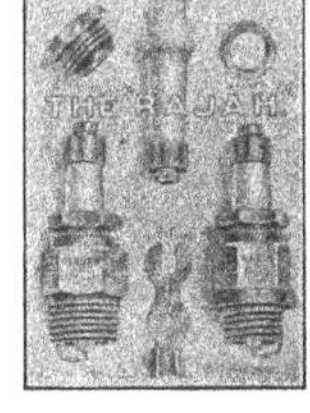

Or the

L'Eclair Spark Plug

Our own Specialty

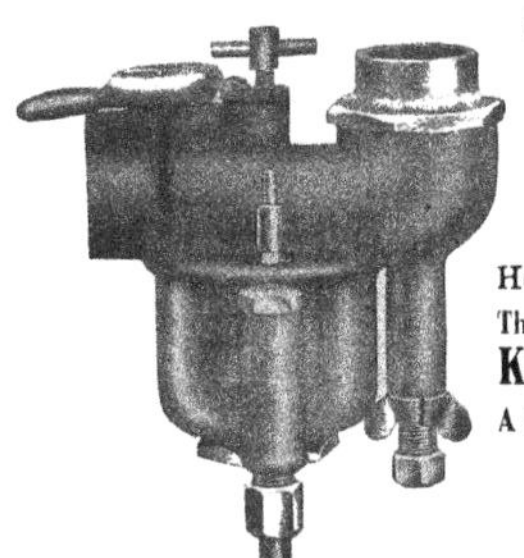

Here is another example,

The

Kingston Carburetor

A Standard line

Or

Columbia Dry Cells

Always wanted
Always dependable
Always in stock by us

We have Coils, Lamps, Horns, Cable, Accumalators, Voltmeters, etc.

Examine the lines at our stands 19, 20 and 21 at Automobile and Sportsmen's Show at Arena, Montreal, April 6 to 13th.

Get our new Catalogue of Accessories. Our new Catalogue of Bicycles and sundries will also interest you.

JOHN MILLEN & SON, Limited

132 Bay St., TORONTO · 321 St. James St., MONTREAL

E. T. WRIGHT & CO. HAMILTON Canada

Manufacturers of Lanterns, Tinware, Bird Cages, Eavetrough and Conductor Pipe, Patent Stove Pipes and Elbows

Since the ***"Victor" Sifter*** was placed on the market, other Sifters have been produced, but none equal to the ***"Victor."*** It stands alone unequalled. Send for sample.

The Anton Carpet Beater

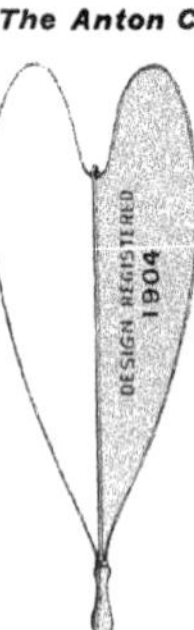

has double the beating surface of any carpet beater on the market. ***Light,*** yet ***strong*** and ***durable; stiff,*** yet ***pliable*** and ***easy to handle.*** Centre rod made of heavy spring steel wire. Outside wires go through the handle and are clinched at bottom end making it impossible for handle to come off; handle filled with lead, thereby making it a perfectly balanced article.

O. G. Eavetrough

(Square Bead)

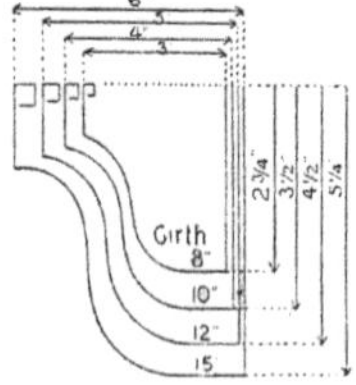

We supply the above in round or half. round, 10-feet lengths. Write for prices.

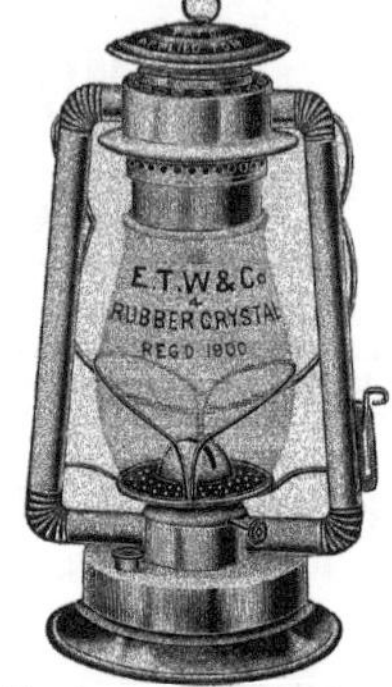

The above cut shows our ***Improved Cold Blast Lantern,*** with side lift. We also make the ***Cold Blast*** with bottom lift.

Plain and japanned.

Balloon Fly Traps

Cleanest and surest mode of catching flies. Packed one dozen in case.

A complete assortment of ***Cages*** always in stock. All sizes, styles and prices. Brass and Japanned. With or without guards. Discount on application.

Cooper's Improved Patent One-piece Stove Pipe Elbows

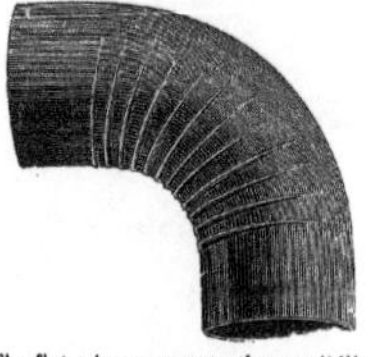

The flat crimp prevents the possibility of dust and dirt accumulating on the elbow. Heaviest and best on the market. Did you ever handle them?

Wright's Mowing Machine Oiler

Made in three styles, plain, coppered and japanned and stencilled. Bent or straight spout. Best oiler on the market. Send for prices.

Galvanized Conductor Pipe

Corrugated or Round, 28 gauge, 10-feet lengths. Write for prices.

A Full Supply of Tinsmiths' Tools

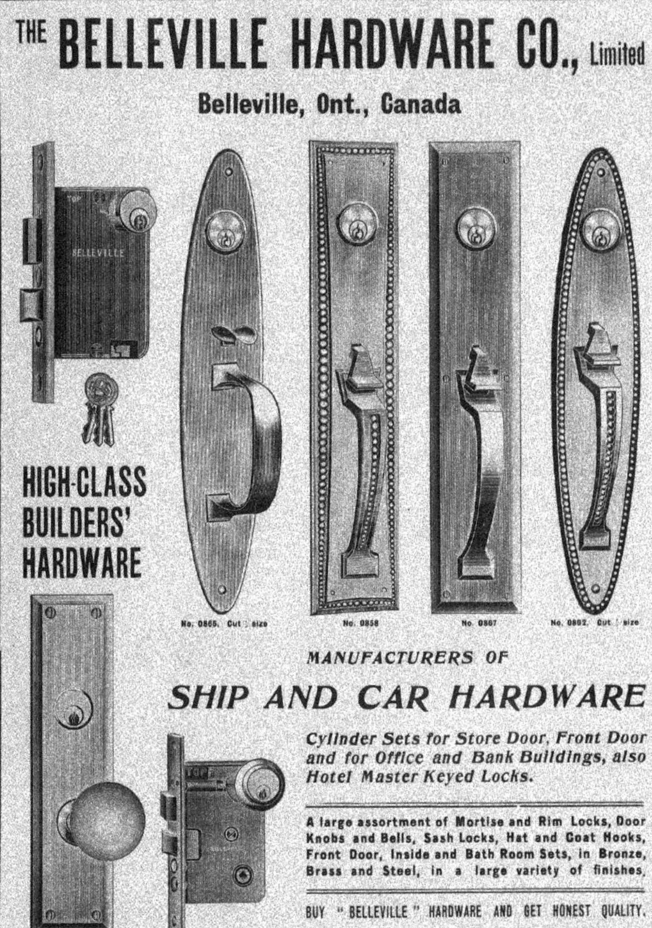
THE BELLEVILLE HARDWARE CO., Limited
Belleville, Ont., Canada
BELLEVILLE
HIGH-CLASS BUILDERS' HARDWARE
No. 0865. Cut size
No. 0858
No. 0867
No. 0862. Cut size
MANUFACTURERS OF
SHIP AND CAR HARDWARE
Cylinder Sets for Store Door, Front Door and for Office and Bank Buildings, also Hotel Master Keyed Locks.
A large assortment of Mortise and Rim Locks, Door Knobs and Bells, Sash Locks, Hat and Coat Hooks, Front Door, Inside and Bath Room Sets, in Bronze, Brass and Steel, in a large variety of finishes.
BUY "BELLEVILLE" HARDWARE AND GET HONEST QUALITY.
A CATALOGUE FOR THE ASKING.

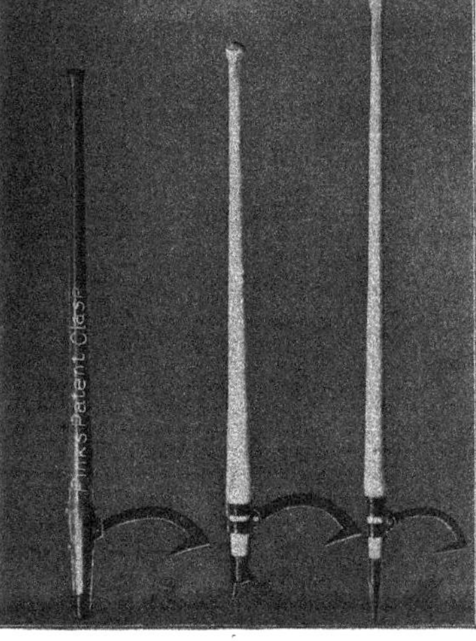
Pink's Patent Clasp

Rabbit No. 1.—"I am done for. My axe is broken."
Rabbit No. 2.—"Why don't you buy a Dundas axe? They stand Hemlock Knots."

Axes may be as good as Dundas Axes are, and yet to have them lead all others as sellers, you must let it be known that you have them. To aid you in this we furnish electros such as appear on this page. If you are amongst the large number that have already ordered, let us hear from you and we will supply what you want, and if you have not yet ordered, mention in your order what electro you want. You want to have the best axe in the market, and you want to let it be known that you have it.

THE DUNDAS AXE WORKS

DUNDAS, ONT.

TRIUMPH ADJUSTABLE STOVEPIPE

PUT TOGETHER.

NESTED.

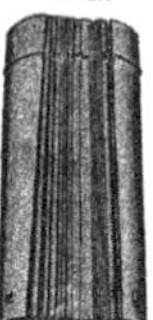

The Best on the Market.

WHY?

BECAUSE IT IS STRONGER. All other patent stovepipes are held together and depend for their strength on one or two very small tabs the same thickness as the sheet. The **Triumph** is held together and depends for its strength on a heavy staple sliding into two loops firmly riveted on.

ADJUSTABLE. The staple can be widened to make the **Triumph** fit larger pipe, or make narrow to fit smaller pipe.

EASIER PUT TOGETHER. The edge slips straight into the fold and the staple is then inserted. A child can put them together.

REQUIRES NO RIVETS OR POUNDING DOWN.

CRIMPED AT ONE END.

SAVES TIME, TEMPER AND FREIGHT. Packed 25 in crate.

Manufactured by

EDWIN CHOWN & SON, Kington, Ont.

1000 MILE AXLE GREASE

HAS NO EQUAL.

NICEST PACKAGE. BEST GREASE.

OUR OTHER SPECIALTIES ARE:

1000 Mile Carriage Oil.	Harness Oil Blacking.
Harness Oil.	Leather Preserver.
Harness Soap.	Magic Hoof Ointment.
Perfection Belt Dressing	Magic Gall Cure.
Matchless Cup Grease.	Olive Floor Oil.

Write for Booklet giving full description of above.

MANUFACTURED BY

The Campbell Manufacturing Co.

Hamilton, Canada.

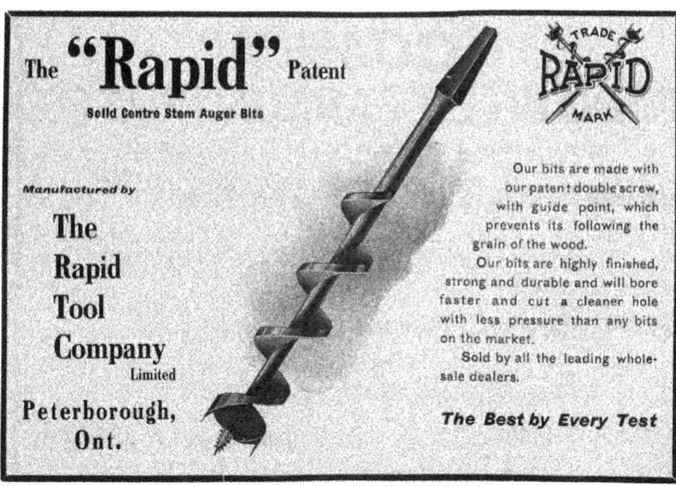

Sovereign Treasure Steel Range

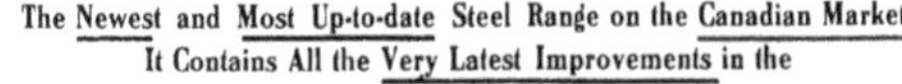

The Newest and Most Up-to-date Steel Range on the Canadian Market.

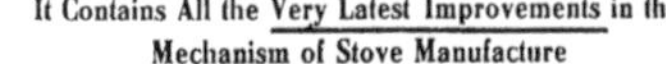

It Contains All the Very Latest Improvements in the Mechanism of Stove Manufacture

An entirely new and original patent "LEVER" for raising front section, for broiling, toasting or coaling fire. Also other new and exclusive features on this range.

"Treasures" are a luxury, because they always please.

The greatest value in stove construction ever offered to the Canadian public.

No competition found with this stove. It sells itself.

Every line agent should have a "Sovereign Treasure" on his floor. Enquiries solicited.

Made either with cast base or nickel plated legs.

Full particulars on request, or from our travelling representatives.

Design registered 1907.

The D. Moore Company, Limited

Hamilton, Ont.

Northwest Distributors:
MERRICK, ANDERSON & CO.
WINNIPEG, MAN.

British Columbia Distributor:
JOHN BURNS
VANCOUVER, B. C.

Canada Stove Works, Harriston, ONT.

The Royal Corona

is a six or four hole Range with many very strong selling features. Has given entire satisfaction. ***No better Range made in Canada.*** Write for prices.

The Sun of Prosperity

is shining very brightly.

During 1906 our business increased over

500 Per Cent.

1907 bids fair to excel it.

The Dealers that are handling the

ROYAL STOVES and RANGES

are

Making Money

The Easy Line to Sell

The Royal Empress

is a six-hole No. 9 Range with copper Reservoir, 20-in. oven, wood and coal linings, cast or steel high shelf. F.O.B. your station (anywhere in Ontario)

$31.00

This Range is a ***Business Builder and is fully guaranteed.***

Let us send you one. If not entirely satisfactory it costs you nothing.

ROYAL EMPRESS

Canada Stove Works, Harriston, ONT.

Steel Ranges

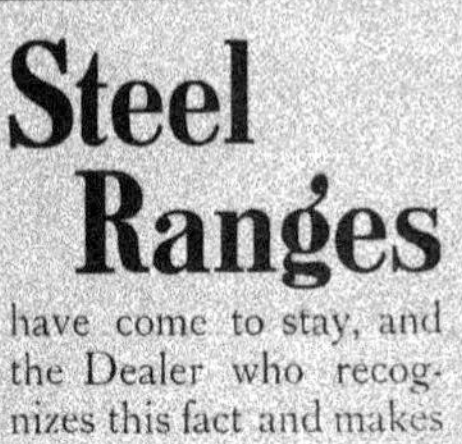

have come to stay, and the Dealer who recognizes this fact and makes them his leaders will have the profitable stove trade of his district.

We make Steel Ranges with right or left hand reservoirs, with cabinet base or on feet, as desired.

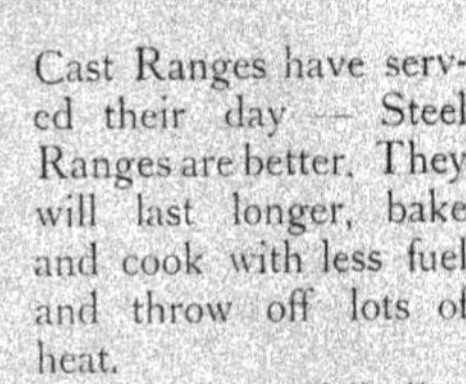

Cast Ranges have served their day — Steel Ranges are better. They will last longer, bake and cook with less fuel and throw off lots of heat.

We make a full line of Family Steel Ranges for city, town and farm trade.

Findlay Bros. Co., Ltd.

CARLETON PLACE.

Branch House, - Winnipeg

e Perfect Idea Cast Range

IS THE ONLY RANGE MADE IN CANADA
THAT IS EQUIPPED WITH A

ent E ectric Sliding Oven Tray

The Latest and Most Convenient Patented Improvement in Ranges.

By buying the **PERFECT IDEA** you secure the agency for the **PATENT ELECTRIC SLIDING OVEN TRAY**

The GREATEST SELLING POINT on any Range made in Canada. Instead of the old-fashioned "Kicker" to open the oven door, which requires the foot to be raised 12 or 15 inches to work it—we have a push button about 5 inches from the floor; by simply pressing this with the foot the door opens and the oven tray slides forward two thirds of the depth of the oven, bringing out whatever is cooking, or to receive anything that is ready for the oven. No need to stoop down to look in the oven. No burning of hands pulling pans out. Just what has been wanted. The door can be opened without the shelf sliding out if desired. GET IN LINE. WRITE US FOR THE AGENCY.

PH OVE CO., Limited

GUELPH, ONT.

The Range With the Selling Points

STEEL RANGE

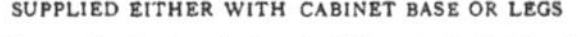

SUPPLIED EITHER WITH CABINET BASE OR LEGS

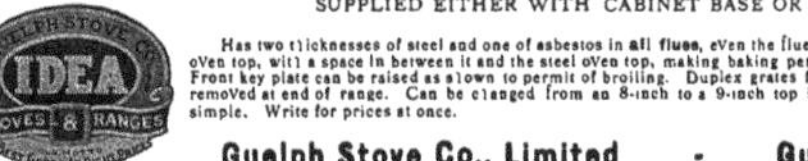

Has two thicknesses of steel and one of asbestos in **all flues**, even the flue **below the oven**. Cast oven top, with a space in between it and the steel oven top, making baking perfect on top and bottom. Front key plate can be raised as shown to permit of broiling. Duplex grates for coal or wood, can be removed at end of range. Can be changed from an 8-inch to a 9-inch top in a few minutes, Very simple. Write for prices at once.

Guelph Stove Co., Limited - Guelph, Ont.

18 Years Practical Use Has Demonstrated

KELSEY EFFICIENCY, ECONOMY, DURABILITY,

THE ALL CAST IRON KELSEY
PRACTICALLY INDESTRUCTIBLE

It is 18 years since the first "Kelsey" Warm Air Generator was made and sold, 3 having been made and sold in 1889.

There were 30,000 pleased users in 1906.

Improvements have been made, but without altering in any way, the many special patented features of construction.

Many imitations, and so-called first-class devices, have come and gone since the Kelsey was introduced, this should be conlusive evidence that **THE KELSEY IS HERE TO STAY.**

Kelsey dealers are not continually up against the "How Cheap" proposition. Why allow your competitor to make prices for you?

Kelsey dealers do not have tc.

Kelsey dealers are given an exclusive agency and protected in their territory.

Kelsey experts always available to assist dealers in interesting intending purchasers and in closing contracts.

The Kelsey agency puts you right up to the front.

Do you wish to get there. If so do not put off writing us.

Sole Kelsey Makers for Canada

THE JAMES SMART MFG. CO., Limited
BROCKVILLE, ONT.

Western Branch—Winnipeg, Man.

J. P. O'SHEA & CO.

We Are in the Front Rank in Glass Work of All Descriptions

All latest designs made in **Ornamental** wheel and **Sand Cut Glass.**
Where we win always, is, that we are direct importers of **British and French Plate Glass.**
We place plate glass insurance.

Manufacturers of Mirrors, &c.

Our Specialties;—Fancy Door Lights for Offices and Railway Work, Grinding, Drilling, Bevelling and Chipping.

Communicate With Us — *Promptness Our Motto*

Tel. Bell E. 450 — ***8 St. Agathe Street, Montreal*** — Tel. Merchants 78

Canada's Best Domestic Range

The Walker Pilot

POINTS OF MERIT:

Large square steel oven.
Balanced oven door at any angle.
Oven thermometer.
Ventilated oven.
Steel high shelf or high closet.
Balanced closet door.
Tea sheives.
Three-ring cover.
Broiler door.
Pouch feed.
Flush top reservoir, controlled by damper.
Porcelain enameled tank.
Cast iron reservoir casing.
Asbestos floor protection.
Large common-sense ash pan.
Ash chutes or guards.
Duplux grates for coal or wood.
Grates removable without disturbing linings.
Oblong fire box.
Proper end draft.
Seymour patent slide draft.
Wood extension.
Rotary check damper.
Asbestos lined.
Reinforced bottom edge.
Ventilated base.
Closely hand-rivetted throughout.
Cast iron flue back.
Simple dampers.
Elegantly nickeled.
Nickel-plated hearth.

"Successful Everywhere"

It possesses all the above "up-to-date" advantages, and yet
Sells at a Competing Price and Makes You Double the Usual Profits.

SEND A POSTAL FOR PARTICULARS

The Walker Steel Range Company, Limited

Grimsby, Ontario

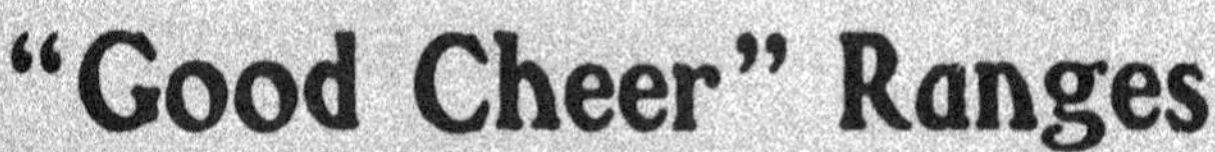

THE "GARNET"

"Money would not buy my 'Garnet' Range if I could not get another just like it." That's what users have said of it.

All loose nickel—readily removed when range is to be polished, and such an easy one it is to keep bright and clean. Square oven, 18 x 18 x 11. Fine firebox, with all cast linings. Grates, etc., removable without disturbing linings or waterfront. A splendid operator in every way.

Improve your range trade by handling the "Garnet."

The JAMES STEWART MFG. CO.

LIMITED

Western Branch:
Foot of James Street, WINNIPEG, MAN.

WOODSTOCK, ONT.

SUMMER SPECIALTIES FOR THE KITCHEN

The work and discomfort of summer cooking can be reduced to the minimum by the use of our **Gas, Gasoline and Oil Stoves, Gas Hot Plates, Refrigerators, Ice Cream Freezers, and other specialties** made to meet this need.

Their great simplicity and economy are considerations with which your customers may not be familiar, and if better understood would be sure to result in many more sales. A little introductory work along this line should prove effective.

Model Wickless Blue Flame Oil Stove

Height—14 inches.

An Ideal Summer Stove. Reliable. Easy to Operate. Easily Cleaned. Handsome Full Cabinet Frame, Nicely Japanned and Ornamented. Removable Safety Oil Tank, Brass Sub-Tank. Automatic Feed. Powerful Brass Burners. Perfect Working Valves.

SUMMER QUEEN (White Flame) OIL STOVE

Without Extension Top

Guaranteed to be better constructed than any other Oil Stove of this style made. Simple, durable, effective.

The Model is a moderate priced refrigerator, but compares favorably with the more expensive lines, and is much better value for the money than many of the so-called high-grade refrigerators.

It is made of hard wood, finished in golden oak; has heavy, solid bronze locks and hinges and is lined with zinc, while the insulation is superior to many of the higher-priced refrigerators.

Made of Hard Wood. Perfectly Constructed. Saves Ice.

Our Summer Goods Catalogue for 1907 has just been completed and mailed. If you have not received a copy please advise us.

The McClary Mfg. Co.

London, Toronto, Montreal, Winnipeg, Vancouver, St. John, Hamilton, Calgary "EVERYTHING FOR THE TIN SHOP."

"Grand Peninsular"

The Ideal Kitchen Range

The "Grand Peninsular" is my ideal of what a perfect cast-iron range should be.

It is the most reliable cooker of its kind—because the oven is built exactly like the oven of an expensive steel range.

Steel radiates heat quicker than cast-iron. Instead of using only a steel BOTTOM —I make the ENTIRE OVEN—top, bottom and sides—of patent Levelled Steel Plate.

That's one reason why the "Grand Peninsular" is such a quick, even cooker—and saves so much fuel.

Then I do away with all guess work about the baking, by putting a thermometer on the oven door. The oven is perfectly ventilated—the air kept fresh and dry—thus preventing soggy baking. I make the Grate Bars large and strong and easily removable.

All the ornamentation—the nickeled parts—can be lifted off the range when cleaning, without loosening a single bolt. You can see Peninsular Ranges at your local dealer's store. 40

CLARE BROS. & CO., Limited	PRESTON, ONT.
CLARE & BROCKEST	WINNIPEG, MAN.
MECHANICS SUPPLY CO.	QUEBEC, QUE.
CUNNINGHAM & WORTH	VANCOUVER, B.C.

This is a sample of our Newspaper Advertising. Let us send you samples of our Illustrated Booklets. We help you sell ***PENINSULAR RANGES.***

THE B. GREENING WIRE CO.
LIMITED

OFFICE AND WORKS:
HAMILTON, ONT.

EASTERN DEPOT:
442 St. Paul St.
MONTREAL

ESTABLISHED 1859

GALVANIZED NETTING

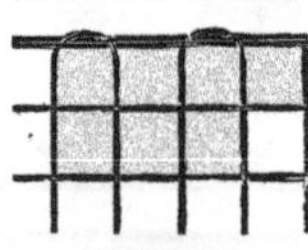
WIRE CLOTH

WIRE ROPE

WIRE DOOR MATS

SOFA SPRINGS

BEE ZINC

A FEW SEASONABLE LINES:

WIRE
Steel, Plain and Galvanized. For hay baling, fencing, coppered spring, coppered soft, tinned mattress and broom, galvanized hard coiled fence wire.

GALVANIZED NETTING
for all purposes. All meshes, widths and strengths in stock.

PAINTED SCREEN WIRE CLOTH
All widths, 18 to 48-in., kept in stock.

WIRE CLOTH
Copper, Brass, Steel and Galvanized. All meshes and strengths, for all purposes.

TRACE CHAINS
Improved quality.

BROWN'S PATENT STEEL WIRE CHAINS
Coil, tie-out, halter and dog chains, etc. Special chains made to order.

GREENING'S PATENT CATTLE CHAINS
and stall fixtures. Samples now ready for fall trade.

WIRE ROPE
Standard and Lang's Lay. For derrick use, passenger and freight elevators, mining tramways and other purposes.

COPPERED STEEL FURNITURE SPRINGS
All standard sizes kept in stock.

STEEL WIRE DOOR MATS
Most sanitary and durable mat on the market. Made either plain or lettered.

PERFORATED SHEET METALS
Zinc, Copper, Brass, Steel, Galvanized Iron. For all purposes.

FOUNDRY SUPPLIES
Riddles, Steel Brushes, Bellows, Shovels, etc.

BEE ZINC
In sheets 36x96-in.

STAPLES
Galvanized Fence and Poultry Netting—Bed and Blind--Electrical, etc.

CHAINS—BROWN'S PATENT

CATTLE CHAINS

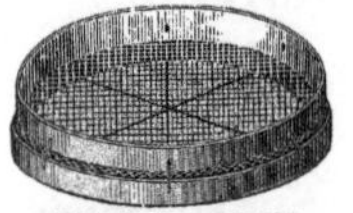
FOUNDRY SUPPLIES

PERFORATED METALS

Wire Rope for Hay Forks, full particulars on application

MATCHLESS EMPIRE

Made in 4 sizes, 14-in., 16-in., 18-in., 20-in. Full sized, very heavy. Built for service, style and power.

RADIANT—EMPIRE—BRILLIANT

For Soft Coal — For Hard Coal

We can't describe these beautiful heaters here; just write us.

MAJESTIC EMPIRE

Made in one size, 22-in. The largest stove of its class made on the North American continent.

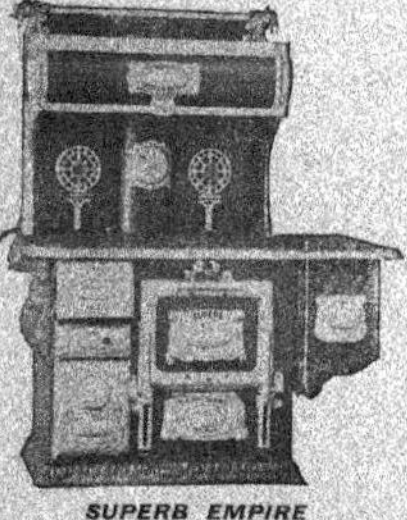

SUPERB EMPIRE

The Perfect Cooker

Made in 3 sizes, 17-in., 19-in. and 21-in. Either 8-in. or 9-in. tops. Not cheap—but unsurpassed.

SUPERIOR EMPIRE

Made in 3 sizes, 16-in., 18-in. and 20-in. Either 8-in. or 9-in. tops. Substantial—price reasonable.

WAYNE

Made in 2 sizes, 16-in. and 18-in. Either 8-in. or 9-in. tops. Attractive—a seller.

IT WILL PAY YOU TO BUY OF US, MR. WESTERN DEALER

"The Western Stove Makers"

W. J. COPP SON & CO

MANUFACTURERS OF

EMPIRE STOVES & RANGES

Fort William, Ontario.

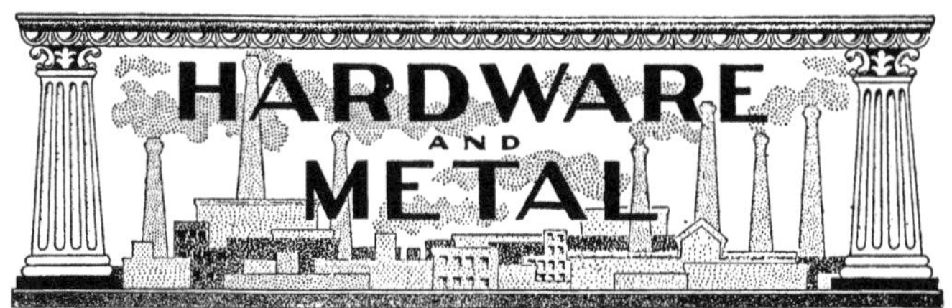

Vol. XIX. MONTREAL, TORONTO, WINNIPEG, MARCH 23, 1907 *No. 12.*

MATTERS OF MUTUAL INTEREST

Just as it often appears that the crest of the wave has been reached in the rising tide of metal prices and to see the wave, gather additional force and rise to even greater heights, so do the publishers of Hardware and Metal each year consider that the Annual Spring Number has reached its limit, only to see the following year far eclipse the mark set in the preceding issue.

The 1906 Spring Number won words of of commendation from many of the best known men in the trade, all agreeing that the issue was creditable to the publishers, of great publicity value to advertisers, besides containing practical reading matter serviceable to every reader. The 1907 Spring Number, now presented to our readers, will, we feel certain, deserve as hearty commendation as any previous issue. The reading matter has been compiled during the past few months by our editorial staff and comprises much practical, educative and authoritative matter which should cause thousands of readers to preserve the issue for future reference.

From an advertising standpoint this year's issue also shows a marked advance, between three and four hundred of the leading manufacturers and jobbers doing business through the hardware trade in Canada using the pages of the issue to make their lines known to the retail trade. The size of the paper has had to be considerably increased in order to include all this valuable matter.

* * *

That the retail trade reads and appreciates the value of the advertisements in this issue has been proven again and again. If the advertisements did not bring results the space used by jobbers and manufacturers would be constantly decreasing instead of steadily increasing. And if the retail trade did not find it of advantage to consult our advertising pages in order to keep posted regarding the latest announcements and lines being introduced by our advertisers, they would not send in the orders which produce the results the advertisers desire.

In reality the advertising pages are equally advantageous to both advertiser and subscriber. It costs less to reach the trade through an advertisement in Hardware and Metal than through circular letters and the results are better. Besides, the retailer prefers to look for the announcements of manufacturers and jobbers in a medium which he reads regularly and which contains the announcements of others in the same line. The Annual Spring Number has been well likened to an annual exhibition wherein the most up-to-date manufacturers and jobbers exhibit their latest lines and present their arguments, outlining why business should be placed with them. An exhibition would be attended by only a small percentage of the trade. The Spring Number will be read by every hardwareman in Canada. Consequently no manufacturer or jobber of standing in the trade can afford to fail to grasp the opportunity of reaching and talking to the trade in this number.

* * *

Another indication of the interest taken in Hardware and Metal by the trade is the increasing number of letters sent to the editor for information regarding the names of manufacturers of lines which merchants do not stock but which they are asked to supply to customers. Much valuable information has been given in the "Letter Box" during the past year, the trade realizing that this "Letter Box" is the one source from which they can secure prompt and impartial information regarding any class of goods sold through hardware stores.

READING MATTER INDEX

The "want ad." department is also being used more liberally by the trade. A merchant who wants to buy or sell a set of tinsmith's tools, wants to secure a clerk or wants to sell his business, or a clerk or traveler who wants to change his position or locate in the west, can find no better medium than Hardware and Metal's "want" column. A. W. Humphries, Parkhill, Ont., recently tried the "want" pages of a leading daily newspaper and got very unsatisfactory results. He then tried Hardware and Metal and transacted the desired business in short order. Mackie & Ryan, Pembroke, secured about seventy replies to a want ad. they recently inserted in our "want" columns at a cost of only 55 cents.

* * *

Probably the most important happenings of the past year, so far as the retail hardware trade in Canada is concerned, have been the organization of the Retail Hardware Associations in several provinces and the successful fight waged against the parcels post c. o. d. legislation proposed in Parliament by Postmaster-General Lemieux, but finally withdrawn as a direct result of the opposition developed by Hardware and Metal and the Retail Hardware Associations.

* * *

A development worthy of mention is the birth of the "Plumber and Steamfitter of Canada" this month. For nearly twenty years Hardware and Metal has been publishing technical articles and news of interest to those engaged in plumbing and steamfitting work in Canada, the aim being to establish a separate plumbing paper as soon as the constituency was large enough to warrant this action. This time has now arrived and the cordial reception given the new paper ensures it a bright future. A large portion of the plumbing and steamfitting work done in Canada, particularly in the smaller cities and towns, is contracted for by hardware merchants and arrangements will be made to allow these to secure more practical information than ever before at a minimum of cost.

The "Plumber and Steamfitter of Canada" has an able staff of writers, including, amongst others, M. J. Quinn, Toronto, who will conduct a department of "Questions and Answers"; C. E. Oldacre, Toronto, a member of the American Society of Heating and Ventilating Engineers, who has a wide reputation as a writer on heating subjects; Wm. Holley, Montreal, Associate in the Royal Sanitary Institute, who has written an able series of articles on the "Advancement of Practical Sanitary Science"; and W. B. Mackay, Toronto, an article by whom on "The Heating of a Hardware Store" will be read with interest by hardwaremen, even though they are not interested in plumbing and heating contracts. These are only a few of the features being arranged for the newspaper, the subscription price for which will be $1 per year.

* * *

An incident which has given us considerable pleasure has been the advance made during the year by one of the contributors to last year's Spring Number. Probably the best and most interesting article in the issue of a year ago was that written by Albert E. Karges, a Woodstock hardware clerk, on the "Interior Arrangement of a Hardware Store." The article excited a great deal of favorable comment, and as a result of the publicity gained by Mr. Karges, he was offered a position as traveler for the James Stewart Manufacturing Company, Woodstock. Mr. Karges went on the road and in calling on customers he was repeatedly asked if he was not the writer of the article on store arrangement. The interest thus established resulted in many orders and after a few months of successful work on the road Mr. Karges was as the beginning of the year promoted to the management of the Winnipeg branch of the business.

The hardware clerk of a year ago is now the western manager of a large manufacturing concern and the change is almost directly due to Hardware and Metal. Mr. Karges had the ability to succeed, but he was modest and his opportunity had not yet come. The editor encouraged Mr. Karges to spend his spare time in studying window-dressing and store arrangement and prepare articles on these subjects. These articles uncovered Mr. Karges' ability and developed his opportunity.

The incident shows clerks a road to success. Use the spare time in increasing the knowledge on store subjects and advancement is certain.

* * *

In response to a letter advising W. J. Illsey, of Winnipeg, that he had been awarded first prize in our Christmas window-dressing contest, Mr. Illsey replied: "The winning of this prize gives me a more earnest desire to increase my capabilities as a window-dresser. I enjoy, and more than formerly, your valuable paper's contents from week to week and endorse what Mr. Ross said in a recent issue: 'Give it to all the clerks.'"

In many ways we endeavor to make Hardware and Metal interesting to the clerk and traveler. We have departments: "What to do Next Month"; "Amongst the Salesmen"; "Window and Interior Display"; "Effective Advertising"; amongst others which are helpful to all in the trade and the jobber and merchant who does not encourage his employes to read the paper is losing an opportunity to make his staff of greater productive value to him.

* * *

One of our Montreal staff recently called on a prominent hardwareman for information regarding market reports, etc., and after the business in hand had been atended to satisfactorily, the conversation shifted round to our market quotations. "You want to be very careful to give the correct quotations as far as possible,' said the gentleman, "because the trade relies so much on the prices you give them, that it would make it very uncomfortable for jobbers and manufacturers, should a wrong price get in there by mistake." In gathering our information about the markets great care is taken and we have to thank both jobbers and retailers for much valuable information received during the past year. We invite correspondence from our readers regarding the quotations we give if it is found at any time that the discounts and net prices we quote are not thoroughly up-to-date.

* * *

Space prevents the reproduction of many testimonial letters, but the three below speak for hundreds received during the past year:

From a retail hardwareman: "I have been taking your Hardware and Metal ever since you started and you are improving with age as it is better to-day than it ever was and a great help to any hardware dealer."—F. A. Campbell, Mitchell, Ont.

From a hardware clerk: "We have always taken your paper in the east and I feel lost here without it. I believe every hardwareman should be a subscriber."—Leroy J. Crown, Edmonton, Alberta.

From a traveler: "You certainly deserve great credit for the progressive way Hardware and Metal is distributed through the trade in Canada. I find it is received by nearly everyone I call on, and, in addition, find that your editorials and comments on trade matters make very interesting reading. It is always a pleasure to pick up your paper whenever I have a few moments to spare waiting for a customer. It must be difficult to make trade reports interesting, but you deserve credit for the successful way you put this matter before the trade."—W. J. Wall, Montreal, Canadian representative of the Crane Company, Chicago, and the J. L. Mott Iron Works, New York.

HARDWARE AND METAL

Established 1888

The MacLean Publishing Co.
Limited

JOHN BAYNE MACLEAN . *President*

Publishers of Trade Newspapers which circulate in the Provinces of British Columbia, Alberta, Saskatchewan, Manitoba, Ontario, Quebec, Nova Scotia, New Brunswick, P.E. Island and Newfoundland.

OFFICES:

MONTREAL, ... 232 McGill Street
Telephone Main 1255
TORONTO, ... 10 Front Street East
Telephones Main 2701 and 2702
WINNIPEG, ... 511 Union Bank Building
Telephone 3726
LONDON, ENG. . . . . . 88 Fleet Street, E.C.
J. Meredith McKim
Telephone, Central 12960

BRANCHES:

CHICAGO, ILL. . . . . . 1001 Teutonic Bldg.
J. Roland Kay
ST. JOHN, N.B. . . . . No. 7 Market Wharf
VANCOUVER, B.C. . . . . Geo. S. B. Perry
PARIS, FRANCE - Agence Havas, 8 Place de la Bourse
MANCHESTER, ENG. . . . . 92 Market Street
ZURICH, SWITZERLAND . . . . Louis Wolf
Orell Fussli & Co.

Subscription, Canada and United States, $2.00
Great Britain, 8s. 6d., elsewhere - 12s

Published every Saturday.

Cable Address { Adscript, London
Adscript, Canada

ENTHUSIASM.

What does enthusiasm mean to the hardware merchant or clerk?

To the one it means increased sales, to the other increased salary.

You must love your goods or you cannot sell them. If you are enthusiastic you can make a possible customer enthusiastic, for there is nothing more contagious than enthusiasm—unless it's indifference.

But enthusiasm is not a made-to-order article. It's not an escape of natural gas or effervescent bluster. To be effective it should be spontaneous—come from the heart as well as the head; it must not be a studied or affected enthusiasm, for there is more than one kind of wireless telegraphy, and men—and women especiallly—will read you like a book.

But how to acquire this enthusiasm: there is only one way.

Know your goods. Read the advertisements in this number, and pay special attention to those which are bristling with enthusiasm, that have the ring of sincerity about them. If a talking point in an advertisement strikes you as being good remember it will probably strike your customers in the same way. Encourage your clerks to read the advertisements and assimilate the selling points of the paints, revolvers, roofing materials and other lines in which they are interested.

Confidence in one's self and in the goods one sells is half the battle.

BENEFITS OF ASSOCIATIONS.

A year ago the retail hardware trade in Ontario was unorganized; the master plumbers' organization in Toronto had been adjudged guilty of criminal conspiracy; the hardware manufacturers had been investigated on charges of unlawful combination, and, with the exception of the strong retail hardware association in western Canada, there seemed to be no ray of hope for the organization of the hardwaremen of Canada.

To-day all is different. The Ontario Retail Hardware Association has come into existence and secured a strong membership. Nova Scotia and British Columbia are partially organized, and practical steps have already been taken to organize a Dominion Retail Hardware Association to enable the retailers in all parts of the Dominion to work together to advantage—just as they have already stood shoulder to shoulder in successfully opposing the parcels post c.o.d. legislation proposed in Parliament by Postmaster-General Lemieux.

The Dominion Government have not yet revised the Criminal Code to allow retail merchants the same rights enjoyed by lawyers, doctors and trades-unionists—who can meet together and jointly agree upon a price at which to sell their various commodities—but on the other hand, their failure to press the charges made against the grocers' guild, the tack manufacturers and others, shows that they see the injustice of the position these business men have been placed in. The grocers' guild are demanding that the Government reimburse them for the legal expense they have been put to, and the master plumbers seem likely to reorganize along lines which will be unobjectionable to any section of the community. Altogether the position of trade organizations in Canada has been vastly improved during the past year.

This is an age of organization, and nothing can stop the getting together of men whose interests are common in opposing certain matters. All the important industries — manufacturing, wholesale and retail—are organized for self protection and advancement, and the retail hardware trade should not be following the rear guard in the procession. They should organize nationally, provincially and in smaller county or district organizations.

Trade organizations have broken down the bars which formerly kept men in the same line of business from meeting in friendly intercourse. They have demonstrated that when competitors get together their illusions that the other fellow wears horns, cloven feet and a forked tail are false, but that he is a good fellow trying to overcome the same problems and make his business pay him a fair profit so that he may live comfortably. Hardwaremen in Canada have met together at conventions, have rubbed shoulders with competitors, and have been better for the experience.

The convention of the Ontario Retail Hardware Association which is to be held in Toronto next Thursday and Friday, March 28 and 29, will be one of the most important gatherings ever held by the trade in Canada. Indications point to a record-breaking attendance, a splendid program comprising business, good fellowship and trade discussion features has been arranged, and the Ontario retail hardwareman who does not avail himself of the opportunity to gain knowledge by hearing the views of some of the most successful men in the trade—well, he's not up-to-date and he's missing a good thing, that's all.

A REFRESHING INCIDENT.

Last month, an example of business integrity was brought to our attention, which was as rare as it was refreshing. We regret that we have been prohibited from mentioning names; it would have given us particular pleasure to have openly expressed the appreciation of the trade for the man who acted with such a high sense of his obligations.

It was the old story of a failure years ago and a struggle ever since to gather together funds to repay creditors, who, in the eyes of the law, had no claim on the bankrupt. It meant the building up from the foundation of a business which has proved a success and which has enabled the owner to pay off all debts with interest to date.

The trade has few such examples before it and this one should serve as a reminder that even in the ordinary course of business, there is something a little higher and finer than the mere adherence to legal obligations.

INDIVIDUALITY OF WESTERN TRADE.

It is often said that the Canadian west is a general store country and easterners who are unacquainted with the actual conditions in the newer part of Canada are apt to accept the statement without question. A trip through the west soon shows that this general statement is only a half truth. There are many general stores in the west and the bulk of some classes of goods is handled in general stores, although in the west, as in the east, a generation ago, the tendency is towards the establishment of specialized stores. But while the general statement may apply to some classes of goods it does not apply to hardware. It would be safe to say that at least 85 per cent of the hardware sold in the west is sold in specialized hardware stores.

A trip through the country or an examination of Bradstreet's or a general business directory will establish this contention beyond the shadow of a doubt. Even the smallest towns have their specialized hardware stores. For the purposes of this article the writer has selected at random five small towns as given in Henderson's Northwest Gazetteer for 1906. Redvers, Sask., with a population of 150 in 1906, had three general stores and two hardware stores; Miami, Man., with a population of 400 in 1906, had two general stores and one hardware store; Saltcoats, Sask., population, 600 in 1906, had three hardware stores and five general stores, Warman, Sask., with a population of 100 in 1906, had three general stores and two hardware stores, and Waskada, Man., with a population of 200, had three general stores and one hardware store. It is, in fact, a difficult matter to find a hamlet anywhere on a line of railway in the west that does not support at least one hardware store. In the newer parts of the country it is the same story as in the old. Along the main line of the Canadian Northern between Winnipeg and Edmonton there are many new towns which had no existence a year ago. The traveler finds in them all at least one hardware store in addition to the general store, handling the usual varied assortment of all classes of merchandise.

The reason is not far to seek. The hardware business is the biggest individual business in the west. The amount of building done each year in a new country is enormous, and straight line hardware stores have been started to supply the demand in builders' and general hardware.

The hardware merchants also have their own retail association quite distinct from the Retail Merchants' Association of Western Canada, an organization of general merchants. They feel that their interests are quite distinct. There would be no separate Retail Hardware Association in the west if the general statement that the west is a general store country applied to the hardware trade.

SHOULD PROVIDE SELLING POINTERS.

Manufacturers can help in educating clerks to increase sales by furnishing them with more information regarding their goods. The maker of a saw or knife or a lock is fully conversant with the article he manufactures. He knows why a certain kind of steel or a brand of iron or a kind of wood is better adapted to the purpose than any other and why the article is made of this particular steel, iron or wood, how a mechanic can best use a tool to the best advantage; why not impart this information to the merchant and his clerk so they can understandingly describe their wares to their customers? How can it be expected that a merchant with hundreds and thousands of different articles on his shelves can talk with knowledge and point out the merits of different articles when the only guide he has is the label on the package giving quantity, size, number and name of article? Manufacturers should use their advertising space in the trade newspapers to provide selling pointers for the merchants and clerks to use when selling the goods.

VALUE OF A BUSINESS EDUCATION.

Every young man entering business with ambition and enterprise as his assets will not be satisfied to stay in a small business. The competitive spirit and the ambitious spirit constrain him to aim towards getting into a big business and become master of it. The thing needed most by such young men to help them to realize their ambitions is a thorough knowledge of those matters which always come up in business transactions.

For instance, how essential is a thorough knowledge of business law! He must know the value of a contract and all that goes to constitute a valid contract. He must know the underlying principles governing large transportation companies, as railways, steamship companies, and express companies. He must know the relations existing between consignor and consignee, between dealer and purchaser.

It is necessary for him to be able to discriminate between good and bad securities, to know what securities represent, the specific laws governing them, the financing of large concerns, the meaning of partnership, and the working of the money market. Too many financially strong merchants have had their business ruined through unwise speculations, not knowing the ins and outs of the stock exchange. A thorough knowledge of the detailed workings of the business concern is indispensable to the young man who is ambitious to better his position, to make it a more lucrative and responsible one. Those who win are the masters of their craft.

WHO PAYS FOR ADVERTISING?

Who pays the cost of advertising? At first sight this question looks easy. One is apt to say offhand that the advertiser pays it. But upon closer investigation it becomes apparent that an advertiser, who advertises properly, gets back all the money he spends in advertising, and a good deal more—else what would be the use of his advertising?

The purchaser doesn't pay it, because he gets the goods as cheaply from an advertising firm as he could from a non-advertising firm. In many cases he buys cheaper. Then who pays the cost of advertising?

The non-advertiser does! By the lack of enterprise in bringing his goods before the attention of the public he loses customers, who buy from advertising people. The profits which are lost by non-advertisers, find their way into the coffers of those who do advertise.

As an example: If $20 spent in advertising brings you in $30 additional profit, your advertising is paid for and you are ten dollars in pocket.

THE MERCHANT'S FINANCES.

Jobbers report that many of their customers have been slow in making their payments this spring, there having been altogether too much money handed over to the promoters of Cobalt mining propositions by merchants who should know better. The money which has been used in gambling in stocks would, as a general rule, earn far more for the merchant by using the funds in buying and taking short term discounts on the many lines which have been advancing in price during recent months. But when a gambling craze is on some suckers must be caught to provide the funds for the promoters' advertising, etc., and the business men, in spite of their supposed knowledge of financial affairs, provide a fair percentage of the cash required.

On the other hand the manufacturers and wholesalers say there is a steadily increasing army of retailers who are getting out of the ancient rut of doing business on a long term basis, fully 60 to 70 per cent. of the hardware trade now taking the discounts allowed for prompt settlement. There is, in fact, a movement springing up amongst the most up-to-date retailers to pay spot cash and demand an extra half per cent. for the accommodation. For instance, a Toronto firm recently bought a $400 bill of goods from a manufacturer and instead of letting the account stand for 30 days and taking off the usual 2 per cent., amounting to $8, he settled at once, and deducted $10, equalling 2½ per cent. The manufacturer expressed himself as being well satisfied to allow the extra discount as he favors encouraging every movement towards prompt settlements. He also pointed out that the man who paid spot cash was not tied to one house, was more independent, and would often receive special favors as a result of his ability and willingness to pay spot cash. In his opinion every dealer should exercise great care in buying so that they would not overload themselves and not have ready cash to meet accounts promptly in order to take the discounts.

We recently heard of one merchant doing a business with an annual turnover of $540,000 who received $24,000 in cash discounts in one year. Would not a dealer doing a business of one-tenth that sum annually find $2,400 a welcome addition to his profits? There are many merchants who are constantly receiving new goods, and to these the discounts should be, and are, a very important item. The dealer who fails to take the discounts is losing money. Go to the bank and borrow the cash if you haven't it on hand. The discounts will more than make it up. It will pay handsomely to pay interest on a loan, when you use the money to take advantage of the discounts. The man who can use his money and make 5 to 10 per cent. on it through discounts is getting the best result.

PRIZES FOR IDEAS.

The editor has been planning to announce a series of competitions in which prizes would be offered for the best letters on subjects of interest to the hardware trade. The idea is to encourage Canadian hardware merchants, travelers and clerks, to take advantage of the facilities offered by Hardware and Metal as a forum in which discussions can take place which will be educational to everyone in the trade.

It is really not necessary to offer prizes, but as mankind loves competitions of a friendly nature and it is necessary to have rules, judges and referees in every contest, it is felt that more interest will be taken if nominal prizes are held up to be striven for. The first subject on which we invite discussion, therefore, is the following:

What is the best plan a hardware merchant can adopt to increase the sale of stoves and kitchen furnishings in the fall and winter season? What methods of display and what system of advertising can be used to the best advantage? Should the dealer canvass his district for business? How can old stoves be disposed of to best advantage and what plan of selling on easy payments brings the best results?

For the best letter answering the above questions received before June 1st, 1907, a prize of $10 will be awarded, and the best letters will be published in order to be of value to the trade in outlining their fall campaign for stove business.

If the number of replies received warrant it this will be followed by other discussions for which prizes will also be awarded. There is no reason why any merchant, traveler or clerk should fail to express their ideas on the question asked.

PUNCTUALITY AN ASSET.

One of the most valuable assets of a business concern is punctuality on the part of the employes. It is always a very admirable and commendable quality, especially in a business man. It shows true appreciation of the value of one's time, and displays a scrupulous sense of honor, self respect, and interest in one's company. Apart from the fact that a dilatory clerk is a drag on a business, punctuality goes far towards increasing the business. A man who lacks this quality, punctuality, to a great extent, it matters not how well he may be equipped, otherwise, will not only suffer many losses in a commercial sense, but will be regarded as unreliable and unsafe in all business transactions involving large sums of money, and as untrustworthy in a moral sense, in matters pertaining to social and civic duty.

It is essential that the employer, also, should be punctual, for an employer who is not punctual in the performance of his own duties is not likely to arouse the enthusiasm of his employes and, as a consequence, his business will suffer in the matter of its management as well as in every other department.

POINTERS FOR CLERKS.

The fellow who is always late in coming in usually the first one to go.

Cheerfulness is catching, and there is always room in the store for a smiling countenance.

Clean hands and clean linen make a favorable impression, while the other kind doesn't.

Jobbers' Relations With Retailers

Selling to Consumers Condemned — Why Retailers Should Support Jobbers — Adjusting Errors in Shipments — Experiences of Two Retailers.

WHY DO RETAIL HARDWARE MERCHANTS FAIL?

By a Prominent Montreal Jobber.

This is a subject of deep importance to both the wholesaler and retailer. Traces of failure in the hardware business are found in a variety of sources.

Starting With Insufficient Capital.

This should never be attempted and is perhaps the main cause of all troubles in the hardware business. Young men with ability, no doubt, believe they can start, these modern days, to do business of $15,000 or $20,000 per year with $1,000 capital. This is utter madness and will only result in the old old story of failure. Great care should be exercised in raising sufficient capital to start with. Under no consideration, start with borrowed capital. This leads to an endless amount of trouble.

Want of Experience.

The wholesale jobber frequently has to refuse credit to young men, who possibly have some capital, but no experience. The two qualifications naturally go hand in hand, and knowledge and thorough training in the hardware trade is more requisite to-day than it has ever been in the past owing to the large variety of goods required to compete an up to date hardware stock. In addition to knowledge, tact and politeness, coupled with caution are absolutely necessary. Remember that this is a time of close and keen competition, and only by those who are strong in the points mentioned, will success be attained.

Tendency to Overbuy.

This is a grevious mistake. The facility of getting goods from the wholesaler to-day has never been equalled, and the retail man does not require to carry the heavy stocks which were formerly necessary. Buying too many goods has probably been the cause of more failures in the hardware trade than anything else we know of. The wholesale jobber, if he is up-to-date, must pay cash, and when the retailer asks for further credit than the usual terms, he is plunging himself into an interest account. Far better, to do a limited business on a carefully purchased stock, than endeavoring to do a large trade requiring an excessively heavy stock. While on this point, it might be mentioned that an up-to-date retailer must keep abreast of the times and be constantly on the lookout for novelties, so as to keep up an attractive display of his goods. Retail hardware stores can be made attractive, and there are many which are a credit to this country, and fully equal to the United States or any other land. There is no reason whatever why a small hardware stock should not be kept in neat order and the goods attractively displayed. This will help materially to draw customers; especially is this so in the larger towns.

Be Careful in Partnerships.

The retail hardware merchant if he enters into partnership should be very careful to select the right man. Want of unanimity amongst partners has often been the cause of failure, the disagreement often leading to a policy of war. To make it still plainer, the writer has known partners who for years hardly ever exchanged a word, the feeling between them had become so bitter; this is wrong and can be easily avoided. Be sure you select a partner thoroughly competent to bear his share of the business. In a partnership where two are concerned, it would be wise to have one an office man and the other a thoroughly competent hardware man. This should make a strong company.

The Credit Problem.

By having a keen oversight over the credits, a retail man may avoid disaster, in fact, he is almost sure to do so. A policy of as near cash as possible, should be adopted by all hardware merchants. Even if their profits appear small success will be attained. Granting of too much credit has always been a great evil, and a continual collecting of accounts and keeping them well within range will add largely to the success of many business and avoid financial distress.

JOBBERS AND RETAILERS MUST STAND TOGETHER

By Weston Wrigley.

It is said that in the New England States there is not a first-class jobbing house—that in fact, there is not such an institution east or north of Philadelphia in the United States. The reason given for this peculiar situation is that jobbers incurred the ill-will of retailers by selling promiscuously to consumers and the retail dealers in retaliation transferred their business to the manufacturers, who in this territory are able to handle the retailers' trade satisfactorily owing to the nearness of the factories to the retail market.

There is a lesson to be drawn by Canadian hardwaremen from this situation. Canadian jobbers—some more than others—are selling too freely to consumers for their own good. The writer knows one Ontario retailer who conducts more than one hardware store and who boasts of buying between 80 and 90 per cent. of his goods direct from manufacturers. The first condition will breed the second. Let the jobbers take action to preserve their hold upon the retailers by eliminating this most important evil. If present tendencies are not checked more and more business will be placed with manufacturers by the retail dealers and the New England experience may be duplicated in Canada.

There is as much need for the jobber as for the retailer, under our method of doing business to-day. Of course business could be done in a sort of "unsight and unseen" method by the large mail order houses selling direct to the consumer, but the objections to this method are too obvious to require enumeration in detail. Who, for instance, would want to buy a ton of coal, a dozen of eggs or a gallon of kerosene by mail order?

A large buyer for several stores may be able to buy most of his goods from manufacturers, but 95 per cent. of the retail hardwaremen in Canada can do business best by using the jobber as an assembler of goods in large quantities to distribute in small lots of infinite variety to hundreds of merchants in all parts of the country.

Any attempt to eliminate the wholesaler is likely to react upon the retailer. The jobbers have carried a large number of retailers in bad years and in prosperous years when the demands for credit by consumers in newly settled districts would have forced the retailers to the wall. The jobbers are men of wide experience and thorough knowledge of market conditions. As a rule they recognize that it is not in their own interest to with one hand sell to legitimate retail customers and with the other hand sell to mail order houses or to the consumers direct. This isn't a square deal—but it is done to a considerable extent.

Without making any comparisons nearer home let us take for an example the case of the recently defunct First National Co-operative Association, otherwise known as the Cash-buyers' Union, of Chicago. This institution did a large mail order trade and claimed to be able to supply anything in the hardware line from "a needle to a threshing machine" from their eight-storey building. When the sheriff took possession it was found that they carried only a $15,000 stock of hardware. But their books showed that they had been doing a big business at cut prices. Where did the goods come from? Was their big stock carried by jobbers or manufacturers? When the concern received an order for a pair of nickle-plated skates and they had only blued cast-steel in stock, was the order passed on to the manufacturer or was it given to a jobber in a direct or indirect way?

It has been clearly shown that the jobbing houses in Chicago and other places supply goods to the mail order concerns who are demoralizing trade for the retailer and now that the mail order problem is developing in Canada (as instance, the proposed parcels post legislation and the subservience of the daily newspapers) it would be well for hardware jobbers and retailers in Canada to take joint action to retain the trade in the regularly established channels. Here is work which the Wholesale Hardware and Retail Hardware Associations could take up with mutual advantage.

While there has certainly been a growing tendency towards eliminating the jobber and retailer in the United States, the former by retail dealers doing business directly with manufacturers, and the latter by the larger mail order houses buying from the manufacturers or manufacturing themselves, there are evidences to show that the movement in this direction is being headed off, one incident which might be quoted being the legal action taken by Montgomery, Ward & Co., Chicago, against the South Dakota Retailers' Association for boycotting the jobbers and manufacturers who supply goods to the Chicago concern.

The disposition to give the jobbers the go-bye and to buy direct from the manufacturer, therefore, not only weakens the retailer in his fight against the catalogue houses, the jobber being a valuable ally in dealing with manufacturers, but it also encourages the chain store idea, which tends to develop a few large firms controlling from two to twenty branch stores, much upon the same plan as the "syndicate" five and ten-cent stores or the "tied" hotels in the large cities controlled by the large brewers.

Looked at from any standpoint it seems clear that the interests of the retailer are bound up with those of the jobber and that the wholesaler is his own worst enemy if he does not do his utmost to eliminate that growing evil—the sale of goods direct to consumers.

EXPERIENCES OF TWO RETAILERS

Two Ontario retail hardwaremen not long ago were induced to outline some of the troubles experienced by the retail dealer in towns remote from the jobbing centres. One from Sturgeon Falls contended that it was the jobers duty, if only from a selfish standpoint, to protect the retailer. He contended that retailers could get along without jobbers much easier than the wholesale houses without retail customers and he gave two instances of ill-treatment at the hands of traveling salesmen for jobbers who go after business direct from consumers.

"A traveler called on me," he said, "and I ordered a quantity of iron. He then called on a customer of mine, the best blacksmith in town, and sold him a year's supply and his shipments came in the same car as mine. The blacksmith owed me $200 at the time and I had to wait about a year for it while the jobber got his cash at once.

"Another instance was even worse. A mill in my town asked me to quote on building paper and I knew that with the quantity I had already bought their order would entitle me to the ten per cent. rebate. I quoted at cost, therefore, but discovered that they had got quotations from two wholesale firms. One of these wrote me saying they had quot, but for me to secure the order as I would get the rebate. The other firm said nothing about it to me but sent their traveler who, when he found I had quoted the same figure, told the millman that I was expecting the ten per cent. rebate, but if the order was given to him the millman could have it himself. The traveler then had the impertinence to call on me for an order and while in my office the phone rang and my wife, who answered the call, was told that the order had been given to this traveler at ten per cent. less than I quoted. My wife repeated the conversation to me and, needless to say, the firm has not received an order from me since.

"There is a great deal of this work going on that we don't hear about and it injures the retailer greatly."

Took a Poor Man's Money.

The other merchant, an Essex County man expressed himself strongly in favor of the retailers organizing a strong association, and outlined his opinions on the troubles of retailers as follows:

"In these days of strenuousness in business a certain amount of leniency is allowable, everyone is pushing to do his best, to do better if possible than the other fellow. I have a case in mind in my own business experience that fully illustrates this.

"I had a call from a traveler representing a wholesale house I did business with, a firm indeed, from whom I bought the most of my general shelf goods and a house that did a good paying business, considered commercially speaking, in the very highest standing; well off, could in fact pick their customers, and enjoyed the reputation of being one of the best all-round houses in the trade.

"After a short time spent in general talk, he informed me he was going to a small town about 9 miles distant (off the railway) where there was no hardware store, but that a party who had been school teaching there, contemplated going into business in that line, having saved by many years of toil, something less than a thousand dollars. Before he went to solicit business, however, he wished to consult me as I might consider it might injure my business to have them sell this man so close by, and that they would be guided by my answer as to the wisdom of opening up an account and that they would not do so unless I was perfectly satisfied.

"I replied to him that it made no difference to me whether they sold to this man or not, that it made no difference to my business and that they could do as they thought fit. I wished, however, before he left to place before him the following facts, viz., that I knew the man in question, that he was perfectly honest; had no knowledge whatever of any business, and that they had had a good general store account in the same village for many years, and I asked him if in view of all these conditions he was justified in selling this man a bill of hardware, take all his money and probably make him lose every cent of it before the year was out.

"The traveler merely smiled, shrugged his shoulders and stated that if he didn't do it somebody else would, and gave me to understand he intended to sell him if he could, if I made no objection. I told him finally that it seemed a disgraceful piece of business that a firm of their standing should have to get down to such small means to capture trade, and that if they cared to be placed on a level with the riff-raff of the trade that had to scrape around for living by such means, he could blaze away so far as I was concerned.

"What was the result? Why it turned out exactly as I predicted, for in about 9 months. the poor man had spent all his money, and as his credit was gone he couldn't do anything but shut up as no one would trust him.

"In this instance it was entirely a case of grab, there was no judgment whatever used, and a house that stood very high in the commercial world had dropped down to the low level of a 3rd or 4th class concern that had no reputation to lose. This happened several years ago, but I have no doubt like cases could be enumerated by retailers all over the country.

"Every excuse seems to be made to make sales, for instance a paint firm with whom I was doing business to some extent, was very anxious (with several others) to sell the paint to cover a very large new building here, and in order to get down low enough to beat all the other fellows he said he couldn't afford to give me a rake-off, and in fact he admitted to me afterwards that he lost about $50 by making the sale. All this forsooth was done so he could say their paint covered the biggest building in the county.

"I could bring forward other cases similar to those, but those two serve to show that the retail hardwaremen have their grievences. Sometimes, of course, these things are done by drummers who do not display any common sense; that it is simply a grab business with them so that they can rake in a little more commission over their regular salary at the end of the year, but as long as the house agree to it the is as bad as the other.

"Now I claim that wholesale houses and manufacturers that can't exist without resorting to such unbusiness like transactions should be out of the business altogether. Competition is keen of course, probably more so at the present day than it has ever been and yet it is not possible for wholesalers and manufacturers to do a clean business? I mean, of course, houses that claim they have an established record that will always bear inspection.

"There are houses in the trade that it is a pleasure to do business with, honest and straightforward in their dealings, courteous and gentlemany in their correspondence, in fact model firms in every respect, but it will not be till the millennium comes that every jobber will have attained to that reputation."

ADJUSTING ERRORS IN SHIPMENTS

The shiper in a jobbing house, like all the rest of us, makes mistakes occasionally and sends to much to one customer and too little to another—and sometimes he sends the wrong articles through error or in substitution for lines which are out of stock and not readily procurable.

In the readjustment of these errors there is opportunity for disagreement, and while hard feeling sometimes results, there is really no occasion for it if both retailer and jobber takes up the matter of difference in the proper spirit. The jobber must, of necessity, be prepared to give and take as he desires to retain the goodwill of his customer. The retailer, on the other hand, too, often takes an arbitrary and inconsiderate position, seeing his own side of the case and refusing to recognize any other.

For instance, a Toronto jobber receives an order for three washing machines of different sizes and of a certain make for shipment to a dealer in the Niagara district. He instructs the manufacturer at Hamilton to forward the order but the latter's shipper sends one machine of a wrong size. The retailer detects the error and immediately ships the wrong machine back to the jobber at Toronto, writing the jobber of his action. The jobber is put to the expense of paying two freight bills in straightening out the order where if a letter had been sent the house explaining the circumstances more prompt and satisfactory attention could have been given the matter.

Every shipment of goods made by a jobber is checked over two or three times and yet mistakes happen. How many times will it be necessary for retailers to check over their incoming shipments to avoid mistakes? And how many devote the attention they should to this important work? It pays to watch this matter closely.

And it pays to be honest about "over" articles received wherever there is a "short" there is almost certain to be an "over." The shipper's assistant might, in checking over a shipment, place an article intended for John Smith in John Brown's pile of goods adjoining.

An instance of this occurred last fall, when a package of pearl handled cutlery was checked off and shipped, but not to the right man. The dealer who did get them had an elastic conscience, as they haven't been returned yet, although, of course, the jobber had the replace the item to the dealer who ordered them in the first place. The jobber lost between $25 and $30 by the mistake and the dealer was in this much by his neglect to report this "over" to the house.

Another incident related is of a dozen plated napkin rings being shipped to the wrong address. In this case, however, the dealer notified the jobber that he had received goods which he hadn't ordered and weren't on his invoice. But he was a wily customer and wrote thanking the wholesaler for his thoughtfulness in sending him such a nice Christmas present.

Inquiry amongst the jobbers brings out the fact that there are few dealers who take advantage of mistakes on the part of the jobbers. There are some, however, who take the view that "if the jobber is fool enough to give them goods for nothing, they are not fools enough to refuse it." And peculiarly enough these are usually the ones who consider the jobbers the least when a shipment contains some "shorts" or wrongly sent goods.

It is an accepted truth that the dealer whose's relations are the most intimate with his traveler and jobber is the one who receives any favor or special attention when it is in the power of the jobber to repay the friendship.

Relations existing between the wholesaler and the retailer should have for their foundation the square deal. Fair and true statements of conditions should also be given by the retailer when asking for credit and fair and square treatment of the retailer when buying goods, the wholesaler giving fair terms and right prices. Mutual trust and mutual dependence result in harmonious relations, with give and take concessions when occasion necessitates.

MR. SARGENT'S EARLY HOURS.

"Some years ago," said a traveling hardware salesman, J. B. Sargent, of New Haven, engaged a young man as private secretary, and asked him to begin work the next morning. 'I am a rather early riser, he said, 'and would like you to get around in good season.'

"The youth was anxious to please, and to make sure that he would be on hand before his employer, he showed up at 7.30 on the following morning. Mr. Sargent was at his desk at work.

"On the second morning he came at 7.00 and found the old gentleman ahead of him. On the third morning he came at 6.30, and found Mr. S. deep in the mail.

"When Mr. Sargent had opened the last letter, he wheeled around and said: 'Young man, I would like to ask you one question.' 'Yes, sir.' 'Would you be good enough to tell me where you spend your forenoons?'"

Merchants and Their Stores

Some Examples of Store Arrangement—Fixtures for Displaying Goods—Business-Getting Suggestions.

ASHDOWN'S STORE AT WINNIPEG

One of the best examples of hardware store construction and arrangement in Canada is the recently constructed building of the J. H. Ashdown Hardware Company, at Winnipeg. Built on a leading corner on Main street, only a block from the city hall and Union Bank building, the location is a favorable one, giving a frontage on three streets, with a width of two stores for window display on Main street. "Canada's Finest Retail Hardware Store," is what W. J. Illsey, manager of the cutlery department of the store claims, he stating this to be the verdict of dozens of traveling men who have visited the store.

Ground Floor Plan of the Ashdown Store at Winnipeg.

The wide frontage and corner location provides five large windows, divided into two main sections, the main door dividing. Here displays are changed frequently, every advantage being taken of this form of advertising. The basement is used largely for storage purposes, the front being occupied by sporting goods, builders' hardware and tools in shelves. The centre section are the nail making quarters, nails being put up in cartons from one to ten lbs. Behind this are lavatories, an elevator shaft and more shelves for tool storage and carriage makers' hardware and blacksmith's supplies. Back of this is the heating apparatus and electric switch-board, next to this being a small tin shop for the

The Housefurnishings Corner in the Ashdown Store.

company's own work. The rear section of all is used for plumbing goods storage and paints, while under the pavement all round the end and side of the building is storage for grate tiling and oil supplies, the "Bowser" pumps being used.

The ground floor is shown in the accompanying diagram. On the right is the sporting goods and in the centre the cutlery department, where sales have averaged about $60 a day. On the left builders' hardware is kept, the goods being sampled on "Bennett" boxes, finished in light oak. Above the drawer section right to ceiling are stored all kinds of surplus goods in the line. This line occupies about half the front section of the store on the left. Beyond come tools of all kinds, machinists', carpenters', moulders', electricians', masons', plasterers', etc. The tool department also occupies a part of the large central show case for finer lines of tools. Back of the elevator is the blacksmith department. Pumps also

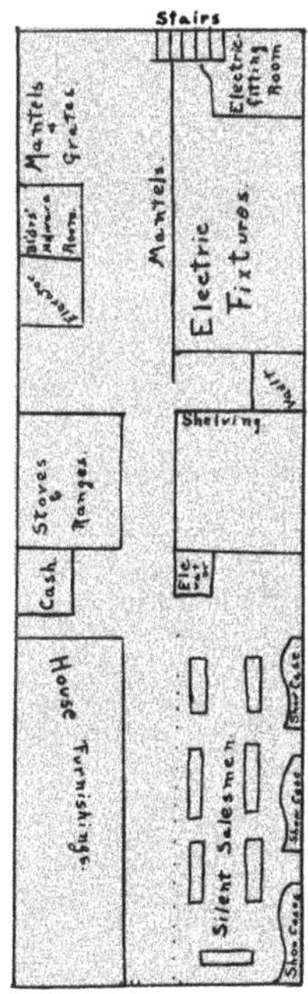

Second Floor of the Ashdown's Winnipeg Store.

figure among this department's outfit. The rear of this ground floor contains the paints and oils and plumbing goods departments. In the rear and opening on Albert street are the delivery rooms and freight elevator.

On the second floor in the front the firm carry a very large stock of enamel and tin and copper ware all displayed on two and three-deck tables, also wire goods and general kitchen utensils. The left front (facing) is filled with silent salesmen cases containing hand-painted china, silverware, fancy Austrian and Doulton and wedgewood china, leather goods, and bronze articles. Along the left wall large showcases contain clocks, silverware, cut glass and china and silver goods. Behind the cash office is the stoves and ranges display on raised platforms, bathroom fixtures, etc. The back half is occupied by mantels and electric fixtures. In a builders' hardware showroom everything new and good may be seen in working order. Unit lock sets, front-door sets, store-door sets and such lines. The third floor is occupied by the general offices and receiving and shipping room. To really understand the interior beauty one must visit it, but the illustrations we produce will give readers who have not seen the store an opportunity of appreciating the excellence of the general arrangement.

Sporting Goods Department on the Ground Floor of the Ashdown Building.

TORONTO'S LIVE RETAIL STORE.

Every hardwareman who visits Toronto should pay a visit to Russill's store, at 126 King street east. A half hour spent in watching the methods of doing business in this store and, if possible, a five-minute chat with Frank M. Russill, the progressive manager of the business, will be time well spent if the visitor is an observing man, capable and willing to learn a few lessons in up-to-date methods of doing business.

About a dozen years ago Mr. Russill was working with his father in a crockery store known as "Russill's in the Market." The store next door fell vacant and it was decided that the premises would be enlarged. A stove salesman happened along and knowing the Russill's to be live business men, he suggested that they put in a line of stoves, offering to supply them on consignment for the first year. One of the clerks had some stove experience and said he could sell stoves, so the move was decided upon. Then some one suggested hardware and a small stock was installed and for seven years the business gradually developed in spite of keen opposition for the farmers' trade.

Paint and Housefurnishings Departments in Russill's Store.

One afternoon five years ago Mr. Russill saw a vacant store across the street and investigated the premises. There was a chance for expansion, he figured, and the new store would provide the space, so a lease was arranged on favorable terms. At that time King street east, near the market, was considered dead from a business standpoint, most city business centering around the departmental stores near Queen and Yonge streets. Old established hardware stores had the bulk of the farmers' business brought by the market also, and Mr. Russill was given three months as the lease of life of his new venture. For a while he catered for the farm trade, but he soon decided that other dealers could handle all the wire fencing, binder twine, wire nails, etc., at small margins, while he would try to keep busy on tools, hardware, paints and other lines which allowed a fair profit. The more business the other fellows did in the unprofitable staple lines the better Mr. Russill was satisfied, as they had less time to compete with him.

By careful advertising on lines which had been proven successful by departmental stores, Mr. Russill gradually developed a splendid business with the city mechanics. He outlived the three months and in three years was prepared to drop the farm trade almost entirely, so satisfactory had the city trade become. Advertising in the country papers was cut out and only city dailies continued. Three delivery wagons and three telephones are kept busy a large percentage of the business being done by telephone and c.o.d. Up to the present the delivery service has been done by contract, the average cost of each wagon, horse and driver per year being about $1,000. Altogether there are over 30 clerks, shippers and other persons employed in the store.

Further expansions are now under

way, $3,200 having just been spent for four of the latest electrical cash registers, the old office and cash carrier system being done away with. Under the old system both clerks and customers were kept waiting for change and there was a chance of error. The cash registers save time all round and will soon pay for themselves, says Mr. Russill, after three months' experience. An automatically printed receipt is given with each purchase, and 2½ per cent. in goods is allowed on $10 worth or over of returned checks. A girl will be able to handle all the cash records from the four machines in two hours' time each day.

Mr. Russill is not an enthusiast on side lines or specialties unless they have proven their utility. "We are distributors, not drummers up," said he, "and the kind of customer we want is the one who will come in and say, 'I want so and so.' We haven't time to do missionary work, and if a man comes in with a new bread knife or a new window latch, we tell him to go and create a demand for it and then we'll buy. I believe in carrying the best lines. For instance, if a man comes in and wants a good paint, I tell him I have S.W.P. and I can recommend it to him as the best on the market. On the other hand, if a man asks for paint and price is a consideration I have a line of my own which I sell at a reasonable price."

The store is organized on departmental lines, each branch being in charge of a manager, who is expected to show results at the end of the year. Mr. Russill is very modest about figures, and declines to say how much his total sales have amounted to, but he admits that each year has seen a substantial gain, and if a department manager is not able to show an increase of from 10 per cent. upward each year, he would consider something wrong. As far as possible buying is done direct from the manufacturer, but jobbers get a fair share of the business. An average of 5 per cent. of the total sales is spent in newspaper advertising. The windows, too, are a great feature, seasonable goods being allowed to remain on display from two to four weeks. Considerable rivalry exists between the clerks regarding the windows and each is anxious to show his goods in order to boost his sales.

All forms of interior advertising is done, every article being marked in plain figures and price tickets, display cards, sample boards, etc., are used wherever possible. "We mark everything and let the goods sell themselves," said Mr. Russill. "See those vises by that display table. A man will come in to buy a hammer and he will see that little vise marked at $3, and he'll buy it also. Another innovation I'm going to adopt is regular meetings of the staff to talk over our business affairs. We will have a 'suggestion box' and any employe making a valuable suggestion will be rewarded. I believe in the clerks educating themselves, and think other manufacturers ought to do as the American Fork and Hoe Company, of Cleveland, has recently done. One of my clerks sent over and got one of the booklets they have issued explaining the talking points of their goods, and I intend sending for enough to hand one to every one on my selling staff."

The basic idea of the arrangement of the Russill store is to give a good impression. The windows are clean and scientifically dressed and the interior, while crowded with goods in every available space, is clean and neat, the silent salesmen and showcases at the front being very attractive. On the right side cutlery and sporting goods are given first place, behind this being builders' hardware and behind the elevator plumbing supplies, heating goods, washing machines, enamelware and general house furnishings. On the left are tool cases and shelf boxes, shovels, forks, horse blankets, sleigh bells, etc., and farther back the paint, oil and brush department, 12 Bowser oil pumps being in use, the Russill Company taking pride in being the first ones in Toronto to instal the system. In the centre are display tables, etc., on which are shown vises, food choppers and mitts, and behind these racks for axes, snow scrapers, etc., and then grindstones and wheelbarrows.

The sketch showing the floor plan will give a general idea of the layout of the store, while the illustration shows the back portion of the store, behind this being a warehouse, shipping office and workroom, 50x58 feet in size, with an entrance on a lane.

MONEY IN HOUSEHOLD SPECIALTIES.

"Is it not an acknowledged fact that all live hardware merchants of to-day are making their most money out of specialties?" writes a washing machine salesman. There is hardly a hardware merchant who has not clerks enough to sell at least 25 per cent. more goods than he does. We will say, for instance, that he is doing a business of $10,000 a year and barely ekeing out a living. This same merchant, by handling a good line of specialties, could easily increase his business to $12,500 a year with no additional expense whatever. He has done an additional business of $2,500 at a profit of not less than 25 per cent. (for there is a good profit in specialties, if properly handled), and the net results is the dealer has made $625 clear money at the same expense.

What are specialties? I call a washing machine a specialty, and am probably better able to talk on the washing machine question than any other. I find hardware dealer who says they can't sell washing machines, and that there are none sold in that locality—he has none on the floor. I leave him, I don't want him to sell our machines, because he can't sell anything. Everything that goes out of his store his customers sell themselves—they come in and ask for five pounds of nails and get them. He don't sell them anything else. Then I go up the street and find a dealer with a couple of washing machines in front of the door, I walk in and find the dealer and clerks are busy. A customer comes in and asks for five pounds of nails, and by the time the clerk has the nails ready he has got the customer interested in a washing machine, and in ten minutes time he has sold him one. He made five cents profit on the nails and two dollars profit on the washing machine. I finally get the dealer's attention for a few minutes, and I find that this dealer sells from 50 to 150 washing machines a year, which is the average sale of an ordinary dealer. This

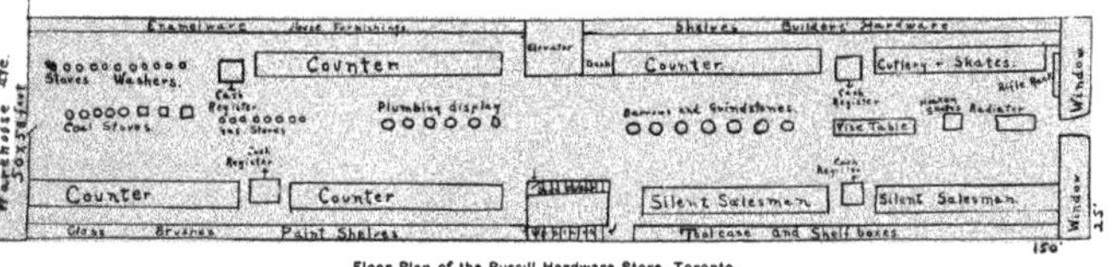

Floor Plan of the Russill Hardware Store, Toronto.

is a contrast you will find in 99 out of every 100 towns in the country. It is the difference between the dead and the live dealer.

This not only refers to washing machines, but to every other specialty as well, and by specialties I mean such things as refrigerators, ice cream freezers, churns, gasoline engines, gasoline gas machines and a hundred and one other things too numerous to mention.

HANDSOME STORE IN ONTARIO TOWN.

Traveling salesmen in western Ontario speak very highly of the new arrangement of Fred Howes' hardware store at Listowel, which was remodeled recently. It is now most up-to-date and in an excellent location for business. The building is two storeys, 19x120 feet, with a full-size cellar, tin shop upstairs and storeroom in rear 36 feet in length. The structure is of cement blocks and the general appearance is excellent.

Interior of Newly Remodeled Store of F. Howes, Listowel, Ont.

Below the sign is nearly 60 square feet of prismatic glass, which provides a splendid light in the interior of the store. The windows are modern and the displays, as will be seen by the illustration on another page, are of a high order. The walls and ceiling of the store are neatly decorated and a skylight provides light for the back of the store.

On the right-hand side are 390 shelf boxes carefully sampled and with nickel-plated pulls. In the centre a large wall showcase has been built for displaying edged tools. On the left side the upper portion of the shelving is used for tin and graniteware, and the lower for mixed paints, a row of hanging lamps and lanterns being also shown. In stove season, rows of stoves are placed on the floor, the surplus stock in the photo partially hiding the handsome counter and showcase. In addition to the rope display and nail bins, as shown in the photo, there is a bolt rack at the rear of the shelving, where each size and kind of bolt, etc., is kept, each being properly labeled to avoid loss of time. Two three-barreled coal oil self-measuring tanks are also installed, Mr. Howes is considering the oil and lamp trade a profitable line to push.

Regarding shelf boxes, Mr. Howes says a hardware store cannot well do without them, as they save space and keep goods clean, besides acting as a silent salesman. Customers see the samples and often buy goods which they had not thought of purchasing.

Mr. Howes was formerly a member of the firm of Anthony & Howes, Georgetown, he having bought out Zilliax & Sarvis, the former owners of the Listowel store, in March, 1906.

ESTABLISH BARGAIN COUNTERS.

There is a lesson for retail hardwaremen in the marvelous success of the "Fairs" and other 5, 10 and 15-cent stores now established on the syndicate basis in all the leading centres of population. They exist largely because of the lack of enterprise on the part of the retail hardwaremen, who might just as well as not have this business as their own if they would make a feature of bargain counters and conduct them in a business way.

In the two illustrations we reproduce suggestions are given of two methods of display adopted, but it will be noticed that everywhere emphasis is put upon the stating of the price in plain figures. The table shown is ten feet long and can be put together in a few hours' time at little expense. The other style

Cheaply Constructed Display Table.

of counter is preferred by some because of the storage space it provides.

To start a bargain counter without expending a cent for fixtures, take large packing boxes and cover them with common wrapping paper. You can do the work yourself and the boxes are probably in your basement now. A few cents will buy wall paper or some fancy cloth which will improve the looks of the counter. Three or four of these boxes put end to end down the centre of your store will make as practical a bargain table as one made of quarter-sawed oak.

SMALL RESULT-BRINGERS.

The few leading and general principles of business are familiar and fairly observed by most business men, but it is in the extra attention to small details that the careful merchant finds his best chance to get ahead and lead the other fellow.

Keep stock clean and well displayed and bring seasonable lines early to the front, before any of your competitors.

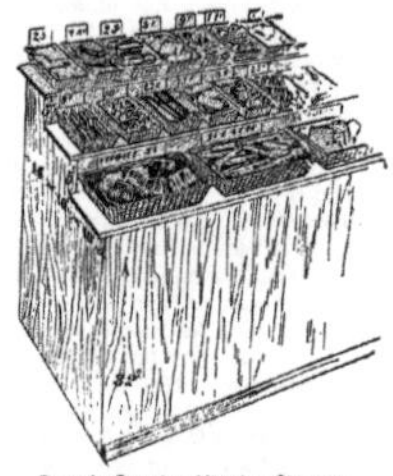

Bargain Counter Allowing Storage.

Let all your present customers know in some way that you appreciate their business and want more of it; also that you are in better shape to serve them than ever before.

Get some good selling specialties which your competitor cannot procure and push them in your store and window display and through your newspaper advertising.

Arrange for a personal business canvass of your locality or send an intelligent representative. See that the names of prospective paint buyers are forwarded to your manufacturer so that he may supplement your work with his advertising matter, and let your newspaper advertising be terse, strong and individual in character.

DISPLAYING STOVES IN HARDWARE STORES

Fred W. Otton, of H. H. Otton & Son, Barrie, has practical ideas regarding the display of stoves, and in a recent letter outlined a few suggestions as follows:

"Where a dealer who has ample floor space and can arrange his stoves so as not to obstruct the shelving, I think it better to have such stoves as base burners and oak stoves elevated along against the wall and have all ranges and cooks on the level below and in front of them, leaving enough space for an aisle between them. In our own store being narrow, only 18 feet wide by 90 feet in depth. It has the advantage of being located on a corner, however, this providing several windows for light and a side entrance to the living apartments above the store. The sketch shows the store in the stove season when every available space is taken up with heating and cooking goods. In the summer season refrigerators take the place of the base burners in front of the office, and lawn mowers, etc., replace the oaks and air-tight heaters. A workshop is located over the back part of the store and there is a storeroom over the second-hand stove room, the front part of this building being a drug store. There is also a fire-proof oil cellar under the stove repair room at the rear of the store, this addition to the store making the building 120 feet in length.

90 ft. Street
18 ft. Window Front Window
Shelves
Display Counter · Silent Salesmen · Ranges · Oaks · Ranges
Air-tight Heaters · Coal and Wood Cooks
Base Burners · Churns, Washers, etc.
Counter · Office
Shelves
Stove Room for opening goods and Repair stoves.
45' Second Hand Stove Show-Room.

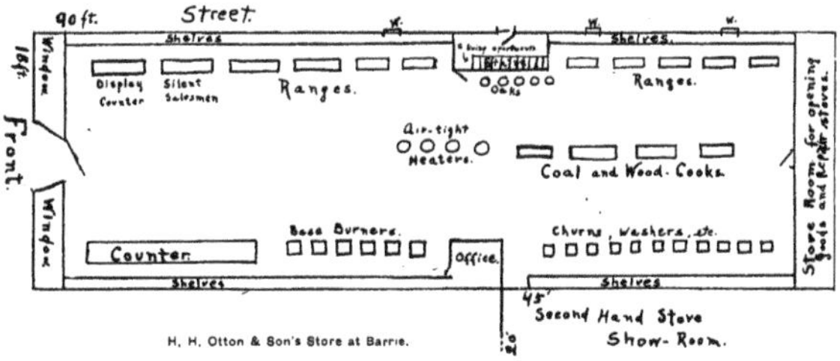

H. H. Otton & Son's Store at Barrie.

store we haven't got the space required for so doing and we use trucks and castors under all, excepting our light heaters.

"We consider it a great advantage to display stoves on red painted trucks or castors where the space is limited, as you can bring your stove out from among the others, turn it around and show it off to better advantage. We use stove trucks mostly our own make, made from 2x4 dressed scantling, with a narrow moulding around the top edge, and use piano castors in them. We find that these run much easier than the stove castor, the wheel being larger and having more bearing, while they are less expensive. We also use castors on our silent salesmen, so we can move it without difficulty to any part of the store for a change, and also make it easy to sweep from under it."

The plan of Otton & Son's store is shown in the accompanying sketch, the

WOULD DISPLAY STOVES ON PLATFORM.

T. H. Latimer, of Latimer & Elliott, Chesley, Ont., says he prefers a long, narrow store to a short, wide one for the display of stoves, as he has done business in one of the latter kind and never could arrange stoves to any degree of satisfaction, the stoves always having a bunched-up appearance. With a long store one can range them up in line according to size, and they not only show off to better advantage but are more convenient to show to customers when making sales. He always uses castors to set his stoves on and likes them better than trucks.

"Were I building a new store," adds Mr. Latimer, "I would cut out the shelves on one side of the store, and have a platform 8 inches high by 4 feet wide placed on floor along the wall, starting at the front show window and extending back as far as I had room to go. I would then place my stoves and ranges on the platform with backs to wall and put on two or three lengths of stovepipe with one elbow so that elbow faces to wall and also put on a pipe collar so that the stove has the appearance of being connected to the chimney.

"On the wall above the stoves I would put up a shelf wide enough to accommodate boilers, dishpans, 5-gallon oil cans, tea kettles, etc. For all other lines of tinware and graniteware I would use double-deck tables placed near the centre of the store where otherwise stoves would stand. By using tables for these lines you can accommodate more than you would with ordinary shelves, and as the ladies can more readily examine the goods, sales would be increased—this being the main point to be considered."

THE BUILDING OF INDIVIDUALITY

Get people's interest aroused in the articles you sell them, not only in the advertising that creates the sale, but in conversation with them, suggests an exchange. It will get them in the habit of regarding their purchases, not as mere things, but as industrial creations full of individual interest and capable of infinite variety of quality. The customer who goes from your store thoroughly interested in some bit of hammer gossip which may be substituted for the usual comments on the weather, will take a special interest in the hammer he has purchased and, if it is a good one, will unconsciously associate its quality with the store from which it came.

There are many interesting facts about

even the smallest mechanical contrivance and by calling a man's attention briefly to something of this sort, the habit is quickly acquired of interesting each thing in the store with individual characteristics. Shelf hardware is no longer so much metal at so many cents a pound, but every piece represents a certain amount of mechanical skill and quality which wins the respect of the purchaser even when the article purchased is of the most trivial kind. This will be found another little blow against the catalogue house that is doing business at too long a range and in too wholesale a manner to individualize. A pound of nails no longer suggests merely the price at which they come, but the process of their manufacture and the mechanical difference between the perfect nail and the shop sweepings that include all sorts of imperfections.

As the habit of close observation grows there comes to the eye beauty of detail that never would otherwise have

Serviceable Rack for Wire.

been noticed but, that, once observed is never forgotten or lost sight of. But do not forget that the customer is usually a busy man. Do not take away from him time for these explanations ; simply put the time he has to give you to good use.

WIRE CLOTH RACK.

The accompanying illustration represents a simple and inexpensive rack for holding wire cloth, which a bright clerk with a knack for using tools can readily construct in a few hours. It is made of 2x3 in. lumber, planed smooth, which may or may not be varnished or painted to harmonize with fixtures or adjacent woodwork. The four upright posts are set about 12 in. apart at the ends, and 20 in. on the sides of the rack. They are braced together at the top and bottom where the projecting arms cross the uprights. There are five sets of these arms, or crosspieces, placed about 1 foot apart and inclining towards the centre so that the rolls of cloth will stay on. The arms are 22 in. long and are fitted together smoothly where their ends meet in the middle of the rack.

Only a few rolls are shown in the cut to avoid obscuring the construction. It will be seen that the ends of the arms are marked with stencil giving the size of the cloth which they hold, 24 to 32 in. being kept on one side of the rack and 34 to 38 on the three upper arms of the other side. The two lower arms on this side are used for brass cloth, while the compartments in the centre contain landscape cloth. One advantage of this rack is that everything it holds can easily be got at from either end, so that if one end is kept clear goods may be stored all around, or it may be built in a corner or against a wall without lessening its convenience.

GETTING TRADE FROM FARMERS.

Catalogue houses, at least the larger ones, depend to a great extent for trade on the rural population, and it is to the farmers that the most of the advertising of these houses is directed. Throughout the United States efforts have been made to enlist the services of rural free delivery carriers, postmasters and others to obtain lists of all farmers and to place catalogues and other advertising matter in their hands.

The rural mail routes that have been established in many sections of the country, have had the effect of making it less necessary for the farmer and his family to go to town. Mail order papers and mail order catalogues are brought direct to his door, so that in many instances, the farmer is really in closer touch with the city establishments, and is better informed as to their offerings, than he is of the goods carried by the merchants in the towns on the rural mail route. The Ad. World says :

"If people don't come in and see what you are doing, it is up to you to go to them and tell them what you have for sale, and why it will be to their advantage to trade with you. You must keep them in mind of you, if you want them to remember that you are still doing business. Don't rely upon the county papers alone, get in direct touch with your possible patrons. Send them a circular or a letter occasionally, telling them about your goods. Do this when you have a lot of new goods come in. Do this whenever you have a bargain to offer.

"If your letter or circular comes in the same mail with a letter or circular from the far away concern, you will at least divide the attention with that concern. People will understand that you are not conceding the trade to the big Chicago mail order concerns, nor to the nearer department stores. They will see that you also claim to sell the most meritorious goods at moderate prices. If you quote prices, and they are as low, or nearly as low in price as the big stores, you will, in most cases, have the first chance of making a sale. Given the chance, it is your own fault if you don't make it.

"The reason that the big concerns are taking away so much of your trade is because they keep hammering people with the assertion that they can give the best values. An assertion made over and over again, without being disputed, is at last accepted as a fact. If you want to hold your trade, go out to your patrons in true mail order style and tell them your side of the story. There is no doubt at all that there are many effective arguments that can be used by the local dealer, and he can quote prices on many goods to back up his claims.

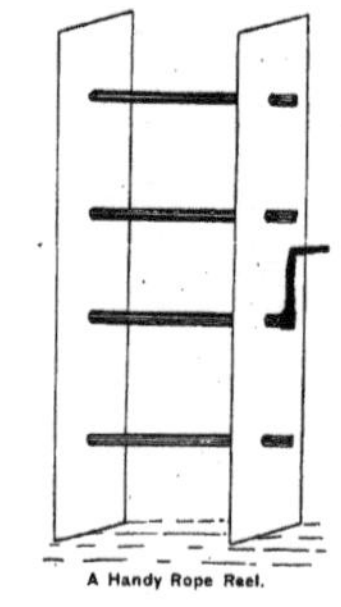

A Handy Rope Reel.

"The form letter is an especially effective way of sending out direct advertising. It would pay many retail dealers to put in a mimeograph or other duplicating device, and send out a letter weekly. Silence is no answer to the persuasive talk of the mail order men. You must meet them on their own ground, with more convincing arguments."

A HANDY ROPE REEL.

The sketch accompanying is of a rope reel made and used by a Wisconsin hardwareman. The reel is made by boring holes in two upright boards fastened against the wall, and running through them round wooden rods. The latter are squared at one end to fit an iron handle which can be made at a cost of a few cents at the blacksmith shop. The reel is a very convenient article, keeping the rope up out of the way, and preventing it from tangling when measuring off.

INDEX AND STOCK CARD SYSTEM

By B. A. Clambeau.

Every retail store should be indexed, very carefully and thoroughly. There are several reasons why this should be done. One is that it preserves that admirable condition expressed by "A place for everything, and everything in its place." It also helps to keep the stock up, preventing the "just out" condition. It helps to keep tab on the rapidity with which any certain stock moves. It preserves a record of prices, frequently settling disputes and often saving money on buying. The hardest part about keeping a card index is the installing. Afterward it is easy, requiring only care. But even the installation is not so hard if properly gone about, the time devoted to it being slack hours.

The card index system is much preferable to the book index system, for several reasons. It is more convenient to install, for it is more elastic. It is more complete in its efficacy, as it permits of the filing of a great deal of data that could not be cared for by the book system. It is neater, and easier to refer to. It is cheaper in the end, because it will outlast a book system, the latter having to be renewed as soon as one department becomes cramped for room, or as soon as the pages are worn out. With the cards, there is no limit to the size to which it may be stretched, if desired, and the wearing out of a card means only to replace it with a new one. Besides, with the book system, old entries are liable to be scratched off. Indeed, that is the only way to dispose of them, while with the card system, cards for stock that is out of date and cards that are too full for further additions may be filed in "defunct" or transfer drawers, where they are easily accessible.

The cards should be four by six inches in size, filed in drawers. Different colors may be used for different departments or classes of stock. The cards should be ruled, both vertically and horizontally, both front and back. No card should be devoted to more than one article.

One side of the card should carry, at the top and left-hand side, the name of the article. On the left-hand side should be placed, in the first column, under the title, the amount in stock on the date the card was filed, entering the date in the second column. Then, thereafter, whenever a new supply arrives, put down the amount, date, name of firm from whom it was purchased, price, discount allowed, price it is to sell for, and, whenever the selling price changes, date of change; condition it came in, whether good or bad. With some articles this last entry is not necessary.

The reverse side of the card should carry at the top the name of the article and its location in the store and in the storeroom. The lower three-fourths may be used as a continuation of the front side.

It is not necessary to have a column for every entry to be made. The date of the change in selling price, for instance, may be noted in the same column with the price only in another color of ink. Same with the condition of arrival when the condition is bad, a cross may be placed just above the price,

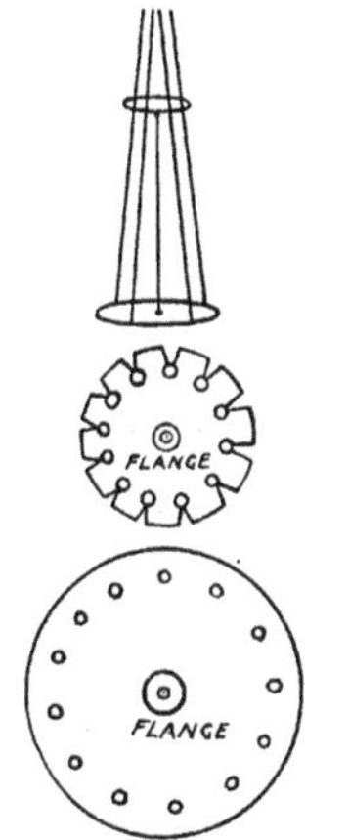

Fishing Tackle Display Stand.

or somewhere else where it is not liable to escape attention.

It is easy to see the importance of this system as far as it is used to denote location of the stock. Many a merchant has forgotten stock piled in remote corners and is ordering new. It saves time and money. The buying record it affords is highly valuable in buying, for from it you may get a line on the last prices you paid, whether the goods are moving rapidly or not, and, in fact, the complete information without which you can not buy to the best advantage for yourself.

The stock and index card system is also of great value in taking an inventory; often shortening that work by several days. And all this cost only a little time in installing, and after that not so much time as you would waste for the lack of it if you didn't have it; and, from ten dollars up, according to the size of index needed.

Never throw a card away. This is important, for you do not know when you will want to refer back to it. Cards should be filed in a transfer drawer as soon as they are filled, and the same should be done with cards of articles no longer in stock. In the case of filled cards, the new or, "live" card in the working case should bear a mark showing that a preceding card has been filed in the "defunct" drawer.

Never, not in a single instance, should the system be neglected once it is installed. A card should be made for everything in stock. Every purchase should be entered. No system is of value unless properly carried out. Do not trust to memory; put it on the card.

Of course, in denoting the location of stock, the store will have to be divided into compartments and each given a designated number or letter. The best method in most cases is to give each show case a double letter, like AA, BB, CC, and so on. The shelf compartments may be lettered and the shelves numbered. Drawers may be treated in the same manner. Then, after this is done, it is easy to tell, when the card says a certain article is at "G-3" that it is in compartment G, shelf 3, or, if it says that it is in "S. R. G. 8," that means that it is in the store-room, in compartment G, shelf 8. The method is very simple.

With this method should be a record on each shelf, wherever such a record would be of assistance, giving a list of the articles belonging on the shelf. This may be placed back out of the customer's sight. It is very convenient in preventing any article from being out of stock, and by its use any person who can read is capable of stocking up the shelves. In the case of drawers, the articles they are supposed to contain may be written on the sides.

FISHING TACKLE DISPLAY STAND

A very useful stand for displaying fishing poles can be made of two circular boards one inch in thickness, as shown in the accompanying drawings. The bottom piece should be bored part of the way through to prevent the butt end of the poles from sliding off, while the top board should be bored through, as illustrated. The two boards are connected by a piece of iron pipe about 3 feet long threaded at both ends and screwed into flanges attached to the circular boards. If desired castors can be added. Any handy clerk can make the fixture in his spare time.

Make Business Produce Profits

Cost of Doing Business – Business that Don't Pay – The Cash System in Operation – The Retailer's Credit – Bookkeeping Suggestions

COST OF DOING A RETAIL BUSINESS

Only four or five in a hundred of the people who go into business succeed. The other ninety-five or six either go to the wall or eke out a bare existence.

Ninety per cent. of the retailers don't know the cost of doing business. It this be true, and the view is strongly held by a great many business men, it is time to make a thorough study of the cost of doing business. The old school definition that profit is the difference between the cost and selling price of an article is wrong. The expense of doing business must be subtracted from the profit.

There are two reasons for this lack of knowledge of the cost of doing business. The first is bad or inefficient methods of bookkeeping; the second a failure to include in the cost of doing business all the items of expenditure that go to make it.

An Example.

To illustrate: Suppose a man is doing $12,000 business per year with $3,000 invested, we would figure the cost something like this:

Interest on capital, 3½ per cent.	$105
Rent	200
Salary to proprietor	800
Wages to employes	850
Insurance	50
Taxes	30
Horse keep, repairing rigs, etc	75
Light and heat	50
Telephone	20
Depreciation in value of fixtures	25
Postage, etc.	25
	$2,230
Now suppose he figured 20 per cent. profit on selling price of goods, $12,000, we have	2,400
Less	2,230
Net profit	$170

The first item is interest on the capital invested. If a man were working on borrowed capital he would pay interest. If a storekeeper borrowed half his capital, the interest on his own money would be as legitimate an item of expense as the interest on what he had borrowed. This should include also interest on the value of stock not paid for. It should include also interest on all outstanding accounts owing him. If a firm has $1,000, or $25,000 open accounts on their books the interest on that is assuredly a legitimate cost of doing business.

No Doubt About These.

Then there are items that leave no room for questioning—rent, taxes, fuel, light, wages, advertising, insurance. If a storekeeper owns his premises he should include rent, or what is the same thing, interest on the value of the property.

If the storekeeper employed a manager, the business would have to pay the manager's salary. If a man manages his own business he should have a salary just the same.

Some people might say "his profits are his salary." Not at all. That is putting him in a worse position than his hired manager. If a business failed, the worst that could befall a hired manager would be the loss of a position. He risks nothing. The storekeeper risks all he puts into the business.

Fixtures Wear Out.

Depreciation of stock and plant is not a myth. Scales wear out, horses get old. The loss of goods in weighing is a case of "many a mickle makes a muckle." It is a very natural tendency in salesmen to appear liberal and the effort to fight against it is not likely to be so strong if one is dealing with some one else's goods. Some people will rasily assert that what is given away in over-weight won't amount to a hill of beans in a year. Suppose that on an average a man gave over-weight of an ounce every time he sold a pound. Many sales would be more than one pound, many would be less. The best cure for over-weight is to have goods weighed up in advance wherever possible.

No one will dispute bad debts as a legitimate item in the cost of doing business account, but some may not have the courage to make a clean breast of it and admit the loss.

Repairs are constantly necessary. Every person utilizing any kind of plant should know the percentage of depreciation and provide for it. If a horse lasts five years, in one year a fifth of his value is used up, has gone into the cost of doing business.

The cost of delivery would surprise some merchants if they would give it careful consideration. Here is how a merchant in an Ontario city doing a business of $125,000 a year figures it out. He keeps eight delivery wagons:—Interest on investment, 6 per cent:

Barn	$1,800	
Rigs	1,000	
Horses	1,200	
	$3,500	$ 210
Wages of drivers		4,000
Keep of horses		1,162
Taxes on stable		33
Repairs		400
Total		5,805

This is just a little over 5 per cent. on the turn-over and shows that it costs this grocer $725 a year to keep one delivery wagon. It will cost the man who keeps only one more.

A man in business is not in business for fun, or for a mere living. He should accumulate a certain sum each year to the good that would keep him from want by the time he is 55 or 60 at any rate.

The temptation to cut prices is the curse of business, and it is ignorance that is at the root of all cutting, the lack of knowledge of the cost of doing business. It is the principal reason for failures. The result of cutting is that a man's family suffers. He is not able to give them a proper education. He is not able to take the recreation due himself.

The proportion of those who succeed in the hardware business is very, very small, probably not over five per cent. It doesn't necessarily follow that all the rest fail; they simply exist.

We place the cost of doing a retail business at 22 per cent. of the total turn-over, and suggest the following as legitimate charges: Interest on investment, interest on open accounts, salary, wages, rent, fuel, light, taxes, advertising, depreciation of stock, loss in overweight, bad debts, delivery, repairs and renewals, paper and twine, incidentals, telephone, etc.

Hardwaremen who have given this subject some thought are invited to write the editor expressing their opinions regarding this estimate and the items included.

WHY I STARTED THE CASH SYSTEM

By A. E. Code, Morris, Man.

It was during the winter of 1904-05 that I first began to investigate the cash system of doing business. During the previous fall and the early part of that winter, I had collected all the accounts that I was liable to get that year, and found that there was a considerable amount still owing to me that I stood a good chance of never getting.

After I paid up the wholesalers for my stock, with the exception of a few odds and ends of accounts, which were not due until February or March, I found that I had more stock on hand than the previous year, and thought I was doing all right. The only thing that bothered me was that there was no visible surplus in the bank. There were also those small accounts to be provided for. There were lots of goods on hand, but the trouble was to turn these goods into money.

It was the case over again of "The Ancient Mariner" thirsting for water in the middle of the ocean. I was literally surrounded with goods, but had very little ready money.

Dealer's Money Paid for Mail Orders.

Just about that time some of my gilt-edged customers came in, bought their goods, and said, "charge it." It was going to be the same old story ; nothing but "Charge ! Chester, charge !" all along the line for another long, weary year. I knew that those customers could afford to pay for what they were getting.

I knew that at that time about one thousand dollars a week was going out of this small town to departmental stores. And whose money was it? Certainly it was my money, or money that should have legitimately come to me in payment for the goods I had sold.

The departmental stores were offering "bargains," in which the sacrifice was generally taken from some hardware line, graniteware, cooking utensils, etc., and they were selling under the cash system.

Now, whose fault was it that some of this money did not come to me ? I take it, it was not the customers'. I was selling under the credit system, and was the customer to blame if he merely fell in line with that system which I must have preferred, else why did I adopt it ?

Three Objectionable Alternatives.

Meanwhile goods were going out on credit and those small accounts with the wholesalers were still unprovided for. How also was I to replace the goods being sold ?

There were, as far as I could see, only three courses open to me. First, I might borrow from the bank ; second, I might endeavor to induce some wholesale house to relax their terms of credit, and supply me with goods until fall ; or, third, I might allow my stock to run down.

Each of these plans was objectionable. If I borrowed from the bank I would have to drop seven or eight per cent. at the outset. Buying on long time from a wholesaler would place me at the mercy of somebody else, who might at any time request that the account be reduced. Besides, this method precluded any possibility of my being able to buy some articles at better prices from other firms. The last method of allowing my stock to run down and thus losing sales, was too objectionable to be seriously considered.

Why should I be compelled to adopt any of these courses ?. Simply that I might procure goods to hand over to people who would not pay the cash for them. Why should I place all I owned at the caprice of a wholesale house, or why should I borrow money in order to lend it against my will ?

This idea seemed to suggest the fact that, we, of the credit system, are really doing a "private banking business," or rather, a "private loaning business," in which we do not take the ordinary security taken by those businesses.

The risks taken in credit sales are too great for the returns. If I buy an article for seventy-five cents and sell it on credit for one dollar, it simply means that I have lent the customer seventy-five cents, and he in effect says to me, "Now if the crop is good, if it is not injured by hail, or rain, or hot winds ; if my profits are greater than my expenses, if the implement men or thresher companies do not get it ahead of you, if everything turns out as well as I expect, and if I remain perfectly honest, I will come in at some future time and give you one dollar."

Of course, some one may say, "That is an extreme case," but those who have been there know that the above type of customer is by no means rare, and if credit sales were confined to those financially strong, there would not be nearly so many entries in the ledger. Those who were well-to-do did not need to buy on credit. They could borrow from the bank, and could make on the deal by buying at a much better price. Let those who needed the money do the borrowing.

The Risk Too Great.

If there were any to whom the bank would not lend the money, why should we take the risk ?

The only solution of the problem seemed to be to insist on cash. But could it be done ? We already sold some goods for cash. Considerable more cash was turned away when we were out of goods, they having been sold on credit, and we could keep a better stock.

I recently asked an adherent of the credit system why he still continued it and what advantages it had, and he said "that it was easier to sell goods," and "better profits could be had." Not very powerful arguments. Let me here enumerate what I consider the advantages to be gained by the cash system.

Advantages of Cash System.

In the outset I would save the amount paid to the bank in interest. I would avoid the risk of being at the mercy of a wholesale house, and I could replace my stock as soon as sold. I would thus be enabled to make several profits on each dollar by keeping it thoroughly alive.

In addition, I would save the price of a new set of book-keeping books—nearly fifty dollars.

I would make two per cent. per month in cash discount, which for the year would amount to twenty-four per cent

I would save the loss by bad debts

I would improve my standing with jobbing houses and thus put myself in line for any snaps that might be going.

I could compete with catalogue houses by adopting their weapons.

I could adopt modern methods, such as bargain sales, etc.

I could sell cheaper.

Instead of wasting time writing up accounts and rendering them, I could use it in improving store displays and in selling.

I could save a waste in postage and stationery.

Each night it would be known just what profit was made during the day, and failure could be entirely avoided by keeping expenses below net profits.

There would be no loss by mistakes in book-keeping and there would be no enemies made by disputed accounts.

I could go after the poor-pay man for business as readily as the others, because I was talking from behind the cash wall. Under the credit system the doubtful customer was rather shunned.

I could cut out the worry, because almost all the worry of business is over accounts or a shortage of money.

A Dream Worth Realizing.

This seemed to me a very alluring picture, a sort of "Elysian fields" for business men, in fact it reminded me of a picture of the "Gates Ajar." I decided at once to adopt the cash system for all time.

I went over my accounts, picked out

those I was anxious about, sought out the men and had a heart-to-heart talk with them. In almost every instance they were in accord with me in the advantages of the new system, for the customer as well as the retailer.

This done, I occupied my time on Good Friday, 1905, with painting some large banners and streamers, red on white. In each of the windows I hung a sign, "We sell for cash only," and farther back in the store another large one proclaimed to all the self-evident truth of "Do not ask for credit and you will not be refused."

They did their work and I found myself finally launched in the new system, the system of cash.

THE CASH SYSTEM IN OPERATION

By A. E. Code, Morris, Man.

When a merchant has finally decided to adopt the "cash system" of business his next step must be to take precautions to guard it. In my own case the first move I made was to render every account on my books. The object of this was to convince the customers that there was more in the new movement than mere idle talk. It certainly had its desired effect. Some people fairly stood aghast at the audacity of the thing. They reasoned: That, while the rendering of accounts in the fall might be pardonable, the sending of them out in March "Well, that seemed too much!" The most surprising result of sending out these accounts was the number of them that were paid. Many were paid that I did not expect till fall.

Realizing that it was considerable of an innovation that I was inaugurating, I was prepared for any number of surprises. As a matter of fact, the change was effected without causing scarcely a ripple. To be sure, some rather shied at the signs in the window—they were somewhat loud—but in a few days they got used to them and the new order of things.

Some of my best friends came in and enquired if it was so what they had heard that I was attempting to run a cash business. On being answered in the affirmative, they gravely shook their heads, looked at me with that compassionate, pitying, "another-good-man-gone-wrong" sort of expression, and assured me I was attempting an impossibility. Especially was it impossible when there was a credit store in opposition. But I found that I retained the trade of a large number of my hitherto credit customers and those, who found it necessary to go elsewhere and buy under the old system, bore no ill-will. During that summer a considerable number of the cash purchases of these people came my way, and in the fall I got an increased share of their trade.

It is pleasing to relate that during the infancy of the new method, I received, I believe, the hearty co-operation and moral support of the rest of the business men of the town, including the opposition hardware. To be sure this was nothing but right, as it was an experiment in which all were equally interested.

It seems to me that the old credit system was ever contrary to the laws of nature. The first law of nature says that "to every action there is an equal and opposite re-action." This truth applied to the commercial world, means, simply, "Value for value."

Let us look at an ordinary credit sale; one in which a note is not taken in payment. The storekeeper hands out the goods, writes out a counter-check slip, and the customer goes out of the store with the goods in one hand, and in the other a receipt showing that he got them. The merchant has certainly given value. What has he got for it? Simply the privilege of making an entry in his day-book, and a vacant space on his shelves. He doesn't get even a receipt from the customer. In short, he does not seem to have gotten "value for value."

I found that my posters and the general circulation of news had spread the tidings pretty thoroughly, and that I had to refuse very few people. Even in this most difficult duty the cash system helped me, and the refusal lost most of its sting. Refusal to sell a man under the "cash system" is merely a pleasant announcement to him that the new method prevents your selling in the old way. That the "cash system" has taken possession of the business and the cash system is practically inexorable.

At this point show the customer some of the goods in which you have reduced prices.. Small articles of well-known value.—The customer readily grasps the advantages and can bear no ill-will, as he realizes that he is up against a "system" in which everyone is treated alike. This is the great point; treat them all alike. Draw the circle complete. In short, maintain the system at all hazards. This done will win the respect of the buying public, and eventually "those who came to laugh" may be induced to "remain to pay." On the other hand, refusal to sell a man goods under the "credit" system is equivalent to telling him that he is dishonest or incapable and is thus practically an insult. This creates an enemy in the district whose presence is a menace to your business.

In actual operation it is less often necessary to refuse customers under the cash than under the credit system. The line had always to be drawn somewhere —between the good and the poor pay— but the trouble generally was that it was not drawn soon enough.

Let it not be supposed that it is an easy thing to guard the new system. More than once my line has been broken down and it requires constant alertness to repair the broken places.

And what a number of different methods are employed by customers to get goods without pay! There is first the man who has "forgotten his purse," and his relative who has "forgotten his check book." After him comes the man who has been sent in for goods by "Mr. Jones," and "Mr. Jones" will pay for them, in fact, "Mr. Jones" is even now hastening down with the money." Next is the man who maintains that "thirty days is cash." Then we have the man who has to see if the article will fit, and closely following him is the man with the "order." Probably the worst of all, though, is the man who deliberately walks off with the goods. I have a case like that now in which a well-to-do foreigner has gone off with a stove. I find it impossible to make him understand the necessity of conforming to my system, and it looks as if I would have to wait until he gets a better grasp of the English language. If he acquires the new language with the same celerity with which he absorbed the stove, I trust to have not long to wait.

It is now nearly two years since I first started the cash system and I am thoroughly satisfied with it. It has done all I expected of it. Many times I have had occasion to be thankful for the protection it gives. Probably a better expression of opinion, as regards the success of the venture can be seen in the fact that after a few months the other stores began adopting it, until now the whole town is cash even the blacksmiths joining it.

Let no one fear to adopt the cash system, in opposition to a credit business. The cash business, if at all well managed, will win every time. From the time, when the first merchant discovered the possibility of running a store for cash the credit system has been doomed. The catalogue house and department store have helped the work. The shortening of terms of credit by wholesales gave the old system a severe blow. In a few short years it will be considered antiquated, and will disappear in that evolution in which "old things pass away and all things shall become new."

BUSINESS THAT DON'T PAY

In spite of the general prosperity of the past year and a marked increase in trade, there are some merchants who, after taking an inventory and totalling up their balance sheets, find that the net profits have not grown proportionately and, in some cases, in spite of an increase in business and a rising market, profits show an actual decrease. Much of this is undoubtedly due to pushing unprofitable lines. In speaking of this fact one hardwareman says in an exchange:

"We are all doing too much business for nothing—selling goods for glory, rather than for profit. Often I used to think that there was business I ought to get, even at a small profit, that would be a help in the course of a year—that could be had without adding to my expense of doing business—and thus would be 'velvet' for me."

"We had a small part of the trade from a certain mill, which amounted to from $35 to $50 a month, in small items —files, saws, paper and emery cloth. The mill is a large one and I knew that these supplies they used in large quantities, purchasing from me only to fill in when entirely out—simply using my stock as a convenience. We charged them fair prices with an average profit of about thirty-five per cent. My prices were never questioned by the purchasing agent, who knew that in buying thus in small lots he had to pay a higher price. Their bill was paid the tenth of each month.

"I figured that if I could get their entire business it would be a big thing for me, but also knew that in such a case I would have to meet close prices. But that the additional business would be velvet for me, I started out for their entire business in my line. We were good friends and they would be glad to give me the business, but I must meet prices. I went into the matter carefully and after some figuring and a good deal of correspondence I secured their next month's supply.

No Money in Big Order.

This first order was about $400. Some goods I had to sell actually at cost. But I knew that by having their trade I would be enabled to purchase larger quantities and felt that thus I would secure lower prices. On this first order of about $400 my profit averaged only about seven per cent. Still I was proud of the order and felt it quite a feather in my hat. From a business of only $35 to $50 I was soon selling this mill about $275 monthly. But my stock increased in these lines. I purchased larger quqntities in order to buy as cheap as possible—and to have the stock on hand when called for. Expenses also increased; I found quite often that I was obliged to wire for some supplies, and in a number of cases had to get the goods by express—as I had to purchase from a manufacturer to get prices, and I found the makers did not ship as promptly as did the jobber.

"Most of my profit of thirty-five per cent., which I had formerly made on the small business, had now been sacrificed to get the large volume—as I found the mill people expected to buy small odd lots or odd sizes of files, etc., at same discount as on the larger quantities. In these various lines I increased my stock from a few hundred to several thousand dollars.

A Lesson Well Learned.

"Some business don't pay, and the sooner a hardwareman can find which it is, the sooner he will become wealthy. It would have been better for me to have held the small pick-up trade of that mill at a fair profit than to have reached out for this larger business which finally tied up a lot of money and which I've really been handling at a loss. In 1906 it actually cost me 13½ per cent. to transact business."

This hardwareman had a brisk business with that mill, but as he himself now figures it, it was a losing trade. Each part of the business done by a merchant should carry its proportion of the expense. Volume of trade without profit enough to cover this expense is of little value to the merchant. Add to the business if possible, make it grow; but add profit-making lines. Cut off "close business" during 1907 and the balance sheet of 1908 will look better and the hardwareman will have more real satisfaction in his business.

BAD BOOKKEEPING AND ACCOUNTING.

Statistics are often quoted which show that only a very small per centage of the men who embark in business on their own account succeed. From three to five per cent., it is usually said, though I question if the average, counting those who achieve a moderate success, is quite as small as that, but all the same the percentage of those who do not make good is very large.

One of the chief reasons for this enormous business death rate can be traced to bad book-keeping and accounting—to the ignorance of the cost of doing business as a merchant, or to the cost of production as a manufacturer. Many merchants accounted shrewd, having no knowledge of accounting themselves, despise it, and utterly fail to appreciate its real purpose. They act on the assumption that any boy or girl fresh from school wh can be hired at the smallest salary, an who is wholly lacking in business train ing, is competent to do their book-keep ing. These green hands might be so i the only function of book-keeping wa to see that sales were properly charge and accounts collected when due. Thes points, of course, are essential, and mus be done correctly if a merchant is t remain solvent, but they are by n means the chief functions of book-keep ing. Books of accounts should be s kept that at any stated period a state ment of the business in each departmen can be presented in detail, and it is no enough that costs should go into genera accounts, but they must be so sub-divid ed that comparisons can be made fron year to year. No accounting system i worthy of confidence which does no classify and locate every dollar that ha passed through the books. That deadl averaging of costs must be eliminated and they must, as far as possible, b wrought out on an accurate and specifi basis. If costs are increasing compari sons will reveal the fact, and if ther are leaks, good book-keeping will detec them, and steps can be taken to sto them.

Work of this kind requires brains an business training, and the money pai for the employing of a competent ac countant is an investment that will yiel large returns in giving to the manage ment facts, instead of guesses, at th costs of running the concern.

Guesswork No Longer Goes.

The demand of the twentieth centur will not admit of guesswork. The man agement of the future must have defin ite, not vague and indefinite, knowledge Success by the rule of thumb process i no longer possible, but can be won onl by exact and definite knowledge. Th cost of running the business is, wit only too many firms, a matter of guess work, and guesswork no longer goes Old Gordon Graham clearly sets fort the changed conditions in his letters t his son when he says: "We started busi ness in a mighty different world; w were all ignorant together, and we did n't have to know fractions to figure pro fits, but now to see profits you have t study astronomy, and when with a pow erful glass you do finally locate them they are away out of the five point deci mal place." Never were truer word spoken, for though now-a-days we thin in large units, there never was a tim when the decimal counted for so much.

It is useless to equip a store wit every modern improvement and then solely through ignorance in the office, t throw away the results, and yet this i just what many business men are doin to-day. The fact is, whatever they may profess in their hearts they look upon book-keeping as unproductive labor, and as the book-keeping of many of them is done they make no mistake in taking this view, for it provides no guide to them in the conducting of their business. Money they will spend on store equipment—they can see its use—but it never occurs to them that money spent on office equipment is just as necessary. The store end of many a business in a general way of speaking now is far away ahead of the business management, looked at from the office end. Many office staffs are past the stage when suggestions of change are welcomed, they may even be past the time when they have a desire or even a capacity for growth. Their special business now is to hold their jobs, and one way to do this is to fight changes, and try to keep things as they are. Such a spirit on the part of the staff can nullify the best accounting system ever installed, simply because it is to be operated by a staff who are bound, to hold their situations, to make it a failure.

THE RETAILER'S CREDIT HIS GREATEST ASSET

An Address by J. D. Campbell, Secretary of the Spokane Jobbers' Association.

No member of your Retail Hardware Association has a more valuable asset, either on your shelves or in the bank, than your credit. Startling as it may seem, it is a fact nevertheless that over 90 per cent. of the business is done on credit. During the year 1905 the manufacturers of the United States did a business of over $13,000,000,000, upon an actual cash investment of $500,000,000, 96 per cent. of all the business done by the manufacturers of the United States was done upon credit. Now that seems startling, and yet sit down some day with a pencil and figure up yourselves the amount of business you do on time—on credit—and the amount you do in cash, and see if it does not startle you.

So I say the most valuable asset you have in your business to-day is your credit; an asset which should be most carefully taken care of; an asset which should be nurtured and not abused. I think also that I can say without fear of contradiction there is no asset you have which is more often abused than that of credit. Not perhaps your own, but the man who buys from you. Because you sell on credit, you must do business on credit. If you do not take care of the credits that are due you, sooner or later you are going to have trouble with those you owe. At least 50 per cent. of the failures reported in one year can be traced directly to bad credits.

Just about a month ago we had occasion to close an unfortunate debtor and take over his business. He had started in business about three years before with an invested capital of $4,000. That was all the cash he had. He has on his books to-day over $7,500, nearly double the amount of his invested capital, standing out on his books. In going over his business for November, 1906, we found his cash business amounted to $1,200 and his credit business $1,500. He had collected on his old accounts during the month $800. So for the month he was $500 behind on his credit business. I said to him, "How soon can you get in the $1,300 of credit in the month of November?" He replied, "I cannot hope to get in a dollar of that until next fall." Now just think of that! How could that man expect to do business and succeed in that manner? He had some notes, it is true, but he had not tried to collect them, and the great majority of his accounts were standing there without drawing any interest and without anything coming from them.

The Retailer's Fault.

I believe that a farmer, or other persons buying goods of you, has no more right to ask you to carry him for a year without interest and without security than you have to go to your home bank and ask them to advance you money to pay your bills without interest and without security, And I believe the credit system is largely the fault of the retailer. I know the farmers can be educated to pay their debts promptly, just as well as any one else. The strange thing to me is that some of the largest accounts on your books are of men who are absolutely good, who can go to the bank and give their note and get the money and pay you? But you have not asked them to do that for fear they will get mad and leave you and go to that man across the street and buy his goods. Isn't that a fact? If this association will pass a hard and fast rule that any man who comes to you for credit should be required to give his note, bearing the legal rate of interest which you have to give at the bank, and make that a bankable note, it would be only a little while until the whole credit system would be changed, and changed for the better, so you could meet your bills promptly and discount them if you pleased by taking the farmers' note to the bank and getting the money.

I have gone over the books of various retailers at various times to find out their condition, and it is astonishing to see the amount of money standing on the books not drawing interest, while you merchants are having interest added up on your past due accounts for them. It is only a question of time when those principles carried out as they have been in the past, are going to involve all of you do a credit business. You cannot help it.

Merchants Should Get Together.

It has been in the past a fact without doubt that if you crowded some of your collections some of your good farmers or neighbors in town, would get huffed and go across the street to buy. But if you all combined and made the same rule when such people go across the street the man over there will hold up a stiff upper lip, and say: "I do not want to sell you on any different terms than Tom Jones would across the street." You would then find out he would be mighty glad to give you his note and pay his bills when they became due. The very system of credit is a matter of education. It is so in all lines of business.

The credit system is more prevalent in the farming districts than in the cities, because the large stores in the cities do practically a cash business. But you who have to carry your neighbor from one year's end to another have to do the credit system; but there is no reason why you cannot inaugurate the system that if you carry a man he must give you good, bankable paper, and that you carry him in that manner.

I believe the wholesaler and retailer should work together to this end. I believe it would be a kindness to the retailers if the wholesalers always insisted on bills being met when due according to the terms of sale. If that was done you would push the man owing you, and he would settle up so you could pay.

What you want is co-operation in this matter; and you must show people dealing with you that you must have your money; that you have your bills to pay, and that if you must carry them to the next fall, you must get a bankable paper so that you can take it to the bank and get your discounts, instead of paying 8 per cent. on your past due accounts.

SELLING GOODS WITHOUT PROFITS.

"I have been behind a retail counter counter for 27 years, and in all my experience I find as a rule the selling of merchandise without profit is largely due to not knowing the cost of doing business," writes C. C. Howell. "How many of you can tell the exact per cent. of cost of selling, or the per cent.

of profits on gross sales of your business last year?

"If you cannot turn your stock at least four times during the year you are not getting the per cent. of profit out of capital invested you ought to. In my experience I have never found one who knew what his cost of merchandise was at the time of delivery to customer. We often find a merchant who says, "That cook stove cost me $19.60 at the factory, and I get $25 for it.' A pretty good profit, he thinks. But let us see when he figures his freight, drayage, blacking, one joint of pipe, damper, delivery and setting up, together with his store expenses, we find his stove cost him nearer $23.50, leaving him $1.50 for profit. This same merchant will wonder where his profits are gone and why he failed in business.

Again, some of us get the idea that we will sell every customer who comes in, regardless of circumstances or price, just to show the fellow down the street that he is not the "only one." A customer comes in and says he can buy a 10-in. strap hinge from Jones for 15 cents a pair. You tell him he certainly is mistaken, that it must have been a 6-in. He assures you that he knows a 6-in. from a 10, and you, after using your best arguments to change his mind sell him a pair of 10-in. for 15 cents, only to find out later that the fellow down the street refused to sell him at that price, and he took this means to 'work you.'"

THE CREDIT MAN'S WORK.

When a credit man represents a concern which carries a large assorted line of goods, and covers an extended territory, he should have a general knowledge of the business, so far as it is associated with the line of credits. He should know the amount of stock on hand, and the stock which it is advisable to move, and amount of profit that the different lines make. He will then be better able to judge whether to accept or reject an order. It is most important that the credit man be familiar with the towns in which his company solicits business, as local conditions vary widely; for illustration, one section may depend upon the farming district for its cash, therefore its cash would only be in circulation during certain months of the year; another town has manufacturing concerns, and, as a rule, pays at least twice a month. Then we have the college town, which only has money during the school term.

The credit man should go over the entire territory at least once a year, and meet personally the customers of the concern, for one can not always judge a man by the letter he writes. He should also see the man's place of business, judge how large a stock he was carrying, the class of goods, and whether they were kept in good order or not, and last, but not least, the opinion of the local banker and fellow townsmen. All of this information would be of particular value, if customers got into trouble.

The salesman's watchword is "Sell Goods!" while the credit man is there to get the money, and he should get his own information. The mercantile agency is good as far as it goes, namely, for a statement of the firm's resources and liabilities, in figures, but they very seldom give any personal information in regard to the man's true character or details of his stock, both of which are most important.

SYSTEM IN THE RETAIL STORE.

When store system is mentioned we should not fall into the error of thinking that it is always some big enveloping thing which covers the whole business like a blanket. Few general merchants can boast of as much store system as that. But nearly every merchant has some system applicable to some particular feature of the business, and it is in comparing notes on these and selecting one here and one there that we finally arrive at a plan for the better handling of the entire business.

The Hardware Trade recently told of a merchant who by checking his freight bills carefully had saved himself in overcharges about $150 in one year. This merchant may be strong on system of this kind and weak on system for other features of the business. Let a merchant take special interest in some particular feature of the business and he will soon evolve a system for handling it which will mark an improvement.

The merchant who saved the $150 on freight overcharges has a penchant for that feature of his business. He has studied freight classification. He has a place for filing his rate records or tariff sheets. He makes sure that a fourth-class item of freight is not elevated to third-class with a proportionately higher rate, by some error-breeding freight clerk in the city freight house. There is much of this done. Thousands of merchants pay these overcharges and never know the difference.

Store system means nothing more than labor-saving ideas and business short cuts. At the same time system furnishes corks to stop up the loss holes. Many stores would be better off if an extra clerk were hired and the proprietor devoted all his time to planning, overseeing and building a system which will save expense and loss.

Out-of-Date Bookkeeping.

It is a wonder that more merchants do not pay more attention to office system. One would think that if there was any place in the store where the merchant would want to see things right at all times it would be in the office--the books. Instead, the big majority of merchants are guilty of carelessness in this most important particular. Many of them seem to think that anything in the shape of bookkeeping system will do. They will spend hours and hours over an out-of-date system when an investment of fifty dollars would save them that much in valuable time every month. Credit men who are on the road a great deal say they strike many antique systems of bookkeeping in retail stores. They do not wonder that so many retailers cannot tell where they are at financially after following the devious paths of such a system.

There are numerous men in every business who seem to cling with a leech-like affection to antique methods. They are the fellows who are sure everything is going to the devil and themselves with it, just because things are not as they were in the days when they made their money more easily. About every ten years a grist of these fellows is run through the hopper and dropped on the dump outside the breastworks, and the trade is benefited that much. That means progress. They take their old accounting ideas along with them.

If nothing else is done it is easy and of primary importance to keep the window glass clean and the show space neat and orderly. Be on the lookout lest any gods or advertising matter in the window, while appearing normal from inside the store, has faded into unsightliness from the street side. Set aside the trimmings of a display for use at a future time. Gradually in this way quite a property outfit will be accumulated. Money spent on crepe paper, ribbons, artificial flowers and other trimmings aids is a good investment.

* * *

The clerk that always has the cash register in balance has a good friend in the head of the house.

How Publicity Helps Retailing

Making the Window Pay Rent — Effective Hardware Advertising— Some Selling Suggestions.

WINDOW DISPLAYS THAT MAKE BUSINESS

By W. J. Illsey, a Winnipeg Hardware Clerk.

Let me in the beginning lay down a few solid rules which should govern so far as possible all displays made. Make a display of one line of goods at a time. Study how best to make each so that all will harmonize. Price cards and uses of articles cards to be used always as much and as elaborate as the merchant may chose. My idea is plain, white, with black lettering. Always study to have a good background more than the filling in front.

Change displays frequently. Some displays will do good work for a longer time than others. Always have the window glass bright and clear as a dirty window detracts rather than attracts. Don't spend too much time on building in a display. Time is too valuable, and the time it will remain is or should be but short.

Next in importance is "which kind of display is the profitable kind to make" No merchant, large or small, but hopes to gain by putting in window displays. If he does it just for a change he'll make by it.

Fancy, catchy displays which attract a big crowd do some good. They make people speak of your store and acknowledge that you are trying to make things go, that you are energetic and enterprising. But in my mind the window display of seasonable goods well exhibited and enough prices given to enable the passer-by to learn what a few of the lines are worth, is the window advertising which brings in the dollars to the till, and that's the object of going into business, ins't it?

Why doesn't the "catch" display bring in just as much? Let's take an example, common in every city or town: Mr. Jones, in walking to business in early Fall, sees in 'Horn and Hill's' window a very fine display of a miniature home. The little house, yard, woodpile, barn, etc., all faithfully carried out in small form. Near the woodpile is an axe, bucksaw and sawhorse, which suggest that someone is in the habit of taking an amount of healthy exercise at this woodpile. No display cards can very well be used in such a window as they would look out of place. Mr. Jones stops, smiles, says to Mr. Smith, who also is looking, "Fine window, that, Mr. Jones." Mr. J. replies in the affirmative and they both pass on and soon forget all about the display. Probably each or both were needing a new axe or saw, but were not tempted to come in and ask the prices. They pursue their course a few blocks when they come to "Blank & Son's" store in whose window is displayed a neat line of bucksaws, saw bucks, and handled axes. Prices and names of each are given and if Mr. Jones or Smith does no buy a saw or axe at this place it's peculiar. They know in an instant what they have to pay for each one and if they do need one just take about 3 minutes to pay the clerk and leave their address and his business is done.

Another example is that of the ab-

WM. J. ILLSEY, WINNIPEG
Who Won First Prize in Hardware and Metal's Recent Holiday Window Display Contest.

sent-minded man. His wife has made him promise a dozen times to bring her home a good pair of shears. He promises and means to keep his promise. Business, however, drives all thought of it from his head and so it goes for days or maybe weeks. But one morning he goes on his way to business and in "Blank & Son's" window he sees shears and scissors of all kinds displayed with cards saying that 50c will buy a good family size of good quality. There's just what he wants and 29 cases out of 29 1-2 he will go in and buy.

In our business here we have such cases almost all the time. And just an instance of what a display of razors, straps, penknifes, etc., will do. Last year I think I had in altogether five such displays and all very similar, and could you but know the extra amount of business we did while they were in (not after so much) you would say, "it pays, doesn't it?"

Too much cannot be said, about window displays of goods. 'Tis the best means of advertising any store can have. The newspapers are good and do good work, but the window as "eyes of the store" show what's inside and invite the public to enter. The people come, too, into the store of the live man who attends to the changing of his window displays and who keeps them interesting to look at.

One point which I will mention in closing, when you wish to show, say, nickel-plated or silver goods, don't show it on yellow or some discordant color. Study harmony of colors and your displays will be vastly beter.

PLANNING A WINDOW.

There are a great many ways of planning a window, writes G. E. MacFadden. One of the best methods is to draw a rough sketch of the framework, or arches, whichever you may decide on using; another way is to go into the window and gradually add to your display until you have completed the trim. Both ways are practised throughout the country by the better trimmers.

At a first thought it might seem best to make a working plan of your window, but sometimes the result will not look so well when completed as the plan promised.

Then, again, if you decide to build your window without any forethought, it is more likely to be a failure than a success.

In my experience, I have always made a rough sketch and have it so firmly impressed on my mind that I can see the finished display before I have really started it.

Working along these lines you often see little points, as you go along, the adoption of which result in a great improvement upon your original plan, and the finished trim may be entirely different from your first sketch; but you started right, working from a fixed plan.

At all times, you should use a centre

piece; this is necessary. It may be an arch, or some other set piece with tools arranged in an artistic way. Circles of different sizes are good to use. Children's rolling hoops and wagon rims are very good articles to arrange tools on, making fan-shaped arches, etc. The sides and back of the window should be covered with plain cloth. Canton flannel is about the best to use, and most inexpensive, and red is one of the best colors for showing up tools.

Large double-headed tacks can be used to good advantage for fastening the different tools to the background. You can first draw out a design, on fan-shape, and tack the various small articles flat against the background.

Don't forget the fact that the eyes of the observer rests on the centre piece of your display. Spend more time and thought on that than on any other part of your window.

This is the reason a small window is often as effective as a larger one. It is true that a large window will hold more goods, but will it make more sales? It all depends upon the arrangement of the articles displayed.

The principal idea of a large window can always be reproduced in a smaller one—this means the centre piece again. It is always the most prominent feature of the window. It must be reduced in size, but the same idea is carried out in both windows. Circles are better than squares or triangles, as the latter are usually stiff and are very much more difficult to trim. Circular steps also make a good centre piece and are much better than triangular or square steps. The pillars or posts that you may use should always be round, as they trim to better advantage. They will support your arches or platforms and can be trimmed in a very artistic way.

As a general rule, the heaviest part of your trim is the centre. Between the centre and the ends of the display should be the lightest part, but this depends considerably on the shape of your window and the character of the trim.

If you have no taste for artistic effect, do not attempt to trim windows.

Do not place your goods too near the window pane; it spoils the effect of the best display. The farther away from the glass the goods are removed, the better they will look, unless they are very small.

Fill tin pans, or wire backets, with small articles, such as tacks, tack hammers, pincers, and various other small articles, and place them at vantage points, and do not forget the price tickets.

If you use signs in the window, see that they are always raised a little from the floor, because they attract attention more readily and are much easier to read than if they are resting on the floor.

If you have any special lines of goods that are not moving fast enough, make a display of them. If you cannot think up an idea, write to me, and I will gladly give you a little help in the matter. If you make frequent changes, you will in a short time, I am sure, see the inestimable value of your window displays.

A HOUSEFURNISHINGS DISPLAY.

In the spring and early fall housefurnishings are a good line to push and one of the most effective methods is to arrange a window display of a novel and striking character and draw attention to the display in the local advertising. If he is asked, the editor of the local paper will be glad to write a reading notice describing the display.

The design of the display shown in the engraving is striking, the arrangement of the spoons in the form of a horseshoe and star having a pleasing effect, which will encourage an inspection of the innumerable household articles shown in the display. Try a similar display, use price cards, and watch results.

Attractive Display of House Furnishings.

When metal letters come off the window, it is easy to cement them on again with a mixture of one part gum mastic, two parts litharge (lead), one part white lead, and three parts linseed oil. Melt together in a homogeneous mass and apply hot. The letters should also be heated to a temperature at least that of the cement.

MECHANICAL DISPLAYS ATTRACT ATTENTION

Mechanical moving displays of one kind and another are becoming more and more popular with merchants. One of these is a farmyard scene with a painted background, a windmill in operation, and cattle, horses, etc., moving around. Another is a fortress apparently engaged with a fleet at sea, while a squadron of soldiers manoeuvre in the foreground. Hunting scenes, with riders following the deer; breweries in operation, and other familiar scenes are also reduced to mechanism and used in window displays.

Popular Mechanics tells how a bicycle can be utilized to make a moving display in the window: Remove the front wheel from the bicycle and in its place fix a seat as at A in the sketch. Take the tire off the back wheel and run a small leather belt around the wheel to transmit power to the dynamo, B, which is connected up, as shown, and causes the little electric car to run when the wheel is operated.

Another novelty is a little mountain made of papier-mache, about two feet square, on the top of which is a little imitation lake made of glass. Around this lake a skater glides without apparent means of locomotion.

Still another one is the hand and blackboard device. It consists of a long shallow box finished in hardwood, between six and eight feet in length and between eighteen inches and two feet high, and so lighted that a blackboard at the back can easily be seen, at the lower right-hand corner of which is a hand. The electric mechanism by which the device is operated, causes this hand to move over to the opposite side of the board, and with a piece of chalk to write rapidly on the blackboard some sentence arranged for by the advertiser. When the sentence has been completed the hand crosses the space with a sponge, erasing what has been written, and then writes a second sentence, which is also erased after time enough has elapsed for it to be read. This device can be placed in the window or other place inside or outside the store in such a position that a spectator can see that no human being operated the hand. The device will write any two sentences continuously.

One of the most striking things in electrical decoration is a cluster light for store illumination, composed entirely of corrugated glass. Readers are well acquainted with the utility of prism lights to reflect and disseminate daylight. These glass prisms are scientifically constructed to produce the same result with artificial light. The reflector mentioned will hold a number of the new high-efficiency lamps, each one in the centre of a corrugated reflector, with a large reflector above which assures that the light rays will be disseminated downward over a considerable space. It is claimed by the manufacturers that this reflector will do the work of an arc light and it certainly has a very handsome appearance. For decorative effects the bulbs can be dipped or striped with colors and the reflection of the prisms will produce a very brilliant and uniform colored pattern, which gives a striking appearance.

A cheap little attachment for the ordinary-sized incandescent lamp is a colored glass cap made to fit over the end of a lamp, being held in place by a spring. This saves the trouble and expense of dipping lamps and produces the same effect.

Another cheap little attachment for an incandescent lamp consists of a cylinder which is divided into different color sections. The cylinder rests on top of a 16-candle power incandescent lamp. The slight heat produced by the lamp causes the cylinder to revolve, flashing different color lights.

One of the most attractive displays is

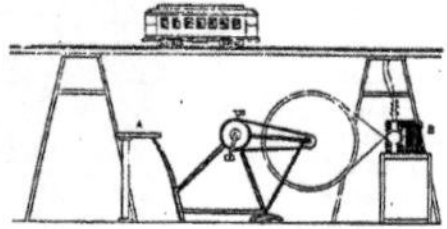

Bicycle Power for Electric Railway.

the large line of shades for covering electric light bulbs. These can be had in numerous shapes, some of the most popular being those representing bunches of grapes, wistaria blossoms and other flowers. A new and pretty incandescent lamp shade is made of oak leaves, both in green and Autumn colors. These leaves are fireproof and semi-transparent.

SOME SELLING SUGGESTIONS.

"Place in one of the show windows every Tuesday morning, some one particular item with a large card stating that on next Monday morning that particular item will be sold for a certain price and the sale continue until the stock is exhausted," suggests a successful retailer. "This must be an absolute bargain and the merchant must figure to lose money on this sale. For instance, a lamp complete that he would usually sell for 25c. or 30c. and which would cost complete say 15c. or 18c. should be sold for 10c. and a lot displayed during the week. The following Monday pense—not an investment. These men have the wrong idea. Advertising is not an expense. Or, if it becomes an expense it is not the right kind of advertising. The proper sort of advertising is that which brings returns commensurate with the money expended. If it does not do that it is not the

Attractive Tool Display.

right kind. And the right kind is not rare by any means.

AN ATTRACTIVE STORE FRONT.

As a rule space is not given to photos of store fronts in Hardware and Metal, but the accompanying illustration has sufficient merit to warrant its reproduction. Above the prismatic glass is an attractive sign below the cement block front of the second story. F. Howes, of Listowel, Ont., whose store is shown, is standing in the doorway.

The displays, both in front of the store and in the windows, are neat and not overdone. A variety of seasonable goods is shown, but taste and good judgment are exercised in the arrangement. Some hardwaremen neglect their window displays for months at a time, while others try to jam as much as possible into their displays. Mr. Howes appears to have struck a happy medium and women buyers can have no objection to entering such an inviting business establishment.

An Attractive Store Front—F. Howes, Listowel, Ont.

ATTRACTIVE TOOL DISPLAY.

In the small illustration we reproduce a very neat display of mechanic's tools is shown. The arrangement of the background are such as can be put together in any ordinary hardware store, and during the next couple of months such a display should bring good results.

ILLUSION ATTRACTED ATTENTION.

A display that brought results was arranged by an enterprising druggist in a Michigan summer resort a year ago. Sporting goods were to be had only from the lone hardware establishment and four or five drug stores; fishing tackle was practically the only goods in demand. The store of the druggist in question was on a prominent street, and he generally gave over one of his small windows to the display of bathing suits, fishing tackle and similar lines. In order to attract particular attention to the window he rigged up an illusion that fooled nearly everyone. A 2x4 wooden beam from the end of which a triangular piece had been sawed was leaned carelessly against the glass window from the inside as if it had fallen from the ceiling. The triangular piece that had been sawed off was glued to the glass on the outside in such a way that it appeared to be a part of the beam extending through the window. To complete the deception, several jagged pieces of glass were glued about the supposed broken place on the inside so that it required a very close inspection to detect that the beam had not fallen against the window and broken through. Almost every passer-by stopped to lok at what seemed to be the peculiar result of an accident, and the drugist's business during the time of the display was noticeably increased.

KEEPING FROST OFF WINDOWS.

An eastern Ontario retailer who had trouble with his windows frosting over overcame the problem in a very simple way. A hole about a foot square was cut, through a pillar, at the upper left-hand corner, and a grate, such as used in a hot air register, placed on the outside. The back of the window was made as near air-tight as possible, so as to keep out the warm air. So long as the temperature inside a window is the same as that of the outside air there is no chance of it frosting over.

In other windows this firm tried boring holes around the sash, but could not get sufficient ventilation. It happened that these windows had transom bars about eighteen inches from the top. The glass above them was put in frames and made to open up. In this way a great amount of cold air can be admitted in

a few moments. Of course it is necessary in this case also to have the back of the window air-tight or almost so.

HOUSECLEANING WINDOW SUGGESTION.

A practical suggestion by a paint manufacturer for a spring window is to place a mop pail, scrubbing brush, and piece of soap on one side of the window and a well arranged display of floor finishes with color cards on the other side. A display card in the background with the words, "Use the scrubbing brush less and the paint brush more," should be a catchy phrase to attract attention and bring results.

NOVEL ADVERTISEMENT.

A United States merchant furnished free to the publishers of the local daily papers all the paper needed for one certain issue if they would allow him to furnish a distinctive color. Without previous announcement the paper came out one day printed on pink paper instead of white. The explanation was given on the reading pages. He also carried a big ad in the advertising columns. He made big sales and was talked about for weeks as the man who "got out the pink paper."

Effective Hardware Advertising

HOW TO PRODUCE IT

By T. Johnston Stewart.

To advertise, or not to advertise, that is the question:
Whether 'tis wiser in a man to suffer
The wants and worries of an unprofitable business;
Or—to sling some ink into a telling ad
And make things hum. To drone—to sleep—
Perchance to dream of dollars when the milliner's bill
Just hangs fire and all the babies in the house
Are crying loud for frocks—when a few words
About the values of our wares would make our stores
The marts of all the people in this town and change
These antique-laden shelves into a silver stream
And boost our bank accounts. This surely is
A consummation devoutly to be wished for. Then,
Let us advertise and keep on advertising.

* * *

To grunt and sweat under a weary life
But that the dread of something not to eat
Makes us keep hustling. The lack of advertising
Explains the weary routine of our days and the game
Of checkers in the back store. Let us shuffle off
Old-fashioned methods and attune our ways
Like unto those of shrewd players of the game
And enterprises of great pith and moment,
In this respect, will no longer turn awry—but
Be what they were meant to be—a source
Of affluence for us and our posterity.
Then,
Let us start advertising and keep on advertising.

* * *

Now wouldn't I be a first-class chump if I were to step to the front of the platform and solemnly declare that the shade of the late Mr. Shakespeare was responsible for the hardwareman's soliloquy. On the contrary, I apologise to the immortal William—not because the parody is inapt or untimely; but because I have turned the idealistic ravings of Denmark's incomparable dreamer into common-sense, twentieth-century, business philosophy. Besides, the only ghost that disturbs our slumbers is—a delicate treasury aad if I can impart a few hints which will be useful to my hardware friends here and now I'll take chances on being able to mollify Brother Bill hereafter.

* * *

The last quarter of the nineteenth century demonstrated beyond the shadow of a doubt that printed salesmanship would be the mightiest lever in the commercial life of the twentieth. Incidentally that just explains why I am an advertising man. This century will revolutionize business methods so thoroughly that the astutest business men of the east—men who have heaped up millions—would be hopelessly outclassed could we fling them unprepared into the year 1920. I tell you there is joy in being alive to-day. Because, it does not matter how insignificant your business may be this vast country of ours will call strong men from every department of commerce and the lesson of all the ages is that the practical, sagacious young man of to-day in any line of endeavor will be the big man and the leader of to-morrow. There can be no doubt about such an assertion because history makes it good. Moreover, I think that Canada's hardwaremen stand to win big prizes during the coming years. Since the Dominion is destined to be the granary of the world that is assured. This immense country cannot be developed without making thousands of wealthy hardwaremen. You can be one of them. Save your dollars, use common sense and advertise—advertise.

* * *

Let me dispel the fears of the man who imagines he cannot "get up an ad." This fact cannot be emphasized too much, viz., you do not need to be a first-class scholar, a master of language to advertise and advertise pretty effectively. The best of us are merely finding out things as far as the advertising game is concerned. The intellect of no one man can comprehend it in all its fulness. We are simply standing on the threshold of a new world which offers huge rewards to men of versatile and creative genius. Set down the advertising man who knows it all as a nonenity in the advertising world.

I intend to reproduce a few ads by amateurs, ads written by hardwaremen and the longer you consider them the more you will be convinced of your own ability to produce good strong advertising copy—copy with a jingle to it—just the kind of copy that sells goods. There are specialists (so-called) who might study the advertising of many country and small city merchants very profitably. You know your business, you know the value of your goods and their selling points, you know the wants of the people in your section and because of that knowledge you can produce good advertising—just the kind of ads which will clear your shelves. Get out your copy, take it to the printer, tell him to arrange it properly and gently hint that if he can help you to make your ad a paying investment you'll be inclined to buy large space regularly.

Material interests sway the business world. The newspaper man dreads the milliner's bill et cetera. Scratch his back and he'll scratch yours. And banish for all time the idea that you cannot produce an ad that will sell goods. You can. Every business man can. At least 99 dealers out of every 100 can do so because there is nothing intricate about retail salesmanship. Try it now. Advertise!

* * *

Running right across the page we have an ad. which simply couldn't keep selling goods. Of course this is granting the truth of practically every statement made in that ad., and although hardwaremen do not belong to even collateral branches of the Washington family my experience has taught me to place implicit faith in their veracity. Now I do not know Taylor Bros. But I do know that the man who is shrewd enough to write an ad. like that is shrewd enough to recognize the value of truth in business. The man who swears fervently that advertising does not pay, because he has tried it, has either wandered from his truth in his advertising, or expected the whole community to make a bee-line to his store after his first ad. or two. Wooing business is pretty much like wooing a girl and we all know enough about coy damsels not to try rushing tactics. You would not drop on your left knee, place your right hand on your heart and pour forth your song of adoration to the "dearest girl on earth," after the first, second, third, or even sixth visit. You cannot make her Mrs. Jones that way and you know it. You see the happy day approaching every time you look into her darkly, deeply, beautifully blue eyes; but there's a glint of resistance still there which says plainer than words, "Don't rush things. Don't expect too much."

Some dealers expect a great deal too much from their advertising and expect it a good deal too soon. Not so with our friends, Taylor Bros. The Carleton Place Herald has been coming in pretty regularly for some time and we notice an ad., right across the top of the editorial page just like the one we have reproduced and, of course, greatly reduced in size, which goes to show that Taylor Bros., not only know how to write an ad., but they also know how to dicker for one of the best locations in any paper. There are few editors who would surrender the whole top of the editorial page to advertising. That's the second best location in any paper published and, when the editor is an interesting writer, it is probably the best. But as a rule the upper right hand corner of the first page is the best possible location—the most valuable space in any daily or weekly. It is always in evidence. It is the first part of the paper to fall under observation and as a rule we glance at that corner just before destroying the sheet.

You can study every item of Taylor Bros. ad. carefully and discover that

Let's take the pretty washerwoman and place her and "Wash Day Necessities" at the extreme right of the ad. Next to the lady placed the bathroom talk and the item on metallic ceilings. That forces the double column talk about the Taylor business principles and the address right into the centre of the ad., and leaves the whole as thoroughly balanced as an ad. could be. Place your little editorial in the centre, gentlemen, have your name and address clearly displayed and it's a certainty that you stand to win in the hardware business. The ad. as it is was bound to sell goods. A little attention to arrangement will make the valuable space more valuable.

* * *

C. Wise & Son's ad. carries out and emphasises our talk on arrangement. It is prettily arranged and the effect is pleasing in the extreme. The eye of the probable customer is captivated at once. There are 'special prices" at Mr. Wise's place and the shrewd buyer de-

forks would please Marie and fill her soul with vain regrets and yearnings for what might have been. This begins to look tragic. And it is. But neither for Bill nor Marie. But for our dear unknown friend, Mr. Wise. Because he has omitted the price. Bill is not impressed. He follows in the wake of Mrs. Smith. And the other fellow does his business.

Mr. Wise I very much fear, that it's 23 for yours—if you don't wake up! It is not enough to produce a pretty ad. It must also be effective. And no retailer's ad. can be effective unless price talks clearly and distinctly all through it.

* * *

You think I am using Mr. Wise pretty roughly. Cheer up! I cut this ad. from an American paper in order that I might give you an object lesson without hurting the feelings of the said Mr. Wise. Few men can swallow a straight-from-the-shoulder criticism in the spirit in which it is given. Mr. Wise lives

4 THE CARLETON PLACE HERALD. JAN 22, 1907

HARD COAL

JACK FROST HAS BEEN KING all week outside. Inside Taylor's Hot Stove has displaced him. He has no place in the homes where our Coal is used

All the Sizes at All Times.

NUT EGG
STOVE GRATE

The quality will always be remembered. Our price is the same as common grades Careful, tidy, delivery

Wash Day Necessities

Our stock is always complete in this line. We have

Washboards, Tubs.
Clothespins,
Clotheslines.

Wringers and Washing Machines The latter from $3 50 up to $7 15.

No trouble to show goods Come in

DID YOU EVER

Think what a privilege and comfort it would be to have a

BATHROOM

in your home? The cost of an outfit is not so much that anyone in ordinary circumstances cannot afford it. We have just completed a first-class job for a farmer. They can easily have a complete Bathroom put in and no trouble with the sewerage.

Information on this subject cheerfully given

We are also Booking Orders for Metallic Ceilings.

This is the modern way of finishing a kitchen.

No plastering required and hence none to fall off.

We have a fully equipped Tin Shop ready to do any work required

Let us figure with you for anything in our line

WE GIVE CREDIT to Responsible parties as freely as ever.

TO THOSE WHO HAVE THE CASH and are willing to part with it the same day as goods are delivered we give

5 p.c. Discount.

This is far more than the money can earn for you in the bank.

This is only one of the many reasons it pays to trade at Taylor's.

CARLOADS

Due to arrive this month

Wire Nails	3 Cars
Barb Wire	1 Car
Black Wire	1 Car
Oil Cake	1 Car
Timothy and Clover	1 Car
Iron and Iron Pipe	1 Car

Besides 10 Cars Coal.

Is not this quantity buying?

AXES

Any Weight. Correct Shape

Take an edge and keep it

For $1.00

Yes, we have some as low as 75c and they stand pretty well, too

To see our line is to buy one.

TO ANYONE PURPOSING BUILDING IN 1907.

We would say be on the move at once

The Hardware prices are advancing so rapidly that in many cases our prices are less than similar goods could be laid down for from Montreal.

It will pay you to get our prices before purchasing elsewhere.

Taylor's Sell for Less.—This we are prepared to prove to any one. By this we mean that anyone who keeps a record of the cost of a year's Hardware will find it less from us than from any other source. We don't have leaders that look good on the surface and then charge exorbitant prices to make it up. We are building a trade on a solid foundation. We set the pace and it keeps others busy to keep it up. The newest goods that we consider reliable are always found here. Quantity buying enables us to sell the same goods for less or better goods for the same price. We thus reasonably expect that 1907 will be the largest year in the Taylor name.

One Price Stores. TAYLOR BROS. & CO. CARLETON PLACE AND ALMONTE

there has been juts enough said in nearly every instance. Jack Frost is a pretty fierce looking gentleman and when you look at him the second time you begin to feel the absolute necessity of a temperature you can regulate, according to your pleasure. There are three strong words about that item which lifts it above common-place advertising. The man who produced the copy considered his customers, and he wrote "careful, tidy delivery." Rest assured that his consideration was amply rewarded. The particular housewife was sure to note the words we've quoted. She drew the attention of her husband to Taylor's way of delivering coal, and you can depend upon it that the Taylor way in this particular was mighty pleasing to all concerned.

I have just one criticism to make about this ad., viz.: It is not arranged nearly as well as it might be and I am sure that Taylor Bros. will agree with me.

cides to read the ad. However, this ad. resembles lots of pretty women. One expects a good deal from a pretty woman, you know, and generally one is greatly disapointed. Because there's nothing decisive about her conversation. This, "How-cute-don't-you-know-gush" makes us weary, and we expected so much from Mr. Wise's ad. It looked so beautiful.

Mrs. Smith wanted a chafing dish. She sees at a glance that Mr. Wise has pretty good chafing dishes. So has the other fellow—and—the other fellow tells the prices just as plainly as printer's ink can. Bill Jones wants to buy a gift for Marie Jenkins on the occasion of her wedding. Marie used to be Bill's steady, and he wants to show her that he has no hard feelings because she changed her mind. Bill feels grateful for this interposition of providence and resolves to do the gift act good and proper. His eyes falls on Mr. Wise's ad. These silver knives and

somewhere in the United States, he's no particular friend of mine, and I am cut loose without putting a drag on my feelings. The point I desire you to assimilate is this, that if there is another dealer in Mr. Wise's town, who cannot ararnge or write ads. nearly as good, but who advertises nevertheless, then that other dealer is reaping the benefit of Wise's advetrising. Mr. Wise is far more philanthropic than any respectable American citizen intends to be.

* * *

Mr. Dealer! You can't afford to advertise for the benefit of the other fellow. Don't do it. There is a certain amount of mutual advantage accruing from all advertising; but when a man writes, arranges, and pays for an ad. which will inevitably be more advantageous to his competitors than himself, then that man is not a wise man and the sooner his son realizes this the better.

The Americans certainly do get out

pretty advertising. They arrange their ads. better and, as a general rule, in a more eye-catching way than either Canadian or British advertisers. But the Americans are short in common sense, as far as their general advertising is concerned. Big American firms are doing immense businesses because of their advertising; but, in many cases, they are not doing all the business their advertising appropriation could land directly or indirectly. The point for the retailer in this is that it is possible to pull just as much business by spending $500 judiciously as by spending $1,000 a little less judiciously. Most retailers could cut out the circular, which they print every now and then because most circulars simply cannot land the business. A mimeograph or type-written letter to all probable customers drawing their attention to your special advertising in the columns of your local paper will be appreciated and prove effective. If your citizens read the local paper then it will pay you best, 99 times out of 100, to place your advertising with them. The only time the man of the house sees a circular is when it it's on fire. He lights his pipe with it. If your circular isn't pretty it has no attraction for a woman. If it is prettily arranged and illustrated, she gives it to the heir-apparent to play with. The court decides in favor of the local press. Use it judiciously and your advertising is bound to pay.

to the catalogue houses, while he talks straight and interestingly to his fellow citizens. He makes a square proposition and the folks of his town are bound to be impressed by that kind of talk. Read Mr. Retzlaff's ad. Read it again. What does it suggest to you? It seems to me that Mr. Retzlaff figured it out this way: "There are carloads of catalogues pouring into this town and district annually. They are mighty good catalogues too—just as pretty as strong talks and nice illustrations about easy-priced, quality goods can make a catalogue. The people of this community will read those catalogues. I don't blame them. They will be impressed—interested up to the buying point again again and again and I can see my profits going to the fellow who can pay a gang of ink-slingers to write up that selling talk for him."

* * *

Mr. Retzlaff reviewed the situation rather dejectedly until that happy idea struck him. And he went on in this cool, argumentive way: "Well, things are not so bad after all. My interest in the game has not petered out yet—not by a jugful. I sell the things those fellows advertise in their dandy catalogues. I can sell at the same price any article in my line that a probable customer may read about in any catalogue. The catalogue houses are simply advertising the stuff I sell and I'm going to let the people of this community know it."

Mr. Retzlaff has struck a happy idea. He realizes that the mail order houses are his keenest competitors; but this knowledge does not seem to worry him in the least degree. He views the situation quite calmly and issues a challenge

Mr. Retzlaff did let the folks in his neighborhood know all about it. He practically says in his ad., "You people know me. You would vouch for my honesty. You saw an article described in Messrs. So-and-So's catalogue and you want it. That description is perfectly correct. Come to my store, bring your catalogue with you and I'll let you have the article you desire at the same price. If I cannot make good then, by all means, get your goods from Messrs.

So-and-So." Are there any retail hardwaremen in Canada who have the grit to print a talk like that and the ability to make it good? Can you afford to do it? Then do it and do it now and let us know the results.

* * *

A sense of fairness compels me to state that Mr. Retzlaff is also an American retailer. Probably you will be surprised to learn that out of 100 hardware ads. sent to me by our clipping bureau, and clipped from Canadian papers published all over the country there was only one which I considered fit for reproduction. Admitting that the young ladies in the clipping bureau overlooked many good ads, yet, I think the fact stands out clearly that there is room for improvement—lots of it.

Now there must be at least 100 hardwaremen in Canada who advertise and advertise effectively every now and then. We would like their ads. forwarded for our mutual benefit. I will promise that my criticism, while remaining effective and profitable, will be as gentle as a summer zephyr playing o'er the bosom of a placid sea. Besides, any ad. we reproduce is bound to come in for a good deal more recommendation than criticism, as a general rule. The moral of all our talk is: Advertise principally in your local papers and keep on advertising.

Samples of ads. are always greatly appreciated.

Meeting Mail Order Competition

Practical Methods which have Proved Successful — Retailer in Small Maryland Town Publishes a House Organ — Iowa Man's Experience — Five and Ten-Cent Stores

WOMAN THE PURCHASING AGENT

Realization of changes necessary before you can hope to compete resultfully with retail mail order houses can hardly show to more immediate advantage in any single direction than in the way you display your goods. And nothing more certainly than his store arrangement indicates whether a merchant appreciates or not that men almost completely have ceased to figure in the day-to-day household buying.

The retail mail order houses show in their every move thorough recognition of the fact that woman has become the "purchasing agent for the American home," says Butler Bros.' drummer, and if you continue to run your store without attempting to find out what women want and the way they want you to offer it, should you complain if they do trade with retail mail order houses and other competitors of yours, who with goods in every-day need cater to those who do the actual every-day buying?

Woman Enjoys Looking.

A woman enters your store as one willing to be interested in your offerings even to the extent of buying things other than those for which she came. The store that makes looking easiest and its display freshest and most varied goes a long way toward inducing her to confine her shopping to that one store. As an up-to-date retailer make sure that the displays in your windows and in the store itself are always such as will interest women. Show a variety with prices clearly indicated in plain figures. Make looking easy and then let shoppers look undisturbed, for in the process they will develop new wants for you to supply.

The retail mail order houses give a hint in this direction. Their immense catalogues are but a display of their wares and can be studied for suggestions as to how the actual goods may be displayed effectively and attractively. You know that though those houses work from a distance, ask payment in advance and talk in print about mere pictures of the goods—you know that, in spite of their disadvantages, retail mail order houses do interest and convince by means of the display in their catalogues. If, working under such disadvantages, the retail mail order houses can create new wants and supply old ones—imitating them, without their disadvantages and working in person among your personal friends, what is not possible for you in this direction?

Go After the Trade.

To sell goods nowadays, it is not enough merely to buy a stock of goods, pile it on shelves behind counters and sit down to wait for trade to come to you. The up-to-date retailer is persistently asking people to buy his goods. That does not mean that he is constantly boring them with the spoken or written question. Words need not be used at all. The display of your goods can be made to ask the question and, perhaps, all the more effective because the asking is wordless.

Frankly criticize all your displays and do not be satisfied with any store arrangement short of that which comes nearest to indicating all the wants you can supply and to suggesting the most new wants to those who enter your store.

Do you let your odd lots accumulate? Or as fast as they appear do you put tickets on them quoting plain figures prices that will make those odd lots part of your bargain ammunition?

Is there a price ticket on every article that can be plainly seen by customers so that lookers get the whole story when they do the looking?

Have you such a thing as a bargain counter in your store?

Do you trim your windows with goods at bargain prices?

Do you run a special sale now and then?

Or, though hard hit by retail mail order houses, are you still running your store in what was the approved way when people came in, asked for what they wanted and went out without looking around at all?

Don't Hide Your Goods.

The ideal store arrangement, remember, is that in which as nearly as possible everything is shown in the attempt to take full advantage of the tendency of people to buy a great many things on sight which they did not have in mind at all when they left home.

In your store people cannot buy on sight unless you make sure that they see your goods. Keep your wares up near the ceiling or under the counters and have you any right to be disappointed because they stay there?

Hundreds of popular priced novelties are on the market and the number is being added to almost daily. Have you ever tried to display some of these novelties on bargain counters in order to learn how freely they sell to women? You may have some of these novelties in your store right now—may even be complaining that they do not sell. But are you justified in expecting them to sell so long as you keep them where they will not be seen?

Get down the old goods, give them a thorough cleaning and then put them where they will be seen by people likely to want them. Go through the entire store with the determination to fix your display as quickly as possible so that people will know without asking that you do have this or that thing. Get things out where they show and to make the story complete, be sure that you fasten price tickets to them. Add as many specialties as you can handle successfully and don't put them so high up that they become shelved goods which is another name for profit-eaters.

Buy and sell for cash. Or if you do give time, include say five per cent. that can be given as a rebate in one form or another to those who do pay cash. Come as near as you can to showing everything you have. Keep changing the display for an atmosphere of newness is most attractive. Study your lines for ability to talk about them so intelligently that people will choose to buy goods you orally describe instead of goods the pictures of which are explained in print by strangers.

Learn Lesson From Opposition.

Study how to make your whole store attractive. Get a reputation for running an attractive store where customers are made to feel at home and to understand that their business is appreciated and you will find that people generally like that way of storekeeping well enough to go blocks out of their way to make even trifling purchases.

Take a lesson from exclusive 5 and 10 cent stores which are seldom advertised by printed matter and in which therefore all the features of up-to-date merchandising are specially emphasized. The 5 and 10 cent stores keep their places bright, clean and attractive. Their goods are always well displayed and their show windows are always kept filled with the best values and the most attractive goods in the store—with prices indicated in plain figures.

All clerks are required to be neat and to attend strictly to business. Trays and merchandise are all wiped or dusted every morning. Window trims are changed weekly or oftener. And in every place and every move you will observe constant studied effort to give prominence to real bargains.

So skilfully does the 5 and 10 cent store use most unmistakable bargains that the average merchant as well as layman takes the suggestion and jumps to the conclusion that the store must own its goods cheaper than other stores own theirs.

Yet when you come to walk up and down the aisles and observe how many articles are priced at 10 cents that you know cost from 48 to 72 cents a dozen, and how many things are marked at 5 cents that you know can be retailed with profit at 3, or even 2 cents—then you begin to realize the power of a few bargains unquestionably big. One can well afford to lose money on a single article if the result is to increase the sales of twenty articles, the other nineteen of which pay a profit of from 33 to 150 per cent.

Turn to any of the retail mail order houses' catalogues and careful study will disclose how identical their merchandising is with that of 5 and 10 cent stores and every other retailer using modern methods. The hints for you in the 5 and 10 cent stores, successful department stores of large cities and the catalogues of retail mail order houses are obvious. Your buying public in general is the same as theirs. And none of them has a monopoly of the courage to pay $1.50 per dozen for goods to retail at a dime nor of that still rarer courage to get a juicy profit on 90 per cent. of his wares.

RETAILER FINDS HOUSE ORGAN PROFITABLE

An interesting address on the subject of House Organs for Retailers, delivered by Charles S. Davis, Oakland, Maryland, at the meeting of the West Virginia Retail Hardware Association last month, was as follows:

Advertising has become a necessity to all classes of business men, so much so that the successful business man is considering every plan and method that will help him place before his trade to the best advantage the various lines he is handling. Catalogues, bulletins and circulars in various forms are now sent broadcast by the thousand to all classes of trade and consumers, so that to-day the man that does not advertise or publish some kind of a circular or bulletin that will advertise his line is not keeping up with the advanced methods that are becoming so prominent throughout our country.

With this thought in view we were led to adopt a system of advertising during the past year which has proven of more than ordinary interest to our friends and to us, publishing a house organ called "The Hardware Bulletin," of 28 pages of regular magazine size, and containing articles of interest to our trade, advertising the various lines we are handling, and at the same time advertising for other merchants some lines that are not handled by us. The plan has been very successful Not only has it increased our business, but it has had the desired effect of offsetting advertising by catalogue houses and keeping at home a large percentage of trade that would otherwise have gone to the catalogue houses and department stores The demand for a publication of this kind has been clearly demonstrated to us during the past year. We have received many requests for sample copies, and quite a number have expressed their intention to adopt a more definite plan of advertising the present year.

How to Publish a House Organ.

The first point to consider is what amount you are willing to spend in advertising. Two per cent. of the amount of goods bought is a rule that many dealers have adopted. This, of course, depends on what plan you adopt, and the access you have to printing offices. So far as possible I believe it would be to your interest to contract with your home printers, especially if you are in a small town. This creates a good feeling and often, as in our case, leads to the copying of articles from your publication in your town or county papers. This not only advertises you to a large extent, but is advertising and creating a demand for your paper free of cost.

The size and style depends on the amount of advertising you intend to do and the lines you expect to advertise. A four-page circular of good size, with letters to the trade on first page and advertising ads. on balance of pages, nicely illustrated by the use of electrotypes that you can secure from the manufacturers whose lines you are selling, has proven to be one of the best and cheapest plans for the small dealer. For the larger dealer a monthly, bi monthly or quarterly publication, nicely illustrated with the various lines you are handling and new goods that you are receiving from time to time, put up in catalogue form and mailed regularly to a carefully selected list of customers and friends will yield a greater influence and go farther to offset catalogue buying than any other form of advertising.

Expense and Mailing List.

Circulars can be furnished by the thousand very cheaply, but a 20 or 30-page magazine or bulletin, including cover and back printed on good paper, weight not to exceed 2 ounces, so that it can be mailed for 1 cent, and size not to exceed 7x9 in., and you furnishing all the data and cuts for illustrating, should be printed, folded and stapled ready to mail for $25 to $35 for the first thousand, and from $15 to $25 for each additional thousand. Where monthly publications are furnished, you will find quite a number of ads., such as of ice cream freezers, screen doors, oil stoves, gas ranges, spring and fall goods, that you will want to reproduce several times or change very little. This is to your advantage and should be taken into consideration by the printer when contracting.

This can be compiled in various ways, and all depends on your location. In small towns where papers are published, the editors can usually furnish a list that will take in a large class of the best people. Mailing list on free mail routes where they can be obtained is a plan that some have used, especially catalogue houses. Another very practical way is to select a leading man or farmer from any section that you want your advertising to cover, and for a small compensation you can usually get a list of post office addresses of the people in that section of the country.

After your own advertising and publication has gone so far as to bring the customer or inquirer to your store, then it is that you want to use every argument possible to make the sale and to place in the hands of your customer printed matter as furnished by the manufacturer to clinch the argument already set forth, and to more forcibly impress on his or her mind the many good qualities of the article in question. Another good and very effective way is to mail with your publication manufacturer's advertisements. This is especially true of new goods, and has the effect of keeping you in the lead on new goods with your trade.

Manufacturers' Co-Operation.

Manufacturers can assist the local merchant by writing ads. of their various lines, placing one or more each week or month in the hardware publications, as they are published and furnishing electrotypes to those who are willing to use same in space they have arranged for in their local papers or organs of their own. Merchants as a rule cannot write their ads, as well as the manufacturer, who is fully acquainted with his line of goods and its many advantages. We are glad to say quite a number of the manufacturers are furnishing the dealers with ads. to publish, but the practice ought to be more general. As a result merchants will not only advertise more largely, but will necessarily have to carry larger and better stocks. This will have the effect of keeping local trade from sending away for many goods now furnished and advertised by the catalogue houses.

This article is based upon our own experience as hardware merchants in a small town, feeling the necessity of some plan to offset the influence of the catalogue houses and city depattment stores, and while the plan has been a success we find it necessary to adopt many of their methods. A good and well-selected stock of goods suited to your trade is required, and in many lines some cheap goods. This is especially true in 5 and 10 cent household goods. This permits you to make compaiisons and to advertise some low-priced articles.

After carefully studying and planning we felt that as our trade was largely with farmers and people of limited means it would be impossible to do a cash business, but we have adopted a plan of putting our store on a cash system. If a party wants to open an account, say, for 30, 60 or 90 days we have an agreement signed in which the party receiving the credit gives us his note payable at a specified time, said note to be credited to an open account less 1 per cent. per month charged against the note until paid for credit allowed. At the same time we give receipt and contract to the party against the note, stating that the note is to be credited on open account less the 1 per cent., and when the note is due if the full amount is not paid we also agree to credit back the difference between book account and note, according to our books. This avoids opening any account, except on a cash basis, and puts your cash and credit customers all on equal basis, with all book accounts payable at a specified time. The plan is new with us, but so far has worked very nicely

USING THE OTHER FELLOW'S WEAPONS

Sidney Arnold in American Artizan

A man whom I have watched pretty closely for the last five years recently started in the hardware business in a fairly good-sized town of Iowa. Previous to that time he had been employed by one of the large hardware jobbers. I recently had an opportunity to interview Charley on the success of his venture and I found him well satisfied. He started in a town rampant with catalogue-house competition, and was successful in fighting this form of competition, so his methods may be considered carefully.

When Charley started in business he looked over the field and discovered a firm belief in his community that the retail catalogue houses offered better values than the local dealers. He looked up the history of the largest of these concerns and noted that their methods were legitimate, using the word in its proper sense. He saw that their growth was to be accounted for by an examination of the laws of nature. In other words, Charley believed that the retail catalogue house system was lawful—as lawful as what is known as the regular method of distributing merchandise. He regarded Beers-Sawbuck & Company as a natural, able and righteous competitor. He put them in the same class as his own competitors in the same town. The classification as such was accepted and the facts taken as they existed. There was no hedging; there were no excuses; there was no denying actual circumstances. The question was not: "How can I drive retail catalogue houses out of business?" but "How can I gain and hold the trde of my community?"

The work of fighting the retail catalogue houses in the matter of supplies, Charley left to what is known as the Joint Committee. He read of their work in the trade papers, but aside from that took little interest in the matter; nor had he much sympathy with their fight. He realized that if the retail catalogue houses feel a demand for certain articles they will find a means of supplying that demand, whether sentiment is against them or not. Charley realized that if a man is starving, he will get food even though the nurse refuses to offer it to him in the form of a milk bottle or a bowl of pap. As will be seen, Charley attempts to minimize campaigns of the retail catalogue houses in his territory rather than to reduce their purchasing power. For all Charley cared, the retail catalogue houses might store their warehouses from basement to ceiling and then build additional ones to hold their goods. It was determined simply that these retail catalogue houses should not empty their warehouses into Charley's community.

How to reduce the retail catalogue houses' sales in his territory was then the question. Charley footed up the balance sheet. To his own credit he placed immediate delivery, personal inspection of goods on the part of the customer, quality, convenience of delivery, proximity of source of sale, unquestioned guarantee, oral rather than written salesmanship. The sole argument for the retail catalogue houses found to be price. Even that was a doubtful advantage.

The premises were laid down and Charley knew where he stood. He could now proceed scientifically. Charley's formal store opening was much as other openings. There was an orchestra hidden behind the conventional palms, a sweet-voiced damsel dispensed biscuits nicely browned in a standard steel range. Mrs. Hogitall of the village gossip shop came in for her share of the advertising novelties and pronounced everything perfect. The youngsters were well taken care of by pocket knives and dolls. Altogether the offer was much as a great many others, only perhaps more successful.

Charley quickly discovered, as he had expected in advance, that the buying public of the community had a most disagreeable tendency toward accepting all the favors to be distributed by local retail stores, to ask them to perform all of the gratuitous, or almost gratuitous, jobs that devolve upon the retailer, and sending to the mail order houses for those goods which netted a good fat profit and which were paid for in cash. This is natural. Charley had noted its presence early in life and realized that people were inclined to help themselves even to the detriment of others.

Sound Logic in This.

Charley did not try to boycott the people of his community who sent their juicy orders to catalogue houses and left him to do the repair work. He aimed not to force his best customers to buy of him but to make them want to buy of him. Consequently he aimed not to show his customers that they ought to buy of him for religious reasons, that the retail catalogue houses were extending and growing tremendously at the expense of the retail merchant, but that he himself, the retail merchanat was growing proportionately faster than the catalogue houses—that he was growing so fast because he was offering better values, that his customers were better satisfied, and that it was up to the people in the community to come to his store and place their hard-earned sawbucks on his counter for their own selfish sakes and not for anybody else's.

He began an advertising campaign that was pursued consistently and scientifically. Every week his customers got a circular letter describing some novelty to be purchased during the succeeding week at a reduced price. The name of the retail catalogue houses was not omitted through fear of advertising. After describing the certain novelty that was offered as a leader for the succeeding week, Charley went on to explain why he could offer this article and all other goods he handled at lower prices than the mail order houses. He commented upon their immense number of clerks, upon the greatly increased freight charges for articles shipped singly rather than in bulk. He referred to the immense sums of money annually thrown away for expensive catalogues and other advertising. He proved that the mail order houses expected a much bigger profit than any retail dealer and showed figures to prove that they are one of he greatest gold mines that financiers have discovered in some years. These and other points were brought up. In addition to this Charley took care to name in the circular letter that while some of the mail order houses offered a good quality, these were exceptions. He mentioned the famous "cheapness at any cost" motto of these concerns. Then he brought up his other points all in a brief way and all hinted at rather than proclaimed broadly.

Common Sense Plans Won.

This work Charley left to an experienced advertising man for the first three or four months. Then, after getting a general idea of what good advertising is like, and having gotten together a kind of guide book, he proceeded to do the work himself, giving it his personal attention. The letters were carefully worded and were the product of perhaps six or eight hours' study at first.

The matter of advertising in the daily papers was done as carefully. Here Charley was helped to a great extent by the manufacturers. These furnished ideas, technical and comical cuts. He was assisted by his advertising man, who sent him proofs of the stock advertising cuts kept on hand by some of the large printing and engraving shops. These advertising cuts can be had at small expense and are, as a rule quite original and satisfactory.

On every occasion Charley distributed souvenirs. His souvenirs were absolutely unique in one particular—they contained absolutely no mark showing who the donor was. Charley had gone to hardware conventions and had received souvenirs from the manufacturers and jobbers who offered displays at these places. In looking over his list of souvenirs one day, the fact struck him that almost all were unavailable for decorative purposes because of the brand or the name of the donor appearing conspicuously on the most prominent part. It occurred to him at the same time that this nullification of beauty was entirely unnecessary. Charley could tell without looking at the trade mark just who had given each of these souvenirs and what that company manufactured. He decided that by giving away a pocketbook his customers would remember that Charley had given it and would remember it longer than if he should have engraved in a prominent part the words Charley's general hardware, Lonesomehurst, Iowa. Charley's argument was good. Try it on yourself. Unless your memory is remarkably poor you will recall who gave you a pearl-handled pocket knife that you carry around, the rule with which you measure short distances, the hand-painted plate which you hang on the wall, or the hundred other things you have received. Moreover, Charley was right in realizing that if one souvenir out of the thousand lacks the name of the donor, the recipient is all the more curious and will himself strive to ascertain the donor and remember it.

All this means that Charley is fighting the retail mail order houses at their own game. Charley realizes that his position is the more logical. He knows that he has more advantages, true advantages, to offer the consuming public in one day than the mail order houses have in a year of fifty-three weeks. Advertising is the strong forte of the catalogue houses. It is a system by which these concerns were started and through which they have grown from ramshackle junk shops into tremendous concerns unrivaled in this or any other country.

Charley is successful because he does not attempt to change nature's laws. He accepts nature as it is and uses it to his own best advantage. While he realizes that the retail mail order houses are here to stay, he feels that his own business is establishd to stick also. And Charley will see that it does stick.

DISPLAY THAT BROUGHT RESULTS.

Concerning the value of window displays in the small town, sporting goods dealers are almost unanimous in declaring that this is their most effective method of gaining publicity for their goods. The experience of one firm will serve to show what benefits are to be derived from this practice. One of the cheapest and most attractive displays heard of recently was put on by a Southen hardware company. It consisted of a mechanical negro's head—one that rolled its eyes from side to side—placed on the stuffed figure of a boy dressed in blue blouse and overalls. The dummy was made very life-like by stiffening the arms and legs with wires which allowed them to be bent in any position. It was placed in the centre of the window with a 6-foot fishing rod fastened in the hands. A pan of water about of five feet long was placed on the floor of the window space so that the line dipped into it, surrounding the pan were rocks, moss and other substances to give the appearance of a river bank. A string of fine, bright hair wire was run from the tip of the rod to the ceiling and thence through a screw eye to some place in the back of the store. By means of this the rod could be raised and lowered as in fishing. Some small catfish were placed in the pan and gave the whole a very natural apearance. It is said that the effect, when the negro rolled his eyes and raised the rod, was very humorous and striking and crowds were gathered before the window during the greater part of the day. That the display was productive of results was proven by the sales which totalled about three times as much as during the previous season. Customers coming into the store frequently asked about the display.

A WASTE OF TIME.

So it is with window trimming. The man who throws a few steel goods or cans of paint into the window and then leave it there until it is fly-specked and covered with dust does not believe in window trimming. He has never received any returns from the trims (?) that he has put in. The only reason he puts them in at all is because his competitors do. He thinks it is a waste of good time.

Contrast this man with the merchant who makes it a point to see that his windows are always clean, are changed frequently—every week anyway, and who sees to it that the displays in the window goods that he is anxious to dispose of—stuff that has dragged or something that is so new to the trade that it needs every sort of pushing to get it started. He will tell you that his windows are the best advereising medium he has—considering the expense. Newspaper advertising is indispensable, but for the outlay a good window will often discount the newspaper ad badly.

After all, like almost everything else, the virtue of window trimming lies in the man who does it. The man who does it because he believes in it, and who puts his heart in the work and takes pride in it will tell you that it is an indispensable adjunct to the business. The man who does it in a haphazard manner, merely because the rest of the merchants do, will say that it is a useless expense and he wished he did not have to waste his time doing it.

With the Men Who Sell

How a Jobbing House is Conducted — Best Method of Posting Travellers — Burning the Candle at Both Ends — Stories of Success and Obstacles to Avoid.

MACHINERY OF A JOBBING HOUSE

By C. C. Fee.

A brief outline of the internal mechanism of the large hardware jobbing house, where goods are sold by the dozen and gross where the retailer sells a single article, should prove of value to quite a number of dealers.

The first thing that crosses the mind of a visitor to one of these large establishments is how on earth things can be kept running smoothly without getting jumbled up. Put the question to any of the employes and they will answer, "System, sir; every man in this house has certain duties to perform, and, by doing his part, he supplies one of the spokes that goes to make the complete wheel."

For the purposes of this article, we will take as our model one of the largest and most up-to-date houses in Canada, as the others are run on the some general principle.

The Buyer's Department.

The man who looks after this important department is a man of wide experience in the hardware business. He must know how any article will sell, so as not to run short, or, more important still, not to have any stock left over at the end of the season. This is the department upon which, practically, the whole success of the firm depends.

The head, with the assistance of two or three "under buyers," interviews all traveling salesmen from the manufacturing houses, and between them keep the stock in the warehouse up-to-date. In this department notebooks are kept for every stock floor of the warehouse, and each morning they are brought around by an office boy and left with the managers of the different floors. In these books the head of each department must enter every day the goods that are beginning to run short in the department under his jurisdiction, and the books are then gathered up in the afternoon by the boy, who brings them all to the buyers. They go over the list of shorts and cross out or place repeat orders according to their judgment, and in this way it's almost impossible for a line to run short unless there's an exceptional demand or the manufacturer is unable to ship the goods in time. Of course, there are always such delays as shortage of cars, congestion of freight, etc., which sometimes put the buyer off the track, but generally this loss of time is allowed for by placing the orders early. This is a point on which many retailers could improve their system, instead of waiting until their shelves were empty before ordering from the wholesale.

The Receipt of Goods.

When a shipment of goods is received at the warehouse it is divided up into lots and sent up to its separate floors, via heavy freight elevators. There are three of these elevators, measuring about 15 by 8 feet; so that goods may be received and sent up on two elevators while the other one is used for orders being sent out. In this way there is no delay by congestion, and the game of "give and take" goes on without interruption.

The Sales Manager.

Goods successfully bought would be of no use unless they were successfully sold. By "successfully," we mean sold at a price that satisfies the customer, and, at the same time, leaves enough profit that the firm may pay its running expenses and have something over to help the banks along. The sales manager is a man who has had wide experience on the road and in the warehouse. He knows about all there is in the art of selling goods, can write beautiful letters, talk like a diplomat, has the stock and prices at his finger tips, keep an enormous army of clerks and stenographers constantly employed, and still find time to see anybody who calls on business. Somewhere round his desk you will find a card saying that social calls may be made after 6 p.m., or some such gentle hint that "gas bags" are out of order. He is in charge of the traveling force, which in any large house will number between 15 and 30. Figure out for yourself whether he has any time to discuss the hockey situation, and next time you call on him judge yourself accordingly.

The Order Department.

The head of this department is over all the warehouse employes, and also looks after the stock and sees that the packers holds no orders over. When an order is received in this department it is handed over to a clerk, who divides it up into sections. Each section is then handed to the managers of the floors on which the different goods are stored. The manager enters the order, number, time, and date received on a form supplied for this purpose, and then hands the order to one of the clerks on his floor to be filled. The goods are laid out and then placed in a truck with the order on top and sent down to the packers, who collect all the goods belonging to one complete order, box them up and send them to the shipping department, where they are delivered to the carters, who ship them off by freight or express, or, if they are city orders, deliver them at their final destination within 24 hours of the time the order was received.

If the dealer on opening his goods finds he has been overcharged, or the goods are short or have been damaged, he writes the sales manager, who turns the correspondence over to the claims clerk. He, in turn, looks into the matter thoroughly: traces the shipment from the time it left the warehouse until it reaches its destination, and finds out upon whom the blame should rest. He then turns the matter over to the financial department, which fixes the matter up with the merchant by either refunding him or placing the overcharge or damage to his credit, to be deducted from his next purchase.

The Sample Room.

Every large wholesale house has a sample room in connection, where prospective buyers may go and look over a sample of almost every line kept in stock. The clerks in this department also make up and keep in order all the travelers' samples, which is no slight undertaking. This department is in charge of a bright clerk, who, himself, is directly under the eye of the sales manager.

The Advertising Man.

The advertising man in a large jobbing house is generally looked upon by the other employes as a being favored by providence. He is, in the first place, only responsible for his actions to the head of the firm, and, on account of the difference between his work and that of a general clerk he is allowed a great deal of freedom which the other employes don't enjoy. He makes up all the advertisements which the firm runs in the different trade papers, besides looking after the printing of catalogues, circulars, etc. His position looks like a golden dream of happiness to the uninitiated, but—well, just get up against it for a while. Far away hills look green.

The Financial Department.

On account of the tremendous amount of bookkeeping necessary in a large house, it would be utterly impossible for one man to look after the books, so everything is done in sections, each clerk having some particular part to look after, for which he is responsible to the head of the department. When a firm has two or three thousand accounts on its books, you can imagine the work it would be to balance them up every month and send out the statements, etc., and this is only a tithe of what is done by this department every month.

The Mail Department.

If the office boy in some small firm were to see the mail that this department gets through every day, he would die of heart disease, but great as the piles of mail are they disappear by magic. Every labor-saving device that can be thought of in this line will be found here. Invoice letters are not addressed by some jobbers, as by folding them in a certain way the name at the top of the invoice shows through the front of the envelope, which has enough cut out of the front of it to show the address, but, at the same time, not enough that the invoice itself can fall out. One of the things that would likely interest a visitor is the envelope sealing machine. This machine is attached to one of the electric light sockets, and is run by its own motor. As fast as a clerk can fed the envelopes into the machine it wets the gum and seals them up.

The Time Keeper.

In a firm employing, say, 150 clerks and warehousemen, it is necessary to keep some kind of a check on their movements, not only in the matter of their arriving in the morning and leaving at night, but also that they don't "duck" out during the day. The warehouse is so constructed that any person entering or leaving must pass a man who has his quarters by the front door. This is the door through which he must records twice a day and reports "lates" and "absents," besides acting as information bureau to any who have business with the firm. He keeps out book agents, pedlars, etc., who, if they got the chance, would overrun the place and keep the different members of the staff from their duties

The above sketch gives some idea of the work in a jobbing house from day to day. Of course there are many things that occur in a day's work that

---

THE COMMERCIAL TRAVELER.

Ella Wheeler Wilcox, in New York Journal.

First in the crowded car is he to offer—
This traveling man, unhonored and unsung—
The seat he paid for, to some woman young
Or old and wrinkled. He is first to proffer
Something—a trifle from his samples may be—
To please the fancy of a crying baby.
He lifts the window and he drops the curtain
For unaccustomed hands. He lends his "case"
To make a bolster for a child, not certain
But its mamma will frown him in the face;
So anxiously some women seek for danger
In every courteous act of any stranger.
Well versed is he in all the ways conducive
To comfort where least comfort can be found.
His little deeds of thoughtfulness abound.
He turns the seat unasked, yet unobtrusive.
Is glad to please you, or to have you please him,
Yet takes it very calmly if you freeze him.
He smoothes the Jove-like frown of the official
By paying the fare of one who cannot pay.
True modesty he knows from artificial;
Will flirt, of course, if you're inclined that way,
And if you are, be sure that he detests you;
And if you're not, be sure that he respects you.
The sorrows of the travelling world distress him:
He never fails to lend what aid he can.
A thousand hearts to-day have cause to bless him,
This much-abused, misused "commercial man."
I do not seek to cast a halo 'round him,
But speak of him precisely as I've found him.

---

go out (or stay in), as no person is allowed to go in or out via the shipping department. This man has his instructions as to who may pass out during business hours. Among the privileged ones are all clerks whose duties consist of outside work, such as banking, buying goods, etc. The ad. man also has a free hand (or foot), and may travel in and out as many times a day as he desires without interruption. All other clerks must show a pass or be reported. The time keeper also goes over the we have not referred to, but the work of the chief departments has been outlined, and, in a general way, some idea given of the work carried on by each.

To polish your show windows, making them as clear as day, use the following paste, applying with a soft rag, and rubbing off with another soft, dry rag:—Take of prepared chalk, nine ounces; white bole, one-half ounce; jeweler's rouge, one-half ounce; water, five ounces; alcohol, three ounces. Mix thoroughly.

BEST METHOD OF POSTING TRAVELING MEN

By W. L. Sanford.

Traveling men are a tremendous factor in the success of all jobbing houses, and to underestimate their worth is a grave mistake. In character and ability they are equal to the men of any profession, and in resource and energy they outstrip them all. The unsuccessful soon drop out, while those who fully measure up to the standard requirements are men of force, sound judgment and broad business views, who in many instances are superior to the men under whom they work. They receive the best pay of all employes, because they earn it--and when there is an important vacancy in the house, it is generally filled by a man who has served a term on the road.

The office man, treading daily the same worn path; performing the same routine of work; surrounded by the same people, and hedged in by the narrow limitations of his little workshop, frequently becomes dwarfed in his views and narrow in judgment—while the traveling man out in the busy world, in constant contact with the representatives of other houses, both in his line and others, meeting people of all kinds and from many sections, with their various and novel ideas, mingling with the consumer, the retailer and the railroader, viewing men and measures from a higher point, becomes himself broader every day of his life.

These facts should be ever borne in mind in dealing with the salesman. Many of his mistakes are from lack of information for which his house, and not himself, is responsible. He is susceptible of business education and is anxious for it, and if he is not well informed, it shows that there is something wrong with the system in the house. He occupies a position different from that of any other employe. While the house shapes a general policy and establishes rules for his guidance, still the details must be left to him, and away from the house he must exercise his own judgment and meet the ever changing conditions as he finds them. The war office plans the campaign, but after all it is the men in the field who win or lose the battle. No matter what stock the house may carry, how great the size and assortment, nor to what advantage it has been bought, it will stay on the shelves and prove a bad investment if the salesman cannot bring these advantages in a convincing manner to the attention of the trade.

Knowledge is Power.

In the sale of goods the posted man is always master of the situation. I found this true in my ten years' experience on the road, as well as in a subsequent and equal experience as buyer and manager of a business. If the buyer knows more of the details of manufacture, of freight rates, and market conditions than the salesman, the latter is at a decided disadvantage, and the opposite is true, when the situation is reversed. It is also a fact that the best posted salesman, all else being equal, invariably controls a larger business than his competitors.

We are all prone to yield our judgment and patronage to those who know more on a subject than we do, and are slow to give a hearing to those whom we think know less. This is especially true in the purchase and sale of goods. Now, if these statements are correct, and I am sure that no one will challenge them, it follows that the thorough posting of salesmen is a matter of vast importance to a house, and so fully am I entrenched in this belief that I devote more time and thought to this work than to any other part of our business.

When freight rates play an important part in the purchase and sale of goods, as is the case in the west especially, the salesmen should be made familiar with both the distributing rates and the shipping weight of goods. The price of an article in Montreal may be 25 per cent. less than the local price, but when the transportation charges are added it may be shown that the local price is 10 per cent. less.

The shrewd and unscrupulous competitor may sell goods for direct shipment from factory, stating that they weigh about so much and take fourth-class rate, when in fact they weigh more than stated and take possibly first-class rate

The salesman should be in a position to show up such false statements, and such unfair competition, and by so doing secure the business to which he is entitled by reason of having a lower delivered price. The profit of a local jobber is practically the differential between carload and less than carload rates, and as this difference is diminished, so is his profit. Therefore, the salesman must be a freight man also, and watch that rates be not lowered by sharp practices.

Know the Goods You Sell.

The salesmen should be given all information possible relative to the manufacture of goods, and the methods employed by manufacturers, and used by competitors to cheapen the cost of goods. If his competitor is selling a forty-five pound vise stamped fifty pounds; or seven ounce wagon covers stenciled ten ounces, or tinware made of eighty-five pounds, instead of 100 pound plate; or patent grass snaths labeled patent bush; or stamped instead of drawn steel rakes, or lighter than standard gauge corrugated iron — he should be able to detect and expose the fraud.

A large number of local salesmen actually believe that large jobbers manufacture the goods bearing their trade mark; while, as a matter of fact, they are the same goods which their own houses are offering for sale, the only difference being in the paper labels which they bear. Furthermore, many old patents have expired and many imitation goods are being placed upon the market constantly, and it is very necessary that the salesmen be kept informed on the subject, otherwise we find him trying to meet with the genuine article the price on the imitation.

This can be accomplished by putting before him whatever information the buyer may gain by his visits to factories, or by conversation with their representatives And here it might be stated

RULES WHICH WILL ENSURE SUCCESS.

Written for Hardware and Metal by H. Styles, Montreal.

If your object is to build up a business very large,
Be certain that politeness rules the salesmen in your charge.
Let your customers be certain they'll be always treated well,
Then you'll find them ever flocking to whatever goods you sell.

If you wish for reputation with the people far and near,
Give good value for their money, let your statement be sincere.
If these lessons you will practise every day with might and main,
You are reasonably certain fame and fortune to obtain.

that a buyer can make no better investment of his time and means than by visiting as often as possible the factories from which he buys. He gains information that cannot be obtained in any other way, and this becomes doubly valuable when imparted to his traveling men, and this should be done at every opportunity.

The salesman of local jobbing houses should be given the actual delivered cost of the goods. I know this policy will meet with much opposition, but I would rather rely upon a man's intelligence than upon his ignorance. It is true that a salesman having the correct cost will more often cut the selling price, but it is also true that in many cases he will fail to secure the business by not having the actual cost. It is just as easy to educate a salesman to get 20 per cent. as 12 per cent. profit, or 30 per cent. as 20 per cent. profit, and when you have educated him to get the profit your business demands, you have accomplished something worth while. You have attained an object that will be of service to the man, and through him a vast benefit to you. Pardon this illustration:

Easier to Lead Than to Drive.

It is a whole lot better and safer to drive a horse with open bridle, having him become accustomed to the cars, than to anchor your safety on the fact that he has on blinds and cannot see the things that might frighten him. It is all in getting a man started right in getting his sights elevated to the proper height. The salesman gets a big profit on razors, not because he doesn't know the cost, but because he is educated to it.

I believe in appealing to a man's pride, intelligence and ambition, in having him feel that he is a part of the business, and that the success of the business depends upon his effort and judgment; and this is quite true so far as his own territory is concerned. I believe in winning his confidence by freely imposing confidence in him. But, if you find a man that is insensible to such an influence, then it is best to let him go. I believe in sitting down with a salesman and having a frank, informal talk, swapping experiences, exchanging ideas, discussing policies and reviewing conditions, getting his views on plans and prices, and then adopting them if they are better than mine, and they frequently are. After such a conference, he will have learned a great deal, and so will the manager if his mind is not clouded with self-importance and self-assertion. A salesman that cannot be entrusted with the actual cost of goods is not worthy of the place he fills, and the quicker you get rid of him, the better.

In this paper I have endeavored to emphasize the matters upon which a salesman should be posted, viz., freight rates and classification, shipping weight of goods; manufacture of goods; light weight and immitation goods; cost of goods.

Just what means shall be employed in conveying this information is a matter of individual preference. In our own business, we supply our salesmen with condensed tariff sheets giving the competitive rates into their territory; we issue weekly change sheets giving the actual delivered cost and selling price, and if the goods on hand have advanced we give both the actual cost, and in parallel columns what the goods would cost to-day. This encourages them to make a profit based on the advance, and at the same time puts them in position to meet competitors who continue selling at the old cost, as many frequently do. We also file in a regular and convenient place for their inspection all letters that we think will be of interest and benefit to them; and these letters they usually look over whenever they come to the house. This puts them in a position to talk intelligently to the trade on all subjects relating to the hardware business. In fact, it educates them, and that is the object toward which we strive.

Other houses have successfully followed the plan of encouraging their traveling salesmen to read the trade papers, some even subscribing for each one on their staff.

NOT WORTH GOING AFTER.

The clown at the circus is the man who makes a fool of himself for the amusement and entertainment of other men, points out the Salesman. Wilder, and ever more reckless grow his pranks as the audience cheers him on. But when at last the show is over—when the lights are out and no one is left to laugh with him—then he is forlorn enough.

How many of us make fools of ourselves by burning the candle at both ends, spurred on by the applause and laughter of the other reckless spenders and "good fellows" in our crowd. But when the show is over—when the candle of income is exhausted—when the fire is out—then we are left forlorn. The crowd of "good fellows" who cheered us on during the burning won't come round afterwards with fire and light when the fun over. Burning the candle at both ends! There's nothing in it that's worth going after.

Burning the Candle at Both Ends.

DICK FAIR'S LUCK.

By C. C. F.

"Pretty nice order, that of Fair's," said the bookkeeper of a large machinery house, to his assistant, speaking of a contract to fit up the plant of a large power house in Western Canada. "Yes," said the assistant, "old Dick's out on top again; he's a lucky beggar."

"Old Dick," was the pet name alloted to the star salesman of the house, not because he was old, but just because.

"I'll admit that Dick certainly is

lucky,' agreed the bookkeeper, "but it's his long head, and eye always on the lookout for business that bring him up against these things. Many of Dick's sales that the uninitiated attribute to luck, are the result of a carefully thought out plan. There's one time though, that luck stepped in and saved Dick his job. He had only been on the road for the house about a year, when on coming back from a trip west, he seemed queer. Used to go round in a kind of a trance, and you'd have to speak to him twice before he'd hear you. Well, one day one of the boys caught him looking at a picture in the back of his watch, and after continual digging, we got out of him that he'd met her in Toronto on his last trip. She was there visiting some cousins or something. Well, anyway, Dick had it bad.

"Next time he went west, was in the Spring. He started off in fine spirits, and we all expected that he wouldn't come back alone, so when one day we got a wire from Dick saying he'd finished his trip and wanted two weeks holidays, with money up-to-date we weren't surprised. Now it just happened that we were pretty busy at the time and the 'old man' said he couldn't spare Dick, and wired him to report at the shop within two days. The only reply we got was 'send my money at once.' We sent him his money and heard nothing more from him when a week had gone by the 'old man' had worked himself up to fever heat, and everybody in the office knew that Dick Fair's goose was cooked.

"About two days leter, we got a big order from him for a complete new set of machinery for a large grist mill in G—. This was Dick's record sale and the chief was tickled to death. Dick landed home about 4 days later and was received with open arms.

"How did you get it, Dick, my boy?" said the chief. What passed between them in the private office, nobody knows, but Dick's salary took a good jump up next week.

"In telling us about his wedding, he said, 'Fellows, that was he narrowest squeak I ever had in my life. If I hadn't got that big order, I'd have been a married man without a job. I knew as much before I started for G—, after getting the 'old man's' wire, but I figured that there were lots of jobs in this broad land waiting for me, but only one girl, so I kissed the job goodbye, and started for the girl. Her daddy took a shine to me first shot and when he heard what my line was, and the scrape I'd got myself in to come up and ask him for his girl, he slapped me on the back so hard that I though I was kicked by a horse and said: 'Young man, you're the right sort. I'll not only give you the girl, but I'll give you such an order, that your boss will sit up and notice things. I own the largest grist mill for miles around here, but the business is so large that the old machinery isn't good enough. I've been thinking for quite a while about ordering a new set, and now you get it.'

"That's what you can call real unadulterated luck fellows. I just ran into that order and it saves me my job, but I wouldn't advise any other chap to do the same thing on prospect, as every nice girl in the land hasn't got a father that owns a mill."

STAND BY THE TRAVELER.

Not long ago several Ontario retail hardwaremen were discussing trade conditions and drawing some comparisons between the present and the past.

One Toronto man spoke of how, in his early days as a retailer, it was necessary to make buying trips to New York and to read United States trade publications in order to keep abreast with the times. "How changed things are now," he said, "by reading our Canadian trade paper I can keep posted, and instead of having to go to New York I can get better satisfaction by doing business with the travelers who call at my store. I know them well and could trust some of them with anything I have. They value my account and realize that the better they serve me the better they are serving themselves."

"Of course," he added, "I could go down to the warehouse but I wouldn't know the young fellows in charge of the sample room as well as the more experienced traveler who calls at my store. Anyway, the jobbers do not seem to encourage buying at their warehouses as their sample rooms are not so complete as in former days and we retailers can get any special discounts that are going by buying through our regular traveler, whereas we are apt to pay a longer price by buying at the warehouse. If we go to New York matters are even worse, as the sample rooms are practically bare and the stocks are carried in some warehouse in the outskirts of Brooklyn or at the factory in Massachusetts or Connecticut. I beileve in standing by the traveler wherever I find he tries to treat me fairly."

Merchants from outside the city, however, said hey found it of advantage to visit the sample rooms at the warehouses two or three times each year, particularly at fair time or during conventions of the Retail Hardware Association when the jobbers look for many of their customers to call and usually have their travelers off the road and in he sample room where they can give personal attention to their customers. One retailer who lived near Toronto told how he had been over-canvassed by traveling salesmen and when one day he received cards from six or eight commercial men, notifying him that they all intended to call on a certain day he made a flank move and run into the city and placed an order at the warehouse of his favorite jobber.

COUNTRY OFFERS BEST CHANCES

"You may talk to me all you please about the advantages of the city," says a man who has clerked in both places, "but I will take my chances in a comparatively small city. I went to work for a big wholesale house in the city once. I had an idea that if a man would get up and hump that his merits would soon be recognized and that he would be pushed right to the front. I rather expected that the head of the establishment would be 'round in a few weeks and ask the foreman who that capable and industrious young man was and that the foreman would introduce me to the head of the establishment, who would commend me and tell me that he had his eye on me and that if I went on as I had started, there was a great future before me.

"Well, I got the hump on myself all right, but the head manager didn't come 'round. I worked for him a year and he never gave the slightest indication that he had ever heard of such a being as me, and what is more, I don't think he knew or cared who I was or where I came from. I didn't even get any commendation from the foreman, but I did succeed in getting several of the rest of the fellows who worked in the same department hot at me. They plainly intimated that I was a fool for doing any more than my regular work called for, and by and by I came to the conclusion that maybe they were right. I felt that I was getting to be simply a little bit of a cog in a great big wheel. If I stayed there long enough I would be worn out just like any other cog that is located in a wheel that is running all the time and then it would be to the scrap pile for me.

"In the store in the small city or town it is different. The head of the concern usually takes a personal supervision of the whole establishment and actually gets to know all the people who are working for him. He notices the ones who are attending strictly to their jobs and pushes them ahead. There is some encouragement for a man to get a hump on himself under those conditions. But in the department store or wholesale house in the great city the individual doesn't amount to much. He has so much work laid out for him to do. If he does it gets no credit, if he fails to do it he gets fired and that is all there is to it."

THE AWAKENING OF BUTTS

F. H. O'Hara Tells How a No-Account Boy Found Success in a Dingy Hardware Basement.

Not many years ago in the Central High school of Detroit, Mich., a lad, called "Butts," in recognition of his small, stout build, was causing his teachers much annoyance. "Butts" was a typical litle sport and used to skip classes daily to smoke cigarettes in the vicinity of the school or to pursue other inclinations quite foreign to study. He was an amiable little fellow, the sort you can't help liking in spite of yourself.

A crisis in the career of "Butts" came when he was a junior. He had made no progress and was a source of irritation to the faculty. Now, the grade principal happened to be one of those good fellow men whom a boy likes naturally. He had played handball with "Butts" a good deal and if there was one person who might have influence with the boy, it was this principal. But it was no use. The teacher tried every method that ever had been employed by him in his long experience. The boy would take everything—scolding, pleading, threats—quietly, even good naturedly; but never mended his ways.

Was Dropped From School.

Then the principal placed him on "schedule." Being placed on schedule is a simple process of elimination. If a student's standings do not become satisfactory in a certain study in one week he drops the study. So with another study the next week, until within a month or six weeks the student has no classes and must bid farewell to the institution.

"Butts" left. His father was a clerk in a hardware store and was provoked with his son. He secured a job for the boy in a large wholesale hardware company, asking that "Butts" be given the "dirtiest job they could find." His request was granted. The lad was placed in charge of a great heap of files in the basement. The place was musty and the light of day never penetrated into the room. Besides, the files were rusty; and, since they were unassorted, when an order same to him from above "Butts" was compelled to rummage through the big pile in order to find the particular files wanted. It was a gloomy job, and the boy felt downcast. But he was resolved to "stick," so he determined to make his position as pleasant as possible. He noted that certain kinds of files were ordered more than others. He found some boards and soon erected a neat case along one side of the dingy room. He labelled each department. Then he placed the files in the case, and found his job less dirty.

One day the manager chanced to pass through the room where "Butts" was working. He noticed the case and the neat arrangement of files, but said nothing. He merely raised "Butts'" wages a trifle—and forgot.

Meanwhile, "Butts" frequently was bothered by orders which he did not understand thoroughly. He did his best. Instead of asking questions, the boy went to the public library and secured a large volume on the subject of files. This he read carefully. Then he rearranged his case and wrote out new labels.

Soon after this, a traveling salesman happened into the office and inquired what was needed in the file line. He was sent down into the basement to investigate. He talked with "Butts" for a time, then went back to the office and exclaimed:

"Where the dickens did you get that boy? Why, he knows ten times more about the file business than I do—the whole thing, from A to Z."

This was in April. They sent "Butts" out on the road selling files. He made good, and more. He was getting a good salary. But in September, shortly after school had re-opened, "Butts" called on his old grade principal. The teacher saw there was something the boy wished to say. 'Butts" left without saying it. However, he returned a few days later.

Went Back to School.

"What's the matter, "Butts?" the principal said. "I can see there's something you want to say."

"Yes, Mr.—," "Butts" began, "there is something. I tried to tell you the last time I was here, but now I'll make a clean breast of it."

"What's the trouble?" inquired the man, thinking that "Butts" was involved in some "scrape.'

Then the boy told him that he wished to return to school.

"What?" cried the principal.

"Well, I'm doing fine now—couldn't hope to do better—but I've been thinking about it, and I believe I know the reason. I've simply been applying myself, and I know I can do that in school."

Then he went on to say that if he could complete his studies and be graduated in June he would remain in school. It meant hard work—seven studies one term and eight the next—but it was arranged, and "Butts" was graduated.

The father was poor, but wanted "Butts" to continue his studies. So "Butts" worked, his father aided as much as possible, and four years later the boy who found success in a heap of files in a dark basement was graduated from the Michigan College of Mines. To-day he gives promise of becoming an exceedingly successful mining engineer.

KEEP TO THE POINT.

The gift of closing a bargain is by no means a common one, and it is left out of the composition of many clerks who are otherwise well equipped. And without this gift all others are of no avail. No matter how skillfully the negotiations may be conducted up to the "sticking point," unless they go beyond that point they are useless. Though indecision on the part of the salesman is the besetting sin which generally causes his failure, there are other elements which contribute to this undesirable end.

One of these is the inability to shut out other matters from the conversation of the moment. Only the unskillful salesman will allow his customer to drag him into general conversation when he is actively engaged in displaying goods and attempting to get his patron's decision. This does not mean that it is not wise to chat pleasantly with customers about social and personal matters, but it does mean that this should not be done when once the business of selling is really in hand. Anything that distracts the mind of the customer from the vital point at issue is always to be avoided. Let all the "visiting" be done before the display of goods has begun.

CONFIDING IN THE CLERK.

There is good opportunity for the employer to exercise diplomacy with his clerks, to keep on good speaking terms and thereby elicit from them goodwill, which goes a long way to make successful and useful clerks. Clerks are human, as are their employers. They are sensitive to praise and blame, and the amount of work they do is greatly governed by the relations existing between them and their employer.

By diplomacy sour temperaments can be easily sweetened, frowning brows can be smoothed out, furrows of care can be erased. On the other hand, self-assertion can be corrected and carelessness in details be corrected. The levity so common with clerks, which always carries an uncomplimentary impression to customers, can, with a little diplomacy, be eradicated. All flaws in the demeanor of clerks may with the least exercise of diplomacy on the part of the employers be corrected and a business institution be put on an unassailable basis.

HOW ONE CLERK SUCCEEDED

By Weston Wrigley

One clerk in a Toronto store recently won promotion by keeping track of the percentage of sales he made in comparison with his wages. He liked clerking and he worked faithfully enough, but he was kept at one counter for nearly two years and was getting discouraged when a bright idea suggested itself to him.

He was thinking one night of looking for another job but he didn't like to leave surroundings which he had grown to like, and in thinking the matter over he commenced to figure out why he hadn't won promotion. He thought he was selling a good lot of goods, but he didn't have any figures to work on, so after considerable thought he decided to stay where he was for a while longer and keep track of his sales. He was earning $9 per week and wanted more and so each day for a week he kept an account of his sales and at the end of the week found that they totalled up to about $170. He divided this up and found that his daily average was less than $30, and his salary figured out at slightly under 5 per cent. of his sales.

With this information he determined to try and increase the percentage. He kept closer tab on his customers, attended to as many as possible, cut out gossiping with other clerks, and with customers, and at the end of a month he found his average nearer six than five per cent. He kept at it, and found that in the past he had merely been taking orders, that by using a little judgment he could develop additional sales. Instead of letting a customer go after she had bought the article asked for, he could induce her to buy other goods by tactfully suggesting and drawing attention to the good qualities and advantages of buying the line now.

The clerk followed his plan up. Every week he compared the figures of his sales with his wages and nearly every week saw an improvement. Inside of two months he had moved away off the counter he had stuck at for two years and in the course of a year he had been promoted three times, he being now working at over double his former salary, with good chances of being promoted into a partnership in the business.

What this clerk did can be done by any wideawake salesman. Too many clerks get discouraged and move from one job to another. It doesn't pay to change too often. In the eyes of almost any business man a salesman doing well in a job is worth almost double as much as one out of a job. Stay with an employer until you have climbed as high as you can. Then, if he is slow to recognize your merit, have a talk with him before looking elsewhere for a position. And take care of the little things. Keep your section of the store as clean as a pin. Have a place for everything and keep everything in its place. Don't give either overweight or underweight. Be patient with customers who don't know exactly what they want. Try and save as much expense as possible, remembering that a dollar saved the store is worth as much or more as a five-dollar sale. Be faithful in your service and endeavor to make friends for the store, and before long your opportunity will come—some opening will develop for you—your employer may find it advisable to take in a partner or some other merchant may be quick to recognize merit.

Attention to business is bound to win, while careless or thoughtless service will not open any doors of opportunity for you. He who wishes to climb the ladder of success must know how far apart the rungs are.

RELATIVES NOT ALWAYS BEST.

As a general thing, the employment of relatives is to be avoided. Traveling salesmen are agreed upon the fact that when they find a storekeeper whose whole force is made up of his sons, daughters, nieces and nephews, he is considered as struggling under a distinct handicap, and his progress is watched with especial care. This is, perhaps, only another way of saying that the man who selects his assistants because they are relatives is not at liberty to make the selection on the broader line of their real qualifications for the work in hand.

HOW BUSINESS IS LOST.

A correspondent in an exchange mentions two cases which occurred to a customer within a short time. In one instance he wished for a fee pieces of sandpaper and went to the hardware store of his neighborhood and paid two cents a sheet for them. The next day he wished a coarser sandpaper and went to another store which was more convenient for him at the time, and handed two cents to the clerk. One cent was given back to him. He remarked: "I thought the price was two cents a sheet," and was answered, "Sandpaper is a cent a sheet, 10 cents a dozen." In the other instance there was a purchase of cheap hooks at still another hardware store. The customer wished one hook and was charged three cents for it. Having decided that the hook was what he wanted, he went to the same store, to the proprietor as it chanced, and was charged 10 cents a dozen for the same hook, the original purchase being shown as the sample. Naturally he complained, and the explanation was not satisfactory. But two cents were returned to him. These are very small matters, but they affect business. Supposing that this customer were to buy a large amount of hardware, or were to recommend a friend who had such a purchase to make, which merchant would be likely to get the business?

CLERKS MUST BE CIVIL.

"It don't always do," said an old hardwareman recently, "to size a man up by his looks. A sporty-looking man with an English accent came into my store and said to a young clerk, 'Say, my boy, can I see some shoes?' 'This ain't no shoe store,' snapped the clerk. 'Can't you see the sign?' 'Is that so?' responded the man. 'I'll look at it from the outside,' and started for the door.

"I had overheard the conversation, unknown to the clerk. I caught the man at the door and asked him to wait a moment. 'Come here, Tom,' I called to the clerk. When he came I said, 'Apologize to this gentleman for your bad manners.' 'I am sorry,' he said. 'I meant it for a joke.' 'One more such joke,' I said, 'and you'll leave this store for good.'

"I opened a conversation with the sporty individual, and found that he was the new owner of our livery stable and wanted horse shoes. I jollied him up, made good friends with him, and in the next five years sold him several hundreds of dollars' worth of goods, which my fool clerk would have thrown to the other store.

"At another time I saw another clerk waiting on a quiet little woman, with a bored manner, as though he had no use for her. I edged up, opened a conversation just as she showed signs of leaving without buying, and found that she was fitting up her home for a boarding-house, as her husband had recently died. I showed my interest in her case, and before she left sold her $87 worth of goods, for which she paid spot cash. She became a steady customer, and is trading with me yet.

"When people come into my store I always show an interest in them and get them to talk about themselves. That's the best way to make friends of strangers. After a man has talked one of my legs off, and I have listened with attention, he begins to feel warm towards me, and would like to do something for me. It's easy enough to sell him goods if he has any notion of buying. If not, he comes around when he needs something. The personal equation is a big thing in the retail hardware business."

Advantages of Profit Sharing

Written for Hardware and Metal by Fred C. Lariviere, Wholesale Hardware Merchant, Montreal

In my endeavours to equip the company of which I am president, with the most modern and up-to-date business methods, I have visited the principal cities of the United States and Canada. In every city I was most cordially received, and every merchant and manufacturer seemed most happy to exchange views. Amongst other things, I was particularly surprised at the great interest and attention paid to the laboring class and the salesmen, everybody seeming to be devising some means of interesting their workers sufficiently to prevent the good ones from searching position from some other house or even paying attention to the tempting offers of competitors.

Objects of Profit-Sharing.

In a recent address on this subject, E. F. DeBrul, secretary if the "Keystone Driller Co.," Beaver Falls, Pa., expresses himself as follows: "We find in Cincinatti, where we have extended the premium system and similar attractions for the men, that by hitting them in their pocket books, we have made unionism an unattractive thing to the average machinist."

The increase of profits has a very good moral effect on the employees, it shows the men the consideration they enjoy from their employer, in placing them above the ordinary salaried employee by becoming interested in the company for whom they are working. They are intimately connected with the welfare of the concern, for the more the company will realize the more the employees will receive, so that this joint interest gives incalculable increased progress. The principal object of profit-sharing is better returns of profits for the employer and higher salaries or compensation for the employe by remunerating properly every moment given for the interest and prosperity of the employer.

Manufacturing Concern Plan.

In the South Metropolitan Gas Company, London, England, profit-sharing has been established seventeen years, the last distribution taking place in July, 1906, and the sum divided amounting to £43,262. By the terms of agreement half of this sum must be employed in buying shares in the company and the other half can be distributed to the beneficiaries or, more properly speaking, to the co-partners, but most of these prefer to invest in the shares of the company, as £41,603 out of £43,262 were placed in 1906.

The shares of profits to the employees has increased 800 per cent. in fourteen years. For 1892 the sum belonging to the employees amounted only to £5,697; eight years ago the accumulated capital which had been exchanged for shares reached £100,000, but to-day the sum is £337,192. The company gives work to 4,998 hands, and the share of each in the accumulated profits is £66.

Crane Co.'s Annual Gift.

Following a custom adopted in 1900, the Crane Company, of Chicago, ge-

FRED. C. LARIVIERE.

nerously shares the fruits of its success with the employees. The basis is a donation at Christmas of ten per cent. of the earnings of each employee. In December, 1906, the distribution amounted to $330,000; this being $25,000 more than in 1905, and making a total of $1,500,000 thus distributed to date. Every employee, irrespective of his position or length of service, receives ten per cent. of his year's earnings in a lump sum, accompanied by a card bearing the good wishes of the company and a bit of wholesome advice as to the wisdom of preparing for the inevitable rainy day. Six thousand employees were the beneficiaries in the last Crane Company's gift.

Keystone Driller Co.'s Plan.

The Keystone Driller Co., Beaver Falls, Pa., has a combined savings deposit and profit-sharing plan which is working to the complete satisfaction of all concerned. It is simple, complete and self sustaining and the benefits have been pronounced. The conditions were as follows:

1. A stock company managed by a regular board of competent directors.
2. Doing a fair business that was capable of some extention.
3. A small amount of treasury stock which could be issued from time to time as desired to workmen.

About $10 was spent in printing blank certificates and the company was ready to introduce the system.

These certificates set forth on its face that it is issued only and exclusively to employees of the company. It is ruled to provide for 20 or more entries, has blank spaces for the computation of earnings, withdrawals, cancellation, and final receipt.

The stub from which the certificate is detached remains in the book and has blanks for corresponding entries, this comprising all the bookkeeping necessary to keep the system running. On the reverse side of the certificate all terms, conditions and limitations are clearly set forth. Briefly and mainly these are:

A Business Proposition.

1. That the owner may withdraw the deposit at any time by giving due notice.
2. That if withdrawn before one month no interest will be paid.
3. If after one and before expiration of four months, interest at four per cent. per annum will be paid.
4. If withdrawn after four months six per cent. will be paid.
5. After the expiration of six months, and if the owner so desires, the deposit shall become profit-sharing and for the time the money is in the hands of the company it will participate upon an exact equality with any other capital invested in the company; the company, however, guaranteeing that the profit shall in no case fall below six per cent.
6. That the amount of profit-sharing

certificates held by any employee at any one time shall be limited to $1,000.

7. That the profit-sharing certificate is at the will of the holder at any time after six months exchangeable for regular corporation stock.

8. That until so exchanged the owner shall have no voice or vote in the management of the company's affairs; but when so exchanged it becomes in all respects equal with the regular corporation stock.

9. That prior to exchange for regular corporation stock it is not transferrable except for purposes of immediate redemption.

10. That all interest or profits, as the case may be, shall be suspended during the existence of any strike in which the holder participates.

11. That the company arbitrarily reserves the right to redeem the certificate at the end of five years, or, in case of the death or disability of the owner, at the end of three years thereafter.

12. That the ownership of $50 or over the profit-sharing stock, other things being equal, shall entitle the holder to preference of employment.

13. That the owner may add to his holding on any regular pay day any amount he sees proper.

Not a Philanthropy.

It will be observed that the plan provides for no so-called charities. It does not propose to offer something for nothing, but, on the other hand, it is a simple, straightforward, business proposition. The interest allowed upon deposits is about what should be paid to the bank for the borrowed money, and for the sake of encouraging thrift and the cultivation of habits of economy, the company can easily afford to favor its employees instead of the bank in this matter. The guarantee of a profit whose minimum shall be at least equal to the current rate of interest is only just, while the depositor has no voice in management. So soon as an employee's certificate amounts to the limit, $1,000, he may surrender and exchange it for regular corporation stock. He is then free to take out a new certificate.

Applied in Jobbing Establishments.

Most of the jobbers seen take as a basis one or two years' record of a traveler, taking into consideration his sales, salary and expenses and the gross profits realized. To make myself clear I submit the following example:

A. sold in two years $200,000 worth of goods, realizing a gross profit of $30,000, or 15 per cent. He received for his salary $4,500 for two years, while his travelling expenses were $3,000, a total of $7,500, or 25 per cent. of the gross profits. In re-engaging this traveler with a view of encouraging and interesting him more, the employer will say: Mr. A., during the last two years you have cost us 25 per cent. of gross profits, or $3,750 per year. We guarantee you this amount as salary and expenses for the coming year, and in addition the difference you will realize between 25 per cent. of the amount of your gross profits and your salary and expenses shall be yours.

I understand that on this basis city travelers are paid 17½ per cent. to 25 per cent., and country travelers from 25 to 40 per cent., according to territory.

Other Plans.

Other employers give a fixed salary and expenses, plus a commission on a certain class of goods, such as cutlery, razors and similar goods carrying a good margin of profit, these bonuses or commissions carying from 2½ per cent. to 7½ per cent, or even 10 per cent. In addition to these commissions, there is also a sharing of profits after the gross profits have covered salary, commissions and traveling expenses, plus say $800 to $1,000 for the employer. After all these are covered the salesmen begins to participate in the profits, on a basis of 40 to 60 per cent.

Still another plan is to guarantee the traveler a fixed salary and expenses with participation in the net profits, varying from 50 to 75 per cent., the traveler's sales bearing its proportion of administration expenses, covering all expenses and losses except salaries of selling staff, interest on capital invested and bad debts. A difference of percentage of expenses must be made for direct shipments from factories and from one's own warehouse.

Applied in Retail Stores.

One method is to establish departments with a manager for each, but in this case the manager is the only one participating in the profits. The profits are made up as follows:

The department is charged with the value of the goods carried in stock. Charges are also made for insurance on the same basis, rent according to location occupied, salary of all clerks in the department, except the manager, salary of travelers making the sales of goods of this department, advertising, donations, telegrams and a certain percentage based on the volume of business to cover the expenses of the office staff, purchasing department, advertising department, fuel, light, stationery, taxes of all kinds, stamps, telephones, delivery of goods and petty expenses.

These expenses are all deducted from the gross profits and the balance divided with proportion of 30 per cent. to 40 per cent. for the manager, from which deduction is made of the amount advanced the manager, and the difference is, his sharing of profits, the employer keeping the 50 per cent. to 70 per cent. as his share and taking the risk of the accounts.

Benefits the Manufacturers.

In the manufacturing industries the first practical result is to avoid conflicts between capital and labor. In these establishments strikes are unknown, for the simple reason that nobody has ever seen strikes amongst capitalists interested in the same enterprise. The employes being shareholders or certificate holders can no longer complain that capital is the exclusive privilege of a few, This fact alone should be sufficient to induce manufacturers to adopt some plan of participation with their employes, but there are other good results.

The employes being directly interested give much better attention to their work. They save waste in supplies and workmanship, devise means of executing the work more quickly, or of cheapening the cost of production. Therefore, it pays the employer dollars and cents to adopt this modern system to say nothing of the high satisfaction of giving a chance to rise to every employe.

In case of dispute the employe is better disposed towards the employer, whom he knows as his friend and is naturally led to discuss the subject with care and good sentiment. An employer who uses his subordinates well is soon known and he never has difficulties of securing good help and always has the most desirable class of workmen.

Jobbers Get Good Results.

Judging from the experience of those who have adopted the premium of profit sharing system the change has been simply wonderful. Salesmen who formerly stayed in the store on Saturdays or Mondays are no more to be found and there is no waste time on the road. Routes are made accordingly, for the salesmen receiving remuneration for every effort and minute used, the prosperity of the house is now his. With this system private detectives are not needed to find out what employes are doing, for the days are not long enough for them to work.

To use the expression of a very able salesmanager of a large jobbing house in the New England states, "Most of us are selfish and the travelling salesman is no exception to the rule." In other cords, place a man in a position that the more he will work, the more he will be interested and the solution of the problem, "How to handle salesmen" is solved.

Of course there would be exceptional cases when it is advisable to pay salesmen a fixed salary, such as working new ground or territory that was tributary to other jobbing centres or when putting out young men who have no connection of established ability as salesmen.

Analysis of the Plans.

I am not in a position to say which of the three manufacturing plans is the best, my studies and experience do not permit me to express an opinion.

For the jobbing houses I do not hesitate to say that the third plan outlined is the best and most equitable, and will try to prove my recommendation. Plans one and two have for a base salary and expenses reduced to percentage on gross profits and should be termed premium propositions. A profit sharing system, to maintain its proper meaning, should place the traveler on the same basis as if he was a partner. His sales and the net profits (not the gross profits) establishing the proportion of his interest. To justify what I have just said, I call attention to the following:

When you subscribe stock in a company, is it not a fact that dividends are only declared on the net profits after expenses are paid? The same rule applies in an ordinary partnership and why should not the same application be made in a proposition with travelers?

The volume of business of a traveler should be charged with a corresponding percentage of the amount of expenses of administration to be deducted from gross profits. To make my views clear I will submit the following example:

Interesting Examples.

C. sells $100,000 on a margin of 14 per cent., or $14,000, gross profits at 40 per cent. profit sharing receives $5,600.

D. is also a country traveler, but devotes all his energies to the sale of the most profitable goods and only secures $50,000 of business at 25 per cent., or $12,500 gross profits, at 40 per cent. receives $5,000. All will admit that D. is the better man, the one that obliges his employer to carry less stock, with less handling, at less cost of administration and with less risk of credit. Still C. would receive $600 more. Is this to the advantage of the employer? I do not hesitate to say no.

On the other hand if you make the "profit sharing" on the net profits the result would not be the same. The traveler selling $100,000 would pay for administration expenses double the amount on the one who sold $50,000, which is only just. But the percentage on the net profits should be larger than that on gross profits and I estimate that a 40 per cent. sharing proposition on gross profits would be equalized by 55 per cent. on net profits, and I now make the following comparison on the basis of gross profits:

A. sold $100,000 worth of goods at 14 per cent., equalling $14,000, at 40 per cent., equalling $5,600, while B. sold $50,000 worth at 25 per cent., equalling $12,500, at 40 per cent., equalling $5,000.

Let us figure on net profit after deducting share of expenses.

A. sold $100,000 at 14 per cent., equalling $14,000, less 7 per cent. for expenses, $7,000, leaving $7,000 at 55 per cent., or $3,850.

B. sold $50,000 at 25 per cent., equalling $12,500, less 7 per cent on $50,000 for expenses, $3,500, leaving $9,000 at 55 per cent., or $4,950.

Does it not appeal to you that the man who brings the better percentage is the best man and should get the greater remuneration?

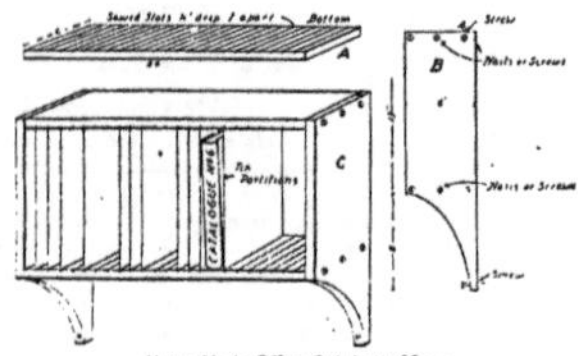

Home-Made Office Catalogue Case.

A difference would have to be made for percentage of expenses on direct shipments from factories, and shipments from one's warehouses. The 7 per cent. referred to includes all expenses of administration except those of the selling staff, whether indoor or outdoor.

The plan submitted for retail stores should give all necessary advantages and very good results, inasmuch as where used it gave very satisfactory returns.

I believe I have proven without the slightest doubt that premium or profit sharing system deserves the very serious attention of all those interested in business, either as manufacturer, jobber or retailer, and if the few remarks I have made are helpful to some I will be more than paid for the trouble I have put myself to in writing on this very important and interesting question of "profit sharing."

CHEAP DISPLAY RACK.

The display rack shown in the accompanying cut is both handsome and prac-

Cheap Display Stand.

tical and can be made for less than half a dollar. The centre rod is made of a bamboo fishing pole, while the cross-pieces are ordinary brass sash curtain rods. By using a 5-16-inch drill bit and boring through the pole close to the joints splitting can be avoided. The upright pole should be gilded. The fixture can be hung from the ceiling or made permanent in the window by fitting in a hole through two 2x4's arranged in X shape.

AN OFFICE CATALOGUE CASE.

Many hardware merchants have too many catalogues, directories and reference books lying around loosely in the office, on top of desks or under shelves. The following sketch shows a simple arrangement for a case to hold such books. Have one of the clerks procure two boards 6 inches wide and 2 or 3 feet long, or just the proper length to fill up some vacant space on your office wall. Square these boards up, commence at one end and with the square draw lines across the boards 1 inch apart. Then saw slots in these boards 1 inch apart and ¼-inch deep, as shown in A. These boards form the top and bottom of the proposed bookcase.

The boss is proud of every clerk who helps him, and soon gets rid of those who don't.

Pioneers of the Hardware Trade

Interesting Career of Winnipeg's Mayor—Other Notable Examples of Long and Faithful Service—Sketches of Men who have Helped to Develop the Hardware Trade in Canada.

WINNIPEG'S HARDWARE MAYOR.

It has been well said that to write the biography of Jas. H. Ashdown would be to write the history of the commercial development of the Canadian West. Although still in the prime of his mental and physical powers, he has seen the development of Winnipeg from the small collection of huts clustering around Fort Garry and the Hudson's Bay trading to its present proud position as the third city of the Dominion and the commercial capital of the Canadian West. All this is in the short space of 41 years. In those years he has built up on of the very largest wholesale hardware businesses in Canada from the smallest beginnings, and has amassed a very large fortune. The tinsmith of 1866 has become the merchant prince of 1907 and mayor of his city by the larges majority ever given in Winnipeg to any candidate for the office.

They who see no romance in business should learn the story of Mr. Ashdown's career. A plain, modest, somewhat retiring man, few would imagine that his is one of the most interesting business careers in Canada. Born in England, he came to Canada when eight years old and he started in life a young friendless boy without a cent to his name and with no prospects except those which were promised by his own native ability and strength of purpose.

When eleven years of age he was serving behind the couner in his father's store in Weston near Toronto. A little later he worked on a bush farm in Brant township and the rough labor of a pioneer lift developed a rugged constitution which has stood him in good stead ever since.

When eighteen years of age, he became a tinsmith's apprentice for John Byrd, of Hespeler, and for three years he worked steadily at his trade in that village. He then went to Chicago and from that city to Kansas. Perhaps it was in Illinois and Kansas that he first began to realize the possibilities of the western half of the continent, but he had not forgotten Canada, and in June, 1866, he entered the Red River settlement.

It was not a prosperous time for the ditsrict which now includes the western metropolis and the Red River Settlement was no place for a faint hearted pioneer. It was the troublous time before the first Riel rebellion, discontent was rife and the country was in financial straits. The story of Mr. Ashdown's connection with that rebellion and of his imprisonment in Fort Garry by the rebels is interestingly told in a quaint old book, Beggs' "Ten Years in Winnipeg" (1869-1879) but space will not permit of an extended account here.

Beggs tells of Mr. Ashdown's financial losses during the rebellion. His shop was closed for the proprietor had no assistant and while he was imprisoned by the rebels his business was at a standstill. It is amusing to read in the quaint old chronicle of the growth of Winnipeg in 1871: "We see about this time an evident reign of prosperity in the business of J. H. Ashdown, for that gentleman, finding his trade increasing so rapidly, had to engage an assistant tinsmith."

JAMES H. ASHDOWN
Mayor of Winnipeg and for 41 years a Hardware Merchant in that City.

From such small beginnings has grown with the development of Winnipeg and the west the big wholesale hardware business of The J. H. Ashdown Hardware Co., and for that growth, although Mr. Ashdown has associated with him many able business men, he himself is mainly resposible. He had always attended strictly to his business and has been in touch with almost every detail of it.

In later years Mr. Ashdown has found time to devote to public duties. As a member of the Transportation Commission he did good work not only for the west but for all Canada. A few months ago he was appointed chairman of the last December, when the city of Winnipeg decided that none but a first-class business man could be entrusted with the control of civic business he was elected Mayor by the largest majority in the history of the city.

THE TRADE'S GRAND OLD MAN.

To see a man eighty-five years of age with a perfectly preserved mentality and physique going about and transacting business with which he has been identified for sixty-eight years and in an establishment with which he has been connected for nearly fifty-one years is assuredly inspiring to all directly or indirectly concerned with the hardware trade. Such a man as above described is S. S. Martin of Rice Lewis & Son, wholesale and retail hardware merchants, Toronto.

Hale and hearty, with a noble and kindly face, which reminds one of William E. Gladstone, Great Britain's grand old man, and with a mind stored with interesting and instructive reminiscences, with ability to re-state them, Mr. Martin is indeed an interesting man to converse with. His reminiscences are not confined alone to the hardware trade of Canada, he having served his apprenticeship in England and spent sixteen years in the trade there. What Mr. Martin doesn't know about the various phases of the hardware trade is hardly worth knowing.

Mr. Martin was born in Lincolnshire, England, in 1822 and at the age of seventeen became an apprentice to a hardware merchant. After serving a short period with this merchant who was rather cynical and little calculated to advance the interests of his employes. Mr. Martin's father thought it wise to change his position. From this time until the year of his coming to Canada, Mr. Martin led a gregarious life. In his youthful wanderings he practically covered the whole of England, working for short periods in Devon, Gloucester, Leicester, Peterborough, Cheltenham, Berkshire and Nottingham, at the last being engaged in taking stocks for valuators. Mr. Martin considers that those years spent in traveling about the old land have been of great value to him in that he thus gained valuable insight into the ways and cross currents of the hardware trade in England.

In 1855 Mr. Martin came to Canada, and on May 4, of the same year he engaged with the Rice Lewis & Son. Mr. Lewis met me at the firm's former site, on the corner of Toronto and King Streets (where the Canadian Northern Railways offices now stand) at 6 o'clock in the morning, and undertook to put me through my facings. He set me at cleaning some hinges on which had accumulated considerable dirt," said Mr. Martin, who accepted the dirty task, pointing out to Mr. Lewis, however, that if he could afford to pay big wages for such work, Mr. Martin could afford to do it. Mr. Martin's annual wage for the first year was $500, which, he gave us to understand, was considerably above the average wage at that time. His sixteen years' experience in England entitled him to higher wages and he got it.

Mr. Martin has been a witness of the evolution of the hardware trade in Canada for over fifty years and his reminiscences of this evolution are particularly interesting. "Very little job-little jobbing," Mr Martin says, "was done in the days when I was first with this firm. We handled a few brass and iron beds made by the Taylor Safe Works, but most of our stock was the regular variety of house furnishings, guns and ammunition, and all the various lines of shelf hardware. Our firm shunned the stove and tinsmithing trade, although this business was done by retailers in the earlier days, there

S. S. MARTIN, SENIOR

For 51 Years with Rice Lewis & Son, Toronto and 68 Years in the Hardware Business.

being no large tinware factories as at present.

In regard to the capacity which Mr. Martin himself filled, he said with a tone of satisfaction: "I had the department that made the money. I bought and sold all that came into the place, never having anything to do with the office." "There were good profits in those days," he added with a sparkling eye.

Fifty years ago practically everything was imported from England, there being almost no home manufacturing with the exception of the Jones firm of Gananoque, who made a line of spades and shovels.

Mr. Martin has never had much faith in paint as a commodity to be handled by wholesale hardware merchants. The company, he said, had about twenty-five years ago taken up the trade in paints, continuing it, however, for only three or four years, the business then being made unprofitable by the competition of the wholesale druggists. Later on, however, the trade reverted to its natural source, the retail hardware trade.

"Rice Lewis & Son made its reputation," said Mr. Martin, by keeping things that no other houses had. I am strongly in favor of a combination,

Sketches of Pioneers Wanted.

A few months ago E. C. Atkins & Co., saw manufacturers, Indianapolis, Ind., drew attention to an employe who had for 41 years served that firm faithfully and well. In looking over the Canadian trade several even more striking examples of sterling worth were noted and it was decided to present to readers of HARDWARE AND METAL a few sketches covering the long careers of usefulness of these honored men in the trade.

While it is doubtful if there are any who can show longer terms of service the editor invites letters telling of any in the trade who have been for 30 years or longer engaged in hardware manufacturing, jobbing or retailing. Information regarding firms which have been in existence for half a century or longer is also solicited. A series of articles on old-time hardware men and business would be interesting to readers of HARDWARE AND METAL.

wholesale and retail business as it is the retail business that allows the firm to carry such a variety and meet the demands of the people."

James Ross, president of the Dominion Coal Company, in conversation with Mr. Martin a year or so ago, asked him to what he attributed the fact of his old age and extraordinary vitality. Mr. Martin said: "I attribute it to the fact that I have always endeavored to live a godly life."

That was sufficient reason. He bears no dishonorable marks of over-indulgence. He has lived a clean, useful life and is at 85 years of age looking forward to the opening of Spring so that he can again enjoy riding down to the store daily on his bicycle. May such a "grand old man" live long enough to pass the century mark and retain his interest in the hardware trade to the last.

A REMARKABLE RECORD.

John Ritchie, whose likeness appears below, celebrated the sixty-first anniversary of his birth. Of these sixty-one years, just forty-five have been spent in the employ of Caverhill, Learmont & Co., Montreal, and by that fact Mr. Ritchie probably holds the Quebec record

JOHN RITCHIE

For 45 Years with Caverhill, Learmont & Co., Montreal.

for continuous service with one firm. Such a career is remarkable for a young country like this, but, when it is mentioned that Mr. Ritchie's father before him, was also an employ of Crathern & Caverhill, pioneers of Caverhill, Learmont & Company's business, a chain of service is linked which to the best of our knowledge cannot be equalled in Canada.

In 1861, when young Ritchie began his long hardware experience, the business was small, in comparison with its present proportions. Almost the whole stock of Crathern & Caverhill was carried in the St. Peter St. warehouse, and Ritchie entered the storeroom there. Later on, it was found necessary to take over a large warehouse on Colborne St., to accommodate the heavy goods and John Ritchie was detailed to this branch as head storeman.

That his duities were faithfully performed, the mere fact of forty-five years' service amply proves. In every deal, he was found to be strictly honest and thorough-going, and no man ever worked harder for his firm. Strong of frame, and robust in health, he was an ideal man for his position. Even in spite of advancing years, he continued to perform his duties in the most faithful spirit, until, a short time ago, he contracted heart trouble, and was sometimes forced to absent himself from his duties. He struggled bravely against his weakness but finally he was compelled to give up, a few short weeks ago and, since then, his health has rapidly failed. He is now confined to his bed, but his friends are hoping that his fine constitution will pull him through to see still further expansion of the business in which he has spent his life.

ARTHUR LADOUCEUR
For 39 Years an Employe of the Canada Horse Nail Company, Montreal.

Mr. Ritchie has always been a strong temperance man, and he is a member of the Fairmount Methodist Church. In his early years, he lived on Ann St., but of late years he has resided in Ontremont, a pleasant suburb of Montreal. "Hardware and Metal" takes pride in showing to its readers so splendid a type of the pioneer hardware employe.

LONG TERM OF MANUFACTURING.

When an employe has remained in the service of one firm and their successors, for such long period as thirty-nine years, that fact should reasonably satisfy most people, first, that the employe was a faithful and deserving person, and second, that the employers were fair and considerate in the treatment of their employes.

We have had lately brought under our notice a worthy example of the good relations which have existed between Arthur Ladouceur, an old employe of the Canada Horse Nail Company, who are the successors in the business of manufacturing horseshoe nails of the firm of W. M. Mooney & Co., who commenced in the year of 1865 at Sault Aux Recollett, eight miles from the city on the north side of the Island of Montral, where water power facilities were obtained. W. M. Mooney being the practical managing partner. The late Honorable James Ferrier, M.L.A., who was at one time mayor of the City of Montreal, and otherwise identified with some of the largest commercial and financial enterprises in Canada, was financially associated with him.

It was a day of small beginnings; the firm commencing with only one machine, and adding to it before the close of the year three others until at present these have been increased five fold. This firm were the first to introduce in Canada the manufacture of horse nails by any mechanical process.

Mr. Ladouceur entered the employ of this company, as a boy of thirteen years of age, and with the exception of two intervals of about one year each time, has been ever since in their service, which represents the long and exceptional period of thirty-nine years, all of which was spent in their forging department, where the nails are hot-forged from Swedish nail rods. Mr. Ladouceur has been a joint foreman in this department for some years, and latterly in entire charge.

Fortunately he has always been of a thrifty disposition and has always managed to save a portion of his earnings, which he wisely invested in productive real estate in the suburbs of Montreal. The returns from this enabled him to decide last September that as his health was not satisfactory he would retire and take a well earned rest from his years of hard work. In accordance with this resolution, he left the employ of the company October 1st. His employers took advantage of the occasion to present him with a "Birks" gold watch and an address in the assembled presence of all the employes of the company.

Such an incident as this, reveals in part, that during the 40 years of its existence the company made more than good horse nails, having made good fellowship amongst its employes.

THE "DEAN" OF THE TRAVELERS

It is doubtful if there is a younger "old man" amongst Canadian travelers than W. H. Dean, a member of the Hamilton Executive Board of the Commercial Travelers' Association, and who for over a quarter of a century has been a traveler for Wood, Vallance & Co., of Hamilton. Judging from the accompanying portrait, forty-five years of strenuous activity as a traveler selling hardware rest remarkably lightly on Mr. Dean's shoulders, and it is safe to say that absence from the road would make him feel like a duck out of water.

Mr. Dean served his apprenticeship to the hardware business with Fuller Bros., of Stratford, sons of the late Bishop Fuller, getting a thorough training and making his first trip in 1862 while in their employ. Leaving them he went to Montreal and engaged with Mulholland & Baker to represent them west and north of Toronto, and other sections at different times for eleven years, afterwards traveling for Frothingham & Workman for two years. For nearly twenty-nine years Mr. Dean has represented in Western Ontario, Wood, Leggat & Co., and their successors, Wood, Vallance & Co.

Mr. Dean has never dissipated and has preserved remarkably well his vigor and vitality, and his general appearance and deportment give one the impression of intellectual muscularity and strong powers of endurance. Mr. Dean has only effected this well-nigh perfect preservation of vitality by having laid down certain and rigorous laws of health when young, and observed them invariably throughout his entire career. Being affable and unassuming in his dealings with his fellow-men, he has won a large number of friends, and, as he himself says, it is a source of pleasure to him in his later days to meet his friends, old and new. In confirmation of the fact that his patrons place great confidence in him, many of his older customers allow him to go behind the counter in their stores and allow him to look over their stock and write out an order for what he thinks they need. Such a privilege accorded to a traveler is enviable to the highest degree. We are sure that we express the sentiments of all who have had dealings with Mr. Dean that we wish him many more happy and useful years. His very name is capable of quite an appropriate application. He is veritably the "Dean" of the hardware trade in Ontario.

W. H. DEAN
Of Wood, Vallance & Co., Hamilton, for 45 Years a Hardware Traveller.

Hardware Retailing in the West

The Hardware Business Stands by Itself and is not a Part of the General Store as Generally Supposed.

It is a common delusion in the east that trade in Western Canada is almost entirely confined to the general store. Whatever may be true of other branches of trade this is not the case with the hardware business, which is conducted almost entirely in special hardware stores. Hardware retailing is a big business in the west, and it attracts to its ranks a class of men who are second in ability to no other body of retailers in any part of the Dominion. Even the smallest and newest hamlet on the newest branch line of railway is pretty sure to have its straight line hardware store established by some enterprising merchant who sees in the new village great possibilities in his own particular line of trade. He is usually a practical hardwareman who has learned the business as a clerk either in eastern Canada or in a store in the older settled portions of Manitoba. In most cases the hardware dealer in the west in a business man in the best sense of the word, and that means much for the successful conduct of his business.

A Specialized Business.

The hardware business is thus specialized in the west and attracts a fine class of business men for the reason that it is perhaps the biggest retail business in the west. In a new country where the incoming settler has everything to build he is of necessity a heavy buyer of hardware. Builders' hardware finds a ready market in the west and the hardware store in the smallest of the new towns finds a profitable trade in this one branch alone.

Settlers Buy Liberally.

Western hardware dealers have the reputation of being liberal buyers. Life in the west gives men big ideas and western business men are apt to buy accordingly. Their business horizon is as wide as the horizon on the prairies and they are accustomed to plan great things. They not only plan; they execute. The western hardware merchant may not have so many customers as his brother hardwareman in the east, but their purchasing capacity is much greater. They want six times as much hardware as they would have wanted had they remained in the east. The western customer is a newcomer who has brought little with him from his former home and he has practically everything to buy. He settles on his quarter or half-section and it has no buildings, no fences, no improvements of any kind. His fencing he is apt to postpone until a later day, but buildings he must have and he is therefore a heavy buyer of builders' hardware; he has brought very little with him in the way of household supplies, and therefore he must buy his supplies of household and general hardware. The newcomer in the west has practically everything to buy, while the customer in the east is pretty well supplied. It is the old story of the difference between business conditions in a new country and in an old.

Must Carry Large Stocks.

The western hardware dealer buys heavily, partly because he has to and partly because he has learned the habit. He is a long distance from his source of supplies and, particularly in the winter months when freights are apt to be uncertain, it is necessary to carry large stocks. Under the best conditions the wholesale centres of Winnipeg, Montreal and Toronto are far away and it takes time to get goods. Orders are accordingly placed farther ahead than is the case in eastern Canada, and they are placed for larger quantities. For this reason, and for the reason that the western dealer finds it absolutely necessary to give long credits—or thinks that it is absolutely necessary to do so, which is the same thing—larger capital is required to conduct a hardware business in Alberta, Saskatchewan or Manitoba than is required in Ontario, Quebec or the Maritime Provinces. Probably it would be quite within the mark to say that at least 50 per cent. more capital is required to finance a hardware store in a Saskatchewan town with an annual turnover of $50,000 per year than is required for a business of the same annual turnover in the east. But if more capital is required, it is also true that the rewards are greater for the amount of capital invested. The western dealer gets bigger profits than his brother in the east.

Not in the Cent Belt.

Because he buys and sells in larger amounts and conducts a growing business in a rapidly growing country it is only natural that the outlook of the western hardware dealer should be widened and his vision enlarged. He is a big man among retail business men, a man of big ideas, a man of buoyant optimism and withal a man of wide business experience and keen business acumen. He is not usually a man to haggle about prices; he is prosperous and liberal and he buys without wasting time to quarrel over a few cents. Until recently the five-cent piece was the coin of least value in circulation in Winnipeg. The advent of a big department store with its 38-cent and 98-cent articles has changed all this and people in the towns farther west now refer with gentle and pitying contempt to the Manitoba capital as a town within the "cent belt" of the effete east. Coppers are dispised in the west and the liberal disposition of the average westerner makes it easy to do business with him. Not that he is in slang parlance an "easy mark." Far from it. Like the man from Missouri, he must always be "shown" that your goods are right, that your proposition is a good one.

No business man in any part of Canada will display a keener insight than will the westerner when he inquires into the merits of your goods or your business proposition. But if you convince him and he wants your goods he will waste little time haggling over a few cents in the price. He makes a good profit himself on all that he sells and he expects you to do the same. He is not in business for his health, and he takes it for granted that you are in business yourself to make money. He expects nothing else. This makes it comparatively easy for a salesman with a good line of samples and representing a reputable well-established house to sell more in the west in a week than he could sell in the east in a month.

SALESMEN'S OPINIONS VALUED.

The commercial traveler's opinion is valued highly by all with whom he comes in contact, for the reason that he has unusual opportunities to acquire information, backed by keen appreciation of current events which enables him to formulate ideas that attract attention, always insuring a respectful hearing.

THE RAILROAD SANDWICH.

The other day an employer asked his traveling man if he always paid fifty cents for his meals on the road. "No," said the traveling man, "we get some for twenty-five cents, but when we are compelled to eat those meals we earn a quarter and are entitled to the extra quarter." And that ended the talk on cheaper meals.

About Goods Hardwaremen Handle

How One Dealer Established a Paint Department—Profits in Heating Contracts—Interesting Furnace Problem—Money in Fishing Tackle—Profitable Side Lines.

THE MAKING OF TURPENTINE

The work in a turpentine orchard is started in the earlier part of the winter with the cutting of the boxes. Until some years ago no trees were boxed of a diameter less than 14 inches; of late, however, saplings under 10 inches in diameter are boxed. Trees of full growth, according to their circumference, receive from two to four boxes, so that the 10,00 boxes are distributed among 4,00 to 5,00 trees on an area of 200 acres.

The boxes are cut from 8 to 12 inches above the base of the tree, 7 inches deep, and slanting from the outside to the interior, with an anle of about 35 degrees. In the adult trees they are 14 inches in greatest diameter and 4 inches in greatest width, of a capacity of about three pints. The cut above this reservoir forms a gash of the same depth and about 7 inches of greatest height. In the mantime the ground is laid bare around the tree for a distance of 2½ feet to 3 feet, and all combustible material loose on the ground is raked in heaps to be burned, in order to protect the tree against danger of catching fire during the conflagrations which are frequently started in the pine forests by design or carelessness. The employment of fire for the protection of the turpentine orchards against the same destructive agency necessarily involves the total destruction of the smaller tree growth, and if left to spread without control beyond the proper limit often carries ruin to the adjoining forests.

During the first days of spring the turpentine begins to flow, and chipping is begun, as the work of scarification is termed, by which the surface of the tree above the box is laid bare beyond the youngest layers of the wood to a depth of about an inch from the outside of the bark. The removal of the bark and of the outermost layers of the wood—the "clipping" or "hacking"—is done with a peculiar tool, the "hacker," a strong knife with a curved edge, fastened to the end of a handle bearing on its lower end an iron ball about 4 lbs. in weight, to give increased force to the stroke inflicted on the tree, and thus to lighten the labor of chipping. As soon as the scarified surface ceases to discharge turpentine freely, fresh incisions are made with the hacker.

The clippings is repeated every week from March to October or November, extending generally over 32 weeks, and the height of the chip is increased about 1½ to 2 inches every month. The resin accumulated in the boxes is dipped into a pail by a flat trowel-shaped dipper, and then transferred to a barrel for transportation to the still. In the first season from six to eight dippings are made. The 10,000 boxes yield at each dip 40 barrels of "dip" or "soft gum," as it is reckoned in Alabama to be, of 240 lbs. net weight. The flow is most copious during the height of the summer (July and August), diminishes with the advent of cooler season, and ceases in October or November. As soon as the exudation of the resin is

How Turpentine Sap Is Secured.

arrested and the resin begins to harden under the influence of a lower temperature it is carefully scraped from the scarified surface and the boxes with a narrow, keen-edged knive attached to a long wooden handle.

The Distilling Process.

In the first season the average yield of dip amounts to 280 barrels, and of the hard gum or scrape to 70 barrels. The first yields 6½ gallons of spirits of turpentine to the barrel of 240 lbs. net, and the latter 31 lbs to the barrel, resulting in the production of 2,100 gallons of spirits of turpentine and 260 lbs. of resin of higher and highest grades. The dipings of the first season are called "virgin dip," from which the finest quality of resin is obtained, graded in the market as water white (WW) and window gloss (WG). In the second year from five to six dippings are made, the crop averaging 225 barrels of soft turpentine and 120 barrels of scrape, making altogether about 1,900 gallons of spirits of turpentine.

The resin, of which about 200 barrels are produced, is of a lighter or deeper amber color, and perfectly transparent, of medium quality, graded as I, H and G. In the third and fourth years the number of dippings is reduced to three. With the flow over a more extended surface, the turpentine thickens under prolonged exposure to the air and loses some of its volatile oil, partly by evaporation and partly by oxidation. In the third season the dip amounts to about 120 barrels, with the scrape to about 100 barrels, yielding about 1,100 gallons of spirits of turpentine and 100 barrels of resin of a more or less dark color, less transparent, and graded F, E and D.

Abandoned After Four Years.

In the fourth and last year dippings of a somewhat smaller quantity of soft turpentine than that obtained the season before and 100 barrels of scrape are obtained, with a yield scarcely realizing 300 gallons of spirits of turpentine and 100 barrels of resin of the lowest quality, classed as C, B and A. After the fourth year the turpentine orchard is generally abandoned. Owing to the reduction in quantity of the raw product, it is not considered profitable by the larger operators to work the trees for a longer time. It is only in North Carolina that the smaller landowners work their trees for ten or more successive seasons, protect the tree against fire, and, after giving them a rest for a series of years, apply new boxes on spaces left between the old chips—"re-boxing."

The process of distillation requires experience and care in order to prevent loss in spirits of turpentine, to obtain the largest quantities of resin of higher grades, and to guard against over-heating. After heating the still somewhat beyond the melting point of crude turpentine, a minute stream of tepid water from the top of the condensing tube is conducted into the still and allowed to run until the end of the process; this end is indicated by a peculiar noise of the boiling contents of the still and the diminished quantity of volatile oil in the distillate. On reaching this point the heating of the still and the influx of water have to be carefully regulated. After all the spirits of turpentine has distilled over the fire it is removed, and the contents of the still are drawn off by a tap at the bottom.

This residuum, the molten rasin, is first allowed to run through a wire cloth, and is immediately strained again through coarse cotton cloth, or cotton batting made for the purpose, into a large trough, from which it is ladled into barrels. The legal standard weight of the commercial package is 280 lbs. The finest grades of resin are largely used in the manufacture of paper, for sizing of soaps, and of fine varnishes; the medium qualities are mostly consumed in the manufacture of yellow soap, sealing wax, in pharmacy, and for other minor purposes! and the lower and lowest quantities are used for pitch in ship and boat building, brewers' pitch, and for the distillation of resin oil, which largely enters into the manufacture of lubricating agents.

INTRODUCING PAINTS IN A HARDWARE STORE

H. A. Johnson tells in The Iron Age how he introduced a line of paints in a country store which had not carried them in stock.

Upon being advised by my clients that they contemplated putting in a line of paint I was requested to suggest plans and methods covering its introduction in all ways. In my preliminary survey of the entire situation I took into consideration the time of the year that was best suited to the subject of paint, and I argued that while summer was the ideal time to use paint, it was not necesarily the best time to introduce the subject. Human nature makes its plans and when ready to act follows beaten paths. Therefore the public mind needed to be disillusioned somewhat. Never before had a hardware store in that town sold paint. Heretofore the paint trade had been in the hands of druggists, decorating establishments and regular paint stores.

Thus I reasoned I would need to inaugurate a campaign of education to rid people of old notions as to who sold paint. In fact, not so much to rid them of old notions as to burn into their minds the fact that paint could be bought in another place. I also reasoned that when the fall sets in people have to stay indoors, and it is then that floors and trim are painted, curtains put up and homes made tidy generally. (That applies more particularly to city homes.) Last, but not least, I felt sure the country people—who once they use it, use lots of paint—did their painting after crops were sold. Their money once in hand they would get ready for winter and their ideas would be centred on getting things snug and dry, and the fall, then, would be the ideal time to hammer a paint proposition at them.

If the truth be told, I cared more about the country trade than I did about about the city trade, because people in the country not only use more paint, but more important fact, there are fewer "renters." Preserving property means a vital something to the man who owns it. To the renter it means nothing—as a rule.

All told, my reflections resulted in the conviction that the best time to bring out the new line was in the fall. That determined, the plans for advertising were laid as are hereinafter set forth.

Using the County Fair.

Even before the paint was in my client's store I learned that the county fair was but a fortnight off, so I immediately urged that space in the exhibition hall be secured. This was done, and a telegram sent off to the paint people telling them a big exhibit was planned for the fair and requesting them to send their exhibition demonstrator to take charge of it. He came forthwith, took full charge of the matter, trimmed the space attractively, so shelving as to display the cans to good advantage, each size of can by itself, and arranging it all in a way that gave the impression of there being far more paint in the space than was actually the case. Among other things, he arranged to have on hand a quantity of boards of various sizes for purposes of practical demonstration.

The week preceding the opening of the fair I saw to it that all the city county newspapers had an announcement to the following effect in a 5-in. single column space, and during the progress of the fair this was printed on cards and distributed from hand to hand on the grounds:

"You are cardially invited to visit our booth at the county fair. We will serve liquid refreshments free. of charge."

I was born and raised in the country, so I knew county fair traits well enough to know all who attended them are looking for souvenirs; in fact, something for nothing. I knew if we advertised something free we would not lack audiences. And my deductions proved true. Naturally, many who came to our booth expected something to eat or drink. Alas! Paint was good for neither purpose. I had promised "liquid refreshments." Was not paint a refreshment? We heard of no annoyance over the town. The majority of people saw the demonstration and asked for color cards. Small sample tins were given out free. Actual orders were booked more than enough to cover the cost of the exhibition and we took the names and addresses of all the onlookers that we could get.

After the fair was over the following advertisement was inserted in local and county papers:

The Liquid Refreshment

that we "served" at the fair occasioned no end of comment. Many thought they were going to get something to eat and we were sorry to disappoint them; many others saw the point and came to know what good refreshment our new line of paint is. Stop to think—paint is refreshment—now, isn't it? This is about the last chance you can have to get your house, fence or property in proper condition to withstand the elemental rigors of winter. When water gets into wood it means rot and rot means that your property decreases in efficiency, strength, looks and value. So paint seems to be just plain horse sense For property indoors, such as floors, trim, etc., we have———, an elastic, transparent, durable Varnish that even heels cannot mar. That sounds like an extravagant claim, but let us prove it and so "show you."

The names and addresses taken at the fair were next carefully transferred to a card and the following leter was sent to every name in the file:

A Personal Letter.

Mr. John Hanson,
Pleasant Plains, Ind.

Dear Sir—We were sorry you could not spare more time while you were in town for the fair. We should like to have shown you even more of the good points of —— paints. But you were good enough to give us your name, so we will explain it more fully by letter.

There was a time, Mr. Hanson, when paint was just white lead and coloring matter, with just enough oil to make it run freely and smoothly. But times have changed, and to-day the paint manufacturer has to go into the chemistry of it, as he has also to go into the chemistry of wood and the changes that the weather brings about it and on it. The good doctor does more than merely bandage a wound. He cleans and antisepticizes it first, before he even thinks of covering it with a bandage. Just so the modern paint maker. The latter must be as thorough.

—— Paint penetrates the pores of the wood. More than simply closing the vent, it solidly fills the pores and worm holes. When —— paint sets the surface of the painted wood is genuinely poreless. This means that snow and and rain can no more injure the wood after it is coated with —— paint than they can harm glass.

We have gone into the subject thus because we know you are alive to the difference between good and poor paint, because we know you have buildings on your place that need painting, because we hope to sell you a bill of paint.

That's why we are inclosing a color-card and price list, showing all discounts. Note the quantity discount. If you can get your neighbors to buy their paint when and where you buy yours, no matter what colors they want, you can all get an extra discount. Should you need putty or glass we will buy it for you and ship it with the paint. Brushes we keep in stock. Hardware, too.

Of course we will ship by freight to ——, which will be a much shorter drive for you than were you to come clear into the city here. Oh, yes, we will pay the freight charges. Your order only can mean a small profit to us. But it'll mean to you the trebling of life and value of every square inch of wood surface that you use it on. Do we get it?

Very truly yours, etc.

One other follow-up goes out a week after the above, containing a cleverly executed carbon copy of the first letter, along with an urgent bit of salesmanship. The letters are doing good work.

Street Car Cards.

Realizing that "he who runs may read," I also planned that he who rode might read as well, the topic being ————paints. So in the street cars of the city were cards of which the following are typical, the display lines being in red, and the text in black ink:

PAINT—A SERMONETTE

Paint is physical insurance. It prevents decay. Property that isn't painted soon deteriorates. The wood decays and its value shrinks. But the wise man uses paint, keeps the wood-pores closed and his property increases in value.

ARE YOU OF THE WISE MAN'S CLASS?

If not, you can join it by getting your paint of

———— ———— ———— ————

JOHNNIE'S "ESSAY" ON PAINT

"Paint is used by men and women. Some use it on the house, others put paint on their faces. My Ma says the women next door paints Her nose is nice and red. (Not Ma's nose). I painted something the other day. He was a dog. He ain't a dog no more, He's dead."

WE DON'T KEEP PAINT—WE SELL IT

The "essay" attracted a good deal of attention, and therefore developed comment, which proves that in street car advertising an ounce of pleasing is worth a good many pounds of instruction. It likewise sold paint and since that is what it was designed to do, it may truthfully be said to have accomplished is purpose.

Keeping it Up.

We have been keeping it up all fall long, striking while the iron was hot, and going after city folks who might have interior woodwork that needed retouching, or outbuildings, fences, etc., that paint might improve. In the meantime we have had a man working on the owners of property, particularly new houses nearing completion. The owner might employ whom he chose to apply the paint. All we wanted was to sell it. We will work that selling plan next year, too, when there is more building going on. It looks good to us.

The opening stock of paint cost about $350, on 90 days' time. It was in stock hardly a month before a good big hole had been made in it. One color, in popular size cans, was gone. And I think that by the time the bill became payable my clients were very willing to meet it. A bill that is willingly met is one that has pleasing thoughts connected with it. And pleasing thoughts nowadays are profits.

THE DESTRUCTIVE POTATO BUG

By W. H. Evans, Canada Paint Company, Montreal.

It is said that a farmer's son at the annual school examination was requested to recite Tennyson's "Queen of the May." Being more practical than poetical and having in mind the "yeoman's burden" that was in store for him, he broke forth as follows:

"You must wake and call me early,
Call me early, mother dear,
For the 'early roses' are all planted
And the potato bug is near."

Eternal vigilance is the only price of safety. Unless repressive measures are taken early by the husbandman his vines will be eaten, and in less than twenty-four hours the field will look as if a Sirocco or hot blast had struck it and blighted the whole potato patch.

The Colorado beetle, or, to give him the Latin term, doryphora decemlineata commonly known as the potato bug, is an undesirable immigrant from "Uncle Sam" who dos not respect the boundary line or the Alien Act, nor does he pay any head tax. Like the lily of the valley, it "toils not neither does it spin"; but unlike the lily, it is not famous for its beauty, and, being a vagrant, he wears ten stripes after the manner of the penetentiary bird. Moreover, the bug hath a voracious appetite, which, as Shakespeare says, "Age does not diminish or custom stale."

Herein is the potato bug's weakness: it devours the tender leaf of the suc-

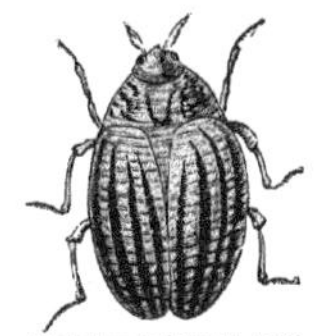

Potato Bug or Colorado Beetle.

culent potato, and at the same time, encompasses its own destruction, provided the vine has been sprinkled with Paris green.

A number of substitutes have been tried, but, so far, no insecticide has been found so efficacious as Paris green, which is a combination of sulphate of copper, arsenic, soda ash and acetic acid. It is sold in a dry powder, and is applied

WM H. EVANS.

by first mixing the Paris green into a paste with water and then placing the green paste into a bucket or pail, and stirring well. It can then be sprayed on the vine or poured by means of a watering can through the perforated spout or nozzle.

In Canada alone about 350 tons of Paris green are sold annually, at an average price of, say, 25 cents per lb. This comes to $175,000. The output in the United States must also be enormous, as Paris green is also largely used to control the destructive cotton boll weevil.

Unchecked, the Colorado beetle commit fearful ravages amongst the potatoes. It was first discovered in 1824 by Thomas Say. It is an oval creature, as may be seen from the cut. The larvae and adults live on the potato plant, and before Paris green was discovered to be a remedy for their depredations, they would sometimes completely destroy the entire potato crop in some sections.

They pass the winter under ground, and sally forth from their hiding places in the beginning of May. The female lays many hundreds of eggs in groups of twelve to twenty on the under side of the potato leaves. The larvae which emerge in about a week are reddish and afterwards orange. They grow quickly, and produce a second and third generation in the same summer.

The home of the Colorado beetle is in the Western States. From Nebraska and Iowa it traveled eastward, favoring Canada with its patronage, until in 1873 it reached, as did Sherman's march to the sea, the eastern shores of America, and like John Brown's soul, the beetle "keeps marching on!"

PROFITS IN FURNACE CONTRACTS

Probably 90 per cent. of the complaints about furnace work are caused by:

1 The fact that competition, architects' specifications or manufacturers' inflated claims of capacity of the furnace caused the dealer to put in a furnace that was too small.

2· Improper proportioning of the cold air supply or hot air pipes or poor workmanship on the job; or,

3. Careless handling of the furnace by the users.

Upstairs StoVe Sa'esroom, Cupar Hardware Company, Cupar, Sask.

For the first the dealer is to a certain extent responsible, if the work was not done under specifications of an architect. If too small a furnace was specified by the architect, it relieves the dealer from blame, provided he calls the architect's or owner's attention to the fact before starting the work.

If the contract is done under an architect's specifications and under a general contract for the completion of the building, the notice should be sent to the general contractor, but if the contract is let by the owner the notice should be sent to the architect, because he drew the specifications.

The second complaint—improper proportioning of the system or poor workmanship—is entirely within the jurisdiction of the dealer. For improper workmanship there is no excuse, and if the dealer hires incompetent help because it is cheap, or fails to give it proper supervision, he deserves all the trouble that results.

The foregoing separates the various complaints into their proper places, and shows which should be charged to the furnace job and which should not.

The cause of the greatest complaint and expense is the incapacity of the plant to heat the house properly. Taking all points into consideration, it would seem to be an easy matter for a contractor to note what relation the year's expenses, figuring and superintending the furnace work and making good the defects for which he is responsible, bears to the business done. This gives the proper percentage to add to the cost of labor and material when figuring new work, and leaves nothing more to add, except the allowance for general expense, to give the actual cost of prospective contracts.

A percentage should be added for profit, not a lump sum, as $100 added

Model Kitchen Display Room.

to a job costing $200 will result in losing the contract, while the same amount added to one costing $1,000 is unreasonably close. If 25 per cent. were added to the small jobs, or 20 or 15 per cent. to the large ones, it would seem more reasonable and would probably result in securing more business and a better average margin of profit. If the conditions suggested are observed this 15, 20, or 25 per cent. is a real profit and not a fictitious profit out of which must come an allowance for general expense, making good defects and just complaints. The allowance is small enough and in some places can probably be increased. The difficulty in a number of places is that the cut rate dealer makes it impossible to secure a reasonable amount of business at a living price.

A Model Pantry.

FURNITURE A GOOD LINE.

The Cupar Hardware Co., Cupar, Sask., whose stove salesroom is shown in the accompanying illustration, are sufficiently discerning and enterprising to realize the great advantage to be gained by combining the hardware and furniture trade. This company have a double store on the ground floor and upstairs have two showrooms, one for stoves and ranges, and the other for furniture. These two lines can be advantageously combined, keeping in mind the fact that newcomers in a town wishing to furnish a home would secure, if they could, their entire furnishings from one company, kitchen range, pantry utensils, dining-room, and bedroom furniture. Morover, home builders, when wishing to buy stoves or furniture, etc., do not wish to look over the stock in the presence of a crowd of customers, and this opportunity afforded by upstairs showrooms to look over the stock privately is a great inducement to them to trade in that store.

The hardware business and the furniture business are becoming combined more and more because they are lines intended for the same purpose—furnishing the home.

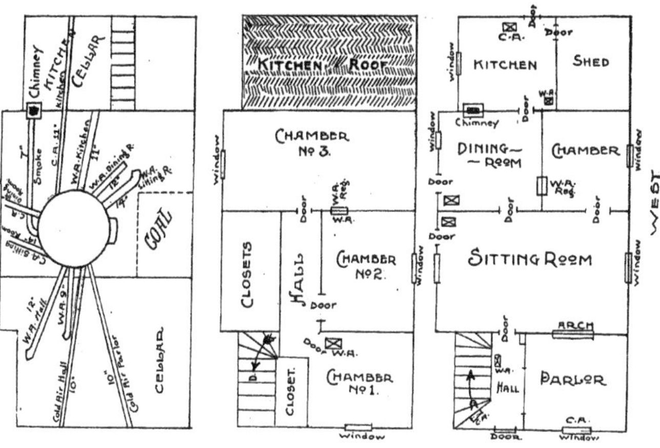

Interesting Heating Problem—How Can Chamber No. 1 be Heated Satisfactorily.

A HOUSE HEATING PROBLEM.

An interesting problem which offers an opportunity for some constructive thought is presented in the design of the house shown on this page.

The bedroom in the upper story, marked No. 1, faces the north, and, owing to the stairway being almost enclosed, the cold air does not travel down to the cold air register at the foot of the stairs. In consequence, when the temperature on the ground floor is 74 degrees, it only averages 50 upstairs in No. 1 chamber.

The firm which installed the furnace has found difficulty in overcoming this trouble and we present the problem to our readers. Stove dealers and travelers for furnace manufacturers are requested to write us, explaining their solution of the problem.

"MODEL" KITCHENS AND PANTRIES.

Although the business done and the store capacity in a large city departmental store is very much larger than that done in a hardware store in a town there are, however, ideas carried out in in the departmental stores which could be carried out in a smaller store very profitably, although, of course, on a smaller scale.

Provided the room over a hardware store is not occupied, it could be very profitably fitted up by the hardware merchant and rooms arranged to represent a kitchen or pantry making it as much like these as possible, completely furnished, if it be a pantry, with shelves and drawers, with all the various kinds of tinware and enamelware arranged on the shelves; if a kitchen furnished with a range, gas stove, sink, etc. Advertise these furnished rooms as special features of the store and arrange for certain days on which ladies may visit and inspect these pantries and kitchens, and the plan will surely prove interesting and remunerative.

PRICE CARD MAKES A SALE.

Not long ago I had on display in my window a steel range with high closet and reservoir, writes Geo. M. Evanson, a western hardwareman. Leaning against the range was a white piece of cardboard upon which I had painted the price $33.50. There was a net profit of $4 in the price for me.

The second day the stove was in the and the price, stopped to carefully ex- window a man and his wife were passing by the store. They noticed the stove amine it, then came inside and the man wanted to know if we warranted that range—"Yes," I replied, "as much as any one can warrant a range for that price."

During a short conversation I discovered that these people were going to buy buy a range and that they had decided to send to a mail order house for it and that seeing the stove in the window and the price being about the same as the one was to cost that they had planned to send for, they thought they would step in and inquire about it.

Beats Out Catalogue House.

I discovered to what catalogue house they were going to send, produced the catalogue, asked the man to show me the stove, and then I had my inning. I picked that catalogue stove so completely to pieces and laid bare to the eyes of these two people all its hidden imperfections, showed them the size of the the oven of the catalogue stove and the size of the oven of my stove, let the man lift the fire box end of my range and then showed him the catalogue stove weighing but 325 pounds. In short I sold that man a range—then and there; why? Will anyone deny that if I had not had a price card on that range I would not have sold it?

Every time I head off an order to a catalogue house I give the item to the local papers and see that all of the publicity posible is given the item. Then the next time I see the farmer I ask him if he will sign a statement of the facts and see to it that every farmer in his neighborhood gets a large postal card with this statement printed thereon.

STOVE SALESMANSHIP.

Good stove salesmanship is rare but invaluable to a large firm. A salesman makes no impression on his customers by simply stating, "this range sells for $45," and then waiting to see this effect on the customer's mind with nothing more to say. To be a successful salesman he must have visited the foundry and carefully inspected every step in the manufacture of the stove, know the materials out of which it is made and be able to point out any special and valuable features.

SPORTING GOODS A BUSINESS BRINGER

By Edward D. Ibbotson.

You may know the story of the man selling soda water fountains. He worked hard to sell one to an old druggist; and could only get he response that the druggist had never had any call for soda water. The up-to-date druggist on the opposite corner put in a soda water fountain and in a few weeks, and during the hot summer days, the profits from it paid the rent of the store. The old, fossilized druggist would not put in the soda water fountain, because he had never had any call for soda water. It's so with many hardware merchants—they do not put in stock, good paying lines, which belong to their business, simply because people do not come and ask for the goods as they do for nails or hammers. The business, however, is there, and it is only ready for the hardwareman to reach out and take up if he knew it.

The story of the druggist and the soda water fountain applies well to many hardware stores with the line of sporting goods—baseball supplies, fishing, tackle, guns and ammunition are just as much hardware as they are anything and belong to the hardware stock and in the hardware store.

Hardware Store a Men's Store.

The best proof that the average hardware dealer has found sporting goods profitable is in the fact that it's very rare that the line is droped, when once taken up, and it's always found to be a positive business bringer. Brings customers into the store—men customers, the buyers of hardware; men buy sporting goods and the hardware store is a man's store. This is one reason why department stores have been and usually are, unsuccessful in selling fishing tackle or other sporting goods; because the department store is not a man's store, and women seldom buy their husband's fishing tackle or his gun goods, although they may buy his shirts or even his neckwear. But when it comes to picking out his fishing outfit or camping goods he draws the line; if he's a sportsman he finds as much pleasure in buying the goods as in using them. So it comes about that men are the buyers of sporting goods—and, as before stated, the hardware store being the man's store, is by far the best store to handle the line.

And so it is, that the hardware dealer will find the handling of sporting goods will bring more men and therefore more prospective hardware customers into his store. The more customers a dealer has coming to his store—no matter for what they may come—the more trade that dealer is going to have in hardware. It may be only a fish line or a box of cartridges, the customer buys to-day. He may formerly have been buying what little pick-up hardware he needed at another store—but if he is well waited on—gets a good fish line or box of cartridges at the right price, sees a good display of hardware—a well arranged and well-kept store—you can rest assured that it's only a question of time that man will become a regular customer of the store. So it's true that the selling of sporting goods, more than any other side line, will prove to be a business bringer for the hardware merchant.

Goods Must Be Displayed.

I went into a store the other day, and, seeing but a very small display of fishing tackle, thought my customer must have sold out well. In reply to my question how the trade had been, he said: "It's been very poor this year; I'm only selling the very cheapest grade of goods." "But where is your stock?" I asked. We were shown the stock, away back in the rear of the store, in a dark corner, tucked back on the shelf. That dealer was complaining about poor trade—said that there had been nothing but cheap goods sold in his town—but right across the street is another store. That merchant said his trade had been good; at first flush it seemed as though he had a big trade—two large cases filled with fishing tackle. "A good display you still have there," I remarked. "Yes, but do you know, right there in those two cases is all the tackle I've got in stock; and I've just had to string it out, and even put those trunk and vest pocket rods, and some cutlery in one of those cases to make a show during the past two weeks. We have had the best season yet, and sold more high-priced stuff and all our fancy tackle." And the order he made up proved what he said was true.

AUTOMOBILE SUPPLIES.

During the past two or three years the automobile industry has developed very extensively. In large centres of population exclusive automobile supply stores have been established to take care of the growing demand. To those closely in touch with the automobile industry it became apparent that in the smaller towns where the exclusive automobile dealer could not command sufficien trade, automobile owners were unable to secure such supplies without inconvenience of ordering same from the nearest supply house, frequently fifty to a hundred miles distant. As hardware dealers are expected to carry in stock almost any article made of metal, they have been approached by automobile owners for the supplies most frequently required, only to be turned away for want of having same in stock. A number of the large hardware jobbers throughout the country have gradually taken up the handling of automobile supplies, and as a consequence, a fair percentage of the dealers have been convinced that they may as well share in the profitable trade that can be secured by catering to the ever-increasing number of automobile enthusiasts.

It may also be noticed that several of the manufacturers of automobile appliences have recently taken steps to directly interest the hardware dealers and therfore it may be expected that in localities where automobile supplies have heretofore been bought to fill specific orders only, will in the near future be found as part of the regular stock of the wide-awake hardware dealers who watch the development of the modern period of motoring. The Hardware Dealers' Magazine suggests that hardware dealers encourage their jobbers to catalogue the most useful and best-advertised automobile appliances so that when it becomes necessary to order same, no time will be lost in lengthy correspondence for the articles wanted. Any line that is profitable to the jobber should be likewise profitable to the retailer.

MONEY IN FISHING TACKLE.

All merchants, no matter what line of business they are engaged in, know that it is not so much the amount of trade that they handle as the profit on their goods that will make them rich. As a corollary to this it may be stated that the more profitable a line of merchandise, the more difficult it is to handle successfully. Goods in which the competition is most keen must be sold on a close margin and consequently they bring the smallest margin of profit. Some goods must be sold close. Much of the stock in any store must be sold with little profit. For example, it would be hardly possible to conduct a hardware store without carrying a stock of nails, and it also would be impossible to sell these goods to large buyers—to contractors or to builders—anywhere in the east, at a profit large enough to pay the running expenses of the store.

This being the case, the wise merchant is the one who is constantly on the lookout for any line which pays a good profit, and at the same time will find a ready sale. The sporting goods and hardware dealers must also realize that each year their fields are being encroched upon by dealers in other lines—the modern house, the general store, the department store, and in some cases the competition is making itself felt from the jobbers in nearby towns. The dealer therefore, must make an effort not only to hold his present trade, but to take advantage of changing conditions.

Profit Making Lines.

"How to enlarge my profits without noticeably increasing my expenses?" This is a question, says the sporting goods dealer, that nearly every thoughtful merchant is asking himself, and this best can be accomplished by gradually

HOW TO ADVERTISE SPORTING GOODS

By T. J. Bowler.

Baseball goods, fishing tackle, lawn tennis goods, footballs, etc., as well as guns and ammunition, all belong legitimately to the hardware trade, and any hardware dealer who neglects this end of the business is doing himself an injustice. If he does not handle this class of goods, his competitor either does or soon will, or perhaps some little fellow will come along with a $200 stock of five and ten cent goods and the first thing he knows he puts in a little line of sporting goods and before long has a nice trade built up. And why? because the the hardware dealer neglected this profitable line and another wide-awake dealer who recently came to town, or moved into the next block, saw where he could increase his line and at the same time make a good profit. As the amount of profit, this depends a good deal on the class of goods you handle. However, the profit should run from 33 1-3 per cent. to 50 per cent.

How to Create Demand.

You could soon create a demand for baseball goods by having your son, your tinner or your clerk join one of the local teams. Get the boys to buy their uniforms from you, even if you are obliged to sell them for absolute cost. They then will buy the rest of the outfit from you. Put up a backstop on the ball grounds with a nice, attractive sign on it setting forth the fact that Smith & Jones Hardware Company are headquarers for all kinds of sporting goods.

You can also create a demand for tennis goods, such as rackets, balls, nets, etc., by having your daughter or your wife organize a tennis club.

Make a nice display of sporting goods in one of your windows early in the season. Get the people of your town talking about what a good time they will have this summr playing lawn tennis, baseball, croquet, fishing, etc. They will know that Smith, the hardwareman, has a nice line and they won't have to send out of town to some catalogue house to get a tennis racket or baseball bat. This is the great difficulty in small towns where dealers do not carry a line large enough to meet these occasional demands and the people get into the habit of sending to the mail order houses.

The dealer might also put an "ad." in the local newspaper. Your jobber will furnish you with a cut of a baseball, catcher's mit, fishing rod or tennis racket. The reputation gained through advertising these goods, though you sell but a few of them, is well worth the price. The people in your neighborhood will soon begin to say, "Well, this man Smith, the hardware dealer, is certainly up-to-date," and a reputation for being up-to-date will hurt no man in business.

You have less competition from catalogue houses on baseball, tennis goods and fishing tackle than any other line that I know of, for the reason that most ball players and tennis players want to see the article they going to buy.

Tobogganing in High Park, Toronto.

It requires but a very little capital to put in a nice, clean stock of well selected sporting goods, outside of guns and ammunition, and I would advise the exclusive use of one window for sporting goods for the months of May, June and July. By this time your customers will all know that you are the headquarters for sporting goods and that Jones is a baseball and tennis enthusiast as well as a good fisherman.

About 75 per cent. of the hardware dealers in the small country towns and about 99 per cent. of the dealers in the large cities are neglecting this branch of the business and letting it drift into other channels, when it rightfully belongs to the hardware trade just as much as stoves or other such hardware.

If this were not a profitable and desirable line, why would the wholesale dealers be carrying such large stocks?

adding profit-making lines to his business.

One such line is fishing tackle. In the hardware business no branch pays a better profit than sporting goods, and nothing in sporting goods can be made as profitable as fishing tackle. This branch of the business has great possibilities. The trade is only in its infancy. Each year the ranks of the old enthusiasts are being added to by a large number of boys and young men, who are for the first time taking up the sport of fishing. Unlike many other outdoor sports or games, fishing is not a fad. A man who has once been a fisherman always remains so.

Another reason for an increase of these goods is because good fishing each year is made more possible. The millions of fish that are being planted every year by the government is having its effect now in making it more easy for a sportsman to find a place to use his tackle. A few years ago if a man was run down from over-work or worry, his physician would recommend a sea voyage. To-day he would be told to take a trip in the woods—to spend a few weeks camping, and this is giving men an opportunity to find the rare sport in fishing and is consequently making an increased and large demand for the goods.

Men Are the Buyers.

Men are buyers of fishing tackle and men are customers of the sporting goods and hardware stores—consequently it is these dealers who can best handle this line of goods, and it is the most natural place for the sportsman to look for his tackle.

It is also true that the larger the assortment and the better the display of fishing tackle that is made, the greater the amount that will be sold. It is possible for a boy with a 5-cent line and a pin fish hook to catch fish. Every man realizes this; also he realizes that the best sport can be had not in the number of fish that he gets as much as going after them and preparing for the trip. So it is that the better display that is made the more goods will be sold, as fishing tackle more than almost any other line of merchandise is largely purchased because the fisherman thinks that with it he will have better sport. If it appeals to the eye it will find a sale and it can not appeal to the eye unless it is well displayed.

In putting in the line of fishing tackle it is of great importance to start right —to have the goods that are best adapted to local trade. A dealer being located in a country where most of the fish are trout would be very foolish to put in a stock of heavy bass or salt water tackle. It would not sell and he would soon become discouraged with the trade. It is, therefore, of great importance not only to have a well assorted line, but to have one that is made up of goods that are used in the locality in which the dealer is. Also in putting in a new line of fishing tackle, like my other class of merchandise, it is of great importance to have the goods at the right prices. Therefore the line should be purchased of a manufacturer who would be able to put the dealer in a position to meet any competition and yet the dealer should not over-buy. Fishing tackle being seasonable goods, the stock should be kept fresh. Each year new articles are brought out and, therefore, the stock should be purchased with care.

HOW TO DEVELOP BUSINESS IN METAL ROOFING

By James G. Lorriman, Advertising Manager the Metal Shingle and Siding Co., Preston, Ont.

A catalogue, and a reasonable amount of push and hustle are the stock-in-trade required to boom the metal roofing business. Moreover, it belongs, logically, to the hardware trade, for hardware dealers are not only in closest touch with building activities, but, as they usually employ a staff of competent sheet metal workers, they are in a peculiarly favorable position to handle the various lines of metal building goods.

The hardware merchant who desires to get the most out of this branch of his business, must keep his eyes and ears open all the time. In this growing time with new buildings going up, and old ones being renovated, in all parts of the country, his opportunities for making sales will be frequent. He should keep himself well informed about what is going on in his town and district, and at the first inkling of a new barn, residence, factory or store, he should hunt up the prospective builder, and begin his campaign.

It is certainly advisable for the hardware dealer, from time to time, to make the round of his district, calling on customers and prospective customers, and learning their needs in the hardware line. It is on these trips that he will be able to find out where new buildings are contemplated, and to work in a few introductory remarks on the metal roofing question. Thus, all the time devoted to "talking up" metal roofing, will be a very small item— in fact, simply incidental to the dealer's regular canvas.

The average man takes a long time to decide how he is going to spend his money in building. He is not inclined to expend a large sum for any purpose, without giving the matter most careful consideration. So the dealer can hardly expect to take his man by storm, and sell him the goods right off the reel. He must nurse his prospect until he has secured his interest and then ask for the ever-ready co-operation of the manufacturer.

This latter item is of immense assistance to the dealer. The wide-spreading advertising, for several years past, has familiarized the public with the advantages of metal roofing, and the missionary stage is over. In addition to large trade catalogues, the leading firms issue attractive and convincing booklets, which are freely distributed among prospective builders, and finally their travelers are always available, to help the dealer close sales.

Considering the fact that the hardwaremen need carry no stock, but invest only his personality, in developing the metal roofing trade. It must be admitted that no branch of his business will better repay him.

SEWING MACHINES.

There is no reason why a hardware merchant should not deal in sewing machines. To the housewife he sells stoves, brooms, sweepers, churns, and kitchen utensils. Sewing machines are in just as good demand, and in such a line there is good opporunity for a wide-awake merchant to make something. Why, then, should he not make use of such an oportunity. They are very saleable articles and afford a fruitful channel for profit.

Discrimination, however, should be exercised by the hardware merchant in choosing the kind of sewing machine he intends to handle. An inferior machine will give him endless trouble, but a first-class machine will afford him pleasure and profit to handle.

The retailer has an advantage over the mail order house. He can show the goods and demonstrate their superior qualities and be in a better position to sell the goods.

PUMPS AS A SIDE LINE.

Opportunity exists for the hardware merchant to do a much larger and much more profitable business than is being done at present. There is business in pumps being done but of a very inferior grade, done mostly by dealers who carry a small stock of the cheaper grades which can be sold like a pail or wash tub when anyone calls for them.

There is room for a larger business, one giving more room for profit. That opportunity presents itself to the merchant doing business on a larger scale. It is a line which will amply repay the efforts made. The greatest difficulty is that hardware merchants are unwilling to push the trade at all; simply care for the trade as it comes to them.

Many manufacturers express regret that the general market has been flooded with so many cheap pumps on which the margin of profit is very small. The retailers can greatly improve this condition if they will. They can co-operate with the manufacturer in handling only first-class goods, enlivening and strengthening the business by coming out from behind the counter and mingling with their customers and personally supervising the installation of the pumps they sell thereby gaining the confidence of their patrons and laying good foundations for future business. The catalogue method of sales would be the easiest and most profitable way of dealing in them.

HANDLING STRAP IRON.

A United States hardware merchant advises that the problem of handling strap and hoop iron has been solved to his satisfaction by running it on reels whenever there is occasion to break bundles. He states that when strung along in the order of their width the various sizes occupy but little space. He also explains that there is no difficulty in making the reels in the store, without giving the job to outside mechanics.

Another dealer says his method of handling hoop and band iron is to separate the bundles on receipt into smaller bundles or lengths of about 25 feet, which are tied up with tarred lath yarn strings, and put into the iron rack, which most dealers in heavy hardware probably have. These small bundles the merchant has found to be most convenient for retailing, as customers will usually take one of them unbroken; otherwise they can be cut with but little trouble.

Two classifications are generally made in referring to this line, a distinction being drawn between the smaller gauges called hoops, and bands, the terms applied to larger sizes. Hoops run from 12 to about 24 gauge, which is the lightest usually carried, while bands include 12 and heavier gauges. Bands, and also hoops down to about 16 gauge, come in scroll bundles about 12 feet long, consisting of several lengths strapped together to make them as compact as possible. Hoops of 16 gauge and smaller come in coils resembling a watch spring, no wider than the iron, and having a large central opening, because small coils could not readily be straightened out when the iron was to be used. Some merchants prefer to order 16 and even 14 gauge hoops in coils, but there is danger that the metal will get waves in it, which must be pounded out at considerable pains before it can be used for some purposes.

An excellent method to accommodate band iron in bundles is to stand it on end against a wall. Most storehouse walls are unfinished, so that there is nothing to scar or damage. Iron rods projecting from the studding or pieces of plank nailed diagonally between the studding and the joints overhead, may be used to form skeleton bins, which should be labeled with the size of iron that they contain. Another form of wall rack adapted to the use of a merchant who carries a small stock of band iron is made by setting up 2x4 joists between floor and ceiling, about 4 feet apart. At intervals of about a foot holes should be bored, in which are driven pieces of round iron to lay the stock on. The projecting ends of the pieces of round iron should be turned up to prevent the iron slipping off. This form of rack is not practicable where a merchant carries more than two or three bundles of a size.

A convenient way of keeping hoop iron is to make a reel, the drum of which is long enough to accommodate a coil of every size carried. This may be constructed roughly, as there is no advantage in having the drum round. Coils should be separated from each other by pegs projecting from the drum, and whenever a coil is cut it should be tied up, to prevent its getting loose or unwinding when the reel is turned for some other coil.

SHOT MAKING.

Manufacturers of shot use a tower about 300 feet in height, in order that the melted lead (which is dropped from the top of the tower) may become congealed before reaching the bottom.

The lead, in the form of "pigs," is taken to the top of the tower and there melted in kettles, and upon becoming liquid is poured over a sieve through which it falls in drops. These drops, as they fall through the air, take a spherical shape, and upon reaching the bottom are received in a vat containing either water or oil. From this vat the pellets are removed to a machine which separates the different sizes. This machine, which somewhat resembles a bureau in appearance, and is so arranged that it will rock, contains a series of shelves, each of which is perforated to allow shot of a certain size to pass through. The arrangement of these shelves is such that the largest shot will stop on the first shelf, the next size on the second, and so on downward until the bottom is reached, where the smallest size is received. The size of the shot can also be governed to a certain extent by the size of the perforations in the sieve at the top of the tower.

KNOW THE TOOLS YOU HANDLE.

At the fifth annual convention of the New York State Retail Hardware Association at Syracuse, N.Y., in February last a very instructive address was given by W. M. Pratt, of Massachusetts, on "Tool Topics." The point he emphasized was the necessity of hardware merchants to know the tools he handles, how they are made, what materials are used, which are cast iron, which malleable, which machinery or Bessemer steel, and which crucible steel. To be well posted on the salient features of the tools they handle, they can't do better than visit the manufacturers themselves, or if they have managers in charge of the tool department they cannot spend a little money to better purpose than by sending these managers to learn how the tools are made and exactly what materials the tools are made of.

IRON FENCING.

Hardware merchants who are enterprising and aggressive and on the lookout for opportunities to expand and strengthen their business will do well to look into the opportunities presented by the iron fencing and ornamental lines. This is a rapidly expanding line and should and does prove profitable to all who take it up.

Quite a number of merchants have already been dealing in tree boxes hitching posts, iron flower vases, and lawn furniture specialties and found such lines attended by excellent financial results. Orders are taken uniformly by catalogue and the specifications sent to the manufacturers. Such a method of dealing in these lines ought to be very profitable and not at all cumbersome.

EVOLUTION OF THE NAIL.

A common nail, it has been said, is an excellent illustration of the difference between old and new methods. Formerly the metal was cut in strips and then forged into shape with hammers, and an expert took about one and one-half minutes for each nail. To-day they are made of steel, and are lighter and stronger. Strips are cut with steam shears and fed into an automatic nail machine. One man tends three machines each machine dropping a nail every second. He turns out a hundred-pound keg of nails in less than two hours, a work that once would have taken him twice as many weeks.

POCKET CUTLERY.

The ambition of every boy with his first pants is to have a pocket and a knife to carry in it. Anything that looks like a knife will do, but as he grows older he yearns for knives with three, four, five or six blades, knives with corkscrews and tweezers in their backs, knives with tortoise-shell and ivory and mother-of-pearl handles, pen knives, file blades, bowie points, big knives and little ones in endless variety. His commercial acumen is early exhibited in knife swapping, his carefulness or lack thereof in the number of knives broken or lost annually. Later in life this character and temperament may be accurately judged by the knife that he carries, whether designed for ornament or service, bright or rusty, sharp or dull. There are day laborers on starvation wages who carry $2 knives in the pockets of 50-cent overals, just as the old-time Texas cowboy cinched a $75 saddle on a $15 pony. And, on the other hand, there are millionaires who, year after year, rust their pockets with knives that no self-respecting street urchin would pick up out of the gutter.

FUN, FACTS AND FICTION

A FIGHT FOR LIBERTY.

A fable by George Ade about sizes up the situation with regard to the dry towns. It is entitled "The Fable of the Single-Handed Fight for Liberty," and is as follows:

"A traveler landed in a blue-law town on Sunday morning and found it as dead as a mackerel. There were only two horses hitched at the Square, and in every window the curtains were down.

"'Why and wherefore this funereal hush?' he inquired of the hotel clerk.

"'The Sunday closers have been at work,' replied the clerk 'You can't get a nip to-day for love or money.'

"'I can't, can't I?' demanded the traveler, indignantly. 'Do the enemies of personal liberty think that they can deprive me of my just rights?' not on your dreamy eyes! Watch me.'

"He cut for an alley and began trying every back door. He would rap three times on a bluff and say, 'It's me,' but there was nothing doing.

"However, he was not to be thwarted. In the absence of the blind pig and the speak easy, he fell back to the prescription gag. Inquiring his way, he walked eight blocks to a physician's residence, and caught the doc. just as he was starting to church. He gave the doc. the Elk's grip, and begged him to save a life. He said he had cramps, and nothing but a large slug of the Scandinavian joy-producer would relieve his agony. Doc. wrote: "Spirts frumenti—Take as directed.' and said it would come to one dollar.

"Then the sufferer went out to find a drug clerk. After a long search he found Mr. Higginson, of the People's Pharmacy, down at Main Street bridge, pushing a baby carriage. At first the ldruggist balked on opening up, but the traveler said he was a dying man, and handed over a good ten-cent cigar.

"At 2 p.m. he went back to the hotel wearing in his pistol pocket a flask of squirrel whiskey the color of kerosene. He was flushed and happy, for he had made a monkey of the law. He invited two other drummers up to 62. They pulled down the curtains and tapped the poison, and nobody could talk for five minutes.

"Two months later the same traveler struck the town one Sunday, and found a baseball team giving a parade.

"'Everything is wide open since the April election,' said the clerk. 'I can get you whatever you want.'

"'All right,' was the reply, 'send up a pitcher of ice water.'

"Moral—Thirst follows the prohibition clause."

* * *

DIDN'T KNOW WHICH BAR.

"Where's the bar?" asked a dirty looking stranger of a waiter at a hotel the other day.

"What kind of a bar?" asked the latter.

"Why the liquor bar, of course; what do you suppose I mean?"

"Well," drawled the boy, "I didn't know but what you might mean a bar of soap."

* * *

FORGOT THE AUTOMOBILE.

"Why do the poor persist in running down the rich?" shouted the new sociologist. "We never heard of the rich running down the poor."

"Gosh, mister," retorted the thin man in faded trousers, "you ain't never seen an automobile, have you?"

* * *

TOOK IT FOR GRANTED.

"I hear the newsboys yelling 'extra!' on the streets. Shall I buy a paper?"

"No, my nerves are in such a condition to-day I couldn't bear to read about a railroad wreck."

* * *

MIGHT HAVE FARED WORSE.

R. J. Cluff, of Cluff Bros., Toronto, tells of the experience of a young eastern woman in the west. She had left the train at a station in New Ontario one afternoon, and asked the only man in sight how she could get to her destination, far out in the country.

"You'll have to wait for the stage in the morning," said the man; "you can't get any rig here."

"But where am I to stop," inquired the young lady; "there's nothing here but the station, and I can't sleep on the floor."

"Guess you'll have to bunk with the station agent,' he replied.

"Sir," she exclaimed, "I am a lady."

"So's the station agent," said the man.

* * *

DIDN'T WAIT FOR THE HATCHET.

Francis J. Torrance, first vice-president of the Standard Sanitary Manufacturing Company, related a funny story at a banquet in Philadelphia recently. Said he: In a certain town in western Pennsylvania there is a house the door of which must be raised a little to be opened, and for this purpose the hatchet is generally used. One night lately a knock came to the door, and a youngster was sent to see who was there.

"Who is there?" he inquired.

"Me," said a voice outside.

The youngster, knowing the voice, shouted back in such a tone that the person outside could hear him:

"It's Mrs. Murphy. Get the hatchet."

Needless to say, Mrs. Murphy didn't wait.

* * *

COULDN'T FOOL THIS CLERK.

J. Walton Peart, of St. Mary's Ont., who sells bird cages and bird seed got a new clerk from the country not long ago. A customer came in shortly afterward and inquired for bird seed. "No you don't, smarty," replied the new clerk, "ye can't joke me. Birds grow from eggs, not seed."

* * *

HE WAS THE BOY.

A Toronto hardwareman directed one of his clerks to hang out a "Boy Wanted' sign at the street entrance a few days ago. The card had been swinging in the breeze only a few minutes when a red-headed litle lad came in with the sign under his arm.

"Say, mister," he demanded, "did you'se hang out this here "Boy Wanted' sign?"

"I did," replied the contractor sternly. "Why did you tear it down?"

Back of his freckles the youngster was gazing in wonder at the man's stupidity. "Hully gee!" he blurted. "Why I'm the boy!" And he was.

Plumbing and Steamfitting

PARAGRAPHS OF INTEREST

By C. E. Oldacre, Toronto

Plumbers who have interested themselves in the septic tank sewage disposal system have opened new fields and found application for their work not otherwise obtainable.

The operations of such tanks are mervelous and of far-reaching benefits to all communities and in application are innumerable.

They are adapted to residences, hospitals, schools, factories, barracks, villages and towns.

Plumbers who have advocated them have added materially to their business, besides conferring great blessings on the public. Those who have not looked into their benefits will find a valuable means of extending their field of operation in old and newly-settled sections.

Septic tanks meet with the highest endorsement of the medical fraternity, and plumbers will find physicians ardent exponents of their use.

Good Rule to Follow.

In single pipe steam work probably fitters have had occasion to note uneven and varying water lines on steam boilers on some heating installations. This is caused in the majority of cases by too small sized mains; that is, either too small flow or too small return, or perhaps both. Where either one or both are present, the water is liable to disappear from the gauge glass. A fairly safe rule to follow in moderate-sized jobs for the size of steam main flow and return is to multiply the diameter of the pipe by itself and this result by 100 and this will give you the number of feet of radiation that can be carried on the main, and then for the return make it one pipe size larger than one-half (1-2) of the diameter of the flow. Example How many feet of radiation can be placed on a 4-inch main ordinarily? Multiply thus, 4x4x100, equal 1,600; showing 1,600 square feet of radiation as the maximum amount of one-half of four bring two, let the return be not less than 2½ inches.

Reducing Size of Pipe.

It is the best practice in steam work to connect the boiler with the full-sized outlet at the boiler, and if any reduction is required do so by means of a reducing elbow at the top end of the nipple leading from the boiler, instead of using a bushing immediately at the boiler. This will reduce friction and give much better results in many cases

In hot water heating, to give equal circulation it will be found to advantage to take the branches off the side of the main, instead of from the top, especially for runs to upper floors.

Are House Traps Dangerous?

Knowing the great effect of currents of pure, fresh air on health-destroying germs, have plumbers fully determined whether the usual "house-trap" is of value? Many contend they are of no value, and their use should be prohibited by law, as it stops these very currents and at once defeats the object for which they are placed in position. Pure air is one of the greatest enemies of infectious germs.

Wall Radiation Best.

Wall radiators give off the greatest amount of heat for every square foot of heating surface of any radiators used to-day. One hundred square feet of wall radiation is practically as good as one hundred and twenty-five feet of radiation of the usual type. They are, however, not compact, and take up too much wall space, and hence are not capable of adaptation in the majority of cases, but where conditions permit will be found very effective.

In a good many sections of this country, as well as other parts of the world, great difficulty is experienced with the domestic water supply from the stoppage of pipes, water-fronts and connections by incrustations. Many cases of clogged, burnt or poor operating hot water connections and water-fronts have been brought to the attention of Canadian fitters. These are due to the solid matter held in solution by the water, and gives to such water the term or appelation "hard. water" The amount of solid matter held in solution varies from 10 to 300 grains per gallon. These solids consist mostly of sulphite of lime and other salts held by the water in solution and are precipitated in various degrees of heat. The temperature for precipitation varies from 170 degrees to 300 degrees Fahrenheit. Where 60 gallons of water containing only 30 grains solid matter per gallon, are used in a day—not a very large amount—for one week, it means nearly 2 pounds of this solid matter that may be deposited. With this going on for any length of time it takes only a short while for the containing pipes, water-front and connections to become thoroughly clogged. How to avoid these incrustations is a very important question and brings to notice the necessity of using liberal-sized connections. Referring to this question from an economical standpoint, it might be interesting to note that it has been stated that London wastes one-third its tea through the hardness of its water. A thousand gallons of London water contains three and one-half pounds of chalk. Calculating from the total population of London it has been figured out that it looses from this cause alone £6,000,000, or approximately, $30,000,000, per annum in extra tea, soap, fuel and labor.

Good Sized Sewers Needed.

Sewers of liberal dimensions are nessary, not only to meet the demands of future conditions, but also to provide ample space for the sewage to be carried away and abnormal quantities of water from sudden down-pours of rain and sudden melting of snow and ice, and as well, that ample space may be provided for a proper circulation of air at all times throughout their entire course. Air is the great destroyer of microbes causing diseases, and perfect ventilation secures plenty of air. The air in the sewers of Paris—some of the largest in the world—is regularly tested, and the investigations show that no disease germs are contained in the air of these sewers. Likewise, the air of the Berlin (Germany) sewers shows that no such amount of disease germs is carried in the air of the sewers as is found in the air of the streets

Early Heating.

It may be interesting to know that the first hotel in the United States to have steam heat was the Eastern Exchange hotel, of Boston, Mass, installed in 1845.

The first hot water radiators were patented by Thomas T. Tasker, Philadelphia, Pa., on March 9th, 1858. The first modern type of cast-iron radiators were patented by Nathaniel H. Bandy, September 22nd, 1874.

WATERWORKS BY-LAWS.

Wetaskiwin, Alta., ratepayers have voted favorably on the following bylaws: To provide for the raising of the sum of $140,000 for the construction of a municipal system of waterworks and sewers. For the raising of the sum of $30,000, for improving and extending the municipal electric light and power plant. For $2,500 for buying a hospital site, and $10,000 for the erection of a hospital.

PROGRESS OF SANITATION

By Wm. Stansfield, in The Sanitary Journal of January 1950.

The advance of sanitation during the past fifty years has been more than sufficient to justify the most sanguine expectations, and the pioneers of the science, who labored during the first decade of the period, would perhaps be as much astonished as gratified could they look around to-day and realize the vast changes and improvements which have taken place within such a short space of time. From comparative statistics we learn that the average span of life has been lengthened by nearly ten years, 9.87 to be exact; that the weight of the people has been increased by about eleven pounds per individual, and the stature advanced by 1¾ inches. The chest measurement has expanded by at least two inches, while the people enjoy robust health invariably to a good old age. During the early part of the century a whole range of zymotic diseases were prevalent: these have disappeared. Consumption was regarded as a fearful scourge—now almost extinct. Cancer research was in its infancy—to-day an isolated case of cancer is a rarity.

If we turn to the newspaper files of 1906, we find that the popular mind was sorely exercised about physical deterioration and decline of the birth rate. Public meetings were held, when a number of little Daniels came to judgment gravely propounding remedies, which, when taken in the light of subsequent events, serve only to provoke a smile.

In a newspaper dated 1899, we read that "out of the 11,000 recruits who offered to enlist in Manchester for the South African war, 8,000 had to be rejected at once on account of physical defects, and of the remaining 3,000, only 1,000 could be passed into the army." This is hard to realize when we look around and reflect that to-day was could scarcely find a man unfit for the service.

And how have these changes, so vast, and so far reaching, been brought about?

So far back as 1900 it was realized that the rising generation of females, especially among the leisured classes, were improving in stature and erectness of carriage, while the men remained stationery; in three more decades the improvement was general and all round. This effect was partly due to the development brought about by the drill exercise in schools—while the vagaries incidental to this improvement need not now concern us, we merely note the result.

About 1920, the last prosecution for smoke nuisance took place, the change being brought about by the storage and transmission of electric force, while improved house construction and artificial heating have almost superseded the domestic fires, the result being that we have now very few fogs, and these not of a dangerous character.

Next came the systematic regulation of the distributive industries and the elimination of competitive waste, with the result that food adulteration disappeared, and the people were enabled to buy commodities at two-thirds of the previous prices. The last prosecution under this head occurred during the thirties.

Coincidently with these changes it had been noted that we appeared to be the dwellings, the people lived in a reeking stench, unavoidably inhaling filth and infection; indeed, the wonder is that they survived at all. This condition of things was operating like the vulture at the nation's vitals, undermining and sapping the vitality of the people, and it is astonishing that while governing bodies paid attention to minor details, this overwhelming cause of sickness and debility should have remained so long in force.

Not that they could plead ignorance, for medical officers were emphatic in denunciation. The explanation seems to be, that along that period the press was in the hands of the capitalists, the capitalists were the property owners—they dreaded a change because of the expense, and hoodwinked the people with the threat of higher rates, although years before it had been ably demonstrated by a sanitary inspector, called ———— (here the name is blurred, but never mind), that water carriage was the cheaper method.

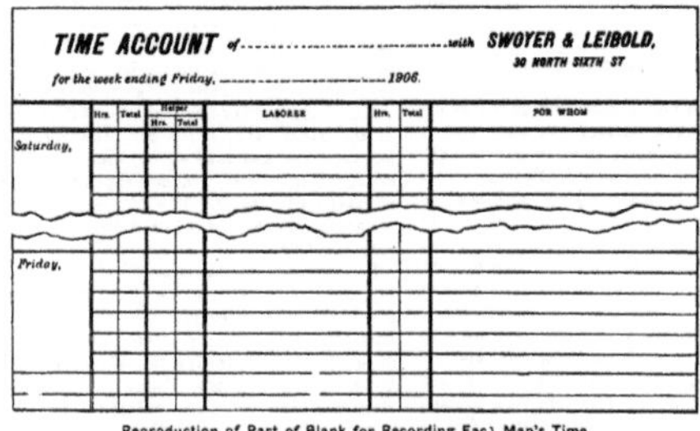

TIME ACCOUNT of .... with **SWOYER & LEIBOLD,**
30 NORTH SIXTH ST
for the week ending Friday, ... 1906.

	Hrs.	Total	Helper Hrs.	Helper Total	LABORER	Hrs.	Total	FOR WHOM
Saturday,								
Friday,								

Reproduction of Part of Blank for Recording Each Man's Time.

breeding two distinct races of children; those in water closet towns being healthy, robust and free from infantile sickness, whilst in those centres where conservancy methods were still retained the juniors showed unmistakable signs of feebleness and decrepitude.

To the present generation it might be necessary to explain the term conservancy. Some thirty years ago it was the custom in some towns to store the human excrement in pails or tubs, and sometimes in pits, called middens. The former were supposed to be emptied weekly, and the latter at less frequent intervals, and as in every case these receptacles were housed at the rear of

At last the state took the matter in hand and dealt with the remaining plague spots in a summary manner, and from that time forward we note a gradual, but marked, improvement in the health of the nation.

All this is now ancient history.

THE PLUMBER AS A BUSINESS MAN.

No more important work confronts the business concern than that of keeping track of the actual cost of doing work cannot be intelligently set. Keeping the cost of doing business means business or of doing any piece of work. Without being informed on both of these costs the price on any piece of keeping a record of everything for which an expenditure is made, including that which does not bring in a direct profit. Such items include insurance, taxes, water rent, horse feed, shoeing the horse, the office boy and a host of other items which will never be included in the time and materials used in putting a heating system in one building or the plumbing system in another. Whether a man keeps his own books or employs a bookkeeper, it is indispensable that a correct account be kept of all these items. Then, if in a year they amount to $1,000 or $10,000 worth of business is being done, it is evident that 10 per cent. must be added to the time and material which enter into the base cost of any job, before the true cost is found and the profit added to form the price to a customer or the bid for contract work. It is equally essential that some system be devised so that whether the workman is doing jobbing or engaged on a piece of contract work his time and the material he uses will be reported to the office. Some benefit may be derived by our readers from an explanation of part of a system used by Swoyer & Leibold, Allentown, Pa., as outlined in the Metal Worker.

SWOYER & LEIBOLD,
Plumbers, Gas and Steam Fitters,
30 N. SIXTH STREET,
ALLENTOWN, PA.

Order № 1755 Date

For whom

Where

Name of Workman

" " Helper

" " Laborer

Description of Work

SWOYER & LEIBOLD,
Plumbers, Gas and Steam Fitters
30 N. SIXTH STREET,
ALLENTOWN, PA.

Order № 1755 Date

Mr.

No. Street

Description of Work

Date		Name	Time

Charge Material

Reproduction of Blank Used for Recording Time and Material.

One blank, it is very evident, is intended to be used in a book with the page perforated, so that a record is kept in the book. From this record the bookkeeper may demand from the workman the other portion within a reasonable time, so that a proper charge can be made. It is arranged to have the date, the name of the customer and where the work is to be done, with space for a brief description of the character of the work at the bottom and for the names of the workmen and help sent to do it. The other portion of the blank provides for the workman a duplicate of the necessary information, with space for him to keep a record of his time and for an enumeration of the materials used or the material returned. A shop which has never had such a system will find some difficulty in putting it into operation when the first effort is made, but as the effort is persisted in the proprietor will soon discover that without its use many items of expense never find their way into the charge as they should, but by their omission effect substantial inroads in the profits of the business.

The second blank will be found invaluable in discovering where the time paid for has been spent. Every conscientious workman is willing to lend his assistance in this work, both for the protection to himself and his employers and customers. We have reproduced this blank only for two days of the week, but the full size blank is 9½ x 12 in., and provides space for every day of the week, with spaces at the bottom for making totals.

Exhibition of a Modern Kitchen and Bath Room at a Pure Food Show at Lewiston, Me.

PAY ATTENTION TO DISPLAY.

Plumbers' establishments, as the majority of them appear at present, can hardly be called business places or stores. More properly speaking they are shops. The word "shop" does not convey to us the idea of size or attractiveness. A large number of plumbers' establishments have fallen into a disreputable condition. No thought is now taken by them for cleanliness, attractive displays, or convenient arrangements. The plumbing business would be put on a stronger and more lucrative basis if more attention were paid to the details and the things of apparently lesser importance.

Plumbers could very easily make their stores (for we should be able to call them by that name) more attractive by arranging a display of their goods; either in rows and on shelves, in the front part of their shops, or by partitioning off a little apartment to represent a bathroom and array as many plumbing goods as possible in this little room. The accompanying illustrations, kindly loaned us by the Domestic Engineering of Chicago, show a very neat and attractive arrangement of goods for a modern kitchen and bathroom. It would involve very little work, and the efforts made would be well repaid by an increase in business and visitors to the store. Plumbers cannot ignore the presence of strong competition, and the fact that the only way to attract customers is to have an attractive display of goods.

PRIVATE 'PHONES.

The hardware dealer in a small country town, where the long distance telephone in the post office or largest hotel is the only one in town, could do something in the line of private 'phones.

There are many prosperous farmers in the surrounding country who would be only too glad to chip in with their nearest neighbor, and install a private wire between their houses. The cost for two instruments, together with wire and batteries enough to work them, is not great, and after the system is once installed the expense, with the exception of renewing the batteries once in a long while, ceases.

There's no reason, that in a town where there is no electrical contractor, the hardware dealer should not make a lot of money in this way. Write any of the electrical supply men whose advertisements appear in this paper for prices and information, then get after the farmers and interest them in the scheme.

Installing such a system is very simple, if you follow carefully the rules sent with the outfit.

You not only make your profit on the telephones, but also what you charge for putting them up.

Sit right down now and write the supply man. If you put it off, you'll very likely forget all about it. Might as well inquire about electric bell outfits while you're at it.

ELECTRIC FLASHLIGHTS.

These articles have passed the novelty stage and have on their merits established themselves as live sellers, for which there is a growing demand.

The hardwareman who will devote a small space in his store to a good variety of them could coax many an elusive dollar out of the main line past his door and sidetrack it in his own till. The electric flashlight is in great demand at summer resorts, where there are times that one would be extremely handy, especially when Mother Moon is on strike.

The writer remembers one particularly dark night at a popular lake shore resort not far from Montreal, when it was impossible to see your hand in front of your face, yet alone pick your way over a very questionable sidewalk. As there was a dance on that evening, many people who would otherwise have been contented with their own front verandahs, found themselves groping along, hanging onto a fence, which in many cases happened to be barbed wire.

One youngster who had a small one dollar flashlight sold it to a man who had to make the late train to town that night for three dollars.

If the hardware dealer near that resort had been an up-to-date man he could have done a pretty nice little business in flashlights.

This is only one instance in which these little articles are useful, but it's worth your while to think the matter over.

A stock will take up very little room and make a good display. They sell from one dollar up.

Attractive Window Display made by a Plumbing Firm at Lewiston, Me.

DISPLAYING LIGHTING FIXTURES

A very attractive display of lighting fixtures is made in J. H. Ashdown's store at Winnipeg. The lighting fixtures department is divided up into little rooms for the purposes of display, the switches being so arranged that the lamps may be shown individually or all together. It is one of the finest displays of fixtures in the west and can be copied to advantage by hardwaremen who go after the electric fixture trade—even if they do not have to carry so elaborate a stock.

Method of Displaying Electric Fixtures in the J. H. Ashdown Hardware Company's Store at Winnipeg.

ELECTRICITY IN THE HOME.

In a $3,000 or $4,000 house, it is a very common matter to wire for electric lights; but by giving the subject a little further attention with your electrical contractor and the architect, it is possible to arrange for socket heating devices at very little extra expense. Such receptacles would be taken off from the lighting wires in a very simple manner. The cost of operating later at lighting rates, would not be large, for the reason that the electric devices which will be used on such a circuit consume only a small amount of electricity and are generally used for only short intervals.

THE POWER OF TO-DAY.

The fact that electricity is fast supplementing steam as motive power is becoming more clearly demonstrated day by day. That electricity is immensely superior to steam for almost all kinds of work is becoming known throughout the length and breadth of this broad land. Electricity in manufacturing plants does not only do away with an immense amount of dirt, but economizes greatly in the way of space. Electric machinery is more compact, and the same horse power can be obtained from a small electric motor, set up on a shelf out of the way, that with a steam plant would necessitate a great engine room with unsightly boilers.

It will take time, no doubt, before our railways adopt electricity in place of steam for the motive power in their locomotives, but that this will be the case nobody seems to deny. The greatest trouble will be the tremendous cost that the change will involve.

Practical tests of electrical apparatus in the leading industrial centres, factories and mines, show that electricity not only does more work, but better work, and with less trouble and expense than is possible with steam.

A GOOD SIDE LINE.

Hardware dealers will do well to keep abreast of the times by getting in a stock of all electrical appliances and get in samples to display and keep a hold on the trade in electrical appliances, as electrically heated ovens, flat-irons, and electric heaters.

The accompanying cut shows that in the summer time the ironing need not be done in the hot, oppressive kitchen: the ironing table may readily be moved to the back porch, or even to a shady spot under the trees nearby. Under such conditions the work will be more satisfactory, both to the laundress and to her mistress.

The mistress of the house sometimes has a little pressing which she can conveniently do herself, or perhaps a dainty bit of lace or embroidery to be ironed, which she does not care to entrust to other hands. In such a case it is not necessary for her to go into the kitchen or laundry. The work may be done in the sewing room, or in her own chamber.

The quality of work done by the electric flat-iron is said to be superior to that done by the ordinary iron. There is no danger of getting soot or smut on the iron, and it does not have to be waxed. There is no difficulty in keeping the iron hot. The heat is easily controlled, and may be regulated to a nicety, insuring first-class work done There is no reason why this electric flat-iron, as well as other simple electrical appliances, should not prove a great success, and hardware dealers will do well to get in this line of goods and endeavor to hold the trade.

ELECTRICITY'S REMARKABLE USES.

In the modern New York house electricity has become domesticated and transformed into an efficient servant who never "talks back." On the electric stove in the kitchen pots begin to simmer and bubble without apparent cause as in the enchanted castle of old. The invisible current sets an electric coffee machine whirring in unison with a meat chopper and both have finished in a jiffy tasks which would have cost the housewife of former days many minutes and much effort. With the perfection of electrical refrigeration carried on in one's own house ice famine and importunate icemen will be but a dream of the past In the laundry, wash tubs, driers, flat-irons, and water heaters, all operated by electrical currents, almost force one to believe in fairies

For the bedroom an electrical comfort is, perhaps, the latest, as it has been introduced within a short time. In appearance this differs not at all from the ordinary bed covering, but beneath the familiar surface is a thickness of asbestos containing a mesh of wire, which, when carrying current provided by a neatly concealed feed wire, heats this electrical cover and makes it efficient in the coldest weather. A pad about two feet square, heated in the same manner, may be used for "cold feet" in place of the troublesome water bag One of these pads is put to an unusual use by a naval officer. On bleak

Ironing by Electricity.

nights while pacing the bridge of his ship he has one neatly tucked beneath his coat, so that the zero winds hold no terrors for him. Because of the fact that this heating pad must be connected to the ship's electrical supply by a wire his brother officers have dubbed him "Monkey on a String"

Other appurtenances of the "electric life" include a clock whose face may be illuminated at night by pressing a button at the head of the bed, a milk warmer which prepares milk so rapidly that a protesting baby is appeased long before its temper is dangerously aroused, a shaving cup, hair drier, and a movable electric light which may be carried about as were candles in bygone days

A house to be built in Schenectady to replace the famous "house without a chimney"—so-called because electrical appliances were used for all heating and cooking purposes—will not only contain the devices enumerated above, but its doors will open and shut mechanically, carpets will be swept, and the lawn mowed by electricity. A burglar breaking into this house will start a general illumination and such a jangling of bells as to cause him to decide that in his case the electric life is not the simple life after all.

ECONOMICAL LIGHTING.

Experiments were recently made in Munich to determine the cost and relative advantages of indirect illumination for school and draughting rooms. The arc lamp was found to give the best results, but intensifying gas burners more economical. This method of lighting has been strongly recommended to the school authorities.

PLUMBING INVENTIONS.

J. A. Frey, Washington, D.C., has secured a patent on an automatic steam and hot water safety cock for use as an attachment to domestic water heaters and steam boilers or generators for relieving pressure of steam when it exceeds the limit of safety. It is adapted for use in the usual way for discharge of water from a heater or boiler to come into action only when the usual turning spigot is closed or adjusted to cut off discharge.

SITUATION AT MORRISBURG.

The article published in Hardware and Metal of March 9, regarding the delay of the Canada Tin Plate and Sheet Steel Co., Morrisburg, in commencing operations, has been contradicted and although the information was received from a supposedly reliable source, it seems to have been not in accordance with the facts.

The delay, it appears, has been entirely beyond the company's control, being due to the failure of the manufacturers of the large driving wheels ordered by the company last June, to deliver them, they having miscalculated the time it would take to turn out the wheels, which are the largest ever cast in Canada, weighing 75 tons and being 30 feet in diameter A statement giving the present situation at Morrisburg will be published in our next issue.

W. R. Hobbs, a prominent hardware merchant of Tillsonburg, died suddenly last week.

W. Illsey, of the Ashdown Hardware Co., Winnipeg, won the $50 cash prize offered by the Asbestos Sad Iron Co. for the best trimmed window with their wares.

Hardware Manufacturing in Canada

New Plants Being Established in all Parts of the Country—Few Lines Not Now Made in Canada.

Half a century ago practically every article sold in Canadian hardware stores was imported from Great Britain, the lines of goods produced in Canada being exceedingly limited and the manufacture of hardware for export having not yet reached large proportions in the United States. Gradually, however, conditions changed, the civil war in the United States and the settling of the Western States aided American manufacturers in developing their industries, while the Confederation of the Canadian Provinces, followed by the construction of the transcontinental railway and the adoption of the National Policy by the Canadian Government, helped to make possible the establishment of numerous hardware manufacturing plants in Ontario, Quebec and the Maritime Provinces.

To-day the situation is that hundreds of plants are in operation in Canada making goods for sale by retail hardwaremen and each year sees many new factories start their wheels in motion as well as many enlargements made to existing concerns. It would be easier now to name the articles not yet produced in Canadian factories than to enumerate those which are manufactured.

A Bright Future.

The development which has been going on for the past half dozen years seems certain to continue for the next quarter of a century, the work of pushing the frontier line of civilization northward and of irrigating the prairie wastes in the west being impeded but slightly by industrial depressions in the United States. Canada's undeveloped lands are the safety valve for present day immigration, and the valve will stay open until the country is filled up. Without attempting to describe every new plant established or to outline the increases which have been made to existing plants, casual reference will be made to a few of the changes made in order to give an idea of the expansion of the hardware manufacturing industry in the older portions of Canada.

Following their well-known aim of supplying "Made-in-Canada" goods wherever possible the Canadian Fairbanks Company have, during the past year, established new factories in both Ontario and Quebec. Jenkins Bros., who have been supplying the Canadian market from their factory at Boston, have found it necessary to manufacture in this country and have extensive works nearing completion at Montreal.

In the same line the Kerr Engine Works, Walkerville; the Penberthy Injector Company, Windsor; the James Morrison Brass Manufacturing Company, Toronto; and the Monarch Brass Works, Port Colborne, have all been extending their operations. The Mann Brass Works have been established at London; two new brass factories are being established at Galt, and Somerville, Limited, are to erect a large factory at Toronto.

Expansion in Heating Trade.

The McClary Manufacturing Company, London, have erected a magnificent new plant at London, and new stove foundries have been established at Welland, Niagara Falls, Ottawa, Morrisburg, and other places. The Acme Can Company, Montreal, have enlarged their plant to a capacity of 150,000 cans per day. The Ontario Steelware, Limited, have entered the field with a full line of kitchen enamelware and it is understood that the Canada Tin Plate Company's plant, now being erected at Morrisburg, will be partially utilized in manufacturing tin and enamelware. E. T. Wright & Co., Hamilton, have added many new lines and are enlarging their plant by one-third.

The Taylor-Forbes Company, Guelph, have made large additions to their plant, particularly in their radiator foundry. The American Radiator Company's plant, at Brantford is now in operation; the Warden King and Son foundry at Montreal has been doubled in size and plans are prepared for their new radiator works at Toronto; the Dominion Radiator Company is moving into a new $250,000 plant; Cluff Bros. and the Pease Foundry Company, have erected fine showrooms in Toronto; the Star Iron Company, Montreal, is enlarging its plant and the Canada Radiator Company, Lachine, have taken over the lines of the Ottawa Furnace Company.

The Montreal Rolling Mills have enlarged their plant and are to instal much new machinery. The Belleville Rolling Mills have been re-opened and greatly enlarged with new furnaces and machinery for making nuts, bolts and horseshoes. New iron industries are also being established at Welland and Fort William with the works at Hamilton being more than doubled in size.

In the lighting field the Montreal companies have all been expanding their business. The R. E. T. Pringle, Robert Mitchell, and Munderloh Companies, have all increased the size of their plant or have made plans to do so. The International Gas Appliance Company, Toronto, have occupied a magnificent new building and the S. F. Bowser Company have established a fine new factory at Toronto.

Several New Paint Factories.

In the paint and oil field there has been an even greater growth, the new Brandram-Henderson white lead works at Montreal being possibly the greatest expansion of the year. The Sherwin-Williams Company have erected a new linseed oil mill at Montreal and are planning new factories at Winnipeg; the Canada Paint Company have taken over the Ontario Lead & Wire Companys' lead business and are otherwise extending their capacity; the Francis-Frost Company have been merged into the big Benjamin-Moore Company, of New York and Chicago, and have erected a magnificent new plant at Toronto Junction; the Blanchite Process Paint Company have established a large business at Toronto; the Canadian Oil Company are enlarging their recently burned works; the Standard Paint Company have erected large works near Montreal, and the Standard Paint and Varnish Works, Windsor; the Imperial Varnish and Color Co., Toronto; A. Ramsay & Son Company, Montreal; the Martin-Senour Company, Montreal; Berry Bros., Walkerville; and P. D. Dods & Co., Montreal, all report increases in their factory output.

Hardware Manufacturing on Increase.

In general hardware a similar development has taken place, a new factory at Windsor being erected by the Lufkin Rule Company; the monster plant of the Plymouth Cordage Company, at Welland; the opening of the McGlashan-Clarke cutlery factory at Niagara Falls; the enlargement of the Canadian Steel Goods Company plant at Hamilton; the doubling in size of the combined Canada Screw and Ontario Tack Companies' factories at Hamilton; the commencement of operations of the Ross Rifle Company, at Quebec; the extensions undertaken by the Sanderson, Harold Co., at Paris; and Ham and Nott at Brantford; the establishment of a branch factory at Montreal; and the erection of a new shop at Preston by the Metal Shingle and Siding Company; the enlargement of the business of the J. H. Still Manufacturing Company, at St. Thomas,' the erection of a new factory by the Brantford Screw Company; and the establishment of a glue making plant at the same place; the enlargement being made by the Dowswell Manufacturing Company, Hamilton; the installation of much new machinery by the Capewell Horse Nail Co., Toronto; the Peterborough Lock Manufacturing Company, Peterborough; Canada Cycle & Motor Company, Toronto Junction; Shirreff Manufacturing Company, Brockville; Cumming Manufacturing Company, Renfrew; Greening Wire Manufacturing Company, Hamilton, and many other concerns indicate the rapid growth of this branch of trade.

We could go on indefinitely describing the developments being made, but the few examples quoted above give some idea of the work that is continually going on. It is safe to say that half the population have no idea of the place Canada is making among the nations of the world. The discontented man who thinks he can do better in the States, is sadly mistaken. Canada at present offers a better chance of success to mankind, young or old, than any land on earth.

MANUFACTURING POSSIBILITIES OF THE WEST

Hardware manufacturing in western Canada is still in its infancy, the investment and output being still very small. Lack of coal and distance from supplies of raw material have combined to retard the growth of this branch of manufacturing, and, although it is, perhaps, foolish to prophesy, it does not seem likely that hardware manufacturing will ever attain to any great importance in the prairie provinces. At Fort William and Port Arthur it seems probable that big industries will be established, and it is not unlikely that British Columbia will soon develop some large concerns engaged in the manufacture of hardware and allied lines; but at present this line of manufacturing is very little developed.

Western branches of eastern enamelware and tinware factories manufacture some of their lines in Winnipeg for sale in the west. The foreign population' supplies unskilled labor, and in some lines of manufacture this can be utilized to advantage. During the last year the Kemp Mfg. and Metal Company have doubled their manufacturing capacity at Winnipeg.

During the last year the Canada Paint Co. have acquired the linseed oil mills in Winnipeg, have added to the plant, and are now manufacturing linseed oil and paints in Winnipeg. The Manitoba Linseed Oil Mills, Limited, has recently been incorporated with a capital of $200,000 to manufacture linseed oil. It is expected that the plant will be in operation to handle the 1907 crop. G. F. Stephens & Co. have enlarged their plant in Winnipeg during the last year and their manufacturing capacity has been greatly increased.

A rolling mill has been built in Winnipeg to make bar iron from the vast supplies of scrap iron now going to waste in the west for lack of an available market, and prospects for its success are considered bright.

Vast Metal Resources of B. C.

Natural conditions favor the development of industries in the line of metal manufactures in British Columbia. Her geographical situation with reference to the rest of the Dominion would also tend to promote such development. Present economic conditions, as relating to manufacturing in the eastern portions of Canada, whence come a large part of the province's supplies, would also further incite interest in the possibilities for manufacturing locally.

A mild climate, rendering the shut-downs forced by more severe weather in other parts, unnecessary to large extent in B.C., would favor the operating of many different industries, especially in which metals are largely used. The existence of natural resources from which may be taken the raw material for manufacture; the possibility of bringing from all the markets of the world raw materials not obtainable locally, and bringing them cheaply by water; the existence of large industries, such as the lumber, canning, fishing and mining industries, demanding large supplies of hardware and metal manufactures; these are natural conditions which favor local production of many articles.

Geographically, British Columbia stands at the western gate of the great Dominion, and is, therefore, furthest away from eastern source of supply. That, and the fact that the territories adjoining her can easily be served from here, should be an immense factor in the production of metal manufactures.

Eastern manufacturers have not used British Columbia at all well; not altogether from neglect, but from inability to handle the enormous amount of business offering in the past few years. Manufacturing in the east has been unable to keep stride with demand, and the demand nearest home has most naturally been given better attention. This should have given greater impetus to local manufactures than it has so far done.

One of the most important considerations acting against the development of local metal manufactures is that these metals are not produced in finished state from the ores being smelted and refined in the province. The only exception to this statement is the lead refining at Trail, the blister copper produced at some of the smelters, and the zinc refining plant at Nelson. Even here the refining is but in its infancy. The greatest lack is, however, in the entire absence of even the mining of the vast ore deposits of Vancouver Island. Iron is still master of the metal world. Vancouver Island on its northern and north-western shores has deposits vast enough to build up an iron and steel industry the equal of the greatest in the world. The coal deposits of the eastern side of the island furnish the necessary complement, and these have been exploited for years, but not yet for the reduction of iron.

Varied Products of Coast Cities.

Notwithstanding all handicaps, there is yet a great deal of manufacturing carried on in the province. The canning of salmon has been responsible for two or three can-making plants and tin plate is a heavy import item. The lumber industry has brought with it the establishment of a saw works, which is thriving and growing with the parent industry. In the city of Victoria a very large plant is operated in manufacture of mixed paints. There is also a large chemical and fertilizer plant there, and the manufacture of excellent stoves and ranges is carried on by the only stove-making plant in the province. The Portland cement plant of the Victoria Portland Cement Co., close to the city of Victoria, is a very large establishment producing an excellent quality of cement.

On the mainland are a number of manufactories tributary to the hardware and allied trades. At Vancouver a can-making factory, and at New Westminster another, supply large quantities of tins for many salmon and other canneries and for syrup and other factories. A glass works is being equipped at New Westminster for the manufacture of bottles in large quantities. Sheet glass will not, as yet, be attempted.

A wire nail works has been located at Vancouver for some years, but, unfortunately, it was burned two years ago, and, though re-established, has suffered from the reverse. Tent and sail factories are prosperous, as both lines are much in demand. Tarred building paper and felt roofing is manufactured by one plant in Vancouver.

One foundry for the making of cast iron pipe, and a small brass foundry, are numbered among the industries supplying the trade. Wire springs and mattresses are largely manufactured locally, one factory in particular turning out large quantities.

The making of wood stave water pipe has been established as an extensive and growing industry in the past three years. The magnificent British Columbia, or Douglas fir, is used, and heavy gauge galvanized steel wire is used for wrapping the sections of pipe. There are now three of these factories competing for the business offering in this line in the west.

Vitrified tile, fire brick, and similar products of the pottery are manufactured by two large plants, one at Victoria and the other at Clayburn, some miles up the Fraser river above New Westminster.

Metal Production Increasing.

The expansion of lode mining in British Columbia is the greatest industrial development of the immediate future. The present progress being made is satisfactory, and the recovering of metal values from the ores is being carried on side by side with the mining. Smelters in the various mining districts run full time and are constantly enlarging their plants. Coal for coking purposes is available in many parts of the province, chiefly the Crow's Nest Pass, the Similkameen, as yet in stage of development, and on Vancouver Island. Electric power, generated by water, is largely applied in the Kootenay and Boundary districts for mining and the operation of smelters.

With the very great increase in metal production which is bound to come and the extension of the refining of these metals within the province, the establishment of many lines of manufacture in which the metals are used may be fairly expected. No use is yet made of copper in any line of manufacture. Lead is only used to manufacture some lead pipe, and that in small way. Silver and zinc, which are both being produced, are not yet used in local manufactures. As stated previously, no iron industry yet exists in which the natural resources of the province are drawn on. Every ton of pig iron used in the foundries of the province is imported. There is no rolling mill, all wrought iron being imported, largely from England. With conditions such as described, it must be the conclusion that following the continued expansion of mining will come the investment of capital in metal manufactures, using the products of the country instead of, as at present, importing even the pig iron, rolled iron, sheet and plate iron and steel, the sheet lead, and largely the lead pipe now consumed. Then the manufacture of tools, of iron, copper and brass work, must engage the attention of those who study the situation with a view to investment of capital in industrial lines. Recently a movement was set on foot for the establishment of a car works, and it is anticipated that the company will be formed. Here, again, the lack of local production of iron is felt.

INDUSTRIAL DEVELOPMENT IN NOVA SCOTIA

By James Hickey, Halifax.

The outlook for a busy season in the hardware and metal lines in this province during the coming season was never brighter. Extensive building operations are planned, and many large manufacturing firms contemplate extending their plants. In Halifax things never looked better for a prosperous season. The erection of the Silliker Car Works, the new roundhouse, which the Government is erecting at a cost of $300,000, and the new machine shops to cost $150,000, will boom things in the building lines. Besides the above large contracts there is a lot of other build-contracts, there are a lot of other buildings proposed, and the possibility of a big electrical power plant to operate an electric railway from Halifax to Bedford is hinted at, although no steps so far have been taken to secure a site. The company has been organized and it is stated propose to start operations this year.

From North Sydney comes a report that the ship chandlers anticipate an exceedingly busy season this year. The enforcement of the Newfoundland Bait Act will very materially help North Sydney. Both the Gloucester and French fishing fleets will be forced to depend largely on North Sydney for their supplies, and the big bait freezer, which has been erected there, will draw the fishermen. Both the American and French fleets are good customers and the supplies which they find necessary to purchase during a season's operations amount to a large sum, and with that port made the headquarters of both fleets, business generally will be greatly benefited.

The Malleable Iron Company of Amherst, N.S., is doing a rushing business and a ready market for their product is found. As fast as the company can turn out the malleable castings are shipped off in carloads, Montreal being a particularly good customer. The company now finds that their present plant is altogether too small for their increasing business, and as soon as the weather permits, they will commence the erection of a new structure. The proposed new building will be 280 feet long.

An Important Merger.

The forward movement in the matter of building up manufacturing establishments has struck the town of Dartmouth, N.S., and there promises to be a boom in the ambitious town as a result of the shareholders of the Starr Manufacturing Company unanimously deciding to amalgamate with the Dartmouth Rolling Mills Company. The decision to unite the two big companies means much for Dartmouth as it is the intention to enter upon new lines of manufacture. The details of the amalgamation have not yet been announced, but sufficient is known to state that the united concern will be operated with a view to future expansion and equipment. The shareholders have decided to increase the authorized capital stock of the Starr Company to $1,000,000. At present the capital is about $100,000, and the capital of the Rolling Mills is about the same. Five hundred thousand dollars' worth of new stock will be placed on the market at once, and the remainder of the new stock retained in the treasury. The directorate of the company has been increased from five to seven. Both companies at the present time have orders booked for months ahead, and are working full time. The rolling mills employ over eighty men, while the Starr works have over one hundred on their pay roll.

Markets and Market Notes

THE WEEK'S MARKETS IN BRIEF.

MONTREAL.

WIRE NAILS—Advanced 5c. per keg.
WIRE — Base price advanced about 5c. per 100 lbs.
TURPENTINE—Advanced 2c. per gal.

TORONTO.

COPPER TACKS AND NAILS — 12½ per cent. advance.
WIRE NAILS—5c. a keg advance.
WIRE—5c. per 100 lbs. advance.

BRIGHT MARKET OUTLOOK.

Business prospects were never brighter with the western hardware trade than they are now at the opening of the spring of 1907. This is the case in spite of an unusually severe winter, with the consequent tie-up of the railways and blockade of traffic of all kinds. It is the case because the prosperity of Western Canada rests upon so sound and substantial a basis that it is not to be seriously affected by temporary drawbacks of the kind indicated and because the programme of railway construction alone is so big as to ensure a busy season for all classes of trade. At the present time there are almost 6,000 miles of railways under contract between the Great Lakes and the Rocky Mountains, the Canadian Northern having 1,500 miles under contract, the Canadian Pacific 1,400 miles, the Great Northern 1,000 miles and the Grand Trunk Pacific 1,900 miles. What this means to the hardware trade and to every other branch of trade in the West can be understood and appreciated to the full only by those who are acquainted with western conditions and have some knowledge of the rapidity with which new towns spring up in the districts through which new lines of railways are laid. In the east the railway follows the villages and towns and is the connecting link between towns and villages already established; but in the West the railway is the pioneer and the new towns soon follow the introduction of the railways. Witness the number of new towns along the main line of the Canadian Northern between Winnipeg and Edmonton. More than one hundred new towns will be founded during 1907 along the line of the Grand Trunk Pacific. This announcement has been made by the officials of the new transcontinental upon which work has been carried on in certain districts during the whole of the winter. Towns have been projected along some 790 miles of the G. T. P. line and it will be the policy of the new road to keep the towns about seven miles apart.

This development means great things for all branches of trade in the West and it accounts for the cheery optimism of the wholesale interests in spite of the hard winter, from the effects of which business has not yet quite recovered.

It means new supplies of all kinds, new building supplies, new builders' hardware. It means new accounts with new merchants, a new territory to be exploited by the wholesale and manufacturing interests. Small wonder that the wholesalers are optimistic; small wonder that they are building new and large warehouses.

In all lines of hardware the market is firm and the last year has seen advances in practically all lines.

Toronto Metal Markets

Office of HARDWARE AND METAL,
10 Front Street East,
Toronto, March 22, 1907

Basic of all markets are the metals and the position of these to-day is one of exceptional strength. During the week no changes have taken place in quotations and as the close of the winter season approaches stocks are rapidly being narrowed down to the minimum, the shortage in black and galvanized sheets in finished materials and in copper ingots being particularly marked. Much trouble is still experienced with the railways, jobbers being put to endless trouble and expense in following up delayed shipments. The railways are being aided, however, by the open weather.

During the past year there has been a tremendous consumption of ingot metals, with very marked advances noticeable on every one. On finished materials the advance is not so large as, while the cost of the raw material has been much higher, the greatly increased output has enabled manufacturers to produce cheaper, the savings in this direction going to keep down the price of the finished article.

The change in the tariff has had little effect on the market, as while duties have been adjusted to keep out all American metals and bring in the British articles on a free basis, this has been counteracted by the higher freight rates necessary to the inland points and the general advance in prices.

A few weeks ago there was much talk of a break in the American iron market. As the year opened there was talk of railroads stopping buying, contracts being cancelled and shipments postponed. Pig iron was to drop heavily, according to the reports, but the market withstood the efforts to break and every week that has gone by has added to the strength of the situation. The iron business is trading on its prospects for the balance of this year, but after that little can be said with confidence. It is safe to say, however, that the railroads, which are the greatest buying factors, cannot stop purchasing, as it will take them years to catch up with the demand for rolling stock required by the wonderful expansion of trade and pioneer work being done in opening up new territory.

A year ago Hardware and Metal predicted that for iron, copper and lead, "higher prices will have to be paid later on," while for tin we said that "present prices will be well maintained." That our view of the situation at that time was very accurate can be seen by the figures we give in the table which follows. It will be seen that there has been a steady advance all along the line, the comparison of the figures ruling to-day, with those of two years ago showing the remarkable jumps which have been made and the reason for the increases in the hardware, paint and plumbing markets. The figures for the past three years for the current week are as follows :

	1905	1906	1907
Summerlee pig iron	$20 00	$25.00	$25 50
Middlesboro do .	21.00	21 50	23 50
Londonderry do.	16.50	21.50	24.00
Bar iron	1 80	2 10	2.30
Ingot tin	.32	.40	.45
Bright coke plates	3 50	3.75	4.25
Ingot copper	16 50	20.00	27 00
Sheet copper .. .	21 00	25 00	35 00
Brass sheets .	23.00	23.00	30.00
Imported pig lead	3.50	4.50	5.60
Zinc spelter	7.00	7.25	7.50
Antimony	9.50	18 50	27.00
Black sheets .	2.70	2.90	[illegible]
Canada plates ...	2 50	2.65	2.90
Galvanized sheets ..	3.90	4 00	4 50
Boiler tubes	8.50	8.70	9 50

With the market at its present high position it requires nerve to predict anything but a decline. We can see no reason, however, to look for a slump and while realizing that some metals, iron for instance, can hardly be expected to maintain its present position for the year 1908, we feel that the various ingot metals will hold firm and there will be not much difference in the prices quoted to-day than those which will rule a year hence. Copper, which has been the great feature of the markets of the past year is very scarce and in steadily increasing demand for electrical purposes. Lead is also scarce, with Germany and Great Britain buying from

Canada and other markets. The tendency of tin is still upward, and antimony is scarce, with the demand on the increase. It is a sign of the times to note that consumers do not raise any serious objection to the advanced prices they have to pay, nor does it appear to be more difficult to effect sales in consequence Possibly this is due to the fact that the boom in the metal trade has been widely published in the trade press, so that almost everyone knows that metals have gone up, both in this country and in foreign markets.

The volume of business being done just now warrants the prediction that 1907 will be a monumental year with the greatest total on record.

Montreal Metal Markets

Office of HARDWARE AND METAL,
232 McGill Street,
Montreal, March 22 1907

When will the end come? Everybody seems to be asking the same question In looking over the metal market reports for the past year, it's quite a surprise to see the actual advances that have taken place It's only natural that this should be so, as the advances have been gradual, and not so noticeable as when we sum them in a bunch. Last year about this time we were quoting copper about 6 cents per pound less than we are at present. This is enough to make anybody who is interested in this metal sit up and take notice, as six cents on every pound bought makes quite a difference in the cost of a large shipment. Copper is not alone either in its mad course, ingot tin holding up its own end, with an advance of 4 or 5 cents during the year. Zinc spelter has also jumped up considerably in the past 12 months and the high figures are well maintained Pig iron last year was also quoted at 3 or 4 dollars lower than at the present time

There are no changes to report this week, most lines remaining firm. Tin fluctuated slightly during the week, but has steadied down again and prices remain as last quoted.

Copper remains firm and unchanged, with the shortage just as marked as ever. The past year has seen an advance of about six cents per pound

Ingot tin has also jumped up considerably during the past year, an advance of 4 or 5 cents per pound being recorded There are no changes to report this week, prices remaining the same, and very firm.

Pig lead also jumped up considerably during the year, present prices showing an advance of about $1.10 per 100 lbs. over the same period twelve months ago Prices this week remain firm and unchanged, although they weakened slightly during the past week.

Toronto Hardware Markets

Office of HARDWARE AND METAL,
10 Front Street East,
Toronto, March 22, 1907.

Hardware trade conditions at present compare well with those of a year ago, demands are just as strong and profits are better. Advances in the prices of all lines have been made; wire nails and copper goods experiencing the greatest advance, mechanics' tools the least advance. Considering advances in price of all lines the average rate would be about 12½ or 15 per cent. Higher prices have not affected the demand and even greater demands are looked for this year than last. The hardware business was the best last year in history, and still a greater volume of trade is confidently expected this year.

The prices in silverware have advanced by 5 to 10 per cent., the advance being caused by a raise in the price of raw material and increased cost of labor. The demand for this line at present is as good, if not stronger than that of a year ago.

Cutlery also has advanced by 5 to 15 per cent.—15 per cent. is a conservative figure, as in one or two particular classes of goods the advance reaches nearly 25 per cent. The demand for this line at present is, if anything, more active than a year ago.

Tinware and enamelware have advanced in price by 10 to 12½ per cent. Copper goods have experienced a big advance, in most cases reaching 25 per cent.

Mechanics' tools have advanced very little, in any case not more than 5 per cent. The demand is strong and should, with the renewal of building operations, be much stronger. All heavy goods manufactured from iron have advanced in price, owing to a raise in duty, averaging about 5 per cent. A year ago there was a shortage in the supply of mechanics' and gardeners' tools, manufacturers being unable to supply the jobbers.

In many cases last year retail dealers repeated orders with the wholesalers, who had repeatedly to send back notices to them that the manufacturers were unable to fill the orders. Jobbers, therefore, had to hold over the orders. This year the situation is more favorable, retailers generally getting in their orders early, thus giving the manufacturers a better chance to get the orders filled.

There is a serious situation in the trade in bolts at present. Large orders are coming in to the jobbers who find that the manufacturers are able to fill only a small part of the orders, owing to the scarcity of raw material. The prices over last year have advanced very little. In some cases there is no change. 7-16 machine bolts a year ago were quoted at a discount of 55 and 5 per cent. The same discount is given at present, the scarcity not having affected the price. The prices of heavy hardware, as vises and anvils, have not changed.

Wringers, washing machines, and churns have been raised in price about 7 per cent. Screen doors have advanced from $6.50 for the walnut stained, 4-inch style, to $7.25, a raise of over 10 per cent. All lines of tinware, such as pails, lanterns, and stovepipes, have been raised in price, galvanized pails 5 per cent., stovepipes, 10 per cent., and lanterns nearly 50 per cent. A year ago the prices in lanterns were cut to pieces during the lantern war.

There has been little advance in the prices on binder twine. The demand is steady, a little stronger than last year, and the trade on the whole is in a very satisfactory condition.

There have been further advances in the price of copper nails and tacks. The new discounts are 30 per cent. and 25 per cent. respectively, making a difference from the discounts a year ago of 22½ to 25 per cent. Wire nails now are quoted at $2.45, making an advance of about 25 per cent. over that of a year ago. Wire has been raised about 10 per cent. over last year's price.

Prices on all lines of hardware have been raised, but the advances have not affected the demand and all prospects point toward an even greater business this year than that of last year, which was the greatest of all previous history.

It has been rumored that a 10 per cent. advance will be made soon in the price of hinges.

Montreal Hardware Markets

Office of HARDWARE AND METAL,
232 McGill Street,
Montreal, March 22, 1907

In looking back over our hardware quotations for the past year, we find almost every line showing advances in prices. One man in a large Montreal jobbing house says that about the "only line in the hardware business that hasn't shown a big increase, is the salary line, which remains firm and unchanged at last year's figures." Most jobbers seem to think that the end is not yet and complain that they can't drive this fact into some retailers' heads. Dealers could often save quite an item on an order, if they would anticipate their wants a little beforehand, instead of waiting till they were completely out of the goods desired.

This week an advance has taken place on all lines of plain wire, the base price having gone up 5 cents per 100 lbs. This also puts up the price of wire nails 5 cents per keg.

Manufacturers are getting ready now to ship their twine orders, which have been mostly all booked for the coming season. Prices remain firm and unchanged.

TORONTO PAINT MARKETS.

Office of HARDWARE AND METAL,
10 Front Street East,
Toronto, March 22, 1906.

The conditions of the paint and oil trade as a whole are at present more satisfactory than they were a year ago. Prices in every line with the exception of glass have advanced and the trade is in a stronger position owing to increased demand, with the prospects of greater building operations the coming summer. A great slump was experienced throughout the trade during January and February of this year, and a big trade must be done during March, April, and May to counterbalance this. The falling off was, of course, due to the enormous booking of orders last fall.

The prices in varnish have been advanced over those of a year ago. Pure orange varnish in barrels a year ago cost from $2.40 to $2.50. At present the prices quoted are from $2.80 to $2.90, an advance of over 16 per cent. The prices of No. 1 orange a year ago were from $2.35 to $2.45. The price at present is $2.50, an advance of about 5 per cent.

Turpentine has advanced in price, and with the scarcity in Georgia and the high cost of labor, there the prospects point toward a still greater advance, one prediction being that prices will reach $1.25 per barrel within twelve months. A year ago the price per single barrel was 98c. a gal., then considered a record figure, but now it is $1.02, an advance of 4 cents. Owing to the great scarcity and advancing cost, painters are looking about for substitutes. As a result, many substitues have been coming on the market, some of them being very inferior in quality and dangerous to handle, as they contain large mixtures of kerosene and benzine.

The prices in linseed oil also have advanced over 16 per cent. For 1 to 4 barrels, raw, a year ago the price was 55 cents, now it is 64 cents. The price of boiled oil has also advanced over a year ago about 10 cents.

A big advance has been made in the price of white lead, reaching almost to $1, over a year ago. With the advance in the price of leads there has been an increased demand for ready-mixed paints.

The tendency of prices on all lines in the paint and oil trade is upward with the exception of glass. The trade in glass is in a very unsatisfactory condition at present. Prices on both sheet and plate glass have declined. If there had been a decline in prices, that alone would have been sufficient to demoralize the trade, as it has done, but the conditions have been aggravated by advances in the cost of material. The selling price of plate glass now is 30 per cent. lower than a year ago, and sheet glass has declined 15 per cent. in price. The cost of material in the European market has advanced on sheet glass 15 per cent. and on plate glass 20 per cent. The reason for the price cutting by the wholesale dealers is that a great deal of friction exists and they seem unable to co-operate. At present, with the high cost of material and the price-cutting, the wholesalers say there is little money in the business.

With the very notable exception of glass, the paint and oil trade at present is in a much better condition than a year ago.

MONTREAL PAINT MARKETS.

Office of HARDWARE AND METAL,
232 McGill Street,
Montreal, March 22, 1907

The past year has been characterized by a buoyancy in almost every line. There is scarcely an article in the paint, color and varnish departments that has not yielded to the lively conditions of trade and stringency or shortage in the primary markets.

Paris green started off with an advance of about 10 cents per pound over the average figures of past years. White lead, the greatest staple in the paint market, was forced upward by the advance in pig lead and changed customs conditions.

Linseed oil pursued the even tenor of its way without notable change for a long time, but owing to the duty being placed on flaxseed, and a sharp increase in the tariff on the oil itself, figures jumped 5 cents per gallon on the oil itself, and this naturally gave a strengthening disposition to liquid paints, varnishes, putty, zinc, ground colors, and all the sundry items into which linseed oil enters.

Trade has been forced to meet these changed conditions Some dealers have been tardy in making their advances, but it is generally conceded that the advances were justified and inevitable.

A fair amount of linseed oil is being shipped and as stocks are light, it is just as well that there is not a heavy call.

Turpentine has advanced 2 cents per gallon, bringing the price of single barrels up to $1 per gallon, and a fair movement is reported.

Window glass, putty, ground white lead and shellac gums and varnishes are all in active demand at this season.

TORONTO PLUMBING MARKETS.

Office of HARDWARE AND METAL,
10 Front Street East,
Toronto, March 22, 1907

On the whole there is very little difference in the conditions of the plumbing trade now from those a year ago. With the advance of about 10 per cent. in the cost of material and labor there has come to counter-balance this increased building operations and demand for a higher class of work.

The most striking feature of the markets this past year has been the increasing scarcity of iron pipe, and if the scarcity continues the trade in a short time will be confronted with a very serious difficulty, as serious almost as a scarcity of food supplies. With the scarcity have come advances in the price of iron pipe. A year ago one-inch black pipe was quoted at $4.37 per 100 feet; at present the price quoted is $5.12, an advance of over 17 per cent. On galvanized pipe a year ago the price quoted for 1-inch pipe was $6.02, at present the price is $6.77, an advance of 12½ per cent. Great difficulty was experienced by Canadian producers last fall in procuring skelp from Pittsburg, and owing to this they are now from 4 to 6 weeks behind with their orders. During the slack months—January, February, and March—the supply houses always endeavor to fill up their stocks in anticipation of the spring trade, but this year they have been unable to secure deliveries and stocks are remarkably low for this season of the year.

Lead pipe also is scarce, and prices have sharply advanced. A year ago the price quoted was 7 cents a pound with 20 per cent. discount. Now, the list price is the same, with only 5 per cent. off.

The price in soil pipe and fittings has changed very little. There has been a good deal of cutting. Owing to the high prices in iron an advance has been made since the first of this year.

Iron pipe fittings have, during the past year, advanced in price. The discount quoted a year ago was 65 per cent., now it is 57½ per cent., a decrease in discount of 7½ per cent. Malleable unions quoted a year ago at 65 per cent. are now quoted at 55 and 5 per cent. discount, making a stiff advance in price.

Galvanized iron range boilers quoted a year ago at $4.75 for the 30-gallon line, are now quoted at $5.00, making an advance of about 5 per cent.

Prices of solder have advanced from 23 cents for half-and-half bar solder in Toronto, to 27 cents, making an advance of about 20 per cent. Wiping has advanced from 20 cents to 23 cents, an advance of 15 per cent.

Advances have been made in enamelware, owing to the increasing cost of iron and other materials as well as the advance made in the duty last fall from 30 per cent. to 35 per cent. A year ago the regular price for 5-foot standard ideal plate E 1 bath tubs was $20.65; now the price quoted at present is $21.35, an advance of about 3 1-3 per cent. The discount on enameled closets and urinals a year ago was 20 per cent., at present the discount quoted is 15 per cent.

On nearly every line of plumbing goods there has been an advance, with the exception of radiators, the prices of which have experienced a good deal of cutting owing to the strong competition resulting from new manufacturers coming into the field. The same condition has prevented hot water boilers from being advanced in keeping with the higher cost of iron. It is said that radiation materials are now being sold at the lowest price in history.

MONTREAL PLUMBING MARKETS.

Office of HARDWARE AND METAL,
232 McGill Street,
Montreal, March, 22 1907.

During the past year, all lines of goods, the plumber has occasion to handle, have jumped to a figure previously unknown in plumbing circles. The prices on the sheets from which iron range boilers are made have advanced about 25 per cent. during the year, not to mention the material from which the copper article is manufactured. Spelter also influences the manufacturing of galvanized boilers, as the galvanizing depends greatly on this material.

In lead pipe there has also been a pretty stiff advance, the prices in 1905, in the English market, being £12 17s. 6d. as against £19 10s., or an advance of about 30 per cent. in two years.

One year ago, iron pipe was quoted at a base discount of 73 1-2 per cent. against 69 per cent. to-day, with prospects of higher prices in the near future. The same advance applies to pipe fittings, soil pipe, etc.

In brass fittings, the advance is even more marked than in the case of iron. A few years ago, brass was selling at 12 cents per pound, to-day it costs 27 cents, and even at this high price there is a decided shortage.

Manufacturers were never so crowded with orders as at present. They are practically independent. It's a case of pay the price or leave it, orders being booked as far as six months ahead.

The advance in iron also affects all lines of heaters, boilers, radiators, etc., and even here the brass valves, etc., cost more than formerly, so it's easy to see that these lines cost the consumer quite a lot more than formerly.

All the sheets used by range makers for the manufacture of steel ranges, camp outfits, etc., have advanced about 25 per cent., or 25 per cent. on the steel sheets, and about 20 per cent. on the tin ones.

The pig iron used for making the castings for enamelware has also advanced about 25 per cent., not to mention the chemicals, etc., entering into enameling process, which have also been boosted up.

We mentioned last week that all lines of iron pipe up to 3 inch had advanced about 4 per cent. The exact figures are not made up, but the advance will not quite reach the 4 per cent. mark, being exactly 3¾ per cent.

Nova Scotia Hardware News

Halifax, March 13.

The stove trade was brisk all the year, and the outlook is excellent. There is a general demand for stoves and ranges. James Hillis, of the firm of James Hillis & Son, who operate a large foundry at Halifax, says that business was never better. The firm has been rushed so that the present plant is not large enough for the business, and they contemplate extending it as soon as possible. The firm has lots of big orders, work being in sight for the best part of the summer. Among the big contracts on hand are iron stairways for the six story Canadian Bank of Commerce, and iron stairways for the four story Chronicle building. In addition, the firm has the contract for the iron work at the new railway roundhouse, a portion of which includes 150 iron columns weighing a ton each.

Jobbers report that enamelware is finding a ready sale on this market. As the quality of the various lines improves, the demand increases.

Reviewing the trade of the past year a leading jobber states that the business broke all records. Trade was good in all lines, the increase in builders' materials being most noticeable. There was a falling off in the sales of mining supplies, but otherwise the business was most satisfactory from every standpoint. Up to the present, no large shipments have been made, but the bookings ahead for the spring give promise of equalling those of last year.

The jobbers are now busy filling orders for lobster supplies, large quantities of which are sent out from Halifax during March and April. Most of the buyers had been holding back for the new price on rope which was made a week ago, and as a consequence, this line of business is a little slower than usual. Rope for lobster traps only is quoted at 14½ cents base, for the best Manila, and to obtain the rope at these figures the seller has to make a declaration to the Consumers Cordage Co. that the rope is for lobster purposes only. The size of the rope is from six-thread to one-half inch. The Consumers Company is forced to quote the above figures to meet the competition of the American rope manufacturers, since the coming into force of the new tariff. The lobster men also use large quantities of ingot tin, which is quoted at 47½c. Pig lead, English, is quoted at $6 per hundred, I. C. 14 x 20 coke is quoted at $4 per box.

It is reported on good authority that the Sydney Foundry and Machine Works anticipate adding to their present works a steel plant for the manufacture of all kinds of steel castings and the putting of the entire works under the head of a joint stock company. This will give employment to a large number of Sydney men.

The electrical business is good. Nearly all new buildings now erected in Halifax are wired for electricity. The severe frosts of the winter have made business brisk for the plumbers. The large number of new buildings which will be erected during the summer will also keep the plumbing trade to the front.

Western Ontario News

London, Ont., March 20, 1907.

Local retail hardware dealers, busy men though they are these days, are not without their trade grievances. Not the least of these is the habit certain jobbers have of selling direct to housekeepers at wholesale prices, thus coming into ruinous competition with the retailer. Discussing the matter with your correspondent, a Dundas Street merchant said: "Almost every housekeeper has a friend in some jobbing house or other, and those who haven't have a friend who possesses such a friend. Consequently when some article is needed in the household a deal is worked with the jobber, directly or indirectly. Why, I was in a certain jobbing house recently, and heard one of the head men shout to a clerk: 'Send to Mrs. ——, —— Avenue, a Bissell carpet sweeper at once,' and—this to the bookkeeper—'Mrs. ——, one Bissell sweeper, $2.' Now, I am forced to take a dozen of these sweepers before I can get them at that price, and have to wait the pleasure of the jobber before they are delivered. And, mind you, these jobbers are far less accommodating to the trade than they are to the private inhividual. Only recently, when I had sold a good order to a customer, and asked the jobber to deliver it to the purchaser instead of my store, he refused to do so. But in the case of a private buyer, the delivery is most prompt. The result has been that so far as I can, I am now buying at wholesale from the manufacturer."

Another grievance aired by this dealer was the methods of a certain city merchant who, besides dealing in notions, carried quite a line of hardware. His plan is to sell certain goods—such, for instance, as nails, lock-sets, etc., at cost price, merely as a bait and an advertisement of other goods. "Oh, I tell you," concluded the merchant interviewed, "this business of ours is not all cream and peaches."

Jobbers, however, are not the only sinners in this respect. A city plumber delights to tell the story of a certain citizen who purchased a force pump from the manufacturer. The latter of course would not undertake to put it up and the buyer himself couldn't do it. So he went to the plumber, who sent up a man to do the work. "A new pump, eh?" remarked the workman. "Where did you get it?" he innocently asked. "At So-and-so's," naming the manufacturer. "And I tell you," grinned the plumber, "that man paid dearly for his pump by the time we got through with him."

Such is business!

There is also a possibility of a rolling mill being started in Amherst by the Rhodes Curry Company. This company uses an enormous quantity of nails during the run of a season, and it is stated that they have for some time past been purchasing as much as eighty tons per week from the rolling mills of Dartmouth.

PUBLIC SPIRITED HARDWARE MERCHANTS

NO. 10

In the recent municipal elections in Ontario there were several well-known hardwaremen elected to help guide the destinies of the different towns and cities. Amongst these was J. J. Coffey, of Barrie, known to many of his fellow-tradesmen on account of his acting as guard of one of the doors at the big convention of the Ontario Retail Hardware Association in Toronto last May. If Mr. Coffey guards the interests of his Barrie constituents as faithfully as he looked after retail association interests at the convention he will do credit to the hardware trade and to himself..

Mr. Coffey guaduated from the hotel business into the hardware store he now conducts, circumstances making it necessary for him to look after an investment he had made and take over an already established hardware business about a year ago. His hotel interests were transferred and he is now devoting all his time to the hardware store. Having learned the advantages of associations in his former vocation, Mr. Coffey lost no time in affiliating with the newly formed retail hardware association.

J. J. COFFEY, Barrie

Elected Alderman in that Town in the recent Ontario Municipal Elections.

HARDWARE TRADE GOSSIP

Ontario.

F. G. Deustadt, Gorrie, has sold out to R. Carson.

H. Ellison, hardware merchant, of Port Stanley, is on the sick list.

J. M. B. Stephens, Bradford, is opening up a hardware store in Newmarket.

The works of the McGregor, Banwell Fence Co., Walkerville, were destroyed by fire this week.

McDonald & Cragg, hardware merchants at Florence, are succeeded by Cragg & Sinclair.

W. O. Greenway, of Montreal, called at the Toronto office of Hardware and Metal this week.

Burglars stole about $100 worth of goods from the hardware store of F. C. McMaster, Havelock.

W. L. Allen, hardware merchant, Cobourg, has awarded the contract for the erection of his new residence.

W. Craig, of the Vokes Hardware Co., Toronto, is leaving for the west to accept a position with a Winnipeg hardware firm.

J. A. French, of Langtree & French, implement dealers at Stirling, will retire from the business, owing to ill-health.

J. A. McDonald, of Tara, has purchased the interest of his late partner, J. H. VanDusen, in the hardware business.

I. Huffman, a prominent citizen of Kingston, and traveler for E. Crown & Sons, of that city, died suddenly on Tuesday.

Frank R. Oliver, formerly of Saskatoon, has been appointed provisional liquidator for the winding-up of the Perrin Plow Co., Smith's Falls.

A. Desmarchais, manager of the plumbing department of Carter Bros.' store, Picton, was in Toronto this week attending the A.O.U.W. convention.

E. B. Fowler has been appointed manager of the Rapid Tool Co., Peterboro. He was formerly secretary-treasurer of the Canada Cordage Co.

H. L. Cameron, of Cameron & Cameron, hardware merchants, Beaverton, is running for the presidency of the Canadian Lacrosse Association. May he win "hands down."

Mayor Ashdown, of Winnipeg, is in Toronto this week on a business trip.

F. R. Oliver, manager of the Perrin Plow Co., Smith's Falls, has purchased the Gould foundry and plant in that town.

The Ontario Plate Glass Importing Co. have commenced business at 20 Wellington street west, Toronto. Chas. Gray, formerly of the Queen City Plate Glass Co., is at the head of the new concern.

Quebec.

G. R. Hutton, hardware merchant and grocer at Richmond, is dead.

F. Malouin's tinshop, at Coaticook, was destroyed by fire last week.

The Damphouse Hardware Co., Louiseville, has been registered.

The Colonial Engineering Co., Montreal, has been incorporated, with a capital of $125,000, for the manufacture of electrical machinery and fixtures. The promoters are: V. E. Mitchell, E. F. Surveyer, A. Chase-Casgrain, J. W. Weldon and S. J. LeHuray, all of Montreal.

Western Canada.

J. L. Evans & Co., hardware merchants, Miniota, Man., are selling out.

Hamilton Bros. & Co., hardware merchants at Tantallon, Sask., have sold to Hooper & Co.

NEW CANADIAN FACTORY.

The Hardy & Dischinger Co., Toledo, wish to start a Canadian branch for the manufacture of paints and lubricating oils. They wish to locate in Toronto, and if the city will give five and a half acres, they will invest $20,000 or $25,000 and employ fifteen or twenty men.

CURRENT MARKET QUOTATIONS.

Mar. 21 1907.

These prices are for such qualities and quantities as are usually ordered by retail dealers on the usual terms of credit, the lowest figures being for larger quantities and prompt pay. Large cash buyers can frequently make purchases at better prices. The Editor is anxious to be informed at once of any apparent errors in this list, as the desire is to make it perfectly accurate.

METALS.

ANTIMONY.

Cookson's .... per lb. 0 27 0 27½

BOILER PLATES AND TUBES.

Plates, ¼ to ½ incl. per 100 lb. 2 50
Heads, per 100 lb 2 75
Tank plates 3-16 inch 2 65
Tubes per 100 feet, 1½ inch . 8 50 ....
„ „ 2 „ 9 00 9 50
„ „ 2½ „ 11 30 11 50
„ „ 3 „ 12 5 13 00
„ „ 3½ „ 16 00 17 00
„ „ 4 „ 20 10 21 50
2 per cent off

BOILER AND T.K PITTS

Plain tinned........} 25 per cent off list.
Spun........

BABBIT METAL.

Canada Metal Company—Imperial, genuine 60c., Imperial Tough, 60c., White Brass 50c.; Metallic, 35c.; Harris Heavy Pressure, 25c.; Hercules, 25c; White Bronze, 15c.; Star Frictionless, 14c; Alluminoid, 10c; No. 4, 9c. per lb.
James Robertson Co.—Extra and genuine Monarch, 60c; Crown Monarch, 50c; No 1 Monarch, 40c; King 30c; Fleur-de-lis, 20c; Thurber, 15c; Philadelphia, 12c; Canadian, 10c; hardware, No 1, 15c; No. 2, 12c; No. 3, 10c. per lb.

BRASS.

Rod and Sheet, 14 to 30 gauge, 25 p c. advance.
Sheets, 12 to 14 in. 0 30
Tubing, base, per lb 5-16 to 2 in ... 0 33
Tubing ½ to 3-inch, iron pipe size.. 0 31
" ½ to 3-inch, seamless.... 0 36
Copper tubing, 6 cents extra

COPPER.

Ingot. Per 100 lb.
Casting, car lots, ... 26 00 27 00
Bars.
Cut lengths, round, ½ to 2 in.... .... 35 00
Sheet.
Plain, 16 oz., 14x48 and 14x60 .... 35 00
Plain, 14 oz. 35 00
Tinned copper sheet, base 38 00
Planished base 43 00
Braziers (in sheets), 4x6 ft., 25 to 30 lb. each, per lb, base . 0 34 0 35

BLACK SHEETS.

	Montreal	Toronto
8 to 10 gauge	2 70	2 70
12 gauge	2 70	2 70
14 "	2 50	2 55
17 "	2 50	2 65
18 "	2 50	2 65
20 "	2 50	2 65
22 "	2 50	2 75
24 "	2 55	2 80
26 "	2 65	2 85
28 "	2 75	3 00

CANADA PLATES.

Ordinary, 52 sheets 2 75 2 90
All bright 3 80 3 90
Galvanized, 52 sheets 4 35 4 45
" 60 " 4 60 4 70

	Ordinary.	Dom. Crown
18x24x52	4 25	4 35
" 60	4 50	4 60
20x28x80	8 50	8 70
" 94	9 00	9 20

GALVANIZED SHEETS

	Fleur-de-Lis	Gordon Crown.	Colborne Crown.
16 to 20 gauge	3 80	3 95	..
22 to 24 gauge .	4 05	4 00	3 75
26 " ..	4 30	4 40	4 30
28 " .	4 55	4 60	4 45

Apollo.
10¾ oz. (American gauge) 4 85
28 gauge........ " 4 45 4 55
26 " .... " 4 15 4 30
24 " .... " 3 85 3 90

	Gorbal's Best.	Comet	Queen's Head.	Bell.
16 to 20 gauge	....	3 60	3 95	....
22 to 24 gauge	....	3 75	4 20	....
26 "	....	4 30	4 45	....
28 "	4 55	4 45	4 70	4 45

Less than case lots 10 to 25c extra.

IRON AND STEEL.

	Montreal	Toronto.
Common bar, per 100 lb	2 15	2 30
Forged iron "	2 40	..
Refined " "	2 55	2 70
Horseshoe iron "	2 55	2 70
Hoop steel, 1½ to 3 in. base		2 80
Sleigh shoe steel "	2 25	2 30
Tire steel	2 40	2 50
Best sheet cast steel........ ...		0 12
B. K. Morton "Alpha" high speed.		0 65
" annealed		0 70
"M" Self-hardening........		0 50
"J" quality, best warranted		0 18
"I" " warranted........		0 14
"B.O" quality		0 09
Colonial black diamond	0 18	0 09
Sanderson's	0 08	0 45
Jessop's	0 12	0 13
Air hardening	0 80	0 65

Conqueror 0 07½ 0 08
Jowett's diamond J 0 06½ 0 07
Jonas & Colver's tool steel .. 0 10 0 20
" "Novo" 0 65
" " annealed 0 65
Jowett & Sons B P L tool steel 0 10½ 0 11

COLD ROLLED SHAFTING.

9-16 to 11-16 inch 0 06
¾ to 1 7-16 " 0 05½
1 7-16 to 3 " 0 05
Montreal 30, Toronto 35 to 40 per cent.

INGOT TIN.

Lamb and Flag and Straits—56 and 28-lb. ingots, 100 lb. $45 00 $45 50

TINPLATES.

Charcoal Plates—Bright
M.L.S., equal to Bradley— Per box.
I C, 14 x 20 base $6 50
I X, 14 x 20 .... 8 00
I XX, 14 x 20 base 9 50
Famous, equal to Bradley—
I C, 14 x 20 base........ .... 6 50
I X, 14 x 20 " 8 00
I X X, 14 x 20 base........ .... 9 50
Raven and Vulture Grades—
I C, 14 x 20 base 5 00
I X " 6 00
I X X " 7 00
I X X X " 8 00
"Dominion Crown Best"—Double Coated, Tissued. Per box.
I C, 14 x 20 base........ 5 50 5 75
I X, 14 x 20 " 6 50 6 75
I XX " x 20 " 7 50 7 75
"Allaway's Best"—Standard Quality.
I C, 14 x 20 base........ .... 4 50
I X, 14 x 20 " 5 25
I XX, 14 x 20 " 6 00
Bright Cokes.
Bessemer Steel—
I C., 14 x 20 base 4 25
20x28, double box 8 50
Charcoal Plates—Terne
Dean or J. G. Grade—
I C., 20x28, 112 sheets 7 25 8 00
IX., Terne Tin 9 50
Charcoal Tin Boiler Plates.
Cookley Grade—
X X, 14x56, 50 sheet bxs.} 7 50
" 14x60, " }
" 14x65, " }
Tinned Sheets.
72x30 up to 24 gauge........ .... 8 50
" 26 " 9 00

LEAD.

Imported Pig, per 100 lb........ 5 50 5 60
Bar, ... 5 75 6 00
Sheets, 2½ lb. sq. ft., by roll ... 0 07½
Sheets, 3 to 6 lb. " 0 07
NOTE.—Cut sheets ½c. per lb., extra. Pipe, by the roll, usual weights per yard, lists at 7c. per lb. and 5 p.c. dis. f.o.b. Toronto.
NOTE.—Cut lengths, net price, waste pipe 8-ft. lengths, lists at 8c.

SHEET ZINC.

5-cwt. casks 8 00 8 25
Part casks 8 25 8 50

ZINC SPELTER.

Foreign, per 100 lb 7 50 7 75
Domestic 7 00 7 25

PLUMBING AND HEATING

BRASS GOODS, VALVES, ETC.

Standard Compression work, 55 per cent
Cushion work, discount 40 per cent.
Fuller work, 65 per cent.
Flatway stop and stop and waste cocks, 60 per cent; roundway, 55 per cent.
J.M.T. Globe, Angle and Check Valves, 45 per cent.
Standard Globe, Angle and Check Valves 55 per cent.
Kerr standard globes, angles and checks, special, 42½ per cent; standard, 47½ p.c
Kerr Jenkins' disc, copper-alloy disc and heavy standard valves, 40 per cent.
Kerr steam radiator valves 60 p.c., and quick-opening hot-water radiator valves, 60 p c.
Kerr brass, Weber's straightway valves, 40 per cent., straightway valves, I B B M, 60 per cent
J. M T. Radiator Valves 50 per cent.
Standard Radiator Valves, 60 per cent.
Patent Quick-Opening Valves, 65 per cent.
Jenkins' Bros Globe Angle and Check Valves 27½ per cent.
No. 1 compression bath cock......net 2 00
No. 4 " " ...... " 1 90
No 7 Fuller's........ .... " 2 25
No. 4½ " .... " 2 35
Patent Compression Cushion, basin cock, hot and cold, per doz., ... $16.20
Patent Compression Cushion, bath cock, No. 2208 .... 2 25
Square head brass cocks, 50 per cent.
" " iron " 60 "
Thompson Smoke-test Machine 25.00

BOILERS—COPPER RANGE.

Copper, 30 gallon........ ...... 33 00
" 35 " 38 00
" 40 " 43 00
15 per cent

BOILERS—GALVANIZED IRON RANGE.

Capacity	Standard.	Extra heavy
30-gallons ...	5 00	7.75
35 "	6 00	8 75
40 "	7.00	9 75

2 per cent., 30 days.

BATH TUBS.

Steel clad copper lined, 15 per cent.

CAST IRON SINKS

16x24, $1; 18x30, $1; 18x36, $1.30.

ENAMELED BATHS.

List issued by the Standard Ideal Company Jan 3, 1907, shows an advance of 10 per cent. over previous quotations.

ENAMELED CLOSETS AND URINALS

Discount 15 per cent.

ENAMELED LAVATORIES.

	1st quality.	Special.
Plate E 100 to E 103 ..	20 & 5 p c.	20 & 10 p c.
" E 104 to E 132....	20 & 10 p c.	30 & 2½ p.c

ENAMELED ROLL RIM SINKS.

	1st quality.	Special.
Plate E 201, one piece.	15 & 2½ p.c.	15 & 10 p.c.

ENAMELED KITCHEN SINKS.

Plate E, flat rim 300, 60 & 10 p.c. 65 & 5 p.c.

HEATING APPARATUS

Stoves and Ranges—Discounts vary from 40 to 70 per cent. according to list.
Furnaces—45 per cent.
Registers—70 per cent
Hot Water Boilers—50 per cent
Hot Water Radiators—50 to 55 p.c
Steam Radiators—50 to 55 per cent
Wall Radiators and Specials—50 to 55 p c.

LEAD PIPE

Lead Pipe, 7c. per pound, 5 per cent. off.
Lead waste, 8c. per pound, 5 per cent. off.
Caulking lead, 6½c. per pound.
Traps and bends, 40 per cent.

IRON PIPE.

Size (per 100 ft.)	Black.		Galvanized
¼ inch........	2 26	¼ inch........	3 08
⅜ "	2 26	⅜ "	3 08
½ "	2 72	½ "	3 57
¾ "	3 67	¾ "	4 72
1 "	5 12	1 "	6 77
1¼ "	6 98	1¼ "	9 23
1½ "	8 37	1½ "	11 07
2 "	11 16	2 "	14 76
2½ "	18 00	2½ "	24 00
3 "	23 50	3 "	31 50
3½ "	30 00	3½ "	40 00
4 "	34 00	4 "	45 00

2 per cent, 30 days.
Malleable Fittings—Canadian discount 30 per cent.; American discount 25 per cent.
Cast Iron Fittings 57½; Standard bushings 57½; headers, 57½; flanged unions 57½, malleable bushings 55; nipples, 70 and 10; malleable lipped unions, 55 and 5 p c.

SOIL PIPE AND FITTINGS

Medium and Extra heavy pipe and fittings, up to 6 inch, discount 60 per cent.
7 and 8-in. pipe, discount 40 and 5 per cent
Light pipe, 50 p.c.; fittings, 50 and 10 p.c.

OAKUM.

Plumbers ....per 100 lb... .. .... 4 75

STOCKS AND DIES.

American discount 25 per cent.

SOLDERING IRONS.

½-lb........ per lb. 0 38
1-lb. or over .... " 0 35

SOLDER.

	Montreal	Toronto
Bar, half-and-half, guaranteed	0 25	0 27
Wiping........	0 22	0 23

PAINTS, OILS AND GLASS

BRUSHES

Paint and household, 70 per cent

CHEMICALS

In casks per lb.
Sulphate of copper (bluestone or blue vitrol) 0 09
Litharge, ground 0 06
" flaked 0 06½
Green copperas (green vitrol) 0 01
Sugar of lead 0 08
Lump olive 0 01½

COLORS IN OIL.

Venetian red, 1-lb. tins pure. 0 09
Chrome yellow " 0 15
Golden ochre " 0 08
French " " 0 08
Marine black " 0 04½
Chrome green " 0 10
French permanent green" 0 13
Signwriters' black " 0 15

GLUE

Domestic sheet 0 10 0 10½
French medal.. .... 0 12 0 12½

PARIS GREEN.

	Berger's	Canadian
600-lb cask	0 25½	0 25½
250 lb. drums	0 25¾	0 25¾
100-lb. "	0 26¼	...
50-lb "	0 26½	0 26½
1-lb. pkgs, 100 in box	0 27½	0 27½
½-lb "	0 29½	0 29½
1-lb. tins, 100 in box	0 28½	0 28½
½ lb pkg	0 30½	0 30

F.o.b Toronto

PARIS WHITE

In bbls 0 90

PREPARED PAINTS.

Pure, per gallon, in tins 1 30
Second qualities per gallon 1 10
Barn (in bbls.)........ 0 65 0 90
Sherwin-Williams paints, 1 gal 1 40
Canada Paint Co's pure 1 30
Standard P. & V. Co.'s "New Era." 1 30
Benj. Moore Co.'s "Ark" B'd 1
" British Navy deck 1
Brandram-Henderson's "English" 1
Ramsay's paints, Pure, per gal. 1
" Thistle, " 1
" Outside, bbls 0 55 0
Martin-Senour's 100 p c. pure, ½ gal. 1
" " " " ¼ gal. 1
" " " " 1 gal. 1
Senour's Floor Paints.....gal. ... 1
Jamieson's "Crown and Anchor" 1
Jamieson's floor enamel...... 1
" barn paints, bbls, per gal. 0
Sanderson Pearcy's, pure 1 25
Robertson's pure paints...... .... 1 20

PUTTY.

Bulk in bbls........ 1 50
Bladders in bbls 1 80
25-lb. tins........ 1 90
Bladders in bulk or tins less than 100 lb 1 85
Bulk in 100-lb. irons........ 1 80

SHINGLE STAINS

In 5 gallon lots 0 75 0 80

SHELLAC.

White........ .... 0 65
Fine orange........ .... 0 60
Medium orange 0 55
F o.b. Montreal or Toronto.

TURPENTINE AND OIL.

Castor oil........ 0 08 0 10
Gasoline 0 22½
Benzine, per gal 0 17 0 20
Turpentine single barrels 1 00 1 02
Linseed Oil, " raw.. 0 60 0 62
" " boiled 0 63 0 65
W. H. & C's English raw oil, bbls .. 0 63
" " boiled, " 0 66

WHITE LEAD GROUND IN OIL. Per 100 lbs

Canadian pure 7 15
No. 1 Canadian 6 80
Munro's Select Flake White.. .... 7 40
Elephant and Decorators' Pure 7 40
Monarch 7 40
Standard Decorator's 7 15
Essex Genuine 6 80
Brandram's B. B. Genuine... .... 8 40
"Anchor," pure 7 00
Ramsay's Pure Lead 6 40
Ramsay's Exterior 8 15
"Crown and Anchor," pure... .... 6 50
Sanderson Pearcy's 7 40
Robertson's C.P., lead........ .... 7 20
W H. & C's matured pure English 8 00 8 25

WHITE AND RED DRY LEAD.

	white	red.
Genuine, 560 lb. casks, per cwt	6 75	6 00
Genuine, 100 lb. kegs, "	7 50	6 50
No. 1, 560 lb. casks, per cwt...	6 25	5 75
No. 1, 100 lb. kegs, per cwt ..	7 00	6 25

WINDOW GLASS.

Size United Inches.	Star	Double Diamond
Under 26	$4 25	$6 25
26 to 40	4 65	6 75
41 to 50	5 10	7 50
51 to 60	5 35	8 50
61 to 70	5 75	9 75
71 to 80	6 25	11 00
81 to 85	7 00	12 50
86 to 90	....	15 00
91 to 95	....	17 50
96 to 100	....	20 50
101 to 105........	....	24 00
106 to 110........	....	27 50

Discount—16-oz, 75 per cent; 21-oz. 80 per cent. per 100 feet. Broken boxes 50 per cent.

WHITING.

Plain, in bbls 0 60
Gilders bolted in bands 0 90

WHITE DRY ZINC.

Extra Red Seal, V.M. 0 07½ 0 08½

WHITE GROUND ZINC.

Pure, in 25-lb. irons 0 08½
No 1, " " 0 07
No. 2, " " 0 05½

VARNISHES.

In 5-gal. lots. Per gal cans
Carriage, No. 1 .... 1 50
Pale durable body........ .... 3 50
" hard rubbing 3 00
Finest elastic gearing 3 00
Elastic oak 1 50
Furniture, polishing 2 00
Furniture, extra........ .... 1 ..
" " No 1........ .... 0 90
" " union........ .... 0 80
Light oil finish........ .... 1 40
Gold size japan 1 80
Brown japan 0 90
No 1 brown japan 0 90
Baking black japan 1 35
No. 1 black japan 0 90
Benzine black japan 0 70
Crystal Damar........ .... 2 80
No 1 " 2 50
Pure asphaltum 1 40
Oilcloth 1 50
Lightning dryer 0 70
Elastilite varnish, 1 gal. can. each.. 2 00
Granitine floor varnish, per gal 2 10
Maple Leaf coach enamels; size 1, 1 20
Sherwin Williams' kopal varnish, gal., 2 50
Canada Paint Co's sun varnish 2 00
"Kyanize" Interior Finish.... .... 2 40
"Flint Lac," coach 1 80
B H. Co's "Gold Medal," in cases 2 00
Jamieson's Copaline, per gal. 2 00

BUILDERS' HARDWARE.

BELLS.

Brass hand bells, 60 per cent.
Nickel, 55 per cent.
Gongs, Sargeants door bells . 5 50 8 00
American, house bells, per lb. 0 35 0 40
Peterboro' door bells, discount 37½ and 10 per cent. off new list.

BUILDING PAPER, ETC.

Tarred Felt, per 100 lb. 2 25
Ready roofing, 2-ply, not under 45 lb. per roll 1 00
Ready roofing, 3-ply, not under 65 lb., per roll 1 25
Carpet Felt ....per ton 60 00
Heavy Straw Sheathing.......per ton 35 00
Dry Surprise.............. 0 40
Dry Sheathing.....per roll, 400 sq. ft. 0 30
Tar " ..... " 400 " 0 45
Dry Fibre......... " 400 " 0 55
Tarred Fibre ...... " 400 " 0 65
O. K. & I. X. L ..... " 400 " 0 70
Resin-sized .... " 400 " 0 45
Oiled Sheathing.... " 600 " 1 00
Oiled " .... " 400 " 0 70
Roof Coating, in barrels.. ...per gal. 0 17
Roof " small packages " 0 25
Refined Tar..............per barrel 5 00
Coal Tar .... " 4 00
Coal Tar, less than barrels....per gal. 0 15
Roofing Pitch....... per 100 lb. 0 80 0 90
Slater's felt ....per roll 0 70
Heavy Straw Sheathing f. o. b. St. John and Halifax 37 50

BUTTS.

Wrought Brass, net revised list.
Wrought Iron, 70 per cent.
Cast Iron Loose Pin, discount 60 per cent.
Wrought Steel Fast Joint and Loose Pin, 70 per cent.

CEMENT AND FIREBRICK.

Canadian Portland 2 00 2 10
Belgium 1 60 1 90
White Bros. English 80 2 05
"Lafarge" cement in wood ... ¼ .. 3 40
"Lehigh" cement, in wood 2 54
"Lehigh" cement, cotton sacks. 2 39
"Lehigh" cement, paper sacks 2 31
Fire brick, Scotch, per 1,000 .... 27 00 30 00
" English 17 00 21 00
" American, low 23 0 25 00
" " high 27 50 35 00
Fire clay (Scotch), net ton 4 95
Paving Blocks per 1,000.
Blue metallic, 9"x4½"x3", ex wharf .. 35 00
Stable pavers, 12"x6"x2", ex wharf .. 50 00
Stable pavers, 9"x4½"x3", ex wharf .. 36 00

DOOR SETS.

Peterboro, 37½ and 10 per cent.

DOOR SPRINGS.

Torrey's Rod........per doz. 1 75
Coil, 9 to 11 in......... " 0 95 1 65
English .... " 2 00 4 00
Chicago and Reliance Coil 25 per cent.

STORE DOOR HANDLES.

Per Dozen 1 00 1 50

ESCUTCHEONS.

Discount 50 and 10 per cent., new list
Peterboro, 37½ and 10 per cent.

ESCUTCHEON PINS.

Iron, discount 40 per cent.
Brass, 45 per cent.

HINGES.

Blind, discount 60 per cent.
Heavy T and strap, 4-in., per lb. net . 0 06½
" " 5-in., " 0 06
" " 6-in., " 0 05¾
" " 8-in., " 0 05½
" " 10-in. and larger .. 0 05¼
Light T and strap, discount 65 p.c.
Screw hook and hinge—
under 12 in.... per 100 lb. 4 75
over 12 in........ " 3 75
Spring, No. 20, per gro. pairs 10 80
Spring, Woodyatt pattern, per gro., No. 5, $17.50 No. 10, $18; No. 20, $10 80; No 12 $20; No. 51, $10; No. 50, $27 50.
Crate hinges and back flaps, 65 and 5 p. c.
Hinge hasps, 65 per cent

SPRING HINGES.

Chicago Spring Butts and Blanks 12½ per cent.
Triple End Spring Butts, 40 and 5 per cent.
Chicago Floor Hinges, 40 and 5 off.
Garden City Fire House Hinges, 12½ p c

CAST IRON HOOKS.

Bird cage........ per doz. 0 50 1 10
Clothes line, No. 61.. " 0 00 0 70
Harness............. " 0 60 12 00
Hat and coat ...... per gro. 1 10 10 00
Chandelier ....per doz. 0 50 1 00
Wrought hooks and staples—
¼ x 5 .... per gross 2 65
5-16 x 5 .... " 3 30
Bright steel gate hooks and staples, 40 p.c.
Hat and coat wire, discount 62½ per cent.
Screw, bright wire, discount 65 per cent.

KNOBS.

Door, japanned and N.P., doz 1 50 2 50
Bronze, Berlin....... per doz. 2 75 3 25
Bronze, Genuine .... " 6 00 9 00
Shutter, porcelain, F. & L. screw .... per gross 1 30 2 00
White door knobs ...per doz 2 00
Peterboro knobs, 37½ and 10 per cent.
Porcelain, mineral and jet knobs, net list

KEYS.

Lock, Canadian dis. 40 to 40 and 10 per cent
Cabinet trunk and padlock American .... per gross 0 60

LOCKS

Peterboro 37½ and 10 per cent
Russell & Erwin steel rim $2 50 per doz
Eagle cabinet locks, discount 30 per cent
American padlocks, all steel, 10 to 15 per cent.; all brass or bronze, 10 to 25 per cent.

SAND AND EMERY PAPER

B. & A. sand, discount, 35 per cent
Emery, discount 35 per cent.
Garnet (Rurton's) 5 to 10 per cent. advance

SASH WEIGHTS.

Sectional..........per 100 lb. 2 00 2 25
Solid .... " 1 50 1 75

SASH CORD.

Per lb....................... 0 31

BLIND AND BED STAPLES

All sizes, per lb.............. 0 07½ 0 12

WROUGHT STAPLES.

Galvanized 2 75
Plain 2 50
Coopers', discount 45 per cent.
Poultry netting staples, discount 40 per cent.
Bright spear point, 75 per cent. discount.

TOOLS AND HANDLES.

ADZES.

Discount 22½ per cent.

AUGERS.

Gilmour's, discount 60 per cent. off list.

AXES.

Single bit, per doz. 5 50 8 50
Double bit, " 10 00 11 00
Bench Axes, 40 per cent.
Broad Axes, 25 per cent.
Hunters' Axes.............. 5 50 6 00
Boys' Axes 6 25 7 00
Splitting Axes 7 00 12 00
Handled Axes 7 00 9 00
Red Ridge, boys', handled.. .... 5 75
" hunters 5 25

BITS

Irwin's auger, discount 47½ per cent.
Gilmour's auger, discount 60 per cent.
Rockford auger, discount 50 and 10 per cent.
Jennings' Gen. auger, net list.
Gilmour's car, 47½ per cent.
Clark's expansive, 40 per cent.
Clark's gimlet, per doz 0 65 0 90
Diamond, Shell, per doz...... 1 00 1 50
Nail and Spike, per gross..... 2 25 5 20

BUTCHERS CLEAVERS

German ....per doz. 6 00 9 00
American............. " 12 00 18 00

CHALK.

Carpenters' Colored, per gross 0 45 0 75
White lump..........per cwt. 0 60 0 65

CHISELS.

Warnock's, discount 72½ per cent.
P. S. & W. Extra, discount 72½ per cent

CROSSCUT SAW HANDLES.

S. & D., No. 3 ....per pair 0 15
S. & D., " 5 .... " 0 22½
S. & D., " 6 .... " 0 15
Boynton pattern.................. " 0 20

CROWBARS.

3½c. to 4c per lb

DRAW KNIVES.

Coach and Wagon, discount 75 per cent.
Carpenters' discount 75 per cent

DRILLS.

Millar's Falls, hand and breast. net list
North Bros., each set, 50c.

DRILL BITS.

Morse, discount 37½ to 40 per cent.
Standard, discount 50 and 5 to 55 per cent.

FILES AND RASPS.

Great Western .... 75 per cent.
Arcade .... 75 "
Kearney & Foot .... 75 "
Disston's .... 75 "
American .... 75 "
J. Barton Smith .... 75 "
McClellan .... 75 "
Eagle .... 75 "
Nicholson .... 66½ "
Globe .... 75 "
Black Diamond, 60, 10 and 5 p c
Jowitt's, English list, 27½ per cent.

GAUGES.

Stanley's discount 50 to 60 per cent.

Wire Gauges.

Winn's, Nos. 26 to 33 ....each 1 65 2 40

HANDLES.

C. & B., fork and hoe, 40 p c, revised list.
American, saw ......per doz. 0 25
American, plane ... per gross 3 15 3 75
Canadian, hammer and hatchet 40 per cent
Axe and cant hook handles, 45 per cent.

HAMMERS.

Maydole's, discount 5 to 10 per cent
Canadian, discount 25 to 27½ per cent
Magnetic tack ....per doz. 1 10 1 20
Canadian sledge ....per lb. 0 07 0 08½
Canadian ball pean, per lb. .. 0 22 0 25

HATCHETS.

Canadian, discount 40 to 42½ per cent.
Shingle, Red Ridge 1, per doz........ 4 40
" 2 .... 4 80
Barrel Underhill 5 05

MALLETS.

Tinsmiths' ....per doz. 1 25 1 50
Carpenters', hickory, " 1 25 3 75
Lignum Vitae........ " 3 85 5 00
Caulking, each 0 60 2 00

MATTOCKS.

Canadian............ per doz. 5 50 6 00

MEAT CUTTERS.

German, 15 per cent
American discount, 33½ per cent.
Gem .... each 1 15

NAIL PULLERS.

German and American........ 0 85 2 50
No. 1 .... 0 85
No. 1573 .... 0 75

NAIL SETS.

Square, round and octagon, per gross 3 38
Diamond 1 00

PICKS.

Per dozen 6 00 9 00

PLANES.

Wood bench, Canadian discount 40 per cent. American discount 50 per cent.
Wood, fancy Canadian or American 37½ to 40 per cent.
Stanley planes, $1 55 to $3 00, net list prices.

TROWELS.

Disston's, discount 10 per cent.
S. & D., discount 35 per cent
Berg's, brick, 924x11 .... 4 00
" pointing, 924x5............ 2 10

FARM AND GARDEN GOODS

BELLS

American cow bells, 63⅓ per cent.
Canadian, discount 45 and 50 per cent.
American, farm bells, each . 1 35 3 00

BULL RINGS

Copper, $1 30 for 2½-inch, and $1.70

CATTLE LEADERS

Nos. 32 and 33 .. ...per gross 7 50 8 50

BARN DOOR HANGERS.

		doz pairs.
Steel barn door..	8 00	10 00
Stearns wood track	4 50	6 00
Zenith	...	9 00
Acme, wood track	5 00	6 50
Atlas	5 00	6 00
Perfect	8 00	11 00
New Milo		6 50
Steel, covered	4 00	11 00
" track, 1 x 3-16 in (100 ft)	....	3 75
" " 1¼ x 3-16 in (100 ft)	....	4 75
Double strap hangers, doz sets	...	6 40
Standard jointed hangers, "	....	6 40
Steel King hangers "	....	6 25
Storm King and safety hangers	....	7 00
" " rail	....	4 25

Chicago Friction, Oscillating and Big Twin Hangers, 5 per cent

HARVEST TOOLS.

Discount 60 per cent.
S. & D. lawn rakes, Dunn's, 40 off
" sidewalk and stable scrapers, 40 off.

HAY KNIVES.

Net list

HEAD HALTERS.

Jute Rope, ⅜-inch....per gross 9 00
" " ½ " .... " 10 00
" " ¾ " .... " 12 00
Leather, 1-inch ......per doz. 4 00
Leather, 1¼ " ...... " 5 20
Web................. " 2 45

HOES.

Garden, Mortar, etc., discount 60 per cent.
Planter.............. per doz. 4 00 4 50

LAWN MOWERS

Low wheel. 12, 14 and 16-inch $2 30
9-inch wheel, 12-inch 2 85
" " 14 " 3 00
" " 16 " 3 12½
High wheel, 12 " 4 05
" " 14 " 4 25
" " 16 " 4 50

SCYTHES.

Per doz. net.................. 8 25 9 25

SCYTHE SNATHS.

Canadian, discount 40 per cent.

SNAPS.

Harness, German, discount 25 per cent.
Lock, Andrews' 4 50 11 00

STABLE FITTINGS

Warden King, 35 per cent

WOOD HAY RAKES.

Ten tooth, 40 and 10 per cent
Twelve tooth, 45 per cent

HEAVY GOODS, NAILS, ETC.

ANVILS.

Wright's, 80-lb. and over.............. 0 10½
Hay Budden, 80-lb and over 0 09½
Brook's, 80-lb. and over 0 11½
Taylor Forbes, handy 0 05
Columbia Hardware Co , per lb 0 09½

VISES

Wright's.............................. 0 13½
Berg's, per lb 0 12½
Brook's................................ 0 12½
Pipe Vise, Hinge, No 1....... .... 3 50
" " " No. 2 ..... . 5 50
Saw Vise 4 50 5 00
Blacksmiths' (discount) 60 per cent.
" parallel (discount) 45 per cent.

BOLTS AND NUTS

Carriage Bolts, common ($1 list .... Per cent.
" " ⅜ and smaller . 60, 10 and 10
" " 7-16 and up .... 55 and 5
" " Norway Iron ($3 list) 55
Machine Bolts, ⅜ and less .... 60 and 10
Machine Bolts, 7-16 and up.... 55 and 5
Plough Bolts .... 55 and 10
Blank Bolts.................... 55
Bolt Ends 55
Sleigh Shoe Bolts, ⅜ and less . 60 and 10
" " " 7-16 and larger 50 and 5
Coach Screws, conepoint , ... 70 and 5
Nuts, square, all sizes, 4c per cent. off
Nuts, hexagon, all sizes, 4¼c. per cent off
Stove Rods per lb., 5½ to 6c.
Stove Bolts, 75 per cent

CHAIN.

Proof coil, per 100 lb., 5-16 in., $4.40, ⅜ in., $3 90; 7-16 in., $3.70, ½ in., $3 50; 9-16 in., $3 45; ⅝ in., $3 35; ¾ in., $3.20; ⅞ in., $3.10; 1 in., $3.10.

Halter, kennel and post chains, 40 to 40 and 5 per cent.
Cow ties .... 40 p.c.
Tie out chains .... 65 p.c.
Stall fixtures .... 35 p.c.
Trace chain .... 45 p.c.
Jack chain iron, discount 35 p.c.
Jack chain, brass, discount 40 per cent.

HORSE NAILS.

'C' brand, 40, 10 and 7½ per cent off list } Oval
M.R.M. Co. brand, 55 per cent. } head
Capewell brand, quotations on application.

HORSESHOES.

M.R.M Co	No. 2. and larger	No. 1 and smaller
Iron	3 60	4 05
Snow	4 05	4 30
Light steel	4 15	4 40
Featherweight, sizes 0 to 4	..	5 75
Toeweight, " 1 to 4	....	7 00

Packing up to 3 sizes in a keg, 10c per 100 lbs ; more than three sizes, 25c. per 100 lbs extra. F.o.b. Montreal, add 15c, Toronto, Hamilton and Guelph.

HORSE WEIGHTS

Taylor-Forbes, 3½c. per lb

NAILS.	Cut.	Wire
2d	3 80	3 45
3d	2 95	3 10
4 and 5d	2 70	2 85
6 and 7d	2 60	2 75
8 and 9d	2 45	2 60
10 and 12d	2 40	2 55
16 and 20d	2 35	2 50
30, 40, 50 and 60d (base)	2 30	2 45

F.o b. Montreal. Cut nails, Toronto 20c higher
Miscellaneous wire nails, discount 75 per cent
Coopers' nails. discount 40 per cent.

PRESSED SPIKES.

Pressed spikes, ⅝ diameter, per 100 lbs, $3 15

RIVETS AND BURRS.

Iron Rivets, black and tinned, 60, 10 and 10.
Iron Burrs, discount 60 and 10 and 10 p.c
Copper Rivets, usual proportion burrs, 15 p c.
Copper Burrs only, net list
Extras on Coppered Rivets, ½-lb packages 1c. per lb ; ¼-lb. packages 2c. lb
Tinned Rivets, net extra, 4c per lb

SCREWS.

Wood, F H., bright and steel, 87½ per cent.
" R. H., bright, dis. 82½ per cent.
" F. H., brass, dis. 80 per cent.
" R. H., " dis 75 per cent
" F. H., bronze, dis. 75 per cent.
" R. H., " dis. 70 per cent.
Drive Screws, dis. 87½ per cent.
Bench, wood ....per doz. 3 25 4 00
" iron .... " 4 25 5 00
Set, case hardened, dis. 60 per cent.
Square Cap, dis. 50 and 5 per cent.
Hexagon Cap, dis. 45 per cent.

MACHINE SCREWS.

Flat head, iron and brass, 35 per cent.
Felister head, iron, discount 30 per cent.
" " brass, discount 25 per cent

TACKS, BRADS, ETC.

Carpet tacks, blued .... 80 and 5
" " tinned............ 80 and 10
" " (in kegs).......... 40
Cut tacks, blued, in dozens only 75 and 10
" ½ weights 60
Swedes cut tacks, blued and tinned—
In bulk......................... 80 and 10
In dozens 75
Swedes, upholsterers', bulk .... 85 and 12½
" brush, blued and tinned bulk 70
Swedes, gimp, blued, tinned and japanned .... 75 and 12½
Zinc tacks 35
Leather carpet tacks............... 40
Copper tacks....................... 25
Copper nails 30
Trunk nails, black 65
Trunk nails, tinned and blued .. 65
Clout nails, blued and tinned .. 66
Chair nails........................ 35
Patent brads....................... 40
Fine finishing..................... 40
Lining tacks, in papers.......... 10
" " in bulk............ 15
" " solid heads, in bulk 75
Saddle nails, in papers.......... 10
" " in bulk............ 15
Tufting buttons, 22 line in dozens only 60
Zinc glaziers' points 5
Double pointed tacks, papers.. 90 and 10
" " " bulk.... 40
Clinch and duck rivets.......... 45
Cheese box tacks, 85 and 5 ; trunk tacks, 80 and 10

WROUGHT IRON WASHERS.

Canadian make, discount 40 per cent.

SPORTING GOODS.

AMMUNITION.

B. B Caps American $2 00 per 1000
C. B Caps American, $2 60 per 1000

CARTRIDGES

Rim Fire Cartridges, 50 and 5 p.c
Rim Fire Pistol, 30 and 5 per cent., American.
Central Fire, Military and Sporting American, 10 per cent. advance. B.B Caps, discount 40 per cent., American.
Central Fire Pistol and Rifle, net Amer. list
Loaded and empty Shells, American 20 per cent. discount. Rival and Nitro, 10 per cent. advance
Empty paper shells American, 10 per cent. advance.
Primers, American $2.05

Wads. per lb.
Best thick brown or grey felt wads, in ½-lb. bags $0 70
Best thick white card wads, in boxes of 500 each, 12 and smaller gauges 0 29
Best thick white card wads in boxes of 500 each, 10 gauge 0 35
Thin card wads, in boxes of 1,000 each, 12 and smaller gauges 0 20
Thin card wads, in boxes of 1,000 each, 10 gauge........ 0 25
Chemically prepared black edge grey cloth wads, in boxes of 250 each— Per M.
11 and smaller gauge 0 60
9 and 10 gauges........ 0 70
7 and 8 " 0 90
5 and 6 " 1 10
Superior chemically prepared pink edge, best white cloth wads in boxes of 250 each—
11 and smaller gauge 1 15
9 and 10 gauges 1 40
7 and 8 " 1 65
5 and 9 " 1 90

SHOT

Ordinary drop shot, AAA to dust $7 50 per 100 lbs. Discount 5 per cent, cash discount 2 per cent, 30 days, net extras as follows subject to cash discount only. Chilled, 40 c; buck and seal, 80c; no 28 ball, $1 20 per 100 lbs.; bags less than 25 lbs., ½c per lb; F O B, Montreal, Toronto, Hamilton, London, St John and Halifax, and freight equalized thereon.

TRAPS (steel.)

Game, Newhouse, discount 30 and 10 per cent.
Game, Hawley & Norton, 50, 10 & 5 per cent.
Game, Victor, 70 per cent.
Game, Oneida Jump (B & L.) 40 & 2½ p. c.
Game, steel, 60 and 5 per cent.

SKATES.

Skates, discount 37½ per cent
Mic Mac hockey sticks, per doz 4 00 5 00

PLANE IRONS.

English ....per doz. 2 00 5 00
Stanley, 2¼ inch, single 24c., double 39c

PLIERS AND NIPPERS.

Button's genuine, 37½ to 40 per cent.
Button's imitation....per doz. 5 00 9 00
Berg's wire fencing 1 72 5 50

PUNCHES

Saddlers ....per doz. 1 00 1 85
Conductor's........ " 3 00 15 00
Tinners', solid........per set 0 72
" hollow......per inch 1 00

RIVET SETS.

Canadian, discount 35 to 37½ per cent.

RULES.

Boxwood, discount 70 per cent.
Ivory, discount 20 to 25 per cent.

SAWS.

Atkins, hand and crosscut, 25 per cent.
Disston's Hand, discount 12½ per cent
Disston's Crosscut ....per foot 0 35 0 55
Hack, complete........each 0 75 2 75
" frame only........each 0 50 1 25

S. & D. solid tooth circular shingle, concave and band, 50 per cent; mill and ice, drag, 30 per cent.; cross-cut, 35 per cent; hand saws, butcher, 35 per cent; buck, New Century, $6 25; buck No. 1 Maple Leaf, $5 25; buck, Happy Medium $4.25; buck, Watch Spring, $1.25, buck, common frame, $4.00.

Spear & Jackson's saws—Hand or rip 26 in., $12 75; 24 in., $11 25; panel, 18 in., $8 25; 20 in., $9; tenon, 10 in., $9.90; 12 in., $10 90; 14 in., $11 50

SAW SETS.

Lincoln and Whiting 4 75
Hand Sets, Perfect 4 00
X-Cut Sets, " 7 50
Maple Leaf and Premiums saw sets, 40 off.
S. & D. saw swages, 40 off.

SCREW DRIVERS

Sargent's ....per doz. 0 65 1 00
North Bros, No. 30 ..per doz. 16 80

SHOVELS AND SPADES.

Bull Dog, solid neck shovel (No 2 pol.) $18 50
(Hollow Back) (Reinforced S Scoop.)
Moose........ $17 50 .... $16 30
Bear........ 15 00 .... 15 30
Fox........ 12 50 .... 14 30
Black Cat.... 10 00 .... 13 30
Canadian, discount 45 per cent.

SQUARES

Iron, discount 20 per cent
Steel, discount 65 and 10 per cent.
Try and Bevel, discount 50 to 52½ per cent.

TAPE LINES.

English, ass skin .....per doz. 2 75 5 00
English, Patent Leather 5 50 9 75
Chesterman's........each 0 90 2 85
" steel........each 0 [illegible] 4 00
Berg's, each........ 0 75 2 50

CUTLERY AND SILVERWARE.

RAZORS. per doz.

Elliot's 4 00 18 00
Boker's 7 50 11 00
" King Cutter 13 50 18 50
Wade & Butcher's........ 3 60 10 00
Lewis Bros.' "Klean Kutter" 8 50 10 50
Henckel's 7 50 20 00
Berg's........ 7 50 20 00
Clauss Razors and Strops, 50 and 10 per cent

KNIVES

Farriers-Stacey Bros., doz 3 50

PLATED GOODS

Hollowware, 40 per cent. discount.
Flatware, staples, 40 and 10, fancy, 40 and 5.

SHEARS.

Clauss, nickel, discount 60 per cent.
Clauss, Japan, discount 67½ per cent.
Clauss, tailors, discount 40 per cent.
Seymour's, discount 50 and 10 per cent.
Berg's........ 6 00 12 00

HOUSE FURNISHINGS.

APPLE PARERS.

Woodyatt Hudson, per doz., net 4 50

BIRD CAGES.

Brass and Japanned, 40 and 10 p c.

COPPER AND NICKEL WARE.

Copper boilers, kettles, teapots, etc. 30 p.c
Copper pitts, 20 per cent.

ENAMELED WARE.

London, White, Princess, Turquoise, Onyx, Blue and White, discount 50 per cent.
Canada, Diamond, Premier, 50 and 10 p c.
Pearl, Imperial Crescent, 50 and 10 per cent
Premier steel ware, 40 per cent.
Star decorated steel and white, 25 per cent.
Japanned ware, discount 45 per cent
Hollow ware, tinned cast, 35 per cent off.

KITCHEN SUNDRIES.

Can openers, per doz 0 40 0 75
Mincing knives per doz 0 50 0 80
Duplex mouse traps, per doz 0 65
Potato mashers, wire, per doz... 0 60 0 70
" wood " .. 0 50 0 60
Vegetable slicers, per doz 2 25
Universal meat chopper No. 0, $1, No. 1, 1 15.
Enterprise chopper, each 1 30
Spiders and fry pans, 50 per cent.
Star Al chopper 5 to 32 .... 1 35 4 10
" " 100 to 103........ 1 35 2 00
Kitchen hooks, bright 0 62½

LAMP WICKS.

Discount, 60 per cent.

LEMON SQUEEZERS.

Porcelain lined...... per doz. 2 20 5 60
Galvanized........ " 1 87 3 85
King, wood........ " 2 75 2 90
King, glass........ " 4 00 4 50
All glass .... " 0 50 0 90

METAL POLISH.

Tandem metal polish paste........ 6 00

PICTURE NAILS.

Porcelain head......per gross 1 35 1 50
Brass head........ 0 40 1 00
Tin and gilt, picture wire, 75 per cent.

SAD IRONS.

Mrs. Potts, No. 55, polished...per set 0 80
" No. 50, nickle-plated, " 0 91
Common, plain........ 4 50
" plated 5 50
Asbestos, per set........ 1 25

TINWARE.

CONDUCTOR PIPE.

2-in. plain or corrugated, per 100 feet, $3.30; 3 in., $4.40; 4 in., $5 80; 5 in., $7.45; 6 in., $9.90.

FAUCETS.

Common, cork-lined, discount 35 per cent.

EAVETROUGHS.

10-inch ....per 100 ft. 3 30

FACTORY MILK CANS.

Discount off revised list, 35 per cent
Milk can trimmings, discount 25 per cent.
Creamery Cans, 45 per cent.

LANTERNS.

No. 2 or 4 Plain Cold Blast....per doz. 6 50
Lift Tubular and Hinge Plain, " 4 75
No 0, safety .... " 4 00
Better quality at higher prices.
Japanning, 50c. per doz. extra.
Prism globes, per doz., $1 20.

OILERS.

Kemp's Tornado and McClary's Model galvanized oil can, with pump, 5 gallon, per dozen 10 92
Davidson oilers, discount 40 per cent
Zinc and tin, discount 50 per cent
Coppered oilers, 20 per cent. off.
Brass oilers, 50 per cent. off.
Malleable, discount 25 per cent

PAILS (GALVANIZED).

Dufferin pattern pails, 45 per cent
Flaring pattern, discount 45 per cent.
Galvanized washtubs 40 per cent.

PIECED WARE.

Discount 35 per cent off list, June, 1899.
10-qt. flaring sap buckets, discount 35 per cent.
6, 10 and 14-qt flaring pails dis. 35 per cent.
Copper bottom tea kettles and boilers, 30 p.c.
Coal hods, 40 per cent.

STAMPED WARE.

Plain, 75 and 12½ per cent. off revised list.
Retinned, 72½ per cent. revised list.

SAP SPOUTS.

Bronzed iron with hooks ....per 1,000 7 50
Eureka tinned steel, hooks 8 00

STOVEPIPES.

5 and 6 inch, per 100 lengths 7 64 7 91
7 inch .... " " 8 18
Nestable, discount 40 per cent.

STOVEPIPE ELBOWS

5 and 6-inch, common........per doz. 1 32
7-inch........ " 1 48
Polished, 15c. per dozen extra.

THERMOMETERS.

Tin case and dairy, 75 to 75 and 10 per cent.

TINNERS' SNIPS.

Per doz........ 3 00 15
Clauss, discount 35 per cent.

TINNERS' TRIMMINGS.

Discount, 45 per cent.

WIRE.

ANNEALED CUT HAY BAILING WIRE

No. 12 and 13, $4; No 13½, $4 10; No. 14, $4 25; No 15, $4 50; in lengths 6' to 11', 25 per cent.; other lengths 20c. per 100 lbs extra; if eye or loop on end add 25c. per 100 lbs to the above.

BRIGHT WIRE GOODS

Discount 62½ per cent.

CLOTHES LINE WIRE.

7 wire solid line, No. 17, $4.90; No. 18, $3.00; No. 19, $2.70; 8 wire solid line, No. 17, $4.45; No. 18, $2.80. No. 19, $2.50. All prices per 1000 ft. measure. F.o.b. Hamilton Toronto, Montreal.

COILED SPRING WIRE

High Carbon, No. 9, $2 90, No. 11, $3 45; No. 12, $3 15.

COPPER AND BRASS WIRE.

Discount 37½ per cent.

FINE STEEL WIRE.

Discount 25 per cent. List of extras: In 100-lb. lots: No. 17, $5 — No. 18, $5.50 — No. 19, $6 — No. 20, $6 65 — No. 21, $7 — No. 22, $7.30 — No 23, $7.65 — No 24, $8 — No. 25, $9 — No. 26, $9 50 — No. 27, $10 — No. 28, $11 — No. 29, $12 — No. 30, $13 — No. 31, $14 — No. 32, $15 — No 33, $16 — No. 34, $17. Extras net—tinned wire, Nos. 17-25, $2—Nos. 26-31, $4—Nos. 32-34, $6. Coppered, 75c.—oiling, 10c.—in 25-lb. bundles, 15c.—in 5 and 10-lb. bundles, 25c.—in 1-lb. hanks, 25c. —in ½-lb. hanks, 38c.—in ¼-lb. hanks, 50c.—packed in casks or cases, 15c.—bagging or papering, 10c.

FENCE STAPLES.

Bright .... 2 75 Galvanized.... 3 15

HAY WIRE IN COILS.

No. 13, $2.60; No. 14, $2 70; No. 15, $2.85; f o b., Montreal.

GALVANIZED WIRE.

Per 100 lb.—Nos. 4 and 5, $3.70 — Nos. 6, 7, 8, $3.15 — No. 9, $2.60 — No. 10, $3 20 — No. 11, $3.25 — No. 12, $2.65 — No. 13, $2.75 — No. 14, $3.75 — No 15, $4.30 — No. 16, $4.30 from stock. Base sizes, Nos. 6 to 9, $2 35 f.o.b. Cleveland. In carlots 12½c. less.

LIGHT STRAIGHTENED WIRE.

Gauge No.	Over 20 in. per 100 lbs.	10 to 20 in.	5 to 10 in
0 to 5	$0.50	$0.75	$1 25
6 to 9	0 75	1 25	2 00
10 to 11	1.00	1 75	2.50
12 to 14	1.50	2.25	3 50
15 to 16	2.00	3.00	4.50

SMOOTH STEEL WIRE.

No 0-9 gauge, $2 30; No. 10 gauge, 6c extra; No. 11 gauge, 12c extra; No 12 gauge, 20c extra; No. 13 gauge, 30c. extra; No 14 gauge, 40c. extra; No. 15 gauge, 55c. extra; No 16 gauge, 70c extra. Add 60c. for coppering and $2 for tinning.

Extra net per 100 lb.—Oiled wire 10c., spring wire $1 25, bright soft drawn 15c., charcoal (extra quality) $1.25, packed in casks or cases 15c., bagging and papering 10c., 50 and 100-lb. bundles 10c., in 25-lb. bundles 15c., in 5 and 10-lb. bundles 25c., in 1-lb hanks, 50c., in ½-lb. hanks 75c., in ¼-lb. hanks $1.

POULTRY NETTING.

2 in mesh 19 w. g., discount 50 and 10 per cent. All others 50 per cent.

WIRE CLOTH

Painted Screen, in 100-ft rolls, $1.62½, per 100 sq. ft.; in 50-ft. rolls, $1 67½, per 100 sq ft. Terms, 2 per cent. off 30 days.

WIRE FENCING.

Galvanized barb........ 2 95
Galvanized, plain twist 3 30
Galvanized barb, f.o.b. Cleveland, $2.70 for small lots and $2 60 for carlots.

WOODENWARE.

CHURNS.

No. 0, $9; No 1, $9; No. 2, $10; No. 3, $11; No. 4, $13; No. 5, $16; f.o.b. Toronto Hamilton, London and St. Marys, 30 and 30 per cent; f o b. Ottawa, Kingston and Montreal, 40 and 15 per cent discount, Taylor-Forbes, 30 and 30 per cent.

CLOTHES REELS.

Davis Clothes Reels, dis. 40 per cent.

LADDERS, EXTENSION.

3 to 6 feet, 11c. per foot; 7 to 10 ft., 12c.
Waggoner Extension Ladders, dis. 40 per cent.

MOPS AND IRONING BOARDS.

"Best" mops........ 1 25
"900" mops 1 25
Folding ironing boards........ 12 00 16 50

REFRIGERATORS

Discount, 40 per cent.

SCREEN DOORS.

Common doors, 2 or 3 panel, walnut stained, 4-in. style........per doz. 7 25
Common doors, 2 or 3 panel, grained only, 4-in., style ....per doz. 7 55
Common doors, 2 or 3 panel, light stair per doz........ 9 55

WASHING MACHINES.

Round, re-acting per doz. 60 00
Square " 63 00
Eclipse, per doz 54 00
Dowswell " 39 00
New Century, per doz 75 00
Daisy........ 54 00

WRINGERS.

Royal Canadian, 11 in., per doz. ... 34 00
Royal American, 11 in. ... 34 00
Ese, 10 in., per doz 37 75
Terms, 2 per cent., 30 days.

MISCELLANEOUS.

AXLE GREASE.

Ordinary, per gross........ 6 00 7 00
Best quality........ 10 00 12 00

BELTING.

Extra, 60 per cent.
Standard, 60 and 10 per cent.
No. 1, not wider than 6 in., 60, 10 and 10 p c.
Agricultural, not wider than 4 in., 75 per cent
Lace leather, per side, 75c.; cut laces, 80c.

BOOT CALKS.

Small and medium, ball ....per M 4 25
Small heel .... " 4 50

CARPET STRETCHERS.

American ....per doz. 1 00 1 50
Bullard's........ " 6 50

CASTORS.

Bed, new list, discount 55 to 57½ per cent.
Plate, discount 52½ to 57½ per cent.

PINE TAR.

½ pint in tins ....per gross ... 7 80
1 " " .... " 9 60

PULLEYS.

Hothouse ....per doz. 0 55 1 00
Axle........ " 0 22 0 33
Screw .... " 0 22 1 00
Awning .... " 0 35 2 50

PUMPS.

Canadian cistern 1 40 2 00
Canadian pitcher spout .. 1 80 3 16
Berg's wing pump, 75 per cent.

ROPE AND TWINE.

Sisal........ 0 10½
Pure Manilla 0 15½
"British" Manilla........ 0 12
Cotton, 3-16 inch and larger... ..0 21 0 23
" 5-32 inch 0 25 0 27
" ⅛ inch 0 25 0 28
Russia Deep Sea 0 16
Jute........ 0 09
Lath Yarn, single 0 10
" double 0 10½
Sisal bed cord, 48 feet........per doz. 0 65
" " 60 feet........ " 0 80
" " 72 feet........ " 0 95

Twine.

Bag, Russian twine, per lb. 0 2
Wrapping, cotton, 3-ply 0 2
" " 4-ply 0 27
Mattress twine per lb. ...... 0 [illegible] 0 45
Staging " " 0 27 0 35

BINDER TWINE

500 feet, sisal 0 09½
500 " standard 0 09½
550 " " manilla 0 10½
600 " " " 0 12½
650 " " " 0 13½
Car lots, ½c. less; 5-ton lots, ¼c. less. Central delivery.

SCALES.

Gurney Standard, 40 per cent.
Gurney Champion, 50 per cent.
Burrow, Stewart & Milne—
Imperial Standard, discount 40 per cent.
Weigh Beams, discount 40 per cent.
Champion Scales, discount 50 per cent.
Fairbanks standard, discount 35 per cent.
" Dominion, discount 55 per cent
" Richelieu, discount 55 per cent.
Warren new Standard, discount 40 per cent.
" " Champion, discount 50 per cent.
" Weighbeams, discount 35 per cent.

STONES—OIL AND SCYTHE.

Washita ....per lb. 0 25 0 37
Hindostan .... " 0 06 0 10
" slip .... " 0 18 0 20
" Axe........ " 0 10
Deer Creek........ " 0 10
Deerlick .... " 0 25
" Axe .... " 0 15
Lily white .... " 0 42
Arkansas........ " 1 50
Water-of-Ayr .... " 0 10
Scythe ....per gross 3 50 5 00
Grind, 40 to 200 lb., per ton.... 20 00 22 00
" under 40 lb., " .... [illegible]
" 200 lb. and over .... [illegible]

ROBERTSON'S MARBLE LAVATORIES

In addition to the goods illustrated, we manufacture and carry in stock a full line of staples, such as Brass Compression, Fuller and Ground Work, Soil Pipe and Fittings, Iron Pipe and Fittings, Engineers', Gasfitters', Steamfitters' and Mill Supplies. A Catalogue for each of these departments will be cheerfully forwarded on application.

The James Robertson Company, Limited

265 King Street West, TORONTO

REPUTATION and PROFIT

You clinch both, Mr. Dealer, by placing your order for

PLUMBERS' SUPPLIES

with a Reliable, Progressive, Twentieth Century House.

Plumbers' Brass Work, Enamelware Closets, Bathroom Woodwork, Iron Pipe and Fittings, Lead Pipe, Lead Traps, etc., etc.

Prompt Deliveries —— *A Specialty*

SOMERVILLE LIMITED

59 Richmond Street East, TORONTO

Persons addressing advertisers will kindly mention having seen this advertisement in Hardware and Metal.

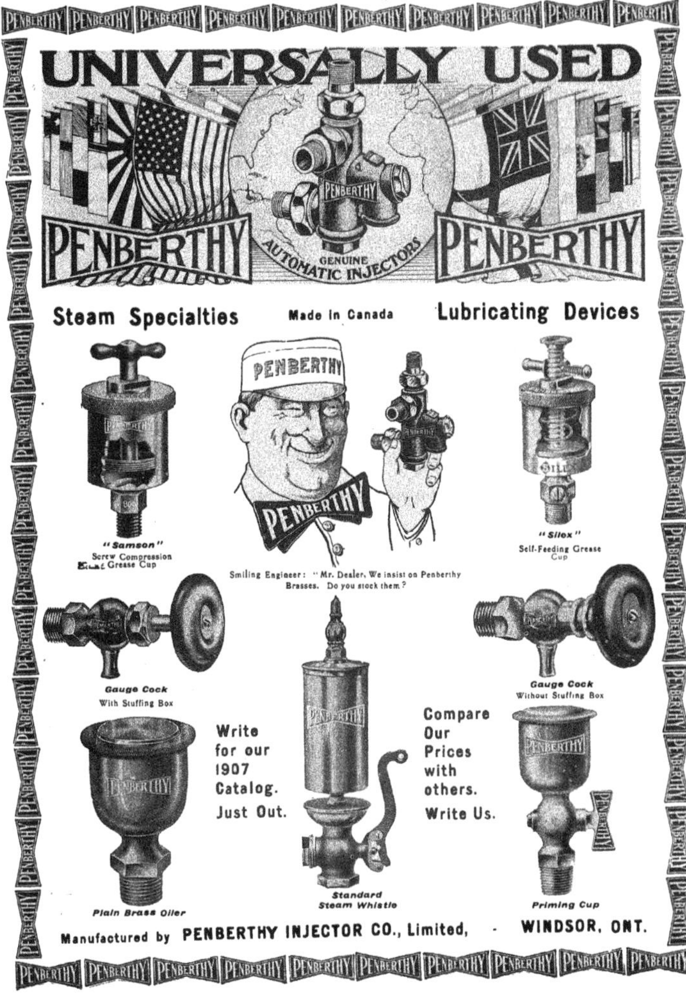
UNIVERSALLY USED
PENBERTHY
GENUINE AUTOMATIC INJECTORS
PENBERTHY
Steam Specialties
Made in Canada
Lubricating Devices
"Samson"
Screw Compression
Grease Cup
Smiling Engineer: "Mr. Dealer, We insist on Penberthy Brasses. Do you stock them?"
"Silex"
Self-Feeding Grease Cup
Gauge Cock
With Stuffing Box
Gauge Cock
Without Stuffing Box
Write for our 1907 Catalog. Just Out.
Compare Our Prices with others. Write Us.
Plain Brass Oiler
Standard Steam Whistle
Priming Cup
Manufactured by PENBERTHY INJECTOR CO., Limited, - WINDSOR, ONT.

The Steamfitters' Worries

¶ Goods ordered weeks ago and not yet delivered—when they arrive they are frequently defective and unsatisfactory your customers are continually asking when their work will be completed.

¶ Isn't that a familiar story?

¶ When you require goods most urgently write us. We carry in stock for your immediate demand the largest assortment of steamfitters' supplies in Canada.

DAISY HOT WATER BOILERS

Wrought Iron Pipe, Cast and Malleable Fittings, Valves, Soil Pipe and Fittings, Lead Traps and Bends, Etc.

CLUFF BROTHERS

Lombard Street, TORONTO

Selling Agents for - - - WARDEN KING & SON, Limited

Let Our Beaver Do Your Work.

You Don't Have to Change Dies.

Cuts 1, 1¼, 1½ and 2" perfect threads, all with one set of chasers

The Hand Stock that Starts Easy and Finishes Easier.

"The New Way"

You Will Find it a sure-enough Beaver.

Write for our special 15-day trial offer.

You cannot afford to take the time to thread pipe by hand in any other way.

Other Sizes of Beavers will be born soon—¼ to ¾—2½ to 4—4½ to 8—8 to 12.

Manufactured by

Borden-Canadian Company

166 Adelaide St., West, Toronto

THE TURNER TORCHES AND FURNACES ARE THE HOTTEST ON EARTH

PROVE IT BY ORDERING A TRIAL SHIPMENT FROM YOUR JOBBER

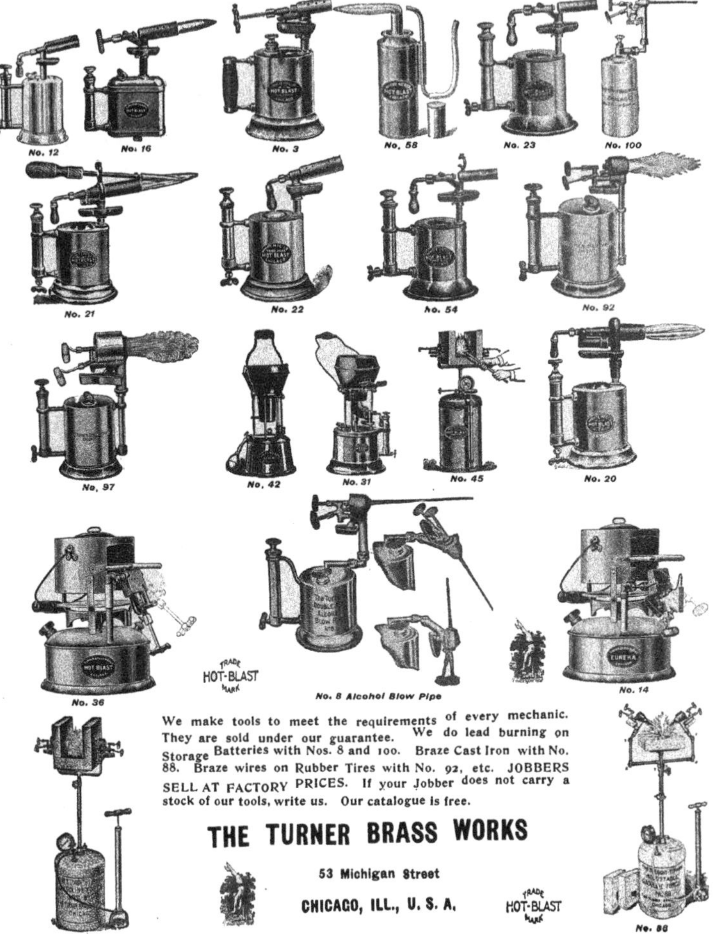

AN IMPORTAN

"WORDS ARE GOOD IF BACKED UP BY DEEDS"

Just remember Roosevelt's famous utterance and if we fail to give you reasonable proofs of the solidarity of our claims, well — pay no attention to this pronouncement.

We manufacture ***Varnish Turpentine,*** and we challenge one and all competitors to place any other turpentine on the market nearly as good.

Our ***Varnish Turpentine*** is produced by a secret and patented process and we guarantee it to be absolutely free from benzine or naphtha. Our

VARNI

TURPEN

dries as quickly as Spirits of Turpentine, by the same process of oxidation and evaporation, and positively cannot leave tacky varnish.

Our ***Varnish Turpentine,*** having a fire test forty degrees higher than benzine, naphtha or kerosene, is absolute and convincing evidence that it contains none of these. It works just like spirits but is infinitely safer.

We guarantee that every barrel of our ***Varnish Turpentine*** will do any work that spirits will do. If not—

RETURN AT OUR EXPENSE

We will mail you a long list of names of men in your business who have used it for years, and ***they are using it to-day.***

A FREE TRIAL—Prove the virtues of our ***Varnish Turpentine at our expense.*** It's heads you win, tails we lose. Can you suggest a fairer way of doing business?

WRITE TO-DAY—IT WILL PAY YOU

The Defiance Mfg. and Supp y o

TORONTO - - - ONTARIO

Advertising that Hits the Mark

0% of ***SHERWIN-WILLIAMS ADVERTISING*** is aimed directly at the paint consumer. We do not shoot into the air and trust to luck that the shot will hit someone interested in paint. We single out each possible buyer and *make sure* of reaching him personally. ***THE SHERWIN-WILLIAMS ADVERTISING*** is to the individual, not to the crowd. It is personal advertising, not just general publicity. Thirty years of active paint advertising have shown us that this direct, personal method is the best for getting business in each community. Of course, we believe in general publicity, too, and each year we are using it more and more; but we first take care in our appropriations of the personal, direct advertising that reaches the real paint buyers in a town. General publicity can never take the place of this kind of work in its effectiveness.

All this leads to the point we want to make and we want you to appreciate—that ***The Sherwin-Williams Agency*** is growing more valuable every year. The dealer who handles our line not only sells the highest standard paints and varnishes, but has behind him all these advertising forces to help him build up a big, profitable business. Good quality goods, backed by our forceful and consistent advertising, have placed us and those who handle our products away ahead of all competition.

Here are the advertising shots that will *hit the mark and land orders for you.* Read these carefully—no other manufacturer is in a position to help you create sales with such advertising and selling helps.

SHOT No. 1— To All Possible Paint Buyers. Just before the painting season we mail to list of property owners furnished by the agent, a Calendar bearing his name. The calendar runs from one painting season to the next, and contains, besides the plates of colored houses, an envelope of paint samples. Each year there is a new design and our agent is given plans and methods for opening the paint season with a rush when the calendars are mailed.

SHOT No. 2— To Those Who are Planning House Painting. All through the year our branch advertising departments in the office nearest the agent send letters, booklets and mailing cards to all names he sends in of persons who intend to paint. The letters are not cheap, but are mailed under two-cent postage. They are a personal appeal and make sales.

SHOT No. 3— To Those Whose Houses Need Painting. The agent frequently sees houses badly in need of paint. The owners don't know it or are putting it off. When he sends their names to us we write them personal letters, and follow them up to induce them to see him about painting. These letters very often go where the agent cannot go himself.

SHOT No. 4— To Painters and Contractors. We want our agent to keep us supplied with lists of painters and contractors. To them we mail direct, good attractive advertising for our painters' specialties. We do this at regular intervals during each spring and fall. Besides we are ready to write personal letters and send literature to the painters at any time of year on any line of our goods that the agent may be anxious to push.

SHOT No. 5— To Architects. For these men, whose help and co-operation the agent so much needs, we publish a monthly magazine. It is artistic, humorous and practical. It's one of the few advertisements that architects really read. Many of them keep every issue on file.

OTHER SHOTS—Besides these five leading "shots," we work personally with our agents in many other ways to help them get "next" to the paint buyers. Here's a partial list:

Two big special paint and varnish campaigns—Spring and Fall—for local use.

The handsomest line of store advertising in the field.

A big assortment of electrotypes for newspaper, hand bill and program advertising.

Posters, fence signs, field signs, etc., for outdoor advertising.

A monthly magazine, brimfull of useful hints for building business.

An 80-page book of catchy and up-to-date window trims.

Special cut-out displays for windows.

The special service of our editorial and art departments in the preparation of any advertising the agent may be doing for any department in his store.

MORE SHOTS—Besides all this (and dozens of other direct advertising features not mentioned), we keep up a big campaign of general publicity—railway bulletins, posters, etc.—to help make The Sherwin-Williams Co. widely known to all people. Within another seven months a national magazine campaign will be added to this.

This describes briefly our advertising helps and how we go after business for you. Write us *at once* for our agency proposition. Let us tell you what an *attractive offer* we can make you and how you can get the paint and varnish business of your locality with *The Sherwin-Williams Line.* Address:

THE SHERWIN-WILLIAMS CO. **LARGEST PAINT AND VARNISH MAKERS IN THE WORLD**

GENERAL OFFICES AND WAREHOUSE: 639 CENTRE STREET, MONTREAL, QUE.

10

A Practical Paint for Practical Painters.

A Profit and Prestige-builder for Live Dealers.

Advertising Methods that Draw Trade to Our Agents.

Give 100 % Satisfaction.

Ask us about our exclusive agency proposition—it is worth looking into.

THE MARTIN-SENOUR CO., Ltd.

"PIONEERS OF PURE PAINTS"

MONTREAL and CHICAGO

The Winnipeg Paint and Glass Co., Ltd. Winnipeg

The Maritime Paint and Varnish Co., Ltd. Halifax, N.S.

It DRIES HARD in one night—Has a beautiful LUSTROUS finish that water will not affect—Won't peel or crack—Cements the cracks and crevices — WEARS LIKE IRON.

THE MARTIN-SENOUR CO., Ltd.

"PIONEERS OF PURE PAINTS"

MONTREAL and CHICAGO

The Winnipeg Paint and Glass Co., Ltd. Winnipeg

The Maritime Paint and Varnish Co. Ltd. Halifax, N.S.

PROMPT DELIVERIES

A Specialty

There you have the difference between success and commercial disaster. We have demonstrated to our own satisfaction and to the entire satisfaction of an army of Hardware Dealers that the next best thing to first quality goods is——**Prompt Shipments.**

Pearcy's Pure Prepared PAINTS

captured the market long since, because of merit and merit alone. It's their sheer merit and superiority over all other brands of prepared paints which **hold the confidence** of of the trade against a host of new-comers.

A Potential Majority of Hardwaremen recommend **Pearcy's Pure Prepared Paints,** because they **assure continuity of trade, profit and reputation.**

Could you desire stronger reasons? Have you our list of

PAINTERS' SUPPLIES?

It will pay you, and pay you will, to place your orders with a thoroughly tried and tested concern—the concern familiarly known as

THE GOOD OLD RELIABLE HOUSE

Sanderson Pearcy & Co., Limited

61, 63, 65 Adelaide St. West, Toronto, Ont.

THE STANDARD PAINT CO.

OF CANADA, Limited

Now having its new Canadian factory near Highland in full operation can now offer to the trade as well as consumer, all of the well-known products so extensively used, almost over the entire Globe. Ruberoid Roofing, the standard for 16 years, durable, elastic and acid-proof as well as fire-resisting, easily applied and always giving full satisfaction.

P. & B. Sheathing and Insulating Papers.
P. & B. Electrical Compounds, Paints and Varnishes
P. & B. Tape for Electrical Insulation.
S. P. C. Paint for Protection of Metal.
Ruberine " " Roofs.

THE STANDARD PAINT COMPANY
OF CANADA, Limited

Room 119 Board of Trade Bldg., MONTREAL.
Selling Agents Ontario and Quebe.
LOCKERBY & McCOMB

SOLVED AT LAST

Your problem is solved at last. The **Cleveland Wall Paper Cleaner** has captured the U.S. markets because of merit and merit alone. There is no other wall paper cleaner nearly as good.

Guaranteed To Keep For Two Years

Not Dangerous. Does Not Contain Acids

The Cleveland Wall Paper Cleaner

does clean wall paper and frescoed walls. It does not streak or spot them, and with ordinary care, almost any one can use it. Let us forward you particulars regarding **your** margin of profit.

Put up in cases of 3 doz. one pound cans
Bulk—10, 25 and 50 pound cans. Prompt Shipments

FRANK U. S. GILBERT
1224 Woodland Ave. CLEVELAND, OHIO

Just try some ordinary green paint on a panel of wood, see how hard it works,—don't cover, lacks body. Then try on another panel

LUCAS IMPERIAL FRENCH GREEN

notice how easy it spreads, how well it covers,—that's body. Makes the cost less for this reason alone, and stays fresh and brilliant. It preserves the wood, keeps the weather out and the color in. It is the **ONE** Green that is free from arsenic or verdigris.

BEWARE OF IMITATIONS

JOHN LUCAS & CO.
NEW YORK, PHILADELPHIA, CHICAGO

THE

Winnipeg Paint and Glass Co.

LIMITED

Our name stands for Quality, Goods, Right Prices, Prompt Shipments, Fair Dealing.
If it's Paint or Paint Specialty Goods we have it. Complete Paint Catalogue just out, write for one.

WE MAKE SHOW-CASES, OFFICE, BANK AND BAR FIXTURES

Calgary — Warehouses — Winnipeg

AL B

is a high-grade proposition.

ALABASTINE

is the most saleable Wall Coating in Canada. "We keep it, but don't sell it," is never said of **ALABASTINE**. You can win and keep the confidence of customers with **ALABASTINE.**

ALABAST

is standard, the demand is increasing, and this season every Canadian home in which a paper or magazine is read will have **ALABASTINE** brought to their notice as never before.

ALABASTINE

obtains new and retains old customers.

Our publicity campaign also includes a liberal supply of tastily printed literature, handsome show-cards, hangers, etc.

In our 32-page pamphlet **"HOMES HEALTHFUL AND BEAUTIFUL,"** we show illustrations in colors of actual **ALABASTINE** decorations. This is without doubt the finest advertising booklet in the trade. Dealers receive copies free, the public pays 10c a copy.

ALABASTIN

is the best, and the best is the cheapest.

Have you ordered yet? The season is now on, and a well assorted stock assures good sales.

The ALABASTINE Company, Lim

Sell Your Paint at a Profit

Refuse to sell paint that you can't make a good profit on.

No use to sell inferior paint, either. This might pay you for a while, but you lose in the long run.

To make a success of your paint department, you must remember one important point at least. That is, to sell paint that leaves you profit enough to pay for your trouble, and at the same time, of such a quality that your customers get their money's worth.

Jamieson's Floor Enamels

We make these enamels in any color desired. Their beautiful finish, wearing qualities, and freedom from that stickiness which is common to most floor **PAINTS**, put them in a class by themselves.

Ask our prices on the following lines:

Varnishes — **Paints**
Japans — **Colors**
Lacquers — **Glues**
Stains — **Bronzes**
Fillers — **Chamois**
Sponges

POLIO

Jamieson's Goods

means satisfied customers, satisfied customers mean more business, more business means more profit (if the goods sold are Jamieson's) and that's what you're in business for. Why not, therefore, sell our paints and reach success by the short line?

Your name and address please?

R. C. JAMIESON & CO.

Manufacturers MONTREAL

MANUFACTURERS OF

Permanent Reds

ens

reens, Blues

FOR AGRICULTURAL IMPLEMENTS, WAGONS, ETC.

Island City Floor Paint Dries in Eight Hours.

G House Paints

MADE FROM THE PUREST MATERIALS. **SEND FOR SAMPLE CARD SHOWING 56 COLORS.**

D CITY

Coach Varnishes.
Railway Varnishes
Furniture Varnishes.
Interior Varnishes.

Coach Colors.
Enamel Paints.
Japans.

Wood Stains. Liquid Aluminum Bronze.
Liquid Gold Bronze.
Pure White Enamel for Dipping or Brush Work.
Enamel Primer for First Coating.

ISLAND CITY Pure White Lead. Pure Decorators' Lead.

OUR ISLAND CITY PURE WHITE PAINT

is non-poisonous, and two coats cover as much as three coats of PURE LEAD.

PURE WHITE ZINC.

THE HOUSE OF QUALITY. Known everywhere for reliability and promptitude.

Our system of advertising helps to sell the goods, and then they are all BACKED BY A GUARANTEE. What more do you want?

DO NOT FORGET THE BRAND "ISLAND CITY."

NOTHING PAYS BETTER.

D & CO

Toronto, Vancouver, Montreal.

LOOK

in our Trade Catalogue No. 10, page 33, and read on the sign

"Imperial Aluminum Paint Does It"

These five words put the story in a nutshell. It does anything Aluminum Paint is supposed to do, and does it so well, we, our agents, and their customers are all satisfied.

MANUFACTURED BY

The Imperial Varnish & Color Co. LIMITED

TORONTO, CANADA.

Stock at
WINNIPEG, MAN., c/o MASON & HICKEY.
VANCOUVER, B.C., c/o DURAND MERCHANDISE CO.

McCaskill, Dougall & Co.

Manufacturers
RAILWAY, CARRIAGE AND BOAT **VARNISHES.**
HIGH-GRADE FURNITURE and HOUSE VARNISHES

MONTREAL.

GLAZIERS' DIAMONDS OF PROVED WORTH

Having a Century's Reputation for Reliability.
MADE BY

A. SHAW & SON, London, Eng.

CANADIAN AGENT

GODFREY S. PELTON, 201 Coristine Building, Montreal

WE SPECIALLY CATER FOR

PAINT GRINDERS.

Our Pure Greens, Permanent Reds, Blues, Lakes, etc. are worth ***attention.*** All qualities, with ***prices that talk!*** ***Write*** to-day for samples to

WM. O. GREENWAY, 13 St. John St., MONTREAL

W. J. COWAN & SONS

Expert Color Manufacturers,
BETHNAL GREEN, LONDON, E. (England)

Are You After More Business?

We have made a Book of Unique Cuts for Advertisers

It contains proofs in color of the quaintest, oddest and most original cuts you ever saw.

Intended for people in search of the unusual or something different that attracts attention and holds the interest.

For the use of advertisers who would appreciate dashy, brilliant and artistic illustrations for pamphlet covers, circulars, mailing cards and newspaper advertisements.

They have been prepared by our Mr. Gibbs who for 10 years designed the world famous National Cash Register advertising.

There are 32 pages to the book, showing about one hundred cuts, together with suggestions for reading matter to use with each cut. The printing of the book alone is a work of art, as it is made in combinations of brilliant colors.

Price 25c (send stamps). It is easily worth ten times that amount for the ideas and suggestions alone that it contains. One advertising man in Chicago saw the rough proofs of this book and ordered $50.00 worth of cuts on the spot. To quote his eXpression:

"I never saw such remarkable cuts in all my life. I can easily use $200 worth. They are not only beautiful and most unusual, but they have strong advertising value."

The cuts sell for $1.00, $2.00, $2.50, $3.00 and $3.50 each in sets of two-color plates. They are all ready for the printer.

Send your 25c for a copy of this book. Get the book even though you may not need the cuts. If you are fond of good advertising and admire the unusual, you will find the 25c you pay for this book an excellent investment.

The Patterson-Gibbs Co.

Advertising Engineers

Heyworth Building, Chicago

ON DUTY

Pure Paint Pays Pleasing Profit

Our paints **are pure paints,** and in handling them you have the satisfaction of pleasing your customers, and at the same time making good money for yourself.

We put nothing but the very best materials into all our lines, with the result that two coats will equal three of any cheaper article. They retain their brilliant, fresh appearance for a long period, and come up to the expectations of your customers.

Many manufacturers **claim** these qualities for their goods, but we would like the chance to **prove** them. Won't you let us have a sample order?

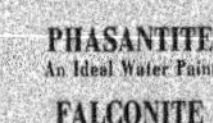

PHASANTITE
An Ideal Water Paint

FALCONITE
White Enamel of High Grade

LACKERITE
Our New Varnish—Stains in 12 Colors

READY MIXED PAINTS — In 3 Grades: FALCON, THREE LEGS and DIAMOND

PETRIFYING LIQUID

FOR DAMP WALLS, HOSPITAL WARDS, ETC.

ESTABLISHED 1796

WILKINSON, HEYWOOD & CLARK LIMITED

300 Notre Dame Street West, Montreal

Branches and Agencies all over the world

FRANCIS C. REYNOLDS
Canadian Manager

McArthur, Corneille & Co.

MONTREAL

Glue and Gelatine

An extensive assortment, to suit all requirements.
WILL BE PLEASED TO SUBMIT SAMPLES AND PRICES

MANUFACTURERS AND IMPORTERS OF . . .

White Lead, Oils and Colors, Prepared Paints, Window Glass, Varnishes, Etc.

SELLING AGENTS IN CANADA

For the GENUINE	And CELEBRATED
Imperial French Green	English Varnishes
of JOHN LUCAS & CO., PHILADELPHIA.	of CHAS. TURNER & SON, LONDON.

Please mention HARDWARE AND METAL when writing.

SPRING TIME IS ALWAYS TIN TIME

And we invite all the big users of cans to get ready early.
There is only one place for everything in tin cans. ***ASK US!***

Sap Buckets, Syrup Cans, round and square,
Friction Top and other Paint Cans
Lead and Color Irons, Putty Boxes and Irons
Jacketted Shipping Cans, all sizes
Square and Oblong Varnish and Oil Cans
Lye Tins and Caustic Drums.

Paint makers and Color workers will profit by dealing with us.
Write or 'Phone Promptitude our motto.

Cans that please, Cans that will outwear any others,
Cans that are made by high-class workers and with the best material.

Tin Can Screws, Tin Chaplets, Roofing Caps.

ACME CAN WORKS

Ontario Street East, - MONTREAL

LIQUID MARINE PAINTS

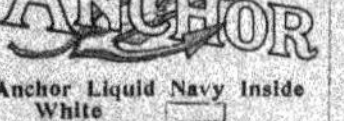

Anchor Liquid Navy Inside White

This flat White Finish is specially prepared for Yacht and Cabin use. A snow white Paint with no gloss.

Anchor Liquid Marine Zinc White

dries hard and has great lustre and covering power. It withstands the action of sea water, and is an absolutely permanent White, unaffected by the gases of the atmosphere.

Anchor Liquid Marine Black

is the glossiest, blackest and most permanent, quick-drying ship's black in use to-day.

Marine Light Green,
Marine Green,
Marine Green Boot-Topping,
Marine Dark Green Verdigris,
Anchor Liquid Marine Greens

constitute the greens that are usually required for Marine purposes. They dry quickly with a fine lustre, and are as durable, both as to wearing properties and as to color, as they can be made.

Anchor Liquid Marine Mast Color

dries quickly, forming the smooth, elastic, durable coating so necessary for this purpose.

Anchor Liquid Marine Seam Paint

The best Composition for filling Vessel seams after being caulked. Being elastic, it will expand and contract with the swelling and shrinking of the wood.

Anchor Aluminum Paint

produces a lasting surface of silvery white. Is being more and more used for marine purposes, being specially adapted to use on ventilators, gas fixtures, furnaces, pipe coils, cabin fixtures and the like. It does not tarnish.

Marine Red Boot-Topping,
Marine Salmon Boot-Topping
Marine Buff Funnel,
Marine P. & B. Funnel,
Marine Plant Indian Red,
Marine Battle Purple,
Marine Lower Hold Composition,
Marine Red Oxide.

ANCHOR MARINE PAINTS are put up in cases of 10-1, 20-½, or 40-¼ Imperial Gallons. 8-Gallon Cases furnished when necessary.

Oxidized Deck Paints

These Paints are made to withstand exposure and the wear of washing.

Pearl,
Light Slate,
Dark Slate,
Light Nile Green,
Red Oxide,
Purple Oxide,
Yellow,
Brown,
Stone.

Marine Varnishes

Smoke Stack Varnish,

Best Spar Varnish,
For the highest class Marine work.

Batavia Damar Varnish,
For Cabins.

Gold Medal Varnish,
For general Marine purposes.

Best Boat Varnish.

Anchor Marine Enamel Whites

Anchor Enamel Yacht White

ANCHOR ENAMEL YACHT WHITE is very elastic and will not chip or peel off. It is manufactured to withstand the action of salt or fresh water and to maintain its gloss.

Anchor Japanese Enamel Cabin White.

BRAND COPPER PAINTS Anti-fouling, for the bottoms of vessels and boats, as a preservative against barnacles and marine grasses, etc.

COPPER PAINT is put up in 10-1's, 20-½s, or 40-¼s.

ANCHOR BRAND BROWN COPPER PAINT	ANCHOR BRAND GREEN COMPOSITION	SQUID BRAND RED COPPER PAINT

BRANDRAM-HENDERSON
LIMITED

Manufacturers - Halifax, St. John - Montreal - Winnipeg

J. J. TURNER & SONS

Peterborough, Ont.

Largest Manufacturers in Canada of Tents, Flags, Awnings, Sails, Waterproof Goods, Horse Blankets, Lap Rugs, Pack Sacks, Sleeping Bags, Bed Ticks, Nose Bags, Coal Bags, Hame Covers and every description of Camping Goods. Also Canoes, Row Boats, Paddles and Oars. Write for Catalogue.

Trade Mark, Registered

BINDER TWINE FOR 1907

The Four Maple Leaf Brands

BETTER THAN EVER

Manufactured only from suitable materials, properly prepared, spun and balled. Clean, bright, strong fibre: round, smooth, even cord. Twine baled in strong sacks, tied with useful laid ropes (no paper or other ballast).

GILT EDGE, 650 feet, Pure Manila — GOLD LEAF, 600 feet, Manila
SILVER LEAF, 550 feet, Standard Manila — MAPLE LEAF, 500 feet, Standard

Made in Canada by

The Brantford Cordage Co., Limited, Brantford, Ont.

Agents wan'ed in localities where we are not at present fully represented. Write for samples and prices.

ROPE AND CORDAGE—A complete stock. Prices right. Orders and enquiries receive prompt attention.

SIMPLE

DURABLE

RELIABLE

STYLE F.

THE LEADER

A Name Familiar to Butter Makers in Connection with the

Highest Grade Churns

ITS STRONG POINTS:

Stiff Steel Frame, with malleable iron fittings, Combination Bow Lever and Pedal Drive. Bicycle Ball Bearings. Bolted Trunnions. Cream Breakers. Barrel easily drained and conveniently detatchable for cleaning.

DEALERS: When you handle this line you handle QUALITY.

The Dowswell Manufacturing Co., Limited

W. L. HALDIMAND & SON, Montreal
EASTERN AGENTS

HAMILTON
ONTARIO

GALVANIZED FENCE HOOK FOR FASTENING WOODEN PICKET ON WIRE FENCES

WIRE NAILS, COILED SPRING, BARB and PLAIN FENCE WIRE, OILED and ANNEALED, CLOTHES LINE WIRE, STAPLES, etc.

THE WESTERN WIRE & NAIL CO., Limited, LONDON, ONT.

PRIEST'S CLIPPERS

BALL BEARINGS

Largest Variety, Toilet, Hand, Electric Power

ARE THE BEST.

Highest Quality Grooming and Sheep-Shearing Machines.

WE MAKE THEM.

SEND FOR CATALOGUE TO

American Shearer Mfg. Co., Nashua, N.H., U.S.A.

Weibusch & Hilger, Limited special New York representatives, 9-15 Murray Street.

Don't Slight Details

The dealer who slights details won't amount to much from a success standpoint. Some men actually imagine that there is not enough money in

BASKETS

Butcher Baskets, Clothes Baskets, Grain and Root Baskets and Patent Strawboard Berry Boxes

to make stocking them worth while. When you stock our Baskets you make your customers lasting friends. Because we guarantee the goods and then there IS a comfortable profit for you besides.

Immediate Delivery a Specialty

The Oakville Basket Co., Oakville, Ont.

MONEY FOR YOU IN HALTERS

All Dealers do not recognize that fact. Most would—if they would stock our leaders

Hercules and Little Giant Halters

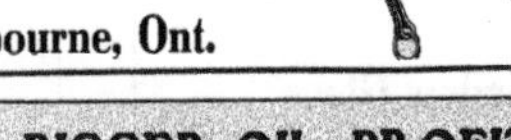

We buy the choicest leather procurable for these halters, and make them just as strong and as good as halters can be made. Our sewed 1 and 1¼ in., and ***"Economy"*** 1 and 1¼ in. should be stocked promptly by every enterprising dealer. Do you carry our line of ***Harness*** and ***Harness parts?*** A post card brings interesting particulars. Write to-day.

G. L. Griffith, - - Melbourne, Ont.

HIGH SPEED STEEL

"NOVO"

Round, Square and Flat Bars.

Milling Cutters, Twist Drills, Reamers, Drill Rods,

Sq Cutters for Tool Holders, Cutter Blanks, Cold Rolled Sheets.

IN STOCK.

WILLIAM ABBOTT

334 St. James Street,
MONTREAL.

BIGGER OIL PROFITS

Are you making any profit on your oil business? Are you making all the profit you ought to make? Or are you losing oil and money through leakage, evaporation, overmeasurement and other causes that always accompany the use of the old-fashioned tin tanks or leaky barrels?

Cut No. 1—Cellar Outfit
For kerosene and similar oils.

You can stop these losses at once and make a good profit on your oil by simply installing a BOWSER SELF-MEASURING OIL STORAGE OUTFIT. With the Bowser these losses are impossible, because the tank is leak and evaporation proof and the pump accurately measures any desired quantity and pumps it directly into the customer's can.

If you want to sell clean oil from a clean store, and at the same time make a good profit, you want a Bowser. Write for free catalogue V, which has full descriptions of the many Bowser Outfits.

S. F. BOWSER & CO., Inc.

66-68 Fraser Ave. Toronto, Can.

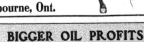

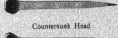

The highest point of perfection is reached, when you adopt the "**Peerless**" style of **Roller Awnings.**

We make them with a reliability and endurance that places our work in the van-guard of the awning trade.

We have other styles also, and make **them** to last—to give general satisfaction.

Catalogue shows how to measure for any style. Easily erected and a pleasure to manipulate.

The Tobin Tent & Awning Co., Ottawa, Ont.

HEAD OFFICE and SHOWROOMS : 201 Sparks St. Phone 3839. **WORKS : 69, 71, 73, 75 Third Ave.**
Long Distance - - - - Phone 432.

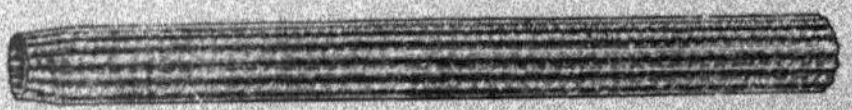

We manufacture Round and Square Corrugated Conductor Pipe, Eavetroughs, Cornices, Skylights, Elbows, Ventilators, Fireproof Metal Windows, etc., and Hot Air Furnaces.

Corrugated Iron Roofing and Siding

STRAIGHT OR CURVED

Goods shipped the same day your order is received.

WHEELER & BAIN

Have you seen our Square Corrugated Pipe?

Toronto

Tents made from the celebrated

"Gourock Tent Duck"

will outwear any others. They are Ideal Tents. Let us tell you why. All styles and sizes kept in stock.

Write for catalogues. *Immediate delivery.*
Special discount to the trade.

Gourock Ropework Export Co.

28 ST. PETER STREET, MONTREAL

Special Typewriter Offer to Hardware Dealers

Retail Price in U.S. $50.00

Special Price to you $40.00 f.o b. Montreal

Get Up-to-date

Like the American.

The New No. 7 Model American is strictly High Grade in every respect and is the first Typewriter ever offered at less than $100.00 possessing all the "Standard" features—universal key-board, ball-bearing carriage, interchangeable steel type bars, printing from ribbon, rapid escapement, unlimited speed, etc., etc. The secret of the low price is in the elimination of parts—other machines 1,500 to 2,000 parts—American 300 parts. Also suitable for travelling men. Weight 11 lbs. Fully guaranteed. Shipped on ten days trial. Write for booklet "The American Way."

KEY AND TYPE ON ONE STEEL BAR

William Stewart & Company

Specialty Department

224 Board of Trade Building, - MONTREAL

Canadian Selling Agents for American Typewriter Co.

SATISFACTION

comes from the use of an article that is made to suit.

A satisfied farrier is always the result of the use of

"M.R.M." HORSE SHOES

The reputation they have earned is due to the strict supervision maintained throughout the process of their manufacture.

MATERIAL used is the best that can be procured for the purpose.

SHAPE required in a perfect shoe has been our special study, and the farrier has less fitting to do in shoeing a horse with "M. R. M." Shoes than is required with most other makes.

CREASING and **PUNCHING** are done in such a way as to insure the proper setting of the Nail.

FINISH of our shoes is clean and smooth, and stamps them as a high-grade article.

"M. R. M." HORSE SHOES are as near perfection as human ingenuity can bring them.

When placing your order for Shoes, specify "M. R. M." and you will get the best.

THIS TRADE MARK  ***STANDS FOR QUALITY.***

The Montreal Rolling Mills Co.

A READY ROOFING that **SELLS** easily, because it has **QUALITY** to back it up, and quality is only obtained by using pure wool felt made under our own inspection, thoroughly saturated with pure asphalt specially selected for this purpose, perfect processes and skilled workmen.

Made in Two Grades.

3 Thicknesses each.

SEE ☞

6 Different Lines, But 1 Quality.

Every Roll bears our "Trade Mark"

"ASPHALT" is coated and finished on both sides with SILICA.

"RUBBER" is coated and finished on both sides with MINERAL RUBBER.

of a Roof with a big "B" in the GABLE.

SEE IT?

The Roofing that will Double Your Sales.

Requires no Coating or Painting.
Absolutely Waterproof.
Practically Fireproof.
Acid and Alkali-proof.
Elastic and Pliable.
Strong and Tough.
Easily Applied.
For Steep or FLAT ROOF.
No Odor—Does not taint water.

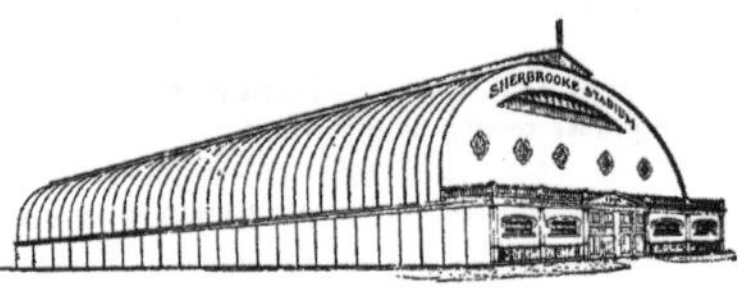

THIS BUILDING IS COVERED WITH IT.

"BRANTFORD-FLOORINE"

As good as Linoleum for floors, but cheaper. 32 inches wide, 36 feet long.

"ASPHALT ROOFING PAINT"

In Special Gallon Cans. 1 doz. per crate.

WRITE FOR PRICES AND SAMPLES.

BRANTFORD ROOFING CO.

LIMITED.

BRANTFORD - CANADA.

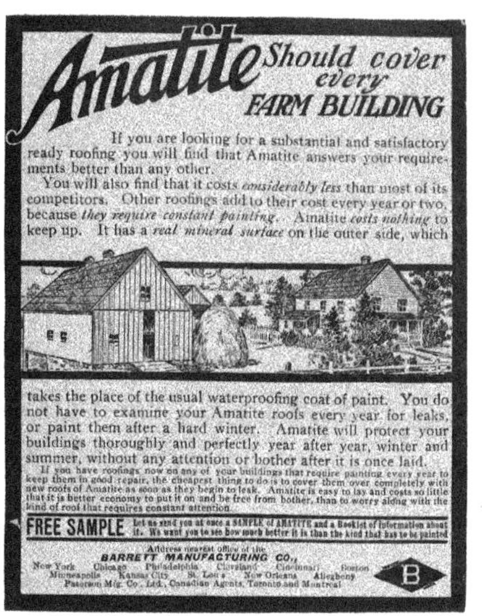

Persons addressing advertisers will kindly mention having seen their advertisement in Hardware and Metal.

Buy the Best.

HERCULES

Sash Cord.

Star Brand Cotton Rope
Star Brand Cotton Clothes Lines
Star Brand Cotton Twine

For Sale by all Wholesale Dealers.

E. T. WRIGHT & CO.

HAMILTON, ONT.

DAIRY PAILS — CALF WEANERS
STRAINER PAILS — MILK CANS
MILK PANS — AND TRIMMINGS
SAP BUCKETS — SAP SPOUTS — SYRUP CANS

Send for our Illustrated Catalogue.

WIRE NAILS
TACKS
WIRE

Prompt Shipment

The ONTARIO TACK CO.
Limited.
HAMILTON ONT.

PAROID ROOFING

The roofing sold on a money-back guarantee

If you are not already selling Paroid Roofing, send for our special proposition. Paroid is the easiest selling ready roofing made. Every roll is sold on a money-back guarantee, and Paroid is the only ready roofing furnished with rust-proof caps.

The best roofing season soon opens and if you are interested, you should write at once. Our advertising plan, which is free to you, will bring customers to your store for other things as well as Paroid. Drop us a postal to-day.

F. W. BIRD & SON, Makers

Established in U.S.A., 1817 **Factory at Hamilton, Ont.**

Offices at Winnipeg, Man., and Hamilton, Ont.

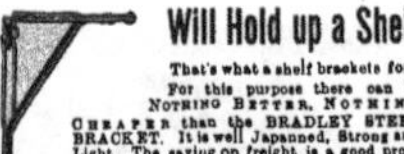

Will Hold up a Shelf

That's what a shelf brackets for.

For this purpose there can be NOTHING BETTER, NOTHING CHEAPER than the BRADLEY STEEL BRACKET. It is well Japanned, Strong and Light. The saving on freight is a good profit aside from the lower price at which the goods are sold. Order direct or through your jobbers

ATLAS MFG. CO.
New Haven, Conn..

COVERT MFG. CO.

Troy, N.Y.

Auto Screw Jack

Harness Snaps, Chain, Rope and Web Goods, etc.

FOR SALE BY JOBBERS AT MFRS. PRICE

STREET PAVING and SIDEWALKS a SPECIALTY

SILICA BARYTIC STONE CO.

OF ONTARIO, Limited.

Water Proof Floors for Malt Houses, Breweries, Slaughter Houses, Cheese Factories, Cellar, Stable Floors, etc.

Head Office
Ingersoll, Ontario.
Walter Mills, General Manage
Ask for quotations for Septic Tanks.

THE JAMES SMART MANUFACTURING CO., LIMITED

BROCKVILLE, ONT. and WINNIPEG, MAN.

HARDWARE DEPARTMENT

Lawn Mowers

The old established and favorably known BROCKVILLE line. In all varieties and sizes. With and without ball bearings. High or low wheels. Three, four or five cutting blades. Ten to twenty inch cut. Unapproached for high finish, durability and perfect working.

Send for Booklets and Quotations

Lawn and Garden Rollers

Four sizes for hand, 250 to 600 lbs. weight. Three sizes for horse, 800 to 1,500 lbs. weight.

"California" Lawn Sprinklers **"Empire" Garden Hose Reels**

Money For You

There's dollars for the shrewd dealer in polishes—but they must be A1 and easy priced. Better make sure. Stock

"Majestic"

brand and you are assured of giving satisfaction. These Polishes are guaranteed to be superior to any other line marketed in Canada.

Send for a test order to-day.

Majestic Polishes, Limited

575 Yonge St.

TORONTO, - CANADA

"Redstone"

High Pressure

Sheet Packing

A packing that will hold. For use in highest pressures for steam, hot or cold water and air. Packs equally well for all.

From actual tests, we believe that this packing is the most durable and satisfactory of any on the market. Try a sample lot and see for yourself.

Manufactured Solely by

THE GUTTA PERCHA & RUBBER MFG. CO.

of TORONTO, LIMITED

HEAD OFFICES,

47 Yonge Street, Toronto.

Branches: Montreal, Winnipeg, Vancouver.

The Peterborough Lock Manufacturing Company, Limited

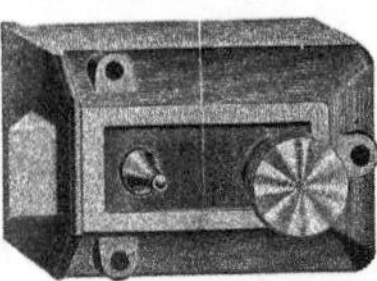

Cylinder Night Latch, No. 108.

Peterborough, Ont.

Manufacturers of all kinds

Rim and Mortise Locks, Inside, Front Door, and Store Door Sets, also full line High-class Builders' Hardware.

Sold by all Leading Jobbers in the Dominion.

THE LONDON FOUNDRY CO., Limited

LONDON, CANADA

SCRAPERS

Wheelbarrows

for Railroad, Dock, Concrete, Farm and Garden

Also

Trucks, for all purposes
Drilling Machines
Forges
Blacksmith Foot Vises, etc.
Wagon Skeins
Bone Cutters
Butter Workers and Shipping Boxes
Cider Mills and Presses

ELECTRIC AND GAS BARGAINS

THESE PRICES TO TRADE ONLY

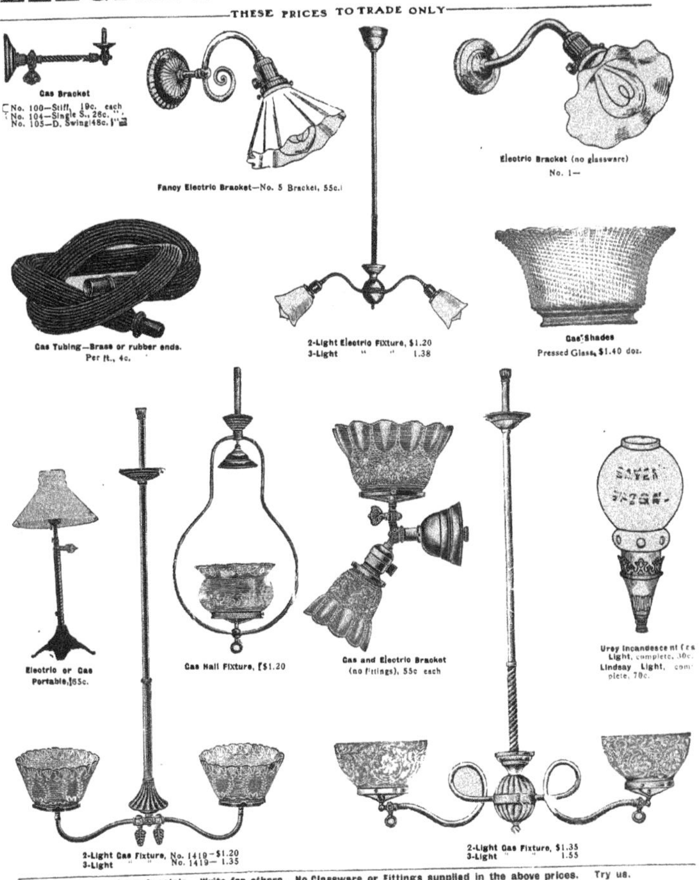

Some of our Specials. Write for others. No Glassware or Fittings supplied in the above prices. Try us.

The SAYER ELECTRIC CO., Everything Electrical and Gas — 10-14 BEAVER HALL HILL, MONTREAL

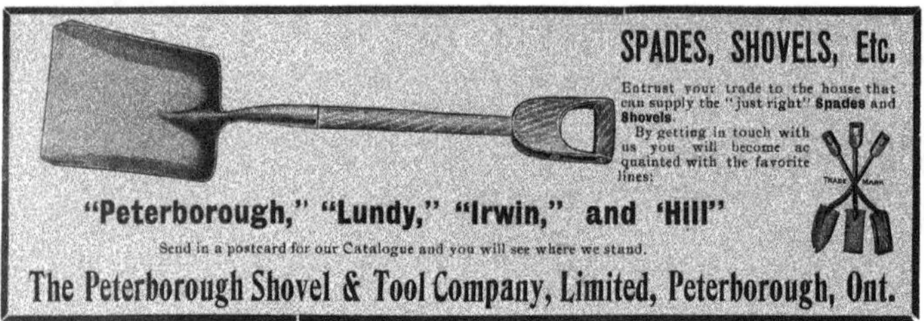

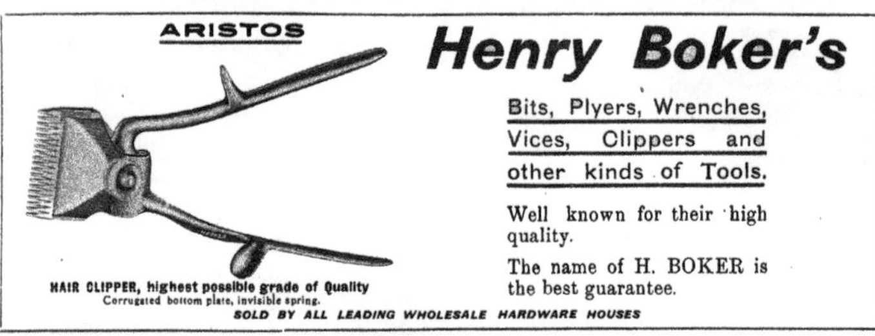

SPECIALISTS

IN AND

MANUFACTURERS and IMPORTERS

OF

GAS AND ELECTRIC

LIGHTING APPLIANCES

FITTINGS FIXTURES GLASSWARE	BEAVER BRAND INCANDESCENT GAS MANTLES	INCANDESCENT COAL OIL LAMPS NO PRESSURE

LIGHTING GOODS ONLY AND A SPECIALTY OF ALL KINDS

INTERNATIONAL GAS APPLIANCE CO., LIMITED

535, 537, 539 Queen St. E., TORONTO, CANADA

THE INDEPENDENT CORDAGE CO.

LIMITED

55 Colborne St.,

TORONTO - CANADA

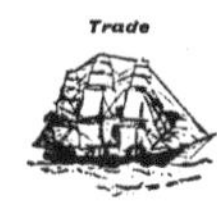

NOW READY

WITH

Plymouth
Pure Manila
Rope

Plymouth
Pure "Prime"
Sisal Rope

and Lath Yarn

ONE CENT

Is it worth one cent to know well not only the paints and brushes you are selling but also those your competitor is selling? A Post Card sent to our advertisers will bring you full information about their goods.

Persons addressing advertisers will kindly mention having seen this advertisement in **Hardware and Metal.**

Wire Cloth

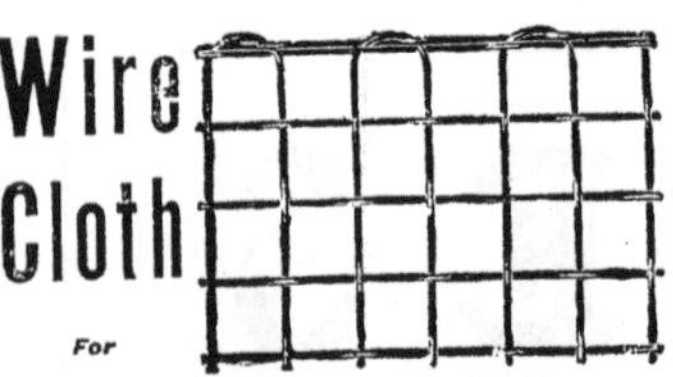

For

Sand Screens, Coal Screens,
Cinder Sifters, Evaporators,
Grain and Seed Cleaning Machinery,
Milk Strainers,
Galvanized and Regalvanized Cloth,
Wire Lathing, Window Protection,
Malt Kiln Floors, Woolen Mills,
Smoke Stacks, Mining Screens,

and for all purposes.

THE B. GREENING WIRE CO., LIMITED

Hamilton, Ont. Montreal, Que.

Sorting the Scrap

Not all scrap makes good bar iron—it must be carefully selected to give satisfactory results.

London Bar Iron is made only from the choicest scrap, selected by experts. Consumers are unanimous that it has the quality.

Orders promptly filled. 62

London Rolling Mills
London, Canada.

Expert Bookkeeping

Investigation and auditing of all classes of accounts undertaken. Trading, profit and loss accounts and balance sheets prepared. Sets of books opened and office systems devised in an up-to-date manner. Collection of accounts a specialty.

Davenport, Pickup & Co.
622 McIntyre Block and 422 Ashdown Block
WINNIPEG MAN

Square your Accounts

INTERNATIONAL PORTLAND CEMENT COMPANY, Limited
OTTAWA, - - ONTARIO.

Manufacturers of strictly high-grade Portland Cement made by the most improved method of manufacture in existence. Each barrel contains three hundred and fifty pounds of pure Portland Cement with no adulteration.

Don't Forget the Name...

NEWMAN'S INVINCIBLE FLOOR SPRINGS

Strong, Quick, Reliable, Effective.

Will close a door against any pressure of wind. Far ahead of ordinary door springs, pneumatic or otherwise. Ask your wholesaler.

W. NEWMAN & SONS, Birmingham.

CLAUSS BRAND DOUBLE SWEDGED BLADE POCKET SCISSORS

Fully Warranted

Hand forged from finest steel. Welded on our Composition Metal. Hardened in water. Tempered by Natural Gas. Full Crocus Finish and nicely nickel plated.

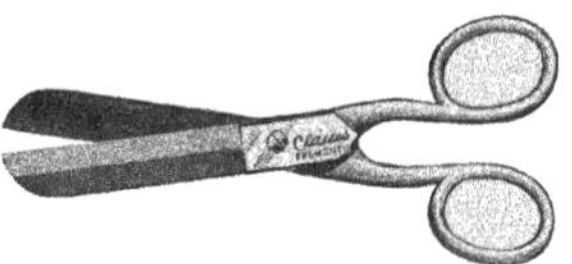

The Clauss Shear Co., - Toronto, Ont.

AN OUNCE OF EXPERIMENT

OFTEN CAUSES A POUND OF REGRET

Let the other fellow do the experimenting while you increase your trade and profits by selling Paterson's Wire Edged Ready Roofing, the only material that has successfully stood the wear and tear of Canada's variable climate for over twenty years.

THE PATERSON MFG. CO., Limited, Toronto and Montreal

QUIT TAKING CHANCES You can't carry on any kind of business campaign to a successful issue—unless you are certain about your base of supplies. When you place an order with us you take out an insurance policy guaranteeing success. Quality—**trade-building** quality is stamped through and through our

READY ROOFING, TARRED FELTS, SHEATHING, WRAPPING AND BUILDING PAPERS

These are our brands

"BLACK DIAMOND" TARRED FELT AND "JOLIETTE" and "CYCLONE" SHEATHING

QUALITY and PRICE always right

ALEX. McARTHUR & CO., Limited, 82 McGILL STREET **MONTREAL.**

F. J. C. COX, Winnipeg, Sole Agent for Northwest Provinces.

We make Electric Fixtures, Sockets, and Cut-Outs

Munderloh & Co

Electrical Supplies of all kinds. — MONTREAL.

Make Your Reputation Solid and Lasting.

There's Just One Sure Way to Do It.

Feature goods which are exclusive goods, with quality stamped all over them.

Goods which mean the come-again customer every time.

The Far-Famed Forsyth Line

familarly known as **THE WISE BUYERS' LINE,** of Household and Hardware Specialties, Bathroom Accessories, Moulding Hooks, Garment Hangers, Bicycle Sundries, etc., etc., have proven themselves sure and certain trade builders. We want permanent agents all over Canada, and are prepared to make our business arrangement a profitable investment for them.

Write to-day. Do not delay.

There are big profits in sight for you.

Write now.

Forsyth Mfg. Co.

BUFFALO, N.Y.

W. F. Canavan, 13 St. John St., Montreal Representative.

Condensed or "Want" Advertisements

Advertisements under this heading 2c. a word first insertion; 1c. a word each subsequent insertion.

Contractions count as one word, but five figures (as $1,000) are allowed as one word.

Cash remittances to cover cost **must** accompany all advertisements. **In no case** can this rule be overlooked. Advertisements received without remittance cannot be acknowledged.

Where replies come to our care to be forwarded, five cents must be added to cost to cover postage, etc.

BUSINESS CHANCES.

HARDWARE and Harness Business for sale in good town. Owner retiring for good reason. Stock about $2,500, in eXcellent shape. This is worth investigating, for particulars. apply Richard Tew, 23 Scott Street, Toronto.

HARDWARE Business; including stoves, tinware and tinsmith tools, in thriving town in West Ontario peninsular, stock about $5,000; building can be leased if desired, dwelling also. Box 583 HARDWARE AND METAL, Toronto. (17)

PLUMBING, Heating, Stoves and Furnishings business for sale—Town of 8,000; first-class farming district; $3,500. Box 595, HARDWARE AND METAL, Toronto. [12]

SITUATIONS VACANT.

BRIGHT, intelligent boy wanted in every town and village in Canada; good pay besides a gift of a watch for good work. Write The MacLean Publishing Company, 10 Front Street E., Toronto. [tf]

EXPERIENCED Hardware Clerk Wanted—One to take change, must be good salesman and stockkeeper. state eXperience and salary eXpected. Box 594, HARDWARE AND METAL, Toronto. [12]

HARDWARE Clerk Wanted — For a retail store; applicant to state age, eXperience and salary eXpected; give name of last employer and how long employed by him, also send references. C. W. Gray, Lethbridge, Alta. [14]

REPRESENTATIVE agent calling upon the wholesale hardware trade in Ontario, wanted to handle a strong line of nails and tacks; only first-class men with connection need apply in confidence to The Bazin Mfg. Co., 94 Arago St., Quebec.

TINSMITH—With 3 or 4 years experience, and with some knowledge of plumbing; state wages; steady job. W. B. Clifton, Alliston, Ont. [14]

WANTED—First-class tinsmith to take charge of shop; must be good furnace man and understand giving estimates. Address Box 593, HARDWARE AND METAL, Toronto. [13]

WANTED—Hardware Clerk, must be good stock keeper, state age, experience and salary expected. Peart Bros., Hardware Co., Limited, Begina, Saskatchewan. [12]

WANTED—EXperienced Hardware Clerk—Must be good stock-keeper, and strictly sober; references required. Hose & Caniff, Kenora, Ont. [13]

WANTED—Immediately, reliable hardware clerk for B.C.; must be good buyer, stockkeeper and salesman of general hardware, etc.; state habits, capabilities, reasons for changing and salary eXpected to Box 596, HARDWARE AND METAL, Toronto.

SITUATION VACANT.

WANTED—First-class tinsmith; married, age 35 to 40; steady work; none other need apply Philip Drench, Walkerville, Ont. [13]

WANTED—EXperienced traveller for builders' supplies; territory, Alberta. Apply Ellis & Grogan, Calgary, Alta.

WANTED—Tinsmith; good general hand for country shop; living rates low, good wages paid weekly; yearly job. Box 600, HARDWARE AND METAL. [14]

SITUATION WANTED

WANTED — By eXperienced hardware clerk and bookkeeper, position as head clerk in retail. Box 598, HARDWARE AND METAL, Toronto.

WANTED—Hardware clerk with 7 years eXperience open for engagement for April 1st; A1 reference; full charge if necessary. Box 599, HARDWARE AND METAL, Toronto. [12]

FOR SALE.

CANADIAN Patent for Sale—Metal "Cramp Plate" for locking horizontal wires to upright stays in wire fence construction; a positive lock; easily manufactured; good investment. P. P. Kee, Brandon, Man. [12]

FOR SALE CHEAP — One hundred gallons paint, well assorted; one St. Thomas acetylene gas machine, with fiXtures for twenty lights. Apply to Box 217, Port Hope.

FOR SALE—Complete set of tinners' tools. Apply Landon & Flesher, Barrie, Ont [11]

FOR SALE—A good set of tinners' tools, including eight-foot brake. Apply Box 597, HARDWARE AND METAL, Toronto. [12]

BOOKKEEPER WANTED

WANTED—A good bookkeeper; stating eXperience, reference and salary. Cunningham Hardware Co., Westminster, B.C. [13]

Persons addressing advertisers will kindly mention having seen this advertisement in Hardware and Metal.

INDEX TO ADVERTISERS.

CLASSIFIED LIST OF ADVERTISEMENTS.

Alabastine.
Alabastine Co, Limited, Paris, Ont

Auditors.
Davenport, Pickup & Co, Winnipeg.

Augers.
Rapid Tool Co, Peterborough

Awnings.
Turner J J, and Sons, Peterborough
Tobin Tent and Awning Co, Ottawa

Axle Grease.
Campbell Mfg Co., Hamilton

Babbitt Metal.
Canada Metal Co., Toronto.
Canadian Fairbanks Co., Montreal
Robertson, Jas Co., Montreal

Baskets.
Oakville, Basket Co, Oakville

Bath Room Fittings
Carriage Mounting Co, Toronto.
Forsyth Mfg. Co., Buffalo, N.Y
Ontario Metal Novelty Co, Toronto

Belting, Hose, etc.
Gutta Percha and Rubber Mfg. Co Toronto.
McLaren, J C, Belting Co, Montreal
Sadler & Haworth Toronto

Bicycles and Accessories.
Johnson's, Iver, Arms and Cycle Works Fitchburg, Mass

Binder Twine.
Brantford Cordage Co, Brantford
Consumers Cordage Co., Montreal.
Independent Cordage Co

Box Strap.
J. N. Warminton, Montreal.

Brass Goods.
Glauber Brass Mfg. Co., Cleveland, Ohio
Kerr Engine Co, Walkerville, Ont.
Lewis, Rice, & Son., Toronto.
Morrison, Jas., Brass Mfg. Co, Toronto
Mueller Mfg Co, Decatur, Ill
Penberthy Injector Co, Windsor, Ont
Taylor-Forbes Co., Guelph, Ont.

Bronze Powders.
Canadian Bronze Powder Works, Montreal.

Builders' Tools and Supplies.
Belleville Hardware Co, Belleville, Ont

Brushes.
Ramsay, A., & Son Co, Montreal.
United Factories, Toronto

Can Openers.
Cumming Mfg Co. Renfrew

Cans.
Acme Can Works, Montreal.

Builders' Tools and Supplies.
Belleville Hardware Co, Belleville
Covert Mfg Co., West Troy, N.Y.
Frothingham & Workman Co, Montreal
Howland, H. S., Sons & Co., Toronto.
Hyde, F., & Co., Montreal.
Lewis Bros. & Co., Montreal.
Lewis, Rice, & Son, Toronto.
Lockerby & McComb, Montreal.
Lufkin Rule Co., Saginaw, Mich.
Newman & Sons, Birmingham.
North Bros Mfg. Co, Philadelphia, Pa.
Stanley Rule & Level Co., New Britain
Stanley Works, New Britain, Conn.
Stephens, G. F., Winnipeg.
Taylor-Forbes Co., Guelph, Ont.

Carriage Accessories.
Carriage Mountings Co., Toronto.
Covert Mfg. Co., West Troy, N.Y.

Carriage Springs and Axles.
Guelph Spring and Axle Co., Guelph.

Carpet Beateas.
Ontario Metal Novelty Co, Toronto.

Cattle and Trace Chains.
Greening, B., Wire Co., Hamilton.

Chains
Oneida Community, Niagara Falls

Churns.
Dowswell Mfg. Co., Hamilton.

Clippers—All Kinds.
American Shearer Mfg. Co., Nashua, N.H

Clothes Reels and Lines.
Hamilton Cotton Co., Hamilton, Ont.

Clutch Nails.
- J. N. Warminton, Montreal

Cordage.
- Consumers' Cordage Co, Montreal.
- Hamilton Cotton Co., Hamilton.

Cork Screws.
- Erie Specialty Co., Erie, Pa.

Cow Ties
- Greening, B, Wire Co., Hamilton
- Oneida Community, Niagara Falls

Counter Check Books
- Carter-Crume Co., Limited, Toronto

Curry Combs
- Eclipse Mfg Co, Ottawa
- Lewis, Samuel & Co, Ltd., Dudley, Eng

Cut Glass.
- Phillips, Geo., & Co., Montreal.

Cutlery—Razors, Scissors, etc.
- Birkett, Thos., & Son Co., Ottawa.
- Clause Shear Co., Toronto
- Dorken Bros. & Co., Montreal.
- Heinisch's, R., Sons Co., Newark. N.J.
- Howland, H. S. Sons & Co., Toronto.
- Hutton, Wm., & Sons, Ltd., London, Eng.
- Lamplough, F. W., & Co., Montreal.
- Phillips, Geo., & Co., Montreal.
- Round, John, & Son, Montreal.

Electric Fixtures.
- Canadian General Electric Co., Toronto.
- Morrison James, Mfg. Co., Toronto.
- Munderloh & Co., Montreal.

Electro Cabinets.
- Cameron & Campbell Toronto.

Emery Wheels
- AjaX Mfg Co., Winnipeg, Man
- BaXter, Patterson Co, Montreal

Enameled Ware
- Kemp Mfg Co, Toronto.

Engines, Supplies, etc.
- Kerr Engine Co., Walkerville, Ont.

Eavetroughs
- Wheeler & Bain, Toronto

Fencing—Woven Wire.
- McGregor-Banwell Fence Co., Walkerville, Ont
- Owen Sound Wire Fence Co., Owen Sound
- Bar-ell HoXie Wire Fence Co, Hamilton.

Files and Rasps.
- Barnett Co. G. & H., Philadelphia, Pa.
- Nicholson File Co., Port Hope

Firearms and Ammunition.
- Hamilton Rifle Co., Plymouth, Mich
- Harrington & Richardson Arms Co., Worcester, Mass.
- Johnson's, Iver, Arms and Cycle Works, Fitchburg, Mass.

Food Choppers
- Enterprise Mfg. Co., Philadelphia, Pa.
- Lamplough, F. W., & Co., Montreal.
- Shirreff Mfg Co, Brockville, Ont.

Galvanizing.
- Canada Metal Co., Toronto
- Montreal Rolling Mills Co, Montreal.
- Ontario Wind Engine & Pump Co, Toronto.

Game Traps.
- Oneida Community, Niagara Falls, Ont

Gas and Electric Fixtures,
- Sayer Electric Co, Montreal.

Glass Ornamental
- O'Shea, J P, & Co., Montreal
- Hobbs Mfg Co., London
- Consolidated Plate Glass Co, Toronto

Glaziers' Diamonds.
- Gibsone, J. B., Montreal.
- Pelton, Godfrey S.
- Sharratt & Newth, London, Eng.
- Shaw, A., & Son, London, Eng.

Grain Scoop.
- Eclipse Mfg Co, Ottawa

Halters.
- Griffith, G L, Melbourne

Handles.
- Still, J. H., Mfg. Co.

Haeness
- Gibson, W. J, Gananoque.

Harvest Tools.
- Maple Leaf Harvest Tool Co, Tillsonburg, Ont.

Hockey Sticks
- Salyerds, E B Preston.
- Still, J. H. Mfg. Co., St. Thomas

Hoop Iron.
- Montreal Rolling Mills Co., Montreal.
- J. N. Warminton, Montreal.

Horse Blankets.
- Heney, E. N., & Co., Montreal.

Horseshoes and Nails.
- Canada Horse Nail Co., Montreal.
- Montreal Rolling Mills, Montreal
- Capewell Horse Nail Co, Toronto

Hot Water Boilers and Radiators.
- Cluff, R. J., & Co Toronto.
- Pease Foundry Co Toronto.
- Taylor-Forbes Co., Guelph

Ice Cream Freezers.
- Dana Mfg. Co, Cincinnati, Ohio
- North Bros. Mfg. Co., Philadelphia, Pa

Ice Cutting Tools.
- Erie Specialty Co, Erie, Pa
- North Bros. Mfg Co., Philadelphia, Pa

Injectors—Automatic.
- Morrison, Jas, Brass Mfg. Co., Toronto.
- Penberthy Injector Co., Windsor, Ont

Iron Pipe.
- Montreal Rolling Mills, Montreal

Iron Pumps.
- Lamplough, F. W., & Co, Montreal.
- McDougall, R., Co, Galt, Ont.

Lanterns.
- Kemp Mfg Co., Toronto.
- Ontario Lantern Co., Hamilton, Ont.
- Wright, E. T., & Co., Hamilton.

Lawn Mowers.
- Birkett, Thos., & Son Co., Ottawa.
- MaXwell, D., & Sons, St. Mary's, Ont.
- Taylor, Forbes Co., Guelph.

Lawn Mower Grinders
- Root Bros. & Co, Plymouth, Ohio

Lawn Swings, Settees, Chairs
- Cumming Mfg. Co., Renfrew.

Ledgers—Loose Leaf.
- Business Systems Toronto
- Copeland-Chatterson Co, Toronto.
- Crain, Rolla L., Co. Ottawa.
- Universal Systems, Toronto.

Linseed Oil
- Canada Linseed Oil Mills, Montreal

Locks, Knobs, Escutcheons, etc.
- Peterborough Lock Mfg. Co., Peterborough, Ont.
- National Hardware Co, Orillia, Ont

Lumbermen's Supplies.
- Pink, Thos, & Co., Pembroke Ont.

Lye
- Gillett, E W, & Co, Toronto

Manufacturers' Agents.
- Fox, C. H., Vancouver.
- Gibb, AleXander, Montreal.
- Scott, Bathgate & Co., Winnipeg.

Metals.
- Canada Iron Furnace Co., Midland, Ont.
- Canada Metal Co., Toronto.
- Eadie, H. G, Montreal
- Gibb, Alexander, Montreal.
- Kemp Mfg. Co, Toronto
- Leslie, A C, & Co., Montreal.
- Lysaght, John, Bristol, Eng.
- Nova Scotia Steel and Coal Co., New Glasgow, N S.
- Robertson, Jas, Co, Montreal
- Roper, J. H., Montreal.
- Samuel, Benjamin & Co, Toronto
- Stairs, Son & Morrow, Halifax, N S
- Thompson, B. & S. H. & Co Montreal

Metal Lath.
- Galt Art Metal Co, Galt
- Metallic Roofing Co., Toronto.
- Metal Shingle & Siding Co., Preston, Ont.

Metal Polish, Emery Cloth, etc
- Oakey, John, & Sons, London, Eng.

Nails Wire
- Dominion Wire Mfg Co, Montreal

Oil Tanks.
- Bowser, S F., & Co, Toronto

Paints, Oils, Varnishes, Glass.
- Brandram-Henderson, Montreal
- Canada Paint Co., Montreal.
- Canadian Oil Co, Toronto
- Consolidated Plate Glass Co, Toronto
- Dods, P. D., & Co, Montreal
- Imperial Varnish and Color Co., Toronto
- Jamieson, R. C., & Co., Montreal.
- Lucas, John & Co, New York
- McArthur, Corneille & Co., Montreal.
- McCaskill, Dougall & Co., Montreal.
- Moore, Benjamin, & Co Toronto.
- Ottawa Paint Works, Ottawa
- Queen City Oil Co., Toronto.
- Ramsay & Son, Montreal.
- Sanderson Pearcy & Co, Toronto
- Sherwin-Williams Co, Montreal.
- Standard Paint Co, Montreal
- Standard Paint and Varnish Works Windsor, Ont.
- Stephens & Co, Winnipeg
- Martin-Senour Co., Chicago.
- Winnipeg Paint & Glass Co., Winnipeg.
- Wilkinson, Heywood & Clark, Montreal

Perforated Sheet Metals.
- Greening, B, Wire Co., Hamilton

Pipe Threading Machines
- Borden-Canadian Co., Toronto.

Plumbers' Tools and Supplies.
- Borden Co., Warren, Ohio.
- Canadian Fairbanks Co., Montreal.
- Cluff, R. J., & Co, Toronto
- Glauber Brass Co., Cleveland, Ohio.
- Jardine, A B, & Co, Hespeler, Ont.
- Jenkins Bros., Boston, Mass.
- Kerr Engine Co, Walkerville, Ont.
- Lewis, Rice, & Son, Toronto.
- Merrell Mfg Co., Toledo, Ohio.
- Mitchell Robert, Co Montreal
- Montreal Rolling Mills Montreal
- Morrison, Jas., Brass Mfg. Co., Toronto.
- Mueller, H, Mfg Co., Decatur, Ill.
- Oshawa Steam & Gas Fitting Co, Oshawa
- Paquet's Maison Jean, Montreal
- Robertson Jas, Co. Montreal
- Robertson, Jas, Co., Limited, Toronto
- Somerville, Limited, Toronto
- Stairs, Son & Morrow, Halifax, N.S
- Standard Ideal Sanitary Co, Port Hope.
- Standard Sanitary Co., Pittsburg.
- Stephens, G F, & Co., Winnipeg, Man
- Turner Brass Works, Chicago.
- Vickery, Orlando, Toronto.

Polishes.
- Majestic Polishes, Toronto

Portland Cement.
- International Portland Cement Co. Ottawa, Ont.
- Hanover Portland Cement Co., Hanover, Ont.
- Hyde, F., & Co., Montreal.
- Thompson B. & S. H. & Co., Montreal

Poultry Netting.
- Greening, B., Wire Co., Hamilton, Ont.

Razors.
- Clauss Shear Co., Toronto

Refrigerators.
- Fabien, C. P., Montreal.

Rivets.
- Parmenter and Bullock Co, Gananoque
- Lewis Samuel & Co., Ltd Dudley, Eng

Roofing Supplies.
- Brantford Roofing Co., Brantford.
- McArthur, AleX., & Co., Montreal.
- Metal Shingle & Siding Co., Preston, Ont.
- Metallic Roofing Co., Toronto.
- Paterson Mfg Co, Toronto & Montreal
- Wheeler and Bain, Toronto

Saws.
- Atkins, E. C, & Co., Indianapolis, Ind
- Shurly & Dietrich, Galt, Ont.
- Spear & Jackson, Sheffield, Eng.

Scales.
- Burrow, Stewart & Milne, Hamilton
- Canadian Fairbanks Co., Montreal.

Screw Cabinets.
- Cameron & Campbell, Toronto.

Screws, Nuts, Bolts.
- Montreal Rolling Mills Co., Montreal.
- Morrow, John, Machine Screw Co., Ingersoll, Ont.

Soil Pipe
- McFarlane, Walter, Glasgow

Sewer Pipes.
- Canadian Sewer Pipe Co., Hamilton
- Hyde, F., & Co., Montreal.

Sharpening Stones
- The Ajax Mfg Co, Winnipeg

Shelf Boxes.
- Cameron & Campbell, Toronto

Shears, Scissors.
- Clauss Shear Co, Toronto

Shelf Brackets.
- Atlas Mfg Co., New Haven, Conn

Shovels and Spades.
- Eclipse Mfg Co, Ottawa
- Peterboro Shovel & Tool Co, Peterboro

Silverware.
- Hutton, Wm, & Sons, Ltd, London, Eng.
- McGlashan, Clarke Co, Niagara Falls, Ont
- Phillips, Geo., & Co., Montreal.
- Round, John, & Son, Sheffield, Eng

Skates.
- Canada Cycle & Motor Co., Toronto.
- McFarlane, Walter, Glasgow

Sprayers
- Cavers Bros., Galt

Spring Hinges, etc.
- Chicago Spring Butt Co, Chicago, Ill.

Sporting Goods
- The Fancy Goods Co. of Canada, Toronto

Stable Fittings
- Dennis Wire & Iron Co, London

Steel Rails.
- Nova Scotia Steel & Coal Co., New Glasgow, N.S

Stove and Furnace Cement
- Sterne, G F, & Sons, Brantford

Stove Pipe.
- Chown, Edwin and Son, Kingston

Stoves, Tinware, Furnaces
- Canadian Heating & Ventilating Co, Owen Sound.
- Canada Stove Works, Harriston, Ont
- Clare Bros. & Co., Preston.
- Copp, W J., Son & Co, Fort William
- Davidson, Thos, Mfg Co., Montreal
- Down Draft Furnace Co., Galt
- Findlay Bros, Carleton Place
- Guelph Stove Co., Guelph.
- Gurney Foundry Co., Toronto.
- Harris, J. W, Co, Montreal.
- Howard, Wm, Toronto
- Joy Mfg Co., Toronto.
- Kemp Mnfg. Co. Toronto.
- McClary Mfg Co London.
- Merrick Anderson, Winnipeg
- Moore D, Co, Limited, Hamilton
- Pease Foundry Co., Toronto.
- Percival Plough & Stove Works, Merrickville.
- Smart, James, Mfg. Co., Brockville
- Stewart, Jas, Mfg. Co, Woodstock, Ont.
- Taylor-Forbes Co., Guelph, Ont
- Telephone City Stoves, Brantford
- Walker Steel Range Co, Grimsby, Ont
- Wright, E. T. & Co., Hamilton.
- 1900 Washer Co, Toronto

Tacks.
- Bazin Mfg Co., Montreal
- Montreal Rolling Mills Co., Montreal
- Ontario Tack Co, Hamilton.

Tents.
- Gourock Rope Work Co, Montreal
- Turner, J J, and Sons Peterborough
- Tobin Tent and Awning Co, Ottawa

Turpentine
- Defiance Mfg Co, Toronto

Twist Drill.
- Abbott, Wm, Montreal

Typewriters.
- Stewart, Wm, and Co, Montreal

Ventilators.
- Harris, J. W, Co., Montreal.
- Pearson, Geo D., Montreal

Wall Paper
- Staunton Limited, Toronto.

Wall Paper Cleaner.
- Gilbert, Frank U S, Cleveland

Washers.
- Parmenter & Bullock Co, Gananoque

Washing Machines, etc
- Dowswell Mfg Co, Hamilton, Ont.
- The Shultz Bros. Co, Brantford.
- Taylor Forbes Co., Guelph Ont
- Woods, Walter, and Co, Hamilton

Wheelbarrows
- London Foundry Co., London, Ont.
- Schultz Bros Co, Ltd, The Brantford.

Wholesale Hardware.
- Amiote Lecours Laveriere Montreal
- Birkett, Thos., & Sons Co., Ottawa.
- Caverhill, Learmont & Co., Montreal.
- Frothingham & Workman, Montreal.
- Hobbs Hardware Co, London
- Howland, H S, Sons & Co, Toronto.
- Kennedy Hardware Co, Toronto
- Lamplough, F. W., & Co, Montreal
- Lewis Bros. & Co., Montreal.
- Lewis, Rice, & Son, Toronto.

Window and Sidewalk Prisms.
- Hobbs Mfg. Co., London, Ont.

Wire, Wire Rope, Cow Ties, Fencing Tools, etc.
- Anthony Wire Fence Co., Walkerville
- Banwell-Hoxie Fence Co, Hamilton
- Dennis Wire and Iron Co., London, Ont.
- Dominion Wire Mnfg. Co, Montreal
- Greening, B., Wire Co, Hamilton
- McGregor Banwell Fence Co, Walkerville
- Owen Sound Wire Fence Co, Owen Sound
- Montreal Rolling Mills Co., Montreal.
- Western Wire & Nail Co, London, Ont

Wrapping Papers.
- Canada Paper Co., Toronto.
- McArthur, Alex., & Co, Montreal.
- Stairs, Son & Morrow, Halifax, N.S.

Wringers
- Connor, J H. & Son, Ottawa, Ont.

Wrenches.
- Baxter Patterson Co., Montreal
- Whitman and Barnes Mfg. Co, St Catharines

VICTOR

SPORTING AND ATHLETIC **GOODS**

The Only Canadian Line

THE ONLY LINE

first, last and always

Satisfactory to the Player
Profitable to the Dealer

These are the marks which are your safeguard.

1907 VICTOR CATALOGUE JUST READY

A card will bring you a copy.

The Fancy Goods Company

OF CANADA, Limited

156 Front Street West - TORONTO

ur Ha

ecia ties

If you are not handling our goods write and obtain Best Prices and Discounts on our well-known lines of **Favorite Churns, Barrows, Bag Trucks** and **Lawn Mowers.** Our Lawn Mowers are further improved for 1907 and lead competition.

They Are Made Better
They Are Finished Better
They Do Better Work
They Sell Better

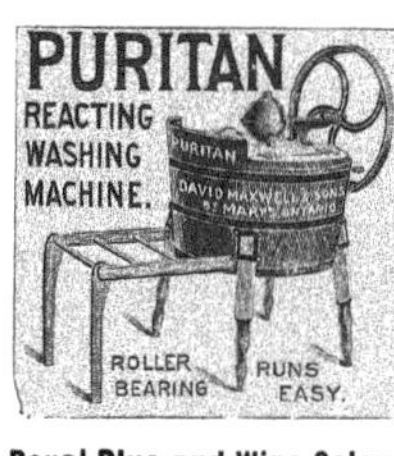

We are now manufacturing a Washing Machine which is in every way superior to anything yet on the market.

It Has New Special Features. Finished in Grained Oak.

Royal Blue and Wine Color, or Silver Aluminum.

If your jobber don't offer **Maxwell** machines don't accept inferior.

WRITE US DIRECT.

MANUFACTURED BY

DAVI MAXWELL & SON

ST. MARYS, ONTARIO

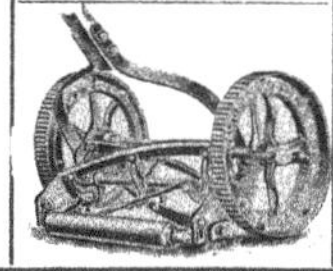

CIRCULATES EVERYWHERE IN CANADA
Also in Great Britain, United States, West Indies, South Africa and Australia.

HARDWARE AND METAL

A Weekly Newspaper Devoted to the Hardware, Metal, Heating and Plumbing Trades in Canada.
Office of Publication, 10 Front Street East, Toronto.

VOL. XIX. MONTREAL, TORONTO, WINNIPEG, MARCH 30, 1907 *NO. 13.*

See Classified List of Advertisements on Page 71.

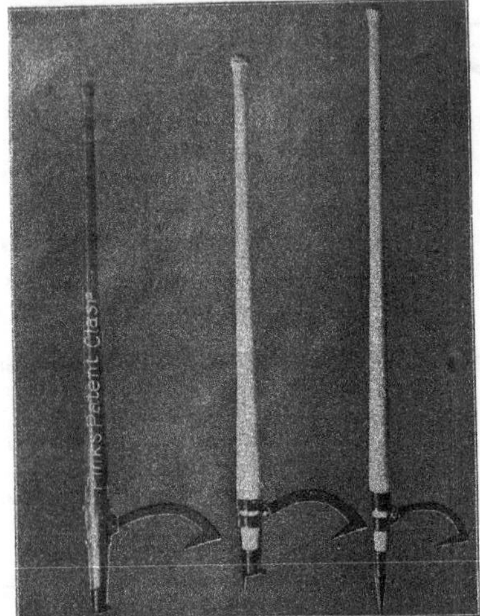

Pink's Lumbering Tools

MADE IN CANADA

Send for Catalogue and Price List

THE STANDARD TOOLS

in every Province of the Dominion, New Zealand, Australia, Etc.

We manufacture all kinds of Lumber Tools

Pink's Patent Open Socket Peaveys.
Pink's Patent Open Socket Cant Dogs.
Pink's Patent Clasp Cant Dogs, all Handled with Split Rock Maple.

These are light and durable tools.

Sold throughout the Dominion by all Wholesale and Retail Hardware Merchants

MANUFACTURED BY

Long Distance Phone No. 87

THOMAS PINK
Pembroke, Ont., Canada.

METALS

ANTIMONY
COPPER
LEAD
TIN
ZINC

Large Stocks **Lowest Prices**

M. & L. Samuel, Benjamin & Co.

TORONTO

A store with attractive weighing machines has an air of reliability about it. The customer has a feeling of confidence in the place and knows that his wants will be supplied with accuracy and promptness.

No. 398, Specially made for the confectionery trade, made in various capacities from 2 to 16 lbs. Has funnel scoop resting on rubber tipped horns. Scoop made of tin or brass.

Howe Scale Co. scales have been renowned for years for their quality, design and finish. They are ornaments in a store, and are accurate and durable.

No. 76, Portable Drop Lever Scale, made in a large range of capacities ; may be had also without drop lever and with iron platform galvanized for salt meat and fish trade.

Do you use **Howe Scale Co. Scales** in your store ?

Do you have them to sell ?

You should both use and sell Howe Scale Co. Scales.

Let us send you a catalogue and quote you prices.

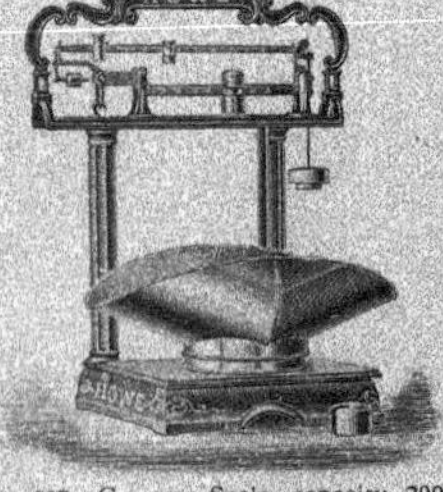

No. 597, Counter Scale, capacity 200 lbs., highly finished in red and gold. The most convenient Counter Scale on the market. Sold with or without scoop.

No. 986 Series, weighs ½ oz. to 40 lbs., made with single or double beam and with tin or brass scoop.

F. & W. Hardware Montreal

F. & W.

FROTHINGHAM & WORKMAN, Limited, MONTREAL, CANADA

THE PRECISION OF

STANDARD DRILLS

Were we asked to tell in two words the secret of the great increase in sales of

STANDARD DRILLS it would be

"Precision" and "Quality"

MADE BY MECHANICS ESPECIALLY SKILLED IN THEIR PARTICULAR BRANCHES, AND EXPERT IN THE USE OF PRECISION GAUGES, WITH WHICH ALL MEASUREMENTS IN THE COURSE OF MANUFACTURE ARE MADE

WITH USING THE HIGHEST GRADE OF MATERIAL AND MODERN MACHINERY AND APPLIANCES THE BEST TO BE HAD IN DRILLS IS THE RESULT.

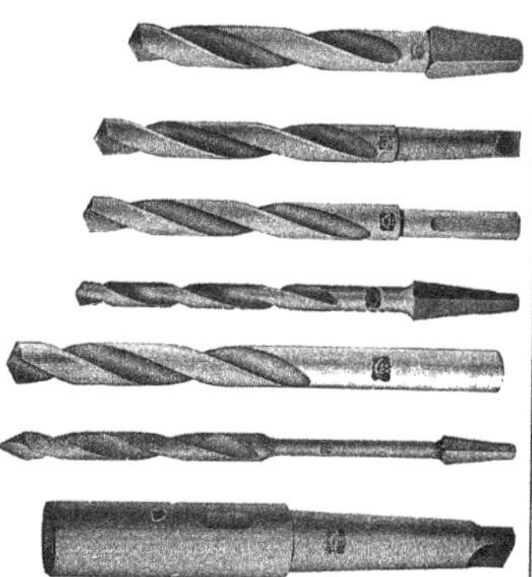

Is it any wonder then that STANDARD DRILLS are now used by the leading

RAILWAYS, CAR MAKERS and LARGE MACHINE SHOPS

When you need Drills let us hear from you; we will gladly co-operate with you with a view to furnishing the best tools adapted to any particular purpose.

Standard Catalogue sent to dealers on request

CAVERHILL, LEARMONT & CO.

Montreal and Winnipeg

Canada Leads the World in the Manufacture of

HIGH-CLASS SAWS

No. 81

The best and cheapest Hand Saw on the market, quality, temper and finish considered. Every saw warranted, Carved applewood handle. Patented, wave finish blade.

No. 1 Maple Leaf Lance Tooth Cross-Cut Saw, and the No. 0 Racer are tempered under our secret process.

No. 0 Narrow Racer Cross-Cut Saw

If you want saws that will sell and give your customer satisfaction, you will stock the Maple Leaf Saws.

Manufactured by **SHURLY & DIETRICH, - GALT, Ontario**

A Good Proposition in Oil Stoves

We have a fine variety for selection. 1, 2, 3 flames; tin or glass basins; nicely Japanned or granite bodies; Cast iron rims and neatly bronzed cast iron stands. Many other patterns.

A postal brings attractive booklet.

F. W. LAMPLOUGH & CO., 9 DeBresoles St., - MONTREAL

Write to-day; it will pay you. This is the Oil Stove time.

SIMPLE

DURABLE

RELIABLE

STYLE F.

THE LEADER

A Name Familiar to Butter Makers in Connection with the

Highest Grade Churns

ITS STRONG POINTS:

Stiff Steel Frame, with malleable iron fittings, Combination Bow Lever and Pedal Drive. Bicycle Ball Bearings. Bolted Trunnions. Cream Breakers. Barrel easily drained and conveniently detatchable for cleaning.

DEALERS: When you handle this line you handle QUALITY.

The Dowswell Manufacturing Co., Limited

W. L. HALDIMAND & SON, Montreal
EASTERN AGENTS

HAMILTON
ONTARIO

GALVANIZED FENCE HOOK FOR FASTENING WOODEN PICKET ON WIRE FENCES

WIRE NAILS, COILED SPRING, BARB and PLAIN FENCE WIRE, OILED and ANNEALED, CLOTHES LINE WIRE, STAPLES, etc.

THE WESTERN WIRE & NAIL CO., Limited, ——LONDON, ONT.

The Peterborough Lock Manufacturing Company, Limited

Cylinder Night Latch, No. 108.

Peterborough, Ont.

Manufacturers of all kinds

Rim and Mortise Locks, Inside, Front Door, and Store Door Sets, also full line High-class Builders' Hardware.

Sold by all Leading Jobbers in the Dominion.

PAROID ROOFING

"THE TIME TESTED KIND"

Plant Steamship Co's, Wharves, Halifax, N.S, covered with Paroid.

Dealers all over Canada and the United States doubled their sales on Paroid last year. Our new advertising plan promises to make a bigger increase for them this year. Paroid is used by the leading railroad systems, manufacturers and farmers throughout Canada. It has stood the test of time and that's the test that tells.

Let us tell you about our advertising plan which will bring customers to your store for other things as well as Paroid. Sign your name to a postal to-day—the best roofing season of the year will soon be here.

F. W. BIRD & SON, Makers

Established in U.S.A. 1817 *Factory at HAMILTON, ONT.*

Offices at Winnipeg, Man., and Hamilton, Ont.

$2 FOR THIS SMALL SUM THE **$2**

MANUFACTURER and SUPPLY MERCHANT

may keep posted on new openings for trade.

The CANADIAN CONTRACT RECORD

reports weekly all projected building and other construction works throughout Canada as well as new business enterprises.

Send your name and address with $2 for a year's subscription to

Canadian Contract Record

Toronto, Montreal, Winnipeg, and Vancouver

$2 **$2**

Auto Spray

Compressed-Air Automatic

Best hand sprayer made. An absolute necessity for every farm and garden. Will repay its cost in one season for potatoes alone. Saves time, labor and material. A boy can do the work. Will run continuously for 6 to 10 minutes.

CAVERS BROS., GALT. ONT.

Sole Agents for the Dominion

The "WAYNE" CARPET BEATER

Handy, Useful and Durable Household Article. "Not a Seasonable Toy"

Is now made in three styles—to retail at popular prices.

Patented Sept. 8, 1903

The "WAYNE" is the best whip upon the market for beating clothes, carpets, rugs and furniture. Constructed of the best coppered steel wire obtainable and mounted on a seasoned hardwood handle.

The wire is fastened to the handle in such a manner that it is impossible to work it loose. A large ferule pressed over the handle prevents the latter from splitting.

The "WAYNE" beater being made of spring steel wire always retains its shape, also eliminates the danger of puncturing or tearing article being cleaned.

This beater is so well balanced that you get full stroke and best results with very little effort. The "WAYNE CARPET BEATER" is a trade winner. Its usefulness is recognized on sight and sells wherever shown. Place your order now.

Manufactured by

Ontario Metal Novelty Mfg. Co.

Limited

111 King Street East, - Toronto, Ont.

Manufacturers of Bath Room Fixtures and Metal Novelties, Dies, Tools, Etc. Look for something new in every week's paper. All our own make. Made in Canada. Send for prices.

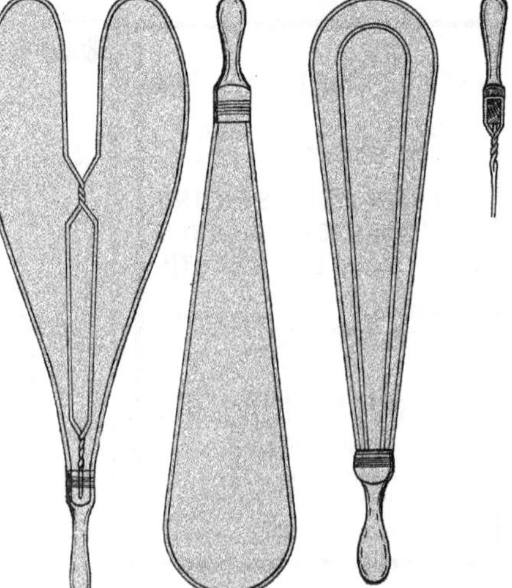

"MAXimum LIGHT GLASS"

Samples and particulars supplied on application.

A residence in the West End of London became vacant because the tenant was not satisfied with the Daylight conditions of ONE ROOM. It was always necessary to burn gas (a continual expense).

The property was vacant for one year. Of the 31 parties who viewed same, 27 complained of the lack of Daylight in this room.

Owner's loss in rent, one year £186 0 0

Someone spoke of the benefits accruing from the use of "Maximum Light Glass."

Argument No. 183

Owner thought of suggestion for one month; Loss in Rent £15 0 0

£201 0 0

Maximum installed as follows:
6 sqrs. 30 x 18 ins., 15 ft. = 1/6 .. .. £1 2 6

This property was let soon afterwards, in fact so soon that the return for rent the second year was £183 0 0

Can you see Argument No. 183?

MAXimum Light Window Glass Co.

Hobbs Manufacturing Co., Limited
Sole Canadian Agents
London Toronto Winnipeg

Dark Interiors ALWAYS Made Light when this Illuminating Glass is used.

The John Morrow Screw, Limited

MANUFACTURERS OF
Set and Cap Screws, Special Milled Work, Engine Studs Etc. Cold Punched Nuts of every variety of finish.

INGERSOLL, ONTARIO

CELEBRATED HEINISCH SHEARS

Tailors' Shears,
Trimmers' Shears,
Tinners' Snips, etc.

ACKNOWLEDGED THE BEST.

R. HEINISCH'S SONS CO. NEW YORK OFFICE, 155 Chambers St. NEWARK, N.J., U.S.A.

SEYMOUR SHEARS

HAVE BEEN THE
Standard for over Half a Century.
"QUALITY UNQUESTIONED."
Each pair of our shears bears the above trade mark.

HENRY T. SEYMOUR SHEAR CO. NEW YORK.
TRADE MARK

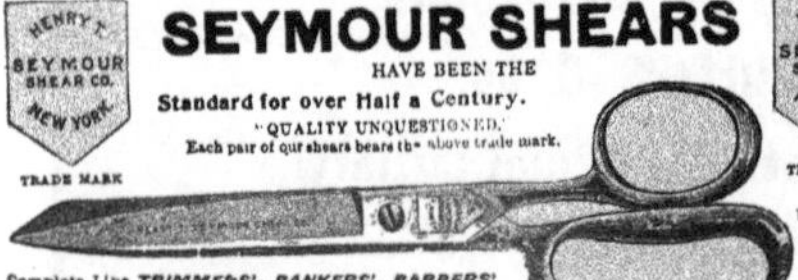

Complete Line *TRIMMERS', BANKERS', BARBERS' and TAILORS' SHEARS, Etc., Etc.*

Latest Catalogue will be sent in exchange for your business card.

Henry T. Seymour Shear Company
WIEBUSCH & HILGER, Limited, NEW YORK, Sole Agents

Covert Mfg. Co.

TROY, N. Y.

Trojan Strap No. 20 — Rein Chains

Harness Snaps, Chain, Rope and Web Goods, etc. For sale by Jobbers, at Manufacturers' prices.

STREET PAVING and SIDEWALKS a SPECIALTY

SILICA and BARYTIC STONE CO. OF ONTARIO, Limited.

Water Proof Floors for Malt Houses, Breweries, Slaughter Houses, Cheese Factories, Cellar, Stable Floors, etc.

Head Office
Ingersoll, Ontario.
Walter Mills, General Manager
Ask for quotations for Septic Tanks.

"The Tool Holder People"

ARMSTRONG BROS. TOOL CO.

Manufacturers of Armstrong Patent Lathe and Planer Tools and other machine shop specialties.

106 N. Francisco Ave. CHICAGO, ILL.

Persons addressing advertisers will kindly mention having seen this advertisement in Hardware and Metal.

Gothic Ceiling Plate, No. 5006.

We want to help you make more money on our

CLASSIFIED Steel Ceilings

They are easier for you to sell than any other ceilings on the market. That's because they are **CLASSIFIED** so as to produce perfect harmony in interior decoration—and sell as cheaply as the meaningless designs of other manufacturers.

Our travellers will be able to boost your ceiling business, if you send them after your prospects. They are at your disposal.

The Metal Shingle & Siding Co.

PRESTON, ONT. Limited

MONTREAL BRANCH FACTORY, Cor. St. Catherine St. and Delorimier Ave.

SALES AGENCIES

ST. JOHN, N.B.	QUEBEC	WINNIPEG	CALGARY
Emerson & Fisher, Ltd.	J. A. Bernard	Clare & Brockest	Ellis & Grogan

Wire Cloth

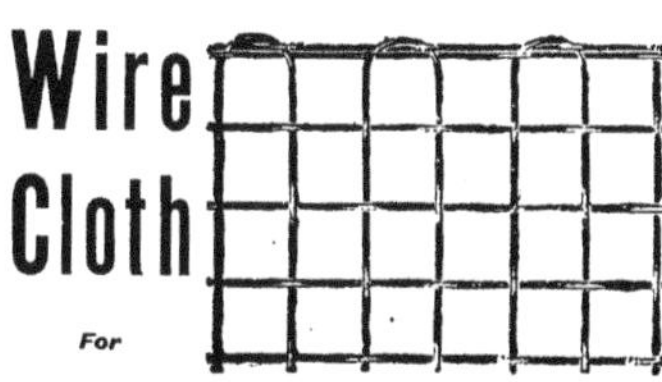

For

Sand Screens, Coal Screens,
Cinder Sifters, Evaporators,
Grain and Seed Cleaning Machinery,
Milk Strainers,
Galvanized and Regalvanized Cloth,
Wire Lathing, Window Protection,
Malt Kiln Floors, Woolen Mills,
Smoke Stacks, Mining Screens,

and for all purposes.

THE B. GREENING WIRE CO., LIMITED

Hamilton, Ont. Montreal, Que.

The best farm fence is one that **stays** after it is put up and doesn't require constant repairs. Strength and wear are sterling qualities in Peerless Woven Wire Fence. Because it demonstrates its ability to fulfill every purpose of a seryiceable fence is the reason that so many miles of it are in use. Wind and weather do not effect it and stock can't go over, under, or break through it for it's made of all No. 9 hard drawn all-steel galvanized wire, locked with the famous

PEERLESS LOCK

the secret of strength in Peerless Fence. It keeps the fence firm and rigid, the wires in place and prevents any openings or breaks being made in it. It allows perfect adjustment to inequalities in ground surface, working equally well on hillsides or level land. The spring in the lateral wires gives it elasticity and there you have it—a perfect fence, no barbs to hurt, no bars to break—a **strong, simple, safe** fence. It's **economical** too—costs little more than plain wire and no expense for repairs. Write for our free fence book—it will interest you.

The Banwell Hoxie Wire Fence Co.,
Dept. J Hamilton, Ont.

BRANTFORD ROOFING

SPECIAL MERITS—

Requires No Coating or Paint.
Acids and Alkali Proof.
Not Affected by Gases or Vapors.
Elastic and Pliable.
Strong and Tough.
Absolutely Water Proof.
Climatic Changes Do Not Affect it.
Both Sides Alike.
Practically Fire Proof.
Non-Volatile.
Easily Applied.
Can be Used on Steep or Flat Surfaces.
Any Workman Can Put it On.
No Odor.
Will Not Shrink or Crack.
Light in Weight.
Does Not Taint Water.

IN SIX VARIETIES

BRANTFORD "ASPHALT." "RUBBER." ROOFING

AT DIFFERENT PRICES

Every dealer in deciding on a line of goods to handle, enquires which is the best ?

invariably they say ***BRANTFORD ROOFING*** leads them all. The first order means a steady customer. We are increasing our sales daily. Carloads of it going out. Carloads raw material coming in.

Samples—Prices—Free. Post Card Will Bring Them.

Brantford Roofing Co., Limited, - Brantford, Can.

LEGAL CARDS

ATWATER, DUCLOS & CHAUVIN
Advocates, Montreal
Albert W. Atwater, K.C., Consulting Counsel for City of Montreal. Chas. A. Duclos. Henry N. Chauvin.

Persons addressing advertisers will kindly mention having seen this advertisement in Hardware and Metal.

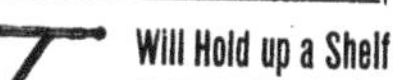

That's what a shelf brackets for.

For this purpose there can be NOTHING BETTER. NOTHING CHEAPER than the BRADLEY STEEL BRACKET. It is well Japanned, Strong and Light. The saving on freight is a good profit aside from the lower price at which the goods are sold. Order direct or through your jobbers

ATLAS MFG. CO.
New Haven, Conn.,

What To Do Next Month

Montreal merchants report the month of March as an off month this year, due, no doubt, to the changeable weather experienced in most parts of Canada. The weather, of course, governs the time when the general spring clean-up shall start, for so long as the snow flies and the "cold north winds do blow" the good housewife considers it winter time, but as soon as Mr. Sun starts to thaw out the frozen world, and the soft breezes begin to travel up from the south, she also begins to throw off her winter inactivity and prepare for the annual renovation.

* * *

Most dealers have ordered and received their spring lines, such as house-cleaning implements, by this time, and they should occupy a prominent place in the show window or store.

* * *

Paints will probably be the most in demand of any article during April. In the cities the exterior, as well as the interior, of most houses, stores, etc., has to receive a new coat of paint, in order to look bright and cheerful, after passing through their storm-swept experiences of the past three or four months. In the country, too, there is a tremendous demand for paints, varnishes and kindred lines, as that buggy must be fixed up and the barn, or house, re-painted. You should handle a good, reliable line of paints if you want to get the lion's share of the trade, as poor paints will begin to blister and peel off shortly after the hot sun gets at them, and this does not tend to make the farmer feel like paying you compliments. You can obtain through your jobber all kinds of advertising matter, explaining the reasons why the paints he handles should be sold, and you should be able to prepare good copy from it for your town paper. Almost any old thing in the paint line will sell for a time, but the dealer who wants to make his paint department pay should handle guaranteed lines only.

* * *

After a woman gets her spring cleaning done, she begins to look round the house for old articles that should be replaced, and the wide-awake hardware merchant can remind her of many articles in her kitchen outfit by preparing a neat circular calling her attention to his line of pots, pans, cutlery, meat and food choppers, bread makers, coffee mills, and the host of things that find their way into every well-regulated household. If you wish to make a success of this scheme, however, you must name prices, as that's the surest way to make a woman sit up and take notice. You may talk all day about great bargains in this and that article without needing any extra help to handle the rush, but give more prices and less "hot air" and you'll get better results.

* * *

You should also watch the papers for notice of any spring weddings about to take place within your vicinity, and then make a bid to supply the happy couple with everything in the hardware of cutlery lines that will be required in the new home. The best way to do this is to make up various-sized lists of kitchen utensils and mark each one "$—— for the lot," but be sure that this price is below what the prices of these goods would total up were they bought separately.

In fact, you could make your list something like this:

PRIZES FOR SUGGESTIONS.

The editor of Hardware and Metal wants additional ideas to help make this department the most interesting and valuable feature in the paper. Travelers for hardware, paint and stove houses, or clerks in hardware stores, are invited to send suggestions for use in retailing the following month, and for every idea accepted for publication a prize of one dollar will be given.

The suggestions should be written on one side of the paper, they should contain from 100 to 200 words, and if sent before April 20 should contain ideas for May retailing, if before May 20, ideas for June retailing, and so on.

Send along the suggestions, help to make this department more interesting, and, incidentally, win one of the prizes.

	Price.	Total.
So many pots	0 00	0 00
So many pans	0 00	0 00
So many knives	0 00	0 00
Grand total	$0 00	$0 00

Show the total cost in good plain figures, and then offer the lot at a nice little reduction to young couples about to be married, and to no others. State the "no others" plainly, because if the bride thinks she is a privileged character she will be eager to accept your offer.

* * *

May first is moving day, and this gives the dealer an excellent opportunity to sell hammers, tacks, picture hooks and wire, step ladders, curtain stretchers, gas or electric fixtures, or anything in the line of goods that is needed by people taking up a new house.

A good idea would be for you to try and get a list of landlords who will have houses vacant soon and send them a letter making a bid to supply them with any goods they need in your line to fix the house up for their new tenants. This may seem a lot of bother to go to on the chance of getting an order, but business is business, and in these days of keen competition the merchant who gets after the orders is the merchant who retires first with a nice little bank account, and his competitors who wait for the orders to come to them envy him and think he's lucky. A retailer, to make a success of his business, must never let a chance slip to make a sale, no matter how small it may be.

* * *

As soon as the frost gets out of the ground people will be getting their gardens ready, and here is another opportunity to work off lawn mowers, hose and all kinds of garden tools. Advertise a package of flower seeds to be given away with every purchase of one dollar or over. It's only a small present, but some women would go a mile to save a few cents, and when they can get something free, no matter how small it is, they will go to the store that gives it, in seven cases out of ten.

Remember, it's not so much the actual sale of a certain article that you must look to in a case like this, but the ultimate outcome from bringing a number of new customers to your store. By treating a customer well and making her feel that it's a pleasure to have her call in at any time to look over your stock, you make a friend of her, and it all depends then on yourself whether you hold her or not.

* * *

Among goods that sell well in the dirty weather are door mats of all kinds, as the winter months generally leave them in pretty poor condition. Wire mats are far ahead of any other kind, as they don't hold the dirt; they also last longer.

* * *

If you have not already been after the secretaries of the sporting clubs in your vicinity you should lose no time in doing so, as we advised about a month ago. You should also make up a sporting goods window, just as soon as the snow begins to melt in the fields. Offer a present of some kind to the boy who buys the first baseball, bat, or any line you may choose. The habit of giving a small premium once in a while causes people to talk about your store, and it doesn't cost you very much, and is one of the very best modes of advertising that you could make use of.

HARDWARE AND METAL

Established 1888

The MacLean Publishing Co.
Limited

JOHN BAYNE MACLEAN - President

Publishers of Trade Newspapers which circulate in the Provinces of British Columbia, Alberta, Saskatchewan, Manitoba, Ontario, Quebec, Nova Scotia, New Brunswick, P.E. Island and Newfoundland.

OFFICES:

Montreal, ... 232 McGill Street, Telephone Main 1255
Toronto, ... 10 Front Street East, Telephones Main 2701 and 2702
Winnipeg, . . . 511 Union Bank Building, Telephone 3726
London, Eng. . . . . . 88 Fleet Street, E.C. J. Meredith McKim, Telephone, Central 12960

BRANCHES:

Chicago, Ill. . . . . 1001 Teutonic Bldg. J. Roland Kay
St. John, N B. . . . . No. 7 Market Wharf
Vancouver, B.C. . . . . Geo. S. B. Perry
Paris, France - Agence Havas, 8 Place de la Bourse
Manchester, Eng. . . . . 92 Market Street
Zurich, Switzerland . . . Louis Wolf, Orell Fussli & Co'

Subscription, Canada and United States, $2.00
Great Britain, 8s. 6d., elsewhere - 12s

Published every Saturday.

Cable Address { Adscript, London / Adscript, Canada

SUCCESSFUL RETAIL CONVENTION

This issue of Hardware and Metal is delayed for a day in order to contain a report of the convention of the Ontario Retail Hardware & Stove Dealers' Association in Toronto on Good Friday. Part of the report has necessarily been held over until our next issue, but sufficient is given to show that the convention proceedings were exceptionally interesting, and any hardwareman who visited the convention was well re-paid by the opportunity of associating for a day or two with several score of the brightest and most up-to-date hardwaremen throughout the province. By a comparison of ideas and discussion of interesting trade topics much valuable information was made available for those in attendance, and several of the delegates were heard to say that the knowledge gained would more than repay them for the expense of attending the convention and payment of a year's membership fee to the association.

The banquet provided an opportunity for retailers and jobbers to discuss some matters of mutual interest, and the three or four hours of recreation in the midst of the convention was thoroughly appreciated. Were there more occasions of this kind the trade would benefit in every way. When men get together at social functions differences which have arisen are apt to be forgotten, and the more fraternal feeling developed is bound to result in more good.

We congratulate the officers and members of the association in the progress which has been made since the formation of the association less than a year ago and we bespeak for the officers elected to serve during the coming year the hearty support of the trade in all parts of the province.

RAILWAY NERVE.

The Grand Trunk Railway in Port Hope passes between the larger portion of the town and the lake front. The business section lies in a somewhat deep valley, and this valley the railway crosses on a high viaduct. The company, without consulting the town, have taken steps to fill in this viaduct and plan to have but two or three narrow archways as a means of communication with the harbor. Canadians everywhere have had ample opportunity to acquaint themselves with the highhandedness of railway corporations, but this is about the limit. Here is one of the prettiest towns in the Dominion—the prettiest, in the belief of every Port Hoper—and the Grand Trunk sets itself unceremoniously to erect an earth wall with a hole in it between the town and the lake. No wonder the town is in arms. It is up to the Railway Commission to stop the depredation.

RETAILER MUST PROTECT HIMSELF.

The retail merchant is the one indispensable link in the chain which connects maker and consumer. Other middlemen may come and go, but in this vast country where the average family wears and uses goods drawn from sources thousands of miles apart, the store near the people is sure to remain.

Yet many makers are deliberately striving to force their goods into the hands of the retailer by advertising direct to the consumer, and then asking the retailer to pay their advertising bills by cutting down his profit.

This is certainly not fair to the retailer, nor does it seem to us good business from the standpoint of the maker. Surely merchants who have the liking and respect of their several communities can do more to make or break a brand of goods so far as their towns are concerned than any other single factor not excepting advertising.

Butler Brothers' Drummer points out that other things being equal, the advertised article wi,l outsell the non-advertised article will outsell the non-adequal if the policy of the maker has furnished the retailer a motive for pushing something else and for keeping the advertised goods in the background.

There is no obligation, moral or legal, on the part of any merchant to handle any certain article. It is his duty to know the quality of his wares and to tell the truth about them. If there is something in the market as good as an advertised article which he can buy at a lower price, there is no reason why he should not cut out the one and handle the other; or, if he stocks both, why he should not put the one in the window and the other on the shelf, out of sight.

Retailers hold themselves too cheap who without protest permit makers to force goods on their shelves which do not pay them a just profit. It costs money to run a retail store, and it is not fair to ask you to handle any article on as thin a margin as 10 to 15 per cent. unless you yourself choose to do so in order to bring more business for yourself in wares that pay better profits.

BE PROGRESSIVE AND UNITED.

The mail order house builds up its business, and continues to extend its territory by being progressive in its ideas and organized in its work. It is thoroughly organized and works in perfect harmony. They are able to take business from the country, not so much because their goods are better in quality or that they fill orders so much more quickly, but because they do not meet with any organized opposition from the rural stores.

The merchant in town and country is self-centred and exclusive and very often not favorably disposed toward the competitors in their own community. Their own knowledge of business is so much further advanced over their competitors and they are so inspired with the idea that no one can teach them, that they are content to remain where they are, to oppose innovations and to defy all attempts at union and organization.

The moment an efficient organization is formed, and more unanimity of ideas is adopted then will the mail order houses meet with some powerful opposition and a damper on their strong rural trade. The purpose of associations is not to provide assemblies of the dealers for expression of goodwill for the time being and some social enjoyment. The great aim of a retail association is the providing of assemblies for the exchange of ideas and mutual education to enable the retail men in rural districts the more intelligently to cope with the growing power and influence of the mail order houses and make possible organized opposition, and do away with the weak and disunited front which the rural districts present.

Ontario Retail Hardware Convention

Best Gathering Since Organization a Year Ago—Large Representation Present—Splendid Set of Officers Elected—The "Question Box" a Decided Success.

"Better than all previous conventions," was the unanimous expression of opinion of the delegates to the Ontario Retail Hardware Convention in Toronto on Thursday and Friday of this week. The attendance was larger than any previous gatherings and the enthusiasm shown indicated the hold which trade organization has taken upon the hardwaremen in Ontario.

The reports of the officers and the doings of the convention, given on the succeeding pages, will show that great progress has been made during the past year, and the delegates who came to Toronto lost little time in dealing with the matters which came before them for action.

About 60 delegates were present and there were almost as many more who wrote, explaining their reasons for

W. G. SCOTT, MOUNT FOREST
Elected President.

their inability to be on hand. The discussions were participated in generally by the delegates and the officers must have been greatly encouraged by the unanimous support they received and for the commendations given them by the delegates for the work they had done during the past year. Great satisfaction was expressed at the progress made in many of the matters taken up and the opinion was repeatedly expressed that although the association had as yet failed to gain its point in several of the matters taken up with the Wholesale Hardware Association, that the fact of a strong organization being in existence was having its effect and there is a steady lessening of the evil complained of—jobbers selling to consumers. In fact, it was pointed out by one or two delegates that there was an increasing number of advertisers in Hardware and Metal who emphasized

BALANCE OF REPORT NEXT WEEK.

Owing to pressure of Space, the report of the Banquet on Thursday night and the Question Box discussion on Friday had to be held over until next week's issue.

Another feature which had to be held over was the correspondence between the Association and the Railway Commission on the Cartage Charges Question.

Watch for the balance of the Report next week.

the point that they were selling to the trade only.

The adoption of the recommendations made by the committee favoring the organization of a Dominion Association should meet with the approval of organized hardwaremen in all parts of Canada. The bringing into existence of such a national organization should result in much benefit to the trade, in the matter of adopting a better system of collecting accounts, the working out of mutual fire insurance and particularly in fighting Dominion and Provincial legislation injurious to the retail trade.

Strong Executive Elected.

The new executive elected is a very strong one and represents all of the older sections of the province, the executive beginning their labors with the prestige of a year's successful work behind them. The steadily growing membership should encourage them in their work of remedying existing trade evils, and it was exceedingly gratifying to see new members enrolled from Kingston, Ottawa, Owen Sound, and many other places, those enrolled being in many cases the strongest firms in the towns and cities represented. As was pointed out by Mr. Purvis of Sudbury, the executive must take up the work of organizing the New Ontario district during 1907. Already there are many members from that district, but there is yet much to be done in the way of enrolling members and taking up the trade evils which affect that large district. The executive also have the advantage of beginning their year with a substantial balance in the treasury. The treasurer's report, showing a balance on hand of $200, considering the heavy expense the association was under during its first year of existence, was a very satisfactory one.

Taken all round, the second annual convention will go down in history as one of the most important and successful gatherings of the retail hardware trade in Ontario. The Question Box discussion was a fitting finale to the gathering, the answers given to the questions asked being listened to with close interest by everyone in attendance. If the present convention could have been improved in any way it would have been by allowing longer time for discussions of this nature. Mention must also be made of the banquet on Thursday evening, every nardwareman present enjoying himself to the limit, and all voted this social feature a decided success.

THURSDAY SESSION.

The convention was called to order at 2.15 p.m., with about half a hundred delegates in attendance, the list of those enrolled being as follows:

The Membership Roll.

(Revised to March 28, 1907.)

Ailsa Craig, J. E. Westcott & Co.; Alliston, W. B. Clifton: Amherstburg,

J. R. HAMBLY, BARRIE
Elected 1st Vice-President.

J. D. Burk; Arthur, D. Brocklebank & Son; Arnprior, J. S. Moir.

Barrie, John Coffey; J. R. Hambly, H. H. Otton & Son; Baysville, Jas. D. Smith; Belleville, W. W. Chown Co.; Hudson Hardware Co., J. W. Walker; Blind River, F. Y. W. Braithwaite; Blenheim, J. W. Fleming & Son; Blyth, McPherson Bros.; Bobcaygeon, A. E. Bottum; Bothwell, G. H. Trott, Colin Reid, Jr. & Bros.; Bolton, O. M. Hodson, Smith & Schaefer; Brantford, Howie & Feely; A. Ballantyne, W. H. Turnbull & Son; Bracebridge, J. L. Fenn, Sons & Co.; Burlington, Jas. S. Allen: Barrie, F. A. Hoar.

Caledonia, McGregor & Co., Clysdale & French; Canfield, John McDonald & Son; Carleton Place, Taylor Bros.; Chesley, Latimer & Elliott.

Dundalk, C. E. Noble; Dundas, G. E. Wilson & Co.; Dunnville, Congdon & Marshall, Jas. Ralston; Dryden, J. D. Hayes.

Embrun, H M Dupuis. Embro, W. J. Geddes; Essex, J. F. Rosebrugh, W. H. Richardson.

Fenelon Falls, Jos. Heard & Son; Fergus, A. E Nichols; Florence, McDonald & Cragg; Forest, Wm. Lochead, Fort William, Gerry Bros.

Glencoe, D. MacLachlan, Jas. Wright & Son; Goderich, Chas. C. Lee, E. P. Paulin; Gore Bay, T. R. Lougheed. Guelph, S. & G. Penfold, T. G. Rudd; Goderich, M. W. Howell.

Hagersville, J. C. Kayser; Hamilton, Alexander Hdwe. Co., A. W. Wright, Wood, Vallance & Co.; Harrow, C. Richardson & Co.; Highgate, J. G. Crosby.

Ingersoll, W. H. Jones, T. N. Dunn, J. T. Norton.

Kenora, Hose & Canniff, A. T. Fife & Co.; Kingsville, Thos. B. McDonald; Kingston, W. A. Mitchell.

Leamington, Greenhill & Moffat; Listowel, Fred Hawes, R. Ross; Little Current,, J. G. Kingsboro; London, Geo. Taylor & Son; Lucknow, Thomas Lawrence.

Markham, A. & H. Wideman; Merlin, W. A. Barr; Midland, F. W. Jeffery; Milverton, J. Rothaermal & Son; Mitchell, Middlemiss & Rankin; Mt. Albert, M. R. Summerfeldt; Monkton, Thos. Fullerton; Mount Forest, Scott & Murphy, C. J. Thornhill; Milton, Clement & Co; Milverton, W. F. Finkbeiner.

Newbury, B. M. Johnston; Newmarket, G. A. Binns; New Hamburg, D. & H. Becker; New Liskeard, G. Taylor Hardware Co.; Niagara Falls, J. H. Clark & Co., Cole & McMurry, J. T. Henderson, Garner Bros., S. E. Boulter.

Oakville, John Kelly; Orono, J. Henry & Son; Orillia, Macnab Bros.; Oshawa, John Bailes & Son; Owen Sound, Christie Bros.; Ottawa, W. G. Charleson.

Parkhill, Brewer & Harrison, A. W. Humphries; Port Arthur, Wells & Emerson, Port Burwell, A. R. Wright; Port Hope, J. B. White; Preston, W. F. Mickus; Providence Bay, W. I. Wagg; Paris, D. Sinclair.

Ridgetown, J. Laing & Son, J. H. Beattie & Co., Wm. F. McMaster; Rodney, D. Mistele.

St. Catharines, S. W. Moore, Watts & Bate, Coy Bros. & Southcott; St. Marys, St. Marys Hdwe. Ltd.; St. Thomas, Geo. W. Brown, R. H. Blackmore, W. J. McMurtry, S. A. Crawford, Ingram & Davey; Shelburne, Button, Spilker & Co.; Sprucedale, Mrs. W. H. Dixon; Stevensville, L. Wale; Stouffville, Silvester Bros.; Stratford, J. R. Myers; Streetsville, James Dandie; Sturgeon Falls, Robert Lillie; Sudbury, Purvis Bros.; Sutton West, W. Holborn.

Tiverton, C. F. Fawcett; Tillsonburg, W. R. Hobbs, E. J. Torrens, C. W. Conn; Teeswater, D. Ferguson; Thamesville, E. S. Hubbell; Thorold, James Wilson, C. N. Stevenson; Tilbury, M. Stewart; Toronto, John Caslor, James Ivory, Geo. Alexander, Prince & Co., E. W. Chard, Geo. R. Plumb, C. F. Moorehouse, J. W. Peacock, H. & G. Jamieson, J. S. Hall, Gurney Oxford Stove & Furnace Company, S. M. Burt; Toronto Junction, R. J. Bruce; Tottenham, R. J. Walkem, Chas. Worrod; Trenton, S. B. McClung & Co., Geo. A. White, Mowat Hardware Co.; Thessalon, Bridge Bros.

Uxbridge, Henry Jones.

Verner, F. A. Ricard.

Walkerville, Walkerville Hdwe. Co.; Watford, Thomas Dodds; West Lorne, McKillop & Ferguson. Welland, J. H. Crow; Whitby, W. M. Pringle; Windsor, Neveaux, Clinton & Baxter; Woodstock, Fred W. Karn; Woodville, V. E. McPherson.

The first order of business was the reading of the president's address and the secretary's and executive committee's reports, these being reproduced in full as follows:

President's Address.

Gentlemen,—It affords me great pleasure to stand before you to-day, as the retiring first president of this association. Only a year ago to-morrow, a handful of us gathered in one of the parlors of the Rossin House, and talked over the matter of organization. "To be or not to be" was the one important question, and after a plain statement of facts to the meeting, by the speaker,

J. WALTON PEART, ST. MARY'S
Elected Second Vice-President.

who had made the call through "Hardware and Metal," it was decided "to be."

However, just as we had nicely arranged things to have a real association of our own, a change came over the spirit of our dreams, for Messrs. Trowern and Brubacher, of the Retail Merchants Association, made the claim that we should become a branch of their association, and as we did not feel that we should speak for the large number of retail hardware and stove dealers of Ontario, we decided to call another meeting, when Mr. Trowern was to be on hand with about 500 hardwaremen who were supposed to be members of his association. Before separating, however, officers were elected temporarily.

On the day appointed, May 16, the meeting was held, when behold, only one man appeared with Mr. Trowern, while a large number were on hand on our side; after considerable discussion, and the wasting of much precious time, a vote was taken, when it was decided, almost unanimously, that we should "go it alone." I have yet to hear of anyone who is sorry for that decision, and the one man who voted against us told me, before leaving the city, that he was sorry he had opposed us, but that he had not fully understood the case as it was later shown to him, or he would not have voted as he did. The officers who had been temporarily elected at the Good Friday meeting, were confirmed at this time.

We had a good meeting that day in St. George's Hall, many matters were brought up for discussion and almost all who were there expressed themselves as being well pleased at having been present. Another good meeting was held at the time of the Exhibition, though that is a hard time to hold a convention; business men who come to buy goods and "see the show" will not attend the meetings. The talk we had from Mr. Lariviere, of Montreal, at this time fully repaid anyone for coming. Now, what have we done between times of meetings? Perhaps a short review of this will not be out of place here.

Your committee having in hand the insurance question, with Mr. Peart as chairman, has made excellent progress, and will have a good report to make. The committee on unjust cartage charges met the Wholesale Association, and talked the matter over with them, but got very little satisfaction. Since that time the committee has been at work on higher ground and will be able to report something more satisfactory at this meeting.

The Postmaster General announced his intention of introducing a bill, with a c.o.d. clause attached, permitting large parcels to be sent by post at a very cheap rate. I need not go into our plans of defeating this bill, but might just say that as president of this association, I sent out a letter to every hardware and stove dealer in Ontario, asking them to see their member of Parliament personally if possible, and if unable to do that, to write him to interview the Postmaster General and protest against the passing of such an unfair measure. This, with the flood of letters which the Postmaster General received regarding the matter about February the 1st, caused him to withdraw his bill, saying that "he could not fight against windmills." So much for the c.o.d. parcel post delivery.

We found that another very prominent daily newspaper was quoting, every Saturday, the wholesale prices of our goods. My attention was called to this by our energetic and efficient secretary, whereupon I wrote the editor of the daily, asking him to call his commercial editor's attention to the matter, as it could do them no good and us much harm. I may say that the quotations have not been published since.

You must not, right at the first, expect too much of us as an association. We are young, we have had to begin to feel our way, but we have done the best we could. If we have not accomplished as much as you expected, we ask you to bear with us and, during the coming year, to go ahead and improve on our past year's work. You will have our experience to build upon and we will be delighted to assist in any way we can, for I feel sure that we have only laid the foundation of our work.

I am anxious to see accomplished this year among other things, a garnishee clause for amounts of $5 and over, allowed in law by the Provincial Legislature; a special rate of express for parcels coming through the express companies into our hands; abolition of the cartage charges at shippers' ends, (I would not object to the abolition of boxing charges, too) ; the establishment of a mutual fire insurance plan, and jobbers prohibited from selling to any but retailers.

Canada is prosperous. Business is booming. Ontario is getting her share, but we must watch our interests and stop the leaks if we hope to make a success of our lines of trade.

Apart from this official and business view of the year's work I wish personally to thank the officers and members of the association for the support they have given me as president, and for the hearty co-operation they have evinced in strengthening the hands of the chief officers. The growth of membership has been most encouraging and there seems to be no doubt but that we shall soon have all the hardware and stove dealers of consequence in the province upon our membership roll.

Good feeling and good fellowship have been strongly marked in our gatherings, and this spirit will serve to wonderfully increase our strength. When we all shove together we should be able to remove large obstacles.

A debt of almost more than gratitude is certainly due to Mr. Wrigley, our secretary, who has been the mainspring and moving spirit of the executive. To him is due the credit for the splendid arrangements for these convention meetings, and for the interesting programmes laid before us.

My most earnest hope and desire is that my successor may be accorded this same warm and encouraging support, and, following my humble efforts, he may lead the association on to better and greater ends. "Keeping everlastingly at it" is sure to bring success.

A. W. HUMPHRIES.

Secretary's Report.

In presenting my report to the second annual convention of the Ontario Retail Hardware Association it is very pleasant to be able to call your attention to the splendid progress made since the first organization meeting was held about eleven months ago. In view of the fact that the association has not had an organizer in the field for more than one month, good work has been done in enrolling upwards of 160 members prior to the date of the convention, and this number is likely to be increased close to the 200 mark before the convention adjourns. During the year expenses of organizing have been heavy, as a large sum has been spent in sending circulars to the trade regarding the parcels post question, notices of conventions and general organization work. No salaries, however, have been paid to any officer, and the executive has freely given its time in attending several special executive meetings held in Toronto. The report of the treasurer should show a substantial balance to the credit of the association.

Progress of Organization Work.

As we are passing the first mile-post in the history of the association it is well to analyze the results of the year's work. First of all it must be admitted that the work of organizing has been well commenced, members being enrolled from Windsor and Kenora in the western portion of the province, to the Ottawa district in the east, as well as from Niagara Falls in the south to Sudbury and New Liskeard in the north. As yet, however, the organization is scattered over too broad a field and there is much work to be done toward filling in the gaps. If a capable organizer could take the field for a couple of months following this convention I feel certain that the province could be thoroughly organized. Already good work has been done by the organization of district associations at Barrie, London and in Middlesex and Lambton districts, while another district association has been proposed for the Niagara Falls vicinity. The London district association has not yet, however, accepted our invitation to join the provincial association in a body. These district associations should be organized in all parts of the province. Another feature of the organization work is the progress which has been made towards bringing into existence a Dominion association, which will unite all hardwaremen in the different provinces into one organization.

WESTON WRIGLEY, TORONTO
Re-elected Secretary

A Good Year's Work.

The greatest success so far secured by our association is in the matter of the parcels post c.o.d. agitation. Our association was the first to take up this matter and in spite of the refusal of the Retail Merchants' Association to co-operate, we conducted the campaign so rigorously that the Postmaster-General was forced to withdraw the legislation aimed to benefit the mail order houses at the expense of the retail trade throughout the country. If for no other reason this work has been sufficient warrant for the existence of the association, as it can safely be said that had the retail hardwaremen in Ontario not organized a year ago the parcels post c.o.d. legislation would now be in force.

The work of the committees on mutual fire insurance and account collections has also been of considerable value to the trade and satisfactory results should follow from the work very shortly. In fighting trade evils the association has also proved the necessity of its existence. The cartage charges question was thrashed out with the wholesale association, and on the jobbers refusing to remedy the evil complained of the association took the matter up with the Railway Commission. The correspondence which will be read to the convention shows that the association is making progress in securing a satisfactory solution of the matter.

Numerous complaints have been made of jobbers selling to consumers and the association has taken up this question with the Wholesale Hardware Association, recommending that they take action to protect the retail trade by adopting a different scale of prices to be used when goods are sold to persons who do not buy them to b? sold again. This matter has not yet been dealt with by the Wholesale Association, and it is one which should receive considerable attention from this convention.

The recommendation of the September convention that enamelware seconds be marked as such in large letters has not yet been adopted by the manufacturers of enamelware. They have, however, assured us that they will give no further cause for complaint in this matter.

We have also taken up the matter of abolition of exemptions from garnishment so as to make it harder for "dead beats" to operate, and the question of securing proper classification on hardware freight has also received some attention.

Takes Time to Produce Results.

The above shows that the association has led an active existence during the past year. It must be remembered, of course, that it takes time to produce results, and I am sure the members of the association realize that the officers have done well in the face of many obstacles. In view of the progress made it would be suicidal for the retail trade to hesitate now. There should be no hanging back as every member added to the fold makes it easier to secure satisfactory results on any work taken up. Our association must secure at least 300 to 400 members in order to be sufficiently strong to command respect from those with whom we have to deal in order to better the retail trade conditions.

What is the most important work for the coming year? During the past twelve months the greatest attention has been given to the cartage charges and parcels post question. The cartage charges question should be pushed to a satisfactory conclusion, but in the secretary's opinion it is far over-shadowed in importance by the greater evil of jobbers and manufacturers selling to consumers. During the coming year this should be the all-important subject under consideration. Until the criminal code is changed there can be no price agreement adopted by our association. There is, however, no reason why arrangements cannot be made to have wholesalers' and jobbers' list dealers adopt a special scale of prices for selling goods to contractors and others who buy to use and not to sell again. Consumers who buy in large quantities are undoubtedly entitled to closer prices than ordinary buyers, but the retail trade should be protected and the existing evil of consumers buying as cheaply as retail dealers should be ended if the retail trade is to continue a factor in commercial life.

Relations With Jobbers.

Our relations with jobbers have been friendly during the past year, although matters of controversy were taken up at the very inception of the retail association. This has probably been the reason why the traveling salesmen have not aided our organization as much as the travelers have aided the retail hardware associations in the United States and western Canada. Jobbers, however, are far-seeing business men and undoubtedly realize that anything the Retail Hardware Association does to better the conditions of the retail trade is bound to have its effect in improving the field for business operations of the wholesale houses. As has been pointed out before, retailers and wholesalers should co-operate together in checking the present tendency of jobbers selling to consumers and retailers buying from manufacturers.

The executive committee are making recommendations on various matters referred to in this report, and it is, therefore, unnecessary to go further into detail than has already been done. The work of the executive has been handicapped somewhat by lack of definite details being furnished with the complaints received regarding jobbers selling to consumers. It is urgent that when a retailer has a specific case that he send full details in order that the association may take definite action.

The question box discussion arranged for at this convention should provide a splendid opportunity to talk over any matters which are injuring the trade throughout the province, as well as making possible any expressions of opinion as to methods whereby the business operations of the members can be improved upon.

In closing this report reference might be made to two suggestions recently made by members of the association. The first one is that the annual convention be held about the end of February, at the time when the annual horse show takes place in Toronto, single fare rates being available for a period covering an entire week. Previous conventions have shown that it is not satisfactory to try and crowd the business of the convention into one or two days. The members would receive better results if the convention occupied at least three days, and ample time was given for the discussion of trade matters through the question box. The second suggestion is that arrangements be made at future conventions for the entertainment of the wives and lady friends of the delegates. These matters should receive the attention of next year's executive committee in arranging for the annual convention in 1908.

WESTON WRIGLEY,
Secretary.

Toronto, March 28, 1907.

Report of the Executive.

The executive committee of the Ontario Retail Hardware Association beg to report as follows:

1. That the campaign outlined at our December 31 meeting against the proposed parcels' post c.o.d. legislation was successful in withdrawal of the same. We acknowledge the co-operation of the Western Canada Retail Hardware Association, and of Hardware and Metal as through its aid the trade in all parts of Canada was encourged to join the campaign.

2. That the Canadian Wholesale Hardware Association have declined to abolish cartage charges at point of shipment and that we have laid the matter before the railway commission, the correspondence herewith submitted showing that the railway companies have no right to refuse to accept shipments delivered by jobbers, and that the matter

JOHN CASLOR, TORONTO
Re-elected Treasurer.

is one for adjustment between shipper and consignee.

3. That the Canadian Wholesale Hardware Association has not yet dealt with our recommendation, "That whenever manufacturers or jobbers sell to buyers outside the trade that sales be made at a price sufficient to protect legitimate hardware dealers. A letter from a member of the association is submitted with the correspondence regarding this matter.

4. That the enamelware manufacturers have not yet complied with our recommendation that "all seconds be stamped as such in plain letters of large size." They have assurred us verbally, however, that we shall have no further cause for complaint in this matter. Correspondence from the Western Canada Retail Hardware Association is submitted showing that the western trade has a similar grievance in that firsts are sold to departmental stores to be sold to consumers at cut prices.

5. That our recommendation that the Division Court Act be amended has not been adopted by the Attorney-General, he stating that no action can be taken at this session of the legislature.

6. That our recommendation that a temporary executive be formed for the proposed Canadian Retail Hardware Association by constituting the presidents and secretaries of the different provincial organizations as the said executive has been adopted by the Western Canada Retail Hardware Association. They have appointed their president and secretary as their representatives and it devolves upon this convention to take such action as will further the organization of the National Association. In this connection we would recommend:

(a) That our newly elected president and secretary be the Ontario representatives on the temporary executive of the national body.

(b) That we suggest that Toronto be the headquarters of the proposed Canadian Retail Hardware Association.

(c) That we suggest that each province be entitled to two representatives.

(d) That we suggest that the national body be financed by an assessment of 25 or 50 cents per member in order to carry on its business and oppose all injurious legislation.

(e) That the proposed temporary executive endeavor to organize a retail hardware association in each province in Canada and by an interchange of experiences strengthen the existing associations.

A. W. HUMPHRIES, Chairman.
WESTON WRIGLEY, Secretary.

Parcels Post Agitation.

The secretary suggested that the report of the executive committee's work should be taken up clause by clause and this was done.

In regard to the parcel posts' matter, the president said that he felt he must say something on this matter. The battle had been won by the secretary and Col. MacLean in the office of Hardware and Metal. He realized that the Colonel was a bashful man, therefore, he (the president) had to speak of the matter. He then went on to tell what his representative in Parliament had said to him regarding what occurred during the fight. This member went to Mr. Lemieux, several times, telling him that a friend of his in the Retail Hardware Association had asked him to aid in the movement for the withdrawal of the measure, and presented various arguments. While he was thus engaged, a dozen others were also trying to secure interviews. Finally Mr. Lemieux asked those present to give him some chance for his life at least. But the next day the same thing happened—not only hardwaremen, but grocers and dry goods dealers. By Feb. 1st, when the hardware dealers' letters poured in by thousands Mr. Lemieux was about beside himself. His usually capacious office was crammed—wagon loads of mail were coming in. "What makes all this rush of mail?" he asked, and the reply was, "Protests." As Mr. Lemieux stated to the House, "he couldn't fight against windmills," and the motion was withdrawn. As president of the association, Mr. Humphries believed that this was the direct work of the MacLean trade journals, and called for an expression of approval from the members. The response was most unanimous.

Discussion on Cartage Charges.

Regarding the addition of cartage charges to the consignee's expenses in shipments, correspondence was read by the secretary from the Wholesale Hardware Association. A letter written to Chairman Killam, of the railway commission, and his reply was then read. As it seemed that the commission had not a clear view of the case, another letter written, was also read to the members. Replying to this, Mr. Cartwright, secretary of the commission, wrote, outlining the agreement between the railways and cartage agents. The secretary said that the executive would make no recommendations, but would leave the matter for the members of the asosciation to deal with as they thought best.

Correspondence on this matter will be published later on the association page in Hardware and Metal.

The president asked for an expression of opinion from the members.

Mr. Noble (Dundalk) thought that the committee had done good work, and had now got to know what is necessary. The matter should, therefore, be left with the committee.

The president explained that the committee was now looking for advice. It wished to know the feeling of the dealers. It would be strengthened by advice and discussion of the members.

Mr. Noble said that the charge was certainly an unjust one. "The matter should be arranged between us and the wholesalers."

The president remarked that he had noticed on one firm's billhead f.o.b. had given place to "ex warehouse." Evidently the jobbers had begun to sit up and take notice.

Mr. Jeffery (Midland) agreed with Mr. Noble.

Mr. Peart (St. Marys) stated that it had been his idea originally to take up the matter with the railway commission. The association had now got the commission to say that the matter is one between them and the shippers. He made the following motion, seconded by Mr. Finkbeiner (Milverton):

Resolution Adopted.

"Be it resolved: that this association being still of the opinion that the sending forward, on the bill of expense of the consignee, of the cartage at shipping point is a gross injustice to the retailer, reaffirm our intention to oppose this unjust condition and authorize the executive to carry the matter again before the railway commission and to take whatever action they deem advisable."

Mr. Taylor (Carleton Place) said that from the correspondence read, the decision given by the Railway Commission appeared to be final. When goods had been delivered to the retailer, there was nothing to prevent his refusing to pay this charge. Let them all stick together and refuse to pay. Inside of six months, he believed, the continual worry and cost of additional correspondence necessitated by this step would tire shippers into compliance with the association's demands.

Mr. Jeffrey (Midland) endorsed Mr. Taylor's position. He said that it was of no use going back to Railway Commission. "If we send back the drafts as fast as they are forwarded, we can soon kill the cartage business."

Mr. Clifton (Alliston) said he had been fighting with a Guelph firm. They told him to go to the Railway Commission. If we stand shoulder to shoulder we can force them to change this. Toronto merchants have goods delivered free of cartage expenses; why shouldn't other dealers throughout the country?

J. S. Hall, (Toronto),—Do Toronto

A. W. HUMPHRIES, PARKHILL
Retiring President—Elected Member of Executive.

dealers buy goods any cheaper than the rural dealers because of not having to pay cartage charges? Will not the outside merchants have to pay more if the charges are done away with?

This question was answered by several members saying they had no objection to paying the full cost of the goods, but the association was fighting for a principle and they wanted the charge made in the proper manner so that it could be included in their costs and figured on in marking the selling price.

J. S. Moir (Arnprior), expressed the opinion that the only way to stop the imposition of cartage charges is to go at the wholesalers. The retailers have no right to pay cartage charges.

"I understand that Montreal houses will deliver goods in Toronto with cartage and boxes free. It is an injustice. In the case of lamp chimneys, for instance, there is an extra charge of 40c. for boxing, which was too often not figured on, the tendency being for dealers to figure their selling price from the cost price given on the invoice."

John Castor (Toronto) said that wholesalers in the city delivered goods to retailers free of cartage charges.

F. Taylor (Carleton Place)—"Just what we want is to get the charge placed directly on the cost of the goods, so that the retailers may be enabled to arrange their prices to cover such expenses. Then we can get our percentage of profit on the cartage charges the same as on all other costs.

W. G. Scott (Mount Forest) expressed his concurrence with Mr Taylor in the matter of cartage charges being made. In the matter of boxes the charge should be put on goods so that we can charge accordingly. He spoke of concerns which have switches into warehouses and the retailers have to pay cartage charges just the same. This is graft of the railway companies at our expense.

Mr. Brocklebank (Arthur) spoke regarding the motion of Mr. Peart. It doesn't make any difference whether we be charged directly or indirectly on the goods delivered so that the merchants could mark their goods at a sufficient advance to defray such expense.

D. A. McNab (Orillia) said that the railways used to pay the cartage themselves. There is no doubt that when we paid cartage in an indirect way we did not feel so sore about it, because we did not know how much we were paying.

Mr. Brocklebank, Jr., (Arthur) expressed the opinion that while the association appreciated very much the work done by the executive committee, he thought that all their efforts were in vain because they were fighting against the winds. He thought it would be well if the association knew what retail firms are paying charges and who are not. A show of hands indicated that about a quarter of those present were not paying cartage charges, either in whole or in part. Continuing, he said: "I've never taken up the matter with wholesalers because I thought it not worth while scrapping about it until we take it up in a body. A good many retailers can't afford to fight the wholesalers and run chances of being turned down.

D. F. McNab (Orillia) expressed his desire that Mr. Peart's motion include the provision that the executive negotiate with the jobbers again.

E. S. Hubbell (Thamesville) expressed his appreciation of the executive's work and favored allowing them to continue their efforts.

F. W. Otton (Barrie) suggested that the question of cartage rates be brought before the wholesale association.

The resolution, as moved by J. W. Peart, was adopted.

Jobbers Selling to Consumers.

The president stated that a certain fence company which had been selling to consumers had realized the error of their way and had fully apologized to the retail trade, and stated that such action would never be repeated.

This brought up the question of the building trade, contractors and others who buy in large quantities. Mr. Howell (Goderich) claimed that this trade was of practically no use to the retailer, for all the orders simply went to retailer-jobbers, such as existed in all the larger cities.

The members were reminded that it

was agreed last year that all complaints of this kind were to be forwarded to the secretary to be dealt with by the executive.

Mr. Child, of Gravenhurst, told how builders in his town bought in Toronto, and was corroborated by Mr. Jeffrey. Other speakers showed how other districts suffered in the same way.

The roofing companies were great offenders in this way, and one member told how all their tenders to the consumer were made uniformly a certain amount lower than the retailer could make it. He suggested that complaints should be written to the secretary, to be supplied to all the members after he had written to those complained of.

Mr. Scott (Mt. Forest) suggested that one way to get hold of the contractor was to go after him in a friendly way, explain the situation to him, and show him just what you can do for him and the advantages of his supporting his own town. His orders with the jobbing retailer will gradually grow smaller, while his little repeat orders will grow larger with you.

The president told how a man from a wholesale house had come to his town, and had been taken by him to a large factory needing tacks in quantity. The salesman took a large order, but he (the president) got no commission out of it. Later on the wholesaler began selling straight to the factory, and still there was no commission. "But since then I have bought nothing from the firm referred to," continued the president, "and now the factory is out of business, so they've lost two customers."

Mr. Scott then moved that all complaints should be sent to the secretary, and that he should notify the members by circular regarding houses complained of, the matter then to be dealt with by the executive. This was seconded by Mr. Caslor.

Mr. Scott also referred to the metal shingle companies, one of which had asked him to be its agent, but when an order had been worked up by him, the manufacturers had covered the barn themselves and allowed no commission.

Another member told him how a large jobbing house had quoted him a price and had then beaten him by five cents a dozen on a tender for several dozen shears for the school board. Later on the jobber secured an opening order for a new store but on having the circumstances explained to him the new dealer promptly turned the order in the way of another firm, and the first firm was told why he had done so.

Mr. Scott's motion was then carried.

Enamelware Seconds.

On the question of the stamping of enamelware seconds as such, the secretary read a letter from the Stamped Metalware Association, and said that he had received a verbal promise that no more trouble would be given.

Mr. Scott believed that perhaps one-half the rural dealers were being shipped seconds—chipped or badly colored stuff, etc., as firsts. He asked that every dealer send back everything about which he had cause to complain. This was what he was doing himself, and it was at least better than doing nothing.

Mr. Brocklebank moved that the question be left with the executive, and that they press for better quality of firsts, and the marking of seconds. The motion was seconded by Mr. Hambly, of Barrie, and carried.

National Organization Recommendations

The convention endorced their former action re the amendment of the Division Court Act, and after adopting the recommendations of the executive in the matter of the proposed Dominion association adopted the report of the executive as a whole.

The convention adjourned shortly after 5 o'clock.

FRIDAY MORNING SESSION.

The convention was called to order at 10 o'clock by the president.

The first subject taken up in the morning session was the report of the accounts collection committee report.

The committee favored the adoption of a similar system of collecting accounts to that in use by the Western Canada Retail Hardware Association, where blank letter forms are supplied the members by the secretary of the association and where necessary additional letters are forwarded to delinquents from the secretary's office. The committee considered, however, that this work could not be undertaken without a paid secretary and the present membership is not large enough to engage such an official, although the members would gain enormously by the establishment of a definite plan of following up delinquents who moved from one place to another or from Ontario to the Western Provinces. It was considered that this matter should be taken up by the National Association when it gets into working order. In the meantime the committee recommended:

D. BROCKLEBANK, ARTHUR
Re-elected Member of Executive.

Collecting Accounts.

"We consider the best method of collecting accounts is to render them in detail and if not promptly collected to call with another statement, securing the cash, a note or a promise to pay at a certain time. In the latter case the call should be promptly made and if not fully collected an installment secured and the matter followed up personally."

The committee urged all hardwaremen to join the association in order that the membership could be increased sufficient to warrant the adoption of the plan proposed.

The report was signed by C.E. Noble (Dundalk) and W. B. Clifton (Alliston) and on motion was adopted.

Mutual Insurance Report.

Following this came the report of the mutual insurance committee, read by J. Walton Peart, the chairman.

"The mutual insurance committee begs to report, having pursued inquiries as instructed at last convention. The replies received from members show great interest and a willingness to carry insurance in a mutual organization. We regret to report, however, that the present insurance laws of our province make it quite impossible at present to organize as we would wish, the law requiring that a joint stock company be formed with a subscribed capital of $300,000 of which 10 per cent. must be paid up before charter will be granted.

"As for the proposed preferential arrangement we have made some inquiries and are informed that in justice to local agents an arrangement only as we suggest is impracticable.

"We suggest that a new mutual insurance committee be appointed; that the trade be requested to join the association and increase our members to such an extent that we will be able to organize a joint stock company, which would be possible were we to have a membership of 500 or 600.

"J. WALTON PEART, Chairman."

The report was followed by the reading of the following letters to the secretary from Lindsay, Lawrence & Wadsworth, Toronto, the solicitors consulted for legal advice in this matter.

Weston Wrigley, Esq., Toronto:

Sir,—I beg to acknowledge your favor of the 25th instant, asking to advise as to organizing a mutual insurance company. To confirm my views I have conferred with Dr. Hunter, Registrar of Insurance Companies, in reference thereto. His statements accord with my own view, that such an insurance company would not be incorporated under the Ontario laws, as they stand at present. The ability of obtaining licenses to carry on mutual fire insurance has been very much curtailed, the law being in a different position now in that regard than it was some years ago, the Government refusing to issue licenses the same as they used to. The only way, therefore, to accomplish your object would be by obtaining incorporation under the Insurance Act as a joint stock company. We presume this would be of no assistance to you, in view of the business you desire to do, being restricted to your own line of business.

A. G. F. LAWRENCE.

Toronto, Jan. 30, 1907.

Weston Wrigley, Toronto:

Sir,—I beg to acknowledge your favor of the 25th inst., asking for information as to the requirements of organizing a joint stock insurance company. In reply I would say that the requirements in effect are as follows:

1. At least five persons must be the applicants, and the application has to receive the written recommendation of the Inspector of Insurance, and the approval of the Minister.

2. Notice has to be published in four consecutive issues of The Ontario Gazette of the intention to apply for incorporation, and in such other papers as may be required. The notice to set out the full names and additions of the applicants, their residences and occupations, the proposed corporate name of the company, the kind of insurance proposed to be transacted, location of head office, the amount of capital stock, number of shares, and the amount of each share.

3. The applicants shall also deliver copies of the proposed by-laws of the corporation.

4. Each director has to be the bona fide holder in his own right to his own use of shares in the capital stock of the company to the amount at least of $1,000 upon which all calls have been duly paid.

5. The capital stock shall not be less than $500,000, and before applying for a license the company shall furnish to the inspector satisfactory evidence that at least $300,000 has been subscribed for and taken up bona fide, and that $30,000 of the said subscribed stock has been paid into some chartered bank of Canada.

These are the chief requirements leading up to the formation of a company. I do not see anywhere in our acts that a subscriber has to be worth a certain amount of capital. He must, of course, pay all calls upon his stock, and at least 10 per cent: of this has to be called up.

I trust the foregoing information will cover the requirements of your association.

A. G. F. LAWRENCE.

Toronto, March 27, 1907.

Various members of the association discussed the subject, the general opinion arrived at being that it would be well nigh impossible to form an insurance company until a larger number of members be enrolled.

A resolution was passed to the effect that the report be laid on the table for further consideration. Messrs. Peart, Brocklebank and Caslor were elected as a committee to act on insurance matters during 1907.

Treasurer's Report.

Treasurer Caslor then gave his report showing receipts $446.71 and expenses for the year $235.71, leaving a balance on hand of $211.

It was moved by E. S. Hubbell and seconded by J. W. Peacock, that the report be accepted. The auditors' report will be published next week.

Great satisfaction was expressed by the delegates at the splendid report made by Mr. Caslor.

Travelers' Affiliation.

A notice of motion, made by W. G. Scott (Mt. Forest), was taken up and discussed. Mr. Scott moved that travelers be admitted as associate members of the association, this being seconded by Mr. Clifton (Alliston). An amendment was made by Messrs. Hambly (Barrie) and Caslor (Toronto), that only retail hardware and stove dealers be admitted.

An interesting discussion was participated in by several members, the preponderance of opinion being that the travelers would prefer to be left alone and that if the doors were opened it would look as though the retailers were trying to enrich their treasury by collecting fees from the travelling salesmen. By a vote of about three to one the amendment was carried and the original motion lost.

Election of Officers.

For president, W. G. Scott was nominated by Mr. Caslor, in a brief speech.

Retiring President A. W. Humphries was nominated for re-election by J. S. Hall, the nomination being seconded by several delegates. Mr. Humphries expressed his regret that circumstances prevented him continuing in the office, and Mr. Scott was elected amidst enthusiasm.

Mr. Coffey nominated Mr. Peart for first vice-president, but Mr. Peart declined in favor of Mr. Hambly, who had ably filled the office of second vice-president during the past year. Mr. Hambly was elected by acclamation.

Mr. Coffey again nominated Mr. Peart as second vice-president, and the election was put through with applause.

For secretary, Mr. Wrigley was re-elected unanimously, Mr. Peart proposing three cheers, which were heartily given.

Mr. Caslor was again made treasurer for next year at the same salary as last year—nothing

Messrs. Peart, Moorehouse and other members suggested that the retiring president be made one of the executive for the coming year and the election of the executive was then proceeded with. Those elected to the executive committee are as follows: A. W. Humphries, Parkhill; D. Brocklebank, Arthur; G. A. Binns, Newmarket; Frank Taylor, Carleton Place; H. Becker, New Hamburg, and W. A. Mitchell, Kingston.

The new president was escorted to the chair by retiring president Humphries, and in a brief address Mr. Scott thanked the members for electing him, expressing confidence that by this time next year the association will have doubled its membership.

Trade Press Commended.

On motion of J. W. Peart (St. Marys) and J. S. Hall, Toronto, the following resolution was carried amidst much hand-clapping:

"That we wish to place upon record our grateful appreciation of the kindness of Lieut.-Col. J. B. MacLean in providing social entertainment for the members of the association and in permitting association matters to occupy such a large and prominent position in his trade paper, Hardware and Metal. Whatever degree of success we have attained is due in no small degree to Lieut.-Col. MacLean and we tender him our heartiest thanks."

Col. MacLean replied briefly, stating that it was his aim to employ specialists to look after every department of the paper in order to make it of value to the subscribers. It had given him great pleasure to hear one member say that he wished his business could be run on the same standard as attained by Hardware and Metal. The manager and editor of the paper were entitled to the credit for the high standard attained and he was sure they would continue to make the paper of increasing value to its readers.

Notice of Motion.

A notice of motion was made by M. W. Howell (Goderich) that the executive have power to change the date of the annual convention. A further suggestion was made by C. F. Moorehouse (Toronto) that the executive revise the whole constitution.

On motion of Messrs. MacNab (Orillia) and Purvis (Sudbury) the sum of $50 was placed in the hands of the executive to provide some suitable souvenir to be tendered the secretary for his active work during the year, the motion being carried in spite of the secretary's protests.

The Question Box.

The business of the convention being concluded about 11.45 a.m. the "Question Box" was opened by J. W. Walton Peart (St. Marys) replying to a query regarding bookkeeping. Mr. Peart was for many years one of the office staff of the D. Maxwell Co., St. Marys, and his talk excited keen interest.

A group photo of delegates was taken at the close of the morning session, copies of the photo being obtainable from the Galbraith Photo Co., 239 Yonge street, Toronto.

Executive Meeting.

The convention adjourned about four o'clock, the members of the executive being then called together by President Scott. It was decided to offer M. W. Howell, of Goderich, the position of organizer of the association, he to commence work at once if he accepts.

While not desiring to establish a precedent, the executive instructed the secretary and treasurer to procure a suitably engraved gold-headed cane to be presented to Retiring President Humphries, as a reminder of the honor he enjoyed in being the first president of the association, and of the executive's appreciation of his work in behalf of the retail trade.

CONVENTION GOSSIP.

Retiring President Humphries had to leave the convention to catch the 4 p.m. train on Friday. The delegates were in the midst of a red hot discussion about oil tanks. but when Mr. Humphries excused himself they interrupted the talk by a hearty chorus of, "He's a jolly good fellow." Mr. Humphries made an exceedingly capable president and in years to come the honor he enjoyed will be even more appreciated than it is to-day.

Markets and Market Notes

THE WEEK'S MARKETS IN BRIEF.

MONTREAL.

COIL CHAIN—Advanced 10c. per 100 lbs.
LINSEED OIL—Advanced 2c. per gal.
TURPENTINE—Advanced 2c. per gal.
SHELLAC VARNISH—Shows advances.

TORONTO.

TURPENTINE—An advance of 2c.
LINSEED OIL—An advance of 1c.
OLD MATERIAL—Several declines.

Montreal Hardware Markets

Office of HARDWARE AND METAL,
232 McGill Street,
Montreal, March 28, 1907

The hardwaremen's rush season is now on here in full swing, and it's a case of "everybody put his shoulder to the wheel and push." Jobbing houses resemble busy bee hives more than anything else, and it's not a good time for anybody to show his face around, if his sole object is to simply view the scenery. Some houses are doing night-work, while others have cut out all concessions such as early closing and Saturday afternoon holidays until further notice. With the prosperous condition of the country, it would be only reasonable to suppose that money should be flowing pretty freely, but reports are to the contrary. Collections are pretty hard to get.

Recent severe snow storms have put the roads here into deplorable condition, and it's impossible to load teams to their full capacity. This naturally has delayed both out-going and in-coming freight to a greater degree than might be expected by the uninitiated. To give some idea of the great volume of business which is being done, one jobber informed our representative that he shipped out four large truck loads of goods the other day before eleven o'clock in the morning. This averages about a team an hour from the time the house opened for business.

Manufacturers of wire nails, wire, etc., report great difficulty in keeping their orders up to date, although they are gradually increasing their plants, and working up to full capacity. No matter what size a plant may be, an order for several thousand kegs of one size of wire nails, such as was received by one Montreal firm last week, makes a fairly large-sized hole in the stock, and naturally makes a shortage in that size that is extremely hard to catch up with, seeing that orders are placed for some time ahead, which test the plant to the utmost.

Market conditions this week show very few changes, the only advance to record being one of 10 cents per 100 lbs. on coil chain. Bolts and nuts remain practically unchanged, with the shortage just as marked as ever. Rivets and burrs, hay wire and wire nails are in good demand at the same old prices, while though cut nails show no change in prices the demand is not exceedingly strong.

A very good business is being done in building paper, but the manufacturers of this line are about the only people we visit who are not rushed to death.

Shot is not in good demand just now, but ammunition is waking up and some orders are now coming in.

Horse nails are moving well at the old prices, but some jobbers are now charging slightly more for their cheaper lines of horse shoes. The advance, however, is not enough to warrant any change in our quotations.

Poultry netting is also in good demand at steady prices.

Orders for binder twine, lawn mowers and green wire cloth are mostly booked for the season, but a few stragglers keep coming in from time to time.

Toronto Hardware Markets

Office of HARDWARE AND METAL,
10 Front Street East,
Toronto, March 28, 1907.

Hardware trade conditions are good, the demand being brisk, with active prices. The few weeks of mild weather have improved the business very much. The demand in a number of lines greatly exceeds the supply, manufacturers repeatedly having to turn down oeders owing to their inability to fill them.

The situation in screens remains the same, with a good demand and limited stocks. The situation is not yet so acute as to warrant an advance in the prices.

The scarcity of wire nails continues, with heavy demand and unsteady prices. The last advance was 5c., making the base price now $2,45. Orders are large, and supply houses find it difficult in some cases to fill them.

Orders for binder twine continue to come in, but have eased off as they have nearly all been booked up. Prices remain firm and unchanged. The price on 500-ft. twine is, if anything, a little weaker.

The demand for mechanics' tools, extension and step-ladders continues strong with firm and unchanged prices. Those who have ordered building paper, washers, wringers, carpet sweepers, lawn mowers and house-cleaning utensils for April 1st are asking for earlier shipment, in anticipation of earlier commencement on house-cleaning operations. Harvest and garden tools have experienced an earlier demand than last year.

The demand for poultry netting continues good.

Bolts, nuts, rivets and burrs are experiencing a strong demand, the supply being limited. Prices continue firm and unchanged. Business in oiled and annealed wire is good, with strong demand and firm and unchanged prices.

The demand for horseshoes and horse nails continues strong. Prices have not yet been affected by the spring demand from new districts and remain firm and unchanged.

Shot is not in much demand now, most of the orders being booked up.

The spring demand for sporting goods is strong, fishing tackle and rifles being particularly active owing to the early opening up of the bay and the approach of the spring shooting season.

Owing to a break-down, the Dominion Iron and Steel Co., Sydney, had to suspend operations for a couple of weeks. This has affected the supply of rods, thus aggravating the scarcity of wire nails. The supply of rods has been limited owing to the delay of the Intercolonial Railway in delivering its shipments to western firms. The damages done at the Sydney works have now been repaired, and the rod, nail and billet mills are now running with full force.

Montreal Metal Markets

Office of HARDWARE AND METAL,
232 McGill Street,
Montreal, March 28, 1907

There is what might be termed a kind of an uneasy atmosphere in metal circles this week, but so far, dealers have not changed their prices. Copper and tin have the honor to be in the spot light as the disturbers of the peace, both these metals having eased off unexpectedly.

Dealers do not expect any great change from present prices in any lines within the near future, unless further fluctuations take place in financial circles.

Copper and tin are both extremely short and no heavy stocks are to be found any place in the city. Dealers report that pig iron is also very scarce and that there is hardly any to be had in town for love or money—it's a case of beg, borrow or steal what you can lay your hands on, and the man with the biggest hand gets the biggest share. This state of affairs is, no doubt, brought about by the severe floods at Pittsburg and along the Ohio River, which have caused the shut-down of some of the busiest mills, the result of which will be to delay still further, shipment of many long overdue orders. If these furnaces have to remain out for any length of time, it will accentuate the situation still more, but it is thought that a week or ten days will again place them in normal condition, but meanwhile, the damage has been done, and the loss of production cannot be made up when it is most needed. It is also said that shipments of Middlesborough irons will not be as great as they have been during the past several months. This state of affairs does not tend to make the outlook for immediate relief very promising.

The copper and tin situation in New York seems to be practically the same as it is here, with the exception that the decline in prices was more marked.

Zinc spelter is about the same as last week, prices remaining firm and unchanged.

Some dealers state that pig lead also weakened slightly during the week, while others report it unchanged.

Antimony shows no special features at present, prices remaining firm and unchanged. The heavy demand for galvanized iron still prevails, and an exceedingly good business is being done in this line.

All other lines, such as boiler tubes, tool steel, cold rolled shafting, etc., remain unchanged. For prices on all lines, see current market quotations at the back of the book.

Toronto Metal Markets

Office of HARDWARE AND METAL,
10 Front Street East,
Toronto, March 28, 1907

Last week closed with sensational declines in the foreign markets, the reduced figures being due to speculation in tin and copper, although other metals such as iron and lead were affected by the general weakness which has followed the serious conditions existing in the stock and money markets. Fluctuations have been entirely due to the tightness in the money market and it is not expected that the decline will hold for any length of time, in fact at the time of writing the market has largely recovered. Factories are as busy as ever and mills are booked ahead for the balance of the year.

The great scarcity of copper is the reason for its extraordinary strength and regardless of the price of stocks it seems assured that the price of the metal itself will hold firm. Under normal circumstances a slump in the value of securities such as recently has taken place would have had a very bad effect upon the copper market. Little excitement has been manifested, however, owing to the general knowledge that the production is not keeping pace with the demand.

Pig iron has fallen off slightly in the Old Country but recent transactions in the American market indicate clearly that there will be no slow down for several months yet. Two months ago there was much talk of canceling booked orders for iron and steel, but there is none of this heard now. Practically all the mills are choked with orders for which specifications have been supplied. The floods in the Pittsburg district have placed many mills and furnaces out of business, and it is probable many orders for pig iron and steel will be considerably delayed in being fulfilled, and dealers who are looking for supplies from this source have this complication to add to previous difficulties in securing shipments. Advices from Great Britain indicate that iron is likely to come forward more slowly than in recent months, and it is interesting to note that stocks of pig iron in Great Britain are a little more than half what they were a year ago, for instance, at Middlesborough a year ago there were about 750,000 tons in stock where today there is considerably less than 500,000 tons.

Business locally is being continued at the high pressure which has been the feature of the market for several weeks past, there being a good volume of trade all through the list of metals. In galvanized iron, tin plates and tinned iron there is a much larger business being done than in any previous year, the call for these goods from dairymen, builders, manufacturers and other consumers being very active. Hoop iron is also in keen demand just now and sheet copper is being bought liberally. All ingot metals are moving freely and despite the fluctuations of outside markets prices on all finished metals are holding firm locally.

Lead, antimony, zinc and spelter are all strong, no weakness having been developed in any of these lines.

The only changes which have taken place are in old material, copper, iron and lead being all subject to slightly reduced quotations. Old rubbers have dropped a cent and they are now subject to new conditions governing the packing. All rubbers will have to be graded better in future, buckles, overshoe tops, etc., being detached before weighing. The quotations are: Copper and wire, 20c. per lb.; light copper, 17½c. per lb.; heavy red brass, 18c. per per lb.; heavy yellow brass, 15c. per lb.; light brass, 11½c. per lb.; tea lead, $4 per 100 lbs.; heavy lead, $4.40 per 100 lbs.; scrap zinc, 4½c. per lb.; No. 1 wrought iron, $12.50; No. 2 wrought, $7.50; machinery cast scrap, $17.50; stove plate, $11.50; malleable and steel, $8.50; old rubbers, 10c. per lb.; country mixed rags, $1 to $1.25 per 100 lbs., according to quality.

London, Eng., Metal Markets.

From Metal Market Report, March 26, 1907.

PIG IRON—Cleveland warrants are quoted at 52s. 10½d., and Glasgow standard warrants at 52s., making prices as compared with last week on Cleveland warrants 1s. 1½d. lower; on Glasgow standard warrants, 1s. 3d. lower.

TIN—Spot tin opened weak at £180 10s., futures at £178 10s., and after sales of 260 tons of spot, and 650 tons of futures, closed firm at £181 12s. 6d. for spot, £179 12s. 6d. for futures, making price, compared with last week £6 12s. 6d. lower on spot, and £6 7s. 6d. lower on futures.

COPPER—Spot copper opened weak at £95 10s., futures £97, and after sales of 800 tons of spot and 1,600 tons of futures, closed firm at £97 for spot, and £99 for futures, making price, compared with last week £8 10s. lower on spot, and £7 15s. lower on futures.

LEAD—The market closed at £19 6s. 3d., making price as compared with last week 7s. 6d. lower.

SPELTER—The market closed at £25 15s., making price as compared with last week 10s. lower.

United States Metal Markets

From the Iron Age, March 28, 1907.

On the whole the buying of pig iron has been rather light, and in at least one case an important inquiry for foundry iron was withdrawn. About 8,000 tons of the foundry iron reported last week to have been sold to a large melter turns out to have been Middlesborough. We understand that during the last week a cargo of close to 5,000 tons of Scotch iron was purchased abroad by an American importer to cover sales here previously made. The Middlesborough market has declined to 53 shillings, 3 pence.

A very curious incident is the offering, for shipment from Shanghai, of a lot of 5,000 tons of basic pig iron made in China. The analysis is a very good one, the iron being guaranteed under 0.05 in sulphur, 0.70 in silicon and 0.30 in phosphorus. It is offered for shipment to arrive here, 2,500 tons in June and 2,500 tons for the third quarter. The price is at a figure which may prove attractive.

In the Central West the scarcity of steel continues. One large steel interest in the Pittsburg district is now in the market for important quantities to cover a shortage in supplies for April and May delivery.

Quite a large tonnage of structural material is in sight. Thus Salt Lake City has work in hand aggregating over 15,000 tons, of which about 3,000 tons has already been contracted for. There is good work coming up, too, from nearly every important Pacific Coast city, and Chicago is expected to call for considerable quantities soon.

Chicago opens bids this week for 12,000 tons of cast iron water pipe. Detroit has just placed 6,000 tons and minor quantities have been sold at Ohio points.

From the Iron Trade Review, March 28, 1907.

New tonnages in iron and steel are not conspicuously active, but there is unmistakable evidence of strength in those materials that were inclined to weakness. This change is most clearly shown in pig iron. It is becoming increasingly difficult to secure spot shipments either from southern or northern furnaces, a fact which has aided in strengthening the market on all deliveries. No price changes have yet appeared other than a firm refusal to shade $23 Birmingham for spot southern No. 2, and an indisposition to go less than $18.50 Birmingham for the last half. Manufacturers of sheets are overwhelmed with orders and specifications, one producer having nothing to offer for the next four to five months, while the best delivery obtainable from many mills is 15 weeks. The movement in structural steel improves with the approach of warm weather, and mills are rapidly filling with specifications. Cast iron pipe was active this week, contracts closed aggregating 7,500 tons. A round tonnage of steel bars has been placed with mills for delivery in the second half. Scrap shows an upward trend.

The movement of wire products continues of an unprecedented nature. Every section of the country is pressing for material, stocks being low and deliveries slow.

The scarcity of semi-finished material has seriously handicaped many of the mills and it will be some time before they can return to their average maximum production. The Humbert, Anderson and Morewood plants of the American Sheet & Tin Plate Co. are closed on account of lack of steel.

British Columbia News

By G. S. B. Perry.

Vancouver, B.C., March 20, 1907.

Application for a water record for 75,000 inches has been made by the Stave Lake Power Co., the record to apply to the Stave River just below the point at which they are now erecting their big power generating station. This application practically means the using over again of the water in the stream and flowing from the lake, after it has passed through the turbines of the power house now being built. As an inch of water—meaning a miner's inch—indicates in British Columbia law, 1.68 cubic feet per minute passing through a pipe or flume, the quantity now being applied for suggests an enormous power development. The Stave Lake Company of which John Hendry, president of the V. W. & Y. Ry. is head, is preparing to enter the field in the coast cities for supply of power to manufacturing plants, and all industries likely to be users of electric energy. The only company now in the field is the B. C. Electric Co., with its subsidiary company, the Vancouver Power Co.

* * *

The day of electric generation from water power seems to have come in British Columbia. The developments on the mainland are but foreshadowing what is to come on Vancouver Island. There the B. C. Electric Railway Company is at present making an effort to secure large additions to present available water power, to provide for expected expansion both in the electrifying of railways on the island and for expansion of many industries. Experts are now out in behalf of the B. C. Electric Company making examinations and reports on every stream available for water power generation. It is persistently asserted that the E. & N. Railway, now owned by the C.P.R., is to be electrified. If this were so, it would be through an arrangement between the C.P.R. and B. C. Electric Railway Co.

* * *

The first carload of coal from the Nicola Valley has been received in Vancouver. The coal was shipped over the new branch line from Coutlee to Spence's bridge, on the C.P.R. main line. The Nicola Valley Coal & Coke Co. shipped the car from their Middlesboro colliery, which has been well developed in anticipation of the railway being completed. It is of excellent quality, and suitable for coking as well as for ordinary steaming and heating purposes. It is asserted that the tests have shown this coal capable of being used for smelting iron ores, and that in this respect it is ahead of other coal mined in the province.

When the C.P.R. extends this Nicola branch to Midway, connecting there with the C. & W. branch and thus reaching the smelters of the boundary district, the haul will be 125 miles shorter than from the Crow's Nest Pass. The comparatively short distance to the coast will also make it possible for Nicola coal to get into the coast market in competition with Vancouver Island mines.

Another marine railway is to be built in Vancouver, to handle vessels of the coasting fleet. Capt. Watts of the Vancouver Shipyards, is preparing plans for a marine railway with three tracks to dock vessels up to fifteen feet draught.

* * *

A Vancouver firm, Evans, Coleman & Evans, secured from the Winnipeg city council recently a large order for cast iron pipe, amounting to some 6,000 tons. This large contract, the largest expected to be open for competition in Canada this year, was secured against competitors representing nearly all the makers of cast iron pipe in Britain as well as the three Canadian pipe foundries. Cast iron has advanced seriously in price and competition has therefore been keen and close.

* * *

Some forty new saw and shingle mills are estimated to be under construction, or contemplated, in the Fraser River district of the lower mainland alone, this year. Of this number about 25 are shingle mills, and of the saw mills many are small portable or semi-portable plants. These are in many cases put in on limits of small area with the expectation of moving them in a few years after cutting out the berth on which they are located. The extreme activity in the lumber market is responsible for the unprecedented development, which is likely to keep up as demand for lumber in the Northwest is keener than ever. Prices, too, show an upward tendency and the report given publicity some time ago that prices would go up another $2 per thousand at April 1st, has not been contradicted, but has really been reaffirmed. The lumber inquiry at Ottawa, now being heard, does not seem to disturb the lumbermen, who claim to be able, if necessary, to amply justify their present price list, as well as to refute the assertions that they have an unlawful combine.

* * *

Included in the cargo of the Canadian-New Zealand liner Bucentaur when she sailed from Vancouver a few days ago, was a consignment of 30 tons of pig lead from the Trail smelter, for New Zealand. This product is refined at the smelter from the silver lead ores of the Slocan. Other consignments in the Bucentaur's cargo were paper, some 200 tons from Eastern Canadian mills, and about a million feet of lumber. The vessels of this line are at present carrying large quantities of lumber and will continue to do so until the general freight offerings increase.

The freight offerings and inquiries for space in the first steamer of the Canadian-Mexican line, which is expected to be sailing from Vancouver in a month, are very encouraging to the promoters of the enterprise. When trade openings in Mexico are exploited by B. C. business houses there is every assurance that the new line of steamers will have all the cargo they can handle. The addition, within twelve months, of two new ocean lines to the shipping of B. C. ports is an indication of growth which is being marked by students of commerce.

N.B. Hardware Trade News

St. John, N.B., March 25.

The hardware dealers here are anticipating a large spring trade. Travelers now on the road are sending in orders for spring lines and the outlook is extremely bright. One prominent merchant told your correspondent that the past three months had been very successful and it was expected that there would be a continuance of good business. There has been practically no change in prices, all lines of goods remaining firm.

* * *

John Young, sales manager of the Canadian Rubber Company, of Montreal, is in the city. He is looking after the Maritime Provinces agencies.

* * *

The quarterly meeting of the Maritime Wholesale Hardware Association will be held in Halifax on Tuesday of this week. The delegates from St. John will be John Keefe, vice-president; T. Carlton Lee, F. W. Murray, W. H. Smith and George A. Lockhart, secretary.

* * *

A. M. Bell, of A. M. Bell & Co., Halifax, was in the city on Tuesday last to attend the funeral of his brother-in-law, the late Dr. A. A. Stockton, M.P.

* * *

The Murchie Lumber Co. has disposed of extensive crown land holdings in the Tobique regions to Stetson-Cutler Co. at a price approximating $108,000. It is said the Murchie Co. will go out of business on the Tobique. The price paid for the property some years ago by the Murchie Co. is said to have been about $80,000, and they have been cutting every year at a good profit and now clear about $28,000 on the deal.

The property comprises about 260 square miles, and is thickly covered by spruce. The purchasers will probably start a pulp mill there.

* * *

A. B. Wetmore, of Chipman, will open a branch of his carriage and farm implement business in Sussex.

* * *

The machinery at the new factory of the Sussex Manufacturing Company's plant is being installed as rapidly as possible. The foundry commenced work last week and the first castings for some time were made.

* * *

William H. Stirling's brass fitting establishment on Water street, was visited by fire on Wednesday night last. It is thought the blaze started in the floor of the second storey, at the furnace flue. Very little damage was done. Mr. Stirling had $2,000 insurance.

* * *

There is considerable interest in the announcement made by Premier Pugsley in the House last week that Mackenzie & Mann may build a railway down the St. John valley, from Woodstock through to St. John. It is believed that this is a part of the scheme for a trans-continental system which this concern is working toward.

Arthur E. Jubien, of Halifax, accountant with the Sydney Cement Company at Sydney, was married in this city on March 12th to Miss Alice M. Armstrong, daughter of Lt.-Col A. J. Armstrong, of this city.

* * *

The Portland Rolling Mills were troubled somewhat last week by a strike among the teamsters. After being out for three or four days the men decided to go back to work, as the company had announced that they would purchase teams if the strikers did not give in.

* * *

Robert W. Graves, Lydia C. Graves, William Miller, of Elizabeth, (N.J.); Frederick A. Young, of St. John, N.B., and Samuel Holt, of Jersey City, are seeking incorporation as the Canadian Mineral Company. The chief place of business will be St. John, and the proposed capital stock is to be $20,000.

MORRISBURG'S NEW INDUSTRY.

Reference was made last week to the progress being made at the Canada Tin Plate and Sheet Steel Company's works at Morrisburg. The company is expecting to commence operations late in April, their machinery being now nearly ready for installation.

Seen by a representative of Hardware and Metal, President Meldrum stated that his company had suffered considerably from local opposition but they were going ahead, and, having failed to secure the duty asked for, would branch out in other fields of manufacture. Mr. Meldrum said: "The delay in getting started is entirely beyond the company's control, and is due to the failure of the manufacturers of the large driving wheels ordered by the company in June last, to make delivery. These wheels are the largest ever cast in Canada, weighing 75 tons each and measuring 30 feet in diameter, and, because of their unusual size, the contractor miscalculated the time in which they could be turned out. Delivery is promised in two or three weeks' time, however, and very shortly thereafter the wheels will be turning in the most modern and well laid out tin plate sheet mills to be found anywhere, and we believe the first to be electrically driven throughout.

"The company's plant is situated just east of Morrisburg, and is connected by a switch with the main line of the Grand Trunk Railway between Toronto and Montreal. It is also intended to run a spur to the river St. Lawrence, about half a mile distant, so that shipments can be handled by water as well as by rail.

"The main building is one of the largest continuous structures in Canada, being about 800 feet in length with an average width of about 90 feet, and is most substantially constructed of steel and stone. When installed the equipment will consist of five hot and four cold mills in the tin plate department; two finishing, one roughing and two cold mills in the sheet steel department: giving a total equipment of 14 mills, with the necessary heating and annealing ovens, pickling and tinning machines, bar and trimming shears, etc. It is the intention to turn out a full line of tin and terne plates, block plates, Canada plates, steel sheets, galvanized sheets, etc., and the mills will have a capacity of about 8,000 boxes per week. There is a large and growing demand for these commodities in Canada at the present time and as under the new tariff the company is permitted to bring in all its machinery and everything entering into the manufacture of its goods free of duty, it should prove an important factor in this trade in the country.

"The municipality of Morrisburg has now completed, at a cost of about $100,000, the power plant for supplying the company with electrical energy, and is under contract to deliver 700 horse power into the company's works free for 24 hours each day. The other concessions the company are to receive include free water and light and exemption from taxation (except for school purposes).

"The ponderous machinery, including the 20-ton electric crane now in operation, will be driven by some 10 or 12 electric motors, the largest of which are two of 650 horse power each for propelling the immense driving wheels coupled with the hot mills. The company will employ about 350 men, most of them skilled mechanics from the U. S., who will be a welcome addition to the population of this country.

"As soon as the present plant is in operation it is the intention to proceed with the erection of three open hearth steel furnaces for producing a supply of steel bars for the tin plate and sheet mills, and an enamelling and stamping plant for making the finished goods will also be added."

A GROWING BUSINESS

The National Hardware Co. was established at Orillia in August, 1905. Employment is given to 70 hands, and to-day orders for goods press in upon the company with such speed that every department is continually taxed to its full capacity. They manufacture a great variety of locks, door knobs and hardware specialties in steel and iron, light castings in iron and brass, and electro-plating in silver, nickel and brass, making a specialty of locks and builders' hardware. So much attention has been given to the perfecting of the company's business in these lines since its establishment that the company is confident that it possesses one of the finest plants for the manufacture of all kinds of fancy brass and iron work in the Dominion.

The machinery and the facilities are of the most modern type, the management being determined to have the best that inventive genius can produce. The officers and directors of the company are: Thos. Venner, president; R. J. Sanderson, vice-president; H. A. Croxall, secretary-treasurer and manager; R. D. Gunn, K.C., R. H. Montgomery, D. C. Thomson.

It would be difficult to find a more certain index of the substantial progress of the country and of her rapidly increasing wealth than is afforded by the rise to sudden prominence of concerns manufacturing along the lines of the National Hardware Company.

LONDON STATIST ON AMERICAN AFFAIRS.

The London Statist says of the American situation: "At such a time as this it would need a bold man indeed who would offer any definite opinion as to the level to which prices are likely to fall. But there is one thing that can be said, there is a general consensus of opinion among the well-informed that trade next year will merely halt in order that capital may accumulate, and if this view be correct, and it certainly coincides with our own view, the companies will not experience much difficulty in maintaining their dividends. The difficulties may arise with companies engaged in considerable construction works which they intend to finance out of capital and who now find difficulty in raising the capital for the purpose. These companies may find it necessary temporarily to reduce their dividends in order to devote a larger portion of their profits to the work of construction."

DEATH OF ALFRED BUEHLER.

In the death last week of Alfred Buehler, vice-president and general manager of the J. H. Ashdown Hardware Co., Winnipeg loses one of her most substantial and most widely respected business men. While the news of Mr. Buehler's death will be learned with sorrow by the hardware trade of Canada it will not occasion any surprise, as it has been known for some time that his days were numbered. Three years ago a prominent physician gave Mr. Buehler two years to live, but he preferred to die in harness and he remained in active management of the business until a very short time before his death.

Mr. Buehler came west in 1879, and after a short experience in a boot and shoe warehouse entered the employ of Mr. Ashdown. He started at the bottom of the ladder and by hard work and first-class ability reached the highest position in the gift of the company.

Personally Mr. Buehler was popular among a wide circle of acquaintances in Winnipeg and the west and his loss is widely mourned.

BUILDING AND INDUSTRIAL.

An American company has bought the Beatty Estate property in Toronto at the corner of Queen and Parliament streets, and will erect there a five-storey departmental store.

The Asbestos Mfg. Co., of London, Ont., has been incorporated with a capital of $25,000 to manufacture and deal in asbestos. The promoters are H. V. Everham, R. Van Z. Mattison, jr., and R. Van Z. Mattison, sr., all of London.

MANITOBA HARDWARE AND METAL MARKETS

Market quotations corrected by telegraph up to 12 a m. Room 511, Union Bank Building, Winnipeg, Man.

With the opening of spring and a consequent improvement of conditions with the railways, the wholesale houses are deluged with orders for present delivery. Building in Winnipeg itself and in most of the western towns seems likely to be much greater than last year and the demand for building supplies and builders' hardware is very active. Values are steady in most lines.

LANTERNS—Cold blast, per dozen, $6.50; coppered, $8.50; dash, $8.50.

WIRE—Barbed wire, 100 lbs., $3.22½; plain galvanized, 6, 7 and 8, $3.70; No. 9 $3.25; No. 10, $3.70; No. 11, $3.80; No. 12, $3.45; No. 13, $3.55; No. 14, $4.00; No. 15, $4.25; No. 16, $4.40; plain twist, $3.45; staples, $3.50; oiled annealed wire, 10, $2.96; 11, $3.02; 12, $3.10; 13, $3.20; 14, $3.30; 15, $3.45. Annealed wires (unoiled) 10c. less; soft copper wire, base, 36c.; brass spring wire, base, 30c.

HORSESHOES—Iron No. 0 to No. 1, $4.65; No. 2 and larger, $4.40; snowshoes, No. 0 to No. 1, $4.90; No. 2 and larger, $4.65; steel, No. 0 to No. 1, $5; No. 2 and larger, $4.75.

HORSENAILS — Capewell brand, quotations on application, No. 10, 22c; No. 9, 24c; No. 8, 24c; No. 7, 26c; No. 6, 28c; No. 5, 30c; No. 4, 36c per lb. Discounts: "C" brand, 40, 10 and 7½ p.c.; "M" brand and other brands, 55 to 60 p.c. Add 15c. per box.

WIRE NAILS—$2.95 f.o.b. Winnipeg, and $2.50 f.o.b. Fort William.

CUT NAILS—Now $2.90 per keg.

PRESSED SPIKES — ¼ x 5 and 6, $4.75; 5-6 x 5, 6 and 7, $4.40; ⅜ x 6, 7 and 8, $4.25; 7-16 x 7 and 9, $4.15; ½ x 8, 9, 10 and 12, $4.05; ⅝ x 10 and 12, $3.90. All other lengths 25c. extra net.

SCREWS.—Flat head, iron, bright, 85 and 10; round head, iron, 80; flat head, brass, 75 and 10; round head, brass, 70 and 10; coach, 70.

NUTS AND BOLTS — Bolts, carriage, ⅜ or smaller, 60; bolts, carriage, 7-16 and up, 50; bolts, machine, ⅜ and under, 50 and 5; bolts, machine, 7-16 and over, 50; bolts, tire, 65; bolt ends, 55; sleigh shoe bolts, 65 and 10; machine screws, 70; plough bolts, 55; square nuts, cases, 3; square nuts, small lots, 2½; hex nuts, cases, 3; hex nuts, small lots, 2½ p.c. Stone bolts, 70 and 10 p.c.

RIVETS — Iron, 60 and 10 p.c.; copper, No. 7, 43c., No. 8, 42½c.; No. 9, 45½c.; copper, No. 10, 47c.; copper, No. 12, 50½c.; assorted, No. 8, 44½c., and No. 10, 48c.

COIL CHAIN—¼ in., $7; 5-16, $5.35; ⅜, $4.75; 7-16, $4.50; ½, $4.25; 9-16, $4.20; ⅝, $4.25; ¾, $4.10.

SHOVELS—List has advanced $1 per dozen on all spades, shovels and scoops.

HARVEST TOOLS—60 and 5 p.c.

AXE HANDLES—Turned, s.g. hickory doz., $3.15; No. 1, $1.90. No. 2, $1.60; octagon, extra, $2.30; No. 1, $1.60.

AXES — Bench axes, 40; broad axes, 25 p.c. dis. off list; Royal Oak, per dozen, $6.25; Maple Leaf, $8.25; Model, $8.50; Black Prince, $7.25; Black Diamond, $9.25; Standard flint edge, $8.75; Copper King, $8.25; Columbian, $9.50; handled axes, North Star, $7.75; Black Prince, $9.25; Standard flint edge, $10.75; Copper King, $11 per dozen.

CHURNS.—45 and 5; list as follows: No. 0, $9; No. 1, $9; No. 2, $10; No. 3, $11; No. 4, $13; No. 5, $16.

AUGER BITS—"Irwin" bits, 47½ per cent., and other lines 70 per cent.

BLOCKS—Steel blocks, 35; wood, 55.

FITTINGS—Wrought couplings, 60; nipples, 65 and 10; T's and elbows, 10; malleable bushings, 50; malleable unions, 55 p.c.

HINGES—Light "T" and strap, 65.

HOOKS—Brush hooks, heavy, per doz., $8.75; grass hooks, $1.70.

STOVE PIPES—6-inch, per 100 feet length, $9; 7-inch, $9.75.

TINWARE, ETC. — Pressed, retinned, 70 and 10; pressed, plain, 75 and 2½; pieced, 30; japanned ware, 37½; enamelled ware, Famous, 50; Imperial, 50 and 10; Imperial, one coat, 60; Premier, 50; Colonial, 50 and 10; Royal, 60; Victoria, 45; white 45; Diamond, 50; Granite, 60 p.c.

GALVANIZED WARE—Pails, 37½ per cent.; other galvanized lines 30 per cent.

CORDAGE — Rope, sisal, 7-16 and larger, basis, $11.25; Manilla, 7-16 and larger, basis, $16.25; Lathyarn, $11.25; cotton rope, per lb., 21c.

SOLDER—Quoted at 27c. per pound. Block tin is quoted at 45c. per pound.

WRINGERS—Royal Canadian, $35; B.B., $39.75, per dozen.

FILES—Arcade, 75; Black Diamond, 60; Nicholson's, 62½ p.c.

LOCKS—Peterboro and Gurney, 40 p.c.

BUILDING PAPER—Anchor, plain, 66c.; tarred, 69c.; Victoria, plain, 71c.; tarred, 84c.

AMMUNITION, ETC.:—Cartridges, rim fire, 50 and 5; central fire, 33½ p.c.; military, 10 p.c. advance. Loaded shells: 12 guage, black, $16.50; chilled, 12 gauge, $17.50; soft, 10 gauge, $19.50; chilled, 10 gauge, $20.50. Shot, ordinary, per 100 lbs., $7.75; chilled, $8.10; powder, F.F., keg, Hamilton, $4.75; F.F.G., Dupont's, $5.

IRON AND STEEL—Bar iron basis, $2.70. Swedish iron basis, $4.95; sleigh shoe steel, $2.75; spring steel, $3.25; machinery steel, $3.50; tool steel, Black Diamond, 100 lbs., $9.50; Jessop, $13.

SHEET ZINC—$8.50 for cask lots, and $9 for broken lots.

PIG LEAD—Average price is $6.

COPPER — Planished copper, 44c. per lb.; plain, 39c.

IRON PIPE AND FITTINGS—Black pipe, ¼ inch, $2.65; ⅜, $2.80; ½, $3.50; ¾, $4.40; 1, $6.35; 1¼, $8.65; 1½, $10.40; 2, 13.85; 2½, $19.00; 3, $25.00. Galvanized iron pipe, ⅜ inch, $3.75; ½, $4.35; ¾, $5.65; 1, $8.10; 1¼, $11.00; 1½, $13.25; 2, inch, $17.65. Nipples, 70 and 10 per cent.; unions, couplings, bushings and plugs, 60 per cent.

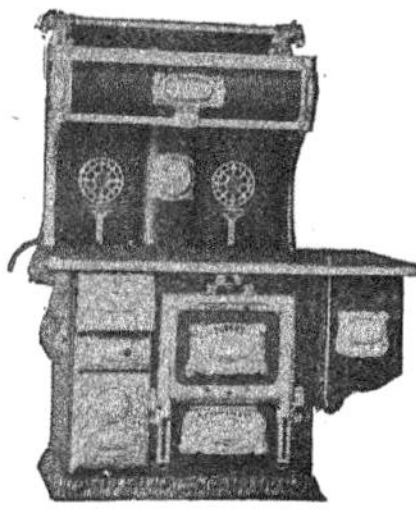

LEAD PIPE—Market is firm at $7.80.

TIN PLATES—IC charcoal, 20x28, box $9.50; IX charcoal, 20x28, $11.50; XXI charcoal, 20x28, $13.50.

TERNE PLATES—Quoted at $9.

CANADA PLATES—18x21, 18x24, $3.40; 20 x 28, $3.65; full polished, $4.15.

LUBRICATING OILS—600W. cylinders, 80c.; capital cylinders, 50c.; solar red engine, 30c.; Atlantic red engine, 29c.; heavy castor, 28c.; medium castor, 27c.; ready harvester, 28c.; standard hand separator oil, 35c.; standard gas engine oil, 35c. per gallon.

PETROLEUM AND GASOLINE — Silver Star in brls., per gal., 20c.; Sunlight in brls. per gal., 22c.; per case, $2.35; Eocene in brls, per gal., 24c.; per case, $2.50; Pennoline in brls., per gal., 24c.; Crystal Spray, 23c.; Silver Light, 21c.; Engine gasoline in barrels, gal. 27c., f.o.b. Winnipeg in cases, $2.75.

PAINTS AND OILS — White lead, Pure, $6.50 to $7.50, according to brand; bladder putty, in bbls., 2¼c.; in kegs, 3¾c.; turpentine, barrel lots, Winnipeg, $1.01; Calgary, $1.08; Lethbridge, $1.08; Edmonton, $1.09. Less than barrel lots 5c. per gallon advance. Linseed oil, raw, Winnipeg, 72c.; Calgary, 79c.; Lethbridge, 79c.; Edmonton, 80c.; boiled oil, 3c. per gal. advance on these prices.

WINDOW GLASS — 16-oz. O.G., single, in 50-ft. boxes — 16 to 25 united inches, $2.25; 26 to 40, $3.40, 16-oz. O.G., single, in 100-ft. cases — 16 to 25 united inches, $4; 26 to 40, $4.52; 41 to 50, $4.75; 50 to 60, $5.25, 61 to 70, $5.75. 21-oz. C.S., double, in 100-ft. cases—26 to 40 united inches, $7.35; 41 to 50, $8.40; 51 to 60, $9.45; 61 to 70, $10.50; 71 to 80, $11.55; 81 to $17.30.

No matter where the store is situated, care of the show windows pays. Anything that pays is worth thought. The show window question cannot be neglected. If a good window makes trade, a poor one drives it away. There is nothing negative about a window; it has an active influence one way or another. If you have not taken up window dressing as a regular feature of your business do it now!

HARDWARE TRADE GOSSIP

Ontario.

F. Clement, of Lynden, was in Toronto this week.

J. S. Allan, hardware merchant, Burlington, has sold out.

A. Henry, of Henry & Son, Orono, was in Toronto this week.

W. Conn, reported to have assigned, is still doing business at Aylmer.

W. Hudson, Dufferin street, Toronto, is putting in a stock of hardware.

M. Ballantyne, of Ballantyne Bros., Kincardine, was in Toronto this week.

Rice & Barnes, hardware merchants, Whitby, have dissolved, Mr. Barnes retiring.

R. Jackson and J. Yorke have opened up a hardware and tinsmith business at Parkhill.

The first annual at-home of the Canada Screw Co.'s employees was held last week at Hamilton.

W. Wickerson, an employee of the steel plant at Hamilton, was killed by a crane last week.

R. J. Walkem, hardware merchant at Tottenham, has recovered from a recent attack of la grippe.

F. J. Taylor has severed his connection with the Western Foundry Co., Wingham, and will be succeeded by M. E. Robson as manager.

J. Cotie, of the Canada Turpentine Co., at Barry's Bay, was caught in some machinery and instantly killed last week.

F. Howe, an employee of the McClary Mfg. Co., London, met with a fatal accident last week, being caught in a belt and killed instantly.

Hardware merchants of Ottawa have petitioned the city council to repeal the early closing by-law, objecting that it is too restrictive upon business.

J. R. Myers, Stratford, whose store was burnt out last December, has again opened up his hardware business. M. Carfield, of Brantford, will take charge of hardware department, Mr. Myers continuing to look after the plumbing and tinsmith department.

J. B. Reade was presented with a Morris chair and several pieces of cut glass on March 16 by the managers and travelers of H. S. Howland, Sons & Co., Toronto, with which company he was connected for about twenty years. He recently accepted the position of buyer for the Kennedy Hardware Co., Toronto.

McKelvey & Birch, hardware merchants, Kingston, whose premises were destroyed by fire last week, have called for tenders for a new building four stories high on the old site. Their net loss will be about $7,000 or $8,000 by the fire. Two stores have been secured near the old stand, in which they will carry on business temporarily. Some twenty-five years ago the firm suffered from a fire even more disastrous than this one. They have been in partnership for over forty years.

Quebec.

A. Huot, Sorel, was in Montreal this week on business.

F. X. Paradis, St. Denis, called on the Montreal jobbers last week.

C. O. Gervais, St. Johns, Que., was in Montreal this week on business.

Morency & Cote, hardware merchants, Montreal, are registered.

P. Cote, Quebec, visited the metropolis this week on business bent.

A. R. Wilson, Sherbrooke, was in Montreal this week on business.

E. Chevalier, Joliette, was in Montreal this week purchasing supplies.

W. C. Webster, Coaticook, was in Montreal this week purchasing supplies.

H. Alguire, Maxville, was in Montreal last week placing orders for spring goods.

J. A. Paquin, St. Eustache, visited some of the Montreal wholesalers during the week.

J. W. Harris, of the J. W. Harris Co., Montreal, has just returned from a trip to Winnipeg.

D. Lavalle, St. Gabriél de Brandon, visited some of the wholesale houses during the week.

Harold G. Eadie, Montreal, is registered as chief agent of the White Fireproof Construction Co.

Mr. Haskell, of the Haskell Lumber Co., Fassett, was among the busy men in Montreal this week.

E. Crepin, Chateaguay, called on the Montreal jobbing houses this week to purchase spring supplies.

Alex. Gibb, one of Montreal's leading manufacturers' agents, has gone south for a holiday, and will visit Atlantic City and ather points.

Mr. McCubbin, New Glasgow, Que., was in Montreal this week on business.

J. H. Ashdown, Mayor of Winnipeg, is on a trip east, and was in Montreal during the week.

F. H. Scott, Montreal, representing several large cutlery and silverware houses of Sheffield, Eng., will move shortly from his present premises in the Temple building to the Coristine building on St. Paul Street.

Jas. Mitchell, Sherbrooke, is putting up a new wholesale warehouse, 125x60 feet, next door to his present store on Wellington street, which will be used in future entirely for retail trade, while the old warehouse down near the tracks will be used exclusively for heavy goods.

Western Canada.

A meeting of the creditors of F. G. Elliott,, hardware merchant, Gainsboro, Sask., was held.

Coade & Hughes, hardware merchants at Carievale, Sask., have dissolved, J. Hughes continuing.

H. Anderson has bought out the tinsmith department of the hardware business of A. V. Gerry, Indian Head.

CATALOGUES AND BOOKLETS

When sending catalogues for review, manufacturers would confer a favor by pointing out the new articles that they contain. It would assist the editor in writing the review.

By mentioning HARDWARE and METAL to show that the writer is in the trade, a copy of these catalogues or other printed matter will be sent by the firms whose addresses are given.

Pedlar's Roofing.

The Pedlar People, Oshawa, have just issued a 128-page booklet which "cancels all previous roofing catalogues," and contains "quotations for immediate acceptance." At the top of every page is the statement "Keeping everlastingly at it brings success." The fifth page contains the terms and conditions of business of the Pedlar people; and the sixth contains some useful items "of mutual interest." The rest of the booklet contains pricel and cuts of the various lines of goods manufactured at their factory. Copies can be had by anyone in the trade who mentions this paper.

Marine Engines.

The Canadian Fairbanks Co., Toronto, have just issued their marine engines catalogue. It is a nicely gotten up booklet of 12 pages containing cuts of their steam and motor engines. Anyone wishing a copy of this catalogue may get one by applying at any of the company's branch offices.

Cooking by Gas.

The Gurney Foundry Co., Toronto, have just issued a splendid booklet, "Cooking by Gas, the Oxford Way." It is is 26-page booklet containing illustrations of their Oxford gas stoves, along with some valuable information regarding the summer evil and how to remedy it. The company states that a special catalogue of these stoves may be had on application and mentioning this paper.

Farm Fences.

The farm fence is one of the most important things about the farm. Except in those districts where neighbors are few and far between and where stock is allowed to roam at will, a good fence is indispensable.

In most cases economy is a factor which forms an important consideration with the farmer. With fences, as well as with other articles, the best economy is to erect a good fence at the start. The cheapest article may have all the appearance of answering the purpose, but when the time and expense necessary for repairs is taken into account, it will be found that it is cheaper in the long run to erect a more serviceable fence and one which stands the wear and tear without so much looking after.

Probably the most practical farm fence yet erected is of the woven wire type. There are a number of different patterns and styles on the market at the present time, and it is well in selecting a fence of this type to be posted in regard to the points of superiority in which a fence of this character should excel, in order to prove of the greatest advantage in farm use. To give the best service a wire fence should possess the qualities of elasticity, tensile strength, firmness and rigidity. It should be so constructed that the upright and lateral wires are securely locked in such a way that they cannot be rooted up from below, shoved down from above or spread sidewise. The wire should be sufficiently heavy and well galvanized, hard enough to give the proper resistance to strain and pressure, but not sufficiently brittle to break under a quick blow. The lock should be of such a character as to allow the fence to be properly erected on uneven as well as level ground, and at the same time hold the wires securely and in their proper relative position.

The Banwell Hoxie Company publish an interesting little folder on the subject of fence erection, which gives valuable information on this important subject, and also gives instructions on the method of making concrete fence posts. A copy may be obtained by addressing the Banwell Hoxie Wire Fence Company, Hamilton, and mentioning this paper.

Original Calendar Design.

The Dominion Cartridge Co., of Montreal, are sending out to their patrons an interesting calendar picture. It represents the Northwest Mounted Police using their cartridges on the prairie in target practice. It is a striking picture and could be used to good advantage for wall or window displays. Anyone wishing a picture will do well to apply to the company.

Bowser's Business Boomer.

S. F. Bowser & Co., Fort Wayne, Ind., and Toronto, issue a neat little fortnightly for their salesmen, the Feb. 20 issue containing a novel primer for Bowser salesmen in alphabetical form. The same issue contained the following reference to one of their Canadian travelers: "Our Thos. Cragg encountered a great deal of trouble in getting over the ground in his Saskatchewan territory during January and February. Nevertheless he is making a very creditable showing. Blizzards have no terror for him, although in order to get out of the way of one he worked a hand-car seven miles one day lately, with the thermometer registering 40 degrees below. It does seem as though there is nothing Bowser salesmen will not do that is honorable to get business. The hand-car racket is the newest one we have heard of."

Morrison's Calendar.

The James Morrison Brass Co., Toronto, have supplied to their customers a large poster-size calendar, and copies will be sent to anyone in the trade on application.

STAR IRON CO.'S CHANGE.

The Maison Jean Paquette, hardware dealers, Montreal, have purchased for a number of years, the right to sell the output of the Star Iron Co., of Montreal, manufacturers of the "Star" boiler.

This was brought about by G. N. Ducharme, the proprietor of the Star Iron Co., wishing to devote more of his time to the manufacturing end of the business, in order to have a more complete line to meet the growing demand for this boiler, and in looking about for somebody to look after the selling end of the business, he hit on Mr. Paquette, as being one of the leading dealers in this line. The proficiency to which Mr. Paquette had brought his own business, showed that he was a competent manager, and this made Mr. Ducharme all the more anxious to close the deal with him.

Mr. Ducharme has lately invested a large amount of capital in the iron industry, and hopes that he will in the near future, be able to give Canada a manufacturing plant that will be a credit to the country at large.

The "Star" boiler is built on a design which is supposed to give the maximum of heat, with the minimum of coal consumption. The sections, instead of being flat, are oval, thereby giving the flame larger surface to cover. This is claimed to have the advantage of heating the water in a great deal less time than it could be accomplished in a flat section boiler.

The clerk who always remembers that he represents the store to the customer is the who who always has a job.

The clerk who passes things to his patrons makes friends for the store, while those that throw or tumble them out drive trade away.

FOUNDRY AND METAL INDUSTRIES

Hutchison Bros. Co., Vancouver, have erected a new foundry in that city to manufacture marine engines and boats.

A number of claims are being staked in the township of Dunnett, west of Sturgeon Falls, where discoveries of copper have been made.

The Hamilton Steel & Iron Co. have secured permits to build new factory buildings and a blast furnace, to cost $300,000, and additions to the open hearth furnaces to cost $50,000.

J. Fogg, who runs a brass foundry on King Street East, Toronto, was sentenced to the Central for three months for receiving brass and lead stolen by his employes from the Gas Works.

Sheet metal shades for automobile storage—that is, sheet metal garages—are becoming a factor in the trade of some sheet metal workers. This material is not only cheap, but it makes a non-inflammable house that can be easily enlarged or removed.

The Gurry Patents Co., Hamilton, has been incorporated with a capital of $40,000, to purchase and use the patent of Edward Gurry for a new pig iron moulding machine. The provisional directors are W. D. Long, P. D. Crerar, and E. Gurry, all of Hamilton.

A patent on a new blast furnace has been secured by H. W. Hixon, Victoria Mines, Ont. The main features of the furnace comprise a lining of refractory material, and an air-jacket, constituting a sunstitute for the water-jacket formerly used, and through which the air passes on its way to the twyers.

The strike of moulders and coremakers which started at the foundry of the Canadian Iron & Foundry Company, Hamilton, on August 29, 1906, has been called off. Only two or three out of 45 strikers have as yet secured positions. At St. Thomas the local union of moulders and coremakers has given up its charter.

Ordinary corrugated iron receives one coat of paint at the rolling mill, the paint usually employed being red oxide of iron thoroughly ground in pure linseed oil, with enough drier mixed in to give it proper drying quality. The first coat of paint is applied by machine, and likely to be imperfect, wherefore the sheets should be painted again after putting them on the building.

NEW METHOD OF TINNING.

The Meaker Company, Chicago, has just perfected a new process for coating iron, steel, copper, and other metals, with tin.

The method is an electric one, and by it an even, bright coat of tin is obtained which retains its lustre and is silvery in appearance. Irregular shaped articles and those that have been machine tooled, are coated without affecting the shape of the article or filling crevices or threads with the metal. Grey iron castings which have not been successfully tinned by the ordinary process, are successfully and economically covered by this method.

Finely perforated brass, copper or steel sheets are now tinned before being perforated. This crude method is necessary, because the tin would fill in the holes were the sheets perforated and then dipped by the hot process. We successfully tin these sheets after they are perforated. The cut edges are also tinned so that on steel sheets no corrosion takes place, and on brass and copper sheets, such as are used in sugar refineries and brewery supply houses, the cut edges being tinned, sulphuric and other acids do not readily attack the metal underneath.

ALZEN, A GERMAN METAL.

The founders of Frankfort, Germany, have, after many experiments, turned out a new metal composed of one part zinc and two parts aluminum, which they have named "alzen." Among the superior qualities claimed for it are that it equals cast iron in strength, is much more elastic, takes a high polish, and does not rust as easily as iron. In casting, the new metal fills out the most minute lines and figures in the forms and molds.

FOR SALE

Have you anything for sale which any Hardware Merchant, Plumber, Stove and Tinware dealer would be interested in ?

Advertise it in our "want ad." column. It will bring results.

Rates—

2c. per word 1st insertion

1c. " " subsequent insertions

HARDWARE and METAL

MONTREAL TORONTO WINNIPEG

Persons addressing advertisers wil kindly mention having seen this advertisement in Hardware and Metal.

BUILDING AND INDUSTRIAL NEWS

HARDWARE AND METAL would be pleased to receive from any authoritative source building and industrial news of any sort, the formation or incorporation of companies, establishment or enlargement of mills, factories, foundries or other works, railway or mining news. All correspondence will be treated as confidential when desired.

Calgary will erect a $45,000 hospital.

The C.P.R. will erect a $20,000 station at Souris.

Regina Baptists will build a new $40,000 church.

Regina's new Y.M.C.A. building will cost $65,000.

A new rink will be built at Ottawa costing $27,000.

Montreal will erect a new postoffice to cost $1,000,000.

A business block, to cost $40,000, will be erected in Lethbridge.

The Canada Glass Co., Montreal, will erect a factory costing $5,000.

Permits aggregating $11,700 were issued in Hamilton, March 18th.

A new Orange Hall is to be erected in Vancouver at a cost of $45,000.

The Manhattan Asbestos Co., Black Lake, Quebec, has been dissolved.

The total of permits issued in Vancouver for March so far is $301,345.

An addition, costing $9,000, is being made to St. Andrew's church, Brantford.

The Meisel Mfg. Co., Port Arthur, will build new factories to cost $120,000.

A new rink will be built in Toronto to cost $300,000, and seating 10,000 people.

Regina will spend $90,000 this summer on new school buildings and grounds.

The Hamilton Parks Board will build a new pavilion in Dundurn Park to cost $8,000.

A new five-storey block will be erected in Regina for F. N. Darke, to cost $100,000.

The National Trust Co., Toronto, is building a branch at Saskatoon to cost $25,000.

The Bank of Montreal has secured a site at Hull for a new branch to cost $35,000.

A new school will be erected at Paris to cost $50,000, the town to pay a fifth of the cost.

The McLaughlin Carriage Co., Oshawa, will build a branch warehouse at Calgary.

The Slingsby Mfg. Co., Galt, will make a $10,000 extension to their plant this spring.

The Canadian Westinghouse Co., Hamilton, will erect a $40,000 addition to its plant.

The Eastern Townships Bank will erect a new building in Montreal to cost $300,000.

The Hotel Dieu, Kingston, is to be remodelled and made into a mica works for Kent Bros.

A new opera house will be erected at Fort William by J. A. Timbers & Son, to seat 800.

A permit has been granted for the erection at Ottawa of a new skating rink to cost $25,700.

Hyslop Bros., Toronto, have secured a permit for the erection of a $30,000 garage at corner of Shuter and Victoria street.

Calgary capitalists are promoting a $600,000 Rocky Mountain Cement Co., with works at Frank.

The Salvation Army will build a new barracks at Toronto Junction with a seating capacity of 500.

Jos. Zuber will build a new four-storey hotel at Berlin, with 180 feet frontage, to cost $75,000.

The contracts for the erection of the Brandon Collegiate have been let. The total cost will be $58,697.

A new school will be erected at Leamington. The cost will be $22,000, less the cost of the old building.

The Canadian Brass Mfg. Co., Galt, will build on the Jackson Wagon Works property in that town.

The village of Alliston has passed a by-law to grant a bonus of $5,000 to the Merner Mfg. Co., of Waterloo.

R. and L. Blackburn, of New Edinburgh, have bought the Thomson property in Ottawa at a cost of $125,000.

Sleeman & Sons' Co., Guelph, will erect a new malt house, to cost $20,000, having a capacity of 75,000 bushels.

Work will be commenced on the Gilson Mfg. Co.'s plant at Guelph at once for the manufacture of gasoline engines.

Gordon, Mackay & Co., Toronto, are erecting a new five-storey factory at corner of Queen and Crawford streets.

The Huber Mfg. Co., of Marion, O., will establish a large manufactory of farm implements at Portage La Prairie.

The Bank of Montreal and the Canadian Bank of Commerce are erecting handsome new branches at Saskatoon.

The contract for the steel pipe for the north Rosedale sewer has been awarded to John Inglis Co., their tender being $4,480.

The Freyseng Cork Co., Toronto, will build a three-storey addition to their factory at the corner of Queen and Sumach streets, to cost $7,000.

The Confederation Life Association, Toronto, has taken out a permit to erect an eight-storey office building on the corner of Queen and Victoria streets to cost $200,000.

R. Locke is to erect six stores on corner of Bloor and Carling streets, Toronto, to cost together $20,000. A. Burns will erect three stores on Dundas street to cost altogether $10,000.

Reiner Bros. & Co., of the village of Wellesley, have been incorporated with a capital of $100,000 to carry on a general store and hardware business. The promoters are: J. G. Reiner, E. K. Reiner, and W. W. Cleghorn.

The Barton Netting Co., of Windsor, has been incorporated with a capital of $40,000 for the manufacture of mantels, tiling and fire place furniture. The promoters are C. J. Netting, C. C. McCloskey and G. F. Trunk, all of Windsor.

OUR LETTER BOX

Correspondence on matters of interest to the hardware trade is solicited. Manufacturers, jobbers, retailers and clerks are urged to express their opinions on matters under discussion.

Any questions asked will be promptly answered. Do you want to buy anything, want some shelving, a silent salesman, any special line of goods, anything in connection with the hardware trade? Ask us. We'll supply the necessary information

Baby Carriage Manufacturers.

A Brantford merchant writes: "Kindly inform us through your journal the names of makers of go-carts and baby carriages in Ontario and Quebec."

Answer.—Manufacturers of the above in Ontario are: American Rattan Co., Walkerton; Anderson Furniture Co., Woodstock; Gendron Mfg. Co., Toronto, and Semmens & Son, Hamilton. In Quebec: H. & F. Giddings & Co., Granby.

U. S. Wringer Manufacturers.

of 500 lbs., $6; ditto, in kegs of 100 lbs.,

Tweeddale & Co., Fredericton, write: "Would you kindly send me some names of wringer manufacturers in the United States?"

Answer.—The following manufacture wringers in the United States: American Wringer Co., New York City; the Boss Washing Machine Co., Cincinnatti, Ohio.

Grinding Machines.

Cameron & Leacock, of Smith's Falls, write: "Can you tell us where we can procure a machine that will grind squaring shear blades, skates and lawn mowers?"

Answer.—The only firm that we know of is Root Bros., Plymouth, Ohio.

The Problem in Discounts.

L. R. Allen, of Bruce, McKay & Co., Summerside, P.E.I., writes: "In reply to your question in your paper, dated March 9th, I beg to submit the following"—

A buys goods at 70 and 10 off list, which is equal to 73 straight discount or net 27 on the dollar.

B buys goods at 60 and 10 off list, which is equal to 64 straight discount or net 36 on the dollar.

From the above A would buy his goods 25 per cent. cheaper than B, and B would pay 33 1-3 per cent. more than A would.

Another Answer to Problem.

A Simcoe, Ont., clerk writes: "I enclose herewith my solution of discount problem in this week's journal."

Suppose, for example, that A and B both buy goods listed at $3.50 per doz.

A buys them at 70 per cent. and 10 per cent. off.

B buys them at 60 per cent. and 10 per cent. off.

A's goods will cost him $3.50 less 70 per cent. plus 10 per cent., which is 94½ cents.

B's goods will cost him $3.50, less 60 per cent. plus 10 per cent., which is $1.26.

Now A will gain on every 94½c. outlay 31½c. more than B, and 31½c. is gain of 1-3 on 94½c., or 33 1-3 per cent.

A therefore buys his goods 33 1-3 per cent. cheaper than B.

Police Goods.

A. J. Walker & Son, Truro, N.S., write: "Would you kindly give us the address of the makers of 'up-to-date' police goods, such as handcuffs, twisters, etc.?"

Answer.—We do not know the address of any such firm, and we would ask any subscriber who knows the address to inform us.

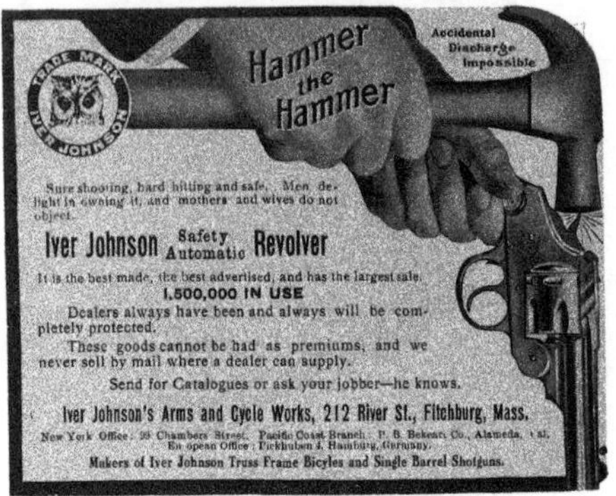

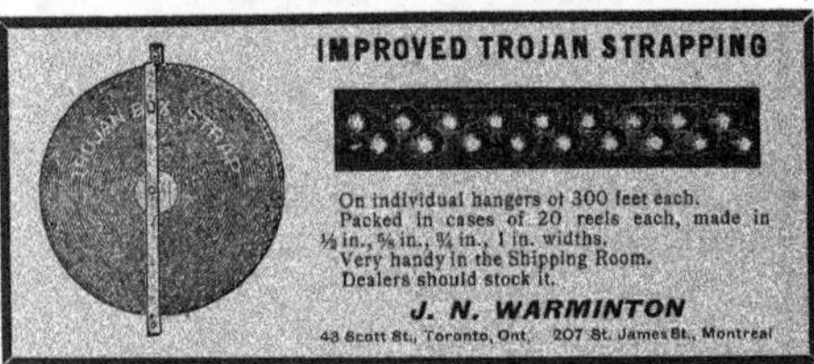

VALUE OF A SHOW WINDOW.

Perhaps the best way to realize the full value of a show window is to hire a few. That which we are accustomed to use daily has a tendency to seem of smaller account than it really is. The show window is of value in several ways; to admit light, as a place to display wares, as a power to draw custom into the store. It should not be forgotten, also, that the merchant is paying part of his rent for it all of the time. Never treat it lightly.

G. S. Simard Co., plumbers, Quebec, have been registered.

CONDENSED OR "WANT" ADVERTISEMENTS.

Advertisements under this heading 2c. a word first nsertion; 1c. a word each subsequent insertion.

Contractions count as one word, but five figures (as $1,000) are allowed as one word.

Cash remittances to cover cost **must** accompany all advertisements. **In no case** can this rule be overlooked. Advertisements received without remittance cannot be acknowledged.

Where replies come to our care to be forwarded, five cents must be added to cost to cover postage, etc.

BUSINESS CHANCES.

HARDWARE and Harness Business for sale in good town. Owner retiring for good reason. Stock about $2,500, in excellent shape. This is worth investigating, for particulars. apply Richard Tew, 23 Scott Street, Toronto.

HARDWARE Business; including stoves, tinware and tinsmith tools, in thriving town in West Ontario peninsular, stock about $5,000; building can be leased if desired, dwelling also. Box 583 HARDWARE AND METAL, Toronto. (17)

FOR SALE.

FOR SALE CHEAP — One hundred gallons paint, well assorted; one St. Thomas acetylene gas machine, with fixtures for twenty lights. Apply to Box 217, Port Hope.

FOR SALE—Complete set of tinners' tools. Apply Landon & Flesher, Barrie, Ont. [14]

WANTED.

WANTED—A house to handle cast-iron tees, elbows, iron fittings and straight-way gate valves of all sizes. HARDWARE AND METAL. [17]

BOOKKEEPER WANTED

WANTED—A good bookkeeper; stating experience, reference and salary. Cunningham Hardware Co., Westminster, B.C. [13]

SITUATIONS VACANT.

BRIGHT, intelligent boy wanted in every town and village in Canada; good pay besides a gift of a watch for good work. Write The MacLean Publishing Company, 10 Front Street E., Toronto. [tf]

HARDWARE Clerk Wanted — For a retail store; applicant to state age, experience and salary expected; give name of last employer and how long employed by him, also send references. C. W. Gray, Lethbridge, Alta. [14]

REPRESENTATIVE agent calling upon the wholesale hardware trade in Ontario, wanted to handle a strong line of nails and tacks; only first-class men with connection need apply in confidence to The Bazin Mfg. Co., 94 Arago St., Quebec.

TINSMITH—With 3 or 4 years experience, and with some knowledge of plumbing; state wages; steady job. W. B. Clifton, Alliston, Ont. [14]

WANTED—First-class tinsmith to take charge of shop; must be good furnace man and understand giving estimates. Address Box 593, HARDWARE AND METAL, Toronto. [13]

WANTED—Experienced Hardware Clerk—Must be good stock-keeper, and strictly sober; references required. Hose & Caniff, Kenora, Ont. [13]

WANTED—Immediately, reliable hardware clerk for B.C.; must be good buyer, stockkeeper and salesman of general hardware, etc.; state habits, capabilities, reasons for changing and salary expected to Box 596, HARDWARE AND METAL, Toronto.

TINSMITH WANTED—Competant man wanted for Brandon, Manitoba; must thoroughly understand his business, especially hot air heating; must be strictly temperate; salary, $1,200 per annum; steady employment to the right man. Apply with references to Brown & Mitchell, Brandon, Manitoba. [14]

PURCHASING AGENT WANTED—We are looking for a young man, a steady worker of intelligence who has had moderate experience in handling or purchasing hardware; some knowledge of factory costs also desirable. Address at once, stating age, experience and salary, Box 602, HARDWARE AND METAL, Toronto. [13]

COST ACCOUNTANT WANTED—Young man with some experience in figuring factory costs; must also have some knowledge of purchasing materials and supplies for factory; give experience and age. Box 601, HARDWARE AND METAL, Toronto. [13]

SITUATION VACANT.

WANTED—First-class tinsmith; married, age 35 to 40; steady work; none other need apply. Philip Drench, Walkerville, Ont. [13]

WANTED—Experienced traveller for builders' supplies; territory, Alberta. Apply Ellis & Grogan, Calgary, Alta.

WANTED—Tinsmith; good general hand for country shop; living rates low, good wages paid weekly; yearly job. Box 600, HARDWARE AND METAL. [14]

SITUATION WANTED

WANTED — By experienced hardware clerk and bookkeeper, position as head clerk in retail. Box 598, HARDWARE AND METAL, Toronto.

Paint, Oil and Brush Trades

HOW TO INCREASE PAINT AND VARNISH SALES THIS SPRING.

There has never been a time in retail paint selling when the merchant with a good line of reputable paints and varnishes had such an opportunity to do a record-breaking business. The prosperity of last year and the prevalent "good times" has put plenty of money in circulation, a great deal of building is going on and the people are becoming educated to a more extensive use of painting materials. The result is a greater demand and a far larger field of output. The business is in evidence; it all depends upon the dealer whether he gets it or not.

In the first place, a certain part of his store should be reserved for his paint stock, and, if possible, be a special department. This will enable him to display the goods to better advantage and to attract more attention to his lines of paints and varnishes. If his shelving is neatly arranged and he has made good and liberal use of the display cards, hangers, posters, color cards, and other store advertising, such as the larger paint manufacturers supply, he will create a good impression when the customer enters the store that will do much toward making sales. A stock slovenly arranged, making it difficult for the clerk to find any particular product, color or package, will not sell goods.

Displaying Paints.

A good feature for store use where the paint stock is not in a prominent place, is a display rack for the aisle, where gallon cans of paint can be shown, or a smaller one that can be set on the counter. Such racks, when rightly used, have proved splendid silent salesmen. They suggest the need of a product to the customer in the store and bring an inquiry about the goods and their prices.

Windows play an important part in bringing trade into the merchant's store, and too much care cannot be exercised in having attractive displays. One company give an eight-page book of window trims well worked-out for the use of their agents. These offer suggestions that can be followed inexpensively, and enable the merchant to change his windows without trouble. One clerk should be delegated to care for the windows, and in the spring two or three paint or varnish displays a week will not be too many. The suggestions you offer through your windows bear fruit in orders later on, and you can well afford to take a little trouble to have good displays. Window display features are also supplied by some paint manufacturers, and advantage can be taken of this to get more distinctive arrangements. These features include "cut-outs," display cards of all kinds, and other handsome advertising matter that secures attention and brings trade.

Newspaper Advertising.

In connection with inside store arrangement and display and window trims, a liberal use should be made of newspaper space. This reaches the majority of people and brings you to their attention. You can secure electros of advertisements ready made up for use from the manufacturer whose line you handle, if he is among the large ones, and thus secure the best kind of display and strong advertisements. If you want to advertise some of your other lines at the same time, however, there are special electros that you can get that will enable you to embody your paint ad, without detracting from the general display. In the spring it is advisable to advertise paint and varnish liberally, as that is the natural painting season and the time to make the strongest bid for business.

The more prominent manufacturers give their agents the advantage of a personal letter service, writing letters to prospective paint buyers for them. One company even provides an extensive spring campaign that includes the mailing of calendars, personal letters to property owners, and follow-up mailing cards. It is an established fact that so strong is this campaign that it seldom fails to land the prospect for the agent. The features are sent out to a list of property owners in his town and vicinity that he believes expect to paint or ought to paint, and which he compiles every year. This service costs him absolutely nothing, except the time necessary to prepare a list of names.

(Continued on page 54.)

SOME day a genius may discover a better process of manufacturing paint than we know; but that day seems far away. Our expert analysts are certain that

NEW ERA PAINT

the most effective "silent salesman" on the market

will hold first place against all comers anywhere in sight. You can recommend New Era with full assurance that your customer will advertise your business right along.

STANDARD PAINT AND VARNISH WORKS CO., LIMITED

WINDSOR - ONTARIO

Most Floor Paints

are very unsatisfactory. By the time they are thoroughly dry, they have begun to wear out.

They always make a sticky floor, and are more or less affected by washing.

These are the paints you have always needed to apologize for.

Want Something New?

Our Floor Enamel

is just what you are looking for.

It gives a mirror-gloss finish to the floor. It dries hard over night, and is never sticky. Morever, it will outwear any floor ***paint*** on the market.

WRITE US FOR COLOR CARD AND PRICES

R. C. JAMIESON & CO.

LIMITED

Sole Manufacturers, MONTREAL

The wise dealer takes no chances. He recommends

Pearcy's Pure Prepared Paints

because he has tried and tested them in every way.

PROMPT DELIVERIES A SPECIALTY

SANDERSON PEARCY & CO., Ltd.

Toronto, Canada

A. RUDD & CO., St. Helens, Lancashire, England

Manufacturers of highest grade

GLAZIERS' DIAMONDS in all patterns. Also **Circle Boards, Beam Compasses, Gauge Glass Cutters, Boxwood Squares and Lathe, Plate Glass Nippers, Atmospheric Soldering Irons, etc., etc., etc.**

Canadian Agent: **I. B. GIBSONE, P.O. Box No. 478 MONTREAL**

GLAZIERS' DIAMONDS OF PROVED WORTH

Having a Century's Reputation for Reliability.

MADE BY

A. SHAW & SON, London, Eng.

CANADIAN AGENT

GODFREY S. PELTON, - 201 Coristine Building, Montreal

BEFORE ordering colors think of **THE VIA GELLIA COLOR CO.**

Raisers and Levigators of Mineral Colors

Mines and works at

Matlock-Bath, Cromford and Bonsall (Eng) and Ovoca (Ireland)

Our specialities are

ENGLISH and TURKEY UMBERS (raw and burnt) TURKEY RED, RED AND PURPLE OXIDES, PURPLE BROWNS, YELLOW OCHRES, ROTTEN STONE (lump and powder) and Superior Ivory Blacks.

These are exclusive lines and are strong money-makers in the trade. Write for samples and prices to

WM. O. GREENWAY, 13 t. John St., Montreal

It will pay you.

Paint and Oil Markets

MONTREAL.

Office of HARDWARE AND METAL,
232 McGill Street,
Montreal, March 28, 1907

A buoyant feeling prevails in paint and varnish circles, and the cry now is for teams and cars to haul the manufactured product away.

Shipments have been somewhat hampered by the state of the roads, this being the transitory time in Montreal between sleighs and wheels.

Prices on all classes of painting material may be classed as being firm, and there is little tendency to recede from quotations, even for round lots, simply because the manufacturers seem all pretty busy.

LINSEED OIL—In good request. There is some see-sawing going on for round lots for importation, but on the whole, jobbing figures are well maintained, at an advance of 2 cents per gallon over last quotations. Prices now are: Raw, 1 to 4 barrels, 62c.; 5 to 9 barrels, 61c.; boiled, 1 to 4 barrels, 65c.; 5 to 9 barrels, 64c.

TURPENTINE—Being shipped very freely, and in view of the high figures now prevailing, stocks are kept as light as possible. There has been an advance of two cents per gallon, and we are now quoting: Single barrel, $1 per gal.; for smaller quantities than barrels, 5c extra per gal. is charged. Standard gallon is 8.40 lbs., f.o.b. point of shipment, net 30 days.

GROUND WHITE LEAD—Bad roads have impeded shipments to some extent, but orders for April and May delivery are of generous proportion. Prices remain: Best brands, Government standard, $7.25 to $7.50; No. 1, $6.90 to $7.15; No. 2, $6.55 to $6.90; No. 3, $6.30 to $6.55; all f.o.b. Montreal.

DRY WHITE ZINC—In fair demand only, with no change in prices, which remain: V.M. Red Seal, 7½c. to 8c.; Red Seal, 7c. to 8c.; French V.M., 6c. to 7c.; Lehigh, 5c. to 6c.

WHITE ZINC GROUND IN OIL—Is inquired for quite freely for decorative purposes. Prices remain as last quoted: Pure, 8½c. to 9½c.; No. 1, 7c. to 8c.; No. 2, 5⅝c. to 6¾c.

PUTTY—The demand is a little stronger this week at the following prices: Pure linseed oil, $1.75 1-5 to $1.85 1-5; bulk in bbls., $1.50 1-5; in 25-lb irons, $1.80 1-5; in tins, $1.90 1-5; bladder putty in bbls., $1.75 1-5.

ORANGE MINERAL—Sluggish without feature. Quoted at: Casks, 8c.; 100 lb. kegs, 8½c.

RED LEAD — Now in active demand, with only light stocks to draw from until navigation opens: Genuine red lead, in casks, $6: in 100-lb. kegs, $6.25; in less quantities at the rate of $7 per 100 lbs.; No. 1 red lead, casks, $5.75; kegs, $6, and smaller quantities, $6.75.

PARIS GREEN—Quiet at present, and it is not thought that there will be much excitement until navigation opens. Quotations remain as follows: In barrels, about 600 lbs., 25½c per lb.; in arsenic kegs, 250 lbs., 25¾c; in 50-lb. drums, 26¼c; in 25-lb. drums, 26¾c; in 1-lb. packets, 100 lbs. in case, 27¼c; in 1-lb. packets, 50 lbs. in case, 27¾c; in ½-lb. packets, 100 lbs. in case, 29¼c; in 1-lb. tins, 28¼c f.o.b. Montreal. Terms, three months net or 2 per cent. 30 days.

SHELLAC GUMS—In consumptive demand, with prices exceedingly steady at the following figures: Bleached, in bars or ground, 46c. per lb., f.o.b. Eastern Canadian points; bone dry, 57c. per lb., f.o.b. Eastern Canadian points; T. N. orange, etc., 48c. per lb., f.o.b. New York.

SHELLAC VARNISH—There is a good call for the various grades of shellac, at somewhat slightly higher figures. Prices now are: Pure white bleached shellac, $3 to $3.25; pure orange, $2.80 to $3; No. 1, orange, $2.60 to $2.80.

PETROLEUM — American prime white coal, 15½c per gallon; American water, 17c per gallon; Pratt's Astral, 19½c per gallon.

WINDOW GLASS—First break, 50 feet, $1.85; second break, 50 feet, $1.95: first break, 100 feet, $3.20; second break, 100 feet, $3.40; third break, 100 feet, $3.95; fourth break, 100 feet. $4.15; fifth break, 100 feet, $4.40; sixth break, 100 feet, $4.95. Diamond Star: First break, 50 feet, $2.30; second break. 50 feet, $2.50; first break, 100 feet. $4.40; second break, $4.80; third break. 100 feet, $5.75; fourth break, 100 feet, $6.50; fifth break, 100 feet, $7.50; sixth break, 100 feet, $7.50; seventh break, 100 feet, $8; eight break, 100 feet, $9. Double Diamond: First break, 50 feet, $3.45; second break, 50 feet, $3.75; first break, 100 feet, $6.75; second break, 100 feet, $7.25; third break, 100 feet, $8.75; fourth break, 100 feet, $10; fifth break, 100 feet, $11.50; sixth break, 100 feet, $12.50; seventh break, 100 feet, $14; eighth break, 100 feet, $16.50: ninth break, 100 feet, $18; tenth break, 100 feet, $20; eleventh break, 100 feet. $24; twelfth break, 100 feet, $28.50. Discount on Diamond Star, 20 per cent.; on Double Diamond, 40 per cent.

TORONTO.

Office of HARDWARE AND METAL,
10 Front Street East,
Toronto, March 23, 1906.

The paint and oil trade has been quiet but should enliven with a few days of mild weather. The demand for all lines is steady, but not particularly strong. Leads have experienced a little demand. Good demand for varnishes, oils and turpentine obtains. It is reported that prices on refined oil in United States have advanced ¼c. This advance has not yet affected Canadian prices, but in all probability it will. Prices on linseed oil and turpentine have advanced owing to the scarcity of raw material. South American seed is scarce, the Canadian supply being diverted to other

purposes. Scarcity of labor and a severe drought in Georgia are the causes of the scarcity and high prices of turpentine.

WHITE LEAD — The demand is steady with firm and unchanged prices. We continue to quote: Pure white, $7.40; No. 1, $6.65; No. 2, $6.25; No. 3, $5.-90; No. 4, $5.65 in packages of 25 lbs. and upwards; ½c. per lb. will be charged extra for 12½ lbs. packages; genuine dry white lead in casks, $7.

RED LEAD—We continue to quote: Genuine in casks of 500 lbs., $6; ditto, in kegs of 100 lbs., $6.50; No. 1, in casks of 500 lbs., $5.75; ditto, in kegs of 100 lbs., $6.25.

DRY WHITE ZINC — There is a little demand for this line. Prices remain firm and unchanged. We continue to quote: In casks, 7½c.; in 100 lbs., 8c., No. 1, in casks, 6½c., in 100 lbs., 7c. Ground in oil—In 25 lb. irons, 8c.; in 12½ lbs., 8½c.

SHELLAC VARNISH — Prices are unchanged with increasing demand: Pure orange, in barrels, $2.70; white. $2.82½ per barrel; No. 1 (orange) $2.50: gum shellac, dry bone, 63c. Toronto. T. N. (orange) 51c. net Toronto.

LINSEED OIL — Another advance has been made in the prices of both raw and boiled oil of 1c. We quote: Raw, 1 to 3 barrels, 65c.; 4 to 7 barrels, 64c.; 8 barrels and over, 63c. Add 3c. to this price for boiled oil, f.o.b. Toronto, Hamilton, London and Guelph, 30 days.

TURPENTINE.—Prices continue variable, the tendency being upward. Another advance of 2 cents has been made. We quote: Single barrels at $1.04, f.o.b. point of shipment, net 30 days; less than barrels, $1.09 per gallon.

PARIS GREEN — There is a little trade in paris green, most of the orders being booked up. Price continues firm and unchanged. We continue to quote English and Canadian at 27½c. base.

PETROLEUM — Trade is improving with increased demand. American prices on the refined products have advanced ½c. Canadian prices have not yet been affected. We continue to quote: Prime white, 13c.; water white, 14½c.; Pratt's astral, 18c.

For additional figures see current quotations at back of paper.

THE BLANCHITE PROCESS.

J. C. B. Blanch, a New York scientist and chemist, discovered in 1901 a new base for paint manufacture which he considered superior to white lead, then in common use. This new essential was white zinc (oxide of zinc.) It is claimed to be a better carrier of oils, more durable, more penetrable, and a non-conductor of electrical currents, which are destroyers of any paints.

On the strength of these points of superiority claimed over white lead in paints, a new process of paint manufacture has developed, called the Blanchite process, after the discoverer. A large factory has been in operation in New York City for five or six years, and in January of this year a plant was opened up on King street west, Toronto. The Blanchite Company are meeting with success in Canada, the demand for their manufactures increasing every week.

There are two special features in the new process of paint manufacture—the oxydization of all oils before going into the mixtures, and the double-grinding of all bases—thereby making the paint finer and more adhesive, for it has been noticed that when the outer coating has been scraped off particles of the paint can be seen filling up the minute cavities of the surface to which it has been applied. The Blanchite paints possess very superior sanitary qualities, being absolutely free from taint and odor.

The Toronto plant, though not as yet very extensive, is very complete, the greater part of the work being done by machinery. In the basement of the plant the power is generated, and the raw material stored. The Blanchite mixes stored here are forced to the upper flats by rotary power pumps, as they are needed. On the main floor is situated the drying kiln at the back, the stock and shipment and labeling rooms, with the offices, at the front. Most of the mills are situated on the second floor—the large double mill mixer, the double, triple and base mixers—and the canning and labeling rooms. The third flat contains the rest of the mixers and triple mill. At the rear of this building is the storehouse for oils.

Great care is taken in thoroughly oxydizing all oils and extracting all acid fats, to purify, as much as possible all ingredients so as to permit of perfect mixture. The present daily capacity of the Toronto plant is 1,200 gallons.

HOW TO INCREASE PAINT AND VARNISH SALES.

(Continued from page 50.)

In addition to these means of getting business, there is out-door advertising, such as large display posters for boards and walls—supplied by the manufacturer—and special features, including circulars, cards, handbills, folders, etc., which the dealer can readily prepare for himself.

The main point to be brought out, however, is this—to increase his paint and varnish sales this spring the merchant must give his heartiest co-operation to the manufacturer, take advantage of all the advertising that he can secure with his line, and then hustle early and late to see that not one possible customer escapes him and that he takes care of the paint needs of his vicinity.

MOVING LINSEED OIL MILL.

The Dominion Linseed Oil Co., Toronto, have decided to move their mill from Elora to Winnipeg, so that they may be more able to supply the demand of the western provinces.

BERRY BROTHERS' VARNISHES

are one grand story of success

BECAUSE THEY ARE THE BEST

They are the safest goods to handle,
and the most reliable goods to use.

Write for Catalogue, Information and Prices.

BERRY BROTHERS, Limited

NEW YORK
BOSTON
PHILADELPHIA
BALTIMORE

Varnish Manufacturers
DETROIT
Canadian Factory—WALKERVILLE ONT.

CHICAGO
CINCINNATI
ST. LOUIS
SAN FRANCISCO

SOME DEALERS WOO DISASTER

100% durability

from a business standpoint simply because they don't discriminate between rational evidence and hot air. Take paint for instance. Every dealer knows that he simply cannot afford to sell or recommend bad paint—or even ordinary paint. He must sell and recommend the best. Make your leader

ARK BRAND PAINT

and you can rest assured of commercial success. We do not say stock **ARK BRAND** because we say it's the best paint. We say you should reccommend and sell **ARK BRAND** because hundreds of practical men have proven its superiority. They say so right over their signatures, and we want to mail the proofs to you.

MURESCO

Well, here's the opinion of practical men

Sarnia, March 12th, 1907

Gentlemen:—Replying to yours of yesterday we have no intention of dropping Muresco. In fact we expect larger sales for it this year than ever. We have sold Muresco for about seven years and have never heard a complaint about it. It gives our customers the best of satisfaction.

Yours truly, (Signed), ROUGVIE BROS.

BENJAMIN MOORE & CO., LIMITED

(Formerly The Francis-Frost Co., Limited)

NEW OFFICES AND WORKS:

TORONTO JUNCTION, CANADA

New York, Chicago, Cleveland, Carteret, N.J.

Dealers in Plumbing Equipment have a Practical Guarantee

of larger profits and greater satisfaction in handling

STANDARD IDEAL CAST IRON PORCELAIN ENAMELED WARE

Made in Canada from designs calculated to meet every requirement of the trade, it is a ware that embodies in itself the durability of iron combined with the rich snowy white opaque finish derived from perfect porcelain enameling.

STANDARD IDEAL PORCELAIN ENAMELED WARE is of non-porous one-piece construction, leaving no joints, turns or crevices for dirt, dust or germs to accumulate. It is absolutely Sanitary.

We make Bath Tubs, Sitz Baths, Shower Baths, Lavatories, Urinal Ranges, Slop Hoppers, Laundry Tubs, Sinks, Closet Ranges, Steamship Supplies, Railway Car Supplies, Hospital Appliances, etc.

WRITE FOR CATALOGUE.

The Standard Ideal Co., Limited

Head Office and Factories, - PORT HOPE, Ont.

Sales Offices and Sample Rooms, 50 Colborne Street, TORONTO, 128 West Craig Street, MONTREAL, and 24 Telfer Block, WINNIPEG

SOLVED AT LAST

Your problem is solved at last. The **Cleveland Wall Paper Cleaner** has captured the U.S. markets because of merit and merit alone. There is no other wall paper cleaner nearly as good.

Guaranteed To Keep For Two Years

Not Dangerous. Does Not Contain Acids

The Cleveland Wall Paper Cleaner

does clean wall paper and frescoed walls. It does not streak or spot them, and with ordinary care, almost any one can use it. Let us forward you particulars regarding **your** margin of profit.

Put up in cases of 3 doz. one pound cans
Bulk—10, 25 and 50 pound cans. Prompt Shipments

FRANK U. S. GILBERT
1224 Woodland Ave. CLEVELAND, OHIO

Plumbing and Steamfitting

WATERFRONT EXPLOSION AT PICTON.

The accompanying photos show the damage done by the explosion of the water front of a range in the Hotel Tecumseth at Picton recently. The lids of the stove were blown through the ceiling and broken to pieces, and the coal was thrown out over the floor. The explosion happened about 4 a.m., waking everyone in the hotel, and the building was saved from destruction only by hard work.

The following description of how the accident was caused and how the repairs were made, written by Arthur Desmarchais, of Picton, Ont., will be read with interest:

"The water front had 133 square inches of heating surface in actual contact with the fire, having sufficient capacity to give good results with a 50-gallon boiler attached to it. But it only had one 30-gallon boiler attached and only one half-inch hot water tap connected to it for the supply of hot water for the hotel and no safety valve attached to the boiler. The grate surface of the range was 165 square inches.

Damage Done by Explosion of Water Front.

"Owing to the small size of the boiler attached to the big water front exposed to the big fire surface it did not take very long for the water to boil in the boiler and this was the only way for the hot water to travel back in the cold water main.

"One cold night the cold water main froze and the hot water having no place to expand something had to go and the water front burst. The water front can be seen on the left of the stove in an "L" shape.

"I have overhauled the job and have reduced the water front one-half the quantity of the old one, and also installed a safety valve on the boiler, and am satisfied that the stove is safe now.

"These photos were taken especially for your paper, and if I have made any mistakes I will appreciate it if other readers will write in your columns and rectify them."

PUBLIC BATHS AT HAWAII.

The following interesting data is quoted from the report on Hawaii of the Commissioner of Labor:

"Upon all the larger plantations public baths are conducted for the use of employes. Sometimes they are partly supported by the employers, and sometimes they are entirely private enterprises, except that the bath house is usually supplied by the plantation. On some plantations hot water as well as the building is provided. Other planters supply free the fuel for heating the water. In other instances the bath house is leased to a contractor, who supplies his own fuel and charges the employes from 26 to 35 cents a month for bathing privileges. The Japanese bathe daily. On some plantations the Koreans have adopted Japanese customs in this respect. Hot water is used and a single large tub—in which both sexes bathe together indiscriminately—suffices for the needs of a number of laborers. Private bath houses, conducted in much the same manner, are common in the Oriental quarters of Honolulu."

A report has been issued as to the hours and wages of the men employed in the public comfort stations in Toronto. There are two men at the corner of Cottingham and Yonge, one working from 6 a.m. to 5 p.m., receiving $12 a week, and the other working from 5 p.m. to midnight, and receiving $8 per week. At Spadina and Queen there are two men working the same hours and receiving the same pay. At Adelaide and Toronto streets there is only one caretaker. He works from 8 a.m. to 6 p.m. week days only, and receives $8.50 per week. On Sundays the men at the Cottingham and Spadina lavatories work from 9 a.m. to 3.30 p.m., and fr 3.30 to 10 p.m., two shifts.

FUNNY STORY FROM WALKERTON.

Thawing out a water pipe is about as exasperating a piece of business as putting up stove pipes, under the most favorable circumstances, and under certain other circumstances it is even worse, says a Walkerton, Ont., paper. Sam Hawthorne struck the very worst case of a frozen pipe the other night, and he lost a whole night's sleep over it. He had been out a little late that night, and when he went to get a drink there was no water. Concluding that it was a case of too much frost, he filled the kettle with water from the cistern and soon had it boiling hot. Then he poured it on the pipe, with no result. There was nothing for it but to fill up the kettle again and give the pipe another dose. This he did, but it still refused to run. This process was kept up steadily till five o'clock the next morning, and at last the water came gushing out. But what made Sam mad was that he learned later on in the day that the frost had nothing to do with the matter. The water had been turned off by the town constable, so as to increase the supply in the reservoir, and he had been operating on the empty pipe.

How the Back of the Range was Smashed.

A. R. Dundas and J. D. Skidmore, of Cobourg, were in Toronto last week.

The employees of W. Mashinter, Adelaide street west, Toronto, returned to work on March 25, agreeing to work for the same wages as they were for-

DISPOSAL OF SEWAGE BY SEPTIC TANKS.

Address by W. B. Gray, at the twenty-fourth annual meeting of the National Association of Master Plumbers, Atlantic City, 1906

The best authorities of to-day predict that, wherever possible to supplant them, common cesspools, of whatever nature, will in due time be things of the past. The septic tank process is the means which will be employed, even in small private installations.

The septic treatment of sewage may be considered a biological rather than a chemical process as its success is dependent upon presenting conditions which favor the rapid growth of certain bacteria. In the complete reduction of sewage by the septic method, bringing it to a harmless state, in the form of nitrates which plant life can take up, two forms of bacteria are employed—anaerobic and aerobic. Air and light retard the multiplication of the first of these. The second requires oxygen and multiplies readily in the open air. The tank or receiver proper is a sort of catch basin, made in form to favor the requirement for the propagation of anaerobic bacteria which reduce the sewage to simple compounds. The tank should be large enough to hold the output of about one day's use of the fixtures discharging into it. Light and air should be excluded. Warmth to a degree is essential, such heat as is common to a pit in the earth, closed at the top, with no unnecessary exposure, together with the heat of waste water and that generated by the action taking place in the sewage itself being sufficient to favor the process in winter weather of quite severe climates. Fifty-four degrees Fahrenheit has been stated to be the minimum temperature permissible in this tank, for little or no septic action can take place at lower temperatures. The waste water of baths and lavatories are not turned into the septic tank alone for the heat they bring, but also to secure dilution of the excrement and matter from other sources, which not infrequently carry too little water to favor the best interests of the process. Both the inlet and outlet of the tank should be arranged to be below the surface of the contents when the tank is full, so that the scum which generally forms on the surface will not be disturbed by entry or exit of matter. This scum, resembling wet ashes, helps to retain the heat and excludes light and air from the mass, all favoring the accomplishment of the purpose The scum may be from a few inches to fifteen to twenty in thickness, according to conditions and nature of the plant.

The contents leaving this initial receptacle having, in it, been brought down from its complex nature to one of simple chemical compounds, principally nitrites, the completion of the reduction process and the change from nitrites to nitrates, is brought about by exposure of the matter to light and air, giving the aerobic micro-organisms a chance to get in their work. This would be accomplished by simply discharging directly into a stream, but a more rapid action is obtained by interposing an open shallow bed of broken stone or slag for the liquid to first flow through so as to break up and bring into contact with the air as large an amount of surface as possible before piping to a stream or otherwise. In this way a more complete reduction is certain before the matter reaches any final source of disposal.

The bacteria necessary to the process are always present in abundance in fresh sewage and no preliminaries are necessary to the operation as described. The resulting product is described as a harmless, colorless and odorless liquid. In this process, admission of air to the tank or lack of sufficient heat or dilution may result in a putrescent state of matter, as found in the common cesspool. As previously indicated, the septic process is not yet widely used, except for town sewage where it is rapidly gaining in favor. Here elaborate methods are adopted to favor the aerobic or oxidizing end of the operation, mostly through filters of special design, all aiming to secure absolute stability and harmlessness of the final discharge from the sewage disposal plant. Better acquaintance will doubtless develop much data bearing on the latitude of conditions under which it will successfully operate. From eight to seventeen days are necessary to set up septic action, according to season and conditions.

SUPPLY MEN LOSE APPEAL.

The long drawn out appeal of the now defunct Central Supply Association against the conviction registered against the organization on the charge of conspiracy with the Master Plumbers' Association has at last been decided, the Court of Appeal upholding Judge Clute's decision.

The fine of $5,000 stands therefore; but as the organization is now out of existence it may never be paid unless the former individual members see fit to contribute to the court or the legal authorities take action to collect the amount.

In dismissing the appeal, Judges Moss, Garrow and MacLaren assent, and Judges Osler and Meredith dissent, they commenting as follows on the case:

A Close Decision.

Justice Garrow in his judgment said: "It is common knowledge that the majority of large operations in manufacturing and dealing in the articles and commodities of commerce are now carried on by joint stock companies. Can it be imputed to the Legislature that the intention in preparing the net was to catch only the small fry? Surely not. Then, again, the section expressly extends to prevent combinations in, among other things, insurance, both life and fire. These subject matters are always carried on by joint stock or mutual companies or corporations. Insurance corporations are not expressly named at all, and, therefore, unless included in the term 'corporation,' found in the first clause of the section, could not under the construction contended for be prosecuted or punished. The theory of the Crown, apparently concurred in by the judge, is that prior to the incorporation there had been such a criminal agreement, first between the individuals comprising both associations before the incorporation of the Plumbers' Association in April, 1905, continued down to the incorporation in September, 1905, of the Supply Association."

Incidentally, Chief Justice Moss asks: "What were the motives that led to the incorporation? The association has no stock in trade, and carries on no business, nor does it buy or sell commodities. It does not profess to be an association formed for general benevolent purposes, and it can scarcely be regarded as a social club."

Justice MacLaren agreed, orally, with the Chief Justice, who said that but for the opinions of his two dissenting brethren he would have experienced little difficulty in coming to the conclusion that the judgment should be affirmed. He, however, sustained Justice Clute's judgment.

Dissenting Opinions.

Justice Osler, who dissents, says: "It cannot be too strongly insisted upon, nor too clearly borne in mind, that the applicants are not responsible, criminally or otherwise, for anything which took place before their incorporation on September 9, 1905. They are not heirs of the sins, or answerable for the misdeeds of their incorporated predecessors. For all these the members of the incorporated association are liable, and have, as we are given to understand, been prosecuted. The Crown had the right to prove the object of the incorporation, and acts done after the incorporation, as tending to prove a conspiracy between the two corporations, but the fact of incorporation alone would not be enough. The prosecution, however, went very far afield, introducing evidence which comprises the bulk of the record of unlawful acts and conspiracy committed by individuals, members of the old incorporated association years before these appellants came into existence. This, in my opinion, was absolutely wrong. In my opinion, the evidence of anything finally done or completed by these defendants after their incorporation is of so slight and flimsy a nature, if any evidence of that kind there be, that the learned trial judge, dealing with a criminal charge of so grave a character, ought to have held that the case was not made out, and should have acquitted the defendants."

For Sins of Others.

Justice Meredith, the other dissenting judge, says: "Bearing in mind the cardinal fact that the appellants were not incorporated until September 6, 1905, it seems to me impossible to avoid the conclusion that they were tried, convicted and sentenced, in a substantial measure at least, for the offences of others, committed long before the appellants had any sort of legal existence, and so were wholly incompetent to commit any crime or do any unlawful act. This is not only apparent from the whole course of the trial, but crops up throughout the reasons given by the trial judge for finding the defendants guilty, and, indeed, it is almost, if not quite, explicitly so stated."

PLUMBING MARKETS

TORONTO.

Office of HARDWARE AND METAL,
10 Front Street East.
Toronto, March 28, 1907

The feature of the Toronto plumbing markets at present is the low prices of radiation. At present radiation is being sold at prices the lowest in history. Trade on the whole has been quiet, owing to the weather, but is now improving with the continued mild weather and prospects point toward a very busy season. With an unprecedented season in the amount of buildings erected there should be a large volume of plumbing business done.

LEAD PIPE—We continue to quote: 5 per cent. off the list price of 7c. per lb. Lead waste, 8c. per lb., with 5 off. Caulking lead, 5¾c. to 6¾c. per lb. Traps and bends, 40 per cent. off.

IRON PIPE—The scarcity in iron pipe continues, with no immediate prospects of repletion. We continue to quote: 1-in. black pipe, $5.12 ; 1-inch galvanized, $6.-77.

IRON PIPE FITTINGS — Prices remain unchanged. We continue to quote: Cast iron fittings, 57½ per cent.; cast iron plugs and bushings, 60 per cent.; flange unions, 60 per cent.; nipples, 70 and 10 per cent.; iron cocks, 55 and 5 per cent.; Canadian malleable, 30 per cent.; malleable unions, 55 and 5 per cent.; malleable bushings, 55 per cent.; cast iron ceiling plates, plain, 65 per cent.; cast iron floor, 70 per cent.; hook plates, 60 per cent.; expansion plates, 65 per cent.; headers, 60 per cent.; hangers, 65 per cent.; standard list.

SOIL PIPE FITTINGS—Demand for these continues to be good. Prices remain unchanged: Medium and extra heavy pipe and fittings, 60 per cent.; light pipe, 50 per cent.; light fittings, 50 and 10 cent.; 7 and 8 in. pipe, 40 and 5 per cent.

RANGE BOILERS—Prices remain firm and unchanged. We continue to quote: Galvanized iron, 30-gal., standard, $5 ; extra heavy, $7.75 ; 35-gal. standard, $6; extra heavy, $8.75; 40-gal., standard, $7 ; 40 gallon, extra heavy, $9.75, net list. Copper range boilers—New lists quote: 30 gallon, $33; 35 gallon, $38; 40 gallon $43. Discounts 5 to 15 per cent.

RADIATORS—The demand continues brisk with firm and unchanged prices. We continue to quote: Hot water, 47½ per cent.; steam, 50 per cent.; wall radiators, 45 per cent.; specials, 45 per cent. Hot water boilers are subject to an open market.

SOLDER—Bar solder, half-and-half, guaranteed, 27c.; wiping, 23c.

ENAMELWARE— Some price-cutting continues in this line. We continue to quote January prices ; Lavatories, first quality, 20 and 5 to 20 and 10 off ; special, 20 and 10 to 30 and 2½ per cent. discount. Kitchen sinks, plate, 300, firsts, 60 and 10 off ; specials, 65 and 5 per cent. Urinals and range closets, 15 off. Fittings extra.

MONTREAL.

Office of HARDWARE AND METAL,
232 McGill Street.
Montreal, March. 28 1907

There is a decided activity noticed just now in plumbing circles, more so in fact than is generally the case at this early season. Wholesalers are having all they can do to keep their work up to date without working overtime, and report that they don't know what they'll do when the real rush comes. This is generally some time in May, when navigation is in full swing, and the railroad and steamship companies give out their new rates. Dealers wishing to get their goods in time, should not delay too long in placing their orders. This refers to almost every line the plumber handles, and in view of the fact that this unusual activity has prevailed since the year started, it goes to show that it is not only spasmodic.

Prices on all lines are very firm, with no changes to report this week.

RANGE BOILERS—Figures on the raw material seem to have reached the high-water mark, as no advances have taken place in this line for some time now. Iron clad, 30-gal., $5 ; 40-gal., $6.50 net list. Copper, 30 gal., $33; 35 gal., $38; 40 gal., $43.

LEAD PIPE—Business is exceedingly brisk with prices firm. Discount is: 5 p.c. f.o.b. Montreal, Toronto, St. John, N.B., Halifax; f.o.b. London, 15c per hundred lbs. extra; f.o.b. Hamilton, 10c per hundred lbs. extra.

IRON PIPE FITTINGS — A marked shortage still prevails in some lines Manufacturers are running their plants full speed, but as they are well stocked up with orders for future shipment there does not seem to be much hope for the present at least, of any relief from the shortage. Discounts on nipples, ¼-inch to 3-inch, 65 per cent.; 3½ to 2 inch, 67½ per cent.

IRON PIPE—Last week's advance of about 4 per cent. on the list is well maintained.

SOIL PIPE AND FITTINGS. —No special features to report. Prices remain: Standard soil pipe, 50 p.c. off list. Standard fittings, 50 and 10 per cent. off list; medium and extra heavy soil pipe, 60 per cent. off. Fittings, 60 per cent. off.

SOLDER—Remains unchanged. Bar solder, half-and-half, guaranteed, 25c.; No. 2 wiping solder, 22c.

ENAMELWARE — We quote: Canadian baths, see Jan. 3, 1907, list. Lavatories, discounts, 1st quality, 30 per cent.; special, 30 and 10 per cent. Sinks, 18x30 inch, flat rim, 1st quality, $2.60; special, $2.45.

The Canadian Brass Mfg. Co., of Galt, has been incorporated, with a capital of $40,000, to manufacture and deal in brass goods. The provisional directors are: W. L. Lefavor, F. E. Brown, and L. J. Osborn.

WARMING WATER FOR SHOWER BATHS.

There are many establishments where steam is available in which a shower bath can be used with advantage by the employes of the plant, for their refreshment in the summer season and for purely bathing purposes at the end of the day's work all the year round.

The device shown in the accompanying illustration was prepared for men working in gas manufacturing establishments.

It is made with a piece of 2-inch or 2½-inch pipe 2½ feet long. At the lower end it is reduced to ¾-inch. Where the cold water enters at the upper end there is a reducing T to ¾-inch. The side opening is where the hot water is taken

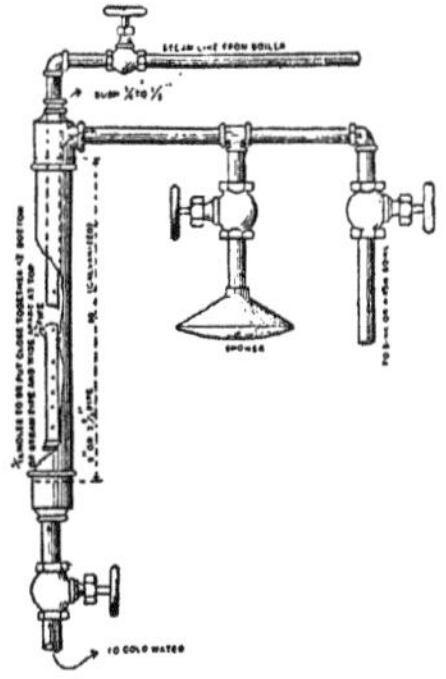

To Warm Water for Shower Baths.

off to the shower or basin. At the top of the T there is inserted a perforated tube of 1½-inch pipe with cap on the lower end. This tube is connected with the live steam line. The user can get the desired temperature of water by adjusting the valves on the cold water and steam inlets. This will be found very handy and a cheap shower for use at the works, or any place where steam is to be had.

Purdy & Mansell Co., Toronto, are moving from their present location on Adelaide street to Bay street.

A plumber inspecting the gas pipes at the Collegiate Institute, Chatham, held a lighted match near a big leak resulting in an explosion, in which he was severely burned.

Everything a Plumber Needs

IS YOUR STOCK COMPLETE?

QUALITY FIRST CLASS

PRICES GOOD

PROMPT DELIVERIES A SPECIALTY

SOMERVILLE LIMITED

59 RICHMOND STREET EAST - TORONTO

WRITE YOUR JOBBER

Mr. Dealer! Here are 3 Standard Lines you Should Stock for Profit

"IXL"
SPRING TOP OILER
Cover is held by spring.
Easily filled.
A good seller.

THE INJECTOR WITH A REPUTATION

"SCREW TOP"
PLAIN BRASS OILER
Made with or without tube inside.
They are safe stock to carry.

PENBERTHY INJECTOR CO., Limited, WINDSOR, ONT.

McDOUGALL PUMPS ARE STANDARD PUMPS

We are going to make the above expression known over the whole civilized portion of this good old world and incidentally educate a few heathen tribes into the bargain. We will do this by "Delivering the Goods" and nothing better in the Pump way is procurable anywhere. We have them for all duties and the price is in your favor.

Quality first, however

THE

R. McDougall Co., Ltd.

GALT, CANADA

Are you interested in any of the lines that are advertised? A Post Card will bring you price list and full information. Don't forget to mention Hardware and Metal.

The Canadian Bronze Powder Works

R. E. THORNE & CO.

The only bronze powder works under the British flag.
High Grade bronze powders and bronze liquids.
Can fill all orders at short notice.

MONTREAL — TORONTO

WORKS AT VALLEYFIELD. *NO ORDER TOO LARGE*

HOTTEST ON EARTH

Our Bonanza Furnaces are known to produce more heat with less fuel than any other make. Jobbers sell at factory prices. Satisfaction guaranteed or money refunded.

No. 38—Price, each, *$7.80* net.
No. 39—One size smaller, *$5.85* net.

THE TURNER BRASS WORKS
53 MICHIGAN STREET, - CHICAGO, U.S.A.

Manufacturers' Agents

CHARLES H. FOX
Vancouver, Canada
MANUFACTURERS' AGENT
Selling to the Hardware Jobbers Only.
Representing Canadian, British and American Manufacturers, correspondence invited. Cable Address, Chasfox Vancouver, P.O.Box 1105.

ALEXANDER GIBB
Manufacturers' Agent and Metal Broker,
13 St. John Street, Montreal
Representing Canadian, British and American Manufacturers. Correspondence invited from firms wishing to be represented.

H. W. MITCHELL
WINNIPEG, MAN.
Correspondence is solicited from manufacturers wishing a representative in Winnipeg.
Travellers call regularly on the wholesale and retail trade in Western Canada.
Highest references and financial responsibility.

Scott, Bathgate & Co.
BROKERS AND COMMISSION MERCHANTS
Cor. Notre Dame St. East and Victoria St., WINNIPEG, Canada
Storage and Warehousing a Specialty.
First-class facilities in a thoroughly modern brick ware house centrally located on the railway track.

PAGE ACME White Fences

Any height to 8 ft. Any length you say. From 16 cents a foot. Gates to match, from $2.25. Last longer. Easy to put up. Get booklet.

PAGE WIRE FENCE CO., Limited

Walkerville - Toronto - Montreal
St. John - Winnipeg 212

Your Interests Are Mine

That explains the increasing volume of my business from month to month. Then, Quality always tells, and, as you know, you can only secure First Quality Enamelware by placing your orders with me. Prompt delivery guaranteed.

ORLANDO VICKERY

178-180 Victoria Street TORONTO, CANADA

There may be better fittings placed on the market some day, but to-day

"Diamond" brand Fittings

hold the market because of their sterling and lasting worth. Remember the brand. Wholesale only.

OSHAWA STEAM & GAS FITTINGS CO., LIMITED
OSHAWA, CANADA

Sharratt & Newth's Glaziers' Diamonds

are unequalled for cutting and wearing qualities.

To be obtained from the principal Dealers in Glass, Hardware, and Painters' Supplies.

Contractors to H. M. Government and the principal English Sheet and Plate Glass Works

World's Best Cutlery

You make no mistake in specifying goods made by **JOSEPH RODGERS & SONS, Limited** when ordering cutlery. A reputation of over 200 years at stake, insures the quality of every article.

Write for catalogues from Sole Canadian Agents

JAMES HUTTON & COMPANY, MONTREAL

Persons addressing advertisers will kindly mention having seen their advertisement in Hardware and Metal.

A Certain Sense of Satisfaction

goes with every MERRELL PORTABLE HAND MACHINE, PIPE THREADING and CUTTING MACHINE. We build them that way.

For doing ALL kinds of threading—under ALL conditions—labor trouble and faulty material included—where speed, durability and ease of control count for something, our word to you is to investigate the MERRELL.

Whether it be stationary or portable—hand or power driven—30 DAYS FREE TRIAL must prove conclusively that your choice should be the MERRELL.

This Portable Hand Machine has encased gears, MERRELL standard quick opening and closing die-head, and the latest improved Cutting-off Knife.

Let us tell you more about this machine—The Chasers, Vise and the large range of work covered.

Catalogue for the asking.
Quick shipments.

THE CANADIAN FAIRBANKS CO.
Sole Agents for Canada Limited
MONTREAL, TORONTO, WINNIPEG, VANCOUVER

Do You Want the Best?

TWO FREE
for every one sent you defective.
Isn't that guarantee enough?

Clauber Brass Mfg. Co.
Expert Makers of Fine Brass Goods
CLEVELAND - - OHIO

Forwell Foundry Co.

BERLIN, ONT.

Manufacturers of

SOIL PIPE, FITTINGS, and CAST IRON SINKS

Ask Jobbers for "F. F. CO." Brand

CURRENT MARKET QUOTATIONS.

Mar. 29, 1907

These prices are for such qualities and quantities as are usually ordered by retail dealers on the usual terms of credit, the lowest figures being for larger quantities and prompt pay. Large cash buyers can frequently make purchases at better prices. The Editor is anxious to be informed at once of any apparent errors in this list, as the desire is to make it perfectly accurate.

METALS.

ANTIMONY.

Cookson's ....per lb. . 0 27 0 27½

BOILER PLATES AND TUBES

Plates, ½ to ½ inch, per 100 lb.	....	2 50
Heads, per 100 lb	...	2 75
Tank plates 3-16 inch	...	2 65
Tubes per 100 feet, 1½ inch	8 50	
" " 2 "	9 00	9 50
" " 2½ "	11 30	11 50
" " 3 "	12 50	13 00
" " 3½ "	16 00	17 00
" " 4 "	20 00	21 50

2 per cent off

BOILER AND T.K. PITTS

Plain tinned .... }
Spun.................. } 25 per cent off list.

BABBIT METAL

Canada Metal Company—Imperial, genuine 60c; Imperial Tough, 60c.; White Brass, 50c; Metallic, 35c; Harris Heavy Pressure, 25c; Hercules, 25c; White Bronze, 15c; Star Frictionless, 14c.; Alluminoid, 10c; No 4, 9c, per lb.

James Robertson Co.—Extra and genuine Monarch, 60c.; Crown Monarch 50c; No 1 Monarch, 40c; King 30c; Fleur de-lis, 20c; Thurber, 15c.; Philadelphia, 12c., Canadian, 10c, hardware, No. 1, 15c; No. 2, 12c; No 3, 10c. per lb

BRASS

Rod and Sheet, 14 to 30 gauge, 25 p c advance

Sheets, 12 to 14 in	0 30
Tubing, base, per lb 5-16 to 2 in ...	0 33
Tubing ½ to 3-inch, iron pipe size..	0 31
" ½ to 3-inch, seamless	0 36

Copper tubing, 6 cents extra

COPPER

		Per 100 lbs
Ingot.		
Casting, car lots...	26 00	27 00
Bars.		
Cut lengths, round, ½ to 2 in.	....	35 00
Sheet		
Plain, 16 oz., 14x48 and 14x60	....	35 00
Plain, 14 oz.	....	36 00
Tinned copper sheet, base	....	38 00
Planished base..	....	43 00
Braziers' (in sheets), 4x6 ft., 25 to 30 lb. each, per lb., base..	0 34	0 35

BLACK SHEETS.

	Montreal	Toronto
8 to 10 gauge	2 70	2 70
12 gauge	2 70	2 40
14 "	2 70	2 55
17 "	2 50	2 65
18 "	2 50	2 65
20 "	2 50	2 65
22 "	2 50	2 75
24 "	2 55	2 80
26 "	2 65	2 85
28 "	2 70	3 10

CANADA PLATES

Ordinary, 52 sheets	2 75	2 90
All bright "	3 80	3 90
Galvanized, 52 sheets	4 35	4 45
" 60 "	4 60	4 70
	Ordinary	Dom. Crown
18x24x52	4 25	4 35
" 60	4 50	4 60
20x28x80	8 50	8 70
" 94	9 00	9 20

GALVANIZED SHEETS.

	Fleur-de-Lis.	Gordon Crown.	Colborne Crown.
16 to 20 gauge	3 80	3 95	....
22 to 24 gauge ..	4 05	4 00	3 75
26 " ..	4 30	4 40	4 30
28 " ..	4 55	4 60	4 45

Apollo.

10¾ oz. (American gauge)........	....	4 85
28 gauge "	4 45	4 55
26 " "	4 15	4 30
24 " "	3 85	3 90

	Gorbal's Best.	Comet	Queen's Head.	Bell.
16 to 20 gauge	....	3 60	3 95	....
22 to 24 gauge	....	3 75	4 20	....
26 "	....	4 30	4 45	....
28 "	4 55	4 45	4 70	4 45

Less than case lots 10 to 25c. extra.

IRON AND STEEL.

	Montreal.	Toronto.
Common bar, per 100 lb...	2 15	2 30
Forged iron "	2 40	
Refined " "	2 55	2 70
Horseshoe iron "	2 55	2 70
Hoop steel, 1½ to 3 in. base...		2 80
Sleigh shoe steel "	2 25	2 30
Tire steel	2 40	2 50
Best sheet cast steel...	...	0 12
B. K. Morton "Alpha" high speed.		0 65
" annealed	....	0 70
"M" Self-hardening	....	0 50
"J" quality, best warranted	....	0 18
"I" " warranted	....	0 14
"B C" quality	....	0 09
Colonial black diamond	0 08	0 09
Sanderson's	0 08	0 45
Jessop's	0 11	0 13
Air hardening	0 60	0 65
Conqueror	0 07½	0 60
Jowett's diamond J	0 06½	0 07
Jonas & Colver's tool steel...	0 10	0 20
" "Novo"	....	0 65
" " annealed		0 65
Jowett & Sons B.P.L. tool steel	0 10½	0 11

COLD ROLLED SHAFTING.

9-16 to 11-16 inch	....	0 06
¾ to 1 7-16 "	....	0 05½
1 7-16 to 3 "	..	0 05

Montreal 30, Toronto 35 to 40 per cent

INGOT TIN.

Lamb and Flag and Straits—56 and 28-lb. ingots, 100 lb. $45 00 $45 50

TINPLATES.

Charcoal Plates—Bright

		Per box.
M.L.S., equal to Bradley—		
I C, 14 x 20 base	...	$6 50
I X, 14 x 20 "	....	8 00
I XX, 14 x 20 base	....	9 50
Famous, equal to Bradley—		
I C, 14 x 20 base	....	6 50
I X, 14 x 20 "	...	8 00
I X X, 14 x 20 base	....	9 50
Raven and Vulture Grades—		
I C, 14 x 20 base	...	5 00
I X "	....	6 00
I X X "	....	7 00
I X X X "	....	8 00
"Dominion Crown Best"—Double Coated, Tissued.		Per box
I C, 14 x 20 base	5 50	5 75
I X, 14 x 20 "	6 50	6 75
I X X x 20 "	7 50	7 75
"Allaways Best"—Standard Quality.		
I C, 14 x 20 base	....	4 50
I X 14 x 20 "	....	5 25
I XX, 14 x 20 "	....	6 00
Bright Cokes		
Bessemer Steel—		
I.C., 14 x 20 base	...	4 25
20x28, double box	....	8 50
Charcoal Plates—Terne		
Dean or J G Grade—		
I C, 20x28, 112 sheets	7 25	8 00
IX., Terne Tin	...	9 50

Charcoal Tin Boiler Plates.

Cookley Grade—
X X, 14x56, 50 sheet bxs. }
" 14x60, " }
" 14x65, " } 7 50

Tinned Sheets.

72x30 up to 24 gauge	....	8 50
" " 26 "	....	9 00

LEAD

Imported Pig, per 100 lb.	5 50	5 60
Bar,	5 75	6 00
Sheets, 2½ lb. sq. ft., by roll ..	....	0 07½
Sheets, 3 to 6 lb. "	....	0 07

Note.—Cut sheets ½c. per lb., extra. Pipe, by the roll, usual weights per yard, lists at 7c. per lb. and 5 p.c. dis. f.o.b. Toronto.

Note.—Cut lengths, net price, waste pipe 8-ft lengths, lists at 8c.

SHEET ZINC.

5-cwt. casks	8 00	8 25
Part casks	8 25	8 50

ZINC SPELTER.

Foreign, per 100 lb	7 50	7 75
Domestic	7 00	7 25

PLUMBING AND HEATING

BRASS GOODS, VALVES, ETC.

Standard Compression work, 55 per cent.

Cushion work, discount 40 per cent.

Fuller work, 65 per cent.

Flatway stop and stop and waste cocks, 60 per cent; roundway, 55 per cent.

J.M.T. Globe, Angle and Check Valves, 45 per cent.

Standard Globe, Angle and Check Valves 55 per cent.

Kerr standard globes, angles and checks, special, 42½ per cent; standard, 47½ p c

Kerr Jenkins' disc, copper-alloy disc and heavy standard valves, 40 per cent.

Kerr steam radiator valves 60 p.c., and quick-opening hot-water radiator valves, 60 p.c.

Kerr brass, Weber's straightway valves, 40 per cent.; straightway valves, I B B M, 60 per cent

J. M. T. Radiator Valves 50 per cent.

Standard Radiator Valves, 60 per cent.

Patent Quick-Opening Valves, 65 per cent.

Jenkins' Bros. Globe Angle and Check Valves 27½ per cent.

No. 1 compression bath cock......net	2 00
No. 4 " " "	1 90
No 7 Fuller's "	2 75
No. 4½, " "	2 35
Patent Compression Cushion, basin cock, hot and cold, per doz., ...	$16.20
Patent Compression Cushion, bath cock, No. 2208	2 25

Square head brass cocks, 50 per cent.
" " iron " 60 "

Thompson Smoke-test Machine $25.00

BOILERS—COPPER RANGE.

Copper, 30 gallon	33 00
" 35 "	38 00
" 40 "	43 00

15 per cent

BOILERS—GALVANIZED IRON RANGE.

Capacity	Standard	Extra heavy
30-gallons ...	5 00	7 75
35 "	6 00	8 75
40 "	7 00	9.75

2 per cent, 30 days

BATH TUBS.

Steel clad copper lined, 15 per cent

CAST IRON SINKS

14x24, $1; 18x30, $1; 18x36, $1.30.

ENAMELED BATHS.

List issued by the Standard Ideal Company Jan. 3, 1907, shows an advance of 10 per cent. over previous quotations.

ENAMELED CLOSETS AND URINALS

Discount 15 per cent

ENAMELED LAVATORIES.

	1st quality.	Special.
Plate E 100 to E 103	20 & 5 p.c.	20 & 10 p.c.
" E 104 to E 132	20 & 10 p.c.	30 & 2½ p.c.

ENAMELED ROLL RIM SINKS.

	1st quality.	Special.
Plate E 201, one piece	15 & 2½ p.c.	15 & 10 p.c.

ENAMELED KITCHEN SINKS.

Plate E, flat rim 300, 60 & 10 p.c. 65 & 5 p.c

HEATING APPARATUS.

Stoves and Ranges—Discounts vary from 40 to 70 per cent. according to list

Furnaces—45 per cent.

Registers—70 per cent

Hot Water Boilers—50 per cent.

Hot Water Radiators—50 to 55 p.c

Steam Radiators—50 to 55 per cent

Wall Radiators and Specials—50 to 55 p.c.

LEAD PIPE

Lead Pipe, 7c. per pound, 5 per cent. off.

Lead waste, 8c. per pound, 5 per cent. off.

Caulking lead, 6½c per pound.

Traps and bends, 40 per cent.

IRON PIPE.

Size (per 100 ft.)	Black.		Galvanized
[illegible] inch	2 26	[illegible] inch	3 08
[illegible] "	2 26	[illegible] "	3 08
[illegible] "	2 72	[illegible] "	3 17
[illegible] "	3 67	[illegible] "	4 72
1 "	5 12	1 "	6 77
1¼ "	6 98	1¼ "	9 23
1½ "	8 37	1½ "	11 07
2 "	11 16	2 "	14 76
2½ "	18 00	2½ "	24 00
3 "	23 50	3 "	31 50
3½ "	30 00	3½ "	40 00
4 "	34 00	4 "	45 00

2 per cent. 30 days.

Malleable Fittings—Canadian discount 30 per cent; American discount 25 per cent.

Cast Iron Fittings 57½; Standard bushings 57½; headers 57½; flanged unions 57½, malleable bushings 55; nipples, 70 and 10; malleable lipped unions, 55 and 5 p.c.

SOIL PIPE AND FITTINGS

Medium and Extra heavy pipe and fittings, up to 6 inch, discount 60 per cent.

7 and 8-in. pipe discount 40 and 5 per cent.

Light pipe, 50 p.c.; fittings, 50 and 10 p.c.

OAKUM.

Plumbers ....per 100 lb... .. .... 4 75

STOCKS AND DIES.

American discount 25 per cent.

SOLDERING IRONS.

½-lb.	per lb.	0 38
1-lb. or over	"	0 35

SOLDER.

Per lb.	Montreal	Toronto
Bar, half-and-half, guaranteed	0 25	0 27
Wiping	0 21	0 23

PAINTS, OILS AND GLASS

BRUSHES

Paint and household, 70 per cent..

CHEMICALS.

	In casks per lb.
Sulphate of copper (bluestone or blue vitrol)	0 09
Litharge, ground	0 06
" flaked	0 06½
Green copperas (green vitrol)	0 01
Sugar of lead	0 08
Lump olive	0 01½

COLORS IN OIL.

Venetian red, 1-lb. tins pure.	....	0 09
Chrome yellow "	....	0 15
Golden ochre "	....	0 08
French " "	....	0 08
Marine black "	....	0 04½
Chrome green "	....	0 10
French permanent green "	....	0 13
Signwriters' black "	....	0 15

GLUE.

Domestic sheet 0 10 0 10½
French medal.. .... 0 12 0 12½

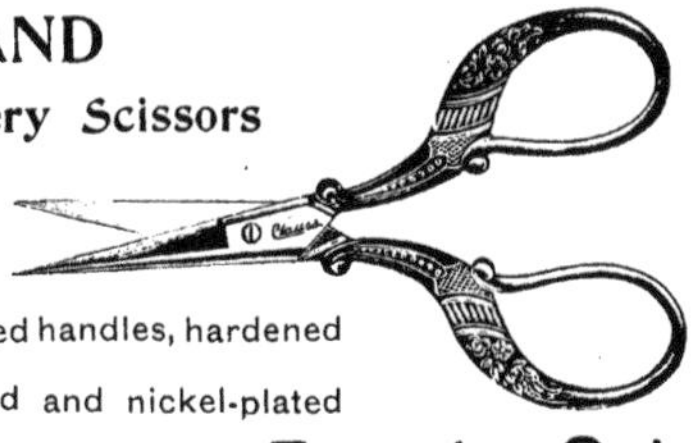

PARIS GREEN.

	Berger's	Canadian
600-lb. cask	0 25½	0 25½
250 lb. drums	0 25½	0 25½
100-lb. "	0 26½	
50-lb. "	0 26½	0 26½
1-lb. pkgs, 100 in box	0 27½	0 27½
½-lb. "	0 29½	0 29½
1-lb. tins, 100 in box	0 28½	0 28½
½-lb pkg	0 30½	0 30

F.o.b. Toronto.

PARIS WHITE.

In bbls 0 90

PREPARED PAINTS.

Pure, per gallon, in tin 1 30
Second qualities per gallon 1 10
Barn (in bbls.) 0 55 0 90
Sherwin-Williams paints, 1 gal 1 40
Canada Paint Co.'s pure 1 30
Standard P. & V. Co.'s "New Era" 1 30
Benj. Moore Co.'s "Ark" B'd 1 25
" British Navy deck 1 50
Brandram-Henderson's "English" 1 45
Ramsay's paints, Pure, per gal. 1 20
" Thistle, ... 1 00
" Outside, bbls 0 55 0 65
Martin-Senour's 100 p.c. pure, ½ gal. 1 50
" " " " ¼ gal. 1 45
" " " " 1 gal. 1 40
Senour's Floor Paints.....gal. 1 20
Jamieson's "Crown and Anchor" 1 20
Jamieson's floor enamel 1 50
" barn paints, bbls. per gal. 0 60
Sanderson Pearcy's, pure 1 20
Robertson's pure paints 1 20

PUTTY.

Bulk in bbls............ 1 50
Bladders in bbls 1 80
25-lb. tins............ 1 80
Bladders in bulk or tins less than 100 lb 1 86
Bulk in 100-lb. irons............ 1 80

SHINGLE STAINS.

In 5 gallon lots 0 75 0 80

SHELLAC.

White 0 65
Fine orange............ .... 0 60
Medium orange 0 55
F o.b. Montreal or Toronto.

TURPENTINE AND OIL

Castor oil............ 0 08 0 10
Gasoline 0 22½
Benzine, per gal 0 17 0 20
Turpentine single barrels 1 00 1 04
Linseed Oil, " raw .. 0 62 0 66
" " boiled 0 65 0 68

WHITE LEAD GROUND IN OIL. Per 100 lbs

Canadian pure 7 15
No. 1 Canadian 6 80
Munro's Select Flake White.. .... 7 40
Elephant and Decorators' Pure 7 40
Monarch 7 40
Standard Decorator's 7 15
Essex Genuine 6 80
Brandram's B. B. Genuine... .... 8 40
" Anchor," pure 7 00
Ramsay's Pure Lead 7 00
Ramsay's Exterior 6 85
"Crown and Anchor," pure. 6 50
Sanderson Pearcy's 7 40
Robertson's C P., lead 7 20

WHITE AND RED DRY LEAD.

	white	red
Genuine, 560 lb. casks, per cwt	6 75	6 00
Genuine, 100 lb. kegs, "	7 50	6 50
No. 1, 560 lb. casks, per cwt ..	6 25	5 75
No. 1, 100 lb. kegs, per cwt ..	7 00	6 25

WINDOW GLASS.

Size United Inches.	Star	Double Diamond
Under 26	$4 25	$6 25
26 to 40	4 65	6 75
41 to 50	5 10	7 50
51 to 60	5 35	8 50
61 to 70	5 75	9 75
71 to 80	6 25	11 00
81 to 85	7 00	12 50
86 to 90	....	15 00
91 to 95	....	17 50

96 to 100 .... 20 50
101 to 105............ .... 24 00
106 to 110............ .... 27 50
Discount—16-oz, 25 per cent; 21-oz. 30 per cent. per 100 feet. Broken boxes 50 per cent.

WHITING.

Plain, in bbls............ .... 0 60
Gilders bolted in bands........ .... 0 80

WHITE DRY ZINC.

Extra Red Seal, V.M... ... 0 07½ 0 08½

WHITE GROUND ZINC.

Pure, in 25-lb. irons 0 08½
No. 1, " " 0 07
No. 2, " " 0 05½

VARNISHES.

In 5-gal. lots. Per gal cans

Carriage, No. 1 .... 1 50
Pale durable body............ .... 3 50
" hard rubbing 3 00
Finest elastic gearing 3 00
Elastic oak 1
Furniture, polishing 2
Furniture, extra............ 1
" " No 1............ 0
" " union 0
Light oil finish............ 1
Gold size japan 1
Brown japan 0
No 1 brown japan 0
Baking black japan 1
No 1 black japan 0
Benzine black japan 0
Crystal Damar............ 2
No 1 " 2
Pure asphaltum............ 1
Oilcloth............ 1
Lightning dryer 0
Elastilite varnish, 1 gal. can, each.. 2
Granitine floor varnish, per gal 2
Maple Leaf coach enamels; size 1, 1
Sherwin-Williams' kopal varnish, gal., 2
Canada Paint Co's sun varnish 2
"Kyanize" Interior Finish.... 2
"Flint-Lac," coach 1
B H. Co's "Gold Medal." in cases 2 50
Jamieson's Copaline, per gal. .... 2 [illegible]

BUILDERS' HARDWARE.

BELLS

Brass hand bells, 60 per cent.
Nickel, 55 per cent.
Gongs, Sargeant's door bells.. 5 50 8 00
American, house bells, per lb. " 35 " 40
Peterboro' door bells, discount 37½ and 10 per cent. off new list.

BUILDING PAPER, ETC.

Tarred Felt, per 100 lb. 2 25
Ready roofing, 2-ply, not under 45 lb. per roll 1 00
Ready roofing, 3-ply, not under 65 lb., per roll 1 25
Carpet Felt ....per ton 60 00
Heavy Straw Sheathing.......per ton 35 00
Dry Surprise............ 0 40
Dry Sheathing.....per roll, 400 sq. ft. 0 50
Tar " ..... " 400 " 0 45
Dry Fibre.......... " 400 " 0 55
Tarred Fibre ...... " 400 " 0 65
O. K. & I. X. L.... " 400 " 0 70
Resin-sized .... " 400 " 0 45
Oiled Sheathing.... " 600 " 1 00
Oiled " .... " 400 " 0 70
Roof Coating, in barrels......per gal. 0 17
Roof " small packages " 0 25
Refined Tar............per barrel 5 00
Coal Tar .... " 4 00
Coal Tar, less than barrels....per gal. 0 15
Roofing Pitch ......per 100 lb. 0 80 0 90
Slater's felt ....per roll 0 70
Heavy Straw Sheathing f. o. b. St. John and Halifax 37 50

BUTTS.

Wrought Brass, net revised list.
Wrought Iron, 70 per cent
Cast Iron Loose Pin, discount 60 per cent.
Wrought Steel Fast Joint and Loose Pin, 70 per cent.

CEMENT AND FIREBRICK

Canadian Portland 2 00 2 10
Belgium............ 1 60 1 90
White Bros. English 1 80 2 05
"Lafarge" cement in wood 3 40
"Lehigh" cement, in wood 2 54
"Lehigh" cement, cotton sacks. ... 2 39
"Lehigh" cement, paper sacks 2 31
Fire brick, Scotch, per 1,000 .... 27 00 30 00
" English 17 00 21 00
" American, low 23 00 25 00
" " high...... 27 50 35 00
Fire clay (Scotch), net ton 4 95

Paving Blocks per 1,000.

Blue metallic, 9"x4½"x3", ex wharf .. 35 00
Stable pavers, 12"x8"x2", ex wharf .. 50 00
Stable pavers, 9"x4½"x3", ex wharf .. 36 00

DOOR SETS.

Peterboro, 37½ and 10 per cent.

DOOR SPRINGS.

Torrey's Rod.........per doz. 1 75
Coil, 9 to 11 in........ " 0 95 1 65
English .... " 2 00 4 00
Chicago and Reliance Coil 25 per cent.

STORE DOOR HANDLES.

Per Dozen 1 00 1 50

ESCUTCHEONS.

Discount 50 and 10 per cent., new list
Peterboro, 37½ and 10 per cent.

ESCUTCHEON PINS.

Iron, discount 40 per cent.
Brass, 45 per cent.

HINGES.

Blind, discount 60 per cent.
Heavy T and strap, 4-in., per lb. net.. 0 06½
" " 5-in., " 0 06
" " 6-in., " 0 05½
" " 8-in., " 0 05½
" " 10-in. and larger.. 0 05½
Light T and strap, discount 65 p.c.
Screw hook and hinge—
under 12 in.... per 100 lb. 4 75
over 12 in....... " 3 75
Spring, No. 20, per gro. pairs 10 80
Spring, Woodyatt pattern, per gro., No 5, $17.50 No. 10, $18; No. 20, $10 80; No. 12 $20; No. 51, $10; No. 50, $27.50.
Crate hinges and back flaps, 65 and 5 p. c.
Hinge hasps, 65 per cent.

SPRING HINGES

Chicago Spring Butts and Blanks 12½ per cent.
Triple End Spring Butts, 40 and 5 per cent.
Chicago Floor Hinges, 40 and 5 off
Garden City Fire House Hinges, 12½ p c.

CAST IRON HOOKS.

Bird cage.............per doz. 0 50 1 10
Clothes line, No. 61.. " 0 00 0 70
Harness.............. " 0 60 12 00
Hat and coat .... per gro. 1 10 10 00
Chandelier ....per doz. 0 50 1 00
Wrought hooks and staples—
¼ x 5 ....per gross 2 65
5-16 x 5 .... " 3 30
Bright steel gate hooks and staples, 40 p.c.
Hat and coat wire, discount 62½ per cent.
Screw, bright wire, discount 65 per cent.

KNOBS.

Door, japanned and N.P., doz 1 50 2 50
Bronze, Berlin....... per doz. 2 75 3 25
Bronze, Genuine .... " 6 00 9 00
Shutter, porcelain, F. & L. screw .... per gross 1 30 2 00
White door knobs ...per doz. 2 00
Peterboro knobs, 37½ and 10 per cent.
Porcelain, mineral and jet knobs, net list.

KEYS.

Lock, Canadian dis. 40 to 40 and 10 per cent
Cabinet trunk and padlock American .... per gross 0 60

LOCKS.

Peterboro 37½ and 10 per cent
Russell & Erwin, steel rim $2.50 per doz
Eagle cabinet locks, discount 30 per cent
American padlocks, all steel, 10 to 15 per cent; all brass or bronze, 10 to 25 per cent.

SAND AND EMERY PAPER.

B. & A. sand, discount, 35 per cent
Emery, discount 35 per cent.
Garnet (Rurton's) 5 to 10 per cent. advance

SASH WEIGHTS.

Sectional...........per 100 lb. 2 00 2 25
Solid .... " 1 50 1 75

SASH CORD.

Per lb............ .... 0 31

BLIND AND BED STAPLES.

All sizes, per lb.............. 0 07½ 0 12

WROUGHT STAPLES.

Galvanized 2 75
Plain 2 50
Coopers', discount 45 per cent.
Poultry netting staples, discount 40 per cent
Bright spear point, 75 per cent. discount.

TOOLS AND HANDLES.

ADZES.

Discount 22½ per cent.

AUGERS.

Gilmour's, discount 60 per cent. off list.

AXES.

Single bit, per doz. 5 50 8 50
Double bit, " 10 00 11 00
Bench Axes, 40 per cent.
Broad Axes. 25 per cent.
Hunters' Axes............ 5 50 6 00
Boys' Axes............ 6 25 7 00
Splitting Axes 7 00 12 00
Handled Axes 7 00 9 00
Red Ridge, boys', handled.. .... 5 75
" hunters 5 25

BITS.

Irwin's auger, discount 47½ per cent.
Gilmour's auger, discount 60 per cent.
Rockford auger, discount 50 and 10 per cent.
Jennings' Gen. auger, net list.
Gilmour's car, 47½ per cent.
Clark's expansive, 40 per cent.
Clark's gimlet, per doz 0 65 0 90
Diamond, Shell, per doz...... 1 00 1 50
Nail and Spike, per gross..... 2 25 5 20

BUTCHERS CLEAVERS

German............per doz. 6 00 9 00
American.......... " 12 00 18 00

CHALK.

Carpenters' Colored, per gross 0 45 0 75
White lump..........per cwt. 0 60 0 65

CHISELS.

Warnock's, discount 72½ per cent.
P. S. & W. Extra, discount 72½ per cent

CROSSCUT SAW HANDLES.

S. & D., No. 3 ....per pair 0 15
S. & D., " 5 .... " 0 22½
S. & D., " 6 .... " 0 15
Boynton pattern.............. " 0 20

CROWBARS.

3½c. to 4c. per lb.

DRAW KNIVES.

Coach and Wagon, discount 75 per cent.
Carpenters' discount 75 per cent

DRILLS.

Millar's Falls, hand and breast. net list
North Bros., each set, 50c.

DRILL BITS.

Morse, discount 37½ to 40 per cent.
Standard, discount 50 and 5 to 55 per cent

FILES AND RASPS.

Great Western .... 75 per cent.
Arcade .... 75 "
Kearney & Foot .... 75 "
Disston's............ 75 "
American .... 75 "
J. Barton Smith .... 75 "
McClellan .... 75 "
Eagle .... 75 "
Nicholson............ 62½ "
Globe .... 75 "
Black Diamond. 60, 10 and 5 p.c.
Jowitt's, English list, 27½ per cent.

GAUGES.

Stanley's discount 50 to 60 per cent.

Wire Gauges.

Winn's, Nos. 26 to 33 ....each 1 65 2 40

HANDLES.

C. & B., fork and hoe, 40 p. c. revised list.
American, saw ...... per doz. 0 25
American, plane ... per gross 3 15 3 75
Canadian, hammer and hatchet 40 per cent
Axe and cant hook handles, 45 per cent.

HAMMERS.

Maydole's, discount 5 to 10 per cent.
Canadian, discount 25 to 27½ per cent.
Magnetic tack ....per doz. 1 10 1 20
Canadian sledge .... per lb. 0 07 0 08½
Canadian ball pean, per lb. .. 0 22 0 25

HATCHETS.

Canadian, discount 40 to 42½ per cent.

Shingle, Red Ridge 1, per doz.		4 40
" " 2, "		4 80
Barrel Underhill		5 06

MALLETS.

Tinsmiths' ... per doz.	1 25	1 50
Carpenters', hickory, "	1 25	3 75
Lignum Vitae	3 85	5 00
Caulking, each	0 60	2 00

MATTOCKS.

Canadian ... per doz.	5 50	6 00

MEAT CUTTERS.

German, 15 per cent.
American discount, 33⅓ per cent.

Gem ... each		1 15

NAIL PULLERS.

German and American	0 85	2 50
No. 1	0 85	
No 1573	0 75	

NAIL SETS.

Square, round and octagon, per gross		3 38
Diamond		1 00

PICKS.

Per dozen	6 00	9 00

PLANES.

Wood bench, Canadian discount 40 per cent. American discount 20 per cent.
Wood, fancy Canadian or American 37½ to 40 per cent
Stanley planes, $1.55 to $3 60, net list prices.

TROWELS.

Disston's discount 10 per cent.
S & D., discount 35 per cent

Berg's, brick, 9¼x11		4 00
" pointing, 9¼x5		2 10

FARM AND GARDEN GOODS

BELLS

American cow bells, 63⅓ per cent.
Canadian discount 45 and 50 per cent.

American, farm bells, each	1 35	3 00

BULL RINGS.

Copper, $1.30 for 2½-inch, and $1 70

CATTLE LEADERS.

Nos. 32 and 33 ... per gross	7 50	8 50

BARN DOOR HANGERS.

	doz.	pairs
Steel barn door	8 00	10 00
Stearns wood track	4 50	6 00
		9 00
Zenith	5 00	6 50
Acme, wood track	5 00	6 00
Atlas	8 00	11 00
Perfect	....	6 50
New Milo	....	11 00
Steel, covered	4 00	11 00
" track, 1 x 3-16 in (100 ft)	....	3 75
" " 1¼ x 3-16 in (100 ft)	....	4 75
Double strap hangers, doz. sets	...	6 40
Standard jointed hangers, "	....	6 40
Steel King hangers	....	6 25
Storm King and safety hangers	....	7 00
" " rail	....	4 25

Chicago Friction, Oscillating and Big Twin Hangers. 5 per cent.

HARVEST TOOLS.

Discount 60 per cent.
S. & D. lawn rakes, Dunn's, 40 off.
" sidewalk and stable scrapers, 40 off.

HAY KNIVES.

Net list

HEAD HALTERS.

Jute Rope, ⅝-inch ... per gross	....	9 00
" " ½ " ... "	....	10 00
" " ¾ " ... "	....	12 00
Leather, 1-inch ... per doz.	....	4 00
Leather, 1¼ " ... "	....	5 20
Web	....	2 45

HOES.

Garden, Mortar, etc., discount 60 per cent.

Planter ... per doz.	4 00	4 50

LAWN MOWERS

Low wheel, 12, 14 and 16-inch	$2 30
9-inch wheel, 12-inch	2 85
" " 14 "	3 00
" " 16 "	3 12½
High wheel, 12 "	4 05
" " 14 "	4 25
" " 16 "	4 50

SCYTHES.

Per doz. net	6 25	9 25

SCYTHE SNATHS.

Canadian, discount 40 per cent.

SNAPS.

Harness, German, discount 25 per cent.

Lock, Andrews'	4 50	11 00

STABLE FITTINGS

Warden King, 35 per cent

WOOD HAY RAKES.

Ten tooth, 40 and 10 per cent.
Twelve tooth, 45 per cent.

HEAVY GOODS, NAILS, ETC.

ANVILS.

Wright's, 80-lb. and over	0 10½
Hay Budden, 80-lb. and over	0 09½
Brook's, 80-lb. and over	0 11¼
Taylor-Forbes, handy	0 05
Columbia Hardware Co, per lb	0 09½

VISES.

Wright's	....	0 13½
Berg's, per lb	....	0 12½
Brook's	....	0 12½
Pipe Vise, Hinge, No. 1	....	3 50
" " " No. 2	....	5 50
Saw Vise	4 50	5 00

Blacksmiths' (discount) 60 per cent.
" parallel (discount) 45 per cent.

BOLTS AND NUTS

	Per cent.
Carriage Bolts, common ($1 list)	
" " ⅜ and smaller	60, 10 and 10
" " 7-16 and up	55 and 5
" " Norway Iron ($3 list)	50
Machine Bolts, ⅜ and less	60 and 10
Machine Bolts, 7-16 and up	55 and 5
Plough Bolts	55 and 10
Blank Bolts	55
Bolt Ends	55
Sleigh Shoe Bolts, ⅜ and less	60 and 10
" " " 7-16 and larger	50 and 5
Coach Screws, conepoint	70 and 5

Nuts, square, all sizes, 4c. per cent. off
Nuts, hexagon, all sizes, 4½c. per cent. off.
Stove Rods per lb., 5½ to 6c.
Stove Bolts, 75 per cent.

CHAIN.

Proof coil, per 100 lb., 5-16 in., $4.50; ⅜ in., $4 00, 7-16 in., $3 80; ½ in., $3 60; 9-16 in., $3 55; ⅝ in., $3.45; ¾ in., $3.30; ⅞ in., $3 20; 1 in., $3.20.
Halter, kennel and post chains, 40 to 40 and 5 per cent.

Cow ties	40 p.c.
Tie out chains	65 p.c.
Stall fixtures	35 p.c.
Trace chain	45 p.c.

Jack chain iron, discount 35 p.c.
Jack chain, brass, discount 40 per cent

HORSE NAILS.

'C' brand, 40, 10 and 7½ per cent. off list } Oval
M R.M. Co. brand, 55 per cent. } head
Capewell brand, quotations on application

HORSESHOES.

M.R M Co	No 2 and larger	No. 1 and smaller
Iron	3 80	4 05
Snow	4 05	4 30
Light steel	4 15	4 40
Featherweight, sizes 0 to 4	...	5 75
Toeweight, " 1 to 4	...	7 00

Packing up to 3 sizes in a keg, 10c per 100 lbs; more than three sizes, 25c per 100 lbs extra. F.o.b Montreal, add 15c, Toronto. Hamilton and Guelph

HORSE WEIGHTS

Taylor-Forbes, 3½c. per lb.

NAILS.	Cut.	Wire.
2d	3 80	3 45
3d	2 95	3 10
4 and 5d	2 70	2 85
6 and 7d	2 60	2 75
8 and 9d	2 45	2 60
10 and 12d	2 40	2 55
16 and 20d	2 35	2 50
30, 40, 50 and 60d (base)	2 30	2 45

F.o.b. Montreal. Cut nails, Toronto 20c. higher.
Miscellaneous wire nails, discount 75 per cent
Coopers' nails, discount 40 per cent.

PRESSED SPIKES.

Pressed spikes, ⅝ diameter, per 100 lbs $3.15,

RIVETS AND BURRS.

Iron Rivets, black and tinned, 60, 10 and 10.
Iron Burrs, discount 60 and 10 and 10 p c.
Copper Rivets, usual proportion burrs, 15 p c.
Copper Burrs only, net list
Extras on Coppered Rivets, ½-lb. packages 1c. per lb.; ¼-lb. packages 2c. lb
Tinned Rivets, net extra, 4c. per lb.

SCREWS.

Wood, F. H., bright and steel, 87½ per cent.
" R. H., bright, dis. 82½ per cent.
" F. H., brass, dis. 80 per cent.
" R. H., " dis. 75 per cent.
" F. H., bronze, dis. 75 per cent.
" R. H., " dis. 70 per cent.
Drive Screws, dis. 87½ per cent.

Bench, wood ... per doz.	3 25	4 00
" iron ... "	4 25	5 00

Set, case hardened, dis. 60 per cent.
Square Cap, dis. 50 and 5 per cent.
Hexagon Cap, dis. 45 per cent.

MACHINE SCREWS.

Flat head, iron and brass, 35 per cent.
Felister head, iron, discount 30 per cent.
" " brass, discount 25 per cent.

TACKS, BRADS, ETC.

Carpet tacks, blued	80 and 5
" " tinned	80 and 10
" " (in kegs)	40
Cut tacks, blued, in dozens only	75 and 10
" ½ weights	60
Swedes cut tacks, blued and tinned—	
In bulk	80 and 10
In dozens	75
Swedes, upholsterers', bulk	85 and 12½
" brush, blued and tinned bulk	70
Swedes, gimp, blued, tinned and japanned	75 and 12½
Zinc tacks	35
Leather carpet tacks	40
Copper tacks	25
Copper nails	30
Trunk nails, black	65
Trunk nails, tinned and blued	65
Clout nails, blued and tinned	65
Chair nails	35
Patent brads	40
Fine finishing	40
Lining tacks, in papers	10
" " in bulk	15
" " solid heads, in bulk	75
Saddle nails, in papers	10
" " in bulk	15
Tufting buttons, 22 line in dozens only	60
Zinc glaziers' points	5
Double pointed tacks, papers	90 and 10
" " " bulk	40
Clinch and duck rivets	45

Cheese box tacks, 85 and 5; trunk tacks, 80 and 10.

WROUGHT IRON WASHERS

Canadian make, discount 40 per cent.

SPORTING GOODS.

AMMUNITION.

B. B. Caps American $2.00 per 1000.
C. B. Caps American, $2.60 per 1000

CARTRIDGES.

Rim Fire Cartridges, 50 and 5 p.c.
Rim Fire Pistol, 30 and 5 per cent., American.
Central Fire, Military and Sporting American, 10 per cent. advance. B B. Caps, discount 40 per cent, American.
Central Fire Pistol and Rifle, net Amer list
Loaded and empty Shells, American 20 per cent. discount. Rival and Nitro, 10 per cent. advance
Empty paper shells American, 10 per cent. advance.
Primers, American $2 05

Wads.	per lb.
Best thick brown or grey felt wads, in ½-lb. bags	$0 70
Best thick white card wads, in boxes of 500 each, 12 and smaller gauges	0 29
Best thick white card wads, in boxes of 500 each, 10 gauge	0 35
Thin card wads, in boxes of 1,000 each, 12 and smaller gauges	0 20
Thin card wads, in boxes of 1,000 each, 10gauge	0 25

Chemically prepared black edge grey cloth wads, in boxes of 250 each—	Per M.
11 and smaller gauge	0 60
9 and 10 gauges	0 70
7 and 8 "	0 90
5 and 6 "	1 10
Superior chemically prepared pink edge, best white cloth wads in boxes of 250 each—	
11 and smaller gauge	1 15
9 and 10 gauges	1 40
7 and 8 "	1 65
5 and 6 "	1 90

SHOT.

Ordinary drop shot. AAA to dust $7.50 per 100 lbs. Discount 5 per cent; cash discount 2 per cent. 30 days; net extras as follows subject to cash discount only; Chilled, 40 c; buck and seal, 80c; no 28 ball, $1 20 per 100 lbs; bags less than 25 lbs., ½c per lb; F.O B. Montreal, Toronto, Hamilton, London, St John and Halifax, and freight equalized thereon.

TRAPS (steel.)

Game, Newhouse, discount 30 and 10 per cent.
Game, Hawley & Norton, 50, 10 & 5 per cent.
Game, Victor, 70 per cent.
Game, Oneida Jump (B. & L.) 40 & 2½ p. c.
Game, steel, 60 and 5 per cent.

SKATES.

Skates, discount 37½ per cent.

Mic Mac hockey sticks, per doz	4 00	5 00

PLANE IRONS.

English ... per doz.	2 00	5 00

Stanley, 2⅝ inch, single 24c, double 39c.

PLIERS AND NIPPERS.

Button's genuine, 37½ to 40 per cent.

Button's imitation ... per doz.	5 00	9 00
Berg's wire fencing	1 72	5 50

PUNCHES.

Saddler's ... per doz.	1 00	1
Conductor's ... "	3 00	15
Tinners', solid ... per set	....	0 85
" hollow ... per inch	....	1 00

RIVET SETS.

Canadian, discount 35 to 37½ per cent

RULES.

Boxwood, discount 70 per cent.
Ivory, discount 20 to 25 per cent.

SAWS.

Atkins, hand and crosscut, 25 per cent.
Disston's Hand, discount 12½ per cent

Disston's Crosscut ... per foot	0 35	0 55
Hack, complete ... each	0 75	2 75
" frame only ... each	0 50	1 25

S. & D. solid tooth circular shingle, concave and band, 50 per cent; mill and ice, drag, 30 per cent; cross-cut, 35 per cent.; hand saws, butcher, 35 per cent.; buck, New Century $6.25; buck No. 1 Maple Leaf, $5.25; buck, Happy Medium $4.25; buck, Watch Spring, $1.25; buck, common frame, $4 00.

Spear & Jackson's saws—Hand or rip 26 in., $12 75; 26 in., $14.25; panel, 18 in., $8 25; 20 in., $9; tenon, 10 in., $9.90; 12 in., $10.90; 14 in., $11 50.

SAW SETS.

Lincoln and Whiting	4 75
Hand Sets, Perfect	4 00
X-Cut Sets "	7 50

Maple Leaf and Premiums saw sets, 40 off.
S. & D. saw swages, 40 off.

SCREW DRIVERS.

Sargent's ... per doz.	0 65	1 00
North Bros., No. 30 ... per doz.	....	16 80

SHOVELS AND SPADES

Bull Dog, solid neck shovel (No. 2 pol.). $18 50

(Hollow Back)		(Reinforced S Scoop)
Moose	$17 50	$16 30
Bear	15 00	15 30
Fox	12 50	14 30
Black Cat	10 00	13 30

Canadian, discount 45 per cent.

SQUARES.

Iron, discount 20 per cent.
Steel, discount 65 and 10 per cent.
Try and Bevel, discount 50 to 52½ per cent.

TAPE LINES.

English, ass skin ... per doz.	2 75	5 00
English, Patent Leather	5 50	9 75
Chesterman's ... each	0 90	2 85
" steel ... each	0 80	8 00
Berg's, each	0 75	2 50

CUTLERY AND SILVERWARE.

RAZORS. per doz.

Elliot's 4 00 18 00
Boker's 7 50 11 00
" King Cutter 13 50 18 50
Wade & Butcher's............ 3 60 10 00
Lewis Bros. "Klean Kutter" 8 50 10 50
Henckel's 7 50 20 00
Berg's 7 50 20 00
Clauss Razors and Strops, 50 and 10 per cent

KNIVES.

Farriers-Stacey Bros, doz 3 50

PLATED GOODS

Hollowware, 40 per cent. discount.
Flatware, staples, 40 and 10, fancy, 40 and 5.

SHEARS.

Clauss, nickel, discount 60 per cent.
Clauss, Japan, discount 67½ per cent.
Clauss, tailors, discount 40 per cent.
Seymour's, discount 50 and 10 per cent.
Berg's............ .... 6 00 12 00

HOUSE FURNISHINGS.

APPLE PARERS.

Woodyatt Hudson, per doz., net 4 50

BIRD CAGES.

Brass and Japanned, 40 and 10 p. c.

COPPER AND NICKEL WARE.

Copper boilers, kettles, teapots, etc. 30 p.c.
Copper pitts, 20 per cent.

ENAMELED WARE.

London, White, Princess, Turquoise, Onyx, Blue and White, discount 50 per cent.
Canada, Diamond, Premier, 50 and 10 p c.
Pearl, Imperial Crescent, 60 and 10 per cent.
Premier steel ware, 40 per cent.
Star decorated steel and white, 25 per cent.
Japanned ware, discount 45 per cent.
Hollow ware, tinned cast, 35 per cent. off.

KITCHEN SUNDRIES.

Can openers, per doz............ 0 40 0 75
Mincing knives per doz 0 50 0 80
Duplex mouse traps, per doz 0 65
Potato mashers, wire, per doz... 0 60 0 70
" wood " .. 0 50 0 60
Vegetable slicers, per doz 2 25
Universal meat chopper No. 0, $1; No. 1, 1 15.
Enterprise chopper, each 1 30
Spiders and fry pans, 50 per cent.
Star A1 chopper 5 to 32 .... 1 35 4 10
" " 100 to 103 .... 1 35 2 00
Kitchen hooks, bright 0 62½

LAMP WICKS.

Discount, 60 per cent.

LEMON SQUEEZERS.

Porcelain lined...... per doz. 2 20 5 60
Galvanized........... " 1 87 3 85
King, wood........... " 2 75 2 90
King, glass........... " 4 00 4 50
All glass .... " 0 50 0 90

METAL POLISH.

Tandem metal polish paste 6 00

PICTURE NAILS.

Porcelain head ......per gross 1 25 1 50
Brass head.......... " 0 40 1 00
Tin and gilt, picture wire, 75 per cent.

SAD IRONS.

Mrs. Potts, No. 55, polished....per set 0 [illegible]
" No. 50, nickle-plated, " 0 [illegible]
Common, plain........................ [illegible]
" plated........................ [illegible]
Asbestos, per set.............. 1 [illegible]

TINWARE.

CONDUCTOR PIPE

2-in. plain or corrugated, per 100 feet, $3.30; 3 in., $4 40; 4 in., $5 8[illegible]; 5 in., $7 45; 6 in., $9.9[illegible].

FAUCETS.

Common, cork-lined, discount 35 per cent.

EAVETROUGHS.

10-inch ....per 100 ft. 3 30

FACTORY MILK CANS.

Discount off revised list, 35 per cent.
Milk can trimmings, discount 25 per cent.
Creamery Cans, 45 per cent.

LANTERNS.

No. 2 or 4 Plain Cold Blast....per doz. 6 50
Lift Tubular and Hinge Plain, " 4 75
No 0, safety.................. " 4 00
Better quality at higher prices.
Japanning, 50c. per doz extra.
Prism globes, per doz., $1.20.

OILERS.

Kemp's Tornado and McClary's Model galvanized oil can, with pump, 5 gallon, per dozen 10 92
Davidson oilers, discount 40 per cent
Zinc and tin, discount 50 per cent.
Coppered oilers, 20 per cent off.
Brass oilers, 50 per cent. off.
Malleable, discount 25 per cent

PAILS (GALVANIZED).

Dufferin pattern pails, 45 per cent.
Flaring pattern, discount 45 per cent.
Galvanized washtubs 40 per cent.

PIECED WARE.

Discount 35 per cent off list, June, 1899.
10-qt. flaring sap buckets, discount 35 per cent.
6, 10 and 14-qt. flaring pails dis. 35 per cent.
Copper bottom tea kettles and boilers, 30 p.c.
Coal hods, 40 per cent.

STAMPED WARE.

Plain, 75 and 12½ per cent. off revised list.
Retinned, 72½ per cent. revised list.

SAP SPOUTS.

Bronzed iron with hooks ....per 1,000 7 50
Eureka tinned steel, hooks 8 00

STOVEPIPES.

5 and 6 inch, per 100 lengths 7 64 7 91
7 inch........ " " 8 18
Nestable, discount 40 per cent.

STOVEPIPE ELBOWS

5 and 6-inch, common........per doz. 1 32
7-inch.......................... " 1 48
Polished, 15c. per dozen extra.

THERMOMETERS.

Tin case and dairy, 75 to 75 and 10 per cent.

TINNERS' SNIPS.

Per doz.......................... 3 00 15
Clauss, discount 35 per cent.

TINNERS' TRIMMINGS.

Discount, 45 per cent.

WIRE.

ANNEALED CUT HAY BAILING WIRE.

No. 12 and 13, $4; No 13½, $4 10; No. 14, $4 2'; No. 15 $4 50; in lengths 6' to 11', 25 per cent; other lengths 20c. per 100 lbs extra; if eye or loop on end add 25c. per 100 lbs. to the above.

BRIGHT WIRE GOODS

Discount 62½ per cent.

CLOTHES LINE WIRE.

7 wire solid line, No. 17 $4.90; No. 18, $3.00; No. 19, $2.70; 6 wire solid line, No. 17, $4.45; No. 18, $2.80. No. 19, $2.50. All prices per 1000 ft. measure. F.o.b. Hamilton Toronto, Montreal.

COILED SPRING WIRE

High Carbon, No. 9, $2 90, No. 11, $3 45; No. 12, $3.15.

COPPER AND BRASS WIRE.

Discount 37½ per cent.

FINE STEEL WIRE.

Discount 25 per cent. List of extras: In 100-lb. lots: No. 17, $5 — No. 18, $5.50 — No. 19, $6 — No. 20, $6 65 — No. 21, $7 — No. 22, $7.30 — No. 23, $7.65 — No 24, $8 — No. 25, $9—No. 26, $9.50—No. 27, $10—No. 28, $11—No. 29, $12—No. 30, $13—No. 31, $14—No. 32, $15—No. 33, $16—No. 34, $17. Extras net—tinned wire, Nos. 17-25, $2—Nos. 26-31, $4—Nos. 32-34, $6. Coppered, 75c.—oiling, 10c.—in 25-lb. bundles, 15c.—in 5 and 10-lb. bundles, 25c.—in 1-lb. hanks, 25c. —in ½-lb. hanks, 38c.—in ¼-lb. hanks, 50c.—packed in casks or cases, 15c.—bagging or papering, 10c.

FENCE STAPLES.

Bright.......... 2 75 Galvanized.... 3 15

HAY WIRE IN COILS.

No. 13, $2 60; No. 14, $2 70; No. 15, $2.85; f.o.b., Montreal.

GALVANIZED WIRE

Per 100 lb.—Nos. 4 and 5, $3.70 — Nos. 6, 7, 8, $3 15 — No. 9, $2.50 — No. 10, $3.20 — No. 11, $3 25—No. 12, $2.65 —No. 13, $2 75—No. 14. $3.75—No 15, $4.30 —No. 16. $4.30 from stock. Base sizes, Nos. 6 to 9, $2.35 f.o.b. Cleveland. In carlots 12½c. less.

LIGHT STRAIGHTENED WIRE.

Gauge No.	Over 20 in. per 100 lbs.	10 to 20 in.	5 to 10 in.
0 to 5	$0 50	$0 75	$1 25
6 to 9	0 75	1 25	2 00
10 to 11	1.00	1 75	2.50
12 to 14	1.50	2.25	3 50
15 to 16	2 00	3.00	4.50

SMOOTH STEEL WIRE.

No. 0-9 gauge, $2 30; No. 10 gauge, 6c. extra; No. 11 gauge, 12c extra; No. 12 gauge, 20c. extra; No. 13 gauge, 30c extra; No 14 gauge, 40c. extra, No 15 gauge, 55c. extra; No. 16 gauge, 70c extra. Add 60c. for coppering and $2 for tinning.

Extra net per 100 lb.—Oiled wire 10c., spring wire $1.25, bright soft drawn 15c., charcoal (extra quality) $1 25, packed in casks or cases 15c., bagging and papering 10c., 50 and 100-lb. bundles 10c, in 25-lb. bundles 15c., in 5 and 10-lb. bundles 25c., in 1-lb. hanks, 50c., in ½-lb. hanks 75c., in ¼-lb. hanks $1.

POULTRY NETTING.

2 in mesh 19 w. g., discount 50 and 10 per cent. All others 50 per cent.

WIRE CLOTH.

Painted Screen, in 100-ft rolls, $1.62½, per 100 sq. ft.; in 50-ft. rolls, $1 67½, per 100 sq ft. Terms, 2 per cent. off 30 days.

WIRE FENCING.

Galvanized barb....... .... 2 95
Galvanized, plain twist 3 30
Galvanized barb, f.o.b. Cleveland, $2.70 for small lots and $2.60 for carlots

WOODENWARE.

CHURNS.

No. 0, $9; No. 1, $9; No. 2, $10; No. 3, $11; No. 4, $13; No. 5, $16.; f.o b. Toronto Hamilton, London and St. Marys, 30 and 30 per cent; f o b Ottawa, Kingston and Montreal, 40 and 15 per cent. discount, Taylor-Forbes, 30 and 30 per cent.

CLOTHES REELS.

Davis Clothes Reels, dis. 40 per cent.

LADDERS, EXTENSION.

3 to 6 feet, 11c. per foot, 7 to 10 ft., 12c.
Waggoner Extension Ladders, dis. 40 per cent.

MOPS AND IRONING BOARDS.

"Best" mops........................ 1 25
"900" mops 1 25
Folding ironing boards......... 12 00 16 50

REFRIGERATORS

Discount, 40 per cent.

SCREEN DOORS.

Common doors, 2 or 3 panel, walnut stained, 4-in. style..........per doz. 7 25
Common doors, 2 or 3 panel, grained only, 4-in., style ....per doz. 7 65
Common doors, 2 or 3 panel, light stair per doz. 9 55

WASHING MACHINES.

Round, re-acting per doz. 60 00
Square " 63 00
Eclipse, per doz 54 00
Dowswell " 39 00
New Century, per doz 75 00
Daisy.................................. 54 00

WRINGERS.

Royal Canadian, 11 in., per doz. ... 34 00
Royal American, 11 in. .. . 34 00
Ese, 10 in., per doz 37 75
Terms, 2 per cent., 30 days.

MISCELLANEOUS.

AXLE GREASE.

Ordinary, per gross 6 00 7 00
Best quality 10 00 12 00

BELTING.

Extra, 60 per cent.
Standard, 60 and 10 per cent.
No. 1, not wider than 6 in., 60, 10 and 10 p.c.
Agricultural, not wider than 4 in., 75 per cent
Lace leather, per side, 75c.; cut laces, 80c.

BOOT CALKS.

Small and medium, ball ....per M 4 25
Small heel 4 50

CARPET STRETCHERS.

American ....per doz. 1 00 1 50
Bullard's............ .... 6 50

CASTORS.

Bed, new list, discount 55 to 57½ per cent.
Plate, discount 52½ to 57½ per cent.

PINE TAR.

½ pint in tins ....per gross ... 7 80
" 9 60

PULLEYS.

Hothouse ....per doz. 0 55 1 00
Axle........ .... " 0 22 0 33
Screw .... " 0 22 1 00
Awning .... " 0 35 2 50

PUMPS.

Canadian cistern 1 40 2 00
Canadian pitcher spout 1 80 3 16
Berg's wing pump, 75 per cent.

ROPE AND TWINE.

Sisal.. 0 10½
Pure Manila 0 15½
"British" Manila.......................... 0 12
Cotton, 3-16 inch and larger.......0 21 0 23
" 5-32 inch 0 25 0 27
" ¼ inch 0 25 0 28
Russia Deep Sea 0 16
Jute.. 0 09
Lath Yarn, single 0 10
" double 0 10½
Sisal bed cord, 48 feet..........per doz. 0 65
" " 60 feet........ " 0 80
" " 72 feet........ " 0 95

Twine.

Bag, Russian twine, per lb. 0 27
Wrapping, cotton, 3-ply 0 25
" " 4-ply 0 29
Mattress twine per lb. 0 33 0 45
Staging " 0 27 0 35

BINDER TWINE.

500 feet, sisal 0 09½
500 " standard 0 09
550 " manilla 0 10
600 " " 0 12
650 " " 0 13½
Car lots, ½c. less; 5-ton lots, ¼c. less. Central delivery.

SCALES.

Gurney Standard, 40 per cent.
Gurney Champion, 50 per cent.
Burrow, Stewart & Milne—
Imperial Standard, discount 40 per cent.
Weigh Beams, discount 40 per cent.
Champion Scales, discount 50 per cent.
Fairbanks standard, discount 35 per cent.
" Dominion, discount 55 per cent.
" Richelieu, discount 55 per cent.
Warren new Standard, discount 40 per cent.
" Champion, discount 50 per cent.
" Weighbeams, discount 35 per cent.

STONES—OIL AND SCYTHE.

Washita ....per lb. 0 25 0 37
Hindostan .... " 0 06 0 10
" slip .... " 0 18 0 20
" Axe.......... " 0 15
Deer Creek............ " 0 10
Deerlick .... " 0 25
" Axe .... " 0 15
Lily white .... " 0 42
Arkansas.............. " 1 50
Water-of-Ayr .... " 0 10
Scythe ....per gross 3 50 5 00
Grind, 40 to 200 lb., per ton..... 20 00 22 00
" under 40 lb., " 24 00
" 200 lb. and over 28 00

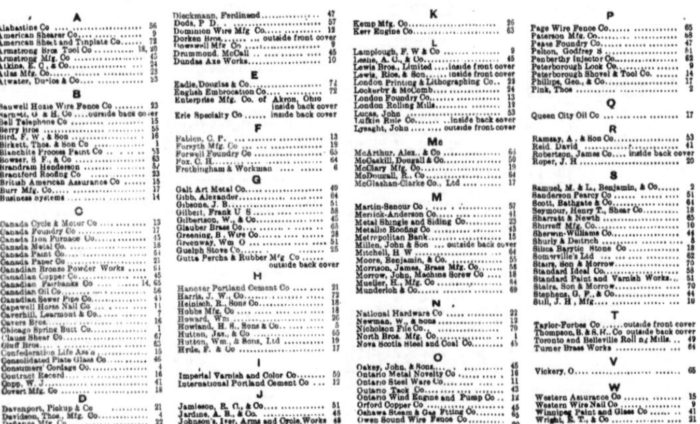

INDEX TO ADVERTISERS.

CLASSIFIED LIST OF ADVERTISEMENTS.

Alabastine.
Alabastine Co, Limited, Paris, Ont

Auditors.
Davenport, Pickup & Co, Winnipeg.

Awnings.
Tobin Tent and Awning Co, Ottawa

Babbitt Metal.
Canada Metal Co, Toronto.
Canadian Fairbanks Co., Montreal.
Robertson, Jas. Co., Montreal.

Bath Room Fittings.
Forsyth Mfg. Co, Buffalo N.Y.
Ontario Metal Novelty Co., Toronto

Belting, Hose, etc.
Gutta Percha and Rubber Mfg. Co. Toronto
Sadler & Haworth Toronto

Bicycles and Accessories.
Johnson's, Iver, Arms and Cycle Works Fitchburg, Mass

Binder Twine.
Consumers Cordage Co., Montreal.

Box Strap.
J. N Warminton, Montreal.

Brass Goods.
Glauber Brass Mfg. Co., Cleveland, Ohio.
Kerr Engine Co., Walkerville Ont.
Lewis, Rice, & Son., Toronto.
Morrison, Jas., Brass Mfg. Co., Toronto.
Mueller Mfg. Co, Decatur, Ill.
Penberthy Injector Co., Windsor, Ont.
Taylor-Forbes Co., Guelph, Ont.

Bronze Powders.
Canadian Bronze Powder Works, Montreal.

Brushes.
Ramsay, A., & Son Co., Montreal.
United Factories, Toronto.

Cans.
Acme Can Works, Montreal.

Builders' Tools and Supplies.
Covert Mfg. Co., West Troy, N.Y.
Frothingham & Workman Co., Montreal.
Howland, H. S., Sons & Co., Toronto.
Hyde, F., & Co., Montreal.
Lewis Bros. & Co., Montreal.
Lewis, Rice, & Son, Toronto.
Lockerby & McComb, Montreal.
Lufkin Rule Co., Saginaw, Mich.
Newman & Sons, Birmingham.
North Bros. Mfg. Co., Philadelphia, Pa.
Stanley Rule & Level Co., New Britain.
Stanley Works, New Britain, Conn.
Stephens, G. F., Winnipeg.
Taylor-Forbes Co., Guelph, Ont.

Carriage Accessories.
Covert Mfg. Co., West Troy, N.Y.

Carriage Springs and Axles.
Guelph Spring and Axle Co., Guelph.

Carpet Beateas.
Ontario Metal Novelty Co, Toronto.

Cattle and Trace Chains.
Greening, B., Wire Co., Hamilton.

Churns.
Dowswell Mfg. Co., Hamilton.

Clippers—All Kinds.
American Shearer Mfg. Co., Nashua, N.H

Clothes Reels and Lines.
Hamilton Cotton Co., Hamilton, Ont.

Clutch Nails.
J. N. Warminton, Montreal.

Cordage.
Consumers' Cordage Co., Montreal.
Hamilton Cotton Co., Hamilton.

Cork Screws.
Erie Specialty Co., Erie, Pa.

Cow Tics
Greening, B., Wire Co., Hamilton

Cut Glass.
Phillips, Geo., & Co., Montreal.

Cutlery—Razors, Scissors, etc.
Birkett, Thos., & Son Co., Ottawa.
Clauss Shear Co., Toronto
Dorken Bros. & Co., Montreal.
Heinisch's, R., Sons Co., Newark, N.J.
Howland, H. S. Sons & Co., Toronto.
Hutton, Wm., & Sons, Ltd., London, Eng.
Lamplough, F. W., & Co., Montreal.
Phillips, Geo., & Co., Montreal.
Round, John, & Son, Montreal.

Electric Fixtures.
Canadian General Electric Co., Toronto.
Morrison James, Mfg. Co., Toronto.
Munderloh & Co., Montreal.

Electro Cabinets.
Cameron & Campbell Toronto.

Enameled Ware
Kemp Mfg. Co., Toronto.

Engines, Supplies, etc.
Kerr Engine Co., Walkerville, Ont

Eavetroughs
Wheeler & Bain, Toronto

Fencing—Woven Wire
Owen Sound Wire Fence Co., Owen Sound.
Page-Hersey Wire Fence Co., Hamilton.

Files and Rasps.
Barnett Co., G. & H., Philadelphia, Pa.
Nicholson File Co., Port Hope

Firearms and Ammunition.
Hamilton Rifle Co, Plymouth, Mich.
Harrington & Richardson Arms Co., Worcester, Mass.
Johnson's, Iver, Arms and Cycle Works, Fitchburg, Mass.

Food Choppers
Enterprise Mfg. Co., Philadelphia, Pa.
Lamplough, F. W., & Co, Montreal
Shirreff Mfg. Co, Brockville, Ont.

Galvanizing.
Canada Metal Co., Toronto.
Montreal Rolling Mills Co., Montreal.
Ontario Wind Engine & Pump Co., Toronto.

Glass Ornamental
Hobbs Mfg Co, London
Consolidated Plate Glass Co., Toronto

Glaziers' Diamonds.
Gibsone, J. B., Montreal.
Pelton, Godfrey S.
Sharratt & Newth, London, Eng.
Shaw, A., & Son, London, Eng.

Handles.
Still, J. H., Mfg. Co.

Harvest Tools.
Maple Leaf Harvest Tool Co, Tillsonburg Ont.

Hockey Sticks
Still, J. H Mfg. Co., St. Thomas.

Hoop Iron.
Montreal Rolling Mills Co, Montreal.
J. N. Warminton, Montreal.

Horse Blankets.
Heney, E. N., & Co., Montreal.

Horseshoes and Nails.
Canada Horse Nail Co, Montreal.
Montreal Rolling Mills, Montreal.
Capewell Horse Nail Co., Toronto

Hot Water Boilers and Radiators.
Cluff, R. J., & Co Toronto.
Pease Foundry Co. Toronto.
Taylor-Forbes Co., Guelph.

Ice Cream Freezers.
Dana Mfg. Co., Cincinnati, Ohio.
North Bros. Mfg. Co., Philadelphia, Pa.

Ice Cutting Tools.
Erie Specialty Co., Erie, Pa.
North Bros. Mfg. Co., Philadelphia, Pa.

Injectors—Automatic.
Morrison, Jas., Brass Mfg. Co., Toronto.
Penberthy Injector Co., Windsor, Ont.

Iron Pipe.
Montreal Rolling Mills, Montreal.

Iron Pumps.
Lamplough, F. W., & Co., Montreal.
McDougall, R., Co., Galt, Ont.

Lanterns.
Kemp Mfg. Co., Toronto.
Ontario Lantern Co., Hamilton, Ont.
Wright, E. T., & Co., Hamilton.

Lawn Mowers.
Birkett, Thos., & Son Co., Ottawa.
Maxwell, D., & Sons, St. Mary's, Ont.
Taylor, Forbes Co., Guelph.

Lawn Mower Grinders
Root Bros. & Co., Plymouth, Ohio.

Ledgers—Loose Leaf.
Business Systems Toronto.
Copeland-Chatterson Co, Toronto.
Crain, Rolla L., Co., Ottawa.
Universal Systems, Toronto.

Locks, Knobs, Escutcheons, etc.
Peterborough Lock Mfg. Co., Peterborough, Ont.
National Hardware Co., Orillia, Ont.

Lumbermen's Supplies.
Pink, Thos., & Co., Pembroke Ont.

Lye
Gillett, E. W., & Co, Toronto

Manufacturers' Agents.
Fox, C. H., Vancouver.
Gibb, Alexander, Montreal.
Scott, Bathgate & Co., Winnipeg.

Metals.
Canada Iron Furnace Co., Midland, Ont.
Canada Metal Co., Toronto.
Eadie, H. G, Montreal
Gibb, Alexander, Montreal.
Kemp Mfg. Co., Toronto
Leslie, A. C., & Co, Montreal.
Lysaght, John, Bristol, Eng.
Nova Scotia Steel and Coal Co, New Glasgow, N.S.
Robertson, Jas, Co., Montreal
Roper, J H, Montreal.
Samuel, Benjamin & Co, Toronto
Stairs, Son & Morrow, Halifax, N.S.
Thompson, B. & S. H. & Co. Montreal.

Metal Lath.
Galt Art Metal Co., Galt.
Metallic Roofing Co., Toronto.
Metal Shingle & Siding Co., Preston, Ont.

Metal Polish, Emery Cloth, etc.
Oakey, John, & Sons, London, Eng

Nails Wire
Dominion Wire Mfg Co, Montreal.

Oil Tanks.
Bowser, S. F., & Co., Toronto.

Paints, Oils, Varnishes, Glass.
Brandram-Henderson, Montreal
Canada Paint Co, Montreal
Canadian Oil Co Toronto
Consolidated Plate Glass Co., Toronto.
Dods, P. D & Co., Montreal
Imperial Varnish and Color Co., Toronto.
Jamieson, R. C. & Co., Montreal.
Lucas John & Co, New York
McArthur, Corneille & Co., Montreal.
McCaskill, Dougall & Co., Montreal.
Moore, Benjamin, & Co Toronto.
Ottawa Paint Works, Ottawa
Queen City Oil Co., Toronto.
Ramsay & Son, Montreal.
Sanderson Pearcy & Co, Toronto
Sherwin-Williams Co., Montreal.
Standard Paint Co., Montreal
Standard Paint and Varnish Works Windsor, Ont
Stephens & Co, Winnipeg.
Martin-Senour Co, Chicago.
Winnipeg Paint & Glass Co, Winnipeg

Perforated Sheet Metals
Greening, B., Wire Co., Hamilton.

Plumbers' Tools and Supplies
Canadian Fairbanks Co., Montreal.
Cluff, R. J, & Co., Toronto
Glauber Brass Co., Cleveland, Ohio.
Jardine, A B., & Co, Hespeler, Ont.
Jenkins Bros., Boston, Mass.
Kerr Engine Co, Walkerville, Ont
Lewis, Rice, & Son, Toronto.
Morrell Mfg. Co., Toledo, Ohio.
Montreal Rolling Mills Montreal.
Morrison, Jas, Brass Mfg. Co., Toronto.
Mueller, H., Mfg. Co., Decatur, Ill.
Oshawa Steam & Gas Fitting Co., Oshaw
Robertson Jas., Co. Montreal.
Robertson, Jas, Co, Limited, Toronto
Somerville, Limited, Toronto
Stairs, Son & Morrow, Halifax, N.S.
Standard Ideal Sanitary Co., Port Hope,
Standard Sanitary Co., Pittsburg.
Stephens, G F., & Co., Winnipeg, Man.
Turner Brass Works, Chicago.
Vickery, Orlando, Toronto.

Polishes.
Majestic Polishes, Toronto

Portland Cement.
International Portland Cement Co. Ottawa, Ont.
Hanover Portland Cement Co., Hanover, Ont.
Hyde, F., & Co., Montreal.
Thompson B. & S. H. & Co., Montreal.

Poultry Netting.
Greening, B., Wire Co., Hamilton, Ont.

Razors.
Clauss Shear Co., Toronto.

Refrigerators.
Fabien, C. P., Montreal

Roofing Supplies.
Brantford Roofing Co., Brantford.
McArthur, Alex., & Co., Montreal.
Metal Shingle & Siding Co., Preston, Ont.
Metallic Roofing Co., Toronto
Paterson Mfg Co., Toronto & Montreal
Wheeler and Bain, Toronto

Saws.
Atkins, E. C., & Co., Indianapolis, Ind
Shurly & Dietrich, Galt, Ont
Spear & Jackson, Sheffield, Eng.

Scales.
Canadian Fairbanks Co., Montreal.

Screw Cabinets.
Cameron & Campbell, Toronto.

Screws, Nuts, Bolts.
Montreal Rolling Mills Co., Montreal.
Morrow, John, Machine Screw Co., Ingersoll, Ont.

Soil Pipe
McFarlane, Walter, Glasgow

Sewer Pipes.
Canadian Sewer Pipe Co., Hamilton
Hyde, F., & Co., Montreal.

Shelf Boxes.
Cameron & Campbell, Toronto.

Shears, Scissors.
Clauss Shear Co., Toronto.

Shovels and Spades
Eclipse Mfg. Co., Ottawa
Peterboro Shovel & Tool Co, Peterboro.

Silverware.
Hutton, Wm., & Sons, Ltd., London, Eng.
McGlashan, Clarke Co., Niagara Falls, Ont.
Phillips, Geo., & Co., Montreal.
Round, John, & Son, Sheffield, Eng.

Skates.
Canada Cycle & Motor Co, Toronto.
McFarlane, Walter, Glasgow.

Sprayers
Cavers Bros., Galt

Spring Hinges, etc.
Chicago Spring Butt Co, Chicago, Ill.

Stable Fittiugs
Dennis Wire & Iron Co, London

Steel Rails.
Nova Scotia Steel & Coal Co., New Glasgow, N.S.

tove Pipe.
Chown, Edwin, and Son, Kingston

Stoves, Tinware, Furnaces
Canadian Heating & Ventilating Co., Owen Sound.
Copp, W. J., Son & Co, Fort William
Davidson, Thos., Mfg. Co., Montreal
Down Draft Furnace Co, Galt
Guelph Stove Co., Guelph.
Gurney Foundry Co., Toronto.
Harris, J. W., Co., Montreal.
Howard, Wm, Toronto
Kemp Mnfg Co. Toronto.
McClary Mfg. Co. London.
Merrick Anderson, Winnipeg
Pease Foundry Co., Toronto.
Smart, James, Mfg Co Brockville
Stewart, Jas., Mfg. Co, Woodstock, Ont.
Taylor-Forbes Co, Guelph, Ont.
Wright, E. T., & Co., Hamilton.

Tacks.
Montreal Rolling Mills Co., Montreal.
Ontario Tack Co., Hamilton.

Tents.
Tobin Tent and Awning Co, Ottawa

Turpentine
Defiance Mfg. Co, Toronto

Ventilators.
Harris, J. W., Co., Montreal.
Pearson, Geo. D., Montreal.

Wall Paper
Staunton Limited, Toronto.

Wall Paper Cleaner.
Gilbert, Frank U. S., Cleveland

Washing Machines, etc
Dowswell Mfg. Co, Hamilton, Ont.
The Shultz Bros Co, Brantford.
Taylor-Forbes Co., Guelph, Ont.

Wheelbarrows
London Foundry Co, London, Ont.
Schultz Bros. Co., Ltd., The Brantford.

Wholesale Hardware.
Birkett, Thos., & Sons Co., Ottawa.
Caverhill, Learmont & Co., Montreal.
Frothingham & Workman, Montreal.
Hobbs Hardware Co., London.
Howland, H. S., Sons & Co., Toronto.
Lamplough, F. W., & Co, Montreal.
Lewis Bros. & Co., Montreal.
Lewis, Rice, & Son, Toronto.

Window and Sidewalk Prisms.
Hobbs Mfg. Co., London, Ont.

Wire, Wire Rope, Cow Tics, Fencing Tools, etc.
Banwell-Hoxie Fence Co., Hamilton
Dennis Wire and Iron Co., London, Ont.
Dominion Wire Mnfg. Co., Montreal
Greening, B., Wire Co., Hamilton.
Owen Sound Wire Fence Co., Owen Sound
Montreal Rolling Mills Co., Montreal.
Western Wire & Nail Co., London, Ont

Wrapping Papers.
Canada Paper Co., Toronto.
McArthur, Alex., & Co, Montreal.
Stairs, Son & Morrow, Halifax, N.S.

Wringers
Connor, J. H. & Son, Ottawa, Ont.

HARDWARE BABBITTS

No. 1, No. 2, No. 3

The wearing quality of our

HARDWARE BABBITTS

is the secret of the large orders the

HARDWARE MERCHANTS

are sending us. If not on our list

GET IN LINE

LIBERAL DISCOUNTS. TERMS RIGHT

The James Robertson Co., Limited

144 William St., MONTREAL, QUE.

PFLUEGER'S Fishing Tackle

Hooks, Flies, Trolls, Spinners, Phantoms, Reels, Furnished Lines.

Everything in Fishing Tackle.

NOTICE—Free to Any Dealer in Sporting Goods, Sent Express Prepaid, 170-Page Illustrated Catalogue No. F25, and Metal Fish Sign in 8-Color Lithograph.

The Enterprise Mfg. Co.

AKRON, Ohio, U.S.A.

WALKER CORK SCREWS

are good holiday souvenirs and live sellers the whole year. Are made well and are made out of good material. That is why we guarantee every one that we put out. When ordering Cork Screws try **Walker's**

Manufactured only by **ERIE SPECIALTY COMPANY,** ERIE, Pa. U.S.A.

LUFKIN MEASURING TAPES

Steel, Metallic, Linen, Pocket, Ass Skin, Pat. Leather, Bend Leather, Etc,

ARE THE BEST AND MOST POPULAR TAPES IN THE WORLD. YOUR STOCK IS NOT COMPLETE WITHOUT THEM.

LUFKIN RULE CO., Saginaw, Mich, U.S.A.

London Office and Warehouse—24 and 26 Holborn. New York City Branch—280 Broadway.

For sale by ALL PROMINENT CANADIAN HARDWARE JOBBERS.

"Redstone"

High Pressure

Sheet Packing

A packing that will hold. For use in highest pressures for steam, hot or cold water and air. Packs equally well for all.

From actual tests, we believe that this packing is the most durable and satisfactory of any on the market. Try a sample lot and see for yourself.

Manufactured Solely by

THE GUTTA PERCHA & RUBBER MFG. CO.

of TORONTO, LIMITED

HEAD OFFICES,

47 Yonge Street, Toronto.

Branches: Montreal, Winnipeg, Vancouver.

SHEET ZINC,

SHEET COPPER,

SPELTER,

INGOT TIN,

BRASS and COPPER TUBES.

For lowest prices send your enquiries to

B. & S. H. THOMPSON & Co.

LIMITED

53 ST. SULPICE STREET, MONTREAL

Lightning Source UK Ltd.
Milton Keynes UK
UKHW021438220119
335636UK00018B/977/P